CIVIL, ARCHITECTURE AND ENVIRONMENTAL ENGINEERING

PROCEEDINGS OF THE INTERNATIONAL CONFERENCE ON CIVIL, ARCHITECTURE AND ENVIRONMENTAL ENGINEERING (ICCAE2016), TAIPEI, TAIWAN, 4–6 NOVEMBER 2016

Civil, Architecture and Environmental Engineering

Editors

Jimmy C.M. Kao
National Sun Yat-Sen University, Kaohsiung, Taiwan, R.O.C.

Wen-Pei Sung
National Chin-Yi University of Technology, Taiping City, Taiwan, R.O.C.

VOLUME 1

CRC Press is an imprint of the
Taylor & Francis Group, an **informa** business
A BALKEMA BOOK

Published by:
CRC Press/Balkema
P.O. Box 447, 2300 AK Leiden, The Netherlands
e-mail: Pub.NL@taylorandfrancis.com
www.crcpress.com – www.taylorandfrancis.com

First issued in paperback 2020

Typeset by V Publishing Solutions Pvt Ltd., Chennai, India

ISBN 13: 978-1-138-02985-9 (set of 2 volumes)
ISBN 13: 978-0-367-73622-4 (Vol 1) (pbk)
ISBN 13: 978-1-138-06583-3 (Vol 1) (hbk)
ISBN 13: 978-1-138-06584-0 (Vol 2)

Visit the Taylor & Francis Web site at
http://www.taylorandfrancis.com

and the CRC Press Web site at
http://www.crcpress.com

Civil, Architecture and Environmental Engineering – Kao & Sung (Eds)
© 2017 Taylor & Francis Group, ISBN 978-1-138-02985-9

Table of contents

VOLUME 1

Structural science and architecture engineering

Building materials and materials science

Construction equipment and mechanical science

VOLUME 2

Environmental science and environmental engineering

Computer simulation & computer and electrical engineering

 ISBN 978-1-138-02985-9

Preface

The 2016 International Conference on Civil, Architecture and Environmental Engineering (ICCAE 2016) was held on November 4-6, 2016 in Taipei, Taiwan, organized by China University of Technology and Taiwan Society of Construction Engineers, aimed to gather professors, researchers, scholars and industrial pioneers from all over the world. ICCAE 2016 is the premier forum for the presentation and exchange of experiences, new advances and research results in the field of theoretical and industrial experience. The conference contained contributions promoting the exchange of ideas and rational discourse between educators and researchers from all over the world.

ICCAE 2016 is expected to be one of the most comprehensive Conferences focused on civil, architecture and environmental engineering. The conference promotes international academic cooperation and communication, and exchanging research ideas.

We would like to thank the conference chairs, organization staff, and authors for their hard work. By gathering together so many leading experts from the civil, architecture and environmental engineering fields, we believe this conference has been a very enriching experience for all participants. We hope all have had a productive conference and enjoyable time in Taipei!

Conference Chair
Dr. Tao-Yun Han
Chairman of Taiwan Society of Construction Engineers

Civil, Architecture and Environmental Engineering – Kao & Sung (Eds)
© 2017 Taylor & Francis Group, ISBN 978-1-138-02985-9

Organizing committee

HONOR CHAIRS

Prof. Ming-Chin Ho, *Architecture & Building Research Institute, Taiwan (Director General)*
Prof. Cheer Germ Go, *National Chung Hsin University, Taiwan*
Prof. Tzen-Chin Lee, *National United University, Taiwan*
Prof. Chu-hui Chen, *China University of Technology, Taiwan*

CONFERENCE CHAIRS

Prof. Jimmy C.M. Kao, *National Sun Yat-Sen University, Taiwan*
Prof. Yun-Wu Wu, *China University of Technology, Taiwan*
Prof. Che-Way Chang, *Chung-Hua University, Taiwan*
Dr. Tao-Yun Han, *Taiwan Society of Construction Engineers*
Prof. Wen-Pei Sung, *National Chin-Yi University of Technology, Taiwan*

CHAIR OF INTERNATIONAL TECHNOLOGICAL COMMITTEES

Prof. Ming-Hsiang Shih, *National Chi Nan University, Taiwan*

INTERNATIONAL TECHNOLOGICAL COMMITTEES

Yoshinori Kitsutaka, *Tokyo Metropolitan University, Japan*
Nasrudin Bin Abd Rahim, *University of Malaya, Malaya*
Lei Li, *Hosei University, Tokyo, Japan*
Yan Wang, *The University of Nottingham, UK*
Darius Bacinskas, *Vilnius Gediminas Technical University, Lithuania*
Ye-Cai Guo, *Nanjing University of Information Science and Technology, China*
Wang Liying, *Institute of Water Conservancy and Hydroelectric Power, China*
Gang Shi, *Inha University, South Korea*
Chen Wang, *University of Malaya, Malaya*

LOCAL ORGANIZING COMMITTEES (TAIWAN)

Wen-der Yu, *Chung Hua University, Taiwan*
Chien-Te Hsieh, *Yuan Ze University, Taiwan*
Ta-Sen Lin, *Taiwan Architects Association, Taiwan*
Hsi-Chi Yang, *Chung Hua University, Taiwan*
Jwo-Hua Chen, *Chienkuo Technology University, Taiwan*
Der-Wen Chang, *Tamkang University, Taiwan*
Cheng Der Wang, *National United University, Taiwan*
Shun-Chin Wang, *Architecture and Building Research Institute, Taiwan*
Yaw-Yauan Tyan, *China University of Technology, Taiwan*
Kuo-Yu Liao, *Vanung University, Taiwan*

Shih-Tsang Chou, *China University of Technology, Taiwan*
Shyr-Shen Yu, *National Chung Hsing University, Taiwan*
Yean-Der Kuan, *National Chin-Yi University of Technology, Taiwan*
Yu-Lieh Wu, *National Chin-Yi University of Technology, Taiwan*
Shih-Heng Tung, *National University of Kaohsiung, Taiwan*
Hsueh-Chun Lin, *China Medical University, Taiwan*
Yao-Chiang Kan, *Yuan Ze University, Taiwan*
Yao-Ming Hong, *Ming Dao University, Taiwan*
P.S. Pa, *National Taipei University of Education, Taiwan*
Shao-Wen Su, *National Chin-Yi University of Technology, Taiwan*
Yi-Ying Chang, *National Chin-Yi University of Technology, Taiwan*
Jun-Hong Lin, *Nanhua University, Taiwan*
Lei Wei, *National Chin-Yi University of Technology, Taiwan*
Ting-Yu Chen, *National Chin-Yi University of Technology, Taiwan*

Civil, Architecture and Environmental Engineering – Kao & Sung (Eds)
© 2017 Taylor & Francis Group, ISBN 978-1-138-02985-9

Acknowledgements

GUIDANCE UNIT

Construction and Planning Agency, Ministry of the Interior, R.O.C.
Architecture and Building Research Institute, Ministry of the Interior, R.O.C.

SPONSORS

Taiwan Society of Construction Engineers

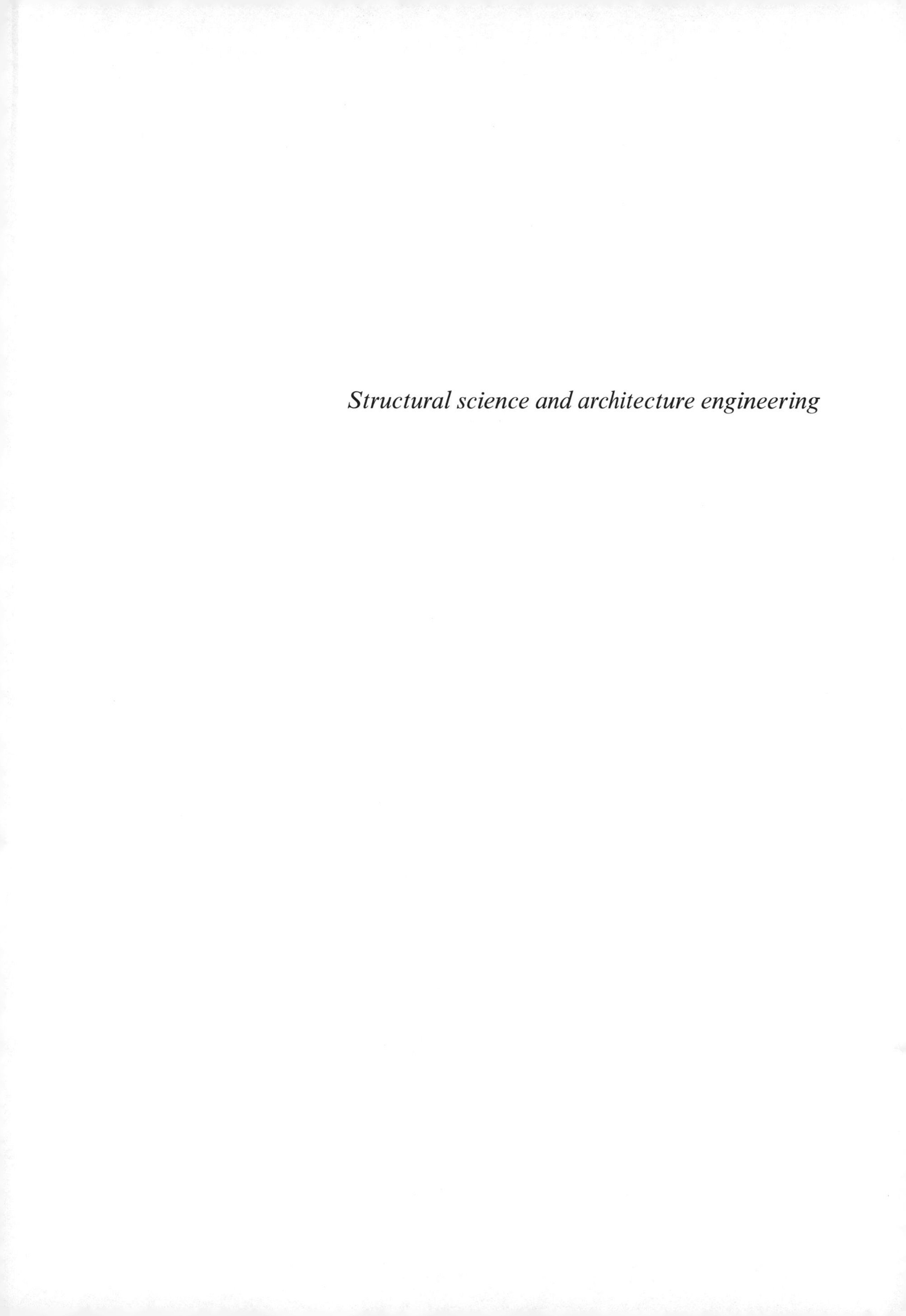

Structural science and architecture engineering

Civil, Architecture and Environmental Engineering – Kao & Sung (Eds)
© 2017 Taylor & Francis Group, ISBN 978-1-138-02985-9

Visual support vector machine-based method for reliability-based design optimization

You Xie & Yizhong Wu
School of Mechanical Science and Engineering, National CAD Supported Software Engineering Centre, Huazhong University of Science and Technology, Wuhan, China

ABSTRACT: In this paper, a method based on Virtual Support Vector Machine (VSVM) and Kriging response surface model is proposed to address Reliability-Based Design Optimization (RBDO) problems for black-box system. In many engineering optimization problems, the number of function evaluations is limited because of the expensive evaluation of the objection function and constraint functions. To get a more precise result, it is necessary to evaluate more samples in a small region, which includes the optimum point. Considering these characters, the RBDO method based on VSVM (RBDO–VSVM) is proposed in this article. RBDO–VSVM uses VSVM classification to approximate the limit state function, which can guide the constraint boundary sampling process. Besides, Trust Region (TR) method is used in the algorithm to downsize the search space to a local range around the constraint boundary, which probably contains the optimum point. Numerical optimization example is posed in the end of the article to validate the efficiency and accuracy of the proposed method.

1 INTRODUCTION

In recent years, RBDO has been studied to design safer and more reliable products, which take the uncertainty of loadings, material properties, environment, and other parameters into consideration. Reliability analysis aims to quantify the uncertainty of the output according to the uncertainty of the system inputs, environment, system itself, and so on. On the contrary, the relationship between inputs and outputs becomes increasingly complicated, and mostly they are black-box systems in the practical engineering application, which makes the aforementioned relationships highly nonlinear and not able to be characterized with explicit functions. Therefore, it is very important do more research on the RBDO problems of black-box systems.

Several reliability analysis methods have been proposed. The First-Order Reliability analysis Method (FORM) is widely used because of its simplicity and efficiency (Du 2008; Yao et al. 2013). To improve the FORM, the Second-Order Reliability analysis Method (SORM) was developed, which can obtain a more precise result. Uncertainty optimization methods aim to find the optimum point by taking the effect of uncertainty into consideration. Traditional uncertainty optimization methods apply a double-loop algorithm, which inserts the reliability analysis into the process of optimization. Methods such as Reliability Index Approach (RIA) and Performance Measure Approach (PMA) belong to the traditional optimization method. A serial Single-Level Approach (SLA) has been developed recently, which decompose the double-loop algorithm into a sequential single-level process. The key of SLA is how to convert the reliability constraints into certainty constraints. The widely used methods include Sequential Optimization and Reliability Assessment (SORA) and reliability index approximate approach. Other typical uncertainty optimization methods such as Single-Loop Single Vector (SLSV) and Single-Level Double Vector (SLDV) aim at uncertainty optimization problems, which insert uncertainty analysis based on the Most Probable Point (MPP) into optimization research.

Meta-Model-Based Design Optimization (MBDO) has been the mainstream method to achieve optimization of the Constrained Black-box system. Typical methods such as Efficient Global Optimization (EGO) emerged as one of the most promising methods for expensive simulators (Jones et al. 1998). EGO is based on the approximation of responses using a Gaussian process model so that the variance of the prediction is available over the whole design space. Expected Improvement (EI) of the object function can also be assessed; hence, the optimum value can be found by maximizing the EI with a global search. Recently, EGO has also developed Efficient Global Reliability Analysis (EGRA) [Bichon BJ et al. 2008]. The EGRA is based on expectation improvement. To improve the efficiency and precision of RBDO, other theories are also developed to address RBDO, such as the Support Vector Machine (SVM). The SVM is a classification method, which constructs only the decision (i.e., limit state) function. Another advantage of this classification method is that SVM can deal with multiple constraints simultaneously. Besides, Virtual SVM (VSVM) (Song, 2013) is developed to improve the accuracy of SVM while maintaining its desirable features, by using the available response function values.

In this paper, the RBDO–VSVM method is proposed to improve the efficiency and accuracy of RBDO. This paper is organized as follows. In Section 1, the mainly situation of the RBDO is reviewed. Section 2 introduces the theoretical background, whereas Section 3 introduces the details of the proposed method. Numerical examples and welded beam design example are demonstrated in Section 4. Conclusion is proposed in the last section.

2 THEORETICAL BACKGROUND

2.1 *EGO and EGRA*

EGO was proposed by Jones [Jones, 2001]. In EGO, a new sampling point, which should be added to the Gaussian process model, is selected by maximizing the amount of improvement in the objective function that can be expected by adding that point. A point could be expected to produce an improvement in the objective function if its predicted value is better than the current best solution or if the uncertainty in its prediction is such that the probability of it producing a better solution is high. Because the uncertainty is higher in regions of the design space with fewer observations, this provides a balance between exploiting areas of the design space that predict good solutions and exploring areas, where more information is needed. EGO was developed in Bichon BJ et al. [2012] to address RBDO problems, which developed the EI into the Expected Feasibility Function (EFF) that can provide an indication of how well the true value of the response is expected to satisfy the equality constraint.

2.2 *Virtual support vector machine*

The basic idea of VSVM is to increase the probability of locating the decision function close to the true limit state function by inserting two opposite signed virtual samples between the given two samples. Virtual samples could be generated from the approximation method using any pair of real samples.

3 RBDO–VSVM

In this section, details of the proposed method will be presented.

The aim of optimization is to find the optimum point; therefore, orientating a small region, which contains the optimum point precisely, is important for optimization the algorithm. The method includes two phases. The aim of Phase 1 is to orientate the region, which contains the optimum point. With the base of Phase 1, the aim of Phase 2 is to refine the meta-model within the trust region, which is near the real optimum point. Detailed information of the two phases is proposed below. The abstract flowchart of the method is depicted in Fig. 1.

3.1 *Phase 1—the coarse search process*

Because the classification approach of VSVM to construct the explicit boundaries is good at dealing with discontinuous and binary responses, both the VSVM and Kriging models are used in Phase 1. In order to improve the efficiency of orientating the small region, which contains the optimum point in Phase 1, EFF and the nearest distance from the existing sample points are applied in the sampling criterion of Phase 1.

The flowchart of Phase 1 is depicted in Fig. 2.

Four steps will be carried out in this phase to find the coarse region, which contains the optimum.

Step 1: Initialize the sample set. In order to obtain the sample set, which meets the uniform distribution, the Latin Hypercube Sampling (LHS) method is used in this step. More than 2^{n+1} samples should be taken as the initial sample set for the problem with n dimensions.

Step 2: Evaluate objective function and constraint functions for each sample and construct/update Kriging models for the objective function and constraint functions based on samples. Besides, the VSVM model is also constructed/updated for constraint functions based on the samples.

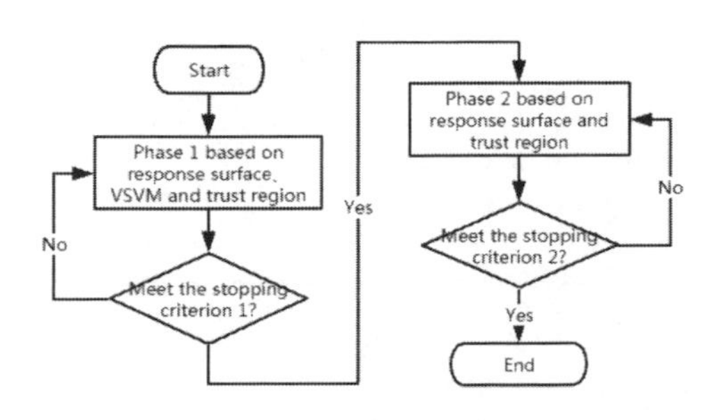

Figure 1. Abstract flowchart of the proposed method.

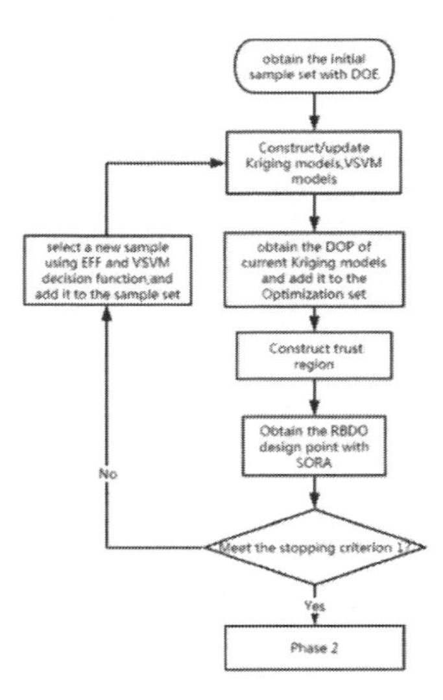

Figure 2. Flowchart of Phase 1.

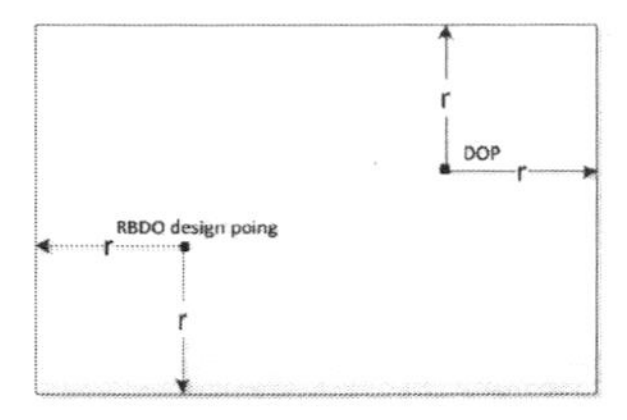

Figure 3. Trust region.

Step 3: Obtain the DOP according to the current object Kriging model and constraints Kriging model with Genetic Algorithm (GA). The optimization problem can be expressed as:

$$\begin{cases} \min \ \hat{f}(\mathbf{x}) \\ \text{s.t} \ \ \hat{g}(\mathbf{x}) = 0 \end{cases}$$

where $\hat{f}(x)$ is the current Kriging model of objection function and $\hat{g}(\mathbf{x})$ is the current Kriging models of constraint functions. This optimization problem is addressed by GA. DOP will be added to the optimization set. The trust region is constructed/updated according to the current DOP $\mathbf{x}_{k+1}$ and the former DOP $\mathbf{x}_k$, as shown in Fig. 3.

If $f(x_{k+1}) < f(x_k)$, which means that the solution can be trusted, then the trust region will be expanded, which means the value of r should be larger; if $f(x_{k+1}) > f(x_k)$, which means that the solution cannot be trusted, then the trust region will be downsized, which means the value of r should be smaller.

If the last two DOP points meet the criterion 1, which is expressed in Eq. (1), go to Step 5; if the last two DOP points do not meet the criterion 1, go to Step 4:

$$\|x_{k+1} - x_k\| \le 2 * r \tag{1}$$

Step 4: Select a new sample based on EFF, SVM, and trust region, which is expressed in Eq. (2), and add it to the sample set. Then, go to Step 2:

$$\begin{cases} \max \text{EFF*D} \\ \text{s.t. } s(x) = 0 \\ x \in \text{current trust region} \end{cases} \tag{2}$$

where D is the minimal distance from the current sample point to the existing sample points. EFF is the expected feasibility function. s(x) = 0 is the VSVM decision function. In the criterion expression, EF is applied to improve the precision of the Kriging model. D is applied to make the sample distribution more uniform. The constraint function of the criterion ensure that the new sample point is located at the SVM decision function and belong to the trust region, which can ensure that the new sample is as close to the optimum as possible.

3.2 *Phase 2—the accurate search process*

The flowchart of Phase 2 is depicted in Fig. 4.

Two steps are included in this phase.

Step 5: Obtain the DOP according to current object Kriging model and constraints Kriging model with Genetic Algorithm (GA) and add it to the Optimization set. The trust region is updated with the method that is used in Step 3. Besides, the RBDO design point is obtained with SORA. If the last two DOP points meet Criterion 2, which is expressed in Eq. (3), the process is complete; if not, go to Step 6:

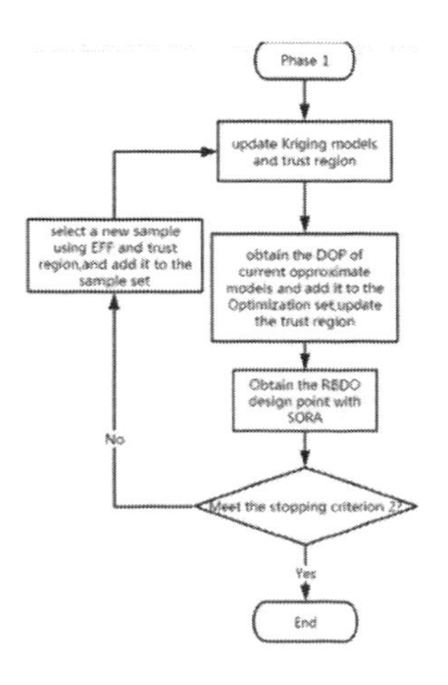

Figure 4. Flowchart of Phase 2.

$$\|x_{k+1} - x_k\| \le \varphi \tag{3}$$

where φ is a small positive number (i.e., 10^{-3})

Step 6: Select a new sample on the basis of EFF and trust region, which is expressed in Eq. (4), and add it to the sample set. Then, go to Step 2:

$$\begin{cases} \max \text{EFF*D} \\ \text{s.t.} \\ x \in \text{trust region} \end{cases} \tag{4}$$

where EFF, D, and the trust region have the same meaning as that in Eq. (2). Kriging models of the constraints can be refined with the criterion within the trust region at each iteration. With the increase in the number of iterations, the accuracy of the Kriging model within the trust region will be increasingly higher.

In the RBDO–VSVM method, because of the uniform distribution of LHS sampling result, the stability of SVM, and taking the minimal distance, D, into consideration, the risk of finding the local optimum point can be reduced to some extent. Besides, the optimum point is obtained with Genetic Algorithm, which can also avoid falling into local optimum point.

4 NUMERICAL EXAMPLE

In this section, a numerical example is proposed to demonstrate the accuracy and efficiency of the proposed method. A comparison between the EGRA and RBDO–VSVM methods is made in this section. This example is the modified Haupt example (Lee and Jung, 2008), which has two design variables, whose objection function and constraint functions are nonlinear.

$$-\begin{cases} \text{Find } \mathbf{X} = [x_1, x_2] \\ \text{Min } f(\mathbf{X}) = (x_1 - 3.7)^2 + (x_2 - 4)^2 \\ \text{s.t.prob}[g(\mathbf{X}) < 0] \le \Phi(-\beta_T^i),\ i = 1,2 \\ g_1(\mathbf{X}) = -x_1 \sin(4x_1) - 1.1x_2 \sin(2x_2) \\ g_2(\mathbf{X}) = x_1 + x_2 - 3 \\ 0.0 \le x_1 \le 3.7,\ 0.0 \le x_2 \le 4.0 \\ \beta_T^1 = \beta_T^2 = 2.0 \\ X^{(0)} = [2.5, 2.5]^T \\ X_n \sim N(d_n, 0.1^2), n = 1,2 \end{cases}$$

In Fig. 5, the feasible samples are denoted by red spots and the infeasible samples are denoted by blue spots. The VSVM decision function, s(x) = 0, is denoted by black lines. The Kriging models of the constraints boundary are denoted by the pink dotted line. The true constraints boundaries are denoted by the green line. The trust region is represented by a red rectangle.

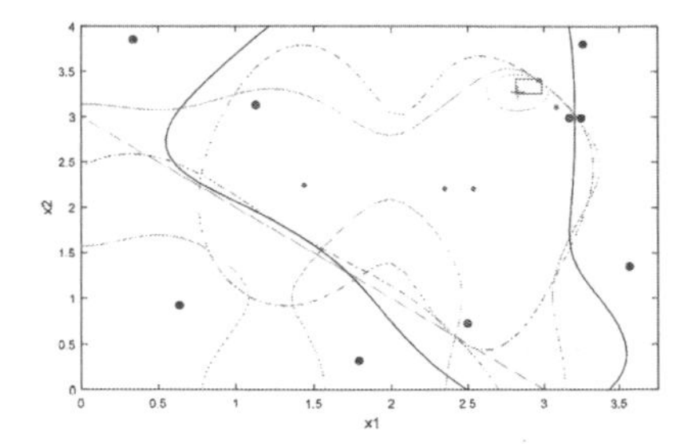

Figure 5. Result of the RBDO–VSVM method.

The final RBDO design point is denoted by the red "+". The RBDO design point obtained by the RBDO–VSVM method is [2.826, 3.268] with 14 samples; the RBDO design point obtained by EGRA is [2.834, 3.251] with 35 samples. To quantify the accuracy of these methods, we adopt the accuracy measure in Eq. (5):

$$\text{error} = \frac{X_d - \hat{X}_d}{X_d} \times 100\% \tag{5}$$

where $\mathbf{X}_d$ is the true optimum design point and $\hat{\mathbf{X}}_d$ is the optimum design point. The norm used in this equation is 2-norm.

The error percent of the RBDO–VSVM method is 0.0487%; whereas that of the EGRA method is 0.468%. Thus, it is evident that the accuracy and efficiency of the RBDO–VSVM method are significantly higher.

5 CONCLUSION AND FUTURE WORK

In this article, the RBDO–VSVM method is proposed to address the black-box system optimization problems. In the RBDO–VSVM method, Kriging models are used to approximate the constraints and objection function, and the VSVM method is used to approximate the limit state function. To improve the efficiency of the optimization process, trust region is used in this article. With the process of iteration, the trust region has an increasingly higher probability to contain the optimum design point so that the Kriging model near the optimum design point can be refined with a higher accuracy.

However, the RBDO–VSVM method still has some limitations in some extreme situation. For example, the method is not good at addressing the optimization problem, whose feasible region is very small, long, and narrow. Under these conditions, it is difficult to find the first sample point within the feasible region, which decreased the accuracy of the VSVM method. Therefore, how to improve the sampling strategy to improve the accuracy of the VSVM still has much room for further research.

ACKNOWLEDGMENTS

This study was supported by the National Natural Science Foundation of China (No. 51575205) and the National High Technology Research and Development Program of China (863 Program) (No. 2013AA041301).

REFERENCES

Bichon, B. J., Eldred, M. S., Swiler, L. P. S., Mahadevan, S., McFarland, J. M. (2008). "Efficient Global Reliability Analysis for Nonlinear Implicit Performance Functions," *AIAA Journal* Vol. 46, No. 10, pp. 2459–2468.

Du, X (2008b). Unified uncertainty analysis by the first order reliability method Journal of mechanical design 130:091401.

Jones, D.R. (2001). "A taxonomy of global optimization methods based on response surfaces." *Journal of Global Optimization* 21(4), 345–383.

Jones, D.R., Schonlau, M. and Welch, W.J. (1998). "Efficient global optimization of expensive black-box functions." *Journal of Global Optimization* 13(4), 455–492.

Lee TH, Jung JJ (2008). A sampling technique enhancing accuracy and efficiency of metamodel-based RBDO: Constraint boundary sampling Computers & Structures 86:1463–1476.

Song, Hyeongjin. "Efficient sampling-based Rbdo by using virtual support vector machine and improving the accuracy of the Kriging method." PhD (Doctor of Philosophy) thesis, University of Iowa, 2013.

Yao W, Chen X, Huang Y, van Tooren M (2013). An enhanced unified uncertainty analysis approach based on first order reliability method with single-level optimization Reliability Engineering & System Safety 116:28–37.

Civil, Architecture and Environmental Engineering – Kao & Sung (Eds)
 ISBN 978-1-138-02985-9

Study on the tunnel group traffic accident risk using a fuzzy comprehensive evaluation model

Bin Ma, Rong Gao & Ying Gao
Xijing University, Xi'an, Shaanxi, China

Xiaojie Wang & Zihan Guo
Xi'an University of Technology, Xi'an, Shaanxi, China

ABSTRACT: This paper aims to evaluate the tunnel traffic accident risk using an Analytic Hierarchy Process (AHP) and the fuzzy comprehensive evaluation methods. The concept of highway tunnel group was put forward for the first time. Three safety factors, including the human factors, vehicle factors, and road factors, were analyzed. The group tunnel traffic accident risk index system was established following the principles of integrity, scientific, credibility, and operable principle. The group tunnel traffic risk evaluation model was put forward using AHP and fuzzy comprehensive evaluation methods. Taking the Luan Chuan Highway Tunnel group traffic accident in Luoyang City as an example, the fuzzy comprehensive evaluation model was used in the application example. The results showed that the fuzzy comprehensive evaluation value of the Luan Chuan Highway Tunnel group traffic was 3.08, indicating that the traffic accident risk of the Luan Chuan Highway Tunnel group was medium.

1 INTRODUCTION

In view of the current serious situation of traffic accidents, this paper evaluates the risk of traffic accidents using the fuzzy comprehensive evaluation method. There are very few research works focusing on the tunnel group. Current references mainly focused on the definition of the tunnel group, the safety evaluation system of tunnel group, the influence factors of the safety of the tunnel group, and the practical safety technology of the tunnel group. However, the definition of the tunnel group and the operation environment of the identification are still not clear. Although the definition of the highway tunnel group is different, it can be mainly divided into two types: static definition and dynamic definition. The static definition of the highway tunnel group is used in the existing traffic environment, given the concept covers tunnel extension attributes, such as the limited value of the fixed distance between two tunnels. The value generally consists of various factors affecting the maximum distance. The dynamic definition of the highway tunnel group is set according to the tunnel design vehicle speed, traffic volume, and variable factors.

Shaofei Wang (2009) defined the tunnel group as the highway, where there are two or more intervals in the shorter tunnel, including the continuous tunnel and the adjacent tunnel. On the basis of the analysis of several factors affecting the role of distance, it is recommended that when the distance between the two tunnels (L) is less than 250 m, it is defined as the adjacent tunnel. When L > 250 m and L < 1000 m, it is defined as the continuous tunnel. The adjacent tunnel and continuous tunnel are two types of the tunnel group. The maximum value of the range of the influence factors is selected as the dividing boundary between the tunnel group and the independent area, which belongs to the static definition. The advantage of the definition is its easy operation, whereas the disadvantage is its poor accuracy. Na Zhou (2010) proposed the concept of relative tunnel spacing and put forward the concept of the critical value of the interaction effect of the tunnel spacing. The tunnel section is a special cover structure of road. The tunnel traffic is still related to the human, vehicle, and road, which fully take into account natural attributes, characteristics of the environment, driver's visual requirements, and vehicle's space requirements (Wang, 2009).

In this paper, the dynamic definition of the tunnel group is used to determine the critical value according to the tunnel group and to determine whether the tunnel group is based on the distance S. Taking the Luan Chuan Highway in Luoyang City as an example, the driving safety characteristics of the highway tunnel group was systematically studied and analyzed. In order to improve the construction and operation of tunnel group, the tunnel group accident risk evaluation model is established and different sections of the actual situation were carried out, which can provide references to deal with the traffic accidents.

2 MATERIALS AND METHODOLOGY

2.1 *Study area and data*

The design speed of Luan Chuan Highway in Luoyang City is 80 km/h, and the running mode is one-way travel. The total distance is 66.58 km, containing 24 tunnels, with a total length of about 10.3 km, accounting for 15% of the total length of the road. The longest tunnel is 2 km, and the rest are short tunnels. The average tunnel spacing is 270 m (Zhang, 2009). The West ditch tunnel group of the Luan Chuan Highway consists of seven tunnels, as shown in Table 1.

2.2 *Establishment of risk evaluation index system*

People (the driver visual characteristics, psychological characteristics), car (vehicle speed, vehicle's own factors), and road (the road itself and environmental factors, including tunnel) are three aspects of the highway tunnel safety factors described in this paper. The evaluation criteria are determined through the concrete analysis of the influence factors of the tunnel group. Then, the risk index system of the highway tunnel group traffic accidents is established (Chao, 2012).

The tunnel group and safety factors are combined with the actual situation to discuss the classification and itemized fixed factors of the tunnel group according to the above principles. The comprehensive safety evaluation system of the highway tunnel group is mainly established from the tunnels of the main structure, traffic engineering facilities, import and mouth and the surrounding environment, electromechanical facilities, emergency rescue system, ventilation and lighting, security management, and maintenance services. According to the reality of the situation of highway tunnel traffic accidents and morphology, Luan Chuan Highway Tunnel group of individual tunnel is divided into three sections. According to the purposes and principles of the highway tunnel traffic accident risk assessment system, the highway tunnel traffic accident risk assessment index system framework is divided into the target layer, the criterion layer, and the index layer. To analyze the key indicators of the evaluation system, the evaluation index value uses five conventional grades according to the various risk indicators assignment.

Table 1. Luan Chuan highway tunnel in Luoyang city.

Tunnel name	Length	Slope (%)	Distance
Taishang tunnel	302	2.9/2.5	1237
Wangyuan tunnel	161	−1.92	1845
Dayangti ditch tunnel	480	+2.45/2.35	73
West ditch tunnel No. 1	95	+2.95	44
West ditch tunnel No. 2	124	+2.95	33
Yangjuan tunnel	753	+2.2/−1.2	691
Yachi ditch tunnel	640		–

2.3 *Safety influence factors of tunnel group*

The analytic Hierarchy Process (AHP) method and the fuzzy comprehensive evaluation method are applied to the established risk evaluation index system and then the evaluation method of the risk assessment model is built.

1. AHP

AHP was first proposed by American operational experts in the mid-20th century proposing the use of the decision-making method. The complex problem is to decompose it into a hierarchical structure. The next level of various factors is relative to the level of each factor for pairwise comparison judgment. The judgment matrix is constructed by the calculation to determine the holding level, with the sorting and the consistency test. Finally, a general arrangement of administrative levels, the combination weights of the factors and sorting, are conducted through the analysis of results to solve the problem (Satty, 1978). This method has the characteristics of a small amount of information and the short-time decision process. Using the analytic hierarchy process, the weight can be determined by the following three steps (Kayastha, 2013):

1. *Establish hierarchical structure*

First, a hierarchical structure model is constructed. The same element has a dominant effect on some elements of the next layer, and it is subject to the domination of the upper layer. In each level, there should not be more than nine elements dominated by the next level of the general; otherwise, it will cause more difficulty to judge.

2. *Structure judgment matrix*

Judgment matrix $A = (a_{ij})_{n \times n}$ has the following properties:

$$a_{ij} > 0, a_{ij} = 1/a_{ij}, a_{ij} = 1(i, j = 1, 2, \cdots, n)$$

In the formula $a_{ij}(i, j = 1, 2, \cdots, n)$, U_i and U_j represent the scale of the importance of a layer of elements relative to its upper layer. The value of the judgment matrix directly reflects people's understanding of the relative importance of each factor, and it is generally used to assign values to the degree of importance and scale. The meanings are shown in Table 2.

Table 2. Judgment matrix scale and its implication.

Scaling	Meaning
1	Compared with the two elements, which is of the same importance.
3	Compared with the two elements, the former is a little more important than the latter.
5	Compared with the two elements, the former is more important than the latter.
7	Compared with the two elements, the former is more important than the latter.
9	Compared with the two elements, the former is more important than the latter.
2, 4, 6, 8	The intermediate values of the two adjacent judgments are indicated.
Reciprocal	The importance ratio of element *j* is $a_{ij} = 1/a_{ij}$.

3. *Calculation of the relative weights of elements under a single criterion and the consistency check*

The maximum characteristic root of A is λ_{max}, and the corresponding characteristic vector is W, which is the weight of the corresponding elements for the relative importance of a certain element at the same level. It is necessary to carry out the consistency test and calculate the consistency test index CI:

$$CI = \frac{\lambda_{max} - n}{n-1}$$

where n is the order of the judgment matrix.

If random consistency ratio $CR = \frac{CI}{RI} < 0.1$, the judgment matrix is satisfied with the consistency; otherwise, it is required to adjust the element value of the judgment matrix. RI value is showed in Table 2.

2. Fuzzy comprehensive evaluation model

In the fuzzy comprehensive evaluation model, the evaluation object is judged according to the specified criteria. The comprehensive evaluation aims at making a comprehensive evaluation of the evaluation object affected by multiple factors. The basic principle of Fuzzy comprehensive evaluation method for risk assessment is as follows: the influence degree of all risk factors is considered, and the weight difference of each factor is set to distinguish the importance difference. By constructing a mathematical model, all possibility degrees of risk and the degree of possibility value are higher for the level of risk to determine the final value. As a kind of concrete application method of fuzzy mathematics, it has been widely used.

First, a set of factors U is set up, $U = \{u1, u2, u3, u4\}$, to determine the value of comprehensive evaluation. Then, we should divide the judgment matrix by using the theory of fuzzy mathematics. The weight of evaluation index is calculated and the results of the fuzzy comprehensive evaluation are finally obtained.

3 RESULTS AND DISCUSSION

According to the established fuzzy comprehensive evaluation model, the evaluation standard, the measured value, the fuzzy switching principle, and the maximum membership principle are used to make a general evaluation of the things or objects, which are restricted by many factors.

1. Set up a set of factors

The factors that affect the traffic accident of tunnel section 1 are set as follows:

$U_1 = \{u1, u2, u3, u4\}$ = {Highway alignment, The alignment of tunnel entrance section, Traffic characteristics of tunnel, Tunnel traffic environment}.

Inside: $U1 = \{c1, c2, c3\}$ = {Highway curve radius, Slope of Expressway, Highway pavement adhesion coefficient}.

$U2 = \{c4, c5, c6\}$ = {Curve radius of tunnel entrance, Longitudinal slope of tunnel portal section, Sight distance}.

$U3 = \{c7, c8, c9\}$ = {Traffic saturation, Cart proportion, Average speed}.

$U4 = \{c10, c11, c12\}$ = {Road adhesion coefficient of tunnel entrance section, Tunnel ventilation, Tunnel lighting}.

The risk index system of traffic accidents of highway tunnel sections 2 and 3 is set up. The set of factors of accidents of sections 2 and 3 was set up as follows:

$V = \{v1, v2, v3, v4, v5\}$ = {Very good, Good, General, Poor, Very poor} = {5, 4, 3, 2, 1}.

2. Comprehensive evaluation of traffic accidents risk of tunnel group

As the different sections of the tunnel are interrelated and influence each other, the comprehensive evaluation value of the tunnel, which is determined by the "compromise generalized criterion", is adopted. The tunnel group is supposed to be consistent with t_i tunnels. The tunnel is set into three sections of the entrance, intermediate, and export. The evaluation values of the tunnel sections are (I = 1, 2 ... J = 1, 2, 3). For a group composed of a plurality of tunnel, "bucket theory" is used to determine the value of comprehensive evaluation of the tunnel group, as well as the tunnel traffic accident risk comprehensive evaluation values for all segments of the tunnel evaluation value minimum value:

$$V = \min(v_{ij})(i = 1, 2, \ldots; j = 1, 2, 3)$$

3. The highway tunnel traffic accident risk assessment

First, the traffic accident risk of the Dayangti ditch tunnel section 1 is assessed. According to the membership function of the fuzzy mathematics, the evaluation matrix of the four subfactors of the criterion layer is obtained from R_1, R_2, R_3, and R_4:

$$R_1 = \begin{pmatrix} 0 & 0 & 0.7 & 0 & 0 \\ 0 & 0.9 & 0 & 0 & 0 \\ 0 & 0 & 0.45 & 0 & 0 \end{pmatrix} R_2 = \begin{pmatrix} 0 & 0 & 0 & 1 & 0 \\ 0 & 0.9 & 0 & 0 & 0 \\ 0 & 0.55 & 0 & 0 & 0 \end{pmatrix}$$

$$R_3 = \begin{pmatrix} 0.49 & 0 & 0 & 0 & 0 \\ 0.55 & 0 & 0 & 0 & 0 \\ 0 & 0 & 1 & 0 & 0 \end{pmatrix} R_4 = \begin{pmatrix} 0 & 0.53 & 0 & 0 & 0 \\ 0 & 0 & 0.47 & 0 & 0 \\ 0.76 & 0 & 0 & 0 & 0 \end{pmatrix}$$

Weights of the four factors are:

$W_1 = (0.0910, 0.4545, 0.4545)$,
$W_2 = (0.6370, 0.1047, 0.2583)$,
$W_3 = (0.1047, 0.2583, 0.6370)$,
$W_4 = (0.6491, 0.0719, 0.2790)$.

A comprehensive evaluation of the four subfactors set is $B_i = W_i \bullet R_i$

$$R = \begin{pmatrix} B_1 \\ B_2 \\ B_3 \\ B_4 \end{pmatrix} = \begin{pmatrix} 0 & 0.4091 & 0.2862 & 0 & 0 \\ 0 & 0.2363 & 0 & 0.637 & 0 \\ 0.1934 & 0 & 0.637 & 0 & 0 \\ 0.21 & 0.344 & 0.034 & 0 & 0 \end{pmatrix}$$

As the weight of factor set U1 is $W = (0.1091, 0.1891, 0.3509, 0.3509)$, the two-level fuzzy comprehensive evaluation vector of Dayangti ditch tunnel section 1 traffic accident is given as follows:

$$B = WgR = (0.1416 \quad 0.2100 \quad 0.2623 \quad 0.2235 \quad 0)$$

As the two-level comprehensive evaluation vectors have been normalized, the weighted-average method is used as $v = \sum_{j=1}^{m} b_j v_j$.

The fuzzy comprehensive evaluation results of three-section traffic accidents of Dayangti ditch tunnel are shown in Table 3.

Similarly, the fuzzy comprehensive evaluation results of West ditch tunnel No. 1, West ditch tunnel No. 2, and Yangjuan Tunnel are shown in Table 4.

On the basis of the above calculation, the final fuzzy comprehensive evaluation result of Luan Chuan Highway Dayangti ditch tunnel group is $v = 3.0819$, which means that the accident risk for Luan Chuan Highway Dayangti ditch tunnel group in Luoyang city is "medium".

Table 3. Fuzzy comprehensive evaluation result of Dayangti ditch tunnel.

Section i	1	2	3
v_i	2.7819	3.1027	2.8536

Table 4. Fuzzy comprehensive evaluation result.

Tunnel	Dayangti ditch tunnel	West ditch tunnel 1#	West ditch tunnel 2#	Yangjuan tunnel
v	3.0819	3.0859	3.1028	3.1153

4 SUMMARY AND CONCLUSION

In this paper, the definition of the highway tunnel group is proposed for the first time. The AHP and fuzzy comprehensive evaluation methods are used to establish the highway tunnel group traffic accident risk evaluation model. The tunnel group safety factors are analyzed from three aspects, including human factors, car factors, and road factors. Taking the tunnel group of the Luan Chuan Highway Tunnel group in Luoyang City as an example, the results showed that the fuzzy comprehensive evaluation result of the Luan Chuan Highway Tunnel group is 3.0819, and traffic accident risk is "medium". The evaluation results have great practical significance for strengthening the safety operation of the tunnel group, which should improve the safety of highway engineering. It thus has good application value in the same projects.

ACKNOWLEDGMENTS

This study was partly supported by the Scientific Research Program Funded by Shaanxi Provincial Education Department (15JK1503, 15JK2168) and Research Funding of Xijing University (XJ150207).

REFERENCES

Chao Feng, Guo Chun, Shi Hongqian, Wang Mingnian. Research on Traffic Accident Risk Index of Impact on Wall in Curved Tunnel [J]. Chinese Journal of Underground Space and Engineering. 2012(S1): 1407–1410.

Kayastha, P, Dhital MR, De Smedt F. Application of the Analytical Hierarchy Process (AHP) for landslide susceptibility mapping: A case study from the Tinau watershed, west Nepal [J]. Computers & Geosciences, 2013, 52: 398–408.

Satty T.L. A scaling method for priorities in hierarchical structures [J]. Journal of Mathematical Psychology, 1978,1(1): 57–65.

Wang Shaofei. Classification of highway tunnel and Discussion on the concept of Highway Tunnel Group [J]. Highway Tunnel. 2009(02): 10–14.

Yi Fu-jun, Han Zhi, Deng Wei. Comprehensive safety evaluation method of highway tunnel group [J]. Journal of Chang'an University (Natural Science Edition). 2012(03): 79–85.

Zhang Yu-chun. Analysis on Risk Causation of Traffic Accidents in highway Tunnel Group [J]. China Safety Science Journal (CSSJ). 2009(09): 120–124.

Zhang Zuohai. Henan geological conditions of Luoyang Luanchuan highway [J]. Resources Environment & Engineering. 2009(S1): 37–40.

Zhou Na. Analysis and evaluation of traffic operation environment of highway Tunnel Group [D]. Chang'an University, 2010.

Civil, Architecture and Environmental Engineering – Kao & Sung (Eds)
 ISBN 978-1-138-02985-9

Applications of new reinforcement methods in city tunnels for shield launching and arriving

Hu Jun, Jiang Baoshi & Wei Hong
College of Civil Engineering and Architecture, Hainan University, Haikou, Hainan, China

Wang Xiaobin
Management Department of Changshu Institute of Technology, Changshu, Jiangsu, China

ABSTRACT: The reinforcement methods at shield shaft will be directly related to the security of shield's departure and reception. Choosing the correct reinforcement method in that phase is crucial. The common reinforcement method at shield shaft is summarized in this paper. The conclusion is as follows: the frequently used reinforcement method for the soft soil area is cement-soil deep mixing pile + high-pressure rotary jet grouting pile. The artificial freezing method is also used when the environment restrains reinforcement and water is the primary method in the soft soil area. When encountering a full-face rock formation, it is not necessary to reinforce and a slice grouting method can be used. A special method may be taken to the special stratum. Meanwhile, three new reinforcement methods are given. Results and conclusions in this research may play a useful role to aid designs of similar projects in the future.

1 INTRODUCTION

The 21st century is the underground engineering age, which is a peak time for the underground space development and utilization of China (Hu, 2013 & 2013). The shield method has been widely applied and rapidly developed, along with sea crossing, hydraulic, hydroelectric, and municipal engineering accompanied with the tunnel construction in the subway. However, the shield method has some construction difficulties and risks, with the entry and exit construction of shield tunnelling being one of the most difficult problems.

Reinforcement methods at shield shaft are important parts of the entry and exit construction technology. Accidents often take place at the end of shield tunnel, so the reinforcement effect of shield shaft is directly related to the safety of the entry and exit construction. Because it plays an important role in the shield tunnel, it is important to select a suitable reinforcement method to ensure that the project is going well (Hu, 2012; Hu, 2011). Based on the common reinforcement methods at the shield shaft and considering the safety, economy, and environment feasibility of the entry and exit construction, in this paper, a special reinforcement method will be used. This may play a useful role to the foundation treatment and soil reinforcement of similar projects in the future.

2 THE COMMON REINFORCEMENT METHOD AT SHIELD SHAFT

There are several reinforcement methods used in the shield shaft, such as the cement-soil deep mixing, high-pressure rotary jet grouting, SMW, artificial freezing, grouting, concrete filling pile, and water lowering methods. A single or a variety of methods can be used for the soil reinforcement (Zeng, 2011; Hu, 2012).

The reinforcement methods at the shield shaft can be divided into the following two categories:

I. Chemical reinforcement,
II. Physical reinforcement.

The cement-soil deep mixing pile + the high-pressure rotary jet grouting pile, SMW pile + high-pressure rotary jet grouting pile, and high-pressure rotary jet grouting pile are often used in the soft soil area. The cement-soil deep mixing pile + high-pressure rotary jet grouting pile has been widely used. The artificial freezing method is used when the environment is restricted on the ground.

SMW pile + high-pressure rotary jet grouting pile and high-pressure rotary jet grouting pile + deep well dewatering could be used in the sand soil area. The reinforcement + artificial freezing method can be used if serious water and sand leakage occurs while drilling or the water can't be pumped by dewatering the well.

A full-face rock formation has a strength far greater than mixing pile or rotary spray pile, which is usually between 0.8 MPa and 1.2 MPa. Therefore, it is not necessary to reinforce when encountering it. However, when fractures develop, a slice grouting method can be used in order to block fissure water. By controlling the grouting pressure to guarantee the effect of grouting seal, a slice-grouting hole is created in a 0.5 m line along the enclosure structure.

When constructing the entry and exit, a particular reinforcement method should be used on the special stratum, including highly weathered soft rock, medium-weathered soft rock, hard clay, and high-density sand layer. For example, the stratum in tunnel portal part is hard plastic silty clay layer between chalukou station and hedingqiao station for the Nanjing subway No. 1 line. Neither mixing pile nor jet grouting pile reinforcement is ideal. Finally, concrete filling pile + rotary spray water stopping are used to solve the problem. According to the actual construction site effects, because there was no leakage or deformation, the reinforcement method was proven to have its own suitability to the special stratum.

3 SEVERAL NEW REINFORCEMENT METHODS FOR SHIELD TUNNEL END

3.1 *Large-diameter cup-shaped horizontal freezing*

Mostly, the cup-shaped horizontal freezing is conducted based on 6.34 m shield diameter. However, the ends of a very large-diameter shield tunnel generally adopt several reinforcement methods. There are few cup-shaped horizontal freeing methods for the ends of the shield tunnel with a large or very large diameter.

This reinforcement method is mainly used for the ends of the shield tunnel with a large or very large diameter. For a shield tunnel with a large excavation diameter of 11.64 m, 122 horizontal freezing holes are designed in this layout. Freezing circle holes with Φ12.3 m are designed along the mouth of the cup body, with a hole pitch with the arc length of 0.76 m. A freezing pipe is often referred to an outer pipe whose length shall be shield host length + 2–3 times the segment width. The freezing circle holes with Φ9.9 m, Φ7.5 m, Φ5.1 m, and Φ2.7 m are designed at the cup bottom along the mouth with hole pitches of 1.11 m, 1.12 m, 1.14 m, and 1.21 m (arc length) respectively. The freezing pipe with circles of Φ9.9 m, Φ7.5 m, and Φ5.1 m along the mouth is called the middle pipe, with a total of 63 pipes. The freezing pipe with a Φ2.7 m circle along the mouth is called the internal pipe, with a total of 7 pipes. A central pipe has one freezing hole at the mouth. The freezing pipe at the cup bottom consists of the middle pipe, inner pipe and central pipe, and the length of freezing hole should ensure a thickness of 2 m of the cup bottom. Fig. 1 refers to the layout of the freezing hole.

One outer pipe, three middle pipes, one inner pipes and one central pile are used in this reinforcement method. The freezing pipe on the cup body is made from the outer pipe. The length of the freezing hole is the shield host length + 2–3 times the segment width. The cup bottom of the freezing pipe consists of the inner pipe, middle pipe and central pipe, with a freezing hole length to ensure a thickness of 2 m of the cup bottom. The 20# low-carbon seamless steel pipe with Φ127 × 5 mm can be used as a freezing pipe. It is connected by welding the outer pipe hoop and using the Φ48 mm × 3.5 mm steel pipe as a feed pipe. This method is used to ensure the formation of the sup-shaped freezing wall with an excavation diameter of 11.64 m at the ends of the shield tunnel with a large diameter, which allows the large-diameter shield tunnelling machine to smoothly get in and out, and to meet the strength and anti-seepage requirements.

3.2 *Combined method of cup-shaped horizontal grout and pipe shed reinforcement*

A combined construction method of the cup-shaped horizontal grout and pipe shed reinforcement is used for the shield tunnel end strata. This

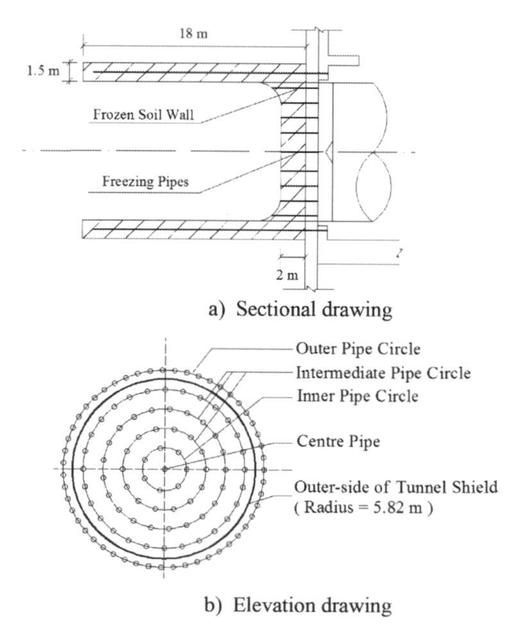

Figure 1. Layout of freezing holes.

method is based on a shield diameter of 6.34 m, with an advanced grouting long pipe shed by a row of 26 Φ108 mm × 5 mm seamless steel pipes 400 mm above the mouth of the working well. The four specifications of the segment length are 1.0 m, 1.5 m, 2.0 m and 3.0 m. Two fabricated interfaces of the pipe shed cannot be assembled at the same section. A hole should never be drilled within 1.0 m of the steel pipe at the retaining structure near the working well. Quincuncial grouting hole is drilled at the other parts of the pipe, with a diameter of 8 mm, 300 mm longitudinal pitch, and 4 rows of even drilling on the pipe body. A box and pin with a drilling casing is used to connect the steel pipe sections. 65 horizontal grouting holes are set in the mouth of working well, and then a Φ45 mm PVC sleeve valve pipe is used in order to form a cup-shaped horizontal grouting reinforcement. The diameters of the circle-grouting hole is 1.5 m (5, 5 m L), 2.7 m (7, 5 m L), 3.9 m (14, 5.2 m L), 5.1 m (14, 12 m L), and 6.1 m (25, 8.6 m L). The circle grouting hole is horizontal at the section of the shield tunnel with diameters 5.1 m and 6.1 m. The lower section is vertical and 13° down from the mouth. A circle grouting hole with a diameter of 3.9 m is vertical and 5° from the mouth. Fig. 2 refers to the layout of the cup-shaped horizontal grouting pipe and pipe shed.

This reinforcement method includes an advanced grouting long pipe shed reinforcement within 160° section at the shield section, as well as 65 horizontal sleeve valve pipes in the mouth for grouting reinforcement. When mixing piles, this reinforcement method can be adopted. While the ground, environment or serious water sand leakage occurs after the chemical reinforcement is used to improve the soil strength, stop water at the shield tunnel end, and to ensure the safety of the shield tunneling machine. This reinforcement method has advantages of a short construction period, low cost, and simple entry and exist operation, compared with reinforcement scheme of "drilling in working well and horizontal freezing" (like cup-shaped horizontal freezing method). Moreover, it can save time, labor, and material cost.

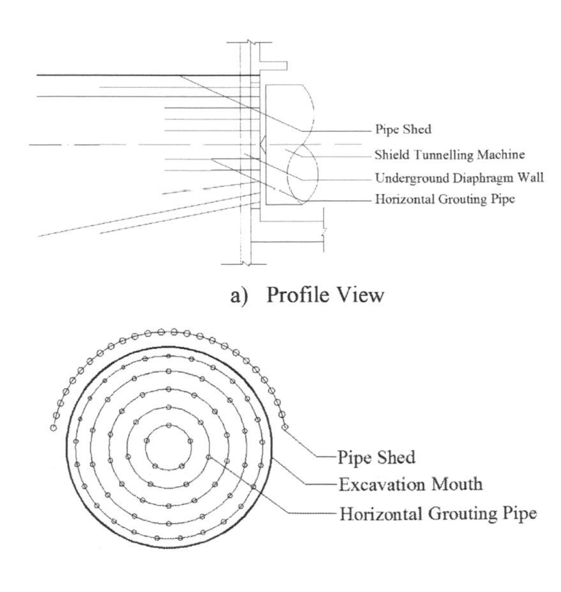

Figure 2. Layout of cup-shaped horizontal grouting pipe and pipe shed.

3.3 *Reinforcement method of glass fiber reinforced polymer diaphragm wall and reinforced concrete box*

Firstly, this reinforcement method completes the major structure of the shield well. It adopts the underground diaphragm wall as the retaining structure. The Glass Fiber Reinforced Polymer (GFRP) is used at the mouth of the slurry wall of the underground diaphragm wall. The shield tunnelling machine breaks through to substitute the bar of the underground diaphragm wall. After that, a reinforced concrete box is constructed before the mouth and sealed with the backfill. At the beginning, the shield tunneling machine is closed with backfill in the reinforced concrete box. Then the reinforced concrete box is full of backfill. Lastly, starting or reaching the direct cutting mouth by shield tunnel machine is completed. The shield tunnel machine directly cuts the mouth in the closed reinforced concrete box to finish starting, as well as to finish reaching.

GFRP can meet safety demands for the foundation ditch during the excavation stage of the shield well due to the glass fiber features. There is no need to set an outer reinforcement lateral to retain the structure at the excavation stage of the shield tunneling machine. Instead, the shield tunnel machine can directly cut the mouth slurry wall to cross the mouth. Not only can this reduce the incidence rate of the collapse accident resulted from the exposed mouth while excavating the mouth, but it can also save time of construction. Using backfill to seal the reinforced concrete box can balance the water and earth pressure in and out of the shield well foundation ditch, as well as avoid the flowing risk of the entry and exit of the shield tunneling machine.

Other similar materials which can substitute the Glass Fiber Reinforced Plastic bars (GFRP) will meet functional requirements of retaining the structure of the shield well, as well as direct cutting mouth demands by the shield tunneling machine after the substitution.

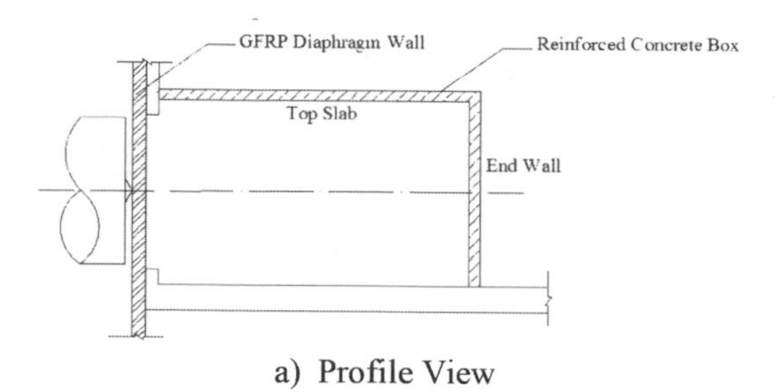

a) Profile View

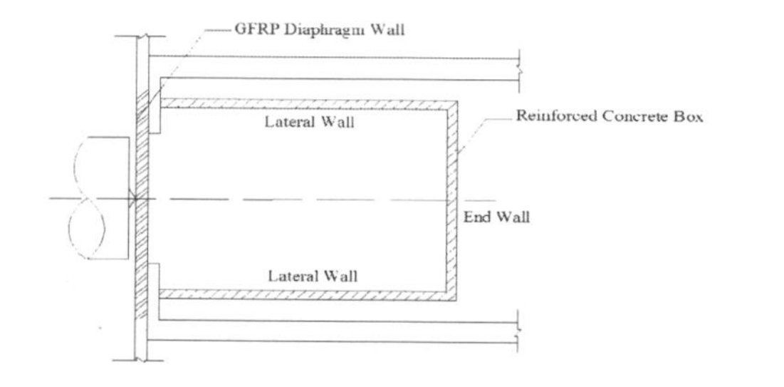

b) Elevation View

Figure 3. Layout of GFRP diaphragm wall and reinforced concrete box.

The reinforced concrete box in this reinforcement method can be substituted by other similar structures, which shall meet demands for balancing the water and earth pressure in and out of the shield well foundation ditch.

This method is applicable for the entry and exit construction of the shield tunneling machine at rock stratum or complex strata, for the hob of the earth pressure balance shield machine can cut the GFRP diaphragm wall. The shield tunnel machine with scraper cannot break the GFRP concrete. In order to avoid serious concrete blocking in the pipe, the slurry balance shield machine has to be controlled.

The layout of GFRP diaphragm wall and reinforced concrete box in this reinforcement method refer to Fig. 3.

4 CONCLUSIONS

1. The cement-soil deep mixing pile + high-pressure rotary jet grouting pile, SMW pile + high-pressure rotary jet grouting pile, and high-pressure rotary jet grouting pile are often used in the soft soil area. When the environment is restricted on the ground, the artificial freezing method is adopted.
2. SMW pile + high-pressure rotary jet grouting pile and high-pressure rotary jet grouting pile + deep well dewatering could be used in the sand soil area. It is essential that pumping tests are taken prior to reinforcement. Generally, the reinforcement should be completed before the dewatering well building in order to avoid high-pressure cement slurry to damage it.
3. It is not necessary to reinforce when encountering a full-face rock formation. However, if a rock fracture develops, a slice grouting method should be used. Some special method may be taken to the special stratum when the common reinforcement methods cannot meet the ideal effect. When the common reinforcement methods can't meet the ideal effects, a special method for the special stratum may be used.
4. The Large-Diameter Cup-Shaped Horizontal Freezing method is used to ensure the formation of the sup-shaped freezing wall at large-diameter shield tunnel end with 11.64 m excavation diameter, to meet requirements for strength and anti-seepage, and to smoothly allow the large diameter shield tunnelling machine to get in and out.
5. This reinforcement method is of short construction period, low cost and simple entry and exit operation when compared with reinforcement scheme of "drilling in working well and horizontal freezing" (like cup-shaped horizontal freezing method). Additionally, it has the ability to save time, labor, and material cost.
6. The advantages of Reinforcement Method of Glass Fiber Reinforced Polymer Diaphragm Wall and Reinforced Concrete Box are obvious. ① The construction period is shortened because no labor is needed to excavate the mouth. ② Engineering costs are reduced because no reinforcement is needed in the lateral retaining structure. ③ Retaining structure can directly stop soil and water at the entry and exist stage of the shield tunneling machine, and reinforced concrete box sealed with backfill can balance the water and earth pressure in and out of shield well mouth, therefore improving the safety of entering and exiting the construction of the shield tunnelling machine.

ACKNOWLEDGEMENTS

This research was supported by the National Science Foundation of China (51368017), the Key Research & Development Science and Technology Cooperation Program of Hainan Province (ZDYF2016226), Science and Technology Program of Hainan Province (ZDXM2015117), Postdoc Research Fund of China (2015M580559), the Scientific Research Project of Education Department of Hainan Province (Hnky2016ZD-7,

Hnky2015-10), the SRF for ROCS, MOHRSS (MOHRSS [2014] 240), the Research Project of Ministry of Housing and Urban Rural Development (2016-k5-060), the Scientific Research Starting Foundation of Hainan University (NO. kyqd1402), the Hainan Natural Science Foundation (20155214, 20155211) and the Academic Innovation Program of Hainan Science Association for the Youth scientific and technological excellence (201505).

REFERENCES

Hu Jun, Yang Ping, Dong Chaowen, et al. Study on Numerical Simulation of Cup-shaped Horizontal Freezing Reinforcement Project near Shield Launching [C]. 2011 International Conference on Electric Technology and Civil Engineering (ICETCE). China: IEEE. 2011, 4: 5522–5525.

Hu Jun, Zeng Hui, Wang Xiaobin. Numerical Analysis of Temperature Field of Cup-shaped Frozen Soil Wall Reinforcement at Shield Shaft [J]. Applied Mechanics and Materials, 2013, 341–342(5): 1467–1471.

Hu Jun, Zeng Hui, Wang Xiaobin. Study on Construction Risk Analysis and Risk Counter- measures of River-crossing Tunnel of Large-diameter Metro [J]. Applied Mechanics and Materials, 2012, 166(5): 2680–2683.

Hu Jun, Zeng Hui, Wang Xiaobin. Experimental Research on the Physi-mechanical Performances of Geosynthetics [J]. Applied Mechanics and Materials, 2013, 341–342(5): 33–37.

Hu, J. "Study on the Reinforcement Methods of Subway Large-diameter Shield Launching in the Sandy Clay with High Water Pressure", Thesis, Nanjing Forestry University, 2012. (in Chinese)

Zeng Hui, Hu Jun, Yang Ping. A Numerical Simulation Study on the Chemical Reinforcement Area at Shield Start Shaft [C]. 2011 International Conference on Electric Technology and Civil Engineering (ICETCE). China: IEEE. 2011, 4: 29–34.

Civil, Architecture and Environmental Engineering – Kao & Sung (Eds)
© 2017 Taylor & Francis Group, ISBN 978-1-138-02985-9

Research on the stress characteristics of concrete carrier pile based on finite element analysis

Jinyun Li
Tianjin College, University of Science and Technology Beijing, Tianjin, China

Aiqing Zhang & Xiangang Han
School of Civil and Environment Engineering, University of Science and Technology Beijing, Beijing, China

ABSTRACT: It is of great significance to describe the stress characteristics of concrete carrier pile in foundation reasonably and accurately for ensuring the safety of superstructures. On the basis of the finite-element numerical simulation method, cast-in-place pile is used as contrast to analyze the change rules of *Q-s* curve of carrier pile by applying axial pressure and lateral pressure repeatedly and verify the correctness of the rules through field test. The result indicates that the stress and displacement of the carrier pile are smaller than those of cast-in-place pile, stress concentration does not exist at the end of carrier pile, displacement change of the soil around the pile end is distributed symmetrically, the displacement at pile end is only 0.07 cm, and there is parabola change between the load increase and subsidence values. Static load test of a single carrier pile was designed in site, the stress reflection wave value of the test pile was measured, *Q-s* curve of the carrier pile is drawn, and the result of the numerical calculation verified the accuracy of numerical calculation.

1 INTRODUCTION

Carrier pile is capable of compacting foundation soil and enlarging the action area of pile end. When soft soil layer exists at the shallow soil layer of the construction site and there is a good soil layer with a certain thickness underneath, it has high application value. In recent years, it has substituted traditional foundation and ground treatment forms in medium and small architectural works as it has high bearing capacity and can adjust the uneven settlement to some extent to show good mechanical properties. According to the principle of the role of vector piles from analysis and research, a concise exposition of the principle of the piles with the force, design, and calculation of bearing capacity was given by Wang (2009). Li (2011) studied the bearing capacity and settlement of the ram-compaction piles with composite bearing. Hou (2015) studied the carrier pile bearing performance and engineering application using the method of field test and theoretical analysis. To explain the stress characteristics of carrier pile, in this paper, we carry out research on the stress of carrier pile and cast-in-place pile, conduct modeling, perform analysis on the bearing process of axial and lateral stresses by the two piles in foundation using large numerical simulation software, and draw the Q-s curve.

2 DESIGN OF THE SIMULATION SCHEME

2.1 *Constitutive model of pile and soil materials*

In this paper, concrete refers to a nonlinear material model with the nonlinear constitutive relation; generally, concrete plastic damage model can well describe the constitutive relation. However, in this paper, especially in pile–soil interaction, elasticity modulus of concrete is much greater than that of soil, and damage rarely occurs in vertical action.

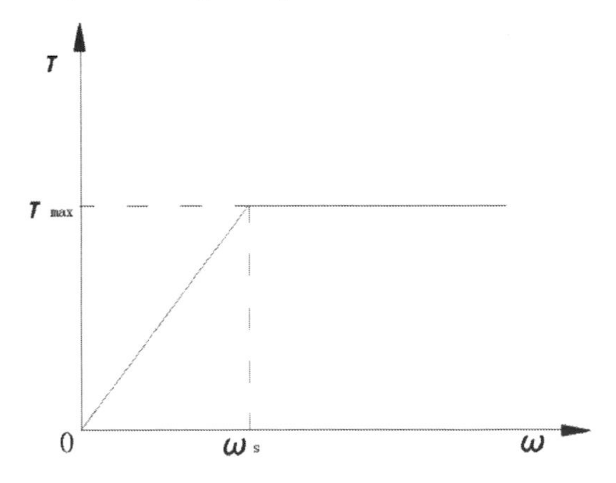

Figure 1. Elastic–plastic Coulomb friction model.

Thus, a linear elastic model is adopted for the constitutive relation of concrete in this paper. The model of soil often refers to Mohr–Coulomb plastic model and Drucker–Prager model. Both of the strength criteria exist in the constitutive relation in an implicit form.

ABAQUS contact simulation refers to point–surface contact analysis based on the surface, and the pile surface is taken as the main surface because of its great rigidity, while soil surface is taken as the auxiliary surface. The point–surface contact unit is composed of coordinate and contact surface. Before using such contact unit, there is no need to know the exact contact position in advance or keep consistent network among the contact surfaces, and large deformation and relative sliding are allowed. Coulomb friction model prevails as the contact constitutive relation, and the relationship between shearing stress and normal stress of the model contact surface is shown in Figure 1.

2.2 *Calculation model*

During modeling, the radius of the soil is 10 times that of the pile cross section and the soil of double pile length is extended below the pile end to minimize the influence of model boundary on the analysis area. The soil body influenced by the carrier of carrier pile is 2–3 m deep in the horizontal direction and 3–5 m deep in the vertical direction. Conduct pile modeling inside Part model, assemble the soil layer model in Assembly, set materials in Property, set analysis step in Step, set pile soil contact surface in Interaction, set birth–death element, set load in Load module, and divide mesh in Mesh module; after completing modeling, submit in Job module for post-treatment process. The

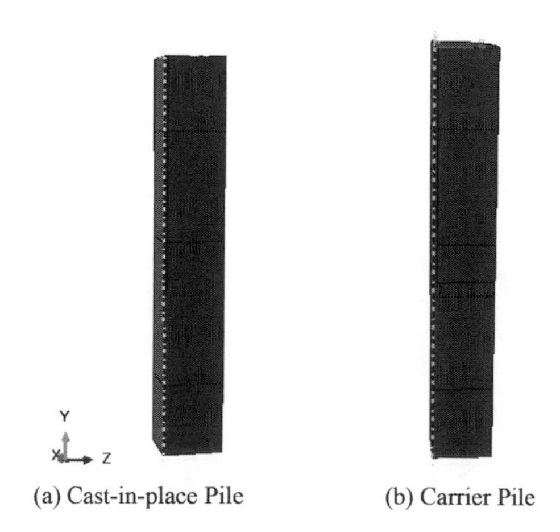

(a) Cast-in-place Pile (b) Carrier Pile

Figure 2. Models of cast-in-place pile soil and carrier pile soil.

Table 1. Summary of soil modeling parameters (1).

Name of the soil layer	Elasticity modulus (MPa)	Density (kg/m³)	Cohesion (kPa)
Plain fill	30.6	1700	25.3
Muddy clay	4.8	1800	19.6
Silty clay	26.6	1900	24.3
Wholly weathered rock	2600	2400	82.0

Table 1. Summary of soil modeling parameters (2).

Name of the soil layer	Internal friction angle (o)	Expansion angle (o)	Poisson's ratio
Plain fill	6.4	3	0.3
Muddy clay	16.5	8	0.3
Silty clay	17.2	8.5	0.3
Wholly weathered rock	28.8	0	0.2

model established is shown in Figure 2. Figure 2 (a) refers to cast-in-place pile soil model and Figure 2 (b) refers to carrier pile soil model.

To ensure that the model conforms to actual engineering, we conduct the survey and sampling of actual engineering geology of the place and obtain the summary of soil parameters required for modeling through indoor test, as shown in Table 1.

The order of soil layer is plain fill, muddy clay, silty clay, and wholly weathered work from the surface, with thicknesses of 6, 9, 12, and 6 m, respectively.

The two models established above are of the same soil-layer structures, and all soil body parameters are obtained from Table 1. The friction coefficient is the same as that of pile body, and the elasticity modulus of the concrete is 2.36×10^4 MPa.

2.3 *Load application*

Apply load to both of them, respectively, by 12 steps, and the applied load is N (measured in kN). To apply the load to the pile evenly, adopt local loading, that is, apply load with the converted P, wherein:

$$P = F / S \tag{1}$$

The loading mode is shown in Table 2.

As for cast-in-place pile, the loading mode is the same as that of concrete carrier pile; to distinguish both of them, apply load by 12 steps, and the loading model is shown in Table 3.

Table 2. Successive load application to concrete carrier pile (kN).

Pile type	1	2	3	4	5	6
Single-pile	250	500	750	1000	1250	1500
Model	62.5	125	187.5	250	312.5	375
P (kPa)	1274	2548	3822	5060	6369	7643
Pile type	**7**	**8**	**9**	**10**	**11**	**12**
Single-pile	1750	2000	2250	2500	2750	3000
Model	437.5	500	562.5	625	687.5	750
P (kPa)	8917	10191	11465	12739	14012	15287

Table 3. Successive load application to concrete cast-in-place pile (kN).

Pile type	1	2	3	4	5	6
Single-pile	50	100	150	200	250	300
Model	12.5	25	37.5	50	62.5	75
P (kPa)	255	510	764	1019	1273	1529
Pile type	**7**	**8**	**9**	**10**	**11**	**12**
Single-pile	350	400	450	500	550	600
Model	87.5	100	113	125	138	150
P (kPa)	1783	2038	2293	2547	2802	3057

2.4 *Calculation result*

Apply load to the carrier pile and cast-in-place pile separately as per the loading mode above, and read the data after each load application. Considering the limit to the paper length, this study only conducts analysis on the stress diagram and displacement diagram of a certain point.

2.4.1 *Stress diagram*

It is evident from Figure 3 (a) and (b) that, under the same vertical load effect, the maximum stress borne by the cast-in-place pile is 904.4 kPa and that borne by the carrier pile is 484.2 kPa. Stress concentration occurs to the end of cast-in-place pile, which indicates that the soil has punching shear destruction, thus being against bearing upper load. While the stress at the end of the concrete carrier pile is distributed evenly and a stress vacuum area appears on the soil layer of the carrier, that is, the stress borne by the carrier pile is much less than that borne by the cast-in-place pile because of the interaction between carrier bottom and soil.

2.4.2 *Displacement diagram*

It is evident from Figure 4 (a) and (b) that, under the same vertical load effect, a certain displacement occurs to the soil around the cast-in-place pile body and the pile end, the displacement distribution of soil around the pile end is uneven, and the pile end displacement is 0.27 cm. Displacement also occurs to the soil around the carrier pile body and the pile end; however, the displacement change of soil around the pile end is symmetrically distributed, and the pile end displacement is only 0.07 cm. It is shown through comparison that the displacement of carrier pile is less than that of the cast-in-place pile.

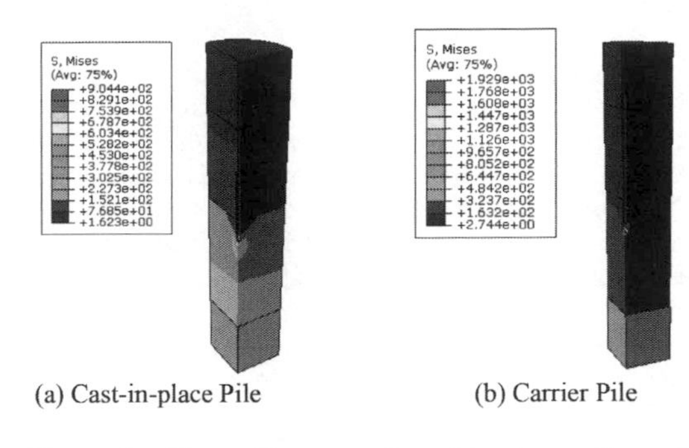

(a) Cast-in-place Pile (b) Carrier Pile

Figure 3. Stress diagram of cast-in-place pile and carrier pile under the 300-kN load effect.

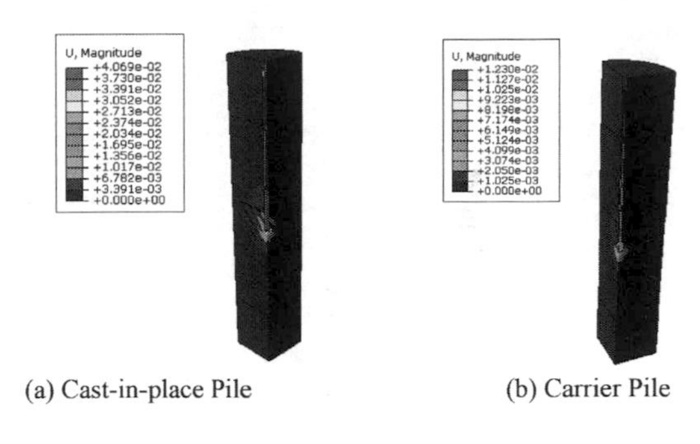

(a) Cast-in-place Pile (b) Carrier Pile

Figure 4. Displacement diagrams of cast-in-place pile and carrier pile under the 300-kN load effect.

2.4.3 *Q-s curve*

Because it is difficult to calculate the pile lateral friction at present, in this paper, we take the settlement curve of pile body as reference during the numerical simulation process, judge the ultimate bearing capacity of pile according to specification, and draw the numerical calculation results to Q-s curve of the carrier pile and the cast-in-place pile, as shown in Figures 5 and 6. During the numerical calculation process, when the cast-in-place pile is loaded to 600 kN, the corresponding Q-s curve has no obvious turning change. Thus, the load is increased to 700 and 800 kN, and the corresponding P values are 3,567 and 4,076 kPa.

It is evident from Figure 5 that the Q-s curve is approximately parabolic and the curve change is relatively gentle. Subsidence value of the carrier pile increases with the increase of load; when the load exceeds 2,500 KN, the soil at the carrier bottom is damaged; then, settlement increases rapidly. Such phenomenon occurs because when bearing load, the carrier pile relies on the friction force between the pile and soil, and the carrier at the bottom of carrier pile restricts the sinking of pile body.

Through observation of Figure 6, Q-s curve is in a steep fall shape. Subsidence value of the cast-in-place pile also increases with the increase of load; before the load reaches 500 kN, the sinkage and load value vary linearly; after the load exceeds 700 kN, the curve decreases sharply. Such phenomenon occurs because that at the beginning of load application, the friction force between the pile and soil is the main source of bearing capacity; after the friction force reaches the limit, the pile end bears the load, but the bearing capacity of pile end is limited, which is less than the bearing load of friction force.

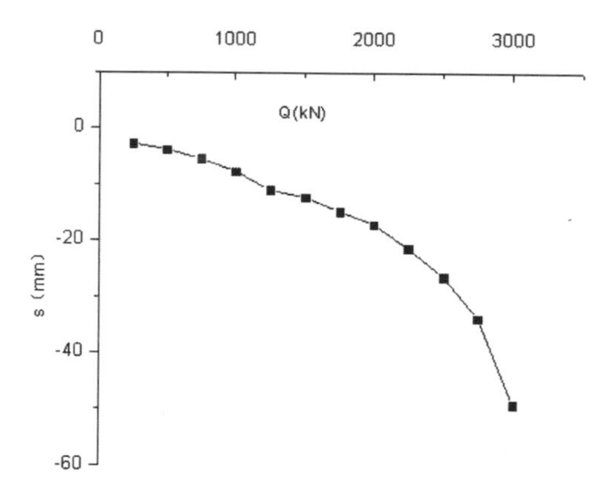

Figure 5. Q-s curve of the carrier pile.

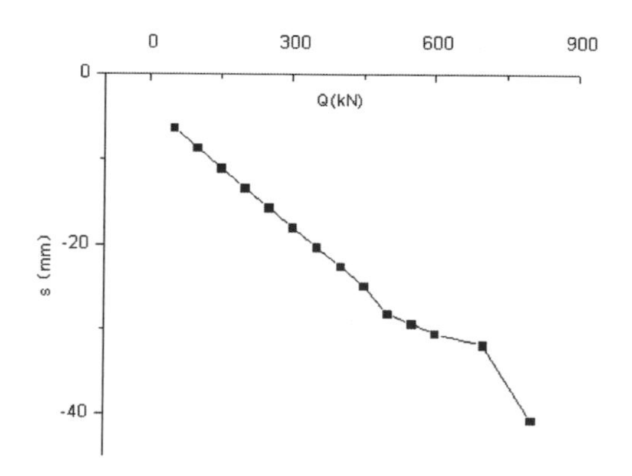

Figure 6. Q-s curve of the cast-in-place pile.

A comparison of Figures 5 and 6 shows that although the subsidence value of both of the piles increases with the increase of load, the variation type of the Q-s curve is different. If the corresponding load values of the same subsidence value are selected for comparison, the load borne by the carrier pile is larger, which indicates that the carrier pile bears a higher load than the cast-in-place pile, and there is parabolic variation rule between the load increase and subsidence value.

3 FIELD LOADING TEST

3.1 *Static load test design of carrier pile*

Select an engineering carrier pile to conduct vertical static load test of single pile and adopt 2h load-keeping method at each step in reference with domestic and international engineering cases. During the testing of 1# test pile, there are 12 steps of load application in total, the first step a 240 kN load is applied, then in each of the subsequent steps, a 160 kN load is applied. The maximum load value is 2,000 kN; after each step of loading, read every 5, 10, 15, 15, 15, 15, 15, 15, and 15 min, and the accumulated time of load keeping is 24h. During the test of 2#, 3#, and 4# test piles, there are 12 steps of load application in total. In the first step, a load of 250 kN is applied; then, in each of the subsequent steps a 250 kN load is applied; after each step of loading, read every 5, 10, 15, 15, 15, 15, 15, 15, and 15 min, and the accumulated time of load keeping is 24h. The load test of the field carrier pile is shown in Figure 7.

3.2 *Static load test results of carrier pile*

The stress wave reflection method is used in site for measuring the pile, which assumes the medium as a one-dimensional section homogeneous bar with continuous elasticity, without taking into account

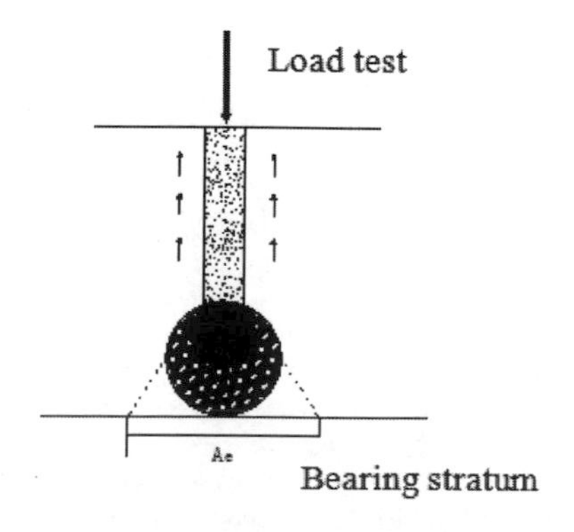

Figure 7. Load test of the field carrier pile.

Table 4. Analysis results of measured curve of reflection wave for test pile (1).

Test pile no.	Construction no.	Pile length (m)	Pile diameter (mm)
1	1	17.00	500
2	2	17.00	500
3	3	17.00	500
4	4	17.00	500

Table 4. Analysis results of measured curve of reflection wave for test pile (2).

Test pile no.	Wave velocity (m/s)	Analysis of pile body integrity
1	3300	III
2	3300	II
3	3300	I
4	3300	I

Table 5. Load–subsidence corresponding analysis of 2#, 3#, and 4# test piles at each step.

Step	Load (kN)	Accumulated subsidence (mm)		
		2# test pile	3# test pile	4# test pile
1	250	0.68	0.64	0.79
2	500	1.78	1.05	1.47
3	750	3.04	1.82	2.07
4	1000	4.63	2.84	2.71
5	1250	7.35	3.48	3.54
6	1500	11.26	4.62	4.56
7	1750	14.78	6.92	5.92
8	2000	20.38	9.83	8.06
9	2250	27.63	12.45	11.33
10	2500	32.54	15.86	17.32
11	2750	37.63	18.46	25.42
12	3000	46.55	21.37	34.92

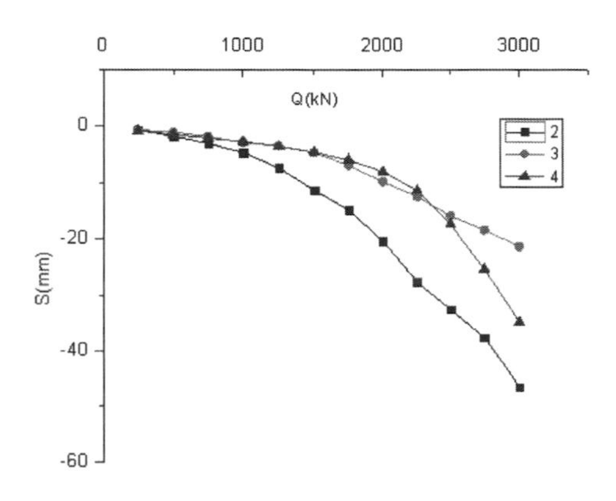

Figure 8. Q-s curve of 2#, 3#, and 4# test piles.

the influence of substances around the medium on the stress wave, which is propagated along the medium. The property of pile body defect and pile end can be exactly judged according to the phase, amplitude, and frequency characteristic of the reflection wave as well as the stratum data, construction record, and practical analysis experience. The measured curve of the reflection wave for test pile and the analysis results of the measured curve of reflection wave for the test pile are shown in Table 4.

It can be known from Table 4 that an obvious defect exists in 1# pile body, which does not meet the specification requirements; thus, the static load test data results of 2#, 3#, and 4# test piles are selected as the test data of the carrier pile. Specific results are shown in Table 5.

The data of Table 5 is drawn as a Q-s curve chart, as shown in Figure 8.

It is evident from Figure 8 that Q-s curve variation trends of the three test piles are the same, all referring to a parabolic change. The variation of 2# test pile is more obvious than that of the 3# and 4# test piles, which is mainly resulted from poor integrity of the pile. However, overall variation rules of the three test piles are the same.

4 CONTRASTIVE ANALYSIS ON FIELD LOAD TEST RESULTS AND NUMERICAL SIMULATION RESULTS

The results of a comparison of the field load test results presented above with the ABAQUS numerical simulation results are shown in Figure 9. It is evident from the figure that the numerical simulation results are similar to the field test results, which indicates that the modeling of numerical simulation and the selection of parameters and loading process are accurate, thus reflecting the actual situation of the field. Therefore, the Q-s curve variation rules of the carrier pile obtained through numerical simulation are correct and authentic.

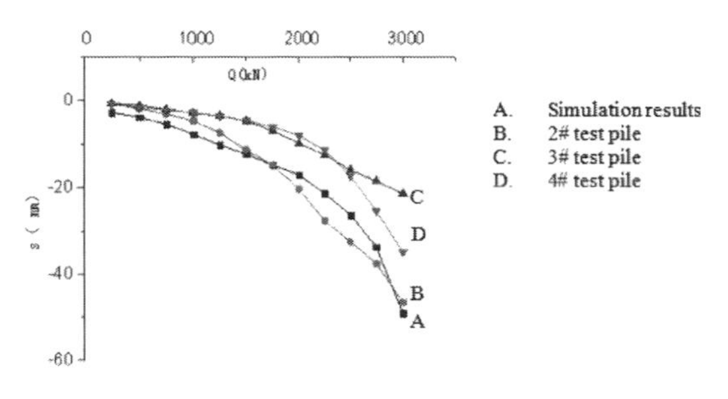

Figure 9. Q-s curve of field test piles and simulation results.

5 CONCLUSION

By comparing the stress, displacement, and Q-s curve of the carrier pile with those of the cast-in-place pile under the same vertical load effect via numerical simulation, it can be concluded that the stress and displacement of the carrier pile are smaller than those of the cast-in-place pile, the displacement change of the soil around the carrier pile end is distributed symmetrically, and the displacement at the pile end is only 0.07 cm, being approximately steady. The carrier pile bears a higher load at the same subsidence value, and there is parabolic variation rule between the load increase and subsidence value. The static load test of a single carrier pile is designed in site, the value of stress reflection wave of the test pile is measured, the Q-s curve of the carrier pile is drawn, and the result agrees with the numerical calculation result. These observations verify that the Q-s curve variation rule of the carrier pile obtained through numerical simulation is accurate.

REFERENCES

Hou Xiaojun. (2015). *Application of Carrier Pile in a Residential Project in Beijing.* Degree. Beijing: China University of Geosciences (Beijing).

Jizhong Wang. (2009). Stress Mechanism and Technical Innovation of Carrier Pile. *J. Engineering Exploration.* 1: 5–8.

Rui Li. (2011). *Research on Composite Foundation of Carrier Pile.* Degree. Xi'an: Xi`an University of Architecture and Technology.

Yachao You & Xiaomeng Li. (2009). Application of Carrier Pile in a Certain Actual Engineering. *J. Architecture.* 2: 119–120.

Yingshe Su. (2011). Application of Carrier Pile Technology in Collapsible Loess Area. *J. Architectural Science.* 27(3): 97–99.

Yongyang Zhang & Xinchao Chen. (2008). Research on Vertical Ultimate Bearing Capacity of Compound Carrier Tamping Extended Pile. *J. Resources Environment & Engineering.* 22(10): 223–225.

Zhuo Zhuang & Xiaochua You n & Jianhui Liao, et al. (2009). *Finite Element Analysis and Application Based on ABAQUS.* Beijing: Tsinghua University Press.

Civil, Architecture and Environmental Engineering – Kao & Sung (Eds)
© 2017 Taylor & Francis Group, ISBN 978-1-138-02985-9

Theoretical analysis of the deformation of shield tunnel segment under fire

Xingbo Han, Yongxu Xia & Lunlei Chai
School of Highway, Chang'an University, Xi'an, China

ABSTRACT: Shield tunnel segment would be exposed to elevated temperature when fire occurs. The elevated temperature will decline the mechanical properties of the segments as well as complicate their deformation and force analysis. The impact of elevated temperature is not fully considered in most studies on the deformation of shield segments. This paper presents a theoretical analysis on the deformation of the segments subjected to elevated temperature. Considering the nonlinear segment section thermal degradation and mechanical property declination, a somewhat generic theoretical calculation model for the segment deformation is established. On the basis of the Free-Form Deformation theory, an analytical solution via the model is then proposed. Moreover, the application of the formulation is illustrated by presenting a case study of Nanjing Metro tunnel segments. The result shows that, when fire lasted 2 h, the maximum deflection reached to 79 mm at the segment crown. The upper half of the segment shows a significant downtrend, whereas its lower half rises. The maximum horizontal deflection occurs at 60 mm down from the crown and reaches to 45 mm at 2 h. Compared with the Chinese code for the design of road tunnel, it would not be safe enough for the escape or rescue of people after the fire lasted 40 min.

1 INTRODUCTION

Tunnel fire is an irresistible rare event, which would bring great damage to both human life and the tunnel structure. The influence of fire on the tunnel lining has been studied wisely. Recently, shield tunnel construction has dramatically increased in China because of its advantage of high mechanization and very little influence on the environment. Thus, the influence of fire on shield tunnel has become a central issue in tunnel disaster release and proof.

Mechanical behavior of a beam subjected to elevated temperature has been widely studied in bridge and construction engineering (Huang et al. 1999, Huang et al. 2000, Bailey 2004). Compared with the normal beam (Heidarpour & Bradford 2009, Bradford 2001, Bradford et al. 2006, Kruppa & Zhao 1995), only few studies have been conducted on the tunnel lining—a curve beam (Pi et al. 2002, Pi & Bradford 2004, Heidarpour et al. 2010). As to the tunnel lining, the thermal degradation along the tunnel section, and lining thickness, the external loads and the boundary condition are more complex than those of a bridge or building beam.

There still have been some studies on the influence of fire on a lining. Several laboratory experiments were carried out by many researchers (Yan et al. 2013, Yan et al. 2015, Lai et al. 2014, Guo 2013, Yan 2007, Choi et al. 2013). Their experiments are mostly concentrated on spalling of the lining concrete or the force of a single segment. Very few studies have been done on the deformation of the whole ring of the segment. Numerical simulation is also well used in studying the influence of fire on lining (Guo 2013, Kodur 2008).

Compared with experiment or numerical simulation, theoretical analysis is relatively weak. A theoretical result of pinned arches under thermal loading was given by Heidarpour (2010). Zhongyou Li (2012) also established a theoretical analysis model of the lining under fire loading. The influence of fire on lining is significantly made clear by this model, but its boundary condition is inconsistent with the real project.

In this paper, we present a mechanical analytical formulation for the deformation of segment ring at elevated temperatures in which the segment ring was treated as a circle. With the symmetry of geometric model and external loads, only half of the ring was analyzed. The calculating model is restrained at its bottom. The cross-section axil force and bending moment are obtained via the free deformation theory by considering external loads such as vertical pressure of soil, lateral pressure of soil, segment self-weight, water pressure, and subground resisting force. Following the segment deformation, regulation under fire situation is obtained from the case study.

2 ANALYTICAL MODEL

The nomenclatures used in this paper are listed as follows:

$\overline{M}$	Bending moment at segment section
$\overline{N}$	Axial force in the segment section
M	Bending moment considering the eccentricity caused by temperature
N	Axial force considering the eccentricity caused by temperature
k	Coefficient of thermal expansion of concrete
E_T	Moduli of elasticity of segment
$T(y)$	Function of segment cross-section temperature distribution
$\alpha_1 \ldots\ldots \alpha_4$	Parameters of the calculating process
y_0	Location of neutral axis of the segment
y	Eccentricity caused by temperature
h	Segment thickness
T_t	Segment temperature of the left-hand side
T_b	Segment temperature of the right-hand side
R	Segment radius
φ	Angle from segment crown to any section
$\theta(\varphi)$	Rotation
$\upsilon(\varphi)$	Radial deflection
$u_x(\varphi)$	Horizontal deflection
$u_y(\varphi)$	Vertical deflection
P_i	Horizontal pressure
q_1	Soil pressure
P_R	Soil resistance
X_i	Unknown redundant force
δ_{ij}	Flexibility coefficient
Δ_{ij}	Unit load deflection
g	Segment self-weight
h	Underground water depth

According to Zhi-guo Yan et al. (2013, 2015), the thermal degradation of a segment section under fire is not a line but a curve, as shown in Fig. 1.

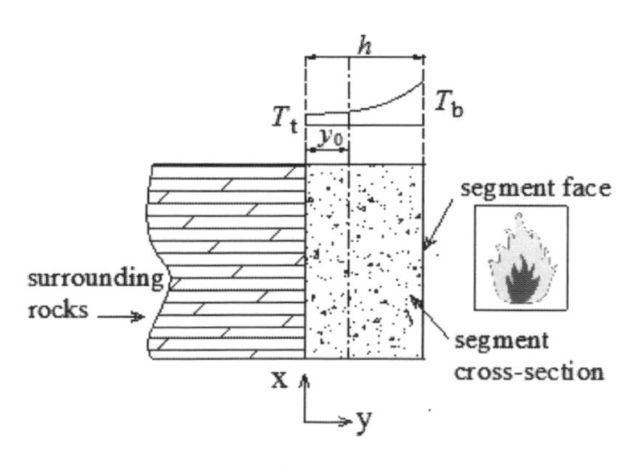

Figure 1. Temperature distribution of lining cross section.

Current experiment shows that concrete elastic modulus decreases and maximum strain increases when subjected to elevated temperature. However, it could still meet the elastic deformation condition before its yield. The concrete could meet the elastic deformation rule before its yield. Hence, the assumptions of this model are:

1. The metric is still in the range of elastic deformation;
2. The section deformation meets the plane cross-section assumption.

As shown in Fig. 1, Tb is the fire close face of the segment, Tt is the fire away face, and h is the segment thickness. According to assumption (1), the strain of segment section can be expressed as:

$$\varepsilon = (y - y_0)\kappa \tag{1}$$

The elevated temperature would lead to a thermal expiation of concrete. The strain caused by thermal expiation is:

$$\varepsilon_T = kT(y) \tag{2}$$

Using Eqs. (1) and (2), the elastic strain of the segment section can be written as:

$$\varepsilon_e = \varepsilon - \varepsilon_T = (y - y_0)\kappa - kT(y) \tag{3}$$

According to assumption (2), ignoring the temperature distribution along tunnel drive direction, the stress of the segment section is:

$$\sigma = E_T\varepsilon_e = E_T\left[(y - y_0)\kappa - kT(y)\right] \tag{4}$$

The axial force (N) in the concrete typical cross section can be obtained by integrating the stresses as:

$$N = \int_0^h E_T\left[(y - y_0)\kappa - kT(y)\right]\mathrm{d}y \tag{5}$$

Eq. (5) leads to a relationship between the total internal axial force and the location of the neutral axis in each part of the cross section as:

$$y_0 = \frac{\kappa\int_0^h E_T y\mathrm{d}y - k\int_0^h E_T y\mathrm{d}y - N}{\kappa\int_0^h E_T\mathrm{d}y} \tag{6}$$

Using Eq. (4), the typical cross-section bending moment can be expressed as:

$$M = \int_0^h ET\left[(y - y_0)\kappa - kT(y)\right]y\mathrm{d}y \tag{7}$$

The curvature (κ) at the cross section can be formulated as a function of the total internal moment and axil force using Eq. (7) as:

$$\alpha_1 M - \alpha_2 N - (\alpha_3 - \alpha_4)k = \kappa \tag{8}$$

where $\alpha_1, \alpha_2, \alpha_3$ and α_4 are defined as:

$$\left.\begin{aligned}
\alpha_1 &= \frac{\int_0^h E_T \mathrm{d}y}{\int_0^h E_T \mathrm{d}y \int_0^h E_T y^2 \mathrm{d}y - \left(\int_0^h E_T y \mathrm{d}y\right)^2} \\
\alpha_2 &= \frac{\int_0^h E_T y \mathrm{d}y}{\int_0^h E_T \mathrm{d}y \int_0^h E_T y^2 \mathrm{d}y - \left(\int_0^h E_T y \mathrm{d}y\right)^2} \\
\alpha_3 &= \frac{\int_0^h E_T y \mathrm{d}y \int_0^h E_T T(y) \mathrm{d}y}{\int_0^h E_T \mathrm{d}y \int_0^h E_T y^2 \mathrm{d}y - \left(\int_0^h E_T y \mathrm{d}y\right)^2} \\
\alpha_4 &= \frac{\int_0^h E_T \mathrm{d}y \int_0^h E_T T(y) y \mathrm{d}y}{\int_0^h E_T \mathrm{d}y \int_0^h E_T y^2 \mathrm{d}y - \left(\int_0^h E_T y \mathrm{d}y\right)^2}
\end{aligned}\right\} \tag{9}$$

The cross-section bend angle can be represented by the sectional curvature (Bradford MA, 2006), which can be computed as:

$$\theta(s) = \int \left[\alpha_1 M - \alpha_2 N - (\alpha_3 - \alpha_4)k\right] \mathrm{d}s \tag{10}$$

As to the circular tunnel segments, Eq. (1) can be further expressed as:

$$\theta(\varphi) = \int \left[\alpha_1 M - \alpha_2 N - (\alpha_3 - \alpha_4)k\right] R \mathrm{d}\varphi \tag{11}$$

The radial deformation can be determined from:

$$v(\varphi) = \int \theta(\varphi) R \mathrm{d}\varphi \tag{12}$$

Further, the horizontal and vertical deformations can be calculated as:

$$\left.\begin{aligned}
u_x(\varphi) &= v(\varphi)\cos(\varphi) \\
u_y(\varphi) &= -v(\varphi)\sin(\varphi)
\end{aligned}\right\} \tag{13}$$

Considering the decrease of elasticity modulus with temperature, the axial force would cause an additional bending moment. Therefore, the axial force and bending moment can be expressed as:

$$N = \overline{N}; M = \overline{M} + \overline{N}\overline{y} \tag{14}$$

where $\overline{y}$ is the location of the neutral axis which is given by:

$$\overline{y} = \frac{\int_0^h E_T y \mathrm{d}y}{\int_0^h E_T \mathrm{d}y} - \frac{h}{2} \tag{15}$$

3 SEGMENT DEFORMATION ANALYSIS

The theoretical segment deformation calculating result is established on the basis of free deformation method and time-dependent material property.

3.1 *Free-form deformation method*

The hypothesis of this calculating result is that the tunnel segment is built in a soft sandy soil or cohesive soil; thus, the soil would not produce a pronounced resistance to the segment. Under this hypothesis, the tunnel segment can be treated as a deformation-free homogeneous circle. The main loads applied on the segments are soil pressure above the segment (q_1), soil horizontal pressures (p_1 and p_2), and soil resistance (p_R).

The segment is a three-time redundant-degree structure, whose internal forces can be solved by force method. The segment structure and its loads are symmetrical with the segment vertical axils; thus, the shearing force along the symmetrical face is 0. Without the shearing force, only two forces are redundant. The basic structure calculating diagrams are shown in Fig. 2. The bottom of the circle is fixed, and the unknown forces, X1 and X2, are moved to the elastic central. The flexibility coefficient δ_{12} equals 0 and then the equation of displacement coordination is expressed as:

$$\left.\begin{aligned}
X_1\delta_{11} + \Delta_{1P} &= 0 \\
X_2\delta_{22} + \Delta_{2P} &= 0
\end{aligned}\right\} \tag{16}$$

The angle between the lining section and the vertical axial force are supposed to equal φ. According to the free-form deformation method, the axial

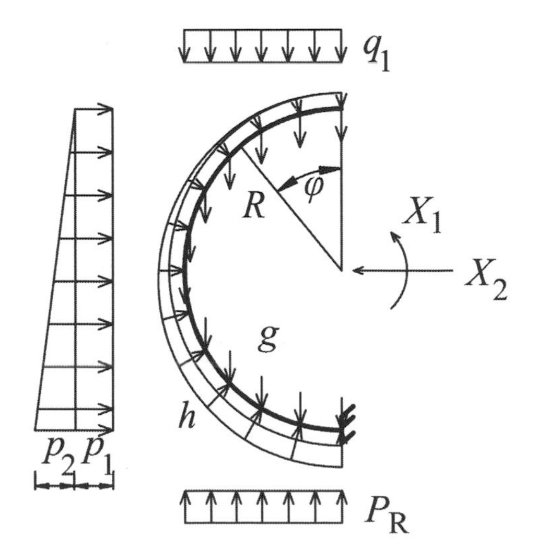

Figure 2. Structure load distribution of the segment.

force and bending moment of any of this section can be solved by:

$$\left.\begin{aligned}\bar{M} &= M_P + X_1 - X_2 R_h \cos\varphi \\ \bar{N} &= N_P + X_2 \cos\varphi\end{aligned}\right\} \tag{17}$$

in which M_P and N_P are expressed by:

$$M_P = \begin{cases} -q_1 R^2 \dfrac{\sin^2\varphi}{2} - p_1 R^2 \dfrac{(1-\cos\varphi)^2}{2} \\ -p_2 R^2 \dfrac{(1-\cos\varphi)^3}{12} R^3 \\ -gR^2(\varphi\sin\varphi + \cos\varphi - 1) \\ +(\cos\varphi + 0.5\varphi\sin\varphi - 1) & \left(0 \le \varphi < \dfrac{\pi}{2}\right) \\ q_1 R^2\left(\dfrac{1}{2} - \sin\varphi\right) - p_1 R^2 \dfrac{(1-\cos\varphi)^2}{2} \\ -p_2 R^2 \dfrac{(1-\cos\varphi)^3}{12} - p_R R^2 \dfrac{(1-\sin\varphi)^2}{2} \\ -gR^2(\varphi\sin\varphi + \cos\varphi - 1) \\ +(\cos\varphi + 0.5\varphi\sin\varphi - 1)R^3 & \left(\dfrac{\pi}{2} \le \varphi < \pi\right) \end{cases} \tag{18}$$

$$N_P = \begin{cases} q_1 R\sin^2\varphi - p_1 R(1-\cos\varphi)\cos\varphi \\ -p_2 R\dfrac{(1-\cos\varphi)^2}{4}\cos\varphi + gR\varphi\sin\varphi \\ +R^2 - 0.5R^2\varphi\sin\varphi - R^2\cos\varphi + hR & \left(0 \le \varphi < \dfrac{\pi}{2}\right) \\ q_1 R\sin\varphi - p_1 R(1-\cos\varphi)\cos\varphi \\ -p_2 R(1-\cos\varphi)^2 \dfrac{\cos\varphi}{4} \\ -p_R R(1-\sin\varphi)\sin\varphi + gR\varphi\sin\varphi \\ +R^2 - 0.5R^2\varphi\sin\varphi - R^2\cos\varphi + hR & \left(\dfrac{\pi}{2} \le \varphi < \pi\right) \end{cases} \tag{19}$$

Eq. (20) resulted from structural mechanics and expressed as:

$$\left.\begin{aligned} X_1 &= -\frac{\Delta_{1P}}{\delta_{11}} = -\frac{1}{\pi}\int_0^{\pi} M_P \mathrm{d}\varphi \\ X_2 &= -\frac{\Delta_{2P}}{\delta_{22}} = \frac{2}{\pi R}\int_0^{\pi} M_P \cos\varphi \mathrm{d}\varphi \end{aligned}\right\} \tag{20}$$

Eq. (20) can be expanded by Eqs. (18) and (19) so that the unknown forces (X_1 and X_2) can be written as a function of applied loads like

$$\left.\begin{aligned} X_1 &= -q_1 R^2\left(\frac{1}{8} - \frac{1}{\pi}\right) + p_1 R^2 \frac{3}{4} + p_2 R^2 \frac{5}{24} \\ &\quad + p_R R^2\left(\frac{3}{8} - \frac{1}{\pi}\right) + gR^2 + \frac{R^3}{2} \\ X_2 &= -q_1 R\frac{1}{3\pi} + p_1 R + p_2 R\frac{5}{16} \\ &\quad + p_R R\frac{1}{3\pi} - \frac{gR}{2} + \frac{3R^2}{4} \end{aligned}\right\} \tag{21}$$

Eq. (17) can be re-expressed using Eqs. (18), (19), and (21) as follows:

$$\bar{M} = \begin{cases} -q_1 R^2 \dfrac{\sin^2\varphi}{2} - p_1 R^2 \dfrac{(1-\cos\varphi)^2}{2} \\ -p_2 R^2 \dfrac{(1-\cos\varphi)^3}{12} - q_1 R^2\left(\dfrac{1}{8} - \dfrac{1}{\pi}\right) + \dfrac{3}{4}p_1 R^2 \\ +\dfrac{5}{24}p_2 R^2 + p_R R^2\left(\dfrac{3}{8} - \dfrac{1}{\pi}\right) \\ +\left(\dfrac{q_1}{3\pi} - p_1 - \dfrac{5}{16}p_2 - \dfrac{p_R}{3\pi}\right)R^2\cos\varphi & \left(0 \le \varphi < \dfrac{\pi}{2}\right) \\ +gR^2\left(2 - \dfrac{1}{2}\cos\varphi - \varphi\sin\varphi\right) \\ -R^3\left(\dfrac{1}{2} - \dfrac{1}{4}\cos\varphi - \dfrac{1}{2}\varphi\sin\varphi\right) \\ -q_1 R^2\left(\sin\varphi - \dfrac{1}{2}\right) - p_1 R^2 \dfrac{(1-\cos\varphi)^2}{2} \\ -p_2 R^2 \dfrac{(1-\cos\varphi)^3}{12} - p_R R^2 \dfrac{(1-\sin\varphi)^2}{2} \\ -q_1 R^2\left(\dfrac{1}{8} - \dfrac{1}{\pi}\right) + \dfrac{3}{4}p_1 R^2 + \dfrac{5}{24}p_2 R^2 & \left(\dfrac{\pi}{2} \le \varphi < \pi\right) \\ +\left(\dfrac{q_1}{3\pi} - p_1 - \dfrac{5}{16}p_2 - \dfrac{p_R}{3\pi}\right)R^2\cos\varphi \\ +p_R R^2\left(\dfrac{3}{8} - \dfrac{1}{\pi}\right) + gR^2\left(2 - \dfrac{1}{2}\cos\varphi - \varphi\sin\varphi\right) \\ -R^3\left(\dfrac{1}{2} - \dfrac{1}{4}\cos\varphi - \dfrac{1}{2}\varphi\sin\varphi\right) \end{cases} \tag{22}$$

$$\bar{N} = \begin{cases} q_1 R\sin^2\varphi - p_1 R(1-\cos\varphi)\cos\varphi \\ -p_2 R\dfrac{(1-\cos\varphi)^2}{4}\cos\varphi \\ -\left(\dfrac{q_1}{3\pi} - p_1 - \dfrac{5}{16}p_2 - \dfrac{p_R}{3\pi}\right)R\cos\varphi & \left(0 \le \varphi < \dfrac{\pi}{2}\right) \\ +gR(\varphi\sin\varphi - \dfrac{1}{2}\cos\varphi) + R^2 \\ -\dfrac{1}{2}R^2\varphi\sin\varphi - R^2\cos\varphi + hR \\ q_1 R\sin\varphi - p_1 R(1-\cos\varphi)\cos\varphi \\ -p_2 R\dfrac{(1-\cos\varphi)^2}{4}\cos\varphi \\ -p_R R(1-\sin\varphi)\sin\varphi \\ -\left(\dfrac{q_1}{3\pi} - p_1 - \dfrac{5}{16}p_2 - \dfrac{p_R}{3\pi}\right)R\cos\varphi & \left(\dfrac{\pi}{2} \le \varphi < \pi\right) \\ +gR\left(\varphi\sin\varphi - \dfrac{1}{2}\cos\varphi\right) + R^2 \\ -\dfrac{1}{2}R^2\varphi\sin\varphi - R^2\cos\varphi + hR \end{cases} \tag{23}$$

The rotated angle is solved by substituting Eqs. (22) and (23) into Eq. (14) and Eq. (8):

$$\begin{aligned}\theta(\varphi) &= \alpha_1 R\left(\int \overline{M} d\varphi + \bar{y}\int \overline{N} d\varphi\right) \\ &\quad - \alpha_2 R\int N\mathrm{d}\varphi - (\alpha_3 - \alpha_4)kR\varphi + C_1\end{aligned} \tag{24}$$

Rewrite Eq. (12) using Eq. (24), the displacement can be determined by:

$$\begin{aligned}\upsilon(\varphi) &= \int\Big\{\alpha_1 R\left(\int \overline{M} d\varphi + \bar{y}\int \overline{N} d\varphi\right) \\ &\quad - \alpha_2 R\int N\mathrm{d}\varphi - (\alpha_3 - \alpha_4)kR\varphi + C_1\Big\}R\mathrm{d}\varphi + C_2\end{aligned} \tag{25}$$

Unknown constants C_1 and C_2 can be worked out by the boundary conditions. The complex functions can be solved by MATLAB or other math-calculating tools.

3.2 *Time-dependent material property*

When a fire occurs, it is reasonable to treat the temperature distribution along the tunnel face homogeneously because of the narrow space of a city metro sector tunnel. A temperature distribution function of the lining section under fire situation was proposed by Qiang Jian (2007) as:

$$T(y) = \alpha\left(T_{f\max,}t\right)e^{\frac{h-y}{\beta(T_{f\max,}t)}} + \gamma\left(T_{f\max,}t\right) \tag{26}$$

where $\alpha(T_{f\max,}\ t)$, $\beta(T_{f\max,}\ t)$, and $\gamma(T_{f\max,}\ t)$ are the parameters correlated with the fire highest temperature and fire lasting time. These parameters can be found in Qiang Jian (2007).

The time-dependent elastic modulus established by Guo Zhenghai & Shi Xudong (2003) can be presented as:

$$E_T = \left[0.83 - 0.0011T_{(y)}\right]E_0 \tag{27}$$

Parameters $\theta(s)$ and $\upsilon(s)$ can be calculated by manipulating Eqs. (24) and (25) using Eqs. (22) and (23). The integration constants of $\theta(\varphi)$ and $\upsilon(\varphi)$ can be worked out by the boundary condition.

4 CASE STUDY

Taking Nanjing Metro as an example, from the in suit test, we knew that the soil pressure is 237 kPa, horizontal pressures p1 and p2 are 131 and 58 kPa, respectively, soil resistance is 228 kPa, segment self-weight is 8.75 kPa, and underground water depth is 9 m.

The thickness of the segment is 0.35 m, racial of segment is 3 m, and the elastic modulus is 29.5 GPa. The recommend highest temperature is approximately 1000–1200°C; hence, 1100°C is taken as the highest temperature in this section. A fire lasting time of 3h is assumed. According to the previously proposed method, the results are shown in Figs. 3–6.

As can be seen from Fig. 3, the whole segment shows a shrinkage trend with the increase in time. The shrinkage of the upper half of the segment is greater than that of the lower half. As shown in Figs. 4 and 5, horizontal shrinkage is smaller than vertical shrinkage. The vertical deflection of the upper half of the segment is greater than that of the lower half. The maximum vertical deflection occurs at the segment crown. It reaches to 41, 61, and 80 mm at times 30 min, 1 h, and 3 h, respectively. As to the lower half of the segment, the maximum vertical deflection occurs when $\varphi = 130°$.

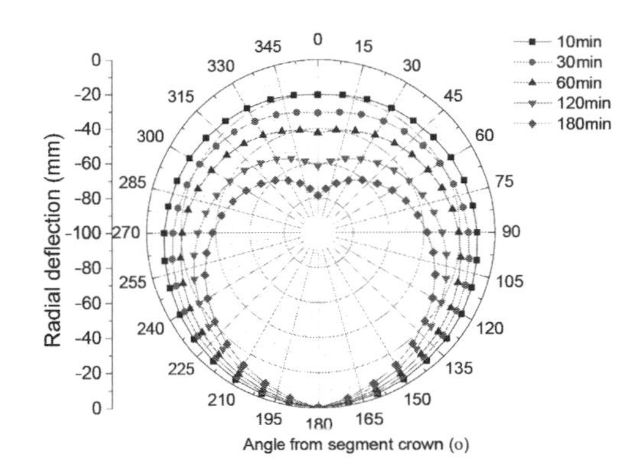

Figure 3. Radial deflection of the segment.

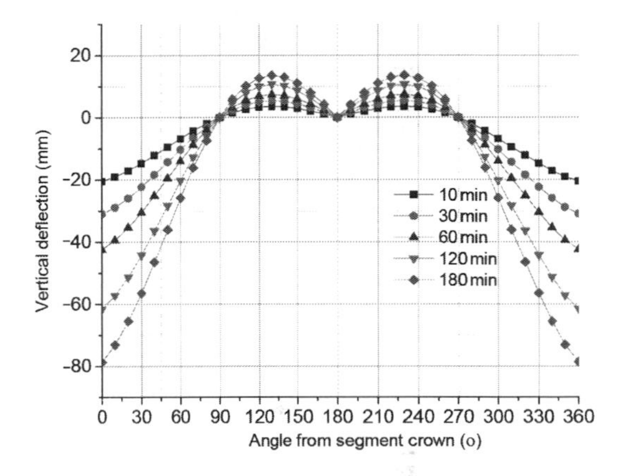

Figure 4. Vertical deflection curve of the segment.

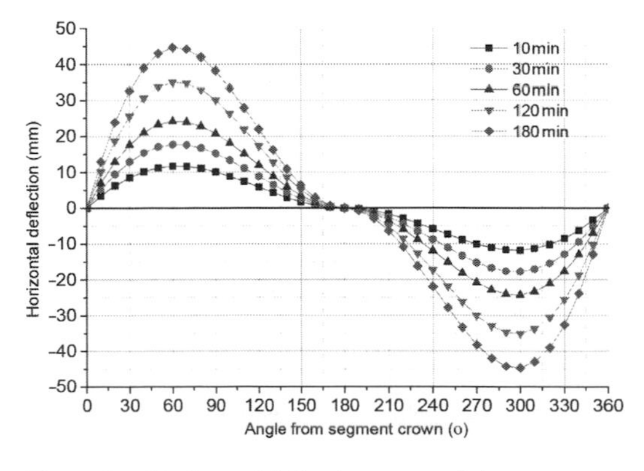

Figure 5. Horizontal deflection curve of the segment.

The maximum horizontal deflection occurs at the position where $\varphi = \pi/3$. It reaches to 17, 24, and 45 mm at times 30 min, 1 h, and 3 h, respectively.

5 CONCLUSION

In this paper, we presented a technique for investigating the shield tunnel lining segment deflection regulation under fire situation. The lining segment was treated as a ring. The nonlinear temperature distribution along the segment thickness and temperature-dependent concrete mechanical properties were considered in this model, which paved a way to solve the deflection of segment at elevated temperature. An analytic formula is established to calculate the segment cross-section bending moment and axil force based on the Free-Form Deformation. The model of this paper is a thermoplastic model. The elastoplasticity of the concrete under fire should be considered further.

According to the case study results, the elevated temperature has a greater influence on the upper half of the segment; hence, more attention should be paid on the upper half when designing as well as repairing after a fire accident.

REFERENCES

Amin Heidarpour, Tung Hoang Pham, Mark Andrew Bradford. (2010). Nonlinear thermoelastic analysis of composite steel-concrete arches including partial interaction and elevated temperature loading. J *engineering structure.* 32, 3248–3257.

Bailey CG. (2004). Membrane action of slab/beam composite floor systems in fire. J *Eng Struct*. 26(12), 1691–1703.

Bradford MA, Ranzi G, Ansourian P. (2006). Composite beams in sub-frame assemblages with longitudinal and transverse partial interaction at elevated temperatures. In: International conference on metal structures. 354–50.

Bradford MA. (2006). Generic model for a composite T-beam at elevated temperature. In: Fourth international workshop on structures in fire. 805–812.

Guo Xingjun. (2013). Experimental Study and Numerical Simulation Analysis of Fire Resistance Performance of Concrete Segment Component of Shield Tunnel D. Changsha: Zhongnan university.

Guo Zhenghai, Shi Xudong. (2003). Behavior of reinforced concrete at elevated temperature and its calculation M. Beijing: Tsinghua University Press. 18–81.

Heidarpour A, Bradford MA. Generic nonlinear modelling of restrained steel beams at elevated temperatures. J *Eng Struct*. 31(11), 2787–2796.

Hong-peng Lai, Shu-yong Wang, Yong-li Xie. (2014). Experimental research on temperature field and structure performance under different lining water contents in road tunnel fire. J *Tunneling and underground space technology*. 43, 327–335.

Huang Z, Burgess IW, Plank RJ. (1999). The influence of shear connectors on the behaviour of composite steel-framed buildings in fire. J *Constr Steel Res*. 51(3), 219–237.

Huang Z, Burgess IW, Plank RJ. (2000). Three-dimensional analysis of composite steel-framed buildings in fire. J *Struct Eng, ASCE*. 126(3), 389–397.

Kodur, V.K.R., Dwaikat, M. (2008). A numerical model for predicting the fire resistance of reinforced concrete beams. J *Cem. Concr. Compos*. 30, 431–443.

Kruppa J, Zhao B. (1995). Fire resistance of composite beams to Eurocode 4 part 1.2. J *Constr Steel Res*. 33, 51–69.

Li Zhongyou, Liu Yuanxue, Liu Shulin, Tan Yizhong Ge Zengchao. (2012). Theoretical Analysis Model of Deformation Behavior of Tunnel Linings Subjected to Fire Load J.*Rock and Soil Mechanics*. 11, 307–310.

Pi Y-L, Bradford MA, Uy B. (2002) In-plane stability of arches. Internat J Solids Structures. 39(1), 105–125.

Pi Y-L, Bradford MA. (2004). Elastic flexural torsional buckling of fixed arches. Quart J *Mech Appl Math*. 57(4), 551–569.

Qiang Jian (2007). A Study onTunnel Lining after Fire Damage to Subway and Evaluation Method Fire Scenarios. D. Shanghai: Tongji university.

Soon-Wook Choi, Junhwan Lee, Soo-Ho Chang. (2013). A holistic numerical approach to simulating the thermal and mechanical behavior of a tunnel lining subject to fire. J *Tunneling and underground space technology*. 35, 122–124.

Yan Zhiguo. (2007). A Study on Mechanical Behaviors and Fireproof Methods of Tunnel Lining Structure During and after Fire Scenarios D. Shanghai: Tongji university.

Zhi-guo Yan, He-hua Zhu, J. Woody Ju. (2013). Behavior of reinforced concrete and steel fiber reinforced concrete shield TBM tunnel linings exposed to high temperatures. J *Construction and building materials*. 38, 610–618.

Zhi-guo Yan, Yi Shen, He-hua Zhu, Xiao-jun Li, Yong Lu. (2015). Experimental investigation of reinforced concrete and hybrid fibre Reinforced concrete shield tunnel segments subjected to elevated temperature. *J Fire safety journal*. 71, 86–99.

Civil, Architecture and Environmental Engineering – Kao & Sung (Eds)
© 2017 Taylor & Francis Group, ISBN 978-1-138-02985-9

Optimization of composition of the claddings of wooden structures

M. Halirova, H. Sevcikova, R. Fabian & E. Machovcakova
Faculty of Civil Engineering, VSB-TU Ostrava, Ostrava-Poruba, Czech Republic

R. Janousek
KNAUF Praha, spol. s r.o., Praha, Czech Republic

ABSTRACT: In this paper, we show a possible method for responsible choice of the optimal composition of the claddings of wooden structures with plasterboard or gypsum board in terms of selected criteria. The proposal unaffected by subjective opinions of solvers will be ensured by comparing the compositions of the claddings of wooden structures.

1 INTRODUCTION

Selection of building materials and technologies can be done in the initial design phase, taking into account the quality of the claddings of wooden structures (Halirova, M., 2015). Multicriterion optimization of selection of individual materials is the selected objective method, which eliminates or minimizes subjective view of solvers and is suitable for the design of claddings and other structures (Takano A., 2014). Six variations of the claddings of wooden structures were assessed using the multicriterion optimization method (Table 1). Seven evaluative criteria were selected (Table 2): thermally technical criteria (heat thermal transmittance value, the lowest indoor surface temperature of the structure, moisture transfer in the structure) (Sevcikova H, 2015), ecological criteria (energy consumption and CO_2 production of building materials), and the economic criterion (cost of materials and assembly).

2 METHODS OF PROCESSING

We determine a set of real material structure variants of cladding and a set of evaluation criteria. Then, we choose the optimal variant using multicriterion optimization (Halirova M, 2013). Requirements for the cladding as well as the requirements

Table 1. Type of claddings with used materials.

Type of cladding	Structure of cladding—building material
A	Fiber gypsum board (12.5 mm) + moisture stop (foil) + squared timber 60 × 140 + mineral insulation (140 mm) + fiber gypsum board (12.5 mm) + polystyrene foam (80 mm) + plaster (7 mm)
B	Fiber gypsum board (12.5 mm) + squared timber 60 × 60 + mineral insulation (60 mm) + moisture stop (foil) + squared timber 60 × 140 + mineral insulation (140 mm) + fiber gypsum board (12.5 mm) + polystyrene foam (150 mm) + plaster (7 mm)
C	Fiber gypsum board (12.5 mm) + squared timber 60 × 60 + wooden fiber insulation (60 mm) + fiber gypsum board (12.5 mm) + squared timber 60 × 140 + wooden fiber insulation (140 mm) + fiber gypsum board (12.5 mm) + wooden fiber board (2 × 80 mm) + plaster (7 mm)
D	Fiber gypsum board (12.5 mm) + moisture stop (foil) + squared timber 60 × 140 + mineral insulation (140 mm) + fiber gypsum board (12.5 mm) + polystyrene foam (80 mm) + plaster (7 mm)
E	Fiber gypsum board (12.5 mm) + squared timber 60 × 60 + mineral insulation (60 mm) + moisture stop (foil) + squared timber 60 × 140 + mineral insulation (140 mm) + fiber gypsum board (12.5 mm) + polystyrene foam (150 mm) + plaster (7 mm)
F	Gypsum board (12.5 mm) + squared timber 60 × 60 + wooden fiber insulation (60 mm) + moisture stop (foil) + squared timber 60 × 140 + wooden fiber insulation (140 mm) + gypsum board (12.5 mm) + wooden fiber board (2 × 80 mm) + plaster (7 mm)

Table 2. Selected evaluation criteria for multicriterion optimization.

Number of criterion	Criterion	Unit of measure
1	Thermal transmittance value U	$W\ m^{-2}K^{-1}$
2	Temperature factor at the internal surface f_{Rsi}	–
3	Annual amount of condensed water vapor $M_{c,a}$	kg/m^2 . year
4	Energy consumption for the production of building material E	MJ/m^2
5	Production of CO_2 during production of building material	g/m^2
6	The price of material in a cladding	CZK/m^2
7	The price of assembly of a cladding	CZK/m^2

for each building material vary. It also depends on the characteristics of the environment in which the building project will be implemented. The method of multicriterion optimization is one of the ways to successfully solve the selection of optimal variant of the cladding according to the specified requirements (Perina Z, 2016).

2.1 *The method of multicriterion optimization*

The optimal variant of the situation is selected using the multicriterion optimization method.

Step 1 – Formulation of goal—usually selection of the optimal variant for specific cladding. Definition of the structure and design of the possible material variants (Perina Z, 2014).

Step 2 – Determination of specific criteria for comparison of various variants. The selection of criteria can be different because of different user or contractor (Halirova M, 2015 & 2016).

Step 3 – Numerical values of individual criteria are obtained here. Each variant and criterion is assessed by the value a_{ij}, and decision matrix (1) is formed. Variants are in columns and criteria are in rows (Lipušček I, 2010). The criterion can be assessed by points from 0 to 10 or from 0% to 100% if the calculation of the criterion is too complicated (Halirova M, 2014).

The decision matrix:

$$\begin{vmatrix} a_{11} & a_{12} & .. & a_{1n} \\ a_{21} & a_{22} & .. & a_{2n} \\ . & . & .. & \\ . & . & .. & \\ a_{m1} & a_{m2} & .. & a_{mn} \end{vmatrix} \begin{vmatrix} f_1 \\ f_2 \\ . \\ . \\ f_m \end{vmatrix} \qquad (1)$$

a_{ij} the value of criterion i, variant j (i-criterion 1 to m, j-variant 1 to n)

f_1–f_m weights, the valid rule $\sum_{i=1}^{m} f_i = 1$

2.2 *Formation of a decision matrix*

Decision matrix is formed of six claddings, which are denoted by alphabets A to F and seven evaluation criteria, which are identified by numbers 1 to 7 (Tables 1 and 2). The values of maximum and minimum (it depends on which value is more useful) can be seen in Table 3.

Step 4 – Determination of criterion weight is the most important step in multicriterion optimization. Each property that is expressed by a criterion has a different significance. It is very important to determine the importance of a certain criterion, and the optimization for different weights of the criterion must be effectively performed (Adekunle, T.O. 2016).

2.3 *Determination of criterion weight*

Determination of weight requires an assessment of the significance of each criterion against other criteria. The weight of a certain property must be preferred. Criteria include an objective element that emphasizes the qualities of claddings and a subjective element, which is the decision-making element. The weight is assigned to individual properties, thereby the seriousness and importance of each criterion is determined (Friedrich D, 2016).

Several different methods can be used for the determination of weights. Each method is more or less suitable for certain criterion. Optimal variant of the solution of all used methods must be evaluated.

The method of quantitative paired comparison belongs among the best methods. This method uses Saaty's matrix highlights priorities of the evaluation criteria. This method is best suited to the elimination of the subjective view of the evaluator. Therefore, this method was selected for further calculations.

2.4 *Method of quantitative paired comparison of criteria*

Scale 1, 2, 3 … 9 and reciprocal values are used for the production of paired comparisons $S = (s_{ij})$. Matrix elements are estimations of proportion of weights of i-th and j-th criteria:

$$S_{ij} \approx \frac{f_i}{f_j}, \quad S_{ii} = 1, \quad S_{ji} = \frac{1}{S_{ij}} \qquad (2)$$
$$i, j = 1, 2, 3, \ldots n$$

Table 3. Decision matrix.

Type of cladding	Max. Min.	Criterion number 1	2	3	4	5	6	7
		Min	Max	Min	Min	Min	Min	Min
A		0.21	0.906	0.0082	430.14	18.28	830.6	761.5
B		0.14	0.955	0.0122	495.19	18.91	1004.3	793.5
C		0.14	0.956	0	992.36	6.86	2478	946.3
D		0.21	0.906	0.0066	448.6	20.06	798.3	749.5
E		0.14	0.955	0.0119	504.42	19.8	965.1	773.3
F		0.14	0.953	0	1066.92	4.83	2277.2	843.5

Table 4. The method of quantitative pair comparison of criteria.

Criterion number	1	2	3	4	5	6	7	S_{ij}	R_{ij}	Weight f_i
1	1	5	4	6	7	2	3	5040	3.38	0.32
2	1/5	1	1/3	3	5	1/5	1/5	0.040000	0.63	0.06
3	1/4	3	1	5	7	1/3	1/3	2.916667	1.17	0.11
4	1/6	1/3	1/5	1	3	1/7	1/7	0.000680	0.35	0.04
5	1/7	1/5	1/7	1/3	1	1/9	1/9	0.000017	0.21	0.02
6	1/2	5	3	7	9	1	2	945	2.66	0.25
7	1/3	5	3	7	9	1/2	1	157.5	2.06	0.20
n = 7									Σ = 10.46	Σ = 1

$$f_1 = \frac{\left(\prod_{j-1}^{n} S_{IJ}\right)^{\frac{1}{n}}}{\prod_{1-1}^{n}\left(\prod_{j-1}^{n} S_{IJ}\right)^{\frac{1}{n}}} \tag{3}$$

2.5 *The method of quantitative pair comparison of criteria*

Saaty´s matrix expresses criteria preferences that contain the equal criteria and slightly and strongly preferred criteria.

Step 5 – The values of the weights of individual criteria are set into the sequence. The sequence of the criteria is made in accordance with the requirements. These values are converted to dimensionless numbers.

a. Cost type is used when the requirement is based on the minimum value (e.g., the cost of energy intensity). Transformation into a dimensionless quantity is as follows: the highest value of max a_{ij} is the lowest value of the evaluation (usually $b_{ij} = 0$); the lowest value of min a_{ij} is the highest value of the evaluation ($b_{ij} = 1$):

$$b_{ij} = \frac{(\max a_{ij}) - a_{ij}}{(\max a_{ij}) - (\min a_{ij})} \tag{4}$$

b. Profit type – is used when the requirement is based on the maximum value (the higher the better).
Transformation is as follows:

$$b_{ij} = \frac{(a_{ij}) - (\max a_{ij})}{(\max a_{ij}) - (\min a_{ij})} \tag{5}$$

Step 6 – unsuitable variant is eliminated. The used criteria are evaluated. The matrix contains variants in columns and criteria in accordance with Step 5 in rows, including weights of each criterion.
Calculation matrix:

$$\left|\begin{matrix} b_{11} & b_{12} & .. & b_{1n} \\ b_{21} & b_{22} & .. & b_{2n} \\ . & . & .. & \\ . & . & .. & \\ b_{m1} & b_{22} & .. & b_{mn} \end{matrix}\right\| \begin{matrix} f_1 \\ f_2 \\ . \\ . \\ f_m \end{matrix}\right| \tag{6}$$

Table 5. Calculation matrix.

Type of the cladding $c_{ij} = b_{ij}.fi.100$	Weight f_i	Criterion number 1	2	3	4	5	6	7	Σ
		0.32	0.06	0.11	0.04	0.02	0.25	0.20	1
A		0	0	3.6	4	0.2	24.5	18.8	51.1
B		32	5.9	0	3.6	0.2	21.9	15.5	79.1
C		32	6	11	0.5	1.7	0	0	51.2
D		0	0	5	3.9	0	25	20	53.9
E		32	5.9	0.3	3.5	0.1	22.5	17.6	81.9
F		32	5.6	11	0	2	3	10.5	64.1

$$b_{ij} \cdot f_{ij} = c_{ij}; \max \sum_{j=1}^{n} c_{ij} = H_i \Rightarrow optimum \quad (7)$$

where:

b_{ij} ... transformed value by Eqs.(4) and (5).

f_i ... criterion weight.

2.6 *Transformation of the decision matrix into the calculation matrix*

The decision matrix is transformed into the calculation matrix. The matrix contains criteria including weights of each criterion in columns and variants of the cladding in rows. The calculation is performed using the calculation matrix. Optimal variant has the highest sum of the products of transformed values and weights.

Step 7 – The result of the calculation is analyzed. The decision-making process can be monocriterial optimization (searching for variant with maximum gain) or multicriterion optimization using different criteria. The result depends on the experience of the researcher and the preferences of the individual criteria. Each researcher (builder, user) prefers other criteria.

3 CONCLUSION—EVALUATION OF MULTICRITERION OPTIMIZATION

The best variant based on the multicriterion optimization is variant E in this case. Variant B is on the second place in this case. This variant has a higher rating for energy consumption for the production of building material and the production of CO_2 during production of building material. The assessment of economic criteria is lower compared to that in variant E. Variant F is on the third place in this case. This variant has lower rating for economic criteria. Variant C has the lowest rating for economic criteria. Variants A and D have the lowest ratings. These variants have the lowest ratings for heat thermal transmittance value and temperature factor at the internal surface. The above mentioned walls have shown that gypsum board has similar, comparable properties and is suitable as a board material for cladding like fiber gypsum board.

ACKNOWLEDGMENT

This study was supported by The Research and Innovation Project for 2016, assigned to VSB-TU Ostrava by Ministry of Education, Youth and Sports of Czech Republic.

REFERENCES

Adekunle, T.O., Nikolopoulou, M. 2016. Thermal comfort, summertime temperatures and overheating in prefabricated timber housing. *Building and Environment: 21–35.*

Friedrich, D., Luible, A. 2016. Assessment of standard compliance of Central European plastics-based wall cladding using Multi-Criteria Decision Making (MCDM). *Case Studies in Structural Engineering: 27–37.*

Halirova, M., Janousek, R., Sevcikova, H., Fabian, R., Machovcakova, E., 2016. Economic Comparison of fire Fighting Measures of Gypsum- based Materials. *Applied Mechanics and Materials. 2015: 467–471.*

Halirova, M., Rykalova, E. 2013. Use of multi-criteria optimization for selection of building materials for reconstruction. *Advanced Materials Research. October 2013: 474–477.*

Halirova, M., Rykalova, E., Perina, Z. 2014. Use of multi-criteria optimization for selection of building materials for new buildings. *Advanced Materials Research:47–50.*

Halirova, M., Sevcikova, H., Rykalova, E., Fabian, R., Janousek, R., 2015. Evaluation of Selected Variants of Claddings on the Basic of CO2 Production and Energy Intensity of Production of Building Materials. *AER- Advances in Engineering Research. November 2015: 43–48.*

Halirova, M., Tymova, P., Perinkova, M. 2015. Sustainable Construction in the Conservation Area. *Advanced Engineering Forum: 132–136.*

Lipuščeк, I., Bohanec, M., Oblak, L., Zadnik Stirn, L. 2010. A multi-criteria decision-making model for classifying wood products with respect to their impact on environment. *International Journal of Life Cycle Assessment:359–367.*

Perina, Z., Halirova, M., Wolfova, M., Fabian, R., Sevcikova, H., Rykalova, E. 2014. Diagnostics of Building Claddings and Design of Technical Solutions with Regards to Energy Performance. *Advanced Materials Research. October 2014:566–572.*

Perina, Z., Sevcikova, H., Fabian, R., Halirova, M., Rykalova, E., 2016. The impact of balcony glazing on the selected building envelope. *Applied Mechanics and Materials. 2015: 628-6315-322.*

Sevcikova, H., Rykalova, E., Fabian, R., Perina, Z., Halirova, M., Wolfová, M. 2015. Mold Issue in Details within the External Walled Structure. *Applied Mechanics and Materials. April 2015: 628–631.*

Takano, A., Hughes, M., Winter, S. 2014. A multidisciplinary approach to sustainable building material selection: A case study in a Finnish context. *Building and Environment:526–535.*

Civil, Architecture and Environmental Engineering – Kao & Sung (Eds)
 ISBN 978-1-138-02985-9

Evaluation of the ultimate strength of RC I-shaped members subjected to a combined action

Pu Wang, Zhen Huang & Quanshen Wang
School of Naval Architecture, Ocean and Civil Engineering, Shanghai Jiaotong University, Shanghai, P.R. China

ABSTRACT: Under complex loads, especially for earthquakes and wind loads, RC I-shaped members are usually subjected to load combinations of axial forces, bending moments, shear forces, and torsion. In this paper, we deduce the relationships between the external forces and the cracking degrees. Then, a model for the evaluation of the ultimate strength of RC I-shaped members is established based on the ultimate equilibrium of twist failure surface. Finally, the model results are compared with the experimental results of former researchers. The model coincides well with the experiment.

1 INTRODUCTION

Nowadays, Reinforced Concrete (RC) I-shaped members have been widely used in construction. RC I-shaped members are mainly designed to undertake the loads of bending moments and shear forces. However, RC I-shaped members in hazards usually fail under different load combinations of axial loads, bending moments, shear force, and torsion. The failure mechanisms of these members are very complex.

Although the theory for the ultimate strength of RC members under axial forces and bending moments is well recognized, there are many different theories for the mechanisms of RC members under shear forces. In the engineering application, equations based on the regression analyses of experimental results are widely used, such as "Code for design of concrete structures" (GB50010-2010).

Empirical equations lack theoretical foundations. Researchers have developed many different theories for the ultimate strength of RC members subjected to combined actions for decades. Collins (1973) established Compressive-Field Theory (CFT) by introducing the compatibility conditions of RC members under shear force. Then, Vecchio & Collins (1986 & 2000) developed CFT into Modified Compressive-Field Theory (MCFT) and Disturbed Stress Field Mode (DSFM). Hsu (1988) established Softened Truss Model Theory by overall consideration of the equilibrium, compatibility conditions, and softened stress–strain relationship of concrete. Pang & Hsu (1996) further developed Fixed Angle Softened Truss Model. These two series of theories could analyze the behavior of RC members under combined actions fairly well by the help of finite-element method.

Huang & Liu (2007) proposed a theoretical model for RC box-section members under combined actions. Wang et al. (2014) proposed a theoretical model for RC rectangular members under combined actions. These two models try to have a clear theoretical basis and be convenient for application. In this paper, we attempt to determine the relationship between the external force and the twist failure surface. Then, a model is established on the basis of the ultimate equilibrium of twist failure surface.

2 MODEL DEDUCING

2.1 *Basic assumption*

The factors affecting the ultimate strength of RC member are very complex. To simply the expression, only main factors are considered and the bellowing basic assumptions are adopted.

1. After cracking, the core concrete is assumed to not undertake any torsion load. All the torsion is assumed to be undertaken by the out-box walls of the concrete.
2. All the reinforcements are assumed to achieve their yield strengths when crossing the warped failure surface.
3. The dowel actions of longitudinal reinforcements are ignored.

2.2 *Shear stress*

Figure 1 shows the section of RC members. A is the area of section, A_{cor} is the area enclosed by the shear flow ($A_{cor} = 2h_1b_{cor} + b_2h_2 = \alpha A$), t is the equivalent wall thickness of the section, and $b_{cor} = b-2t$, $h_{cor} = h-2t$. According to the Bredt thin-walled theory, the shear flow on the out-box wall can be expressed as a constant (q). The shear flow (q) is express as:

$$q = \frac{T}{4qh_1b_{cor} + 2qb_2h_2} = \frac{T}{2A_{cor}} \tag{1}$$

where T is the torsion on the member and α is the ratio of A_{cor} to A.

Figure 2 shows the warped failure surface. The left, right, and bottom sides are denoted by 1, 2, and 3, respectively.

The shear force on each side wall due to the torsion (T) can be expressed as:

$$V_{T1} = qh_2 = \frac{Th_2}{2\alpha A}; V_{T2} = qh_2 = \frac{Th_2}{2\alpha A};$$
$$V_{T3} = qb_{cor} = \frac{Tb_{cor}}{2\alpha A}; V_{T4} = qb_{cor} = \frac{Tb_{cor}}{2\alpha A} \tag{2}$$

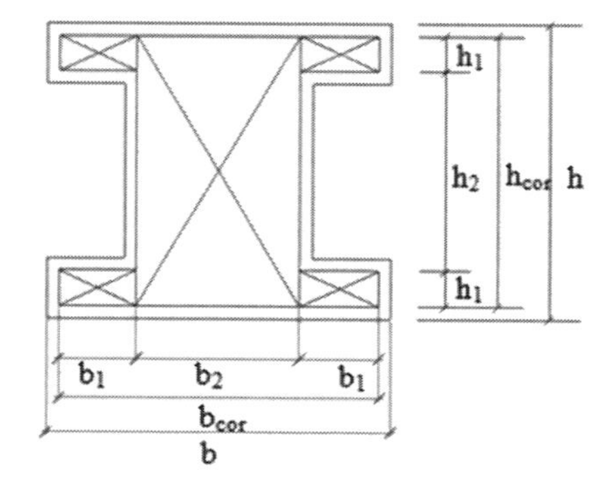

Figure 1. Section of RC members.

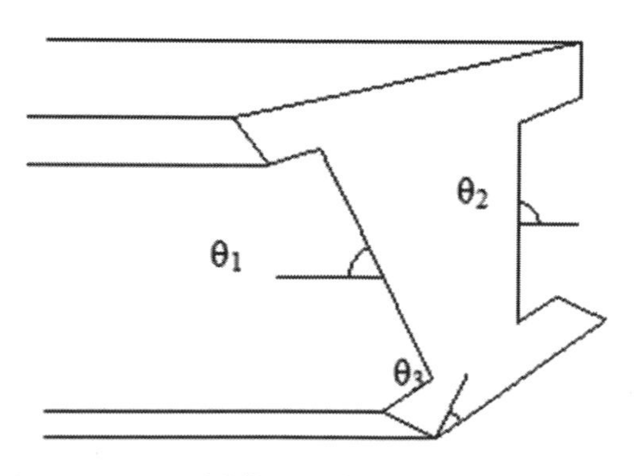

Figure 2. Warped failure surfacer.

The shear stress due to the external shear force (V) can be approximately regarded as a constant value ($\tau_v = V/A$). Then, the shear forces on the left and right side walls can be expressed as:

$$V_{V1} = V_{V2} = \tau_v h_2 t = \frac{Vh_2t}{A} \tag{3}$$

By adding the shear forces from Eqs. 1 and 2, the total shear force on each side wall under the torsion and shear force is equal to:

$$\begin{aligned} V_1 &= V_{T1} + V_{V1} = \frac{Th_2}{2\alpha A} + \frac{Vh_2t}{A} \\ V_2 &= V_{T2} - V_{V2} = \frac{Th_2}{2\alpha A} - \frac{Vh_2t}{A} \\ V_3 &= V_{T3} = qb_{cor} = \frac{Tb_{cor}}{2\alpha A} \\ V_4 &= V_{T4} = qb_{cor} = \frac{Tb_{cor}}{2\alpha A} \end{aligned} \tag{4}$$

The shear force on each side wall is undertaken by the concrete and transvers reinforcements. On the basis of this assumption, transverse reinforcements achieve the yield strength when approaching the failure. Therefore, the shear force on each side wall can be expressed as:

$$\begin{aligned} \tau_1 &= \frac{V_1 - n_1A_{sv1}f_{yv}}{th_2} = \frac{T}{2\alpha At} + \frac{V}{A} - \frac{n_1A_{sv1}f_{yv}}{th_2} \\ \tau_2 &= \frac{V_2 - n_2A_{sv1}f_{yv}}{th_2} = \frac{T}{2\alpha At} - \frac{V}{A} - \frac{n_2A_{sv1}f_{yv}}{th_2} \\ \tau_3 &= \frac{V_3 - n_3A_{sv1}f_{yv}}{tb_{cor}} = \frac{T}{2\alpha At} - \frac{n_3A_{sv1}f_{yv}}{th_2} \end{aligned} \tag{5}$$

where A_{sv1} is the area of a single stirrup, f_{yv} is the yield strength of transverse reinforcements, and n_1, n_2, and n_3 are the numbers crossing the failure surface.

The relationships between n_1, n_2, n_3, and cracking degrees are:

$$n_1 = \frac{h_{cor}\cot\theta_1}{s}; n_2 = \frac{h_{cor}\cot\theta_2}{s}; n_3 = \frac{b_{cor}\cot\theta_3}{s} \tag{6}$$

where θ_1, θ_2, and θ_3 are the inclined degrees of warped failure surface and s is the space of stirrups.

2.3 *Normal stress*

The mean normal stress on the section is $\sigma = N/A$. On the left side wall of the member, the principal stresses can be calculated as:

$$\sigma_1 = \frac{\sigma + \sqrt{\sigma^2 + 4\tau_1^2}}{2}; \sigma_2 = \frac{\sigma - \sqrt{\sigma^2 + 4\tau_1^2}}{2} \tag{7}$$

where σ_1 and σ_2 are the principle stresses on the section and τ_1 is the shear stress on the section.

Figure. 3 shows the failure criterion of Tasuji–Slate–Nilson. On the basis of the failure criterion, it can be deduced that:

$$\tau_1 = \frac{\sqrt{f_c^2 + (1-\beta) f_c \sigma - \beta\sigma^2}}{1+\beta} \tag{8}$$

where f_c is the compress strength of the concrete, f_t is the tensile strength of the concrete, and β is defined as f_c/f_t.

The shear strength of the concrete will decrease with the development of cracks. Considering the decreasing, a factor γ is added to the right-hand side of Eqs. 7 and 9 as:

$$\tau_1 = \frac{\gamma\sqrt{f_c^2 + (1-\beta) f_c \sigma - \beta\sigma^2}}{1+\beta} = \Phi(N) \tag{9}$$

where the equation is defined as $\Phi(N)$ and γ is a factor considering the strength lowering after cracking.

Substituting Eq. 9 into Eq. 5, we obtain Eq. 10:

$$n_1 = \left(\frac{T}{2\alpha At} + \frac{V}{A} - \Phi(N)\right)\frac{th_2}{A_{sv1} f_{yv}} \tag{10}$$

Similarly,

$$n_2 = \left(\frac{T}{2\alpha At} - \frac{V}{A} - \Phi(N)\right)\frac{th_2}{A_{sv1} f_{yv}}$$
$$n_3 = \left(\frac{T}{2\alpha At} - \Phi(N)\right)\frac{tb_{cor}}{A_{sv1} f_{yv}} \tag{11}$$

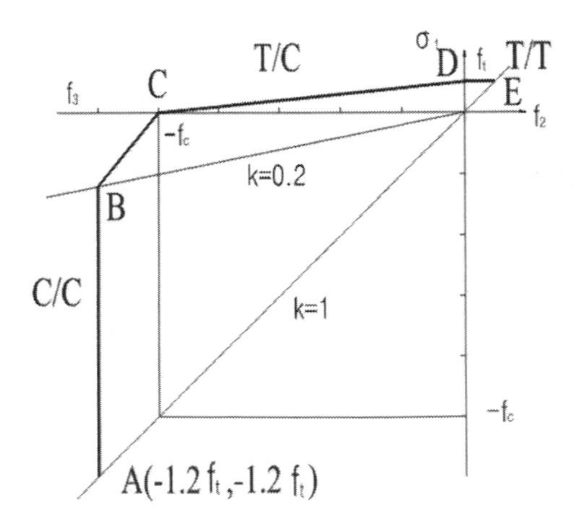

Figure 3. Failure criterion of Tasuji–Slate–Nilson.

Substituting Eq. 6 into Eqs. 10 and 11, we obtain Eq.12:

$$\cot\theta_1 = \left(\frac{T}{2\alpha At} + \frac{V}{A} - \Phi(N)\right)\frac{th_2 s}{h_{cor} A_{sv1} f_{yv}}$$
$$\cot\theta_2 = \left(\frac{T}{2\alpha At} - \frac{V}{A} - \Phi(N)\right)\frac{th_2 s}{h_{cor} A_{sv1} f_{yv}} \tag{12}$$
$$\cot\theta_3 = \left(\frac{T}{2\alpha At} - \Phi(N)\right)\frac{ts}{A_{sv1} f_{yv}}$$

2.4 *Model formula*

On the basis of the equilibrium of moment to the center of compressive reinforcement, Eq. 13, $\Sigma M_z = 0$, can be established:

$$\begin{aligned} M + N\frac{h_{cor}}{2} &= f_y A_s h_{cor} \\ &- n_1 A_{sv1} f_{yv}\left(\frac{h_{cor}\cot\theta_1}{2} + \frac{b_{cor}\cot\theta_3}{2}\right) \\ &- n_2 A_{sv1} f_{yv}\left(\frac{h_{cor}\cot\theta_2}{2} + \frac{b_{cor}\cot\theta_3}{2}\right) \end{aligned} \tag{13}$$

where A_s is the area of longitudinal reinforcements in the tensile area and f_y is the tensile strength of longitudinal reinforcements.

Substituting Eqs. 10, 11, and 12 into Eq. 13, we obtain Eq. 14:

$$\frac{N}{2f_y A_s} + \frac{M}{f_y A_s h_{cor}} + \frac{h_2 st^2 V^2}{h_{cor} f_y A_s A^2 f_{yv} A_{sv1}} + \frac{h_2 s(b_2 + h_2)(T - T')^2}{4\alpha^2 h_{cor} f_y A_s A^2 f_{yv} A_{sv1}} = 1 \tag{14}$$

Define

$$\begin{aligned} &N_0 = 2f_y A_s; M_0 = f_y A_s h_{cor}; \\ &V_0^2 = \frac{h_{cor} f_y A_s A^2 f_{yv} A_{sv1}}{h_2 st^2}; \\ &T_0^2 = \frac{4\alpha^2 h_{cor} f_y A_s A^2 f_{yv} A_{sv1}}{h_2 s(b_2 + h_2)}; \\ &T'' = T - T'; T' = 2\alpha At\Phi(N) \end{aligned} \tag{15}$$

Equation 14 becomes Eq. 16:

$$\frac{N}{N_0} + \frac{M}{M_0} + \frac{V^2}{V_0^2} + \frac{T''^2}{T_0^2} = 1 \tag{16}$$

3 VALIDATION OF THE PROPOSED MODEL

Because of the scarcity of studies on the ultimate strength for RC I-shaped members subjected to the

Table 1. Reinforcements and concrete strength.

	Top		Bottom		Concrete
	A'_s *(mm²)*	f_y' *(Mpa)*	A_s *(mm²)*	f_y*(Mpa)*	f_c *(Mpa)*
IRC-1	*1519.8*	*397*	*2026.8*	*720*	40.3
IRC-3	1519.8	397	2026.8	720	44.0
IPC-1	1519.8	397	2026.8	720	41.8
IPC-2	1519.8	397	2026.8	720	49.3
IPC-3	1519.8	397	2026.8	720	45.0
IPC-4	1519.8	*397*	*3564.8*	*730*	43.2

Table 2. Comparison of model and experimental results.

	Experiment		Model		Ratio
	M(kN·m)	V(kN)	M(kN·m)	V(kN)	(Mod/Exp)
IRC-1	755.6	559.7	551.5	408.5	0.730
IRC-3	849.2	629.0	658.3	487.6	0.775
IPC-1	954.2	706.8	691.5	512.2	0.724
IPC-2	869.4	644.0	663.7	491.6	0.763
IPC-3	840.6	622.7	651.6	482.7	0.775
IPC-4	1005.8	745.0	1047.7	776.0	1.042
Average					0.802
Standard deviation					0.120
Variation coefficients					14.96%

combined actions of axial force, bending moment, shear force, and torsion, it is very difficult to find the experimental data. In order to verify the validation of the proposed model, experimental results of RC I-shaped members subjected to bending moment and shear force by Sudhira et al. 2008 are referred and compared with the proposed model. Table 1 and Figure.4 are obtained from Sudhira et al. (2008). The stirrup space is 225 mm for IPC-2 and 125 mm for others. The yield strength is 353 MPa for IPC-3 and 438 MPa for others.

Table 2 shows the comparison of model and experimental results. The model results are safer and more conservative than the experimental results. From Sudhira et al. (2008), the first five members failed when the tensile reinforcements were ruptured. Therefore, the ultimate strength of reinforcement should be used in the calculation of ultimate strength of members. However, the model adopts the yield strength of reinforcements in calculation. As a result, the values from the model are less than the experimental results. If we use the ultimate strength of longitudinal reinforcements in the calculation of model, the ratios of model calculation to experimental results are 1.06, 1.07, 1.01, 1.06, and 1.09, respectively, for the first five members. Considering this, the model results can be regarded to coincide well with the experimental results.

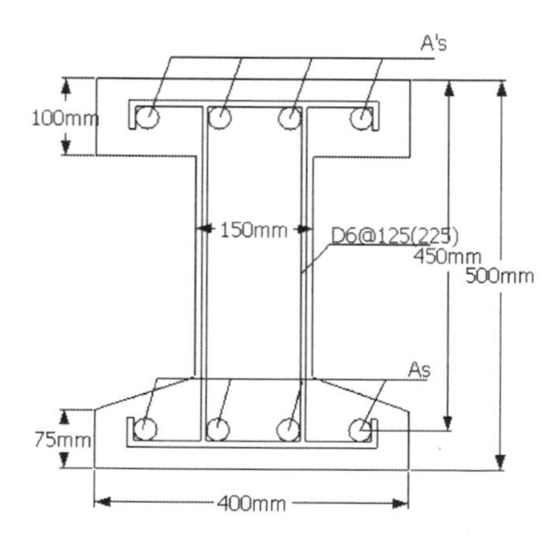

Figure 4. Cross section of experimental beams.

4 CONCLUSION

The study leads to the following main conclusions:

1. Compared with the experiment data, the validation of the proposed model is verified. The physical concept of the model is clear, and the expression of the model is concise. Therefore, the proposed model is convenient to be used in the application.

2. In the consideration of the requirement of torsion, the width of web plate should be confined with a minimum value, which should be further studied. In order to use the flange plates of I-shaped members to bear torsions, it is recommended that the overhang length of flange plates should be less than three times of its thickness.
3. In the text, the transverse reinforcements are assumed to achieve the yield strength. However, in fact, the transverse reinforcements may not yield in many situations. It needs to be further studied.
4. The experiments are very difficult for RC I-shaped members under combined actions of axial force, bending moment, shear force, and torsion. As a result, the experimental data are very rare in the aspect. The experimental studies are much needed in order to study the ultimate strength of RC I-shaped members under combined actions.

ACKNOWLEDGMENTS

The authors acknowledge the National Natural Science Foundation of China for its financial support to this research project (No. 51178265).

REFERENCES

Collins M.P. (1973). Torque-twist characteristics of rein-forced concrete beams [J]. Inelasticity and Nonlinearity in Structural Concrete. 1973. 211–231.

Hsu T.T.C. (1988). Softened Truss Model Theory for Shear and Torsion [J]. ACI Structural Journal. 1988. 85 (6): 624–635.

Huang Zhen, & Liu Xila (2007). Unified approach for analysis of box-section members under combined actions. [J]ASCE Journal of Bridge Engineering, 2007, Vol. 12, No.4:494–499.

Pang XB, & Hsu T.T.C. (1996). Fixed Angle Softened Truss Model for Reinforced Concrete [J]. ACI Structural Journal. 1996. 197–207.

Sudhira De Silva, Hiroshi Mutsuyoshi, & Eakarat Witchureangkrai (2008). Evaluation of Shear Crack Width in I-Shaped Prestressed Reinforced Concrete Beams. Journal of Advanced Concrete Technology Vol.6, NO.3, 443–458, October 2008.

Vecchio F.J., & Collins M.P. (1986). The modified compression field theory for reinforced concrete elements subjected to shear [J]. ACI Journal. 1986.83(2): 219–231.

Vecchio F.J. (2000). Disturbed stress field model for reinforced concrete: formulation [J]. Journal of Structural Engineering. 2000. 126 (9): 1070–1077.

Wang Pu, Huang Zhen, & Sun Liang (2014). A unified model of ultimate capacity of RC members with a rectangular section under combined actions, 2014 International Conference on Sustainable Development of Critical Infrastructure IC-SDCI 2014, 204–212.

Civil, Architecture and Environmental Engineering – Kao & Sung (Eds)
© 2017 Taylor & Francis Group, ISBN 978-1-138-02985-9

Three-dimensional centrifuge and numerical modeling of pile group due to twin tunnelling in different depth

Shaokun Ma, Yu Shao, Ying Liu, Junyu Chen & Jie Jiang
College of Civil Engineering and Architecture, Guangxi University, Nanning, China

ABSTRACT: Tunnelling activity inevitably induces change of soil stress and hence may cause adverse effects on nearby pile foundations. In this study, a three-dimensional centrifuge model test and a series of numerical analyses were carried out to investigate the effects of pile group in clay due to twin tunnelling in different depth. The springline of twin tunnels in the test were located at half a tunnel diameter above the pile toe (Test FF). Numerical back-analysis of centrifuge model test and parametric study were also carried out to get more insight of twin tunnels-pile interaction. Two more configurations, i.e., twin tunnels located at mid-depth of the pile (Simulation MM) and below the pile toe (Simulation BB), were investigated in the numerical analyses. The relative elevation between the twin tunnels and pile group has a significant effect on the pile settlement and the pile bearing capacity. Apparent loss of the pile bearing capacity (i.e., 90.2%) and the largest pile settlement (i.e., 5.5% of pile diameter) due to twin tunnelling are identified in Simulation BB. The maximum value of the loss of the pile bearing capacity was 2.7 times of the minimum value (Simulation MM). Additionally, the maximum transverse tilting of pile group induced by twin tunnelling is also observed when C/D = 3.7 (i.e., Simulation BB).

1 INTRODUCTION

In congested urban areas, tunnel construction inevitably induces ground deformation and stress redistribution in the soil. This may cause adverse effects on adjacent pile foundations. The effects of pile foundations due to tunnel excavation have been investigated by many researches.

Numerical and analytical studies have been carried out to investigate the effects of tunnelling on pile foundations. Marshall (2012) investigated on the effects of tunnelling on end-bearing piles located above the tunnel by using a new analytical method based on cavity expansion theory. It was found that pile failure occurred when the load-carrying capacity reduced below 80% of its initial value due to stress redistribution by tunnelling. Three-dimensional coupled-consolidation analyses were adopted by Soomro et al. (2015) to investigate the responses of pile group subjected to single tunnelling located at different depths. The results show that the largest bending moment in the piles occurs when a tunnel is excavated at a cover-to-diameter ratio (C/D) of 1.5.

Centrifuge model tests have also been carried out to investigate the tunnel-pile interaction mechanism (Jacobsz et al. 2004, Loganathan et al. 2000, Marshall 2009, Marshall & Mair 2011). Loganathan et al. (2000) investigated the responses of a 2 × 2 pile group with working load in stiff clay subjected to a single tunnel excavation. The induced bending moments and settlement of pile group were significantly influenced by the relative positions between the tunnel and the pile toe.

Recently, more and more twin tunnels are being constructed near piles with the development of metro system in cities. However, only limited research was carried out to study the effects of twin tunnelling on piles. In the centrifuge tests reported by Ng & Lu (2014), they investigated the responses of a loaded pile subjected to twin tunnelling at different depths located on both sides of the pile. It was revealed that the effects of the tunnelling sequence on the pile settlement and the loss of pile capacity were significant.

In order to comprehend the responses of pile group induced by twin tunnelling, a centrifuge model test and the displacement controlled finite element model (DCM) based on ground loss were applied to investigated the responses of pile group under working load in clay to twin tunnelling with different depth. The effects of the three-dimensional tunnel excavation process were simulated in-flight and investigated by the numerical analysis. In addition, pile group settlement, apparent loss of pile capacity and transverse tilting of pile cap obtained from the two approaches were compared with each other.

2 CENTRIFUGE MODELLING

2.1 *Experimental program and setup*

The centrifuge model test was carried out at the Geotechnical Centrifuge Facility of Hong Kong

University of Science and Technology. According to the size of the model tunnel and model pile group, the centrifugal acceleration of 40 g (where g is the acceleration due to earth gravity) was adopted in the test. Fig. 1 shows the cross section of Test FF, where the springline of the tunnels was located half tunnel diameter above the pile toe. The diameter of each model tunnel (D) was 152 mm (6.08 m in prototype scale) and the cover-to-diameter ratio (C/D) was 2.5. The pile group was centered between the tunnels, so that the horizontal distance between the tunnel axis and the closest pile was 0.75D.

2.2 *Simulation of tunnel excavation*

In this study, the tunnel excavation was simulated using the equivalent volume loss method. This was achieved by draining out 2% of water in the rubber bag. Fig. 2(a) shows the model tunnels and the pile group. Each model tunnel consists of five cylindrical rubber bags with de-aired water corresponding to five advances of tunnel construction. The three-dimensional tunnel excavation process was simulated in-flight by draining away a controlled amount of water from each rubber bag.

2.3 *Model pile group and instrumentation*

All the model piles were fabricated from aluminum tubes, 20 mm in diameter and 600 mm in length. The spacing between two piles was $3.5d_p$ (pile diameter), 70 mm in the model. The size of the pile group is often applied in building engineering and bridge engineering. As shown in Fig. 2(b), each pile was equipped with semiconductor Strain Gauges (SGs) bonded at 10 levels, with a spacing of 60 mm,

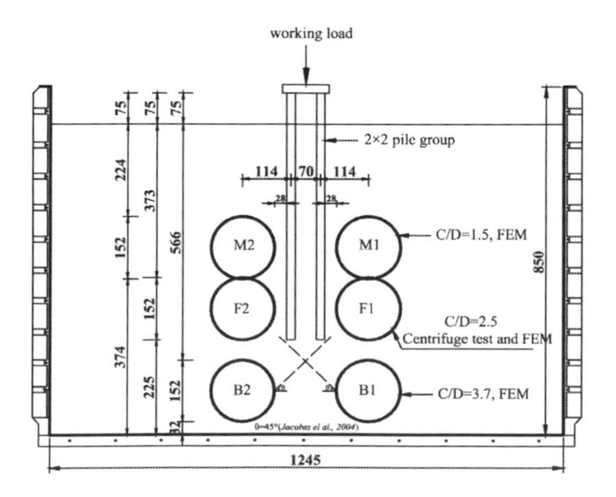

Figure 1. Elevation view of the centrifuge model test and the other two more configurations (dimensions in millimetres).

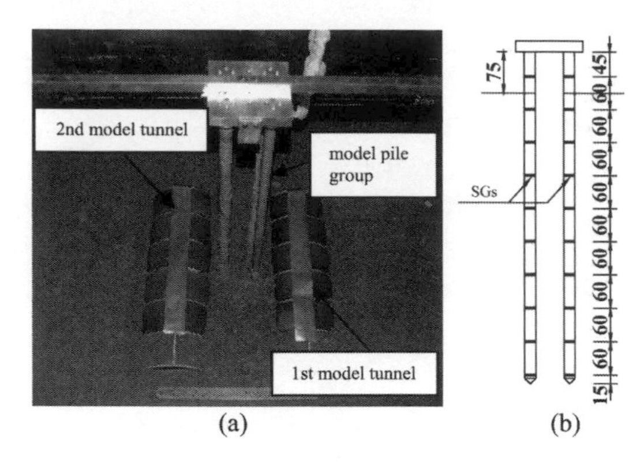

Figure 2. (a) Model tunnels and model pile group; (b) The distribution of SGs (dimensions in millimetres).

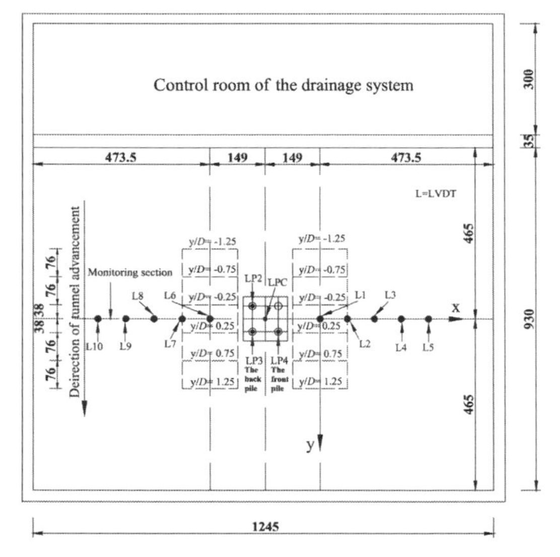

Figure 3. Plan view of the centrifuge model test (dimensions in millimetres).

along the external surface to record the axial force and bending moments. The SGs were protected by a thin layer of epoxy coating uniformly pasted to the shaft of each pile. The Young's modulus (E_p) of the aluminum model pile was 70 GPa. Thus, the model piles are equivalent to a 0.8 m diameter (in prototype scale) concrete pile with flexural stiffness and bending moment capacity of 721 MNm^2 and 800 kNm, respectively. In the paper, the effects of pile installation were not considered, and the pile group was "wished-in-place".

Fig. 3 illustrates the plan view of the model pile group, model tunnels and instruments. 10 Linear Variable Differential Transformers (LVDTs) were installed to measure ground surface settlement. Four LVDTs (LP2 to LP4 and LPC) were installed on the pile cap to measure the settlement of the pile group.

Table 1. Physical and mechanical parameters of soil.

Free expansion ration	Maximum dry density	Plastic limit	Liquid limit	Plasticity index	Optimum water content
59%	1880 kg/m^3	25%	53%	28	12.9%

Table 2. Testing procedure.

Events	Testing time (min)
Centrifuge acceleration spun up from 1 g to 40 g	10
Apply working load to the pile until 1500 N was reached	240
Excavate the first and second tunnel	20
Pore water pressure dissipated gradually	1200
Centrifuge acceleration reduced from 40 g to 1 g	30

2.4 *Model preparation and Testing procedure*

The clay used in the tests was obtained from Nanning, Guangxi Province, China. The mechanical properties of the clay are summarized in Table 1. The clay was dried in the laboratory and then crushed and sieved at 2 mm, and mixed uniformly with a certain amount of water to reach a water content of 18%, then cured for 24 h.

The testing procedure of the test is shown in Table 2.

3 3D FINITE ELEMENT NUMERICAL SIMULATION

In order to provide more insights into the effects of twin tunnelling with different C/D ratios on a pile group, the finite element software ABAQUS was used in this study. A displacement controlled method (DMC) based on ground loss was used to back-analyze the centrifuge Test FF. After verifying the numerical model, a numerical parametrical study was performed. As shown in Fig. 1, two C/D (C/D = 1.5 and 3.7) ratios were adopted in the numerical parametric study.

3.1 *Constitutive model and model parameters*

In this study, the Modified Cam Clay model was adopted for the soil. The model parameters obtained by oedometric test and triaxial test were summarized in Table 3. The model pile group was modelled as a linear elastic material with Young's modulus, Poisson ratio and unit weight of 33 GPa, 0.2 and 25 kN/m^3, respectively. The soil–pile interface was simulated according to the Coulomb friction law included in ABAQUS. According to the test results obtained from Peng (2012), the two parameters i.e. interface friction coefficient (μ) and limiting displacement (γ_{lim}) were assumed as 0.5 and 5 mm, respectively.

Table 3. Model parameters of soil.

Parameter	Value
Soil density ρ_d (kg/m^3)	1800
Initial void ratio e_0	0.9
Coefficient of earth pressure at rest K_0	0.68
Poisson ratio v	0.35
Gradient of normal compression line of soil λ	0.09
Gradient of swelling line κ	0.015
Gradient of critical state line M	0.9

3.2 *Finite element mesh and boundary conditions*

The mesh had dimensions of 50.6 m × 26 m × 34 m in prototype scale. Eight-node brick elements were adopted for both the soil and the pile group. The boundary conditions assumed for the bottom and vertical sides were pin and roller supports, respectively. In numerical analysis, the piles were modelled using "wished-in-place" method. This was the same as that in the centrifuge model test.

3.3 *Numerical modeling procedures*

Each numerical analysis followed the same procedures as those in the centrifuge test. Firstly, set up the 3D finite element mesh and the initial boundary conditions. Then applying working load to the pile group, 160 kN at a time until 2.4 MN is reached. The displacement boundary of the tunnel sections to be excavated should be restrained before killing the soil elements inside the tunnel. According to the construction sequence of the centrifuge test, apply the target displacement boundary conditions to reach the target volume loss (i.e. 2%). The displacement pattern of the tunnel cross sections adopted in this study was suggested by Park (2004). It is important to note that the restrained displacement boundary should be released during applying the target displacement boundary conditions. Finally, impose the same displacement boundary conditions to the other tunnel sections by repeating the step stated above to simulate the three-dimensional tunnelling process.

4 INTERPRETATION OF MEASURED AND COMPUTED RESULTS

All the results presented in this paper are in prototype scale unless stated otherwise.

4.1 *Determination of the axial load-carrying capacity of the pile*

In order to evaluate the security of the pile group and the quantitative effects on the pile group due to twin tunneling excavation, first of all, a simulation was carried out to determine the bearing capacity of the pile group. In the simulation, the pile group was loaded incrementally, with each loading increment of 800 kN (500 N in model scale), up to the final load of 24 MN (15000 N in model scale).

Fig. 4 shows the load settlement relationship of the pile group obtained from the numerical simulation. In the figure, settlement of the pile group (S_p) was normalized by the pile diameter (d_p).

In this study, the failure criterion (i.e., 10% of pile diameter) proposed by IMMMFE (1985) and the failure criterion proposed by Ng et al. (2001) for large diameter piles (i.e., $d_p > 0.8$ m) were used to determinate ultimate bearing capacity of the pile group, the equation of the displacement failure criterion proposed by Ng et al. (2001) as follows:

$$\delta_{ph,\max} \cong 0.045 d_p + \frac{1}{2}\frac{P_h L_p}{A_p E_p} \tag{1}$$

where $\delta_{ph,max}$ is the pile settlement used to define the ultimate load, L_p is the length of the portion of pile inserted in soil, P_h is the pile head load, d_p is the pile diameter, E_p is the elastic modulus, A_p is the cross-sectional area of the pile. As was showed in Fig. 4, based on the failure criterion proposed by Ng et al. (2001) and IMMMFE (1985), the ultimate bearing capacity of the pile group were deduced as 9.3 MN and 15.7 MN, respectively. It is known that the working load of 1500 kN (2.4 MN in prototype) was much smaller than the ultimate bearing capacity of the pile group deduced by Ng et al. (2001) and IMMMFE (1985). Thus, the pile group with the load of 2.4 MN was secure.

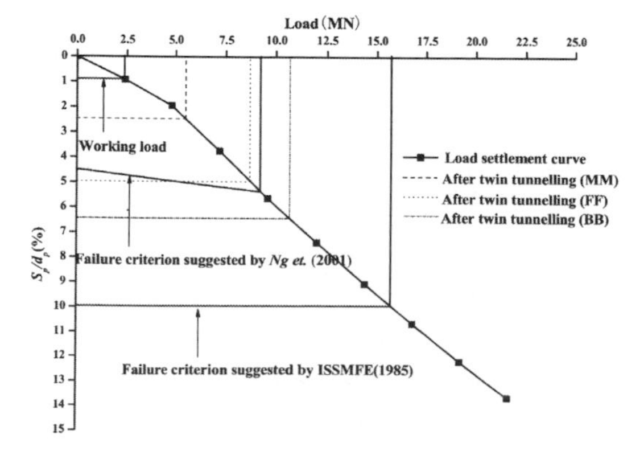

Figure 4. Load-settlement curve obtained from the result of simulation.

4.2 *Settlement of pile group*

Fig. 5(a) and (b) show the development of measured (Tests FF) and calculated (Simulations MM, FF and BB) pile settlement. The pile settlement was normalized by the pile diameter (d_p). As was shown in Fig. 3, the y value which was normalized by the tunnel diameter (D) is the distance between the tunnelling face and the monitoring section. In addition, settlement data was obtained from the LPC in Fig. 3.

It can be seen that the pile group settled linearly during the first tunnelling in Simulation MM in Fig. 5(a). In contrast, non-linear settlement characteristics of the pile group were observed in Simulations FF and BB and Tests FF. The settlement of pile group in the cases of Simulation FF calculated by the DCM was fitted well with the results measured from the centrifuge test. In the Simulation FF, BB and Test FF, the settlement of pile increased rapidly as the tunnel face approached the monitoring section, i.e., when the first tunnel advanced from y/D = –0.25 to 0.25. However, the pile settlements increased at a lower rate as the tunnel face continued to advance beyond the monitoring section. Similar settlement characteristics of the pile group subjected to twin tunnelling in sand were observed in the centrifuge tests reported by

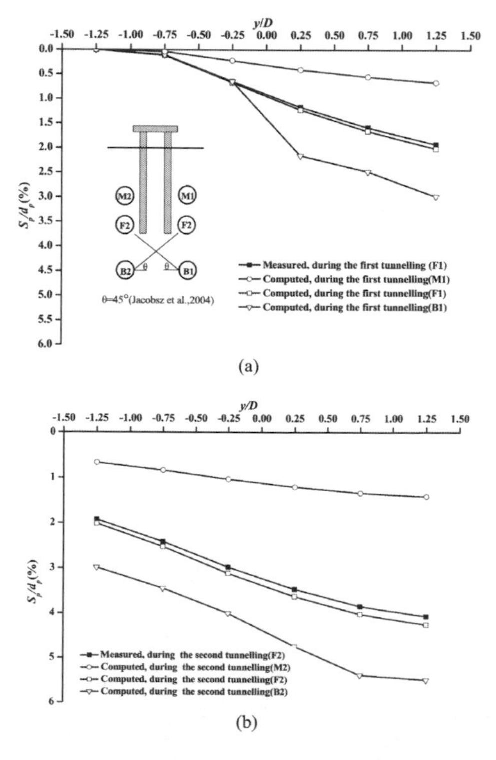

Figure 5. Settlement of pile group due to twin tunnelling: (a) during the first tunnelling; (b) during the second tunnelling.

Ng et al. (2014). After the first tunnelling, in the cases of Simulation MM, Test FF, Simulation FF and Simulation BB, the settlement of the pile group were 0.67%$_p$, 1.93% d_p, 2.02% d_p and 2.98% d_p, respectively. It can be seen that the maximum settlement of the pile group induced by the first tunnelling occurred in Simulation BB. This was because the pile group was completely located within the major influence zone (as was shown in the inset of Fig. 5(a)), as proposed by Jacobsz et al. (2004).

It can be seen from Fig. 5(b) that the settlement characteristics of the pile group subjected to the second tunnelling were similar to that induced by the first tunnelling. In addition, the settlements of the pile group induced by the second tunnelling were only 112%, 111%, 110%, 84% and 85% of that induced by the first tunnelling, respectively. In addition, the final maximum settlement of pile group induced by twin tunnelling were 1.42%d_p, 4.07% d_p, 4.26% d_p and 5.51% d_p in Simulation MM, Test FF, Simulation FF and Simulation BB, respectively. Based on the centrifuge and computed results, the maximum settlement of pile group induced by twin tunnelling increased with the increase of C/D. However, from the parametric study, it can be concluded that the effect of excavation of twin tunnels on pile group settlements can be ignored when C/D is higher than 5.5.

4.3 *Apparent loss of pile group capacity due to twin tunnelling*

The pile bearing capacity is often quantified using the pile settlements. In this study, the relative loss of the pile bearing capacity (P_{Loss}) was used to quantify the apparent loss of pile group capacity due to twin tunnelling. P_{Loss}, and was defined as:

$$P_{Loss} = \frac{N - N_w}{N_0} \cdot 100\% \tag{2}$$

where N is the load deduced from the load-settlement curve (see Fig. 4), N_w is the working load, and N_0 is the ultimate bearing capacity of pile group.

As shown in Fig. 4, in Simulation MM, the settlement of the pile group due to working load was 0.91% d_p (i.e. 7.3 mm), and the induced pile group settlement due to twin tunnelling was 1.6% d_p (i.e. 12.8 mm), which was equivalent to apply an additional load of 3.1 MN on the pile cap (based on the load-settlement curve shown in Fig. 4). In the Simulation MM, the ultimate bearing capacity of the pile group (N_0) was 9.3 MN deduced by the failure criterion proposed by Ng et al. (2001). Hence, the loss rate of pile group capacity due to twin tunnelling (P_{Loss}) was 33.7%. In the Simulation FF and BB, the values of P_{Loss} were 68.7% and 90.2%, respectively. In addition, the value N_0 is 15.7 MN if the failure criterion proposed by IMMMFE (1985) be used. Hence, in Simulation MM, FF and BB, the values of P_{Loss} were 19.7%, 40.1% and 52.7%, respectively. It can be seen that, no matter which criterion was adopted, the maximum value of P_{Loss} was induced in Simulation BB, while the minimum value of P_{Loss} was induced in Simulation MM. This was consistent with the centrifuge tests reported by Ng et al. (2014). The maximum value of P_{Loss} was 2.7 times of the minimum value.

4.4 *Transverse tilting of pile cap*

Fig. 6 (a) and (b) shows the transverse tilting of the pile cap during the first and second tunneling (i.e. Simulation MM, Test FF, Simulation FF and Simulation BB). The tilting was determined as the ratio of differential settlements measured from the central points of the front pile (LP4) and the rear pile (LP3) to the distance between them. A positive value means the pile cap tilts towards the first tunnel and vice versa.

It can be seen from Fig. 6(a) that the maximum calculated tilting of the pile cap (0.116%) occurred in Simulation BB. This was due to the stress reduction induced by tunneling in Simulation BB, which was the largest among the three configurations.

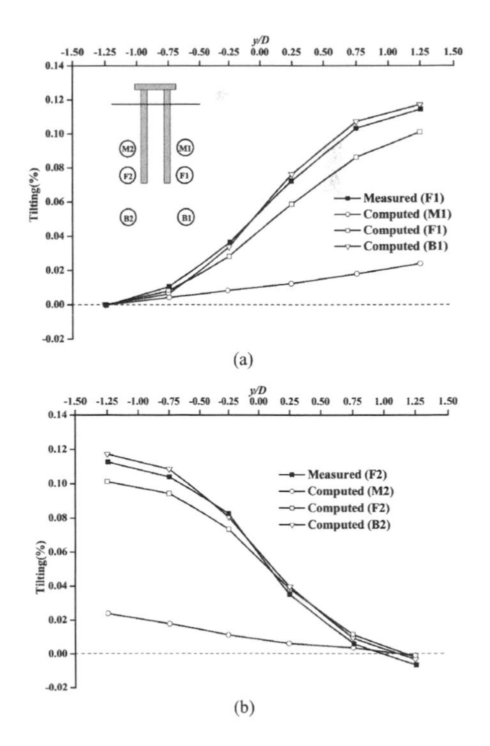

Figure 6. Transverse tilting due to twin tunnelling: (a) during the first tunnelling; (b) during the second tunnelling.

Similar to the settlement of the pile group, the increment of tilting of the pile cap also increased as the tunnel face approached the monitoring section and decreased as tunnel face continued to advance beyond the monitoring section. Different from the test or the simulations in this study, the relation between tilting of the pile cap induced by the first tunnelling and second tunnelling in Simulation MM and the process of tunnelling was linear. Comparing Fig. 6(a) with Fig. 6(b), the value of tilting was reduced gradually and close to zero during the second tunnelling. It implies that the plastic strains generated adjacent to the two tunnels did not overlap with each other.

The maximum tilting of pile cap induced by the first tunnelling in Simulation MM, Test FF, Simulation FF and Simulation BB were 0.024%, 0.114%, 0.100% and 0.116%. It can be seen that the maximum value of tilting was lower than the tilting limit suggested by Eurocode 7 (2001) for buildings and Technical code for building pile foundations (China industry standards for buildings, JGJ 94-2008) (i.e. 0.2%).

5 CONCLUSIONS

In this study, a three-dimensional centrifuge test and a series of numerical analyses were carried out to investigate the effects of pile group under working load in clay due to twin tunnelling in different depth. A centrifuge model test, i.e, Test FF (i.e., C/D = 2.5) was carried out. In addition, Simulation MM (i.e., C/D = 1.5), FF (i.e., C/D = 2.5) and BB (i.e., C/D = 3.7) were investigated using numerical analysis. Based on the results obtained from centrifuge model test and numerical analysis, the following conclusions can be drawn.

1. In all the cases, the pile settlement increased as the tunnel face approached the monitoring section and then decreased as the tunnel face continued to advance beyond the monitoring section. The maximum settlement of the pile group induced by twin tunnelling occurred in the case of C/D = 3.7 (i.e., Simulation BB).
2. The maximum loss of pile bearing capacity induced by twin tunnelling occurs when the whole pile group is located within the major influence zone due to tunnelling proposed by Jacobsz (2001). In the cases of C/D = 1.5, C/D = 2.5 and C/D = 3.7, the loss rate of the pile bearing capacity (P_{Loss}) are 33.7%, 68.7% and 90.2%, when the displacement failure criterion proposed by Ng et al. (2001) was used.
3. The maximum transverse tilting of pile group induced by twin tunnelling is also observed in the case of C/D = 3.7 (i.e., Simulation BB). The maximum tilting of the pile cap is lower than the tilting limit suggested by Eurocode 7 (2001) for buildings and Technical code for building pile foundations (China industry standards for buildings, JGJ 94-2008) (i.e. 0.2%).

ACKNOWLEDGEMENTS

The authors would like to acknowledge the financial support from National Natural Science Foundation of China(No. 51508113; No. 41362016), Guangxi Key Laboratory of Geomechanics and Geotechnical Engineering(13-KF-02)

REFERENCES

IMMMFE (1985). Axial pile loading test-part 1: static loading. *Geotechnical Testing Journal*, **8**: 79–90.

Jacobsz, S.W., Standing, J.R., Mair, R.J., Hagiwara, T., & Sugiyama, T. (2004). Centrifuge modelling of tunnelling near driven piles. *Soils and Foundations*, **44**(1): 49–56.

Loganathan, N., Poulos, H.G., & Stewart, D.P. (2000). Centrifuge model testing of tunnelling-induced ground and pile deformations. *Géotechnique*, **50**: 283–294.

Marshall, A. (2009). Tunnelling in sand and its effect on pipelines and piles. Ph.D. dissertation, Cambridge: University of Cambridge.

Marshall, A.M. (2012). Tunnel-Pile Interaction Analysis Using Cavity Expansion Methods. *Journal of Geotechnical and Geoenvironmental Engineering*, **138**(10): 1237–1246. doi: 10.1061/(ASCE)GT.1943-5606.0000709.

Marshall, A.M. & Mair, R.J. (2011). Tunnelling beneath driven or jacked end-bearing piles in sand. *Canadian Geotechnical Journal*, **48**(12): 1757–1771. doi: 10.1139/t11-067.

Ng, C.W.W. & Lu, H. (2014). Effects of the construction sequence of twin tunnels at different depths on an existing pile. *Canadian Geotechnical Journal*, **51**(2): 173–183. doi: 10.1139/cgj-2012-0452.

Ng, C.W.W., Yau, T.L.Y., Li, J.H.M., & Tang, W.H. (2001). New Failure Load Criterion for Large Diameter Bored Piles in Weathered Geomaterials. *Journal of Geotechnical and Geoenvironmental Engineering*, **127**(6): 488–498. doi: 10.1061/(ASCE)1090-0241(2001)127:6(488).

Park, K.H. (2004). Elastic solution for tunnelling-induced ground movements in clays. *International Journal of Geomechanics*, **4**(4): 310–318. doi: 10.1061/(ASCE)1532-3641(2004)4:4(310).

Peng, S. (2012). Influence of stress relief due to deep excavation on capacity of pile foundations. PhD thesis, Hong Kong University of Science and Technology, Hong Kong.

Soomro, M.A., Hong, Y., Ng, C.W.W., Lu, H., & Peng, S. (2015). Load transfer mechanism in pile group due to single tunnel advancement in stiff clay. *Tunnelling and Underground Space Technology*, **45**: 63–72. doi: 10.1016/j.tust.2014.08.001.

Standards, P.R.O.C. (2008). Technical code for building pile foundations. China Architecture and Building Press.

Civil, Architecture and Environmental Engineering – Kao & Sung (Eds)
© 2017 Taylor & Francis Group, ISBN 978-1-138-02985-9

Index selection and weight analysis of the underground development of urban municipal public facilities

Lu Ye
College of Architecture, Soochow University, Suzhou, Jiangsu, China

Liming Zhu
Suzhou Institute of Architectural Design Co. Ltd., Jiangsu, China

ABSTRACT: Municipal public facilities underground development is one of the most important contents of urban underground space planning. In this paper, we study the major factors affecting the underground development of public facilities in the urban planning, including locational factor, geological environment, economy, engineering technology conditions, policies and management mechanisms, and disaster prevention safety. The subjective and objective weighting methods are adopted to analyze the weights of these factors. Weight analysis lays the foundation for the establishment of evaluation model and provides scientific and reasonable evidences for establishing the "feasibility and appropriateness of underground development of municipal public facilities" evaluation system and the resolutions of the underground development of municipal public facilities.

1 INTRODUCTION

Underground development of municipal public facilities is one of the major functions of urban underground space development and utilization. Underground municipal public facilities have great advantages in upgrading the safety, stability, and post-stage maintenance of infrastructure. In large cities like Shanghai, Beijing, Shenzhen, and Guangzhou in China, the underground development of municipal public facilities has achieved certain development, for example, large-scale underground transformer substation, underground sewage treatment site, and underground comprehensive pipe rack.

The development of urban underground space facilities shall be closely combined with urban planning. With the development of the city and social economy, it shall consider multiple factors when making resolutions on whether the facility shall be developed underground. Even if the development conditions and construction technologies are feasible, only when the expected "benefits" of underground development of municipal public facilities are more obvious than the ground development, it would present the specific enforceability. However, "benefits" do not only include the direct economic ones resulted from land savings generated from underground development, but also bring more social and environmental benefits, such as the contribution to urban transportation and the environmental improvement. Meanwhile, municipal projects are the life projects mainly invested by the government. It shall be planned in advance for intensive utilization of land source, low-carbon sustainable development planning, and improving the comprehensive benefits of the engineering.

In this paper, we study the influential factors of urban underground space development level and specific indicator factors of underground development from macro and micro perspectives, respectively.

2 LOCATIONAL FACTOR

Locational factor can be analyzed from macro, medium and micro levels. For the construction location, it judges from the municipal public facility's requirement level on the underground development according to the requirements of urban planning.

Macro and medium locational factors represent the development level of urban underground space indirectly. According to the current practices of Chinese cities, the management development of urban underground space is mainly taken charge by the civil air defense departments. The civil air defense level of these cities affects the development force for underground space to a large extent. Therefore, in the macro locations, the air defense level and population of cities directly decide the importance of the underground space in their urban development.

Medium location mainly represents the position of cities in the surrounding city clusters. For instance, cities in the core positions of the city clusters are superior to those not in the core positions in terms of the urgency of underground space development.

Micro location factor mainly evaluates the influences of the micro environment of different locations in the city on the development conditions and approaches of municipal public facilities. For example, the feasibility for carrying out underground development in the city center is completely different from that in the suburb. The soundness of micro location conditions affects the benefits resulted from the underground development directly. As urban area in China is expanding continuously, many municipal public facilities in the suburb previously gradually move to the city center and the value of the land is becoming increasingly higher. Therefore, underground development enjoys an increasingly higher feasibility. To sum up, the level-2 factors under locational factor are divided into three items: macro location of macro factor, medium location, and the micro location of micro factor.

3 ECONOMIC FACTOR

3.1 *Influential mechanism*

In accordance with the actual status, China is still in the stage of prioritizing the economic development. Therefore, when it comes to the underground development of municipal public facilities, economic strength and input–output ratio are the biggest concerns of decision makers.

The main factors hindering underground development of municipal public facilities under the current status are as follows: the financial income of the city is too low to have capacity to make the underground development of municipal public facilities; underground development calls for huge investment. For example, for a 110 kV transformer substation, the capital consumed by its underground development and ground development is 1.5–5.1. The capital for development is available, but the benefits brought by underground development are not obvious.

Economic factor comes first, which is applicable for not only municipal public facilities but also other urban facilities. However, it is common that there are multiple plans for municipal public facilities. Among the evaluation standards, the weight of economic factor varies.

3.2 *Evaluation indicator selection*

Economic factor is mainly divided into three level-2 indicators: Per capita GDP, per capita financial income, and per capita disposable income.

3.2.1 *Per capital GDP*

Per capita GDP mainly represents the economic development level of the city and affects the strength and model of underground space development. It facilitates us to master the stage of urban underground space development from the overall perspective. Widely recognized "threshold theory" holds: When the per capita GDP reaches US $3000, it indicates that the city has entered the large-scale underground development stage (Shu 2002a, b, Tong 1994).

3.2.2 *Per capita financial income*

In the current stage, a major model for Chinese municipal public facility development is the allocation from government finance. Therefore, the financial income of the city affects the ability of municipal public facility development to a large extent, especially when the underground development increases the financial investment.

3.2.3 *Per capita disposable income*

Per capita disposable income is one of the important factors evaluating the living standards of urban residents. Cities featuring high per capita disposable income have strong economic dynamics, and their residents have high recognition and acceptance on the new facilities. If the residents of a city are still striving to meet daily necessities and its government is developing new municipal public facilities, it would receive great resistance from residents and leads to unsound social effects.

4 POLICY MECHANISM FACTOR

4.1 *Influential mechanism*

The current municipal public facility planning face the following problems: lack of large-scope regional municipal public facility planning; municipal public facilities are separate and lack connection with each other; traditional land supply model leads to the shortages of intensive land use of municipal public facilities; municipal public facility regulations are old and there are contradictories among them; and the integration with other public service facilities is insufficient. One of the advantages of underground development of municipal public facilities is intensive land use and land saving. However, because of the factors of governmental system or policy mechanism, the underground intensive development for some municipal public facilities faces great difficulties. The most collective representation is in the construction and operation of integrated trenches. Integrated trenches have been developed for several years abroad. It has been proved to be a sound measure in solving the underground pipe problem. Since the development and operation of Zhangyang Road of Pudong District in Shanghai nearly 20 years ago, there has been no significant progress. The development of some integrated trenches could hardly make ends meet when it enters the operation stage with suffering losses. The major reason is that the integrated trench joining-responsible units failed to clarify the responsibility ratios of their departments. The integrated trench, which had been proved to have sound benefits in foreign countries, came across many problems in the domestic development. The fundamental reason derives from the problem of Chinese municipal public facility management system and management mechanism. In 2015, 69 cities throughout China initiated the underground integrated pap rack development, and the construction scale was about 1000 km. In March 2016, it was pointed out that the urban planning construction management shall be strengthened in the government working report and proposed the clear target of over 2000 km underground integrated pap rack. It is foreseeable that urban underground integrated pap rack would become the important content for urban infrastructure development in the future. [8] This indicates that the changes of policy mechanism have huge influences on the development of urban municipal public facility.

4.2 *Evaluation indicator selection*

It shall follow the line of comprehensive planning, integrative planning, and establishing indicators to study the underground development of municipal public facilities. Standardized municipal public facilities and planning system are beneficial for the underground development of municipal public facilities. On the basis of the above mentioned analysis, level-2 indicators under government factor are mainly divided into the optimal level of related policies and regulations and the implementation force of government in driving the underground development of municipal public facilities.

5 URBAN CONSTRUCTION FACTOR

The urbanization process of cities has significant influences on their public facility development. The higher the urbanization level, the larger the burden for the urban life engineering system and the higher the requirements and scale for construction. Therefore, urbanization ratio is a macro factor in evaluating the underground development level of the city in this model.

Urban annual completion ratio is the objective indicator in evaluating urban construction strength. The completion level presents the construction capacity and construction efficiency of the city. Therefore, two level-2 indicators would be set under the first indicator of urban construction: urbanization rate and urban annual completion ratio.

Meanwhile, from a micro perspective, plot ratio, population density, land using property, construction land level, and seismic grade of the region also serve as factors in analyzing whether the underground development is worthy.

6 CONSTRUCTION BASIS OF UNDERGROUND SPACE DEVELOPMENT

If the construction basis is weak, it would encounter the following resistances in considering the underground development of facilities: shortage of the engineering geological documents of the city, shortage of the underground development capacity, shortage of government decision-making experience, and shortage of residents' acceptance. The main level-2 indicators for underground space development foundation are per capita underground space area and underground space construction ratio.

7 GEOLOGICAL AND ENGINEERING TECHNOLOGY FACTOR

Analyzing engineering technology condition and geological condition together, they represent the difficulty in implementing the municipal public facilities in the project implementation stage from subjective and objective aspects. If the economic condition policy mechanism is permitted, the underground development could not be realized without the guarantee of mature engineering technology. The mutuality of the geological conditions decides whether the underground development could be carried out smoothly for this facility. They have close relation with the economic factor. The investment of the underground development of a project is in inverse proportion to the engineering technology and local geological situations. If engineering technology conditions in Beijing and Shanghai are the same, the costs for developing underground transformer substation are different because of the geological condition. In some regions of Beijing, they do not need foundation pit dewatering and enclosure. In comparison, the expense of enclosure of a large foundation pit in Shanghai is approximately 40–50% of that of the whole project.

In the same way, Shanghai now has the capacity to make underground development of 500 kV transformer substation, which is developed on the strong engineering technology condition and economic foundation. As the engineering technology condition is becoming more mature, the geological conditions of a region are by no means the decisive factor anymore. For example, the underground space of Shanghai stands at the cutting edge and serves as the model for domestic cities under the support of construction capacity and capital of the government.

8 ENVIRONMENTAL FACTOR

Environmental factor mainly analyzes the influences of underground municipal public facilities in pre-setting areas on the surrounding environment.

First, underground development of municipal public facilities could significantly reduce pollutions of the surrounding area, such as the underground development of transformer substation and sewage treatment plant. Meanwhile, the underground development of these facilities could alleviate the psychological depression of surrounding residents. Greening implementation on the land of underground development could increase the greening ratio of this area and improve air quality in the surrounding areas.

Furthermore, Chinese society is confronting with increasingly large pressure from the international community in terms of carbon reduction and environmental protection. How to maximize the environment-friendly and low-carbon development strategies in the underground space planning and whether the above mentioned targets could be fulfilled through underground development of ground facilities have become the research hot spots for people in the industry.

Urban municipal public facility is the foundation for urban development, serving as a crucial point in guaranteeing the sustainable development of cities. "The connotation of low carbon development of municipal public facility is the same with that of low-carbon economy, that is, improve energy efficiency and minimize the carbon emission. The final target is to reduce environmental pressure and realize sustainable development; the core is low-carbon technology innovation, system innovation and the transformation of the development concept" (Zhang, B. et al. 2010).

It is apparent that carbon emission mainly exists in the last three stages among planning, designing, construction, operation, and dismantles stages of municipal public facilities. However, the soundness of planning and designing is extremely important for the emission volume in the last three stages. These two stages are the key for carbon emission reduction. Therefore, the underground development of municipal facilities is apparently beneficial for the whole lifecycle of the municipal public facilities.

9 WEIGHT ANALYSIS OF FACTORS

To sum up, Tables 1 and 2 show the evaluation indicators and influential factors. After analyzing and selecting major factors, defining the weight of these factors is an important procedure in defining whether the underground development is applicable for this facility by using mathematic model as "indicator + weight + analysis model → analysis result".

9.1 *Methods for the weight of micro factors*

In this paper, micro factors are mainly presented as major planning control indicators. Therefore, it could adopt the objective weighting method to redistribute the weight of these factors.

9.2 *Methods for the weight of macro factors*

Analytic Hierarchy Process is the decision-making method that breaks down decision-related factors to hierarchies of target, standard, and plan and

Table 1. Macro influencing factors.

Indicator (level 1)	Indicator (level 2)	Unit
Location	Urban location (macro)	——
	Urban location (medium)	——
Economy	Urban per capita GDP	10,000 RMB
	Per capita financial income	10,000 RMB
	Per capita disposable income	10,000 RMB
Urban construction	Urbanization ratio	——
	Urban annual completed area	10,000 m^2
Underground space	Underground space status	10,000 m^2
	Facility construction volume of underground space	10,000 m^2
Policy mechanism	Government implementation force, soundness of government-related policies	—— ——

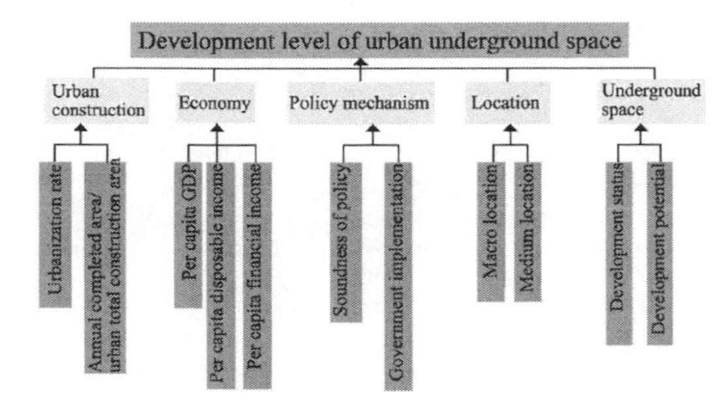

Figure 1. Analysis hierarchy process model.

Table 2. Micro influencing factors.

Evaluation indicators	Unit
Land use property	——
Locational value (micro)	——
Greening ratio	%
Population density	10,000 men/km^2
Land development intensity	——
Construction density	——
Construction land grade	——
Building antiseismic grade	——

Table 3. Micro factor weights.

Micro factors	Unit	Weight
Land use property	——	0.125
Locational value (micro)	——	0.125
Greening ratio	——	0.125
Population density	10,000	0.125
Land development intensity	men/km^2	0.125
Construction density	——	0.125
Construction land grade	——	0.125
Building antiseismic grade	——	0.125

Table 4. Macro factor weights.

Macro factor	Unit (RMB)	Weight (ranking)
Per capita GDP	10,000	0.1705 (1)
Government implementation force	——	0.1584 (2)
Macro location	——	0.1197 (3)
Per capita disposable income	10,000	0.1000 (4)
Underground space construction rate	——	0.0942 (5)
Urbanization ratio	——	0.0905 (6)
Policy optimization level	——	0.0712 (7)
Per capita financial income	10,000	0.0587 (8)
Medium location	——	0.0538 (9)
Per capita underground space area	——	0.0423 (10)
Annual completion rate	——	0.0407 (11)

then makes qualitative analysis and quantitative analysis on this basis. In this paper, we adopt yaahp, the Analytic Hierarchy Process software, to analyze the macro factors and establish target hierarchy, standard hierarchy, and plan hierarchy (Figure 1).

Four major procedures in establishing a model in the Analytic Hierarchy Process are as follows: establish the hierarchical structure model on the basis of the mentioned analysis: highest hierarchy (target hierarchy), middle hierarchy (standard hierarchy), and bottom hierarchy (plan hierarchy); construct all judgment matrixes in different hierarchies, that is, defining weights of all factors by comparing the importance of different indicators of the same hierarchy. In this paper, we constructed a judgment matrix through yaahp software; separated hierarchy sequencing, made computing sequencing on factors of relatively high hierarchy. The overall hierarchy sequencing and re-sequencing weights of targets for all factors in this model, especially in the plan of the bottom level, were used to obtain the final weight of the analysis.

In evaluating the weight of factors subordinated to "urban underground space development level", it shall first compare the important level of factors in the standard level and then compare all factors, especially in the plan level, and finally find the weight of macro factors in the plan level through software in Table 4.

10 CONCLUSIONS

In this paper, we analyzed and selected indicators affecting the underground development of urban municipal public facilities and adopted corresponding weighting approach to make weight analysis on various indicators. It lays a foundation for establishing "feasibility and applicability of the underground development of municipal public facilities" evaluation indicator system.

After defining weights for corresponding indicators, it could select appropriate constitutive model to establish "feasibility and applicability of the underground development of municipal public facilities" evaluation indicator system. The applicability of underground development of municipal public facilities deduced through this method makes the primary exploration for introducing quantitative calculation to the decision of underground development of urban facilities.

ACKNOWLEDGMENTS

This work was financially supported by the National Natural Science Fund (51508359).

REFERENCES

Deng, X. & Li, J. & Zeng, H. 2012. Research on Computation Methods of AHP Wight Vector and Its Applications. *Mathematics in Practice and Theory* 42(7):93–100.

Shu, Y. 2002. *The development And The Uses of Underground space resources*. Shanghai: Tongji University Press.

Tong, L.X. 1994. *Underground Architecture*. Ji'nan: Shandong Science & Technology Press.

Tong, L.X. & Zhu, W.J. 2009. *The city underground space resources evaluation and exploitation and utilization planning*. Beijing: China Architecture Industry Press.

Yan, P. & Dai, S.Z. 2010. Integration of Infrastructure Facilities Under the Background of Intensive Land Use. *City Planning* 186(1):109–115.

Zhang, B. & Huang, Y. & Cheng, Y. 2010. The Low carbon and development path of underground municipal public utilities. *Construction Economy*. 335(9):97–100.

Zhu, D.M. 2004. The basic law of the city underground space development. *Underground Space* 24(3):365–367.

Zhang, Y. & Liu, C.L. & Yang, J.S. 2016. Reseach on Sustainable Development of Promotion the Construction of Urban Underground Integrated Pap Rack. [J]. *Water & Wastewater Engineering* 42(6):1–3.

Civil, Architecture and Environmental Engineering – Kao & Sung (Eds)
© 2017 Taylor & Francis Group, ISBN 978-1-138-02985-9

Risk assessment of pipelines in operation based on AHP and application of MATLAB

Lei Chen, Guixuan Wang & Jie Zhao
Civil Engineering Technology R&D Center, Dalian University, DaLian, LiaoNing, China

ABSTRACT: The pipeline risk evaluation model is established on the basis of the analysis of risk factors encountered in the operating period of pipeline. Moreover, subfunction program of the model is written using MATLAB. Applying the model to the risk evaluation of a real oil pipeline in Dalian, which can identify the risk safety grades and the risk response measures in the operating period, the calculated result is consistent with the actual one. Examples show that using the risk evaluation model established by the combination of Analytic Hierarchy Process (AHP) and MATLAB can not only draw the specific quantitative indicators to carry out the objective assessment of pipeline risks, but also simply and efficiently provide a reliable basis to the risk assessment and management of the operating pipeline.

1 INTRODUCTION

With the increase of urbanization, the demand for oil, natural gas, and tap water is increasing, which promotes the rapid development of pipeline transportation. It is expected that, in the period of "Thirteen Five", the total length of China's oil, gas, and water supply pipelines will reach 200,000 km, and the government will build the inter-regional oil and natural gas supply network system and urban water supply pipe network system (ZHANG Zhihong, 2012). With the passage of time, a number of pipelines are in "sick" operating state; with the increasing use of pipelines, it is necessary to perform health diagnosis and risk management for those pipelines.

In recent years, scholars specialized in pipeline fields in China have made a preliminary exploration and research on the risk of pipeline and obtained some fruitful results. Liu Jiaming (2014) determined that the weight of pipeline corrosion, equipment, third-party damage, and natural environment are, respectively, 0.49, 0.31, 0.12, and 0.08. By introducing several improvements in AHP, it obtained the pipeline health index, which is 0.707, by fuzzy comprehensive evaluation, which is in potentially unsafe levels. Wang Junfang, Yu Qianxiu (2009) evaluated 10 pipe sections of an oil pipeline using the unascertained measure model. The evaluation results were consistent with the actual ones, which indicates that unascertained measure model can effectively evaluate the oil and gas pipelines. Wang Kai (2009) carried out qualitative and quantitative analyses for long-distance pressure pipeline by using fault tree analysis method to identify the weak sections of pipeline and the risk factors and the degree of risk, which provides specific data to operation of pipeline and improves the safety. All these evaluation methods lack a procedural calculation model. The calculation and analysis process still needs to be simplified and the computational efficiency also needs to be improved.

In this paper, we conducted a research on risk factors encountered in the operation of pipeline on the basis of AHP, wrote the subfunction program of AHP using MATLAB, and built the pipeline risk evaluation model. This model can not only carry out quantitative assessment of pipeline risk, but also make the corresponding risk management program according to assessment results, which can provide a reliable judgment basis for the follow-up risk management of pipeline. Besides, the calculation and analysis process are simple and it can work efficiently.

2 AHP STRUCTURE EVALUATION MODEL

2.1 *The basic principles and steps of AHP*

Analytic Hierarchy Process is a hierarchical weight decision analysis method put forward by Professor Saaty from the United States in the early 1970s (SU Yi, 2011). This method lists all elements related to the target issues and then establishes goal, rule, and index layer so as to establish a multilevel structural analysis model. It is a modeling, quantitative, and systematic process referred to the thought process of decision makers on complex systems. Furthermore, it reflects the subjective importance of each factor and thus expands the quantitative analysis and research, which can be divided into the following steps: establishing a hierarchical structure analysis model; constructing judgment matrix; calculating the weight vector of judgment matrix; testing the consistency; and making judgment.

First, the analytic hierarchy process is to construct a hierarchical analysis model with issue principled. Second, the analytic hierarchy process stresses the upper and lower subordinate relationships strictly and uses the 1–9 scaling method to carry out a quantitative comparison of the importance of this level of factors with the previous layer and establish the judgment matrix:

$$A = \left(a_{ij}\right)_{m \times n} \tag{1}$$

where a_{ij} is the important degree of the i factor relative to the j factor and the matrix A is a reciprocal matrix, which satisfies a_{ij}, $a_{ij} = 1/a_{ji}$, $a_{ii} = 1$. For hierarchical structure, the number of judgment matrix is the same as the number of affiliations from the top layer to the bottom layer. After the establishment of the judgment matrix, the most important thing is to obtain the maximum eigenvalue and the corresponding eigenvector, that is, the weight vector. Usually, the characteristic polynomial is used to obtain the eigenvalue of matrix, which can be expressed as:

$$\det\left(A - \lambda I\right) = 0 \tag{2}$$

where I is the unit vector, λ is the eigenvalue, from which the characteristic vector can be obtained.

Practice shows that with the increase in the complexity and the structures of the problem, the calculation and analysis process of AHP will be very complex, especially for the solution of the matrix eigenvalue and eigenvector. In fact, the eigenvalues and eigenvectors can be easily obtained and the consistency test of judgment matrix by writing a function program in MATLAB, which greatly simplifies the calculation.

2.2 *Pipeline risk index evaluation system*

The number of risk factors that affect the operation of pipeline is high; therefore, the selection of evaluation index will have an important effect on the accuracy and reliability of the final evaluation results. Referring to the research of pipeline damage in China and abroad (JIANG Delin, 2011; Thomas jones, 1992 & Benkherouf A, 1988) and pipeline integrity specifications, we selected 12 representative risk indicators affecting the safety operation of pipeline to build a hierarchical evaluation model, which can more fully reflect the risk characteristics of the pipeline operation period, as shown in Figure 1.

2.3 *Classification of pipeline risk safety grade*

Currently, most of the pipeline condition assessment using fuzzy evaluation method, such as pipeline operation safety, exists some security risks and other fuzzy qualitative evaluation, and there is no clear pipeline operation risk safety grade classification method and corresponding risk response measures. With reference to the four-level method used in pipeline security status evaluation in industrial pipeline, we determine to use the four

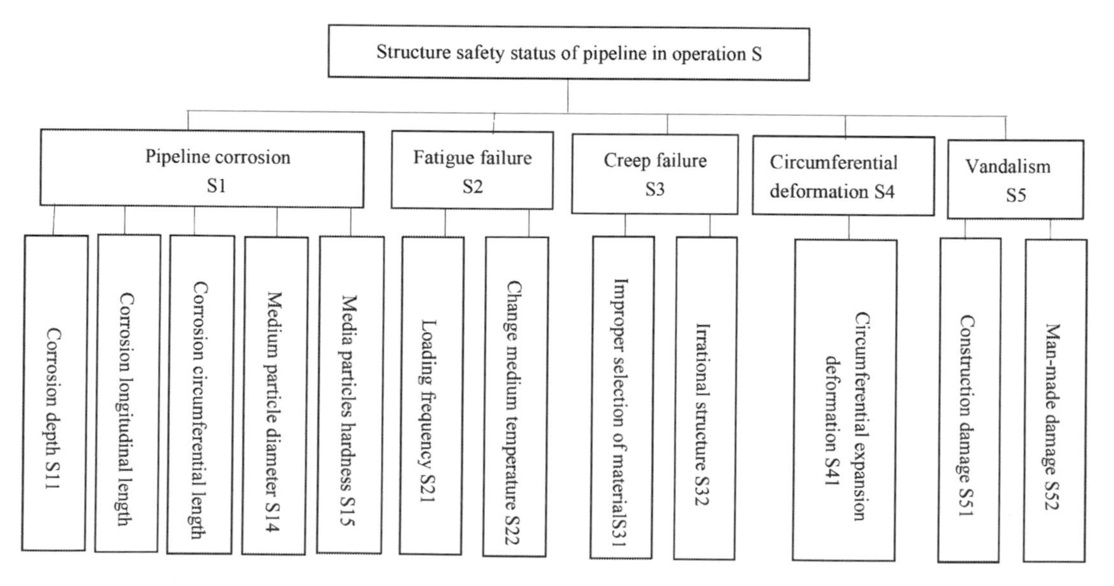

Figure 1. Safety assessment system of operating pipelines.

Table 1. Risk level and response mechanisms of operating pipelines.

Risk level	Response mechanisms
First level of risk	(Weight greater than or equal to 0.2; extremely dangerous): take immediate measures
Second level of risk	(Weight ranged from 0.15 to 0.2; severely damaged): master the risk dynamic and make preparations to cope
Third level of risk	(Weight ranged from 0.1 to 0.15; moderate damage): exclude potential risks within 7 days
Fourth level of risk	(Weight less than 0.1; slight damage): strengthen risk management to eliminate risk as soon as possible

division methods as the operational pipeline risk safety level classification method, and according to the weights of risk factors and the experience of experts and site management personnel, it established the pipeline risk security level and response mechanisms, as shown in Table 1.

3 PROGRAMMING MODEL DESIGN OF AHP

We use MATLAB programming language to write the subfunction of Analytic Hierarchy Process (AHP) to achieve the calculation and analysis procedure. The procedure is that when inputting the judgment matrix, seeking single-level sorting, total sorting, consistency checking, and eventually getting the total ranking weight of all risk factors will be carried out automatically in the procedure (SONG Fei, 2008).

3.1 *Solving normalized eigenvectors*

Calling function *eig* in MATLAB is to solve the maximum eigenvalue and the corresponding eigenvector of matrix. The call format is $[V, D] = eig(A)$, where V is the eigenvector matrix and D is eigenvalue matrix. Defining function maxeigvalvec.m to implement.

function [maxeigval, w] = maxeigvalvec (*A*); % Seeking maximum eigenvalue and the corresponding normalized eigenvectors of judgment matrix A

[*v*, *d*] = eig (*A*); % *v* is eigenvector matrix and *d* is eigenvalue matrix

eigenvalue = diag (*d*);

maxeigval = max (eigenvalue); % Seeking maximum eigenvalue maxeigval

maxeigvec = v (:, 1); % Seeking eigenvector corresponded to the maximum eigenvalue

w = maxeigvec ./sum(maxeigvec); % Eigenvector maxeigvec normalization

end

3.2 *Calculation of single-level sorting and consistency test*

When constructing judgment matrix, because of the diversity of risk the pipe encountered and the understanding limitations of inspectors for each risk factor, the consistency of the judgment matrix will inevitably be damaged and result in the biased judgment result. In order to avoid the error, emergence of A is more important than B, B is more important than C, while C is more important than A; in the same layer, the consistency of judgment matrix is checked by the obtained maximum eigenvalue λ_{max}. Therefore, it is necessary to test the consistency of the judgment matrix. Defining function examine.m to test the consistency of matrix:

function [CI, RI] = examine (maxeigval, *A*); % Consistency test of single-level sorting weight vector, maxeigval is maximum eigenvalue, *A* is judgment matrix

N = size(*A*,1);

RI = [0.0 0.0 0.58 0.90 1.12 1.24 1.32 1.41 1.45 1.49 1.51 1.54 1.56 1.58 1.59]; % *RI* is consistency index of average random (He Furong, 2014)

RI = RI(*n*);

CI = (maxeigval-*n*)/(*n*-1); % *CI* is consistency index

CR = CI/RI; % *CR* is consistency ratio

if CR < 0.1

disp ('Judgment matrix *A* passes the consistency test')

else

disp ('Judgment matrix *A* does not pass the consistency test, please readjust the judgment matrix')

end

3.3 *Calculation of total-level sorting and consistency test*

Defining function tsw = tolsortvec (utw, dw, CIC, RIC); % Seeking total-level sorting weight vector and testing its consistency, tsw is the total ranking weight of risk factors in the second layer, utw is the total ranking weight row vector of factors in the upper layer, dw is the single-sorting weight vector matrix of the lower-layer factors for the upper-layer factors, CIC is the consistency index column vector, and RIC is the index column vector of random consistency.

tsw = dw * utw;

CR = utw .* CIC / (utw .* RIC);

if CR > = 0.1

```
disp ('total-level sorting does not pass the consistency test, please readjust the judgment matrix')
else
disp ('total-level sorting passes the consistency test')
end
```

4 ENGINEERING APPLICATIONS

4.1 *Project overview*

A tank farm is located in the Ganjingzi district of Dalian, the pipeline of which operated for more than 10 years. The length of the oil pipeline is about 600 m, the diameter of pipe is 200 mm, and the thickness of wall is 6 mm. The pipeline adopted an overhead layout scheme; the bottom of the pipe is about 30 cm above the ground. According to the site investigation and inspection records of staff, we know that both the inner surface and the outer surface of the pipe have a different degree of corrosion; it is accompanied by a certain degree of micro deformation and damage caused by human factors.

According to the hierarchy structural diagram in Figure 1, the safety state S of the pipeline is the target layer, the pipeline corrosion (S1), fatigue damage (S2), creep rupture (S3), circumferential deformation (S4), and vandalism (S5) are the criterion layer, and the index layer is specifically divided into 12 influencing factors. We use 1–9 scale methods to carry out the pairwise comparison of risk factors so as to establish the judgment matrix of criterion layer and index layer, as shown in Tables 2–6.

For the circumferential deformation, the influence factor is only expansive deformation, so the judgment matrix is S4 = [1].

Table 2. Judgment matrix (S).

S	Pipeline corrosion S1	Fatigue failure S2	Creep failure S3	Circumferential deformation S4	Vandalism S5
Pipeline corrosion S1	1	7	6	4	2
Fatigue failure S2	1/7	1	1/2	1/5	1/6
Creep failure S3	1/6	2	1	1/3	1/5
Circumferential deformation S4	1/4	5	3	1	1/2
Vandalism S5	1/2	6	5	2	1

Table 3. Judgment matrix (S1).

S1	Corrosion depth S11	Longitudinal length S12	Circumferential length S13	Particle diameter S14	Particles hardness S15
Corrosion depth S11	1	3	2	7	5
Longitudinal length S12	1/3	1	1/2	4	3
Circumferential length S13	1/2	2	1	5	4
Particle diameter S14	1/7	1/4	1/5	1	1/3
Particles hardness S15	1/5	1/3	1/4	3	1

Table 4. Judgment matrix (S2).

S2	Loading frequency S21	Medium temperature S22
Loading frequency S21	1	3
Medium temperature S22	1/3	1

Table 5. Judgment matrix (S3).

S3	Improper selection of material S31	Irrational structure S32
Improper selection of material S31	1	1/2
Irrational structure S32	2	1

Table 6. Judgment matrix (S5).

S5	Construction damage S51	Manmade damage S52
Construction damage S51	1	3
Manmade damage S52	1/3	1

4.2 *Write calculation program*

4.2.1 *Input judgment matrix*

```
Clear; % Clear all variables
  S = [1 7 6 4 2;1/7 1 1/2 1/5 1/6;1/6 2 1 1/3 1/5;1/4 5 3 1 1/2;1/2 6 5 2 1];
  S1 = [1 3 2 7 5;1/3 1 1/2 4 3;1/2 2 1 5 4;1/7 1/4 1/5 1 1/3;1/5 1/3 1/4 3 1];
  S2 = [1 3;1/3 1];
  S3 = [1 1/2;2 1];
  S4 = [1];
  S5 = [1 3;1/3 1];
```

4.2.2 *Calculation of single-level sorting and consistency test*

```
[ maxS, wS ] = maxeigvalvec (S);
  [maxS1, wS1] = maxeigvalvec (S1);
  [maxS2, wS2] = maxeigvalvec (S2);
  [maxS3, wS3] = maxeigvalvec (S3);
  [maxS4, wS4] = maxeigvalvec (S4);
  [maxS5, wS5] = maxeigvalvec (S5);
  [RIS, CIS] = examine (maxS, S);
```

4.2.3 *Calculation of total-level sorting and consistency test*

```
dw = zeros (12, 5);
  dw (1:5,1) = wS1; dw (6:7,2) = wS2; dw (8:9,3) = wS3; dw(10,4) = wS4; dw(11:12,5) = wS5;
  CIC = [CIS1; 0;0;0;0];
  RIC = [RIS1; 0;0;0;0];
  tsw = tolsortvec (wS, dw, CIC, RIC)
```

By default, the order of judgment matrix S2, S3, S4, S5 is less than 3, which completes consistency and does not need to carry out the consistency test, so RIS2 = 0, CIS2 = 0; RIS3 = 0, CIS3 = 0; RIS4 = 0, CIS4 = 0; RIS5 = 0, CIS5 = 0.

4.3 *The operation result of program*

Running the above procedures, the interface will pop up two times "judgment matrix passes the consistency test" and a time "total-level sorting passes the consistency test", which explains that judgment matrix S, S1, and the total-level sorting all pass the consistency test and will obtain the total-level sorting weight of each risk factor in the second layer at the same time: tsw = [0.1968 0.0782 0.1231 0.0205 0.0381 0.0320 0.0107 0.0218 0.0436 0.1604 0.2061 0.0687].

According to the 1.3 section of safety level of pipeline risk, by comparing the calculated results, we can know that the weight of construction damage or vandalism is 0.2061, which belongs to the first level of risk; the weights of corrosion depth and circumferential expansion deformation are 0.1968 and 0.1604, respectively, which belong to the second level of risk; the weight of circumferential corrosion length is 0.1231, which belongs to the third level of risk; and the weight of transmission medium, loading frequency, pipeline structure, and material are all less than 0.1, which belong to the fourth level of risk. By combining with the operating conditions and daily management records of site management staff, it is clear that the construction of surroundings and man-made structures accidentally damage and cause a lot of adverse effects on the operation of the pipeline. Besides, because of the external natural environment and internal medium, the pipe also faces the risk of corrosion. As a result, the calculation results are in agreement with the actual condition of pipeline. Finally, we determine the corresponding risk response measures according to the calculated risk level so as to ensure the safe operation of the pipeline.

5 CONCLUSION

Here, we realized the calculation and analysis steps of AHP by MATLAB programming language and established the pipeline risk evaluation model. The model was applied to the risk evaluation of an actual pipeline risk assessment; by comparing the calculated results with the actual condition of pipeline, we can draw the following conclusions:

1. According to the fundamental principles of AHP, pipeline risk factors, and industrial pipeline security status level, the risk hierarchy assessment model of pipeline structure in operation period was established. The use of MATLAB programming language to write the AHP calculation procedure realizes the procedure of calculation and analysis and reduces the quantitative calculation to derive more accurate results.
2. Application of the programmed calculation model of AHP to the risk assessment of an actual oil pipeline can quickly obtain the weight value of each risk factor the pipe has faced in the operation period and can determine the risk level and take the corresponding risk

response measures. After the above procedure, the problem which primary seemed vague and difficult to judge became clear. Therefore, it is a good method to deal with the fuzzy problem and has strong generalization and practicability in the future.

REFERENCES

Benkherouf A, Auiodina A Y. Leak detection and location in gas pipelines. *IEE Processing,* 1988, 135 (2): 142–148.

He Furong. Research and application of analytic hierarchy process in construction bidding. Southwest Jiao Tong University, 2014, ChengDu.

Jiang Delin, Wang Zhijie and Sun Xia. Failure mode and preventive measures of pressure pipeline. *Chemical Technology and Development,* 2011(40)5: 53–54.

Liu Jiaming. The application of fuzzy comprehensive evaluation method in the safety of oil and gas pipelines. *Chemical Engineering and Equipment,* 2014(1): 201–204.

Q/SY1180-2009, Specification for Pipeline Integrity Management.

Song Fei, Zhao Fasuo. The application of analytic hierarchy process and Matlab program for risk analysis of Underground engineering. *Journal of Earth Sciences and Environment*, 2008(30)3: 292–296.

Su Yi, Zhang Hong. Research on the risk of credit business of commercial banks based on analytic hierarchy process. *New West,* 2011(13): 46–47.

Thomas jones, Harold Berger. Thermographie detection of impact damage in graphite-epoxye composites. *Materials Evaluation*, 1992 (11): 1446–1453.

Wang Junfang, Yu Qianxiu. Uncertain measurig model applied to safety evaluation of oil and gas pipeline systems. *Oil & Gas Storage and Transportation,* 2009(28)5: 14–18.

Wang Kai. Study on Risk Assessment Method Based on Fault Tree Analysis in Long Pipelines. *Wuhan University of Technology,* 2009, WuHan.

Zhang Zhihong, Wang Lijuan, LI Kefu. Development Trend of oil & Gas Storage and Transportation Technology in China. *Petroleum Science and Technology Forum,* 2012, 31(1): 1–6.

Civil, Architecture and Environmental Engineering – Kao & Sung (Eds)
© 2017 Taylor & Francis Group, ISBN 978-1-138-02985-9

Effect of ground co006Editions on design parameters for bridge foundations

M.T.A. Chaudhary
Department of Civil Engineering, Kuwait University, Kuwait

ABSTRACT: Ground conditions (geotechnical properties and geological setting) influences the near surface response of a strata subjected to seismic excitation. Geotechnical parameters required for computation of dynamic impedance of bridge foundations include damping ratio (β), shear wave velocity (V_s) and soil shear modulus (G). Values of these parameters are sensitive to the level of non-linear strain induced in the strata due to seismic ground motion. This paper attempted to investigate the effect of variation in soil properties like Plasticity Index (PI), Over Consolidation Ratio (OCR), effective stress (σ'), depth of soil strata over bedrock (H) and Impedance Contrast Ratio (ICR) on seismic design parameters (β, V_s and G) for two soil types (C and D in AASHTO code). It was found that variation in soil properties influenced seismic design parameters in soil type D more profoundly than soil type C.

1 INTRODUCTION

Influence of site conditions on seismic design spectra is well recognized in current codes. AASHTO code (2010) caters for these effects by characterizing the site conditions into six classes based on shear wave velocity in the top 30 m depth (V_{s30}) or alternative procedures based on average SPT N values or undrained shear strength of top 30 m strata. Design spectrum for a bridge is constructed based on the mapped PGA of the site, short period spectral acceleration, S_s and long period spectral acceleration, S_1 along with site modification factors F_{PGA}, F_a and F_v. Effect of variation in soil properties like Plasticity Index (PI), Over Consolidation Ratio (OCR), effective stress (σ'), depth of soil strata over bedrock and variation in stiffness of the bedrock are currently not included in the code based site characterizing and design spectrum construction processes. This study examined the influence of these geotechnical and site parameters on spectral acceleration, SA, used for seismic design of bridges.

2 METHODOLOGY

2.1 *Background*

Site conditions are classified into six categories (A to F) in the AASHTO code based on V_{s30}. Site Classes A and B are rock sites with V_{s30} more than 760 m/s. Shallow spread footings are commonly used for bridges in these site classes. Site class C represents very hard soil or soil-rock with V_{s30} between 360 and 760 m/s. Shallow spread foundations are suitable in this site class for upper range of V_{s30} while deep pile foundations can be used for the lower values of V_{s30}. Site class D has V_{s30} from 175 m/s to 360 m/s. Pile foundations are commonly used for bridges in this site class. This study focused on the most commonly occurring soil classes C and D which are suitable for both shallow as well as deep foundations and are characterized by a wide variation in V_{s30} (175–760 m/s), PI (0–60), OCR (1–10) and σ' (20–1500 kPa). It is to be noted that AASHTO site classes C and D roughly corresponds to EC-8 soil classes B and C respectively (CEN, 2003).

2.2 *Procedure*

Sensitivity of seismic design parameters (β, V_s and G) to variation in V_{s30}, PI, OCR, σ', depth of soil strata and variation in bedrock stiffness; characterized by variation in V_{sR} based on CSIR classification for rocks (Bieniawski 1974) was undertaken in this study. The seismic bedrock characteristics considered in this study are classified as rock classes I to V in which class I is a very good rock ($V_s > 3353$ m/s) and class V a poor rock ($V_s = 600$ m/s). Five generic soil profiles falling within the V_{s30} range for soil classes C and D were selected from Douglas et al (2009) as depicted in Fig. 1. Variations in PI, OCR, σ', strata depth and Impedance Contrast Ratio (ICR) between soil strata and bedrock considered in the study are summarized in Table 1. Twenty actual far-field ground motions varying in PGA from 0.036 to 0.47 g were selected from ATC-63 (2008) and Chaudhary (2016a) to perform one-dimensional non-linear seismic site response analysis. Acceleration response spectra of these strong motions is depicted in Fig. 2. Table 2 lists the salient features of the selected ground motions as well as group these

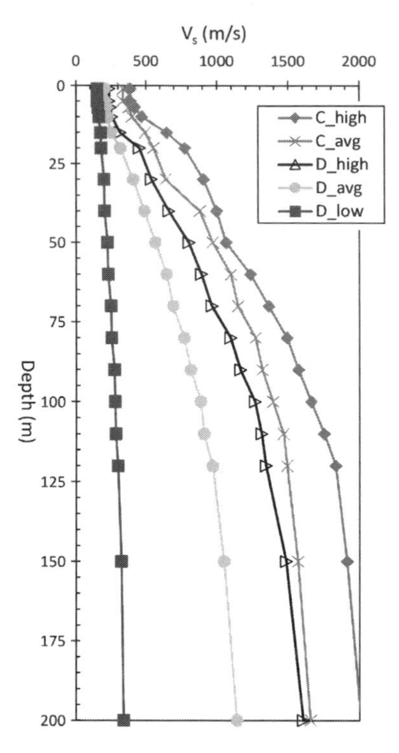

Figure 1. Shear wave velocity profile for various site classes.

Table 1. Variation in soil parameters.

Parameter	Range
V_{s30} (m/s)	600 (C_high) 475 (C_avg) 350 (D_high) 275 (D_avg) 175 (D_low)
PI	0, 15, 60
σ′ (atm)	2, 4
Strata depth (m)	40, 110
$V_{s\ bedrock}$ (m/s)	600, 760, 1350, 2251, 3353

into Design Basis Earthquakes (DBE), Functional Evaluation Earthquakes (FEE) and Maximum Credible Earthquake (MCE) based on a design PGA of 0.2 g. Ground motions 18 to 20 were scaled down to match MCE for the design PGA.

More than 2400 analysis were carried out using 1-D site response analysis software STRATA (Kottke and Rathje, 2008) for various combinations of soil profiles, soil properties, strata depth and ICR. STRATA performs a 1-D linear/non-linear seismic response analysis of the soil column in the time domain and incorporates strain dependent non-linear shear modulus reduction and damping curves from a number of sources. In this study, modulus reduction and damping curves developed by Darendeli (2001) were used for the soil classes included in the study.

The choice of PI values as listed in Table 1 were based on the generally accepted limit for sand, medium plastic and highly plastic soils. OCR was kept constant (= 1) for all soil profiles based on the observations of Guerreiro et al. (2012). Two values of confining pressure were considered: 2 atm for sandy soils (PI = 0) and 4 atm for soils with PI = 15 and 60. The choices of confining pressure were made based on the observations of Guerreiro et al. (2012) and to limit the number of simulation cases to a realistic number. Salient results of 110 m deep strata analysis are presented herein. Refer to Chaudhary (2016b) for other cases.

Table 2. Median values of β (%) – 110 m strata.

	DBE (PGA < 0.27g)			FEE (0.27g < PGA < 0.36g)			MCE (0.38g < PGA < 0.46g)		
Soil	PI 0	15	60	0	15	60	0	15	60
C_high	2.44	1.63	1.57	4.31	2.63	2.07	5.42	3.17	2.29
C_avg.	4.34	2.66	2.10	6.17	3.86	2.73	8.56	5.18	3.44
D_high	7.68	4.27	2.91	9.21	7.14	4.88	11.46	8.46	6.08
D_avg.	10.92	6.72	4.28	12.68	8.97	5.77	15.01	12.43	8.21
D_low	11.50	9.42	6.63	11.80	9.58	7.10	13.71	12.07	9.52

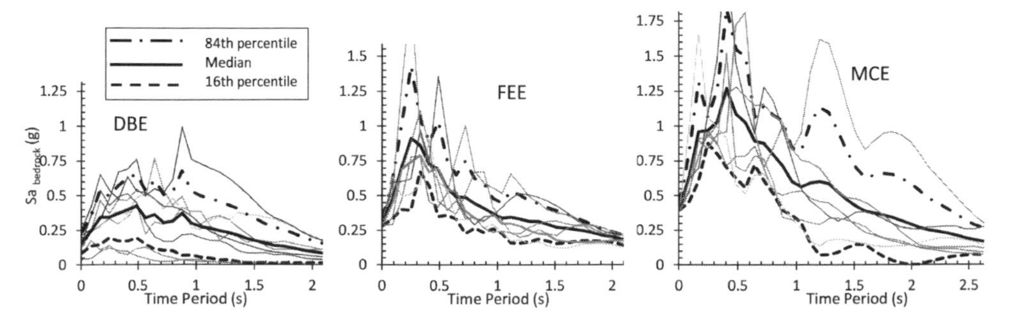

Figure 2. Spectral acceleration of selected input ground motions for DBE (PGA < 0.27 g), FEE (0.27 g < PGA < 0.36 g) and MCE (0.38 g < PGA < 0.46 g).

3 RESULTS & DISCUSSIONS

3.1 *1-D site analysis results*

Variation in damping ratio, shear wave velocity and shear modulus along strata depth for various analysis cases as outlined in Table 1 were determined by conducting 1-D site response analysis for the chosen ground motions. Figure 3 presents typical variation in these parameters along depth for Soil type D-avg for 110 m deep strata. Median and 16 & 84 percentile values of these parameters are marked in the figure with solid bold and dashed bold lines respectively.

In the design of deep bridge foundations, soil parameters in the top 30 m depth are mostly used. Therefore, weighted damping ratio in the top 30 m depth is computed by the following relationship for a particular earthquake in an analysis case:

$$\beta_{30\mathrm{m}} = \frac{\sum d_i \beta_i}{30} \quad (1)$$

Time averaged V_{s30} is the shear wave velocity based on the time for the shear wave to travel from a depth of 30 m to the ground surface and was computed as:

$$V_{s30} = \frac{30}{\Sigma\left(\frac{d_i}{V_{si}}\right)} \quad (2)$$

Variation in the average values of damping ratio, shear wave velocity and shear modulus in top 30 m depth with respect to PGA, PI and ICR are presented in Figure 4 for soil type C_avg as an example. Tables 2 and 3 list the median values of β and V_{s30} for 110 m strata. The following observations were made:

1. Damping ratio increases while V_{s30} and G decrease with an increase in PGA.
2. Damping ratio decreases with an increase in PI. This means that non-plastic (sandy) soils exhibit more damping than plastic (clayey) soils. Whereas V_{s30} and G increase with increasing PI. This means that non-plastic (sandy) soils have smaller shear wave velocity compared to clayey soils.
3. Damping ratio increases with a decrease in V_{s30}. That is, weaker soils exhibit more damping. The reverse is true for V_{s30} and G, which decrease as the soil becomes weaker.
4. An increase in ICR within a soil type and for the same PI value, increases the damping ratio. Whereas, increase in ICR within a soil type and for the same PI value, decreases the shear wave velocity and soil shear modulus.
5. The impact of ICR on variation in damping ratio, shear wave velocity and shear modulus is more in soil type D compared to soil type C.

3.2 *Effect of ICR on β & V_{s30}*

Effect of ICR at the bedrock-soil interface on seismic wave propagation and design parameters like β & V_{s30} was investigated in the study. ICR varied between 1 to 6.75 for type C soils and between 1 to 18 for type D soils. Difference in β & V_{s30} for various ICR values within the same soil type and PI with respect to the median values in the group (as reported in Tables 2 and 3) were used to delineate the impact of ICR on these parameters.

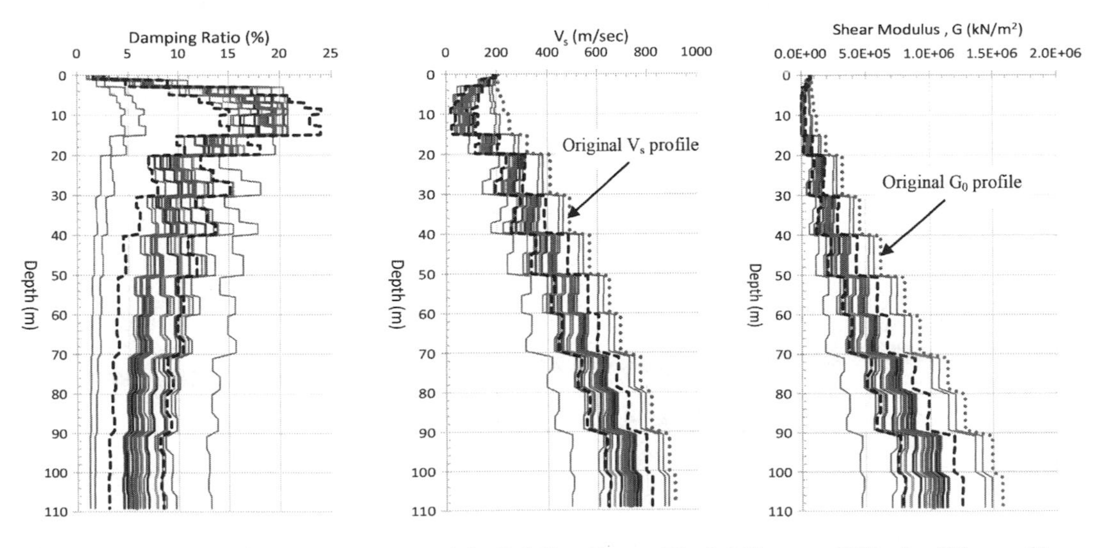

Figure 3. Variation in β, V_{s30} and G with depth for Soil Class D_avg., PI = 0 & $V_{s\,bedrock}$ = 3353 m/s – 110 m profile.

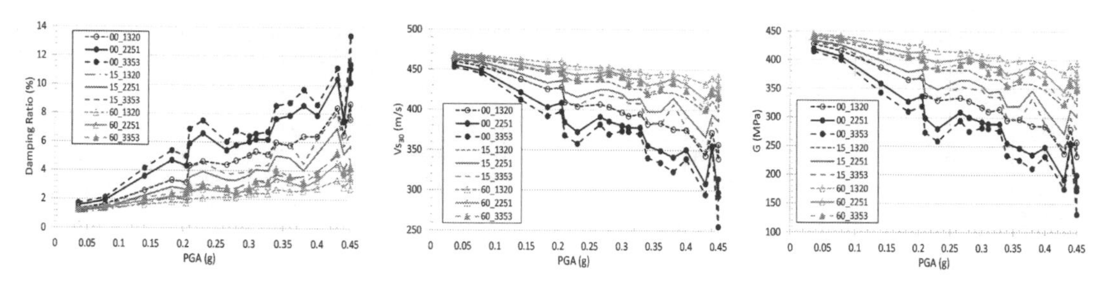

Figure 4. Variation in β, V_{s30} and G with PGA in top 30 m depth for Soil Class C_avg. – 110 m profile depth.

Table 3. Median values of V_{s30} (m/s) – 110 m strata.

	DBE (PGA < 0.27g)			FEE (0.27g < PGA < 0.36g)			MCE (0.38g < PGA < 0.46g)		
Soil	PI 0	15	60	0	15	60	0	15	60
C_high	558.73	578.01	589.54	516.68	554.11	576.74	491.42	543.02	571.84
C_avg.	408.65	436.72	453.75	377.94	415.96	442.21	340.20	394.01	430.54
D_high	247.52	302.06	329.45	211.27	253.37	297.23	164.52	225.79	278.09
D_avg.	160.91	210.95	242.06	144.59	185.94	225.91	101.75	140.32	201.93
D_low	103.46	116.60	136.79	100.36	115.77	133.89	84.10	98.31	118.58

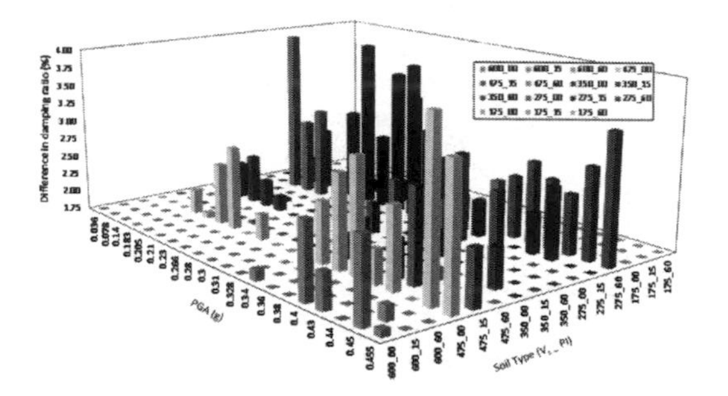

Figure 5. Effect of ICR on damping ratio.

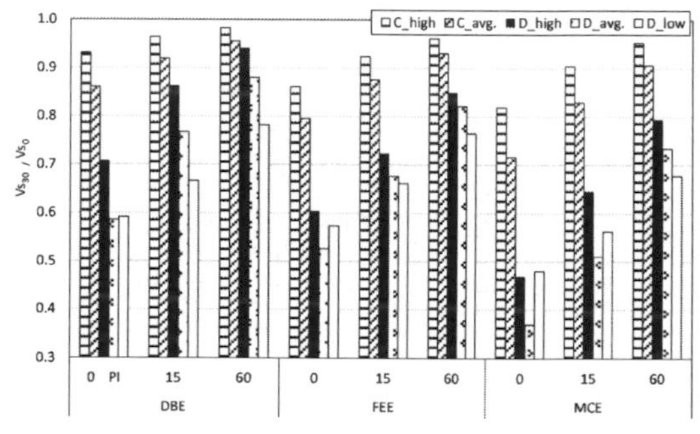

Figure 7. Reduction in V_{s30} with earthquake intensity.

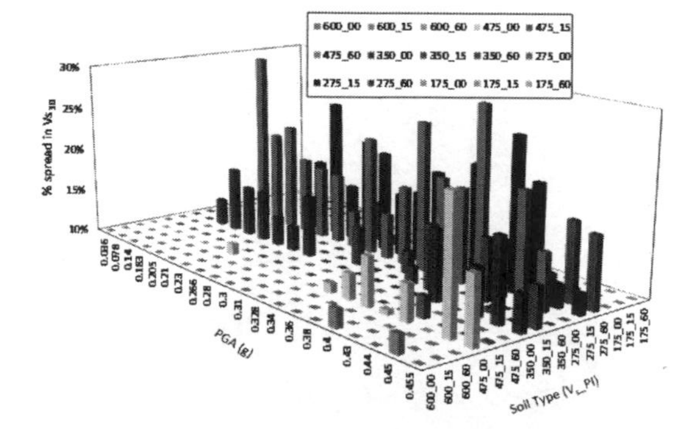

Figure 6. Effect of ICR on shear wave velocity.

A difference in β of more than 1.75% was taken as the threshold for considering the influence of ICR. Whereas a spread of more than 10% between V_{s30} for extreme ICR cases with respect to the median V_{s30} for the group was taken as the threshold for shear wave velocity. Figures 5 and 6 depicts the results for critical cases for damping ratio and shear wave velocity respectively for 110 m deep strata. It can be observed that variation in ICR has most effect of soil type D for both β and V_{s30}.

3.3 *Reduction in V_{s30} as compared to V_{s0}*

Figure 7 depicts the reduction in final median value of V_{s30} as compared to V_{s0} for a soil type. Reduction in V_{s30} can be noted with increasing PGA as well as reducing PI values. Type D soils displayed more reduction than type C.

4 CONCLUSIONS

The following conclusions are drawn from this study:

i. β, V_{s30} and G are strongly influenced by PGA, soil type and PI.
ii. ICR influences β, V_{s30} and G more profoundly in soil type D.
iii. V_{s30} reduces with PGA and the reduction is very significant (up to 60%) in soil type D with PI = 0.

ACKNOWLEDGEMENT

This work was supported by Kuwait University, Research Grant No. EV01/15.

REFERENCES

AASHTO (2010). AASHTO LRFD Bridge Design Specifications. American Association of State Highway and Transportation Officials, Washington, DC.

ATC-63 (2008). Quantification of Building Seismic Performance Factors. ATC-63 Project Report prepared by the Applied Technology Council for the Federal Emergency Management Agency, Washington, DC.

Bieniawski, ZT. (1974). Geomechanics Classification of Rock Masses and its Application in Tunnelling, Proceedings of the 3rd International Congress on Rock Mechanics, ISRM, Denver, 2(2), 27–32.

CEN (2003). European Committee for Standardization TC250/SC8/, Eurocode 8: Design Provisions for Earthquake Resistance of Structures, Part 1.1: General rules, seismic actions and rules for buildings, PrEN1998-1.

Chaudhary, M.T.A. (2016a). Seismic response of bridges supported on shallow rock foundations considering SSI and pier column inelasticity, KSCE Journal of Civil Engineering, DOI:10.1007/s12205-016-0352-5, 2016, pp. 1–11.

Chaudhary, M.T.A. (2016b). Implication of soil and seismic ground motion variability on dynamic pile group impedance for bridges, Research Project Report # EV01/15, Kuwait University, Kuwait.

Darendeli, M.B. (2001). Development of a new family of normalized modulus reduction and material damping curves, PhD thesis, University of Texas, Austin, 394 pp.

Douglas, J., Gehl, P., Bonilla, L.F., Scotti, O., Régnier, J., Duval, A.M. and Bertrand, E. (2009). Making the most of available site information for empirical ground-motion prediction, Bulletin of the Seismological Society of America, 99(3), 2009, pp. 1502–1520.

Guerreiro, P., Kontoe, S., and Taborda, D. (2012). Comparative study of stiffness reduction and damping curves, 15th World Conference on Earthquake Engineering, Lisbon, CD ROM, pp. 1–10.

Kottke, A.R., and Rathje, E.M. (2008). Technical Manual for Strata. PEER Report 2008/10. University of California, Berkeley, California, http://nees.org/resources/strata.

Civil, Architecture and Environmental Engineering – Kao & Sung (Eds)
© 2017 Taylor & Francis Group, ISBN 978-1-138-02985-9

Research on multisource information fusion of tunnel geological prediction based on evidence theory

Zonghui Liu, Yetian Wang, Hui Ke, Heng Wu & Dong Zhou
College of Civil Engineering and Architecture, Guangxi University, Nanning, China
Key Laboratory of Disaster Prevention and Structural Safety of the Ministry of Education, Guangxi University, Nanning, China

ABSTRACT: On the basis of a great amount of multivariate obtained information (geological information, geophysical prospecting data, etc.), tunnel geological comprehensive prediction is the procedure to predict the unexcavated geological situations. According to the characteristics of the current tunnel geology comprehensive prediction method, D-S evidence theory is introduced to construct the multisource information fusion model of tunnel geological comprehensive prediction in this paper. The evidence obtained in each step is superposed layer by layer following the comprehensive prediction process from macroscopic prediction to the geological analysis in tunnel and then comprehensive geophysical prospecting (long distance and short distance). With more sufficient evidence, the unfavorable geological information is much more clear for predicting the unfavorable geology of tunnel barrel, which can be narrowed down to a specific location. Citing the instance of comprehensive geological prediction of the left line in the LingJiao Tunnel, the proposed application processes and results of multisource information fusion model of tunnel geological comprehensive prediction based on the evidence theory are introduced in detail. The prediction of four anomalies are in good agreement with the prediction of the expert groups and the actual excavation results, which shows that the proposed comprehensive prediction model can fuse various information and is capable of simulating multiexpert decision-making process without reducing the accuracy.

1 INTRODUCTION

Geological comprehensive prediction of tunnel is a very complicated process and the geological predicting workers have to face large amounts of geological materials and field data, which should be organized and analyzed effectively to obtain correct predicting results. Recent years have seen some comprehensive predicting models put forward (Ge et al. 2010, Xu et al. 2013, Xu et al. 2011, Yuan et al. 2011), which can integrate multiaspect information and have achieved good results in some engineering cases. There are still some shortcomings and limitations in these studies; for instance, the fuzzy evaluation method in the existing comprehensive predicting model mainly reflects the consistency but no clear description of conflict between each evaluation index. Meanwhile, as a complicated engineering system, the geological prediction of tunnel produces several uncertainties, but the current fuzzy evaluation methods are difficult to fully reflect the uncertain characteristics of the information.

The evidence theory is a generalization of the probability theory, which has a weaker axiom system and a more rigorous reasoning process than the probability theory, and can reflect the uncertainty of objects more objectively (Fan & Zuo. 2006; Tan & Xiang. 2008). As a method of uncertainty reasoning, it can express the uncertainty caused by randomness and fuzziness. Dempster evidence combination rule can integrate evidences without complete information simply and effectively and reflects the conflict of different evidences to some degree. The evidence theory is widely used in the fields of information fusion and expert system, and it is a relatively frontier research topic worldwide (Du et al. 2014, Jia et al. 2014). The multisource information fusion method of tunnel geological comprehensive prediction proposed in this paper aims at maximum elimination of the multisource results of the single predicting method and improvement of the accuracy of prediction.

2 OVERVIEW OF D-S EVIDENCE THEORY

2.1 *Frame of discernment*

The set Θ, in which any result that people care about can be found, is a set of mutually exclusive proposition of problems that need to be determined, and the determination often depends on the level of knowledge of the decision maker. Shafer calls Θ as the frame of discernment, and the formula is expressed as follows:

$$\Theta=\{F_1, F_2,\cdots F_n\} \tag{1}$$

where F_n is an element of frame of discernment.

2.2 *Basic function*

To describe and deal with the uncertainty after establishing the frame of discernment, the concepts of basic probability assignment function, belief function, and likelihood function are introduced.

The set 2^Θ composed of all subsets of Θ is the power set of Θ. If the function set m: $2^\Theta \rightarrow [0, 1]$ satisfies

$$m(\phi)=0 \tag{2}$$

$$\sum_{A\subseteq\Theta} m(A)=1 \tag{3}$$

then m could be named as the Basic Probability Assignment (BPA) of Θ, which expresses the trust of evidence in subset A of the frame of discernment. Formula (2) demonstrates that a null proposition does not generate any probability, and formula (3) shows that the sum of the probability assignment of all subsets is equal to 1. A which satisfies $m(A) > 0$ is called the focal element. If subset A obtained only one element, it could be called a single-element focal element.

The belief function of D-S evidence theory is also named as the lower function, which indicates the trusting level in an assumed set and its value is defined as the sum of the current basic probability assignment of all subsets, and is expressed as follows:

$$Bel(A)=\sum_{B\subseteq A} m(B) \tag{4}$$

The plausibility function can be described as the upper function, reflecting the extent of doubt about A, that is, how we do not doubt A or A is reliable, and the formula is given in the following expression:

$$Pl(A)=\sum_{B\cap A\neq\phi} m(B) \tag{5}$$

The trust interval, [Bel(A), Pl(A)], is made up of the belief function and the plausibility function, to measure the degree of certainty of some hypotheses.

2.3 *The Dempster evidence combination rule*

The Dempster combination rule is the essence of the D-S evidence theory, which reflects the combined action among evidences. Using the rule, a new belief function could be figured out from several belief functions based on different evidences, which does not conflict completely, in the same frame of discernment. Therefore, the new belief function can express the combined actions of those evidences.

Assuming that there are two independent evidences in the same frame of discernment whose focal elements are B_i and C_i ($i = 1, 2, \ldots, m$; $j = 1, 2, \ldots, n$), where m and n, respectively, represent the number of the two evidence focal elements, then m1 and m2 are the corresponding two BPAs, and the formula of the synthetic BPA $m_\oplus$: $2^\Theta \rightarrow [0, 1]$ is as follows:

$$m_\oplus(A)=\frac{\sum_{B_i\cap C_j=A} m_1(B_i)\, m_2(C_j)}{1-K}, A\neq\phi \tag{6}$$

$$m_\oplus(A)=0, A=\phi \tag{7}$$

where $\oplus$ is the combination mark and $K=\sum_{B_i\cap C_j=\phi} m_1(B_i)\, m_2(C_j)$ is the conflict coefficient. When the number of evidence is more than two, they can be combined one by one according to the formula as Dempster rule meets commutative law and associative law. Dempster rule can integrate multi-information and reflect the conflict condition between different information or evidences. Generally, a larger value of K means a greater conflict between two evidences. However, some studies show that K is not enough to describe the conflict, which should combine other factors such as the distance of evidence and evidence similarity measure (Bi et al. 2016). They are not discussed in detail.

3 D-S EVIDENCE INTEGRATION MODEL OF TUNNEL GEOLOGICAL COMPREHENSIVE PREDICTION

3.1 *D-S evidence integrate model of tunnel geological comprehensive prediction*

According to the characteristics of the current comprehensive prediction method of tunnel geology, D-S evidence theory is introduced to construct the multi-source information fusion model of tunnel geological comprehensive prediction in this paper.

The evidence obtained in each step is superposed layer by layer following the comprehensive prediction process from macroscopic prediction to the geological analysis in tunnel and then comprehensive geophysical prospecting (long distance and short distance).

With the more sufficient evidence, the unfavorable geological information is much more clear for predicting the unfavorable geology of tunnel barrel, which can be narrowed down to a specific location.

The four most common geophysical prospecting methods included in this model are the Tunnel Seismic Prediction (TSP), the Landsonar (LDS), the Transient Electromagnetic (TEM), and the Ground Penetrating Radar (GPR). Other geophysical methods can be also incorporated into this system in a similar way.

According to the characteristics of various predicting methods system, the tunnel geological prediction information fusion process consists of three stages, and the model structure is shown in Figure 1.

The first stage is information sorting and preprocessing, which then extracts corresponding characteristic parameters to construct some evidence subspaces in which all evidence reflects the tunnel geological condition from an independent aspect.

For example, the main sources of the basic geological information of the tunnel macro-geological prediction come from the existing survey and design results, regional geological data, and complementary surface geological survey results.

What comes in first is the extraction of the stratigraphic lithology, geological structure, topography and geomorphology, climate conditions and surface water features, and other indicators to construct the subspace of the macro-geological predicting evidence.

The second stage is the local information fusion, namely evidences obtained from each prediction method are input onto the constructed local

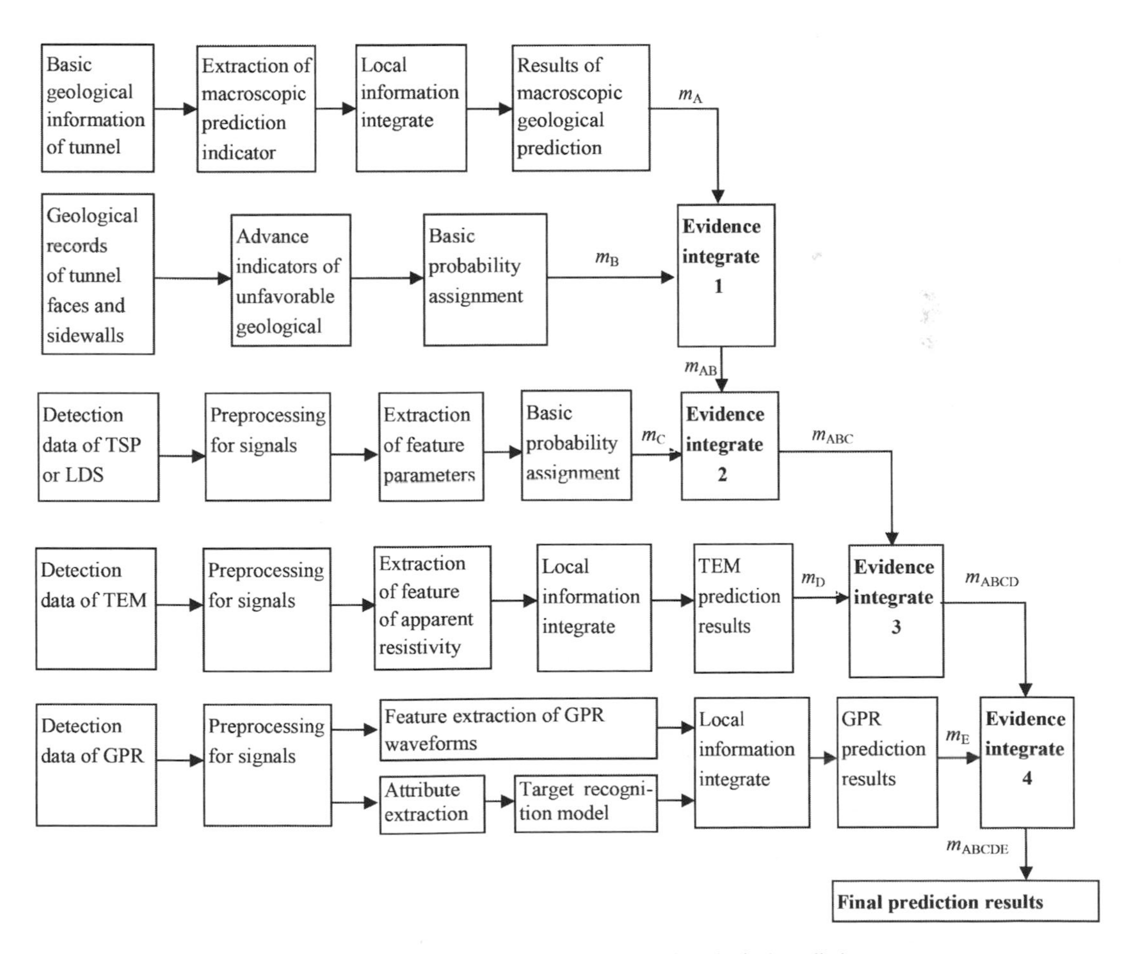

Figure 1. Structure frame of the information fusion model for tunnel geological prediction.

information fusion submodel. This paper consists of three submodels of local information fusion: macro geological prediction, TEM data interpretation, and GPR target recognition.

With step-by-step integration of framework shown in Figure 1, the third stage is transferring the predictions of each submodel to new evidences. Then, the final prediction, F_c, can be drawn as according to the following rules:

Rule No.1: $Bel(F_c) = \max\{Bel(F_j), F_j \subseteq \Theta\}$

Rule No.2: $Wm(\Theta) < \varepsilon_1$

Rule No.3: $Bel(F_c) - Bel(F_j) > \varepsilon_2$

Rule No.4: $Bel(F_c) > Wm(\Theta)$

where $[Bel_j, pl_j]$ is the belief interval of evidence to all propositions in the frame of discernment Θ and $Wm(\Theta)$ is the uncertainty of evidence. Rules 1 and 2 are the basic conditions to ensure that the predicting results can be sufficiently distinguished. Rules 3 and 4 guarantee the advantage of reliability of the determined predictions, in which ε_1, ε_2 should be chosen based on the actual situations. If the above four rules can not be satisfied at the same time, the prediction would not be determined. Two reasons may explain this situation: (1) the prediction does not include in the frame of discernment; therefore, it is necessary to re-identify the frame; (2) the evidence selection is not reasonable; therefore, it is necessary to select again or further selecting more evidences for fusion calculation.

4 THE GENERATION METHOD OF THE BPA

Because of the complexity and diversity in application of D-S evidence theory, it is always difficult to construct the BPA of all evidence in practice, which should be constructed according to the characteristics of the research issue, as there is no unified mode (Lin et al. 2007, Xiang et al. 2015).

Some scholars have already proposed several methods to different problems, such as the BPA generation method based on fuzzy membership, the BPA converted from the measure value based on pattern recognition (Jiang et al. 2008); the determination of BPA according to the types of target and the weighting coefficient of environment (Han et al. 2014); and the establishment of the BPA based on the Euclidean distance of the interval.

Geological analysis and various geophysical methods are involved in the geological comprehensive prediction of tunnel. Different BPAs are adopted in different methods for which they have different principles according to their own characteristics. In this paper, the three adopted methods are as follows.

- The formation of the BPA of all evidence based on the Euclidean distance of interval number (Cao et al. 2015). The quantitative value of the jth evidence is $Q_j = [x^l, x^u]$, and the evaluating interval of the jth evidence corresponding to the ith prediction is $F_i = [y^l, y^u]$. According to the definition of Euclidean distance, the Euclidean distance between interval $Q_j = [x^l, x^u]$ and interval $F_i = [y^l, y^u]$ is:

$$D_2(\tilde{Q}_j, \tilde{F}_i) = \left[(x^l - y^l)^2 + (x^u - y^u)^2 \right]^{0.5} \tag{8}$$

Using formula (8), all Euclidean distance between the jth evidence and each possible prediction F_i can be calculated. A smaller Euclidean distance means a higher supporting degree of the jth evidence to F_i. As the Euclidean distance can reflect the support degree of evidences to the prediction, there is further determination of the BPA of the prediction corresponding to the jth evidence:

$$m[A_j(F_i)] = \frac{1/D_2(\tilde{Q}_j, \tilde{F}_i)}{\sum_{i=1}^{n} \frac{1}{D_2(\tilde{Q}_j, \tilde{F}_i)}} \tag{9}$$

- The conversion of pattern recognition results to the BPA. Using Ground Penetrating Radar (GPR), for example, to identify the unfavorable geological body, the corresponding attribute parameters (such as spectral attributes and instantaneous attributes) based on the measured GPR data are extracted first (Forte et al. 2012). Obtained by pattern recognition method, the recognition measure of the unfavorable geological body in the frame of discernment can be finally converted to the BPA after normalization. This method embraces the objectification of the BPA and the avoidance of the complexity when constructing the BPD.
- The acquisition of the BPA on the basis of the measured data by the assistance of experts experience. The interpretation of TEM, LDS, and other geophysical methods being difficult to be quantified and the artificial intelligence methods, like pattern recognition, being difficult to be applied, experts in relevant fields are often needed to interpret the measured data and give the corresponding BPA.

5 FIELD APPLICATION

5.1 *Field description and geological condition*

Lingjiao Tunnel located in Cenxi, Guangxi Province, China, is a separated highway tunnel (length = 790 m, maximum burial depth = 130 m). The distance between the center lines of the two openings is 30 m, whereas the net distance is 17 m.

The tunnel is located in the northeastern foot of Yunkai Mountain, an erosional-denudation landform of high hills. The main litho-stratigraphic stratum of the tunnel is weathered migmatite with regional metamorphism. Overall, the rock mass has a broken and inlaid cataclastic structure with weak interlayer bonding.

There are two regional faults, F1 and F2, in the tunnel location, whose intersection causes many complicated conditions in geology.

To guarantee safety during construction, comprehensive geological prediction conducted by Censhui highway headquarters for the preceding geology is undertook by several research institutes using multiple predicting means as well as comprehensive analyses about the prediction by experts.

On the basis of the comprehensive geological prediction of the left line in the LingJiao Tunnel, the application process and effect are introduced in detail for the method proposed in this paper as multisource information fusion based on evidence theory for tunnel geological comprehensive prediction.

Geological analysis process, field geophysical process, and the corresponding geophysical result involved in this paper can be found in Zhou et al. (2015). Because of the limited length of this paper, no further details are included.

5.2 *Comprehensive geological prediction processes and results based on evidence theory*

The following is the prediction of geology circumstance in front of the tunnel face FK18+230 by using the proposed information fusion model based on the evidence theory, geological analysis, and geophysical prospecting results.

5.2.1 *Constructing the frame of discernment*

According to the macroscopic geological predicting conclusion, unfavorable geology in fault fracture zones mainly includes: fractured rock mass (fractured rock mass with mud zone), soft layer (filled cavity) mainly filled with soft plastic and flow plastic clay, and fracture cleft (cavity) without filling material. Therefore, the frame of discernment can be expressed as: $\Theta = \{A, B, C, D\}$.

Where A is the relatively intact surrounding rock; B is the fractured rock mass (fractured rock mass with mud zone), C is the soft layer (filled cavity) mainly filled with soft plastic and flow plastic clay, and D is fracture cleft (cavity) without filling material.

5.2.2 *The basic probability assignment*

1. Section FK18+228~FK18+202

Evidences of tunnel geological prediction in this section include the result of four prediction methods: geological analysis, LDS detection, TEM detection, and GPR detection. Therefore, the evidence set can be expressed as: $P = \{P_1$ (geological analysis), P_2 (LDS), P_3 (TEM), P_4(GPR)$\}$.

Geological experts believe that the geological analysis can only determine that the above three types of unfavorable geologies are all likely to exist when entering section FK18+220 in the fault fracture zone; however, the uncertainty of specific location and the reason for their anomalies make it almost impossible to be a intact rock. Consequently, the basic probability of the evidence m_1 can be assigned as $m_1(A) = 0.25$, $m_1(B,C,D) = 0.75$.

Detecting results of the LDS method show that the range of abnormal area is obviously reduced and there are several faults and cavities, in which, however, the material can not be determined and there may be false anomalies because of far detecting distance. The BPA of LDS detection, which has higher accuracy than geological analyses, and the location of the anomalies of LDS are unified as $m_2(A) = 0.2$ and $m_2(B,C,D) = 0.8$.

The Transient Electromagnetic Method (TEM) can only determine the region where low resistance exists of the tunnel space. The detection results show that the section FK18+213~FK18+207 is obvious in the area of low-resistance anomalies, where there is a high possibility of low-resistivity, water-bearing formations, compared to other areas whose relatively high apparent resistivity means little possibility of low-resistivity, water-bearing formations. Therefore, the BPA of the TEM detection are $m_3(A,B,D) = 0.2$, $m_3(C) = 0.8$ in obvious low-resistivity abnormal zone and $m_3(A,B,D) = 0.8$, $m_3(C) = 0.2$ in other regions. The targets A, B, and D are all regarded as high-resistance areas.

Using the attribute extraction and target recognition technology of GPR described in Forte et al. (2012), we can divide abnormal zones and distinguish the types of the unfavorable geology. Consequently, attributes, instantaneous amplitude, instantaneous phase, amplitude spectrum, and phase spectrum are extracted. Table 1 shows the result of BPA converted from pattern recognition of three abnormal regions in Fig. 7 of Zhou et al. (2015). According to the pattern recognition result, the abnormal regions and the types of unfavorable geology can almost be determined; however, the superiority in reliability of the abnormal types is not obvious.

2. Section FK18+202~FK18+150
Evidences of tunnel geological prediction in this section include the result of geological analysis, TEM detection, and LDS detection. Therefore, the evidence set can be expressed as $P = \{P_1$ (geological analysis), P_2 (LDS), P_3 (TEM)$\}$.

The detection result of TEM and LDS shows the existence of a large cavity in the upper right of section FK18+188~FK18+183. Synthesis of evidences is mainly used to determine the type of filling materials.

The BPA of geological analysis results can be also assigned as $m_1(A) = 0.25$, $m_1(B,C,D) = 0.75$.

Although the detection results of LDS detected inside and outside both show that there is a cavity, and the filling material still remains unknown. Therefore, the BPA is $m_2(A) = 0.1$, $m_2(B) = 0.1$, $m_2(C,D) = 0.8$.

Meanwhile, TEM can show obvious existence of low-resistivity anomalies and the BPA is $m_3(A,B,D) = 0.2$, $m_3(C) = 0.8$.

5.2.3 *Combination of multiple evidence*

Using D-S evidence fusion, the above four anomalies are fused separately, and their results are shown in Table 2.

5.2.4 *Conclusions of prediction*

According to the reliability-dominance judging rule in Section 3, the above four types of abnormal geological condition can be determined. Section FK18+220~FK18+219 is the fracture rock mass (fracture rock mass with mud zone); there is a cavity without filler in the right wall of section FK18+220; section FK18+213~FK18+207 is the soft layer (filled cavity) mainly filled with soft plastic and flow plastic clay, and the right side of section FK18+188~FK18+183 is a filled cavity whose filling material is loose and mixed rock soils with the saturated water.

It can be seen from the above processes of evidence fusion that a single prediction method can not determine the specific type of the unfavorable geology. With the fusion of multiple evidences, the state of uncertainty declines continuously while the superiority of reliability of the detection target was observed simultaneously; finally, the type of anomalies can be judged clearly. For instance, the GPR can only show the existence of a cavity in the right wall of section FK18+220, with no information of whether filled or not for its strong interference. Fortunately, this problem is addressed after the fusion of TEM result.

5.3 *Expert predictions and actual excavating situations*

To guarantee the accuracy, the prediction of geological conditions ahead of the FK18+230 is made after discussion, organized by the command, of experts in geological and geophysical fields. The agreement of prediction between experts and the evidence fusion theory demonstrates that the evidence fusion theory proposed in this paper can fuse multiaspect information to simulate the decision-making process of experts.

Excavation was completed by using advanced support, and it was performed in the fault fracture zone. The excavation shows that the main anomalies lie in section FK18+212~FK18+206, which has cavities, and is a fault fracture with mud zone mainly filled with yellow soft plastic and flow plastic clay. A small scale of mud burst occurred when the excavation came to section FK18+18 but without large-scale disaster because of early prevention. Precisely, the excavation is in good agreement with the prediction.

Table 1. BPA of abnormal areas of the GPR.

	Mass function			
Abnormal regions	$m_1(A)$	$m_1(B)$	$m_1(C)$	$m_1(D)$
FK18+220~FK18+219	0.151	0.452	0.296	0.101
FK18+213~FK18+207	0.054	0.285	0.439	0.222
The right wall of FK18+220	0.070	0.126	0.402	0.402

Table 2. Synthesis results of the mass function.

	Mass function			
Abnormal regions	$m_{1234}(A)$	$m_{1234}(B)$	$m_{1234}(C)$	$m_{1234}(D)$
FK18+220~FK18+219	0.021	0.706	0.116	0.158
FK18+213~FK18+207	0.002	0.126	0.775	0.098
The right wall of FK18+220	0.011	0.198	0.158	0.633
The upper right of FK18+188~18+183	0.010	0.024	0.774	0.194

6 CONCLUSIONS

Information fusion is a process of combination or comprehensive processing of multisource data and information, and the information fusion technology is actually a function simulation of complex problems handled in the human brain, in order to obtain more accurate and reliable conclusion than single information. The geological comprehensive prediction of tunnel is a typical information fusion process, which aims at predicting the unexcavated tunnel geological situations according to massive obtained information (geological information, geophysical prospecting data, etc.). Its ultimate goal is to eliminate the multiplicity of the single predicting method and improve the predicting accuracy.

According to the characteristics of the current comprehensive prediction of tunnel geology, D-S evidence theory is introduced to construct the multisource information fusion model of tunnel geological comprehensive prediction in this paper. The evidence obtained in each step is superposed layer by layer following the comprehensive prediction process from macroscopic forecast to the geological analysis in tunnel and then comprehensive geophysical prospecting (long distance and short distance). With the more sufficient evidence, the unfavorable geological information is much more clear for predicting the unfavorable geology of tunnel barrel, which can be narrowed down to a specific location.

Suffered from the wide fractured zone originated by the intersection of two regional faults and thick water dykes formed by magmatic intrusion into the fault zone, the geological conditions of the Lingjiao Tunnel are very complicated, which may contain many types of unfavorable geologies. On the basis of this project, the application process and result of multisource information fusion model of tunnel geological comprehensive prediction based on the evidence theory are introduced in detail. Predictions of four anomalies in this project are in good agreement with the prediction of the expert groups and actual excavating results, which shows that the proposed comprehensive predicting model can fuse various information and is capable of simulating multiexpert decision-making processes without reducing the accuracy.

REFERENCES

Bi, W.H., Zhang, A., Li, C. (2016). "Weighted evidence combination method based on new evidence conflict measurement approach." *Control and Decision*, 31(1), 73–78.

Cao, W.G., Yang, W.K., Qu, Y.C. (2015). "Combination evaluation method for the classification of rock mass quality based on d-s theory of evidence." *Journal of Human University (Natural Sciences)*, 42(5), 86–91.

Du, X.L., Zhang, X.F., Zhang, M.J., et al. (2014). "Risk synthetic assessment for deep pit construction based on evidence theory" *Chinese Journal of Geotechnical Engineering*, 36(1), 155–161.

Fan, X., Zuo, M.J. (2006). "Fault diagnosis of machines based on D–S evidence theory." *Pattern Recognition Letters*, 27(5), 366–385.

Forte, E., Pipan, M., Casabianca, D., et al. (2012). "Imaging and characterization of a carbonate hydrocarbon reservoir analogue using GPR attributes." *Journal of Applied Geophysics*, 81, 76–87.

Ge, Y.H., Li, S.C., Zhang, Q.S., et al. (2010). "Comprehensive geological prediction based on risk evaluation during tunneling in karst area." *Chinese Journal of Geotechnical Engineering*, 32(7), 1124–1130.

Han, D.Q., Yang, Y., Han, C.Z. (2014). "Advances in DS evidence theory and related discussions." *Control and Decision*, 29(1), 1–11.

Jia, Y.P., Lü, Q., Shang, Y.Q., et al. (2014). "Rock burst prediction based on evidence theory." *Chinese Journal of Geotechnical Engineering*, 36(6), 1079–1086.

Jiang, W., Zhang, A., Yang, Q. (2015). "Fuzzy approach to construct basic probabil ity assignment and its application in multi-sensor data fusion systems." *Chinese Journal of Sensors and Actuators*, 21(10), 1717–1720.

Lin, H., Lin, C., Weng, R.C., et al. (2007). "A note on plat's probabilistic outputs for support vector machines." *Machine Learning*, 68(3), 267–276.

Tan, Q., Xiang, Y.H. (2008). "Application of weighted evidential theory and its information fusion method in fault diagnosis." *Journal of Vibration and Shock*, 27(4), 112–116.

Xiang, Y.H., Zhang, G.Q., Pang, Y.X. (2015). "Multi-information fusion fault diagnosis using SVM & improved evidence theory." *Journal of Vibration and Shock*, 34(13), 71–77.

Xu, S.C., Chen, J.P., Zuo, C.Q., et al. (2011). "Application of the fuzzy comprehensive evaluation method to karst forecasting in tunnel construction." *Modern Tunnelling Technology*, 48(5), 76–81.

Xu, S.C., Zhang, S.L., Chen, J.P. (2013). "The application of fuzzy-analytic evaluation in geological disaster assessment for tunnel construction." *Chinese Journal of Underground Space and Engineering*, 9(4), 946–953.

Yuan, X.S., Zhang, Q.S., Xu, C.H., et al. (2011). "Analytic hierarchy process based optimization for tunnel comprehensive geologic prediction." *Journal of Engineering Geology*, 19(3), 346–351.

Zhou, D., Liu, Z.H., Wu, H., et al. (2015). "Application of the comprehensive geological prediction techniques in the lingjiao tunnel." *Modern Tunneling Technology*, 52(5), 171–177.

Civil, Architecture and Environmental Engineering – Kao & Sung (Eds)
© 2017 Taylor & Francis Group, ISBN 978-1-138-02985-9

Pushover analysis of modularized prefabricated column-tree steel frame

Z.P. Guo, A.L. Zhang & X.C. Liu
College of Architecture and Civil Engineering, Beijing University of Technology, Beijing, China

ABSTRACT: A prefabricated steel frame with Z-shaped cantilever beam splicing is proposed, which can realize a good seismic performance through the slip of the splice. The moment-rotation relationship model of the splice is constructed through force analysis about various states of loading process of splice. A simulation method utilizing connecting element of ABAQUS to realize the constitutive relation of the splice in the overall structure is proposed. Then modal analysis and pushover analysis are conducted on the modularized prefabricated column-tree steel frame and rigid frame under the same condition, and the result shows that there is a very small difference of the natural periods of vibration of the two frames and the performance points under seismic action. The column-tree frame can dissipate seismic energy by utilizing slip of the splice, and compared with the rigid frame, the column-tree frame achieves the purpose of multiaspect seismic fortification.

1 INTRODUCTION

Quality of welding joint can be guaranteed as welding of beam and column is completed in the factory for column-tree frame. What's more, high-strength bolt connection has been conducted on the site to quicken installation. Therefore, column-tree frame structure system has been widely applied (Astaneh-Asl 1997). The seismic behavior of column-tree frame is analyzed and the improvement measures are put forward. Mcmullin et al. (2003) have conducted theoretical and experimental researches about seismic performance of semi-rigid column-tree steel frame and put forward that column–tree construction can utilize splice assemblies to provide stable energy dissipation during lateral movement. Chen & Lin (2006 & 2013) have researched seismic performance of reinforced column-tree connection, put forward that the specimens can successfully develop ductile behavior with no brittle fracture by forming the plastic hinging of the beam away from the beam–column interface. Oh et al. (2014) have conducted quasi-static test for column-tree connections with weakened beam splices, and put forward the moment resisting capacity of the specimens doesn't obviously decrease. Also, their energy dissipation capacity is better than that of the specimen following the full-strength design principle.

Above all, most of traditional column-tree connections follow the full-strength design principle. The seismic damage of the weld between beam and column can't be effectively reduced and the number of bolts is too much. One effective way to resolve brittle failure of beam column connection is to shift away plastic hinge of beam to prevent beam column connection from brittle fracture in the earthwork via energy dissipation of shift-away plastic hinge (Kim et al. 2002, Lee & Kim 2007, Popov et al. 1998, Uang et al. 2000). In this paper, the splice is designed as semi-rigid and shift-away of plastic hinge is achieved by using slip energy dissipation of the splice. What's more, modular plate installation has not been formed and assembly degree is not high because placing and maintenance are conducted for concrete floor on the construction site. Therefore, a kind of column-tree frame that is applicable to modularized prefabricated steel structures is proposed on the basis of traditional column-tree structure system. The frame is assembled through the splicing of connection on the construction site after the beam and slab are assembled in the factory. Assembly experiment has been conducted for sample engineering of two-floor modularized prefabricated steel structure frame as shown in Figure 1. The experiment proves that the structure has advantages of simple on-site installation and quick assembly.

As stated above, it is necessary to research the seismic performance of modularized prefabricated column-tree steel frame (Liu et al. 2015a, b, Zhang & Liu 2013). In this article, aimed at the modularized prefabricated steel frame with Z-shaped cantilever beam splicing, a method of utilizing connecting element of ABAQUS finite element software to simulate the beam and column connection with Z-shaped cantilever beam splicing of the overall structure is proposed, and the reliability of using the connecting element method in the overall structure is verified. On this basis, modal analysis and pushover analysis are conducted on

Figure 1. Assembly experiment of two-floor structure.

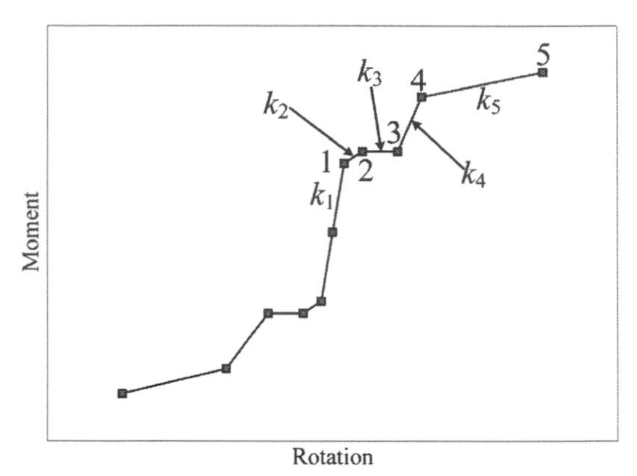

Figure 2. Moment-rotation relationship of the splice.

Figure 3. Failure modes of specimen.

the overall structure of prefabricated column-tree steel frame, and the result is compared with the rigid frame under the same condition.

2 CONNECTING ELEMENT METHOD

The moment-rotation relationship model of the splice is constructed through force analysis about various states of loading process of splice. ABAQUS finite element software connecting element is employed to simulate the constructive relation of splice in the overall structure, and the moment-rotation curve of beam and column connection simulated with this method is compared with the completed static loading curve and skeleton curve.

2.1 *Moment-rotation relationship model of the splice*

Taking the splice as the research object, the moment-rotation relationship model of the splice is constructed according to the mechanical theory and finite element calculation result.

According to the analysis on the overall loading process, the states of loading process of the splice can be divided into four stages: elastic stage, slip stage, bearing capacity strengthening stage and elastic-plastic stage. Each stage is analyzed to determine the moment, rotation and rotation stiffness of the splice in each stage, and the moment-rotation relationship model finally determined of the splice is shown in Figure 2.

2.2 *Verification of connecting element method*

Figure 3 shows the test picture of column-tree connection with Z-shaped cantilever beam splicing, and static loading and quasi-static loading tests are conducted to this connection, with the column being rolled square steel tube RHS200 × 10 and beam H300 × 200 × 8 × 12, both of which are made of Q235B steel, and during the test 10.9 grade high-strength bolt of M 20 is employed, which is made of 40 Cr steel, and anti-slip coefficient of specimen is 0.51 through test.

In the overall structure model, the constructive relation of the splice under the static load is simulated by defining the nonlinear connecting element, the connection property is defined as HINGE, which is composed of translation connection property JOIN (restriction u1, u2, u3) and rotational connection property REVOLUTE (restriction ur2, ur3), and rotation is only allowed in the direction of ur1, the moment-rotation relationship of the splice is entered as the connection property in this direction. Based on the completed static loading and quasi-static loading tests and by reference to the actual boundary condition in the test, both the bottom and top of the square steel tube in the finite element simulation are simulated as hinge, and the lateral support in the test is simulated by setting a lateral restriction in the

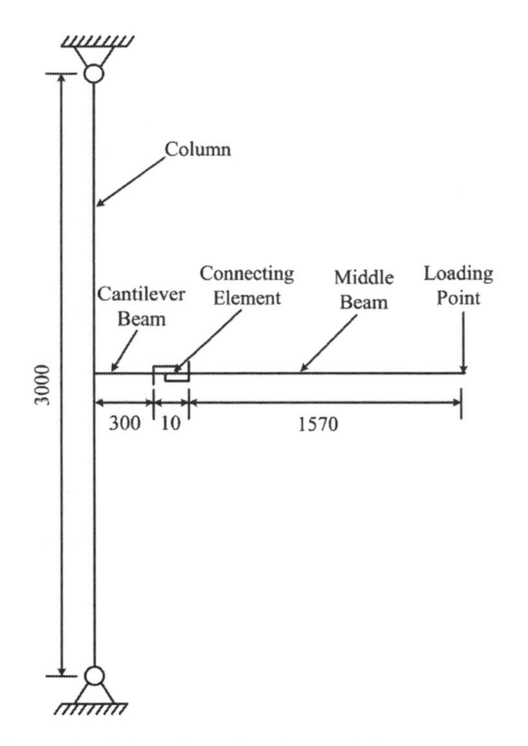

Figure 4. Finite element analysis model.

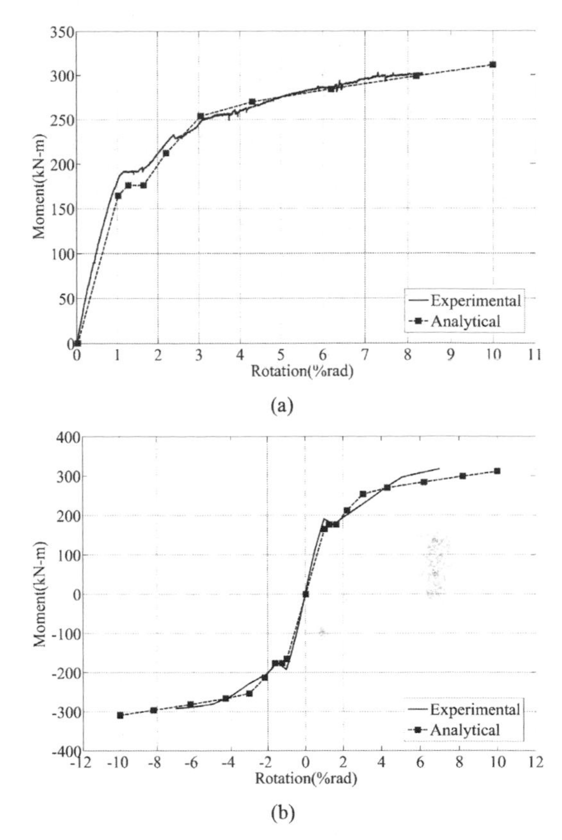

Figure 5. Comparison of moment-rotation curves between experiments and FEM analysis. (a) Static test. (b) Quasi-static test.

I-shape steel beam. A finite element analysis model is constructed with the same size as the test piece, beam and column are simulated with beam element, and the splice is simulated with nonlinear connecting element, as shown in Figure 4.

Static loads are applied to the loading point in both the positive and negative directions, and the moment-rotation curve at beam end obtained from the finite element analysis is compared with the moment-rotation curve at beam end obtained from the static test and skeleton curve obtained from the quasi-static test, as shown in Figure 5. Due to only one-way loading in the static loading test, the moment-rotation curve at beam end obtained from the static test only has the positive part. Figure 5 shows that the finite element analysis result coincides with the test result well, so it can be considered that the connecting element method is feasible to the application in the overall structure.

3 FINITE ELEMENT MODEL OF OVERALL STRUCTURE

3.1 *Overview of overall model*

A calculation example of 3 × 8-span steel frame residence building of nine-story is designed in this section, which is proposed to be built in a region of 8 degree of fortification intensity, with a plane size of 28.8 m × 12 m and it has 8 spans transversely, with a span of 3.6 m and 3 spans longitudinally, with a span of 4 m. The overall structure is 27 m high, with 3 m of each floor. The structure is designed to a rigid structure with the structural design software, with column being rolled square steel tube RHS 200 × 10 and beam H 300 × 200 × 8 × 12. This structure is named as frame RF, and then the rigid joint of the frame RF is changed as column-tree joint, and the sectional size of the beam and column remains unchanged, the column-tree frame named as frame CTF is constructed, as shown in Figure 6. The beam and column are simulated with beam element B32 for which the shear deformation can be considered, and the floorslab is simulated with shell element S4R. The detailed size of splice connection in the frame CTF is shown in Figure 7, modal analysis and pushover analysis are conducted of the rigid structure and column-tree steel frame respectively, and the difference of frame RF and frame CTF

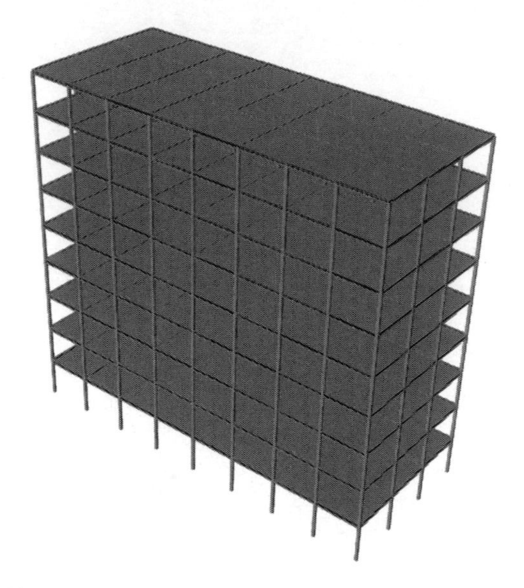

Figure 6. Schematic of three-dimensional frame model.

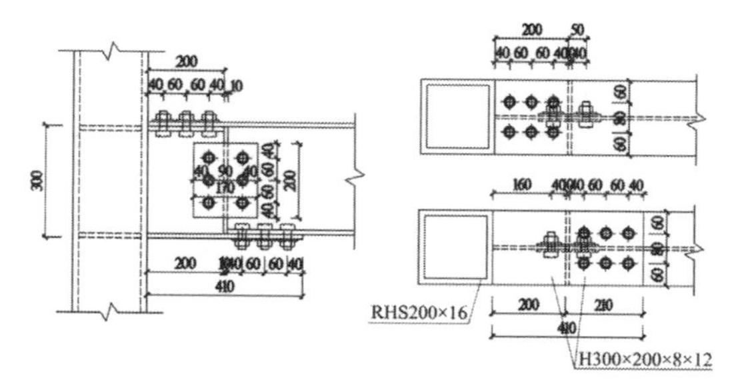

Figure 7. Details of splice connection.

is compared in terms of seismic performance in details.

3.2 *Definition of material parameters*

Both the beam and column are made of Q235B steel, and the multilinear isotropic hardening model is employed. The floorslab concrete has a strength grade of C 30, and based on the consideration of elastic material, the elastic modulus is 3.0×10^4 MPa. The representative value of gravity load is calculated by "$1.0 \times$ dead load $+0.5 \times$ live load", which is considered by increasing the density of floorslab during modeling.

3.3 *Boundary condition and load application*

In the structural model, the bottom of the column is fixedly connected, and all 6 degrees of freedom are restricted, the seismic force is applied according to the inverted triangle model.

$$F_i = \frac{G_i H_i}{\sum_{j=1}^{n} G_j H_j} V_b \tag{1}$$

Where, F_i represents the horizontal load in floor i, G_i and H_i represent the representative value of gravity load and calculated height of floor i, and V_b represents the total seismic shear at the bottom of the structure.

4 COMPARISON OF STRUCTURAL CHARACTERISTICS

A linear perturbation analysis step is set in ABAQUS for modal analysis and to obtain the structural frequency. Lanczos method is selected for the solver of eigenvalue, and the first 12-order frequency and natural period of vibration are calculated respectively for frames RF and CTF, and by observing the vibration mode in each order of the two frames, it is found that frame CTF has its mode of vibration consistent with that of RF in each order, i.e. the change of rigid joint into column-tree joint does not change the original mode of vibration of the structure. The first 6-order natural period of vibration of the structure is shown in Table 1. According to Table 1, compared with the traditional frame, frame CTF has its rigidity reduced slightly with a very small a amplitude, and the difference of natural period of vibration is very small, only 1.71% maximally, so it can be considered that the change of the rigid joint into column-tree joint has a very small influence on the natural period of vibration, and it can be neglected basically.

5 PUSHOVER ANALYSIS RESULT OF THE STRUCTURE

In this article, a pushover analysis is made on two steel frames RF and CTF, and capacity spectrum

Table 1. Natural vibration period comparison of RF and CTF.

	T_{RF}*	T_{CTF}*	
Vibration mode	s	s	Difference
1	1.71%	1.58	1.60
2	1.50%	1.53	1.56
3	1.61%	1.37	1.39
4	1.13%	0.53	0.54
5	1.53%	0.52	0.53
6	1.54%	0.45	0.46

* T_{RF} refers to natural vibration period of frame RF, T_{CTF} refers to natural vibration period of frame CTF.

method is adopted. The response of the two frames under frequent earthquake and rare earthquake has been analyzed, so as to illustrate the characteristics of seismic performance of the overall structure of column-tree steel frame.

5.1 *Capacity spectrum method*

Pushover analysis method is to apply a monotonic increasing horizontal load to the structure according to a certain horizontal loading model, push the structure to a given target displacement or a ultimate status gradually, analyze the mechanical behavior of the structure after entering the plastic status and judge whether the bearing capacity and deformation of the structure and members can meet the requirements of seismic design.

The analysis steps of capacity spectrum method are as follows:

1. Solve the internal force of the structure under the action of vertical load, and then apply the horizontal pushover force. The inverted triangle loading model is employed in this article.
2. Carry out pushover analysis to obtain the pushover curve. Then convert the pushover curve into acceleration spectrum-displacement spectrum relation curve. The conversion formula is:

$$S_a = \frac{V_b}{M_1^*}, S_d = \frac{U_n}{\Gamma_1 \phi_{n1}} \tag{2}$$

Where, Γ_1 and M_1^* are respectively the participation coefficient of the first mode of vibration of the structure and effective mass; V_b and U_n are respectively the base shear and top displacement.

3. Establish the elastic-plastic demand spectrum curve: first, convert the elastic response spectrum curve with damping ratio of 5% into an elastic demand spectrum curve, and the conversion formula is:

$$S_d = \left(\frac{T}{2\pi}\right)^2 S_a \tag{3}$$

Then, preset the acceleration values as a_p and displacement value as d_p on the capacity spectrum, and according to the principle of equal energy, the capacity spectrum is double linearized to obtain the acceleration value a_y and displacement value d_y at the yield point, as shown in Figure 8.

Then calculate the hysteretic damping ratio β_0 based on formula (4), and reduce the elastic demand spectrum to obtain the elastic-plastic demand spectrum.

$$\beta_0 = \frac{1}{4\pi}\frac{E_D}{E_{SO}} \tag{4}$$

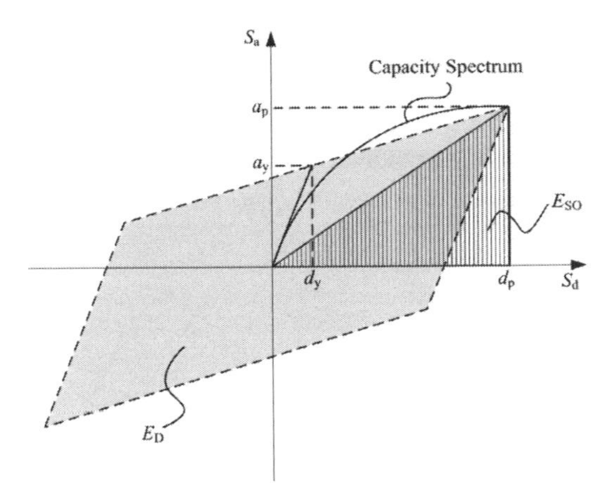

Figure 8. Response spectrum reduction with damping ratio.

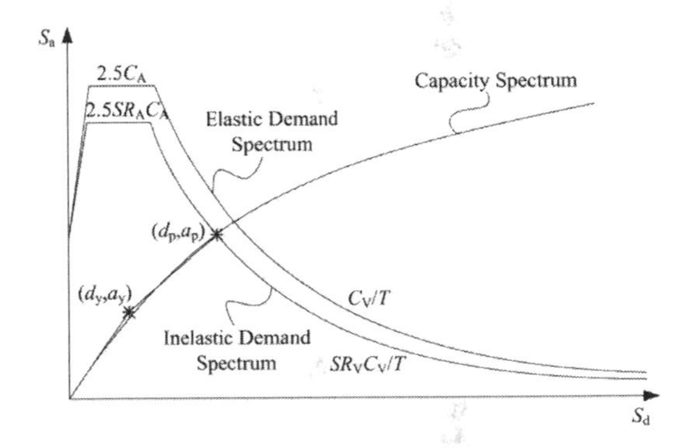

Figure 9. Demand spectrum and capacity spectrum.

Where, E_D is the energy consumption of hysteretic damping, E_{SO} is the maximum strain energy, d_p and a_p are respectively the maximum displacement and the corresponding acceleration of the equivalent single degree of freedom system; d_y and a_y are respectively the corresponding displacement and acceleration at yield of equivalent single degree of freedom system. The sum of hysteretic damping ratio β_0 and inherent damping ratio β can be expressed with equivalent damping ratio β_{eff}, as follows:

$$\beta_{eff} = \kappa\beta_0 + \beta \tag{5}$$

Where κ is the adjustment coefficient of equivalent damping, used to consider that the structure with different energy consumption capacities has different hysteretic damping ratios.

The equivalent damping ratio β_{eff} can be used to calculate the reduction factors SR_A and SR_V of the response spectrum. In Figure 9, SR_A and SR_V respectively represent the reduction coefficient of the constant acceleration zone and constant speed zone. C_A and C_V are the seismic factors, C_A represents the effective peak acceleration, C_V represents the speed response spectrum of the structure system with damping ratio of 5%.

4. Determination of performance point. If the error between the intersection of capacity spectrum and elastic-plastic demand spectrum and the trial point is within 5%, it is considered that the intersection is the performance point, otherwise, this intersection should be used as a new trial point for calculation in the same way, until the point solved is within the permissible error range.

5.2 *Calculation of structural performance point*

After determination, the performance point is converted into the top displacement of the structure, i.e. the target displacement. The seismic performance of the structure can be evaluated by analyzing the plastic zone and story drift angle at the moment when the structure reaches the target displacement. If there is no intersection between the capacity spectrum and demand spectrum, the structure does not have anti-collapse capacity. According to the calculation steps above, the frames RF and CTF are respectively calculated, and the results is as shown in Figure 10.

According to the figure above, the performance point of the frame is as shown in Table 2, it can be seen that the two frames have a very small difference in performance point, and under the frequent earthquake and rare earthquake, the two frames have a very small difference of top displacement, indicating that under the condition of achieving prefabrication, the frame CTF does not have a great difference from the fame RF in terms of seismic performance of overall frame.

5.3 *Evaluation on structural seismic performance*

Under the frequent earthquake, there is no plasticity occurring in each component of the frames RF and CTF when the structure reaches the state of performance point, and both frames are in an elastic state. Under the rare earthquake, some beams of frame RF are in plastic state, and there is no plastic deformation of the column. Some connecting elements of the frame CTF are in nonlinear state, and without plastic deformation of column and beam. This shows that under the rare earthquake, the frame RF can utilize the plastic deformation of beams to dissipate energy, and the frame CTF can utilize the slip of splices to dissipate energy. With the development of plasticity in frame RF, plastic deformation occurs to the column gradually, and the yield mechanism of the structure is: frame beam-frame column. With the development of plasticity in frame CTF, plastic deformation occurs to the beam and column respectively, and the yield mechanism of the structure is: splice—frame beam-frame column. The yield mechanisms of the two frames show that both frames conform to the seismic requirements of strong column and weak beam, and frame CTF realizes the energy dissipation of the splice, and compared with frame RF, it achieves the purpose of multiaspect seismic fortification.

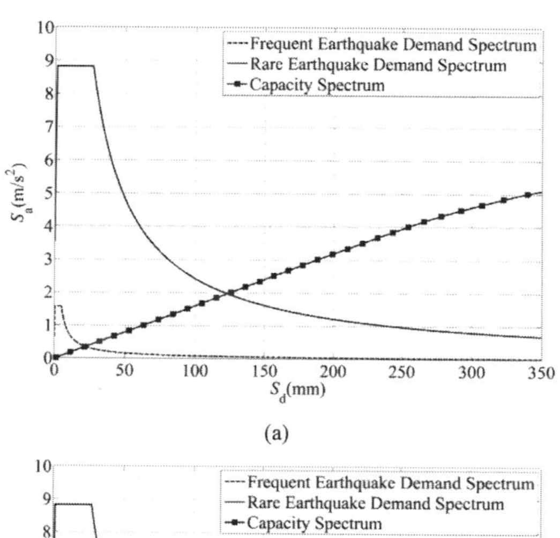

(a)

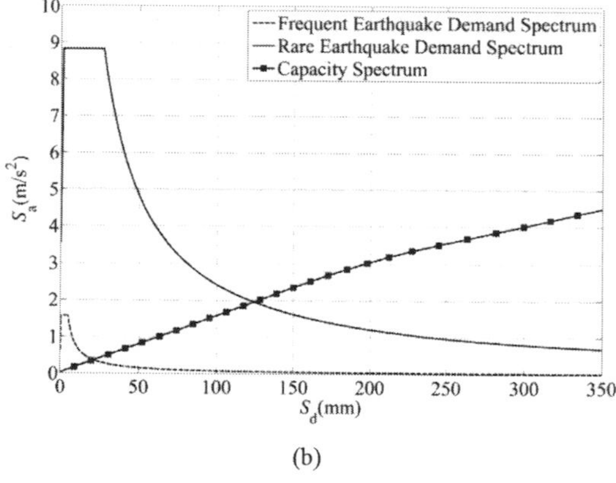

(b)

Figure 10. Analysis result of pushover. (a) Frame RF. (b) Frame CTF.

Table 2. Performance points of RF and CTF.

Frame	Earthquake	S_d mm	S_a mm	Top displacement m/s²
RF	Frequent	22.3	0.34	28.6
	Rare	123.0	1.97	157.7
CTF	Frequent	21.4	0.36	27.4
	Rare	124.0	1.94	159.0

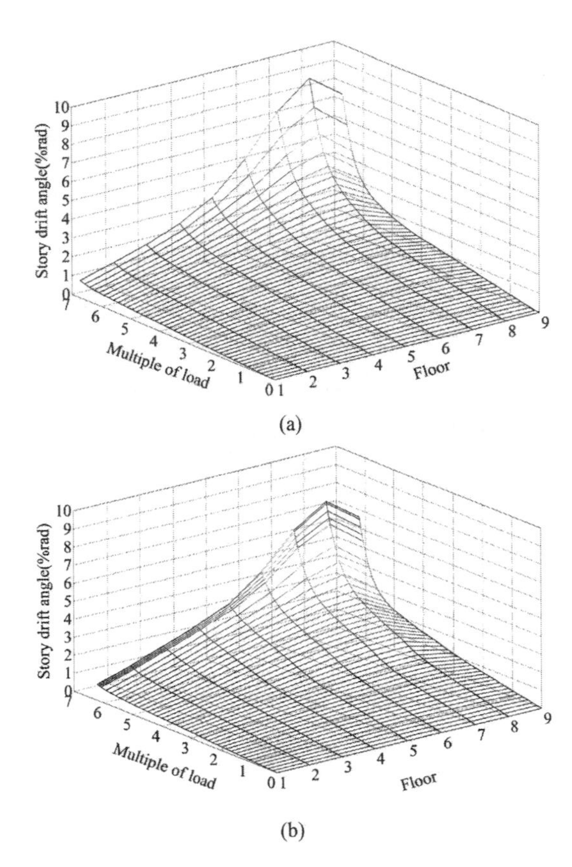

Figure 11. Story drift ratio of the frames. (a) Frame RF. (b) Frame CTF.

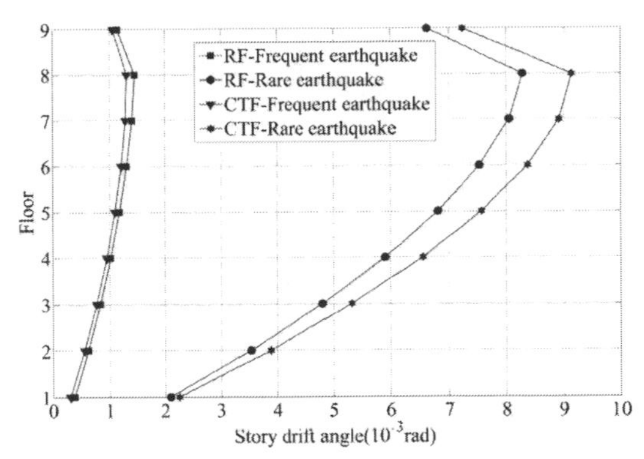

Figure 12. Story drift ratio of the frames in frequent earthquake and rare earthquake.

The curve of change of story drift angle of the two frames with the number of floors and multiple of load is extracted, as shown in Figure 11. It shows that with the increase of load, the story drift angle of each floor increases, and on the first 8 floors, with the increase of floor, the story drift angle increases, and on the 9th floor, it decreases. The maximum story drift angle of both frames RF and CTF occurs on the 8th floor under the frequent earthquake and rare earthquake, the maximum story drift angles of frames RF and CTF under the frequent earthquake are respectively 1/689 and 1/764, both less than 1/250, meeting the requirements of seismic code. The maximum story drift angle of frames RF and CTF under the rare earthquake are respectively 1/121 and 1/109, both less than 1/50, meeting the requirements of seismic code.

Figure 12 shows the distribution of story drift angle of the structure along the height under frequent earthquake and rare earthquake, and it can be seen that under the frequent earthquake, the structure has a relatively even deformation, and under the rare earthquake, there is a sharp increase deformation on the 8th floor, which is the weak layer in the structure, and during design, it is required to mainly check whether its story drift angle meets the seismic requirements.

6 CONCLUSIONS

In this article, ABAQUS is employed to build two steel frame models, and the modal analysis and pushover analysis are respectively conducted on the overall structures. The calculation results of rigid frame and column-frame steel frame are compared and analyzed in details, and it is concluded that:

1. The moment-rotation curve of the beam column joint obtained by the connection element method and the experimental results have been compared and analyzed. The results show that the connection element method can simulate the response of the tree column frame under the static load accurately.
2. The column-tree frame has its mode of vibration consistent with that of rigid frame, and the difference of natural period of vibration is very small, only 1.71% maximally. The change of the rigid joint into column-tree joint has a very small influence on the natural period of vibration.
3. The two frames have a very small difference in performance point under the frequent earthquake and rare earthquake, indicating that under the condition of achieving prefabrication, the column-tree frame does not have a great difference from the rigid frame in terms of seismic performance of overall frame.
4. The yield mechanisms of the two frames show that both frames conform to the seismic

requirements of strong column and weak beam, and the column-tree frame realizes the energy dissipation of the splice, and compared with the rigid frame, it achieves the purpose of multi-aspect seismic fortification.

5. The maximum story drift angle of both column-tree frame and rigid frame occurs on the 8th floor under the frequent earthquake and rare earthquake and the 8th floor is the weak layer in the structure.

ACKNOWLEDGMENTS

The writers gratefully acknowledge the support for this work, which was funded by the National Natural Science Foundation of China (51278010) and the Key Project of Natural Science Foundation of Beijing (8131002).

REFERENCES

Astaneh-Asl A. 1997. Seismic design of steel column-tree moment-resisting frames. *Structural Steel Educational Council*, Berkeley, California.

Chen, C. C., Lin, C.C., & Lin, C.H. 2006. Ductile moment connections used in steel column-tree moment-resisting frames. Journal of Construct*ional Steel Research, 6*2(8), 793–801.

Chen, C.C., & Lin, C.C. 2013. Seismic performance of steel beam-to-column moment connections with tapered beam flanges. *Engineering Structures, 48, 588–601.*

Kim, T., Whittaker, A.S., Gilani, A.S.J., Bertero, V.V., & Takhirov, S.M. 2002. Experimental evaluation of plate-reinforced steel moment-resisting connections. *Journal of Structural Engineering*, 128(4), 483–491.

Lee, C.H., & Kim, J.H. 2007. Seismic design of reduced beam section steel moment connections with bolted web attachment. *Journal of Constructional Steel Research*, 63(4), 522–531.

Liu, X.C., Pu, S.H., Zhang, A.L., Xu, A.X., Ni, Z., & Sun, Y., et al. 2015. Static and seismic experiment for bolted-welded joint in modularized prefabricated stee*l structure. Journal of Constructional S*teel Research, 115, 417–433.

Liu, X.C., Xu, A.X., Zhang, A.L., Ni, Z., Wang, H.X., & Wu, L. 2015. Static and seismic experiment for welded joints in modularized prefabricated steel structure. *Journal of Constructional Steel Research*, 112, 183–195.

Mcmullin, K.M. & Astaneh-Asl, A. 2003. Steel semirigid column–tree moment resisting frame seismic behavior. *Journal of Structural Engineering*, 129(9), 1243–1249.

Oh, K., Li, R., Chen, L., Hong, S.B., & Lee, K. 2014. Cyclic testing of steel column-tree moment connections with weakened beam splices. *International Journal of Steel Structures*, 14(3), 471–478.

Popov, E.P., Yang, T.S., & Chang, S.P. 1998. Design of steel MRF connections before and after 1994 Northridge earthquake. *Engineering Structures*, 20(12), 1030–1038.

Uang, C.M., Yu, Q.S., Noel, S., & Gross, J. 2000. Cyclic testing of steel moment connections rehabilitated with RBS or welded haunch. *Journal of Structural Engineering*, 126(1), 57–68.

Zhang, A.L., & Liu, X.C. 2013. The New Development of Industrial Assembly High-Rise Steel Structure System in China. *Pacific Structural Steel Conference, Sentosa, Singapore,* pp. 976–981.

Civil, Architecture and Environmental Engineering – Kao & Sung (Eds)
© 2017 Taylor & Francis Group, ISBN 978-1-138-02985-9

Study on reuse of heritages of an old industrial area in Harbin Gongyijie area against the backdrop of urban renewal

Xuezhu Shan & Wen Cheng
School of Architecture, Harbin Institute of Technology, Harbin, Heilongjiang, P.R. China

ABSTRACT: The urban old industrial area, because of its geographical location and close spatial relationship with the urban land, has become an important aspect of urban renewal. Against the backdrop of decreasing new land development in China, there is a huge demand of transforming the old industrial area. To study the transformation style of the urban old industrial area and to promote implementation of the transformation project is a major approach to enliven the urban old industrial area and increase the environmental quality of the urban material space. In this paper, we adopt the transformation planning of the old industrial area in Harbin Gongyijie Area as an example and summarize characteristics and problems facing the old industrial area. Combining the characteristics and specific transformation requirements of urban renewal in the Area, the author hopes that this study can provide technical support and references for planning and implementation of heritage reuse in the urban old industrial area through the functional integration of the urban space, the ecological environmental construction, the historical heritage protection, and the space culture re-shaping.

1 INTRODUCTION

Currently, China's urbanization and industrialization has entered a new development period. With constant adjustment of the urban space structure and the industrial layout, a large number of industries have been migrated because of their singular functions and serious pollution. The remaining industrial land is closely related to the urban space. Being convenient in transportation, the old industrial area has been a major target for urban construction against the backdrop of urban renewal. However, as the urban construction level uplifts and the new urban land development decreases, the large-scale extended construction and the old urban area-leveled reconstruction have gradually shown their defects, and the urban development mode is seeking for adjustment (Li, R. et al. 2014).

Through urban renewal, the regional economic development and construction can be enhanced, the public service facility and infrastructure construction level can be lifted to a new level, and more attention will be paid to protecting ecological environment, reshaping space culture, and protecting and reusing industrial heritages. Transformation of the urban old industrial area on the basis of the backdrop of urban renewal can improve urban functions, alleviate pollution, protect the ecological environment, sustain the history and culture of the old industrial area, and improve both economic efficacy and social and ecological efficacy (Li, D. 2009).

2 CHARACTERISTICS AND TRANSFORMATION DEMANDS OF URBAN OLD INDUSTRIAL AREAS

2.1 *Major characteristics of urban old industrial areas*

Because of historical transformation and urban industrial structural adjustment, old industrial areas of Harbin are mainly located on the edge of downtowns and feature the traditional manufacturing industry enterprises. Its disadvantages include a small industrial development space and serious environmental pollution, and its advantages are favorable geographical conditions, profound industrial cultural deposit, and strong space plasticity.

To sum up, the old industrial area in Harbin Gongyijie Area has the following characteristics: 1) Favorable geographical conditions: Because of historical reasons, urban industrial areas are often located in expensive areas such as the main urban area. With the acceleration of urbanization, improvement of the urban space layout, and the constant adjustment of the industrial structure, old industries are moved out, leaving a space with a high added-value area. Such areas have favorable basic conditions for redevelopment. 2) Carrier of history and culture: The material space and environment peculiar to the urban old industrial area, including the industrial texture, the large-scale production, construction and space and the structures, and its industrial production atmosphere, including production slogans and sounds of machine operation—all these form a part of a city's history and its characteristics. They are basic elements of the indispensable industrial slogans during the urban renewal and transformation. 3) Strong plasticity of the space environment: The texture of the urban old industrial area is highly orderly. The industrial buildings are huge in size, high in floors, and large in space. Their spatial integrity is strong. Therefore, the urban old industrial area shows a high degree of flexibility and plasticity in terms of space structure, layout mode, industrial building transformation, and facade shaping.

2.2 *Transformation demands of urban old industrial areas*

On the one hand, urban old industrial areas boast advantages in terms of geographical location and transportation; on the other hand, they are faced with challenges, including industrial decline, waste of resources, and environmental pollution. Against the backdrop, renewal and transformation of urban old industrial areas have put forward more specific requirements in terms of reasonability of the regional industrial structure, supplement of the city functions, protection and utilization of regional industrial heritages, and improvement of regional space environment.

Specific requirements are as follows:

1. Adjustment of the industrial structure: In the current stage, the urban industrial structural adjustment mainly aims to develop the tertiary industry, stimulate new demands of land use, and put forward more reasonable requirements of the current urban space and the land use structure. Generally speaking, because of the fact that the industrial land use takes up a huge percentage in the urban old industrial area and lack of land for the tertiary industry, it is imperative to supplement relevant land use types and adjust the land structure for the convenience of upgrade and transformation of the overall industrial structure and stimulation of the regional economic development.
2. Improvement of the city functions: Transformation of urban old industrial areas calls for supplementation and improvement of the urban functional space in these areas. Realization of a city's economic and ecological functions relies on renewal and transformation of the urban land utilization structure, optimization of the urban industrial space layout, and enhancement of the urban greening construction. At the same time, the relationship among residence, public services, and infrastructure construction should be coordinated, the residents' living environment should be improved, the public service level should be upgraded, and the infrastructure construction should be strengthened so as to address problems such as the outdated infrastructure and lack of public services in urban old industrial areas. Besides, reusing of heritages of urban old industrial areas has a significant effect on the function of urban culture, for example, doing tourism exploitation is a principal means to enhance the regional cultural functions.
3. Improvement of the comprehensive urban environment quality: Urban old industrial areas mainly feature traditional industrial enterprises. Current industries, such as the ordinary machine manufacturing industry and the boiler industry, are mostly traditional industries with an extensive development mode, backward equipment conditions, and a huge influence on ecological environment. They pose a great obstacle to improvement of the urban ecological environment.
4. Protection of history and culture: Industrial heritages are a record of historical traces with precious cultural value. Therefore, to dig the historical and cultural deposit of industrial heritages and to protect them have become key development tasks of a city. (Li, H. & Xiao, Y. 2014) In the current era, as a large number of industries move out, buildings, industrial process, and other cultural information left behind to date need carriers to further preserve them. While endowing land with new functions and value, urban renewal can maximally meet a city's requirements of making use of industrial areas and protecting industrial heritages.

3 PROBLEMS EXISTING IN HARBIN GONGYIJIE OLD INDUSTRIAL AREA

3.1 *Overview*

Harbin Gongyijie Industrial Area is located in the Harbin Urban-Rural Continuum. The Gongyijie Area is close to Majiagou River on the east. The major research area is the area enclosed by Gongyijie Street, Huaneng Railway, and Majiagouhe River, which covers an area of 0.3 km^2 (Fig. 1).

There are dozens of plants in Gongyijie Area (Table 1). The boiler industry, the manufacturing industry, and other relevant industries are dominating in the area. The remaining are processing links at the bottom of the industrial chain.

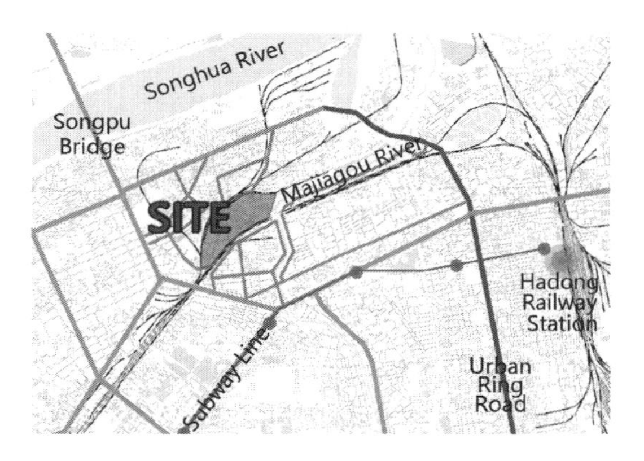

Figure 1. Local map.

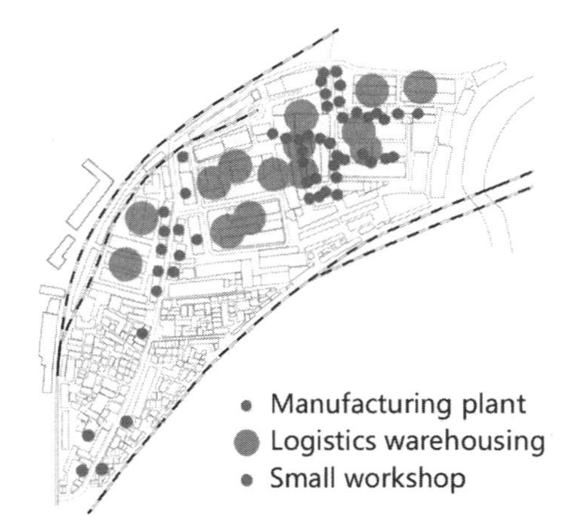

Figure 2. Distribution map of plants.

Table 1. List of industrial types in Gongyijie Area.

Industrial type	Industrial institutions
Plantation	Material companies
Fishery	Fish farms
Manufacturing	Furniture plants, asbestos tile plants, boiler plants, building material companies, steel manufacturers, electricity plants, special equipment plants, machinery plants of light industry, oxygen plants, boiler plants, fuel companies, and material plants
Building	Cement plants
Catering	Fishing villages
Wholesales and retails	Steel markets and goods yards
Logistics warehousing	Ceramic warehouses

3.2 *Main problems*

3.2.1 *The single industrial structure*

In Harbin, the tendency of industries' outsourcing and core industries' hollowing gradually appears in Gongyijie Area. The overall industrial structure is imbalanced: In the secondary industry, the priority is given to the general machinery manufacturing industry and boiler industry, which account for a large proportion but are at the low level. Besides, the developmental model of industrial chain is extensive and the technical foundation is weak, which has a little effect on the overall economic development. Moreover, the proportion of tertiary industry is small, with its development lying behind (Fig. 2).

3.2.2 *Waste of land resources*

With the increasing expansion of Harbin urban land, the geographical position of Gongyijie Area has gradually been the downtown area, which was originally the industrial area at the edge of city. Now, its geographical condition has obvious advantages, and the land value is significantly promoted. As for the present internal situation of this area, the priority is given to the industrial land with scattered usage and mixed organization in which the lands of industries, residents, and corporate facilities are mixed. In addition, because of various reasons such as history and economic development, the majority of its internal factories have been abandoned without production or in the half-production state, but they still occupy most of land in this area, which leads to a waste of land resources in land use (Fig. 3). Here, the industrial land use accounts for 50.97% of the total land area.

3.2.3 *Disappearance of distinctive features*

The Gongyijie is near Majiagou, and its good ecological landscape resources have not been effectively utilized. At the same time, the constructional situation and quality in jurisdictional area of Gongyijie is comparatively poor: The industrial landscape, such as the original large workshops and the private sidings of railway transportation, has been lacking maintenance for many years, The phenomena of private and arbitrary construction could be seen everywhere, which forms a large number of negative space and landscape, and seriously affects the style of urban landscape. In addition, the evaluation system in Table 2 is used to evaluate the status of the construction quality, and quality evaluation of the existing industrial heritage is shown in Fig. 4.

3.2.4 *Serious environmental pollution*

The industries and enterprises that contribute to serious environmental pollution in Gongyijie district include asbestos tile factories, sewage treatment plants, boiler factories, and special equipment

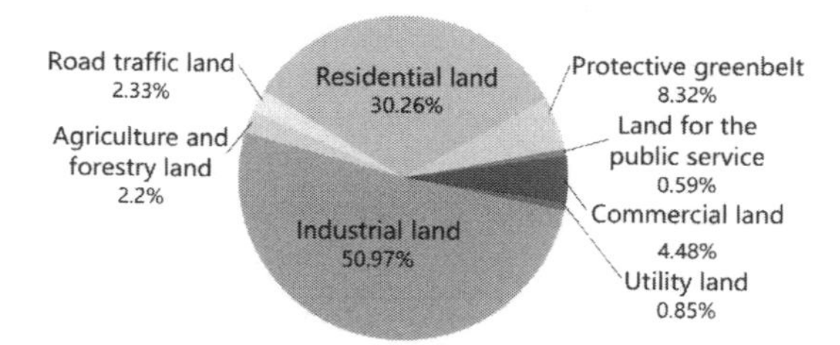

Figure 3. Land use ratio map.

Table 2. List of industrial heritage evaluation index.

First-level index	Factor	Second-level index	Factor
Representative	62.95%	Historical value	28.87%
		Artistic value	6.27%
		Technology value	4.05%
		Social value	9.16%
		Economic value	14.61%
Authenticity	12.64%	Plane function	7.02%
		Facade appearance	3.18%
		Structure equipment	1.22%
		External environment	1.22%
Integrity	6.84%	Material form	5.13%
		Nonphysical form	1.71%
Regional	17.57%	Regional traditional technology	2.85%
		Regional characteristics	2.52%
		Regional rarity	5.54%
		Mark local industry beginning	6.66%

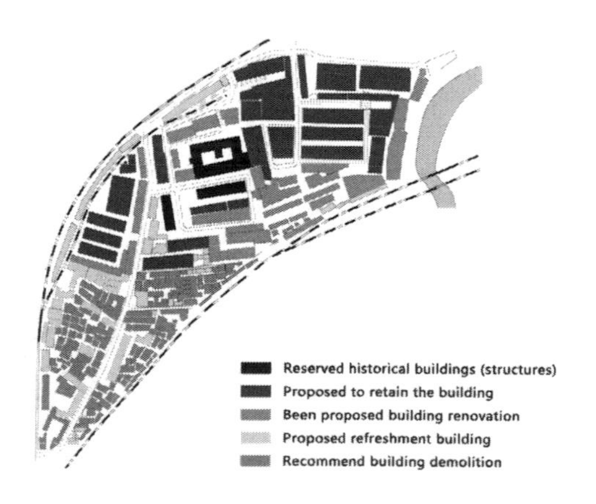

Figure 4. Construction quality evaluation map.

factories. In the production process, soot, sewage, toxic or harmful gases, solid wastes, etc. have been emitted for a long period of time. Meanwhile, this type of industries and enterprises has always been the traditional industries with outdated equipment and lack of technological innovation, leading to problems such as high energy consumption and serious environmental pollution. They seriously interfere with the surrounding residents' daily production and living and threaten the urban ecological environment that needs to be governed.

3.2.5 *Incomplete coverage of public service and backward infrastructure construction*

The public service facilities in Gongyijie Area are in the lack of system with the incomplete type of internal corresponding public service facilities, low constructional standard, and poor constructional quality. The service radius of the only schools is too large with small radiation scope. The current urban construction is seriously backward. For example, the roads have no specific scales, and the system is in the chaos. The hardening of road surface has not been finished, and there is no bus passing through the area. Besides, the corresponding infrastructure set of water and energy supply is insufficient, which does not match with the further development of city, the improvement of living standard, and the modernization of lifestyle.

4 THE RECONSTRUCTION STRATEGY OF THE RENEWAL OF INDUSTRIAL HERITAGES IN GONGYIJIE AREA

According to the goal of regional development, the present situation and problems, and the updating demand, the reconstruction of industrial heritages preservation in Gongyijie Area adjusts the regional positioning and industrial structure, reshapes the spatial environment, and revitalizes the regional energy. The specific strategies and working paths of the renewal of industrial relic preservation in Gongyijie Area are analyzed from the perspectives of industrial development, urban context, urban function, urban material environment, and urban spatial environment.

4.1 *Functional positioning of Gongyijie area in Harbin*

As for the update and reconstruction of Gongyijie Area, the actual demands of its industrial heritages in Harbin's future urban development should be comprehensively considered in order to achieve the overall development of city. Analyze the relative developmental strategies of Harbin at the current stage with gradually achieving "the secondary industry's exist and the tertiary industry's entrance" and outsourcing the second industries under the backdrop of heavily developing the tertiary industries; combine Gongyijie Area with the present situation of its surroundings and relative planning;

comply with the actual situation of regional economic development; balance the land use structure and ratio in Gongyijie Area; adjust industrial structure; put the Gongyijie Area on the position of vice center of external Harbin that is guided by modern service; and possess a comprehensive area of being open, dynamic, ecological, and livable.

4.2 *Specific strategies of the renewal planning in Gongyijie area*

4.2.1 *Adjust the original industrial structure and encourage the development of the tertiary industry*

The original industrial structure in Gongyijie Area is imbalanced with the secondary industry occupying a large proportion. First, the industrial layout should be adjusted. The companies with limited special equipment, and the factories with occupying large land, consuming much energy, and heavily polluting environment such as steel plants, asbestos plants, and cement plants should be relocated or shut down. At the same time, capital investment should be increased, the relevant supporting policies should be perfected, and the great efforts should be made in developing tourism, finance, information, and community service. Besides, the ecological tourism along the river, industrial tourism, and other modern service industries with low energy consumption, no pollution, and high added value in the tertiary industry should be introduced in order to achieve the upgrade and transformation of industrial structure in Gongyijie Area.

4.2.2 *Extend the urban context and reasonably protect and utilize the industrial heritages*

Industrial old area records the history of urban development. Thus, the principle of historical preservation should be adhered to in the reconstruction planning. The industrial remaining buildings and structures with certain memorable significance, such as blast furnace, chimney, tower crane, and railway and transport equipment, inside the base should be protected and utilized (Li, L. 2012). Taking the workshop of cement factory that possesses the industrial atmosphere in Gongyijie Area as an example, its function should be replaced by introducing multiple functions of business, resident, culture, and so on, and making a reconstruction design of its space corresponding to the functional demand in order to protect the industrial relics and inject the new vigor into Gongyijie Area.

4.2.3 *Perfect the urban functions, and complement public service and basic infrastructure construction*

The original land use model in Gongyijie Area lacks planning of the public service facilities and attention to the infrastructure construction, and the overall function lacks a systematic planning. The updating process should be based on the public service facilities of Gongyijie Area and its surroundings as well as the functional layout of business and finance, culture and entertainment, sports and health, education and scientific research, and so on to supplement the categories of facilities and improve the service level. On the one hand, the original scale of education facilities and medical facilities within the jurisdictional area should be improved, and the social welfare facilities and sports facilities should be supplemented so as to complete the public service system. On the other hand, the water supply facilities ought to be repaired, and the level of infrastructure services should be improved to perfect the urban functions from all aspects.

4.2.4 *Strengthen the construction of green system and improve the urban environment by using the ecological restoration measures*

In the need of thoroughly improving the ecological environment in Gongyijie Area, the ecological and livable environment should be built within the internal Gongyijie Area in which a green system has not been planned at the current stage. Now, there still exists more public space to be developed in Gongyijie Area such as the green space with ecological protection (such as isolating noise) and ecological restoration function (such as easing land pollution) (Nan, X. 2013). For example, Majiagou River with good ecological landscape and osmosis should be first developed as the public green space. The priority should also be given to restore the industrial land that is seriously polluted by the original cement factory as the public green space so as to partly construct the green system and curb the pollution and negative effect, construct an urban landscape with high quality, and improve the urban environmental quality (Zhou, Q. et al. 2009).

5 CONCLUSION

Through the way of urban renewal and on the basis of regional functions, the planning should optimize the allocation of land resources and adjust the land using type and proportion. Besides, the planning should guide the adjustment of industrial structure, strengthen the public service facilities and infrastructure construction, perfect the urban functions, protect and utilize industrial heritages, and improve the quality of urban environment so as to enable the urban old industrial areas to adjust to the demands of regional social and economic development, protect historical context, and restore ecological environment. Because the renewal of urban old industrial area has an extensive involvement and is a complex system, this study on the strategic renewal of urban old industrial areas by taking the Gongyijie Area in Harbin as an example, basing on its current conditions and existing problems, is certainly one-sided. With the gradual deepening of relative studies in the future, the renewal and reconstruction strategies of urban old industrial areas would be continually and systematically improved so that it can provide a more complete and specific guideline for the reuse of different types of old industrial areas.

REFERENCES

Li, D. (ed.) 2009. *Principles of urban planning.* Beijing: China Building Industry Press.

Li, H. & Xiao, Y. 2014. Conservation and renewal of urban old industrial areas under the guidance of cultural planning. *Planners.* 07:40–44.

Li, L. 2012. *Study on the protection and reuse of industrial historic buildings in the industrial area of Hanyang*. Wuhan: Huazhong University of Science and Technology.

Li, R., Chi, X. & Lin, G. 2014. Renewal strategy of old industrial blocks in the perspective of new urban development concept. *Planners.* 30(3):208–212.

Nan, X. 2013. *Analysis and optimization strategy on the transformation of the old industrial zone of Shenzhen overseas Chinese town under the ecological perspective*. Shenzhen: Harbin Institute of Technology.

Zhou, Q., Lei, H. & Chen, X. 2009. Explore the harmonious way of the transformation of the old industrial areas of the city. *City Planning Review* 3:67–70 + 86.

Civil, Architecture and Environmental Engineering – Kao & Sung (Eds)
© 2017 Taylor & Francis Group, ISBN 978-1-138-02985-9

Research on lateral resistant behavior and design of multicover plates connection for prefabricated steel plate shear wall with beam-only-connected infill plate

X. Zhang, A.L. Zhang & X.C. Liu
College of Architecture and Civil Engineering, Beijing University of Technology, Beijing, China

ABSTRACT: In this paper, the design principle for the connection of prefabricated steel plate shear wall with beam-only-connected infill plate is investigated. Expressions for determining single-bolt demands for the purposes of design are developed, and a multicover plates connection for prefabricated steel plate shear wall with beam-only-connected infill plate is designed. The lateral resistant behavior for both welded connections and multicover plates connections for steel plate shear wall with beam-only-connected infill plate are analyzed and compared by using ABAQUS finite-element software. The effects of changes in the bolt pre-tightening force for multicover plates connection are investigated as well. Then, the FEM simulation results are verified by monotonic loading test of a multicover plates connection with 6 mm infill plate, the lateral resistant behavior and the deformation of cover plates for both FEM simulation results and test results are compared. The results show that the lateral resistant capability of beam-only-connected steel plate shear wall with prefabricated multicover plates connection is less than that with welded connection; in tension field, severe tearing happens on the bolt holes, and the loss of bolt pre-tightening force is large; the tearing of bolt holes and the loss of bolt pre-tightening force get smaller when they are away from the tension field; the buckling of infill plate can cause deformation in cover plates. There was no significant difference on force transmission performances between welded connections and prefabricated multicover plates connections.

1 INTRODUCTION

After entering the 21st century, China's construction is facing an urgent need for industrial technology improvement. Combining with the highest crude steel output ranks of the world and the current situation of oversupply of steel, there are opportunities for the development of prefabricated steel structure. Prefabricated steel structure has many advantages such as high quality, short construction period, energy conservation, and clean production and construction (Zhang 2013). Comparing with concrete structure, prefabricated steel structure system is more suitable for industrialization. This is because steel has better machinability and it is more suitable for industrial production and processing; steel structure is lighter and more suitable for transportation and assembly; steel structure is suitable for the high-strength bolt connection and much easier for assembly, maintenance, and replacement (Zhang 2014). The A.L. Zhang research group of Beijing University of Technology earlier proposed and studied the industrial prefabricated steel structure. However, the different lateral resistant systems for prefabricated high-rise steel structure still need to be further researched.

Steel plate shear wall system is a type of efficient lateral resistant system, which was developed in the 1970s. It consists of infill steel plate and boundary frame members. The infill steel plate can be connected with the boundary frame members by bolts or welding (Guo 2009). Infill steel plate is a lateral resistant member in steel plate shear wall system, which provides the primary lateral strength and energy dissipation (Clayton, Berman & Lowes 2015). It protects the main frame structure and damages before the main frame structure as the first line of defense and improves the collapse-resistant capacity of the structure.

In most steel plate shear walls, infill plates are connected to both the beams and columns. However, connecting the infill plates to the beams only has been proposed as a means of reducing boundary frame demands and mitigating infill steel plate damage (Xue 1994). Steel plate shear wall with beam-only-connected infill plate can also provide convenience for structural openings and lateral stiffness adjustment (Choi & Park 2009). Because bolted connection is more suitable for assembly, steel plate shear wall with bolted connection is more suitable for prefabricated high-rise steel structure than the welded one.

At present, most of the researches mainly focus on different infill steel plates and boundary frame members, and only few researchers investigate different welding connections between infill steel plates and boundary frame members. Studies on bolted connections are almost none. In this paper, expressions for determining single-bolt demands for the purposes of design are developed first. Then, a multicover plates connection for prefabricated steel plate shear wall with beam-only-connected infill plate is designed and analyzed.

2 DESIGN FOR PREFABRICATED STEEL PLATE SHEAR WALL WITH BEAM-ONLY-CONNECTED INFILL PLATE

2.1 *Design principle*

The earliest design principle of steel plate shear wall is to ensure that the shear buckling of the plate is not earlier than the shear yield. This means that the buckling of the infill plate is not allowed. Therefore, the shear buckling stress of the infill steel plate is designed as the ultimate bearing capacity. The later studies show that the infill steel plates can form a tension field after buckling (Thorburn 1983). When tension field develops, the structure still has efficient lateral stiffness and shear capacity. Ever since then, many studies are focused on the tension field, the results of which show that rational utilization of tension field can provide not only the good seismic performance and energy dissipation capacity but also low steel quantity, light weight, and small seismic response (Hosseinzadeh & Tehranizadeh 2014).

The strain of tension field is large and the stress is greater than the yield strength of infill steel plate. Therefore, steel plate shear wall needs to be designed by using the ultimate tensile strength of the tension field. For frictional high-strength bolted connection, the slippage between infill plate and fishplate should be avoided in the design. The shear capacity of single bolt should not be less than the ultimate tensile strength of the tension field it takes.

2.2 *Expressions for determining single-bolt demands*

As Figure 1 shows, take the diameter of bolt hole as d. Assuming that the spacing between two bolt holes is the minimum:

$$b = 2d \tag{1}$$

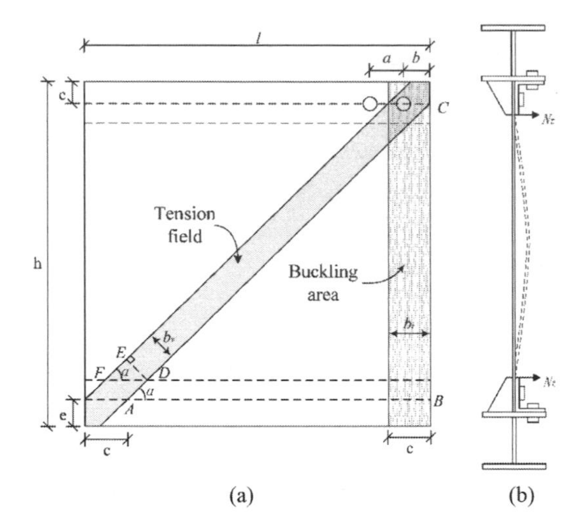

Figure 1. Model for design calculation. (a) Model for prediction of inclination angle of tension field. (b) Buckling of infill plate.

$$a = 3d \tag{2}$$

$$e = 2d \tag{3}$$

$$c = b + \frac{a}{2} = 2d + \frac{3d}{2} = 3.5d \tag{4}$$

where a, b, e, and c are shown in Figure 1. If the angle of tension field is α, then:

$$\tan\alpha = \frac{L_{BC}}{L_{AB}} = \frac{h-2e}{l-c} = \frac{h-4d}{l-3.5d} \tag{5}$$

$$\sin\alpha = \frac{L_{ED}}{L_{FD}} = \frac{b_v}{c} = \frac{b_v}{3.5d} \tag{6}$$

The width of the tension field for single bolt (b_v) is given in equation (7):

$$b_v = 3.5d \cdot \sin\left(\tan^{-1}\left(\frac{h-4d}{l-3.5d}\right)\right) \tag{7}$$

The cross-sectional area of tension field for single bolt (A_v) is given in equation (8):

$$A_v = 3.5d \cdot \sin\left(\tan^{-1}\left(\frac{h-4d}{l-3.5d}\right)\right) \cdot t \tag{8}$$

The ultimate tensile strength of tension field for single bolt (N_v) is given in equation (9):

$$N_v = 3.5d \cdot \sin\left(\tan^{-1}\left(\frac{h-4d}{l-3.5d}\right)\right) \cdot t \cdot f_u \tag{9}$$

As Figure 1(b) shows, buckling of infill steel plate will cause out-of-plate load on bolts, the magnitude of which can reach 10% of the yield strength of tension field. If the out-of-plate load is ignored, the bolt hole will slip easily. Therefore, the out-of-plate load should be considered while designing. The out-of-plate load for single bolt(N_t) can be design using equation (10):

$$\begin{aligned} N_t &= \left(A_t \cdot f\right) \times 10\% \\ &= \left(b_y \cdot t \cdot f\right) \times 10\% \\ &= 0.35 d \cdot t \cdot f \end{aligned} \tag{10}$$

For frictional high-strength bolted connection, the shear capacity of single bolt $\left(N_v^b\right)$ can be calculated by equation (11):

$$N_v^b = k_1 \cdot k_2 \cdot n_f \cdot \mu \cdot P \tag{11}$$

where μ is the friction coefficient. For ordinary steel structure, take k_1 as 0.9; for large bolt hole, take k_2 as 0.85; the number of friction surface (n_f) should be 2 in multicover plates connection, then the expression for shear capacity of single bolt can be simplified as equation (12):

$$N_v^b = 0.9 \times 0.85 \times 2 \times \mu \times P = 1.53 \mu \cdot P \tag{12}$$

Table 1. Assembled connection design sheet.

l	H	t	d	η	α	N_v	N_t	i
mm	mm	mm	mm			kN	kN	
2700	2700	6	30	0.5	0.78	163	13.5	1.02
2700	2700	5	30	0.5	0.78	136	11	0.85
2700	2700	4	30	0.5	0.78	109	9	0.68
900	900	6	30	0.5	0.78	164	13.5	1.03
900	900	5	30	0.5	0.78	137	11	0.86
900	900	4	30	0.5	0.78	110	9	0.69

The design value of tensile capacity for a single bolt can be calculated by equation (13):

$$N_t^b = 0.8P \tag{13}$$

When frictional high-strength bolt undertakes shear and tension simultaneously, the bearing capacity should satisfy equation (14):

$$i = \frac{N_v}{N_v^b} + \frac{N_t}{N_t^b} \leq 1 \tag{14}$$

The multicover plates connection for prefabricated steel plate shear wall with beam-only-connected infill plate can be designed by using the design sheet shows in Table 1, when Q235B steel and level 10.9 M24 high-strength bolts are used.

2.3 *Design of multicover plates connection for prefabricated steel plate shear wall with beam-only-connected infill plate*

From the calculated results in Table 1, the difference between ultimate tensile strength of tension field for single bolt in full-size model and the one in reduced scale model is less than 1%. For the convenience of finite-element analysis and test, the reduced scale model has been designed.

The characteristics of prefabricated multicover plates connection of prefabricated steel plate shear wall with beam-only-connected infill plate are: 1) large bolt holes are arranged at the top and bottom of the infill steel plate in order to increase the allowable value of machining error and make the installation more convenient; 2) multiple welded angle steel are used for connecting the infill plate and boundary beam. There are normal bolt holes on both sides of the cover plates. In this paper, there are three normal bolt holes on each side of the cover plate and a 10 mm space between each cover plate; 3) the bolt holes on beam flange are large; 4) there are stiffeners at both ends of the bottom plate to avoid deformation caused by buckling

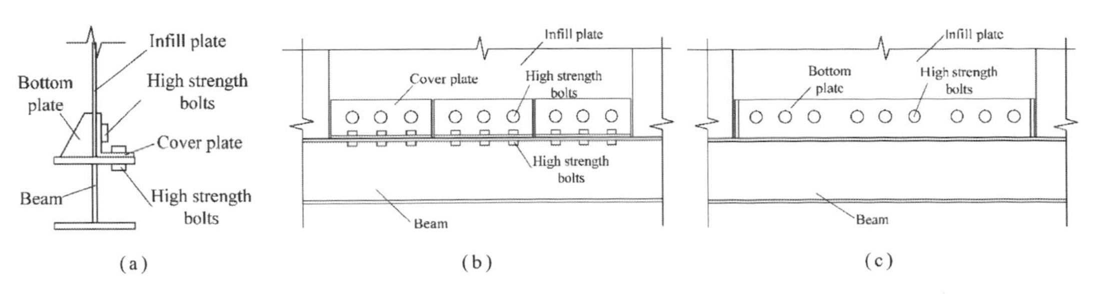

Figure 2. Design layout for multicover plates connection. (a) Cross section of connection, (b) front view of connection, and (c) back view of connection.

of infill steel plate. The design layout of multi-cover plates connection of prefabricated steel plate shear wall with beam-only-connected infill plate is shown in Figure 2.

3 FINITE-ELEMENT ANALYSIS

3.1 *Introduction of the FE model*

In this paper, six models are analyzed by using ABAQUS software. The details of the six models are shown in Table 2. All models are made by Q235B steel. H-section beams of H200 × 200 × 8 × 12 are used as the boundary beam.

There is no connection between infill plate and columns. To avoid the lateral stiffness of columns, the simplified analysis model is shown in Figure 3.

In order to simplify the finite-element model further and make the calculation more convenient, columns are modeled as rigid wire element of RB3D2, and the other parts are modeled as 3D solid element of C3D8R. The models are under displacement-controlling loading, and the loading point is at one end of the top beam. The models are fixed at both ends of bottom beam, and they make sure that there is no out-of-plane displacement of the top beam. The finite-element models are shown in Figure 4.

3.2 *Lateral capacity analysis*

Figure 5 shows the loading-displacement curve of six finite-element models. By comparison, the lateral capacities of multicover plates connections are lower than those of the welding connections; however, the trends of the curves are similar. As can be seen from the deformation of the models, the effective height of infill plate with welding connection is smaller because the welded part between infill plate and bottom plate has nearly no deformation. This means the effective height of infill plate with welding connection should be the distance between upper and lower bottom plates, which is smaller than the one with multicover plates connection.

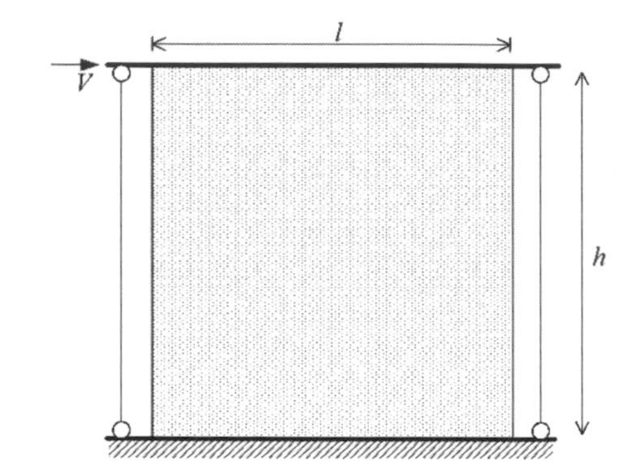

Figure 3. Model for simplified calculation.

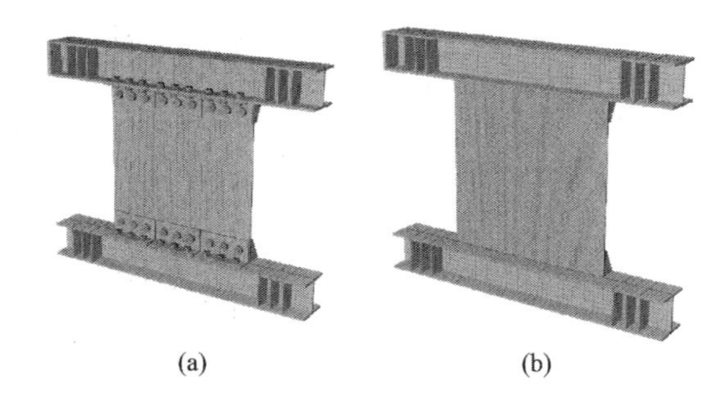

Figure 4. Model for finite-element analysis. (a) Assembled connection. (b) Welded connection.

As can be seen from the curves, when the lateral loading reaches a certain value, every curve has its own descent part. The thinner the infill steel plate, the larger the reduction. The reduction of lateral loading is caused by the buckling of infill steel plate. When the infill steel plate starts to buckle, the tension field has not developed, the lateral loading will decrease until the tension field is formed. For models A5, A4, W5, and W4, the tension fields are obvious by the end of the simulation, but the tension fields of models A6 and W6

Table 2. Details of finite-element model (mm).

	Assembly connection			Welding connection		
Model	A6	A5	A4	W6	W5	W4
Bolt diameter	24	24	24	24	24	24
Height of infill plate	900	900	900	900	900	900
Width of infill plate	900	900	900	900	900	900
Infill plate thickness	6	5	4	6	5	4
Cover plate thickness	10	10	10	10	10	10
Bottom plate thickness	10	10	10	10	10	10

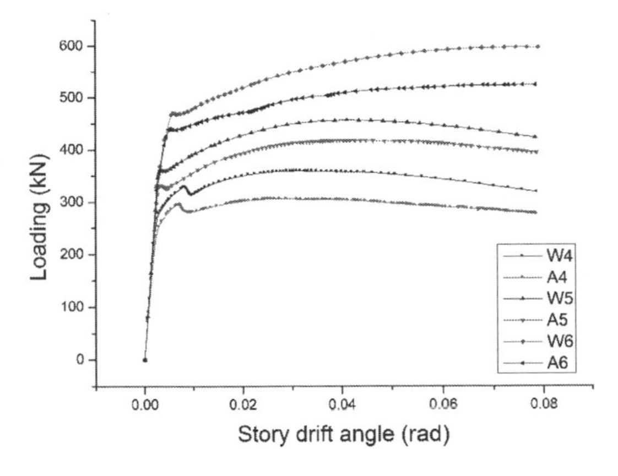

Figure 5. Loading-story drift angle curve for FE models.

Figure 7. Loading device.

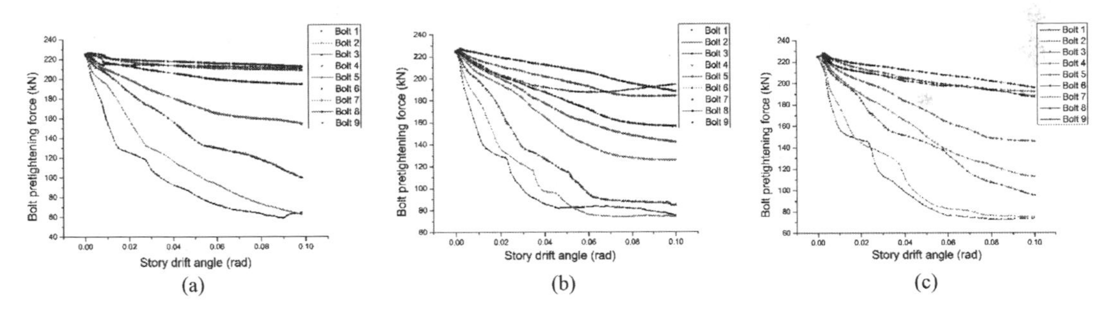

Figure 6. Bolt pre-tightening force-story drift angle curves. (a) Model A4. (b) Model A5. (c) Model A6.

are still developing. This means models A6 and W6 do not belong to thin steel plate shear wall.

3.3 *Bolt pre-tightening force analysis*

Figure 6 shows how the bolt pre-tightening force changes in models A4, A5 and A6 when they are under lateral load. The bolts are the bottom connection bolts on the infill plate, and the bolt numbers are 1–9 from the left to right. This means bolts 1 and 9 are the first and last bolts in the tensile zone.

By comparison, from bolt 1 to bolt 9, the losses of bolt pre-tightening force are decreasing. The largest loss happens on bolt 1, which is about 70% of the total bolt pre-tightening force. The farther the bolt from the tensile zone, the less loss of the bolt pre-tightening force. Only about 10% of total bolt pre-tightening force was lost until the end of simulation. This loss is caused by the buckling of infill plate.

From the deformation of the FE models, the bolt hole of bolt 1 slips and tears first, then the bolt 2, and so on. The farther the bolt from the tensile zone, the smaller the slippage of the bolt hole. There is nearly no slippage at bolts 7, 8, and 9.

4 MONOTONIC LOADING TEST

4.1 *Introduction of the test*

In order to verify the results of FE analysis, a test specimen is designed on the basis of the FE model. A prefabricated steel plate shear wall with 6-mm-thick infill plate has been tested under monotonic loading. The loading device is depicted in Figure 7. Because of the stroke limit of the loading device, the test specimen was pulling at the end of the top beam instead of pushing.

4.2 *Comparative analysis of lateral capacity*

Figure 8 shows the comparison of the loading-story drift angle curve for both test results and FE results. The lateral capacity of the test member is about 8% less than the FE results, but the trends of both curves are similar. The test results prove the validity of the finite-element modeling.

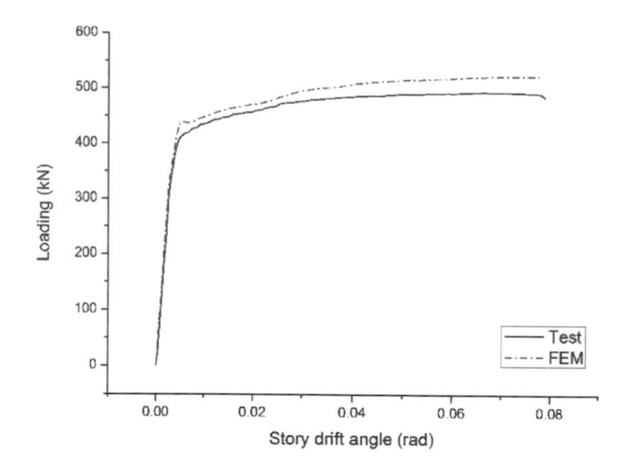

Figure 8. Loading-story drift angle curve for FEM and test specimen.

4.3 *Comparative analysis of deformation*

When the story drift angle is 0.005, the first bolt holes in the tensile zone of both the test specimen and the FE model start to slip until the story drift angle reaches 0.02. Then, the bolt hole touches the bolt shank and starts to deform.

Buckling of the infill plate happens at nearly the same time when the infill plate slips. Figure 9(b) shows the deformation of test specimen at the end of pulling. Comparing with the deformation of FE model under pushing, they are close to completely symmetrical. Therefore, the FE modeling is verified.

The beams have obvious shear deformation at the tensile end of the connection. The cover plates are slightly deformed because of the buckling of the infill plate. Therefore, the suggestion is that the

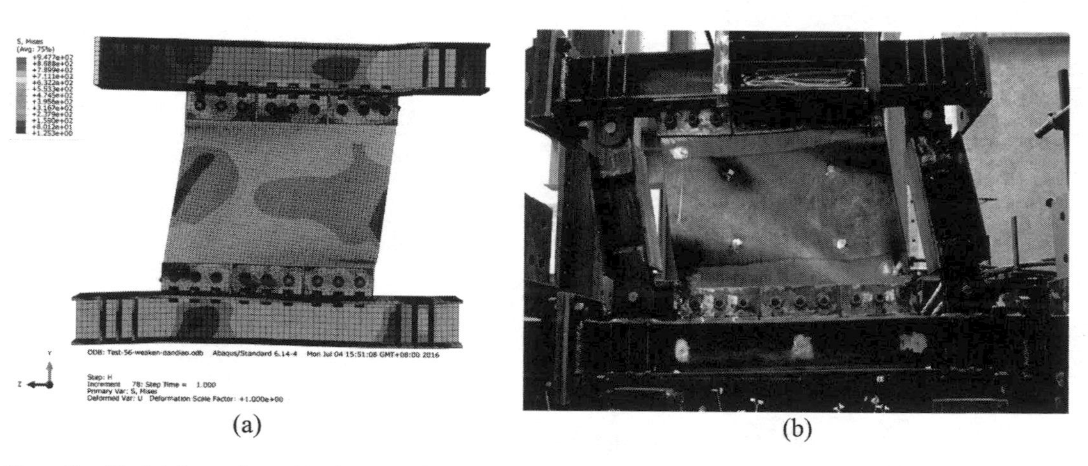

(a) (b)

Figure 9. Final deformations. (a) Finite-element model. (b) Test specimen.

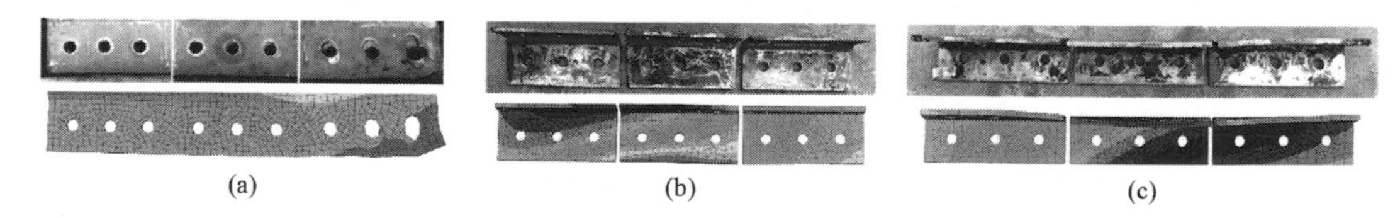

(a) (b) (c)

Figure 10. Comparison of residual deformation. (a) Infill plate connection zone. (b) Side view of cover plates. (c) Top view of cover plates.

stiffening rib is arranged on the beam at the end of the connection.

4.4 *Comparative analysis of residual deformation*

Figure 10(a) shows the comparison of residual deformations on infill plate after loading. The bolt hole in tension is teared, and the hole at the other end is unwounded. Figure 10(b) and (c) shows the cover plates, which are removed after test; both sides of the cover plates have slightly deformed. The bolt holes are all unwounded. The deformation of cover plates is slightly smaller than that of the FE model, because after the pulling, the test specimen has been pushed to the original position before dismantling.

5 CONCLUSION

In this paper, first, the design principle for the connection of prefabricated steel plate shear wall with beam-only-connected infill plate is investigated, expressions for determining single-bolt demands for purposes of design are developed, and a multicover plates connection for prefabricated steel plate shear wall with beam-only-connected infill plate is designed. Assuming that the spacing between two bolt holes is the minimum, the angle of tension field is related to not only the dimension of infill plate but also the diameter of bolt holes. Together with the development of the tension field, the angle of it will increase.

In order to study the influence of lateral capacity on welded connection and bolted connection of steel plate shear wall, six FE models have been analyzed by using ABAQUS finite-element software. By comparison, the lateral capacities of multicover plates connections are lower than the welding connections, but their trends are similar. The effective height of the infill plate with welding connection is smaller than that of the multicover plates connection. The bolt pre-tightening force will decrease during loading; the closer the tension field, the larger the loss of bolt pre-tightening force. The deformation and slippage of tension field are serious.

Then, the FEM simulation results are verified by monotonic loading test. The results show that the multicover plates connection has a good force transmission performance. The beams have obvious shear deformation at the tensile end of the connection. To ensure that the steel plate shear wall gets damaged first, the stiffening rib at the end of the connection on the beam is suggested.

REFERENCES

Choi, I.R. & Park, H.G. (2009). Steel plate shear walls with various infill plate designs. *Journal of Structural Engineering*, 135(7), 785–796.

Clayton, P.M., Berman, J.W. & Lowes, L.N. (2015). Seismic performance of self-centering steel plate shear walls with beam-only-connected web plates. *Journal of Constructional Steel Research*, 106, 198–208.

Guo, Y.L. & Zhou, M. (2009). Categorization and performance of steel plate shear wall. *Journal of Architecture and Civil Engineering*, 26(3):1–13. (in Chinese)

Hosseinzadeh, S.A.A. & Tehranizadeh, M. (2014). Behavioral characteristics of code designed steel plate shear wall systems. *Journal of Constructional Steel Research*, 99(8), 72–84.

Thorburn, L.J., Kulak, G.L. &Montgomery, C.J. (1983). *Analysis of steel plate shear walls*. Edmonton: University of Alberta.

Xue, M. & LU, L.W. (1994). Interaction of infilled steel shear wall panels with surrounding frame members. *Proceedings of Structural Stability Research Council Annual Technical Session*, Bethlehem, PA, 339–354.

Zhang, A.L. (2014). The key issues of system innovation, drawing up standard and industrialization for modularized prefabricated high-rise steel structures. *Industrial Construction*, 44(8):1–6. (in Chinese)

Zhang, A.L. & Liu X.C. (2013). The new development of industrial assembly high-rise steel structure system in China. *Proceeding of 10th Pacific Structural Steel Conference*. Singapore.

Civil, Architecture and Environmental Engineering – Kao & Sung (Eds)
© 2017 Taylor & Francis Group, ISBN 978-1-138-02985-9

Research on sensitivity and control criteria of cable length error in a double-strut cable dome

C. Sun & A.L. Zhang
College of Architecture and Civil Engineering, Beijing University of Technology, Beijing, China

ABSTRACT: Cable dome is a kind of lightweight and efficient large-span tensegrity structure. The manufacturing error of cable length will seriously affect the pre-stress distribution of cable dome and deviate structural performance from the design goal. With a new type of double-strut cable dome as calculation example, the sensitivity of cable length error to structural behavior is analyzed, under the construction method of "fixed length cable". Then, using reliability theory and nonlinear programming method, the control criteria of cable length error are obtained for this structure. The accuracy of control criteria is verified by the Monte Carlo method. The results show that for the double-strut cable dome, the length errors of ridge cables, inner diagonal cables, and outer hoop cables have an obvious effect on the internal forces of other cables. The method used in this paper can get reliable control criteria of cable length error. However, the excessive allowable error in specification will lead the structure unable to meet the design requirements.

1 INTRODUCTION

In recent years, tensegrity systems have attracted much attention of engineers and become popular as roofs for arenas and stadiums because of their lightweight and architectural impact. Cable dome is the first civil structure inspired by the tensegrity principle (Richard 1962), and its engineering applications include two forms: Geiger type (Geiger et al. 1986, Zhang et al. 2012) and Levy type (Levy, 1994), as shown in Figure 1.

As a tensegrity structure, the pre-stress of cable dome depends on the change of cable length during the tensioning process. To control the pre-stress distribution conveniently, the construction method of "fixed length cable" is always adopted as shown in Figure 2, which makes the anchor end of cable unadjustable so that we can stretch the cable from its original length to the connection directly through jack equipment (Guo et al. 2011). The construction method of "fixed length cable" is economical and efficient, avoiding time and labor consumption to adjust cable force repeatedly. However, if there is a big error in the original cable length, the pre-stress distribution will seriously deviate from the design value after completing construction, which will impact subsequent mechanical properties under the external load. Therefore, it is necessary to study the influence of cable length error and the control criteria.

At present, the research of tensegrity structure mainly concentrates on form finding (Juan & Tur 2008, Tibert & Pellegrino 2011) and structural behavior (Fu 2005, Ge et al. 2011); however, the research on the sensitivity of cable length error and control criteria is rare. Quirant et al. (2003) studied the effect of cable length error on tensegrity gird by the Monte Carlo method. Gao (2004)

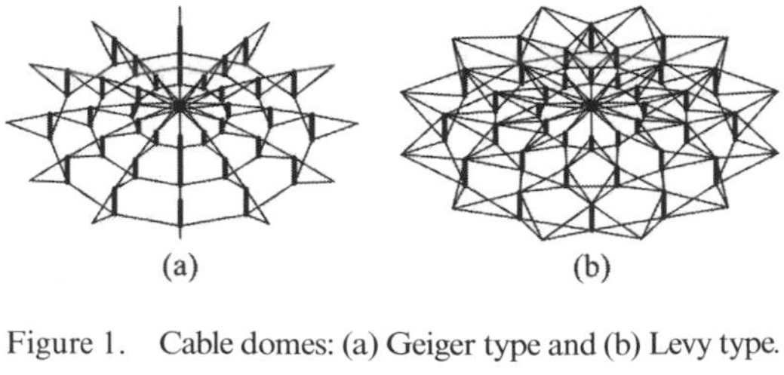

(a) (b)

Figure 1. Cable domes: (a) Geiger type and (b) Levy type.

Figure 2. Unadjustable anchor end of cable.

and Zhang (2011) adopted the orthogonal test to study the sensitivity of manufacturing error for suspen-dome and cable dome. Tian (2011) studied construction tolerance of spoke structure through reliability theory and experiment test. Both the American standard ASCE/SEI 19-96 (1997) and Chinese specification JGJ 257-2012 (2012) provided allowable error according to the manufacturing level, without considering the requirement of structural behavior.

In this paper, we study the sensitivity and control criteria of cable length error for a new type of double-strut cable dome. First, under the construction method of "fixed length cable", the sensitivity of cable length error to structural behavior is analyzed. Then, reliability theory and nonlinear programming method are used to obtain the control criteria of cable length error. Finally, the accuracy of error limit is verified by the Monte Carlo method.

2 DOUBLE-STRUT CABLE DOME STRUCTURE

2.1 *Structural features*

Double-strut cable dome is composed of ridge cables (RC1, RC2, and RC3), diagonal cables (DC1, DC2 and DC3), hoop cables (HC1 and HC2), central strut (ST1), and diagonal struts (ST2 and ST3), as shown in Figures 3–4. In Geiger and Levy cable domes, only one strut is connected between the top and bottom nodes. However, in double-strut cable dome, except for central strut, each node is connected to two diagonal struts, so diagonal struts are continuously arranged in circular direction.

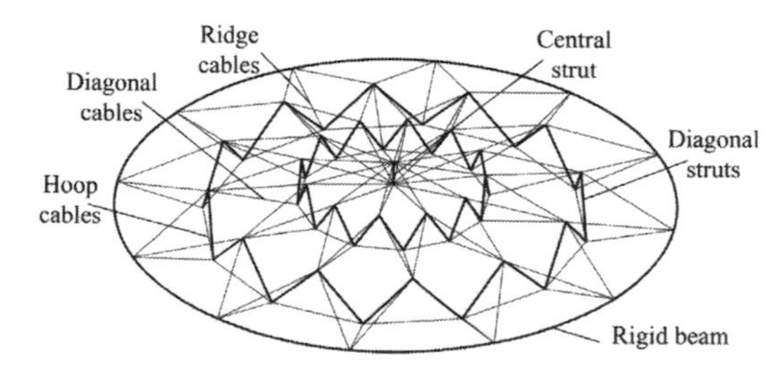

Figure 3. Perspective view of double-strut cable dome.

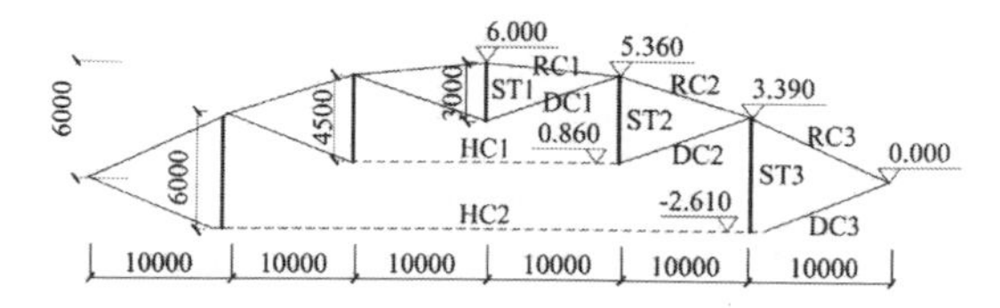

Figure 4. Member arrangement and profile view of double-strut cable dome.

Double-strut cable domes (Zhang et al. 2016) have better structural stiffness than the Geiger type because of the presence of triangular nets. Furthermore, the number of diagonal cables is only half that of the Levy type; thus, it needs fewer scaffolds and jack equipment to apply pre-stress through tensioning diagonal cables.

2.2 *Calculation model*

The calculation model is a double-strut cable dome with 60 m span and 12 divisions in circular direction. To ensure a reasonable structural design and safety reserves, optimization was carried out to guarantee that no cable relaxes and no strut buckles under the design load (1.05 kN/m^2), and the cable force does not exceed 50% of breaking force. The optimal design results of all components under the initial pre-stress state and design load are shown in Table 1.

Numerical analyses were carried out using ANSYS, a nonlinear finite-element software package. In the finite-element model, element Link180 is applied to simulate both cables and struts, and the elasticity modulus of cables and struts are 1.6×10^5 and 2.06×10^5 MPa, respectively. The paper only discusses the impact of cable length error under the initial pre-stress state, without considering gravity and the external load.

3 SENSITIVITY ANALYSIS OF CABLE LENGTH ERROR

Sensitivity analysis of cable length error was carried out under the pre-stress state, as shown in Table 1. Taking the cable force on axis 1 as the

Table 1. Design of components.

	Section	F_0* kN	F_P* kN	F_u* kN	Safety factor
ST1	Φ95 × 4	−148.2	−111.9	−168.6	1.14
ST2	Φ83 × 4	−30.2	−41.2	−45.4	1.10
ST3	Φ168 × 5	−119.0	−189.0	−198.8	1.05
RC1	Φ20	193.4	7.5	405.1	2.09
RC2	Φ18	142.7	26.6	329.8	2.31
RC3	Φ26	323.1	207.6	677.5	2.10
DC1	Φ12	53.8	38.7	150.7	2.80
DC2	Φ24	204.4	259.5	578.9	2.23
DC3	Φ46	656.4	990.7	2090.8	2.11
HC1	Φ32	397.4	504.5	1019.8	2.02
HC2	Φ65	1300.0	1965.0	4153.6	2.11

* F_0 is the internal force under initial pre-stress state, F_P is the internal force under design load, and F_u is the ultimate capacity of components.

reference value, a micro error of the initial length was assigned for each cable from axis 1 to 12 successively to obtain the changes of reference value. If the inner ridge cable (RC1) on axis i has an initial cable length error as δ_{RC1}, and it leads to the cable force changes of RC1 on axis 1 as $\Delta F_{RC1,RC1(i)}$, then we define:

$$s_{RC1,RC1(i)} = \Delta F_{RC1,RC1(i)} / \delta_{RC1} \tag{1}$$

where s = sensitivity to cable length error, which reflects the impact of unit length error of a cable on axis i on the internal force of a cable on axis 1. By repeating the above process, all values of sensitivity s in the double-strut cable dome can be calculated as shown in Figures 5–8.

Figure 5 shows that the cable force of ridge cables on axis 1 is affected by the length error of ridge cables and diagonal cables on all axes. Because of structural symmetry, the impact of ridge cables' (RC1, RC2, and RC3) length error is shown on the right-hand side, and the impact of diagonal cables' (DC1, DC2, and DC3) length error is shown on the left-hand side.

It can be seen from Figure 5 that the length error of RC1 and RC2 has a significant impact on RC1 and RC2 of the same axis and RC3 of adjacent axes, but has few impact on the ridge cables far away from axis 1. If the initial length error of RC1 on axis 1 is 5 mm, the cable force variation of RC1 on the same axis is 5.08% and that of RC2 is 2.85%, while the cable force variation of RC1 on axis 2 is only 1.84% and that of RC2 is 1.01%. Therefore, the impact weakens rapidly from axis 1.

The length error of DC1 has a significant impact on RC1 and RC2 of the same axis. If the length error of DC1 on axis 1 is 5 mm, the cable force variation of RC1 on the same axis is 1.49% and that of RC2 is 0.96%. The other diagonal cables have little impact on the ridge cables force.

Figure 6 shows that the cable force of diagonal cables on axis 1 is affected by the length error of ridge cables and diagonal cables on all axes. It can be seen from Figure 6 that only DC1 is sensitive to the length error of DC1 and ridge cables of the same and adjacent axes. If the length error of DC1 on axis 1 is 5 mm, the cable force variation of DC1 on the same axis is 9.99%, and that on the adjacent

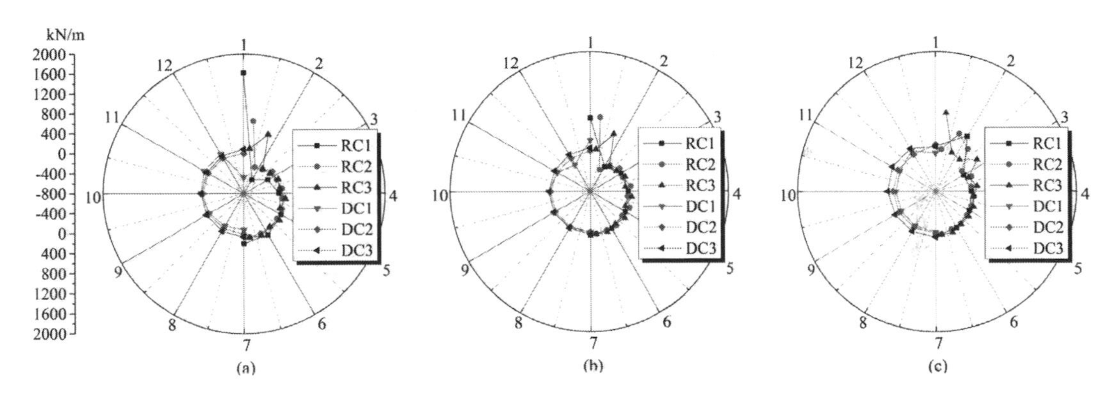

Figure 5. Sensitivity of ridge cables force: (a) RC1, (b) RC2, and (c) RC3 to cable length error.

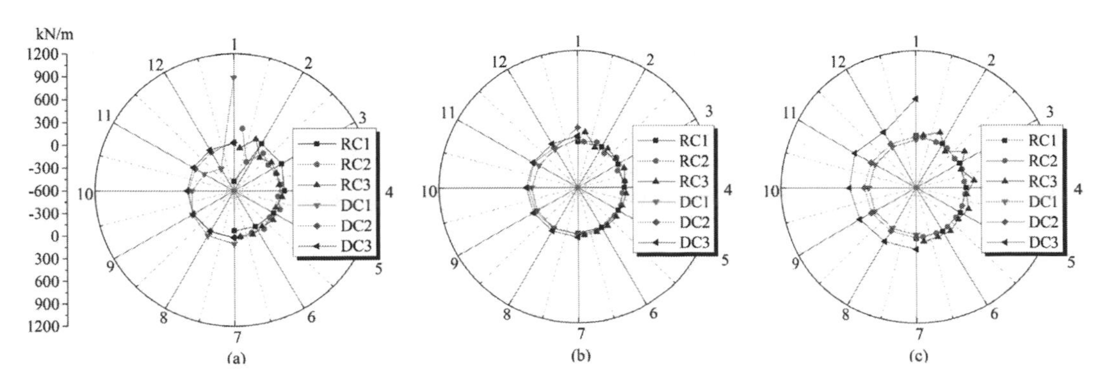

Figure 6. Sensitivity of diagonal cables force: (a) RC1, (b) RC2, and (c) RC3 to cable length error.

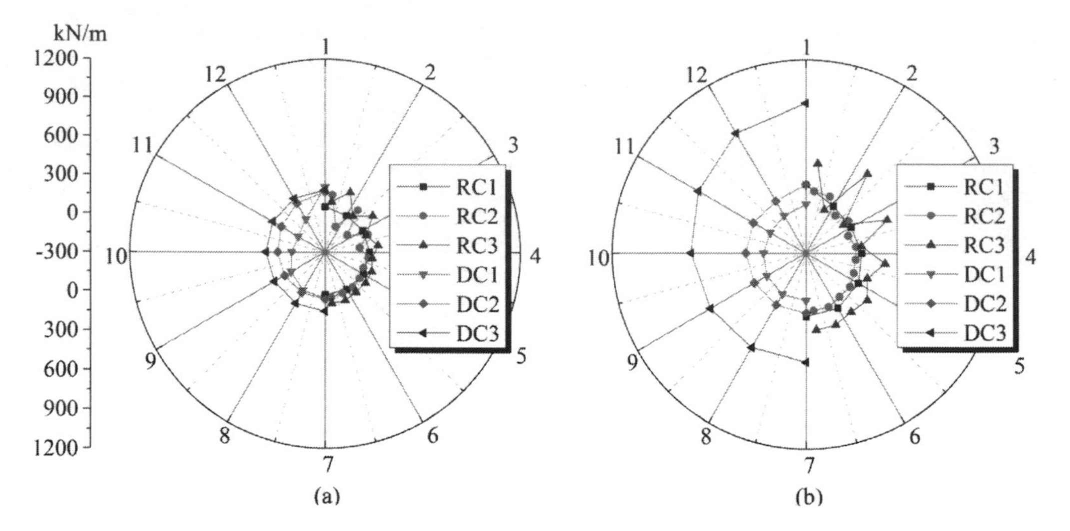

Figure 7. Sensitivity of hoop cables force: (a) HC1 and (b) HC2 to cable length error.

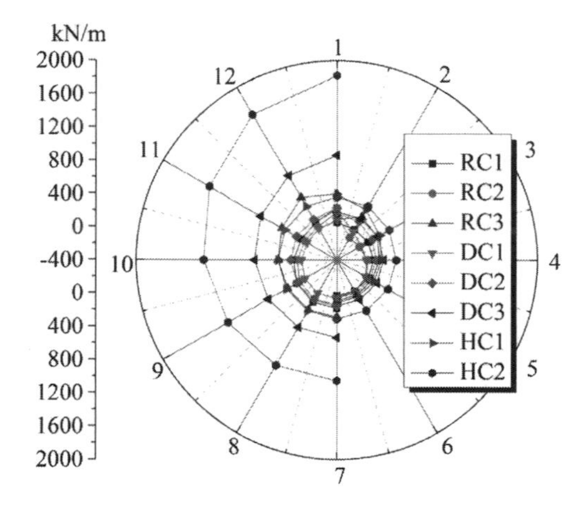

Figure 8. Sensitivity of all cables force to the length error of hoop cables.

axis 2 is 3.01%. However, both DC2 and DC3 are not sensitive to the length error of ridge cables and diagonal cables. If the length error of DC3 on axis 1 is 5 mm, the cable force variation of DC3 on the same axis is only 0.48%.

Figure 7 shows that the cable force of hoop cables on axis 1 is affected by the length error of ridge cables and diagonal cables on all axes. The figure also shows that hoop cables are not sensitive to the length error of ridge cables and diagonal cables. When the length error of HC1 on axis 1 is 5 mm, the cable force change of HC2 on the same axis is only 0.18% and that of HC2 is only 0.29%.

Figure 8 shows that all cables force is affected by the length error of hoop cables. The impact of HC2's length error is shown on the left-hand side, and the impact of HC1's length error is shown on the right-hand side. It can be seen that HC2 has a significant impact on all cables, much greater than that of HC1. If HC2's length error on axis 1 is 5 mm, then the average cable force variation of each group of similar cables would exceed 0.50%.

In conclusion, the length error of ridge cables and internal diagonal cables has a significant impact on the internal force of adjacent local cables. And that of the outer hoop cables has a significant impact on the overall cable force.

4 CONTROL CRITERIA OF CABLE LENGTH ERROR

4.1 *Constraint inequations of error variance based on reliability theory*

Let the structural resistance be R and load effect be S. Then, the structural state equation is $G=R-S$. In the case of $G>0$, the structure is safe. If the structural state equation is expressed as a linear form, as shown in (2), then the structural reliability index, β, can be expressed as (Nowak 2012):

$$G = g(X_1, X_2, \dots, X_n) = a_0 + \sum_{i=1}^{n} a_i X_i \tag{2}$$

$$\beta = \frac{a_0 + \sum_{i=1}^{n} a_i \mu_{X_i}}{\sqrt{\sum_{i=1}^{n} \left(a_i \sigma_{X_i}\right)^2}} \tag{3}$$

Let ΔF_{RC1} be the total force change of RC1. According to the sensitivity analysis in Section 3, we can obtain ΔF_{RC1} by adding the impact of all cables as (4). Repeating the similar process, the force change of all cables write as (5).

Let $[\Delta F_{RC1}]$ be the force deviation limit of RC1 under initial pre-stress state. On the basis of the reliability theory, the state equation of RC1 is expressed as (6). Set δ_{RC1} to satisfy a normal distribution with mean 0 and variance σ^2_{RC1}, namely $\delta_{RC1} \sim N(0, \sigma^2_{RC1})$. According to (2) and (3), we can obtain the reliability of the RC1 as (7).

inside the surface is a valid combination of the variance of different cables' length error. The points closer to the original point present a higher reliability, while those farther from the original point present a lower reliability. A proper variance combination shall meet the following two conditions: First, its distance to the original point shall be as far as possible to relax the request of manufacturing accuracy. Second, the value of each σ_i shall be as close as possible to take similar control criteria for different cables.

$$\Delta F_{RC1} = \sum_{i=1}^{n} s_{RC1,RC1(i)}\delta_{RC1} + \ldots + \sum_{i=1}^{n} s_{RC1,DC1(i)}\delta_{DC1} + \ldots + \sum_{i=1}^{n} s_{RC1,HC2(i)}\delta_{HC2} \tag{4}$$

$$\begin{pmatrix} \Delta F_{RC1} \\ \vdots \\ \Delta F_{DC1} \\ \vdots \\ \Delta F_{HC2} \end{pmatrix} = \begin{pmatrix} s_{RC1,RC1(1)} & \cdots & s_{RC1,RC1(n)} & \cdots & s_{RC1,HC2(1)} & \cdots & s_{RC1,HC2(n)} \\ \vdots & \vdots & \vdots & \vdots & \vdots & \vdots & \vdots \\ s_{DC1,RC1(1)} & \cdots & s_{DC1,RC1(n)} & \cdots & s_{DC1,HC2(1)} & \cdots & s_{DC1,HC2(n)} \\ \vdots & \vdots & \vdots & \vdots & \vdots & \vdots & \vdots \\ s_{HC2,RC1(1)} & \cdots & s_{HC2,RC1(n)} & \cdots & s_{HC2,HC2(1)} & \cdots & s_{HC2,HC2(n)} \end{pmatrix} \begin{pmatrix} \delta_{RC1} \\ \vdots \\ \delta_{DC1} \\ \vdots \\ \delta_{HC2} \end{pmatrix} \tag{5}$$

$$g_{RC1}(\delta) = [\Delta F_{RC1}] - \sum_{i=1}^{n} s_{RC1,RC1(i)}\delta_{RC1} - \ldots - \sum_{i=1}^{n} s_{RC1,DC1(i)}\delta_{DC1} - \ldots - \sum_{i=1}^{n} s_{RC1,HC2(i)}\delta_{HC2} \tag{6}$$

$$\beta_{RC1} = \frac{[\Delta F_{RC1}]}{\sqrt{\sum_{i=1}^{n} (s_{RC1,RC1(i)}\sigma_{RC1})^2 + \ldots + \sum_{i=1}^{n} (s_{RC1,DC1(i)}\sigma_{DC1})^2 + \ldots + \sum_{i=1}^{n} (s_{RC1,HC2(i)}\sigma_{HC2})^2}} \tag{7}$$

$$\sum_{i=1}^{n} (s_{RC1,RC1(i)})^2\sigma^2_{RC1} + \ldots + \sum_{i=1}^{n} (s_{RC1,DC1(i)})^2\sigma^2_{DC1} + \ldots + \sum_{i=1}^{n} (s_{RC1,HC2(i)})^2\sigma^2_{HC2} \le \left(\frac{[\Delta F_{RC1}]}{[\beta]}\right)^2 \tag{8}$$

where n = the total number of similar cables. For RC2 and RC3, $n = 24$; and for other cables, $n = 12$.

The structural reliability must satisfy certain tolerance value $[\beta]$, namely $\beta \ge [\beta]$. By substituting (7) to $\beta \ge [\beta]$, we get (8). Following the same process, eight critical state equations and reliability expressions of all ridge cables, diagonal cables, and hoop cables can be obtained successively. Finally, a set of inequalities can be obtained and simplified as (9), where $m = 8$.

$$\begin{aligned} &a_{11}\sigma_1^2 + a_{12}\sigma_2^2 + \ldots + a_{1m}\sigma_m^2 \le 1 \\ &a_{21}\sigma_1^2 + a_{22}\sigma_2^2 + \ldots + a_{2m}\sigma_m^2 \le 1 \\ &\ldots\ldots \\ &a_{m1}\sigma_1^2 + a_{m2}\sigma_2^2 + \ldots + a_{mm}\sigma_m^2 \le 1 \end{aligned} \tag{9}$$

4.2 *Calculation of error limit based on nonlinear programming*

The set of inequalities in (9) can enclose a high-dimensional curved surface. Each point $(\sigma_1, \sigma_2 \ldots \sigma_n)$

Therefore, the maximum value of σ_i arithmetic product is chosen as the selection principle of variance combination. Set $x_i = \sigma_i^2$ and this issue can be transformed to a nonlinear programming problem. The objective function is: max $\{\Pi\ x_i\}$. The constraint conditions are:

$$\begin{aligned} &a_{11}x_1 + a_{12}x_2 + \ldots + a_{1m}x_m \le 1 \\ &a_{21}x_1 + a_{22}x_2 + \ldots + a_{2m}x_m \le 1 \\ &\ldots\ldots \\ &a_{m1}x_1 + a_{m2}x_2 + \ldots + a_{mm}x_m \le 1 \\ &x_1 \ge 0, x_2 \ge 0, \ldots, x_m \ge 0 \end{aligned} \tag{10}$$

Use the double-strut cable dome structure mentioned above as an example. Select the allowable deviation of cable force as 10% of the designed value; thus, $[\Delta F]$ can be calculated. Set the allowable structural failure probability to 10^{-5} and the corresponding reliability index $[\beta]$ to 4.27. Combining with the sensitivity analysis results in Section 3, all the coefficients a_{ij} (mm^{-2}) in (10) can be calculated through (4)–(8), and the

Table 2. Control criteria of cable length error.

Component	Original length l_0 m	Error limits δ_C mm	δ_C/l_0
RC1	9.971	2.471	0.025
RC2	10.791	3.105	0.029
RC3	12.283	5.799	0.047
DC1	10.237	1.339	0.013
DC2	9.937	18.654	0.188
DC3	9.622	19.828	0.206
HC1	5.337	6.803	0.127
HC2	10.683	9.592	0.090

Table 3. Error limits of cable length in JGJ 257-2012.

Cable length m	Allowable error mm
≤50	±15
50 < L ≤ 100	±20
>100	±L/5000

final results are shown in (11). Then, MATLAB is adopted to solve the nonlinear programming issues to obtain the optimal variance combination of length error.

If the cable processing satisfies 3σ quality management standards, then all cables' length error control criteria (δ_C) is shown in Table 2. Table 3 also presents the control criteria in Chinese specification JGJ 257-2012. It can be seen through comparison that, for most cables in double-strut cable dome, the error limits of cable length calculated by the method proposed in this paper are much less than 15 mm. Therefore, the allowable error of JGJ 257-2012 may lead to an unacceptable deviation of cable force and fail to guarantee the structural behavior.

5 VERIFICATION OF CONTROL CRITERIA BASED ON THE MONTE CARLO METHOD

To verify the validity of the error limits of cable length δ_C in Table 2, 1200 samples of random length error satisfying $\delta \sim N(0,(\delta_C/3)^2)$ were extracted using the Monte Carlo method (Olsson et al. 2003) and introduced into the finite-element model. All the statistics data of cable force for double-strut cable dome structure under the initial pre-stress state are shown in Figure 9.

Curves a, b, and c in Figure 9 present the maximum, minimum, and average values of different cable forces. It can be seen that while considering the impact of random cable length error, the average value of cable force coincided with the design value, which demonstrates that the quantity of samples is sufficient. When the error limits satisfy δ_C in Table 2, cable force of DC1 deviates from the design value most obviously. The maximum force deviation is 6.54%, followed by RC1

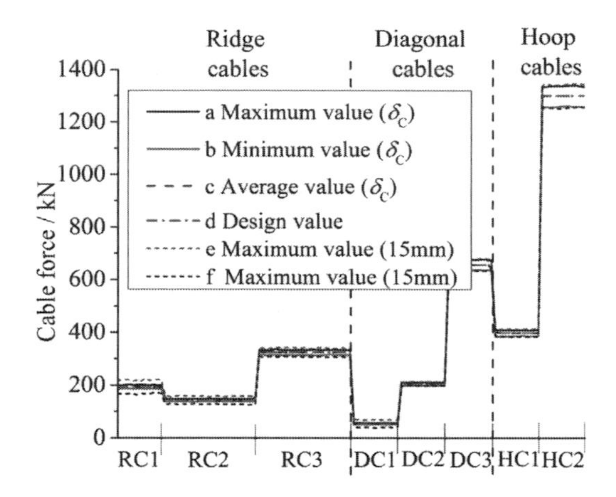

Figure 9. Statistics data of cable force.

$$a = \begin{Bmatrix} 162.196 & 55.769 & 32.374 & 14.230 & 0.106 & 2.303 & 0.380 & 9.521 \\ 51.256 & 71.394 & 33.877 & 8.294 & 0.731 & 2.417 & 4.047 & 10.086 \\ 5.796 & 6.599 & 14.823 & 0.456 & 0.618 & 2.988 & 1.642 & 11.911 \\ 184.983 & 117.196 & 33.108 & 634.038 & 3.233 & 2.819 & 24.430 & 12.010 \\ 0.095 & 0.712 & 3.107 & 0.223 & 2.121 & 2.371 & 4.532 & 9.610 \\ 0.201 & 0.230 & 1.462 & 0.019 & 0.231 & 4.385 & 0.870 & 16.222 \\ 0.090 & 1.040 & 2.183 & 0.447 & 1.199 & 2.364 & 4.728 & 9.500 \\ 0.212 & 0.245 & 1.485 & 0.021 & 0.239 & 4.134 & 0.891 & 16.081 \end{Bmatrix} \quad (11)$$

and RC2, with the maximum force deviations of 4.27% and 4.50%, respectively. The force deviation of all cables has not exceed 10% of the design value, which demonstrates that control criteria in the paper are accurate and reliable.

Figure 9 also presents the statistical data of cable force when we adopt the control criteria of JGJ 257-2012. The error limit of cable length is 15 mm, and the random sample satisfies $\delta \sim N(0,5^2)$, as shown by curves e and f. The maximum force variations of DC1, RC1, and RC2 are 30.5%, 16.7%, and 12.8%, respectively, higher than the 10% of the design value. Therefore, the excessive allowable error of cable length in JGJ 257-2012 would cause the structural performance failure to satisfy the design requirements.

6 CONCLUSIONS

On the basis of the research on the sensitivity and control criteria of cable length error for a double-strut cable dome, the following conclusions were drawn:

1. For the double-strut cable dome, the length error of ridge cables (RC1, RC2, and RC3) and internal diagonal cables (DC1) has a significant impact on the internal force of adjacent local cables, and that of outer hoop cables (HC2) has a significant impact on the overall cable force.
2. The constraint inequations of the variances of cable length error have been derived on the basis of the reliability theory. Then, on the basis of the results of sensitivity analysis and nonlinear programming method, the control criteria of cable length error were obtained from the optimal combination of variances. This approach can be adopted to study the manufacture control of cables for other tensegrity structures.
3. The accuracy and reliability of the control criteria in this paper were verified by the Monte Carlo method. To guarantee the structural behavior, the error limits of most cables in double-strut cable dome shall be much less than the Chinese specification's value. The excessive allowable error of JGJ 257-2012 would lead to an unacceptable deviation of cable force and fail to meet the design requirements.

REFERENCES

American Society of Civil Engineers. 1997. *ASCE/SEI 19-96 Structural applications of steel cables for buildings*. Washington, D.C.: ASCE Press.

Fu, F. 2005. Structural behavior and design methods of tensegrity domes. *Journal of Constructional Steel Research*, 61(1), 23–35.

Gao, B.Q., & Weng, E.H. 2004. Sensitivity analyses of cables to suspen-dome structural system. *Journal of Zhejiang University Science*, 5(9), 1045–1052.

Ge, J. et al. 2012. Analysis of tension form-finding and whole loading process simulation of cable dome structure. *Journal of Building Structures*, 33(4), 1–11.

Geiger, D.H., Stefaniuk, A., & Chen, D. 1986. The design and construction of two cable domes for the Korean Olympics. *In Proc. of the IASS Symposium on Shells, Membranes and Space Frames*, 2, 265–272.

Guo, Y. et al. 2011. Tensioning experiment on spoke structural roof of Bao'an Stadium. *Journal of Building Structures*, 32(3), 1–10.

Juan, S.H., & Tur, J.M.M. 2008. Tensegrity frameworks: static analysis review. *Mechanism and Machine Theory*, 43(7), 859–881.

Levy, M.P. 1994. The Georgia Dome and beyond: achieving lightweight-long span structures. *In Spatial, Lattice and Tension Structures*, 560–562.

Ministry of Housing and Urban-Rural Development of the People's Republic of China. 2012. *JGJ 257-2012 Technical specification for cable structures*. Beijing: China Architecture & Building Press.

Motro, R., & Nooshin, H. 1984. Forms and forces in tensegrity systems. *In Proceedings of the Third International Conference on Space Structures*, 180–185.

Nowak, A.S. & Collins, K.R. 2012. *Reliability of structures*. CRC Press.

Olsson, A., Sandberg, G., & Dahlblom, O. 2003. On Latin hypercube sampling for structural reliability analysis. *Structural safety*, 25(1), 47–68.

Quirant, J., Kazi-Aoual, M.N., & Motro, R. 2003. Designing tensegrity systems: the case of a double layer grid. *Engineering structures*, 25(9), 1121–1130.

Richard, B.F. 1962. Tensile-integrity structures. U.S. Patent 3,063,521 Nov.13 1962.

Tian, G. et al. 2011. Experiment on sensitivity to construction tolerance and research on tolerance control criteria in spoke structural roof of Bao'an Stadium. *Journal of Building Structures*, 32(3), 11–18.

Tibert, A.G., & Pellegrino, S. 2011. Review of form-finding methods for tensegrity structures. *International Journal of Space Structures*, 26(3), 241–255.

Zhang, A.L., Sun, C., & Jiang, Z.Q. 2016. Calculation method of prestress distribution for levy cable dome with double struts considering self-weight. *Engineering mechanics*. doi: 10.6052/j.issn.1000-4750.2016.01.0056

Zhang, G. et al. 2012. Design and research on cable dome structural system of the National Fitness Center in Ejin Horo Banner, Inner Mongolia. *Journal of Building Structures*, 33(4), 12–22.

Zhang, J.H., Wang, Z.Q., & Zhang, Y.G. 2011. Sensitivity Analysis of Manufacture Errors for Cable Dome. *In Advanced Materials Research*, 163, 822–827.

Civil, Architecture and Environmental Engineering – Kao & Sung (Eds)
© 2017 Taylor & Francis Group, ISBN 978-1-138-02985-9

Structural design of a prefabricated steel frame structure with inclined braces

Xuechun Liu, Hexiang Wang & Zhipeng Guo
Beijing Engineering Research Center of High-rise and Large-span Prestressed Steel Structures, Beijing University of Technology, Beijing, China

ABSTRACT: A new modular prefabricated steel structure that is characterized by high construction efficiency, low labor intensity, convenient quality control, and consumption of excess capacity of the steel industry is used in this paper. The structural design and in-depth research are performed in this paper using a finite-element method. The strength, rigidity, and stability of the structure and relevant components are calculated. The results show that all indices are compliant with the standards and regulations. Through pushover analysis and finite-element analysis of the joints, the mechanical characteristics of the system under the design load and the failure mode under the ultimate load are obtained, together with the elastoplastic development pattern; yield failure mode and potential mechanisms were further explored. The results also show that the structural system is consistent with the design philosophy of "strong joint, weak component". The system is basically in the elastic stage with reasonable yield mechanisms in case of major earthquakes. Therefore, both the elastic and elastoplastic performances comply with structural design requirements. This paper presents an innovative design method that suits the system and provides a technical reference for the design of similar structures.

1 PROJECT OVERVIEW

Let us assume a building containing 28 floors above ground with a total building height of 99.35 m. The 1st through 3rd floors are the annex with a height of 5.5 m, length of 144 m, width of 56.1 m, and total area of 21,503 m^2. The main functions of the annex include offering dining, leisure, conference, and cultural exchange facilities. Meanwhile, the floors above the 4th floor are divided into two towers, with a length of 42.9 m, width of 46.8 m, and total area of the main building of 88,345 m^2. The height of the tower's 4th floor is 3.6 m, the height of the tower's 5th floor through 28th floor is 3.3 m, and both are mainly used as office space. The building also contains three basements, with a height of 3.9 m, length of 188.7 m, width of 64.9 m, and total area of 36,126 m^2. They are mainly used for parking and equipment rooms. The total floor area of the project is 147,025 m^2. The building design is depicted in Figure 1.

Figure 1. The building design sketch.

Two seismic joints are set between the two towers and the middle annex building, which actually divide the superstructure into three independent structural units, as shown by the architectural elevation in Figure 2. The figure shows that the basement and middle annex building are constructed using reinforced concrete, whereas the towers adopt the prefabricated steel frame structures with inclined braces. Because the two towers are left–right symmetric, only the left tower is chosen for structural calculation and analysis in this paper.

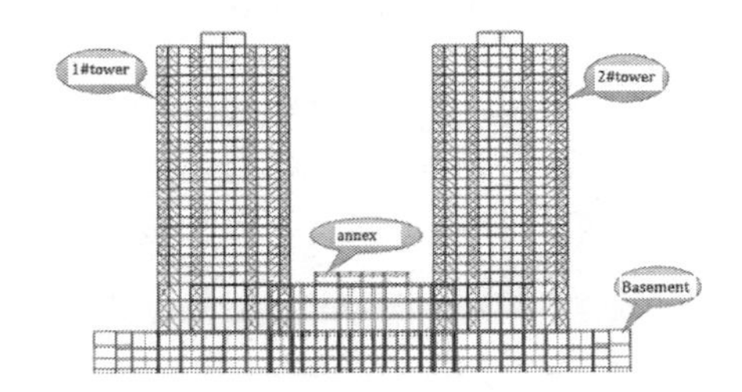

Figure 2. The architectural elevation.

2 STRUCTURAL SYSTEM

A new type of steel structural system, namely prefabricated high-rise steel frame structural system with inclined braces, is applied to the project. The prefabricated structural system is characterized by high construction speed, industrialized production, and low labor intensity, with its industrialization level exceeding 90%. The structure is composed of a mainboard and lean-brace column. As shown in Figure 3, the mainboard consists of the column base, profiled steel concrete floor, and truss. The beam is a welded steel truss with channel steel, angle steel, and a steel plate, and is commonly used to facilitate the laying of pipelines. In a factory, the beams and floor are prefabricated into mainboards that are used as modules for the assembly of water, heat, and electrical facilities. Before the mainboards are shipped from the factory, the decorative layer of the floor slabs and the placement of pipelines for water, heat, and electrical facilities are completed; the interfaces of the connections between the modules are deliberately reserved for a later step.

As shown in Figures 4–6, the lean-brace column consists of a column and lean brace that are welded at approximately one-third of the length of the column and bolted to the truss at approximately 45° to the other end. Therefore, the lean brace works together with the beam and column. During the construction process, the prefabricated mainboards and lean-brace columns are connected by high-strength bolts via flanges at both ends of the column and the mainboard column base. The connection between mainboards is a semi-rigid connection that connects the end of the truss to the column base by a high-strength bolt. The lean brace allows the connection of the beam and column to become more compact and strengthen the joint area. Meanwhile, the lean brace and beam-column frame jointly carry the load, which helps solve the common problem that the joints of a traditional frame are easily damaged during

Figure 3. Mainboard.

Figure 4. Diagonal braces column.

Figure 5. Diagonal braces joint.

earthquakes. In addition, the installation process is convenient. Figure 6 shows the assembly process of the whole structure.

The modular prefabricated steel structure is a new type of steel structural system. Compared with conventional structures such as reinforced concrete structures, the modular prefabricated steel structure is characterized by a short construction period, low self-weight, low labor intensity, and high industrialization level. Therefore, it is a green building (Zhang A. L, 2013 & Liu W. L, 2012). The design standards of this type of structure are fixed, and the components are fabricated by automatic

Figure 6. Assembly of the whole structure.

production lines. At the construction site, only connecting components such as bolts, self-tapping screws, and sealing materials are used for dry assembly, and the practice of wet operation is not required. As a result, the construction process is convenient and fast (Xiaoyun G, 2001), and it has been widely applied in developed countries and regions including Europe, the United States, Japan, and Canada (Jaillon L, 2009). However, the steel structures are only applied to multistory buildings, especially to low-rise frame structures currently (Wang W, Dao T. N, Teh L. H, 2012 & Zha X. X, 2010). By contrast, studies on the applications of prefabricated steel structures for high-rise frame structures are rare (Ye Z. H, 2012; Case F, 2014 & Festjens H, 2013). In this paper, we propose solutions to the key problems of the design process of prefabricated high-rise steel frame structures with lean-brace joints on the basis of computational analysis and empirical research on a modular prefabricated high-rise steel frame structure with lean-brace joints used by Zhuhai International Business Service Base, which provides technical support for the design of similar frame structures.

3 COMPUTATIONAL MODEL

ETABS, the software for the finite-element design and analysis, is applied in the present study for the elastic analysis. It includes static analysis, response spectrum analysis, and structural design. Because ETABS is based on the method of finite-element analysis, it could provide linear analysis, nonlinear analysis, dynamic seismic analysis, and static pushover analysis on the basis of the Chinese code for steel frame structures. Through repeated trial calculations and adjustments, the final analytical model of the project is determined, as shown in Figure 7.

To be consistent with the true load-bearing conditions of the structure, beam element modeling is applied to all components. Meanwhile, webs and

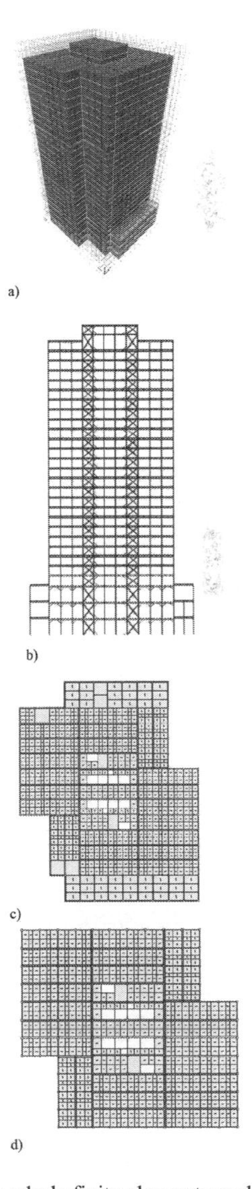

Figure 7. The whole finite-element model: a) isometric view, b) elevation, c) plan of floors 1–3, and d) plan of a standard floor.

chords of the trusses are modeled independently, and their joints are rigid connections. The floor slabs are modeled by profiled steel plate models, which are integral within a slab sector. Gaps are reserved between the slab sectors to conform to the actual situations. The test results show that the connection of the truss and column is a semi-rigid connection, which is simulated by the spring element built at the joints of the beam and the column; the rigidity of the joints is obtained from the test results. The material of the jack truss and keel is Q235B, and the remaining component is Q345B. Table 1 provides the component section sizes.

4 LOAD AND BOUNDARY CONDITIONS

The design reference period of the project is 50 years, and the safety class of the building structure is Class II. The design category of the foundation is Category I; the seismic fortification intensity is Degree 7, Class III site; the design earthquake grouping is Group I; the characteristic period is 0.45 s; and the classification of the seismic fortification is Class C (Class B for the skirt building). The basic wind pressure for 50 and 10 years are 0.85 and 0.50 kN/m^2, respectively. The floor load is chosen based on the GB50009-2012 Load Code

Table 1. Component section size and specifications.

No. of Sections	Position	Section size (mm)
3-section (1st–3rd floor)	Main beam chord	C160 × 80 × 12
	Secondary beam chord	C80 × 43 × 5 × 8
	Middle beam chord	C80 × 80 × 10
	Main beam web	2L63 × 10
	Secondary beam web	L50 × 5
	Middle beam web	L63 × 8
	Column	400 × 400 × 40
	Brace foot	C140 × 140 × 12
	Knee-brace	C150 × 130 × 16
	Lean-brace	H300 × 400 × 30 × 30
2-section (4th–18th floor)	Main beam chord	C160 × 80 × 10
	Secondary beam chord	C80 × 43 × 5 × 8
	Middle beam chord	C80 × 80 × 8
	Main beam web	2L63 × 10
	Secondary beam web	L50 × 4
	Middle beam web	L63 × 8
	Column	400 × 400 × 35 (4th–8th floor) 400 × 400 × 30 (9th–13th floor) 400 × 400 × 25 (14th–18th floor)
	Brace foot	C140 × 140 × 12 (4th–13th floor) C140 × 140 × 10 (14th–18th floor)
	Knee-brace	C150 × 130 × 12 (4th–13th) C150 × 130 × 10 (14th–18th floor)
	Lean-brace	H300 × 400 × 25 × 25 (4th–12th floor) H300 × 400 × 20 × 20 (13th–18th floor)
1-section (19th–28th floor)	Main beam chord	C160 × 80 × 8
	Secondary beam chord	C80 × 43 × 5 × 8
	Middle beam chord	C80 × 80 × 6
	Main beam web	2L63 × 10
	Secondary beam web	L50 × 4
	Middle beam web	L63 × 8
	Column	400 × 400 × 20 (19th–23rd floor) 400 × 400 × 15 (24th–28th)
	Brace foot	C140 × 140 × 8
	Knee-brace	C150 × 130 × 8
	Lean-brace	H300 × 400 × 15 × 15 (19th–28th)
Lift well	Main beam	H300 × 300 × 12 × 12
	Secondary beam	H300 × 150 × 6.5 × 9

for Design of Building Structures and JGJ3-2010 Technical Specification for Concrete Structures of Tall Buildings. Two methods, namely the response spectrum method and nonlinear pushover analysis are applied to consider the impacts of earthquakes on the prefabricated steel structure. The seismic spectrum analysis considers bidirectional horizontal earthquakes. More details of the load case and combinations are shown in Table 2. Overall, 19 load combinations that consider seismic and wind direction cases simultaneously were evaluated.

5 ANALYSIS RESULTS

5.1 *Modal analysis*

The 30-order modal is computed, and the modal mass participation ratio reached more than 90%, which meets the requirements of the code. The developed modal is shown in Figure 8, and the 1st- and 2nd-order modal of the structure is horizontal vibration, with a period T_1 = 2.594 s and T_2 = 2.424 s, respectively. The 3rd-order modal is the torsional vibration modal, with a period T_3 = 1.684 s. The period ratio of torsional to translation modal is 0.65 < 0.90, which conforms to the code requirements.

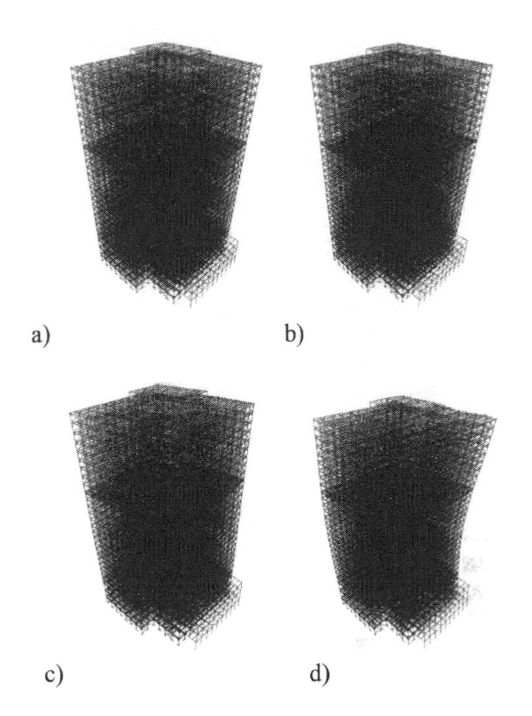

Figure 8. Modal shape: a) 1st order (T_1 = 2.594 s), b) 2nd order (T_2 = 2.424 s), c) 3rd order (T_3 = 1.684 s), and d) 4th order (T_4 = 0.847 s).

5.2 *Analysis of story drift*

For the case of common earthquakes, the maximum story drift angle along the X-direction under the influence of X-directional earthquakes is 1/995 and that along the Y-direction under the influence of Y-directional earthquakes is 1/1471, which is far less than the limit value of 1/250 stipulated by the reference. Under the influence of wind loads, the displacements of various floors change uniformly, indicating that the change of structural rigidity is uniform. The maximum story drift angle along the X-direction is 1/862 and 1/680 along the Y-direction, which are lower than the limit value of 1/400 stipulated by the reference.

Table 2. Combinations of the design load.

Combination type	Combination formula
Gravity load	1.35 Dead + 1.4(0.7) Live
	1.2 Dead + 1.4 Live
Gravity load + wind load	1.2 Dead + 1.4 Live ± 1.4(0.6) Wind
	1.2 Dead + 1.4(0.7) Live ± 1.4 Wind
	1.0 Dead + 1.4(0.7) Live ± 1.4 Wind
Gravity load + wind load + horizontal earthquake	1.2 (D + 0.5 L) ±1.3 R ± 1.4(0.2) W

5.3 *Shear–weight ratio*

Under the influence of X-directional earthquakes, the minimum shear–weight ratio is 0.019, which is larger than 0.015. Under the influence of Y-directional earthquakes, the minimum shear–weight ratio is 0.019, which is larger than 0.015. Therefore, both cases are in compliance with the relevant standards.

5.4 *Strength and stability of components*

The aspects of computing the component strength and stability, the strength and stability of columns, chord, web of the truss, and lean brace are all calculated using press and bend components model. The computational results show that the equivalent stress ratios under various load combinations are all less than 1.0, which are in accordance with the requirements of the corresponding strength and stability.

6 PUSHOVER ANALYSIS

As shown in Figure 9, when developing an overall ANSYS finite-element model, a beam is modeled

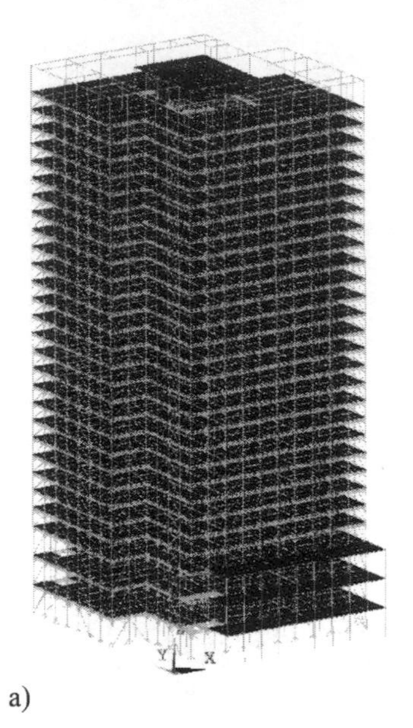

a)

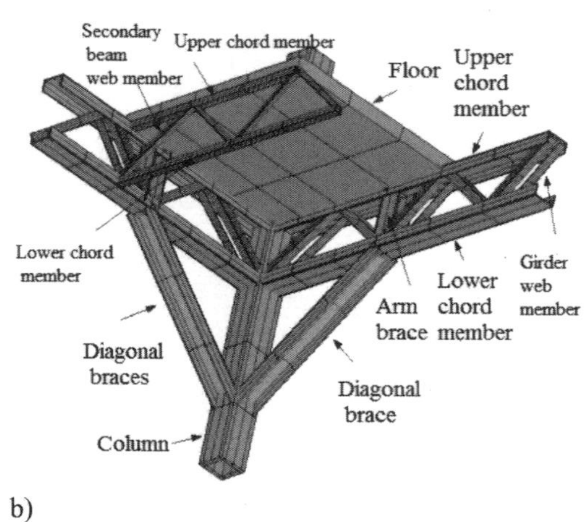

b)

Figure 9. Finite-element model: a) complete model and b) structural detail.

using a beam 189 element and a floor slab is modeled using a shell 181 element. The model has 1,190,391 joints and 872,827 elements.

A pushover analysis is performed for the finite-element model. First, the gravity equivalent load is applied, and 0.2 times the total base reaction force under common earthquakes is applied to each step until the results are divergent with damage occurring to the whole or a portion of the structure. Because the shape is almost the same along the vertical direction and the stiffness varies smoothly along the vertical direction, only the influence of the first vibration mode is considered. The total seismic reaction force distributed in an inverted triangle is applied to the structural model. The structure is damaged when the total X-horizontal reaction is 5.81 times the reaction of common earthquakes. Additionally, the structure is damaged when the total Y-horizontal seismic reaction is 5.6 times the reaction of common earthquakes. As shown in Figure 10, the maximum ultimate story drift angle occurs on the 10th floor, and the story drift angle between the 7th and 14th floor is larger, with the maximum value of 1/200, which is less than the limit value of 1/50 stipulated by the code. Therefore, the lateral rigidity of the structural system is larger, and the story drift angle meets the requirements of the standards. However, it also shows that this structure has poor ductility.

The load displacement curve of the top point of the structure is shown in Figure 11. It is evident from the figure that during the whole loading process, the load displacement curves along the X- and Y-directions are straight lines, which indicates that the structure is in the elastic state. Additionally, under the ultimate load, large-scale yield of the structural components has not been observed, and the yield positions are mainly concentrated on the webs and some chords at the outer edges of the structural plane. Therefore, the yield mechanism of the structural system is first the beam web is damaged, and then, the beam chord is damaged, while, all of the column stress is lower than the yield

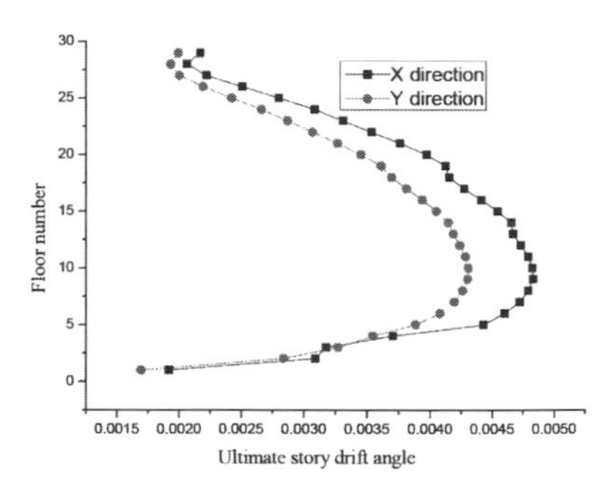

Figure 10. Ultimate story drift angle.

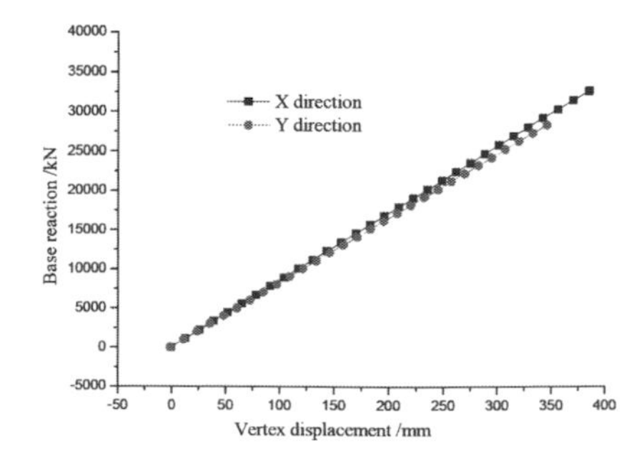

Figure 11. Curve of the base reaction-vertex displacement.

strength. The yield mechanism of the structure is reasonable; the structure is basically in the elastic state for major earthquakes, so no collapse events would occur at this time.

The capacity spectrum proposed by American ATC-40 is applied in the capacity spectrum analysis of the load displacement curve of the base reaction-vertex displacement. The capacity curve is converted into the capacity spectrum by using formulas (1) and (2):

Ordinate of the capacity spectrum: $S_{ai} = \frac{V_i / G}{\alpha_1}$ (1)

Abscissa of the capacity spectrum: $S_{di} = \frac{\Delta_{i,\text{roof}}}{\gamma_1 \varphi_{1,\text{roof}}}$ (2)

The mass participation factor of the first vibration mode is:

$$\alpha_1 \frac{\left[\sum_{i=1}^{N}(m_i \varphi_{i1})\right]^2}{\sum_{i=1}^{N} m_i \left[\sum_{i=1}^{N}(m_i \varphi_{i1}^2)\right]} \quad (3)$$

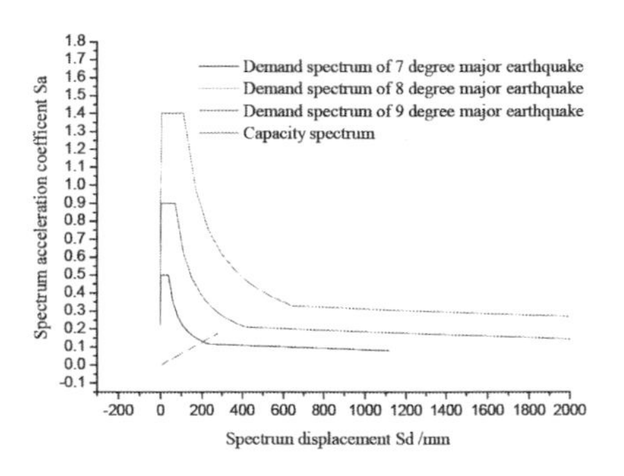

a)

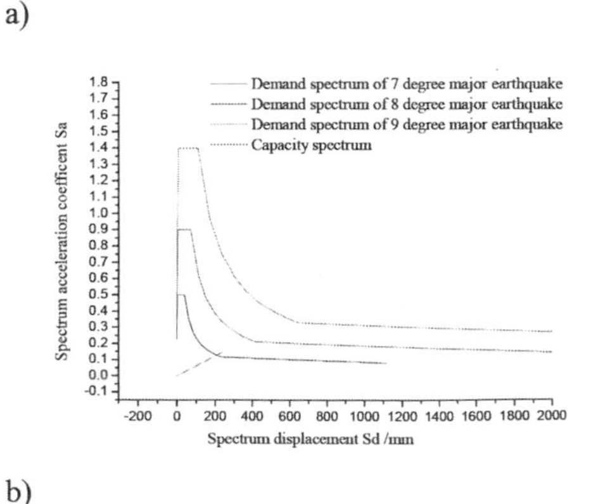

b)

Figure 12. Capacity spectrum and demand spectrum: a) X-direction and b) Y-direction.

The participation factor of the first vibration mode is:

$$\gamma_1 \frac{\sum_{i=1}^{N}(m_i \varphi_{i1})}{\sum_{i=1}^{N}(m_i \varphi_{i1}^2)} \quad (4)$$

$\Delta_{i,\text{roof}}$—— Vertex displacement of the structure

$\varphi_{1,\text{roof}}$—— Vertex amplitude of the first vibration mode

Formula (5) is used to convert the standard response spectrum in the Building Seismic Code into an ADRS spectrum. The abscissa of the demand spectrum is the same as the standard response spectrum. The ordinate is:

$$S_{di} = \frac{T_i^2}{4\pi^2} S_{ai} g \quad (5)$$

The structure is basically in the elastic state before damage occurs. Therefore, an elastoplastic reduction of the demand spectrum is not necessary. The curves of the capacity spectrum and demand spectrum are shown in Figure 12.

In Figure 12(a), in the curve of the capacity spectrum and demand spectrum in the X-direction, the vertex displacement corresponding to the performance points for degree 7 major earthquakes is 301 mm, and the total base reaction force is 25,851 kN. Under these loads, the damage rates of the Q235 steel components, Q345 steel

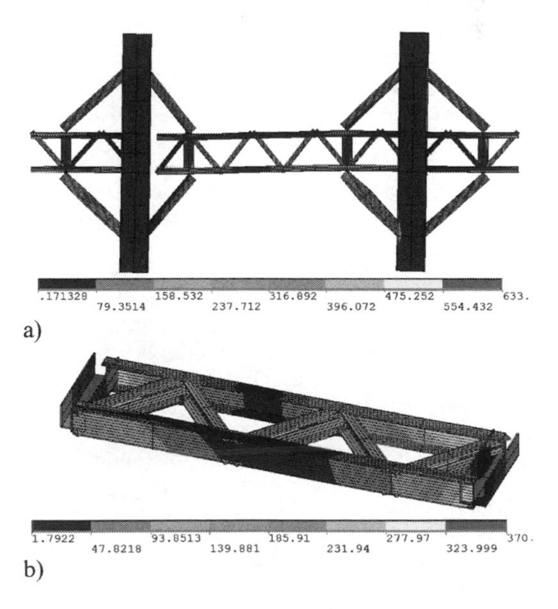

Figure 13. Stress of members under ultimate load: a) stress in the frame and b) stress in the truss.

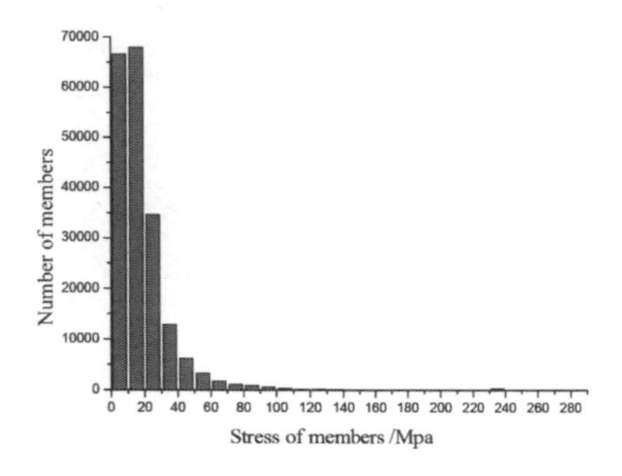

a)

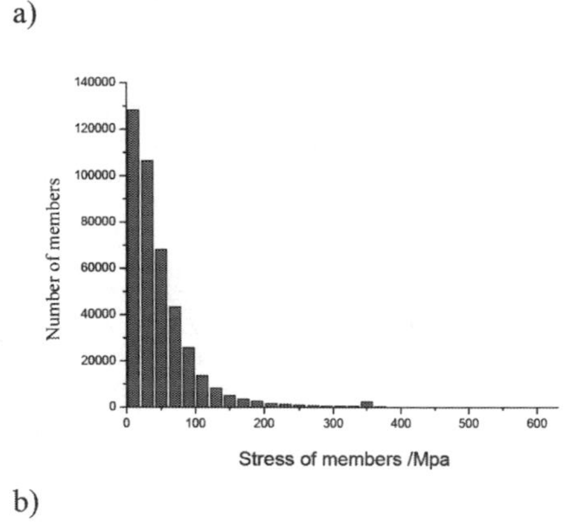

b)

Figure 14. Maximum component stress diagram under the limit load a) Q235 steel and b) Q345 steel.

components, and concrete floor slab are 0.02%, 0.23%, and 0.37%, respectively. Therefore, all structures in the elastic state display low damage rates. By contrast, the main damaged positions are webs and some chords at the building outer edges. In Figure 12(b), in the curve of the capacity spectrum and demand spectrum in the Y-direction, the vertex displacement corresponding to the performance points for degree 7 major earthquakes is 294 mm, and the total base reaction force is 24,280 kN.

The structure loses bearing capacity under loads that are 5.81 times the common earthquake forces in the X-direction. Under this ultimate load, the maximum stress is 633 MPa, and the yielding components are mainly distributed along trusses at the outer surfaces of the floor plane positions. As shown in Figure 13, under the ultimate load, the stress of columns is smaller; the maximum stress occurs at the truss web and the first chord close to the inclined brace. Because the ductility of the shear failure is lower than the bending failure, the web section should be increased during the design process to assure that the bending failure first occurs within the chord of the truss, followed by failures of the web, and finally column failures. This process assures a reasonable yield mechanism of the structure, and the performance of structure in the Y-direction is the same as that in the X-direction.

To study the failure state and safety after failure, a statistical analysis is performed on the stress of each member under the ultimate load, as shown in Figure 14. The result shows that the stress of only 418 members of the Q235B steel exceed the yield strength in the X-direction load because the Q235B steel is mainly used for secondary trusses that bear gravity loads but no horizontal loads. The stress of only 2544 members of the Q345B

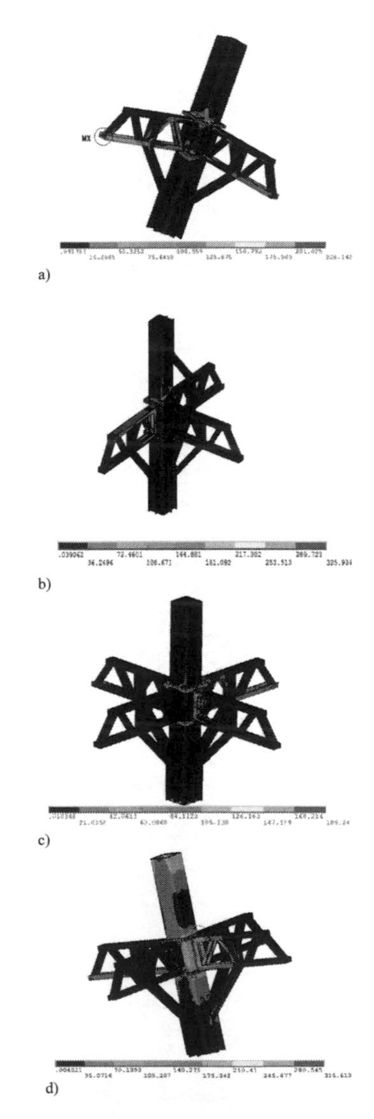

Figure 15. Equivalent stress under the design load: a) L-type connection, b) T-type connection, c) cross-type connection I, and d) Cross-type connection II.

steel exceeded the yield strength. The damage rate is only 0.23% because most of other members are in the elastic state, which indicates that the failure of the structure is due to the failure of a few members including a truss, but not including a column and brace, so the structure would not collapse. The structure could continue to be used after the broken members are repaired or replaced. The performance of the structure in the Y-direction load is similar to that in the X-direction. Therefore, the failure mechanism could be summarized as follows: the web and chord members fail first by an axial force and end bending moment, the column may or may not fail next, and the brace may or may not fail last; the whole structure will not collapse, but local failure will occur.

7 FINITE-ELEMENT ANALYSIS OF THE JOINTS

To study the performance of the joints, finite-element models of four typical joints are established in this paper. The loads for the most unfavorable load cases are extracted from the ETABS overall model and applied to the joint. The mechanical characteristics of the joints under either the design load or ultimate load are analyzed, and the mechanical characteristics of joints under the design load are shown in Figure 15. Through an ultimate bearing capacity analysis, the ultimate load and ultimate failure mode are determined. The safe reserves of the joints are obtained, and the design safety of the structure is guaranteed.

As shown in Figure 14, all four typical joints are in the elastic state without component yielding or buckling under the design load. Meanwhile, no gaps and relative slippage appear between the bolted gusset plates and columns, which suggest that the internal forces between the beam and column joints are very small because of the existence of the inclined braces. Under the design load, the size and arrangement of the bolts are reasonable, and all joints and beam–column components are well connected in a safe load-bearing state.

As shown in Figure 16, the load-displacement curve of the end of truss, four typical joints would not reach the ultimate state until the actual load force is 3.08–3.59 times the design load (i.e., the most unfavorable load case), so the safety reserve is large enough. The failure positions are all on the trusses outside the inclined braces that are designed to protect these joints. In general, the structural system complies with the design principle of *strong joint, weak component*.

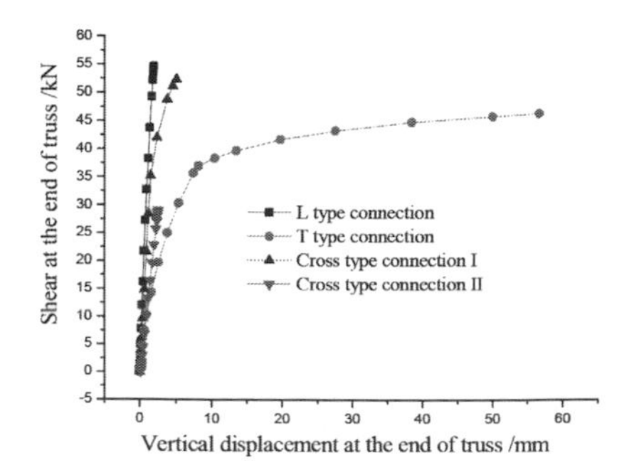

Figure 16. Load-displacement curve of the end of truss.

8 CONCLUSIONS

In this paper, we introduced the design method of a modular prefabricated high-rise steel frame structure with inclined braces used by Zhuhai International Business Service Base Project. Using joint analysis in an ANSYS finite element and elastoplastic pushover analysis of major earthquakes, the following conclusions are drawn:

1. The modal, story drift, shear–weight ratio, stability, and strength of the components of the structure are in accordance with the design control indices of the relevant standards and requirements, and the structural system is both safe and reliable.
2. Four typical joints under the design load are in the elastic state without the occurrence of component yielding. The ultimate load of the typical joints is 3.08–3.59 times the design load; therefore, there is a large safety reserve.
3. The ductility of the structural system is worse. The maximum elastoplastic story drift angle is only 1/200, and the maximum elastic floor displacement is 1/416. Therefore, compared with conventional structures, the design of the structure needs to limit the story drift angle more strictly.
4. The load bearing and displacement of the structure meet the requirements of major earthquakes. For major earthquakes, only a few structural components are damaged, and the whole structure generally displays linear elastic properties. Therefore, the safety reserve of the entire system is relatively high. Meanwhile, after rare earthquakes, the structure could continue to be used upon replacement of partially damaged components.
5. According to the requirement of a performance-based design, the bearing capacity of the structural system meets the requirement of performance standard I, and the story drift could meet the requirement of performance standard II.

6. All failure positions are on the truss outside the lean brace, indicating that the lean brace has served to protect the joints. Additionally, the entire structural system conforms to the design principle of *strong joint, weak component*.

ACKNOWLEDGMENTS

This work was supported by the National Natural Science Foundation of China (51278010), the Science and Technology Plan of Beijing Municipal Commission of Education (KM201610005012), and the Beijing Natural Science Foundation (8131002).

REFERENCES

Case, F., Beinat, A., Crosilla, F., & Alba, I. M. (2014). Virtual trial assembly of a complex steel structure by Generalized Procrustes Analysis techniques. Automation in Construction, 37, 155–165.

Dao, T. N., & van de Lindt, J. W. (2012). Seismic performance of an innovative light-frame cold-formed steel frame for midrise construction. Journal of Structural Engineering, 139(5), 837–848.

Festjens, H., Chevallier, G., & Dion, J. L. (2013). A numerical tool for the design of assembled structures under dynamic loads. International Journal of Mechanical Sciences, 75, 170–177.

GB50011-2010, "Code for seismic design of buildings," China Planning Press, 2010 (in Chinese).

Jaillon, L., & Poon, C. S. (2009). The evolution of prefabricated residential building systems in Hong Kong: A review of the public and the private sector. Automation in Construction, 18(3), 239–248.

JGJ99-1998, "Technical specification for steel structure of tall buildings," China Planning Press, 1998 (in Chinese).

Liu, W. L. (2012). Research on Design and Application of Steel Prefabricated House. M.S. thesis, Shandong University, Jinan, China, 2012.

Teh, L. H., & Gilbert, B. P. (2012). Net section tension capacity of cold-reduced sheet steel channel braces bolted at the web. Journal of Structural Engineering, 139(5), 740–747.

Wang, W., Chen, Y. Y., Yu, Y. C., Tong, L. W., Yang, J. X., Liu, D. W., & Furumi, K. (2012). Floor-by-floor assembled steel braced structures for prefabricated buildings. Building Structure, 42(10), 48–52.

Xiaoyun, G., & Hong, Y. (2001). Discussion on application of prefabricated houses in foreign countries [J]. Industrial Construction, 8, 005.

Ye, Z. H. (2012). Study on the Present situation and countermeasure of the domestic prefabricated steel structure. M.S. theiss, Nanchang University, Nanchang, China, 2012.

Zha, X. X., Wang, L. L., & Zhong, S. T. (2010). The method of constructing mutil-storied used shipping container buildings and the deduction of practical formula about structural security. Building Structure, 40(6), 462–465.

Zhang, A. L., & Liu, X. C. (2013, October). The new development of industrial assembly high-rise steel structure system in China. In Proceedings of the 10th Pacific Structural Steel Conference (PSSC'13), 976–981.

Civil, Architecture and Environmental Engineering – Kao & Sung (Eds)
© 2017 Taylor & Francis Group, ISBN 978-1-138-02985-9

A study on the design of an ungated community

Fan Jia
Internal Trade Engineering Design and Research Institute, Beijing, P.R. China

Renxiang Wang
Institute of Information Engineering, CAS, China

Fuli Ma
College of Architecture and Civil Engineering, TYUT, China

ABSTRACT: Gated community is a very common residential area in Chinese mainland. Now, it is time to think about the gated lifestyle and the current situation of a city. The ungated community can help a city to reduce the number of traffic jam events and let citizens enjoy more open landscapes and more convenient public facilities. Establishing an ungated community needs open-road network system, density architectural forms, friendly urban interface, and open landscape.

1 INTRODUCTION

From ancient to modern times, the traditional mode of dwelling in mainland China is the quadrangle courtyard, which is the basic unit of the court–alley–street pattern. While time has developed, our family patterns and lifestyles have changed a lot. Quadrangle courtyard changed to multistorey residential, even high-rise buildings, and the courtyard walls changed to residential walls and railings. These changes are all based on the closed form of the residential quarters and happen inside the community, despite the various aspects of the Chinese communities have got huge progress in the last several decades.

On October 1, 2014, *Green Residential Area Standards* was formally implemented, which was hosted and compiled by the China Real Estate and Housing Research Association. The standards clearly put forward the concept of open residential district and block in the general principles. The standards also advocated opening the residential area to the city, sharing green spaces and service facilities with citizens, constructing a harmonious and unified development system of urban communities. Today, the design of open community has exceeded the simple concept of breaking the wall of the residential area. It stresses that the community is an integral and organic part of the whole city instead of a small independent city in the urban space.

2 LITERATURE REVIEW

Since the release of the *CPC Central Committee and State Council on further strengthening urban planning and construction management*, it quickly sparked a heated debate. Two views of management raises high concerns, which are *China will no longer build closed residential area in principle and Residential quarters and units that have been built will compound to gradually open*. The management has made architectural experts to study it carefully. The most famous expert for the research of ungated community is Christiande Portzamparc. He not only put forward the concept of an ungated community, but also put this idea into practice in his design, which creates a method of reinterpreting city streets and shaping the urban space. According to him, the urban district is full of urban vitality, pluralistic, and open. In China, the research of ungated community is affected by Chinese policy. There are two tendencies to study the ungated community. The first one is research about foreign ungated community theory, for instance, *The Planing Concept of Open Block and Its Enlightenment on the Urban Residential Construction in China/ Yong YU, Zhitao Li*. The other one is the analysis about the practice of ungated community, for example, *Discussing open community based on the Aga Khan Award winners-Aranya Community/Shanshan Wu, Gangyi Tan*. As there has been no research about exploring how to design the ungated community to date, it will be discussed and studied in this paper, and the conclusion may provide the guide and reference to build the ungated community.

3 CONCEPT AND CONNOTATION ANALYSIS

3.1 *Residential district*

A residential area (50–100 ha) or residential quarter (10–35 ha) and even smaller scales of place for living. The connotation of the core is the space for living, recreation, and daily activities within a certain geographical area.

3.2 *Community*

Generally defined as a social community, people of which have common value view and close relationship. The basic elements of community are certain number of population, certain extent of the region, certain scale of facilities, certain characteristics of the culture, and certain type of organization.

3.3 *Gated community*

Generally defined as a residential area with a strict access control. Its internal parks, trails, sports fields, and even the public spaces like river and the sea are owned by residents. The most obvious feature is closed and forbidden or restricted to enter.

3.4 *Ungated community*

In contrast to the gated communities, there is no entrance guard, wall, or fence. Public spaces (roads, service facilities, green spaces, etc.) are all open to citizens. Its core is connectivity, sharing, and mixing.

Connectivity: Integration of community and the surrounding urban areas. Integration of the road network of community and the city road system. Integration of the community neighborhood green and the city green system.

Sharing: Landscape architecture, recreation facilities, roads, gardens, and other public venues in the open community are shared by urban residents.

Mixing: Building types in the open community not only meet the needs of urban functional diversification, but also accommodate the needs of residents of different classes, races, and ages.

4 ANALYSIS OF ADVANTAGES AND DISADVANTAGES OF A GATED COMMUNITY

4.1 *Advantages*

1. Gated community has good privacy and security of living space for the residents. First, effective community boundary creates a small scope of territory, which conforms to the human psychological demand of security. Then, because of the limit of the external citizens and vehicles, children can play safely and elderly persons can move around freely. The management of gated community is safe and easy and also suitable for traditional living habits.
2. Gated community has good community attachment. The creation of community attachment needs appropriate population and the similar value orientation. It is easy to produce a sense of attachment in gated community in certain boundary.
3. Gated community has rich and varied interior space and landscape environment. Despite the walls to the urban space that gated community gives, the interior residential space is very rich. It is another garden world within the cold enclosure, where there is beautiful and gorgeous views, many activity spaces, and different and useful facilities.

4.2 *Disadvantages*

1. Gated community causes traffic jams. Gated community, especially large ones, generally use walls to make several public roads of a few blocks privatization and to prevent external vehicles to travel. The road network within the walls can not continue the city road system, forming a circular or flexural road network. It forms a large scale of city road network outside the walls of community so that there is not so many roads for vehicles to choose, and then, they gather in those several major roads, which naturally leads to congestion.
2. Gated community occupies public resources. The public spaces in gated community are not shared by the citizens. Public resources in different communities have been probably duplicated. The phenomenon of gated community is the symbol of urban pathology. The reduction of social public space is the plunder of the poor class.
3. Gated community disappears the street life, which is a virtuous circulatory community's necessary space state. However, the walls of gated community weaken the fun of strolling along the street and reduce business activity of the street. The function of the street is only traffic.

Gated community is an isolated place, just like heavily guarded fortress. It makes the urban space depressed. It takes security as an excuse to digest the information like that of the traditional city. When public service, public facilities, and public roads are private, community responsibility ends in front of the wall, where shall we to talk about the so-called equal democracy? It reduces the motivation of residents to communicate and the chances of social intercourse. Where there is no social intercourse, the nation will be lazy to fulfill its social responsibilities.

The mode of gated community destroys the overall planning of the whole city and runs in the opposite direction with economic benefit, social benefit, and environmental benefit.

5 THE SIGNIFICANCE OF BUILDING AN UNGATED COMMUNITY

5.1 *Urban perspective: The sustainable development of residential area and urban area*

On the road system, it is different with the gated community road system that ungated community gives road space back to the city so that much more space changes into city road system. The way people can choose is more by the net road system, which helps to solve the problem of traffic jam.

On the landscape system, greens and facilities of ungated community have more accessibility and are easier to share. It helps to have more and larger open space of the city, to reduce the same facilities and improve the utilization of the existing ones, to avoid unnecessary waste, and to achieve a sustainable development.

5.2 *Residents perspective: Diversity of living environment and experience*

Ungated community mixes functions of dwelling, commerce, working, and so on. It offers a more convenient physical environment. Interface of continuous blocks integrated with service facilities is easier to create a vibrant street life, which provides a safe environment for working and increases the communication between citizens.

5.3 *Developer perspective: Flexible methods of developing*

The way that divides land into small scales and large numbers is beneficial to flowing of fund and phased development of large-scale projects. Smaller block is a flexible layout mode, which can be combined or divided according to the city plan. In addition, ungated community provides a large number of street interface, which has a huge commercial value. Through the reasonable arrangement of the space of streets, developers can gather popularity, and the new development projects can be quickly transformed into a mature area.

6 EXPLORATION OF THE DESIGN OF AN UNGATED COMMUNITY

6.1 *Establishing ungated residential district space*

1. Establishing ungated residential district space requires building density with certain features. It needs an appropriate scale to control block, as a space carrier to bear the interactive mode of life. The building density with certain features ensures the level of opening and closing, and it also creates living atmosphere that belongs to the block. In addition, it will make certain that people will obtain plenty of experience level of vision, and hence it is easier to make people approve residential environment.
2. Establishing ungated residential district space requires open-road network system. It will be fully considered that keeping the good connection between the road organization in dwelling districts, which belongs to cities, so ungated residential district will promote traffic smoother both in dwelling districts and cities. It will make appropriate road network density, and the most important is that ungated residential district will solve the increasingly serious traffic problems in modern urban design.
3. Establishing ungated residential district space requires friendly urban interface. Urban interface of open community needs to be coordinated with the urban street landscape. We should not only focus on elevation and skyline, but also take into consideration that space element and details should penetrate into residential district. Public area of community must always be located in the urban interface, and there is an outgoing space characteristic. This area is the part that people in community and city can share and exist. The space of ungated residential district is mainly reflected in the opening of public area.
4. Establishing ungated residential district space requires designing open landscape.

6.2 *Creating ungated community life*

Ungated community life is mainly reflected in psychological experience of neighborhood association and community activity. Planners and designers find that creating community life will be affected positively by focusing on the following points.

1. *Mixed plentiful living experience*

What we most importantly focused on is the plentiful and mixed living experience, and it is helpful to diversify the formation of residents. Diverse residents come from several residential products. Apart from residential products, there will be a substantial proportion of public buildings and facilities, and it will not only attract other community residents to communicate here, but also will save land and energy. In addition, residential district with good green landscape environment design will make community attract people stay longer and go sightseeing more interestingly, so it will form a mixed living experience.

2. *Interactive communication lifestyle*
Ungated community provides a wealth of space of communication and includes a composite function of the public buildings and open public space. Diverse demographic composition is main part of Interactive communication. It can arouse people's curiosity more and also facilitate communication that took place. The style of multiple residents living is not only an important way to reach living in harmony in the long run, but also to put an end to the situation of "elite" exclusive advantages of resources and satisfy the psychological needs of all residents. Therefore, it is important to make it obtain a positive and inclusive attitude of mind, promoting social harmony.

3. *Planning model of shared resource*
Supporting open public service facilities not only conform to the law in the operation, but it is also beneficial to the sharing of resources. Public services can be determined according to a reasonable size and positioning of the entire region, to avoid duplication and homogeneous competition. Meanwhile, the facility can be gathered by leveraging economies of scale to attract a large number of community people, to ensure that all kinds of traffic facilities, is conducive to survive and avoid unnecessary waste.

Community green space can be united with urban parks and green resources as system. Residents can enjoy the different resources in their own communities and share resources belonging to other communities through the convergence of urban space.

7 CONCLUSION

It is rational to exist that closed community is the current resident situation in the stage when urbanization rate grows rather fast in Chinese mainland. However, in the long run, single closed communities restrict the sustainable development of cities. Open community is a kind of practice mode, which has more positive impact on urban space because of its open-space structure. For the current situation of China, we should neither negate closed community completely nor give up the exploration of new styles of residence. When the open communities and closed communities coexist, China's residential environment and the urban quality can progress continuously.

REFERENCES

Christiande Portzamparc. Open Street — A European-style residential road and Massena New Area. Urban Environment Design. 2015, 93(6).

Shanshan Wu, Gangyi Tan. Discussing *open community* based on the Aga Khan Award Winners-Aranya Community. Community Design. 2013(1):137–141.

Yong YU, Zhitao Li. The Planing Concept of *Open Block* and Its Enlightenment on the Urban Residential Construction in China. Planners. 2006, 22(02): 101–104.

Civil, Architecture and Environmental Engineering – Kao & Sung (Eds)
© 2017 Taylor & Francis Group, ISBN 978-1-138-02985-9

Protection and utilization of Gaichun Garden ruins of the Summer Palace

Di Zhao & Long Zhang
Tianjin University School of Architecture, Tianjin, China

ABSTRACT: The Summer Palace enjoys the last and most complete large-scale royal garden in Chinese feudal society. Gaichun Garden, located in the Longevity Hill, was built in Qianlong Period. As a garden located in mountainous region, with gullies in front and cliff on the back wall, Gaichun Garden contains profound artistic conception, where multiple cultures such as Confucianism, Taoism, and Buddhism converge, reflecting the ideology of unification of Emperor Qianlong and the saint, and is of profound cultural connotation and historical research value. The surviving gardens and buildings have suffered destruction too many times to reflect the historical appearance. In this paper, research will be carried out on the remaining buildings, rockery and stone settings, ancient trees, and visitor utilization to understand cultural relics content, analyze existing problems and safety hazards, and propose relevant protection strategies, playing a positive role on the overall protection of the Summer Palace, display, and value cognition.

1 INTRODUCTION

The construction of Qingyi Garden was started in the 15th year reign of Emperor Qianlong (1750) and completed in his 30th year reign (1765), with the basic framework constituted by the Longevity Hill and Kunming Lake, known as one of "Three Hills and Five Gardens" of the Royal Gardens in the northwestern suburbs of Beijing. In the 10th year reign of Emperor Xianfeng (1860), the Anglo-French Allied Force incinerated and plundered many gardens in the suburb of Beijing, due to which Qing-Yi Garden also suffered serious damage. In the 12th year reign of Emperor Guangxu (1886), Empress Dowager CiXi decided to repair Qingyi Garden and renamed it as the Summer Palace. However, the repair was only concentrated on the main sceneries in the first part of the Longevity Hill and Kunming Lake as well as partial sceneries in the second part of the Longevity Hill and Kunming Lake due to the economic and political situation, and no repair was done to gardens within the Garden, containing the most compact designs and the richest architectural forms. Remains are still present in Gaichun Garden, Gouxuxuan, Qiwangxuan, Kanyunqshi, and Huacheng Pavilion, which serve as a rich wealth of historical information and cultural connotations.

The river area of the second part of the Longevity Hill and Kunming Lake is located in the north part of the Summer Palace, accounting for 12% of the total area, where the mountain terrain and water system remain almost the same as that in Qingyi Garden, but are artificially created, able to shape a profound quiet canyon view, and a perfect place to build scattered, profound small garden.

Garden architectures are shaded in the mountains and plants by these scattered gardens within the Garden, not in pursuit of the architectural scale but focusing on details and artistic conception. Besides, Gaichun Garden, located at the gentle slope, with three tablelands piled by human, combines the royal garden feature of elegance and wealthy and mountainous feature of natural wilderness. The overall architectural layout of Gaichun Garden has skillfully made use of mountainous region and palisades to successfully shape varied spaces, in line with the artistic conception in the book of *Yuan Ye*, Jicheng (1988) Mountainous region, which is the best place to build garden as the terrain will be high and low, twist and turn, and precipitous and flat, naturally.

2 OVERVIEW OF GAICHUN GARDEN

Gaichun Garden was initially built in the 23th year reign of Emperor Qianlong (1758) as an early garden architecture complex, located on the slope beside Taohua Gully in the west of the second part of the Longevity Mountain, composed of Gaichun Garden and Weixianzhai. Architectures inside the garden buildings like Yunzhenchangqie Palace, Zhuyu, Qingkexuan, Xiangyanshi, and Liuyun Tower have taken advantage of natural rock gully to form a mountainous garden through adjustment based on local conditions (Fig. 1). Gaichun Garden and Weixianzhai were places for Emperor Qianlong to study, rest, play the zither, brew tea, worship Buddha, and where multiple cultures such as Confucianism, Taoism, and Buddhism converge, reflecting the ideology of unification of Emperor Qianlong and the saint. Cong Yipeng & Zhao di (2015) Fresh and elegant decoration and local plants and trees embody leisure life in mountain. According to the record in the poems written by Emperor Qianlong, he had been to Gaichun Garden for 60 times and composed 98 poems to show his affection towards it. Unfortunately, Gaichun Garden did not escape from the damage when the Anglo-French Allied Force incinerated and plundered many gardens in the suburb of Beijing. During the reign of Guangxu, Gaichun Garden did not receive any repair and thus the residual architectures gradually dilapidated. In the 1950s, Clock Pavalion and the Gate of Qingkexuan were repaired. Summer Palace Management Office(ed) (2006) in the 1990s, the Management Office of the Summer Palace, cleaned, protected, and displayed the ruins.

3 GARDEN CREATION AND FEATURES

Emperor Qianlong named this garden as "Gaichun Garden", literally meaning "the garden is full of the beauty of spring", indicating that there were plenty of flowers and plants in this garden where spring was in the air. The second part of the Longevity Hill is mostly quiet natural scene, and Emperor Qianlong did not mention any kind of plants or flowers in the poetry, so he might be inspired by the beauty of spring. Gaichun Garden and Weixianzhai are, respectively, located on both sides of the west part of Taohuagou, connected by corridors on which Clock Pavalion was located, 84 m from east to west and 60 m from north to south. Gaichun Garden was built on hills with multilayer tablelands and 15 m elevation difference, with each layer linked by corridors or stairs (Figs. 2, 3). Climb the corridors to the top terraces and overlook the mountains; Qingkexuan, Liuyun Tower, and Xiangyanshi are located along the cliffs. Weixianzhai at the first layer in the lowest terrain is a closed courtyard. The tableland at the second layer is higher than courtyard walls where you can enjoy scenery outside of the garden. Yunzhenshangqie Palace is also built on the tableland, bamboo open hall, called Zhuyu, on the east side, and corridor on the west side to Weixianzhai. Climb the corridors to the third tableland and overlook the mountains, where Qingkexuan, Liuyun Pavilion, and Xiangyanshi are built along the cliffs, and the back walls of Qingkexuan and Liuyun Tower as well as large rocks beneath are natural (Fig. 4). Liuyun Tower is built on the rocks and some of the buildings are outstanding, mentioned as the pavilion of Yongji Temple of Jinling in the poems written by Emperor Qianlong. Xiangyanshi is a stone cave artificially created using natural caves to worship Guanyin and a place for Qianlong's study and self-cultivation.

Figure 1. Restored scene of Gaichun Garden and Xianwei study.

4 GAICHUN GARDEN STATUS OVERVIEW AND ANALYSIS

Repair record in the reign of Guangxu showed that the surviving buildings in Gaichun Garden

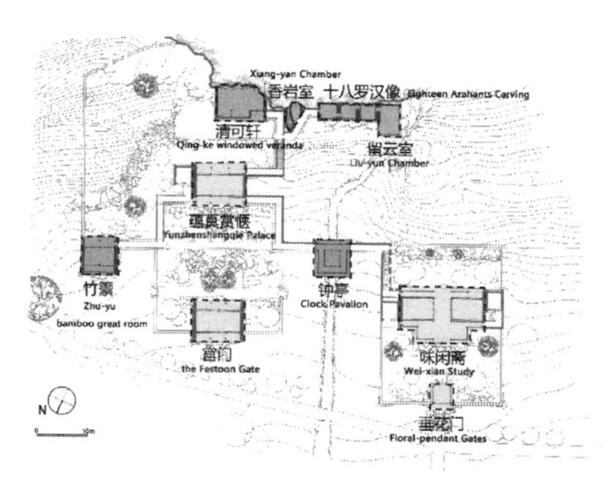

Figure 2. Gaichun Garden plan.

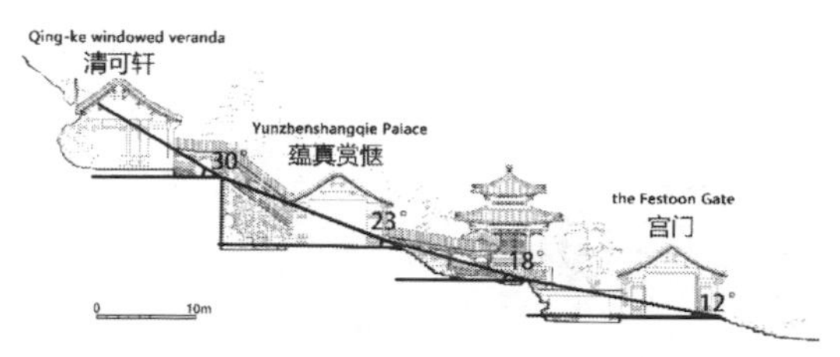

Figure 3. Map of tableland slope in Gaichun Garden.

were the Gate, the Festoon Gate of Weixianzhai, Weixianzhai, Qingkexuan, Yunzhenchangqie Palace, Zhuyu, and Clock Pavalion. In the 12th year reign of Guangxu (1886), Empress Dowager CiXi decided to rebuild the Summer Palace, but Gaichun Garden was not repaired and gradually reduced to ruins. Hence, only a restored gate of Gaichun Garden is available now, whereas others became ruins, such as architectural platform, plinth, damaged walls, and partially-damaged components, among which architectural platform and plinth are relatively well-preserved; therefore, we can get architectural plane measurement, column distribution and architectural frontal width and depth based on the platform measurement and plinth distribution (Fig. 5).

4.1 *Building and construct remains*

In the 1990s, the Clock Pavalion and the Gate were renovated. Now, the building bases are still distinguishable in Gaichun Garden ruins, artificial hills and stones are well-preserved, and there are still some palisades left at the back walls of Qingkexuan. Emperor Qianlong's inscription and the josses of the 18 Arhats are still distinguishable, but the remaining building bases and damaged constructs are incomplete and suffer weathering and shifting (Figs. 6, 7). For example, architectural platforms become flexible and landslide, trees grow on some platforms, and architectural components are missing, damaged, and repaired by other material. Plenty of pavements, both indoor and outdoor, are missing, and remaining pavements suffer serious weathering; trees are grown here and there on the ground.

Figure 4. Cut national rock of the back wall of the Qingkexuan ruins.

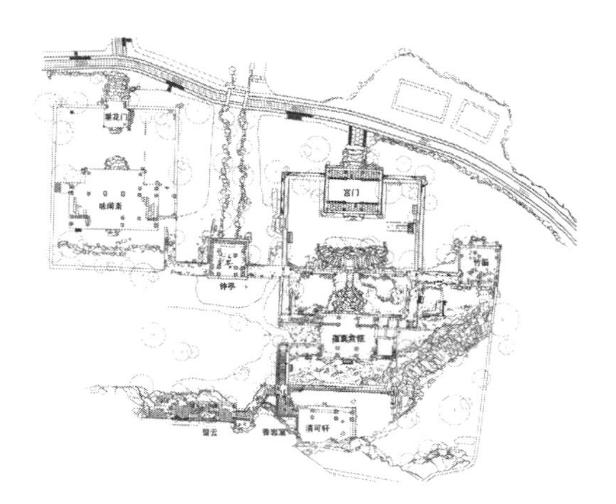

Figure 5. Surveying and mapping of Gaichun Garden ruins.

4.2 *Stone inscription*

The remaining artificial hills and stones can be classified as chiseled cliffs and piled artificial hills or as yellow stone and green stone by material. Chiseled cliffs imitate natural extraordinary power and highlight the history, whereas yellow and green stones are heavy and dense, in line with the fun of mountain garden. In general, the forms of groups of artificial hills and cliffs in Gaichun Garden are well preserved, but the stones suffer weathering and some artificial hills crumble down, having negative impact on sight view. Emperor Qianlong's poems and inscriptions on the cliff of the back wall of

Figure 6. Weixianzhai ruins.

Figure 7. Difference between the repair material and original architectural material for Clock Pavalion.

Qingkexuan are partly distinguishable. As for the josses of the 18 Arhats at Liuyun Pavilion, their heads are damaged during wars and by human, but their dress lines are distinguishable, although there are some weathering impact (Figs. 8, 9).

4.3 *Ancient trees and vegetation*

Wang Qiheng et al. (2009): According to the description in the poems written by Emperor Qianlong, the vegetation in Longevity Hill are transplanted, mainly evergreen pines and cypresses, also with many deciduous trees. As for the second part of the mountain, Xuan and Tang are scattered, landscapes have twists and turns but quiet, so there are many pine trees but also some other plants, like peach tree, apricot tree, *Acer truncatum*, locust tree, and *Koelreuteria paniculata*, to highlight seasonal changes. There remain seven second-level ancient trees in Gaichun Garden, that is, Chinese junipers and *Platycladus orientalis*, planted at the period of Qingyi Garden. Besides, there are several Chinese pines and cypresses outside of the garden, reflecting the air of emperor. In the poems written by Emperor Qianlong, there are 36 poems related to plants, describing the plant species, configuration mode, and the artistic conception, to express the aesthetic taste of Emperor Qianlong for plants. In the poems, more than 10 plants are clearly mentioned, like the pine, plum, willow, bamboo, rattan, lotus and lotus moss as well as quiet grove, flower rain, and field scenes.

There are remains of three ancient cypress trees and four *Platycladus orientalis* in the garden and remains of *Pinus tabulaeformis*, *Platycladus orientalis*, and Chinese junipers outside of the garden. Others are mainly wild species grown in recent years, basically common trees of North mountain, broadly in line with the species and the garden environment, plant varieties totally up to 11 (91 trees), three species of evergreen, and eight species of deciduous plants, including *Acer truncatum*, *Koelreuteria paniculata*, Chinese scholar trees, *Platycladus orientalis*, Chinese junipers, *Amygdalus davidiana*, *Broussonetia papyrifera* and elm, but the plant species and configuration method cannot reflect the original appearance of Qingyi Garden. Plant landscape restoration should reflect the artistic conception of gardens, increase space level, highlight seasonal changes, protect ecology and water environment, and strengthen the role of display.

Figure 8. Stone sculpture of 18 Arhats.

Figure 9. Stone sculpture's mapping of 18 Arhats.

4.4 *Water and soil erosion*

Because of changes in water resource in the west mountain area of Beijing City, in the second part of the Longevity Hill, there are some surface water only in the rainy season, but short of water apparently in the remaining seasons. The barren area in Gaichun Garden accounts for 30% of the whole garden, especially in the upstream of Taohuagou and the east area of Qingkexuan (Fig. 10), where there are no vegetation cover but only few trees. As the mountain topography belongs to slope with

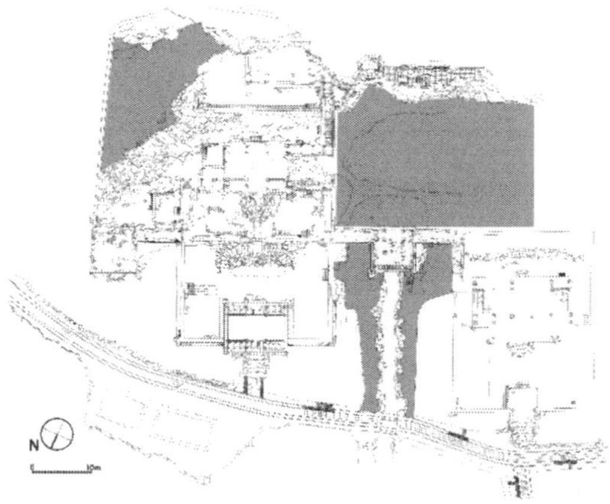

Figure 10. Serious water and soil erosion area.

loose soil surface and the soil erosion is obvious after rainfall; therefore, it is speculated that the soil decline is more than 20 cm. In this area, there are two ways: (1) to cover shrubs and to plant taller trees and (2) to restore the artistic concept of the garden and conserve water and soil. The areas with tableland and stones to keep water and soil have better surface condition, as the soil is blocked to a certain degree, it is speculated that the soil decline is within 10 cm, and the recovery focuses on protecting buildings and stacked stone from erosion.

4.5 *Management and display*

In the 1990s, the Management Office of the Summer Palace has cleaned, preserved, and presented the ruins of Gaichun Garden. It has set up stone signs in the ruins to briefly introduce each building and display area inside the gate of Gaichun Garden to introduce historical changes and cultural connotations and to show remains and the full-recovery model (Fig. 11). However, there are some safety concerns in the remain; it is not fully open to visitors, about 200 visitors each day to Gaichun Garden and Weixianzhai, mostly for short stay to rest and enjoy the scenery, relics, or wall inscription. The visitor questionnaire statistics showed that about 86% of visitors to Gaichun Garden were mostly those accidentally passing by, majority of whom had no idea about the history and culture of Gaichun Garden and surrounding garden sites, but showed great interest in the history evolution, site ontology show, and garden culture and aesthetic; however, this kind of publicity and display facilities obviously cannot meet the demand.

5 GAICHUN GARDEN RELICS PROTECTION AND UTILIZATION

The ruins at the second part of the Longevity Hill, as part of the Summer Palace, have an important

Figure 11. Renovated palace gate display and commercial services.

historical and cultural significance. The research on such ruins should adhere to the approach of "overall protection, scientific renovations, and rational use", and the core of explicit protection is refurbishment and use. On the basis of their own characters, we should explore history feature and the overall aesthetic effect from the viewpoint of sustainable development and consider the layout and visual integrity combined with the surrounding environment. The sites, including yamagata, drainage, vegetation, and architectural sites, must be protected to reflect the original style and features when Qingyi Garden, which was in a period of great prosperity, and must be restored and renovated on the basis of full and accurate historical information to protect and use the sites in a sustainable way. Now, we should give priority to site preservation and protection, and the subsequent restoration and presentation must be conducive to reflect the historical authenticity of the Summer Palace.

In order to ensure the minimum interference to the relics and maintain the landscape to the greatest degree, those existing but in good condition or those able to use after conservation treatment will maintain the status quo; most natural stone, rockery, and stacked stone will continue to be used after weathering treatment. The dislocated building components should return to the seat and components, which can be clearly identified, and added according to the site situation. Inscriptions and poems of Qianlong Emperor of Qing dynasty remain on the back wall of Qingkexuan and carvings of 18 Arhats, which are of great historical value, and protection components can be added to prevent further weathering and destruction.

Meanwhile, ruins information publicity should be enhanced to show the public Chinese traditional culture and garden activities, giving full play to its historical, artistic, and scientific value and educational role. The publicity and display should reflect the original state of Gaichun Garden as real as possible and deliver accurate historical information. The original artistic conception, design techniques, garden artistic concept, history, and other information about Gaichun Garden should be displayed through panels, poems, paintings, and models. Inside building remains is evident the technology of Chinese architectural traditional carpentry and masonry, as well as ruin protection method. Gardening activities like Qianlong reading, playing the zither, brewing tea, and worshiping Buddha in Gaichun Garden should be reproduced; illustrated by poems, calligraphy, and paintings; and displayed. For protecting buildings and building sites, ancient trees should be paid more attention for conservation, and garden plant landscape should be repaired and garden artistic concept at Qingyi Garden should be reproduced.

6 CONCLUSION

Relevant historical documents collection and collation, site collection of spatial information, and the systematic analysis of gardening ideas, art, and cultural connotation can not only fulfill the gaps of the study of the Gaichun Garden, but also enrich the overall research achievements about the Summer Palace and deepen the value. The conservation of Gaichun Garden sites should focus on the sustainable development of the garden and surrounding ecological environment, improve overall quality and landscape values, realize "authenticity protection, rational utilization and sustainable development" of the world cultural heritage, and emphasize the importance of continuity, maintenance, and preservation in the protection planning. The case study of Gaichun Garden can provide basic information for the Summer Palace site protection, promote the present relics protection, and display the progress.

ACKNOWLEDGMENT

Project 51308381 was supported by the National Natural Science Foundation of China.

REFERENCES

Cong Yipeng. & Zhao di. 2015. An Aesthetic Analysis of Plants Landscape in Gaichun Garden during Qingyi Garden Period. *China Landscape Architecture* 11: 82–85.

(Ming) Jicheng (original), Chen Zhi (notes). 1988. *The explanation of Yuan Ye*: 58. Beijing: Chinese architectural industry press.

Summer Palace Management Office(ed.). 2006. *The Summer Palace Journal*. Beijing: Chinese Forestry Publishing House.

Wang Qiheng et al. 2009. Collocation Analysis of Plant Historic Landscapes in the Summer Palace. *Journal of Tianjin University (Social Sciences)* 06:504–508.

Civil, Architecture and Environmental Engineering – Kao & Sung (Eds)
© 2017 Taylor & Francis Group, ISBN 978-1-138-02985-9

Development of a web-based hybrid BIM cost-estimating system for fire safety engineering

Ming-Hui Wen
Department of Commercial Design and Management, National Taipei University of Business, Taiwan

Wei-Chin Huang & Yun-Wu Wu
Department of Architectural Design, China University of Technology, Taiwan

ABSTRACT: In this study, a building project cost estimation system is developed using the Web-based technology. Also known as the Web-based hybrid BIM cost-estimating system, it incorporates the Building Information Management (BIM) with external cost estimation data to facilitate cost estimation in the early design stage and increase the efficiency of project implementation. The system is mainly composed of the following modules: "Basic Project Information", "BIM System", "Fire Safety Equipment Property Database", and "Web-based System Management Interface". Through its BIM components, it can provide efficient estimation and references for decision making. To verify the feasibility and benefits of the system, a case study is conducted by using the Web-based hybrid BIM cost-estimating system and the traditional 2D CAD method to evaluate the same case and then compare the cost evaluation accuracy and efficiency of these two methods. According to the comparison results, the system developed in this study is better than the 2D CAD method in terms of both cost estimation accuracy and efficiency.

1 INTRODUCTION

1.1 *Research background*

In project planning, cost estimation about the budget, materials, and all the other required resources is a fundamental task for policy making (Lee, Kim & Yu 2014). However, in the early design stage, cost estimation can be made more challenging because of the restrictions of insufficient time and information (Sonmez 2004). For project managers, both accuracy and efficiency of cost estimation are very important in order to ensure the overall efficiency of the following work.

Cost estimation is one of the critical activities for construction projects (Zhiliang et al. 2011). In the past, project managers often depended on subjective judgments based on prior experiences in their cost estimation (Yu 2014). However, this method is mainly based on personal experiences and cannot produce accurate cost estimation due to the differences between the estimated prices and actual market prices of certain items.

In recent years, with the emergence of Informational Communication Technology (ICT), project managers can use CAD 2D computer software for assistance in their cost estimation. However, the components in the CAD 2D design drawing may not be fully comply with the on-site operation. In addition, most of the CAD 2D software lacks information consistency and data connections of the operation and, therefore, cannot help project managers to achieve efficient cost estimation.

BIM is a type of digital information model that can integrate all the architecture design graphics and documents (Chen & Hou 2014; Chien, Wu & Huang 2014; Porter et al., 2014; Meža, Turk & Dolenc 2014). It can help project managers to quickly access detailed design information and find out the required quantities of materials by automatically calculating related figures. However, without sufficient information or data from cost estimation, BIM will not be able to provide comprehensive references for project managers.

To help project managers to achieve rapid and accurate cost estimation in the early design stage, the Web-based hybrid BIM cost-estimating system is developed in this study. In addition to the characteristic features of a traditional BIM system, the system developed in this study can provide geometric and nongeometric information from its 3D modeling for project managers. Moreover, in this system, there is a cost database and cost-estimating model established for each construction project component. This mechanism can enable the BIM system developed in this study to have more accurate cost estimation of the construction (Jarde & Alkass 2001). Moreover, together with the rapid development of the construction industry, construction projects nowadays are growing in size and complexity and require participation of more professionals (Choi, Choi & Kim 2014).

1.2 *Research purpose*

This study is intended to develop a Web-based system. Through the systematic and synchronized management of project information on the Web, specialists and professionals from different fields can work together to estimate the costs of a project whenever and wherever they are needed. To verify the feasibility and benefits of the BIM system, a case of fire safety system inside a factory building is used in this study to explore the cost estimation accuracy and efficiency of the system.

The following discussion in this paper is divided into four chapters (with the Introduction above as the first chapter). The second chapter is the literature review, covering issues about difficulty in cost estimation, BIM system, Web-based technology, database, and fire safety equipment. The third chapter introduces the BIM system developed in this study, including its framework structure and operation procedure. The fourth chapter is the comparison of the cost estimation accuracy and efficiency of the BIM system developed in this study and the traditional 2D CAD method by using these two methods to evaluate the case of fire safety system deployment in a factory building. The last chapter discusses the research restriction of this study, benefits of the BMI system developed in this study, and directions of related future research to promote its application.

2 LITERATURE REVIEW

2.1 *Difficulty in cost estimation*

In the stage of early planning and design, it is important to consider the required amount of capital investment in the construction project first. Therefore, the accuracy of cost estimation is the key to successful operation and future profitability of the project. However, if the cost estimation itself is time-consuming and difficult to implement, it will cause extra costs instead. Therefore, this study is mainly intended to explore how to establish an efficient cost estimation system in order to solve the above-mentioned problems.

The cost estimation for construction projects is subject to difficulties caused by uncertainties in architecture information and requirements. To achieve fast, economic, and accurate cost estimation within limited time, computer software can be used to help obtain the cost information of facilities and equipment for the architecture (Kim, Seo & Hyun 2012).

In the existing method of construction design and planning, CAD 2D graphics are used. However, manual estimation and calculation is still required to determine the costs and quantities of building elements and required equipment. The failure to automatically generate quantity information of the required equipment and materials results in extra cost estimates in the early design and lack of accurate quantity information of required materials for cost estimation.

2.2 *Building Information Model (BIM)*

The greatest change brought by the rapid development of ICT for the construction industry is the application of BIM (Bryde, Broquetas & Volm, 2013). The concept of visualization, communication, and integration of BIM (Ding, Zhou & Akinci 2014; Grilo & Jardim-Goncalves 2011; Park et al. 2011) enables architects and professional technicians to work together on the same platform, sharing and exchanging information (Isikdat 2012). This can help reduce conflicts of wire, pipe, and duct configuration as well as omissions in quantity calculation of building elements. In addition, it can allow the users to preview the design results before the project is implemented, promote effective team communication and cooperation (Chen & Luo 2014; Qiu 2011; Cheng & Wang 2010; Popov et al. 2006; Han, Yue & Lu 2012), integrate drawings and documents, reduce conflicts (Chen & Luo 2014), and shorten lead time (Kim, Kim & Seo 2012). It can also solve the problems of wire, pipe, and duct collision/interference and calculation errors in CAD 2D configuration and material estimation caused by its lack of 3D specification presentation. In addition, the integral database of BIM can be customized to directly display the WBS component information and quantity information. Without manual calculation, it can help reduce man-made errors and improve design quality (Miettinen & Paavola 2014). However, BIM itself does not contain sufficient cost information or data for cost evaluation while it cannot integrate or analyze data from similar cases. Such problems can be solved by integrating the BIM system with external cost estimation data.

2.3 *BIM database*

The database can help BIM to realize the storage, exchange, and sharing of data (Vanlande, Nicolle & Cruz 2008). Through the Database Management System (DBMS) software, users can create projects, operate the system as well as maintain, manage, and inquire data. In addition, it can help ensure data safety and integrity as well as enable data transmission among multiple users. With the DBMS software, manual management of project data is no longer needed and all the project data can be centrally stored and accumulated. The information/data of cost estimation items and unit price analysis are integrated in the Relational DataBase Management System (RDBMS), which can provide detailed cost estimation information/data in the early design stage and then improve the accuracy of cost estimation. By incorporating the database, the BIM system developed in this study can effectively facilitate information sharing/exchange among engineers and solve the problem of data integration and interoperability among different construction projects (Isikdag 2012).

2.4 *Web-based technology*

The major purpose of incorporating the Web technology into the BIM system in this study is to provide a communication and cooperation platform (environment) for users, promote organizational efficiency, and enhance the values of BIM (Vanlande, Nicolle & Cruz 2008). Through the information connection, contribution, and exchange on the WWW or central server (Chou & O'Connor 2007; Turban et al. 2007), engineers can quickly access and exchange different types of required information and knowledge and then work on the same project even if they are located in different places. This can solve the problems of inconvenient access to historical data and lack of communication among engineers in the traditional manual and CAD 2D cost estimation. Therefore, in this study, interactive and dynamic Web pages are developed, and the open-source and free software of LAMP (LINUX + Apache + MySQL + PHP) is used to establish the server database and processing program. Users can access and use this Web-based system through any interface device that can open Web browsers. For cost evaluators, the web-based system can enable more flexibility of time and space, save time and costs, and promote cooperation and communication among users and peers to work more seamlessly. With ICT, all the data in the evaluation process can be fully recorded to provide valuable information for the evaluators for their future analysis and improvement of the cost evaluation method.

2.5 *Fire safety equipment*

Fire safety equipment is intended to prevent buildings from the potential threat of fire hazards and, more importantly, give people more time of evacuation in a fire so as to reduce causalities (Kobes et al. 2010). Poor design, maintenance, and use of fire safety equipment due to consideration of cost reduction pose a huge threat to people's safety and property. Therefore, to ensure reasonable investments in fire safety equipment, it is necessary to develop an effective method of cost estimation to prevent both excessive and insufficient budgets for fire safety equipment installation and maintenance. By introducing BIM into the fire safety equipment deployment design in a building's spatial structure (Isikdag, Underwood & Aouad 2008), it is possible to manage the design through 3D model visualization and integrate with other digital monitoring systems. For future use and management of the design, it can also help prevent spatial conflicts and modification costs during the installation process, reduce personnel costs, and improve the efficiency of fire safety equipment management and maintenance (Godager 2011).

3 DEVELOPMENT OF THE WEB-BASED HYBRID BIM COST-ESTIMATING SYSTEM

3.1 *System introduction*

The Web-based hybrid BIM cost-estimating system developed in this paper is composed of the following modules: "Basic Project Information", "BIM System", "Fire Safety Equipment Property Database", and "Web-based System Management Interface". Through the system, project managers can have efficient access to data and information required to make decisions and judgments in cost evaluation. The following is a brief introduction to each module and the process of using this system. The conceptual flowchart of the proposed system is shown in Figure 1.

Basic project information: initial collection and analysis of basic information of the evaluation factors in the project, including basic information investigation of the fire safety equipment legally required for buildings, definition of each item in cost evaluation as well as investigation scope and restrictions.

BIM system: providing the project model, property information of fire safety equipment in building elements as well as quantity and drawing information of each component.

Fire Safety Equipment Property Database: transferring information from pervious projects and providing information about fire safety equipment classification budge list, detail price list, and unit cost required for the connections between the BIM model and the Web-based interface.

Web-based system management interface: binding the information inquiry module with the other modules and providing an easy-to-use interface for

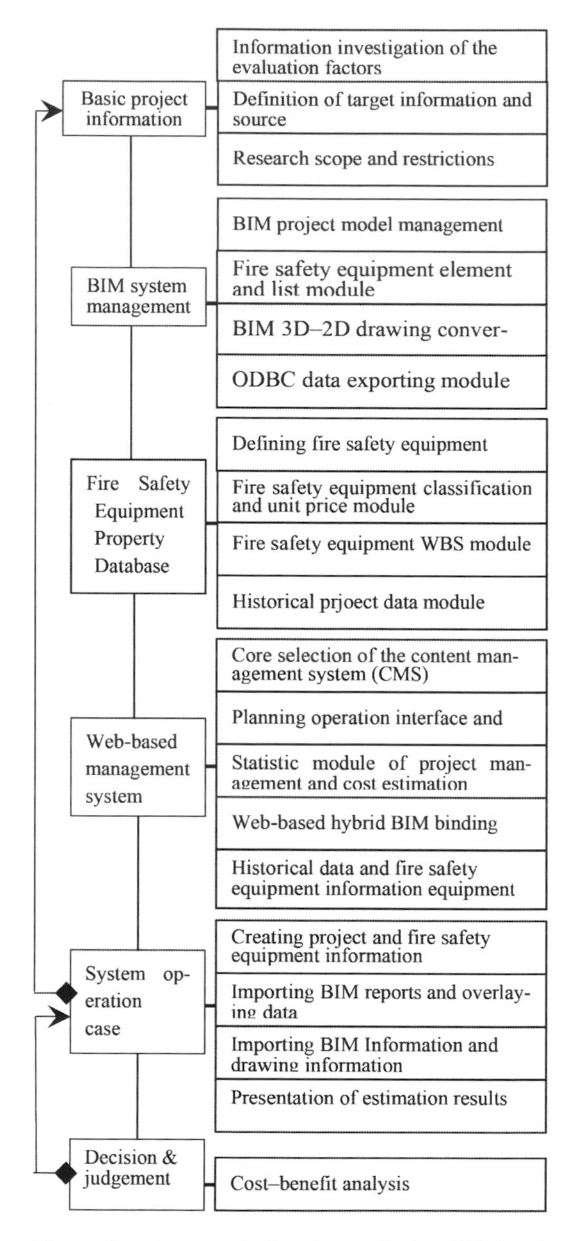

Figure 1. Conceptual flowchart of the Web-based hybrid BIM cost-estimating system.

users, allowing them to create projects, budget lists, detailed price lists for the project cost estimation.

System operation and case analysis: First, a project is created on the Web-based system management interface together with its budget list and detailed price list. Then, the fire safety equipment elements of the BIM model and the detailed quantity reports are imported into the Web-based interface. By overlaying the binding module and the Fire Safety Equipment Property Database, the Web-based hybrid BIM cost-estimating system can generate accurate estimation results and produce the budget list and detailed price list after receiving the quantity reports of the BIM project. Finally, the BIM 3D models are imported and converted into 2D drawings to fully present the documents required for cost estimation.

Decision and judgment: producing the cost estimation results for the evaluators to conduct cost–benefit analysis and find references for decision makers.

3.2 *Work Breakdown Structure (WBS) and unit cost estimation of components*

For the code classification of the fire safety equipment, budget lists, and detailed price lists in the case of this study, references are drawn from the cost estimation of similar cases by some engineering consultancies as well as the listing prices of material/equipment of the suppliers to quantify the estimation basis for each item. In addition, references are also drawn from the Public Construction Computer Estimate System (PCCES) and the code classification of the Public Construction Work Guideline Codes developed by the Public Construction Commission under the Executive Yuan in Taiwan based on the CSI Master Format.

The codes for the BIM system in this study can be divided into two types: work codes (10-digit codes) and resource codes (11-digit codes for materials and miscellaneous items with the prefix code and 10-digit code; and 13-digit codes for labor and equipment with the prefix code and 12-digit code):

- XXXXX: Public Construction Work Guideline Codes (totally five digits)
- □□□□□□: Code for function or specification (four digits for materials and miscellaneous items and six digits for labor and equipment)
- Δ: Code for pricing unit (one digit)
- L: Code for labor (prefix code)
- E: Code for equipment (prefix code)
- M: Code for materials (prefix code)
- W: Code for miscellaneous items (prefix code)

The code classification structure developed for the BIM system in this study can help project managers to understand the type of each work related to the fire safety equipment, work structure, connections among components, and unit price of each component in the project (Choi, Choi & Kim 2014; Chou & O'Connor 2007). The WBS of the fire safety equipment work in the project is mainly divided into eight categories (Table 1). Each category represents a type of the most fundamental work in the project, and each category is composed of several subcategories. By comparing with the historical data of fire safety equipment in 10 other projects, the components and their unit costs in the case of this study show no significant differences from those data in more than 70% of the compared cases, indicating that the WBS categorization developed in this study can be used as a standard.

3.3 *Work categorization and relational matrix of the work components*

The fire safety equipment work in this study is categorized into "fire extinguishing equipment", "fire alarm equipment", "refuge and escape equipment", and "necessary equipment for fire escape" in accordance with the categorization in the "Standard for Installation of Fire Safety Equipment Based on Use and Occupancy" promulgated by the National Fire Safety Agency of the Ministry of the Interior in Taiwan.

According to related research and the basic requirements in governing legal regulations, fire safety equipment generally include fire extinguishers, indoor hydrants, outdoor hydrants, automatic sprinklers, manual fire alarm equipment, emergency broadcasting equipment, direction indicating equipment, evacuation equipment, emergency lighting equipment, water supplying pipes, fire water reservoirs, and exhaust equipment. Therefore, the content of the Web-based hybrid BIM cost-estimating system can satisfy the demand for cost estimation of at least 60% of the fire safety equipment used in buildings in Taiwan.

The amount of required manpower and working hours is evaluated on the basis of the working progress to estimate the unit price and labor ratio of each category of fire safety equipment and then calculate the installation cost of each category of equipment on the basis of the algorithm rules in the database. The relational matrix of the work components is illustrated in Table 2.

Table 1. Work breakdown structure of fire safety equipment project.

Code	Classification
EA	Equipment and apparatus
TL	Transportation & lifting
TW	Tubing and wiring
TC	Testing costs
IC	Installation costs
DF	Documentation fees
MD	Maintenance and diagnostics
FV	Fire technician visa fees

3.4 *User interface flow of the Web-based hybrid BIM cost-estimating system*

After the input of the required information, such as quantity and unit cost, of the fire safety equipment is completed, the Web-based hybrid BIM cost-estimating system exports all the data in the form of a .txt file to the ODBC database.

Table 2. Relational matrix of the work components in the project.

Work category	Item no	Work description	EA	TL	TW	TC	IC	DF	MD	FV
Firefighting equipment										
	M1052011106	Fire Extinguisher	√	√			√			
	M1052011203	Fire sand	√	√						
	M1392012108	Indoor fire hydrant	√	√	√	√	√	√	√	
	M1393100004	Automatic sprinkler	√	√	√	√	√	√	√	√
Fire Alarm device										
	M1385100008	Automatic fire alarm	√	√	√	√	√	√	√	√
	M1678100008	Emergency radio	√	√	√	√	√	√	√	
Evacuation equipment										
	M1653020006	Direction indica. Lamps	√		√					
	M1052032104	Automatic descend.	√		√	√	√	√	√	
Fire rescue Necessary equipment										
	M1583200004	Exhaust equipment	√	√	√	√	√	√	√	
	M1052042108	Emergency power sockets	√		√	√				
Maintenance/Diagnosis										
	L000002A10004	Regular diagnosis	√							
	L000002A20004	Regular inspection	√	√		√				
	L000002A30004	Non-regular inspection	√	√		√				

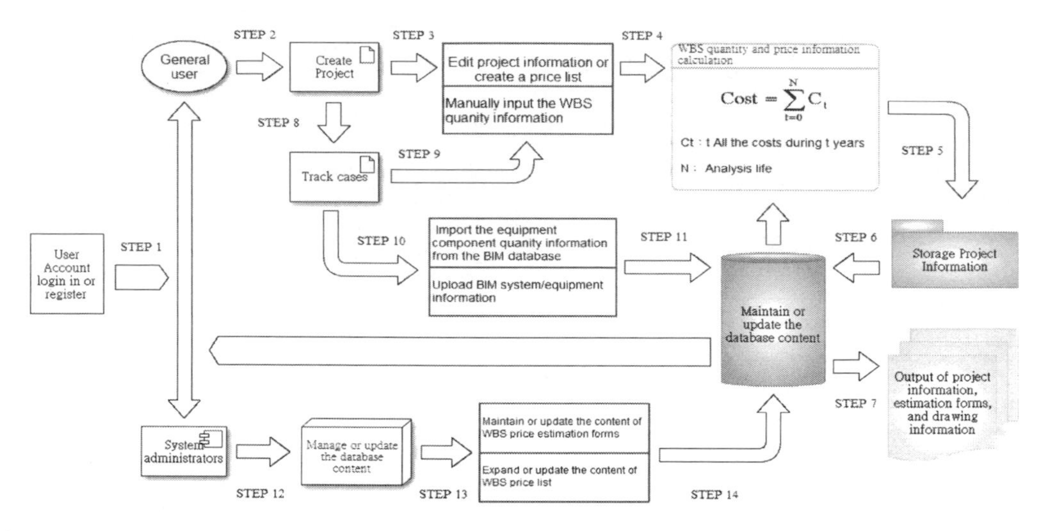

Figure 2. User interface flow of the web-based hybrid BIM cost estimating system.

When the system is connected to the Internet, users can view and modify their projects via the Web browser anytime and anywhere on devices or platforms connected to the Internet. The framework of the BIM Web-based cost estimation system is illustrated in Figure 2, and the following is an introduction to its function flows respectively for nonadministrators and administrators:

Function flow A: Nonadministrators

- Step 1: Log into the system as general users by entering their passwords;
- Step 2: Create a new project (continue with Step 3) or modify an existing project (jump to Step 8);
- Step 3: Input the information and data required for project such as basic information of the building, building site information, basic information of fire safety equipment of different categories, budget list, and detailed price list. Then, the system will provide cost estimation examples and updated unit price information about each category of fire safety equipment for the references of the users;
- Step 4: The input data will be imported into the cost estimation module to calculate the required quantities;
- Step 5: Complete the newly created project and obtain the initial cost estimation;
- Step 6: The data of the newly created project are stored in the Fire Safety Equipment Property Database for future modification or addition of any subproject;
- Step 7: After the creation of the project is complete, the system can generate the budget list and detailed price list as well as provide related BIM 2D drawings and documents;

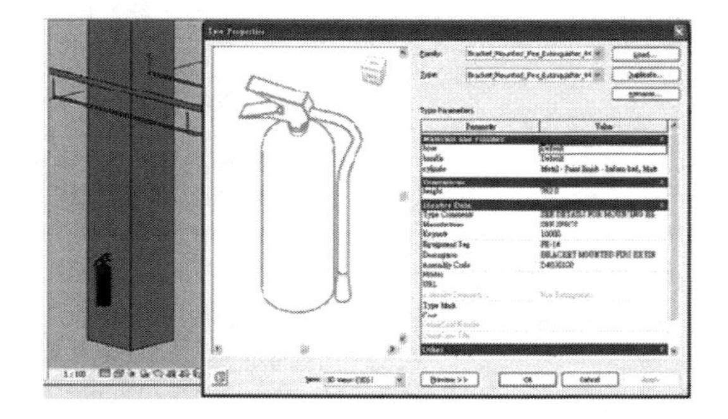

Figure 3. Models of fire safety system component and building element.

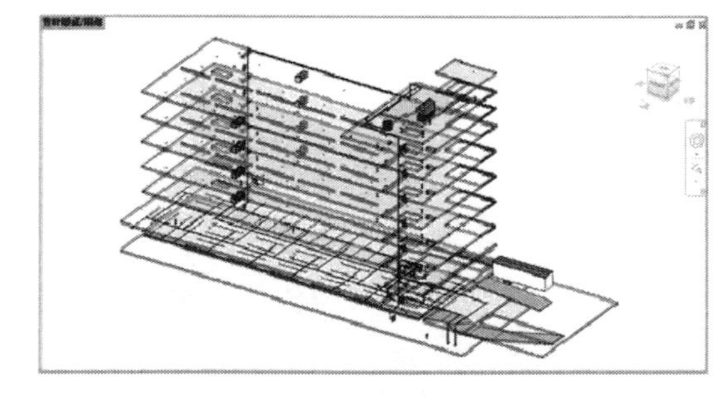

Figure 4. Quantity information of the fire safety system components.

- Step 8: After the completion of the new project, the users can track and modify the project (Step 9) or import the .txt file of the BIM model quantity report into the binding module of the Web-based hybrid BIM system to update the quantity information stored in the BIM system toward the quantity of fire safety equipment actually used in the project (Step 10) or upload the BIM 2D drawings (Step 11);

Function flow B: Administrators

- Step 12: Log into the system as administrators to maintain and update information in the system on a regular and irregular basis;
- Step 13: Use of the administrator privileges, including: (a) maintain the Web-based hybrid BIM cost-estimating system and manage the general user access; (b) expand, maintain, and update the unit cost, detailed price list and total budget list of the fire safety equipment as well as the content of historical projects; (c) change, maintain, and update the content and format of the detailed price list, total budget list, and BIM 2D drawings;
- Step 14: Updated data are stored in the Fire Safety Equipment Property Database for future maintenance and updating of the content in the system.

4 EMPIRICAL STUDY

4.1 *Case description*

The case in this study is a factory building (six floors above ground and one floor below) with a total floor area of 7,780 m^2. The total project cost of this building is NT$ 67,690,000 (~US $2,256,000). Then, in accordance with governing legal regulations, 3D BIM models of the following fire safety equipment in a building must have fire extinguishing equipment, fire alarm equipment, refuge and escape equipment, and necessary equipment for fire rescue.

4.2 *Web-based hybrid BIM cost-estimating system*

The Web-based hybrid BIM cost-estimating system is used as a communication platform for drawing and cost estimation. First, the Web-based hybrid BIM system is used for the fire safety equipment design and cost estimation of the above-mentioned case. Second, the 2D CAD model is used for the fire safety equipment and cost estimation of the same case. Then, the time and costs used in the two cost estimation methods are compared. The process of using the Web-based hybrid BIM cost-estimating system for cost estimation is shown below:

- Step 1: Build the 3D model of the building;
- Step 1.1: Build the models of parametric building elements;
- Step 2: Generate the drawings of fire safety system components and set up their properties;
- Step 3: Generate the model of the fire safety system (Figure 3);
- Step 4: The content of the inbuilt list-making tool of the BIM system is used to automatically generate the quantity information of the required fire safety system components;
- Step 5: Import the BIM quantity information of the fire safety system components to the BIM cost-estimating system Figure 4);
- Step 6: Conduct unit price analysis and create the detailed price list;
- Step 7: Create the budget list.

In general, the 3D BIM models created by the Web-based hybrid BIM cost-estimating system in this study can fully represent the actual conditions of the fire safety equipment in all aspects and then generate the detailed price list and budget list for cost estimation. By quantifying the evaluation criteria for each fire safety system component and evaluating the amount of required manpower and working hours on the basis of the working progress, the BIM system can estimate the unit price and labor rate of each component to calculate the costs of establishing different categories of the fire safety equipment in this case.

4.3 *2D CAD cost evaluation method*

The 2D CAD cost evaluation method is also used in this study to evaluate the same case as a comparison against the Web-based hybrid BIM system to find out the differences in the costs and benefits of these two cost evaluation methods. The process of using the 2D CAD cost evaluation method is shown below:

- Step 1: Create the 2D CAD drawings;
- Step 2: Create the "BLOCK" symbol drawings that represent the fire safety system components;
  - Step 2.1: Create the shop detail drawing of the fire safety equipment;
- Step 3: Design the fire safety system and create the disposition plan of the fire safety system;
  - Step 3.1: Create the riser diagram of the fire safety system;
- Step 4: Manually estimate the quantity of each component in the fire safety system and building element;
- Step 5: Conduct unit price analysis of the fire safety equipment and component;
- Step 6: Create the detailed price list for the cost estimation;
- Step 7: Create the budget list.

4.4 *Comparison and discussion*

By comparing the working hours required to evaluate the same case using the traditional 2D CAD method and the Web-based hybrid BIM system in this study (see Table 3 for comparison results), it is found that the 3D BIM system takes 19.6% and 22.2% less working hours than the 2D CAD method, respectively, to create the building (drawings) models and component (blocks) models. In the estimation of the quantity of the system and component, the BIM system takes 75% less working hours than the 2D CAD method because of its automatic output of the component quantity information. However, the 3D BIM system requires 25% more working hours for the system and equipment layout than the 2D CAD method for the former to adjust the height of each water pipe, electrical wire, and/or water duct to ensure its connection with the corresponding fire safety system component while the 2D CAD method just uses the schematic drawing with the shop detail drawing and the rise diagram as supplements to indicate 3D specifications. Nevertheless, in terms of the overall working hours, the 3D BIM method still requires 22.2% less time than the 2D CAD method. It is because the 2D CAD method requires manual estimation of the quantity, length, and height of the system, equipment, and component while the automation function of the 3D BIM model can significantly reduce time required for these estimation activities.

The above-mentioned case analysis and comparison results reflect the fact that the construction industry is a highly fragmented industry with a very detailed and complex division of work (Ahmad, Russel & Abou-Zeid 1995; Cheah & Chew 2005). With the integration of work information through the BIM system, the problems with complex division of work can be solved. With the BIM system, design modifications and conflict detection can be done in the design stage, saving both time and money (Miettinen & Paavola, 2014). The BIM system is also helpful for the construction industry to improve its design development efficiency, reduce working hours, prevent construction material waste, and effectively improve productivity and design quality (Park et al. 2011). In the early design stage, the cost estimation using the BIM system can help architects and building owners to analyze the feasibility of building projects and effectively control the costs of the projects (Cheng, Tsai & Hsieh 2009).

By comparing the cost estimation of the 2D CAD method and the Web-based hybrid BIM system for the same case (see Table 4 for comparison results), it is found that, with the same fire safety equipment categories and quantities, the cost of water pipes, electrical wires, and air ducts as well as the miscellaneous cost estimated by the Web-based hybrid BIM system are 5.7% less than those estimated by the 2D CAD method, whereas the labor cost estimated by the BIM system is 2% less than that estimated by the 2D CAD method. In terms of the overall cost, the estimate of the Web-based hybrid BIM system is 2.2% less than that of the 2D CAD method.

As indicated by the comparison results, the Web-based hybrid BIM system has more accurate estimates of the pipe/wire/duct cost, miscellaneous cost, and labor cost, whereas its estimate of the total cost is 2.2% less than that of the 2D CAD method. With more accurate cost estimation, the BIM system is helpful for effective cost control.

By comparing the working hours of the Web-based hybrid BIM system and the traditional 2D

Table 3. Working hour comparison between the 2D CAD method and the Web-based hybrid BIM system.

Item	2D CAD method (working hour)	Web-based hybrid BIM system (working hour)	Cost–benefit analysis (%)
Creating building (drawings) models	56	45	19.6
Creating component (blocks) models	36	28	22.2
Layout of the system and equipment	8	10	–25.0
Quantity estimation of the system and component	8	2	75
Total	108	84	22.2

Table 4. Comparison of cost estimations of the 2D CAD method and the Web-based hybrid BIM system.

Item	2D CAD method (US $1,000)	Web-based hybrid BIM system (US $1,000)	Cost–benefit analysis (%)
Equipment cost	54.5	54.5	0
Pipe, wire, and duct cost	29.3	27.6	5.7
Miscellaneous cost	10.3	9.7	5.7
Labor cost component	66.9	65.7	2.0
Estimate of total cost	161.0	157.5	2.2

CAD method for the same case, we obtained some main results. First, the BIM system requires 75.0% less working hours than the 2D CAD method in the quantity estimation of the system and component, 41.8% less working hours in the Creating building (drawings) models & creating component (blocks) models, and 22.2% less time in the total working hours.

The result is contributed by the following reasons: (1) The system can allow different experts to work together whenever and wherever they need and, therefore, promote more efficient work cooperation. In addition, because of the component quantity data stored in the database, manual evaluation is not needed, which also helps to reduce time for design modification in the design planning stage. (2) In the Web-based hybrid BIM cost-estimating system, the basic information/data about the evaluation items, component cost, component unit price, and so on, are stored in the database, which saves time to collect public information about the price of each component or to visit suppliers to inquire the price. (3) The system also saves time for the production of the shop detail drawing and riser diagram for each component of the system.

Second, by comparing the cost estimation of the Web-based hybrid BIM cost-estimating system and the 2D CAD method, it is found that the total cost estimate of the former is 2.2% less than that of the latter. It is probably because of the following reasons: (1) The 2D CAD method only uses the schematic drawing with the shop detail drawing and the rise diagram as supplements to indicate 3D specifications. Therefore, it is difficult for the 2D CAD method to have accurate estimates of the lengths of the pipes, wires, or ducts because of the lack of accurate 3D specifications. In addition, the measure standard may vary from people to people, which may also result in inconsistent estimation results. (2) The estimated lengths of the pipes, wires, and ducts are positively correlated with the labor cost. With longer pipes, wires, and ducts to install, more labor is needed, which will also increase the total cost.

5 CONCLUSION

In this study, the Web-based hybrid BIM cost-estimating system is developed with the intention of providing assistance to the cost estimation work in the early design stage of a building project and promoting the efficiency of the project implementation. To verify the feasibility and performance of the system, a case study is conducted by using the BIM system and the traditional 2D CAD method to evaluate the same case, and then the working hours and cost estimates of the BIM system and the 2D CAD method are compared. The comparison results indicate that the Web-based hybrid BIM cost-estimating system is better than the 2D CAD system in terms of estimation efficiency and accuracy.

It is probably because the Web-based hybrid BIM cost-estimating system is built on the basis of work cooperation while its inbuilt database can directly display relevant information and quantity of each component. Therefore, the statistic information/data of the BIM system about the layout, relative position to building structure, and 3D specifications of the components are more accurate and helpful for effective cost control than those of the 2D CAD method. In addition, the BIM model system can incorporate the dimension, position, 3D specifications, and other information of the equipment as 3D models inside the building structure; therefore, it can obtain accurate quantification data of the equipment. Finally yet importantly, on the basis of the requirements of difference projects, the equipment quantity information can be modified and then directly exported via the database of the BIM system or exported as a .txt report and then printed out. To conclude, the Web-based hybrid BIM cost-estimating system uses its database to integrate all the information/data and, as a result, makes it more convenient and fast for cost estimation.

Moreover, by incorporating the fire safety equipment into the 3D building structure model, the BIM system can accurately represent the spatial configuration of the power line and cable tray (power supply), water pipes (water supply), and air ducts (exhaust system). Such a visualized design is helpful in eliminating problems caused by uncertain factors (3D spatial configuration and quantity of required materials) in the initial cost estimation of a building project.

The restriction of this study lies in its exclusion of economic factors, such as net values, internal rate of return, price index, inflation rate, and interest rate, in the cost estimation analysis. Therefore, a further analysis of sensitivity toward these factors is required. In addition, the influence of the lifecycle cost of a building project for different types of relevant entities (such as owner, investor, developer, building administrator, and building operator) is also a topic that requires further exploration in future research on the economic benefits of fire safety equipment for buildings.

To promote future application of the Web-based hybrid BIM cost-estimating system, the technology acceptance model can be applied to determine the requirements and preferences of fire safety professionals in their cost estimation operation and explore their possible interest in the

BIM system, providing references for the future online application and digital learning of this system.

REFERENCES

Ahmad, I.U., Russell, J.S., & Abou-Zeid, A. (1995). Information Technology (IT) and integration in the construction industry. *Construction Management and Economics, 13*(2), 163–171.

Bradley, D.C., Mascaro, M., & Santhakumar, S. (2005). A relational database for trial-based behavioral experiments. *Journal of neuroscience methods, 141*(1), 75–82.

Bryde, D., Broquetas, M., & Volm, J.M. (2013). The project benefits of Building Information Modelling (BIM). *International Journal of Project Management, 31*(7), 971–980.

Cheah, C.Y., & Chew, D.A. (2005). Dynamics of strategic management in the Chinese construction industry. *Management Decision, 43*(4), 551–567.

Chen, H.M., & Hou, C.C. (2014). Asynchronous online collaboration in BIM generation using hybrid client-server and P2P network. *Automation in Construction, 45*, 72–85.

Chen, L., & Luo, H. (2014). A BIM-based construction quality management model and its applications. *Automation in construction, 46*, 64–73.

Cheng, B., & Wang, Y. (2010). BIM's content and its application in contemporary architectural design. In *Management and Service Science (MASS), 2010 International Conference on* (pp. 1–4). IEEE.

Cheng, M.Y., Tsai, H.C., & Hsieh, W.S. (2009). Web-based conceptual cost estimates for construction projects using Evolutionary Fuzzy Neural Inference Model. *Automation in Construction, 18*(2), 164–172.

Chien, K.F., Wu, Z.H., & Huang, S.C. (2014). Identifying and assessing critical risk factors for BIM projects: Empirical study. *Automation in Construction, 45*, 1–15.

Choi, J., Choi, J., & Kim, I. (2014). Development of BIM-based evacuation regulation checking system for high-rise and complex buildings. *Automation in Construction, 46*, 38–49.

Chou, J.S., & O'Connor, J.T. (2007). Internet-based preliminary highway construction cost estimating database. *Automation in Construction, 17*(1), 65–74.

Coronel, C., & Morris, S. (2016). Database systems: design, implementation, & management. Cengage Learning.

Ding, L., Zhou, Y., & Akinci, B. (2014). Building Information Modeling (BIM) application framework: The process of expanding from 3D to computable nD. *Automation in construction, 46*, 82–93.

Godager, B.A. (2011). Analysis of the information needs for existing buildings for integration in modern BIM-based building information management.

Grilo, A., & Jardim-Goncalves, R. (2011). Challenging electronic procurement in the AEC sector: A BIM-based integrated perspective. *Automation in Construction, 20*(2), 107–114.

Han, N., Yue, Z.F., & Lu, Y.F. (2012). Collision detection of building facility pipes and ducts based on BIM technology. In *Advanced Materials Research* (Vol. 346, pp. 312–317). Trans Tech Publications.

Isikdag, U. (2012). Design patterns for BIM-based service-oriented architectures. *Automation in Construction, 25*, 59–71.

Isikdag, U. (2012). Design patterns for BIM-based service-oriented architectures. *Automation in Construction, 25*, 59–71.

Isikdag, U., Underwood, J., & Aouad, G. (2008). An investigation into the applicability of building information models in geospatial environment in support of site selection and fire response management processes. *Advanced engineering informatics, 22*(4), 504–519.

Jrade, A., & Alkass, S. (2001). A conceptual cost estimating computer system for building projects. *AACE International Transactions*, IT91.

Kim, H.J., Seo, Y.C., & Hyun, C.T. (2012). A hybrid conceptual cost estimating model for large building projects. *Automation in Construction, 25*, 72–81.

Kim, I., Kim, J., & Seo, J. (2012). Development of an IFC-based IDF converter for supporting energy performance assessment in the early design phase. *Journal of Asian Architecture and Building Engineering, 11*(2), 313–320.

Kobes, M., Helsloot, I., De Vries, B., & Post, J.G. (2010). Building safety and human behaviour in fire: A literature review. *Fire Safety Journal, 45*(1), 1–11.

Lee, S.K., Kim, K.R., & Yu, J.H. (2014). BIM and ontology-based approach for building cost estimation. *Automation in construction, 41*, 96–105.

Meža, S., Turk, Ž., & Dolenc, M. (2014). Component based engineering of a mobile BIM-based augmented reality system. *Automation in construction, 42*, 1–12.

Miettinen, R., & Paavola, S. (2014). Beyond the BIM utopia: Approaches to the development and implementation of building information modeling. *Automation in construction, 43*, 84–91.

Park, J., Kim, B., Kim, C., & Kim, H. (2011). 3D/4D CAD applicability for life-cycle facility management. *Journal of computing in civil engineering, 25*(2), 129–138.

Popov, V., Mikalauskas, S., Migilinskas, D., & Vainiūnas, P. (2006). Complex usage of 4D information modelling concept for building design, estimation, sheduling and determination of effective variant. *Technological and economic development of economy, 12*(2), 91–98.

Porter, S., Tan, T., Tan, T., & West, G. (2014). Breaking into BIM: Performing static and dynamic security analysis with the aid of BIM. *Automation in Construction, 40*, 84–95.

Qiu, X. (2011). Building Information Modelling (BIM) adoption of construction project management based on Hubei Jingzhou bus terminal case. In *2011 International Conference on Business Computing and Global Informatization.*

Sonmez, R. (2004). Conceptual cost estimation of building projects with regression analysis and neural networks. *Canadian Journal of Civil Engineering, 31*(4), 677–683.

Turban, E., Leidner, D., McLean, E., & Wetherbe, J. (2007). *Information technology for management: transforming organizations in the digital economy.* John Wiley & Sons, Inc.

Vanlande, R., Nicolle, C., & Cruz, C. (2008). IFC and building lifecycle management. *Automation in Construction, 18*(1), 70–78.

Zhiliang, M., Zhenhua, W., Wu, S., & Zhe, L. (2011). Application and extension of the IFC standard in construction cost estimating for tendering in China.*Automation in Construction, 20*(2), 196–204.

Civil, Architecture and Environmental Engineering – Kao & Sung (Eds)
© 2017 Taylor & Francis Group, ISBN 978-1-138-02985-9

Test and parametric analysis of postfire seismic performance of SRC column–RC beam joints

Guohua Li, Long Xie & Dongfu Zhao
School of Civil and Transportation Engineering, Beijing University of Civil Engineering and Architecture, Beijing, China

ABSTRACT: In order to study the hysteretic behavior of SRC column–RC beam joints after fire, fire and pseudo-static tests were carried out. On the basis of the experimental results, the numerical simulation was developed to study the influence of parameters such as heating time, axial compression ratio, concrete strength, and profiled steel ratio on the hysteretic behavior of the joint after fire. The results show that the bearing capacity and stiffness of joints will degrade after fire. With the increase of axial compression ratio, concrete strength, and profiled steel ratio of column, the bearing capacity and stiffness of the SRC column–RC beam joints were improved.

1 INTRODUCTION

Building fire is not only a serious threat to people's lives and property, but also it causes damage to the building itself. Sometimes, although building collapses are not caused by fire, the structure performance of building must be estimated to decide whether it can be used continuously. Because of the superiority of high bearing capacity, good earthquake resistance, and high stiffness, the Steel Reinforced Concrete (SRC) structure is widely used in high-rise buildings. At present, most of the studies on the column–beam joints are mainly focused on the seismic behavior at ambient temperature, and only few studies have been concerned with the seismic behavior after fire.

In this study, three pseudo-static tests were conducted, and the influence of heating time on the seismic behavior of the SRC column–RC beam joints after fire was determined. Then, the influence of bearing capacity and stiffness under cyclic loading was studied by numerical simulation.

2 EXPERIMENTAL PROGRAMS

2.1 *Test specimens*

Three specimens were used in the test: A, B, and C. The main purpose of the test is to investigate the influence of heating time on the seismic behavior of the SRC column–RC beam joints, and three specimens are of the same. The joints 1960 × 300 × 300 mm for columns and 2870 × 300 × 300 mm for beams were designed on the basis of the relate Chinese codes (JGJ. 1996, JGJ. 2001, YB. 2007). The H-shaped steel section encased in column is HW175 × 175 × 7 × 10, which is fabricated from Grade Q345. The longitudinal reinforcement, 18mm diameter bar, and transverse reinforcement, 10mm diameter bar, were used for both the columns and beams. The yield strength of bars is 345 MPa, and the concrete with grade C40 for the specimens was constructed in specimens.

2.2 *Test method*

In this study, two types of test were conducted: fire test and pseudo-static test. Figure 1 shows a fire furnace and setup of the pseudo-static test. Specimen A was the control one, and pseudo-static test was developed directly without fire condition. Specimens B and C were exposed to fire before the pseudo-static test. Specimens B and C were fire-tested under ISO834 standard fire (GB/T. 2008), whose heating times are 60 and 90 min, respectively. The fire conditions were simulated in a burning furnace with dimensions 3000 × 3000 × 1000 mm in the China Academy of Building Research.

Figure 1. The furnace and test setup.

After fire test, pseudo-static test was conducted, and three column–beam joint specimens were tested on a special steel girder frame for the pseudo-static test. Equal and opposite displacements were imposed at the beam ends by hydraulic jacks, and the top and bottom of the column were pinned against translation. An axial load of 0.35 column axial load capacity was applied on the top of columns. The test was displacement-controlled and the specimens were subjected to an increasing cyclic displacement up to its failure.

2.3 *Hysteretic loops and envelopes*

Table 1 shows the test results. The yield load (P_y), maximum load (P_{max}), and ultimate load (P_u) decreased with the prolongation of heating time. The bearing capacities of J2 and J3 are 85.2%, 72.9% of those of J1. Ductilities of the joints are 2.59, 2.53, and 2.0, respectively, which shows that the bearing capacities of the specimens reduced significantly when the heating time exceeds 60 min. Hysteretic loops and envelopes of the specimens are shown in Figure 2. An obvious pinch effect was observed in the hysteretic loops, primarily due to the slip between the reinforced bars, profiled steel, and concrete. The area surrounded by the hysteresis loops of Specimen A is larger than that of Specimens B and C, showing that when undergoing fire, the energy dissipation capacity would decrease. From the envelopes, the stiffness of the joint also decreased with the prolongation of heating time.

Table 1. Test results.

	P_y	Δ_y	P_{max}	Δ_{max}	P_u	Δ_u
Specimen	kN	mm	kN	mm	kN	mm
A	67.7	14.8	100.1	33.4	84.6	38.3
B	55.9	16.3	85.2	36	72.3	41.2
C	53.9	18.5	72.9	32	64.5	37.1

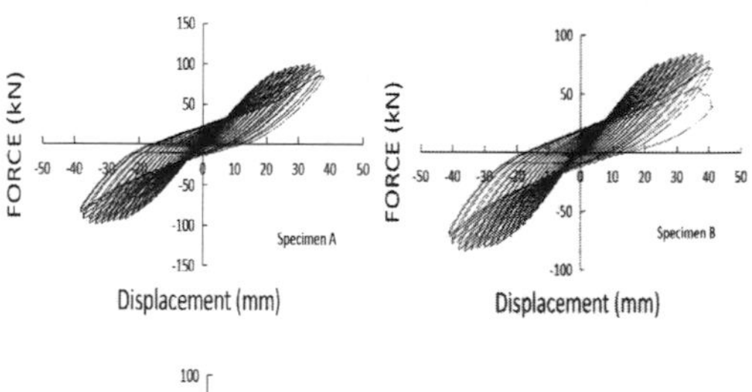

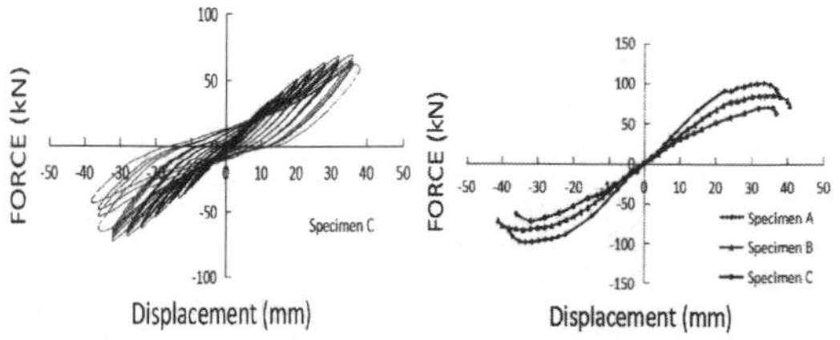

Figure 2. Load-displacement hysteretic loops and envelopes.

3 FINITE-ELEMENT ANALYSES

On the basis of the pseudo-static test after fire, parameter analysis was carried out to study the influence of parameters such as heating time, axial compression ratio, concrete strength, and steel ratio on the hysteretic behavior of joint after fire. A total of 10 finite-element models of SRC column–RC beam joint were established by using the general finite-element package, ABAQUS/Standard v6.10, as shown in Figure 3. The values of the parameters are shown in Table 2. For validation, Specimens J1, J2, and J3 are test pieces simulated by the finite-element method.

Thermo-structural analysis was conducted to simulate the behavior of the SRC column–RC beam joints after fire. For thermal analysis, concrete and profiled steel were modeled with the element DC3D8, which was replaced by the element C3D8R in structural analysis. Element DC1D2 was used to model the reinforcement bar and replaced by the element T3D2 during structural analysis. ISO834 standard heating curve is used as the heating curve

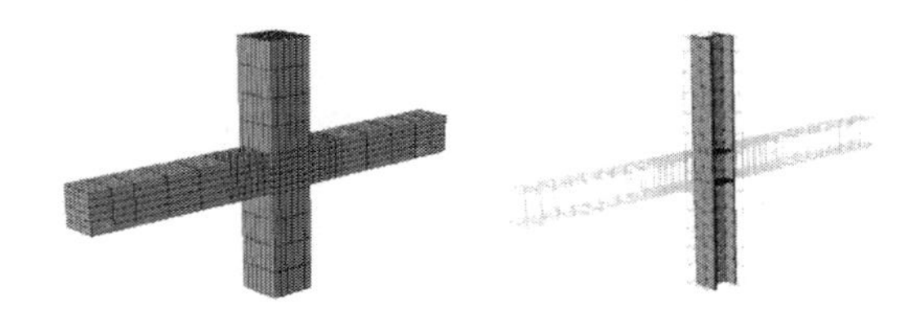

Figure 3. Model of the finite analysis.

Table 2. The parameters.

No.	Heating time min	Axial compression ratio	Concrete strength MPa	Steel ratio %
J1	–	0.35	40	4.96
J2	60	0.35	40	4.96
J3	90	0.35	40	4.96
J4	120	0.35	40	4.96
J5	60	0.45	40	4.96
J6	60	0.55	40	4.96
J7	60	0.35	30	4.96
J8	60	0.35	50	4.96
J9	60	0.35	40	4.23
J10	60	0.35	40	6.72

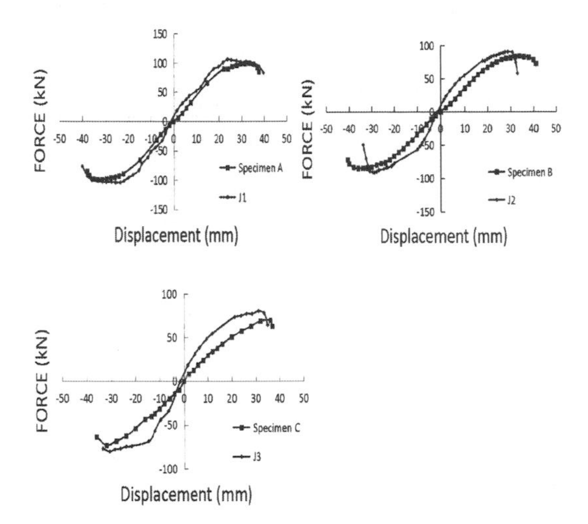

Figure 4. Envelopes of the test and numerical simulation.

and the cooling process is considered to be a straight line. The heat radiation and heat convection are considered in heat analysis. Because of the analysis focusing on the postfire behavior of the joints, as the temperature field of the joints was known, a special program was used to extract the maximum temperature of each element in the whole process of heating and cooling. On the basis of the maximum temperature of the element, the correct constitutive model of concrete was chosen in structural analysis.

For structural analysis, the concrete and steel are considered as isotropic materials, and the bond slip between concrete and profiled steel and the slip between concrete and reinforcement bars are not considered. The relationship between stress and strain of concrete at high temperature was adopted in accordance with Zhoudao Lu (Zhoudao Lu et al. 1993), and the constitutive relation model of steel after high temperature recommended by Linhai Han (Linhai Han. 2007) is used as the material constitutive relation. In order to better simulate the damage process of concrete, the concrete damage plasticity model is used, which could simulate the concrete cracking and crushing conveniently.

Figure 4 compares envelopes of the test with numerical simulation. It is evident from the figure that the bearing capacity gives a close prediction of the test results. The stiffness value obtained from numerical simulation is larger than that of the test, which is mainly because of some simplification in the numerical model, such as homogeneous material of concrete, no slip between concrete and profiled steel, no slip between concrete and reinforced bars, and the support restraint ideally. The difference between test results and numerical simulation results is in the control scope and acceptablity.

4 PARAMETRIC STUDY ON BEARING CAPACITY AND STIFFNESS OF JOINT AFTER FIRE

4.1 *Heating time*

Figure 5 illustrates envelopes of the joints at different heating times. It can be seen that the bearing capacity of the specimen decreases gradually with the prolongation of heating time. The bearing capacities of J2, J3, and J4 are 85.2%, 75.6%, and 54.2% of those of J1, respectively. The stiffness values obtained from the simulation of J1–J4 are 3.71, 3.15, 2.59, and 1.69 kN/mm, respectively, for three specimens, which show that the prolongation of the heating time would lead to the stiffness degradation. The material properties deteriorate at high temperature, which causes degradation of the bearing capacity and stiffness. When heating time exceeds 120min, the bearing capacity and stiffness decreased sharply, about half of those of specimen J1.

4.2 *Axial compression ratio*

According to specification regulated data in the *Technical Specification for Steel Reinforced Concrete Composite Structures*, the axial compression ratios were assigned as 0.35, 0.45, and 0.55, respectively. Figure 6 shows envelopes of J2, J5, and J6. It is found that the influence of axial compression ratio on the bearing capacity of the joints is very small. The bearing capacities of the three specimens are 90.3, 93.6, and 101.39 kN. And the simulation gives stiffness values of 3.15, 3.25, and 3.43 kN/mm, respectively. Therefore, the axial compression ratio increase can lead to the stiffness increase, but the amplitude is very small.

4.3 *Concrete strength*

For this investigation, concrete strengths C30, C40, and C50, are selected, which are commonly used

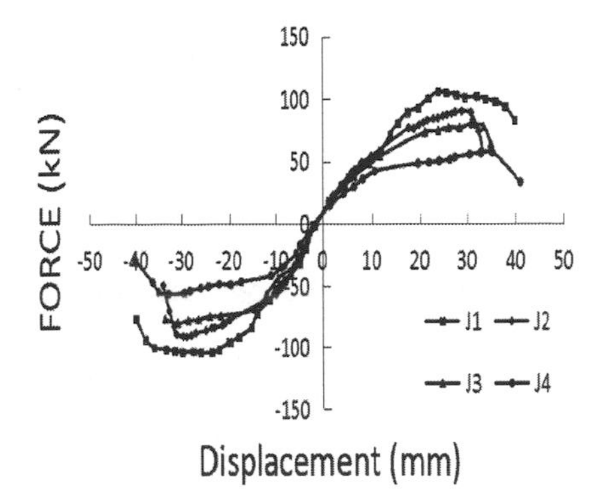

Figure 5. Effect of heating time on strength and stiffness.

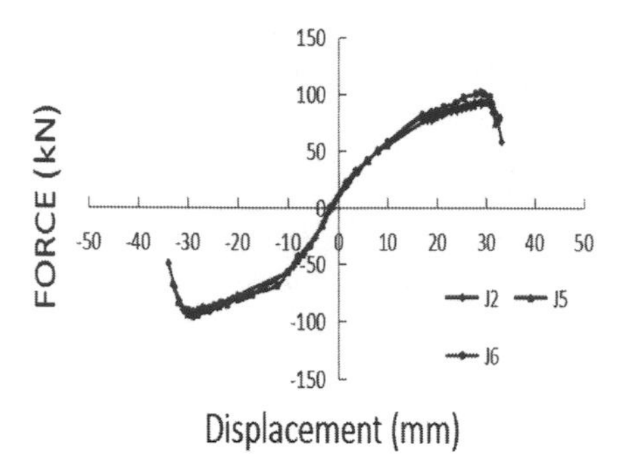

Figure 6. Effect of axial compression ratio on strength and stiffness.

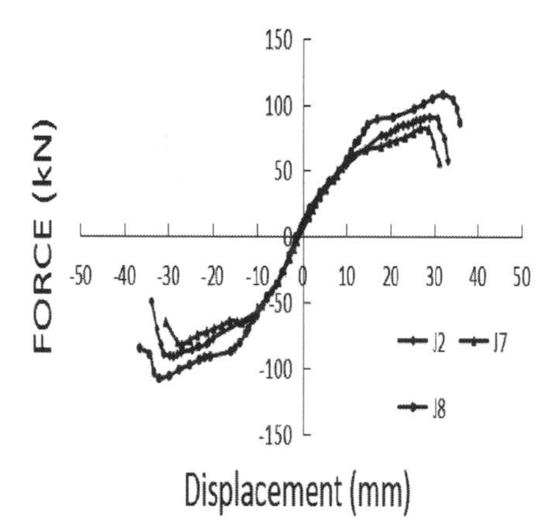

Figure 7. Effect of concrete strength on bearing capacity and stiffness.

in engineering practice. The envelopes, as shown in figure 7, of Specimens J7, J2, and J8 illustrate that the bearing capacity and stiffness of the joint increased effectively when the strength of concrete is improved, as well as the stiffness, which are 3.07, 3.15, and 3.36 kN/mm, respectively. The bearing capacities of the three specimens are 82.1, 90.3, and 107.6 kN, respectively.

4.4 *Profiled steel ratio*

The influence of profiled steel ratio was examined by changing the cross section of steel. The type of profiled steel is based on the regulations of the *Technical Specification for Steel Reinforced Concrete Composite Structures*, and the profiled steel ratios of J9, J2, and J10 are 4.23%, 4.96%, and 6.72%, respectively. The results in Figure 8 indicate that the bearing capacity and stiffness are greater when the profiled steel ratio is improved. The values of stiffness are 2.25, 3.15, and 3.71 kN/mm, respectively.

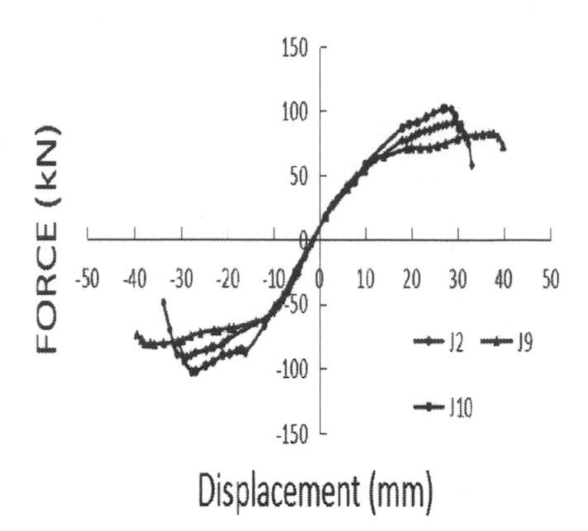

Figure 8. Effect of profiled steel ratio on strength and stiffness.

5 CONCLUSIONS

The fire test and pseudo-static test were conducted to study the seismic performance of SRC column–RC beam joints. Then, on the basis of the test, the finite-element analysis was developed to conduct parameter study. The test and numerical simulation results indicate that the bearing capacity and stiffness will degrade after fire, and the longer the heating time, the more is the deterioration effect. The increase of axial compression ratio, concrete strength, and profiled steel ratio of column can lead to the significant improvement of bearing capacity and stiffness of the SRC column–RC beam joints, except for the axial compression ratio, which has a small influence on the bearing capacity.

REFERENCES

GB/T. 2008. Fire-Resistance Tests-Elements of Building Construction. Beijing: Standard Press of China, 9978.8-2008.

JGJ. 1996. Specification of Testing Methods for Earthquake Resistant Building. Beijing: China, Architecture & Building Press, 101–96.

JGJ. 2001. Technical Specification for Steel Reinforced Concrete Composite Structures. Beijing: China, Architecture & Building Press, 138–2001.

Linhai Han. 2007. Concrete filled steel tube structure-theory and practice. 2nd Edition. Beijing: Science Press.

YB. 2007. Technical Specification of Steel-Reinforced Concrete Structures. Beijing: Metallurgical Industry Press, 9082–2006.

Zhoudao Lu, et al. 1993. The Research of reinforcement and repair on reinforced concrete beam. In *The Proceedings of State Key Laboratory for Disaster Reduction in Civil Engineering*, Tongji University, 152–162.

Civil, Architecture and Environmental Engineering – Kao & Sung (Eds)
© 2017 Taylor & Francis Group, ISBN 978-1-138-02985-9

Using agent-based simulation approach for estimating the efficiency of the building project design team

M.H. Tsai
Department of Civil Engineering, Tamkang University, Taipei City, Taiwan

ABSTRACT: The Project Team Collaborative Efficiency Simulator (PTCES) proposed by Tsai and Huang (2015) was an agent-based simulation tool helping project managers to estimate the workflow efficiency resulted from the collaborative behaviour of the team members. Applying the PTCES, this study continuously extended the practice of efficiency estimation to the architectural design project team and facilitated the architectural design project efficiency evaluation. A case of four-floor nursery school building design was studied in this paper including verification, validation and sensitivity analysis experiment steps. According to simulation results of the case study, the efficiency bottleneck activities of the design process were revealed, and the collaboration willing threshold, communication unit time and the exception probability were found to be the most significant factors influencing the design process efficiency (EC) and the project design duration.

1 INTRODUCTION

In many project tasks, teamwork plays a vital role in getting things done and the efficiency of the results. Effective teamwork is one of the predictors of organisational success since it can cause rapid information exchange and increase responsiveness (Farhangian et al. 2013). The design is a complex process which always involves many participants from different disciplines to work together during the design process (Girard & Robin 2006). Therefore, design collaboration becomes a crucial element in the design process and has a significant effect on the final design performance (Yin et al. 2011). Moreover, large scale and multidisciplinary architectural and engineering projects (e.g., the design of a commercial building) are often complex. They usually involve many interdependent activities and require intensive collaboration among actors (i.e., designers) to deal with activity interdependencies (Jin & Levitt 1996). To make such projects more effective and efficient, managers need to understand where collaboration requirements are and whether the collaborations could be finished efficiently and effectively by the current project team. Consequently, a large amount of research has paid attention to improving collaborative design performance (Girard & Robin 2006; Bstieler 2006; Busi & Bititci 2006; Lahti et al. 2004; Easley et al. 2003; Jin & Levitt 1996).

Jin and Levitt (1996) successfully developed the Virtual Design Team (VDT), a computational model of project organisations, to analyse how activity interdependencies raise coordination needs and how organisation design and communication tools change team coordination capacity and project performance. Based on the VDT foundation, Levitt (2012) illustrated the method applying agent-based simulation, such as VDT model, to help managers design the work processes and organisation of project teams engaged in a great, semi-routine but complex and fast-paced projects. Although the large numbers of organisational and individual level behavioural parameters available in the VDT model can potentially represent cultural phenomena, the social factor is one of the essential perspectives for developing the next generation simulation model (2012). To enhance the social collaborative capability of the project team in the simulation model, Tsai and Huang (2015) adopted the social network philosophy and the Agent-Based Modelling and Simulation (ABMS) approach to develop the Project Team Collaborative Efficiency Simulator (PTCES), which breaks the project team into individual agents and modeling the behaviors and the social network to simulate the collaborative interactions and performance.

The simulation approach provides a useful way to evaluate the dynamic status of a project team efficiency. With repeatable scenario experiments, we can access not only the collaboration performance but also the efficiency improvement strategies by using the simulation approach. Following this idea, this study focused on the architectural design project team and continuously used the PTCES to facilitate the collaborative efficiency analysis of social-network-based and agent-based simulation.

2 METHOD

2.1 *Problem description*

The social-network-based Project Team Collaboration Efficiency Simulation (PTCES) model has been developed for the performance analysis of project teams by Tsai and Huang (2015) using the ABMS approach and has also appropriately used to simulate the construction project duration and cooperation efficiency of the project team. Since the PTCES is an ongoing research project aiming at estimating the project team performance with different organisational structures and collaboration mechanisms, this study adopts the PTCES to evaluate the collaborative efficiency of the project design team for validating the soundness of the PTCES and even discovering the essential impact factors of design performance.

The design process itself is the result of collaboration developed by the designers during the design phrase (Girard & Robin 2006). A building design process composed of the designing activities can be revealed by extracting the functional requirements of the owner, and executed by the corresponding actors of the design team. The PTCES crumbled the activities of design processes into single-actor-performed ones to bring the simulation model into a micro-level perspective so that users can model the detail behaviors of designers with their interdependencies and relevant information/skills.

Following the ABMS philosophy, team members are modeled as agents possessing individual characteristics and social network properties, kept performing assigned activities not only dealing with information but by also communication with one another through his/her existing connections in both the formal and informal networks. That is, team members need to communicate with the neighbors via their collaboration networks for their insufficient information or skills.

2.2 *Project team collaborative efficiency simulation*

The PTCES (Tsai & Huang 2015) refers to the information-processing view of organisations in the VDT model (Jin & Levitt 1996; Anon 1999; Fridsma 2003) to calculate the design process collaboration efficiency. Accordingly, the following mathematical relations referred to Jin and Levitt (1996) are applied to be the foundation of efficiency calculations of the PTCES.

For a given activity of a building design project, TW presents the total work volume of it, which is the sum of the Primary Work volume (PW) and the Collaboration Work volume (CW) as shown in Equation (1).

$$TW_i = PW_i + CW_i \tag{1}$$

where the TWi, PWi and CWi are the corresponding TW, PW, CW of the activity i.

Besides, the PW consists of originally-planned Production Work (PWo) and the production for rework (PWr) resulted from the failure of the original production work as shown in Equation (2).

$$PW_i = PWo_i + PWr_i \tag{2}$$

where the PWi, PWoi and PWri are the corresponding PW, PWo, PWr of the activity i.

From Equation (1) and (2), TW is the sum of the PWo, PWr and CW (Equation 3); the spent time of PW is the sum of the processing time of PWo, PWr and CW (Equation 4).

$$TW_i = PWo_i + PWr_i + CW_i \tag{3}$$

$$t_{TWi} = t_{PWoi} + t_{PWri} + t_{CWi} \tag{4}$$

According to Equation (4), for management viewpoint, we could say the project team is perfectly well performed if the time of TW (t_{TW})equals to the time of PWo(t_{PWo}). Therefore, the ratio (EC, Efficiency coefficient) of t_{PWo} and t_{TW} is the presentation of activity efficiency. As Equation (5) shows, lower EC implies that more PWr time (t_{PWr}) or CW time (t_{CW}) has occurred during the project performed and vice versa.

$$EC_i = \frac{t_{PWo_i}}{t_{TW_i}} \tag{5}$$

$$EC_p = \frac{\sum_{i=1}^{n} t_{PWo_i}}{\sum_{i=1}^{n} t_{TW_i}} \tag{6}$$

where EC_i ($0 \leqq EC_i \leqq 1$) presents the efficiency coefficient of activity i, and EC_p ($0 \leqq EC_p \leqq 1$) is the efficiency index of the whole project performed by the team members.

Besides, since PTCES assumes one communication task takes one-unit time in the simulation model, the CW time spent by an actor for one activity depends on the total communication frequencies between him and his coworker.

Unlike CW time, as shown in Equation (7) and (8), PWo time and PWr time will be calculated according to actor's information processing speed and the work volumes (PWo and PWr).

$$\mathrm{RPS}_{ji} = [x \sim \mathrm{poisson}(\mathrm{APS}_{ji})] \tag{7}$$

$$t_{PWo(ij)} = \frac{PWo_i}{RPS_{ji}} \tag{8}$$

$$t_{PWr(ij)} = \sum_{t=1} \frac{PWo_i(t) + PWc_i(t)}{RPS_{ji}} \quad (9)$$

where RPS_{ji} is actor j's Random Processing Speed for action i; APS_{ji} is actor j's Average Processing Speed for action i; $t_{PWo(ij)}$ is the PWo time of activity i performed by actor j; $t_{PWr(ij)}$ is the PWr time of activity i performed by j; the $PWo_i(t)$ is primary production work volume of activity i at time t, and $PWc_i(t)$ are the correction production work volume of activity i as exception occurred to activity i at time t.

Based on Equation (7), (8) and (9), the PWo time and PWr time can be calculated according to Actor j's RPS which is randomly generated by the Poisson distribution function of Actor j's APS. In this study, the values of actors' APS need to be surveyed with the project team members. According to the aforementioned mathematical model, Figure 1 illustrates the primary function of PTCES.

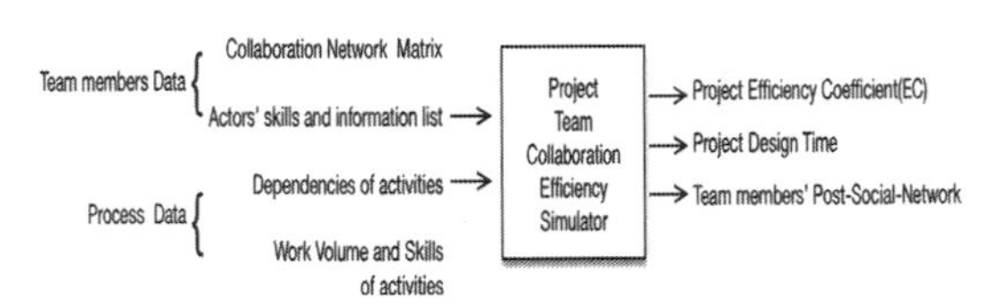

Figure 1. Inputs and outputs PTCES (Tsai & Huang 2015).

2.3 *Data collection of case study*

This study applied a real building design project case for testing the soundness of the PTCES and discovering the potential strategy of design efficiency enhancement. The historical data of the study case are collected by deep interviews with the project managers and observation surveys for the design actors of the architectural and planning firm. Table 1 shows the 43 activities of the four-floor nursery school building design case, which are the process data inputted to the PTCES. The actual design time in Table 1 are compared with the simulation results to calibrate the simulation model. Figure 2 reveal communicative social network of the team members that will be transferred to the collaborative design network matrix and actors' skills and information list as shown in Figure 1.

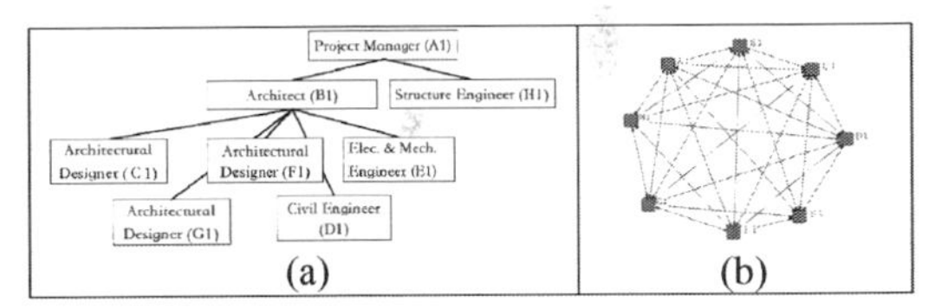

Figure 2. (a) Project team structure (formal relationships); (b) Collaborative design network of project team.

Table 1. Design activity schedule of the study case.

Design Phase	Activity Name	Activity ID	Actors	TW (amount of design documents)	Necessary Inputs (# of information)	Outputs	Successor Activity	Actual Design Time (day)
Preparation	Site location identification	AEC01	A1	1	H01 H03 H08	G01	AEC02	1
	Context of the site determination	AEC02	A1	1	H02 H08 G01	G02	AEC03	1
	Urban planning	AEC03	A1	1	H01 H02 H08 G02	G03	AEC04	1
	Usage requirements planning	AEC04	B1	7	H01 H06 G03	G04	AEC05	6
	Budget Evaluation	AEC05	A1	1	H01 H16 H17 H19 G04	G05	AEC06	3
Regulation Survey	Land use approval document	AEC06	A1	2	H02 H16 H17 G05	G06	AEC07	2
	General layout planning and technical standards evaluation	AEC07	A1	7	H02 H16 H17G06	G07	AEC08	5
Basic Design	General façade design	AEC08	B1	10	H01 H04 H09 H12 H15 H18 G07	G08	AEC09 AEC19	7
	Ground floor plan design (for certification)	AEC09	B1	7	H01 H02 H09 H10 H12 H14 H18 H19 G08	G09	AEC10	5
	Ground floor plan design (for constructor)	AEC10	C1	8	H01 H09H12 H18 H19 G09	G10	AEC11	6
	First floor plan design (for certification)	AEC11	B1	7	H01 H04 H09 H10 H12 H14 H18 H19 G09	G11	AEC12 AEC13	6
	First floor plan design (for constructor)	AEC12	D1	14	H01 H09 H12 H18 H19 G11	G12	AEC21	5
	Second floor plan design (for certification)	AEC13	B1	7	H01 H04 H09 H10 H12 H14 H18 H19 G11	G13	AEC14 AEC15	6
	Second floor plan design (for constructor)	AEC14	E1	14	H01 H09 H12 H18 H19 G13	G14	AEC21	10
	Third floor plan design (for certification)	AEC15	B1	7	H01 H09 H10 H12 H14 H18 H19 G13	G15	AEC16 AEC17	6
	Third floor plan design (for constructor)	AEC16	F1	14	H01 H09 H12 H18 H19 G15	G16	AEC21	5
	Forth floor plan design (for certification)	AEC17	B1	7	H01 H09 H10 H12 H14 H18 H19 G15	G17	AEC18	5
	Forth floor plan design (for constructor)	AEC18	G1	14	H01 H09 H12 H18 H19 G17	G18	AEC21	10
	Landscape design (a)	AEC19	A1	7	H01 H04 H09 H10 H12 H18 G08	G19	AEC20	8
	Landscape design (b)	AEC20	C1	7	H01H04 H09 H10 H12 H18 G19	G20	AEC21	6
	Building sections and elevations design	AEC21	D1	7	H01 H12 H13 H18 G12 G14 G16 G18 G20	G21	[illegible]	3
Facility Planning and Design	Wnidows arrangement and design	AEC22	E1	4	H01 H09 H10 H12 H14 H15 H18 G21	G22	AEC34	3
	Stairs design (a)	AEC23	B1	5	H01 H02 H12 H14 H15 H18 G21 G23	G23	AEC27 AEC28	4
	Stairs design (b)	AEC24	F1	5	H01 H02 H12 H14 H15 H18 G21 G23	G24	AEC27 AEC28	4
	Carriageway design (a)	AEC25	A1	7	H01 H02 H09 H12H15 H17 H18 G21 G26	G25	AEC31 AEC32	7
	Carriageway design (b)	AEC26	G1	7	H01 H02 H09 H12H15 H17 H18 G21 G25	G26	AEC31 AEC32	7
	Ventilating analysis and design (a)	AEC27	B1	3	H01 H02 H12 H15 H18 G23 G24 G28	G27	AEC29 AEC30	3
	Ventilating analysis and design (b)	AEC28	H1	3	H01 H02 H12 H15 H18 G23 G24 G27	G28	AEC29 AEC30	2
	Detailed façade design (a)	AEC29	B1	15	H01 H02 H04 H09 H12 H15 H18 G27 G28 G30	G29	AEC35 AEC36	11
	Detailed façade design (b)	AEC30	C1	15	H01 H02 H04 H09 H12 H15 H18 G27 G28 G29	G30	AEC35 AEC36	12
	Structural design (a)	AEC31	A1	14	H01 H02 H07 H10 H12 H15 H17 G25 G26 G27 G28 G32	G31	AEC37 AEC38	14
	Structural design (a)	AEC32	H1	14	H01 H02 H07 H10 H12 H15 H17 G25 G26 G27 G28 G31	G32	AEC37 AEC38	10
	Enclosure design	AEC33	D1	7	H01 H02 H07 H09 H12 H15 H17 H18 G21	G33	AEC37 AEC38	3
	Planting Design	AEC34	E1	7	H04 H09 H10 CH12 H15 H16 H17 G22	G34	AEC39 AEC40	5
	Lifts design (a)	AEC35	B1	7	H01 H02 H12 H14 H15 G29 G30 G36	G35	AEC39 AEC40	7
	Lifts design (b)	AEC36	H1	7	H01 H02 H12 H14 H15 G29 G30 G35	G36	AEC39 AEC40	5
	Water supply and drainage design (a)	AEC37	A1	7	H01 H02 H12 H15 G31 G32 G33 G38	G37	AEC41	7
	Water supply and drainage design (b)	AEC38	D1	7	H01 H02 H12 H15 G31 G32 G33 G37	G38	AEC41	3
	HVAC facility arrangement (a)	AEC39	B1	7	H01 H02 H12 H15 G34 G35 G36 G40	G39	AEC42 AEC43	7
	HVAC facility arrangement (b)	AEC40	E1	7	H01 H02 H12 H15 G34 G35 G36 G39	G40	AEC42 AEC43	5
	Lightning arrester design	AEC41	A1	1	H01 H02 H12 H15 G37 G38	G41		1
	Barrier-free space design (a)	AEC42	B1	7	H01 H02 H12 H14 H15 G39 G40 G43	G42		6
	Barrier-free space design (b)	AEC43	F1	7	H01 H02 H12 H14 H15 G39 G40 G42	G43		6

2.4 *Simulation plan*

To facilitate the collaborative performance analysis of the building design teams by using ABMS and social network philosophy, this study continues applying the PTCES above as the tool for simulation experiments. Four steps, namely, (1) calibration, (2) verification, (3) validation and (4) sensitivity analysis, are included in the simulation plan.

3 RESULTS

3.1 *Calibration*

While running simulations in PTCES, the tunable parameters of the PTCES includes (1) exception (failure) probability (Pe), (2) failure for rework probability (Pr), (3) unit time of communication (Cu), (5) correction volume ratio(Cr), and (6) collaborative willing threshold ($\hat{W}$) (Tsai & Huang 2015). The values of all parameters vary from project to project. Accordingly, we can only estimate the value range of each parameter by interviewing the staff of the studied project team and surveying the historical data. Since users need to set up the default values of all parameters in the PTCES, this study tunes the case model based on the surveyed value ranges to approach the actual state of the project. Table 2 shows the calibration result and the corresponding parameter settings which fit the actual performance mostly. The validation and the following experiments could be performed with the same parameter settings in Table 2.

Table 2. Calibration Results with 100 experiments (300 simulations per experiment).

Parameter settings after calibration:
Pe = 30%, Pr = 40%, Pc = 40%, Cr = 40%, Wthr = 0.1, Cu= 1.2

Output factors	Process duration (Days)	PWo Time (Days)	PWr Time (Days)	CW Time (Days)	EC
Values after calibration	119.39 (118*)	66.73	15.26	46.5	0.54

*actual project design duration.

3.2 *Verification*

The statistical verification, including the discrete degree and the t-test approaches, is applied to ensure the soundness of the simulation model of the case study and the PTCES (Xiaorong Xiang et al. 2005; Sargent 2012). Table 3 shows the discrete degrees of the case study simulations (100 experiments and 150 simulations per experiment). According to the standard deviation and mean absolute deviation of each estimated value, since they are relatively small, we could say that the stability of the PTCES is acceptable.

3.3 *Validation*

After calibration and verification, the outputs of the conceptual model of the PTCES and the case study are verified to be correct, consistent, and stable. In this step, we further compared the simulation outputs with the actual performance in the real project to validate the reasonability of the addressed conceptual model (Xiaorong Xiang et al. 2005).

Figure 3 shows the result of each design activity simulated by the PTCES of the case study. The average value of EC of the whole project is 0.54 and the project design total time is 120 days. The histogram diagram in Figure 3 shows the time spent in each activity, including PWo time, PWr time and CW time; the line chart illustrates the EC

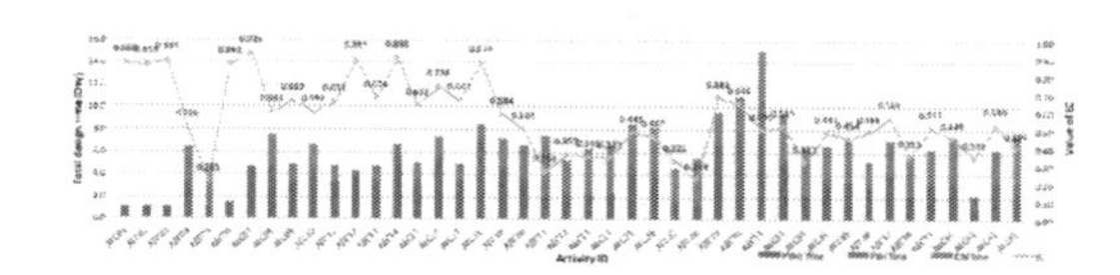

Figure 3. Design time and EC value of each design activities (estimated by PTCES).

Table 3. The stability of the PTCES outputs.

Factors	Project Duration	PWo Time	PWr Time	CW Time	Average EC (0≤EC≤1)
Average	118.67 day	66.81 day	15.11 day	46.24 day	0.54
SD	1.628	0.759	0.722	1.151	0.024
MAD	1.364	0.617	0.607	0.971	0.020

SD: Standard Deviation, MAD: Mean Absolute Deviation.

value of each activity. Ideally, we need to validate the results with the real values of the project; however, it almost impossible to collect the exact time spent for each activity. Accordingly, this study can only verify the results in Figure 3 with the roughly-estimated values by interviewing all project staff.

The distribution of the output values in Figure 3 match the estimation of the interviewed staff. Taking activity AEC21 (Building sections and elevations design) as an example, since the architect needs to collaborate with the actors "C1", "E1", "F1" and "G1" to integrate the design results, the efficiency of the activity AEC21 is below 30% in average. Accordingly, the simulated efficiency of the activity AEC21 is 26.3% which is reasonably within the range estimated by the architect.

Meanwhile, based on the time distributions and the efficiency values of all activities, we can also discover the bottlenecks of the efficiency and the project duration (total design time). Figure 4 shows the distribution of each activity with its TW and EC values. Taking EC = 0.5 and TW = 6.2 day (average time) as the thresholds, the coordinate can be divided into four zones (Zone "A", "B", "C" and "D"). For the company, the 11 activities in the zone "D" would be the critical performance bottlenecks of the workflow because of very low EC values and long durations.

Summarily, the model after the verification and the validation provides the quantitative and meaningful information inside the design workflow. Managers could discover the potential efficiency bottlenecks and causes related to the social networks of the project team.

3.4 *Sensitivity analysis*

According to the results of the sensitivity analysis, the highly-influencing parameters of the project efficiency can be determined. Meanwhile, since the all the parameters in the PTCES, i.e., Pr, Pe, Cr, Ŵ, and Cu, are holistic factors of a project, the project

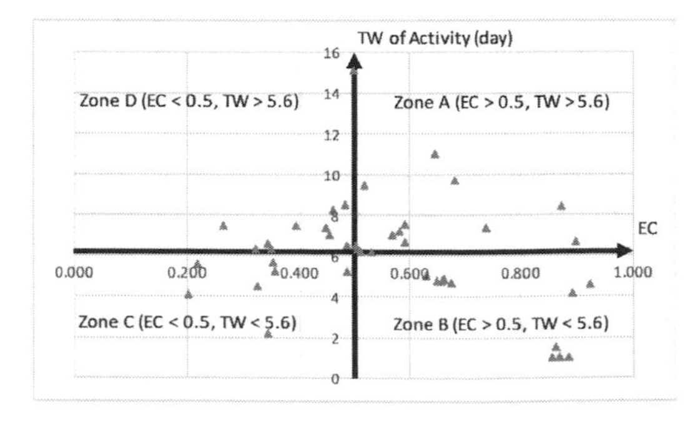

Figure 4. Activity EC—TW distribution diagram.

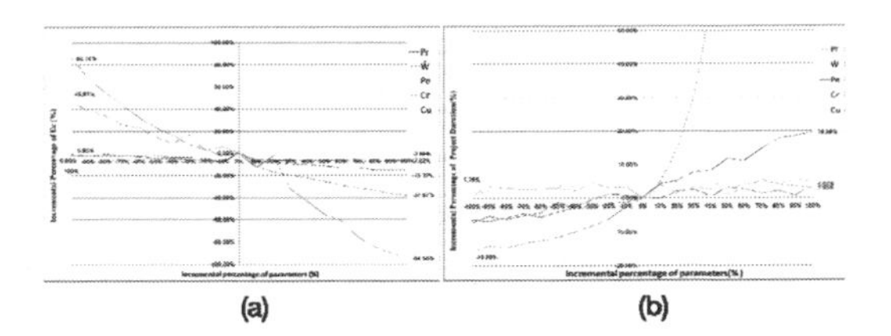

Figure 5. Sensitivity analysis for (a) EC and (b) total project design time (total duration of the architectural design project).

managers could find the way improving the total project efficiency from the global perspective.

Figure 5a shows the liner adverse effect on the EC of each parameter, in which the collaborative willing threshold (Ŵ) provides the highest influence on the project efficiency as the incremental percentage is positive, and the unit time of communication (Cu) is the factor with the most significant impact in the negative incremental percentage area. In Figure 5b, Ŵ is still the most significant factor for the project design time, while the Pe is the secondary one. Therefore, how to reduce Ŵ and Cu values could be the vital issue for enhancing the efficiency, and Ŵ and Pe are primary factors controlling the total project design time.

4 CONCLUSIONS AND FUTURE WORKS

By using the PTCES proposed by Tsai and Huang (2015), this study facilitated the architectural design project efficiency evaluation. A small but complete building project design case was addressed to be the test model in this study. Following the framework of the PTCES, the communicative design network of the project team, the design process with all activities and their corresponding information were collected to build a simulation input data model. Through the calibration, verification, and validation experiment steps, we can grasp the estimated quantitive efficiency information of the global project and the one in the manner of micro perspective.

Based on the simulation, the potential efficiency bottleneck activities can be revealed, and the cause could be analysed by tracing back to the social network features and the interdependency relationship between activities. Moreover, with the sensitivity analysis results, the collaborative willing threshold (Ŵ) and the communication unit time (Cu) are the highest influence factors on the project efficiency.

Summarily, in this study, the architectural design project with a small design team was modelled and simulated successfully with the PTCES. However, due to the limitations and model assumptions of the PTCES, some specific and unusual features of the studied case were ignored or simplified. How to modify the model of the PTCES to match the behaviours of the project team in AEC industry is the critical issue in the future.

ACKNOWLEDGEMENTS

The authors gratefully acknowledge the financial support of the Ministry of Science and Technology of Taiwan, R.O.C. through its grants MOST 104 - 2221 - E - 032 - 047 –

REFERENCES

Anon, 1999. Simulating project work processes and organizations: Toward a micro-contingency theory of organizational design. Management Science, 45(11), pp. 1479–1495.

Bstieler, L., 2006. Trust Formation in Collaborative New Product Development*. Journal of Product Innovation Management, 23(1), pp. 56–72.

Busi, M. & Bititci, U.S., 2006. Collaborative performance management: present gaps and future research. International Journal of Productivity and Performance Management, 55(1), pp. 7–25.

Easley, R.F., Devaraj, S. & Crant, J.M., 2003. Relating Collaborative Technology Use to Teamwork Quality and Performance: An Empirical Analysis. Journal of Management Information Systems, 19(4), pp. 247–268.

Farhangian, M. et al., 2013. Modelling the Effects of Personality and Temperament in Small Teams. In Coordination, Organizations, Institutions, and Norms in Agent Systems IX. Lecture Notes in Computer Science. Cham: Springer International Publishing, pp. 25–41.

Fridsma, D.B., 2003. Organizational Simulation of Medical Work: An Information-Processing Approach. Stanford University.

Girard, P. & Robin, V., 2006. Analysis of collaboration for project design management. Computers in Industry, 57(8–9), pp. 817–826.

Jin, Y. & Levitt, R.E., 1996. The virtual design team: A computational model of project organizations. Computational and Mathematical Organization Theory, 2(3), pp. 171–195.

Lahti, H., Seitamaa-Hakkarainen, P. & Hakkarainen, K., 2004. Collaboration patterns in computer supported collaborative designing. Design Studies, 25(4), pp. 351–371.

Levitt, R.E., 2012. The Virtual Design Team: Designing Project Organizations as Engineers Design Bridges. Journal of Organization Design, 1(2), pp. 14–41.

Sargent, R.G., 2012. Verification and validation of simulation models. 7(1), pp. 12–24.

Tsai, M.-H. & Huang, S.-R., 2015. Agent-based Project Team Collaboration Behavior Simulation. In International Conference on Innovation, Communication and Engineering. Xiangtan, Hunan, P.R. China.

Xiaorong Xiang et al., 2005. Verification and Validation of Agent-based Scientific Simulation Models. In Agent-Directed Simulation Conference. pp. 47–55.

Civil, Architecture and Environmental Engineering – Kao & Sung (Eds)
© 2017 Taylor & Francis Group, ISBN 978-1-138-02985-9

Field assessment of window daylighting with prismatic glazing

Z. Tian & D. Xue
School of Architecture, Soochow University, Suzhou, China

Y.P. Lei
Suzhou Institute of Building Science Group, Suzhou, China

ABSTRACT: Prismatic glazing has the potential to improve indoor luminance levels and luminous comfort. Through field measurements and the comparison with Desktop Radiance simulation results, this study shows that using prismatic glazing at side windows can improve indoor luminance levels and luminance uniformity, especially for inner spaces. The technology can work effectively in both sunny and cloudy days with better performance in the former. Prismatic glazing systems refract light into inner spaces and thus can provide better luminous comfort environment for building occupants.

1 DAYLIGHTING MEASURES

Daylight is an effective resource to improve the health and energy savings of indoor occupants (HesChong Mahone Group Inc. 2003). In general, there are two methods for daylight: skylight daylight and side window daylight, with the latter having much wider applications. Limited by the aperture parameter factors such as size, shape, and position, providing good daylight for large depth spaces through side-windows presents a challenge.

To improve daylight area, two approaches are often utilized: light shelf on the window or a solar tube on the roof or above the basement. Light shelf can reflect daylight to the ceiling and then reflect to the inner spaces. The solar tube has a similar function to skylight but through more modular structure and intensive light accumulation.

Because of dust accumulation on the panels, it is necessary to maintain routine housekeeping to clean the light shelf so that the light reflective system can work effectively. Meanwhile, adding the light shelf may bring hurdles to elevation maintenance.

Basically, solar tubes are utilized in the top floor or the first floor of basement, and the utilization of solar tubes in other cases is very rare.

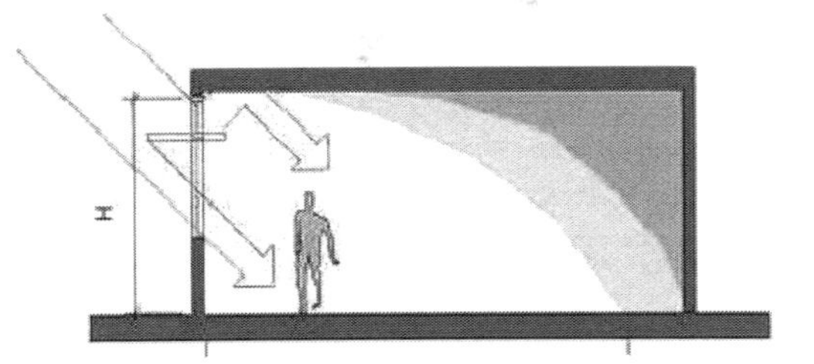

Figure 1. Light shelf on the windows.

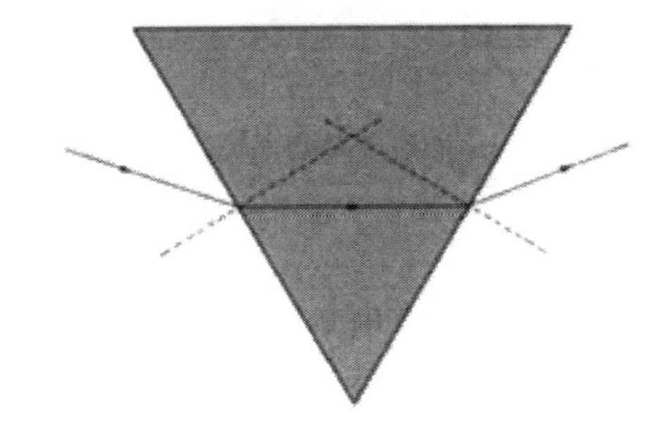

Figure 2. Light transmittance through prismatic film.

2 DAYLIGHT WITH PRISMATIC GLAZING

The mechanism of prismatic glazing is adding a sawtooth prismatic film between double-glazing (Jiangsu HURD, 2013) so that the light passing through prismatic film will be majorly refracted upward to the ceiling (Figure 2) and then reflected downward to the inner spaces. The prismatic film works similarly to the light shelf, but no light will be blocked by the light shelf panels.

Prismatic glazing has wide applications. It can be utilized in elevation side windows to increase daylit areas and spaces as well as to improve luminous comfort by luminance uniformity (Figure 3). Prismatic glazing can also be utilized in skylight in atrium of courtyard to increase the daylit hours (Figure 4).

As a new light directive guidance system, prismatic glazing systems can reduce the excessive daylight close to the window areas by reflect light to the inner spaces and thus can improve the daylit area and luminance uniformity. As light is refracted in the prismatic film, the appearance of prismatic glazing is translucent. In many cases, it is utilized in the clearstory and combined with conventional glazing in the low windows so as to accommodate both the visual and daylighting functions in office buildings (Figure 5).

3 FIELD ASSESSMENT CASE STUDY

The ground floor of Sumin Decoration Company Office Building was chosen as the field assessment study case. The exterior and interior of the building are shown in Figures 6 and 7. The floor plan and field measurement spots grid set according to Liu (2010) and Standard GB/T5699 (GAQSIQ, 2008) are illustrated in Figure 8.

Field measurements are conducted throughout for a year. For simplification, the Spring Equinox, Summer Solstice, Autumn Equinox, and Winter Solstice are the typical times for field measurement (Yun, 2007).

Although the field measurement results were collected through each hour from 9 AM to 5PM, according to the LEED requirement (USGBC, 2009), field measurement results were presented as 9 AM, 12PM, and 15PM for both sunny days and cloudy days in this paper.

4 FIELD MEASUREMENT RESULTS VS. SIMULATION DATA

The field measurement results of luminance levels within the ground floor were compared with Desktop Radiance (LBNL, 2016) simulation data for analysis. The sunny summer season results and cloudy autumn season were presented.

4.1 *Summer sunny day's data*

From Table 1, in sunny days at 9 AM, the exterior direct sunlight is majorly on the east side. The luminance levels on the west side are higher than the simulated ones with conventional glazing, especially at the areas close to west side windows.

At 12PM, the luminance levels on the west side are overwhelmingly higher than simulated data

Figure 3. Prismatic glazing on the side windows (Suzhou linshine optronics building).

Figure 4. Prismatic glazing on the atrium (Kushan Hushi electronics office building).

Figure 5. Combined prismatic and conventional glazing (Shanghai green building technology incubator building).

Figure 6. Exterior of Sumin Decoration Office.

with conventional glazing, especially at the inner spaces where the luminance levels increased from 72% to 81%.

At 3PM, the luminance levels with prismatic glazing have a much better luminance uniformity distribution. Simulated data showed that the luminance may reach 33137 Lux with direct sunlight and strong solar radiation at the space 4 m from west side windows, while only around 278 Lux at the space 20 m from windows. This may cause luminous discomfort and glare issues. In the field measurement, the corresponding luminance levels are 3089 and 490 Lux and thus the luminance contrast is reduced from 119 to 6.3 by using prismatic glazing (the cells in Table 1 marked as "X" mean there is a column so that no measurement was conducted at this point).

Figure 7. Interior of ground floor of Sumin Decoration Office.

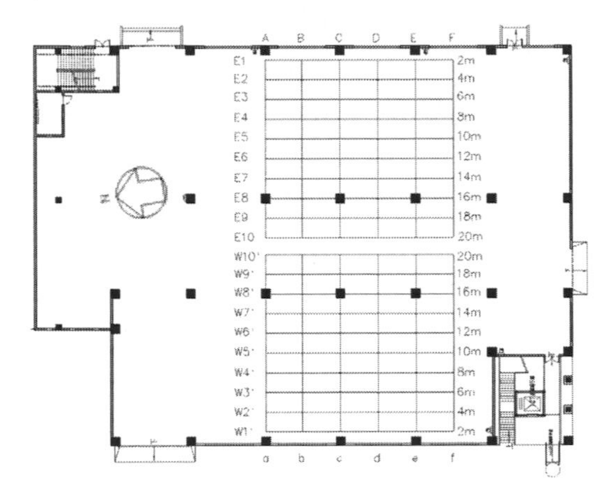

Figure 8. Field measurement spots grid.

4.2 *Autumn cloudy day's data*

The data of autumn season cloudy days are presented in Table 2. The data are field measurement of the luminance levels on the east side of the building with prismatic glazing and simulated data with conventional glazing.

At 9 AM, the luminance levels with prismatic glazing are higher than the data with corresponding conventional glazing, except the data at spots 4 m from window, where the prismatic reflect light to the inner spaces and increase the luminance levels.

At 12PM, at spots 20 m from window with prismatic glazing, the luminance levels are 100% higher than the data with conventional glazing.

At 3PM, the luminance difference with prismatic glazing and conventional glazing is not significant. Although the luminance with prismatic glazing at spaces within 10 m to window

Conventional Glazing (Lux)						Date: July 8		Prismatic Glazing (Lux)					
						Time: 9:00							
341	302	321	279	284	292	W10'	20m	322	352	277	284	320	270
326	276	286	264	252	279	W9'	18m	300	341	253	300	272	254
X	301	X	335	X	263	W8'	16m	X	363	X	320	X	241
272	276	278	280	269	271	W7'	14m	291	379	286	331	283	267
317	328	302	325	304	299	W6'	12m	315	414	290	360	289	274
311	302	312	347	304	284	W5'	10m	358	398	325	404	314	284
372	354	349	343	411	311	W4'	8m	391	440	380	418	350	300
396	419	403	381	471	352	W3'	6m	440	495	408	476	374	363
424	435	420	442	398	427	W2'	4m	580	569	560	713	525	555
342	430	322	420	318	398	W1'	2m	724	1108	582	870	468	729
a	b	c	d	e	f	Points	Distance	a	b	c	d	e	f
						Time: 12:00							
191	194	180	196	213	241	W10'	20m	332	324	334	355	368	420
211	210	185	235	209	238	W9'	18m	310	356	332	367	351	410
X	249	X	225	X	253	W8'	16m	X	374	X	385	X	401
264	258	269	277	278	266	W7'	14m	410	405	418	405	452	427
305	340	290	322	321	295	W6'	12m	450	446	490	449	446	430
331	322	336	388	322	329	W5'	10m	524	538	555	553	524	488
400	386	389	376	460	367	W4'	8m	584	600	645	630	592	565
472	505	462	451	530	443	W3'	6m	720	700	745	745	692	602
573	585	558	573	548	574	W2'	4m	980	1100	930	1090	900	890
414	555	389	560	396	551	W1'	2m	1001	1447	990	1472	860	1360
a	b	c	d	e	f	Points	Distance	a	b	c	d	e	f
						Time: 15:00							
270	278	254	259	249	288	W10'	20m	480	490	488	520	442	430
280	328	317	321	276	307	W9'	18m	500	569	510	541	446	483
X	415	X	399	X	320	W8'	16m	X	626	X	682	X	525
510	495	514	539	470	398	W7'	14m	900	824	836	750	745	600
674	592	632	579	537	475	W6'	12m	972	980	936	875	800	675
753	727	720	736	714	601	W5'	10m	1108	1069	1078	940	922	801
929	896	962	861	832	793	W4'	8m	1327	1256	1138	1044	1054	960
1106	1217	1102	1102	1069	1041	W3'	6m	1403	1456	1290	1300	1158	1070
1300	33137	1283	33053	1258	33017	W2'	4m	1786	1914	1742	1687	1432	1612
891	1073	881	1021	814	965	W1'	2m	1960	3089	1808	2690	1413	2548
a	b	c	d	e	f	Points	Distance	a	b	c	d	e	f

Table 1. Luminance levels with conventional and prismatic glazing (summer sunny days, west side).

Conventional Glazing (Lux)						Date: Nov. 12		Prismatic Glazing (Lux)					
						Time: 9:00							
a	b	c	d	e	f	Points	Distance	a	b	c	d	e	f
184	357	200	349	243	389	E1	2m	362	651	286	702	342	669
469	534	490	536	538	585	E2	4m	326	387	336	397	370	432
309	351	361	360	369	414	E3	6m	232	256	249	244	274	294
214	233	240	287	280	298	E4	8m	209	242	245	242	225	276
164	194	211	198	222	235	E5	10m	207	211	216	210	196	253
121	137	148	162	165	172	E6	12m	186	180	184	194	187	231
124	121	131	135	152	172	E7	14m	164	151	175	177	180	217
X	99	X	111	X	147	E8	16m	X	142	X	165	X	185
72	86	77	101	84	117	E9	18m	122	137	133	138	121	160
77	84	83	85	90	107	E10	20m	114	128	125	126	111	139
						Time: 12:00							
a	b	c	d	e	f	Points	Distance	a	b	c	d	e	f
191	304	195	314	247	369	E1	2m	402	698	427	764	445	689
466	533	543	559	532	609	E2	4m	376	475	382	509	485	550
315	347	367	379	377	408	E3	6m	268	300	298	346	340	350
217	246	248	254	310	293	E4	8m	262	278	284	294	327	326
168	183	189	194	215	241	E5	10m	236	228	256	240	318	317
144	136	151	157	182	192	E6	12m	205	210	237	212	274	326
104	126	138	141	150	174	E7	14m	171	194	229	235	276	258
X	106	X	121	X	141	E8	16m	X	150	X	205	X	230
82	96	85	101	95	137	E9	18m	153	171	194	192	209	233
81	86	86	94	98	101	E10	20m	160	167	175	179	202	219
						Time: 15:00							
a	b	c	d	e	f	Points	Distance	a	b	c	d	e	f
191	304	195	314	247	369	E1	2m	227	320	222	342	259	358
466	533	543	559	532	609	E2	4m	187	232	193	199	209	231
315	347	367	379	377	408	E3	6m	140	143	148	156	180	203
217	246	248	254	310	293	E4	8m	145	134	141	143	200	223
168	183	189	194	215	241	E5	10m	130	132	129	133	167	219
144	136	151	157	182	192	E6	12m	111	111	120	109	159	184
104	126	138	141	150	174	E7	14m	102	108	112	119	156	151
X	106	X	121	X	141	E8	16m	X	104	X	112	X	154
82	96	85	101	95	137	E9	18m	106	105	112	17	123	143
81	86	86	94	98	101	E10	20m	103	114	105	111	114	127

Table 2. Luminance levels with conventional and prismatic glazing (autumn cloudy days, east side).

are lower than the data with conventional glazing, at the inner space 20 m from windows, the luminance levels are still around 30% higher than the luminance with conventional glazing at the east side of the building. Tables 1 and 2 indicate that even prismatic glazing systems have better luminous performance on luminance level and luminance uniformity improvement than conventional glazing in sunny days; the system still work effectively in cloudy days, especially at the inner spaces.

5 CONCLUSIONS

This paper introduced the properties and advantages of prismatic glazing system. By comparing the indoor luminance levels measurement with prismatic glazing to Desktop Radiance simulation results with conventional glazing, we summarized the following features with prismatic glazing:

1. Prismatic glazing can provide better luminance levels and luminance uniformity for interior spaces, especially for inner spaces.
2. Prismatic glazing can avoid direct sunlight with conventional glazing by refracting light into inner spaces and thus can provide better luminous comfort environment for building occupants.
3. Both the incident light angle and intensity will affect the performance of prismatic glazing system for side window daylight.
4. Side window prismatic glazing systems can significantly improve the luminance levels for spaces over 10 m from windows.
5. Prismatic glazing can improve indoor luminance levels for both sunny days and cloudy days while the improvement of luminance levels is more obvious and effective in sunny days.

ACKNOWLEDGMENTS

This work was funded by the Priority Academic Program Development of Jiangsu Higher Education Institutions (PAPD) and Jiangsu Nature Science Foundation (BK20161215).

REFERENCES

GAQSIQ. 2008. *Method of Daylighting Measurements, GB/T5699–2008*. Beijing: Standards Press of China.

HesChong Mahone Group Inc. 2003. Daylighting and Retail Sales[R]. California Energy Commission. Sacremento, CA.

Jiangsu HURD. 2013. *A Guide to Green Building Application Technology*. Nanjing: Phoenix Science Press.

LBNL. Desktop Radiance. www.radsite.lbl.gov/deskrad/

Liu, X. 2010. *Building Physics*. Beijing: China Architecture and Building Press.

USGBC. 2009. *Leadership in Energy and Environmental Design Reference Guide*. US Green Building Council.

Yun, P. 2007. *Lighting Environment Simulation of Building*. Beijing: China Architecture and Building Press.

Civil, Architecture and Environmental Engineering – Kao & Sung (Eds)
© 2017 Taylor & Francis Group, ISBN 978-1-138-02985-9

Mechanical analysis of rockburst considering the "locked in" stress in the driving face of the roadway

Xiao-Bin Yang, Xin-Xing Han, Hao Wang & Tian-Bai Zhou
School of Resource and Safety Engineering, China University of Mining and Technology (Beijing), Beijing, China

ABSTRACT: In the driving face of rock roadway in underground engineering, local dynamic phenomena or dynamic disasters often occur, such as rockburst, pressure bump, shock bump, etc, in which the occurrence of rockburst in the driving face is more frequent. In order to study its formation mechanism, a stress inclusion was assumed to exist in front of the driving face of rock roadway based on the "locked in" stress hypothesis and the mechanical model of driving roadway was simplified under consideration of "locked in" stress. Then based on the basic theory of elastic mechanics, the stress distribution laws of the rock around a stress inclusion under the combined stress of the overlying rock and the inclusion was deduced. By analyzing the stress distribution of the surrounding rock of the inclusion, a conclusion is obtained that when there is a certain distance from the driving head to the inclusion, the surrounding rock would be destroyed along the direction of the maximum principal stress. When the excavation work is close to the inclusion, the rock between destroy weak surfaces will be thrown to the excavation space causing dynamic disasters under the combined stress of the overlying rock and the inclusion and left the crater section of "V" type. The mechanical analysis results of this paper can provide theoretical support for the prevention of rockbursts in the excavation process of underground geotechnical engineering.

1 INTRODUCTION

Local dynamic phenomena or dynamic disasters such as rockburst, pressure bump and shock bump often take place in the driving face of rock roadway of underground engineering, in which, especially, rockburst occurs frequently. Rockburst is one of geological hazards about dynamic instability in the mining process of underground cavity under high geostress, the cave wall will arise stress differentiation due to unloading function caused by the excavation of surrounding rocks, the elastic strain energy stored in rock mass is released suddenly and at the same time the rock mass will generate burst loose, peeling off, ejection and even throwing phenomenon (Xu & Wang 2000). After the occurrence of rockburst, the crater section of "V" type often appears (Tan 1989, Feng et al. 2012, Chen et al. 2012, Zhou et al. 2015). Rockburst seriously affects the construction schedule of underground excavation works, damages the support and construction equipment and endangers the safety of the workers (Wang et al. 2003). However, due to the complexity of the rockburst phenomenon, it is still one of the worldwide difficult problems of rock mechanics (Zhang & Fu 2008). So far, many experts and scholars have studied the occurrence mechanism of rockburst, and made important contributions. Xie (1993) used the method of fractal geometry to study the occurrence mechanism of rockburst. Ge and Lu (1999) simulated rockburst behavior and preliminarily discussed the rockburst phenomenon in the tunnel based on the concept of DDA (Discontinuous Deformation Analysis) and artificial jointing. Chen et al. (2009) simulated the failure process of rock under different conditions by laboratory experiments and put forward the index of judging the occurrence of rockburst from the point of view of energy. Qin et al. (2009) expounded the source, destination and the release reason of the energy of the deformation rockburst from the point of view of energy and put forward the principle of the energy storage of the rock mass. However, these research results can not explain the localization of dynamic disasters about rock and the formation of the pit of "V" type.

Considering the non homogeneity of rock mass and its complicated physical and mechanical environment, the "locked in" stress hypothesis proposed by Academician Tan in 1979 is an important theory about the occurrence of localization, the hypothesis thinks that the "locked in" stress is the cause of rock engineering disasters (Tan 1979, Tan & Kang 1980, Tan & Kang 1991). Academician Qian et al. discussed the existence of "locked in" stress from the point of view of microstructure and deformation of rock (Qian et al. 2004, Qian & Zhou 2013). Academician Wang (2009) also considered that rock contained unreleased sealing stress. Yue (2014, 2015) put forward and tried to demonstrate the fluid inclusion was a specific, real, measurable and computable stress inclusion. He thought the inclusion was a specific form of existence and function of "locked in" stress and strain energy and further analyzed and discussed the fluid inclusions in the rock were the tensile or expanding volumetric force sources of the dynamic disasters of rock in underground engineering.

This paper thought that the inclusion in rock mass led to the occurrence of the localization disaster of rock, the combined function of the stress closed in inclusions and tectonic stress resulted in the local energy concentration in the rock mass. When excavating to a certain distance, the balance of the energy of the inclusions and the bearing capacity of the surrounding rock will be broken, the release of energy will form a dynamic phenomenon or dynamic disaster. Based on these, the paper simplified mechanical model of rock containing an inclusion in the driving face, theoretically deduced the stress distribution of rock mass under the combined action of "locked in" stress and overburden stress and verified the reason for the existence of the pit of "V" type after rockburst.

2 MECHANICAL MODEL OF THE ROCK MASS CONTAINING AN INCLUSION IN THE DRIVING FACE OF THE ROADWAY

In order to explain the localization characteristics of rockburst survival and the occurrence of the pit of "V" type after rockburst, a stress inclusion containing "locked in" stress was assumed to exist in the rock in front of the driving face in the process of advancing the roadway. The mechanical model was simplified under considering the interaction effect of the inclusion and the surrounding rock on the heading face. The stress distribution of rock containing an inclusion in the driving face of roadway was obtained by applying the basic theory of elastic mechanics.

Without consideration of the influence of the driving force and the horizontal stress, the mechanical model of rock mass containing an inclusion in the driving face of rock roadway was simplified as shown in Fig.1. In the model, we symmetrically chose a REV (Representative Elementary Volume) with a certain size containing a inclusion, the size of the REV was much larger than that of the inclusion. In consideration of the infinite length dimension(compared to the width of the roadway) perpendicular to the excavation direction of the roadway, the model was simplified as a plane strain model, the vertical direction of the model was influenced by the vertical stress of the overlying strata and the size of the stress was assumed to be P; The front of the roadway was considered to be infinitely extended along the driving direction and the driving head was a free surface, so it was approximately considered that the rock in front of working face was not affected by horizontal stress. The inclusion in rock mass was regarded as a sphere, which was a circle in the plane problem. The "locked in" stress in the conclusion is uniformly distributed internal pressure, which interacted with the surrounding rock and formed a mechanical model of relative equilibrium. Without taking into account the deformation and failure process of the inclusion, the effect of the "locked in" stress contained in the conclusion on the surrounding rock mass and the effect of the overlying strata on the REV were only considered. For this model, a stress conclusion with a radius of a and uniformly distributed internal pressure of q was assumed to exist in the middle of it; we took the center of the stress inclusion as the coordinate origin of the model, the direction horizontal to the right (the driving side of roadway) as the positive direction of x axis, the vertical downward direction as the positive direction of y axis.

Under consideration of the function of the "locked in" stress in the circular inclusion, the problem of the rectangular coordinate was converted into that of the polar coordinate by using the basic theory of elastic mechanics. The line boundary outside was transformed into the circle one. With the origin O as the center, a big circle with a radius of b much larger than a was made. The stress boundary of the model is: $\sigma_x = 0$, $\sigma_y = -P$, $\tau_{xy} = 0$ (σ_x, σ_y and τ_{xy} are the horizontal stress, vertical stress and shear stress respectively).

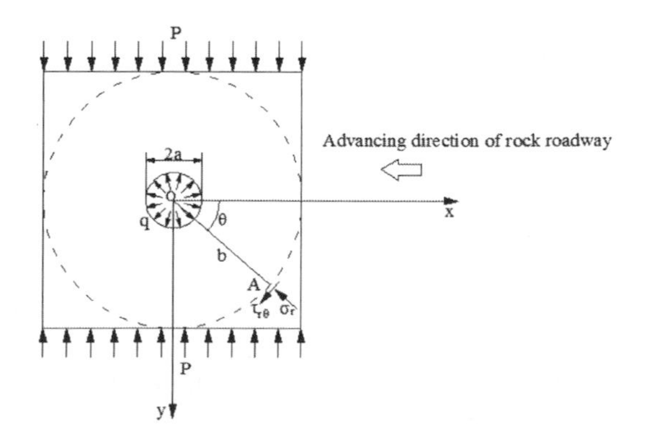

Figure 1. Simplified mechanical model of rock around the inclusion in the driving face of rock roadway.

By using the coordinate transformation formulas, the stress components of A in polar coordinates can be expressed as

$$\begin{cases}\left[\begin{array}{l}(\sigma_r)_{r=b}=\dfrac{\sigma_x+\sigma_y}{2}+\dfrac{\sigma_x-\sigma_y}{2}\cos 2\theta+\tau_{xy}\sin 2\theta= \\ -\dfrac{P}{2}+\dfrac{P}{2}\cos 2\theta\end{array}\right] \\ (\tau_{r\theta})_{r=b}=-\dfrac{\sigma_x-\sigma_y}{2}\sin 2\theta+\tau_{xy}\cos 2\theta=-\dfrac{P}{2}\sin 2\theta\end{cases} \tag{1}$$

where σ_r is the radial stress; $\tau_{r\theta}$ is the shear stress; θ is the polar angle; r is the pole diameter.

The mechanical model in the polar coordinate was changed into a model of thick wall cylinder with a inner diameter of a and a outer diameter of b, the stress boundary condition can be expressed as

$$\begin{array}{l}\text{inner boundary:}\begin{cases}(\sigma_r)_{r=a}=-q \\ (\tau_{r\theta})_{r=a}=0\end{cases} \\ \text{outer boundary:}\begin{cases}(\sigma_r)_{r=b}=-\dfrac{P}{2}+\dfrac{P}{2}\cos 2\theta \\ (\tau_{r\theta})_{r=b}=-\dfrac{P}{2}\sin 2\theta\end{cases}\end{array} \tag{2}$$

According to the stress situation of the model and based on the superposition principle of elastic mechanics, the mechanical model of thick wall cylinder in polar coordinates was split into model I and model II, which were shown in Fig. 2 and Fig. 3 respectively.

The stress boundary condition of the mechanical model I can be expressed as

$$\begin{array}{l}\text{inner boundary:}\begin{cases}(\sigma_r)_{r=a}=-q \\ (\tau_{r\theta})_{r=a}=0\end{cases} \\ \text{outer boundary:}\begin{cases}(\sigma_r)_{r=b}=-\dfrac{P}{2} \\ (\tau_{r\theta})_{r=b}=0\end{cases}\end{array} \tag{3}$$

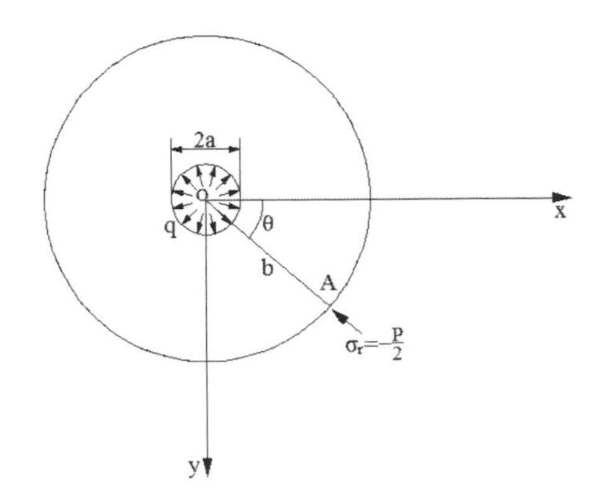

Figure 2. Schematic diagram of mechanical model I.

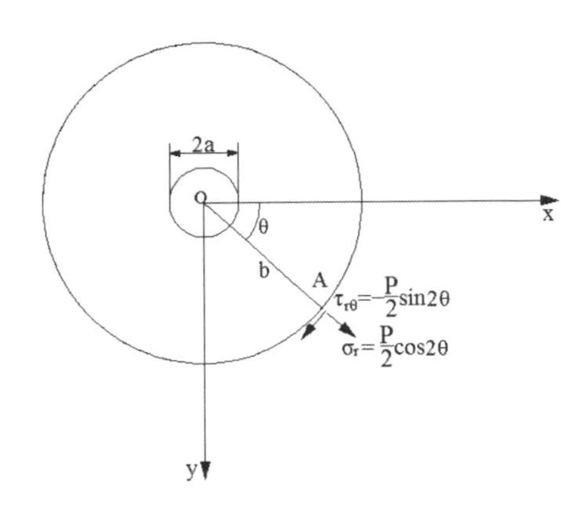

Figure 3. Schematic diagram of mechanical model II.

The stress boundary condition of the mechanical model II can be expressed as

$$\begin{array}{l}\text{inner boundary:}\begin{cases}(\sigma_r)_{r=a}=0 \\ (\tau_{r\theta})_{r=a}=0\end{cases} \\ \text{outer boundary:}\begin{cases}(\sigma_r)_{r=b}=\dfrac{P}{2}\cos 2\theta \\ (\tau_{r\theta})_{r=b}=-\dfrac{P}{2}\sin 2\theta\end{cases}\end{array} \tag{4}$$

3 SOLUTION OF THE MECHANICAL MODEL OF THE DRIVING FACE CONTAINING AN INCLUSION

The model I and II in the polar coordinate abovementioned were solved respectively, and the results were superimposed to obtain the stress situation of the original model.

1. *Solution of the mechanical model I*

The solution of model I is a axisymmetric plane problem, so set stress function as

$$\varphi=A\ln r+Br^2\ln r+Cr^2+D \tag{5}$$

where φ is a stress function; A, B, C, D are undetermined constants.

The stress components can be expressed as

$$\begin{cases}\sigma_r=\dfrac{1}{r}\dfrac{d\varphi}{dr}=\dfrac{A}{r^2}+B(1+2\ln r)+2C \\ \sigma_\theta=\dfrac{d^2\varphi}{dr^2}=-\dfrac{A}{r^2}+B(3+2\ln r)+2C \\ \tau_{r\theta}=\tau_{\theta r}=0\end{cases} \tag{6}$$

where σ_r is the radial stress; σ_θ is the circumferential stress; $\tau_{r\theta}$ and $\tau_{\theta r}$ are shear stress.

The mechanical model I is a axisymmetric model, in which $B = 0$. Substituting $B = 0$ into Eq.(6) yields

$$\begin{cases} \sigma_r = \frac{A}{r^2} + 2C \\ \sigma_\theta = -\frac{A}{r^2} + 2C \\ \tau_{r\theta} = \tau_{\theta r} = 0 \end{cases} \tag{7}$$

Combining Eq.(3) and Eq.(7), the solution can be expressed as

$$\begin{cases} A = \frac{(P-2q)a^2b^2}{2(b^2-a^2)} \\ 2C = \frac{2qa^2 - Pb^2}{2(b^2-a^2)} \end{cases} \tag{8}$$

Substitution of Eq.(8) into Eq.(7) yields

$$\begin{cases} \sigma_r = \frac{(P-2q)a^2}{2\left(1-\frac{a^2}{b^2}\right)}\frac{1}{r^2} + \frac{2q\frac{a^2}{b^2} - P}{2\left(1-\frac{a^2}{b^2}\right)} \\ \sigma_\theta = -\frac{(P-2q)a^2}{2\left(1-\frac{a^2}{b^2}\right)}\frac{1}{r^2} + \frac{2q\frac{a^2}{b^2} - P}{2\left(1-\frac{a^2}{b^2}\right)} \\ \tau_{r\theta} = \tau_{\theta r} = 0 \end{cases} \tag{9}$$

Because of $b >> a$, Eq.(9) can be simplified as

$$\begin{cases} \sigma_r = \left(\frac{P}{2} - q\right)\frac{a^2}{r^2} - \frac{P}{2} \\ \sigma_\theta = \left(q - \frac{P}{2}\right)\frac{a^2}{r^2} - \frac{P}{2} \\ \tau_{r\theta} = \tau_{\theta r} = 0 \end{cases} \tag{10}$$

2. *Solution of the mechanical model II*

Stress function can be assumed as

$$\psi = f(r)\cos 2\theta \tag{11}$$

where ψ is a stress function; r is polar radius; $f(r)$ is only a function of r.

Compatible equation can be written in the following form

$$\nabla^4 \psi = \left(\frac{\partial^2}{\partial r^2} + \frac{1}{r}\frac{\partial}{\partial r} + \frac{1}{r^2}\frac{\partial^2}{\partial \theta^2}\right)^2 \psi = 0 \tag{12}$$

Substituting Eq.(11) into Eq.(12) produces

$$f(r) = Ar^4 + Br^2 + C + D\frac{1}{r^2} \tag{13}$$

where A, B, C, D are undetermined constants.

Substituting Eq.(13) into Eq.(11) yields

$$\psi = \left(Ar^4 + Br^2 + C + D\frac{1}{r^2}\right)\cos 2\theta \tag{14}$$

The stress components can be expressed as

$$\begin{cases} \sigma_r = \frac{1}{r}\frac{\partial \psi}{\partial r} + \frac{1}{r^2}\frac{\partial^2 \psi}{\partial \theta^2} = -\left(2B + \frac{4C}{r^2} + \frac{6D}{r^4}\right)\cos 2\theta \\ \sigma_\theta = \frac{\partial^2 \psi}{\partial r^2} = \left(12Ar^2 + 2B + \frac{6D}{r^4}\right)\cos 2\theta \\ \tau_{r\theta} = -\frac{\partial}{\partial r}\left(\frac{1}{r}\frac{\partial \psi}{\partial \theta}\right) \\ \quad = \left(6Ar^2 + 2B - \frac{2C}{r^2} - \frac{6D}{r^4}\right)\sin 2\theta \end{cases} \tag{15}$$

Combining Eq.(4) and Eq.(15), and make $a/b = 0$, the solution can be expressed as

$$\begin{cases} A = 0 \\ B = -\frac{P}{4} \\ C = \frac{Pa^2}{2} \\ D = -\frac{Pa^4}{4} \end{cases} \tag{16}$$

Substitution of Eq.(16) into Eq.(15) generates

$$\begin{cases} \sigma_r = \frac{P}{2}\left(1 - \frac{a^2}{r^2}\right)\left(1 - \frac{3a^2}{r^2}\right)\cos 2\theta \\ \sigma_\theta = -\frac{P}{2}\left(1 + \frac{3a^4}{r^4}\right)\cos 2\theta \\ \tau_{r\theta} = -\frac{P}{2}\left(1 - \frac{a^2}{r^2}\right)\left(1 + \frac{3a^2}{r^2}\right)\sin 2\theta \end{cases} \tag{17}$$

By adding Eq.(10) and Eq.(17), the stress solution of the original model can be expressed as

$$\begin{cases} \sigma_r = \left(\frac{P}{2} - q\right)\frac{a^2}{r^2} \\ \quad -\frac{P}{2} + \frac{P}{2}\left(1 - \frac{a^2}{r^2}\right)\left(1 - \frac{3a^2}{r^2}\right)\cos 2\theta \\ \sigma_\theta = \left(q - \frac{P}{2}\right)\frac{a^2}{r^2} - \frac{P}{2} - \frac{P}{2}\left(1 + \frac{3a^4}{r^4}\right)\cos 2\theta \\ \tau_{r\theta} = -\frac{P}{2}\left(1 - \frac{a^2}{r^2}\right)\left(1 + \frac{3a^2}{r^2}\right)\sin 2\theta \end{cases} \tag{18}$$

4 RESULT ANALYSIS OF THE STRESS SOLUTION

In the absence of excavation or the working face is still far from the stress inclusion, the stress inclusion and the surrounding rock are in a stable equilibrium state. From the point of view of long-term stability and statistical significance, it is approximately considered that the stress q of the stress inclusion is equal to the vertical stress P of the original rock(Yue 2014, Yue 2015), so Eq.(18) can be simplified as

$$\begin{cases} \sigma_r = -\dfrac{P}{2}\dfrac{a^2}{r^2} - \dfrac{P}{2} + \dfrac{P}{2}\left(1-\dfrac{a^2}{r^2}\right)\left(1-\dfrac{3a^2}{r^2}\right)\cos 2\theta \\ \sigma_\theta = \dfrac{P}{2}\dfrac{a^2}{r^2} - \dfrac{P}{2} - \dfrac{P}{2}\left(1+\dfrac{3a^4}{r^4}\right)\cos 2\theta \\ \tau_{r\theta} = -\dfrac{P}{2}\left(1-\dfrac{a^2}{r^2}\right)\left(1+\dfrac{3a^2}{r^2}\right)\sin 2\theta \end{cases} \tag{19}$$

Considering the stress situation of the rock in the driving face, combining the dynamic phenomenon of rockburst with the spatial failure structures of rock after disasters, based on the stress distribution of rock mass containing an inclusion, the occurring mechanism of the dynamic phenomenon of rock mass was analyzed. This paper used the maximum principal stress criterion to determine whether the rock mass containing an inclusion was broken and the rupture direction.

The calculation formula of maximum principal stress about the surrounding rock containing a inclusion in the polar coordinate can be expressed as

$$\sigma_{\max} = \frac{\sigma_r + \sigma_\theta}{2} + \sqrt{\left(\frac{\sigma_r - \sigma_\theta}{2}\right)^2 + \tau_{r\theta}^{\,2}} \tag{20}$$

Substituting Eq.(19) into Eq.(20) yields

In Eq.(21), σ_{max} is a function of r and θ, The angle of the extreme value of the principal stress of radius r can be solved when $\partial\sigma_{\max}/\partial\theta = 0$. Making $r=ka$, $\partial\sigma_{\max}/\partial\theta = 0$ can be simplified as Eq.(22).

Because Eq.(22) is a complex implicit function of k and θ, the curve of the implicit function when $k\in[1, 20]$ and $\theta[-90°,90°]$ can be obtained by applying MATLAB software. Combining with Eq.(21), the data points of the curve of the maximum principal stress can be found and the relation between the angle of the maximum principal stress and the radius of surrounding rock was shown in Fig.4 through using Origin software. With the center of the stress inclusion as the origin O, The failure trace of surrounding rock close to the side of the mining space ($\theta\in[-90°,90°]$) was shown in Fig.5.

From Fig.4, we can see that the angle of the maximum principal stress of the surrounding rock close to the inclusion($r\leq 1.54a$) is located in the direction of ±90°, with the increase of the radius, the angle θ of the maximum principal stress gradually decreases and the reduced amplitude of the angle gradually decreases until the angle is approximate constant. When there is a certain distance from the driving head to the inclusion body, according to the maximum principal stress criterion, the surrounding rock containing a inclusion will be destroyed along the direction of the maximum principal stress under the combined stress of the overlying rock and the inclusion, the extended and perforative cracks will form a structural section of "V" type in the upper and lower part of the inclusion which was shown in Fig.5.

When the excavation work is close to the inclusion, due to the loss of the binding effect of the rock close to the side of the mining space, the rock between destroy weak surfaces would be thrown to the excavation space causing dynamic disasters under the combined stress of the overlying rock and the inclusion and left the blasting crater of "V" type, which is consistent with the appearance of "V"-shaped section after rockburst in the process of spot excavation.

$$\sigma_{\max} = \left\{\left[\frac{P\cos 2\theta\left(\dfrac{3a^4}{r^4}+1\right)}{4} - \frac{Pa^2}{2r^2} + \frac{P\cos 2\theta\left(\dfrac{a^2}{r^2}-1\right)\left(\dfrac{3a^2}{r^2}-1\right)}{4}\right]^2 + \frac{P^2\sin^2 2\theta\left(\dfrac{a^2}{r^2}-1\right)^2\left(\dfrac{3a^2}{r^2}+1\right)^2}{4}\right\}^{\frac{1}{2}} - \frac{P}{2} + \frac{P\cos 2\theta\left(\dfrac{a^2}{r^2}-1\right)\left(\dfrac{3a^2}{r^2}-1\right)}{4} - \frac{P\cos 2\theta\left(\dfrac{3a^4}{r^4}+1\right)}{4} \tag{21}$$

$$\left\{\left[\sin 2\theta\left(\frac{6}{k^4}-\frac{4}{k^2}+2\right)\right]\left[\frac{1}{2k^2}-\frac{\cos 2\theta}{4}\left(\frac{6}{k^4}-\frac{4}{k^2}+2\right)\right] + \cos 2\theta\sin 2\theta\left(\frac{1}{k^2}-1\right)^2\left(\frac{3}{k^2}+1\right)^2\right\} \Big/ 2\left\{\left[\frac{\cos 2\theta}{4}\left(\frac{6}{k^4}-\frac{4}{k^2}+2\right)-\frac{1}{2k^2}\right]^2 + \frac{\sin^2 2\theta}{4}\left(\frac{1}{k^2}-1\right)^2\left(\frac{3}{k^2}+1\right)^2\right\}^{\frac{1}{2}} + \frac{2\sin 2\theta}{k^2} = 0 \tag{22}$$

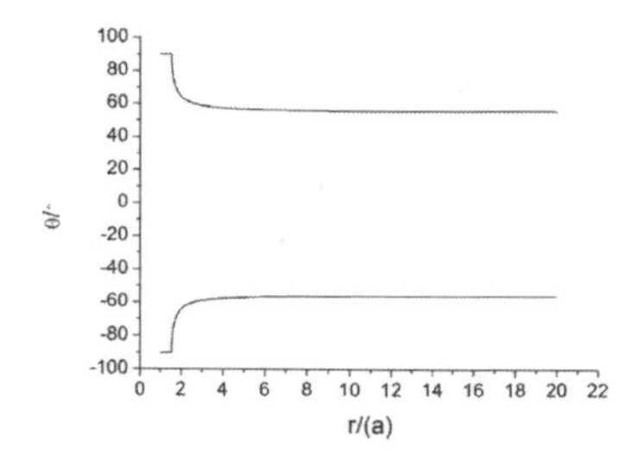

Figure 4. Relation between the angle of the maximum principal stress and the radius of surrounding rock.

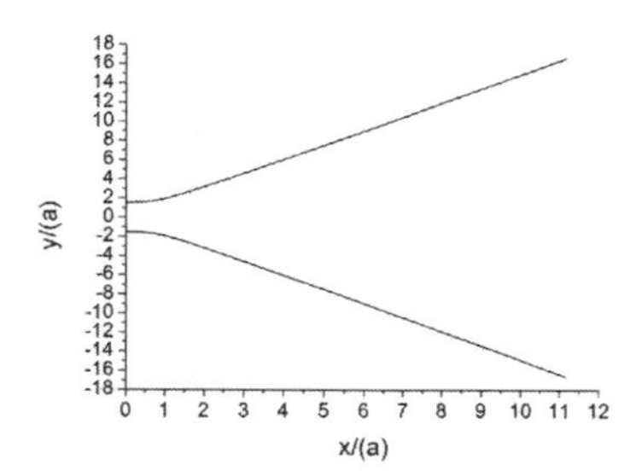

Figure 5. Failure trace line of surrounding rock.

5 CONCLUSIONS

Considering the localization problem of dynamic disasters and the formation of the pit of "V" type after disasters in the excavation process of rock roadway, based on the "locked in" stress hypothesis and the basic theory of elastic mechanics, Through theoretical analysis, the following conclusions were obtained:

1. Assuming that an inclusion is existed in the rock mass in the driving face, combining with the stress situation of the surrounding rock containing a inclusion, the mechanical model of the surrounding rock containing a inclusion is established through theoretical analysis.
2. Based on the basic theory of elastic mechanics, the stress distribution law of the surrounding rock containing an inclusion is obtained by theoretical derivation.
3. By applying the maximum principal stress criterion, the distribution law of the maximum principal stress and the destruction direction and trajectory about the surrounding rock containing an inclusion are gained, which explain the formation of the blasting crater of "V" type after the occurrence of the dynamic disaster.

ACKNOWLEDGEMENT

The authors are grateful to Natural Science Foundation of China (No. 50904071, 51274207), the Fundamental Research Funds for the Central Universities (2010QZ01) for the financial support to this work.

REFERENCES

Chen, B.R., Feng, X.T., Ming, H.J., et al., 2012. Evolution law and mechanism of rockburst in deep tunnel: time delayed rockburst. *Chinese Journal of Rock Mechanics and Engineering* 31(3): 561–569.

Chen, W.Z., Lv, S.P., Guo, X.H., et al., 2009. Reserch on unloaning pressure tests and rockburst criterion based on energy theory. *Chinese Journal of Rock Mechanics and Engineering* 28(8): 1530–1540.

Feng, X.T., Chen, B.R., Ming, H.J., et al., 2012. Evolution law and mechanism of rockbursts in deep tunnels: immediate rockburst. *Chinese Journal of Rock Mechanics and Engineering* 31(3): 433–444.

Ge, D.Z., Lu, J.F., 1999. Discontinuous numerical simulation of the rockburst behavior. *Chinese Journal of Rock Mechanics and Engineering* 18(Supp.): 936–944.

Qian, Q.H., Qi, C.Z., Wang, M.Y., 2004. Rock dynamics under strong loading// Wang Sijing. Century achievements of China rock mechanics and engineering. *Nanjing: Hehai University Press.*

Qian, Q.H., Zhou, X.P., 2013. Effects of incompatible deformation on failure mode and stress field of surrounding r*ock mass. Chinese Journal of Rock Mechanics* and Engineering 32(4): 649–656.

Qin, J.F., Zhuo, J.S., 2009. Study on the energy storage ability criterion in a kind of rockburst and applic*ation. Chinese Journal of Computational Mechanics* 26(3): 318–323.

Tan, T.K., 1979. Vice-president address note// Proceedings of Congress on Rock Mechanics of International Society for Rock Mechanics. *Montreux, Suisse (Switzerland).* 3: S253–254.

Tan, T.K., Kang, W.F., 1980. Lock in stress, creep and dilatancy of rocks and constitutive equation. *Rock Mechanics and Rock Engineering* 13(1): 5–22.

Tan, T.K., Kang, W.F., 1991. On the locked in stress, *creep and dilatation of rocks and th*e constitutive equations. Chinese Journal of Rock Mechanics and Engineering 10(4): 299–312.

Tan, Y.A., 1989. The mechanism researc*h of rockburst. Hydrogeology and Engineering Geo*logy (01): 34–38.

Wang, Q.H., Li, X.H., Gu, Y.L., et al., 2003. Rockburst hazard and its forcast and treatments in underground engineering. *Journal of Chongqing University* 26(7): 116–120.

Wang, S.J., 2009. Geological nature of rock and its deduction for rock mechanics. *Chinese Journal of Rock Mechanics and Engineering* 28(3): 433–450.

Xie, H.P., 1993. Fractal characteristics and mechanism of rocks. *Chinese Journal of Rock Mechanics and Engineering* 12(1): 28–37.

Xu, L.S., Wang, L.S., 2000. Study on the rockburst type classification. *Journal of Geological Hazards and Environment Preservation* 11(3): 245–247.

Yue, Z.Q., 2014. Gas inclusions and their expansion power as foundation of rock "locked in"stress hypothesis. *Journal of Engineering Geology* 22(4): 739–756.

Yue, Z.Q., 2015. Expansion power of compressed micro fluid inclusions as the cause of rockburst. *Mechanics in Engineering* 37(3): 287–294.

Zhang, J.J., Fu, B.J., 2008. Rockburst and its criteria and control. *Chinese Journal of Rock Mechanics and Engineering* 27(10): 2034–2042.

Zhou, H., Meng, F.Z., Zhang, C.Q., et al., 2015. Effect of structural plane on rockburst in deep hard rock tunnels. *Chinese Journal of Rock Mechanics and Engineering* 34(4): 720–727.

Civil, Architecture and Environmental Engineering – Kao & Sung (Eds)
© 2017 Taylor & Francis Group, ISBN 978-1-138-02985-9

Study on the identification of major safety defects in port engineering

Jianhua Peng
China Academy of Transportation Sciences, Beijing, China

ABSTRACT: At present, no regulation or standard has been developed for the identification of major safety defects in port engineering in China, as well as there is no common cognition for the identification of major safety defects in port engineering, which varies widely. In this paper, three research methods (accident deduction, risk assessment of safety defects, and expert consultation) are proposed for the identification of major safety defects in port engineering, which provide technical support to the development of standard for the identification of major safety defects in port engineering. By using the above three methods, the list of major safety defects in port engineering is put forward.

1 INTRODUCTION

The impacts of different safety defects on the safe production are not the same. In order to distinguish them and to find the focus of prevention and control of safety defects, it is usually needed to classify the safety defect and identify the major safety defect. However, there is no regulation or standard developed for the identification of major safety defect in port engineering in China, and there is no common cognition on the identification of major safety defect in port engineering, which varies widely. To a certain extent, it has affected the investigation and management of safety defect of port engineering in China. Therefore, it is necessary to study the identification method of major safety defect and propose the list of major safety defects.

2 DEFINITION OF MAJOR SAFETY DEFECT IN PORT ENGINEERING

From the viewpoint of accident statistic data of China port engineering in recent years (see Table 1), the number of deaths per year is generally no more than 10. Major safety defect is often defined as the safety defect, which may cause more than 10 deaths. Therefore, the definition of major safety defect in port engineering needs to be different from the former.

With reference to classification of safety defect in the "Interim provisions on the management of safety defect", major safety defect in port engineering can be defined as the safety defect for which hazard is great, rectification is difficult, all or part of the construction should stop, and rectification needs a period of time. Great hazard refers to the death or significant property damage. Difficult rectification refers to the rectification requiring high costs or a long time.

Table 1. Accident statistic data of China port engineering (2005–2013).

Year	Number of accidents	Number of deaths
2005	2	2
2006	6	8
2007	3	3
2008	2	2
2009	4	4
2010	2	2
2011	3	3
2012	1	8
2013	1	2

3 METHOD FOR THE IDENTIFICATION OF MAJOR SAFETY DEFECT IN PORT ENGINEERING

3.1 *Accident deduction*

An accident is caused by safety defect. By means of accident deduction, analyzing the cause of accidents (major accidents) is an effective method to identify the major safety defect. Select 24 port engineering accidents (2005–2013) to analyze their causes and perform statistical analysis of deductive results. Distribution of safety defect in port engineering is shown in Table 2.

In addition, both highway engineering and port engineering are traffic construction projects, which are more common in safety management. Therefore, select some typical highway engineering major accidents to analyze their causes, such

Table 2. Distribution of safety defect in port engineering (24 port engineering accidents).

Type of safety defect in port engineering	Occurrence frequency
Workers do not use protective equipment	7
Violation of operating procedures	6
Violation in commanding	4
Risk-taking operation	3
Absent responsibility of the guardianship	3
Employees do not receive safety education and training	2
Special operation personnel do not obtain the certificate of operation qualification	2
Absent responsibility of safety technical disclosure	2
No regular safety check	2
Wire rope defect	2
Operate or stay under lifting object	2

as Dongjiashan tunnel project "12.22" major gas explosion accident, Tuojiang bridge "8.13" special major collapse accident, Jianshan bridge collapse accident, and Meliet interchange ramp bridge falsework collapse accident. According to the accident deduction data, the main safety management defects are shown below:

- Disordered safety management;
- Illegal subcontracting;
- Absent responsibility of supervision personnel;
- Without a license;
- Without training;
- Unauthorized alteration of construction scheme;
- Unreasonable schedule;
- Absent responsibility of supervisor.

3.2 *Risk assessment of safety defect*

Risk assessment of safety defect is a direct research method for the identification of major safety defect. Risk assessment of safety defect from two aspects of the possibility of the accident and the possible consequence is expressed as:

$$R = f(p,s) \tag{1}$$

where R is the risk of safety defect; p is the possibility of the accident; and s is the possible consequence.

The specific risk assessment methods include risk matrix, LEC, and so on. The risk matrix model of safety defect is shown in Figure 1.

Safety defect in the first quadrant, which can easily lead to an accident and huge loss, can be identified as the major safety defect. Of course, this applies only to a rough analysis. It is more scientific to draw the risk contour lines on the "risk matrix". The upper right area of contour in Figure 1 is the distribution of major safety defect.

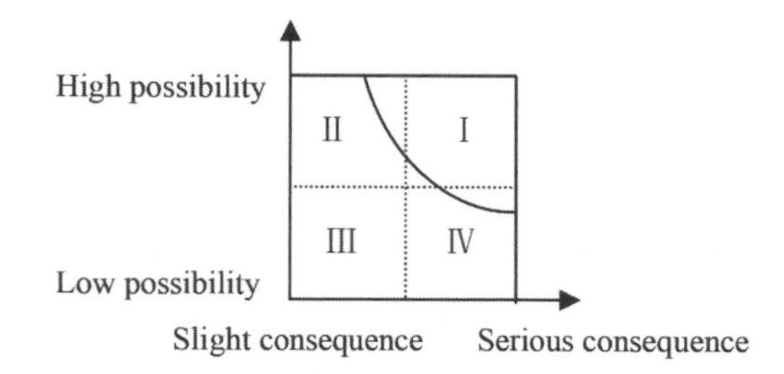

Figure 1. Risk matrix model of safety defect.

On the basis of the risk of safety defect, considering the difficulty of the rectification, the major safety defect is determined as:

$$M = f(p,s,d) \tag{2}$$

where M is the decision value of safety defect; p is the possibility of the accident; s is the possible consequence; and d is the difficulty of the rectification.

At this point, the risk matrix is extended to 3D risk map, where p, s, and d are located in x, y, and z axes, respectively, and risk contour lines are extended to contour plane.

However, considering the three indicators make it is difficult to determine the major safety defect, a simplified method is used to determine the risk of safety defect (R) through coupling of p and s first and then the decision value of safety defect (M) through coupling of R and d (see Equation 3):

$$M = f(p,s,d) = f(p \times s,d) = f(R,d) \tag{3}$$

where M is the decision value of safety defect; p is the possibility of the accident; s is the possible consequence; d is the difficulty of the rectification; and R is the risk of safety defect.

Rating criteria of p, s, and d are shown in Table 3.

The numerical distribution of $R\ (p,s)$ is shown in Table 4.

Classification criteria of $R\ (p,s)$ is shown in Table 5.

Taking into account the correlation between the accident and safety defect, the risk of safety defect should be priority to the rectification difficulty of safety defect. Therefore, when the risk of safety defect is very high, regardless of the difficulty of rectification, it should be regarded as a major safety defect. The risk matrix of R and d is shown in Table 6.

The lower right corner of Table 6 shows the distribution of major safety defect, including 10 types

Table 3. Rating criteria of p, s, and d.

Numerical value[1]	p[2]	s[3]	d[4]
1	Very low possibility	Slight consequence	Very easy
2	Low possibility	Commonly consequence	Easy
3	Medium possibility	Great consequence	Medium
4	High possibility	Severe consequence	Difficult
5	Very high possibility	Extraordinarily severe consequence	Very difficult

Notes: [1]Values 1, 2, 3, 4, and 5 refer to different degrees of state. [2]Very low possibility = less than 10%; Low possibility = 10–30%; Medium possibility = 30–70%; High possibility = 70–90%; Very high possibility = more than 90%. [3]Slight consequence = minor injury but not be in hospital, or property damage below 20,000 RMB; Common consequence = minor injury and be in hospital, or property damage 20,000–100,000 RMB; Great consequence = seriously injured but not disabled, or property damage 100,000–200,000 RMB; Severe consequence = seriously injured and disabled, or property damage 200,000–500,000 RMB; Extraordinarily severe consequence = death or property damage more than 500,000 RMB. [4]Very easy = rectification time less than 3 days, or rectification fund less than 20,000 RMB; Easy = rectification time 3–7 days, or rectification fund 20,000–50,000 RMB; Medium = rectification time 7–15 days or rectification fund 50,000–100,000 RMB; Difficult = rectification time 15–30 days or rectification fund 100,000–200,000 RMB; Very difficult = rectification time more than 30 days or rectification fund more than 200,000 RMB.

Table 4. Numerical distribution of $R\ (p,s)$.

	s				
p	1	2	3	4	5
1	1	2	3	4	5
2	2	4	6	8	10
3	3	6	9	12	15
4	4	8	12	16	20
5	5	10	15	20	25

Table 5. Classification criteria of $R\ (p,s)$.

Numerical value	R (Folding point)	Description of risk
(1,2,3,4)	(1')	Very low risk
(5,6,8,9,10,12)	(2')	Low risk
(15,16)	(3')	Medium risk
(20)	(4')	High risk
(25)	(5')	Very high risk

Table 6. Risk matrix of R and d.

	d				
R	1	2	3	4	5
1'	—	—	—	—	—
2'	—	—	—	—	—
3'	—	—	—	3' × 4	3' × 5
4'	—	—	4' × 3	4' × 4	4' × 5
5'	5' × 1	5' × 2	5' × 3	5' × 4	5' × 5

Table 7. Determination of major safety defect (an example).

Safety defect	Without use of personal protective equipment (safety belt) in the process of aerial work	
$p = 5$ (very high possibility)	$s = 5$ (extraordinarily severe consequence)	d=1 (very easy)
$R\ (p \times s) = (25)/(5')$		$d = 1$ (very easy)
$M = (R, d) = 5' \times 1$		

of state: (3' × 4), (3' × 5), (4' × 3), (4' × 4), (4' × 5), (5' × 1), (5' × 2), (5' × 3), (5' × 4), and (5' × 5). Take the safety defect "Without use of personal protective equipment (safety belt) in the process of aerial work" as an example. It can lead to an accident easily, so the possibility of the accident is very high and the value of p is 5. It can lead to casualties, the possible consequence is extraordinarily severe, and the value of s is 5. It is easy to rectify, the value of d is 1. Its decision value in the major safety defect determination is shown in Table 7.

M is located in the area of major safety defect. Therefore, the safety defect "Without use of personal protective equipment (safety belt) in the process of aerial work" is determined as a major safety defect.

3.3 *Expert consultation*

Through consulting with experts in the field of port engineering construction and safety production extensively and understanding their views on the major safety defect in port engineering, the list of major safety defect is revised constantly. In the application of this method, designing some questionnaires or conducting in-depth communication on specific issues can ensure that expert consultation is more efficient.

4 LIST OF MAJOR SAFETY DEFECT IN PORT ENGINEERING

By using the above three methods, the list of major safety defect in port engineering (Table 8) is put forward.

Table 8. List of major safety defect in port engineering.

Type	Major safety defect
Human unsafe behavior	1. Without use of personal protective equipment in aerial work, water operation, etc.
	2. Violation of operating procedures or violation in commanding in water and underwater operation, blasting operation, etc.
	3. Risk-taking operation, such as working at height when the wind degree is greater than 6
Unsafe physical condition	1. No or improper protective measures
	2. Defect of equipment, facilities, accessories, and material
	3. No obvious safety warning signs are set up.
Adverse environmental conditions	1. Inadequate lighting in excavation and support of foundation pit, lifting operation, etc.
	2. The construction resident set in debris flow area, landslide, etc.
Defects in management	1. Disordered safety management
	2. The safety management personnel of the construction unit has not passed the examination
	3. Special operations personnel do not obtain the certificate of operation qualification
	4. Illegal subcontracting
	5. Unauthorized alteration of construction scheme
	6. Unreasonable schedule
	7. The safety production cost is less than 1.5% of the engineering cost
	8. Absent responsibility of safety technical disclosure

5 CONCLUSIONS

1. From the viewpoint of accident statistic data of China port engineering in recent years and with reference to classification of safety defect in the "Interim provisions on the management of safety defect", major safety defect in port engineering is defined.
2. Three research methods (accident deduction, risk assessment of safety defect, and expert consultation) for the identification of major safety defect in port engineering are proposed.
3. By using the above three methods, the list of major safety defect in port engineering is put forward, including four types of human unsafe behavior, unsafe physical condition, adverse environmental conditions, and defects in management.

REFERENCES

China Academy of Transportation Sciences (CATS). 2007. *Research on key technologies of highway and waterway engineering*. Beijing: CATS.

China Academy of Transportation Sciences (CATS). 2008. *Study on supervision mode and measures of safety production of traffic construction project*. Beijing: CATS.

Peng, Jianhua. 2008. Identification on hazard and safety defect of construction project. *Journal of Safety Science and Technology* 4(4):126–128.

Zhang, Jinglin & Cui, Guozhang. 2002. *Safety system engineering*. Beijing: Coal Industry Press.

Civil, Architecture and Environmental Engineering – Kao & Sung (Eds)
 ISBN 978-1-138-02985-9

Investigating the impact of greenery on the driver's psychology at a freeway tunnel portal

D.Q. Xiao, Z.W. Shen & X.C. Xu
School of Civil Engineering and Mechanics, Huazhong University of Science and Technology, Wuhan, Hubei, China

ABSTRACT: The great environment difference inside and outside freeway tunnels is the major cause for a higher accident rate than that in other tunnel sections. To assess the psychology of drivers because of load and distract factors form the traffic environment, investigation and analysis of heart rate of drivers was carried out in this paper. By using the illumination meter and dynamic electrocardiograph, based on a mass of experimental data, we found the quantitative relationship between the greenery form and the drivers' heart rate changes. The results showed that: following the deepening process of greenery level, illuminance gap narrowed and greenery on the tunnel portals can effectively reduce the illumination difference inside and outside the tunnel. Drivers are mentally released because of the greening settings when traveling in a tunnel section.

1 INTRODUCTION

With the rapid development of freeway construction in China, traffic accidents occur everywhere especially at the tunnel portal. According to statistics in China, the frequency of accidents near the entrance and exit of tunnels is much higher than that in the interior parts, as the amount of accidents that happened within 200–400 m from the entrance accounts for 70% of the total (Ma Z.L. et al. 2009). This type of crash tends to be more severe, increasing the probability of fatalities or incapacitating injuries.

The cause of freeway tunnel accidents is extremely complex and can be generally summarized to be caused by three factors: the driver, the vehicle, and the environment. From a subjective aspect, the driver, as an operator of the vehicle, a road user, and an experiencer of the environment, is the primary factor affecting the traffic safety. The vehicle condition and outer environment stimulation are also important for a secure trip.

From the viewpoint of objective conditions, freeway tunnel is narrow and relatively closed. Compared with the general freeway sections, the main variation of tunnel sections is the sudden change in illumination. Significant contrast between darkness and brightness increases the visual and psychological burden on drivers. In the daytime, drivers need to adapt to sudden darkness when entering tunnels and sudden brightness when leaving tunnels. At night, drivers need to suffer from bright adaptation when entering and dark adaptation when leaving tunnels. To overcome this visual adaption issue, in addition to narrowing the illumination gap inside and outside the tunnel by tunnel lighting, reasonable greenery configuration can also play an important buffering role in sudden change of light; however, this method was rarely applied in tunnel landscape.

Therefore, the aim of this study is to investigate the impact of greenery on driver's psychology located at freeway entrance/exits. With the implementation of illumination meter and dynamic electrocardiograph into abundant tests, the driver's physiological and psychological indicators and environmental indicators were achieved; the impact of greenery on drivers' psychology was investigated via quantitative analysis methods; and the application of greenery to reduce tunnel accidents was discussed.

2 LITERATURE REVIEW

The variations of physiological and psychological characteristics of drivers describe the impact of environment changes on drivers when approaching and leaving the tunnels. Many studies have investigated this problem all over the world.

Initially, Groeger and Rothengatter (1998) studied the cognitive process and social psychology of drivers. Byung Chan Min et al. (2002) found that subjects feel tension under the high-speed driving condition, and as the speed of a car increased, the sympathetic nervous system of passengers became more highly activated. Hoogendoorn (2010) evaluated the mental load of drivers in the perception process of traffic accidents with the application of the driver's physiological indices such as breathing and heart rate. Bucchi et al. (2012) investigated the construction process of driver personality with psychological methods. In this way, the authors provided useful indications for the design of safe roads.

Other research efforts have been made to tunnel lighting and road greening. The lighting specification of China (2000) stipulated detailed parameters of luminosity of tunnels in daytime but ignored the fact that drivers would suffer from different visual disturbances between day and night. International Commission on Illumination (2004) published "Guide for the Lighting of Road Tunnels and Underpasses" in 2004. Kircher and Ahlstrom (2012) investigated how driving performance was influenced by tunnel design factors and proposed that tunnel design and illumination have some influence on the drivers' behavior, but visual attention given to the driving task is the most crucial factor. Wei et al. (2012) explored the relationship between night lighting and driving safety in long express tunnels. On the basis of real road experiments and fuzzy assessments, this study showed that night luminosity was too high and should be controlled below 9 Lx.

In China, tunnel entrance and exit are regarded as black points of traffic accident. Wang et al. (2010) analyzed the illumination design in tunnels, especially at entrance and exit. Du et al. (2007, 2013) investigated the pupillary change of the driver at the entrance and exit of highway tunnel. The purpose of these studies was to analyze the phenomena of visual turbulence during light and dark adaption and the conversion duration for visual turbulence. From the viewpoint of content and methods, this study attributed pupil area variation completely to drivers' nervousness and ignored the impact of illumination and dark adaptation time. Most of the developed countries have their own technical standards on highway greening design in highway engineering. In 1965, the United States enacted the Highway Beautification Act, and Japanese government issued the Greening Technology Benchmark in 1976. As mentioned in Flexibility in Highway Design (2000) published by the federal highway administration, an important aspect of highway landscape design is the handling of the trees. The AASHTO (1996) put forward three suggestions about landscape design and plant configuration, the first and most important one was that drivers should benefit from the beauty and safety vegetation.

To sum up, previous studies have showed that the driver's visual adaptation difficulty caused by sharp variation of luminance is the prime reason of accident black spot at tunnel entrance. Although some of them mentioned the improvement of light environment by greening, the support of driving experiments is lacked (Pan X.D. et al. 2001, Xiao D.Q. 2011, Zhang Q & Chen Y.R. 2005).

3 METHODOLOGY

3.1 *Experiment design*

Because of the particularity of tunnel sections and the differences in tunnel illumination, length, and design, drivers will suffer visual adaption generated by physical and psychological reactions.

3.1.1 *Vehicle selection*

We select Volvo XC-60 as the experimental vehicle, which can fully meet the experimental demand for its performance. Besides, drivers are familiar with the vehicle's condition, which could lower the possible heart rate fluctuation caused by the experimental car.

3.1.2 *Tunnel selection*

The experimental tunnel sections belong to the Hurong Freeway in Hubei Province of China (also known as Huyu Freeway, Hubei). About 30 typical tunnels were selected, whose lengths are between 186 and 9,000 m. The speed limits of these tunnels range from 60 to 80 km.h^{-1}. All the tunnels are one-way traffic without interference from vehicles in the other way, and all the measurement was done in a normal traffic flow.

3.1.3 *Participants*

Two male drivers, aged 34 and 22, with different driving experience and a corrected visual acuity above 5.0 were selected. Both of them are not familiar with the experiment sections.

3.1.4 *Apparatus*

In this experiment, Qingxin L8 Dynamic Electrocardiograph (Shenzhen Qingxin Electronics co., LTD.) is selected to record drivers' physiological data (mainly heart rate data) when traveling through tunnel sections, UNIT UT382 Series Illumination Meter produced by UNI-TREND Technology (China) Limited for illumination data, and Driving Recorder for the driving environment.

3.2 *Experimental measurement factors*

3.2.1 *Tunnel illuminance*

The illumination device is capable of storing and analyzing data in real time; moreover, it can transfer data to a PC for analysis by specialized software UT382 Interface Program. The experiment was carried out in daytime on a cloudy day.

Illuminance Coefficient here is defined to assess the illuminance variation, and the specific equation is as follows:

$$\Delta I = \frac{I_{out} - I_{in}}{T} \tag{1}$$

where ΔI is the illuminance Coefficient; I_{out} is the mean illumination value over time interval (5 s) before the vehicle entered the tunnels; I_{in} is the mean illumination value over time interval (5 s) after the vehicle entered the tunnels; and T is a time interval (10 s) before and after the vehicle entered the tunnels.

3.2.2 *Driver's electro cardio data*

During the driving cognitive process, the driver would generate psychological and physiological changes due to external environmental stimuli. When the environment is relatively simple and the information density is low, the driver will be in a relatively relaxed state with a regular heart rate. Contrarily, the driver will be more intense and the heart rate is relatively high under a complex environment. Therefore, continuous heart rate detection can be used to assess the workload of drivers.

Compared with Heart Rate (HR), Heart Rate Growth (HRG) is a more practical index to evaluate drivers' psychological changes. The specific calculation is shown in the following formula:

$$HRG = \frac{HR1 - HR2}{HR2} \times 100\% \quad (2)$$

where HRG is the heart rate growth; HR1 is the heart rate when driving; and HR2 is the heart rate under normal condition. This paper focused on the impact on drivers when driving in a particular tunnel road, so the formula (1) can be improved as follows:

$$N_j = \frac{n_j - \bar{n}}{\bar{n}} \times 100\% \quad (3)$$

where N_j is the heart rate growth when driving in moment j (%); n_j is the heart rate when driving in moment j (bpm); and $\bar{n}$ is the average heart rate when driving in the freeway (bpm).

4 DATA COLLECTION AND PROCESSING

The experimental tunnels were classified into three types to perform a better analysis: nongreenery tunnels, unilateral greenery (or high cut) tunnels, and bilateral greenery (or high cut) tunnels. Thus, we selected 12 eligible tunnels in three different greenery forms from 30 experiment tunnels.

Average Heart Rate Growth (AHRG), used to describe the impact of the greenery, is defined as the average heart rate growth during this specified 20 s: 10 s before and after the moment when the HRG reaches a peak. The definition of AHRG made the heart rate changes of drivers more obvious and easier to describe during the adaption period.

4.1 *Illumination variation*

In general, the illuminance outside the tunnel can reach tens of thousands of lux, but only dozens of lux inside on a sunny day. Three representative tunnels in different forms were selected. The luminosity variation pattern could be seen in Figure 1. Graph (a) is the illumination variation curve of nongreenery tunnel, graph (b) belongs to unilateral

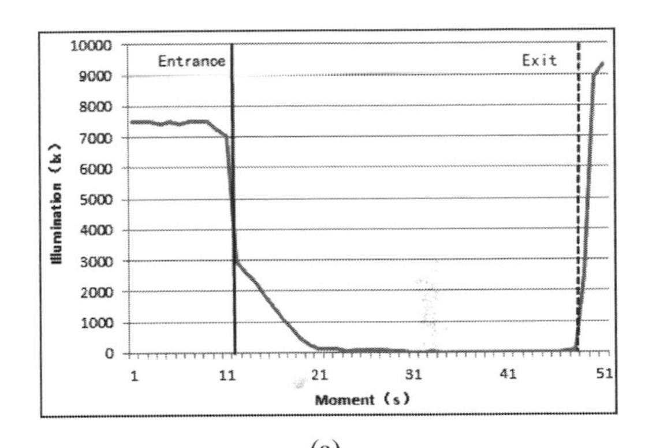

(a)

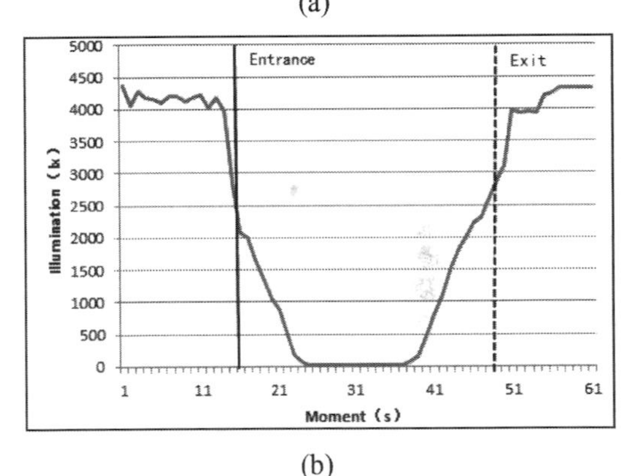

(b)

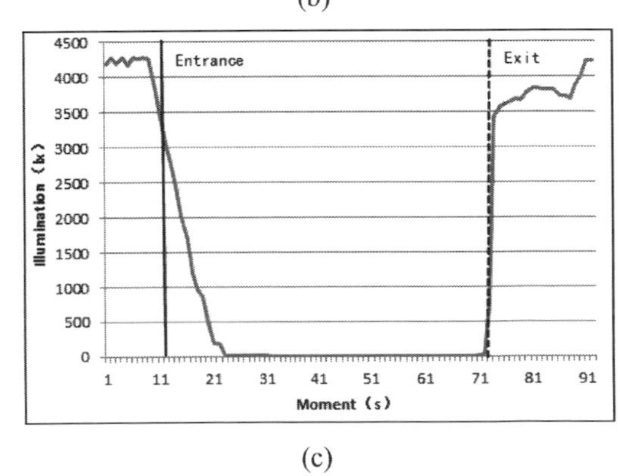

(c)

Figure 1. Illumination variation curve of three types of tunnel.

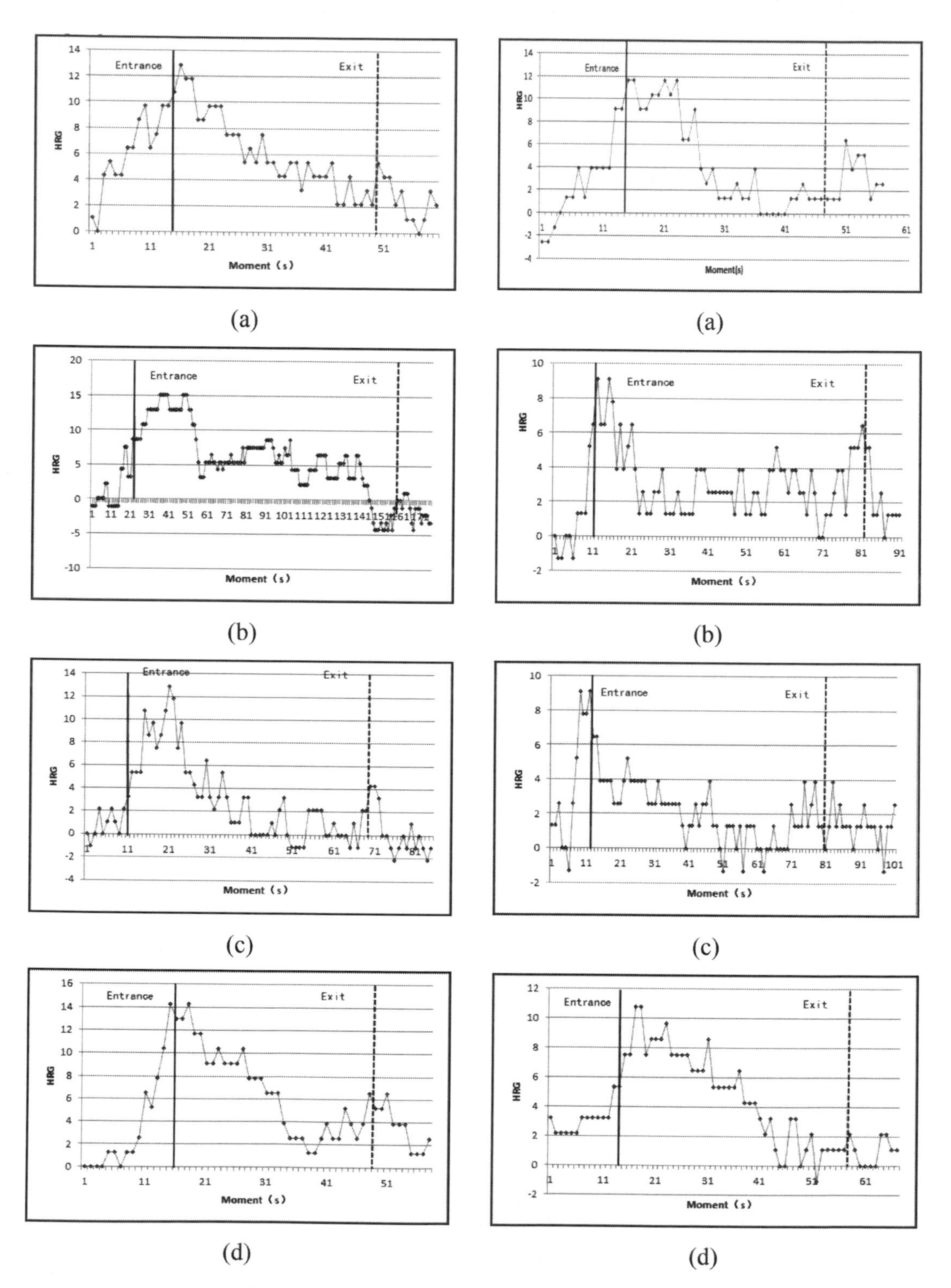

Figure 2. Drivers' psychology variation in nongreenery tunnels.

Figure 3. Drivers' psychology variation in unilateral greenery tunnels.

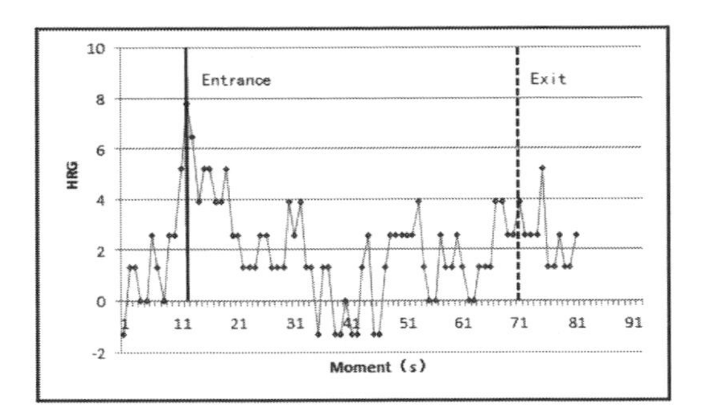

(a)

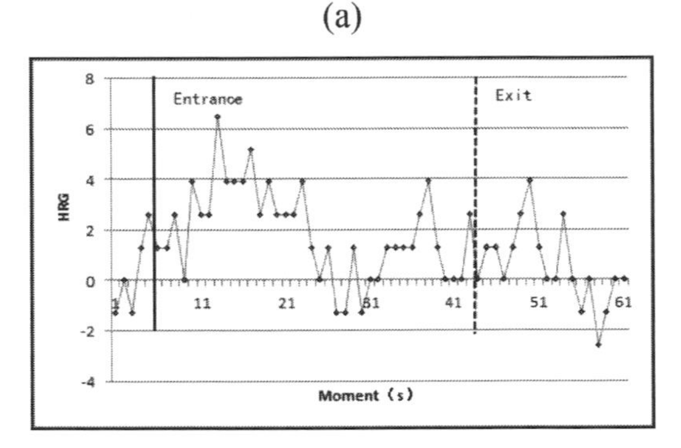

(b)

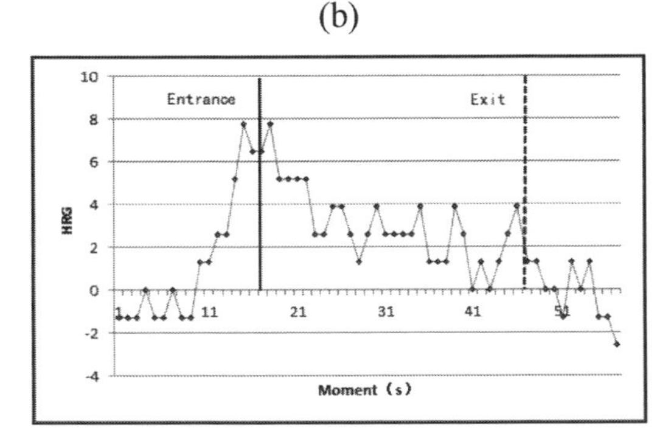

(c)

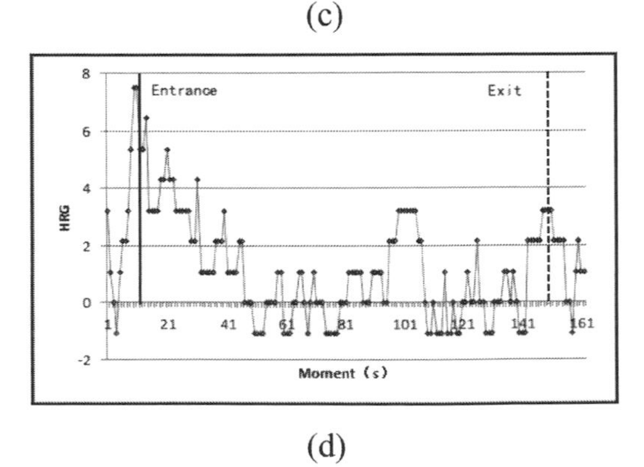

(d)

Figure 4. Drivers' psychology variation in bilateral greenery tunnels.

greenery tunnel, and graph (c) corresponds to bilateral greenery tunnel.

Seen from the three graphs above, all of the three tunnels have the similar illumination variation. The luminosity inside the tunnel was much lower than that at both ends. The illumination was enhanced in entrance and exit, which could reach to several thousands.

4.2 *Drivers' psychological variation*

We selected 12 eligible tunnels in three different greenery forms and drew the diagram on how drivers' heart rate changes over time. As shown in Figures 2, 3, and 4, the solid line represents the location of the tunnel entrance, and the dotted line represents the tunnel exit. Average heart rate of driver A is 77 bpm, whereas that of driver B is 93 bpm. Graphs (a) and (b) show the heart rate variation curves of driver A, and graphs (c) and (d) show the corresponding figures of driver B.

Driver's heart rate increase rapidly when approaching the tunnel entrance and then reaching a HRG peak (Figs. 2, 3, and 4). After leaving the tunnel, the HRG also reached a peak, which is lower than the first peak when entering. The peak HRG of three types of tunnels varies from each other. For nongreenery tunnels, all of the peak HRG values are over 12% (Fig. 2). For unilateral greenery tunnels, the peak HRG is more than 8% but less than 12% (Fig. 3) and for bilateral greenery tunnels the peak HRG is more than 6% but less than 8% (Fig. 4).

5 RESULTS

Illuminance coefficient of different types of tunnels can be figured out and exhibited in Table 1. By comparison, we found that the illuminance coefficient decreases with the development of greenery at entrance, particularly in unilateral greenery tunnels where the illuminance factor drops by 56.6% compared with nongreenery tunnel; in the bilateral greenery tunnel, the illuminance factor declines by 68.3%.

The AHRG indicators are figured out in Table 2. As the greenery develops, both drivers' AHRG show a downward trend, drivers are mentally released when traveling in a tunnel section with better greening. Besides, while approaching the entrance of a tunnel, the drivers' heart rate rises rapidly and levels out after a peak; interestingly, the heart rate increases less near the exit. There is a tendency in each curve that the HRG descends over time. It indicates that drivers are more intense at entrance than at exit, as dark adaption is more difficult than light adaption.

Table 1. Illuminance coefficient of different tunnel forms.

Tunnel form	Luminance outside I_{out} (lx)	Luminance inside I_{in} (lx)	Luminance coefficient ΔI
Nongreenery	7361	2247	511.4
Unilateral greenery	3846	1625	222.1
Bilateral greenery	4023	2400	162.3

Table 2. Drivers' AHRG indicator when travelling in tunnels.

Tunnel form	Driver	Tunnel length (m)	AHRG (%)
Nongreenery	A	707	9.03
		2722	6.34
	B	1141	4.36
		672	8.05
Unilateral greenery	A	286	7.21
		1182	4.09
	B	1570	3.96
		890	6.61
Bilateral greenery	A	1080	3.18
		751	2.99
	B	724	3.38
		900	3.55

6 CONCLUSION AND FUTURE WORK

This study investigated the impact of greenery freeway tunnel on driver's psychology, and statistics of heart rate and environmental illumination at tunnel portals has been recorded and quantitatively analyzed. With the implementation of illumination meter and dynamic electrocardiograph into abundant tests, we collected the driver's physiological and psychological indicators and environmental indicators, achieved the effect of studying the impact of greenery on drivers' psychology via quantitative analysis methods, and discussed the possibility of the application of greenery to reduce the number of tunnel accidents. The following conclusions are drawn:

1. Greenery in tunnel portals can decrease the illuminance variances, moderate the illuminance difference, and help drivers be easier to adapt to dark and light variation; with the deepening of greenery, the tunnel illuminance coefficients of three greenery forms decreased gradually; the illuminance factor of nongreenery tunnel has reduced 68.3% compared with the data of bilateral greenery tunnel.
2. Greenery in portal dramatically alleviate drivers' nervousness and ensure traffic safety; with the greenery promoting, the heart rate of two drivers rise less, especially in bilateral greenery tunnel, where the AHRG has fallen by 52.1% averagely. On the basis of the drivers' psychological analysis in three different greenery tunnel, we could learn that greenery at portal can effectively ease the anxiety of drivers through the tunnel, thereby improving traffic safety.

In this paper, the information of illumination and psychology in expressway tunnels was obtained according to Chinese traffic condition and characteristics of drivers. The conclusion was verified by sufficient driving tests in Hurong Freeway. In fact, on the one hand, the greenery in expressway tunnel entrance and exit is not only a type of road landscape design or ecological restoration design, but it can also play a role like sunshade or light-abatement grille, which could remove dark and light adaptation. For example, under the premise of not shielding the sightline of driving, tall and leafy arbors with a large crown diameter should be planted in a higher density in the tunnel entrance and exit sections. The planting space and height change in a transitionary way, that is, along the direction of the vehicle travelling toward the tunnel, gradually narrowing planting space and improving planting height. On the other hand, green plants can relieve tension while driving, thus making the driver feel safer. Therefore, for the sections that could make drivers tense and nervous such as the tunnel sections, greenery should be used to eliminate or reduce the fear and anxiety of drivers to a certain extent, thus improving traffic safety.

However, it would be better to include other traffic information such as speed, flow, and lane-changing behaviors because illumination and greening may not be the only reason for increasing the heart rate. And as a result, we should have compared crashes with greening to find out how greening affects crash occurrences. Therefore, in future studies, we will conduct more tests to take other factors, such as speed, flow, and position of the sun with respect to the tunnel entrance, into account.

ACKNOWLEDGMENT

This work was supported by the Natural Science Foundation of China (No. 51308242).

REFERENCES

Alberto B, Sangiorgi C, & Vignali V. 2012. Traffic Psychology and Driver Behavior. *Procedia—Social and Behavioral Sciences* 53:973–980.

Byung C.M, Soon C.C, Se J.P, Chul J.K, Mi-Kyong S, & Kazuyoshi Sakamoto. 2002. Autonomic responses of young passengers contingent to the speed and driving mode of a vehicle. *International Journal of Industrial Ergonomics* 29:187–198.

CIE 88: 2004, *Guide for the Lighting of Road Tunnels and Underpasses*, International Commission on Illumination, Vienna, Austria.

Du Z.G., Pan X.D, & Guo X.B. 2007. Visual adaptation index for driving safety at entrance and exit of highway tunnel. *Journal of South China University of Technology (Natural Science Edition)* 35(7):15–19.

Du Z.G, Huang F.M, Yan X.P, & Pan X.D. 2013. Light and dark adaption time based on pupil area variation at entrance andexit Areas of highway tunnel. *Journal of Highway and Transportation Research and Development* 30(5):98–102.

Du Z.G, Pan X.D, Yang Z, & Guo X.B. 2007. Research on visual turbulence and driving safety of freeway tunnel entrance and exit. *China Journal of Highway and Transport* 20(5):101–105.

Flexibility in Highway Design. 2000. Washington D.C., USA: Federal Highway Administration (FHWA).

Groeger J.A, & Rothengatter J.A. 1998. Trafficpsychology and behavior. *Transportation Research Part F*1:1–9.

JTJ 026. 2000. 1-1999 *Specifications for Design of Ventilation and Lighting of Highway Tunnel*. Beijing: People's Education Press.

Kircher K, & Ahlstrom C. 2012. The impact of tunnel design and lighting on the performance of attentive and visually distracted drivers. *Accident Analysis and Prevention* 47:153–161.

Ma Z.L, Xu C.F, & Zhang S.R. 2009. Characteristics of traffic accidents in Chinese freeway tunnels. *Tunnelling and Underground Space Technology* 24:350–355.

Pan X.D, Gotou Jun'ichi, & Makoto Y. 2001. Studies on Evaluation of forest roadssurface by driver's psychological and physiological responses. *Applied Forest Science* 10(2):27–30.

Raymond H, Hoogendoorn S.P, Brookhuis K, & Daamen W. 2010. Mental Workload, Longitudinal Driving Behavior, and Adequacy of Car-Following Models for Incidents in Other Driving Lane. *Transportation Research Record* 2010:64–73.

The Roadside Design Guide. 1996. Washington D.C., USA: American Association of State Highway and Transportation Officials (AASHTO).

Wang Y.Z. Guo Y, & Liao Z.G. 2010. Safety analysis for illumination design at tunnel entrance and exit. *Inproceedings of the international conference on intelligent computation technology and automation (ICICTA '10)* 3:255–259.

Xiao D.Q. 2011. *Study on optimization of freeway greening engineering based on three objectives* [PhD Thesis]. Changan: Changan University.

Zhao W.H, Chen H, Yu Q, Liu H.X, & Wang W.H. 2012. Study on optimization of night illumination in expressway long tunnels. *Discrete Dynamics in Nature and Society* 2012 November: 1–9.

Zhang Q, & Chen Y.R. 2005. Analysis of relations between planting and safety of high-type highways. *Highway* 2005(11):202–207.

Civil, Architecture and Environmental Engineering – Kao & Sung (Eds)
© 2017 Taylor & Francis Group, ISBN 978-1-138-02985-9

Modeling two-dimensional rubble mound breakwater using dolos at armor layer and geotube at the core layer

O.C. Pattipawaej & H.J. Dani
Civil Engineering Department, Universitas Kristen Maranatha, Bandung, Jawa Barat, Indonesia

ABSTRACT: Coastal protection structure is used to protect the coast against erosion that is usually determined by the availability of materials at or near the job site, the sea conditions, water depth, and the availability of equipment for the implementation of the work. The purpose of this study is to conduct experimental research in the laboratory to obtain the rubble-mound breakwater design optimum and tools before it is applied directly in the field to reduce the risk of failure of construction. This study is conducted for laboratory testing of a two-dimensional rubble-mound breakwaters model with dolos at armor layer and geotube at core layer using regular wave and three variations of slope in front of the structure facing seaward, that is, 1:1.5, 1:2, and 1:2.5. The water level varies for the conditions of non-over-topping, over-topping, and submerged. The two-dimensional rubble-mound breakwaters model with the slope in front of the structure facing seaward 1:2.5 shows the most stable rubble-mound breakwater model.

1 INTRODUCTION

Indonesia as an archipelago country does not escape the impact of global warming due to the sea level rise. The impact of sea level rise results in the coastline change due to erosion and/or abrasion. In addition to global warming, forces of nature such as waves, currents, tides, and construction errors caused major damage to the existing breakwater. In general, construction fault lies in the tilt angle of the building, a thick layer, and heavy rock, in addition to other factors. This can lead to the collapse of the breakwater.

Breakwater is a structure on the coast to protect the coast against erosion. The type of breakwater structure used is generally determined by the availability of materials in or near the work site, the sea conditions, water depth, and the availability of equipment for the implementation of the work. Rubble-mound breakwaters use structural voids to dissipate the wave energy. Rock or concrete armor units on the outside of the structure absorb most of the energy, while gravels or sands prevent the wave energy's continuing through the breakwater core.

Construction of a breakwater on the field will take considerable time and cost. A laboratory test using a physical model is proved to be effective in avoiding such costs before constructing breakwater in the field. In this study, we will discuss the optimization of two-dimensional model of the rubble-mound breakwater using dolos at the armor layer and geotube at the core layer, because it has a high permeability that balances the water pressure. Wave run-up will be included in determining the peak elevation breakwater.

2 COASTAL PROTECTION STRUCTURE

Coastal structures are intended to protect shoreline or navigation channels from the effects of waves and other hydrodynamic force (Fith, et al., 2014). Design of coastal structures must consider a range of wave heights and periods combined with water level variations. The most common types of coastal structures are breakwaters. Breakwater is a structure constructed on coasts as part of coastal protection from the effects of both weather and longshore drift. Breakwater reduces the intensity of wave action in inshore waters and thereby reduces coastal erosion. Breakwaters may also be small structures designed to protect a gently sloping beach and placed at a distance of 1–300 feet offshore in relatively shallow water. Breakwaters can be constructed with one end linked to the shore, in which case they are usually classified as sea walls; otherwise, they are positioned offshore from as little as 100 m up to 300–600 m from the original shoreline. The types of breakwaters are rubble-mound structures, impermeable sloping structures, vertical wall structures, composite

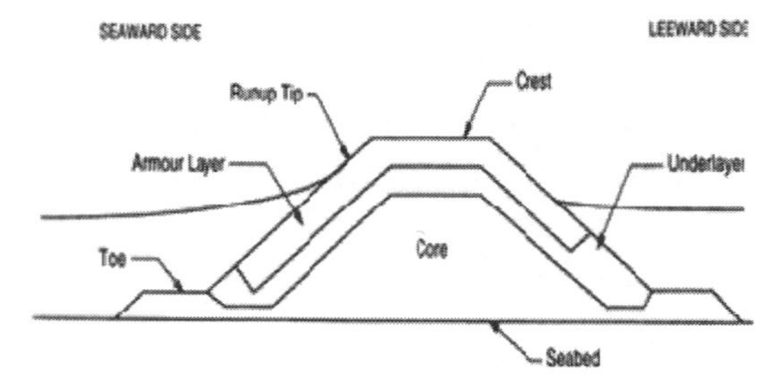

Figure 1. Cross section of a typical rubble-mound breakwater.

structures, floating structures, and pneumatic and hydraulic breakwaters (Hughes, 1993).

Rubble-mound structure is used extensively for breakwaters, jetties, revetments, seawalls, and wave absorber. Rubble-mound structures have a core of quarry-run stones, sand, slag, or other suitable materials protected from waves action by one or more stone under layers and a core layers of relatively large, selected quarry stones or specially shaped concrete armor units (Hudson, 1959). Rubble-mound breakwaters use structural voids to dissipate the wave energy. Rock or concrete armor units on the outside of the structure absorb most of the energy, while gravels or sands prevent the wave energy's continuing through the breakwater core. The integrity of a rubble-mound structure is primarily a function of the stability of the individual armor units that form the seaward face of the structure. In shallow water, revetment breakwaters are usually relatively inexpensive. As water depth increases, the material requirements and, hence, costs increase significantly. Figure 1 shows a typical cross section of a rubble-mound breakwater.

3 STABILITY OF RUBBLE-MOUND BREAKWATER

An important aspect in the design of a rubble mound is its stability to wave attack. Three aspects of the effect of waves on rubble-mound breakwater are wave run-up, overtopping, and transmission (Figure 2). Run-up is defined as the vertical height above still water level to which waves incident upon a structure can be expected to travel up the face of the structure. Wave run-up is important in defining both the amount of wave energy transmitted over and through permeable rubble-mounds and also the quantity of water that may be expected to overtop the structure (Hur, Lee, and Cho, 2012).

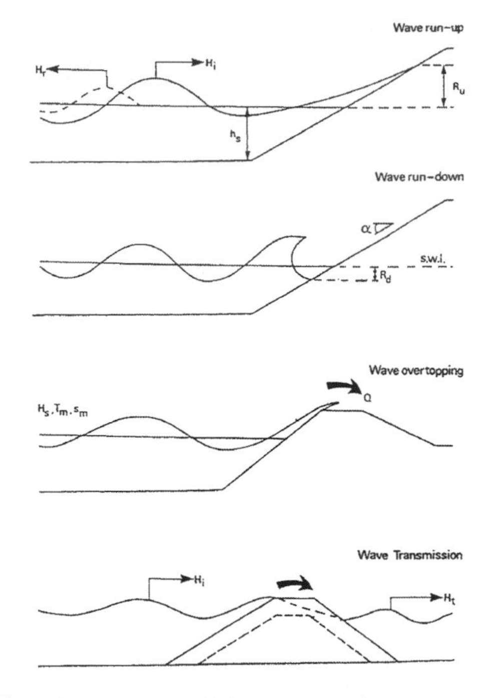

Figure 2. Waves on rubble-mound breakwater.

4 EXPERIMENT SETUP AND PROCEDURE

The experimental test was conducted on the wave flume at Laboratory Balai Pantai, Ministry of Public Works in Buleleng, Bali, Indonesia. The wave flume is 40 m long, 0.6 m wide and 1.2 m high. It is equipped with a piston-type wave generator that produces regular wave. Figure 3 shows three wave probes that are installed in the wave flume.

Simulation of the wave generated by the wave maker uses regular wave model. The model of rubble-mound breakwater is constructed with the seaward face at slopes 1:1.5, 1:2, and 1:2.5 using an armor layer composed of individual units of weight W and density ρ. The armor units use dolos with the average weight of 265 g, relative density of 2.209 g/cm^3, and K_d of 8. The dolos is randomly placed at the armor layer (CIRIA, 2007). The core layer is set to be stable by using geotube. Three variations of slopes facing the seaward side of the breakwater model are applied: 1:1.5, 1:2, and 1:2.5. The wave period is 2 s. The wave height for different slopes of breakwater models can be seen in Table 1.

The water level tests are conducted at four different water levels: 38 cm (mean low water level), 54 cm (mean high water level), 63 cm (crest elevation), and 70 cm (submerged). The structure is assumed to be founded on an impermeable bed, and the duration of exposure to waves is not considered.

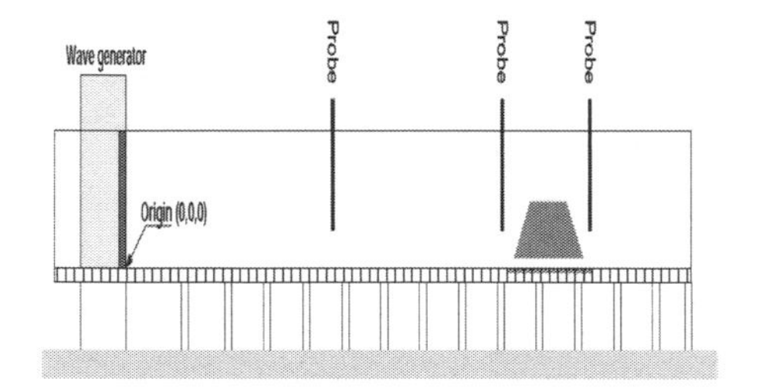

Figure 3. Side view of the flume.

Table 1. Wave heights at different slope of breakwater model.

Slope	Wave height (cm)
1:1.5	12.42
1:2	13.66
1:2.5	14.72

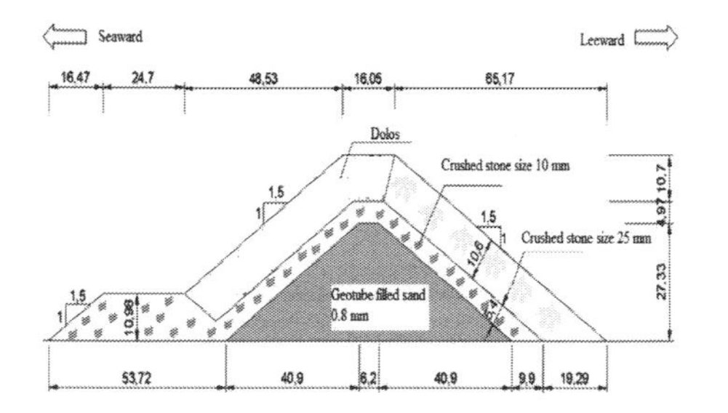

Figure 4. Side view of rubble-mound breakwater model with slope of 1:1.5.

The rubble-mound breakwater model is designed on the basis of the stability of rubble-mound structure using the Hudson formula (Hudson, 1959). The similarity of the rubble-mound breakwater model is 1:10. Figure 4 presents the side view of the rubble-mound breakwater model with the slope of 1:1.5.

5 RESULT AND DISCUSSION

Wave direction, either normally incident or oblique, refers to the direction of wave travel with respect to the breakwater axis. Tables 2–4 present the results of wave height measured using zero upcrossing statistic in front of rubble-mound structure model with slopes 1:1.5, 1:2, and 1:2,5 facing the seaward side, respectively. The highest wave height occurred at crest elevation.

The significant wave height, runup, and rundown at 38 cm of water elevation can be seen in Table 5. It can be concluded that $H_{1/3}$ and $Ru - R_d$ at the mean low water level show the lowest values at the slope 1:2.5 of the rubble-mound breakwater model. It means that the highest energy of wave absorption occurred at the slope of 1:2.5 facing the seaward side of the rubble-mound breakwater model.

Table 2. Wave heights based on zero upcrossing statistics in front of rubble-mound structure with slope 1:1.5 facing the seaward side.

Water level (cm)	H_{max} (cm)	$H_{1/3}$ (cm)	$H_{1/10}$ (cm)	$H_{average}$ (cm)
70	18.0	15.5	16.7	14.4
63	27.0	25.8	26.3	23.9
54	16.2	15.4	15.9	14.1
38	15.5	13.7	14.5	11.5

Table 3. Wave heights based on zero upcrossing statistics in front of rubble-mound structure with slope 1:2 facing the seaward side.

Water level (cm)	H_{max} (cm)	$H_{1/3}$ (cm)	$H_{1/10}$ (cm)	$H_{average}$ (cm)
70	18.5	16.2	16.7	13.4
63	19.9	18.5	19.1	16.7
54	14.3	14.0	14.1	13.1
38	15.9	14.5	15.0	13.0

Table 4. Wave heights based on zero upcrossing statistics in front of rubble-mound structure with slope 1:2.5 facing the seaward side.

Water level (cm)	H_{max} (cm)	$H_{1/3}$ (cm)	$H_{1/10}$ (cm)	$H_{average}$ (cm)
70	17.9	15.7	16.3	14.9
63	18.1	17.3	17.7	16.0
54	17.6	16.7	17.1	15.4
38	13.8	12.6	13.1	11.3

Table 5. Significant wave height, runup, and rundown at 38 cm of water elevation.

Format of slope	$H_{1/3}$ (cm)	Runup, R_u (cm)	Rundown, R_d (cm)	$R_u - R_d$ (cm)
1:1.5	13.7	32.0	8.5	23.5
1:2	14.5	28.0	8.0	20.0
1:2.5	12.6	20.0	6.0	14.0

When the water level is 54 cm or at the mean high water level, only rundown occurred. The rundown for each different slope of the rubble-mound breakwater model is shown in Table 6. The wave absorber gives the highest value for the slope of 1:2.5 for the rubble-mound breakwater model.

The elevation of the crest should be minimum at which overtopping occurs. This should be based on maximum wave runup (Palmer and Christian,

Table 6. Rundown at the mean high water level.

Format of slope	Rundown, R_d (cm)
1:1.5	14.5
1:2	17.0
1:2.5	22.0

Table 7. Wave transmission coefficient at the water level 70 cm.

Slope	H_t(cm)	H_i (cm)	K_t
1:1.5	13.1	15.5	0.85
1:2.0	9.0	16.2	0.56
1:2.5	8.3	15.7	0.53

1998). Unfortunately, the overtopping at 63 cm of water elevation cannot be recorded due to the instrument's problem during the experiment.

The wave transmission coefficient (K_t) is defined as the ratio of the transmitted wave height (H_t) at the leeside to the incident wave height (H_i) at the breakwater seaward, as follows:

$$K_t = \frac{H_t}{H_i} \tag{1}$$

The value of K_t indicates the effectiveness of a rubble-mound breakwater at submerged condition to attenuate waves (Yuliastuti and Hashim, 2011, Zhang and Li, 2014). The value of K_t varies between 0 and 1. A value of zero implies that there is no transmission, whereas 1 means no reduction in wave height (there is no barrier in front of the wave). The result of wave transmission coefficient is presented in Table 7 at the water level of 70 cm (submerged).

On the basis of the result of wave transmission coefficient, slope 1:2.5 of the rubble-mound breakwater model can reduce the wave height more significantly than the slopes 1:1.5 and 1:2 of the breakwater model.

6 CONCLUSIONS

The function of rubble-mound breakwater is to protect the coastal area. Because the rubble-mound breakwater needs to be stable, the two-dimensional physical model of rubble-mound breakwater is conducted using three different slopes, 1:1.5, 1:2, and 1:2.5, facing the seaward side. The water levels are applied at the mean low water level, the mean high water level, the crest elevation, and submerged using the flume with regular wave. The rubble-mound breakwater model is designed on the basis of stability of this breakwater using the Hudson formula. The armor units of the rubble-mound breakwater model are dolos placed randomly facing the seaward side. The core is set stable using geotube. The wave absorber, transmission wave, and displacement of armor units give the best result for the slope of 1:2.5 of the rubble-mound structure model.

The results of two-dimensional physical testing for this rubble-mound breakwater model will certainly be helpful in making the optimum slope of rubble-mound breakwater facing the seaward side in coastal areas of Indonesia with erosion and/or abrasion condition. The results of the two-dimensional rubble-mound breakwater model will be very useful if it can proceed to the next stage of the three-dimensional physical test by using regular and irregular waves. In addition, test results of physical model of this structure can also improve the existing rubble-mound breakwater, especially damaged rubble-mound breakwater.

ACKNOWLEDGMENTS

The authors thank DIPA Kopertis Wilayah IV, Ministry of Research, Technology, and Higher Education, the Republic of Indonesia, for providing the grant accordance with the Letter Agreement of Implementation Research Grant No. DIPA-023.04.1.673453/2015, on November 14, 2014, and the first revision on March 3, 2015. They also thank Balai Pantai, Ministry of Public Works, for providing the research facilities to undertake the research.

REFERENCES

CIRIA, CUR, CETMEF. 2007. *The Rock Manual*. The use of rock in hydraulic engineering, 2nd edition, C683, CIRIA, London.

Fith, L.B., et al. 2014. Between a rock and hard place: Environmental and engineering considerations when designing coastal defense structure. *Coastal Engineering*. 87:122–133.

Hudson, R.Y. 1959. Laboratory Investigation of Rubble mound Breakwaters. *Journal of the Waterways and Harbors Division, American Society of Civil Engineers* 85(WW3): 93–121.

Hughes, S.A. 1993. Physical Models and Laboratory Techniques in Coastal Engineering. *Advanced Series on Ocean Engineering* 7. River Edge, NJ: World Scientific Publishing Co. Pte. Ltd.

Hur, D.S., Lee, W.D. and Cho, W.C. 2012. Characteristics of wave run-up height on a sandy beach behind dual-submerged breakwaters. *Ocean Engineering* 45: 38–55.

Palmer, G., and Christian, C.D. 1998. Design and construction of rubble mound breakwaters. *IPENZ Transactions*, 25(1).

Yuliastuti, D.I. and Hashim, A.M. 2011. Wave transmission on submerged rubble mound breakwater using L-blocks. *2nd International Conference on Environmental Science and Technology IPCBEE* 6, Singapore: IACSIT Press.

Zhang, S.X. and Li, X. 2014. Design formulas of transmission coefficients for permeable breakwaters. *Water Science and Engineering* 7(4): 457–463.

Civil, Architecture and Environmental Engineering – Kao & Sung (Eds)
© 2017 Taylor & Francis Group, ISBN 978-1-138-02985-9

Shear stress calculation of rubber asphalt overlay and stress-absorbing layer

Jizong Tan
Guangxi Key Laboratory of Road Structure and Materials, Nanning, China
Guangxi Transportation Research Institute, Nanning, China

Jianguo Wei
Guangxi Key Laboratory of Road Structure and Materials, Nanning, China
Changsha University of Science and Technology, China

Chan Pan
Guangxi Key Laboratory of Road Structure and Materials, Nanning, China
Guangxi Transportation Research Institute, Nanning, China

Kaixi Huang
Guangxi Communications Professional Technology Institute, Nanning, China

ABSTRACT: By using BISAR mechanical calculation software to calculate and analyze the rubber asphalt compound pavement shear stress of different interlayer adhesive states and under the condition of stress-absorbing layer thickness, we know that the adhesive states of stress-absorbing layer, rubber asphalt surface layer, and underlying cement pavement have a direct impact on the distribution of shear stress of asphalt overlay and stress-absorbing layer, and the maximum shear stress of asphalt layer bottom will first decrease and then increase with the increase of friction parameters. Therefore, the bottom layer should have good adhesion to prevent the premature entry of shear stress growth period.

1 PREFACE

The stress-absorbing layer is arranged on the general cement concrete deck and asphalt surface layer, which can effectively disperse the stress concentration of cement concrete joints or cracks under load effect in order to delay or prevent the appearance of reflective crack of asphalt pavement. The commonly used stress-absorbing layer types mainly include the layer distribution method stress absorbing layer (such as SAMI), modified asphalt mixture stress absorbing layer, the compound stress absorbing interlayer, geosynthetics, and synchronous macadam transition layer, among which the first three have better impermeable, crack resisting, and bonding properties, whereas the effect of the latter two in the practical application is not significant. The cost of modified asphalt stress absorbing layer and compound stress absorption interlayer engineering is high, and the difficulty of construction is relatively high. By contrast, the layer distribution method stress absorbing layer is relatively more economical, reasonable, and convenient for construction. Rubber powder-modified asphalt with high elasticity, toughness, and high deformation is more applicable to the layer distribution method stress absorbing layer. In addition, the rubber powder-modified asphalt stress absorbing layer is also an effective measure of waste tires.

The layer distribution method for rubber powder-modified asphalt stress absorption layer is widely used in compound pavement engineering; however, some technical problems still remain to be solved, including setting rubber powder-modified asphalt binder stress-absorbing layer asphalt overlay design theory and method of structure, the rubber asphalt binding material to improve the technical performance, and the construction technology and other aspects. In light of this, the above rubber asphalt stress-absorbing layer shear characteristics are studied in this paper, in order to further improve the technology of rubber powder-modified stress absorbing layers.

2 ANALYSIS OF THE INTERLAMELLAR SHEAR STRESS

The combination form of general rubber asphalt compound pavement structure is shown in Fig. 1.

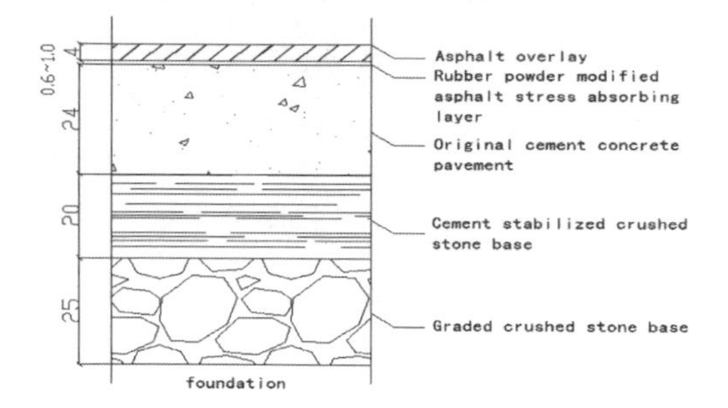

Figure 1. Pavement structure combination.

Table 1. Pavement structure parameters.

Structure layer	Thickness (cm)	Modulus (MPa)	Poisson's ratio
Asphalt overlay	4.0	1200	0.35
Absorbing layer	0.6~1.0	400	0.35
Cement concrete	24.0	30000	0.15
Cement-stabilized stone base	20.0	1500	0.20
Graded stone	25.0	350	0.35
Soil matrix	–	100	0.40

In addition to the rubber powder-modified asphalt stress-absorbing layer structure of each layer of compressive resilient modulus and Poisson's ratio, we refer to "asphalt pavement design specification" -JTG D50-2006 value. However, the rubber powder-modified asphalt stress-absorbing layer refers to the modified asphalt stress-absorbing layer parameters. The values of the pavement material parameters for each layer are shown in Table 1.

Using BISAR pavement structure in Table 1 in the interlayer under BZZ-100 standard axle load conditions, shear stress is calculated to analyze the thickness of the shear stress distribution, contact conditions, the interlayer shear stress, and the effects of stress-absorbing layer thickness on shear stress.

2.1 *Distribution of shear stress in the lower layers of different adhesive states*

In this analysis, the stress-absorbing layer and the upper and lower positions under different conditions of friction parameters ∂ each point along the driving direction (0,1a,2a,3a) at the XX and YY normal stress component analysis asphalt surface and the bottom of the maximum stress absorbing layer shear stress distribution. The results are shown in Figures 2–6.

As can be seen from Figures 2–6, the maximum shear stress distribution of the stress-absorbing compound pavement overlay and the stress absorbing underlying layer can be changed with different adhesive statuses, and the main changes are as follows:

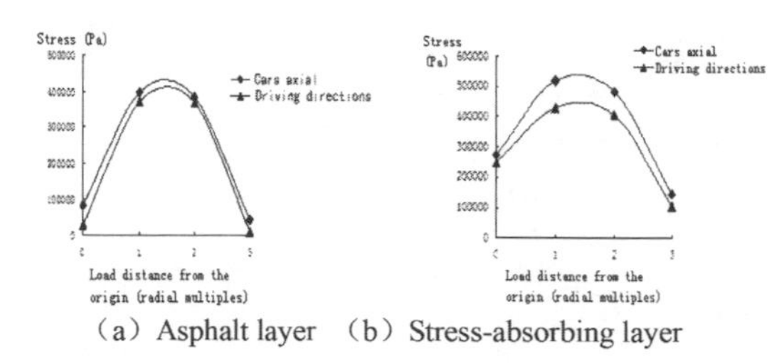

(a) Asphalt layer (b) Stress-absorbing layer

Figure 2. Shear stress at the underlying layer under the condition of complete continuity between two layers.

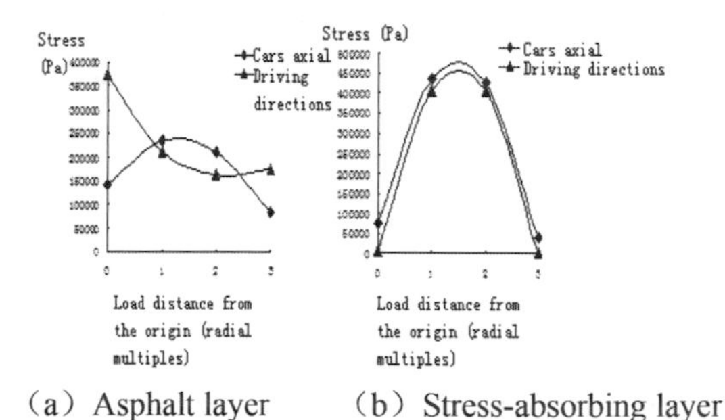

(a) Asphalt layer (b) Stress-absorbing layer

Figure 3. Shear stress at the underlying layer when the upper layer is 0.5.

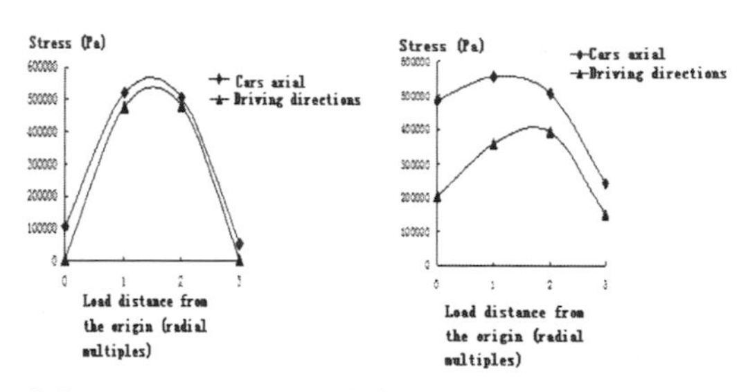

(a) Asphalt layer (b) Stress-absorbing layer

Figure 4. Shear stress at the underlying layer when the upper layer is completely sliding.

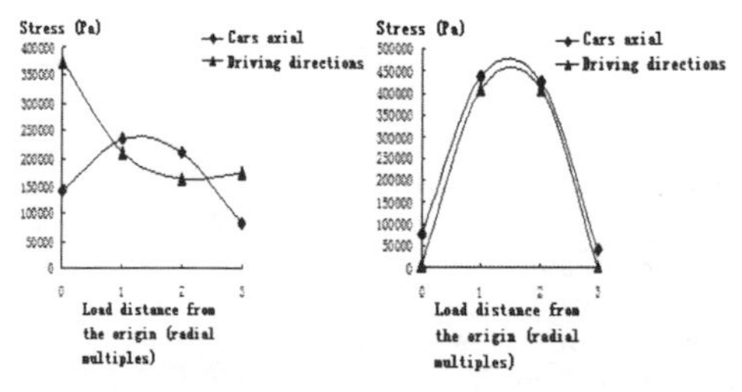

(a) Asphalt layer (b) Stress-absorbing layer

Figure 5. Stress absorbing layer friction parameter is 0.5.

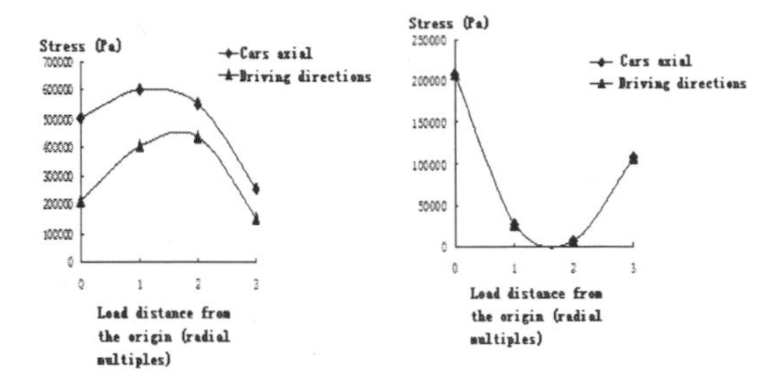

(a) Asphalt layer (b) Stress-absorbing layer

Figure 6. Shear stress at the underlying layer when the absorbing layer is completely sliding.

Table 2. Shear compliance under different friction parameters.

Adhesive layers	Friction parameters	Modulus (MPa)	Poisson's ratio	AK (N/m²)
Asphalt	0	1200	0.35	0
overlay	0.25			3.99E-11
and stress-	0.5			1.20E-10
absorbing	0.75			3.59E-10
layer	0.99			1.19E-08
Stress-	0	400	0.35	0
absorbing	0.25			1.20E-10
layer with	0.5			3.59E-10
the original	0.75			1.08E-09
concrete pavement	0.99			3.56E-08

1. When every level of compound pavement is completely in continuous state, asphalt layer and stress-absorbing layer stress was basically consistent along the direction of travel distribution, and auto axial and driving direction stress are all parabolas, and the axial stress is always greater than the direction of traffic stress.
2. When an adhesive status between the asphalt surface layer and the stress-absorbing layer changes, the distribution of stress in asphalt overlay takes a significant change along the directions of traffic. When the friction parameter is in the case of 0.5, the axial stress and stress variation of the direction of travel are exactly opposite, and the maximum shear stress is travel direction stress in load center, which proves that the interlayer adhesive status affects the stress distribution of the asphalt surface layer and the maximum shear stress direction. At the same time, under three friction parameter conditions, absorbing layer stress changing rules along the direction of traffic are not obvious, but there still exists possibility of the presence of the influence of different friction parameters on stress variation.
3. When the stress-absorbing layer and the original cement coagulation pavement adhesive status change, the variation of asphalt overlay and stress-absorbing layer of stress were significantly changed. When the friction parameter is 0.5, that is to say, when the stress-absorbing layer and the cement concrete road surface are not fully bonded, the stress distribution is basically equal to the effect generated the asphalt layer—the changing adhesive status of the stress-absorbing layer. Changes in stress distribution stress-absorbing layer occurs when incomplete bonding is not large; however, when the layers completely coincide, the driving direction and the distribution of the axial stress are exactly opposite with fully bonded case, namely the maximum shear stress is seen as a parabola, which decreases first and then increases.
4. From the distribution of stress in different layers and the variation of bonded conditions, it can be seen that the interlayer shear stress distribution varies, and this effect will change with the variation of friction parameters.

2.2 *Variation of maximum shear stress between the layers with different bond states*

BISAR is used to make mechanical analysis for stress-absorbing layer and the upper and lower positions, respectively, under different conditions of friction parameters. The friction parameters were 0, 0.25, 0.5, 0.75, and 0.99, in which the friction parameter 0 is for complete adhesion and 0.99 is for nearly complete slide. The variation of the maximum shear stress under different conditions of friction parameters is shown in Figs. 7 and 8.

As shown in Figures 7 and 8, the change of the bonding condition between stress absorption layer and the upper and lower layers affects the maximum shear stress between itself and the bottom layers of asphalt directly. With the increase in friction parameters between asphalt layer and stress-absorbing layer, the maximum shear stress

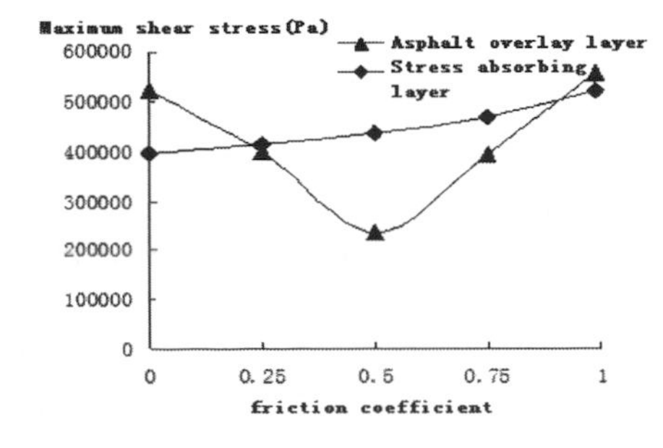

Figure 7. Maximum shear stress at the bottom of the layer under the condition of friction parameter of different asphalt overlay stress-absorbing layer.

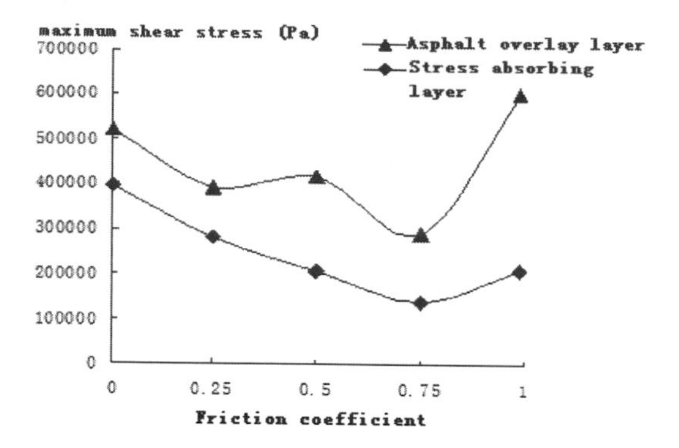

Figure 8. Maximum underlying shear stress between cement pavement and stress-absorbing layer under the condition of different friction parameter.

of rubber asphalt overlay will first decrease and then increase with the increase of friction parameters, while the maximum shear stress of the bottom stress absorption layer increases with it. The good bonding state of asphalt overlay layer can reduce the maximum shear stress of the stress absorption layer and prevent the premature failure of the stress absorbing layer. Although the change trend of asphalt overlay was first reduced and then increased, reducing the maximum shear stress of asphalt overlay by reducing the bonding degree between layers is not suitable, because at the beginning of building, the initial bonding state between the layers is not ideal. At the same time, over time, the bonding condition of interlayer will be gradually reduced, so it should not be the measures to prevent premature to enter the maximum shear stress growth. The bonding condition between the stress-absorbing layer and the cement pavement has a significant impact on the maximum shear. The maximum shear stress of rubber asphalt overlay and stress absorption layer bottom will first decrease and then increase with the increase of friction parameters. And for the asphalt layer bottom, completely sliding and in relatively complete bonding condition, the maximum shear stress increases by 15.23%, which also proves that we cannot reduce the maximum shear stress by reducing the adhesion of the layers.

2.3 *Effect of different stress-absorbing layer thicknesses on the maximum shear stress of the bottom layer*

BISAR is used for the mechanical analysis of rubber asphalt overlay (stress-absorbing layer) under different conditions of stress-absorbing layer thickness.

The thicknesses were 0, 0.6, 0.7, 0.8, 0.9, 1.0 and 1.5 cm. It can be found from calculation that when we do not set stress-absorbing layer, the maximum shear stress of the asphalt overlay bottom layers is 5.959 × 105, and the calculation results of bottom shear stress when settings with other different thickness of stress absorption layer are as shown in Figure 9.

As it can be seen from Figure 9, with the increase in the thickness of stress-absorption layer, the maximum shear stress between asphalt overlay and the bottom stress-absorbing layers decreases continuously, the maximum shear stress decreases by 12.59% when stress absorbing layer thickness is set at 0.6 cm. When the thickness is 1.0 cm, it decreases by 23.55%, and when the thickness is 1.5 cm, under the maximum shear stress, it decreases by 32.01%, which proves that the thickness effect is the effective measure to reduce the maximum shear stress of asphalt pavement; however, this effect increases with the thickness gradually. At the same time, for rubber powder-modified asphalt stress-absorbing layer, its thickness is 0.6–1.0 cm, and in the actual use of the process, because of the influence of Webster effect, the lower layer of asphalt will form a certain thickness of the oil-rich layer, the actual reduction effect of the maximum shear stress is greater than the one of calculation results. Therefore, the rubber powder-modified asphalt stress-absorbing layer is an economical and reasonable anticracking measure. In addition, from the variation of the maximum shear stress of absorption bottom layer in different thickness, it can be seen that increasing thickness cannot effectively reduce the maximum shear stress of stress-absorption layer. When the thickness of the stress-absorbing layer increases from 0.6 to 1.0 cm, the maximum shear stress decreases by only 3.44%. When the thickness of the stress-absorbing layer increases to 1.5 cm, the maximum shear stress decreases by only 7.5%. Therefore, the increasing of thickness of the stress-absorbing layer cannot effectively reduce the maximum shear stress by itself, which requires the stress-absorbing layer material's good elastic recovery, and the rubber powder asphalt has

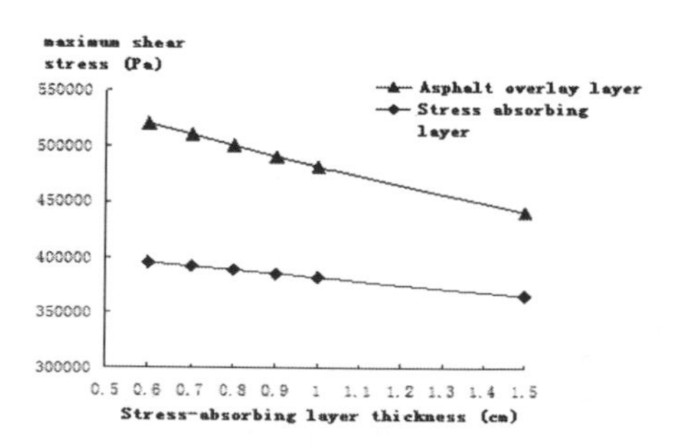

Figure 9. Bottom layer maximum shear stress under different thicknesses of stress-absorbing layer.

a high modulus and a high viscoelasticity trait, which is in line with this request.

3 CONCLUSION

1. By using BISAR mechanical calculation software, we calculated and analyzed the rubber asphalt compound pavement shear stress of different interlayer adhesive states, and under the condition of stress-absorbing layer thickness, we knew that the adhesive states of stress-absorbing layer, rubber asphalt surface layer, and underlying cement pavement have a direct impact on the distribution of shear stress of asphalt overlay and stress-absorbing layer, and the maximum shear stress of asphalt layer bottom will first decrease and then increase with the increase in friction parameters. Therefore, the bottom layer should have good adhesion to prevent the premature entry of shear stress growth period.
2. With the increase in the thickness of the stress-absorbing layer, the maximum shear stress of asphalt overlay is significantly reduced, but the effect of the maximum shear stress of the stress-absorbing layer is not obvious.

ACKNOWLEDGMENT

Guangxi Key Lab of Road Structure and Materials 2013 Open Subject, 2013gxjgclkf-004.

REFERENCES

Department of Transportation Highway Science Research Institute. Thin Layer Asphalt Concrete Surface Layer Technology Research [R]. Beijing: 2004.

Guan Chang Yu, Wang Zheren, Guo Dazhi. Study on the Bonding State of Pavement Structure Layer [J]. Chinese Journal of Highway, 1989, 2(1): 70–80.

Huang Xiaoming, Liao Gongyun. Application of ABAQUS Finite element software in road engineering [M]. Nanjing: Southeast University press, 2008.

Ruymbeke E. van, Keunings R., V. Stephenne. Determination of the Molecular Weight Distribution of Entangled Linear Polymers from Linear Viscoelasticity Data. Journal Non Newtonian Fluid Mwch. 2002.

The Ministry of transport of the people's communist country. Specifications for design of highway asphalt pavement (JTG D50–2006). Beijing; People's Communication Press, 2006.

Thomas Bennert, Ali Masher, and Joseph Smith. Evaluation of Crumb Rubber in Hot Mix Asphalt. Final Report[R]. Department of Civil and Environmental Ehgineering, Rutgers University. 2004.

Civil, Architecture and Environmental Engineering – Kao & Sung (Eds)
© 2017 Taylor & Francis Group, ISBN 978-1-138-02985-9

Experimental study on beam-column joints of new prefabricate assembly frame structures

W.Z. Zhu & J.C. Zhang
School of Civil Engineering, Guangzhou University, Guangzhou, China

ABSTRACT: The seismic performance of four new-type beam-column joints and two ordinary cast-in-site beam-column connections was compared with low-cycle loading tests. The study is mainly on the failure modes and mechanisms, hysteresis loops, ductility, and stiffness degradation of the joints. It is demonstrated that the specimens, belonging to classical bending failure, meet the seismic requirement of strong-shear and weak-bending for beam-column joints. There are no cracks on the surface of the joint region, and no relative slips along the interfaces between the cast-in-site column and composite beams for the bonds raised with concrete keys and reinforced bars. The seismic resistance performance of the new-type beam-column joints is almost equal to that of the ordinary connections. The construction measures of the prefabricate assembly frame structure can guarantee the safety and reliability of the novel joints.

1 INTRODUCTION

Construction measures, including composite beam and concrete shear key, used in New Prefabricate Assembly Frame Structure (NPAFS), could ameliorate troubles in traditional assembly structure system, such as low precision and poor structural integrity, and effectively improve construction quality and structural performance, shorten the construction period, reduce environmental pollution, and save resources and energy (Zhang, 2013). The NPAFS, good in integrity and seismic performance, and high production efficiency, is applicable for multi-layer and small high-rise structure, and is significant to the development of urbanization construction and housing industrialization of China.

Most related experiment researches at home and abroad assume that the NPAFS can achieve the same or similar seismic performance of the cast-in-site structures (Li, 2006). By the full-scale model tests, Zhao Bin et al. (2005) had studied the seismic behaviors of cast-in—site high strength concrete beam-column sub—assemblage, precast concrete beam-column sub—assemblage with cast-in-site high strength concrete joint and precast concrete beam-column sub—assemblages with fully assembled joints, such as the failure pattern, hysteresis characteristic, skeleton curve, strength and stiffness deterioration, shear deformation of the joint core zone, rotation of the beam end and column end. Through testing the seismic behavior of precast layered slab and beam to column exterior joints under high axial loading and low—cycle reversed loading, Yan Weiming et al (2010) indicated that the earthquake resistance performance of this type of connection is almost equal to that of monolithic connections. At present, connections assembled with cast-in-site column and composite beams and monolithic connections are frequently used in Assembly integral frame joints in Chinese (2004). Luo Qinger et al. (2009) conducted contrast tests of one precast monolithic R.C. frame interior column joint and one cast-in-site R.C. frame interior column joint. Additionally, connecting techniques of precast columns used presently are not ideal to structural integrity, only suitable to multi—layer structures (CECS43:92).

In the paper, a novel beam column joint, composed of composite beams and cast-in-site column, was proposed. As shown in Fig. 1, the main construction measures of the joint include composite beams, protruding bars of the beams welded in the joint region, and concrete shear keys at the beam—ends. Concrete shear keys are used to improve the shear-resistance capacity of the joint.

In the actual construction process, protruding bars are unable to be accurately aligned for deviation of component fabrication, positioning

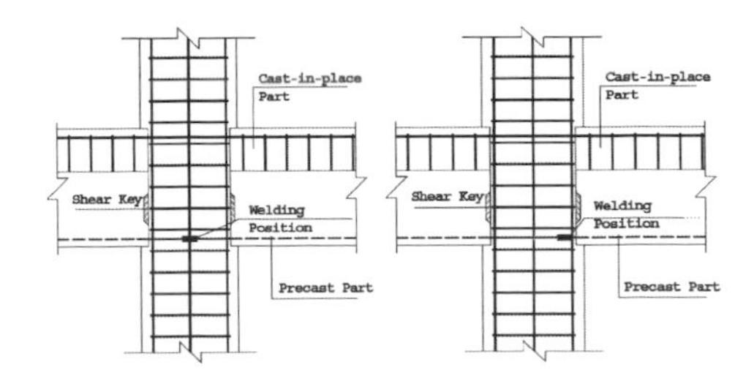

Figure 1. New-type joint for reinforced frames.

inaccuracy, and space restriction in the node, so it is difficult to be welded together. While welding the bars in edge region of the node can reduce the difficult in adjusting them.

In order to ensure the safety and reliability of the construction measures, low-cycle reverse loading tests were conducted on the novel frame beam column joints for comparing the lading process, failure mode, hysteretic ductility, energy dissipation, and stiffness degradation of the joints with cast-in-site beam column joints.

2 SPECIMEN DESIGN

A total of six full-scale specimens of interior RC beam-column joints from an assembled integral residence was made and tested. Four of the specimens are the novel beam-column joints presented in the paper, only different in welding positions of the bottom steel bars, in specimens 1 and 2 (PC-1-1 and PC-1-2) the bottom bars welded at the column center-line, and in specimens 3and 4 (PC-2-1 and PC-2-2) at the edge region of the joints. Specimens 5 and 6 (RC-1-1 and RC-1-2) are ordinary cast-in-site concrete connections. The numbers and sizes of the components are shown in Tables 1 and 2, and material performance of the steel and concrete used is demonstrated in Tables 3 and 4.

For these specimens, the beam and column ends were the points of inflection under the applied lateral load. The beam length was measured as

Table 1. Types and numbers of the tested joints.

Component type	prefabricated components		Cast-in-site members
	Bars welded at column center-line	Bars welded at the edge region	
Number	PC-1	PC-2	RC-1
Quantity	2	2	2

Table 2. Dimensions of the tested joints/mm.

Specimen	Column section	Beam section		
		Prefabricated part	Cast-in-site part	Total cross section
PC-1-1	500 × 500	300 × 350	300 × 200	300 × 550
PC-1-2				
PC-2-1				
PC-2-2				
RC-1-1	500 × 500	–	300 × 550	300 × 550
RC-1-2				

Table 3. Parameters of concrete material.

Specimen		Cubic compressive strength/MPa	Modulus of elasticity $E_c/10^4$ MPa	Maturity
PC-1-1	Prefabricated part	62.32	4.714	28
	Cast-in-site part	59.03	4.709	28
PC-1-2	Prefabricated part	60.55	4.714	28
	Cast-in-site part	63.38	4.709	28
PC-2-1	Prefabricated part	56.62	4.525	28
	Cast-in-site part	59.03	4.808	28
PC-2-2	Prefabricated part	60.55	4.881	28
	Cast-in-site part	63.38	4.902	28
RC-1-1		56.45	4.926	28
RC-1-2		62.28	4.908	28

Table 4. Properties of steel.

Type	Grade	Yielding strength/MPa	Ultimate strength/MPa	Modulus of elasticity × 10^5/MPa
D22	HRB335	372	577	2.0
D25	HRB335	368	573	2.0

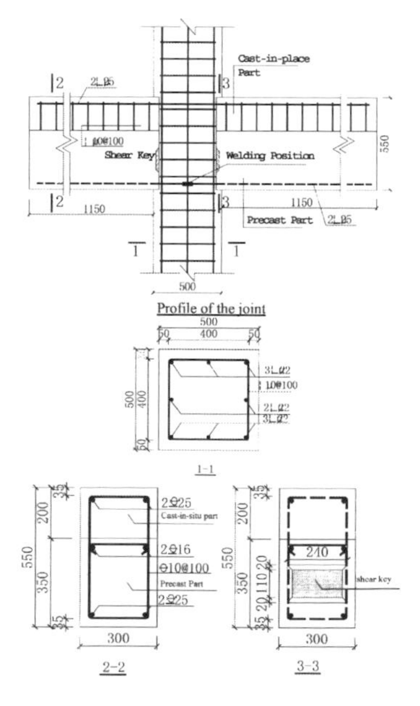

Figure 2. Geometry and reinforcement layout of the joints/mm.

1.4 m from the column axis to the point of inflection. The materials of the joints are similar to the prototype structure, made of C60 concrete, and symmetrically reinforced with HRB335 bars. The detailed geometry and reinforcement of the joints are shown Fig. 2.

3 TEST LOADING AND DESCRIPTION

3.1 *Loading system*

Considering the effect of earthquake, vertical load is subjected on the beam ends of the joints by two vertical actuators. To simulating the actual mechanical behavior of the beams, the loading positions are 1.1 m away from the edge of the column.

Axial compression ratio, u, is an important parameter to ensure the failure modes of the concrete column in earthquakes being large eccentric compression and ductility damage (GB50011). Previous studies believe that when $u \geq 0.3$, the angles of seismic waves have little effect on the section ductility of column; and when $u \leq 0.3$, the influence of concrete strength grade on the section ductility may be neglected (Xu, 2012). So a constant axial load of $0.3 f_c A_g$, in which f_c and A_g are design value of compression strength of concrete and sectional area of the column, respectively, was subjected on the top of the column, and maintained in the testing.

The loading system selected is a load—displacement hybrid control (STMERB, 1997), as shown in Fig. 3. The estimated cracking load f_{cr} of the composite beam is 80 kN, and the yield load f_y is 160 kN. Before yield of steel bars in the beam, the tests were executed with force control, and one cycle for each load level. The first 3 load increments are 50%, 30%, and 20% of the cracking load, respectively. When the beam cracking, the load increment applied is switched to 20% of the yield load, and changed to 10% of the yield load when the loading approaching the yield load.

When the beam reaches its yield strength, the loading system was switched to the displacement control. Each load level was subjected on the specimens for three cycles gradually to achieve the displacement ductility factor $\Delta/\Delta_y = \pm 1.5, \pm 2.0, \pm 2.5$, …, where the yield displacement Δ_y was measured during the test. The ultimate deformation of the specimens is determined corresponding to the beam-end displacement when the bearing capacity reduced to 80% of the maximum load. The setup of the loading test are shown in Fig. 4.

The stress values in the specimens were measured with DH3815N static strain instrument, with measuring range of ±20000 με, system error of 0.5% ± 3 με, resolution of 1 με, and drift of 4 με/4 h. Deflection was collected with an automatic displacement acquisition system. Load was read with pressure sensors. All instruments were calibrated in accordance with specified criteria before tests, and a cyclic vertical load of 5 kN was applied on the beam-ends for 2 times to check whether the instruments performing normally or not. The arrangement of steel and concrete strain gauges for the test components is shown in Figs. 5 and 6.

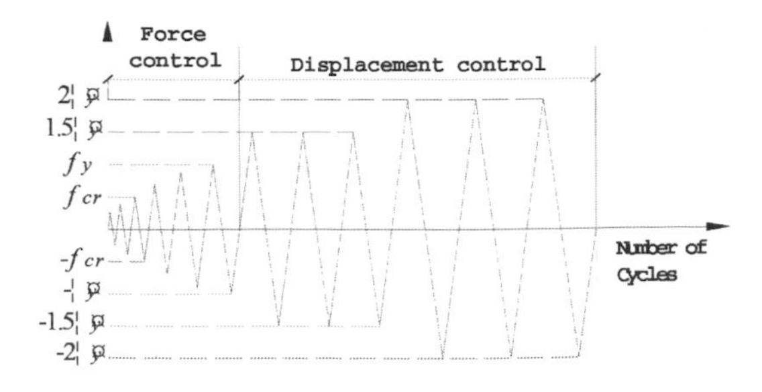

Figure 3. Loading system of the test.

Figure 4. Setup of the loading test.

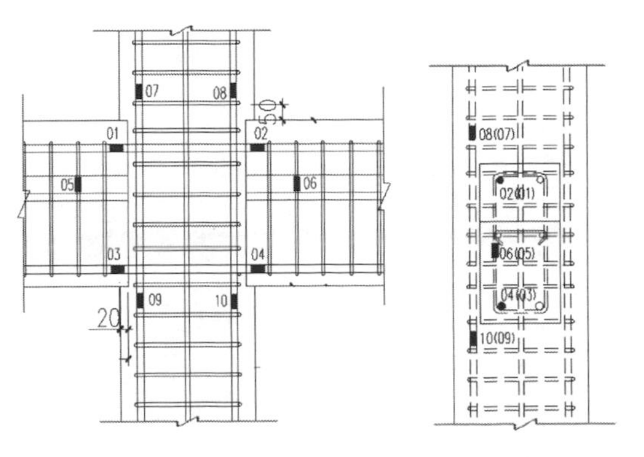

Figure 5. Arrangement of the steel-stress gauges.

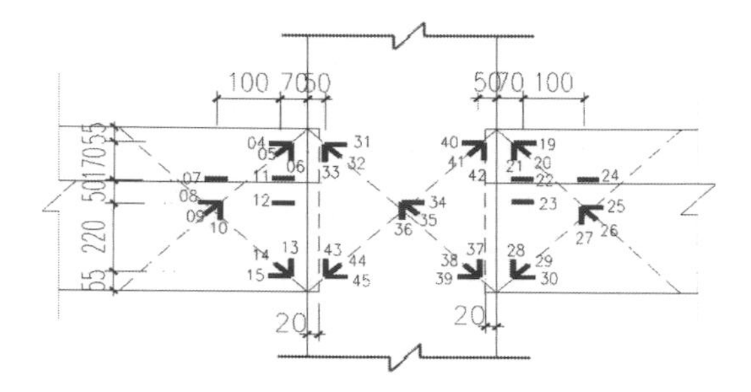

Figure 6. Arrangement of the concrete-stress gauges.

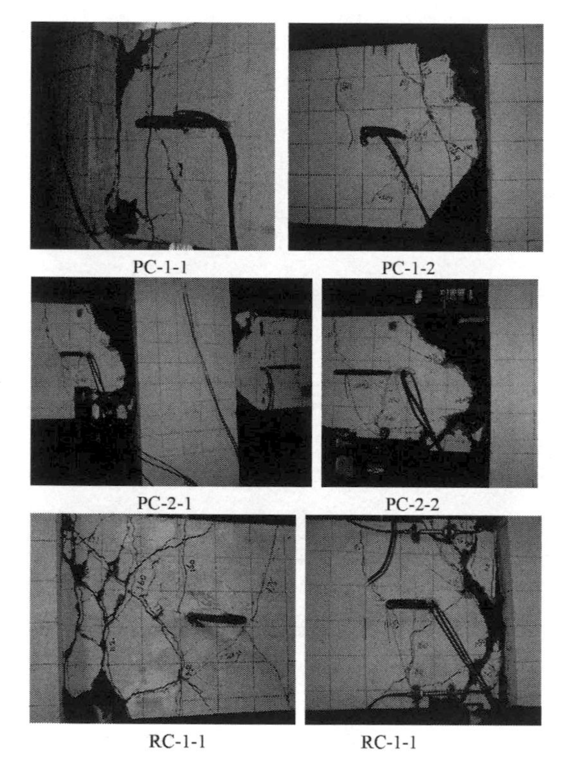

Figure 7. Patterns for the failure modes of the joints.

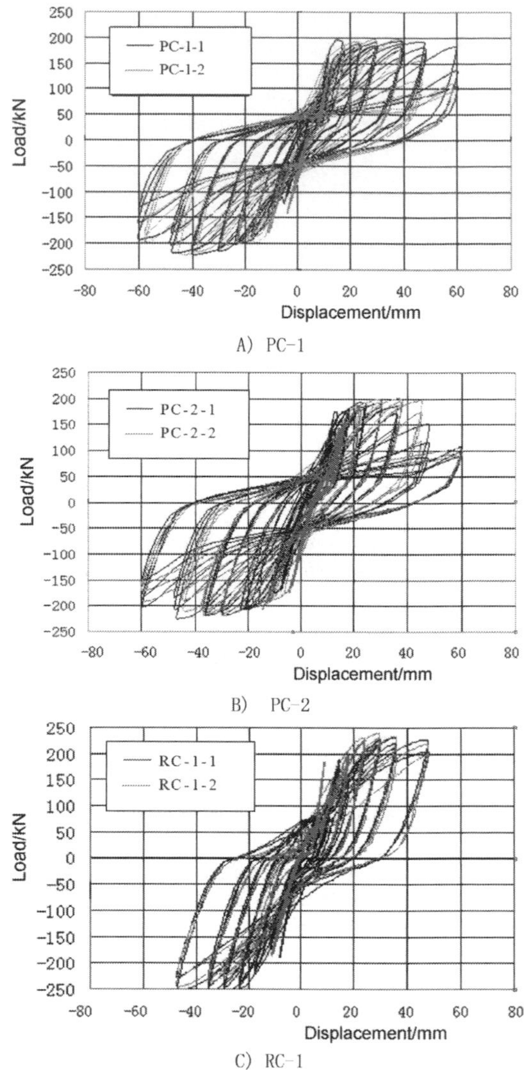

Figure 8. Load-displacement hysteretic loops.

3.2 *Test process*

The failure modes of the components tested are shown in Fig. 8. All specimens exhibit bending failure modes under cyclic loading, plastic hinges fully developed in beam body near the joint. For the specimens having present basically similar failure features, the loading process was divided into three phases.

In the first phase, concrete cracks in these specimens are firstly observed in the beams, and the width, depth and distribution range develop with load increasing. In the second phase, new cracks appear constantly and the original cracks are gradually developing into crossing cracks. In the last phase, cracks develop rapidly, with concrete crushed in compression zone, plastic hinge formed in the beam, and loading capacity of the column decreased. In the process, no cracks were observed on the surface of the joint core while reinforced bars show no brittle failure. The observed crack patterns of the specimens are shown in Figure 7.

At the third loading stage, there are no obvious changes on the surface of the PC-1 specimens, while measured strains show that the beam concrete has reached cracking degree. At the fourth loading stage, hair cracks are observed at the bottom of the left beam in the region near the joint. At the fifth stage, about 144 kN, vertical cracks with different width appear on beams, with the largest length of 170 mm and width of 0.06 mm. To the sixth stage, the vertical cracks have developed in varying degree, and form diagonal cracks gradually. At the seventh stage, crossing cracks are formed at different extent, and the beams yield. The corresponding beam-end displacements are 12.0 and 13.0 mm, respectively, for PC-1-1 and PC-1-2. In the displacement control phase, the width, length and the distribution range of the cracks expand continuously with the displacement increasing gradually, and plastic hinge formed at the beam-end. Finally, when the deformations reach $5\Delta_y$, the hysteresis curves become unstable, and the specimens are failure.

At the third loading stage, hair cracks were observed at the top of the right beam of PC-2 specimens in the region near the column. At the fourth loading stage, cracks are observed at the top of both beams. At the fifth stage, vertical cracks with different width appear on the bottom of beams, and origin cracks develop continuously. At the sixth stage, origin cracks extend further, and form diagonal cracks gradually. At the seventh stage, the original cracks have various degree of development with a main diagonal crack formed obviously. The corresponding beam-end displacements are 12.3 and 13.7 mm, respectively, for PC-2-1 and PC-2-2. The appearance in the displacement control phase is similar to PC-1specimens.

The demonstration of RC-1 specimens during 3 to 6 load stages is similar to that of PC-1 basically. While reinforced bars of PC-1 are not yield at the 7th load stage, 10% of the estimated yield load of the beam is still used as the load increment in the 8th stage. The corresponding beam-end displacements are 12.6 and 13.4 mm, respectively, for RC-1-1 and RC-1-2. In the displacement control phase, the hysteresis curves become unstable when the deformations reach $4\Delta_y$, and the specimens are failure.

4 RESULTS AND DISCUSSION

4.1 *Hysteresis behavior*

The hysteretic responses of specimens, shown in Figure 8, are presented in the form of displacement versus corresponding vertical applied load on beam-ends. Pinching is obviously observed in the hysteresis curves (Mansour, 2001). Skeleton curves, shown in Fig. 9, were obtained from the curves. Some results inferred from the experimental phenomena and Figs. 9 and 10 are given below.

1. The hysteretic behavior of the specimens is depended on the rotation capacity of the beam-end plastic hinge. The reduction in strength of the specimens is mainly corresponding to the yield of the reinforced bars and concrete crushing in the beams.
2. The ultimate bearing capacity of the ordinary beam-column connections is greater around 10% than that of the novel beam-column joints.
3. The novel beam-column joints with different welding position of the reinforced bars at beam bottom are similar in hysteresis curves and skeleton curves. Welding position of the reinforced bars has little effect on hysteretic behavior of the specimens.

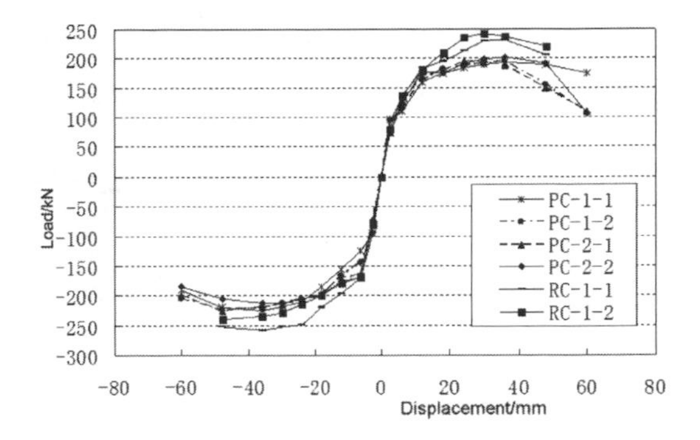

Figure 9. Comparison of load-displacement skeleton curves.

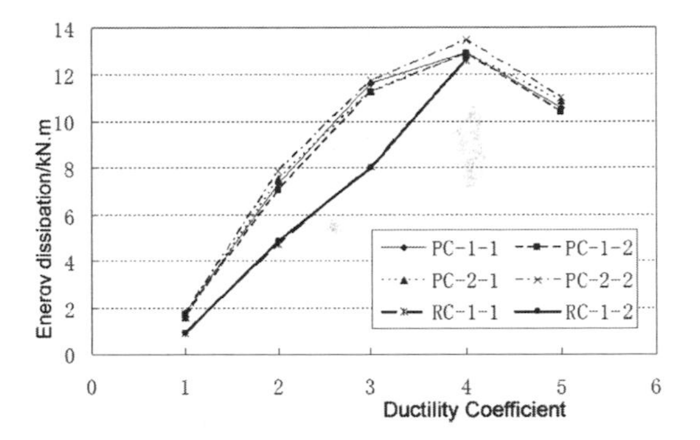

Figure 10. Comparison of energy dissipation.

4.2 *Ductility coefficient of beam*

Ductility Coefficient is ratio of extreme deformation to yield deformation, and reflects plastic deformation capacity of structural components (Cai, 2005). In structural seismic analysis, ductility coefficient is usually used to represent the ductility of structural elements, and can be calculated by equation (1).

$$\beta_D = \frac{\Delta_u}{\Delta_y} \quad (1)$$

Table 5. Displacement ductility coefficient of the joints.

Specimen	P_y/kN	Δ_y/mm	P_{max}/kN	Δ_{max}/mm	P_u/kN	Δ_u/mm	β_D
PC-1-1	176.2	12	219.9	40.3	186.9	59.9	4.99
PC-1-2	178.5	13	223.6	47.4	190.1	48.1	3.70
PC-2-1	179.1	12.3	229.0	29.3	203.2	47.6	3.87
PC-2-2	179.6	13.7	216.6	37.3	184.1	52.9	3.86
RC-1-1	193.2	12.6	258	47.9	221	46.4	3.68
RC-1-2	195.4	13.4	242	45.2	219	45.2	3.37

where, β_D is ductility coefficient, Δ_u is extreme deformation, and Δ_y is yield deformation.

Ductility coefficient calculated of each specimen, shown in Table 5, is between 3.37 and 4.99, and meets the requirements of seismic ductility[12]. Ductile deformation capacity of the novel beam-column joints is greater than that of the ordinary beam-column connections with equivalent reinforcement and concrete strength. Additionally, no obvious effect of welding position of the reinforced bars is observed on ductility performance of the novel beam-column joints.

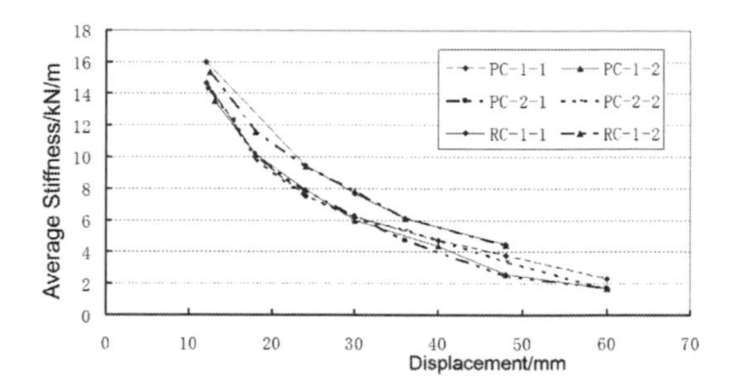

Figure 11. Comparison of stiffness degradation.

4.3 *Energy dissipation capacity*

Area of hysteresis loop can be used to evaluate the energy dissipation capacity of the component in earthquake. Average energy dissipation of each specimen under each levels of deformation for each specimen is shown in Fig. 10.

The figure indicates that plastic hinges of the novel beam-column joints appear earlier than those of the ordinary beam-column connections, and the energy dissipation capacity of the joints inclines when ductility coefficient greater than 4.0, demonstrating that plastic hinge start to destroy. Additionally, the energy dissipation capacity of the novel beam-column joints is greater than that of the ordinary beam-column connections, and two types of the novel joints are more or less similar in energy dissipation.

4.4 *Stiffness degradation*

Stiffness degradation is that stiffness of structural members decreases with their deformation increasing under cyclic loading. Rate of the degradation, used for evaluating seismic capacity of the joints, can be demonstrated with average stiffness of each circulation at all levels:

$$K_j = \sum_{i=1}^{n} \frac{\left(\left|P_j^{i+}\right| + \left|P_j^{i-}\right|\right)}{\left(\left|\Delta_j^{i+}\right| + \left|\Delta_j^{i-}\right|\right)} \bigg/ n \quad (2)$$

In which, P_j^{i+}, P_j^{i-} are the positive and negative peak load of the ith cycle in jth loading stage, respectively, $\Delta_j^{i+}, \Delta_j^{i-}$ are deformation values corresponding to P_j^{i+} and P_j^{i-}, respectively, n is the number of cycles.

Deductions on the stiffness degradation of each specimen, shown in Fig. 11, are given below.

1. The stiffness degrades with deflection increasing after beam yield. However, the deterioration trends tend to mitigation.
2. Stiffness degradation of the novel beam-column joints is similar.
3. In the process of stiffness degradation, the stiffness of the ordinary beam-column connections is greater than that of the novel beam-column joints.

5 CONCLUSION

Based on the low-cycle loading tests of four novel beam-column joints and two ordinary cast-in-site beam-column connections, we can draw the following conclusions.

1. All the joints exhibit typical bending failure modes, and plastic deformation capacity of the beams is fully developed. The construction measures, such as concrete keys in the joint and steel bars welded, meet the ductility requirement of strong-shear and weak-bending for concrete frame joints.

2. The ultimate strength and stiffness of the novel concrete frame joints are slightly lower than those of the ordinary beam-column connections, but their ductility and energy dissipation capability are higher, so two types of joints are similar in seismic performance.
3. Welding positions of the bottom steel bars of the beams have little effect on the mechanical properties of the tested specimens, for two kinds of the novel joints having similar stiffness degradation, ductility and energy dissipation.

In conclusion, the novel frame joints of assembling integral reinforced concrete frame structure presented in this paper can reach the mechanical level of ordinary beam-column concrete connections, and the construction measures for the novel frame joints can ensure the safety and reliability of buildings.

ACKNOWLEDGMENTS

This study has been supported by the Collaborative Innovation Major Project of Guangzhou City under grant Number (201604020071).

REFERENCES

Cai Jian, Zhou Qing, Fang Xiaodan. Study on Seismic Dis placement Ductility Factor of Reinforced Concrete Frame Structure [J]. *Earthquake Resistant Engineering and Retrofitting*, 2005, 27(3): 2–6 (in Chinese).

CECS43:92. Design specification for joints and connections of precast monolithic reinforced concrete frames [S]. 1992, 11 (in Chinese).

Zhang Jichao, Chen Jiefeng, XuYong, et al. Discussion of the new prefabricate assembly integral frame structure technology [J]. *Building Structure*, 2013, 43(s1):1357-03 (in Chinese).

GB 50011-2010. Code for design of buildings [S]. Beijing: China Architecture & Building Press, 2010, 8.

Li Zhenbao, Dong Tingfeng, YanWeiming, et al. Study on Seismic Performances of Hybrid Precast Concrete Beam-to-column Connections [J]. *Journal of Beijing University of Technology*, (2006), 32(10):898–900 (in Chinese).

Luo Qinger, Wang Yun, Weng Yuhui, et al. Experimental study of beam-column joint of precast monolithic R.C. Frame [J]. *Industrial Construction*, 2009, 39(2): 80–83 (in Chinese).

Mansour M, Lcc J, Hsu T. Cyclic stress-strain curves of concrete and steel bars in membrane elements [J]. *Journal of Structural Engineering*, 2001, 127(12): 1402–1411.

STMERB, Specification of testing methods for earthquake resistant building[S]. Beijing: China Architecture & Building Press, 1997 (in Chinese).

Xu Yidong, Wang Qiwen, Chen Yunxia. Research on limiting value of axial compression ratio ductility brhavior of RC square column [J]. *Journal of Building Structures*, 2012, 33(6):103–109 (in Chinese).

Yan Weiming, Wang Wenming, Chen Shicai, et al. Experimental study of the seismic behavior of precast con crete layered slab and beam to column exterior joints [J]. *China Civil Engineering Journal*, 2010, 43(12):56–61.

Zhao Bin, Lu Xilin, Liu Haifeng. Experimental study on seismic behavior of precast concrete beam-column subassemblage with cast-in-situ monolithic joint [J]. *Journal of Building Structures*, 2004, 25(6):22–28 (in Chinese).

Zhao Bin, Lü Xilin, Liu Lizhen. Experimental study on seismic behavior of precast concrete beam-column sub-assemblage with fully assembled joint [J]. *Earthquake Engineering and Engineering Vibration*, 2005, 25(1):81–87(in Chinese).

Civil, Architecture and Environmental Engineering – Kao & Sung (Eds)
© 2017 Taylor & Francis Group, ISBN 978-1-138-02985-9

Seismic analysis of large span spatial steel frame with isolation system

W.Z. Zhu
School of Civil Engineering, Guangzhou University, Guangzhou, China

ABSTRACT: Base isolation technique was employed in the E Block of the Guangdong Science Center (GSC) to protect the huge span frame from damage under external loads, especially typhoon and earthquakes. In this application, isolating bearings are directly placed at the base of the main columns of the building. Non-linear dynamic analysis and shaking table tests were conducted to study the effect of the base isolation on vibration modification, evaluate the improvement of the seismic-resistance capability, and estimate the deflection compatibility of the isolation. A 1/35th scale model was constructed and tested on shaking table under a series of base excitation with gradually increasing acceleration amplitudes. The results show that the experimental test setup can simulate the vibration characteristics of the prototype structure perfectly, and the stiffness simulation for isolating bearings is successful. With the comparison of the natural vibration characteristics and the dynamic responses, such as accelerations and displacements, of the structure for the cases with or without base isolation, it was demonstrated that base isolation can remarkably improve the seismic-resistance capability of the building and improve the torsion performance of structure with irregular plan. The investigation indicates that the isolation design for GSC is rational and is a useful effort for isolation design in huge civil engineering.

1 INTRODUCTION

Base isolation is an effective measure to alleviate earthquake disasters to structures. Since the 60's of last century, extensive researches on isolation technique have been conducted in practical engineering. It has been indicated that base isolation is very effective in keeping the utilization functions of buildings under severe earthquakes (Skinner, 1993; Su, 2001; Masahiko, 2006). Topkaya (2004) conducted a parametric study using three-dimensional finite elements with geometric and material nonlinearities, and identified the important geometric specimen properties that influence the measured shear modulus. Providakis (2009) discussed the aseismic performances of two actual Reinforced Concrete (RC) buildings with various Lead-Rubber Bearing (LRB) isolation systems. Kilar etc. (2009) analyzed the positive and negative effects of different bearing distributions on displacements and rotations of the superstructure as well as the base isolation system and tried to determine the most favorable distribution of isolators that is able to balance the effects of introduced eccentricities. Goda etc. (2010) assessed the cost-effectiveness of seismic isolation technology from the lifecycle cost-benefit perspective. All of the above studies indicated that seismic isolation can reduce vibration in superstructures significantly, especially the unfavourable torsion effect and can be cost-effective in mitigating seismic risk. The study reported by Jangid (2007) found that the optimum yield strength of the LRB is in the range of 10%–15% of the total weight of the building under near-fault motions. Dicleli etc. (2007) indicated that the accuracy of the equivalent linear analysis results is affected by the ratio of peak ground acceleration to peak ground velocity of the ground motion as well as the intensity of the ground motion relative to the characteristic strength of the isolator, and the effective damping equation currently used in the design of seismic-isolated structures must incorporate the effective period of the structure and frequency characteristics of the ground motion for a more accurate estimation of seismic response quantities.

Although the isolation technique has been used in civil engineering for more than 40 years, such as U.S. Court of Appeals, San Francisco Airport International Terminal, with isolation systems mainly consisting of sliding bearings (Charleson, 2008), Lead-Rubber Bearings (LRB) used in large span structures are seldom reported.

Guangdong Science Center (GSC), a large span steel frame consisting of special shaped components, is new landmark architecture of Guangdong Province, China. The potential fatalness to the Block is of great seriousness for its extremely complex internal force state. To ensure the safety of the building and people, devices and exhibitions in it under external loads, especially typhoons and strong earthquakes, the base isolation system of LRB was employed in the E Block of GSC.

Since isolation system was used in such a large span steel framework for the first time, whether the huge columns directly mounted on the isolating bearings could work compatibly or not, liking the case with a uniform isolating layer, is curious. To achieve optimum design and favorable seismic effect, a shaking table test and nonlinear dynamic analysis were conducted to compare the seismic performance of the base isolated structure with that of the base-fixed one. In this study, a 1/35th-scale shaking table specimen was constructed and tested to evaluate the seismic characteristics.

2 PROJECT OVERVIEW

GSC is located at Wanzuitou, west of Xiaoguwei Island in Panyu district of Guangzhou. The construction area is at the northern of the Cenozoic era sedimentation basin of Pearl River Delta of China, 1.1 km from a nearest fault line. The area has a geological condition of high hazard level, and many potential geological disasters, such as liquefaction, earthquake-induced landslides. The fortification intensity for the construction site is categorized as 7. The type of site-class is specified as type II corresponding to the equivalent shear-wave velocity and the overlaying thickness of the soil profile, and the predominant period of the ground motion is taken as 0.45 s.

The main structure consists of seven Blocks, A, B, C, D, E, F, and G, as shown in Fig. 1. Among them, E Block is the biggest exhibition hall, liking a ship sailing on the sea, used for exhibiting valuable space instrumentations, and its structural style is the newest large span steel frame with many special shaped components ever used in China.

The E Block has three stories with span length of 40 m, and the supporting system consists of six huge truss columns with cross sectional area of 4 m × 9 m. Huge trusses of 15 m high, linking the huge columns along the longitudinal direction, are installed at the 2nd storey and reached out to support the ship's bow. Lattice beams are connected to the huge columns along the transverse direction on each floor. Then huge trusses, lattice beams and huge columns are made up of the huge irregular frame structure, and the lateral and vertical force resisting system. Junior trusses, 3 m in height and 6 m in space distance, are installed on the 2nd and 3rd floor. The structural plan and the elevation of E Block are shown in Figs. 2 and 3.

Figure 1. Building model of GSC.

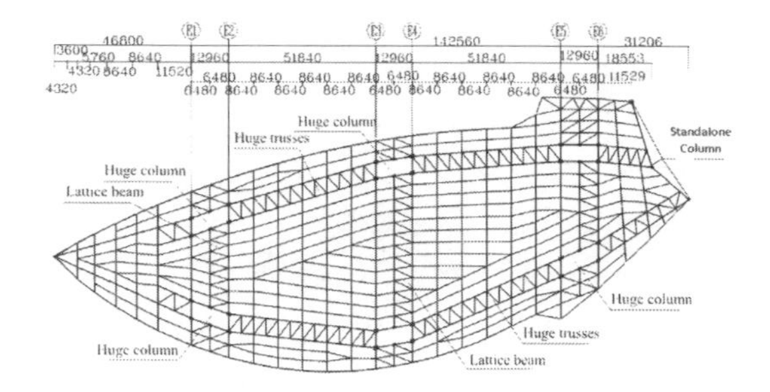

Figure 2. Structural plan of Block E.

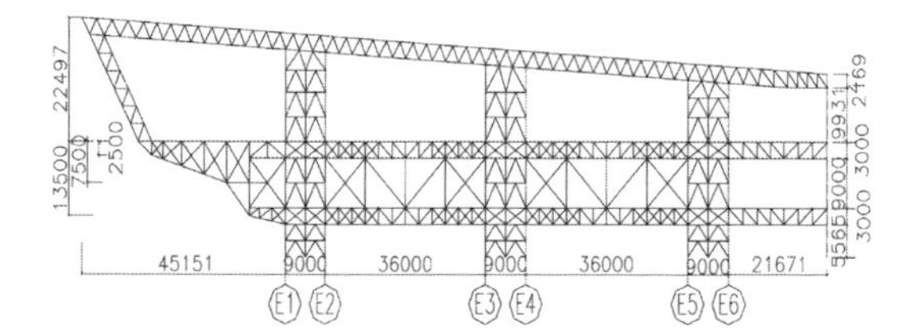

Figure 3. Sketch map of main structure.

3 ISOLATION DESIGN

It is very difficult to design base isolation in the E Block, for the span length of about 40 m. If a uniform isolation layer was chosen, the layer would be about 1.5 m high, and the isolating bearings should be embedded about 2 m into soil with the level of the structure unchanged. While the ground-water level is only 0.4 m below existing surface grade, the bearings would be immersed into water. If the mounting height of the bearings was higher than the existing ground, the level of the building should be lifted up about 2 m. Therefore isolating bearings are placed directly at the base of each huge column, taking advantage of the rigid characteristic of the huge columns to keep all the bearings working compatibly. LRBs (Lead Rubber Bearings) were chosen as isolating bearings for the well-rounded research and the received computational model.

Table 1. Parameter of the bearings /mm.

Item	GZY-1100	GZY-1000	GZY-800	GZP-500	GZY-500*
Diameter	1100	1000	800	500	500
Height	347	339	325	194	347
Thickness of rubber layers	6×27	6×27	5×32	4.68×19	6×27
Thickness of steel plates	3.1×26	2.8×26	2.8×31	2.5×18	3.1×26
Diameter of the lead plug	220	200	160	–	220
First shape coefficient	45.8	41.7	40	24.5	12.1
Second shape coefficient	6.8	6.2	5	5.5	3.15

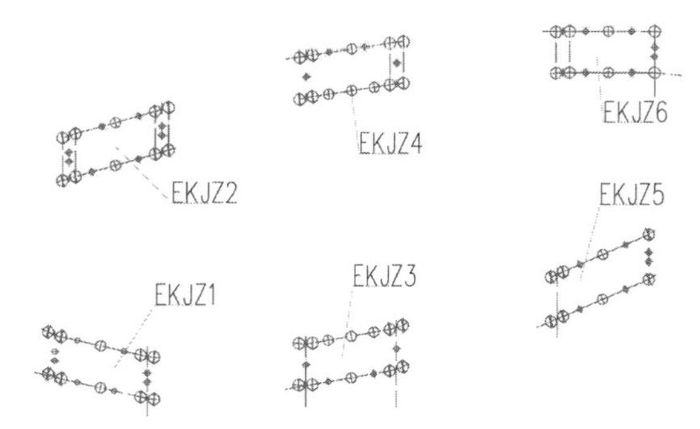

Figure 4. Arrangement of bearings under huge columns.

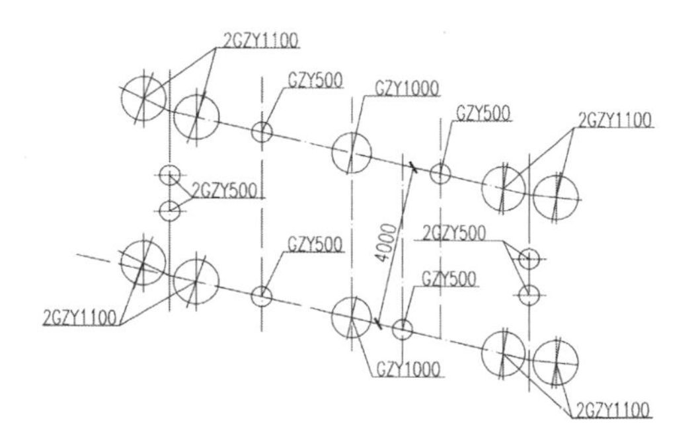

Figure 5. Arrangement of bearings under EKJZ1 column.

To secure the deformation harmony of the isolating bearings with different sizes and satisfy the requirement of the practical engineering, isolating bearings, with the diameters of 1100, 1000, 800 and 500 mm, are selected for bearing stability under extreme earthquakes. The 1100 mm bearings were the biggest bearings ever used in China. The parameters of the bearings are shown in Table 1, in which, GZY500* is a damping bearing for wind resistance formed by the commonly used GZP500 bearing through increasing the cross sectional area of lead bars. The arrangement of bearings under the huge columns is shown in Figs. 4 and 5.

In addition, there are three other standalone steel columns shown in Fig. 2. Considering the smaller cross sectional area of the columns, a partial isolation layer was used to connect all the smaller columns to two huge columns nearby. Then the seismic deformation compatibility of the smaller columns with the whole building is achieved.

4 FEM SIMULATION

4.1 *Analytical model*

In the analytical model, isolating bearings were modeled with two-directional coupled viscous damping element models. The non-linear relationship of the two-directional coupled force and displacement can be written as

$$\begin{cases} f_{u2} = m_2 k_2 d_{u2} + (1 - m_2) F_{y2} z_2 \\ f_{u3} = m_3 k_3 d_{u3} + (1 - m_3) F_{y3} z_3 \end{cases} \tag{1}$$

where, k_2 and k_3 are the lateral stiffness of the isolating bearing in two directions respectively [$kN{\cdot}m$]; m_2 and m_3 are the ratios of the post yield stiffness to the elastic stiffness in two directions; F_{y2} and F_{y3} are the yield force in two directions; z_2 and z_3 are hysteresis variables of the isolating bearing, under the condition of $\sqrt{z_2^2 + z_3^2} \le 1$ and the yield surface is $\sqrt{z_2^2 + z_3^2} = 1$. The initial values of z_2 and z_3 are zero, and meet the differential equations:

$$\begin{Bmatrix} \dot{z}_2 \\ \dot{z}_3 \end{Bmatrix} = \begin{bmatrix} 1 - a_2 z_2^2 & -a_3 z_2 z_3 \\ -a_3 z_2 z_3 & 1 - a_3 z_3^2 \end{bmatrix} \begin{Bmatrix} k_2 F_{y2} / d_{u2} \\ k_3 F_{y3} / d_{u3} \end{Bmatrix} \tag{2}$$

In which,

$$a_2 = \begin{Bmatrix} 1 & d_{u2} z_2 > 0 \\ 0 & d_{u2} z_2 \le 0 \end{Bmatrix}, a_3 = \begin{Bmatrix} 1 & d_{u3} z_3 > 0 \\ 0 & d_{u3} z_3 \le 0 \end{Bmatrix}.$$

The steel trusses, girders and columns were modeled with linear elastic beam-column elements, and the floors and curtain walls as shell elements. The analytical model of the prototype structure is shown in Fig. 6, in which there are shell element of 2578, beam-column element of 12907 and viscous damping element of 78.

According to the Chinese Code for seismic design of buildings (CCSDB, GB 50011-2010), at least 2 sets of strong earthquake records and one set of artificial acceleration time-history curve should be selected based on the intensity, the design seismic group and site-class, and the maximum value of the calculated results should be chosen as the final result of the time-history analysis method. Considering the spectral density properties of Type-II site soil, El Centro (1940) and Taft (1952) ground motion records were selected as the ground motions for the model tests, for the frequency spectrum feathers of the two waves being similar to those of the site soil. The time history of artificial acceleration is developed considering the motion design spectrum for Type-II site soil. The spectrum characteristics of the wave are shown in Figs. 7 and 8. The peak ground accelerations of the selected records were scaled to 0.055, 0.1 and 0.31 g, corresponding to seismic intensity of frequent, basic and rare earthquakes, respectively.

The artificial time history of acceleration is 1-D wave, while El Centro and Taft ground motion records are all 2-D waves. If the main direction is X, the peak acceleration in the direction Y can be calculated by:

$$a_y = 0.85 \cdot a_x \tag{3}$$

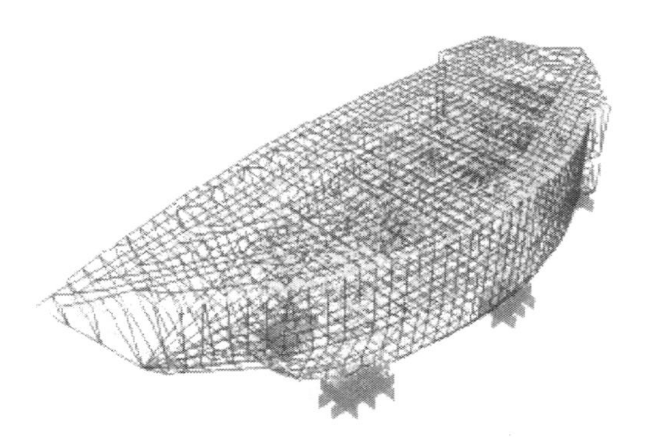

Figure 6. FEM model of the structure.

where, a_x, a_y represent the peak acceleration in direction X and Y, respectively. The factor of 0.85 was regulated by CCSDB.

4.2 *Analysis of FEM results*

4.2.1 *Natural periods and vibration modes*

The vibration periods and modes calculated for the base-isolated and base-fixed buildings are presented in Table 2, Table 3, Fig. 9, and Fig. 10, respectively. The results illustrate that the isolation system prolongs the building's periods of vibration, and the vibration energy concentrates to

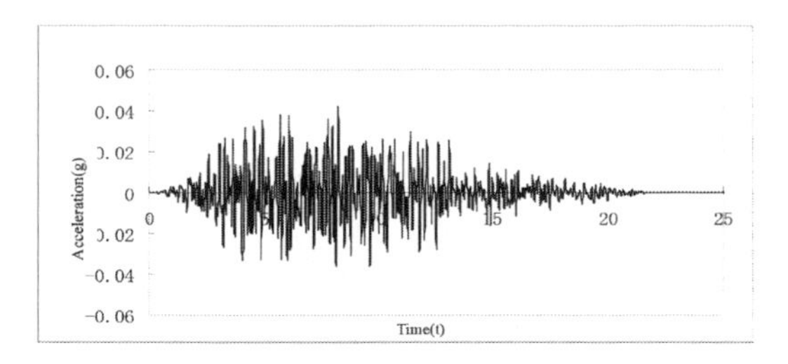

Figure 7. Time history curve of artificial acceleration.

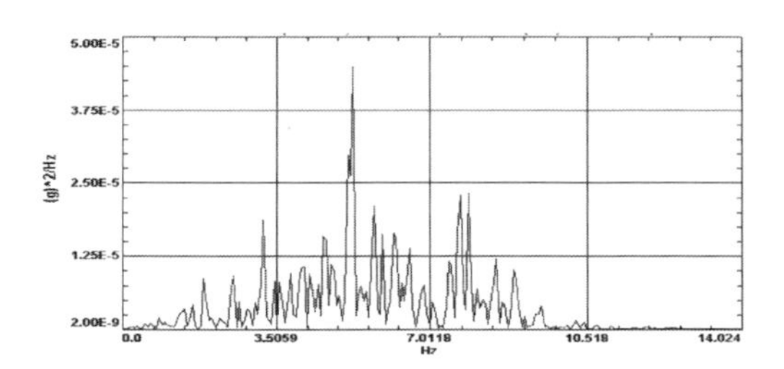

Figure 8. Spectrum of the artificial acceleration.

Table 2. Comparison of periods and modes from FEM.

Type	Mode	Period (s)	Direction
Base-fixed structure	1	0.854	Y
	2	0.717	Torsion
	3	0.590	X
Isolated structure	1	1.819	Y
	2	1.710	X
	3	1.495	Y

Table 3. Participating mass ratios of vibration modes and accumulated values for the isolated structure.

Mode	UX translation of X-Dir	UY translation of Y-Dir	UZ translation of Z-Dir	ΣUX_i	ΣUY_i	ΣUZ_i	RX Rotation of X-Dir	RY Rotation of Y-Dir	RZ Rotation of Z-Dir
1	0.03307	0.86	0.0000038	0.03307	0.86	0.0000038	0.25	0.002043	0.89
2	0.96	0.03816	0.0000051	0.99	0.9	0.0000089	0.01077	0.06025	0.01648
3	0.004791	0.09566	0.0000055	1	0.99	0.0000149	0.0482	0.0005402	0.08844

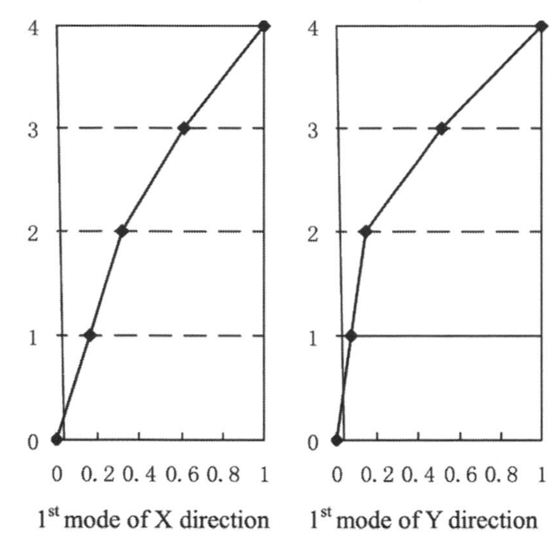

Figure 9. Vibration modes of base-fixed building from FEM.

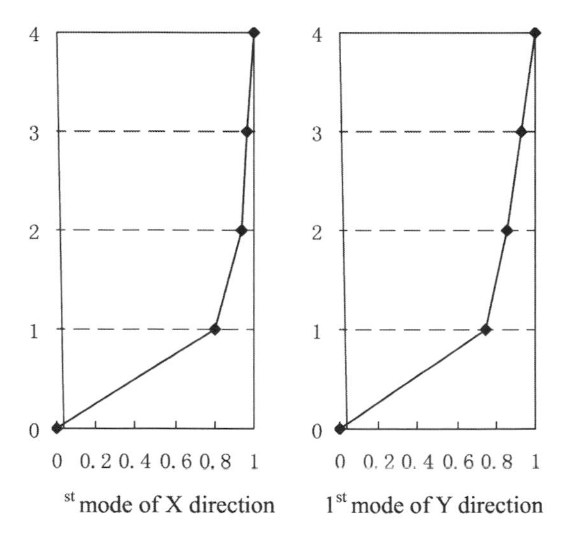

Figure 10. Vibration modes of isolated building from FEM.

the first three modes. (In Fig. 9 and Fig. 10, the number "0" denotes the surface of the shaking table, and "1" denotes the first floor). In the base-isolated building, the isolation system has large deformation but the superstructure behaves essentially rigid, while the base-fixed one behaves like a flexural beam. With isolation system, the main modes change from bent or/and torsion vibrations to translations. Isolation system can upgrade the vibration modes of the irregular superstructure, and make the seismic response smooth and regular.

4.2.2 *Acceleration response*

The maximum acceleration responses for each floor of the buildings subjected to the 1940 El Centro ground motion record, scaled to 0.31 g PAG, in two directions are shown in Table 4. It is shown that the maximum accelerations on the roof of base-fixed building are 0.496 and 0.412 g in x and y directions, with corresponding amplification factors of 2.65 and 1.87, respectively. The maximum accelerations on the roof of base-isolated building are 0.061 and 0.076 g in x and y directions, with amplification factors of approximately 0.30. The amplification factors for the roof of isolated building are far below those of the base-fixed one, with a reduction of about 85%.

4.2.3 *Displacement response*

Analytical results indicate that with isolation system, the structural deformation converges to the isolating bearings, and the maximum deformation of the bearings is about 107 mm for the base-isolated building subjected to the 1940 El Centro record with rarely intensity. The ratios of the maximum storey drift to the ground motions, scaled to frequent, basic and rare intensity, along X and Y direction are shown in Table 5. According to CCSDB, the analytical results in Table 5 are all less than the limit value of elastic story drift rotation for multi-story and tall steel structures (1/250), which means that both buildings are in elastic range.

Table 4. Maximum accelerations of all floors from FEM/g.

	Response of X-Dir			Response of Y-Dir		
Floor	Isolated	Base-fixed	Isolated/Base-fixed	Isolated	Base-fixed	Isolated/Base-fixed
4	0.061	0.496	0.123	0.076	0.412	0.184
3	0.060	0.291	0.206	0.067	0.287	0.234
2	0.055	0.213	0.258	0.066	0.229	0.288
1	0.055	0.187	–	0.066	0.220	–
0	0.187	0.187	–	0.220	0.220	–
Maximum amplification factors	0.30	2.65		0.33	1.87	

Table 5. Analytical ratios of the maximum inter-storey drifts from FEM.

Seismic intensity	Base-fixed structure (1/θ)		Isolated structure (1/θ)		Reduction ratio	
	X-Dir	Y-Dir	X-Dir	Y-Dir	X-Dir	Y-Dir
Frequently	5213	2726	13581	7976	0.62	0.66
Basic	1719	1205	5643	3876	0.70	0.69
Rarely	1077	726	3418	2183	0.68	0.67

Comparisons between the base-fixed and the based-isolated buildings in Table 5 show that the isolation system reduces the storey torsion drift and the internal forces of the base-isolated building, while maximum elastic storey drift for the isolated building are probably only one third of that for the base-fixed one.

5 MODEL TEST

5.1 *Description of the shaking table*

The shaking table test was conducted on MTS shaking table system at Guangzhou University. The table can input three-dimensional and six degree-of-freedom motions. The dimension of the table is 3 m × 3 m, and the maximum payload is 150 kN. The shaking table can vibrate with two maximum horizontal direction accelerations of 1.0*g* and a maximum acceleration of 2.0*g* vertically, with the frequency ranging from 0.1 Hz to 50 Hz.

5.2 *Model manufacture*

The model was designed by scaling down the geometric and material properties from prototype structure (Lu, 2007; Deyin, 1996). The basic model similitude rules were established from the scaling theory (Sabnis, 1983), which is generally known and therefore there is no need for further explanation of the theory. Since the dynamic behavior of a structure is fully described by means of three basic quantities, only three independent parameters can be selected when designing a model. Subsequently, other parameters are expressed in terms of the basic scale factors chosen. Considering the capacity and the size of the shaking table to be used, the dimension scaling parameter (S_L) was chosen as 1/35, so the weight of the model was kept less than 200 kN, the maximum load of the shaking table.

Steel structural elements were modeled with grade Q235 steel pipe, which is used as main materials in the prototype structure. Its Young's modulus scaling parameter (S_E) is 1.0.

Table 6. Similitude scale factors for the test model.

Controls parameters	Symbols	Calculation Equations	Values
Length	S_L	L_m/L_p	1/35
Young's modulus	S_E	E_m/E_p	1.00
Acceleration	S_a	$S_a = S_E S_L^2/S_m$	2.439
Time	S_t	$S_t = \sqrt{S_L/S_a}$	0.11
Frequency	S_f	$S_f = 1/S_t$	9.26
Mass	S_m	Q_m/Q_p	1/3000
Displacement	S_d	$S_d = S_L$	1/35
Strain	S_ε	$S_\varepsilon = 1$	1.00
Stress	S_σ	$S_\sigma = S_E$	1.00
Stiffness of isolating bearings	S_k	$S_k = S_m S_a/S_L$	1/35

Subsequently, the third parameter (*Sa*) concerning the horizontal acceleration can be determined. Since the peak value of the noise is close to or larger than the amplitude of seismic waves under frequent occurrences intensity (the maximum acceleration is 35 cm/s^2), errors would occur in the test data acquisition. Amplified accelerations are demanded to prevent measurement from being distorted by the noise. According to the instrument capacities, *Sa* was selected to be 2.439. The test program was designed on the basis of the final similitude scaling relationship. The main scaling parameters are presented in Table 6. It is noted worthy that the following equation should be established according to the similitude rule:

$$\alpha_m/\alpha_p = g_m/g_p \tag{4}$$

where a_m is the horizontal acceleration of the model structure, α_p is the horizontal acceleration of the prototype structure, g_m is the gravitational acceleration of the model structure, and g_p is the gravitational acceleration of the prototype structure. However, Eq. 4 is no longer feasible unless the model is excited in a centrifuge. In this case, adoption of the artificial mass becomes the only choice to fulfill the similitude requirements. In practical test, this is achieved by the addition of suitably distributed weights, which are attached to the model in a manner that does not change the strength and stiffness of the floor beams and slabs.

To ensure an effective transmission for the earthquake motions to the base of the test model, the base plate of the model was mounted on the shaking table through two methods, bolt connections and isolation connections. The arrangement of the isolation bearings on the shaking table for the experimental model is shown in Fig. 11, while the performance param-

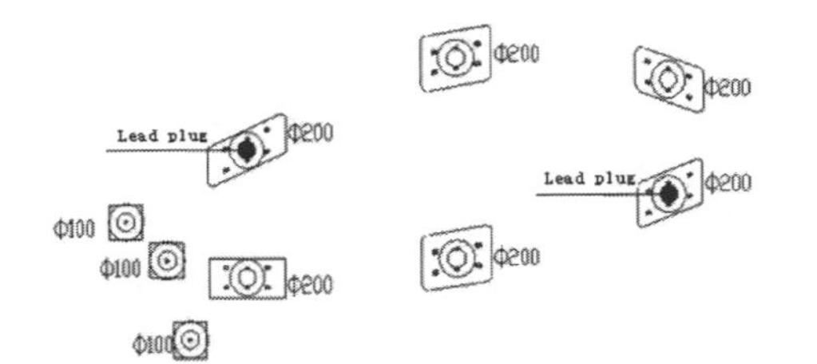

Figure 11. Arrangement of isolating bearings on the shaking table.

Table 7. Performance parameters of the bearings in shaking table test setup.

Type	GZP200G6		GZY200G6	
Lateral deformation	50%	100%	50%	100%
Yield stiffness (kN/mm)	0.39	0.37	1.97	1.32
Yield load (kN)	0.437		0.758	
Effective lateral stiffness (kN/mm)	0.38	0.33	1.86	1.20
Effective damping ratio (%)	5	5	8	10

Figure 12. Test model on the shaking table.

eters are presented in Table 7. Fig. 12 shows the constructed test setup. The displacements of the bearings were measured with laser displacement transducers. It should be pointed out that the designed input acceleration amplitudes differ from the measured amplitude on shaking table, to some extent due to the effects of the control system and mechanical system of the shaking table. So the acceleration amplitude on shaking table should be measured.

The ground motions applied to the test setup with increasing amplitudes are the same as the records selected for previous dynamic analysis. According to the similitude factors shown in Table 6, the frequency ratio 9.26 means that the time ratio is 1/9.26. The incrementally scaled excitations are input successively in a manner of time-scaled earthquake waves. After different series of ground acceleration are inputted, white noise is scanned to determine the natural frequencies of the model structure.

5.3 *Results analysis for model test*

5.3.1 *Natural periods and vibration modes*

According to similarity law (Zhou, 2005; Huang, 1997), the vibration characteristics of the prototype building are shown in Table 8, Fig. 13 and Fig. 14.

Table 8. Comparison of natural periods from the analytical and experimental studies (s).

	Direction of mode	Base-fixed structure	Isolated structure
Analytical value	x	0.590	1.710
	y	0.854	1.819
Experimental value	x	0.67	1.54
	y	1.02	1.77
Error	x	11.54%	–9.94%
	y	16.27%	–2.59%

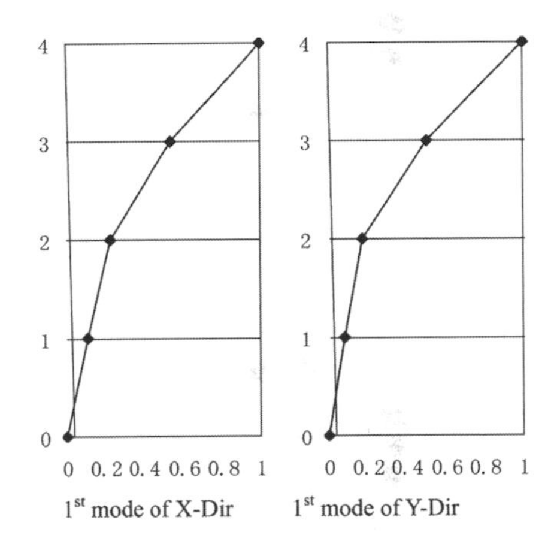

Figure 13. Vibration modes of base-fixed building from model test.

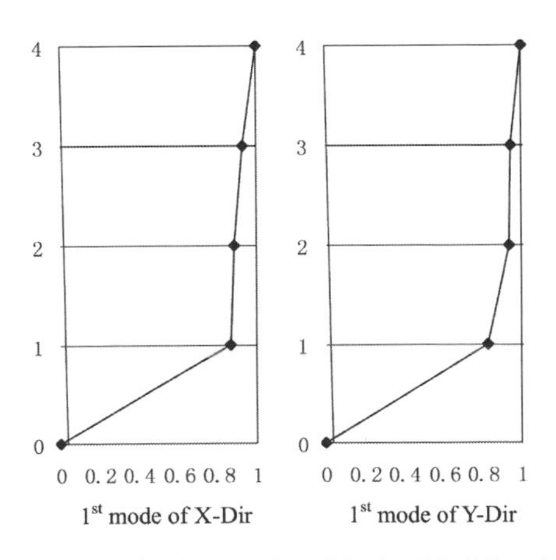

Figure 14. Vibration modes of isolated building from model test.

The fundamental period of the base-isolated building from the experimental data is 1.54 s, which is within 9.94% of the analytical results, while the fundamental period of the base-fixed building from the experimental studies is 0.67 s, with a difference of 11.54%. It is indicated that the fundamental periods of the building from experimental studies meet well with the analytical results. The shapes of the vibration modes for the test specimen with base-fixed and base-isolated are similar to the analytical modes for both building, respectively. It reveals that the test specimen can capture the vibration characteristic of the prototype structure perfectly, and the stiffness simulation for isolation bearing is successful.

5.3.2 *Acceleration response*

According to the Eq. 5, the peak acceleration of the prototype structure can be calculated from the experimental results:

$$a_{pi} = K_i a'_m / S_a \tag{5}$$

where a_{pi} is the peak acceleration for the i th floor of the prototype structure (g); K_i is the magnification factor of the acceleration at the i th floor; a'_m is the peak acceleration of the applied ground motion (g); S_a is the similitude scale factor for acceleration, 2.439, as shown in Table 6.

Subjected to ground motions scaled to 0.31 PGA, the peak accelerations of each floor of the prototype structure are shown in Table 9. It indicates that the acceleration response of the base-fixed structure is increasing as the height increased, and the magnification factors of the acceleration on the roof are approximately 2.10 and 1.57 in x and y directions, respectively. The acceleration response on the roof of the isolated building is only 0.056 and 0.064 g in x and y directions, with the magnification factor of approximately 0.30. The acceleration magnification factors for the roof of the base-isolated building are about 6 times smaller than those for the base-fixed one, meet well with the analytical results.

5.3.3 *Displacement response*

The peak displacement for the prototype structure can be calculated from the experimental results using Eq. 6:

$$D_{pi} = \frac{a'_m}{a_t} \frac{D_{mi}}{S_d} \tag{6}$$

where D_{pi} is the peak displacement for the i th floor of the prototype structure (mm); D_{mi} is the peak displacement for the i th floor of the test model (mm); a_t is the peak acceleration measured on the shaking table corresponding to a'_m (g); S_d is the similitude scale factor for displacement, 1/35, as shown in Table 6.

The ratios of the peak drifts on each storey to the ground motions are presented in Table 10. It is indicated that the ratio of the maximum inter-story drift rotation for the base-fixed building to rare intensity earthquakes is 1/785, and the ratio for the isolated building is 1/2339, less than the limit value of elastic story drift rotation for multi-story and tall steel structures (1/250). The average storey drifts of the isolated building are less than

Table 9. Maximum accelerations of all floors from experimental studies /g.

	Response of X-Dir			Response of Y-Dir		
Floor	Isolated	Base-fixed	Isolated/Base-fixed	Isolated	Base-fixed	Isolated/Base-fixed
4	0.056	0.393	0.14	0.064	0.345	0.19
3	0.041	0.268	0.15	0.062	0.193	0.17
2	0.038	0.192	0.20	0.053	0.191	0.28
1	0.053	0.187	–	0.056	0.220	–
0	0.187	0.187	–	0.220	0.220	–
maximum magnification factor	0.30	2.10		0.29	1.57	

Table 10. Ratios of the maximum inter-story drifts of the prototype building from experimental studies.

Earthquake intensity	Base-fixed building (1/θ)		Isolated building (1/θ)		Reduction ratio	
	Y-Dir	X-Dir	X-Dir	Y-Dir	X-Dir	Y-Dir
Frequently	5593	2880	14269	8318	0.61	0.65
Basic	1852	1395	6035	4322	0.69	0.68
Rarely	1134	785	3440	2339	0.67	0.66

Table 11. Maximum deformation responses of the prototype building from experimental studies (mm).

Input direction	Wave	X-Dir	Y-Dir
X-Dir	El Centro	111.65	
Y-Dir	El Centro		73.5
X+Y-Dir	Artificial	43.4	39.9
	El Centro	95.2	71.75
	Taft	63.7	57.05

those of the base-fixed one by reducing the peak story drifts by a reduction of 61–69%, which is within 10% of the analytical drifts.

The peak displacement of the isolation layer for the prototype structure estimated from the experimental results of the test specimen subjected to the rare intensity ground motions is shown in Table 11. The peak displacement in the case of the 1940 EI Centro record reaches to 111.65 mm, consisting well with the previous analytical results, 107 mm.

6 CONCLUSION

The E-Block of Guangdong Science Center is a complex large span steel frame. Isolation system was introduced in the mega structure for the first time in the China. Through the non-linear dynamic analysis and shaking table test, the following conclusions are the main finding of the study:

1. Base isolation system lengthens the fundamental period of the building, and the basic vibration modes change from bent or/and torsion to translation, upgrade the vibration modes of the superstructure, and make the seismic response smooth and regular.
2. The natural characteristic of the prototype structure calculated form the experimental results matches the FEM analytical results well, so the experimental test setup can simulate the dynamic properties of the prototype structure preferably.
3. During the ground motions scaled to 0.31 g PGA, the acceleration magnification factors for the base-isolated building are about 6 times smaller than those of the base-fixed one. The modification effect of the isolation system is remarkable.
4. The storey drifts of the base-isolated building are less than those of the base-fixed one greatly, and the reduction ratios are about 0.61–0.69. The deviation to the analytical values is less than 10 percent.
5. Subjected to the ground motions scaled to 0.31 g PGA, both the base-fixed and the base-isolated buildings are in elastic stage, while the internal forces in the base-isolated building is probably one third of those in the base-fixed one. The peak deformations of the isolating layer in the case of the 1940 EI Centro record are in the range of 39.9–111.65 mm, which consists with the analytical value.

On the whole, the isolation system can improve the seismic performance of the structure, especially for the structure with an irregular plan. Besides, because of the stiffness of the huge columns, the response of each huge column of the base-isolated building is harmony basically. The study indicates that the isolation design of GSC is rational, and is a helpful explore to the seismic design measure for huge civil engineering.

ACKNOWLEDGEMENTS

This study has been supported by the Collaborative Innovation Major Project of Guangzhou City under grant Number (201604020071).

REFERENCES

Cem Topkaya (2004). "Analysis of specimen size effects in inclined compression test on laminated elastomeric bearings". *Engineering Structures* 26(5):1071–1080.

Charleson A.W. (2008). "Reflections on Aspects of New Zealand's Seismic Resilience: Comparisons with California Practice". *NZSEE Conference*. No.59.

China Ministry of Construction. "Code for seismic design of buildings (GB 50011-2010) (2010)". Beijing: China Architecture and Building Press.

Goda, K., C.S. Lee, H.P. Hong (2010). "Lifecycle cost–benefit analysis of isolated buildings". *Structural Safety* 32(1): 52–63.

Huang Ximing, Zhou Fulin (1997). "Research on modal test of structure system and Computer simulation". *Engineering Mechanics*. 123(1): 44–48.

Jangid R.S., (2007). "Optimum lead–rubber isolation bearings for near-fault motions". *Engineering Structures* 29 (10) 2503–2513.

Li Deyin, Wang Bangmei (1996). "Model Test of Structure". Beijing: Science Press.

Masahiko Higashino, Shin Okamoto (2006). "Response control and seismic isolation of buildings". Taylor & Francis Group.

Murat Dicleli, Srikanth Buddaram (2007). "Comprehensive evaluation of equivalent linear analysis method for seismic-isolated structures represented by sdof systems". *Engineering Structures* 29(8):1653–1663.

Providakis C.P. (2009). "Effect of supplemental damping on LRB and FPS seismic isolators under near-fault ground motions". *Soil Dynamics and Earthquake Engineering* 29(1) 80–90.

Sabnis GM, Harris HG, White RN, Mirza MS (1983). "Structural Modeling and Experimental Techniques". Prentice-Hall: Englewood Cliffs, NJ.

Shen Zuyan, Hunag Kuisheng, Chen Yiyi, Tong Jun (2006). "Shaking Table Test Model with Steel Braced-Frame Structure About Main Plant of Large Thermal Power Plant". *Journal of Architecture and Civil Engineering*. 23(4):1–5.

Skinner RI, RobinsonWH, McVerry GH (1993). "An Introduction to Seismic Isolation". John Wiley & Sons Ltd.

Su Jingyu, Zeng Demin (2001). "Research and Application of Seismic Isolation System for Civil Engineering Structures in China". *Earthquake Engineering and Engineering Vibration,* 21(4):94–101.

Vojko Kilar, David Koren (2009). "Seismic behaviour of asymmetric base isolated structures with various distributions of isolators". *Engineering Structures*, 31 (4): 910–921.

Xilin Lu, Yin Zou, Wensheng Lu (2007). "Shaking table model test on Shanghai World Financial Center Tower". *Earthquake Engng Struct. Dyn.* 36:439–457.

Zhou Fulin, Zhang Jichao (2005). "Guangdong science center isolation and energy dissipation structure model to simulate an earthquake shake table test study (ST-0501-189)". Guangzhou: Earthquake Engineering Research & Test Center of Guangzhou University.

Civil, Architecture and Environmental Engineering – Kao & Sung (Eds)
© 2017 Taylor & Francis Group, ISBN 978-1-138-02985-9

Effect of material ingredients on the cement-based self-leveling material

Jun Liu, Jiajian Hu, Runqing Liu & Yuanquan Yang
School of Materials Science and Engineering, Shenyang Jianzhu University, Shenyang, China

ABSTRACT: The effect of water-reducing agents, fly ash, and re-dispersible latex powders on the flexural strength, compressive strength, and fluidity of the cement-based self-leveling material was investigated. The results show that the addition of fly ash can improve the fluidity of self-leveling mortar; however, excessive fly ash leads to a higher water requirement. Fluidity of the system decreases when additional fly ash is added at a constant water–cement ratio, which decreases the compressive strength and flexural strength. Water-reducing agents can improve the fluidity of self-leveling mortar; however, their compressive strength decreases after 1–7 days. Addition of re-dispersible latex powders leads to more water absorption and unfavorable fluidity.

1 INTRODUCTION

Compared with developed countries, the research on the ground leveling materials started late in China, which was proposed about at the end of the 1980s and began about the 1990s. Originally, the development speed, popularization, and promotion are relatively slow (Dong Sufen, 2009). Because of the rapid development of China's reform and opening-up policy and rapid development of economy, high quality of foreign re-dispersible latex powder, water-reducing agent, and other materials were introduced (Ren Zeng zhou, 2010). Therefore, the quality of these materials has been greatly improved in China, and their price is relatively low too. Thus, it is important to provide a good environment for further research so that self-leveling materials undergo fast development.

Cement-based self-leveling mortars have advantages of good fluidity, fast strength development, simple construction, and good leveling, and they are often used in concrete ground of office buildings, apartments, shopping malls, schools, hospitals, factories, parking lots, and ship (Gao Shujuan, 2013 & Huang Tianyong, 2015). They have unique advantages compared with conventional mortar in floor construction. However, there are many problems in their popularization and application. Their high price is a particularly serious problem; therefore, it is very important to control the cost of self-leveling mortar. Because the total amount of the emission of fly ash from China's coal-fired power plant annually increased, the pressure on the environment became severe. Use of fly ash can not only reduce the price, but also reduce the enormous pressure on the ecological environment and improve the self-leveling mortar's fluidity at a certain extent.

Recently, researchers have studied the preparation and properties of self-leveling mortar in China. Gu Jun (2008) studied polycarboxylate superplasticizer-modified self-leveling cement and pointed out that the polycarboxylate was suited for self-leveling cement mortar and is highly cost-effective. However, polycarboxylate superplasticizer may change air entraining performance slightly, so bubble problem should be attended. Wu Jie and Guo Qiang found that the complex retarder can make up for the deficiency of a single retarder, which has the advantages of each retarder and improves the comprehensive performance of the retarder. Hassan et al. used POFA and PBC in self-leveling cement mortar; POFA and PBC can prevent the separation of the mortar with a high-efficiency water-reducing agent. The paper by Canbaz M described that the setting time of cement increased with the increase of the additive. In Self-Compacting Mortar (SCM), Gneyisi et al. used fly ash instead of natural fine aggregate, in which the work of self-leveling mortar was improved effectively.

It is very important to study the effect of material composition on the physical properties of cement-based self-leveling material, considering that the composition of the cement-based self-leveling material was decided by the mixture ratio. According to the different mixture ratios, different properties of self-leveling materials resulted. Fly ash, water-reducing agent, and re-dispersible latex powder were critical factors and hence their effects on the mechanical properties of leveling material were investigated. It is significant to guide the production and preparation of self-leveling materials. The experimental research program in this paper is designated to change fly ash, water-reducing agent, and re-dispersible latex powder content, aiming to study the influence of mixture ratio on compressive strength and flexural strength of cement-based self-leveling mortar.

2 EXPERIMENT

2.1 *Materials*

Portland cement used in this study is provided by Dalian Xiao Ye Tian cement plant. Fly ash, water-reducing agent, re-dispersible latex powder, calcium hydroxide, and sand are also used in this experiment.

2.2 *Mix ratio design*

The proportions of mortar mixture were 20:40:40:0.6:0.35:1:1 by mass of ordinary Portland cement, fly ash, water-reducing agent, re-dispersible latex powder: retarder: calcium hydroxide, respectively. Fly ash, water-reducing agent, and re-dispersible powder were selected as the three main variables. Fly ash was used at 20%, 25%, 30%, 35%, and 40% replacement by weight for cement while others were kept constant. Re-dispersible latex powder was used at 0.3%, 0.35%, 0.4%, 0.45%, 0.5%, and 0.55%, and water-reducing agent was used at 0.5%, 0.55%, 0.60% 0.65%, 0.70%, and 0.75%, respectively.

2.3 *Methods*

In accordance with the proportion of raw materials, the materials were weighted and then poured into a mixing container. Then, low-speed stirring was performed for 1 min, stopped for 30 s, continued at high speed for 1 min, stopped for 5 min, and then continued at high speed for 15 s. If there are bubbles, stirring was stopped for 1 min. The specimens were demolded and cured in lime-saturated water at 23 ± 2°C until testing.

The specimens were kept in molds for 24 h at the room temperature of 23 ± 2°C, and the relative humidity was 50. Wind speed of the test area was less than 0.2 m/s. All experimental methods were in accordance with the JC/T985-2005, "ground with cement-based self-leveling mortar".

3 RESULT AND DISCUSSION

3.1 *Effect of addition of fly ash on the performance of self-leveling mortar*

According to the experiment, the mixture ratio of fly ash was changed. The fly ash was used at 20%, 25%, 30%, 35%, and 40%, while others were kept constant. The compressive strength and flexural strength of the mortar were determined at 1 and 7 days; the test results are presented in Figure 1(a) and Figure 1(b). Fluidity of the self-leveling material was determined, and the test results are presented in Figure 1(c).

Fly ash has a certain effect on fluidity in the cement-based self-leveling material, and the fluidity

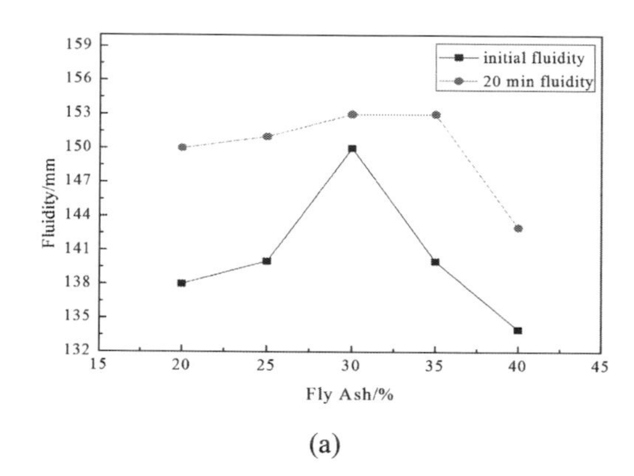

(a)

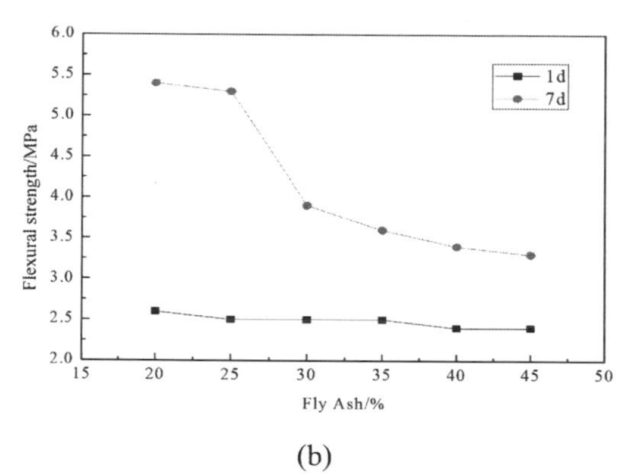

(b)

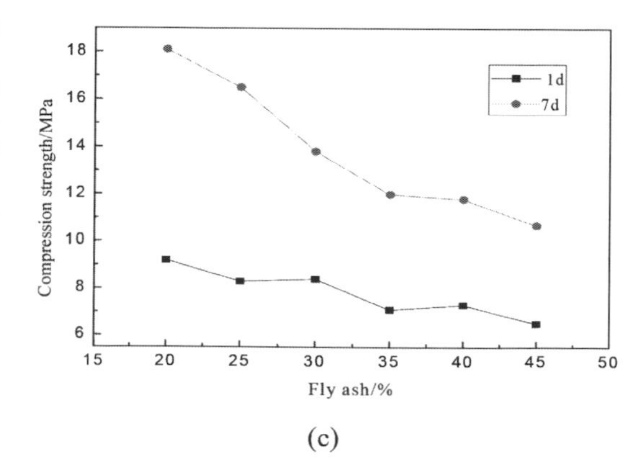

(c)

Figure 1. Effect of addition of fly ash on the strength and fluidity of the self-leveling mortar.

of fly ash can be improved by appropriate addition of fly ash. Figure 1(a) shows the effect of fly ash on the fluidity of the self-leveling mortar, the initial fluidity and 20-min fluidity of self-leveling mortar are more than 130 mm in accordance with the national standard. The fluidity at 20 min is higher than the initial fluidity, which is due to the effect of retarder. The cement flocculation structure is destroyed by the retarder, and more free water is released so that the fluidity increases. When the fly ash is 30%, the fluidity of fly ash reaches its maximum. The initial fluidity and 20-min fluidity are 150 and 153 mm, respectively. With the increase of fly ash content, the fluidity of self-leveling mortar first increases and then decreases. The result may be attributed to the existence of a large amount of glass beads from fly ash. Because of the particle size effect and water-reducing performance from these glass beads, the fluidity of self-leveling mortar increases. When fly ash content is more than 30% and the water cement ratio is kept constant, a large amount of water is needed, so fluidity of self-leveling mortar decreases. It is important to choose an appropriate amount of fly ash to improve the fluidity of self-leveling mortar.

Figure 1(b) shows the effect of fly ash content on the flexural strength of self-leveling mortar, and the amount of fly ash has a great influence on the flexural strength at 7 days. When the amount of fly ash is less, its flexural strength is higher, the flexural strength at 1 day reaches 2.6 MPa, and that at 7 days reaches 5.4 MPa. With the increase in fly ash content, the flexural strength of the self-leveling mortar at 1 day decreased slowly and remains almost unchanged. This result may be attributed to the pozzolanic effect of fly ash. There is almost no hydration reaction for fly ash at 1 day, during which period fly ash just plays the role of a micro aggregate and a filler. When the amount of the fly ash is 20–25%, mortar flexural strength at 7 days decreased slowly. However, when the amount of the fly ash is 25–30%, the flexural strength at 7 days decreased rapidly, by about 26.4%. When the amount of the fly ash mortar is 30–45%, the flexural strength at 7 days tends to be low. With the increase of fly ash content, the flexural strength of self-leveling mortar showed an overall downward trend. On the one hand, the rate of cement hydration decreases by low activity of fly ash. On the other hand, calcium silicate hydrate is reduced, which results in a gradual decrease in the strength of the self-leveling mortar.

Figure 1(c) shows the effect of fly ash content on the compressive strength of self-leveling mortar. When fly ash content is 20%, the compressive strength reaches its maximum value, that is, 18 MPa. With the increase of fly ash, strength at 1 and 7 days were gradually decreased. Compared 45% with 20%, self-leveling mortar's compressive strength at 1 and 7 days were decreased by 40.9% and 29.4%, respectively. When the amount of fly ash is more, the pozzolanic effect of fly ash is very low and reaction cannot fully take place, so the strength decreases by a large margin. Considering the influence of fly ash content on the fluidity and compressive strength, fly ash content of 35% is the best.

3.2 *Effect of addition of water-reducing agent on the performance of self-leveling mortar*

According to the experiment, the mixture ratio of water-reducing agent was changed. The water-reducing agent was used at 0.5%, 0.55%, 0.60% 0.65%, 0.70%, and 0.75%, respectively, while others were kept constant. The compressive strength and flexural strength of mortar were determined at 1 and 7 days, and the test results are presented in Figure 2(a) and Figure 2(b). Fluidity of self-leveling material was determined, and the test results are presented in Figure 2(c). High-performance water-reducing agent is the primary material of self-leveling mortar, and it is the most important influential factor on the fluidity and strength.

Figure 2(a) presents the effect of addition of fly ash water-reducing agent on the fluidity of self-leveling mortar. The effect of water-reducing agent on the fluidity is obvious. When water-reducing agent is 0.5%, fluidity of self-leveling mortar is the minimum. Initial fluidity and 20-min fluidity are 120 and 135 mm, respectively, but the initial fluidity does not meet the national standard. The national standard requested the initial fluidity to be more than 130 mm. With the increase of water-reducing agent, the fluidity increases obviously. When the content of water-reducing agent is low, liquidity is not large. This is because water-reducing agent is not enough to disperse cement particles, and the thickness of adsorbed water film on the surface of cement was reduced. When the amount of water-reducing agent is large, the fluidity of self-leveling mortar is gradually increased. This is because the water-reducing agent is mixed into fresh cement paste and cement particles dispersed by water-reducing agent become more and more obvious. The flocculation structure of cement particles was damaged more significantly, thus more free water from flocculation structure was released. Therefore, the fluidity of cement paste increased. Fluidity at 20 min did not decrease; however, it was higher than the initial fluidity, which is mainly caused by the combined effect of retarder and water-reducing agent during the hydration.

Figure 2(b) presents the effect of water-reducing agent on the flexural strength of self-leveling

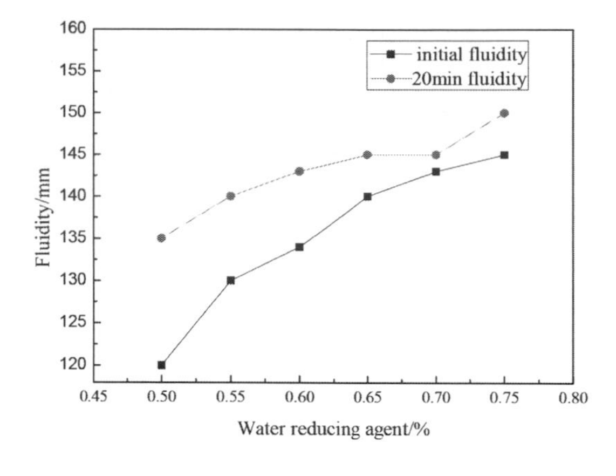

(a)

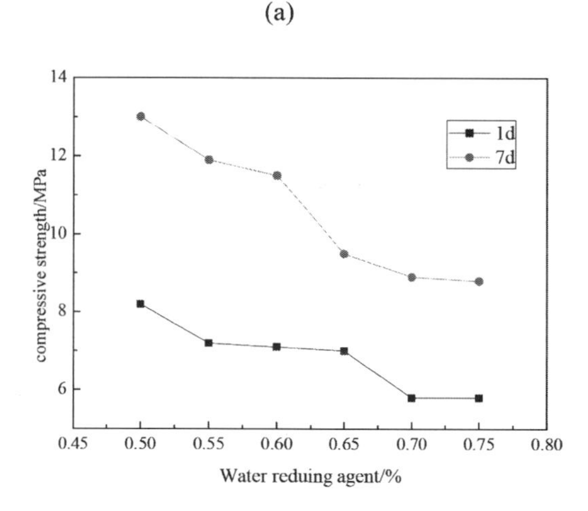

(b)

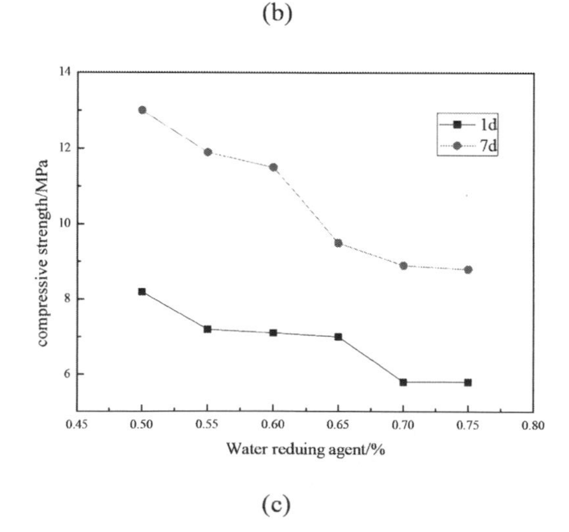

(c)

Figure 2. Effect of addition of water-reducing agent on strength and fluidity of the self-leveling mortar.

mortar. The influence of water-reducing agent on the flexural strength at 1 and 7 days is different. With the increase of water-reducing agent, the flexural strength at 1 day first increases and then decreases. When the amount of water-reducing agent is 0.60%, the flexural strength at 1 day is the best. At the low content of water-reducing agent, cement particles were dispersed by water-reducing agent more homogeneously. Mortar specimens become more compacted, so strength at 1 day increases. When the content of water-reducing agent increases, $Ca(OH)_2$ of the hydration products become plate from the block, hardening structure of the mortar become loose, so the appropriate water-reducing agent has strong effect on the early strength of the self-leveling mortar. However, with the increase in the content of water-reducing agent, the mortar flexural strength at 7 days decreases. The result may be attributed to the hydration. With water-reducing agent, more plate calcium silicate hydrates were produced, and the mortar structure was more loose, thereby reducing the strength.

Figure 2(c) presents the effect of water-reducing agent on the compressive strength of self-leveling mortar. The effect of water-reducing agent at 1 and 7 days is the same. With the increase in the content of water-reducing agent, the compressive strengths at 1 and 7 days are gradually reduced, 0.5% was compared to 0.75%, compressive strengths at 1 and 7 days were decreased by 29.3% and 32.3%, respectively.

3.3 *Effect of addition of re-dispersible latex powder on the performance of self-leveling mortar*

According to the experiment, the mixture ratio of re-dispersible latex powder was changed. The re-dispersible latex powder was used at 0.3%, 0.35%, 0.4%, 0.45%, 0.5%, 0.55%. The fluidity, compressive strength, and flexural strength are measured.

Figure 3(a) presents the effect of re-dispersible latex powder on the compressive strength of self-leveling mortar. The influence of re-dispersible latex powder with water-reducing agent and fly ash is different. When re-dispersible latex powder is 0.3%, the initial fluidity reached 143 mm, 20-min fluidity reached 145 mm; in other words, 20-min fluidity is not lost. When re-dispersible latex powder is 0.5%, the initial fluidity is 128 mm, 20-min fluidity is 135 mm, 0.5% compared with 0.3%, the initial fluidity and 20-min fluidity decreased by 11.7% and 6.9%, respectively. With the increase of re-dispersible latex powder, the fluidity of mortar is gradually reduced, which is due to the function of water retention. When re-dispersible latex pow-

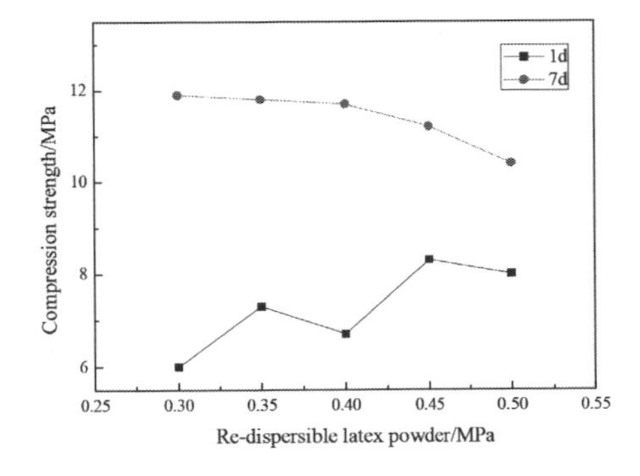

(a)

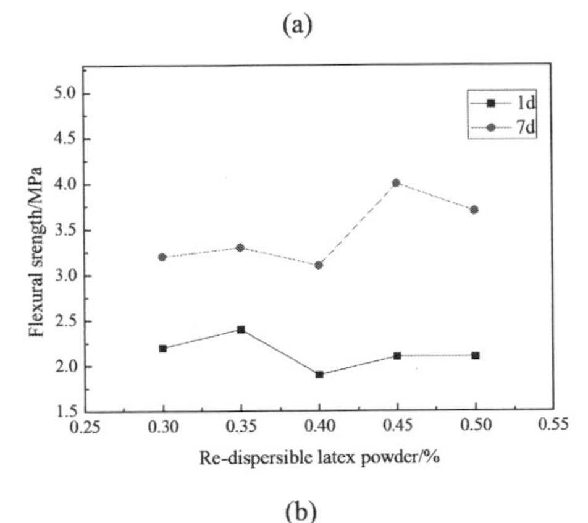

(b)

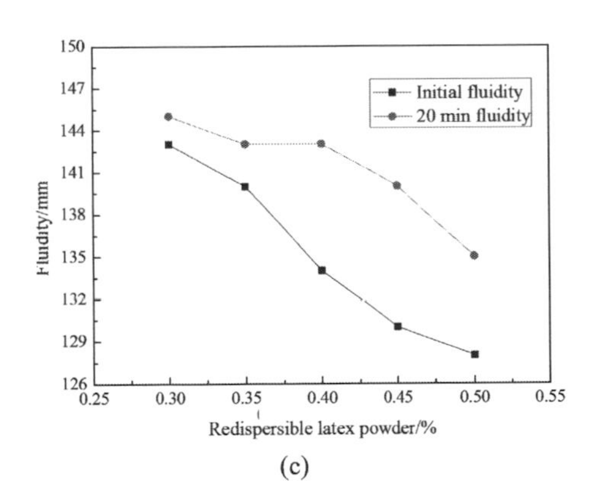

(c)

Figure 3. Effect of addition of re-dispersible latex powder on strength and fluidity of the self-leveling mortar.

der was excessive, cohesion of the new self-leveling mortar became so large that fluidity was reduced. Therefore, the amount of emulsion powder should not be too high.

Figure 3(b) presents the effect of re-dispersible latex powder on the flexural strength of self-leveling mortar. When re-dispersible latex powder is 0.4%, the flexural strength at 1 and 7 days are the lowest, 1.9 and 3.1 MPa respectively, and the flexural strength at 1 day does not meet the national standard. When the powder is 0.35%, the maximum flexural strength at 1 day is 2.4 MPa. When the powder is 0.45%, the maximum flexural strength at 1 day is 4.0 MPa.

Figure 3(c) presents the effect of re-dispersible latex powder on the compressive strength of self-leveling mortar; with the increase of powder, compressive strength at 1 day increased gradually. When the powder content is 0.3%, the compressive strength at 1 day reached 6 MPa. When the powder content is 0.5%, compressive strength at 1 day reached 8.3 MPa and the strength was increased by about 27.7%. Because of the initial hydration, the water content is further reduced by the hydration of cement or dry powder. The polymer particles gradually condensed, and a thin film was formed. The film has a higher ability of deformation, and the hydration products and cement aggregate comingle, resulting in an increase of compressive strength. However, the compressive strength at 7 days is different from that at 1 day. With the increase in the amount of powder, the compressive strength decreased from 11.9 to 10.4 MPa gradually, which is due to the destruction of the original structure by hydration product, so the mortar strength at 7 days did not increase but decrease. Therefore, the amount of the re-dispersible latex powder should be controlled in a certain range.

4 CONCLUSIONS

1. In a certain extent, fly ash can increase the fluidity of self-leveling mortar, early strength of the mortar almost is not influenced; however, fly ash is not unfavorable to mortar compressive strength and compressive strength at 7 days.
2. Water-reducing agent can improve the fluidity of self-leveling mortar; however, it is unfavorable to the compressive strength of self-leveling mortar at 1 and 7 days.
3. Re-dispersible latex powder can increase the water absorption and raise the corresponding consistency. With the increase of re-dispersible latex powder, the fluidity of mortar is gradually reduced, the strength at 1 day was increased, and the compressive strength at 7 days was decreased.

ACKNOWLEDGMENT

The authors acknowledge the financial support from Major projects of the national science and technology ministry “12th Five-Year” science and technology support program (2014BAL03B01-04).

REFERENCES

Dong Sufen. Study on Preparation and properties of cement based self-leveling mortar [D]. Chongqing University, 2009.

Gao Shujuan, Liu Wenbin. Preparation of cement based self-leveling mortar [J]. Sichuan building materials, 2013, v.39; No. 17606: 47–48.

Gu Jun. Application of poly carboxylic acid series water reducing agent in cement based self leveling [C]. 2008. Production and application of third (China) International Building dry mixed mortar technology conference proceedings, 277–280.

Huang Tianyong, Chen Xufeng, Zhang Yinxiang, Yan Peiyu. The mechanism of Cement based self leveling mortar [J]. Portland bulletin, 2015, V.34; No.22910:2864–2869.

Ren Zeng zhou. Preparation and application of high performance cement based self-leveling mortar [D]. Chongqing University, 2010.

The article, Zhong Chao Ming, et al. Research on size change rate of CAC based self leveling mortar [J]. concrete, 2013 (6): 97–99.

Civil, Architecture and Environmental Engineering – Kao & Sung (Eds)
© 2017 Taylor & Francis Group, ISBN 978-1-138-02985-9

Rebuilding the city parks: How great is the effectiveness of environment-friendly constructions?

R.Y. Tallar & A.F. Setiawan
Maranatha Christian University, Bandung, Indonesia

J.P. Suen
National Cheng Kung University, Tainan, Taiwan, R.O.C

ABSTRACT: Recently, the number of ecological restoration projects at waterbodies have increased significantly and applied in many areas. Many researchers have been working on large-scale restoration projects with some complex variables and considerations involved, but only few are focused on small-scale projects. The purposes of this study are: (1) to evaluate and monitor the ecological restoration projects at Barclay City Park, Tainan City, Taiwan R.O.C, by using Post Project Appraisals (PPAs) and compare with several city parks in Bandung, Indonesia; (2) to promote an ecological approach that incorporates public perceptions into city parks restoration projects in Bandung City, Indonesia, to determine the environment-friendly constructions that provide long-term benefits to the ecosystem. The data were collected by kick samplings and used following the Before-After-Control-Impact (BACI) method. We sampled and analyzed seven water quality parameters, including temperature, electrical conductivity, salinity, pH, dissolved oxygen, and so on, at several sections, and then identified the aquatic habitat within. The results showed that water quality condition was still unclear with limited ecological value. The study also identified that the degraded water quality condition has increased the public awareness. Therefore, the application of sustainable ecological projects in urban drainage systems could also improve the aesthetic value in the context of urban landscape and gain growing public interest due to its positive effects on water quality and quantity issues. Furthermore, the study recommended water quality monitoring and public hearings programs. Further recommendation is the principles of ecological effectiveness, methodological and economic efficiency, and sociocultural engagement, which should be interwoven in the application of the guidelines and the framework for planning and implementing ecological restoration.

1 INTRODUCTION

1.1 *Background*

Recently, the number of ecological restoration projects at waterbodies has increased significantly and applied in many areas. Many researchers are working on large-scale restoration projects with some complex variables and considerations involved, but only few are focused on small-scale projects. Restoring degraded ecosystems or creating new ones has become a huge global concern. In Indonesia, for instance, many cities are rebuilding their city parks to achieve a sustainable ecological development through environment-friendly construction. This patchwork movement to rebuild the city parks is well supported by the government. Such projects are, moreover, likely to become far more common as the cities rapidly urbanize, turn to green infrastructure, or environment-friendly construction to address problems like climate change, flood control, and pollution within cities. However, hardly anyone does a proper job of measuring the results, and when people do, it generally turns out that ecological restorations seldom function as intended.

Therefore, an appropriate approach was introduced, that is, Post Project Appraisals (PPAs). PPAs involves the evaluation of the effectiveness of restoration projects on the basis of systematic data collection (Tompkins and Kondolf, 2007). PPAs are not just the records of project assessment, but rather are the references to improve future projects.

The purposes of this study were: (1) to evaluate and monitor the ecological restoration projects at Barclay City Park, Tainan City, Taiwan R.O.C, using Post Project Appraisals (PPAs) and compare with several city parks in Bandung, Indonesia; (2) to promote an ecological approach that incorporates public perceptions into city parks restoration projects in Bandung City, Indonesia, to determine the infrastructure components or environment-friendly constructions that provide long-term benefits to the ecosystem.

2 METHODS

Several methods were used in this study. In order to analyze by using PPAs, the first methods were facies mapping and water quality sampling for the first study areas. In the second set, we used statistical analysis by SPSS 17.0 and a set of questionnaire prepared for the respondents in the second study. A simple and short presentation was also prepared to introduce the purpose of this questionnaire. In the questionnaire, the respondents were asked to rank their satisfactions about the features at city parks related to environment-friendly construction. The respondents were asked to rate each statement by marking one of five boxes that most exactly/closely matched their perspectives. The responses were later coded on a five-point numerical scale for analysis: 1 = Dissatisfied, 2 = Somehow dissatisfied, 3 = Neither satisfied nor dissatisfied, 4 = Somehow satisfied, and 5 = Satisfied. Then, they were asked to respond to written items on the basis of their suggestions, if they have any.

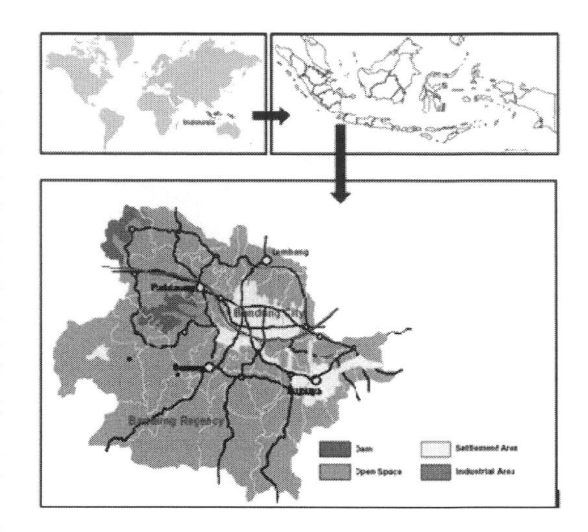

Figure 1. Location of Bandung City, Indonesia.

3 STUDY AREA

In this study, two areas with several sites were selected. The first locations were Barclay Memorial Park and Tainan Athletic Park in Tainan City, Taiwan R.O.C. The second locations were Lansia Park and Maluku Park in Bandung City, Indonesia. The Barclay Memorial Park was located in the southern Tainan City with Zhu Creek flowing across and its site is about 3.3 ha. The site was abandoned in 1994 due to the land dispute between the Tainan City and Tainan County Governments causing garbage and safety problems until the reconstruction in 2001 (Tainan Business Culture and Art Foundation, 2008). The second site was Tainan Athletic Park, which was enclosed by Jiangkang road to the north, Datong road to the east, Dalin road to the south, and Nanmen road to the west. The park is about 130 acres and contains facilities for all types of sporting activities. Meanwhile, the second locations were conducted in two city parks at Bandung City, Indonesia. Bandung City is counted as the capital of West Java province in Indonesia (Fig. 1). It is Indonesia's third largest city by population, with over 2.4 million persons from 30 districts. Geographically, Bandung City has an area of about 16.729,65 ha. In 2014, the average temperature ranged between 22.5 and 23.7°C. The average annual rainfall precipitation in 2014 is 198.8 mm. Meanwhile, Bandung City is reported to have around 600 parks, with the city government targeting the adoption of thematic concept in 30 parks. The major purpose of the city parks in Bandung City is recreation intended for improving ecological environment with harmoniously formed landscapes and waterscapes of natural features.

4 RESULTS AND DISCUSSIONS

On the basis of the water quality sampling result, DO varies inversely with the temperature; increasing temperature (warm water) can hold less dissolved oxygen because a portion of oxygen converts from liquid state to a gas, and decreasing temperature (cold water) can hold more dissolved oxygen. It was indicated in the last survey in Barclay Memorial Park on December 2013 that the value of DO was higher than that in the previous survey because in the data collected during the beginning of the winter season, DO was higher than the DO measured in March 2010, which was in the spring season. It was also known that DO in the upstream is higher than in the downstream and it was very poor, which was attributed to the nonflowing water. Nonflowing water occurs because there was no elevation on the surface and no aeration naturally of creek's water. Low DO due to large amounts of floating plants consumes oxygen and algae. When organic matter, such as animal waste or improperly treated wastewater, enters a body of creek, vegetation and algae growth increases and DO decreases and is decomposed through the action of aerobic bacteria.

This study also identified that Fragrant Water Lilies (*Nymphaea odorata*) dominate as the floating plant in Barclay Memorial Park. Basically, this plant is desirable because they enhance the aesthetics of waterbody, create nice flowers, and are essential to a healthy ecosystem. Besides, they can serve as a food source for organisms in the waterbody, hiding places for fish and insects, and nutrient sinks. However, they can overrun the waterbody and become undesirable if not controlled, as happened in the Barclay Memorial Park. These floating aquatic weeds can cause severely low DO levels because they physically block the transfer of oxygen from the atmosphere to the water's surface. They can also block light and prevent photosynthesis (oxygen production) by submerged plants. Moreover, death and decay of floating plants provides the conditions for a further depletion of DO levels.

At Barclay Memorial Park, the DO level was poor with a value of 3.54 mg/l on upstream, 8.39 on middle stream, and 1.57 mg/l on downstream. At Tainan Athletic Park, the dissolved was also poor with a value of 4.44 mg/l on upstream, 5.89 on middle, 4.66 on downstream, indicating neighborhood sewage and no elevation on the surface of water. At levels below 5 mg/l, some fish become stressed. Large fish usually have a greater need for oxygen than smaller fish of the same species. On the basis of visual investigation, small fish have adapted to waters that were chronically low in DO level. At downstream of Barclay Memorial Park, fish cannot survive because of the DO level lower than 1.57 mg/l.

At Barclay Memorial Park and Tainan Athletic Park, pH condition in water was neutral (7.44–7.71). Stream water usually ranges from a pH of 6.5 to a pH of 8.5; this range is considered to be an optimal range for most aquatic life. The natural pH varies from river to river, but the pH range of a river will generally remain stable. The conductivity in Barclay Memorial Park was higher than that in Tainan Athletic Park because of the higher temperature in the former. However, the conductivities in Barclay Memorial Park and Tainan Athletic Park were below the SPC minimum standard for freshwater lakes/streams, which is 2 μS/cm (range: 2–100 μS/cm). By knowing that the conductivity of Zhu Creek was below the minimum standard, it does not mean that the Zhu Creek was not polluted because again on the basis of visual investigation, there were some fish like *Java tilapia* that can survive in polluted waterbody. Conductivity in waterbody was affected primarily by the geology of the area through which the water flows. The stream that passed through Tainan Athletic Park consisted of concrete bedrock; that is why the streams tend to have a lower conductivity because concrete was composed of more inert materials that did not ionize (dissolve into ionic components) when washed with water. On the contrary, streams that run through Barclay Memorial Park areas with clay soils tend to have a higher conductivity because of the presence of materials that ionize when washed with water. A failing sewage system would increase the conductivity because of the presence of chloride, phosphate, and nitrate; an oil spill would also lower the conductivity (http://water.epa.gov). Besides, the conductivity depends on temperature. During higher winter flows, the lower saline surface waters dominate, resulting in a decrease in in-stream conductivity.

Salinity of Barclay Memorial Park and Tainan Athletic Park were in the low range (around the Minimum Amount of Salinity of Brackish Water = 0.5 ppt). The low amount of salinity depends on the low amount of conductivity. The higher conductivity indicates higher salinity. Some solutions were offered, which can partly remove floating plants (Fragrant Water Lilies—Nymphaea Odorata) to increase dissolved oxygen content, water flow, and aeration.

On the contrary, the study also used public perceptions through satisfaction survey for the second study area in Bandung City, Indonesia. The data obtained from the questionnaires were processed using the SPSS 17.0 program for statistical analysis. Specifically, Multivariate Analysis of Variance (MANOVA) was used, and this was a way to test hypotheses in which one or more independent variables or factors were proposed to have an effect on a set of two or more dependent variables (Mathew 1989). MANOVA was used in this work to test whether different types of respondents and locations of city parks affect the perceptions of the respondents with regard to a measure of aesthetic value based on various perceived environment-friendly construction attributes (bioretention, buffer strips, permeable pavements, canopy trees, and others). In SPSS, there are four multivariate measures: Wilks' lambda, Pillai's trace, Hotelling's trace, and Roy's largest root. The difference between the four measures is the way in which they combine with the dependent variables in order to examine the amount of variance in the data. The results of the multivariate analysis cannot be shown in this paper due to the space limitation. However, in Wilks' lambda criteria, the intercept showed Wilks' $\lambda = 0.003$, $F(5, 155) = 9008.768$, $p < 0.001$, and the parks showed Wilks' $\lambda = 0.444$, $F(21, 1316) = 57$, $p < 0.001$.

The results indicated that most visitors satisfied with the available environment-friendly construction features in targeted city parks. Canopy is the most preferable environment-friendly construction feature according to public perceptions, whereas bioretention is the least preferable one. Some comments from respondents mentioned that they agreed and supported environment-friendly construction to be applied in public city parks. Most of them also stated that it is necessary to improve the aesthetic value by applying those features in study area.

5 CONCLUSIONS

The study also identified that the degraded water quality has increased public awareness. Therefore, the application of sustainable ecological projects in urban drainage systems could also improve the aesthetic value in context of urban landscape and gained growing public interest, as a result of its positive effects on water quality and quantity issues. Furthermore, the study recommended the necessary of continuing water quality monitoring and public hearings programs. Further recommendation is that the principles of ecological effectiveness, methodological and economic efficiency, and sociocultural engagement should be interwoven in the application of the guidelines and the framework for planning and implementing ecological restoration.

The outcomes showed that the public city parks are one of the most effective application areas for green infrastructure concept design. Green infrastructure investments in the study area boost the economy, enhance community health and safety, and provide recreation, wildlife, and other benefits that can increase the level of public satisfaction. This study can guide local planners and decision makers from the other cities to create and develop their public city parks on the basis of green infrastructures concept design. As with other forms of infrastructure, green infrastructure requires sustainable management and maintenance arrangements to be in place if it is to provide benefits and services in the long term. Arrangements for managing green infrastructure and funding its management over the long-term should be identified as early as possible when planning green infrastructure and factored into the way that it is designed and implemented.

REFERENCES

Iojă C.I., Rozylowicz L., Pătroescu M., Niță M.R., Vânau G.O. (2011) Dog walkers' vs. other park visitors' perceptions: The importance of planning sustainable urban parks in Bucharest, Romania. Landscape and Urban Planning 103 (1):74–82.

Mathew T. (1989) MANOVA in the multivariate components of variance model. Journal of Multivariate Analysis 29 (1):30–38.

Meitner MJ (2004) Scenic beauty of river views in the Grand Canyon: relating perceptual judgments to locations. Landscape and Urban Planning 68 (1):3–13.

Pettebone D, Newman P, Lawson SR, Hunt L, Monz C, Zwiefka J (2011) Estimating visitors' travel mode choices along the Bear Lake Road in Rocky Mountain National Park. Journal of Transport Geography 19 (6):1210–1221.

von Essen L.M., Ferse S.C.A., Glaser M., Kunzmann A. (2013) Attitudes and perceptions of villagers toward community-based mariculture in Minahasa, North Sulawesi, Indonesia. Ocean & Coastal Management 73:101–112

Yamashita S (2002) Perception and evaluation of water in landscape: use of Photo-Projective Method to compare child and adult residents' perceptions of a Japanese river environment. Landscape and Urban Planning 62 (1):3–17.

Yao Y, Zhu X, Xu Y, Yang H, Wu X, Li Y, Zhang Y (2012) Assessing the visual quality of green landscaping in rural residential areas: the case of Changzhou, China. Environ Monit Assess 184 (2):951–967.

Civil, Architecture and Environmental Engineering – Kao & Sung (Eds)
© 2017 Taylor & Francis Group, ISBN 978-1-138-02985-9

Renaissance with recycled materials: Reconstruction of Guifeng Temple

Y.S. Gao
School of Human Settlements and Civil Engineering, Xi'an Jiaotong University, Xi'an, Shaanxi, China

J. Xu
School of Human Settlements and Civil Engineering, Xi'an Jiaotong University, Xi'an, Shaanxi, China
Institute of Heritage Sites and Historical Architecture Conservation, Xi'an Jiaotong University, Xi'an, Shaanxi, China

ABSTRACT: Green building materials represent not only the material itself, but also it is a representation of a use policy, covering the whole life cycle of a construction. In view of this, when it achieves some success in energy saving, nonpollution, and recycling of building materials, it provides new ideas for the green building materials research. In the reconstruction of Guifeng Temple in Mount Zhongnan, the designers considered the future development trend of the region, especially taking three factors of the reuse of building materials in account, selected of the structure and installment construction, and finally made the plan and design systematically with local building materials. More than protecting the traditional architectural style, meanwhile, it diversifies the development of the temple, leading to a preliminary mode of reconstruction, which adapts to the local economy and culture.

1 INTRODUCTION

This design is a part of historical area planning design of Guifeng Temple, located in Mount Zhongnan, over 30 km away from Xi 'an at 667 m above sea level. The climate here is warm and semi-humid with clear four seasons, known as continental monsoon climate with warm spring, hot and rainy summer, moderate moist autumn, and cold and dry winter.

There is a tradition of seclusion for literati and Buddhist, with natural environment in good condition. "Guifeng Moonlight" is the most famous sight in temple so far. Besides, it is rich in natural resources, and several natural villages connect with each other into a town at the foot of the mountain, where strawberry plants have been extensively cultivated as the main economic crop. Buildings in the town and structures in the field are all built up with the local materials in various and flexible ways.

After communicating with presiders many times, we arrived at two main contradictions, the first is building materials can only rely on local human transportation, so we chose the construction process as simple as possible and the other is building insulation design. Resolution strategies are grouped into two measures, one is building materials are all set in a standard size within two men's luck (the longest edge no more than 3.3 m), and the construction technologies limiting to stitching or masonry local mature technologies; the other is to choose the unified and closed architectural layout in north China, thus the whole building forms several organic, closed yard as well as half-covered space for outdoor activities to adjust the micro climate.

2 RECONSTRUCTION OF GUIFENG TEMPLE

2.1 *Local green building materials*

An investigation was put forward before design, simply about the local building materials.

Agriculture is the pillar industry of this region, thousands of cement columns stand in fields as supports of strawberry cultivation, which are now being replaced by metal industries. Such cement columns share a common size about 1.2 m high and a section size of 150*150 (mm), for they are produced by pouring cement into the template until solidification. Without internal reinforcement, their bending strength is very low.

In addition, there are serious problems among houses in the villages around and most of the original buildings in the temple, especially structure security and thermal properties. After evaluating their cultural value, we planned to rebuild all those prefabricated and adobe houses, gaining a large amount of bricks from demolition.

The selection of building materials goes through an assessment by "resource recycling", for the reference factors include local source of construction waste, the strength of stock, transportation cost, safety, and sustainable aspects of comparison. Finally, we used the cement column in strawberry fields, clay brick from demolition, coal cinder from production or living, and steel from prefabricated houses.

2.2 *Various structure*

As building materials are diverse, different structures are used. Houses in villages are masonry structures in forms of multistories, holding a strength high enough until the villagers start to build additional contractures on roof, which greatly reduced the structure security. Moreover, some villagers misuse prefabricated houses as permanent buildings, which causes a great damage to both thermal insulation and safety performance.

Besides, since the prefabricated houses are steel-framed structures while the common house are masonry structures, connections between the two parts are extremely weak.

In the existing constructions of Guifeng Temple, masonry structure and brick–concrete structure take the main part. Although they are made from bricks, the connection materials or masonry constructions are simply mud or mixed with cement mortar. Thus, these constructions are easily damaged and easy to dismantle.

2.3 *Installment construction*

The first reconstruction in Guifeng Temple was a hovel divided into three bedrooms for monks, followed by an expansion of one prefabricated house serving as a monastery and apartments since the monks increased. Along with the temple's further development, several brick houses were built for the monastery, which have a library and a meditation room attached to them. Soon afterward, one more brick house was built for reception. After the recovery of "law" in the temple, a bathroom and laundry room in masonry structure was added.

The "law" requires more strict patterns to the layout of the temple, so an entrepreneur among the believers decided to donate a new reconstruction, thus the design team is invited.

As a heritage ontology of the Qing dynasty, reparation of the main hall varies from the using requirements; not only the construction needs to be repaired, but also lights or other technical equipment need to be added. In the latest design, because the wood rot of the hall cracked, it needs a replacing wood, whose raw material is donated by devotees fundraising and other temples' old material. Benefitting from the gradual accumulation of those old material, construction realized a condition of "repair the old as old".

It is obvious that compared with one-time construction, an installment construction is more suitable for the continuous development as well as a stable reparation.

3 OPTIMIZATION THEORY SUPPORT

3.1 *Green building materials*

On the basis of Lynn Froeschle's book, Environmental Assessment and Specification of Green Building Materials, green building materials perform well especially in resource efficiency, affordability, and so on. In this process, green building materials represent a use policy throughout the whole life cycle of a construction.

Resource efficiency can be accomplished by utilizing recycled or renewable, natural, durable, local available, and plentiful materials.

Affordability means building costs are comparable to those of conventional materials or within a project-defined percentage of the overall budget throughout its life cycle. From another perspective, it also means lower operating costs associated with changing space configurations and greater design flexibility.

3.2 *Composite structure*

Composite structure has high flexibility to cope with differences in building materials and the demand of the architectural space. The last few decades have witnessed that composite materials not only revolutionized traditional design concepts but also made possible an unparalleled range of possibilities, for composite structures have expanded its range of application to nearly all types of combination. Therefore, a cantilever structure can composite with a masonry structures.

Composite structures offer benefits of using materials in various sizes and reducing cost by combining large components and small components. While structures composed of large components are usually better than a structure composed of a large number of smaller components more easily without damage and dismantling, it is recommended to use large components in a composite structure.

Connection mode within a composite structure also has an important effect on reducing cost for easy dismantling, so materials adopt the mechanical connection method involving wood or steel structures.

3.3 *Installment construction*

Installment construction has been popular among several construction units because of the gradual collection mode of recycled material from demolitions in pace with updates of villages nearby.

According to a project management, the main factors in the installment construction are size, capital, planning, risk, costs, using requirements project promotion, and reasonable tax avoidance. Temples' demand changes constantly; learning the lessons of the temple overextended in early "Reform and Opening up", requirements of the reconstruction of temple need to be tested by time, and can only be realized gradually rather than overnight. In terms of saving money, the most effective way of controlling cost is reasonable tax avoidance.

The administration of taxation on real estate development enterprises on the management of land value-added tax liquidation of circular (Guoshuifa [2006] NO.187) regulation states: "the land value-added tax should be calculated according to departments of the state in real estate projects approved, while in the installment construction, the land value-added tax should be carried out cooperating with each stage."

Conventionally, an estate feasibility study report should be submitted when applying for a land approval, including the development schedule. Companies tend to pick up the land overall, but construct in installment. Because of the different land range, a discount in tax is obvious.

4 MORE SYSTEMATIC RECONSTRUCTION MODE

4.1 *Recycling resources from surrounding villages*

Local materials in most stock that can be conveniently transported are recycled clay brick from demolition. Bricks are used to build up walls whose thickness is 240 mm, the most common mode of local wall (see Figure 1(a)).

The second most common stock and the most convenient one to be transported is cement column in strawberry fields. According to section size and length, cement columns will be piled up in the orthogonal direction alternately, without any adhesive material (see Figure 1(b)), beneath which waterproofing materials are set at the interface with steel beams, forming the roof construction.

To make it easier to be disassembled and built, steel beams are applied as structure, which come from prefabricated houses or old buildings. All the qualified beams are set at a length of 3.6 m, considering the size of clay brick and cement column.

Coal cinder from production or living is used to build square and road (see Figure 1(c)). After the compaction process, the surface can gain enough leveling and compaction.

As mentioned above, "resource recycling" in this design is locally available, plentiful, and renewable, forming a building strategy as an implementation of green building materials.

4.2 *Designed for disassembly*

Spaces with different functions in the temple require different forms or sizes of a space, leading to differences among structures.

Small-span spaces like monk's apartment, guest room, and bathroom use the masonry structure, for the local traditional building method is to pile up the bricks, with special mud masonry.

Long-span spaces like the main hall, meditation room, and monastery use large cantilever structures. In this case, only one column is needed to support a whole square (see Figure 1(d)), less than a frame structure, in which four columns are needed in every square.

Besides, to realize a recycle of resources both in constructing this temple and making it recyclable for other construction in the future, it is designed

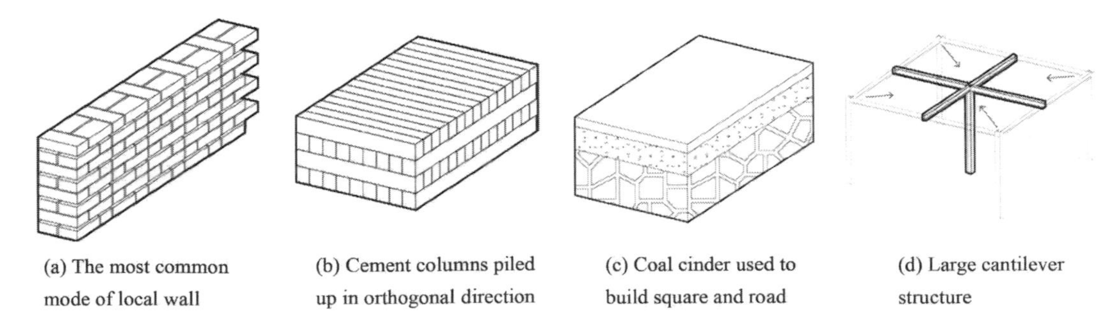

(a) The most common mode of local wall

(b) Cement columns piled up in orthogonal direction

(c) Coal cinder used to build square and road

(d) Large cantilever structure

Figure 1. Structures used in the reconstruction.

for disassembly at the beginning. The steel components are connected by mechanical joint, and the cohesion between curtain wall and brick wall is applied to a construct site.

Finally, for the convenience of construction and disassembly, decorative components are avoided as much as possible, and all detail structures use dry hanging method rather than the wet homework.

4.3 *Installment construction in pace with local development*

Throughout the reconstruction of the Guifeng Temple, it is realized systematically in pace with the temple demands to build more constructions, not scaled up sharply. Neither the capital volume of Buddhist support nor the probability of one-time funds from a big benefactor is the focus, but the economical construction in every few years itself was supported by clear function and demands. In this way, it avoids a massive rapid construction, causing a waste of resources, and ensures adequate following funds for reconstruction, which helps realize an update in the life cycle.

It has always been a tradition, whenever it is, temples accept believers' donation, which flows to construct the temple. This type of capital circulation pattern results in a persistent update and becomes a reference for the latest reconstruction. Although the latest capital volume can push the temple achieve a rapid expansion, both the public who donated and monks prefer to regard the mass recovery as long-term planning, which should be realized in a staggered way, attaching a fund's plan divided into construction funds and operating funds. In this plan, gradual capital represents a capital using patterns, advocating to guarantee a continuous supply of operating funds.

Except from gradual collection of materials and the actual requirement of temple, what is more important in the necessities of installment construction is to keep pace with the local development, that is, the town at the foot of the mountain.

Temple in essence belongs to the cultural construction, taking on a function of presenting the social culture, whose service object is the public. In terms of the Guifeng Temple, a sharp expansion may limit the temple's normal operation because of the inability of the local foundation facilities to meet the demand; or it causes an uncontrolled impact on local development because of a sudden change on land use mutation, which closely related to the local public, the main force of the temple protection.

In view of this, the designers considered the relationship between the reconstruction of Guifeng Temple and the development of the towns in a systematic and careful way. Finally, we put forward a plan covering issues about waste management, scenic area development, support facilities, and people's livelihood improvement, which adapts to the development of the local economy and culture.

5 SUBTOTAL

The reconstruction of Guifeng Temple is a renaissance with recycled materials, exploring a systematic using policy complying with the principles of green building materials. All the building materials applied in the design are not only local available but also expanding the source from local development and the agricultural facilities. On the one hand, the material is not limited to present conditions, but in combination with regional development trend, so the abandoned clay bricks from demolition of houses are used. On the other hand, it is not limited to materials served as buildings neither, but using agricultural facilities engineering materials as well, such as strawberries in the field of concrete column and coal production and living. Considering material sources are collected in installment, construction is designed to be realized in stages. Meanwhile, the use of materials for building maintenance operation stages is well planned.

In short, in pursuit of the green building material using policy, material selection, structure, construction, and operation management methods are all considered in an integrated way. Such a material using policy leads to a preliminary mode of reconstruction adapting to local economy and culture, whose background is thus a more universal solution, which remains to be explored by concluding more experience in projects.

ACKNOWLEDGMENT

The authors acknowledge the support of the State Natural Science Fund of China (No.51278416).

REFERENCES

Froeschle, L.M. (1999). Environmental Assessment and Specification of Green Building Materials. Salt Lake City: The Construction Specifier.

Jacobs, J. (2006). The death and life of great American cities. Jiangsu: Yilin Press.

Spiegel, R. & Meadows, D. (1999). Green Building Materials: A Guide to Product Selection and Specification. New York: John Wiley & Sons Inc.

Xiaolei, G. (2010). Research on Construction and dismantling and material recycle technology. Tianjin: Tianjin University.

Xudong, W. (2011). Protection and Development of Ancient Villages in Shaanxi Province from "Organic Renewal". Xi'an: Xi'an building university of science and technology.

Yigang, P. (2010). Theory of Architectural Space Combination. Beijing: China Building Industry Press.

Civil, Architecture and Environmental Engineering – Kao & Sung (Eds)
 ISBN 978-1-138-02985-9

Research on safety management of construction projects based on game theory

J.X. Zhao, M.M. Wang, J.H. Sun & K. Li
School of Management, Qingdao University of Technology, Qingdao, China

ABSTRACT: Shortage of construction project safety management has been one of the problems that hinder the development of construction industry. Strengthening the construction safety management can not only promote the development of the construction industry, but also promote the social stability. This paper uses game theory to study the relationship between the main parties of construction project from the macro perspective. In this paper, we first use the static game theory of mixed strategy to establish the game model between supervision units and contractors, then use principal-agent system to establish the game model between owners and supervision units, and finally use the first model into the second model to construct an extensive form between owners and supervision units and contractors. Furthermore, in this paper, we obtain the construction project safety equilibrium point and put forward the safety management strategy for the supervision unit and contractor. Finally, the application of the model is introduced by a case study.

1 INTRODUCTION

At present, in China, increasing attention has been paid to the security problem of construction, and if we cannot effectively carry out safety management in construction project, it will result in a wide range of security incidents, such as casualties, machinery and equipment damage, and project entity damage, which resulted in huge economic losses to project participants and caused serious negative impact on society. Therefore, increasing attention has been paid to the safety of construction projects, and various theories and methods for the safety management of construction projects have emerged in an endless stream.

As early as 1996, Blair put forward that all project stakeholders (owners, design units, contractors (subcontractors), government, and insurance companies, and so on) are responsible for the safety of the project (Blair 1996). Ngowi and Rwelamila (1997) pointed out that in the construction industry, safety and health are the only responsibilities of the contractor, which has been regarded as one of the main causes of frequent accidents (Ngowi et al. 1997).

In the process of safety management of construction project, the problem of responsibility and benefit distribution has a direct impact on the effect of safety management. Therefore, scholars such as F.L. Zhang introduced the game theory into the construction safety management and set up a game model of safety management between project safety supervision authorities and contractors to determine the proportion of the supervisory organ carrying out sampling of the contractors and the punishing the contractors for unsafe construction (Zhang et al. 2002). On the basis of that, P.H. Tu also concluded the relationship between security costs, punishment, and the probability of inspection (Tu 2004). J.M. Wang and other scholars further concluded the relationship between safety construction costs, regulatory costs, penalties, and regulatory efficiency (Wang 2007). The above studies were largely limited to the game relationship between the safety supervision authorities and contractors and less consideration of the impact of owners and supervision units on the construction safety management. In view of this, in this paper, we established the noncooperative game model between supervision units and contractors, and owners and supervision units.

2 THEORETICAL ANALYSIS

Game theory mainly studies decision-making parties, which present a conflict of interest to make a set of strategies and behaviors that are interdependent in the process of mutual antagonism and competition (Wang et al. 2015). All participants in the premise of following a certain rule analyze the strategies other people may take, to choose decisions that meet their own interests, thereby obtaining a certain interest. Game theory involves the most basic elements such as participants, strategies, and benefits.

In the process of project construction, the involved subjects include design units, owners, supervision units, supply units, and contractors, and they are mutually independent and self-financing economic entities. The actions and interests of each participating subject have a certain effect on the formation of construction safety. Although the overall goal of all participation units is consistent, as various units consider their own benefits, there will be a certain exclusion between them, so we cannot exclude the behaviors that participants do not follow the contract to construction and not seriously examine, which led to the construction project safety problems (Wang, 2013). Game theory happens to be a kind of method to study the behavior of decision-making parties, so the introduction of the game theory into the construction project safety management restricts the occurrence of adverse behavior. This paper mainly uses the mixed strategy of static game theory to establish the game model between supervision units and contractors, and then uses principal-agent system to establish the game model between owners and supervision units.

3 GAME MODEL BETWEEN SUPERVISION UNITS AND CONTRACTORS

3.1 *Introduction of mixed-strategy game*

1. A set of participants in a game $i, i \in (1,2,..,n)$;
2. Strategic space for each participating party $S_i, S_i = \{s_{i1},...,s_{ik}\}$;
3. Benefit function of each participating party $u_i(S_1,S_2,...,S_n)$;
4. Random selection of each participating party in k optional strategies with probability distribution $p_i = (p_{i1},...,p_{ik})$, where $0 \le p_{ij} \le 1, j = 1,...,k$ and $p_{i1} + ... + p_{ik} = 1$.

The participating parties involved in the mixed-strategy game model are the supervision units and contractors. The supervision unit's strategic space is:

$S_1 = [s_{11}, s_{12}, s_{13}]$ = [Supervision and inspection and the conclusion is correct, Supervision and inspection but the unsafe construction is wrong to see as a safe construction, No supervision and inspection]

and the contractor's strategic space is:

$S_2 = [s_{21}, s_{22}]$ = [Carry out construction according to the contract, Does not carry out construction according to the contract]

3.2 *Model hypothesis*

1. The total cost at which contractors carry out safe construction in accordance with the provisions of the safety is defined as C_1; the total cost at which contractors do not carry out safe construction in accordance with the provisions of the safety is defined as C_2, and $C_1 > C_2$. The probability that the supervision units find and punish contractors who do not carry out safe construction in accordance with the provisions of the safety is defined as $\gamma(0 \le \gamma \le 1)$, and it reflects the efficiency of supervision and inspection of the supervision units, and the contractors should pay a fine, which is defined as a. If the supervision units do not find the illegal acts of contractors, then the supervision unit should bear certain economic responsibility, which is defined as f. The cost of supervision and inspection by the supervision units is defined as k, and if the supervision units do not carry out supervision and inspection, then they do not need to pay any cost.
2. The probability that the supervision unit conducts supervision and inspection is p_{11}, and the probability of not conducting supervision and inspection is p_{12}, where $p_{11}=1-p_{12}$. The probability that the contractor conducts safety construction according to the contract is p_{21}, and the probability that the contractor does not carry out safety construction according to the contract is p_{22}, where $p_{22}=1-p_{21}$.

3.3 *Model construction and solution*

According to the model hypothesis, the payoff matrix between the supervision unit and contractor was obtained, as shown in Table 1.

In the case of the mixed strategy, the expected payoff of the supervision unit is:

$$\begin{aligned} u_1 &= p_{11}\gamma\left[-kp_{21} + (a-k)(1-p_{21})\right] + p_{11}(1-r)\left[-kp_{21} + (-k-f)(1-p_{21})\right] - (1-p_{11})f(1-p_{21}) \\ &= ap_{11}r - ap_{11}rp_{21} - p_{11}k + fp_{11}r - fp_{11}rp_{21} - f + fp_{21} \end{aligned}$$

According to the mixed-strategy Nash equilibrium, it is necessary to select an appropriate value

Table 1. Payoff matrix between the supervision unit and contractor.

Strategic space of the supervision unit	Strategic space of contractor	
	$s_{21}(p_{21})$	$s_{22}(1-p_{21})$
$s_{11}(p_{11}*\gamma)$	$-k, -C_1$	$a-k, -C_2-a$
$s_{12}(p_{11}*(1-\gamma))$	$-k, -C_1$	$-k,-f,-C_2$
$S_{13}(1-p_{11})$	$0,-C_1$	$-f, -C_2$

for p_{11}, which makes the u_1 maximum; therefore, the optimal first-order condition is:

$$\frac{\partial u_1}{\partial p_{11}} = a\gamma - a\gamma p_{21} - k + f\gamma - f\gamma p_{21}$$

To make the above expression equal to 0, that is:

$$p_{21}* = 1 - \frac{k}{a\gamma + f\gamma} \quad (1)$$

the expected payoff of contractor is:

$$u_2 = p_{21}\left[-C_1 p_{11}\gamma - C_1 p_{11}(1-\gamma) - C_1(1-p_{11})\right] + (1-p_{21})\left[(-C_2 - a)p_{11}\gamma - C_2 p_{11}(1-\gamma) - C_2(1-p_{11})\right] = -p_{21}C_1 - C_2 - ap_{11}\gamma + p_{21}C_2 + p_{21}ap_{11}\gamma$$

The partial derivative of u_2 to p_{21}: $\frac{\partial u_2}{\partial p_{21}} = -C_1 + C_2 + ap_{11}\gamma$

To make the first-order derivative equal to 0, we draw the simplification as:

$$p_{11}^* = \frac{C_1 - C_2}{a\gamma} \quad (2)$$

It can be concluded that the mixed-strategy Nash equilibrium for the game is $(p_{11}^* = \frac{C_1 - C_2}{a\gamma}, p_{21}^* = 1 - \frac{k}{a\gamma + f\gamma})$. From the model, it can be seen that the value of Nash equilibrium (p_{11}*, p_{21}*) is related to C_1, C_2, a, γ, k, and f.

The probability that the supervision unit conducts supervision and inspection is proportional to the cost difference between the cost at which the constructor conducts safety construction according to the contract and the cost at which the contractor does not conduct safety construction according to the contract; the greater the difference in cost, the greater the probability of supervision and inspection. At the same time, p_{11} is inversely proportional to a and γ; the more fine and the higher the regulatory efficiency, the lower the probability of supervision and inspection.

Probability p_{21} is inversely proportional to k; the less the cost of supervision and inspection, the greater the probability that the contractor will carry out safety construction. Probability p_{21} is proportional to a, γ, and f; the more fine and the higher the regulatory efficiency and the greater the economic responsibility of supervision units when the supervision fails, the greater the probability that the contractor will carry out safety construction. Therefore, in order to urge the construction unit to carry out safety construction, not only do the supervision units increase the penalties to the contractor who does not carry out safety construction, but also improve their regulatory efficiency continuously and reduce the cost of supervision and inspection.

4 ESTABLISHMENT OF GAME MODEL BETWEEN OWNER AND SUPERVISION UNIT

4.1 *Principal-agent theory*

The relationship between principal and agent is an important part of modern economic research, which is often referred to as the "principal agent theory". Its core content is a two-side dynamic game. The principal's interests is closely related to the agent's behavior, but the principal cannot directly control the agent's behavior and even difficult to supervise the agent's work, and can only influence indirectly agent's behavior by remuneration.

Principal-agent model is divided into four categories. The first category is the principal-agent model without uncertainty. It refers to the agent's work without uncertainty, indicating that the agent's output is a deterministic function of the degree of effort. The second category is the uncertain but can be supervised, that is, the agent's efforts results are uncertain, but the principal can supervise the agent completely. The third category is the uncertain and non-supervised principal-agent model, that is, the agent's efforts results are uncertain, and the principal cannot supervise the work of agent. The fourth category is the principal-agent model of selecting continuous reward and continuous effort level, whose efforts results are uncertain and not be supervised, but the principal can choose reward function, and the agent's effort level is continuously changing.

4.2 *Model hypothesis*

1. When the safety performance of construction project is up to the standard, the owner's income is $R(g)$, and the supervision unit's remuneration is w, where $w>k$, $w>f$.
2. When the safety performance of construction project is not up to the standard, the owner's income is $R(b)$, and the supervision unit's remuneration is $w - f$.
3. If the owner does not carry out the commission, then the payoffs of the owner and the supervision unit are 0.
4. If the supervision unit refused to commission, the payoffs of the owner and the supervision unit are 0.

4.3 *Establishment and solution of the uncertain and non-supervised principal-agent model*

The relationship between owners and supervision units is the same as that between the principal and the agent. The supervision unit's work not only has uncertainty, but also there are certain difficulties in owner's supervision to supervision unit; therefore, in this paper, we use the third category of principal-agent model, that is, the uncertain and non-supervised principal-agent model.

The extensive form between the owner and the supervision unit is shown in Figure 1.

In Figure 1, digital 1 represents the owner and digital 2 represents the supervision unit. Because of the uncertainty, the "natural" game player'0' is introduced. In the square brackets, the former symbol is on behalf of the owner's benefit, and the latter symbol is on behalf of the supervision unit's benefit.

When the supervision unit works hard and the safety performance of the construction project is up to the standard, the benefit of owner is $R(g) - w$ and that of the supervision unit is $w - k$ or $w - k + a$. When the supervision unit works hard and the safety performance of the construction project is not up to the standard, the benefit of the owner is $R(b) - w + f$, and the benefit of the supervision unit is $w - k - f$. When the supervision unit does not work hard but the safety performance of the construction project is up to the standard, the benefit of the owner is $R(g) - w$, and the benefit of the supervision unit is w. When the supervision unit does not work hard and the safety performance of the construction project is not up to the standard, the benefit of the owner is $R(b) - w + f$, and the benefit of the supervision unit is $w - f$.

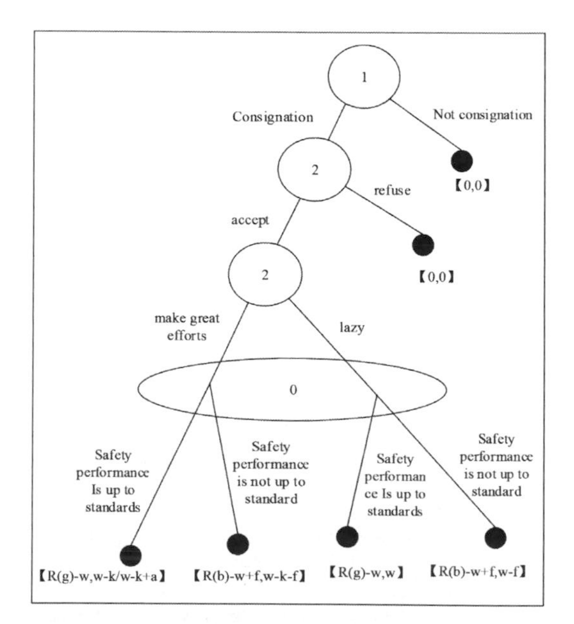

Figure 1. Game extensive form between owner and supervision unit.

Complete and perfect information dynamic game needs to use the backward induction method for analysis. In the third stage, because the safety performance of the construction project is related to contractor and combined with the mixed-strategy game model among the supervision unit and the contractor, we further improve the principal-agent model between the owner and the supervision unit by adding contractor into the third stage, as shown in Figure 2.

In Figure, digital 3 represents contractor. In the third stage, we used the backward induction method to analyze the incentive compatibility constraint that the supervision unit works hard, which is the expected benefit of efforts is more than lazy, as shown below:

$$p_{21}(w-k)+(1-p_{21})[\gamma(w-k+a)+(1-\gamma)\times(w-k-f)]>(1-p_{21})(w-f)+p_{21}war\times(1-p_{21})+fr(1-p_{21})-k>0$$

$$p_{21}<1-\frac{k}{ar+fr} \tag{3}$$

When the probability that the contractor carries out safety construction is less than $1-\frac{k}{ar+fr}$, the supervision unit will choose to work hard. This is consistent with the result of static game theory of mixed strategy. On the contrary, when the

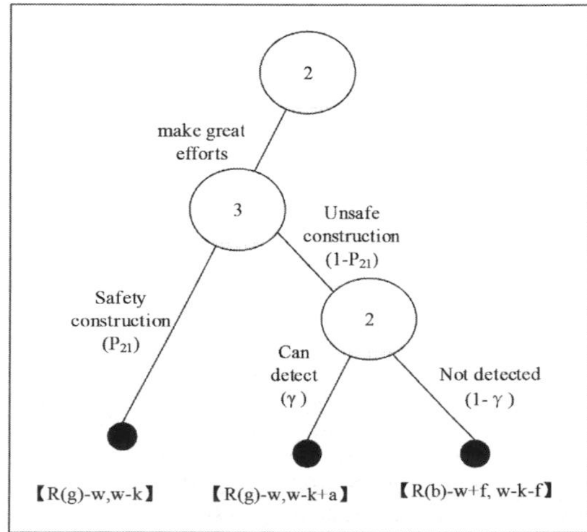

Figure 2. The extensive form that the supervision unit works hard.

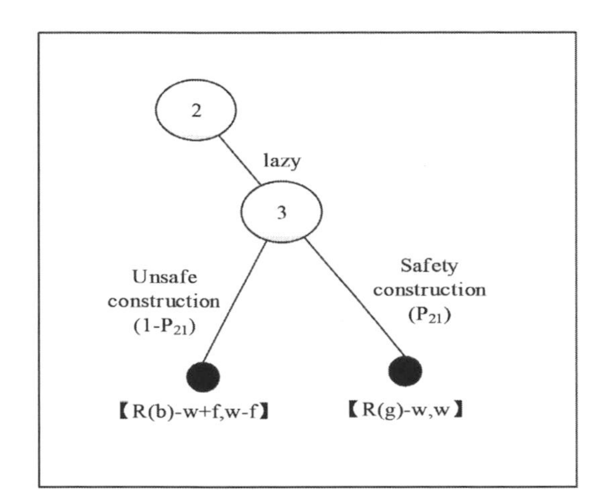

Figure 3. The extensive form that supervision unit does not work hard.

probability that the contractor carries out safety construction is more than $1-\frac{k}{ar+fr}$, the supervision unit may choose to be lazy.

Back to the second stage, the participation constraint of supervision units is the expected benefit of accepting a commission more than lazy. Regardless of the state of supervision unit selection efforts, and its benefits are greater than 0, the supervision unit must choose to accept the commission.

Similarly, push back to the first stage, the expected benefit that the owner selects to commission is greater than that without commission, so the owner unit will also choose to commission.

5 CASE ANALYSIS

Construction unit "A" develop and construct a residential area of two residential buildings, the construction area is 40,000 m^2. Now the construction project will be entrusted to the supervision company "B", and contractor "C" obtains the construction project by bidding. The cost at which contractor "C" carries out safe construction in accordance with the provisions of the safety is 20,000 yuan more than the cost at which contractor "C" does not carry out safe construction in accordance with the provisions of the safety, and the probability that the supervision company "B" conducts supervision and inspection is 1, the efficiency of supervision and inspection of the supervision company "B" is 0.9. Then, according to the mixed-strategy Nash equilibrium: $p_{11}^* = \frac{C_1-C_2}{a\gamma}$. Therefore, the contract should be specified that the fine of contractor C's unsafe acts is more than 220,000 yuan. Keeping the efficiency of supervision and inspection unchanged, if the probability of supervision and inspection is reduced to 0.5, then in the contract, contractor "C" shall be fined more than 440,000 yuan for unsafe acts. If the provisions of the contract for contractor "C" of a fine of 500,000 yuan, and the efficiency of supervision and inspection is raised to 0.95, then the probability that the supervision company "B" conducts supervision and inspection can be reduced to 0.42. Thus, if supervision units want to reduce the probability of inspection and the probability of occurrence of safety accident, they should improve their efficiency of supervision and inspection and increase the fine for contractors not carrying out safe construction in accordance with the provisions of the safety. The model can be used to make a reasonable amount of fines, thus restricting the contractor's unsafe construction behavior.

Similarly, if the contract stipulates k for 50,000 yuan, γ for 0.9, f for 200,000 yuan, and a for 200,000 yuan, then according to the mixed-strategy Nash equilibrium: $p_{21}^* = 1-\frac{k}{a\gamma+f\gamma}$. It can be concluded that the probability of contractor "C" for the safe construction is 0.86. If the cost of supervision and inspection is reduced to 40,000 yuan and the other conditions are unchanged, then the probability of contractor "C" for the safe construction is 0.889. It can be seen that by reducing the cost of supervision and inspection, the probability that contractors carry out construction in accordance with the provisions of the safety can be improved.

6 CONCLUSIONS

1. The study shows that safety problems of construction project is closely related to supervision units' efficiency of supervision and inspection and construction quality of contractors. Therefore, the owners shall take comprehensive consideration to select the supervision units with high efficiency and contractors with good construction quality.
2. Supervision units shall improve the quality of employees, strengthen their sense of responsibility, and reduce or avoid the mistakes in work or dereliction of duty to improve the efficiency of supervision and inspection and to make full use of modern information management technology, such as visual management of the project to reduce the cost of supervision and inspection.
3. Through the above game model, we can make a reasonable amount of punishment to restrict the behavior that contractors do not carry out construction in accordance with the provisions of the safety. If contractors do not carry out safe construction, then the supervision units shall consider the comprehensive use of a variety

of ways to punish them, for example, warning, fines, and downtime rectification.

REFERENCES

Blair, E.H. 1996. Achieving a Total Safety Paradigm through Authentic Caring and Quality Professional Safety. *J. Journal of American Society of Safety Engineers* 6: 18–20.

Ngowi, A.B. & Rewlamila, P.D. 1997. Holistic Approach to Occupational Health and Safety and Environmental Impacts. In: Haup, Theo C, Rwelamila P.D, eds. Proceedings of Health and Safety in Construction: Current and Future Challenges. *J. Pentech: Cape Town* 8: 20–29.

Tu, P.H. 2004. Game analysis on safety management of engineering construction project. *Journal of wuhan university of science and engineering* 17(8):90–91.

Wang, J.M. & Feng, Y.H. 2007. The application of game theory in the safety management of construction project. *Shaanxi Coal* 26(5): 74–75.

Wang, J. & Cai, Y.Q. 2015. Analysis and construction for threshold signature scheme based on game theory. *Journal on Communications* 26(5): 74–75.

Wang, X.Y. 2013. Research on quality supervision of construction project based on Evolutionary Game Theory. *Project Management Technology* 36(5): 1–8.

Zhang, F.L. & Liu, l. 2002. The Application of Game Theory In Project Safety Management. *Systems Engineering* 20(6): 33–37.

Civil, Architecture and Environmental Engineering – Kao & Sung (Eds)
© 2017 Taylor & Francis Group, ISBN 978-1-138-02985-9

Prestress distribution calculation method of a ridge tube cable dome with annular struts

A.L. Zhang
Beijing Engineering Research Center of High-rise and Large-span Prestressed Steel Structure, Beijing University of Technology, Beijing, China
Key Laboratory of Earthquake Engineering and Structural Retrofit of Beijing, Beijing University of Technology, Beijing, China

Y. Bai
Beijing Engineering Research Center of High-rise and Large-span Prestressed Steel Structure, Beijing University of Technology, Beijing, China

X.C. Liu
Beijing Engineering Research Center of High-rise and Large-span Prestressed Steel Structure, Beijing University of Technology, Beijing, China
Key Laboratory of Earthquake Engineering and Structural Retrofit of Beijing, Beijing University of Technology, Beijing, China

C. Sun
Beijing Engineering Research Center of High-rise and Large-span Prestressed Steel Structure, Beijing University of Technology, Beijing, China

ABSTRACT: Cable dome structure is a large-span spatial structure with high structural efficiency, which is suitable for large-span roof structure. In order to compensate the weaknesses of traditional cable dome and enrich cable dome structure schemes, a new type of cable dome, ridge tube cable dome with annular struts, was proposed by the authors of this paper. Compared with the traditional cable domes, the structure avoids the disadvantage of ridge cable easy relaxation, has greater whole rigidity and good static performance, and relatively less construction difficulty. The cable dome structure and the whole rigidity are mainly determined by the prestress distribution, so force-finding analysis is the key to the new cable dome research. In this paper, the redundant constraint was replaced by constraint force, and according to the node equilibrium relationship and considering self-weight, the prestress distribution calculation formulas of the ridge tube cable dome with annular struts were deduced. The calculation method reduced the operation cost, made the horizontal force and vertical force transfer in the internal structure clear, and made the calculation convenient for engineers both by hand and by computer. Simulation results using ANSYS finite-element iteration method verified the accuracy of the calculation formulas. Using derived calculation formulas, the prestress calculation table of the ridge tube cable dome with annular struts, with self-weight, and under different constraint force level was obtained to provide reference for engineering design and study the mechanical properties.

1 INTRODUCTION

The cable dome is a full-tension space structure with light form and high efficiency (Lu et al. 2005), where components are prestressed to provide the whole rigidity. The traditional Geiger and Levy cable domes have been successfully applied in many domestic and foreign large-span structures, such as the 1985 Seoul Olympic Games gymnastics and fencing arenas (Geiger et al. 1986) and the 1996 Atlanta Olympic Games stadium (Levy 1994). In 2010, the first cable dome building in mainland China—National Fitness Center in Eijin Horo Banner, Ordos City, Inner Mongolia, was established (Zhang, Liu, Ge, Zhang et al. 2012). However, several limitations existed in the traditional Geiger and Levy cable domes: (1) under high roof vertical load, ridge cables are prone to slackening, causing stiffness, sudden decline, and large displacement; (2) a large number of components are flexible cables, and the whole rigidity of the structure is not adequate and sensitive to the eccentric load; and (3) in the flexible hinge cable system, it is difficult to control component size precision and construction deviation and results in tension forming scheme increasingly difficult. Therefore, scholars worldwide have put forward the new cable dome structure scheme, which mainly includes the layout change of ridge cables (Yuan & Dong 2005, Bao 2007), key member replacement of the structure (Li 2007), or combination forms of the structure (Dong et al. 2010, Zhang & Liu 2010, Lu et al. 2015), thus improving the weakness of the traditional cable force characteristics or relatively lowering the difficulty of construction, which promoted the construction of cable dome structure.

Because of the facts that the cable dome structure has almost no natural rigidity and the whole rigidity is provided by prestress tension, the initial prestress distribution determination according to the topology of cable dome structure, namely the force-finding analysis, is the key in tension forming of the cable dome structure. Methods of force-finding analysis mainly include the singular value decomposition method of the main balance matrix (Pellegrino 1993), integrity-feasible prestressing method (Yuan & Dong 2001), the nonlinear finite-element method (Dong Z.L. et al. 1995), and the node equilibrium method (Wang et al. 2010, Dong et al. 2009). However, methods above are mostly based on the theory of complex matrix, which require professional software analysis and fail to clear the internal structure force mechanism; some do not usually consider member and node weight, only to calculate the initial prestress, while in fact cable dome is the equilibrium state of the prestress under gravity, so ignoring self-weight does not conform to the reality.

In order to further improve the limitations of the traditional cable domes and enrich the cable dome structure schemes, the author presented a new full-tension structure system—ridge tube cable dome with annular struts. The new system avoids the shortcomings of ridge cable easy slackening, the whole rigidity of the structure is greater than the traditional cable dome, and the construction difficulty is lowered. In this paper, the structural characteristics of ridge tube cable dome with annular struts were introduced. Aiming at the form finding, the redundant constraint was replaced by constraint force; according to the node equilibrium relationship and considering self-weight, the prestress distribution calculation formulas of ridge tube cable dome with annular struts was deduced. The formulas reduced the operation cost, made the horizontal force and vertical force transfer in the internal structure clear, and made the calculation convenient for engineers both by hand and by computer. Simulation results using ANSYS finite-element iteration method verified the accuracy of the calculation formula. Using derived calculation formulas, the prestress calculation table of ridge tube cable dome with annular struts, which considering self-weight and under different constraint force level, was obtained, to provide reference for the study of engineering design and mechanical properties.

2 RIDGE TUBE CABLE DOME WITH ANNULAR STRUTS

Ridge tube cable dome with annular struts replaced ridge cables of the traditional cable dome with rigid tube, that is, a seamless pipe. The rigid tube topology was triangular, and struts were connected serially and annularly, namely horizontal projection and hoop cable found in coincidence. The three-dimensional structure schematic diagram of the ridge tube cable dome with annular struts is shown in Figure 1.

Compared with the traditional cable dome, struts were arranged annularly, the number of components connected to the upper and bottom nodes were seven and five, respectively, and a triangular mesh formed by the struts and hoop cable enhanced the stability of the structure significantly, which resulted in a greater whole rigidity and high convenience for installation and positioning. The number of diagonal cables was half of that of the traditional Levy type cable dome, so material consumption and tension construction workload were significantly reduced. The ridge tubes prevented ridge cables' slackening, which avoided causing sudden decline of stiffness and large displacement, with higher structure bearing capacity. In addition, round steel tubes were cheaper than high-strength cables, with more price advantage and reduced construction cost.

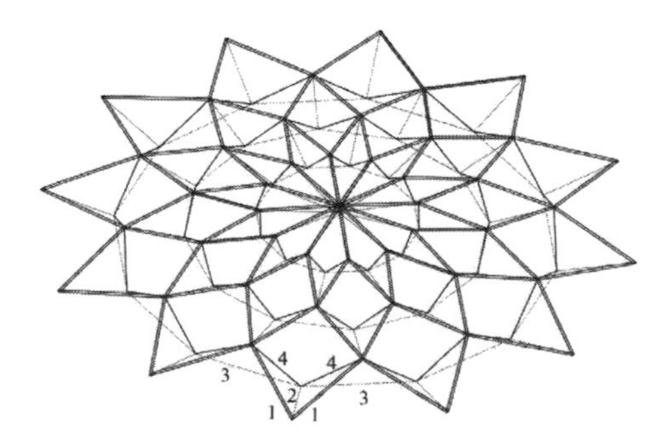

Figure 1. Three-dimensional structure schematic diagram of ridge tube cable dome with annular struts. Specification: 1–ridge tubes; 2–inclined cables; 3–hoop cables; 4–struts.

3 PRESTRESS DISTRIBUTION CALCULATION FORMULAS WITH SELF-WEIGHT

The 1/n unit with single strut as the center of ridge tube cable dome with annular struts was chosen, as shown in Figure 2. The upper node number of struts from the inside to the outside were 1, 2, ..., m, and the corresponding bottom node numbers were 1', 2',..., m, where m and n represented radial and circular section numbers, respectively. In morphological analysis of the ridge tube cable dome with annular struts, states of self-equilibrium stress mode were 19, mode of mechanism displacement was 0, so the cable dome was geometric stability system with no mechanism displacement. Considering the symmetric conditions and the uniaxial deformation of cables and struts, the cable dome had only one integral feasible prestress mode.

The radial horizontal constraint of the central strut upper node was set as the redundant constraint, which was then replaced by the constraint force. According to the moment equilibrium of isolated bodies in each circle, vertical force and radial force transferred by ridge tubes, inclined cables were calculated from inside and out.

3.1 *Vertical force transfer process*

The vertical load of cable dome was transferred through ridge and inclined cables, from the inside out by ring to the outer ring. Horizontally arranged hoop cables did not transfer the vertical load. The weight of elements and nodes were supposed as the equivalent vertical node load (G_i) at upper nodes, as shown in Figure 2. Vertical transfer force from the center node to the second node was calculated as:

$$G_{1\to 2} = G_1 = \frac{G_0}{n} \tag{1}$$

$$\text{when } i \geq 2, \quad G_{i\to(i+1)} = \frac{1}{2}\sum_{j=1}^{i} G_j$$

where G_j was the vertical load at the j node from the center. When single strut as the center, one-piece vertical load was $G_{1=}\ G_0/n$, where G_0 is the equivalent vertical load at the center node; $G_{i\to(i+1)}$ was the vertical transfer force from node i to node $(i + 1)$.

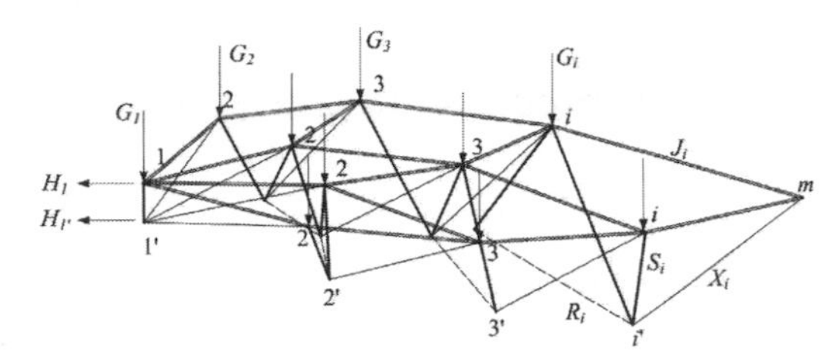

Figure 2. 1/n unit with single-center strut.

3.2 *Horizontal force transfer process*

Upper and bottom nodes were subjected to the radial horizontal forces H_i and $H_{i'}$, respectively, gradually transferred to the outside. Select typical isolated body with struts inside, as shown in Figure 3, according to moment equilibrium, and the bottom radial horizontal force was calculated as:

$$H_{1'} = \frac{G_{1\to 2}L_1 + H_1 h_1^J}{h_1^X} \tag{2}$$

$$\text{when } i \geq 2, \quad H_{i'} = \frac{2G_{i\to i+1}L_i + H_i h_i^J}{h_i^X \cos\frac{\pi}{n}}$$

where h_i^J and h_i^X represent vertical heights from the upper and bottom nodes of the i strut to the horizontal level of the $(i + 1)$ node, respectively, and L_i represents the horizontal projection distance between the upper node i to the line $(i + 1) - (i + 1)$:

$$L_1 = r_1 \cos\frac{\pi}{n}, \text{ when } i \geq 2, \quad L_i = r_{i+1}\sin\frac{\pi}{n}\cot\varphi_{i,(i+1)}$$

The upper node radial horizontal force (H_i) was obtained using formula (2), so the bottom node radial horizontal force ($H_{i'}$) could be calculated.

Analyzing the relationship of the node radial horizontal force by rings, as shown in Figures 3 and 4, the recurrence formula of the node radial horizontal force was obtained using formula (3). To illustrate this, the third-ring nodes were taken for example. The third node radial horizontal force was transferred through two ridge tubes in the third ring to the fourth node, the bottom 3' node

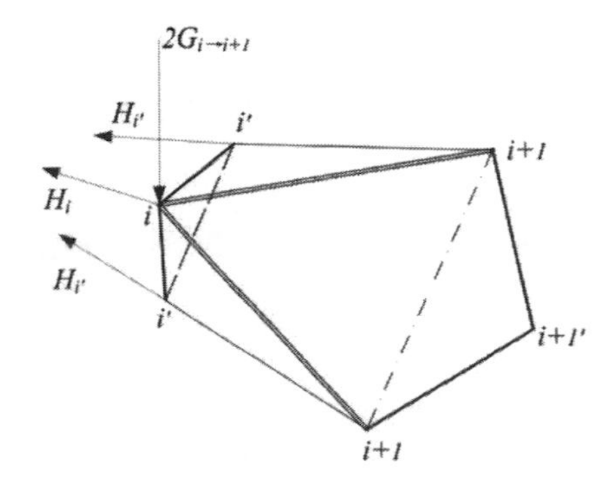

Figure 3. Typical isolated body.

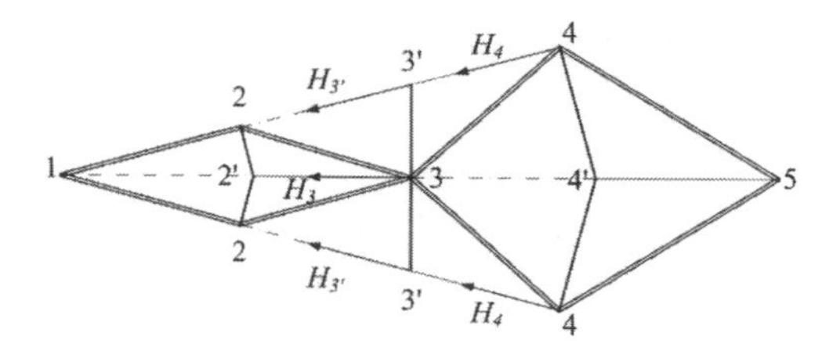

Figure 4. Upper node radial horizontal force.

radial horizontal force was transferred through one inclined cables in third ring to the fourth node; the fourth node radial horizontal force was transferred through two ridge tubes and one inclined cable in third ring to the third node. Provided directions of different node radial horizontal force, $H_{i'}$ was calculated as:

$$H_2 = H_{1'} + H_1 \tag{3}$$

$$\text{when } i \geq 3, \quad H_i = H_{i-1'} + H_{i-1}\frac{\cos\varphi_{i,(i-1)}}{\cos\varphi_{(i-1),i}}$$

where $\varphi_{(i+1),\,i}$ is the angle between the horizontal projection of ridge and inclined cables at node $(i + 1)$, $\varphi_{i,(i+1)}$ is the angle between the horizontal projection of ridge tubes and radial lines at node i.

$$\varphi_{(i+1),i} = \arctan\left(\frac{r_i \sin\frac{\pi}{n}}{r_{i+1} - r_i\cos\frac{\pi}{n}}\right)$$

$$\varphi_{i,(i+1)} = \arctan\left(\frac{r_{i+1}\sin\frac{\pi}{n}}{r_{i+1}\cos\frac{\pi}{n} - r_i}\right)$$

3.3 *Prestress distribution with self-weight considered*

The initial prestress of the cable dome was composed of two parts: one part was the ideal initial prestress when the constraint force was P and weight was neglected; the other part took self-weight into consideration and the constraint force was zero.

First part: When the constraint force was P, and the weight was neglected, radial horizontal constraint of the center strut upper node was replaced by the constraint force P, namely, $H_I = P$, $G_i = 0$. According to the derived formulas (1)–(3), each node radial horizontal force formula could be obtained from the inside to the outside when weight is neglected:

$$H_{1'} = \frac{h_1^J}{h_1^X}P \tag{4}$$

$$H_2 = H_{1'} + H_1 = \frac{h_1}{h_1^X}P \tag{5}$$

$$H_{2'} = \frac{H_2 h_2^J}{h_2^X\cos\frac{\pi}{n}} = \frac{h_1 h_2^J}{h_1^X h_2^X\cos\frac{\pi}{n}}P \tag{6}$$

$$H_3 = H_{2'} + H_2\frac{\cos\varphi_{3,2}}{\cos\varphi_{2,3}} = \left(\frac{h_1 h_2^J}{h_1^X h_2^X\cos\frac{\pi}{n}} + \frac{h_1}{h_1^X}\frac{\cos\varphi_{3,2}}{\cos\varphi_{2,3}}\right)P \tag{7}$$

$$H_{3'} = \frac{H_3 h_3^J}{h_3^X\cos\frac{\pi}{n}} = \left(\frac{h_1 h_2^J}{h_1^X h_2^X\cos\frac{\pi}{n}} + \frac{h_1}{h_1^X}\frac{\cos\varphi_{3,2}}{\cos\varphi_{2,3}}\right)\frac{h_3^J}{h_3^X\cos\frac{\pi}{n}}P \tag{8}$$

From the formulas above, the ideal initial prestress varied linearly with P, so an increase in constraint force enhanced cable domes' overall prestress level.

Second part: When the constraint force was zero and the weight was considered, the radial horizontal constraint of the center strut upper node was $H_I = P$, $G_i \neq 0$. According to the derived formulas (1)–(3), each node radial horizontal force formula could be obtained from the inside to the outside under gravity:

$$H_{1'} = \frac{G_{1\to2}L_1}{h_1^X} \tag{9}$$

$$H_2 = H_{1'} \tag{10}$$

$$H_{2'} = \frac{2G_{2\to3}L_2 + H_2 h_2^J}{h_2^X\cos\frac{\pi}{n}} \tag{11}$$

$$H_3 = H_{2'} + H_2\frac{\cos\varphi_{3,2}}{\cos\varphi_{2,3}} \tag{12}$$

$$H_{3'} = \frac{2G_{3\to4}L_3 + H_3 h_3^J}{h_3^X\cos\frac{\pi}{n}} \tag{13}$$

Adding the two-part prestress together, on the basis of the geometric relationship of components, supposing the internal force of the cable tension as positive and the struts compression as negative, internal force formulas of each component types could be deduced.

Force of ridge cables was calculated as:

$$J_1 = \frac{H_1}{\cos\alpha_1} \tag{14}$$

when $i \geq 2$, $J_i = \dfrac{H_i}{2\cos\alpha_i \cos\varphi_{i,(i+1)}}$

Force of inclined cables was calculated as:

$$X_i = \frac{H_{i'}}{\cos\beta_i} \tag{15}$$

Force of struts was calculated as:

$$S_1 = n\tan\beta_1 H_{1'} - G_1^X \tag{16}$$

when $i \geq 2$, $S_i = \dfrac{\tan\beta_i H_{i'} - G_i^X}{2\sin\theta_i}$

Force of hoop cables was calculated as:

$$R_i = \frac{1}{2}H_{i+1'}\left(\sin^{-1}\frac{\pi}{n} + \frac{\tan\beta_{i+1} - G_{i+1}^X}{\tan\theta_{i+1}}\right) \tag{17}$$

where α_i represents the angle between ridge cables and the horizontal plane, β_i represents the angle between inclined cables and the horizontal plane, and θ_i represents the angle between struts and the horizontal plane:

$$\alpha_i = \arctan\left(\frac{h_i^J \sin\varphi_{(i+1),i}}{r_i \sin\frac{\pi}{n}}\right)$$

$$\beta_i = \arctan\left(\frac{h_i^X}{r_{i+1} - r_i \cos^{-1}\frac{\pi}{n}}\right)$$

$$\theta_i = \arctan\left(\frac{h_i}{r_i \sin\frac{\pi}{n}}\right)$$

4 CASE VERIFICATION

In order to verify the correctness of the derived formulas, the results of internal force distribution using ANSYS finite-element modeling and the prestress distribution calculation method were compared. Case: a ridge tube cable dome with annular struts; span, 60 m; height, 4 m; circumferential segment number, 12; with two hoop cables; and the angle between inclined cables and the horizontal level was 15°. The calculation diagram is shown in Figure 5, where JG, XS, HS, and CG represent the ridge tubes, inclined cables, hoop cables, and struts, respectively, and 1–3 are ring numbers from the inside to the outside. Q345B seamless steel tubes were adopted as ridge tubes and struts, whose yield strength was 345 MPa and the elastic modulus was 2.06 × 105 MPa. High-strength steel wires were adopted as cables, with the tensile strength of 1670 MPa and the elastic modulus of 1.85 × 105 MPa. Component specification from the inside to the outside: the ridge tubes of Φ245X8, Φ203X8, Φ168X7, inclined cables of Φ14, Φ28, Φ40, struts of Φ159X8, Φ102X4, Φ168X10, hoop cables of Φ42, Φ75.

The finite-element model of ridge tube cable dome with annular struts was established, applied with the ideal initial prestress and under gravity, structure internal force distribution, and the maximum vertical displacement under the effect of gravity were obtained. While the initial prestress level remained unchanged, the last obtained prestress distribution was taken as the new initial prestress into the next analysis. After several iterations, when the internal force distribution under the gravity was stable and the maximum vertical displacement was minimum, the iteration of prestress distribution was terminated. In this way, after seven iterations, the final prestress distribution of ridge tube cable dome with annular struts was obtained, and the iterative process is shown in Table 1.

By using the simple calculation formulas proposed, considering the gravity of the structure, the weight was transformed into the equivalent vertical node load (G_i) acting on the upper nodes, constraint force P as 500kN. By substituting the

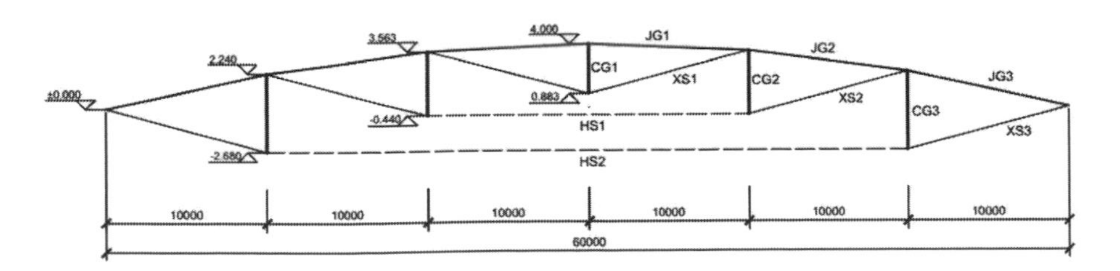

Figure 5. Structure calculation diagram.

Table 1. Finite-element iterative process.

Components		Initial value	1st	2nd	3rd	4th
		Iterative data/kN				
Ridge tubes	J_1	−262.259	−254.026	−295.015	−289.813	−290.265
	J_2	−49.631	−48.996	−56.888	−55.915	−56.005
	J_3	−173.725	−174.031	−202.043	−198.613	−198.933
Inclined cables	X_1	500.000	430.211	508.708	499.190	500.088
	X_2	334.910	293.469	346.017	339.596	340.194
	X_3	633.574	580.188	680.300	668.014	669.155
Struts	S_1	84.441	82.190	95.384	93.697	93.842
	S_2	308.216	308.703	358.088	351.970	352.535
	S_3	848.202	872.233	1011.536	994.517	996.116
Hoop cables	R_1	601.321	601.852	698.192	686.258	687.360
	R_2	1702.458	1747.099	2026.525	1992.384	1995.590
Displacement/mm	U_{max}	0.000	10.097	1.599	0.021	0.002

Components		5th	6th	7th	Formula results	Error
Ridge tubes	J_1	−290.210	−290.214	−290.213	−289.478	−0.25%
	J_2	−55.995	−55.996	−55.996	−56.015	0.03%
	J_3	−198.899	−198.902	−198.902	−198.970	0.03%
Inclined cables	X_1	499.994	500.002	500.001	500.478	0.10%
	X_2	340.130	340.135	340.135	340.277	0.04%
	X_3	669.032	669.042	669.041	669.283	0.04%
Struts	S_1	93.824	93.825	93.824	93.588	−0.25%
	S_2	352.474	352.479	352.479	352.579	0.03%
	S_3	995.946	995.961	995.959	996.309	0.04%
Hoop cables	R_1	687.241	687.251	687.250	687.450	0.03%
	R_2	1995.250	1995.280	1995.276	1995.981	0.04%
Displacement/mm	U_{max}	0.000	0.000	0.000	0.000	–

Table 2. Prestress reference table of ridge tube cable dome with annular struts (kN).

Component force	$Gi \neq 0$ $P = 0$	$Gi = 0$ $P = 10$	$Gi \neq 0$ $P = 50$	$Gi \neq 0$ $P = 100$	$Gi \neq 0$ $P = 200$	$Gi \neq 0$ $P = 300$	$Gi \neq 0$ $P = 500$
J_1	26.968	5.250	53.219	79.470	131.972	184.474	289.478
J_2	6.336	0.994	11.304	16.272	26.207	36.143	56.015
J_3	25.078	3.478	42.467	59.857	94.635	129.413	198.970
X_1	0.000	10.010	50.048	100.096	200.191	300.287	500.478
X_2	5.047	6.705	38.570	72.093	139.139	206.185	340.277
X_3	35.102	12.684	98.520	161.938	288.774	415.610	669.283
S_1	9.066	1.690	17.518	25.970	42.875	59.779	93.588
S_2	44.069	6.170	74.920	105.771	167.473	229.175	352.579
S_3	147.296	16.980	232.198	317.099	486.901	656.704	996.309
R_1	85.554	12.038	145.744	205.933	326.312	446.692	687.450
R_2	291.894	34.082	462.303	632.712	973.529	1314.346	1995.981

results of formulas (4)–(13) into formulas (14)–(17), the internal force value of all types of components were calculated, and the formula calculation results are shown in Table 1.

As Table 1 indicates, the results of the simple calculation formulas proposed and the finite-element modeling results after several iterations were basically the same, while the force maximum error was less than 0.3%, which indicated that the theoretical formulas proposed in this paper did not require the finite-element software calculation, reduced the computational cost of iterative process, and determined the cable dome's prestress distribution with self-weight accurately.

For convenience of structure design for technicians, by adopting prestress distribution calculation formulas proposed in this paper, the prestress reference table of ridge tube cable dome with annular struts in different constraint force level and under gravity is shown in Table 2. According to the engineering requirement, selecting the corresponding constraint force level and the prestress distribution under gravity, the initial prestress distribution with self-weight of ridge tube cable dome with annular struts was calculated rapidly.

5 CONCLUSIONS

In order to determine the initial prestress distribution of ridge tube cable dome with annular struts, this paper set the radial horizontal constraint of the central strut upper node as the redundant constraint, which was then replaced by the constraint force. According to the moment equilibrium of isolated bodies in each circle, vertical force and radial force transferred by ridge tubes and inclined cables were calculated from inside and out. Thus, the calculation formulas of prestress distribution with self-weight were derived. The initial prestress of the cable dome was composed of two parts: one part was the ideal initial prestress when the constraint force was P and weight was neglected; the other part took self-weight into consideration and the constraint force was zero. Compared with the finite-element iteration method, the calculation formula results were accurate, which cleared the horizontal and vertical force transmission relationship in the internal structure and was convenient for technicians. In this paper, the prestress reference table of ridge tube cable dome with annular struts in different constraint force level and under gravity was given. According to the engineering requirement, selecting the corresponding constraint force level and the prestress distribution under gravity, the initial prestress distribution with self-weight of ridge tube cable dome with annular struts was calculated rapidly.

ACKNOWLEDGMENTS

This work was supported by the National Natural Science Foundation of China (51278010) and the Science and Technology Plan of Beijing Municipal Commission of Education (KM201610005012) and the Beijing Natural Science Foundation (8131002).

REFERENCES

Bao, H.Z. 2007. The theoretical and experimental research of bird-nest cable dome. *Doctoral dissertation, Zhejiang University*.

Dong, S.L., Wang, Z.H. & Yuan, X.F. 2009. A simplified calculation method for initial prestress of levy cable domes with the consideration of self-weight. *Engineering Mechanics,* 26(4), 1–6.

Dong. S.L., Wang, Z.H. & Yuan, X.F. 2010. Static behavior analysis of a space structure combined of cable dome and single-layer lattice shell. *Journal of Building Structures*. 31(03): 1–8.

Dong, Z.L., He, G.Q. & Lin, C.Z. 1995. Finding of equilibrium states of tensegrity systems. *Journal of Building Structures,* 20(05): 24–28.

Geiger, D.H., Stefaniuk, A. & Chen, D. 1986. The design and construction of two cable domes for the Korean Olympics. *Shells, Membranes and Space Frame, Proceedings IASS Symposium. Madrid, Spain: IASS*. 265–272.

Ge, J.Q., Zhang, A.L., Liu, X.G., Zhang, G.J., Ye, X.B., Wang, S. & Liu, X.C. 2012. Analysis of tension form-finding and whole loading process simulation of cable dome structure. *Journal of Building Structures*. 33(04):1–11.

Levy, M.P. 1994. The Georgia dome and beyond achieving lightweight-long span structures. *Proceedings of IASS-ASCE International Symposium. Madrid, Spain: IASS*. 560–562.

Li, Z.Q. 2007. Behavior analysis and system improvement of cable domes. *Doctoral dissertation, Zhejiang University.*

Liu, X.C., Zhang, A.L., Liu, Y.J., Ge, J.Q. & Zhang, G.J.2012. Experimental research and performance analysis of new-type cable-strut joint of large-span cable dome structure. *Journal of Building Structures*.33(04):46–53.

Lu, C.L., Zhang, A.L. & Zhang, G.J. 2005. The development and deepening of the study on prestressed steel structures. *Progress in Steel Building Structures,* 8(2):41–48.

Lu, J.Y., Wu, X.L., Zhao, X.L. & Shu, G.P. 2015. Form finding analysis of cable-strut tensile dome based on tensegrity torus. *Engineering Mechanics*. S1:66–71.

Pellegrino, S.1993. Structural computations with the singular value decomposition of the equilibrium matrix. *International Journal of Solids and Structures*. 30(21): 3025–3035.

Wang, Z. H., Yuan, X.F. & Dong, S.L. 2010. Simple approach for force finding analysis of circular geiger domes with consideration of self-weight. *Journal of Constructional Steel Research,* 66(02), 317–322.

Yuan, X.F. & Dong, S.L. 2001. Application of integrity feasible prestressing to tensegrity cable domes. *China civil engineering journal*. 34(02): 33–37.

Yuan, X.F. & Dong, S.L. 2005. New forms and initial prestress calculation of cable domes. *Engineering Mechanics*. 22(02):22–26.

Zhang, B.K. & Liu, W.C. 2010. Design and analysis of hybridized self-balance prestressed latticed structure composed of shell and cable dome. *Journal of Shenyang Jianzhu University (Natural Science)*. 26(03):452–457.

Zhang, A.L., Liu, X.C., Li. J. Ge, J.Q. & Zhang, G.J. 2012. Static experimental study on large-span cable dome structure. *Journal of Building Structures*.33(04):54–59.

Zhang, G.J., Ge, J.Q., Wang, S., Zhang, A.L., Wang, W.S., Wang, M.Z. & Xu, K.R. 2012. Design and research on cable dome structural system of the National Fitness Center in Ejin Horo Banner, Inner Mongolia. *Journal of Building Structures*. 33(04):12–22.

Civil, Architecture and Environmental Engineering – Kao & Sung (Eds)
© 2017 Taylor & Francis Group, ISBN 978-1-138-02985-9

A new multiple-model analysis method considering structural uncertainty

Fengqi Zhu
Southeast University, Nanjing, China

ABSTRACT: Because of the corrosion of the main body structure and connection, difference between construction and design, and so on, structure parameters cannot be known according to the original design or existing data. Thus, the structural uncertainty is generated, which widely exists in the actual structure. It is very necessary that uncertainty analysis is carried out on the structure. However, classical Finite-Element (FE) method could only establish a certain FE model and could not react and transmit uncertainty of key parameters. From a practical engineering perspective, this paper provides the intuitivism apprehension of uncertainty of probability and statics by multiple-model probability FE method. The specific method of improved multiple model was that multiple FE models were established on the basis of the model of the key parameters of Markov Chain Monte Carlo (MCMC) sampling first. Thus, it was considered, transmitted the influence of uncertainty, and reflected the actual structure character. Then, the second-order response surface method was brought to fit the structural dynamic and static response results to improve the efficiency of multiple-model modeling. Through the above methods, multiple models could be applied to the engineering practice. In a numerical example, the parameters of girder elasticity modulus, section height, and support spring stiffness were analyzed, multiple models were built, and reliability was assessed.

1 INTRODUCTION

In general, structures would deteriorate over time; especially, the degeneration of structural critical areas might lead to monolithic catastrophic failure. In the inspection of a structure, the phenomena of crack and corrosion of the main structure and the padding and failure of connection parts were discovered (Shinae 2013). They are the main sources of structural uncertainty for structural performance evaluation; therefore, it is of great importance to figure out the way of quantitatively evaluating their influences on structure performance.

Many scholars have done a lot of work in this research field. Subrata Chakraborty et al. updated the finite-element model through response surface method; Ching and Beck gave the initial key parameter a certainty probability and proceeded finite-element model update through Bayesian theory combining structural monitoring data. Both of them treated model updating as an optimization problem. Through different kinds of optimistic algorithm, they searched for an optimal model that keeps with the field test data. Koh et al. utilized multiple models to partially solve this problem. First, they relegated the key parameter by PCA method and then by using K-means method to cluster the models. Jianzhang et al. proposed utilizing Markov chain Monte Carlo sampling method to calibrate the uncertainties. The process utilizes high-quality sampling algorithm to generate multi-finite-element model library and then utilizes multi-finite-element model library to assess structural response. However, because of the high cost, it is hardly executed at practical engineering structure. Zong et al. used RSM to optimize finite-element model to find the best single model. In this paper, response surface technology and multimodel idea are used to analyze structure uncertainty. Because of the influence of uncertainty, the key parameters of the model is difficult to decide in the process of establishing the finite-element model of engineering structure; at the same time, utilizing the usual model correction technique is not reliable. Therefore, in this paper, a multimodel idea was used to sample the key parameters of structure, which can provide the structural probabilistic information to engineers. Furthermore, the structural probability response can be predicted by sampling. In order to improve the computational efficiency, this paper will inlet response surface model in the computational model. By using the response surface model to replace multiple models, we can greatly improve the computational efficiency, which could make the idea of multimodel possible in structural engineering.

This paper is organized as follows. First, a multiple-model method is proposed using the

MCMC sampling and RSM technologies. Then, ANSYS batch tool is used to operate ANSYS in background under MATLAB control to achieve multiple-model method. Finally, a multimodel was built and reliability assessment was studied with a numerical example.

2 MULTIPLE-MODEL METHOD

Multiple-model method is used to sample the key parameters of a structure, which can provide the structural probabilistic information for engineers. At the same time, in order to improve the computational efficiency, response surface model is brought in the computational model.

Introduction of the Markov Chain Monte Carlo (MCMC) sampling method can rapidly converge to the objective function, $\pi(x)$, under the framework of Bayesian theory. The MCMC sample can realize the chain sample and overcome the impact of high-dimensional parameters. The process is a recursive procedure, involving the following steps:

1. Randomly generated initial parameters θ_0.
2. Assuming that θ_{i-1} is the kth parameter of the Markov chain, candidate parameter θ_c is generated by proposal distribution $q(x)$:

$$\theta_c = q(\theta_c / \theta_{k-1}) \tag{1}$$

3. Acceptance probability of candidate parameter θ_c is:

$$\alpha(\theta_{k-1}, \theta_c) = \min\left(1, \frac{\pi(\theta_c) q(\theta_c / \theta_{k-1})}{\pi(\theta_{k-1}) q(\theta_{k-1} / \theta_c)}\right) \tag{2}$$

4. If θ_c is accepted, then $\theta_k = \theta_c$; otherwise, repeat steps 2 and 3 until θ_c is received.
5. Make $k = k + 1$, repeat steps 2–4 until the process converges.

In a multiple-model framework, objective function π is elected as the model probability conditioned by known structural information G; proposed distribution q is elected as the prior distribution of the key unknown parameters; the key unknown parameter θ in the studied structure is a combination of elasticity modulus, stiffness of expansion joint, bearing longitudinal stiffness, and bearing vertical stiffness:

$$\pi(\theta) = P(M(\theta)|G); \quad q = P(M(\theta)) = \prod_1^4 P(\theta_j) \tag{3}$$

where $P(\theta_j)$ is the prior distribution of the jth parameter.

According to Bayesian theory, $P(M/G) = \frac{P(G/M)P(M)}{\int P(G/M)P(M)}$. Assuming $\int P(G/M)P(M)$ is constant and combining (2) and (3), we have:

$$\alpha(\theta_{i-1}, \theta_c) = \min\left(1, \frac{P(G / M_c(\theta_c))P(M_c)}{P(G / M_{i-1}(\theta_{i-1}))P(M_{i-1})}\right) \tag{4}$$

Assuming that the error of the model belongs to normal distribution and independent from each other:

$$P\left(G|M_i(\theta_i)\right) = \frac{1}{\sqrt{2\pi\sigma^2}} exp\left(-\frac{obj^2}{2\sigma^2}\right) \tag{5}$$

To further improve the sampling efficiency, an extended M-H algorithm is adopted because of its high efficiency. Two novel ideas, Adaptive Metropolis (AM) and Delayed Rejection (DR), are successfully combined to improve the computational efficiency of the M-H algorithm. The reader is advised to refer to ref. [Haario *et al.* (2006)].

Response Surface Method (RSM) is used to approximate implicit functions between response value and design parameter by explicit response surface model. RSM is very good at copping with the issue with implicit multiresponse object function for its satisfied accuracy and good efficiency. Because the amount of sample set used to reconstruct the response surface is small (Zhao and Qiu 2013), the second-order RSM is reconstructed by Central Composite Design (CCD) method, F test of variance analysis is used to identify the parameter of significance, and the second-order polynomial is the following fit response surface:

$$y = \beta_0 + \sum_{i=1} \beta_i x_i + \sum_{i=1} \sum_{j=1} \beta_{ij} x_i x_j + \sum_{i=1} \beta_i x_i^2 \tag{6}$$

3 NUMERICAL EXAMPLE OF MULTIPLE-MODEL METHOD

Through a numerical example of nonuniform beam of the MM method proposed in this paper, the high precision and the whole performance of this method is verified for the natural frequencies

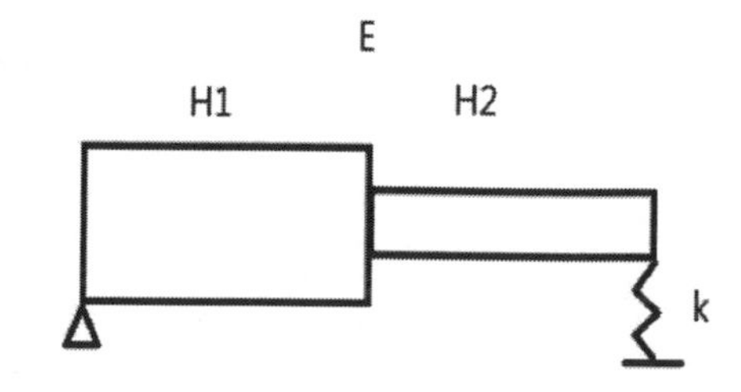

Figure 1. The numerical example.

Table 1. Parameters of real value and initial value.

	Elasticity modulus	Height of Section 1-1	Height of Section 2-2	Spring stiffness
Real value	3.5×10^{10} Pa	0.6 m	0.3 m	1×10^5 Pa
Initial value	3.1×10^{10} Pa	0.4 m	0.2 m	0.5×10^5 Pa
Distribution	Normal	Uniform	Uniform	Uniform

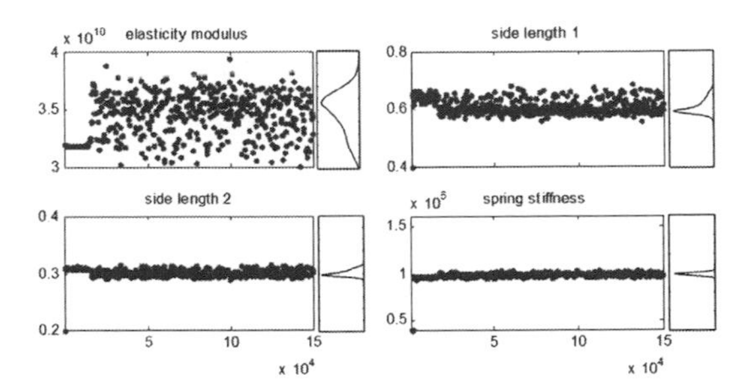

Figure 2. Four-parameter MCMC results.

and formation of the structure and the displacement and strain under the action of static loads. The numerical example is shown in Fig. 1 and the real values of the parameters are as below: the elasticity modulus of the non-uniform beam is 3.5×10^{10}; section of the beam is square; the heights of Sections 1-1 and 2-2 are 0.6 and 0.3 m, respectively; and the spring stiffness of the bearing is 1×10^5 Pa (Table 1).

The parameters of elasticity modulus, height of Section 1-1, height of Section 2-2, and spring stiffness are elected to consider uncertainty. Extraction of the first five-order natural frequency and formation are displacement and strain under two load cases: loading of 100 kN at mid-span and 1/4 span of the beam, respectively. On the basis of the modal and static response result constructing the objective function (Eq. 7), 15,000 samplings were carried out using MCMC sampling method and response surface technology (Fig. 2):

$$\pi(\theta) = \sum_{i=1}^{m} \frac{|\omega_{ai} - \omega_{ei}|}{\omega_{ei}} + \sum_{i=1}^{m} (1 - MAC_i) + \sum_{i=1}^{n} \frac{|\delta_{ai} - \delta_{ei}|}{\delta_{ei}} + \sum_{i=1}^{n} \frac{|\varepsilon_{ai} - \varepsilon_{ei}|}{\varepsilon_{ei}} \tag{7}$$

where ω_{ai} and ω_{ei} are analytical and experimental natural frequencies of the ith mode, respectively, and Φ_{ai} and Φ_{ei} are the analytical and experimental modal shapes of the ith mode. MAC value can be calculated from Φ_{ai} and Φ_{ei}.

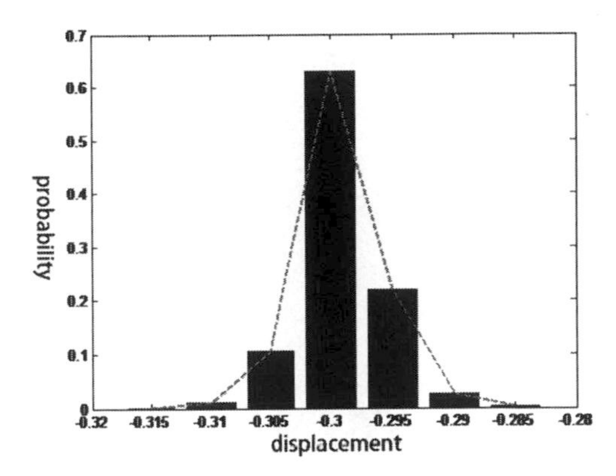

Figure 3. Displacement of spring support.

As shown in Fig. 3, on the basis of the above established multiple-model library of the variable cross-sectional beam displacement of spring support forecasting 0.3 m, the probability follows normal distribution. It equals the real displacement value of 0.3 m.

4 CONCLUSION

1. A multiple-model method through MCMC sampling and response surface technology is proposed in this paper, which is used in structure performance evaluation.
2. Using the response surface technology, the uncertainty problem of large finite element in real structure can be solved by combining the large FE technology with the multiple-model idea.

REFERENCES

Ching J.Y. and James L. Beck (2004). New Bayesian Model Updating Algorithm Applied to a Structural Health Monitoring Benchmark. *Structural Health Monitoring*, Vol. 3.

Haario H., M. Laine, M. Mira, E. Saksman (2006). DRAM: Efficient adaptive MCMC. *Statistics Computing.* Vol. 16.

Koh H.M. , W. Park and H.J. Kim (2013). Recent activities on operational monitoring of long-span bridges in Korea. *The 6th International Conference on Structural Health Monitoring of Intelligent Infrastructure Hong Kong*, December 9–11.

Shinae Jang, J. Li, Billie F. Spencer Jr (2013). Corrosion Estimation of a Historic Truss Bridge using Model updating. *Journal of Bridge engineering*, July.

Subrata Chakraborty, Arunabh Sen (2014). Adaptive response surface based efficient Finite Element Model Updating. *Finite Elements in Analysis and Design*, Vol. 80.

Zhang J., C.F. Wan and Tadanobu Sato (2013). Advanced Markov Chain Monte Carlo Approach for Finite Element Calibration under Uncertainty. *Computer-Aided Civil and Infrastructure Engineering*, Vol. 28.

Zhao W.T., Z.P. Qiu (2013). An efficient response surface method and its application to structural reliability and reliability-based optimization. *Finite Elements in Analysis and Design*, Vol. 67.

Zong Z.H., M.L. Gao, Z.H. Xia (2011). Finite element model validation of the continuous rigid frame bridge based on structural health monitoring Part 1: FE model updating based on the response surface method. *China Civil Engineering Journal*, Vol. 44, No. 2.

Civil, Architecture and Environmental Engineering – Kao & Sung (Eds)
© 2017 Taylor & Francis Group, ISBN 978-1-138-02985-9

Promotion of transportation on urban spatial distribution and industrial development in Beijing

Xiao Peng
China Urban Sustainable Transportation Research Center (CUSTReC), China Academy of Transportation Sciences, Beijing, China

ABSTRACT: The relationship between urban transportation and urban spatial layout and industry is the eternal topic of urban development. On the basis of the mechanism analysis of urban transportation infrastructure construction and urban industry and spatial layout, the solution and relative suggestions of the coordinated development of urban transportation, urban industry, and spatial layout are put forward. The research results are helpful to change the transportation infrastructure, which restricts the city industrial development and urban space layout and realizes the benign interaction of the transportation and urban industry and spatial layout in Beijing. The research result will also provide technical support for the traffic infrastructure construction in Beijing and promote the adjustment and upgrading of industrial structure to guide the harmonious coexistence of industry between center city and new town.

1 INTRODUCTION

The relationship between urban transportation infrastructure and urban industrial and spatial layout is interaction and reaction. It is not only a simple single direction but also a complex cycle. The adjustment of industrial development and urban spatial layout is an important characterization of urban economic and social development. Realizing the interaction mechanism of traffic infrastructure construction and the development of industry and urban spatial layout are the important basis to formulate transportation policy and solve complex urban traffic problems. It helps to solve the efficiency problems in limited land resources.

1.1 *Urban industrial development and urban spatial layout affect the trip mode choice and travel demand distribution*

In urban areas, besides the living district, the main travel demand and trip distribution come from the workplace. All kinds of urban industrial layout are the main components of workplace and also an important part of urban spatial layout. Eventually, a different urban spatial layout affects the urban industrial distribution and position, which results in different trip generation in different areas. By the way, different urban industrial and spatial layouts may lead to different trip directions and routes, which directly determines the distribution of whole urban traffic.

1.2 *Traffic infrastructure distribution makes different accessibilities in different spaces and affects the urban industry layout*

Accessibility is one of the important elements for selecting the location of industrial areas. The location quality is the main reason directly affecting the industry development. As accessibility is caused by the different traffic infrastructure distribution, the ultimate factor affecting the urban industry layout is the urban traffic infrastructure construction and distribution.

It is necessary to spread the industry in space if it is wanted to be developed. Transport can play a role in the spatial diffusion of the urban industry. The construction of traffic infrastructure will stimulate the industrial development along the line. The population and main industrial activities gradually increase the urban transportation line, thereby forming industrial centers. At the same time, a good urban traffic infrastructure will reduce the cost of trip and product transportation and provide the possibility to expand the industrial agglomeration and diffusion.

The practical experience worldwide shows that the actual shape and structure of the city are the result of the interaction between the transportation

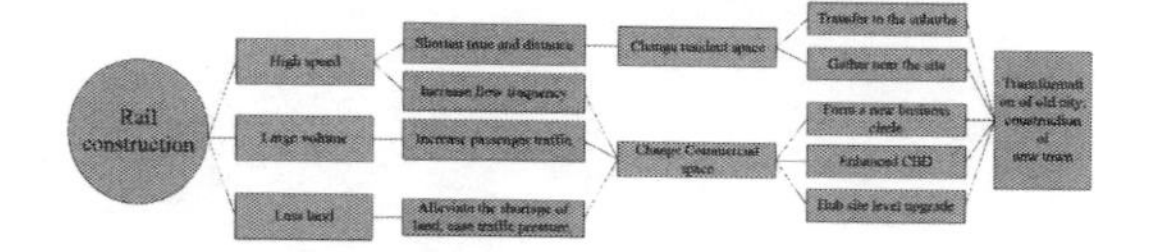

Figure 1. Traffic infrastructure promotes and guides the urban spatial layout.

development strategy and the spatial development strategy. Among all types of transport mode, rail transit will bring great changes in the passenger flow, improve the accessibility, and change the location condition along the rail lines. Hence, the urban land use and the industrial spatial layout will be affected profoundly, followed by the urban spatial structure and morphology. As the most important component, rail transit will promote city renewal and urban spatial layout restructure.

2 CURRENT SITUATION ASSESSMENT

2.1 *Beijing urban transportation supply*

In recent years, the increasing transportation construction investment has resulted in the rapid growth of urban roads in Beijing. Taking 2005 as the base year, the total mileage of urban streets reached 6223 km at the end of 2015, with an average annual increase of 11%. The total mileage of highway reached 21885 km at the end of 2015, with an average annual increase of 9%. However, compared with the developed cities, the current urban road density and land use rate of Beijing are still significantly lower. The total mileage of urban buses and tram vehicles reached 21716 km by the end of 2015.

At the end of 2015, the motor vehicle and private car ownership reached 5.62 and 4.4 million in Beijing, respectively, which are more than 0.81 million and 0.66 million from the corresponding figures in 2010. In line with international practices of 10 cars per 100 households, on average, per 100 households in Beijing have 25.8 motor vehicles and 14.6 private cars. It means that Beijing has begun to enter the automobile society. In the future, Beijing will face more and more serious transport pressure. In addition, mobile population is also increasing rapidly. A number of private cars are used for commuting, which results in the surging of motor travel demand.

At the end of 2015, the city's highway mileage was 21876 km, with an increase of 27.2 km over 2014 and 762 km over 2010. Among them, the expressway mileage was 982 km, in agreement with that of the previous year and an increase of 79 km over 2010. In 2015, the urban road mileage is 6435 km, an increase of 9 km over 2014 and 80 km over 2010.

Table 1. Various indicators of bus and tram in 2011–2015.

Year	2011	2012	2013	2014	2015
Routes (number)	740	779	813	877	876
Length (km)	19338	19547	20575	20347	20315
Vehicles (units)	21575	22146	22486	24083	24347

Table 2. Vehicle ownership in 2011–2015.

Year	Motor vehicle ownership (10,000)	Private vehicle ownership (10,000)	Private car ownership (10,000)
2011	498.3	389.7	286.2
2012	520	407.5	298.2
2013	543.7	426.5	311
2014	559.1	437.2	316.5
2015	561.9	440.3	316.5

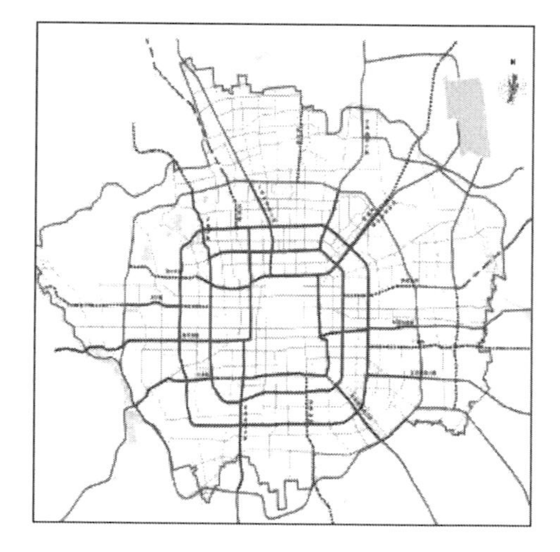

Figure 2. Urban road network of Beijing.

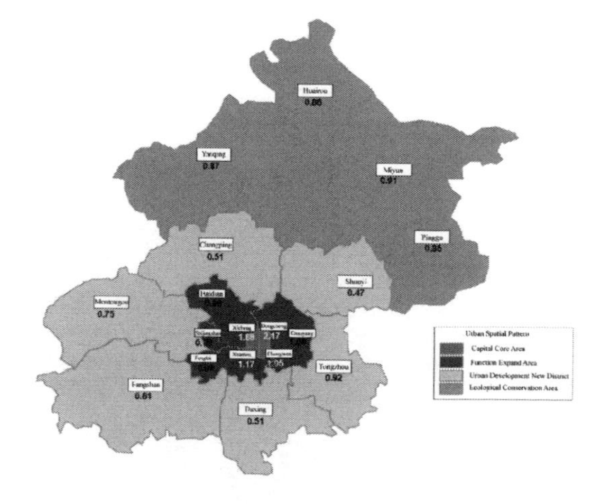

Figure 3. Layout of urban functions.

2.2 *Beijing urban industrial space layout*

In accordance with the latest urban planning, Beijing, under the jurisdiction of 16 areas, can be divided into four parts: the Capital Core Area, the Function Expand Area, the Urban Development Area, and the Ecological Conservation Area.

The main features of the urban industry in Beijing are: (a) The spatial structure is evolved from the single center to the suburbanization, which means Beijing is in the direction of the development phase to the multicenter layout. (b) The six areas (the original eight districts) in the city are still the main bearing area of the urban population and industry. (c) The economic growth of the core area of the capital function and urban function development zone is the fastest, and the population growth in the urban function development zone and the urban development zone are important. (d) Industrial agglomeration effect is gradually emerging, and development zones and industrial parks have become the important carrier of economic development in the capital. (f) A number of different functions and industrial positioning of different professional parks or bases have begun to emerge.

2.3 *Main problems*

a. The single-center urban spatial structure led to the central area of traffic travel and tidal traffic congestion.

The construction of Beijing city is still a layout pattern of the old city as the center, surrounded by the edge. The old city, set in a 62 km^2 in the second ring, as the center of the political, economic, and cultural centers, sums up three commercial centers of Xidan, Wangfujing, and the front door and the Financial Street business district. In addition, it also has more than 20 central ministries, 100 bureau level authorities, and 250 units of the Beijing Municipal Committee and government owned units, while the other is the residential area. With the fourth and fifth ring roads constructed, the Beijing pie continues spreading, which is called "where is the ring road, where is the pie spreading". This city development mode moves residents out of the old city, while all types of social activities are still in the old downtown. At the same time, strengthening the construction of functional areas in the fifth ring and the construction of three new key city is sluggish, and the functions of population living, industrial development, public service in the core area, and development area are still concentrated, and the new cities are weak to ease the downtown population; instead, they possess a large proportion of foreign population, which failed to effectively undertake the center area population ease.

b. The unbalance of economy and industry increases traffic congestion.

There is a big gap per capita GDP between the south and north of Beijing. The state administrative organs, universities, and other cultural and educational institutions and large and medium-sized hospitals most concentrated in the north of Beijing. The important resource in north Beijing is more than that in south Beijing, which results in more traffic density in north Beijing than south Beijing. The high-density resource allocation in north Beijing crowds out the traffic infrastructure space, increasing the difficulty to alleviate the traffic flow by using the transportation methods. The back end of south Beijing exacerbates the traffic flow between south Beijing and north Beijing.

c. The "closed courtyard" or "large-scale design" planning reduces the traffic capacity and public transportation attraction.

Traditional courtyard culture and large-scale design planning in Beijing is not suitable for the current city development. On the one hand, due to walk out of the courtyard to take public transport, bigger courtyard means that the walk distance and time are larger, which reduces the attraction of public transportation. On the other hand, although the self-structure of the courtyard is relatively complete, it cuts the urban spatial structure result in the lack of branch road and impedes microcirculation, which weakens the traffic organization ability of main urban road.

d. "Demand following" traffic development mode lags behind urban development

Traffic construction is always lagging behind the development of urban land use, which results in passivation following the urban construction.

At present, the traffic construction standards of all types of building and facilities are relatively low.

It cannot meet the travel demand when the building is completed. The main features lack enough parking spaces and do not have building for public transportation facilities. In addition, the traffic impact evaluation mechanism is not perfect, and public transportation construction always lacks behind the urban land development.

3 TREND AND DEMAND

3.1 *Population of Capital Core Area is further reduced*

Because of the differential rent and other economic factors, the population of Capital Core Area will continue to transfer to the Function

Expand Area. The industrial spatial structure of Capital Core Area will change little. Economic activity in this area is still along the traditional "two axis" distribution. The main industry includes administrative office, commercial, tourism, and culture. In the meantime, education and medical service functions will expand outside Capital Core Area by stimulating Function Expand Area, and the attraction of Capital Core Area will decrease.

3.2 *Function Expand Area will become the agglomeration area of industry and population*

The population of Function Expand Area will increase. However, because of the limited land capacity, the growth rate will decrease. Because various high-quality resources are still mainly distributed in the central area and the traditional concept, Function Expand Area will remain the most active area of economy in Beijing, which is also affected by the traditional concept. Various function parks will develop rapidly. The traditional center, such as CBD and Zhongguancun Science and Technology Park will become more and more mature, and the Olympic Center, Lize financial center, Dahongmen business district will get into the fast lane.

3.3 *The urban fringe community is gradually improved*

The population of 10 marginal groups will further expand. However, the area of 10 marginal groups will not continue to expand through the control of land development and transportation infrastructure. Affected by the development of new towns, the industry in marginal groups will limit development. The main industry includes the living facilities of education, health care, entertainment, and business. The population is mainly composed of two parts: one is the relocation residents in Capital Core Area and the other is new Beijing people who work at Function Expand Area.

3.4 *The city develops along the rail track of beads*

Because of the driving effect from rail traffic, the land development density surrounding the rail station will further increase. The population will also gather surrounding the rail station. As the land use in downtown is limited, the leading role of rail traffic in this area is limited, which is less than that outside downtown. Urban development will be along the rail track of beads between downtown and new towns.

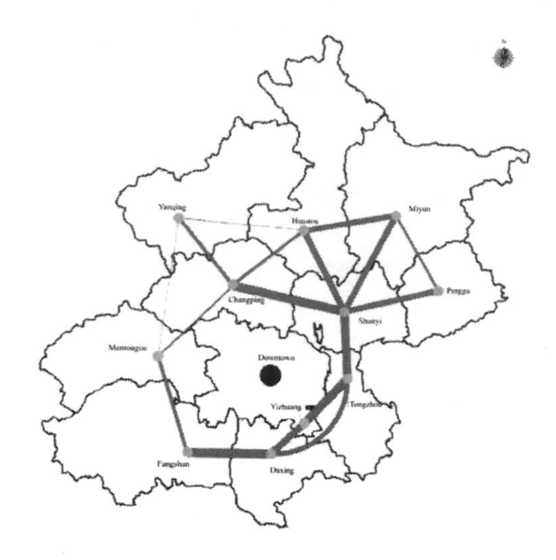

Figure 4. Traffic connection strength between new towns.

3.5 *Population and industry in new town grow rapidly*

As the land resources in new town is abundant and there are some encouragement policies and measures, such as the investment policies about urban infrastructure and other various industrial development, the new town will enter a period of rapid growth. Especially the four new towns, namely Shunyi, Tongzhou, Yizhuang, and Daxing, will form a certain scale first.

3.6 *Travel demand in downtown and new town tends to be stable and that in new town and between new towns enhances*

In the future, new town will reach a certain scale. Travel demand internal new town and between new towns will enhance. Therefore, the traffic infrastructure in those areas will be strengthened. Along with the formation of antimagnetic system in new town, the travel demand between downtown and new town will become stable gradually.

4 SCHEME AND MEASURES

In the future, the overall development ideas of Beijing are designed. The first is for the downtown. The main task is to improve the transportation system and enhance the quality of public transport services. The second is for the new town. The main task is to adhere to the Transit Oriented Development mode to build the new town. According to the urban comprehensive planning of Beijing, the buildings of Daxing, Tongzhou, Yizhuang, and Shunyi will be focused. The third is for the inter-city traffic. The main task is to build Beijing–Shenyang high-speed rail, huge logistics base, and other traffic infrastructure to promote the metropolitan development. The important tasks are as follows.

4.1 *To continuously increase public transport to guide the construction in new towns*

According to the development needs of traffic demand in new towns, three main tasks need to be carried out: (a) building the BRT system timely to connect different function areas in new town, which alleviates the pressure of rail traffic; (b) according to the urban space development in new town, the public transportation should be built in the same time; and (c) promoting the construction of rail transit and guiding urban industry and spatial layout from downtown to new towns. It is the key period of building rail transit in Beijing currently. Therefore, the development ideas of connecting new towns and easing downtown should be adhered. The rail transit corridor construction will continue to be enhanced to guide the urban industry and urban spatial layout.

4.2 *To increase the capacity of the road network and the service level of public transport*

As the Function Expand Area is the largest urban development area and the most active area, two main tasks need to be carried out: (a) to expand the capacity of road network through opening the broken road, dredging transport congestion points, increasing branch network construction, and other measures; (b) to improve the accessibility and comfort of public transportation through additional rail and BRT construction in the Function Expand Area.

4.3 *To build a diversified and high-quality public transportation in Capital Core Area*

The transportation system should be built unswervingly with public transportation-oriented walking and bicycle traffic. During promoting the rail system, a small electric bus service should be built according to the alley system; free bicycle system should be promoted through the integration of the existing bicycle rental network. At the same time, the congestion charging system should be studied and designed in order to ease the road congestion within the fourth ring road.

4.4 *To improve transportation infrastructure in the urban fringe area*

In the future, the urban fringe area will become more and more mature. Therefore, it is necessary to improve the connecting transportation facilities level with downtown and other neighbor functional area. A BRT system connecting urban fringe area and neighbor functional area should be built. In the same time, the BRT system connecting urban fringe area and downtown should be built timely according to the subway operation. The aim is to alleviate the pressure of subway.

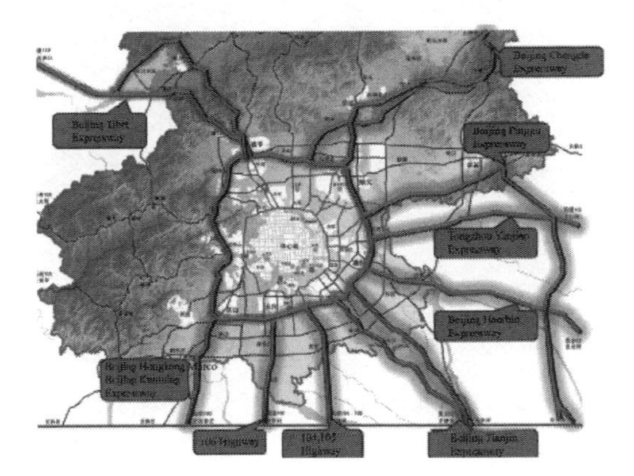

Figure 5. Freight corridor scheme in Beijing.

4.5 *To develop the traffic corridor between the new towns*

With the development of the new town, the traffic demand is bound to increase.

It is necessary to establish complex rail transport and road transport corridors between the new towns. The six ring road toll is proposed to make the road as a urban passenger corridor. Rail S6, which is proposed to connect the new town, is proposed to be extended as a passenger corridor.

4.6 *To improve the traffic corridor of big Beijing metropolitan area*

With the development of the new town, the six ring road will bear more and more passenger traffic. Therefore, it is very necessary to build a new freight corridor to reduce the pressure. The big city freight corridor surrounding Beijing is proposed to be built. The corridor, which will connect Zhuozhou, Guan, Langfang, Xianghe, Sanhe, Pinggu, and Dachang, will intersect Jinggangao, Jingkun, Jingkai, Jingtai, Jingjin, Jingjintang, Jingha, Tongyan, and Jingping expressways. It will improve the freight flow capacity in the Jingjinji city groups and reduce the traffic pressure of Beijing transit transportation.

5 CONCLUSION

The method of guiding the development of urban industry and urban spatial layout by the traffic infrastructure is studied in this paper. The proposals and measures for the superlarge cities are put forward to realize the integrated development of transportation and city. The relative research result can provide experience and reference for growing upto superlarge cities. Then, it is possible to promote the cities in China to experience a more sustainable development path.

ACKNOWLEDGMENT

This research was funded by the National Natural Science Foundation Project (project number: 41471459).

REFERENCES

Adelheid Holl. Twenty Years of Accessibility Improvements: The Case of the Spanish Motorway Building Programme. Journal of Transport Geography, 2007, 15(4):286–297.

Deng X, Huang J, Rozelle S, et al. Growth, Population, Industrialization and Urban Land Expansion of China. Journal of Urban Economics, 2008, 63(1):96–115.

Kim, T.J., Rho, J.H., Suh, S., 1989. Integrated Urban Systems Modeling: Theory and Applications. Kluwer Academic Publishers, Dordrecht, The Netherlands.

Li Shi-ming. Politic, Economy, Culture: The Triple Impetus Variation of Chinese Urban Development. Modern Urban Research, 2006(6):23–29.

Southworth, F., 1995. A Technical Review of Urban Land Use-Transportation Models as Tools for Evaluating Vehicle Reduction Strategies, Research Report ORNL-6881, OakRidge National Laboratory, OakRidge, TN.

Wegner, M., Fuurst, F., 1999. Landuse transport interaction: state of the art. Deliverable of the project TRANSLAND of the 4th RTD Framework Programme of the European Commission.

Yang Shao-hui, Ma Lin, Chen Sha. Urban and Urban Transportation: Development Courses and Interactions. Urban Transport of China, 2009, 7(4):1–6.

Yang Shao-hui, Ma Lin, Chen Sha. Interaction of Spatial Structure Evolution and Urban Transportation. Urban Transport of China, 2009, 7(5):45–48.

Civil, Architecture and Environmental Engineering – Kao & Sung (Eds)
 ISBN 978-1-138-02985-9

Selection of interruptive protection methods in rapid transit underground construction

Hsi-Chi Yang
Department of Construction Management, Chung Hua University, Hsinchu, Taiwan

Chen-Chung Liu
Department of Civil Engineering, Chung Hua University, Hsinchu, Taiwan

ABSTRACT: Since the MRT construction is mainly shield tunneling and deep excavation, it is important to select proper interruptive protection method to prevent the impact of MRT construction on the existing structures and vice versa. The purpose of this research is to select the proper interruptive protection method to be used along the MRT lines to ensure the safety of existing structures and tunneling excavation. This research first determines the criteria and sub-criteria to be used in the initial assessment framework for the protection of the existing structures and tunneling excavation, and then it establishes the final assessing framework by using Delphi method. Furthermore, the Analytic Hierarchical Process is used to determine the relative weights of elements of each level in the hierarchical structure of assessment framework. Finally, the selection process is presented by using a study case in the Nan-Gang line of the Taipei Metropolitan Rapid Transit system.

1 INTRODUCTION

Taiwan is heavily populated in most of metropolitan areas. Since the MRT demand has been increased drastically, there is a sharp increase of rapid transit routes. However, a lot of structures, either nature or man-made, have been existed along the planned routes. It is generally required to protect the safety of the nearby existing structures when the routes are constructed. The MRT construction is mainly shield tunneling and deep excavation. When its adopted preventive strategies cannot reduce the external impact on the existing structures, some supplemental protective methods have to apply to the existing structures. The generally used supplemental protective methods can be classified into two categories, the interruptive protection methods and the existing structure reinforcement methods. Generally speaking, it is highly unlikely to reinforce the existing structures, especially when they are privately owned. Thus, it is important to select proper interruptive protection method to cut off the impact of existing structures on the MRT construction and vice versa. When neglecting the impact from either tunneling excavation or existing structures along the MRT line, disasters can occur. The Taipei MRT tunneling projects had encountered a lot of existing structures including airport facilities, historical monuments, viaducts and rivers. This research tries to investigate how to select the proper interruptive protection method to be used when the protection of existing structures becomes necessary. The selection of a proper interruptive protection is a decision making process influenced by a lot of criteria. The criteria to be used are different based on the nature of a project. Pan (2006) had determined the criteria to be used in deep excavation. Yang and Deng (2012) had determined the criteria to be used in supplementary grouting.

The purposes of this research are: (1) Establish the assessment framework for the selection of interruptive protection methods and determine the criteria and sub-criteria to be used in the framework, and (2) Use the decision mechanism in the framework to select the proper protection method for an underground MRT project.

2 METHODOLOGY

In this study, foreign and domestic literatures were reviewed and experts were interviewed to formulate the criteria and sub-criteria to be used in the initial assessment framework. The final assessment framework was determined by using the

Delphi method. The relative weights of criteria and sub-criteria were identified by using the Analytic Hierarchical Process (AHP). The Delphi and AHP questionnaires were conducted with a panel of 15 senior engineers and professionals in foundation engineering. They had 14 to 30 years of working experience with an average of 20 years.

The Delphi method is designed to explore opinions of a group of knowledgeable persons in order to gain a consensus on a particular topic without bringing the group together (Uhl 1983). The Delphi method pools expert judgment in an iterative process that involves anonymity and opportunity to reflect on and respond to other experts' opinions. Questionnaires are mailed to a group of expert panelists, soliciting their opinion on a topic of interest. Researchers then synthesize the results and distribute them to the panelists in additional waves for reflection and comment.

The AHP developed in the early 1970's by T. Saaty. There are six procedures to run AHP: (1) analyzing problem and displaying evaluative factors, (2) constructing the hierarchy, (3) establishing pair-wise comparison matrices, (4) computing eigenvectors and eigenvalues, (5) testing inconsistency of pair-wise comparison matrices, and (6) computing the relative weight of each factor (Lee 1999). Chen et al. (2012) had used the above two methods to study the application of green architecture to residential buildings. Yang and Lo (2016) had used them to study green energy management.

3 INTERRUPTIVE PROTECTION METHODS

Five commonly used interruptive protection methods in Taipei Metropolitan Rapid Transit construction are considered in this research. They are (1) Pile wall method, (2) Slurry wall method, (3) Pipe roofing method, (4) Freezing method and (5) Grouting method (Sun 2010).

4 ASSESSMENT FRAMEWORK

4.1 *Establishment of assessment framework*

The assessment framework was established first using the Delphi Method. Through literature reviews and expert interviews, an initial assessment framework was formulated with four major criteria, including Safety, Environment, Time and Cost along with a total of 18 sub-criteria for the four major criteria (see Table 1). In order to validate its integrity and completeness, three rounds of expert questionnaires had been conducted when the opinions of the respondents were converged. The four-level (**A**, **B**, **C**, **D**) hierarchical diagram of the final assessment framework is shown in Fig. 1. The 18 sub-criteria in the initial framework have been cut to 13. The alternatives (Level 4) are the above five interruptive protection methods.

4.2 *Identification of relative weights of elements*

After having establishing the elements of each level in the hierarchical diagram of the assessment framework, the AHP method was used to identify the relative weights of elements of each level. Three types of pair-wise comparisons matrices had been established: (1) the evaluation matrix of level 2 criteria with respect to the goal (A); (2) the evaluation matrix of sub-criteria with respect to the related criterion; and (3) the evaluation matrix of alternatives with respect to each sub-criterion. At level 2, the evaluation matrix of criteria to the goal (**A**) is shown in Table 2. At level 3, four evaluation matrices were required to perform. The evaluation matrices of sub-criteria with respected to the four criteria are shown respectively in Table 3, Table 4, Table 5 and Table 6. At level 4, a total of 13 evaluation matrices were required to perform so that the relative weights of the alternatives (shown in the following study cases) with respect to each of the 13 sub-criteria could be obtained.

4.3 *Interruptive protection methods scoring table*

Based on the relative weights for the four major criteria at Level 2 and the 13 sub-criteria at Level 3, an interruptive protection method scoring table can be established (see Table 3). Among the four major criteria, "Safety" is the most important criterion with a weight of 54.2%; followed by Cost, Time and Environment. Among the sub-criteria, geological conditions, underground water, planned structure conditions and protected structures conditions in "Safety" criterion are the top four sub-criteria to be considered first. Then, the contractor can consider the cost effect. The relative weights in the table reflect what really happens in the construction site. If the top four sub-criteria have more or less been ignored, it may require more than the original estimated cost later on. The purpose of establishing a scoring table is that it can be used as a basis for planning personnel to select a proper interruptive protection method.

4.4 *Priority weights for alternatives*

From the relative weights at Level 4 and using the relative weights in the scoring table, the priority weights for the alternatives could be obtained (shown in the following study cases) and the proper interruptive protection method determined.

Table 1. Criteria and sub-criteria of the initial assessment framework.

	Sub-criteria	Descriptions
Safety	Geologic conditions	Soil stratum properties and N values
	Underground water	Water head & soil permeability
	Construction technique	Construction technique used in the interruptive protection method
	Project structure conditions	Location and position of the structure to be build and its construction technique (Ex. Pipe Jacking, Shield Tunneling, NTAM, etc.)
	Protected structures conditions	The type (building, airport, historical monument or itself, etc.), area, height of structures to be protected; Relative distance between the planned and protected structures
	Drift woods	The presence of drift wood in soil
Environment	Interrupted objects	Objects to be interrupted include force, displacement, water and vibration, etc. and their locations
	Environmental impact	Environmental pollution (Ex. Noise, vibration, grout water pollution, etc.) caused by the interrupter construction
	Traffic impact	Traffic impact on users and surrounding facilities when the interrupter is going to build.
	Constructibility	Construction position and available operational space for the interrupter to be built
Time	Machine & manpower	Machine availability and maneuver; Available laborers and Coordination of different laborers
	Construction duration	The time required to complete a project when the interruptive protection method chosen is used.
	Construction progress	Daily work load; Off-day construction
	Weather conditions	Rainfall, typhoon & earthquake
	Construction method characteristics	The construction characteristics of the chosen protection method will affect the time to complete the interrupter.
Cost	Construction cost	Direct cost and indirect cost
	Benefit cost	Cost difference among using different interruptive protection methods
	Loss cost	Tangible and intangible cost due to construction accidents

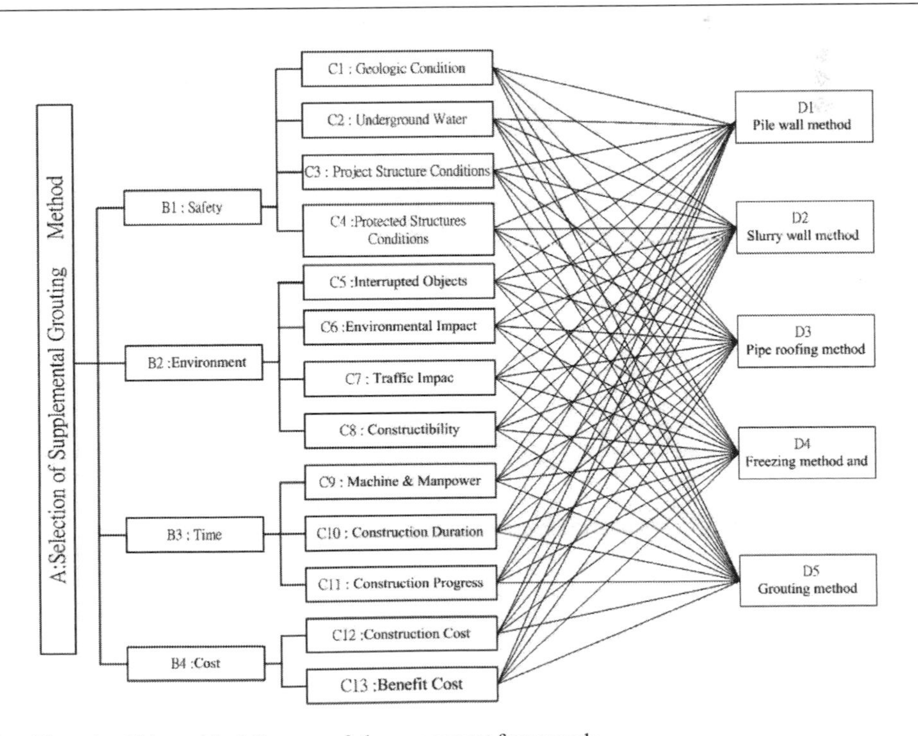

Figure 1. Three-level hierarchical diagram of the assessment framework.

Table 2. Evaluation matrix of level 2 criteria with respect to the goal (A).

A	B1	B2	B3	B4	Weights
B1	1	5.261	3.285	2.606	0.542
B2	0.190	1	1.127	0.969	0.141
B3	0.304	0.887	1	1.138	0.156
B4	0.384	1.032	0.879	1	0.161

λmax = 4.06; C.I. = 0.02(≒ 0, ok); C.R. = 0.02(<0.1, ok).

Table 3. Evaluation matrix of sub-criteria with respect to the safety criterion (B1).

B1	C1	C2	C3	C4	Weights
C1	1.000	1.810	1.232	1.204	0.315
C2	0.552	1.000	1.490	1.423	0.256
C3	0.812	0.671	1.000	1.584	0.237
C4	0.831	0.703	0.631	1.000	0.192

λmax = 4.11; C.I. = 0.04(≒ 0, ok); C.R. = 0.04(<0.1, ok).

Table 4. Evaluation matrix of sub-criteria with respect to the environment criterion (B2).

B2	C5	C6	C7	C8	Weights
C5	1.000	1.844	1.300	1.045	0.312
C6	0.542	1.000	1.171	1.164	0.230
C7	0.769	0.854	1.000	0.998	0.223
C8	0.957	0.859	1.002	1.000	0.236

λmax = 4.05; C.I. = 0.02(≒ 0, ok); C.R. = 0.02(<0.1, ok).

Table 5. Evaluation matrix of sub-criteria with respect to the time criterion (B3).

B3	C9	C10	C11	Weights
C9	1.000	1.966	1.151	0.425
C10	0.509	1.000	0.732	0.233
C11	0.869	1.366	1.000	0.343

λmax = 3.01; C.I. = 0(≒ 0, ok); C.R. = 0(<0.1, ok).

Table 6. Evaluation matrix of sub-criteria with respect to the cost criterion (B4).

B4	C12	C13	Weights
C12	1.000	0.939	0.484
C13	1.065	1.000	0.516

λmax = 2; C.I. = 0(≒ 0, ok); C.R. = 0(<0.1, ok).

5 A CASE STUDY: CB420 BID

CB420 Bid is a shield tunneling construction passing through Sung-Sun Airport. A schematic diagram of CB420 Bid is shown in Fig. 2(a). The CB420 bid, a section in Nei-hu line, includes two tunnels passing through the aircraft stands, taxiways and runways of Sung-Sun airport first, and then No. 1 National Highway and Keelung River. Since this section of shield tunneling construction has to pass through operational Sung-Sun airport, the key points of design and construction are to maintain its regular operation and flight safety. During the design stage, two very stringent control values for the runway and taxiway were set, ±2.0 cm for warning and ±2.5 cm for taking action. These values are far above the requirement

Table 7. Interruptive protection methods scoring table.

A	Criteria	Sub-criteria	Weights	Rank	Score
Selection of Interruptive Protection Method	B1: Safety (54.2%)	Geologic conditions	17.1%	1	
		Underground water	13.9%	2	
		Planned structure conditions	12.9%	3	
		Protected structures conditions	10.4%	4	
	B2: Environment (14.1%)	Interrupted objects	4.39%	9	
		Environmental impact	3.24%	12	
		Traffic impact	3.14%	13	
		Constructibility	3.33%	11	
	B3: Time (15.6%)	Machine & manpower	6.63%	7	
		Construction period	3.63%	10	
		Construction progress	5.34%	8	
	B4: Cost (16.1%)	Construction cost	7.8%	6	
		Benefit cost	8.3%	5	

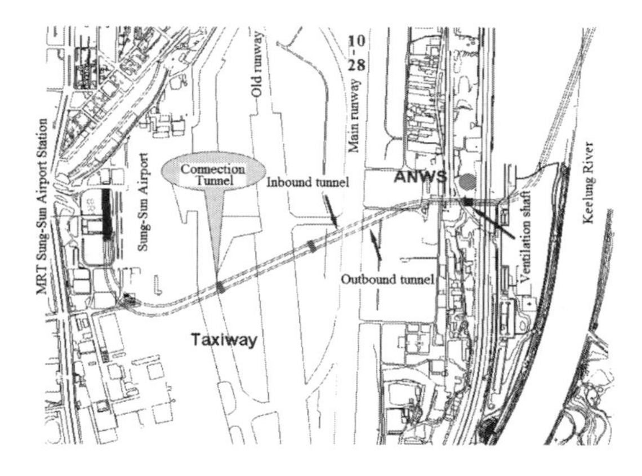

Figure 2. A schematic diagram of CP7.

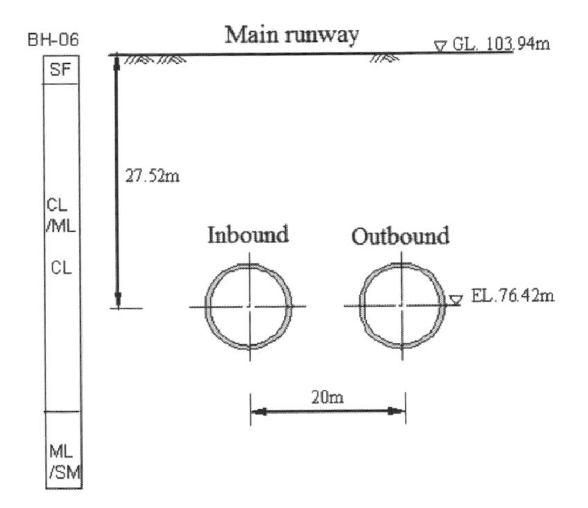

Figure 3. Excavated driftwood.

Table 8. CB420 priority weights for alternatives.

A	Weights	D1	D2	D3	D4	D5
B1	17.1%	0.166	0.207	0.207	0.071	0.349
B2	13.9%	0.117	0.211	0.118	0.190	0.364
B3	12.9%	0.165	0.430	0.168	0.067	0.170
B4	10.4%	0.148	0.324	0.202	0.084	0.242
B5	4.39%	0.199	0.261	0.275	0.094	0.172
B6	3.24%	0.302	0.184	0.174	0.126	0.214
B7	3.14%	0.163	0.116	0.218	0.147	0.356
B8	3.33%	0.226	0.207	0.137	0.112	0.318
B9	6.63%	0.387	0.230	0.111	0.112	0.160
B10	3.63%	0.220	0.243	0.193	0.140	0.204
B11	5.34%	0.289	0.189	0.152	0.128	0.242
B12	7.8%	0.366	0.122	0.119	0.087	0.307
B13	8.3%	0.259	0.197	0.204	0.112	0.228
Priority weights		0.211	0.246	0.177	0.12	0.27
Rank		3	2	4	5	1

for general shield tunnel construction. Also, in order to minimize the amount of settlement, the distance between the inbound and outbound tunnels had been increased to reduce the magnifying settlement effect of two tunnels being too close together. The analytic shield tunneling section is shown in Fig. 2(b). In addition to design, to satisfy the stringent control values, the interruptive protection method has to be used to control the settlement caused by shield tunneling excavation. It was decided to use the grouting method.

After having conducted questionnaire on Level 4 for CB420, the relative weights of alternatives with respect to the 16 sub-criteria were obtained. Table 8 shows how to use the relative weights in Level 3 and Level 4 to determine the priority weights for the alternatives. The grouting method (D1) with a weight of 0.27 was perceived to be the best alternative for interruptive protection.

6 CONCLUSIONS

Since the environment of each underground project is not quite the same, it is not easy to select a proper interruptive protection method. If a wrong method is used, the risk of having a failure project will increase. However, a lot of uncertainty in the selection process has made the selection even more difficult. Therefore, it is not easy to properly select a safe, fast and economical method. This research established the criteria and sub-criteria to be used in the assessment framework for selecting a proper interruptive protection method by Delphi method. Then, AHP was used to help the planning personnel effectively evaluate and select the proper interruptive method. The selection process and scoring table established had been verified by a study case. It has proved that the selection process presented is suitable to help decision makers select the proper interruptive protection method.

ACKNOWLEDGMENT

This work was partially supported by National Science Council, Taiwan, under grant NSC101-2622-E-216-004-CC3, which is greatly acknowledged. The authors also gratefully acknowledge the helpful comments and suggestions of the reviewers, which have improved the presentation.

REFERENCES

Chen, R., H.C. Yang and H.C. Chang (2012). The Application of Green Architecture to Residential Building Development. *Applied Mechanics and Materials* 193–194, 34–39.

Lee, Z.R. (1999), The Device Variables of Tea Web Server Design. *Journal of Agriculture and Forest* 48(2), 85–101 (in Chinese).

Pan, N.F. (2006). Selection of Earth Retaining Methods for Deep Excavation in Kaohsiung Area Using Fuzzy AHP. *Journal of Architecture* (Architectural Institute of Taiwan) 58, 19–40 (in Chinese).

Sun, G.N. (2010), Introduction to the Construction Methods and Progresses of the Civil Construction Projects along the Hsinchung-Luzhou Line. *Journal of Rapid Transit Technolog* 42, 41–68 (in Chinese).

Uhl, N.P. (1983). Using the Delphi Technique in Institutional Planning. *New Directions for Institutional Research* 37, 81–94.

Yang, H.C. and J.W. Deng (2012). The Selection of Suppleental Grouting Methods for Disaster Prevention in Rapid Transit Underground Tunnel Excavation. *Disaster Advances*, 5(4), 391–397.

Yang, H.C. and Y.H. Lo (2016). Establishment of Green Energy Management Evaluation Framework in Sustainable Buildings. *ICIC Express Letters, Part B: Applications* 7(3), 563–570.

Civil, Architecture and Environmental Engineering – Kao & Sung (Eds)
© 2017 Taylor & Francis Group, ISBN 978-1-138-02985-9

Damage mechanism and dynamic analysis of tunnel lining structures under internal blast loading

Changyu Chen, Yibo Jiang & Jun Yang
PowerChina RoadBridge Group Co. Ltd., Beijing, China

Lifu Zheng & Zhipeng Li
University of Science and Technology Beijing, Beijing, China

ABSTRACT: Considering the intense threats of explosion caused by gas leaks or terrorist activities in highway tunnels, a Three-Dimensional (3D) Finite-Element (FE) method was used to study the dynamic response and damage mechanism of tunnel lining structures subjected to internal blast loading. Motivated by the reality that explosion in tunnels damages not only the lining structures but also lives and properties of people, a study based on the gas explosion in Luodai Tunnel is carried out. The coupled fluid–solid interaction was considered in this study, and the explosive–gas–lining system was modeled by using AUTODYN. The result of the numerical study shows that the part of inverted arch close to the hypocenter damages first under the impact load of gas explosion, followed by the sidewall nearby two arch springing exhibiting various degrees of damage. Eventually, the entire tunnel lining was divided into pieces of concrete blocks by the longitudinal and circumferential cracks, and the carrying capacity of the lining structure was degenerated. In addition, the deformation caused by a direct blast impact can also increase steadily because of the continuous fluctuation of the explosion shock wave due to inertia.

1 INTRODUCTION

With the sustainable and rapid development of highway, railway, and urban transportation system, tunnels are widely used, and are becoming an inextricable part of the modern civil infrastructure. In recent decades, tunnel explosion accidents have resulted from gas leaks or terrorist attacks, which are proved to be a great threat to human society. Explosion inside a tunnel may not only damage the tunnel lining structures, but also lead to further loss of lives and properties of people. The blast behaviors of tunnel structures will be distinct from those of other structures caused by their features, especially (1) high longitudinal to cross-sectional dimension ratio, (2) pressures reflected from the tunnel boundary, (3) completely confined by the surrounding ground, and (4) coupled behavior of air blast inside the tunnel and wave propagation through the surrounding ground. On the basis of the above reasons, unique requirements, which differ from the design strategies for other structures, are essential to designing a tunnel to endure a potential blast; therefore, it is imperative to understand its dynamic response under the blast loading. Because of sociopolitical issues, it is normally difficult to carry out the experimental determination of tunnels' dynamic response. Hence, applying the advanced numerical analysis to practical tunnel problems is very important.

Analyses of the behaviors of tunnels subjected to the blast loading have been conducted by researchers worldwide. On the basis of 3D numerical analysis methods, Chille et al. (1998) studied the dynamic response of underground electric plant under internal explosive loading. They took coupled fluid–solid interaction into account, but not considered the nonlinearity and failure of rock and concrete as well as the interaction between different solid media. For traffic tunnels, Choi et al. (2006) carried out 3D FE analyses to investigate the blast loading and deformation in lining structures. They found that blast loading on lining structures was different from the normal one obtained by using CONWEP. The above two studies focused on underground structures in rock mass, as we all know, which are more resistant to internal blast loading than in soils. Dynamic analysis of circular lined tunnels under external blast loading was studied by Lu et al. (2005) and Gui and Chien (2006) using the FE program. Then, a difference method to analyze underground tunnels and cavities under blast loadings was adopted by Feldgun et al. (2008). Liu (2009) analyzed subway tunnels subjected to blast loading using FE method and modeled blast loading using CONWEP, while the high strain rate behavior of soils under blast loading was not considered. A similar study considering this soil characteristic was carried out by Higgins et al. (2012). However, their research only considered elastic stress–strain response of lining structures. And the explosive was simulated using the JWL equation of state. The performances of different shock absorbing foam materials, steel, and concrete tunnel linings subjected to explosion were compared by Chakraborty et al. (2013). They calculated blast loading by using a coupled fluid dynamics simulation. Nevertheless, using the JWL equation of state to carry out rigorous 3D simulations of lining tunnels in rock mass with properly simulated explosive load is rather unused in the literature because of the complex practical problems.

The specific objective of the present study is to use a 3D nonlinear FE analysis to understand the dynamic response and damage of lining structures under blast loading; meanwhile, researchers and practicing engineers will have more interests to study the behavior of tunnels under non-nuclear explosions so that this field can be understood further. With these purposes, numerical simulation was used to study such problems by using the 3D Finite Element program, AUTODYN. The numerical study was based on the Luodai Tunnel in Chengdu, which encountered an explosion caused by gas leaks in February 2016, according to the scale of explosion in Luodai, the equivalent explosive is 800 kg TNT. Considering its importance as well as its representativeness, the coupled fluid–solid interaction was adopted and the pressure was applied as impulse loading at appropriate positions in the tunnel lining structures. In addition, the possible effects of the reflection and superposition of pressure were discussed.

2 PROJECT BACKGROUND

Town Luodai is located in Chengdu, Sichuan Province, China. As the passenger volume increases, extension of the Luodai to Wufeng Line has been planned. The extension involves two tunnel drives, with a total length of 4924 m, route length of 11.056 km, and total investment of 1.5478 billion yuan. Luodai Tunnel is the one pass through the Town Luodai, with a total length of 2915 m, which is judged to be a high gas tunnel.

Explosion caused by a gas leak occurred at the entrance of the Luodai Tunnel on February 24, 2015. Two detonation points were determined on the right track and one on the left, causing different degrees of damage to the structures of the primary lining, secondary lining, and invert arch. Longitudinal cracks and widespread damage appear in the secondary structure, with the most serious fracture mainly concentrating at the section near the explosion and the maximum fracturing area up to 28 m^2. In the primary lining section, concrete damage and localized rock collapses occurred and part of the steel was twisted. Also, longitudinal cracks appeared in the invert arch section at the entrance of the tunnel.

3 EXPLOSION DAMAGE CONDITIONS AT LUODAI TUNNEL

Above all, the specific damage situation of lining structures in Luodai was investigated and summarized below, mainly including four aspects: primary lining structure damage, secondary lining structural damage, inverted arch damage, and the situation of seepage.

3.1 *Primary lining*

Section ZK2+812~ZK2+925 of the left tunnel is a primary lining structure. Because of the impact of the explosion, the shotcrete has been cracked and flaked off; the uneven and rough phenomenon on the surface is much more evident, as shown in Figure 1(a). Steel at the arch section has also been cracked, causing large deformations, part of the concrete fell on the ground, and there are also some depression areas, as can be seen at the top of the lining structures in Figure 1(b). Seven large cavities came into sight in section ZK2+815~ZK2+895, and many concrete blocks and broken steel fell off and piled up on the pavement.

Section K2+818~K2+925 of the right tunnel is also a primary lining structure. Similar damage occurred on the surface of the shotcrete. Steel arch in section K2+820~K2+870 cracked, part of the concrete fell off, thereby causing large deformations. Five large cavities can be seen in the section K2+820~K2+858. In addition, section K2+818~K2+925 was excavated by phases, the sidewall on both sides and arch lining were also badly damaged. Many concrete blocks and broken steel fell off and piled up on the pavement.

(a) Deformed steel (b) Concrete fracture

Figure 1. Primary lining damage in Luodai Tunnel.

3.2 *Secondary lining*

Secondary lining damage mainly occurs due to concrete cracks, water seepage, and partial breakage. Longitudinal cracks mainly distributed in the vault and the lining near the waist, slit width mostly ranges from 0.2 to 3 mm, individuals near the detonation point with a width greater than 3 mm, most of the depth of cracks penetrate the secondary lining thickness, as shown in Figure 2(a). Water seepage phenomenon appeared in the partial fracture position. Partially broken area localized mainly in the vault near detonation point, and the maximum crushing area is up to 28 m^2, as shown in Figure 2(b).

Statistics indicate that there are 10 cavities in the left tunnel, with six in the vault and four on the sidewall. The right tunnel involves two cavities, both seriously damaged and located in the vault, whose specific conditions are shown in Table 1.

3.3 *Inverted arch*

Tunnel inverted damage was mainly in the form of radial cracks and concentrated in the section of the entrance of the tunnel. As shown in Figure 3, the cracking position coincides with the longitudinal concrete construction joints, and the specific distribution is shown in Table 2.

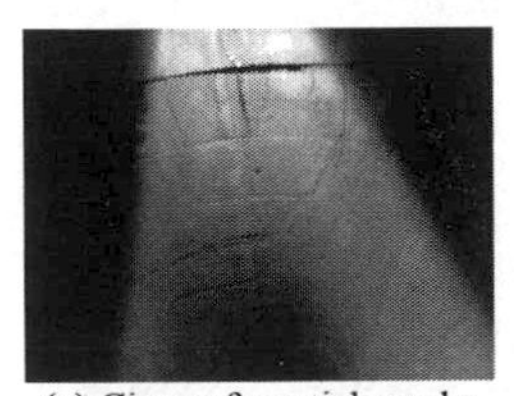

(a) Circumferential cracks

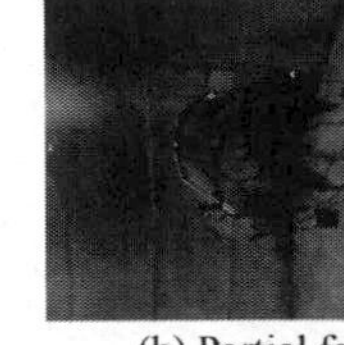

(b) Partial failure

Figure 2. Secondary lining damage in Luodai Tunnel.

Table 1. Cavities in secondary lining structures.

No	Location	Section	Size m
1	Vault	ZK2+062	0.2*0.3
2	Vault	ZK2+235~ZK2+239	4.0*2.5
3	Vault	ZK2+281	0.4*0.8
4	Vault	ZK2+341	1.0*3.5
5	Vault	ZK2+437~ZK2+439	2.0*2.0
6	Vault	ZK2+807~ZK2+809	2.0*6.0
7	Vault	K2+292~K2+300	7.0*4.0
8	Vault	K2+381~K2+384	3.0*3.0
9	Sidewall (left)	ZK2+766~ZK2+768	2.0*2.0
10	Sidewall (left)	ZK2+798~ZK2+800	2.0*2.0
11	Sidewall (left)	ZK2+803~ZK2+806	3.0*2.0
12	Sidewall (right)	ZK2+800~ZK2+803	3.0*2.0

Figure 3. Longitudinal cracks on the inverted arch at the entrance of the tunnel.

Table 2. Cracks in inverted arch.

No	Direction	Section	Width mm
1	Longitudinal	ZK2+035~ZK2+115	3–10
2	Oblique	ZK2+580~ZK2+600	3–5
3	Longitudinal	ZK2+730~ZK2+750	3–5
4	Longitudinal	K2+040~K2+105	3–10
5	Oblique	K2+459~K2+469	3–5
6	Longitudinal	K2+533~K2+535	7

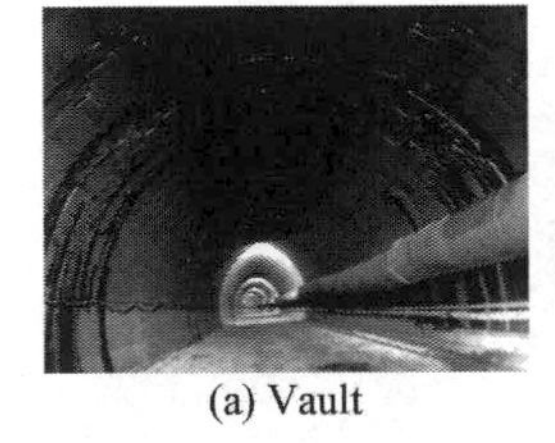

(a) Vault

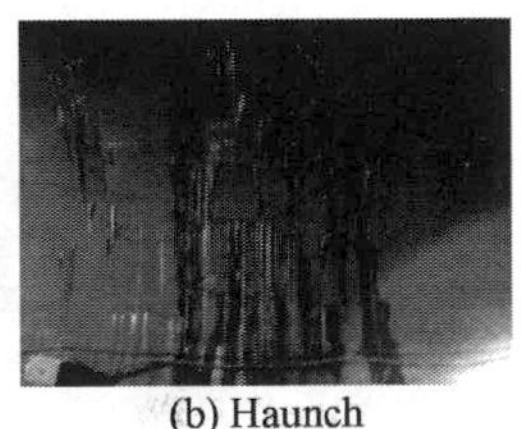

(b) Haunch

Figure 4. Actual situation of water leakage in Luodai Tunnel.

3.4 *Water seepage*

Section ZK2+040~ZK2+170 pavements in the left tunnel has local waterlogging problems. There are even some water flows from both sides of the pattern outside the tunnel. A similar case can be seen in section ZK2+230~ZK2+290. Figure 4(a) and 4(b) show the occurrence of dome seepage and haunch water seepage in section ZK2+240 and ZK2+370, respectively. The left sidewall in section ZK2+760~ZK2+790 and the right pattern in section ZK2+773~ZK2+787 were filled with puddles. Section K2+040~K2+190 in the right tunnel has partial waterlogging problem on the pavement as well as some water flows from both sides of the pattern outside the tunnel. The left side of the haunch of the section K2+147 emerges slight seepage. Sections K2+320 and K2+350 have serious water seepage along the circumferential of the lining, which has obvious waterlogging on the road, with other sections having no significant ponding. On the right side of the pavement, section K2+625~K2+635 has a slight water log.

4 THREE-DIMENSIONAL FINITE-ELEMENT MODELING

Sophisticated analytical simulations must be required to assess the effect of explosions on the underground facilities properly. In addition, several factors, such as explosion, structure, and ground, need to be considered for the application of numerical analyses (Du et al. 2009). Current traditional approaches for determining the effect of underground internal explosion often seem too simple to get reliable results. The coupled Euler–Lagrange nonlinear dynamic analysis is a better analysis method, which has been widely applied in dynamic or earthquake analyses. Euler–Lagrange coupling is especially fit for simulating procedure of the explosion and the interaction with the underground structure (Ngo et al. 2008). It is a powerful feature to deal with fluid–structure and gas–structure interaction problems.

AUTODYN is an entirely integrated analysis program, which is particularly designed for nonlinear dynamics problems. This computer program is used to study underground internal explosion by conducting 3D coupled Euler–Lagrange nonlinear dynamic analysis. AUTODYN-3D is a finite-difference, finite-volume, and finite-element-based calculation program, which has been well developed and validated for a wide range of impact penetration and blast problems (Autodyn, 2003). It is designed to simulate contact problems, such as nonlinear dynamics, large strain and deformation, fluid–structure interaction, explosion, shock and blast wave, impact, and penetration, and it has been widely utilized in gas and oil industry, aerospace, national defense, nuclear power, automotive, chemical, and other dynamic related fields.

The FE models in this paper were established on the basis of double single-line tunnels in Luodai, Chengdu, by using AUTODYN-3D. The inner diameter of this circular tunnel was 5.5 m, with a 50-cm-thick concrete liner. As presented in Figure 5(a), the model is a half model layout, with a symmetrical boundary set along the longitudinal direction and 800 kg of TNT equivalent detonating at 1.5 m above the tunnel pavement. With a strong foundation paved by the Euler–Lagrange coupling method, serial analysis of 3D nonlinear dynamic explosion–lining–ground interaction was carried out to study the response to internal explosion of tunnel lining structure.

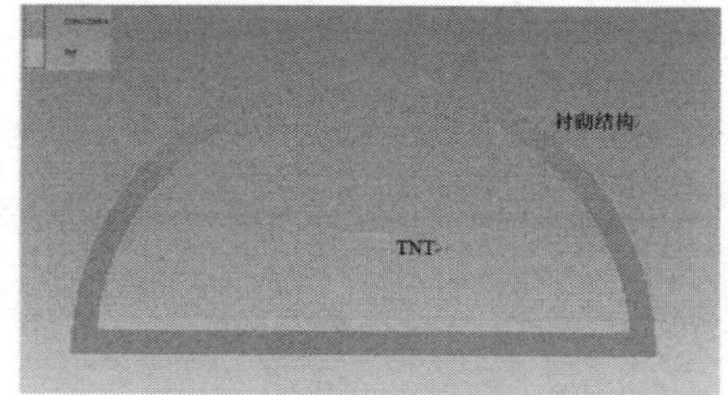

(a) The model layout

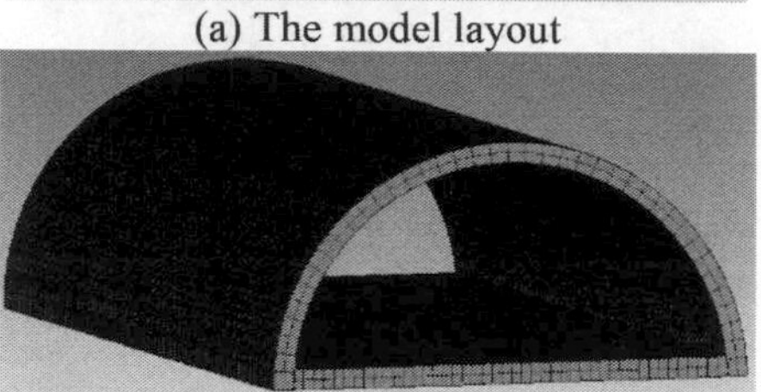

(b) The FE mesh of lining

Figure 5. 3D finite element model.

4.1 *Lagrangian finite-element modeling of concrete lining*

The 3D FE model of the tunnel in rock mass was developed by using AUTODYN with the Lagrangian analysis option. Figure 5(b) shows the FE mesh of the tunnel lining structures, while both mesh convergence and boundary convergence studies were not considered at this stage. A tunnel with a length of 30 m was modeled in rock, and the thickness of concrete lining is 50 cm.

Involving the dynamic response of concrete materials, a strain-rate-dependent concrete model, namely RHT concrete model, is introduced to the analysis in this paper, which is a modular strength model for brittle materials that contain several features in common with many similar constitutive models (Li et al. 2006). It can represent the material characteristics of concrete better and reasonably describe the entire process, from elastic to plastic, until crushed by introducing three different failure surfaces, as shown in Figure 6. It is proved to be a suitable model to simulate the dynamic behavior of concrete material under blast loading.

4.2 *Material properties for concrete*

The grade of concrete in the tunnel lining model has been set as M35, and the properties of concrete are listed in Table 3, utilizing the concrete damage plasticity model, whose yield function was given by Lubliner et al. (1989). The peak tensile strain of concrete under quasi-static load is 0.0002. Taking the softening stage into consideration, the strain when the concrete completely fails is reasonably valued 0.0006. Under blast loading, concrete strain

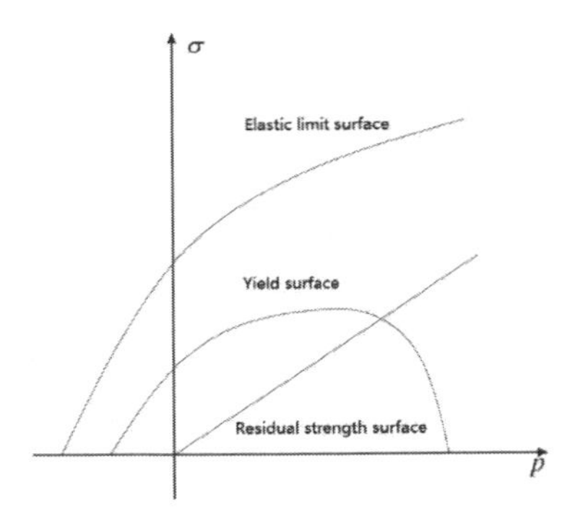

Figure 6. Curve of elastic limit surface, yield surface, and residual strength surface.

Table 3. Concrete material properties.

	Material
Properties	Concrete (M30)
Density (kg/m^3)	2750
Yield strength (MPa)	35
Tensile strength (MPa)	3.5
Shear strength (MPa)	6.3
Shear modulus (GPa)	16.7
Poisson's ratio	0.22

rate ranges from 10 to 10^3 s^{-1}, correspondingly, and the dynamic increase factors can reach the point 5.0 or more (Carreira et al. 1986). While the failure strain enhanced slower than the amplification of strength, the concrete failure tensile strain should take 0.001 when RC structural damage analysis was carried out under blast loading.

4.3 *Eulerian finite-element modeling of explosive*

The explosive material has been modeled in AUTODYN using the Eulerian modeling technique. A typical FE mesh of explosive is shown in Figure 5(b). Eulerian continuum three-dimensional eight-node reduced integration elements have been used to simulate Eulerian explosive material and the surrounding air domain inside the tunnel (TM5-1300). Through the general contact defined between explosive, air, and tunnel lining surface, the Eulerian and Lagrangian elements can interact with each other. Free outflow boundary condition has been defined at the boundary of air domain. Thus, blast pressure propagates freely out of the air domain without any reflection when it reaches boundaries of the air domain. The rationality of Eulerian elements mesh determines the efficiency of capturing the propagation of blast wave through air and the surrounding concrete lining.

Table 4. JWL material properties for TNT explosive.

	Material
Properties	TNT
Density (kg/m^3)	1630
Detonation wave speed (m/s)	6930
C_1 (GPa)	373.7
C_2 (GPa)	3.747
r_1	4.15
r_2	0.9
ω	0.35

Jones Wilkens Lee (JWL) equation of state is used to simulate the pressure–volume relation of explosives (Zukas et al. 2003). In this model, the pressure (Ps)–volume (v) relationship can be represented as the sum of functions given by:

$$p = C_1\left(1-\frac{\omega}{r_1 v}\right)e^{-r_1 v} + C_2\left(1-\frac{\omega}{r_2 v}\right)e^{-r_2 v} + \frac{\omega\theta}{v} \quad (1)$$

C_1, r_1, C_2, r_2, and ω are material constants in which C_1 and C_2 have dimensions of pressure and the remaining constants are dimensionless.

During the numerical calculation, air is assumed to be an ideal gas, with its properties generally described by using the linear polynomial equation of state (Welch et al. 1997), according to the Gama equation, which can be expressed as:

$$p_\alpha = (\gamma - 1)\frac{\rho}{\rho_\alpha}E_\alpha \quad (2)$$

where p_α, γ, ρ, ρ_α, and E_α are gas pressure, ratio of specific heats, the current air density, initial air density, and the internal energy of gas per volume, respectively.

4.4 *Material properties for TNT*

In the JWL equation of state (equation (1)), the former two exponential terms are high-pressure terms, whereas the latter is a low-pressure term, which deals with the high volume of cloud due to explosion (Veyera et al. 1995). Table 4 lists the material properties used for TNT explosive.

5 RESULTS AND DISCUSSION OF NUMERICAL SIMULATION

5.1 *Dynamic response*

The explosive–gas–lining structure system exhibiting a similar dynamic response under the internal blast loading though the magnitudes distinctly varied from different tunnel, ground, or explosive parameters (Goel et al. 2011). Both the velocity and displacement of structures under explosion are regarded as important factors to measure the extent of damage (Goel et al. 2012). According to the specific damage situation in the Luodai Tunnel, the responses of the system when subjected to a blast loading of 800 kg TNT explosion are studied in this section. Defining upward and toward the left is positive, downward and toward the right is negative.

The lining structures velocity, which was accompanied with extreme lining, vibration increased dramatically under the blast loading. Figure 7 shows the change of velocity at the measuring points in the lining with time. The velocity in the inverted arch increased immediately after explosion and reached the peak value (34.8 m/s) at about 2 ms, followed by a fluctuation with overall decreasing magnitude. The velocity in vault and haunch exhibited a similar trend, originally, although their peak value is less than that in the inverted arch and was followed by a fluctuation with increasing magnitude.

Figure 8 shows the displacement at measuring points in the lining structures with internal explosion increased continuously, and they were found to be different at different locations. The large displacements concentrated at the section close to inverted arch right under the explosion point, with measuring points in vault and haunch both move upward. To sum up, under the blast loading of 800 kg TNT, the tunnel lining structures exhibited an expansion deformation in the vertical direction. The measuring points located on the left haunch moves to the left, similarly, the right moves to the right, which means tunnel lining also exhibited an expansion deformation in the horizontal direction. Lining structures undergo deformation subjected to the direct blast impact, and the deformation will continue to increase, due to the continuous fluctuation of the explosion shock wave after attenuation due to inertia. By simulating and observing the displacements of the measuring points in the lining, the result shows that failure occurred only at limited location of the lining in this case under blast loading, which is consistent with the reality of the damage situation in Luodai Tunnel.

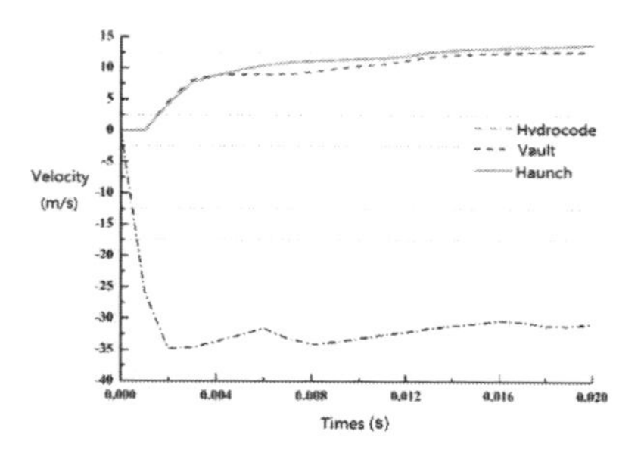

Figure 7. Velocity at measuring points subjected to 800 kg TNT.

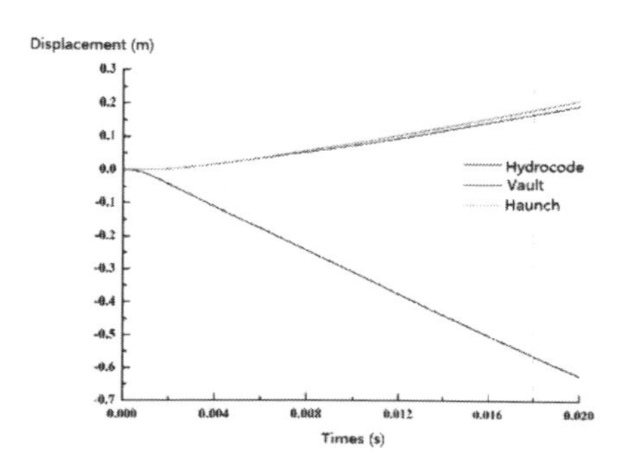

Figure 8. Displacement at measuring points subjected to 800 kg TNT.

5.2 *Damage mechanism*

The damage distribution in the tunnel lining at 2, 4, 6, 8, 10, and 13 ms are shown in Figure 9. Several pictures are taken in the Luodai Tunnel, as shown in Figure 10, which reflect the actual damage at the accident site. The analysis for damage mechanism and the comparative analysis between numerical simulation and actual situation are discussed in this subsection.

Under the gas explosion impact load, the damage first appears in the inverted arch and vault, which are in the vertical direction of the explosion point, then expands outward, eventually causing a burst

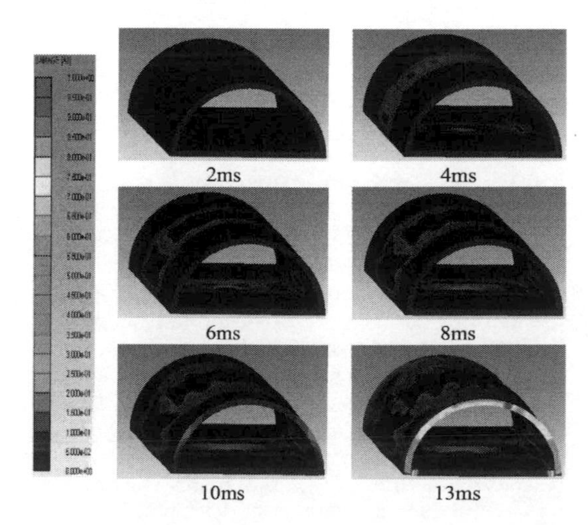

Figure 9. Process of damage in concrete tunnel lining subjected to 800 kg TNT.

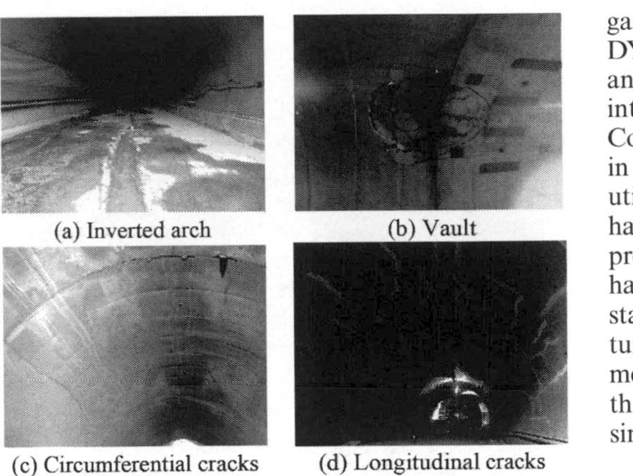

Figure 10. Actual damage of concrete lining structure.

pit of radius 4 m right under the explosion point, leading to the inverted arch pavement damage. As Figure 10(a) shows, many cracks travel through inverted arch have arisen. The part of arch springing damage earlier due to the tensile stress is caused by vertical deformation of inverted arch. Then, the sidewall nearby two arch springing, especially at the two sides close to the hypocenter, exhibits different degrees of damage subjected to the blast loading. Meanwhile, under the blast loading, the lining structure damaged range keeps evolving, as shown in Figure 10(b), and the tunnel vault lining is locally broken and a cavity formed. As the shock wave spreads to the entire inner wall of the tunnel lining, the initial compression waves first act on the wall and then turn to tension wave after reflection leading to tension along the surface of the inner ring. There are many longitudinal cracks exhibited in the hypocenter, arch springing, haunch 45° and 60°, and vault parts. Meanwhile, a certain distance from the blast center has cracks along the inner ring, and the specific actual damage situation is shown in Figure 10(c),(d). These cracks divided the entire tunnel lining structures into pieces of concrete blocks, and with the blast undergoing multiple reflections inside the tunnel, the damage and cracks keep aggravating (Bischoff et al. 1991). Although the shock wave gradually decays, the internal structure, because of the inertia, will continue to fluctuate and further deepen the damage on tunnel lining, eventually leading to the loss of the carrying capacity of the lining structure, which is also consistent with the actual damage situation in Luodai Tunnel.

6 CONCLUSIONS

According to the gas explosion and the specific damage situation in Luodai Tunnel, the explosive–gas–lining system was modeled by using AUTODYN. In this paper, 3D nonlinear finite element analyses of highway tunnels in rock mass under internal blast loading have been simulated. The Coupled Eulerian–Lagrangian (CEL) analysis tool in finite element software AUTODYN has been utilized for research purpose. The explosive TNT has been modeled using the Eulerian elements. The pressure–volume relationship of TNT explosive has been performed by using the JWL equation of state. Considering fluid–structure and gas–structure interactions, dynamic response and damage mechanism of the lining structures are discussed in this paper. On the basis of the extensive numerical simulations, the following conclusions are drawn:

1. Subjected to the internal explosion, the velocity of lining structures increased heavily, accompanying with extreme lining vibration. The velocity in the inverted arch increased dramatically and reached the peak velocity after explosion immediately, which was followed by a fluctuation with overall decreasing magnitude. The velocity in vault and haunch exhibited a similar trend, originally, however, their peak values were less than the 1 in the inverted arch, and were followed by a fluctuation with increasing magnitude.
2. Subjected to the blast loading, the displacement in the lining structures with internal explosion increased continuously. To sum up, the tunnel lining structures exhibited an expansion deformation in both the vertical and horizontal directions. It is found that lining structures undergo deformation subjected to the direct blast impact, and the deformation will continue to increase, resulting from the continuous fluctuation of the explosion shock wave after attenuation due to inertia.
3. Under the gas explosion impact load, part of inverted arch and arch springing were damaged earlier. Then, the sidewall nearby two arch springing, especially at the two sides close to the hypocenter, exhibits different degrees of damage subjected to the blast loading. The tunnel vault lining is locally broken and a cavity is formed. The longitudinal and circumferential cracks divided the entire tunnel lining into pieces of concrete blocks, and the damages and cracks keep aggravating, with the blast undergoing multiple reflections inside the tunnel. Although the shock wave gradually decays, the internal structure, because of the inertia, will continue to fluctuate, to further deepen the damage on tunnel lining, eventually leading to the loss of the carrying capacity of lining structure. On the basis of this conclusion, it can be seen that more attention should be paid to those critical sections while evaluating blast-resistance of tunnel structures.

4. By simulating and analyzing, it is found that the results of numerical simulation are in good agreement with the actual damage situation in Luodai Tunnel, which means that the simulation can reasonably and effectively reflect the dynamic response process and reveal the damage mechanism of the lining structure of highway tunnel subjected to the blast loading. This paper offers a reference for the antiknock performance evaluation and postdisaster rehabilitation, ultimately accomplishing the purpose of guiding and helping the repair work in Luodai Tunnel. Similar principles could also be applied for designing new highway structures and other structures, which are recommended to take blast resistance into account.

REFERENCES

Bischoff, P.H. & Perry, S.H. Compressive behaviour of concrete at high strain rates. *Materials and Structures, Materiaux et Constructions* 24:425–450 (1991).

Carreira, D.J. & Chu, K. Stress strain relationship for reinforced concrete in tension. *ACI Journal, January-Febuary* 83(3):21–28 (1986).

Chakraborty, T. & Larcher, M. & Gebbeken, N. Comparative performance of tunnel lining materials under blast loading. *3rd International Conference on Computational Methods in Tunnelling and Subsurface Engineering*, Ruhr University Bochum (2013).

Chille, F. & Sala, A. & Casadei, F. Containment of blast phenomena in underground electrical power plants. *Advances in Engineering Software* 29:7–12 (1998).

Choi, S. & Wang, J. & Munfakh, G. & Dwyre, E. 3D nonlinear blast model analysis for underground structures. Geo Congress 2006 ASCE:1–6 (2006).

Du, H. & Li, Z. Numerical analysis of dynamic behaviour of rc slabs under blast loading. Trans. Tianjin University 15:061–064 (2009).

Feldgun, V.R. & Kochetkov, A.V. & Karinski, Y.S. & Yankelevsky, D.Z. Internal blast loading in a buried lined tunnel. *International Journal of Impact Engineering* 35:172–183 (2008).

Goel, M.D. & Matsagar, V.A. & Gupta, A.K. & Marburg, S. An abridged review of blast wave parameters. *Defence Science Journal* 62:300–306 (2012).

Goel, M.D. & Matsagar, V.A. & Gupta, A.K. Dynamic response of stiffened plates under air blast. *International Journal of Protective Structures* 2:139–155 (2011).

Gui, M.W. & Chien, M.C. Blast resistant analysis for a tunnel passing beneath Taipei Shongsan airport—a parametric study. Geotechnical and Geological Engineering 24:227–248 (2006).

Higgins, W. & Chakraborty, T. & Basu, D. A high strain-rate constitutive model for sand and its application in finite element analysis of tunnels subjected to blast. *International Journal for Numerical and Analytical Methods in Geo mechanics* 37:2590–2610 (2012).

Horsham. *Autodyn Theory manual* [J]. UK: Century Dynamics Ltd, 2003.

Li, Z.X. & Liu, Y. & Tian, L. Dynamic Response and Blast-resistance Analysis of Double Track Subway Tunnel Subjected to Blast Loading within One Side of Tunnel [J]. *Journal of Beijing University of Technology* 32(2):173–181 (2006).

Liu, H. Dynamic analysis of subway structures under blast loading. *Geotechnical and Geological Engineering* 27:699–711 (2009).

Lu, Y. Underground blast induced ground shock and its modelling using artificial neural network. *Computers and Geotechnics* 32:164–178 (2005).

Lubliner, J. & Oliver, J. & Oller, S. & Onate, E.A. Plastic damage model for concrete. *International Journal of Solids and Structures* 25:299–329 (1989).

Ngo, T. & Mendis, P. Modelling reinforced concrete structures subjected to impulsive loading using concrete lattice model. *Electronic Journal of Structural Engineering* 8:80–89 (2008).

TM5-1300 Structures to resist the effects of accidental explosions. US Departments of the Army and Navy and the Air Force (1990).

Veyera, G.E. & Ross, C.A. High strain rate testing of unsaturated sands using a split-hopkinson pressure bar. *3rd International Conference on Recent Advances in Geotechnical Earthquake Engineering and Soil Dynamics* 1.11 (1995).

Welch, C.R. In-tunnel air blast engineering model for internal and external detonations[C]//*Proceedings of the 8th International Symposium on Interaction of the Effects of Munitions with Structures*. 195–208 (1997).

Zukas, J.A. & Walters, W.P. Explosive effects and applications. *Springer-Verlag New York*, (2003).

Civil, Architecture and Environmental Engineering – Kao & Sung (Eds)
© 2017 Taylor & Francis Group, ISBN 978-1-138-02985-9

The seismic responses of near-fault frame structure cluster on basin induced by the rupture of reverse fault

W. Zhong
Chongqing College of Electronic Engineering, Chongqing, P.R. China

ABSTRACT: The objective of this article is to study the effects of basin on seismic responses of near-fault frame structure cluster during an earthquake. An integrated simulation method, which can simultaneously implement wave propagation in structures and earth medium, is used to obtain seismic responses of superstructures due to the rupture of reverse fault. The seismic responses of near-fault frame structures for two cases without basin and with basin are simulated during a hypothetical M_w 6.0 earthquake. The existence of basin can increase greatly the shear forces of structural members and change earthquake risk positions of frame structures. It is observed that there exists the trend of beating for the case with basin. The geological structure of basin should be considered in simulating seismic responses of near-fault structure cluster.

1 INTRODUCTION

It is known that the basin has great effects on seismic ground motions, which is also named as basin effects such as ground-motion amplification, ground-motion duration extension and so on. Many researchers studied the basin effects and suggested that the subsurface geology of basin needs to be known in detail to simulate ground motions (Pitarka & Irikura 1996, Morales et al. 1996, Graves et al. 1998, Graves & Wald 2004, Graves 2008, Day et al. 2008). The basin regions that they concerned are San Fernando basin, the Los Angeles basin, Granada basin and San Bernardino basin, which include many building clusters in actual cities. Their research works have neglected the existence of structures on the surface of the basins.

In fact, the earthquake response of the structures in urban region is one of the main reasons for resulting in the severe seismic hazard. So the reasonable computational model should not only include earth media and seismic sources but also include structure clusters in city.

Since 1996, some researchers have combined building clusters in urban area at the surface of basin with half-space earth media and incident plane waves, which include plane P wave, plane SV wave and plane SH wave, to study far-field site-city interaction or earthquake responses of buildings. There are three different types of building models they used: (1) the buildings simplified as blocks (Wirgin & Bard 1996, Tsogka & Wirgin 2003a, b, Groby et al. 2005, Groby & Wirgin 2008, Semblat et al. 2002, Semblat et al. 2004, Kham et al. 2006, Semblat et al. 2008), (2) the single oscillator models of buildings (Guéguen 2002, Boutin & Roussillon 2004, Boutin & Roussillon 2006, Uenishi 2010, Uenishi 2013), (3) FEM models of frame structures (Lombaert & Clouteau 2006, Lombaert & Clouteau 2009).

Apart from the far-field cities as mentioned above, there are also many other near-fault cities during an earthquake. For the near-fault cities, the rupture process of causative fault produces the characteristics of near-fault ground motions such as hanging-wall effect, directivity effect, near-fault velocity pulses, and static offsets, which can often result in complicated seismic responses and severe earthquake disaster of near-fault superstructures in city. Therefore, the rupture process of causative fault should be considered in studying site-city interaction or earthquake responses of building clusters in near-fault cities.

In recent years, a few researchers used finite-fault seismic sources and building clusters in city situated on the basin to study the site-city interaction during a near-fault earthquake. Guidotti et al. (2012) studied the site-city interaction during the 22 February 2011 M_w 6.2 Christchurch earthquake. The concerned region is Christchurch Central Business District, which is situated on the surface of the alluvial basin and is composed of around 150 buildings. The research results showed that the presence of the city changes considerably ground motions inside the city of Christchurch. Isbiliroglu et al. (2015) studied coupled soil-structure interaction effects of building clusters during earthquakes. They simulated the coupled responses of multiple simplified building models located within the San Fernando Valley during the 1994 Northridge earthquake.

In the research works mentioned above, a building or a set of buildings on urban area is simplified as a low-velocity block and then is connected to substratum. All blocks and half-space medium underneath the free surface constitute heterogeneous continuum. This type of block model of building is first proposed by Wirgin & Bard (1996). However, this kind of simplified building model can not be used to obtain the actual earthquake responses of structures, e.g. the internal forces (bending moments, shear forces and axial forces of structural members) and the deformations of structure as well as the structural members.

In our previous work, Liu & Zhong (2014) introduced an integrated system consisting of frame structure clusters, half-space viscoelastic earth medium and causative fault. Based on this integrated system, an integrated numerical simulation method was developed for achieving flexural wave propagation in frame structure clusters and wave propagation in viscoelastic earth medium as well as bidirectional wave propagation between structure cluster and earth medium due to a near-fault earthquake.

In this article, I shall use the integrated numerical simulation method in the literature (Liu & Zhong 2014) and construct the computational model including one building cluster, earth media (with basin and without basin) and finite-fault seismic source to simulate seismic responses of frames in structure clusters during a hypothetical M_w 6.0 earthquake induced by the rupture of reverse fault. I shall study the shear forces of structures in cluster, earthquake risk positions of frame structures and snapshots of displacement of all structures.

2 THE NUMERICAL METHOD FOR SEISMIC WAVE PROPAGATION IN THE INTEGRATED SYSTEM

Figure 1(a) describes the integrated system including plane frame structure cluster, half-space viscoelastic earth medium with basin and causative fault. In order to achieve integrated simulation for seismic wave propagation in near-fault frame structure clusters and viscoelastic earth medium due to rupture of the causative fault, we have constructed three types of the investigated lumps (Liu et al. 2012, Liu & Zhong 2014, Zhong & Liu 2016): (1) the investigated lump in plane frame structure, (2) the investigated lump in viscoelastic earth medium and (3) the investigated lump for structure-soil connection. The schematic illustrations about these three types of investigated lumps

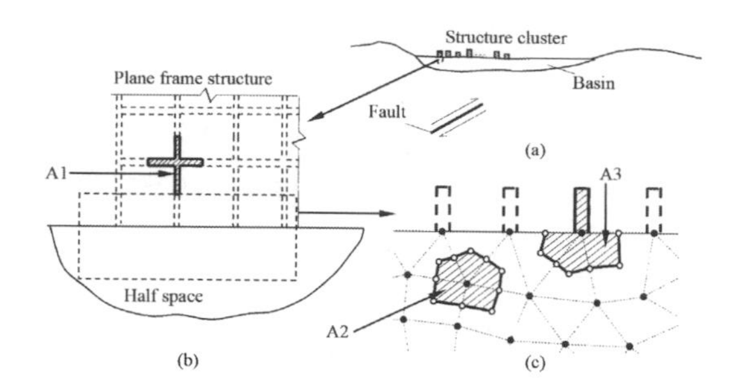

Figure 1. Schematic illustration of the investigated lumps in the integrated system. (a) Simplified diagram of the integrated system including structure cluster, half-space earth medium with basin and fault. (b) The investigated lump (shaded part A1) in the plane frame structure in structure cluster shown in figure 1(a). (c) The investigated lump (shaded part A2) in earth medium and the investigated lump (shaded part A3) for structure-soil connection.

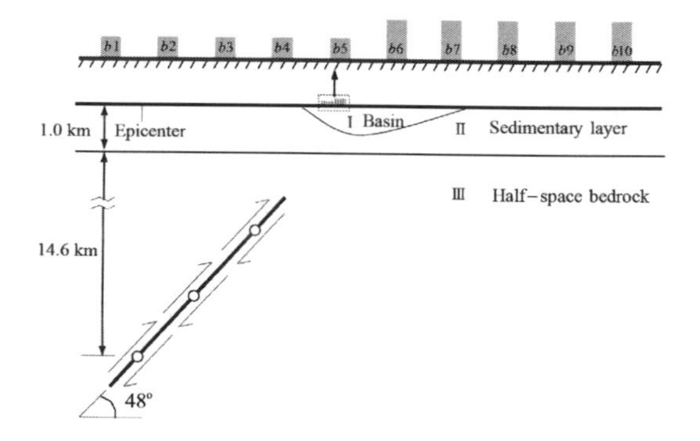

Figure 2. Computational model of the integrated system. The three circles denote three subfaults. The "model 1" of earth media considers the basin. The "model 2" neglects basin and the material property of basin region in "model 1" is the same as sedimentary layer. The epicentral distance of center of the structure cluster is 4.18 km. The *b1–b5* are identical 3-bay 3-story plane frames. The *b6–b10* are identical 3-bay 6-story frames.

in integrated simulation for wave propagation are respectively shaded parts A1, A2 and A3 shown in Figures. 2(b), (c).

In our previous work, Liu & Zhong (2014) have established governing equations of the above three types of the investigated lumps and given an integrated numerical simulation method, which can implement simultaneously flexural wave propagation in plane frame structures and viscoelastic wave propagation in earth medium as well as bidirectional wave propagation between the plane frame structure and earth medium during an near-fault earthquake. The key of implementation of the integrated method is the connection between structures and earth medium.

It is simple for program implementation of this integrated numerical simulation method. On the one hand, the structures model and half-space earth media are built independently. On the other, the method will be implemented by only telling the computer which investigated lumps in ground floor columns of frame structures and at the surface of half-space earth medium belong to the same investigated lump for structure-soil connection.

In this article, I shall use the integrated simulation method to obtain seismic responses of structure cluster due to the rupture of causative fault and study the influences of basin on earthquake responses of plane frame structures in structure cluster.

3 NUMERICAL SIMULATION

In this section, in order to study the influence of basin on the earthquake responses of structure cluster during a M_w 6.0 hypothetical near-fault earthquake, two different models of earth media will be constructed. I will obtain the seismic responses of near-fault structure cluster on hanging wall during a M_w 6.0 hypothetical earthquake by using the integrated simulation algorithm. The shear forces of structure cluster and earthquake risk positions of all frame structures for the cases without basin and with basin as well as snapshots of simulated displacement of all structures in cluster with basin are given during a reverse fault earthquake.

3.1 *Computational model*

For the seismic source, finite-fault earthquake model used here is the same as the literature (Liu & Zhong 2014). For a hypothetical earthquake of M_w 6.0, the optimal size of each subfault is about 2.5 km, and then the number of subfaults along the dip-slip direction is 3. The final slip displacement for each subfault and the average rise time are respectively 0.35 m and 2.0 s. The focal depth and dip angle are set about 15.6 km and 48°, respectively.

For the earth media, two different geological configuration models are constructed to obtain earthquake responses of structure cluster on the surface of earth media. The models are referred to hereafter as "model 1" and "model 2". In "model 1", as is shown in Figure 2, the geological structure includes basin, sedimentary layer and half-space bedrock. The length and the maximum depth under free surface of basin is 3.1 km and 450 m, respectively. The epicentral distance of the center of basin is about 5.18 km. Depth of the uniform interface between sedimentary layer and the half-space bedrock is 1.0 km. In "model 2", comparing to "model 1", the interface between basin and sedimentary layer is neglected and "model 2" is only composed of sedimentary layer with flat free surface and the half-space bedrock. The interface of them is also 1.0 km in depth. For the "model 1", the material properties are shown in Table 1. For the "model 2", the material properties of only II (sedimentary layer) and III (half-space bedrock) in Table 1 are used.

For the frame structure cluster, as is shown in Figure 2, one structure cluster is considered, the center of the cluster is 4.18 km far from the epicenter. The cluster consists of 5 RC plane frame structures of 3-bay 3-storey (see *b*1–*b*5 in Figure 2) and 5 RC plane frame structures of 3-bay 6-storey (see *b*6–*b*10 in Figure 2) distributed with equal interval of 27.6 m. The spans of all plane frame structures are 6.9 m. C25 concrete is used in all structural members. Elastic modulus and density of concrete are respectively 2.8×10^{10} N/m^2 and 2500 kg/m^3.

3.2 *Simulation results*

Figure 3 shows time history curves of the shear forces of structure clusters without basin and with basin during a hypothetical M_w 6.0 near-fault earthquake. The outputs are the middle sections of the first column (from left to right) in the first story of the structures *b*1–*b*10. Compared with the case without basin, the increase rate of peak values of the shear forces is about 11.8% (*b*6)~146.3% (*b*1) for the case with basin. It is shown that the presence of basin can increase greatly the shear forces of plane frame structures in structure cluster at the surface of the basin.

Figure 4 shows the positions where peak column shear force and peak beam-end bending moment of each structure in structure cluster is maximum value due to rupture of causative fault for the cases without basin and with basin. These positions are named as earthquake risk positions of frame structures.

Table 1. Material properties of "model 1".

No.	thk. (km)	Velocity (km/s)		Density (kg/m^3)	Quality factor	
I	0.45	Vs	1.04	2000	Q_s	20
		V_p	1.80		Q_p	30
II	1.0	V_s	1.73	2500	Q_s	173
		V_p	3.00		Q_p	260
III	∞	V_s	3.46	2700	Q_s	346
		V_p	6.00		Q_p	519

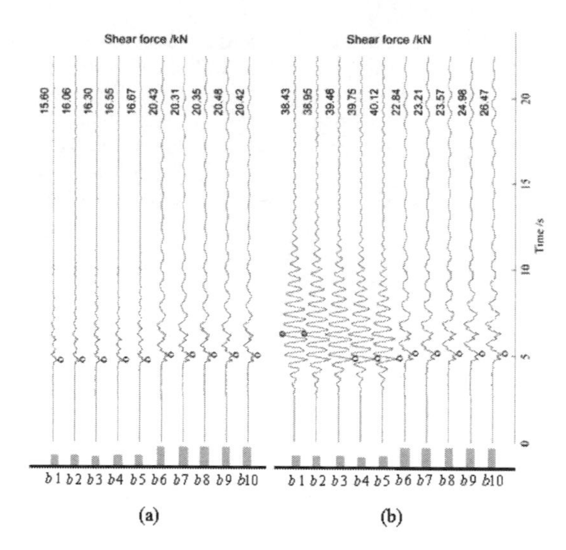

Figure 3. The shear forces of structure cluster (a) without basin and (b) with basin during a hypothetical M_w 6.0 earthquake. The outputs are the middle sections of the first column (from left to right) in the first story of the structures *b*1-*b*10. The symbol circle denotes the position of peak shear force of the output for each structure.

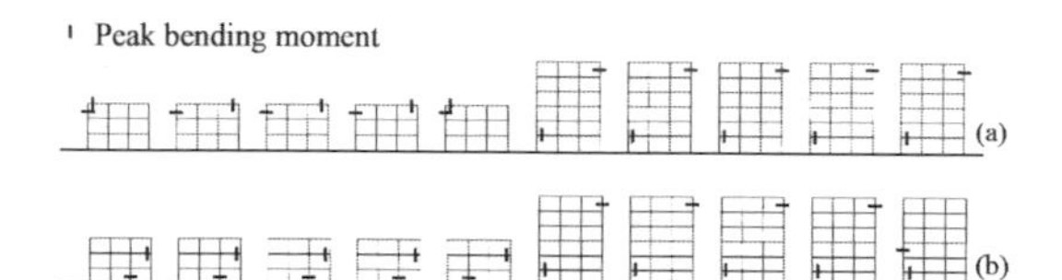

Figure 4. Earthquake risk positions of the structures in structure cluster during a M_w 6.0 hypothetical earthquake for the cases without basin (a) and with basin (b).

For the case without basin, earthquake risk positions of beam ends and columns in 3-story structures are all located on the third story, earthquake risk positions of beam ends and columns in 6-story structures are respectively located on the first story and the sixth story.

From the Figure 4(b), for the case of 3-story structures with basin, earthquake risk positions of beam ends and columns are respectively located on the second story and the first story. For the case of 6-story structures with basin, earthquake risk positions of beam ends are located on the first story, earthquake risk positions of columns are located on the sixth story (*b*6–*b*9) and the third story (*b*10).

The simulating results show that the existence of basin changes the earthquake risk positions of frame structures.

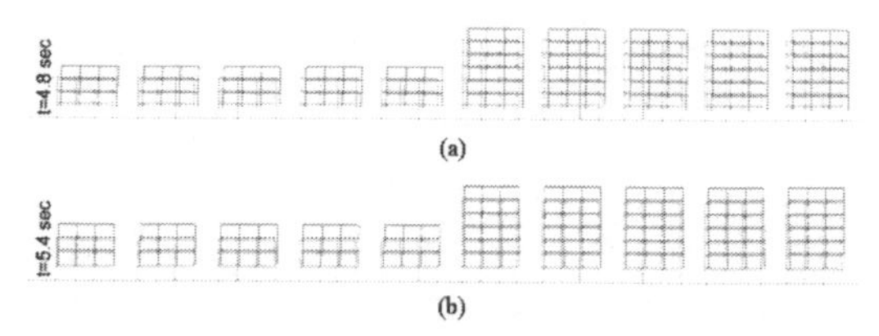

Figure 5. Snapshots of displacements of all structures in structure cluster at time 4.8 s (a) and 5.4 s (b) after initial rupture of the causative fault for the case with basin. The deformations of the structures are magnified here by a factor 50.

Figure 5 shows snapshots of simulated displacement of all structures in cluster at time 4.8 s and 5.4 s after initial rupture of the causative fault for the cases with basin. The simulated displacements includes displacements of ground motion and displacements resulting from structure deformation. Following the figure, the vertical displacement of each structure are apparently larger than the horizontal displacement. It is observed that the structures *b*5 and *b*6 at time 5.4 s have the trend of beating with basin.

4 CONCLUSIONS

An integrated numerical simulation method is used to study the influence of basin on earthquake responses of near-fault structure cluster induced by the rupture of reverse fault. By constructing two different models of earth media without basin and with basin, the seismic responses of near-fault structure cluster on the hanging wall are obtained during a hypothetical M_w 6.0 earthquake due to the rupture of reverse fault. And then the shear forces of structure cluster and earthquake risk positions of all structures are given for the cases without basin and with basin. The snapshots of displacement of structure cluster at different times are illustrated for the case with basin. Some numerical simulation results and conclusions are obtained as follows:

1. The presence of basin can increase greatly the shear forces of structure cluster at the surface of the basin.
2. The existence of basin have a great influence on the earthquake risk positions of frame structures.
3. The trend of beating for the case with basin is observed from snapshots of simulated displacement of structure cluster.

In view of the results and conclusions mentioned above, the geological structure of basin need to be known in detail and should not be neglected when simulating seismic responses of near-fault structure cluster induced by the rupture of causative fault.

ACKNOWLEDGMENTS

Project is Supported by Scientific and Technological Research Program of Chongqing Municipal Education Commission (Grant No. KJ1503001).

REFERENCES

Boutin, C. & Roussillon, P. 2004. Assessment of the urbanization effect on seismic response. *Bulletin of the Seismological Society of America* 94(1): 251–268.

Boutin, C. & Roussillon, P. 2006. Wave propagation in presence of oscillators on the free surface. *International Journal of Engineering Science* 44: 180–204.

Day, S.M. Graves, R.W. Bielak, J. Dreger, D. Larsen, S. Olsen, K.B. Pitark, A. Ramirez-Guzman, L. 2008. Model for basin effects on long-period response spectra in Southern California. *Earthquake Spectra* 24(1): 257–277.

Graves, R.W. Pitarka, A. & Somerville P.G. 1998. Ground-motion amplification in the Santa Monica area: effects of shallow basin-edge structure. *Bulletin of the Seismological Society of America* 88(5): 1224–1242.

Graves, R.W. 2008. The seismic response of the San Bernardino basin region during the 2001 Big Bear Lake earthquake. *Bulletin of the Seismological Society of America* 98: 241–252.

Graves, R.W. & Wald, D.J. 2004. Observed and simulated ground motions in the San Bernardino basin region for the Hector Mine, California, earthquake. *Bulletin of the Seismological Society of America* 94: 131–146.

Groby, J.P. Tsogka, C. & Wirgin, A. 2005. Simulation of seismic response in a city-like environment. *Soil Dynamics and Earthquake Engineering* 25(7–10): 487–504.

Groby, J.P. & Wirgin, A. 2008. Seismic motion in urban sites consisting of blocks in welded contact with a soft layer overlying a hard half-space. *Geophysical Journal International* 172(2): 725–758.

Guéguen, P. Bard, P.Y. Chávez-García, F. 2002. Site-city seismic interaction in Mexico city–like environments: an analytical study. *Bulletin of the Seismological Society of America* 92(2): 794–811.

Guidotti, R. Mazzieri, I. Stupazzini, M. Dagna P. 2012. 3D numerical simulation of the site-city Interaction during the 22 February 2011 M_W 6.2 Christchurch earthquake. *Electronic proceedings of the 15th World Conference of Earthquake Engineering, 15th WCEE*, 24–28. Portugal: Lisbon.

Isbiliroglu, Y. Taborda, R. & Bielak, J. 2015. Coupled soil-structure interaction effects of building clusters during earthquakes. *Earthquake Spectra* 31(1): 463–500.

Kham, M. Semblat, JF. Bard, PY. & Dangla, P. 2006. Seismic site-city interaction: main governing phenomena through simplified numerical models. *Bulletin of the Seismological Society of America* 96(5): 1934–1951.

Liu, T. Luan, Y. Zhong, W. 2012. Earthquake responses of clusters of building structures caused by a near-field thrust fault. *Soil Dynamics and Earthquake Engineering* 42: 56–70.

Liu, T. & Zhong, W. 2014. Earthquake responses of near-fault frame structure clusters due to thrust fault by using flexural wave method and viscoelastic model of earth medium. *Soil Dynamics and Earthquake Engineering* 61–62: 57–62.

Lombaert, G. & Clouteau, D. 2006. Resonant multiple wave scattering in the seismic response of a city. *Waves in Random and Complex Media* 16: 205–230.

Lombaert, G. & Clouteau, D. 2009. Elastodynamic wave scattering by finite-sized resonant scatters at the surface of a horizontally layered halfspace *Journal of the Acoustic Society of America* 125: 2041–2052.

Morales, J., Singh, S.K. & Ordaz, M. 1996. Analysis of the Granada (Spain) earthquake of 24 June, 1984 (M = 5) with emphasis on seismic hazard in the Granada basin. *Tectonophysics* 257: 253–263.

Pitarka, A. & Irikura K. 1996. Basin structure effects on long-period strong motions in the San Fernando valley and the Los Angeles basin from the 1994 Northridge earthquake and an aftershock. *Bulletin of the Seismological Society of America* 86(1B): S126–S137.

Semblat, J.F. Kham, M. Guéguen, P. Bard, P.Y. & Duval, A.M. 2002. Site-city interaction through modifications of site effects. *On 7th U.S. Conference on Earthquake Engineering*. Boston.

Semblat, J.F. Kham, M. Bard, P.Y. & Guéguen, P. 2004. Could "site-city interaction" modify site effects in urban areas? *In Proceedings of the 13th World Conference on Earthquake Engineering, Canadian Association for Earthquake Engineering, Ed., Vancouver, British Columbia, Canada, August 2004*. International Association for Earthquake Engineering, Paper 1978.

Semblat, J.F. Kham, M. Bard, P.Y. 2008. Seismic-wave propagation in alluvial basins and influence of site-city interaction. *Bulletin of the Seismological Society of America* 98(6): 2665–2678.

Tsogka, C. & Wirgin, A. 2003a. Simulation of seismic response in an idealized city. *Soil Dynamics and Earthquake Engineering* 23: 391–402.

Tsogka, C. & Wirgin, A. 2003b. Seismic response of a set of blocks partially imbedded in soft soil. *Comptes Rendus Mecanique* 331(3): 217–224.

Uenishi, K. 2010. The Town Effect: Dynamic Interaction between a Group of Structures and Waves in the Ground. *Rock Mechanics and Rock Engineering* 43(6): 811–819.

Uenishi, K. 2013. "Unexpected" failure patterns and dynamic collective behaviour of an assembly of buildings subjected to horizontal impact. *Engineering Failure Analysis* 35: 125–132.

Wirgin, A. & Bard P.Y. 1996. Effects of buildings on the duration and amplitude of ground motion in Mexico city. *Bulletin of the Seismological Society of America* 86(3): 914–920.

Zhong, W. & Liu, T. 2016. Application of an investigated lump method to the simulation of ground motion for Beichuan town during the Wenchuan earthquake. *Journal of Earthquake and Tsunami* 10(1): 1650002-1-1650002-14.

Civil, Architecture and Environmental Engineering – Kao & Sung (Eds)
© 2017 Taylor & Francis Group, ISBN 978-1-138-02985-9

Wave propagation FEA and experimental validation of aluminum plate using piezoelectric elements

S. Yan, S. Zhang, Z.Q. Wang, P.T. Zhao & Y. Dai
School of Civil Engineering, Shenyang Jianzhu University, Shenyang Liaoning, China

ABSTRACT: In this paper, we aim to develop a new Finite-Element Method (FEM) to analyze a coupled aluminum plate structure pasted with a pair of Piezoelectric ceramic (PZT) patches as transducers for Structure Health Monitoring (SHM) and validate the efficiency of the proposed method. The finite-element software ABAQUS is applied directly to the coupled SHM system by using piezoelectric elements. Piezoelectric ceramic patches are used as either generating or receiving signal device. The piezoelectric elements are directly used for the PZT material. It is considerably focused on the influence law of PZT polarization direction and detection wave spectral characteristics for detection effectiveness. The whole process of generating, propagating, and receiving of the detection signal is numerically simulated and the results are analyzed. In order to verify the efficiency of the proposed method, a model test of an aluminum plate structure under the same condition is carried out, and the results are compared to that of the finite-element analysis. The results show that the proposed finite-element model and the finite-element detection method match well with the experimental results.

1 INTRODUCTION

As a kind of common structure, aluminum plate is widely used in engineering, especially in the manufacture of aircraft, vessels, and many other structures in various fields. It has a very important significance to monitor the working state of aluminum plate structures over time. At present, ultrasonic Lamb wave Nondestructive Testing (NDT) technology is widely applied in the health monitoring of aluminum plate structures (Wang 2007, Liu 1999). Recently, more and more people use finite-element software for the simulation of Lamb wave detection of aluminum plate (Xu 2015); however, in the present simulation analysis, most of the analysis methods are directly exciting and extracting the Lamb wave displacement signal on the structure surface, which still have a certain difference with the actual experiment using the sensor to detect the aluminum plate, especially the influences of adhering layers on the transformation between electric voltage and structural movement, resulting in a great difference between numerical simulation using equivalent displacement input and output and experimental results. Accordingly, in this paper, on the basis of the research of the existing active Lamb wave detection (Yan & Zhang & Meng 2010, Cao 2008, Yan et al 2013), using finite-element software ABAQUS and considering the influence of PZT polarization direction and the detection wave spectrum characteristics of the detection result, a finite-element method is proposed for the detection of aluminum plate by directly using the piezoelectric element, and the results are verified by experiments. This method can utilize the piezoelectric effect of the piezoelectric element to directly excite and receive the voltage signal, which is closer to the actual experiment than the previous simulation method.

2 NUMERICAL SIMULATION MODEL

2.1 *Excitation signal selection*

Lamb wave propagation is a dispersive and multimode method. The modes of Lamb waves excited by different frequency bands are also different. Through the frequency dispersion curve of Lamb wave propagation in aluminum plate, it can be found that when the frequency and the plate thickness of the product are less than 1 MHz and 1 mm, there are only two modes of guided waves of S_0 and A_0, which is convenient for the analysis of the signal. Therefore, the excitation signal is selected as the five peak wave narrowband signal modulated by Hanning window function, whose center frequencies are 150, 200, and 300 kHz, respectively, and the amplitudes are all 1V, as shown in Figures 1–3, respectively.

From the figures, we can find that when the frequencies are increasing in time domain, the amplitudes of the spectra are decreasing and the frequency band is expanding in frequency domain.

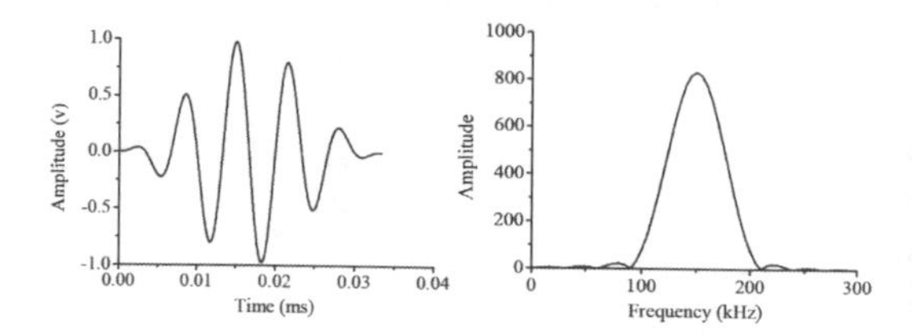

Figure 1. 150 kHz excitation signal.

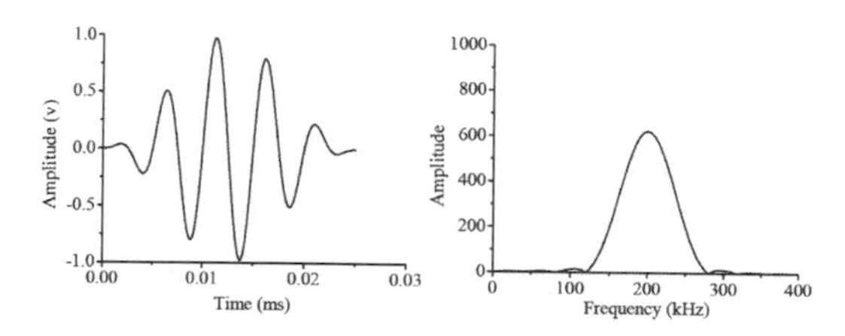

Figure 2. 200 kHz excitation signal.

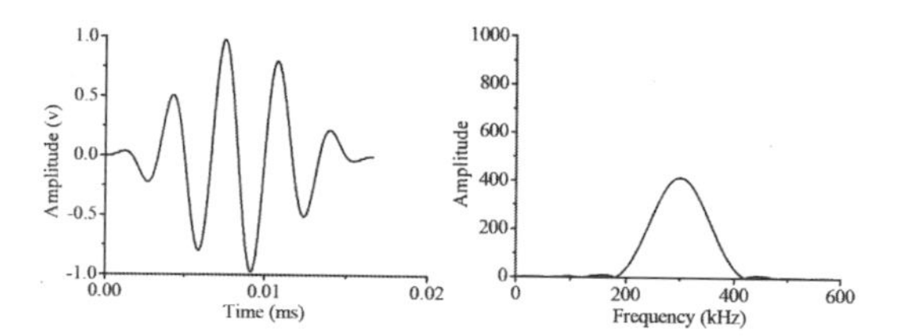

Figure 3. 300 kHz excitation signal.

2.2 *Model establishment*

The aluminum plate and PZT patch models provided by ABAQUS and used in this paper are the shell elements. They use the solid element and the piezoelectric element, respectively. The size of the aluminum plate model is 80 mm × 80 mm × 1 mm, and the size of the piezoelectric patch element is 12 mm × 6 mm × 1 mm. The distance between driver (PZT1) and receiver (PZT2) is 480 mm, and the distance between the exciter and the left boundary is 160 mm along the horizontal direction, and the distance between the exciter and the right boundary is also 160 mm, as shown in Figure 4. The parameters of the materials are shown in Tables 1–2.

2.3 *Boundary conditions*

The polarization directions of PZT1 and PZT2 are set to be perpendicular to the thickness direction of the plate from the upper surface to the lower surface. The periodic potential of 100V is applied on the upper surface of the PZT1 along thickness direction, and the lower surface is set as the zero-potential-energy surface. At this point, the electric potential direction is parallel to the polarization direction, and PZT1 will generate a periodic vibration along the thickness direction of the plate. Meanwhile, the lower surface is set as the zero-potential-energy surface along the thickness direction of the receiving piezoelectric element PZT2, which can extract the voltage change caused by the deformation in the thickness direction. Similarly, when the polarization direction and the electric potential direction of the piezoelectric element are set along the length direction, the displacement and voltage signals along the length direction of the piezoelectric element will be excited and received.

Figure 4. Structure assembly.

Table 1. Parameters of the materials.

Settings	Density kg/m^3	Young's modulus Pa	Poisson's ratio
Aluminum	2800	7.5E+10	0.33
Piezoelectric element	7600		

Table 2. Parameters of the PZT materials.

Dielectric constant F/m	Elastic constant N/m^2	Piezoelectric constant C/m^2
D11 8.11E-09	D1111 1.21E+11	e2 23 12.3
D22 8.11E-09	D1122 7.54E+10	e3 11 –5.4
D33 7.35E-09	D2222 1.21E+11	e1 13 12.3
	D1133 7.52E+10	e3 22 –5.4
	D2233 7.52E+10	e3 33 15.8
	D3333 1.11E+11	
	D1212 2.26E+10	
	D1313 2.11E+10	
	D2323 2.11E+10	

* e1 11 e1 22 e1 33 e1 12 e2 12 e2 13 e1 23 e2 11 e2 22 e2 33 e3 12 e3 13 e3 23 are zero.

2.4 *Time incremental step selection and size of element*

The element size and time increment of FEM have a great influence on the convergence and accuracy of numerical solution. In general, the smaller the incremental step, the finer the grid division and higher the accuracy of the model. In order to ensure the accuracy of the calculation and to consider the influence of the computational efficiency, the cell grid size is set to 2 mm, and in order to improve the efficiency of the operation, the difference grid is adopted and the grid of plate boundary section is rough. And on the basis of ensuring that the incremental step size is less than 1/20 of the corresponding period of the maximum frequency wave, the time increment step is set to 0.1 μs. Figure 5 shows the mesh division of the model.

2.5 *Simulation result analysis*

During simulation results analysis, the potential along the length direction of the sensor, caused by the S_0 mode, and the electric potential along the thickness direction of the sensor, caused by the A_0 mode of the reference point, are extracted. According to the frequency dispersion curve and the corresponding frequency, the S_0 and A_0 modes are identified by checking and comparing with the group velocity, and the time between the peak values in the signal wave packet is taken as the arrival time of the signal center frequency group. The detection signals in time domain are shown in Figures 6–11.

The Figures show that both the S_0 and A_0 mode Lamb waves have energy attenuation phenomenon in the propagation process, and the energy attenuation of mode A_0 is more serious than that of mode S_0. When reaching the boundary, both the mode signals have reflections and phase transition. With the increase of the excitation frequency, the energy attenuation effects of the two mode signals are more serious in the propagation process. In the above-mentioned figures, the first wave packet

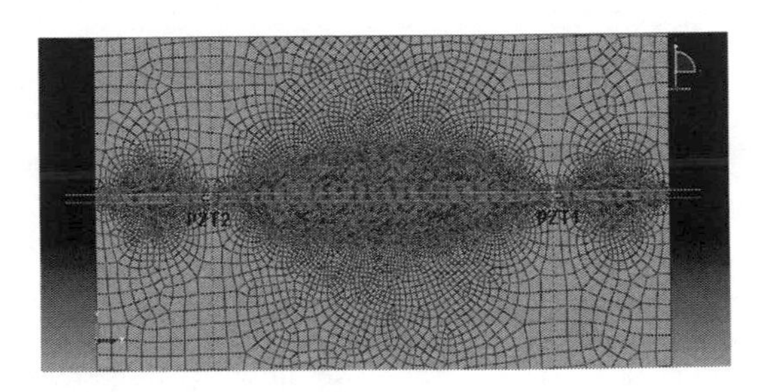

Figure 5. Mesh division of model.

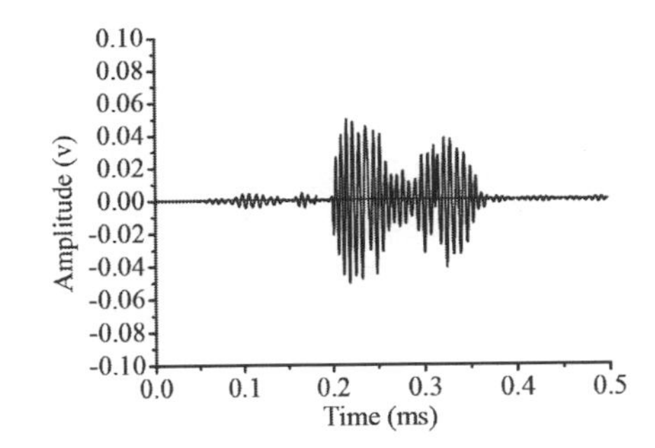

Figure 6. 150 kHz A_0 mode received signal.

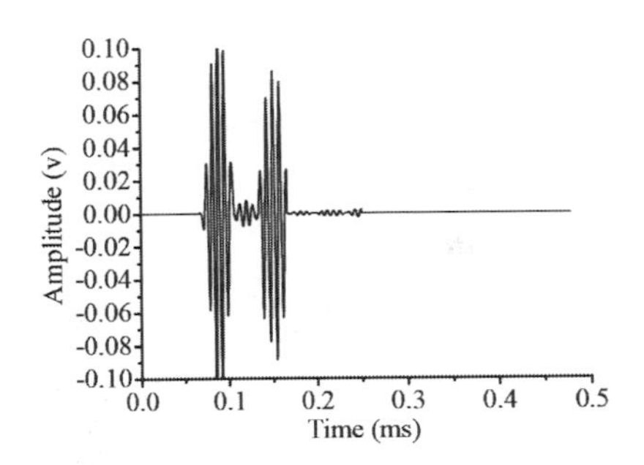

Figure 7. 150 kHz S_0 mode received signal.

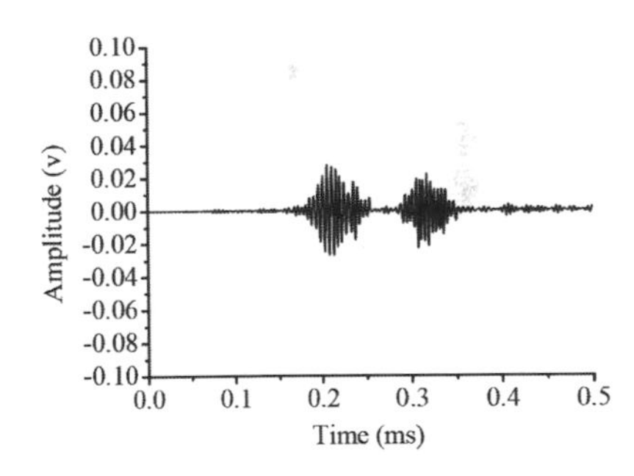

Figure 8. 200 kHz A_0 mode received signal.

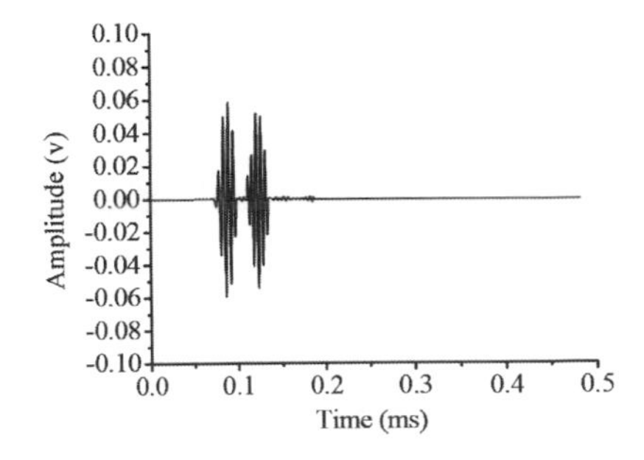

Figure 9. 200 kHz S_0 mode received signal.

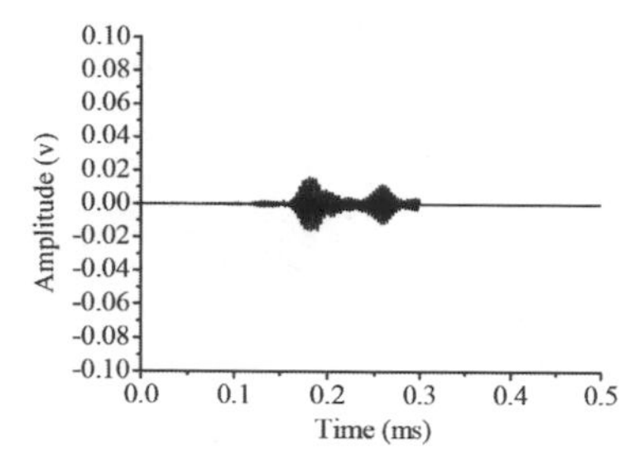

Figure 10. 300 kHz A_0 mode received signal.

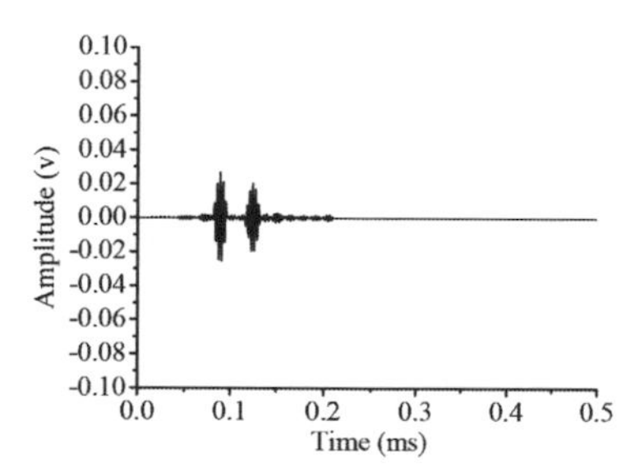

Figure 11. 300 kHz S_0 mode received signal.

from the left is the initial received signal, and the second wave packet is the boundary reflection signal.

3 EXPERIMENTAL VERIFICATION

3.1 *Experimental setup*

To validate the efficiency of the numerical simulation, a model test of an aluminum plate pasted with a pair of PZT transducers are performed. The size of the aluminum plates used for test specimens is 800 mm × 800 mm × 1 mm (length × width × thickness). The piezoelectric ceramic patches with a square single-side electrode as sensor and actuator are pasted on aluminum plate surface. The system equipment is mainly composed of a DG1022 arbitrary waveform transmitter (signal generator), an aluminum plate, and a DS1102E digital oscilloscope. The excitation signal used in the experiment is a five-peak wave signal, and the central frequencies of the excitation signal are 200 and 300 kHz.

3.2 *Experimental analysis*

The central frequencies of 200 and 300 kHz modulation five-peak wave signals are selected as the excitation voltage, and the voltage amplitude is 100V. The experimental data are shown in Figures 12 and 13, respectively.

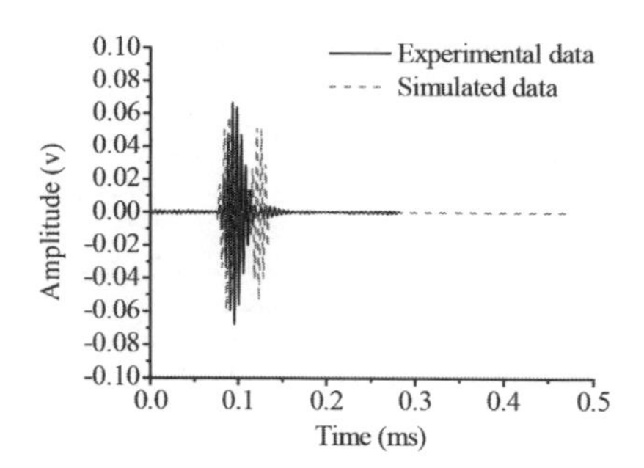

Figure 12. 200 kHz received signal.

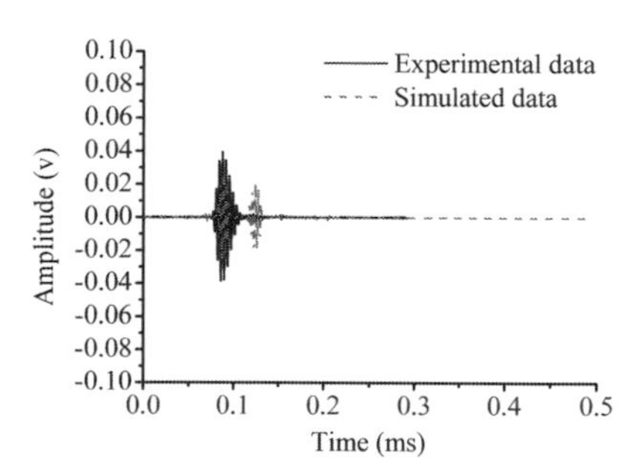

Figure 13. 300 kHz received signal.

Because the piezoelectric plate used in the experiment is in the direction of vibration along the length, the electric potential signal is mainly based on the S_0 mode. In the experiment, the boundary of the aluminum plate is adhered to the rubber mud, which is used to absorb the reflection wave of the boundary, and the interference of the boundary reflection signal is too small to be measured. In the above figures, the first wave packet is the initial received signal, and the second small wave packet is the boundary reflection signal. By comparison, the waveform and the amplitude of the two central frequency detection signals in both FEM simulation and experiment match well, indicating the validation of the efficiency for the proposed method.

4 CONCLUSION

Using ultrasonic Lamb wave to detect the structure of an aluminum sheet has very important significance, according to the basic principle of the piezoelectric effect and the frequency dispersion curve of Lamb wave in aluminum plate, and a new method for the detection of aluminum sheet based on the piezoelectric element is proposed by using finite-element software. The results were compared with the experiment results. This new method takes voltage as the excitation and the extraction signal, and it is more applicable to the actual situation than the commonly used simulation method, which takes displacement as excitation and extraction signal. In this paper, the element properties, time increment, and mesh division of the model used in finite-element software are introduced, which may provide some references for the future research on the piezoelectric simulation of ABAQUS.

ACKNOWLEDGMENTS

This study was partially supported by the National Natural Science Foundation of China with grant 51278313, Liaoning Natural Science Foundation of China with grant 2015020595, and Science Foundation of Liaoning Education Department of China with grant LJZ2016029.

REFERENCES

Cao, Zheng Min. 2008. Plotting and Experimental Validation of Lamb wave Dispersion Curve. *PTCA* 44(9): 482–484.

Liu, Zhen Qing. 1999. Ultrasonic Lamb Waves in Nondestructive Testing. *NDT* 21(9): 409–413.

Wang, Du. & Zheng, Xiang Ming. 2007. Experimental Study for Nondestructive Testing of Steel Sheet by Lamb Waves. *NDT* 29(4): 193–196.

Xu, Ye Dong. 2015. Research on Lamb Wave Excited by Flanging Piezoelectric Wafer. *Journal of Applied Acoustics* 34(6): 547–553.

Yan, Shi. & Zhang, Hai Feng & Meng, Yan Yu. 2010. Numerical Calculation and Experimental Validation for Lamb Wave Dispersion Curves. *J. HUST* 27(1): 1–4.

Yan, Shi. Zhao, Naizhi. Wanf Qiujing and Song, Gangbing. 2013. Dynamic Mechanical Model of Surface-bonded PZT Actuator: Theory and Experiment[J]. *Applied Mechanics and Materials*, 303–306: 1732–1735.

Civil, Architecture and Environmental Engineering – Kao & Sung (Eds)
© 2017 Taylor & Francis Group, ISBN 978-1-138-02985-9

Numerical simulation on seismic behavior of subway station structure under near-fault ground motions

Baizan Tang, Su Chen & Xiaojun Li
Institute of Geophysics, China Earthquake Administration, Beijing, China

Haiyang Zhuang
Civil Engineering and Earthquake Disaster Prevention Center of Jiangsu Province, Nanjing, China

ABSTRACT: Based on the finite method of ABAQUS software, the numerical simulation on seismic response of the soil-subway station structure interaction were performed to reveal the obvious nonlinear properties of soil and the effect of soil-structure interaction under near-fault ground motions. In the numerical simulation, an advanced numerical model was crafted to consider the nonlinear static and dynamic coupling interactions between the soil and the underground structure. Dynamic behavior of the underground structure, including the lateral displacement and acceleration response of the structure, acceleration response of soil-subway station structure, were analyzed, respectively. The research results could provide references for the seismic design and construction of underground structure.

1 INTRODUCTION

With rapid development and urbanization in China, exploitation and utilization of underground space has become a major concern. By the end of 2015, subways have been built or approved for construction in 37 Chinese cities with a total operating mileage 2933 km and 1947 operating subway station approximately. As a result, the structural styles of the subway underground structure become more and more complicated. In recent years, the strong earthquake disaster investigations indicate that underground structures suffered severe damage in near-fault area, such as the 1994 Northridge earthquake, 1995 Kobe earthquake, 1999 Kocaeli earthquake and 1999 Chi-Chi Earthquake (Samata et al. 1997, Kalkan et al. 2006, Corigliano 2007, Cunha et al. 2014). But the studies on the effect of near-fault ground motion on the seismic behavior of underground structures are limited (Chen et al. 2010, Tao et al. 2011, Corigliano et al. 2011, Davoodi et al. 2013, Zhao et al. 2015). Selected a set of near-fault ground motions of Loma Prieta and Coalinga records as seismic inputs, Chen et al. (2010) carried out the simulation analysis for the Double-layer vertical overlapping metro tunnels based on finite element software ABAQUS to investigate the seismic response of subway underground station. It was shown that the relative horizontal displacement induced by near-fault ground motion was greater than that induced by far-field ground motion in metro tunnels. At present, the subway station model with double-layer three-span is widely used in China, accordingly, this paper focused on analyzing the seismic behavior of double-layer three-span subway station structure under near-fault ground motions with a commercial Finite Element Model (FEM) software. To simulate the dynamic properties of soil, a modified Martin-Seed-Davikendov viscoelastic constitutive model was applied.

2 DEFINITION AND CHARACTERISTICS OF NEAR-FAULT GROUND MOTION

Near-fault ground motion is the motion that is typically assumed to be restricted to within 20 km of a fault (Mavroeidis & Papageorgiou 2003). However, this definition is not universal because near-fault effects attenuate as distance increases, which, in turn, leads to a greater effect of factors such as magnitudes and local site conditions on ground motion. The special characteristics of near-field ground motions are directly related to the earthquake source mechanism, rupture direction relative to the site, and slip direction of the rupture fault. The distinguishing characteristic of near-fault ground motion is the pulses generated by the directivity effect and fling-step effect. These pulse-type ground motions often contain one or more distinct pulses in the acceleration, velocity and displacement time histories, most frequently in velocity. Meanwhile, many new typical characteristics including hanging wall effect and vertical effect are also gradually recognized (Li &

Xie 2007). The ground motions are characterized by pulse-type wave shape, long pulse period, abundant long-period components, and sometimes large permanent ground displacement.

3 NUMERICAL MODEL

3.1 *Soil foundation and underground station structure*

A 2-dimensional subway station numerical model with regular cross-section performed by ABAQUS was shown in Figure 1. The nonlinear static and dynamic coupling model was developed for the soil-underground structure dynamic interaction. In this numerical model, CPE4R (a 4-node plan strain, reduced integration) elements were used to mesh the soil foundation and the subway station structure; B21 (a 2-dimensional linear beam) elements were embedded into the concrete to simulate the rebar. The CPE4R element size is 0.2 m × 0.2 m for the subway station structure. Based on Liao's study (2013), the maximal height of element $h_{\max}$ in the case of shear wave propagation in the soil should be determined by:

$$h_{\max} = \left(\frac{1}{75} - \frac{1}{160}\right) \cdot V_S / f_{\max} \tag{1}$$

where V_S means the shear wave velocity. $f_{\max}$ denotes the maximal vibration frequency of the input motion. Hence, the maximum element height of the soil foundation from the top to the bottom of the ground ranges from 0.8 to 2.3 m.

3.2 *Material properties*

The stress-strain relationship of soil shows characteristics of nonlinearity, hysteresis, and accumulative deformation under cyclic loading. The constitutive model for soil used in the paper is a modified Martin-Seed-Davikendov viscoelastic model (Chen et al. 2005), which is based on the Davikendov skeleton curve of a one dimensional dynamic stress-strain relation. The dynamic shear stress-strain curves were constituted by Masing rules (Martin & Seed 1982). This soil constitutive model was recently implemented into the commercial software ABAQUS (Chen et al. 2011). The Davikendov skeleton curve was corrected by sectional functions, and the upper limit of failure shear strain amplitude (γ_{ult}) was applied as the sectional point. The modified skeleton curve could approach the upper limit of failure shear stress (τ_{ult}) when the shear strain amplitude approached infinite.

The Davidenkov skeleton curve is given as follows:

$$\tau(\gamma) = G \times \gamma = G_{\max} \times \gamma\left[1 - H(\gamma)\right] \tag{2}$$

where

$$H(\gamma) = \left\{\frac{(\gamma/\gamma_0)^{2B}}{1 + (\gamma/\gamma_0)^{2B}}\right\}^A \tag{3}$$

The actual shear stress and strain relationship of soil is $\gamma \to \infty$, $\tau(\gamma) \to \tau_{ult}$. Therefore, the piecewise function method was used to describe the skeleton curve, and the Davidenkov model was modified as follows:

$$\tau(\gamma) = \begin{cases} G_{\max} \times \gamma\left[1 - H(\gamma)\right] & \gamma_c \le \gamma_{ult} \\ G_{\max} \times \gamma_{ult}\left[1 - H(\gamma_{ult})\right] & \gamma_c > \gamma_{ult} \end{cases} \tag{4}$$

$$\tau_{ult} = G_{\max} \times \gamma_{ult}\left[1 - H(\gamma_{ult})\right] \tag{5}$$

The equations used to calculate the damping ratio can be derived. The hysteretic curve of the modified shear stress-strain relationship and

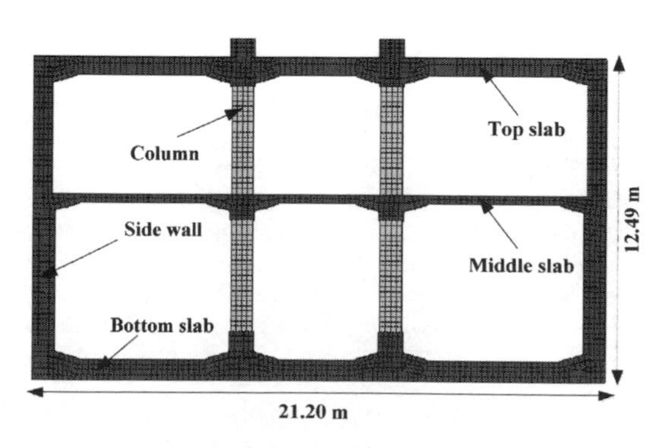

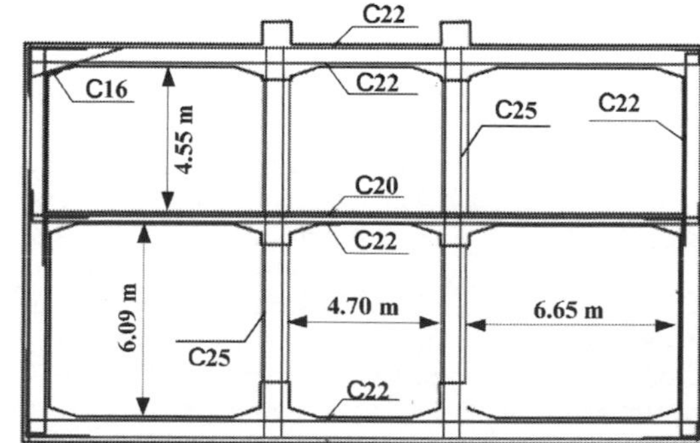

Figure 1. The cross-sectional dimensions and distributed rebar of the subway station.

calculation flow diagram were shown in Figure 2. Table 1 presents the physical and mechanical parameters of the soil foundation.

The concrete damaged plasticity constitutive model presented by Lee & Fenves (1998), which was adopted to simulate the material plasticity and continuum damage mechanics behavior of concrete. The strength level of concrete adopted in the subway station is No. C30, and its material properties are summarized in Table 2.

3.3 *Boundary condition and input motion*

The static and dynamic coupling boundary condition of the soil foundation adopted the artificial constrained boundary. According to the method, the lateral boundary is constrained in the

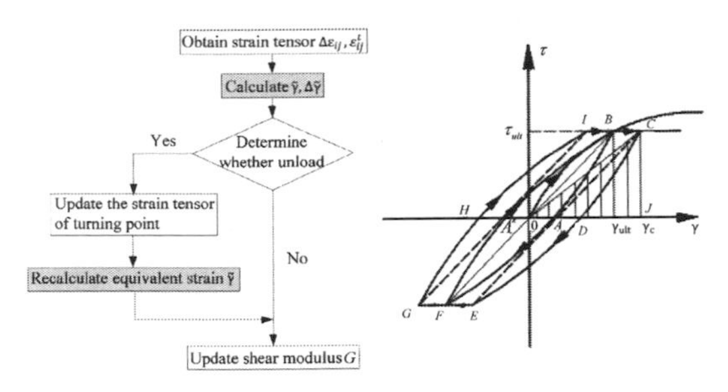

Figure 2. Hysteretic loop of shear stress-strain of modified Davikendov model and calculation flow diagram.

horizontal direction, and the horizontal reaction forces (RF) at the nodes on the lateral boundary output at the end of static analysis. Prior to the dynamic analysis, the lateral boundary condition is released in the horizontal direction with a horizontal RF that is reserved by the artificial loading method and constrained in the vertical direction. The conversion process of boundary condition is shown in Figure 3. To weaken the reflected wave effect on the lateral boundary of soil foundation, the horizontal distance from the lateral boundary to the subway station was set as 89.4 m, and the damping factor of the elements near the boundary was amplified fivefold (Zhuang et al. 2015).

Prior to the dynamic simulation, the interaction system is subjected to gravity loading and hydrostatic pressure. In the ABAQUS/Standard analysis, these loads are specified in two consecutive static steps. To examine the effects of near-fault ground motions on the dynamic response of the subway station, three representative near-fault ground motion recordings with forward directivity effect were selected as seismic input from the Next Generation Attenuation Project database (PEER NGA Database 2005), which were recorded at NO.07-SN El Centro strong motion station during the Ms 6.5 Imperial Valley earthquake on Oct 15, 1979 in California, United States, and NO. TCU075-EW, TCU129-EW during the Ms 7.6 Chi-Chi earthquake on Sep 21, 1999 in Taiwan. The near-fault

Table 1. Properties of soils surrounding the underground structure.

Soil profiles	Tickness (m)	Unit weight $(KN \cdot m^3)$	Average shear-wave velocity (m/s)	Poisson's ratio (v)	Damping factor α_1	The angle of shearing resistance (°)
Silty clay	3.0	19.4	100	0.49	0.001	14.90
Clay	20.5	19.4	210	0.49	0.001	14.90
Silty-fine sand	20.0	19.4	320	0.49	0.001	14.90
Clay	8.0	20.0	355	0.49	0.001	14.90
Clay	28.5	21.5	400	0.49	0.001	19.50

Table 2. Damaged plasticity model parameters of the concrete C30.

Parameters	Value	Parameters	Value
Elastic modulus E (GPa)	30.00	Initial compressive yield stress σ_{c0} (MPa)	13.00
Poisson's ratio ν	0.15	Limited compressive yield stress σ_{cu} (MPa)	24.10
Density ρ (Kg/m³)	2450	Initial tensile yield stress σ_{t0} (MPa)	2.90
Dilation angle ψ (°)	36.31	Compression stiffness recovery parameter ω_c	1.00
Damping ratio	0.05	Tensile stiffness recovery parameter ω_t	0.00

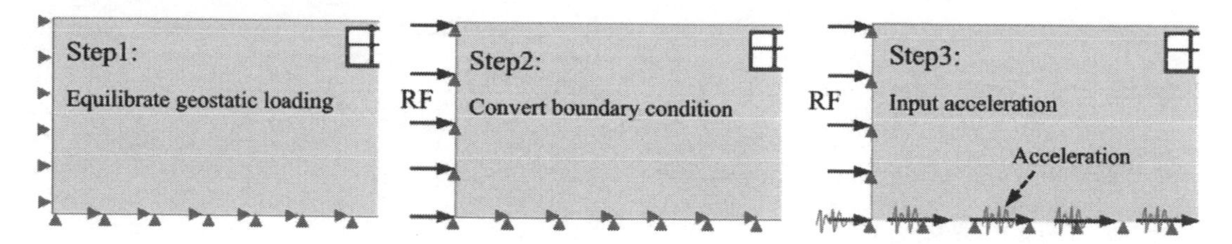

Figure 3. Conversion process of boundary condition in analysis.

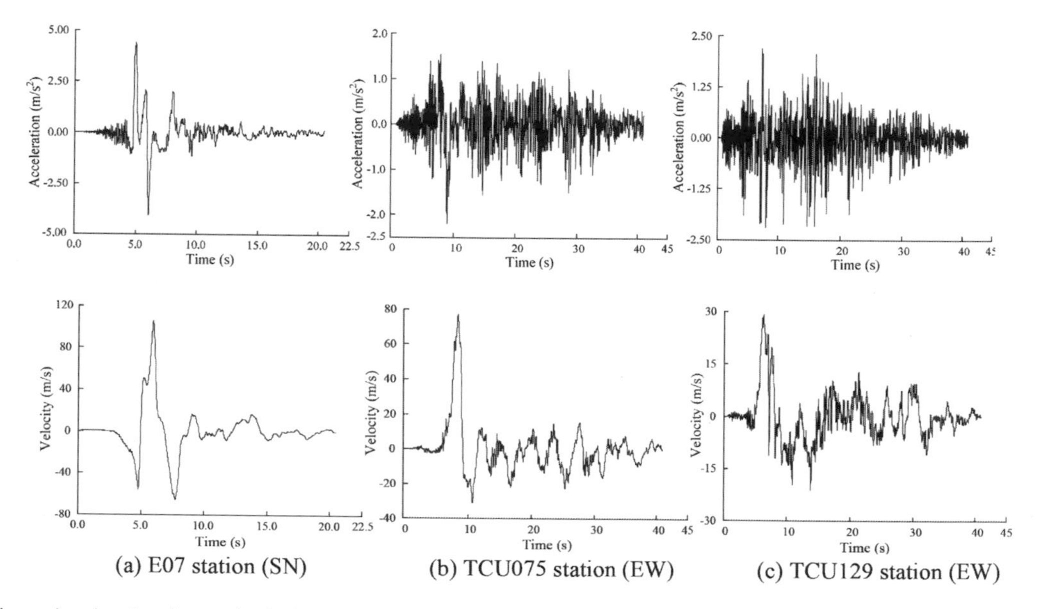

Figure 4. Acceleration and velocity time histories of input near-fault ground motions.

ground motions were oriented in the fault-normal direction with rupture distance of 0.56, 0.91 and 1.84 km, respectively. The acceleration and velocity time histories of three near-fault ground motions are shown in Figure 4.

4 RESULTS AND ANALYSIS

4.1 *Acceleration responses of the soil foundation*

The peak acceleration amplification factors along the soil depth under the three ground motions are shown in Figure 5. Note that the peak acceleration amplification factor increases from the bottom to the top of the soil foundation, especially visible from the bottom slab of subway station to the ground surface. The result means that the soil had different effect on the propagation of waves with different frequency components and the presence of the subway structure had a strong effect on the response of the soil foundation. It's clear that the peak acceleration amplification factor is almost larger than 1, and the curve shapes were basically similar for the forward directivity effect of near-fault ground motions. The values for the PGA amplification factors at the ground surface are much larger than 1.50 (Chen et al. 2015), ranging from 2.28 to 2.65. It indicates that the softening behavior of the topsoil was more sensitive to the near-fault ground motion with characterize of long pulse period and abundant low frequency components.

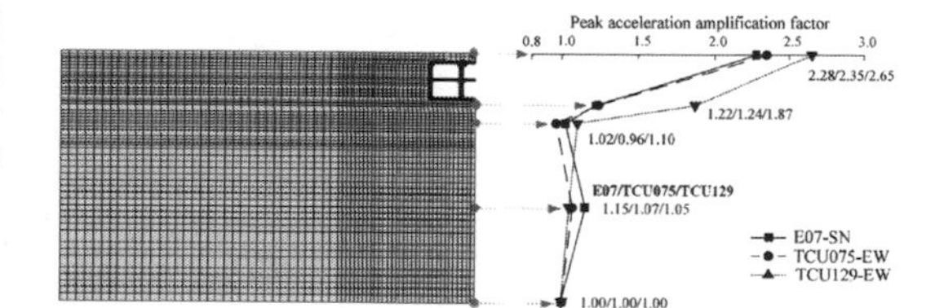

Figure 5. Peak acceleration amplification factors of the model soil.

4.2 *Acceleration responses of the subway station structure*

Figure 6 presents the peak acceleration amplification factor at the subway station slabs subjected to near-fault ground motions. Note that the peak acceleration amplification factor at the middle slab of the subway station is smaller than that of the bottom slab, and then it increases obviously at the top slab. This finding is obviously inconsistent with the acceleration response rule for the general ground structure, and the effect of the soil-underground structure interaction should be the main factor. In addition, the peak amplification factor at the middle slab induced by the input motion of TCU129 record is the smallest with the same inputting PGA = 0.23 g and it reaches the largest for the input motion of E07 record due to the different frequency spectrum characteristics of input near-falult ground motion.

4.3 *Lateral displacement responses of the subway station*

Figure 7 plots the relative displacement distribution of measuring points along the structural height subjected to near-fault ground motions. It's quite clear that the lateral displacements of the subway station have positive correlation with the height of the subway station, and the peak relative lateral displacements on the top slab range from 32.99 to 34.89 mm. The maximum story drift was 20.40 mm on the middle slab under input motion

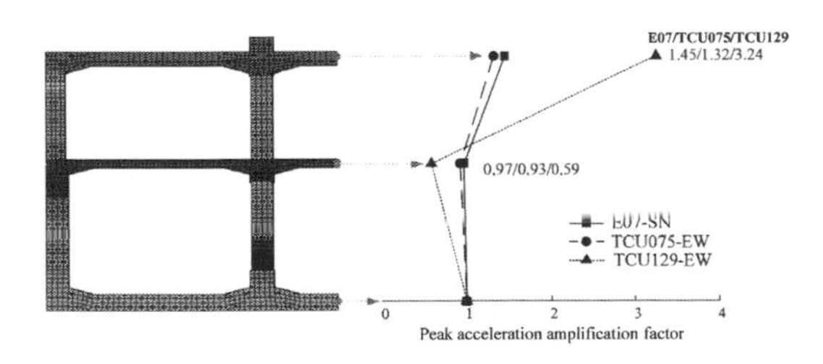

Figure 6. Peak Acceleration amplification factor for the slabs of the subway station.

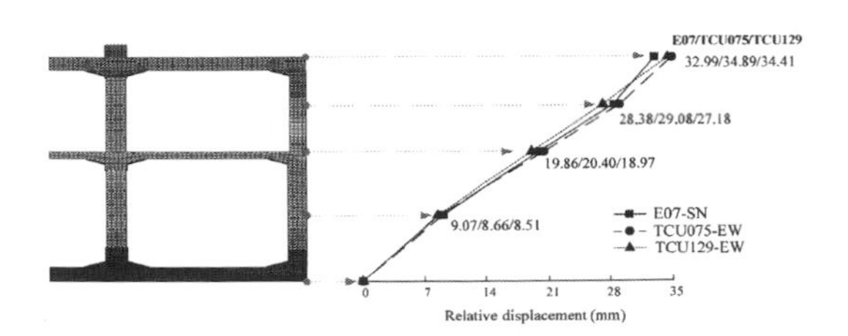

Figure 7. Relative displacement of the subway station.

of TCU075 record, and 15.44 mm on the top slab under input motion of TCU129 record with PGA = 0.23 g. Meanwhile, the inter-story drift reached 1/356 at the middle slab and 1/339 at the top slab, respectively. The result shows that frame-type subway station was regular on the vertical direction, and also had excellent integrity. According to the code for seismic design of buildings (GB 50011–2010) and recent research (Zhuang et al. 2015), the ultimate value of elastic inters-tory drift for Reinforced Concrete (RC) frame structure should range from 1/550 to 1/430. It indicates that the structural members of subway station may be damaged under the near-fault ground motions. In addition, the curve shapes show that the deformation form of subway station structure is basically shearing type.

5 CONCLUSIONS

The FEM simulation model was crafted to model the soil foundation and the underground structure. Acceleration responses of the soil-underground structure system and the lateral displacement response of the subway station were analyzed. The following concluding remarks and recommendations could be drawn:

1. The softening behavior of the topsoil was more sensitive to the near-fault ground motion with character of long pulse period and abundant low frequency component.
2. The frequency spectrum characteristics of the input near-fault ground motion had significant effect on the dynamic response of the subway station structure.
3. The structural members of subway station may be damaged under the near-fault ground motions, and the deformation form of subway station structure is basically shearing type.

ACKNOWLEDGEMENTS

The authors gratefully acknowledge the financial support provided by the National Natural Science Foundation of China (No. 51421005; No. 51508526; No. 51278246) and the Project of Construction and Development of Innovative Teams for Universities and Colleges in Beijing (No. IDHT20130507).

REFERENCES

Chen, G.X. & H.Y. Zhuang (2005). Developed nonlinear dynamic constitutive relations of soil based on Davidenkov skeleton curve. *Chinese Journal of Geotechnical Engineering*. 27(8), 860–864 (in Chinese).

Chen, G.X., L. Chen, L.P. Jing, & H. Long (2011). Comparison of implicit and explicit finite element methods with parallel computing for seismic response analysis of metro underground structures. *Journal of the China Railway Society*. 33(11), 111–117 (in Chinese).

Chen, G.X., S. Chen, X. Zuo, X.L. Du, C.Z. Qi, & Z.H. Wang (2015). Shaking-table tests and numerical simulations on a subway structure in soft soil. *Soil Dynamics and Earthquake Engineering*. 76, 13–28.

Chen, L., G.X. Chen, & L.M. Li (2010). Seismic response characteristics of the double-layer vertical overlapping metro tunnels under near-field and far-field Ground Motions. *China Railway Science*, 31(1), 79–86.

Corigliano, M. (2007). Seismic response of deep tunnels in near-fault conditions. *PhD dissertation*. Politecnico di Torino, Italy, p. 222.

Corigliano, M., L.Scandella, C.G. Lai, & R. Paolucci (2011). Seismic analysis of deep tunnels in near fault conditions: a case study in Southern Italy. *Bulletin of Earthquake Engineering*. 9(4), 975–995.

Cunha, A., E. Caetano, P. Ribeiro, & G. Müller (2014). Effects of Near-Fault Ground Motions on the Seismic Response of Concrete Gravity Dams. *Proceedings of the 9th International Conference on Structural Dynamics*. Porto, Portugal, 30 June–2 July, Paper No. 305–310.

Davoodi, M., J M.K. afari, & N. Hadiani (2013). Seismic response of embankment dams under near-fault and far-field ground motion excitation. *Engineering Geology*. 158, 66–76.

Kalkan, E. & S.K. Kunnath (2006). Effects of fling step and forward directivity on seismic response of buildings. *Earthquake spectra*. 22(2), 367–390.

Lee, J. & G.L. Fenves (1998). Plastic-damage model for cyclic loading of concrete structures. *Journal of engineering mechanics*. 124(8), 892–900.

Li, S. & L.L. Xie (2007). Progress and trend on near-field problems in civil engineering. *Acta Seismologica Sinica*. 20 (1), 105–114.

Liao, Z.P. (2013). *Introduction to wave motion theories in en—gineering*. Beijing: Science Press.

Martin, P.P. & H.B. Seed (1982). One-dimensional dynamic ground response analyses. *Journal of the geotechnical engineering division*. 108(7), 935–952.

Mavroeidis, G.P. & A.S. Papageorgiou (2003). A mathematical representation of near-fault ground motions. *Bulletin of the Seismological Society of America*. 93(3), 1099–1131.

National Standard of the People's Republic of China (2010). *GB50011-2010 Code for seismic design of buildings*. Beijing: China Architecture & Building Press.

PEER NGA Database. Pacific Earthquake Engineering Research Center, University of California at Berkeley.

Samata, S., H. Ohuchi, & T. Matsuda (1997). A study of the damage of subway structures during the 1995 Hanshin-Awaji earthquake. *Cement and Concrete Composites*. 19(3), 223–239.

Tao, L.J., W.P. Wang, B. Zhang, Q.H. Wang & Y.J. Wei (2011). Effects of rupture forward directivity and fling step of near-fault ground motions on dynamic responses of representative subway station structure. *Journal of Earthquake Engineering and Engineering Vibration*. 31(6), 38–44.

Zhao, W.S. & W.Z. Chen (2015). 1558. Effect of near-fault ground motions with long-period pulses on the tunnel. *Journal of Vibroengineering*. 17(2), 841–858.

Zhuang, H.Y., Z.H. Hu, & G.X. Chen (2015). 1555. Numerical modeling on the seismic responses of a large underground structure in soft ground. *Journal of Vibroengineering*. 17(2), 802–815.

Zhuang, H.Y., Z.H. Hu, X.J. Wang, & G.X. Chen (2015). Seismic responses of a large underground structure in liquefied soils by FEM numerical modelling. *Bulletin of Earthquake Engineering*. 13(12), 3645–3668.

Civil, Architecture and Environmental Engineering – Kao & Sung (Eds)
© 2017 Taylor & Francis Group, ISBN 978-1-138-02985-9

Finite—infinite element analysis of soil dynamic response under moving load

Chenqi Tian & Liqun Yuan
School of Civil Engineering, Shandong Jianzhu University, Jinan, Shandong, China
School of Architecture and Civil Engineering, Liaocheng University, Liaocheng, Shandong, China

ABSTRACT: In recent years, the rail transit in China has been developed rapidly. But the noise and vibration caused by rail transit have serious impact on people's working and living. The propagation of vibration wave in the foundation soil is the key problem for the reasearch of environmental vibration. In this paper, different element stiffness matrices and group sets are obtained by using three dimensional, one direction bidirectional, and three direction mapping infinite element methods. Dynamic responses of semi-infinite soil are calculated under different loads. The characteristics of vibration for the semi-infinite soil space under surface load are also analyzed and summarized.

1 INTRODUCTION

After the emergence of the concept of infinite element, the domestic and foreign scholars have put forward various kinds of static, dynamic, wave frequency domain or infinite element in the time domain, which can simulate the infinite domain. Infinite element method is developed to solve the effectively. Researchers have proposed different types of infinite element and shape function. But most of the methods are based on the unite shape function, which is multiplied by an attenuation factor, to represent the amplitude of wave decay gradually when the wave extends to infinite element in the relevant direction. Although at present theoretical researchers for the coupling method between finite elementh and infinite element have made some progress, infinite elementh method is a new kind of special method compared with the finite element method. Theory and application of infinite element method are limited. Many questions in practical engineering applications still exist. In this paper, the coupling finite element and infinite element method is used to explore the dynamic problem of the three-dimensional elastic semi infinite soil under moving load.

2 ESTABLISHMENT OF THE MODEL

In this paper, we mainly consider the dynamic response of the arch at the end of the tunnel. So the formation parameters are calculated by stratum soils selection. Detailed parameters of the materials used in the numerical calculation model are shown in Table 1.

Table 1. Material parameters of the finite element model.

Gravity density γ/kN/m³	Elastic modulus E/MPa	Poisson ratio v	cohesion C/kPa	internal friction angle ϕ/°
19.2	4.0	0.30	30	21°

3 FINITE ELEMENT BOUNDARY CONDITION

Dynamic calculation, the first to calculate the characteristic value of the model. The key to the eigenvalue calculation is to define the elastic boundary condition. Model boundary is adopted in the form of curved spring, and the spring coefficient is calculated according to the following formula.

Vertical ground reaction coefficient: $k_v = k_{v0} \times \left(\frac{B_V}{30}\right)^{-0.75}$

Horizontal ground reaction coefficient: $k_h = k_{h0} \times \left(\frac{B_h}{30}\right)^{-0.75}$

The calculation formula of the damping ratio is as follows:

P-wave: $C_p = \rho \times A \times \sqrt{\frac{\lambda+2G}{9.81\times\gamma}} = c_p \times A$

S-wave: $C_s = \rho \times A \times \sqrt{\frac{G}{9.81\times\gamma}} = c_s \times A$

Through the calculation of the model: $k_x = 139.4$ kN/m², $k_y = 116.8$ kN/m², $k_z = 139.4$ kN/m², $c_p = 101.9$ kN•sec/m, $c_p = 54.5$ kN•sec/m

4 FINITE ELEMENT ANALYSIS

Finite element model of moving load calculation y, x, z directions of the dimensions are 10 m, 15 m, 5 m, the outer and the bottom of the infinite element. The loading is divided into two kinds, respectively for moving load and moving harmonic load.

4.1 *Moving force load*

Moving force load size as $P = 1$ N,moving along y axis, speed is V.

In Figure 1 and the thick solid line is the load moving trajectory.

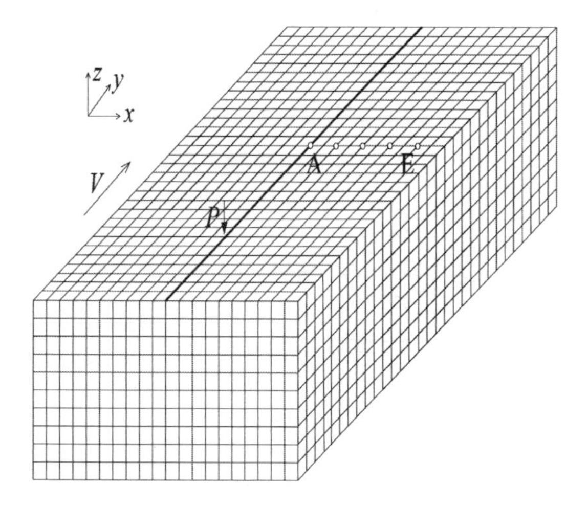

Figure 1. Schematic diagram of subsoil model.

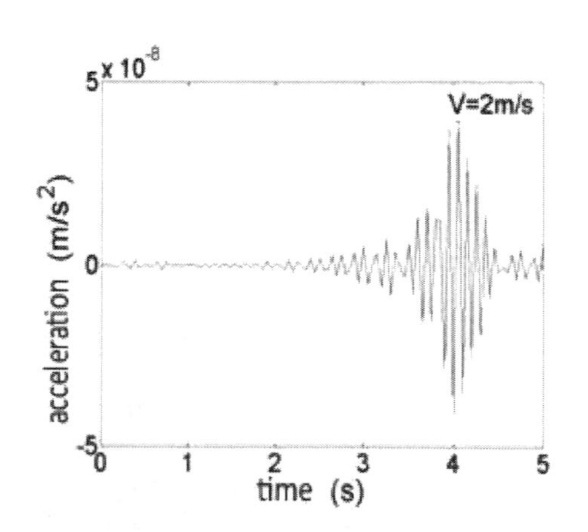

Figure 2. Time curves of vertical acceleration (V = 2 m/s).

Figure 2 to Figure 5 shows the dynamic response of the node E at different speeds. The position of the node E distance from the moving load 4 m.

Through the calculation, it is known that the acceleration of the node E is larger when the

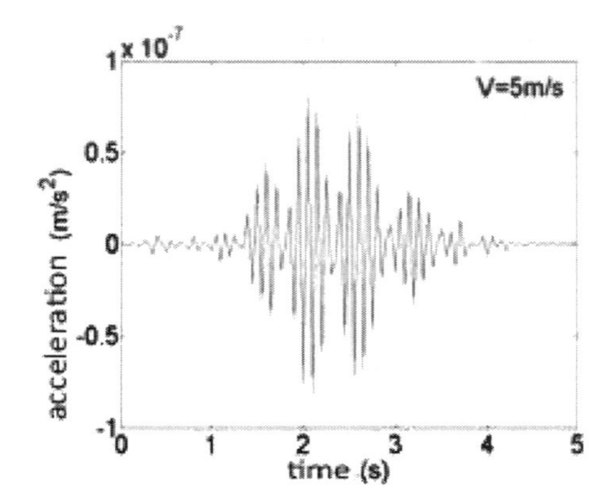

Figure 3. Time curves of vertical acceleration (V = 5 m/s).

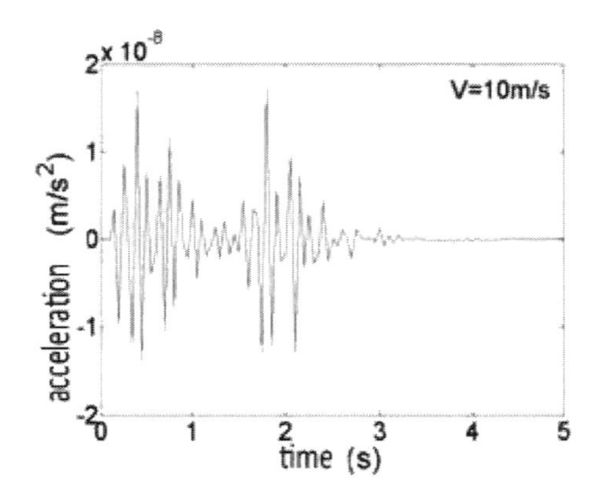

Figure 4. Time curves of vertical acceleration (V = 10 m/s).

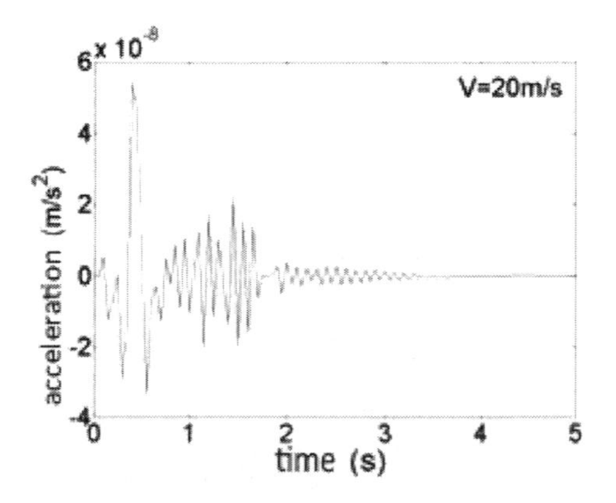

Figure 5. Time curves of vertical acceleration (V = 20 m/s).

moving load speed is 10 m/s. When the velocity is 0~10 m/s, the root mean square value of the acceleration increases with the speed. When the velocity is greater than 10 m/s, the root mean square value of the acceleration decreases.

4.2 *Moving harmonic load*

Different frequency moving harmonic loads are applied on the ground and loads velocity is 5 m/s. Table 2 gives effective value of acceleration for the different frequency loads of the node A and E.

In order to study the dynamic response characteristics of the foundation soil, the acceleration response of the node E under different load frequencies is analyzed by Fourier transform, which are shown in Figures 6~11.

By comparing the acceleration time history curves of the joints under different loading frequencies, it is known that the dynamic response

Table 2. Effective value of acceleration (10^{-8} m/s^2).

Frequency (Hz)	0.16	0.8	1.6	5	10	16	20	30
A	196.3	287.1	598.9	233.8	4064	2017	7028	5070
E	0.200	0.201	0.2134	1.462	69.239	44.42	36.13	39.85

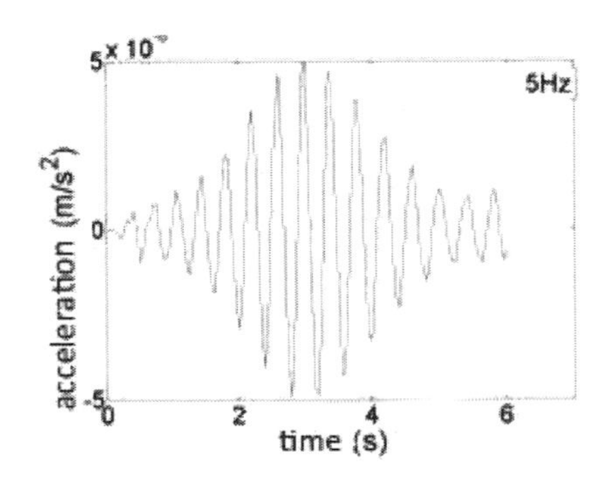

Figure 6. Time curves of vertical acceleration (5 Hz).

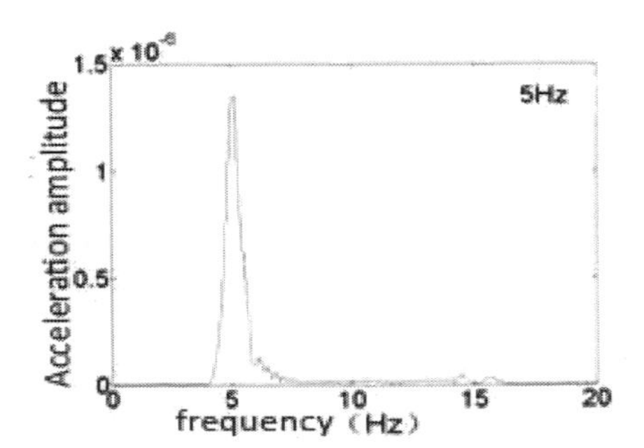

Figure 7. Spectrum analysis.

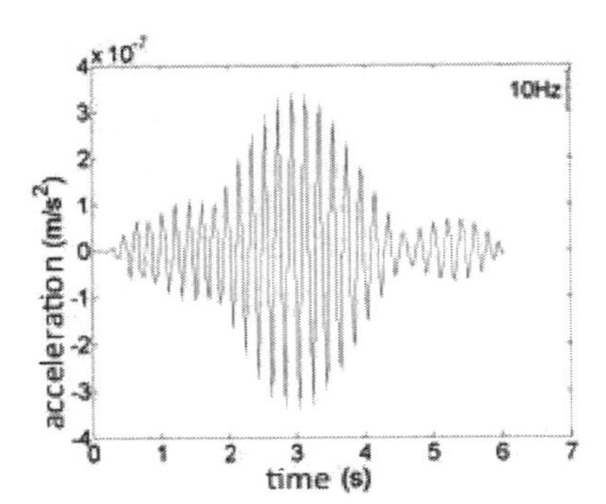

Figure 8. Time curves of vertical acceleration (10 Hz).

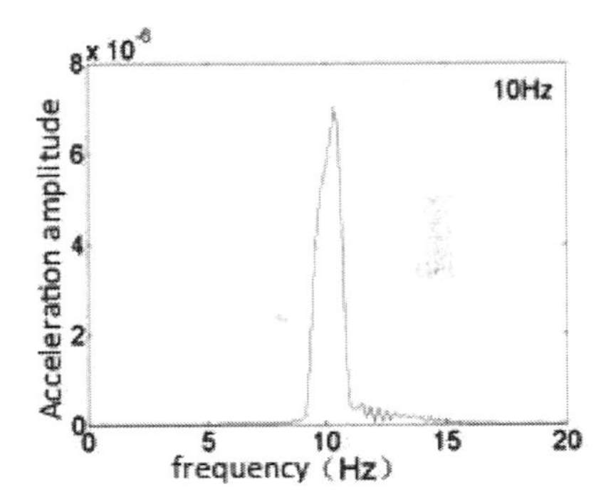

Figure 9. Spectrum analysis.

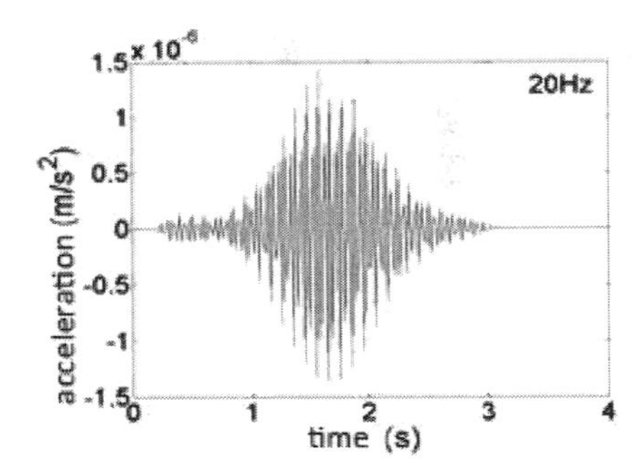

Figure 10. Time curves of vertical acceleration (20 Hz).

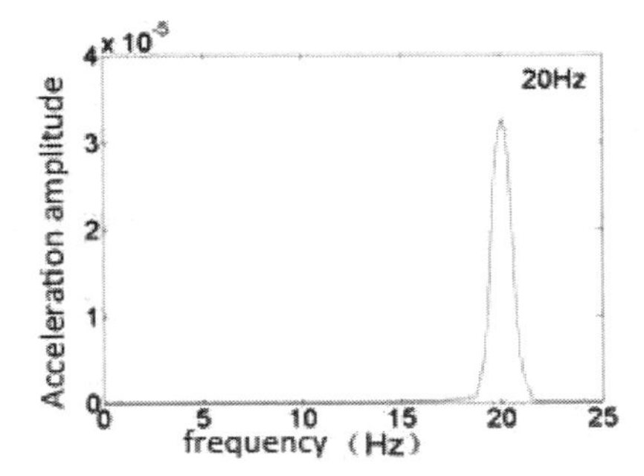

Figure 11. Spectrum analysis.

of the foundation increases with the increase of the load frequency, but the attenuation of the dynamic response of the foundation is accelerated. At the same time, by means of the Fourier transform of the acceleration of the joint, the frequency value of the Fourier transform is the frequency of the load.

5 CONCLUSIONS

In this paper, the three-dimensional finite element and infinite element method are employed to simulate the soil dynamic response. Some valuable results are found. The results obtained are of great reference value for the analysis of formation dynamic response under traffic load.The research results not only have a certain academic value for the infinite element method and soil mechanics, but also provide technical support for the important transportation infrastructure construction in the future. The results can be also a good reference for the study of underground space security.

ACKNOWLEDGEMENTS

This work was financially supported by Liao cheng university Science Foundation (318051524) and National Natural Science Foundation of China (41502277).

REFERENCES

Andersen, L. S.R.K. Nielsen. Reduction of ground vibration by means of barriers or soil improvement along a railway track [J]. Soil Dynamics and earthquake Engineering, 2005, 25: 701–716

Degrande G, Lombaert G. High-speed train induced free field vibrations: in situ measurements and numerical modeling [A]. Wave 2000 [C], 2000, 29–42.

Degrande G, Schillemans L. Free field vibrations during the passage of a Thalys high-speed train at variable speed [J]. Sound and vibration, 2001, 247(1): 131–144.

Xia He, et al. Traffic-induced vibrations and their influeces on surrounding environments [A]. MCCI'2000[C]. Beijing, 2000, 123–130

Yang, Y.B., and Hung, H.H. (2009). Wave Propagation for Train-Induced Vibrations—A Finite/Infinite Element Approach. Singapore, World Scientific.

Yuan L, Liu H, Men Y. Research advances in the interaction among stratum-ground fissure-tunnel under subway dynamic loading [C]. Shanghai, China: Trans Tech Publications, 2012.

Yuan Liqun, Liu Nina and Yang Mi. Soil Pressure Analysis of Subway Tunnel Across Ground Fissures[J]. Advances in Engineering Research, 2016, 65:73–76

Civil, Architecture and Environmental Engineering – Kao & Sung (Eds)
© 2017 Taylor & Francis Group, ISBN 978-1-138-02985-9

Evaluation of the processing suitability of rural domestic sewage treatment, taking Fangshan in Beijing as a case study

Yong Wang
College of Resources and Environmental Science, China Agricultural University, Beijing, China
Fangshan District Beijing Water Authority, Beijing, China

ABSTRACT: In order to understand the adaptability and the effect of multiple rural domestic sewage treatment techniques in rural areas, their technical suitability could be evaluated by using the entropy fuzzy matter-element model and the AHP analysis method by taking the rural sewage treatment station in Fangshan District, Beijing, China, as an example. The result showed the high repeatability of the two evaluation methods. The sequence of processing suitability was: Constructed Wetland (CW) > MBR > Solar Anaerobic Biological Filter (SABF).

1 INTRODUCTION

The features of rural domestic sewage are of small scale and dispersed, which have large difference for the water quantity and water quality of different regions. In fact, it was difficult to collect and had high organic pollution concentration and low processing ratio (LI Juan, 2012; WANG Shou-zhong, 2008). For the quality, the COD and BOD_5 were higher than the city domestic sewage but with good biochemical reactivity (WANG Shou-zhong, 2008; HOU Jing-wei, 2012). According to the specialties, the common treatment technologies were frequently used as constructed wetlands, underground soil permeability system, aerobic biological treatment system, and anaerobic biological treatment systems (WANG Baoxue, 2015).

China is an agricultural country with a large population, consisting of 634,000 villages and 22,000 towns. According to statistics, 50% of the total displacement came from the counties, towns, and villages. The county wastewater treatment rate was 11%, whereas the corresponding figures of villages and towns were less than 1%. The promulgation of "water pollution prevention action plan" will drive RMB 2 trillion yuan of investment in water pollution governance. However, the fault management shows that a number of sewage treatment stations have low efficiency of operation. The inadequate analysis and evaluation of planning, designing, construction, operation, maintenance, management, and so on have resulted in insufficient understand of the applicability and reliability of different treatments in different areas (Liang Hanwen, 2011). Therefore, suitability analysis is really important for rural domestic sewage treatment.

Recently, Fangshan District has invested more than 300 million RMB for the rural domestic sewage control and constructed 115 wastewater treatment stations in 20 towns and more than 80 villages, were including Changyang Town, Qinglonghu Town, and Zhoukoudian Town.

This paper aims to provide a reference for the rural sewage treatment processing construction and management by discussing different treatment technologies and carry out adaptability investigation and evaluation of actual running rural sewage treatment station in Fangshan District.

2 METHODS

2.1 *Investigation*

The specific information of Fangshan District wastewater treatment station's processing type, investment, designed quality of inflow and outflow, operation management, and pollutant treatment effect was gathered by questionnaires, investigation, and data collection.

2.2 *Monitoring*

The evaluated stations were sampled and monitored for water quality five times from January to April, 2016. The monitored indictors were water temperature, COD, BOD_5, NH_4^+-N, and TP.

2.3 *Evaluation method*

At present, the widely adopted evaluation methods are contrast method, Analytic Hierarchy Process (AHP), and causal analysis and comprehensive evaluation method (Li Jicheng, 2007). For this study, the entropy weight fuzzy matter-element model (HAO Rui-xia, 2013) (EWFM) and AHP method (GUO Jin-yu, 2011 & Bottero M, 2011) were applied.

3 RESULTS AND DISCUSSION

3.1 *Selection of evaluation stations*

During the analysis and research, 115 treatment stations were classified into 11 processing types (whose position and process distribution are shown in Figure 1).

The representative potential stations were selected according to the topographies, technology penetration, and operation situation. After discussing with the local management department, five processing types were chosen: Membrane Bioreactor (MBR), Constructed Wetland (CW), Pretreatment + Constructed Wetland (PCW), Solar Anaerobic Biological Filter (SABF), and Multistage Vertical-flow Constructed Wetlands (MVCW). In addition, the representativeness of plain, hills, and mountainous areas should be considered, so finally six stations were evaluated on the research. The names and process types of the selected stations are shown in Table 1.

3.2 *Evaluation by entropy weight fuzzy matter-element model*

First, the investment and operating cost was uniformed to price of per ton. The station serviced population was in accordance with the equivalent population: a calculation of the resident population and the tourist population who had same pollution discharge with resident population per day. The Fangshan tourism resource abundance value was 26% (Li Jun, 2007). For qualitative and semi-qualitative indicators, the standard of evaluation was established for quantitative evaluation. The quantitative evaluation used 10 division systems, whose scoring standards are shown in Table 2.

According to the investigation and water quality monitoring results, the researchers established recombination fuzzy matter-element and uniformed data, whose results are shown in Table 3. Judgment matrix was uniformed for the value of optimal subordinate degree. It was a nondimensional process for the various elements, which achieved element comparability.

The calculated definition entropy (S_j) of each evaluation indicators and built vector S.

$$S^T = \begin{pmatrix} A & B & C & D & E \\ 0.987 & 0.979 & 0.987 & 0.968 & 0.981 \\ F & G & H & I & J \\ 0.980 & 0.989 & 0.989 & 0.984 & 0.982 \\ K & L & M & & \\ 0.981 & 0.980 & 0.974 & & \end{pmatrix}$$

The calculated weight of each evaluation indictor (ρ_i) as vector ρ.

$$\rho^T = \begin{pmatrix} A & B & C & D & E \\ 0.056 & 0.087 & 0.055 & 0.132 & 0.080 \\ F & G & H & I & J \\ 0.085 & 0.048 & 0.047 & 0.066 & 0.074 \\ K & L & M & & \\ 0.080 & 0.083 & 0.107 & & \end{pmatrix}$$

From the results of entropy weight, its average was 0.070, the value of the annual operating rate was the highest, followed by ammonia nitrogen removal rate, actual running scale/design scale, and the investment cost per ton. The weight of emergency response capability was the lowest and significantly lower than other indicators. Calculated Euclid Approach Degree (EAD) of evaluation stations (φ_i) and built vector (R_φ). The results are as follows:

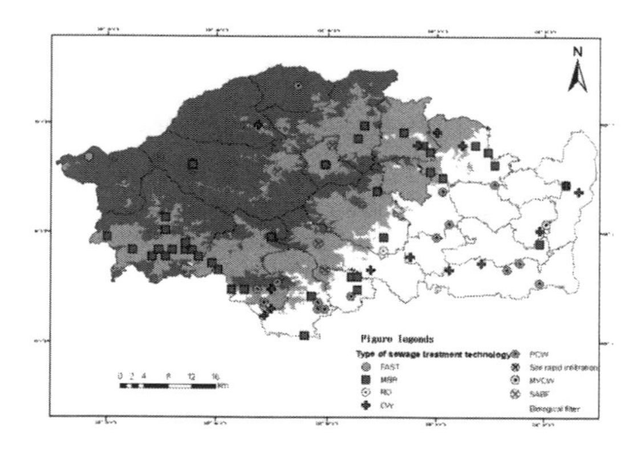

Figure 1. Position and process distribution.

Table 1. Names and process types of evaluation stations.

Number	M1	M 2	M 3	M 4	M 5	M 6
Station name	Nan He (NH)	Miao Er Gang (MEG)	Si Ma Ta (SMT)	Huang Tu Po (HTP)	Long Men Tai (LMT)	Hou Shi Yang (HSY)
Process	MBR		CW	PCW	SABF	MVCW
topography	Plain		Mountainous	Hills	Mountainous	Plain

Table 2. Quantization value of the evaluation indicators.

Evaluation indicators	Function score				
	10–8	8–6	6–4	4–2	2–0
Supervisory ability	Very strong	Strong	Ordinary	Weak	Weaker
Operating journal	Very detailed	Detailed	Ordinary	Rough	No
Emergency response operation	Very strong	Strong	Ordinary	Weak	No
Inspection frequency	>10	10–6	6–3	3–0	0

Table 3. Fuzzy matter-element.

Item	NH	MRG	SMT	HTP	LMT	HSY
Emergency ability (A)	6.76	5.73	4.70	6.27	5.73	6.27
Equipment condition degree (B)	5.00	1.00	5.00	3.00	3.00	5.00
Water quantity change adaptability	9.00	9.00	9.00	9.00	9.00	9.00
Water quality change adaptability	9.00	9.00	9.00	9.00	9.00	9.00
Equipment operation difficulty (C)	3.00	3.00	5.00	3.00	3.00	3.00
Annual operating rate (D)	1.00	1.00	0.27	0.27	0.99	0.99
Service population (person/m^2) (E)	3.89	11.12	2.24	1.71	12.45	1.57
Actual running scale/design scale (F)	0.67	0.83	0.17	1.00	0.33	1.00
Operation cost (¥/t) (G)	1.62	1.62	1.28	2.09	0.85	1.47
Investment cost (10^4¥/t) (H)	1.34	1.18	0.90	1.20	1.20	1.50
SS removal rate (I)	0.62	0.93	0.77	0.97	0.68	0.79
COD removal rate (G)	0.59	0.59	0.95	0.89	0.27	0.90
BOD_5 removal rate (K)	0.88	0.60	1.00	0.94	0.26	0.95
TP removal rate (L)	0.69	0.39	0.99	0.93	0.19	0.91
NH_3–N removal rate (M)	0.39	0.03	0.99	0.97	0.10	1.00

Table 4. EAD value.

Station	Process	EAD	Difference from the EAD
HSY	MVCW	0.70	–
NH	MBR	0.63	–0.07
HTP	PCW	0.58	–0.05
SMT	CW	0.56	–0.02
MRG	MBR	0.51	–0.05
LMT	SABF	0.46	–0.05

$$\mathbf{R}_{\varphi}^{T} = \begin{pmatrix} M_1 & M_2 & M_3 & M_4 & M_5 & M_6 \\ 0.63 & 0.51 & 0.56 & 0.58 & 0.46 & 0.70 \end{pmatrix}$$

According to the calculated results, sorted by the EAD of evaluation station treatment process, the results are shown in Table 4.

The order of stations according to the advantages of the station treatment process suitability was HSY (MVCW) > NH (MBR) > HTP (PCW) > SMT (CW) > MRG (MBR) > LMT (SABF).

Combined with the distribution map, the EAD was lower in the mountainous area. There was a general lack of management issues in the stations in mountains area, increasing the equipment maintenance cost and reducing the actual scale/design scale. These lead to the low adaptability to the environment, which caused the low EAD value. Although MRG is in the plain area, the equipment has high operating cost, and it could not immediately put into service.

3.3 *Evaluation by analytic hierarchy process*

The data in AHP kept same with entropy weight fuzzy matter-element model; partially quantitative data were marked on a scale of 0–10, scoring followed rules:

The bigger optimal indicators:

$$\nabla_{ji} = \frac{v_{ji}}{\text{Max } v_j} \times 10$$
$$1 \le j \le n$$

The smaller optimal indicators:

$$\nabla_{ji} = \frac{\text{Min } v_j}{v_{ji}} \times 10$$
$$1 \le j \le n$$

The scoring matrix C of qualitative and quantitative indicators is obtained as follows:

$$C = \begin{pmatrix} A & B & C & D & E & F & G & H & I & J & K & L & M \\ 7 & 5 & 3 & 10 & 3 & 7 & 5 & 7 & 6 & 6 & 9 & 7 & 4 \\ 6 & 1 & 3 & 10 & 9 & 8 & 5 & 8 & 10 & 6 & 6 & 4 & 0 \\ 5 & 5 & 5 & 3 & 2 & 2 & 7 & 10 & 8 & 10 & 10 & 10 & 10 \\ 6 & 3 & 3 & 3 & 1 & 10 & 4 & 8 & 10 & 9 & 9 & 9 & 10 \\ 6 & 3 & 3 & 10 & 10 & 3 & 10 & 8 & 7 & 3 & 3 & 2 & 1 \\ 6 & 5 & 3 & 10 & 1 & 10 & 6 & 6 & 8 & 9 & 10 & 9 & 10 \end{pmatrix}$$

The building judgment matrix is shown in Table 5.

Matrix biggest characteristic root was $\lambda_{max} = 14$, CI = CR = 0, the corresponding characteristic vector unit was the unit vector $\boldsymbol{\omega}_1$:

$$\boldsymbol{\omega}_1^{T} = \begin{pmatrix} A & B & C & D & E \\ 0.047 & 0.057 & 0.066 & 0.066 & 0.066 \\ F & G & H & I & J \\ 0.057 & 0.075 & 0.066 & 0.066 & 0.075 \\ K & L & M & & \\ 0.075 & 0.075 & 0.057 & & \end{pmatrix}$$

The comprehensive scores of the evaluation stations could come from element matrix, multiply characteristic vector, and the corresponding matrix's biggest characteristic root, and the results are shown in Table 6.

Table 5. AHP judgment matrix.

	A	B	C	D	E	F	G	H	I	J	K	L	M
A	1	5/6	5/7	5/7	5/7	5/6	5/8	5/7	5/7	5/8	5/8	5/8	5/6
B	6/5	1	6/7	6/7	6/7	1	3/4	6/7	6/7	3/4	3/4	3/4	1
C	7/5	7/6	1	1	1	7/6	7/8	1	1	7/8	7/8	7/6	7/6
D	7/5	7/6	1	1	1	7/6	7/8	1	1	7/8	7/8	7/8	7/6
E	7/5	7/6	1	1	1	7/6	7/8	1	1	7/8	7/8	7/8	7/6
F	6/5	1	6/7	6/7	6/7	1	3/4	6/7	6/7	3/4	3/4	3/4	1
G	8/5	4/3	8/7	8/7	8/7	4/3	1	8/7	8/7	1	1	1	4/3
H	7/5	7/6	1	1	1	7/6	7/8	1	1	7/8	7/8	7/8	7/6
I	7/5	7/6	1	1	1	7/6	7/8	1	1	7/8	7/8	7/8	7/6
J	8/5	4/3	8/7	8/7	8/7	4/3	1	8/7	8/7	1	1	1	4/3
K	8/5	4/3	8/7	8/7	8/7	4/3	1	8/7	8/7	1	1	1	4/3
L	8/5	4/3	8/7	8/7	8/7	4/3	1	8/7	8/7	1	1	1	4/3
M	6/5	4/3	6/7	6/7	6/7	1	3/4	6/7	6/7	3/4	3/4	3/4	1

Table 6. Evaluation results by AHP.

Station	Process	Score of AHP
HSY	MVCW	7.4
SMT	CW	7.0
HTP	PCW	6.9
NH	MBR	6.5
MEG	MBR	6.2
LMT	SABF	5.7

Table 7. Comparison of the results of AHP and EWFM.

Station (process)	Score of AHP	EAD × 10
HSY (MVCW)	7.4	7.0
NH (MBR)	6.5	6.3
HTP (PCW)	6.9	5.8
SMT (CW)	7.0	5.6
MEG (MBR)	6.2	5.1
LMT (SABF)	5.7	4.6

3.4 *Comparison of the two evaluation models*

Compared the results of the EWFM and AHP methods, the Euclid approach degree was 10 times as that of the EWFM model score (Table 7), and two evaluation methods showed a highly repeatability.

The entropy weight matter-element model reduced the fuzziness of evaluation indicators and avoided the subjective judgment influence. The AHP represented the experts' experienced judgments. The result showed that the two evaluation methods showed a good repeatability, and their advantages were slightly different.

4 CONCLUSION

1. Using the EWFM model and AHP method to evaluate the technological suitability of rural sewage treatment station, the evaluation results had good repeatability.
2. In the EWFM model, the entropy weight of the annual operating rate, NH_3–N removal rate, and actual scale/design scale were high, which were the main basis for the selection of wastewater treatment processes.
3. CW processing type had obvious advantages in mountainous areas. Also, the MVCW and MBR processing types had advantage in plain areas.

REFERENCES

Bottero M, Comino E, Riggio V. Application of the Analytic Hierarchy Process and the Analytic Network Process for the assessment of different wastewater treatment systems [J]. Environmental Modelling & Software, 2011, 26(10):1211–1224.

Guo Jin-yu et al. Study and Applications of Analytic Hierarchy Process [J]. China Safety Science Journal. 2008, 18(5):148–153.

Hao Rui-xia et al. Comprehensive Evaluation and Analysis on the Operation of Wastewater Treatment Plant Based on Matter-element Model Combined With Entropy Weight [J]. Journals of Beijing University of Technology. 2013(3):445–451.

Hou Jing-wei et al. Review about Characteristics of Rural Domestic Wastewater Discharge [J]. Journal of Anhui Agri. Sci. 2012, 40(2): 964–967.

Liang Hanwen et al. Investigation and analysis of rural wastewater discharge characteristics *in thr*ee typical areas of China [J]. Chinese Journal of Environmental Engineering. 2011, 9(5):2054–2059.

Li Jicheng, L iHongcheng, Zhang Xiance et al, Overview on Investing and Constructing Post Project Evaluation Methods and Development Progress [J]. TUHA OIL& GAS, 2007, 12(1): 86–91.

Li Juan. Discussion on status and technology of sewage treatment during the "Twelfth Five-Year Plan" [J]. Shanxi Architecture. 2012.12, 38(34):156–157.

Li Jun. An Appraisal of Beiiing' Tourism Environmental Carrying Capacity and Its Potentiality [D]. Capital University of Economics and Business. 2007.

Wang Baoxue. Discussion on the Domestic Sewage Treatment Technology for rural Areas of China [J]. Hydropower and New Energy. 2015, 3:67–69.

Wang Shou-zhong, Zhang Tong. Characteristics and Prevention Countermeasures of Chinese Rural Water Pollution [J]. China Water & Waste-water. 2008, 9(24): 1–4.

Civil, Architecture and Environmental Engineering – Kao & Sung (Eds)
 ISBN 978-1-138-02985-9

Influence analysis of the incremental launching of steel box girder and local stability control

Fu-Jun Xie
Central South University, Changsha, Hunan, P.R. China
Transport Bureau of Hengyang City, Hengyang, Hunan, P.R. China

ABSTRACT: Incremental launching method is generally used for steel box girders of long-span self-anchored suspension bridge during construction. First, the mechanical characteristics of the pushing process and the three bending moment equations were introduced. Second, in Guihua Bridge, design parameters influence analysis of the incremental launching process of steel box girder was carried out. The results of the analysis showed that the maximal displacement at the front of guide beam, the bending moment at steel guide beam root, and the maximum tensile stress of steel box girder as well as the compressive stress of steel box girder periodically change along with the construction process. Meanwhile, elastic modulus and Poisson's ratio have a little influence on the stress and deformation of the steel box girder during incremental launching process. The length of the steel guide beam and its average line weight density have an important influence on the stress and deformation of the steel box girder during incremental launching process. Finally, on the basis of the elastic thin plate theory of small deflection, local stability analysis of steel box girders during the process of incremental launching was carried out. Theoretical calculation showed that the local stability of the steel box girder does not meet the requirements before stiffening, but after stiffening, the local stability fulfills the requirements. By considering the constraint factors of plate groups, material nonlinearity, initial geometric imperfection as well as residual stress, finite-element analysis showed that the stress by self-weight has a little influence on critical buckling stress, which can be neglected. Installing a diaphragm plate to improve the effect of local buckling of the steel box girder is obvious. The analysis results show that the stress of local buckling for the steel box girder is less than the structural maximum working stress during the process, and the risk of instability would be presented.

1 INTRODUCTION

At present, construction of long-span suspension bridge is a trend. In order to reduce the number of anchor constructions and balance the tension of main cable, the main cable is directly anchored on a stiffening girder, which is a self-anchored suspension bridge (Dong & Li 2015). Regardless of the type of self-anchored suspension bridge, the main girder is built inevitably by the methods of support construction method and incremental launching method. In most cases, with regard to the limited space and cost control, incremental launching method is more used for building bridges.

Existing analysis of the incremental launching method is more focused on concrete girders (Ding & Liu 2002) and less focused on steel girders (Beaney & Martin 1993), especially for the local stability analysis of the construction process of the incremental launching method (Li & Shao 2011). It is much less reported (Guo & Wei 2009). By taking Guihua Bridge as an example, the influence parameters of steel box girders were analyzed by incremental launching method and the local stability was evaluated.

2 INTRODUCTION TO INCREMENTAL LAUNCHING METHOD

The incremental launching method is a bridge construction method (Li & Zuo 2006). The specific construction process is as follows. First, beam bodies are built using cast-in-place technique or assembled on the bridge end. Second, the beam bodies get through temporary sliding supports on the top of piers by pushing jack action. Finally, the beam bodies are in place. The main characteristics of a continuous beam bridge under dead load determined by the incremental launching method are the same as those of the main girder segment presented above, and internal force of every section in the whole bridge is a cycle with repetitive change (Qin & Chen 2013).

Finite-element method is generally used for internal force calculation during construction under dead load by incremental launching method (Rosignoli 1995). Solving steps and methods are briefly introduced in Figure 1.

1. The basic structure is shown in Figure 1(b). M_1, M_2, and M_i represent superfluous moments at intermediate support.
2. According to the following general formula, three bending moment equation (i) for each bearing is taken as:

$$M_{i-1}\frac{l_i}{I_i}+2M_i\left(\frac{l_i}{I_i}+\frac{l_{i+1}}{I_{i+1}}\right)+M_{i+1}\frac{l_{i+1}}{I_{i+1}}=-6\left(\frac{B_i^{\phi}}{I_i}+\frac{A_{i+1}^{\phi}}{I_{i+1}}\right) \quad (1)$$

where l = span length, I = Cross section of the bending moment, and A^{ϕ} and B^{ϕ} are the reaction forces of the virtual beam.
3. By solving simultaneous equations, M1, M2… are obtained; then, every span's internal force and deformation are calculated according to a single span beam.
4. Fixed end and cantilevered end dispose.

After the structure in Figure 1(a) changes to that shown in Figure 1(b), repeat steps (1)–(3) to obtain the problem solutions.

In order to avoid solving simultaneous equations, for designing of a constant section, such as two spans to five spans continuous beam, a relevant engineering handbook can be consulted and bending moment can be obtained directly from that (Shi & Lin 2012).

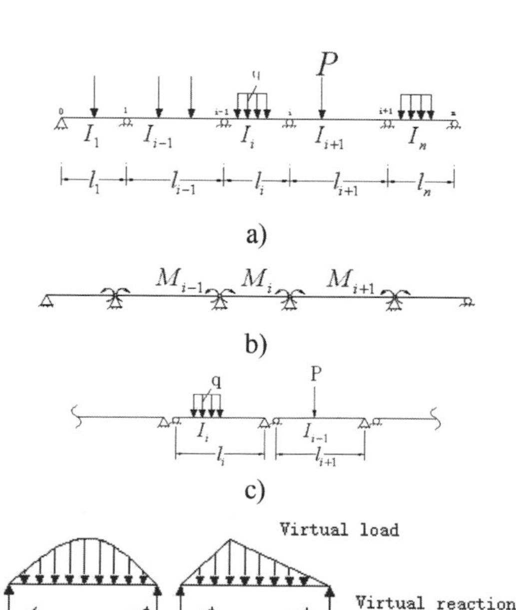

Figure 1. Three bending moment equations solving steps.

3 INTRODUCTION OF STEEL BOX STIFFENING GIRDER WITH INCREMENTAL LAUNCHING METHOD

Guihua bridge is a three-span self-anchored suspension bridge with twin towers in Shaoyang City. The main span arrangement is 60 m+120 m+60 m. Its steel box stiffening girder is built by incremental launching method; after main girders pushing, the main cable can be erected by PPWS method; next, a suspender can be installed, eventually into a whole bridge.

A temporary pier is set across 1#–2# span in Guihua Bridge. During steel box stiffening girders launching construction, a variable stiffness guide beam is used. Its height decreased gradually from root to end. The connection between guide beam and steel box girders uses big box structures, whereas double-H structure is used in other places. Between the two-beam and the double-H structure, Φ 140 × 4 steel tube trusses can be used as supports. The steel guide beam in the connection parts, such as bottom slab and web and top slab, were thickened and strengthened.

4 PARAMETER INFLUENCE ANALYSIS

According to the characteristics of the incremental launching method, the Elastic modulus (E), Poisson's ratio (μ), and Length (l) as well as the average line weight density of the guide beam are chosen as parameters for launching construction influence analysis. Standard parameters of the influence analysis are shown in Table 1. The results are shown in Figures 2–5.

Table 1. Standard parameters of incremental launching influence analysis.

Location	Parameters	Value
Steel box girder	E	2.06×10^{11} Pa
	α	1.2×10^{-5}/°C
	γ	76.98 kN/m^3
	μ	0.3
Guide beam	E	2.06×10^{11} Pa
	α	1.2×10^{-5}/°C
	γ	76.98 kN/m^3
	μ	0.3
	Average line weight density	44.44 kN/m

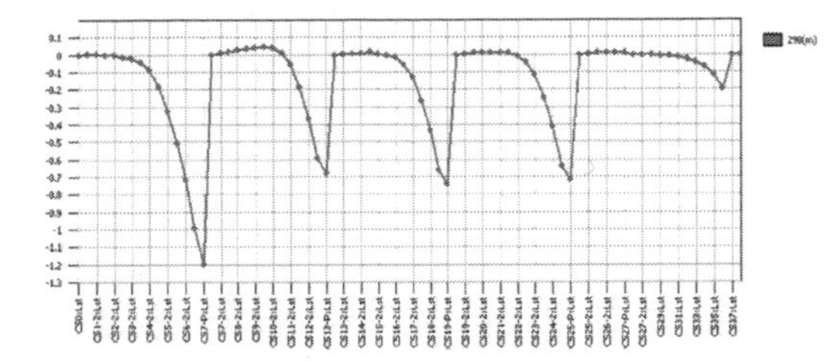

Figure 2. Deflection at the front of guide beam with construction steps.

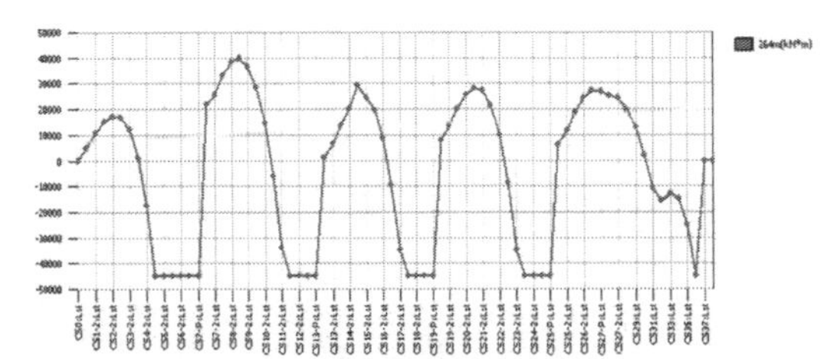

Figure 3. Bending moment at guide beam root with construction steps.

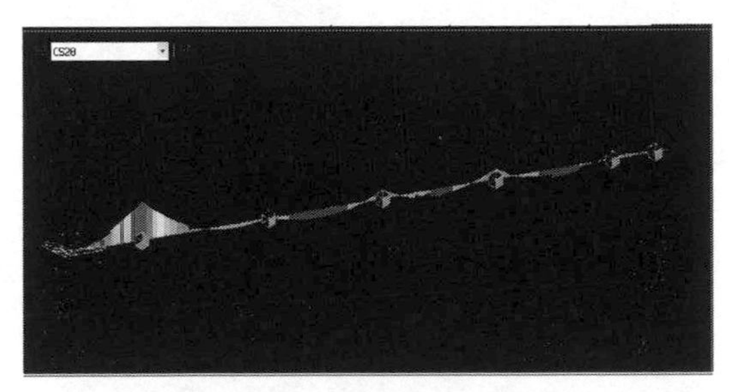

(CS28,Max=351.951MPa)

Figure 4. Maximum tensile stress during incremental launching process at the top of the main girder.

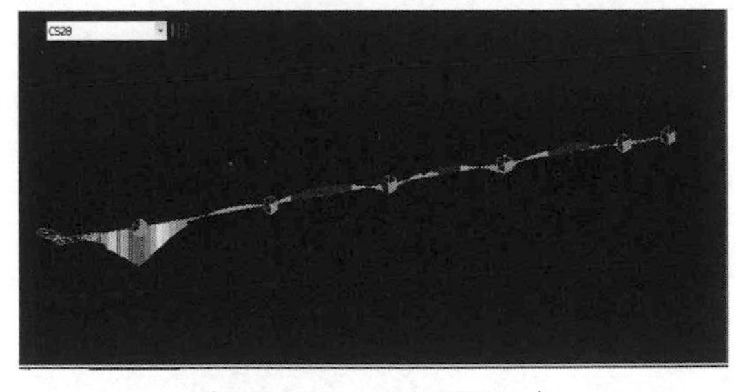

(CS28,Max=351.951MPa)

Figure 5. Maximum compressive stress during incremental launching process at the bottom of the main girder.

As the steel box girder is pushed forward, the maximum deflection at the front of guide beam presents a periodic change (Figure 2) and appears in the first span at the maximum cantilever state, whose maximum value is 1.20113 m.

As the steel box girder is push forward, the bending moment at guide beam root presents a periodic change, and there is a numerical value stable stage in every push cycle (Figure 3). The analysis shows that the theoretical maximum bending moment is $0.5ql^2 = 44995.5$ kN.m at the maximum cantilever state, considering the actual pushing length of the steel box girder; the simulation result of the maximum bending moment is 45.01728 MN. m.

From Figures 4–5, the maximum tensile stress of the top plate and compressive stress of the bottom plate in the steel box girder appear in the CS28 state, and the maximum stress is 351.951 MPa, which is less than 380 MPa (allowable stress) and meets the requirements.

4.1 *Elastic modulus influence analysis*

Several steel structural commonly used elastic moduli of guide beam are taken as examples for steel box girder launching construction elastic modulus influence analysis. The maximum tensile stress and compressive stress and the maximum bending moment at guide beam root as well as displacement at the front of guide beam are chosen for computing objects, and the results are shown in Table 2 and Figures 6–7.

Table 2. Analysis results for guide beam changes with the elastic modulus.

E: $\times 10^{11}$ Pa	(1) MPa	(2) MPa	(3) MN.m	(4) m
1.95	351.951	351.951	44.9959	1.20521
2.0	351.951	351.951	45.0076	1.20327
2.05	351.951	351.951	44.9954	1.20142
2.06	351.951	351.951	45.0172	1.20113
2.1	351.951	351.951	44.9964	1.19966

*(1) = maximum tensile stress of steel box girder, (2) = maximum compressive stress steel box girder, (3) = bending moment at guide beam root, (4) = deflection at the front of guide beam. The same below.

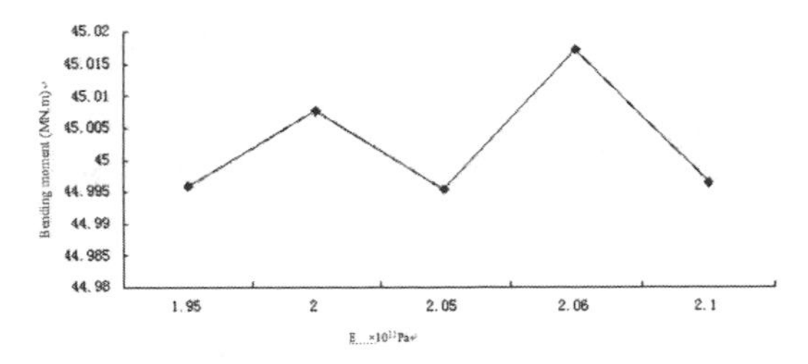

Figure 6. Variation of the maximum bending moment at guide beam root with elastic modulus.

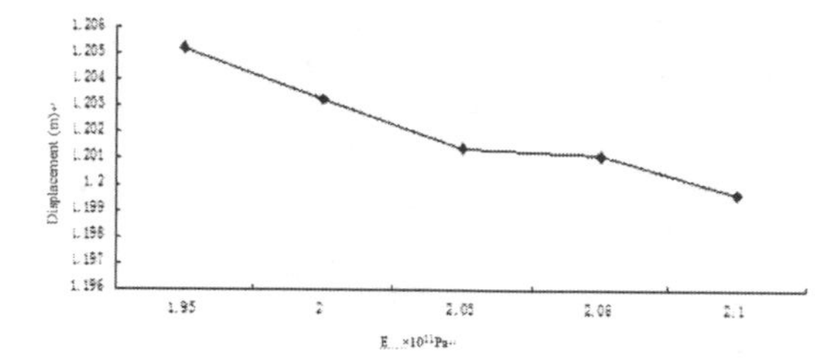

Figure 7. Variation of the maximum deflection at the front of guide beam with elastic modulus.

With the increase of elastic modulus, the maximum bending moment at guide beam root presents periodic changes (Figure 6). From Table 2, its extreme difference is 12.2 kN. m, less than 1% of the average value, which meets the engineering precision requirement. Carefully analysis is required, as computer rounding-off errors may be caused during the rigidity integration.

With the increase of elastic modulus, the maximum deflection at the front of guide beam decreases. Both show an approximate linear relationship (Figure 7 and Table 2). The ratio of elastic modulus variation to displacement variation is 26.96×10^{11}Pa/m. With the change of elasticity modulus, the maximum tensile stress and compressive stress of the steel box girder are constants.

4.2 *Poissn's ratio influence analysis*

Poisson's ratio of 0.2, 0.25, and 0.3 are adopted for the guide beam are taken as examples for steel box girder launching construction Poisson's ratio influence analysis. The maximum tensile stress, the maximum compressive stress, and the maximum bending moment at guide beam root as well as displacement at the front of guide beam are chosen for computing objects, and the results are shown in Table 3.

From Table 3 and Figures 8–9, with the increase of Poisson's ratio, the maximum bending moment at guide beam root increases. Both show a nonlinear relationship. With the increase of Poisson's ratio, the maximum deflection at the front of guide beam increases; both show a linear relationship, and the ratio of Poisson's ratio variation to displacement variation is 714.2857/m.

Through the analysis, it was found that the maximum tensile stress and compressive stress of steel box girders that are constants appear in a double cantilever state of continuous beam.

4.3 *Length of guide beam influence analysis*

The length of the guide beam is 45 m, the longest span is 60 m during launching construction, and their ratio is 0.75. According to "Code for construction and quality acceptance of bridge works in city" (CJJ2-2008), the length of the guide beam should be 0.6–0.8 times of the launching construction span (Chen & Tan 2015). On this basis, guide beam's length and analysis results are shown in Table 4.

From Table 4 and Figures 10–11, with the increase in the length of the guide beam, the maximum bending moment at the guide beam root increases. Both show a linear relationship, and the ratio of length variation to bending moment variation is 0.54775 m/MN.m.

Table 3. Analysis results of guide beam changes with Poisson's ratio.

μ	(1) MPa	(2) MPa	(3) MN.m	(4) m
0.2	351.951	351.951	44.99551	1.20099
0.25	351.951	351.951	44.99552	1.20106
0.3	351.951	351.951	45.01728	1.20113

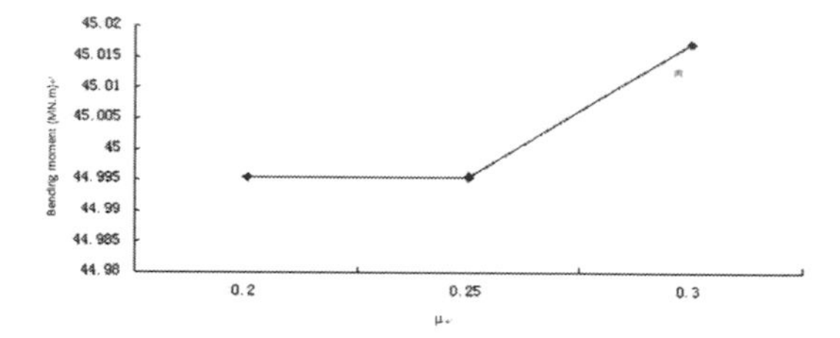

Figure 8. Variation of maximum bending moment at guide beam root with Poisson's ratio.

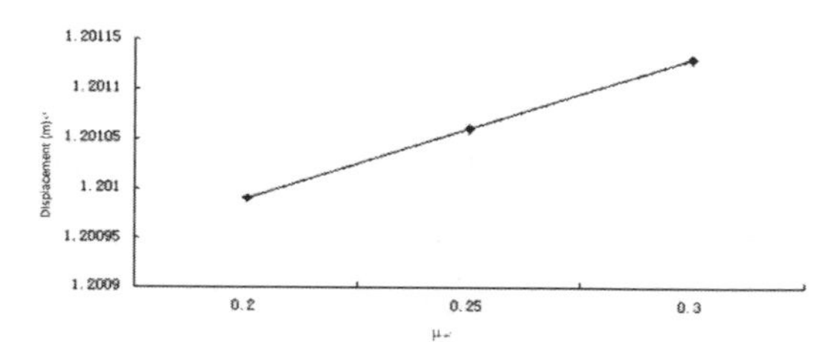

Figure 9. Variation of maximum deflection at the front of guide beam with Poisson's ratio.

Table 4. Analysis results of guide beam changes with length.

L (m)	Ratio	(1) MPa	(2) MPa	(3) MN.m	(4) m
34.255	0.571	351.951	351.951	27.23041	1.19688
37.255	0.621	351.951	351.951	31.24699	1.00936
40.255	0.671	351.951	351.951	35.5520	1.12296
42.5	0.708	351.951	351.951	40.1349	1.07826
45	0.75	351.951	351.951	45.01728	1.20113
48	0.8	351.951	351.951	51.200	1.12275
51	0.85	351.951	351.951	57.80115	1.06975

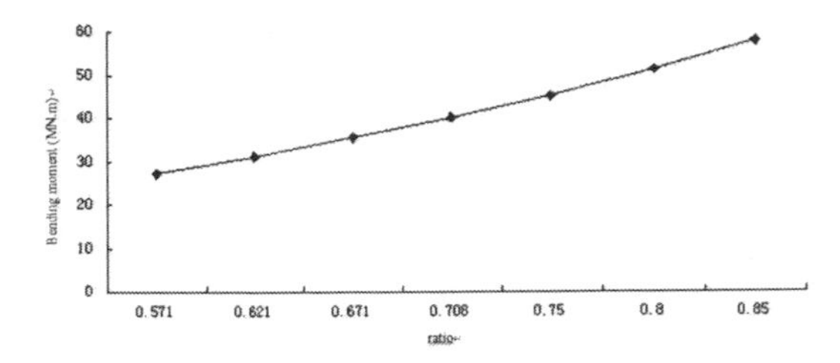

Figure 10. Variation of maximum bending moment at guide beam root with length.

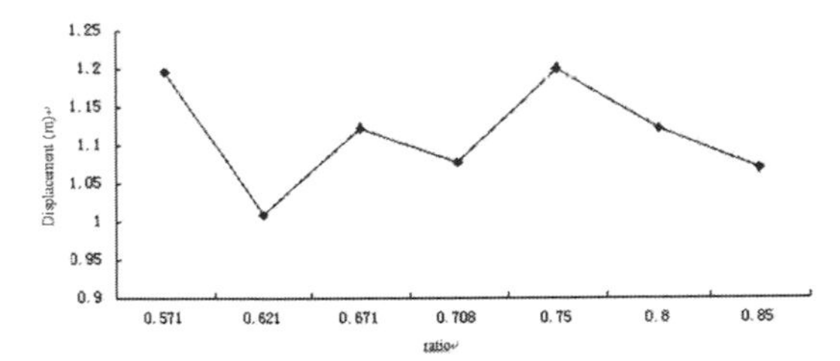

Figure 11. Variation of maximum deflection at the front of guide beam with length.

With the change of length of the guide beam, the maximum deflection at the front of the guide beam presents periodic changes. The reason for this phenomenon is that when the length of guide beam is less than the span length, the maximum deflection appears in the maximum cantilever state. With the increase in the length of the guide beam, the deflection increases gradually. When the length of beam is larger than the span length, the maximum deflection appears in the double cantilever state of continuous beam with two spans; with the increase in the length of the guide beam, the displacement increases gradually.

Analysis results show that the maximum tensile stress and compressive stress of the steel box girder that are constants appear in double cantilever state of continuous beam.

4.4 *Influence analysis of average line weight density of guide beam*

The designed average line weight density of guide beam for the bridge is 44.44 kN/m. Now, at the same length, change the average line weight density of guide beam between 0.6 times and 1.1 times, Analysis results are shown in Table 5.

From Table 5 and Figures 12–13, we know that the maximum bending moment at guide beam root and maximum deflection at the front of guide beam increase gradually with the increase of average line weight density of guide beam. Both present a linear relationship. The ratios of average line weight density variation to bending moment variation and deflection variation are 0.9779 kN.m/MN.m and 53.49284 kN.m/m, respectively.

Analysis shows that with the change of equivalent average line weight density, the maximum tensile and compressive stress of steel box girder, which are constants appear in a double cantilever state of continuous beam.

Table 5. Analysis results of guide beam changes with average line weight density.

Average line weight density	(1) MPa	(2) MPa	(3) MN.m	(4) m
44.44×0.6	351.951	351.951	26.99729	0.87178
44.44×0.7	351.951	351.951	31.49684	0.95412
44.44×0.8	351.951	351.951	35.99637	1.03645
44.44×0.9	351.951	351.951	40.49597	1.11879
44.44×1.0	351.951	351.951	45.01728	1.20113
44.44×1.1	351.951	351.951	49.49508	1.28305

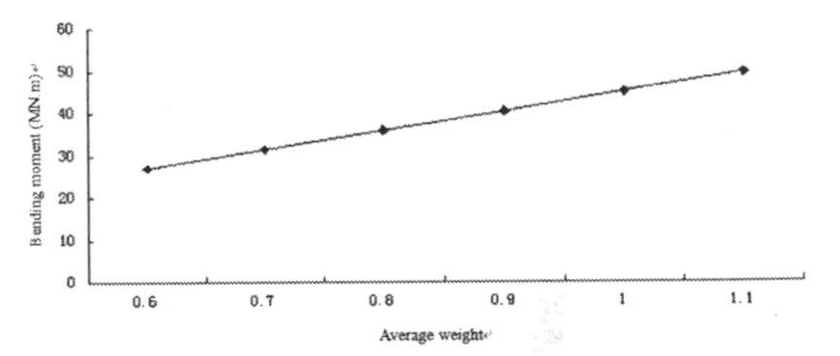

Figure 12. Variation of maximum bending moment at guide beam root with average line weight density.

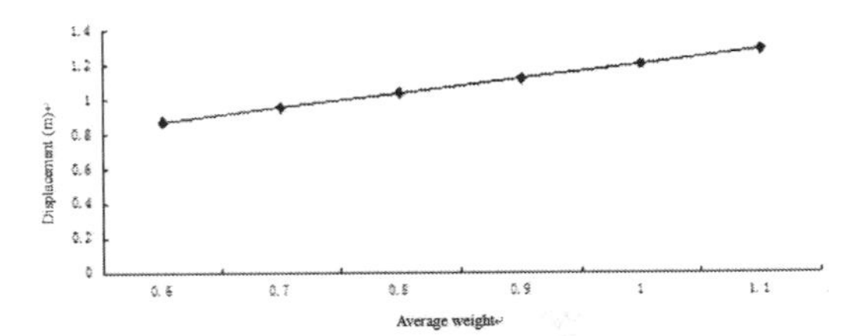

Figure 13. Variation of maximum deflection at the front of guide beam with average line weight density.

5 LOCAL STABILITY CONTROL OF STEEL BOX GIRDER PUSHING

According to the small deflection theory of elastic thin plate, the plate only bears lateral loads and the deflection is small (Yang & Chen 2001). In this case, all point displacements in neutral surface parallel to the surface are not considered. An elastic curved surface of the thin plate is a neutral surface, whose stretching and shear strains are zero.

The elastic thin-plate element is analyzed as shown in Figure 14.

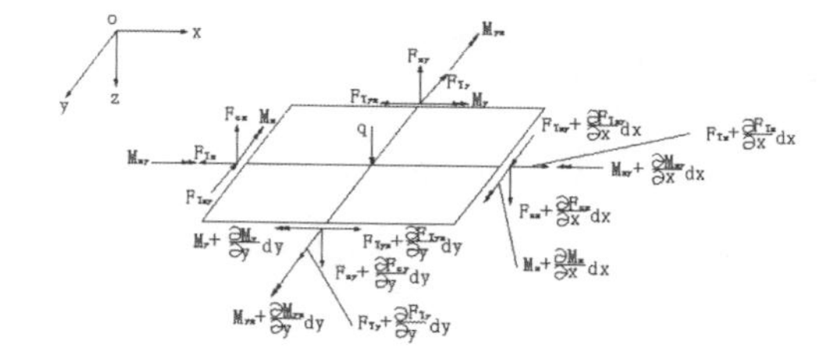

Figure 14. Balanced infinitesimal body of elastic thin plate.

From the analysis, the stability equation of thin plate (formula 2) is found.

$$D\nabla^4\omega-\left(F_{Tx}\frac{\partial^2\omega}{\partial x^2}+2F_{Txy}\frac{\partial^2\omega}{\partial x\partial y}+F_{Ty}\frac{\partial^2\omega}{\partial y^2}\right)=q \quad (2)$$

During incremental launching construction, plates at different positions are chosen for analysis, its lateral load $q=0$, which is comparable to the internal force of structure is small and can be neglected. In this case, according to constraint conditions and load conditions, the corresponding longitudinal load Fx can be solved by the (heavy) trigonometric series.

Simply supported at four edges, a pair of opposite edges bearing uniform distributed pressure Fx (acting on the edge of b length) is given by:

$$F_x=\frac{\pi^2a^2D}{m^2}\left(\frac{m^2}{a^2}+\frac{n^2}{b^2}\right)^2=k\frac{\pi^2D}{b^2} \quad (3)$$

where k is the stability coefficient of plate, given by $k=\left(\frac{mb}{a}+\frac{n^2a}{mb}\right)^2$, n = 1, m for any natural number, $k_{min}=4$.

Simply supported at four edges, one pair of opposite edges bearing uniform distributed pressure Fx (acting on the edge of b length) and another pair of opposite edges bearing uniform distributed pressure $Fy=\alpha Fx$ (acting on the edge of a length) are given by:

$$F_x=\frac{\pi^2D}{a^2}\frac{\left(m^2+\frac{n^2a^2}{b^2}\right)^2}{m^2+\alpha\frac{n^2a^2}{b^2}}=k\frac{\pi^2D}{a^2} \quad (4)$$

where k is the stability coefficient of the plate, given by $k=\frac{\left(m^2+\frac{n^2a^2}{b^2}\right)^2}{m^2+\alpha\frac{n^2a^2}{b^2}}$, n = 1, m = 1, $k=\min\left(\frac{2\alpha-2}{\alpha},\frac{4\alpha-4}{\alpha^2}\right)$.

When Fy denotes tension, α is negative.

Simply supported at three edges, and one edge parallelled to pressure is free, a pair of simply supported edges bearing uniform pressure Fx (acting on the edge of b length) is given by:

$$F_x=k\frac{\pi^2D}{b^2} \quad (5)$$

where k is the stability coefficient of the plate, the solution expression as a four-order coefficient determinant is zero. Because of space constraints, the concrete expressions are not listed. Analysis shows that when a >> b, $k_{min}=0.425$. In other cases, k is associated with Poisson's ratio and a/b.

A pair of opposite edge is fixed, and the other pair of edges is simply supported by bearing uniform distributed pressure Fx (acting on the edge of b length). The expression of critical load Fx (formula 5) is found. Analysis shows that when a/b is approximately equal to 0.7, $k_{min}=6.97$; in other cases, k is associated with a/b.

A pair of opposite edge is fixed, and the other pair of edges id fixed and simply supported by bearing uniform distributed pressure Fx (acting on the edge of b length). The expression of critical load Fx (formula 5) is found. Analysis shows that $k_{min}=5.42$; in other cases, k is associated with a/b (Zhangi 2003).

A pair of opposite edge is fixed, and the other pair of edges is fixed and free from the bearing uniform distributed pressure Fx (acting on the edge of b length). The expression of critical load Fx (formula 5) is found. Analysis shows that $k_{min}=1.28$; in other cases, k is associated with Poisson's ratio and a/b (Zhang & Yuan 2014).

Thus, the following expression is given for critical buckling stress:

$$\sigma=\frac{F_x}{t}=k\frac{\pi^2D}{b^2t}=\frac{k\pi^2E}{12(1-\mu^2)}\left(\frac{t}{b}\right)^2 \quad (6)$$

The steel box girder that is axial-bending component during pushing construction bears compressive and bending stresses. The standard section is 5 m long, with two pieces of diaphragm plate at 1.67 m interval in it. Analysis shows that the maximum compressive stress is 351.951 MPa during pushing construction.

A top plate (5 m × 2.8 m × 22 mm) is chosen for analysis, without considering the local stress change of plate near the pushing jack action. According to Saint Venant principle, using formula (5), Fx = 14.70 MPa << 351.951 MPa. It shows that local buckling under pressure will occur and stiffening measures need to be taken. The reinforcement design shown in Figure 15 is adopted, and a

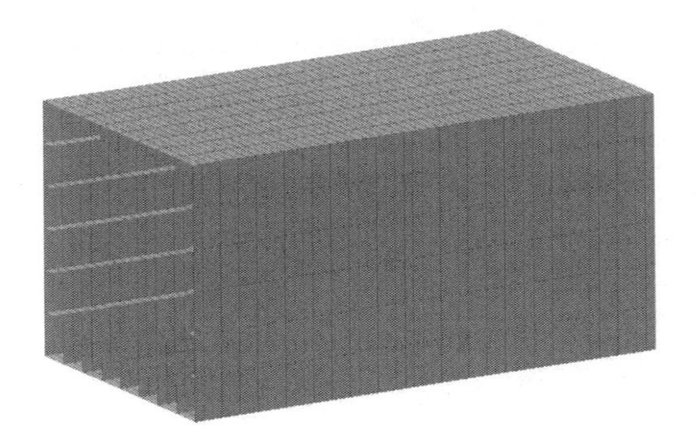

Figure 15. Local stability calculation model for steel box girder.

top plate between two longitudinal stiffeners (5 m × 0.35 m × 22 mm) is taken as an example. Using formula (5), Fx = 940.64 MPa >> 351.951 MPa. It shows that the structure meets the local instability requirements.

In a similar way, a bottom plate (5 m × 2.8 m × 24 mm) is chosen for analysis. Using formula (5), Fx = 17.49 MPa << 351.951 MPa. It shows that local buckling under pressure will occur and stiffening measures need to be taken. The reinforcement design shown in Figure 15 is adopted, and a bottom plate between two longitudinal stiffeners (5 m × 0.35 m × 24 mm) is taken as an example. Using formula (5), Fx = 1119.4 MPa >> 351.951 MPa. It shows that the structure meets the local instability requirements.

A web plate (5 m × 2.5 m × 16 mm) is chosen for analysis. Using formula (5), Fx = 9.75 MPa << 351.951 MPa. It shows that local buckling under pressure will occur, and stiffening measures need to be taken. The reinforcement design shown in Figure 15 is adopted, and a web plate between two longitudinal stiffeners (5 m × 0.415 m × 16 mm) is taken as an example. Using formula (5), Fx = 353.88 MPa > 351.951 MPa. It shows that the structure meets the local instability requirements. However, its safety reserve is too small, so the construction process should be controlled strictly to ensure the safety.

The constraint factors and material nonlinearity and initial geometric imperfections as well as residual stresses are also considered in the local stability analysis of the plate. With consideration of the above factors, the buckling stress is smaller than that of formula (5). Meanwhile, the finite-element analysis software is used for the analysis. The local buckling analysis model of steel box girder in Guihua Bridge and its results are shown in Figure 16 and Table 6.

The analysis shows that, without considering the influence of diaphragm plate, the buckling of plate around the box section occurs under axial pressure, and its buckling mode is shown in Figure 16(a). Considering the influence of the diaphragm, the first part of local buckling is the web, whose buckling mode is shown in Figure 16(b).

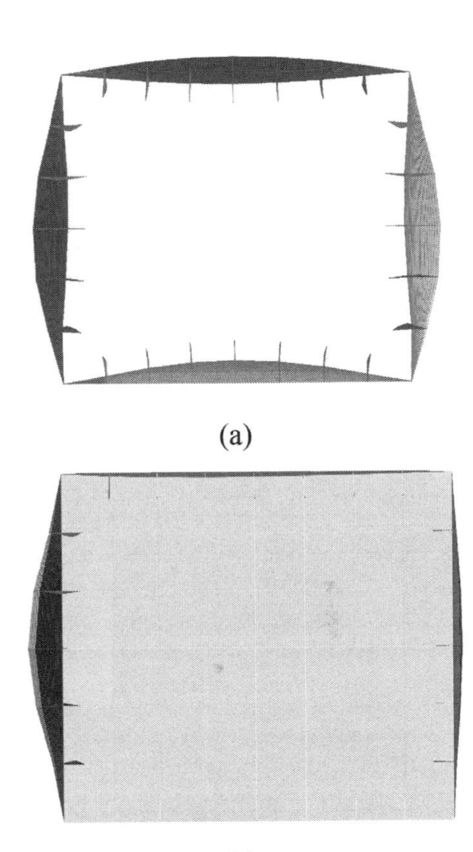

Figure 16. Local buckling instability mode.

Table 6. Local buckling analysis result for steel box girder (kN, MPa).

	Critical axial load	Buckling stress of top plate	Buckling stress of web plate	Buckling stress of bottom plate
(1)	13232.13	69.04	82.70	63.29
(2)	13306.65	69.43	83.17	63.65
(3)	42421.60	221.34	265.15	202.92

*(1) Do not consider the influence of diaphragm plate and structural weight; (2) Consider structural weight but do not consider the influence of diaphragm plate; (3) Consider the influence of diaphragm plate and structural weight.

From Table 6, without considering the influence of the diaphragm plate and considering structural weight, the critical axial load is 13306.65 kN. Without considering the influence of diaphragm plate and structural weight, the critical axial load is 13232.13 kN, which is slightly less than the critical load of considering the structural weight value (0.9944 times). It shows that the influence of bending moment caused by structural weight on critical buckling stress is very small and can be neglected. Considering the influence of diaphragm plate and structural weight, the critical axial load is 42421.60 kN, which is 3.188 times that without considering the influence of diaphragm plate and considering structural weight. It shows that the application of the diaphragm plate to improve the effect of local buckling of steel box girder is obvious. At the same time, it is not difficult to see that the maximum stress of local buckling of steel box girder is smaller than the structural working stress. Again, the instability risk of launching construction may happen.

6 CONCLUSIONS

The maximum deflection at the front of guide beam and maximum bending moment at guide beam root and maximum tensile stress of steel box girder as well as its maximum compressive stress have changed periodically during launching construction.

The change of guide beam elastic modulus of guide beam has little influence on stress and deformation of launching construction. With the increase of elastic modulus, the maximum bending moment at guide beam root presents periodic changes. With the increase of elastic modulus, the maximum deflection at the front of guide beam decreases, and both show an approximate linear relationship. The ratio of elastic modulus variation to displacement variation is 26.96×10^{11} Pa/m. With the change of elasticity modulus, the maximum tensile stress and compressive stress of steel box girder are constants.

The change of Poisson's ratio of guide beam also has little influence on stress and deformation of launching construction. With the increase of Poisson's ratio, the maximum bending moment at guide beam root increases. Both show a nonlinear relationship. With the increase of Poisson's ratio, the maximum deflection at the front of guide beam increases. Both show a linear relationship, and the ratio of Poisson's ratio variation to displacement variation is 714.2857/m. The maximum tensile stress and compressive stress of steel box girders that are constants appear in double cantilever state of continuous beam.

The change of the length of guide beam and its average line weight density have a great influence on stress and deformation of launching construction. With the increase in the length of the guide beam, the maximum bending moment at guide beam root increases. Both show a linear relationship, and the ratio of length variation to bending moment variation is 0.54775 m/MN.m. With the change in the length of guide beam, the maximum deflection at the front of the guide beam presents periodic changes. The maximum tensile stress and compressive stress of the steel box girder, which are constants, appear in the double cantilever state of continuous beam.

With the increase in average line weight density of guide beam, the maximum bending moment at guide beam root and maximum deflection at the front of guide beam increase gradually. Both present a linear relationship. The ratios of average line weight density variation to bending moment variation and deflection variation are 0.9779 kN.m/MN.m and 53.49284 kN.m/m, respectively. The maximum tensile and compressive stresses of steel box girder, which are constants, appear in a double cantilever state of continuous beam.

Theoretical calculation shows that the local stability does not meet the requirements before stiffened in steel box girder. After stiffening, local stability meets the basic requirements, but its safety reserve is too small, so the construction process should be controlled strictly to ensure the safety.

Analysis by using the finite-element model shows that the influence of bending moment caused by structural weight on critical buckling stress is very small, and can be neglected. The application of the diaphragm plate to improve the effect of local buckling of steel box girder is obvious. Analysis shows that the maximum stress of local buckling of the steel box girder is smaller than the structural working stress, thereby increasing the possibility of occurrence of instability risk of launching construction.

REFERENCES

Beaney, N J & Martin, J M (1993). Design and construction of the Dornoch Firth Bridge. *Proceedings of Institution Civil Engineerings, Transportation*, 100(7):145~156.

Chen, Shuangqing & Tan, Guangyao (2015). Supporting pier stress analysis and optimization research with direct incremental launching. *Hunan communication science and technology*, 41(3):92–95.

Ding, Yang & Liu, Xiliang (2002). Design Method of Local Buckling and Post-buckling Strength for Welded I Beams. *Journal of Building Structures*, 23(3):52–59.

Dong, Chuanwen & LI, Chuanxi, Zhang Yuping & Zhang, Jianren (2015). Single-step modulus search and composition method for determining the scheme of support elevation adjustment during girder launching with vertically varied curvatures. *China Civil engineering Journal*, 48(1):101–111.

Guo, Qingfeng & Wei, Min & Qi, Liang (2009). Local stability analysis of steel box girder of Liede bridge in incremental launching process. *Sichuan architecture*, 29(4):202–203.

Li, Chuanxi & Zou, guisheng. 2006. Local stability calculation methods review for axial compression steel box girder. Journal of China & Foreign Highway, 26(3):129–133.

Li, Lifeng & Shao, Xudong (2007). Model Test on Local Stability of Flat Steel Box Girder. *China Journal of Highway and Transport*, 20(3):60–65.

Liu, gang (2011). Local stability comparison of seismic provisions between chinese and american standard. Steel Construction, 26(114):42–45.

Qin, Lin & Chen, Jinfen (2013). Study of incremental launching schemes for closure of Beipanjiang River bridge. *China Civil engineering Journal*, 41(3):43–46.

Rosignoli Macro. 1995. Prestressing schemes for incrementally launched bridges. *Journal of Bridge Engineering*, 5(2):107~115.

Shi, Gang & Lin, Cuocuo (2012). Experimental study on local buckling of high square box section stub columns under Axial Strength Steel Compression. *Industrial Construction*, 42(1):18–25.

Yang, Huxiang & Chen, Xianglin (2003). The maturation of the incremental launching technology of prestress concrete continuous bridge. *Highway*, 3(3):10~13.

Zhang, Xiaodong (2003). The incremental launching technology of bridge. *Highway*, 9(0): 45~51.

Zhang, Yupiny & Yuan, Peng (2014). Analysis on Local Stability and Strength of Steel Box Girder in Construction Process of Jiaxing Shaoxing Bridge. *Journal of Highway and Transportation Research and development*, 31(2):47–53.

Civil, Architecture and Environmental Engineering – Kao & Sung (Eds)
© 2017 Taylor & Francis Group, ISBN 978-1-138-02985-9

Study on active strategies for thermal environment regulation in the zero-energy houses of Solar Decathlon, China

Feng Shi, Wei Jin, Weiwei Zheng & Tao Zhuang
School of Architecture and Civil Engineering, Xiamen University, Xiamen, China

ABSTRACT: Active strategies for thermal environment regulation are the key point to create a comfortable building environment and important in building energy efficiency. In this paper, we take the zero-energy buildings in Solar Decathlon competition as examples to analyze the active strategies in solar houses and sort them out according to their technical characteristics. Energy-consuming active strategies such as the HVAC systems, which are used to control the building thermal environment and air quality, are analyzed in the paper. And their application in SDC2013 has been discussed to provide references for contemporary green building design.

1 INTRODUCTION TO SOLAR DECATHLON, CHINA

The international solar decathlon competition (Solar Decathlon) is a solar building technology competition for global universities, which is called "The Olympic Games in the field of solar building", and is launched and hosted by the U.S. Department of Energy. It challenges collegiate teams to design, build, and operate solar-powered houses that are comfortable, livable, and sustainable living spaces. Solar energy is the only energy source of each house with an area of about 80 m^2. These solar houses must meet the requirements of the various functions of residence and be operated actually for test contests. Five sessions of the competition are hosted in the United States (SDA2002 SDA2005, SDA2007, SDA2009, and SDA2011), two sessions in Spain (SDE2010, SDE2012), and one session in China for the first time (SDC2013). SDC2013's participants include 20 teams from 35 universities with students from over 35 nationalities in 13 countries on six continents, which reflect that China attaches a great importance to the application and promotion of new energy and performs the action of building energy conservation and emissions reduction, and also greatly expands the influence of the competition.

Zero-energy solar houses exhibited in SD reflect different universities' progress in the field of building energy conservation and solar energy application (Peng et al. 2015). In the contests, indoor air temperature and air humidity were measured during the schedule time in the competition, and HVAC system design, ventilation design, and air quality were also evaluated by experts. Therefore, active strategy for thermal environment regulation was one of the key points for each team in their solar house design.

2 ACTIVE DESIGN STRATEGY

Every competition has strict requirements for the building indoor temperature and humidity, according to the scoring criteria of the competition. For example, indoor temperature should be controlled between 22 and 25°C and humidity within 60% in SDC2013. If outdoor climate is comparatively awful, the use of active environmental control equipment to control the indoor environment is extremely important. How to combine HVAC system with building design to achieve the purpose of energy saving and take full advantage of the solar energy attract significant attention of each competition team.

3 AIR-CONDITIONING SYSTEM

The competition site for all sessions of Solar Decathlon are chosen in places that have an abundance of solar energy, and the competition time is always in summer or early autumn; therefore, during the competition, an air-conditioning system is generally needed. There are various types of air-conditioning systems in different teams' house, such as common split-type air-conditioning system, central air-conditioning system, and other special types of air-conditioning systems (Real et al. 2014).

3.1 *All air-conditioning system*

All air-conditioning systems use outdoor air as the cold and heat source. The outdoor air will be cooled and dehumidified by a centralized processing unit in the system. Because of the small building size, the pipelines of all air-conditioning system will not be complicated, which could deal with sensible heat and latent heat load in the meantime, and the effect of temperature and humidity control is so good that there is no need to set a fresh air system. For these reasons, all air-conditioning systems have been widely used. This system's processing unit has a large volume, which needs a large equipment room in the house. Its air duct dimensions are larger than those of the cryogen-type air-conditioning system, which means it needs higher ceiling.

For example, Wellington Victoria University took part in SDA2011. Their house "First Light" adopted the all air-conditioning system. The schematic diagram of their processing unit and supply and return pipes are shown in Fig. 1. In order to handle the fresh air, Mitsubishi's heat recovery new fan is used in this system, which can recycle the heat of exhaust air to save energy.

3.2 *VRF air-conditioning system*

VRF air-conditioning system is a frequency conversion air-conditioning system with high energy-saving efficiency by using one outdoor unit to connect several indoor units. Air cooling heat transfer method is used in outdoor units, which use refrigerant as transmission medium adapting to the changes of air-conditioning load by changing the refrigerant flow rate. Heat recovery VRF air-conditioning system can do better; it controls not only the refrigerating capacity of different indoor units but also two parts of indoor units for cooling and heating, fitting the different needs of different space thermal environment regulation.

Xiamen University's house "Sunny Inside" in SDC2013 uses VRF air-conditioning system, connecting one outdoor unit to three indoor units to satisfy the cooling and heating demand of the living room, dining room, and bedroom. Because of the different service conditions of these three rooms, VRF system can meet the different needs of each room. Meanwhile, system efficiency can be higher when part of indoor unit is working after the system works at full capacity, which saves energy.

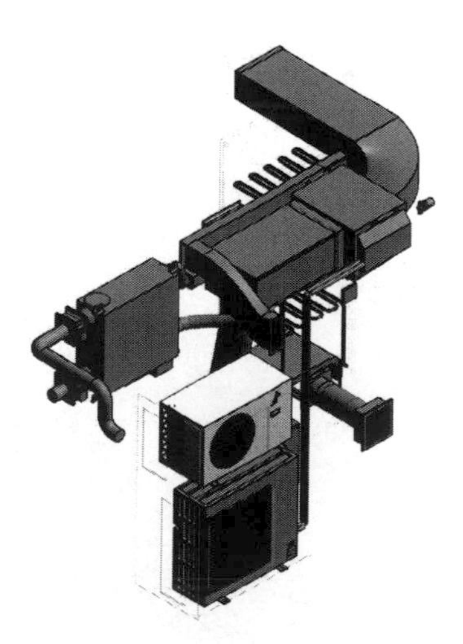

Figure 1. Schematic diagram of air-conditioning system in "First Light".

3.3 *Solar energy air-conditioning system*

Solar energy air-conditioning system is a new type of air conditioning system that uses solar thermal collector to get hot water to drive absorption or absorbing-type refrigeration unit to get the required refrigerant water. This system is mostly used in large and medium-sized central air-conditioning systems, in which the higher the water temperature in the collector, the higher the cooling efficiency.

However, Tel Aviv University's house "Living Patio" in SDC2013 uses a special small solar energy air-conditioning system with a TIGI honeycomb solar thermal collector made of transparent insulation material, which and can collect high-temperature hot water (110°C) to heat the refrigerant to be gas and drive the compressor to refrigerate (Fig. 2). This system greatly reduces the electricity consumption of the compressor by saving a lot of energy.

A night radiation cooling system is also used in "Living Patio". Water radiant coils are set on the north side of the building using the sky long wave radiation cooling effect, which can open the circulating water to make and collect cold water in a cold water tank at night. When there is little cloud in Datong, cold water with temperature 2–3°C lower than air temperature will be guided into phase change materials storage tank to exchange heat with phase change materials, whose phase transition temperature is around 18°C as auxiliary cold source of the air-conditioning system (Fig. 3).

3.4 *Temperature- and humidity-independent control air-conditioning system*

Temperature- and humidity-independent control air-conditioning system with a sensible heat disposal unit and a latent heat disposal unit can control indoor temperature and humidity, respectively, to achieve the goal of energy saving. Because of

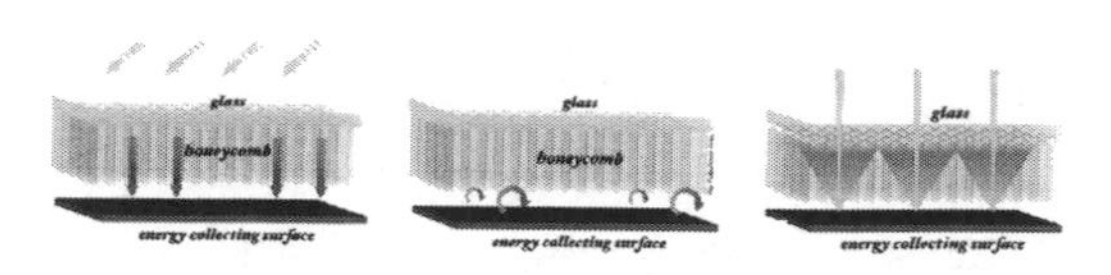

Figure 2. TIGI honeycomb solar thermal collector in "Living Patio".

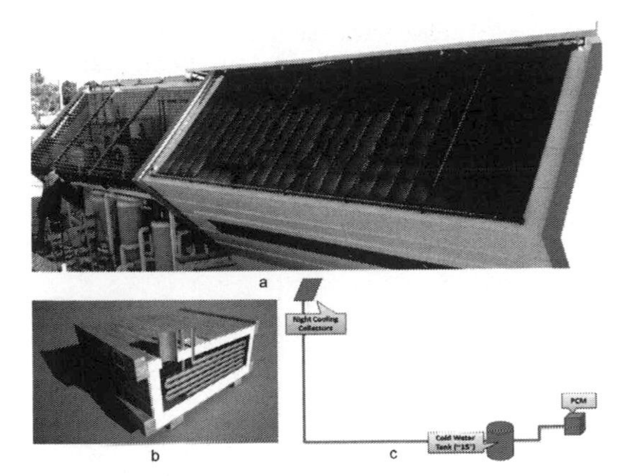

Figure 3. Night radiation cooling system in "Living Patio".
a: water radiant coils on the north side of the building
b: Phase change energy storage tank in the system
c: system diagram.

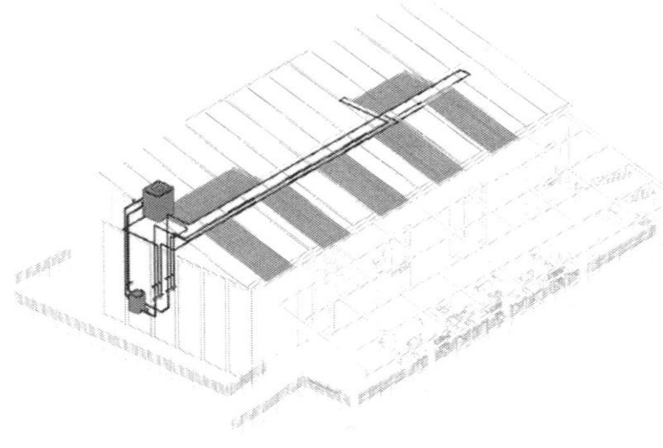

Figure 4. Capillary net radiation air-conditioning system in "Qi Ju".

the sensible heat disposal unit control-independent indoor temperature, there is no need to use refrigerant at low temperature for condensation dehumidification. On the contrary, refrigerant at higher temperature is preferred. It can improve the performance coefficient of refrigerant and save energy at the same time. Higher-temperature refrigerant also solves the problem of moisture condensation preventing the growth of bacteria on wet surfaces, which can improve indoor air quality effectively. Latent heat disposal unit, in other words dehumidification system, is mostly set with fresh air system that supplies dry fresh air, which undertakes the task of dehumidification and ensuring air quality. Dehumidification system can adopt these methods, such as cooling dehumidification, solution type dehumidification, and wheel-type dehumidification.

Xi'an University of Architecture and Technology's house "Qi Ju" use capillary net radiation at the end of system to control indoor temperature. The temperature of circulating water in the capillary, which is at the end of air conditioning system, is 16°C, and when the system is on, the temperature of architecture radiation surface can reach 17–19°C, with refrigerating efficiency of 85 W/m^2. Meanwhile, this system is quiet and has a strong ability of heat storing as well as self-adjusting to balance (Fig. 4).

4 HEATING SYSTEMS

The climate characteristics at the competition site during the competition and one year's climate characteristics at the permanent place for the house must be considered when designing houses for Solar Decathlon. The heating system that regulates thermal environment should also be taken into consideration. Cold and warm air-conditioning system that can provide refrigeration and heating at the same time could be used in places that do not need heating in winter. On the contrary, in places that need the heating, the independent heating system is necessary for the need of heating.

On account of the fact that hot water item occupies one of the 10 scoring items, which highlights the importance of hot water, teams often combine hot water system with the need of heating in design to ensure the supply of hot water. Southeast University's house "Solark" in SDC2013 uses floor radiation heating system with a water tank to store hot water, which is connected to the hot water supply system. This way of heating is good for the indoor temperature uniform distribution and has a strong heat storage capacity to improve the comfort of heating.

5 PHASE CHANGE MATERIALS VENTILATION SYSTEM

Phase change materials ventilation system is a special ventilation system that makes use of the large temperature difference between day and night during the competition time, storing cooling capacity of cold air in the night by phase change materials and releasing it for refrigeration in the daytime (Lin et al. 2014). The phase transformation point of phase change materials in the system is very important, which should be chosen according to the local average temperature in day and night.

For example, there are four phase change materials energy storage ducts under the courtyard of "Sunny Inside" of Xiamen University. Phase change materials whose phase transformation point is 23°C and phase change latent heat value is

187 kJ/kg packaged with aluminum foil bag are put into the ducts with a fan in each duct. There is also a duct on the ceiling of living room with phase change materials packaged by stainless steel tube using the material whose phase transformation point is 19°C for one half and 23°C for the other half. Air inlet, outlet, and electric air valve are set on each duct to control ventilation module of the duct. The temperature in Datong at night during the competition is around 16°C and 30°C at noon, which means the temperature difference between day and night is large. At night, open the outdoor air vents using cold air outside to cool the phase change materials, shut down the outdoor air vents, open indoor air vents when needed in the daytime to cool indoor air, which can partly replace the air conditioning.

6 INTELLIGENT CONTROL SYSTEM

Intelligent control system is widely used in Solar Decathlon, in which sensors are set to monitor the climate parameters in and out of the house and target controlling all types of devices in the building according to the feedback parameters. Through the intelligent control method, all types of devices in the building are connected together for automatic control. And the system can regulate the operation mode of various devices according to the needs of energy conservation. Different technical strategies can be reasonably integrated in this way, promoting each other to achieve the effect of 1 + 1 > 2.

Intelligent control system is also used in Tsinghua University and Florida International University's house "O-House" (Zhang et al. 2014), which can control air-conditioning system, the angle of the photovoltaic panel bracket, and so on. It is strict for the control requirement of temperature and humidity in the competition that relying on the automatic control of air conditioning is a risk that the indoor temperature may easily go out of control between 22 and 25°C. In consequence, an air-conditioning control system that can automatically adjust the operation of air conditioner according to the indoor temperature and humidity is set in "O-House" in order to keep the indoor temperature within the prescribed scope (Rodriguez-Ubinas et al. 2014). Finally, their team won the first place in the test item of temperature and humidity with this system in SDC2013 (Fig. 5).

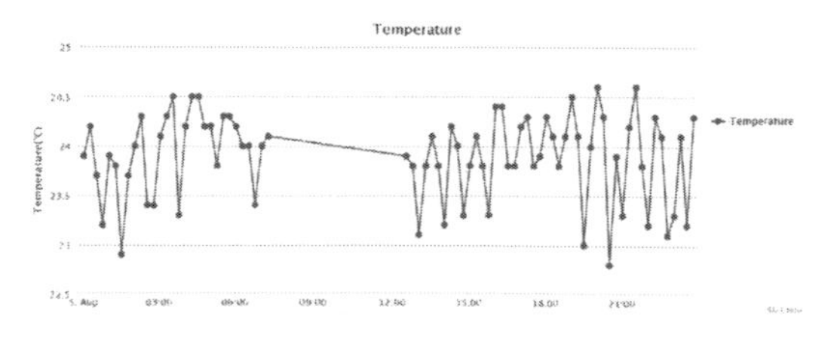

Figure 5. Result of temperature and humidity test of "O-House" on August 5, 2013.

7 CONCLUSION

For green building, passive strategies could be used initially in the regulation of building thermal environment and make the most use of natural conditions to improve the local thermal environment. However, when the passive measures cannot be used to meet the requirements of indoor thermal environment, the active strategies can be combined with passive measures to reduce energy consumption and improve the performance of some equipment. Among this, the progress of technology intelligent control system can be used to monitor climate parameters and optimize the working performance of all the equipment, which integrate all types of technical strategies reasonably according to the different applicability to achieve the goal of energy-saving firmly.

ACKNOWLEDGMENTS

This study was supported by the National Natural Science Foundation of China (Grant No. 51308481) and the Fundamental Research Funds for Central Universities (Grant No. 20720150102).

REFERENCES

Lin, W., Ma, Z., Sohel, M.I., & Cooper, P. (2014). Development and evaluation of a ceiling ventilation system enhanced by solar photovoltaic thermal collectors and phase change materials. Energy Conversion & Management, 88, pp. 218–230.

Peng, C., Huang, L., & Huang, J. (2015). Design and practical application of an innovative net-zero energy houses with integrated photovoltaics: a case study from solar decathlon china 2013. Architectural Science Review, 58(2), pp. 144–161.

Real, A., García, V., Domenech, L., Renau, J., Montés, N., & Sánchez, F. (2014). Improvement of a heat pump based hvac system with pcm thermal storage for cold accumulation and heat dissipation. Energy & Buildings, 83, pp. 108–116.

Rodriguez-Ubinas, E., Ruiz-Valero, L., Vega, Sergio. and Neila, J. (2012) Applications of Phase Change Material in Highly Energy-Efficient Houses. Energy and Buildings, 50 (7), pp.49–62.

Zhang, H., Li, J., Dong, L., & Chen, H. (2014). Integration of sustainability in net-zero house: experiences in solar decathlon china. Energy Procedia, 57, 1931–1940.

Civil, Architecture and Environmental Engineering – Kao & Sung (Eds)
© 2017 Taylor & Francis Group, ISBN 978-1-138-02985-9

Research and enlightening of ecological infrastructure-oriented "multiple planning integration" based on Germany's spatial order and structure planning—illustrated by the example of Dujiangyan

Yanjin Li & Fei Fu
Southwest Jiaotong University, Chengdu, Sichuan, China

Yang Yi
Sichuan Branch of Chongqing Planning and Designing Institute, Chengdu, Sichuan, China

ABSTRACT: Spatial order and structure planning based on overall planning of "multiple planning integration" plays a guidance role for other spatial plannings. In this paper, we analyze Germany's spatial planning system and explore highest-level spatial order planning and spatial structure planning on the basis of spatial order planning. Dujiangyan is one of the second-batch pilot cities of "multiple planning integration" in Sichuan Province. On the basis of absorbing the experience of Germany's spatial order and structure planning, Dujiangyan innovatively constructs spatial base map of ecological infrastructure-oriented "multiple planning integration" and specifies ecological infrastructure-oriented spatial order and structure planning. The above innovation practice aims to strengthen the overall planning and coordination of the citywide space and to provide reference experience to other practice of other cities' "multiple planning integration".

1 INTRODUCTION

Integrated Reform Plan for Promoting Ecological Progress published by the Communist Party of China (CPC) Central Committee and the State Council in September, 2015, asks for solving the outstanding problems of ecological environment, safeguarding national ecological security, improving environmental quality, improving the efficiency of resource utilization, and promoting the formation of the new pattern of modernization of harmonious development between human and nature. Therefore, this paper proposes the consideration of ecology-oriented "multiple planning integration". In the field of spatial planning, Germany's spatial planning system restricts and guides the inferior spatial planning through superior spatial order and spatial structure planning, which has overall consideration. Duiiangyan is one of the second-batch pilot cities of "multiple planning integration" in Sichuan Province. Its pilot work analyzes spatial order and spatial structure planning of Germany's spatial planning and absorbs successful experience, combining with its own rich ecological infrastructure to be used for reform practice of ecological infrastructure-oriented "multiple planning integration".

2 GERMANY'S SPATIAL PLANNING SYSTEM

Germany's spatial planning system (Turowski, Gerd, 2012) has the explicit level relationship and appropriate legal basis. Each level of Germany's spatial planning system coordinates with each other to achieve spatial sustainable development. Germany's spatial planning system is based on Spatial Planning Act. Germany's spatial planning system emphasizes coordination among all levels. On the basis of spatial order and structure planning, Germany's spatial planning system coordinates different spatial influences and probable conflicts among different levels of spatial planning system and proposes solutions and preventive measures for protecting independent spatial function and land use to coordinate social sustainable development, contradictions in economic construction, and the relationship between land use and ecological protection of land to achieve sustainable development.

Comprehensive spatial planning at Federal Level does not focus on coordination but makes spatial planning principles to implement the vision and to provide actions of saving space for possible foreseeable trends. Comprehensive Spatial Planning at State level includes comprehensive spatial planning at land level and regional planning. Comprehensive spatial planning at land level launches spatial structure planning to provide sustainable development goals for respective planned territory. Regional planning establishes relationships for spatial structure, plans functional relationships between regional and urban land use, and constructs legal framework for planning bureaus and projects such as residence, industry, and infrastructure. Comprehensive spatial planning at municipality level is dominated by urban land use planning. Urban land use planning, the core tool of urban development, specifies the purpose of land in the process of urban development (like commercial purposes and public purposes), determines floors of highest buildings, maximum proportion of limits of construction land and relevant additional conditions and measures during the construction to realize urban sustainable development, to ensure suitable living environment for human beings, to provide natural foundation, and to maintain a fair land use. Sectoral planning is different from other plannings. It aims at special planning like natural conservation, forest, land conservation, water resource management, agriculture, transportation, and communication to establish the system procedures of development measures. Collaborative planning at public institutions level does not participate in the actual spatial planning, but it requires corresponding public institutions to provide suggestion and information for the positive development of urban space in the coordination with comprehensive spatial planning at Federal State, municipalities level, and sectoral planning.

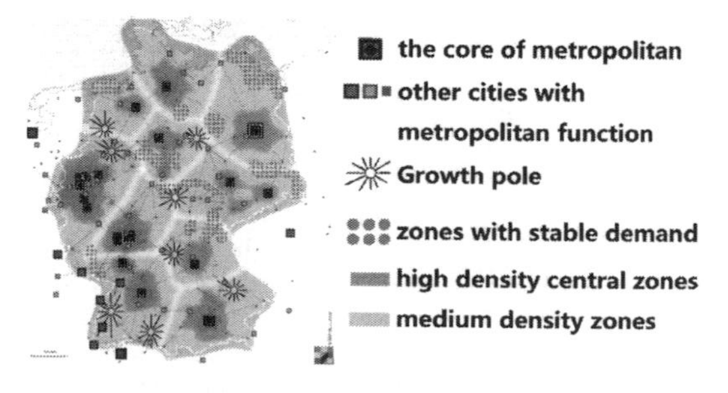

Figure 1. Germany's spatial order planning.

3 SPATIAL ORDER AND STRUCTURE PLANNING OF GERMANY'S SPATIAL PLANNING

3.1 *The-highest-level spatial order planning*

Spatial order planning is launched by comprehensive spatial planning at Federal Level. First, space is divided into three parts: metropolitan zones, radiation zones of countrysides, and transition zones among metropolitan areas (flexible development space zones). Second, spatial order includes two parts: high-density central zones and medium-density zones. Third, zones with stable demand are determined in different density zones. Thus, comprehensive spatial planning at Federal Level determines the principle of national spatial development (Min Xie, 2011).

3.2 *Spatial structure planning based on superior spatial order planning*

Comprehensive spatial planning at land level (Turowski, Gerd, 2012) follows policy requirements and requirements of spatial order planning. Comprehensive spatial planning at land level launches spatial structure planning on the basis of spatial order planning and land use status, economic development, population size, and so on. Spatial structure planning determined central-place systems and axes of spatial development axis.

Table 1. Central-place systems of Germany's spatial structure planning (Turowski, Gerd, 2012).

Characteristic	Classification	Role
To ensure that all parts of the national territory are supplied with both public and private sector services they require, and that employment is available in all parts of the country	Lower-order center (open zones)	To supply the basic everyday needs solely of the local population
	Intennediate-order center (relative high-density zones)	To supply the less everyday needs of the population of a larger area
	Higher-order center (high-density space)	To supply the "higher" or more specialized needs of the population of a wider ("higher-order") region

4 SPATIAL ORDER AND STRUCTURE PLANNING OF ECOLOGICAL INFRASTRUCTURE-ORIENTED "MULTIPLE PLANNING INTEGRATION" — ILLUSTRATED BY THE EXAMPLE OF DUJIANGYAN

4.1 *Spatial base map of ecological infrastructure-oriented "multiple planning integration"*

4.1.1 *Background*

Germany's spatial planning based on inter-regional planning achieves coordination of each region to safeguard sharing and using of spatial function and resourcc with location advantage and special geographical environment. Dujiangyan is one of the second-batch pilot cities of "multiple planning integration". Its work based on overall city planning innovatively pays attention to space division of ecological infrastructure, which is changed from resource factors. Paying attention to evaluation of the present situation transforms into evaluation of intensive optimization potential of ecological infrastructure evaluation. And spatial development pattern of advantage resources transforms into intensive optimization development pattern of ecological infrastructure of regional space.

4.1.2 *Concept*

Dujiangyan's spatial base map based on the direction of the main body function, environmental carrying capacity of resources, urban development potential, cognition of the base map of ecological infrastructure resources, and spatial development vision pattern has two characteristics: static presentation of ecological infrastructure and adapting to the spatial development pattern. Spatial base map is the objective reflection of the spatial elements and generally forms three types of space: urban space, village ecological infrastructure space, and ecological protcction space, which can be expressed by various types of "polygon-chain". Village ecological infrastructure (Liang Mao, 2012) is the material basis for residents' existence and development and for the normal operation of urban residents. It provides natural service for village residents and sustainable development of urban; village ecological infrastructure space includes ecological productive land in village, such as natural or artificial forest land, farmland, grassland, river wetland, coastal beach, green space, and water body in the village built area and living space for residents. Spatial development pattern changes with the appreciation of space resource to form a different spatial pattern.

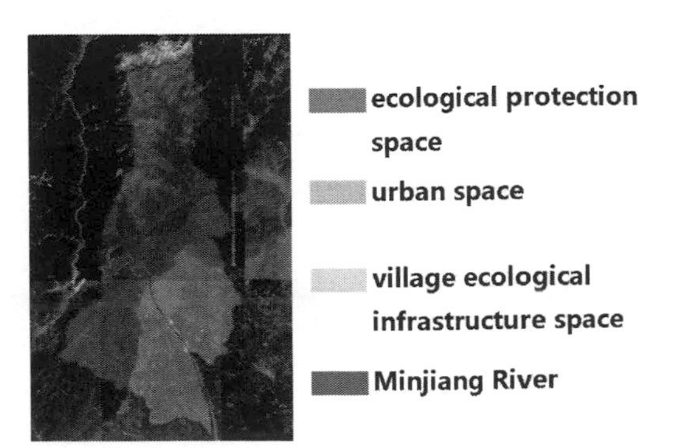

Figure 2. Space division of Dujiangyan's "multiple planning integration".

4.2 *Improving spatial order and structure planning based on spatial base map of ecological infrastructure-oriented "multiple planning integration"*

Dujiangyan's "multiple planning integration" by analogy with Germany's spatial order planning determined the spatial structure on the basis of three types of space: high-density central city zone, medium-density village ecological infrastructure space zones, and low-density ecological protection zone. Satellite towns, key towns, general towns, and regions with a stable demand are further determined in different density zones.

Germany's spatial structure planning based on spatial order planning determines the principles and objectives of state spatial coordinated development and structural relationship between dense space and open space. On the basis of overall planning, Dujiangyan preliminarily forms space pattern of "one city and four town" to achieve coordination of whole-region space and to strengthen the relationship between urban space and ecological infrastructure. On the basis of space pattern of

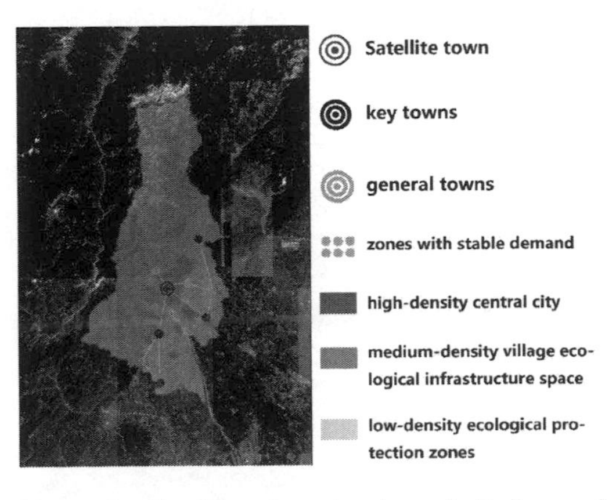

Figure 3. Spatial order planning of Dujiangyan's "multiple planning integration".

Table 2. Central-place systems of spatial structure planning of Dujiangyan's "multiple planning integration".

Classification	Corresponding region
Lower-order center (open zones)	General towns of village ecological infrastructure space
Intennediate-order center (relative high-density zones)	Key towns of village ecological infrastructure space
Higher-order center (high-density space)	Central city

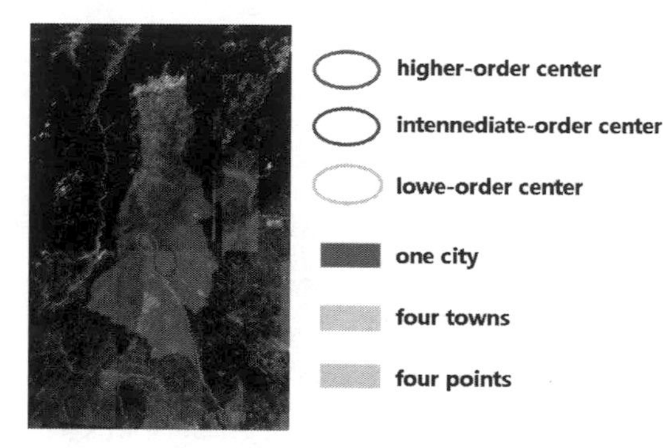

Figure 4. Central-place systems and "one city, four towns and four points" structures of Dujiangyan's "multiple planning integration".

overall planning and by analogy with central-place systems of Germany's comprehensive spatial planning at land level, central city (high-density space) is treated as higher-order center. Key towns (relative high-density space) are treated as higher-order centers. General towns (open space) are treated as intennediate-order centers. Finally, spatial structure of "multiple planning integration" of "one city, four towns and four points" is formed.

On the basis of spatial structure planning, Germany's spatial planning links up the densely populated zones and the more peripheral zones through existing transport and public infrastructure to form the communication axes to vest with the location advantage and for structure development. Besides, Germany's spatial planning linearly connects proper living center to form the settlement spatial development axes to safeguard ordered settlement development and to protect open space (Turowski, Gerd, 2012). Dujiangyan's "multiple planning integration" based on spatial structure of "multiple planning integration" of "one city, four towns and four points", relying on the existing transportation and ecological infrastructure, determines the spatial development axes to strengthen the coordination of whole-region space: riverside development axes (settlement spatial development)

Figure 5. Riverside development axes of Dujiangyan's "multiple planning integration".

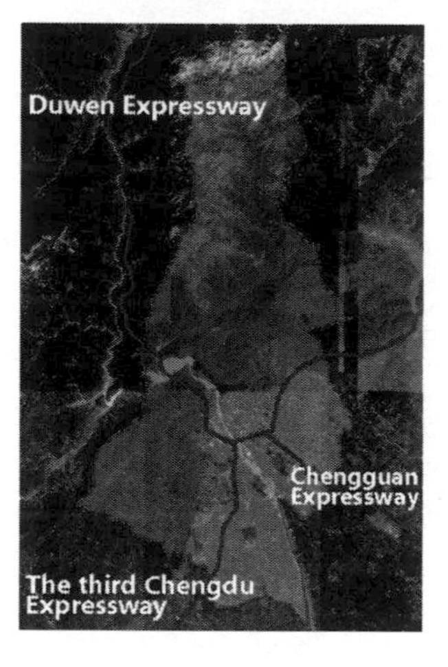

Figure 6. Three comprehensive development axes of Dujiangyan's "multiple planning integration".

and three comprehensive development axes (communication axes). Three comprehensive development axes, relying on Chengguan Expressway, Duwen Expressway, the third Chengdu Expressway, link central city, "four towns and four points", and the more peripheral zones of the urban space, village infrastructure space, ecological protection space to highlight the advantages of the central city and to strengthen the relationship between the spatial structures. Three comprehensive development axes divide ecological protection space into two zones (north and south). Riverside development axes maintain the orderly development of living space. Riverside development axes link urban space and village infrastructure space and divide urban space and village infrastructure space into four regions. Relying on spatial development axes and rich tourism and culture resource along the mountains, tourism service industry and leisure creative agriculture can be developed to inspire development for mountainous areas and to open ecological protection space. On the basis of spatial structure of "multiple planning integration" of "one city, four towns and four points", the spatial structure development model of "one core, two axes, four regions and two zones" is formed.

5 CONCLUSION

Germany's spatial planning uses spatial order and structure planning as principles, which has confinement effect and shows their scientific rationality. The practice of Dujianyan's "multiple planning integration" put forward a spatial base map of ecological infrastructure-oriented "multiple planning integration" and ecological infrastructure-oriented spatial order and structure planning. The above innovation practice aims to strengthen the overall planning and coordination of the citywide space and provides reference experience to other practice of other cities' "multiple planning integration".

ACKNOWLEDGEMENT

The content of this paper is based on a project funded by the technology research and development project of science and technology benefit of Chengdu Municipal Science and Technology Bureau—《Research on agricultural village ecological infrastructure based on GIS platform》(2015-HM01-00356-SF).

REFERENCES

Christoph Riegel. Infrastructure resilience through regional spatial planning prospects of a new legal principle in Germany[J]. Int. J. Critical Infrastructures, 2014, 10(1):17–21.

Integrated Reform Plan for Promoting Ecological Progress[Z]. The Communist Party of China (CPC) Central Committee and the State Council, 2005.

Liang Mao. Study on rural ecological infrastructure in western Liao Ning province from the perspective of ecological productive land[D]. Harbin: Harbin Institute of Technology, 2012.

Min Xie, Lijun Zhang. Explanation of the ideas of German Spatial Planning[J]. Land and Resources Information, 2011, (7):9–12.

Turowski, Gerd. Spatial planning in Germany: Structures and concepts, Akademie für Raumforschung und Landesplanung[M]. Hannover: Akademie für Raumforschung und Landesplanung (ARL), 2002.

Civil, Architecture and Environmental Engineering – Kao & Sung (Eds)
© 2017 Taylor & Francis Group, ISBN 978-1-138-02985-9

Homogeneous generalized yield function and limit analysis of structures with rectangular sections

Wei Ou, LuFeng Yang & Wei Zhang
Key Laboratory of Engineering Disaster Prevention and Structural Safety of China Ministry of Education, School of Civil Engineering and Architecture, Guangxi University, Nanning, China

ABSTRACT: A problem for the Elastic Modulus Adjustment Procedure (EMAP) is that the limit load solution varies with the initial load when the Generalized Yield Function (GYF) is employed. To overcome this problem, we use a Homogeneous Generalized Yield Function (HGYF)-based Elastic Modulus Reduction Method (EMRM) for structures with rectangular section. The procedure for selecting fitting points from the yield surface was determined by the original generalized yield function and is presented here for the formulation of the relevant HGYF. The HGYF for rectangular section is presented based on regression analysis. The Element Bearing Ratio (EBR), the reference EBR and the uniformity of EBR for rectangular section are defined on the basis of the HGYF, the strategy of elastic modulus adjustment is proposed for the evaluation of the lower bound limit load based on deformation energy conservation principle. Numerical examples show that the proposed method can provide a promising result for limit analysis of the structures with rectangular section.

1 INTRODUCTION

The limit load is an important index for measuring the overall safety of an engineering structure. The limit analysis method for complex structures can be classified into two categories: experimental and numerical methods. Although experimental methods are intuitive, they are time-consuming and expensive. The numerical analysis method has been widely employed because it overcomes the experimental method's aforementioned deficiencies.

The EMAP is a promising numerical method for determining limit loads using a series of linear elastic finite element analysis iterations. In the past 20 years, a number of EMAPs have been proposed, including generalized local stress–strain method (Seshadri & Babu 2000), elastic compensation method (Mackenzie *et al.* 2000), linear matching method (Chen & Ponter 2001), modified elastic compensation method (Yang *et al.* 2005) and m_α–Tangent (Seshadri & Hossain 2009). Most EMAPs are based on the state of stresses and strains; therefore, a refined finite element analysis model that is discretized with solid elements is usually adopted. The limit analysis using this model is inconvenient for calculations involving large and complex structures. To overcome this problem in the limit analysis of constructions, Generalized Yield Functions (GYFs) have been introduced into the EMAPs (Shi *et al.* 1996; Hamilton *et al.* 1996; Hamilton 2002; Marin-Artieda & Dargush 2007; Pisano & Fuschi 2011). Yang *et al.* (2012, 2014) developed the GYF-based Elastic Modulus Reduction Method (EMRM) for determining the limit load. The EMRM improves the efficiency and accuracy of EMAPs by developing the Element Bearing Ratio (EBR), the reference EBR—a dynamic criterion for identifying highly stressed components—and the strategy for adjusting the elastic modulus.

Research has shown that limit load solutions vary with the initial load when the structure is under a combination of internal forces. This variation occurs because most of the GYFs are non-homogeneous functions that cannot maintain the proportional relationship between GYFs and the external load claimed by the EMAPs. A solution to this problem for structures with angle steel proposed preliminarily by Yang *et al.* (2014). However, there is still a need for HGYF-based EMRM for structures with rectangular section.

In this paper, the principle of point selection for establishing HGYF is presented considering the curvature and shape characteristics of GYF. The HGYF is then established for rectangular section. The key parameters for limit analysis are defined; these parameter include EBR, reference EBR and uniformity of EBR. The strategy for the elastic modulus adjustment procedure is given in terms of the conservation of energy. Based on the above analysis, a HGYF-based EMRM is provided for the limit analysis of structures with rectangular section. Numerical examples demonstrate that the provided HGYF-based EMRM can perform promising limit analyses of structures with rectangular section.

2 ELASTIC MODULUS REDUCTION METHOD FOR LIMIT ANALYSIS

2.1 *Generalized Yield Criterion and EBR*

A Generalized Yield Criterion (GYC) that is expressed by internal forces is used to determine the limit state of a cross-section from the elastic stage to the plastic stage. The general formula of GYC under a combined axial force and biaxial bending moments is as follows:

$$f(n_x,m_y,m_z)\le 1, \tag{1}$$

where $f(\cdot)$ denotes the GYF, which is determined with respect to different section; $f(\cdot)$ equals 1 and represents the plastic yield of a cross-section. $n_x = N_x/N_{px}$, $m_x = M_x/M_{px}$, $m_y = M_y/M_{py}$, where lowercase n_x, m_y and m_z are dimensionless parameters. N_x, M_y and M_z are internal forces, and the variables with subscript p denote the corresponding fully plastic section forces that depend on the geometric shape of the cross-section and the strength of the material.

The EBR is developed as a governing parameter of the EMRM. Based on the GYF, EBR can be expressed as follows:

$$r_k^e = \sqrt[N_1]{f(n_x,m_y,m_z)}. \tag{2}$$

where N_1 is the highest power of the GYF.

2.2 *Strategy for elastic modulus adjustment*

The deformation energy conservation principle and the linear elastic finite element method were adopted to develop the strategy of elastic modulus adjustment:

$$E_{k+1}^e = \begin{cases} E_k^e \dfrac{2\left(r_k^0\right)^2}{\left(r_k^e\right)^2+\left(r_k^0\right)^2}, & r_k^e > r_k^0 \\ E_k^e, & r_k^e \le r_k^0 \end{cases} \tag{3}$$

where E_k^e and E_{k+1}^e are the elastic moduli for element e in the kth and $(k+1)$th iterations, and r_k^0 is the Reference EBR (REBR) that provides the threshold for determining whether the elastic moduli are reduced. To improve the accuracy and efficiency of the calculations, the REBR r_k^0 was defined by employing the uniformity of the EBR as follows:

$$r_k^0 = r_k^{\max} - d_k\times\left(r_k^{\max}-r_k^{\min}\right), \tag{4}$$

where $r_k^{\max}$ and $r_k^{\min}$ are the maximum and minimum of the EBRs in the kth iteration, respectively, and d_k is the uniformity of the EBR, which can be defined as follows:

$$d_k = \frac{\overline{r}_k + r_k^{\min}}{\overline{r}_k + r_k^{\max}}, \tag{5}$$

where $\overline{r}_k$ is the mean value of the EBRs in the structure, and d_k with a dynamic value range of (0, 1] reflects the uniformity of the EBR.

The parameters, $r_k^{\max}$, $r_k^{\min}$, $\overline{r}_k$ and d_k, reflect the distribution condition of the EBR for each iteration step and are the characteristic distribution parameters of the EBR. The REBR is defined in expression (4) and provides dynamic adaptive criteria for identifying the highly stressed element with an EBR greater than the REBR. The failure evolution and plastic limit state of engineering structures are simulated by reducing the elastic moduli of highly bearing elements in the iterative process.

2.3 *Limit load solution*

According to the linear elastic finite element, the limit load P_L^k in the kth iteration can be determined by

$$P_{\mathrm{L}}^k = P_0 / r_k^{\max}, \tag{6}$$

where P_0 is the initial load.

The above iterative process is repeated until the limit load of two adjacent iteration steps meets the following criterion of convergence:

$$\left|\left(P_{\mathrm{L}}^k - P_{\mathrm{L}}^{k-1}\right)/P_{\mathrm{L}}^{k-1}\right| \le \varepsilon, \tag{7}$$

where ε is the prescribed admissible error; 0.001 is adopted in this paper.

If the convergence criterion is met in the mth iteration, the lower bound of the structural limit load P_{L} is obtained as follows:

$$P_{\mathrm{L}} = \max(P_{\mathrm{L}}^1, P_{\mathrm{L}}^2, \cdots, P_{\mathrm{L}}^m). \tag{8}$$

3 HOMOGENEOUS GENERALIZED YIELD FUNCTION BASED ELASTIC MODULUS REDUCTION METHOD FOR STRUCTURES WITH RECTANGULAR SECTIONS

3.1 *GYF for rectangular section*

The GYF for rectangular section derived by Duan & Chen (1990) is introduced here and then employed in a HGYF formulation in the EMRM.

The GYF for rectangular section using combined n_x, m_y and m_z can be expressed as follows (Duan & Chen 1990):

$$f = m_y^{\alpha} + m_z^{\alpha} - (1 - n_x^2)^{\alpha} + 1, \tag{9}$$

where $\alpha = 1.7 + 1.3n_x$.

According to equation (9), the GYF for plane structure with a rectangular section can be deduced as follows:

$$f = n_x^2 + m_y. \tag{10}$$

The N_{px}, M_{py} and M_{pz} values of rectangular section can be expressed as follows:

$$N_{px} = \sigma_s bh,\ M_{py} = \frac{1}{4}\sigma_s bh^2,\ M_{pz} = \frac{1}{4}\sigma_s b^2 h, \tag{11}$$

where b and h are the width and height of the cross-section, respectively.

3.2 *Limitation of the GYF and its improvement*

The GYFs mentioned above have been introduced into the EMAPs using sparse mesh for improving calculation efficiency. equation (6) shows that the limit load P_L^k can be obtained from the maximum EBR of $r_k^{\max}$ in the kth iteration, where GYF is employed in $r_k^{\max}$ by the basic EMRM. However, results may vary with different initial loads because there is a non-proportional relationship between the external load P_0 and the GYF of f.

To overcome the nonproportional problem, the homogeneous formula for GYF, i.e. HYGF, can be expressed as follows:

$$\overline{f}(n_x, m_y, m_z) = \sum_{i=1}^{H} a_i n_x^q m_y^g m_z^{N_2-q-g} \\ (q = 0,\ldots,N_2;\ g = 0,\ldots,N_2 - q), \tag{12}$$

where H, N_2 and a_i are the number of polynomial terms, power and undetermined coefficients of the HGYF, respectively.

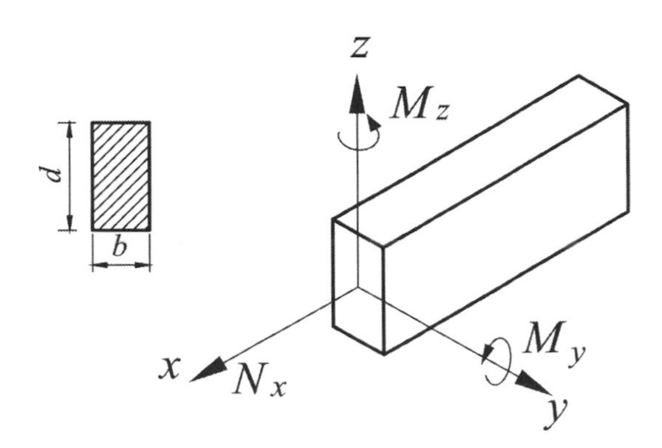

Figure 1. Internal forces in rectangular section.

3.3 *Selection principle of points from GYF*

To give the HGYF shown in equation (12), the representative points from the GYF must first be selected. For better illustrating the features of the GYF, the selection principle of points for HGYF regression analysis is given as follows:

1. Determination of the range of all the dimensionless parameters n_x, m_y and m_z.
2. Calculation of the extremum points of $f = 1$ using the method of the extremum of function.
3. Selection of representative points in a principle of reflecting the shape of yield surface, considering its global and local features.
4. Provision all the sample points for fitting the HGYF.

For the rectangular section defined in equation (9), the detailed process of points selection is as follows:

1. First, the range of n_x, m_y and m_z can be determined as [0, 1] according to equations (1) and (9). The range of m_y is then determined as $[0, 1-n_x^2]$ by considering equation (10).
2. Second, when $n_x = 0$ and $n_x = 1$, the GYF of $f = 1$ is found at its extremum points.
3. Third, m_y can be obtained from the formula of $a(1-n_x^2)$, in which the values of a and n_x are incrementally selected in a step of 0.01 at the range of [0, 0.05] and [0.95, 1.00] to reflect local features of the GYF, while the values are evenly selected in a step of 0.05 at the range of [0.05, 0.95] to reflect global features of the GYF.
4. At last, m_z can be calculated according to equation (9), and all 841 sets of sample points $(n_{x,j}, m_{y,j}, m_{z,j})$, $j = 1, 2 \ldots 841$, are determined for fitting the HGYF.

3.4 *HGYF for rectangular section*

According to the aforementioned principles and steps for regression analysis, the 4th power HGYF for rectangular section is obtained as follows:

$$\begin{aligned}\overline{f}_4 = {} & 1.02n_x^4 + 0.89n_x^3 m_y + 1.41n_x^3 m_z + 4.30n_x^2 m_y^2 \\ & -2.05n_x^2 m_y m_z + 3.55n_x^2 m_z^2 - 0.12n_x m_y^3 \\ & -1.21n_x m_y^2 m_z - 0.69n_x m_y m_z^2 + 0.03n_x m_z^3 \\ & +0.99m_y^4 + 0.38m_y^3 m_z + 2.29m_y^2 m_z^2 \\ & +0.33m_y m_z^3 + 0.99m_z^4.\end{aligned} \tag{13}$$

3.5 *HGYF based EMRM for limit analysis*

The r_k^e can be rewritten according to the HGYF formulated in equation (13):

$$r_k^e = \sqrt[N_2]{f_{N_2}(n_x, m_y, m_z)}. \tag{14}$$

Equation (14) overcomes the problem of variation of limit load P_0 result with different initial loads because r_k^e is proportional to P_0. The HGYF-based EMRM for limit analysis is then established by replacing equation (2) with equation (14).

4 NUMERICAL EXAMPLES

A user routine in a commercial FE package of ANSYS 8.0 purchased by the Guangxi University was employed for constructing FE models and computation in the following examples.

4.1 *Multi-story multi-span frame*

In this example, we considered a four-span, four-story frame structure under a concentrated load and a uniformly distributed load, as shown in Figure 2. The span length L is 4.0 m and the story height of column H is 2.0 m. The rectangular section aforementioned are used here, and their geometry parameters are given in Table 1. The initial elastic modulus is 210 GPa, and the initial yield strength is 310 MPa. Every member is discretized as four elements of beam189 in the FEM model. The HGYF-based EMRM with aforementioned $\bar{f}_4$ is employed to determine the limit load of the frame. The results calculated by the HGYF-based EMRM with rectangular section are compared to the results from the EMRM with the original GYF f and the Elasto-Plastic Incremental Analysis (EPIA), as listed in Table 2. The iteration process of the limit load calculated is shown in Figure 3.

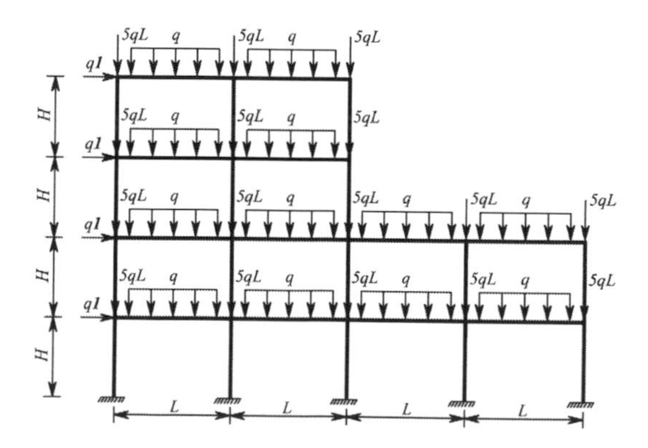

Figure 2. Multi-story multi-span frame.

Table 1. Geometry parameters.

Geometry parameters (cm)	
b	d
4.00	7.60

Table 2. Limit load of frame structure (kN).

	EMRM					
	Original GYF f			4th power HYGF $\bar{f}_4$		
EPIA	$P_0 = 1$	$P_0 = 2$	$P_0 = 3$	$P_0 = 1$	$P_0 = 2$	$P_0 = 3$
186.97	169.08	149.63	138.15	183.91	183.91	183.91

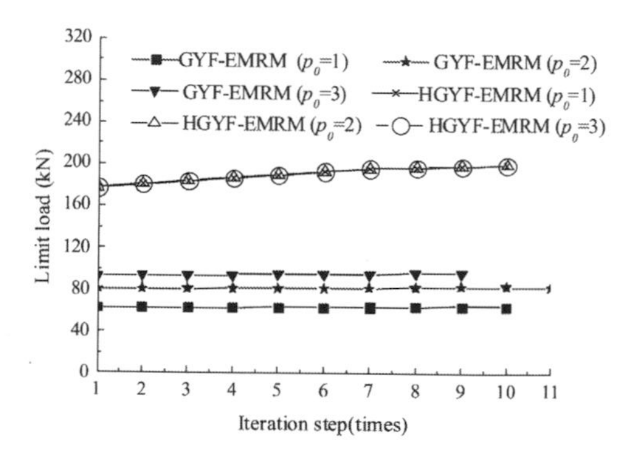

Figure 3. Iterative process of limit load.

The results yielded by the HGYF-based EMRM agree with the results from the EPIA, whlie the results yielded by the GYF-based EMRM vary with different initial loads due to nonhomogeneity.

4.2 *Spatial frame*

This example considers a three-story spatial frame under some concentrated loads, as shown in Figure 4. The geometry parameters of rectangular section are the same as the previous example. The initial elastic modulus is 210 GPa, and the initial yield strength is 310 MPa. Every member is discretized as four elements of beam189. The HYGF-based EMRM with aforementioned $\bar{f}_4$ is employed to perform the limit analysis of this three-story spatial frame structure. The results calculated by the HGYF-based EMRM with the rectangular section were compared to the results from the EMRM with the original GYF f and EPIA, as listed in Table 3.

The results of the EMRM with f vary with different initial loads and are inaccurate because of the nonhomogeneity of f, whereas the limit loads calculated by the HGYF-based EMRM was not affected by the initial load P_0 and agreed well with the results calculated by the EPIA.

The iteration process of the limit load calculated is shown in Figure 5. It can be seen that a calculation result without varying with initial load, yielded by the HGYF-based EMRM, can

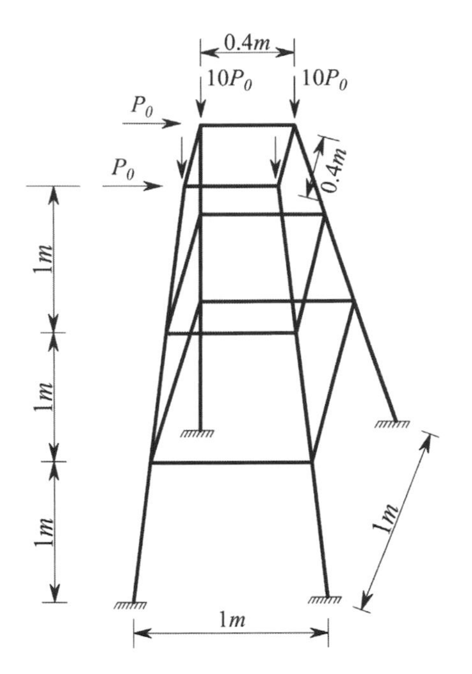

Figure 4. Three-story spatial frame.

Table 3. Limit load of spatial frame structure (kN).

	EMRM					
	Original GYF f			4th power HYGF $\overline{f}_4$		
EPIA	$P_0 = 1$	$P_0 = 5$	$P_0 = 20$	$P_0 = 1$	$P_0 = 5$	$P_0 = 20$
50.97	57.88	53.56	47.88	50.54	50.54	50.54

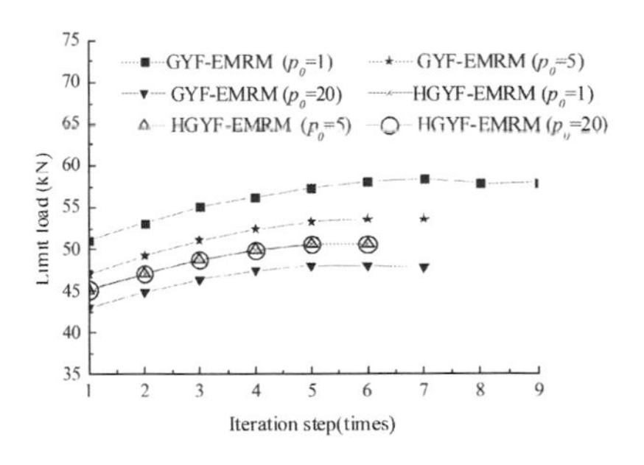

Figure 5. Iterative process of limit load.

be obtained after a few times of iterative analysis, whereas the results of the GYF-based EMRM show that application of the original GYF will result in variability of results with different initial load and iteration steps.

The results of this example verified that the HGYF-based EMRM is a promising limit analysis method for overcoming the nonhomogeneity problem that existed in EMAPs.

5 CONCLUSIONS

In this paper, the HGYF-based EMRM for structures with rectangular section is presented. The HGYF for rectangular section was determined precisely using the selection principle of points from the GYF; the 4th power HGYF $\overline{f}_4$ was elaborately provided. Numerical examples show that the HGYF-based EMRM is an accurate and efficient calculation method for the limit load analysis of structures with rectangular section that overcome the nonhomogeneity problem that existed in EMAPs.

REFERENCES

Chen, H.F. & Ponter, A.R.S. (2001). Shakedown and limit analyses for 3-D structures using the linear matching method. *Int J Press Vessels Pip. 78*, 443–451.

Duan, L. & Chen, W.F. (1990). A yield surface equation for doubly symmetrical sections. Engng Struct. 12, 114–119.

Hamilton, R. & Boyle, J.T. (2002). Simplified lower bound limit analysis of transversely loaded thin plates using generalised yield criteria. *Thin-Walled Struct. 40*, 503–522.

Hamilton, R. & Mackenzie, D., Shi, J., Boyle, J.T. (1996). Simplified lower bound limit analysis of pressurised cylinder/cylinder intersections using generalised yield criteria. *Int J Press Vessels Pip. 67*, 219–226.

Mackenzie, D. & Boyle, J.T., Hamilton, R. (2000). The elastic compensation method for limit and shakedown analysis: a review. *J Strain Anal. 35*, 171–188.

Marin-Artieda, C.C. & Dargush, G.F. (2007). Approximate limit load evaluation of structural frames using linear elastic analysis. *Engng Struct. 29*, 296–304.

Pisano, A.A. & Fuschi, P. (2011). Mechanically fastened joints in composite laminates: Evaluation of load bearing capacity. *Compos Part B: Eng. 42*, 949–961.

Seshadri, R. & Babu, S. (2000). Extended GLOSS method for determining inelastic effects in mechanical components and structures: Isotropic materials. *J Pressure Vessel Technol. 122*, 413–420.

Seshadri, R. & Hossain, M.M. (2009). Simplified limit load determination using the mα-tangent method. *J Pressure Vessel Technol. 131*, 021213.

Shi, J., Boyle, J.T., Mackenzie, D. & Hamilton, R. (1996). Approximate limit design of frames using elastic analysis. *Comput Struct. 61*, 495–501.

Yang, L.F., Li, Q., Zhang, W., Wu, W.L. & Lin, Y.H. (2014). Homogeneous generalized yield criterion based elastic modulus reduction method for limit analysis of thin-walled structures with angle steel. *Thin-Walled Struct. 80*, 153–158.

Yang, L.F., Yu, B. & Ju, J.W. (2014). Incorporated Strength Capacity Technique for Limit Load Evaluation of Trusses and Framed Structures under Constant Loading. *J Struct Eng. 150*, 04015023.

Yang, L.F., Zhang, W., Yu, B. & Liu, L.W. (2012). Safety evaluation of branch pipe in hydropower station using elastic modulus reduction method. *J Pressure Vessel Technol. 134*, 041202.

Yang, P., Liu, Y., Ohtake, Y., Yuan, H. & Cen, Z. (2005). Limit analysis based on a modified elastic compensation method for nozzle-to-cylinder junctions. *Int J Press Vessels Pip. 82*, 770–776.

Civil, Architecture and Environmental Engineering – Kao & Sung (Eds)
© 2017 Taylor & Francis Group, ISBN 978-1-138-02985-9

Numerical simulation analysis of the influence of shield tunneling on an adjacent tunnel

Meide He
Beijing Municipal Engineering Research Institute, Beijing, China
Beijing Construction Engineering Quality Third Test Co. Ltd., Beijing, China

Jun Liu
Beijing University of Civil Engineering and Architecture, Beijing, China

ABSTRACT: In the case of a shield tunnel construction in Beijing, on the basis of the interaction mechanism of neighborhood tunnels during tunnel excavating, a 3D nonlinear numerical simulation model was set up. The developing laws of the segment lining stress, ground deformation around tunnels, segment lining deformation, and changing bending moment of segment lining in different tunnel spaces are analyzed in detail. The results show that the stress growth rate of segment lining and tunnel space are linearly descending related to the situation of the immediately adjacent shielding. The ground settlement form is similar to the case of the single-tunnel construction due to space constraint. Furthermore, it is found that the form is single mode. Under horizontal compression and vertical tension, the segment lining deformation is in ellipse deformation trend. The response of the left existing tunnel to adjacent shielding has hysteresis effect. The result shows that 3D numerical model is a useful method, which can directly display the interaction law of neighborhood tunnels during construction. The study results can be used to guide the construction of immediately adjacent shield tunnels.

1 INTRODUCTION

With the continued construction of the urban subway engineering, shield is increasingly used in the underground engineering; increasingly acute problems are met in the construction of neighboring tunnels. He Meide (2014) studied a certain large-section pedestrian upper passageway crossing the running shield tunnel of Beijing subway. The vertical displacements of shield tunnel structure, the horizontal convergence displacements of shield tunnel structure, and the vertical displacements of track bad structure were monitored and analyzed. The analysis result of this project shows that grouting from floor of passageway can significantly control upfloating, which can be applied in similar projects. On the basis of a constructing running shield tunnel subway side-through close to a high building (He Meide, 2010) using the finite-element method calculation model and site-monitoring method, the foundation characteristics of the high building affected by the constructing shield tunnel are studied. The settlements of the monitoring points reach the maximum value when the shield is 10.0 m away from the monitoring surface. The simulation and the site monitoring results agreed well. Through the analysis of the ground movement law resulted by three typical location relations during the excavation process of the adjacent tunnel with the boundary element method, (Liao Shaoming et al. (2006) drew the conclusion that the tunnel depth and the relationship between the adjacent tunnels has a great impact on the ground displacement. Bi Jihong et al. (2005) carried out numerical simulation on the two-lane tunnel engineering with the shield spacing transiting from 7.2 to 3.0 m constructed by shield tunneling with the finite-element method and analyzed the impact of excavation of right line works on the left tunnel. Hu Wei and Zeng Dongyang (2007) studied the distribution and variation law of surface lateral settlement tank and vertical settlement–uplift curve caused by the construction of parallel shield tunnel with the double-lane tunnel spacing of 6.0 m, 12.0 m with three-dimensional finite-element method, and used three-dimensional curves to draw the spatial distribution variation law of surface settlement–uplift variance curves during the parallel shield tunneling process. In the case of the proximity-overlapped shield tunnel construction in the second phase of Shanghai Mingzhu Transit. Zhang Haibo (2005) simulated the stress and deformation of the old tunnel caused by the post-built tunnels in the case of short-distance overlap with three-dimensional nonlinear finite element. Yamaguchi et al. (1998) summarized and analyzed the interaction of the four parallel tunnels near Tokyo from the angle of design and construction, respectively.

Because of the importance of the interaction during the construction process of the immediately adjacent tunnel, the domestic research of the interaction on the adjacent tunnel is focused on the variation law of the surface settlement and the surrounding layer movement caused in the process of tunnel excavating, and little analysis is done on the influence of the succeeding tunnel on the existing tunnel. Consequently, the development law of segment lining stress, ground deformation around tunnels, segment lining deformation, and changing bending moment of segment lining in different tunnel spaces were analyzed with the 3D numerical simulation method based on the interaction mechanism of neighborhood tunnels during tunnel excavating in this paper, and the research result was used to guide the engineering practice.

2 ENGINEERING SITUATION

Earth pressure balance shields were used in the construction of the two-lane tunnel project. The two shields were advancing in the left and right tunnels with a certain distance apart with the left shield in the front. The tunnel segment lining had an external diameter of 6.0 m and internal diameter of 5.4 m; the precast reinforced concrete segment lining with a ring width of 1.2 m and thickness of 0.3 m was used. The whole ring was divided into six parts, including three standard A-type segment linings, two adjacent capped block B-type segment linings, and a capped block C-type segment lining. Precast reinforced concrete segment lining was C50 grade concrete with the impermeability grade of S10; M24 high-strength bolts were also adopted in the vertical and circumferential connecting bolts for the lining. The minimum tunnel clearance was 1.68 m; the section with the clearance between two lines of less than 2.0 m was about 300 m long, which was the immediately adjacent construction.

The formation conditions of the shield construction section from top to bottom were as follows: miscellaneous fill layer ①, mixed fill layer ①$_1$; silt layer ③, silty clay layer ③$_1$; silty clay layer ⑥, silt layer ⑥$_2$; medium-coarse sand layer ⑦$_1$, fine sand layer ⑦$_2$; silty clay layer ⑧; pebble gravel layer ⑨, medium-coarse sand layer ⑨$_1$; silty clay layer ⑩, clay layer ⑩$_1$. The layers crossed by the tunnel are mainly silt clay layer ⑥ and silt layer ⑥$_2$; the formation was mixed up with a small amount of fine sand layer ⑦$_2$. The physical and mechanical parameters of the main formation are shown in Table 1.

3 THREE-DIMENSIONAL NUMERICAL MODEL

Analysis was conducted with three-dimensional continuum fast Lagrange finite difference software FLAC3D, and Mohr–Coulomb model was adopted for the soil mass in the calculation; shield segment lining was considered as a linear elastic material. The main parameters of the segment lining were: elastic modulus, 34.5 GPa; weight, 25.0 $kN \cdot m^{-3}$ and; Poisson's ratio, 0.2. In view of the constantly changing distance between the left and right tunnels, the cross-sectional size of the established model was 90 m × 52 m (in the x- and y-axis direction), the vertical dimension was 204 m (in the z-axis direction), namely the spacing between the left and right tunnels was 4.77 m at the 1131st ring from one end of the model, from which the spacing between the left and right tunnels was gradually decreased until the minimum spacing at the 1280th

Table 1. Physical and mechanical parameters of stratum.

Layer number	Layer name	Layer thickness/m	Unit weight $\gamma(kN \cdot m^{-3})$	Modulus of compressibility $E_S\ (MPa)$	Poisson's ratio u	Cohesion $C\ (kPa)$	Angle of internal friction $\varphi\ (°)$
①	Miscellaneous Fill	2.4	16.6	–	–	10.0	8.0
③$_1$–③	Silt	5.4	19.3	9.65	0.30	30.8	15.7
④	Silty clay	4.0	20.1	7.75	0.31	35.0	9.0
⑥	Silty clay	4.4	19.8	10.10	0.29	46.0	15.0
⑥$_2$	Silt	6.0	20.2	13.20	0.30	33.0	28.0
⑦$_1$–⑦$_2$	Silty sand	3.6	19.9	35.00	0.28	0.0	32.8
⑦	Round gravel	1.4	21.1	60.00	0.25	0.0	40.0
⑧	Silty clay	5.0	19.7	11.40	0.30	54.0	9.0
⑨	Fine sand	1.6	20.6	35.00	0.27	0.0	30.0
⑩	Silty clay	6.2	19.7	10.90	0.29	20.0	20.0

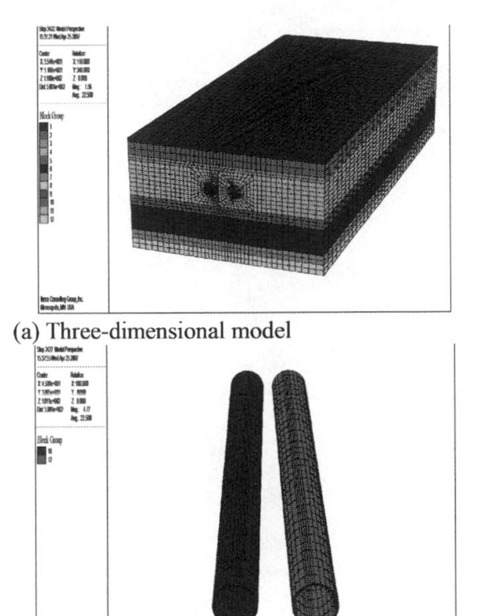

(a) Three-dimensional model

(b) Location relationship between the tunnels

Figure 1. 3D numerical calculation model.

ring section in the left line reached 1.62 m. Then, the tunnel spacing gradually increased again until the spacing between the left and right tunnels was increased to 1.88 m at the 1300th ring section of the left line at the other end of the model. The tunneling construction process of the shield was considered to be advanced gradually; the advance step of the shield in this model was set as the width of five segments of 6.0 m. The optimized model was divided into 47,872 units and 67,027 nodes.

In order to eliminate the displacement produced during the generation of the initial stress field, the left tunnel was further advanced after the displacement was set to zero after the formation of the initial stress field. The view of three-dimensional model and the location relationship between the left and right tunnels is shown in Figure 1.

4 ANALYSIS OF RESULTS

4.1 *Stress of segments*

During the advance process of the shield, the stress of segment lining in the left tunnel was closely related to the damage of segment lining, and the maximum principal stress (σ_1) of the segment lining was an important indicator of its stress state. Therefore, the maximum principal stress was extracted in the result to analyze the change and development status of the stress of the segment. During the advance process of the shield in the right tunnel, the maximum principal stress of the segment was also constantly changing, as shown in Figure 2.

During the construction process in the right tunnel, because of the compression of the shield tunneling on the left tunnel, the segment of the left existing tunnel showed the horizontal compression state in the x-axis direction and stretched state in the y-axis direction so that the left tunnel assumed an oval development trend. As seen from the calculation result of the overall model in Figure 2, the maximum principal stress of the segment of the left tunnel was −7.44 MPa after the two-lane tunnel was completed, which was 32.15% greater

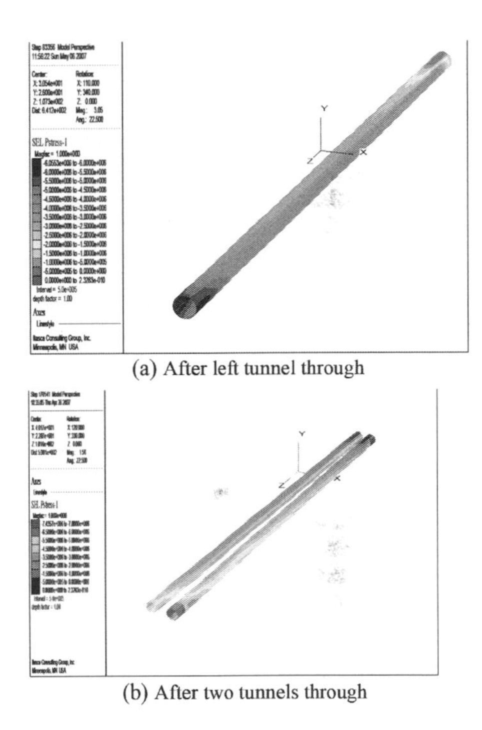

(a) After left tunnel through

(b) After two tunnels through

Figure 2. Contour of the maximum principal stress of segment lining during shield tunneling.

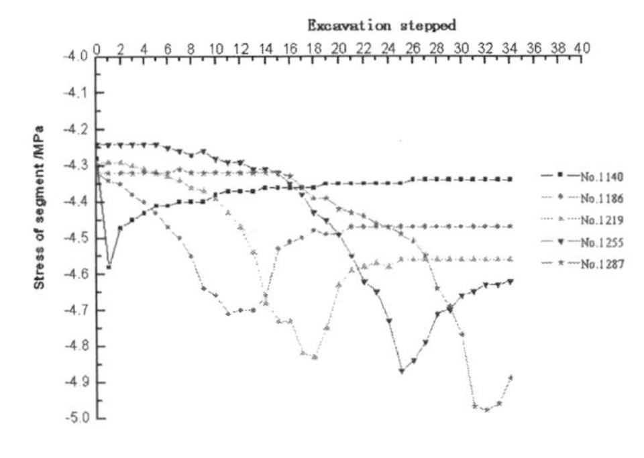

Figure 3. Curve of the maximum stress of segment lining in left tunnel monitoring sections.

Table 2. Stress of segment lining in left line monitoring sections.

	No. 1140	No. 1186	No. 1219	No. 1255	No. 1287
Tunnel spacing (m)	4.53	3.35	2.63	1.73	1.65
Maximum stress of segment after left through (MPa)	–4.28	–4.32	–4.29	–4.24	–4.32
Maximum stress of segment after right through (MPa)	–4.58	–4.71	–4.83	–4.87	–4.98
Maximum stress corresponding to excavation steps (step)	1	11	18	25	32
Monitoring sections where excavation steps (step)	2	12	18	25	32

than the maximum principal stress of the segment of –5.63 MP when the left tunnel was completed.

As could be seen from the curve of the maximum stress of segment lining in monitoring sections of the left tunnel in Figure 3, before the shield in the right line reached the monitoring segments, the stress of the monitoring sections had changed. Because both the 1140th and 1287th rings were located adjacent to the two ends of the model, the segment stress curve was not complete. However, according to the curve variation law of the remaining three sections, it could be seen that the changing trend of the curve was consistent, that is, before the arrival of the shield, the stress of the segment began to increase; the stress of the segment was the largest when the shield reached the monitoring section or the shield advanced into the monitoring section. It could be seen that the segment region with significant changes of stress was in the range of –D to 0. The stress result of the monitoring sections of the segment is shown in Table 2.

In the stress results of the monitoring sections in Table 2, the segment showed the compression state in the 1140 ring with the largest clearance between the left and right tunnels. The maximum stress was increased by 7.00% from 4.28 to 4.58 MPa; in the 1287th ring, with a clearance between the left and right tunnels of only 1.65 m, the maximum stress was increased by 15.28% from 4.32 to 4.98 MPa, and the compression trend of the segment was further strengthened.

As was seen from the step at which the monitoring sections of the segments achieved the maximum stress, both the 1140th and 1186th rings in the immediately adjacent section of the left and right tunnels reached their respective maximum stress one advance step before the shield's arrival, with stress growth rates of, respectively, 7.00% and 9.03%. As was seen from the 1219th ring of the proximity section to the 1255th ring, the 1287th ring in the basically parallel section of the left and right tunnels and the segment lining stress in the three monitoring sections reached their respective maximum stress with the segment stress growth of the three sections of, respectively,

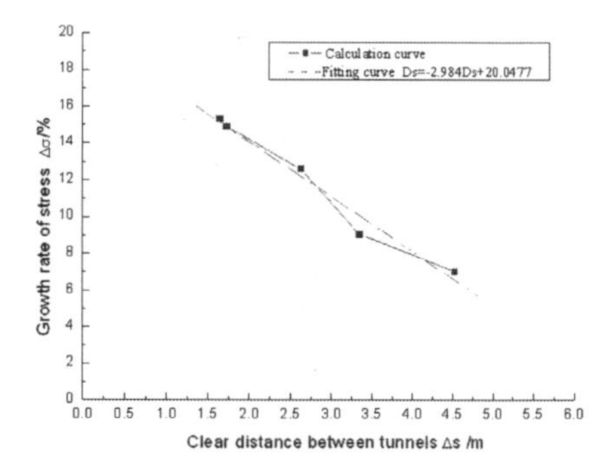

Figure 4. Relationship between clearance and stress growth rate of segment lining.

12.59%, 14.89%, and 15.28%. According to the relationship between the segment stress growth rate of five monitoring sections and the tunnel spacing, the linear fitting curves of the stress growth rate $\Delta\sigma = -2.984\Delta s + 20.0477$ (where $\Delta\sigma$ is the segment stress growth rate and Δs is the clearance of the two-lane tunnel) reflected that the segment stress growth rate was basically linearly descending with a clearance between the left and right tunnels, as shown in Figure 4.

4.2 *Ground deformation around the tunnel*

During the construction process of the immediately adjacent tunnels, besides the great impact on the left existing tunnel, the surrounding stratum would certainly be disturbed in a certain degree. As there was a very small spacing between the left and right tunnels, the surface subsidence distribution caused by the construction of two tunnels was very similar to that caused by a single tunnel, as shown in Figure 5. In the vertical axis location of center line of the left and right tunnels, the surface sedimentation was the greatest, the surface sedimentation tank was very similar to that of single lane, and the sedimentation tank was also a "single peak" rather than the "twin-peak" sedimentation tank caused by the general construction of left and right tunnels.

4.3 *Deformation of segments*

Because of the compression on the left existing tunnel during the shield tunneling process in the right tunnel, the segment showed the deformation trend

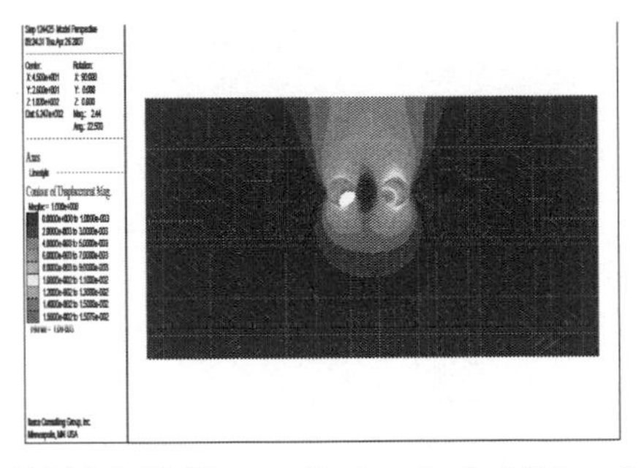

(a) Right half of the tunnel boring after the left through

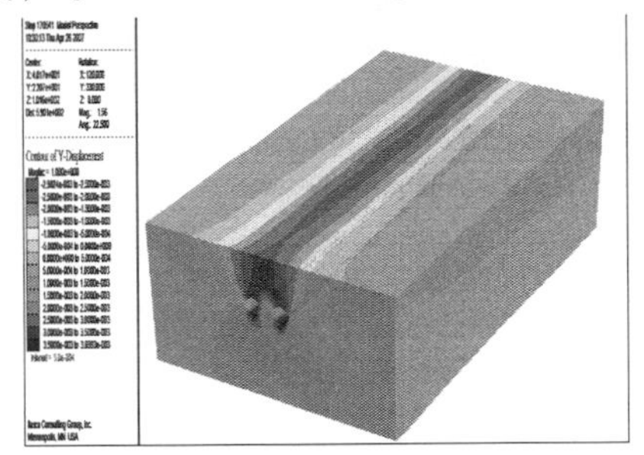

(b) After two tunnels through

Figure 5. Contour of deformation in surrounding rocks.

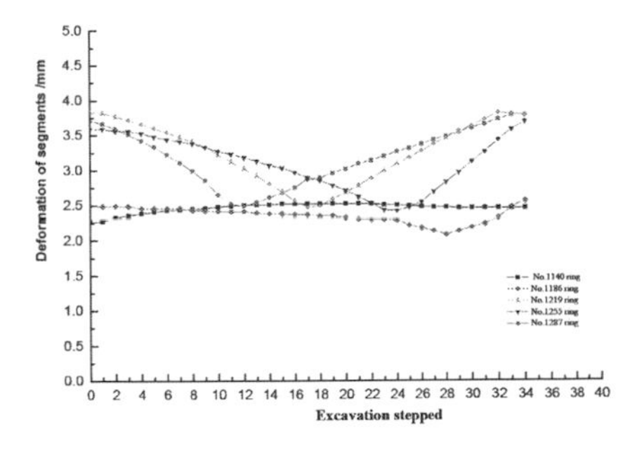

Figure 6. Duration curve of maximum deformation of segment lining in monitoring section.

of horizontal compression and vertical tension. The left shield tunnel was also shifted as a whole compared with the new right tunnel; during the construction process of the right tunnel, the deformation of the segment was also affected by "space effect". The duration curve of maximum deformation of segment lining in Figure 6 showed that when the shield approached the five monitoring sections and passed through the right tunnel segment, the overall deformation of the segments had the slowing trend with varying degrees; after the shield passed some excavation steps, the maximum deformation of the segments began to stabilize after some moderate growth, and the final deformation was slightly larger than the initial deformation. Through comparison of the deformation of the segment in the monitoring sections before and after the passing of the shield, the deformation was increased to 2.46 mm in the 1140th ring with the largest increment of 8.85%; the deformation was increased to 3.79 mm in the 1219th ring with the smallest increment of only 1.30%.

4.4 *Bending moment of segments*

During the different construction stages of the tunnel, there were some differences among the maximum plus–minus bending moments of the segment, as shown in Table 3. Through the comparison of the plus–minus segment bending moment (M_x), it could be seen that during the excavation process of the left tunnel, there was a small difference between the maximum plus–minus bending moments, the largest difference between their absolute values was not more than 0.2%; during the excavation process of the right tunnel, the absolute difference between the maximum plus–minus bending moments of the left and right tunnel was 5.03%. As was seen from the bending moment of the segment, during the excavation process of the left tunnel, the largest difference between their absolute values was 55.53%; during the excavation process of the right tunnel, the difference between the absolute value of the maximum bending moment of the left and right lines of the tunnel was up to 276.32%. It could be seen that the segment bending moment (M_x) was apparently greater than M_y at the same construction phase, and the absolute

Table 3. Statistical comparison of bending moment in segment lining during different construction stages.

		Excavation of the left half tunnel	Left tunnel through	Excavation of the right half tunnel	Right tunnel through
M_x $(kN \cdot m)$	Sagging moment	33.91	33.63	38.17	40.02
	Hogging moment	−33.97	−33.70	−40.09	−39.82
M_y $(kN \cdot m)$	Sagging moment	4.43	5.24	7.18	8.40
	Hogging moment	−6.89	−6.83	−27.02	−28.39

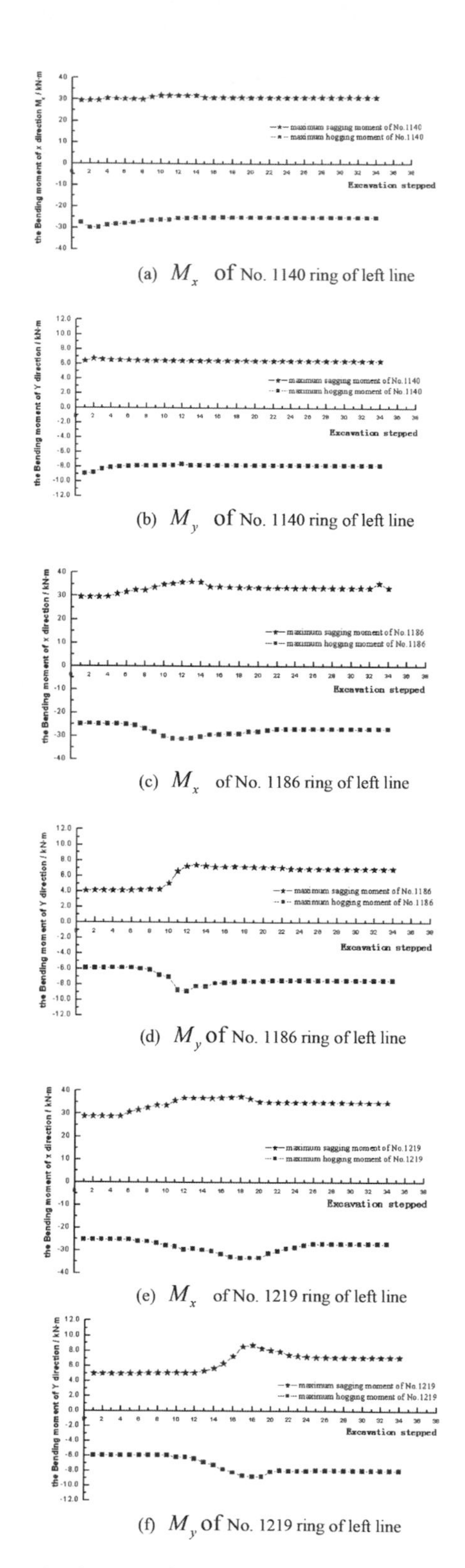

(a) M_x of No. 1140 ring of left line

(b) M_y of No. 1140 ring of left line

(c) M_x of No. 1186 ring of left line

(d) M_y of No. 1186 ring of left line

(e) M_x of No. 1219 ring of left line

(f) M_y of No. 1219 ring of left line

Figure 7. (*Continued*).

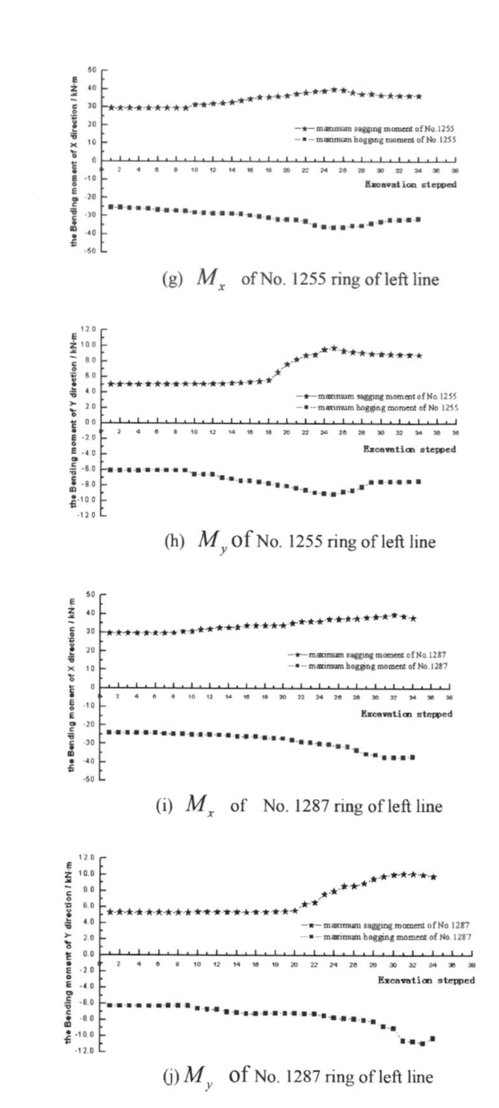

(g) M_x of No. 1255 ring of left line

(h) M_y of No. 1255 ring of left line

(i) M_x of No. 1287 ring of left line

(j) M_y of No. 1287 ring of left line

Figure 7. Duration curve of bending moment in monitoring section segment lining.

value of the maximum plus–minus bending moment of M_y was significantly greater than that of M_x. It could be seen in the shield tunneling process of the right tunnel, through comparison of the bending moment M_x and M_y on the same section, M_y is more affected by the existing tunnel than M_x.

During the shield tunneling in the right line, the bending moment of five monitoring sections in the left line followed the same change rules, as shown in Figure 7. Before the shield arrived at the monitoring section, the plus–minus bending moment of M_x and M_y of the ring segments began to increase at different levels; after the shield passed the corresponding sections in the right line, the plus–minus bending moments of M_x and M_y of the ring segments began

Table 4. Statistical comparison of maximum bending moment of segment lining in different monitoring sections.

		No. 1140	No. 1186	No. 1219	No. 1255	No. 1287
M_x $(kN \cdot m)$	Sagging moment	32.12	36.23	37.57	39.92	39.90
	Hogging moment	−29.91	−31.15	−33.35	−36.27	−37.31
M_y $(kN \cdot m)$	Sagging moment	6.70	7.40	8.73	9.76	10.15
	Hogging moment	−8.96	−8.87	−8.81	−9.09	−10.88

to reduce at different levels until it becomes stable. The monitoring sections of the moment, both the maximum positive moment and maximum negative moment, withstood by the same ring showed the trend of symmetrical change, which was consistent with the "ellipse" deformation trend of the segment. After the shield passed the sections in the right line, the bending moments of the monitoring sections were increased at varying degrees. As could be seen from the comparison between the maximum bending moments of the segments in Table 4, the change of bending moment of the segment was also related with the distance between the left and right tunnels, namely the smaller the clearance between the left and right lines, the greater change of the bending moment withstood by the segment.

5 CONCLUSIONS

In the case of construction of an immediately adjacent shield in Beijing, in this paper, the influence of shield construction on the immediately adjacent tunnel is analyzed through establishing a 3D nonlinear elastic–plasticity numerical model of the construction of immediately adjacent shield tunnels. The calculation result shows that:

1. In the light of the analyses of four indexes, namely the stress variation of the segment lining, the deformation law of the ground around the tunnel, deformation of the segment itself, and the variation law of the bending moment of segment lining reflect the sensitive degree of the clearance between the left and right tunnels at different levels. Because the segment itself has a higher rigidity, the deformation development of the segment itself is less influenced by the existing tunnels. The process of tunnel excavating has certain hysteresis effect on the influence of immediately adjacent left existing tunnel.
2. The stress growth rate of the adjacent segments caused by the tunnel excavating is basically linearly descending related to the situation of tunnel space.
3. The ground settlement form caused by the construction of double-line tunnel is similar to that caused by the single-tunnel construction due to the little space between the left and right tunnels, which is single-mode.
4. Development of the maximum plus–minus bending moments (M_x and M_y) of the segment is in a trend of symmetrical variation, which is consistent with the ellipse deformation trend of the whole segment of the left existing tunnel under horizontal compression and vertical tension.

ACKNOWLEDGMENTS

This study was supported by the Beijing Outstanding Talents Training funded projects (No. 2015754154700G160) and the Beijing Energy Conservation and Emission Reduction Key Technology Collaboration Innovation Center (02058416003).

REFERENCES

Bi Ji-Hong, Jiang Zhi-Feng, Chang Bin. 2005. Numerical simulation for constructions of metro tunnels with short distance between them[J]. *Rock and Soil Mechanics*, 26(2):277~281.

He Meide, et al. 2010. Study of impact of shield tunneling side-crossing on adjacent high building[J]. *Chinese Journal of Rock Mechanics and Engineering*, 29(3):603–608.

He Meide, et al. 2014. Deformation analysis for shield tunnel upper crossing by close range large section passageway [J]. *Chinese Journal of Rock Mechanics and Engineering*, 33(Supp2):3682–3691.

Hu Wei, Zeng Dong-Yang. 2007. Research on the influence of ground-movement caused by parallel shield tunnel construction [J]. *Journal of Railway Engineering Society*, 2007 (3):50~55.

Liao Shao-Ming, Yu Yan, Bai Ting-Hui, et al. 2006. Distribution of ground displacement field owing to two overlapped shield tunneling interaction[J]. *Chinese Journal of Geotechnical Engineering*, 28(4):485–490.

Yamaguchi I, Yamazaki I, Kiritani Y. 1998. Study of ground-tunnel interactions of four shield tunnels driven in close proximity, in relation to design and construction of parallel shield tunnels [J]. *Tunnelling and Underground Space Technology*, 13(3):289–304.

Zhang Hai-Bo, Yin Zong-Ze, Zhu Jun-Gao. 2005. Numerical simulation of influence of new tunnel on short distance overlapped old tunnel during shield tunneling [J]. *Rock and Soil Mechanics*, 26(2) 282~286.

Civil, Architecture and Environmental Engineering – Kao & Sung (Eds)
© 2017 Taylor & Francis Group, ISBN 978-1-138-02985-9

Development and validation of the effective stress algorithm of the Davidenkov constitutive model using the Byrne pore pressure increment model

Bin Ruan, Guoxing Chen & Dingfeng Zhao
Institute of Geotechnical Engineering, Nanjing Tech University, Nanjing, China
Civil Engineering and Earthquake Disaster Prevention Center of Jiangsu Province, Nanjing, China

ABSTRACT: In this paper, the initial stress–strain skeleton curve was simulated using the Davidenkov model, and simplified loading and unloading rules were proposed instead of the "upper boundary" rule of the extended Masing criteria. Through a large number of resonant column tests of undisturbed Nanjing fine sandy soil samples, the fitting parameters of the small shear modulus (G_{max}) and reference shear strain (γ_0) formulations, including the effect of confining pressure, are given. A new effective stress method using the integration algorithm of the modified Davidenkov constitutive model and the Byrne pore water pressure increment model was implemented in ABAQUS. Using the drained cyclic triaxial test data of the saturation Nanjing fine sand, the parameters of the Byrne pore model are given. Undrained cyclic triaxial tests and its three-dimensional numerical simulation for the saturation Nanjing fine sand have been performed, and the results of the measurement in tests and the calculation almost agreed. The results indicated that the effective stress algorithm proposed in this paper can well simulate the pore pressure increasing process of the liquefiable sand, and the scientific rationality and validity of the integration algorithm were also verified.

Keywords: effective stress algorithm; modified Davidenkov constitutive model; Byrne incremental model; numerical simulation; cyclic triaxial tests

1 INTRODUCTION

Earthquake-induced liquefaction is a common phenomenon. The root cause of liquefaction is that the pore water pressure increase and then soil layers liquefy under the ground motion. In recent years, with the development of computer technology, many scholars are using numerical simulation method to simulate the development of pore pressure and liquefaction characteristics during the vibration of the engineering site. Its primary problem is to use the reasonable soil dynamic constitutive model and the vibration pore pressure model for describing the phenomenon of sand liquefaction and its programming.

A large number of experimental studies have shown that the use of ideal Davidenkov constitutive model with three parameters can describe the nonlinear dynamic characteristics of various types of soil, and a large amount of experimental data has been accumulated (Chen Guo-xingm 2005 & Rong Mian-shui, 2013). Because the amplitude of actual ground motions are often unequal, additional provisions need to be added to correct the irregular loading and unloading stress–strain hysteresis curve of the Davidenkov model. "Extended Masing Rule" builds the theoretical framework of soil stress–strain curves under irregular loading and unloading. On the basis of previous research, this paper proposes the modified irregular loading and unloading criteria, which follows the Davidenkov "Backbone Curve Rule", the "N Times Method" proposed by Pyke (Pyke R M, 1979), and the "Upper Skeleton Curve Criteria". On the basis of the explicit module of ABAQUS, this paper developed explicit subroutine modules.

Seed and Martin et al. (Seed H B, 1975 & 1976) proposed a stress model to calculate the average pore water pressure according to the dynamic triaxial experiment data of isotropic consolidation sand under undrained conditions. Finn and Lee et al. (1977) amended the Seed and Martin model considering the effect of initial shear stress on dynamic pore pressure. Martin and Finn et al. (1975) established a seismic vibration strain model of pore water pressure on the basis of the consistency condition of saturated soil volume change. Byrne (1991) utilized the experimental data to establish the experience function between accumulated volume strain and the incremental volume strain on the basis of Martin and Finn's model, which made the model more simple and practical.

On the basis of the work above, in this paper, we will describe the Davidenkov constitutive model, which was modified with irregular loading–unloading criterion and the Byrne pore pressure increment model. It describes the nonlinear characteristics of soil liquefaction by the subroutine in ABAQUS. Undrained cyclic triaxial tests of Nanjing fine sand and the corresponding three-dimensional numerical simulation were carried out, and the target of simulating pore water pressure growth process of liquefied foundation was achieved. The initial verification of the three-dimensional effective stress subroutine and the feasibility analysis of the results were also analyzed in ABAQUS.

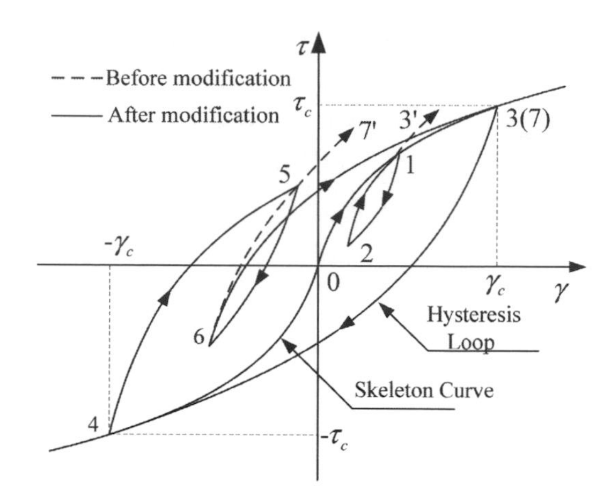

Figure 1. Stress–strain curves of the Davidenkov model with irregular loading and unloading rules.

2 MODIFIED DAVIDENKOV CONSTITUTIVE MODEL

2.1 *The loading–unloading criterion of the Davidenkov skeleton curve under irregular loading and unloading situation*

Martin and Seed (Martin P P, 1982) proposed the skeleton curve formula of the Davidenkov constitutive model:

$$\tau = G \cdot \gamma = G_{\max} \cdot \gamma \cdot \left[1 - H(\gamma)\right] \tag{1}$$

where

$$H(\gamma) = \left\{ \frac{(\gamma / \gamma_0)^{2B}}{1 + (\gamma / \gamma_0)^{2B}} \right\}^{A} \tag{2}$$

where τ is the shear stress, γ is the shear strain, $G_{\max}$ is the initial shear modulus, and A, B, and γ_0 are the experimental parameters of soil.

To construct the hysteresis curve of the Davidenkov constitutive model under irregular cyclic stress, the Masing rules suitable for constant amplitude loading need to be modified, as shown in Figure 1. The loading–unloading criterion can be amended as follows:

1. The initial loading curve goes along the skeleton curves, such as 0→1 curve.
2. If the loading and unloading curve encounters skeleton curves before the steering, it should follow the extended Masing rules and the "upper backbone curve" rule, that is, the subsequent stress–strain curve goes along skeleton curve, 2→1→3 curve before correction should be corrected to 2 → 1 → 3 curve.
3. After stress is up to the steering point, the subsequent stress–strain curve goes along the direction from the current inflection point to the maximum (minimum) point in history 6→7' curve before correction should be corrected to 6→7 curve. On the basis of the "N-Times Law" proposed by Pyke, at this time, the stress–strain hysteresis curve obeys the following relationship:

$$\tau - \tau_c = G_{\max} \cdot (\gamma - \gamma_c) \cdot \left[1 - H\left(\frac{|\gamma - \gamma_c|}{2n}\right)\right] \tag{3}$$

where γ_c is the strain at loading and unloading inflection point.

According to criterion (3), in addition to recording the point of most value in history, when stress is up to the steering point, this criterion just needs to remember the current strain and parameter N at the inflection point, that is, determine the direction of stress–strain curve. This is an effective solution to the problem of large amount memory of to expand the "Masing rule" at inflection points.

2.2 *Modification of G_{max} and γ_0 considering the initial effective confining pressure*

When using the average modulus of the soil layers in seismic response analysis, most of the high-frequency component of ground motion cannot propagate to the surface through the site. By modifying shear modulus and the reference strain with the initial effective confining pressure, we can fix this problem, so that the seismic response analysis results are more in line with the actual situation. Guoxing Chen (Zhuang H, 2015) summarized a number of experimental studies, which showed that: For noncohesive soils, in the same shear strain amplitude, the greater the initial effective confining pressure, the slower the movement of the shear modulus attenuation curve. Parameter γ_0 of the curve of dynamic shear modulus ratio plays a major role in the performance of the attenuation. The higher the value of γ_0, the slower the attenuation of the curve, that is, the greater the initial effective confining pressure, the greater the value of γ_0. Therefore, we need to add the correction formula of G_{max} and γ_0 considering the initial effective confining pressure in the subroutine to meet the needs of different users.

Through abundant resonant column tests of undisturbed soil samples of Nanjing fine sand, 43 groups of test data with initial effective confining pressure, soil shear modulus, and parameter γ_0 were selected. Using Origin to fit data, variation curves in which G_{max} and γ_0 increase with the increase of σ_c', as shown in Figure 2. At the same time, according to empirical formulas in the literature (Zhang J & Zhang J, 2005), G_{max} and γ_0 were calculated as:

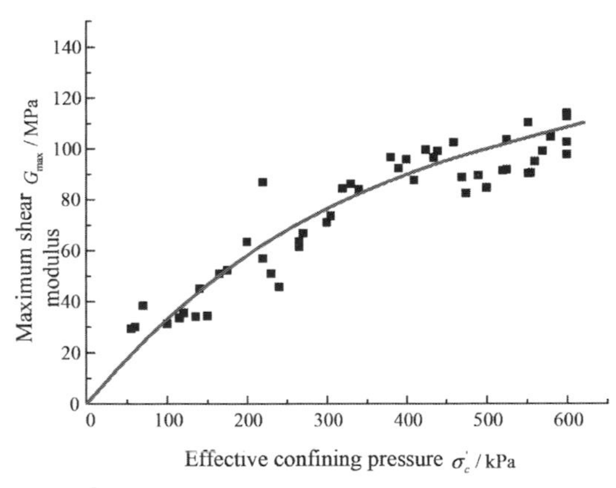

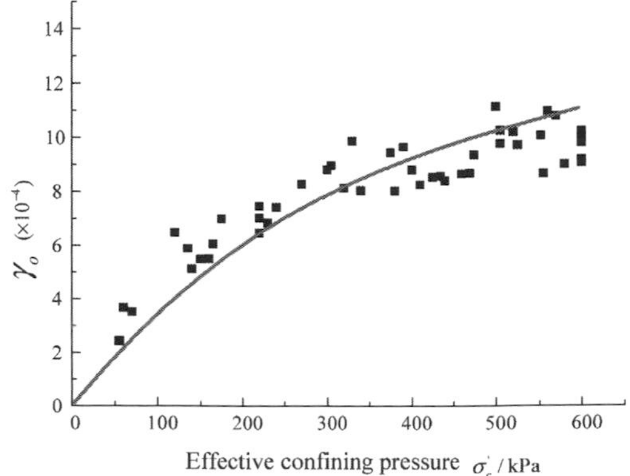

Figure 2. Fitted curve of tests.

$$G_{max} = G_{max}^{ref} \cdot \left(\frac{\sigma_c'}{\sigma_{ref}'} \right)^a \tag{4}$$

$$\gamma_0 = \gamma_0^{ref} \cdot \left(\frac{\sigma_c'}{\sigma_{ref}'} \right)^a \tag{5}$$

where G_{max}^{ref} and γ_0^{ref} are experimental values under confining pressure σ_{ref}' and a is the fitting parameter associated with soil characteristics.

3 BYRNE PORE WATER PRESSURE INCREMENT MODEL

On the basis of the consistency condition of volumetric change of saturated soil during earthquakes, Martin and Finn established the basic formula to obtain the increment of pore water pressure of saturated soil. As the decrease of effective stress will cause the rebound of soil volume, the volume permanent compression is induced by cyclic shearing and the pore water discharge and soil volume rebound are induced by the decrease of effective normal stress to meet the compatibility criterion. Assuming positive volume change of soil compression and pore water discharge:

$$\Delta\varepsilon_{v,d} + \Delta\varepsilon_{v,r} = \Delta\varepsilon_{v,f} \tag{6}$$

where $\Delta\varepsilon_{v,d}$ is the soil compression volume induced by shearing at each cycle, $\Delta\varepsilon_{v,r}$ is the soil volume increment induced by the decrease of effective normal stress, and $\Delta\varepsilon_{v,f}$ is the discharge volume of pore water at each cycle induced by shearing. Assuming that at the moment the earthquake occurs, there is no drainage in saturated sand, that is, $\Delta\varepsilon_{v,f} = 0$. Then, equation (6) can be written as:

$$\Delta\varepsilon_{v,d} = -\Delta\varepsilon_{v,r} \tag{7}$$

Assuming E_r as resilient modulus of saturated sand, Δu as the increment of pore water pressure, and $\Delta\sigma$ as the increment of effective normal stress, we get:

$$\Delta\varepsilon_{v,r} = \frac{\Delta\sigma}{E_r} = -\frac{\Delta u}{E_r} \tag{8}$$

Substituting equation (8) to (7), we get:

$$\Delta u = E_r \Delta\varepsilon_{v,d} \tag{9}$$

Byrne (1991) obtained the relation between cumulative volume strain ($\varepsilon_{v,d}$) and volumetric strain increment ($\Delta\varepsilon_{v,d}$) using experimental data of Martin and Finn (1975):

$$\frac{\Delta \varepsilon_{v,d}}{\gamma_a} = C_1 \exp(-C_2 \frac{\varepsilon_{v,d}}{\gamma_a}) \quad (10)$$

$$E_r = \frac{(\sigma_c' - u)^{1-m}}{mk(\sigma_c')^{n-m}} \quad (11)$$

In equation (10) and equation (11), γ_a is the shear strain amplitude of the Nth stress cycle, and σ_c' is the initial effective normal stress, which are represented by octahedral shear strain amplitude (γ_{oct}) and initial octahedral effective normal stress(σ_{oct}), respectively, in two-dimensional and three-dimensional conditions, u is the excess pore water pressure, and C_1, C_2, m, n, and k are parameters related to soil characteristics.

Byrne used the experimental data of Martin and Finn (1975) and Tokimatsu and Seed (1987) (Tokimatsu K, 1987) to obtain the empirical formula of coefficients C_1 and C_2 represented by the sand relative density (D_r) or the modified blow counts of standard penetration test N_1:

$$C_1 = 0.076 D_r^{-2.5} \text{ or } C_1 = 8.7 N_1^{-1.25} \quad (12)$$

$$C_2 = \frac{0.4}{C_1} \quad (13)$$

The generation of excess pore water pressure will lead to the decrease of soil shear modulus, in order to consider the influence induced by the rise of pore water pressure on the mechanical properties of the soil, the formula of dynamic pore water pressure, soil shear modulus, and reference shear strain was established according to the experiment (Liyanathirana D S, 2002):

$$\mathrm{G}_{\max} = G_{\max}^{ref} \cdot \left(\frac{\sigma_c'}{\sigma_{\mathrm{ref}}'}\right)^a \cdot \left(\frac{\sigma_c' - u}{\sigma_c'}\right)^b \quad (14)$$

$$\gamma_0 = \gamma_0^{ref} \cdot \left(\frac{\sigma_c'}{\sigma_{ref}'}\right)^a \cdot \left(\frac{\sigma_c' - u}{\sigma_c'}\right)^b \quad (15)$$

where u is the pore water pressure and b is the fitting parameter related to soil characteristics.

4 IMPLEMENTATION OF EFFECTIVE STRESS ALGORITHM IN ABAQUS

4.1 *Secondary development platform in ABAQUS*

ABAQUS modules support constitutive model programs written by FORTRAN language, users can realize the subroutine development through the interface provided by ABAQUS UMAT/VUMAT (implicit/explicit), and the main program will automatically call the constitutive model specified by the user in the calculation process.

The development involved the interface program, which is the material custom programs of VUMAT, is briefly introduced as follows:

The main task of the VUMAT subroutine is: (1) the stress increment ($\Delta\sigma$) should be achieved on the basis of the incoming strain increment ($\Delta\varepsilon$) from ABAQUS; and (2) the state variables need to be transferred in the process of updating and solving.

The simulation of soil pore pressure needs coupling calculation with both static and dynamic steps, which can be achieved by multitasking. However, the multitasking calculation method is complicated to be adopted. In order to facilitate the employment of the subroutine, the constitutive model program was divided into two computing stages, including the static stage and the dynamic stage. The static calculation stage used the original in situ stress balance calculation method in ABAQUS. The soil dynamic stress–strain relationship during dynamic calculation phase relied on the modified Davidenkov constitutive model based on irregular loading and unloading criterion, and the Byrne pore water pressure increment model presented above is embedded in the constitutive model.

4.2 *Calculation procedure of the constitutive model*

For large-scale numerical simulation problems, explicit algorithm can ensure the calculation accuracy, while providing higher computational efficiency (Chen Guo-xing, 2011). Therefore, explicit algorithm subroutine was compiled through ABAQUS VUMAT interface. Figure 3 shows the

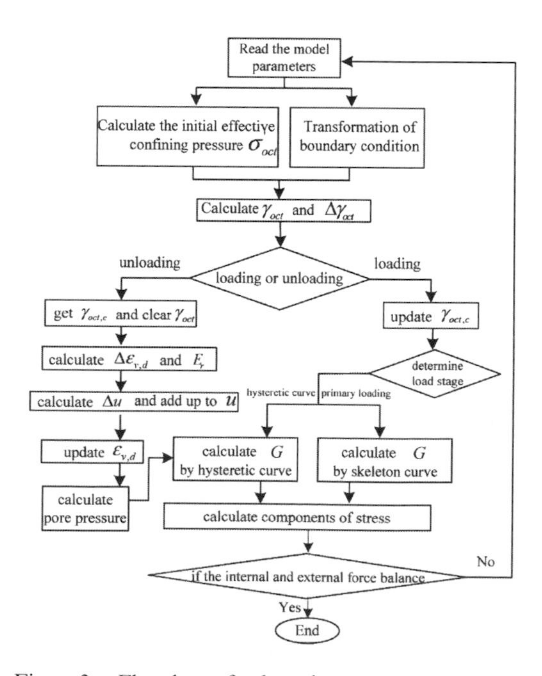

Figure 3. Flowchart of subroutine.

subroutine calculation process. First, the subroutine obtains the initial octahedral effective confining pressure (σ'_{oct}) in static analysis phase and the transition of the boundary conditions. Second, in the stage of dynamic analysis, the initial loading section should be judged and the shear modulus (G) is obtained using the initial load curve. Then, the turning point of loading and unloading must be judged. If there is a turning point, then update the octahedral shear strain amplitude and calculate pore water pressure increment (Δu) and the volumetric strain increment ($\Delta\varepsilon_{v,d}$), update pore water pressure (u) and the accumulated volume strain ($\varepsilon_{v,d}$) after accumulation, and update the stress.

5 VERIFICATION OF THE SUBROUTINE MODULE

5.1 *Determination of parameters in Byrne pore pressure increment model*

Saturated drained cyclic triaxial tests of remolded Nanjing fine sand were carried out. The sample diameter was 39.1 mm, the height H was 80 mm, and the relative density was 50%. The basic physical parameters are shown in Table 1. The layered wet ramming method and saturated vacuum extraction method were applied to remold the soil. The saturation time was more than 24 h. The sinusoidal shear strain load was applied on the specimen with amplitudes of 0.1%, 0.2%, and 0.3% at 1Hz.

The relation curve between volumetric strain and cyclic number (N) was achieved by tests mentioned above, as shown in Fig. 4. The experimental results of crystal silica sand were represented by dotted lines from Byrne P M (1991). Data points of different shear strain amplitudes were selected in Fig. 4. According to formula (10), the modified relation curve between volumetric strain increment ($\Delta\varepsilon_{v,d}/\gamma_a$) and volumetric strain ($\varepsilon_{v,d}/\gamma_a$) was fitted out, as shown in Fig. 5. The Byrne pore pressure increment model parameters of Nanjing fine sand are $C_1 = 0.730$ and $C_2 = 0.475$. Besides, the dotted lines are the fitting curves of crystal silica sand obtained from Byrne P M (1991).

As shown in Figs. 4 and 5, the development trends of relation curve between volumetric strain ($\varepsilon_{v,d}$) and cyclic number (N) of Nanjing fine sand and crystal silica sand were basically consistent. Under the cyclic loading of same shear strain amplitude, volumetric strain increased as the cyclic number increases. At the same certain cycle, with the increase of shear strain amplitude, the volumetric strain increased too; the development trends of relation curve between the modified volumetric strain increment ($\Delta\varepsilon_{v,d}/\gamma_a$) and volumetric strain ($\varepsilon_{v,d}/\gamma_a$) were basically coincident, the values of parameters are relatively close.

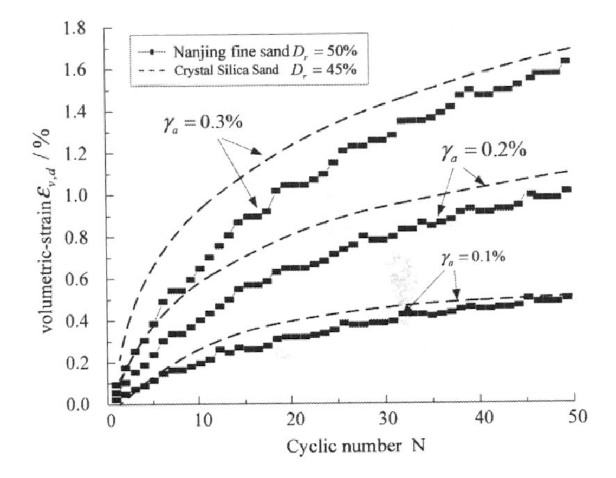

Figure 4. Volumetric strains from constant amplitude cyclic simple tests.

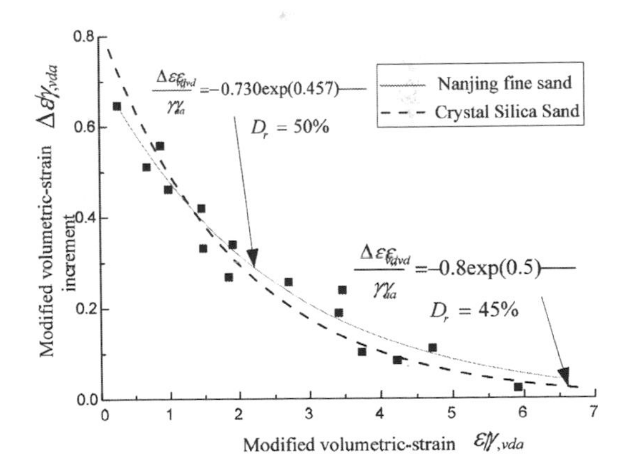

Figure 5. Normalized incremental volumetric strains.

Table 1. Basic physical properties of the Nanjing fine sand.

Mean diameter d_{50} (mm)	Uniformity coefficient $c_u = d_{60}/d_{10}$	Curvature coefficient $c_c = d_{30}^2/(d_{10}\,d_{60})$	Specific gravity G_s	Maximum void ratio e_{max}	Minimum void ratio e_{min}	Relative density D_r(%)
0.12	2.31	1.07	2.72	1.15	0.62	50

Notes: Effective diameters d_{10}, d_{30}, and d_{60} mean that 10%, 30%, and 60% (by weight) of particles are smaller than the number, respectively.

These conclusions illustrated the validity of the test results with each other.

5.2 *Undrained cyclic triaxial tests of saturated Nanjing fine sand*

Saturated undrained triaxial tests of remolded samples of Nanjing fine sand were carried out. The sample diameter D = 39.1 mm, height H = 80 mm, relative density $D_r = 50\%$, and the basic physical parameters are shown in Table 1. When pore pressure ratio $\mu_N/\sigma_c' = 1$, the test was stopped. The dynamic sinusoidal stress of 1Hz was exerted on the sample, and N_c represents the experienced cyclic number.

Time history curves of pore water pressure and hysteretic curves of soil samples are shown in Figs. 6 and 7 from triaxial tests, where initial effective stress $\sigma_c' = 100$ kPa and dynamic cyclic shear stress ratio $\tau_d/\sigma_c' = 1.05$.

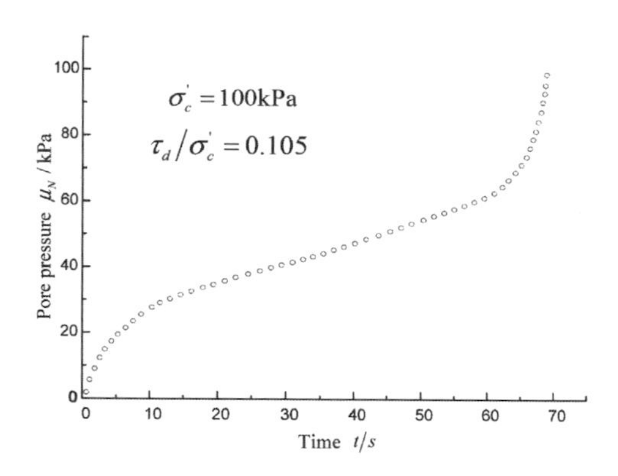

Figure 6. Time–history curve of pore pressure.

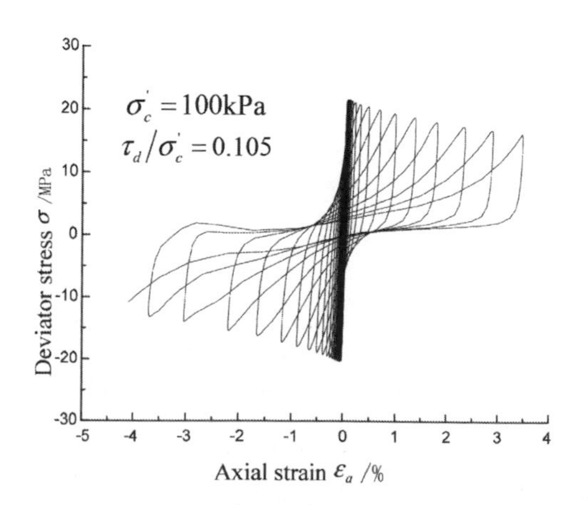

Figure 7. Strain–stress curve in the test.

5.3 *Numerical simulation of undrained cyclic triaxial tests of saturated Nanjing fine sand*

Undrained cyclic triaxial tests of Nanjing fine sand were compared with the corresponding three-dimensional numerical simulation, which was utilized to verify the correctness of the calculation results with effective stress algorithm in subroutine modules under simple stress condition. The sample model was made of an eight-node hexahedron reduced integration unit (C3D8R), whose unit number was 96. Two rigid blocks with same cross section without weight were set on the top and bottom of the model, which made the load uniformly distributed under cyclic load. The rigid block on the bottom was fixed. Cyclic load was applied on the upper rigid block. Stress artificial boundary was applied around the model, as shown in Fig. 8. Parameters for numerical calculation are shown in Table 2.

The relation curves between pore pressure ratio (μ_N/σ_c') and cyclic number ratio (N/N_c) of undrained cyclic triaxial tests of Nanjing fine sand under different isotropic consolidation conditions are described in Fig. 9, with the corresponding numerical simulation results. The figure shows that with the gradual increase of cyclic number, the pore pressure finally developed to the initial effective confining pressure. From the overall shape of the curve, we can see that, under the same initial effective confining pressure, with the increase of axial dynamic shear stress ratio, the curve shape developed from S-type to hyperbola. The conclusion was in agreement with Binghui Wang's (Wang Bing-hui, 2011) summary of the transition about

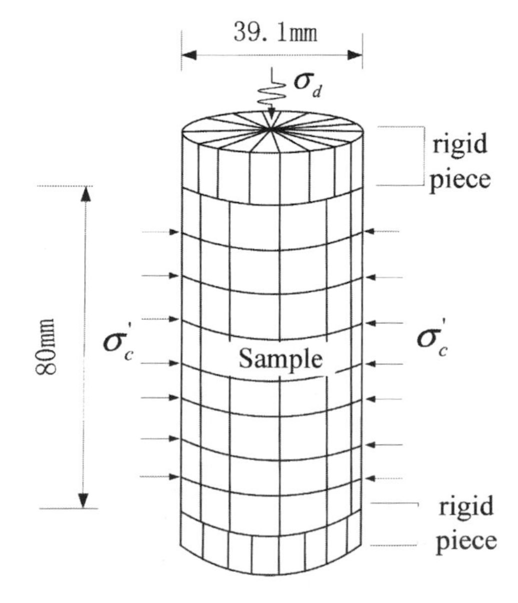

Figure 8. Numerical mode of cyclic triaxial test.

Table 2. Parameters of numerical calculation.

Model type	Parameter	Parameter values	Model type	Parameter	Parameter values
Modified Davidenkov model parameters	Unit weight	21.2 kN/m³	Pore pressure model parameters	u	0.444
	shear wave velocity	205.8 m/s		C_1	0.730
	A	1.02		C_2	0.475
	B	0.36		m	0.43
	u	0.49		n	0.62
	r_0	0.00038		k	0.0025
	a	0.5		b	0.5

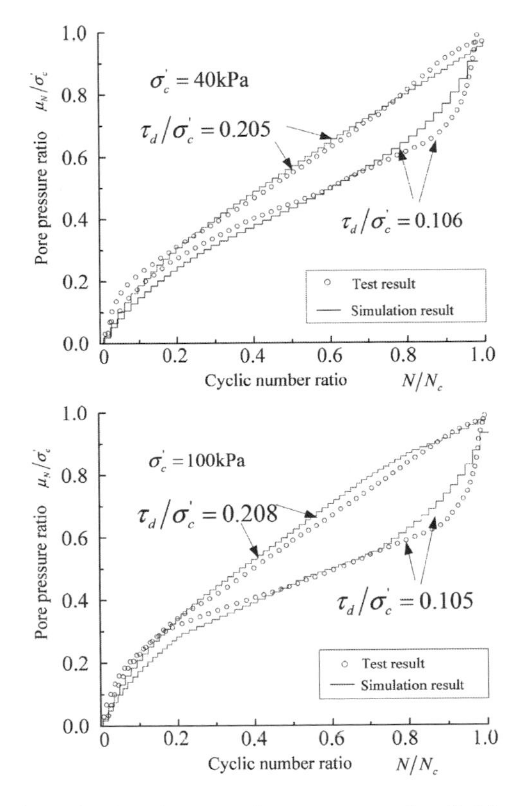

Figure 9. Development of pore pressure of cyclic triaxial test and numerical simulation.

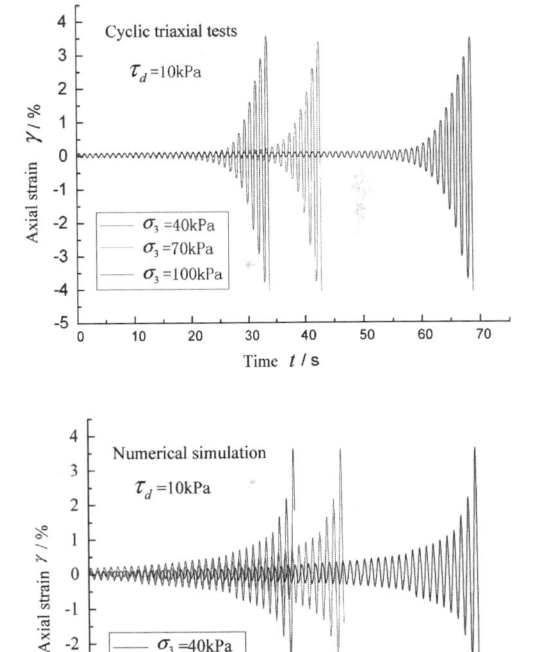

Figure 10. Time–history curves of strain in different initial confining pressure.

pore pressure ratio development from type B to type A. With the increase of the initial effective confining pressure, pore pressure development regulation under the same (τ_d/σ_c') of development basically remained unchanged.

Fig. 10 shows the axial strain curves of cyclic triaxial tests and numerical simulation under cyclic loading at constant amplitude with different initial effective confining pressures. Furthermore, with the increase of the initial effective confining pressure, the samples took longer time to reach the liquefaction state. Sample failures of cyclic triaxial tests and numerical simulation were, at the same time, 35, 45, and 70 s, respectively. Both results were in accordance with the trumpet-shaped development regulation. The numerical simulation results in the initial stage of cyclic load are not almost unchanged as the triaxial test results, but slowly grew. In strain mutation stage, the simulation results were not as obvious as the results of triaxial tests. The simulation liquefaction timing and the strain amplitude of the sample model basically agreed with the results of triaxial tests.

6 CONCLUSION

- On the basis of the Davidenkov constitutive model modified by irregular loading and unloading criterion and the introduction of Byrne-modified Martin and Finn pore pressure model, as well as a large number of resonant column test data fitting, this paper proposed the modified formula of G_{max} and γ_0 considering the different initial effective confining pressure. On account of the explicit module of ABAQUS platform, it realizes the subroutine development of the modified Davidenkov constitutive model with irregular loading and unloading criterion and the Byrne incremental model of pore water pressure. In order to get the parameters of Byrne pore pressure model of Nanjing fine sand, the saturated Nanjing fine sand drained cyclic triaxial tests were carried out. Undrained cyclic triaxial tests were also conducted for the same sand, and three-dimensional numerical simulation was conducted for the tests. The following conclusions are drawn:

1. The trends of relation curves between volumetric strain and cyclic number (N) of Nanjing fine sand and crystal silica sand are consistent basically. Under the cyclic loading of same shear strain amplitude, volumetric strain increases with the number of cycles. At the same cycle, the volumetric strain increased, while the shear strain amplitude increased.
2. As the cyclic number ratio increased, the pore pressure ratio (μ_N / σ_c') increased gradually, and the pore pressure (μ_N) finally developed to the initial effective confining pressure (σ_c'). Under the same initial effective confining pressure, the shape of μ_N / σ_c' curve changed from S-type to hyperbola with the increase of axial dynamic shear stress ratio (τ_d / σ_c'). At the same cyclic shear stress ratio τ_d / σ_c', as the initial confining pressure (σ_c') increased, the pore pressure development was essentially constant, and the pore pressure development trend of simulation results and triaxial test results were basically the same.
3. Under the condition of invariable amplitude of cyclic loading, as σ_c' increased, the soil sample took more time to liquefy. The simulation liquefaction time basically agreed with the result of triaxial tests, and both results accorded with the trumpet-shaped development regulation. The axial strain in the initial stage of cyclic load in numerical simulation is not the same as the corresponding results in triaxial tests, which remain almost unchanged, but slowly grows; the strain mutation is not as obvious as the result of triaxial tests.

REFERENCES

Byrne P M. A cyclic shear-volume coupling and pore pressure model for sand [C]. Mssouri, USA: 1991:47–56.

Chen Guo-xing, Chen Lei, Jing Li-pin. Comparision of implicit and explicit finite element methods with parallel computing for seismic response analysis of metro underground structures[J]. Journal of the China Railway Society, 2011, 33(11): 111–118.

Chen Guo-xing, Zhuang Hai-yang. Developed nonlinear dynamic constitutive relations of soils based on Davidenkov skeleton curve[J]. Chinese Journal of Geotechnical Engineering, 2005, 27(08): 860–864.

Chen Guo-xing. Geotechnical Earthquake Engineering [M]. Beijing:science press,2007.

Finn W D L, Lee K W, Martin G R. An effective stress model for liquefaction[J]. Journal of the Geotechnical Engineering Division, ASCE, 1977, 103(6): 517–533.

Liyanathirana D S, Poulos H G. Numerical simulation of soil liquefaction due to earthquake loading[J]. Soil Dynamics and Earthquake Engineering, 2002, 22(7): 511–523.

Martin G B, Finn W D L, Seed H B. Fundamentals of liquefaction under cyclic loading[J]. Journal of Geotechnical Engineering, ASCE, 1975, 101(5): 423–438.

Martin P P, Seed H B. One-dimensional dynamic ground response analyses[J]. Journal of the Geotechnical Engineering Division, ASCE, 1982, 108(7): 935–952.

Pyke R M. Nonlinear soil models for irregular cyclic loadings[J]. Journal of Geotechnical Engineering, ASCE, 1979, 105(6): 715–726.

Rong Mian-shui, Li Hong-guang, LI Xiao-jun. Applicability of Davidenkov model for soft soils in sea areas[J]. Chinese Journal of Geotechnical Engineering, 2013, 35(S2): 596–600.

Seed H B, Martin G R, Lysmer J. Pore-Water pressure changes during soil liquefaction [J]. Journal of the Geotechnical Engineering Division, ASCE, 1976, 102(4): 323–346.

Seed H B, Martin G R, Lysmer J. The generation and dissipation of pore water pressures during soil liquefaction[R]. earthquake engineering research center, university of california, berkeley, 1975.

Tokimatsu K, Seed H B. Evaluation of settlements in sands duo to earthquake shaking[J]. Journal of Geochemical Engineering, ASCE, 1987, 113(8): 861–878.

Wang Bing-hui, Chen Guo-xing. Pore water pressure increment model for saturated Nanjing fine sand subjected to cyclic loading[J]. Chinese Journal of Geotechnical Engineering, 2011, 33(02): 188–194.

Zhang J, Andrus R D, Juang C H. Normalized shear modulus and material damping ratio relationships[J]. Journal of Geotechnical and Geoenvironmental Engineering, 2005, 131(4): 453–464.

Zhuang H, Hu Z, Wang X, et al. Seismic responses of a large underground structure in liquefied soils by FEM numerical modelling[J]. Bulletin of Earthquake Engineering, 2015, doi:10.1007/s10518-015-9790-6.

Civil, Architecture and Environmental Engineering – Kao & Sung (Eds)
© 2017 Taylor & Francis Group, ISBN 978-1-138-02985-9

Urban ITS framework and essential parts of the ITS frame structure

Meng-qi Li, Zhen-hua Wang, Wei Wan & Ze-tao Jing
China Aerospace System Engineering Company, Beijing, China

ABSTRACT: Urban ITS systems have been constructed by different departments batch by batch and stage by stage at present; hence, there are many problems among them, such as mutual independence, difficulty of information sharing, and lack of top-level design. To resolve these problems, the method of urban ITS framework and important content of its construction is presented in this paper, which is based on the advanced information technologies such as big data. The framework includes six parts, including the data center, the efficient dispatching system, the convenient services, publishing system, and so on. The framework provides a reference and guideline for the urban ITS system design from the overall view of the city.

1 INTRODUCTION

Since the 1990s, China has made several achievements in the urban ITS construction in many cities, but not without problems. On the one hand, because of the lack of overall planning and top-level design, most ITS systems have so many problems, such as mutual independence, self-containment, lack of information sharing, and business collaboration. On the other hand, the new generation technologies, such as IOT, cloud computing, and big data, have developed rapidly. Meanwhile, new demands and requirements of the urban ITS construction are put forward. Therefore, not only how to build a more seamless-connected ITS systems, but also how to integrate the existing intelligent transportation systems and information resources to mining and applying have become new challenges. A new framework of the urban ITS system and its important construction contents are presented on the basis of the aerospace system engineering and the advanced information system architecture design method in developed countries. The objective of this paper is to provide a reference and guidance for the urban ITS construction of cities in China.

In the 1980s and 1990s, the United States, Japan, and the European Union have proposed the TIS framework (Wang, 2004). Meanwhile, the ITS framework of China is also proposed in the "Eleventh Five" period (Liu, 2004; Zhang, 2005), and the system framework effectively promoted the development of the ITS in various countries. However, because of the rapid development of the information technology, the urban intelligent transportation system framework has been difficult to adapt to the requirements based on new technologies and businesses. The urban ITS systems primarily constructed by government are the fastest growing and more mature e-government systems. With the large-scale construction of e-government and the development of new technologies, the United States, Britain, Australia, and other European countries have proposed different methods of e-government architecture design, such as Federal Enterprise Architecture (FEA) in the United States, the UK E-Government Interoperability Framework (E-GIF), the Australian Government architecture (AGA), the U.S. Global Information Grid (GIG) and other methods (Jin, 2008; Fang, 2007). In these methods, the United States has first begun to study the architecture design method of e-government, and is also the most advanced and mature one. Frameworks of other countries are referenced from ideas and methods of the FEA more or less. The above advanced information system design methods are referenced also by the paper.

2 URBAN INTELLIGENT TRAFFIC INFORMATION SYSTEM STATUS AND PROBLEMS

With the development of the information technology, integration of the urban intelligent transportation construction has become a major challenge and a strategic issue.

Information integration includes four aspects: information collection, information processing, information mining, and information publishing operations. Its aim is to achieve traffic information to optimize the allocation of resources to broaden the application field of information resources and maximize the value of information mining.

2.1 *Traffic information collection*

Traffic information is composed of static and dynamic traffic information. Specifically, dynamic traffic information currently collects operational data including highways, public transportation information, long-distance passenger statistics, and rail and air data. Static traffic information focuses on the basic geographic information. Traffic information collection has some problems such as limited scope of the acquisition, lack of integration of information processing, and noncomprehensive information collection.

2.2 *Processing and handling traffic information*

Traffic information processing in accordance with the processing methods can be divided into two sites: collected traffic information within their decentralized processing application system and collected traffic information is aggregated to the data processing center for centralized processing. Traffic information system processes have the following problems: data sharing limitations, insufficient processing depth, lack of standards and process specifications, inadequate public service and interdepartmental applications, and lack of appropriate mechanisms and funding.

2.3 *Use of traffic information*

The traffic control department achieves several management functions by using processed traffic information such as transport command and control, traffic monitoring, traffic enforcement, and traffic information services. The current e-transport information mining degree is not deep enough, has information use limitations, and is a single source of information.

2.4 *Traffic information release and operations*

Traffic management departments can provide traffic information through a variety of ways of public traffic information, such as website, radio, mobile client, and roadside traffic guidance. This type of the problem, including traffic information services market, is not mature, lacks legal protection, and the operating system is not perfect.

3 THE URBAN IT FRAMEWORK

Urban Intelligent Transportation System mainly consists of many aspects: road traffic management, emergency roads, rails, transportation companies, traffic governments, e-charges, and so on. The definition of the level and relationship in urban ITS will play an important role for the future building. On this basis, this paper proposes a more scalable and advanced urban intelligent transportation system framework, namely "one data center, and five application systems" (the red by category data centers need to exchange information, shown in Figure 1), from the perspective of the overall city.

3.1 *Construction of the urban traffic data center*

A comprehensive data center can effectively coordinate the relationship between data centers and data exchange platform to achieve acquisition integration, intensive treatment, and the depth of excavation. Other modules can use sharing data in time from five system modules storage in a traf-

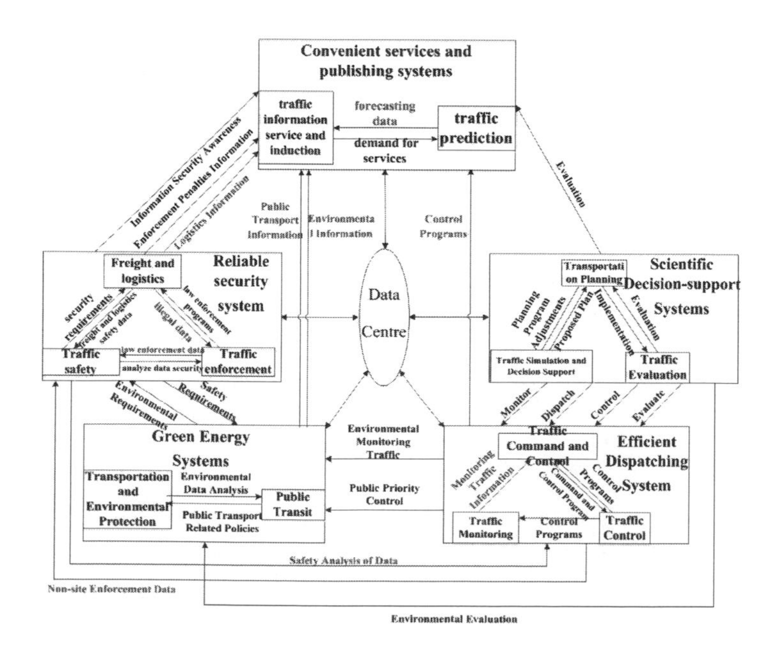

Figure 1. Top-level framework of urban IT.

fic data center. Meanwhile, the other data do not have to put in a data center, so modules can take full advantage of the resources within the system to process traffic information as well as reduce the pressure on the comprehensive traffic data center.

We constructed an urban transportation data center that meets the demands of national transport planning, construction, operation, management, and public transport information service. This data center integrates urban transportation data resources to achieve the purposes of storage, management, search, share, analysis, and publicizing of national traffic data resources. Furthermore, advanced ITS construction method will enhance the service quality of the city traffic management.

3.2 *Efficient dispatching system*

Urban road network has its own characteristics. Although increasing the construction of transport infrastructure can solve part of the traffic problem, it does not solve the problem fundamentally. Under the limited growth conditions of the road network infrastructure and minimal traffic operating costs, building an efficient dispatching system can improve the capacity of the largest possible pass through and rapidly respond to emergencies. A blameless and efficient dispatching system should include daily dispatch and emergency dispatching for major urgent events.

The relationship between dispatching system and external system is shown in Figure 2. Traffic departments draft control plan for the transportation program has to use comprehensive information. After the implementation of traffic control plans, the management needs to get evaluation data in time. On the basis of the above evaluations, the traffic control department can understand the method effectiveness. Traffic control plans also need to agree with the traffic information forecast and traffic information service and issue specific traffic control measures to the public by ways of forecasting and services.

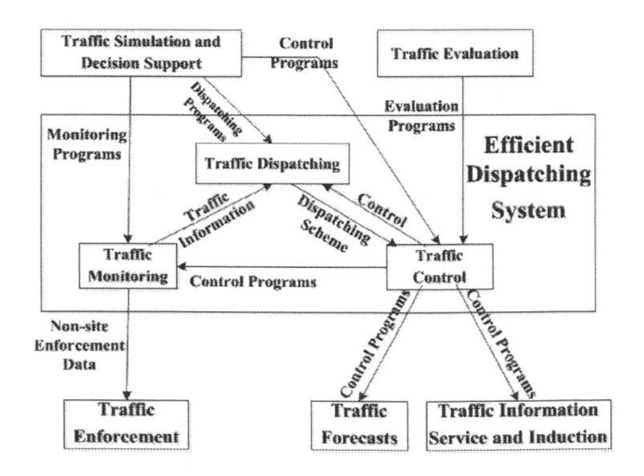

Figure 2. Efficient dispatching system.

3.3 *Convenient services and publishing system*

Convenient services and publishing system can provide convenient travel information services for the public by road-side traffic guiding panels or application programs. Meanwhile, this system can guide public to avoid traffic congestion.

3.4 *Scientific decision support system*

Scientific decision support system includes traffic simulation and decision support, transportation planning, and traffic evaluation. This system can not only propose quantitative assessment analysis of traffic scheduling management program, but also give analysis and mining report on historical traffic data, providing decision-making support for the basis of transportation infrastructure planning and construction or other studies.

3.5 *Reliable security system*

The number of traffic accidents increasing with the increasing number of vehicles in cities. Improving road safety and establishing a reliable security system is a long and complex process.

To ensure the operation coordination between all systems, the interlayer and intralayer data transmission require a unified specification in the Intelligent Transportation System. These specifications and system operation management together constitute the essential standards system. In addition, the unauthorized use of the Intelligent Transportation System, which is an information system that has a direct effect on the transport system, may seriously affect the normal operation of transportation systems or even cause inestimable confusions. Therefore, the entire Intelligent Transportation System should be subjected to comprehensive and strict control under the security system to prevent data theft, illegal system intrusion, and other issues.

3.6 *Green energy system*

Green environmental protection system can improve the quality of the environment and the satisfaction of urban transportation. This system is the inherent requirement of the whole society to achieve sus-

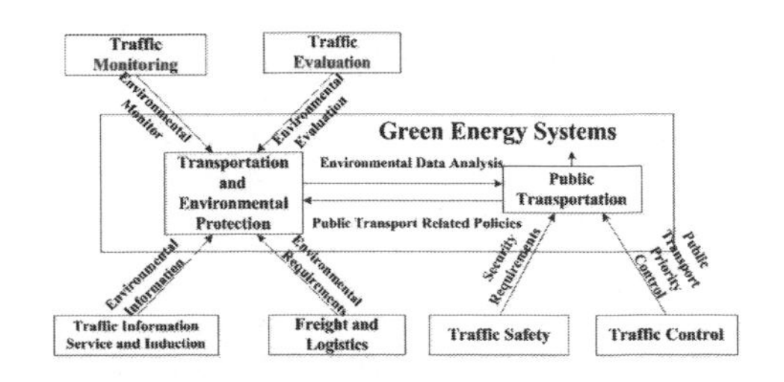

Figure 3. Green energy system.

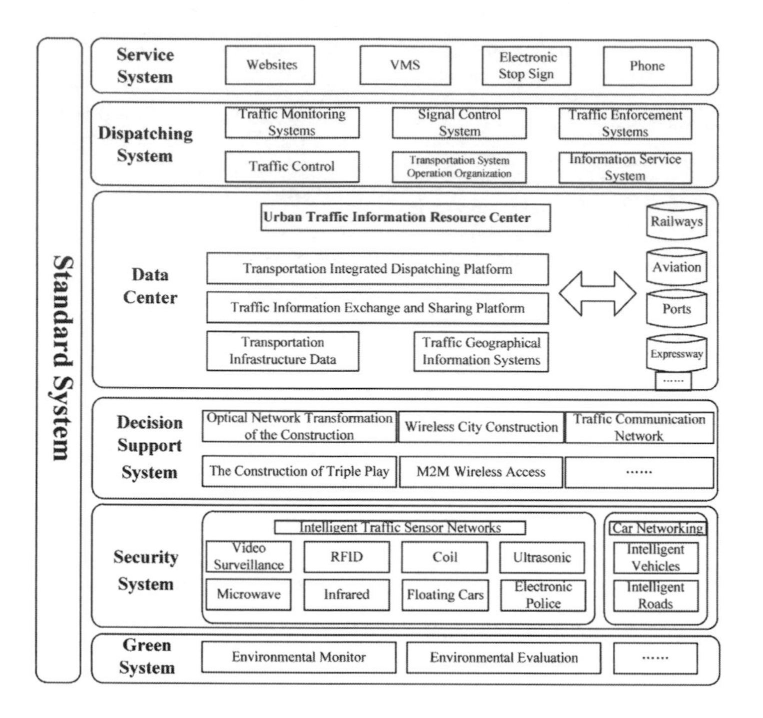

Figure 4. Application of urban ITS framework.

tainable development. Creating energy-saving and environmental protection systems should build an appropriate detection system that will use big data technologies to summarize the causes of pollution, put forward recommendations of environmental protection, and promote the construction of environmentally friendly transportation by means of information.

Transportation and environmental protection is responsible for the traffic pollution monitoring and supervision, mainly for monitoring emissions compliance, noise pollution, and dust pollution. Public transportation administration uses the surveillance to introduce the policy of traffic and environmental protection, set reasonable bus routes, and reduce rail traffic noise pollution. This system can effectively improve the quality of urban environment and create favorable conditions for the city's construction.

3.7 *The application of urban ITS framework*

Combining with the city's characteristics, the urban intelligent framework described in this paper has been utilized in its traffic system. It includes one data center and five systems (shown in Figure 4). The framework has been applied to intelligent traffic plan, design, and construction works in the city. It can effectively solve the problem presented in the second chapter.

4 CONCLUSIONS

To resolve the urban ITS construction problems, the new urban intelligent transportation system framework is proposed in this paper. The framework provides a reference and guidance to further research, plan, design, and construction of urban intelligent transportation system. However, due to space constraint, only the urban intelligent transportation system framework has been described in brief, and details of ITS system architectural design method and construction method based on this framework need further research.

REFERENCES

Fang C, Journal of Chongqing University of Posts and Telecommunications (Natural Science), 19(2), 249–252 (2007).

Jin JJ, Wei ZJ, Information Construction, 06, 47–49 (2008).

Liu DM, Beijing University of Technology (2004).

National ITS Architecture Team National ITS Architecture 4.0 (Washington DC, USA, 2002).

Wang XJ, Qi TY, Cai H, *Principle and application of Intelligent Transportation Systems (ITS) framework and architecture* (China railway press, Beijing, 2004).

Zhang K, Qi TY, Liu DM, Wang CY et al, Journal of Transportation Systems Engineering and Information Technology, 5(5), 6–11 (2005).

Civil, Architecture and Environmental Engineering – Kao & Sung (Eds)
© 2017 Taylor & Francis Group, ISBN 978-1-138-02985-9

Development and implementation of a product module division system based on the function–principle–structure

Jing Chen
College of Mechanical and Control Engineering, Guilin University of Technology, Guilin Guangxi China

Limin Wang
College of Information Science and Engineering, Guilin University of Technology, Guilin Guangxi China

Liangliang Yang
College of Mechanical and Control Engineering, Guilin University of Technology, Guilin Guangxi China

ABSTRACT: According to the theory of modular design and module partition key technology, we design and develop a product module partition system, which is based on the included system management, product management, expert evaluation, and module partition. Taking the gearbox for example, here, we use the module division principle and the method proposed in this paper to divide its main parts into several modules. The experimental results have verified the feasibility and validity of this method, and we have preliminary realized the engineering application of the product module partition method on the basis of function–principle–structure.

1 INTRODUCTION

Modular design takes the module as object and mainly includes module division, module planning, and module combination. Modular design technology first appeared in the early 20th Century. Since the concept of modular design was put forward in the 1950s, it has been applied in many fields, such as furniture and machine tools. After the 1970s, the modular design method in Chinese machine tool industry has gradually emerged and has been widely used in other industries (Gong, 2007). Pahl and Beitz (Pahl, 1996) thought that the modular design is used to complete the mapping from the demand domain to the functional domain and then complete the mapping from the functional domain to the domain based on the consideration of the performance of the modules. Li Yupeng (Li, 2011) aimed at modeling the difficulties of traditional module identification method, computed complexity, could not effectively deal with the evaluation data problem of fuzzy and uncertainty, and put forward the product module clustering method based on rough set. The evaluation data are transformed into the form of rough numbers to fuzzy cluster analysis. In the cluster graph, the module is divided according into upper and lower threshold values, which makes the data processing more objective. On the basis of the simulated annealing algorithm, Shan quan (Quan, 2007) integrated the population of the genetic algorithm and proposed the module division method on the basis of improved simulated annealing algorithm, which combines the advantages of both, thus avoided the following shortcomings: the simulated annealing algorithm is difficult to operate, the efficiency is not high, and the genetic algorithm is easy to fall into the local optimal solution.

This paper takes the composition of the product components for the module object and considers the number of assembly and the relationship among the parts. With a gearbox as an example, we analyzed the Function–Principle–Structure (F-P-S) on the basis of the product module partition method and introduced the system of product module partition on the basis of F-P-S implementation.

2 PRODUCT MODULE PARTITION

2.1 *Product module partition principle and overall planning*

The most basic principle of module division is independent of function and structure. On the basis of functional requirements, it takes structure as the carrier to meet customers' needs and ultimately ensures that the strong correlation among the internal modules and the weak coupling among the modules. Another basic principle is the changing needs of customers, including the higher needs for quantity and quality. These higher needs will create products with more functions. The existing module partition method (Quan, 2012) mainly includes two types: one is the function analysis method, which emphasizes the function of the module, and it is mainly used in the phase of requirement analysis and conceptual design; the other is to emphasize the independence of the module structure of the parts analysis method; this method is carried out mainly through the analysis of product components and structural correlation between parts and combined with other factors to complete the division of modules. Product module division process is summarized as: the establishment of the existing product structure parts information management database. According to the total function of product design code, the model is decomposed into subfunctions by F-P-S mapping model and the establishment of the functional structure of the product. The principle of each subfunction and its realization structure are analyzed, and the principle of division based on F-P-S is obtained. The function–principle–structure of the characteristics of indicators or related characteristics is quantified as a module division of the comprehensive evaluation of the data. Finally, through the fuzzy clustering method to achieve the cluster analysis of product components, we get the final module partition scheme. The process of product module partition is shown in Figure 1.

2.2 *Product module partition method based on cluster analysis*

The integrated correlation matrix is transformed into fuzzy equivalent matrix by using the transitive closure method (Wang, 2003). On the basis of the fuzzy equivalent matrix, the parts of the product are clustered by the method of calculating the maximum correlation degree among the components.

First, the threshold of cluster analysis is determined. In the cluster analysis, the threshold value can be artificially set; however, there is some uncertainty in the selection of threshold. In this paper, by calculating the average value of the correlation distance among parts, as the threshold of cluster analysis, the formula for calculating the average correlation degree is:

$$K = \frac{\sum_{i=1}^{n}\sum_{\substack{j=1 \\ i \neq j}}^{n} d_{ij}^{*}}{n(n-1)} \tag{1}$$

Second, a pair of components with the largest correlation degree is calculated, and they are gathered together as a module. Each line represents a correlation between one part and other parts of the n–1, and $d_{ii}^{*} = 1(i = 1,\ldots,n)$. Therefore, by calculating the largest element in the fuzzy equivalent matrix D*, the formula to calculate the parts x_p, x_q with the largest correlation degree is:

$$T_{(p,q)} = \max\{d_{ij}^{*}(i = 1,2,\ldots,n; j = 1,2,\ldots,n; i \neq j)\} \tag{2}$$

Because the parts x_p, x_q correlation is the largest, the module is denoted as M(p,q).

The processing method for the original matrix is: First, paddle the p,q line and the p,q column in the matrix D*. Second, calculate the relationship

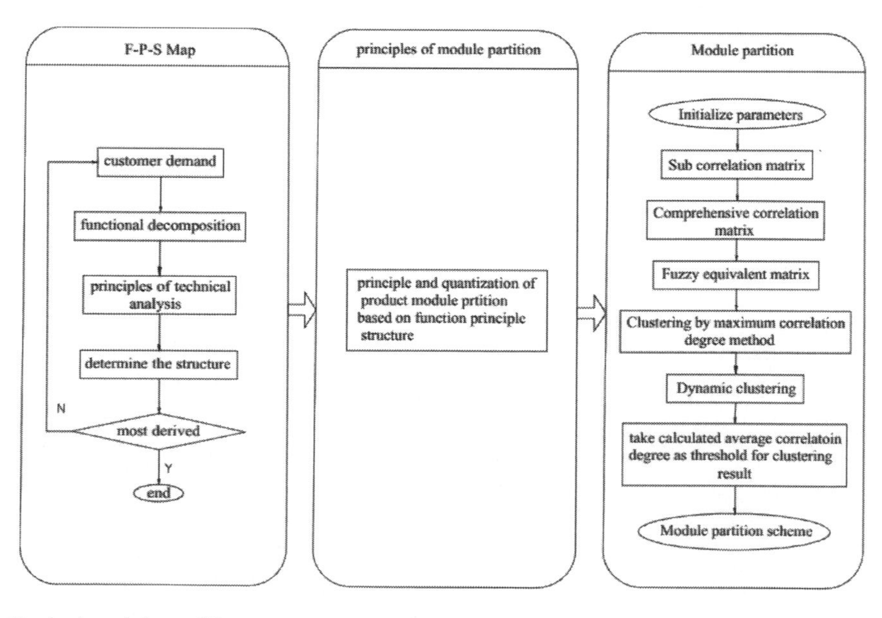

Figure 1. Product module partition process.

distance between the module M(p,q) and the rest of the parts as:

$$d_{iM}^* = \max\{d_{ip}^*, d_{iq}^*(i = 1,2,...,n; i \neq p \neq q)\} \quad (3)$$

Third, add the module M(p,q) to the matrix and take the relationship distance d_{iM}^* between the module M(p,q) and the remaining parts as an element that the module M(p,q) is in the matrix. Therefore, the order of the new matrix after processing the original matrix is n–1.

Then, use the second step of the algorithm to calculate the largest correlation degree of a pair of components or modules and put them into a module. In the second step, the processing for n–2 order matrix is determined by using the original matrix.

Finally, with the average correlation degree K as threshold, according to the second step and the third step cycle operation, until the largest correlation degree is less than the average correlation matrix K, terminate the process of clustering algorithm. The resulting matrix is the result of cluster analysis as well as module partition.

3 THE SYSTEM DEVELOPMENT BACKGROUND AND THE OVERALL FRAMEWORK OF THE SYSTEM

For system implementation, the choice of development platform is very important. The development platform can realize the algorithm and play a powerful role in computer-aided product design. The design in this paper, based on the F-P-S of the product module partition of the system algorithm, mainly uses Microsoft's mainstream development tool Microsoft Visual studio 2005. This tool has superior application range and execution efficiency. System development platform and environment selection are as follows:

The hardware configuration: core T5870; 1024 MB of memory; 256 MB of memory

Software configuration: Windows 7 Operating System; Microsoft Visual studio 2005

Management module includes system management submodule and product submanagement module. System management module mainly carries out the management tasks of the personnel, the system involves the process of expert score, the system management module achieves the expert's account number, password, and permissions management, and product management module is mainly composed of parts of the product information management, including the entry, query, delete, and update function. The management module is to provide the necessary information and data support for the main module.

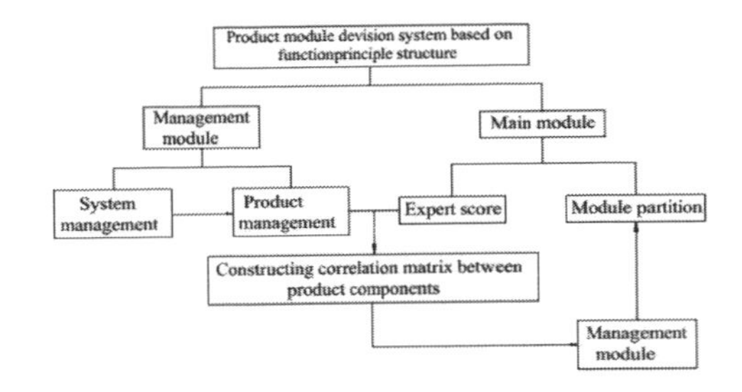

Figure 2. System integrated framework.

Main modules include expert evaluation submodule and module partition submodule. Expert evaluation module scores mainly through the experts on the correlation between parts, and factor score includes three aspects: the function-principle-structure score results in real-time display, and data can be queried, modified, deleted. The submodule mainly provides expert evaluation results and clustering analysis of parts using fuzzy clustering algorithm, and displays the final module partition results.

4 SYSTEM FUNCTION REALIZATION

In order to facilitate the evaluation of information between parts statistics and calculation, in this paper, with the aid of computer-aided methods and combining with the proposed product module of the F-P-S based on the developed theoretical knowledge and the F-P-S product module partition system, the system mainly consists of management module and the main module. The main module is mainly used for the cluster of the product module. The management module mainly manages the data in the process of product division. Both seamless integration and information sharing in the whole system, to a large extent, improve the efficiency of product modular design.

4.1 *Management module*

The management module includes system management submodule and product management submodule. The system management module includes two parts: user login and personnel management. User login part includes user authority identification and determines user's permissions; personnel management part allows system users to add, delete, and modify permissions and passwords. Product management submodule mainly manages the relevant information of the components of products. The add part is mainly to add parts' name which is composed of the structure of the parts, parts drawing number, component properties, parts material, procurement company information, and so on.

Module partition sub module

Fuzzy equivalent matrix:

	01	02	03	04	05	06	07	08	09	10	11	12	13	14	15	16	17
01	1	0.384	0.384	0.384	0.384	0.384	0.384	0.384	0.384	0.384	0.384	0.384	0.384	0.384	0.384	0.384	0.384
02	0.384	1	0.384	0.384	0.384	0.384	0.384	0.384	0.384	0.384	0.384	0.384	0.384	0.384	0.384	0.384	0.384
03	0.384	0.384	1	0.384	0.384	0.384	0.384	0.384	0.384	0.384	0.384	0.384	0.384	0.384	0.384	0.384	0.384
04	0.384	0.384	0.384	1	0.648	0.384	0.384	0.648	0.486	0.486	0.648	0.384	0.648	0.486	0.486	0.486	0.648
05	0.384	0.384	0.384	0.648	1	0.384	0.384	0.648	0.486	0.486	0.648	0.384	0.648	0.486	0.486	0.486	0.648
06	0.384	0.384	0.384	0.384	0.384	1	0.384	0.384	0.384	0.384	0.384	0.384	0.384	0.384	0.384	0.384	0.384
07	0.384	0.384	0.384	0.384	0.384	0.384	1	0.384	0.384	0.384	0.384	0.384	0.384	0.384	0.384	0.384	0.384
08	0.384	0.384	0.384	0.648	0.648	0.384	0.384	1	0.486	0.486	0.648	0.384	0.648	0.486	0.486	0.486	0.648
09	0.384	0.384	0.384	0.486	0.486	0.384	0.384	0.486	1	0.7	0.486	0.384	0.486	0.7	0.7	0.738	0.486
10	0.384	0.384	0.384	0.486	0.486	0.384	0.384	0.486	0.7	1	0.486	0.384	0.486	0.738	0.762	0.7	0.486
11	0.384	0.384	0.384	0.648	0.648	0.384	0.384	0.648	0.486	0.486	1	0.384	0.648	0.486	0.486	0.486	0.648
12	0.384	0.384	0.384	0.384	0.384	0.384	0.384	0.384	0.384	0.384	0.384	1	0.384	0.384	0.384	0.384	0.384
13	0.384	0.384	0.384	0.648	0.648	0.384	0.384	0.648	0.486	0.486	0.648	0.384	1	0.486	0.486	0.486	0.648
14	0.384	0.384	0.384	0.486	0.486	0.384	0.384	0.486	0.7	0.738	0.486	0.384	0.486	1	0.738	0.7	0.486
15	0.384	0.384	0.384	0.486	0.486	0.384	0.384	0.486	0.7	0.762	0.486	0.384	0.486	0.738	1	0.7	0.486
16	0.384	0.384	0.384	0.486	0.486	0.384	0.384	0.486	0.738	0.7	0.486	0.384	0.486	0.7	0.7	1	0.486
17	0.384	0.384	0.384	0.648	0.648	0.384	0.384	0.648	0.486	0.486	0.648	0.384	0.648	0.486	0.486	0.486	1

Average correlation degree:

k=0.492

Gearbox module is divided into:
Module 1:1.2.3.6.7.12
Module 2:4,5,6,11.13,17
Module 3:9.10,14.15,16

Figure 3. Module cluster analysis interface.

4.2 *Main module*

The main module of product module division system based on F-P-S consists of two parts: expert evaluation submodule and module partition submodule.

4.2.1 *Expert evaluation submodule*

After the expert raters choose two parts, there comes the need to score, the score of each of the two parts between the function correlation, structure correlation, and principle correlation. Evaluation data would be displayed in real time by reading documents based on F-P-S module division principle and quantitative methods. Experts determine the product function, the principle, and the structure in to evaluate weight. In the construction of the integrated correlation region enter function, principle, and structure of weight value, the system will automatically calculate the two parts between the integrated correlation results of real-time display. Then, the cycle is completed by evaluating each two component and the final comprehensive correlation matrix is obtained.

4.2.2 *Module partition submodule*

The system applies the method of transitive closure, transforms the comprehensive correlation matrix into fuzzy equivalence matrix, and finally displays the transformation, thereby completing the matrix transformation process. To calculate the fuzzy equivalence matrix element, the average correlation degree is obtained. The matrix interface is shown in Figure 3.

5 CONCLUSION

In this paper, we introduced the product module division system, design platform, development tools, management module, and main module on the basis of function-principle-structure. Furthermore, the relationship between parts and components based on function, principle and structure was evaluated. On the basis of the cluster analysis of the comprehensive correlation matrix, the division of product modules based on function-principle-structure is realized. The effectiveness of this method is verified by the example of the division of the gearbox. The product module partition method can provide support to product design on the basis of modular, and the further research works will focus on the evaluation of the product module partition results.

REFERENCES

Haijun Wang, Baoyuan Sun. Analysis of Product Modular Formation Process Based on Fuzzy Clustering. Computer Integrated Manufacturing System, 2003, 9(12):123–12.

Pahl G, Beitz W. Engineering Design: a systematic process. London: Springier, 1996.

Quan Shan, Guangrong Yan, Yi Lei. Research and Application of Improved Simulated Annealing Algorithm in Module Partition. Computer Engineering, 2007,(12):208–210+213.

Technology of Rapid Design Based on Product Platform: [PhD thesis]. Nanjing: Nanjing University of Science and Technology, 2007.

Xiaobin Chen. Research and application of green module partition method for electromechanical products: [Master Dissertation]. Zhejiang University, 2012.

Yupeng Li, Xuening Zhu, Xiuli Geng etal. Module Clustering Method and Its application Based on Rough Set. Mechanical Design and Research. 2011,(04):1–5+12 W.

Zhibing Gong, Research and Implementation of Key.

Civil, Architecture and Environmental Engineering – Kao & Sung (Eds)
 ISBN 978-1-138-02985-9

Parameter optimization of space truss and aerospace load viscoelastic damping composite structure

Zhenwei Zhang & Peng Wang
School of Mechanics, Northeastern University, Shenyang, China

Haitao Luo
Shenyang Institute of Automation, Chinese Academy of Sciences, Shenyang, China

ABSTRACT: The acceleration response value of space truss and aerospace load structure is exceeded in the acceptance level vibration test. The method of applying the viscoelastic damping layer is used to reduce the vibration of the space truss and the aerospace load structure. The test results show that the damping effect is obvious, and it is needed to determine the structure parameters and material parameters of the viscoelastic damping layer. The structural parameters include the location of the viscoelastic damping layer, the thickness of the damping layer, and the thickness of the constrained layer. The material parameters include the shear modulus of the damping layer, damping factor of the damping layer, and the selection of the constrained layer. With index of damping factor and the resonance frequency of vibration, the influence of structural parameters and material parameters on the damping effect of viscoelastic damping materials is obtained by the optimization and comparison of damping factors.

1 INTRODUCTION

According to the practical need, the method of applying the viscoelastic damping layer (Shen, 2006) to the aerospace load box structure is used to reduce the vibration. In this paper, the constrained damping layer is adopted, which is composed of damping layer and constrained layer. Viscoelastic damping layer is in the form of simple, light weight, easy to paste, excellent damping performance, does not need to change the existing structure, the use of less material will be able to achieve a larger damping effect (Wang, 1990).

For a viscoelastic damping scheme, it needed to determine the structural parameters and material parameters of viscoelastic damping layer. The structural parameters include the location of the viscoelastic damping layer, the thickness of the damping layer, and the thickness of the constrained layer. The material parameters include the shear modulus of the damping layer, the damping factor of the damping layer, and the selection of the constrained layer. In general, the material of the general constrained layer is the same as that of the structure layer. For the space truss and aerospace load structure, the 2A12 plate is selected as the constrained layer. The existence of the viscoelastic damping layer, on the one hand, has a great influence on the quality and stiffness of the structure, thus affecting the natural frequency of the structure. On the other hand, due to the viscoelastic damping layer is damping material, so the damping characteristic of the whole structure will be affected, and the resonance response peak and damping factor of the whole structure will be greatly affected.

2 EVALUATION INDEX OF VIBRATION REDUCTION EFFECT

The fundamental frequency of the vibration reflects the vibration characteristics of the structure. High vibration frequency shows that the structure stiffness is large, then the vibration is easy to decay. Therefore, the fundamental frequency of vibration from one side reflects the damping effect of the structure is good or bad. The damping factor (Yang, 2011) as an index to evaluate the damping effect because viscoelastic damping vibration reduction method is based on the form of no significant change in structure to increase the damping of the structure, so the damping level can accurately reflect how the damping effect of the scheme. The resonance response peak value of the structure is another index of evaluating the effect of vibration reduction because the resonant response of the structure is greatly influenced by the damping. The greater the damping is, the smaller the resonance response is, the smaller the damping is, the greater the resonance response is. In this paper, the vibration frequency and damping factor is the evaluation index of vibration reduction effect.

3 CALCULATION OF DAMPING FACTOR

Damping factor is obtained firstly for simple beam and plate structure through theoretical analysis (Ross, 1959; Mead, 1969). Johnson (Johnson, 1982) proposed using modal strain energy method to calculate damping factor, which makes it possible to use the finite element software to calculate the damping factor. The method of finite element software to calculate the damping factor has a direct frequency response method, modal strain energy method (Ren, 2004; Zhao, 2013) and complex eigenvalue method. In this paper, the complex eigenvalue method (Rikards, 1993) is used to calculate the damping factor. By using the complex eigenvalue calculation function provided by MSC/NASTRAN, it can be more convenient to determine the damping factor of the structure. By using the complex modulus model, the constitutive relation of viscoelastic materials is:

$$\begin{aligned}\sigma_0 &= E^*(\omega)\varepsilon_0 = \left[E'(\omega) + iE''(\omega)\right]\varepsilon_0 \\ &= E'(\omega)\left[1 + i\eta(\omega)\right]\varepsilon_0\end{aligned}$$

$$\eta(\omega) = \frac{E''(\omega)}{E'(\omega)}$$

In the formula, σ_0 and ε_0 represent the time-varying harmonic stress respectively. $E''(\omega)$ is complex modulus of materials. $\eta(\omega)$ is material damping factor. The free vibration equation of the composite structure is:

$$\begin{aligned}M\ddot{u}^* + K^*u^* &= 0 \\ K^* &= K + iK'\end{aligned} \tag{1}$$

In the formula, M is mass matrix. K^* is complex stiffness matrix. K and K' represent the real part and imaginary part of the complex stiffness matrix respectively. u^* is displacement. To make $u^* = \phi^* e^{iw^* t}$, formula (1) can be transformed into:

$$K^*\phi^* = \lambda^* M\phi^* \tag{2}$$

In the formula, ω^* is complex frequency. ϕ^* is complex feature vector. λ^* is complex eigenvalue. After solving formula (2) can be obtained:

$$\lambda^* = \lambda + i\lambda'$$

Thus, the n-th order modal damping factor η^n:

$$\eta_n = \frac{\lambda_n'}{\lambda_n}$$

This method is more intuitive and clear, and can describe the dynamic mechanical properties of viscoelastic materials very well (Bagley, 1979).

4 THE COMPOSITION OF SPACE TRUSS AND AEROSPACE LOAD STRUCTURE

The simulation model of space truss and aerospace load is mainly composed of the following parts, as shown in Figure 1. They are: box structure, long tube, short tube, connector, connection box, constrained damping layer and tooling.

Here, the exterior of the box structure is plate structure, and the interior is a hollow structure. The box structure is composed of the upper and lower cover plate, left and right cover plate, the front and back cover plate. Each two faces are connected by screws. The viscoelastic damping layer on the outer wall of the box structure is a constrained damping layer, as shown in Figure 2. The box structure is connected with the four connection boxes at the top of space truss through screws. In addition to the four surfaces in contact with the box structure, it has a diameter of 18 mm and a depth of 2 mm circular recess at central position of each remaining surface, for positioning the connector, as shown in Figure 3.

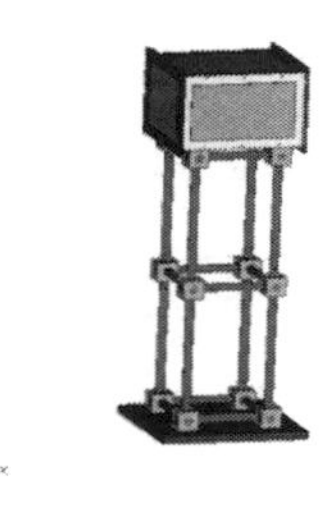

Figure 1. Finite element model of space truss and aerospace load structure.

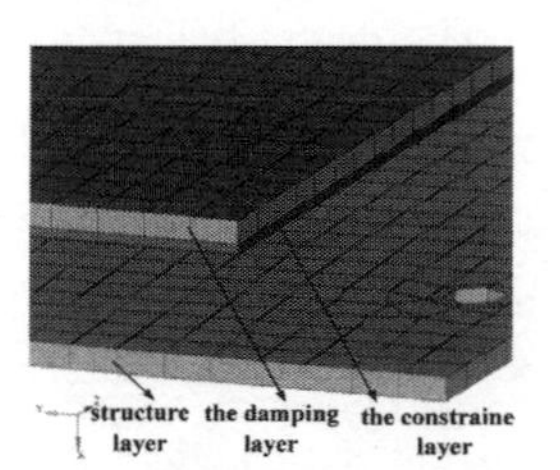

Figure 2. Constrained damping layer.

Figure 3. Connection box and connector.

5 PARAMETER OPTIMIZATION OF VISCOELASTIC DAMPING COMPOSITE STRUCTURE

In order to study the influence of damping layer thickness on the damping effect, take six groups of thickness values in the range of 0.25 mm–2.5 mm. The resonance frequency and damping factor increment obtained by the Nastran complex eigenvalue method are shown in Table 1. List the first six orders damping factor values. In order to express the influence of the damping layer thickness on the resonant frequency and damping factor, draw data from the table into a graph. Wherein the horizontal axis is the resonance frequency, and the vertical axis is the damping factor increment.

It can be seen from the figure that the resonance frequency decreases with the increase of the thickness of the damping layer. It can be seen from the data in table, when the damping layer thickness is 0.25 mm, the fundamental frequency is 23.7 Hz, the thickness is 2.5 mm, the fundamental frequency is 23.1 Hz. By referring to the second order to sixth order data, we can see that with the increase of the thickness of the damping layer, the decrease of the resonance frequency is not large. When the thickness of the constrained layer is constant, with the increase of the thickness of the damping layer, the damping factor increases. The reason is that the thickness of the damping layer increases and the energy consumption in the process of vibration increases. Then study the influence of the thickness of the constrained layer on the vibration damping effect, take six groups of thickness values in the range of 0.5 mm–3 mm, and the thickness of the damping layer is constant.

It can be seen from the figure that the resonance frequency decreases with the increase of the thickness of the constrained layer. It can be seen from the data in table, when the damping layer thickness is 0.5 mm, the fundamental frequency is 24.8 Hz, the thickness is 3 mm, the fundamental frequency is 22.8 Hz. By referring to the second order to sixth order data, we can see that with the increase of the thickness of the constrained layer, the decrease of the resonance frequency is not large. When the thickness of the damping layer is constant, with the increase of the thickness of the constrained layer, damping factor firstly increases and then decreases, and the damping factor reaches the maximum when the thickness of the constrained layer is 2 mm. The reason is that the constrained layer thickness increases, and the restraining ability of the constrained layer is increased, damping layer deformation is bigger, the energy consumed in the vibration process is more, and the damping factor increases. With the increase of the thickness of the constrained layer, the mass increases and plays a leading role, and the damping factor decreases.

Next, the influence of damping layer material on the damping effect is studied, and the shear modulus of the damping layer are chosen as the material property parameters. Viscoelastic damping materials

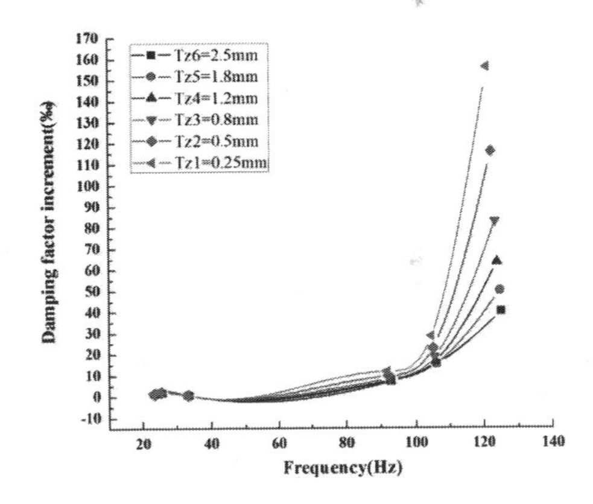

Figure 4. The curve of the damping factor increment and the resonance frequency with the change of the thickness of damping layer.

Table 1. Damping factor increment and the resonance frequency with the change of the damping layer thickness.

	$T_{z1} = 0.25$ mm		$T_{z2} = 0.5$ mm		$T_{z3} = 0.8$ mm		$T_{z4} = 1.2$ mm		$T_{z5} = 1.8$ mm		$T_{z6} = 2.5$ mm	
Order number	Frequency (Hz)	Damping factor increment (‰)	Frequency (Hz)	Damping factor increment (‰)	Frequency (Hz)	Damping factor increment (‰)	Frequency (Hz)	Damping factor increment (‰)	Frequency (Hz)	Damping factor increment (‰)	Frequency (Hz)	Damping factor increment (‰)
1	23.7	1.12	23.6	1.14	23.5	1.16	23.4	1.24	23.2	1.44	23.1	1.70
2	25.6	1.72	25.6	1.76	25.5	1.78	25.3	1.93	25.2	2.23	24.9	2.63
3	33.7	0.52	33.7	0.54	33.5	0.57	33.4	0.62	33.3	0.71	33.1	0.85
4	92.9	6.94	92.8	7.19	92.7	7.27	92.5	7.98	92.2	9.50	91.8	11.62
5	106.3	15.17	106.1	15.24	105.8	16.26	105.6	18.47	105.1	22.60	104.6	28.32
6	125.3	39.96	124.8	49.87	124.0	63.41	123.4	82.81	122.3	115.78	121.1	155.79

Table 2. Damping factor increment and the resonance frequency with the change of the constrained layer thickness.

Order number	T_{y1} = 0.5 mm Frequency (Hz)	T_{y1} = 0.5 mm Damping factor increment (‰)	T_{y2} = 1 mm Frequency (Hz)	T_{y2} = 1 mm Damping factor increment (‰)	T_{y3} = 1.5 mm Frequency (Hz)	T_{y3} = 1.5 mm Damping factor increment (‰)	T_{y4} = 2 mm Frequency (Hz)	T_{y4} = 2 mm Damping factor increment (‰)	T_{y5} = 2.5 mm Frequency (Hz)	T_{y5} = 2.5 mm Damping factor increment (‰)	T_{y6} = 3 mm Frequency (Hz)	T_{y6} = 3 mm Damping factor increment (‰)
1	24.8	0.48	24.4	0.80	23.9	1.04	23.5	1.16	23.1	1.10	22.8	1.06
2	26.9	0.80	26.4	1.28	25.9	1.63	25.5	1.78	25.0	1.72	24.6	1.68
3	34.9	0.15	34.4	0.46	34.0	0.52	33.5	0.57	33.1	0.51	32.8	0.47
4	94.5	3.60	93.9	4.81	93.2	6.48	92.7	7.27	92.1	7.07	91.5	6.87
5	108.4	11.84	107.4	13.10	106.5	14.50	105.8	16.26	105.2	16.10	104.5	15.50
6	132.1	31.10	129.3	38.00	126.5	53.40	124.0	63.41	122.2	61.6	120.4	60.70

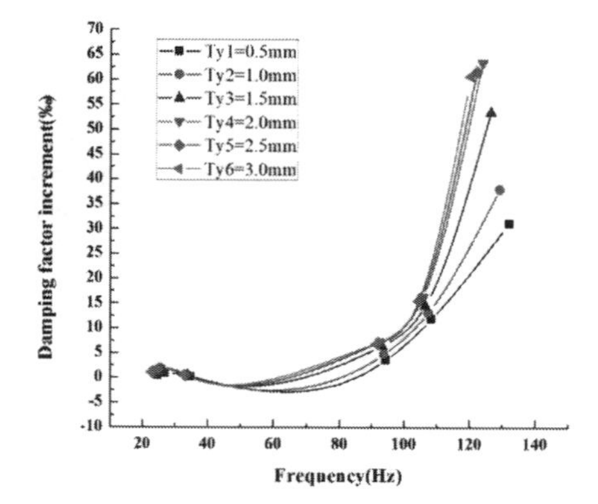

Figure 5. The curve of the damping factor increment and the resonance frequency with the change of the thickness of constrained layer.

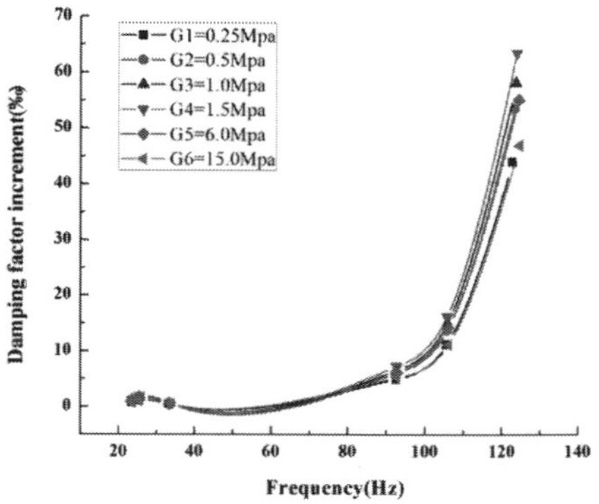

Figure 6. The curve of the damping factor increment and the resonance frequency with the change of the shear modulus of damping layer.

Table 3. Damping factor increment and the resonance frequency with the change of the shear modulus of damping layer.

Order number	G_1 = 0.25 (MPa) Frequency (Hz)	G_1 = 0.25 (MPa) Damping factor increment (‰)	G_2 = 0.5 (MPa) Frequency (Hz)	G_2 = 0.5 (MPa) Damping factor increment (‰)	G_3 = 1 (MPa) Frequency (Hz)	G_3 = 1 (MPa) Damping factor increment (‰)	G_4 = 1.5 (MPa) Frequency (Hz)	G_4 = 1.5 (MPa) Damping factor increment (‰)	G_5 = 6 (MPa) Frequency (Hz)	G_5 = 6 (MPa) Damping factor increment (‰)	G_6 = 15 (MPa) Frequency (Hz)	G_6 = 15 (MPa) Damping factor increment (‰)
1	23.5	0.68	23.5	0.92	23.5	1.06	23.5	1.16	23.5	1.02	23.6	0.69
2	25.5	1.05	25.5	1.43	25.5	1.59	25.5	1.78	25.6	1.58	25.7	1.09
3	33.5	0.41	33.5	0.54	33.5	0.55	33.5	0.57	33.5	0.54	33.6	0.46
4	92.6	4.97	92.6	6.05	92.6	6.55	92.7	7.27	92.7	6.17	92.8	5.24
5	105.7	11.15	105.8	13.73	105.8	14.99	105.8	16.26	106.0	14.26	106.2	11.17
6	122.9	43.94	123.4	53.41	123.8	58.07	124.0	63.41	124.5	055.11	124.8	46.93

are generally a kind of high molecular polymer, mainly rubber and plastic, shear modulus is less than 20 MPa. Take six groups of data in the range of 0.25 Mpa–15 Mpa for research and analysis.

It can be seen from the figure that the resonance frequency increases with the increase of the shear modulus of damping layer. It can be seen from the data in table, when the shear modulus of damping layer is 0.25 Mpa, the fundamental frequency is 23.5 Hz, the shear modulus is 15 Mpa, the fundamental frequency is 23.6 Hz. By referring to the second order to sixth order data, we can see that with the increase of the shear modulus of damping layer, the increase of the resonance frequency is not large.

With the increase of the shear modulus of damping layer, damping factor firstly increases and then decreases, and the damping factor reaches the maximum when the shear modulus of damping layer is 1.5 Mpa. The reason is that the shear modulus of damping layer increases, on the one hand, the stiffness of space truss and aerospace load structure is gradually increased, the relationship between the frequency and the stiffness of the structure shows that the fundamental frequency of the structure increases gradually. On the other hand, the stiffness of the damping material is gradually increasing, and the deformation of the damping layer decreases gradually. The damping factor of the structure is related to the strain energy of the damping layer, and the strain energy is proportional to the stiffness and strain. Therefore, under the common influence of the increasing stiffness and the decreasing of the strain, the damping factor appears to increase first and then decrease.

The placement of constrained damping layer is an important aspect of constrained damping layer optimization design. Because the constrained damping layer is to rely on the energy dissipation of the damping layer to achieve the vibration of the structure. The constrained damping layer is applied to the different locations to generate different modal strain energy distribution, its energy dissipation is also different, which will affect the vibration reduction effect. In this paper, the constrained damping layer is pasted on five different locations as shown in Figure 7.

From position 1 to position 5, the resonance frequency decreases first and then increases, and the damping factor increases at first and then decreases. The damping factor reaches the maxi-

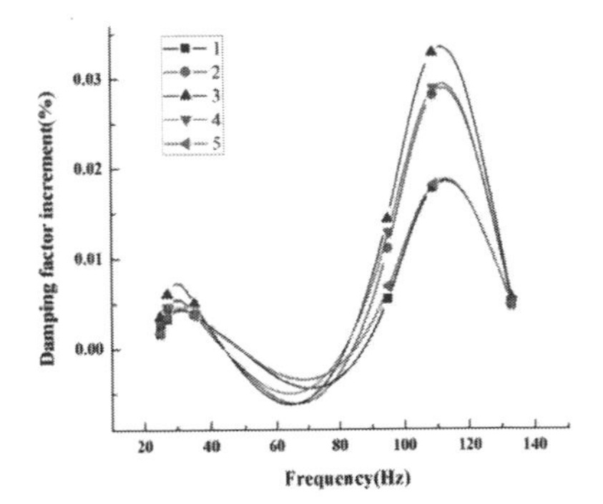

Figure 8. The curve of the damping factor increment and the resonance frequency with the change of paste position of constrained damping layer.

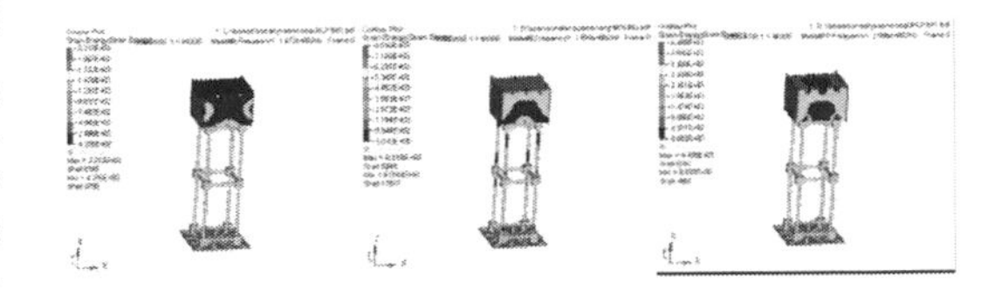

Figure 9. Modal strain energy distribution of eight order to ten order.

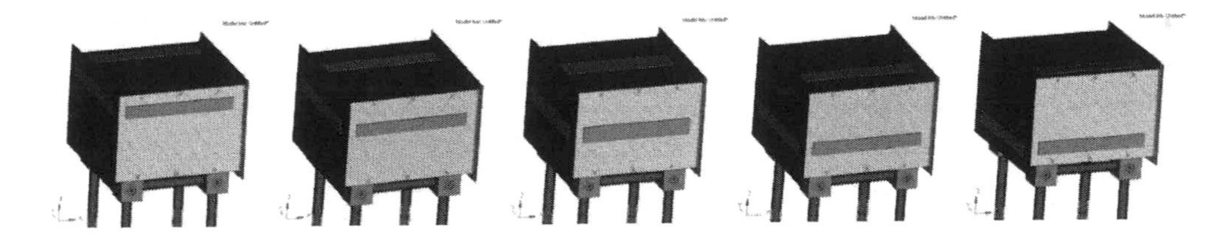

Figure 7. Paste position of constrained damping layer.

Table 4. Damping factor increment and the resonance frequency with the change of paste position of constrained damping layer.

	Position 1		Position 2		Position 3		Position 4		Position 5	
Order number	Frequency (Hz)	Damping factor increment (‰)	Frequency (Hz)	Damping factor increment (‰)	Frequency (Hz)	Damping factor increment (‰)	Frequency (Hz)	Damping factor increment (‰)	Frequency (Hz)	Damping factor increment (‰)
1	25.2	0.0016	25.2	0.0025	25.0	0.0035	25.2	0.0027	25.2	0.0017
2	27.3	0.0032	27.3	0.0044	27.1	0.0060	27.3	0.0046	27.3	0.0034
3	35.3	0.0037	35.3	0.0038	35.2	0.0050	35.3	0.0038	35.3	0.0037
4	94.9	0.0054	94.9	0.0110	94.8	0.0143	95.1	0.0128	95.2	0.0068
5	109.0	0.0177	109.0	0.0281	108.9	0.0328	109.2	0.0288	109.3	0.0181
6	133.0	0.0047	133.0	0.0050	132.9	0.0055	133.0	0.0052	133.0	0.0048

mum in position 3. In order to analyse the reasons, the modal strain energy distribution is investigated. at first. According to the results of modal analysis, in the first ten orders modes, the vibration mode of the space load box is in the order of eighth to tenth. The modal strain energy distribution of the three order is listed.

It can be concluded from the variation law of damping factor at different position of constrained damping layer and position diagram of modal strain energy: The greater the modal strain energy is, the greater the damping factor is. The reason is that the constrained damping layer is applied to the position where the modal strain energy is larger, and the shear deformation of the damping layer is large in the course of the vibration of the structure, the energy consumption is large, the damping factor of the structure is big, the resonance response peak is small, and the effect of vibration reduction is good. Therefore, the position where the modal strain energy is larger should be chosen to paste constrained damping layer in the optimization process.

6 CONCLUSION

1. When the thickness of the constrained layer is constant, the thickness of the damping layer is in the range of 0.25 mm–2.5 mm, with the increase of the thickness of the damping layer, the damping factor increases.
2. When the thickness of the damping layer is constant, the thickness of the constrained layer is in the range of 0.5 mm–3 mm, and damping factor increases first and then decreases with the increase of the thickness of the constrained layer.
3. The shear modulus of the damping layer gradually increased from 0.25 MPa to 15 MPa. The damping factor of the structure is related to the strain energy of the damping layer, and the strain energy is proportional to the stiffness and strain. Therefore, under the common influence of the increasing stiffness and the decreasing of the strain, the damping factor appears to increase first and then decrease.
4. The constrained damping layer is applied to the position where the modal strain energy is larger, the more energy is consumed by the damping layer, the greater the damping factor is, the more obvious the effect is.

REFERENCES

Bagley R. L. Applications of Generalized Derivatives to Viscoelasticity[R]. AIR FORCE MATERIALS LAB WRIGHT-PATTERSON AFB OH. 1979.

Johnson C. D., Kienholz D. A. Finite element prediction of damping in structures with constrained viscoelastic layers [J]. AIAA journal, 1982, 20(9): 1284–1290.

Mead D. J., Markus S. The forced vibration of a three-layer, damped sandwich beam with arbitrary boundary conditions [J]. Journal of sound and vibration, 1969, 10(2): 163–175.

Ren Zhigang, Lu Zhean, Lou Menglin. Calculation of frequency and loss factor of composite sandwich structures [J]. Earthquake engineering and engineering vibration, 2004, 24 (2):101–106.

Rikards R. Finite element analysis of vibration and damping of laminated composites [J]. Composite structures, 1993, 24(3): 193–204.

Ross D., Ungar E. E., Kerwin E. M. Damping of plate flexural vibrations by means of viscoelastic laminae [J]. Structural damping, 1959, 3: 44–87.

Shen Zhichun, Liang Lu, Zheng Gangtie, et al. The vibration suppression of a certain type of satellite payload bracket [J]. Journal of Astronautics, 2006, 27 (3): 503–506.

Wang Mozhai. Application of viscoelastic rubber damping materials in aerospace equipment [J]. Aerospace materials technology, 1990(4): 69–71.

Yang Jiaming, Zhang Yichang, Wu Lijuan. Optimization of damping composite structure performance of multilayer viscoelastic [J]. aviation journal, 2011, 32 (2):265–270.

Zhao Xiaochun, Li Xiangning, Li Kai, et al. Vibration characteristics and optimization of sandwich plate. Journal of Ship Research [J].Chinese, 2013, 8 (4).

Civil, Architecture and Environmental Engineering – Kao & Sung (Eds)
© 2017 Taylor & Francis Group, ISBN 978-1-138-02985-9

A unified theory of thermodynamically consistent microplane elastoplastic damage, and elastoplastic damage models

W.S. Li
Electric Power Research Institute of Guangdong Power Grid Co. Ltd., Guangzhou, China

J.Y. Wu
State Key Laboratory of Subtropical Building Science, South China University of Technology, Guangzhou, China

ABSTRACT: Microplane theory is a class of general approach to establish the three-dimensional constitutive relation of materials from the simpler one- or two-dimensional behavior on the generic planes with predefined spatial orientations. Although successfully implemented, verified, and applied to almost all types of materials, the theoretical aspects of microplane theory is still an open issue and further investigations are necessary. Within the framework of irreversible thermodynamics, we proposed a unified microplane theory in this paper. A detailed description of the general formulation, including evolution laws of the involved internal variables derived from the principle of maximum dissipation, was presented. As illustrative examples, the microplane elastoplastic, damage, and combined elastoplastic damage models were derived as the special cases of the proposed unified theory. The predicted results under cyclic uniaxial compression and biaxial compression show that the proposed theory is capable of describing the nonlinear behavior of concrete under dominant compression. This paper is the first step of the research work aiming to develop a thermodynamically consistent microplane model for concrete.

1 INTRODUCTION

Microplane theory is a class of general approach to construct a three-dimensional constitutive relation of materials from the simpler one- or two-dimensional behavior on the generic planes with predefined spatial orientations. Although the word "microplane" was coined by Bažant and Oh (1983) to refer in particular to the kinematically constrained models, the microplane theory can be traced back to the classical elastoplastic concept of Mohr–Coulomb-type failure envelope for a generic plane (Mohr 1900). The slip theory of plasticity for crystalline metals (Taylor 1938, Batdorf & Budiansky 1949) and the multilaminate model for fractured rocks (Zienkiewicz & Pande 1977) can be viewed as the special cases of the generalized microplane theory, although the static constraint was adopted in these models.

Although the (generalized) microplane theory has been successfully implemented, verified and applied to almost all types of ductile, brittle, or quasi-brittle materials, the theoretical aspects of the microplane theory is still an open issue and further investigation is required. The interrelations between microplane models and the classical disciplines of solid mechanics, such as plasticity and damage mechanics, have been discussed in detail by Carol et al. (1991, 1997). It was only recently that the earlier versions of microplane models have been found to be thermodynamically inconsistent, and during some loading cases, negative dissipations would be produced (Carol et al. 2001). To eliminate this deficiency, a new thermodynamically consistent microplane damage model was proposed by Kuhl et al. (2001) and further elaborated in Wu (2009); however, the capability of fitting the test data of concrete cannot be guaranteed.

Noticing the above facts, we aim to develop a new microplane model, which is not only capable of good data fitting but thermodynamically consistent. This paper is the first step of this work, in which a unified microplane theory was proposed within the framework of irreversible thermodynamics. The general formulation, including the consistent evolution laws of the involved internal variables, was presented in detail. As illustrative examples, by adopting different microplane unloading/reloading stiffness, the microplane elastoplastic, damage, and combined elastoplastic damage models were derived as the special cases of the proposed theory. Finally, the proposed model was applied to concrete under cyclic uniaxial compression and biaxial compression, respectively, and the numerical predictions were also presented to preliminarily verify its validity.

2 GENERAL FORMULATION

2.1 *Kinematic and kinetic settings*

Let the configuration occupied by a solid with reference position vector x in $\mathbb{R}^{ndim}$ be denoted by $\mathcal{B} \subset \mathbb{R}^{ndim}$. The boundary $\partial\mathcal{B}$ is subdivided into two disjoint parts: $\partial\mathcal{B} = \partial\mathcal{B}_u \cup \partial\mathcal{B}_t$ with $\partial\mathcal{B}_u \cap \partial\mathcal{B}_t = \varnothing$. In solid $\mathcal{B}$, the displacement fields are denoted by $\boldsymbol{u}: \mathcal{B} \subset \mathbb{R}^{ndim}$. The macroscopic stress and symmetric strain tensors are introduced as $\boldsymbol{\sigma}: \mathcal{B} \rightarrow \mathbb{R}^{ndim \times ndim}$ and $\boldsymbol{\varepsilon}: \mathcal{B} \rightarrow \mathbb{R}^{ndim \times ndim}$, respectively. Upon the assumption of infinitesimal strains, the macroscopic strain tensor $\boldsymbol{\varepsilon}$ is defined as the symmetric part of the displacement gradient $\nabla \boldsymbol{u}$:

$$\boldsymbol{\varepsilon} = \nabla^{sym} \boldsymbol{u} \tag{1}$$

The macroscopic strain tensor $\boldsymbol{\varepsilon}$ is usually decomposed into the following additive form:

$$\boldsymbol{\varepsilon} = \boldsymbol{\varepsilon}_{vol} + \boldsymbol{\varepsilon}_{dev} \tag{2}$$

with the macroscopic volumetric and deviatoric strain tensors, $\boldsymbol{\varepsilon}_{vol}$ and $\boldsymbol{\varepsilon}_{dev}$, respectively, given by

$$\boldsymbol{\varepsilon}_{vol} = \mathbb{I}_{vol} : \boldsymbol{\varepsilon}, \qquad \boldsymbol{\varepsilon}_{dev} = \boldsymbol{\varepsilon} - \boldsymbol{\varepsilon}_{vol} = \mathbb{I}_{dev} : \boldsymbol{\varepsilon} \tag{3}$$

where $\mathbb{I}_{vol}$ and $\mathbb{I}_{dev}$ denote the corresponding symmetric fourth-order volumetric and deviatoric projection operators:

$$\mathbb{I}_{vol} = \frac{1}{3} \boldsymbol{I} \otimes \boldsymbol{I}, \qquad \mathbb{I}_{dev} = \mathbb{I} - \mathbb{I}_{vol} = \mathbb{I} - \frac{1}{3} \boldsymbol{I} \otimes \boldsymbol{I} \tag{4}$$

According to the kinematic constraint, the volumetric and deviatoric strain vectors on a specific microplane (represented by its unit normal vector $\boldsymbol{n}$), $\boldsymbol{e}_{vol}$ and $\boldsymbol{e}_{dev}$, are determined by:

$$\boldsymbol{e}_{vol} = \boldsymbol{n} \cdot \boldsymbol{\varepsilon}_{vol} = e_V \boldsymbol{n}, \quad \boldsymbol{e}_{dev} = \boldsymbol{n} \cdot \boldsymbol{\varepsilon}_{dev} = e_D \boldsymbol{n} + \boldsymbol{e}_T \tag{5}$$

with the microplane volumetric strain e_V, deviatoric strain e_D, and tangential strain $\boldsymbol{e}_T$ resolved as the projections of the macroscopic strain tensor:

$$e_V = \boldsymbol{V} : \boldsymbol{\varepsilon}, \quad e_D = \boldsymbol{D} : \boldsymbol{\varepsilon}, \quad \boldsymbol{e}_T = \boldsymbol{T} : \boldsymbol{\varepsilon} \tag{6}$$

where the projection tensors $\boldsymbol{V}$, $\boldsymbol{D}$, and $\boldsymbol{T}$ are expressed as:

$$\boldsymbol{V} = \frac{1}{3} \boldsymbol{I}, \quad \boldsymbol{D} = \boldsymbol{n} \otimes \boldsymbol{n} - \frac{1}{3} \boldsymbol{I}, \quad \boldsymbol{T} = \boldsymbol{n} \cdot \mathbb{I} - \boldsymbol{n} \otimes \boldsymbol{n} \otimes \boldsymbol{n} \tag{7}$$

It is more convenient to express the microplane tangential strain vector $\boldsymbol{e}_T$ as the product of its magnitude e_T and its unit vector $\boldsymbol{1}_T$:

$$\boldsymbol{e}_T = e_T \boldsymbol{1}_T, \quad e_T = |\boldsymbol{e}_T| = \sqrt{\boldsymbol{e}_T \cdot \boldsymbol{e}_T}, \quad \boldsymbol{1}_T = \frac{\boldsymbol{e}_T}{e_T} \tag{8}$$

The macroscopic stress tensor ($\boldsymbol{\sigma}$) is specified by the constitutive equation $\boldsymbol{\sigma}(\boldsymbol{\varepsilon}, \boldsymbol{\alpha})$ and satisfies the balance of linear momentum at equilibrium:

$$\text{div}\, \boldsymbol{\sigma} + \boldsymbol{f} = \boldsymbol{0} \tag{9}$$

for the body force $\boldsymbol{f}$ in the domain $\mathcal{B}$, with $\boldsymbol{\alpha}$ denoting the set of internal variables (including the damage and plastic internal variables in this paper). In contrast to microplane strain components, the conjugate microplane stress components on the microplane cannot in general be equal to the projections of the macroscopic stress tensor $\boldsymbol{\sigma}$ if the microscopic strain components represent the projection of macroscopic strain tensor $\boldsymbol{\varepsilon}$. Thus, static equivalence or equilibrium between the micro and macro levels, here called the micro-macro homogenization formulation, has to be enforced in the weak form.

2.2 *Elastoplastic damage model in strain space*

In the formulation of strain space plasticity, the stress tensor $\boldsymbol{\sigma}$ is expressed as the difference between the reversible elastic part σ^e and the permanent plastic part $\boldsymbol{\sigma}^p$:

$$\boldsymbol{\sigma} = \boldsymbol{\sigma}^e - \boldsymbol{\sigma}^p \tag{10}$$

where the elastic stress $\boldsymbol{\sigma}^e$ is related to the strain tensor $\boldsymbol{\varepsilon}$ by the generalized Hooken law:

$$\boldsymbol{\sigma}^e = \mathbb{E} : \boldsymbol{\varepsilon}, \quad \boldsymbol{\varepsilon} = \mathbb{C} : \boldsymbol{\sigma}^e \tag{11}$$

with $\mathbb{E}$ and $\mathbb{C} = \mathbb{E}^{-1}$ being the unloading stiffness and compliance tensors, respectively. For a pure elastic degradation (fracturing, damage, etc.) material, we have, $\mathbb{E} = \mathbb{E}^{sec}$ with $\mathbb{E}^{sec}$ representing the secant stiffness. However, for a pure elastoplastic material, the unloading stiffness $\mathbb{E}$ is the same as the elasticity tensor $\mathbb{E}^0$:

$$\mathbb{E}^0 = \frac{E^0}{1 - 2\nu^0} \mathbb{I}_{vol} + \frac{E^0}{1 + \nu^0} \mathbb{I}_{dev} \tag{12}$$

where E^0 is Young's modulus and ν^0 is Poisson's ratio.

The differentiation of stress $(11)_1$ to time yields the following rate form constitutive relation:

$$\dot{\boldsymbol{\sigma}} = \mathbb{E} : \dot{\boldsymbol{\varepsilon}} + \dot{\mathbb{E}} : \boldsymbol{\varepsilon} - \dot{\boldsymbol{\sigma}}^p = \dot{\boldsymbol{\sigma}}^{co} - \dot{\boldsymbol{\sigma}}^{cr} \tag{13}$$

where $\dot{\boldsymbol{\sigma}}^{co}$ is the continuum stress rate, $\dot{\boldsymbol{\sigma}}^{d}$ is the degradation stress rate, and $\dot{\boldsymbol{\sigma}}^{cr}$ is the cracking stress rate, which are defined by:

$$\dot{\boldsymbol{\sigma}}^{\rm co} = \mathbb{E} : \dot{\boldsymbol{\varepsilon}}, \quad \dot{\boldsymbol{\sigma}}^{\rm d} = -\dot{\mathbb{E}} : \boldsymbol{\varepsilon}, \quad \dot{\boldsymbol{\sigma}}^{\rm cr} = \dot{\boldsymbol{\sigma}}^{\rm d} + \dot{\boldsymbol{\sigma}}^{\rm p} \tag{14}$$

Herein, the continuum stress rate ($\dot{\boldsymbol{\sigma}}^{\rm co}$) is the stress rate that would be obtained by preventing the microcracks from evolution further, whereas $\dot{\boldsymbol{\sigma}}^{\rm cr}$ is the decrease or relaxation of the incremental elastic behavior due to the microcracks evolution, consisting of two parts: (i) the degradation stress rate ($\dot{\boldsymbol{\sigma}}^{\rm d}$) that is recoverable and (ii) the irreversible plastic strain rate ($\dot{\boldsymbol{\sigma}}^{\rm p}$). In the 1-D case, the above definitions of the continuum, damage, and cracking stress rates are illustrated in Fig. (1). It is worth noting that Eq. (13) is rather similar to the counterpart in the strain space elastoplasticity. However, the constant elastic isotropic stiffness ($\mathbb{E}^0$) in the later formulation is replaced by the variable stiffness tensor ($\mathbb{E}$) herein, which means $\dot{\boldsymbol{\sigma}}^{\rm e} \neq \dot{\boldsymbol{\sigma}}^{\rm co}$.

Because of the evolution of microcracks, the concrete-like quasi-brittle material exhibits both the stiffness degradation and the permanent irreversible deformations or the elastoplastic coupling referred in the literature (Dafalias 1977; Han & Chen 1986). Therefore, to complete the entire model, we have to also give the evolution laws for the stiffness. This would be a rather difficult job if a tensor invariant-based macroscopic constitutive model (e.g., plasticity, damage mechanics, or their combinations) is used. Comparatively, within the framework of microplane theory, because the macroscopic properties of material can be determined as the integral or weighted summation of the contributions from all the predefined microplanes over, we can easily solve this problem.

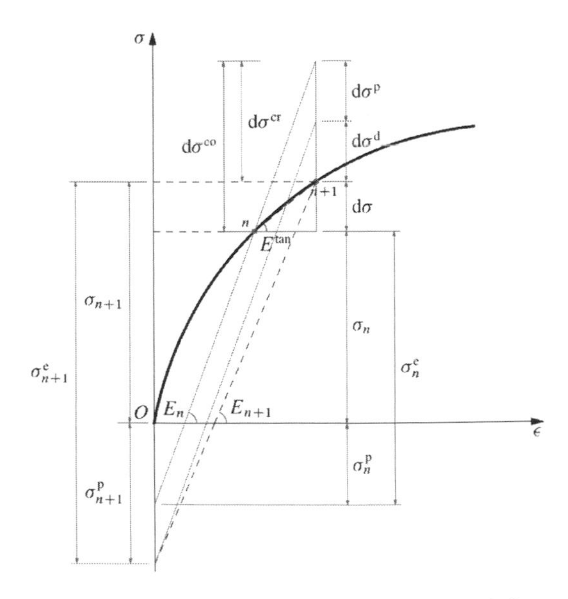

Figure 1. Definitions of the continuum, degradation, and cracking stress rates under 1-D case.

2.3 *Thermodynamics description*

For any isothermal and pure mechanical process, the second law of thermodynamics reads the following Clausius–Duhem inequality:

$$D = \boldsymbol{\sigma} : \dot{\boldsymbol{\varepsilon}} - \dot{\bar{\psi}} = \dot{\psi} - \boldsymbol{\varepsilon} : \dot{\boldsymbol{\sigma}} \geq 0 \tag{15}$$

where D is the energy dissipation during the process to be considered; the Gibbs Free Energy (GFE) (ψ) is related to the Helmholtz Free Energy (HFE) ($\bar{\psi}$) by the classical Legendre transformation:

$$\bar{\psi} + \psi = \boldsymbol{\sigma} : \boldsymbol{\varepsilon} \tag{16}$$

A standard assumption is the existence of a macroscopic GFE potential of material in isothermal conditions:

$$\psi = \psi(\boldsymbol{\sigma}^{\rm e}, \phi, \boldsymbol{q}) = \psi^{\rm co}(\boldsymbol{\sigma}^{\rm e}, \phi) - \psi^{\rm cr}(\boldsymbol{q}) \tag{17}$$

where the continuum GFE potentials ($\psi^{\rm co}$ and $\psi^{\rm cr}$) are assumed to be decoupled, where ϕ is the set of damage variables that characterize the stiffness degradation and $\boldsymbol{q}$ is a given set of stress-like internal variables that fully describe the irreversible plastic deformations. For the hyperelastic material, the continuum GFE potential of the material ($\psi^{\rm co}$) is equal to the continuum HFE potential ($\bar{\psi}^{\rm co}$):

$$\psi^{\rm co} = \frac{1}{2}\boldsymbol{\sigma}^{\rm e} : \mathbb{C}(\phi) : \boldsymbol{\sigma}^{\rm e} = \frac{1}{2}\boldsymbol{\varepsilon} : \mathbb{E}(\phi) : \boldsymbol{\varepsilon} = \bar{\psi}^{\rm co} \tag{18}$$

In the microplane theory, the continuum HFE potential ($\bar{\psi}^{\rm co}$) (or the equivalent GFE potential $\psi^{\rm co}$) can be expressed as the integral of its microplane counterpart $\bar{\psi}^{\rm co}$:

$$\begin{aligned} \psi^{\rm co} &= \bar{\psi}^{\rm co} = \frac{3}{2\pi}\int_{\Omega} \bar{\varphi}^{\rm co}\, {\rm d}\Omega \\ \bar{\varphi}^{\rm co} &= \frac{1}{2}(\phi_{\rm V} E^0_{\rm V} e_{\rm V} \cdot e_{\rm V} + \phi_{\rm D} E^0_{\rm D} e_{\rm D} \cdot e_{\rm D} + \phi_{\rm T} E^0_{\rm T} e_{\rm T} \cdot e_{\rm T}) \end{aligned} \tag{19}$$

where Ω is the surface of a unit hemisphere representing the set of all possible microplane orientations; $E^0_{\rm V}$, $E^0_{\rm D}$, and $E^0_{\rm T}$ are the microplane elastic moduli to be determined; and $\phi = \{\phi_{\rm V}, \phi_{\rm D}, \phi_{\rm T}\}$ represents the set of microplane volume, deviatoric, and tangential damages, respectively. Similarly, we further assume that the cracking GFE potential ($\psi^{\rm cr}$) can be expressed in the integral of its microplane counterpart ($\varphi^{\rm cr}$) over all possible spatial orientations:

$$\psi^{cr}(\boldsymbol{q}) = \frac{3}{2\pi}\int_\Omega \varphi^{cr}(\boldsymbol{q})\,d\Omega \tag{20}$$

where $\boldsymbol{q} = \{q_V, q_D, q_T\}$ denotes the set of stress like internal variables on each microplane that fully characterizes the microplane volumetric, normal deviatoric, and tangential deviatoric plastic flows. Therefore, the rate of the macroscopic GFE potential is given by:

$$\dot{\psi} = \frac{\partial \psi^{co}}{\partial \boldsymbol{\sigma}^e} : \dot{\boldsymbol{\sigma}} + \frac{1}{2}\boldsymbol{\sigma}^e : \dot{\mathbb{C}} : \boldsymbol{\sigma}^e - \frac{3}{2\pi}\int_\Omega \boldsymbol{\kappa}\cdot\dot{\boldsymbol{q}}\,d\Omega \tag{21}$$

where $\boldsymbol{\kappa} = \partial\varphi^{cr}/\partial\boldsymbol{q}$ denotes the set of strain-like internal variables conjugated to $\boldsymbol{q}$.

Substituting Eq. (21) into the Clausius–Duhem inequality (15) and making use of the standard arguments along with the additional assumption that the unloading is recoverable, we obtain the same constitutive relation as in Eq. (11):

$$\boldsymbol{\varepsilon} = \frac{\partial \psi^{co}}{\partial \boldsymbol{\sigma}^e} = \mathbb{C} : \boldsymbol{\sigma}^e, \quad \boldsymbol{\sigma}^e = \frac{\partial \bar{\psi}^{co}}{\partial \boldsymbol{\varepsilon}} = \mathbb{E} : \boldsymbol{\varepsilon} \tag{22}$$

In accordance with Eq. (22), the continuum HFE potential defined in Eq. (19) leads to:

$$\boldsymbol{\sigma}^e = \frac{3}{2\pi}\int_\Omega (\boldsymbol{V}\cdot s_V^e + \boldsymbol{D}\cdot s_D^e + \boldsymbol{T}^T\cdot s_T^e)\,d\Omega \tag{23}$$

where the microplane elastic stress components are given by:

$$s_V^e = \phi_V E_V^0 e_V, \quad s_D^e = \phi_D E_D^0 e_D, \quad s_T^e = \phi_T E_T^0 e_T \tag{24}$$

and the stiffness tensor $\mathbb{E}$ is expressed as:

$$\mathbb{E} = \frac{3}{2\pi}\int_\Omega (\phi_V E_V^0 \boldsymbol{V}\otimes\boldsymbol{V} + \phi_D E_D^0 \boldsymbol{D}\otimes\boldsymbol{D})\,d\Omega + \frac{3}{2\pi}\int_\Omega (\phi_T E_T^0 \boldsymbol{T}^T\cdot\boldsymbol{T})\,d\Omega \tag{25}$$

Upon initial virgin state, substituting the relation $\phi_V = \phi_D = \phi_T = 1$ into Eq. (25) and comparing the obtained result to Eq. (12), we obtain the definitions for the microplane elastic moduli:

$$E_V^0 = \frac{E^0}{1-2\nu^0}, \quad E_D^0 = E_T^0 = \frac{E^0}{1+\nu^0} \tag{26}$$

where, as discussed in Carol et al. (1991), the deviatoric and tangential elastic moduli, $E_D^0 = E_T^0$, are assumed to take the same value herein.

Therefore, the continuum stress rate ($\dot{\boldsymbol{\sigma}}^{co}$) in Eq. (13) can be expressed in the form similar to Eq. (23):

$$\dot{\boldsymbol{\sigma}}^{co} = \frac{3}{2\pi}\int_\Omega (\boldsymbol{V}\cdot\dot{s}_V^{co} + \boldsymbol{D}\cdot\dot{s}_D^{co} + \boldsymbol{T}^T\cdot\dot{s}_T^{co})\,d\Omega \tag{27}$$

where the microplane volumetric, deviatoric, and tangential continuum stress rates are defined in an incremental elastic form:

$$\dot{s}_V^{co} = \phi_V E_V^0 \dot{e}_V, \quad \dot{s}_D^{co} = \phi_D E_D^0 \dot{e}_D, \quad \dot{s}_T^{co} = \phi_T E_T^0 \dot{e}_T \tag{28}$$

In accordance with the second law of thermodynamics, the constitutive relation derived above has to satisfy the following energy dissipation inequality:

$$D(\boldsymbol{\Gamma}) = \boldsymbol{\varepsilon} : \dot{\boldsymbol{\sigma}}^p - \frac{1}{2}\boldsymbol{\varepsilon} : \dot{\mathbb{E}} : \boldsymbol{\varepsilon} - \frac{3}{2\pi}\int_\Omega \sum_i \kappa_i\cdot\dot{q}_i\,d\Omega \ge 0 \quad \text{with } \kappa_i = \frac{\partial\psi^{cr}}{\partial q_i} \tag{29}$$

where and thereafter, to make the algebraic manipulations more convenient, the compact notation $\boldsymbol{\Gamma} := \{\boldsymbol{\varepsilon}, \boldsymbol{\kappa}\}$ is introduced for the strain-like variables constrained within a closed convex (not necessarily smooth) admissible domain in the strain space ($\mathcal{E}$):

$$\boldsymbol{\Gamma} \in \mathcal{E} := \{\boldsymbol{\Gamma} \in \Re^{n\dim \times n\dim} \times \Re^q \mid F_i(\boldsymbol{\Gamma}) \le 0\} \tag{30}$$

where $F_i(\boldsymbol{\Gamma}) \le 0$ are the microplane volumetric, deviatoric, and tangential loading surfaces.

2.4 *Evolution laws for internal variables*

The evolution laws for the stiffness tensor ($\mathbb{E}$) and the stress-like variables ($\boldsymbol{\Sigma} := \{\boldsymbol{\sigma}^{cr}, \boldsymbol{q}\}$) can be obtained from the postulate of maximum energy dissipation: for given admissible strain state variables $\boldsymbol{\Gamma}$, the rates $\dot{\mathbb{E}}$ and $\dot{\boldsymbol{\Sigma}} = \{\dot{\boldsymbol{\sigma}}^{cr}, \dot{\boldsymbol{q}}\}$ are those values that lead to a stationary point of the energy dissipation D:

$$\boldsymbol{\Gamma} = \arg\max[D] \in \mathcal{E} \tag{31}$$

To find the solution of the above constrained optimization problem (31), the following Lagrangian function is generally introduced:

$$\mathcal{L}(\boldsymbol{\Gamma}) = -D(\boldsymbol{\Gamma}) + \frac{3}{2\pi}\int_\Omega \sum_i \dot{\lambda}_i F_i(\boldsymbol{\Gamma})\,d\Omega \tag{32}$$

where $\dot{\lambda}_i$ denotes the Lagrangian multipliers constrained by the Karush–Kuhn–Tucker conditions:

$$F_i \le 0, \quad \dot{\lambda}_i \ge 0, \quad \dot{\lambda}_i F_i = 0 \tag{33}$$

Although other forms of functions can be postulated, herein, we assume that the individual cracking surfaces are simply expressed in the following same form:

$$F_i = f_i(e_i) - \kappa_i(q_i) \leq 0, \quad f_i = |e_i| \tag{34}$$

where f_i are the non-negative microplane loading functions. The cracking stresses increase for virgin loadings, whereas they remain constant for unloading and reloading. The case of virgin loading corresponds to the actual value of the cracking strain-like internal variables, as it is larger than the maximum values reached so far, that is, the cracking thresholds κ_V, κ_D, and κ_T. Once further cracking process occurs, the cracking thresholds are updated as:

$$\kappa_i = \max_{t_0 \leq t \leq t_n}[0, |e_i|_t] \tag{35}$$

From the necessary optimality conditions we obtain the associated evolution laws (normality rule) for the strain-like internal variables $\boldsymbol{\Gamma}$:

$$\begin{aligned} \dot{\boldsymbol{\sigma}}^{cr} &= \frac{3}{2\pi}\int_\Omega (\boldsymbol{V} \cdot \dot{s}_V^{cr} + \boldsymbol{D} \cdot \dot{s}_D^{cr} + \boldsymbol{T}^T \cdot \dot{s}_T^{cr})\, d\Omega \\ \dot{q}_i &= -\dot{\lambda}_i \frac{\partial f_i}{\partial \kappa_i} = \dot{\lambda}_i \end{aligned} \tag{36}$$

for the microplane cracking stress rates:

$$\begin{aligned} \dot{s}_V^{cr} &= \dot{\lambda}_V \frac{\partial f_V}{\partial e_V} = \dot{\lambda}_V \text{sign}(e_V) \\ \dot{s}_D^{cr} &= \dot{\lambda}_D \frac{\partial f_D}{\partial e_D} = \dot{\lambda}_D \text{sign}(e_D) \\ \dot{s}_T^{cr} &= \dot{\lambda}_T \frac{\partial f_T}{\partial e_T} = \dot{\lambda}_T \frac{\partial f_T}{\partial e_T} \boldsymbol{l}_T = \dot{\lambda}_D \boldsymbol{l}_T \end{aligned} \tag{37}$$

When the material lies in the elastic domain, that is, $F_i(\boldsymbol{\Gamma}) < 0$, from Eq. $(33)_3$, we have $\dot{\lambda}_i = 0$; upon virgin loading state, the Lagrangian multipliers $\dot{\lambda}_i > 0$ are determined by the following consistency conditions:

$$\dot{\lambda}_i \dot{F}_i(\boldsymbol{\Gamma}) = 0 \quad \Rightarrow \quad \dot{F}_i(\boldsymbol{\Gamma}) = 0 \tag{38}$$

from which we can solve the Lagrangian multipliers as follows:

$$\begin{aligned} \dot{\lambda}_V &= H_V \text{sign}(e_V)\dot{e}_V \quad \Rightarrow \quad \dot{s}_V^{cr} = H_V \dot{e}_V \\ \dot{\lambda}_D &= H_D \text{sign}(e_D)\dot{e}_D \quad \Rightarrow \quad \dot{s}_D^{cr} = H_D \dot{e}_D \\ \dot{\lambda}_V &= H_T \dot{e}_T \quad \Rightarrow \quad \dot{s}_T^{cr} = \boldsymbol{H}_T \cdot \dot{\boldsymbol{e}}_T \end{aligned} \tag{39}$$

where the second-order tensor $(\boldsymbol{H}_T)$ is defined as $\boldsymbol{H}_T = H_T(\boldsymbol{l}_T \otimes \boldsymbol{l}_T)$, and $H_i = \partial q_i / \partial \kappa_i$ denotes the hardening/softening functions.

For the continuum stress rate expressed in Eq. (27) and the cracking stress rate given in Eq. (36a), we can rewrite the stress rate (13) as:

$$\begin{aligned} \dot{\boldsymbol{\sigma}} &= \frac{3}{2\pi}\int_\Omega (\boldsymbol{V} \cdot \dot{s}_V + \boldsymbol{D} \cdot \dot{s}_D + \boldsymbol{T}^T \cdot \dot{s}_T)\, d\Omega = \mathbb{E}^{\tan} : \dot{\boldsymbol{\varepsilon}} \\ \dot{s}_V &= \dot{s}_V^{co} - \dot{s}_V^{cr}, \quad \dot{s}_D = \dot{s}_D^{co} - \dot{s}_D^{cr}, \quad \dot{s}_T = \dot{s}_T^{co} - \dot{s}_T^{cr} \end{aligned} \tag{40}$$

where the tangent stiffness tensor $(\mathbb{E}^{\tan})$ is expressed as:

$$\begin{aligned} \mathbb{E}^{\tan} = \mathbb{E} &- \frac{3}{2\pi}\int_\Omega (H_V \boldsymbol{V} \otimes \boldsymbol{V} + H_D \boldsymbol{D} \otimes \boldsymbol{D})\, d\Omega \\ &- \frac{3}{2\pi}\int_\Omega (\boldsymbol{T}^T \cdot \boldsymbol{H}_T \cdot \boldsymbol{T})\, d\Omega \end{aligned} \tag{41}$$

2.5 *Hardening/softening functions*

In accordance with Eq. (39) we have $H_i = 0$ during the unloading/reloading regimes. Therefore, we only need to determine the hardening/softening functions H_i of virgin loading regions, which can be derived if the microplane stress–strain curves are known. On the basis of the previous experiences of microplane models, in the presented paper, we use the following secant stress–strain curves for the virgin loading:

$$s_V = \omega_V E_V^0 e_V, \quad s_D = \omega_D E_D^0 e_D, \quad \boldsymbol{s}_T = \omega_T E_T^0 \boldsymbol{e}_T \tag{42}$$

where ω_i are the appropriate functions in terms of the corresponding internal variables κ_i:

$$\omega_i = \hat{\omega}_i(\kappa_i) = \hat{\omega}_i(|e_i|) \tag{43}$$

3 EXEMPLIFIED MODELS

In accordance with the selection of damage variables ϕ_i, which characterize the stiffness degradation upon unloading/reloading/reloading states, different microplane models, elastoplastic model, damage model, and combined elastoplastic damage model can be postulated.

3.1 *Microplane elastoplastic model*

We first consider the case that under both tensile and compression states, all the microplane volumetric, deviatoric, and tangential behaviors follow the classical elastoplasticity. That is, the microplane unloading branches always have the initial slopes E_i^0:

$$\phi_i = 1 \tag{44}$$

Therefore, the stiffness tensor $(\mathbb{E})$ in Eq. (25) becomes the elasticity tensor $(\mathbb{E}^0)$, and the stress relaxation ($\dot{\boldsymbol{\sigma}}^{cr}$) is entirely contributed from the plastic flows on each microplane.

The microplane elastoplastic model obtained above is very similar to the strain space elastoplasticity, which can deal with strain softening. The theoretical and numerical comparisons between this strain-based microplane elastoplastic model with kinematic constraint, the stress-based one with kinematic constraint (Brocca and Bažant 2000; Kuhl et al. 2001), and the one with static constraint (Batdorf and Budiansky 1949; Carol and Bažant 1997) will be presented elsewhere.

3.2 *Microplane damage model*

Comparatively, we assume that under both tensile and compression states, all the microplane volumetric, deviatoric, and tangential tensile behaviors follow the pure elastic degradation (damage) model, namely the unloading–reloading branches point to the origin point:

$$\phi_i = \omega_i \tag{45}$$

Under such cases, the obtained microplane damage model is entirely the same as the one proposed in Kuhl et al. (2001).

3.3 *Microplane elastoplastic damage model*

To realistically describe both stiffness degradation and plastic flows, we adopt the following simple rule: the volumetric and deviatoric tensile behavior is assumed to be pure elastic degradation (damage), whereas the volumetric, deviatoric compressive behavior and the tangential performance are assumed to follow the classical plasticity.

More specifically, the volumetric and deviatoric unload–reloading branches are assumed to be straight lines with a certain slope. The compressive behavior is assumed to be elastoplastic, that is, the unloading branches are assumed to always have the initial slopes E_V^0 and E_D^0, that is, $\phi_V = \phi_D = 1$, and the origins of the tensile parts of the diagrams always shift to the points in which the unloading compressive branches intersect the horizontal (strain components) axis. On the contrary, the unloading–reloading behavior in tension is assumed to be a straight line pointing to the origin of that curve, that is, $\phi_V = \omega_V$ and $\phi_D = \omega_D$.

For the tangential unloading–reloading behavior, we assume a differential rule. For the unloading branch with initial stiffness E_T^0, zero stress is assumed when the horizontal axis is reached during unloading. For the reloading branch, a straight line pointing to the maximum point reached so far is followed. Thus, a reloading branch with a slope smaller than the initial stiffness (E_T^0) is postulated herein.

3.4 *Illustrative numerical results*

Let us adopt the expressions for the nonincreasing function ω_i (Carol et al. 1992):

$$\hat{\omega}_V(e_V) = \begin{cases} \exp[-(|e_V|/a_1)^{p_1}] & \text{Iff } e_V \geq 0 \\ (1+|e_V|/a)^{-p} + (|e_V|/b)^q & \text{Iff } e_V < 0 \end{cases}$$

$$\hat{\omega}_D(e_D) = \begin{cases} \exp[-(|e_D|/a_1)^{p_1}] & \text{Iff } e_D \geq 0 \\ \exp[-(|e_D|/a_2)^{p_2}] & \text{Iff } e_D < 0 \end{cases}$$

$$\hat{\omega}_T(|e_T|) = \exp[-(|e_T|/a_3)^{p_3}]$$

where a, b, p, and q are parameters to control the volumetric compressive behavior; parameters p_1, p_2, and p_3 are exponential parameters determining the curve shape; and a_1 and a_2 are positive parameters that affect the peak strength of the individual stress component. To model the macroscopic confinement on the behavior of the tangential stress, the value of parameter a_3 is assumed to depend on the macroscopic confinement represented by the volumetric strain e_V, that is, $a_3 = a_3^0 + \xi\langle -e_V \rangle$, where ξ is an additional parameter. Seven parameters can be generally assumed to have the same values for most concrete: a = 0.005, b = 0.225, $p = 0.25$, $q = 2.25$, $p_1 = 1.0$, $p_2 = p_3 = 1.5$. Moreover, Poisson's ratio is fixed as $\nu_0 = 0.18$.

The first example corresponds to the cyclic behavior of a concrete specimen subjected to a uniaxial compression. In this example, the uniaxial strain (ε_{xx}) unloads at 0.0015, 0.0024, 0.0035, 0.0045, and finally increases to the final value of 0.006. The predicted stress–strain curves are shown in Fig. 2, with the parameter values being E^0 = 26.8 GPa, $a_1 = 7.0 \times 10^{-5}$, $a_2 = 2.0 \times 10^{-3}$, $a_3^0 = 1.7\times10^{-3}$, and $\xi = 0.0$. It is remarked that because of the many possible combinations of loading/unloading/reloading on each microplane, both the macroscopic envelop curve and the hysteretic loops can

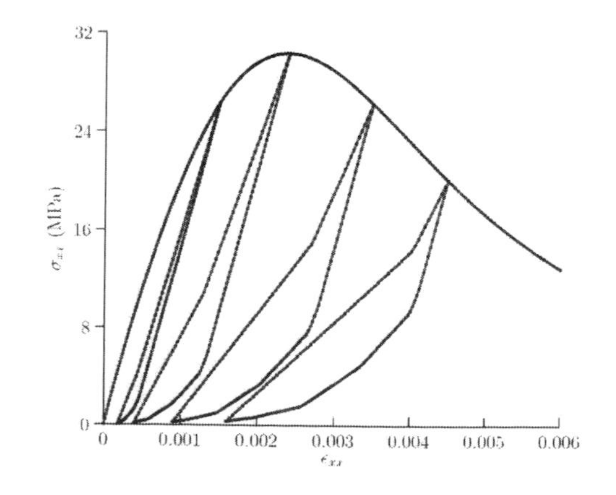

Figure 2. Cyclic uniaxial compression.

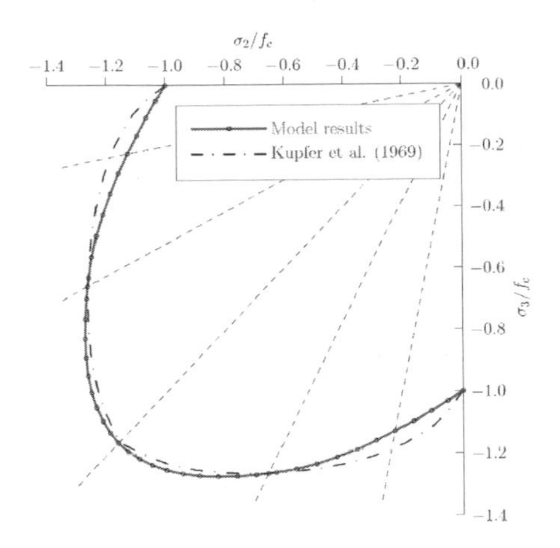

Figure 3. Comparison of strength envelop with biaxial compressive test by Kupfer et al. (1969).

be obtained realistically, although the latter is not a specific purpose of this work.

In the second example, we consider the behavior of concrete under biaxial compression, with the parameters being E^0 = 30 GPa, $a_1 = 5.0 \times 10^{-5}$, $a_2 = 1.7 \times 10^{-3}$, $a_3^0 = 1.5\times10^{-3}$, and $\xi = 0.88$. The parameters yield the uniaxial compressive strength of 32.4 MPa. We fixed the first principal stress σ_1 as zero during all the simulations and changed the ratio between the second and third principal stresses (σ_2/σ_3). The predicted strength envelope under biaxial compression is illustrated in Fig. 3, which agrees with the test data (Kupfer et al. 1969) well.

4 CONCLUSIONS

In this paper, a unified microplane theory was proposed within the framework of irreversible thermodynamics. A detailed description of the general formulation, including the consistent evolution laws of the involved internal variables, was presented. As illustrative examples, the microplane elastoplastic, damage, and combined elastoplastic damage models were derived as the special cases of the proposed microplane theory. The numerical predictions show that the proposed theory is capable of realistically describing the nonlinear behavior of concrete.

It is important to note that the proposed model does not eliminate the inherent deficiencies exhibited by the kinematically constrained microplane theory, that is, the pathological tensile behavior. Moreover, because the associated evolution laws of internal variables were used, the deviatoric-induced dilatancy, typical for quasi-brittle materials, cannot be realistically described by the proposed model. It is the next step of this research work to solve these two problems.

ACKNOWLEDGMENTS

This study was supported by the Science and Technology Project (GDKJ00000030) of China Southern Power Grid Co. The second author (J.Y. Wu) acknowledges support from the State Key Laboratory of Subtropical Building Science (2015ZB24, 2016KB12) and the Fundamental Research Funds for the Central University (2015ZZ078).

REFERENCES

Batdorf, S.H. & Budiansky, B., 1949. A mathematical theory of plasticity based on the concept of slip. *National Advisory Committee for Aeronautics* (N.C.C.A), Technical Note No. 1871, Washington, DC.

Broacca, B. & Bažant, Z.P., 2000. Microplane constitutive model and metal plasticity *Appl. Mech. Rev., ASME*, 53(10): 265–281.

Bažant, Z.P. & Oh, B.-H., 1983. Microplane model for fracture analysis of concrete structures. *Proc. Symp. Interact. Non-Nucl. Munitious Struct.*, US Air Force Academy, Springs. Co.; 49–55.

Carol, I. & Bažant, Z.P., 1997. Damage and plasticity in microplane theory. *International Journal of Solids and Structures*, 34(29): 3807–3835.

Carol, I., Bažant, Z.P. & Prat, P.C., 1991. Geometric damage tensor based on microplane model. *Journal of Engineering Mechanics, ASCE*, 117(10): 2429–2448.

Carol, I., Jirásek, M. & Bažant, Z.P., 2001. A thermodynamically consistent approach to microplane theory. Part I. Free energy and consistent microplane stresses. *International Journal of Solids and Structures*, 38: 2921–2931.

Carol, I., Prat, P.C. & Bažant, Z.P., 1992. New explicit microplane model for concrete: Theoretical aspects and numerical implementation. *International Journal of Solids and Structures*, 29(9): 1173–1191.

Dafalias, Y.F., 1977. Elasto-plastic coupling within a thermodynamic strain space formulation of plasticity. *Int. J. Non-linear Mech.*, 12: 327–337.

Han, D.J. & Chen, W.F., 1986. Strain-space plasticity formulation for hardening-softening materials with elastoplastic coupling. *International Journal of Solids and Structures*, 22: 935–950.

Kuhl, E., Steinmann, P. & Carol, I., 2001. A thermodynamically consistent approach to microplane theory. Part II. Dissipation and inelastic constitutive modelling. *International Journal of Solids and Structures*, 38: 29322952.

Kupfer, H., Hilsdorf, H.K. & Rüsch, 1969. Behavior of concrete under biaxial stresses. *ACI Journal*, 66: 656666.

Mohr, O., 1900. Welche Umstande bendingen der Bruch und der Elastizitätsgrenze des maaterials. *Z. Vereins Deutscher Ingenieure*, 44: 1–12.

Taylor, G.I., 1938. Plastic strain in metals. *J. Inst. Metals*, 62: 307–324.

Wu, J.Y., 2009. An alternative approach to microplane theory. *Mechanics of Materials*, 41: 87–105.

Zienkiewicz, O.C. & Pande, G.N., 1977. Time-dependent multi-laminate model of rocks—a numerical study of deformation and failure of rock masses. *Int. J. Num. Anal. Meth. Geomech.*, 1: 219–247.

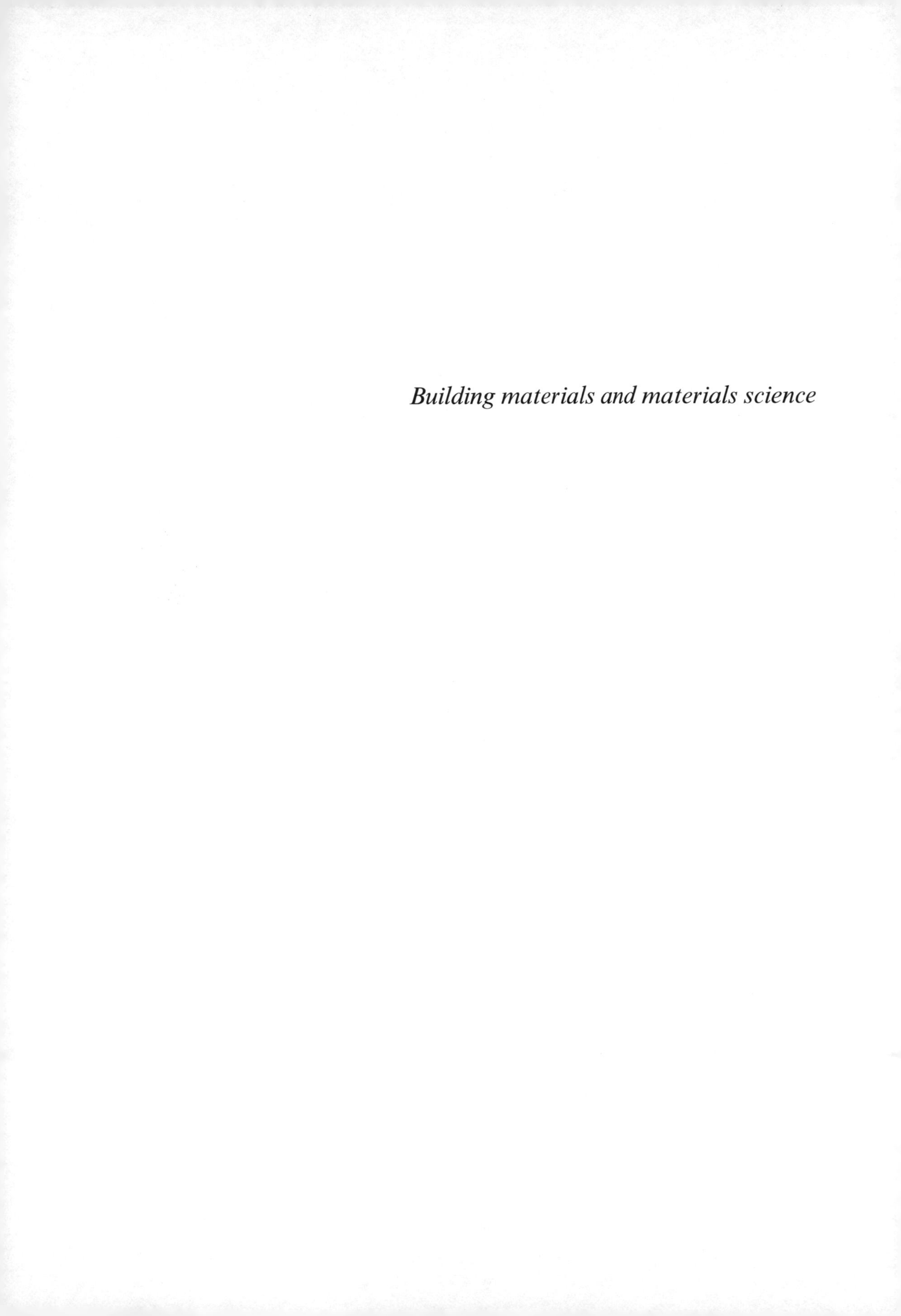

Building materials and materials science

Civil, Architecture and Environmental Engineering – Kao & Sung (Eds)
© 2017 Taylor & Francis Group, ISBN 978-1-138-02985-9

Preparation of sebacic acid by a phenol-free method

Ling Xu
College of Chemistry and Chemical Engineering, Inner Mongolia University for Nationalities, Tongliao Inner Mongolia, China
Inner Mongolia Key Laboratory of Castor Breeding, Inner Mongolia Collaborate Innovation Cultivate Center for Castor, Inner Mongolia Industrial Engineering Research Center of Universities for Castor, Tongliao, China

Lili Mao, Yiming Feng, Jingchao Zhang, Meng Wang, Tingting Zhao, Xiaojie Gao & Rui Ren
College of Chemistry and Chemical Engineering, Inner Mongolia University for Nationalities, Tongliao, China

ABSTRACT: At present, wastewater containing phenol has been produced due to the addition of phenol or cresol as diluent in the preparation of sebacic acid via the traditional high-temperature cracking castor oil method, which resulted in severe pollution. Recently, an environment-friendly method has been introduced to prepare sebacic acid by cracking castor oil. In this process, liquid paraffin is used as diluent and iron oxide is the catalyst. The optimal reaction temperature for preparing sebacic acid is 553 K, and the alkali concentration is 14 mol/L in the alkaline hydrolysis process.

1 INTRODUCTION

Sebacic acid is named from Latin sebaceous, which means the production of fat. It is freely soluble in alcohols, ethers, and ketones and slightly soluble in water (Diamond, 1965; Liang, 1996). Sebacic acid is an important chemical raw material, mainly used in the preparation of sebacic acid ester. In addition, sebacic acid is one of the main reagents for manufacturing nylon engineering plastics, resins, and fibers, which can also be used for the preparation of polyamides, polyurethanes, lubricating oil, and lubricating oil additives (Zhao, 2006; Ma, 2007). Therefore, the manufacturing technique of sebacic acid has attracted increasing attention in the past several decades. Many methods have been proposed to prepare sebacic acid, such as high-temperature alkali fusion cracking of castor oil, and nitrate epoxidation of decane ring decanol.

It is reported that sebacic acid is one of China's important traditional export products, whose production has reached more than 10000 ton/year, meeting about one-third of the global demand (Chen, 1990). The main raw material for the production of sebacic acid is castor oil. However, China is one of the three largest castor-planting countries, and Tongliao City accounts for a quarter of the national castor-cultivated area. Therefore, using castor oil hydrolytic cleavage to produce sebacic acid at Tongliao City has the advantage of raw materials. Especially in recent years, castor industries have been listed as essential development project. The pace of modernization and industrialization of castor was constantly accelerated due to the opportunity of the Western big development in the Inner Mongolia Autonomous Region. In addition, it is significant to use castor oil as raw material to produce chemical products in Tongliao, which highlights the local characteristics and also promotes the development of regional economy.

In the traditional process, sebacic acid was prepared by cracking castor oil using phenol or cresol as diluent, and the yield of sebacic acid was comparatively low, that is, 35.7–44.0% (Stephen, 1973). At present, a large amount of wastewater containing phenol was produced in the process of preparing sebacic acid from castor oil by the traditional high-temperature alkali fusion cracking method. Using this method, phenol content was 2000–3000 mg/L by water quality analysis and 120 million tons of wastewater containing phenol was produced each year. It is generally known that phenol or cresol is poisonous to all living individuals. Especially, it can mouse into organisms through skin, and contacts with protoplasmic protein to form insoluble protein, which makes cells lose their activity. Especially, it has a greater affinity to the nervous system and causes lesions to the nervous system. Therefore, developing a phenol-free method for the production of sebacic acid is highly recommended.

In addition, Wang et al. (Wang, 2012) reported that the yield of sebacic acid can reach 67.3% by cracking castor oil under the condition of alkali by using liquid paraffin as diluents and Pb_3O_4 as catalyst. However, lead oxide is harmful to the environment. To solve this problem, research on phenol-free method has been carried out to the preparation of sebacic acid. The poisonous phenol diluents and lead oxide catalyst were replaced by green paraffin oil and iron oxide, respectively. It can reduce the cost and avoid the pollution of phenol.

2 EXPERIMENT

2.1 *Materials and instruments*

NaOH, Fe_2O_3, castor oil, hydrochloric acid, and liquid paraffin used in this experiment were analytically pure.

PD-10 precision pH meter (Beijing sartorius Instrument System Co. Ltd.), DHT-type stirring electric heating sleeve (Shandong Juancheng Hualu Electic Instrument Co. Ltd), and FT-IR Spectrometer (Thermo Nicolet Corporation) were used.

2.2 *Preparation of sebacic acid*

2.2.1 *Saponification castor oil*

Castor oil (30 mL) and NaOH (120 mL, 3 mol/L) were mixed in a three-neck flask, and then the mixture was heated to boil under stirring for 30 min. After cooling, the solution was adjusted to pH 6 by adding 6 mol/L HCl, and was allowed to standing for 20 h.

2.2.2 *Alkali cracking of castor oil*

Liquid paraffin was added to a 500 mL three-neck flask, and then a certain concentration of NaOH was added to the flask. When the temperature reached the reaction temperature, a certain amount of Fe_2O_3 was added into the solution. At the same time, the saponification castor oil and 40% NaOH solution dropped slowly into the mixed solution. After that, the reaction was controlled at a certain temperature and kept for 1 h.

2.2.3 *Product separation*

At the end of the reaction, three times volume of distilled water was added into the mixture when the temperature dropped below 333 K. Then, the mixture was heated to 363 K and adjusted to pH 6 by adding 6 mol/L HCl. The mixture separated was separated into three layers, with the upper layer consisting of liquid paraffin, middle water, and the bottom layer white precipitate. The middle water layer was adjusted to pH 2, and the white precipitate was collected in aqueous phase. Finally, the crude product of sebacic acid was obtained by the filtering and drying process.

2.2.4 *Product purification*

About 15 times of water was mixed to the crude sebacic acid and the crude sebacic acid products and then heated to boiling for 3 h to increase the soluble content of sebacic acid in the water. After natural cooling, the crystal obtained was purified of sebacic acid (Li, 2009).

3 RESULTS AND DISCUSSION

3.1 *Temperature of the alkali cracking*

Figure 1 shows the FT-IR spectra of the crude product of sebacic acid, which was prepared at different temperatures.

When the temperature was below 533 K, the purity of sebacic acid was relatively low. With increasing alkali cracking temperature, the purity of sebacic acid was improved. When the cracking temperature reached 553 K, the main peaks of FT-IR spectrum were consistent with the standard spectrum of sebacic acid (Zhou, 2014). The bands at 2927 and 2855 cm^{-1} were assigned to saturated C–H stretching vibration. The bands at 1697, 1421, and 1353 cm^{-1} were for the C = O stretching vibration, the O–H bending vibration in the dimer as well as the C–O stretching vibration, and the C–H bending vibration in CH_2, respectively. The bands at 1239–1186 cm^{-1} correspond to the characteristic absorption of C–O bond in the monomer. The peak at 930 cm^{-1} was O–H bending vibration of sebacic acid dimer. When the temperature was higher than 553 K, carbonation phenomenon of the reagent and product occurred.

Figure 2 shows the FT-IR spectrum of purified sebacic acid at a cracking temperature of 553 K.

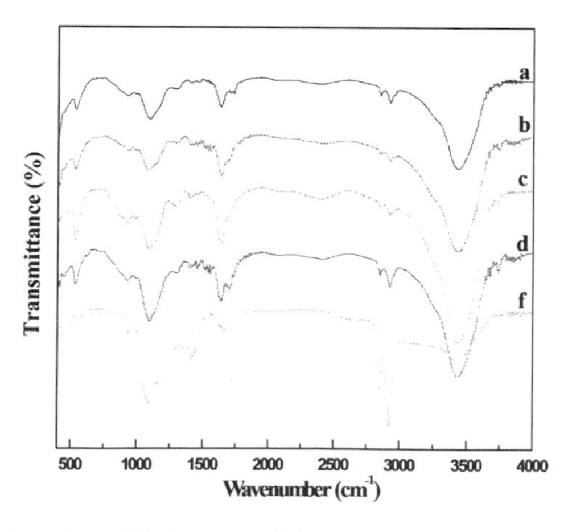

Figure 1. FT-IR spectra of the crude product of sebacic acid under different reaction temperatures. a: 473 K, b: 493 K, c: 513 K, d: 533 K, and f: 553 K.

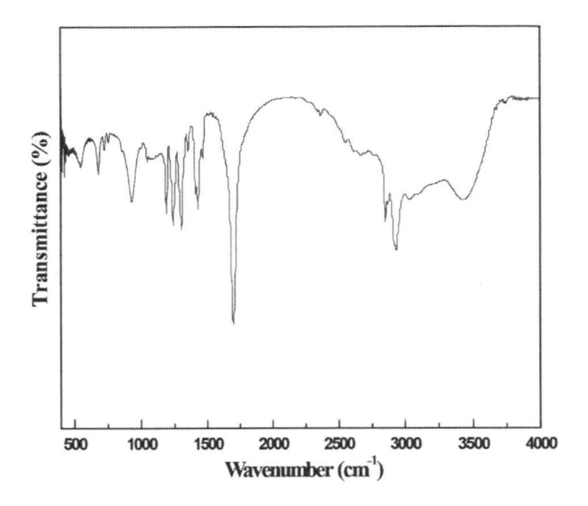

Figure 2. FT-IR spectrum of purification of sebacic acid at 553 K.

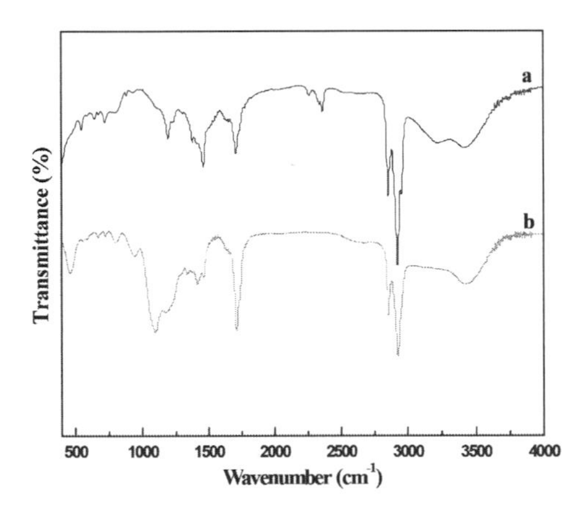

Figure 3. FT-IR spectra of crude sebacic acid with different alkali concentration. a: 7 mol/L, b: 14 mol/L.

The relative intensities of infrared diffraction peaks at 1697, 1421, 1353 cm^{-1} were significantly enhanced. The result shows that crystallization and purification of sebacic acid was beneficial.

3.2 *Alkali concentration in alkali cracking process*

The effect of different concentrations of NaOH on the alkali cracking castor oil to prepare sebacic acid has been investigated. The FT-IR result is shown in Figure 3. When the reaction temperature was controlled at 553 K, the amount of castor oil and liquid paraffin was 30 mL and 90 mL, respectively, the dosage of catalyst was 1% of the quality of castor oil, the alkali concentration was 14 mol/L, and the purification and crystallization of sebacic acid was optimal (as shown in Figure 3).

3.3 *Effect of different catalysts on the yield of sebacic acid*

Without catalyst, the yield of sebacic acid was less than 20%. After the addition of various metal oxide catalysts, the yield of sebacic acid increases. Compared with Al_2O_3 and ZnO, Fe_2O_3 exhibits a higher catalytic activity. Hence, Fe_2O_3 was selected as the catalyst for the preparation of sebacic acid.

4 CONCLUSION

Sebacic acid was prepared by cracking castor oil by a phenol-free method. The parameters for the optimal manufacturing technique are as follows: temperature, 553 K; alkali, 14 mol/L NaOH; catalyst, Fe_2O_3; and diluent, paraffin oil. This method can reduce environmental pollution because of the paraffin oil substitution of poisonous phenol, and is a potentially industrialized method for the preparation of sebacic acid.

ACKNOWLEDGMENTS

The authors acknowledge the National Natural Science Foundation of China (21561024) and the open foundation financial support of Inner Mongolia Industrial Engineering Research Center of Universities for Castor of Inner Mongolia University for Nationalities (MDK2016005 and BMYJ2015-05).

REFERENCES

Chen Guanrong et al., Beijing, chemical industry [M]. chemical industry press, 1, 654, 1990.

Diamond M.J. Binder, R.G.J., Am. Oil. Chem. Soc., 1965, 42: 882.

Li Ying. Sebacic acid solubility determination [J]. Sichuan chemical industry, 2009, 12 (4): 40–43.

Liang Liang, Liang Yiceng. Application of microwave radiation technology in organic synthesis [J], chemical bulletin, 1996, 3: 26.

Ma Jiancheng, Xia Qing, Zhang Fengbao. Castor oil cracking of sebacic acid production process of [J]. Chemical industry and engineering, 2007, 24: 362.

Stephen N, Purification of polybic acids [P], US 3746755, 1973-07-17.

Yanxiong Wang, Xiaoli Zhang, Hongya Li, The study on the process of cleaner production related to the preparation of sebacic acid by cracking of castor oil. Industrial Catalysis, 2012, 20: 68–71.

Zhao Tianbao chemical reagent, Chemical Handbook [M] Beijing Chemical Industry Publishing House, 2006.892.

Zhou Wanji. Research on Preparing Sebacic Acid by Activated Carbon Loaded Iron Oxide Catalytic Cracking of Castor Oil [D]. 2014.

Civil, Architecture and Environmental Engineering – Kao & Sung (Eds)
© 2017 Taylor & Francis Group, ISBN 978-1-138-02985-9

Effects of electrospinning parameters on the pore structure of porous nanofibers

Zhaoyang Sun & Lan Xu
National Engineering Laboratory for Modern Silk, College of Textile and Engineering, Soochow University, Suzhou, China

ABSTRACT: Electrospinning provides a simple and convenient method for generating polymer fibers which have been widely applied to produce porous nanofibers. The fibers produced by electrospinning have several excellent properties such as high surface-area-to-volume ratio, surface functionality and outstanding mechanical performance. The effects of collect distance and spinning voltage on the pore length distribution as well as the variation trend were explored and researched.

1 INTRODUCTION

The Electrospinning is regarded as one of the most simple and powerful technique for fabricating continuous and thin fibers. It has been widely utilized for preparation of polymer, ceramic, metal and composite fibers with high specific surface area (Dai, 2011), porosity and diameters ranging from tens of nanometers to a few micrometers (Panthi, 2015). In the past years, electrospinning technique has become an effective method for fabricating multifunctional nanometer materials from various polymers and composites (Zhang, 2015). One of the advantages of using electrospinning to fabricate porous nanofibers is its potential for control of pore structure. Xu et al. added sodium alginate in the hole of the nanofibers to make the wound dressing (Xu, 2016). Zhang et al. found as absorbent, the porous CeO_2 nanofibers adsorbed the MO were not only determined by the specific surface area, but closely related to the pore size (Zhang, 2016).

Due to its low cost, mechanical strength, and minimal inflammatory response, Polylactic Acid (PLA) has been widely studied for use in the biomedical industry (Sun, 2016). Although widely used in tissue engineering applications, biocompatible PLA electrospun meshes have displayed a high degree of shrinkage (Rowe, 2015).

In this paper, we fabricated porous PLA nanofibers with high specific surface area via the electrospinning method and investigated systematically the effects of electrospinning parameters, such as collect distance and spinning voltage, on the pore structure of PLA nanofibers.

2 EXPERIMENTAL SECTION

2.1 *Materials*

The base material for electrospun fibers was Polylactic Acid (PLA) with a molecular weight of 100,000 g/mol. Chloroform (CF) was obtained from Shanghai Chemical Reagent Co. Ltd, and N, N-dimethylformamide (DMF) was purchased from Guoyao Chemical Reagent Co. Ltd. All materials were used without any further purification.

All concentration measurements were done in weight by weight (w/w). PLA solutions were prepared at a concentration of 7 wt% by using mixture of CF and DMF with the weight ratio 9/1. The obtained solutions were magnetically stirred at 25°C for 3 h.

2.2 *Instrumentation*

The electrospinning setup consisted of a syringe, a needle, a grounded collecting plate, a flow meter and a variable DC high-voltage power generator (0–30 KV, DW-P303-1ACF0, Tianjin DongWen high-voltage power generator Co, LTD). The needle tip was connected to a DC high-voltage generator via an alligator clip.

2.3 *Electrospinning process*

The PLA solution was dropped into a 10 ml syringe connected with a metal needle that was controlled by a syringe pump at a constant flow rate 0.6 mL/h. The collect distance ranged from 8 cm to 14 cm. The spinning voltage ranged from 10 kv to 30 kv. All the experiments were carried out at 25 ± 3°C with the relative humidity of 50 ± 5%.

3 RESULTS AND DISCUSSION

Fig. 1 and Table 1 showed SEM images and pore length distribution under the influence of different spinning voltages. The collect distance was 15 cm. It could be seen that the distribution of pore length of the electrospun porous fibers was more nonuniform with the increase of the applied spinning voltage. In this process, when the values of the flow rate and the collect distance were constants, higher voltage meant higher value of the jet speed. With increasing

Table 1. The relationship between the spinning voltage and the pore length.

Spinning voltage	Average length (nm)	Minimum length (nm)
10 kv	77.97	65.19
15 kv	86.38	84.95
20 kv	99.78	96.28
25 kv	96.09	76.80
30 kv	90.81	83.05

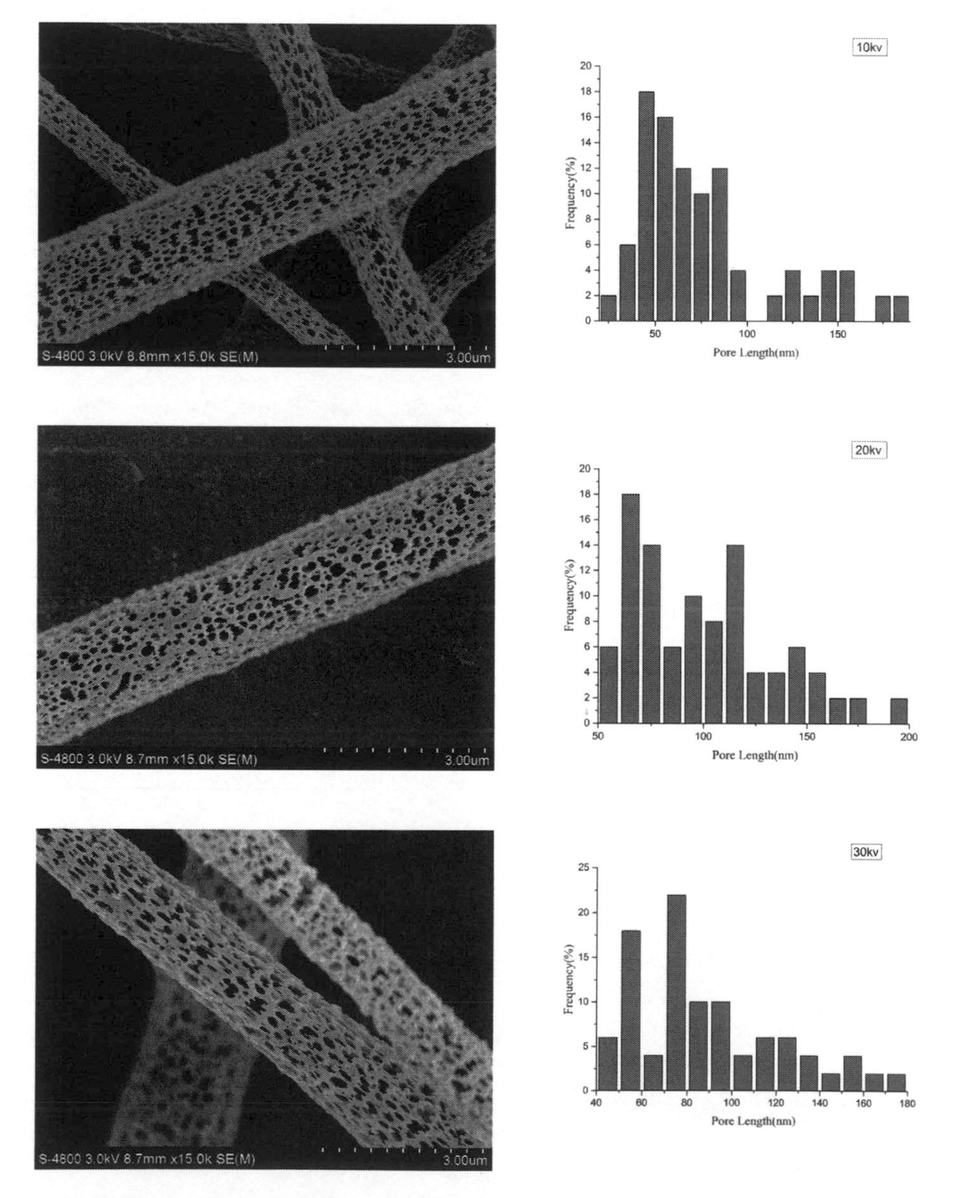

Figure 1. SEM images and pore length distribution under the influence of different spinning voltage.

the jet speed, the bending instability of jets could be exacerbated, as a result, both the distribution of pore length became nonuniform. That meant the most optimized spinning voltage was 10 kv.

Fig. 2 and Table 2 showed SEM images and pore length distribution under the influence of different collect distance. The applied spinning voltage was 15 Kv. It showed that the uniformity of pore distribution was improved firstly and then deteriorated with the increase of the collect distance. When the values of the flow rate and the applied

Table 2. The relationship between the collect distance and the pore length.

Collect distance (cm)	Average length (nm)	Minimum length (nm)
8 cm	79.53	68.59
11 cm	86.45	84.70
14 cm	78.19	78.58
18 cm	80.04	75.12
20 cm	92.96	80.85

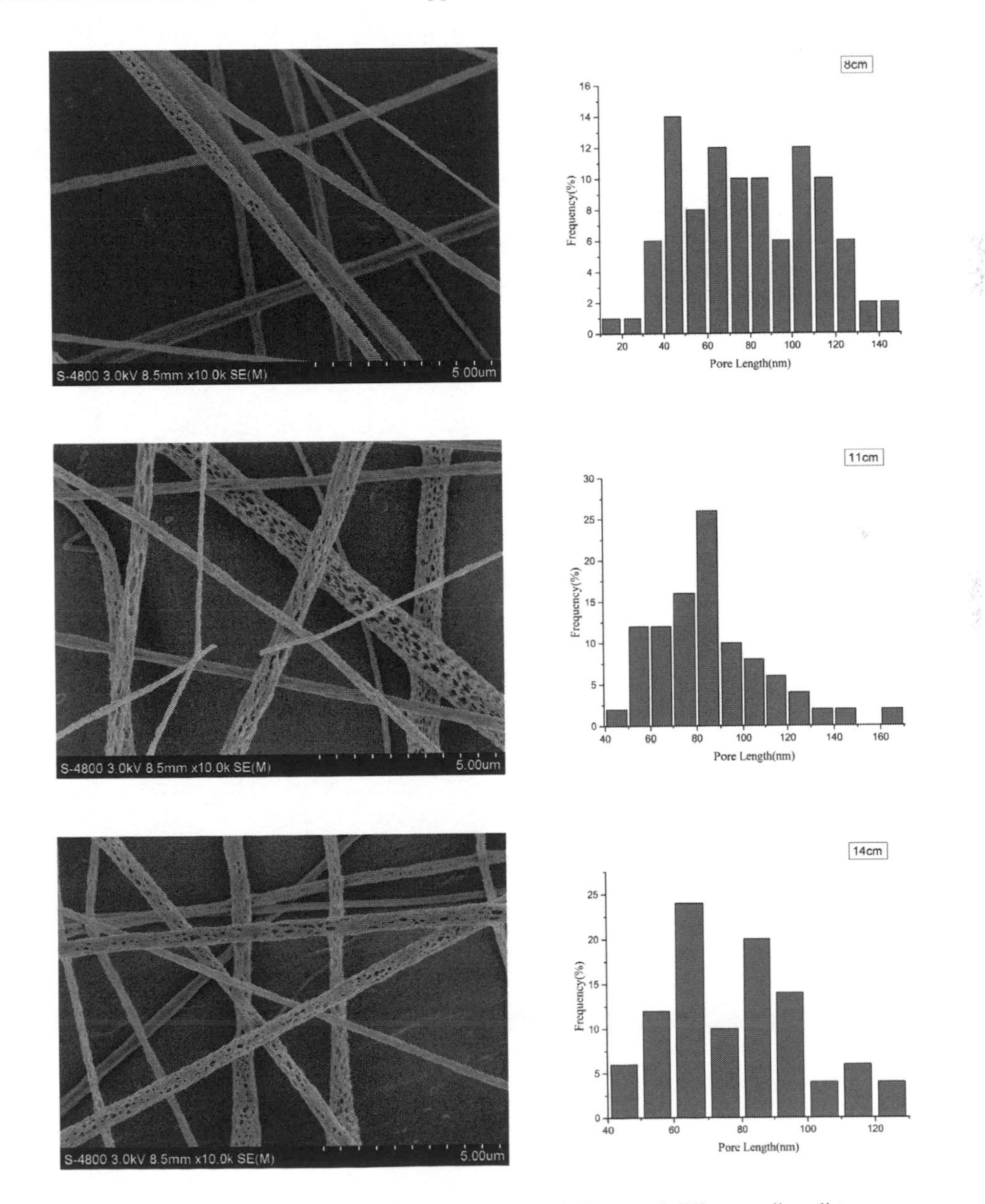

Figure 2. SEM images and pore length distribution under the influence of different collect distances.

voltage were constants, with the increase of the collect distance the charged jet could be more easily accelerated to the higher speed before it was collected. Higher value of the jet speed resulted in the exacerbated bending instability of jets and the nonuniformity of pore distribution. In addition, when the collect distance was too small the formation of pores on the jets would be hindered due to inadequate volatilization in the electrospinning process. Therefore, the most optimized collect distance was 11 cm.

4 CONCLUSION

In this paper, electrospun porous fibers were prepared by adjusting electrospinning parameters, such as collect distance and spinning voltage. And the pore distribution of obtained porous fibers was investigated. The results showed that electrospinning parameters played an indispensable and crucial role in electrospinning process, which directly affected the porous structure having uniform but tunable lengths could be controlled by adjusting electrospinning parameters in the process.

REFERENCES

Dai Y.Q., W.Y. Liu, E. Formo, Y.M. Sun, Y.N. Xia, Polym. Adv. Technol., **22**, 7, 326–338(2011).

Panthi G., M. Park, H.Y. Kim, S.Y. Lee, S.J. Park, Ind. Eng. Chem., **21**, 26–35(2015).

Rowe M.J., Kamocki K., Pankajakshan D.J. Biomed. Mater. Res. **104**, 3(2015).

Weihong Xu, Renzhe Shen, Yurong Yan, J. Serb. Chem. Soc, **65**, (2016).

Zhang H., Q. Niu, N.T. Wang, Eur. Polym. **71**, 440–450(2015).

Zhang Y., R. Shi, P. Yang. Ceram. Int, **42**, 12, 14028–14035(2016).

Zhaoyang Sun, Chenxu Fan, Xiaopeng Tang. Appl. Surf. Sci, **387**, 828–838(2016).

Civil, Architecture and Environmental Engineering – Kao & Sung (Eds)
© 2017 Taylor & Francis Group, ISBN 978-1-138-02985-9

Preparation of orderly nanofibers using bubble-electrospinning

Zhong-Biao Shao & Lan Xu
National Engineering Laboratory for Modern Silk, College of Textile and Engineering, Soochow University, Suzhou, China

ABSTRACT: Aligned nanofibers fabricated by electrospinning have extensive applications. Bubble-electrospinning as a mass production technology has some advantages compared with other electrospinning techniques to prepare nanofibers. The highly aligned PAN nanofibers were successfully fabricated by the simple rapid method which was a bubble-electrospinning with the parallel micropipette electrodes. The effects of spinning voltage on the degree of nanofiber alignment and diameter distribution were investigated.

1 INTRODUCTION

Electrospinning provides a simple and convenient method for generating nanofibers (Dai, 2011), which has been applied to prepare nanofibers made of organic polymers, ceramics, and polymer/ceramic composites (Zhao, 2016). Due to the wide application of orderly nanofibers in tissue engineering (Mehrasa, 2015), drug deliver (Goh, 2013), reinforcements (Terao, 2010), filtration membranes (Kaur, 2012), and so on, a number of methods have been proposed to fabricate aligned nanofibers (Zhang, 2012; Heidari, 2013). But all these methods have respective defectiveness, accompanied with the problems of easily jammed and inefficiency. Preparation of highly aligned nanofibers has received much attention in recent years. In this regard, bubble-electrospinning technique as an interesting and effective alternative to the classical electrospinning technique are explored to prepare aligned nanofibers (He, 2014).

In this paper, we modified a bubble-electrospinning with the parallel micropipette electrodes was successfully developed for mass production of highly aligned nanofibers for a long spinning time(Shao,2010). And also, the morphology and the degree of alignment of nanofibers with the effect of voltages of polymer solution were researched.

2 EXPERIMENTAL SECTIONS

2.1 *Materials*

Polyacrylonitrile (PAN) with a molecular weight of 150,000 g/mol was supplied Beijing lark branch co., LTD. N,N-Dimethylformamide (DMF) was obtained from commercial source and used as solvent. All materials were used without any further purification.

2.2 *Instrumentation*

The bubble-electrospinning setup included a bubble generator, a spinneret, an air pump, a pair of parallel electrodes, and a variable DC high-voltage power generator (0–100 KV, DW-P303-1ACF0, Tianjin DongWen high-voltage power generator Co, LTD).

2.3 *Bubble-electrospinning process*

All concentration measurements were done in weight by weight (w/w). A control amount of PAN powder was dissolved in the DMF with the weight ratio 1:9 (PAN: DMF). The mixture was magnetically stirred at 80°C for 4 hours in an electromagnetism stirrer (Angel Electronic Equipment (Shanghai) Co., LTD), stirring speed from 950 to 1200 rounds/min. The mixture was dropped into in a 10-ml syringe which was mounted in a syringe pump. A pair of parallel electrodes was placed vertically upper to the spinneret, and the distance was 15 cm. The applied voltage connected to the generator varied from 25 KV to 45 KV. All processes were carried out at room temperature (25°C) in a vertical spinning configuration at a relative humidity of 45%.

3 RESULTS AND DISCUSSION

Fig. 1 and Table 1 showed SEM images and diameter distribution of aligned nanofibers under the influence of different spinning voltages. As could be seen, when the applied voltage was 25 KV, the diameter distribution of nanofibers was sparse and

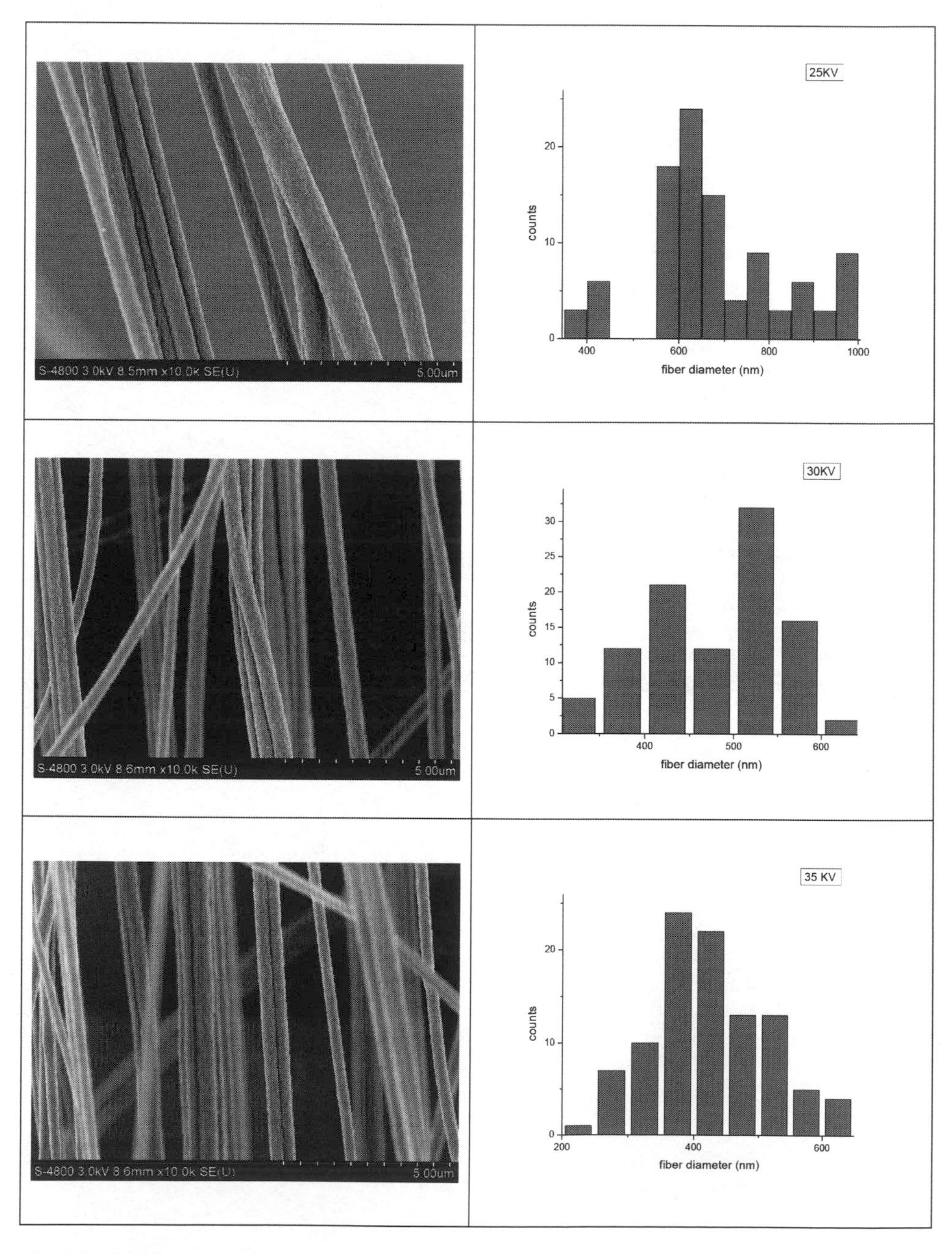

Figure 1. (*Continued*).

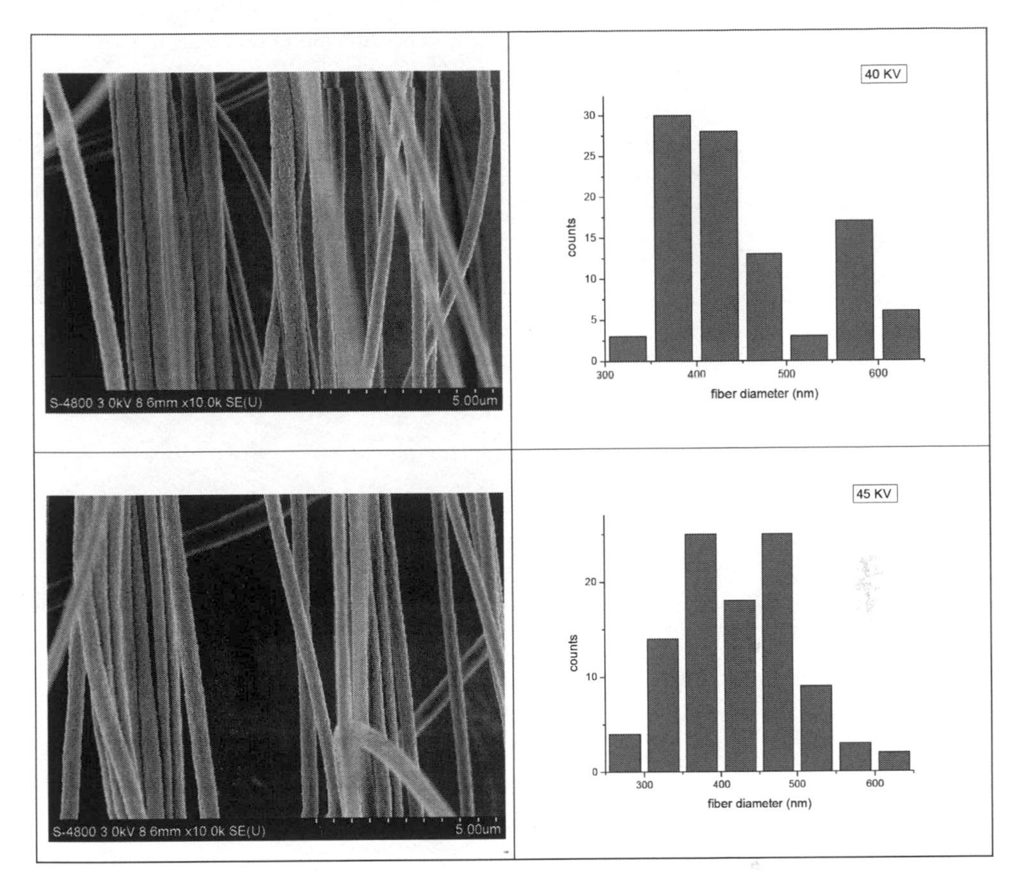

Figure 1. SEM images and diameter distribution of aligned nanofibers under the influence of different spinning voltage.

Table 1. The relationship between spinning voltage and nanofiber diameter.

Voltage	Average diameter (nm)	Minimum (nm)
25 KV	680	379
30 KV	480	318
35 KV	754	550
40 KV	464	335
45 KV	445	275

nonuniform. Moreover, with the increase of the applied spinning voltage, the diameter distribution of nanofibers became dense and more uniform. However, when the applied voltage was 45 KV, the bending instability of jets was exacerbated, the diameter distribution of nanofibers became sparse again. That meant the most optimized spinning voltage was 30–40 KV.

4 CONCLUSION

In this work, the aligned PAN nanofibers have been successfully fabricated by the bubble-electrospinning with the parallel micropipette electrodes. And the effects of spinning voltage on the nanofiber alignment and diameter distribution were researched. The results showed the optimized spinning voltage ranged from 25 KV to 45 KV. A large number of highly aligned nanofibers produced by the modified bubble—electrospinning would have promising applications of fiber-reinforcement, fiber-oriented liquid crystal, and tissue engineering.

ACKNOWLEDGEMENT

The work is supported financially by PAPD (a project funded by the Priority Academic Program Development of Jiangsu Higher Education

Institutions), Jiangsu Provincial Natural Science Foundation of China (Grant No. 11672198), Natural Science Foundation of the Jiangsu Higher Education Institutions of China (Grant No. 14 KJA130001), and Suzhou Science and Technology Project (Grant No. SYG201434).

REFERENCES

Dai Y.Q., W.Y. Liu, E. Formo, Y.M. Sun, Y.N. Xia, Polym. Adv. Technol. **22**, 7, 326–338 (2011).

Goh Y.F., I. Shakir, R. Hussain, J. Mater. Sci. **48**, 3027–3054 (2013).

Heidari I., M.M. Mashhadi, G. Faraji, Chem. Phys. Lett. **590**, 231–234 (2013).

He J.-X., K. Qi, Y.-M. Zhou, S.-Z. Cui, Polym. Int. **63**(1), 288–294 (2014).

Jianghui Zhao, Hongying Liu, lanXu, Mater.Des. **90**, 1–6 (2016).

Kaur S., D. Rana, T. Matsuura, S. Sundarrajan, S. Ramakrishna, J. Membr. Sci. **390**, 235–242 (2012).

Mehrasa M., M.A. Asadollahi, K. Ghaedi, H. Salehi, A. Arpanaei, Int. J. Biol. Macromol. **79**, 687–695 (2015).

Terao T., C.Y. Zhi, Y. Bando, M. Mitome, C.C. Tang, D. Golberg, J. Phys. Chem. C **114**, 4340–4344 (2010).

Zhang Q.C., L.H. Wang, Z.M. Wei, X.J. Wang, S.R. Long, J. Yang, J. Polym. Sci. Pol. Phys. **50**, 1004–1012 (2012).

Zhongbiao Shao, Therm. Sci. (to be published).

Civil, Architecture and Environmental Engineering – Kao & Sung (Eds)
© 2017 Taylor & Francis Group, ISBN 978-1-138-02985-9

Environmental radioactivity measure and dose estimation of ^{90}Sr in some area

Li Ma
Medical Protection Laboratory, Naval Medical Research Institute, Shanghai, China

Xin Ren
Graduate School, National Defense University, Beijing, China

Kai Zhang
Medical Protection Laboratory, Naval Medical Research Institute, Shanghai, China

ABSTRACT: Objective To measure the current environmental radioactivity levels of ^{90}Sr in the area, and to estimate the doses to local residents. Methods The water samples, surface siol and biologic samples were collected and analyzed by radiochemstry techniques. Results Based on the date, the annual effective dose rate of ^{90}Sr to the local people was 1 μSv per year, only 1‰ of per year limited by national standards. Conclusion the annual effective dose rate of ^{90}Sr to the local people is far less than the dose of national standards.

1 INTRODUCTION

In the 1960 s, the Ministry of health has established environmental radioactivity monitoring network in the national scope, which has played an important role in the effect evaluation of the impact of previous nuclear tests and foreign major nuclear accidents on the regions of our country, and the valuable information on environmental radioactivity level in our country has been accumulated. With the development of the national nuclear power industry and the establishment of military nuclear base and facilities, the radiation environment investigation in the surrounding areas has also been included in the routine monitoring. According to the requirements of environmental protection work, a large number of studies on the investigation and evaluation of radionuclides background in environmental medium have been carried out since 2003 in this unit.

As the artificial radionuclides in the environment, radioactive strontium has a total of 17 isotopes (the atomic number is from ^{81}Sr to ^{97}Sr), and they are β radiation source and ^{89}Sr and ^{90}Sr have the most significant meaning of Toxicology. ^{90}Sr has a long half-life, with severe damage to agriculture and animal husbandry, and it can enter the human body through the food chain. Consequently, ^{90}Sr pollution has attracted people's attention. The present situation investigation and evaluation of nuclear radiation environment in a certain area has been carried out by the unit in 2012, the radioactive levels of ^{90}Sr in water, soil and biological samples within the scope of evaluation have been investigated and monitored, and the accumulated effective dose is calculated using the corresponding dose conversion factor recommended by the ICRP (International Commission on Radiological Protection) and it is compared with the relevant dose limit in order to evaluate the adult personal annual effective dose caused by ^{90}Sr in this area.

2 INSTRUMENTS AND METHODS

Equipments are as follows: F6000-60 ashing furnace, Thermo Fisher Company, America. BH1216 III type two channel low background alpha and beta measuring instrument, Beijing nuclear instrument factory. XDB030301 type undisturbed soil sampler, Kardinal Technology and Trade Co., Ltd.

2.1 *The principle of layout and area*

The evaluation scope is 20 km area of taking nuclear facilities as the center (10 km area is the key investigation scope). The concentric circle is used to dispose the dot in different directions within the scope of the evaluation range, and the development of layout scheme and selection of measuring point are assigned combined with the terrain features of the surrounding and the situation of population distribution.

2.2 *Samples acquisition and measurement*

In this investigation, the 4 the water samples, 12 biological samples and 7 soil samples were collected.

2.2.1. *Water samples*

The drinking water of residents including reservoir water and tap water was collected, the sampling methods are used according to the radiation environmental monitoring technical specifications of HJ/T61-2001, and the radiochemical analysis method of strontium-90 in the water 2-(2-ethylhexyl) phosphate extraction chromatography of GB 6766-86 is used as the methods for measuring.

2.2.2 *Soil samples*

The soil sampling points are selected in a relatively open, domesticated and less activities of livestock zones, the soil below 10 cm of surface is taken, and the sampling methods are used according to the HJ/T 61-2001 radiation environmental monitoring technical specifications and the general rules and regulations of EJ 428-1989 environment nuclear radiation monitoring in the collection and preparation of soil samples, and the methods of measurement use the strontium 90 analysis method of EJ/T 1035-1996 in the soil.

2.2.3 *Biological samples*

The local representative samples of wheat, sweet potatoes, pork, spinach, scallops, noodle fish and seaweed etc. are collected. The sampling methods are used according to the HJ/T 61-2001 radiation environmental monitoring technical specifications and the the basic rules of biological sampling in environmental radiation monitoring of EJ 527-1990, and the methods of measurement use strontium-90 radiochemical analysis method in biological samples of GB11222.1-89-(2-ethylhexyl) phosphate extraction chromatography method.

2.3 *Efftive dose evaluation method of ^{90}Sr for residents in a year*

2.3.1 *Transfer coefficient*

The transfer coefficient P_{ij} is used to describe the environmental behavior of the nuclide and calculate the dose as the basic parameters in the calculation model. Transfer coefficients are used to describe the relationship between the cumulative concentration or dose in a series of environmental compartments, such as the transfer relationship from I compartment to J compartment. The migration pathway and transfer coefficient of radionuclides commonly used in the environment dose evaluation by UNSCEAR can be shown in Figure 1. For example, P_{34} is the value that the activity concentration of the nucleus is divided by the time integral concentration of the nucleus in the food.

2.3.2 *Evaluation of radiation dose through the food*

Becuase the residents personal annual effective dose of ^{90}Sr is mainly caused by the diet (water), the residents annual intake of various types of food and drinking water in this area must be considered in the calculation. The average personal annual drinking water intake of adult is 730 L/year, the average annual individual consumption of food can be shown in Table 1.

To evaluate the exposure dose based on the biological sample (including water) radionuclide monitoring data:

$$D_a = \sum_{i,j} D_{ai,j} = \sum_{i,j} C_{ai,j} \upsilon_{ai} P_{aj45} \quad (1)$$

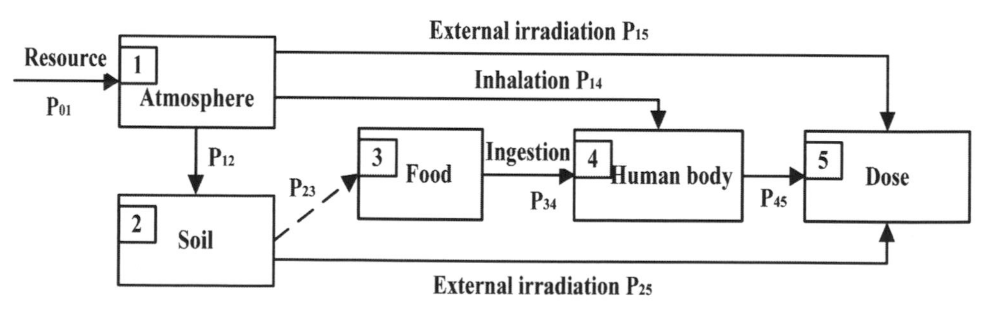

Figure 1. Land transfer pathway of radionuclide and the dose effect on human body.

Table 1. Food consumption of adults in a region (kg/year).

Food	Vegetable	Fruit	Grain	Livestock	Poultry meat	Egg	Milk	Fresh water produc	Sea food	Tea
Adult	133.7	7.8	210.6	8.2	3.2	7.6	4	1	3	0.3
Self sufficiency in the evaluation area	100	80	100	100	100	100	100	100	100	100

Among them, D_a is the radiation dose of personal annual ingestion, Sv. D_{ai} is the individuals annual ingestion of the ith food, Sv. $C_{ai,j}$ is the quality activity of radionuclides j in the ith food, Bq/kg. v_{ai} is the annual intake of the ith food, kg. p_{aj45} is the dose caused by the intake of unit radionuclide radioactivity j, Sv/Bq, and the value of ^{90}Sr is taken for 28 nSv/Bq. The irradiation dose of intaked food is evaluated based on the outdoor radionuclide monitoring data of soil:

$$D_a = \sum_i D_{ai} = \sum_i 150 * A_i * p_{i2345} * (1-\delta) \quad (2)$$

Among them, D_a is the radiation dose of personal annual ingestion, nSv. D_{ai} is the individuals annual ingestion of the ith food, nSv. A_i is the quality activity of the ith radionuclide in soil, Bq/kg. p_{ai2345} is the dose caused by intaking the jth unit radioactivity radionuclide, nSv/Bq·m^2, and the value of ^{90}Sr is taken for 53nSv/Bq. δ is the containing water in soil, %.

3 RESULTS

3.1 *Measurement of water samples*

The analysis results of artificial radionuclides ^{90}Sr in drinking water samples around the area are shown in Table 2. As can be seen from Table 2, the activity range of artificial radionuclide ^{90}Sr in the water samples around the investigate area is from 2.7 mBq/L to 5.1 mBq/L, the average value was 4.1 ± 0.2 mBq/L, and the maximum value was about 1.5 times of the control point.

3.2 *Measurement of biological samples*

The results of artificial radionuclide ^{90}Sr in the sample of biological samples around the area are shown in Table 3. It can be seen as follows from Table 3: the activity of ^{90}Sr in the scallops, pork, tea, wheat and chicken is relatively higher, the activity of ^{90}Sr in milk, kelp, spinach, sweet and potato is relatively lower, and the activity of ^{90}Sr in noodle fish, crucian carp, and apple is the lowest. In the above mentioned biological samples, the content of artificial radionuclide ^{90}Sr is lower than the limit of national standard.

3.3 *Measurement of soil sample*

The analysis results of artificial radionuclides ^{90}Sr in the soil samples around the area are shown in Table 4. As can be seen from Table 3, the activity range of artificial radionuclide ^{90}Sr in the soil samples around the investigate area is from 0.7 mBq/kg·fresh to 2.43 mBq/kg·fresh, the average value was 1.57 ± 0.05 mBq/kg·fresh, and the maximum value was about 4 times of the control point.

3.4 *Evaluation of personal annual effective dose*

Considering that the uncertainty of dose evaluation can be effectively reduced by reducing the migration process of radionuclides in the environment and the human body, the ingestion exposure dose assessment value of food is used based on the full access of radionuclide concentration monitoring data in the biological samples. According to the nuclide dose calculation model shown in Figure 1 and provides the food consumption of

Table 2. Radiochemical analysis results of the water samples.

Name of samples	reservoir water of number one	Tap water of number one	Tap water of number two	Well water of number one	Control point
^{90}Sr (mBq/L)	2.7 ± 0.1	4.5 ± 0.2	5.1 ± 0.3	4.2 ± 0.2	3.0 ± 0.1

Table 3. Radiochemical analysis results of the biological samples.

Name of samples	Wheat	Apple	Sweet potato	Spinach	Scallop in Shell	Noodle fish
^{90}Sr (mBq/kg·fresh)	108.3 ± 2.8	13.8 ± 3.0	68.1 ± 2.4	81.5 ± 4.7	129.0 ± 3.0	21.2 ± 5.0
Name of samples	Pork	Chicken	Kelp	Milk	Tea	Crucian carp
^{90}Sr (mBq/kg·fresh)	155.3 ± 5.8	107.8 ± 2.6	92.5 ± 6.6	94 ± 4.6	128.1 ± 6.1	35 ± 5.2

Table 4. Radiochemical analysis results of the soil samples.

Name of samples	Point 1	Point 2	Point 3	Point 4	Point 5	Point 6	Point 7
^{90}Sr (mBq/kg·fresh)	1.31 ± 0.05	1.77 ± 0.06	2.43 ± 0.06	2.04 ± 0.05	1.02 ± 0.05	0.70 ± 0.05	1.72 ± 0.05

adults in average and personal drinking water in a year shown Table 2, the resident personal annual dose of ^{90}Sr is about 38.7 Bq and it is far below the national standards dose limit of 3.8×10^4 Bq. The calculation value of personal annual effective dose of ^{90}Sr is 1 μSv according to the formula (1), and it is only 1 per thousand of national standard annual dose limit value.

4 DISCUSSION

The radioactive contents of ^{90}Sr in the water, soil and biological samples around are measured through the investigation of environmental radioactivity level in a certain area, the content of ^{90}Sr in the biological samples is lower than the national standard limit values, and the content of ^{90}Sr in the water and soil samples is slightly higher than the control trace, but the calculations results show that the personal annual effective dose is rarely little, and it is only 1 per thousand of national standard annual dose limit value, which indicate that the public exposure caused by ^{90}Sr is minimal in human activities around the area. In this investigation, the exposure of ^{90}Sr on the human body is produced by beta ray, the annual effective dose estimation measurement data of ^{90}Sr is mainly from the measurement results of drinking water and food, and the external radiation caused by soil will not be considered. The results of this investigation and evaluation can provide an effective basis for the radioactive level scientific evaluation of ^{90}Sr in the surrounding environment of nuclear facilities and the individual annual effective dose and collective effective dose of the public in the region caused by ^{90}Sr.

REFERENCES

GB 14882–1994. The standards for the limiting concentrations of radioactive substances in the food [S].

Haijun Wang, Yifang Yang, Yongjie Lu. Application of thermoluminescent in the investigation of the environmental gamma radiation levels in the [J] Navy Medical Journal, 27(2), pp. 110–112, 2006.

Maoxiang Zhu. health effects and the promotion measures of radionuclide [J]. Canceration distortion mutation, 23(6), pp. 468–472, 2011.

Sichang Qin, Zhengping Fan, Chengxiang Lei, etal. Study on the method of radioactive background in environmental media [J]. Navy Journal of medicine, 27(1), pp. 8–10, 2006.

UNSCEAR 2000 [R]. Ionizing radiation sources and its effects [M]. Taiyuan: Shanxi Science and Technology Press. pp. 27, 2002.

Yongjie Lu, Zhengping Fan, Yifang Yang, etal. Preparation and implementation of the background investigation on environmental gamma radiation [J]. Navy Medical Journal, 27(1), pp. 5–8, 2006.

Yongmei Xie, Zhenming Zhuang, Yongzhong Song. Investigation on environmental electromagnetic radiation level in suburban counties of Nanjing city [J]. Chin radiation health, 18(4), pp. 461–462, 2009.

Ziqiang Pan. Monitoring and evaluation of ionizing radiation environment [M]. Beijing: Atomic Energy Press. pp. 54, 2007.

Civil, Architecture and Environmental Engineering – Kao & Sung (Eds)
© 2017 Taylor & Francis Group, ISBN 978-1-138-02985-9

Effect of humic acid on the adsorption of levofloxacin to goethite

Xiaopeng Qin, Hong Hou, Long Zhao, Jin Ma, Zaijin Sun & Wenjuan Xue
State Key Laboratory of Environmental Criteria and Risk Assessment, and Soil Pollution Effect and Environmental Criteria, Chinese Research Academy of Environmental Sciences, Beijing, P.R. China

Fei Liu & Guangcai Wang
School of Water Resources and Environment, and Beijing Key Laboratory of Water Resources and Environmental Engineering, China University of Geosciences (Beijing), Beijing, P.R. China

ABSTRACT: Adsorption of a widely used antibiotic Levofloxacin (LEV) to goethite was studied using batch experiments. The adsorption of LEV to goethite increased proportionally with the increase of Humic Acid (HA) concentration, therefore the additional LEV adsorbed can be attributed to its adsorption to LEV. The amounts of LEV adsorbed to NOM are consistent with the literature values for adsorption of antibiotics with similar structures on free (*i.e.* in the absence of goethite) HA and Fulvic Acid (FA). The Excitation-Emission Matrix (EEM) spectra results indicated the interactions (*i.e.* adsorption) between HA and LEV, which become strong at neutral conditions, and much weaker at pH 3.0 and 9.0. The presence of goethite and HA will affect the environmental fate of antibiotics.

1 INTRODUCTION

Antibiotics have been widely used in human and veterinary treatments, and they are frequently detected in the environment (Hu et al., 2010). The presence of antibiotics may change the microbial community and the antibiotic resistance genes (Zhang et al., 2013). The environmental fate of antibiotics is influenced by their adsorption to soil minerals and Natural Organic Matter (NOM) (Tolls, 2001).

Adsorption of antibiotics to minerals and to NOM has been studied previously. In this respect, a range of minerals (*i.e.* goethite, hematite, and kaolinite) and NOM have been investigated. Cation exchange and surface complexation were considered as the major mechanisms in their adsorption on respectively clay minerals (Figueroa et al., 2004) and metal (hydr)oxides (Gu and Karthikeyan, 2005). Contradictory results have been reported regarding the effects of NOM on antibiotics adsorption to minerals. Some studies have shown that presence of NOM promoted antibiotic adsorption to goethite (Zhao et al., 2011). However, different results were reported as well. Peng et al. (2012) observed that the addition of HA (5–50 mg/L) suppressed the adsorption of norfloxacin on Titanium Oxide (TiO_2). Similarly, inhibition of norfloxacin adsorption to soils by small organic acids was also reported (Zhang and Dong, 2008). In addition, several authors found that NOM had no significant influence on the adsorption of antibiotics to soil minerals (Yan et al., 2012). Clearly, a consistent picture regarding the effect of NOM on antibiotics adsorption to minerals is still lacking.

In this work, a common iron mineral goethite (α-FeOOH) was used as the adsorbent, and a widely used antibiotic Levofloxacin (LEV) was used as an example. The aim is to investigate its adsorption to goethite in the presence of HA, and the complexation between HA and LEV.

2 MATERIALS AND METHODS

2.1 *Materials*

LEV (> 98.0%, Tokyo Chemical Industry Co.) was used without further purification. The commercial HA was purchased from Sinopharm Chemical Reagent Co. Acetonitrile (HPLC grade, Honeywell Burdick & Jackson) was obtained. Goethite was prepared using the common method (Hiemstra et al., 1989; Antelo et al., 2005). Other chemicals were analytical reagent grade or higher. MilliQ water was used in the experiments. The properties of LEV, HA, and goethite were reported in our previous studies (Qin et al., 2012; Qin et al., 2014b).

2.2 *Adsorption experiments*

Adsorption isotherms of LEV to goethite at pH 5.0 in the absence or presence of PHA at two concentrations (3.63 and 7.25 mg C/L PHA) were obtained from results of batch experiments.

During the experiments, the 50 mL polyethylene centrifuge tubes that contained the suspensions were flushed with N_2 gas to minimize the influence of CO_2. Both the LEV-NOM and goethite stock solutions were prepared in the background of 0.01 M NaCl. The final goethite concentration was 0.8 g/L. The total LEV concentration was 1–10 μM, and the final NOM concentration was 0, 3.63, or 7.25 mg C/L. The pH of all suspensions was adjusted to 5.0 using either 0.01 M HCl or 0.01 M NaOH solutions. All suspensions were shaken for 24 h at 175 rpm in the dark under room temperature. After equilibration, the end pH of each suspension was measured with a pH meter (Sartorius PB-10, Germany). Then the suspensions were centrifuged at 12000 rpm for 20 min, and the supernatants were analyzed immediately. All treatments were carried out in triplicates.

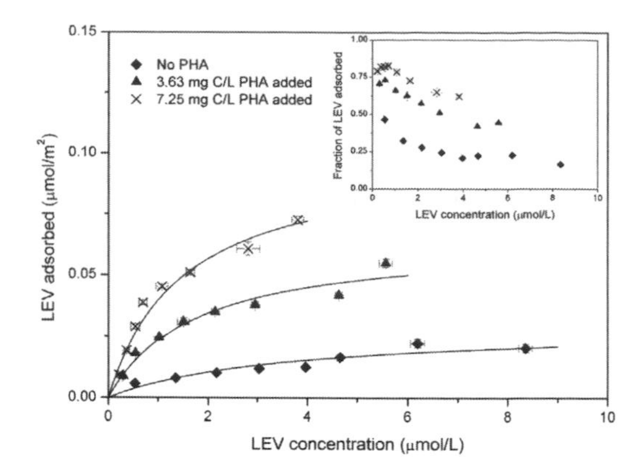

Figure 1. Adsorption of LEV on goethite in the absence and presence of HA at pH 5. The concentration of goethite is 0.8 g/L. Error bars (±1 Standard deviation, n = 3) are shown in the figures.

2.3 *Chemical analyses*

The concentration of LEV in solutions was determined using the High Performance Liquid Chromatography (HPLC) method as reported (Qin et al., 2014a). The concentration of LEV adsorbed to goethite was calculated from the differences in the concentrations of LEV in solutions before and after adsorption.

The Excitation-Emission Matrix (EEM) spectra of PHA or (and) LEV was measured with a fluorescence spectrophotometer (Hitachi F-7000, Japan). The emission wavelength and excitation wavelength were in the range of 250–600 nm and 200–500 nm, respectively. The concentrations of PHA and LEV were 5 mg C/L and 10 μM, which were diluted 25 times during the EEM spectra measurements.

3 RESULTS AND DISCUSSION

3.1 *Effect of HA on LEV adsorption*

The adsorption of LEV to goethite was measured at pH 5.0 at a range of LEV concentrations in the absence and presence of 3.63 and 7.25 mg C/L HA (Figure 1). It seems that an adsorption plateau for LEV has been reached at 0.021 μmol/m^2 in the absence of NOM, which indicates a relative weak adsorption. The presence of PHA promoted LEV adsorption significantly even when only 3.63 mg C/L PHA was added. Increasing HA concentration to 7.25 mg C/L increased LEV adsorption further. More than 16%, 42%, and 62% of the total LEV were removed from the solution in the absence and presence of 3.63 and 7.25 mg C/L PHA, respectively. According to the preliminary experiments, nearly all (100% and 96%) of the added HA was adsorbed to goethite in our study, which is equivalent to a PHA loading on goethite of 0.047 mg C/m^2 and 0.089 mg C/m^2.

The adsorption isotherms of LEV were well fitted to the Langmuir model ($R^2 > 0.84$) as shown in Table S1. The fitted maximum adsorption (q_m) of LEV increased from 0.029 (without PHA) to 0.099 μmol/m^2 (7.25 mg C/L PHA). The fitted affinity constants (K) in the presence of PHA were about 2 times larger than that without PHA. The results show that both the adsorption affinity and the maximum binding capacity have increased due to the presence of PHA.

Under different pH, the ratios between the various LEV species in solution change due to the protonation or deprotonation of their functional groups (*i.e.* carboxyl group, amine group). At pH 5.0, most of LEV molecules will be cationic, because under this pH most of the piperazinyl groups and carboxyl groups will be protonated. Goethite surface will also be positively charged at pH 5.0 (PPZC = 9.0–9.3). The electrostatic repulsion between LEV molecules and the surface of goethite is part of the reasons for the small amount of LEV adsorbed to goethite (< 0.025 μmol/m^2).

The adsorption of LEV to goethite increased significantly in the presence of HA added at two concentrations (Fig. 1), which can be attributed to two reasons, (1) synergistic electrostatic effects from PHA adsorbed, and (2) co-adsorption of LEV to mineral-bound HA. NOM is usually negatively charged under common pH in natural systems. With sufficient amount of NOM adsorbed on the mineral surface, the NOM-mineral assemblages may become negatively charged, and even when the mineral surface is intially positively charged, a charge reversal can take place (Weng et al., 2005;

Kumpulainen et al., 2008). The positively charged antibiotics molecules (at pH 5.0 in solutions) can be adsorbed more easily to the negative charged NOM-mineral assemblages due to electrostatic attraction. On the other hand, it has been shown that antibiotics are also strongly adsorbed on NOM (Carmosini and Lee, 2009; Ding et al., 2013). Consequently, when PHA is adsorbed, some LEV will be co-adsorbed with PHA to goethite.

From the fitted adsorption maximum (q_m), we can derive that the qm increased by 0.036 μmol/m^2 at 3.63 mg C/L HA, and by 0.070 μmol/m^2 when PHA was doubled. These results indicate an increase in adsorption capacity propotional to the amount of PHA added, which suggests that the electrostatic intereaction plays probably a smaller role than the co-adsorption mechanism in determining the effect of NOM on LEV adsorption to goethite. However, because the NOM loading at both HA concentrations (0.047 and 0.089 mgC/m^2) is relatively low, the electrostatic effect may still be in the range that is almost propotional to the amount of NOM present. Therefore based on the data we cannot make definitive conclusion regarding the mechanims of NOM effect on LEV adsorption.

If we assume that the additional amount of LEV adsorbed when PHA was added is solely due to LEV adsorption to adsorbed PHA, we can calculate the amount of LEV adsorbed to HA (excluding LEV adsorbed to goethite in the absence of HA). The calculated amounts of LEV adsorbed to HA are in the range of 90–820 mmol/kg C, which are in the same order of magnitude as the reported values for ciprofloxacin adsorption on LHA, PPHA, and PPFA (200–650 mmol/kg C) (Carmosini and Lee, 2009), norfloxacin adsorption on a coal HA (260–380 mmol/kg C) (Zhang et al., 2012), ofloxacin adsorption on a sediment HA (150–720 mmol/kg C) (Pan et al., 2012), and tetracycline adsorption on LHA and AHA (30–45 mmol/kg C) (Ding et al., 2013). In these literatures, the equilibrium dialysis technique, fluorescence quenching method, and Solid Phase Extraction (SPE) method were used to measure antibiotics adsorption at different concentrations (2.5–79 mg C/L) of NOM in the absence of minerals. The advantages of using the mixtures of NOM and minerals in our current work are first of all that the NOM-mineral assemblages are more representative for the adsorbents in the natural systems; and secondly, in the NOM-mineral-antibiotic ternary experiment, the soluble and adsorbed antibiotics can be separated much more easily compared to when only NOM is present.

3.2 *EEM spectra measurements of HA and LEV*

In order to investigate the interactions between NOM and LEV, the EEM spectra contour maps of HA, LEV, and their mixtures were determined (Fig. 2). In all the contour maps, there are two red lines. The left one is the Rayleigh scattering peak ($\lambda_{em} = \lambda_{ex}$), and the right one is the second-order Rayleigh scattering peak ($\lambda_{em} = 2\lambda_{ex}$). There are three peaks in the contour maps of HA and LEV, and four peaks in their mixtures. The positions and fluorescence intensities of the peaks change with the solution pH (Table 1).

As shown in the EEM spectra of HA under different pH (Figure 2a, b, and c), three peaks at around E_x/E_m of 225/335 nm, 325/465 nm, and 285/485 nm are observed (Table 1), which are respectively related to aromatic protein, model HA

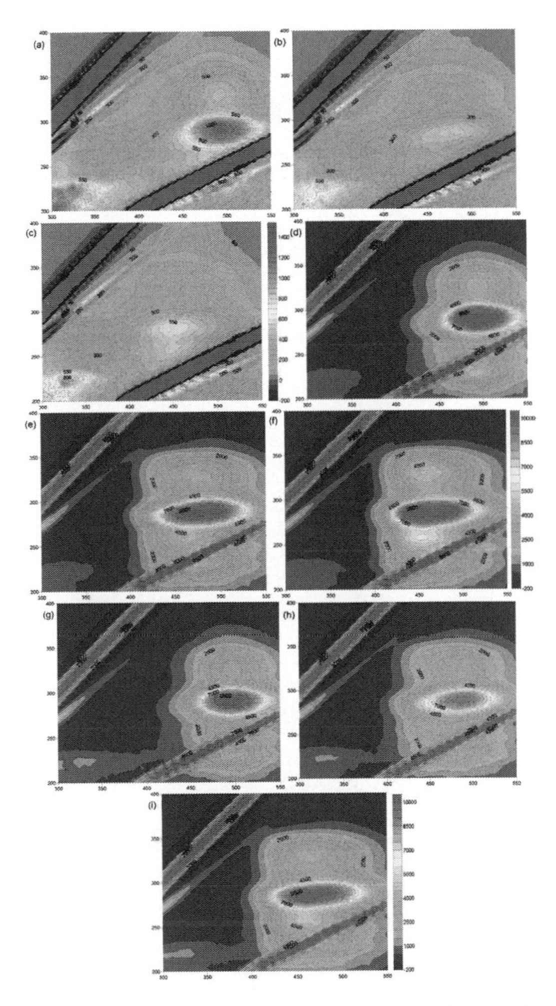

Figure 2. The EEM spectra contour maps of HA (a, b, and c, 0.2 mg C/L), LEV (d, e, and f, 0.4 μM), and the mixtures of HA and LEV (g, h, and i) at pH 3.0, 6.0, and 9.0. The positions and intensities of the peaks are shown in Table 1.

Table 1. Three-dimensional fluorescence characteristics of HA, LEV, and their mixtures.

System	Peak 1 ($\lambda_{em}/\lambda_{ex}$)	Intensity	Peak 2 ($\lambda_{em}/\lambda_{ex}$)	Intensity	Peak 3 ($\lambda_{em}/\lambda_{ex}$)	Intensity	Peak 4 ($\lambda_{em}/\lambda_{ex}$)	Intensity
HA (pH 3.0)	220/330	1016	325/490	469.4	290/495	1272	ND[a]	ND
HA (pH 6.0)	225/335	828.4	325/465	203.4	285/485	533.7	ND	ND
HA (pH 9.0)	220/325	977.5	330/450	272.4	280/445	604.0	ND	ND
LEV (pH 3.0)	ND	ND	325/490	4297	290/495	9999.9[b]	230/500	4980
LEV (pH 6.0)	ND	ND	330/470	4226	290/480	9999.9	230/495	4224
LEV (pH 9.0)	ND	ND	330/455	5065	290/465	9999.9	255/450	6154
LEV-HA (pH 3.0)	225/325	1221	325/495	3805	290/495	9999.9	225/485	4378
LEV-HA (pH 6.0)	225/325	1120	330/480	3372	290/490	9427	225/490	3615
LEV-HA (pH 9.0)	225/330	1124	330/465	4230	285/470	9999.9	225/480	4108

[a]ND means not detected, or no peaks are found.
[b]The value exceeds the detection limit of the fluorescence spectrophotometer.

polymers, and HA-like regions (Chen et al., 2003). The fluorescence intensities of the peaks are the highest at acidic conditions (pH 3.0). The λ_{em} values of peak 2 and 3 decrease with the increasing pH. These results support that pH affects the morphology of NOM.

There are a main peak (290/480 nm) and two nearby small peaks in the EEM spectra of LEV (Figure 2d, e, and f). At high pH, the λ_{em} values of the three peaks also decrease, and the intensity is the highest. This is because under different pH conditions, LEV molecule is mainly cationic, zwitterionic (or neutral), or anionic.

Four peaks are observed in the mixtures, which seems to be the simple sum of the EEM spectra of PHA and LEV (especially for peak 1), and the intensities of the peaks should be larger than that of the single one. However, as shown in Table 1, the intensities of the other three peaks are much smaller than the sum values, even smaller than that of LEV along, especially at pH 6.0. As a result, the interactions (*i.e.* adsorption) between HA and LEV occur, which become strong at neutral conditions, and much weaker at pH 3.0 and 9.0.

4 CONCLUSIONS

Adsorption isotherms at different concentrations of PHA were well fitted to the Langmuir model, and additional amount of LEV adsorbed is proportional to the amount of PHA added, which indicates that the main mechanism of increased LEV adsorption is probably co-adsorption with NOM adsorbed. The Excitation-Emission Matrix (EEM) spectra results indicated the interactions (*i.e.* adsorption) between HA and LEV, which become strong at neutral conditions, and much weaker at other conditions.

REFERENCES

Antelo, J., Avena, M., Fiol, S., López, R. & Arce, F. 2005. Effects of pH and ionic strength on the adsorption of phosphate and arsenate at the goethite–water interface. *J. Colloid Interface Sci.* 285, 476–486.

Carmosini N. & Lee L.S., 2009. Ciprofloxacin sorption by dissolved organic carbon from reference and bio-waste materials. *Chemosphere* 77(6), 813–820.

Chen W., Westerhoff P., Leenheer J.A. & Booksh K., 2003. Fluorescence excitation-emission matrix regional integration to quantify spectra for dissolved organic matter. *Environ. Sci. Technol.* 37(24), 5701–5710.

Ding Y., Teppen B.J., Boyd S.A. & Li H., 2013. Measurement of associations of pharmaceuticals with dissolved humic substances using solid phase extraction. *Chemosphere* 91(3), 314–319.

Figueroa R.A., Leonard A. & MacKay A.A., 2004. Modeling tetracycline antibiotic sorption to clays. *Environ. Sci. Technol.* 38(2), 476–483

Gu C. & Karthikeyan K.G., 2005. Sorption of the antimicrobial ciprofloxacin to aluminum and iron hydrous oxides. *Environ. Sci. Technol.* 39(23), 9166–9173.

Hiemstra, T., De Wit, J.C.M. & Van Riemsdijk, W.H., 1989. Multisite proton adsorption modeling at the solid/solution interface of (hydr)oxides: a new approach: II. Application to various important (hydr) oxides. *J. Colloid Interface Sci.* 133, 105–117.

Hu X., Zhou Q. & Luo Y., 2010. Occurrence and source analysis of typical veterinary antibiotics in manure, soil, vegetables and groundwater from organic vegetable bases, northern China. *Environ. Pollut.* 158(9), 2992–2998.

Kumpulainen S., Von Der Kammer F. & Hofmann T., 2008. Humic acid adsorption and surface charge effects on schwertmannite and goethite in acid sulphate waters. *Water Res.* 42(8–9), 2051–2060.

Pan B., Liu Y., Xiao D., Wu F.C., Wu M., Zhang D. & Xing B., 2012. Quantitative identification of dynamic and static quenching of ofloxacin by dissolved organic matter using temperature-dependent kinetic approach. *Environ. Pollut.* 161, 192–198.

Peng H., Feng S., Zhang X., Li Y. & Zhang X., 2012. Adsorption of norfloxacin onto titanium oxide: effect of drug carrier and dissolved humic acid. *Sci. Total Environ.* 438: 66–71.

Qin X., Liu F. & Wang G., 2012. Fractionation and kinetic processes of humic acid upon adsorption on colloidal hematite in aqueous solution with phosphate. *Chem. Eng. J.* 209, 458–463.

Qin X., Liu F., Wang G., Li L., Wang Y. & Weng L., 2014b. Modeling of levofloxacin adsorption to goethite and the competition with phosphate. *Chemosphere* 111, 283–290.

Qin X., Liu F., Wang G., Weng L. & Li L., 2014a. Adsorption of levofloxacin onto goethite: effects of pH, calcium and phosphate. *Colloids Surf. B.* 116:591–596.

Tolls J., 2001. Sorption of veterinary pharmaceuticals in soils: a review. *Environ. Sci. Technol.* 35(17), 3397–3406

Weng L., Koopal L.K., Hiemstra T., Meeussen J.C.L. & Van Riemsdijk W.H., 2005. Interactions of calcium and fulvic acid at the goethite-water interface. *Geochim. Cosmochim. Acta* 69(2), 325–339.

Yan W., Hu S. & Jing C., 2012. Enrofloxacin sorption on smectite clays: effects of pH, cations, and humic acid. *J. Colloid Interface Sci.* 372(1), 141–147.

Zhang J. & Dong Y., 2008. Effect of low-molecular-weight organic acids on the adsorption of norfloxacin in typical variable charge soils of China. *J. Hazard. Mater.* 151(2–3), 833–839.

Zhang Q., Zhao L., Dong Y. & Huang G., 2012. Sorption of norfloxacin onto humic acid extracted from weathered coal. *J. Environ. Manage.* 102, 165–172.

Zhang Y., Xie J., Liu M., Tian Z., He Z., van Nostrand J.D., Ren L., Zhou J. & Yang M., 2013. Microbial community functional structure in response to antibiotics in pharmaceutical wastewater treatment systems. *Water Res.* 47(16), 6298–6308.

Zhao Y., Geng J., Wang X., Gu X. & Gao S., 2011. Adsorption of tetracycline onto goethite in the presence of metal cations and humic substances. *J. Colloid Interface Sci.* 361(1), 247–251.

Civil, Architecture and Environmental Engineering – Kao & Sung (Eds)
© 2017 Taylor & Francis Group, ISBN 978-1-138-02985-9

Calcium and manganese affect ethanol fermentation by *Pichia stipitis* in cadmium-containing medium by inhibiting cadmium uptake

Min-Tian Gao
Shanghai Key Laboratory of Bio-energy Crops, School of Life Sciences, Shanghai University, Shanghai, China
Energy Research Institute of Shandong Academy of Sciences, Jinan, China

Xingxuan Chen, Wei Zheng, Qingyun Xu & Jiajun Hu
Shanghai Key Laboratory of Bio-energy Crops, School of Life Sciences, Shanghai University, Shanghai, China

ABSTRACT: In this study, the effect of Cadmium (Cd) on ethanol fermentation by *Pichia stipitis* was investigated. *P. stipitis* was sensitive to Cd, which reduced cell growth and diminished ethanol production. The inhibitory effect of Cd on ethanol fermentation decreased upon the addition of Calcium (Ca) and Manganese (Mn); the role of Mn was less significant compared to that of Ca. These results correlated with the inhibitory effect of the metals on the uptake of Cd by *P. stipitis*, with Ca having a greater effect on the uptake of Cd than Mn. However, Mn could also increase the inhibitory effect of Ca on the uptake of Cd. Simultaneous addition of both Ca and Mn lowered cellular Cd level and resulted in a greater degree of fermentation, demonstrating that this treatment can be used to improve ethanol production from Cd-contaminated biomass.

1 INTRODUCTION

Bioethanol is considered an alternative to petroleum. It can be easily produced from agricultural products (Nikolić et al., 2010), such as corn meal and sugarcane. However, bioethanol production from such biomass competes with food suppliers and thereby causes food shortages and price increases. Lignocellulosic materials, such as forest and agricultural residues, are a potential alternative feedstock for the production of bioethanol because their use would not affect the food supply. However, lignocellulosic biomass contains a greater amount of lignin and complex polysaccharides than starch and monosaccharides. These features make lignocellulosic biomass difficult to be hydrolyzed to fermentable sugars that can then be bioconverted into ethanol, resulting in high costs for ethanol production.

To reduce the high cost of ethanol production from lignocellulosic biomass, research has focused on biorefinery improvements such as efficient production of cellulase, improvement of pretreatment, and process development for ethanol production. Although improvements in the efficiency of biorefinery processes could lead to a decrease in the cost of ethanol production, such efforts would be insufficient for substituting ethanol for petroleum since the price of ethanol would still be much higher than petroleum. Therefore, alternate research strategies should be considered.

Environmental pollution by toxic metals has been increasing worldwide because of industrial progress (Xue et al., 2014). Cd is toxic at low levels; it can enter the food chain, accumulate in the body through repeated exposure, and can exert irreversible effects (Abadin et al., 2007). Crops grown on heavy metal-contaminated soils can absorb heavy metals. Cd absorbed by plant roots and transported to ground tissues of plants poses a potential threat to human health through the food chain (Kuboi and Yazaki, 1986). Chronic exposure to Cd causes a variety of health problems, such as Itai-Itai disease. Xu et al. (2013) studied the uptake and distribution of Cd in sweet maize grown on contaminated soils and demonstrated that the Cd concentration in sweet maize decreases from its highest levels in the sheath to its lowest levels in the fruit. Therefore, soil remediation with crops (phytoremediation) could be a promising technology. In the future, the demand for soil remediation would increase markedly, resulting in the generation of a potentially huge market. However, fruits contaminated with Cd cannot be used as food and the contaminated straws of crops cannot be used as fodder. The disposal of metal-containing biomass has been a significant limitation of this technology. However, since metal-contaminated biomass contains starch, cellulose, and hemicelluloses, it could be repurposed as raw materials for biorefinery. The combination of biorefinery and soil remediation can not only ease environmental problems, but can also reduce both costs for biorefinery and soil remediation. Furthermore, problems concerning the disposal of metal-containing biomass could potentially be solved and the released heavy metals could be removed from the food chain.

In the past decade, a few teams have studied the response of *S. cerevisiae* to Cd. Oliveira et al. (2011) found that the use of vinasse minimized the negative effects of Cd on cell mass concentration, cell viability, and budding rate. Gharieb and Gadd (2004) showed that a *S. cerevisiae* strain deficient in Glutathione Synthase Hydroxylase (GSH) displayed a higher sensitivity to Cd than its wild-type strain. Mielniczki-Pereira et al. (2011) reported that the detoxification of Cd is mainly dependent on ion binding with GSH in *S. cerevisiae*. GSH detoxification of Cd also occurs in other organisms (Rehman and Anjum, 2010). Moreover, metal ions can affect uptake of Cd and subsequently reduce its toxicity (Zhao et al., 2004). This detoxification could be linked to competitive transport between metal ions due to the similar radii of Ca and Cd (Long et al., 2012); in support of this, Ca^{2+} -ATPases contribute to Cd tolerance in *S. cerevisiae* (Mielniczki-Pereira et al., 2011).

Despite numerous studies on toxicity of Cd towards *S. cerevisiae*, no study has investigated its effects on ethanol production by the yeast strain *P. stipitis*. In particular, there has been no study on the effect of Cd on the metabolism of xylose for ethanol production. The objective of this work was to study the effects of Cd on the metabolism of glucose and xylose in order to enhance detoxification and increase fermentation efficiency in *P. stipitis*, a highly efficient xylose-fermenting strain. Accordingly, the effects of Ca and Mn on Cd uptake by *P. stipitis* were investigated.

2 MATERIALS AND METHODS

2.1 *Fermentation*

The yeast strain *Pichia stipitis* (CICC 1960) used for ethanol fermentation was obtained from Shanghai Industrial Microbiology Institute Tech. Co., Ltd. Fermentations were carried out in shaking test tubes with 5 mL medium containing 40 g/L glucose, 20 g/L xylose, 8 g/L peptone, 2 g/L yeast extract, 2 g/L $(NH_4)_2SO_4$, 1 g/L $MgSO_4{\cdot}7H_2O$, and 1.5 g/L KH_2PO_4. Fermentations were started at an initial OD_{600} of 1.0 and incubated at 30°C and 200 rpm. For the investigation of the effect of metal ions on ethanol production, the concentration of $CdCl_2$ ranged from 25 to 100 μmol/L. In preliminary experiments, the optimum molar ratios of Ca and Mn to Cd were found to be 100:1 and 1:1, respectively, and were therefore used in this study.

2.2 *Cell dry weight and enzyme activity*

After 24-h fermentation, cells were collected for measurements of cell dry weight and enzyme activity. For cell dry weight measurement, cells were washed with water followed by drying at 105°C overnight. For enzyme activity measurement, cells were ruptured with glass beads (425–600 μm) by vortexing at 30-s intervals and were then kept on ice for 30 s. The resulting suspensions were centrifuged at 12,000 rpm, 4°C for 10 min and the supernatants were used for enzyme activity measurement. Xylitol Dehydrogenase (XDH) activity was measured by monitoring the reduction in $NAD(P)^+$ at 340 nm and 30°C in a reaction mixture containing: 71 mM Tris-HCl buffer (pH 8.2), 0.5 M xylitol, and 1.2 mM NAD+/NADP+. Xylose Reductase (XR) activity was measured by monitoring the oxidation of NADPH at 340 nm and 30°C in a reaction mixture containing: 71 mM Tris-HCl buffer (pH 7.2), 2 M xylose, and 1.2 mM NADPH. One unit of enzyme activity was defined as the amount of enzyme that reduced or oxidized 1 μmol $NAD(P)^+$ or NAD(P)H per min.

2.3 *Measurement*

Samples were taken periodically and centrifuged at 15,000 rpm for 5 min. Glucose, xylose and ethanol concentration in the fermentation broth were analyzed on an Exformma EX1600 HPLC system (Shanghai Wufeng Scientific Instruments Co., Ltd., Shanghai, China) using an Aminex HPX-87H column (300 × 7.8 mm, Bio-Rad, USA). The separation of sugars and ethanol was performed at 0.6 mL/min and at 65°C with 5 mM H_2SO_4 as the mobile phase.

The concentration of $NAD(P)^+$ or NAD(P)H was measured with a spectrophotometer (UV-2102C, Unico Instrument Co., Ltd, Shanghai, China) at 340 nm. The concentration of Cd was measured with an atomic absorption spectrophotometry (AH-670, Shimadzu Co., Ltd, Japan).

3 RESULTS AND DISCUSSION

Since *P. stipitis* can ferment both glucose and xylose, fermentation was carried out in media containing both glucose and xylose, and Cd concentration was varied from 25 to 100 μmol/L; fermentation without Cd was used as a reference. As shown in Fig. 1, at 25 μmol/L Cd, both glucose and xylose were completely consumed within 48 h, resulting in 17.6 g/L ethanol. Compared to fermentation without Cd, fermentation with 25 μmol/L Cd had much lower consumption rates of glucose and xylose, and consequently lower titer of ethanol, indicating the inhibitory effect of Cd on ethanol fermentation by *P. stipitis.* The inhibitory effect became more severe with increasing Cd concentration. At Cd concentrations above 50 μmol/L, xylose consumption at 48 h markedly decreased, resulting in a sharp decrease in ethanol titer (Fig. 1C, D). The lesser consumption of xylose could result from much low consumption rate of glucose (Fig. 1B). It should be noted that OD 600 also decreased proportionally with increasing Cd concentration (Fig. 1 A). However, the decrease in OD 600

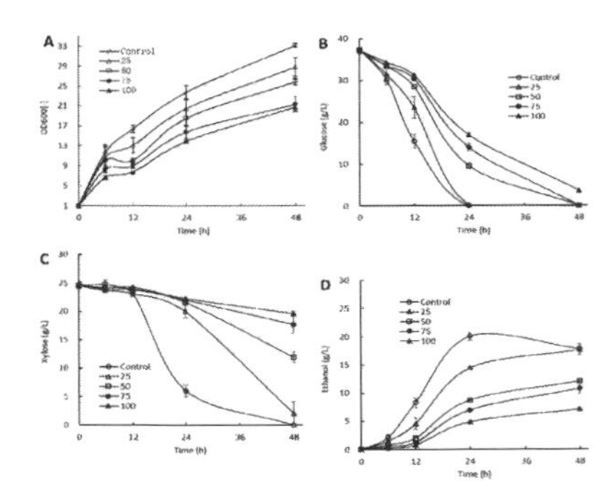

Figure 1. Effect of Cd on ethanol fermentation. The concentrations of cell mass (A), glucose (B), xylose (C) and ethanol (D) change over time during fermentation in the presence of Cd at increasing concentrations (25, 50, 75, 100 μmol/L) compared to that in the absence of Cd. All values are expressed as the mean ± standard deviation (SD) for two experiments (n = 2).

was lower than that in ethanol production; e.g., the maximum OD 600 at 100 μmol/L was 1.6-fold lower than that without Cd, while the maximum concentration of ethanol at 100 μmol/L was 2.9-fold lower than that in the absence of Cd. These results demonstrated that Cd had higher influence on ethanol production than on cell growth.

Cd toxicity towards plants and microorganisms has been widely reported (Perfus-Barbeoch et al., 2002). Its toxicity is mainly related to induction of oxidative stress and interference with intracellular signaling and DNA repair. Other metals (e.g., zinc (Zn), magnesium (Mg), Mn and Ca) can affect plant uptake of Cd and subsequently reduce toxicity of Cd to the plants (Zhao et al., 2004, Long et al., 2012). The effects of Mn, and Ca on ethanol fermentation are shown in Figure 2. In these experiments, Cd concentration was set at 50 μmol/L. Among the metals, Zn and Mg had little effect on the tolerance of *P. stipitis* towards Cd (data not shown). In contrast, Ca and Mn had apparent effects on the tolerance of *P. stipitis* towards Cd (Fig. 2). When Ca or Mn was added, the increase in OD 600 was much higher than that with Cd alone at 12 h (Fig. 2 A). In the earlier period (0–12h) of the fermentations, the effect of Ca on the growth rate was similar to that of Mn, and after 24 h, the effect of Ca was higher than that of Mn (Fig. 2 A). The addition of Ca or Mn also had positive effects on sugar consumption, as both the consumption rates of glucose and xylose increased sharply relative to those where only Cd was present (Fig. 2B, C). However, there were little differences in the effects of Ca and Mn on glucose consumption, while Ca had more significant effect on the consumption of xylose than that of Mn (Fig. 2B). The effects of Ca and Mn on sugar consumption can be explained by the uptake of Cd. As shown in Fig. 3, when Ca or Mn was not present, the Cd concentration in cells increased markedly at 24 h. The maximum Cd concentration was 297.8 μg/g of dried cell. This uptake of Cd was limited by the addition of Ca or Mn. When Ca or Mn was present, the Cd concentration in cells increased very little by 12 h. Afterwards, Cd concentration in cells maintained a low level relative to Cd alone even though there were increases in Cd concentration along with time, indicating the inhibitory effects of Ca and Mn on the uptake of Cd. The lower uptake of Cd would be responsible for the higher rates of sugar consumption. The inhibitory effects of Ca and Mn on the uptake of Cd could be linked to competitive transport between metal ions, owing to their similar atomic radii. Because cellular Cd concentrations were maintained at a relative low level in the presence of Ca or Mn in the earlier periods of fermentation, minimal inhibitory effect of Cd on glucose consumption was observed. In contrast, when xylose began to be consumed, large amounts of Cd accumulated in cells, e.g., at 24 h the cellular Cd concentrations were 157.9 μg/g of dried cell and 237.9 μg/g of dried cell, for Ca and Mn additions, respectively. The accumulation of Cd in cells would

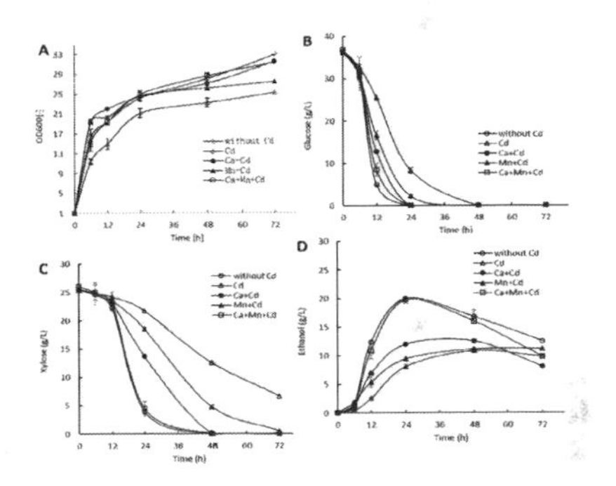

Figure 2. Effects of Ca and Mn on ethanol fermentation with Cd. The concentrations of cell mass (A), glucose (B), xylose (C) and ethanol (D) change over time during fermentation with addition of metals (Ca, Mn, both Ca and Mn) in the presence of Cd compared to that in the absence of Cd. The results represent the mean of two independent experiments.

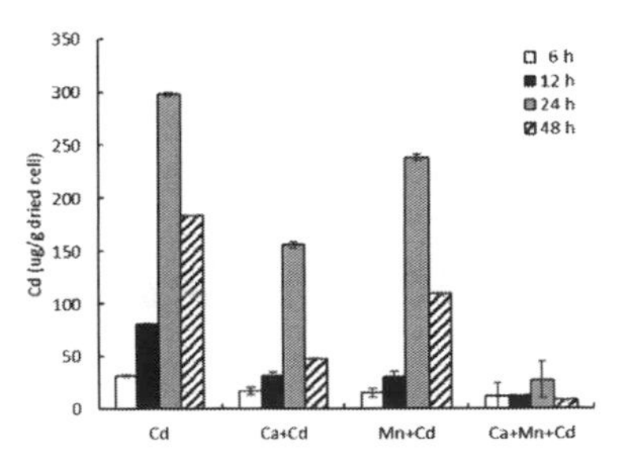

Figure 3. Effects of Ca and Mn on Cd uptake by *P. stipitis*. The intracellular Cd content changes over time during fermentation with addition of metals (Ca, Mn, both Ca and Mn) in the presence of Cd. Bars indicate means (n = 2).

Table 1. Effects of Cd on XR and XDH activities.

	Relative enzyme activities (%)	
	XR	XDH
Without Cd	100	100
With Cd	76.22 ± 3.78	32.81 ± 2.21

result in lower consumption rates of xylose. Enzyme activity assays reveal that Cd significantly inhibited xylose metabolism (Table 1). Inhibited metabolism of xylose could be attributable to the low conversion rate of xylose into ethanol in the presence of Cd. Since Ca had a marked effect on Cd uptake compared to Mn, the consumption rate of xylose during fermentation was higher in the presence of Ca than that in the presence of Mn.

We found that the addition of ions had a positive effect on the production of ethanol at 12 h (Fig. 2D), while ethanol was hardly produced after 24 h even though the consumption of xylose was improved by the addition (Fig. 2C). This result demonstrates that the production of ethanol was linked to the accumulation of Cd in cells, and under Cd-induced stress, the conversion of xylose into ethanol decreased, which could explain why Cd had marked influence on ethanol production compared to its effects on cell growth. After 24 h, Cd release was apparent (Fig. 3), which may be owing to destruction of cell membrane by Cd accumulation.

Taken together, it was clear that the addition of Ca or Mn was not sufficient to cope with Cd toxicity towards *P. stipitis,* even though the addition could limit the uptake of Cd. Surprisingly, the use of both Ca and Mn significantly inhibited the uptake of Cd (Fig. 3). The Cd concentration in cells was maintained at a very low level and did not drastically change during fermentation. The reason for these effects is not clear; however, ethanol fermentation was not inhibited by Cd in the presence of both Ca and Mn (Fig. 2D). Under these conditions, the production rate of ethanol, the consumption rates of sugars were no different from those in media with only Cd (Fig. 2B,C), while the maximum titer of ethanol was 4-fold higher (Fig. 2D).

4 CONCLUSIONS

In this study, we found inhibitory effects by Ca and Mn on the uptake of Cd by *P. stipitis.* The tolerance of *P. stipitis* towards Cd increases upon joint application of Ca and Mn, an effect that would be beneficial to the development of *P. stipitis* as a key player in cellulosic ethanol production. However, the exact mechanism through which Ca and Mn together limit the uptake of Cd by *P. stipitis* remains unknown; further, how *P. stipitis* utilizes xylose for cell growth under Cd stress is unclear. A more detailed study is therefore required to investigate the mechanism of the effect of the joint application of Ca and Mn on the uptake of Cd by *P. stipitis.*

ACKNOWLEDGMENTS

This research was supported by Special Fund Agroscientific Research in The Public Interest (No. 201503135–14), by the Scientific Research Projects of Shanghai Science and Technology Committee (No. 14540500600 and NO.16391902000), by Shanghai Municipal Education Commission (No. 14ZZ091), and by National Natural Science Foundation of China (No. 21307093).

REFERENCES

Abadin, H., et al. 2007. *Toxicological Profile for Lead,* Agency for Toxic Substances and Disease Registry (US).

Gharieb, M. M. & Gadd, G. M. 2004. Role of glutathione in detoxification of metal(loid)s by Saccharomyces cerevisiae. *Biometals An International Journal on the Role of Metal Ions in Biology Biochemistry & Medicine,* 17, 183–8.

Kuboi, T. & Yazaki, J. 1986. Family-dependent cadmium accumulation characteristics in higher plants. *Plant & Soil,* 92, 405–415.

Long, Z., et al. 2012. Endogenous nitric oxide mediates alleviation of cadmium toxicity induced by calcium in rice seedlings. *J Environ Sci,* 24, 940–8.

Mielniczki-Pereira, A. A., et al. 2011. New insights into the Ca 2+ -ATPases that contribute to cadmium tolerance in yeast. *Toxicology Letters,* 207, 104–111.

Nikolić, S., et al. 2010. Production of bioethanol from corn meal hydrolyzates by free and immobilized cells of Saccharomyces cerevisiae var. ellipsoideus. *Biomass & Bioenergy,* 34, 1499–1456.

Oliveira, R. P. D. S., et al. 2011. Response of Saccharomyces cerevisiae to cadmium and nickel stress: the use of the sugar cane vinasse as a potential mitigator. *Biological Trace Element Research,* 145, 71–80.

Perfus-Barbeoch, L., et al. 2002. Heavy metal toxicity: cadmium permeates through calcium channels and disturbs the plant water status. *Plant Journal for Cell & Molecular Biology,* 32, 539–48.

Rehman, A. & Anjum, M. S. 2010. Multiple metal tolerance and biosorption of cadmium by Candida tropicalis isolated from industrial effluents: glutathione as detoxifying agent. *Environmental Monitoring & Assessment,* 174, 585–95.

Xu, W., et al. 2013. Uptake and distribution of cd in sweet maize grown on contaminated soils: a field-scale study. *Bioinorganic Chemistry & Applications,* 2013, 959764–959764.

Xue, J. L., et al. 2014. Positive matrix factorization as source apportionment of soil lead and cadmium around a battery plant (Changxing County, China). *Environmental Science & Pollution Research International,* 21, 7698–707.

Zhao, Z. Q., et al. 2004. Effects of forms and rates of potassium fertilizers on cadmium uptake by two cultivars of spring wheat (Triticum aestivum, L.). *Environment International,* 29, 973–978.

Civil, Architecture and Environmental Engineering – Kao & Sung (Eds)
© 2017 Taylor & Francis Group, ISBN 978-1-138-02985-9

Performance comparison of Biological Aerated Filters packed with plolyurethane sponge and ceramic particles

G.L. Yu, X.J. Yan, H. Chen, C.Y. Du & Y.J. Fu
School of Hydraulic Engineering, Changsha University of Science and Technology, Changsha, Hunan, China
Key Laboratory of Water-Sediment Sciences and Water Disaster prevention of Hunan Province, Changsha, China

J.X. Zhang
Changjun High School, Changsha, Hunan, China

ABSTRACT: Biological Aerated Filters (BAFs) are a promising biological oxidation technology for wastewater treatment and reuse. The characteristics of packing media in BAFs considerably affect the removal of pollutants. This study aims to compare and investigate the pollutant removal performance of two identical bench-scale BAFs, BTF 1 and BTF 2, to evaluate the feasibility of employing polyurethane sponge as BAF media. BTF 1 and BTF 2 were packed with polyurethane sponges and ceramic particles, respectively, and evaluated for COD and NH_4^+-N removal at various hydraulic loadings. When the hydraulic loading was varied from 5 L/h to 20 L/h, the overall COD removal efficiencies of BAF 1 and BAF 2 reached 82.8% and 78.5%, respectively. BAF 1 performed better in COD removal than did BAF 2 under low hydraulic loadings of 5 and 10 L/h. The removal efficiency for COD and NH_4^+-N rose with increasing media height in both BAF systems. The majority of COD and NH_4^+-N removal occurred within a media height of 0.4 m. The BAF 1 system exhibited a higher and steadier treatment performance for COD and NH_4^+-N removal because of the large specific surface area and high porosity rate of the polyurethane sponges. These results indicate that polyurethane sponges have potential as media in BAFs.

1 INTRODUCTION

Biological Aerated Filters (BAFs) are a promising biological oxidation technology for wastewater treatment and reuse (Bao et al., 2011; Qiu et al., 2010). A BAF is a fixed film biological process used for the simultaneous removal of carbonaceous matter, ammonia, and suspended solids in a single unit primarily by interception, biological metabolism, and adsorption (Ryu et al., 2008). The advantages of this system include high efficiency, small footprints, ease of operation and management, and modular construction; the system also does not require a secondary clarifier (Osorio and Hontoria, 2001). Given these merits, BAFs serve as an effective alternative to the traditional activated sludge process commonly used in wastewater treatment. BAF systems are currently widely used for secondary and tertiary treatment of wastewater or reclaimed water reuse. Improving the treatment efficiency and operation stability of BAFs necessitates further study of impact factors such as packing media and operation optimization (Shen et al., 2009; Qiu et al., 2010).

The packing materials in BAFs act as solid interceptors, solid/liquid separators, and biofilm carriers. The characteristics of the packing media used in BAFs considerably affect wastewater treatment performance. Moreover, packing media determine capital construction investment and operating cost (Han et al., 2009; Qiu et al., 2010). The media used in BAFs fall under two categories. The first comprises conventional packing media, such as ceramic particles, granular zeolites, sands, shales, and carbonate media (Qiu et al., 2010). Extensive research has been done on this category. Although these mineral materials show excellent biological performance, they are prone to attrition and easily form clogging in media by biomass or attached high-density solids, thereby leading to frequent backwashing (Kent et al., 1996). The other category is synthesis media, which include plastic materials and carbon foams (Bao et al., 2011). They have very low attrition levels but exhibit inferior performance (Mendosa-Espinosa and Stephenson, 1998). Therefore, the development of novel and effective packing materials for BAF is a crucial requirement.

As a synthesis media, polyurethane sponge has a high porosity and external surface area, rough surface, and uniform open-pore reticulated construction. However, no report has been published on the performance of BAF systems that use polyurethane sponge as packing media. The present study aims to compare the performance of two BAFs packed with the polyurethane sponge and ceramic particles under the same operational conditions to evaluate the suitability of polyurethane sponge as BAF media.

2 MATERIALS AND METHODS

2.1 *Experimental set-up*

Two identical bench-scale biofilters, designated as BAF 1 and BAF 2, were packed with polyurethane sponge and ceramic particles, respectively. Fig. 1 shows a schematic of the BAFs. The reactors are made of transparent Plexiglas. Each reactor is cylindrical with an overall height of 1.2 m and a diameter of 0.1 m. The media within a reactor measures 0.8 m with a total effective volume of 6.3 L. Aside from a sample port placed at the base of each reactor, four liquid sampling ports are located at 0.2 m intervals along the height of each reactor. A sieve plate divides each of the reactors into two sections. The bottom section (height, 0.15 m) of the column serves as a reservoir for influent water and provides uniform water distribution. The upper section (height, 0.1 m) of the column houses a gravel support layer, which is located beneath a media layer. The support gravel protects the water nozzle from clogging by smaller media and improves the distribution of gas bubbles over the cross-sectional area of the media. To provide backwash water, a storage tank collects effluent. A water pump injects the liquid influent at the base of each column and oxygen is supplied by an air blower that injects air through porous stones located at the bottom of each reactor. An air flow meter controls the airflow according to operational requirements.

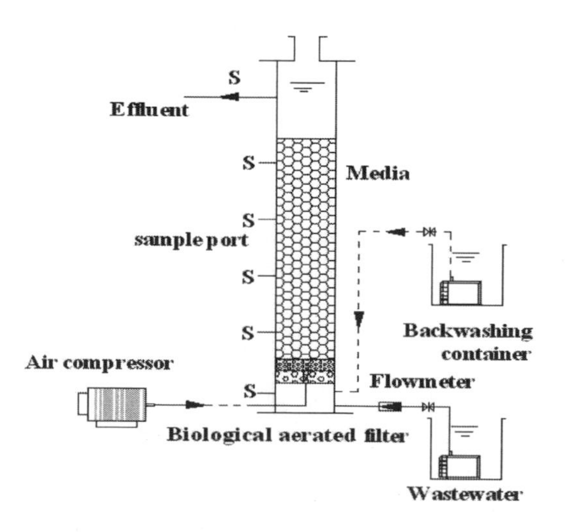

Figure 1. Schematic diagram of the BAFs.

Table 1. Characteristics of polyurethane sponge and ceramic particles.

Media	Shape	Size (mm)	Surface area (m^2/m^3)	Porosity rate
Polyurethane	Cylindrical	$\varphi 100 \times 100$	≥ 950	95%
Ceramic particles	Granular	4–6	≤ 400	30%

Table 2. Main ingredients of the simulated domestic wastewater.

$C_6O_6H_{12}$ (mg/L)	NH_4Cl (mg/L)	KH_2PO_4 (mg/L)	$CaCl_2$ (mg/L)	$FeCl_3$ (mg/L)	$MgSO_4$ (mg/L)
300 ± 1.0	80 ± 0.5	20 ± 0.5	25 ± 0.5	3.7 ± 0.05	2.5 ± 0.05

For the experiment, both BAFs were operated in co-current modes with gas and wastewater flowing upward. The BAFs were backwashed every 48 h. Combined air–water backwashing was employed, with the procedures implemented in sequence as follows: air scouring for 4 min, air and water washing for 6 min, and water rinsing for 10 min. The water and air flushing strength were set at 8 and 10 (L/m^2·s).

Two BAF reactors were packed with polyurethane sponge and ceramic particles. Open-pore reticulated polyurethane sponges with a pore size of 10 pores per cm (Shenzhen Jiechun Filter Material Co., Ltd., Guangdong, China) were used as the filter bed media. Table 1 shows the characteristics of the two media.

2.2 *Synthetic wastewater*

The two BAFs were fed with synthetic wastewater throughout the experimental period. Table 2 shows the composition and concentration of synthetic domestic wastewater.

2.3 *Analytical methods*

Samples from the influent, effluent, and sampling ports were obtained regularly at different filter heights. The concentrations of Chemical Oxygen Demand (COD), ammonia-N (NH4+-N), and Suspended Solids (SS) were analyzed according to the Standard Methods for the Examination of Water and Wastewater (APHA, AWWA, WEF, 2005). Air and water temperatures, pH, and Dissolved Oxygen (DO) were routinely monitored daily during the evaluation period. Flow rates to the BAFs were measured with a glass rotameter. Head loss was measured using a piezometric tube. All the samples were measured in triplicate.

3 RESULTS AND DISCUSSION

3.1 *Startup stage*

To accelerate biomass growth, the polyurethane sponge was marinated and inoculated with activated sludge from a secondary sedimentation tank at Changsha Guozhen Wastewater Treatment Co., Ltd (Hunan, China). The concentration of mixed liquor suspended solids in the original activated sludge was 3500 mg/L. The two BAFs were inoculated simultaneously.

The startup of the BAFs was carried out with the bioreactor operated in batch mode and seeded with sludge and synthetic wastewater for three days. After this, the operation mode was shifted to continuous flow mode. The synthetic wastewater flow rate adopted for 4 days was 3 L/h. During this startup period, the operational parameters were kept at a hydraulic retention time of 2.5 h and a Dissolved Oxygen (DO) concentration higher than 2 mg/L at room temperature. The accumulation process lasted for one week, after which a yellowish-brown biomass was observed to have accumulated on the packing materials. Figs. 2 and 3 present the removal performance of polyurethane sponge and ceramic particles during the start-up. As shown in the figures, COD and NH_4^+-N removal during the start-up did not differ significantly between the two BAFs. Over the first 7 days of the start-up, the volumes of effluent COD and NH4+-N in the two BAFs declined with running time. On day 8, the first backwashing was carried out. COD and NH_4^+-N removal was relatively steady as indicated by the effluent quality. This result shows that a successful startup was conducted for the two BAFs. The BAFs were then switched to normal operations to evaluate treatment performance.

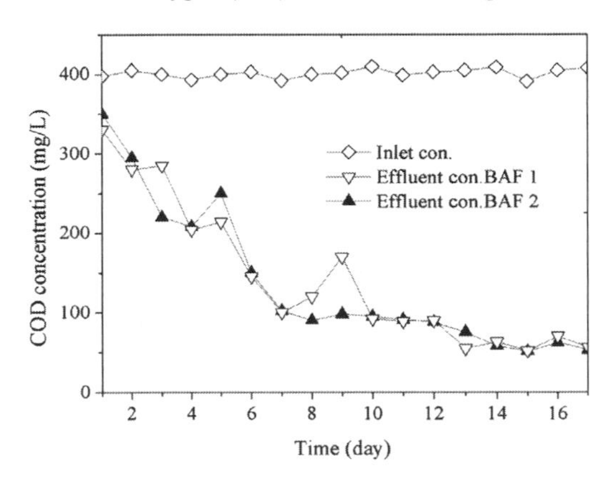

Figure 2. COD removal performance in BAFs with polyurethane sponge and ceramic particle during start-up.

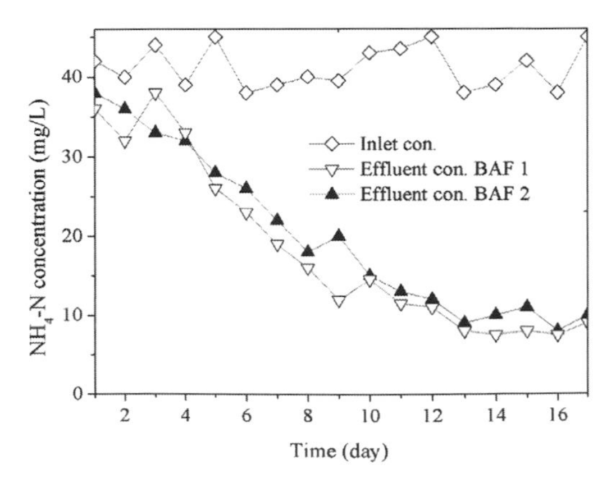

Figure 3. NH_4^+-N removal performance in BAFs with polyurethane sponge and ceramic particle during start-up during start-up.

3.2 *COD removal performance of the BAFs under various media heights*

Fig. 4 shows the COD removal performance of the two reactors under different hydraulic loadings with media height. The results illustrate promising COD treatment performance for the two materials. At hydraulic loadings of 5, 10, 15, and 20 L/h, the COD concentration and removal efficiency of both reactors showed a similar tendency. The total average removal efficiencies of BAF 1 and BAF 2 were 82.8% and 78.5%, respectively. Even under a hydraulic loading of 20 L/h, the COD removal efficiencies of the two BAFs were 78.0% and 80.6%, respectively. The corresponding residual COD concentrations were 56.9 and 50.2 mg/L for BAF 1 and BAF 2, respectively [Fig. 4(d)]. Therefore, the effluent quality satisfies the requirement for reuse.

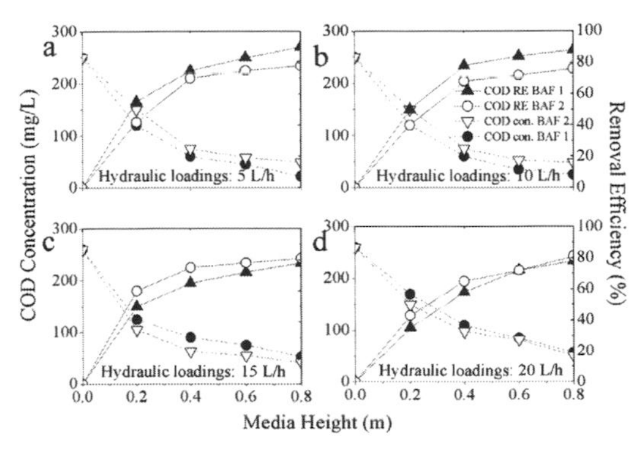

Figure 4. Profile of COD removal in BAFs with different media under hydraulic loadings of 5, 10, 15 and 20 L/h (●) COD concentration in polyurethane sponge BAF; (▽) COD concentration in biological ceramic BAF; (▲) COD removal efficiency in polyurethane sponge BAF; (○) COD removal efficiency in biological ceramic BAF.

Moreover, Fig. 4 shows that the majority of the COD content in the BAFs can be removed within a 0.4 m media in the upflow direction under various hydraulic loadings. At a media height of 0.4 m, the COD removal efficiency of the BAFs reached more than 63% and 56%. The removal efficiency depended primarily on the accumulation of activated biological film in the filter media and mass transfer efficiency. Organic matter degradation occurred primarily in the inlet area because of the competitive advantage of heterotrophic bacteria. The biofilm attached from the lowermost to the uppermost section media contributed to the degradation of organic matter, whereas only the biofilm in some sections close to the inlet played an important role in the aforementioned process. The lowermost surface of the sponge media contributed significantly to COD removal for the upflow BAFs because of the accumulation of a more active biomass in this region. The concentrations of organic compounds and DO were sufficient at the bottom of the filters, which favored the growth of heterotrophic bacteria.

BAF 1 performed better for COD removal than did BAF 2 under low hydraulic loadings (5 and 10 L/h; Fig. 4). The difference in removal efficiency between the two BAFs is more than 10%. By contrast, BAF 2 exhibited higher removal efficiency under high hydraulic loadings, but its performance did not differ sharply from that of BAF 1. The differences in the surface characteristics, specific surface areas, and morphological characteristics of the media were the principal factors that influenced COD removal. The polyurethane sponges used in BAF 1 presented a considerably more regular shape, which allows for even biofilm distribution on the BAF. This uniform distribution resulted in a high mass transfer coefficient under low shear by water. However, an increase in hydraulic loadings resulted in the low removal efficiency for biofilm runoff from the media.

3.3 *NH_4^+-N removal performance of the BAFs under various media heights*

Fig. 5 shows the average removal efficiency for and variations in concentration of NH_4^+-N in the BAFs under hydraulic loadings of 5, 10, 15, 20 L/h. Overall, BAF 1 performed slightly better than did BAF 2 for ammonia removal under the same hydraulic loadings. The BAF functioned as a push-flow reactor, in which competition between carbonation heterotrophic and autotrophic nitrifying bacteria exists. In addition, organic matter inhibited the growth of autotrophic nitrifying bacteria. Therefore, organic matter should generally be degradable first along the flow direction for NH_4^+-N removal to gradually increase. In this study, however, the tendency of concentration and removal efficiency for NH_4^+-N with media height in the two BAFs was the same as that for COD. The removal efficiency for NH_4^+-N increased with increasing media height. Most of the NH_4^+-N reductions also occurred at a media height of 0.4 m (Figs. 4 and 5).

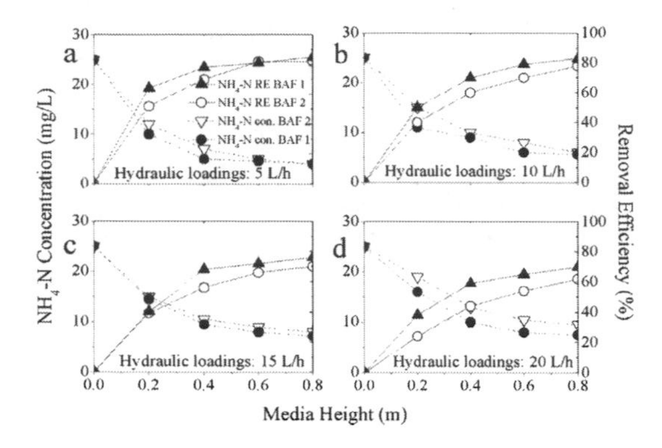

Figure 5. Profile of NH_4^+-N removal in BAFs with different media under hydraulic loadings of 5, 10, 15 and 20 L/h. (●) NH_4^+-N concentration in polyurethane sponge BAF; (▽) NH_4^+-N concentration in biological ceramic BAF; (▲) NH_4^+-N removal efficiency in polyurethane sponge BAF; (○) NH_4^+-N removal efficiency in biological ceramic BAF.

Three possible factors explain the phenomenon of most of the NH_4^+-N reductions within 0.4 m media. The first is influent wastewater quality. Glucose and ammonium chloride, which are easily decomposed and absorbed by microbes, were used as carbon and nitrogen source, respectively. There was adequate DO within a media height of 0.4 m in the two BAFs, thereby promoting the possibility of coexistence between carbonation heterotrophic and nitrifying autotrophic bacteria. In addition, as carbonation heterotrophic bacteria degrade organic matter, they consume some of the nutritional substances including NH_4^+-N.

The second factor is the manner by which the media was configured. The microbes in the BAF can be divided into the biofilm attached on the media and the activated sludge suspended in the media gaps. The morphological characteristics of the ceramic particles thickened and stabilized the biofilm in the inner surface and concavity of the ceramic particles. The biofilm was also minimally affected by hydraulic conditions. Therefore, the ceramic particles are suitable for the growth of bacteria with long generation times. The convexity or angularity of ceramic particles, which were subjected to collision and attrition, as well as the shear stress on water flow and air bubbles, which had a high concentration of organic matter and DO, promoted the growth and rapid reproduction of heterotrophic bacteria. Therefore, the configuration of the ceramic particles induced carbonation, which caused heterotrophic and nitrifying autotrophic bacteria to occupy different locations, and reduced the competition for space between them. Polyurethane sponge has a large specific surface area and high porosity rate, which allowed for sufficient space for the carbonated heterotrophic and nitrifying autotrophic bacteria.

The third factor is once the autotrophic nitrifying bacteria that were cultivated in the biofilm matured, they were able to grow steadily (Ohashi et al., 1995). During the stable operation of the BAFs, the maturation of the nitrifying bacteria in the biofilm weakened and eventually eliminated the competitive inhibition between carbonated heterotrophic and nitrifying bacteria.

Fig. 5 shows that the concentration of effluent NH_4^+-N increased when the hydraulic loading was increased from 5 L/h to 20 L/h. The hydraulic loading imposed a more obvious effect on NH_4^+-N than on COD removal. In particular, the effect on NH_4^+-N removal efficiency was sharper under high hydraulic loadings (10 and 20 L/h). These results differ from those of a previous study conducted with a single-stage Biofor system: nitrification efficiency was suggested to improve with increased hydraulic loading (Pujol et al., 1998). The effect of hydraulic loading on BAF performance is dual: high hydraulic loading improved the substrate, DO, and mass transfer rates between the environment and biofilm, but it also meant a short retention time for the substrate. The other factor that may have led to this dual effect is the competition between the heterotrophic and nitrifying bacteria for oxygen and space.

4 CONCLUSION

BAFs packed with polyurethane sponge or ceramic particles were employed for the treatment of synthetic wastewater under the same conditions. The overall COD removal efficiencies of the BAFs were an average of 82.8% and 78.5%. BAF 1, packed with polyurethane sponge, exhibited better performance for COD removal than did BAF 2, packed with the ceramic particles, under low hydraulic loadings of 5 and 10 L/h. The removal efficiency for COD and NH_4^+-N increased with increasing media height in both BAFs, while most of the reductions in COD and NH_4^+-N occurred within a media height of 0.4 m. The polyurethane sponges with large specific surface areas and high porosity rates provided sufficient space for the carbonated heterotrophic and nitrifying autotrophic bacteria, which resulted in the higher removal efficiency for NH_4^+-N.

Compared with the BAF 2 system, the BAF 1 system exhibited a higher and steadier treatment performance for COD and NH_4^+-N removal. Therefore, a BAF system for wastewater treatment can extensively use polyurethane sponge as its packing media.

ACKNOWLEDGEMENTS

Financial support from the National Natural Science Foundation of China (Grant No.: 51308069), the Scientific Research Fund of Hunan Provincial Education Department (Project Contract No.: 14C0034), the Open Research Fund Program of Key Laboratory of Water-Sediment Sciences and Water Disaster prevention of Hunan Province (Project Contract No.: 2012SS04), and the Water Conservancy Science and Technology Project of Hunan Province (Project Contract No.: [2015]186–13) is highly appreciated.

REFERENCES

APHA, AWWA, & WEF (2005). *Standard Methods for examination of water and wastewater*, 21st ed. Washington DC.

Bao, Y., L. Zhan, C.X. Wang, Y.L. Wang, W.M. Qiao, & L.C. Ling (2011). Carbon foams used as packing media in a biological aerated filter system. *Mater. Lett.* 65 3154–3156.

Han, S., Q. Yue, & M. Yue (2009). Effect of sludge-fly ash ceramic particles (SFCP) on synthetic wastewater treatment in an A/O combined biological aerated filter. Bioresource Technol. 100, 1149–1155.

Kent, T., C. Fizpatrick, & S. Williams, (1996). Testing of biological aerated filter (BAF) media, Water Sci. Technol. 34, 363–370.

Mendosa-Espinosa, L., & T. Stephenson (1998). A process model to evaluate the performance of a biological aerated filter, *Biotechnology Terbniqzrar* 12, 373–375.

Ohashi, A., D.G. Virai de Silva, B. Mobarry, J.A. Manem, D.A. Stahl, & B.E. Rittmann (1995). Influence of substrate C/N ratio on the structure of multi-species biofilms consisting of nitrifiers and heterotrophs, *Water Sci. Technol.* 32, 75–84.

Osorio, F., & E. Hontoria (2001). Optimization of bed material height in a submerged biological aerated filter, *J. Environ. Eng.* 127, 974–978.

Pujol, R., H. Leemel, & M. Gousailles (1998). A keypoint of nitrification in an upflow biofiltration reactor, *Water Sci. Technol.* 38, 43–49.

Qiu, L.P., S.B. Zhang, G.W. Wang, & M.A. Du (2010). Performances and nitrification properties of biological aerated filters with zeolite, ceramic particle and carbonate media. *Bioresource Technol.* 101 7245–7251.

Ryu, H.D., D. Kim, H.E. Lim, & S.I. Lee (2008). Nitrogen removal from low carbon-to nitrogen wastewater in four-stage biological aerated filter system, *Process Biochem.* 43, 729–735.

Shen, J.Y., R. He, & H.X. Yu (2009). Biodegradation of 2, 4, 6,-trinitrophenol (picric acid) in a biological aerated filter (BAF), *Bioresource Technol.* 100, 1922–1930.

Civil, Architecture and Environmental Engineering – Kao & Sung (Eds)
© 2017 Taylor & Francis Group, ISBN 978-1-138-02985-9

Preparation and evaluation of the Ni-Al catalyst derived from LDHs synthesized on γ-Al_2O_3 support

Xiaokang Zhao, Yuqing Li, Yan Xu & Xihua Du
School of Chemistry and Chemical Engineering, Xuzhou Institute of Technology, Xuzhou, China

Jingjing Xu
Jiangsu Zhongneng Polysilicon Technology Development Co. Ltd., Xuzhou, China

ABSTRACT: In this paper, Ni-Al catalyst derived from LDHs was prepared. CH_4-CO_2 reforming was chosen as the probe reaction to evaluate the catalytic activity. The traditional catalyst with alike composition (Ni-Al-IMP) was also prepared as the reference. At the temperature range of 650–850, the Ni-Al-LDH catalyst exhibited a higher activity, and the side reaction of reverse water–gas reaction was inhibited. The related samples, including catalyst support, catalyst precursor, and catalyst, were characterized by XRD, SEM, and FT-IR. The results showed that the Ni-Al-LDH with CO_3^{2-} serving as the interlayer anion was successfully synthesized on the γ-Al_2O_3 support and presented in the layer structure. Ni existed as the divalent cation in this LDH and was present in the atomic-level dispersion, resulting in a smaller crystal size, better dispersion, and more active sites. It can be concluded that the highly dispersed Ni crystal played a key role in the enhanced catalytic activity of Ni-Al-LDH catalyst.

1 INTRODUCTION

With the rapid development of society and economy, fossil energy has been widely used. And the resulting experimental issues have aroused widespread attention in recent years (Feely et al. 2011, Wang et al. 2015, Saidi et al. 2015). Moreover, the finite-energy resource is difficult to meet the increasing energy demands of the society in the long run. Thus, the sustainable global development of energy and environment is a critical and strategic challenge.

Although CO_2 has been considered to be the main human-related greenhouse gas (Liu et al. 2016), it is also a kind of carbon resource. By means of chemistry and chemical engineering, converting CO_2 and CH_4 to syngas (a mixture of H_2 and CO), which can be converted to liquid fuel, is an effective way to realize the carbon cycling, relieving the experimental and energy crisis.

Ni-based catalyst has been recognized as the strongest candidate for the industrial application of CH_4–CO_2 reforming because of its lower price and higher catalytic activity. As CH_4–CO_2 reforming ($CH_4 + CO_2 = 2H_2 + 2CO$) is strongly endothermic, a higher temperature is required to achieve better conversion. However, the Ni catalyst can be easily deactivated by the sintering and agglomeration of Ni atoms. Furthermore, the carbon content of this system is relatively high (compared with the commercial CH_4–H_2O reforming system), and the deposited carbon generated by side reactions may cover and wrap the Ni atoms, thereby deactivating it. It is reported that the size of Ni crystal has a close relationship with the generation of deposited carbon, more Ni atoms are needed to promote the side reaction ($CH_4 = C + 2H_2$) at high temperature. Thus, a catalyst with better Ni dispersion, smaller Ni crystal size, and stronger interaction between Ni and support can be produced by CH_4-CO_2 reforming (Han et al. 2014).

Layered Double Hydroxides (LDHs) are lamellar mixed hydroxides containing positively charged main layers and undergoing anion exchange chemistry. Researches revealed that LDHs have high specific surface area, inherent metal atom dispersion, and alkalinity, which make them possible to become a good candidate catalyst material (Romero et al. 2014, Rezende et al. 2015, Zhang et al. 2014, Montañez et al. 2014). Ni-Mg-Al catalyst derived from LDHs has been prepared and used in the autothermal reforming of CH_4 by Katsuomi Takehira (2004), and a higher catalytic activity was obtained by comparing with the conventional catalyst prepared by the impregnation method. Alak Bhattacharyya et al. (1998) synthesized a hydrotalcite clay catalyst via CH_4–CO_2 reforming and gained prospective excellent performances. In our previous work, we prepared Ni-Co-Al catalyst with LDH precursor for CH_4–CO_2

reforming and found that the catalyst with Ni/Co ratio of 8/2 achieved the highest conversions and best stability in 100 h testing (Long et al. 2013).

Several studies focused on catalysts derived from pure LDH materials. However, the mechanical strength of this type of catalyst is poor, which makes it not well suited for the CH_4–CO_2 reforming. In this paper, LDHs were in-site synthesized on a porous γ-Al_2O_3 support. And its catalytic activity was evaluated and compared with the reference catalyst prepared by impregnation method. XRD, SEM, and FT-IR were used to analyze its microstructure.

2 EXPERIMENTAL

2.1 *Preparing Ni-Al catalyst derived from LDHs*

Details of the preparation process of Ni-Al catalyst derived from LDHs were described as follows: first, 1.783 g of $Ni(NO_3)_2 \cdot 6H_2O$ and 0.737 g of CH_2N_4O (the molar ratio of CH_2N_4O/NO_3^- was 1) were mixed and dissolved in 5 ml of deionized water to make a solution. Second, 2.64 g of γ-Al_2O_3 particles (the diameter was about 0.38–0.83 mm) serving as support and Al^{3+} source were added to the prepared solution and impregnated for 2 h. Then, the excess liquid was removed, and the obtained particles were aged for 6 h at 130°C to form the LDH structure followed by filtering, washing to neutral pH, and drying for 12 h at 70°C in sequence. The obtained product was the precursor of the Ni-Al catalyst, that is, Ni-Al-LDHs/γ-Al_2O_3. The precursor was then calcined (air atmosphere, 550°C, 5 h) and reduced (H_2 atmosphere, 700°C, 2 h) to obtain the target catalyst, Ni-Al-LDH.

The reference catalyst was prepared by the common impregnation method described as follows: first, 1.783 g of $Ni(NO_3)_2 \cdot 6H_2O$ was dissolved in 5 ml of deionized water. Then, 2.64 g of γ-Al_2O_3 particles was put into the solution and impregnated for 4 h followed by drying, calcination, and reduction under the same conditions mentioned above to obtain the reference catalyst noted as Ni-Al-IMP.

2.2 *Evaluation of catalyst*

The catalytic activity was evaluated in a fixed-bed continuous-flow reactor. The reactor, made of quartz with 6 mm internal diameter, was mounted horizontally inside a tubular furnace. The catalyst (200 mg) was charged in the quartz tube reactor. The reaction temperature varied within the temperature range of 650–850°C, and the reactant feed was $CH_4/CO_2 = 2/3$ with the GHSV of 30 L/(g·h). The effluent was passed through a trap to condensate residual water and then analyzed by an in-line gas chromatography (GC-102) equipped with a TCD detector. The conversions of CH_4 and CO_2 and the selectivities of H_2 and CO were calculated.

2.3 *Characterization of catalyst*

X-Ray Diffraction (XRD) analysis used to speculate the average Ni crystal size and Ni dispersion of catalyst was operated on the DX-1000 equipment with Cu Kα radiation source at 40 kV and 25 mA in the diffraction range of $2\theta = 5$–$80°$. The sweep rate was 0.08°/min. The microstructure of the sample was observed using JSM-6340F Scanning Electron Microscopy (SEM) developed by JEOL company at 0.5–30 keV with an amplification of 50–200000 X. Fourier transform IR (FT-IR) spectra of samples were obtained by Thermo Scientific™ Nicolet™ iS™50 FT-IR spectrometer to confirm the interlayer exchangeable anions. The scan region was 400–4000 cm^{-1}.

3 RESULTS AND DISCUSSION

3.1 *Evaluation results*

The catalytic performance of Ni-Al-LDH and Ni-Al-IMP catalyst is shown in Figure 1. It can be seen clearly that the Ni-Al-LDH catalyst has obviously better catalytic performance. It is worth noting that the Ni-Al-IMP showed higher H_2 selectivity but lower CO selectivity, which may be caused by the side reaction of reverse water-gas-shift reaction ($CO + H_2O = H_2 + CO_2$). The better performance of Ni-Al-LDH catalyst should be ascribed to the more active sites produced by the atom level dispersion of Ni atoms in the LDH precursor.

3.2 *Characterization of catalyst*

Figure 2 shows the XRD patterns of different samples, where there are visible peaks at $2\theta = 10.4°$, 20.63°, 33.77°, and 61.44° on the Ni-Al-LDH precursor, which are the typical characters corresponding to the diffraction of LDHs at crystal faces of 003, 006, 009, and 110, respectively (Manukyan et al. 2015, Khan et al. 2016, Mikulová et al. 2007). It can be deduced that the LDHs have been successfully synthesized on the γ-Al_2O_3 support.

These peaks disappeared once the precursor was calcined and reduced. Comparing the XRD pattern of Ni-Al-LDH to that of Ni-Al-IMP, it can be discovered that there is a relatively sharp Ni crystal peak at $2\theta = 76.1°$ on the pattern of Ni-Al-IMP catalyst, indicating that the Ni crystal agglomerated to some extent. There is no obvious peak on the pattern of Ni-Al-LDH catalyst, which means a better dispersion and a smaller crystal size were obtained.

The images of the microstructure of γ-Al_2O_3 support and Ni-Al-LDH precursor obtained by SEM are shown in Figure 3. Apparently, the surface of γ-Al_2O_3 support is tough and irregular, and there are a number of small pores, creating favorable

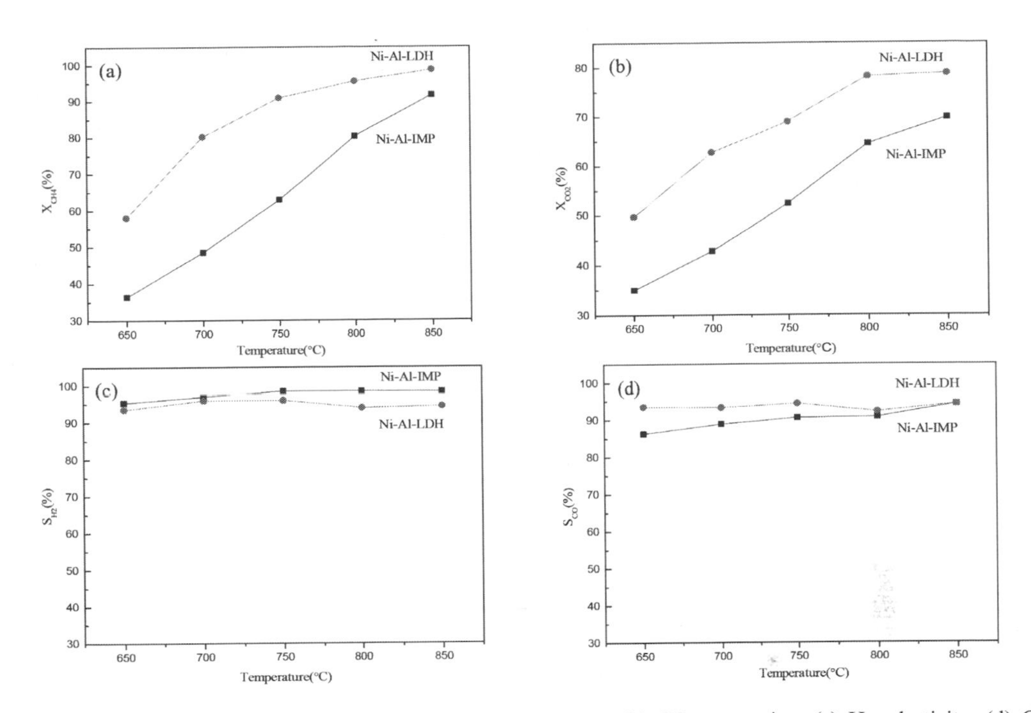

Figure 1. Catalytic performance of catalyst: (a) CH_4 conversion; (b) CO_2 conversion; (c) H_2 selectivity; (d) CO selectivity.

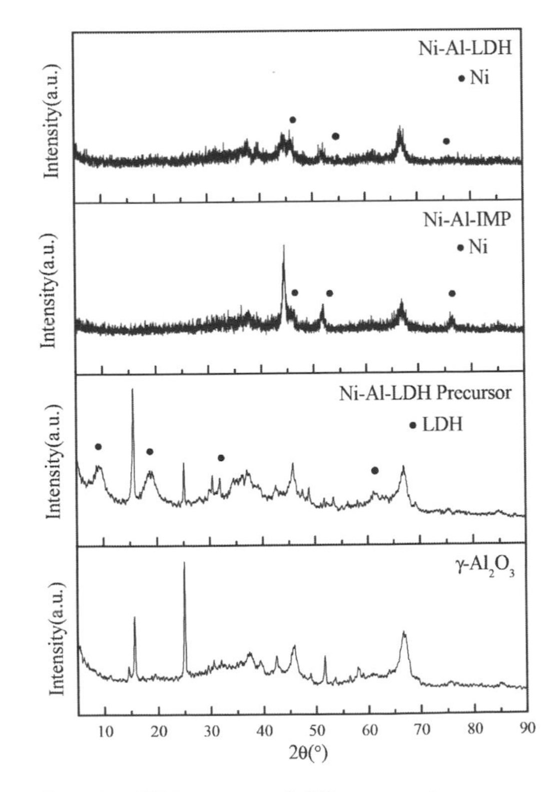

Figure 2. XRD patterns of different samples.

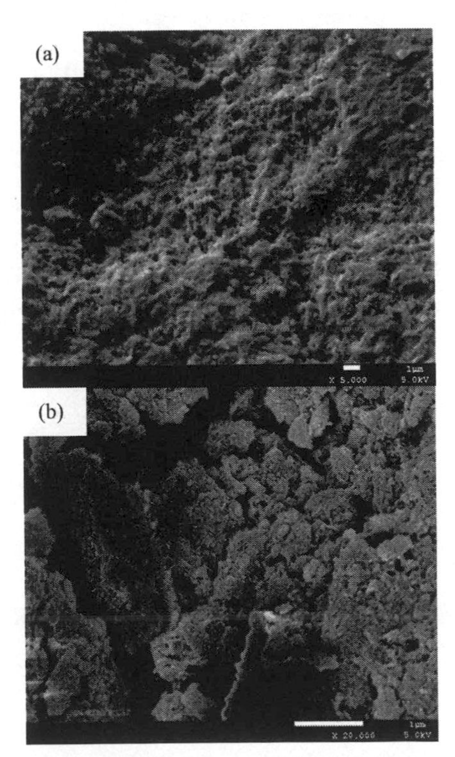

Figure 3. Images of different samples: (a) γ-Al_2O_3 support; (b) Ni-Al-LDH precursor.

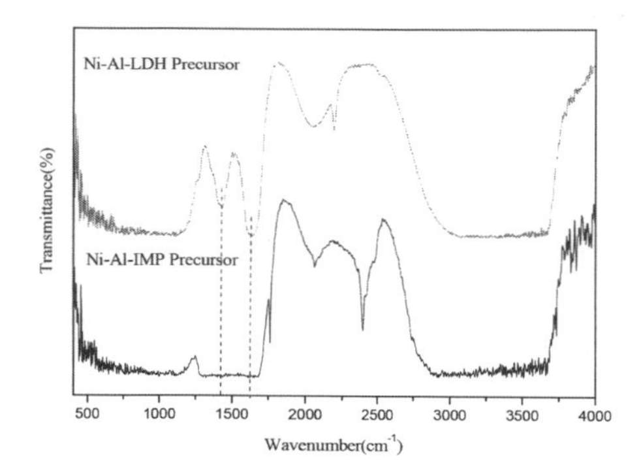

Figure 4. FT-IR spectra of catalyst precursor.

properties for serving as the catalyst support. The Ni-Al-LDH precursor exhibited the expected layer structure shown in Figure 3(b), which also proved the formation of LDHs on the γ-Al_2O_3 support.

As there are two types of anions (NO_3^- and CO_3^{2-}) in the LDH preparation system, the FT-IR spectrums of Ni-Al-LDH and Ni-Al-IMP precursors were obtained to certain the interlayer anion. As shown in Figure 4, the peaks at 1430 and 1640 cm^{-1} belong to the vibration of the carbonate species and the bending vibration of interlayer water (Qi et al. 2008, Zhang et al. 2011), respectively. Therefore, it can be concluded that the LDHs synthesized on the γ-Al_2O_3 support with CO_3^{2-} serving as the interlayer anion, the crystal water exists in the synthetic LDHs.

4 CONCLUSIONS

LDHs with the interlayer anion of CO_3^{2-} were synthesized on the γ-Al_2O_3 support, and the resulting Ni-Al-LDH catalyst was evaluated by CH_4-CO_2 reforming. Compared with the Ni-Al-IMP catalyst, the Ni-Al-LDH showed an enhanced catalytic activity, which is ascribed to the unique layer structure of LDHs. Ni existed as a divalent cation in LDHs and was of the atomic level dispersion, resulting in smaller Ni crystal size, better Ni dispersion, and more active sites, leading to a better catalytic activity.

ACKNOWLEDGMENTS

This study was supported by the Natural Science Foundation of Jiangsu Higher Education Institutions of China (Grant number: 4100716090) and the Research Project of Xuzhou Institute of Technology (Grant number: XKY2015308).

REFERENCES

Bhattacharyya A, Chang V W, Schumacher D J. (1998). CO_2, reforming of methane to syngas: I: evaluation of hydrotalcite clay-derived catalysts. *Applied Clay Science*. 13, 317–328.

Feely R A, Doney S C. (2011). Ocean Acidification: The Other CO_2 Problem. *Aslo Web Lectures*. 3, 1–59.

Han J W, Kim C, Park J S, et al. (2014). Highly Coke-Resistant Ni Nanoparticle Catalysts with Minimal Sintering in Dry Reforming of Methane. *Chemsuschem*. 7, 451–456.

Khan S B, Khan S A, Asiri A M. (2016). A fascinating combination of Co, Ni and Al nanomaterial for oxygen evolution reaction. *Applied Surface Science*. 370, 445–451.

Liu C, Gu Q, Ye J, et al. (2016). Perspective on catalyst investigation for CO_2 conversion and related issues. *CIESC Journal*. 67, 6–13.

Long H, Xu Y, Zhang X, et al. (2013). Ni-Co/Mg-Al catalyst derived from hydrotalcite-like compound prepared by plasma for dry reforming of methane. *Journal of Energy Chemistry*. 22, 733–739.

Manukyan K V, Cross A J, Yeghishyan A V, et al. (2015). Highly Stable Ni-Al_2O_3 Catalyst Prepared from a Ni-Al Layered Double Hydroxide for Ethanol Decomposition toward Hydrogen. *Applied Catalysis A General*. 508, 37–44.

Montañez M K, Molina R, Moreno S. (2014). Nickel catalysts obtained from hydrotalcites by coprecipitation and urea hydrolysis for hydrogen production. *International Journal of Hydrogen Energy*. 39, 8225–8237.

Mikulová Z, Čuba, P, Balabánová J, et al. (2007). Calcined Ni-Al layered double hydroxide as a catalyst for total oxidation of volatile organic compounds: Effect of precursor crystallinity. *Chemical Papers*. 61, 103–109.

Qi T, Reddy B J, He H, et al. (2008). Synthesis and infrared spectroscopic characterization of selected layered double hydroxides containing divalent Ni and Co. *Materials Chemistry & Physics*. 112, 869–875.

Rezende S M D, Franchini C A, Dieuzeide M L, et al. (2015). Glycerol steam reforming over layered double hydroxide-supported Pt catalysts. *Chemical Engineering Journal*. 272, 108–118.

Romero A, Jobbágy M, Laborde M, et al. (2014). Ni(II)-Mg(II)-Al(III) catalysts for hydrogen production from ethanol steam reforming: Influence of the Mg content. *Catalysis Today*. 470, 398–404.

Saidi K, Hammami S. (2015). The impact of CO_2, emissions and economic growth on energy consumption in 58 countries. *Energy Reports*. 1, 62–70.

Takehira K, Shishido T, Peng W, et al. (2004). Autothermal reforming of CH_4 over supported Ni catalysts prepared from Mg-Al hydrotalcite-like anionic clay. *Journal of Catalysis*. 221, 43–54.

Wang G, Chen X, Zhang Z, et al. (2015). Influencing Factors of Energy-Related CO_2 Emissions in China: A Decomposition Analysis. *Sustainability*. 7, 14408–14426.

Zhao Q, Chang Z, Lei X, et al. (2011). Adsorption Behavior of Thiophene from Aqueous Solution on Carbonate—and Dodecylsulfate-Intercalated ZnAl Layered Double Hydroxides. *Industrial & Engineering Chemistry Research*. 50, 10253–10258.

Zhang F, Li M, Yang L, et al. (2014). Ni-Mg-Mn-Fe-O catalyst derived from layered double hydroxide for hydrogen production by auto-thermal reforming of ethanol. *Catalysis Communications*. 43, 6–10.

Civil, Architecture and Environmental Engineering – Kao & Sung (Eds)
© 2017 Taylor & Francis Group, ISBN 978-1-138-02985-9

Synthesis and hydrodesulfurization performance of NiMo sulfide catalysts supported on an Al-Si mixed oxide

Qun Yu
CNOOC Zhongjie Petrochemicals Co. Ltd., Hebei, China

Jingcheng Zhang, Jun Nan, Haibin Yu, Xuefeng Peng, Han Xiao, Guoliang Song, Shangqiang Zhang, Shuai Wang, Yuting Zhang & Guohui Zhang
CNOOC Tianjin Chemical Research and Design Institute, China

ABSTRACT: This work presents a comparative study of the textural, superficial and catalytic properties of NiMo/Al_2O_3-SiO_2 sulfide catalysts during the Hydrodesulfurization (HDS) of complex organic sulfur compounds, such as DBT and Cn-DBT. The Al-Si mixed oxide support was synthesized by the sol-gel method, the catalysts were synthesized by the co-impregnation method using an atomic ratio of Ni/[Ni/(Ni+Mo)] = 0.42. The materials were characterized by N_2 physisorption, XRD, SEM and HRTEM. The hydrogenation activity was investigated on a 100 ml high-pressure hydrogenation unit, using high-nitrogen catalytic cracking diesel oil as feedstock. This catalyst exhibited the best pore size and high specific surface area, coupled with the presence of Ni and Mo species in octahedral coordination, as well as good morphological properties.

1 INTRODUCTION

As a consequence of the increasing concern about environmental pollution, more stringent legislation to limit the sulfur content in transportation fuels has been introduced throughout the world (C. Song, 2003 & P. Raybaud, 2007). One of the most cost-effective ways is to improve the HDS activity of catalysts. Conventional NiMo catalysts are usually prepared by depositing molybdenum and nickel oxides on the surface of alumina. The activity of conventional supported catalyst is limited by the support, so it is difficult to have a more substantial increase. It is now known that one of the main factors that affect catalyst activity is the interaction between the active components and the support, since metal-support interactions influence not only the dispersion of the active species, but also their reducibility and sulfidability. The strong interaction between Mo oxide and the support leads to the difficult and incomplete reduction-sulfidation of Mo^{6+} to Mo^{4+}. Extensive research studies have been dedicated to develop improved catalysts for the process in order to meet with legislations. In this sense, different materials have been studied as catalyst supports, specially the mixed oxide supports, such as the Al_2O_3-SiO_2(-TiO_2), since the presence of SiO_2 in alumina could facilitate redox processes for the active phases of Mo (J.R. Grzechowiak, 2001 & S.K. Maity, 2006) and W and therefore facilitate the formation of active octahedral type sites of the Mo and W oxide species. Due to the fact that this type of sites exhibits higher HDS activity, it is expected to obtain better catalysts.

The objective of this work is to contribute to elucidate their catalytic activity differences through a comparative study of the textural and superficial properties of NiMo sulfide catalysts supported on an Al-Si mixed oxide during the Hydrodesulfurization (HDS) of complex organic sulfur compounds, such as DBT and Cn-DBT.

2 EXPERIMENTAL

2.1 *Support preparation*

The Al-Si mixed oxide was synthesized by the sol-gel method, using 5 wt% of SiO_2. The required amounts of alkoxide precursors $Al(OC_4H_9)_3$ and $Si(OC_2H_5)_4$ were added to 150 ml of isopropanol at 60°C and kept stirred constantly for an hour, later, the system was cooled down to low temperature (3°C). Separately, a hydrolysis solution was prepared (water, ethanol, isopropanol and nitric acid using 13:8:5:0.5 ml, respectively). This solution was added drop-wise to alkoxides in solution to form a gel. The obtained gel was aged and then dried at room temperature. Finally, the materials was extruded, dried at 120°C for 12 h and calcined at 550°C (10°C/min) for 3 h under air flow. The γ-Al_2O_3 (S_{BET} = 292 $m^2 \cdot g^{-1}$) support was prepared, used as the reference support.

2.2 *Catalysts preparation*

NiMo sulfide catalysts were synthesized using the co-impregnation method with atomic ratio Ni/[Ni/(Ni+Mo)] = 0.42. As precursors we used: Ammonium Thiomolybdate (TMA), which were synthesized in the laboratory by previously reported methods and $Ni(NO_3)_2 \cdot 6H_2O$. These thiosalts and nickel nitrate were separately dissolved in MEA/water and then mixed to obtain a solution. After that the support was impregnated, dried at 120 °C for 2 h, then calcined at 400 °C for 3 h under N_2 flow. Finally, the catalysts were cooled down to room temperature under nitrogen gas flow. The catalysts were labeled as NiMoS/Al_2O_3-SiO_2. The NiMoS/γ-Al_2O_3 catalyst was prepared, used as the reference catalyst.

2.3 *Materials characterization*

2.3.1 *N_2 physisorption*

The textural properties for the oxide support and sulfide catalysts were determined from the adsorption-desorption isotherms of nitrogen at –196°C, recorded with a Quantachrome Autosorb 1. The specific surface areas were calculated by means of the Brunauer, Emmett and Teller (BET) model. Previous to their measurements, the samples were treated at 250°C for 3 h under vacuum.

2.3.2 *Scanning Electron Microscopy (SEM)*

The sulfide catalysts were studied by Scanning Electron Microscopy using a Phillips XL30 (ESEM) microscope with an energy dispersive X-ray spectroscopy (EDX) attachment. Samples were suspended in isopropanol by sonication and then deposited on carbon coated copper grids.

2.3.3 *XRD and HRTEM*

The X-Ray powder Diffraction (XRD) determinations were carried out in a Panalytical X' Pert Pro MPD diffractometer, equipped with a curved graphite monochromator, using Cu-Kα radiation (λ = 1.5406 Å) operating at 45 kV and 40 mA at a scanning rate of 5 min–1. HRTEM was performed with a JEM-2100 analytical microscope operated at 200 kV.

2.3.4 *Catalytic activity measurement*

The activity evaluation was carried out in a fixed bed system, using FCC diesel as feedstock. The operating conditions were set as follows: reaction temperature = 320, 340, 360°C, reaction pressure = 6.5 MPa, LHSV = 1.5 h^{-1} and H_2/Oil rate = 500:1. Prior to the activity test, the catalysts were presulfided in situ. The sulfur concentration was measured in a Multi EA3100 trace S/N analyzer of German Jena Company.

3 RESULTS AND DISCUSSION

3.1 *N_2 physisorption*

Figure 1 and Figure 2 show the N_2 adsorption-desorption isotherms for the supports and the catalysts. The supports and catalysts exhibit a type IV isotherms (IUPAC classification), which indicates that they are mesoporous materials (R. R. Xu, 2004). However, it possible to observe different type of hysteresis, since the Al_2O_3-SiO_2 oxide support and the corresponding catalyst show a H1 type of hysteresis, indicating that these materials have regular mesoporous channels. On the other hand, the γ-Al_2O_3 and the NiMo catalyst exhibit a clear H2 type of hysteresis loop, characteristic of materials with irregular mesoporous chan-

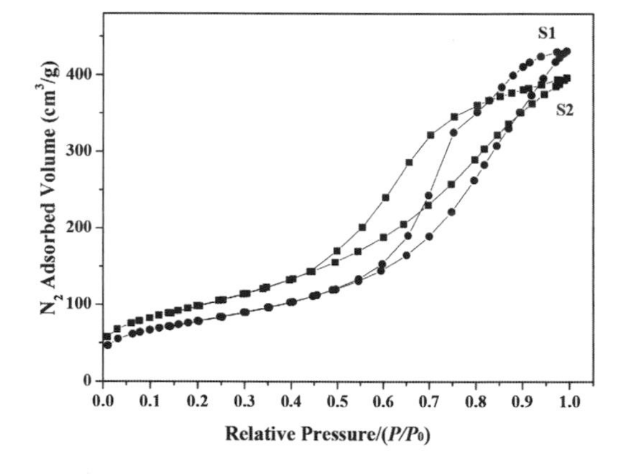

Figure 1. N_2 adsorption-desorption isotherms for the Al_2O_3-SiO_2 oxide support and γ-Al_2O_3.

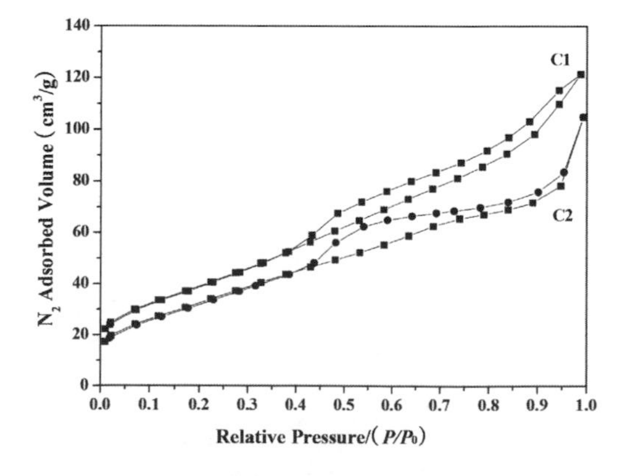

Figure 2. N_2 adsorption-desorption isotherms for the NiMoS/Al_2O_3-SiO_2 and NiMoS/γ-Al_2O_3 catalysts.

Table 1. BET surface area and pore volume of Al_2O_3-SiO_2 and γ-Al_2O_3 support.

Sample	BET surface area ($m^2 \cdot g^{-1}$)	Pore volume ($cm^3 \cdot g^{-1}$)
Al_2O_3- SiO_2	335	0.71
γ-Al_2O_3	292	0.65
NiMoS/Al_2O_3-SiO_2	192	0.40
NiMoS/γ-Al_2O_3	165	0.36

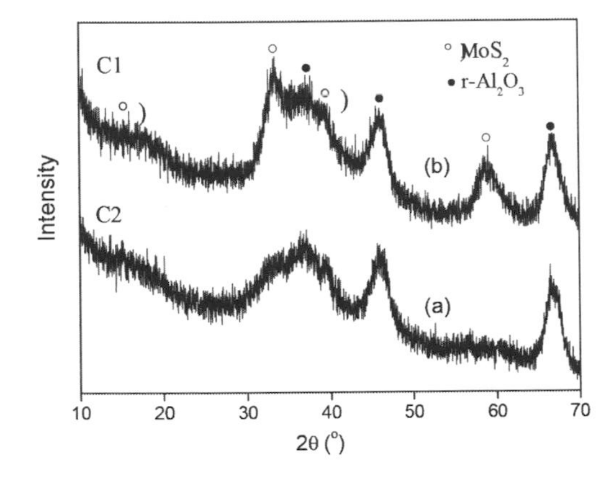

Figure 3. XRD patterns of the NiMoS/Al_2O_3-SiO_2 and NiMoS/γ-Al_2O_3 catalysts.

nels, which suggest that pores occlusion probably takes place. By comparison (Table 1), the NiMoS/Al_2O_3-SiO_2 catalyst presents higher surface area (192 m^2/g) and pore volume (0.40 mL/g), which is obviously larger than that of NiMoS/γ-Al_2O_3 (165 m^2/g and 0.36 mL/g, respectively). The higher specific surface area and pore volume mean more active sites, so that high catalytic activity is to be expected.

3.2 *XRD Analysis*

In Fig. 3, the XRD patterns of the NiMoS/Al_2O_3-SiO_2 and NiMoS/γ-Al_2O_3 catalysts are displayed. Four weak reflections at 14.4, 33, 40, and 58° are observed which are the characteristic reflections (002), (101), (103) and (110) of a poorly crystalline 2H-MoS_2 structure. No reflection of nickel sulfide phases is detected for all the catalysts, suggesting that Ni atoms replace the Mo atoms on the edge or corner sites and form so-called Ni-Mo-S phase.

3.3 *SEM Analysis*

The surface morphology of the support materials prepared via different methods are shown in Fig. 4 A and 4B. The main particles of Al_2O_3-SiO_2 oxide materials are comprised of sphere-like nano-

Figure 4. SEM images of Al_2O_3- SiO_2 and γ-Al_2O_3 support, S1(A) and S2(B).

particles, and the distribution of the particles seems to be practically uniform, with the average size of about 500 nm. The three-dimensional porous structure can supply with more channels for participating molecules to contact the active sites. The images of γ-Al_2O_3 materials presents that the materials are composed of massive particles with irregular shape and severely agglomerated. It is obvious that the utilization of active components is lower. During the co-precipitation process, the introduce of Si influenced the form of the primary grains and the particle growth is controlled, which result in higher surface area and pore volume. It is possible to improve the textural properties for the catalyst.

3.4 *HRTEM Analysis*

The HRTEM images of the NiMoS/Al_2O_3-SiO_2 and NiMoS/γ-Al_2O_3 catalysts are shown in Fig. 5a and 5b. HRTEM is a powerful technique for studying the changes in the morphology of active phases

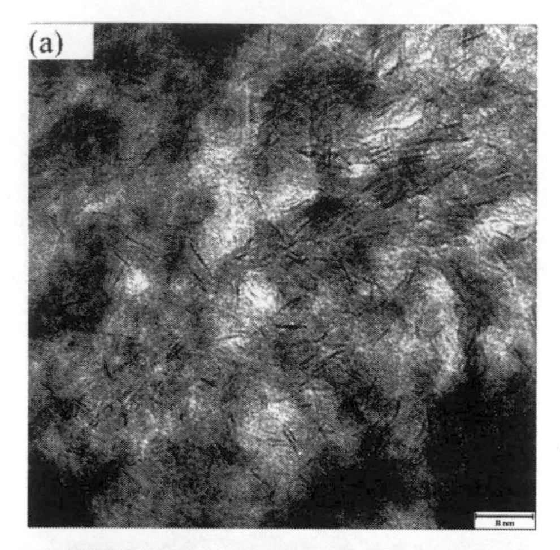

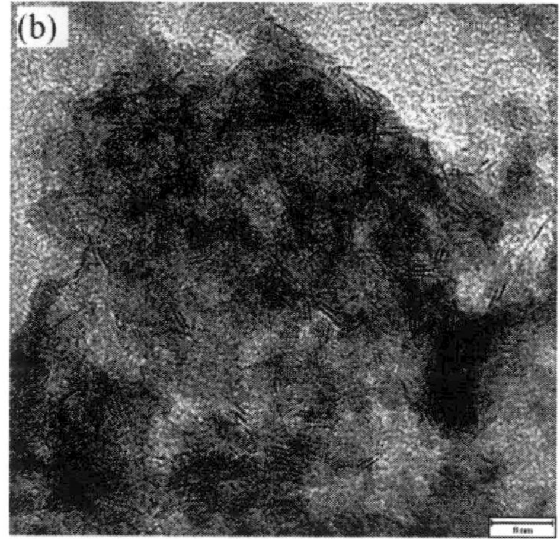

Figure 5. SEM images of $NiMoS/Al_2O_3-SiO_2$ and $NiMoS/\gamma-Al_2O_3$ catalysts, C1(a) and C2(b).

and has been applied in many studies of sulfide catalysts. The black thread-like fringes correspond to the MoS_2 or Ni-Mo-S crystallites. It is clear that the MoS_2 slabs are homogeneously dispersed on the support and no exceptionally large aggregates of MoS_2 are observed. In all catalysts, only few are single slabs and most of the MoS_2 slabs contain two or three layers.

3.5 *Catalytic activity*

The hydrotreatment of FCC diesel was studied under conditions of 6.0 MPa and 1.5 h^{-1}, which are close to those used in industrial applications. Table 2 summarizes the activity results of FCC diesel on the $NiMoS/Al_2O_3-SiO_2$ and $NiMoS/\gamma-Al_2O_3$ under steady-state conditions after 5 h of reaction. As can be seen from the table, the $NiMoS/Al_2O_3-SiO_2$ shows an improvement of activity. This enhancement of higher activity observed in the $NiMoS/Al_2O_3-SiO_2$ catalyst is more closely related to the textural and superficial properties, since this catalyst exhibited the largest pore size, as well as the best HDS precursor species, such as Mo and Ni species in octahedral symmetry (S. Eijsbouts, 2007).

Table 2. The activity evaluation results of FCC diesel on $NiMoS/Al_2O_3-SiO_2$ and $NiMoS/\gamma-Al_2O_3$ catalysts.

Catalyst	Sample	S content, μg/g	N content, μg/g	Density, g/ml
	FCC diesel	2483	1040	0.9263
C1	330°C	122.5	50.3	0.8960
	350°C	42.8	13.1	0.9111
	360°C	15.2	9.3	0.8902
C2	330°C	138.4	69.1	0.8966
	350°C	51.3	23.2	0.9114
	360°C	22.5	13.6	0.8906

4 CONCLUSIONS

We shown that the Ni-Mo catalyst, using the Al-Si mixed oxide as support increases the HDS activity compared to $NiMoS/\gamma-Al_2O_3$ confirming the beneficial effect of using the mixed oxide as support for the preparation of highly active catalysts. This change in metal dispersion and the decreased interaction of the molybdate with the support probably leads to a slight improvement in the sulfidation of the Ni and particularly of the Mo phase.

REFERENCES

Eijsbouts, S., S.W. Mayo and K. Fujita. Appl. Catal. A: Gen Vol. 322 (2007), p. 58.

Grzechowiak, J.R. J. Rynkowski, I. Wereszczako-Zieliñska, Catalytic hydrotreatment on alumina–titania supported NiMo sulphides, Catalysis Today, 65 (2001) 225–231.

Maity, S.K., J. Ancheyta, M.S. Rana, P. Rayo, Alumina–Titania Mixed Oxide Used as Support for Hydrotreating Catalysts of Maya Heavy CrudeEffect of Support Preparation Methods, Energy & Fuels, 20 (2006) 427–431.

Raybaud, P. Appl. Catal. A: Gen., 322 (2007), p. 76.

Song C. and X. Ma: Appl. Catal. B: Environ Vol. 41 (2003), p. 207.

Xu, R.R., W.Q. Pang and J.H. Yu, in: Molecular Sieve and Porous Materials Chemistry, Beijing: Science Press (2004).

Civil, Architecture and Environmental Engineering – Kao & Sung (Eds)
© 2017 Taylor & Francis Group, ISBN 978-1-138-02985-9

Study on solidification of Ningbo mucky soil

Jianyang Lu & Riqing Xu
Research Center of Coastal and Urban Geotechnical Engineering, Zhejiang University, Hangzhou, China

Qing Yan
Zhejiang Provincial Institute of Communications Planning, Design and Research, Hangzhou, China

Xu Wang
Research Center of Coastal and Urban Geotechnical Engineering, Zhejiang University, Hangzhou, China

ABSTRACT: Response Surface Methodology (RSM) was utilized to optimize the admixtures and ascertain the quantitative equation. For Ningbo soil with the natural water content of 89.5% and organic matter content of 17.1%, the optimum ratio of four additives was obtained as 12: 8.87: 3.96: 1.28, which is named ZJU-15. Through another set of response surface tests, a quantitative equation was gained, which relates the 28-day unconfined compressive strength of soil about three parameters of organic matter content, water content and ZJU-15 content. The influence of the three factors on the strength of solidified soil was determined by the single factor and mutual-influence analysis.

1 INTRODUCTION

Ningbo soil is the typical case of soil of coastal areas and inland cities in China, which has the characteristic of high contents of water and organic matter. Before constructions were launched, the soil stabilization must be carried out first on those grounds (Wang, 2012; Huang, 2012; Feng, 2007).

This study employs Response Surface Methodology (RSM) to study on the solidification of Ningbo soil (Ragonese, 2002). RSM is a combination of statistical and mathematical techniques, and we have some experience on its application on soil stabilization. Furthermore, we choose the 28-day unconfined compressive strength as the response values, taking into account the 28-day strength as an important indicator of the actual project.

So far, some scholars have done a lot of study related to either the water content or the organic matter content (Tang, 2000; Chen, 2005). However, there is hardly any research considering both of them. On the other hand, many scholars have devoted themselves to the optimization of admixtures to enhance the solidification effect of cement (Chang, 2014; Paulo, 2013; Jacob, 2012).

This study is divided into two stages. Firstly, the undisturbed mucky soil was taken as object, and its physical parameters can be tested by geotechnical test. RSM was utilized to optimize the admixtures and ascertain the quantitative equation. Secondly, based on the quantitative equation which was ascertained at the first stage, we varied the contents of organic matter, water and ZJU15 in order to verify their influence on the strength of solidified soil.

2 MATERIALS AND METHODOLOGY

2.1 *Soil samples*

The soil samples used in this investigation were obtained from Ningbo, China. Basic index properties (see Table. 1) were determined according to the American Society for Testing and Materials (ASTM) standards.

2.2 *Stabilized materials*

The cement used in this study was ordinary Portland cement produced by Qianjiang Cement Company, China. Gypsum was produced by Hangzhou Gypsum Company, China.

The slag, fly ash and triethanolamine were produced by Shanghai Jiangfu Company, China.

Table 1. Physical and mechanical properties of soil.

soil	w %	r $kN{\cdot}m^{-3}$	d_s	e	W_p %	W_L %	W_0 %
clay	89.5%	16	2.73	1.67	54.2	26.1	7.1

Note: w is content of water, r is the unit weight, d_s is specific gravity of soil, e is porosity ratio of soil, W_p is plastic limit of soil, and W_L is liquid limit of soil. $W0$ is content of organic matter.

2.3 *Unconfined Compressive Strength tests (UCS)*

In this study, the UCS values of 28-day was selected to establish response values. UCS tests were conducted after the following process according to the ASTM standards. At first stage, the undisturbed mucky soil, cement, slag, fly ash and triethanolamine were added into a container with certain mix proportions, and then the mixture was stirred until it acquired a uniform consistency. The mixture was poured into cube samples of 70.7 mm × 70.7 mm × 70.7 mm, which were to be cured in a temperature-controlled room at 20 ± 2°*C*. After a fixed period (28D), the cube samples were taken out and UCS test was carried out.

2.4 *Single doped experiment*

Based on the research (LI, 2012), when the cement content is 15%, there is a turning point on the curve of cement soil strength for the soil with 6% humic acid content. When the humic acid content is increased, its adverse impact on the strength of cement-stabilized soil will tend to be stable. In addition, the strength of cement-soil will increase with the initial water content increasing. At the point of peak strength, the corresponding water content approaches the soil liquid limit.

We chose slag, fly ash and triethanolamine as additives for the following reasons: slag and fly ash have preferable chemical activity, which can promote the hydration reaction; triethanolaminecan could regulate the overall curing effect.

According to the test results, the range of additives for the next phase could be decided as: 15% cement, 5 ~ 7% slag, 5 ~ 7% fly ash, and 1.4 ~ 1.6% triethanolamine (additive percentages are that of the soil).

2.5 *Response surface method*

Based on the result from the single doped experiment, we adopted the Central Composite Rotatable Design (CCRD) response surface method.

RSM was applied to the experimental data by using a statistical software. Regression coefficients were obtained by fitting the experimental data into a second order polynomial model. The generalized second-order polynomial model proposed for the response surface analysis is as follows:

$$Y = \beta_0 + \Sigma\beta_i\ X_i + \Sigma\beta_{ii}\ X_{i2} + \Sigma\Sigma\ \beta_{ij}\ X_i\ X_j \qquad (1)$$

where β_0, β_i, β_{ii}, β_{ij}, are regression coefficients for constant, linear, quadratic and interaction terms, respectively. X_i and X_j are the coded values of the independent variables.

3 RESEARCH ON THE COMPOSITE ADDITIVE EQUATION

This stage of the study comprised of cement content of W_c (%), slag content of W_s (%), flyash content of W_f (%), and triethanolamine content of W_t (%), corresponding to the three code levels and the actual content as presented in Table 2. The results

Table 2. Independent variables and their levels for CCRD.

		Coded levels				
Variables	Coded	−2	−1	0	1	2
W_c(%)	X_1	0	12	15	18	21
W_s(%)	X_2	0	3	6	9	12
W_f(%)	X_3	0	3	6	9	12
W_t(%)	X_4	0	0.8	1.6	2.4	3

Note: X_1 = (Wc-15)/3, X_2 = (Ws-6)/3, X_3 = (Wf-6)/3, X_4 = (Wt-1.6)/0.8.

Table 3. CCRD experimental design matrix with 28d experimental results of UCS.

No.	X_1	X_2	X_3	X_4	Y_{28}/kPa
1	−1	−1	−1	−1	182.51
2	1	−1	−1	−1	1428.31
3	−1	1	−1	−1	771.92
4	1	1	−1	−1	2292.09
5	−1	−1	1	−1	201.29
6	1	−1	1	−1	1339.36
7	−1	1	1	−1	521.14
8	1	1	1	−1	2120.55
9	−1	−1	−1	1	118.10
10	1	−1	−1	1	1194.42
11	−1	1	−1	1	613.59
12	1	1	−1	1	2254.59
13	−1	−1	1	1	289.86
14	1	−1	1	1	1685.05
15	−1	1	1	1	748.81
16	1	1	1	1	2292.50
17	−2	0	0	0	0.00
18	2	0	0	0	2257.39
19	0	−2	0	0	238.87
20	0	2	0	0	1395.48
21	0	0	−2	0	1422.45
22	0	0	2	0	1494.91
23	0	0	0	−2	1559.45
24	0	0	0	2	1363.32
25	0	0	0	0	1380.83
26	0	0	0	0	1382.13
27	0	0	0	0	1383.36
28	0	0	0	0	1382.09
29	0	0	0	0	1382.10
30	0	0	0	0	1382.15

Table 4. ANOVA for 7d regression model 28d experimental results of UCS.

Variable	Quadratic sum	Mean square	F value	P value
X_1	5.965E + 006	5.965E + 006	768.74	<0.0001
X_2	1.362E + 006	1.362E + 006	175.51	<0.0001
X_3	5781.20	5781.20	0.75	0.4016
X_4	66.97	66.97	8.630E-003	0.9272
X_1X_2	76456.40	76456.40	9.85	0.0068
X_1X_3	1357.37	1357.37	0.17	0.6817
X_1X_4	849.87	849.87	0.11	0.7453
X_2X_3	25785.13	25785.13	3.32	0.0883
X_2X_4	167.51	167.51	0.022	0.8851
X_3X_4	64230.57	64230.57	8.28	0.0115
X_1^2	97345.21	97345.21	12.54	0.0030
X_2^2	3.886E + 005	3.886E + 005	50.08	<0.0001
X_3^2	317.05	317.05	0.041	0.8425
X_4^2	420.65	420.65	0.054	0.8190
Model	$R^2 = 0.9856$			

obtained from the four-factor (X_1, X_2, X_3, X_4) and five-level (−2, −1, 0, 1, 2) Central Composite Rotatable Design (CCRD) are summarized in Table 3. Depending on the Design Expert software, it is convenient to obtain the response surface regression analysis of the results in Table 3. The regression model coefficient and variance analysis results are shown in Table 4.

This research adopts the F distribution to study the regression analysis results for a more rigorous analysis. In the analysis process, firstly, we need to determine the significance level α. In Table 4, if the P value is lower than 0.05, it indicates that the variable is statistically significant. Otherwise, it must be removed in the optimization analysis just for its not being significant. In this study, the significance level (α) is 0.05.

Accordingly, excluding the non-significant items X_3, X_4, X_1X_3, X_1X_4, X_2X_3, X_2X_4 in the 28d regression model, we obtained the regression equation after adjusting as:

$$Y_{28} = 1049.21 + 503.75X_1 + 240.42X_2 + 85.13X_1X_2 + 62.13X_3X_4 - 69.36X_1^2 - 102.73X_2^2 \quad (2)$$

This calculation showed that the coefficient of determination R^2 of the adjusted regression equation is 0.9856, which indicates that Equation (2) fits well with the actual situation. The further optimization analysis by Design Expert software revealed that the 28d strength of the stabilized soil reaches its maximum value when the contents of cement, slag, fly ash and triethanolamine are 12%, 8.87%, 3.96% and 1.28%, respectively. Thus, the composite additive ratio is W_c: W_s: W_f: W_t = 12: 8.87: 3.96:1.28. And the predicted value of Y_{28} is 2.403*MPa* against the actual value of 2.388*MPa*. This optimum ratio is named ZJU-15 in this paper.

Form the variance analysis, we get the linear effect of X_1 and X_2 on the strength of solidified soil. The linear effect and surface effect are both significant. At the same time, the interaction effect of X_1X_2, X_3X_4 are significant.

4 SOLIDIFICATION SCHEME

4.1 *Soil samples*

The soil samples used in this investigation were obtained from Ningbo, China. Basic index properties (see Table 1) were determined according to the American Society for Testing and Materials (ASTM) standards. The natural water content of these soils is higher than their liquid limit. The soil samples were conserved in the oven with the temperature ranging 105~110°C for more than 8 hours. After the soils have been smashed completely, they were put through a 2 mm sieve, and sealed to be spare.

The research (Aslan, 2008) has shown that, humus is the main component of the organic matter in soil, and humic acid is the main ingredient that affects the solidification effect of cement.

Humic acid was produced by Changsheng Industry Co., Ltd., China. Its purity is 90%. Considering the test requirements of soil samples with different contents of humic acid, we made trial mixes with different contents of humic acid beforehand.

4.2 *Experimental design and results*

As introduced above, this stage of the study comprised of organic matter content of Wo (%), water content of Ww (%), ZJU-15 content of W_Z (%), corresponding to the three code levels and the actual content as presented in Table 5. The results obtained from the three-factor (X5, X6, X7) and five-level (−1.682, −1, 0, 1, 1.682) Central Composite Rotatable Design (CCRD) are summarized in Table 6. Depending on the Design Expert software, it is convenient to obtain the response surface regression analysis of the results in Table 6. The regression model coefficient and variance analysis results are shown in Table 7.

As the research on the composite additive equation, we determine the significance level α as 0.05. After excluding the non-significant items X_5X_6, X_6X_7 in the 28d regression model, we obtained the regression equation after adjusting, as:

$$Y'_{28} = 1186.77-88.52X_5-86.89X_6+73.71X_7-41.14X_5X_7-20.65X_5^2-72.67X_6^2+9.95X_7^2 \quad (3)$$

Table 5. Independent variables and their levels for CCRD.

Variables	Coded	Coded levels				
		−2	−1	0	1	2
$W_O(\%)$	X_5	0.795	2.5	5	7.5	9.205
$W_W(\%)$	X_6	26.36	40	60	80	93.64
$W_Z(\%)$	X_7	1.272	4	8	10	14.728

Note: $X_5 = (W_O-5)/2.5$, $X_6 = (W_W-60)/20$, $X_7 = (W_Z-8)/2$.

Table 6. CCRD experimental design matrix with 28d experimental results of USC.

No.	X_5	X_6	X_7	Y_{28}'/kPa
1	−1	−1	−1	1150.42
2	1	−1	−1	1087.45
3	−1	1	−1	982.33
4	1	1	−1	870.76
5	−1	−1	1	1388.52
6	1	−1	1	1113.89
7	−1	1	1	1210.12
8	1	1	1	981.11
9	−1.682	0	0	1293.67
10	1.682	0	0	978.12
11	0	−1.682	0	1134.65
12	0	1.682	0	842.91
13	0	0	−1.682	1102.36
14	0	0	1.682	1342.57
15	0	0	0	1186.34
16	0	0	0	1186.34
17	0	0	0	1186.34
18	0	0	0	1186.34
19	0	0	0	1186.34
20	0	0	0	1186.34

Table 7. ANOVA for 28d regression model.

Variable	Quadratic sum	Mean square	F value	P value
X_5	1.07E+05	1.07E+05	534.7	<0.0001
X_6	1.03E+05	1.03E+05	515.19	<0.0001
X_7	74202.4	74202.4	370.78	<0.0001
X_5X_6	1.11	1.11	5.55E-03	0.9421
X_5X_7	13538.35	13538.35	67.65	<0.0001
X_6X_7	677.12	677.12	3.38	0.0957
X_5^2	6147.11	6147.11	30.72	0.0002
X_6^2	76096.72	76096.72	380.25	<0.0001
X_7^2	1427.92	1427.92	7.14	0.0234
Model	$R^2 = 0.9766$			

The calculation shows that the coefficient of determination R2 of the adjusted regression equation is 0.9766, which indicates that Equation (3) fits well with the actual situation.

The quantitative equation (3) shows the relationship between strength (Y'28) and organic matter content (X5), water content (X6), ZJU-15 content (X7). X5 and X6 could be determined by geotechnical test. As long as the strength need is designed, the volume of ZJU-15 can be calculated accurately, instead of empirical estimation or lunching a new research. Therefore, the quantitative equation is very practical and has great application prospect.

4.2 *Single factor and mutual-influence analysis*

Form Equation (3) it is easy to find out that X_5, X_6 and X_7 have significant linear effect on the strength of solidified soil. The mutual-influence of X_5 and X_7 is also significant.

The first derivative of Y'_{28} with respect to X_5:

$$d(Y'_{28})/d(X_5) = (-88.52-41.14X_7) -41.3X_5 \quad (4)$$

The value $X_5 = -(88.52 + 41.14X_7)/41.3$ is an Extreme Point (EP). It means when W_O is greater than extreme point, the negative effect increases with the increase of content on the strength. However, the value of ZJU-15 content (X_7) determines $EP(X_5)$. The increase of X_7 will reduce $EP(X_5)$. It means that the increase of ZJU-15 content will effectively eliminate the negative effect of organic matter on strength of solidified soil.

The first derivative of Y'_{28} with respect to X_6:

$$d(Y'_{28})/d(X_6) = -86.89-145.34X_6 \quad (5)$$

$EP(X_6) = -0.598$, the corresponding true water content is 48.04%. It means when $X_6 < -0.598$, the increase of water will promote the strength of solidified soil. All the substances involved in the curing reaction depend on the water environment. When $X_6 > -0.598$, if there is too much water, it will lead to too low ion concentration. Furthermore, the curing reaction is weakened, and the formation of crystals and the coupling effect with the soil particles will be greatly reduced.

The first derivative of Y'_{28} with respect to X_7:

$$d(Y'_{28})/d(X_7) = (73.71-41.14X_5) + 19.9X_7 \qquad (6)$$

$EP(X_7) = (41.14X_5-73.71)/19.9$, the organic matter content (X_5) determines $EP(X_7)$. When $X_7 < EP(X_7)$, ZJU-15 can not have a positive effect on the strength of solidified soil. When $X_7 > EP(X_7)$, the increase of ZJU-15 will promote the strength evidently.

5 CONCLUSION

For Ningbo soil with the natural water content of 89.5% and organic matter content of 17.1%, the optimum ratio of cement, slag, fly ash and triethanolamine is 12: 8.87: 3.96: 1.28.

The quantitative equation (3) shows the relationship between strength and organic matter content, water content and ZJU-15 content. It is very practical and has great application prospect.

According to the synthesis of single factor and mutual-influence analysis, ZJU-15 can effectively eliminate the negative effect of organic matter on strength of solidified soil, and more than 48.04% of water content will reduce the effect of ZJU-15.

REFERENCES

Aslan, N. [J]. Powder Technology, 2008, 185(1): 80–86.

Chang S., Xu R., Li X., Liao B., Wang X. [J]. Rock and Soil Mechanics, 2014, 01:105–110.

Chen H., Wang Q. [J]. Chinese Journal of Rock Mechanics and Engineering, 2005, 24(s2): 5816–5821.

Feng Z., Zhu W., et al. [J]. Chinese Journal of Rock Mechanics and Engineering, 2007, 26(s1): 3052–3057.

Huang Y., Zhu W, Zhou X. et al. [J]. Rock and Soil Mechanics, 2012, 33(2): 2923–2928.

Jacob J. Sauer, et al. T. [J]. Journal of Geotechnical and Geoenvironmental Engineering, 2012, 138(8): 968–980.

Li X., Xu R., Rong X. Journal of Central South Univ—ersity. 2012. 19: 2999–3005.

Paulo J. Venda O., et al. [J]. Journal of Geotechnical and Geoenvironmental Engineering, 2013, 139(5): 810–820.

Ragonese R, Macka M, et al. (2002) J. Pharm. Biomed. Anal. 27: 995–1007

Tang Y., Liu H. [J]. Chinese Journal of Geotechnical Engineering, 2000, 22(5): 549–554.

Wang D., Xu W. [J]. Rock and Soil Mechanics, 2012, 33(12): 3659–3664.

Civil, Architecture and Environmental Engineering – Kao & Sung (Eds)
© 2017 Taylor & Francis Group, ISBN 978-1-138-02985-9

A study on the utilization of iron and steel industrial solid waste

Y.B. Zhang
School of Metallurgical and Ecological Engineering, University of Science and Technology Beijing, Beijing, China
Central Research Institute of Building and Construction Co. Ltd., MCC Group, Beijing, China

R. Zhu & Y. Wang
School of Metallurgical and Ecological Engineering, University of Science and Technology Beijing, Beijing, China

C.S. Yue & X.J. Piao
Central Research Institute of Building and Construction Co. Ltd., MCC Group, Beijing, China

ABSTRACT: As by-products of iron and steel industry, blast furnace slag has been widely used as supplementary cementitious material in concrete. However, the utilization rate of steel slag is still relatively low and its open-air stacking occupies more space. Corrosion of reinforcement is regarded as one of the most important indexes of concrete durability. In this paper, we report the effect of slag admixture on the corrosion of reinforcement on the basis of the investigation of pore liquid pH, electric flux, and corrosion behavior of the reinforcement under dry–wet circulation conditions. Experimental results show that concrete with ground granulated blast furnace slag has better resistance of chloride-ion penetration but lower level of liquid-phase alkalinity. The pore liquid alkalinity of concrete with ground steel slag keeps at a high level, but the density of concrete is lower because of the slower growth of concrete strength in the early period. Compared with single-doped slag, ground iron and steel slag helps the concrete maintain a high pore liquid alkalinity as well as good resistance of chloride-ion penetration, which would provide a better protection to fixture within concrete.

1 INTRODUCTION

Production of crude steel has been increasing since 2000 worldwide. From 2013, China's crude steel output has occupied nearly half of the global production (see Figure 1), yielding large amounts of by-products as iron slag, steel slag, fly ash, and so on. In recent decades, several studies have been conducted all over the world, especially in the field of utilization as construction materials because of its feasibility and market space.

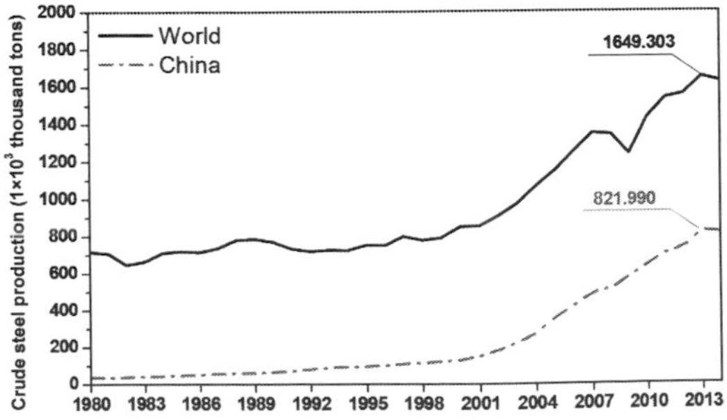

Figure 1. Global crude steel production statistics.

Steel slag is generated from the process of steel production, which has cementitious activity due to its composition of a certain amount of C_3S, C_2S, *etc.* Steel slag is often called as "Over-burnt Portland Cement Clinker", different from the cement clinker, with calcination temperature of about 1450°C and formation temperature of about 1650°C. Relatively, minerals crystallized from steel slag are more complete and compact, leading to its slow hydration.

Blast furnace slag comes from iron-making sections. As a result of water quenching treatment, the mineral structure remains vitreous in a thermodynamically unstable state, remaining potential hydraulic cementitious activity. Reactive SiO_2 and Al_2O_3 are the main factors influencing slag activity under an alkaline stimulating agent, $Ca(OH)_2$. In the progress of the reaction of minerals and water, strength is derived from hydrated calcium silicate, hydrated calcium aluminate, and so on.

In addition to saving energy and resources and reducing environmental impact, slag powders are adopted in concrete to improve its working performance (Zheng 2005). With the active mineral admixture, concrete would improve particle size distribution and pore structure, which leads to good performance in liquidity, secretion of water, setting time, resistance to chloride ion permeability (Kyong 2005, Erhan 2008, Hüseyin 2007), and so on.

Durability is an essential factor in the life and safety of concrete structures (Ann 2009). Corrosion of reinforcement that is usually caused by carbonation or chloride is considered the most important and direct reason of concrete durability failing. During concrete hydration process, calcium hydroxide [$Ca(OH)_2$], calcium silicate hydrate [$3CaO{\cdot}SiO_2{\cdot}3H_2O$], unhydrated tricalcium silicate, and dicalcium silicate are the main carbonized materials reacted with carbon dioxide, which diffuses into the concrete from atmosphere. $CaCO_3$, $Al(OH)_3$, and other products cause the formation of loose and non-cementitious state in concrete structures, which finally results in the destruction of concrete structures (Feng 2009). Another issue is that chloride ions would penetrate into the concrete reaching the steel surface to cause electrochemical action. Steel surface passivation film is destroyed in the first place. Therefore, the formation of pitting gradually spread throughout the steel surface (Ann 2007). Corrosion products bring about an increase in volume leading to concrete cracking, delamination, and flaking. The destruction facilitates the passage of water and chloride intrusion, accelerating corrosion development even more.

2 MATERIALS AND METHODS

2.1 *Raw materials*

Ground granulated Blast Furnace slag (BF) and ground steel slag (SS) used in the experiment were both sourced from a large iron and steel corporation in Jiangxi province, China. As seen from Table 1, composition difference between the two kinds of slag mainly concentrates in siliceous and aluminum constituents. Slag were ground by a ball mill to the specific surface area of 400 ± 5 m^2/kg, when density of blast furnace slag was 2900 kg/m^3 and density of steel slag was 3350 kg/m^3. Basic physical property of slag is shown in Tables 2~3 (according to Chinese standard GB/T 18046–2008 and GB/T 20491–2006). Standard cement was ordinary Portland cement supplied by Jidong Cement Co., Ltd, China.

2.2 *Experimental apparatus*

As strong alkaline substances, concrete's range of pH remains 12.5–13.0. Carbonation mainly affects the pH value of pore liquid within the concrete. The alkalinity of the concrete gradually lowered from surface to interior. The passivation film on the steel surface would become unstable and lose

Table 1. Chemical composition of slag (%).

	CaO	SiO_2	Al_2O_3	Fe_2O_3	FeO	MgO	P_2O_5	MnO
SS	41.35	12.07	1.88	11.67	14.32	9.12	2.41	1.26
BF	38.56	31.24	12.61	10.71	–	7.77	–	–

Table 2. Flexural and compressive strengths of cement mortar with SS.

Ratio(%)		Specific surface area (m^2/kg)	Fluidity (mm)	Flexural strength (MPa)		Compressive strength (MPa)	
Standard cement	SS			7d	28d	7d	28d
100	/	308	254	8.5	11.2	45.0	58.1
70	30	400	262	6.6	9.0	30.9	46.9

Table 3. Flexural and compressive strengths of cement mortar with BF.

Ratio(%)		Specific surface area (m^2/kg)	Fluidity (mm)	Flexural strength (MPa)		Compressive strength (MPa)	
Standard cement	BF			7d	28d	7d	28d
100	/	327	223	5.7	8.0	29.4	47.0
50	50	400	247	5.0	7.7	24.2	50.9

the protective effect on reinforcement when the pore liquid pH around reinforcement is up to about 11.5. Besides, chloride ions reaching the surface of steel may also cause electrochemical effect; thus, corrosion of reinforcement occurs.

Therefore, through the pore liquid pH tests, carbonation performance, and electric flux of the concrete with slag admixtures and corrosion behavior of the reinforcement within the concrete under dry–wet circulation conditions, the effect of slag admixture on the corrosion of reinforcement would be determined.

2.2.1 *Pore liquid alkalinity test*

Cement paste with 30% or 40% (mass ratio, same mentioned below) SS or BF and ordinary Portland cement under standard curing conditions (temperature 20°C±2°C; relative humidity, 95%). Tests were carried out at specified curing periods (1, 3, 7, 28, 60, 100, 140, 170, and 200 days).

2.2.2 *Electric flux test*

Chloride ion permeability of concrete would be reflected by the electric flux values. The greater the electric flux values, the better the chloride ion permeability in the concrete. As test objects, ground iron and steel slag (IS) are composed of SS and BF at certain ratio of 2:8, 3:7 or 4:6. According to Chinese standard GB/T 50082–2009, concrete was prepared with a mineral admixture (SS-30%, BF-30%, IS-30%, IS-40%, and IS-50%). Tests were conducted at a specified curing period (28 days).

2.2.3 *Corrosion behavior test*

In the corrosion of reinforced concrete, alternative wet and dry condition is the most adverse environmental effect. The situation of steel corrosion can be reflected directly by measuring the mass loss of steel corrosion. According to Chinese standard GB/T 50082–2009, the samples were prepared with mineral admixture (SS-10%, SS-30%, SS-40%, BF-30%, BF-50%, BF-70%, IS-10%, IS-30%, and IS-40%). Tests were carried out at specified periods (15 and 20 cycles). Mass loss rate of steel corrosion was calculated by the following formula:

$$L_W = \frac{w_0 - w - \dfrac{(w_{01} - w_1) + (w_{02} - w_2)}{2}}{w_0} \times 100$$

L_W – mass loss rate of steel corrosion (%);
W_0 – initial weight of steel (g);
W – mass of pickled corroded steel (g);
W_{01}, W_{02} – initial mass of standard steel (g);
W_1, W_2 – mass of pickled standard steel (g).
Accuracy of mass loss results should be 0.01%.

3 RESULTS AND DISCUSSION

3.1 *Pore liquid pH value*

During the 200-day test period, pore liquid pH values of samples mixed with steel slag (SS-30% and SS-40%) and pure cement sample (CE) develop with the same trend, where pH of early age (3 days) is low and then returns to normal along with gradual decrease over time. pH values reach the lowest point at 140 days. However, pH values of samples mixed with blast furnace slag (BF-30% and BF-40%) fluctuate, decreasing over the test period, as shown in Figure 2.

From the overall perspective, the pore liquid alkalinity of pure cement sample remains in the highest state, followed by SS and BF samples. Also, the impact shows a positive correlation with dosage of admixture content due to the alkalinity order from high to low in the cement, steel slag, and blast furnace slag.

As for CE and SS samples, the same trend shows that ground steel slag has no significant impact on the alkalinity characteristics of the cement paste. In another word, as alkaline substances, the cementitious property of steel slag depends on its minerals such as C_2S and C_3S, which is similar to that of Portland cement, and the hydration process was almost similar.

However, for BF samples, as low-alkalinity slag powder, the blast furnace slag has potential hydraulic cementitious activity due to its reactive SiO_2 and Al_2O_3 activated by the alkaline agent, $Ca(OH)_2$. As a result, alkaline substances are produced and consumed at the same time, forming a fluctuating trend. Because of the lower level of alkalinity caused by blast furnace slag, dosage of the blast furnace slag in concrete engineering should be emphasized and controlled strictly (Song 2006).

3.2 *Electric flux*

As is shown in Figure 3, electric flux test results reflect the resistance to chloride-ion penetration

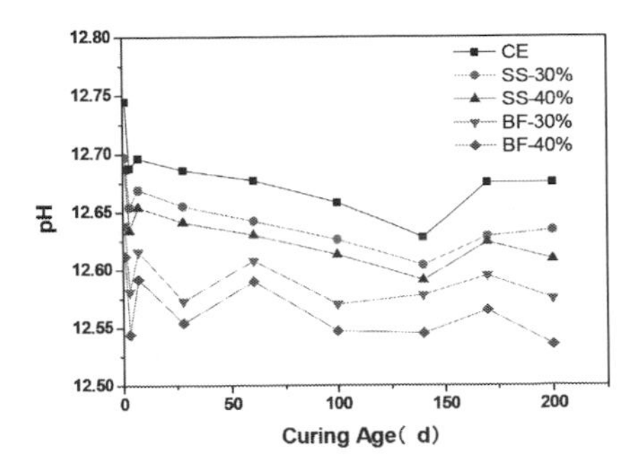

Figure 2. pH values of cement paste pore liquid.

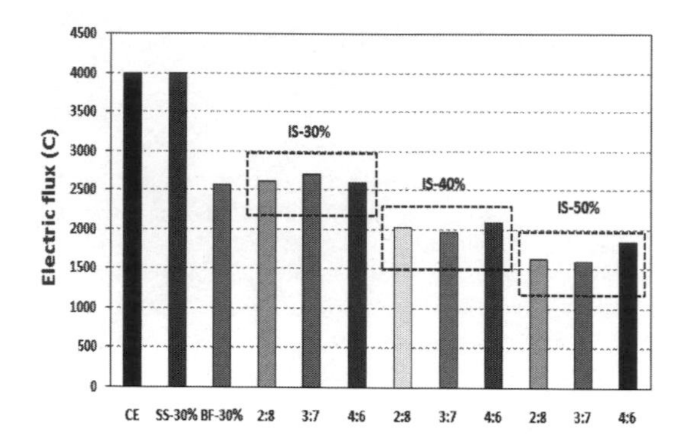

Figure 3. Electric flux of concrete samples.

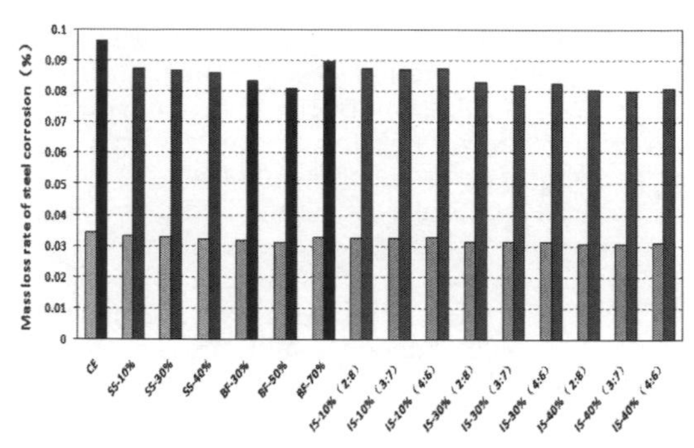

Figure 4. Mass loss rate of steel corrosion.

of concrete primarily related to compactness. Test results of pure cement sample (CE) and 30% steel slag sample (SS-30%) are beyond the scope of the instrument test, both of which were recorded as 4000 C. As a result of the slow pace of hydration of steel slag, SS-30% displays no significance. Studies have shown that the relative diffusion coefficients of chloride ion of concrete mixed with 30% ground steel slag in curing ages of 28 and 90 days are 5.22×10^{-12} and 1.45×10^{-12} m^2/s, respectively (Li 2003), indicating that as the curing period extends, the slag powder can sufficiently hydrate and improve the density of concrete.

Compared with sample CE, sample BF-30% shows an obvious advantage, as the ground blast furnace slag has a short term of activating and reacting. Furthermore, pore structures of concrete are filled with hydration products, making the structure denser and improving the impermeability of concrete.

When it comes to the IS-series samples, it is found that incorporated slag powders demonstrate good performances. Influences show positive relation with slag content. Especially in the samples IS-50% (2:8 and 3:7), electric fluxes fall more than 50% margin compared with sample CE's, which evidence that the admixture composed of both ground steel slag and ground blast furnace slag can show improvement in concrete compactness, surpassing single-doped slag powder. As for the results of different ratio of IS, 2:8 and 3:7 are almost at the same level, but addition of ground blast furnace slag leads to weaken the effect, especially when the percentage of IS reaches 50%.

3.3 *Steel corrosion behavior within concrete*

Mass loss rate of steel corrosion within concrete gives a direct visual comparison between different samples, as shown in Figure 4, where light-colored bars refer to the test of 15 cycles and dark ones represent that of 20 cycles. It is found that all the concrete samples result in significant differences between 15 and 20 cycles.

For results of 15-cycle samples, the component of concrete has a little effect; however, data of 20-cycle samples show difference to some degree. For the samples mixed with ground steel slag (SS series), because of the longer test period, the cementitious property of SS plays a role in ensuring the alkalinity of the liquid within concrete, meanwhile improving its compactness, which effectively reduce steel corrosion. For the samples added by blast furnace slag (BF series), steel corrosion reduced significantly, which mainly depend on the hydration products of BF, playing the role of concrete and improving the pore structure of the concrete. However, it is noted that the dosage of BF reached 70% and the steel corrosion rate recovered, which is consistent with the above findings. It should be taken to control of the content when BF is used as admixture in concrete. For the incorporation of the slag powder sample (IS series), with the increase of the slag powder content, the effect tends to improve significantly, while the composition of the slag powder has little effect on the results.

4 CONCLUSION

As inevitable by-products of iron and steel industry, metallurgical slag needs to be utilized as a reasonable resource through technical means to avoid dumping, occupation of land, and environmental pollution. Mineral composition causes cementitious activity or potential activity in steel slag and blast furnace slag, which is an important issue in the application in the field of construction materials. Steel slag can help the concrete alkalinity keep a higher level, but concrete strength can hardly increase fully at early age because of the slow pace of steel slag hydration. Blast furnace slag can improve the density of concrete by quick reaction of products; however, because of the possible low alkalinity, the slag cannot adequately protect steel, so its dose should be controlled strictly. When the two types of slag mentioned above were mixed at a certain ratio, it can ensure the strength and compactness of concrete as well as keep the pore liquid alkalinity at a higher level. In these regards, this research would shed light on improving the concrete durability, reducing steel corrosion rate, and effectively extending the service life of reinforced concrete.

REFERENCES

Ann K.Y., Song H. W (2007). Chloride threshold level for corrosion of steel in concrete. *Corrosion Science. 49*, 4113–4133.

Ann K.Y., Jung H.S., Kim S.S., et al (2009). Effect of calcium nitrite-based corrosion inhibitor in preventing corrosion of embedded steel in concrete. *Cement and Concrete Research. 36*, 530–535.

Erhan G., Mehmet G (2008). A study on durability properties of high-performance concretes incorporating high replacement levels of slag. *Materials and Structures. 41*, 479–493.

Feng N.Q., Xing F (2009). Concrete and durability of concrete structure. *Beijing: China Machine Press.*

Hüseyin Y., Halit Y., Serdar A (2007). Effects of cement type, water/cement ratio and cement content on sea water resistance of concrete. *Building and Environment. 42*, 1770–1776.

Kyong Y.Y., Eun K.K (2005). An experimental study on corrosion resistance of concrete with ground granulate blast-furnace slag. *Cement and Concrete Research. 35*, 1391–1399.

Li Y.X (2003). Study on composition, structure and properties of cement and concrete with steel-making slag powder mineral additive, *China Building Materials Academy*.

Song H.W., Velu S (2006). Studies on the corrosion resistance of reinforced steel in concrete with ground granulated blast-furnace slag-An overview. *Journal of Hazardous Materials B. 138*, 226–233.

Zheng K.R., Sun W., Jia Y.T., et al (2005). Effects of slag dosage on hydration products and pore structure of cement paste at high water to binder ratio. *Journal of the Chinese Ceramic Society. 33*, 520–524.

Civil, Architecture and Environmental Engineering – Kao & Sung (Eds)
© 2017 Taylor & Francis Group, ISBN 978-1-138-02985-9

Experimental study of the drainage and consolidation characteristics of tailings discharge and accumulation by gradual height rising under the action of drainage material

Jian Su, Xiaoduo Ou & Kaiwen Hou
Key Laboratory of Disaster Prevention and Structural Safety of Ministry of Education, Guangxi University, Nanning, China
Key Laboratory of Disaster Prevention Mitigation and Engineering Safety, Guangxi University, Nanning, China

ABSTRACT: A tailing drainage-consolidated physical model with discharge and accumulation process by gradual height rising, in which geotechnical composite drainage material is used as vertical drainage channel, was established. At the same time, the drainage-consolidated process was simulated by a self-developed test device, which was designed on the basis of the physical model described. The results showed that: (1) the rate of drainage is in accordance with the autoregressive integrated moving average model: $ARIMA(2,1,8) \times (0,0,0)$ on the condition of drainage-consolidated process of tailings with gradual height rising. The results predicted by the model show that: the growth trend of maximal drainage rate becomes gradually higher, which is much less than the actual peak rate, and the drainage performance fitness has a large scope for its development; (2) excreting dry upper water accumulation in time provides the essential conditions for consolidation of tailings under the effect of geotechnical composite drainage material, the average stress degree of consolidation can reach 62.32–70.81%, and the average strain degree of consolidation can reach 61.27–68.91% after the experiment of 104 days, which clearly reveals the promotion of consolidation effect.

1 INTRODUCTION

The resource of bauxite is abundant in Guangxi, China, with reserves reaching about 5.15 hundred million tons. Tailings reservoir for washing tailings storage is an indispensable part of mineral mining and processing. The hydraulic transmission and wet storage are used for the bauxite tailing of Guangxi. The tailing slurry, with low permeability, high void ratio, and high water content, has the maximum moisture content rate up to more than 300% (OU Xiao-duo 2014). Because of the lack of effective drainage channel in the tailings accumulation body, the accumulation of water in the reservoir is severe. With plenty of rainfall and extremely developed karst landform, overtopping and underground karst leakage accident of Guangxi bauxite tailings dam occur frequently (LI jie-quan 2012).

Therefore, there are a lot of research about the characteristics of sedimentation–consolidation (Ofori 2011, Shamsai 2007, YIN Guang-zhi 2011), the drainage consolidation mechanical model (Hann 2013, ZHANG Nan 2013), and the drainage consolidation technology (Suits 2010, Wickland 2010) of tailings slurry that are carried out by many scholars. However, because the Guangxi bauxite tailing has unique characteristics (DU Chang-xue 2006), the achievements of these studies have not been effective for practical application yet. On the basis of the typical bauxite tailings heap environment, we put forward to produce a geotechnical drainage material with good drainage performance and effective filtration of fine-grained tailing as vertical drainage channels. By equal proportion model test, the drainage mechanism and soil consolidation effects of tailings under the influence factors, such as side pressure, soil particle clogging effect, and the consolidation of the soil and water, have been stimulated. Combined with the model test results, the drainage performance and the consolidation effect of tailings slurry in the step piling environment were analyzed. On the basis of summarizing and analyzing the change of water displacement and the characteristics of the drainage rate, the mathematical model of drainage rate has been established by using SPSS and combined with the pore water pressure and sedimentation data to calculate the mud consolidation and makes the contrastively analysis.

2 TEST OF DRAINAGE-CONSOLIDATED MODEL OF TAILINGS BY GRADUAL WET STACK

2.1 *Drainage-consolidated physical model for tailings*

Tailings slurry is discharged to the tailings reservoir by hydraulic transportation after a special grinding, choosing, and washing process. The resulting product is a type of completely disturbed suspended slurry with extremely high content of clay particles. The suspended state of mud eventually formed structural clays with certain intensity after settlement and self-weight consolidation under the action of vertical drainage water. The sedimentation consolidation process of grading a heap of discharge slurry is not only the sedimentation consolidation process of this pile row mud, but it is also affected by the completed pile row mud, which is also a process of interaction. By comprehensive consideration of consolidation characteristics of tailings and operation characteristics of tailings, the drainage-consolidated physical model of model test was established:

1. The tailing slurry pile row, step-by-step continuous drainage;
2. Sedimentation consolidation process was carried out with the help of gravity, that is, it adopts the gravity sedimentation consolidation of the lower drainage method;
3. The top is the draining profile, the bottom in addition to the drain is the draining profile, and the rest is not the draining profile. The surrounding of the cylinder is not the draining profile.

The drainage-consolidated model of tailings by gradual wet stack is shown in Figure 1.

2.2 *Model test design*

2.2.1 *Model design*

In order to give full play to the drainage performance of geotechnical composite drainage material and fully understand the changing situation of tailings consolidated with geotechnical composite drainage material as the vertical drainage channels. The design and manufacture of test equipment is based on the equal-ratio physical model of drainage consolidation. The test device is mainly composed of a base, a test chamber, and a sensor mounting bracket (as shown in Figure 2). The test chamber is a thickening cylindrical stainless steel water tank with the bottom diameter, height, and thickness of 2.0 m, 1.8 m, and 7 mm, respectively. There are mounting brackets on both ends of the test chamber, which were independent of the test chamber instrument, in case they do not deformation due to the test load. The test chamber adopted a concrete pouring special drain at the bottom according to the requirement of drainage. The test used geotechnical composite drainage material as vertical drainage material, which consists of two parts: thin fabric and plastic core material. The high-strength support of core material ensures adequate drainage section, at the same time, outsourcing membrane, as the filter prevents the particles from moving into the blocking drainage channels and ensures that the drainage material can be unobstructed along the horizontal direction. The drainage material bottom was fixed in a special drain of the test chamber. And because of the pre-pull of the fixed device at the top, the drainage material can be in a state of uniform elongation. After the geotechnical composite drainage material installation was complete, sand was chosen to be laid in the port as a filter layer based on the situation of fine clay grain size distribution, by which the chamber used to discharge filter membrane water.

2.2.2 *Tailings slurry preparation and heap row*

The selected tailings mud came from the branch of China aluminum industry in Guangxi 2# tailings. Through indoor test, its basic properties are shown in Table 1.

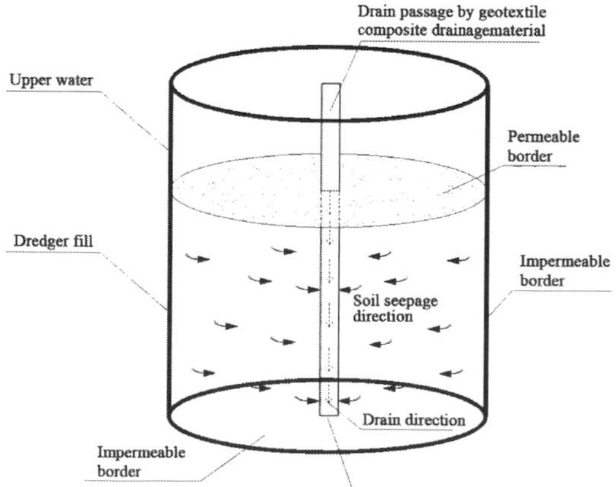

Figure 1. Experimental model of drainage consolidation.

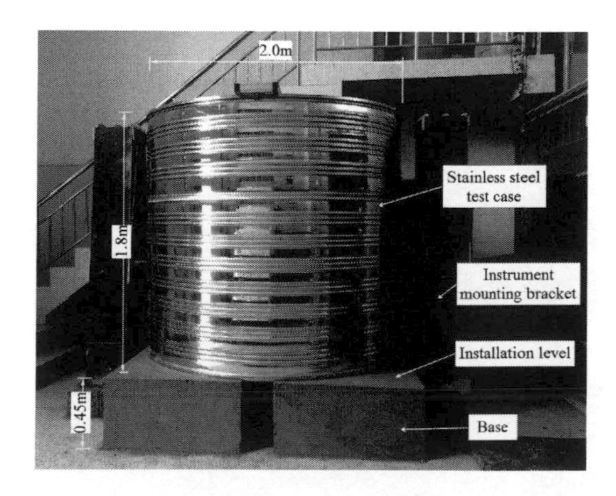

Figure 2. Test device.

Table 1. Formatting sections and subsections.

Proportion (g/cm^3)	Plastic index (%)	Liquidity index (%)	Permeability coefficient (m/s)	Slurry concentration (%)
2.83	23.5	5.97	2.44×10^{-6}	11–25

Figure 3. The status in the middle/last stage of experiment.

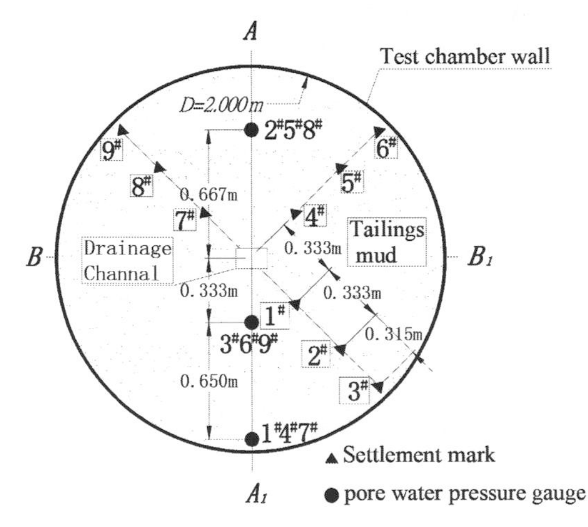

Figure 4. Layout plan of sensors test.

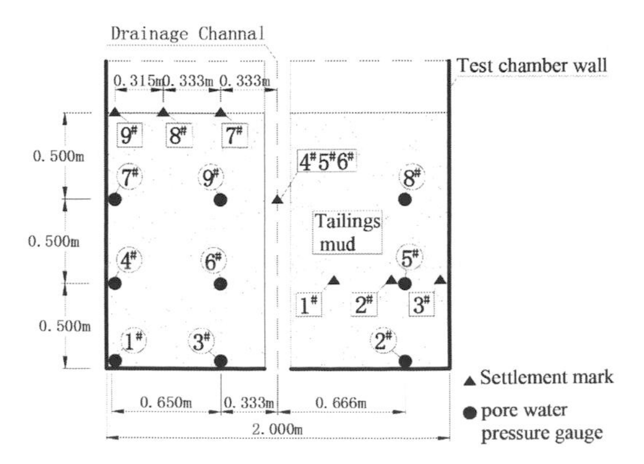

Figure 5. Layout profile of sensors test.

Tailings soil samples became dispersed mud evenly after the process of "dry—preliminary crushing—fully soaked—stirring", and then the mixed mud began to heap row after being diluted and stirred until the concentration reached 20% (row rate and per level pile row height is based on the branch of China aluminum industry in Gangxi 2# tailings in nearly 2 years to simulate the actual situation). The test lasted 21 days, with a total pile of 21 times and row height of 0.070 m. Eventually, mud surface height is 1.475 m and the water surface height is 1.747 m (as shown in Figure 3).

2.2.3 *Test index test*

The test design consists of monitor displacement, velocity, pore water pressure, layered settlement, and other indicators (sensor layout as shown in Figures 4–5).

Collect displacement by 4–8 h one time, and after that record the displacement in 2 min. Because the test chamber volume is large enough, displacement of drainage material in 2 min basic remains the same. Therefore, it can be thought that short-term water collecting data reflect the instantaneous velocity.

Because of the frequent fluctuation of the water surface height in the heap row period of the text and the gradual change of the last stage of test drainage tailing parameter, the frequency to monitor the pore water pressure monitoring data is divided into two periods: before trial begins to upper water draining, test one time in every 8 h; and the upper water drained until the end of test, one time in every 2 days.

Taking into account the impact of the heap row mud on the sedimentation scale test is large, the test of subsidence starts when heap row is complete; test frequency is once a week.

3 DRAINING PERFORMANCE ANALYSIS OF GEOTECHNICAL DRAINING MATERIAL

3.1 *Analysis of the influence factors on displacement*

To figure out the influence of the thickness of the mud and the overlying water on the displacement of geotechnical composite draining material, the 17 valid piling up data were analyzed, whose displacement is shown in Figure 6.

It can be seen from Figure 6 that the growth trend of daily average displacement is obvious, but changes in growth between the early piling up and the late piling up is large, which can be divided into two phases: linear growth and nonlinear growth phases. In the linear growth phase (which is one to

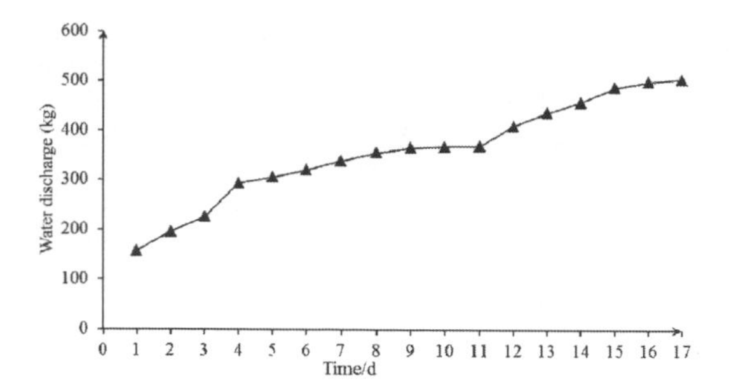

Figure 6. Total water discharge of discharge and accumulation process per day.

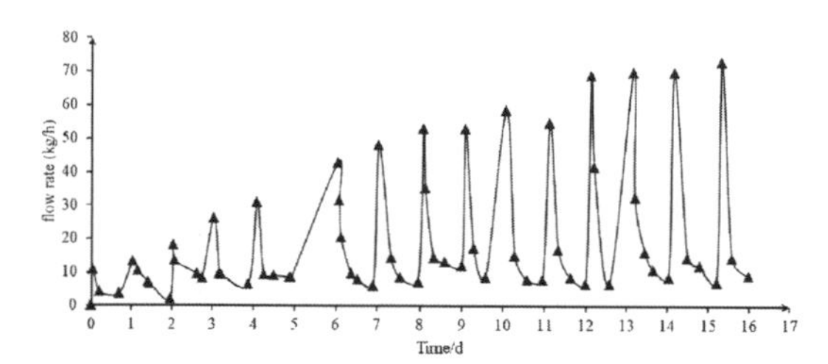

Figure 7. Flow velocity of discharge and accumulation process of tailings.

four times piling up), the growth of displacement is large and the increment is average, the average increment is around 50 kg every time; in the phase of linear growth (which is 5–17 times piling up), growth of displacement significantly slows, the average increment is only around 12 kg every time.

3.2 *Analysis of the characteristics of the drainage rate*

In the experiment, test of flow was conducted in different times of a day (e.g., before piling up, after piling up, and any two time between them), and instantaneous flow values were obtained in four different time points each day, as shown in Figure 7.

Figure 7 illustrates that drainage rate has the following characteristics:

1. Time correlation: because the test was conducted by the manner of piling up gradually, there is a lot of relevance between the velocity distribution of later period and the early piling up, and the velocity shows the great characteristics of the time series.
2. Long-term trend: along with the tailings, slurry pile up continuously, maximum velocity shows an obvious increasing trend, and its growth and changing law of flow is consistent. When it is one to four times piling up, the velocity shows almost a linear increase; and when it is 5–17 times piling up, the growth slows significantly. The minimum flow rate has also increased, but the growth is small. The reason is that when the water level goes down to a certain degree, membrane filter and mud seriously hindered the drainage rate.
3. Cycle variability: because of the intermittent of piling up, the flow rate shows obvious cycle changes; in a pile and idle period for one cycle, the velocity level changes cycle.

3.3 *Drainage rate model establishment and prediction*

The velocity values above reflect the velocity variation characteristics under the environment of particular pile up, but it is difficult to obtain quantitative values of the drainage rate in the late piling up stage. For this purpose, this paper uses autoregressive integrated moving-average model by SPSS software to simulate the drainage rate of mathematical modeling. It does not consider the causal relationship between the variables, but considers the laws of changes and developments of the variable in time. It can be used in the quantitative forecast and analysis in the late velocity. The model mathematical expression from the software is as follows:

$$\Psi(B)\Phi(B^s)w_t = \theta(B)\Theta(B^s)u_t, \tag{1}$$

formula:

$$\Phi(B^s) = 1 - \Phi_1 B^s - \Phi_2 B^{2s} - \cdots - \Phi_p B^{ps},$$

$$\Theta(B^s) = 1 - \Theta_1 B^s - \Theta_2 B^{2s} - \cdots - \Theta_p B^{ps},$$

where $w_t = \nabla_S^D \nabla_{z_t}^d$, ∇ is the difference operator, B the is backward-shift operator, $\Psi(B)$ and $\theta(B)$ are established functions, and u_t is the random residual term.

$AR(p)$ is the order-p autoregressive process and $MA(q)$ is the moving-average process. Fixing both of them, the autoregressive moving-average model $ARMA(p,q)$ was established. By using d times differential transform to stabilize it, the model can also be referred to: $ARIMA(p,d,q) \times (P,D,Q)$; parameters p, d, q, P, D, and Q need to set up when simulating.

3.4 *Establishment of the autoregressive moving-average model*

For model building, the continuous date of time unit 1 is essential, but the in-practice the data could not be collected continuously, so it is necessary to fill the missing values to form a complete

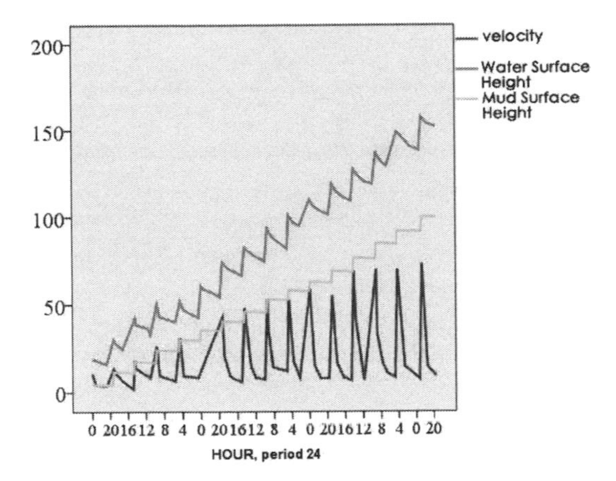

Figure 8. Sequence chart of flow velocity, thickness of tailings, and thickness of overlying water.

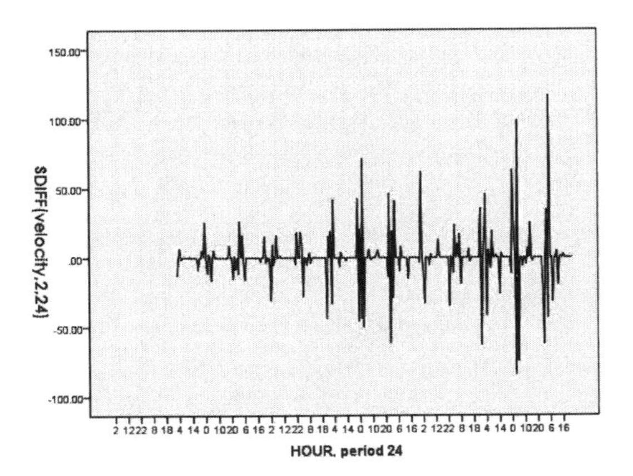

Figure 9. Sequence chart of velocity of stabling.

time series of the observed data. The missing value is completed with a linear interpolation; then the thickness of the mud and water surface elevation, both corresponding to the velocity as the independent variable condition in model, are introduced; time variables are defined by using "day/24 hour"; finally, the time sequence is drawn in Figure 8. Because the time series models are based on smooth sequence, the mean condition has no change over time, variance does not change with time, and the relevant number only relates to the time interval and has nothing to do with time point. Therefore, the second-order difference and the seasonal difference were used to smooth the data, which is shown in Figure 9.

After the parameters are put into the model, the model parameters could be analyzed as *ARIMA* (2,1,8) × (0,0,0). Smooth coefficients (R^2 and R^2) of the model were 0.964 and 0.990, respectively, which shows that the model fits very well, and the trend and seasonal volatility had a good performance.

3.5 *Model prediction*

By using the above model, the drainage rate of 17–23 times piling up has been forecast. Because the model is built on two independent variable conditions of clay surface height and water height, using piling up rate to calculate relevant and putting the data into the model, the projections can be obtained as shown in Figure 10.

The prediction shows that the maximum velocity prediction is relatively reasonable, but the predictive value of the minimum velocity is significantly large. The reason is that the model is just introduced in the mud surface height and height of the water as a condition of the independent variables, but the velocity is also affected by

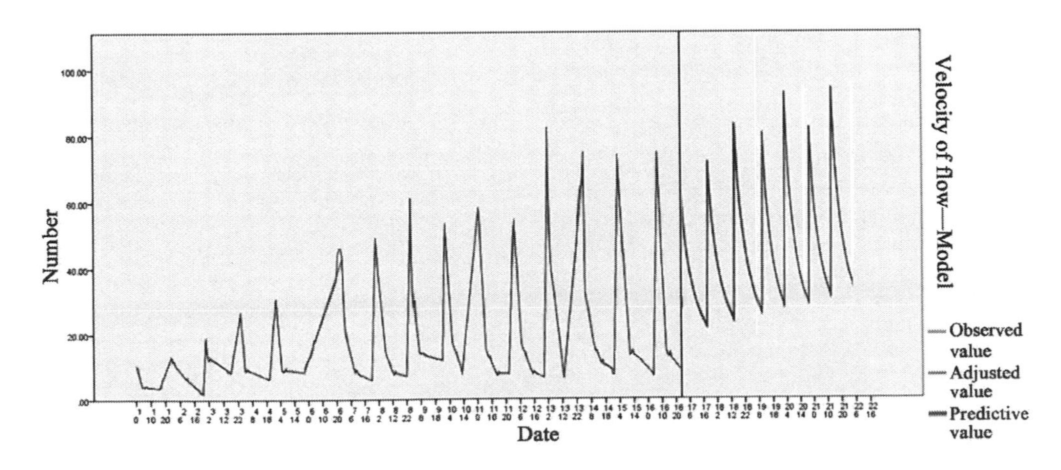

Figure 10. Predicting outcomes of discharge and accumulation process for 17–23 times.

cortical thickness, mud consolidation degree, and the overlying water turbidity factors. From its change trend, as tailings pile up, the thicknesses of both the mud and the overburden layer increase, the maximum displacement increases gradually, far from peak flow in capacity of composite drainage material, and drainage performance has much space.

4 ANALYSIS OF THE CONSOLIDATION EFFECT OF TAILINGS SLURRY

4.1 *Analysis of stress degree of consolidation*

The trial has installed nine pore water pressure meters ($1^{\#}$–$9^{\#}$), and the pore water pressure monitoring last 125 days. Among them, the pile row period was 21 days and the stopping discharge period was 104 days. According to Terzaghi's one-dimensional consolidation theory, stress degree of consolidation in the arbitrary point of tailings is defined as effective stress and total super hydrostatic pressure value in that point. The degree of dissipation of the pore water pressure can be used to define the degree of consolidation when the consolidation degree was calculated according to the measured pore water pressure data:

$$U_{\sigma} = \left(1 - \frac{\Delta u_t}{\sum \Delta u}\right) \times 100\%, \tag{2}$$

where Δu_t is the t time pore water pressure increment value and $\sum \Delta u$ is the cumulative pore water pressure increment.

According to the measured curve of pore water pressure, the stress degree of consolidation of $1^{\#}$–$9^{\#}$ pore water pressure gauge measuring point in 21 days, calculated by formula (2), is shown in Figures 11–13.

By contrast, Figures 11–13 show that the change of stress degree of consolidation of the tailings has the following characteristics:

1. In terms of stress degree of consolidation at a different depth, the average degree of consolidation of the underlying (67.51%) is slightly higher than that of the middle layer (62.32). Because the upper layers are greatly influenced by factors such as surface evaporation and sunshine, the average degree of consolidation (70.81%) is significantly larger.
2. In terms of the stress degree of consolidation at the same height and at different distances from the center of the drainage, the stress degree of consolidation value increase gradually with the distance, but the growth rate decreases in the same period of time.

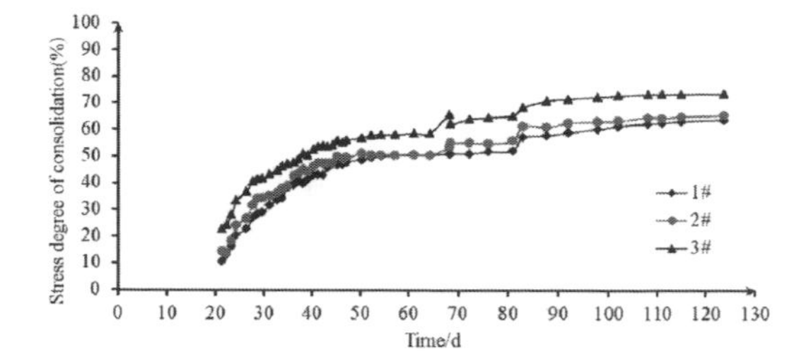

Figure 11. Stress degree of consolidation curve for measured No. $1^{\#}$–$3^{\#}$ points of pore water pressure.

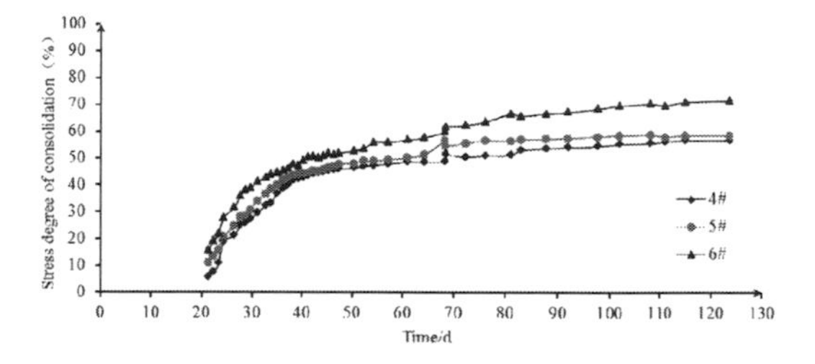

Figure 12. Stress degree of consolidation curve for measured No. $4^{\#}$–$6^{\#}$ measured points of pore water pressure.

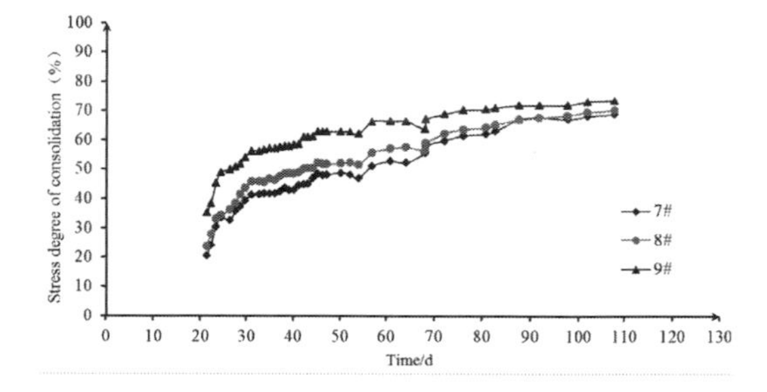

Figure 13. Stress degree of consolidation curve for measured No. $7^{\#}$–$9^{\#}$ measured points of pore water pressure.

4.2 *Analysis of strain consolidation degree*

The trial has installed nine settlement marks ($1^{\#}$–$9^{\#}$), and the settlement deformation monitoring last 117 days. Among them, the pile row period was 14 days, and the stopping discharge period was 104 days. The degree of consolidation of tailings is quantitatively analyzed by the degree of strain consolidation through the change in the volume of tailings. As tailings continue in heap row, post-stack tailings discharge mud can be regarded as the load of the upper part of the lower tailings, so the expression of its strain consolidation is:

$$\overline{U} = \frac{s_t - s_d}{s_{\infty} - s_d}, \tag{3}$$

where $\overline{U}$ is the strain consolidation degree, s_t from the *T-S* curve takes any *t* time settlement, s_d is the instantaneous settlement, and s_∞ is the final settlement.

The settlement s_t can obtain settlement–time curves of each moment when the degree of consolidation is calculated. Through formula (3), it is known that only the instantaneous settlement s_d and the final settlement s_∞ need to be calculated to calculate the strain consolidation degree of tailings. In this paper, the method of exponential function was used to calculate the consolidation degree of tailings, and the expression of average consolidation degree of the soil layer is:

$$\overline{U} = 1 - \alpha e^{-\beta t}, \tag{4}$$

where $\overline{U}$ is the average degree of consolidation, α, β are the parameters of the theoretical solution under different drainage conditions, α is the theoretical value, and β is the undetermined parameters.

Choosing any three times, t_1, t_2, and t_3, after the load stopped from the measured settlement–time (i.e., s–t) curve when calculating, and t_3–t_2 = t_2–t_1. According to formula (4), three equations can be written as follows:

$$s_\infty = \frac{s_3(s_2 - s_1) - s_2(s_3 - s_2)}{(s_2 - s_1) - (s_3 - s_2)}, \tag{5}$$

$$s_d = \frac{s_t - s_\infty(1 - \alpha e^{-\beta t})}{\alpha e^{-\beta t}}, \tag{6}$$

$$\beta = \frac{\ln\left(\dfrac{s_2 - s_1}{s_3 - s_2}\right)}{t_3 - t_2}, \tag{7}$$

On the basis of the measured data of settlement, settlement of 1# 9# standard strain curve of degree of consolidation of measuring points can be calculated from formulas (5)–(7), as shown in Figures 14–16.

By contrast, Figures 14–16 show that each settlement test point has the following characteristics:

1. In terms of different depths of strain degree of consolidation, the lower degree of consolidation, is higher than the upper value. At the test ending, the average strain consolidation degree of 1#–3# settlement measuring point (0.5 m stack depth) is 68.91%, the average strain consolidation degree of 4#–6# settlement measuring point (1.0 m stack depth) is 63.87%, and the average strain consolidation degree of 7#–9# settlement measuring point (tailings surface) is 61.27%.
2. In terms of the strain degree of consolidation in the same height, strain consolidation degree decreases with the increase of the distance from drainage channels; however, damping decreases in the same period of time.

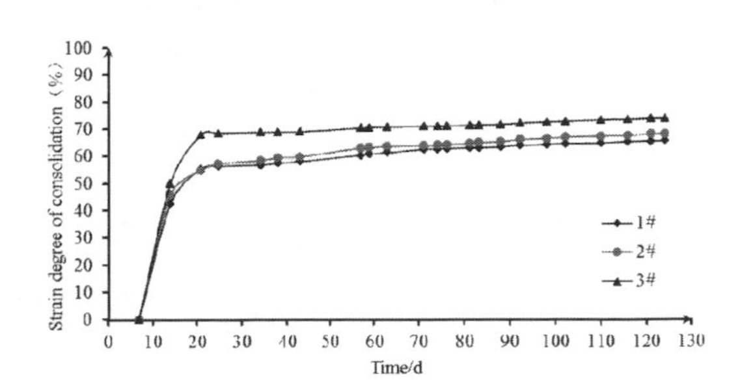

Figure 14. Strain degree of consolidation curve for measured No. 1#–3# measured points of settlement.

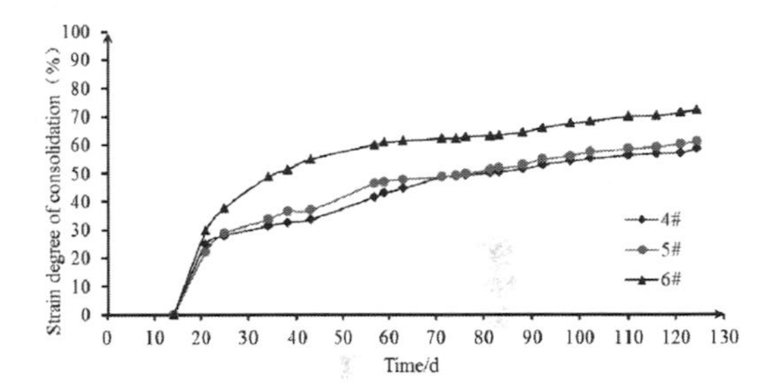

Figure 15. Strain degree of consolidation curve for measured No. 4#–6# measured points of settlement.

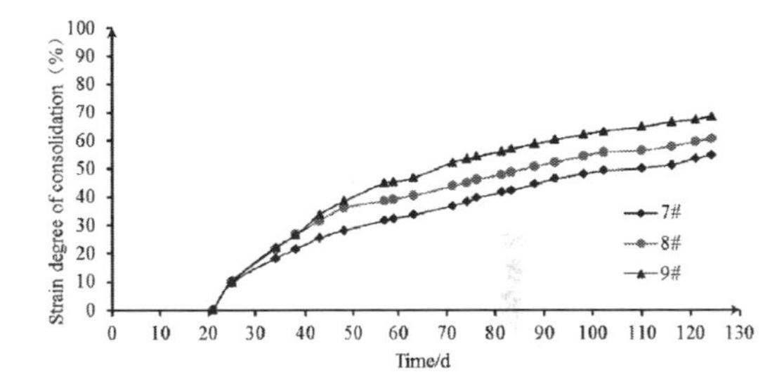

Figure 16. Strain degree of consolidation curve for measured No. 7#–9# measured points of settlement.

4.3 *Comparative analysis between stress degree of consolidation and strain consolidation degree*

The classical consolidation theory adopts small strain and linear hypothesis and regards stress and strain consolidation degree as equivalent. However, engineering practice and laboratory tests showed that the soil constitutive relationship is definitely not a simple linear relationship, especially due to the high compressibility of soft U_σ and U_ε, there are obvious differences. On the basis of the characteristics of pore water pressure and settlement mark arrangement, the average stress consolidation degree and average strain consolidation degree of the 0.5 m height (corresponding to the pore water pressure meter number: $4^{\#}$–$6^{\#}$ and corresponding settlement standard serial number: $1^{\#}$–$3^{\#}$) and 1.0 m height of the cabinet (corresponding to the pore water pressure meter number: $7^{\#}$–$9^{\#}$ and corresponding settlement standard serial number: $4^{\#}$–$6^{\#}$) were compared. The results are shown in Figures 15 and 16. Considering that the sedimentation will depress with tailings, the pore water pressure gauge was respectively fixed at 0.5 and 1.0 m, so the actual test position of the average strain consolidation degree is less than the average stress consolidation degree.

As Figures 17 and 18 show, the calculation results of two types of consolidation degree have minor variations, and these variations have the following characteristics:

1. Both the strain and stress average consolidation degrees at the depth of 1.0 m of tailings were better than those at the depth of 0.5 m. To average strain consolidation degree, the growth of the largest interval is before 21 days (the pile row), which means the upper load has a great influence on the calculation of strain consolidation degree; however, to average stress consolidation degree, the growth of the largest interval is 20–40 days, which illustrates that the drainage has more influence on the calculation of stress consolidation degree.

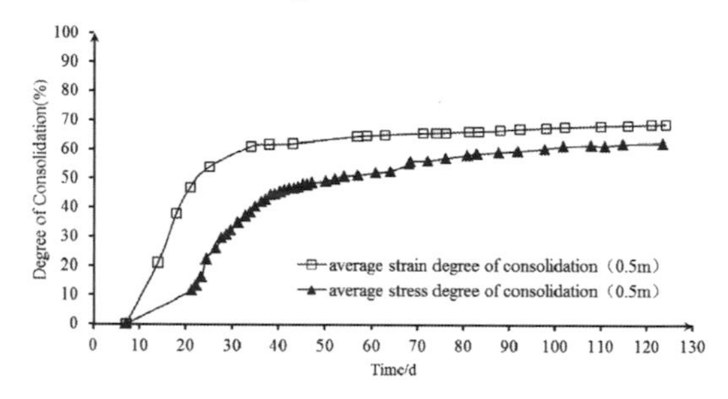

Figure 17. Average value of strain degree of consolidation and stress degree of consolidation for tailings depth of 0.5 m.

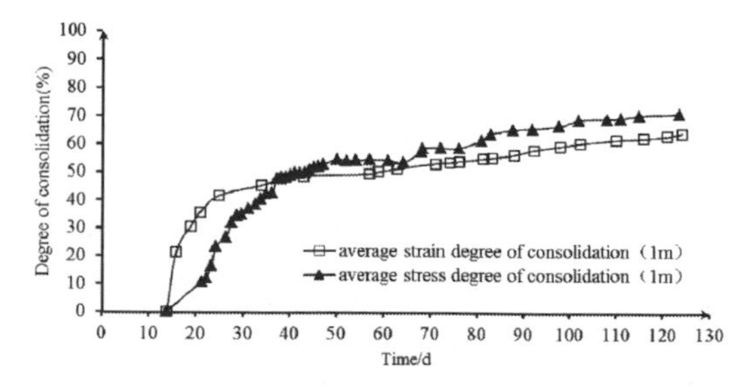

Figure 18. Average value of strain degree of consolidation and stress degree of consolidation for tailings depth of 1.0 m.

2. By comparing the average strain consolidation degree and the stress consolidation degree of the tailings at 0.5 m depth, the former is larger than the latter. The study (WEI Ru-long 1993) shows that the time lines of strain consolidation degree and stress consolidation degree were connected from beginning to end. Although there is difference in the middle of the process, the strain consolidation degree was always larger than the consolidation stress degree. The results of the experiment are consistent with those in Figures 22 and 23.

The average strain consolidation degree and stress consolidation degree of tailings at 1.0 m depth are still larger than the stress consolidation degree in 35 days. Then, it is just the opposite. After 35 days, the overlying water drained. As the distance between the tailings surface to the pore water pressure gauge at 1.0 m depth is only 0.3 m, the stress consolidation degree calculation results are larger than those of the strain consolidation degree, due to external factors such as evaporation, sunshine, and others.

5 CONCLUSION

This paper begins with the study of the drainage consolidation of tailings reservoir as the target area, bauxite tailings reservoir in the west of Guangxi, China. It adopts geotechnical composite drainage material as vertical drainage channels and simulated the mud consolidation effect by indoor pile model test and then reaches following conclusions:

1. In the condition of tailings slurry step piling, the drainage rate of geotechnical composite drainage material has the characteristics of time correlation, long-term trend, and periodic cycle, which is consistent with the characteristics of the autoregressive moving-average model. The model was established by SPSS and the computational fitted are $ARIMA(2,1,8) \times (0,0,0)$. Smooth coefficient of the model of R^2 and R^2 were 0.964 and 0.990. It shows that model fits very well. Model prediction analysis shows that the maximum water displacement has a tendency of gradually increase, so the performance of the geotechnical composite drainage material still has a large space.
2. The varying curve of pore water pressure has the following characteristics: lower-layer pore water pressure was significantly higher than that of the upper layer; for each additional 0.5 m on the tailings thickness, the pore water pressure value average increases about 5 kPa. The pore water pressure value of same height layout increases

gradually with the increase in the distance from the drainage channel; however, the growth rate decreases gradually. During the heap period, pore water pressure and height of water layer show good synchronization with the gradual increase of water level fluctuation wave. During the stop-discharge period, the pore water pressure value generally showed a decreasing trend, but at about 35 days, an obvious inflection point appeared, and the speed of decline also decreased significantly.

3. For the same height layout sedimentation test points, sedimentation value decreases gradually with the increase of the distance from drainage channels, but the decrease slows down gradually. The sedimentation curve can be divided into two stages: during the heap period, the lower part of the tailings soil is under staging load of upper part and the sedimentation value increased considerably; and in the stop-discharge period, the sedimentation value increased significantly and decreased.
4. After tailings mud stop heap 104 days, variation range of the average stress consolidation degree of each depth becomes 62.32–70.81%, and the average strain consolidation range of the variation is 61.27–68.91%. The time curves of strain consolidation degree and stress consolidation degree were connected at the beginning and end, but they were different in the middle of the process. The strain consolidation degree was always larger than the stress consolidation degree in the process.

REFERENCES

Du Chang-xue (2006). Study of Physical Mechanics Property and Solidified Technology of Bauxite Mine Slime. *D. Central South University*.

Hann E M & Miils S V, et al (2013). Refining estimates of foundation settlements due to consolidation of fine-grained tailings deposits. *C. Geotechnical and Geophysical Site Characterization 4-Proceedings of the 4th International Conference on Site Characterization 4,2*: 1707–1714.

Li Jie-quan & Ou Xiao-duo, et al (2012). An experimental study on self-weight consolidation of the red mud tailings placed in *karst regions. J. Journal of Guangxi U*niversity, 37(1):152–159.

Ofori P & AV Nguyen, et al (2011). Shear-induced flock structure changes for enhanced dewatering of coal preparation plant ta*ilings. J. Chemical Engineering Journal*, 172(2): 914–923.

Ou Xiao-duo & Liao You-fang, et al (2014). Large-strain consolidation calculation model on both horizontal and vertical drainage of wet filling t*ailings. J. Advances in Science and Technology of Water Resourc*es, 34(5): 28–34.

Shamsai A & Pak A (2007). Geotechnical characteristics of copper mine tailings: a case study. J. Geotechnical and Geological *Engineering, 2007, 25(5): 591–602.*

*Sui*ts LD & TC Sheahan, et al (2010). Laboratory Study of Soft Soil Improvement using Lime Mortar (Well Grade*d) Soil Columns. J. Geotechnical Testing Journal, 33(3)*: 225–235.

Wei Ru-long (1993). Derivation for coefficient of consolidation from settlement observation. *J. Chinese Jounal of Geotechnical Engineering, 15(2)*: 12–19.

Wickland B E & Wilson G W (2010). Hydraulic conductivity and consolidation response of mixtures of mine waste rock and tailings. *J. Canadian Geotechnical Journal, 47(4)*: 472–485.

Yin Guang-zhi & Zhang Qian-gui, et al (2011). Experimental study of nonlinear characteristics of deformation evolution for meso-scopic structure of tailings. *J. Chinese Journal of Rock Mechanics and Engineering, 30(8)*:1604–1612.

Zhang Nan & Zhu Wei (2013). Study of sedimentation and consolidation of soil particles in dredged slurry. *J. Rock and Soil Mechanics, 34(6):* 1681–1686.

Civil, Architecture and Environmental Engineering – Kao & Sung (Eds)
© 2017 Taylor & Francis Group, ISBN 978-1-138-02985-9

LQR control of across-wind responses of tall buildings using composite tuned mass dampers

Y.M. Kim & K.P. You
Department of Architecture Engineering, Long-Span Steel Frame System Research Center, Chonbuk National University, Jeonju, Korea

J.Y. You
Department of Architecture Engineering, Songwon University, Gwangju, Korea

S.Y. Paek & B.H. Nam
Department of Architecture Engineering, Chonbuk National University, Jeonju, Korea

ABSTRACT: Composite Tuned Mass Damper (CTMD) is a vibration control device, which consists of an active–passive tuned mass damper supported on the main vibrating structure. The performance of CTMD in suppressing wind-induced vibration of tall building is investigated. Optimum tuning frequency and damping ratio of a single Passive Tuned Mass Damper (PTMD) for minimizing the variance response of the damped main structure under random loads derived by Krenk are used for the optimum parameters for CTMD. Optimum parameters of CTMD with different mass ratios of an Active Tuned Mass Damper (ATMD) to PTMD are 0.01, 0.03, 0.05, 0.1, 0.3, and 0.5, which are chosen as the parameters of CTMD. The control force generated by the actuator of ATMD is estimated by a Linear Quadratic Regulator (LQR) controller. Fluctuating across-wind load, considered as a stationary random process, was simulated numerically using the across-wind load spectrum by Kareem. Comparing the controlled across-wind responses of a tall building with CTMD with those of an original tall building, the reduction rate of rms responses is 15–30%. Therefore, CTMD system is effective in mitigating excessive wind-induced vibrations of a tall building.

1 INTRODUCTION

Modern tall building are more lighter and slender with a lower natural frequency and damping ratio, so these tall buildings are more sensitive to wind-induced vibration. One way of mitigating such an excessive wind-induced vibration of tall buildings is by using vibration control devices. Passive Tuned Mass Damper (PTMD) is a classical vibration control device, which consists of a mass, a spring, and a damper supported on the main vibrating structure (Den Hartog 1956). The original idea of PTMD was proposed by Frahm in 1909, who invented a vibration control device called a vibration absorber using a spring-supported mass without damper (Frahm 1909). It was effective when the absorber's natural frequency was close to the excitation frequency. However, it was difficult tuning the absorber's natural frequency to the excitation frequency. This control device was improved by introducing a damper in the spring-supported mass (Ormondroyd et al. 1928). Later, Den Hartog derived optimum tuning frequency and damping ratio for the undamped main structure under harmonic load (Den Hartog 1956). While Den Hartog considered harmonic loading only, Warburton and Ayroinde derived optimum parameters of PTMD for the undamped main system under harmonic and white noise random excitations (Ayoringde 1980, Warburton 1981, 1982). Krenk derived the optimum parameters of PTMD for the damped main structure under random excitations with the condition that the mass ratio is small and the main structure's damping ratio is less than that of PTMD (Krenk et al. 2008). And a number of PTMDs have been installed in tall buildings to suppress wind-induced vibrations of tall buildings (McNamara 1977, Housner et al. 1997). The Center Point Tower in Sydney is one of the first PTMDs in a building, and a 400-ton PTMD has been installed for the Citicorp Center in New York. Another PTMD has been designed in the John Hancock Tower in Boston (Housner et al. 1997). All of the PTMDs have been installed to mitigate wind-induced vibrations. At that time, it was accepted as a fact that the performance of PTMD could be improved by incorporating a feedback controller through the use of an actuator as an active control force in the design of PTMD, which was called Active Tuned Mass Damper (ATMD) (Housner et al.1997). In 1972, Yao introduced the modern control theory into vibration control of a civil engineering structure, and many auxiliary control devices combining with modern optimal control theory have been developed for mitigating the excessive wind-induced vibrations of tall buildings (Yao 1972). One way of ATMD design for reducing wind-induced vibration of tall building using assumed deterministic harmonic wind loading was presented by Chang and Soong in 1980 (Chang et al. 1980). That was the first active control study for mitigating wind-induced vibration of tall buildings with ATMD using a Linear Quadratic Regulator (LQR) controller. Since then, many studies have been advanced for obtaining optimal control force for reducing wind-induced vibration of tall buildings on the basis of the modern optimal control technique (Ankireddi et al. 1996, 1997). However, the fact that ATMD is superior to PTMD reducing wind-induced vibration of tall buildings is still question (Ricciardelli et al. 2003). From the viewpoint of the modern optimal control theory, fluctuating across-wind load acting on a tall building can be treated as a stationary random process, and a constant power spectrum could be considered as a system noise. Then, many advanced studies for mitigating wind-induced vibration of tall buildings on the basis of the modern optimal control theory, including Linear Quadratic Gaussian (LQG), H_2 and $H_\perp$ have been developed, and a number of tall buildings are currently implemented with active control device systems (Yang et al 2002, 2003). Nishimura and Wang et al. pointed out that the disadvantage of a single PTMD is its error in tuning the natural frequency of PTMD to that of the main structure and fitting the optimum damping ratio of PTMD. The size restriction of PTMD limits the vibration control effect (Nishimura et al.1994, 1998, Wang et al. 1999).

In overcoming such problem of PTMD, an active–passive composite tuned mass damper, that is, a Composite Tuned Mass Damper (CTMD), was proposed by Nishimura and Wang et al. (Nishimura et al. 1994, 1998, Wang et al. 1999). CTMD system consists of ATMD, which is attached to the PTMD supported on the main vibrating structure, as shown in Figure 1.

In this study, the performance of CTMD for suppressing across-wind induced vibrations of tall building is investigated. The control force generated by the actuator of ATMD is estimated by a Linear Quadratic Regulator (LQR) controller. Fluctuating across-wind load was simulated numerically using the across-wind load spectrum proposed by Kareem (Kareem 1982). Dynamic across-wind responses of tall buildings with CTMD are estimated and compared with the responses of a single PTMD and the original tall building without CTMD. The controlled rms responses with CTMD is reduced about 15–30% rms response of the original tall building without CTMD, which is close to the controlled response with PTMD. Therefore, CTMD systems are effective in mitigating excessive wind-induced vibrations of tall building.

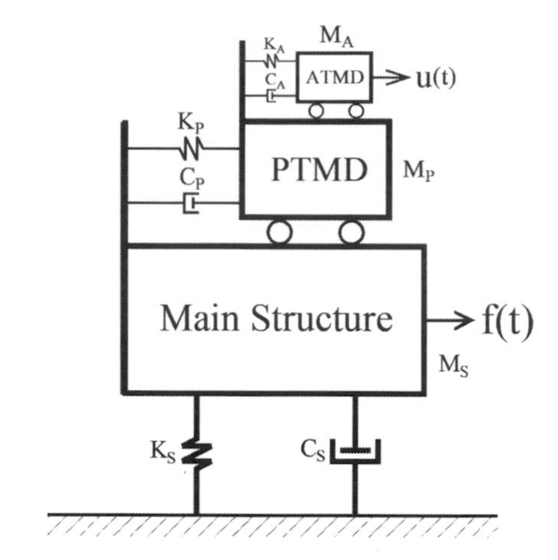

Figure 1. Tall building CTMD model.

2 EQUATIONS OF MOTION

Dynamic response analysis procedure can be simplified if the contribution of higher modes of the tall building is ignored, so the response is represented by the motion of the first mode (Kareem et al. 1995, Ankireddi et al. 1996). Therefore, tall building-CTMD system can be modeled as the first-mode generalized SDOF/CTMD system, as shown in Figure 1.

The linear dynamic equations of motion of the CTMD system can be written as:

$$M_S\ddot{X}_S + C_S\dot{X}_S + K_SX_S - C_P\dot{X}_P - K_PX_P = f(t) \quad (1)$$

$$M_p\ddot{X}_p + C_p\dot{X}_p + K_pX_p - C_p\dot{X}_A - K_AX_A + M_P\ddot{X}_S = u(t) \quad (2)$$

$$M_A\ddot{X}_A + C_A\dot{X}_A + K_AX_A + M_A\ddot{X}_S + M_A\ddot{X}_P = -u(t) \quad (3)$$

These equations of motion can be rewritten in the state space variable representation as:

$$\dot{X} = AX + Bu(t) + Ff(t) \quad (4)$$

where

$$X = \begin{bmatrix} X_S & X_P & X_A & \dot{X}_S & \dot{X}_P & \dot{X}_A \end{bmatrix} \quad (5)$$

$$A = \begin{bmatrix} 0 & 0 & 0 & 1 & 0 & 0 \\ 0 & 0 & 0 & 0 & 1 & 0 \\ 0 & 0 & 0 & 0 & 0 & 1 \\ -\frac{K_S}{M_S} & \frac{K_P}{M_S} & 0 & -\frac{C_S}{M_S} & \frac{C_P}{M_S} & 0 \\ \frac{K_S}{M_S} & -\left(\frac{K_P}{M_P}+\frac{K_P}{M_S}\right) & \frac{K_A}{M_P} & \frac{C_S}{M_S} & -\left(\frac{C_P}{M_P}+\frac{C_P}{M_S}\right) & \frac{C_A}{M_P} \\ 0 & \frac{K_P}{M_P} & -\left(\frac{K_A}{M_A}+\frac{K_A}{M_P}\right) & 0 & \frac{C_P}{M_P} & -\left(\frac{C_A}{M_A}+\frac{C_A}{M_P}\right) \end{bmatrix} \tag{6}$$

$$B = \begin{bmatrix} 0 & 0 & 0 & 0 & \frac{1}{M_P} & -\left(\frac{1}{M_A}+\frac{1}{M_P}\right) \end{bmatrix}^T \tag{7}$$

$$F = \begin{bmatrix} 0 & 0 & 0 & \frac{1}{M_S} & -\frac{1}{M_S} & 0 \end{bmatrix}^T \tag{8}$$

where M_S, C_S, and K_S are generalized mass, damping, and stiffness coefficients corresponding to the first mode of the tall building;

M_p, C_p, and K_p are the mass, sdamping, and stiffness coefficients of the PTMD;

M_A, C_A, and K_A are the mass, damping, and stiffness coefficients of the ATMD;

X_S and $\dot{X}_S$ are the main structural displacement and velocity;

X_P and $\dot{X}_P$ are the relative displacement and velocity of the PTMD with respect to the main structure;

X_A and $\dot{X}_A$ are the relative displacement and velocity of the ATMD with respect to the PTMD;

u(t) is active control force;

f(t) is the generalized fluctuating across-wind load associated with the first mode.

3 OPTIMUM PARAMETERS OF PTMD

While the basic concept of PTMD for reducing vibrations of the main structure has been well established, the optimum parameters of PTMD could be different for different structures and external loading conditions (Ayoringde (1980), Warburton (1981, 1982)). Warburton derived the optimum parameters for tuning the natural frequency and damping ratio of PTMD for the undamped main structure under stationary random load (Warburton 1982). Krenk derived the optimum parameters of PTMD for the damped main structure under the condition of mass ratio is small and the primary structure's damping ratio is less than that of PTMD as (Krenk et al. 2008):

$$f_{opt} = \frac{1}{1+\mu} \tag{9}$$

$$\xi_{opt} = \frac{\sqrt{\mu}}{2} \tag{10}$$

where f_{opt} is the optimum tuning frequency ratio, ξ_{opt} is the optimum damping ratio, and μ is the mass ratio.

4 OPTIMUM PARAMETERS OF CTMD

The optimum parameters of ATMD for minimizing rms responses of the main structure are similar to those of PTMD. It was known that tuning the frequencies of PTMD to the fundamental natural frequency in the main structure is more effective than tuning it to different natural frequencies (Kareem et al. 1995). Accordingly, the natural frequency and applied damping ratio of ATMD are tuned to the natural frequency and damping ratio of PTMD as in Eq. (9) and Eq. (10) with small mass ratios.

5 LINEAR QUADRATIC REGULATOR CONTROLLER

The LQR control method is a widely used modern optimal control technique in structural vibration control problems (Chang et al. 1980). In LQR control law, all continuous time state-space variables are available, and linear dynamic equations of motion of the system can be written in terms of the state-space formulation as in Eq. (4). The external force term in Eq. (4) can be treated as a system noise input and hence Eq. (4) can be written as (Dorato et al. 1995):

$$\dot{X}(t) = AX(t) + Bu(t) \tag{11}$$

The objective of LQR control law is to find out a state-feedback optimal control force, u(t), that minimizes the deterministic cost functional J, maintaining the state close to the zero state. The cost functional (J) is given by:

$$J = \int_0^\infty (X(t)^T QX(t) + u(t)^T Ru(t))dt \quad (12)$$

where Q is a positive semi-definite state weighting matrix and R is a positive definite control weighting matrix. The term X(t)T QX(t) in Eq. (12) is a measure of control accuracy and the term u(t)TRu(t) is a measure of control effort (Dorato et al. 1995, Lewis et al. 2012). Minimizing J with keeping the system response and the control effort close to zero needs appropriate choice of the weighting matrices Q and R (Suhardjo et al. 1992). If it is desirable that the system response be small, then large values for the elements of Q should be chosen by selecting the matrix Q to be diagonal and to make the large value of the diagonal element for any respective state variable to be small (Suhardjo et al. 1992). If the control energy needs to be small, then large values of the elements of R should be chosen. The state-feedback optimal control force, u(t), is derived as (Dorato et al. 1995, Lewis et al. 2012):

$$u(t) = -KX(t) \quad (13)$$

where $K = R^{-1}B^TP$

In Eq. (13), K is called an optimal controller gain and P is the unique, symmetric, positive semi-definite solution to the Algebraic Riccati Equation (ARE) given by Dorato et al. (1995) and Lewis et al. (2012):

$$A^TP+PA-PBR^{-1}B^TP+Q = 0 \quad (14)$$

Then, the closed-loop system using the optimal control force, u(t), becomes:

$$X = (A-BK)X(t)= AcX(t) \quad (15)$$

where Ac is the closed-loop system matrix

In LQR control law, the cost functional J keep minimizing, indicating that a larger value of state weighting matrix Q makes the state X(t) smaller, that is, the poles of the closed-loop system matrix Ac is further left in the s-plane so that the state X(t) decays faster to zero (Lewis et al. 2012).

6 NUMERICL SIMULATION OF A FLUCTUATING ACROSS-WIND LOAD

The dynamic along-wind response of tall buildings can be estimated reasonably by a gust factor approach (Solari 1993). However dynamic across-wind response cannot be estimated by a gust factor approach. The complex nature of the across-wind loading, resulting from an interaction of incident turbulence, flow separation, vortex-shedding, and unsteady wake development, has prevented theoretical prediction from estimation (Kareem 1982). The fluctuating across-wind load can be treated as a stationary random process, which can be simulated numerically in the time domain using the across-wind load power spectral density data. That is particularly useful for some response estimations that are more or less narrow-banded random processes, such as the across-wind response of tall buildings (Shinozuka 1987). The numerical simulation procedure presented in this work is taken from Shinozuka (Shinozuka 1987):

$$f(t) = \sum_{k=1}^{N} \sqrt{S_F(\omega_1)\Delta\omega}\cos(\omega_k t + \phi_t) \quad (16)$$

where $S_F(w_1)$ is the value of the spectral density of across-wind load corresponding to the first modal resonant frequency.

$\Delta\omega=(\omega_u-\omega_1)/N$;

$\omega_k = \omega_l + (k-1/2)\Delta\omega$;

ω_u = upper frequency of $S(\omega)$;

ω_l = lower frequency of $S(\omega)$;

Φ_t = uniformly distributed random numbers between 0 and 2π;

N = number of random numbers.

The across-wind load power spectral density used in Eq. (16) is that proposed by A. Kareem

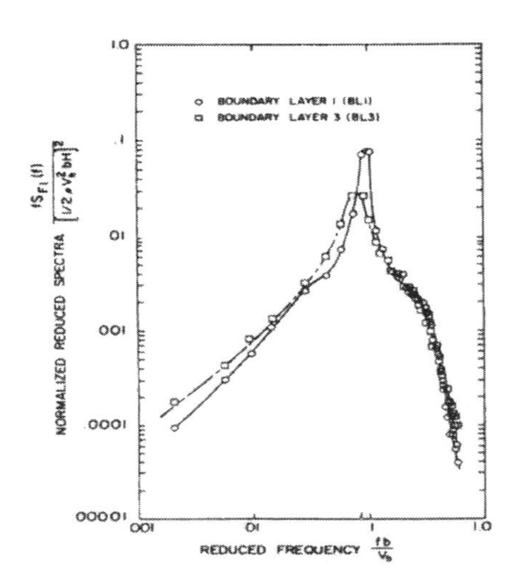

Figure 2. Normalized reduced spectra of across-wind load in a suburban (BL1) and urban (BL3).

(Kareem 1982). Kareem used a 5 sq in. (127 mm^2) square, 20-in. (508 mm^2) tall prism model and eight pressure transducers, integrating eight simultaneously-monitored channels of pressure data on a building model's surface and obtained the normalized across-wind load spectra of the across-wind forcing function on a square cross section tall building that is exposed to urban and suburban atmospheric flow conditions as shown below:

7 NUMERICL EXAMPLE

This numerical example is taken from "Numerical Example" by A. Kareem (Kareem 1982). A tall building was chosen, whose height (H) is 180 m, width (B) is 30 m, depth (D) is 30 m, natural frequency (f_1) is 0.2 Hz, critical damping ratio is 0.01, air density is 0.973 kg/m^3, hourly mean wind speed at the building height (V_h) is 24.4 m/s, the reduced velocity (V_h/f_B), corresponding to the mean hourly wind speed, is 4.0, generalized mass of a first mode shape is 10,942,500 kg, and $S_f(f_n)$ is 3.149×10^8 kg^2/Hz. The optimum parameters of CTMD are considered to have the same values as for the PTMD. The optimum parameters of PTMD are: mass ratio (μ) = 0.01, tuning frequency (f_{opt}) = 1.0, and damping ratio (ξ_{opt}) = 0.05. The numerically simulated across-wind load and response without CTMD are shown in Figures 3 and 4. The rms displacement response without CTMD shown in Figure 4 is 0.0047 m, which is a good approximation to that of Kareem's closed-form response of 0.0040 m.

7.1 *Across-wind responses with PTMD*

The rms response with a single PTMD with optimum parameters is shown in Figure 5, which shows that the controlled rms response is reduced with reduction ratio of 28% comparing with that of the original tall building without PTMD.

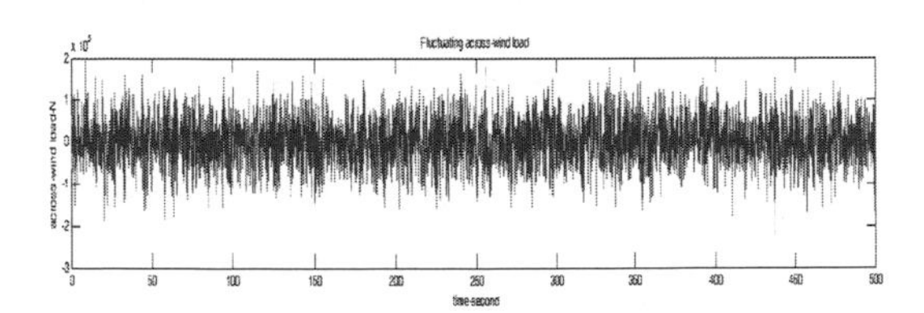

Figure 3. Simulated across-wind load in time domain.

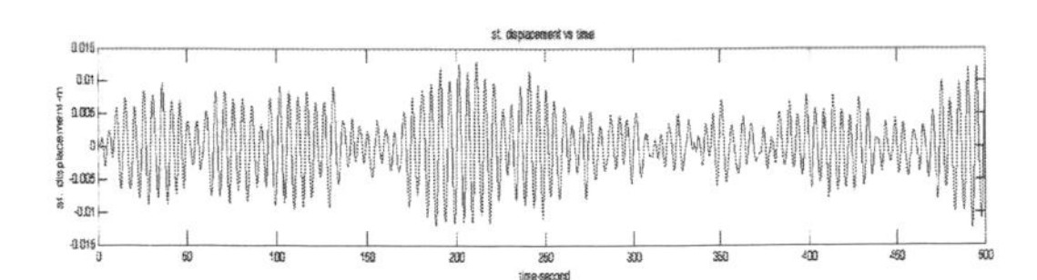

Figure 4. Across-wind response without CTMD (rms = 0.0047 m).

7.2 *Across-wind responses with CTMD*

For estimating LQR controller, the weighting matrices Q and R are selected as:

$$Q = 1.0 * 10^8 * \begin{bmatrix} 100 & 0 & 0 & 0 & 0 & 0 \\ 0 & 1 & 0 & 0 & 0 & 0 \\ 0 & 0 & 1 & 0 & 0 & 0 \\ 0 & 0 & 0 & 100 & 0 & 0 \\ 0 & 0 & 0 & 0 & 1 & 0 \\ 0 & 0 & 0 & 0 & 0 & 1 \end{bmatrix} \quad R = [1.0 * 10^{-42}]$$

The dynamic across-wind responses of tall building with CTMD, which have a different mass ratio (μ_{AP}) of ATMD to PTMD are 0.01, 0.03, 0.05, 0.1, 0.3, and 0.5, which are presented in Figures 6–11.

As shown above, comparing the controlled rms responses with those of the original tall building, about 15–30% reduction effect was presented, which shows close to the reduction effect of using PTMD. And the effectiveness of CTMD is increased as the mass ratio of ATMD to PTMD is increased within an allowable limit. That is, as

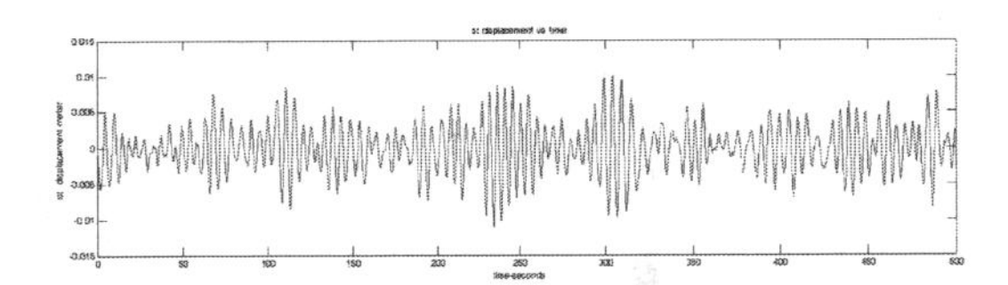

Figure 5. Across-wind response with PTMD (rms = 0.0034 m).

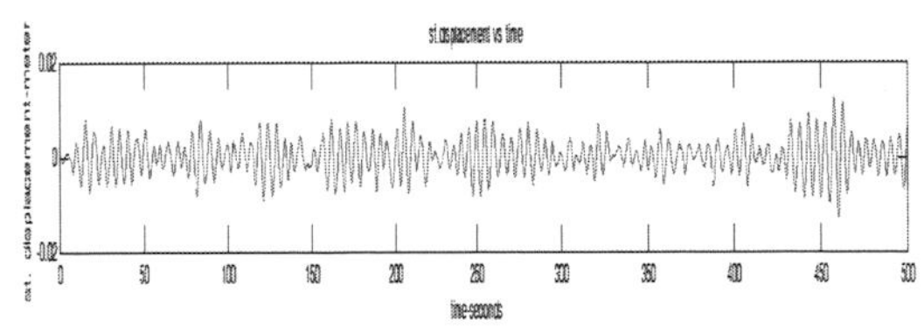

Figure 6. Across-wind responses with CTMD (μ_{AP} = 0.01, rms = 0.0037 m).

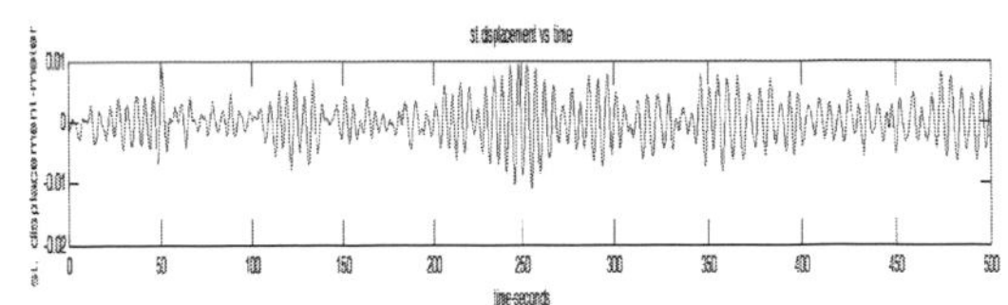

Figure 7. Across-wind responses with CTMD (μ_{AP} = 0.03, rms = 0.0034 m).

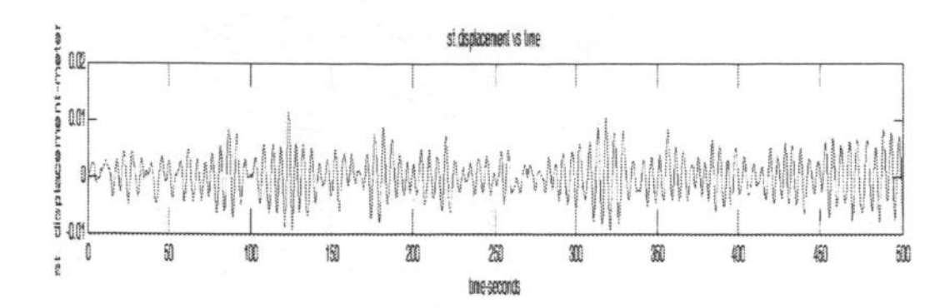

Figure 8. Across-wind responses with CTMD ($\mu_{AP} = 0.05$, rms = 0.0035 m).

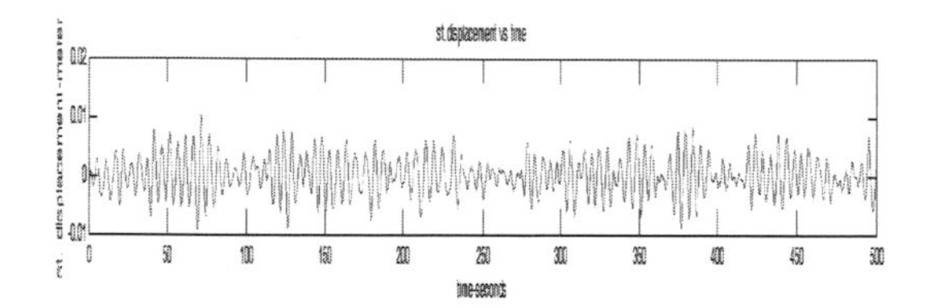

Figure 9. Across-wind responses with CTMD ($\mu_{AP} = 0.1$, rms = 0.0033 m).

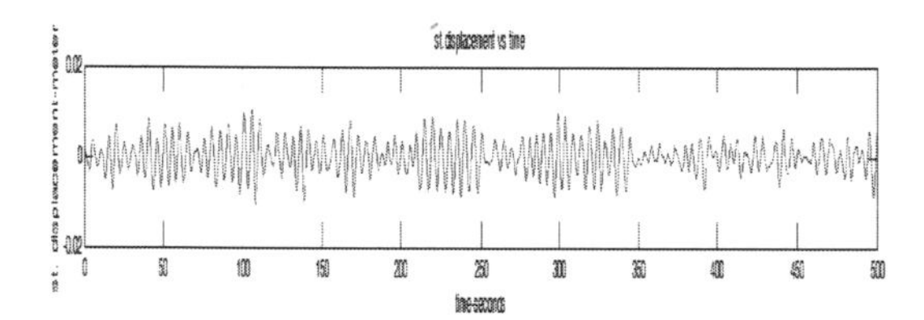

Figure 10. Across-wind responses with CTMD ($\mu_{AP} = 0.3$, rms = 0.0037 m).

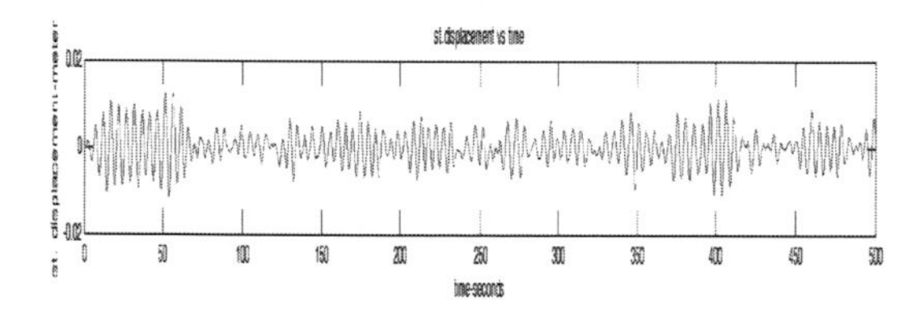

Figure 11. Across-wind responses with CTMD ($\mu_{AP} = 0.5$, rms = 0.0040 m).

the mass ratio of ATMD to PTMD is increased 10 times with the optimum tuning frequency ratio and damping ratio, a reduction effect of 10% is presented. However, an adverse effect of 8–15% increased response is obtained when the mass ratio is increased to 0.3 and 0.5. Therefore, CTMD system is effective for suppressing wind-induced vibrations of tall buildings.

8 CONCLUSIONS

The performance of Composite Tuned Mass Dampers (CTMDs) with LQR controller for mitigating across-wind responses of tall buildings is investigated. Optimum values of tuning frequency ratio, damping ratio, and mass ratio for PTMD were used for the optimum parameters for ATMD of CTMD. The fluctuating across-wind load was simulated numerically using the across-wind load spectrum by Kareem. Comparing the rms response with CTMD with that of the original tall building without CTMD, the rms responses of a CTMD is reduced about 15–30% than that of the original tall building, which is close to that of PTMD. Therefore, CTMD system with an LQR controller, which has optimum parameters, is effective in mitigating excessive wind-induced vibration of tall buildings.

ACKNOWLEDGMENT

This work was supported by the National Research Foundation of Korea (NRF) grant funded by the Korea government (MEST) (No. 2011-0028567).

REFERENCES

Ankireddi, S. & Yang, H.T.Y. 1996. Simple atmd control methodology for tall building subject to wind loads. *Journal of Structural Engineering,* ASCE 122: 83–91.

Ankireddi, S. & Yang, H.T.Y. 1997. Multiple objective lqg control of wind-excited building, *Journal of Structural Engineering*, ASCE 123: 943–951.

Ayoringde, E.O. & Warburton, G.B. 1980. Minimizing structural vibrations with absorbers. Earthquake Engineering and Structural Dynamics 8: 219–236.

Chang, J.C.H. & Soong, T.T. 1980. Structural control using active tuned mass dampers, *Journal of Engineering Mechanics* ASCE 106: 1091–1098.

Den Hartog.J.P. 1956. Mechanical Vibration. Fourth Edition, McGraw-Hill. New York. (Reprinted by Dover, NewYork, 1985.)

Dorato, P. Abdallah, C. & Cerone, V. 1995. Linear-Quadratic Control. *Prentice-Hall, New Jersey*.

Frahm, H. 1909. Device for damped vibration of bodies. U.S. Patent No. 989958. October 30.

Housner, G.W. Bergman, L.A. Caughey, T.K. Chassiakos, A.G. Claus, R.O. Masri, S.F. Skelton, R.E., Soong, T.T., Spencer, B.F. & Yao, J.T.P. 1997. Structural control: past, present, and future. *Journal of Engineering Mechanics (ASCE)* 123(9): 897–971.

Kareem, A. & Kline, S. 1995. Performance of multiple mass dampers under random loading. *Journal of Structural Engineering* ASCE 121(2): 348–361.

Kareem, A. 1982. Across response of buildings, *Journal of the SD, ASCE* 108(ST4): 869–887.

Krenk, S. & Hogsberg, J. 2008. Tuned mass damped structures under random load. *Probabilistic Engineering Mechanics* 23: 408–415.

Lewis, F.L. & Vrabie, D.L. Syrmmos, V.L. 2012. Optimal Control, Third Edition, *John Wiley & Sons.*

McNamara, R.J. 1977. Tuned mass dampers for buildings, *Journal of the Structural Division* 103: 1785–1798.

Nishimra, I. Sakamoto, M. Yamada, T. Koshika, N. & Kobori, T. 1994. Acceleration feedback method applied to active—passive composite tuned mass damper. *Journal of Structural Control* 1(1–2): 103–116.

Nishimra, I. Yamada, T. Sakamoto, M. & Kobori, T. 1998. Control performance of active-passive composite tuned mass damper. *Journal of Smart Materials & Structures* 7(5): 637–753.

Ormondroyd, J. & Den Hartog, J.P. 1928. The theory of the dynamic vibration absorber, *Transactions of ASME*, Vol.50 (APM-50–7): 9–22.

Ricciardelli, F. Pizzimenti, A.D. & Mattei, M. Passive and active mass damper control of the response of tall buildings to wind gustiness. 2003. *Journal of Engineering Structures*. 25: 1199–1209.

Shinozuka, M. 1987. Stochastic fields and their digital simulation, edited by Schueller, G.I. & Shinozuka, M. Martinus Nijhoff Publishers, *Stochastic Methods in Structural Dynamics* 93–133.

Solari, G. 1993. Gust buffeting, *Journal of Structural Engineering, ASCE* 119(2): 383–398.

Suhardjo, J. Spencer, B.F. & Kareem, A. 1992. Active control of wind-excited buildings: A frequency domain based design approach. *Journal of Wind Engineering and Industrial Aerodynamics*, 41–44:1985–1996.

Wang, C.M. Yan, N. & Balendra, T. 1999. Control on dynamic structural response using active-passive composite-tuned mass dampers. *Journal of Vibration and Control* 5: 475–489

Warburton, G.B. 1981. Optimum absorber parameters for minimizing vibration response, Earthquake Engineering and Structural Dynamics 9: 251–262

Warburton, G.B. 1982. Optimum absorber parameters for various combinations of response and excitation parameters, Earthquake Engineering and Structural Dynamics 10: 381–401.

Yang, J.N, Lin, S. & Jabbari, F. 2003. H_2-based control strategies for civil engineering structures, *Journal of Structural Control* 10: 205–230.

Yang, J.N. Lin, S. J.H.Kim. & Agrawal, A.K. 2002. Optimal design of passive energy dissipation systems based on H and H_2 performances, *Journal of Earthquake Engineering and Structural Dynamics* 31: 921–936.

Yao, J.T.P. 1972. Concept of structural control, *Journal of SD*, ASCE 98: 1567–1574.

Civil, Architecture and Environmental Engineering – Kao & Sung (Eds)
© 2017 Taylor & Francis Group, ISBN 978-1-138-02985-9

Time-dependent reliability analysis of steel fiber-reinforced concrete beams under sustained loads

Yi Li
College of Resources and Civil Engineering, Northeastern University, Shenyang, China

Kiang Hwee Tan
National University of Singapore, Singapore

Shi-meng Yu
College of Resources and Civil Engineering, Northeastern University, Shenyang, China

ABSTRACT: This study was aimed at examining the time-variable flexural capacity of Steel Fiber Reinforced Concrete (SFRC) sections subjected to long-term loading. Based on tests results of two series of steel fiber reinforced concrete beams subjected to sustained loading over a period of ten years, the time-variable reliability indices for flexural strength were obtained considering the time-dependent resistance of SFRC beams under applied loads. Results show that under a specified sustained load, the reliability index is significantly higher in SFRC beams than in beams without steel fibers. With the same steel fiber content, the larger the applied load, the smaller is the reliability index. For a service life of 100 years, the reliability index of SFRC beams was found to first increase and then decrease; with the reliability index reaches the maximum value when $t = 2a$ and steel bars begin to rust, and the reliability index significantly decreased when $t = 37.8a$

1 INTRODUCTION

Steel Fiber-Reinforced Concrete (SFRC) has wide applications in the field of civil engineering as it can effectively improve the cracking and toughening characteristics, thus eliminating the inherent defects of concrete. On the long-term behavior of structural elements under service loads, Goistseonc et al. (2009) had studied the time-dependent characteristics of the flexural capacity, curvature and neutral axis depth of reinforced concrete beams placed in sodium chloride solution with a concentration of 5%, and subjected to sustained loads for 70 days. Lee et al. (2013) studied the middle span displacement, section stiffness, and stresses and strains in reinforced concrete beams subjected to repeated loads at frequencies of 0.2 and 1 Hz for 80 days. Creep was found to influence the mechanical properties and structural response of long-span concrete-filled steel tube arch bridges (Ma et al. 2015). The effect of creep and shrinkage on the deflection and structural response of steel-concrete composite bridges after 300 days of service under different ambient temperature and humidity has also been studied (Hui et al. 2015).

Based on flexural tests, the deformation and cracking characteristics of the steel-polyester fiber beams under sustained load over 17 months was determined by Emilia et al. (2013). The reliability index of in-service pre-stressed concrete bridge was calculated by Jin (2013) by using neural network with sampling methods. The time-dependent reliability of prestressed sleeper has been analyzed by Monte Carlo Method (Saeed et al. 2011).

Previous studies dealt with concrete elements subjected to sustained loads sustained over a relatively short period. Adding steel fibers makes the material more complex, it is difficult to ensure the stability in sustaining loads for a long period. Thus, for SFRC beams, the change in reliability index as a result of long sustained load period being influenced by many factors has not been reported.

This paper is based on flexural test results from the SFRC beams subjected to sustained loads over a period of ten years. The time-variable load effect and resistance level of SFRC beams under long-term load is considered for two groups of beams; Group I beams with the same sustained load level but with different steel fiber content, and Group II with the same fiber content but with different sustained load levels. The time-dependent flexural resistance of SFRC sections is established for the evaluation of time-dependent reliability.

2 STATISTICAL PARAMETERS

2.1 *Tensile strength of concrete*

The average and standard deviation of time—dependent tensile strength for concrete grade C10~C110 are respectively (Faxing et al. 2004):

$$\mu_{f_t(t)} = 0.24\mu_{f_{cu}(t)}^{2/3} \tag{1}$$

$$\sigma_{f_t(t)} = 0.24\sigma_{f_{cu}(t)}^{2/3} \tag{2}$$

where $f_{cu}(t)$ = time-dependent axial compressive strength of concrete cube.

2.2 *Effective cross-section area of reinforcement*

Due to corrosion caused by carbonation, the reduced steel reinforcement area at time t is taken as (Zhenping et al. 2009):

$$A_s(t) = \begin{cases} A_{s0} & , \ (t \le t_1) \\ A_{s0}[1 - \frac{\lambda}{r_0}(t - t_1)]^2 , & (t > t_1) \end{cases} \tag{3}$$

where A_{s0} = initial area of rebar; λ = rebar corrosion rate; r_0 = initial radius of rebar; t_1 = initial corrosion time, calculated according to Yi et al. (2009).

2.3 *Bond strength between rebar and concrete*

With the extended service time, $\varphi(t)$ is used to account for the loss in bond strength between steel and concrete; $\varphi(t) = 0$ when the steel bars are not corroded, and 0.95 while the steel began to rust.

2.4 *Uncertainty coefficient of calculation model*

The uncertainty coefficient of calculation model P can be expressed in general as:

$$P = \frac{K_s}{K_j} \tag{4}$$

where K_s = actual resistance of the structure and K_j = resistance calculated by code using real geometric parameters and mechanical properties of materials.

3 TIME-DEPENDENT RELIABILITY ANALYSIS OF FLEXURAL CAPACITY

The flexural capacity of SFRC can be calculated as:

$$M_{fu} = f_{fc}bx\left(h_0 - \frac{x}{2}\right) + f'_y A'_s (h_0 - a'_s) - f_{ftu}bx_t\left(\frac{x_t}{2} - a\right) \tag{5}$$

where x is the depth of concrete compression zone, given by:

$$f_{fc}bx = f_y A_s - f'_y A'_s + f_{ftu}bx_t \tag{6}$$

in which M_{fu} = design value of the flexural capacity; f_{fc} = design value of axial compressive strength; f_y, f'_y = design value of tensile and compressive strength of longitudinal reinforcement, respectively; A_s, A_s' = total area of tensile and compressive bars, respectively; b = section width; h_0 = section effective height; a = distance from the centroid of longitudinal tension reinforcement to the tension edge of the section; a_s' = distance from the centroid of longitudinal reinforcement in compression zone to the edge of the section; x_t = height of equivalent stress in tensile zone; f_{ftu} = tensile strength of equivalent stress block of steel fiber reinforced concrete in tensile zone.

3.1 *Height of tensile zone of concrete*

The tensile stress distribution is simplified as an equivalent rectangle stress block, with height as (CECS38-2004):

$$x_t = h - \frac{x}{\beta_1} \tag{7}$$

where x = height of equivalent rectangular stress pattern in compression zone

3.2 *Tensile strength of tensile zone*

Tensile strength of the equivalent rectangular tensile stress block can be calculated as:

$$f_{ftu} = f_t \beta_{tu} \lambda_f \tag{8}$$

where β_{tu} = effect of steel fiber on tensile strength; λ_f = characteristic value of steel fiber content. Over time, considering the reduced strength of material and the reduced effective cross section of the concrete and reinforcing steel bars, along with Eqs. (1) - (8), the time-dependent flexural capacity is:

$$R(t) = p \cdot \varphi(t) \begin{bmatrix} f_{fc}(t)bx(t)\left(h_0 - \frac{x(t)}{2}\right) + f'_y A'_s(t)(h_0 - a'_s) \\ -f_t(t)\beta_{tu}\lambda_f b\left(h - \frac{x(t)}{\beta_1}\right)\left(\frac{h - \frac{x(t)}{\beta_1}}{2} - a\right) \end{bmatrix} \tag{9}$$

where x can be calculated as:

$$x(t)=\frac{f_y A_s(t)-f_y' A_s'(t)+f_t(t)\beta_{tu}\lambda_f bh}{f_{fc}(t)b+\frac{1}{\beta_1}f_t(t)\beta_{tu}\lambda_f b} \tag{10}$$

in which $f_{fc}(t)$ = axial compressive strength of SFRC; $A_s(t)$, $A_s'(t)$ = area of longitudinal tensile reinforcement and compression bars, respectively; $f_{ftu}(t)$ = tensile strength of equivalent rectangular stress block of SFRC in tensile zone; $f_t(t)$ = tensile strength of steel fiber reinforced concrete; at time t.

The time-dependent load effect is expressed by $S(t)$, and the time-dependent limit state function of SFRC beams is

$$Z(t)=p\cdot\varphi(t)\times\begin{bmatrix} f_{fc}(t)bx(t)\left(h_0-\frac{x(t)}{2}\right)+f_y' A_s'(t)(h_0-a_s') \\ -f_t(t)\beta_{tu}\lambda_f b\left(h-\frac{x(t)}{\beta_1}\right)\left(\frac{h-\frac{x(t)}{\beta_1}}{2}-a\right)\end{bmatrix}-S(t) \tag{11}$$

4 EXAMPLE CALCULATION

4.1 *Test specimen design*

Ten RC beams were subjected to sustained flexural loading over a period of 10 years (Tan et al. 2005), using hanging weights to ensure the stability of long term load. The sustained load level was 0.5 times the ultimate load capacity, and the content of steel fiber was varied in Group I specimens (Table 1). Whereas, the content of steel fiber was 0.1% and the load level was varied in Group II specimens.

The beams measured 100 mm × 125 mm × 2000 mm with an effective span of 1800 mm. Concrete consisted of ordinary portland cement, natural sand and granular granite stone with maximum size of 10 mm. The steel fiber has a length of 0.5 mm, width of 30 mm and hook ends. Longitudinal tensile reinforcement consisted of two 10-mm diameter bars; transverse reinforcement comprised 6-mm links at 75 mm spacing. Concrete grade was C40.

Table 1. Beam designation.

Group	Test specimen	Steel fiber content ρ_f / %	Sustained load level
I	A-50	0	0.50
	B-50	0.5	0.50
	C-50	1.0	0.50
	D-50	1.5	0.50
	E-50	2.0	0.50
II	C-35	1.0	0.35
	C-50	1.0	0.50
	C-59	1.0	0.59
	C-65	1.0	0.65
	C-80	1.0	0.80

*Sustained load level = applied load Ps divided by design ultimate load Pu; Pu = 23.3 kN (based on Beam A-50).

4.2 *Test instrumentation*

The beams were loaded at quarter points. Deflections at mid-span were measured at 10 mm from the front and back faces and averaged. Crack width was measured by hand-held microscope, with an accuracy of ±0.02 mm.The deflection and crack widths were measured at 1d, 50d, 138d, 230d, 370d, 2284d (6.25 years) and 3678d (10 years).

4.3 *Time-dependent reliability*

The statistical parameters were determined using the deterioration model for each mechanical property. Using the first-order second-moment method, the time-dependent reliability of flexural capacity of SFRC beams was obtained. The time-dependent reliability indices for beams under 10-year sustained load are shown in Figure 1(a) and Figure 2(a). The time-dependent reliability indices of beams under 100-year sustained load were predicted as shown in Figure 1(b) and Figure 2(b).

1. As shown in Figure 1(a), when the external load was kept constant ($0.50P_u$) and the steel fiber content varied (0, 0.5%, 1.0%, 1.5%, 2.0%), the reliability index of SFRC beams is always higher than Beam A-50 without steel fiber at any time. The reliability index of Group I specimens in descending order is:

$$\beta(B-50)\rangle\beta(C-50)\rangle\beta(D-50)\rangle\beta(E-50)\rangle\beta(A-50)$$

Therefore, the incorporation of steel fiber can significantly improve the flexural capacity of the beam. The optimal steel fiber content was 0.5% (Beam B-50).

2. As shown in Figure 2(a), for Group II beams with the same steel fiber content but with different sustained loads ($0.35P_u$, $0.50P_u$, $0.59P_u$, $0.65P_u$, $0.80P_u$), the reliability index in descending order is:

$$\beta(\text{C-35})>\beta(\text{C-50})>\beta(\text{C-59})>\beta(\text{C-65})>\beta(\text{C-80})$$

Indicating that the larger the sustained load, the smaller the reliability index. Also, β(C-59) and β(C-65) are close to each other, indicating that the closer the applied loads, the closer are the reliability indexes, and that the results are consistent with the experimental results.

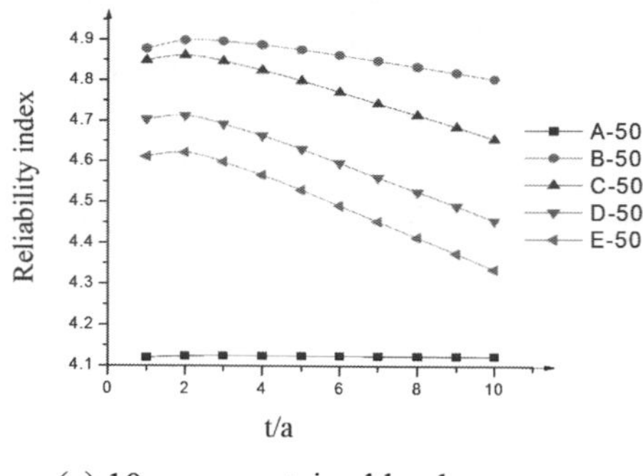

(a) 10 years sustained load

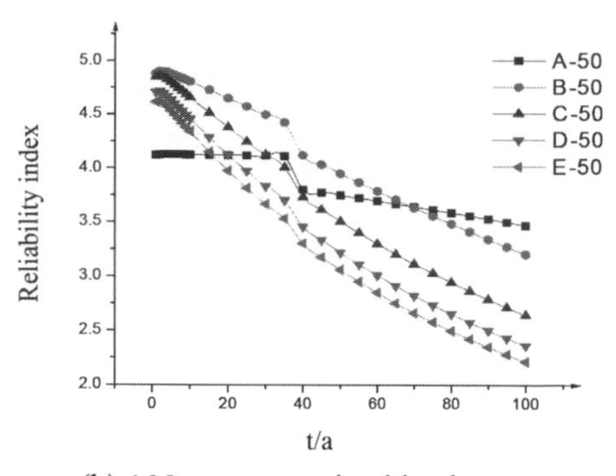

(b) 100 years sustained load

Figure 1. Time-dependent reliability of Group I beams.

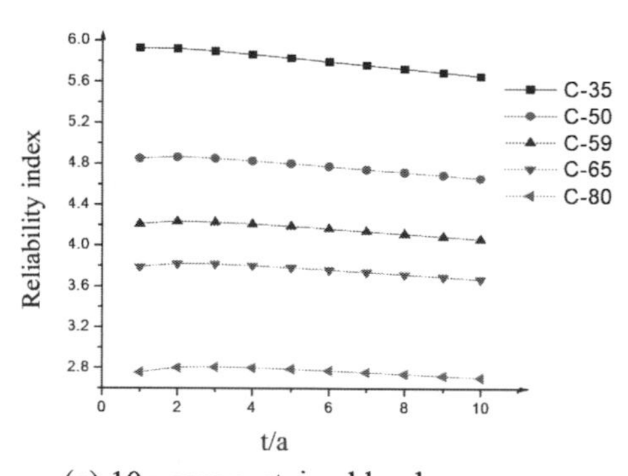

(a) 10 years sustained load

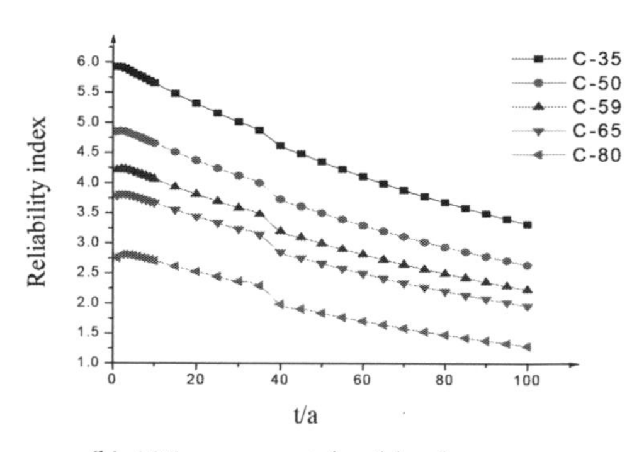

(b) 100 years sustained load

Figure 2. Time-dependent reliability of Group II beams.

3. As shown in Figures 1 and 2, the reliability index of SFRC beams first increased and then decreased during the service period. The reliability index of the test specimens reaches the maximum value when $t=2a$; this phenomenon may be due to the increase in early strength of the beam. With increasing time of service, the performance of concrete and reinforcing steel bar is deteriorated, the bond strength is weakened, and the reliability index is gradually reduced. According to the reliability evaluation criteria, when $t = 50a$, Group I: E-50 and Group II: C-59, C-65, C-80 do not meet the requirement of flexural capacity. When $t = 100a$, Group I: C-50, D-50, E-50 and Group II: C-50, C-59, C-65, C-80 do not meet the requirement of flexural capacity.
4. According to the Zheng et al. (2009), the steel bars begin to rust when $t_1 = 37.8a$. As shown in Figures 1(b) and 2(b), the reliability index curves of all the specimens show a sharp drop at this point. This is because the carbonation depth has reached the surface of the steel bars. As a result, the steel bars begin to rust which leads to a significant decrease in the yield tensile strength, and consequently a significant reduction in the flexural capacity of the beams.

5 CONCLUSION

The reliability index of SFRC beams is significantly higher than a beam without steel fibers. For Group I beams, with the same applied load but different steel fiber content, the reliability index in descending order is:

$$\beta(B\text{-}50) > \beta(C\text{-}50) > \beta(D\text{-}50) > \beta(E\text{-}50) > \beta(A\text{-}50)$$

For the same steel fiber content, the larger the sustained load, the smaller is the reliability index. The reliability index of Group II beams in descending order is:

$$\beta(\mathrm{C}\text{-}35) > \beta(\mathrm{C}\text{-}50) > \beta(\mathrm{C}\text{-}59) > \beta(\mathrm{C}\text{-}65) > \beta(\mathrm{C}\text{-}80)$$

For 100 years of service, the reliability index of SFRC beams first increases and then decreases, with the reliability index reaching a maximum value while $t = 2a$. The steel bars begin to rust, resulting in a significant reduction in the reliability index when $t = 37.8a$.

ACKNOWLEDGEMENTS

This work presented is jointly supported by New Century Excellent Talents in University (No. NCET-12-0107), Program for Liaoning Excellent Talents in University (LR2015024).

REFERENCES

Ding, F. & Yu, Z. 2004. Unified calculation method for tensile properties of concrete. *Journal of Huazhong University of Science and Technology (City Science Edition)* 21 (3) : 29–35. (in Chinese)

Dong, Z. & Niu, D. 2006, 38 (2) : 204–209, Liu Xifang. Method for calculating the time of initiation of reinforcement in the general atmosphere. *Journal of Xi'an University Of Architecture And Technology (Natural Science Edition)*, 2006, 38 (2) : 204–209. (in Chinese).

Emilia, V., Francesco, M. & Maria, A.A. 2013. Long term behavior of FRC flexural beams under sustained load. *Engineering Structures* 56 (6): 1858–1867.

Goitseone, M., Pilate, M. & Mark, A. 2009. Behaviour of RC beams corroded under sustained service loads. *Construction and Building Materials* 23(4): 3346–3351.

Hui, Y.B., Brian, U, Sameera, W.P. 2015. Time- dependent behaviour of composite beams with blind bolts under sustained loads. *Journal of Constructional Steel Research* 112 (3): 196–207.

Jin, C. 2013. Serviceability reliability analysis of prestressed concrete bridges. *KSCE Journal of Civil Engineering* 17(2): 415–425.

Lee, H, John, P.F. & Anne, N. 2013. Behaviour of cracked reinforced concrete beams under repeated and sustained load types. *Engineering Structures* 56(3): 457–465.

Ma, Y.S. & Wang Y.F. 2015. Creep influence on structural dynamic reliability. *Engineering Structures* 99(5): 1–8.

Saeed, M. & Ehsan, V. 2011. Time-dependent reliability analysis of B70 pre-stressed concrete sleeper subject to deterioration. *Engineering Failure Analysis* 18(1): 421–432.

Tan, K.H. & Saha, M.K. 2005. Ten-Year Study on Steel Fiber-Reinforced Concrete Beams Under Sustained Loads. *ACI Structural Journal* 102 (3) : 472–480.

Technical specification for fiber reinforced concrete structures[S]. (CECS38–2004). *China Planning Press*. 2004, 11. (in Chinese).

Zheng, Y., Liu, M. & Shen X.. 2009. Calculating models of time-dependent reliability for existing reinforced concrete beam. *Journal of Traffic and Transportation Engineering* 9 (2) : 45–49. (in Chinese).

Civil, Architecture and Environmental Engineering – Kao & Sung (Eds)
© 2017 Taylor & Francis Group, ISBN 978-1-138-02985-9

Research on the heat transfer characteristics of rock and soil under the effects of vertical double U-type buried-pipe heat exchanger

Xiao-duo Ou, Xin Pan, Heng Dai, Ying-chun Tang & Zi-yan Liu
Key Laboratory of Disaster Prevention and Structural Safety of Ministry of Education, Guangxi University, Nanning, China
Key Laboratory of Disaster Prevention Mitigation and Engineering Safety, Guangxi University, Nanning, China

ABSTRACT: This study based on a vertical double U-type buried-pipe heat exchanger of ground source heat pump engineering project in Nanning through an in situ test monitored the temperature changes in horizontal and vertical directions of the buried-pipe surrounding rock layer and explored the geotechnical heat transfer characteristics. Our results showed that in the deepest measuring point of the monitored hole, the temperature progressively drops in the unit run-time, which accumulated a drop of 5.2°C in 3 years. However, the temperature did not change during the running of the unit stop. This research also revealed that the change of temperature at other measuring points has a correlation with the operation of the unit. The measuring point temperature decreased during the unit run time, whereas the measuring point temperature increased but is still lower than the initial ground temperature when the unit was not operating. It is also found that the increasing degree of the temperature at each measuring point diminished with the increasing depth of each point. For example, the accumulation of temperature at 5 m deep, 1# hole, amounted to 2.4°C in 2009, whereas at a depth of 15 m, it amounted to 0°C. The research further showed that the temperature of each measuring point dropped every year and that the variation degree of temperature decreased with the increase of the depth of the measuring point, which suggests that the ground source heat pump has a cumulative effect on the ground temperature.

1 INTRODUCTION

The sustainable development of society requires the development and utilization of energy to coordinate with the development of environment protection and ecological balance, minimize the damage to our environment in the process of utilizing energy, exploit renewable and clean energy, and to increase the efficiency of energy utilization (Yu Zhongyi 2008). Geothermal energy, one kind of renewable resources, has drawn researchers' attention because of its sustainability and wide applicability across regions. In order to make better use of geothermal energy, many scholars have studied the characteristics of geotechnical heat transfer and obtained some achievements. Wang Wei (2010) performed a large number of tests for determining the properties of permafrost heat transfer. Wang Tie-hang (2010) summarized previous studies on water heat coupling in the field of loess, frozen soil, expansive soil, soil science, and so on. He further pointed out the limitations of research on the water heat coupling function. Sutton et al. (2005) worked out the drilling soil thermal resistance analytic expression when groundwater flows around. As a widely used ground source, the ground source heat pump also produces many research results. Relevant literature (Michel A B 2001 & 2002, Yavuzturk C et al 2001 & 2002, Michel P. A 2001) established the mathematical model of the buried-pipe heat exchanger and the numerical simulation, and North America choose IGSHPA model (Ball D A et al. 1983) method as the standard method to determine the size of the underground buried-pipe heat exchanger. Studies (ZHANG Guozhu et al. 2012, Yingchun Tang et al. 2011, Yan Ren 2010, Chen Ying et al. 2009, Hu Yingning et al. 2009) have shown the characteristics of soil heat transfer under the influence of ground source heat pump. However, in terms of different hydrogeology and engineering geology, soil heat transfer has different characteristics. As the ground source heat pump is finding wider and wider application in Guangxi, in view of the Guangxi Nanning basin characteristics in geotechnical layer, buried pipe heat exchanger under the action of soil heat transfer experiment research is necessary.

2 SUMMARY OF OBSERVATION STATION

The surface of the field is flat and belongs to the north shore of Yong Jiang River of the Nanning

Table 1. Parameters of different soil layers.

Number	Depth (m)	Rock and soil layers	Moisture content (%)	Specific weight (kN/m^3)	Void ratio	Saturability	Thermal conductivity w/(m·K)	Specific heat kJ/(m^3·K)K)]	Thermal diffusivity (mm^2/s)
1	5.0	Clay Silty	24.2	19.9	0.68	98.1	1.575	2.716	0.573
2	7.0	Clay Silt	22.5	20.3	0.61	100.0	1.874	2.959	0.672
3	12.0		23.5	20.4	0.68	97.4	2.367	3.171	0.771
4	30.0	Mudstone	16.7	20.1	0.56	81.3	1.981	3.625	0.565

basin II level terrace. The field has two layers of groundwater within the drilling depth. The first layer is surface water and perched water, assigned product in fish ponds, depression, and tillage soil, mainly from atmospheric precipitation and surface water, and the volume of water in this layer is relatively minor. The second layer is pore water, whose static level was buried deeply, 5.80–8.50 m, in the exploration period. Figure 1 shows the geological histogram of the field based on the geological survey report. The result of the thermal physical property test and the physical and mechanical properties test of the geotechnical layer sampling of the field are shown in Table 1.

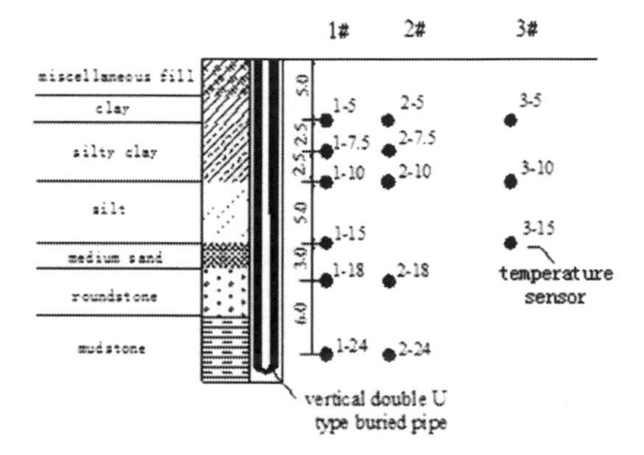

Figure 1. Profile of sensor layout.

3 TEST DESIGN OF OBSERVATION STATION

This research was based on the ground source heat pump engineering of student apartments of Guangxi University Xingjian College of Science and Liberal Arts. The heat pump is a ground-source heat pump, using 60 double U vertical buried-pipe heat exchangers at an embedment depth of 25 m and spacing of 5 m. We selected an engineering well at the edge for monitoring. The temperature sensors were arranged around the monitoring well. In the horizontal direction, the distance between the monitoring holes and the heat exchange tube were 0.5, 1, and 2 m. In the vertical direction, temperature sensors were embedded at different depths in each monitoring hole. The temperature sensors were buried underground at depths of 5, 7.5, 10, 15, 18, and 24 m in the 1# monitoring hole. The temperature sensors were also buried at the same depth (except 15 m deep outside) in the 2# monitoring hole. The temperature sensors were buried at depths of 5, 10, and 15 m in the 3# monitoring hole. In sum, there were 3 holes and 14 temperature sensors to monitor the temperature of the surrounding rock layer of the engineering well. The specific arrangement of sensors is shown in Figures 1 and 2 (temperature sensor number: X–Y means the temperature sensor is embedded in X# monitoring hole and depth Y)

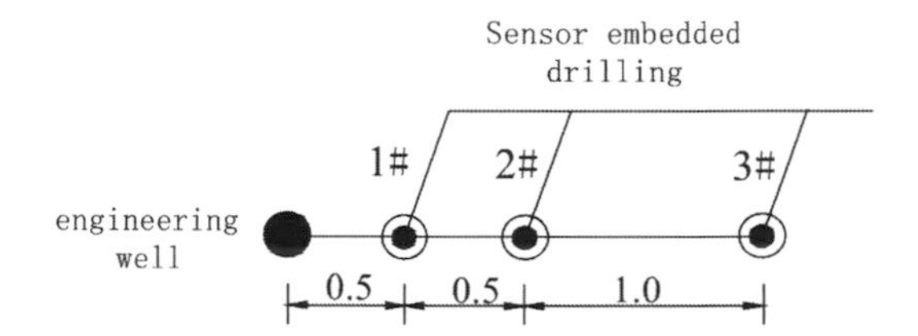

Figure 2. Arrangement plan of sensor.

4 IN SITU TEST PROCESS

After all the heat pump engineering wells have been completed, we selected a well at the edge as the research object. The monitoring holes were drilled surrounding the well according to the test design, and temperature sensors were buried in each monitoring hole in order to monitor geotechnical temperature changes of the well. First, we tied the temperature sensor to the PPR pipe on the corresponding position according to the experimental design depth and then put the pipe and temperature sensor into the monitoring hole and refilled the hole with sand. After all the sensors were embedded, we monitored the change of rock layer temperature surrounding the buried pipe.

1. Before the running of heat pump units, we detected the temperature of various rock layers

using the temperature checking instrument twice a day for a week, taking its mean value as the initial temperature of a rock layer.

2. In the early operation of heat pump units, our monitoring frequency is twice a day. After the temperature change of rock layer became relatively stable, our monitoring frequency reduced to once a day. Finally, according to the monitoring results of temperature change of geotechnical layer during early stages, our monitoring frequency gradually became once every 2 weeks.

5 TEST RESULTS AND ANALYSIS

5.1 *Soil Initi0061l temperature test*

Before the running of heat pump units, we detected the temperature of various rock layers twice a day for a week, taking its mean value of 14 times as the rock layer's initial temperature. The test results are shown in Table 2 and Figure 3.

As Figure 3 shows, the lowest geotechnical layer temperature is at the depth of 10 m, and the average temperature is 23.1°C. The soil temperature within the depth range of 0–10 m reduced as the depth increases; however, the soil temperature below the depth of 10 m increased as the depth increases.

Cause analysis: during the whole test period, the outside temperature is higher than the soil temperature. The soil temperature nearing the ground surface was affected by the external temperature to a great extent. The ground surface and 5.80–8.50 m deep soil layer contain groundwater, which can transfer heat as it flows and thus result in a decrease of soil temperature within that depth. Therefore, soil temperature decreased as depth increased from 0 to 10 m. On the contrary, there is no groundwater in soil layers of 10–24 m depth range. Its temperature is less affected by the external temperature, and the heat diffusion to the ground is less possible, which resulted in an increase of rock layer temperature with an increase of its depth.

Table 2. Initial soil temperature of observatory.

Depth/m Temperature/°C Number	5.0	7.5	10	15.0	18.0	24.0
1# hole average	24	23.3	23.1	23.2	23.5	23.6
2# hole average	24	23.4	23.1	—	23.5	23.7
3# hole average	23.9	—	23.1	23.2	—	—
Each hole average temperature	24.0	23.4	23.1	23.2	23.5	23.6

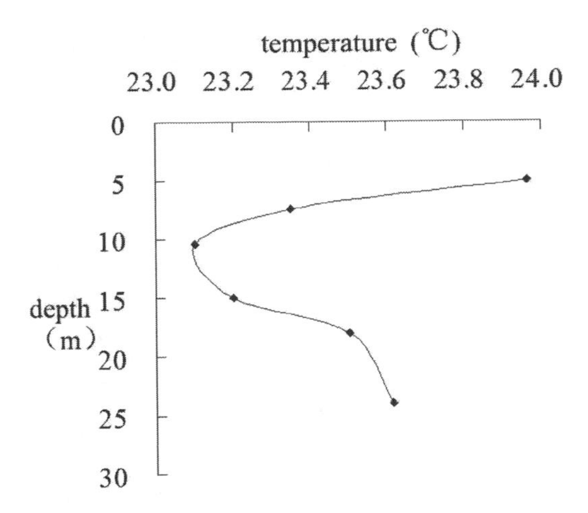

Figure 3. Original temperature profile.

Table 3. Running state of ground source heat pump.

Number	Date	Days	Unit operation condition
1	2008–10–18	215	Run
2	2009–5–21	157	Stop
3	2009–10–25	218	Run
4	2010–5–31	147	Stop
5	201010–25	204	Run
6	2011–5–17	154	Stop
	2011–10–18		

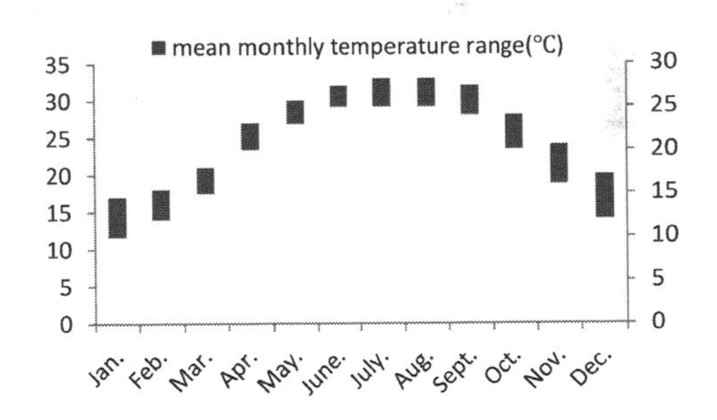

Figure 4. Nanning produced historical average temperature profile.

5.2 *Soil temperature change in the monitoring hole at different depths*

Our analyses of soil temperature change were based on the monitoring data during the 3-year time span from 18 October 2008 (beginning of the research) to 18 October 2011. The unit's running condition during this period is shown in Table 3. The unit's stop running period represents the convalescence of surrounding rock-layer temperature.

Figure 4 lists the average temperatures of Nanning. As can been seen from the graph, the average air temperature is higher during the period of the unit's nonrunning period from May to October, which was above 20°C. The average temperature is lower during the period of the unit's running period from November to April, which was between 10 and –20°C.

In order to understand the geotechnical vertical heat transfer condition, we compared temperature changes of different soil layers of each monitoring hole as time changes and listed the temperature decrease degree of every year, as shown in Figures 4 and 5. Points 1–24, 2–5, 2–24, 3–5, 3–10, and 3–15 were destroyed in 2011so part of the data is missing.

As shown in Figure 5 (a), the temperature of point 1–24 (1# hole buried depth 24 m, the same below) continued to decline during the unit's running period, which accumulated at 5.2°C in 3 years. However, the temperature remained unchanged when the unit stopped running. Except for point 1–24, the temperature changes of all the other monitoring points has linked with the unit running condition. The temperature kept declining when the unit was running, whereas the temperature started to increase when the unit

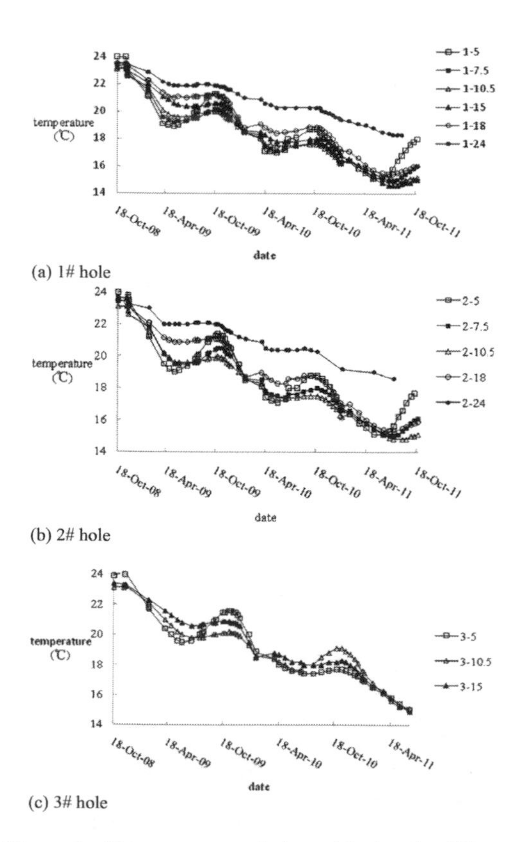

Figure 5. Temperature variation with time in different strata at each hole.

stopped running. The temperature of each point reached the maximum in October (before unit running). In 2009, the highest degree of temperature increase was monitored in point 1–5, increasing gradually from 18.9 to 21.3°C with a total increase of 2.4°C. The lowest degree of temperature increase was in point 1–15, increasing gradually from 20.6 to 20.4°C and then up to 20.6°C, with an accumulated increase of 0°C. In 2010, the highest degree of temperature increase was monitored in point 1–5, increasing from 17 to 18.7°C with a total increase of 1.7°C. The lowest degree of temperature increase was in point 1–10, increasing from 17.6 to 17.5°C and then up to 17.6°C with a total increase of 0°C. In 2011, the highest degree of temperature increase was monitored in point 1–5, increasing from 15.1 to 18°C with a total increase of 2.9°C. The lowest degree of temperature increase was in point 1–10, increasing from 15.5 down to 14.9°C and then up to 15.2°C with a total increase of –0.3°C. The temperature of all monitoring points decreased every year, and failed to return to the initial temperature. The temperature of some points dropped first and then rose while the unit stopped running, but both the decrease and increase of temperature was relatively small within no more than 1°C. The temperature change of each point has a similar tendency that the degree of temperature change decreased as the depth of monitoring point increased. As Figure 4 (b) and (c) show, the temperature change of each monitoring points of 2# hole and 3# hole was similar to that of the corresponding points of 1# hole. However, as there were fewer points in 3# hole and because of the data loss in 2011, this tendency of temperature change is not that obvious.

As Figure 6 (a) shows, the difference between the highest and initial temperature of each monitoring points in 1# hole during the unit's stop running period is as follows: In 2009, the maximum detectable temperature change was 3.1°C with points 1–7.5 and 1–10; In 2010, the maximum detectable temperature change was 5.6°C with point 1–7.5. In 2011, the maximum detectable temperature change was 8.2°C with point 1–15. The figure also shows that the difference between the highest and initial temperatures of each monitoring point in 1# hole during the unit's stop running period was increasing every year. The difference between the highest and initial temperatures of each monitoring points in 2# hole and 3# is shown in Table 4. As Figure 5 (b) and (c) show, the highest difference between the highest and initial temperatures of each monitoring points in 2# hole and 3# hole during the unit's stop running period was also increasing every year. Figure 6 (c) shows that the temperature change of both points 3–5 and 3–15 in 3# hole was higher

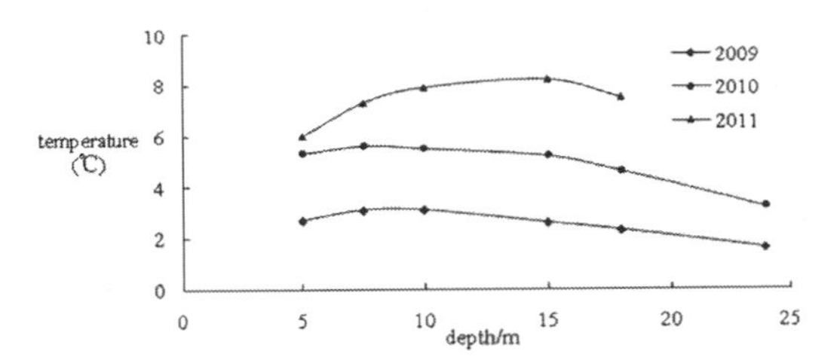

(a) 1# hole

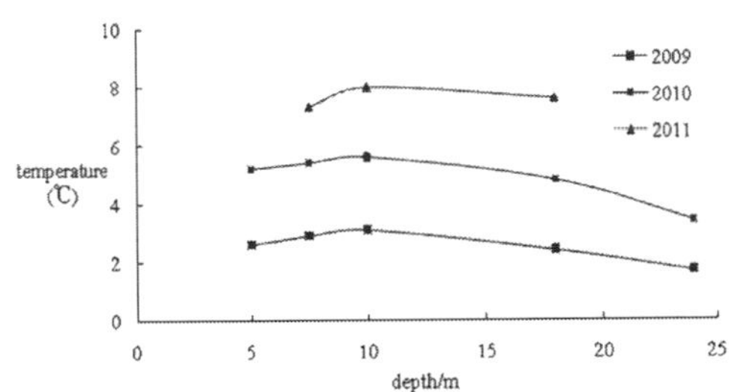

(b) 2# hole

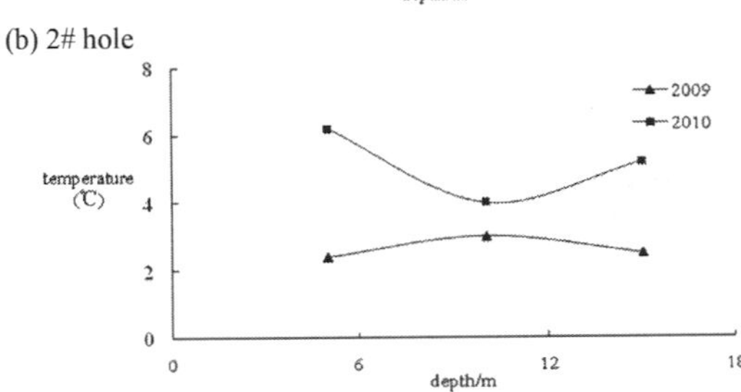

(c) 3# hole

Figure 6. Variation of maximum temperature and initial.

Table 4. Shows that 2# and 3# holes have the highest difference between the temperature in the unit's stop running period and the initial temperature.

	2009	2010	2011
2# hole the most difference point and change value	1–10, 3.1°C	1–10, 5.6°C	1–10, 8°C
3# hole the most difference point and change value	1–15, 2.5°C	1–5, 6.2°C	

than that in point 3–10 in 2010. Nevertheless, because of inadequate monitoring points in 3# hole, the finding is not conclusive and requires a further investigation.

5.3 *Soil temperature change at monitoring points of same depth in different holes*

In order to understand the geotechnical lateral heat transfer, we compared the soil temperature change of monitoring points at the same depth as time changes. The results are shown in Figure 7.

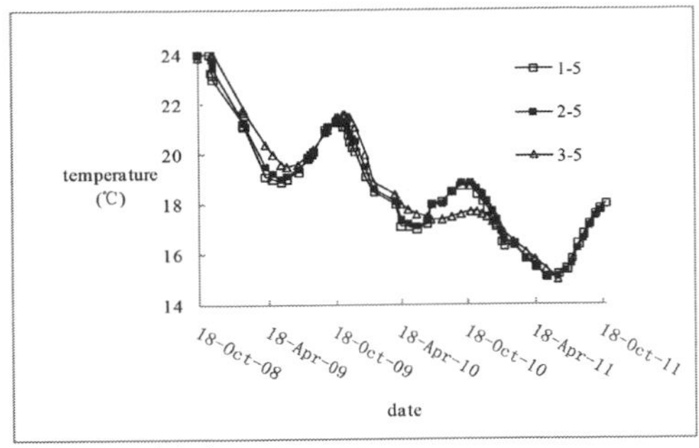

(a) 5m at each hole

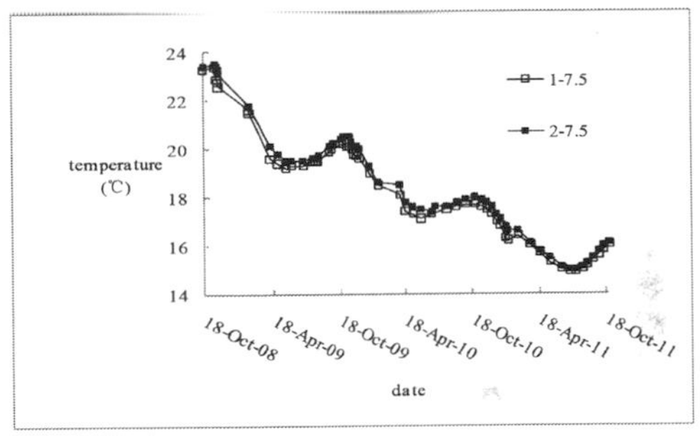

(b) 7.5m at each hole

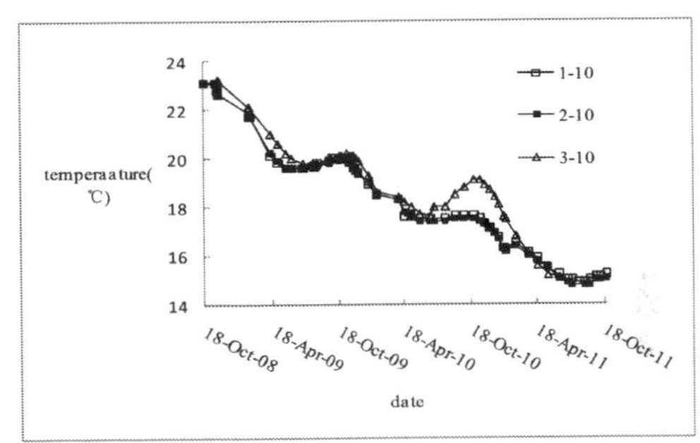

(c) 10m at each hole

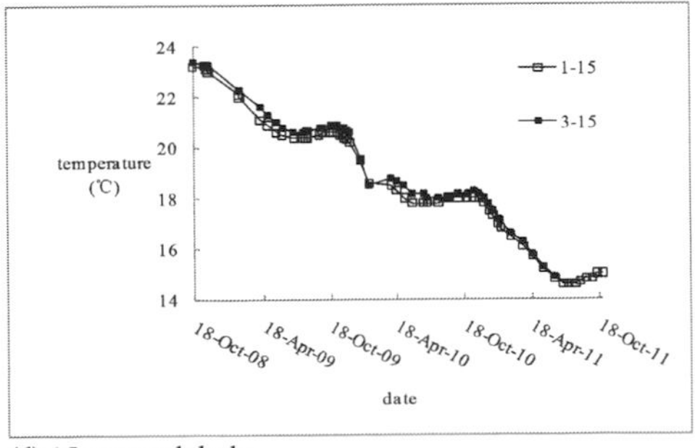

(d) 15m at each hole

Figure 7. (*Continued*)

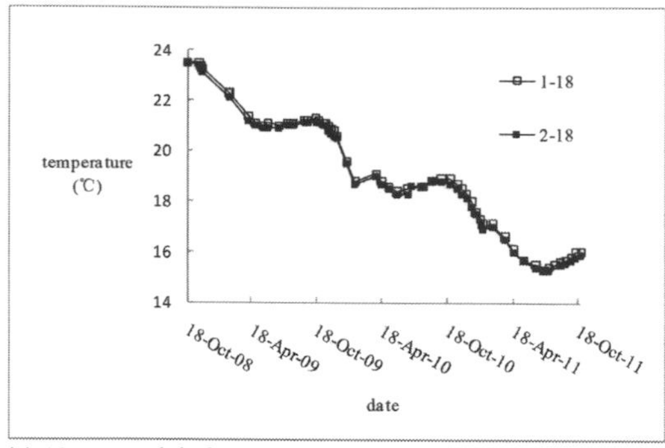

(e) 18m at each hole

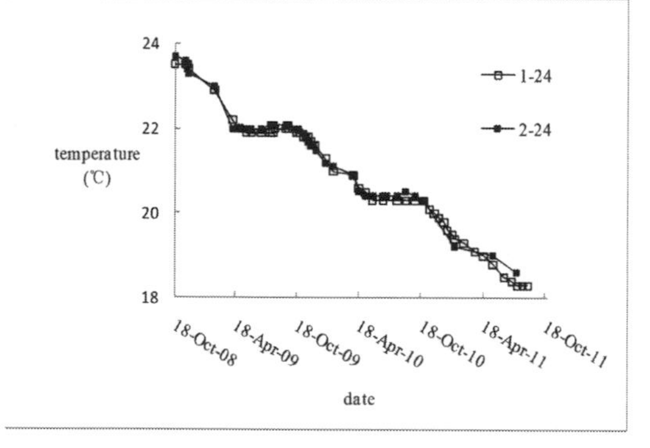

(f) 24m at each hole

Figure 7. Temperature variation with time in the same depth of different hole.

As Figure 7 (a) shows, the temperature of monitoring point at a depth of 24 m in each hole dropped when the unit was running, but the temperature did not recover when the unit stopped running. As Figure 6 (a)–(f) show, the temperature change of monitoring points at the same depth in each hole showed a similar tendency and change values. The temperature change of monitoring points has correlation with the unit's operation condition. The temperature of each monitoring point was dropping when the unit was running in winter, whereas the temperature was rising when the unit stopped running in summer. Additionally, the temperature change of each monitoring point was decreasing in degree with the increase of depth. During the unit's running period for each year, the closer it is to the heat pump, the earlier the minimum temperature will emerge. The minimum temperature first appeared in 1# hole, and then in 2# hole and finally in 3# hole. The deeper the monitoring point, the later the minimum temperature will emerge. For example, the minimum temperature at 5 m deep point generally appeared in May, but the temperature at 10 m deep point and 15 m deep point continued to drop until the minimum temperature appeared in July. The time span for temperature dropping to the minimum was longer.

Cause analysis: Figure 1 shows that the soil is clay at 5 m deep, silty clay at 7.5 m deep, silt at 10 m deep, sand at 15 m deep, round stone at 18 m deep, and mudstone at 24 m deep. As Table 1 shows, the specific heat capacity of the soil increases with depth, for example, the specific heat capacity of clay is 2.716 kJ/(m^3. K) at 5 m deep and 3.625 kJ/(m^3. K) of mudstone at 24 m deep. This shows that the temperature change needs to absorb or release more heat with the increase of soil depth, thus the temperature change of soil fluctuated less with the increase of soil depth. On the contrary, with the continuous effect of the higher temperature of outside air, the shallow soil will exchange energy with the outside air and lead to an increase of soil temperature, which may explain the larger temperature fluctuation of shallow soil when the unit stopped running.

6 CONCLUSION

Through the establishment of in situ observation station, we have obtained long-term observation data. By analyzing the soil temperature variation surrounding the engineering well, we concluded that:

1. Soil layer initial temperature distribution: the temperature of soil layer at the depth range of 5–10 m decreased with the increase of depth; however, at the range of 10–24 m the temperature increased with the increase of depth. The average temperature of the minimum temperature point (10 m deep) was 23.1°C.
2. The temperature of the deepest monitoring point decreased to some extent when the unit was running, but it did not recover when the unit stopped running. The temperature changes of other monitoring points have correlation with the operation of the unit. The temperature of monitoring points was dropping when the unit is in operation, whereas it began to increase when the unit is off operation but did not recover to the initial temperature. The 3# hole, 2 m horizontal distance from the engineering well, was the furthest one from the well, but showed temperature variation when the unit was on and off operation. This result suggested that the engineering well's thermal radius of influence is more than 2 m.
3. The variation in temperature of each monitoring point in the same monitoring hole was similar, so was the variation of temperature of each monitoring hole at the same depth. The degree of temperature change decreased with the

increase of the depth of monitoring points. The main reason for this tendency is that the shallow soil layer was influenced by the temperature of the outside world and that the heat capacity of the soil increased with the increase of depth. The difference between the highest and initial soil temperatures when the unit stopped running increased every year, which suggests the cumulative effect of soil temperature.

REFERENCES

Ball D A, Fischer R D & Hodgett D L (1983). Design methods for ground-source heat pumps. J. ASHRAE Transactions, 89(2): 416–440.

Chen Ying, Yang Mei & Shi Baoxing (2009).Experi-mental investigation on soil temperature restorat-ive characteristics for soil source heat pump in intermittent cooling operation. J. ACTA Energiae Solaris Sinica, 30(10): 1193–1197.

Hu Yingning, Lin Jun & Yin Xiangming (2009).Study on heat exchange performance of ground exchangers in the water-rich soil areas. J. ACTA Energiae Solaris Sinica, 30(1): 5–11.

Michel A B (2001). Ground-coupled heat pump system simulation. J. ASHRAE Transactions, 107(1): 605–616.

Michel A B (2002). Uncertainty in the design length calculation or vertical ground heat exchangers. J. ASHRAE Transactions, 108(1): 939–943.

Michel P (2001). A simplified tool for assessing the feasibility of ground-source heat pump projects. J. ASHRAE Transactions, 107(1): 120–129.

Sutton M G, Nutter D W, Couvillion R J (2003). A ground resistance for vertical bore heat exchangers with groundwater flow [J]. Journal of Energy Resources Technology, 125(3): 183–189.

Wang Tie-hang, Li Ning & Xie Ding-yi (2005). Necessity and means in research on soil coupled heat-moisture-stress issues. J. Rock and Soil Mechanics. (3): 488–493.

Wei Wang (2010). The ExPeriments Researeh on Thermal Conductivity of Frozen soil. D. Jilin University.

Yan Ren (2010). Performance characteristic analysis and experimental study of moderate buried double-U tube ground-source heat pump system. D. Beijing University of technology.

Yavuzturk C & Chiasson A D (2002). Performance analysis of U-tube, concentric tube, and standing column well ground heat exchangers using a system simulation approach. J. ASHRAE Transactions, 108(1): 925–938.

Yavuzturk C & Spitler J D (2001). Field validation of a short time step model for vertical ground-loop heat exchangers. J. ASHRAE Transactions, 107(1): 617–625.

Yingchun Tang & Xiaoduo Ou & Baotian Wang (2011). Experimental Study on the Heat Transfer in Rock Layers with the Vertical Downhole Heat Exchanger. Advanced Materials Research. C. Vols. 168–170:2243–2248.

Yu Zhongyi (2008). Research on Heat Transfer Characteristic of Vertical Ground Heat Exchangers in Ground Source Heat Pump Systems. D. Huazhong University of Science & Technology.

Zhang Guozhu, XIA Caichu1 & MA Xuguang, et al (2012). Rock-soil thermal response test of tunnel heating system using heat pump in cold region. J. Chinese Journal of Rock Mechanics and Engineering. (1): 99–105.

Civil, Architecture and Environmental Engineering – Kao & Sung (Eds)
 ISBN 978-1-138-02985-9

Nitrogen removal performance of ANAMMOX-PVA granules immobilized by different preparation methods

Y.M. Han
College of Mechanical Engineering, Dalian University, Dalian, China
R&D Institute of Fluid and Powder Engineering, Dalian University of Technology, Dalian, China

ABSTRACT: PVA (Polyvinyl Alcohol) was used to encapsulate ANAMMOX (ammonia oxidation bacteria) in order to decrease microbial loss in ANAMMOX and its related process for wastewater treatment. Different types of granules with ANAMMOX sludge immobilized in PVA gel of different mass concentrations 6%, 10%, 15%, and 20%, were prepared by cross-linking in calcium chloride solution and deep-freezing method with 10% PVA solution. The results of nitrogen removal experiments showed that 10% was the best concentration for ANAMMOX sludge to be immobilized in PVA gel when better nitrogen removal performance and longer life could be achieved; the granules prepared by chemical cross-linking method had better performance than those prepared by freezing cross-linking method. The granules prepared using the former method removed nitrite in a shorter time period (after 24 h), whereas those prepared using the latter method reacted in 7 days.

1 INTRODUCTION

PVA is a type of macromolecule organic used widely in pharmaceuticals industry and biological field for its nontoxic and excellent biocompatibility organic compounds. S. Okabe (2011) indicated that ANAMMOX processes were promising for nitrogen removal and recognized them as cost-effective with low energy and no extra organic carbon. Although there are several running wastewater treatment plants via ANAMMOX and its related process in Europe, the practical engineering applications of the ANAMMOX process are limited and still in laboratory process because of the low growth rate of ANAMMOX bacteria in many other countries. Because of this limitation, one of the main challenges in implementing the ANAMMOX process is to ensure bacterial cells' retention inside the reactors. In practical engineering applications of ANAMMOX process, microbial loss is always a problem due to aeration and stirring. For this reason, immobilization of microbial cells has received increasing interest in wastewater treatment to minimize the risk of biomass washout from the reactors and provide stabilized treatment reports. Most immobilization methods were carried out by attachment on the surface of carriers forming biofilm or forming granular biomass directly. Entrapment of the microbial biomass into gel pellets has been reported in recent years.

The use of synthetic polymers such as urethane, Polyethylene Glycol (PEG), and Polyvinyl Alcohol (PVA) for the entrapment of microorganisms was reported as advantageous in the field of wastewater treatment. There were two techniques reported to immobilize microorganisms using PVA; the first step in both cases was to combine ANAMMOX sludge with a PVA solution, and then the mixture was polymerized. In order to solidify the mixed-solution PVA and ANAMMOX sludge, some chemical solution was used in one technique and freezing method was used in the other. In the chemical cross-linking method, the mixed gel solution was shaped into spherical gel beads first and then dropped into a saline solution by G.L. Zhu (2009). Physical cross-linking through the freezing method was the other preparation technique reported by T.H. Hsia (2008), who immobilized ANAMMOX in polymer gel by the physical cross-linking method. The lowest temperature reported in freezing cross-linking was –20°C from A. Magrí (2012), who successfully immobilized swine wastewater nitrifying sludge in PVA by freezing at –8°C.

This study was carried out in order to find out the best PVA concentration for better nitrogen removal and mechanical strength through the chemical cross-linking method and then contact the effectiveness of entrapping and activity recovering period for ANAMMOX cross-linking through chemical method and freezing method in this concentration. Our experiments included preparation of PVA gel granules by cross-linking in calcium chloride solution and nitrogen removal process; preparation of

PVA gel granules through the aforementioned two methods and nitrogen removal process.

2 MATERIAL AND METHOD

2.1 *Immobilization procedure*

2.1.1 *Seed sludge and pretreatment*

ANAMMOX seed sludge was derived from a running ANAMMOX biofilm reactor, which was stably operated for more than 400 days with Nitrogen Loading Rate (NLR) of 2.0 $kg\text{-}N{\cdot}m^{-3}{\cdot}d^{-1}$ and Nitrogen Removal Rate (NRR) of more than 1.7 $kg\text{-}N{\cdot}m^{-3}{\cdot}d^{-1}$. The ANAMMOX seed sludge solution kept static for 30 min to cause the sludge deposit in the first; then, the liquid supernatant was removed and the remaining sludge was washed thrice with PBS (pH 7.2); the sludge was put into the centrifugal machine and centrifuged for 10 min at 1000 rpm, and the supernatant liquid was removed.

2.1.2 *Preparation of PVA solution*

The PVA polymer used in this study was in powder form (average molecular weight was about 10000 and purity 94%). The PVA powder was sprinkled on warm water and mixed by hands to form a polymer suspension. The suspension was then heat preserved to 55°C and stirred on a magnetic stirrer for 2 h to obtain a complete PVA dissolution. After cooling to room temperature, the PVA solution is ready. The mass concentration of PVA solution was set to 6%, 10%, 15%, and 20% to investigate the concentration effect on the nitrogen removal capability of the gel granules.

2.1.3 *Preparation of gel granules*

① Cross-linking in calcium chloride solution

PVA solutions (30 mL) prepared in Section 2.1.2 at different concentrations were mixed with the same volume of sludge. The solution was pumped by a peristaltic pump with a soft pipe of inner diameter 4.8 mm installed and dropped into the calcium chloride solution with a mass concentration of 4% and was kept at 4°C before immobilization. The ANAMMOX–PVA solution drops solidified to form granules with diameter between 4 and 6 mm. Then, the floating granules and calcium chloride solution were kept at 4°C together. After 24 h, the granules in calcium chloride solution were put out.

② Cross-linking in freezing

The same volume and scale solution with ① were mixed and packed into a sheet; then, the ANAMMOX–PVA solution was kept under low-temperature condition (–10°C). After 2 h, the whole gel pieces were put out and cut to squares with sides 4 cm.

Table 1. Components of the influent.

Composition	$(NH_4)_2SO_4$	$NaNO_2$	$KHCO_3$	KH_2PO_4
Concentration/ $mg \cdot L^{-1}$	100–300		187.5	162

2.2 *Composition of synthetic wastewater and analytical procedure*

The concentrations of NH_4^+–N and NO_2^-–N in the influent and effluent were analyzed to determine the NRR of the total inorganic nitrogen. The nitrogen removal rate was calculated by the difference of the total nitrogen concentrations between the influent and effluent. The concentrations of NH_4^+–N and NO_2^-–N were measured by ordinary method. The dissolved oxygen in the reactor was measured using a DO meter (JB-2, Iinesa, China), and the pH of the effluent was measured using a pH meter (METTLER TOLEDO FE20, USA).

2.3 *Nitrite removal performance*

The gel granules were put into conical flasks with an effective volume of 250 mL after washing thrice with PBS. Synthetic wastewater (250 mL) with components in Table 1 was then fed and the neck was sealed. The flasks were placed into the orbital shaker incubator and cultivated under the conditions of 10 rpm and 28°C. The cultivation period was 25 days, and the concentrations of NH_4^+–N and NO_2^-–N were detected once every 2 days.

3 RESULTS AND DISCUSSION

3.1 *PVA concentration effect*

The biological sludge immobilization would affect the growth of bacteria, so the nitrogen removal performance must be tested. In addition, the mechanical property of the gel granules was another important influence factor for the wastewater treatment because the purpose of biological sludge immobilization was to decrease the loss of biomass in the reaction process. The period of wastewater treatment was always several months or years, so the life span of the gel granules would play an important role on the test performance. The concentration of PVA solution was the key influence factor to the degradation period and mechanical property of the gel granules.

The nitrogen removal experiments were carried out for almost 20 days, and NH_4^+–N and NO_2^-–N were detected once every 2 days. The results are shown in Figs. 1 and 2.

Obvious nitrogen removal could be achieved in all the samples, and the nitrogen removal efficiency was

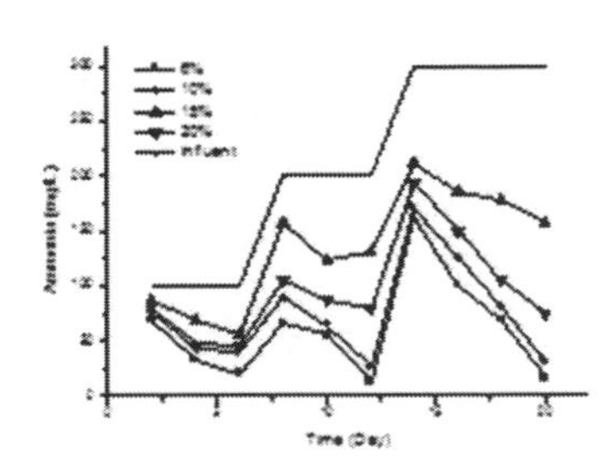

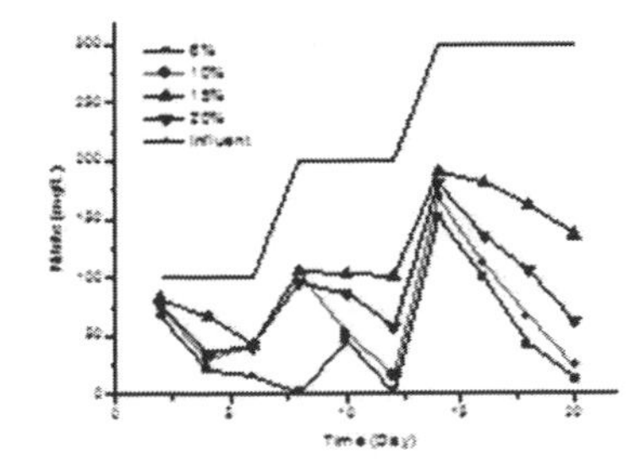

Figure 1. Ammonia and nitrite concentration changing profile.

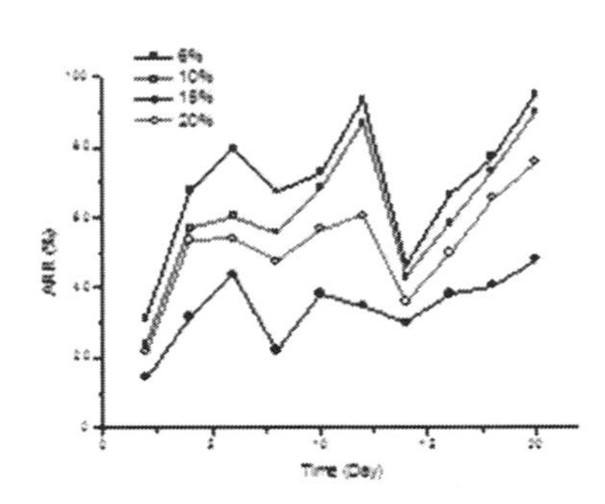

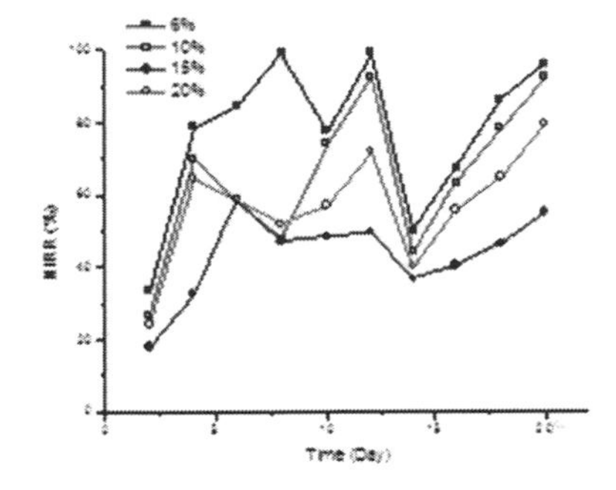

Figure 2. ARR and NIRR changing profile.

almost in inverse proportion with the PVA concentration. The highest Ammonia Removal Rate (ARR) and Nitrite Removal Rate (NIRR) about 97% was achieved in the PVA–ANAMMOX sample at 6% concentration on the 12th and 20th days, whereas the ARR and NIRR of all the other samples were lower. It was also indicated that the lowest ARR and NIRR about 50% appeared in the sample of 15% PVA concentration but not in the sample with 20% concentration. That was because the PVA–ANAMMOX solution with 20% PVA was so sticky that it was difficult to prepare the granules, and many sheets, sticks, and other out-of-shape granules were formed. The abnormal granules had worse mechanical property and shorter degradation period, so there were no obvious granules existing after 7 days and the flocculent remainder increased ARR and NIRR up to 80%.

Sample with 15% PVA had the lowest nitrogen removal efficiency (only 50% after 48 h) for the gel with a higher PVA concentration, which blocked the mass transfer although the granules had the better shape even on the 20th day.

3.2 *Contrast nitrogen removal experiments by two different preparation methods*

The PVA–ANAMMOX gel granules were prepared by the two methods described in Section 2.1.3 with a PVA concentration of 10%. The gel granules were put into conical flasks after cross-linking and then the flasks were placed into the orbital shaker

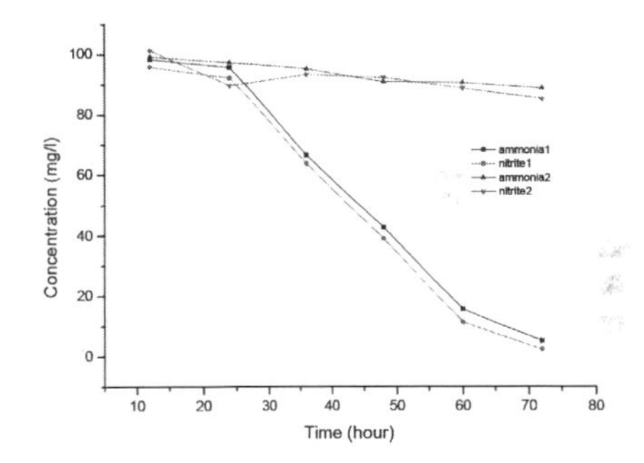

Figure 3. Nitrite removal performance in 72 h. ① Chemical cross-linking ② Freezing cross-linking.

incubator and cultivated under the conditions of 10 rpm and 28°C. The concentration of ammonia and nitrite in the influent was 100 mg/L. The cultivation period was 72 h, and the concentrations of NH_4^+–N and NO_2^-–N were detected once every 12 h. The experimental results are shown in Fig. 3. The profiles indicated that there was obvious nitrogen removal performance in the flask containing sample prepared through method ① after 24 h, and most of the gel granules in sample ① came up to the water surface and there are also plenty of white lucid bubbles on the water surface instead of sinking at the bottom at the time of the reaction beginning. The concentrations of ammonia and nitrite were below 40 mg/L after 48 h, and almost all of the gel granules came up to the water surface. The gel granules came up because there was some

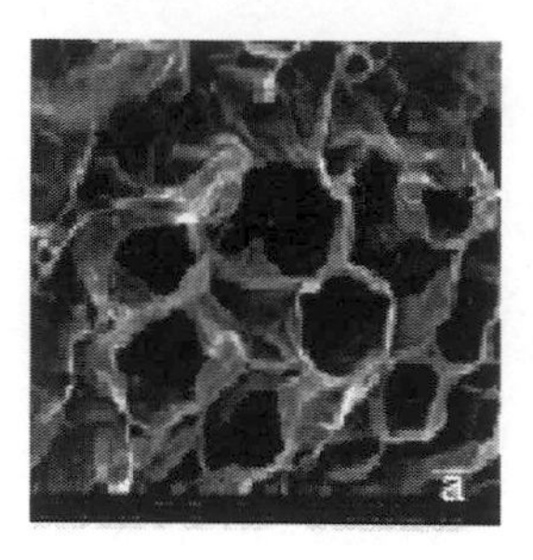

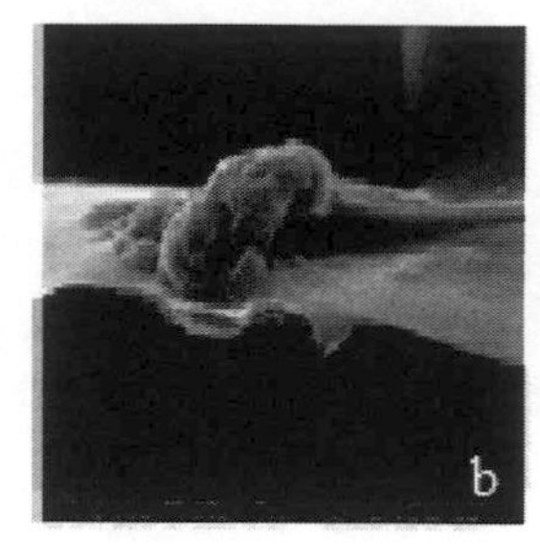

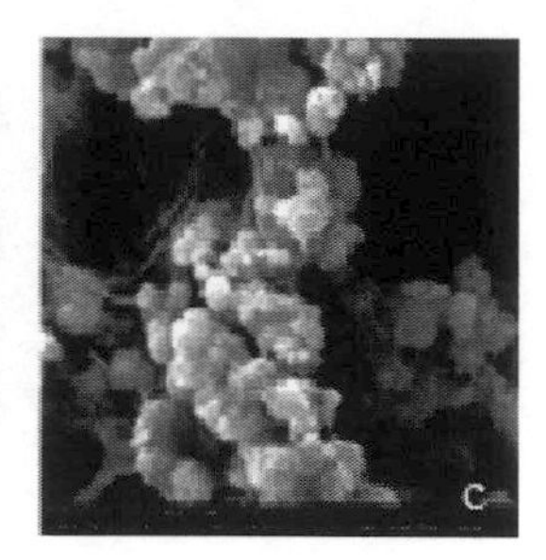

Figure 4. SEM images of the granule's section.
a) × 200; b) × 5000; c) × 10000.

nitrogen inside the gel granules, and the average density of the granules was reduced. The nitrogen removal rate increased continually after then on.

The sample prepared using method ② had distinct performance from sample ①. No obvious nitrogen removal could be achieved in 72 h, and the gel pieces still sank at the bottom of the flask. ARR and NIRR increased slowly from the 4th day and reached to 70% after 7 days.

It was indicated from the contrastive analysis that the gel granules by method ① were kept at 4°C for only 24 h, so the activity of ANAMMOX was restrained in a short time and they could be recovered in 24 h. The gel pieces by method ② were kept under a low-temperature condition at −10°C for 2 h so the ANAMMOX bacteria came into a short dormancy and the recovery period would be much longer than that of method ①. The preparation method: cross-linking in calcium chloride solution was superior, and cross-linking in freezing of most wastewater treatment process for quick setup operation was necessary.

3.3 *SEM images*

SEM imaging was used to observe the structure and characterize the morphology of the gel granules after pretreatment of the samples on the 10% PVA concentration by method ①. The gel samples were frozen for 1 h initially after washing thrice with distilled water and then vacuum-dried (the pressure less than 30 Pa) for 3 h in a freeze-dryer. Then, they were put under SEM to observe after being sliced up and metal-sprayed. Figure 4 shows the SEM images. The cellular structure is shown in Fig. 4 a), acquired at × 5000 magnification, and all the holes inside of the granule formed a channel through which nutrition could be shipped. ANAMMOX cenobium was immobilized on the thick wall of gel from Fig. 4 b) acquired at × 5000 magnification, and the structure was proved to be useful for ANAMMOX growth. Images of Fig. 4 c) were acquired at × 10000 magnification, which clearly showed that there were plenty of spherical bacteria.

4 CONCLUSION

There was obvious nitrogen removal in all the immobilized gel samples prepared by PVA and hence the immobilization method was successful and could be used to decrease the biomass loss in wastewater treatment. The preparation method of gel granules had a great impact on nitrogen removal: the ANAMMOX–PVA gel granules prepared by chemical cross-linking recovered in 24 h and NRR reached 60% in 48 h, whereas those produced by physical cross-linking could not remove any nitrogen in 72 h, so the method of cross-linking in calcium chloride solution was suitable for reactor setup phase. The nitrogen removal capability of ANAMMOX–PVA gel granules was under influence and inversely proportional to PVA concentration (10%), which was higher than the other three.

REFERENCES

Furukawa, K., Y. Inatomi, S. Qiao, et al. (2009). Innovative treatment system for digester liquor using anammox process. Bioresour. Technol. 100, 5437.

Hsia, T.-H., Y.-J. Feng, et al. (2008). J. PVA-alginate immobilized cells for anaerobic ammonium oxidation (anammox) process Ind. Microbiol. Biotechnol. 35, 72.

Isaka, K., Y. Date, T. Sumino, et al. (2006). Growth characteristic of anaerobic ammonium-oxidizing bacteria in an anaerobic biological filtrated reactor Appl. Microbiol. Biotechnol. 70, 47.

Magrí, A., M.B. Vanotti, A.A. Szögi (2012). Anammox sludge immobilized in polyvinyl alcohol (PVA) cryogel carriers Bioresource Technology. 114, 231.

Okabe, S., M. Oshiki,et al. (2011). Development of long-term stable partial nitrification and subsequent anammox process Bioresource Technology. 102, 6801.

Sliekers, A.O., N. Derwort, et al. (2002). Completely autotrophic nitrogen removal over nitrite in onesingle reactor. Water Res. 36, 2475.

Zhu, G.L., Y.Y. Hu, Q.R.Wang (2009). Nitrogen removal performance of anaerobic ammonia oxidation co-culture immobilized in different gel carriers. Water Sci. Technol. 59, 2379.

Civil, Architecture and Environmental Engineering – Kao & Sung (Eds)
 ISBN 978-1-138-02985-9

Seismic response of a long-span cable-stayed bridge with slip-shear metal damper

Changke Jiao
Engineering General Institute of Shanghai Construction Group, Shanghai, China

Xin Dong
Tongji Architectural Design (Group) Co. Ltd., Shanghai, China

ABSTRACT: The relative displacement between floors is utilized by a Shear Metallic Damper (*abbr. SMD*) to dissipate energy during earthquake and has been used widely in buildings to protect the main structure from damage under strong seismic motions. *SMD* can be driven by the relative displacements between girder and towers or piers to dissipate energy for long-span cable-stayed bridge with floating or half-floating structural style. Hazardous internal force may be induced by implementing an *SMD* between girder and tower or piers directly, as the normal displacement requirements for temperature or vehicles are needed for long-span cable-stayed bridge with floating or half-floating structural style, and it may damage the superstructure of the substructure of bridges. In this paper, a strategy is presented to eliminate the above hazardous internal force by implementing a limit groove above *SMD* with initial gaps in moving direction, named as Slip-shear Metal Damper (*abbr. SSMD*). The initial gaps can satisfy the daily displacement requirement, and the *SMD* can dissipate energy once the displacement between girder and tower or pier exceeds the initial gaps. In this study, the implementation of a hysteric model for *SSMD* is presented and the seismic reduction of *SSMD* for long-span cable-stayed bridge is investigated.

1 INTRODUCTION

Shear Metallic Damper (abbr. *SMD*) has been widely used in building structures to reduce the seismic response due to the excellent performance of dissipating energy during earthquake. The yield displacement of *SMD* is very small as the allowed displacements between adjacent floors of buildings are small, and the stiffness of *SMD* is relative large in the deformed plane. The initial stiffness can be strengthened by extending the steel plate of the damper in its own plane, the energy dissipating ability can be increased by changing the geometry of the steel plate of the damper, which was proposed by Li (2006), and quasi-static tests of five types of *SMD* were carried out. A novel steel shear panel damper called a buckling restrained shear panel damper (abbr. *BRSPD*) was proposed by *Deng* (2014, 2015), in which two restraining plates were used to limit the out-plane displacement of energy dissipation plate, quasi-static tests of five specimens were carried out to investigate the performance of the *BRSPD*s, and the dampers were used to reduce the seismic response of the bridge. However, the expansion caused by temperature was not considered. The hysteric model for *SMD* and Buckling-Restrained Brace (*abbr. BRB*) can be simplified by bilinear model (Lanning, 2016). Hazard internal force may be caused by installing *SMD*s or *BRB*s between girder and piers or tower directly for bridge, in which expansion movements are required by bridges usually. The dynamic characteristic of the bridge may change dramatically by installing *SMD*s or *BRB*s directly, which may influence the seismic response of bridge. *BRB* were used in long-span bridges for seismic mitigation by Lanning (2016), and Vincent Thomas Bridge was taken as a case study. Several *BRB*s were installed between girder and towers directly, which causes the first three mode frequency to increase about 22%, 30%, and 4%, respectively. The effects of bilinear spring on the seismic response of multispan cable-stayed bridges were determined by *Okamoto* (2011), and bilinear springs were very effective in reducing the dynamic displacements and bending moments of the towers. However, the effects of the bi-linear springs on the expansion caused by temperature were not considered. The effect of E-shape metal damper on seismic reduction of longitudinal displacement of Bosporus bridge was carried out by *Apaydln* (2010), and it was pointed out that the application of damper affects the distribution of internal loads because the E-shape damper connected deck and tower directly.

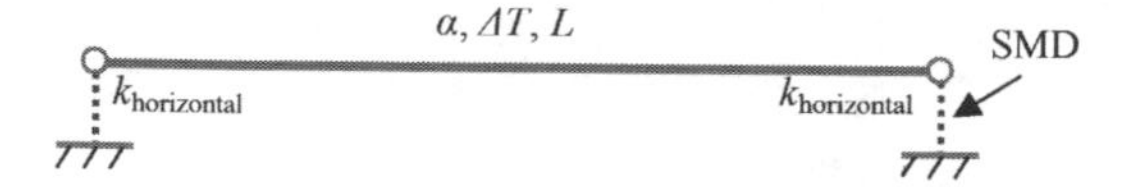

Figure 1. Schematic of the simplified girder model.

SMD can be driven by relative displacements between girder and towers or piers for long-span cable-stayed bridge with floating or half-floating supporting systems. There is a significant difference between the building structures and bridges, and expansion movements are required by bridges between girder and towers or piers, which may be caused by temperature or moving vehicles. For example, temperatures may cause a significant internal force in the girder. A simply supported steel bridge is taken as an example to demonstrate the effect of temperature on the expansion, and the diagram is shown in Fig. 1. The length of the bridge (L) is 500 m, the section area (A) is 0.9 m^2, and the temperature change (ΔT) is 20°C; the gravity of the bridge is ignored. The elastic module (E) and temperature coefficient (α) of steel are 2.06×10^8 kPa and 1.2×10^{-5} respectively. *SMD* was installed directly between girder and supports, with the lateral stiffness of *SMD* $k_{horizontal}$ of 2e5 kN/m and the free expansion of each end of 0.06 m (calculated by $\alpha \Delta T L/2$), and the axial force is 7795.4 kN (calculated by $F = 0.5\alpha\Delta T/[1/EA+1/(Lk_{horizontal})]$). It is obvious that the internal force caused by temperature is very high.

Improvements are proposed to ordinary *SMD* by adding a U-shaped groove on the top of *SMD*, with an initial gap y_{slip}, named as *SSMD*, and the hazard internal force (such as temperature internal force) can be avoided by the initial gaps. *SSMD* dissipates energy when the displacements between tower and girder or pier and girder exceed y_{slip}. In this paper, slip model for *SSMD* is present in the finite-element model, and the influence of *SSMD* on the seismic response of long-span cable-stayed double-deck bridge is investigated.

2 HYSTERIC MODEL OF SSMD

The main parameters of the bilinear model (Fig. 2(a)) of *SMD* are initial stiffness (k_1), yield displacement (y_d), and enhanced stiffness (k_2). The U-shaped groove above *SMD* can be modeled as a nonlinear elastic spring (Fig. 2(b)) with slip gap y_{slip} and big stiffness k_∞. The hysteric model of *SSMD* can be simulated by a combination of bilinear spring and the nonlinear elastic spring (Fig. 2 (c)), and it will be demonstrated by a single degree of freedom with *SSMD*. The configuration of *Sdof* is shown in Fig. 3(a), and the ground motion (u_g)

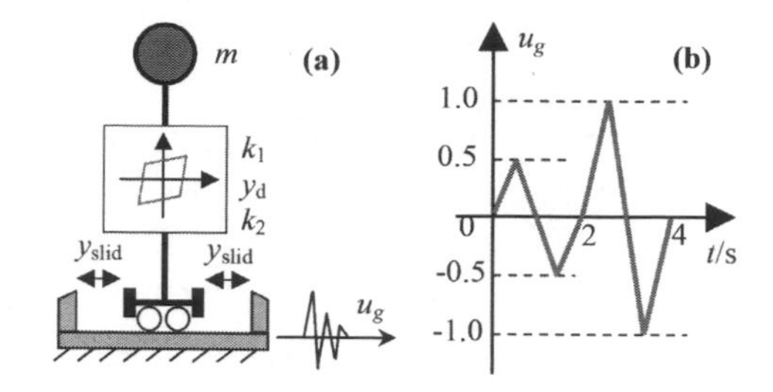

Figure 2. Hysteric model for dampers (a. bilinear model, b. limit groove, and c. bilinear model with slip gap).

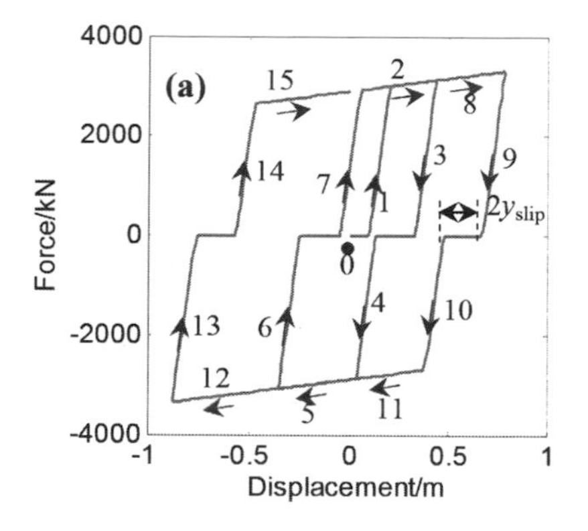

Figure 3. Configuration of *Sdof* (a) and the applied displacement history (b).

Table 1. Parameters of *Sodf*.

Parameter	Value
k_1/kN/m	3×10^4
y_d/m	0.1
k_2/kN/m	600
y_{slip}/m	0.1
k_∞/kN/m	3×10^8
m/t	100

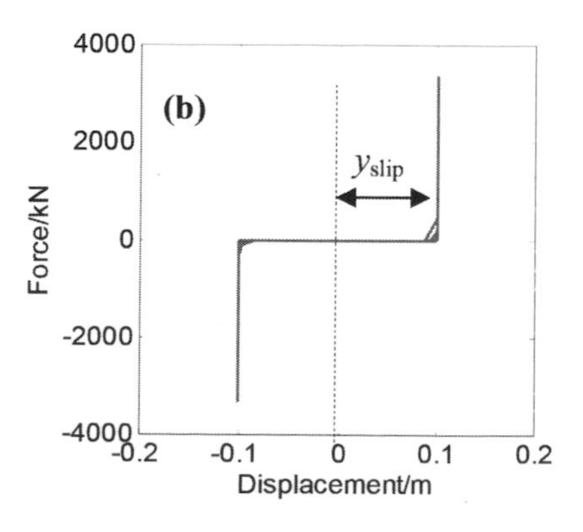

Figure 4. Hysteric curves for *SSMD* (a) and the sliding-stop slot (b).

is depicted in Fig. 3(b). The parameters of *SSMD* are listed in Table 1. Automatic dynamic increment time step is adopted because the stiffness suddenly changes at the maximum slip displacement (y_{slip}), and the minimum step allowed in the analysis is set to 1×10^{-5} s.

Hysteretic curves for *SSMD* and the groove are depicted in Fig. 4, and the developments of SSMD are the same as the one shown in Fig. 1(c). From the sudden changes happened to the nonlinear elastic spring at the maximum displacement of y_{slip}, it is clear that the combination of bilinear and nonlinear elastic spring can be used to simulate the behavior of *SSMD* properly.

3 FE MODEL

The FE model for Shanghai Minpu Bridge was established with beam and shell elements (Fig. 5 Jiao 2013). The bridge has a span of 1212 m and the arrangement is 4 × 63 m + 708 m + 4 × 63 m. There are eight lanes on the top deck and six on the bottom deck. The girder of mid-span is a combination of orthotropic steel truss and bridge decks. The side-span is a combination of concrete decks and steel truss web members. The web members include oblique, side-oblique, and vertical ones. The H-shaped reinforced concrete tower has two transverse beams, and the height of the tower is 210 m. Cables are simulated with a single truss element with no compression. Initial stresses were set in cables before analysis. The frequencies of the first lateral, vertical, and torsional vibration modes are 0.08275, 0.2575, and 0.28547 Hz, respectively.

Four *SSMD*s are arranged between towers and girder, and the parameters for *SSMD*s are shown

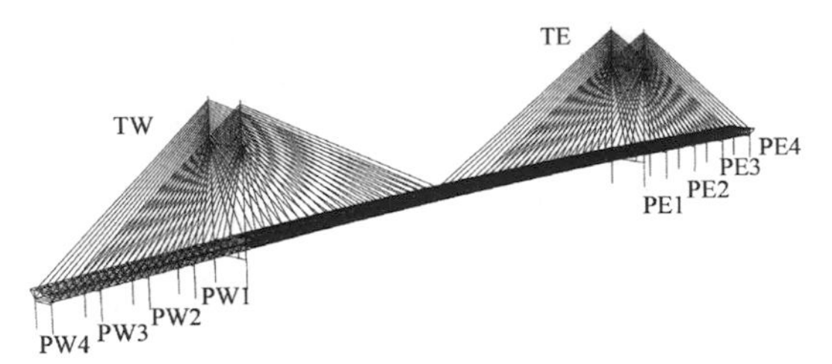

Figure 5. FE model of minpu bridge.

Table 2. Parameters of *SSMD*.

Parameter	Value
k_1/kN/m	9×10^5
y_d/m	0.02
k_2/kN/m	0.02 k_1
y_{slip}/m	0.05 and 0.08

Table 3. Effects of temperature on the relative displacement between girder and tower and pier.

	Negative (−30°C)		Positive (+30°C)	
	PW4	PE4	PW4	PE4
W/o SMD	195.7	−195.6	−195.6	195.7
With SMD	177.4	−177.4	−175.7	175.8
Difference/mm	−18.3	18.3	19.9	−19.9
Difference/%	−9.4	−9.3	−10.2	−10.2

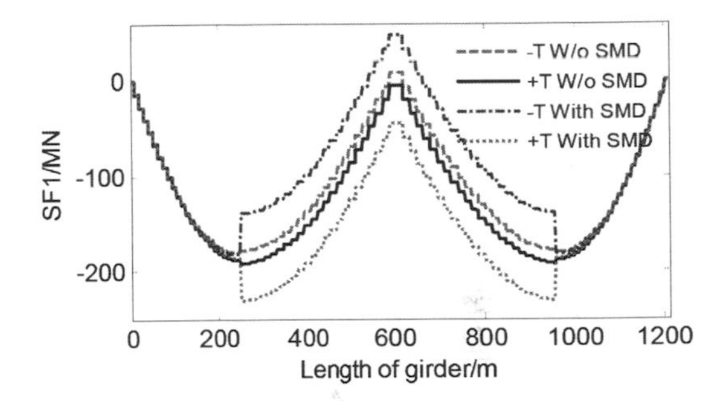

Figure 6. Effect of temperature on axial force of girder.

in Table 2. Two groups of slip gap are considered during the nonlinear dynamic time history analysis. The ground motion or *Tianjin* is applied at the bottom of the tower and piers, and the magnitude of ground motion is adjusted to 0.3 g.

Temperature effects on the static displacements between girder and pier (PW4 and PE4) at the end of the girder are shown in Table 3, in which negative (−30°C) and positive (+30°C) values indicate that the temperature of structure rises and falls 30°C from zero. Fig. 6 illustrates the effect on the axial force of girder (identified as SF1), when *SMD*s arc installed directly between towers and girder. It is can be found that the variety of axial force of girder are dramatically up to 25%, which may induce a harmful influence on the bridge and can be avoided by *SSMD* with initial gaps.

4 SEISMIC RESPONSES ANALYSIS

Seismic responses of the bridge with *SSMD*s are shown in Figs. 7 and 8. The hysteric curves of *SSMD*s are divided into two parts by the distance of $2y_{slip}$ (Fig. 7(a)). It is indicated that the *SSMD* may dissipate less energy than *SMD* during the ground motion because *SMD*s dissipate energy all the time while slip motions happen to *SSMD*s without dissipating energy. The time history of longitudinal displacements between girder and

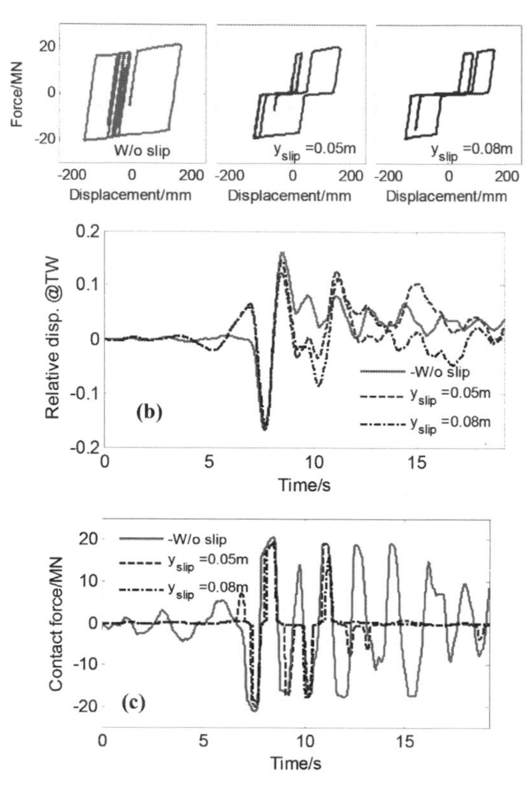

Figure 7. Hysteric curves of *SSMD* (a), history of relative displacement between girder and tower (b), and contact force between groove and SMD (c).

tower is totally different from that without slip gaps (Fig. 7(b)), and the connection strength between towers and girder by *SSMD*s is weaker than that by *SMD*s. The time history of contact forces between groove and *SMD*s is demonstrated in Fig. 7(c). The contact forces exist all the time when *SMD*s are installed directly between girder and towers during the ground motion and change with the relative displacement between girder and towers. However, zero contact force appears for *SSMD*, when *SMD*s are separated from the above grooves.

Envelop of shear forces (identified as SF2) and bending moments (identified as SM1) of tower column are shown in Fig. 8(a). Shear and moment below the transverse beam of tower are reduced about 20% by *SSMD*s compared to that of *SMD*s. It is due to the fact that as some of inertia force of girder is not translated to the transverse beam of tower by *SSMD*s, it has little influence on the shear and moment for column above transverse beam. Envelopes for axial forces of oblique (identified as XYXFG) and vertical (identified as XYSFG) web members are depicted in Fig. 8(b), and *SSMD*s can also reduce the maximum axial forces for some

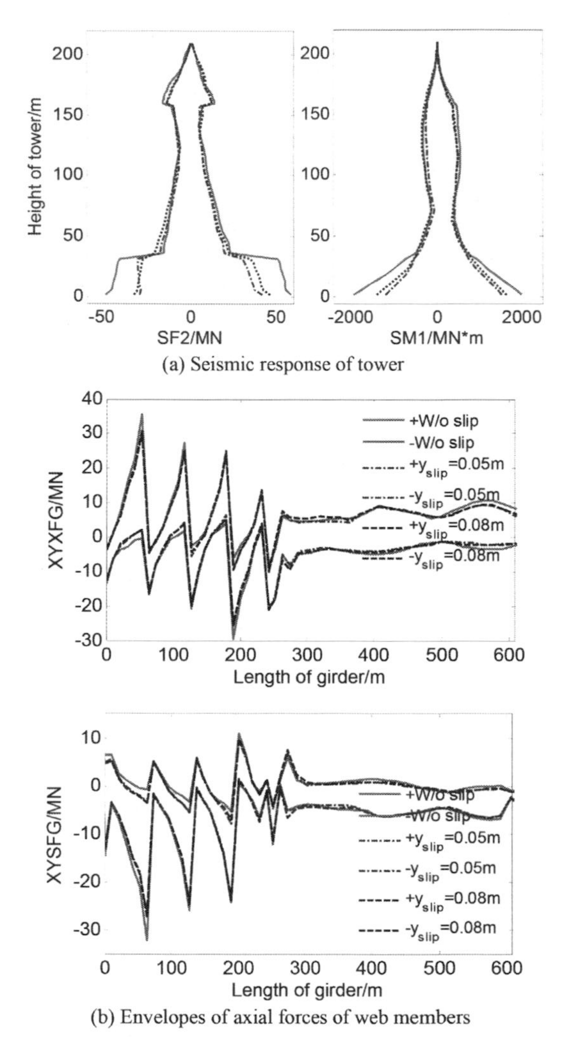

Figure 8. Seismic response of tower (a) and the envelopes of axils forces of web members (b).

web members especially near piers, which may be up to 20%.

5 CONCLUSIONS

The disadvantage of installing *SMD*s between gird and piers or towers directly may arise due to internal restrain, and a strategy is presented in this paper to eliminate the above hazardous internal force by implementing a limit groove with initial gaps between *SMD* and the groove, namely *SSMD*. The initial gaps can satisfy the daily displacement requirement, and the *SMD* can dissipate energy once the displacements between girder and tower or pier exceed the initial gaps. The implementation of hysteric model for *SSMD* is present, and a case study is performed in detail. The results show that:

1. The strategy is feasible and the implementation of the hysteric model is reasonable.
2. *SSMD* can overcome hazardous internal forces caused by increase in the decrease of temperature or motions of vehicles.
3. Compared with ordinary *SMD*s, *SSMD*s can reduce the connection stiffness between girder and piers or towers, and the displacement response may increase slightly and internal forces reduce significantly when proper parameters of *SSMD*s are chosen.
4. It is recommended that the ordinary *SMD* should not be installed directly on bridge to dissipate energy, and an *SSMD* with proper parameters may be a good alternative option.

ACKNOWLEDGMENTS

This study was supported by the National Natural Science Foundation of China under Grant No. 51408355; National Natural Science Found for Young Scholars of China under Grant No. 51408353; and Plan of Morning Star for Young Scholars of Shanghai under Grant No. 15QB1404800.

REFERENCES

Apaydln N.M (2010). Earthquake performance assessment and retrofit investigations of two suspension bridges in Istanbu. *Soil Dynamics and Earthquake Engineering*, 30: 702–710.

Corte G.D., M. D'Aniello & R. Landolfo (2013). Analytical and numerical study of plastic overstrength of shear links. *Journal of Constructional Steel Research*, 82: 19–32.

Gang Li & Li Hongnan (2006). Study on vibration reduction of structure with a new type of mild metallic dampers. *Journal of Vibration and Shock*, 25(3): 66–72.

Jiao Changke & Aiqun Li (2013). Research on random seismic response of long-span cable-stayed bridge under multi- excitation. *Journal of Vibration Engineering*, 26(5): 707–714.

Kailai Deng, Pan Peng & Li Wei, *et al* (2015). Development of a buckling restrained shear panel damper. *Journal of Constructional Steel Research*, 106: 311–321.

Kailai Deng, Pan Peng & Yukun Su, *et al* (2014). Development of an energy dissipation restrainer for bridges using a steel shear panel. *Journal of Constructional Steel Research*, 101: 83–95.

Lanning J., G. Benzoni & C.M. Uang (2016). Using buckling-restrained braces on long-span bridges rull-scale testing and design implications. I: full-scale testing and design implications. ASCE, *Journal of Bridge Engineering*, 04016001.

Lanning J., G. Benzoni & C.M. Uang (2016). Using buckling-restrained braces on long-span bridges. II: feasibility and development of a near-fault loading protocol. ASCE, *Journal of Bridge Engineering*, 04016002.

Okamoto Y. & S. Nakamura (2011). Static and seismic studies on steel/concrete hybrid towers for multi-span cable-stayed bridges. *Journal of Constructional Steel Research*, 67(2): 203–210.

Civil, Architecture and Environmental Engineering – Kao & Sung (Eds)
© 2017 Taylor & Francis Group, ISBN 978-1-138-02985-9

Research on the interfacial bond behavior between CFRP sheet and steel plate under the static load

Long Zhang & Shuang-Yin Cao
School of Civil Engineering, Southeast University, Nanjing, China

ABSTRACT: The interfacial bond behavior between CFRP and steel plate plays an important role in the strengthening effects of steel structure reinforced by CFRP. In this paper, experimental studies based on double-shear specimen were conducted under static load, which considered the influence of different bond width and layer number. This paper presented the failure modes, bond strength, strains of CFRP, and bond–slip relationship. Test results showed that the bond strength of the specimen could be significantly improved by the way of bonding layers of CFRP sheets or increasing the bond width although the ductility was reduced because of layers of CFRP. The bond–slip curve between the interface of CFRP sheet and steel plate had an approximately bilinear shape.

1 INTRODUCTION

Fiber-Reinforced Polymer (FRP), especially Carbon Fiber-Reinforced Polymer (CFRP) has gained more and more attention in strengthening engineering structures for its high strength-to-weight ratio, stiffness-to-weight ratios, and corrosion resistance. The study of steel structure strengthening by CFRP started relatively late and was less compared to that on the strengthening of concrete structure by CFRP. Recent studies have found that external bonding with CFRP can effectively improve the flexural (Deng et al. 2004, Linghoff & Al-Emrani 2010), tensile (compressive) (Shaat & Fam 2004, Sawulet et al. 2012), and antifatigue (Bocciarelli et al. 2009, Wu et al. 2012) performance of the steel structure.

As the most critical part, the interfacial bond behavior directly determines the strengthening effect of steel structure reinforced by CFRP. At present, the research on the interface bond mechanism and the peeling properties of steel structure reinforced by CFRP is relatively less (Yu et al. 2012). In this paper, experimental studies considering different bond width and layer number were conducted under static load on the basis of six double-shear specimens. The failure modes, bond strength, strains of CFRP, and bond–slip relationship were analyzed, which would provide a basis for the design of the steel structure reinforced by CFRP.

2 EXPERIMENTAL WORKS

2.1 *Material property*

The ordinary CFRP sheets (MKT-CFC300) and matched epoxy resin adhesive were supplied by NJMKT Corp. According to the manufacture's catalog, the mechanical properties of the CFRP sheets (MKT-CFC300 with a thickness of 0.167 mm) were Young's modulus (210 GPa) and tensile strength (3750 MPa). The mechanical properties of the adhesive were shear strength (16.2 MPa) and Young's modulus (2041 MPa). The steel plates with a thickness of 5 mm were made of Q235 steel. Young's modulus and yield strength were 201GPa and 263 MPa, respectively.

2.2 *Specimen preparation*

Six double-shear specimens were designed as shown in Fig. 1. The figure shows that two steel plates with a gap of 2 mm were connected by pasting CFRP sheets onto both sides. In order to observe the whole process of debonding failure, the specimen was designed to consist of two parts: one with a bond length of 200 mm called the testing part and the other one with a bond length of 230 mm called strengthening part, which could make sure that the debonding failure occurred at the testing part first. Before pasting CFRP sheets, the surface of the steel plates should be cleaned first by using an abrasive disk for removing the dust and then scrubbed with absolute ethanol. The specimens were put into a temperature and humid-

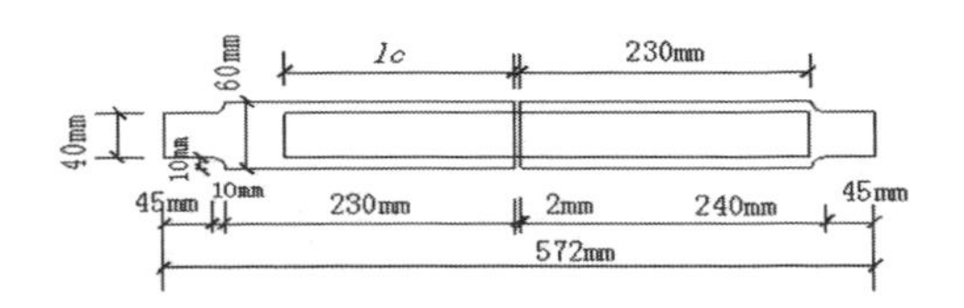

Figure 1. Schematic of specimen.

ity chamber for curing for 7 days when CFRP sheets were pasted onto the surface of steel plates. The test control parameters are shown in Table 1.

2.3 *Arrangement of measuring points and testing process*

For each specimen, some strain gauges were installed on the surface of testing part in order to measure the strain distribution on the surface of CFRP sheets. One strain gauge was installed at a distance of 5 mm from the middle span while the rest of them were installed with a gap of 15 mm from each other along the longitudinal axis. Particularly, strain gauges were installed in a double row on S120–55–1 for discussing the regularities of transverse strain distribution as shown in Fig. 2. The static strain recorder (DH3816) was used to collect the static strain values.

All specimens were tested by a universal testing machine (CMT5105) with a maximum capacity of 100 kN, as shown in Fig. 3. The displacement control mode was adopted with a loading rate of 0.2 mm/min in this test. The test was stopped when the CFRP sheets stripped from the steel plate completely.

Table 1. Specimen number and control parameters.

Specimen no.	Bond length/mm	Bond width/mm	Number of layers
S200-40-1	200	40	1
S200-40-2	200	40	2
S200-40-3	200	40	3
S200-30-1	200	30	1
S200-55-1	200	55	1
S120-55-1	120	55	1

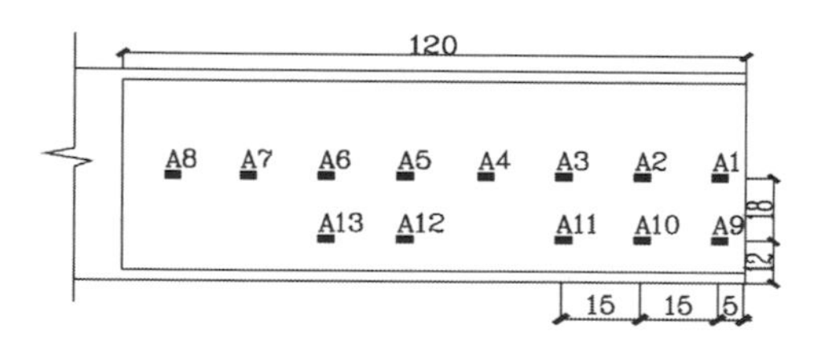

Figure 2. Demonstration of mounted strain gauges on S120-55-1.

Figure 3. Test setup.

3 TEST RESULTS AND DISCUSSIONS

3.1 *Fatigue test phenomenon*

In the initial stage, there was no obvious change in the appearance of the specimens, which were in elastic phase. After a certain period of loading, a slight fabric tearing sound could be heard, indicating CFRP strips started peeling from the middle span. Peeling of CFRP sheets propagated steadily with the increment of load, while the peeling developed rapidly with scratching noise after load reached 80% of the ultimate load. CFRP sheets stripped from the steel plate completely with a loud noise while the specimen reached the ultimate load, and then the test stopped.

3.2 *Failure mode*

Recent studies (Zhao & Zhang 2007) have shown that the common failure modes in a CFRP bonded steel system subjected to a tensile force include: (a) steel and adhesive interface failure; (b) cohesive failure; and (c) mixed failure including the first two modes. In this paper, only mode (a) and mode (c) appeared in the tests, which are shown in Fig. 4.

a. Mixed failure (S200-40-1)

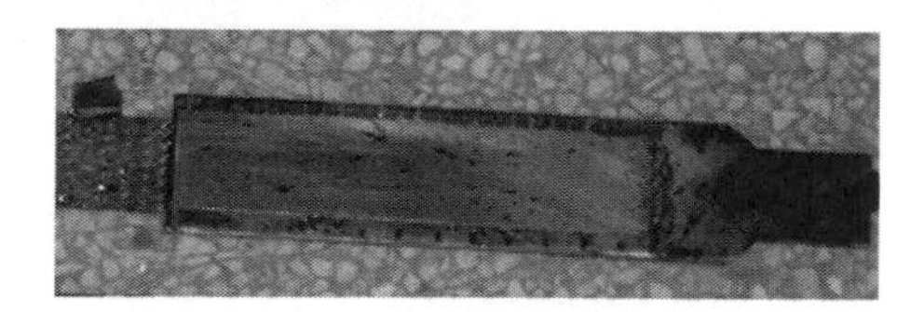

b. Steel and adhesive interface failure (S200-40-3)

Figure 4. Two failure modes.

Table 2. Test results.

Specimen no.	Bond strength/kN	Major failure mode
S200-40-1	14.72	Mode c
S200-40-2	22.81	Mode a
S200-40-3	23.08	Mode a
S200-30-1	11.01	Mode a
S200-55-1	28.22	Mode c
S120-55-1	28.01	Mode c

As shown in Table 2, the bond width and layer number have a directly influence on the failure mode of specimens. Mixed failure occurred in S200-40-1, whereas S200-40-2 and S200-40-3 are subjected to steel and adhesive interface failure. Meanwhile, the former shows a better ductility than the latter through analysis, as failure of the latter are controlled by the performance of steel-adhesive interface, which is a weaker region. Also, the failure modes of the specimens turn from mode (a) to mode (c) with the increase of bond width. This is because the bond area is enlarged with the increase of bond width, which can reduce bond stress and improve stress performance of the interface.

3.3 *Bond strength*

In Table 2, the bond strength of S200-40-2 increased as much as 55% compared with S200-40-1, whereas the bond strength of S200-40-3 has no significant improvement as S200-40-2. The results indicate that the bond strength is largely influenced by the layers of CFRP sheets within certain limits. And also, the bond strength of specimens rapidly increased with the increment of bond width, but the relationship was not linear. A conclusion that the bond length that exceeds the effective bond length (about 45 mm) has hardly caused any infection to the bond strength can be reached through the result of S120-55-1 compared with others.

3.4 *Strain distribution of CFRP*

The strains of S120-55-1 that have almost identical value along the transverse section are shown in Table 3, which indicates that the CFRP sheets are stressed uniformly along transverse section when the specimens are subject to axial tension.

Typical strain distribution of CFRP along the bond length under the static loading is shown in Fig. 5. It can be seen that the changing processes of strains include two phases: the strain distribution locate mainly in the small area of 25 mm nearby the middle span before the peeling and gradually decrease to zero with an increment of distance from the middle span. The strains increase with the increment of load, and the shear stress transfer of the interface is only completed by the friction after peeling. At the moment, from Fig. 5, there is a long straight segment, which coincidently equals the peeling length for almost all curves. The area of delamination enlarges with the increment of load, and develops gradually to the load-end until final failure occurs.

3.5 *Bond–slip relationship*

The bond–slip curve reflects the constitutive relation of the local interfacial bond performance. The average shear stress between the two strain gauges mounted on the longitudinal axis of the CFRP sheet were calculated using the following relationship:

$$\tau_i = \frac{(\varepsilon_{i+1} - \varepsilon_i) \times t_i \times E_i}{\Delta l_i} \tag{1}$$

where Δl_i is the distance between strain gauges i and i+1; ε_i is the strain in the CFRP sheet at strain gauges i; and E_i and t_i are the elastic modulus and thickness of the CFRP sheet, respectively. The local slip can be derived from equation (2):

$$s_i = \int \varepsilon(x)\,dx \tag{2}$$

The relationship between local slip of CFRP-steel and shear stress of CFRP at each measuring

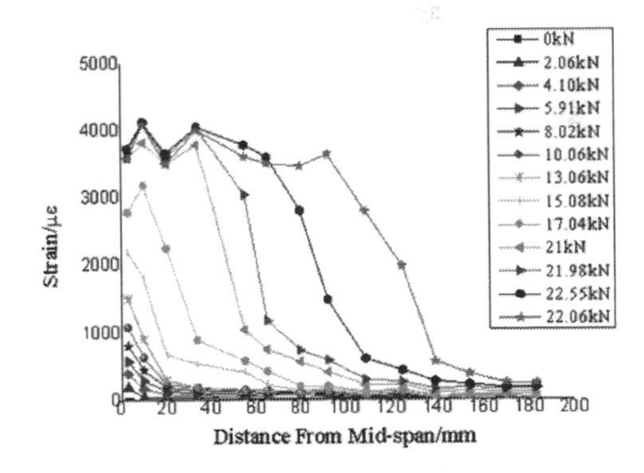

Figure 5. Strain distribution of S200-40-2.

Table 3. Strain of S120-55-1.

	Strain along longitudinal axis			Strain of the row sideways			Ratio		
Load	A1	A3	A6	A9	A11	A13	A9/A1	A11/A3	A13/A6
5.05 kN	292	70	71	309	73	72	1.06	1.04	1.01
6.99 kN	439	104	104	470	105	104	1.07	1.01	1.00
11.29 kN	887	182	180	963	181	175	1.09	0.99	0.97
14.83 kN	1462	267	256	1573	262	247	1.07	0.98	0.96
17.18 kN	1877	322	299	1988	314	290	1.06	0.98	0.97
20.79 kN	2753	434	392	2910	419	373	1.06	0.97	0.95

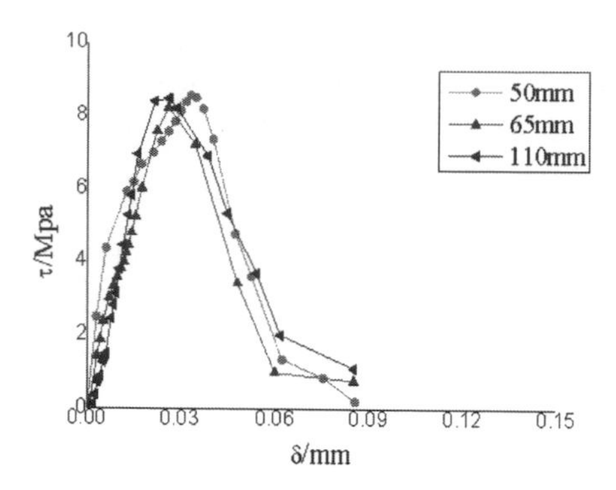

Figure 6. Bond–slip relationship of S200-40-3.

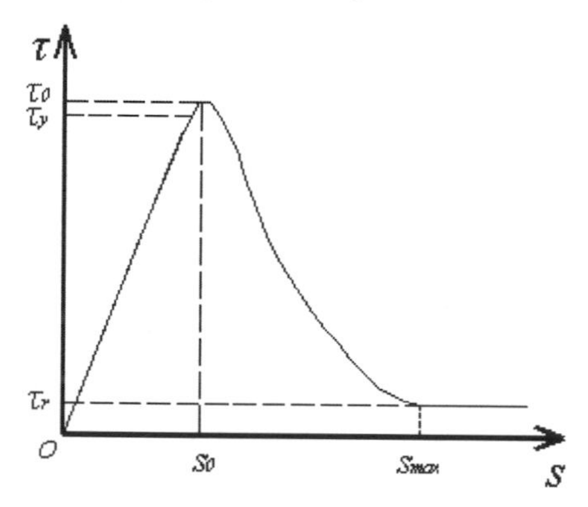

Figure 7. Approximate mode of bond–slip relation.

point can be calculated by Eqns. (1) and (2). Fig. 6 shows the typical bond–slip relationship curve, in which the horizontal and vertical axes represent the calculated slip and local bond stress, respectively, and the curve generally shows a bilinear shape. Also, previous study by Xia & Teng (2005) indicated that a bilinear model, which is similar to that for concrete, could be adopted for the CFRP-bonded steel system. Combining with the results of experiment and existing studies, the curve in Fig. 7 can approximately represent the interfacial bond–slip relationship of CFRP-steel.

It can be seen that the curve consists of an ascending elastic stage and a descending softening stage. In the elastic stage, the shear stress increases linearly with the increment of slip, while the specimen will enter the softening stage after the shear stress reaches maximum shear stress (τ_0). Then, the shear stress decreases with the increment of slip until τ_r, which is caused by the interfacial friction and test error. It should be noted that τ_r equals zero in an ideal conditions.

4 CONCLUSIONS

In this paper, experimental studies considering different bond width and layer number were conducted under static load based on six double-shear specimens. The following conclusions can be drawn:

1. Although the bond strength can be significantly improved by increasing the layers of CFRP sheets, the failure mode changes from mixed failure to steel and adhesive interface failure, which is a relatively brittle failure mode. However, the bond strength of the specimen rapidly increase with the increment of bond width, and the stress performance of the interface can be well improved. The bond length, which exceeds the effective bond length, can hardly cause any infection to the bond strength.
2. The CFRP sheets are stressed uniformly along the transverse section when the specimens are subject to axial tension. The strains and the area of delamination, which develops gradually to the load-end, increase with the increment of load until final failure occurs.
3. The bond–slip relationship of CFRP-steel generally emerges as a bilinear shape, which is similar to that for concrete and consists of an ascending elastic stage and a descending softening stage.

REFERENCES

Bocciarelli, M.et al. (2009). Fatigue performance of tensile steel members strengthened with CFRP plates. *Composite Strutures, 87(4),* 334–343.

Deng, J., Lee, M., Moy, S. (2004). Stress analysis of steel beams reinforced with a bonded CFRP plate. *Composite structures, 65(2),* 205–215.

Linghoff, D., Al-Emrani, M. (2010). Performance of steel beams strengthened with CFRP laminate–Part 1: laboratory tests. *Composites Part B: Engineering, 41(7),* 509–515.

Sawulet, B. et al. (2012). Experimental study on FRP fast anti-buckling strengthening technique for axial compressive steel members. *Engineering Mechanics,* 105–113 (in Chinese).

Shaat, A., Fam, A. (2004). Strengthening of short HSS steel columns using FRP sheets. *Proceedings of the 4th international conference on advanced composite materials in bridges and structures.* 2004.

Wu, G. et al. (2012). Experimental Study on the Fatigue Behavior of Steel Beams Strengthened with Different Fiber-Reinforced Composite Plates. *Journal of Composites for Construction, 16(2),* 127–137.

Xia, S., H., Teng, J., G. (2005). Behaviour of FRP-to-steel bonded joints. *Proceedings of the international symposium on bond behaviour of FRP in structures,* 419–26.

Yu, T. et al. (2012). Experimental study on CFRP-to-steel bonded interfaces. *Composites Part B: Engineering, 43(5),* 2279–2289.

Zhao, X., Zhang, L. (2007). State-of-the-art review on FRP strengthened steel structures. *Engineering Structures. 29(8),* 1808–1823.

Civil, Architecture and Environmental Engineering – Kao & Sung (Eds)
© 2017 Taylor & Francis Group, ISBN 978-1-138-02985-9

Influence of thick and loose sediments on the vacuum load transfer

Jiaxing Weng
Geotechnical Engineering Institute of Southeast University, Nanjing, Jiangsu Province, P.R. China

ABSTRACT: Field experiments were carried out to determine the influence of thick and loose sediments on the vacuum load transfer in vacuum preloading. The results indicate that it is greatly affected by the loose sediment when vacuum load is transferred to Prefabricated Vertical Drain (PVD) through sand layer and not influenced when the transfer is through a pipe system because of its good airtightness. Thus, pipe system is more suitable as the horizontal transfer medium of vacuum load than sand layer in the case of overlying loose and thick soil layer. Besides, vacuum pressure in the PVD (VP_{PVD}) decays rapidly in the loose sediment layer, especially in the unsaturated zone, and the decay rate increases with the increase of VP_{PVD} in this layer. The settlement analysis validates the transfer rule of vacuum load indirectly. Soil compression occurs mainly in the loose sediment layer because most of vacuum energy is consumed in this layer.

1 INTRODUCTION

Vacuum consolidation was proposed in the early 1950 by Kjellmann (1952). The studies and promotions of vacuum-induced consolidation have been continuing to date (Holtz, 1975; Choa, 1989; Cognon et al., 1994; Bergado et al., 1998; Tang & Shang, 2000; Chai & Miura, 2000; Mohamedelhassan & Shang, 2002; Indraratna et al., 2004, 2005; Chai et al., 2005, 2006, 2007, 2008; Yan & Chu, 2005; Walker & Indraratna, 2006, 2009; Rujikiatkamjorn & Indraratna, 2007; Rujikiatkamjorn et al., 2007, 2008; Saowapakpiboon et al., 2008a, b, 2009, 2010). Vacuum preloading technique is a soft ground improvement method that has been successfully applied to accelerate the rate of consolidation and to eliminate the instability problem. The engineering practice has proved that the vacuum preloading developed based on the soil consolidation theory is a relatively economic, convenient, and reliable method for soil improvement.

Generally speaking, vacuum preloading technique is most suitable to reinforce the reclaimed soft foundation and not applicable to the foundation with an overlying thick (not less than 3 m) and loose soil layer. It greatly limits the application of the vacuum preloading in engineering because thick and loose sediments in quaternary are widespread around the world (Guorui, 1991).

The transfer of vacuum load has a significant influence on the reinforcement effect of vacuum preloading technique. Thus it is necessary to determine the influence of thick and loose sediments on the vacuum load transfer in the vacuum preloading for the further improvement of the method.

2 FIELD EXPERIMENT

2.1 *Experimental site*

The field experiments were adopted here because it was difficult to simulate the thick and loose sediments in the laboratory. The experimental site was selected in the Chongwan zone of the west dyke of Grand Canal (pile No. 49+580-50+335) (Fig. 1) after extensive investigation. According to the engineering investigation data, the soil profile here mainly includes three layers. The upper layer is a loose sediment in quaternary with thickness of about 5 m. The second layer below is clayey sludge with the almost same thickness. And the lower layer is silty sandy loam. The geological conditions of the site meet the requirements of the experiment well.

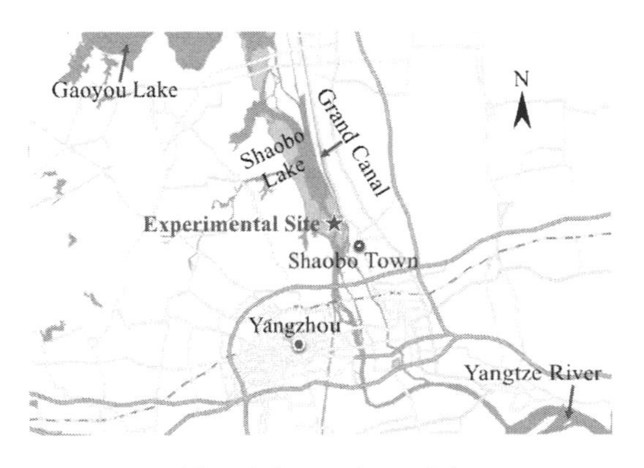

Figure 1. Position of the experimental site.

The plasticity chart of soil samples is shown in Figure 2. In accordance with the Unified Soil Classification System (USCS), loose sediment is low plasticity clay (CL), clayey sludge is high plasticity clay (CH), and silty sandy loam is low plasticity silt (ML).

2.2 *Experimental scheme*

A sand layer is generally used as the horizontal transfer medium of vacuum load in the vacuum preloading method (Fig. 3). Vacuum load generated by vacuum pump is transferred into Prefabricated Vertical Drains (PVDs) through the sand layer. Then, vacuum load is transferred into surrounding soil layers through PVDs. Both of the transfer chains above may be affected by the overlying loose sediment, which has direct interfaces with the sand layer and PVDs.

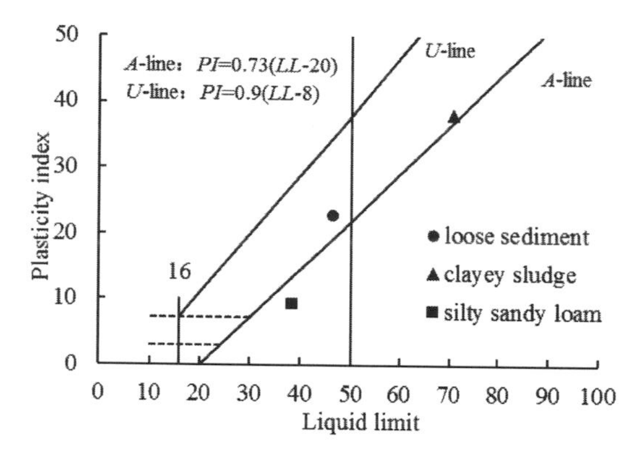

Figure 2. Plasticity chart of soil samples.

In order to determine the influence of the overlying thick and loose sediments on each transfer chain of vacuum load in vacuum preloading, a newly developed type of horizontal transfer medium of vacuum load was introduced here (Saowapakpiboon et al., 2008; Saowapakpiboon et al., 2011; Liang et al., 2013; Long et al., 2015). In this method, airtight pipe system is adopted to replace the sand layer (Fig. 4). PVDs are connected to the vacuum pump through an airtight pipe system to form a whole by a prefabricated fitting cap for each individual PVD. Thus, vacuum load generated by vacuum pump can be transferred into the surrounding soil layers through PVDs directly, and the overlying thick and loose sediments can only affect the vacuum load transfer in the PVDs. For convenience, the vacuum preloading methods with sand layer and pipe system are, respectively, referred to simply as SVP and PVP below. Schematic diagrams of SVP and PVP are shown in Figures 3–4, respectively.

2.3 *Implementation process*

A plain layout of the trial site is presented in Figure 5. PVP was applied in zone I, and SVP was applied in zone II. Major implementation processes (Fig. 7) are summarized as follows.

- Leveling the trial site and removing the sharp objects.
- Installing PVDs on a square grid with a spacing 1.0 m into an average depth of 15.0 m (Fig. 8)
- SVP: Layering the sand layer with medium-coarse sand, with mud content not more than 5%. A sand layer of 20 cm thickness was laid firstly. Then, perforated pipes were placed on the top of the sand layer to collect the water, and

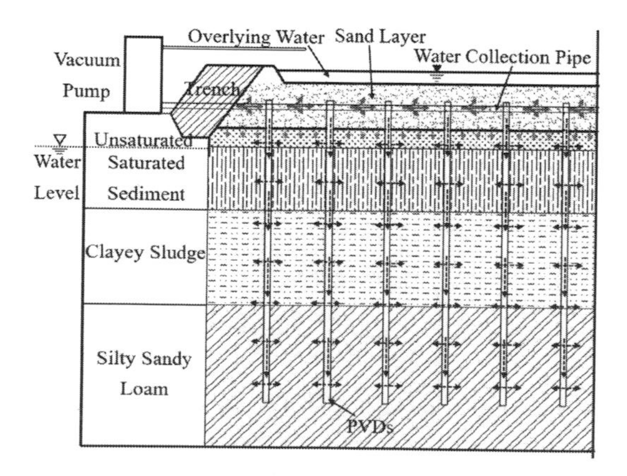

Figure 3. Schematic diagram of vacuum load transfer of SVP.

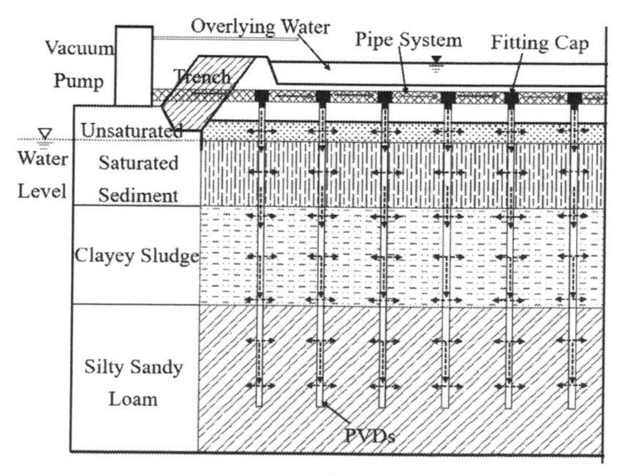

Figure 4. Schematic diagram of vacuum load transfer of PVP.

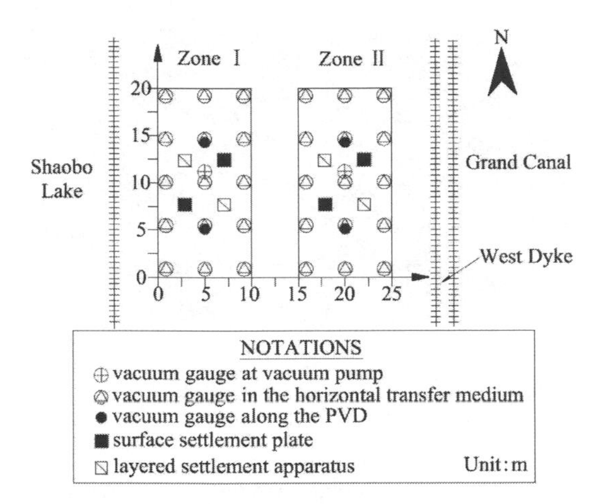

Figure 5. Plain layout of the trial site and monitoring instruments.

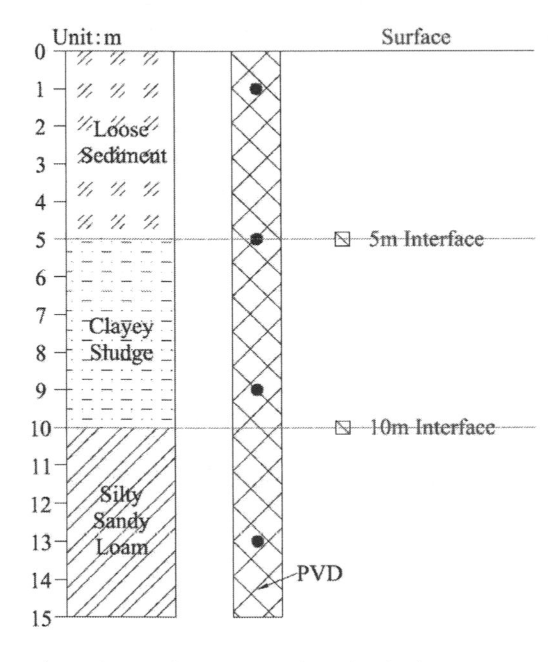

Figure 6. Profile diagram of monitoring instruments.

exposed parts of PVDs were put on the surface of the perforated pipes. Finally, laying of the sand layer was continued to 50 cm (Fig. 9).

- PVP: Assembling PVDs with fitting caps and reinforced pipes of 25 mm diameter. Then, connecting the branch pipes with the PVC main tubes to form a whole by fitting joints (Fig. 10).
- Installing monitoring instruments including vacuum gauges, surface settlement plates, and layered settlement apparatuses according to the plain layout and profile diagram of monitoring instruments (Figs. 5–6).

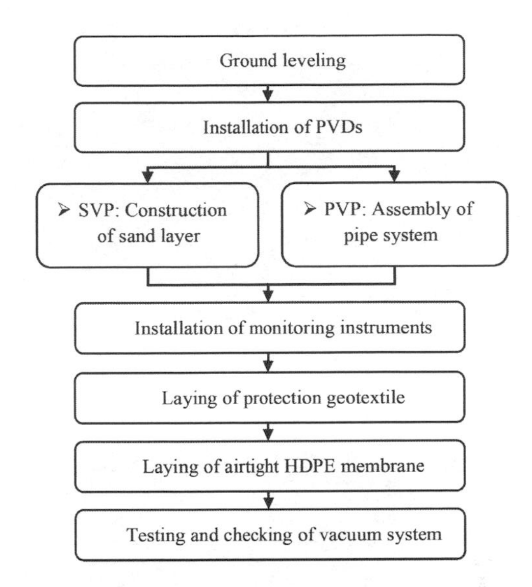

Figure 7. Implementation processes.

Figure 8. Installation of PVDs.

Figure 9. Construction of the sand layer.

- Laying down a layer of nonwoven geotextile with 350 g/m^2 to protect the High-Density Polyethylene (HDPE) membranes against punching from sharp materials (Fig. 11).
- Laying down three layers of the airtight HDPE membranes with 1.5 mm thickness on the top of protection geotextile (Fig. 12). Then, the edges

Figure 10. Assembly of pipeline system.

Figure 11. Laying of protection geotextile.

Figure 12. Laying of airtight HDPE membrane.

Figure 13. Excavated trench.

Figure 14. Testing and checking of airtightness.

of membranes were embedded in the excavated trench with about 1.5 m depth and backfilling with in situ clay and bentonite slurry (Fig. 13).

- Starting vacuum pumping for testing and checking of airtightness (Fig. 14). Then, covering the membrane with water to prevent aging of the membrane.

3 RESULTS AND DISCUSSION

3.1 *Vacuum pressure at vacuum pump*

Each zone was controlled independently by one vacuum pump with a vacuum gauge. Vacuum pressure at Vacuum Pump (VP_{VP}), which is shown in the vacuum gauge, can reflect the working performance of the vacuum pump and airtightness of the trial zones. The magnitude and change rate of VP_{VP} are plotted against time in Figures 15 and 16, respectively.

It can be seen that change law for VP_{VP} of the two vacuum pumps is almost the same. VP_{VP} rises to 80 kPa rapidly in the first 10 days, then increases slowly between 10 and 25 days, and finally stabilizes at about 85 kPa. This is because air under the airtight membrane is pumped out rapidly at an early stage of vacuum pumping and then the change rate of VP_{VP} decreases with the decreasing of air density (Fig. 16). It indicates that the trial zones are well sealed and the initial conditions of the two trial zones are basically the same.

3.2 *Vacuum pressure in the horizontal transfer medium*

Vacuum gauges were installed at different locations of the horizontal transfer medium to monitor the distribution of the vacuum pressure in the horizontal transfer medium (VP_{HTM}). VP_{HTM} of different locations are shown in Figure 17. VP_{HTM} of different locations is almost the same whether in

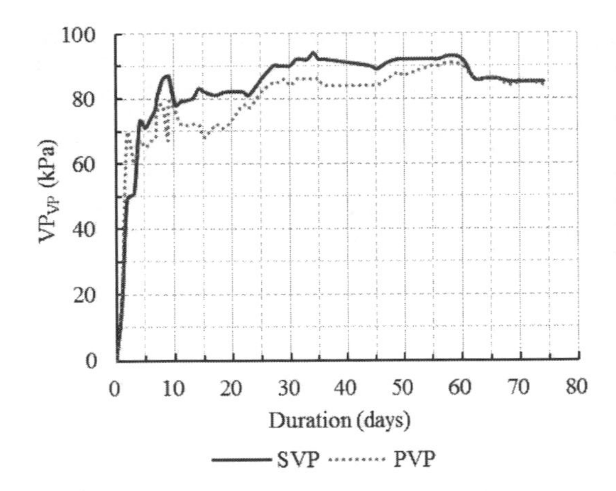

Figure 15. Vacuum pressure at vacuum pump.

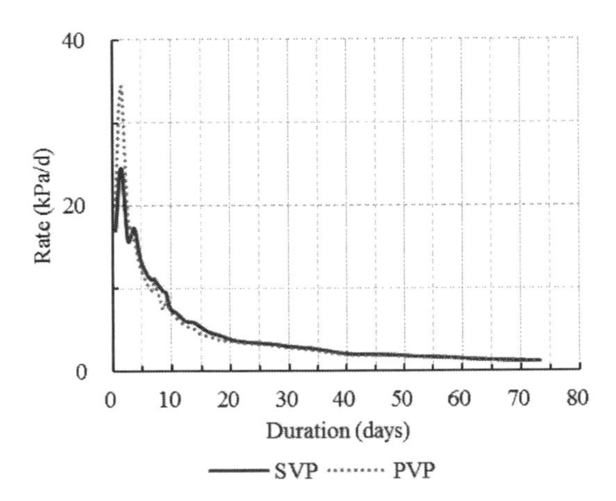

Figure 16. Change rate of vacuum pressure at vacuum pump.

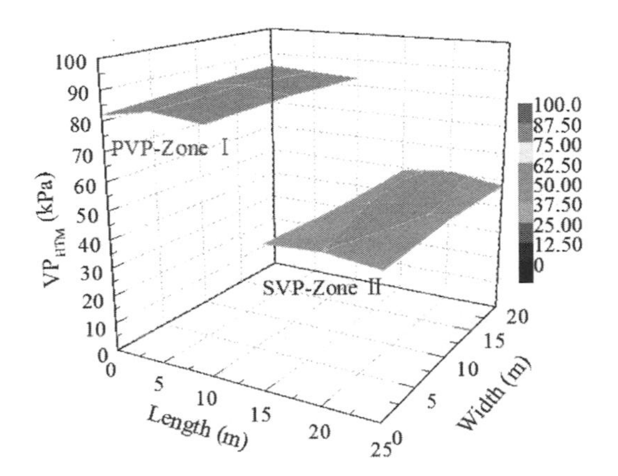

Figure 17. Distribution of the vacuum pressure in the horizontal transfer medium.

Figure 18. Status of the overlying loose sediment.

the sand layer or pipe system. VP_{HTM} in the pipe system is about 85 kPa, which is almost the same as VP_{VP}. VP_{HTM} in the sand layer is only about 50 kPa, which is 35 kPa lower than that in the pipe system.

However, the engineering practice proved that VP_{HTM} in the sand layer could also achieve 80–90 kPa (Shang, 1998) under good sealing condition. Therefore, it is most likely due to the influence of the loose sediment. It belongs to the point-to-point contact that pipe system directly connects with the PVDs. It is not influenced by the loose sediment when vacuum load is transferred to the PVDs through the pipe system because of its good airtightness (Fig. 4). To the sand layer, it belongs to the surface-to-surface contact that the sand layer directly overlays on the loose sediment. It may be greatly affected by the loose sediment when vacuum load is transferred to the PVDs through the sand layer because both of them are loose and porous. According to the engineering geologic investigation, buried depth of underground water level is about 1.0 m. Thus, the loose sediment layer can be divided into unsaturated loose sediment (about 0.0–1.0 m) and saturated loose sediment (about 1.0–5.0 m) further. Large numbers of plant roots and fissures in the unsaturated loose sediment (Fig. 18) can also be the vacuum transfer channels besides PVDs. Therefore, large amounts of vacuum energy can be consumed on the interface between the sand layer and the unsaturated loose sediment or in the soil cracks (Fig. 3).

Therefore, it can be inferred that the pipe system is more suitable as the horizontal transfer medium of vacuum load than the sand layer in the case of overlying loose and thick soil layer.

3.3 *Vacuum pressure in the PVD*

Vacuum pressure along the PVD is a key parameter of vacuum preloading technique. Thus, vacuum pressure was measured at the different depths of PVD to identify the effect of loose sediment on the vacuum load transfer in the PVD. Vacuum pressure at the top of the PVD (VP_{TOP}) is approximately equal to the average of VP_{HTM}. Vacuum pressure in the PVD (VP_{PVD}) after the stabilization of VP_{VP} is plotted against depth in Figure 19. It can be seen that VP_{PVD} has a clear loss along the depth direction of the PVD. And VP_{PVD} values of PVP at different depths are significantly greater than those of SVP at the corresponding depths. VP_{PVD} of PVP can be transferred up to 13.0 m in PVP, whereas that of SVP cannot be detected at the depth of 13.0 m all the time. This indicates that the transfer depth of vacuum load is less than 13.0 m in the SVP zone.

Besides, it can also be observed that the decay of VP_{PVD} is obviously influenced by the properties of the surrounding soil layers. VP_{PVD} decays rapidly in the loose sediment layer, especially in the unsaturated zone. It decays almost linearly and more slowly in the soil layers beneath the loose sediment. The VP_{PVD} curve of "Translational SVP" in Figure 19 is obtained by translating the curve of "SVP" until the two VP_{PVD} points at the depth of 5.0 m coincide. Comparing with two curves, the decay rate of VP_{PVD} increases with the increase of VP_{PVD} in the loose sediment layer. However, in the soil layers beneath the loose sediment, the decay rate of VP_{PVD} of SVP and PVP is almost the same with a constant being about 3–4 kPa/m. It indicates that the decay rate of VP_{PVD} is irrelevant to the magnitude of VP_{PVD} in the soil layers beneath the loose sediment.

3.4 *Surface and layered settlement*

Surface settlement plates and layered settlement apparatuses were installed to monitor the settlement of ground surface settlement and different soil layer interfaces, respectively. The settlement data can be used to validate the transfer rule of vacuum load indirectly because the compression of soil layers is caused by the vacuum load. The surface and layered settlement of SVP and PVP is plotted against time in Figures 20 and 21, respectively. In the SVP zone, the compression of the loose sediment layer is about 27.9 cm, which is almost equal to the surface settlement. The settlement of the 5 and 10 m interface is not detected. In the PVP zone, the compression of the loose sediment layer is about 26.5 cm, which accounts for about 78% of the surface settlement (34.0 cm). The compression of the clayey sludge layer is about 7.5 cm and that of the silty sandy loam is almost zero. Accordingly, soil compression caused by vacuum load occurs mainly in the loose sediment layer. It is in accordance with the analysis results

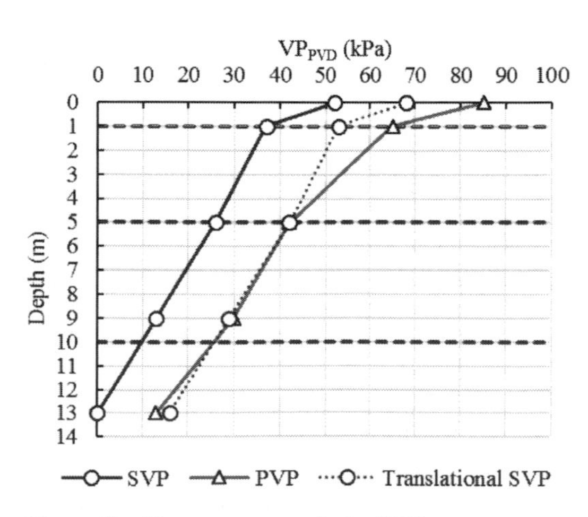

Figure 19. Vacuum pressure in the PVD.

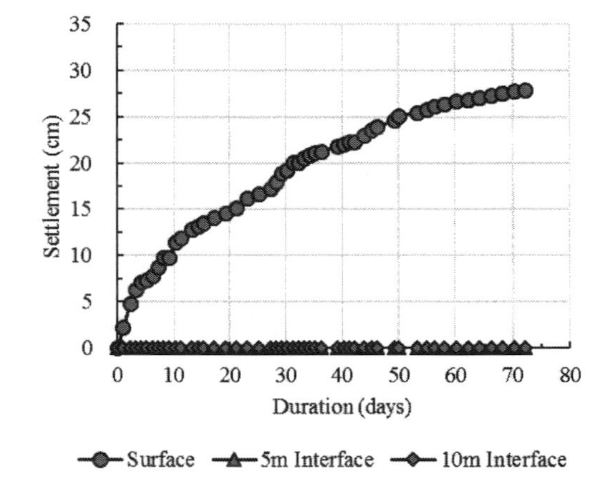

Figure 20. Surface and layered settlement of the SVP zone.

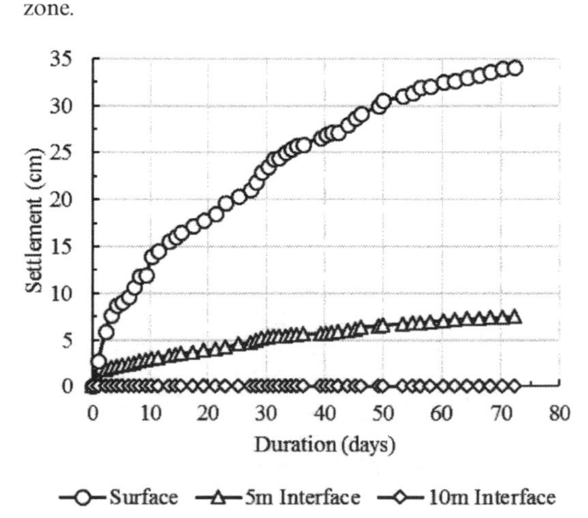

Figure 21. Surface and layered settlement of the PVP zone.

above that most of vacuum energy is consumed in the loose sediment layer.

Besides, it can also be seen that the compression of the loose sediment layer of the SVP zone is greater than that of the PVP zone although the surface settlement of the SVP zone is less. This also confirms the front analysis results that more vacuum energy is consumed on the interface between the sand layer and the unsaturated loose sediment or in the soil cracks because of more vacuum transfer channels including plant roots, fissures, and PVDs (Figs. 3, 18).

4 CONCLUSION

Field experiments have been carried out to determine the influence of thick and loose sediments on the vacuum load transfer. On the basis of the analysis results, the following conclusions can be made:

1. It is greatly affected by the loose sediment when vacuum load is transferred to the PVDs through the sand layer because both of them are loose and porous. And it is not influenced by the loose sediment when vacuum load is transferred to the PVDs through the pipe system because of its good airtightness.
2. Pipe system is more suitable as the horizontal transfer medium of vacuum load than sand layer in the case of overlying loose and thick soil layers.
3. Vacuum pressure in the PVD decays rapidly in the loose sediment layer, especially in the unsaturated zone, and the decay rate increases with the increase of vacuum pressure in the PVD in this layer. While it decays almost linearly and more slowly in the soil layers beneath the loose sediment, the decay rate is irrelevant to the magnitude of vacuum pressure in the PVD.
4. The settlement analysis validates the transfer rule of vacuum load indirectly in the case of overlying loose and thick soil layers. Soil compression caused by vacuum load occurs mainly in the loose sediment layer because most of vacuum energy is consumed in this layer.
5. The compression of the loose sediment layer of the SVP zone is greater than that of the PVP zone although the surface settlement of the SVP zone is less. It is mainly due to more vacuum transfer channels, including plant roots, fissures, and PVDs.

ACKNOWLEDGMENTS

This study was supported by the National Key Technology Support Program of China During the "12th Five-Year Plan" of Ministry of Science and Technology of the People's Republic of China (Award Number 2015BAB07B06). The authors thank Jiangsu Hongji Water Conservancy Construction Engineering Co. Ltd., Yangzhou, China, for its assistance in the implementation of the field experiments. Liu Zhizhong and Shen Minyi are highly appreciated for their assistance in data acquisition.

REFERENCES

Bergado, D.T., Chai, J.C., Miura, N., & Balasubramaniam, A.S. (1998). PVD improvement of soft Bangkok clay with combined vacuum and reduced sand embankment preloading. *Geotechnical Engineering Journal 29 (1)*, 95–121.

Chai, J.C., Carter, J.P., & Hayashi, S. (2005). Ground deformation induced by vacuum consolidation. *Journal of Geotechnical and Geoenvironmental Engineering, ASCE 131 (12)*, 1552–1561.

Chai, J.C., Carter, J.P., & Hayashi, S. (2006). Vacuum consolidation and its combination with embankment loading. *Canadian Geotechnical Journal 43 (10)*, 985–996.

Chai, J.C. & Miura, N. (1999). Investigation of factors affecting vertical drain behavior. *Journal of Geotechnical and Geoenvironmental Engineering 125 (3)*, 216–226.

Chai, J.C. & Miura, N. (2000). A design method for soft subsoil improvement with prefabricated vertical drain. *Proceedings of International Seminar on Geotechnics in Kochi, Japan*, 161–166.

Chai, J.C., Miura, N., & Bergado, D.T. (2008). Preloading of clayey deposit by vacuum pressure with cap-drain: analyses versus performance. *Geotextiles and Geomembranes 26 (3)*, 220–230.

Chai, J.C., Miura, N., & Nomura, T. (2007). Experimental investigated on optimum installation depth of PVD under vacuum consolidation. *Proceedings 3rd China-Japan Joint Geotechnical Symposium, China*, 87–92.

Choa, V. (1989). Drains and vacuum preloading pilot test. Proceedings 12th International Conference on Soil Mechanics and Foundation Engineering, Riode Janeiro, Brazil, 1347–1350.

Cognon, J.M., Juran, I., & Thevanayagam, S. (1994). Vacuum consolidation technology: principles and field experience. *Proceedings Vertical and Horizontal Deformations of Embankments (Settlement'94) 40 (2), ASCE Special Publication,* 1237–1248.

Guorui, G. (1991). The distribution, formation and engineering properties of quaternary regional sediments in china. *Geological Society London Engineering Geology Special Publications 7 (1)*, 491–501.

Holtz, R.D. (1975). Preloading by vacuum: current prospects. *Transportation Research Record 548*, 26–69.

Indraratna, B., Bamunawita, C., & Khabbaz, H. (2004). Numerical modeling of vacuum preloading and field applications. *Canadian Geotechnical Journal 41 (6)*, 1098–1110.

Indraratna, B., Sathananthan, I., Rujikiatkamjorn, C., & Balasudramaniam, A.S. (2005). Analytical and numerical modelling of soft soil stabilized by PVD

incorporating vacuum preloading. *International Journal of Geomechanics 5 (2)*, 114–124.
Kjellmann, W. (1952). Consolidation of Clay Soil by Means of Atmospheric Pressure. *Proceedings on Soil Stabilization Conference, Boston, USA*, 258–263.
Liang, A. Z., Jianwei, B. F., Yunzeng, C. G., & Fengjie, D. H. (2013). Application of new "straight line" vacuum preloading method in soft soil foundation treatment. *Proceedings of the 7th International Conference on Asian and Pacific Coasts Bali, Indonesia*, 785–789.
Long, P. V., Nguyen, L. V., Bergado, D. T., & Balasubramaniam, A. S. (2015). Performance of PVD improved soft ground using vacuum consolidation methods with and without airtight membrane. *Geotextiles & Geomembranes 43 (6)*, 473–483.
Mohamedelhassan, E. & Shang, J.Q. (2002). Vacuum and surcharge combined one-dimensional consolidation of clay soils. *Canadian Geotechnical Journal 39 (5)*, 1126–1138.
Rujikiatkamjorn, C. & Indraratna, B. (2007). Analytical solutions and design curves for vacuum-assisted consolidation with both vertical and horizontal drainage. *Canadian Geotechnical Journal 44 (2)*, 188–200.
Rujikiatkamjorn, C., Indraratna, B., & Chu, J. (2007). Numerical modelling of soft soil stabilized by vertical drains, combining surcharge and vacuum preloading for a storage yard. *Canadian Geotechnical Journal 44 (3)*, 326–342.
Rujikiatkamjorn, C., Indraratna, B., & Chu, J. (2008). 2D and 3D numerical modeling of combined surcharge and vacuum preloading with vertical drains. *International Journal of Geomechanics 8 (2)*, 144–156.
Saowapakpiboon, J., Bergado, D. T., Hayashi, S., Chai, J. C., Kovittayanon, N., & De Zwart, T. P. (2008). Ceteau PVD vacuum system in soft Bangkok clay: a case study of the Suvarnabhumi airport project. *Lowland Technology International 10 (1)*, 42–53.
Saowapakpiboon, J., Bergado, D. T., Voottipruex, P., Lam, L. G., & Nakakuma, K. (2011). PVD improvement combined with surcharge and vacuum preloading including simulations. *Geotextiles & Geomembranes 29 (1)*, 74–82.
Saowapakpiboon, J., Bergado, D.T., Chai, J.C., Kovittayanon, N., & De Zwart, T.P. (2008a). Vacuum-PVD combination with embankment loading consolidation in soft Bangkok clay: a case study of the Suvarnabhumi Airport Project. *Proceedings 4th Asian Regional Conference on Geosynthetics, Shanghai, China*, 18–27.
Saowapakpiboon, J., Bergado, D.T., Hayashi, S., Chai, J.C., Kovittayanon, N., & De Zwart, T.P. (2008b). CeTeau PVD vacuum system in soft Bangkok clay: a case study of the Suvarnabhumi Airport Project. *Lowland Technology International 10 (1)*, 42–53.
Saowapakpiboon, J., Bergado, D.T., Thann, Y.M., & Voottipruex, P. (2009). Assessing the performance of Prefabricated Vertical Drain (PVD) with vacuum and heat preloading. *Geosynthetics International 16 (5)*, 384–392.
Saowapakpiboon, J., Bergado, D.T., Youwai, S., Chai, J.C., Wanthong, P., & Voottipruex, P. (2010). Measured and predicted performance of Prefabricated Vertical Drains (PVDs) with and without vacuum preloading. *Geotextiles and Geomembranes 28 (1)*, 1–11.
Shang, J. Q., Tang, M., & Miao, Z. (1998). Vacuum preloading consolidation of reclaimed land: a case study. *Canadian Geotechnical Journal 35 (5)*, 740–749.
Tang, M. & Shang, J.Q. (2000). Vacuum preloading consolidation of Yaogiang Airport Runway. *Geotechnique 50 (6)*, 613–653.
Walker, R. & Indraratna, B. (2006). Vertical drain consolidation with parabolic distribution of permeability in smear zone. *Journal of Geotechnical and Geoenvironmental Engineering, ASCE 132 (7)*, 937–941.
Walker, R. & Indraratna, B. (2009). Consolidation analysis of a stratified soil with vertical and horizontal drainage using the spectral method. *Geotechnique 59 (5)*, 439–449.
Yan, S.W. & Chu, J. (2005). Soil improvement for a storage yard using the combined vacuum and fill preloading method. *Canadian Geotechnical Journal 42 (4)*, 1093–1104.

Civil, Architecture and Environmental Engineering – Kao & Sung (Eds)
© 2017 Taylor & Francis Group, ISBN 978-1-138-02985-9

Enhancement on fireproof performance of construction coatings using calcium sulfate whiskers prepared from wastewater

Tsung-Pin Tsai
Department of Civil Engineering, Chung Hua University, Hsinchu, Taiwan

Hsi-Chi Yang
Department of Construction Management, Chung Hua University, Hsinchu, Taiwan

ABSTRACT: This work is dealing with the fireproof performance of $CaSO_4$ composite painting for fire passive protection in building construction. An efficient microwave-assisted method is adopted to fabricate high-crystalline $CaSO_4$ whiskers from wastewater. The as-prepared $CaSO_4$ whiskers display one-dimensional structure with a high aspect ratio of 40. The thermal resistive behavior of $CaSO_4$-containing paints are investigated by using thermo-gravimetric analyzer, differential scanning calorimetry, and direct flaming test. The addition of $CaSO_4$ whiskers not only improves the anti-flammability but also reduces the ignition temperature of construction painting. Accordingly, the $CaSO_4$ whiskers can be considered as an effective fire retardant additive for improving the fireproof ability of construction coatings.

1 INTRODUCTION

With the crisis of resources and environmental issues becoming serious, the development of eco-materials is still taken into account as a global challenge. One strategy to counter this is to use the materials from sustainable sources, which would be renewable and economically competitive, capable of providing sustainable percentage of the resources for human utilization (Jiang J.X, 2015). Recently, there is an increasing demand to rationalize industrial wastewater treatment methodologies, including chemical precipitation, ion exchange, liquid-phase adsorption, reverse osmosis, and electrochemical methods (Deng L.C, 2013). The goal of the methods is to accommodate any reductions in discharge limits or step towards zero-discharge industries (Baltpurvins K.A, 1996). It is generally known that many industrial wastewaters contain high concentrations of sulfate, mainly originated from mining and mineral process. The sulfate concentration usually ranges from 250 to 2000 mg/L, which surpasses the secondary drinking water standard of 250 mg/L. To reduce the amount of wastewater, herein this present work proposes an efficient microwave method to recycle the sulfate species from acidic wastewater, opening a "green" gate for making calcium sulfate ($CaSO_4$) whiskers as fireproofing construction materials.

Recently, $CaSO_4$ whiskers have gradually earned much attention in scientific and technological applications owing to their superior properties. The $CaSO_4$ whiskers, single crystals of calcium sulfate with a high aspect ratio, display excellent performance in mechanical strength, thermal stability, chemical resistance, and compatibility (Mao X.L, 2014). Thus, the $CaSO_4$ whiskers exhibit great potential as reinforcing agents in plastics, ceramics, paper, and so on. Within the above scope, the target of the present study is to examine an optimal ratio of $CaSO_4$ whiskers on the improved fireproof performance of construction painting. Both thermal stability and flammability of $CaSO_4$-containing paintings are systematically investigated. This study proves that the usage of $CaSO_4$ whiskers is very beneficial for fire retardant in terms of enhanced fire-barrier performance, low cost and waste utilization.

2 EXPERIMENTAL SECTION

Both Thermo-Gravimetric Analyzer (TGA, Perkin Elmer TA7) and differential scanning calorimetry (DSC, TA Instrument Q20) were adopted to inspect the thermal stability and calorimetric change of $CaSO_4$ whiskers. The TGA analysis was conducted under an oxygen atmosphere with a heating rate of 30°C/min, ramping between 50°C and 900°C.

To examine the fireproof performance, one water-soluble painting mixed with different amounts of $CaSO_4$ whiskers were prepared to coat over calcium silicate plate. Herein the calcium silicate ($CaSiO_3$)

plates were carefully cut into an area of 5×5 cm^2, and its thickness was around 0.8 cm. Four recipes were set at 0, 5, 10, and 15 wt% in the weight ratios of $CaSO_4$ whiskers to water-soluble painting. For good uniformity, the $CaSO_4$-containing paintings were blended with a three-dimensional mixer using zirconia balls for 10 min. The as-prepared paintings were then pasted on the substrates with a doctor blade and then dried at 80°C in an oven overnight. The flame retardancy of as-prepared paintings on calcium silicate plates was tested by using a high-performance flamethrower. The distances between the paintings and the top of flame were set at 1, 5, and 10 cm, where the surface temperatures on the calcium silicate plates were 1100, 570, and 150°C, respectively. In the fireproof test system, three thermocouples (*K* type) were adopted to detect surface temperatures of the plates.

3 RESULTS AND DISCUSSION

TGA has been considered as a common technique that offers quantitative decomposition information on a composite material and can be used to investigate degradation kinetics and char formation. Herein the paintings mixed with different ratios of $CaSO_4$ whiskers were subjected to a controlled temperature regime of 50–900°C at a heating rate of 30°C/min. The weight loss for each sample was monitored throughout the chemical decomposition in air, as shown in Fig. 1.

Without the addition of $CaSO_4$ whiskers, the TGA curve of original painting consists of three weight-loss stages: (i) 100–200°C (dehydration), (ii) 200–350°C (ignition and decomposition), and (iii) 650–800°C (pyrolysis and degradation). The main weight loss takes place during the stage (ii) for all samples, liberating gases and volatiles from the paintings. The effectiveness of flame retardant can be evaluated on the basis of volatiles with increasing temperature, depended on the amount of residue produced. It can be found that the weight loss is a decreasing function of the amount of $CaSO_4$ whiskers.

In comparison, the residual weight of original painting is approximately 68.6 wt.% after thermal oxidation at 900°C, whereas the maximal residue can reach as high as 77.6 wt.% for the painting mixed with 15 wt.% $CaSO_4$ whiskers. Accordingly, the introduction of $CaSO_4$ whiskers displays a positive effect on the anti-flammability of construction painting. This improved retardancy is attributed to the fact the $CaSO_4$ crystals offer a high melting temperature (i.e., 1460°C), allowing its use for high-temperature operation.

To support this argument, DSC was adopted to examine the heat flux as a function of temperature under air flow starting from 25°C to 300°C, as depicted in Fig. 2.

For comparison, the DSC curve of pure $CaSO_4$ whiskers is also provided. As observed from Fig. 2(a), two endothermic peaks at 90°C and 180°C can be found. The former can be attributed to the departure of 1.5 mol H_2O from dihydrated calcium sulfate, whereas the later originates from the departure of 0.5 mol H_2O from hemihydrated calcium sulfate obtained after the incomplete dehydration of dihydrated calcium sulfate and from the

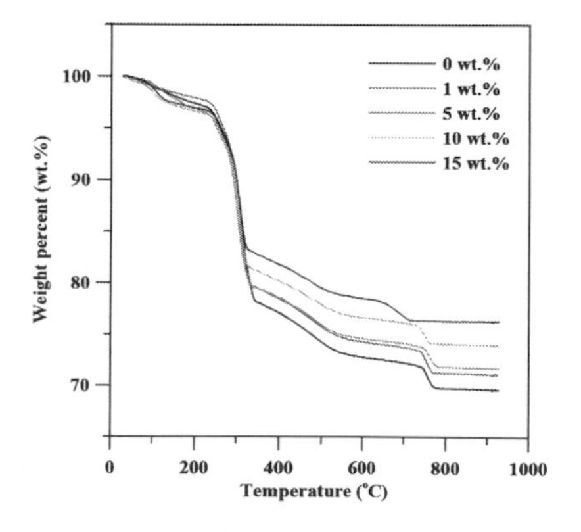

Figure 1. Typical TGA curves of the painting with different amounts of $CaSO_4$ whiskers.

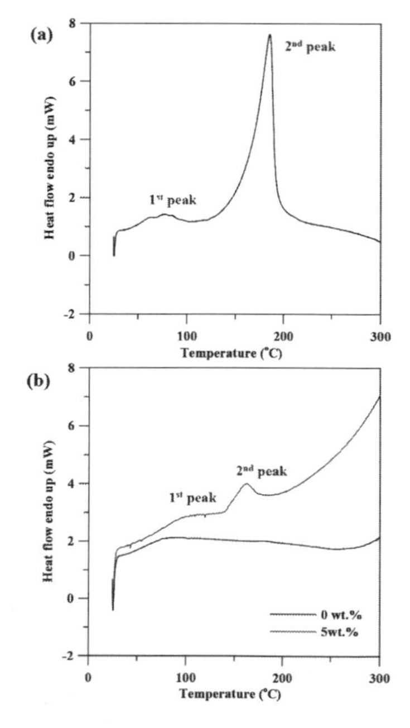

Figure 2. Typical DSC curves of (a) as-prepared $CaSO_4$ whiskers and (b) the painting without and with 5 wt.% $CaSO_4$ whiskers.

β-hemihydrate of plaster (i.e., a phase transformation) (Abidi S, 2015). This result reveals that the heat transfer through the $CaSO_4$-containing painting can be impeded until the endothermic steps are totally completed (Kolaitis D.I, 2014). Therefore, the thermal decomposition of $CaSO_4$-containing paint delivers a trailing effect, resulting in the delay of ignition temperature. On the basis of experimental results, the $CaSO_4$ whiskers can serve as an effective fire retardant additive for tailing away the decomposition reactions.

4 CONCLUSIONS

This work presented an efficient microwave-assisted method to fabricate high-crystalline $CaSO_4$ whiskers from wastewater. The introduction of $CaSO_4$ whiskers displayed a positive effect on the anti-flammability and the reduced ignition temperature of construction painting, demonstrated by TGA and DSC analyses. This result could be attributed to the fact that the heat transfer through the $CaSO_4$-containing painting is seriously impeded until the endothermic reaction steps (i.e., dehydration of gypsum ($CaSO_4{\cdot}2H_2O$) and crystalline phase change of β-hemihydrated plaster) were totally completed. On the basis of experimental results, the $CaSO_4$ whiskers could serve as an effective fire retardant additive for improving the fire protection and thermal resistive coatings. The present study shed some lights on an eco-environmental pathway to prepare "green" materials (i.e., $CaSO_4$ whiskers) for high-performance retardant additive in construction applications.

ACKNOWLEDGEMENTS

The authors are very grateful for the financial support from the Ministry of Science and Technology (Taiwan) under the contract NSC 101-2221-E-216-031-040.

REFERENCES

Abidi, S. 2015. Impact of perlite, vermiculite and cement on the thermal conductivity of a plaster composite material: experimental and numerical approaches. *Compos. Pt. B-Eng.*, 68: 392–400.

Baltpurvins, K.A. 1996. Heavy metals in wastewater: modelling the hydroxide precipitation of copper (II) from wastewater using lime as the precipitant. *Waste Manage.*, 16: 717–725.

Deng, L.C. 2013. Reactive crystallization of calcium sulfate dihydrate from acidic wastewater and lime. *Chin. J. Chem. Eng.*, 21: 1303–1312.

Jiang, J.X. 2015. Effect of flame retardant treatment on dimensional stability and thermal degradation of wood. *Constr. Build. Mater.*, 75: 74–81.

Kolaitis, D.I. 2014. Fire protection of light and massive timber elements using gymsum platerboards and wood based panels: A large-scale compartment fire test. *Constr. Build. Mater.*, 73: 163–170.

Mao, X.L. 2014. Effects of metal ions on crystal morphology and size of calcium sulfate whiskers in aqueous HCl solutions. Ind. *Eng. Chem. Res., 53:* 17625–17635.

Civil, Architecture and Environmental Engineering – Kao & Sung (Eds)
© 2017 Taylor & Francis Group, ISBN 978-1-138-02985-9

Simulation of nonlinear ultrasonic waves in anisotropic materials using convolution quadrature time-domain boundary element method

T. Saitoh
Department of Civil and Environmental Engineering, Gunma University, Gunma, Japan

T. Maruyama
Department of Civil Engineering, Tokyo University of Science, Tokyo, Japan

A. Furukawa
Department of Civil and Environmental Engineering, Tokyo Institute of Technology, Tokyo, Japan

ABSTRACT: In recent years, a new ultrasonic nondestructive testing method that utilizes nonlinear ultrasonic behavior generated in cracks or at a bi-material interface has emerged. However, the mechanism of generating subharmonics and higher harmonics is theoretically still not well understood. In this research, to simulate higher harmonics, dynamic contact problems of a crack face with contact boundary conditions in anisotropic materials are investigated by the Convolution Quadrature Time-Domain Boundary Element Method (CQBEM). Numerical results show that the cracks with contact boundary conditions in anisotropic materials excite higher harmonics as well as those in isotropic ones.

1 INTRODUCTION

Nonlinear ultrasonic testing has shown promise as a nondestructive method for testing materials used in nuclear power plants and in other important industries. In general, nonlinear ultrasonic testing uses subharmonics or higher harmonics, which include the frequency components corresponding to half harmonics or the harmonic series of an incident wave. Nonlinear ultrasonic testing has, however, not been applied in practical applications because the mechanism of generating these harmonics is theoretically still not well understood. Several researches on nonlinear ultrasonic testing have been done (Solodov, 2011) to investigate the mechanism of generating subharmonics and higher harmonics. Numerical analyses of nonlinear ultrasonic waves have been implemented by Saitoh et al. (2010), Hirose et al. (2014) and Maruyama et al. (2014), using the Convolution Quadrature time-domain Boundary Element Method (CQBEM). The CQBEM has been researched by Schanz (2001), Saitoh et al. (2009) and Maruyama et al. (2016) for transient wave analysis. In the CQBEM formulation, the convolution integrals of the time-domain boundary integral equation are numerically approximated by the Convolution Quadrature Method (CQM). Application of the CQM to time-domain BEM improves the numerical stability of the classical time-domain BEM, even for small time step sizes, which are not allowed in the classical time-domain BEM. Therefore, the CQBEM is particularly helpful for the simulation of nonlinear ultrasonics, which requires a time step as small as possible for high precision solutions. In this paper, simulation of nonlinear ultrasonic waves in anisotropic materials is carried out using the CQBEM. In the following sections, the CQBEM for 2-D anisotropic elastodynamics is presented. Next, contact boundary conditions on the crack face are discussed. Finally, nonlinear ultrasonic waves generated by a crack with the contact boundary Conditions are demonstrated.

2 ANISOTROPIC ELASTODYNAMIC PROBLEMS

2.1 *BEM formulation*

We consider the scattering of an incident wave $u_i^{\text{in}}(\boldsymbol{x},t)$ by a crack with surface S in an anisotropic solid D, as shown in Figure 1. The small indices used throughout this paper, such as $(\ )_i$ range from 1 to 2 unless otherwise stated. Additionally, summation over repeated subscripts is implied throughout this paper. The equation of motion at point x and time t is given as follows:

$$C_{ijkl}u_{k,lj}(\boldsymbol{x},t)+f_i(\boldsymbol{x},t)=\rho\ddot{u}_i(\boldsymbol{x},t) \tag{1}$$

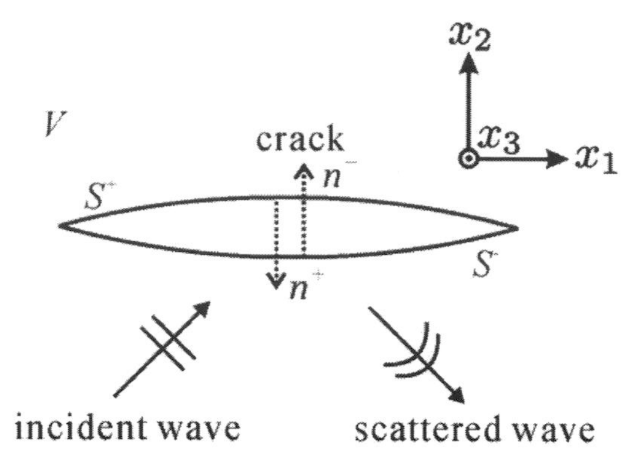

Figure 1. Scattering of an incident wave by a crack.

where $u_i(\boldsymbol{x},t), \sigma_{ij}(\boldsymbol{x},t)$, and $f_i(\boldsymbol{x},t)$ represent the displacement, stress, and body force component, respectively. The dot notation $(\cdot)$ and the symbol $(\ )_{,i}$ denote the partial derivative with respect to time t and space x_i, respectively. ρ is the density of the anisotropic elastic material V. In addition, the constitutive equation is given by

$$\sigma_{ij}(\boldsymbol{x},t) = C_{ijkl}u_{k,l}(\boldsymbol{x},t) \tag{2}$$

where C_{ijkl} is the elastic constant. The boundary conditions on traction free cracks are given by

$$t_i^{\pm}(\boldsymbol{x},t) = 0 \text{ on } S^{\pm},\ \phi_i(\boldsymbol{x},t) = 0 \text{ on } \partial S \tag{3}$$

where $t_i(\boldsymbol{x},t)$ is the traction corresponding to the displacement $u_i(\boldsymbol{x},t)$, and $\phi_i(\boldsymbol{x},t)$ is the crack opening displacement defined by $\phi_i(\boldsymbol{x},t) = u_i^+(\boldsymbol{x},t) - u_i^-(\boldsymbol{x},t)$. Moreover, ∂S denotes the edge of the crack, and S^+ and S^- show the positive and negative sides of the crack face, respectively. The regularlized hypersingular integral equation is written as follows:

$$\begin{aligned} &t_p^{\text{in}}(\boldsymbol{x},t) \\ &= C_{pckl}n_c(\boldsymbol{x})\Big[-\rho\int_S U_{ik}(\boldsymbol{x},\boldsymbol{y},t) * \ddot{\phi}_i(\boldsymbol{y},t)n_l(\boldsymbol{y})dS_y \\ &+ e_{db}e_{lj}\int_S C_{ijna}U_{nk,a}(\boldsymbol{x},\boldsymbol{y},t) * \phi_{i,b}(\boldsymbol{y},t)n_d(\boldsymbol{y})dS_y\Big] \end{aligned} \tag{4}$$

where $*$ is the convolution integral with respect to time t and $U_{mj}(\boldsymbol{x},\boldsymbol{y},t)$ is the fundamental solution for 2-D anisotropic elastodynamics, and $n_k(=n_k^+)$ represents the outward unit normal vector component. The zero initial condition is assumed in the boundary integral equation. In addition, $t_p^{\text{in}}(\boldsymbol{x},t)$ is the traction component of an incident wave $u_p^{\text{in}}(\boldsymbol{x},t)$ and e_{st} is the permutation symbol. Normally, the time-domain boundary integral equation (4) is discretized by the timestepping scheme. However, the well known timestepping scheme sometimes produces numerical errors if we use small time step size Δt. To overcome the difficulty, the CQM is utilized for the time discretization of the boundary integral equation (4).

2.2 *Time discretization by the CQM*

Discretizing the regularized hypersingular integral equation (4) using a piecewise constant approximation of the unknown crack opening displacement $\phi_i(\boldsymbol{y},t)$ and using the CQM for the convolutions of equation (4) yield the discretized boundary integral equation as follows:

$$\begin{aligned} t_p^{\text{in}}(\boldsymbol{x},n\Delta t) = \sum_{\alpha=1}^{M}\sum_{k=1}^{n}\left[A_{pi}^{n-k,\alpha}(\boldsymbol{x}) + B_{pi}^{n-k,\alpha}(\boldsymbol{x})\right] \\ \phi_i(\boldsymbol{y}^{\alpha},n\Delta t) \end{aligned} \tag{5}$$

where M is the number of discretized boundary elements. In addition, $A_{pi}^{m,\alpha}(\boldsymbol{x})$ and $B_{pi}^{m,\alpha}(\boldsymbol{x})$ are the influence functions, which are defined by

$$\begin{aligned} &\mathrm{A}_{pi}^{m,\alpha}(\boldsymbol{x}) = \rho C_{pckl}n_c(\boldsymbol{x}) \\ &\times\frac{\mathcal{R}^{-m}}{L}\sum_{l=0}^{L-1}\left[-\int_{S_\alpha} n_l(\boldsymbol{y})s_l^2\hat{U}_{ik}(\boldsymbol{x},\boldsymbol{y},s_l)dS_y\right]e^{-\frac{2\pi iml}{L}}, \end{aligned} \tag{6}$$

$$\begin{aligned} &\mathbf{B}_{pi}^{m,\alpha}(\boldsymbol{x}) = C_{pckl}C_{ijna}n_c(\boldsymbol{x})e_{lj} \\ &\times\frac{\mathcal{R}^{-m}}{L}\sum_{l=0}^{L-1}\left[\int_{\partial S_\alpha}\frac{\partial}{\partial y_a}\hat{U}_{nk}(\boldsymbol{x},\boldsymbol{y},s_l)d(\partial S_y)\right]e^{-\frac{2\pi iml}{L}} \end{aligned} \tag{7}$$

where $\hat{U}_{ik}(\boldsymbol{x},\boldsymbol{y},s)$ and $d(\partial Sy)$ are the Laplacedomain fundamental solution for 2-D anisotropic elastodynamics and sides of each boundary element S_α, respectively. S_l denotes the Laplace parameter. Moreover, $\mathcal{R}$ and L are the CQM parameters (Lubich, 1998). Equation (5) can be solved with initial and boundary conditions from the first time step to the last time step.

3 CONTACT BOUNDARY CONDITIONS

In a usual analysis of elastic wave scattering by a crack, it is assumed that the boundary condition on the crack face is traction free in all time steps. In this paper, however, we consider the contact boundary conditions on the crack face proposed by Hirose (Hirose, 1994). The summary of the contact boundary conditions are described in this section.

3.1 *Separation state*

The separation state shows the usual boundary condition imposed on the crack faces for dynamic

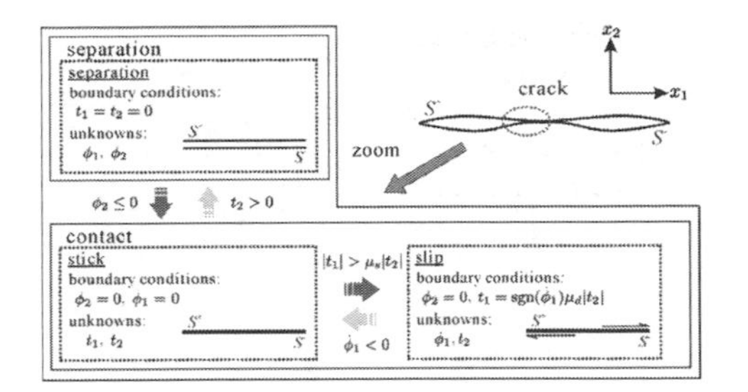

Figure 2. Contact boundary conditions.

crack problems. The crack faces S^+ and S^- are separated from each other as shown in Figure 1. Therefore, the traction free boundary condition $t_i = 0$ is considered at this state.

3.2 *Stick contact state*

In the stick contact state, the relative velocity of the horizontal crack opening displacement $\dot{\phi}_1(\boldsymbol{x},t)$ and the vertical crack opening displacement $\phi_2(\boldsymbol{x},t)$ must be zero as shown in the lower left of Figure 2.

3.3 *Slip contact state*

In the slip contact state, we can assume the boundary condition given by

$$\phi_2 = 0, \quad t_1 = \mathrm{sgn}(\dot{\phi}_1)\mu_d|t_2| \tag{8}$$

where μ_d and sgn( ) are the crack face frictional coefficient and signum function, respectively. The difference between stick and slip contact states is whether the crack opening velocity $\dot{\phi}_1(\boldsymbol{x},t)$ can be allowed or not. In each time step, the stick, slip, or separation contact boundary condition is determined by the crack opening displacement and stress states. The transition from one state to the others is summarized in Figure 2.

4 NUMERICAL RESULTS

Some numerical results are shown in this section. We assume the material V with a crack as CFRP with the following material constants:

$$C_{pq} = \begin{bmatrix} 45.91 & 1.84 & 41.87 & 0 & 0 & 0 \\ & 3.98 & 1.84 & 0 & 0 & 0 \\ & & 45.91 & 0 & 0 & 0 \\ & & & 1.0 & 0 & 0 \\ & \text{sym.} & & & 2.02 & 0 \\ & & & & & 1.0 \end{bmatrix} \tag{9}$$

Note that the material constants are normalized by the C_{66}.

4.1 *Elastic wave scattering by non-contacting crack in anisotropic solid*

First, in order to check the validity of the CQBEM, we consider the scattering of elastic waves by a noncontacting crack as shown in Figure 3(a). The incident plane qP wave is given by

$$u_i^{\mathrm{in}}(\boldsymbol{x},t) = u_0 d_i \frac{1}{2}\left[1-\cos(2\pi f h)\right] \left\{H(h) - H\left(h-\frac{1}{f}\right)\right\} \tag{10}$$

where u_0 is the amplitude, d_i is the polarization vector component, H is the Heaviside step function,

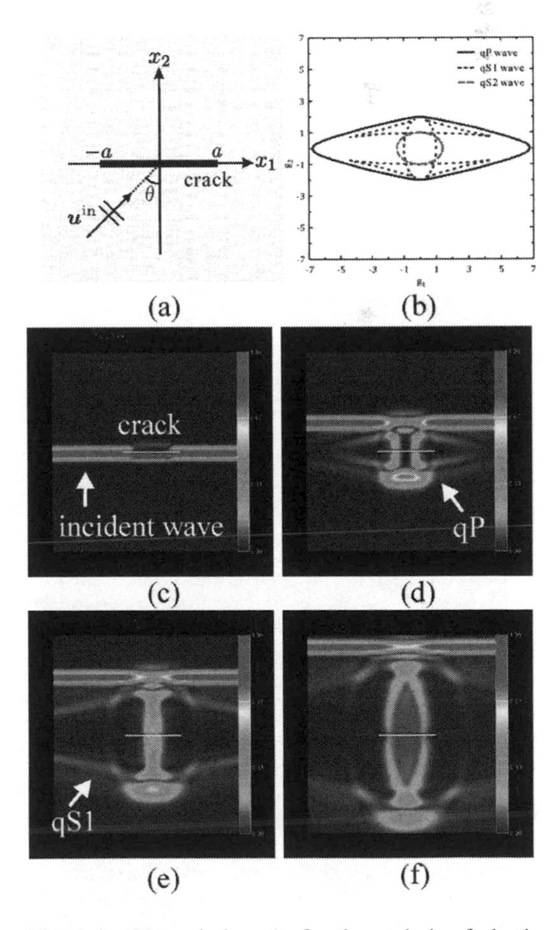

Figure 3. Numerical results for the analysis of elastic wave scattering by non-contacting crack. (a) analysis model (b) group velocities of the CFRP, (c)-(d) absolute value of the total wave fields around the crack.

and f is the frequency of the incident wave, respectively. In addition, c^{in} is the wave velocity for the incident wave propagation direction. The function h is defined by

$$h = t - \frac{x_1 + a}{c^{\text{in}}}\sin\theta - \frac{x_2}{c^{\text{in}}}\cos\theta \tag{11}$$

In this analysis, we set the parameters as $\boldsymbol{d} = (0,1)$, $\theta = 0$, $c^{\text{in}}\Delta t / a \simeq 0.1$, and $N = L = 128$. The incident wave length λ^{in} is given by $\lambda^{\text{in}} / a = 1.0$. The group velocity curves of the CFRP with the material constants given in equation (9) are shown in Figure 3(b). The group velocity of qP wave propagating in the x_1 direction is faster than that in the x_2 direction. Figures 3(c)-(f) show the absolute value of the total wave field around the crack at $c^{\text{in}}t / a = 0.4, 1.89, 3.39, \text{and } 4.89$. The scattered qP and qS1 waves are generated by the interaction between the incident wave and the crack. The wave fronts of these scattered waves concord with the group velocity curves. Therefore, it is confirmed that the obtained numerical results are valid from the physical point of view.

4.2 *Simulation of nonlinear ultrasonic waves*

Next, the scattering of an incident plane qP wave by a crack with the contact boundary conditions analized by the CQBEM. In this analysis, the following incident qP wave is considered:

$$u_i^{\text{in}}(\boldsymbol{x},t) = u_0 d_i \sin(2\pi f h) H(h) H\left(\frac{10}{f} - h\right) \tag{12}$$

where

$$h = t - \frac{x_1 + a}{c^{\text{in}}}\sin\theta - \frac{x_2}{c^{\text{in}}}\cos\theta. \tag{13}$$

In this analysis, we set the parameters as $\boldsymbol{d} = (0,1)$, $c^{\text{in}}\Delta t / a \simeq 0.1$, $N = L = 128$. The incident wave length λ^{in} is given by $\lambda^{\text{in}} / a = \pi$. Figure 4(a) shows the time histories of the total displacements u_2 / u_0 at the point $(x_1, x_2) = (0.0, -4.0a)$. The total displacement obtained by considering the contact boundary conditions is distorted compared with the non-contacting result shown by dashed line in Figure 4(a). Figure 4(b) depicts the Fourier spectrum of the total displacements u_2 / u_0 of Fig. 4(a). As shown in Figure 4(b), the scattered wave by the non-contacting crack has only one spectral component at $k^{\text{in}}a = 2.0$, which is corresponding to the fundamental frequency of the incident wave. However, the scattered wave by the crack with contact boundary conditions has not only the fundamental frequency but also the higher frequency component at $k^{\text{in}}a = 4.0$ and $k^{\text{in}}a = 6.0$, which are the two and three times of the fundamental frequencies and called the second and third order harmonics, respectively. The use of Fourier spectrums of the scattered waves may have a potential to detect a closed crack in anisotropic materials which cannot be detected by the linear ultrasonic testing based on the acoustic impedance mismatch.

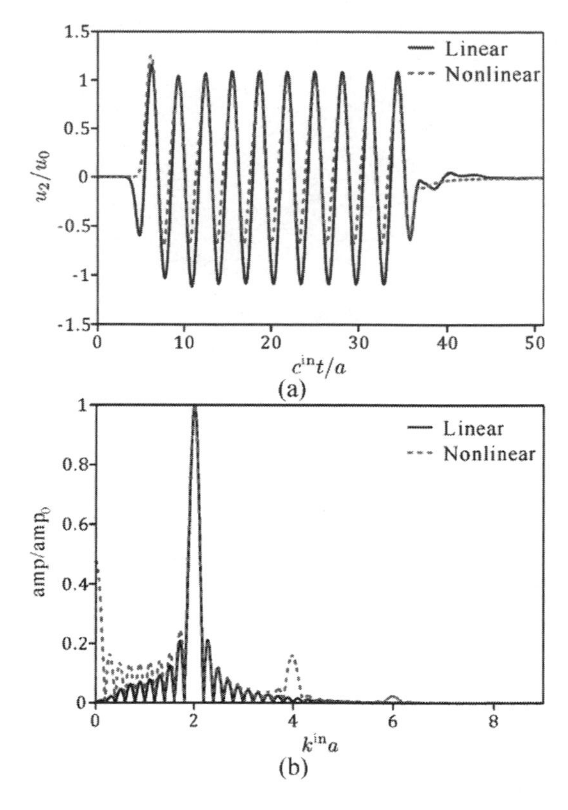

Figure 4. (a) Total displacements u_2/u_0 and (b) their Fourier spectra.

5 CONCLUSIONS

In this paper, the simulation of nonlinear ultrasonic waves is implemented by the CQM. In the future, we will extend this 2-D simulation to 3-D one.

ACKNOWLEDGEMENT

This work was supported by Ono Acoustics Research Fund.

REFERENCES

Furukawa, A., Saitoh, T. & Hirose, S. (2014). Convolution quadrature time-domain boundary element method for 2-D and 3-D elastodynamic analyses in general anisotropic elastic solids, *Eng. Anal. Bound. Elem. 39*, 64–74.

Hirose, S. (1994). 2-D scattering by a crack with contact-boundary conditions, *Wave Motion 19*, 37–49.

Hirose, S. & Saitoh, T. (2014). Numerical simulation of nonlinear ultrasonic waves due to bi-material interface contact, *J. Phys., Conf. Ser. 19*, 520 012001, 37–49.

Lubich, C. (1998). Convolution quadrature and discretized operational calculus I, *Numer. Math. 52*, 129–145.

Maruyama, T., Saitoh, T. & Hirose, S. (2014). 3-D BEM simulation of nonlinear ultrasonic waves due to contact acoustic nonlinearlity, *Proc. 1st Int. conf. comput. eng. and sci. safety and environ. prob.*, 756–759.

Maruyama, T., Saitoh, T., Bui, T. Q. & Hirose, S. (2016). Transient elastic wave analysis of 3-D large-scale cavities by fast multipole BEM using implicit Runge-Kutta convolution quadrature, *Comput. Method. Appl. M. 303*, 231–259.

Saitoh, T., Chikazawa, F. & Hirose, S. (2014). Convolution quadrature time-domain boundary element method for 2-D fluid-saturated porous media, *Appl. math. model. 38*, 3724–3740.

Saitoh, T., Nakahata, K. & Hirose, S. (2010). Improved time-domain BEM analysis for a solid-solid interface with contact boundary conditions, *Theor. Appl. Mech. Japan. 58*, 9–17.

Saitoh, T., Zhang, Ch. & Hirose, S. (2009). Large-scale multiple scattering analysis of SH waves using time-domain FMBEM, *Adv. Bound. Elem. Tech. 4*, 159–164.

Schanz, M. (2001). Wave propagation in viscoelastic and poroelastic continua, A Boundary Element Approach, *Springer*.

Solodov, I. Yu., Doring, D., Busse, G. (2011). New opportunities for NDT using non-linear interaction of elastic waves with defectsk, *J. Mech. Eng. 57–3*, 169–182.

Civil, Architecture and Environmental Engineering – Kao & Sung (Eds)
© 2017 Taylor & Francis Group, ISBN 978-1-138-02985-9

Study on the layout method of newly-added feeder bus lines coordinating with bus rapid transit

Zhong Guo, Pinghong Wei, Xiao Peng & Chengzhi Chang
China Urban Sustainable Transport Research Centre (CUSTReC), Beijing, China

Lu Tong
School of Traffic and Transportation, Beijing Jiaotong University, Beijing, China

Juan Miguel Velásquez
WRI Ross for Sustainable Cities, World Resources Institute, Washington, USA

ABSTRACT: With a rapid development of Bus Rapid Transit (BRT) system in China, many studies on planning and design of BRT have been conducted; however, there is only little research on the feeder system of the BRT system, especially for the layout of the newly added feeder lines. In this paper, we first put forward that the reasonable influence area of BRT and the layout of feeder lines should be determined on the basis of questionnaire surveys of passengers close to the BRT stations. Second, the selection principle of newly added feeder lines and the feeder stops were proposed. Then, a mathematical model with the connection benefits (including travel time cost and operation cost) as a minimum objective function has been conducted for the first time, an improved genetic algorithm was designed to calculate the model, and the layout scheme of newly added feeder lines was solved. Finally, the methodology was applied to the case of Beijing BRT line 1. With the goal of 80% passengers' interest, the reasonable influence radius of Beijing BRT line 1 is 3.89 km. After the calculating the model, eight newly added feeder bus lines were designed linking to the Beijing BRT line 1. The results reveal that the layout method of newly added feeder bus lines coordinating with BRT can be well used to support the feeder system design of Beijing BRT.

1 INTRODUCTION

Bus Rapid Transit (BRT) can be defined as that with a large capacity and high speed, similar to those of rail transit. It is also as flexible and cheap as regular bus transit, so it is widely used throughout the world (2013). Between 2004 and 2014, 23 major cities in China established and implemented 40 BRT corridors, reaching a total length of 2761 km (2014).

However, the development of BRT is still in an early stage. It has not formed network operation and is still insufficient in service scope, so it is necessary for regular bus lines to collect and distribute passengers and provide feeder service, thus improving the operational efficiency of bus rapid transit and cultivating passenger flow for newly added bus rapid transit lines. Therefore, it is particularly important to carry out the research on the method of laying newly added feeder buses to BRT lines.

A utility model is developed and a quantitative study on reasonable influence area of BRT and feeder lines is conducted in this paper. The study also designed a genetic algorithm to calculate the model and give the layout scheme on new-added feeder lines.

2 LITERATURE REVIEW

In recent years, with the rapid development of rail and BRT, many studies have begun to focus more on feeder bus lines. The methodologies on feeder lines layout are of two types: benefit analysis and network optimization. The objective function of benefit analysis is to find the optimal interest and analyze the interests of feeder bus system on the basis of public transit demand spatial and temporal distribution function.

Among the international studies on the layout of feeder bus lines, Vuchic (2005) provided internationally recognized methodologies to determine the structure of feeder bus network according to the maximum passenger transport demand. Martins and Pato (1998) established the optimal number of feeder bus lines for urban rail transit systems and the methods or algorithms for laying them. Kuan et al. (2006) introduced the heuristic genetic algorithm to solve the complex N-P problem concerning feeder bus network optimization. Verma and Dhingra (2005) formulated an optimization model of line layout to account for the integration of rail transit and feeder line operations. Wiasinghe

(1977, 1980), Kuah et al. (1988), Schonfeld et al. (1998), and MD. Shoaid Chowdhury (2001) established the profit analysis model from different perspectives and carried out studies on the possible maximum benefit of feeder bus system. Shrivastava (2006, 2001) and S.N. Kuan (2005) studied the application of different heuristic algorithms in bus network optimization model.

However, because the construction and operation of BRT systems is relatively later in China, there is little research regarding the layout of BRT feeder lines. Among the existing research, it is important to highlight the work of Dong (2011) to research methodologies to add new feeder bus lines on the basis of land use characteristics, for example, of universities, hospitals, and other densely populated areas. Fang (2013) also explored feeder line layout on the basis of the shortest path method and the optimal searching algorithm to lay feeder bus lines one by one. In 2009, Zhu (2009) put forward a method on the time-value model and network coordination of regular bus network adjustment on the basis of coordination theory of BRT and regular network. In 2011, Bi (2011) did malicious attack experiment on BRT network coordinated with regular bus network, testing the robustness of BRT network and verifying the importance of BRT linked to regular bus.

Because most of the above-mentioned studies investigated the layout of feeder bus lines for rapid rail transit, this paper aims to propose a methodology specific to BRT, using genetic algorithms, which will consider improving access to public transport while minimizing passenger and operation costs. Section 2 will depict the methodology to determine the influence area of BRT lines, section 3 will discuss the methodology proposed to determine the layout of feeder lines, including a genetic algorithm. Section 4 presents an application for the case of Line 1 in Beijing, China, and Section 5 includes the paper conclusions.

3 METHODOLOGY

3.1 *Definition of reasonable influence area of BRT*

The first step in the methodology is to determine the reasonable influence area of BRT lines. We define influence area as the area around BRT lines where passengers will have access to the BRT system. This influence area should take into account passengers who might walk as well as those transferring from other modes such as cycling or feeder buses.

The influence area of a station can be divided into reasonable influence area and potential influence area (2013). The definition of the reasonable influence area is based on the reasonable distance that most passengers are willing to travel to access the BRT line, including passengers who walk, cycle, or drive and use park and ride facilities (2011). The concept that corresponds to a reasonable influence area is the potential attraction, which refers to the distribution of some areas that still attract passenger flow in addition to the stations that reasonably attract passenger flow.

The reasonable influence area of a BRT station includes the influence area of bicycles and the influence area of regular bus transit; the emphasis of this paper is the reasonable influence area of bus rapid transit lines to conventional public transport passenger flow, which is used to determine the layout scope of feeder bus lines. According to the concept of reasonable influence area, it is a quantity value aiming for meeting most passengers' benefits and not wasting transport resources; therefore, on the basis of GIS network information and station questionnaires and with the goal of considering the benefits of a certain proportion of passengers, a buffer analysis can be carried out, to determine the reasonable influence area of bus rapid transit lines to the passenger flow of regular bus transit.

3.2 *Selection of feeder stations*

3.2.1 *Selection of BRT stations*

Not all BRT stations need a feeder bus, and their need should be judged according to the size of the feeder or the existing station passenger volume. If the feeder passenger volume is too small, the feeder bus line will probably be not cost-effective to increase new feeder bus lines; if a rapid transit station already has a heavy passenger flow volume, and newly added feeder passenger flow will lead to exceeding the platform's capacity, then it is not suitable to add new feeder bus lines to this station.

3.2.2 *Selection of regular bus stops*

If it is determined that adding feeder lines is suitable, the station selection of regular bus should be conducted within the identified reasonable influence area of the BRT line. Stations should be divided into two categories: one category is the stations for passengers to get on regular buses. These can be obtained by looking at passenger transfers from regular bus stops to BRT stations in the Origin-Destination (OD) survey. The other category is the regular bus transit stops that do not allow passengers to transfer to BRT. When selecting the above two types of regular bus stops, the size of the feeder passenger flow volume should also be considered, and if the resulting feeder passenger flow volume is too small, it is not economical to add new feeder bus lines. Moreover, feeder stations can be added according to specific needs; for example, new feeder stations can be considered at special locations, such as scenic spots, stations, residential areas, and public central areas with potential passenger demand.

4 MODEL ESTABLISHMENT

4.1 *Analysis of influencing factors*

There are many factors that affect the layout of feeder bus lines; therefore, it is necessary to make simplifications and assumptions. The basic assumptions of this model are as follows:

1. The feeder passenger flow volume at each regular bus stop that needs the layout of feeder bus lines is known, and the distance between these stops and the BRT station are known.
2. The bus capacity (passenger/bus) and operating speed of each vehicle running on feeder bus lines are known.
3. The feeder buses stop at every feeder lines' stops, the dwell time at stops is constant.
4. There is only one feeder bus line passing through every feeder regular bus stop.
5. All feeder lines must start or end at a BRT station, and they can only be connected to one BRT station.
6. Depending on passenger volume, two or more feeder bus lines can be laid for a bus rapid transit station.

4.2 *Analysis of influencing factors*

On the basis of the above requirements, a layout model for newly added feeder bus lines can be established. In this model, the basic parameters are defined as shown in Table 1:

The layout model can be established as follows, and the objective function is:

Table 1. Parameter description.

Name	Meaning	Name	Meaning
N	Total Quantity of regular bus transit stations that need the layout of feeder bus lines	v_0	Operating speed of the vehicles on feeder bus lines
T_I	Total in-vehicle travel time of all the passengers	t_0	Stopping time of vehicles at each station on feeder bus lines
k_n	Ridership at the nth feeder regular bus transit station	c_1	Unit in-vehicle travel time cost of passengers
d_n	Length of feeder bus line from the nth feeder regular bus transit station to the bus rapid transit station	S_n	Quantity of stations through which the line between the nth feeder regular bus transit station and the bus rapid transit station has passed
T_W	Total waiting time of all the passengers	c_2	Unit waiting time cost of passengers
h_l	Departure interval of the nth feeder regular bus transit station on the lth feeder bus line	F_l	Fleet scale of the lth feeder bus line
N_l	The quantity of regular bus transit stations through which the lth feeder bus line has passed	d_l	Length of the lth feeder bus line
L	Total quantity of feeder bus lines required to be laid	c	Operating cost per vehicle per unit time on a feeder bus line
d_{max}	Upper limit value of the length of a feeder bus line	d_{min}	Lower limit value of the length of a feeder bus line
h_{max}	Upper limit value of the departure interval of a feeder bus line	h_{min}	Lower limit value of the departure interval of a feeder bus line
P	Rated passenger capacity of a feeder bus	$Colmax$	Maximum operating cost of the lth feeder bus line that can be provided by enterprises
q_{max}	Maximum section passenger volume of a feeder bus line	R	Maximum feeder passenger flow volume that can be accepted by the bus rapid transit stations
D_l	Spatial straight-line distance between the origin and end stops of the lth feeder bus line	ρ	Constant, nonlinear coefficient

$$\min \quad C_T = C_1 \sum_{n=1}^{N} k_n \left(\frac{d_n}{v_0} + s_n t_0 \right) + C_2 \sum_{n=1}^{N} \frac{1}{2} k_n h_l + \sum_{l=1}^{L} \frac{2 d_l / v_0 + t_0 \cdot N_l}{h_l} \cdot c \tag{1}$$

where the first part refers to the total in-vehicle travel time cost of all the passengers transported by feeder buses, the second part refers to the total waiting time cost of all the passengers who are waiting for feeder buses at regular bus stops, and the third part refers to the feeder operation cost based on Xiong et al. (2013).

The following constraints should be considered:

1. Line length constraint: Long feeder bus lines can lead to complicated line functions, and short lines can lead to insufficient passenger flow; thus, the line cannot be operated normally.

$$d_{min} \le d_l \le d_{max} \tag{2}$$

2. Departure headway constraint: The departure headway of feeder bus lines is determined by comprehensively considering feeder lines' passenger demand per unit of time and operation costs:

$$h_{min} \le h_l \le h_{max} \tag{3}$$

$$h_{min} = \frac{c \cdot \left(2 d_l / v_0 + t_0 \cdot N_l \right)}{C_{olmax}} \tag{4}$$

$$h_{max} = \frac{P}{q_{max}} \tag{5}$$

3. BRT station capacity constraint: If a BRT station has a large passenger volume, and newly added feeder passengers will cause the platform to exceed its capacity, then it is not suitable to set up a feeder bus line at the station:

$$\sum_{l=1}^{L} \sum_{n=1}^{N_l} k_n \le R \tag{6}$$

4. Nonlinear coefficient constraint:

$$\frac{d_l}{D_l} \le \rho \tag{7}$$

In conclusion, the above layout model can be written as:

$$\min C_T = C_1 \sum_{n=1}^{N} k_n \left(\frac{d_n}{v_0} + s_n t_0 \right) + C_2 \sum_{n=1}^{N} \frac{1}{2} k_n h_l + \sum_{l=1}^{L} \frac{2 d_l / v_0 + t_0 \cdot N_l}{h_l} \cdot c \tag{8}$$

$$s.t. \begin{cases} d_{min} \le d_l \le d_{max} \\ h_{min} \le h_l \le h_{max} \\ \sum_{l=1}^{L} \sum_{n=1}^{N_l} k_n \le R \\ \frac{d_l}{D_l} \le \rho \end{cases}$$

5 DESIGN OF GENETIC ALGORITHM

In this paper, we propose the following process to create feeder lines: First, calculate the distance between each regular bus stop within the influence area and the BRT station, then choose two BRT stations with the shortest distance as the potential stations to connect with the feeder line. Boolean variables 0–1 can be defined as control variables for the final selection of the two stations: "0" indicates the nearest BRT station to the regular bus stop regular bus, and "1" indicates the next-nearest BRT station regular bus. Then, select value 0 or 1 in the control variables for all the regular bus stops according to a probability of 50%.

Finally, on the basis of the selection results of the control variables for all the feeder regular bus station and aiming at each BRT station, connect the potential feeder regular bus stations to the BRT station with feeder lines, starting with the closest in distance regular bus, thus generating a scheme of feeder bus lines as an individual. It is important to note that a BRT station may not be connected to any feeder line.

To improve the process of creating feeder lines, an improved genetic algorithm to solve the model can be designed as follows.

5.1 *Chromosome coding*

The chromosome coding should be conducted by using binary encoding, that is, using 0–1 to indicate the connection between each feeder regular bus station and a BRT station.

For example, there are n regular bus stops numbered from 1 to n and four BRT stations (A, B, C, and D). If we assume that the nearest and next-nearest BRT stations of regular bus stop 1 are A and B, respectively, the two-numbered feeder regular bus station's nearest and next-nearest bus rapid transit stations are B and C, respectively, and the three-numbered feeder regular bus station's nearest and next-nearest bus rapid transit stations are C and D, respectively, then the individuals (1 0 1...) indicate that among the eventually-generated set of feeder bus lines, the one—and two-numbered feeder regular bus stations are connected to the bus rapid transit station B and the three-numbered feeder regular bus station is connected to the bus rapid transit station D.

5.2 *Population initialization*

The population should be initialized through the chromosome coding of all feeder regular bus stations by selecting control variable 0 or 1 at the probability of 50%.

5.3 *Fitness function*

In order to avoid superindividual (having a fitness value that is considerably greater than the average fitness value of the population and leads to the problem of early convergence in genetic algorithm) in the population, this paper transforms the fitness function as follows:

$$F(x)=\left[\frac{f(b)-f(x)}{f(b)-f(a)}\right]^2 \tag{9}$$

where f(a) and f(b) indicate the minimum and maximum values among the objective function values, respectively; f(x) indicates the objective function value; and F(x) indicates the fitness value.

5.4 *Selection strategy*

Selection strategy adopts a method that combines the optimal individual conservation, the fitness proportionate selection, and the roulette wheel selection. Specific steps are as follows: first, sort the individuals in the population according to the fitness values, retain 2% of total individuals in the population number that have higher fitness values, and directly copy them to the next generation. Then, calculate the fitness proportion, that is, the probability of each individual to be selected. Finally, calculate the cumulative probability of each individual in accordance with their selection probability and generate a random number, then the corresponding individual should be selected for the crossover if the random number fell into a region of the cumulative probability, and in order to select the individual for the crossover, it is necessary to carry out several rounds of selection.

5.5 *Crossover strategy*

As for the crossover operation, two-point crossover should be adopted, that is, after randomly setting up two crossover points in two matching individual coding strings, exchange part of the chromosome between two crossover points set up by two individuals, and the crossover probability is p_cross = 0.4.

5.6 *Mutation strategy*

As for the mutation operation, the site mutation should be adopted, and the mutation probability is p_muta = 0.2. In terms of a binary-encoded individual, if the original site is 0, then it will become 1 through the mutation operation and vice versa.

6 CASE STUDY

6.1 *Survey design and data collection*

The layout of newly added feeder buses to the Beijing BRT line 1 is analyzed as an example. A questionnaire survey was carried out at 17 stations along the lines at morning peak 7:00–9:00 during two working days, thus obtaining the departure points of the passengers that are transferring from regular buses to BRT. A total of 3,076 valid samples are obtained and 83 regular bus lines and 225 stops are involved; according to a statistics based on regular bus stations, the quantity of the passengers that need feeder buses to bus rapid transit lines at each regular bus station can be obtained. In order to meet the research needs, it is necessary to map BRT lines and their station information as well as the information about 225 regular bus stops in ArcGIS, which stores regular bus stop names, the quantity of passengers with feeder needs, and latitude and longitude coordinates among other attributes.

6.2 *Influence area definition*

When selecting the layout scope, it is necessary to fully consider the needs of passengers to transfer from regular bus transit to bus rapid transit and the efficient use of newly added feeder bus lines, thus considering most passengers' travel demands and making no waste of transport resources. On the basis of the above analysis, this paper selects the reasonable influence area of bus rapid transit lines to the passenger flow of regular bus transit as the layout scope of feeder bus lines. By using the method in Section 2 and by considering 75%, 80%, and 85% of the passengers' interests respectively as the goals, the reasonable influence area of the Beijing BRT line 1 can be determined, as shown in Figure 1. This paper uses the value of 80% to layout feeder lines.

With the goal of 75%, 80%, and 85% passengers' interest, respectively, reasonable influence area of Beijing BRT line 1 (as shown in Figure 1) and the influence radius under different goals (as shown in Table 2) are identified.

6.3 *Selection of feeder stops*

The selection of feeder stops includes the selection of BRT stations and the selection of regular bus stops. The regular bus stations should be selected according to the quantity of the passengers at each station, estimated from the survey, who need feeder buses to connect to BRT lines. This paper chose 29 stations with relatively large demand for feeder (passenger volume is more than 20 people), and new feeder bus lines are added to these stations to meet the constraint of large passenger demand for feeder. The selection of BRT stations is based on the size of feeder passenger volume at stations. The 29 regular bus stops are numbered 1–29, five bus rapid transit stations are numbered A–E, and the station information is initialized, as shown in Table 3. The distances between stations are calculated through latitude and longitude coordinates in the ArcGIS.

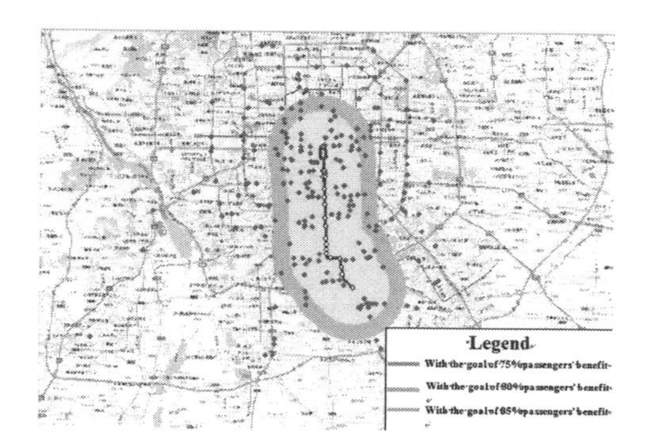

Figure 1. Reasonable influence area of bus rapid transit lines under different goals.

Table 2. Influence radius of BRT with different goals.

Indicators	75% passengers' interest	80% passengers' interest	85% passengers' interest
Reasonable influence area			
Influence radius (m)	3430	3890	4940

Table 3. Passenger quantity at each feeder regular bus station (Unit: person).

No.	Passenger quantity	No.	Passenger quantity	No.	Passenger quantity
1	32	11	27	21	34
2	23	12	26	22	25
3	53	13	47	23	23
4	31	14	42	24	47
5	27	15	36	25	32
6	34	16	33	26	41
7	30	17	41	27	30
8	35	18	34	28	28
9	52	19	39	29	35
10	38	20	36	–	–

6.4 *Feeder lines layout*

1. Parameter value

The model's parameter values are shown in Table 4:

2. Calculation

By using the Matlab2013b (as shown in Figure 2), the layout scheme consisting of a total of eight feeder bus lines were generated, including: Route1:1–3-4–7-A; Route2:2–6-5-A; Route3:10–11–8-9-B; Route4:12–14–13–15–16-D; Route5:17–18-A; Route6:20–22–23–21–19-A; Route7:25–24-B; and Route8:29–28–27–26-D. The layout of each line is shown in Figure 3.

Table 4. Model's parameter values.

Parameter name	Unit	Value	Parameter name	Unit	Value
Stopping time of feeder buses at each station	sec	20	Operating speed of feeder buses	km/h	15
The operating cost per feeder bus per unit time	Yuan	1,500	Unit in-vehicle travel time cost of passengers	Yuan/min Yuan/min	0.2
The maximum operating cost of each feeder bus line that can be provided by operators	Yuan	4,500	Unit waiting time cost of passengers	Yuan/min Yuan/min	0.4
Rated passenger capacity of a feeder bus	Person	80	The upper limit value of the length of a feeder bus line	km	8
The upper limit value of the departure headway of a feeder bus line	min	8	The lower limit value of the length of a feeder bus line	km	2
The lower limit value of the departure headway of a feeder bus line	min	5	–	–	–

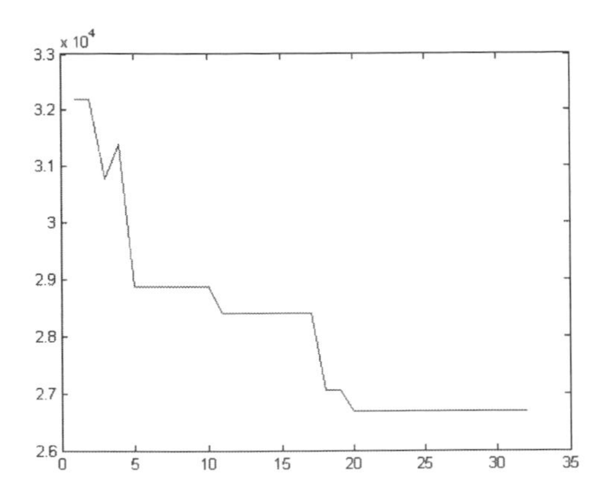

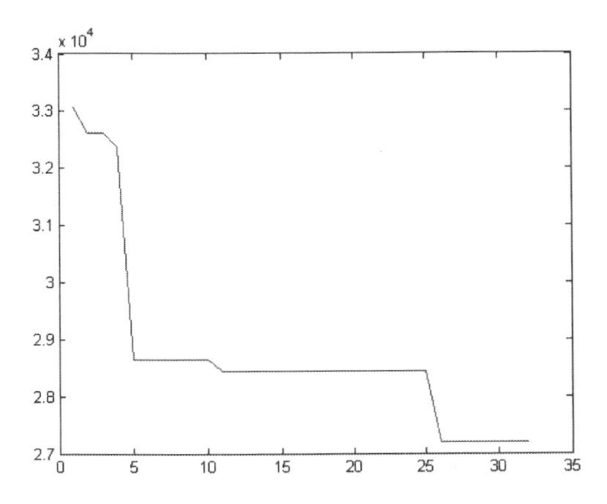

Figure 2. Iterative process during the running time of algorithm.

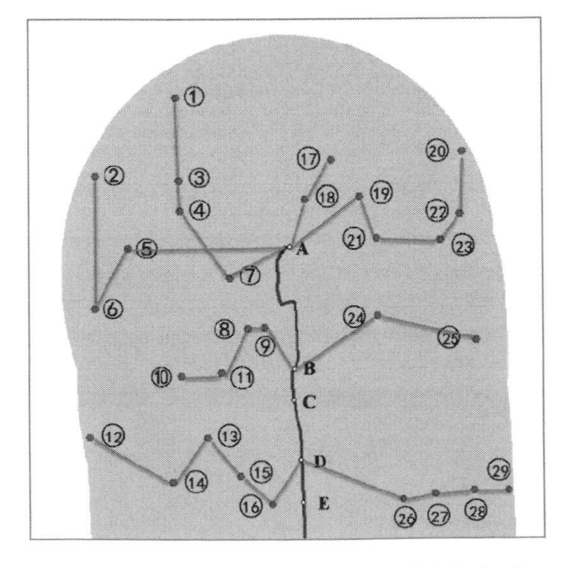

Figure 3. Layout scheme of newly added feeder bus lines.

7 CONCLUSIONS

The layout of newly added feeder buses to BRT lines is an important issue that relates to increase the efficiency of the whole city's public transportation system. In this paper, we discussed the method to determine the reasonable influence area of BRT and the method to lay feeder bus lines within the influence area, proposed a mathematical model for feeder bus line layout, put forward an improved genetic algorithm to solve it, and conducted a case study for adding feeder buses to the Beijing BRT line 1. The following conclusions can be summarized from this study.

First, through the survey and GIS analysis on Beijing BRT line 1, in reasonable passenger attract area aspects, because of sparse stops at the suburban district, the trip distance of passengers is much longer with a high average speed for regular bus, so it is more attractive for regular bus passengers. Conversely, the influence area is smaller in the central district. With the goal of 80% passengers' interest, the reasonable influence radius should be 3.89 km.

Second, this paper proposed the selection principle of newly added feeder lines and the feeder stops. A mathematical model with the connection benefits (including travel time cost and operation cost) as a minimum objective function has been conducted for the first time. Using genetic algorithm to calculate the model, the layout scheme of newly added feeder lines was planned. Applying to the Beijing case, eight feeder bus lines were designed linking to the Beijing BRT line 1, and the results prove the validity of the model.

It is important to note that although in this study the model involved the BRT capacity constraint issues, the coordination of departure frequency of BRT and feeder lines was not taken into account. The newly added feeder lines should be designed considering the operating coordination as well as the reasonable influence area, which will improve the efficiency of the feeder system and will be the directions for our future work.

ACKNOWLEDGMENTS

This study was funded by the Volvo Research Educational Foundation (VREF), an National Science Foundation (NSF) under grant No. 41471459, and the Fundamental Research Funds for the Central Universities (Beijing Jiaotong University) under grant No. 2015RC089-C. The authors gratefully acknowledge Juan Miguel Velasquez, Cristina Albuquerque, and Pablo Andrés Guarda Rosas from ALC-BRT for their contributions to this paper.

REFERENCES

Carrigan, A., R. King, J.M. Velasquez, M. Raifman, and N. Duduta. Social environmental and economic impacts of BRT systems. Bus rapid transit case studies from around the world. Washington D.C., United States of America, Carrigan, 2013.

Chen S, Schonfeld P. Joint optimization of a rail transit line and its feeder bus system. Journal of Advanced Transportation, 1998, 32(3): 253–284.

Dong, X. Smooth connection between bus rapid transit and regular bus transit. Chengdu: Southwest Jiaotong University, 2011.

Fang, X. Research on the method of laying and optimizing feeder bus lines for urban rail transit. Chengdu: Southwest Jiaotong University, 2013.

Kuah, Perl. Optimizition of feeder bus routes and bus-stop spacing. Journal of Transportation Engineering, 1988, 114(1): 341–354.

Kuan, S.N., H.L. Ong, and K.M. Ng. Solving the feeder bus network design problem by genetic algorithms and ant colony optimization. Advances in Engineering Software, Vol. 37, No. 6, 2006, pp. 351–359.

Kuan, S.N., H.L. Ong, K.M. Ng. Solving the feeder bus network design problem by genetic algorithms and ant colony optimization. Advances in Engineering Software, 2005(10): 351–359.

Martins, C.L., and M.V. Pato. Search strategies for the feeder bus network design problem. European Journal of Operational Research, Vol. 106, No. 2, 1998, pp. 425–440.

National Report on Urban Passenger Transport Development. Ministry of Transport, P.R China, 2014.

Niu, S. Research on the Optimization of Feeder Bus Network in Coordination with Urban Rail Transit. Chengdu: Southwest Jiaotong University, 2011.

Pabhat Shrivastava, Margaret O'Mahony. A model for development of optimized feeder outes and coordinated chedules-A genetic algorithms approach. Transportation Policy, 2006, 13(5): 413–425.

Shoaid Chowdhury, Steven Chien. Optimization of Tranfer Coordination for Intermodal Transit Network. Transportation Research Board 80th Annual Meeting, 2001, January 7–11.

Shrivastava, Dhingra. Development of feeder routes for suburban railway station using heuristic approach. ASCE Journal of Transportation Engineering, 2001, 127(4): 334–341.

Verma, A., and S.L. Dhingra. Feeder bus routes generation within integrated mass transit planning framework. Journal of Transportation Engineering, Vol. 131, No. 11, 2005, pp. 822–834.

Vuchic, V.R. Urban Transit: Operations, Planning and Economics. John Wiley & Sons, New Jersey, United States of America, 2005.

Wang, S., L. Sun, and J. Rong. Research on influence area of rail transit stations in Beijing City. Journal of Transportation Systems Engineering and Information Technology, Vol. 13, No. 3, 2013, pp. 184–185.

Wirasinghe, Herdle, Newell. Optial parameters for a coordinated rail and bus transit system. Transport Science, 1977, 11(4): 359–374.

Wirasinghe. Nearly optimal Parameters for a rail feeder bus system on a rectangular grid. Transport Research Part A, 1980, 14(1): 33–40.

Xiaoguo Zhu. Research on coordination of bus transit rapid and ragular bus. Changsha: Changsha University of Science and Technology, 2009.

Xiaoying Bi, Xuewu Chen, Zimu Li. Study on characteristics of BRT network based on L-topological spaces. Journal of Transportation System Engineering and Information Technology, 2011(5): 173–180.

Xiong, J., W. Guan, and L.Y. Song. Optimal routing design of a community shuttle for metro stations. Transportation Engineering, Vol. 139, 2013, pp. 1211–1223.

Civil, Architecture and Environmental Engineering – Kao & Sung (Eds)

Design of a campus view of a virtual roaming system based on VR

Ling Yu
Art and Design School, Dalian Polytechnic University, Dalian, China

ABSTRACT: Virtual reality is a new and advanced technology that produces a three-dimensional space by computer simulation to provide users with visual, auditory, tactile, and other sensory simulations. In this paper, by comprehensively using Vega Prime and Visual Studio software, we construct a virtual campus roaming system. The main research work of this paper is as follows: 1. On the basis of early use of three-dimensional modeling of MultiGen Creator on the campus of Nanchang University, we use Vega Prime and Visual Studio software to build a virtual campus roaming system. 2. We design the virtual campus roaming system database, which covers the data of various functional departments of the school, in order to facilitate the user's inquiry and understanding of the school. 3. We study the key technologies of virtual simulation technology: roaming technology and collision detection technology. The design improves the flexibility, immersion, and interaction of the system.

1 INTRODUCTION

Before the 21st Century, because of the less development of computer hardware and computer technology, people's understanding and description of the real world were still in the form of binary. The binary description of the real world is quite different from the actual real world, which cannot reflect various types of images of the three-dimensional world. Therefore, under many circumstances, it will affect people's understanding of the real world. With the rapid development of computer technology, virtual reality technology (referred to as VR, also translated as the spirit and reality) is increasingly widely used. This is an integrated technology, relating to the field of computer graphics, simulation technology, multimedia technology, human–computer interaction technology, sensor technology, artificial intelligence, and so on. Its core idea is to simulate a real world as realistic as possible in a computer model. People are roaming, driving, and training in this virtual model so as to achieve the equal effect as it is in the real world. As a result, it is widely used in commercial, military training, and fields with emergency and uncertainties (such as reproduction of the scene of the traffic accident, fire rescue simulation, aerospace, and micro simulation). VR has three main features: immersion, interaction, and imagination.

2 THE OVERALL DESIGN OF THE VIRTUAL CAMPUS ROAMING SYSTEM

The design of the virtual campus roaming system is divided into three parts: the establishment of a virtual campus model, using Prime Vega to quickly customize the virtual campus roaming system ACF files and using Studio2005 Visual to write a virtual campus roaming system on the basis of MFC (Biocca F, 2013).

2.1 *System environment configuration*

Real-time rendering three-dimensional virtual scene and large-scale spatial data have high requirements on the system hardware. In the following, it describes the hardware configuration and software used in the system.

The hardware configuration of the system is as follows:

CPU: Intel (R) Core2 Q9500 2.83 GHz
Memory: 2GB
Hard disk: 500GB
Graphics: 1GB independent memory
Operating system: Windows XP
Three-dimensional modeling software: MultiGen Creator 3.2
Virtual campus roaming configuration software: Vega Prime 2.2.1
Virtual campus roaming system development platform: Visual Studio 2005
Database: Microsoft Access 2003

In the early implementation of the project, the computer configuration is low, resulting in when using Gen Creator Multi 3.2 software modeling, it cannot normally display the texture image and thus unable to continue to carry on this project. We suggest using 1GB or more independent video card so that the system can smoothly carry out interactive roaming, query, and so on.

2.2 *Overall design of virtual campus roaming system based on MFC*

2.2.1 *ACF file of virtual campus roaming system*

The development of visual simulation application program is usually divided into the establishment of virtual 3D model and the design of virtual roaming engine. In the design of virtual roaming engine, if we do not use ACF file, then it can only use the program code to replace the ACF file, create an instance of all objects, and proceed relationship configuration of these objects of a class. If the virtual scene is more complex, not using the ACF file will cause lots of problems and inconveniency for configuration and modification of the scene. As a consequence, in this paper, we use the ACF file in the design of virtual roaming engine. If there are changes in the virtual scene, it can be modified in the ACF file. In the framework of virtual campus roaming system, the function of Vega Prime is written in C++ language, and the developing environment of roaming engine is Vega Prime 2.2.1, so we use Visual Studio 2005 as the development platform of the system; choose application program based on MFC (Microsoft Foundation Classes), so it can be liberated from the implementation details of the cumbersome framework program in the study, and put the main focus on the study of simulation function and interactive function of the system. To sum up, the construction of virtual campus roaming system is divided into three steps, shown as follows.

1. On the basis of collecting the campus data in the previous period, we make use of Gen Creator Multi to build the virtual campus three-dimensional model.

   The campus map file *.dwg is stored in the South CASS7.0 mapping software as CAD R12/LT2 DXF *.dfx Auto format. File *.dfx is input into the Gen Creator Multi, namely the upcoming map is input into the Gen Creator Multi modeling software. On the basis of the two-dimensional map in Gen Creator Multi, respectively, we model buildings, roads, trees, streetlights, lakes, grasslands, gates, and so on.

   We use Photoshop software to process the photos taken in the early stage and produce texture image. Virtual campus 3D model will be used for Prime Vega visual simulation, and the size of the texture must be 2n*2n, so the size of the texture is set to 1024*1024 pixels. Make the texture on the surface of the three-dimensional model, and then the establishment of three-dimensional virtual campus modeling is completed.
2. Load the modeling file *.flt in step (1) into Prime Vega, the virtual campus, and proceed the preliminary design of engine of virtual campus roaming system, such as pipeline configuration, window configuration, rendering of virtual environment, motion mode, and navigation path configuration, so as to get the *.acf file.
3. Load *.acf file obtained in step (2) into the MFC framework built to complete various functions of the system, such as walking and driving mode, fixed-path roaming mode, environment control, improvement of query, and modification of attribute.

2.2.2 *Development of virtual campus roaming system based on MFC framework*

MFC is a fairly mature library. Developing application program having a good graphical user interface under Windows platform based on MFC library is the most convenient way. Therefore, a good graphical user interface based on MFC is the most convenient way to develop visual simulation program.

The development of virtual campus roaming system based on the MFC framework needs to solve two problems. One is how to render the three-dimensional scene of virtual campus to MFC view window. The other is how to proceed frame cycle of the virtual scene. First, vp Window provided by Vega Prime comes from vs Window, which is generated from vr Window that provides a set Parent (Window win) function. It is seen that it only needs to transfer the kind of MFC view window to the vp Window, namely vp Window *vp Win = *vp Window: begin (); vp Win->setParent (m_h Wnd); only by doing like this can the virtual campus 3D scene be rendered to view window (Earnshaw R. A, 2014). Second, the system frame cycle can be achieved in two types of methods. The first one is to set a timer, namely after application program completing, run the initialization, definition, and configuration, setting a timer, and then proceed frame cycle in the message processing function. The second method is to create an additional Windows thread and complete Vega Prime application in the thread.

3 DETAILED DESIGN OF SYSTEM FUNCTION

3.1 *Specific design of system function*

Consistency of style is very important. Note the spacing, punctuation, and caps in all the examples below.

3.1.1 *Walking mode*

In Vega Prime, use vp Motion to define abstract motion model, while the specific motion patterns are as follows: Motion Drive, Motion Fly, Motion Walk, Motion UFO, and so on, all of which inherit the class of vp Motion. vp Motion can specify input device as the method of controlling motion mode, such as keyboard and mouse. If no input device is specified, the motion model will enable default of their devices as the input of the system. The control of default in the motion mode is packaged in the various classes of motion mode. For example, in the driving mode, Source Boolean controls the acceleration and deceleration of the vehicle model, and Source Float controls the left turn and right turn of the vehicle model.

In the navigation interface of the system in Visual Studio, add a Radio type button and a Slider type controlling button, respectively, to set up whether choose walking mode or not and the speed of walking mode, and the control node of walking control code mode is as follows:

Vp Transform *Transform_walk = vp Transform: find ("Transform_walk").

3.1.2 *Automatic roaming mode*

In the realization of the automatic roaming mode function, the tool to be used is Path in the LynX Prime. Path Tool is a tool used to create path control points and navigation files in the automatic roaming of a system. Fixed path in Vega Prime has two meanings: the first is a series of control points of the navigation path set by the users, which are stored in the .way file. The file stores the 3D coordinates of each control point and the attitude of each point; which is followed by a navigation device. Navigation file provides a data structure for each control point. Information stored in the data structure contains the connection method from current control point to the next control point, moving speed from the current control point to the next control point and so on. These messages are stored in the .nav file, and the navigator, through the interpretation of the data in the .nav file, controls the motion of object or the observer's state.

In the automatic roaming mode, five automatic roaming paths are set up, which basically cover the whole campus scene. Five school bus models are added in the virtual scene of the campus, each of which uses an automatic roaming route. In the navigation system interface, add a Radio button to control the automatic roaming mode selection, and a dropdown list box to determine which path the user should choose. The realization code of the two controllers is shown as follows.

```
Realization process of Radio button is:
If Radio button is pressed, the automatic path is
   set up as the first path.
Int nCount = m_Combo Auto.Get Count();
if (n Count > 0)
{
m_Combo Auto.Set Cur Sel(0);
On Selchange Combo Auto ();
}
vp Transform *tran = vp Transform:
find("Transform_car0");
Drop-down list box control implementation
   process is:
int n Sel = m_Combo Auto.Get Cur Sel ();
if (n Sel ! = –1)
{
// obtain the selected path and set up it
vp Transform *Trans = (vp Transform*) m_
Combo Auto.Get Item Data (n Sel);
   if (Trans)
(*vp Observer::begin())->set Look from(Trans);
}
```

In order to prevent the emergence of the car model in the mountain drill, it is necessary to add tripod Isector detector whose target is the terrain. Tripod Isector detector is composed of line segment that is, respectively, perpendicular to the Isector (X-Width/2, Y-Length/2), (X+Width/2, Y-Length/2), and (X, Y, +Length/2); these three coordinates, coordinate, and X and Y represent the current position of Isector. The method is that three line segments and the points intersect with the terrain form a triangle, output the triangle center value of Z and pitch value and roll value corresponding to the center Z and the current direction of the intersect vector, so the car and terrain model can be tightly attached together, but not to drill into the mountain, and the effect is shown in Figure 1.

3.2 *Application of system database*

3.2.1 *Establishment of tree structure and modification of nodes*

In order to enhance the interactivity and immersion of the whole system, users can browse the virtual campus scene better and have a better understanding of the whole virtual 3D scene, add a

Figure 1. Tripod detection effect diagram.

set of tree structure and attribute display bar at the interface of the system, and all objects in the three-dimensional scene has a corresponding node in the tree structure. As shown in Figure 2, the three-dimensional scene is the Information Engineering Institute of the campus. In the tree structure, there is a corresponding node called "Information Engineering Institute".

The structure of the tree building method is as follows: first, to create a global database object and connect to the database each time the system boots; second, when the database connection is successful, to create the tree structure by using a recursive function according to the relationship of nodes in the Observer Position table.

When we use the right mouse button to click the node in the tree structure, the management menu will pop up. When the root node is clicked, the deleting item node and the modified position item will become gray, which is unable to be used. When the other node is right-clicked, all options are available. When the user clicks the adding node option, it will pop up a dialog box, and then the user can choose the type of node and add a subnode. For example, when the school builds a few teaching buildings, the school has introduced a number of new teachers, and in the new term school has recruited some freshman, able to add subnodes through the user's choice. Therefore, the realization of this function has a significant impact on the school management, planning, and user's perception of the school. The method to add a subnode is first adding a subnode for the current node in the tree structure, then setting up the value of new nodes in Observer Position table, and finally inserting the attribute data in the attribute table for the new nodes. The method to delete a node is as follows: first, to find the brother node or parent node of this node, to ensure that after deleted, it still exists the selected node, and then delete the node and all subnodes. Finally, delete the information in response to the database.

Figure 2. Tree structure.

3.2.2 *Realization of modifying location and querying function*

When the user opens the tree and right-clicks on a node, it will pop up a menu, in which there exists an option called "modify the location", whose function is to record the current position of the view and store it in a database. The implementation of this function is convenient for users to modify the scene position in the database and the major node to achieve the function as shown below (Gan, Q, 2015):

```
observer->get Position (&x, &y,&z);
observer->get Rotate (&heading, &patch, &roll);
// obtain the current selected point
HTREEITEM itm = this->Get Selected Item ();
Observer Position * obp = (Observer Position*)
  this->Get Item Data(itm);
// store the data in the database
sz SQL.Format("UPDATE observerposition
  SET x = %lf, y = %lf, z = %lf, heading
  = %lf, patch = %lf, roll = %lf WHERE
  id = %d",
x, y, z, heading, patch, roll, obp->id);
p SYSDB->Execute SQL (so SQL);
Update the data in the system
obp->x = x;
obp->y = y;
obp->z = z;
obp->heading = heading;
obp->patch = patch;
obp->roll = roll;
```

The function of query is that when the user does not know the position of a node in the virtual scene, it can double-click the node in the tree structure and render the 3D virtual scene on the left-hand side of the screen, and the view will be cut from the view of the current node to the view of the node that the user double-clicks. That is to say, "fly" to the destination from the current position, and complete the inquiry work. As a result, this function improves the user understanding of 3D virtual scene and greatly enhances the interactivity and immersion of virtual campus roaming system. The design idea of the function: when the user double-clicks a node, it will first calculate the six parameters, X, Y, Z, H, P, and R, of the current view and the difference value of the view of the destination node, and then the difference value is averagely divided into several parts, to get the increment of each flight, and finally open a timer (Jiao, N, 2014). In each period of time, complete a viewpoint change of flight increment. Destruct the timer and complete query function until six freedom parameters of the current viewpoint is equal to the six freedom parameters of the viewpoint that the user double-clicks The main code of this function is as follows:

```
// Calculate the difference value between the two
  view points
```

```
double x Len = m_pos End.x - m_pos Begin.x;
double y Len = m_pos End.y - m_pos Begin.y;
double z Len = m_pos End.z - m_pos Begin.z;
double h Len = m_pos End.heading - m_pos
    Begin.heading;
double p Len = m_pos End.patch - m_pos
    Begin.patch;
double r Len = m_pos End.roll - m_pos Begin.
    roll;
Calculate the increasing amount in each flying
m_pos Delt.x = x Len/n Times;
m_pos Delt.y = y Len/n Times;
m_pos Delt.z = z Len/n Times;
m_pos Delt.heading = h Len/n Times;
m_pos Delt.patch = p Len/n Times;
m_pos Delt.roll = r Len/n Times;
// Set the position of view point
vp Observer *observer = *vp Observer::begin ();
observer->set Position(m_pos Begin.x,m_pos
    Begin.y,m_pos Begin.z);
observer->setRotate(m_pos Begin. heading, m_
    pos Begin. patch, m_pos Begin.roll).
```

3.2.3 *Voice information broadcasting function*

Music added in the virtual scene usually has three modes: the first one is to simulate a real sound in life, which usually considers the factors that the sound reduces due to the propagation in the medium and Doppler effect, so it is true. For example, the roar of aircraft in flight and driving sound will be weakened as the car and the airplane objects get away from the observer; the second kind is background music, whose sound does not change with the location of the observer; the third is speech information, namely voice introduction. The sound will not change with the change of the location of the observer.

Essential node objects in the tree structure of the object system are added with the voice information. The voice information, through special recording, has been added to the sound value of each node objects in the database, and users can also use the menu item to manually add voice information for the node object. When the node object in the tree structure is double-clicked, the speech information will automatically play so that the user can increase understanding of 3D virtual campus.

4 CONCLUSION

People's understanding and accepting ability of visual, acoustic, tactile, and other sensory information is far greater than that of numbers in such kind of abstract information. Along with the rapid development of computer technology, visual simulation is more and more widely used in military training, emergency, simulation of driving, city planning, and many other areas.

This paper completes the development of the virtual campus roaming system based on MFC framework, in which the research is on environment rendering technology, so users can browse through the mouse to control to roam in virtual campus. The research work is summarized as follows. First, use GPS and total station to measure the whole area of the campus, then get a two-dimensional map of the campus, and finally use Multigen Creator to make a three-dimensional model of the scene of the campus (Jun, T, 2015). In addition, by in-depth study using Vega Prime software technology, import the virtual campus 3D model into Prime Vega software and configure the relevant parameters to get the ACF file, namely the initial and simple visual simulation of the virtual campus.

The Access is chosen as the virtual campus roaming system of external database to realize data query, modify, and carry out other functions. The model in the scene is corresponding to the attribute information model, to complete the query function of the virtual campus scene model, and be convenient for users' understanding and cognition of the 3D virtual campus; add data model and node of the tree structure for the system to modify the attributes of the nodes to enhance the system scalability and performance management; add voice information broadcasting function for each data node in the system to improve the immersion and interactivity of the system, so as to increase users' awareness of the system. Three-dimensional virtual campus technology plays a very important role in the external propaganda, enrollment, and the appearance of the campus, and it is an important part of campus information construction. Therefore, the development of the virtual campus roaming system is of great significance to the development of visual simulation system in the future.

REFERENCES

Biocca, F., & Levy, M.R. (Eds.). 2013. Communication in the age of virtual reality. Routledge.

Earnshaw, R.A. (Ed.). 2014. Virtual reality systems. Academic press.

Gan, Q., & Li, W. 2015. VRML-based virtual roaming system of Pingdingshan museum. Journal of Terahertz Science and Electronic Information Technology, 4, 023.

Jiao, N., Liu, S., Li, L., Wang, Q., Zhao, J., & Zhao, P. 2014. Design and Research of Lunar Rover Human-Computer Interactive Roaming System. InProceedings of the 13th International Conference on Man-Machine-Environment System Engineering (pp. 293–300). Springer Berlin Heidelberg.

Jun, T., Fengjun, L., & Ping, W. 2015. National museum research based on virtual roaming technology. In 2015 Seventh International Conference on Measuring Technology and Mechatronics Automation (pp. 683–688). IEEE.

Civil, Architecture and Environmental Engineering – Kao & Sung (Eds)
© 2017 Taylor & Francis Group, ISBN 978-1-138-02985-9

Analysis on sustainable financial framework and fare adjustment strategy of China's urban public transit

Sa Xu, Cheng Li, Xianglong Liu & Yunke Du
China Academy of Transportation Sciences, Ministry of Transport, Beijing, China

ABSTRACT: This paper aims to design a feasible financial framework and fare adjustment strategy to intensify the financial sustainability of China's urban public transit service, therefore to support the policy of transit priority and the development of transit metropolises in China. The fiscal problems in China's public transit, the skyrocketing operating costs of public transit companies, unsustainable urban public financial subsidies, decreasing attraction of low fares to the passenger flow, are analyzed by the data on the operating costs, fares, subsidies and other information collected from Beijing, Shenzhen, Zhengzhou and other China's typical urban public transit companies. The relationship among transit service quality, costs, fares and government subsidies is analyzed systematically based on the demands of three major interest subjects in city public transit, namely passengers, businesses and governments. Based on the principle of "acceptable for the public, sustainable for businesses, and affordable for government finance", the paper proposes a feasible service quality-oriented and sustainable financial framework of urban public transit operation for China which clarifies the responsibilities of all the stakeholders in the framework. Finally, the paper offers specific fare adjustment strategy, in which the public transit fare adjustment coefficient is playing the role in the trigger mechanism.

1 INTRODUCTION

With China's urbanization and mechanization pressing ahead, cities continue to expand, and urban population and traffic volume increase rapidly. As a result, traffic congestion, security issues, environmental pollution and energy depletion are increasingly prominent, becoming the bottleneck of urban sustainable development and imposing great pressure to the construction of a resource-conserving and environment-friendly society. Urban public transit possesses advantages such as high holding volume and efficiency, and low energy consumption and pollution. Learning from Europe, Japan, Singapore and South America, China is striving to prioritize public transit to reduce environmental pollution and energy consumption, and to achieve low carbon and sustainable development.

Chinese government has attached great importance to the development of urban public transit, and issued many important documents, including Guiding Opinions of the State Council on Giving Priority to Public Transportation in Urban Development, to further specify national prioritization of public transit development. Pushed by the accelerating urbanization and guided by the national transit priority policies, China's urban public transit has come a long way. As shown in the annual report of China urban passenger transportation development, (MoT, 2011–2013), by the end of 2013, the annual passenger capacity and daily capacity of China's urban public transit have amounted to 128.335 billion person times and 351 million person times respectively. There are 41,783 operating routes for buses and trolleybuses nationwide, totaling 749,000 km. Metro construction is moving forward steadily. 15 cities in China have 81 rail transit lines with a total length of 24,079 km. An urban rail transit network covering major cities like Beijing, Shanghai, Guangzhou and Shenzhen have taken shape. 22 cities have operated the BRT (Bus Rapid Transit) system with the operating routes amounting to about 2,752 km. Cities like Jinan, Changzhou and Zhengzhou have built a preliminary online operating system for BRT which is gradually playing a major role in urban transport systems.

To implement the transit priority policies, governments in many cities have issued a series of relevant policies and measures. One important measure is to combine low fares with large financial subsidies. In other words, governments purchase public services from businesses through financial subsidies. The measure has promoted the utilization and share of public transit by reducing the traveling costs of the public. Due to deficient support in relevant policies, however, urban public transit faces problems on the sustainability of finance, such as the ever-increasing operating costs, the difficulties of public transit businesses in operation, the overwhelmed financial subsidies and so on. Those problems have affected, to some extent, the healthy and sustainable development of China's urban public transit.

2 REVIEW OF LITERATURES AND INTERNATIONAL EXPERIENCES

Methods of fare and subsidy adjustment is a common issue in many cities worldwide and there are many valuable experience in their practices.

In 1998, Singapore adopted the price-cap model for the regulation of public transit fares. The fare adjustment cap formula of this model was "CPI+X", among which CPI represented change in the Consumer Price Index over the preceding year, and "X" was set to compensate the operators for net cost (after considering wages and productivity) increases beyond inflation. When economy growth slowed down or passengers' capacity to pay fares declined markedly, the Public Transport Council would launch this model to adjust fares. This formula was optimized in 2005, further balancing the interest of public transport operators and passengers. Some macroeconomic indexes such as CPI and WI were involved. The optimized formula was: fare adjustment cap = price index = 0.5CPI + 0.5 WI-X. Between 2005 and 2012, X was set at 1.5%.

In South Korea, the Seoul Metropolitan Government (SMG) is authorized by the Seoul Metropolitan Council (SMC) to adjust the basic fare of public transit every two years. The requirement for adjustment is usually proposed by public transport businesses and associations to relevant authorities who will, after examination and approval, report to the President. The decision is approved for execution by the mayor of Seoul. Besides, SMG will also solicit opinions from SMC.

In Hong Kong, bus companies have been allowed to promote fares with the increase of inflation. The government has a Fare Adjustment Arrangement for Franchised Buses (FAA): operators can decide by themselves when to submit a fare adjustment application, and the government also has the right to lower fares. In every season, the Hong Kong Transport Department (HKTD) will calculate the supportable fare adjustment rate based on a formula: supportable fare adjustment rate = 0.5 x Change in Wage Index for the Transportation Section + 0.5 x Change in CCPI – 0.5 x Productivity Gain.

Chinese scholars have conducted broad studies on the financial sustainability of public transport in recent years.

He and Ye (2012) pointed out the problems in existing subsidies, including ambiguous targets (cannot guarantee that the limited subsidies are used by the most needy ones), a lack of diverse forms (mainly direct financial subsidies for state-owned public transit companies) and limited capital resources (all come from local finance). They proposed that a clear line should be drawn between policy-related losses and operational losses, routes making profits and ones suffering losses; subsidies for producers and those for consumers, infrastructure construction and public transit operation, and between two purposes, namely providing universal services and guiding the way to travel.

Mao and Sun (2011) pointed out that to address problems facing the operation of public transit companies, the government should, first, innovate its subsidy mechanism for public transit and build a long-term mechanism, and second, create a reasonable linkage pricing mechanism and properly introduce market regulation,

Li, Song and Ren (2009) provided many suggestions. First, completing subsidy procedures to ensure that subsidies are provided and utilized in a scientific and efficient way. Second, building an information disclosure mechanism to provide institutional guarantee for the efficient use of subsidies. Third, strengthening social evaluation and supervision of public transit subsidies to ensure timely feedback, and making service quality an important evaluation indicator of financial subsidies. Forth, supporting IT application in public transit and introducing a competition mechanism.

Yang (2011) pointed out three problems existing in China's low fare policy. First, as people with their own cars are not sensitive to public transport fares, it is difficult to promote the utilization of public transport. Second, as low fares have attracted many cyclists and pedestrians, public transport means have become increasingly crowd, forcing some passengers to travel by their own cars. Third, targeted measures are lacked in subsidizing the traveling of vulnerable groups. Yang suggested that more subsidies should be used to aspects that can promote public service quality, such as shortening traveling distance and passengers' waiting time, reducing transfer times, and shortening passengers' distance to bus stations.

3 ANALYSIS ON FINANCIAL SUSTAINABILITY OF CHINA'S PUBLIC TRANSIT

The policy of combining low fares with large financial subsidies in urban public transit service has promoted the utilization and share of public transit markedly in China. Simultaneously, the urban public transit is facing problems on the sustainability of finance, which has affected, to some extent, the healthy and sustainable development of China's urban public transit.

The first problem is the ever-increasing operating costs. The operating costs of public transit companies can be divided into the following categories: labor cost, fuel cost, depreciation cost of fixed assets, maintenance cost, management cost, financial expense, sales tax and additional expenses. According to a survey on the operating costs of urban public transit businesses in 18 municipalities directly under Henan Provincial Government (Figure 1), labor cost and fuel cost together, totaling almost 65% of the overall costs, form the largest part in the cost structure of public transit businesses.

As global economy has a fast growing demand for energy, global oil prices have increased steadily in the past decade, so have the gasoline and diesel prices in China. In 2014, global oil prices declined markedly, but decline in China's domestic gasoline prices has been much shallower than that in the international market. Figure 2 shows the growth of China's gasoline prices between 2003 and 2013.

With rapid economic growth, national income and per capita wage have increased at a rather high speed, driving up the labor costs in almost every industry. As shown in Figure 3, according to the Statistics data (National Bureau of Statistics of China, 1990–2012), the annual salaries of urban employees in China between 1990 and 2013.

According to above analysis, the major reason for the rising operating costs for urban public transit companies lies in the fast growth of price index, energy cost, labor cost and other costs (see Figure 4).

The second problem China's urban public transit facing is the low fares and no fare adjustment mechanism. Guided by transit priority policies, many provinces and cities have regarded public transit as a commonweal cause and kept the low fare policy for a long time. A survey conducted in six cities, namely Beijing, Shanghai, Shenzhen, Guangzhou, Zhengzhou and Jinan (see Table 1) shows that these cities haven't changed public transit fares for years and do not have a stable and complete fare adjustment mechanism.

In these cities, the implementing fares (fares that passengers actually pay) are lower than cost prices, and the part of price higher than cost prices is compensated by the fuel subsidies of the Central Government and public financial subsidies from local governments. A survey shows that the ratio of implementing fares to cost prices are less than 50% in Beijing, Shanghai and Shenzhen, and that in Beijing is even lower than 20%. This means Beijing Municipal Government provides a financial subsidy that is four times of the implementing fares for public transit companies to support the daily operation of the public service system.

The third problem is the difficulties of public transit businesses in operation in China. According to the above data and analysis, the fast increasing price index, labor cost and fuel prices have driven up the operating costs of China's public transit businesses. Fares have dropped away from costs for a long time, leading to severe losses of public transit companies. Most of them are counting on public financial subsidies for operation. Meanwhile, under the strategy of prioritizing public transit development, governments have asked these companies to

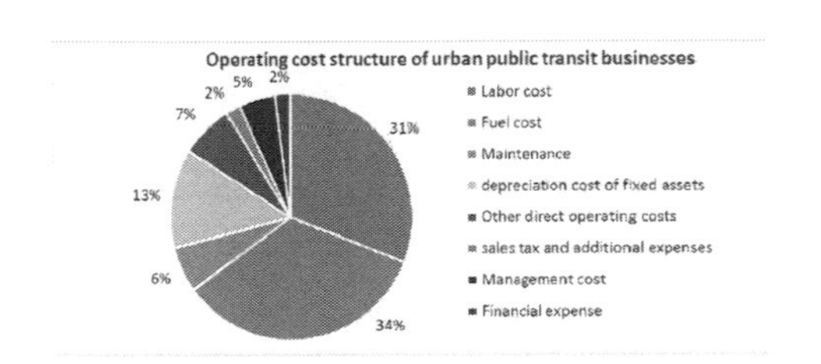

Figure 1. Operating cost structure of urban public transit businesses in Henan Province.

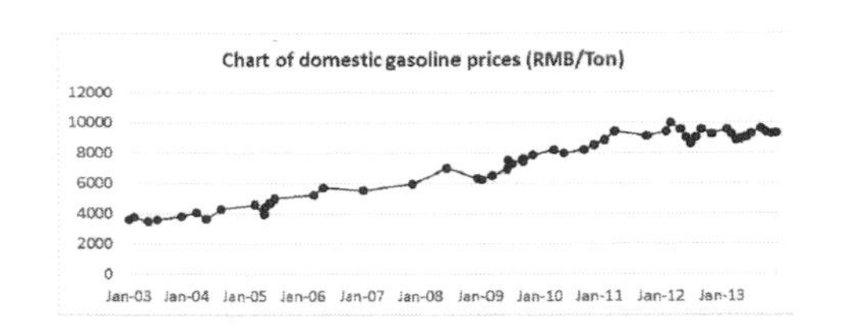

Figure 2. Chart of domestic gasoline prices (2003–2013).

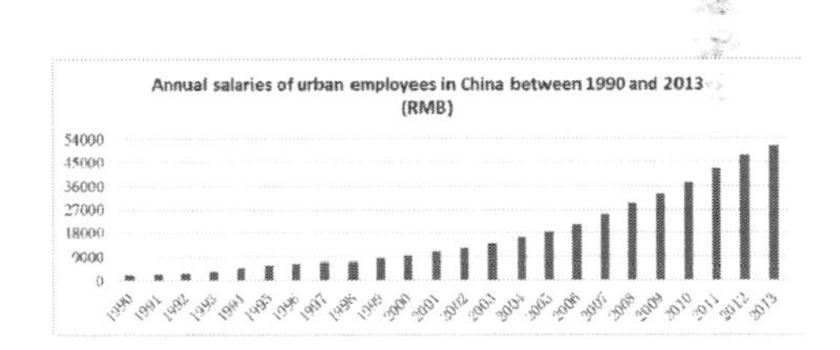

Figure 3. Annual salaries of urban employees in China between 1990 and 2013.

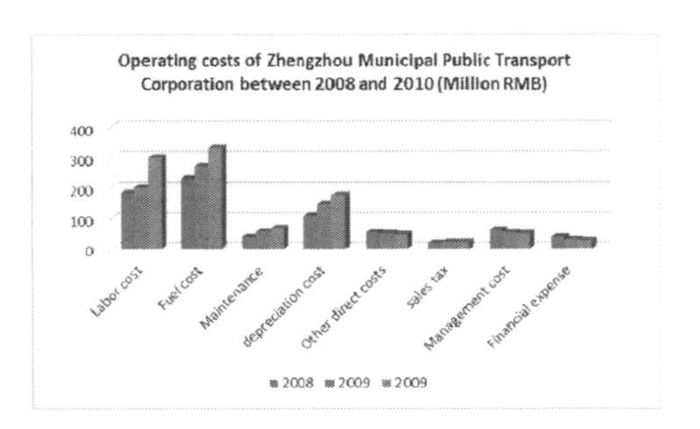

Figure 4. Operating costs of Zhengzhou Municipal Public Transport Corporation between 2008 and 2010.

Table 1. Public transit fares in 6 Chinese typical cities (2013).

City	Base fare (RMB)	Years without adjust	Price discount	Fare comparing to cost
Beijing	Base fare: 1	6 years	Have	Lower than cost
Shanghai	Aver. 1.8 /ride	14 years	Have	Lower than cost
Shenzhen	Aver. 2.02 /rid	7 years	Have	Lower than cost
Guangzhou	Base fare: 1 or 2	14 years	Have	Lower than cost
Zhengzhou	Flat price: 1	10 years	Have	Lower than cost
Jinan	Base fare: 1 or 2	14 years----	Have	Lower than cost

Table 2. The ratio of implementing fares to cost prices in 3 Chinese typical cities (2011).

	Amount and ratio of cost for each ride		
City cost price (RMB)	Implementing fare	Subsidy from central government	Subsidy from municipal government
Shenzhen	49%	16%	35%
Beijing	0.40	0.29	2.46
	13%	9%	78%
Shanghai	1.80	0.42	1.38
	50%	12%	38%

add new routes and increase frequencies. Some of the newly added routes have little passenger flow and thus little fare revenue, imposing more pressure on the policy-related losses of these companies. Besides, these companies indicate that financial subsidies have fallen short of their policy-related losses. Governments have had no auditing methods to check the authenticity of the financial information submitted by companies, and while authorizing subsidies, they would simply lower the subsidies applied by companies, leading to greater losses of the companies. Long-term losses have led to high working intensity of drivers, low salaries, and bad vehicle conditions, lowering the service quality and attraction of urban public transit and affecting the healthy operation of public transit companies. Citizens are quite dissatisfactory towards the vicious cycle in public services.

Overwhelmed financial subsidies is the fourth problem in current China's urban public transit. Under the background of China's transit priority policies, many cities such as Beijing and Shenzhen have started to subsidize low fares of public transit to make up for the policy-related losses of public transit businesses.

Beijing started to adopt low fares and financial subsidies in 2007. Bus routes adopted a single fare of 1 RMB. Passengers with an ordinary card could get a discount of 60% off, and those with student cards, 40% off. Only a small amount of routes adopted distance-graduated fares. A flat fare of 2 RMB was needed for all metros except the airport line. Since the implementation of low fare policies, the subsidies provided by Beijing Municipal Government skyrocketed from 1.81 billion RMB in 2005 to 20 billion RMB in 2013, an average growth rate of 38% in 8 years. The proportion of public transit subsidies in urban budget revenue increased from 2.24% to 5.89% (Figure 5).

Shenzhen also started to adopt low fares and financial subsidies in 2007. Since the implementation of low fare policies, the subsidies provided by Shenzhen Municipal Government skyrocketed from 1.033 billion RMB in 2008 to 6.3 billion RMB in 2013, an average growth rate of 44% in 6 years (Figure 6).

According to the above data, all cities carrying out low fare policies and public transit subsidies see a remarkable increase in their subsidies and the proportion of subsidies in local budget revenue and expenditure, and suffer from extreme pressure on public finance.

The fifth problem is the decreasing attraction of low fares. Low fares have greatly saved people's traveling costs, and promoted the utilization and share of public transit at the initial stage. In recent years, however, companies suffering from skyrocketing operating costs and low fares have made it harder to guarantee or improve service quality. As the public has become less interested in low fares and asked for higher service quality, and families have been buying more cars, the single preferential policy of low fares can no longer attract more people to use public transit, thus leading to the declining utilization rate of public transit. Figure 7 shows changes in the proportions of various transportation modes in Beijing's urban traffic from 1986 to 2011. Although the overall proportion of public transit including rail transit had been increasing due to the fast growth of passenger flow in urban rail transit, the contribution rate of public buses and trolleybuses declined from 28.9% in 2009, two years after ticket price reform, to 28.2% in 2011.

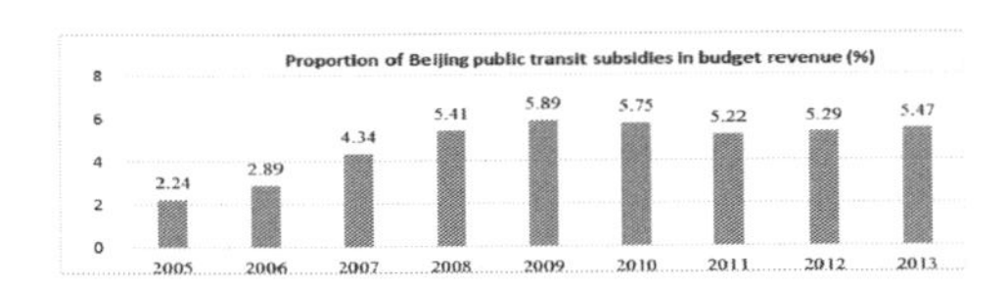

Figure 5. Proportion of Beijing transit subsidies in budget revenue between 2005 and 2013 (%).

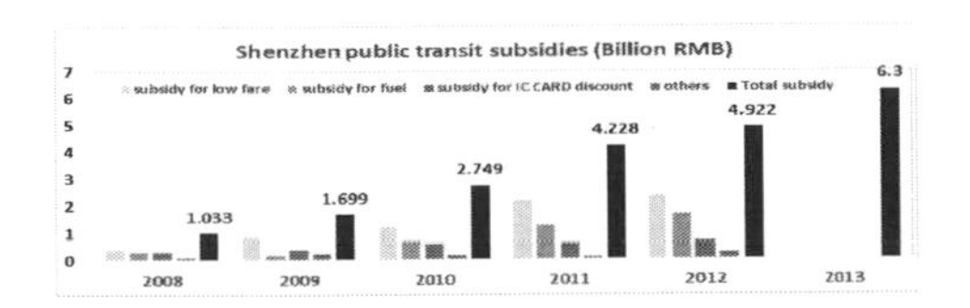

Figure 6. Shenzhen public transit subsidies between 2008 and 2013.

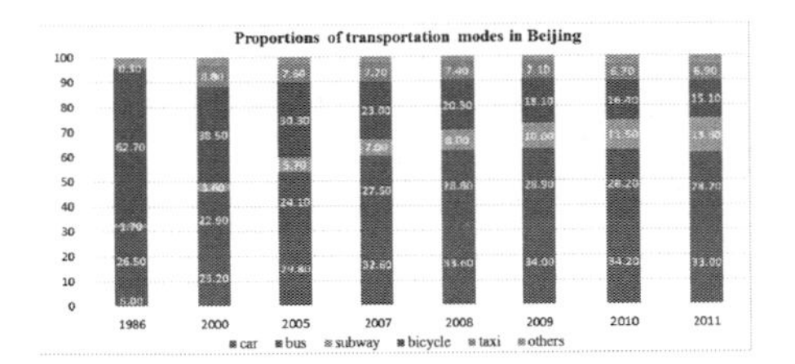

Figure 7. Proportions of transportation modes in Beijing from 1986 to 2011.

4 SPECIFIC STRATEGY OF FARE ADJUSTMENT MECHANISM

Only when costs change will the proposed framework launch a fare adjustment, and the adjustment range should take into consideration changes in costs and profits (involving subsidies) and citizens' acceptance and affordability. Improving service quality is the precondition for promoting fares.

4.1 *An analysis of key factors influencing operating costs of urban public transit companies with cost composition of a few public transportation companies as examples*

The operating costs of urban public transportation, which include labor cost, fuel cost, maintenance, depreciation cost of fixed assets, financial expense and other cost items directly related with operation, are core factors influencing fares of urban public transportation. Different cost items have a varying degree of influence over the total operating cost, among which labor cost and fuel cost take about 70% or even higher of the total operating cost and thus have biggest impact on the total operating cost.

Within a certain period of time, there are many factors influencing cost levels. Among them, labor cost is affected by the overall economic level in a city and the supply and demand dynamics of human resources in the industry, fuel cost and maintenance cost are affected by price indexes and depreciation and financial expense are mainly affected by the scale and speed of asset expansion of public transportation companies. On the basis of the analysis of key operating cost items of public transportation companies and the factors influencing each cost item, a fare adjustment model of urban public transportation is put forward. Meanwhile, along with social and economic progress, while experiencing rising operating costs, public transit companies also benefit from an increase in productivity. Therefore, in the fare adjustment model, we shall take into account some modifying factors by learning the experiences of Singapore and Hong Kong so that the public can share the benefits of economic and social progress while the public transit companies interests can be protected too.

Singaporean government developed the first fare adjustment formula for urban public transportation in 1998 and has since optimized and adjusted its formula along with social and economic development. The formula developed in 1998 was as follows: The ceiling of fare adjustment = CPI + X. CPI here refers to the change of CPI of the previous year and X refers to sharing of productivity rise. In 2004, Singaporean government optimized this formula to better balance the interests of public transport providers and commuters. The new formula was as follows: the ceiling of fare adjustment = Price index – X = 0.5 x CPI + 0.5 x WI - X. In this formula, X was 1.5% during 2005 to 2012.This formula incorporated macroeconomic indicators such as CPI and Wage Index (WI) which is an index for national monthly average wage change. In 2014, the formula was fine-tuned as: the ceiling of fare adjustment = price index – X = 0.4 x CPI + 0.4 x WI + 0.2 x EI - 0.5%. The key change here is the incorporation of energy index into the formula. Obviously, along with the change in economic level, industrial scale and energy mix, the key factors influencing the operating costs of public transit companies have evolved. Besides CPI which is a key economic indicator, wage level and energy price index have a growing impact on the operating costs.

Hong Kong has also established a fare adjustment mechanism for public transit. The adjustment can be upward or downward. Bus companies have the discretion to decide when to submit a fare increase application while the government has the authority to launch the downward adjustment mechanism. Transport Department derives a range of fare change every quarter on the basis of the following formula: the range of fare change = 0.5 x wage index change in the transport sector + 0.5 x CPI change – 0.5 x productivity increase.

Wage level, CPI, energy price index are still the key factors influencing the operating costs of urban public transit companies. However, with the implementation of low fare policy and a growing demand for public transport, under the condition that the fiscal and taxation support system for public transit is still underdeveloped by all levels of government, urban public transit companies are faced with capital shortfall in general and borrowing to make ends meet is quite common, therefore, financial cost becomes a key factor. With capital shortfall, overheads can not be neglected in the cost structure.

In a nutshell, a fare adjustment formula is put forward that is well suited to the features of the urban public transit industry in China. We did a trial run on the formula using Qingdao Bus Group as an example. Through analyzing the operating cost data of Qingdao Bus Group in 2012–2014, we estimated the weighting coefficient and key indicators of each factor in the formula.

4.2 *Trigger mechanism of fare adjustment*

4.2.1 *Considerations*

- Computing results of the fare adjustment formula.
- Changes in operating costs and profits since last adjustment.
- Prediction of future costs, earnings and returns.
- Reasonable rates of return of public transit companies.
- Acceptance and affordability of citizens,
- Service level of public transit.

4.2.2 *Conditions*

Public transit fare adjustment coefficient K:

$$K = C_1 \times \left(L_n / L_0 \right) + C_2 \times \left(E_n / E_0 \right) + C_3 \times CPI_n + C_4 \times \left(F_n / F_0 \right) + C_5 - 0.5 \times X$$

If K exceeds the threshold (such as 5%), operators and government departments can launch an application for adjustment.

4.2.3 *Notes for K*

- n: the year of fare adjustment;
- 0: base/reference year (the time of the last adjustment);
- C1: weight coefficient of salaries in the overall costs;
- C2: weight coefficient of fuel cost in the overall costs;
- C3: weight coefficient of other operating costs that are closely related to the overall price index (such as cost in tires and repair cost) in the overall costs;
- C4: weight coefficient of other major changes in operating costs (such as tax change) in the overall costs;
- C5: weight coefficient of other operating costs that are slightly related to the overall price index (such as depreciation cost in fixed assets) in the overall costs;
- X: Absorptivity of businesses through their own efforts, the proportion of cost increase when companies promote productivity through technological innovation and management optimization.

4.2.4 $C_1 + C_2 + C_3 + C_4 + C_5 = 1$

4.2.5 *Basis for factors in the formula of K*

- L represents the average salary in the city;
- E represents the average fuel price in the city;
- CPI is the Consumer Price Index of the city released authoritative institutions;
- F represents the weighted average of three-five years lending rate published by the people's bank of China (the people's bank of China's Announcement);
- Absorptivity is settled by the government in advance.

4.2.6 *Design idea of the formula*

As mentioned in section 2, labor cost and fuel cost account for a major part (about 70%) in the operating costs of China's urban public transit, and thus C1 and C2 should be calculated separately. To simplify the calculation, expenses closely related to CPI can be calculated together as C3, and changes that are less related to CPI but will greatly affect operating costs (such as tax change) should be calculated as C5. Any normal companies should reduce costs through technological innovation and management optimization, and thus absorptivity is involved to encourage the self-improvement of the companies.

4.3 *The decision mechanism of fare adjustment*

The computing result of K should serve as the basis for fare adjustment, but it shall not determine the adjustment range. The final range shall be recalculated based on the computing result of the fare adjustment formula, changes in operating costs and profits since last adjustment, prediction of future costs, earnings and returns, reasonable rates of return of public transit companies, and acceptance and affordability of citizens. It is suggested that the recalculation be conducted by an independent third-party, such as a professional consulting agency, entrusted by the city financial departments.

Acceptance and affordability of citizens is a crucial consideration. Price policies in public transit directly affect the interests of the public especially the low income group. While investigating the affordability of the public, China can refer to the achievements of cities like Singapore, and set an affordability indicator, such as an upper limit in the proportion of public transit spending in monthly family income for typical families. As an important standard measuring whether ticket price is reasonable, the affordability indicator should follow closely the following indexes: 1) the proportion of monthly average public transit spending in total family income; 2) the proportion of monthly average public transit spending in total family spending. Under the precondition of a stable ratio between fares and citizens disposable income and stable passenger flow, the fare can be adjusted based on the increase of family income, and a diverse fare system can be built based on the principle of high quality and competitive price to provide differentiated public transit services and meet the various demands of different groups.

Fare adjustment does not only mean the increase or decline in fares, but also refers to the adjustment in the fare system which serves as an expression and adjustment method of fare level and price relations. At present, fare systems can be divided, based on factors such as travel time, distance and areas, into flat fare, metered fare, time sharing fare and partition fare. One form of fare adjustment is that cities can design its fare system by comprehensively taking into consideration its passenger transport structure, features of passenger flow and systemic conditions, and build a multi-level, differentiated price system to strengthen the attraction of public transit.

In the proposed framework, improving service quality is the precondition for promoting fares. While applying for promoting fares, public transit companies should promise to enhance their services. Assessment on service level should follow standard procedures to ensure its fairness and accuracy.

As shown in Figure 8, detailed procedures for fare adjustment are:

1. Public transit companies submit a fare adjustment application and an introduction report to city de- partments of transport and finance.
2. City departments of transport and finance work together to decide, based on the launching formula of fare adjustment, whether to accept the application or not.
3. After the application is accepted, the reasonable adjustment range would be recalculated. This is suggested to be done by an entrusted independent third party such as professional consulting institutions.
4. The city public transit consulting agency consising of relevant government departments, companies, experts and citizen representatives would study the adjustment plan after the recalculation.
5. The city public transit consulting agency consising of relevant government departments, companies, experts and citizen representatives would assess the exiting service level of public transit, and make the assessment result the precondition for promoting fares. Public transit companies shall promise to provide better services after the adjustment.
6. Relevant departments disclose the reasons and range of fare adjustment to the public based on relevant laws and regulations, and organize hearings.
7. Relevant departments make the decision of fare adjustment.

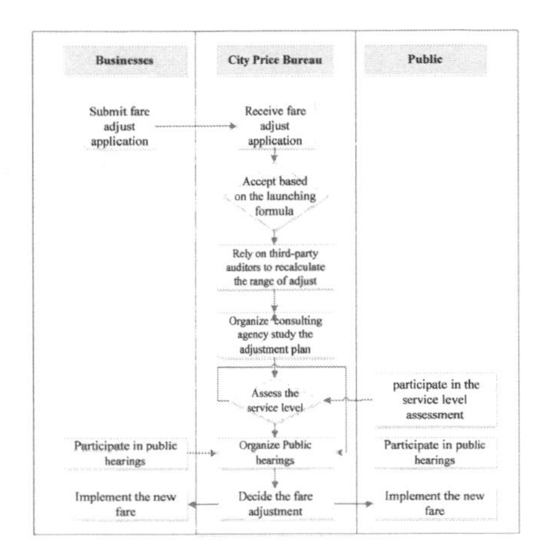

Figure 8. Flow chart of fare adjustment.

When companies do not apply for adjustment, relevant government departments shall recalculate fares on a regular basis, for example, every three years. Based on the recalculation, and considering the economic and social development, the government shall decide whether to conduct fare adjustment or not.

5 THE FARE ADJUSTMENT CAST STUDY-QINGDAO

Average adult base cash fare level of Qingdao Bus transportation is CNY1.00 with no adjustment from 2000 to now. By analyzing the operating cost data of Qingdao Public Transportation Group in 2012–2014 and statistics of Qingdao government, the factors of fare adjustment formula are as followings:

According to the fare adjustment formula, $K_{qingdao2014} = 1.113$, the Qingdao government should trigger the fare adjustment mechanism and the upper limit of new adjusted average adult base cash fare level should be 1.113($K_{qingdao2014} * P_{2012} = 1.113*1$). Even Qingdao government adjust the fare level up to CNY1.113, the new fare level would be much lower than the real cost fare.

Table 3. The factors of fare adjustment formula-Qingdao case study.

Factors	Value
L_{2014}/L_{2012}(a)	1.296
E_{2014}/E_{2012}	0.942
$CPI_{(2014.12)}$(b)	1.026
F_{2014}/F_{2012}	0.920
C_1(c)	0.598
C_2(c)	0.238
C_3(c)	0.078
C_4(c)	0.026
C_5(c)	0.060
X	0.113

(a) Average wage of employed persons in urban units. Available at http://www.stats-qd.gov.cn.
(b) Available at http://www.stats-qd.gov.cn.
(c) Average weight coefficient in the overall costs from 2012 to 2014.

Table 4. Fare level comparison-Qingdao case study.

Average adult base cash fare	Fare level (CNY)
Operating cost fare	2.845
Implementing fare	1.000
Adjusted fare	1.113

6 CONCLUSIONS

Sustainable finance is essential for the stable development of public transit companies, efficient utilization of government subsidies, and better service quality in public transit. Focusing on the three major stakeholders of public transit, namely the government, businesses and the public, this paper deeply analyzed issues related to the financial sustainability of China's urban public transit. Based on the principle of "acceptable for the public, sustainable for businesses, and affordable for government finance", the paper proposed a service quality-oriented framework featuring the interaction among operating costs, financial subsidies and fare adjustment and offered specific strategy in fare adjustment, providing a reference for cities in China and other countries to build a scientific fare and subsidy adjustment system and to foster the healthy development of public transit.

ACKNOWLEDGEMENT

Thanks for Volvo Research and Educational Foundations (VREF) funded the research. Authors are also grateful that the data set used in this study was kindly provided by Transport Department of Henan Province, Zhengzhou Bus Communication Corporation, Shenzhen Bus Group Co. Ltd and Qingdao Bus Group.

REFERENCES

Beijing transportation research center (2008–2012), Annual report of beijing transportation development, Beijing, Beijing transportation research center.

China Urban Sustainable Transportation Research Center (Custrec), China Academy of Transportation Sciences, Mot, P.R. China (CATS) (2012), Chinese Urban Public Transportation Fare Adjustment Mechanism Study Report.

Hongkong Ministry of Transport. References in pu bic bus service regulations (Chapter 230): Applying for price raising from Jiulong Bus Co. Ltd. & Longyun Bus Co. Ltd. [EB/OL]. (2014–11–06)[2015–03–16]. Http:// www.td.gov.hk.

Huiling Mao & Hongdi Sun. (2011). Innovative mechanism of fare and subsidy to promote sustainable development of the transit. J. Transport Business China. 6: 32–35.

Jianping LI et al. (2009). Views of construction of the urban public transport subsidies mechanism. J. Economic Forum. 1: 28–32.

Jiayao He & Zhenxiang Ye. (2012). The development direction of beijing public transport subsidies refine finance. J. Economic research guide. 5: 12–15.

Ministry of Transport of the People's Republic of China (2011). Introduction to urban public transport management, Beijing, China communications press.

Ministry of Transport of the People's Republic of China (2011–2013). Annual report of China urban passenger transportation development, Beijing, China communications press.

National Bureau of Statistics of China (1990–2012), China statistical yearbook, Beijing, China statistical press.

Public Transport Council. Public Transport Council. 1.0% Fare Adjustment for 2011 with Full-day Concession, Lower Fares on NEL/CCL For Senior Citizens and No Increase in All Child/Student Fares. [EB/OL]. (2014–08–11)[2015–03–05]. http:// www.ptc.gov.sg

Tong Liu et al. (2007). Urban public transportation fare reform experiences from Singapore public transportation fare making model. J. Price theory and practice. 10: 46–47.

Xiande Shen. (2006). Urban public transportation fare making system. J. Zhejiang Statistics. 7: 21–22.

Xianglong Liu et al. (2014). Sustainable financial issues on urban public transportation. J. China Road Transport. 11: 23–26.

Yang Yang. (2010). Analysis on the efficiency of urban public transport subsidies under the transit priority policy. J. Comprehensive transportation. 10: 5–10.

Civil, Architecture and Environmental Engineering – Kao & Sung (Eds)
© 2017 Taylor & Francis Group, ISBN 978-1-138-02985-9

Sanitation improvement and the bottleneck analysis of equity in China: From the perspective of specialists

Xiaolong Li
School of Economics, Peking University, Beijing, China
China National Health Development Research Center, NHFPC, Beijing, China

Yanqing Miao
China National Health Development Research Center, NHFPC, Beijing, China

ABSTRACT: Accessible improved sanitation is critical to child health, and inequities in improved sanitation can be interpreted as health inequities across socio-economic groups. This study assesses the equity of sanitation improvement in China and finds out the bottleneck problems on equity across provinces. Based on United Nations Children's bottleneck analysis framework and practical situation, we established a scale questionnaire, which included four dimensions—policy environment, supply, demand and quality. A new equity-score model was generated to review sanitation improvement. We regarded questions with the three lowest scores as the bottleneck of sanitation improvement within province. The bottleneck problem of policy environment, demand, quality across provinces were Question 11, Question 23, Question 29 and 30. Questions on supply dimension were much better. As for the equity score, Question 6,9,25,27,28 were the lowest scores and indicated the inequity on sanitation improvement. The bottleneck problems of equity derive from relevant propaganda campaign, training and following interventions. Economic condition was the primary factor influencing sanitation improvement. The key work should be more effectively promoting the sanitation improvement on demand dimension. Health Departments should highlight the announcement of health-related knowledge.

1 INTRODUCTION

Globally, approximately 2.4 million deaths (4.2% of all deaths) could be prevented each year if everyone practiced appropriate hygiene and had good, reliable sanitation and safe drinking water (Prüss-Üstün et al. 2008). Most of these deaths are children in developing countries and result primarily from diarrhea and subsequent malnutrition, as well as other diseases that are attributable to malnutrition. It has been acknowledged worldwide that good sanitation plays an important role in reducing domestic 0-5-year-old child mortality (Hutton 2013). The World Health Organization estimates that there are 74,000 children under 5 years old die each year in China due to diarrhea (Boschi-Pinto et al. 2008). The keys to diarrhea disease control are access to safe water, improved sanitation and regular hygienic practices.

Sanitation improvement in rural China is worthy of close attention because China is one of the largest developing countries, and has a rural population of 971 million people. The rural population accounts for approximately 72% of the total population (National Bureau of Statistics of China 2012). The Chinese government has attached importance to sanitation improvement for over the last decade and has incorporated the improvement of sanitation in rural areas into its national 5-year plan since the 1990 s. In 2009, to promote public health service levels and equity, the Chinese government launched a 3-year health reform program and set sanitation improvement as one of the major public health services. Opinions of the Communist Party of China (CPC) Central Committee and the State Council on expanding medical and health system reforms indicated that government input on health services would be gradually increased (Zhang et al. 2013). In recent 3 years, the Chinese government invested a total of 4.448 billion yuan (0.74 billion US dollar) for rural sanitation improvement (Ministry of Finance 2012). In addition, the coverage rate of improved sanitation in rural China rose from 50.9% in 2003 to 71.7% in 2012 (Ministry of Health P.R. China 2013).

In terms of the disparity in sanitation coverage rates among different regions, by the end of 2011, the highest was 98% in Shanghai and the lowest was 40.9% in Guizhou Province (Laiyun 2012). Clearly the sanitation coverage rate varies greatly among different regions due to substantial gaps in economic development levels, efficiency of government execution, and other factors. However, macro-data on the national level often concentrates on economy, education and temperature etc. which are essential but not easy to change in a short term. Therefore, some influential factors encountered in the process of public health service program may be more visible and practical. It will be considerably beneficial to improve sanitation and health equity. Based on the current situation and equity, the objective of this study was to better understand the sanitation conditions in rural China and identify the factors that influence sanitation improvement and its equity. In addition, the implications of the findings can be provided as references for policy-makers.

2 DATA AND METHODOLOGY

2.1 *Questionnaire design and data collection*

In accordance with United Nations Children's bottleneck analysis framework and practical situation, we established a scale questionnaire, which included four dimensions-policy environment, supply, demand and quality. The questionnaire had 31 questions. Specifically, Question 1–3 were basic information. Question 3–14, 15–18, 19–26, 27–31 were questions on environment, supply, demand and quality respectively. Each question's score range was from 1 to 5. We chose Patriotic Health Campaign Committee Offices with more than 3 staff to fill out the questionnaire in each province and used average values to represent the score.

109 questionnaires were handed out in Anhui, Beijing, Chongqing, Fujian etc. 29 provinces and all were taken back. The number of questionnaires every province were not less than 3, and 14 questionnaires was taken back in Gansu province.

2.2 *Methodology*

We summarized the methods of previous scholars to quantify influential factors and generated a new equity-score model to review sanitation improvement:

$$SC_i = \sum_{j=1}^{N} (1-\mu_j)\cdot(ASP_{ij} - \overline{ASP_i})$$

$$ASP_{ij} = SP_{ij}\cdot \mu_j$$

$$\overline{ASP_i} = \frac{\sum_{j=1}^{N} ASP_{ij}}{N}$$

where SC_i is theequity score of Question i for national sanitation improvement. μ_j is the sanitation coverage rate in province *j*, using the data in 2012. ASP_{ij} is the absolute score of Question *i* in province *j*. SP_{ij} is the original score of Question *i* in province j in the questionnaire. $\overline{ASP_i}$ is the average absolute score of question i.

This model was constructed as follows:

The score in the questionnaire just reflected the situation within province. If we want to standardize the results, we should adjust for the score consistent with overall situation in each province. Here we used the SP_{ij} to multiply μ_j, then we could get the standardized score across province;

Not the overall level of questions but differences among provinces should be focused on if we aimed to find out the equity and relevant bottleneck. For example, every province had a low score of Question A and provinces with higher sanitation coverage owned higher scores of Question B while those with lower sanitation coverage had lower score of Question B, then we should regard Question B not A as the bottleneck question of equity;

We used $ASP_{ij} - \overline{ASP_i}$ to eliminate national common information and just keep difference information. However, as for the information $ASP_{ij} - \overline{ASP_i}$, we should add weight value for further adjustment. For example, two provinces with different sanitation coverage had the same value of $ASP_{ij} - \overline{ASP_i}$. But the province with lower sanitation coverage would bring about worse health results, so we used $(1-\mu_j)\cdot(ASP_{ij} - \overline{ASP_i})$ to obtain the adjusted score.

Total score of Question i was calculated by summarizing all scores of every province. The score of SC explained the equity of sanitation improvement and low scores indicated which were bottleneck problems.

3 RESULTS

3.1 *Score distribution and bottleneck problems*

We regarded questions with the three lowest scores as the bottleneck of sanitation improvement within province. The bottleneck problem of policy environment across provinces was mainly Question 11- "Sanitation Management department is independent on Health department". Questions on supply dimension were much better. As for the dimension of demand, Question 23- "Local living standard has reached the level of being fairly well off" was the bottleneck problem. The bottleneck problems of quality dimension were Question 29 (most local government equips sanitation-improvement workers with excrement-processed vehicles) and Question 30 (excrement exposure is monitored and recorded regularly and sustainably).

Finding out the dimensions where bottleneck problems existed was essential to policy-making and direction improvement. So we calculated the average score of each dimensions, as was shown in Table 1. We can see that the dimension of demand had the lowest score in most provinces (21 out of 29). 6 provinces out of 29 had the lowest score in the quality dimension. It was very evident that the bottleneck problems in dimension of demand were worst. Therefore, demand dimension should be specially concentrated on in future work to further improve the sanitation coverage.

The equity score in each question are shown in Table 2. Questions with the lowest scores were Question 6 (The sanitation institute's level is high within province, −307.29), Question 19 (Areas without sanitation improvement yet can be fully aware

Table 1. Score of each dimension in each province.

Province	Environment	Supply	Demand	Quality
Anhui	3.63	3.90	2.78	3.08
Hebei	3.58	4.25	3.25	3.27
Heilongjiang	3.36	3.08	3.00	3.07
Henan	3.08	3.83	3.42	2.87
Hubei	4.58	4.33	3.25	3.60
Hunan	4.08	4.92	3.71	3.53
Jiangsu	4.33	4.92	4.75	4.27
Jiangxi	3.89	4.08	3.79	3.27
Jilin	4.47	4.42	3.58	3.67
Liaoning	4.42	4.50	4.42	4.00
Neimenggu	4.03	4.00	3.04	3.40
Beijing	4.29	4.69	4.50	4.60
Ningxia	3.43	3.54	2.95	3.06
Qinghai	4.17	4.42	3.25	4.00
Shandong	4.25	4.25	3.71	4.00
Shanxi	3.81	4.58	3.75	3.40
Shanxi (Xi'an)	3.81	4.17	3.08	4.13
Sichuan	3.75	4.00	3.38	3.80
Tianjin	3.31	3.08	2.79	3.07
Xinjiang	4.50	5.00	4.42	5.00
Chongqing	3.64	3.83	2.96	3.13
Yunnan	3.21	3.81	2.81	3.05
Zhejiang	3.77	3.69	3.44	3.75
Fujian	3.94	3.92	2.96	3.73
Gansu	3.43	3.54	2.91	3.36
Guangdong	3.39	3.33	3.75	3.47
Guangxi	4.22	4.25	4.04	4.07
Guizhou	3.89	4.58	3.21	3.60
Hainan	3.94	3.42	2.50	3.00
Average Score	3.87	4.08	3.43	3.59

Table 2. Equity score in each question.

Dimension	Questions and equity scores							
Environment	3	4	5	6	7	8	9	10
	−246.41	−223.33	−274.52	−307.29	−296.01	−286.42	−244.53	−262.99
	11	12	13	14				
	−127.45	−260.41	−287.27	−271.08				
Supply	15	16	17	18				
	−269.05	−198.14	−231.57	−238.55				
Demand	19	20	21	22	23	24	25	26
	−335.16	−249.67	−269.56	−240.73	−239.05	−294.04	−311.19	−270.36
Quality	27	28	29	30	31			
	−300.82	−298.41	−218.41	−224.27	−270.82			

of the expense, –335.16), Question 25 (Residents' knowledge, attitudes and behaviour were improved in recent year and understand the harmful consequence of no sanitation, –311.19), Question 27 (All family who have had sanitation improvement can use toilet consistently, without going back to the situation of excrement exposure, –300.82) and Question 28 (Training on sanitation knowledge and improvement is fully and consistent, –298.41). Question 11 (Sanitation Management department is independent on Health department, –127.45) had the highest equity score among all questions.

4 DISCUSSION

The analysis is divided into two steps: the first step is to establish the questionnaire and get the absolute score, which in fact quantify the potential problems in public health service work and directly reflect weak aspects within province and on the national level; the second step is to find the disparities between provinces through equity score model and therefore bottleneck problems of equity can be found.

According to the research results, Economic condition was the primary factor influencing sanitation improvement. The key work in the foreseeable future should be more effectively promoting the sanitation improvement on demand dimension. Specifically, government should advance levels of people's daily life. Besides, Health Departments should highlight the announcement of health-related knowledge.

The bottleneck problems of equity derive from relevant propaganda campaign, training and following interventions, which also prove the right measures of human resources, financial and material investment on sanitation improvement in previous healthcare reform. However, the quality of service and training should not be ignored and it will generate enormous improvement effect on the sanitation equity in China.

Equity in sanitation improvement is difficult to measure, so we put forward a method to calculate the equity score. It is also the first time a study has systematically evaluated sanitation improvement during these years in China. Sanitation improvement is in fact influenced by many other factors, including climate and nationality (Whittington et al. 1993; Chen J et al. 2013, Hossain 1996). How these factors influence the equity of sanitation improvement and any bottlenecks in further equity improvement must be included in further study. It is also important to consider variables such as Knowledge, Attitude and Practice (KAP) which cannot be arrived at directly. We believe there is a marked effect of disparity in KAP on sanitation improvement in China and can strongly influence the ratio of improved sanitation. In addition, our study is based on provincial panel data for China, which means that the imparities we discussed are among provinces, not within each province. To determine provincial imparities, more detailed data are needed, such as data at the municipal level or county level.

The ultimate goal of evaluating equity in sanitation improvement is to eliminate inequity, which requires determining what caused the inequity so the bottleneck and any constraints on sanitation improvement can be removed. This is the common perspective of many researchers and related departments and an important research goal.

REFERENCES

Boschi-Pinto, C., Velebit, L. & Shibuya, K. 2008 *Estimating child mortality due to diarrhoea in developing countries. Bulletin of the World Health Organization* 86, 710–717.

Chen J, Li Z, Gao X, Du H, Yu L, Ren H, et al. 2013 T*oilet retrofit in rural areas of China:* impact factors and effect analysis. *Chin Rural Health Serv Adm* 002:181–3.

Hossain SI. 1996 *Making an equitable and efficient education:* The Chinese experience. China: Social Sector Expenditure Review, World Bank;

Hutton, G. 2013 *Global costs and benefits of reaching universalcoverage of sanitation and drinking-water supply. Journal of* Water *& Health* 11, 1–2.

Kakwani N, Wagstaff A, Van Doorslaer E. .1997 *Socio-economic inequalities in health: measurement, computation, and statistical inference. J Econ*77(1),87–103.

Laiyun S. 2012 *China Statistical Yearbook* 2012. China Statistics Press, Beijing.

Ministry of Finance 2012 *Report on the implementation of central and local budgets for* 2012. Xinhua, Beijing. Available from: http://news.xinhuanet.com/english/china/2012-03/16/c_131472175.htm.

Ministry of Health P.R. China 2013 *China Health Statistics Yearbook 2013.* Peking Union Medical College Press, Peking.

National Bureau of Statistics of China 2012 *China Statistical Yearbook* 2012. China Statistics Press, Beijing.

Prüss-Üstün, A., Bos, R., Gore, F. & Bartram, J. 2008 Safer water,better health: c*osts, benefits and sustainability of interventions to protect and promote health.* World Health Organization:Geneva.

Zhang, X., Xiong, Y., Ye, J., Deng, Z. & Zhang, X. 2013 *Analysis of government investment in primary health-care institutions to promote equity during the three-year health reform program in China. BMC Heal Serv Res.* 13, 114.

Civil, Architecture and Environmental Engineering – Kao & Sung (Eds)
© 2017 Taylor & Francis Group, ISBN 978-1-138-02985-9

Ultimate capacity of uniaxially compressed perforated steel plates strengthened by using CFRPs

Xin Tao & Shuangyin Cao
Key Laboratory of Concrete and Prestressed Concrete Structures of Ministry of Education, Southeast University, Nanjing, China

ABSTRACT: The aim of the present study is to investigate the ultimate capacity of perforated plates strengthened by using CFRPs under uniaxial compression along its longitudinal direction. A new test rig combined with an existing universal hydraulic testing machine has been designed to achieve simply supported along all edges. Some 36 plates were tested. Test results showed that if a CFRP is pasted on the convex surface of initial deflection, it will improve the capacity of uniaxially compressed perforated steel plates very well; otherwise, it will almost be of no effect. It also shows that the use of CFRP-strengthened perforated plates exhibit better results in a big slenderness plate with a small hole and a medium slenderness plate with a big hole. And with the use of CFRP-strengthened perforated plates with a certain plate slenderness ratio (b/t), such as b/t = 66, no matter how the size of the hole' diameter changes, the CFRP-strengthening effect remains stable. Also, it can be seen that the pasting of a CFRP cannot increase the stiffness before buckling, but it can improve the mechanical property afterwards.

1 INTRODUCTION

Fiber-Reinforced Polymers (FRPs) have a high strength to weight ratios and excellent resistance to corrosion and environmental degradation. It is also very flexible and forms all kinds of shapes, and it is easy to handle during construction (Alsayed et al., 2000, Moy, 2001, Teng et al., 2002); and so, it has been widely used in the strengthening of steel structures recently (Zhao and Zhang, 2007, Teng et al., 2012).

Steel plate elements are the main structural components of the desk and platforms on ships and ship-shaped offshore structures, box girder bridges and architecture structures. Openings are usually introduced in such plates for various reasons (Cheng and Zhao, 2010). The openings can significantly reduce the ultimate load of the plates. The ultimate strength of plates with openings has then been studied in the literature, e.g. Narayanan and Rockey, 1981, Narayanan and Chow, 1984, Paik, 2007. Cheng and Zhao, 2010 study the opening and strengthening of perforated steel plates that are subjected to uniaxial compressive loads and four types of stiffeners are mainly discussed; FEM has been employed to analyze the elastic and elasto-plastic buckling behaviors of strengthened and un-strengthened perforated plates.

It is clear from the above description that FRP is good at strengthening steel structures, many perforated plates need to be strengthened, and several ways to strengthen the perforated plates have been provided, but no one had tried to strengthen the perforated steel plates by pasting CFRP, and no research has been carried out on the ultimate capacity of uniaxially compressed perforated steel plates strengthened by CFRP.

In this study, the ultimate capacity of uniaxially compressed perforated steel plates strengthened by CFRP has been studied.

2 PROBLEM DEFINITION

Ultimate capital of perforated steel plates strengthened by CFRP subjected to uniaxial compression along its longitudinal direction is considered in the study. About 36 plates were designed, and all of them possessed the following properties: (i) square shape; (ii) four edges simply supported; (iii) circular hole as if it has an opening; (iv) one layer of transverse CFRP inside and one layer of longitudinal CFRP outside in signal or both sides of the plate if it was strengthened. A perforated steel plate strengthened by CFRP is shown in Figure 1. The length and width of the plates

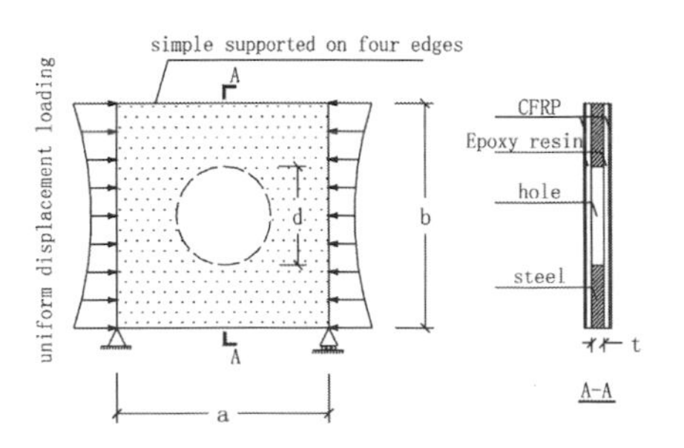

Figure 1. Geometrical conditions of specimens.

are represented by a and b. The thickness of the steel plate and CFRP are given as t and t_f, respectively.

In this paper, some variations have been considered as follows:

1. The situation of the CFRP pasting: one-third was plates without CFRP pasting, the other one-third was single-side CFRP pasting and the rest was plate with double-side CFRP pasting.
2. Plate slenderness ratio b/t, which is an important parameter that governs the plate strength: in this study, the plate slenderness ratio b/t was 92, 66 and 43, respectively, corresponding to the steel plate's width t of 2.61 mm, 3.64 mm and 5.58 mm.
3. The hole diameter ratio d/b is unlikely to be larger than half the width of the plate in practice, and so it is confined to half the width of the plate in this paper. In this study, the hole diameter ratio d/b was 0, 0.1, 0.3 and 0.5, respectively.

3 METHOD OF ANALYSIS

The ultimate capacity of uniaxially compressed perforated steel plates strengthened by using a CFRP was studied with experimental methods, and a test rig was fabricated in the laboratory (see Figure 2); the test rig was used in combination with an existing universal hydraulic testing machine. The applied loads were measured by using the dial of the universal hydraulic testing machine and static strain testing systems, which were composed by using a force sensor, DH3816, and computer. The displacement and deformation of the plates are achieved by using a displacement meter. The

Figure 2. Overall picture of the test rig.

Table 1. Details of the plate specimen.

Group	t (mm)	b/t	D (mm)	d/b	Specimen no. Without CFRP	d_0/b	With single-side CFRP	d_0/b	With double-side CFRP	d_0/b
1	2.61	92	0	0	CP92-1	0.933	SFSCP92-1	−0.895	FSCP92-1	0.483
			24	0.1	CP92-2	0.424	SFSCP92-2	−0.794	FSCP92-2	0.465
			72	0.3	CP92-3	0.949	SFSCP92-3	1.181	FSCP92-3	0.641
			120	0.5	CP92-4	0.678	SFSCP92-4	−0.682	FSCP92-4	1.050
2	3.64	66	0	0	CP66-1	0.200	SFSCP66-1	−0.266	FSCP66-1	0.263
			24	0.1	CP66-2	0.236	SFSCP66-2	−0.511	FSCP66-2	0.220
			72	0.3	CP66-3	0.150	SFSCP66-3	0.132	FSCP66-3	0.311
			120	0.5	CP66-4	0.454	SFSCP66-4	0.310	FSCP66-4	0.288
3	5.58	43	0	0	CP43-1	0.086	SFSCP43-1	0.058	FSCP43-1	0.105
			24	0.1	CP43-2	0.062	SFSCP43-2	0.133	FSCP43-2	0.184
			72	0.3	CP43-3	0.110	SFSCP43-3	0.139	FSCP43-3	0.051
			120	0.5	CP43-4	0.169	SFSCP43-4	−0.152	FSCP43-4	0.043

Table 2. List of material properties.

Steel plate	t (mm)	Young's modulus (MPa)	Yield stress (MPa)	Tensile strength (MPa)
	2.61	206	228.7	327.8
	3.64	206	253.2	369.3
	5.58	206	287.9	434.1
CFRP	t_f (mm)	Young's modulus (MPa)	Fiber mass per unit area (g/m^2)	Tensile strength (MPa)
	0.167	260	300	3492.5
Epoxy resin	–	Tensile shear strength (MPa)	Compressive strength (MPa)	Tensile strength (MPa)
	–	19.6	87.9	48.4

details of the plate specimen are shown in Table 1, d_0 is the initial deflection of the steel plate, and if δ_0/b is positive, it means that the CFRP was pasted on the convex surface of initial deflection, if δ_0/b is negative, it means that the CFRP was pasted on the concave surface of the initial deflection. The material properties are listed in Table 2.

4 RESULTS AND DISCUSSION

4.1 *Failure patterns*

Figure 3 shows the typical failure pattern of these plates. The reason why it is not so obvious to see the final out-of-plane deflection is that the experiments were ended soon after the bearing capacity declines. From Figure 3, we can see that the failure mechanism consists of four yield lines which start from the four corners and end near the hole edges, and also the steel plates and CFRP have a good coordinated working performance throughout the experiment.

4.2 *Ultimate load*

In Table 3, the ultimate load of each specimen is presented. It can be easily seen that the ultimate load of specimens with two sides CFRP pasting (FSCP) is bigger than specimens with no CFRP pasting (CP); however, the ultimate load of specimens with one-side CFRP pasting (SFSCP) was not stable. Considering the initial deflection in Table 1, it seems that if the CFRP is pasted on the convex surface of initial deflection, it will improve the capacity of uniaxially compressed perforated steel plates very well; otherwise, it will almost be of no effect. This result can be explained as the CFRP has a good tensile behavior, but its compressive behavior is very poor.

Figure 4 shows the influences of the plate slenderness ratio (b/t) on the ultimate load. P_y is the yield load of the steel plate. A comparison of Figure 4a and b shows that the trends of perforated plates strengthened by using the CFRP are very similar to those curves of perforated plates without CFRP strengthening, that is, P_u/P_y degrades as b/t is gradually enlarged. Figure 4c shows that when the holes are small, such as d/b = 0 and d/b = 0.1, the CFRP strengthening effect increases as the plate slenderness ratio (b/t) is gradually enlarged. When the holes are large, such as d/b = 0.3 and b/t = 0.5, the CFRP strengthening effect first increases and then degrades as the plate slenderness ratio (b/t) is gradually enlarged. It means that the use of CFRP-strengthened perforated plates yield better results in a big slenderness plate with a small hole and a medium slenderness plate with a big hole.

Figure 5 shows the influences of the hole diameter ratio d/b on the ultimate load.

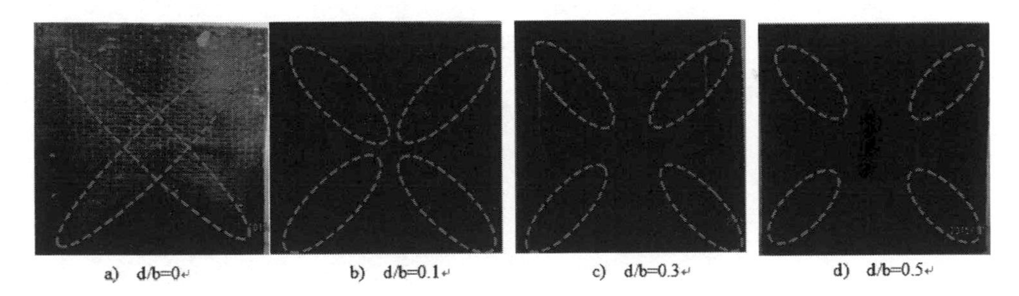

Figure 3. Pictures of typical failure patterns of specimens.

Table 3. List of experimental ultimate loads.

Group	d/b	Specimen no.	P_u kN	Specimen no.	P_{sfu} kN	Specimen no.	P_{fu} kN
1	0	CP92-1	94.02	SFSCP92-1	101.86	FSCP92-1	126.67
	0.1	CP92-2	90.11	SFSCP92-2	96.64	FSCP92-2	113.90
	0.3	CP92-3	79.66	SFSCP92-3	92.72	FSCP92-3	94.02
	0.5	CP92-4	74.44	SFSCP92-4	78.35	FSCP92-4	87.50
2	0	CP66-1	154.10	SFSCP66-1	135.81	FSCP66-1	215.49
	0.1	CP66-2	148.00	SFSCP66-2	154.10	FSCP66-2	195.89
	0.3	CP66-3	130.59	SFSCP66-3	168.46	FSCP66-3	178.91
	0.5	CP66-4	114.92	SFSCP66-4	144.95	FSCP66-4	160.89
3	0	CP43-1	283.38	SFSCP43-1	304.27	FSCP43-1	282.07
	0.1	CP43-2	267.40	SFSCP43-2	309.50	FSCP43-2	306.89
	0.3	CP43-3	229.84	SFSCP43-3	258.57	FSCP43-3	275.54
	0.5	CP43-4	168.73	SFSCP43-4	197.19	FSCP43-4	247.79

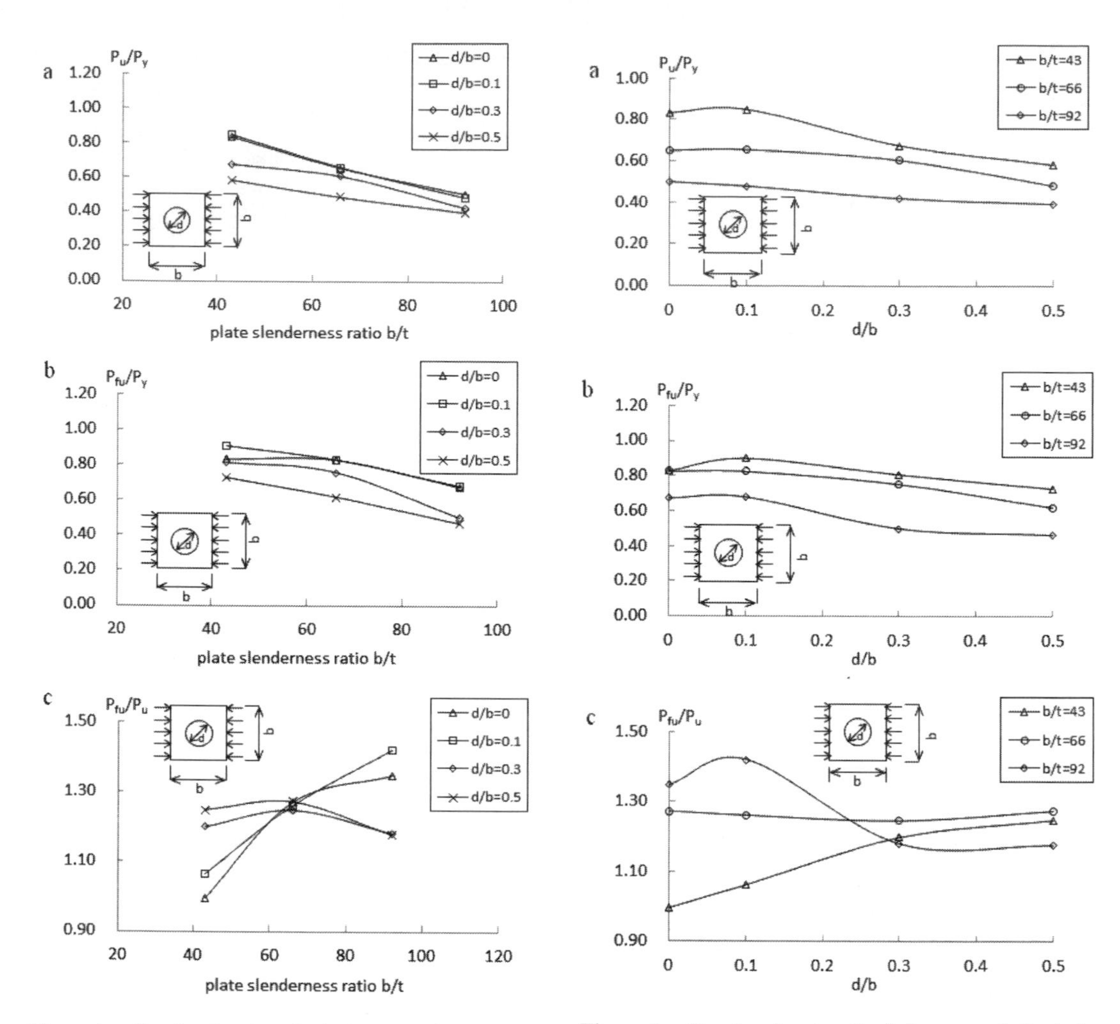

Figure 4. Graphs showing the influences of plate slenderness ratio.

Figure 5. Graphs showing the influences of the hole diameter ratio.

The results of perforated plates strengthened by CFRP are also very similar to those curves of perforated plates without CFRP strengthening, that is, P_u/P_y first remains unchanged and then degrades as d/b is gradually enlarged. Figure 5c shows that when plate slenderness ratio (b/t) is small, such as b/t = 43, the CFRP strengthening effect increases as d/b is gradually enlarged. When the plate slenderness ratio (b/t) is medium, such as b/t = 66, the CFRP strengthening effect remains unchanged as d/b changes. When the plate slenderness ratio (b/t) is high, such as b/t = 92, the CFRP strengthening effect first increases and then degrades as d/b is gradually enlarged. It means that, with the use of CFRP-strengthened perforated plates with a certain plate slenderness ratio (b/t), such as b/t = 66, no matter how the size of the hole' diameter changes, the CFRP strengthening effect remains stable.

4.3 *Load–deflection curves*

Some typical relationships between compressive load and in-plane displacement of the plates have been plotted in Figure 6. Some conclusions can be obtained as follows:

1. By comparing Figure 6a and b, we can see that the pasting of CFRP cannot increase the stiffness before buckling, but it can improve the mechanical property afterwards.
2. From Figure 6c, d and e, it can easily be seen that the stiffness of the specimens decreases along with an increase in the hole's diameter.
3. From Figure 6c, d and e, we also know that the greater the slenderness ratio b/t, the more

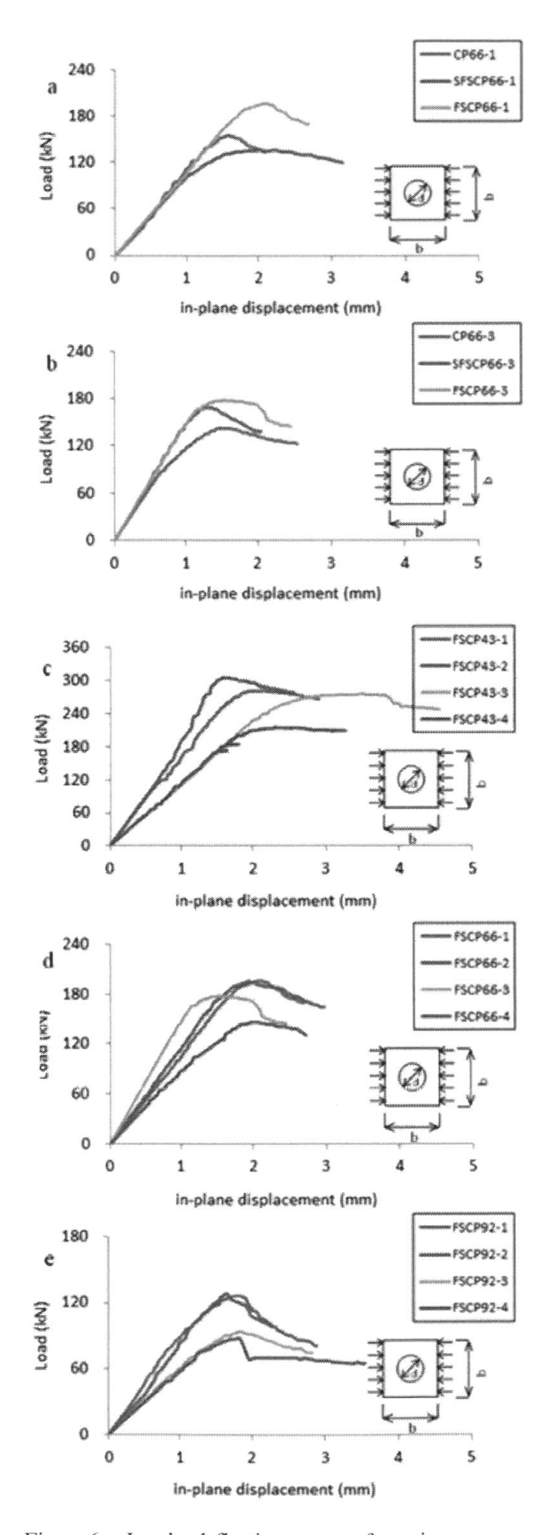

Figure 6. Load—deflection curves of specimens.

instable the compressive load and in-plane displacement curve.

5 CONCLUSIONS

In this study, the perforated steel plates strengthened by using the CFRP that are subjected to uniaxial compression are studied. Some general conclusions can be summarized as follows:

1. The failure mechanism of perforated plates strengthened by using the CFRP consists of four yield lines which start from the four corners and end near the hole's edges.
2. If the CFRP is pasted on the convex surface of initial deflection, it will improve the capacity of uniaxially compressed perforated steel plates very well; otherwise, it will be almost of no effect.
3. The use of CFRP-strengthened perforated plates yields better results in a big slenderness plate with a small hole and a medium slenderness plate with a big hole.
4. With the use of CFRP-strengthened perforated plates with a certain plate slenderness ratio (b/t), such as b/t = 66, no matter how the size of the hole' diameter changes, the CFRP strengthening effect remains stable.
5. The pasting of the CFRP cannot increase the stiffness before buckling, but it can improve the mechanical property afterwards.

REFERENCES

Alsayed, S., Al-Salloum, Y. & Almusallam, T. 2000. Fibre-reinforced polymer repair materials—some facts. *Proceedings of the Institution of Civil Engineers-Civil Engineering*. Thomas Telford Ltd, 131–134.

Cheng, B. & Zhao, J. 2010. Strengthening of perforated plates under uniaxial compression: Buckling analysis. Thin-Walled Structures, 48, 905–914.

Moy, S.S. 2001. FRP composites*: life extension and strengthening of metalli*c structures, Thomas Telford.

Narayanan, R. & Chow, F. 1984. Ultimate capacity of uniaxially compressed *perforated plates. Thin*-Walled Structures, 2, 241–264.

Narayanan, R. & Rockey, K. 1981. Ultimate load capacity of plate girders with webs containing circ*ular cut-outs. Proceedings of the institu*tion of civil engineers, 71, 845–862.

Paik, J.K. 2007. Ultimate strength of steel plates with a single circular hole under *axial compressive loadi*ng along short edges. Ships and Offshore Structures, 2, 355–360.

Teng, J., Chen, J. & Smith, S. 2002. FRP: Strengthened RC Stru*ctures. Journal of Composites for Construction, A*SCE, 6, 232–245.

Teng, J., Yu, T. & Fernando, D. 2012. Strengthening of steel structures with fiber-reinforc*ed polymer composites.* Journal of Constructional Steel Research, 78, 131–143.

Zhao, X.-L. & Zhang, L. 2007. State-of-the-art review on FRP strengthened steel structur*es. Engineering Structures, 29*, 1808–1823.

Civil, Architecture and Environmental Engineering – Kao & Sung (Eds)
 ISBN 978-1-138-02985-9

Research on the measurement of the permeability coefficient of porous asphalt pavement

Guoqi Tang
Southeast University, Nanjing, China

Lei Gao, Tianjian Ji & Jianguang Xie
Nanjing University of Aeronautics and Astronautics, Nanjing, China

ABSTRACT: The drainage capacity of the porous asphalt pavement is related with the permeability of the porous asphalt concrete. On the basis of drainage mechanism of the asphalt pavement, the horizontal permeability coefficient and vertical permeability coefficient of the drainage pavement can be measured by the experimental test. The relationship between the water permeability coefficient and permeability coefficient tested by the pavement seepage meter was established in this research. Finally, the measurement method of the permeability coefficient of the drainage asphalt pavement was proposed. It is proved that the permeability coefficient of the porous asphalt pavement can be measured in a rapid way by the pavement seepage meter.

1 INTRODUCTION

The permeable asphalt mixture is used for the porous asphalt pavement as the surface layer. The porous asphalt pavement can take the rainfall into the drainage function layer, and laterally discharge the rainwater through the inside layer. It significantly improves the driving safety and comfort of the rainy days. Discharging the rainwater fallen on the road surface is the primary function of the porous asphalt pavement, and its drainage capacity is related with the permeability of porous asphalt concrete (Pisciotta M. 2010). The permeability coefficient is the parameter of the penetrating quality of the representation material. The permeability coefficient is mainly influenced by the composed particle size, gradation, and void content. Carman, Kozeny and Carman, established the aggregate permeability coefficient equation through the theory of capillary and the hydraulic radius (Kozeny J. 1927 & Carman P C. 1956). Walsh, J B and Brace, W F introduced the void shape distribution parameters, and revised Kozeny-Carman equation (Walsh J B. et al. 1984). Masad et al. proposed the simple equation to approximately calculate the asphalt concrete permeability coefficient through the theory of capillary and the hydraulic radius (Masad E. et al. 2004). Cooley, L et al. designed the asphalt concrete mixture ratio in accordance with the method of Superpave, and made the asphalt concrete penetration test (Cooley L A. et al. 2002). It was considered that the permeability of drainage asphalt road was closely related to the concrete density of the coarse graded asphalt. Kanitpong Kunnawee et al. considered that the permeability of asphalt concrete is not only significantly influenced by the void content, but also by the thickness of the specimen, aggregate shape and grading and other factors (Kanitpong K. et al. 2001). Therefore, in order to assess the drainage of pavement permeability, it is necessary to measure the permeability coefficient of road surface drainage. The permeability coefficient of the drainage asphalt pavement can be rapidly determined by the pavement seepage meter.

2 ANALYSIS ON THE DRAINAGE MECHANISM OF THE POROUS ASPHALT PAVEMENT

The rainwater falls on the surface of the road and then it is permeated to the inside of the pavement through the air voids. Firstly, the initial moisture void is filled with water. Then, the water is accumulated in the structure layer. Finally, the seepage is happened. After absorption and accumulation of the pavement to the moisture, it begins producing seepage. But due to the rainfall intensity and rainfall duration and other condition, the rainfall could completely laterally infiltrate to discharge to the pavement through the drainage surface. Part of the water may infiltrate into the drainage layer and the surface runoff could form, which leads to the adverse effect on the performance of the pavement.

The difference between the runoff producing situations of the drainable pavement is that the drainage surface void content of the drainable pavement is large while the thickness is thinner. The bottom of the draining-layer is a water-proof (sealing) layer, but the moisture is only accumulated and infiltrated in the drainage layer. Moreover, with the different moisture content, the capacity of the transmitting water of the draining-layer material has been changed. Therefore, there are large differences in runoff producing between the drainable pavement and the soil.

The rainwater is vertically downward infiltrated into the drainage asphalt surface which is the two-phase process. The first stage is the external control stage. If the external eater supply rate does not exceed infiltration capacity, then the infiltration rate is equal to the Inflow rate. The second phase is the drainage layer control phase. With the drainage surface saturation degree increasing, the capacity of the surface water conduction is reduced. When it is less than external water supply rate, the rainwater begins forming the runoff to discharge it to the surface of the pavement through the comprehensive slope of the pavement surface. At this time, the infiltration rate is equal to the infiltration capacity of the surface layer. At last, the infiltration capacity reduces to the saturated hydraulic conductivity of the draining-layer.

The air void content of the drainage asphalt pavement is 20% or above, which is much larger than the soil. It is not easy to form a super runoff. Due to the limited thickness of the porous structure and its impermeable bottom, the rain that falls to the road surface mostly discharges through lateral seepage.

The reasons for road surface runoff formation saturation is not only related to the rainfall intensity or the rainfall duration, but also related to the drainage condition, drainage in the surface layer, the ditch drainage condition and the side ditch drainage conditions. Especially, the road surface runoff is affected by the drainage of permeation performance of the pavement material (Zhu Y. et al. 2004). Therefore, the foremost precondition of research on the drainage performance of the drainage pavement is to study the permeation performance of the drainage pavement surface.

3 THE TEST OF THE PERMEABILITY COEFFICIENT OF THE DRAINAGE SURFACE LAYER

3.1 *Test device of the permeability coefficient*

The permeability coefficient is significantly affected by the gradation of the material and the air void content. For a specific material, the air void content is more important. The void content of the asphalt concrete material is relatively easier to determine and in the drainage asphalt pavement construction, the drainage performance of the pavement is also ensured by the void content. And the relationship between the void content and the permeation performance of the porous asphalt concrete can be determined by the penetration test of drainage asphalt mixture (Ma X. et al. 2004).

According to the principle of penetration test of soil mechanics, two sets of tester for vertical permeability and horizontal permeability of the asphalt concrete were designed as shown in Figure 1 and Figure 2. In the vertical penetration test, the big Marshall specimen without remolding was directly installed in the penetrant test device. The rutting plate specimen without the two side baffles was used for the horizontal permeability test. In order to ensure the precision of test results, the constant water head was used for the permeability test, which makes the penetrant test conforming to the range of application of Darcy law.

Figure 1. Test device of vertical permeability coefficient.

Figure 2. Test device of lateral permeability coefficient.

Table 1. The mix designs of the three kinds of drainage asphalt mixtures.

Sieve size (mm)		19	16	13.2	9.5	4.75	2.36	1.18	0.6	0.3	0.15	0.075	Oil stone ratio (%)
PAC-10	A	100	100	100	99.4	30.8	21	14.9	11.5	10.2	9.2	7.5	5.2
	B	100	100	100	99.4	25.9	16.3	10.8	7.7	6.6	5.8	4.6	5.1
	C	100	100	100	99.3	20.0	11.5	7.4	5	4.2	3.6	2.9	5.0
	D	100	100	100	99.3	15.1	7.7	4.9	3.4	2.9	2.5	2.0	4.8
PAC-13	E	100	100	88.3	59.9	24.1	21.1	18.5	12.6	9.3	7.6	6.0	4.6
	F	100	100	88.6	60.1	18.1	15.8	13.9	9.6	7.2	6.0	4.7	4.6
	G	100	100	87.2	55.5	11.7	10.2	9.3	7.1	5.9	5.2	4.3	4.6
	H	100	100	89.4	61.2	7.8	6.6	6.4	5.0	4.2	3.8	3.2	4.6
PAC-16	I	100	99.1	89.6	53.7	23.5	19.7	14.0	9.1	7.5	6.9	6.5	4.5
	J	100	99.0	86.8	46.0	17.7	15.0	10.8	7.3	6.1	5.7	5.4	4.5
	K	100	98.9	83.6	36.6	9.0	7.5	5.2	3.3	2.7	2.5	2.3	4.5
	L	100	98.8	81.1	31.0	7.1	6.2	4.9	3.8	3.5	3.3	3.3	4.5

3.2 *Test method of the PAC permeability coefficient*

At present, the commonly used drainage asphalt mixture structure layer in China is PAC-10, PAC-13 and PAC-16 mixture. The penetrant test is mainly used to test the permeability performance of these several materials. First of all, according to the gradation composition of the asphalt mixture and the former engineering experience, the pass rate of the different void content situation of these several kinds of asphalt concrete can be determined. The dosage of each grade material and the theoretical density should be calculated. For the vertical penetration test, the specimen height of the big Marshal specimen under different void content should be calculated according to the theoretical density. The big Marshal Marshall specimen with the different void contents should be the same static pressure molding. For the horizontal seepage, the quality of specimen in the different void content shall be calculated, to form the rutting plate of the different void contents.

According to the material composition characteristics of the drainage asphalt concrete and the related construction experience, the material compositions of the several kinds of asphalt concrete were determined. The mix proportions of the three kinds of drainage asphalt mixture are shown in Table 1. The test results of the vertical permeability coefficient and the lateral permeability coefficient of rolling rutting plate by using static pressure molding specimens were described in Table 2.

Table 2 shows that with the increase of void content, the vertical permeability coefficients and lateral permeability coefficients of the three kinds of drainage asphalt mixture are increased, and the lateral permeability coefficient and vertical permeability coefficient of the same void content are almost the same. This is consistent to the research conclusions of Zhang Fan and other scholars on the relationship between the vertical permeability coefficient and lateral permeability coefficient of the drainage asphalt mixture (Zhang F. et al. 2009).

Table 2. The vertical permeability coefficient and lateral permeability coefficient of the three kinds of the drainage asphalt mixtures.

Type (%)		Void content	Vertical permeability coefficient (cm/s)	Void content (%)	Lateral permeability coefficient (cm/s)
PAC-10	A	16.1	0.06	16.3	0.07
	B	20.8	0.26	20.9	0.28
	C	24.5	0.43	24.7	0.46
	D	26.6	0.51	26.7	0.59
PAC-13	E	15.4	0.04	15.6	0.05
	F	19.2	0.20	18.9	0.20
	G	23.1	0.36	23.9	0.44
	H	26.4	0.51	25.9	0.54
PAC-16	I	16.1	0.05	16.5	0.09
	J	19.8	0.23	20.6	0.27
	K	25.1	0.44	25.2	0.51
	L	27.1	0.57	26.7	0.58

4 THE SIMPLIFIED DETERMINATION TECHNOLOGY OF THE PERMEABILITY COEFFICIENT OF DRAINAGE ASPHALT MIXTURE

4.1 *The method of indoor testing permeability coefficient of the pavement seepage meter*

The pavement seepage meter can be used for the lateral water penetration ability of the drainage asphalt mixture specimen and the test shall be made according to the site water permeability test method in the Japan drainage pavement guideline (scheme), and to calculate and determine the seepage coefficient. But due to the varying-head is used for the test method of pavement seepage meter, which the height of the water head is from 100 ml to 100 ml. So, the flowing of water flow within the specimen is non-laminar, and the permeability coefficient is not able to be calculated by the Darcy law. Therefore, the pavement seepage meter is used to make the Permeable coefficient of determination of taking off side mode. The rutting plate is shown as Figure 3. In the test process, through the method of external water pipe, the height of water head is kept at 100 ml scale line.

The surface height of control pavement seepage meter constantly determines the water discharge. Then, according to the related theory and through the seepage meter, and based on the formula (1), to calculate the lateral permeability coefficient K:

$$K = \frac{Q}{\rho_w tAi} = \frac{QL}{\rho_w tA(h_1 - h_2)} \quad (1)$$

In the formula, Q is the quality of water which the Infiltration is passing test specimen in the t time period, g; ρ_w is the density of water; g/cm^3; L is the effective flow length in the test specimen *cm*. A is the effluent cross sectional area of the specimen, cm^2; h_1, h_2 are respectively water head height and the seepage water head height under the constant water head, cm.

The PAC-13 rutting plates with different void contents of 18%, 20%, 22% and 24% are used to test the constant water head of pavement seepage meter. The test results are as shown in Figure 4.

Figure 4 shows that with the increase of void content, the permeability coefficient of rutting plate also gradually increasing. And, permeability coefficient numerical value and lateral permeability coefficient of the rutting plate which is using dismantling the two side baffles along the rolling direction are basically the same. By using the pavement seepage meter with a suitable hydraulic gradient, the permeability coefficients of porous asphalt mixtures of the same gap are basically the same, which lays the foundation for rapidly testing the water seepage coefficient.

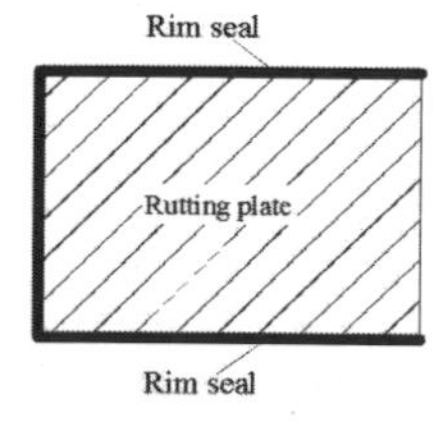

Figure 3. The lateral permeability coefficient of the rutting plate measured by the pavement seepage meter of constant water head.

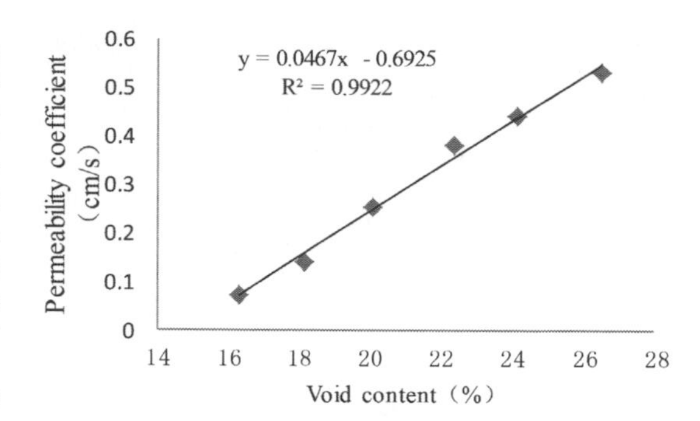

Figure 4. The relationship between the void content of the rutting plate and permeability coefficient tested by the pavement seepage meter.

4.2 *The rapid test of the pavement seepage meter in the construction site*

The water surface in the seepage meter was declined from 500 ml to 400 ml scale. With the water yield of 15 s, the permeable coefficient of road surface can be calculated. The method is simple in operation and distinguish the conditions of water seepage of road surface well, but is unable to accurately determine the void content and permeability coefficient of road surface. If through a lot of test to determine the relationship between the void content and the seepage time first, and according to the relationship between the void content and permeability coefficient, then we can through the time tested by the seepage meter to determine the relationship between void ratio and permeability coefficient of the road surface more accurately. The pavement seepage meter was used to test the water seepage time of the rutting plate with the above different void contents. The relationship between void content and the water seepage time is as shown in Figure 5. Based on the void content as the medium, to determine the relationship between rutting plate seepage time tested by the pavement seepage meter and the lateral permeability coefficient is shown in Figure 6.

Figure 5 shows that there is good curvilinear relationship between the void content and the water seepage time of the drainage pavement. R^2 is 0.991. Therefore, in the drainage asphalt pavement construction acceptance, according to the Figure 5,we can quickly determine whether the void content of this section pavement can meet the requirements or not, so we can speed up the construction progress. The relation curves of the seepage time of the rutting plate and the horizontal permeability coefficient has been established in Figure 6. In the construction, the permeability coefficient of this pavement can be calculated according to the seepage time, so we can evaluate the water seepage performance of the drainage asphalt road surface.

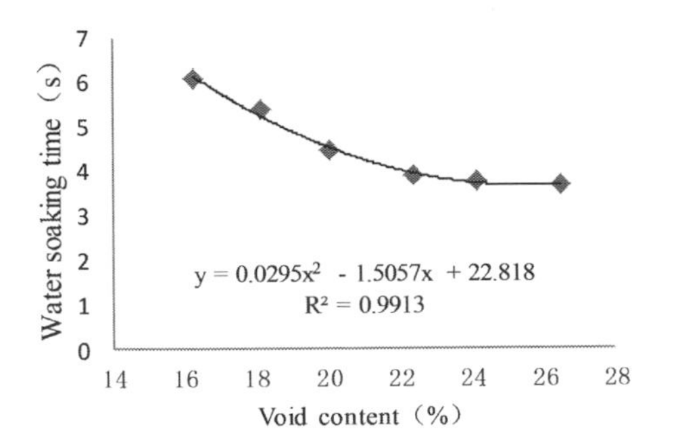

Figure 5. The relationship between void content of the rutting plate and the seepage time tested by the pavement seepage meter.

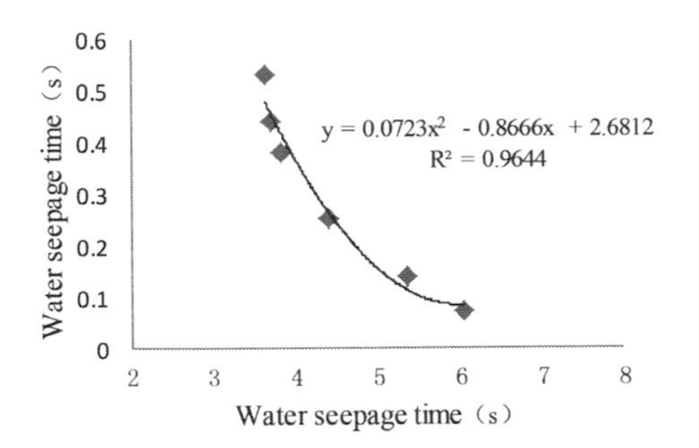

Figure 6. The relationship between the seepage time of the rutting plate tested by the pavement seepage meter and the lateral permeability coefficient.

5 CONCLUSION

In order to evaluate the permeability performance of drainage asphalt mixture, a large number of indoor experiments were conducted to test the permeability coefficient of drainage asphalt mixture. It comes to the following conclusions:

1. Based on the simple operation of the device of the vertical permeability and horizontal permeability of the seepage drainage asphalt mixture, which has the seepage principle. The test result showed good representative and stability, which can be used to accurately test the permeability coefficient of drainage asphalt mixture.
2. The vertical permeability coefficient measured by the big Marshall specimen of not mold releasing and the lateral permeability coefficient measured by the lateral permeability apparatus are almost the same, which shows that drainage asphalt mixture has good isotropic properties, and it can be considered that in engineering applications. The big Marshall compaction specimen is relatively convenient, and can be used to determine the vertical permeability.
3. The seepage meter can be used to make simple experiment to test the permeability coefficient of drainage asphalt mixture, through the rutting plate specimen of the remolding overhead frame. The lateral seepage meter shows there are good linear correlations between the seepage time of the different void content and the permeability coefficient measured. Through this relationship, the seepage time tested in the field can be used to calculate the permeability coefficient of drainage asphalt mixture. It is helpful to establish drainage asphalt pavement construction quality control and maintenance countermeasures.

ACKNOWLEDGEMENTS

This research is supported by National Science and Technology Support Program (2015BAL02B02) and Primary Research and Development Plan of Jiangsu Province (BE2015349). All the authors of the following references are much appreciated.

REFERENCES

Carman P C. 1956. Flow of gases through porous media. *Academic Press.*

Cooley Jr. L A, Prowell B D. & Brown E R. 2002. Issues pertaining to the permeability characteristics of coarse-graded superpave mixes. Colorado Springs, CO, United states: Association of Asphalt Paving Technologist.

Kanitpong K, Benson C H & Bahia H U. 2001. Hydraulic conductivity (permeability) of laboratory-compacted asphalt mixtures. *Transportation Research Record* 2001(1767): 25–32.

Kozeny J. 1927. Über kapillare Leitung des Wassers im Boden: (Aufstieg, Versickerung u. Anwendung auf die Bewässerg). *Gedr. mit Unterstützg aus d. Jerome u. Margaret Stonborsugh-Fonds*. Hölder-Pichler-Tempsky, A.-G. [Abt.:] Akad. d. Wiss.

Ma X, Ni F. & Wang Y. 2009. Test and analysis on permeability of porous asphalt mixture. *Journal of Building Materials* 12(2): 168–172.

Masad E, Birgisson B. & Al-Omari A. 2004. Analytical Derivation of Permeability and Numerical Simulation of Fluid Flow in Hot-Mix Asphalt. *Journal of Materials in Civil Engineering* 16(5): 487–496.

Pisciotta M. 2010. Functional characteristics of double draining layer flexible pavements. *Transportation Research Record* 2010(2101): 88–97.

Walsh J B. & Brace W F. 1984. Effect of Pressure on Void Content and the Transport Properties of Rock. *Journal of Geophysical Research* 89(B11): 9425–9431.

Zhang F, Chen R. & Ni F. 2010. Techniques of permeability testing for porous asphalt pavement mixture. *Journal of Southeast University (Natural Science Edition)* 40(6): 1288–1292.

Zhu Y, Chen R. & Ni F. 2004. Research on evaluation method of drainage performance of the porous asphalt pavement. *Journal of Highway and Transportation Research and development* 21(8): 9–11.

Civil, Architecture and Environmental Engineering – Kao & Sung (Eds)
© 2017 Taylor & Francis Group, ISBN 978-1-138-02985-9

Mechanical performances of icing atop cement concrete surfaces coated with emulsion wax curing agents

Jia-liang Yao
School of Traffic and Transportation Engineering, Changsha University of Science and Technology, Changsha, Hunan, China
Key Laboratory of Road Structure and Material of Ministry of Transport (Changsha), Changsha, Hunan, China

Min-jie Qu
School of Traffic and Transportation Engineering, Changsha University of Science and Technology, Changsha, Hunan, China

ABSTRACT: In road engineering, Emulsion Wax Curing Agents (EWCAs) are mainly used for cement concrete construction curing or as the bond breaker between the cement concrete pavement surface layer and the lean concrete base layer. Inspired by the lotus leaf effect, in this research work, a series of the laboratory tests were performed to evaluate the mechanical performances of the icing atop the cement concrete surface coated with an EWCA. The test results indicate that friction and bonding between the icing and the concrete surface coated with an EWCA have been reduced, and the icing is easier to be broken and removed. Thus, spraying EWCA atop the cement concrete surfaces might improve the efficiency of the mechanical deicing process on the cement concrete surface.

1 INTRODUCTION

Icing atop pavements tends to disrupt the entire transport network, thereby leading to traffic paralysis and huge economic losses. Therefore, deicing and skid resistance of pavement surfaces have been a crucial part of pavement curing and maintenance work in winter, as well as a menace for the road management departments in winter.

At present, various approaches including chemical, mechanical, thermal and coating are used to deice pavements in winter. The mechanical deicing approach is performed by using the mechanical devices to remove snow and ice atop pavements. In spite of the low efficiency of deicing, the mechanical approach is environment-friendly, thereby doing less harm to the vegetation and the environment around. Consequently, the mechanical approach is widely used. However, under low temperature conditions, with an increase in the bonding between the ice and the pavements, the efficiency of the mechanical deicing approach alone usually reduces to an unacceptable level. What is worse is that the pavements may be damaged by the violent deicing.

The coating approach has been used in some asphalt pavement projects. Ma et al. (2014) reported that the silicone hydrophobic material which is used as the surface coating of the asphalt pavement could effectively reduce the bonding between the asphalt pavement and the icing atop. However, the coating approach is rarely reported to be used for deicing in cement concrete pavement projects. Therefore, the authors of this paper wanted to explore the mechanical performances of the icing atop the surface of cement concrete coated with an Emulsion Wax Curing Agent (EWCA) through a series of laboratory tests. The test results were expected to be used as a reference for improving the efficiency of the mechanical deicing process with the help of EWCAs.

2 METHODOLOGIES

2.1 *Emulsion Wax Curing Agents (EWCAs)*

In cement concrete pavement projects, the cement concrete pavement is likely to be bonded to stabilized or lean concrete subbases. Thus, it is a common practice to place a bond-breaking medium atop the subbase surface. The common practice in the United States is to place two coats of a wax-based curing compound on the subbase surface as the bond breaker layer (Okamoto 1994).

The major components of the EWCA used in this research work are paraffin emulsion and compound emulsifiers that are mixed in a given proportion which is determined by the projects' conditions, including weather, structure types, and the depth of the EWCA layer to be formed. EWCAs have been applied extensively in various areas, including textile, paper, leather, fiberboard, polishing, fruit preservation, architecture, gardening, ceramics and otherwise. In road engineering, EWCAs have been used with success worldwide to cure several thousand kilometers of cement concrete pavements, and cement-treated or lean concrete subbases. The corresponding technical specifications for EWCAs have been established by the American Society of Testing Materials C309 (ASTM 2007) and American Association of State Highway Officials M148 (AASHTO 2005). Note that the material requirements for EWCAs in both AASHTO M 148 and ASTM C309 are identical.

The choice of the EWCA used in this research was primarily determined by its curing function and bond-breaking function, with inspiration from the lotus leaf effect. In this research, the solid content of the EWCA was no less than 25%, with the mass ratio of polymer materials to paraffin being 1:2 and a qualified film-forming or solidifying performance. A special device was used to spray the EWCA so as to ensure uniformity of spraying and to control the spraying volume.

2.2 *Tests for shear failure stress and integrated coefficient of friction*

In order to better evaluate the mechanical performances of the icing atop the pavements, the shear failure stress and friction between the icing and the concrete pavement with or without EWCA coating were tested. In this research work, a shear test apparatus was self-developed. And the specimens for the direct shear test were casted in PVC pipe molds, as shown in Figure 1.

Figure 1. Picture showing specimens for shear tests.

The molded specimens were cylinders that measured 110 mm in diameter, 100 mm in height, and 2.3 kg in weight. After the specimens were cured under the standard curing conditions and reached 30–35 MPa compressive strength, the specimens were divided into two groups. EWCAs were sprayed on the top-ends of one group with the spraying volume set at 0.2 L/m^2 while the control group of specimens was not sprayed with the EWCA. After the sprayed EWCA solidified, cold water was poured into the molds of the two groups of specimens to form a 5-cm-thick icing atop the specimens by placing the molds with specimens in a freezer at −18°C for 12 hours. After the molds were removed, the specimens with icing atop were prepared, as shown in Figure 2.

When the specimens were fully prepared for tests, the centers of the jack, the measuring force ring, and the specimen were arranged in a horizontal line. The load pushed the jack slowly until the bond between the icing and the cement specimens was completely broken, and the value of the measuring force ring at that time was recorded. This value was regarded as the maximum horizontal thrust.

After the maximum horizontal thrusts were measured, the maximum shear failure stresses and the integrated coefficients of friction could be calculated.

2.3 *Pull-out tests*

The pull-out tests were carried out to measure the pull-out forces to pull out the icing from the cement concrete surface, so as to determine the pull-out normal failure stress of the icing.

To perform the pull-out test, first, the cement concrete specimens were casted in cylindrical steel molds that measured 110 mm in diameter (the bottom molds). After the specimens were cured under the standard curing conditions and reached 30–35 MPa compressive strength, the specimens were divided into two groups. An EWCA were

Figure 2. Picture showing cement concrete specimens with icing atop.

sprayed on the top-ends of one group with the spraying volume set at 0.2 L/m^2 and was solidified while the control group of specimens were not sprayed with an EWCA. Next, another set of steel molds of the same diameter (the top molds) were attached to the bottom molds, and the joints were sealed. The connected top molds and bottom molds with specimens were placed in a freezer at −18°C for 1 hour, and then cold water was poured into the top molds and the whole set was placed back into the freezer to form the 50-mm-thick icing atop all specimens.

When the specimens were fully prepared for tests, pull-out devices were used to draw the top molds and the bottom molds to the opposite directions. At the time when the icing and the concrete were broken apart, the pull-out force P was the maximum and was recorded.

2.4 *Tests for impact on icing*

The tests for impact on the icing were performed by following the drop-weight method with a set of self-made test apparatuses.

Cement concrete specimens were prepared by means of machine vibrations, with dimensions of 300 mm by 300 mm and thickness of 50 mm, cured under the standard curing conditions and reached 30–35 MPa compressive strength. The ready specimens were divided into two groups with one sprayed with EWCA of 0.2 L/M^2 on the top-ends and the other without EWCA as the control group. Next, the top surfaces of two groups of specimens were covered with ice and the icing was 10 mm thick. Afterwards, a ball weighing 200 grams was allowed to fall down through the hole in the fixed bracket from a height of 100 cm. The center of the hole in the fixed bracket and that of the specimens were in a vertical line. The impacts were repeated to observe the cracking of the icing atop the specimens and the related data were recorded, including two sets: the cycles of impact (N1) for the initial cracking (when the first crack on the icing appeared, the cycles of impact taken by the icing); the cycles of impact (N2) for the final fracture of the icing.

3 RESULTS

3.1 *Results of shear failure stress and integrated coefficient of friction*

In order to obtain the shear failure stresses and the integrated coefficients of friction between the icing and the cement concrete surface of two groups of specimens, the maximum horizontal thrusts to break the icing from the concrete surface were measured. The measured maximum horizontal thrusts of two groups of specimens are listed in Table 1.

The maximum shear failure stress can be determined by the ratio of the maximum horizontal thrust T_{max} to the shear area A ($1/4\pi \times 110^2$ mm^2):

$$\tau = T_{max} / A \quad (1)$$

According to equation (1), when the specimens are coated with an EWCA, the shear failure stress is 0.14 MPa in average, while the ice shear stress for the control group of specimens is 0.37 MPa in average. It can be seen that the shear failure stress of the specimens with an EWCA coating is decreased by 62%, which indicates that the EWCA tends to reduce the bonding of the icing to the concrete surface, which makes it easier for the icing to be broken apart from the concrete surface under the horizontal thrusts.

With respect to the integrated coefficient of friction f_h, it is the ratio of the maximum horizontal thrust of specimen T_{max} to the dead weight of the specimen m:

$$f_h = T_{max} / m \quad (2)$$

According to equation (2), when the specimens are coated with EWCA, the integrated coefficient of friction is 57 in average, while the integrated coefficient of friction for the control group of specimens is 155 in average. It can be seen that the integrated coefficient of friction of the specimens with EWCA coating is reduced by 63%. Such a reduction indicates that the EWCA coating has great effects on the interface condition between the icing and the cement concrete surface, with the interface shifting from bonding to semi-bonding. This shift makes it easier for the icing to be removed away from the concrete surface under the horizontal thrusts.

3.2 *Results of pull-out tests*

The results of the pull-out forces to pull out the icing from the concrete surface are shown in Table 2.

Table 1. Results of maximum horizontal thrusts to break the icing from a concrete surface.

	Specimens with EWCA coating	Specimens without EWCA coating
Maximum horizontal thrust P/KN	1.6	3.2
	1.0	3.4
	1.3	3.6
Average value P/KN	1.3	3.5

Table 2. Results of pull-out forces.

	Specimens with EWCA coating	Specimens without EWCA coating
Pull-out forces P/KN	1.5	2.2
	1.4	2
	1.4	2.5
Average value P/KN	1.4	2.3

The pull-out normal failure stress σ can be defined as the ratio of the maximum pull-out force P_{max} to the cross-sectional area A ($1/4\pi \times 110^2$ mm^2) of the specimen for the pull-out tests:

$$\sigma = P_{max} / A \qquad (3)$$

According to equation (3), when the specimens are coated with EWCA, the pull-out normal failure stress is 0.15 MPa in average, while the pull-out normal failure stress for the control group of specimens is 0.24 MPa in average. It can be seen that the pull-out normal failure stress of the specimens with EWCA coating is reduced by 38%, which indicates that the EWCA tends to lessen the bonding of the icing to the concrete surface, which makes it easier for the icing to be pulled out from the concrete surface under the pull-out forces.

3.3 *Results of tests for impact on icing*

Through the tests for impact on icing, the cycles of impact for the initial cracking of the icing (N1) and the cycles of impact for the final fracture of the icing (N2) were recorded, as shown in Table 3.

Table 3 shows that the value of N1 for the specimens coated with EWCA is 60% less than N1 for the control specimens without EWCA coating; while the value of N2 for the concrete surface coated with EWCA is 52% less than N2 for the control specimens without EWCA coating. And the difference between N1 and N2 for the specimens with EWCA coating is 6, while the difference between N1 and N2 for the control specimens without EWCA coating is 12, which indicates that the icing atop EWCA coating is more prone to breaking after the initial cracking. The underlying reason might be that the EWCA coating has great effects on the interface condition between the icing and the cement concrete surface, with the interface shifting from the bonding to the semi-bonding state. This shift makes it easier for the icing to be broken under the impacts.

Table 3. Results of cycles of impact on icing.

	Specimens with EWCA coating	Specimens without EWCA coating
Cycles of impact for the initial cracking (N1)	2	5
	2	6
	2	5
Average value of N1	2	5
Cycles of impact for the final fracture (N2)	7	15
	8	18
	7	17
Average value of N2	8	17

4 CONCLUSIONS

This research explored the feasibility to improve the efficiency of the mechanical deicing process on the cement pavement surfaces by spraying the EWCA atop the cement concrete surfaces, through a series of laboratory tests (including the tests for shear failure stresses and integrated coefficients of friction, the pull-out tests, and the tests for impact on the icing) to test the mechanical performances of the icing atop the cement concrete surface with or without the EWCA coating. By comparing the test results of the specimens coated with EWCA with those of the control specimens without EWCA coating, the following conclusions can be drawn:

1. By spraying the EWCA coating atop the cement concrete surface, the latter is changed from a hydrophilic surface into a hydrophobic one, which is conducive for the good drainage of rainwater, and thus helps to prevent the concrete surface from icing.
2. If rainwater or snow on the cement concrete surfaces turns to be the icing layer, by spraying the EWCA coating atop the concrete surface, the bonding between the icing and the concrete surface can be reduced greatly, and the same is true for the cycles of impact to break the icing. The underlying reasons might be that the EWCA coating has great effects on the interface condition between the icing and the cement concrete surface, with the interface shifting from the bonding to the semi-bonding state. This shift makes it easier for the icing to be de-bonded from the concrete surface, and to be cracked and broken.

Thus, the results of the laboratory tests mentioned in this research indicate that spraying the EWCA atop the cement concrete surface might improve the efficiency of deicing through the mechanical approach greatly. Further studies will be performed to confirm the above conclusions through a series of field tests.

ACKNOWLEDGMENTS

This research was supported by the National Natural Science Foundation of China under grant no. 51578080, grant no. 51178064, and grant no. 51302020. And it was also (Project No. 2013-01) funded by the Transportation Department of Hunan Province, China.

REFERENCES

AASHTO (2005). American Association of State Highway Officials (AASHTO). Standard Specification for Liquid Membrane-Forming Compounds for Curing Concrete. AASHTO Designation: M 148-05.

ASTM (2007). ASTM C 309. Standard Specification for Liquid Membrane-Forming Compounds for Curing Concrete. American Society of Testing Materials, Philadelphia, PA.

Ma, H., R. C. Yang & S. Z. Qian (2014). Research on asphalt concrete pavement deicing technology. *Journal of Southeast University*, 30(3), 336–342.

Okamoto, P.A, P.J. Naussbaum, K.D. Smith, M.I. Darter, T.P. Wilson, C.L. Wu, & S.D. Tayabji (1994). Guidelines for Timing Contraction Joint Sawing and Earliest Loading for Concrete Pavements, Volume 1: Final Report. FHWA-RD-91-079. FHWA, U.S. Department of Transportation, February 1994.

Civil, Architecture and Environmental Engineering – Kao & Sung (Eds)
© 2017 Taylor & Francis Group, ISBN 978-1-138-02985-9

A study on residential area's adaptability design for the aged in China based on open building theory

Xue Meng
Harbin Institute of Technology, Harbin, China

Benchen Fu
The Architectural Design and Research Institute of HIT, Harbin, China

ABSTRACT: Through the analysis of serious aging problems in China, this paper aimed to explore the value of Open Building Theory for guiding residential area's adaptability design for the aged in China. The main concepts involved in open building theory, including overall concept, hierarchy control, showing respect to individual interests and dynamic adjustment to need, were highly associated with residential area's adaptability design for the aged by the combination of theoretical research and case study. Based on that, this paper proposed two aspects of adaptability design techniques, including the design of residential planning and dwelling units, providing references for future rational design of residential areas for the aged in China.

1 INTRODUCTION

1.1 *Open building theory*

1.1.1 *The development of open building theory*

In 1961, N. John Habraken published *Support: An Alternative to Mass Housing,* in which he proposed to separate housing design and construction into two parts- "Supports" and "Detachable Units". This reflects his thinking between housing's stability and change in deal with the built environment (Kendall 2009). The research aims to advocate residents to participate in the design and construction process and then realize the diversity and individuality in mass housing construction (Jiang & Jia 2014). As the Director of SAR (Stichting Architect Research), he led the research institute to search in this field and formed the prototype of open building theory (Miao et al. 2016).

In 1980s, open building was proposed by Stephen Kendall (Fig. 1), who studied with John Habraken in MIT. After that, this theory went through a long-term complement investigation and gained the great influence via OHI (Open House International) and Open Building Foundation's promotion. In the last tens of years, housing construction in many countries got boost based on open building theory.

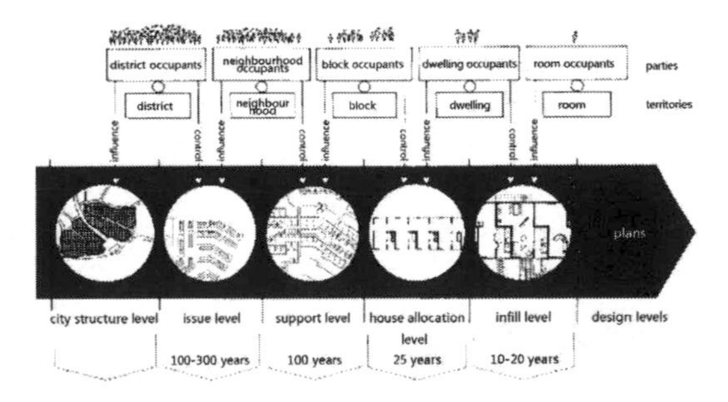

Figure 1. Environmental levels in Open Building (Kendall 2009).

1.1.2 *The main concepts involved in open building theory*

Open building theory emphasizes several main concepts, including overall concept, hierarchy control, showing respect to individual interests and dynamic adjustment to needs (Habraken & Teicher 2000). The proposal's original intention was to deal with contradictions between urgent needs for mass housing and diversity of social housing after World War II's rapid urban development (Jiang & Jia 2014). However, with the continuous evolution of open building theory in the following few decades, it has been more than a design theory. In a broad way, "Open Building" in this theory stands for "open-type building", implying a way to achieve dynamic, adjustable and adaptive design methods for architecture, which would help realize the sustainable and better urban living environment. It has already got rid of the initial limitations of housing design and construction technology. From urban environment and collectivity to living unit and individual, open building theory covers the extensive ranges, becoming a significant component of sustainable urban built environment.

1.2 *Problems of residential area's adaptability for the aged in China*

1.2.1 *Aging population in China*

Since 1999, China has entered an aging society with increasingly aging degree. By the end of 2015, the elderly (>60 years old) population would have increased to 221 million, reaching 16% of the total population according to the statistics of China State Council (2011). By 2030, the amount of older adults would double to 400 million and China would be faced with the world's highest degree of aging population (2011). Challenges brought out by aging population are everywhere.

1.2.2 *Problems of aging at home*

With the development of undertakings for the aged in China, pension mode gradually improves. The national implementation of "9073" Policy provides various pension choices for the elderly. Nearly 90% residents would adopt the manner of "aging at home". Compared with the other two types, aging at home is "tangible and relevant" (Senior Homes 2016). It is in accordance with residents' conventional ideas and long-term living habits, which is accepted by more families (Liu et al. 2015). However, this brings about enormous pressure of aging at home. In addition, more problems still remain, including demographic changes, relatively backward residential service facilities, etc.

1.2.3 *Problems of residential areas*

As the space carriers of aging at home, residential areas become the key to implementation of the national pension policy. Figure 2 shows the continuous and vigorous development of newly started housing area in China (NBSC 2015). While the quality of the living environment could hardly meet the dynamic needs of residents.

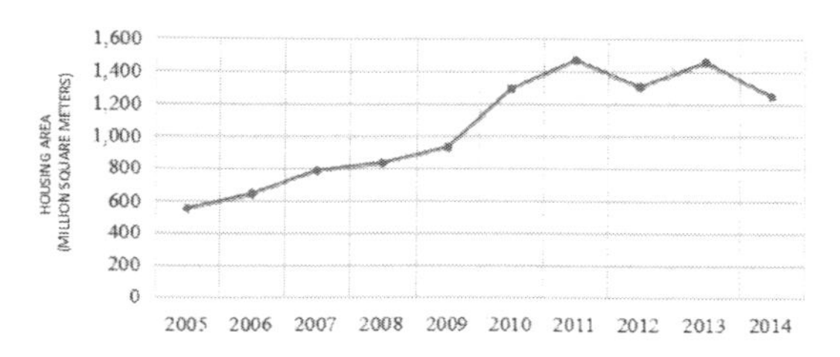

Figure 2. Development of housing area newly started by real estate enterprises.

In the face of aging crisis, the so-called "settlements for the aged" is far from enough. What we need is the residential area where we could generally live for long until old age according to Elderly Housing Construction in Japan (2011). This reflects the residential adaptability for the aged.

In conclusion, the design and construction of residential areas equipped with adaptability is imperative. Thus, open building theory was introduced into this study to explore the rational design concept. Then two aspects of design techniques (planning and dwelling units) were proposed, providing approaches to better achieve dynamic, long-term and sustainable living environments.

2 OPEN BUILDING THEORY'S REFERENTIAL VALUE

2.1 *The practical application of open building theory in housing*

Based on open building theory and its expansion, many countries have set up their own construction system and standards according to their national conditions. Japan experienced the development of KEP (Kodan Experimental Housing Project), NPS (New Planning System), CHS (Century Housing System), etc. Since 21th century, KSI (Kikou Skeleton Infill) system obtained a comprehensive implementation. KSI emphasizes the separation of skeleton and infill, indicating the structure's durability and living pattern's flexibility for long (Qin et al. 2014). In 1980s, universal design concept was proposed in the US in order to create safe, convenient, healthy and comfortable environment for all. According to this concept, the particular needs of various people were taken into consideration, including the elderly, the disabled, etc. Then based on that, *Universal Design in House* presented the housing's separation into structural section and non-structural section, achieving a high degree of residential adaptability. The Building and Construction Authority in Singapore enacted *Universal Design Guide*, giving consideration to different group's pension and housing needs. Through the references of Japan's KSI and western countries housing experience, China achieved the establishment of CSI (China Skeleton Infill) in 2010, giving response to better residential adaptability, quality improvement and extension of life-cycle (Li 2010).

2.2 *Open building theory's introduction into residential area's adaptability design*

Faced with serious aging problems and stress on aging at home, residential area's adaptability for the aged, which aims to provide inclusive, flexible and sustainable living environment, is imperative.

As a complex living system, residential areas contain a series of components including residential buildings, residential exterior space, public service facilities, etc. And the main concepts involved in open building theory provide rational approaches to realize the long-term sustainability. At this time, from the perspective of open building theory, a residential area is regarded as a whole system. Overall concept emphasizes the system's long-term integrity. Hierarchy helps to achieve relatively independent and stable subsystems. Showing respect to individual interests and division of decision-making are conducive to instant feedbacks. Dynamic adjustment provides guarantee of flexibility and adaptability in the long term. All these four aspects guide and constraint to ensure the sustainable development of residential areas. In this case, residential adaptability design for the aged may be suitable for everyone.

3 DESIGN TECHNIQUES

Based on open building theory, buildings and its surroundings are never static. They should always make adjustment according to the development of society and technology. Thus interaction between people, time, building and environment should be established to realize the high-quality and sustainable living environment (Lan & Wei 2015). In general, there are mainly two aspects (Table 1). First, in residential planning level, the function and scale of subsystems (e.g. exterior space, service facilities) should be adjusted and optimized according to the population, behaviors and needs of elderly people. Second, in unit level, the interior space ought to be alterable and flexible to meet the living habits of the elderly.

3.1 *Residential planning*

Specifically, the guiding value of open building theory mainly reflects in the following aspects.

Table 1. Adaptability design for the aged.

Concepts of open building theory	Design techniques
□ Overall concept	Planning: adjustment in function and scale
□ Hierarchy	
□ Showing respect to individual interests	Unit: flexibility of space separation and utilization
□ Dynamic adjustment	

3.1.1 *Flexible configuration of public service facilities*

Based on the planning and construction standards, the configuration (e.g. type, scale, inner function) of facilities should be adjusted according to residential population structure, regions and urban location (Wu et al. 2012). In addition, either residential objective conditions (e.g. construction time, resource allocation) or residents' subjective preferences would make a difference on the demands of service facilities. Thus relatively elastic and flexible configuration are appropriate for the implementation of sustainable living environment (Fig. 3).

3.1.2 *Reservation land for future extension and replacement*

Reserving part of land for future extension and replacement is encouraged in response to residential dynamic requirements. The reservation area could first be utilized as residential external public space. Along with the growing aging population, the shortage of required facilities gradually worsens. Then the reservation land could be converted into supplementary service facilities to fulfill the greater demands of the elderly.

3.1.3 *Utilization and renovation of the remaining*

More attention ought to be paid to focus on the present living status and residents' behavior habits. With aggravating trend of aging problems and demographic change, the existing standards of planning is no more applicable. Some buildings may stand empty as the present needs for some facilities such as kindergartens and primary schools are not that much. In addition, other idle facilities (e.g. factory buildings) could also be renovated to serve for the elderly (Zhan & Wu 2014). We should make full use of idle and inefficient status of facilities and balance the function utilization. Either function replacement or ratio adjustment would be beneficial to ease the supply imbalances of residential service for the aged.

3.2 *Dwelling units*

The main problem involved in residential adaptability design for the aged is that changes in family

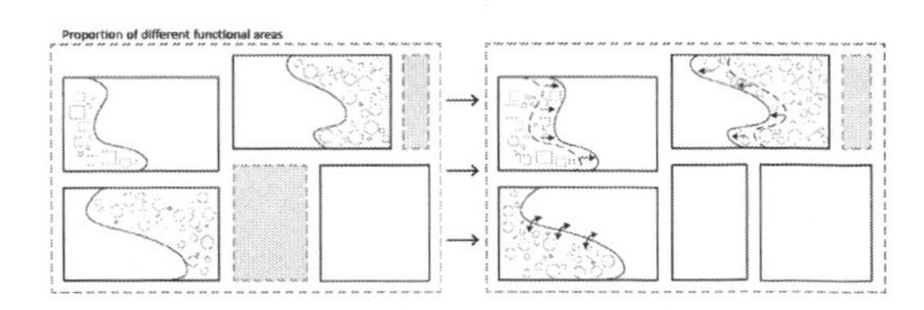

Figure 3. Flexibility of configuration in residential planning.

structure and members' physical conditions would lead to variation of functional layout and space details (Pan & Ding 2014). Conventional housing's integrated construction methods lead to the lack of flexibility and adaptability. While the separation design and construction of "Supports" and "Infills", which is one of the most significant design principles of open building theory, provides a guarantee for unit's orderly optimization and dynamic adaptability (Wang & Mei 2014). First, "Supports" (e.g. structural system) are regarded as the basic skeletons for residential interior space segregation. The durability of "Supports" also extends the utilization life of residential dwellings. Then, the flexible segregation of "Infills" (e.g. light partition walls) enables the instant adaptability, satisfying the diverse needs of occupants at different ages. It realizes the richness and practicability of space utilization. This is the expression of "maintaining the status quo" under hierarchy control, reflecting the typical characteristics of residential adaptability for the aged, namely spatial versatility and variability. The combination of "Supports" and "Infills" ensures the long-term utilization during life cycle as well as improving living comfort (Qin et al. 2014). For instance, Figure 4 shows Tila housing block whose apartments have no space segregation except for 2 baths.

Figure 4. The apartment with no space segregation except for 2 baths (CIB W104 2015).

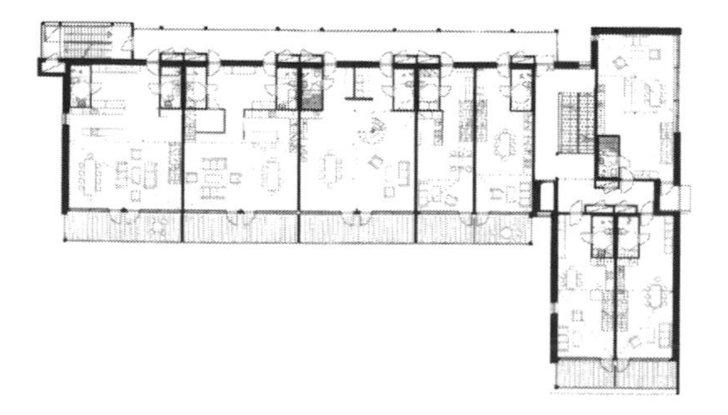

Figure 5. Occupants' various preferences for space division (CIB W104 2015).

The occupants could divide the inner space regarding their daily behaviors, preferences and willingness (Fig. 5). And adjustments would also be made to adapt to changing ages.

4 CONCLUSIONS

With the intensification of aging situation in China, residential area's adaptability design for the aged is required. The introduction of open building theory would help guide and promote the residential area's construction and renovation. In this way, the disadvantages of conventional design methods could be avoided. The theory emphasizes building is not one-time but long-term (Jia 2009). The hierarchy control enables the overall system as well as independent subsystems to be stable and durable. And the more open decision-making provides positive ideas for diversity, stability and sustainability of residential area's construction in China. Meanwhile, we could see that the some attempts have been made to pursue residential area's adaptability design for the aged, such as Summit's Guanghe Yuanzhu project in Beijing based on SI system (Qin et al. 2015). These projects help confirm the referential value of open building theory and provide practical experience for future construction of livable, healthy and sustainable residential areas in China.

ACKNOWLEDGEMENTS

We would like to express our gratitude to National Natural Foundation of China (NSFC), Project Number 51578174, for funding this study.

REFERENCES

China State Council. (2011). *"The Twelfth Five-Year Plan" Development of Aging Undertakings in China.* [Online] Available from: http://www.gov.cn/zwgk/2011-09/23/content_ 1954 782.html [Accessed 17th September 2011].

CIB W104. (2015). *Report on the Tila Open Building Project in Helsinki.* [Online] Available from: http://www.open-building. org/ob/cases.html [Accessed 28th May 2015].

Elderly housing Consortium. (2011). A Handbook of Elderly Residence Design. M. Beijing: China Building Industry Press.

Habraken, N. J. & Teicher, J. (2000). The structure of the ordinary: form and control in the built environment. J. Architectural Review, 2000, 205 (1225): 88–89.

Jia, B. (2009). Defining the Interface of the Private and the Shared—The Cultivation of Citizenship in the Economical Housing. J. New Architecture, 03, 4–15.

Jiang, Y. & Jia, B. (2014). Industrialization and Customization in the Housing Industry of Japan. J. Architectural Journal, s1, 120–125.
Kendall, S. (2009). *Open Building Concepts.* [Online] Available from: http://open-building.org/ob/concepts.html [Accessed 19th May 2009].
Lan, X. & Wei, H. (2015). Discussion on the Approach of Housing Industrialization Base on the "Open Building" Theory. J. Architecture & Culture, 07, 192–193.
Li, A. (2010). CSI System Will Become the Foundation of Hundreds years housing—An Interview with Liu Meixia of Housing Industry Development Center. J. China Concrete, 12, 12–14.
Liu, D. et al. (2015). A Study on Universal Design of Housing with Home-Based Care for the Aged. J. Architectural Journal, 06, 001–008.
Miao, Q. et al. (2016). Architectural design and construction features of European residential buildings under the concept of open architecture. J. Housing Industry, 04, 18–25.
National Bureau of Statistics of China (NBSC). (2015). *Statistics of Real Estate Development.* [Online] Available from: http://data.stats.gov.cn/easyquery.htm?cn = C01 [Accessed 10th August 2016].
Pan, H. & Ding, W. (2014). The Edification of New Trends of Proper Aging Collective Housing Design in Japan on Housing Design for the Elderly and Living Facilities for the Aged in China. J. Architecture & Culture, 12, 160–161.
Qin, S. et al. (2014). A Study of the Development of Japanese Infill Components System from KEP to KSI. J. Architectural Journal, 07, 017–023.
Qin, S. et al. (2015). Practicing Public Housing of Adaptability to the Aged—on the Project of Summit's Yuanzhu in Beijing. J. Architectural Journal, 06, 28–31.
Senior Homes. (2016). *Aging at Home: How to Prepare.* [Online] Available from: http://www.seniorhomes.com/p/ag-ing-at-home/ [Accessed 9th August 2016].
Wang, M. & Mei, H. (2014). Design Strategies of Rural Houses in Frigid Areas of Northeast China Based on The Open Building Theory. J. Urbanism and Architecture, 28, 110–112.
Wu, Q. et al. (2012). Research on the Development Dilemma and Optimization Strategy of Public Service Facilities in Old Community—A Case Study of Lujiazui Area. J. Shanghai Urban Planning Review, 01, 049–054.
Zhan, Y. & Wu, F. (2014). Planning Strategies for Aging Megacities: A Case Study of Shanghai. J. Urban Planning Forum, 06, 38–45.

Civil, Architecture and Environmental Engineering – Kao & Sung (Eds)
© 2017 Taylor & Francis Group, ISBN 978-1-138-02985-9

Safety management strategies for public infrastructure projects

C.W. Liao
China University of Technology, Taipei, Taiwan

T.L. Chiang
Occupational Safety and Health Administration under the Ministry of Labor of Taiwan, New Taipei City, Taiwan

ABSTRACT: Labor inspection services in many countries are not capable of carrying out their roles and functions. In Taiwan, the Lowest Bid Tendering method (LBT) has been adopted by most public entities. As a result, occupational safety management for public construction projects faces challenges such as diminishing budgets, a reduction in staff, and the continued growth of the workforce, and thus it is getting increasingly difficult. This article examines management strategies for occupational safety in the construction industry. Several strategies were conducted to improve the safety performance of a public construction project in Taiwan—the widening project of the Wugu-Yangmei section of National Freeway No.1. A questionnaire method is used to analyze different strategies in order to explore the understandability and effectiveness of each strategy. The findings identified in this article provide a direction for more effective safety management strategies and injury prevention programs.

1 INTRODUCTION

The Government Procurement Law in Taiwan has two categories of contract award methods: the Lowest Bid Tendering method (LBT) and the most favorable Tendering method (MFT), a type of multicriteria bid evaluation method (Wang et al., 2006). LBT features simple tendering procedures and non-controversial issues, making it one of the most popular contract-awarding methods (Tzeng et al., 2006). In Taiwan, LBT has been adopted by most public entities (project owners) for a long time. As contractors compete solely on the bidding price, the LBT method may result in awarding contracts to unreasonably low bidders (Tzeng et al., 2006). Such cases imply that the winner may cut corners during construction to maintain profits, especially in occupational safety and health issues (Wang et al., 2006). Multi-level contracting and illegal sub-contracting in the construction industry may worsen this situation. As a result, occupational safety management for public construction projects faces challenges such as diminishing budgets, a reduction in staff, and the continued growth of the workforce, and thus it is getting increasingly difficult to manage.

In addition, the owner of a public project (government agencies) typically takes an overseeing rather than a proactive role in safety management for public construction projects. This situation implies that safety for public construction projects is entirely determined by the supervisory practices carried out by the supervisors and the spontaneous activity of the contractors. If the supervisor does not carry out supervision and constrain the contractor actively, or if the contractor lacks a sound approach to construction management and safety, accidents are highly likely to occur.

In Taiwan, occupational safety is the responsibility of the employer, while the occupational safety and health laws are enforced by the Occupational Safety and Health Administration under the Ministry of Labor. The inspectors are mandated to carry out site visits without prearranged appointments to inspect work situations, working hours, construction safety, or any aspect of accident risk (Yränheikki and Savolainen, 2000). Generally, inspectors seek to modify worker behavior by raising the penalties for non-compliance with the regulations but they may also take a more flexible approach. For example, they may aid in helping workers adapt work systems to conform to production demands while at the same time correct compliance problems. Nevertheless, labor inspection services in many countries are not capable of carrying out their roles and functions (Eisenbraun, 2013; International labour office, 2006; Weil, 2008). They are often understaffed, underequipped, undertrained, and underpaid. A labor inspectorate must develop its strategies and deploy its resources in the most effective way. Ranking management strategies from worst to best is perhaps the most straightforward way for safety management on construction sites.

In this article, a public infrastructure project was selected for a single case study. The Occupational Safety and Health Administration were taken as a leading role, and the owner, the supervisors, and the contractors were united to explore effective safety management strategies. This article examines management strategies for occupational safety in the construction industry. Several strategies were conducted to improve the safety performance of a public infrastructure project in Taiwan—the widening project of the Wugu-Yangmei section of National Freeway No. 1.

2 SAFETY MANAGEMENT STRATEGY

Several safety studies have investigated occupational accidents within the construction domain. Much has been discussed about the factors that drive safety performance. Organizational safety policy is essential to safety performance on construction projects. It can ensure that proper safety management should not be seen as a burden or something rigid to be isolated from other aspects of management (Sawacha et al., 1999; Hinze and Wilson, 2000; Jaselskis et al., 1996). Since effective training of workers can greatly reduce unsafe acts, worker training is the most influential factor that improves safety performance (Hinze and Wilson, 2000; Huang and Hinze, 2003). At a construction project, more formal and informal safety meetings with supervisors or specialty contractors are important for achieving better safety performance (Jaselskis et al., 1996). Construction accidents occasionally result because the safety equipment necessary to perform the job safely is not present on construction sites, or workers cannot effectively use the safety equipment that is provided for their use (Chi et al., 2005; Toole, 2002). Safety inspections can be used to monitor safety compliance on sites. Increasing the number of site safety inspections has a positive impact on safety performance (Hinze and Wilson, 2000; Jaselskis et al., 1996). The use of safety incentives and penalty schemes can effectively increase safety awareness to mitigate site casualties. Safety incentives and penalties should be applied to promoting safety practices (Jaselskis et al., 1996; Tam and Fung, 1998). Workers' safety attitude is an important factor contributing to occupational accidents. If positive attitudes of workers towards safety can be reinforced and embedded within a group, successful safety management can then be achieved (Abdelhamid and Everett, 2000; Aksorn and Hadikusumo, 2008; Toole, 2002). As the high turnover rate makes it more difficult for workers to adjust to varying work conditions with different demands set by different employers, it is associated with higher injury rates (Harper and Koehn, 1998; Hinze and Gambatese, 2003; El-Mashaleh, et al., 2010).

Based on the above, there are eight important influence factors: organizational safety policy, worker training, safety meetings, safety equipment, safety inspections, safety incentives and penalties, workers' safety attitude, and labor turnover rates. For the factors, workers' attitude towards safety and labor turnover rates, occupational safety and health administration has no direct influence. Checks on safety equipment and implementation of safety incentives and penalties are typically merged into safety inspections. Therefore, factors to reduce construction accidents that are available to the competent authorities can be divided into four main types: organizational safety policy, safety training, safety meetings, and safety inspections. Further, On the basis of the four main factors for occupational safety management, some applicable strategies and activities have to be identified.

Through in-depth interviews with the responsible employees from the labor inspectorate, project owner and supervisors, and regard for current labor safety management practices in Taiwan, we develop related strategies to prevent construction accidents. Then the corresponding safety activities were progressively presented for the entire project during 2009–2011, as shown in Table 1.

With regard to the seminars, as soon as construction accidents or near misses occur, using an internet platform to immediately disseminate this information as well as gathering relevant cases and developing

Table 1. Safety management activities for this project.

Activity	Time	Code	Description
Regular			
Joint inspections between project owner and supervisors	Once a month	SI 2.3	
Training of health and safety engineers from project owner and supervisors	Twice a year	ST 3.1ii	
General safety seminars for bridge construction	Ten times a year	ST 3.1	

(Continued)

Table 1. (*Continued*).

Activity	Time	Code	Description
Irregular			
Form strategic partnerships between project owner, supervisors, and contractors	Oct 2009	SP 1.1	
Establish inspection programs	Dec 2010	SI 2.1	
Generate safety advocacy action on bridge foundation excavation and pre-construction	Mar 2010	ST 3.1i	
Generate safety advocacy action on precast segmental method	Mar 2010	ST 3.1i	The two earliest projects used the precast segmental method
Organize demonstrations for deck working	Jun 2010	ST 3.2	Since the area of construction is insufficient, a large amount of working platforms and construction roads have to be used
Establish a management system for health and safety supervision	Jun 2010	SM 4.3	
Place safety illustrations in work entrances on freeway interchanges, ramps, and hard shoulders	Jul 2010	ST 3.3	In response to the location of a large number of work areas on freeway interchanges and ramps
Generate advocacy action on collapse prevention during bridge construction	Nov 2010	ST 3.1	Incident involving collapse of interchange scaffolding occurred in October 2010
Chinese New Year safety inspections	Dec 2010	SI 2.1	
Organize on-site safety diagnosis for precast segmental method	Dec 2010	SM 4.4	
Personnel carry out control and inspection programs on-site	Feb 2011	SI 2.2	
Establish an internet platform for exchange of information	Feb 2011	SM 4.2	
Equipment inspections for bridge piers	Mar 2011	SI 2.1	Incident involving collapse of steel staircase occurred in March 2011.
Organize safety seminars for bridge operation vehicle incidents	Mar 2011	SM 4.1	Incident involving part of an operation vehicle falling off in January 2011
Organize safety seminars for work operations next to the freeway	Mar 2011	SM 4.1	1. Incident involving crane collapsing on freeway occurred in March 2011 2. Incident involving collapse of steel staircase onto freeway occurred in March 2011.
Establish strengthened inspection programs	Apr 2011	SI 2.1	Increase the frequency of inspections and level of sanctions in response to frequent construction accidents
Produce construction safety illustrations and checklists for the ten major hazards	Apr 2011	ST 3.3	
Assessment of major hazards and revisions to standard operating procedures	May 2011	SM 4.1	
Organize demonstrations for collapse prevention of bridge pier rebar	May 2011	ST 3.2	Incident involving collapse of bridge pier rebar occurred in May 2011
Organize safety seminars for collapse prevention of bridge pier rebar	Jun 2011	SM 4.1	Incident involving collapse of bridge pier rebar occurred in May 2011
Organize safety seminars for incidents involving bridge operation vehicles and work operations next to the freeway	Aug 2011	SM 4.1	Incident involving steel rail pile hitting freeway fence occurred in July 2011
Organize on-site safety diagnosis for projects with the poorest safety records	Aug 2011	SM 4.4	Contractors who have had repeated incidents
Generate advocacy action on construction health and safety self-management	Sep 2011	ST 3.1	
Training for crane operators	Oct 2011	ST 3.1ii	Incident involving crane collapse occurred in August 2011
Strengthen weekly joint inspections	Nov 2011	SI 2.3	
Organize accident prevention forum	Dec 2011	SM 4.1	
Organize demonstrations for pier table bracket damage	Dec 2011	ST 3.2	

countermeasures, a seminar is also organized. This aims to ensure that related construction personnel are immediately aware of the risk of similar incidents and can jointly develop countermeasures.

3 PROJECT PROFILE

The widening Project of the Wugu-Yangmei section of National Freeway No.1, approximately 40 km in length, starts from the Xizhi-Wugu viaduct and ends north of the Yangmei toll station. A separate viaduct is adopted as the main construction type of the Widening Project. The section between Taishan and Zhongli intermediate transfer connections is designed as a 3-lane viaduct while the other section is a 2-lane bridge according to the traffic demand. Taishan intermediate transfer connection and Zhongli intermediate transfer connection serve as traffic exchanges between the ground section of National Freeway No.1 and the elevated bridge of the Widening Project. One interchange is also established to connect National Freeway No.2 to access Taoyuan International Airport. The total budget of the project is about USD 2.72 billion, making it Taiwan's highest per-unit cost road to date (Taiwan Area National Expressway Engineering Bureau, 2014).

The project was taken over by the Taiwan Area National Expressway Engineering Bureau on February 4, 2009, and opened to traffic in April 2013 (Taiwan Area National Expressway Engineering Bureau, 2014). The entire route is divided into 12 bids, and it is elevated, making construction work very dangerous. The elevated road was very high, with many of the columns more than 30 meters in height. In addition, construction took place very close to the existing Freeway No. 1 and neighboring houses. During the pre-construction phase, it was needed to demolish houses and acquire land, lay out a large number of construction roads, and build gantries and other construction facilities. It was necessary to make multiple crossings of freeways, expressways, and local roads during construction. This article examines the influence of several management strategies for occupational safety on public construction projects in Taiwan using this project for a single case study.

4 CONCLUSIONS

In this article, we examine management strategies for occupational safety in the construction industry. Several strategies were conducted to improve the safety performance of the widening project of the Wugu-Yangmei section of National Freeway No.1. Subsequently a questionnaire method will be used to analyze the advantages and disadvantages of different strategies in order to understand the effectiveness of each strategy. The final part of the study will use the results of questionnaire survey to provide a direction for more effective safety management strategies.

REFERENCES

Abdelhamid, T.S., Everett, J.G., 2000. Identifying root causes of construction accidents. Journal of Construction Engineering and Management 126, 52–60.

Aksorn,T., Hadikusumo, B.H.W., 2008. Critical success factors influencing safety program performance in Thai construction projects. Safety Science 46, 709–727.

Chi, C.F., Chang, T.C., Ting, H.I., 2005. Accident patterns and prevention measures for fatal occupational falls in the construction industry. Applied Ergonomics 36, 391–400.

Eisenbraun S. (Eds.), 2013. "Taiwan 2013 human rights report", Country reports on human rights practices for 2013, United States Department of State-Bureau of Democracy, Human Rights and Labor, Washington, DC.

El-Mashaleh, M.S., Rababeh, S.M., Hyari, K.H., 2010. Utilizing data envelopment analysis to benchmark safety performance of construction contractors. International Journal of Project Management 28, 61–67.

Harper, R.S., Koehn, E., 1998. Managing industrial construction safety in Southeast Texas. Journal of Construction Engineering and Management 124, 452–457.

Hinze, J., Gambatese, J., 2003. Factors that influence safety performance of specialty contractors. Journal of Construction Engineering and Management 129, 159–164.

Hinze, J., Wilson, G., 2000. Moving toward a zero injury objective. Journal of Construction Engineering and Management 126, 399–403.

Huang, X., Hinze J., 2003. Analysis of construction worker fall accidents. Journal of Construction Engineering and Management 129, 262–271.

International labour office, 2006. "Strategies and practice for labour inspection", Governing Body, 297th Session, Geneva.

Jaselskis, E.J., Anderson, S.D., Russell, J.S., 1996. Strategies for achieving excellence in construction safety performance. Journal of Construction Engineering and Management 122, 61–70.

Sawacha, E., Naoum, S., Fong, D., 1999. Factors affecting safety performance on construction sites. International Journal of Project Management 17, 309–315.

Taiwan Area National Expressway Engineering Bureau (TANEEB), Available at: http://gip.taneeb.gov.tw/ct.asp?xItem = 23777&ctNode = 3887 (5/12/2015).

Tam, C.M., Fung I.W.H., 1998. Effectiveness of safety management strategies on safety performance in Hong Kong. Construction Management and Economics 16, 49–55.

Toole, T.M., 2002. Construction Site Safety Roles. Journal of Construction Engineering and Management 128, 203–210.

Tzeng, W.L., Li, J.C.C., Chang, T.Y., 2006. A study on the effectiveness of the most advantageous tendering method in the public works of Taiwan. International Journal of Project Management 24, 431–437.

Wang, W.C., Wang, H.H., Lai, Y.T., Li J.C. C, 2006. Unit-price-based model for evaluating competitive bids. International Journal of Project Management 24, 156–166.

Weil, D. (2008). "A strategic approach to labour inspection." International Labour Review, 147, 349–375.

Civil, Architecture and Environmental Engineering – Kao & Sung (Eds)
© 2017 Taylor & Francis Group, ISBN 978-1-138-02985-9

The influence of the horizontal load on the stress–strain relationship of an asphalt layer

Yonghong Wang
School of Transportation, Wuhan University of Technology, Wuhan, China

Ximing Tan
CCCC Second Highway Consultants Co. Ltd., China

Xiongjun He
School of Transportation, Wuhan University of Technology, Wuhan, China

Yu Wang
Xingda Luqiao Hubei Company, China

Cheng Wang
CCCC Investment Co. Ltd., China

Tao Liu
Xianning Anda Highway Maintenance Engineering Co. Ltd., China

ABSTRACT: With the popularization and development of the highway network, many heavy duty vehicles run on every grade highways. The continuous braking, starting and steering of the vehicle in the position of the intersection of the highway bring about common problems such as the passage of the asphalt pavement and the cracking of the vehicle (Wang Jun, 2010 & Roque R, 2000). In addition, the thawing environment such as high temperature in summer or winter freezing will reduce the shear resistance of the asphalt mixture, resulting in track, elapse, crowd etc. These early problems have seriously affected vehicle comfort and have hindered the normal operation of the road, and have caused huge losses to the national economy. Therefore, it is significant to master the influence law of the asphalt pavement structure under horizontal load conditions.

1 INTRODUCTION

Because of the popularization and development of the highway network, more and more duty vehicles are on every grade highways. High temperatures in summer or winter freezing temperatures will reduce the shear resistance of the asphalt mixture. It also will result in track, elapse, crowd etc. So, it is important to master the influence law of the asphalt pavement structure under horizontal load conditions.

2 THE ESTABLISHMENT OF THE FINITE ELEMENT MODEL

2.1 *The force characteristic analysis of the asphalt layer when the horizontal force is acting on the asphalt surface layer*

The main design method of the asphalt pavement is to place the wheel load as equivalent with double circular vertical loads for the structure design on the pavement thickness. This method only considers the vertical force of the wheel load without considering the horizontal force and vertical force on the road. But the asphalt pavement is not only influenced by the vehicle, but also will be influenced by a horizontal force, such as friction etc. Therefore, when the vehicle is on the road surface, the vertical force and the horizontal force of the vehicle should be considered together.

2.2 *Establishing the finite element model structure*

When the asphalt pavement is loaded by using a car, the rear wheel of the vehicle will generate force that includes the vertical force and the horizontal tangential force. The horizontal tangential force of the asphalt pavement is mainly caused by the friction between the tire and the road surface. The pavement structure and parameters' values are listed in Table 1.

Table 1. Pavement structure and parameters.

Structure layer	Thickness (cm)	Modulus (MPa)	Poisson's ratio
Asphalt layer	4	1200	0.25
PMC layer	28	21000	0.18
Subgrade	500	40	0.35

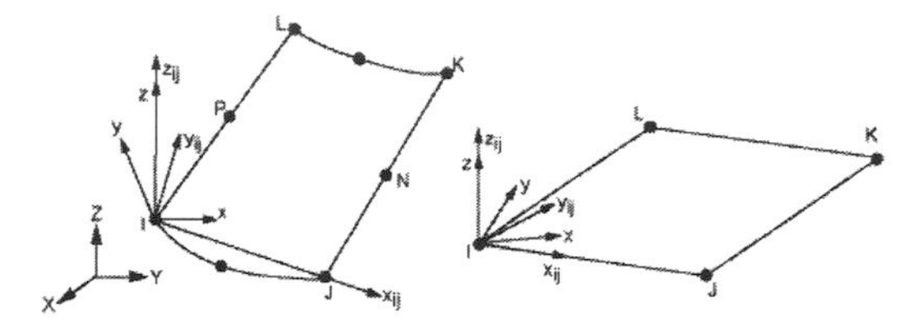

Figure 1. Schematic diagram of the surface effect element SURF154 geometry.

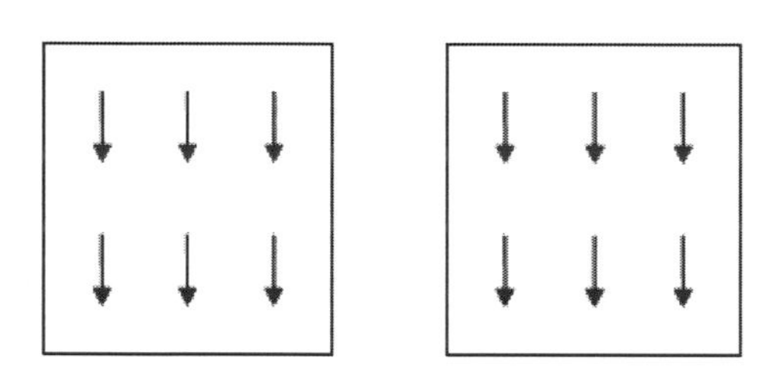

Figure 2. Schematic diagram of horizontal force loading along the driving direction.

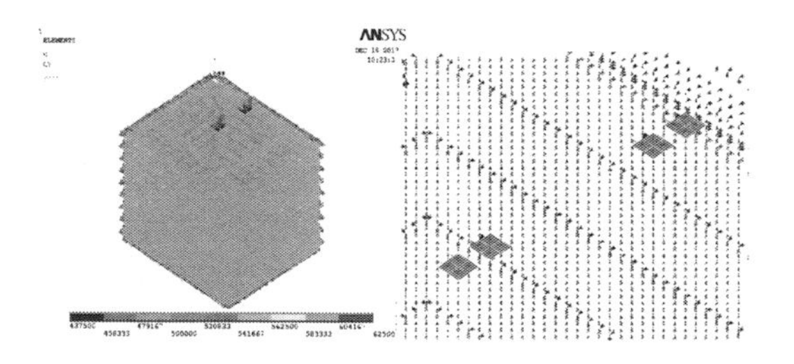

Figure 3. The loading pattern of the horizontal force in finite element calculation.

The main method is to use the ANSYS finite element software to analyze the loading level of the pavement structure and the selection of the asphalt surface effect unit is SURF154 (Jiang Yi, 2009). Asphalt surface effect element SURF154 can simulate the vertical force and horizontal force together on the 3D cell surface. The SURF154 geometry of the surface effect unit is shown in Figure 1. At the same time, the vertical load and horizontal load are added to the SURF154 unit to simulate the vertical force and horizontal force generated by the vehicle load on the pavement structure. The schematic diagram of horizontal force loading along the driving direction is shown in Figure 2. In the finite element calculation, the loading diagram of the horizontal force along the driving direction is shown in Figure 3.

3 CONSIDERING THE EFFECT OF THE HORIZONTAL LOAD ON STRESS AND STRAIN OF THE ASPHALT LAYER

3.1 *The effect of the bonding state between the asphalt layer and PMC layer on the stress and strain of the asphalt layer*

The effects about the incorporation conditions on the maximum shear stress, shear strain, and tensile stress of the asphalt layer are listed in Tables 2–4.

As can be seen from the Tables 2–4, it indicates that the absolute smooth state of the stress and strain is greater than the completely continuous state. It can be explained that the asphalt layer and PMC layer between the continuous ratio of the pavement structure under the load is more favorable. Taking into account the actual situation for the large gap between the PMC and asphalt layers, the surface is very rough, and so the two layer interface is nearly continuous and the calculation of the layer between the combined states are set to completely continuous states.

Table 2. The effects about the incorporation conditions on the maximum shear stress.

Incorporation conditions	Thickness (cm)	Maximum shear stress (MPa)
Absolutely smooth	0	0.437
	2	0.461
	4	0.452
Completely continuous	0	0.385
	2	0.392
	4	0.304

Table 3. The effects about the incorporation conditions on the shear strain.

Incorporation conditions	Thickness (cm)	Maximum shear strain
Absolutely smooth	0	0.450
	2	0.480
	4	0.471
Completely continuous	0	0.450
	2	0.480
	4	0.471

Table 4. The effects about the incorporation conditions on the tensile stress.

Incorporation conditions	Thickness (cm)	Tensile stress (MPa)
Absolutely smooth	0	0.215
	2	0.237
	4	0.284
Completely continuous	0	0.080
	2	0.091
	4	0.104

3.2 *Effect of transverse force on the stress and strain of the asphalt layer*

When the vehicle is in an emergency braking, climbing or turning state on the asphalt pavement, the horizontal force coefficient is about 0.5, and its maximum value is approximately equal to the friction coefficient of the road surface. Therefore, it is important to analyze the impact of the transverse force on the stress and strain of the asphalt layer. The analysis of the transverse force coefficient is 0.1, 0.3, 0.5, and 0.7 and the asphalt thickness is 8 cm.

With an increase in the horizontal force coefficient, the maximum shear stress increases to 41.5% and 32.8% when the horizontal force coefficient increases from 0.1 to 0.7 on the road surface and the depth is 2 cm. It will be normal running of the vehicle when the horizontal force coefficient is about 0.2, and so the level of load generated by the normal running of the vehicle on the road affects the road surface to a little extent. When the car is in the uphill, emergency braking or turning will have a greater impact on the road.

When the horizontal force coefficient is less, the horizontal shear stress becomes smaller, but the value and the maximum shear stress become larger. When the horizontal force coefficient is large, the horizontal shear stress is large. Its value is very close to the maximum shear stress value. The larger the lateral force coefficient, the more prone it is to horizontal shear damage. Figure 4 shows the tensile stress of the asphalt surface.

As can be seen from Figure 4, the maximum tensile stress of the asphalt surface appears in front of the wheel, which is due to the action of the backward horizontal force. With an increase in the transverse force coefficient, the tensile stress also exhibits an obvious increase. When the transverse force coefficient increased from 0.1 to 0.7, the maximum tensile stress increased by more than two times. The horizontal displacement of the asphalt layer also increases with an increase in the lateral force coefficient, and it will have a significant impact on the horizontal displacement when the

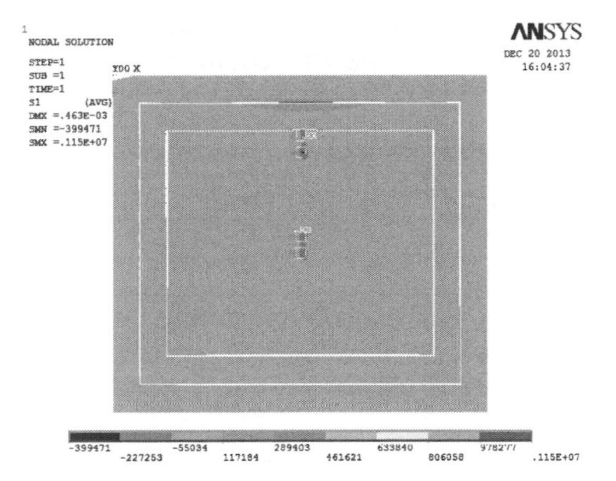

Figure 4. Tensile stress of the asphalt surface.

lateral force coefficient is larger. The production of horizontal displacement can induce the disease of the asphalt layer, and so it is more likely to occur in an area in which the lateral force coefficient of the vehicle load is large in the intersection.

3.3 *Influence of the asphalt layer thickness on the stress and strain of the asphalt layer in summer and winter*

An asphalt mixture is regarded as a special temperature sensitive material, in which the mechanical index and pavement performance of the asphalt mixture will also undergo a significant change with the change of temperature of the external environment (Chen Fengfeng, 2007 & Ma Xin, 2008). In a high temperature environment such as summer, the asphalt modulus becomes small and the mixture shows a softening flow in the high temperature. And so, it will reduce its own shear performance. If the traffic in the asphalt pavement shear performance is less, the asphalt pavement will form the possibility of permanent deformation failure which is quite big. In winter, the stress relaxation performance of asphalt is decreased, the modulus and brittleness increase, and low temperature cracking is the main failure mode. Therefore, it is necessary to study the temperature distribution of the asphalt pavement in the pavement structure when the asphalt pavement is driving at high temperature. The modulus of asphalt in summer is 450 MPa and the modulus of asphalt in winter is 3400 MPa. The force coefficient can be assumed as 0.3 when calculating the transverse.

The vertical shear stress of the asphalt layer will increase with an increase in the depth; but when the depth is more than 6 cm, the vertical shear stress (shear strain) will no longer increase, and slowly decrease. The increase of the asphalt layer thickness will increase the vertical shear stress (shear strain) in the asphalt layer. With an increase in the depth of the road surface, the maximum tensile stress gradually decreases and the amplitude increases with an increase in the asphalt layer thickness, and the maximum tensile stress of the pavement increases with an increase in the thickness.

According to the chart, the maximum shear stress and the maximum tensile stress in summer and winter are not quite different. When the asphalt layer is less than 6 cm, the maximum shear stress is larger in summer than in winter, but when the asphalt layer thickness is greater than 6 cm, the stress value of summer and winter is basically the same. At the same time, the maximum tensile stress values of summer and winter are also different and the value of tensile stress in summer is slightly larger than that in winter.

It can be obtained through the contrast analysis in the winter and summer, in which low temperature and high temperature asphalt layers are under the action of the vertical force and the horizontal force is nearly the same. The shear capacity of the asphalt mixture is greatly lower under the high temperature than the low temperature conditions. It is very important to improve the high temperature performance of the asphalt concrete mixture on resistance of the asphalt pavement shear failure.

4 REASONABLE TYPE AND THICKNESS RANGE OF ASPHALT LAYERS

The effect of the horizontal force has a certain effect on the stress of the asphalt layer. By the analysis of the different layers of the state, it can be seen that the maximum shear stress of the PMC layer changes little when the asphalt layer is under the absolute smooth and completely continuous conditions, but the maximum tensile stress of the absolutely smooth state is two times larger than the completely continuous state.

The different asphalt layer thickness stress analysis results show that the maximum shear stress (shear strain) location is the asphalt surface layer. When the depth is more than 2 cm, the shear stress (shear strain) will decrease with an increase in the depth. It is important to improve the shear performance above the asphalt layer to prevent the shear failure. At the same time, the change of the asphalt layer thickness has little effect on the internal shear stress (shear strain). When the thickness of the asphalt layer is more than 6 cm, the maximum shear stress can be increased by an increase in thickness. All of these indicate that the location of the asphalt near the pavement is the most unfavorable position of the asphalt layer in PMCCP. It is important to improve the high temperature performance of the asphalt concrete mixture on the resistance of the asphalt pavement shear failure.

Under different vehicle driving conditions, the lateral force of the road surface is different, and the lateral force size is characterized by the lateral force coefficient (Ahmed Abdelazim Eltahan, 1996 & Yang Xueliang, 2007). The asphalt layer stress changes in the different lateral force coefficients show that the horizontal force coefficient remained at a low level, the shear stress (strain), tensile index, and horizontal displacement will not produce significant changes with horizontal force coefficient changes. When the value of the transverse force coefficient is larger, the magnitude of the reduction is more obvious, while the normal travel of the car or the prediction of the lateral force coefficient of the brake is generally no more than 0.3. But on the ramp or city road intersection, the horizontal force of the vehicle caused is large. And so, necessary measures should be taken to improve the performance of the asphalt layer to resist the lateral force caused by shear stress and tensile stress. From the above analysis, it can be observed that the reasonable thickness of the asphalt layer that is based on the shear resistance can be recommended as 4 to 8 cm.

5 CONCLUSIONS

1. Establish the finite element model of the pavement structure and determine the parameters of the pavement structure layer, element type, and calculation parameters. The direction of the horizontal force is parallel to the driving direction, which is used to simulate the horizontal force generated by the vehicle acceleration, braking, or uphill.
2. The influence of the horizontal stress on the stress and strain of the asphalt layer when the PMC and the asphalt surface are in different contact states (i.e., completely continuous and absolutely smooth). Different depths of the absolute smooth state shear stress tensile stress is greater than that of the completely continuous state, which can be used to explain that between the asphalt layer and the PMC layer such that the continuous state is more favorable than the smooth state. The pavement structure will generate a stress response when the asphalt mixture is considered by using the horizontal force in the summer and winter.
3. The different thicknesses of the asphalt layer on the asphalt layer shear stress (shear strain) and tensile stress have little effect on the asphalt layer. It set up a thick layer of asphalt that may lead to internal shear stress (shear strain) that is increased by a certain extent. It can be obtained from the analysis of the influence of vehicle load on the asphalt pavement stress that the asphalt concrete pavement should not set the asphalt layer thickness.

REFERENCES

Ahmed Abdelazim Eltahan. A Mechanistic-Empirical Approach to The Development of A Stochastic Reflection Crack Prediction Model for Flexible Pavement Overlays[D]. Texas A&M University, PHD, December 1996.

Chen Fengfeng, Huang Xiaoming. Mechanical response of rigid base asphalt pavement under heavy load [J]. Highway traffic science and technology, 2007, 24 (6): 41–45.

Jiang Yi. Shop mechanics analysis and crack control measures of [D]. and the old asphalt pavement Changsha: Hunan University, 2009.

Ma Xin, Guo Zhongyin, Yang Group. Mechanical analysis of asphalt pavement damage mechanism considering horizontal loads [J]. Highway engineering, 2008.33 (5): 74–76.

Roque R. Evaluating measured of tire contact stresses predict pavement response and performance [J]. Transportation Research Record. 2000, 1716:73–81.

Shen Jinan. Road performance of asphalt and asphalt mixture[M]. Beijing: People's Communications Press. 2001.

Wang Jun. Finite element analysis of the structural finite element analysis of cement concrete pavement [J]. North traffic. 2010. 12–15.

Yang Xueliang, Liu Boying. Finite element analysis of temperature field and structure coupling of asphalt pavement [J]. Highway traffic science and technology, 2007, 23 (11): 1–4.

Civil, Architecture and Environmental Engineering – Kao & Sung (Eds)
© 2017 Taylor & Francis Group, ISBN 978-1-138-02985-9

A study on the protection type of flow-guided porthole dies with twin cavities for semi-hollow Al-profiles

Rurong Deng & Peng Yun
Guangzhou Vocational College of Science and Technology, Guangzhou, China

ABSTRACT: Through an actual example, a new structure with twin cavities in a die for semi-hollow Al-profiles was introduced. A structure named protection type of flow-guided porthole die to solve the cantilever strength is presented. And the selection and the optimization of structure parameters are described in detail. The design of the portholes, the determination of the bridge, and the selection of bearing belt are included. It is shown that the structure with twin cavities for semi-hollow profiles is effective, and the structure can greatly improve the production efficiency and reduce the cost. It is worth promoting. The purpose of this work is to provide reliable data and reference for further research and development of this technology on the extrusion die with multi-cavities in a die.

1 INTRODUCTION

Semi-hollow Al-profiles are quite common in the production of aluminum profiles extrusion dies. The key to the production of a semi-hollow section is the die. The die design of semi-hollow profiles is more difficult and complicated, and the main problem to be solved is the strength of the die. In order to solve the problem, domestic and foreign scholars, experts and engineering and technical personnel have carried out a wide range of exploration and research. In addition to the study of mold materials and heat treatment, more studies are required to explore the die structure, by changing the die structure and changing the stress condition of the cantilever to improve the die strength. This way is particularly effective. In recent years, with the rise of land, labor and energy costs, the overall cost of enterprises rose significantly. In such an economic environment, it is not realistic for enterprises to depend on increasing the number of equipment, the expansion of the plant or to human-intensive and other traditional investments to expand the production scale. The best way to improve productivity and reduce costs is to rely on technological innovation and the use of innovative ways to drive. The extrusion technology of multi-cavities in a die is one of the best ways and methods. The key factor in this extrusion technology is the die, especially for a wide variety of semi-hollow sections, because this type of material usually forms a relatively small section. Therefore, it is more practical to use a model of multi-cavities extrusion. Through an actual example, a new structure named protection type of flow-guided porthole with twin cavities in a die for the semi-hollow Al-profiles will be introduced for peer reference. And to provide reference data for the development of more and more effective die structures for similar profiles.

2 DETERMINATION OF DIE STRUCTURE PARAMETERS

2.1 *The formation of structures*

The semi-hollow sections of the flow guide protection type die structures are formed, that is, for some of the profile sections with a length to width ratio of more than three of the narrow and long sections, according to the structural characteristics of the profile. The porthole bridges are used to cover and protect the cantilever, and there is a certain stress space between the bridge and cantilever. In the extrusion process, even if the porthole bridge bends downward flexibly under force, it would not touch the cantilever instead of bearing the force, and the cantilever will not bear the extruded metal's positive pressure. In the die, the cantilever can be considered as a cantilever beam, and the bridge is a simple supported beam under a uniform load and the strength of the simple supported beam is greatly improved under the same loading conditions. Obviously, the change or shift of the force conditions can improve the strength of the mold. And so, as long as the bridge has sufficient strength, the strength of the cantilever will be ensured. At the same time, as the cantilever beam, this bending moment that the end of the beam exerts on the root is the biggest, and so in order to improve the stress state and metal supply of the cantilever end, the guide grooves must be set in the place of the corresponding bridge, and the end of the cantilever is completely under the protection of the bridge; through the bridge, the metal will perform welding and diversion actions into the sites, thereby reducing the extrusion force of the end, so as to improve the strength of the cantilever.

2.2 *The arrangement of die holes*

A typical section of the semi-hollow profile is shown in Figure 1.

The maximum tongue ratio of the section was 6.3, and the ratio of the width to the length was 3.1. In the layout of the die with twin cavities, the die strength and smooth extrusion must be considered, but more important is taking into account the full use of the potential of the extrusion machine, thereby making it conducive to the adjustment of the metal flow and saving costs. The layout of the die is shown in Figure 2.

The layout shown in Figure 2(a) is a symmetrical layout of the left and right, but its requirement of the size of the die is more larger, and the parts of the die hole that are close to the edge of the container and away from the extrusion center are not conducive to shape, so that the layout cannot fully play the capacity of the container. The scheme shown in Figure 2(b) is just the opposite; the lack is that the die holes are arranged up and down, and the

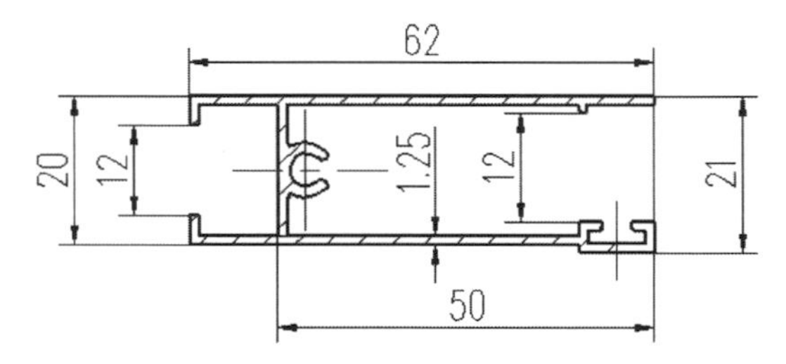

Figure 1. Schematic showing the signal of the semi-hollow section.

surface of the two profiles will come in contact in the extrusion. But according to the surface quality requirements and the final selection of the profile surface treatment and comprehensive consideration, the scheme shown in Figure 2(b) will be more appropriate. By this way shown in Figure 2(b), if changing the site of the positioning pin in the die, the way of a symmetrical layout of the left and right can be realized similarly; the way shown in Figure 2(b) is similar to the symmetry of the left and right, and it is more conducive to solve the balanced distribution of the metal in the design.

2.3 *The design of feeder holes and the bridge*

2.3.1 *The design of feeder holes*

On the basis of the layout shown in Figure 2(b), the arrangement of portholes is carried out, which mainly includes the determination of the area and size of the portholes, the feeder ratio, and the area relationship between the portholes. According to the experience, the preliminary programs should be completed. With the help of CAD software, a mature arrangement of portholes by experience will be selected to establish a three-dimensional model under the environment of UG, and the models must be saved as a fixed pattern into the extrusion simulation software. Simulation, calculation, and observation were carried out, and the results were compared and analyzed, and combined with the experience to make correction, in particular, the effect of the communal porthole S1 on the metal flow and deformation, simulation and correction were repeated, until the final determination of the best program was completed. The latter program is shown in Figure 3.

Its main parameters are as follows:

1. The feeder ratio is 16.
2. The bridge size is of width 22 mm and thickness 85 mm.
3. The area relationship between the porthole S1 and S2 is shown in Equation 1.

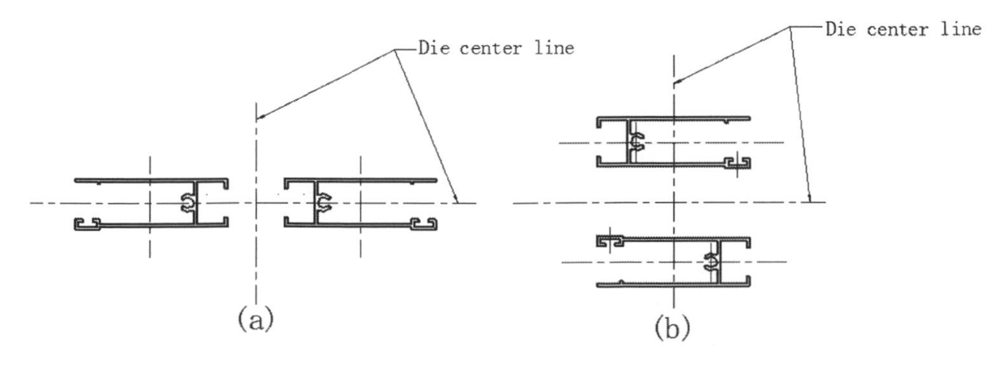

Figure 2. Schematic of the signal of die holes arrangement.

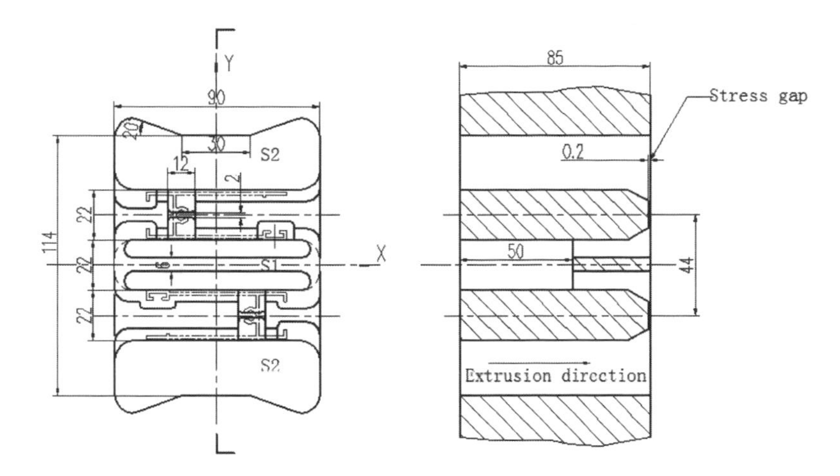

Figure 3. Schematic of the signal of portholes arrangement.

$$S1 = 0.8S2 \quad (1)$$

4. The result of the simulation shows that when the area relationship meets certain conditions as shown in Equation 2, it is easy to maintain consistency of the flow rate of the metal. The deformation degree of the cantilever can be almost neglected.

$$S1 = (0.70 \sim 0.85)S2 \quad (2)$$

In addition, the common porthole is the form that one divides into two; its advantage is that it is easy to adjust the distribution of the metal; more important is that the metal flow of two portholes does not affect each other ahead of the allocation. If the separation that one porthole divides into two is not carried out, on one hand, it will form a rigid zone and dead zone of the metal flow in the central part of the common porthole, which will make the corresponding profile surface appear as coarse grains or a bright band, which will affect the surface quality and performance of the profile. On the other hand, the sensitivity to the manufacturing error will increase, which will affect the synchronization of the extrusion. But the length of the original flow of the metal into the first porthole named as first stage depth can be obtained by simulation, when the metal reaches the depth. It is divided into two strands of the metal into two port holes respectively, and from the simulation, it is found that when the depth of the first stage is greater than 25 mm, the depth will have almost no effect on the metal flow.

2.3.2 *The structure of the bridge*

It is mainly used to determine the appropriate distance from the bottom edge of the bridge to the die hole and the size of flow guiding groove of the corresponding end of the cantilever. The structure of the bridge is shown in Figure 4.

From the simulation, it is found that the distance of 3 mm–5 mm from the bottom edge of the bridge to the die hole is more suitable. If the distance is too small, the metal flow becomes complicated, the extrusion force will increase, and if the distance is too big, the deformation of the cantilever will increase.

2.3.3 *The stress gap*

In order to ensure that the bridge does not touch the cantilever in the extrusion process, the stress gap should be set up to prevent the bridge from touching the cantilever, the gap value of the experience can be taken from 0.15 mm to 0.25 mm.

2.3.4 *The selection of bearing*

In the die with twin cavities, the choice of the bearing can be selected according to the ordinary single cavity. When one is selected, the other one is the same as it.

For the flow guide protection type die structure, because the central part of the cantilever end is completely under the protection of the bridge, through the bridge, the metal will make a welding and diversion into the site, and therefore it is difficult for the metal to reach the site. Corresponding to the site, the length of the bearing is the shortest, and the choice of the bearing must be based on the point. The selection of the bearing based on experience can be modified through the simulation operation. The latter bearing is shown in Figure 5.

2.4 *The die structure composition*

The die is composed of the male die and the female die, as shown in Figure 6.

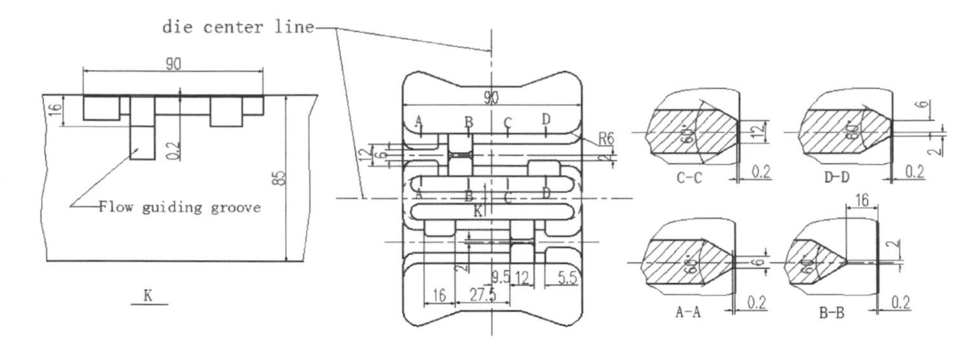

Figure 4. Schematic of the signal of the bridge structure.

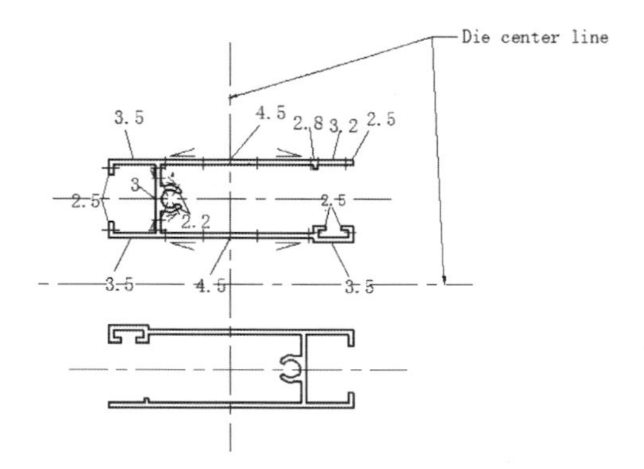

Figure 5. Schematic of the signal of the bearing.

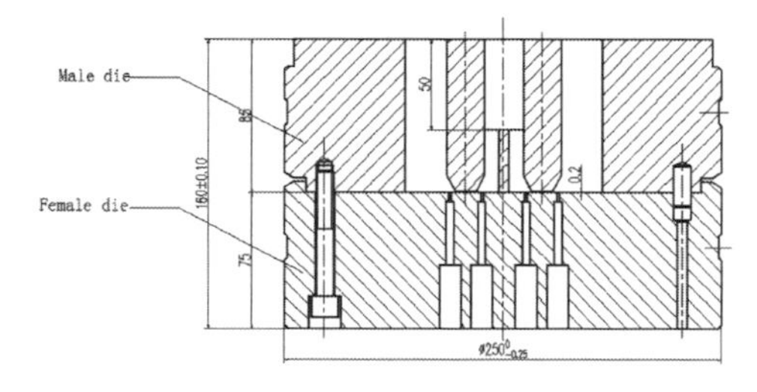

Figure 6. Schematic of the signal of the die structure composition.

Taking the cost of the die, extrusion coefficient and die strength into account, we chose an 18MN extrusion machine, whose container has an inner diameter of 185 mm. After a calculation, the extrusion coefficient is determined as 63, the die size of the selection is that the diameter is 250 mm and its thickness is 160 mm.

3 CONCLUSIONS

When the design of porthole die, computer simulation, revise of experience and test of extrusion and use tracking were prepared, the results show that the die is successful in one time. And with repeated use after being nitrided, the die extrusion production life reached 24.2 tons. Thus, it can be shown that the new structure named flow guided protection type with twin cavities in a die is effective, and it can improve production efficiency and reduce cost; at the same time, we can see that it is a structure worth promoting. This provides reliable practical data for the further exploration and research of multi-cavities extrusion die, in particular, for the semi-hollow section and accumulates valuable experience.

REFERENCES

Deng Rurong, Huang, Xuemei. Design of the extrusion die of semi-hollow aluminum profile [J]. Light Alloy Fabrication Technology, 2015, 43(4):51–54.

Hu Dongpo, Wang Leigang, Huang Yao. Steady-state simulation and die improvement on the extrusion of prolate aluminum profile [J]. Forging & Stamping Technology, 2015, 40 (4): 69–73.

Kuang Weihua, Chen Biaobiao. Research on design and structure of extrusion die for cantilever aluminum profile [J]. Hot Working Technology, 2013, 42(21): 136–138.

Liu Jingan. Die design, manufacture, application and maintains for aluminum profiles extrusion [M] Beijing: Metallurgical Press, 1999: 181–183.

Sun Xuemei, Zhao Guoqun. Fake porthole extrusion die structure design and strength analysis for cantilever aluminum alloy profiles [J]. Journal of Mechanical Engineering, 2013, 49 (24): 39–44.

Wang Liwei. Optimization design of extrusion die for the bigger slenderness ratio half hollow aluminum profile [J]. Die and Mould Manufacture, 2011(4): 61–64.

Xie Jianxie, Liu Jingan, Die design, manufacture, application and maintains for aluminum profiles extrusion [M]. Beijing: Metallurgical Press, 2012: 133–138.

Xu Yongli, Huang Shuangjian, Pang Zugao, et al. Failure analysis of extrusion die and optimization of heat treatment process for aluminum alloy circular tube [J]. Forging & Stamping Technology, 2015, 40(2): 116–122.

Yu Mingtao, Li Fuguo. Simulation extrusion process of the sketch hollow aluminum profile based on infinite volume method [J]. Die and Technology, 2008(4): 40–43.

Civil, Architecture and Environmental Engineering – Kao & Sung (Eds)
© 2017 Taylor & Francis Group, ISBN 978-1-138-02985-9

A study on covering and protection type porthole dies for semi-hollow Al-profiles

Rurong Deng & Peng Yun
Guangzhou Vocational College of Science and Technology, Guangzhou, China

ABSTRACT: A new structure named covering type for the semi-hollow Al-profiles is presented. Through a common actual case, the determination of structure parameters is introduced in detail. The arrangement of portholes, the structure design of the chamber, and the selection of the bearing are included. The method of checking the die strength is introduced. According to the extrusion results, the structure of the traditional solid die and the covering type structure are compared. The characteristics of the latter structure are simple and easy to process. The practical application shows that the new die structure can enhance the die life, improve the production efficiency, and reduce the cost. High precision and the surface brightness of the profiles were obtained. The structure is worth promoting. The aim is to provide reliable data and reference for further research and development on the structure of semi-hollow Al-profiles.

1 INTRODUCTION

With the development of modern manufacturing technology as well as the understanding and research of aluminum alloys, aluminum alloy profiles have been widely used, and the market demand is huge. In the huge market demand, the proportion of the half hollow section is quite common. And in the semi-hollow section of the production, the mold is the key element. This is because the mold design of the semi-hollow profile is of a higher grade, and the strength of the mold is the key. Therefore, vigorously developing a suitable semi-hollow profile extrusion die structure has a very important significance. In this paper, through the common actual examples, a covering type porthole die structure is introduced specially for semi-hollow aluminum profiles to offer reference.

2 THE COVERING AND PROTECTION TYPE PORTHOLE DIE STRUCTURE

The covering and protection structure is a kind of structure which is used with a porthole structure with covering and protection instead of the traditional solid structure for semi-hollow profiles. At present, the porthole structure commonly used in the semi-hollow section is the cutting type porthole die, but the structure can easily produce raised wire and convex flanges on the surface of profiles. The covering and protection type porthole die structure is constructed in such a manner that the center part of the male die is used as a cover or shelter to protect the cantilever, so that the cantilever does not bear the positive pressure of the metal during the extrusion process, and the cantilever is completely protected. And in the female die, the cantilever must protrude or raise upward, but it has no contact with the male die, and there is a gap named stress gap between the top of the bulge and male die; as a result of the gap, even if the male die is bent downward by force in the extrusion process, the male die cannot touch the cantilever, and the cantilever will be protected effectively, which improves the stress state of the cantilever and reduces the force area of the cantilever, so as to greatly improve the strength of the cantilever and the die. And the porthole die structure is used to make full use of the bridge as the protective shield of the cantilever, avoid the cantilever and its end in particular, thereby bearing directly the positive pressure of the metal extrusion, so as to improve the strength of the die.

3 DETERMINATION OF DIE STRUCTURE PARAMETERS

3.1 *The design of portholes and the bridge*

The design of the portholes includes the determination of the feeder ratio, the size of the port holes, and the structure of the bridge. This is the key. The section shown in Figure 1 is a common section of civil building doors and windows, which is a typical semi-hollow section.

The purpose of using the new structure is to protect the cantilever from the direct impact of the metal. Therefore, it is important to avoid the impact of the end part of the cantilever directly to withstand the impact of the metal, as far as possible by the use of the bridge for shelter. According to the characteristics of the profile shown in Figure 1, it is more suitable to use two portholes and a bridge, as shown in Figure 2. And so, one can do what the cantilever is arranged for under the protection of the porthole bridge, and the use of the structure and the area of portholes can be taken larger to increase the feeder ratio, which can reduce extrusion pressure, lighten the cantilever force, and improve the cantilever strength.

Because the cantilever must protrude or raise upward, the structure of the bridge should be designed according to the protruding boundary dimension of the corresponding cantilever. The structure of the bridge is shown in Figure 3.

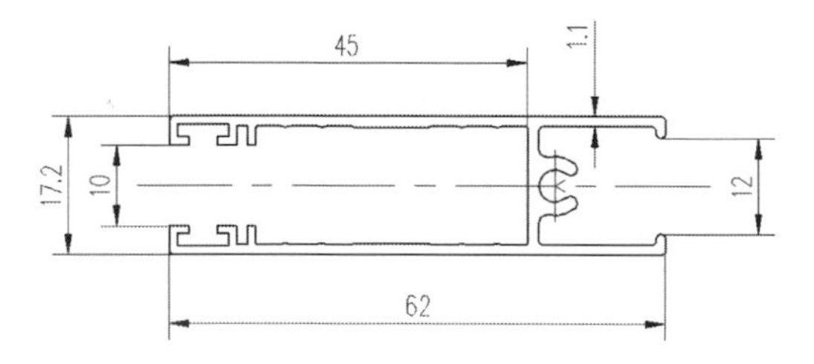

Figure 1. Schematic of the signal of a typical semi-hollow section.

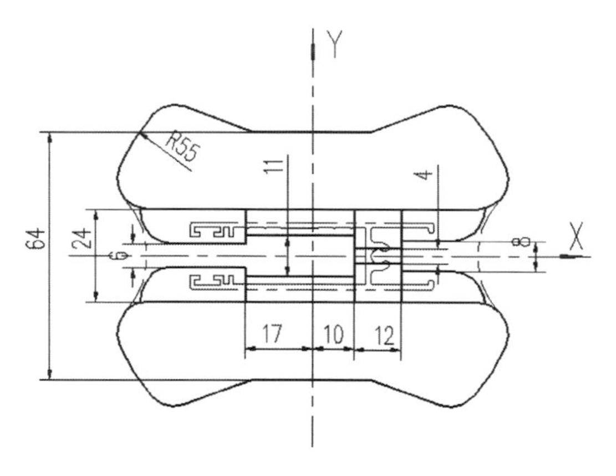

Figure 2. Schematic of the signal of the portholes arrangement.

3.2 *The welding chamber and the bearing*

The shape of the welding chamber is determined in terms of the outer contour of the portholes. The key is to determine the size of the raised boundary of the cantilever, which is the key to the structure of the covering and protection type. The size relation between the raised part and the die hole of the cantilever is shown in Figure 4.

The principle is that the distance from the edge of the protruding part to the edge of the die hole of the cantilever can be determined to be in the range from 2 mm to 3 mm. At the same time, with a stress gap between the top of the raised part of the cantilever and the bottom of the bridge, for ease of processing, the gap is built by cutting the top of the cantilever, the experience value shows that the gap is a range from 0.5 mm to 1.2 mm, the bridge does not produce a force to a cantilever, and so the value of the gap is taken as 1.0 mm. The welding chamber structure and bearing are shown in Figure 5.

3.3 *The die structure composition*

The die is composed of the male die and the female die, as shown in Figure 6.

Taking the cost of the die, extrusion coefficient, and die strength into account, we chose a 10MN extrusion machine, with a container with an inner diameter of 135 mm. After a calculation, the extrusion coefficient is determined to be 61.2, the feeder ratio is 20.1, and the die size of the selection is that the diameter is 200 mm and its thickness is 130 mm.

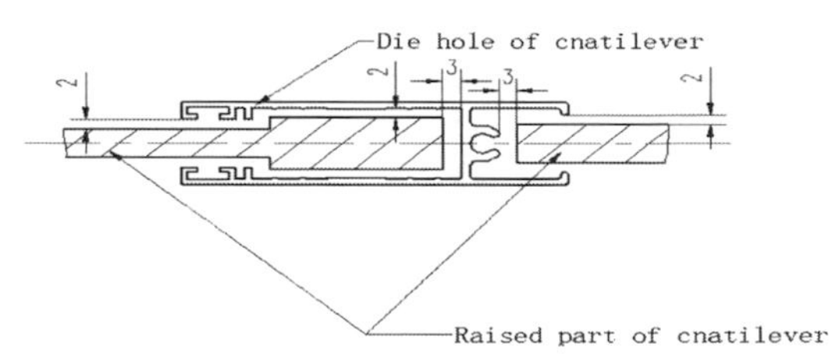

Figure 4. Schematic of the signal of the raised part of the cantilever.

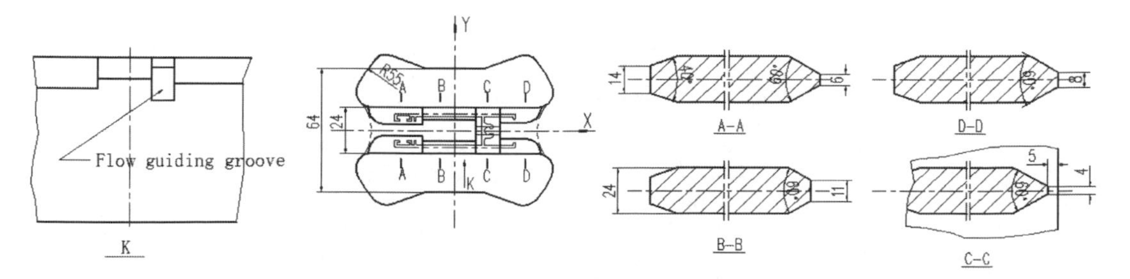

Figure 3. Schematic of the signal of the structure of the bridge.

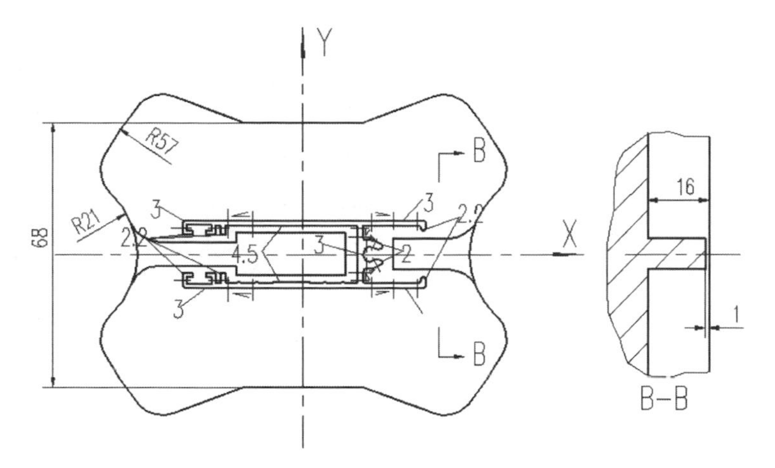

Figure 5. Schematic of the signal of the welding chamber and the bearing.

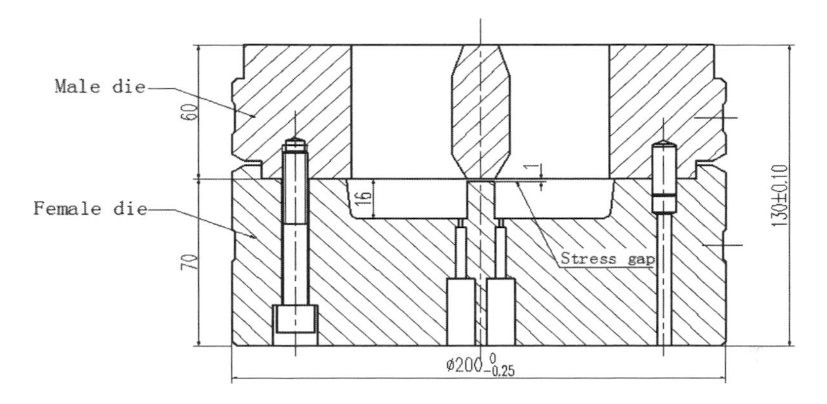

Figure 6. Schematic of the signal of the die structure composition.

4 CHECK THE STRENGTH OF THE DIE

Although the cantilever of the semi-hollow profile extrusion die is equivalent to a cantilever beam, because of the protection of the bridge, the bridge bears positive pressure directly instead of the cantilever during the extrusion process, and so checking the bridge strength instead of the cantilever is allowed. In the process of extrusion, the bridge is subjected to a cyclic stress, and so it is more effective to use an empirical formula to check the strength of the bridge in the actual process, as shown in Equation 1.

$$n = \frac{h \times b \times 2 \times [\sigma]}{s \times p} \tag{1}$$

where

1. n is the safety factor (the porthole die is greater than three and the solid die is greater than two).
2. *b is* the thickness of the bridge, mm.
3. s is the total area of the pressure on the bridge, mm^2.
4. p is the extrusion press, MPa.
5. $[\sigma]$ is the bending stress of the die material in the working area, and the value is 1150 Mpa.

The strength was calculated according to Equation 1.

$$n = \frac{60 \times 24 \times 2 \times 1150}{14 \times 80 \times 690} = 4.3$$

The results show that the mold has sufficient strength.

But in practice, the covering and protection type porthole die structure for the semi-hollow section material must be supported by using a special die support; otherwise, the die life will be greatly reduced. The thickness of the special die support should not be less than 70 mm.

5 EXTRUSION RESULTS' COMPARISON

According to the extrusion tracking for the profile shown in Figure 1, the traditional solid and the

Table 1. A comparison of the die structure and extrusion results.

Die structure	Die life (tons)	Wall thickness deviation	Extrusion marks and brightness	Dimension precision level	Change of opening size	Processing difficulty	Die material
Traditional solid	Less than 1.2	Obvious	Deep and rough	General	Big	Big	62 kg
Covering and protection type	26.7	Not obvious	Low and shining	High	Medium	Simple	35 kg

covering and protection type of the porthole die structures were used, and the results of the comparison are given in Table 1. The results show that the new structure has the obvious advantage.

6 CONCLUSIONS

When the test of extrusion and use tracking were completed, the results show that the die is successful in one time. And by the repeated use after being nitrided, the die extrusion production life reached 26.2 tons. Thus, it can be shown that the new structure named covering and protection type is effective and the new structure can improve production efficiency and reduce cost. The new structure is worth promoting. At the same time, we can see that the design of portholes, the bridge structure, the die welding chamber structure, and the selection of the bearing are key and important to the new structure.

REFERENCES

Deng Rurong, Huang, Xuemei. Design of the extrusion die of semi-hollow aluminum profile [J]. Light Alloy Fabrication Technology, 2015, 43(4): 51–54.

Hu Dongpo, Wang Leigang, Huang Yao. Steady-state simulation and die improvement on the extrusion of prolate aluminum profile [J]. Forging & Stamping Technology, 2015, 40(4): 69–73.

Kuang Weihua, Chen Biaobiao. Research on design and structure of extrusion die for cantilever aluminum profile [J]. Hot Working Technology, 2013, 42(21): 136–138.

Liu Jingan. Die design, manufacture, application and maintains for aluminum profiles extrusion [M] Beijing:Metallurgical Press, 1999: 181~183.

Sun Xuemei, Zhao Guoqun. Fake porthole extrusion die structure design and strength analysis for cantilever aluminum alloy profiles [J]. Journal of Mechanical Engineering, 2013, 49(24): 39~44.

Wang Liwei. Optimization design of extrusion die for the bigger slenderness ratio half hollow aluminum profile [J]. Die and Mould Manufacture, 2011(4): 61–64.

Xie Jianxie, Liu Jingan, Die design, manufacture, application and maintains for aluminum profiles extrusion [M]. Beijing: Metallurgical Press, 2012: 133~138.

Xu Yongli, Huang Shuangjian, Pang Zugao, et al. Failure analysis of extrusion die and optimization of heat treatment process for aluminum alloy circular tube [J]. Forging & Stamping Technology, 2015, 40(2): 116–122.

Yu Mingtao, LI Fuguo. Simulation extrusion process of the sketch hollow aluminum profile based on infinite volume method [J]. Die and Technology, 2008(4): 40–43.

Civil, Architecture and Environmental Engineering – Kao & Sung (Eds)
© 2017 Taylor & Francis Group, ISBN 978-1-138-02985-9

A study on protection type porthole dies of semi-hollow Al-profiles with four cavities

Rurong Deng & Peng Yun
Guangzhou Vocational College of Science and Technology, Guangzhou, China

ABSTRACT: A new porthole die structure named protection type with four cavities in a die for semi-hollow Al-profiles is presented in this article. Through a practical case, the determination of parameters for the new structure were described, mainly including the structure composition that it is made up of three parts in replacement of the traditional two piece structure; the design of portholes and the structure of chamber in the female die and the selection of the bearing were also included. The method of checking the die strength was introduced. And according to the extrusion results, the structure of the traditional porthole die in a common code, the cutting type of porthole die with single cavity and twin cavities and the protection type of four cavities were compared. It indicated that the characteristics of the latter structure with four cavities are simple and easy to process. The practical application shows that the new structure can greatly improve the efficiency of the extrusion machine and significantly prolong the die life. Furthermore, the high precision and the surface brightness of profiles are obtained. The structure is worth promoting. The purpose is to promote the technology, provide reliable reference data for further development of the porthole die with multi-cavities in a die for semi-hollow profiles.

1 INTRODUCTION

The porthole extrusion die with four cavities, in particular for the semi-hollow section, is a multi-cavity die extrusion technology. It is designed particularly for semi-hollow profiles, with four cavities in a die, where four identical profiles will be extruded synchronously at a time. It is a good method to rely on technological innovation to cope with the continued rise of cost of human resources and energy costs and the lack of land supply. Especially in an extrusion machine equipped with a trend of large scale, it can solve the problem since a small machine has too many orders to complete, and a large machine has not enough orders to work on. In the extrusion of multi-cavities, the die is the key. Research on this technology exhibits an ascending trend in our country; especially in recent years, the majority of experts, scholars, and engineering and technical personnel carry out lot of research and exploration. Through an actual case, a new kind of four cavities protection type structure will be introduced, which is specially designed for the semi-hollow section, as a reference to designers.

2 THE SECTION ANALYSIS AND DIE STRUCTURE

Figure 1 shows a typical civil profile used in building doors and windows.

The section of the profile is provided with a hollow part and a typical semi-hollow part, the area of the section is 291 mm^2, and the maximum tongue ratio of the semi-hollow part is 6.5.

Compared with the determination principle of the semi-hollow section, the ratio of the tongue is more than the maximum value allowed in the opening, and so the part can be determined as a semi-hollow section. In practice, if the die structure is still an ordinary structure, then the strength of the cantilever in the extrusion process is very dangerous and prone to fracture, thereby resulting in premature failure of the die. Therefore, this section should be ruled as a semi-hollow section with a hollow part. For single-cavity dies, in order to solve the strength of the cantilever in semi-hollow profile extrusions, more structures are usually used in a cutting type porthole die, and its cantilever is cut into two parts, and one part is separated as the core of the male die. In the die with single

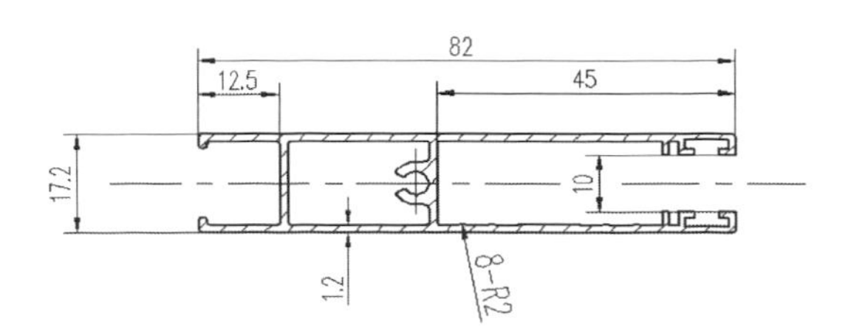

Figure 1. Schematic of the signal of the section of the profile.

cavity, the cores can be placed in the center region of the container, the core force is in a state of equilibrium, and it will not be affected by an additional force with the pressure difference formed around the core. But in a die with multi-cavities, with an increase in cores, it cannot achieve this arrangement. Because of a pressure gradient that is formed across the section of the container, the pressure gradient makes the core take an elastic offset in the extrusion process, the phenomenon of raised wire and convex flange is produced in the porthole die with single cavity by the cutting type porthole die, but the problem will be more serious in the porthole die with multi-cavities. At the same time, with an increase in the number of cores cut from the cantilevers, the manufacturing difficulty and error will increase and the difficulty to realize the synchronization of the extrusion will increase. In order to eliminate the phenomenon, in the die with multi-cavities, especially for the dangerous site of the cantilever, we can set a fake core in the male die to protect the cantilever with the help of the feeder bridge. In the extrusion, the fake core cannot touch the cantilever, and the cantilever will be protected effectively by the fake core and the bridge, which will improve the stress state of the cantilever and reduce the force area of the cantilever, so as to greatly improve the strength of the cantilever and the die. The die structure is shown in Figure 2.

An extrusion machine of capability 35MN is selected and the die dimensions are as follows: the diameter is 400 mm and thickness is 230 mm. In order to facilitate the processing of the die and the adjustment of metal flow velocity, the traditional structure type of the porthole die is changed, and a front feeder plate is added. For the determination of the number of cavities in portholes, on one hand, the equipment condition to realize the extrusion should be considered to improve the efficiency of the equipment, the larger in capability the machine is, the higher the efficiency of the equipment is, and the lower the cost will be. But it is more important to consider the difficulty of the die design and manufacture to ensure the synchronization and operability of the extrusion. On the other hand, it is necessary to select a reasonable extrusion coefficient, which is advantageous to the extrusion forming and production processes. If the extrusion coefficient is not reasonable, the waste will increase, and the rate of finished products will reduce. Under a comprehensive consideration, experience shows that the extrusion coefficient is more appropriate in the range from 60 to 80. A comparison of the extrusion program and the die structure for the section is shown in Figure 1 and given in Table 1.

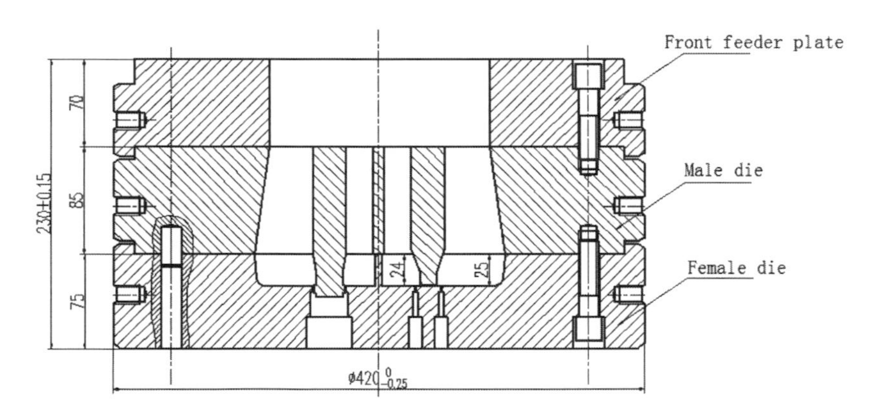

Figure 2. Schematic of the signal of the die structure.

Table 1. A comparison of the die structure and extrusion program.

Die structure	Capability of machine	Inner diameter of container	Number of cavities	Extrusion coefficient	Die dimension (diameter and thickness)	Operator number
Cutting type single cavity	10MN	130 mm	1	45.6	200 × 130	5
Cutting type twin cavities	18MN	185 mm	2	46.2	250 × 160	5
Protection type four cavities	35MN	300 mm	4	63	400 × 230	5

3 PARAMETERS' DETERMINATION OF THE DIE STRUCTURE

3.1 *The die hole arrangement*

It must be considered to make full use of the capacity of the machine and give play to the potential of the container in the arrangement of the die hole. It is important to avoid as far as possible to make the die lie close to the inner wall of the container, because this is not conducive to metal forming and will increase the size of the die. At the same time, the arrangement should make the holes in a symmetrical arrangement as far as possible; it is conducive to realize consistency of the flow of the metal in the extrusion process and ease of adjustment and balance of the flow of the metal. The die holes layout is shown in Figure 3.

3.2 *The front feeder plate*

The traditional porthole die consists of a male die and a female die. However, for a die with multi-cavities used in a large machine, if the male die thickness is too large, when the strength is sufficient, it will increase the difficulty of the die-making process and reduce quenching of the die in the heat treatment. And an addition of the front plate can ensure predistribution of the metal before entering the male die with a larger feeder ratio, which will help to achieve synchronization and reduce the pressure of extrusion. During the determination of the die holes layout, the front feeder plate can be designed. The metal supply of the two die holes can be separated from the left and the right or up and down. Its structure is shown in Figure 4.

3.3 *The design of port holes*

The design of port holes include hole layout, the area of the port holes, and determining the feeder ratio of port hole area and the size of false core is more important. The arrangement of the port holes can be in many forms. Based on the layout in Figure 3, the design of the portholes is carried out, according to the experience, the preliminary programs should be finished With the help of CAD software, two or

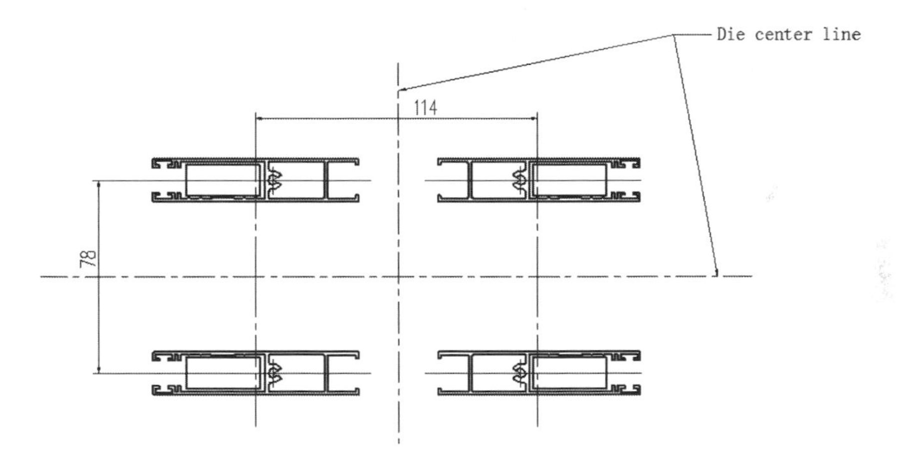

Figure 3. Schematic of the signal of die hole arrangement.

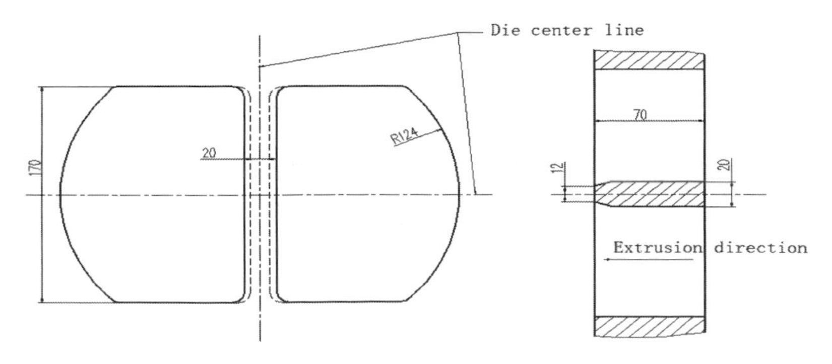

Figure 4. Schematic of the signal of the front plate.

three kinds of mature arrangement of portholes by experience will be selected to establish a three-dimensional model under the environment of UG, and the models must be saved as a fixed pattern into the extrusion simulation software, the simulation and calculation and observation are carried out, and the results were compared and analyzed, and combined with the experience to make correction, repeating simulation and correction, until the final determination of the best program. The arrangement of the port holes is shown in Figure 5.

The advantage of this layout is that, in the simulation, it is found that consistency of the metal flow velocity can be easily achieved. Each die hole is supplied separately with metal from two portholes. As long as the area of the porthole S1 is determined, the area of the porthole S2 is relatively easy to adjust, and the adjustable range is large, and the layout of the bridge and the die hole need not be modified or changed. On the other hand, the arrangement can enable the cantilever to be placed under the protection of the bridge, and the metal will not directly impact the cantilever, especially the end of the cantilever, so that the force of the cantilever is minimal. And so, the strength of the cantilever is the best. More important is that, from the simulation it can be observed that the change of the area of the portholes has an effect on the cantilever.

Its main parameters are as follows:

1. The feeder ratio is 17.4.
2. The dimensions of the bridge are as follows: width is 24 mm and thickness is 85 mm.
3. The area relationship between the portholes S1 and S2 are shown in Equation 1.

$$S1 = 0.81S2 \tag{1}$$

4. The result of the simulation shows that when the area relationship meets certain conditions, as shown in Equation 2, consistency of the flow rate of the metal can be easily achieved.

$$S1 = (0.75 \sim 0.85)\ S2 \tag{2}$$

5. The largest circumcircle of the porthole is 270 mm.
6. The structure of the bridge is similar to dripping of water and its feeder position is a trapezoid, which is conical in shape in the outlet part.
7. The stress space is taken in false core, the stress gap is measured according to the experience of 1.0 mm, and so the height of the false core is 24 mm.

In the above-mentioned die structure parameters, determination of the size of the false core is very important. Experience and simulation show that the minimum distance from the fake core to the cantilever's opening is 3 mm and to other die hole edge is 2 mm, and the shape of the rules must be designed as far as possible in order to facilitate processing. The size of the protection type false core is shown in Figure 6

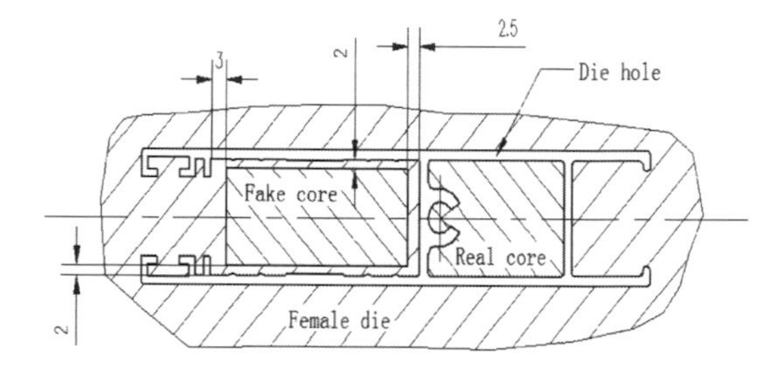

Figure 6. Schematic of the signal of the fake core.

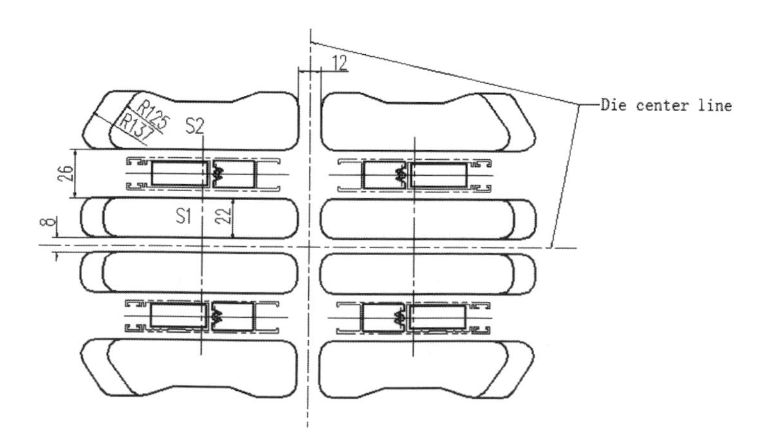

Figure 5. Schematic of the signal of port hole arrangement.

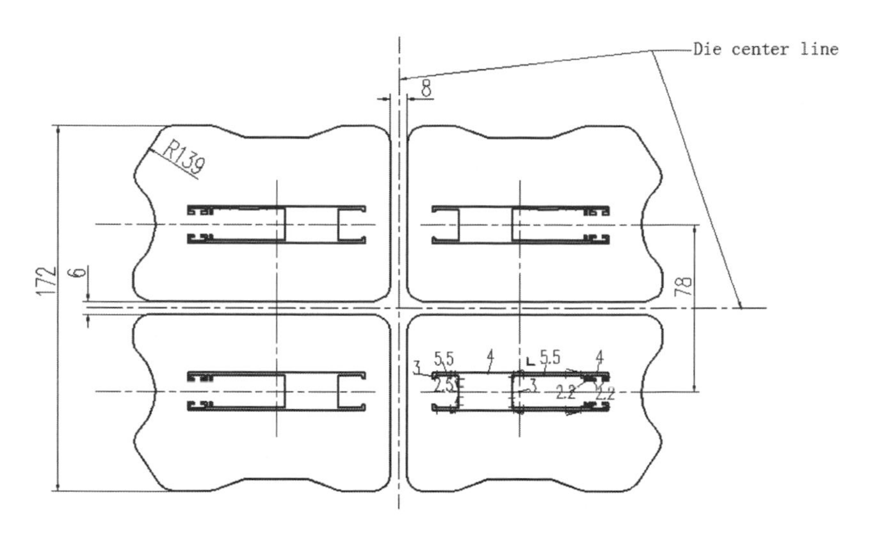

Figure 7. Schematic of the signal of the chamber and the bearing.

Table 2. A comparison of the die structure and extrusion results.

Die structure	Capability of machine	Wall thickness deviation	Extrusion marks and brightness	Dimension precision level	Die life (tons)	Die material
Ordinary single cavity	10MN	Obvious	Deep and rough	General	Less than 1	52
Cutting type single cavity	10MN	Not obvious	Low and shining, raised wire	General	8.5	65
Cutting type twin cavities	18MN	Not obvious	Low and shining, raised wire	General	21.7	105
Protection type four cavities	35MN	NO	Low and shining	High	More than 73.2	290

3.4 *The chamber of the female die and the bearing*

In order to eliminate the effect of manufacturing errors on the extrusion synchronization and to avoid the rigid region of metal flow in the central part of the cavities to cause the coarse grain in the profile, the welding chamber must be designed to be independent respectively. A wall must be set between the welding chamber, with a desirable width of 6–8 mm and the height is 8–10 mm. For the choice of the bearing is selected according to the single cavity. When one is selected, the other one is the same as it is. The welding chamber and the bearing are shown in Figure 7.

4 A COMPARISON OF EXTRUSION RESULTS

According to the section shown in Figure 1, the traditional ordinary porthole die with single cavity, the cutting type porthole die with single hole, the cutting type porthole die with twin cavities, and the protection type porthole die with four cavities were separately used. We have carried out the test of extrusion and extrusion production tracking, and the results and comparison are shown in Table 2.

It can be seen from the results that the protection type die of the porthole die with four cavities in a die for the semi-hollow section can significantly improve the efficiency of extrusion and greatly improve the life of the die, as well as reduce costs.

5 CONCLUSIONS

The structure of the protection type porthole with four cavities is used in the section, as shown in Figure 1. The results show that the die life of this kind of structure can reach 73.2 tons. It can be seen that this structure is effective and it can significantly improve the efficiency and reduce the cost of the production. If we use more number of dies with multi-cavities in a die for extrusion production, we can completely solve the problems in that small machines have too many orders to complete, and large machines have not enough orders to work on and significantly improve the efficiency. In the die structure with four cavities, especially for the semi-hollow profiles, the determination of the capability of the machine, the design of portholes, and the welding chamber structure and bearing is very important. But it is more critical to select a reasonable and effective die structure to reduce the extrusion force and to ensure the strength of the cantilever and to make it easy to process. This structure is worth promoting. For further research and development of the new die structure with multi-cavities for the semi-hollow section, we have accumulated experience and reliable practice data.

REFERENCES

Deng Rurong, Huang, Xuemei. Design of the extrusion die of semi-hollow aluminum profile [J]. Light Alloy Fabrication Technology, 2015,43(4):51–54.

Hu Dongpo, Wang Leigang, Huang Yao. Steady-state simulation and die improvement on the extrusion of prolate aluminum profile [J]. Forging & Stamping Technology, 2015,40(4):69–73.

Kuang Weihua, Chen Biaobiao. Research on design and structure of extrusion die for cantilever aluminum profile [J]. Hot Working Technology, 2013,42(21):136–138.

Liu Jingan. Die design, manufacture, application and maintains for aluminum profiles extrusion [M]. Beijing: Metallurgical Press, 1999:181–183.

Sun Xuemei, Zhao Guoqun. Fake porthole extrusion die structure design and strength analysis for cantilever aluminum alloy profiles [J]. Journal of Mechanical Engineering, 2013,49(24):39–44.

Wang Liwei. Optimization design of extrusion die for the bigger slenderness ratio half hollow aluminum profile [J]. Die and Mould Manufacture, 2011(4):61–64.

Xie Jianxie, Liu Jingan, Die design, manufacture, application and maintains for aluminum profiles extrusion [M]. Beijing: Metallurgical Press, 2012:133–138.

Xu Yongli, Huang Shuangjian, Pang Zugao, et al. Failure analysis of extrusion die and optimization of heat treatment process for aluminum alloy circular tube [J]. Forging & Stamping Technology, 2015,40(2):116–122.

Yu Mingtao, Li Fuguo. Simulation extrusion process of the sketch hollow aluminum profile based on infinite volume method [J]. Die and Technology, 2008(4):40–43.

Civil, Architecture and Environmental Engineering – Kao & Sung (Eds)
© 2017 Taylor & Francis Group, ISBN 978-1-138-02985-9

Numerical and experimental study on stability of aluminium alloy riveted and stiffened panels under in-plane shear

Dawei Wang, Zhenkun Lei, Ruixiang Bai & Yu Ma
State Key Laboratory of Structural Analysis for Industrial Equipment, Dalian University of Technology, Dalian, China

ABSTRACT: Shear experiment on aluminium alloy riveted and stiffened panel was conducted to obtain the shear buckling mode, buckling load, load capacity, and failure mode of the structure. Using a combination of fringe projection method and strain gauge sensor, the shear buckling morphology and instability evolution process were detected in real-time. Finite Element Method (FEM) was used to analyze the buckling performance and load capacity. Experimental and simulation results show that the shear instability of the structure forms the skin local buckling between stiffeners.

1 INTRODUCTION

Stiffened aluminium alloy panel structure with a low weight can significantly improve the buckling critical load, which had been widely used in aircraft wing and fuselage structure. The riveting technology avoids the residual stress problem caused by high temperature in laser welding technology. The stiffened panel as a typical aircraft structure has a common failure mode of buckling under compression, shear, torsion and bending. The stiffened panel still has a considerable post-buckling load capacity, which can be used to increase structure loading capacity.

There are many theoretical and experimental works on the buckling of stiffened panels in recent years. Rossini et al. (2016) has studied the pre-buckling and post-buckling behavior of adhesively-bonded and riveted and panels subjected to in-plane compression by numerical simulation and experiments. It showed that the bonded panels have a superior performance than the riveted panels in buckling load and buckling mode. Villani et al. (2015) measured the buckling load, collapse load and failure mode of bonded panels, which agreed to FEM simulations.

However, the selection of element, mesh, and material modeling have a important role for the nonlinear FEM analysis of riveted panels under uniform shear loading (Murphy, 2005). Cricrì et al. (2014) designed a novel multi-hinged test fixture to provide a pure shear load for stiffened panels, and the experimental results showed that there are different buckling modes in comparison with that of a traditional picture frame test fixture. Therefore, the nonlinear FEM analysis of riveted panels has influenced by the selection of boundary conditions and must be validated by experiments.

The manufacturing defects, contact and damage problems are inevitable, which need to be considered in the nonlinear FEM analysis. Anyfantis et al. (2012) studied the progressive post-buckling and the final failure response of stiffened composite panels based on the structural nonlinear FEM. Orifici et al. (2008) combined the post-buckling deformations and several composite damage mechanisms to successfully predict the damage growth and collapse behaviour of composite stiffened structures under compression. The buckling behavior influenced by the type of stiffeners, panel orthotropic ratio, tensile stiffness and shear stiffness ratios of stiffener and skin (Jain, 2010).

In this paper, a shear experiment on the aluminium alloy riveted and stiffened panel was conducted to investigate the buckling mode and failure, which were compared with the FEM stability analysis. The experimental and simulation results provide a reference for engineering application of the structure.

2 EXPERIMENTS

2.1 *Specimen description*

As shown in Fig. 1(a), the specimen was made of high strength aluminium alloy, which consists of skin and 4 Z-shaped stiffeners, and the skin and stiffeners were riveted. The shear load was applied to the specimen by a diagonal tension, as shown in Fig. 1(b). The strain gauges were pasted to the symmetrical and back-to-back positions (shown as the "+" symbol in Fig. 1) on the skin to monitor the initial buckling and post-buckling failure.

2.2 *Test procedure*

The shear experiment was carried out in a four-column tester (CSS-100T, Changchun tester Co.), as shown in Fig. 2(a). The diagonal tensile load was applied to the specimen until a large deformation failure. A loading rate was 6 kN/min and the final failure load was 471.5 kN, as shown in Fig. 2(b).

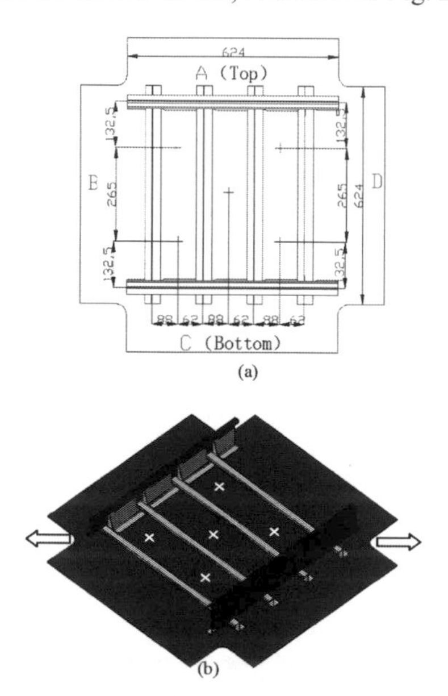

Figure 1. (a) Specimen geometry and (b) diagonal tensile shear.

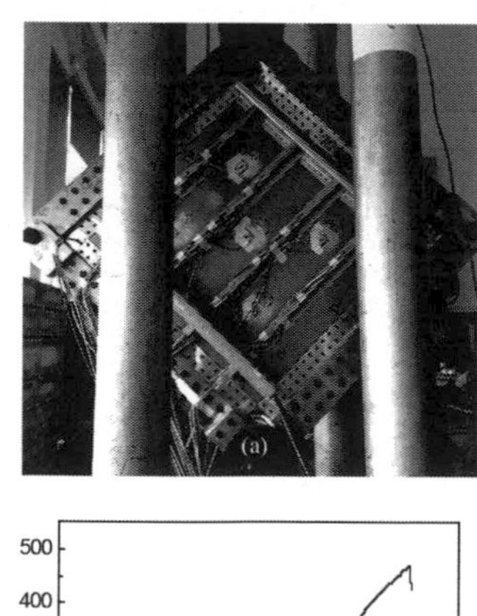

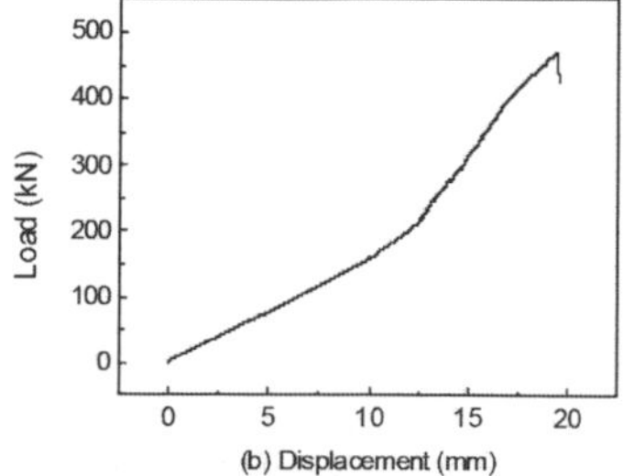

Figure 2. (a) Tensile shear test and (b) loading history.

3 RESULTS AND DISCUSSIONS

3.1 *Strain measurement*

The strain history for typical positions is shown in Fig. 3. An obvious strain bifurcation phenomenon appeared at the moment of 10 min., which implies that the initial skin buckling appeared at the time. Then, the stiffeners bear the main load and the post-buckling deformation occurs with the increase of strains until to the maximum load of stiffener instability. At this time, the panel structure cannot afford a more load, namely the failure load.

3.2 *Optical measurement*

A multi-frequency fringe projection method was adopted to measure the buckling mode of skin at different time (Lei, 2015). Four sets of sinusoidal and phase-shifted fringes with the fringe frequencies of 1, 4, 16 and 64 were projected by a projector (TLP-X2000) onto the skin in turns, and the distorted fringe patterns were simultaneously grabbed by a CCD camera (F-080B). An image acquisition frame rate was 4 fps and the image size was 1024×768 pixels.

Then a four-step phase-shifting method and a multi-frequency phase unwrapping technique were combined to obtain the phase of surface topography. The phase difference before and after deformation was converted to the skin deflection (Lei, 2016).

The full-field deflections of buckling wave were given by optical measurement, as shown in Fig. 4, which is more vivid than the electrical strain measurement (Fig. 3). Some peaks and troughs of local buckling appeared in the skin between the stiffeners and there are 4 buckling half waves on the skin. The direction of the wires is along the direction of the stiffeners. The horizontal direction of the buckling waveforms is consistent with the direction of diagonal loading.

3.3 *FEM simulation*

Numerical simulation on the shear behaviour of the stiffened panel was implemented by ABAQUS/

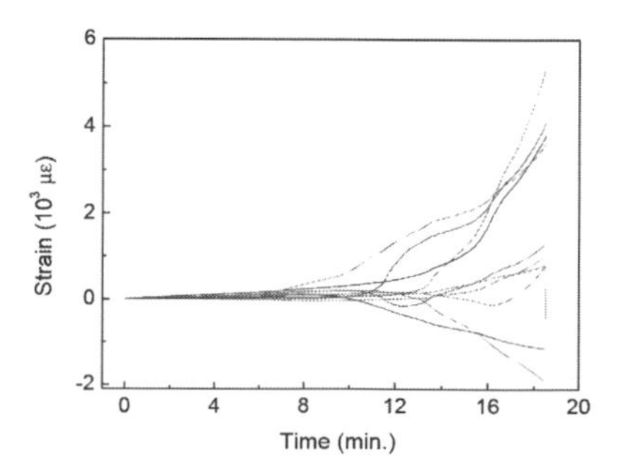

Figure 3. Bifurcation phenomenon of strains.

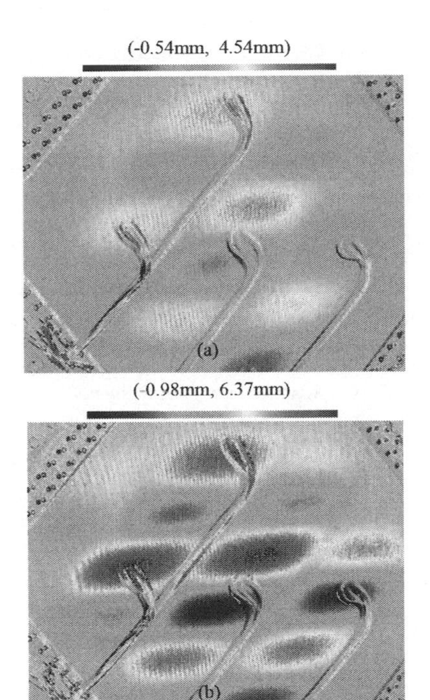

Figure 4. Buckling modes at back of the specimen under different loads, (a) 300 kN and (b) 400 kN.

Standard module. The finite element mesh, loading and boundary conditions of the stiffened panel are shown in Figs. 5(a) and 5(b). A 4-node doubly curved thin or thick shell element of S4R was chosen from the ABAQUS element library according to the basic assumption of Kirchhoff thin plate theory.

In order to improve the computational efficiency and accuracy, a 2-node linear beam element "B31" has been chosen to discrete clamps and the MPC constraint was used to simulate the pin connection between the clamps. A hard contact without friction was adopted. A fixed support constraint was applied to a diagonal point of the finite element model. A tensile displacement load was applied to the other corner of the diagonal and the out-of-plane displacement of the four-side loading regions were constrained.

Before analyzing the load capacity of the stiffened panel, a linear buckling analysis should be performed firstly to obtain the buckling modes. The initial geometric imperfections of the panel were assumed to be a linear buckling of a certain order of deformation. The post-buckling load capacity analysis was performed by editing the INP file for introduction of the initial defect. The result is shown in Fig. 5.

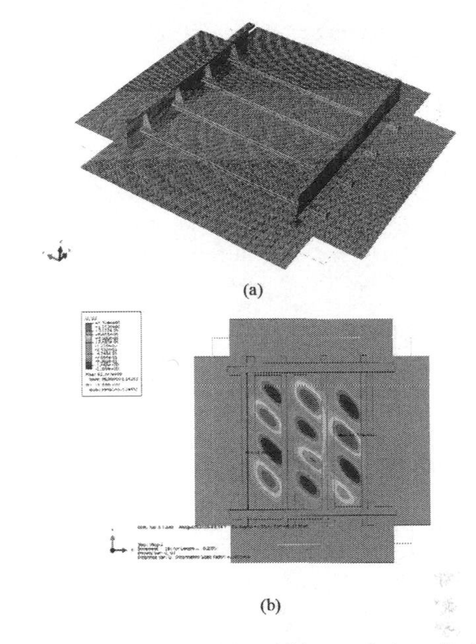

Figure 5. (a) FEM model and (b) out-of-plane displacement under the critical buckling load.

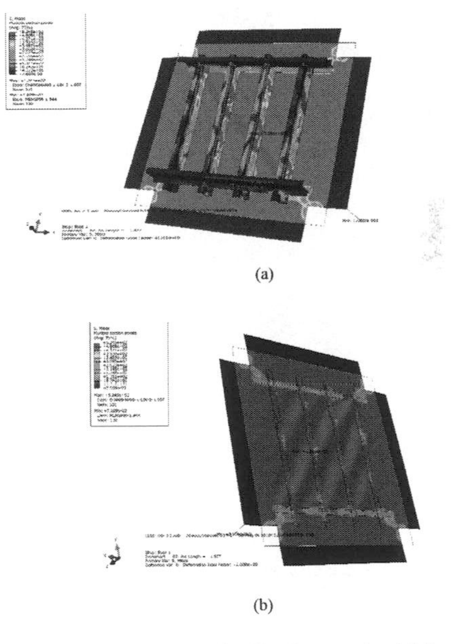

Figure 6. Misses stress distribution on the (a) front and (b) back of stiffened plate under failure load.

Fig. 6 is the Misses stress distribution of the structure under the ultimate load. It shows that the failure mode of the structure is the skin deformation between the stiffeners, and the plastic deformation is very large near the diagonals, which are in agreement with the experiment.

The connection strength of the riveting area in the actual structure is determined by the rivet strength. Through the experiment, it is found that the failure mode of the stiffened panel is the rivet failure. In this paper, the connection mode of the riveting is simplified, and the effect of the rivet failure is ignored. It is believed that the structural loss of load capacity is mainly caused by the plastic deformation of the skin and stiffeners, which is the main reason for the calculation error.

4 CONCLUSIONS

The riveted and stiffened aluminium alloy panel under in-plane shear test was conducted. The experimental result shows that the failure mode of the stiffened panel under in-plane shear includes the skin local buckling between the stiffeners, the skin plastic deformation, the stiffener bending and torsion deformation, and the partial rivet connection failure. Then, the FEM model was established and successfully predicted the buckling load and the failure load, which are close to the experimental ones.

ACKNOWLEDGMENTS

The authors thank the National Basic Research Program of China (2014CB046506), the National Natural Science Foundation of China (Nos. 11472070, 11572070), the Research Subject of State Key Laboratory of Structural Analysis for Industrial Equipment (Nos. S14207, S15202), and the Natural Science Foundation of Liaoning Province of China (No. 201502014).

REFERENCES

Anyfantis KN, Tsouvalis NG. Post Buckling progressive failure analysis of composite laminated stiffened panels. Applied Composite Materials. 2012, 19(3–4): 219–236.

Cricrì G, Perrella M, Calì C. Experimental and numerical post-buckling analysis of thin aluminium aeronautical panels under shear load. Strain. 2014, 50(3): 208–222.

Jain HK, Upadhyay A. Buckling behavior of blade-, angle-, T-, and hat-stiffened FRP panels subjected to in-plane shear. Journal of Reinforced Plastics & Composites. 2010, 29(24): 3614–3623.

Lei ZK, Bai RX, Tao W, Wei X, Leng RJ. Optical measurement on dynamic buckling behavior of stiffened composite panels under in-plane shear. Optics and Lasers in Engineering. 2016, 87: 111–119.

Lei ZK, Wang CL, Zhou CL. Multi-frequency inverse-phase fringe projection profilometry for nonlinear phase error compensation. Optics and Lasers in Engineering. 2015, 66: 249–257.

Murphy A, Price M, Lynch C, Gibson A. The computational post-buckling analysis of fuselage stiffened panels loaded in shear. Thin-Walled Structures. 2005, 43(9): 1455–1474.

Orifici AC, Alberdi IODZ, Thomson RS, Bayandor J. Compression and post-buckling damage growth and collapse analysis of flat composite stiffened panels. Composites Science and Technology. 2008, 68(15–16): 3150–3160.

Rossini MB, Donadon MV. Numerical and experimental analysis of adhesively bonded stiffened panels subjected to in-plane compression loading. Advanced Materials Research. 2016, 1135: 140–152.

Villani ADPG, Donadon M, Arbelo MA, Rizzi P, Montestruque CV, Bussamra F, Rodrigues MRB. The post-buckling behavior of adhesively bonded stiffened panels subjected to in-plane shear loading. Aerospace Science & Technology. 2015, 46: 30–41.

Civil, Architecture and Environmental Engineering – Kao & Sung (Eds)
© 2017 Taylor & Francis Group, ISBN 978-1-138-02985-9

Process optimization for the new nano-complex internal coating and its anti-corrosion and drag reduction properties in natural gas pipes

Baohong Hao, Bin Sun, Zhihong Liu, Yulong Han & Longfei Zhao
School of Mechanical Engineering, Beijing Institute of Petrochemical Technology, Beijing, China

ABSTRACT: A new nano-composite coating is developed by using the blending method for use inside the pipe of natural gas. The "systematic effect" and the "cascading effect" produced by the hybrid composition of nanoparticles and the matrix are used to achieve optimization for the complex coating on functions such as anti-corrosion and drag reduction. The difficult-to-coordinate bottleneck problem which is an "internal and external trouble" existing inside the pipe will be solved at the same time. The optimized composite coating, on one hand, can exert a "super bonding force" with the wall of the pipe which is not easy to peel; on the other hand, the coating can show a "super alienation" behavior to the medium which is not easy to scale. An acceleration simulation method has been used to evaluate the performance indicators of anti-corrosion and drag reduction through experiments. The results of experiments show that the composite coating added with nano-additives shows a decline of surface roughness and visible latent function of drag reduction. The results of the experiments also show that the corrosion rate has declined by an average of more than one time.

1 INTRODUCTION

By referring the literature, we know that (Luo, 2010; Ma, 2012; He, 2013; Fan, 2008) the most prominent problems in our country's existing coating are high surface roughness and poor durability of coatings. Firstly, the roughness of the pipe directly affects the transport efficiency. Secondly, poor durability is a potential security risk which will affect the secure transportation of natural gas. And the roughness is a precipitating factor for the decline of durability. Especially for the occasion where too much H_2S and CO_2 are present in the natural gas, the corrosive environment of the internal wall of the pipe will become more severe, and a little increase in the surface roughness of the pipe will directly cause a significant decline in the durability. Therefore, it is imperative that a composite coating needed to be developed to deal with the durability of coating and the roughness of the surface of coating synchronously. Developing a new, long-lasting and efficient nano-composite coating can alleviate fundamentally the coating's peeling inside the pipe. When the temperature and pressure of the medium change and wear is caused by the impurity of the medium etc., an efficient and long-lasting protection is realized for the internal wall of the natural gas pipe, thereby resulting in control of corrosion, reduction of wear, prolonging of service life and improvement of durability so as to ensure the safe operation of natural gas pipelines in the West–East Gas Transportation Project (Zhu, 2006).

2 EXPERIMENTAL MATERIALS AND EXPERIMENTAL METHODS

X65 (micro-alloyed high-strength low-alloy steel) is a model of the test specimen of the pipe; the additive is nano-aluminum dioxide and nano-titanium dioxide; the matrix coatings are epoxy resin paint, alkyd anti-rust paint and synthetic thinner. The testing device applied is the CS electrochemical workstation; the instruments involved are ORION420A type pH meter, X-ray diffractomer, and scanning electron microscope.

3 ANALYSIS FOR EXPERIMENTAL RESULTS

3.1 *Mechanism of drag reduction and contrastive analysis of experimental figures*

Fig. 1 shows the schematic diagram of the vortex produced when the natural gas flows through the pipe.

From Fig. 1, we can see that when the internal wall of the pipe is not coated, the height (roughness) of embossments on the internal wall is high and the vortex zones formed behind these are large in size; therefore, the drag loss produced is also big; When the internal wall is coated, as the fineness of the internal wall surface is increased, the vortex zones produced behind the embossments are smaller in size, and the drag loss therefore becomes small. From Fig. 1(c), we can see that the fineness of the internal wall of the pipe (boundary) has a big impact on the drag coefficient; that is to say that, we

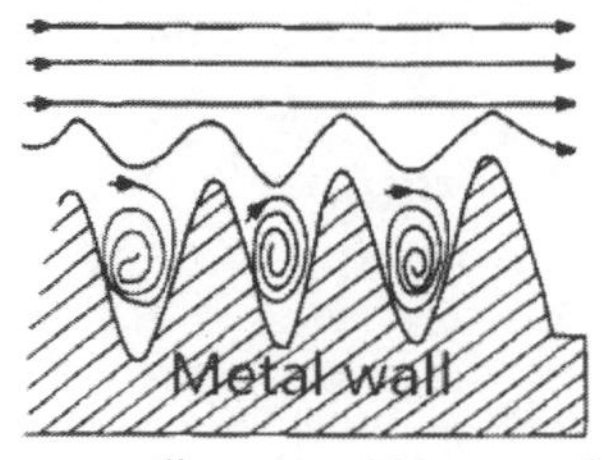

(a) Vortex zone diagram without coating

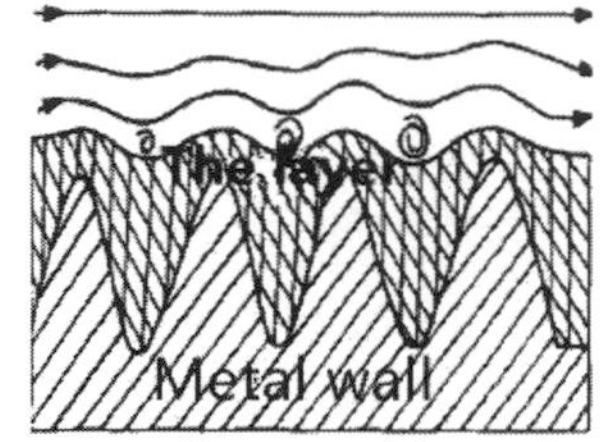

(b) Vortex zone diagram with coating

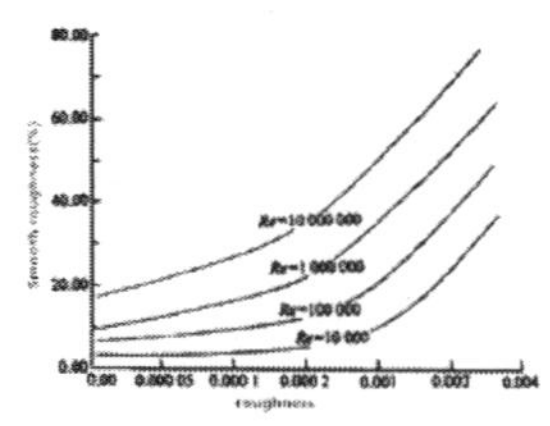

(c) Circular pipe smooth drag reduction rate

Figure 1. Schematic diagram of the vortex zone and effect chart of fineness and drag reduction rate.

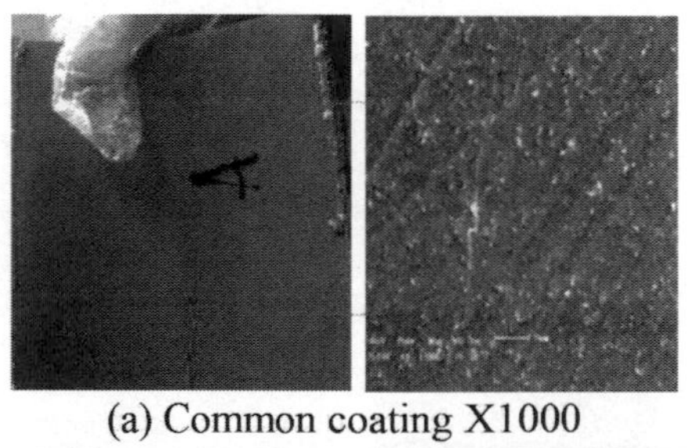

(a) Common coating X1000

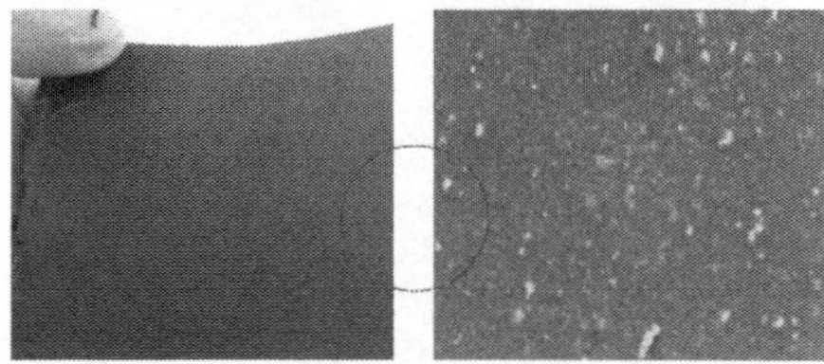

Figure 2. Visual and microscopic analysis figures for the surface roughnesses of nano-composite coating and common coating.

can reduce the drag in the later transportation of natural gas by improving the fineness of the composite coating. Fig. 2 shows the visual and microscopic analysis comparison figures of the surface appearances, which are enhanced with additives and without additives.

From Fig. 2, we can see directly that the composite coating added with nano-additives can exhibit a higher level of fineness; the fine, smooth, and flat surface predicts a potential of drag reduction when the fluid flows through, while for the composite coating added without nano-additives, whose surface is rough and has grooves on it, it is easy for the liquid to attach to produce channeling or vortex and form the fluid drag force. Through scratch test, it was confirmed that the nano-composite coating possesses a higher scratch resistance.

3.2 *Anti-corrosion self-cleaning effect and analysis for the simulated accelerated corrosion test*

Fig. 3 is the principle diagram of the lotus leaf effect.

From Fig. 3, we can see that the natural nanostructure of the lotus leaf determines its self-cleaning

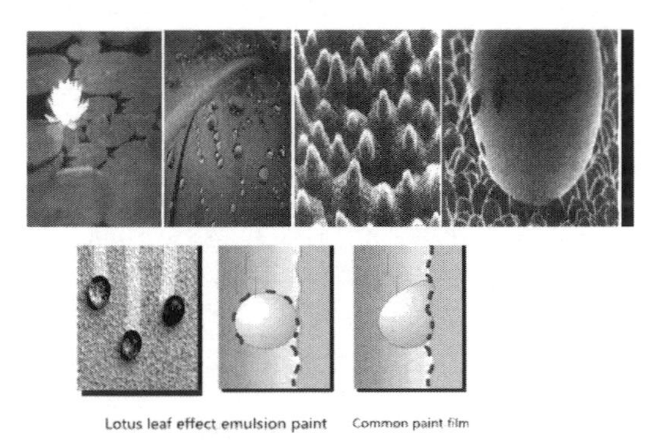

Figure 3. Schematic diagram of the self-cleaning function of the nano-composite coating.

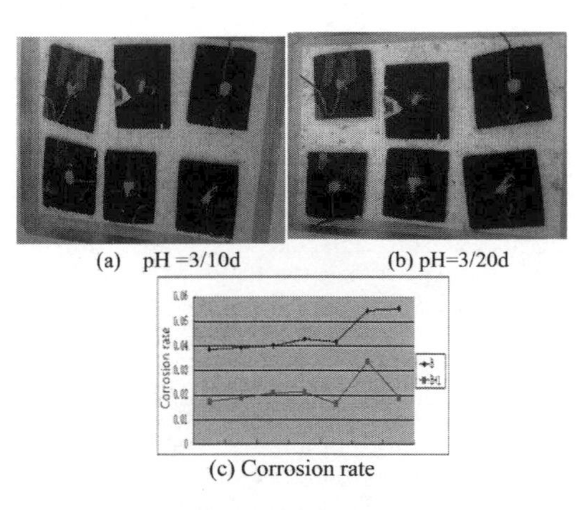

(a) pH =3/10d (b) pH=3/20d

(c) Corrosion rate

Figure 4. Simulated accelerated corrosion test and chart of the corrosion rate of electrochemical detection.

function, which means that we can adjust the structure of the composite coating to make it nanocrystallized through method of composite modification and realize the super alienation effect between the material and the outside world. Fig. 4 shows intuitionistic figures of the corrosion test and a comparative analysis chart of the corrosion rate of the electrochemical detection result for the coatings with nano-additives and without nano-additives.

From Fig. 4, we can see that after accelerated corrosion for 22 days, the test specimen of the nano-composite coating in the corrosive bath in a strong acid environment (pH = 3) can be observed with only little signs of corrosion, with no bubbling phenomenon and no peeling phenomenon. However, for the common coating, we can see that the visible corrosion products are formed and a certain degree of deposits occurred on the surface. Through electrochemical detection, it shows that the corrosion rate of the nano-composite coating is obviously lower than the common coating and it will be stable basically along with the time of corrosion and exhibits a rising trend just at the later stage of the corrosion; however, the corrosion rate of the common coating exhibits a rapid growth trend. From the contrastive analysis, we can see that the corrosion rate of the composite coating added with nano-additives is only half of the corrosion rate of the common coating, which indicates that the nano-additive can reduce the corrosion rate of the test specimen of the composite coating exponentially and improve the corrosion resistance exponentially. Fig. 5 shows EDAX spectra to prove the existence of nano-additives.

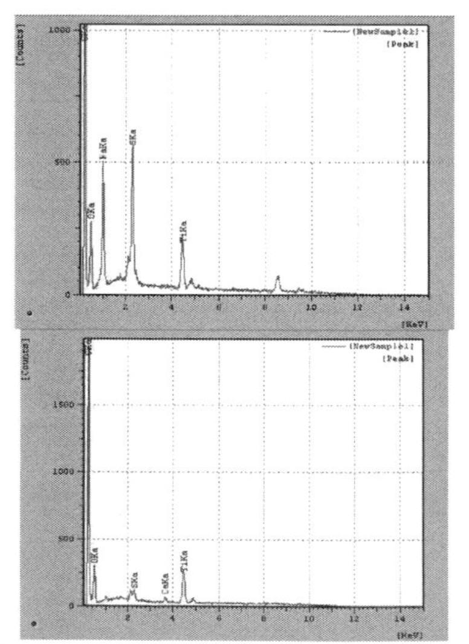

(a) Energy spectrum of the common coating and
(b) Energy spectrum of the nano-composite coating

Figure 5. EDU contrastive figures of the nano-composite coating and composite coating.

From Fig. 5, we can see that when the composite coating is enhanced with nano-additives, the X-ray diffraction peaks show a visible epoxy characteristic peak of the matrix coating, after adding with nano-additives, as the nanoparticles have initiated a chemical reaction with the matrix, and so it weakens the strength of the characteristic peak. This proves the existence and true effectiveness of the nano-additive in the composite coating.

4 CONCLUSION

In conclusion, after a visual inspection for macro-appearance and micro-observation for local appearance, the nano-composite internal coating owns a higher surface fineness which can reduce the drag force exerted during the transportation effectively. Through the simulated accelerated corrosion and the electrochemical detection, it indicates that the nano-composite coating exhibits a better "self-cleaning" function and "anti-corrosion function". The existence of the nano-additives in the composite coating and the effectiveness that it can improve the corrosion resistance to the test specimen has been verified by energy spectrum detection and analysis. The conclusion is summarized as follows:

1. From the visually observed corrosion situation of the test specimen, we can see that the nano-composite coating in the process of stimulated accelerated corrosion shows a strong acid resistance and corrosion resistance. The maximum pH that can be resisted is 3, and the tolerance time can reach to 20 days such that there is nearly no corrosion occurred on it. It is evident that the nano-composite coating has better strong acid resistance when compared to the common coating.
2. Nano-composite coating has a higher level of surface fineness and a potential of drag reduction. The addition of nanoparticles can, on one hand, act as a "nail" going deep into the coating to form a relatively strong bonding force with the wall of the pipe and prevent blisters and folds; on the other hand, it can suspend on the surface of coating and play a role in shimming, compacting, and repairing the surface.
3. The anti-corrosive function of the coating enhanced with nanoparticles also shows a certain degree of self-repair function; the related mechanism is to be studied and discussed later.

ACKNOWLEDGMENTS

The authors gratefully acknowledge the subsidization obtained from the city level URT project of Beijing (Project Nos: 16032082001/036 and 2016 J00036).

REFERENCES

Fan Zhaotin, Yuan Zongming, Liu Jia etc. Corrosion Mechanism and Protection Research of H_2S and CO_2 on Pipeline [J]. Oil-Gas Surface Engineering, 2008, 27(10):39–40.

He Renyang, Tang Xin, Zhao Xiong etc. Research Progress of Corrosion Protection Technology for Oil and Natural Gas Pipeline [J]. Process Equipment & Piping, 2013, 50 (1): 53–54.

Liu Guangwen, Qian Chengwen, Yu Shuqing etc. Drag Reduction Technology of Internal Coating in Gas Transmission Pipeline [M]. Petroleum Industry Press, 2003: 2–3.

Luo Peng, Zhao Xia, Zhang Yiling etc. Analysis of Direct Assessment on Internal Corrosion of External Gas Pipelines [J]. Oil & Gas Storage and Transportation, 2010, 29(2): 137–140.

Ma Weifeng, Luo Jinheng, Cai Ke etc. The Situation and Existing Problem of Resistance—Drug Reduction Technology of Internal Coating in Gas Transmission Pipeline [J]. 2012, 9: 18–20.

Zhu Xin. Corrosion of Long-distance Pipeline and Protection [J]. Petrochemical Corrosion and Protection, 2006, 23(1): 30–35.

Civil, Architecture and Environmental Engineering – Kao & Sung (Eds)
© 2017 Taylor & Francis Group, ISBN 978-1-138-02985-9

An improved gray wolf optimization algorithm and its application in iron removal from zinc sulfate solution using goethite

Tiebin Wu
Hunan University of Humanities, Science and Technology, Loudi, Hunan, China

Jianhua Liu
Hunan University of Technology, Zhuzhou, Hunan, China

Yunlian Liu, Qiaomei Zhao & Xinjun Li
Hunan University of Humanities, Science and Technology, Loudi, Hunan, China

ABSTRACT: The Gray Wolf Optimization (GWO) algorithm has disadvantages including low solution accuracy and being easily trapped in the local optimum. In order to overcome these disadvantages, an Improved Gray Wolf Optimization (IGWO) algorithm is proposed. The research applies chaos theory to generate the initial population to ensure that more individuals are distributed in the searching space. A non-linear convergence factor described based on the logarithmic function is proposed to replace the linear decreasing convergence factor so as to coordinate the exploration and development capacities of the algorithm. In order to avoid occurrence of the premature convergence of the algorithm, a mutation operator with adaptable mutation probability is introduced. The numerical experimental results of four benchmark test functions and the data simulation results in the removal of irons from zinc sulfate solution using goethite show that the IGWO algorithm has a favorable optimization performance.

1 INTRODUCTION

The swarm intelligent optimization algorithm, as a global optimization method derived from the simulation of the behavior mechanism of biotic communities in nature, has developed rapidly in recent years. Gray Wolf Optimization Algorithm (GWO) is a new swarm intelligent optimization algorithm proposed by Mirjalili et al. (Mirjalili, 2014) in 2014. Originating from simulating the class mechanism and predation of gray wolf groups, the algorithm can be realized easily through programming, needs little parameters, and shows a strong global searching ability (Mirjalili, 2014). Since its proposal in 2014, the GWO algorithm has been widely used in various fields, including power flow optimization (El-Fergany, 2015) and optimum control of direct current motors (Madadi, 2014). However, similar to other swarm intelligent optimization algorithms, the GWO algorithm also has disadvantages such as low solution accuracy and premature convergence.

In order to solve these disadvantages of the standard GWO algorithm, some improvement methods are proposed. By using four benchmark test functions and parameter optimization of iron removal from zinc sulfate solution through goethite, the optimization performance of the algorithm proposed in this study is verified.

2 STANDARD GWO ALGORITHM

Gray wolves in a gray wolf population are divided into head wolf α, deputy leader β, common wolves δ, and underclass wolves ω according to the class mechanism. In addition, the lower the class is, the larger is the number of individuals. To capture a prey, other individuals besiege the prey in an organized manner under the leadership of head wolf α. In the GWO algorithm, the individual with the highest fitness in the group is defined as α and those whose fitness is ranked second and third are defined as β and δ. In addition, others are described as ω. Moreover, the position of the prey is defined as the global optimal solution of an optimizing problem. In a D-dimensional searching space, if the position of the ith wolf is $X_i = (X_i^1, X_i^2, \cdots, X_i^D), X_i^d$ represents the position of the ith wolf in the d-dimensional space.

For the position of the ith wolf in the d-dimensional space, the behavior of the gray wolf gradually approaching and besieging a prey is described by Equation (1).

$$X_i^d(t+1)=X_p^d(t)-A_i^d\left|C_i^d X_p^d(t)-X_i^d(t)\right| \tag{1}$$

where t represents the current iteration times and $X_p=\left(X_p^1,X_p^2,\cdots,X_p^D\right)$ describes the position of the prey. $A_i^d\,|\,C_i^d X_p^d(t)-X_i^d(t)\,|$ stands for the besieging step and A_i^d and C_i^d are defined as follows:

$$A_i^d=a\cdot(2\cdot rand_1-1) \tag{2}$$

$$C_i^d=2\cdot rand_2 \tag{3}$$

where $rand_1$ and $rand_2$ represent random variables in the range of [0, 1]. Variable a is the convergence factor, which plays an important role in coordinating exploration and development capacities of the algorithm. a decreases linearly from 2 to 0 with an increase in the iterations during the evolution, as shown in Equation (4).

$$a=2-t/t_{\max} \tag{4}$$

where $t_{\max}$ refers to the maximum iteration times.

While solving optimization problems, there is no priori knowledge concerning the global optimal solution beforehand. In addition, the gray population updates their positions according to their distances to the positions of α, β and δ, namely, X_α, X_β and X_δ.

$$\begin{cases} X_{i,\alpha}^d(t+1)=X_\alpha^d(t)-A_{i,1}^d\,|\,C_{i,1}^d X_\alpha^d(t)-X_i^d(t)\,| \\ X_{i,\beta}^d(t+1)=X_\beta^d(t)-A_{i,2}^d\,|\,C_{i,2}^d X_\beta^d(t)-X_i^d(t)\,| \\ X_{i,\delta}^d(t+1)=X_\delta^d(t)-A_{i,3}^d\,|\,C_{i,3}^d X_\delta^d(t)-X_i^d(t)\,| \end{cases} \tag{5}$$

$$X_i^d(t+1)=\sum_{j=\alpha,\beta,\delta} w_j\cdot X_{i,j}^d(t+1) \tag{6}$$

where $w_j\,(j=\alpha,\beta,\delta)$ represents the weight coefficients of α, β, and δ, and $d=1, 2, \ldots, D$.

$$w_j=\frac{f\left(X_j(t)\right)}{f\left(X_\alpha(t)\right)+f\left(X_\beta(t)\right)+f\left(X_\delta(t)\right)} \tag{7}$$

where $f(X_j(t))$ refers to the fitness of the jth wolf in the tth generation. $f(X_\alpha(t))$, $f(X_\beta(t))$, and $f(X_\delta(t))$ represent the fitness values of head wolf α, deputy leader β, and common wolves δ in the tth generation, respectively. The detailed pseudo codes of the GWO algorithm can be found in the literature (El-Fergany, 2015).

3 IMPROVED GRAY WOLF OPTIMIZATION ALGORITHM

3.1 *Population initialization based on the chaotic method*

It can be obtained from the mechanism of evolutionary algorithms that the quality of the initial population has a great influence on the convergence, searching efficiency, and stability of evolutionary algorithms. The individuals of the initial population need to be uniformly distributed within the whole searching apace as much as possible.

Chaos, as a non-linear phenomenon characterized by ergodicity and randomness, can traverse all states without repetition within a certain range according to its own law. For this characteristic, it can be applied in optimization algorithms to initialize populations so as to ensure diversity of individuals. A transcendental function is selected in the study, as shown in Equation (8).

$$x_{k+1}=\beta\cdot\sin(\pi x_k) \tag{8}$$

where β is a non-negative real number. Given an initial value x_0, a random sequence within the range of $[-\beta,\beta]$ is produced. Given that the optimal individual in a certain generation of a population is x_p and selected individuals who are about to enter into the next generation are described as x_i. And then, the Euclidean distance d_{ip} of each individual from the optimal individual is calculated by using Equation (9).

$$d_{ip}=\sqrt{\sum_{j=1}^{n}(x_{pj}-x_{ij})^2},\quad j=1,2,\ldots,n \tag{9}$$

where $i=1,2,\ldots,N$, and N represents the population size. A threshold value σ is set. If $d_{ip}<\sigma$, then the population $x_i(i=1,2,\ldots,N)$ is initialized by using a chaos sequence, which is produced by employing Equation (8).

3.2 *Non-linear convergence factor*

For swarm intelligent optimization algorithms, only when the exploration and development capacities are coordinated, a strong robustness and quick convergence rate can be obtained.

By improving Equation (4), the equation for the non-linear convergence factor is updated based on a logarithmic formula, as shown in Equation (10).

$$a(t)=a_{initial}-\log\left(1+\lambda\cdot\frac{t}{t_{\max}}\right) \tag{10}$$

where t represents current iteration times and t_{max} refers to the maximum iteration times. In addition, $a_{initial}$ stands for the initial value of the convergence factor a whose value is two in this study. $\lambda \in [0,1]$ is called the regulatory factor and $\lambda = 0.7$.

It is obtained from Equation (10) that the convergence factor a non-linearly decreases with an increase in the number of iteration times and therefore effectively coordinates the exploration and development capacities of the GWO algorithm.

3.3 *Mutation mechanism with adaptable mutation probability*

The standard GWO algorithm is short of the mechanism of getting rid of the local optimum, and so premature convergence is likely to occur in the later period of the optimization.

Therefore, a mutation mechanism with adaptable mutation probability is introduced. That is to say that, mutation occurs in the process of evolution and the mutation probability is shown in Equation (11).

$$p_m = \phi(f) \times p_{m0} + (1 - \phi(f)) \times p_{m1}$$

where p_{m0} and p_{m1} represent small and large mutation probabilities $(p_{m0} \in [0.01, 0.10]$ and $p_{m1} \in [0.2, 0.6)$, respectively. In addition, $\phi(f)$ stands for the function relating the objective function value of the GWO algorithm, as shown in the following equation.

$$\phi(f) = \begin{cases} 0 & if \quad \frac{f_{avg}}{f_{max}} > h_f \\ 1 & if \quad \frac{f_{avg}}{f_{max}} \le h_f \end{cases}$$

where f_{avg} represents the average value of objective functions of all gray wolves (suppose that all objective function values are more than or equal to zero). In addition, f_{max} refers to the objective function value corresponding to head wolf and h_f stands for the coefficient ($h_f \in [0.85,1.0)$). Obviously, if f_{max} and f_{avg} are close, it shows that the diversity of the population decreases. That is to say that, the population activities decrease, so it calls for mutation with great probabilities to improve the population activities.

4 NUMERICAL EXPERIMENTS AND ANALYSIS

4.1 *Test functions*

In order to prove the validity of the Improved Gray Wolf Optimization (IGWO) algorithm based on the logarithmic function proposed in this study, four benchmark test functions are widely used across the world from the literature (Mirjalili, 2014) and are utilized to conduct numerical experiments. The dimensions of the four test functions are all set as 30.

4.2 *Comparing IGWO with GWO and HGWO algorithms*

By using the IGWO algorithm proposed in the study, the four benchmark test functions are solved. In addition, the IGWO algorithm is compared with the standard GWO algorithm and Hybrid GWO algorithm with differential evolution (HGWO) (Zhu, 2015). In order to ensure fairness, the parameters of the three algorithms are set as follows: the population size and and maximum iteration times are set as N = 30 and 500 separately. Each function independently operates for 30 times with above parameters and their optimal values, average values, worst values, and standard deviations are recorded. Additionally, the comparison results are shown in Table 2. In which, the results of the HGWO algorithm are directly excerpted from the literature (Zhu, 2015). The bold figures in Table 2 represent the best results in the comparison.

Table 1. List of four benchmark test functions.

Function name	Dimension	Function expressions	Searching regions	f_{min}	Convergence precision
Sphere	30	$f_1(x) = \sum_{i=1}^{n} x_i^2$	[–100,100]	0	1×10^{-10}
Schwefel 2.22	30	$f_2(x) = \sum_{i=1}^{n} \lvert x_i \rvert + \prod_{i=1}^{n} \lvert x_i \rvert$	[–10,10]	0	1×10^{-10}
Schwefel 1.2	30	$f_3(x) = \sum_{i=1}^{n} \left(\sum_{j=1}^{i} x_j \right)^2$	[–100,100]	0	1×10^{-10}
Schwefel 2.21	30	$f_4(x) = \max_i \{\lvert x_i \rvert, 1 \le x_i \le n\}$	[–100,100]	0	1×10^{-10}

Table 2. Optimization results of the four functions using the three algorithms.

Function	Algorithm	Optimal value	Average value	Worst value	Standard deviation
f_1	GWO	8.42E-29	8.08E-28	1.71E-27	6.09E-28
	HGWO	2.92E-34	1.12E-32	8.95E-32	2.32E-32
	IGWO	**1.57E-56**	**8.76E-54**	**4.23E-53**	**1.47E-53**
f_2	GWO	1.77E-17	1.97E-16	6.66E-16	1.82E-16
	HGWO	1.65E-20	9.33E-20	3.60E-19	6.92E-20
	IGWO	**2.96E-33**	**2.66E-32**	**6.75E-32**	**2.53E-32**
f_3	GWO	2.53E-09	2.82E-05	3.44E-04	1.25E-04
	HGWO	6.07E-11	3.18E-08	3.08E-07	6.55E-08
	IGWO	**1.09E-17**	**1.52E-12**	**6.41E-12**	**2.22E-12**
f_4	GWO	1.34E-07	7.96E-07	2.03E-06	6.76E-07
	HGWO	5.81E-09	4.17E-08	2.39E-07	4.56E-08
	IGWO	**1.98E-16**	**7.99E-15**	**2.08E-14**	**6.73E-15**

Table 3. A comparison of oxygen consumption.

Working conditions	Oxygen additive amount optimized using the IGWO algorithm	Oxygen additive amount optimized in the literature (Xiong, 2013)	Oxygen additive amount in the actual working operation
1	517.97	529.80	646
2	539.23	550.19	674
3	551.33	552.91	708
4	577.31	591.31	732

As shown in Table 2, the IGWO algorithm proposed in the study is obviously superior to the other two algorithms.

5 APPLICATION IN INDUSTRIAL OPTIMIZATION

Zinc sulfate solution leached in the process of zinc hydrometallurgy contains many iron ions. In order to ensure the safe and stable operation of subsequent procedures, the iron ions in the solution need to be removed to a proper content. Removing irons by using goethite is to oxidize ferrous irons in the solution gradually into ferric irons by adding oxygen into five successive reactors. The optimization model of the iron removal process by using goethite is established in the literature (Xiong, 2013). In addition, four typical working conditions proposed in the literature (Xiong, 2013) are optimized by applying the IGWO algorithm proposed in the study. The overall optimal values of the oxygen additive amount in the five reactors are given in Table 3.

From Table 3, it can be seen that under the four typical working conditions of precipitating iron using goethite, the oxygen additive amounts optimized using the IGWO algorithm were separately saved by 19.82%, 19.99%, 22.12%, and 21.13% when compared with those in the practical commercial process. The optimized oxygen additive amounts were 1.82%, 1.59%, 0.22%, and 1.93% less than those optimized using the double population coevolution algorithm, respectively.

6 CONCLUSIONS

The simulation experiment results of four benchmark test functions show that, when compared with other swarm intelligent optimization algorithms, the IGWO algorithm is more competitive. The simulation results of the industrial process of iron removal using goethite show that the IGWO algorithm is applicable to the parameter optimization of the complex industrial process.

ACKNOWLEDGMENTS

This work was partially supported by the National Science Foundation of China (61503131), Loudi Science and Technology Project (No. 16), Scientific Research Fund of Hunan Provincial Education Department (No. 14B097 and No. 15c0722), and Natural Science Foundation of Hunan Province

(2016 JJ3079) and Hunan Institute of Humanities Science and Technology University Youth Project (2009QN04).

Corresponding author: Liu Jianhua, E-mail: jhliu0615@163.com

REFERENCES

El-Fergany A, Hasanien H. Single and multi—objective optimal power flow using grey wolf optimizer and differential evolution algorithms[J]. Electric Power Components and Systems, 2015, 43(13): 1548–1559.

Hydrometallurgy [J]. Control and Decision, 2013, 28(4): 590–594.

Madadi A, Motlagh M. Optimal control of DC motor using grey wolf optimizer algorithm [J]. Technical Journal of engineering and Applied Science, 2014, 4(4): 373–379.

Mirjalili S, Mirjalili S M, Lewis A. Grey wolf optimizer [J]. Advances in Engineering Software, 2014, 69(7): 46–61.

Xiong Fuqiang, Gui Weihua, Yang Chunhua, et al. A double population co-evolution algorithm for process of zinc.

Zhu A, Xu C, Li Z, et al. Hybridizing grey wolf optimization with differential evolution for global optimization and test scheduling for 3D stacked SoC [J]. Journal of Systems Engineering and Electronics, 2015, 26(2): 317–328.

Civil, Architecture and Environmental Engineering – Kao & Sung (Eds)
© 2017 Taylor & Francis Group, ISBN 978-1-138-02985-9

Determination of multi-element in abstergent by ICP-MS with microwave digestion

BeiLei Wu, ZhenXin Wang & JiaMei Ye
China Certification & Inspection Group NingBo Co. Ltd., NingBo, P.R. China

ZhenXing Lin, Hao Wang & Chuan Luo
Ningbo Entry-Exit Inspection and Quarantine Bureau, NingBo, P.R. China

ABSTRACT: An efficient method was developed for the determination of the multi-element in abstergent by ICP-MS with microwave digestion. Using the pressure control device to monitor the process of digestion and to explore the conditions of digestion, the parameter of ICP-MS and appropriate inner internal standard element were optimized and chosen in terms of the amount of digestion, digestion system and the procedure of microwave digestion. The detection limit was between 0.002~0.362 ng/ml. The relative standard deviations were less than 10%. The spiked recoveries were 90.8%~109.6%.

1 INTRODUCTION

The detergent is one kind of inseparable chemical in daily life. Benefiting from the new technology, detergent producers may add some harmful ingredients during the production process in order to obtain better performance for their products, which will lead to the environment deterioration. For example, the phosphorus detergent will make rivers and lakes become eutrophic, which can lead to serious fish deaths, red tide, etc. and heavy metals accumulating in the environment. The heavy metals will be transferred into the animal and human body through the food chain, finally causing teratogenic and carcinogenic diseases.

Since the 1960s, Europe and the United States have successively promulgated a number of directives and standards for detergent specifications. The EU released 2011/382/EU and 2011/263/EU. The USA released GS-8 and GS-11. China has promulgated and implemented several national and industrial standards, such as GB/T 9985-2000 and HJ 458-2009 standards, which standardized the composition and dosage of the detergent. According to those directives and standards, there are 14 kinds of heavy metals having been severely restricted, but the existing detection technology is still stuck in traditional chemical detection method. It is necessary to have a quick, effective and accurate detection method. Therefore, it is necessary to study the efficient detection method for the 14 kinds of heavy metals.

Currently, pre-treatment methods are dry ashing, reflux wet digestion, catalytic wet digestion and extraction. Those methods have disadvantages, such as waste of reagents, long-time consuming, easy sample contamination, low efficiency, poor accuracy, etc. Microwave Digestion is a new sample digestion technology in recent years, with small reagent consumption, shorter operation time, uneasy to pollution, great precision and other advantages (Chen, 2009; Liu, 2005; Chen, 2006; Luo, 2013; Luo, 2015). The microwave digestion-inductively coupled plasma mass spectrometry is used to determinate multi-element in abstergent.

2 EXPERIMENTAL

2.1 *Instruments*

Agilent 7700x ICP-MS (Agilent Technologies Co. Ltd, USA), replace the injection set and torch set to resistant to hydrofluoric acid special set. Milli-Q water purification system (EMD Millipore Co., Ltd USA); MARS microwave digestion system (CEM Technologies Co. Ltd, USA), with pressure control device.

2.2 *Reagents*

All reagents were of analytical-reagent grade or higher. Nitrate acid (71%) and Hydrogen peroxide (30%) (Suzhou Crystal Clear Chemical Co., Ltd, China); Hydrofluoric Acid (Sinopharm Chemical

Table 1. ICP-MS optimization parameter for determination multi-element.

Parameter	Value	Parameter	Value
RF Power	1550 w	Spray chamber temperature	2°C
Sample depth	8.0 mm	Assistant gas flow	0.13 L/min
Carrier gas flow	1.0 L/min	Reaction gas flow (He)	0.43 L/min

Reagent Co., Ltd, China); Multi-element Calibration standard 2 A (Agilent Technologies Co. Ltd, USA) contains 27 elements, including the 9 kinds of elements measured in this article. Concentrations were 10 mg/kg; Single standard of Tin (1.0 mg/mL) and Mercury (1.0 mg/mL) (Central Iron and Steel Research Institute, China). Single standard and Multi-element Calibration standard 2A was diluted by 5% nitric acid solution, with the configuration of 100 ng/mL, 50 ng/mL, 20 ng/mL, 10 ng/mL and 1 ng/mL standard working solution standbys.

2.3 *Analytical procedures*

0.2 g (accurate to 0.1 mg) abstergent sample was weighed and placed at PTFE microwave digestion barrels. Then 6 mL nitric acid was added. After 3 hours, 2 mL hydrogen peroxide was added, the lid was closed, and the barrels were put into the microwave digestion system. After the program of digestion completed, there was a natural cooling process, followed by transferring the solution to a 50 mL calibrated flask. Then the solution was diluted to a vessel with ultrapure water. Reagent blanks were included in each series of digestions. Sample solution was analysed by ICP-MS.

The parameter of the ICP-MS was shown in the Table 1.

3 RESULTS AND DISCUSSION

3.1 *Amount of sample*

There are variety kinds of detergents. Most of the detergents contain volatile organic compounds. Therefore, the digestion process will release a lot of gas when the sample weight is too heavy, resulting in insecurity. However, it will reduce accuracy of the determination of certain elements when the sample weight is too light. By choosing different sample weight in the experiments, the digestion tank pressure would be controlled in the safe range when the sample weight is 0.93 g, as shown in Figure 1. It is recommended to control the sample

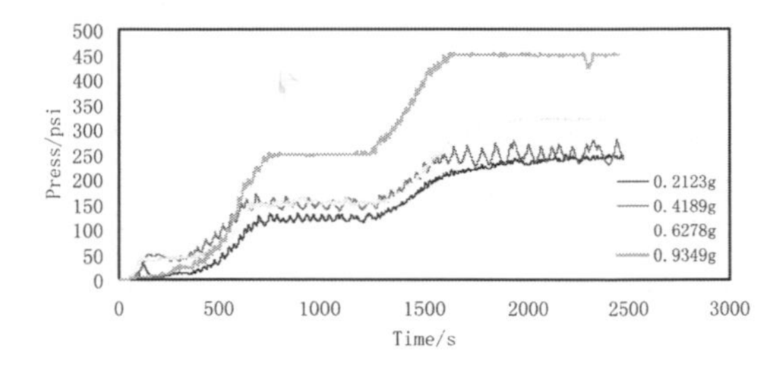

Figure 1. Effect of the sample weight on pressure of the microwave digestion system.

weight of microwave digestion in the range of 0.5~1.0 g to avoid experiment accidents.

3.2 *The digestion solution*

Nitric acid was used as digestion reagents in experiments. In order to determine a proper amount of nitric acid, 4 mL, 6 mL and 8 mL HNO_3 were used during washing supplies microwave digestion sample test. When the amount of nitric acid reached 6 mL, the sample could be digested well with transparent digestion solution. Peroxide hydrogen was added to help reduce stress and promoting digestion. Tests showed that for the detergent sample with weight of 0.5 g, 2 ml peroxide hydrogen was needed. Microwave digestion with medium pressure led to gentle reaction which conduced to shorten experiment time and safe operation.

3.3 *The program of microwave digestion*

Microwave digestion process was studied. Detergents can have several states, such as shower gel is in good mobility as a liquid, cleanser presents paste and detergent is in powder state. Therefore, to study digestion procedures for different samples must use different experimental temperature. The affection of temperature, time and temperature program were studied by a segmented temperature program, heating up to 120°C in the first 5 minutes, holding 2 minutes, then raising temperature to 170°C in next 5 min, holding 2 minutes, finally raising temperature to 180°C in 2 min, holding 20 minutes. The heating power of the whole digestion process is set up to 1200 W as constant power.

3.4 *Linear equation and detection limit*

Linear equation was created by the five different concentrations of solution; the detection limit of each element was calculated by the linear equation. Linear equation and detection limit of each element were shown in the Table 2.

Table 2. Linear equation and detection limit of each elements.

Element	Mass	Linear equation*	r	Detection limit, ng/g
V	51	Y = 0.0352x + 0.000209	1.0000	0.002357
Cr	52	Y = 0.0471x + 0.0040	1.0000	0.04334
Mn	55	Y = 0.0265x + 0.0048	1.0000	0.3509
Fe	56	Y = 0.0419x + 0.0930	0.9999	0.2681
Ni	60	Y = 0.0235x + 0.0083	1.0000	0.08324
Cu	63	Y = 0.0647x + 0.0335	1.0000	0.0746
Zn	66	Y = 0.0112x + 0.0159	1.0000	0.3625
As	75	Y = 0.0024x + 0.00007602	1.0000	0.04471
Se	78	Y = 0.0000972x + 0.00001032	1.0000	0.3185
Cd	111	Y = 0.0013x + 0.0000151	1.0000	0.008807
Sn	118	Y = 0.0033x + 0.0004836	1.0000	0.04664
Ba	137	Y = 0.0007375x + 0.0001442	1.0000	0.1134
Hg	202	Y = 0.0006519x + 0.00001294	0.9997	0.02475
Pb	208	Y = 0.0082x + 0.0034	1.0000	0.01491

*: X: standard concentration; Y: signal intensity.

Table 3. Accuracy and recovery of the method of the Shampoo.

Element	RSD, %, (n = 7)	Recovery, %, (n = 5) (Add 50 ng standard)
V	4.3	97.9
Cr	5.4	96.4
Mn	4.8	98.8
Ni	2.2	96.5
Cu	1.2	91.9
Zn	3.0	107.5
As	2.4	93.9
Se	3.4	90.8
Cd	4.4	100.4
Sn	2.3	105.8
Ba	4.6	103.5
Hg	3.6	108.5
Pb	4.3	104
V	4.3	97.9
Cr	5.4	96.4
Mn	4.8	98.8

3.5 *Accuracy and recovery of the method*

The accuracy of the developed method was tested with recovery experiments by spiking standard solution into the different kinds of sample solution. The analytical results and recoveries are shown in

Table 4. Accuracy and recovery of the method of the Shaving cream.

Element	RSD, %, (n = 7)	Recovery, %, (n = 5) (Add 50 ng standard)
V	3.9	104.3
Cr	5.3	106.7
Mn	4.1	104.8
Ni	5.0	103.7
Cu	3.0	96.0
Zn	2.9	95.1
As	2.1	103.5
Se	2.7	96.2
Cd	5.8	94.4
Sn	7.0	106.6
Ba	5.8	93.6
Hg	5.1	105.5
Pb	1.2	106.6
V	3.9	104.3
Cr	5.3	106.7
Mn	4.1	104.8

Table 5. Accuracy and recovery of the method of the Cleaning liquid.

Element	RSD, %, (n = 7)	Recovery, %, (n = 5) (Add 50 ng standard)
V	3.7	105.5
Cr	7.1	108
Mn	0.4	109.6
Ni	5.0	106.8
Cu	2.3	104
Zn	8.2	99.8
As	3.4	107.3
Se	4.3	105.3
Cd	2.9	96.6
Sn	5.2	103.4
Ba	3.1	93.4
Hg	3.1	103.3
Pb	3.5	104.6
V	3.7	105.5
Cr	7.1	108
Mn	0.4	109.6

Table 3, Table 4 and Table 5 with different sample. The relative standard deviations were 0.4% ~ 8.2%. The spiked recoveries were 90.8%~109.6%.

4 CONCLUSIONS

A rapid method was established for the determination of 14 kinds of harmful elements in detergent by using nitric acid and peroxide hydrogen for sample digestion and using ICP-MS for simultaneous determination of the harmful elements with a wide linear range and better accuracy. Microwave digestion process is monitored by a pressure control device, meanwhile the amount of solvent for digestion, acid digestion system and procedures have been optimized so that the digestion effect of various sample is good. Optimizing the parameters of ICP-MS, the accuracy of this method was judged by standard addition recovery methods. Reproducibility of this method was validated by the repeated measurement samples.

ACKNOWLEDGEMENTS

This work was financially supported by General Administration of Quality Supervision, Inspection and Quarantine of the People's Republic of China (Fund No.: 2013IK303) and the Ningbo social science development research project (Fund No.: 2012C50049).

REFERENCES

2011/263/EU Commission Decision of 28 April 2011 on establishing the ecological criteria for the award of the EU Ecolabel to detergents for dishwashers (Notified under document C (2011) 2806) (1) [S].

2011/382/EU Commission Decision of 24 June 2011 on establishing the ecological criteria for the award of the EU Ecolabel to hand dishwashing detergents (notified under document C(2011) 4448) (1) [S].

Chen Guoyou, Du Yingqiu, Ma Yonghua. Chinese journal of analytical chemistry, VOL. 37. No. 10: A058 (2009).

Chen Yuhong, Zhang Hua, Shi Yanzhi, etc. Environmental chemistry VOL 25. No. 4: 520 (2006).

Chuan Luo, Shujun Dai, Jiamei Ye, Advances in Energy Science and Equipment Engineering, ICEESE 2015, 30 MAY 2015, Guangzhou, China.

GB/T 9985-2000 Detergents for hand dishwashing [S].

GS-11 Paints and Coatings [S].

GS-8 Household Cleaning Products [S].

HJ 458-2009 Technical for environmental labeling products Household detergents [S].

Hygienic Standard for Cosmetics (2007 Edition).

Liu Conghua, Huang Lina, Yu Yidong. Chinese Journal of Analysis laboratory VOL. 24. No. 2:66 (2005).

Luo Chuan, BeiLei Wu, Ying Zhang. Asian Journal of Chemistry VOL. 25, No 12, 6993 (2013).

Civil, Architecture and Environmental Engineering – Kao & Sung (Eds)
© 2017 Taylor & Francis Group, ISBN 978-1-138-02985-9

Fabrication and application of superhydrophobic-superoleophilic porous Cu sponge

Yingjie Xing & Zhengjin Han
Key Laboratory for Precision and Non-Traditional Machining Technology of Ministry of Education, Dalian University of Technology, Dalian, China

Jinlong Song
Collaborative Innovation Center of Major Machine Manufacturing in Liaoning, Dalian University of Technology, Dalian, China
Key Laboratory for Precision and Non-Traditional Machining Technology of Ministry of Education, Dalian University of Technology, Dalian, China

Jing Sun, Jie Luo, Ziai Liu, Jiyu Liu, Liu Huang & Xin Liu
Key Laboratory for Precision and Non-Traditional Machining Technology of Ministry of Education, Dalian University of Technology, Dalian, China

ABSTRACT: Superhydrophobic-superoleophilic materials arises more and more attention because of their possible application on oil/water separation. Here, chemical oxidation and palmitic acid modification were used to fabricate superhydrophobic-superoleophilic surface on porous Cu sponge. Scanning Electron Microscope (SEM), Energy Dispersive Spectrometer (EDS) and Intelligent Roman Spectroscopy (DXR) were used to detect the surface micromorphology, chemical composition and wettability of the Cu sponge. The results show that the nanometer-scale structures with the composition of CuO and Cu_2O were covered on the whole 3D skeleton of Cu sponge after chemical oxidation and the resulted surfaces show superhydrophilicity and superoleophilicity. After palmitic acid modification, the surface wettability was transferred into superhydrophobicity with contact angle of 156° and tilting angle of smaller than 10°, but still showing superoleophilicity. Based on the fabricated superhydrophobic-superoleophilic porous Cu sponge, pouring-type, gravity driven oil/water separation and surface-tension-driven, gravity assisted oil/water separation was realized with a high separation efficiency.

1 INTRODUCTION

1.1 *Type area*

Water and oil are the important resource of our daily life. The qualities of water and oil affect numerous aspects of the human productivity. The waste water in oil seriously affects the quality and service efficiency of the oil. Research on oil/water separation technology has great significance in terms of improving the quality of oil and water. Researchers have showed keen interest in the special performance of the extreme wettability in some materials in the nature (Barthlott W & Neinhuis C 1997). Thanks to the different applications in the special wettability of water and oil, achieving oil/water separation has been in the center of public concern (Sanhu Liu et al. 2015). Porous materials like fibrous membranes (Viswanadam G 2013), filter paper (Zhang M et al. 2012), metal net (Lee C H et al. 2011, Wang C X et al. 2009) and polymeric sponge (Yang Y et al. 2014) are all applied to the research of superhydrophobic surface preparation and oil/water separation. Although the aforementioned materials have favorable extreme wettability, there are limitations in the separation efficiency, separation velocity, pressure resistance, and stability of resisting water attack.

With the characteristics of high porous rates, large specific surface area and high specific strength, porous Cu sponge has been a new object of study in the oil/water separation area. Porous Cu sponge is a three-dimensional and multi-functional material with a large amount of connected or non-connected holes in the Cu substrate. It presents good machinability and processing properties and has excellent electromagnetic shielding and noise elimination ability. Simultaneously, porous Cu sponge still retains the properties of Cu-like good electrical conductivity, excellent thermal conductivity, ductility and strong base solution in reserve, thus having a wide range of application prospects such as electronics, medical, mechanics and national defense (Wen Zhang et al. 2015). In this paper, a simple method that combines chemical oxidation and low surface energy modification was used to construct nano-roughness and achieve the superhydrophobicity. The influence of oxidation time on the surface structure was studied in this paper, and the extreme wettability, water pressure resistance, oil/water separation efficiency of the porous Cu sponge were studied at the same time.

2 EXPERIMENTS

2.1 *Materials*

Porous Cu sponge (99.9%, pore size 160 μm) was purchased from Kunshan dessco Industrial Co., Ltd. acetone, HCl (37%), H_2O_2 (30%), ethanol, palmitic acid, hexane, hexadecane, mineral oil was purchased from Tianjin Kemiou Chemical Reagent Co., Ltd. Corn oil was purchased from Luhua Co.

2.2 *The preparation of superhydrophobic–superoleophilic porous Cu sponge*

Firstly, porous Cu sponge was cut into 20 mm × 20 mm squares and ultrasonically cleaned by acetone, 1 mol/L dilute hydrochloric acid and deionized water in sequence for two minutes. To prevent rusting of porous Cu sponge after cleaning, its moisture was mostly absorbed by absorbent paper immediately and porous Cu sponge was then put into the oven with 60°C to dry and for later experiments. After that, the samples were heated in the boiling water, and three-or-four drops of 30% H_2O_2 water solution were dropped into water every ten minutes for four times. After 40 minutes, the samples were taken out and cleaned before put in the oven with 60°C. Then, the dried porous Cu sponge was put in 0.1 mol/L ethanol solution of palmitic acid at room temperature for 30 minutes. Finally, the superhydrophobic-superoleophilic porous Cu sponge could be obtained after cleaning by absolute ethanol.

2.3 *Characterization*

Scanning Electron Microscope (SEM) was employed to observe the surface micro topography of the samples. Energy Dispersive Spectrometer (EDS) was used to characterize surface elements of the samples. Raman spectrometer was employed to qualitatively analyze ingredients of the samples.

The contact angle and rolling angle measuring instruments were used to characterize wettability of porous Cu sponge at room temperature. When measuring the static water contact angle, 5 μL water droplets were dropped on one sample surface for 5 different places, and the average value was regarded as the measured data. During measurement of the rolling angle, the samples were rotated, and the average of 5 measurements values with different rolling direction were taken as the rolling angle of the sample surface.

Home-made gravity-driven oil/water separator and surface tension-gravity dual driven oil/water separator (Song J L et al. 2015) were used to conduct the oil/water separation experiments of four types of oil with less dense than water. The oils used are hexane, hexadecane, mineral oil, corn oil. The corresponding surface tensions are 17.9 mN/m, 27.5 mN/m, 33.4 mN/m and 32.15 mN/m. The surface tension of water is 72.1 mN/m.

3 RESULTS AND DISCUSSION

3.1 *The construction of nanostructure in the surface of porous Cu sponge*

Figure 1(a-f) shows the SEM images of porous Cu sponge oxidized in the boiling water for different time. It can be observed that the surface morphology of the porous Cu sponge changes obviously with the oxidation time. When the oxidation time was relatively shorter (10 min), the nanoparticles on the surface of porous Cu sponge formed gradually, their sizes were quite different, and the distribution of them was uneven (Figure 1(b)). When the oxidation time was 20 min, porous Cu sponge surfaces were evenly covered by a layer of nanoparticles

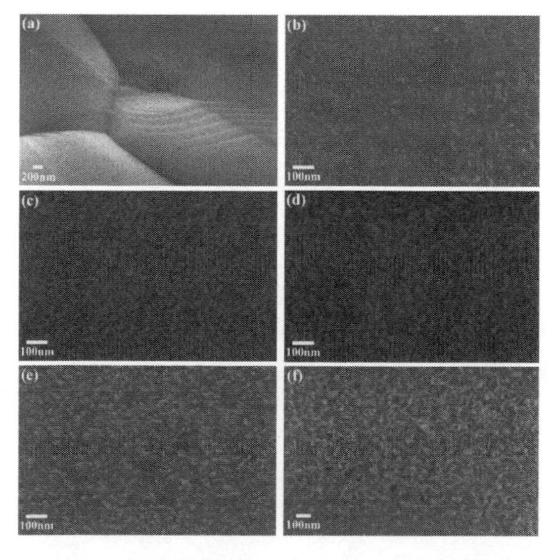

Figure 1. SEM images of porous Cu sponge at different oxidation times: (a) 0 min; (b) 10 min; (c) 20 min; (d) 30 min; (e) 40 min; (f) palmitic acid modification.

with the uniform size, and the diameter of the nanoparticles was about 10 nm (Figure 1(c)). When the oxidation process lasted for 30 min, the nanoparticles became larger, and a dense and even oxide layer was formed on the porous Cu sponge surface with the diameter of nanoparticles being 20 nm (Figure 1(d)). When the oxidation time was 40 minutes, the nanoparticles developed upward and became column-like with the height of 30 nm (Figure 1(e)). As shown in Figure 1(f), after the porous Cu sponge was oxidized for 40 minutes followed by immersing in 0.1 mol/L palmitic acid ethanol solution for 30 minutes, a lot of even flocs with the length of 100 nm appeared on the surface. This demonstrates that the porous Cu sponge formed nanoscale oxide layers when being boiled, and that rough nanostructures were fabricated by the self-assembly reaction during the modification process of palmitic acid ethanol solution.

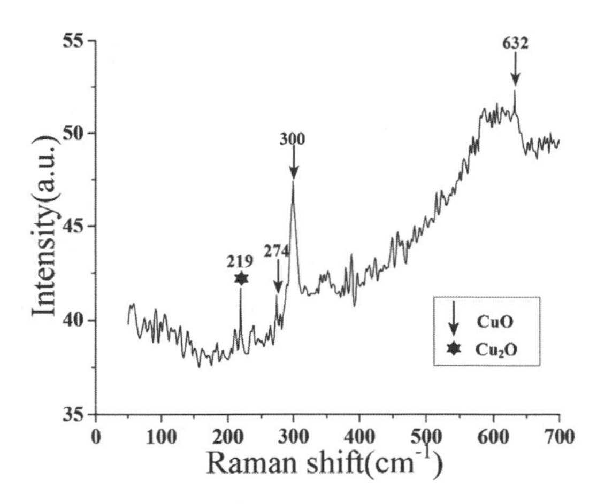

Figure 3. DXR spectrum of porous Cu sponge after oxidation.

3.2 *Analysis of the composition of porous Cu sponge surface*

EDS was employed to detect the surface elements of samples, as shown in Figure 2. Before the oxidation process, porous Cu sponge just has Cu element (Figure 2(a)). After oxidization for 40 min, there was O element on the surface of porous Cu sponge (Figure 2(b)), thus further confirming the oxidation reaction between Cu and the oxygen from the H_2O_2 in the boiling water. Figure 2(c) shows the EDS result of the porous Cu sponge surface after 40 minutes oxidation and 30 minutes modification of palmitic acid ethanol solution, it can be observed that C element appeared on the surface of porous Cu sponge and the content was 18.26wt%. In addition, the content of oxygen element also increased, which demonstrates the self-assembly process of palmitic acid ethanol solution.

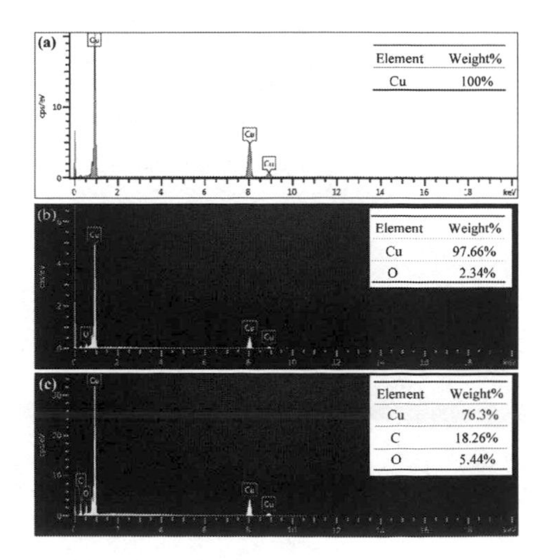

Figure 2. EDS spectrum of porous Cu sponge: (a) ordinary porous Cu sponge; (b) oxidation treatment 40 minutes; (c) palmitic acid modification.

The ingredients of the samples surface were qualitatively analyzed by Raman spectroscopy. Figure 3 shows the spectrum of the porous Cu sponge surface after 40 minutes oxidation. The peaks around 274, 300 and 632 cm^{-1} are attributed to CuO (Masudy-Panah S et al. 2015), while the one at 220 cm^{-1} was contributed by Cu_2O (Abdel-fatah M et al. 2016). During the modification process of palmitic acid ethanol solution, CuO and palmitic acid reacted and generated Cu palmitate, a low surface energy substance, which has a low enough surface energy to be superhydrophobic (Sheng Lei. 2014).

3.3 *Wettability of the porous Cu sponge surface*

Figure 4 is the digital photo of static water droplets on porous Cu sponge which have been processed differently. Oxidation and modification process have great influence on the wettability of porous Cu sponge surface. For a clean ordinary porous Cu sponge surface, water droplet can be semi spherical on it and the static contact angle of water is 118° (Figure 4(a)). Even if overturning the porous Cu sponge, water droplet will not fall off from the surface. Palmitic acid is used to reduce the surface energy of porous Cu sponge. As a result of the slowly self-assembly reaction between palmitic acid and porous Cu sponge, Cu palmitate which has low surface energy generated on the surface. After modification by palmitic acid, the wettability of porous Cu sponge surface was reduced significantly and the static contact angle of water

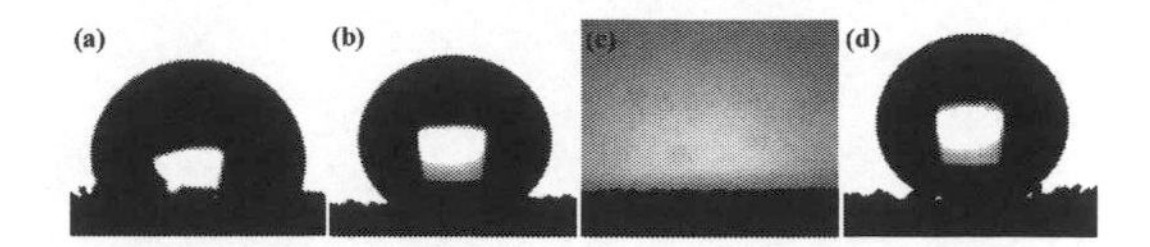

Figure 4. The static contact angles of water droplets on porous Cu sponge: (a) ordinary porous Cu sponge; (b) after modification; (c) after oxidation; (d) after oxidation and modification.

could be 134° (Figure 4(b)), showing hydrophobicity. However, the reaction between palmitic acid and Cu is very slow and the porous Cu sponge surface cannot obtain superhydrophobicity in a short time. Figure 4(c) shows the porous Cu sponge that has been oxidized in the boiling water for 40 minutes. Uniform nanostructures are formed on the surface which shows superhydrophilicity and water droplet can spread rapidly on it with a static contact angle of 0°. The superhydrophilic porous Cu sponge can obtain superhydrophobicity after modification by palmitic acid which has a static contact angle of 156° with a tilting angle smaller than 10°. The water pressure resistance of the porous Cu sponge was measured by using a homemade pressure-resistant measuring device and the porous Cu sponge can withstand 1.1 kPa pressure. The results clearly indicate that the superhydrophobicity of the porous Cu sponge surface is owing to the joint action of oxidation and low-surface energy material modification.

3.4 *Oil/water separation characteristics of the porous Cu sponge*

Oil/water separation methods include biodegradable methods (Qiqi Shi et al. 2009, Huan Liu & Yongming Sun. 2014), chemical treatments (New Japan Chemical Co Ltd. 1982, Zhonghua Yao. 1993) and traditional physical treatments (Eiichi Kobayashi et al. 2011, Yongge Wang. 2011). However, these methods have a great number of drawbacks: biodegradation methods have a high demand to the environment; chemical treatments are easy to cause secondary pollution; traditional physical treatments are cumbersome and the final recovery of oil usually needs to take further separation. The extreme wettability of the surfaces can make oil and water show two different kinds of wettability. Using these materials can avoid the disadvantages of the conventional oil/water separation methods.

3.4.1 *Pouring type gravity driven oil/water separation*

In the pouring type gravity-driven oil water separation process, due to the fact that the light oil firstly contacted and wetted the filter, the after reaching water could not pass through the superhydrophobic filter and formed a barrier below the light oil which hinders the contact of the light oil. Superhydrophobic filter could not use this method to separate light oil water mixture. Similarly, dry-superhydrophilic filter could not use this method to separate light oil water mixture either. When using the superhydrophilic filter which was pre-wetted by water, due to the microstructure of the filter screen was filled with water, light oil could not pass through the filter. In contrast, water would contact the filter due to its greater density to light oil (Song, J.L. 2015). Therefore, the superhydrophilic porous Cu sponge pre-wetted by water in pouring gravity-driven water separation was used, as shown in Figure 5. Figure 6 is the oil/water separation efficiency of n-hexane, hexadecane, mineral oil and corn oil. The maximum of separation efficiency can be up to 99.4%. Separation efficiency could maintain 90% after repeating the separation

Figure 5. Oil/water separation using pouring type gravity driven method.

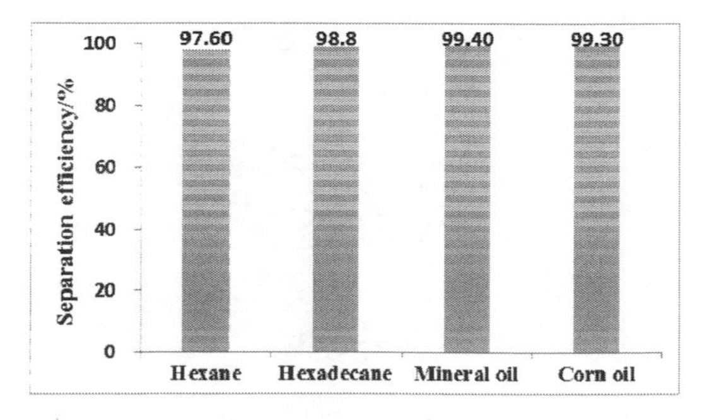

Figure 6. The oil/water separation efficiency of pouring type gravity driven method.

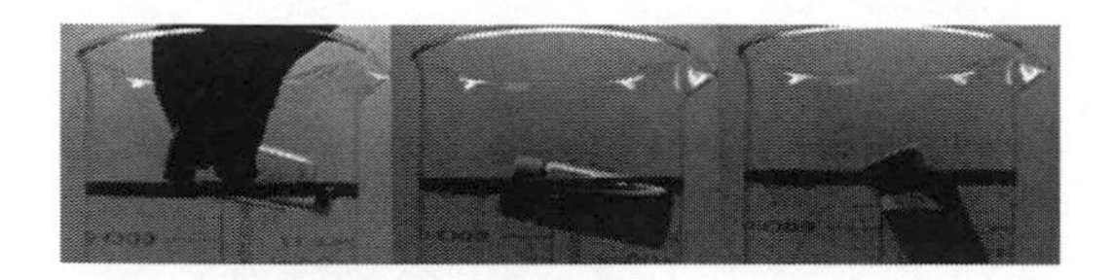

Figure 7. The oil/water separation process driven by both surface tension and gravity.

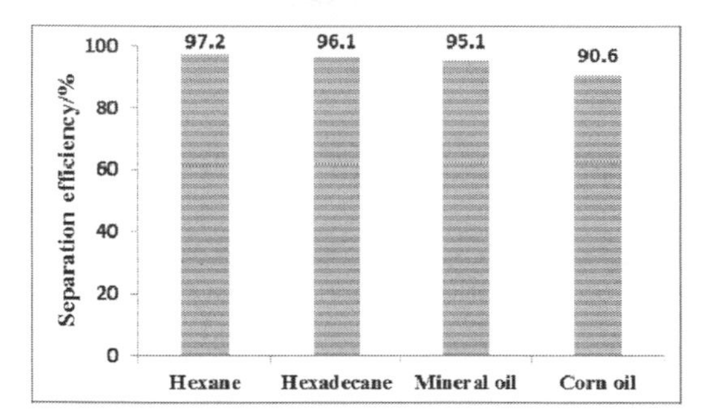

Figure 8. Oil/water separation efficiency of surface tension-gravity dual driven method.

experiment by 10 times, showing a good oil/water separation recycling capacity.

3.4.2 *Oil/water separation driven by both surface tension and gravity*

According to the reference paper, we installed the prepared superhydrophobic porous Cu sponge in the glass tub which has an opening at the top to constitute an oil skimmer. During the oil/water separation and the collection of the oil slick, the skimmer was placed in the mixture of oil and water. The oil skimmer can float on the water due to buoyancy and part of the superhydrophobic porous Cu sponge could contact with the oil slick and water below. Owing to the superhydrophobic porous Cu sponge has different wettability to water and oil, oil slicks could penetrate the Cu when the oil slick contacted with porous Cu sponge. Then oil flowed into the bottom of skimmer, and water was prevented from the skimmer. The process of using an oil skimmer to collect the oil slick was shown in Figure 7. Figure 8 shows the oil water separation efficiency of collecting n-hexane, hexadecane, mineral oil and corn oil mixture using this method, which shows a high oil/water separation efficiency more than 90%.

4 CONCLUSION

A simple chemical oxidation process to construct nano-roughness on the surface of porous Cu sponge was suggested. The surface morphology of porous Cu sponge was gradually changed from a smooth surface to a rough surface covered with Cu_2O and CuO nanoparticles with the increasing of the oxidation time. At last, the surface becomes a homogeneous nano-columnar structure, forming nanometer-scale rough structures. With the help of nano-roughness, the Cu sponge shows superhydrophilicity and superoleophilicity. After palmitic acid modification, the surface wettability was transferred into superhydrophobicity with contact angle of 156° and tilting angle of smaller than 10°, but still showing superoleophilicity. The fabricated porous Cu sponge can be used in the oil/water separation with separation efficiency more than 90%.

ACKNOWLEDGEMENTS

This project was financially supported by National Natural Science Foundation of China (NSFC, Grant No.51605078 and No.51275071) and the Fundamental Research Funds for the Central Universities (DUT15RC(3)066).

REFERENCES

Abdelfatah M et al. 2016. Fabrication and characterization of low cost Cu2O/ZnO Al solar cells for sustainable photovoltaics with earth abundant materials. *Sol Energ Mat Sol C*; 145: 454–61.

Barthlott W & Neinhuis C. 1997. Purity of the sacred lotus, or escape from contamination in biological surfaces. *Planta*, 202(1):1–8.

Eiichi Kobayashi et al. 2011. Development of oil Boom System for Stormy Weather. The 11th, 2001. International Offshore and Polar Engineering Conference. *Stavanger, Norway*: 366~373.

Huan Liu & Yongming Sun. 2014. Marine oil spill processing device Overview. *Guide of Sci-tech Magazine*, 06:253+276.

Lee C H et al. 2011. The performance of superhydrophobic and superoleophilic carbon nanotube meshes in water-oil filtration. *Carbon*. 49(2):669–676.

Masudy-Panah S et al. 2015. Optical bandgap widening and phase transformation of nitrogen doped cupric oxide. *J Appl Phys*; 118.

New Japan Chemical Co Ltd. 1982. Gelling agent for oil recovery form water. *JP*: 57 195 174.

Qiqi Shi et al. 2009. Off shore oil spill collection device concept design. *Ship Engineering*, S1:136–139.

Sanhu Liu et al. 2015. Advances in superhydrophobic water separate materials. *Chemical Research*, 06:561–569.

Song J L et al. 2015. Barrel-Shaped Oil Skimmer Designed for Collection of Oil from Spills. *Adv Mater Interfaces*; 2.

Song J L. 2015. Fabrication and Application of Extreme Wettability Surfaces on Engineering Metal Materials. Dalian University of Technology.

Sheng Lei. 2014. Preparation of the superhydrophobic Cu mesh and oil-water separation performance research. Nanchang Hang Kong University.

Viswanadam G. 2013. Water-Diesel Secondary Dispersion using Superhydrophobic Tubes of Nanifibers. *Separation and Purification Technology*, 104: 81–88.
Wen Zhang et al. 2015. Nano Cu porous Cu sponge needle superhydrophobic surface electrochemical Construction technology and properties of oil-water separation. *Journal of Southeast University (Natural Science)*, 01: 69–73.
Wang C X et al. 2009. Facile approach in fabricating superhydrophobic and superoleophilic surface for water and oil mixture separation. *ACS Applied Materials & Interfaces*, 1(11): 2613–2617.
Yang Y et al. 2014. Multifunctional foams derived from poly (melamine formaldehyde) as recyclable oil absorbents. *J Mater Chem A*, 2(2): 9994–9999.
Yongge Wang. 2011. Adherent oil spill recovery equipment design research. Dalian Maritime University.
Zhang M et al. 2012. Fabrication of coral-like superhydrophobic coating on filter paper for water-oil separation. *Appl Surf Sci*, 261(15): 764–769.
Zhonghua Yao. 1993. Preparation and Properties of sorbitol type gelling agent. *Shanghai Environmental University*, 12(11): 24–25.

Civil, Architecture and Environmental Engineering – Kao & Sung (Eds)
 ISBN 978-1-138-02985-9

Preparation of porous ceramics using $CaCO_3$ as the foaming agent

Jin Li & Yanlin Chen

School of Material and Chemical Engineering, Hubei University of Technology, Wuhan, China
Hubei Provincial Key Laboratory of Green Materials for Light Industry, Hubei University of Technology, Wuhan, China
Collaborative Innovation Center of Green Light-weight Materials and Processing, Hubei University of Technology, Wuhan, China

ABSTRACT: Porous ceramics were synthesized by using oily soil and waste glass as the main materials, calcium carbonate as the foaming agent, and sodium silicate as the fluxing agent by controlling the sintering process. The samples were characterized by using XRD and SEM. The influence of the quantity of oily soil used on the porosity and flexural strength of samples were further evaluated. The results show that when the quantity of the oily soil increased, the porosity increased first and then decreased regularly, and the flexural strength decreased first and then increased. When the quantity of the oily soil was 40 wt%, the corresponding highest porosity was 69.44% and the lowest flexural strength was 2.16 MPa.

1 INTRODUCTION

The oil wells are used to drill for oil. In recent years, with the rising demand for oil, a large number of oil wells was established. This leads large quantities of oily soil becoming solid waste. People are not paying attention to this, and oil wells caused large quantities of pile and soil in the atmosphere and water and soil pollution (Cao & Yu 2005, Jia et al. 2009). C. M. Kao et al. (2001) published that oil pollution is one of the world pollution causes. At present, soil pollution in the oil wells is mostly treated by using biological technology processing (Deng et al. 2012, Lösekann et al. 2007, Jones et al. 2008). And for the use of problem is almost a blind spot. There is less reported literature in this arena. Therefore, these have become important problems to be solved, such as how to deal with the waste oil soil and how to recycle waste.

The development of the porous inorganic material materials began in the 1930s. Saint-Gobain company was first developed with calcium carbonate as the foaming agent of porous glass in 1935 (Zhang et al. 2010). At present, China and the United States are the main production agent countries of porous ceramics (Tian et al. 2014). With the improvement of people environmental protection consciousness, people began to attach importance to the sustainable development cycle use of solid wastes. Porous ceramics is prepared by using ceramics as the main raw material, by adding foaming agent and a porous material of various kinds of modified additives (Gao et al. 2014). Its interior is filled with uniform connected or closed pores; it is a light–weight material, and has high strength, flame retardancy and good sound absorption and water retention properties (Qiao et al. 2009). Therefore, the porous ceramics have extremely broad prospects for development in building energy efficiency, noise reduction, and green engineering areas (Zhang & Zeng 2006).

This study describes a method for manufacturing porous ceramics. With oily soil and waste glass as the main raw materials, calcium carbonate as the foaming agent, and sodium silicate as the fluxing agent, preparation of porous ceramics is carried out. In this article, oily soil and waste glass and the relationship between the porosity and flexural strength of samples are studied.

2 EXPERIMENTAL

2.1 *The raw materials of experiments*

The oily soil used in experiment was derived from Dongying Shengli Oil Field. Waste glass was obtained from Wuhan Changli Glass Co., Ltd. The foaming agent calcium carbonate and flux of sodium silicate were purchased from Sinopharm Chemical Reagent Co., Ltd. The main chemical compositions of oily soil and waste glass are given in Table 1.

2.2 *Preparation of porous ceramics*

Oily soil was placed in a ventilated area to dry to achieve half dryness. The oily soil and waste glass

Table 1. Chemical compositions of oily soil and waste glass (wt%).

Raw materials	SiO_2	Al_2O_3	Fe_2O_3	MgO	CaO	K_2O	Na_2O	LOI
Oily soil	62.50	21.00	7.39	2.59	2.47	1.88	1.51	8.42
Waste glass	69.59	5.18	0.53	4.42	5.59	1.00	13.69	0.44

Table 2. Quantity of raw materials provided by oily soil and waste glass in different experimental plans.

	Main raw material ratio (wt%)			
Sample	Oily soil	Waste glass	$CaCO_3$	Sodium silicate
A1	30	45	20	5
A2	35	40	20	5
A3	40	35	20	5
A4	45	30	20	5
A5	50	25	20	5

were placed in two ball milling tanks; water was added to the mixture and grinding balls were prepared such that the ratio of the weight of materials to the ball to the water is 1:2:1. After that, the mixture was milled in a ball mill for 2 hours and the mixture was placed in an oven to dry and ground to powder. Oily soil and waste glass raw materials were made to pass through a 100 mesh screen. According to Table 2, oily soil, waste glass, and calcium carbonate were mixed in different proportions and the mixture was pressed into a substrate at a pressure of 10 MPa. It was placed in a muffle furnace for sintering in an air atmosphere. The sintering system is given as follows: first the speed was increased at the rate of 7°C/min up to reaching 400°C, and then was held for 30 min. The furnace temperature was then increased at the rate of 15°C/min up to reaching the foaming temperature of 850°C, and then was held for 20 min. Finally, the speed was reduced at the rate of 15°C/min down to 600°C, and then was held for 30 min. Samples were removed from the furnace and allowed to cool.

2.3 *The test method of the sample*

The elementary composition and content of the sample were examined by using a scanning X-ray fluorescence spectrometer XRF-1800. The crystalline phases were analyzed by using a D/MAX-RB1 X-ray diffraction instrument. The surface morphology could be observed by using JSM 6390 filament lamps for a scanning electron microscope. The porosity of the sample was measured according to National Standard GB/T 1966-1996 Standards. The flexural strength of the sample was detecting by using an INSTRON-1195 electronic Universal Material Testing Machine (0.5 mm/min loading speed) according to National Standard GB/T 1964-1996.

3 RESULTS AND DISCUSSION

3.1 *SEM analysis*

It illustrates the microstructural changes with an increase of oily soil in Fig. 1. The pore sizes increases with an increasing in oily soil reaching maximum values (251 μm) at the 40 wt%. Further, increasing the content of oily soil at 45 wt% (132 μm) and 50 wt% (47 μm) caused a decrease in the pore size. The formation of crystalline phases and the extent of crystallinity, as well as their dependence on the starting composition and sintering temperature, are other important factors that determine the structural evolution of porous ceramics were studied (Fernandes et al. 2009). The preparation of desirable porous ceramics requires ceramics with low crystallizing tendency (Spiridonov & Orlova 2003). Crystals influence the viscosity of the glass and its foaming behavior. Furthermore, crystals affect the structural integrity of the foam because of the mismatch of thermal expansion coefficients. This can cause opening or cracking of the lamellae between adjacent pores, which enhances further coalescence of the pores during the foaming process (Hasheminia & Yekta 2012). Thus, the different contents of oily soil changed the phase composition of the resulting porous ceramics by an increased rate of the crystallization process. For the samples with lower contents of oily soil, as the crystallization of glasses increases with addition of the oily soil, pore sizes increase and become less homogeneous due to the increasing number of pores (Fig. 1b and c). However, the excess amounts of oily soil tend to limit the foaming ability of the system. This effect is also attributed to the formation of more crystallites on the surface of the glass particles, which would increase the viscosity of the system, thereby resulting in increasing densification. Therefore, the contents of samples (45 wt% and 50 wt%) decrease again. Due to the higher content of crystallites, the pores of 50 wt% are not homogeneous in nature (Fig. 1e).

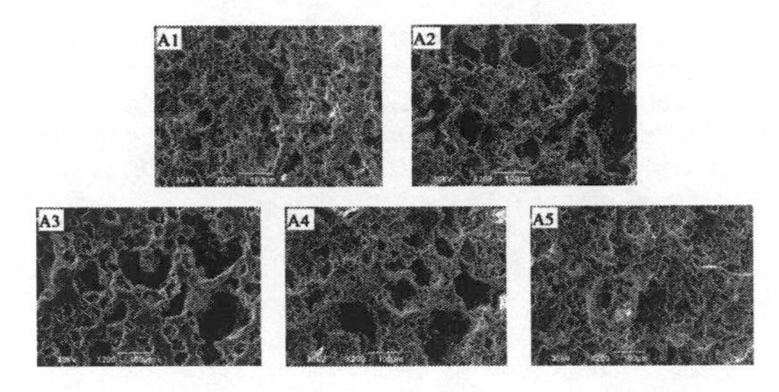

Figure 1. SEM images of samples: (a) A1 (30 wt% oily soil); (b) A2 (35 wt% oily soil); (c) A3 (40 wt% oily soil); (d) A4 (45 wt% oily soil); and (e) A5 (50 wt% oily soil).

3.2 *XRD analysis*

The crystalline phases formed in the samples with different contents of oily soil after heat treatment at 850°C are presented in Fig. 2. The XRD pattern of the content of 30 wt% consists of quart and calcium silicate characteristic peaks. More extensive crystallization of $CaAl_2O_4$ and augite ($Ca(Mg, Fe)Si_2O_6$) phases occurred for the higher added quantities of oily soil and the quartz phase gradually disappears. When the oily soil reached up to 50 wt%, the content is the most crystalline according to the XRD pattern. It is suggested that the contents of oily soil has a great influence on the crystalline phase types of samples since the relative intensity of the peaks change dramatically with an increase in the quantity of oily soil.

3.3 *Porosity and flexural strength analysis*

As shown in Fig. 3, when the content of oily soil was about 30 wt%, 35 wt%, 40 wt%, 45 wt%, and 50 wt%. The porosities of samples were 39.48%, 50.36%, 69.44%, 41.31%, 27.71%, respectively. The flexural strengths were 3.93 MPa, 2.57 MPa, 2.16 MPa, 3.53 MPa, 5.49 MPa, respectively. The porosity was in the range of 27.71%–69.44%. Results show that, at first, an increase in the quantity of oily soil only slightly increased the porosity (69.44%), thus reaching maximum values at the 40wt% oily soil. Further increasing the quantity of oily soil to 50 wt% caused a sharp decrease in the porosity. The Table 3 shows open porosity and closed porosity evolutions of different quantities of oily soil. The content of oily soil has no big influence to open porosity, but it is obvious that the closed porosity first increased and then decreased. When the content of oily soil is less than 40 wt%, as the contents increased, increasing the viscosity of the system has a good effect on the reservation of bubbles, thus the closed porosity increased. However, when the content of oily soil is more than 40 wt%, because of the formation

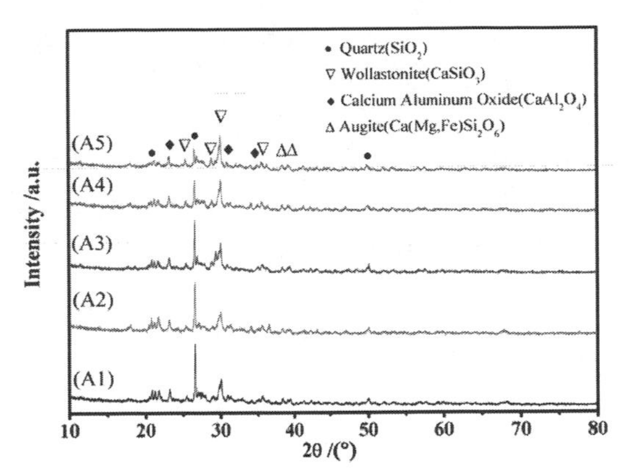

Figure 2. XRD patterns of samples: (a) A1 (30 wt% oily soil); (b) A2 (35 wt% oily soil); (c) A3 (40 wt% oily soil); (d) A4 (45 wt% oily soil); and (e) A5 (50 wt% oily soil).

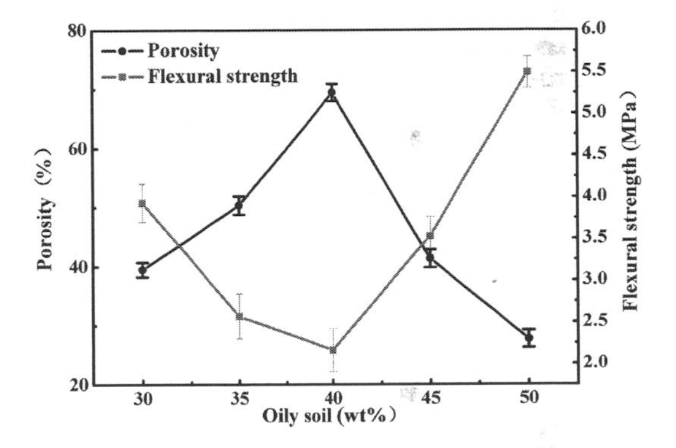

Figure 3. Graph showing porosity and flexural strength evolution for different quantities of oily soil.

Table 3. Open porosity and closed porosity evolutions of different quantities of oily soil.

Oily soil (wt%)	30	35	40	45	50
Open porosity (%)	21.53	22.83	24.15	21.24	22.57
Closed porosity (%)	17.95	27.53	45.29	20.07	5.14

of crystals, the viscosity is too big to be bubbles, and thus the closed porosity decreased. This is in correspondance with the microstructure shown in Fig. 1.

Generally, the porous materials' size and thickness of walls play an important role in the flexural strength (Bernardo & Albertini 2006). As shown in Fig. 2, the smaller pore size and their homogeneous distribution in 30 wt% of oily soil (72 μm) resulted in a flexural strength (3.93 MPa) that is significantly higher than that obtained for 35 wt% of oily soil (2.57 MPa). The higher oily soil quantity of 45 wt% and 50 wt% showed higher values of flexural strength (3.53–5.49 MPa) due to a denser microstructure and smaller porous structure. More extensive crystallization phases occurred for higher added quantities of oily soil. This also increased the flexural strength of porous ceramics. The results of the present study demonstrated that the porosity and flexural strength of porous ceramics presented a negative correlation.

4 CONCLUSION

In this study, with oily soil and waste glass as the main materials, calcium carbonate and sodium silicate were used to synthesize porous ceramics of different properties by controlling the sintering process and changing the ratio of raw materials. We provide a feasible solution to make full use of the oily soil. The porosity increased first and then decreased regularly, and the flexural strength decreased first and then increased with an increase in the quantity of oily soil used. These present a negative correlation. When the amount of the oily soil was 40 wt%, the corresponding highest porosity was 69.44% and the lowest flexural strength was 2.16 MPa.

REFERENCES

Bernardo, E., & Albertini, F. (2006). Glass foams from dismantled cathode ray tubes. *J. Ceramics International. 32(6)*, 603–608.

Cao, G. & Yu, H. (2005). Oil pollution and governance. *J. Coastal Enterprises and Science & Technology. (3)*, 92–94.

Deng, S. Y., Xue-Yi, X. U., & Qiu, Q. H. (2012). Present situation and prospect of research of remediation in oil-contaminated soil. *J. Northern Horticulture. (14)*, 184–190.

Fernandes, H. R., Tulyaganov, D. U., & Ferreira, J. M. F. (2009). Preparation and characterization of foams from sheet glass and fly ash using carbonates as foaming agents. *J. Ceramics International. 35(1)*, 229–235.

Gao, X., Xing-Jun, L. V., & Wang, L. (2014). Research status and prospects of foam glass. *J. Research & Application of Building Materials. 4*, 6–8.

Hasheminia, S., Nemati, A., & Yekta, B. E. (2012). Preparation and characterisation of diopside-based glass–ceramic foams. *J. Ceramics International. 38(3)*, 2005–2010.

Jia, J., Liu, Y., & Li, G. (2009). Contamination characteristics and its relationship with physicochemical properties of oil polluted soils in oilfields of China. *J. Journal of the Chemical Industry & Engineering Society of China. 60(3)*, 726–732.

Jones, D. M., Head, I. M., & Gray, N. D. (2008). Crude-oil biodegradation via methanogenesis in subsurface petroleum reservoirs. *J. Nature. 451(7175)*, 176–180.

Kao, C. M., & Prosser, J. (2001). Evaluation of natural attenuation rate at a gasoline spill site. *J. Journal of Hazardous Materials. 82(3)*, 275–289.

Lösekann, T., Knittel, K., & Nadalig, T. (2007). Diversity and abundance of aerobic and anaerobic methane oxidizers at the haakon mosby mud volcano, barents sea. *J. Applied & Environmental Microbiology. 73(10)*, 3348–3362.

Qiao, Y. U., Jing, Y., & Wang, C. (2009). Foam glass and cyclic utilization of solid offal. *J. Materials Review. 23(1)*, 93–96.

Spiridonov, Y. A., & Orlova, L. A. (2003). Problems of foam glass production. *J. Glass and Ceramics. 60(9)*, 313–314.

Tian, Y. L., Zhan, M, & Sun, S. B. (2014). A Review on Technological Development of Foam Glass and Designation of Generation of Production Lines. *J. Glass & Enamel. 42(3)*, 26–32.

Zhang, J., Yong-Sheng, W. U., & Zhang, X. (2010). Research development of foam glass producing technique. *J. Materials Review. 28(4)*, 25–30.

Zhang, X., & Zeng, Z. (2006). Application of eco-friendly material—foamed glass. *J. Journal of Building Materials. 9(2)*, 177–182.

Civil, Architecture and Environmental Engineering – Kao & Sung (Eds)

Spacing optimization modeling and simulation of thermo-optic VOA gold thin film interconnect

Shanting Ding, Lei Xu, Fei Huang & Menglan Gong
School of Mechanical Engineering, Hubei University of Technology, Wuhan, China

ABSTRACT: Along with increase of interconnects in thermo-optic devices, its overall size is becoming larger and larger. Aiming at the challenges of miniaturization and integration, a method to optimize the spacing of thin film interconnects in thermo-optic devices was proposed. Based on the working principle, the relationship between the distance and interaction as well as the temperature distribution and the refractive index were analyzed by finite element simulation and optical simulation software respectively. The simulation results were compared with those of reliability experiments. It is obvious that the method of reducing the interconnect spacing is feasible for the 100 μm interconnect spacing, and it is meaningful for device design and manufacture.

1 INTRODUCTION

With the rapid development of optical communication industry, themo-optical devices with the characteristics of small size, low driving power, low cost and easy manufacture, represented by Variable Optical Attenuator (VOA) and optical switch, have been widely applied (Xingjian Lv et al. 2004, Junxing Yu 2014). However, the device size has being growing obviously along with the fast increasing of channels. For example, a typical commercial available thermo-optical VOA has 16 channels for the early products. When the channel number is improved to 20, the overall size of the device has increased 1.5 times. According to the working principle of thermo-optic devices, each channel requires a film interconnect as the role of heater to establish a temperature gradient and achieve light attenuation. The requirements for high integration and miniaturization of the device are needed in the size and spacing of the thin film interconnects. At present, there are a lot of researches on the problems of size optimization and reliability of the thin film interconnects, but there is little research on the spacing optimization.

To investigate the influence of interconnect spacing on the interaction of the optical waveguides, the temperature distribution of polymer optical waveguides under the heating of gold thin film interconnects has been analyzed. On this basis, the interaction of multiple heating interconnects are studied, and the influence of interconnects spacing on the device operation is clarified.

2 WORKING PRINCIPLE OF THERMO-OPTIC VOA

Thermo-optic VOA is based on the Thermo-optic effect to achieve the attenuation. In general, the production of thermo-optic VOA requires the use of micro-processing technology in the formation of thin film interconnect as a heating electrode on the optical waveguide. After the electrode is energized, heat will be transferred to the optical waveguide and the temperature field distribution will be generated. Finally changes the refractive index of the waveguide and attenuates the optical signal. Figure 1 gives the structure of a thermo-optic VOA. The input and output single-mode waveguides are connected to the optical fiber and they have the same matching coefficients. The heating electrode is at an angle β to the plane of the multimode waveguide so that the attenuation ratio of the optical signal is conformed to the design claim. After the electrode is energized, the refractive index of the waveguide decreases as the temperature

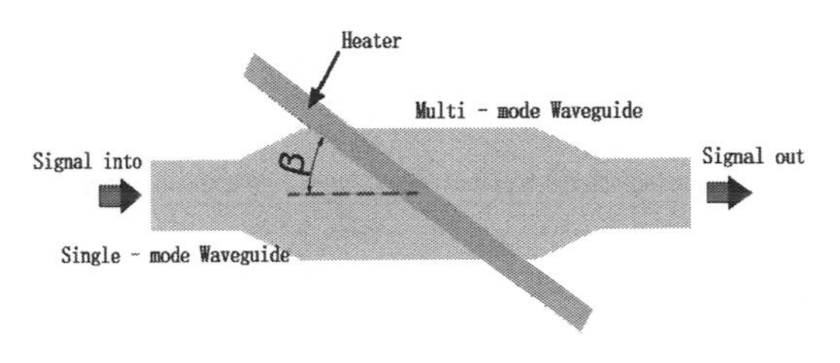

Figure 1. Structure of the thermo-optic VOA.

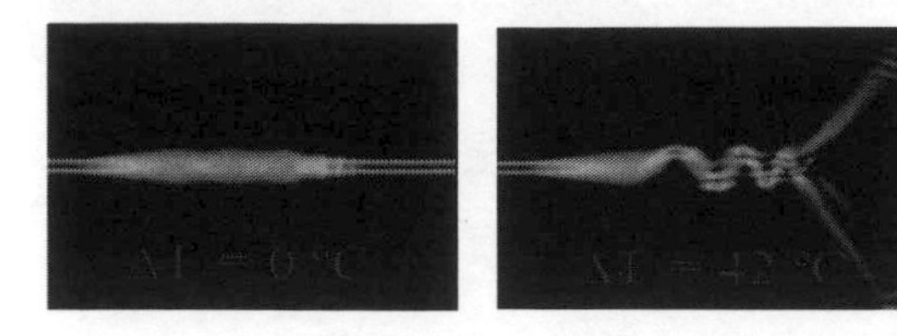

Figure 2. Optical signal attenuation of the thermo-optic VOA.

increases, the direction of the refractive index gradient forms an angle β with the beam, and then the beam in the multimode region is reflected at 2β. When the reflection angle is greater than the transmission angle, the reflected light beam enters the high-order mode and is attenuated. Figure 2 shows the attenuation of optical signal. ΔT is the rising temperature of the waveguide core relative to the substrate. When ΔT equals 0°C, the refractive index of the waveguide does not change and the input optical signal is not attenuated. When ΔT equals 42°C, the refractive index of the waveguide core layer is smaller than than that of the surroundings and the optical signal is totally refracted.

3 THERMO-OPTIC EFFECT AND THERMAL FIELD ANALYSIS

3.1 *Thermo-optic effect*

Thermo-optic effect is the optical properties of the optical medium. When the temperature gradient of the material exists, it will cause the changes of the refractive index of the material corresponding to the temperature gradient distribution changes (Yuliang Liu et al. 1996). Therefore, the change of refractive index n with temperature T can be attributed to the change of n with density ρ and the change of polarization rate with temperature. The expression of thermo-optic coefficient dn/dT can be written as (Edmond J. Murphy 1999):

$$\frac{dn}{dT} = -\left(\rho\frac{\partial n}{\partial \rho}\right)_T \gamma + \left(\frac{\partial n}{\partial T}\right)_\rho \tag{1}$$

γ refers to the expansion coefficient of the material. According to the Lorentz-Lorenz (Lorentz) equation, $(\rho\partial n/\partial\rho)_T$ can be derived from:

$$\left(\rho\frac{\partial n}{\partial \rho}\right)_T = (1-\Lambda_0)\frac{(n^2+2)(n^2-1)}{6n} \tag{2}$$

Λ_0 is the strain polarization constant, which indicates the density change effect in the material caused by atomic polarization. For polymer materials, the intermolecular effect is very weak and much smaller than 1, such as Polymethyl-Methacrylate, Λ0 = 0.15. Since the refractive index of most polymer materials is close to 1.5 and the expansion coefficient is 2×10^{-4}/°C in glassy state (Sihua Zeng 2004), the following equations can be obtained according to equations (1) and (2).

$$-\left(\rho\frac{\partial n}{\partial \rho}\right)_T \gamma \sim -10^{-4}/°\text{C} \tag{3}$$

For polymers of constant density, the refractive index varies with temperature is about 10^{-6} per degrees Celsius. Therefore, the value of equation (1) is mainly determined by the first term.

$$\frac{dn}{dT} \sim -\left(\rho\frac{\partial n}{\partial \rho}\right)_T \gamma \sim -10^{-4}/°\text{C} \tag{4}$$

That is, the refractive index decreases by 10^{-4} per degree of temperature rise.

3.2 *Thermal field analysis*

The steady-state heat conduction equation can be solved by a two-dimensional temperature field model established through a buried channel waveguide (Ying Li 2004). The heating model of the thermo-optic VOA is given in Figure 3. The upper layer is the Au thin film interconnect; the lower layer is the polymer optical waveguide; the middle green is the waveguide core layer which has the highest refractive index to realize the transmission of optical signal in it. When electric current is applied to the electrode, heat is generated and transferred to the lower polymer optical waveguides. Thereby a temperature field distribution is formed. The thermal conductivity of polymeric materials is in the range of 0.17 to 0.22, which is very close and isotropic. So the core and cladding of the waveguide can be considered to have the same thermal conductivity. Since the thermal conductivity is much greater than the thermal radiation and thermal convection in this model, the steady-state analysis only takes into account the

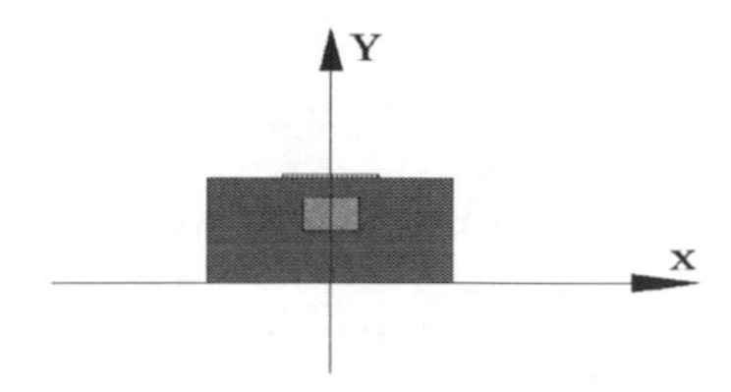

Figure 3. Heating model of the thermo-optic VOA.

heat transfer problem. In addition, gold interconnects are so thin (only a few hundred nanometers) that it can be regarded as a thin wall and all the heat is passed to the lower layer. Compared to the polymer, the silicon material under the polymer has good thermal conductivity. So the temperature of the silicon substrate is considered to be room temperature.

Thus, the basic thermal physical model has been established. For the two-dimensional steady-state heat conduction problem with constant heat source (Jianmin Liu 2006), the heat conduction equation is presented by:

$$\frac{\partial^2 T}{\partial x^2}+\frac{\partial^2 T}{\partial y^2}+\frac{\dot{q}}{\lambda}=0 \tag{5}$$

The domain of the equation is partitioned into rectangular finite difference meshes with step sizes Δx, Δy, and x = iΔx, y = iΔy. According to the known second-order central difference formula, the finite difference of the second partial derivative of the temperature in equation (5) can be shown as:

$$\left.\frac{\partial^2 T}{\partial x^2}\right|_{i,j}=\frac{T_{i-1,j}-2T_{i,j}+T_{i+1,j}}{(\Delta x)^2} \tag{6}$$

The above two equations are substituted into equation (5) to obtain the finite difference scheme of two-dimensional steady-state heat conduction equation with internal heat source.

$$\frac{T_{i-1}-2T_{i,j}+T_{i+1,j}}{(\Delta x^2)}+\frac{T_{i,j-1}-2T_{i,j}+T_{i,j+1}}{(\Delta y^2)}+\frac{\dot{q}_{i,j}}{\lambda}=0 \tag{7}$$

where $T_{i,j}$ and $\dot{q}_{i,j}$ are the values of the temperature and the internal heat source at node (x,y), respectively.

4 SIMULATION ANALYSIS

4.1 *Thermo-optic VOA interconnect simulation model*

According to the relevant parameters of the thermo-optic VOA, the geometric model is established in Figure 4. The upper layer is gold thin film interconnects and the lower layer is polymer optical waveguide. In order to make the simulation results more accurate, the arc structure of the interconnect is neglected during modeling. The interconnect structure is designed as follows: length equals 1000 um, width equals 10 um, height equals 0.3 um and the spacing between interconnects is constant. The thickness of the polymer material is 15 μm.

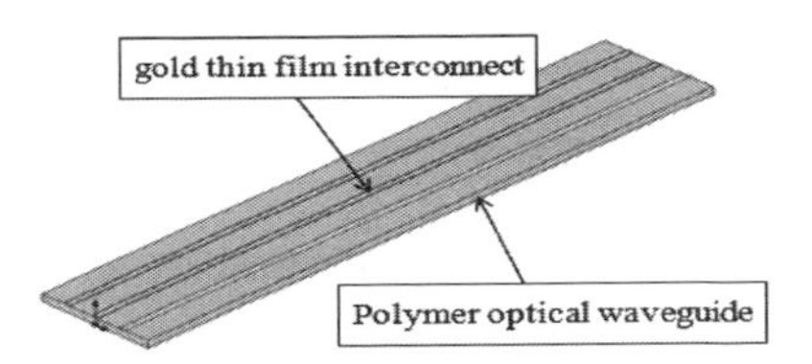

Figure 4. Thermo-optic VOA multiple interconnects model.

Table 1. Material parameters.

Material	Thermal conductivity W/m.K	Resistivity Ω.m
Gold	315.5	2.20E-08
Polymer	0.18	1.00E+15

4.2 *Model parameter setting*

By means of the finite element analysis software of ANSYS, the thermo-optic VOA model with different interconnect spacings is simulated to study the influence of interaction of multiple interconnects. The material parameters involved in the simulation are shown in Table 1.

4.3 *Simulation results*

In the simulation, the ambient temperature is set to 25°C, the current of gold film interconnects is 0.07 A and the interconnect spacing is 100 um. Contours are drawn from eight fixed and equant points through Y direction as shown in Figure 5. It can be seen from the figure that there is a minimum temperature SMN = 25°C at the bottom of the polymer and a maximum temperature SMX = 149.742°C on the most intemediate interconnect. The temperature decreased along the interconnects toward the polymer. When ΔT = 42°C, the refractive index of the waveguide core is less than the refractive index of the cladding so that all the light can be attenuated. From the contour map it can be determined that the position of the waveguide core is between D and E.

From equation (4), we know that the change of refractive index can be known by temperature distribution. In order to clearly show the correspondence, the ANSYS simulation results were imported into Rsoft optical simulation software. The change of temperature distribution and refractive index are presented in Figure 6, where ΔT is the temperature

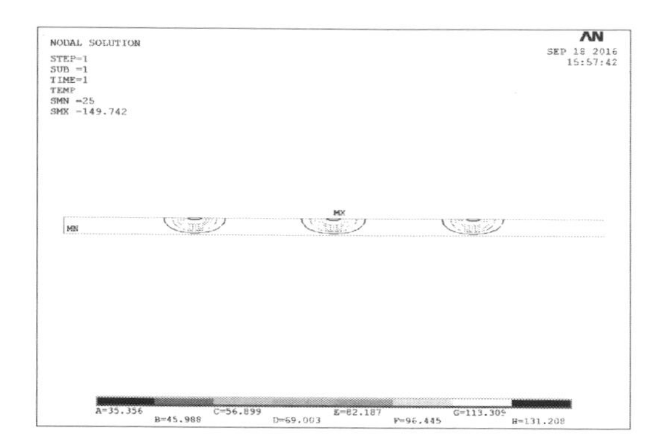

Figure 5. Thermo-optic VOA cross-section temperature counters.

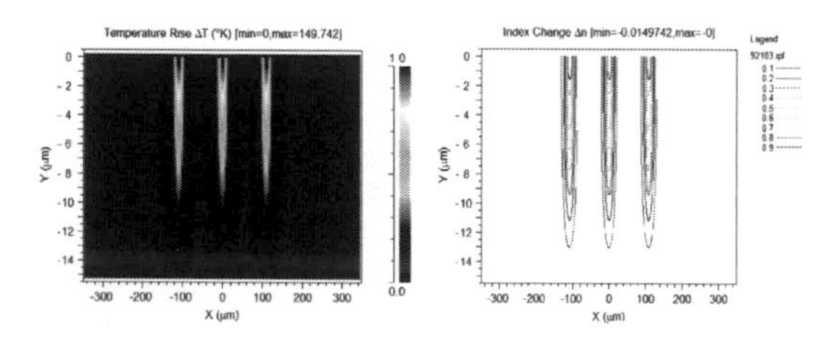

Figure 6. Temperature distribution and refractive index.

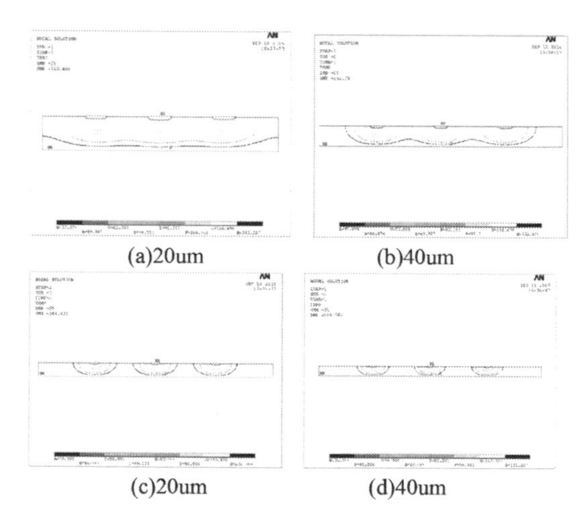

Figure 7. The cross-sectional temperature distribution of VOA with different interconnect spacing.

rise of the position relative to the substrate, Δn is the change of the corresponding refractive index.

To explore the relationship between the interconnection distance and interaction, the four models of 20 um, 40 um, 60 um and 80 um are simulated respectively, and the temperature distribution is shown in Figure 7.

5 DATE ANALYSIS AND RESULTS DISCUSSION

As the temperature distribution data are close, difference method is used to study the relationship between the size of the interconnect spacing and the interaction. The temperature distribution at each spacing is compared with the temperature distribution at 100 μm, which can be seen in Figure 8. It can be seen that the temperature of the waveguide increases gradually from point 1 to point 8, and the increasing rate gradually decreases. When the interconnect spacing is 80 um or 60 um, the value of Δ°C is very small and close. When the distance is 20 um, Δ°C sharply increases and the value of the eighth equalization point even reaches 10°C that would have a great impact on the life and reliability of the device. Taking into account the interconnect and the electrode is connected through the arc, 20 um size is too small and is not conducive to the design of the device layout. Contrasting the four curves, it was found that the interaction was enhanced with the reduction of the interconnect spacing.

According to the spacing optimization simulation results, the trial sample of 80 um interconnect

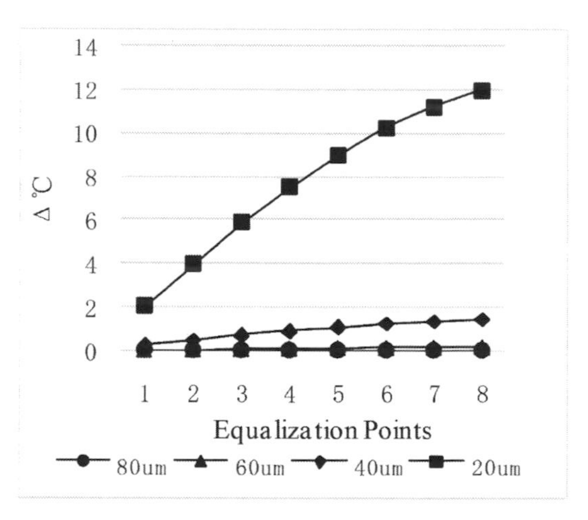

Figure 8. The curve of temperature difference under different spacing.

Figure 9. 80 um interconnect spacing test sample.

spacing is made and the experimental result is shown in Figure 9. The reliability experiment results show that the layout of interconnect spacing is feasible.

6 CONCLUSION

In order to reduce the overall size of the device, the finite element simulation software is used to analyze the interaction of interconnects at different spacings, and the results show that the interaction increases gradually with the decrease of the interconnect spacing. The test samples with 80 um interconnect spacing are fabricated by the spacing optimization design. The reliability test results show that the simulation results are correct and the distance can be optimized, which can be used as a reference for the miniaturization and integration of devices in optoelectronic industry.

ACKNOWLEDGEMENT

The authors are grateful to the support Natural Science Foundation of Hubei Province (2014CFB178), Hubei Provincial Department of Education Research Project (Q20111405), Hubei Provincial Department of Education Key Project (D20131407).

REFERENCES

Edmond J. Murphy (1999). Integrated optical circuits and components design and applications. *New York Basel Marcel Dekker, Inc.* 245–247.

Jianmin Liu (2006). Principle of heat and mass transfer and its application in electric power. *M. Beijing: China Electric Power Press.*

Junxing Yu (2014). Interconnect Failure Analysis and Reliability Modeling of the Thermo-optic Effect VOA. *D. Hubei University of Technology.*

Sihua Zeng (2004). Study on Polymeric Waveguide Optical Attenuator. *D. Huazhong University of Science and Technology.*

Xingjian Lv, Jun Yang, Ming Hai (2004). Polymer Waveguide Device's Latest Development. *J. Journal of Atmospheric and Environmental.* 17(6).

Ying Li (2004). Study on Thermo—Optical Variable Attenuator of Organic Polymers. *D.* Zhejiang University.

Yuliang Liu, Enke Liu & Jinsheng Luo (1996). Thermo-optic effect in the whole silicon optical waveguide switch. *J. Chinese Science Bulletin*, 6:575–576.

Civil, Architecture and Environmental Engineering – Kao & Sung (Eds)
 ISBN 978-1-138-02985-9

Growth and characterization of thick Ge epilayers on Si and Silicon-on-Insulator (SOI) substrates with low temperature Ge buffers

Zhiwen Zhou
School of Electronic Communication Technology, Shenzhen Institute of Information Technology, Shenzhen, Guangdong Province, P.R. China

ABSTRACT: High-quality thick Ge epilayers on Si and Silicon-on-Insulator (SOI) substrates with smooth surface and low threading dislocation densities are critical for device performance. In this work, Ge epilayers on Low Temperature Ge (LT Ge) buffer layers were grown utilizing two-step approach by ultrahigh vacuum chemical vapor deposition and characterized by atomic force microscope, x-ray diffraction, transmission electron microscopy, counting etch pit density and photoluminescence measurements. The results showed that the surfaces of LT Ge buffers were rough, but the misfit strains were nearly fully relaxed (strain relaxation reached 95%). However, the rough LT Ge surface was smoothed by subsequent growth at elevated temperature. It suggested that LT Ge buffer was not essential to be atomically flat as known for the high quality Ge growth. On proper LT Ge buffer, 1-μm thick Ge epilayers with narrow symmetric diffraction peak (full width at half maximum of about 90 arcsec), smooth surface (root mean square roughness is about 3 nm), low Threading Dislocation Density (TDD, $6 \times 10^7/cm^2$), and strong photoluminescence spectra at 1.55 μm wavelength were obtained both on Si and SOI substrates.

1 INTRODUCTION

The heteroepitaxy of thick Ge epilayers on Si and Silicon-on-Insulator (SOI) substrates with ideal characteristics, i.e., flat surface morphologies, few misfit dislocations and thin intermediate buffer layers, has attracted great attention due to the high compatibility of Ge with mature Si complementary metal oxide semiconductor processes and the extensive applications in optoelectronics for high-volume and large-scale electronic–photonic integration, such as efficient photodiodes operating in the infrared wavelength range of 1.3–1.55 μm (Wang 2015), high mobility field effect transistors (Wu 2015), ultralow energy electro-absorption modulators (Liu 2008), innovative direct bandgap light emitting diodes (Sun 2009), and even monolithic lasers (Liu 2010). Moreover, Ge is the preferred template for the integration of Sn (Wirths 2015) and III-V compound semiconductors (Liu 2011) on Si. However, the greatest challenge to grow high quality thick Ge layers on Si is the 4.2% lattice mismatch between the two elements, which could cause high surface roughness resulting from Stranski–Krastanov growth mode and high density threading dislocations due to plastic strain relaxation. Various growth techniques and treatments have been put forward to solve the problems. These include compositionally graded SiGe buffer layers (Thomas 2003), two thin SiGe intermediate buffer layers (Luo 2003), and two-step approach (Liu 2008, Sun 2009, Liu 2010). In two-step approach, a Ge film is initially deposited at low temperature (LT: 300–400°C) in first step, which constitutes a buffer whose thickness lies between 30 and 100 nm and enables plastic strain relaxation without the nucleation of any 3D islands. In second step, the growth temperature is increased to high temperature (HT: 500–850°C) for the formation of additional thicker Ge layer with higher crystalline quality and growth rate.

In this paper, we reported the growth of thick Ge epilayers on Si and SOI substrates by Ultrahigh Vacuum Chemical Vapor Deposition (UHVCVD) and characterizations by Atomic Force Microscope (AFM), X-Ray Diffraction (XRD), High-Resolution Transmission Electron Microscopy (HRTEM), counting Etch Pit Density (EPD) and Photoluminescence (PL) measurements.

2 EXPERIMENTAL DETAILS

All samples were grown by a single-wafer cold wall UHVCVD system. The base pressure is 5×10^{-8} Pa. Pure Si_2H_6 and GeH_4 were used as precursors. The substrates were 4 inch N-type Si (100) and SOI (100) wafers with resistivity in the range of 0.1~1 Ω•cm. After baking the wafer at 850°C and the growth of 300 nm Si buffer at

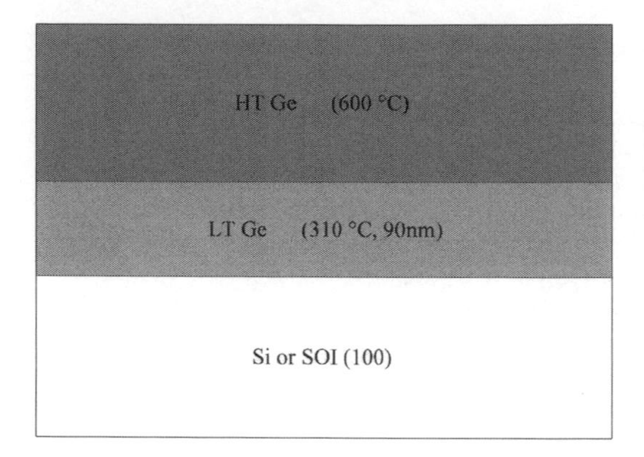

Figure 1. Ge deposition diagram.

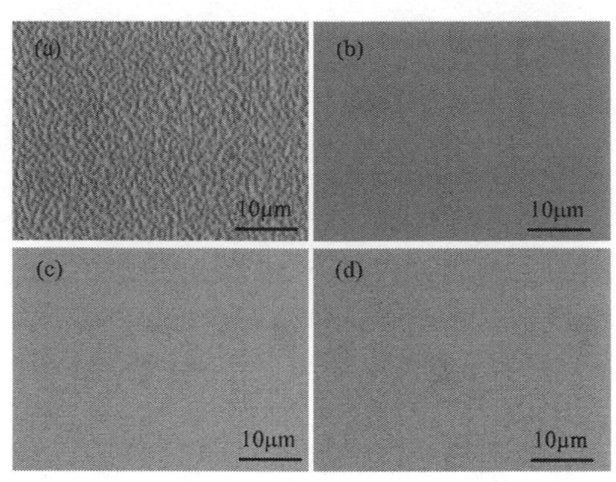

Figure 2. Optical images of samples (a) A, (b) B, (c) C, and (d) D.

750°C, it's ready for Ge growth. The growth of Ge epilayers proceeded in two steps. First, thin LT Ge buffer layers were grown at the temperature of 310°C with the thickness of 90 nm. Second, thick HT Ge epilayers were grown at fixed 600°C with various thicknesses. Figure 1 shows the Ge deposition diagram. The chamber pressure is about 2×10^{-2} Pa during deposition. The growth rate is 0.5 nm/min and 1.2 nm/min for LT Ge and HT Ge, respectively. In the present study, we focus on four samples (A to D). Sample A to C were grown on Si wafers, but sample D was grown on SOI wafer. Sample A is the thin LT Ge buffer with thickness of 90 nm. Sample B and C are the thick HT Ge epilayers on the LT Ge buffer. The thickness of LT Ge buffer is 90 nm, but the thickness of HT Ge is 210 nm and 910 nm for sample B and C, respectively. Sample D is a replica of sample C except that the wafer is SOI.

The surface morphologies and roughnesses of all samples were analyzed by atomic force microscopy (Seiku Instruments, SPI4000/SPA-400) in a tapping mode, and Nomarski optical microscope. The strain status and crystal quality of Ge layers were evaluated by double crystal x-ray diffraction measurements (Bede, D1 system), using a Cu $\kappa_{\alpha 1}$ ($\lambda = 0.15406$ nm) as the x-ray source. The threading dislocation densities of Ge films were characterized by counting pits formed by selectively chemical etching and high-resolution transmission electron microscopy. The photoluminescence spectra were measured at room temperature using an Ar^+ laser emitting at 488 nm.

3 RESULTS AND DISCUSSION

Figure 2 shows the large-scale surface morphologies of four samples viewed by Nomarski optical microscope. For sample A of LT Ge buffer, the surface is rather rough, but samples B to D with HT Ge epilayers on LT Ge buffers have flat surfaces. The detailed surface morphologies were measured by AFM and shown in Figure 3. The surface root mean square (rms) roughness for sample A is 16.9 nm. After the growth of HT Ge epilayers, the rough surfaces were smoothed; the rms roughness is 2.3, 3.4 and 2.9 nm for sample B to D, respectively. The rms value was largely reduced. It is interesting to note that a typical cross-hatch surface pattern, due to strain fields arising from

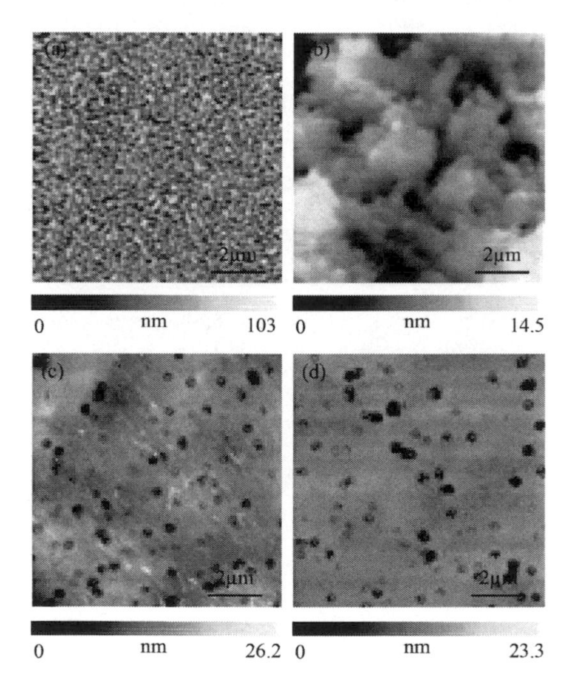

Figure 3. AFM images of samples (a) A, (b) B, (c) C, and (d) D.

inhomogeneous distribution of misfit dislocations, associated with thick SiGe layers grown on Si substrates is not visible in these samples.

Figure 4 shows the XRD results of these four samples along with the theoretical positions of bulk Si and Ge. Sharp peaks from Si substrates and relatively broad peaks from Ge epilayers are clearly observed. For sample A of LT Ge buffer, Ge diffraction peak is much broader and asymmetric, which are signs of some imperfection of a crystal quality, like mosaicity, due to the presence of defects, and the non-uniformly distributed strain. Based on the peak positions, the calculated degree of strain relaxation reached 95% for sample A. For sample B to D with HT Ge epilayers, Ge peaks become symmetric and narrower. The Full Width at Half Maximum (FWHM) for sample A to D is 1200, 490, 90 and 95 arcsec, respectively. The modified Ge diffraction peaks suggest the improvement of crystal quality. Based on the peak positions of the HT Ge epilayers (samples C and D), which are much closer to that of the Si substrate in comparison with the bulk Ge one (−5640 arc sec), the Ge epilayers are tensily strained. The tensile strain in Ge epilayers is calculated to be 0.16%. The tensile strain is induced by the thermal expansion coefficient mismatch between Ge and Si.

Figure 5 shows cross-sectional HRTEM images of sample C. Although we cannot evaluate the exact TDD by HRTEM for its localized view, the TDD distribution can still be outlined by analyzing HRTEM images. It can be seen that most of misfit dislocations are confined at the LT-Ge and Si interface as shown in the inset. Some of them thread upward and generate threading dislocations and then some of the threads meet and annihilate in a region at the beginning of the HT-Ge layer.

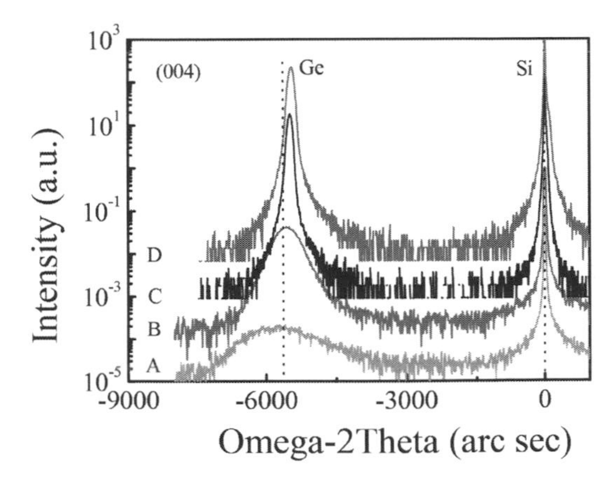

Figure 4. Omega-2Theta XRD scans around (004) order for samples A, B, C, and D. Straight dashed line indicates the theoretical peak position of bulk Si and Ge (−5640 arc sec).

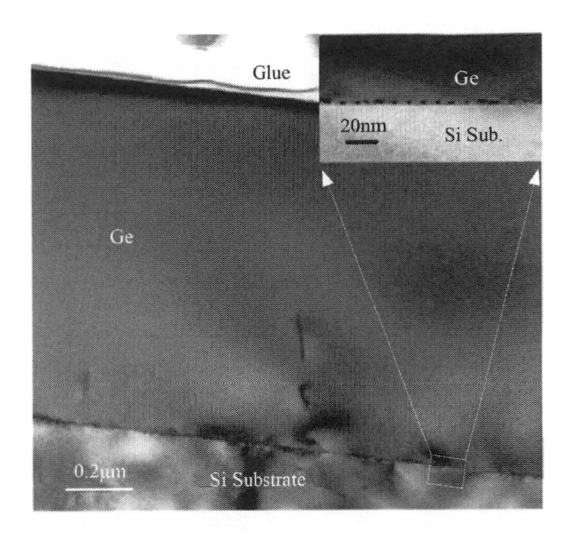

Figure 5. Cross-sectional images of HRTEM of sample C.

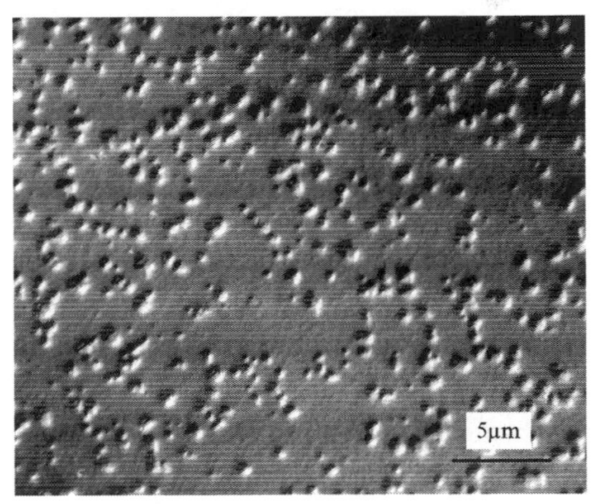

Figure 6. Optical image of sample C after chemical etching.

The typical optical image of the etched Ge surface for threading dislocation measurement is shown in Figure 6, and the etch pit densities are averaged to be 6×10^7 cm^{-2}. TEM result is in fair agreement with the EPD measurements.

The PL spectra of the Ge epilayers on Si and SOI wafers (samples C and D) are shown in Figure 7. The PL peak of the sample C with the Ge layer on Si wafer is at about 1.56 μm, which shifts to longer wavelength compared to 1.52 μm for bulk Ge. This red shift should be due to the tensile strain-induced Γ point conduction band down forward shift. The tensile strain induced Ge band gap shrinkage is shown in Figure 8. For sample D with the Ge epilayer on SOI wafer, the PL spectrum is

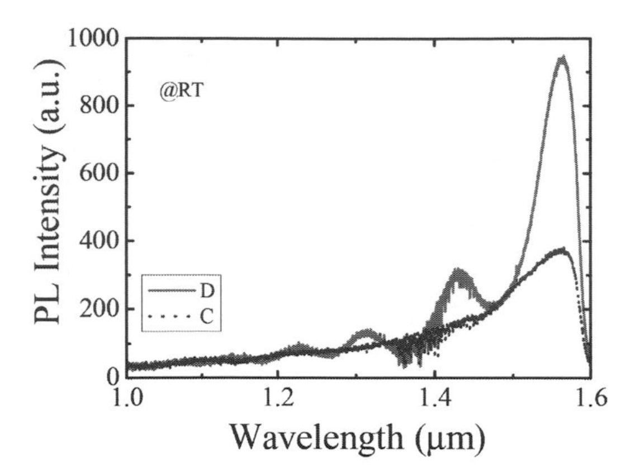

Figure 7. PL spectra of samples C and D.

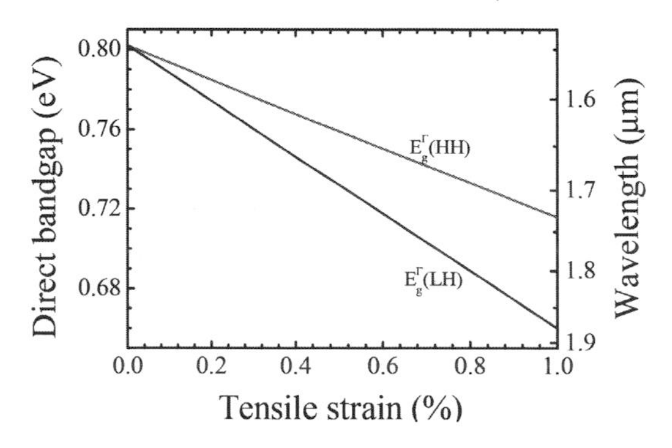

Figure 8. Direct bandgap of Ge vs. tensile strain.

modulated by Fabry-Perot cavity formed between the surface of Ge and the interface of buried SiO_2. The enhancement of the luminescence peak intensity of the Ge layer on SOI wafer in comparison with the Si wafer is about 2.5 fold and the FWHM of the maximum peak is reduced from 200 to 76 nm. What more interesting is that the integration over all spectra of the Ge layer on SOI wafer has a factor of 1.5 fold increases compared to that on Si wafer. This result implies that the absolute amount of light emitted in the Ge on SOI substrate is significantly increased.

4 CONCLUSION

We have grown high quality thick Ge epilayers on Si and SOI substrates utilizing low temperature Ge buffers by ultrahigh vacuum chemical vapor deposition. It has been demonstrated that LT Ge buffers were rough, but the misfit strains were fully relaxed. On 90 nm thick LT Ge buffer, high quality 1 μm thick of Ge epilayers were obtained both on Si and SOI substrates. XRD peaks were narrow and symmetric with the FWHM of about 90 arcsec. AFM showed the smooth surfaces with rms roughness of about 3 nm. TEM and counting etch pit methods indicated the Ge epilayers had a low TDD of $6 \times 10^7/cm^2$. Strong photoluminescence spectra at 1.55 μm wavelength were shown.

ACKNOWLEDGEMENTS

This work was partially supported by Foundations for the Excellent Youth Scholars of Educational Commission of Guangdong Province (YQ20140123).

REFERENCES

Liu, H. et al 2011. Long-wavelength InAs/GaAs quantum-dot laser diode monolithically grown on Ge substrate. *Nature Photon.* 5: 416–419.

Liu J., X. Sun, R. Aguilera, L. C. Kimerling, J. Michel 2010. Ge-on-Si laser operating at room temperature. *Opt. Lett.* 35(5): 679–681.

Liu J., M. Beals, A. Pomerene, S. Bernardis, R. Sun, J. Cheng, L. C. Kimerling, J. Michel 2008. Waveguide-integrated, ultralow-energy GeSi electro-absorption modulators. *Nature Photon.* 2: 433–437.

Luo G. L., T. H. Yang, E. Y. Chang, C. Y. Chang, K. A. Chao 2003. Growth of High-Quality Ge Epitaxial Layers on Si (100). *Jpn. J. Appl. Phys.* 42: L517-L519.

Sun X., J. Liu, L. C. Kimerling, J. Michel 2009. Room-temperature direct bandgap electroluminesence from Ge-on-Si light-emitting diodes. *Opt. Lett.* 34(8): 1198–1200.

Thomas S. G., et al, 2003. Structural Characterization of Thick, High-Quality Epitaxial Ge on Si Substrates Grown by Low-Energy Plasma-Enhanced Chemical Vapor Deposition. *J. Electron. Mater.* 32(9): 976–980.

Wang C., C. Li, J. Wei, G. Lin, X. Lan, X. Chi, C. Lu, Z. Huang, C. Chen, W. Huang, H. Lai, S. Chen 2015. High-Performance Ge p-n Photodiode Achieved With Preannealing and Excimer Laser Annealing. *IEEE Photonics Technology Letters* 27(14): 1485–1488.

Wirths S. et al 2015. Lasing in direct-bandgap GeSn alloy grown on Si. *Nature Photon.* 9:88.

Wu H., M. Si, L. Dong, J. Gu, J. Zhang, P. Ye 2015. Germanium nMOSFETs With Recessed Channel and S/D: Contact, Scalability, Interface, and Drain Current Exceeding 1 A/mm. *IEEE Transactions on Electon Devices* 62(5): 1419–1426.

Civil, Architecture and Environmental Engineering – Kao & Sung (Eds)
© 2017 Taylor & Francis Group, ISBN 978-1-138-02985-9

Effect of $ZnSO_4$ and $SnSO_4$ additions on the morphological of α-Al_2O_3 flakes for pearlescent pigment

Huan Qin, Ling Wang & Yan Xiong
School of Materials Science and Engineering, Hubei Province Key Laboratory of Green Materials for Light Industry, Hubei University of Technology, Wuhan, P.R. China

ABSTRACT: The flaky α-Al_2O_3 is one of desirable substrate materials for pearlescent pigments. In the present work, the flaky α-Al_2O_3 with aspect ratio larger than 100 were prepared via a novel routine combining the sol-gel process with molten-salt growth method. Furthermore, $ZnSO_4$ or/and $SnSO_4$ were selected as the morphology modifiers and the effects of the ratio and amount of the modifiers on the morphologies of α-Al_2O_3 platelets were investigated by Scanning Electron Microscopy (SEM). The results showed that the flaky α-Al_2O_3 with an average size of ~30 μm and ~0.2 μm in thickness were prepared using 2 wt% $ZnSO_4$ and 0.2 wt% $SnSO_4$ as the modifiers by fired at 1100°C for 5 h.

1 INTRODUCTION

The flaky α-Al_2O_3 with the special two-dimensional structure not only has good performance like mechanical strength, hardness and high heat conductivity as general α-Al_2O_3, it is also equipped with double peculiarities of micro powder and nano-powder properties, appropriate surface activity, preferable adhesion and ability of reflecting rays, remarkable shielding effect and etc. Therefore, the flaky α-Al_2O_3 has extensive application in areas of cosmetics industry, fire-proofing (Cemail (2002), Karlsson (2006)), pearlescent pigment (Egashira (2006), Teaney (1999) & Sharrock (2000)), toughening ceramics (Matthew (1997), Zhang (2010)). For example, ceramic matrix composite enhanced with the flaky α-Al_2O_3 has higher breaking tenacity, compared to that enhanced with alumina granules (Shukla (2007), (2011), Bonderer (2009)). In addition, the alumina platelets with high-quality is the ideal high-grade pearlescent pigment substrate. As a pearlescent pigment base material, it has a high surface reflectivity and a high reflectance capability over other powdery materials. And it improved the color brightness, hue and saturation greatly, and produced special effects of color vector angle. The flaky α-Al_2O_3 is the ideal high-grade pearlescent pigment substrate (Wu (2014), Zhang (2014)).

At present, many studies are focusing on adjusting the thickness and size of the flaky α-Al_2O_3 (Tan (2012), Zhu (2011) & Su (2011), among others), but some of them are faced with the problem of poor stability and repeatability (Jin (2004), Su (2011) & Zhu (2011)). In this work, a method of molten salt growth which combined the sol-gel method was adopted, and that can prepare the flaky α-Al_2O_3 with characteristic of large radius-thickness ratio, stability of physical and chemical performance and good repeatability.

2 EXPERIMENT PROCEDURE

2.1 *Raw materials*

The $Al_2(SO_4)_3 \cdot 18H_2O$ and Na_2CO_3 were used as raw materials. K_2SO_4 and Na_2SO_4 were used as molten salts. $ZnSO_4$ or/and $SnSO_4$ were added as morphology modifiers. All of the materials above were produced in Sinopharm Group Chemical Reagent Co., Ltd. and were 99.7% in purity.

2.2 *Sample preparation*

$Al_2(SO_4)_3 \cdot 18H_2O$ was dissolved in de-ionized water to form an Al^{3+} solution with 1 mol/L in concentration, K_2SO_4 and Na_2SO_4 were added separately into the $Al_2(SO_4)_3$ solution with the molar ratio of 1.6:2.4:1. The Na_2CO_3 solution which was prepared with the concentration of 3 mol/L was dropped in the speed of 15 ml/min into the $Al_2(SO_4)_3$ solution at 65°C until the pH reached about 6.8. And the intermediate product was aged at 95°C for one day to obtain a translucent colloid after stirred rapidly for 15 min. Then dried at 110°C and sintered at 1100°C for 5h to obtain a precursor. Finally the sintered product was dispersed in a sulfuric acid solution, repeatedly washed with distilled water and alcohol, and dried to obtain a flaky α-Al_2O_3 sample.

2.3 *Characterization*

The crystal structure of the samples were characterized by X-ray diffraction pattern (XRD, Bruker D8 Advance diffractometer) with monochromatized Cu Kα radiation (λ = 1.5418 Å) in the 2θ range of 10–90 deg. The morphologies of the prepared samples were observed by field-emission Scanning Electron Microscopy (SEM, S4800 Hitachi, Japan) with a voltage of 80 kV.

3 RESULTS AND DISCUSSION

The effect of $ZnSO_4$ content on the morphology of flaky α-Al_2O_3 was investigated by SEM (Fig. 1). With the addition of $ZnSO_4$, the average size of the sample was also increasing, but its thickness was gradually thickening. When the amount of $ZnSO_4$ was 2.0 wt%, the average size of the alumina can be up to 25 μm or more and the thickness was about 0.2 μm.

The SEM micrographs of alumina platelets in Fig. 2 (a–f) shows the effect of $SnSO_4$ content on the morphology of platelets. The addition of small amounts of $SnSO_4$ had a large effect on the sample morphology. The increased addition amount of $SnSO_4$ led to the decrease of the size of the flaky

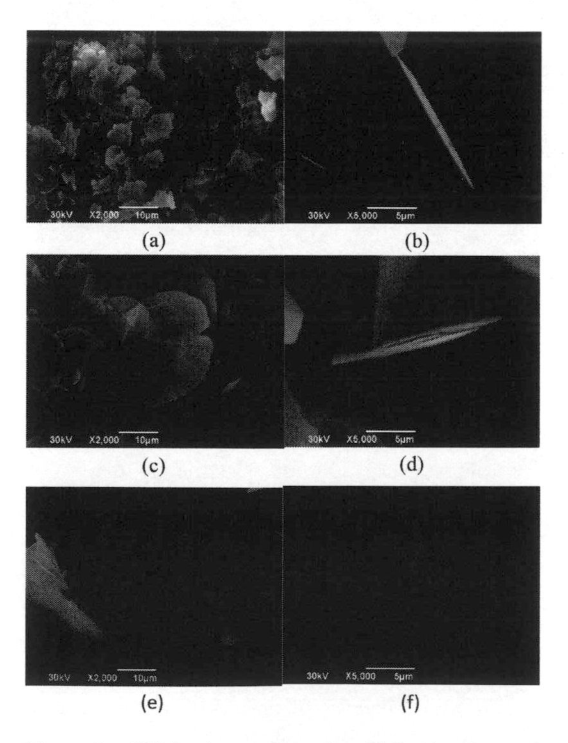

Figure 1. SEM micrographs of α-Al_2O_3 platelets prepared through a 5 h calcination at 1100°C. (a), (b) 1 wt%, (c), (d) 2 wt%, (e), (f) 3 wt% of $ZnSO_4$.

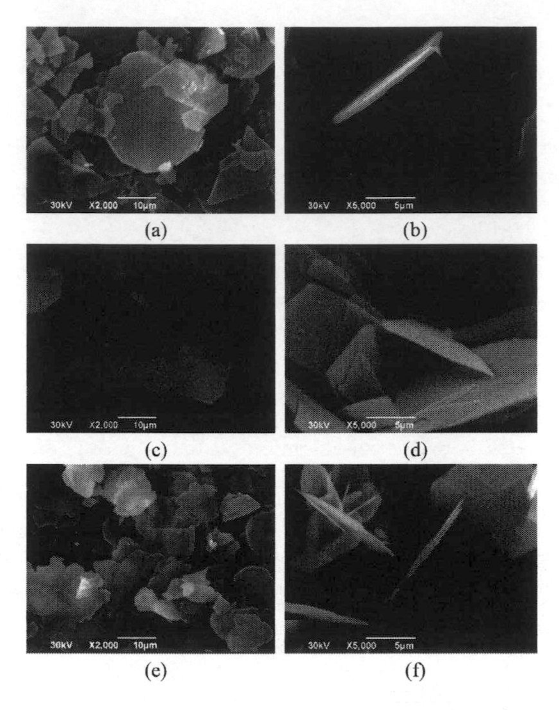

Figure 2. SEM micrographs of α-Al_2O_3 platelets prepared through a 5 h calcination at 1100°C. (a), (b) 0.2 wt%, (c), (d) 0.4 wt%, (e), (f) 0.8 wt% of $SnSO_4$.

α-Al_2O_3 and the corresponding reduction of thickness. When the amount of $SnSO_4$ was 0.4 wt%, the average size of the alumina can be 30 μm and the thickness was about 0.4 μm.

Fig. 3 (a–f) presents the SEM micrographs of alumina platelets with various amounts of both $ZnSO_4$ and $ZnSO_4$. The addition of the two materials exerts a competitive effect on the morphology of the flaky α-Al_2O_3 particles. When the addition amount of $ZnSO_4$ and $SnSO_4$ was 2 wt% and 0.2 wt%, the average size of the alumina can be 30 μm or more and the thickness was about 0.2 μm, which to achieve the best.

In present study, the samples added additives would get flake-like structure, may be the additives could inhibit the growth of the longitudinal crystal, so that it presents a special two-dimensional plane structure. And the larger size and thinner thickness of the sample, the more beautiful streamer color it has. That may as a result of its translucent sheet structure that occurred in the interference effect of light.

The samples obtained after calcination at 1100°C for 5h at a heating rate of 5°C/min were analyzed by XRD in Fig. 4. The sharp diffraction peaks appeared at 2θ of 35.2°, 57.56°, 43.4°, 25.66° and 37.9°, which were consistent with the diffraction peak positions of standard spectra. Thus it

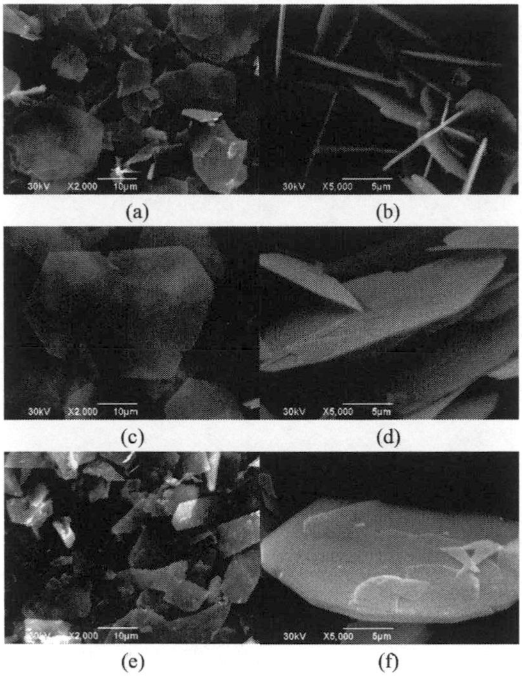

(a) (b)

(c) (d)

(e) (f)

Figure 3. SEM micrographs of α-Al_2O_3 platelets prepared through a 5 h calcination at 1100°C. (a), (b) 2 wt% /0.8 wt%, (c), (d) 2 wt%/0.2 wt%, (e), (f) 3 wt%/0.8 wt% of $ZnSO_4$/$SnSO_4$.

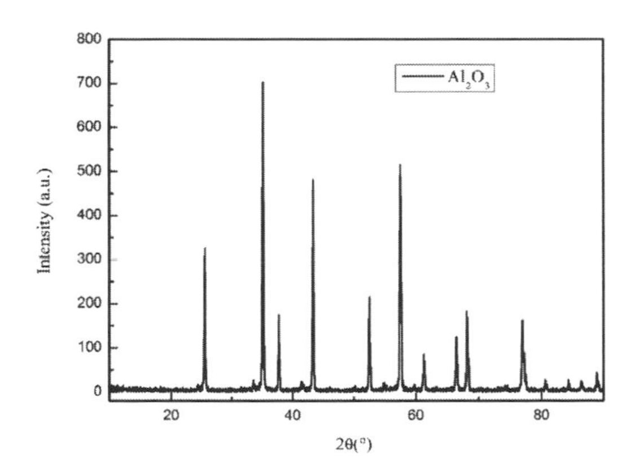

Figure 4. XRD pattern of Al_2O_3 particles.

can be concluded that the flaky α-Al_2O_3 prepared by this method was α-Al_2O_3, which belongs to the trigonal system, and has good crystallinity.

4 CONCLUSIONS

The flaky α-Al_2O_3 prepared by the novel routine which combined the sol-gel process with molten-salt growth method could be used as substrate materials for pearlescent pigments successfully. During the reaction, individual additions of $ZnSO_4$ and $SnSO_4$ had adverse effects on the morphology of the platelets. The particles of α-Al_2O_3 would be large and thick as the amount of $ZnSO_4$ increased. $SnSO_4$ had the opposite effect on the morphology which resulted in the formation of small and thin α-Al_2O_3 flakes. But when the mixtures of the $ZnSO_4$ and $SnSO_4$ were used as morphology modifiers, the average size and the thickness were both improved. And the flaky α-Al_2O_3 with the average size ~30 μm and ~0.2 μm in thickness under the condition of adding 2 wt% $ZnSO_4$ and 0.2 wt% $SnSO_4$.

REFERENCES

Bonderer, L.J., Studart, A.R. & Woltersdorf, J. (2009). Strong and ductile platelet reinforced polymer films inspired by nature: microstructure and mechanical properties. J. *J. Mater. Res.* 24 (9), 2741–2754.

Cemail, A. (2002). The role of fine alumina and mullite particles on the thermomechanical behavior of alumina-mullite refractory materials. J. *Mater. Lett.* 57: 708–714.

Egashira, M., Utsunomiya, Y. & Yoshimoto, N. (2006). Effects of the surface treatment of the Al_2O_3 filler on the lithium electrode/solid polymer electrolyte interface properties. J. *Electrochim. Acta.* 52: 1082–1086.

Jin, X.H., Gao, L. (2004). Size control of alpha-Al_2O_3 platelets synthesized in molten Na_2SO4 flux. J. *J. Am. Ceram. Soc.* 87, 533–540.

Karlsson, P., Palmqvist, A.E.C. & Holmberg, K. (2006). Surface modification for aluminium pigment inhibition. J. *Adv. Colloid Interface Sci.* 121–134.

Matthew, M., Ingrid, H. (1997). Messing, Texture development by templated grain growth in liquid-phase-sintered α-alumina. J. *J. Am. Ceram. Soc.* 80: 1181–1187.

Sharrock, S.R. & N. Schuel (2000). New Effect Pigments Based on SiO_2 and Al_2O_3 Flakes. J. *Eur. Coat.* 12: 20–23.

Shukla, D.K., V. Parameswaran. (2007). Epoxy composites with 200 nm thick alumina platelets as reinforcements. J. *J. Mater. Sci.* 42 (15), 5964–5972.

Shukla, D.K., Srivastava, R.K. (2011). Effect of alumina platelet reinforcement on dynamic mechanical properties of epoxy. In *Proceedings of the World Congress on Engineering.*

Su, X.H., J.G. Li (2011). Low temperature synthesis of single-crystal alpha alumina platelets by calcining bayerite and potassium sulfate. J. *J. Mater. Sci. Technol.* 27 (11), 1011–1015.

Tan, H. (2012). Influence of calcium oxide on alumina phase transition and morphology development in NaCl-KCl flux. J. *Mater. Technol.* 27 (2), 165–168.

Tan, H.B. (2011). Preparation of plate-like α-Al_2O_3 particles in Na_2SO_4 flux using coal fly ash as starting materials. J. *Mater. Technol.* 26 (2), 87–89.

Teaney, S., Pfaff, G. & Nitta, K. (1999). New Effect Pigments Using Innovative Substrates. J. *Eur. Coat.* (4): 90–96.

Wu, N., Chen, Q.H. & Zhou, W.M. (2014). An Exploration of Factors Affecting the Preparation of SiO_2-Coated α-Al_2O_3 Pearlescent Pigment. C.//Advanced Materials Research. Trans Tech Publications. 983, 26–29.

Xu, G.G., H.Z. Cui & G.Z. Ruan. (2013). Synthesis of highly dispersed α-Al_2O_3 platelets in KCl-K_2SO_4-Na_2SO_4 composite flux. J. *Mater. Res. Innov.* 17, 224–227.

Zhang, J., Cheng X. & Zeng L. (2014). Preparation of Flaky Aluminum Oxide for Pearlescent Pigment. J. *China Ceramic Industry*. 6, 001.

Zhang, L., J. Vleugels (2010). Fabrication of textured alumina by orienting template particles during electro phoretic deposition. J. *J. Eur. Ceram. Soc.* 30: 1195–1202.

Zhu, L.H., R.R. Tu & Q.W. Huang (2011). Molten salt synthesis of α-Al_2O_3 platelets using $NaAlO_2$ as raw material. J. *Ceram. Int.* 38, 901–908.

Zhu, L.H., Q.W. Huang (2011). Morphology control of α-Al_2O_3 platelets by molten salt synthesis. J. *Ceram. Int.* 37, 249–255.

Civil, Architecture and Environmental Engineering – Kao & Sung (Eds)
© 2017 Taylor & Francis Group, ISBN 978-1-138-02985-9

A facile cost-effective method for preparing superhydrophobic carbon black coating on Al substrate

P. Wang, B. Sun, T. Yao, X. Fan, C. Wang, C. Niu, H. Han & Y. Shi
School of Energy, Power and Mechanical Engineering, North China Electric Power University, Baoding, China

ABSTRACT: In this research, a simple technique for fabrication of superhydrophobic coating was prepared via spraying hydrophobic carbon nanoparticles suspension on aluminum alloy surface. The static water contact angle and sliding angle on the silica nanoparticle depostied aluminum alloy surface were 160 ± 1° and 5°, respectively. In addition, the as-prepared superhydrophobic surface exhibited excellent self-cleaning property. This approach may open a new way in the fabrication of superhydrophobic surfaces with self-cleaning property.

1 INTRODUCTION

As is well known, aluminum and its alloys have achieved extensive industrial applications owing to its abundance in nature, good ductility, low specific weight, excellent electrical conductivity and excellent machinability (Zheng et al. 2016). However, aluminum and its alloys are prone to contamination in damp environment. Therefore, it is a needed to form a surface layer on Al and its alloys which protect them from a wide range of contamination. One promising approach is to transform the hydrophilic nature of Al and its alloy surfaces to be superhydrophobic.

Superhydrophobic surfaces with static water contact angles higher than 150° and sliding angles lower than 10° have attracted more and more attention because of their numerous applications in self-cleaning (Wang et al. 2015), anti-icing (Ganne et al. 2016), anti-corrosion (Zhang et al. 2016, Zhang et al. 2014), oil-water separation (Liu et al. 2015), and so forth. It is well known that the superhydrophobic property is the result of a combination of suitable surface roughness and low surface energy (Du et al. 2007). A great number of artificial superhydrophobic surfaces have been developed by various methods such as layer-by-layer deposition (Zhai et al. 2004), template-based methods (Chang et al. 2013), sol-gel process (Latthe et al. 2014), lithography (Park et al. 2012), etc. However, these artificial surfaces are prone to lose their self-cleaning properties when they are subjected to impact of water (Deng et al. 2012). Moreover, fluoropolymers such as perfluorooctanoic acid and perfluorooctyltrichlorosilane are conveniently utilized to prepare superhydrophobic surface. Recent researches revealed that long-chain polyfluorinated compounds have a documented ability to bioaccumulation and potential adhverse effects on human offspring (Li et al. 2016). Therefore, there is an increasing demand to prepare non-fluorinated superhydrophobic materials for their nontoxic and low-cost benign.

In this paper, we presented a simple, cost-effective and environmental-friendly way to prepare an aluminum alloy sheet with superhydrophobicity and self-cleaning properties. Painting is intrinsically cheap, low-expertise route to mass production and applicable to almost any substrate types (Wang et al. 2016). Here, the suspension was simply prepared by one-pot process at room temperature. Through painting the as-prepared suspension, the aluminum alloy sheet shows a contact angle of 160 ± 1° and a sliding angle of 2°. It also shows perfect self-cleaning property. When the water droplet is dropped on the nano silica deposited aluminum alloy sheet covered with graphite powder, it roll off the surface and carried the graphite powder away.

2 EXPERIMENTAL PROCEDURE

Carbon black (VXC 72R) was supplied by Cabot. Trimethoxypropylsilane (PTMS, assay: 97%) and ethanol were supplied by Sinopharm Chemical Reagent Co., Ltd. Acetic acid (assay: 36%) was purchased from Tianjin Fuchen Chemical Reagents Factory.

Carbon black was first dispersed in ethanol, and PTMS was added dropwise under stirring at room temperature. The concentration of carbon black and PTMS was 20 and 50 mg/mL, respectively. The solution was further adjusted at PH = 3.5 with acetic acid. In the following experiments, the solution was painted onto the Al 6061 substrate using a disposable Pasteur pipette (1 ml), and cured at room temperature for 24 h.

FE-SEM (Field-emission scanning electron microscope) images were obtained on a Zeiss Supra 55 instrument at 5–10 kV. Prior to FE-SEM measurements, a thin Au layer (ca. 5 nm) was deposited on the specimens by sputtering. The water contact angle and sliding angle were measured with a SL100B apparatus at ambient temperature. The volume of the individual water droplet in all measurements was 5 μL. The average contact angle and sliding angle values were obtained by measuring the same sample at least in five different positions.

3 RESULTS AND DISCUSSION

The synthesis of hydrophobic carbon nanoparticles was a key part of our strategy for the fabrication superhydrophobic coating. During the preparation of coating solution, the hydrolysis reaction of PTMS could be described as Figure 1, where R is $CH_2CH_2CH_3$. Furthermore, the hydrophilic carbon nanoparticles were functionalized through the attraction of hydrogen bond formation by pairs of Si-OH groups.

Figure 2a and b showed the SEM top-images of carbon nanoparticle deposited Al 6061 substrate with low and high magnifications, respectively. From the low magnification SEM images (Figure 2a), it could be founded that carbon nanoparticles uniformly cover the paper substrate. The high magnification SEM (Figure 2b) showed nanoscale porous structure. During the evaporation of ethanol, the distance between nearby nanoparticles is much greater than several nanometers (Li et al. 2015). Thus, the interparticle forces display attractive force rather than replusive force which lead to the nanoparticle aggregation. Furthermore, as the nanoparticles come closer, one nanoparticle tended to combine with another nanoparticle through van der Waals force and hydrogen bonds. Therefore, after the evaporation of ethanol solvent, the porous structure with large scale homogeneity but local irregularity was spontaneously formed.

$$H_3CO{-}\underset{OCH_3}{\overset{CH_2CH_2CH_3}{Si}}{-}OCH_3 + H_2O \xrightarrow{\text{Hydrolysis}} HO{-}\underset{OH}{\overset{R}{Si}}{-}OH \quad (1)$$

$$C(OH)_n + HO{-}\underset{OH}{\overset{R}{Si}}{-}OH \xrightarrow{\text{Dehydration}} R{-}Si{-}O{-}C{-}O_3{\equiv}Si{-}R \quad (2)$$

Figure 1. Scheme of creating a carbon nanoparticle superhydrophobic coating.

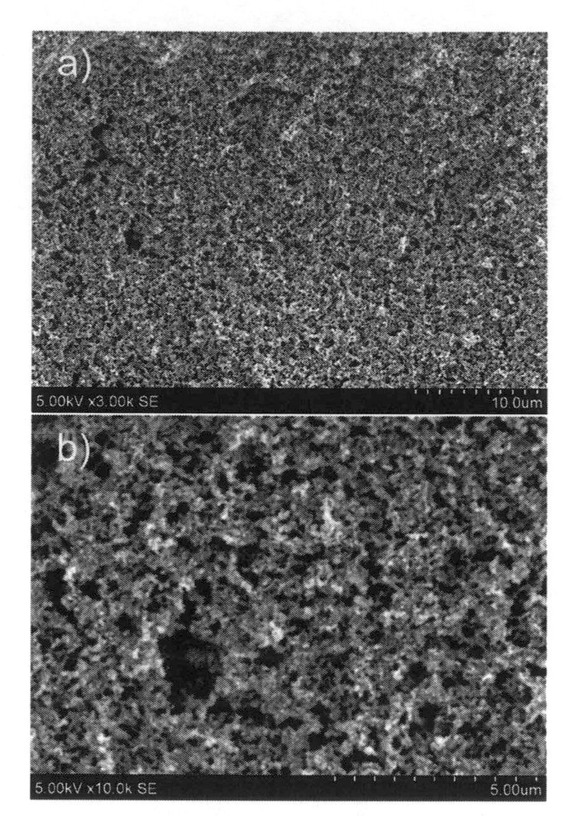

Figure 2. Top-view SEM images of carbon black deposited aluminum alloy surface with low magnification (a) and high magnification (b), respectively.

As shown in Figure 2b, the pore size ranged from nanoscale to sub-microscale.

To study the surface wettability of carbon nanoparticle superhydrophobic coating, the water contact angle and sliding angle of the as-prepared sample were measured, as shown in Figure 3. The carbon nanoparticle deposited superhydrophobic surface showed a high water contact angle of $160 \pm 1°$ (Figure 3a). Meanwhile the water droplets don't come to rest on the surface when the sliding angle is 5°, suggesting low adhesion between water droplet and carbon nanoparticle deposited surface (Li et al. 2012). As is known to all, the surface wettability can be described by the Wenzel (Wolansky et al. 1999) and Cassie-Baxter models (Johnson et al. 1964). Both two wettability states can lead to a high contact angles. However, only the Cassie-Baxter model can result in a very low sliding angle (Liu et al. 2013). Herein, Cassie-Baxter theory was employed to explain the mechanism of superhydrophobic behavior on the carbon nanoparticle superhydrohobic coating. On the as-prepared surface, the apparent solid-liquid contact

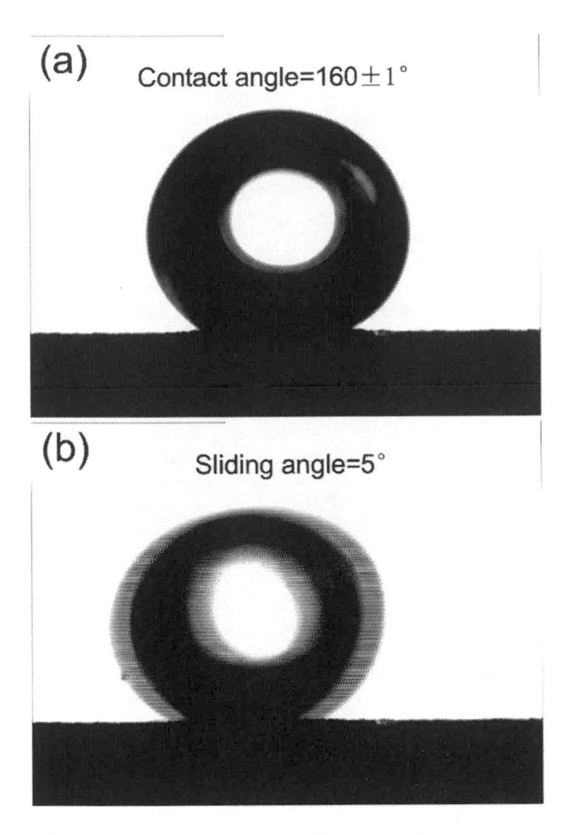

Figure 3. Water droplets on the nano silica deposited aluminum allogy substrate with (a) a contact angle of 160 ± 1° and (b) a sliding angle of 5°.

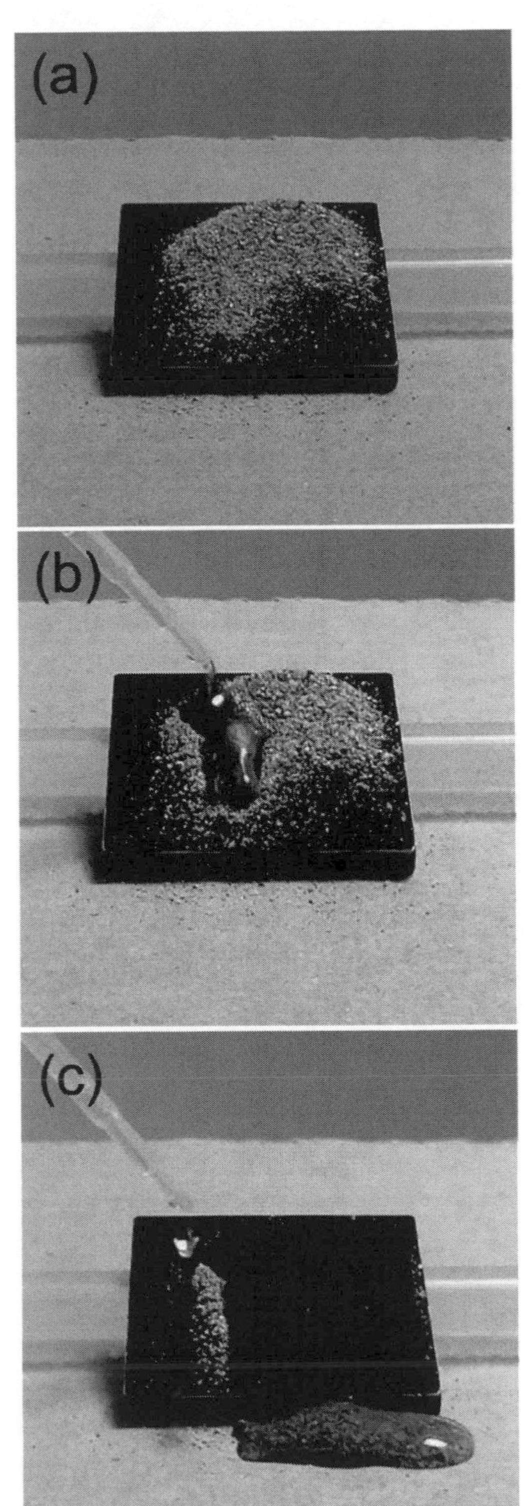

Figure 4. (*Continued*).

is a real composite contact of solid-liquid-gas due to the hierarchical structures. In this composite state, air layer can be trapped in the micro cavities and nanopores between silica nanoparticles, and then water droplets will be suspended due to the "air cushion" underneath water droplets. Thus, the surface adhesion between droplets and superhydrophobic silica nanoparticle deposited surfaces is extremely low and the water drops can slide off with a slight sliding angle.

The silica nanoparticle deposited aluminum alloy surface also shows excellent self-cleaning property. The self-cleaning performance of the carbon nanoparticle superhydrophobic coating was investigated using sand as contaminant dust particles (Figure 4). First, a thin layer of sand was sprinkled onto the as-prepared superhydrophobic coating (Figure 4a). Then, water droplets were gently placed on this contaminated surface. The dust particles were immediately adsorbed on the surface of the water droplet as soon as they came into contact with the water droplet (Figure 4b and 4c). After several sliding and rolling processes,

Figure 4. Self-cleaning process on the carbon nanoparticle deposited aluminum alloy surface: (a) the surface with sand as a model of contaminant; (b-c) the contaminated surface with one water drop on it; (d) the contaminated surface after the sliding of water droplets.

water droplets brought away the contaminants completely and left a clean surface (Figure 4d), which confirmed the excellent self-cleaning ability of the prepared graphene coating.

4 CONCLUSION

In conclusion, we have developed a facile and cost-effective spray-coating method to prepare carbon nanoparticle superhydrophobic coating. The carbon nanoparticle deposited aluminum alloy surface exhibited high water contact angle (160 ± 1°) and low sliding angle (5°). Furthermore, the as-prepared superhydrophobic aluminum surface exhibited excellent self-cleaning property. Thus, the aformentioned great properties make this method a promising candidate to fabricate a superhydrophobic surface on an aluminum sheet for wide application.

ACKNOWLEDGEMENTS

This work was supported by the National Nature Science Foundation of China (No.51607067), the Natural Science Foundation of Hebei Province (No. E2015502023) and Fundamental Research Funds for the Central Universities (No. 2015 MS124).

REFERENCES

Chang, K. et al. (2013) Nanocasting technique to prepare lotus-leaf-like superhydrophobic elecroactive polyimide as advanced anticorrosive coatings, *ACS Appl. Mater. Interfaces 5*, 1460–1467.

Deng, X. et al. (2012) Candle soot as a template for a transparent robust superamphiphobic coating, *Science 335*, 67–70.

Ganne, A. et al. (2016) Combined wet chemical etching and anodic oxidation for obtaining the superhydrophobic meshes with anti-icing performance, *Colloids Surf. A 499*, 150–155.

Johnson, R. et al. (1964) Contact angle hysteresis. III. study of an idealized heterogeneous surface, *The J. Phys. Chem. 68*, 1744–1750.

Latthe, S. et al. (2014) Development of sol-gel processed semi-transparent and self-cleaning superhydrophobic coatings, *J. Mater. Chem. A 2*, 5548–5553.

Li, F. et al. (2015) Transparent and durable SiO_2-containing superhydrophobic coatings on glass. *J. Appl. Poly. Sci. 132*, 41500.

Li, J. et al. (2012) One-step process to fabrication of transparent superhydrophobic SiO_2 paper, *Appl. Surf. Sci. 261*, 470–472.

Li, Y. et al. (2016) One-step spraying to fabricate nonfluorinated superhydrophobic coatings with high transparency, *J. Mater. Sci. 51*, 2411–2419.

Liu, C. et al. (2015) Versatile fabrication of the magnetic polymer-based graphene foam and applications for oil-water separation, *Colloids Surf. A 468*, 10–16.

Liu, W. et al. (2013) Fabrication of the superhydrophobic surface on aluminum alloy by anodizing and polymeric coating, *Appl. Surf. Sci. 264*, 872–878.

Park, H. et al. (2012) Superhydrophobicity of 2D SiO_2 hierarchical micro/nanorod structures fabricated using a two-step micro/nanosphere lithography, *J. Mater. Chem. 22*, 14035–14041.

Qu, M. et al. (2007) Fabrication of superhydrophobic surfaces on engineering materials by a solution-immersion process, *Adv. Mater. 17*, 593–596.

Wang, P. et al. (2016) Transparent and abrasion-resistant superhydrophobic coating with robust self-cleaning function in either air or oil, *J. Mater. Chem. A 4*, 7869–7874.

Wolansky, G. et al. (1999) Apparent contact angles on rough surfaces: the Wenzel equation revisited, *Colloids Surf. A 156*, 381–388.

Zhai, L. et al. (2004) Stable superhydrophobic coatings from polyelectrolyte multilayers, *Nano Lett. 4*, 1349–1353.

Zhang, X. et al. (2014) Preparation of superhydrophobic zinc coating for corrosion protection, *Colloids Surf. A 454*, 113–118.

Zhang, Z. et al. (2016) Mechanically durable, superhydrophobic coatings prepared by dual-layer method for anti-corrosion and self-cleaning, *Colloids Surf. A 490*, 182–188.

Zheng, S. et al. (2016) Development of stable superhydrophobic coatings on aluminum surface for corrosion-resistant, self-cleaning, and anti-icing applications. *Mater. Design 93*, 261–270.

Civil, Architecture and Environmental Engineering – Kao & Sung (Eds)
© 2017 Taylor & Francis Group, ISBN 978-1-138-02985-9

Preparation of multifunctional flake alumina ceramics

Qianrui Wang, Chong Liu, Kaixuan Du & Yan Xiong
School of Materials Science and Engineering, Hubei Province Key Laboratory of Green Materials for Light Industry, Hubei University of Technology, Wuhan, P.R. China

ABSTRACT: In the field of inorganic non-metallic materials, flake Al_2O_3 is one of the hot spots of research due to its unique two-dimensional structure and the larger aspect ratio. $Al_2(SO_4)_3$, Na_2SO_4, K_2SO_4, Na_2CO_3 and Na_3PO_4 are used as raw materials. And a novel routine method obtained by combining the sol–gel process with molten-salt growth is used to prepare flake Al_2O_3. Meanwhile, the effects of additive content and sintering temperature on its aesthetic properties and micro-structure are also investigated. The results show that the best aspect ratio and the excellent aesthetics property of flake Al_2O_3 are obtained by using 0.0136% $SnSO_4$, 0.1000% $TiOSO_4$, and 0.0900% $ZnSO_4$ as the additives and by sintering at 1200°C for 7 h.

1 INTRODUCTION

In recent years, flake Al_2O_3 has emerged into a kind of composite materials with excellent performance. It has extensive applications in different fields, including cosmetics industry, refractory (Cemail (2002), Karlsson (2006)), pearlescent pigments, toughening ceramics (Matthew (1997), Zhang (2010)), and other fields. Generally, flake Al_2O_3 grain size is 5–50 μm, and the thickness is 100–500 nm. The well-developed crystal particles of flake Al_2O_3 also show a regular hexagonal morphology (Yang (2004), Pei (2010)). Flake Al_2O_3 ceramics with excellent performance are widely used in expensive automobile basic paint materials, high-grade pearlescent pigments and cosmetics (Egashira (2006), Teaney (1999), Sharrock & Schuel (2000)), and advanced abrasives (for the electronics industry, mono-crystalline silicon substrate polishing) (SU (2004)). But these raw materials are mainly dependent on import, such as Germany's Merck and BASF companies. Although, there are some reports in the Chinese literature, flake Al_2O_3 mentioned in the report is not big in size, the aspect ratio is small and the stability and repeatability of preparation are poor. However, in China, automotive-related industries have a huge demand of flake Al_2O_3, providing broad prospects for flake Al_2O_3. When compared with the traditional molten salt method (Jin & Gao (2004), Hsiang (2007), Su & Li (2011)) to prepare ceramic powder, we have adopted a novel routine method by combining the sol–gel process with molten-salt growth (Hill & Supancic (2002), Zhou (2003), Akkaya (2014)), the flake Al_2O_3 prepared by this method has a large aspect ratio, stable physical and chemical properties, and good repeatability.

2 EXPERIMENTAL PROCEDURE

2.1 *Raw materials and reagents*

All of the following materials were produced in Sinopharm Group Chemical Reagent Co. Ltd. and were 99.7% pure:

$Al_2(SO_4)_3{\cdot}18H_2O$, Na_2SO_4, K_2SO_4, $SnSO_4$, Na_2CO_3, $ZnSO_4{\cdot}7H_2O$, $TiOSO_4$, and Na_3PO_4.

2.2 *Preparation process*

The molar ratio of the molten salt and Al^{3+} is 4:1. The concentration of 0.5295 mol/L $Al_2(SO_4)_3{\cdot}18H_2O$, 0.8469 mol/L K_2SO_4, and 1.2787 mol/L Na_2SO_4 are prepared respectively, thereby composing solution 1. And then, a certain percentage of $SnSO_4$, $TiOSO_4$, and $ZnSO_4{\cdot}7H_2O$ are added to solution 1. Next, 0.3396 mol/L Na_2CO_3 and 0.0018 mol/L Na_3PO_4 are prepared as the second solution, namely 2. The solution 2 is dropped at a speed of 5 ml/min into solution 1 at 65°C and is kept under continuous stirred at a rate of 400 r/min for 15 min by using a peristaltic pump until the reaction is completed to obtain a gel mixture of the two solutions. The temperature of the reaction container is raised to 90°C and aged for 24 h. The gel is evaporated to dryness. And then, the dried precursor powder is placed in a corundum crucible and compacted and placed in a muffle furnace. The dried precursor powder is heated respectively to 1000°C, 1100°C, 1200°C and 1300°C at the rate of 9°C/min. The solid agglomerate obtained after calcination is dispersed with concentrated sulfuric acid and then repeatedly washed with a large amount of distilled water and alcohol to remove impurities and finally dried to obtain a test sample.

2.3 *Properties and characterization of the sample*

The morphology analysis instrument of flake alumina employs a JSM-5610 LV scanning electron microscope, made by Japanese Electronics Co., LTD, by using high-energy electron beam, on the surface of flaky alumina powder through scanning frame by frame, inspiring different physical signals, which are saved as printing results. The microstructure phase analysis of samples uses an X-ray Microanalysis System (EDS) made by the American General Company, as the attachment configuration of scanning electron microscopy in a JSM-5610 LV Scanning Electron Microscope (SEM). A Phoenix spectrometer is a spectrum analysis system with high resolution images and component diagram.

3 RESULTS AND DISCUSSION

3.1 *Surface morphology characterization and mechanism analysis*

Figure 1 shows the XRD pattern for flake aluminium oxide.

According to the XRD pattern of the sample, the crystal shape of flaky aluminium oxide is α-Al_2O_3. Through the contrast with the standard peak, the location of the characteristic peak is conformed to the standard. And so, the crystal type of the final samples is the alpha type of the crystal. α-Al_2O_3 is a kind of nanometer alumina. The size and specific surface area of α-Al_2O_3, which appears in a white fluffy powder state, are respectively 20 nm and 50 m^2/g or higher. In view of the existence of defects, semiconductor and metal composite materials, etc. present a rare superiority instead. Addition of inhibitors inhibits the grain's longitudinal growth, thereby bringing out the special two-dimensional plane structure on flake alumina.

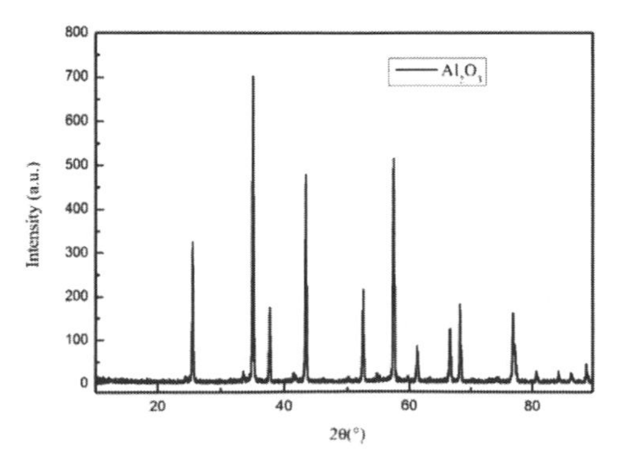

Figure 1. XRD pattern of flake Al_2O_3.

Figure 2 shows the SEM image of the analysis of the crystal microstructure of flake alumina, which is prepared under the condition of aging at 95°C for 36 h and sintering at 1000–1300°C, at the rate of 9°C/7 min for 7 h. Results indicate that, at 1200°C, the grain size and thickness of the flake alumina are the best and the aspect ratio is the largest. Meanwhile, the samples possess a translucent shape and the macroscopical morphology has a beautiful streamer color. Because of its translucent sheet-like structure, the light interference effect was observed through the light path through reflection. Due to the anisotropy of the crystal, the addition of appropriate additives can inhibit the growth of the crystal in the longitudinal direction, so that it presents a special two-dimensional planar structure.

From the SEM images shown in Figure 2, we can observe that most of the α-Al_2O_3 exhibit a flaky structure and stay in a translucent crystalline state when not dispersed. On this basis, we measured the size and thickness of grain and compared the aspect ratio through the chart, in order to explore a suitable formula to produce a low-cost and high quality semi-transparent flaky alumina matrix.

Tables 1–3 provide the analysis of the sample of the flake alumina shown in Fig. 2. $ZnSO_4$, $SnSO_4$, and $TiOSO_4$ were added at 65°C, and then aged

Figure 2. Thick contrast SEM images of flake alumina with grain diameter.

Table 1. The morphology of the sample obtained at 1100–1300°C by adding $ZnSO_4$ during sintering of alumina.

Sintering temperature (°C)	Biggest size (μm)	Average thickness (μm)	Particle size (μm)	Aspect ratio
1000	31.04	0.43	20.30	47.21
1100	20.13	0.40	13.60	34.00
1200	28.40	0.18	27.70	153.89
1300	31.00	0.20	24.40	122.00

Table 2. The morphology of the sample prepared at 1100–1300°C by adding $SnSO_4$ during sintering of alumina.

Sintering temperature (°C)	Biggest size (μm)	Average thickness (μm)	Particle size (μm)	Aspect ratio
1000	25.71	0.58	21.50	37.09
1100	44.96	0.20	33.80	169.00
1200	49.02	0.86	40.80	47.44
1300	26.75	0.29	24.20	83.45

Table 3. The morphology of the sample prepared at 1100–1300°C by adding $TiOSO_4$ during sintering of alumina.

Sintering temperature (°C)	Biggest size (μm)	Average thickness (μm)	Particle size (μm)	Aspect ratio
1000	36.89	0.30	26.12	87.07
1100	34.46	0.25	25.50	102.00
1200	26.55	0.21	15.60	74.29
1300	25.68	0.22	20.23	91.95

at 95°C for 36 h, and sintered for 7 h at the rate of 9°C/min at 1000–1300°C; the aspect ratios of different additives at different temperature points vary greatly; Table 1 aims to explore the impact of temperature on its molding, which additive is added and the quantity of additives.

It can be seen from Table 1 that under the same conditions of additive $ZnSO_4$, sintering temperature, holding time, and so on, the samples sintered at 1200°C are very thin. SEM images show that the samples are translucent and thin and their grain sizes have reached the best level, and also the diameter to thickness ratio is relatively large, but the maximum size is not perfect when compared to other temperatures.

It can be seen from Table 2 that under the same conditions of additive $SnSO_4$, sintering temperature, holding time, and so on, the maximum size and grain size of samples sintered at 1200°C are very large, but the average thickness is the largest when compared to the other samples. Therefore, it can be concluded that addition of $SnSO_4$ can increase the grain size of flaky alumina, but it will also greatly increase the thickness of the sample.

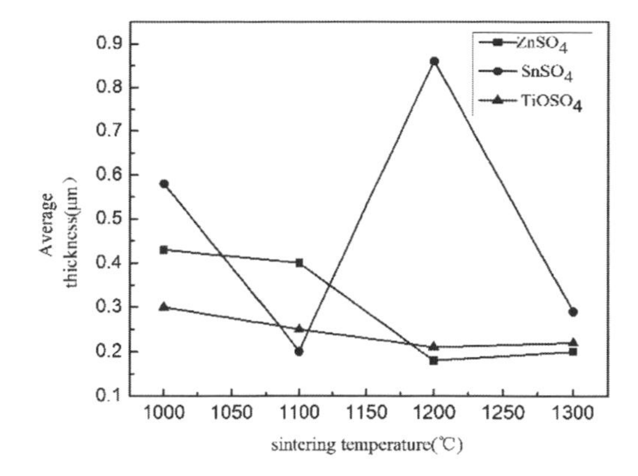

Figure 3. Chart of the average thicknesses obtained by adding $ZnSO_4$, $SnSO_4$, and $TiOSO_4$ at different temperatures.

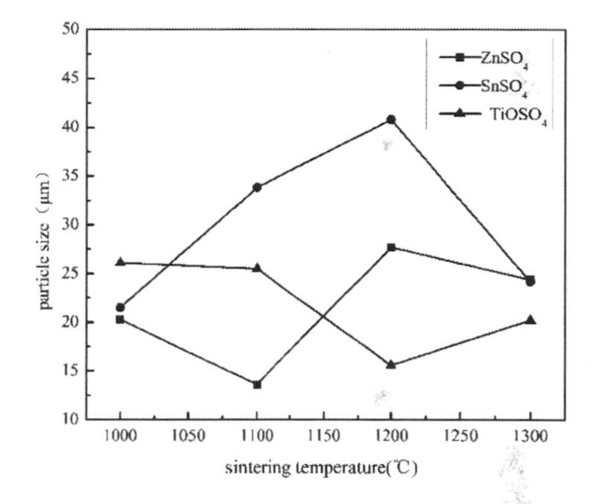

Figure 4. Chart showing the particle size distribution of samples prepared by adding ZnSO4, SnSO4, and $TiOSO_4$ at different temperatures.

It can be seen from Table 3 that under the same conditions of additive $TiOSO_4$, sintering temperature, holding time, and so on, the average thickness of samples sintered at 1200°C is smaller than that of other samples. But the maximum size and grain size of the sample are the largest among the several temperature points. Therefore, it can be concluded that the addition of $TiOSO_4$ can reduce the thickness of the flaky alumina, but it will also increase the grain size of the sample.

It can be seen from Table 4 that under the conditions of adding additive $ZnSO_4$, $SnSO_4$, and

Table 4. The morphology analysis of sintering powder obtained at 1200°C and 7 h by adding $ZnSO_4$, $SnSO_4$ and $TiOSO_4$ at different temperatures.

Additive dosage	Diameter to thickness ratio
1.77% $ZnSO_4$ + 0.13% $SnSO_4$	124.33
0.21	
26.11	
1.77% $ZnSO_4$ + 0.26% $SnSO_4$	54.67
0.39	
21.32	
0.89% $ZnSO_4$ + 0.13% $SnSO_4$	49.35
0.60	
29.61	
0.09% $ZnSO_4$ + 0.0136% $SnSO_4$ + 0.10% $TiOSO_4$	168.00
0.20	
33.60	

$TiOSO_4$ together, the average thickness and particle size of the samples sintered at 1200°C are much better than those obtained by using the individual additives.

4 CONCLUSION

The composite molten salt method is adopted to prepare flake alumina with the raw materials of $Al_2(SO_4)_3$, Na_2SO_4, K_2SO_4, Na_2CO_3, Na_3PO_4, and additives such as $ZnSO_4$, $SnSO_4$, $TiOSO_4$ to ensure that samples have a more perfect microscopic appearance. The results show that when the sintering temperature is 1200°C, sintering time is 7 h, the amount of $SnSO_4$, $ZnSO_4$, $TiOSO_4$ is 0.0136%, 0.1000% and 0.0900%, respectively, the flaky alumina obtained has the largest diameter to thickness ratio and a beautiful streamer color. It can also meet the needs of applications market for being employed in expensive automobile basic paint materials and high-grade pearlescent pigments.

REFERENCES

Akkaya U.O. & F Z Tepehan. (2014). Influence of Al_2O_3: TiO_2 ratio on the structural and optical properties of Al_2O_3-TiO_2 nano-composite films produced by sol gel method. *J. Compos. Part B: Eng. 58,* 147–151.

Cemail, A. (2002). The role of fine alumina and mullite particles on the thermomechanical behavior of alumina-mullite refractory materials. *J. Mater Lett 57,* 708–714.

Egashira M. et al. (2006). Effects of the surface treatment of the Al_2O_3 filler on the lithium electrode/solid polymer electrolyte interface properties. *J. Electrochim. Acta. 52,* 1082–1086.

Hill R F & Supancic P H. (2002). Thermal conductivity of planet-filled polymer composites. *J. Joumal of the American Ceramic Society. 85(4)*, 851–857.

Hsiang H I et al. (2007). Synthesis of alpha-alumina hexagonal platelets using a mixture of boehmite and potassium sulfate. *J. J. Am. Ceram. Soc. 90*, 4070–4072.

Jin X H & Gao L. (2004). Size control of alpha-Al_2O_3 platelets synthesized in molten Na_2SO_4 flux. *J. J. Am. Ceram. Soc. 87,* 533–540.

Karlsson P et al. (2006). Surface modification for aluminium pigment inhibition. *J. Adv. Colloid Interface Sci.* 121–134.

Matthew M et al. (1997). Messing, Texture development by templated grain growth in liquid-phase-sintered α-alumina. *J. J. Am. Ceram. Soc. 80,* 1181–1187.

Pei Xinmei & Rong lan. (2010). Preparation of Flaky Alumina by Molten Salt Method. *J. Materials Review. 24(16),* 152–153.

Sharrock S R & Schuel N. (2000) New Effect Pigments Based on SiO_2 and Al_2O_3 Flakes *J. Eur. Coat. 12,* 20–23.

Su X. & Li J. (2011). Low temperature synthesis of single-crystal alpha alumina platelets by calcining bayerite and potassium sulfate. *J. J. Mater. Sci. Technol. 27,* 1011–1015.

Su Zhou. (2004). Molten salt synthesis flaky alumina powder. *D. Changsha: Central South University.*

Teaney S et al. (1999). New Effect Pigments Using Innovative Substrates. *J. Eur. Coat. (4),* 90–96.

Xin T.S.J. et al. (1997) *P.* CN Pat: CN1150165A.

Yang Y. et al. (2004). Journal of Process Engineering. *(Z1),* 279–283.

Zhang L. et al. (2010). Fabrication of textured alumina by orienting template particles during electro phoretic deposition. *J. J. Eur. Ceram. Soc. 30,* 1195–1202.

Zhou Z.J. et al. (2003). *Silicate bulletin. 22(4),* 11.

Civil, Architecture and Environmental Engineering – Kao & Sung (Eds)
© 2017 Taylor & Francis Group, ISBN 978-1-138-02985-9

Study on the deformation of plastic films by virtual particle method

X.H. Yang, Z. Yang, Y.X. Wang, C.Y. Lv & X. Cheng
School of Mechanical Engineering, Shandong University of Technology, Zibo, China

ABSTRACT: Sorting of waste plastics by wind is a kind of high-efficient and energy-saving processing method. However, the deformation of plastic films in the wind sorting process seriously affects the theoretical analysis accuracy. A novel virtual mass method has been introduced in this paper. The mathematical model for film deformation has been created and the line unit equation has been deduced. Based on the method, the deformation values at the wind speeds of 0.2 m/s and 0.35 m/s have been obtained. It coincides well with the analytical results from dynamic meshes and fluid solid coupling by the Ansys Fluent software. The wind sorting accuracy can be improved by achieving the deformation rules of plastic films. It can be used to guide the optimization of wind sorting machines for plastic films.

1 INTRODUCTION

With the urbanization goes forward, domestic waste increases year by year. According to the statistics, waste plastics occupy 10% of total waste in large and medium-sized cities (Zhu 2010, Li 2013). The wind sorting of waste plastics has been attracted increasingly attentions because it is a kind of non-pollution and high-efficient sorting method. The optimal dimensions of wind nozzles have been identified for accurate sorting of waste plastics by wind sorting devices (Hu 2015). The wind blowing machine's optimal wind entrance angle has been identified for horizontal wind sorting of domestic wastes (Li 2012). The efficiency of separation and sorting has been enhanced by developing a separation and sorting system for plastic waste (Gao 2009). These studies provide references for further studies of plastic sorting.

The plastics in domestic wastes are composed of plastic films. Therefore, the sorting accuracy and efficiency will be affected seriously due to the deformation of films in the sorting process. The novel virtual particle method is introduced in the paper to study the deformation of plastic films in the separating and sorting process. The mathematical model for film deformation has been created and the deformation rules have been analyzed. The method has been evaluated by the fluid solid coupling analyses using the Ansys Fluent software. It lays solid calculating and analysis foundations for enhancing the accuracy and efficiency of sorting.

2 VIRTUAL PARTICLE METHOD

There are limited plastic deformation studies in sorting processes since it is very difficult to model and calculate the plastic films. The consequences are that the errors between theoretical and experimental results are very large. Aiming at this problem, a novel virtual particle method is proposed to study the deformation of plastic films in the sorting process.

By the virtual particle method, limited uniformly distributed particles are applied to describe the film. The mass of the film is lumped at the particle. The particle is connected by line cell without masses. Positions of line cells are calculated by position functions. The schematic view of the simplified model is shown in Figure 1 with mesh structures. The internal and external forces bearing by the films are described the equivalent concentrated forces of the particles. The overall deformation and spatial positions of the films are defined by position vectors of particles.

The mass of the particle can be described as,

$$M_a = \frac{M}{n} \tag{1}$$

where, M_a is the mass of the point, M is the mass of the film, and n is the particle number in the model.

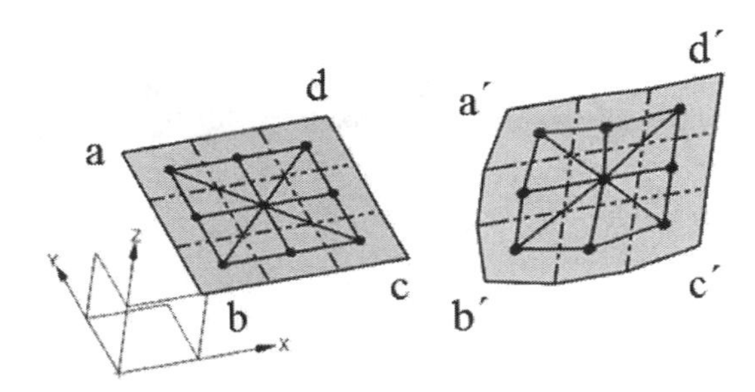

Figure 1. Film models described by particles.

In order to solve the internal forces of the unit, the pass unit concept is introduced by neglecting the effects of unit geometrical changes in variations of films positions. By the virtual particle method, film behaviors are described by the motion trajectories of all particles under loads. The motion and deformation of films are time functions of particle position vectors. Then, the time trajectory of the particle can be solved, where the time trajectory is also called the path.

As shown in Figure 2, given the starting and ending time as t_0 and t_f for particle A, y_0 and y_f is the corresponding position vector, respectively. The time function y(t) is the position vector at time t. The path of the particle between time t_0 and t_f is divided into three segments. The segments between time t_a and t_b is defined as path cell. The time function is continuous inside one path cell. The path cell abides by motion laws and Hooke law while the rigid body motions are excluded.

The virtual reverse motion concept is proposed in order to obtain the deformation value from overall displacement by excluding the rigid body displacement after the deformation of films. Inside the path cell between time t_a and t_b, the geometrical structure of the film at the starting time t_a is selected as the reference. The rigid body movement of line cell at the time between t_a and t will be estimated. The translation is defined as the displacement value of arbitrary point at the line cell. The rotational motion is defined as the spinning value of the line cell between time t to t_a. According to the estimated values, the line cell makes a virtually inverse rigid body movement at time t, which is called the virtual inverse movement. Inside one path cell, the difference between the formation after virtually reverse movement and the actual one is the tiny deformation and rigid body movement, which can be calculated by micro strain and engineering stress. Therefore, the node force can be solved. The force direction will be changed after the positive rigid body movement since the node force is a group of balance forces. Therefore, the internal force inside the cell can be obtained by vector rotation of the node force direction.

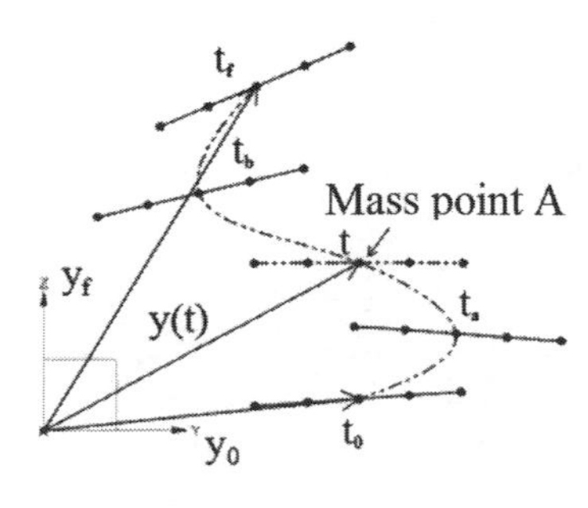

Figure 2. The path cell ($t_a \leq t \leq t_b$).

3 MATHEMATIC MODELLING AND ANALYSIS

3.1 *Deformation of line cells*

A standard bar cell (1_a 2_a) is selected from the standard path cell between time t_a and t_b. Its cross section area is A_a and the Young's modulus is E. The bar axis coordinate (x') is set up with the superimposition of node 1_a and original point, where the x' axis is parallel to the bar axis. Then, the position vector Xa on the bar at time ta and t can be calculated by,

$$\begin{cases} x_a = (1-s)x_{1a} + sx_{2a}, s = \dfrac{x'_a}{l_a} \\ x = (1-s')x_1 + sx_2, s = \dfrac{x'_a}{l_a} \end{cases} \tag{2}$$

The cross section of the bar remains unchanged and s is equal to s'. Therefore, the node deformation is,

$$(x - x_a) = (1-s)(x_1 - x_{1a}) + s(x_2 - x_{2a}) \tag{3}$$

namely,

$$u = (1-s)u_1 + su_2 \tag{4}$$

where, u is the displacement of arbitrary point on the bar, u_1 and u_2 are the node displacements of the rod cell.

The relative displacement between the two points is,

$$u - u_1 = \Delta s(u_2 - u_1), \Delta s = \frac{\Delta x'_a}{l_a} \tag{5}$$

where, $\Delta x'_a$ is the distance between arbitrary two points on the bar at time t_a.

3.2 *Internal forces of line cell*

The bar cell takes node 1 as the fixed reference point. The rotational angle is $-\theta$. The position variation between two arbitrary points on the bar is $\Delta\eta^r$. If the relative position variation is $\Delta\eta_2^r$ between node 2 and node 1 on the bar, then,

$$\Delta\eta^r = \Delta s \eta^{2r} \tag{6}$$

Given the deformations of two nodes on the bar as η_1^d and η_2^d, the following equation can be got since $\eta_1^d = 0$.

$$\eta^{2d} = (u_2 - u_1) + \eta^{2r} \tag{7}$$

After rotation, the cell and the bar cell are collinear. The geometrical relationship satisfies the hypothesis of mechanics of materials for calculating axial deformations. Under the coordinate,

$$\eta^{2d} = (l - l_a)e'_{ax} \tag{8}$$

The relative displacement in virtual formation is,

$$\Delta\eta^d = (\Delta x' - \Delta x'_a)e'_{ax} \tag{9}$$

It is tiny deformation and rotation of the bar cell. The geometrical variation can be neglected. The tiny strain ε and it value ε' are,

$$\varepsilon = \varepsilon' e'_{ax} = \frac{\Delta\eta^d}{\Delta x'_a} = \frac{l - l_a}{l_a} e'_{ax} \tag{10}$$

Under the defined coordinate, the internal force vector f and axial stress σ between two points on the bar are,

$$\sigma = \frac{f}{A_a} = \frac{f}{A_a} e'_{ax} = \sigma' e'_{ax} \tag{11}$$

The bar cell is elastic. Based on Hooke law, there are linear relationships between stress and strain at the time t_a and t. Namely,

$$\sigma' - \sigma'_a = E_a \varepsilon' \tag{12}$$

Therefore,

$$f = f' e'_{ax} = (f_a + A_a E_a \varepsilon) e'_{ax} \tag{13}$$

Given relative virtual displacement $\delta(\Delta\eta^d)$, virtual deformation power is,

$$\Delta(\delta U) = f\delta(\Delta\eta^d) \tag{14}$$

The deformation virtual power of the bar cell can be calculated after the integration of the overall bar cell.

$$\delta U = \left[f_a + \frac{A_a E_a}{l_a}(l - l_a) \right] \delta(l - l_a) \tag{15}$$

Since the axial force of the bar cell is in balance and $\Sigma F = 0$, then $f_1 = f_2$ and the virtual power is,

$$\delta W = f'_2 \delta \eta_2^d \tag{16}$$

Based on virtual power theory, δU = δW and the internal force of the node is,

$$f'_1 = f'(-e'_{ax}),\ f'_2 = -f'_1 \tag{17}$$

After obtaining the internal force of nodes, positive motions are processed. Since the force is in balance for two node point, the value remains unchanged after translation. Only the rotation makes the force direction change. Then, the internal force between two nodes under the coordinate is,

$$f'_1 = f'(-e'_x),\ f'_2 = -f'_1 \tag{18}$$

3.3 *Motion control function of particles*

Particles abbey the second Newton law. The motion function of a single DOF is,

$$m\frac{d^2x(t)}{dt^2} = F(t),\ t \geq 0 \tag{19}$$

where, m is a constant, x(t) and F(t) are functions of time.

Central difference equation is adopted to solve the function and the acceleration can be described as,

$$\ddot{x} = \frac{d^2x(t)}{dt^2} = \frac{1}{h^2}(x_{n+1} - 2x_n + x_{n-1}) \tag{20}$$

where, h is the time increment step and is a constant.

The difference equation at time t can be as follows.

$$x_{n+1} = \frac{F^n}{m}h^2 + 2x_n - x_{n-1} \tag{21}$$

Then,

$$\begin{cases} x_{-1} = x_0 - h\dot{x}_0 + \dfrac{h^2\ddot{x}_0}{2} \\ x_1 = x_0 + h\dot{x}_0 + \dfrac{h^2F^0}{2m} \end{cases} \tag{22}$$

After obtaining the value when n = 1, the difference equation can be used to solve the function step by step forward and the corresponding displacement of each time step can be achieved.

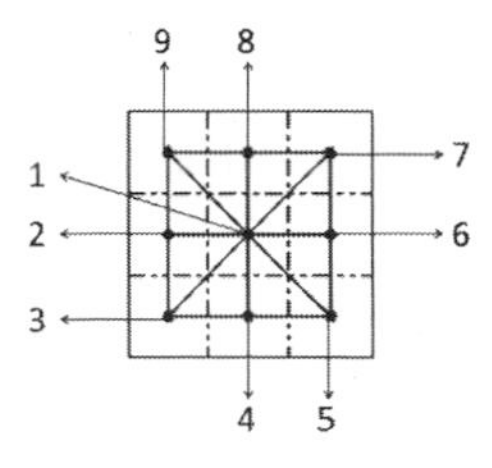

Figure 3. Schematic view of the virtual particle model of the film.

3.4 *Model calculation*

The film model is created as shown in Figure 3 by the virtual particle method. The model is composed of nine particles and the lines connecting these points. The size and the thickness of the cell film are 12 mm × 12 mm and 0.05 mm, respectively. Since the film is distributed symmetrically, 1/4 of upper right corner of the film is used to calculate the model. Given the film properties for middle density PE material, the elastic modulus $E = 172$ N/mm^2, poisson ratio $\nu = 0.439$, and the density $\rho = 0.00139$ g/mm^3. The film is bearing the wind force, windage, and gravity perpendicular to the film surface. The wind speed is given as 0.2 m/s and 0.35 m/s.

3.5 *Solving results*

The software Matlab is used to program and calculate the function based on virtual particle method. The force bearing vs. displacement variation is shown in Figure 4 for particle 1 while the wind speed is 0.2 m/s. With the increase of the force bearing, the displacement of particle 3 becomes larger and larger. The maximum value is 3.2969×10^4 mm.

4 SIMULATION AND EVALUATION

In order to evaluating the above calculated results, the FEM software Ansys Fluent is used to mode and calculate the two-way fluid solid coupling for films. The model is shown in Figure 5 with 2.62 million mesh cells and 2.07 million nodes.

Figure 6 shows the simulation results while the wind speed is 0.2 m/s, which coincides with the calculated one and the introduced mathematical modeling method has been evaluated.

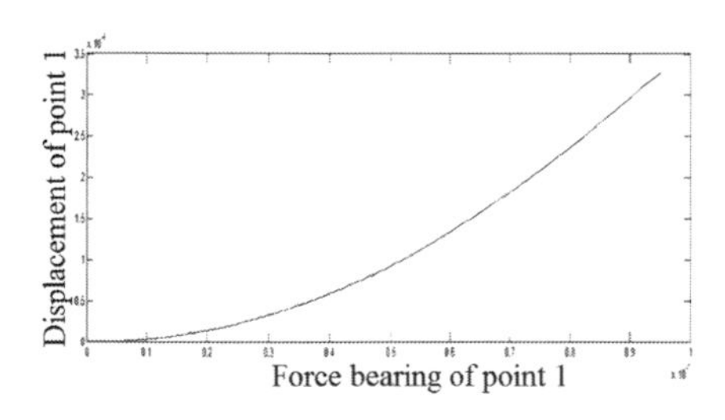

Figure 4. The calculating result of particle 1.

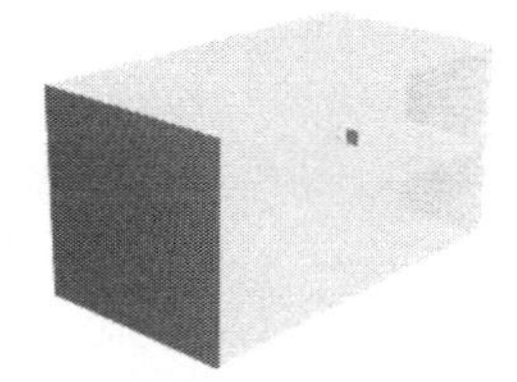

Figure 5. FEM model.

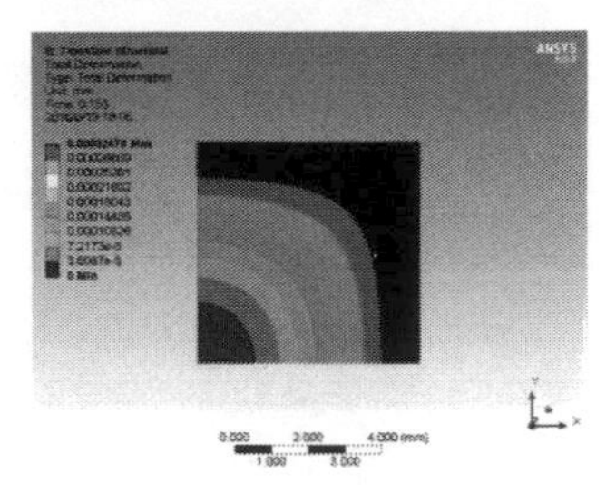

Figure 6. Simulation results.

5 CONCLUSIONS

1. A novel mathematical calculating method, virtual particle method, has been introduced for modeling the deformations of plastic films.
2. The two-way fluid solid coupling method is used to build the FEM model for plastic films by the software Ansys Fluent. The simulation results coincide with that from the mathematic model.
3. The deformation rules can be solved in the near future based on the proposed method. It will be applied to design optimizations of wind sorting equipments of waste plastic films for enhancing the sorting accuracy.

ACKNOWLEDGEMENT

The authors wish to express their appreciation for the financial support by the Natural Science Foundation of Shandong Province (ZR2013EEM011, ZR2014EEM046) and the Key R&D Project of Shandong Province (2015GGB01328).

REFERENCES

Gao, G.S., J.F. Wang, H.Y. Zhang & L. Feng (2009), Wind sifting system for plastics from municipal solid waste: Light Ind. Mach., 27(2), 104–106.

Hu, B., S.T. Wang, G.L. Tang, L. Ren & S.H. Zhang (2015), Design research of waste plastic accurate sorting device based on fluent software: Chinese J. Appl. Mech., 32(6), 1075–1081.

Li, B., Z.Y. Dong, Y.C. Zhao, Y.H. Chen & J.L. Zhu (2012), Study on characteristics of the horizontal air separation for municipal solid waste: Env. Eng., 30, 96–101.

Li, C.H., X.H. Yang & X. Peng (2013), Study on the movement dynamics of waste plastic film in the separation chamber: Mech. Sci. Tech. for Aero. Eng., 32(5), 744–750.

Zhu, X.H. & L.N. Yu (2010). Sorting technology of waste plastic in urban garbage: *Recy. Res.* 3, 33–37.

Civil, Architecture and Environmental Engineering – Kao & Sung (Eds)
 ISBN 978-1-138-02985-9

An experimental study of installing-holes distances parameter optimization for metallic inserts bonded into CFRP laminates

Jing Liu, Hui Zhang & Kun Hou
China Aero-Polytechnology Establishment, China

ABSTRACT: Metallic inserts are one kind of aerospace fasteners, which are usually installed in metallic components by using an interference fit. However, when metallic inserts are installed in the CFRP laminates by using the traditional installing method, delamination and low efficiencies are troublesome. Therefore, excellent quality and cost-effective installing of metallic inserts into the CFRP laminates remains a challenge. In this paper, a series of experiments were carried out to study the installing-holes distances parameter optimization for metallic inserts bonded into CFRP laminates.

1 INTRODUCTION

Being an advanced material, CFRP laminates are characterized by properties such as high specific strength, high specific stiffness, and good corrosion resistance, which have been widely used in manufacturing of aeronautics structure components. Meanwhile, high rigidity and brittleness of CFRP laminates bring about new technical problems and difficulties in assembling (Mccarthy, C. T., & Mccarthy, M. A. 2005). Especially for metallic inserts, the traditional interference fit is not adapted to these new types of materials, which would easily lead to stress concentration and delamination at the fastener hole of the CFRP laminates (Camanho, P. P., Fink, A., Obst, A., & Pimenta, S. 2009). This results in reduced strength against fatigue and compromised structure integrity (Hopkins, M. 2006). Therefore, it is important to develop new techniques to improve the quality and efficiency of mechanically fastened metallic inserts in CFRP laminates (Camanho, P. P., Tavares, C. M. L., Oliveira, R. D., Marques, A. T., & Ferreira, A. J. M. (2005.) Over the years, some researchers have studied the bonding to fasten the metallic inserts in CFRP laminates, the method of which is considered as a good method to satisfy the joint strength and avoid the assembly defect (Camanho, P. P., & Matthews, F. L. 2000). In this paper, a series of experiments were performed to study the installing-holes distances parameter optimization for metallic inserts bonded in CFRP laminates.

2 EXPERIMENTAL SETUP AND PROCEDURE

2.1 *Setup and conditions*

The CFRP laminates bonded with metallic inserts are shown in Figure 1 and there are two metallic inserts in each sample. The test parameters for the experiments are summarized in Table 1. Figure 2(a) shows a computer-controlled electronic universal testing machine, which is used to test the push-out values of metallic inserts installed in the CFRP laminates and mainly consists of a loading spindle, a samples fixture, a worktable, a power supply system, and a computer. The maximum loading and feed speed are 20 kN and 500 mm/min, respectively. Figure 3(a) shows a torsion testing machine, which is used to test the breakaway torque values of metallic inserts installed in the CFRP laminates and these mainly consist of a torsion spindle, a samples fixture, a feed direction worktable, a power supply system, and a computer. The maximum torque and spindle speed are 50 N.m and 500 rpm, respectively.

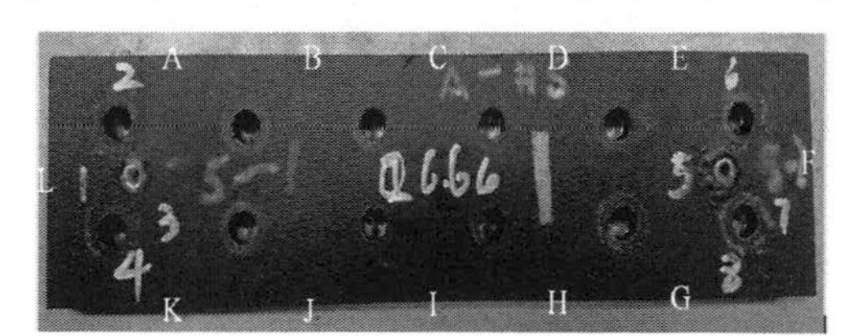

Figure 1. Picture showing the sample of a CFRP laminate bonded with metallic inserts.

Table 1. Test parameters for the experiments.

Parameters	Sample
Outer diameter of the metallic insert	Ø3 mm
Material of the metallic insert	Steel
Thickness of the CFRP laminate	5 mm
Adhesives	Epoxy Resin Adhesive
Material of mandrel	A286
Installing-holes distances type	A<B<C<D
Injection pressure	(0/60/80/100)PSI

2.2 *Experimental procedure*

Four type installing-holes distances samples are prepared to test the push-out and breakaway torque values of metallic inserts bonded into CFRP laminates. Firstly, each sample is prepared to finishing injection pressure for 0/60/80/100PSI. Secondly, the push-out and breakaway torque tests are carried out. The test conditions are summarized in Table 2.

The push-out test process is illustrated in Figure 2. The computer-controlled electronic universal testing machine applies axial load on the metallic insert that is bonded into CFRP laminates via a mandrel until the metallic insert looses in an axial direction.

The breakaway torque test process is illustrated in Figure 3. The torsion testing machine applies radial torque on the metallic insert bonded into CFRP laminates via a mandrel until the metallic insert looses in the radial direction.

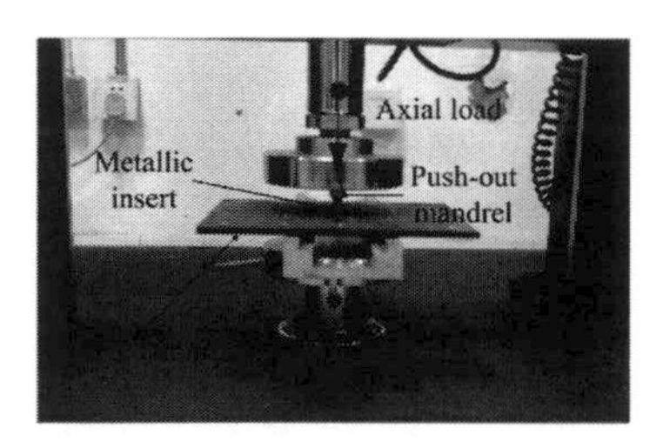

(a) Computer-controlled electronic universal testing machine

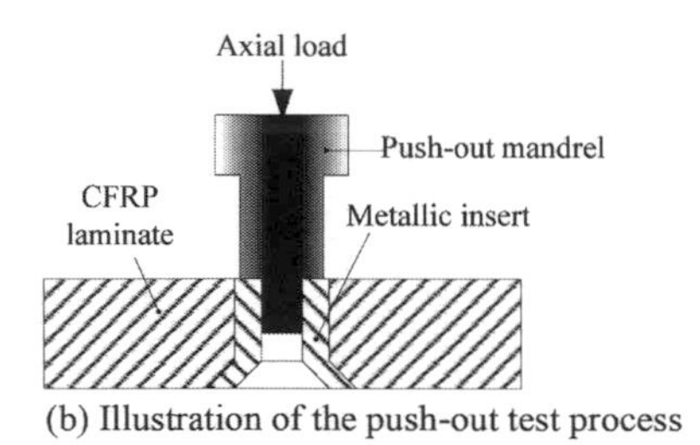

(b) Illustration of the push-out test process

Figure 2. Picture and schematic of the push-out test.

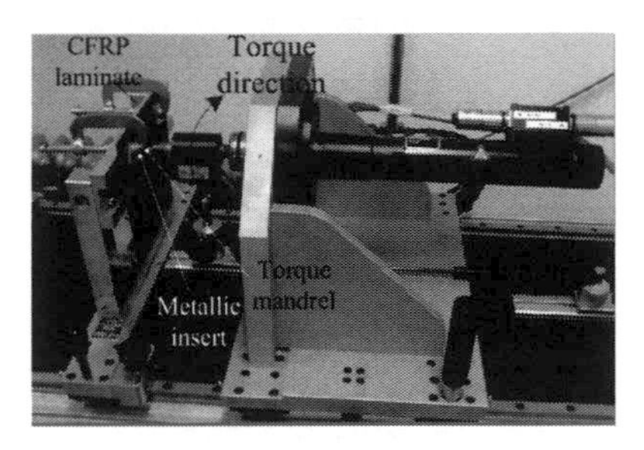

(a) Torsion testing machine

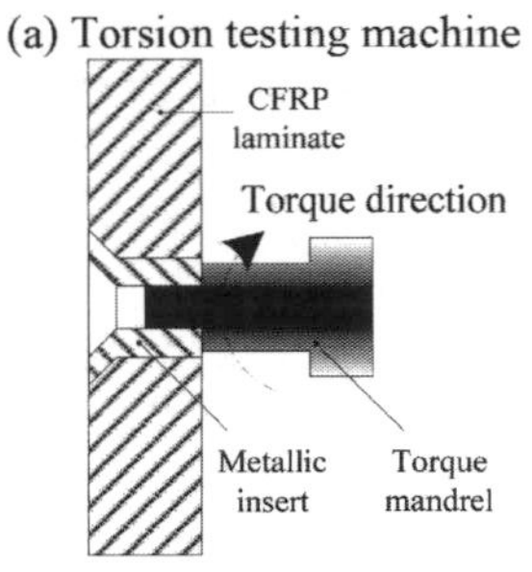

(b) Illustration of the breakaway torque test process

Figure 3. Picture and schematic of the breakaway torque test.

Table 2 Test conditions

Samples	Injection pressure	Metallic inserts	Tests	Quantity
A	0 PSI	SD-A-1#-1~SD-A-1#-2	Push-out/breakaway torque	1/1
	60 PSI	SD-A-2#-1~SD-A-2#-2	Push-out/breakaway torque	1/1
	80 PSI	SD-A-3#-1~SD-A-3#-2	Push-out/breakaway torque	1/1
	1000 PSI	SD-A-4#-1~SD-A-4#-2	Push-out/breakaway torque	1/1
B	0 PSI	SD-B-1#-1~SD-B-1#-2	Push-out/breakaway torque	1/1
	60 PSI	SD-B-2#-1~SD-B-2#-2	Push-out/breakaway torque	1/1
	80 PSI	SD-B-3#-1~SD-B-3#-2	Push-out/breakaway torque	1/1
	1000 PSI	SD-B-4#-1~SD-B-4#-2	Push-out/breakaway torque	1/1
C	0 PSI	SD-C-1#-1~SD-C-1#-2	Push-out/breakaway torque	1/1
	60 PSI	SD-C-2#-1~SD-C-2#-2	Push-out/breakaway torque	1/1
	80 PSI	SD-C-3#-1~SD-C-3#-2	Push-out/breakaway torque	1/1
	1000 PSI	SD-C-4#-1~SD-C-4#-2	Push-out/breakaway torque	1/1
D	0 PSI	SD-D-1#-1~SD-D-1#-2	Push-out/breakaway torque	1/1
	60 PSI	SD-D-2#-1~SD-D-2#-2	Push-out/breakaway torque	1/1
	80 PSI	SD-D-3#-1~SD-D-3#-2	Push-out/breakaway torque	1/1
	1000 PSI	SD-D-4#-1~SD-D-4#-2	Push-out/breakaway torque	1/1

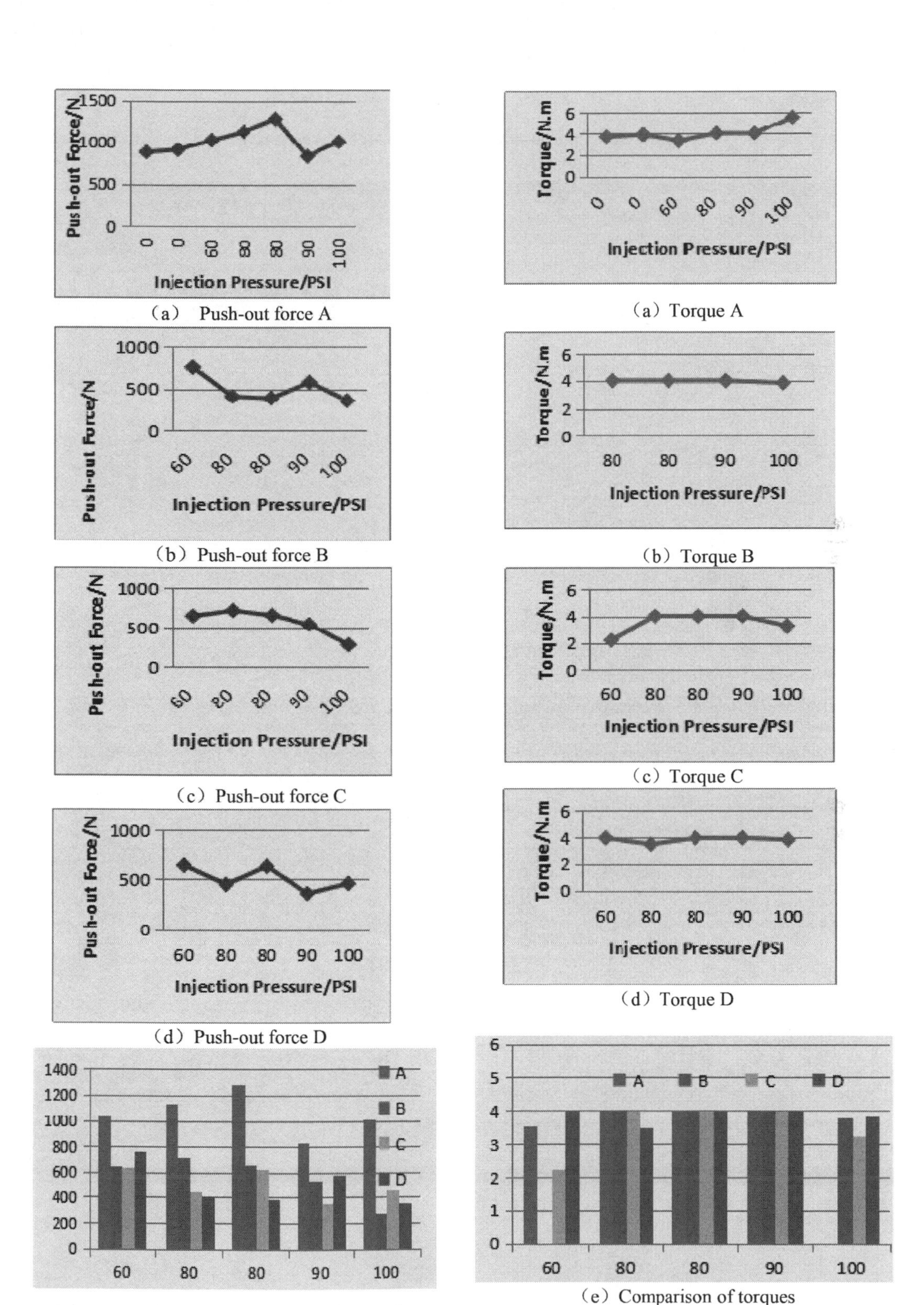

(a) Push-out force A

(b) Push-out force B

(c) Push-out force C

(d) Push-out force D

(e) Comparison of push-out force distributions

Figure 4. Graphs showing results of the push-out test.

(a) Torque A

(b) Torque B

(c) Torque C

(d) Torque D

(e) Comparison of torques

Figure 5. Graphs showing results of the breakaway torque test.

3 EXPERIMENTAL RESULTS AND DISCUSSION

3.1 *Push-out test results and discussions*

The push-out test results are shown in Figure 4. The minimum push-out value of the A type sample is 835 N and the maximum fluctuation in amplitude of the push-out is 449 N under the four different types of injection pressure. The minimum push-out value of the B type sample is 280 N and the maximum fluctuation in amplitude of the push-out is 440 N under the four different types of injection pressure. The minimum push-out value of the C type sample is 360 N and the maximum fluctuation in amplitude of the push-out is 280 N under the four different types of injection pressure. The minimum push-out value of the D type sample is 360 N and the maximum fluctuation in amplitude of the push-out is 400 N under the four different types of injection pressure. Figure 4 shows that the fluctuation in amplitude of the push-out for four types samples is not according to the regulation. Therefore, it is mainly attributed to the different mixed ratios for the adhesive to cause different adhesive strengths. Moreover, the installing-holes distances of metallic inserts play a less important role in adhesive strength than that of the different mixed ratios for adhesives.

3.2 *Breakaway torque test results and discussions*

Under the four different types of injection pressure, the breakaway torque test results are shown in Figure 5. In the test, the threshold value of stopping the breakaway torque is set for 4 N.m. In these tests, all the inserts are not loosened. However, due to the fact that the strength of mandrels is not enough to lose the bonded metallic inserts, a less number of mandrels were broken when the torque value is less than 4.0 N.m. The minimum torque value of the A type sample is 3.55 N.m. The minimum torque value of the B type sample is 3.83 N.m. The minimum torque value of the C type sample is 2.23 N.m. The minimum torque value of the D type sample is 3.52 N.m. The torque curves during the test of four types of installing-holes distances are displayed in Figure 5(e). Because most of the breakaway torque values can reach the threshold value, it is considered that the performance of the bonded joint is stable.

4 CONCLUSIONS

In this paper, a series of experiments were carried out to study the installing-holes distances parameter optimization for metallic inserts bonded into CFRP laminates. By performing the push-out and breakaway torque test for the four types of installing-holes distances samples, the installing-holes distances parameter optimization experiments are studied and the following conclusions are drawn:

1. For different types of installing-holes distances, the push-out bearing capacity of metallic inserts bonded into CFRP laminates is not decreased obviously. The performance of the bonded joint is stable.
2. For different types of installing-holes distances, the breakaway torque bearing capacity of metallic inserts bonded into CFRP laminates is not decreased significantly. The performance of the bonded joint is stable.
3. The joint strength of metallic inserts bonded into the CFRP laminates is affected largely by the adhesive property than that of the installing-holes distances, and so the adhesive parameters and installation process should be strictly controlled.

REFERENCES

Camanho, P. P., Fink, A., Obst, A., & Pimenta, S. (2009). Hybrid titanium–cfrp laminates for high-performance bolted joints. *Composites Part A Applied Science & Manufacturing, 40*(12), 1826–1837.

Camanho, P. P., & Matthews, F. L. (2000). Bonded metallic inserts for bolted joints in composite laminates. *Proceedings of the Institution of Mechanical Engineers Part L Journal of Materials Design & Applications, 214*(1), 33–40.

Camanho, P. P., Tavares, C. M. L., Oliveira, R. D., Marques, A. T., & Ferreira, A. J. M. (2005). Increasing the efficiency of composite single-shear lap joints using bonded inserts. *Composites Part B Engineering, 36*(5), 372–383.

Hopkins, M., Dolvin, D., Paul, D., Anselmo, E., & Zweber, J. (2006). Structures technology for future aerospace systems. *Aiaa/asme/asce/ahs/asc Structures, Structural Dynamics, and Materials Conference and Exhibit* (pp.76).

Mccarthy, C. T., & Mccarthy, M. A. (2005). Three-dimensional finite element analysis of single-bolt, single-lap composite bolted joints: part ii—effects of bolt-hole clearance. *Composite Structures, 71*(2), 159–175.

Civil, Architecture and Environmental Engineering – Kao & Sung (Eds)
© 2017 Taylor & Francis Group, ISBN 978-1-138-02985-9

Formation and test methods of the thermo-residual stresses for thermoplastic polymer matrix composites

Changchun Wang, Guangquan Yue, Jiazhen Zhang & Jianguang Liu
Beijing Aeronautical Science and Technology Research Institute, Commercial Aircraft Corporation of China Ltd. (COMAC), Beijing, China

Jin Li
Key Laboratory of Ningxia for Photovoltaic Materials, Ningxia University, Yinchuan, China

ABSTRACT: Fiber reinforced thermoplastic composites have been widely applied in the civil aircraft design for their excellent advantages, which were produced under the conditions of high temperature and cooling rate. The thermo-residual stresses in the composites will be formed due to the mismatch in thermal expansion coefficients of the fiber and matrix, anisotropy of the adjacent fabric layers, and the temperature gradient distribution. The thermo-residual stresses have an important role to play in the mechanical properties of the thermoplastic parts and need to be considered in the design. It will be beneficial to study the mechanism and test methods of thermo-residual stresses to control the residual stress level of the thermoplastic composites, which are the main contents of this study and the formation reasons of the residual stresses were discussed from three levels. The test methods of the thermo-residual stresses were also studied by using the non-destructive and destructive methods, and their advantages and disadvantages are also discussed.

1 INTRODUCTION

Fiber reinforced thermoplastic composites have been widely used in aerospace for the high toughness and short molding cycle (J. Bernhardson, 2000 & G. Schinner, 1996). Thermo-residual stresses will be created when the parts are cooled to service temperature for the mismatch of the thermal expansion coefficient of the fiber and thermoplastic matrix (P.P. Parlevliet, 2006 & 2007). The factors such as temperature gradient, crystalline shrinkage, and interaction of the tool and composites will cause the thermo-residual stresses and affect the mechanical properties of the composites, which need to be considered to obtain qualified parts.

The formation of thermo-residual stresses is estimated from three levels in the present study to control and decrease the stress levels in the thermoplastic composites. In addition, test methods of the stresses were discussed based on the non-destructive and destructive methods as well.

2 FORMATION AND EFFECTS ON THERMO-RESIDUAL STRESSES

2.1 *Formation mechanism*

Nairn et al. tested the thermo-residual stresses of the carbon fiber reinforced polyether sulfone by using the photo-elastic method (J.A. Nairn, 1985). The effect of the temperature on thermo-residual stresses of fiber composites was also studied by Favre (J.A. Barnes, 1994). The thermo-residual stresses can be discussed from three levels, i.e. constituent materials, laminate plates, and structure parts. The interaction between the fiber and matrix is shown in Figure 1 and the classification of thermo-residual stresses is presented in Table 1.

2.2 *Constituents level thermo-residual stresses*

2.2.1 *Effect of the fiber on thermo-residual stresses*

The carbon fiber has been widely used in the aerospace industry and the thermal expansion

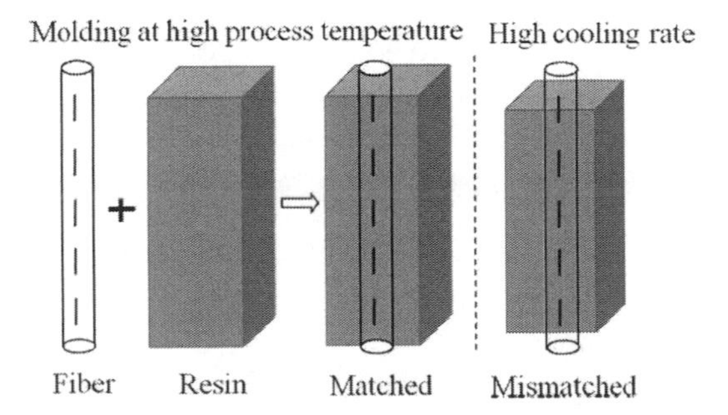

Figure 1. Diagram of the interaction of the fiber and matrix for thermoplastic composites.

Table 1. Classification of the thermo-residual stresses.

Levels	Mechanism analysis
Constitute materials	Deformation inconsistency of the fiber and matrix in the cooling process for the difference of the thermal expansion coefficient
Laminate plates	Mismatch of the thermal expansion coefficient of the fabric layer along the radial and transverse direction of the composites, including that of the transverse crystalline layer
Structure parts	Temperature gradient in the cooling process for the thermoplastic composites having a certain thickness

interaction along the axial and vertical directions of the fiber causes the thermo-residual stresses in the cooling process and leads to fiber deformation; however, it is smaller than that of the polymer matrix in the whole process of thermoplastic composite materials. The effect of the thermo-residual stresses caused by the fiber can be neglected based on previous studies.

2.2.2 *Effect of the matrix on thermo-residual stresses*

The thermoplastic matrix exhibits visco-elastic properties when the process temperature is cooled to service temperature, in which the thermo-residual stresses will be released when to the molecular thermal motion has sufficient energy. The matrix has the thermal elasticity properties when the temperature decreases, and the thermo-residual stresses at this stage will exist in the final composites. The cooling rate plays an important role on the thermo-residual stresses, which degrades the thermo-residual stress relaxation at a higher cooling rate (K.S. Kim, 1989). These two mechanisms coexist and compete with each other. The thermo-residual stresses were also affected by the elastic modulus of the matrix and and a decreasing trend with a decrease in the elasticity modulus.

2.3 *Plate level thermo-residual stresses*

All the influence factors of the constituent materials level will also act at this level; but the effect of the layup pattern on the thermo-residual stresses was mainly concerned, including the angle-ply information and fiber structure. The thermo-residual stresses increase with an increase in the difference of the stress-free temperature and the service temperature, which can be determined by using the heating method based on the thermoplastic composites until the deformation disappears (S.L. Gao, 2000). During the crystallization process, the thermo-residual stresses of the obtained composites caused by the temperature gradient of the laminates will be accumulated.

2.4 *Structural level thermo-residual stresses*

The thermo-residual stresses will be formed when the composites attain a certain thickness, and it is affected by the cooling process. The thermo-residual stresses in the thermoplastic composites different from each other due to the different cooling processes, though with the same crystallinity degree. The annealing method can be used to release the thermo-residual stresses and decrease the residual stresses gradient. It improves the crystallinity level of the thermoplastic composites; but it will increase the process time and the energy consumption.

Mould materials used for molding the composites also have an important effect on the thermo-residual stresses, which play a role on the thermal and mechanical properties (E.M. Asloun, 1989). The thermal interaction will obviously affect the formation of the thermo-residual stresses, which is determined by the mould materials' properties and the cooling rate of the parts' surfaces. The mechanical interaction is caused due to the mismatch of the thermal expansion coefficient of the tool and composites. It will degrade the compressive residual stresses on the surface plies of the composites near the mould and has a transverse stress distribution in the laminates.

3 TEST METHODS OF THE THERMO-RESIDUAL STRESSES

The thermo-residual stresses of the thermoplastic composites formed during the molding process have the important effect on its mechanical properties, and it will be beneficial to obtain the residual stresses level in the designing process of the laminates. The commonly used test methods of non-destructive and destructive methods for obtaining the thermo-residual stresses are presented in Table 2.

Table 2. Test methods of the thermo-residual stresses.

Test patterns	Detailed methods		
Non-destructive testing methods	Constituent material	Embedding sensors	Warpage
Destructive testing methods	Removal layer	Blind-hole	Grooving

3.1 *Methods of the constituent material*

The properties of the constituent material will be changed as the composites exhibit change in their thermo-residual stresses, which can be measured by using the photo-elastic, Raman spectrum and conductivity methods. Photo-elasticity studies are conducted by using the static stress analysis, in which a transparent matrix is required to obtain the stress field (A. Pawlak, 2001). The molecular orientation distribution alters when there are thermo-residual stresses that can be calculated with the light intensity and phase difference. This method can be used to determine the thermo-residual stress distribution; however, it can be only used in the laminates of the low fiber volume fraction.

The Raman spectroscopy technique is widely used to obtain the thermo-residual stress of the composites by monitoring the change of the Raman spectrum peak position, which is highly dependent on the internal stress level (A.S. Nielsen, 2002). In the test, the plots of the spectrum peak and the thermo-residual stresses of the fiber in air and the matrix should be firstly constructed and studied to determine its intrinsic features. The glass fiber shows a tiny Raman response and the aramid fibers can be added into the thermoplastic matrix as the sensors due to the strong Raman responses. The results of this method exhibit high accuracy; however, it needs to take the matrix peaks' contribution into account to avoid errors in the test process.

The electrical conductivity of the composites will be affected by strain, damage, and temperature, which can be used as indicators without an additional sensor that will change the original thermo-residual stresses in the thermoplastic composites. Its resistivity will also be affected by the pressure in the molding (S.K. Wang, 2001). Moreover, poor adhesion of the adjacent layers will increase the electrical resistance as well, which means more efforts are needed to obtain the thermo-residual stresses in the test.

3.2 *Methods of embedding sensors*

The sensors embedded into the thermoplastic composites can be used to monitor the stress with the results of the sensors, which commonly include the resistance strain gauge, fiber optic sensors, and metallic particles with X-rays. The strain gauges have been used to obtain the residual stress change of the carbon fiber/PEEK laminates by placing into the surface layer in the heating and cooling processes (R.H. Fenn, 1993).

The fiber optical sensors have been widely used as the strain gauges to obtain the thermo-residual stresses of the composites for their high accuracy, which includes the fiber Bragg grating and the Fabry–Perot interferometric sensors (B.P. Arjyal, 1998). However, the stresses concentration will be created when the fiber optical sensors are perpendicular to the fiber direction.

The X-ray diffraction method is achieved by placing the metallic particles into the matrix, by monitoring the deflection of the peak angle and crystal lattice caused by thermo-residual stresses. The particles can be silver, copper, and aluminum materials. It can be used to measure the intra-laminar and inter-laminar thermo-residual stresses; but it can only give the surface properties of the specimens with the thickness lying between 0.3 and 0.5 mm.

3.3 *Methods of the deformation measurement*

The light wave's interference principle can be used to obtain the thermo-residual stresses of the composites by measuring its deformations. It was caused by the Moiré effect as the light through the two gratings grids at a small angle. The residual stresses can be obtained with the in-plane and out-of-plane deformations. However, it can only show the macroscopic residual stress information.

The thermo-residual stresses can also be studied with the warpage of the unsymmetrical lay-up laminates. Nairn et al. studied the thermo-residual stresses in the composites by using the warpage method. The residual stresses increase with the positive proportional increase of the laminate warpage of a certain thickness. However, the results' accuracy may be degraded due to the errors of the test results.

Residual stress of the curved member plate can be obtained with the deflection chord length and curvature of the specimen. The deflection is obtained with the microscope and the chord size is obtained with the ruler. This method is easy to manipulate; but it cannot provide the residual stresses along the thickness direction. It also cannot be used to study the residual stress of the symmetric laminates.

3.4 *Methods of the destructive technology*

The thermo-residual stresses of the thermoplastic composites can also be determined with the destructive techniques, which are achieved by removing the part of the laminate to make the residual stresses release. These stresses can be obtained by using the methods of first-layer removal and blind-hole by testing the caused deformation.

The first-layer removal method is widely used in testing the thermo-residual stresses by removing the layer of the thermoplastic laminate and testing the released strain. It has a poor precision, and the thermal and micro-cracks in the removal process will affect the residual stress (M. Eijpe, 1997). The process simulation method was used to reduce the effect of the removal process by placing a thin film into the adjacent layers. It separates the adjacent layers and reduces the interaction, the validity of which has been proved with the test results.

The blind-hole method is the stresses relaxation technique in drilling holes in the composites and monitored the strain changes to determine the thermo-residual stresses. However, the obtained strain might be released in the drilling-hole process (A. Makino, 1994). In addition, the crack buckling method and the notch technique are also used in testing of the thermo-residual stresses according to the previous studies, which can be used to obtain the strain, and then the residual stress is obtained with the change of the crack length before and after the damage process.

4 CONCLUSIONS

The thermo-residual stresses may be caused by the mismatch of the thermal expansion coefficient of the fiber and matrix and the interaction of the adjacent layers for the anisotropy properties, as well as the temperature gradient of the laminates. The formation of the thermo-residual stresses of the thermoplastic composites were discussed from three different levels in the present study. Moreover, the destructive and non-destructive methods of measuring the thermo-residual stresses were also studied. The advantages and disadvantages of each method were evaluated, which can be used to produce the thermoplastic composites with a suitable residual stress level.

ACKNOWLEDGMENTS

The authors would like to acknowledge the financial support of the Beijing Municipal Science and Technology Project for this work, Grant No. Z151100002815021.

REFERENCES

Arjyal, B.P., C. Galiotis, S.L. Ogin and R.D. Whattingham, Residual strain and Young's modulus determination in cross-ply composites using an embedded aramid fiber strain sensor, Composites Part A, 29 (1998) 1363–1369.

Asloun, E.M., M. Nardin, J. Schultz, Stress transfer in single-fiber composites: effect of adhesion, elastic modulus of fiber and matrix, and polymer chain mobility, J. Mater. Sci. 24 (1989) 1835–1844.

Barnes, J.A., G.E. Byerly The formation of residual-stress in laminated thermoplastic composites, Compos. Sci. Technol. 51 (1994) 479–494.

Bernhardson, J., R. Shishoo, Effect of parameters on consolidation quality of GF/PP commingled yarn based composites, J. Thermoplast Compos. 13 (2000) 292–313.

Eijpe, M.P., C. Powell, Residual stress evaluation in composites using a modified layer removal method, Compos. Struct. 37 (1997) 335–342.

Fenn, R.H., A.M. Jones, G.M. Wells, X-ray diffraction investigation of triaxial residual stresses in composite-materials, J. Compos. Mater. 27 (1993) 1338–1351.

Gao, S.L., J.K. Kim, Cooling rate influences in carbon fiber/PEEK composites. Part 1. cystallinity and interface adhesion, Composites Part A, 31 (2000) 517–530.

Kim, K.S., H.T. Hahn, R.B. Croman, The effect of cooling rate on residual stresses in a thermoplastic composite, J. Compos. Tech. Res. 11 (1989) 47–52.

Makino, A., D. Nelson, Residual stress determination by single-axis holographic interferometry and hole drilling-Part I: Theory, Exp. Mech. 34 (1994) 66–78.

Nairn, J.A., P. Zoller, Matrix solidification and the resulting residual thermal stresses in composites, J. Mater. Sci. 20 (1985) 355–367.

Nielsen, A.S., R. Pyrz, A novel approach to measure local strains in polymer matrix systems using polarized Raman microscopy, Compos. Sci. Technol. 62 (2002) 2219–2227.

Parlevliet, P.P., H.E.N. Bersee, A. Beukers, Residual stresses in thermoplastic composites-A study of the literature-Part I: Formation of residual stresses, Composites Part A, 37 (2006) 1847–1857.

Parlevliet, P.P., H.E.N. Bersee, A. Beukers, Residual stresses in thermoplastic composites-A study of the literature-Part II: Experimental techniques, Composites Part A, 38 (2007) 651–665.

Parlevliet, P.P., H.E.N. Bersee, A. Beukers, Residual stresses in thermoplastic composites-A study of the literature-Part III: Effects of thermal residual stresses, Composites Part A, 38 (2007) 1581–1596.

Pawlak, A.P., Zinck, A. Galeski and J.F. Gerard, Photo-elastic studies of residual stresses around fillers embedded in an epoxy matrix, Macromol. Symp. 169 (2001) 197–210.

Schinner, G., J. Brandt, H. Richer, Recycling carbon-fiber reinforced thermoplastic composites, J. Thermoplast Compos. 9 (1996) 237–245.

Wang, S.K., Z. Mei, D.D.L. Chung, Interlaminar damage in carbon fiber polymer-matrix composites studied by electrical resistance measurement, Int. J. Adhes. Adhes. 21 (2001) 465–471.

Civil, Architecture and Environmental Engineering – Kao & Sung (Eds)
© 2017 Taylor & Francis Group, ISBN 978-1-138-02985-9

Effects of lithium slag from lepidolite on Portland cement concrete

Qi Luo
School of Materials Science and Engineering, Tongji University, Shanghai, China
School of Materials Science and Engineering, Nanchang University, Jiangxi, China

Yufeng Wen, Shaowen Huang, Weiliang Peng, Jinyang Li & Yuxuan Zhou
School of Materials Science and Engineering, Nanchang University, Jiangxi, China

ABSTRACT: The effects of lithium slag from lepidolite on the workability, mechanical properties, and parts of the durability of concrete were investigated. SEM was used to observe and analyze the interface structure of lithium slag concrete. The results show that adding a moderate amount of lithium slag can improve the workability and mechanical properties of concrete; and its anti-impact property will be enhanced significantly. Adding 20% lithium slag can improve the impact resistance of concrete by 45%. And the study of the interface structure shows that lithium slag can consume calcium hydroxide effectively, bring more amount of gels, inhibit the formation of ettringite, and enhance the density of concrete.

1 INTRODUCTION

The associated lepidolite content in the Tantalum–Niobium Ore resource in Jiangxi province is extremely abundant (Zeng Chuanlin 2012, Zeng Qingling 2014). However, extracting lithium from lepidolite will produce massive solid waste slag (hereinafter referred to as lithium slag) in industries. The main treatment methods of lithium slag are landfills, dam construction, and piling at present, which will cause serious pollution and land wastage. Strengthening the study of comprehensive utilization of extracting lithium from lepidolite has a great significance for economizing the resource and protecting the environment.

Using lithium slag as the cement concrete admixture is one of the important ways of using lithium slag. At present, there are many studies about using lithium slag from spodumene as a cement concrete admixture in China (Yang Hengyang 2012, Zhang Lanfang 2008, Zhang Lanfang 2012, Hu Zhiyuan 2008). However, very few studies concentrate on using lithium slag from lepidolite as a cement concrete admixture. By studying the effect of lithium slag from lepidolite on the workability, mechanical properties, anti-impact property, anti-chloride ion permeability, and interface meso-structure of concrete, we can provide theoritical reference for the application of lithium slag from Yichun lepidolite in the concrete field.

2 EXPERIMENTAL SECTION

2.1 *Raw materials*

Cement: P.O42.5R cement was produced by East Asia JiangXi Cement Co. Ltd. The apparent density of the cement is 3.07 g/cm^3. Lithium slag: lithium slag from lepidolite was produced by Jiangxi Ganfeng Lithium Co. Ltd. The specific surface area of lithium slag is equal to 550 m^2/kg and its middle diameter is equal to 12 μm. The chemical constituents of lithium slag are given in Table 1. Fine aggregates: sand was obtained from Gan River with a fineness modulus of 2.5 and apparent density of 2.7 g/cm^3. Coarse aggregate: continuous grading limestone artificial gravels were obtained with the maximum grain size of 20 mm. Water reducing agent: the polycarboxylate superplasticizer (powder) was obtained with a water reducing ratio of 26%.

2.2 *Experimental method*

The mixture proportion of lithium slag and cement concrete is given in Table 2; the water–binder ratio is 0.47 and the mixing value of the water reducer is 0.4% invariably. Workability of cement concrete: according to GB/T 50080-2002 Standards for test method of performance on ordinary fresh concrete. Mechanical properties of cement concrete: according to GB/T 50081-2002 Standards for test method of mechanical properties on ordinary concrete to test flexural and compressive strength of cement concrete. Anti-impact property of cement concrete: according to the falling weight test results. The impact hammer is a hemispherical cylinder (height is 150 mm and diameter is 150 mm). The weight of the impact hammer is 4.5 kg. The height of the impact hammer (the distance between the bottom of the impact hammer and specimen surface) is 1 m. When the first crack appears in the specimen, the

Table 1. The chemical constituents of lithium slag.

Materials types	SiO_2	Al_2O_3	CaO	Fe_2O_3	MgO	SO_3	Na_2O	K_2O	TiO_2	Loss	Sum
Lithium slag	47.62	21.56	2.02	0.48	0.12	0.03	10.68	3.05	3.46	0.14	90.63

Table 2. The mixture proportion of lithium slag cement concrete (kg/m^3).

Sample	Water	Cement	Lithium slag	Fine aggregate	Coarse aggregate
L0	234	500	0	640	1139
L1	234	450	50	640	1139
L2	234	400	100	640	1139
L3	234	350	150	640	1139

impact times (denoted by n) can be recorded as the maximum times of impact resistance. With an increase in n, the impact resistance of cement concrete becomes stronger. Interface structure: the interface structure of cement concrete was studied by using a QUANTA-200FEG Field Emission Environment Scanning Electron Microscope (FEI, America).

3 RESULTS AND DISCUSSION

3.1 *Effect of lithium slag on the work performance of cement concrete*

The test results of the slump of lithium slag concrete are shown in Fig. 1.

As test results are shown in Fig. 1, while mixing an increased quantity of lithium slag, the slump of ordinary concrete showed a decreasing trend. The experimental results also indicated that the normal concrete without lithium slag has poor cohesiveness and water retention, and there is bleeding at the bottom while testing the slump. After adding 10% lithium slag, the slump of cement concrete is slightly improved. Meanwhile, the cohesiveness and water retention of cement concrete are improved obviously. The aggregate is firmly wrapped by cement paste and no apparent water bled from cement concrete. When the mixing amount of lithium slag is more than 10%, the slump of cement concrete began to decrease. When it is increased to 30%, the slump of cement concrete collapsed to 132 mm. It indicates that mixing a little lithium slag can improve the workability of cement concrete to a certain amount. However, when adding more than 10% lithium slag, with an increment in the mixing amount, the water demand of cement concrete increased, and the flowability decreased. Therefore, the mixing amount of lithium slag should be controlled to 13% in practical applications to ensure that the slump of cement concrete is in accordance with the design requirements.

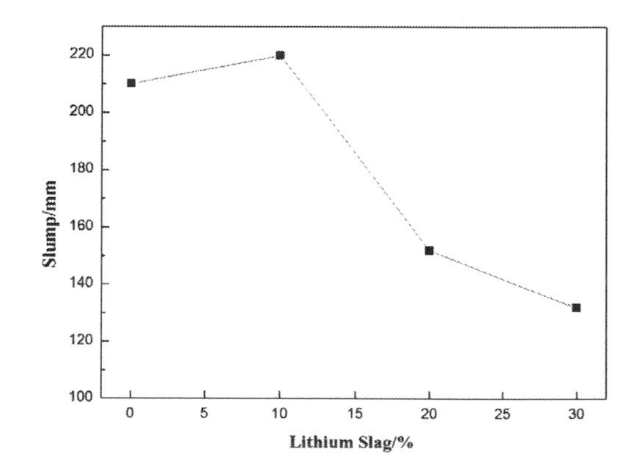

Figure 1. Effects of lithium slag on the work performance of cement concrete.

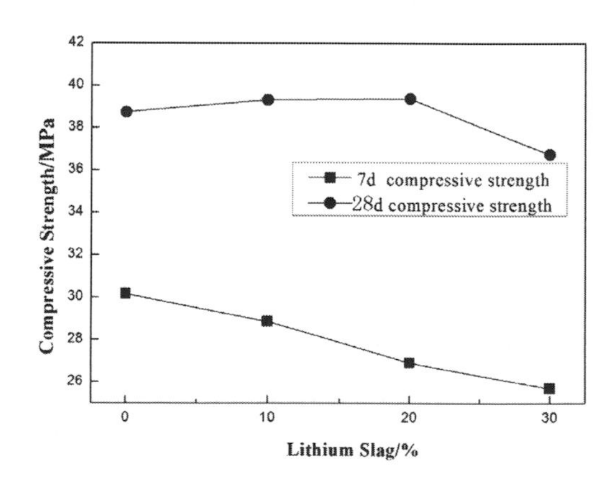

Figure 2. Graph showing the effect of lithium slag on the concrete compressive strength.

3.2 *Effect of lithium slag on concrete compressive strength*

Fig. 2 shows the relationship between compressive strength and the amount of lithium slag added. Adding within 20% lithium slag can increase the strength in 28 days dramatically. The compressive strength ratio can reach 95% in 28 days while adding 30% lithium slag. But adding lithium slag has a negative effect on the strength in 7 days. With addition of lithium slag after the mixing amount reaches 10%, the compressive strength on 7 days decreases slowly. While adding 30% lithium slag, the strength ratio reaches 85%. The result shows that adding lithium slag can increase the compressive strength in the middle-later period and decrease the prophase strength in low amplitude.

3.3 *Effect of lithium slag on concrete flexural strength*

The test results of the flexural strength of lithium slag concrete are shown in Fig. 3. As depicted in Fig. 3, the effect of lithium slag on concrete's flexural strength and the effect on the compressive strength are approximately same. But the negative effect of the flexural strength reduced significantly in 7 days. When the mixing amount of lithium slag is less than 30%, the flexural strength is about 5.9 MPa in 7 days, and ranged from 5.7 MPa to 6.1 MPa. However, in this amount, the incorporation of lithium slag played a positive role in the flexural strength. Even if the mixing amount reached 30%, the flexural strength still exceeds that of the sample without lithium slag in 28 days. These may be due to the fact that after the cement hydration reached a certain extent, lithium slag can undergo a secondary hydration reaction with products from cement hydration, which will make the concrete more compact and improve the interface structure of concrete, thereby increasing the flexural strength of concrete. However, when the mixing amount of lithium slag is in excess, the hydration products will be reduced. As lithium slag cannot hydrate by itself, residual lithium slag can only fill in concrete, which decreases the strength of concrete gradually.

3.4 *Effect of lithium slag on anti-impact property of concrete*

Depending on the test results of impact resistance of lithium slag concrete (see Fig. 4), we can see that with the incorporation of lithium slag, the impact resistance of concrete can be improved significantly. When compared to the specimen without lithium slag (L0), the impact resistance of specimens with 10%, 20%, and 30% lithium slag improved at different degrees. Concrete with 20% lithium slag (L2) has the strongest impact ability that is 45%, which is higher than L0. This result has a strong relativity with the compressive strength and flexural strength of lithium slag concrete. It shows that the incorporation of lithium slag can improve the mechanical properties of concrete.

3.5 *Effect of lithium slag on the interfacial micro-structure of cement concrete*

Figs. 5 and 6 are SEM images of interfaces of cement concrete without lithium slag and with 10% lithium slag, respectively after 28 days. As shown in Fig. 5, without lithium slag, a lot flocculent C-S-H gels accompanied with many acicular ettringites appear in the interfaces of cement concrete. As depicted in Fig. 6, the interface structure of cement concrete with 10% lithium slag is more tight and the hydration product mainly is flat C-S-H gels. The amount of ettringite used was significantly increased when compared with the cement concrete without lithium slag. There were few ettringites in the sample in Fig. 6. It shows that, in the doped cement concrete of lithium slag, more $Ca(OH)_2$ is consumed and the production of ettringite is reduced due to its pozzolanic effect (Wang Ling 1998). Meanwhile, as a result of adding lithium slag, the amount of C-S-H gels is increased and the density of cement concrete is improved, which is an important reason for strengthening of cement concrete.

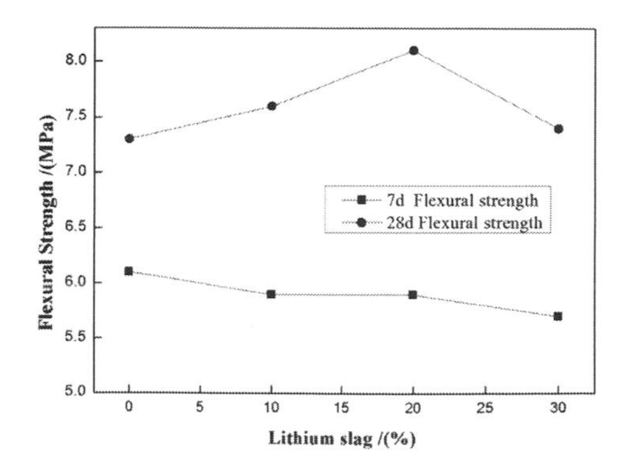

Figure 3. Graph showing the effect of lithium slag on concrete's flexural strength.

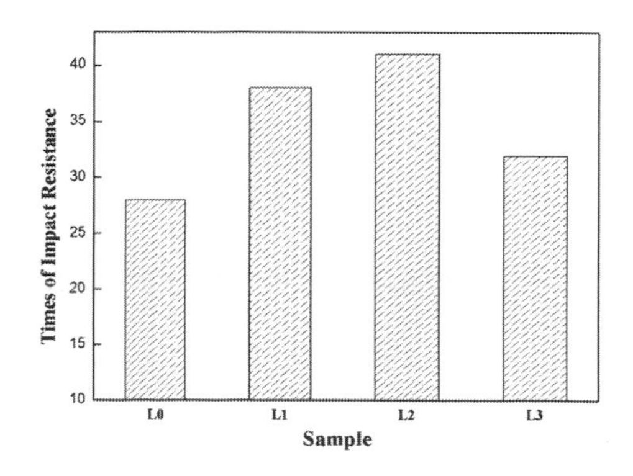

Figure 4. Graph showing the effect of lithium slag on the anti-impact property of concrete.

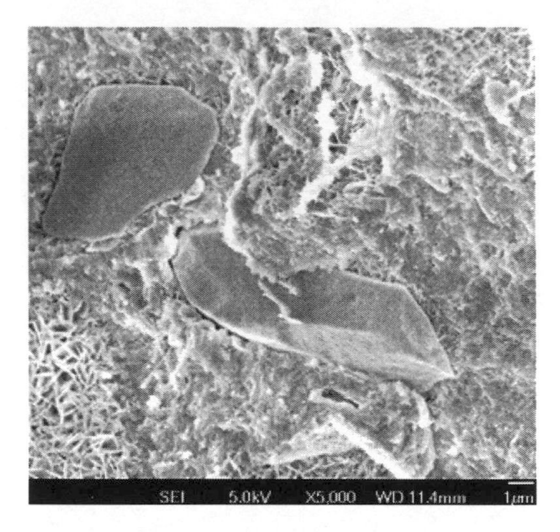

Figure 5. SEM image of interfaces of L0 after 28 days.

Figure 6. SEM image of interfaces of L1 after 28 days.

4 CONCLUSIONS

As a concrete admixture, lithium slag from lepidolite has a higher activity index than traditional volcanic materials. While mixing 30% lithium slag, the compressive strength ratio after 7 days and 28 days reach 85% and 95%, respectively, which shows that lithium slag is a high quality material as a concrete admixture.

Impact resistance of concrete can be improved significantly when mixing lithium slag. Adding 20% lithium slag can improve the impact resistance of the concrete by 45%. And adding little lithium slag can also improve the working performance of concrete.

SEM observation shows that adding lithium slag can improve the interface structure, inhibit the formation of ettringites, and enhance the density of the concrete.

REFERENCES

Hu Zhiyuan (2008). *Studyon concrete mixing lithium Slag with other mineral admixtures*. Chongqing, China: Chongqing University.

Wang Ling (1998). Ralation between the C_3A content and the sulphate resistance property of concrete. *Cenment Technology, 4,* 28.

Yang Hengyang, Zhou Hailei, Shi Kebin, et al (2012). Experimental study on the cracking resistance at early ages of high performance concrete added with lithium slag and fly ash. *Concrete, 1,* 65–67.

Zeng Chuanlin (2012). *Study on Preparing Lightweight Ceramsite Using Leaching Residual Slag of Lepidolite ore*. Nanchang, China: Nanchang University.

Zeng Qingling (2014). *Research on White Composite Portland Cement Using Lithium Slag From Lithium Mica By Chlorine Salt Autoclaving Technology*. Nanchang, China: Nanchang University.

Zhang Lanfang (2008). Study on performance of concrete mixed with lithium slag. *Concrete, 4,* 44–46.

Zhang Lanfang, Chen Jianxiong, Yue yu, et al (2012). Experimental study on the high strength concrete added with lithium slag. *New Building Materials, 3,* 29–31.

Civil, Architecture and Environmental Engineering – Kao & Sung (Eds)
© 2017 Taylor & Francis Group, ISBN 978-1-138-02985-9

Sorption properties of TEMPO-oxidized natural cellulose to iron ions

Cunzhen Geng, Yinghua Lou & Dongyun Yan
College of Environmental Science and Engineering, Qingdao University, Qingdao, P.R. China

Zhihui Zhao, Zhixin Xue & Yanzhi Xia
Co-Innovation Center for Marine Biomass Fibers, Materials and Textiles of Shandong Province, Qingdao, P.R. China
Research Institute of Marine Fiber Novel Materials, Qingdao University, Qingdao, P.R. China

ABSTRACT: The aim of this paper was to investigate the adsorption behavior of iron ions in wastewater with a novel type of adsorbent, 2,2,6,6-tetramethylpiperidine-1-oxyl radical (TEMPO)-oxidized natural cellulose, which is synthesized in a special oxidation system with raw natural cellulose. The effects of adsorbent dosage, pH, temperature, initial concentration, and adsorption time on the adsorption efficiency were explored. The results demonstrated that the removal efficiency of iron ions with TEMPO-oxidized cellulose could be achieved as high as 99% under certain conditions. Adsorption isothermal studies also revealed that the adsorption process of iron ions was well-fitted with the Freundlich and Langmuir-type adsorption isotherms. Our results indicated that TEMPO-oxidized cellulose may have wide-ranging applications in the sequestration or removal of metal ions and many other wastewater treatment industries as a new adsorption material in future.

1 INTRODUCTION

Modern industries discharge a certain quantity of wastewater containing large amounts of heavy metals every day. Because of their toxicity, bioaccumulation, and their various forms, these metals do great harm to the ecological environment and survival of human beings after being discharged into the environment. A number of methods are available to remove heavy metals from water or wastewater, including chemical precipitation, ion exchange, adsorption, membrane process, reverse osmosis, and so on (H. Xu & Y. Liu. (2008), T. A. Kravchenko et al. (2009), Y. Y. Xiong et al.(2014), G. Ferraiolo et al.(1990), V.K. Gupta et al. (2004), V. K. Gupta et al.(2005)). Among these technologies, adsorption is considered to be the most effective and efficient approach for dealing with large volumes of wastewater in recent years (B. Volesky(2001), A. T. Paulino et al. (2008)). And the adsorption process has been applied as one of the most popular treatment processes due to various advantages, such as high efficiency, low cost, wide applicable scope, *etc*. Cellulose, which is a kind of inexpensive material and the main constituent of plants, is the most abundant biodegradable natural polymer biomass on the earth and has been utilized in many applications such as pulp, paper, filters, film, and the textile industries (D. Klemm et al.(2005)). To use cellulose as a heavy metal ion adsorbent, however, chemical modifications such as esterification, etherification, and oxidation targeting the hydroxyl group present in cellulose are essential (V. A. G. Leandro et al. (2008), W. O. David et al. (2008), L. V. A. Gurgel et al. (2008)). Furthermore, up to now, few reports can be found on the adsorption and reduction mechanisms of iron ions onto TEMPO-oxidized natural cellulose except literature about sorption TEMPO-oxidized cellulose hydrogel to some heavy metal ions (N. Isobe et al.(2013)).

In this paper, a new adsorption material was prepared by using the chemical-modified natural cellulose, namely TEMPO-oxidized natural cellulose. And ferric ion-sorption experiments were also carried out in order to study the adsorption properties of the new materials. Natural cellulose has great superiority as the new experimental material, since cellulose is the most abundant and renewable biopolymer and is one of the promising raw materials available in terms of cost for preparing various functional materials. And so, it will provide strong support for the application of this new type of adsorbent. With the single variable controlling method, all kinds of factors affecting the adsorption efficiency were investigated, and the optimum conditions for the new removal of iron ions with TEMPO-oxidized natural cellulose were determined.

2 MATERIALS AND METHODOLOGY

2.1 *Materials*

Iron ions solution was prepared by using iron sulfate. TEMPO-oxidized natural cellulose was synthesized in our laboratory. Deionized water was used for all experiments. Other reagents used were of analytical grade in all experiments.

2.2 *Synthesis of the new adsorbent*

Natural cellulose was oxidized in an oxidation system constituted by 2,2,6,6-tetramethylpiperidine-1-oxyl radical (TEMPO)/NaBr/NaClO under certain conditions (A. Isogai & Y. Kato (1998), T. Saito et al. (2005)). The new adsorbent prepared is a long chain molecule with lots of hydroxyl groups and some carboxyl groups, and its chemical structure is illustrated in Fig. 1. The details of the synthesis of TEMPO-oxidized natural cellulose are as follows: designed weight Natural Cellulose (NC) was suspended in purified water which contained TEMPO (0.1 mmol/g NC) and sodium bromide (1.0 mmol/g NC). And then, a designed amount of the 10% NaClO solution (5.0~10.0 mmol/g NC) was added to the slurry dropwise, and the mixture was stirred by mechanical means at room temperature. And meanwhile, the pH of the mixture system was maintained between 10.5 and 11.0 by using a pH stat until no 0.5 M NaOH consumption was observed in the reaction process. Oxidation was quenched by adding ethanol (ca. 10 mL). Oxidized cellulose was washed thoroughly with purified water and then ethanol by filtration. The wet product was dried by lyophilization and then followed by vacuum-drying at 50°C for about 48 h. And then, the synthesis compounds were milled into powder after drying for the following adsorption experiments.

2.3 *Adsorption experiments*

Adsorption was carried out according to the following steps: (1) 20 mL of different concentrations of wastewater was placed into small beakers; (2) a certain amount of adsorption materials was filled into the solution after its pH was adjusted with sodium hydroxide and hydrochloric acid; (3) all the suspended solutions were placed in a constant temperature oscillator for a designed duration of time; (4) the concentration of iron (III) ions was determined with adjacent dinitrogen spectrophotometry after stopping, oscillating, and standing for a period of time. Therefore, the optimum adsorption conditions were determined on the basis of setting the adsorption conditions by using the orthogonal principle.

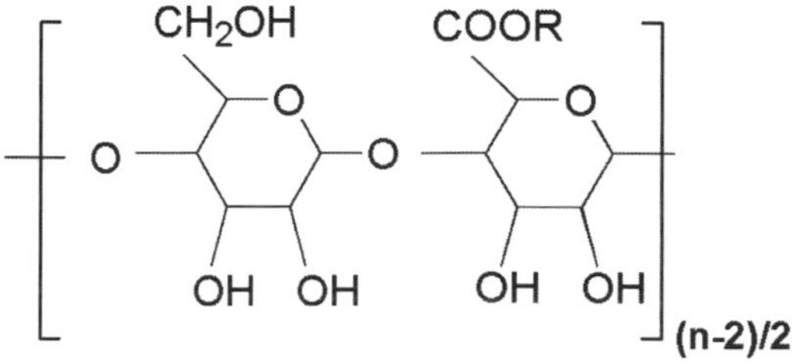

Figure 1. Structure of oxidized natural cellulose (R, H, or Na).

2.4 *Calculation method*

The effect of adsorbing iron ions can be formulated by adsorbing capacity (q_e) and adsorption efficiency (R%).

Adsorbing capacity (q_e) refers to the quality of Fe^{3+} ions adsorbed into the adsorbent (1 g mass) when the adsorption system achieves adsorption equilibrium. And it was calculated according to Equation (1).

$$q_e = \frac{(C_0 - Ce) \times V}{m} \tag{1}$$

where q_e ($mg\,g^{-1}$) is the adsorbing capacity, C_0 ($mg\,l^{-1}$) is the initial concentration of Fe^{3+}, C_e ($mg\,l^{-1}$) is the equilibrium concentration of Fe^{3+}, V (l) is the total volume of wastewater, and m (g) is the mass of adsorbent.

The removal efficiency (R%) refers to the specific value between the quality of Fe^{3+} ions adsorbed into the adsorbent and the initial quality of Fe^{3+} ions when the adsorption system achieves adsorption equilibrium. It is defined as Equation (2).

$$R\% = \frac{(C_0 - C_e)}{C_0} \times 100\% \tag{2}$$

3 RESULTS AND DISCUSSION

3.1 *SEM results*

The morphologies of natural cellulose and NOCs were studied by SEM analysis in order to make a comparison more directly before and after synthesis of NC. Two typical images at two magnifications of NC and TEMPO-oxidized natural cellulose are shown in Fig. 2a and b. Evidently, both of them are characterized by a level of inhomogeneity owing to natural growth. And the dimension of TEMPO-oxidized natural cellulose is 4–8 nm in width. But there many mangy sags and crests on the surface of TEMPO-oxidized natural cellulose, which increases undoubtedly the surface area of the adsorbents.

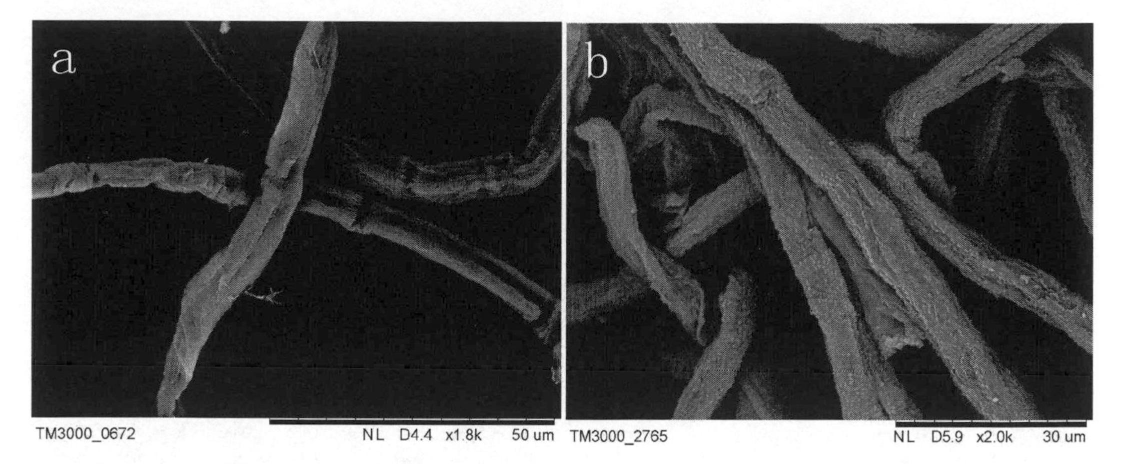

Figure 2. SEM micrographs of natural cellulose (a) and TEMPO-oxidized natural cellulose (b).

3.2 *Adsorption experiments results*

3.2.1 *The influence of adsorbent dosage on the removal efficiency*

The influence of adsorbent dosage on the removal efficiency of iron ions is shown in Fig. 3. The removal efficiency of iron ions increased significantly with an increasing amount of adsorbent in the range of 0.1–0.15 g, and then it began to fall down slightly when the adsorbent dosage was larger than 0.15 g. It could be inferred that the removal efficiency of iron ions was the optimum and adsorption process was saturated when the adsorbent dosage was about 0.15 g.

The behavior of efficiency dependency on the absorbent dosage could be explicated such that the removal efficiency of the heavy metal increased along with an increase in absorbent quality, and that was to say with an increase in the number of active adsorption sites (Z. k. Luan & H. X. Tang (1993)). Moreover, due to an uneven distribution of oxidized cellulose surface positions, adsorption sites had strong and weak points naturally. The surface active points enabled competitive adsorption among the adsorption sites when the adsorbent dosage was in excess. In addition, owing to the hydrogen-bond interaction, cross-linking among the high molecules of the adsorbent would be formed in this case. And so, it would lead to the result that surface active points cannot be made full use of and were not conducive to the adsorption. Thereby, it would result in a decrease in the efficiency.

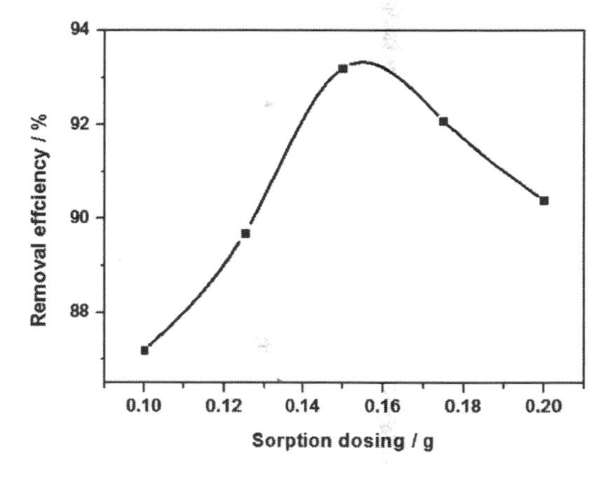

Figure 3. The influence of adsorbent dosage on the removal efficiency of Fe^{3+} ions.

3.2.2 *The influence of adsorption time on the removal efficiency*

In our research, we found that adsorption time was one of the important influence factors. From Fig. 4, it can be observed that, in the beginning, the concentration gradient between the solution and the surface of TEMPO-oxidized natural cellulose was very high and hence the driving force was strong too; therefore, the resulting adsorption speed was very high. The removal efficiency of iron ions had reached more than 90% at the time of 40 min, and then the removal efficiency still increased despite the decrease in the concentration gradient and the adsorbent tending to saturation gradually. It is not difficult to find that the adsorption efficiency was reaching the maximum and the adsorbent was reaching saturation at 60 min from Fig. 4. As time goes on, desorption began to appear and resulted in the declining of the removal efficiency.

3.2.3 *The influence of the initial concentration of metal ions on the removal efficiency*

The initial concentration of metal ions had an effect on the occurrence of ion exchange and

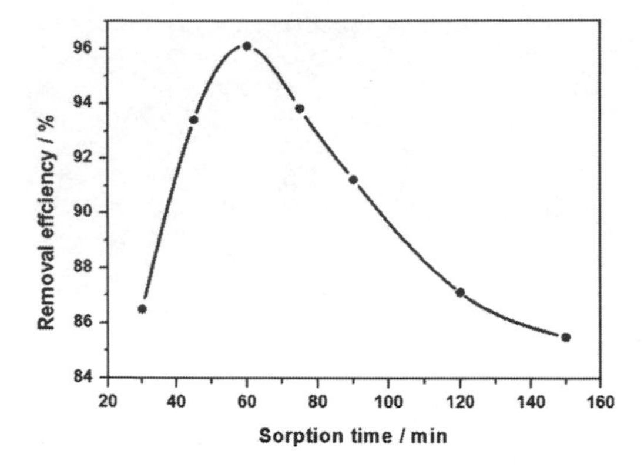

Figure 4. The influence of adsorption time on removal efficiency of Fe^{3+} ions.

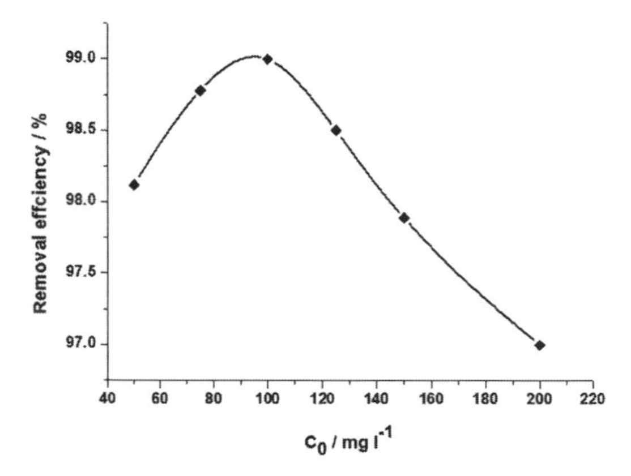

Figure 5. The influence of the initial concentration of metal ions on the removal efficiency of Fe^{3+} ions.

complexation in the suspended solution. The dependency of the removal efficiency of iron ions on the initial concentration of metal ions is shown in Fig. 5. Meanwhile, the graph presented a curve rising at first and declining in the studied range, while the best initial concentration of the oxidized cellulose adsorption of iron ions is 100 mg·L^{-1}. This may be due to the fact that, when the initial concentration increased in the adsorption process, the concentration gradient increased between the solution and the surface of TEMPO-oxidized natural cellulose, thereby leading to an increase in the driving force. The adsorption capacity of TEMPO-oxidized natural cellulose increased rapidly at first and then went back into equilibrium. When it surpassed the equilibrium level, the high concentration solution reduced the chance of functional atoms combining with iron ions, due to the fact that the amount of iron ions in the solution exceeded the maximum adsorbing capacity of the adsorbent at this time. Therefore, the adsorption quantity no longer increased. And so, the excessive increase of the initial concentration led to a reduction in the adsorption efficiency. It indicated that the initial concentration had a fitting range.

3.2.4 *The influence of temperature on the removal efficiency*

Temperature was one of the critical factors affecting adsorption. It was mainly due to the high sensitivity of the solubility of copper to temperature in a certain temperature range, which promoted adsorption. However when the temperature exceeded a certain limit, high temperatures would lead to excessive thermal motion of the molecules and ions, which was not conducive to form stable chemical bonds, and therefore led to a decrease in the adsorption efficiency ultimately.

The influence of temperature on the removal efficiency of Fe^{3+} ions is revealed in Fig. 6. The adsorption temperature experiments were performed in the range of 15–35°C. The adsorption efficiency of iron ions increased slowly with an increase in the temperature. When the temperature was about 25°C, the adsorption efficiency of oxidized cellulose was the highest (more than 99%). And then, it slightly reduced with an increase in temperature. In general, the impact of temperature on the adsorption efficiency of iron ions with TEMPO-oxidized natural cellulose as the adsorbent was not obvious, and the adsorption efficiency of iron ions was above 98% in a wide range (15–30°C). Therefore, it was sufficient to indicate that there would be a wide range of available temperatures for the sorbent of oxidized cellulose in the wastewater treatment process.

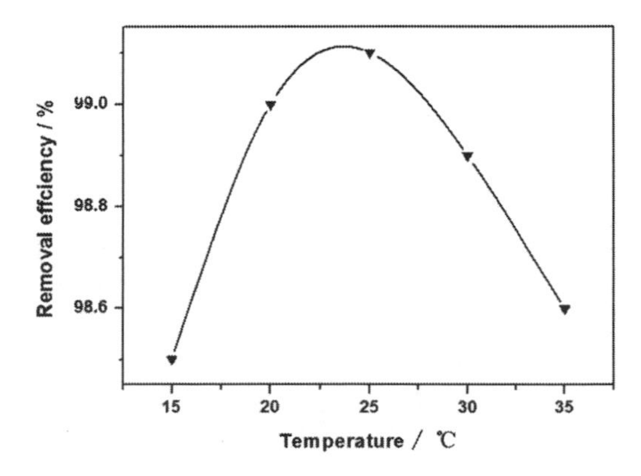

Figure 6. The influence of temperature on the removal efficiency of Fe^{3+} ions.

3.2.5 *The influence of pH on the removal efficiency*

The pH of the solution was another important factor that affects the adsorption process. Considering that the pH value of wastewater is usually below 7, the simulated wastewater was treated with acid solution (pH ranged from 2 to 7) in our experiment. The influence of the pH value on the removal efficiency of iron ions is shown in Fig. 7.

The adsorption efficiency was low at low pH and improved obviously with an increase in the pH. The efficiency reached more than 90% when the pH ranged from 4 to 7. Especially, under the pH value of 4–5, the highest removal efficiency was achieved, close to 93%. The removal efficiency then began to decrease with an increase in pH. It can be expected that the removal efficiency would be lesser when pH > 7, because iron ions and OH^- groups can easily cause precipitation in the alkali solution.

When pH was low, both effects of pH were disadvantageous to the efficiency. On one hand, more hydronium ions were in the solution, which would compete with iron ions at the active site on the surface of the adsorbent; on the other hand, the adsorbent was in a cationic atmosphere, and this would decrease the number of the active sites at which adsorption may take place. When the pH was high, two competing factors determined the efficiency. Fe^{3+} and OH^- ions formed deposits more easily, and the metal ions were surrounded by anions and were not easily combined with adsorbent. At the same time, pH could also affect the acid–base properties of the adsorbent, which led to a change of its structure. Therefore, there was an optimal pH for the efficiency. The condition of partial acid was more desirable for copper removal especially when the pH is 4–5.

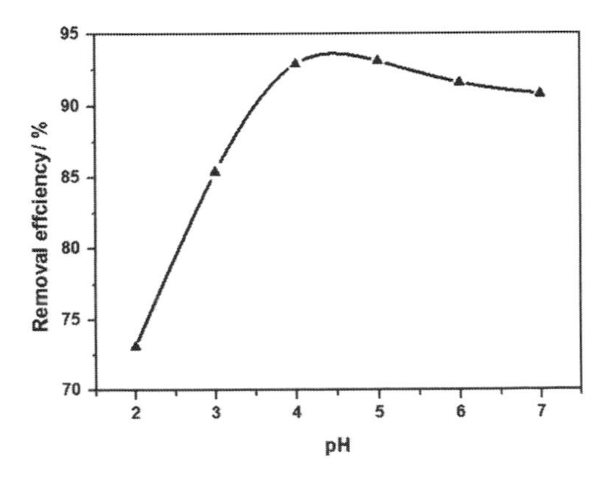

Figure 7. Graph showing the influence of the pH value on the removal efficiency of Fe^{3+} ions.

3.2.6 *Adsorption isotherms*

The capacity of the adsorbent can be estimated through adsorption equilibrium studies. It is usually described by adsorption isotherms. And the experimental data were applied to the Langmiur and the Freundlich isotherm equations, respectively. Two equations (Equation (3) and Equation (4)) were obtained as follows:

Langmuir equation:

$$\frac{1}{q_e} = \frac{0.0069}{C_e} + 0.0052 \quad \left(R^2 = 0.9991\right) \tag{3}$$

Freundlich equation:

$$\ln q_e = 0.9484 \ln C_e - 1.7755 \quad \left(R^2 = 0.9996\right) \tag{4}$$

where C_e (mg·L^{-1}) is the equilibrium concentration of iron ions and q_e (mg g^{-1}) is the adsorption capacity of iron ions at equilibrium.

According to the above-mentioned two equations, the high correlation coefficients indicate that the adsorption process of iron ions by TEMPO-oxidized natural cellulose corresponded well with the Freundlich and Langmuir model. It also demonstrated that the adsorption process is a physical–chemical mixed model which needs further detailed study.

4 CONCLUSIONS

TEMPO-oxidized natural cellulose is a new type of adsorbent, which can be synthesized by natural cellulose under an oxidized condition. We found that the material had a great effect on the adsorption of iron ions from wastewater. The optimum conditions of adsorption from the experiments were obtained. The removal efficiency of iron ions could reach as high as 99% under the conditions where the temperature was 25°C, the iron ions concentration was 100 mg·L^{-1} and the value of pH was 4–5 in the solution, dosing 0.15 g adsorbents and adsorption time of 60 min. Kinetic studies also showed that the adsorption isotherm of Fe^{3+} corresponded with the Freundlich and Langmuir models. Other adsorption mechanisms such as adsorption thermodynamics will be further studied subsequently. In conclusion, the present work predicts that the new adsorption material will have wide-ranging application in the sequestration or removal of metal ions and many other industrial processes in the future owing to the advantages such as raw materials easily available with many sources, large adsorption capacity, ease of operation, and low cost.

ACKNOWLEDGMENTS

The authors are grateful to the National Natural Science Foundation of China (No. 51303089) and Special Fund for Self-directed Innovation of Shandong Province of China (No. 2013CXB80201) for financial support.

REFERENCES

David, W.O., B. Colin & F.O. Thomas. 2008. Heavy metal adsorbents prepared from the modification of cellulose: a review, Bioresour.Technol. vol. 99: 6709–6725.

Ferraiolo, G., M. Zilli & A. Converti. 1990. Fly-ash disposal and utilization, J Chem Technol Biotechnol vol. 47: 281–306.

Gupta, V.K., P. Singh & N. Rahman. 2004. Adsorption behavior of Hg (II), Pb (II) and Cd (II) from aqueous solution on duolite C-433: a synthetic resin, J Colloid Interf. Sci. vol. 275: 398–403.

Gupta, V.K., V.K. Saini & N. Jain. 2005. Adsorption of As (III) from aqueous solutions by iron-oxide coated sand, J Colloid Interf. Sci. vol. 288: 55–61.

Gurgel, L.V.A., R.P. Freitas & L.F. Gil. 2008. Adsorption of Cu(II), Cd(II), and Pb(II) from aqueous single metal solutions by sugarcane bagasse and mercerized sugarcane bagasse chemically modified with succinic anhydride, Carbohydrate Polymers, Vol. 74 (4): 922–929.

Isobe, N., X.X. Chen, U.J. Kimb, S. Kimura, M. Wada,T. Saito & A. Isogai. 2013. TEMPO-oxidized cellulose hydrogel as a high-capacity and reusable heavy metal ion adsorbent, Journal of Hazardous Materials vol. 260:195–202.

Isogai, A. & Y. Kato. 1998. Preparation of polyuronic acid from cellulose by TEMPO-mediated oxidation, CELLULOSE vol. 5: 153–164.

Klemm, D., B. Heublein, H.P. Fink & A. Bohn. 2005. Cellulose: fascinating biopolymer and sustainable raw material, Angew. Chem. Int. Ed. vol. 44: 3358–3394.

Kravchenko, T.A., L.L. Polyanskiy, V.A. Krysanov, E.S. Zelensky, A.I. Kalinitchev & W.H. Hoell. 2009. Chemical precipitation of copper from copper-zinc solutions onto selective sorbents, Hydrometallurgy vol. 95:141–144.

Leandro, V.A.G., K.J. Osvaldo, P.F.G. ossimiriam & F.G. Laurent. 2008. Adsorption of Cu (II), Cd (II), and Pb (II) from aqueous single metal solutions by cellulose and mercerized cellulose chemically modified with cuccinic anhydride, Bioresour. Technol. vol. 99(8): 3077–3084.

Luan, Z.K. & H.X. Tang. 1993. Study on surface properties of ore tailing particles and adsorption of heavy mental ions II. Adsorption of heavy metals onto ore tailing particles, Environmental Chemistry vol. 12 (5): 356–364, (in Chinese).

Paulino, A.T., L.B. Santos & J. Nozaki. 2008. Removal of Pb^{2+}, Cu^{2+}, and Fe^{3+} from battery manufacture wastewater by chitosan produced from silkworm chrysalides as a low-cost adsorbent, Reactive and Functional Polymers vol. 68(2): 634–643.

Saito, T., I. Shibata, A. Isogai, N. Suguri & N. Sumikawa. 2005. Distribution of carboxylate groups introduced into cotton linters by the TEMPO-mediated oxidation, Carbohyd. Polym., vol. 61 (4):414–420.

Volesky, B. 2001. Detoxification of metal-bearing effluents: biosorption for the next century, Hydrometallurgy vol. 59: 203–207.

Xiong, Y.Y., Z.T. Fu & W.S. Huang. 2014. Research on Treatment of Simulated Wastewater Containing Copper by Chemical Sedimentation Method, Environmental Protection Science vol. 40: 35–39, (in Chinese).

Xu H. & Y. Liu. 2008. Mechanisms of Cd^{2+}, Cu^{2+} and Ni^{2+} biosorption by aerobic granules, Separation and Purification Technology vol.58: 400–410.

Civil, Architecture and Environmental Engineering – Kao & Sung (Eds)
© 2017 Taylor & Francis Group, ISBN 978-1-138-02985-9

Research on the precision of grinding experiments in an iron ore

G.M. Shi
School of Resource and Environmental Engineering, Jiangxi University of Science and Technology, Ganzhou, P.R. China
Jiangxi Key Laboratory of Mining Engineering, Jiangxi University of Science and Technology, Ganzhou, P.R. China

L.P. Jiang, J.M. Zhao, R.Q. Zhao, Y.C. Zhou & S.C. He
School of Resource and Environmental Engineering, Jiangxi University of Science and Technology, Ganzhou, P.R. China

ABSTRACT: In order to obtain better grinding homogeneity and lower grinding cost, the optimization experiments of grinding were carried out in the fourth series of the Meishan iron ore concentrator. When compared with the results before application, the results of the test showed the following: the average size of the secondary overflow was similar, but the qualified fraction increased to 0.6 percentage point; the fraction of −10 μm decreased to 1.05 percentage point while decreased to 15.56 percentage points in magnetic separation tailings; the average size increased to 0.06 mm. Simultaneously, the power consumption reduced to 1.68 kw•h/t, the ball consumption dropped to 0.1 kg/t, the noise of grinding dropped to 2~6 dB, the temperature of pulp decreased to 0.5~2°C, the life of primary mill liners extended to 6 months and the life of secondary mill liners extended to 1 year. The grinding costs reduced to 1.70 RMB/t and the effect of energy conservation was very significant. Results of the fourth series could be applied in the same type grinding operations of other series.

1 INTRODUCTION

Mostly, the iron ore was the magnetite, martite, hematite, siderite, and pyrite and silicate minerals in Meishan. The structure of the ore mainly comprised of dense patches, disseminated, breccia and mottled structure. Currently, Meishan iron ore dressing concentrator adopted a two stage closed-circuit grinding process. The grinding operation power consumption accounted for 46% of the total power consumption. It exerted great pressure for the settlement, emissions, and comprehensive utilization of tailings (Liu, 2005). Shi Guiming (2013) studied the optimization of particle size obtained through grinding on the tungsten polymetallic ore grinding system and improved the metal distribution characteristics. Xiao Qingfei (2007) reviewed the development of the grinding medium and optimization research. Wang Peng (2010) studied the shape and size selection principle of the new type of grinding media. Li Jian (2010) discussed the quality of mill grinding, and thought it can improve the quality of grinding, save energy, and reduce consumption. In order to improve the uniformity of the grinding process and reduce costs, in this paper, the fourth series grinding system in Meishan is studied.

2 GRINDING SYSTEM STATUS BEFORE THE INDUSTRIAL TEST

The fourth series closed-circuit grinding system of the Meishan iron ore concentrator was made up of two sets of a 2.7 × 3.6 m ball mill and a 1.6 m double helix grading unit. The mill speed was 21.7 r/min, filling rate was 38%, and grinding concentration was above 88%.

The process investigation deciphered the following problems before the industrial test: (1) the primary stage spiral classifier has no underflow, with classifier efficiency approaching 0%, which did not have any screening effect; (2) the primary stage steel ball size was 120 mm and the secondary stage ball size was 80 mm, which led to the −10 microns productivity of the primary stage classification overflow being 11.75%, and the −10 microns productivity of the secondary stage classification overflow being 15.39%, thereby increasing the over-crushing. On the other hand, it would impact the mill cylinder liner and reduce the service life of the lining board. At the same time, the remaining amount of useless energy was too much and was transferred into heat; (3) the liberation degree of the Fe particle size range was 0.3 mm~38 microns and the liberation degree of the S particle size range was 76~10 microns, the liberation scope of which was different and therefore, Fe was suitable for coarse grinding and S was suitable for fine grinding; (4) the –10 microns products of secondary classification overflow was in high grade that the Fe grade was 30.26%, S grade was 0.77%, and the metal mass loss was greater which leads to the recovery of Fe. S was only 80.28% and 86.59% in qualified products of 0.3 mm~10 microns fraction.

3 INDUSTRIAL TEST PLAN

According to the research results of the laboratory's high-precision grinding and practice experience, the major industrial test under the premise of without changing the existing mill capacity is shown as follows:

1. The grinding concentration was adjusted, the concentration and overflow fineness of the first stage grinding classification process were classified, the amount of sand return was increased, and the grinding process was improved. At the same time, a new ball was added in accordance with the method Φ 100: Φ 80 = 60%:40%.
2. The grinding concentration was adjusted, the concentration of the secondary grinding classification process was classified, and the classifier overflow fineness was stabilized. At the same time, a new ball was added in accordance with the method Φ 60 = 100%.

4 INDUSTRIAL TEST RESULTS

4.1 *Particle size distribution in the grinding-classification circuit*

When compared with the effect of the industrial test, the fourth series were in contrast with the third series. The fourth series has implemented a new industrial test plan since October 1, 2013. It obtained sample and analysis data after stable operation from November. Particle size distributions of the third series production and the fourth series grinding-classification circuit are listed in Table 1.

It can be seen from Table 1 that after the high-precision grinding process, although the secondary classification overflow failed to meet –0.074 mm which accounted for 63%, the average particle size of secondary classification overflow products was the same, the content of the 0.3~10 microns qualified fraction was higher and the content of –10 microns over the crushed fraction was lighter. Moreover, the product quality of the secondary classification overflow of the fourth series was better than that of the third series overall.

4.2 *Product quality of overflow in secondary classification*

During the industrial experiment, the comparison results of the product quality of secondary classification overflow in the third and fourth series are shown in Table 2.

It can be seen from Table 2 that after the utilization of high-precision grinding technology, the better product quality of the secondary classification overflow was obtained under the same processing capacity of the mill.

4.3 *Sorting effect of secondary classification overflow*

To test the quality improvement of the secondary classification overflow in fourth series, the contrast experiments for the third or fourth series were carried out in a laboratory. The dressing process and parameters were the same with the actual process

Table 1. Particle size distribution of the product of the grinding-classification circuit/%.

Indicator		Primary grinding	Primary sand return	Primary overflow	Secondary grinding	Secondary sand return	Secondary overflow
$\gamma_{-76\,\mu m}$	3	35.45	0	36.55	30.50	13.38	64.15
	4	28.20	13.16	32.78	32.93	13.84	60.55
$\gamma_{-10\,\mu m}$	3	9.28	0	9.63	3.83	3.36	15.20
	4	7.63	3.93	8.60	7.43	3.38	14.15
$\gamma_{0.3-10\,\mu m}$	3	55.45	0	54.37	74.32	58.28	80.15
	4	43.72	19.5	48.73	75.05	53.14	80.75

Table 2. Comparison results of the product quality of the final overflow in different series/%.

Indicator	$\gamma_{-0.076\,mm}$	$\gamma_{0.3\,mm-10\,mm}$	D/mm	$\gamma_{-10\,\mu m}$
3#	64.15	80.15	0.10	15.20
4#	60.55	80.75	0.10	14.15
Increased/decreased amplitude	–5.61	+0.75	0	+6.91

of the Meishan iron ore concentrator. The dressing effects of the secondary classification overflow in 3# and 4# series are shown in Table 3.

It can be seen in Table 3 that after the utilization of high-precision grinding technology, the productivity and grade of the iron concentrate were the same, but the loss of Fe and S also reduced because of the improvement of the particle size characteristics. The recovery of iron reduced to 0.02 percentage point and the recovery of sulfur reduced to 0.23 percentage point.

In order to further investigate the separation influence of magnetic tailings, the separation indexes of magnetic separation tailings in the 3# and 4# series are listed in Table 4.

It can be seen from Table 4 that after the utilization of high-precision grinding technology, the −10 microns productivity of magnetic separation tailings reduced to 15.56 percentage point, and the average particle size increased to 0.06 mm.

4.4 *Saving energy and reducing consumption*

After the utilization of high-precision grinding technology, the grinding efficiency and quality were greatly improved. Saving energy and reducing consumption of the grinding process were significant. The energy-saving and consumption-reducing indicators are listed in Table 5.

It can be seen from Table 5 that the mill power consumption reduced to 1.68 kW•h/t, steel consumption reduced to 0.1 kg/t, mill noise dropped to 2~6 dB, grinding pulp temperature reduced to 0.5~2°C, the service life of the first grinding mill lining board prolonged for 6 months, and the second stage grinding can be extended to more than

Table 3. The dressing effect of the secondary classification overflow in different series/%.

	3#					4#				
		Grade/%		Recovery/%			Grade/%		Recovery/%	
Product	Productivity/%	TFe	S	TFe	S	Productivity/%	TFe	S	TFe	S
Sulfur rough concentrate	3.69	40.60	21.09	3.15	72.00	3.31	42.02	23.26	2.91	60.32
Sulfur middling	1.21	39.21	7.75	1.00	8.71	2.30	41.34	9.71	1.99	17.53
Iron ore concentrate	84.52	52.31	0.18	93.03	14.09	84.32	52.40	0.26	92.31	17.18
Magnetic separation tailings	10.58	12.68	0.53	2.82	5.20	10.07	13.30	0.63	2.80	4.97
Undressed ore	100.00	47.53	1.08	100.00	100.00	100.00	47.87	1.28	100.00	100.00

Table 4. The separation indexes of magnetic separation tailings in different series/%.

Indicator	Productivity	Grade	Recovery	$\bar{D}$ / mm	$\gamma_{-10\,\mu m}$
3#	10.58	12.68	2.82	0.04	60.98
4#	10.07	13.30	2.80	0.10	45.42
Increased/decreased amplitude	−4.82	+4.89	−0.71	+150	−25.52

Table 5. Energy-saving and consumption-reducing indicators of 4# series before and after the industrial application (%).

	Power (kwh/t)	Steel consumption (kg/t)		Noise (dB)		Pulp temperature (°C)		Life of the lining (month)	
Indicators	1+2	1	2	1	2	1	2	1	2
Before	13.79	0.5	0.45	95–99	94–96	33~34	34	−8	24
After	12.31	0.45	0.4	92–94	91–93	31~32	33.5	12	36
Increased/decreased	+12.12	−10	−11.11	3~5dB	3dB	2°C	0.5°C	4mon	1a

1 year. Saving energy and reducing consumption were very significant.

4.5 *Benefits of saving energy and reducing consumption*

In the 4# series, according to the price of the industry, the cost of electricity was 0.6 RMB/(kW•h), the cost of steel ball was 4000 RMB/t, the price of the manganese steel lining board was 160000 RMB/set, the price of the magnetic lining board was 210000 RMB/set, and the annual total cost of grinding reduced by 1.70 RMB/t.

5 CONCLUSIONS

1. The appropriate grinding condition is very important for the improvement of the grinding quality.
2. After the introduction of a high-precision grinding technological process, the quality of the secondary classification overflow and magnetic separation tailings was significantly improved. The qualified fraction increased to 0.6 percentage points and the −10 microns content reduced to 1.05 percentage points in the secondary classification overflow. The −10 microns content reduced to 15.56 percentage points, and the average particle size increased to 0.06 mm in the magnetic separation tailings.
3. After the introduction of the high-precision grinding technological process, the energy-saving effect was very significant. The mill power consumption reduced to 1.68 kW•h/t, the steel consumption reduced to 0.1 kg/t, the mill noise reduced to 2~6 dB, the grinding pulp temperature reduced to 0.5~2°C, and the grinding cost reduced 1.70 RMB/t.

ACKNOWLEDGMENTS

The authors gratefully acknowledge the financial assistance of the Foundation of Jiangxi Educational Committee

(GJJ150651) and The Key Project of Jiangxi University of Science and technology (NSFJ2015-K02).

REFERENCES

Li Jian, Zhang Wei, Zhang Xiaoyu. Discussion of improving the quality of mill grinding [J]. Mining machinery, 2010, 38 (15): 97–99.

Liu Anpin, NI Wen, Zhang Zugang. Experimental study of flocculation deep cone enrichment for Meishan tailings [J]. Metal mine, 2005 (10): 30 ~ 33.

Shi Guiming, Wu Caibin, Xiao Liang, etc. Laboratory study of optimizing grinding system for tungsten polymetallic [J]. Non-ferrous metal science and engineering, 2013, 4 (5): 79–84.

Wang Peng. Application research of grinding medium in dressing [J]. Metal mine, 2010 (1): 132–134

Xiao Qinfei, SHI Guiming, Duan Xixiang. Progress and optimization of the grinding media [J]. Mining machinery, 2007, 35 (1): 28–32.

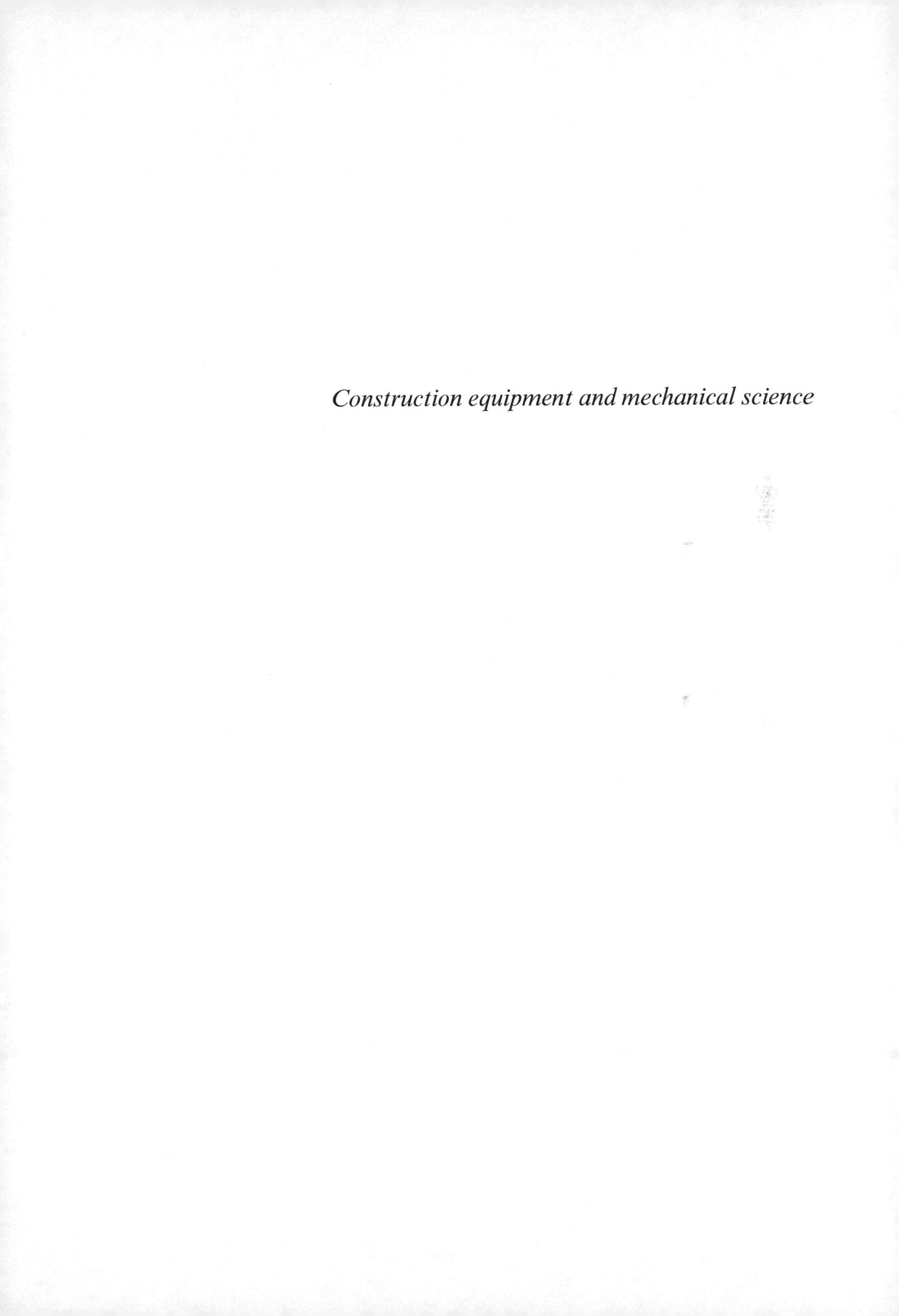

Construction equipment and mechanical science

Civil, Architecture and Environmental Engineering – Kao & Sung (Eds)
© 2017 Taylor & Francis Group, ISBN 978-1-138-02985-9

An experimental study on the dynamic characteristics of compact transmission lines bundles following ice shedding

Kunpeng Ji, Bin Liu, Xueping Zhan & Xiangze Fei
China Electric Power Research Institute, Xicheng District, Beijing, China

Chun Deng & Liang Liu
Jibei Electric Power Research Institute, Xicheng District, Beijing, China

ABSTRACT: Ice shedding is one of the main reasons threats the mechanical security of overhead electric transmission lines. This study attempts to gain the dynamic response of conductor bundles following sudden ice shedding, by conducting a series of ice shedding tests on a reduced scale single span conductor bundle physical model in a laboratory. The influence of ice load, lumped mass number, and location was systematically studied, and the time histories and maximum values of displacement at different key points were measured. With a total of 30 ice shedding scenarios in a reduced scale conductor bundle physical model, these indicate that the ice load, the number of lumped mass, location, and number of released lumped mass significantly affects the dynamic response of the conductor bundle. It is found that the maximum displacement of measured points increases with the ice deposit mass and the maximum displacement along the span in all the scenarios always occurs at the midpoint. Also, the conductor vibrates mainly following the first order mode and it is affected by the particular ice shedding arrangement. This study helps to understand the dynamic response of conductor bundles, which is significant to the design of overhead transmission lines in cold regions.

1 INTRODUCTION

Atmospheric icing is one of the main natural hazards that threaten the safety of overhead transmission lines. It is reported that atmospheric icing and ice shedding have caused serious loss to electric power networks in countries like America, Canada, China and Russia (Farzaneh, 2008). The accidents resulting from icing can be divided into two catalogues, one is electrical accidents, such as flashover and short circuit, and the other is mechanical accidents, such as strands or conductor breakage, insulator rupture, cross-arm deformation, or tower collapse (McClure, 2002).

Ice shedding is a phenomenon in which ice chunks are shed off the conductor or ground wires suddenly due to high temperature, strong winds, or mechanical de-icing (Morgan, 1964; Jamaleddine, 1993). When ice shedding happens, the conductor will jump dramatically and the tension of the conductor will change largely; the gravitational and strain potential energy of ice and conductor were released and converted into kinetic energy following ice shedding.

The compact transmission lines have superior economic and social benefits to ordinary transmission lines, because these need less corridor space, cause less environmental pollution, but increase the natural transmission power and reduce the wave impedance, by reducing the phase conductor distance and increasing the sub-conductor number and conductor diameter. However, it has higher risk to ice shedding accidents due to the closer phase distance.

A series of ice shedding tests were conducted by Morgan and Swift on a five-span 230 kv transmission line, from which the equation of the conductor jump height after ice shedding was deduced and the effects of different insulator assembly methods were studied (Morgan, 1964). Jamaleddine *et al.* carried out several ice shedding tests on a two-span reduced scale line and was the first to use non-linear commercial finite element software to study the dynamic response of the conductor following ice shedding. It is proved that the numerical simulation results agree well with the model tests (Jamaleddine, 1993). Roshan Fekr and McClure simulated 21 ice shedding scenarios, by changing conductor density, to study the influence of ice thickness, span length, and elevation difference on the ice shedding response (Fekr, 1998). Meng *et al.* conducted an ice shedding test on a real scale transmission line and studied the influence of conductor damping (Meng, 2012). Li *et al.* study the towers–lines coupled effect and the anti-vibration measure for towers during the ice shedding process (Li, 2008). Yang studied the dynamic loads on Ultra-High Voltage (UHV) towers following ice shedding (Yang, 2014). Kollar *et al.* was the first who systematically studied the dynamic characteristics of bundled conductors in different ice shedding scenarios, such as whole span shedding, large chunks shedding and un-zipping shedding, and the FE modeling method was validated by using a reduced scale twin sub-conductor bundle (Kollar, 2008; László, 2013). In the study, the influence of ice shedding from certain sub-conductors, sub-conductor distance, and sub-conductor number were studied. Shen et al. studied the influence of different FE models, the equivalent single conductor model, and the bundle model (Shen, 2012). Ji *et al.* considered the induced ice shedding effects in dynamic analysis, and so the time-varying characteristics of the system can be considered in FE analysis, which is more realistic (Ji, 2015).

Although the ice shedding phenomenon has been studied in previous studies, there is a lack of ice shedding tests on bundle conductors, which are widely used in extra and UHV transmission lines. Thus, in this paper, a reduced scale physical transmission line model of the conductor bundle is designed to obtain its dynamic characteristics in different ice shedding scenarios. This study will offer more experimental data for transmission line designs in cold regions and is helpful for the mechanical safety of transmission lines and the operation of electric power networks.

2 EXPERIMENTAL SETUP

The reduced scale physical model is a single span bundle with eight sub-conductors and no spacer is installed for simplicity. A high strength buckle was used to connect the bundle to the dead end at the left end and the right end of the bundle is connected to a tension sensor first by using a buckle, and then the tension sensor is connected to a screw, which is used to adjust the conductor tension. Five laser sensors are set underneath the conductor bundle on the test platform. The lumped mass was hung on the bundle with a rope to model the ice load and the ice shedding phenomenon is modelled by suddenly cutting the rope. The experimental setup is shown in Fig. 1.

In the tests, laser sensors produced by Panasonic were used to detect the real time displacement of measured points and force sensors produced by Tecsis were used to measure the tension of sub-conductors.

A total of 30 scenarios of ice shedding tests were studied, by changing the total mass of the ice deposit, lumped mass number and ice shedding locations. The combinations of ice shedding scenarios are listed in Table 1 and shown in Fig. 2 (only the five lumped loads scenario is illustrated).

In Table 1, 5%, 10%, and 15% refer to the fact that the total ice load is 5%, 10%, and 15% of the conductor bundle mass, respectively. The total ice load was modeled by 1, 3, and 5 lumped masses in different scenarios.

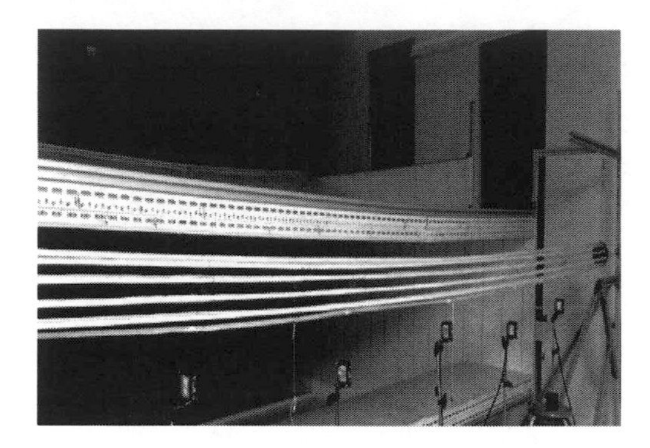

Figure 1. Picture of an ice shedding experimental setup.

Table 1 Details of ice shedding scenarios.

Ice load	Lumped mass number	Ice shedding location
5%	1	M_1
	3	M_1, M_2, $M_1 + M_2$, $M_1 + M_3$
	5	M_1, M_2, M_3, $M_2 + M_3$, $M_2 + M_3 + M_4$
10%	1	M_1
	3	M_1, M_2, $M1 + M_2$, $M1 + M_3$
	5	M_1, M_2, M_3, M_2+ M_3, $M_2 + M_3 + M_4$
15%	1	M_1
	3	M_1, M_2, $M_1 + M_2$, $M_1 + M_3$
	5	M_1, M_2, M_3, $M_2 + M_3$, M_2+M_3+M_4

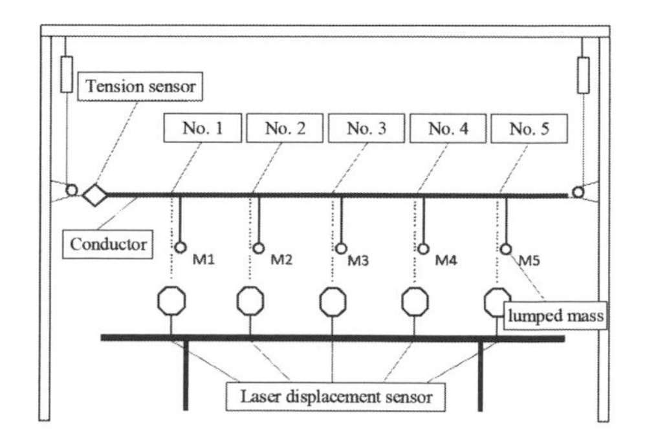

Figure 2. Schematic of the ice shedding scenario arrangement (with five lumped loads).

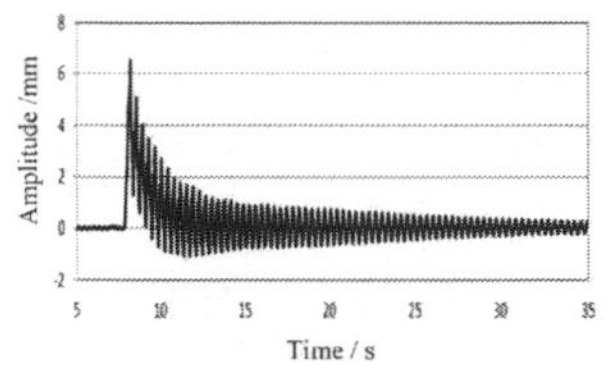

(a) Displacement time history of point 3 (M2 was released)

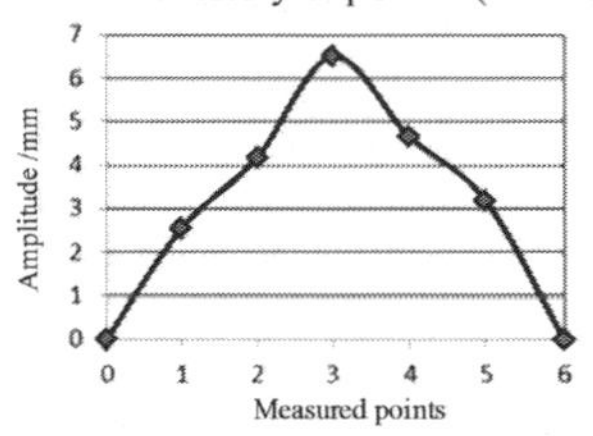

(b) Maximum jump height of the five measured points (M2 was released)

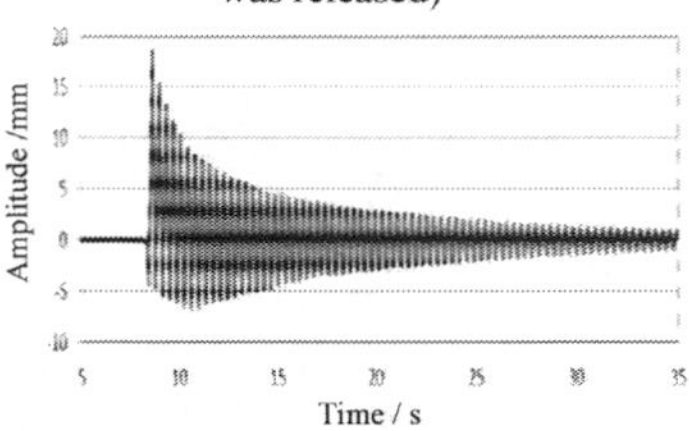

(c) Displacement time history of point 3 (M1+M2 were released)

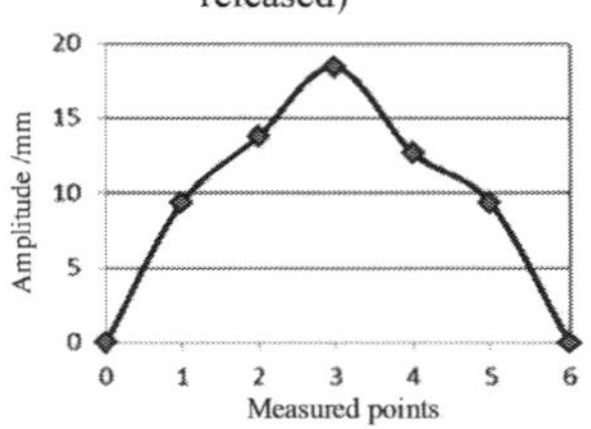

(d) Maximum jump height of the five measured points (M1+M2 were released)

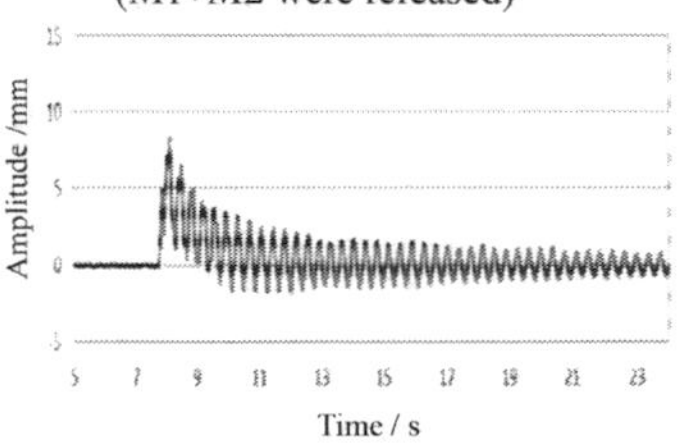

(e) Displacement time history of point 3 (M1+M3 were released)

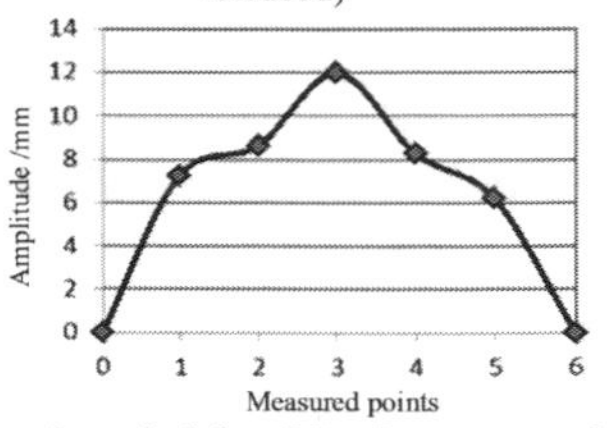

(f) Maximum jump height of the five measured points (M1+M3 were released)

Figure 3. Results of the test where the ice load is simulated by three lumped masses.

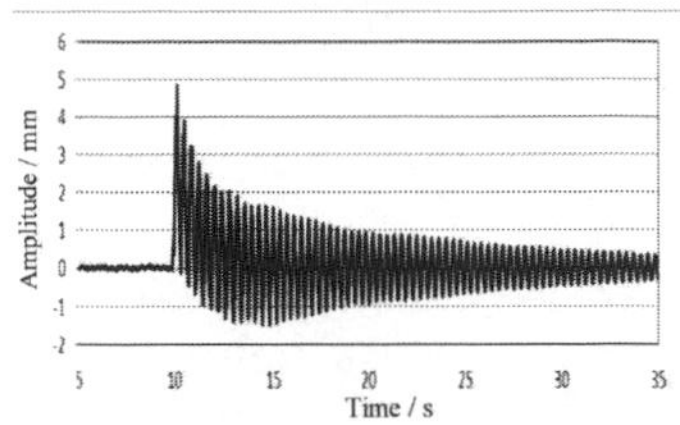

(a) Displacement time history of point 3 (M3 was released)

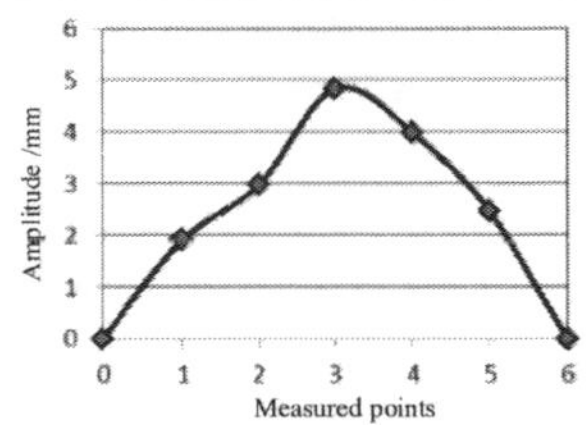

(b) Maximum jump height of the five measured points (M3 was released)

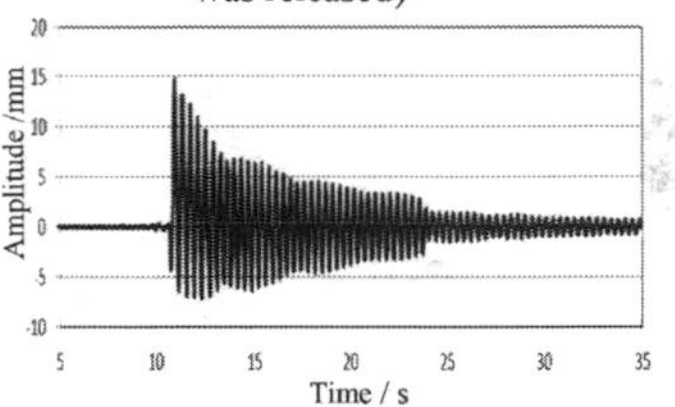

(c) Displacement time history of point 3 (M2+M3 were released)

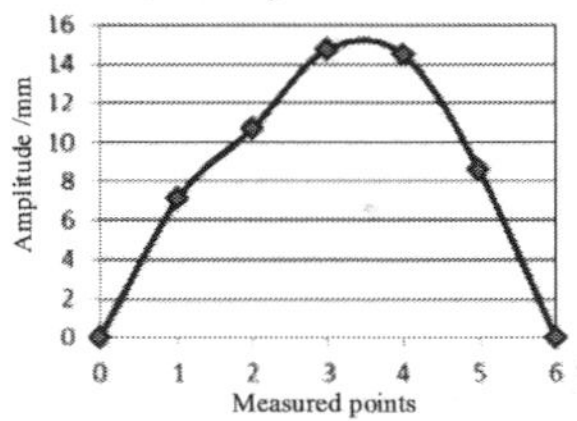

(d) Maximum jump height of the five measured points (M2+M3 were released)

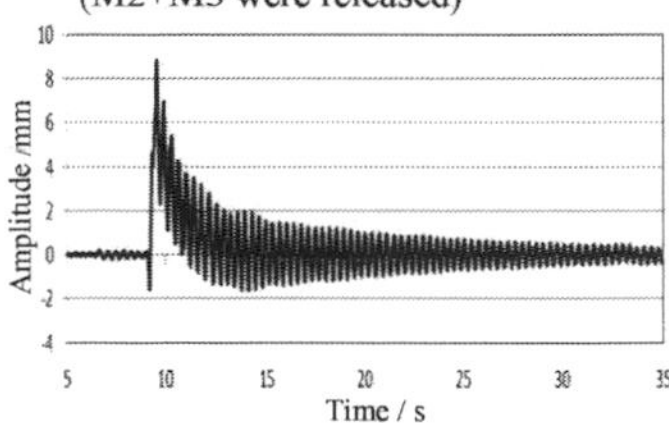

(e) Displacement time history of point 3 (M2+M3+M4 were released)

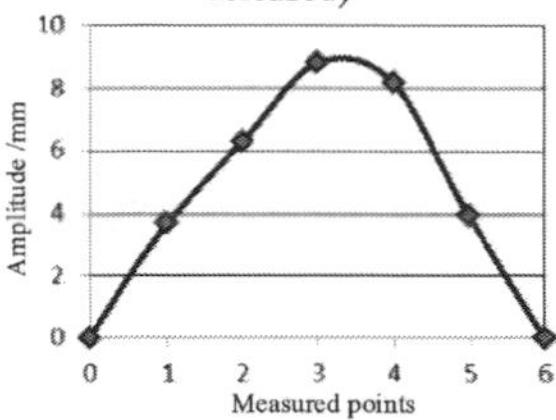

(f) Maximum jump height of the five measured points (M2+M3+M4 were released)

Figure 4. Results of the test where the ice load is simulated by the five lumped mass.

3 DYNAMIC RESPONSE ANALYSIS OF BUNDLED CONDUCTORS IN THE EXPERIMENTS

In the tests, the displacement of the five measure points is recorded. Figures 3 and 4 show the displacements time history of the midpoint (point 3) and maximum displacements of every measured point in different test scenarios, where the total ice load is 15% of the cable mass. For the sake of paper length, only the scenarios with three (Figure 3) and five (Figure 4) lumped masses are listed in this article.

It indicates from the figures that the dynamic response of the conductor bundle is significantly influenced by the number of lumped mass, location, and number of released lumped mass. The maximum displacement along the span in different scenarios always occurs at the midpoint and the displacement time history decays due to the damping effect. Among all the tests, the maximum jump height of the midpoint appears in the scenario with only one lumped mass. In addition, the conductor vibrates mainly following the first order mode and is affected by the particular ice shedding arrangement, and the second order mode may be excited when uneven and non-synchronous ice shedding happens.

Besides, the comparison of 5%, 10%, and 15% ice load scenarios shows that the maximum displacement of measured points increases with the ice deposit mass, and the vibration modes with the same number and arrangement of lumped mass are similar when the ice load varies.

4 CONCLUSIONS

A reduced scale single span conductor bundle physical model was established to study the dynamic response of the bundle following sudden ice shedding. By changing the ice load, lumped mass number, and releasing the arrangement of the lumped mass, it obtained the displacement time history and maximum displacement of the measured points. The test results show that the maximum displacement always occurs in the midpoint of the span and increases with total ice load. The main vibration mode was excited in the tests is the first order mode with one loop. The number and release location of the lumped mass have significant influence on the dynamic response of the conductor bundle. The study and findings in this paper are useful for the design of overhead transmission lines in cold regions.

ACKNOWLEDGEMENT

This research work was financially supported by the State of Grid of China's scientific project—*the Applicability and Anti-Disaster Ability Improvement of Compact Transmission lines* (No. SGTYHT/14-JS-188).

REFERENCES

Farzaneh, M., C. Volat, A. Leblond, Anti-icing and deicing techniques for overhead lines. (Springer, Netherlands, 2008).

Jamaleddine, A., G. McClure, J. Rousselet, R. Beauchemin, *Comput. Structures* **47**, 523 (1993).

Ji, K., X. Rui, L. Li, C. Zhou, C. Liu, G. McClure, Shock Vib. **2015**, Article Number 635230, (2015).

Kollar L.E. and M. Farzaneh, IEEE Trans. Power Del. **23**, 1097 (2008).

Li, L., Z. Xia, G. Fu, Z. Liang, J. Vib. Shock **27**, 32 (2008). (In Chinese)

László E Kollar and Masoud Farzaneh, IEEE Trans. Power Del. **28**, 604 (2013).

McClure, G., K.C. Johns, F. Knoll, G. Pichette, *in Proc. 11th Int. Workshop Atmospheric Icing Structures*, Paper 9 (Brno, Czech Republic, 2002).

Meng, X., L. Hou, L. Wang, M. MacAlpine, G. Fu, B. Sun, Z. Guan, W. Hu, Y. Chen, Mech. Syst. Signal Process. **30**, 393 (2012).

Morgan, V.T., D.A. Swift, *Proc. Insti. Elect. Eng.*, **111**, 1736 (1964).

Roshan M. Fekr and G. McClure, Atmos. Res. **46**, 1 (1998).

Shen, G., L. Xu, X. Xu, W. Lou, B. Sun, Power Syst. Tech. **36**, 201 (2012). (In Chinese)

Yang, F., J. Yang, H. Zhang, J. Cold Reg. Eng. (ASCE) **28**, 1 (2014).

Civil, Architecture and Environmental Engineering – Kao & Sung (Eds)
© 2017 Taylor & Francis Group, ISBN 978-1-138-02985-9

A single-lens optimal imaging system based on a liquid crystal spatial light modulator and micro-scanning optical wedge

Ning Xu, Zhi-Ying Liu & Dong Pu
School of Opto-Electronic Engineering, Changchun University of Science and Technology, Changchun, Jilin, China

ABSTRACT: A simple and high-resolution single-lens imaging system is generated by using a liquid crystal spatial light modulator to correct wavefront distortion and reconstructing a high resolution image with micro scanning by using an optical wedge. A zygo interferometer is used to measure of wavefront data of single-lens focusing infinite distance; ZEMAX is used to simulate to obtain wavefront data of finite distance and use Zernike Polynomials to describe it. Based on the relationship between the phase and grayscale of a liquid crystal spatial light modulator, a conjugate wavefront of distortion wavefront is obtained for completing static wavefront correction before assessing the correction effect by Imatest software. In order to improve the image resolution, an optical wedge is used to perform multiple 2 × 2 micro-scannings of a same scene for imaging. A high-resolution static scene image is reconstructed by using a fast high-resolution algorithm based on the Keren registration method and template convolution.

1 INTRODUCTION

With the development of science and technology, demands for optical systems with a small volume and high resolution imaging have become increasingly intense. In an optical system, single lens imaging can be employed to reduce its volume effectively. As a result, the structure and price of this system are substantially simplified and brought down respectively. However, single lens imaging is provided with certain wavefront distortions that are able to give rise to imaging blurring for such an optical system. Therefore, the solution to the wavefront distortion is the core of using single lens imaging.

In recent years, application of adaptive optics into wavefront distortion correction has become a hotspot due to the development of liquid crystal display technology and relevant techniques as well as abundant liquid crystal materials. As a wavefront corrector, the liquid crystal spatial light modulator is featured with high resolution, low energy consumption, small volume, ease of control, low price, etc. (Cai et al, 2008). In this paper, defocus and aberration are compensated by virtue of the liquid crystal spatial light modulator. A 1272 × 1024 pixels pure phase type liquid crystal spatial light modulator developed by the Hamamatsu Company in Japan is adopted in combination with Zemax simulation and Zygo interferometer test to carry out fitting for wave surfaces so as to achieve an effect of compensation. In addition, Imatest is utilized to grade it based on images that are processed by using the liquid crystal spatial light modulator or otherwise. Thus, wavefront distortion correction effects of the modulator are evaluated.

Additionally, a method of employing a CMOS Sensor for integration on a focal plane is also used in this paper. CMOS staggers half of the pixel along the direction of the linear array to perform twice imaging. In other words, the method of sub-pixel imaging is adopted. Moreover, an approach of 2 × 2 micro-scanning is also applied to fill the infinitesimal displacement to improve the sampling frequency so that the original CMOS sampling frequency can be increased by times. In this way, the imaging quality is further improved.

2 DEFOCUS AND ABERRATION CORRECTION BY USING THE LIQUID CRYSTAL SPATIAL LIGHT MODULATOR

2.1 *Construction of experimental optical path*

A K9 single lens with a focal length of 35 mm and a diameter of 25.4 mm is selected for experiments in this work. The distance from a front-end test card to a lens is 500 mm, while it is 800 mm from a back-end test card to the lens. Figure 1 shows an overall structure block. Due to limitations of the field of view, ISO 12233 reflective test cards of 0.5X are chosen. They should be placed up and down without overlaying. In addition, a read monochromatic source of 633 nm is adopted. While the

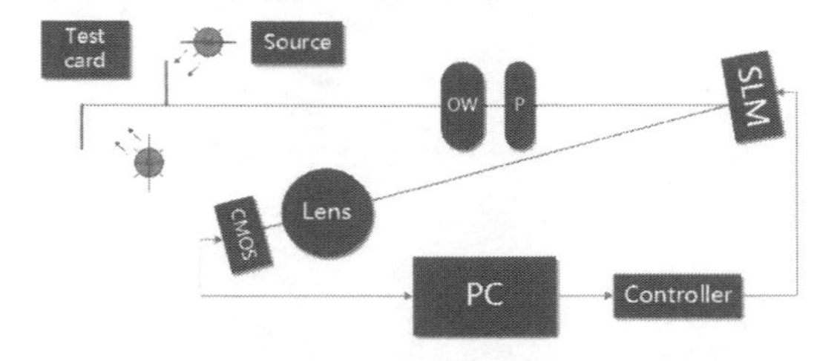

Figure 1. The overall structure diagram.

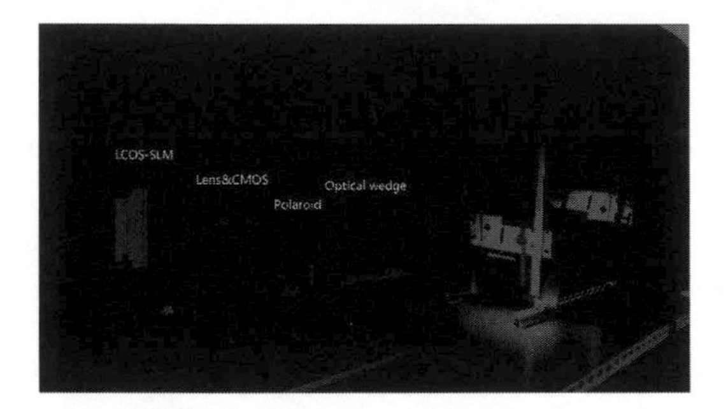

Figure 2. Picture of the overall structure of appearance.

micro-scanning optical wedge is aimed at achieving high resolution imaging implementation, the polarizing film only plays a part in horizontal polarized light because of the adoption of a liquid crystal spatial light modulator. Furthermore, it is under the control of this modulator that controls it through the PC. The relevant overall structure is presented in Figure 2.

2.2 *Simulation and calculation of defocus*

Defocus usually takes place in an optical system that is constructed by using a single lens. Therefore, it is assumed that there are two ISO12233 test cards at the front and the back ends to evaluate this system. The corresponding focal length is adjusted to make the front-end test card just situated at the focal plane. At this time, a defocus phenomenon appears correspondingly as long as the test card at the back end is moved. Thus, defocus of the optical system can be simulated.

As given in Figure 3, the focal plane is located at a test card plane at 500 mm. Clearly, we almost cannot see a test card at 800 mm nearly, thereby signifying that a great distorted wavefront exists at 800 mm. In the case that MTF curves under a half field of view with an object height of 500 mm are simulated by Zemax, a giant aberration exists in a single lens system and the edge field of view is also dramatically different from the central field of view, etc., under the circumstance that no additional elements are added, as shown in Figure 4.

Zemax can be utilized to obtain wavefront data under conditions of a finite element object distance. In order to acquire a clear image on a CMOS sensor, the front-end test card is adjusted in the first place. And then, the distance from the back-end test card to the lens can be figured out, so as to further simulate the corresponding coefficient of Zernike Polynomials by virtue of Zemax, as presented in Figure 5. In detail, item 4 of this

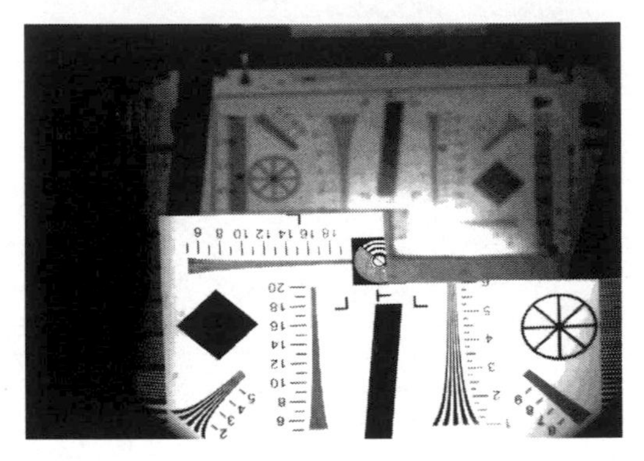

Figure 3. Picture showing the original images.

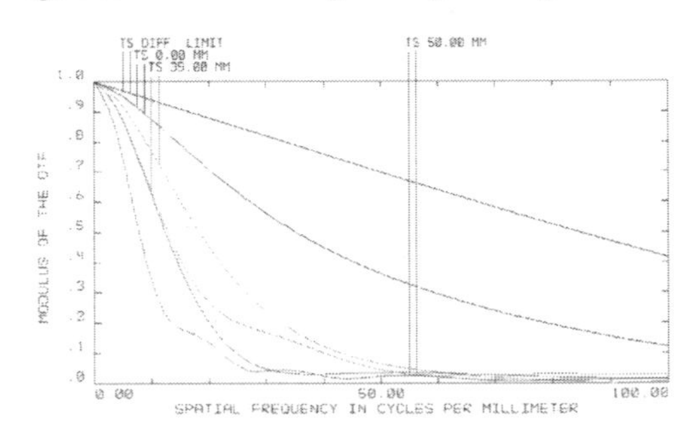

Figure 4. MTF curve with 500 mm object.

```
Z  1   583.46412714 : 1
Z  2     0.00000000 : (p) * COS (A)
Z  3     0.00000000 : (p) * SIN (A)
Z  4   965.67738563 : (2p^2 - 1)
Z  5     0.00000000 : (p^2) * COS (2A)
Z  6     0.00000000 : (p^2) * SIN (2A)
Z  7     0.00000000 : (3p^2 - 2) p * COS (A)
Z  8     0.00000000 : (3p^2 - 2) p * SIN (A)
Z  9   482.10638288 : (6p^4 - 6p^2 + 1)
Z 10     0.00000000 : (p^3) * COS (3A)
Z 11     0.00000000 : (p^3) * SIN (3A)
Z 12     0.00000000 : (4p^2-3) p^2 * COS (2A)
Z 13     0.00000000 : (4p^2-3) p^2 * SIN (2A)
Z 14     0.00000000 : (10p^4 - 12p^2 + 3) p * COS (A)
Z 15     0.00000000 : (10p^4 - 12p^2 + 3) p * SIN (A)
Z 16   155.96612089 : (20p^6 - 30p^4 + 12p^2 - 1)
Z 17    -1.02439574 : (p^4) * COS (4A)
Z 18     0.00000000 : (p^4) * SIN (4A)
Z 19     0.00000000 : (5p^2 - 4) p^3 * COS (3A)
Z 20     0.00000000 : (5p^2 - 4) p^3 * SIN (3A)
Z 21     0.00000000 : (15p^4 - 20p^2 + 6) p^2 * COS (2A)
Z 22     0.00000000 : (15p^4 - 20p^2 + 6) p^2 * SIN (2A)
Z 23     0.00000000 : (35p^6 - 60p^4 + 30p^2 - 4) p * COS (A)
Z 24     0.00000000 : (35p^6 - 60p^4 + 30p^2 - 4) p * SIN (A)
Z 25    96.79109300 : (70p^8 - 140p^6 + 90p^4 - 20p^2 + 1)
Z 26     0.00000000 : (p^5) * COS (5A)
Z 27     0.00000000 : (p^5) * SIN (5A)
Z 28    -1.18552559 : (6p^2 - 5) p^4 * COS (4A)
Z 29     0.00000000 : (6p^2 - 5) p^4 * SIN (4A)
Z 30     0.00000000 : (21p^4 - 30p^2 + 10) p^3 * COS (3A)
Z 31     0.00000000 : (21p^4 - 30p^2 + 10) p^3 * SIN (3A)
Z 32     0.00000000 : (56p^6 - 105p^4 + 60p^2 - 10) p^2 * COS (2A)
Z 33     0.00000000 : (56p^6 - 105p^4 + 60p^2 - 10) p^2 * SIN (2A)
Z 34     0.00000000 : (126 p^8 - 280p^6 + 210p^4 - 60p^2 + 5) p * COS (A)
Z 35     0.00000000 : (126 p^8 - 280p^6 + 210p^4 - 60p^2 + 5) p * SIN (A)
Z 36    71.61581051 : (252p^10 - 630p^8 + 560p^6 - 210p^4 + 30p^2 - 1)
Z 37    53.81510563 : (924p^12 - 2772p^10 + 3150p^8 - 1680p^6 + 420p^4 - 42p^2 + 1)
```

Figure 5. Zenike coefficient simulated by Zemax.

coefficient refers to spherical aberration and aberration items of the wavefront fitted.

2.3 *Aberration measurement*

As for the imaging at infinity, a digital wave phase-shifting interferometer is employed to precisely measure wave aberration of the single lens. Thus, wavefront matrix data are acquired.

The digital wave phase-shifting interferometer is used to precisely measure the wavefront of an optical system and provides data required by the subsequent corrections. The experimental optical path for measurement is shown in Figure 6. GPI-XPIV, a digital wave phase-shifting interferometer manufactured by Zygo, is adopted. To be specific, its principle is a Fizeau interferometer. Owing to the accuracy of this apparatus itself, we only need to place the polarizing film, optical system and reflector before such an interferometer (Wang & Yu, 2008). In addition, phase-shifting interference is employed by it. Concerning its major advantages, uncertainty can be up to 1/50 wave length in spite of a high measurement precision. Besides, data processing software mounted in the interferometer can export binary data files of the wavefront and also provides an exe program that can be utilized to convert such binary files into wavefront XYZ values stored in ASCII code. Based on a concrete wavefront data matrix, the wavefront can be accurately corrected and the relevant measurement process is shown in Figure 7. As for its experimental results, they are given in Figure 8.

Figure 8 shows that wavefront patterns can be displayed in the gray scale in the MetroPro after measurement by virtue of the Zygo interferometer.

Figure 6. Picture of the experimental optical path structure.

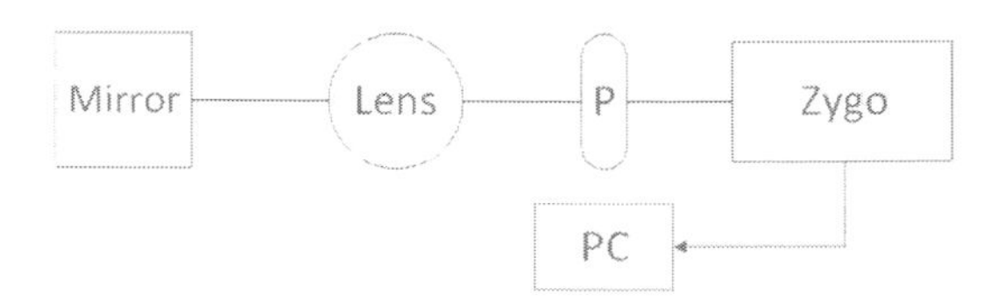

Figure 7. Measurement structure diagram.

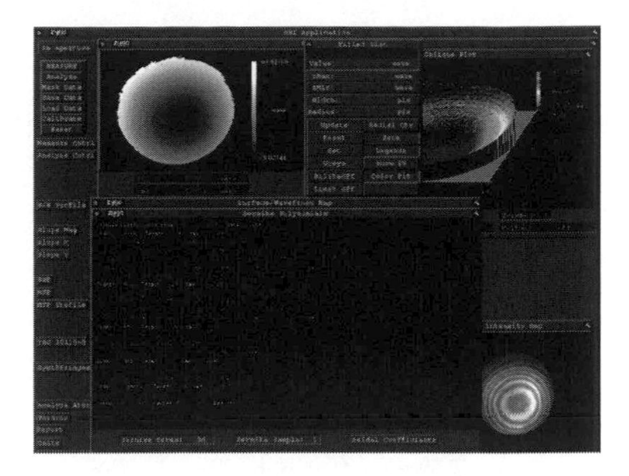

Figure 8. Zygo interferometer output interface.

In addition, 36 coefficients of the Fringe Zernike Polynomials are also given. The wavefront of an optical system or optical elements is always continuous and smooth. Considering this, the continuous function is used to represent wave aberration or the surface shape of an optical system. In normal cases, a series of orthogonal linear combinations of the Zernike Polynomials are adopted for this representation (Zhang et al, 2015). Although the Zernike Polynomials are provided with a considerably high accuracy, wavefront data exported by MetroPro directly are more accurate. Notably, the Zernike Polynomials will be employed to describe these uniformly for the convenience of the unification between defocus and aberration.

2.4 *Wave aberration correction for liquid crystal spatial light modulator*

The core component used for wave aberration correction is the liquid crystal spatial light modulator, a device performing spatial light modulator for spatial distribution of optical waves (Wyant et al, 1992). Generally, a spatial light modulator is constituted by multiple independent units that are arranged into one-dimensional or two-dimensional arrays spatially. Every unit can be controlled by optical or electric signals independently. Hence, its optical properties are altered according to such signals so as to modulate optical waves lighting on it (Love, 1997).

A distorted wavefront can be represented by using a series of linear combinations of orthogonal Zernike Polynomials, which is able to describe complex wavefront phase information and generate sufficient precision at the same time (Born & Wolf, 1978). The Zernike Polynomials in a circle is usually expressed in 2-dimensional polar coordinates as follows (Nam & Rubin, 2005):

$$Z_n^m(\rho,\theta)=\begin{cases}N_n^m R_n^{|m|}(\rho)\cos(m\theta) & \mathrm{m}\geq 0,\\ -N_n^m R_n^{|m|}(\rho)\sin(m\theta) & \mathrm{m}\leq 0,\end{cases} \tag{1}$$

$$N_n^m=\sqrt{\frac{2(n+1)}{1+\delta_{m0}}},\quad \delta_{\mathrm{ij}}=\begin{cases}1 & \mathrm{i}=\mathrm{j},\\ 0 & \mathrm{i}\neq \mathrm{j},\end{cases} \tag{2}$$

$$R_n^m(\rho)=\sum_{s=0}^{n-|m|/2}\times\frac{(-1)^s(n-s)!}{s!\left(\frac{n+|m|-s}{2}\right)!\left(\frac{n-|m|-s}{2}\right)!}\rho^{n-2s}, \tag{3}$$

In the equations, angular and radial frequency numbers of the polynomials are provided respectively. They are important parameters reflecting the spatial frequency of the Zernike Polynomials (Cai et al, 2008).

The liquid crystal spatial light modulator should carry out phase modulation under a condition of polarized light. Therefore, the polarizing film is installed in the optical path. After this modulator is electrified, the liquid crystal correspondingly rotates according to a gray-scale map loaded so as to change the phase distribution on an incident wave surface. In the case that the coefficient of the Zernike Polynomials is obtained, control signals of the Zernike pixel should be determined. In other words, the compensated coefficient of the Zernike Polynomials is found. Popularly speaking, a conjugate coefficient of the original Zernike coefficient is taken. At that time, Matlab is utilized in conjunction with phase modulation characteristics of the liquid crystal spatial light modulator (Cai et al, 2007) to write the wave aberration fitting code and further generate gray-scale maps for compensated defocus and aberration. Subsequently, gray-scale signals are transferred to the drive circuit of the liquid crystal spatial light modulator through the PC. It is important to note that a monochromatic source of 633 mm is employed during the experiment. Under such a circumstance, the phase value (0–2)π of the X10468 series liquid crystal spatial light modulator manufactured by Hamamatsu is linearly correlated to the gray level (0–255). It means that there is no need to recalibrate the relation between the phase and gray scale under such a waveband, as shown in Figure 9.

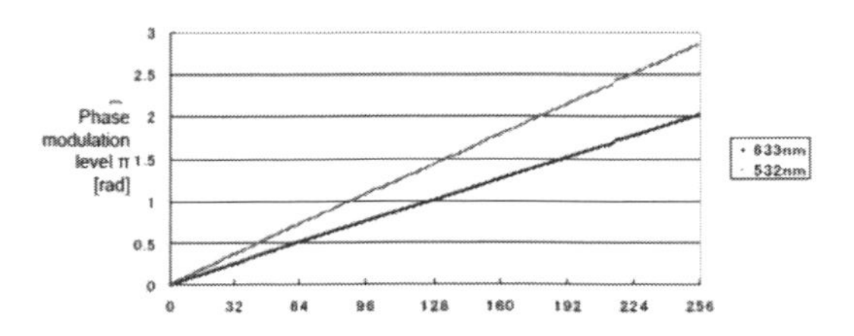

Figure 9. Phase modulation curve of the X10468 series modulator.

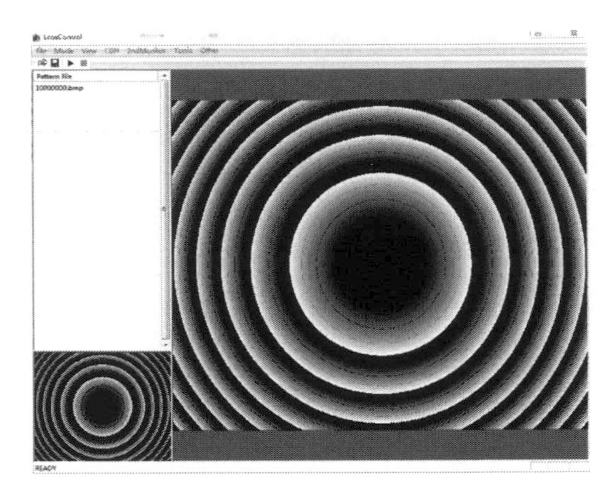

Figure 10. Modulator driver software to load the Fresnel grayscale image.

To expand the scope that can be adjusted by using the liquid crystal spatial light modulator, the gray-scale map portrayed is converted into a Fresnel gray-scale map. As a result, the modulation range of such a modulator can be enlarged. However, its modulation range cannot be extended infinitely. As far as the X10468 series is concerned, the maximum phase that it can modulate is at 25π. And then, the gray-scale map portrayed is loaded onto software of the liquid crystal spatial light modulator (Fig. 10).

2.5 *Estimation on the correction effect of liquid crystal spatial light modulator*

Imatest is digital image evaluation software developed by the Imatest LCC in America and it has been extensively applied. The overall system of this software is established based on Matlab, which consists of multiple function modules, such as SFR, Colorcheck, Stepchart, etc. In addition, Imatest takes advantage of the Spatial Frequency Response (SFR) which requires lower cost to measure the MTF curve. This is a very simple approach. The required MTF curves can be acquired only by analyzing black and white slanted edges with a certain angle of inclination (Chen et al, 2014).

3 HIGH RESOLUTION RECONSTRUCTION FOR MICRO-SCANNING

A CMOS sensor of 1.3 million pixels is adopted during the experiment. The pixel size of the CMOS is 5.2 μm. Dependent on the scheme of liquid crystal spatial light modulator, the CMOS sensor becomes a factor restricting the improvement

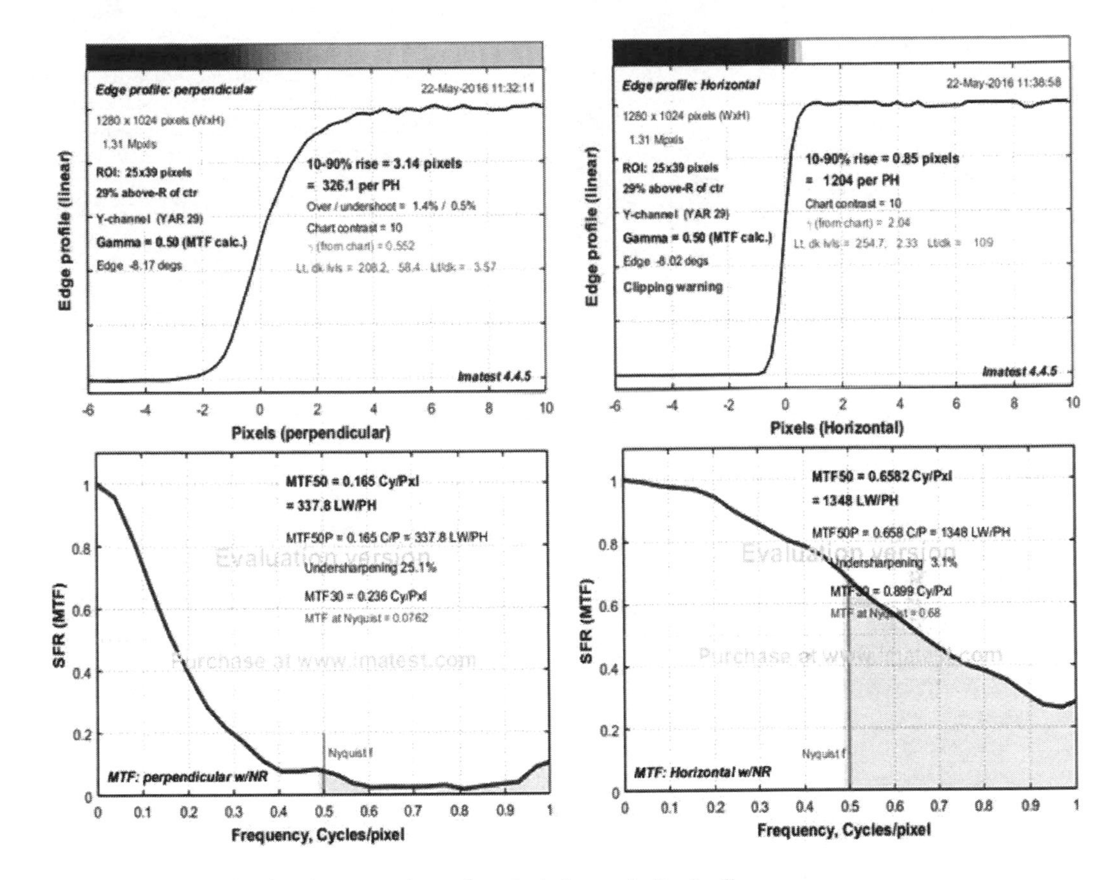

Figure 11. Graphs showing the comparison of results before and after loading.

of image quality. In order to perfect the resolution of the imaging on the premise of changing no CMOS sensor, high resolution imaging is realized by means of micro-scanning.

Micro-scanning is an over-sampling imaging process that takes place multiple times for an identical scene and each sampling is referred to as sub-sampling. According to the micro-scanning process, images achieved by sampling are noted down each time and then images obtained by each subsampling process are recombined in line with their sequences of micro-scanning. As a result, high resolution images are obtained through subpixel processing. What needs to be pointed out is that the relevant scenes and fields of view are required being static or moving slightly in the processes of sampling and imaging. In other words, micro-scanning can acquire a high spatial resolution by sacrificing the temporal resolution (Zhang et al, 1999).

Theoretically, the larger the number of scanning is, the better the reconstruction effect will be. But, a 2 × 2 micro scanning pattern is the most reasonable in terms of this experiment for the purpose of efficiency improvement. To be specific, four pictures are captured by rotating the micro-scanning optical wedge and these are integrated into a high resolution image based on the corresponding algorithm. In a word, the 2 × 2 micro-scanning method is selected for use, according to the below-mentioned equations:

$$\delta = (n-1)\alpha \quad (4)$$

$$\Delta = \delta f_0 \quad (5)$$

In line with equations for deviation angle δ and displacement Δ, the wedge angle of the optical wedge is 21" and the associated machining tolerance is ±2", as shown in Figure 12, which is a schematic diagram of the micro-scanning structure. If the optical wedge rotates around a ray axis to perform azimuthal rotation, the focus point of the converged light beams can form a circumference on the focal plane and this circumference uses the original image point as the center and Δ as the radius. In the case that the rotating optical plate respectively stays at four positions that are

90° away from each other, that is 45°, 135°, 225°, and 315°, for under sampling imaging (as the high resolution image reconstruction principle given in Figure 13), micro displacement of adjacent sampling points in a detector can be expressed as P = L/2 (Kim et al, 2000). In this equation, L is the pixel pitch and the value of micro-scanning displacement is (√2 L)/4.

With an aim to improve the possibility of real-time application of the high resolution reconstruction technique for images and enhance its tolerance of the registration error, a fast robust high-resolution image reconstruction algorithm based on Keren registration and interpolation is used. According to this algorithm, low resolution images after registration are projected onto a high

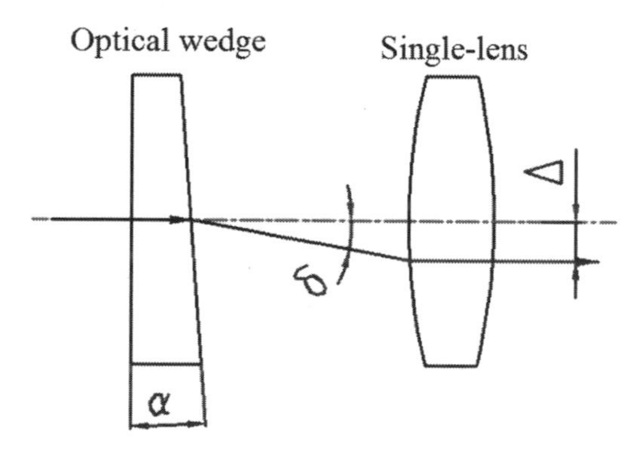

Figure 12. Schematic of the micro-scanning structure principle.

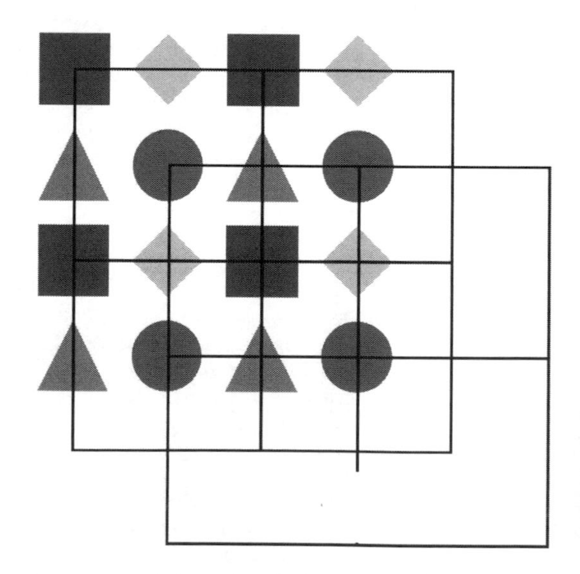

Figure 13. Schematic of the High-resolution image reconstruction principle.

resolution grid in conformity with transformation parameter. And then, template convolution is utilized to iteratively fill missing pixel values so as to reconstruct a high resolution image. Additionally, such an algorithm is compared with another four high resolution image reconstruction algorithms, including the non-uniform interpolation method, the convex set mapping method, the robust iterative backward mapping method and the structurally adaptive normalized convolution method. Experimental results indicate that this algorithm is insensitive to registration errors within a certain range of precision, although it is provided with some advantages in terms of speed and reconstruction effects. Totally speaking, it is an effective, robust and rapid high-resolution multi-frame image reconstruction algorithm. Hence, a fast high-resolution algorithm based on Keren registration and template convolution is implemented and used for the resolution reconstruction of static scene image sequences. In line with Table 1, this algorithm can be employed to estimate pixel points with a difference of 0.0001. In other words, superiorities of the registration based on the Keren algorithm possess characteristics such as high speed, high precision, etc. (Fortin et al, 1996). After completing registration based on the Keren algorithm, Structure-Adaptive Normalized Convolution (SANC) is used to integrate images together. Pictures with four frames (m = 4) and two amplification factors (q = 2) are selected for irregular interpolation by virtue of twice-normalized convolutions. In this manner, high resolution reconstruction can be implemented. Results after high resolution reconstruction of the above-mentioned two algorithms are shown in Figure 14.

Table 1. Keren algorithm registration results.

	Rotations	Shifts x	Shifts y
LR_1	0	0	0
LR_2	0.0003	–0.2682	0.0470
LR_3	–0.0004	–0.2439	0.3153
LR_4	0.0030	0.0529	0.2622

Figure 14. Partial results of the high resolution image.

4 CONCLUSIONS

An optimal single lens imaging system is constructed with wave aberration correction by using the liquid crystal spatial light modulator and the high-resolution by using a micro-scanning optical wedge. Firstly, a high precision digital interferometer is utilized to measure the aberration of the single lens together with the acquisition of wavefront data of out-of-focus Zernike Polynomials by means of Zemax simulation. And then, Matlab software is adopted to accurately portray a conjugate gray-scale map that is loaded onto the liquid crystal spatial light modulator later. Thus, wave aberration of the optical system can be corrected depending on the linear phase modulation characteristics of the modulator. In addition, effect improvement generated by the employment of the liquid crystal spatial light modulator is tested by Imatest. Moreover, 2 × 2 micro-scanning imaging is made use of to perform an oversampling of multiple times and an identical scene. With respect to the micro-scanning imaging, images obtained through each sampling are recorded in the first place; and then, images achieved by each subsampling are recombined otherwise according to their sequences of micro-scanning. These are processed into high-resolution images by sub-pixels. This method is featured with high precision, good adaptability, etc. As the liquid crystal technology progresses, the cost of a modulator in the future can be further brought down and it will have a smaller volume. Therefore, a preferable solution will be provided for obtaining high-resolution images by simplifying the optical system.

REFERENCES

Born M., Wolf E (1978) *Principles of Optics* Beijing: Science Press, P610.

Cai D.M., Ling N., Jiang W.H. (2007) *SPIE* 6457 64570P-1.

Cai Dongmei, Ling Ning, & Jiang Wenhan (2008). Analysis on Zernike Aberration Performance Fitting based on Pure Phase Liquid Crystal Spatial Light Modulator. *Optical Technique*, 57(2), 897–903.

Chen Jiaxin (2014). A Study on Mobile Terminal Camera Performance Evaluation System based on Imatest. Shanghai Jiaotong University.

Jame C. Wyant, & Katherine Creath (1992). Basic Wavefront Aberration Theory for Optical Metrology. *New York, Academic Press*, 28–35.

Jean Fortin, & Paul Chevrere (1996). Realization of a fast microscanning device for infrared focal plane arrays. *SPIE*. 1996, 2743, 185–196.

Kim, H.S., W.K. Yu, Y.C. Park, et al. Compact MWIR camera with × 20 zoom optics. *SPIE*. 2000, 4369, 673–679.

Love, G.D. (1997), Wavefront correction and production of Zernike modes with a liquid crystal SLM, *Appl. Optics* 36 (I March 1997).

Nam J, Rubin S J (2005) *J.Opt.Soc.Am.A* 22 1709.

Wang Zhihua, & Yu Xin (2008). A Study on Phase Modulation Measurement and Wavefront Correction of Liquid Crystal Spatial Light Modulator. *Acta Physica Sinica*, 02, 196–199+201.

Zhang Haitao, & Zhao Dazun (1999). A Study on Mathematical Principle and Implementation of Photoelectronic Imaging System Frequency Spectrum Aliasing Reduction by Microscanning. *Acta Optica Sinica*. 19(9), 1263–1268.

Zhang Hongxin, Zhang Jian, Qiao Yujin, Si Junshan, & Ma Wei (2013). A Study on Wavefront Stimulation and Error Compensation of Liquid Crystal Spatial Light Modulator. *Journal of Optoelectronics Laser*, 05, 838–842.

Civil, Architecture and Environmental Engineering – Kao & Sung (Eds)
© 2017 Taylor & Francis Group, ISBN 978-1-138-02985-9

A geometry-based path optimization approach for motion control of the NC grinding machine

Fan Zhang, Jie Gong, Yijun Liu, Wen Xia, Maofang Huang & Yunfei Fu
Agricultural Product Processing Research Institute at Chinese Academy of Tropical Agricultural Sciences, Chinese Agricultural Ministry Key Laboratory of Tropical Crop Products Processing, Zhanjiang, P.R. China

ABSTRACT: A Geometry-Based Path Optimization Approach (GBPOA) is proposed to solve the motion control of the NC grinding machine which contains four processing procedures. The GBPOA firstly analyzes the structure of the DXF graphical file and develops an Analyzing Algorithm to obtain the data from the DXF file based on Java; and then a Zooming Algorithm is proposed to solve the motion path of the tool of the NC grinding machine according to the value of grinding; next a Path Compensation Algorithm is proposed to solve the path offset problem caused by tool wear in the process of extract grinding and polishing; finally, the NC codes of four processing procedures are generated automatically. A CNC grinding system proves the feasibility and effectiveness of the above-mentioned method, which can improve the machining accuracy of parts and company production efficiency.

1 INTRODUCTION

A grinding machine, which is also called grinder, is used for processing components by using four procedures of kibbling, fining grinding, extract grinding and polishing. A grinder can be widely used in the areas of glass, stone, metal components, ceramic tile, and so on. It is suitable for grinding the inclined plane of the metal strip with different sizes and thicknesses, such as round edge, straight edge, and duck mouth-shaped edge type. DXF (Drawing Exchange Format) is developed by Autodesk for enabling data interoperability between AutoCAD and other programs. By reading and identifying the DXF file of the workpiece, the machine can generate the track of processing automatically.

It is a typical engineering problem that changes the data of DXF files to the NC code, which has appealed to many researchers. D Sreeramulu presented an automated feature recognition system, which was intended to extract the geometric information of rotational parts from the DXF file, and utilized this information to recognize the turning features (D Sreeramulu, 2011). Chen Shumin analyzed DXF files and extracted the coordinate data, and then generated NC codes for the cutting path of layout results (2012, Chen Shumin). Ying Bai studied a motion control system by obtaining the graphic data from the DXF file, and then generated the processing data and sent the data to the designed microcontroller to drive a stepper motor (Ying Bai, 2013). Huibin Yang proposed a graphic information extraction method to transform the graphic information identified from the DXF file into the bottom motion controller's codes, which can drive the engraving machine (Huibin Yang, 2015). Gabriel Mansour proposed an expert CNC system, which analyzed the DXF file and optimized the machine path process by deciding the optimum path (Gabriel Mansour, 2013). Kovacic constructed an autonomous and intelligent CAD/CAM programming system for the cutting device controller based on an evolutionary genetic algorithm (Kovacic, 2005). In short, studying the DXF file is advantageous to the second development of drawing software. The above-mentioned studies have been used to automate the tool path of the CNC (Computerized Numerical Control) machine and to improve the processing efficiency. However, it is not easy and flexible for the grinding machine to change the processing path spontaneously when the feeding amount is adjusted. Furthermore, in the procedures such as extract grinding and polishing, the tool abrasion may lead to deformation of the part. And so, it is necessary to develop an open interface for the grinding machine, which can generate all the NC codes for four processing procedures.

A Geometry-Based Path Optimization Approach (GBPOA) is devised to solve the motion control of the NC grinding machine. The GBPOA firstly discusses the structure of the DXF and develop a graphical recognition interface based on Java, which can read the graphical information, and can translate the graphical data into coordinate or line information (Section 2); Nextly, a zooming algorithm is proposed to solve the motion path of the tool of the NC grinding machine according to the value of grinding, and presents a path compensation algorithm to solve the

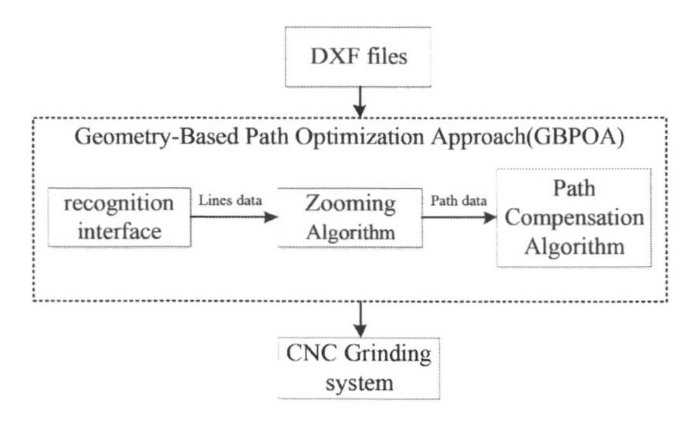

Figure 1. The framework of GBPOA.

path offset problem caused by tool wear in the process of fine grinding and polishing (Section 3). Last but not least, a CNC grinding system, which contains the NC code of four procedures, is totally developed to work directly with the DXF file (Section 4). The framework of the GBPOA is shown in Fig. 1.

2 ANALYZING ALGORITHM FOR THE DXF FILE

As a CAD data file format, DXF has the function of data interoperability between AutoCAD and other programs. The complete structure of DXF includes seven segments called SECTION, which contains HEADER, CLASSES, TABLES, BLOCKS, ENTITIES, OBJECTS, and EOF. Each section has its corresponding function, while the ENTITIES covers all the information of the drawing entities, including the name of each primitive layer, the name of each line, the basic geometry data of each line, and so on. And so, by analyzing the ENTITIES from DXF file, we could obtain the graphical information effectively.

The ENTITIES section is made up of group codes and associated values, which contains all the graphic element objects including LINE, LWPOLYLINE, CIRCLE, ARC, ELLIPSE, and SPLINE. The graphic element objects cover all the geometry data for drawing lines. For example, the LINE has the group code of 10, 20, and 30 corresponding to x, y, and z of the starting point, while the group code of 11, 21, and 31 corresponding to x, y, and z of the terminal point. The other ENTITIES have the same format as that of a group code corresponding to a value.

According to the characteristic of the DXF file, an analyzing algorithm is proposed to extract the graphic data based on Java. In order to store the necessary information, a parametric variable named *LineList,* whose object type is ArrayList, is used with all the information of each feature such as the name, the start and end point, the central point of the circle, the radius, and so on. When a graphic element object is analyzed, the result will be added to the end of *LineList*. The implementation of the analyzing algorithm can be described as follows:

Analyzing(*fileList*, *lineString*)
Parameters:
fileList: all string information of DXF file
lineString: string of each row
1) ***for*** each *lineString* of itemList ***do***
2) if *lineString* = 'ENTITIES'
3) *lineString*=fileList.ReadLine( )
4) obtain the data from *lineString*
5) add the data to *lineList*
6) ***end for***
7) ***return*** *lineList*

After analyzing all the primitive data, we sort the data, so that these can be read and converted more efficiently. And in order to correspond with NC code, in this paper, a set of data storage rules is also developed, which can be described as follows: firstly, 'G01' is taken as the storage symbol of LINE, and then the starting point and end point follow alongwith. Secondly, 'G02' or 'G03' is taken as the storage symbol of CIRCLE or ARC, and among these 'G02' represents the clockwise arc, while the other represents the counterclockwise arc, and then the central point, the start angle, and the end angle follow alongwith. Particularly, it is difficult for the NC grinding machine to process the ELLIPSE and SPLINE line, and so in the range of acceptable accuracy of the project, the ELLIPSE and SPLINE line are discretized into a multi-segment line. And so, we use 'GPL' to represent the LWPOLYLINE and other multi-segment lines, which is integrated with lots of points. From then, the type of *lineList* contains two parts content, one of which is the code of the line whose index is 0, and the other is the data of line. Table 1 shows the storage rules of analyzed DXF graphical data.

Table 1. The storage rules of analyzed DXF graphical data.

Code	Type of ENTITIES	Data character	Index
G01	LINE	Start point (x, y)	(1,2)
		End point (x, y)	(3,4)
G02 or G03	ARC and CIRCLE	Central point (x, y)	(1,2)
		radius	3
		start angle and end angle	4,5
GPL	LWPOLYLINE, ELLIPSE and SPLINE	The coordinate of the *i*-th point (x, y)	(2*i*-1,2i) (*i* = 1,2,...)

3 ZOOMING ALGORITHM AND PATH COMPENSATION ALGORITHM

Traditionally, the final size and shape of a component are always given so that we can get the processing path easily. However, several questions are ignored such as the irregular shape of the blank workpiece, the orientation of the workpiece on the grinding machine, and the abrasion of the grinding wheel. As for this, during the practical manufacture process, the DXF file of the blank component and the grinding amount would be designated, where the workpiece is processed with a principle of using a small grinding amount one time but many times. It means that the processing path will be zoomed in or zoomed out several times. Therefore, after setting the grinding amount, a zooming algorithm is proposed to compute the processing path as follows:

Zooming(*lineList, gAmount, inOrOut*)
Parameters:
lineList=**Analyzing**(*fileList,lineString,lineList*)
gAmount: the grinding amount of single processing
inOrOut: specified external grinding or internal grinding
pathList: aggregation of each processing path
1) ***for*** each item *line* of lineList ***do***
2) calculate the two paralled lines *pLine* of line according to the *gAmount*
3) *isIn* = whether *pLine* is interior or exterior of the figure
4) ***if*** *isIn* equals to *inOrOut*
5) *pathList*.add(*pLine*)
6) ***end for***
7) ***for*** each item *pLine* of *pathList* ***do***
8) *nextPLine* = the next item of *pLine* in the *pathlist*
9) *node* = compute the node of the *pLine* and the *nextPLine*
10) update the data of each *pLine* of *pathList* according to the *node*
11) ***end for***
12) ***return*** *pathList*

After calculating the processing path according to the zooming algorithm, this result can be used to generate the NC code when grinding the workpiece in the procedures of kibbling and fining grinding. However, tool wear must be considered when grinding the component in the procedures of extract grinding and polishing because of the requirement of grinding accuracy. The orientation offset of the processing workpiece caused by tool wear is shown in Fig. 2. Therefore, in order to

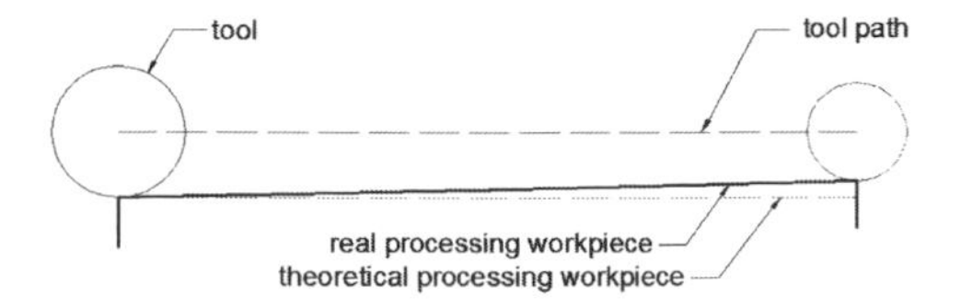

Figure 2. Schematic of the orientation offset of the processing workpiece.

avoid the distortion of the processing workpiece, a path compensation algorithm is proposed to compute the compensated path. The implemented course of the algorithm can be revealed as follows:

PathCompensation(*pathList, wearRate*)
Parameters:
pathList = Zooming(lineList,gAmount,inOrOut)
wearRate: the wear rate of grinding tool for fixed-distance *fd*
cPathList: the path after tool compensation
cPointList: all the point for Arc compensation, whose size is Arc' scircumference in *fd*
1) ***for*** each item *path* of *pathList* ***do***
2) *if path* is Line or Lwpolyline
3) calculate the compensating line *cPath* according to *wearRate*
4) ***if*** *path* is Arc
5) get all *cPointList* of *path*
6) ***for*** each step *i* of *cPointList'* s size ***do***
7) cp_i = *cPcPointLis*.get(*i*), cp_{i+1} = *cPcPointLis*.get(*i*+*1*)
8) calculate the compensation point cp_i' of cp_i and cp_{i+1}' of cp_{i+1} according to *wearRate*
9) calculate the compensation *cPath* whose start point is cp_i', end point is cp_{i+1}'
10) set radius that is the same of *path*
11) *cPathList*.add(*cPath*)
12) ***end for***
13) ***return*** *cPointList*

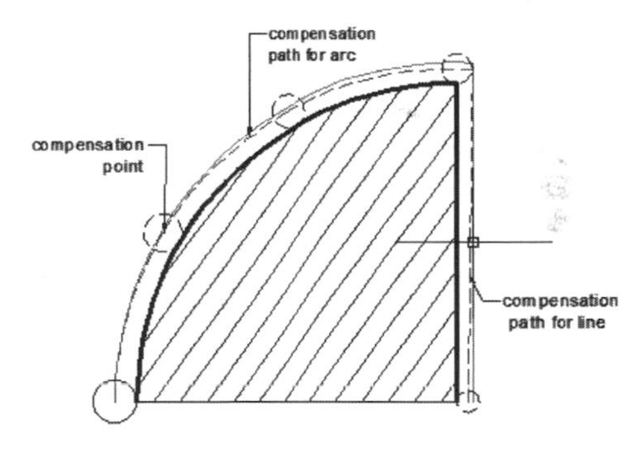

Figure 3. Schematic of the case for the path compensation algorithm.

Line 3 calculates the offset slope of the processing line, so as to obtain the compensating line for the line segment. For the arc segment, a subsection compensation method is presented in line 4–line 10. Firstly, we define some parameters of the arc segment such as the radius r, the central point P_c, the start point P_0, the end point P_{end}, and the wear rate wr. Secondly the compensation points P_i $(i = 1,2,..n)$ of the arc segment according to the wear rate, thereupon the arc can be described as $(P_0, P_1, P_2, \ldots, P_i, \ldots, P_n, A_{end})$. When the tool goes to P_i, the wear of the tool became $i * wc$, and then P_i will be dealt with the point P_i' that is the intersection between the circle, whose radius is $(r - i * wc)$ and central point is the P_c, and the line segment whose start point is P_i and end point is P_c.

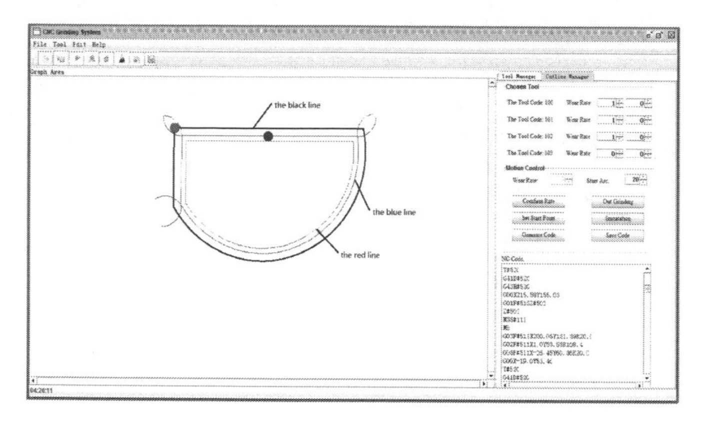

Figure 4. Screenshot with schematic of the CNC grinding system.

```
NC Code:
N400
M191
G28G91Z0
G90G54
G00X2.0 Y53.25
G01F#141X2.83
Y130.39
G03F#511X-16.95Y150.6R20.0
G00X2.83Y150.39
T#530
G41D#530
G43H#530
G00X-17.17Y150.39
G01F#513Z#502
```

Figure 5. Part of the NC code of the presented graph.

And then, the curve of (P_{i-1}, P_i) will become the curve of (P_{i-1}', P_i). In order to fit the rule of the grinder that only line and arc segments can be processed, the curve of (P_{i-1}', P_i) will be dealt with the arc whose radius is r. Therefore, the compensating path of the arc segment is going to be (Arc (1), Arc (2), …, Arc (end). A case in point is shown in Fig. 3.

4 THE CNC GRINDING SYSTEM

Based on the Geometry-Based Path Optimization Approach (GBPOA), a CNC grinding system, which can generate the NC codes of the four procedures automatically, is totally developed to work directly with the DXF file. The CNC grinding system developed by Java is shown in Fig. 4.

From Fig. 4, it can be observed that the graph area displays the processing of the part. The black line presents the blank part, the red line shows the shape of the product and the blue line shows the tool path of the grinder. The cutline manger pane shows the operation mode. After setting all the parameters, the NC code will be generated in the NC code area.

Fig. 5 shows one part of the NC code of the presented graph for the grinding machine.

5 CONCLUSION

In this paper, a novel approach, such as the GBPOA, is presented to solve the motion control of the NC grinding machine. The three main steps of this approach are as follows: analyzing the DXF graphical file, performing tool path calculation, implementing path compensation; these are critical to the GBPOA. The CNC grinding system based on the GBPOA is developed to handle automatically the processing path of the parts for four procedures, and to generate the NC codes for the NC grinder. With more intelligence, the system can be quite practically used.

Future work will be focused on the path optimization for more parts and more constraints.

REFERENCES

Chen Shumin & Liu Qiang (2012). DXF applications to layout cutting. *J.* Computer Applications and Software, 29(5), 143–187.

Gabriel Mansour, Apostolos tsagaris (2013). CNC Machining Optimization by genetic algorithms using CAD Based system. *J.* International Journal of Modern Manufacturing Technologies. 5:75–80.

Huibin Yang (2015). DXF File Identification with C# for CNC Engraving Machine System. *J.* Intelligent Control and Automation, 6, 20–28.

Kovacic M., Brezocnik M. (2003). Evolutionary programming of a CNC cutting machine. *J.* The International Journal of Advanced Manufacturing Technology. 22(1):118–124.

Sreeramulu, D CSP R (2011). A new methodology for recognizing features in rotational parts using STEP data exchange standard. *J.* International Journal of Engineering, Science and Technology, 3(6), 102–115.

Ying Bai (2013). Design of Motion Control System Based on DXF Graphic File. *J.*, Information Technology Journal, 12(15), 3096–3102.

Civil, Architecture and Environmental Engineering – Kao & Sung (Eds)
© 2017 Taylor & Francis Group, ISBN 978-1-138-02985-9

A study on carbon electrode-based photoelectrochemical-type self-powered high-performance UV detectors

Qi Meng, Xiao Fan, E'qin Wu & Liang Chu
Advanced Energy Technology Center and School of Science, Nanjing University of Posts and Telecommunications (NUPT), Nanjing, P.R. China

ABSTRACT: The carbon nanoparticle film was applied as the counter electrode for photoelectrochemical-type UV detectors for the first time, in which the ultraviolet light-response layer was the TiO_2 nanoparticle film. Under zero bias conditions, the UV detector was sensitivity under pulse UV light, and showed rectangular square-wave signals. The response time and recovery time were 0.06 s and 0.17 s, respectively. The carbon electrode-based self-powered UV detector has potential commercial application because of low cost and high sensitivity.

1 INTRODUCTION

The UltraViolet (UV) detection technology has attracted wide attention due to its applications in food disinfection, binary switch, flame detection, optical communications, switch of UV lights, *etc* (Z. H. Lin, 2014 & H. Zhou, 2015). Nowadays, different types of UV detectors have been achieved, such as Schottky contact (H. Chen, 2012), P–N junction (Y. Q. Bie, 2011), *etc*. However, most of those have complicated preparation processes and low sensitivity. Particularly, they need external power as a force field to separate the excited electrode–hole pairs, which waste energy.

Z. L. Wang has proposed the "self-powdered" concept, which means the devices or systems are wireless devices to be self-powered without using battery (Z. L. Wang, 2006 & 2012). Photoelectrochemical-type UV detectors can work under zero bias, which are self-powered ones. After the devices absorb UV light, the electron–hole pairs are excited, and separated by the redox reaction to achieve detection function without external energy. For now, the counter electrodes of the photoelectrochemical-type UV detectors are mainly Pt and Au (S. M. Hatch, 2013 & T. Dixit, 2016 & S. Lu, 2014), which are noble metals and expensive (W.-J. Lee, 2011). It is well-known that carbon materials have numerous advantages, such as cheap, broad application prospects and high stability. In addition, some members in our group previously studied that carbon electrodes were used commendably for dye-sensitized solar cells (W. Ahmad, 2015).

In this work, we developed a photoelectrochemical-type self-powered UV detector, which was based on carbon electrodes for the first time. The UV sensitive layer was the TiO_2 film (Z. Song, 2016 & Z. Liu, 2015 & X. Li, 2012). Under zero bias voltage conditions, the detector showed high sensitivity and stable signals. The response time and recovery time were 0.06 s and 0.17 s, respectively, and exhibited a rectangular square wave signal in pulsed UV irradiation. Due to the advantages of low cost, signal stability, high sensitivity, and self-powered character, the as-prepared UV detector has great potential in business applications.

2 EXPERIMENTAL DETAILS

2.1 *Chemicals and reagents*

Fluorine-doped Tin Oxide (FTO, ~7 Ω/2) glasses were purchased from Nippin Sheet Glass Co., Ltd. P25 (TiO_2, 21 nm) and carbon nanoparticles (25 nm) were obtained from Degussa. Ethyl cellulose (200CPS), TiCl4 (99.0%), terpineol, acetic acid (99.0%), pethanol (99.7%), and acetone (99.5%) were purchased from Sinopharm Chemical Reagent Co. Ltd. Guanidinium thiocyanate (GuSCN, 99.0%) was acquired from Amresco. Lithium iodide (LiI, 99.999%), iodine (I2, 99.99%), 1-methyl-3-propylimidazolium iodide (PMII, 98%), 4-Tert-Butylpyridine (4-TBP, 96%) and tert-butyl alcohol (99.5%) were obtained from Aladdin. Acetonitrile (99.8%) and valeronitrile (99%) were purchased from Alfa Aesar.

2.2 *Fabrication of TiO_2 and carbon electrodes*

The cut FTO glasses were washed with detergent under running water, and sonicated in deionized water, acetone, and ethanol for 20 min, respectively. After dried, the well-cleaned FTO glasses were soaked in 0.04 M TiCl4 solution at 70°C for 30 min to form a compact TiO_2 layer (L. Chu, 2015 & J.S. Zhong, 2015). P25 powder (1 g) or carbon powder (1 g) was mixed with 0.2 g acetic acid, 0.5 g ethyl cellulose, 3 g terpineol and some ethanol under continuous stirring by using a magnetic stirrer. And then, the mixture was grinded for about 20 min, and finally ultrasonically treated for about 10 min to make the expected pastes. Subsequently, the TiO_2 paste and carbon paste were printed onto the $TiCl_4$-treated and clean FTO glasses, respectively. And then, the TiO_2 layer was heated at 125°C for 15 min, at 325°C for 5 min, at 375°C for 5 min, and at 450°C for 30 min, and the carbon layer was heated at 125°C for 15 min and at 325°C for 30 min.

2.3 *Fabrication of the UV detector*

The electrolyte was comprised of 0.6 M PMII, 0.05 M LiI, 0.03 M I_2, 0.1 M GuSCN, and 0.5 M 4-TBP in the solution of acetonitrile and valeronitrile (V: V = 85: 15). For UV detectors, the carbon electrode was buckled on the TiO_2 photoelectrode, separated, and sealed by using a 50 μm plastic sheet, where the internal space was filled with above-mentioned liquid electrolyte.

2.4 *Characterization*

The surface morphologies of the TiO_2 and carbon electrodes were characterized by using Scanning Electron Microscopy (SEM, FEI NOVA NanoSEM 450). The phase identification was examined by using an X-Ray Diffractometer (XRD) (Bruker D8 Advance X-ray diffractometer). The ultra-high pressure mercury lamp (Model CHF-XM500, Beijing Changtuo Technology Co., Ltd.) provided 365 nm UV light (7 mW/cm^2). An electrochemical workstation (AUT84315, Netherlands) was used to obtain the current–voltage curve and current response under pulsed UV irradiation.

3 RESULTS AND DISCUSSION

Fig. 1a and b show SEM images of TiO_2 and carbon film electrodes. The results show the typical mesoporous films with uniform space and without agglomeration of nanoparticles. The nanoparticles propose an enormous specific surface area for sufficient contact with the electrolyte and the uniform space is beneficial for the injecting of the electrolyte. Thus, the TiO_2 and carbon nanoparticle mesoporous film could be used as an efficient UV sensitive layer and counter electrode, respectively. The XRD patterns of TiO_2 and carbon electrodes in Fig. 1c and d reveal that the solvents and thickening agent in the pastes disappeared completely during annealing. P25 was mixed phases of anatase and rutile, carbon was the amorphous phase, and the phase of FTO substrates has the peaks of S(110), (101), (200), and (211) in the patterns.

Fig. 2 is the curve of current versus voltage under dark and UV irradiation conditions. When the device was irradiated by using UV light, current was significantly enhanced. More importantly, at zero bias voltage, the device has a significant current with UV irradiation. The results indicate that the UV detector is a self-powered device, which can work without external energy.

In fact, this UV detector is a photoelectrochemical type, whose schematics and work principle are shown in Fig. 3a and b. Firstly, electron–hole pairs were generated after absorbing ultraviolet light of

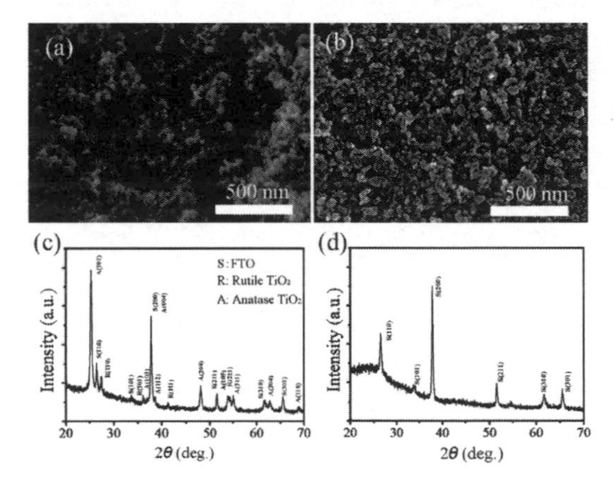

Figure 1. SEM images of **a**: TiO2 photoelectrodes and **b**: carbon electrodes. Representative XRD patterns of **c**: TiO2 photoelectrodes and **d**: carbon electrodes.

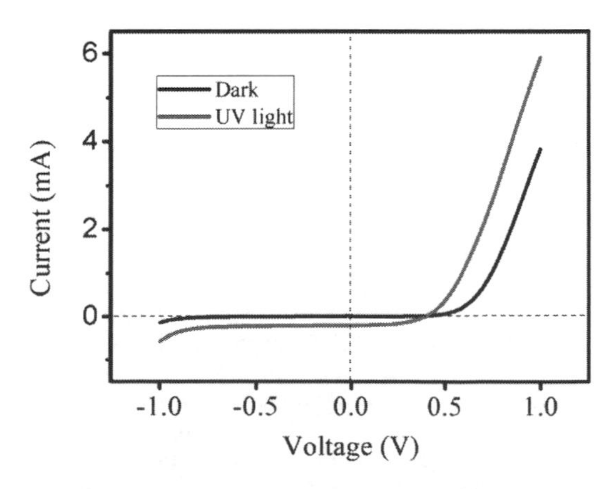

Figure 2. Current–voltage curves of the detector with or without UV illumination.

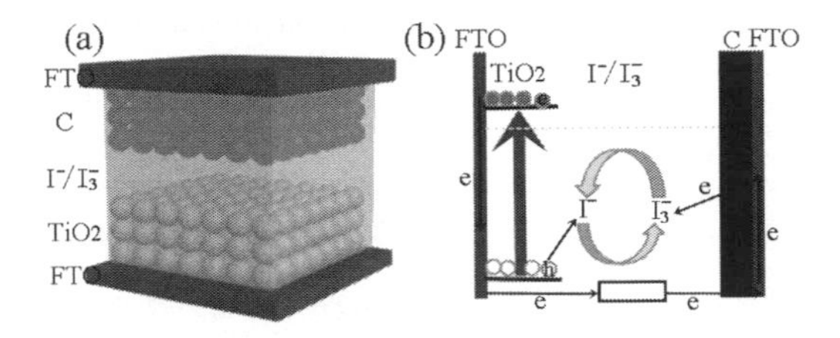

Figure 3. (a) Schematic illustration of the self-powered UV detector; and (b) work principles of the self-powered UV detector.

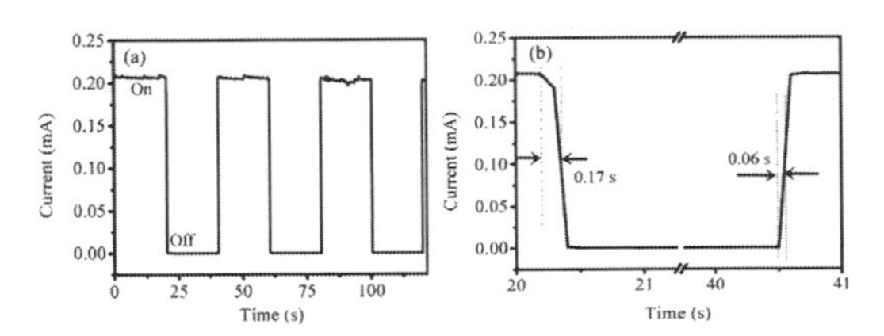

Figure 4. (a) The response and recovery currents under pulse UV illustration and zero bias; (b) the response and recovery times.

TiO_2 nanoparticles; and then, the electrons flow to conductive substrate through the TiO_2 layer, and the holes flow to the electrolyte. Meanwhile, the carbon electrode exhibits catalytic activity for electrons. The electrodes and holes carry out oxidation and reduction reactions between redox couples to ensure the continued operation of the self-powered UV detector.

Fig. 4a shows the current response of the UV detector under pulse UV irradiation at zero bias voltage. The device has a rectangular square wave signal with or without UV irradiation. The result indicates the self-powered UV detector with high responsibility and stability. To reveal the response and recovery time more clearly, the enlarged response and recovery currents are shown in Fig. 4b. The response and recovery times are the times needed to reach (1–1/e) and to drop to 1/e of the maximum photocurrent (J.J. Hassan, 2012), respectively, which are 0.06 and 0.17 s. The results indicate that the self-powered UV detector has high sensitivity and stability. As a consequence, the self-powered UV detector has a simple preparation process, low prices, high sensitivity, good stability, and high performance, which is a desired kind of the commercial UV detector.

4 CONCLUSION

In summary, the carbon electrode was first used in photoelectrochemical-type UV detectors, which has advantages of simple preparation process, low cost, high performance, and stability. More importantly, the UV detector is self-powered. In light of the simple fabrication, zero energy consumption, and outstanding UV-sensitive performance, we look forward to great potential applications in various fields, such as flame detection, UV radiation surveillance, binary switch, optical communications, and switch of UV light.

ACKNOWLEDGMENTS

The authors acknowledge the Natural Science Foundation of Jiangsu Province, China (BK20150860), the Seed Project Funded by Introducing Talent of NJUPT, China (NY215022), and the Science and Technology Innovation Training Program (STITP) of NJUPT (XYB2015303) for financial support.

REFERENCES

Ahmad, W., L. Chu, M.R. Al-Bahrani, X. Ren, J. Su, Y. Gao 2015. P-type NiO nanoparticles enhanced acetylene black as efficient counter electrode for dye-sensitized solar cells Mater. Res. Bull. 67, 185–190.

Ahmad, W., L. Chu, M.R. Al-bahrani, Z. Yang, S. Wang, L. Li, Y. Gao 2015. Formation of short three dimensional porous assemblies of super hydrophobic acetylene black intertwined by copper oxide nanorods for a robust counter electrode of DSSCs RSC Adv. 5, 35635–35642.

Bie, Y.Q., Z.M. Liao, H.Z. Zhang, G.R. Li, Y. Ye, Y.B. Zhou, J. Xu, Z.X. Qin, L. Dai, D.P. Yu 2011. Self-powered, ultrafast, visible-blind UV detection and optical logical operation based on ZnO/GaN nanoscale p-n junctions Adv. Mater. 23, 649–653.

Chen, H., L. Hu, X. Fang, L. Wu 2012. General fabrication of monolayer SnO2 nanonets for high-performance ultraviolet photodetectors Adv. Funct. Mater. 22, 1229–1235.

Chu, L., Z. Qin, J. Yang, X.a. Li 2015. Anatase TiO2 Nanoparticles with exposed {001} facets for efficient dye-sensitized solar cells Sci. Rep. 5, 12143.

Dixit, T., I.A. Palani, V. Singh 2016. Hot holes behind the improvement in ultraviolet photoresponse of Au coated ZnO nanorods Mater. Lett. 181, 183–186.

Hassan, J.J., M.A. Mahdi, S.J. Kasim, N.M. Ahmed, H. Abu Hassan, Z. Hassan 2012. High sensitivity and fast response and recovery times in a ZnO nanorod array/p-Si self-powered ultraviolet detector Appl. Phys. Lett. 101, 261108.

Hatch, S.M., J. Briscoe, S. Dunn 2013. A self-powered ZnO-nanorod/CuSCN UV photodetector exhibiting rapid response Adv. Mater. 25, 867–871.

Lee, W.-J., M.-H. Hon 2011. An ultraviolet photo-detector based on TiO2/water solid-liquid heterojunction App. Phys. Lett. 99, 251102.

Li, X., C. Gao, H. Duan, B. Lu, X. Pan, E. Xie 2012. Nanocrystalline TiO2 film based photoelectrochemical cell as self-powered UV-photodetector Nano. Energy. 1, 640–645.

Lin, Z.H., G. Cheng, Y. Yang, Y.S. Zhou, S. Lee, Z.L. Wang 2014. Triboelectric nanogenerator as an active UV photodetector Adv. Funct. Mater. 24, 2810–2816.

Liu, Z., F. Li, S. Li, C. Hu, W. Wang, F. Wang, F. Lin, H. Wang 2015. Fabrication of UV Photodetector on TiO2/Diamond Film Sci. Rep. 5, 14420.

Lu, S., J. Qi, S. Liu, Z. Zhang, Z. Wang, P. Lin, Q. Liao, Q. Liang, Y. Zhang 2014. Piezotronic interface engineering on ZnO/Au-based schottky junction for enhanced photoresponse of a flexible self-powered UV detector ACS Appl. Mater. Inter. 6, 14116–14122.

Song, Z., H. Zhou, P. Tao, B. Wang, J. Mei, H. Wang, S. Wen, Z. Song, G. Fang 2016. The synthesis of TiO2 nanoflowers and their application in electron field emission and self-powered ultraviolet photodetector Mater. Lett. 180, 179–183.

Wang, Z.L., J.H. Song 2006. Piezoelectric nanogenerators based on zinc oxide nanowire arrays Science. 312, 242–246.

Wang, Z.L., W. Wu 2012. Nanotechnology-enabled energy harvesting for self-powered micro-nanosystems Angew. Chem. Int. Edit. 51, 11700–11721.

Zhong, J.S., Q.Y. Wang, X. Zhu, D.Q. Chen, Z.G. Ji 2015. Sovolthermal synthesis of flower-like Cu3BiS3 sensitized TiO2 nanotube arrays for enhancing photoelectrochemical performace Alloy. Compd. 641, 144–147.

Zhou, H., P. Gui, Q. Yu, J. Mei, H. Wang, G. Fang 2015. Self-powered, visible-blind ultraviolet photodetector based on n-ZnO nanorods/i-MgO/p-GaN structure light-emitting diodes J. Mater. Chem. C. 3, 990–994.

Civil, Architecture and Environmental Engineering – Kao & Sung (Eds)
 ISBN 978-1-138-02985-9

A scheme study of CFB power generation by using pet coke

H.T. Liu, L. Xue & D.F. Xu
SEPCO Electric Power Construction Corporation, Jinan, Shandong Province, China

ABSTRACT: High sulfur pet coke power generation through circulating fluidized bed technology is discussed in this paper. When compared with coal, pet coke has the following characteristics such as high carbon content, high sulfur content, high heat value, low ash content, and low volatile matter. CFB power generation technology is considered to be a good method to use high sulfur pet coke. A six times 300 MW sub-critical unit (6 × 300 MW) is proposed in this project.

1 INTRODUCTION

Pet coke is a by-product produced during heavy/residual oil's delayed coking process. Its main component is the element C, accounting for more than 80 wt%. The rest of the elements are H, O, N, and metallic elements. It has characters such as high heat value, high carbon content, low ash content, low volatile matter, etc. With respect to the sulfur content existing in the pet coke, content more than 3% is called high sulfur pet coke (Zhu 2012). High sulfur pet coke is normally used as fuel, and therefore, it is also called fuel grade pet coke (Miao et al. 2014). However, due to the large amount of pollutant emission, such as SO_x and NO_x, the large-scale use of high sulfur pet coke is limited. The ideal way to use high sulfur pet coke should consider both efficiency and cleanliness. IGCC power generation technology and CFB power generation technology are both considered to be good options to use high sulfur pet coke (Ding 2014, Xu et al. 2011, Wang & Leng 2009). When compared with IGCC technology, CFB power generation technology has advantages such as high combustion efficiency, high availability, large load adjustment range, and less corrosion to equipment and nature. There are already some commercial pet coke-fired CFB plants operating in the world.

Saudi Arabia is the world's biggest oil producer in contrast to most of the world's oil reserves. Its pet coke production grows continuously along with the growth of crude oil production. However, since the quality of crude oil declines continuously, the rate of production of high sulfur pet coke grows rapidly. Commissioned by one Saudi Arabia oil company, our company is carrying out research on the CFB power generation scheme by using high sulfur pet coke produced in this oil company's refinery. The aim of this paper is to provide reference for related research.

2 DESIGN INPUT

Pet coke used in this project is adopted from two nearby refineries. Each refinery can produce 6000 tons pet coke per day. Elements analysis and other analyses are shown in Tables 1 and 2.

When compared with coal, pet coke studied in this project has the following characteristics such as high carbon content, high sulfur content, high heat value, low ash content, and low volatile matter.

According to local emission requirements, the plant emission shall satisfy World Bank Targets, which is given in Table 3.

Table 1. Elements' analysis of pet coke (count as per air received base) wt%.

C_{ar}	H_{ar}	O_{ar}	N_{ar}	S_{ar}	M_{ar}	A_{ar}
80.1	3.2	0.1	1.3	6.7	8.5	0.1

Table 2. Others' analysis of pet coke.

Item	Result	Item	Result
Volatile matte	Max. 12%	Density	881 kg/m
High heat value	35631 kJ/kg	HGI	Min. 40
Vanadate	867 ppmv	Vanadate plus Nickel	1775 ppmv

Table 3. Baseline Emission Targets (DA-Degraded Airshed-Region with poor air quality).

Item	Upper limit
Excess O_2 in dry flue gas	3%
Particulate Matter (PM)	30 mg/Nm^3
SO_2	200 mg/Nm^3
NO_2	200 mg/Nm^3

3 PET COKE FIRED CFB POWER GENERATION

3.1 *Process description*

The process of CFB power generation by using pet coke is roughly as follows: pet coke and limestone are crushed and then fed by using a feeding device to a combustion chamber. Pet coke is heated by using cycled high temperature materials and is burned soon. Limestone is used as a desulfurizing agent and supplement as the bed material at the same time. In the combustion chamber, the bed material, with the fuel, is fluidized with primary air which turbulently transports the solids up to the entire height of the combustion chamber. Combustion of the fuel takes place as it rises and heat is transferred to the membrane water-wall tubing that forms the walls of the combustion chamber. The hot combustion gases with the entrained solids are released at the top of the combustion chamber into the cyclone separator. The separator separates the solids from the combustion gases and returns the solids, including any unburned solid fuel, through the non-mechanical loop seal to the combustion chamber where they mix with incoming fresh fuel. The long solids residence time at combustion temperature and the retention and continuous recirculation of the solids ensure high combustion efficiencies for a wide range of fuels and result in an ideal system for the mixture of fine limestone with the fuel for efficient SO_2 retention in solid ash form.

Limestone is allowed to react continuously and therefore, it is necessary to continuously feed limestone with the fuel. The sulfation reaction requires that there is always an excess amount of limestone present. The amount of excess limestone that is required is dependent on a number of factors, such as the amount of sulfur in the fuel, the temperature of the bed, and the physical and chemical characteristics of the limestone. The ideal desulphurization efficiency can reach more than 90%. According to the situation, flue gas desulphurization measures can also be considered to be added onto the tail flue channel to satisfy SO_2 emission requirements.

Provisions are made for primary and secondary air supply to the combustion chamber. The primary air is supplied through the wind box to the fluidizing grid and provides the main fluidization air flow. The secondary air provides a staged combustion effect to ensure high combustion efficiencies and to minimize NO_x production. In addition, the Selective Non-Catalytic Reduction (SNCR) denitration system can also be adopted to meet NO_x emission requirements.

Flue gas and some fine size particulate matter leave the separator and pass through the convection section, which contains superheat, reheat, and economizer banks, along with an air preheater. The flue gas then enters a particle collector, where particulate matter is removed in compliance with environmental regulations. Clean flue gas is discharged to the stack via an induced draft fan.

Feedwater is sent to the CFB and heated to produce steam. Steam with high temperature and pressure is then sent to a steam turbine and obtains the power from the steam turbine generator. Details of the process are shown in Figure 1. Pet coke-fired CFB power generation technology has the following characteristics such as large load adjustment range, high availability, less corrosion to equipment and nature. Power and steam can be released as output at the same time.

3.2 *Research progress*

CFB technology using pet coke as fuel has been developed rapidly since 1980s. Till now, there are already dozens of power plants that are being commercial operated. Usage of pet coke in CFB boilers can be mainly divided into two ways, one is pure burning of pet coke and another is mixed burning pet coke with coal (Dong et al. 2014). At present, the main problems of pure pet coke burning methods are as follows: high carbon content in fly ash, little amount of cycle material, instability of bed temperature, high temperatures of the dense–phase zone, limited boiler load, slag of the bed material in the return feeder and cyclone separator, high

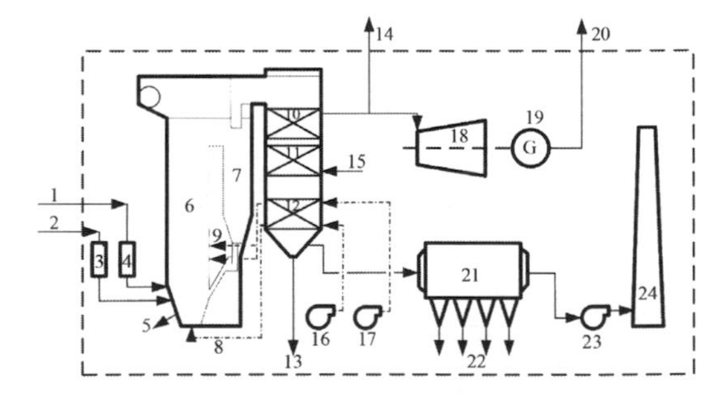

Figure 1. Process diagram of CFB power generation by using pet coke.
1-Pet coke; 2-limestone; 3-limestone feed device; 4-pet coke feed device; 5-bottom ash; 6-CFB combustion chamber; 7-cyclone separator; 8-primary air; 9-secondary air; 10-boiler superheater; 11-boiler economizer; 12-air preheater; 13-ash discharge; 14-steam; 15-feedwater; 16-primary air fan; 17-secondary air fan; 18-steam turbine; 19-steam turbine generator; 20-power; 21-particle collector; 22-fly ash; 23-induced draft fan; 24-stack.

temperature corrosion at the separator center pipe, ash deposited on the heating surface, etc. (Pei et al. 2013, Zheng & Chen 2004). The high carbon content in fly ash can be controlled through measures such as strict control of the material particle size, thereby choosing an appropriate material circulation ratio and adjusting the secondary air volume. Problems, such as little amount of cycle material, instability of bed temperature, high temperature of the dense–phase zone, and limited boiler load, can be dealt with through measures such as improving the separation efficiency of the cyclone separator, adding sand in the bed material, and adjusting the ratio of primary and secondary air. Currently, there are no effective measures to overcome other problems (Pei et al. 2004, Li et al. 2005, Tang & Li 2004, Xiang & Yuan 2004, Liu et al. 2002, Yuan & Liu 2002). Further studies are still needed to proceed with.

4 SCHEME OF THIS PROJECT

Based on the pet coke production of two refineries and the heat value of the pet coke, it is assumed that thermal efficiency of the whole power plant is about 40%; it can be preliminarily evaluated that the scale of this power plant is around 1980 MW. Considering the mature mode of commercial products (steam turbine and CFB boiler), a six times 300 MW unit (6 × 300 MW) is proposed in this project. Related parameters of the steam turbine and CFB boiler are described as follows:

4.1 *Steam turbine*

As provided by the supplier, the information of the steam turbine is as follows: sub-critical, once reheated, four cylinders with one single flow High Pressure turbine (HP), one single flow Intermediate Pressure turbine (IP) and one double flow Low Pressure turbines (LP), with single axial, single back pressure condensing, water cooling condensing mechanisms. Main parameters of the steam turbine are given in Table 4.

4.2 *CFB Boiler*

To match with the steam turbine's capacity, six sub-critical CFB boilers are selected. The Boiler Maximum Continuous Rating (BMCR) is set to be the same as that of the turbine Valve Wide Open (VWO) operating condition and is equal to 1.05 times that of the Turbine Maximum Continuous Rating (TMCR) operating condition. The information of the boiler provided by the supplier is as follows: sub-critical, single steam drum, natural circulation, balanced draft, autoignition, and combustion-supporting. Relevant details are given in Table 5.

4.3 *Other systems*

Other auxiliary systems, such as the combustion system, steam and water system, pet coke handling system, limestone handling system, ash removal system, water supply system, chemical water treatment system, waste water treatment and drainage system, firefighting system, air conditioning and ventilation system, electrical system, instrumentation and control system, environmental protection, etc., are all designed according to related standards and design inputs to balance the plant.

Table 4. Main parameters of the steam turbine (TMCR operating conditions).

Item	Unit	Parameter
Mode		N300-16.67/565/565
Rated output (TMCR + 0%)	MW	300
Maximum output (VWO + 0%)	MW	315
Rated steam pressure (main steam)	MPa	16.67
Rated steam temperature (main steam)	°C	565°C
Rated steam pressure (reheat steam)	MPa	4.16
Rated steam temperature (reheat steam)	°C	565
Rated main steam flow	t/h	925
Maximum steam flow	t/h	975
Rated exhaust pressure	kPa	10
Feed water temperature	°C	273.2
Heat rate	kJ/kWh	8017.7
Rated speed	r/min	3600
Regenerative system		3 HP heaters, 4 LP heaters and 1 deaerator

Table 5 Main parameters of the boiler (BMCR operating condition).

Item	Unit	Parameter
Maximum steam flow	t/h	975
Superheat steam outlet pressure	MPa	16.67
Superheat steam outlet temperature	°C	565
Reheat steam flow	t/h	860.5
Reheat steam inlet pressure	MPa	4.75
Reheat steam outlet pressure	MPa	4.37
Reheat steam inlet temperature	°C	372.6

5 CONCLUSION

High sulfur pet coke is normally used as fuel. When compared with coal, pet coke studied in this project has the following characteristics: high carbon content, high sulfur content, high heat value, low ash content, and low volatile matter. CFB power generation technology is considered to be a good method to use high sulfur pet coke. It has advantages such as high combustion efficiency, high availability, large load adjustment range, and low corrosion to equipment and nature. The scale of this project is evaluated around 1980 MW. A six times 300 MW sub-critical unit (6×300 MW) is proposed. Mode of the steam turbine is N300-16.67/565/565. Six sub-critical CFB boilers, under BMCR operating conditions that are equal to 1.05 times that of TMCR operating conditions, are selected to match with the steam turbine.

REFERENCES

Ding, S.R. 2004. Use research of pet coke. *Lubes & Fuels* 14(5/6): 26–29.

Dong, J., Chen, X.D., Peng, R.Y. & Yang, L.H. 2014. Engineering application of petroleum coke-fired CFB boilers. *Power Equipment* 28(5): 328–331.

Li, J.F., Lv, J.F., Zhang, J.S., Liu, Q., Yang, H.R. & Yue, G.X. 2005. Application of high sulphur content petroleum coke burned in circulating fluidized bed boiler. *Boiler Technology* 36(2): 37–42.

Liu, D.C., Yuan, G.C., Zhang, C.L., Zhang, S.H., Chen H.P. & Lin, G. 2002. Some problems on the design of petroleum coke circulating fluidized bed boilers. *Boiler Technology* 33(6): 20–22, 27.

Miao, C., Yang, W.J. & Wang H. 2014. Situation and trends in petroleum coke supply & demand, by region of the world—Highlights for Argus Asia petroleum coke summit 2014. *International Petroleum Economics* 22(5): 15–20.

Pei, Y.M. & Feng, B.Y. 2013. Research of blending combustion of CFB boiler coal and petroleum coke. *Power System Engineering* 29(4): 10–11.

Tang, B.X. & Li, P.N. 2004. The fundamental researching of the bed temperature of 75t/h CFB boiler where only petroleum coke is combusted. *Petrochemical Safety Technology* 20(3): 30–43.

Wang, J.H. & Leng, B. 2009. Rational utilization of high sulfur petroleum coke and analysis of its pollution control measures. *Environmental Protection Circular Economy* 29(4): 27–29.

Xiang, X.B. & Yuan, G.C. 2004. Design and operation analysis of 75t/h petroleum coke circulating fluidized bed boiler. *Energy Conservation* 23(8): 43–46.

Xu, K.Z., Miao, C. & Song, A.P. 2011. Simply analysis of clean utilization technologies in high sulfur petroleum coke. *Guangzhou Chemical Industry* 39(20): 25–28, 42.

Yuan, G.C. & Liu, D.C. 2002. Application of combustion technology on petroleum coke fired circulating fluidized bed. *Energy Engineering* 2: 45–47.

Zheng, W.J. & Chen J. 2004. Existing problems and resolutions in CFB. *Journal of Sinopec Management* 35(2): 11–14.

Zhu, H.Y. 2012. Analysis on petroleum coke market supply situation in China. *Refining and Chemical Industry* 23(6): 44–46.

Civil, Architecture and Environmental Engineering – Kao & Sung (Eds)
© 2017 Taylor & Francis Group, ISBN 978-1-138-02985-9

Design of the pilot-scale double effect evaporator for high-concentration sodium hydroxide production

Le Gao
Department of Chemical Engineering, University of Minnesota Twin Cities, Minnesota, USA

ABSTRACT: The objective was to use the pilot plant double effect evaporator in order to concentrate 75,000 gallons of aqueous sodium hydroxide from 10 wt% to 40 wt%. The purpose of the experiment was to determine the steady-state operating temperatures and pressures in order to calculate the overall heat transfer coefficients, U, and determine steady-state cooling water rates to the condenser. In order to accomplish the experimental purpose of running the pilot plant at steady-state operation, the accumulation rates of mass and heat were calculated and plotted over time. The closest time step to steady state was at 4500 ± 6 seconds from the start of data collection. The mass accumulation for this data set and the energy accumulation were obtained from the experimental data. The heat transfer coefficient for the first effect, second effect and condenser were compared to each other, which were found to be within the expected literature values. The cooling water rate provided to the condenser was calculated. Using these results, a design was proposed with a feed rate of 0.8 kg/s of the 10 wt% NaOH into the first effect and a cooling water flow rate of 0.004 kg/s. This resulted in an output flow rate of 0.2 kg/s 40 wt% solution, with a steam consumption of 1.35 kg/s and a steam economy of 0.4 kg vapor/kg steam.

1 INTRODUCTION

Analysis of steady-state operation of double effect evaporator and prediction were made on the process of sodium hydroxide solution. 1. Double effect distillation and thermal integration applied to the ethanol production process. The ethanol production process was simulated using Aspen Plus® software. The process included double effect distillation technology which achieved the reduction in the steam consumption. The study assessed the double effect distillation in ethanol production process, and its impact on energy consumption and electricity surplus production in the cogeneration system. The economic assessment was also done based on equipment cost. As a result, the reduction in steam consumption was calculated. Figures: process flow diagrams of conventional and double effect processes steam consumption and cost. 2. Prediction of energetic and exergetic performance of double effect absorption system. The topic was about solar double effect absorption system. The performance of double effect absorption system was analyzed and optimized using a computer program which has been developed by Engineering Equation Solver software to describe the mathematical model of the used absorption system. Also, the study predicted the Coefficient of Performance (COP) and exergetic efficiency of the system, and formulated them in equations as functions of the operating parameters of the cycle. Figures: thermodynamic performance based on temperature of high pressure generator. 3. Design and computer simulation on multi-effect evaporation seawater desalination system using hybrid renewable energy sources in Turkey. The multi-effect evaporation was simulated with Visual Basic programming language based on data obtained by 18 stations for ten years. The comparison of the result with the literature values showed the accordance. Also, it compared the solar and wind energy as the energy source of evaporator. Multiple effect units; energy figure: process flow diagrams of multi-effect; flowchart. 4. Thermodynamic analysis of a trigeneration system consisting of a micro gas turbine and a double effect absorption chiller. A mathematical model of double effect absorption chiller was built using a MATLAB programming language to estimate the thermodynamic performance. The trigeneration system was evaluated at different operating conditions: ambient temperatures, generation temperatures and microturbine fuel mass flow rate. More about microturbine than double effect. 5. Simulation and Control of a Commercial Double Effect Evaporator: Tomato Juice The commercial double effect tomato paste evaporation system was simulated. The open loop simulation showed that the sample evaporator provided stable process. Backward feeding. 6. Dynamic Simulation of a Nonlinear Model of a Double Effect Evaporator. The commercial double effect tomato paste evaporator was simulated using ACSL. Feed flow rate and product solids concentration from the simulation were compared with the experimental measurements. 7. Dynamic Response of a Double Effect Evaporator. A simple mathematical model was built with unsteady state material and energy balances. Agreement achieved between experimental response of the evaporator and the predicted response.

The purpose of the experiments was to determine the steady-state operating temperatures and pressures, in order to calculate the overall heat transfer coefficient of each of the heaters used in the double effect evaporator system. Steady state was measured as having zero mass or energy accumulation. Radiative and convective heat losses were also calculated for use in the energy balance and the design problem.

2 EXPERIMENT AND MODEL

2.1 *Experimental procedure*

2.1.1 *Start up*

Before the experiment, all valves were closed. Blow-down valves of pressure gauges P1-P4 were opened to blow out gases initially. The filling of evaporator started from 1st effect unit. Before filling, LLC was activated by opening valve A-14 to supply compressed air to LLC. After LLC was connected to the 1st effect, the water was fed to 1st heater passing through rotameter R1. The 1st disengaging chamber was filled to steady state operating level. Before filling 2nd effect unit, the water supply to vacuum pump was opened and then the vacuum pump was started to create vacuum in condenser. The vent valve, valve of condensate receiver to drain were closed to ensure that condenser unit was not open to atmosphere. To fill 2nd effect unit, the valve connecting 1st and 2nd effect was opened to feed water from 1st effect to 2nd effect unit. The circulating pump was started when filling 2nd effect, which forced a circulation between heater and disengaging chamber in 2nd effect. Then the cooling water was fed to condenser After the 2nd disengaging chamber was half filled, the steam was fed to the 1st heater. The gauge pressure of steam supply was controlled to be around 170 Kpa (Sharqawy, 2010). After the steam was supplied to the system, the cooling water started to be fed to the condenser. The flow rate of cooling water was controlled to obtain 20°C temperature rise in cooling water. Condensate collected in condensate receiver was drained before it reached the liquid level of 0.70 m. After the operation of DEE system, the valve of steam calorimeter was opened to measure the condition of steam before and after adiabatic expansion. The measurement of calorimeter was taken until the temperature of T8 and T9 were almost equal, which meant the steam constantly passed through the calorimeter and got ready for data collection.

2.1.2 *Attainment of steady state*

To achieve steady state, two important independent variables, liquid level heights of 1st and 2nd disengaging chamber were controlled to reach constant. The liquid level of 1st disengaging chamber was controlled by the liquid level controller, since the sensor automatically closed the feed valve as the liquid level increased. The liquid level of 2nd disengaging chamber was controlled manually through turning the feed valve to 2nd effect. Through the calculation and analysis in section 3.2 and 4, the system took around 4400 s to achieve the steady state of mass and energy. The rate of cooling water was changed around 3:30 PM to obtain 20°C temperature rise of cooling water passing condenser.

2.1.3 *Data collection*

In week 1, the double effect evaporator was filled with proper operation and all procedures, valves, and dimensions of equipment were recorded. In week 2, the filling of the double effect evaporator started at 1:40 PM. Operations such as filling and feeding steam were finished before 3:20 PM. Data collection started at 3:25 PM, and measurements were taken around every 10 min.

Measurements of the experiment included temperature of different streams or positions (T1-T13, TD4, TD5), pressure of different streams or positions (P1-P5), flow rate of different streams (R1-3, steam trap 1–2), and liquid level height of two disengaging chambers. Systematic errors with the equipment were recorded as well.

2.1.4 *Shut-down [13]*

The shut-down followed the procedures in handout [13] to ensure safety. The inlet of steam and water to the whole system was first stopped. Valve (S-11) for steam supply was closed and the remaining steam was vent through calorimeter. Valves for water feed to 1st and 2nd effect units were closed. The vacuum to condenser was opened fully and the condensate rotameter (R3) was bypassed. The condensate in receiver was drained through the vacuum pump. The circulating pump was stopped. The condenser and 2nd effect was vented once the temperature of 1st effect decreased to 80°C (175°F). The vacuum pump was shut off, and the feed water from main and water supply to vacuum pump were stopped by closing valves WM1 and VP-2. All drains and interconnecting valves were opened to drain water remained in the evaporator. Air supply to LLC was stopped by closing valve A-14. Blow-down valves of pressure gauges and drain valve of vacuum pump were opened to drain water.

2.1.5 *Precaution*

Water was used in the experiment instead of sodium hydroxide solution. Basic rules such as safety goggles, long pants, and close-toed shoes were required in the laboratory. Special precautions for Double Effect Evaporator experiment were avoiding contacting hot pipes and hot equipment. Also, to make a safe experiment when operating the equipment, the water supply for vacuum pump was opened. The liquid level of disengaging chamber and condensate receiver was monitored to avoid overfilling. The shutdown procedures in handout was followed carefully after experiment.

2.1.6 *Observation and Improvement*

Liquid levels of two disengaging chambers were observed to be in ±0.06 m, which were roughly determined to be mass steady state in 1st and 2nd effect units. Cooling water flow rate was one of the important independent variables of the experiment. In the experiment, the flow rate of water feed to condenser was so fast at first that the temperature raise only 10°C in cooling water. The cooling water flow rate was decreased after 4000 s of operation (around 4:30 PM) to obtain a temperature rise over 20°C in cooling water, which caused difficulty to achieve steady state. All settings should be controlled before operating and data collecting. The change of conditions during the operating may break the trend and take longer time to reach steady state.

2.2 *Modeling*

2.2.1 *Double effect evaporator*

The function of the double effect evaporator is to concentrate a solution by evaporation of the solvent. In each effect, steam is used to heat the liquid stream, which is separated in a disengaging chamber into vapor and condensate. In a feed-forward system, the vapor from the first effect enters the second effect heater and is used to heat the condensate from the first effect. The second effect is kept at a lower pressure than the first effect by the vacuum pump so that it operates at a lower temperature. To characterize steady state, Equations 1 and 2 were used to describe the mass and heat balances, respectively.

$$\Sigma m_{in} - \Sigma m_{out} = acc \quad (1)$$

$$\Sigma Q_{in} - \Sigma Q_{out} = acc \quad (2)$$

min and Qin represent the mass and heat terms entering the system, m_{out} and Q_{out} represent the mass and heat terms exiting the system, and acc is the accumulation term, which is defined to be zero at steady state. The system was solved with overall mass and heat balances, as well as individual mass and heat balances on each effect.

Using steady-state values for temperature and heat transferred in each effect, the overall heat transfer coefficient describing the effectiveness of the heater can be calculated from equation 3:

$$U_i = Q_i/\Delta T_{m,i}A_i \quad (3)$$

U_i is the overall heat transfer coefficient the heater, Q_i is the heat transferred from the steam in the heater, $\Delta T_{m,i}$ is the mean temperature difference between the steam and the boiling point of the disengaging chamber, and Ai is the total area of heat transfer. The subscript "i" is used to denote that the values are specific to the heater in the i th effect.

Heat in the liquid and vapor streams were calculated with the following equations:

$$Q_{liquid} = m_L * H_L \quad (4)$$

$$Q_{vapor} = m_v * \lambda_V \quad (5)$$

m_L and m_V represent the mass flow rate of the liquid or vapor streams, respectively, H_L represents the enthalpy of the liquid stream, and λ_V represents the latent heat of vaporization of the vapor stream. The enthalpy of a liquid stream containing a solute could be found as a function of concentration and temperature from enthalpy-concentration charts. The assumption for the vapor streams was that they remained at the boiling point of the liquid in the disengaging chamber and exchanged heat by converting between liquid and vapor state. Latent heats of vaporization were found from temperature from the following correlation:

$$\lambda_V = 2.501 * 106 - 2.369 * 103 * T + 2.678 * 10{-}1 * T^2 - 8.103 * 10{-}3 * T^3 - 2.079 * 10{-}5 * T^4 \quad (6)$$

This correlation was found to be valid for pure liquid water within 0.01% for a temperature range between 0–200°C when fit to experimental data [14].

3 RESULTS AND DISCUSSION

3.1 *The measurements of the double effect evaporator*

The definition of steady state is a point at which the accumulation of mass and energy within the system approaches zero. It was important to determine at which point the pilot system began to operate at steady state. The mass accumulation terms for the system are displayed in in Figure 2(a-c). The mass flow rate data from the system was significant to calculate the rate of accumulation of mass in the overall system, the first effect, and the second effect.

To calculate the rate of accumulation of energy in the system, it was necessary to find how much steam first entered the system. This was found through the use of the steam calorimeter. The steam calorimeter data and steam quality value were used to estimate the energy loss during the process. It was also necessary to calculate the amount of energy dissipated out of the system as shown in Figure 3. It was found that convective energy loss was negligible compared to that of radiative energy loss, while the radiative energy loss was significant, due to the significant sources of radiative energy loss determined by the comparisons in different curves of Figure 3.

The energy over time values for liquid and vapor streams associated with the pilot system were calculated using the methods were presented. The accumulation of energy in the system at various times of operation are further characterized in the following chapter.

Using the energy over time values in Figure 3, it was possible to calculate the overall heat transfer coefficients for the heat exchangers in the pilot system. The overall heat transfer coefficients were calculated at the time at which the system was thought to be closest to steady state operation. The values for the overall heat transfer coefficients (as well as the parameters important to calculate them) are: 8200 ± 502 W (m^2 K)$^{-1}$) for the first effect heater, 2950 ± 57 W (m^2 K)$^{-1}$) for second effect heater and 135 ± 6 W (m^2 K)$^{-1}$) for condenser as shown in Table 1.

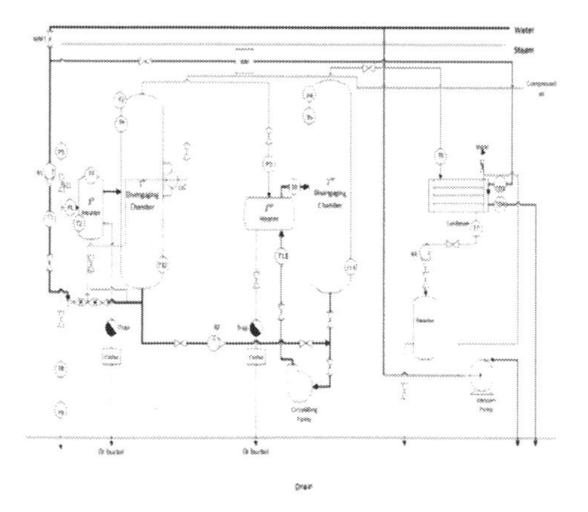

Figure 1. Process flow diagram of the pilot plant double effect evaporator.

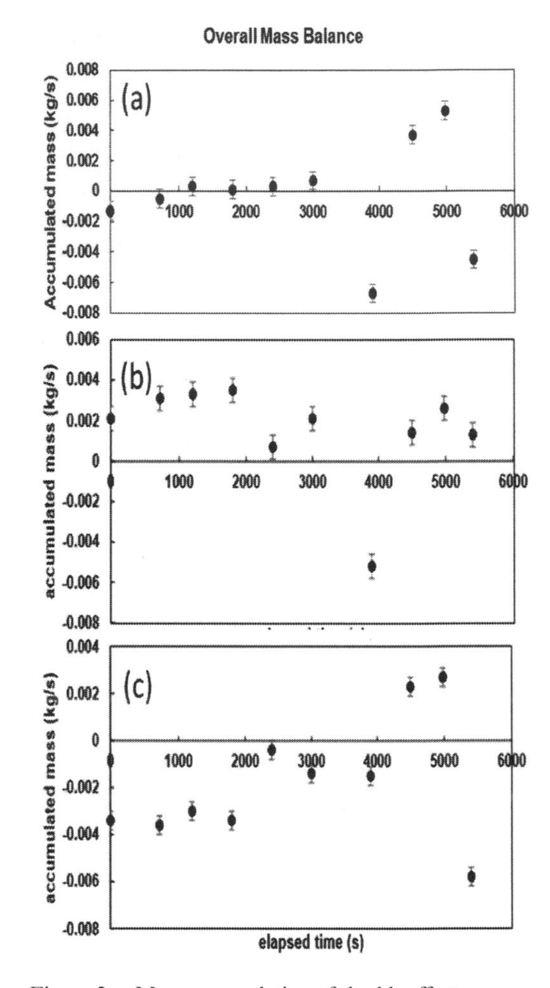

Figure 2. Mass accumulation of double effect evaporator: the accumulation of mass in the system over time for the (a) overall mass balance; (b) first effect accumulated mass and (c) second effect accumulated mass.

The general equation for each energy balance was Equation 4. The energy balance equations were used to monitor the rate of accumulation of energy in the system over time. It was approximated that the steam exiting each effect would be a completely saturated liquid. Because of this, the rate of energy input into the system by the steam. Knowing these values, the rate of accumulation of energy in the system was then calculated using the values rate of energy transferred to and from the system from Figure 3, Trial 8 and radiative energy loss values from Trial 6 because they were taken at similar times. The overall rate of accumulation of energy values and the values of the terms of which it was comprised of are featured in Figure 4 and Table 2.

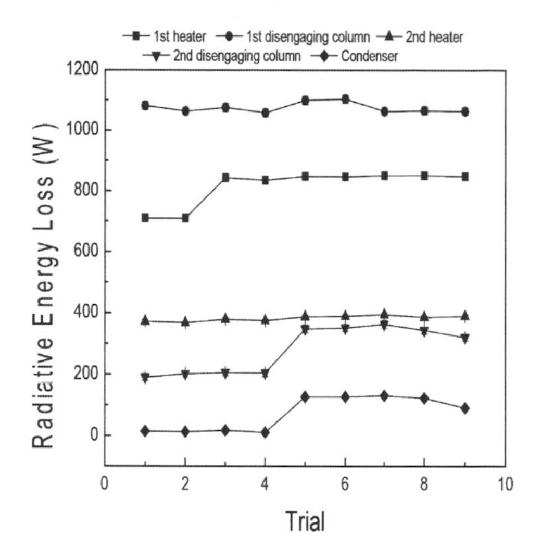

Figure 3. Radiative energy loss data: Contains significant radiative energy loss values from equipment in the pilot system. The method for calculating these values and determining which values were significant is detailed in the model section.

Table 1. Heat transfer coefficient.

1st Effect unit (± 502 W(m^2 K)$^{-1}$)	2nd Effect unit (± 57 W(m^2 K)$^{-1}$)	Condenser (± 6 W(m^2 K)$^{-1}$)
8200	2950	135

3.2 *Design of sodium hydroxide concentration procedure*

The stated objective of the Figure 5 was to use the same pilot plant double effect evaporator system to concentrate 75,000 gallons of aqueous sodium hydroxide solution from 10 wt% to 40 wt%. This could be achieved with only two small modifications to the system (from Figure 1). First, the main water line was disconnected from the feed to the first effect heater and replaced with a feed line connecting to the supply of 10 wt% NaOH solution. The rotameter R1 would be used to set an upper limit on the feed mass flow rate, which is also controlled by the LLC to maintain steady-state in the 1st effect disengaging column. Additionally, a new pipe would need to be attached to the 2nd disengaging chamber to allow the concentrated solution to connect to other units for further processing. This flow rate would be controlled by the valve connected to rotameter R4. Start-up of the system could be done in a similar manner to the experimental procedure, described in experimental section. The first disengaging chamber would be allowed to fill to the steady-state level controlled by the level controller. Next, the second disengaging

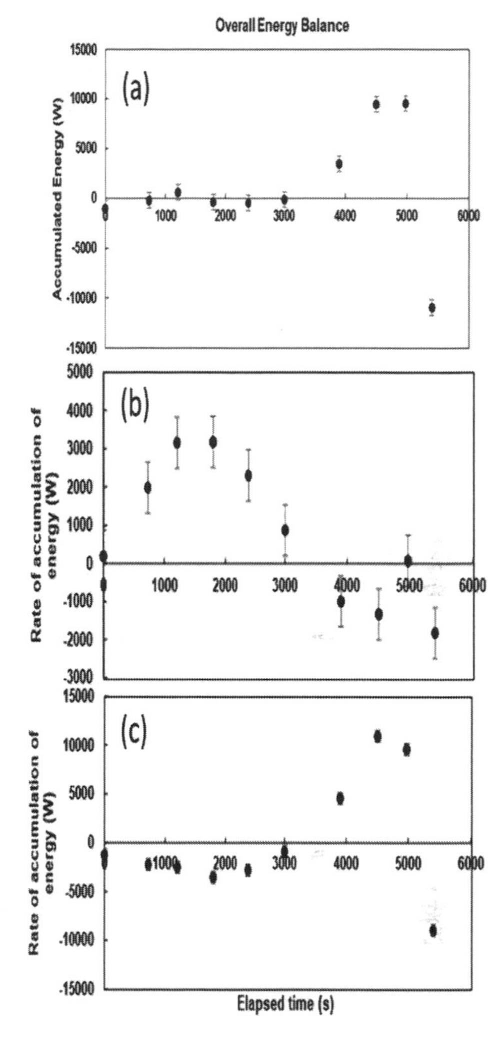

Figure 4. Energy balance of double effect evaporator: rate of accumulation of energy values vs. elapsed time for the (a) overall energy balance; (b) first effect energy balance and (c) second effect energy balance.

chamber would be filled from the first disengaging chamber by opening the valves through R2 and the circulating pump. The circulating pump would be powered on until the 2nd disengaging chamber was approximately half filled. The vacuum pump would be switched on, but the rotameter valve R4 would not be opened until the system had reached approximately steady state operation.

The derivations of Equations 4 were used to calculate the correct feed mass flow rate and cooling water flow rates from the mass balances, heat balances, and the capacity 30 equation for each effect. Assumptions were used to define the system and allow for a steady-state solution. The system

Table 2. Energy of streams in double effect evaporator.

Elapsed Time (±5 s)	Steam (±570 W)	Feed water (W)	Condensed steam leaving condenser (±42 W)	Steam leaving 2nd disengaging column (kW)	Steam leaving 1st disengaging column (±267 W)	Liquid transferred from 1st effect to 2nd effect	Steam entering 1st heater (±271 W)	1st effect accumulation rate of energy (W)	2nd effect accumulation rate of energy (W)	Overall energy accumulation rate (W)
0	34.3	1700 ± 221	4840	29900 ± 515	30700	3300 ± 162	25900	200 ± 696	−1200 ± 604	−1100 ± 801
720	34.3	1718 ± 116	4570	29400 ± 515	29100	3200 ± 162	24600	2000 ± 670	−2200 ± 604	−300 ± 78
1200	34.1	1700 ± 116	4320	28400 ± 516	27500	3200 ± 161	23200	3200 ± 670	−2500 ± 604	600 ± 778
1800	34.1	1700 ± 115	4310	29400 ± 516	27500	3200 ± 162	23200	3200 ± 670	−3500 ± 605	−400 ± 778
2400	34.0	1700 ± 115	4240	29400 ± 516	27000	4400 ± 161	22800	2300 ± 670	−2800 ± 604	−500 ± 778
3000	34.5	1900 ± 109	4660	29200 ± 516	29700	4000 ± 161	25000	900 ± 669	−900 ± 604	−200 ± 778
3900	34500	1800 ± 167	5000	25000 ± 516	32100	3300 ± 160	27000	−1000 ± 669	4600 ± 607	3500 ± 788
4500	33700	2000 ± 112	4930	18400 ± 519	31300	3800 ± 160	26400	−1300 ± 669	11000 ± 607	9500 ± 780
4980	32600	1900 ± 110	4540	17700 ± 519	28600	3900 ± 163	24100	100 ± 669	9600 ± 608	9500 ± 780
5400	33300	1500 ± 115	4920	38000 ± 516	31022	3600 ± 162	26100	−1800 ± 670	−9000 ± 606	−10900 ± 779

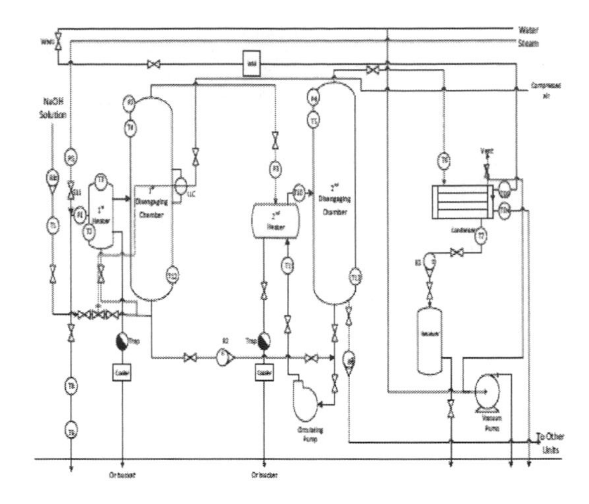

Figure 5. Modified process flow diagram of pilot plant double effect evaporator used in design problem.

was solve using a trial-and-error method in Excel. The feed was assumed to start at room temperature (20°C), with a mass flow rate of 0.8 kg/s. This mass flow rate was chosen with the assumption that slower mass flow rates would waste more energy through radiative heat losses. The flow rate through the circulating pump was chosen such that the linear velocity through the pump was at least 1 m/s in order to prevent fouling effects in the more concentrated second effect, as suggested by (Geankoplis, 2003) (page 299). The steam entering the first effect heater was assumed to start at 100°C (rounded up from the values reported in Table 2), and it was assumed to have the same quality value of 0.996 (from Table 2). The enthalpy of the steam entering was calculated from steam quality, and it was assumed to exit with the same enthalpy as the saturated liquid. Convective heat losses were assumed to be negligible compared to radiative heat losses. Radiative heat loss in piping was assumed to be negligible compared to losses in the heaters, columns, and condenser.

Cooling water flow rates were chosen to give a temperature drop as close to 20°C as possible. Cooling water was assumed to enter at 13°C, similar to what was measured in Table 4. The procedure for determining steady state from the spreadsheet was as follows: 1. Assume a mass flow rate. 2. Change boiling temperature for the first effect along with the enthalpy at the outlet (from Figure 8.4-3 in (Geankoplis, 2003)) until the concentration exiting the first effect is within the range of 0.1~0.4.3. Determine boiling point rise from Figure 8.4-2 (Geankoplis, 2003) and saturation pressure from steam Tables. 4. Change boiling temperature in 2nd effect along with the enthalpy at the outlet (Figure 4 (Geankoplis, 2003)) until

Table 3. Steam quality.

Enthalpy of saturated steam at system pressure (236972 Pa) (±0.66 kJ kg^{-1} K^{-1})	Enthalpy of saturated liquid at system pressure (236972 Pa) (±2.02 kJ kg^{-1} K^{-1})	Enthalpy of superheated steam at barometric pressure (97697 Pa) (±1.2 kJ kg^{-1} K^{-1})	Steam quality (±0.001)
2714.29	527.74	2705.53	0.996

Table 4. Design of the double effect evaporator for the NaOH concentration process.

Parameters	1st EFFECT	2nd EFFECT	Condenser	
Liquid flow rate in (kg/s)	0.8	2.9	CW flow rate in (kg/s)	0.004
Solids content in (wt%)	0.1	0.4	CW temp. in (°C)	13.0
Liquid flow rate out (kg/s)	0.257	0.200	CW temp. out (°C)	21.2
Solids content out (wt%)	0.312	0.400	e (Pa) 6000	6000
Boiling temperature (°C)	99.2	70.2		
Saturation pressure (Pa)	50000	6000	Result	
Boiling Point Elevation (°C)	20	30	Steam consumption (kg/s)	1.35
Evaporation rate (kg/s)	0.543	0.057	Steam economy (kg vapor/kg steam)	0.4
Steam/vapor flow rate (kg/s)	1.353	0.543		
Steam/vapor temperature (°C)	100	79.21		
Recycle stream flow rate (kg/s)	N.A.	2.7		

the concentration exiting the 2nd effect is 0.4. 5. Calculate mass flow rate of product produced, verify mass balances and calculate steam economy.

4 CONCLUSIONS

A double effect pilot system initially ran with a cooling water flow rate through the final condenser that caused a temperature increase in the cooling water which was below the operating criteria of 20°C. The system ran at this condition for approximately the first 4000 seconds of data collection. At approximately 4000 seconds, the cooling water flow rate to the condenser was lowered from 0.5783 kg s^{-1} to 0.0421 kg s^{-1}. This caused change in temperature of the cooling water through the condenser to increase from 10 ± 0.6°C to 30 ± 0.6°C. Because of this, data from the pilot system that can be used to characterize steady state must be taken after this correction to the cooling water flow rate to the final condenser was made. Three data sets were taken after this correction. These were taken at elapsed times of 4500 ± 5 seconds, 4980 ± 5 seconds, and 5400 ± 5 seconds. Analysis of the accumulated mass and energy was performed to determine which of these data sets gives the best portrayal of the pilot system at steady state operation.

Using these results, a design was proposed with a feed rate of 0.8 kg/s of the 10 wt% NaOH into the first effect and a cooling water flow rate of 0.004 kg/s. This resulted in an output flow rate of 0.2 kg/s 40 wt% solution, with a steam consumption of 1.35 kg/s and a steam economy of 0.4 kg vapor/kg steam.

REFERENCES

Bird, Byron R., Stewart, Warren E., Lightfoot, Edwin N., Klingenberg, Daniel J. *Introductory Transport Phenomena*. 1st ed. Wiley, United States of America, 2014.

Geankoplis, John Christie. *Mass Transfer and Separation Process Principles*. 4th ed. Prentice Hall, Englewood Cliffs, NJ, 2003.

Sandler, Stanley. Chemical, *Biochemical, and Engineering Thermodynamics*. 4th Ed. Wiley, United States of America, 1999.

Sharqawy, Mostafa H., John H. Lienhard, and Syed M. Zubair, Thermophysical properties of seawater: a review of existing correlations and data, *Desalination and Water Treatment* 2010, 16(1–3), 354–380.

Steam Traps. Encyclopedia of Chemical Engineering Equipment.http://encyclopedia.che.engin.umich.edu/Pages/SeparationsMechanical/SteamTraps/SteamTraps.html. Retrieved on May 3rd.

Civil, Architecture and Environmental Engineering – Kao & Sung (Eds)
© 2017 Taylor & Francis Group, ISBN 978-1-138-02985-9

Early fault diagnosis of wind turbine gearboxes based on the DSmT

Zhouqun Liu & Guochu Chen
Electric Engineering School, Shanghai DianJi University, Shanghai, China

ABSTRACT: Aiming at the problem that early faults of the gear boxes are difficult to be identified effectively and the conventional DST-based fault diagnosis method requires no conflict between the evidence, a fault diagnosis system based on a DSmT wind turbine gearbox is proposed. The DSmT fusion method is used to fuse the conflicting evidences from each evidence source. An analysis of the gear box fault reliability value of numerical changes and determination of the fault type is carried out in this work. The results show that the DSmT is applied to the early fault diagnosis of the wind turbine gear box, which exhibits better fusion results than the DST method.

1 INTRODUCTION

A gearbox is the key equipment that connects the low speed shaft and the high speed shaft. Due to long-term bearing of noise, high temperature, vibration and other complex factors, it is prone to faults. Since wind farms are generally designed in mountainous areas that are far from residential areas or offshore far from the land, maintenance is extremely difficult when it fails. Therefore, to master the early fault characteristics of the wind turbine gear box, and according to the characteristics of early troubleshooting, one can effectively improve the maintenance efficiency and reduce the loss to a large extent.

2 DSMT THEORETICAL BASIS

DSmT (Dezert–Smarandache Theory) is an effective theory that is used to fuse the conflict evidence proposed by Dezert and Smarandsche in 2002, which mainly deals with high uncertainty, high conflict, and imprecise information source evidence (Florentin Smarandache, 2011). DSmT can deal with the information fusion problem of any type of independent information sources represented by using the reliability function. In the subsequent development of DSmT, the combination rules of PCR1, PCR2, PCR3, PCR4, PCR5, and PCR6 are proposed. Among these, PCR6 (Florentin Smarandache, 2006) is a more accurate conflict allocation method. DSmT is able to handle complex, static or dynamic fusion problems without being restricted by the DST framework.

2.1 *DSmT theoretical basis*

Let $\Theta = \{\theta_1, \ldots\ldots, \theta_n\}$, called the hyper-power set D^Θ [3] in the DSmT frame, which is defined as the set of all composite propositions built from elements of Θ with $\cup$ and $\cap$ operators is such that:

1. $\varnothing, \theta_1, \ldots\ldots, \theta_n \in D^\Theta$.
2. if $A, B \in D^\Theta$, then $A \cap B \in D^\Theta$ and $A \cup B \in D^\Theta$.
3. No other elements belong to D^Θ, except those obtained by using rules 1 or 2.

When DSmT is applied to the wind turbine gearbox fault diagnosis, θ_i represents the failure mode of the fan gear box. The wind turbine gear box failure mode mainly wears away, gets pitting, gets scratches, and so on.

DSmT presents a new method for describing and analyzing problems. The hyper-power set in DSmT is based on the DST power set to relax the exclusive constraints. The proposition θ_i in frame Θ composes a set of finite propositions that expresses all the features of the fusion problem (Xusheng Zhai, 2012).

2.2 *Gearbox of wind turbine conflict evidence fusion rules*

The PCR6 rule is one of the more accurate fusion methods in DSmT, and it can obtain better fusion results than PCR5 when there are more than two sources of evidence. Under the generalized recognition frame Θ, there is a mapping of $m(\cdot): D^\Theta \to [0,1]$, which satisfies the following three conditions:

1. $M(\theta) = 0$, θ is an empty set;
2. $\sum_{A \in D^\theta} m(A) = 1$, where $A \in D^\theta$;
3. $0 \le m(A) \le 1$, $A \in D$^θ;

And then, m (A) is the generalized basic confidence distribution function of A (Florentin Smarandache, 2011).

Assuming that m1(·) and m2(·) denote the generalized basic confidence distribution functions provided by two independent and reliable sources, the classical combination rule (DSmC) and PCR6 rules for two information sources are as follows:

The classical combination rule (DSmC) is given as follows:

$$m_{DSmC}(Y)=\sum_{\substack{A,B\in D^{\Theta}\\ A\cap B=Y}} m_1(A)m_2(B) \vee / Y\in D^{\Theta} \tag{1}$$

Proportional Conflict allocation Rule (PCR6):

$$m_{PCR6}(\varnothing)=0$$
$$m_{PCR6}(Y)=m_{12\cdots s}(A)+\sum_{i=1}^{S} m_i(A)^2 \sum_{\substack{\cap_{k=1}^{s-1} Y_{\sigma_i(k)}\cap A\equiv\varnothing\\ (Y_{\sigma_i(1)},\cdots Y_{\sigma_i(s-1)})\in(G^{\theta})^{s-1}}} \left(\frac{\Pi_{j=1}^{s-1} m_{\sigma_i(j)}\left(Y_{\sigma_i(j)}\right)}{m_i(A)+\Sigma_{i=1}^{s-1} m_{\sigma_i(j)}\left(Y_{\sigma_i(j)}\right)}\right) \tag{2}$$

Among $m_i(A)+\sum_{j=1}^{s-1} m_{\sigma_i(j)}\left(Y_{\sigma_i(j)}\right)\neq 0$, and a detailed explanation of this equation can be found (Florentin Smarandache, 2006).

When comparing the classical combination rule with the proportional conflict allocation rule, we can see that the PCR6 method is more accurate. The conflict reliability between X and Y is composed of the conflict between X in information source 1 and Y in information source 2 and the conflict between Y in information source 1 and X in information source 2. With PCR6, one redistributes the partial conflicting mass to A and B proportionally with the masses $m_1(A)$ and $m_2(B)$ assigned to A and B, respectively and also the partial conflicting mass to A and B proportionally with the masses $m_2(A)$ and $m_1(B)$ assigned to A and B, respectively; thus, one gets now two weighting factors in the redistribution for each corresponding set A and B.

Since there are more than two information sources in this paper, PCR6 is used as the fusion rule in this paper.

3 EARLY FAULT DIAGNOSIS SYSTEM OF THE WIND TURBINE GEAR BOX

3.1 *Wind turbine gear box's early fault diagnosis structure*

The vibration sensor is arranged at several key positions of the fan gearbox. The data measured by each sensor are firstly anti-aliased and then EMD is decomposed (Florentin Smarandache, 2004) and information entropy extraction is performed to obtain the corresponding eigenvector. The BP neural network is used to obtain the confidence distribution of each source of evidence, and finally PCR6 fusion is used to acquire the fault type.

3.2 *Extraction of the feature vector*

There are two steps to extract the eigenvector, which are EMD and IMF energy entropy extraction.

3.2.1 *Empirical Mode Decomposition (EMD)*

EMD decomposition can change a frequency irregular wave into a single frequency of a wave + residual wave form. For a given signal y(t), the EMD decomposition process is as follows:

1. Find all the maximum points of the sequence y(t) and fit the upper envelope of the original data with the cubic spline interpolation;
2. Find all the minimum points of the sequence y(t) and fit the lower envelope of the original data with the cubic spline interpolation;
3. The mean value of the upper and lower envelopes is denoted by using ml, the original sequence y(t) is subtracted from the average envelope ml to obtain a new data sequence hl;
4. If hl satisfies the IMF (Lu Shen, 2010) criterion, it is denoted as the first IMF. If there are negative local maxima and positive local minima, this is not an Intrinsic Mode Function and needs to be continued. Filter until the first IMF component is obtained.
5. Calculate the residual signal $r_1(t)=y(t)-c_1(t)$ as a new signal, repeat the previous steps, until all the IMF components are found.
6. After the decomposition of each component, the original signal can be expressed as follows:

$$\text{Original wave}=\sum IMF_S+\text{Residual wave}$$

The main purpose of EMD is to smoothly process the signal and obtain a series of the IMF component from high-frequency to low frequency, in order to prepare for the extraction of information entropy.

3.2.2 *Fault feature vector extraction based on EMD energy entropy*

Information entropy is a subject which was formed and developed with the development of communication technology. In 1948, Claude Shannon first proposed the concept of information entropy. That is, the amount of information contained in a signal source is called information entropy. EMD energy entropy-based fault feature vector extraction steps are given as follows (Xusheng Zhai, 2012):

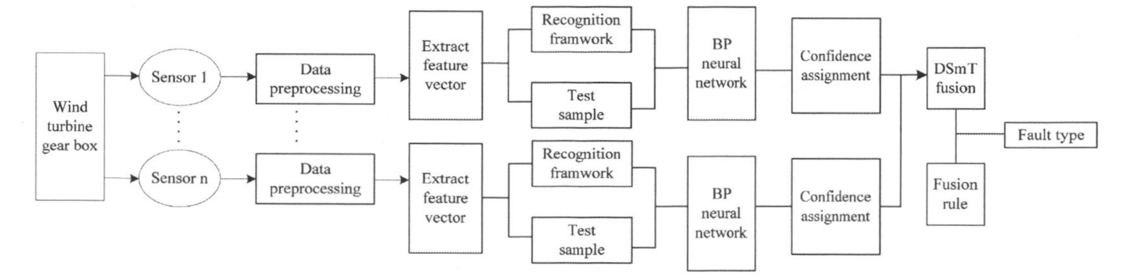

Figure 1. Schematic of the gearbox's early fault diagnosis structure.

1. For EMD decomposition of the vibration signal to be processed, select the main fault features including the first n IMF components.
2. Find the total energy of each IMF component E_i.

$$E_i = \int_{-\infty}^{+\infty} |C_i(t)|^2 \, dt \quad i = 1, 2, \cdots, n \tag{4}$$

3. Construct an eigenvector T with energy as an element.

$$\mathrm{T} = [E_1, E_2, \cdots E_n] \tag{5}$$

4. The vector is normalized

$$\mathrm{E} = \left[\Sigma_{i=1}^{n} |E_i|^2 \right]^{1/2} \tag{6}$$

$$T' = \left[\frac{E_1}{E}, \frac{E_2}{E}, \cdots \frac{E_n}{E} \right] \tag{7}$$

And then, T' is the fault feature vector.

3.3 *Wind turbine gear box early fault diagnosis data processing*

1. Place the vibration sensor at an appropriate position on the test bed;
2. Filter the measurement data;
3. Obtain the eigenvector according to equation (2.2);
4. Through BP neural network training, we can obtain the output $\{p_i(1), p_i(2), \cdots, p_i(q)\}$ of the ith sub BP neural network. If there are n fault patterns in the diagnosis system, then $q = n + 1$.
5. DSmT decision fusion is carried out, where the ith input mi of DSmT is the confidence distribution of the output of the ith sub-BP neural network, which is calculated as follows:

$$\mathrm{m}_i(j) = \frac{p_i(j)}{\sum_{j=1}^{q} p_i(j)} \tag{8}$$

$$m_i = \{m_i(1), m_i(2), \cdots, m_i(q)\} \tag{9}$$

where x = 1, 2, …, q.

4 FAULT DIAGNOSIS OF THE WIND TURBINE GEARBOX

This test uses an experimental platform which includes a two horsepower motor, a torque sensor, a power meter, and electronic control equipment to simulate the fault of the wind turbine gear box. The tested bearing supports the motor shaft. A single point of failure was placed on the 6205-2RS JEM SKF bearing using EDM technology with a fault diameter of 0.007 inches. An experimental platform is shown in Figure 2.

In the experiment, the acceleration sensor was used to acquire the vibration signal, and the sensor was placed on the motor housing by using the magnetic base. The acceleration sensor is mounted near the support housing of the drive end spindle of the motor housing. The vibration signal is collected by using a 16-channel DAT recorder and processed later in the MATLAB environment. The digital signal sampling frequency is 12000S/s, and the sampling frequency of the drive side bearing fault data is 48000S/s. This study focuses on the motor drive end bearing fault research. The speed is 1797 rpm and the load is zero.

Figure 2. Picture of the test-bed physical graphics.

Table 1. Diagnosis case 1 of the wind turbine gearbox early fault.

The sensor number	BP network output	Basic confidence assignment	DST fusion	PCR5 fusion	PCR6 fusion
Number 1	[0.020 0.861 0.019 0]	$m(A_1) = 0$ $m(A_2) = 0.957$ $m(A_3) = 0.021$ $m(N) = 0.022$			
Number 2	[0.874 0 0.129 0]	$m(A_1) = 0.871$ $m(A_2) = 0$ $m(A_3) = 0$ $m(N) = 0.129$	$m(A_1) = 0$ $m(A_2) = 0$ $m(A_3) = 0$ $m(N) = 1$	$m(A_1) = 0.4172$ $m(A_2) = 0.5672$ $m(A_3) = 0.0003$ $m(N) = 0.0153$	$m(A_1) = 0.2770$ $m(A_2) = 0.7139$ $m(A_3) = 0.0003$ $m(N) = 0.0088$
Number 3	[0.011 0.889 0 0]	$m(A_1) = 0$ $m(A_2) = 0.988$ $m(A_3) = 0$ $m(N) = 0.012$			

Table 2. Diagnosis case 2 of the wind turbine gearbox early fault.

The sensor number	BP network output	Basic confidence assignment	DST fusion	PCR5fusion	PCR6 fusion
Number 1	[0 0.373 0 0.529]	$m(A_1) = 0$ $m(A_2) = 0.414$ $m(A_3) = 0$ $m(N) = 0.586$			
Number 2	[0.232 0 0.567 0]	$m(A_1) = 0.290$ $m(A_2) = 0$ $m(A_3) = 0.710$ $m(N) = 0$	Not applicable	m(A1) = 0.0575 m(A2) = 0.1587 m(A3) = 0.5125 m(N) = 0.2713	m(A1) = 0.0735 m(A2) = 0.1022 m(A3) = 0.6397 m(N) = 0.1846
Number 3	[0.114 0 0.679 0]	$m(A_1) = 0.144$ $m(A_2) = 0$ $m(A_3) = 0.856$ $m(N) = 0$			

Firstly, the sensor data from the nearest fault source are selected as the training samples, and 12000 samples are taken as a unit to measure the vibration samples of the vibration system of the bearings under the conditions of outer ring damage, inner ring damage, sphere damage, and normal state. And three sub-neural networks were trained on the basis of the samples. The vibration data of three sensors were taken as test samples to test the actual running state of the gearbox.

The recognition frame of DSmT is $\{A_1, A_2, A_3, N\}$, which is composed of the sphere fault, inner circle fault, and outer circle fault, where A_1, A_2, and A_3 represent the three failure modes and N represents the normal state. The vibration signals collected by using the three sensors in different parts of the test bed are processed by EMD decomposition, and the first nine IMF components are taken. And the eigenvectors are calculated by using equations (4), (5), (6), and (7). And then, the eigenvectors are fed as input into the BP neural network, and the outputs of the BP network are transformed into confidence distributions by using equation (8) and (9), and then merged according to the DSmT decision rule or fusion rule.

As in the early weak stage, the wind turbine gear box fault is not particularly evident. As can be seen from Tables 1 and 2, some sensors detect a failure mode, and some are not detected, and part of the detected information between the sensors conflict with each other. In Diagnostic Case 1, the probability of the failure mode N measured by the three sensors is very small, but the DST fusion result shows that the failure mode is N, which is obviously illogical. However, the difference between the fault modes A1 and A2 is small, and it is obvious that the results of the PCR6 rule are more obvious than the PCR5 rule. In Diagnostic Case 2, the DST method cannot be applied, while PCR5 and PCR6 can diagnose the final failure mode as "outer-ring failure (A3)". However, it is clear that the PCR6 diagnostic method is more accurate.

5 CONCLUSIONS

The characteristics of the early fault of a wind turbine gear box are weak, and it is difficult to identify these effectively. In this paper, an early fault diagnosis system based on DSmT for the wind turbine gearbox is designed. Based on the basic idea of information fusion, the sensor is arranged in several key parts of the wind turbine gear box, and the vibration information is obtained under the weak characteristics of the fault feature. And then, the EMD decomposition and IMF energy entropy extraction are used to obtain the eigenvector, combined with BP neural network and PCR6 fusion rule in DSmT thereby making the final judgment about the fault condition.

1. The traditional DST and other methods cannot effectively identify the weak faults of the wind turbine gearbox early, and DSmT can be used to synthesize the conflict evidence effectively, which is suitable for the early fault diagnosis of the wind turbine gearbox.
2. Reasonable results can be obtained when the PCR5 fusion rule is used for fault decision, but the PCR6 fusion rule is more accurate.

ACKNOWLEDGMENT

This work was supported by the Natural Science Foundation of Shanghai Municipal Education Commission of China (Grant no.13YZ140).

REFERENCES

Florentin Smarandache, Jean Dezert. Advances and Applications of DSmT for Information Fusion [M]. New Mexico: American Research Press. 2004:3–31.

Florentin Smarandache, Jean Dezert. Advances and Applications of DSmT for Information Fusion [M]. New Mexico: American Research Press. 2006:13–15.

Florentin Smarandache, Jean Dezert. Advances and Applications of DSmT for Information Fusion [M]. New Mexico: American Research Press. 2006:49–50.

Florentin Smarandache, Jean Dezert. DSmT Advances and Applications of DSmT for Information Fusion [M]. National Defend Industy Press. 2011:39–45.

Florentin Smarandache, Jean Dezert. DSmT Advances and Applications of DSmT for Information Fusion [M]. National Defend Industy Press. 2011: VIII.

Lu Shen, Fuchun Yang,Xiaojun Zhou. Gear Fault Feature Extraction Based on Improved EMD and Morphological Filtering [J]. Journal of Vibration and Shock, 2010, 29(3).

Xusheng Zhai, Jinhai Hu, Shousheng Xie. Fault Diagnosis in Early Vibration of Aeroengine Based on DsmT [J]. Journal of Aerospace Power, 2012, 27(2).

Xusheng Zhai, Jinhai Hu, Shousheng Xie. Fault Diagnosis in Early Vibration of Aeroengine Based on DsmT [J]. Journal of Aerospace Power, 2012, 27(2).

Civil, Architecture and Environmental Engineering – Kao & Sung (Eds)
© 2017 Taylor & Francis Group, ISBN 978-1-138-02985-9

A study of an improved method for horizontal alignment based on the line element

Y.H. Li, Y.J. Liu & Y.L. Li
Dalian University of Technology, Dalian, China

ABSTRACT: Based on the building-blocks method and taking line element as the foundation unit, an improved method for horizontal alignment is presented. It can solve the problem that the building-blocks end position cannot be controlled effectively. There are four independent parameters in this method, and the rest of the parameters can be calculated by themselves. Multiple types of horizontal alignment can be displayed by using the improved method, which can increase the selectivity.

1 INTRODUCTION

Methods of horizontal alignment are composed by using the straight-line and curve method. The straight-line method is the easiest method to use, because the calculation is relatively simple. However, it is difficult to make full use of the circular curve and transition curve. The straight-line method cannot deal with the complicated line effectively. The curve method improves the straight-line method, which is setting the wire before the curve. Considering the specifications and the actual requirements, taking the line element as the foundation unit can make the line more fluent. The curve method, according to the requirements of the different line layout, can be divided into the fitting method, building-blocks method, synthetic method, chord tangent method, etc. The idea that the complex horizontal alignment can be composed by several line elements dates back to the 1990s, when the famous studies on this subject were published (Li 1993, Wu 1965). This research field is still quite active (Xu 1997) and the controlled segmentary element design method has been investigated. More recently, Miao et al. (2001) proposed a method for the highway's horizontal alignment based on the segmentary element. In order to add the new line element, we need to carry out three steps by using the building-blocks method: first, we should determine the curve type and then determine the line element control parameters. According to the dynamic formulation of the end position, the rest of parameters can be calculated.

2 PROBLEMS IN LINE BUILDING-BLOCKS METHOD

2.1 *Difficulty in controlling the end point coordinates*

The building-blocks method is often difficult to accurately meet the requirements of controlled conditions. Generally, end point coordinates of the new design does not coincide with the reference coordinates.

2.2 *Single type of line element*

Building-blocks based on the line element can only display the preset line element type (circular curve, transition curve, and tangential straight line), thereby making the design lack selectivity.

3 IMPROVMENT OF THE HORIZONTAL ALIGNMENT DESIGN

3.1 *Parameters*

In the traditional building-blocks method, the line element is used as the basic unit of the route design. In this article, the improvement is based on the traditional building-blocks method, and it emphasizes that the coordinate position has an effect on the linear position. There are eight basic parameters in common use, which are starting point coordinates B, end point coordinates E, starting tangential angle α_B, end tangential angle α_E, starting point radius R_B, end point radius R_E, line element deflection angle β,

Table 1. List of the types of line element.

Type of line element	Abbreviation
Tangential straight line	TSL
Circular curve	CC
Forward holonomic transition curve	FHTC
Reverse holonomic transition curve	RHTC
Forward non-holonomic transition curve	FNHTC
Reverse non-holonomic transition curve	RNHTC

and the length of the line element L. There are four independant parameters at most, which are starting point coordinates B, end point coordinates E, starting point radius R_B, and end point radius R_E.

3.2 *The type of line element*

3.2.1 *The design steps*

Determine the starting point parameters of the line, including the starting point coordinates, starting tangential angle, as well as the starting point radius. For the first line element, it requires setting starting point parameters. In other cases, the end point of the previous line is the starting point of the next line, without additional settings.

Take the dynamic drawn position as an end point of line element. After determining the end point's position, we can calculate the basic parameters of six types, which can provide convenience for the designer to choose.

4 CALCULATION OF PARAMETERS

In this section, we briefly summarize the calculated steps of the line element, referring for a detailed description to the deflection angle.

The line element deflection angle and other parameters of a circular curve can be calculated by using the analytical method. As for all kinds of transition curves, an iterative method is the best choice. In this paper, we take the reverse holonomic transition curve as an example.

In order to obtain a reasonable horizontal alignment, the calculation of the reverse holonomic transition curve is divided into six steps.

Step one: establish a relative coordinate system $O'X'Y'$. Take the tangential direction of the end point E as $+X'$, and anticlockwise 90° as $+Y'$. Apparently, the end point E and the relative coordinate origin should be overlap.

Step two: calculate the direction angle of vector **BE** in the relative coordinate system.

Starting tangential angle α_B is known, we can set α_B in equation (1) to calculate the direction angle of vector **BE** in the relative coordinate system.

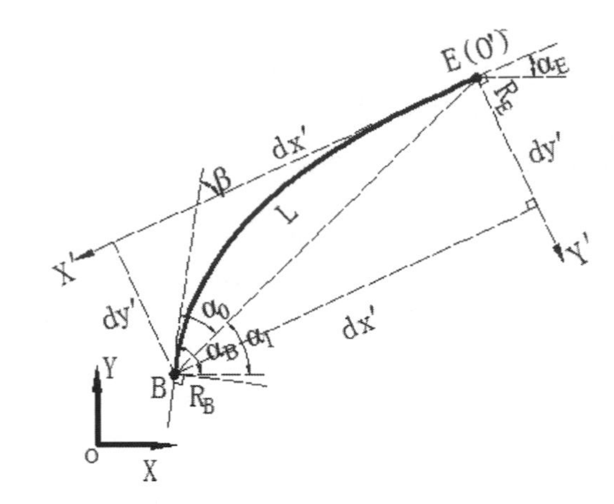

Figure 1. Graph showing the reverse holonomic transition curve.

$$\alpha_O = \alpha_1 - \alpha_B, \alpha_O \in (-\pi, +\pi), \alpha_O \neq 0 \tag{1}$$

where α_1 is the vector direction angle in the cartesian coordinate system.

Step three: calculate the line element deflection angle. The reverse holonomic transition curve calculates the line element deflection angle by using an iterative method and the bisection method. Firstly, we should calculate the deflection angle by dichotomy:

$$\beta > \beta_{\min} = |\alpha_o|, \beta < \beta_{\max} = \pi \tag{2}$$

$$\beta_0 = (\beta_{\max} + \beta_{\min})/2 \tag{3}$$

Direction angle of the end point is given as follows:

$$\alpha_E = \alpha_B + \delta_0 \beta_0 \tag{4}$$

where δ_0 is the coefficient. If the line element is turned left, $\delta_0 = +1$, else $\delta_0 = -1$.

Substituting equation (4) into equation (5), the tangent direction angle can be drawn as follows:

$$\alpha'_{O'} = \alpha_E + \pi \tag{5}$$

where $\alpha'_{O'}$ is the tangent direction angle of the cyclotron line origin O'.

Since the tangent direction angle, coordinates of origin O' and starting point are provided, the line element deflection angle β can be worked out by using the following iterative equation:

$$\mathrm{d}x' = 2R\beta f_x(\beta) \tag{6}$$

where dx' is the abscissa of the starting point B in the relative coordinates and $f_x(\beta)$ is a function about the line element deflection angle β.

$$dy' = 2R\beta^2 f_y(\beta) \tag{7}$$

where dy′ is the co-ordinate of the starting point B in the relative coordinates, $f_y(\beta)$ is a function about the line element deflection angle β.

$$f_x(\beta) = \sum_{i=1}^{\infty}(-1)^{i-1}\frac{\beta^{2i-2}}{(2i-2)!(4i-3)} \tag{8}$$

$$f_y(\beta) = \sum_{i=1}^{\infty}(-1)^{i-1}\frac{\beta^{2i-2}}{(2i-1)!(4i-1)} \tag{9}$$

$$k' = {dy'}/{dx'} = \beta {f_y(\beta)}/{f_x(\beta)} \tag{10}$$

where k' is the ratio of the relative coordinates.

$$\beta = k' {f_x(\beta)}/{f_y(\beta)} = k' f(\beta) \tag{11}$$

Under the condition of the first trial, the line element deflection angle β, which is the tangent direction of the starting point B relative to the tangent direction of the origin O', can be obtained. Set β in equation (12) to determine whether it is real.

$$\Delta\beta = \beta_0 - \beta \tag{12}$$

where β_0 is the result of the first trial calculation.

Apply the following rules:

$$|\Delta\beta| < \xi = 10\mathrm{e}^{-6} \tag{13}$$

Line element deflection angle β is exact.

Else,

$$|\Delta\beta| > \xi = 10\mathrm{e}^{-6}, \Delta\beta > 0, \beta_{\max} = \beta \tag{14}$$

$$\Delta\beta < 0, \beta_{\min} = \beta \tag{15}$$

Return to the equation (3) and enter the second trial calculation to obtain the second computation of deflection angle β. Set β in equation (12). Circulatory computation like this does not come to an end until the difference satisfies the inequality equation (13). The line element deflection angle β is exact. Step three is completed.

Step four: calculate relative coordinates of the starting point (X'_{BE}, Y'_{BE}).

Based on the accurate solution of the deflection angle, we can calculate the tangent angle of transition curve origin.

Calculate the relative coordinates (X'_{BE}, Y'_{BE}) of the starting point in the relative coordinate system, according to equation (16).

$$\begin{Bmatrix} X'_{BE} \\ Y'_{BE} \end{Bmatrix} = \begin{bmatrix} \cos(\alpha'_{O'}) & \sin(\alpha'_{O'}) \\ -\sin(\alpha'_{O'}) & \cos(\alpha'_{O'}) \end{bmatrix} \begin{Bmatrix} X_B - X_E \\ Y_B - Y_E \end{Bmatrix} \tag{16}$$

Step five: calculate the tangent angle α_E of the end point E.

Calculate tangent angle α_E of the end point E by setting β in equation (17):

$$\alpha_E = \alpha_B + \delta_0\beta \tag{17}$$

Step six: calculate the rest of the parameters of the reverse transition curve.

$$R_B = \frac{X'_{BE}}{2\beta f_x(\beta)} \tag{18}$$

Substitute equation (18) in equation (19), and we will obtain the length:

$$L = 2R_B\beta \tag{19}$$

where L is the length of the line element.

5 CALCULATED EXAMPLE

As shown in Figure 2, in the cartesian coordinate system, the starting point coordinates are (200, 200), the starting tangential angle α_B is 40°, and the end point coordinates are (400, 300).

We can use the above-mentioned steps to calculate the rest parameters. The result is shown in Table 2, which presents the end tangential angle,

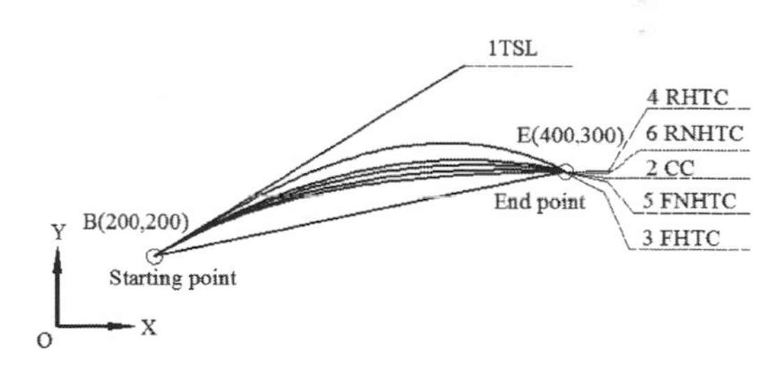

Figure 2. Schematic of the example of the improved method.

Table 2. Calculation of the basic line element.

No.	Types	α_E	R_B	R_E	β	L
1	TSL	40.00°	∞	∞	0	223.60
2	CC	−1.52°	344.86	344.86	−36.49°	219.07
3	FHTC	−15.04°	∞	116.84	−55.04°	224.47
4	RHTC	12.72°	228.57	∞	−27.28°	217.59
5	FNHTC	−4.98°	650.00	179.81	−44.98°	221.15
6	RNHTC	8.59°	270.00	756.54	−31.41°	218.16

starting and end point coordinates, line element deflection angle, and the length of the line element. Six basic line elements can also be drawn, which can facilitate the coordination of routes and increase selectivity.

6 CONCLUSIONS

1. When compared with the traditional building-blocks method, the coordinates of the end position can be effectively controlled.
2. There are three independent parameters in the tangential straight line and circular curve. The rest have four independent parameters. The typical independent parameters are the starting and end point coordinates and the starting and end radii.
3. For the continuous design, parameters related to starting point are known; we only need to determine the end point coordinates to obtain the rest parameters of the line element.
4. Line element deflection angle β is the key to the calculation of the untypical independent parameters. The circular curve uses the analytical method, while the other four types use an iterative method.
5. We can draw six basic line elements, which can increase the selectivity.

REFERENCES

Andrzej, K. (2014). New Solutions for General Transition Curves. *Surveying Engineering* 140(1): 12–21.

Cai, H.H. & Wang, G.J. (2010). Construction of cubic C-Bézier spiral and its application in highway design. *Zhejiang University* 2010(1): 68–74.

He, J.L & Zhang, X.S. (2004). Optimal Design on Geometric Compounds of A Highway. *Forest Engineering* 20(6): 58–59.

Jiang, Q.P. (2002). Study of the new types of transition curve of road. *China Journal of Highway and Transport* 15(2): 16–18.

Li, F. (1993). Building-Blocks for interchange Ramps Plane Alignment Layout. *Southeast University* 23(4): 25–31.

Miao, K. & Zhan, Z.Y. (2001). Study of interactive design method for highway's horizontal alignment based on segmentary element. *Highway and Transport* 14(3): 25–29.

Wu, G.X. (1994). A New Curve Method for Designing Highway Horizontal Alignment—Comprehensive Method. *ChongQing JiaoTong Institute* 15(suppl): 24–31.

Xu, J.L. (1997). The Controlled Segmentary Element Design Method of Interchange. *Xi'an Highway University* 17(4): 5–9.

Zhao, J. (2013). Based on the Weft to Interchange Design Aided Design Functions of Software Development. *Chang'an University*.

Civil, Architecture and Environmental Engineering – Kao & Sung (Eds)
© 2017 Taylor & Francis Group, ISBN 978-1-138-02985-9

Theoretical analysis of mechanical properties and mechanical design for fabricated zero initial cable force friction dissipation

Q.X. Ye & A.L. Zhang
College of Architecture and Civil Engineering, Beijing University of Technology, Beijing, China

ABSTRACT: A type of Fabricated Zero initial cable force Friction Dissipation Re-centering Brace (FZFDRB) is proposed. The theoretical analysis of mechanical properties was carried out and mechanical design for FZFDRB was also constructed. The results show that, the brace has a simple structure. There is no stiffness degradation in the loading process, and the brace has a full hysteretic curve and shows stable energy dissipation law. When the brace stops re-centering at the residual displacement, releasing the high-strength bolts of the brass–slot steel friction plates can let the brace continue re-centering to zero point. This indicates that FZFDRB has an excellent re-centering function.

1 INTRODUCTION

As a component used to resist lateral force, brass can take on part of the lateral load in the structure, so as to reduce the damage degree of the structure. Due to its various forms and ease of processing, steel braces are widely researched and applied. From research at home and abroad, the steel brace mainly includes five types: ordinary steel brace, attached energy dissipator brace, buckling-restrained brace, prestressed cable brace, and re-centering function brace. Ordinary steel brace, attached energy dissipator brace, and buckling-restrained brace may produce irreversible residual deformation during major earthquakes, and the main structure could not be effectively protected, the structural seismic performance in the subsequent aftershocks will be severely affected (Wakabayashi, M. & Nakamura, T. 1973, Yoshino, T. & Kano, J. 1971, Christopoulos et al. 2003). Research shows that when the residual deformation angle is more than 5‰, the maintenance cost may even be greater than the cost of reconstruction after the quake (Erochkol 2011). A pre-stressed cable brace is often the cross layout in the structure. It can improve the lateral stiffness of the framework, and make it have a certain ability of re-centering. But the pre-stressed cable brace has no energy dissipation capacity. Structures have to damage the main framework to dissipate energy (Zhang 2014). With the improvement of attached energy dissipator brace and buckling-restrained brace, the re-centering function brace takes the advantages of buckling-restrained in compression and energy dissipation. It can avoid the damage, and eliminate or reduce residual deformation of the structure. Zhu et al. (2006) proposed a kind of Self-centering Friction Damping Brace (SFDB) and performed tests accordingly. Miller et al. (2011) connected stranded SMA wires to ordinary cables to provide the restoring force, utilized them in the mild steel energy-dissipative buckling-restrained brace with three-layer steel pipes to make a new type of the Self-Centering Buckling-Restrained Brace (SC-BRB), and conducted pseudo-static testing on a large-scaled specimen thereof. Liu et al (2012) placed an energy-dissipative mild steel into the above-mentioned specimen and below gaps between inner and outer steel pipes of equal length, allowed pre-stressed strands to pass through the inner and outer steel pipes and fixed them on both ends of the steel pipes to create a novel type of Self-Centering Buckling-Restrained Brace (SCBRB), and then tested its seismic performance via pseudo-static testing. Chen et al. (2014) proposed three kinds of large sizes of the self-centering brace based on SMA bars or plates, and gave the brief description to the construction. The author (2016) summarizes the advantages and disadvantages of currently research about the re-centering function brace, and points out the key problems needed to study in the future.

Previous studies showed that the re-centering braces still have the following shortcomings:

1. The re-centering material has to be pre-stressed, which makes the construction work more complicated. The pre-stressing force of pre-stressed cables inevitably loses after long-term placement. Tandem cables can improve the deformability, but reduce the cable stiffness by half. Due to the relatively high cost and difficulties in anchorage connection, the application of the

shape memory alloy as the re-centering material is not yet mature.

2. At present, the re-centering braces normally adopt the structure of inner and outer dual pipes or even triple pipes and require the braces to be self-centered. According to this design concept, the overall brace structure is quite complicated. Moreover, the machining precision of the pipe length could also make a big difference to the load-bearing capacity of the brace.

To solve the inadequacies of the re-centering brace, a new type of the Fabricated Zero initial cable force Friction Dissipation Re-centering Brace (FZFDRB) is proposed; the structure and mechanical design are elaborated and the theoretical studies of mechanical properties are also made.

2 STRUCTURE, OPERATING PRINCIPLE AND THE RESTORING FORCE MODEL OF THE FZFDRB

2.1 *Structure of the FZFDRB*

The FZFDRB is comprised of the core part and outer holding pipes, of which, the outer holding pipes are two channel steels and the structure is as shown in Fig. 1. The main components are as follows: 1. loading end, 2. front anchoring part, 3. dowel bar, 4. small cover plate, 5. high-strength strain bolt (5a. bolt 1 and 5b. bolt 2), 6. fixed channel steel, 7. damper of the brass–slot steel friction plates (the part in the red circle in Fig. 1 consists of 7a. guard plate, 7b. brass friction plate, and 7c. slot steel friction plate), 8. pressure transducer and anchorage, 9. rear anchoring part, 10. 1st group of cables, 11. 2nd group of cables, and 12. fixed end.

The assembly process is as follows: the cross plate of the loading end is passed through the cross cut holes of the front anchoring part and then welded to the left connecting plat of the dowel bar. Let high-strength bolts pass through the corresponding preserved holes of the damper of the brass–slot steel friction plates, rear anchoring part, and the right connecting plate of the dowel bar to form the friction dissipation part of the FZFDRB. The 1st group of cables are passed through the preserved hole on the left connecting plate of the dowel bar and then fixed to the front anchoring part and the right end of the dowel bar, respectively by using the anchorages. The 2nd group of cables are passed through the preserved hole on the right connecting plate of the dowel bar and then fixed to the rear anchoring part and left end of the dowel bar, respectively by using the anchorages to form the core part. The assembled core part is fastened by using the two fixed channel steels, high-strength bolts, and small cover plates.

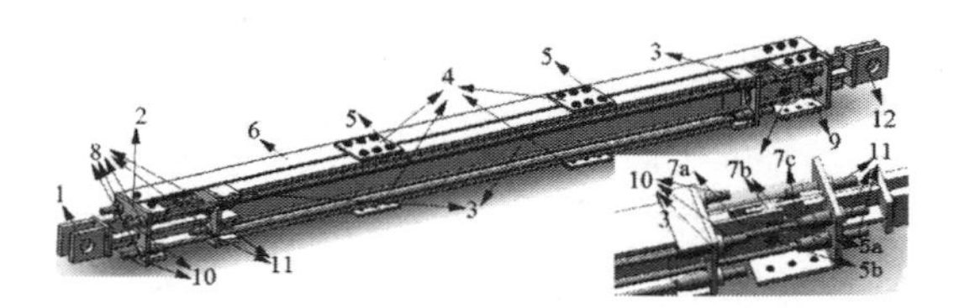

Figure 1. Schematic of the structure diagram of the FZFDRB.

2.2 *Operating principle of the FZFDRB*

Fig. 2 shows the simplified schematic diagram of the FZFDRB. One end of the cables is connected to the brass friction plate, and the other end is connected to fixed end. The slot steel friction plate is also fixed. And so, the moving of the brace is equal to sliding around of the brass friction plate on the slots steel friction plate. It is assumed that, when under pressure, the brace is equivalent to a brass friction plate with movement to the right, and when under tension, the brace is equivalent to the brass friction plate with movement to the left. When the loading end of the FZFDRB drives the dowel bar to move right, the damper of brass–slots steel plates has friction energy dissipation under compression. At this point, the brace is equivalent to as shown in Fig. 4 and brass friction plates have right distance to the slots steel friction plate. The first group of cables are tensioned, and the second group of cables with no internal force, it naturally

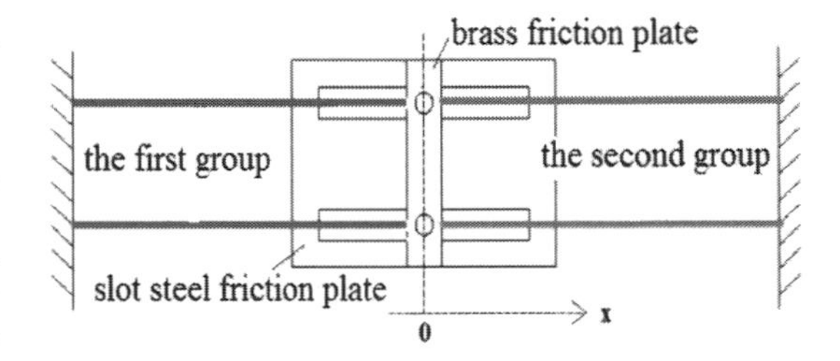

Figure 2. Simplified schematic diagram of the FZFDRB.

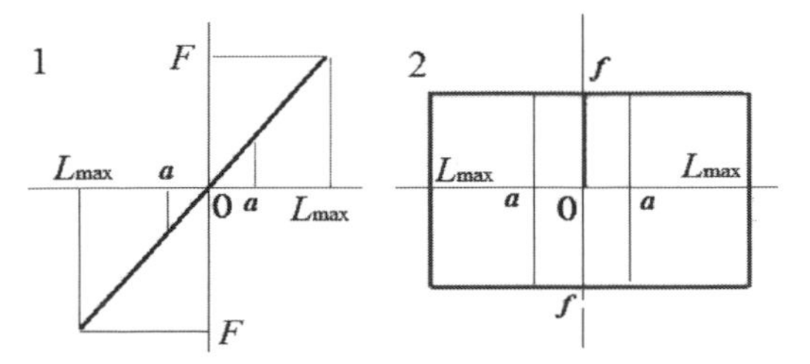

Figure 3. Schematic of the composition of the FZFDRB restoring force model.

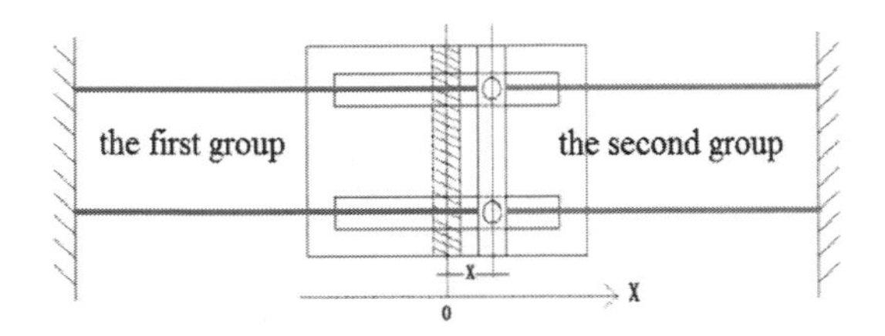

Figure 4. Schematic of the positive pressure diagram of the FZFDRB.

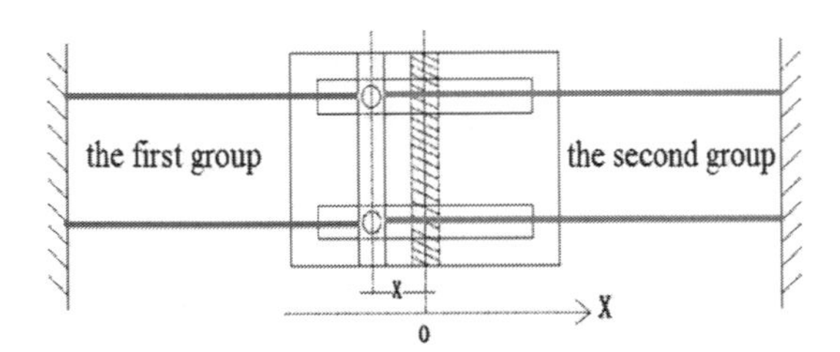

Figure 5. Schematic of the negative tension diagram of the FZFDRB.

prolapse. When the loading end of the FZFDRB drives the dowel bar to move left, the damper of brass–slots steel plates has tensile friction energy dissipation that is equivalent as shown in Fig. 5. The brass friction plates have left distance to the slots steel friction plate. The second group of cables are tensioned, and the red cables naturally prolapse with no internal force.

2.3 *Restoring force model of the FZFDRB*

As shown in the operating principle of the FZFDRB, the restoring force model of the brace consists of two parts: 1) one is provided by the steel cables, i.e. elasticity, 2) the other is provided by the damper of brass–slot steel friction plates, as shown in Fig. 3. If only think about the force balance of the FZFDRB itself, a shows the residual displacement when the friction force is equal to the cable forces in Fig. 3, and the displacement of the brace can be calculated according to equation (1):

$$a = f / k \tag{1}$$

where, f is the sliding friction and k is the stiffness of the cables; Controlling the size of a can limit the brace deformation, control the residual deformation of the structure to less than 5 ‰ and guarantee the building process in a repairable area. After the brace has stopped re-centering and artificially relaxing the high strength bolts of the damper of brass–slot steel friction plates, the re-centering force of the brace is equal to the friction force, and the re-centering function of the brace can be realized.

The restoring force model of the FZFDRB can be deduced by using the force mechanism of the brace. Taking the brass friction plate (shadow) center as the initial position to establish the equation and right is positive.

Positive, $X > 0$, as shown in Fig. 4: the 1st group of cables is tensioned, while the 2nd group of cables is slacked, and X extends from 0 to L_{max}

$$F_1 = K_1 \cdot x + f \tag{2}$$

Negative, $x > 0$, as shown in Fig. 4: the 1st group of cables is tensioned, while the 2nd group of cables is slacked, and x extends from L_{max} to 0

$$F_2 = K_1 \cdot x - f \tag{3}$$

Negative as shown in Fig. 5, $X < 0$: the 2nd group of cables is tensioned, while the 1st group of cables is slacked, and X extends from 0 to $-L_{max}$

$$F_3 = K_2 \cdot x - f \tag{4}$$

Positive, $x < 0$, as shown in Fig. 5: the 2nd group of cables is tensioned, while the 1st group of cables is slacked, and x extends from $-L_{max}$ to 0

$$F_4 = K_2 \cdot x + f \tag{5}$$

K_1 is the stiffness of the 1st group of cables,

$$K_1 = E \cdot A_1 / L_1$$

K_2 is stiffness of the 2nd group of cables,

$$K_2 = E \cdot A_2 / L_2 \tag{6}$$

A_1 is the effective area of the 1st group of cables;

A_2 is the effective area of the 2nd group of cables;

L_1 is the calculated length of the 1st group of cables;

L_2 is the calculated length of the 2nd group of cables; and

f is the sliding friction, calculated by using the equation: $f = n_1 \cdot \mu \cdot N$, wherein, n_1 is the number of friction surfaces, μ is the friction coefficient, measured from the test as shown in Fig. 7, for the brass–slot steel friction plates, $\mu = 0.26$, N is the pressure value of the friction surface, with bolt connection, $N = n \cdot P$, where, n is the number of bolts, and P is the pre-tightening force of high-strength bolts.

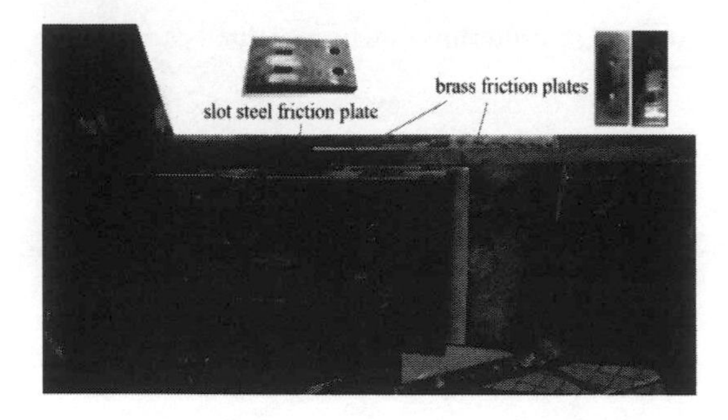

Figure 6. Picture showing the experiment of friction plates.

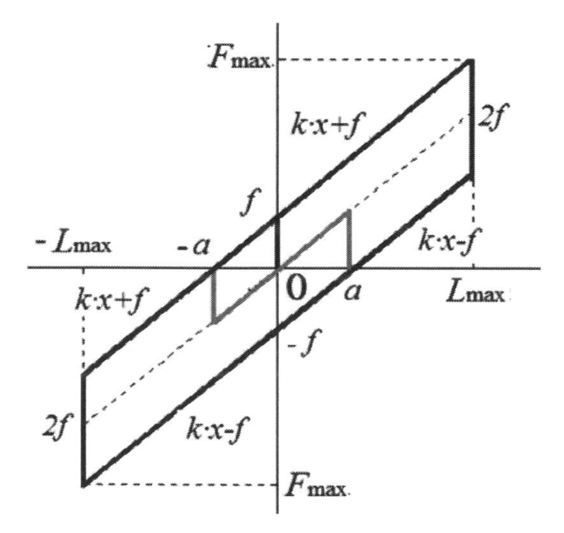

Figure 7. Graph showing the restoring force model of the FZFDRB.

The same cable length and cross section are taken, so that the two groups of cable stiffness are also the same. The restoring force model of the brace is shown in Fig. 7. The broken line shows the process that when the cable force is equal to the friction force, the residual displacement is *a* mm and the brace stops re-centering. Loosen the high strength bolts at this time and the friction to zero, the brace continues to re-center to zero with the re-centering force that is equal to the friction force.

3 DESIGN OF THE STRUCTURE SIZE OF THE FZFDRB

Currently, the connection of the frame and brace usually uses a large number of bolts or welding to achieve rigid connection or half rigid connection. In such cases, when the structure has a displacement, the brace will be the bend member under an external load, and the end of the brace may be easily buckling or even rupturing failure. In view of the above reasons, the FZFDRB ends adopt the form of articulation, thereby making the brace always as the axial stress components under lateral load. And so, the design of the brace can only consider the axial load. When some components of the brace yield or are buckling under loading, that will need extra re-centering force; therefore, the design requirements of the brace components always stay flexible. Fig. 8 shows the internal structure of the FZFDRB.

Under tension, as shown in Fig. 9a, the loading end, dowel bar, guard plate, and the cables shown in Fig. 9a have the same axial displacement Δ to the left. The axial force of components is shown in Fig. 9b-e, the design formula can be received by using the force equilibrium equation. Known from the analysis of Fig. 9e, the force is entirely transmitted to the fixed end from the loading end. And so, the internal force between the front anchoring part, channel steels, and rear anchoring part is zero, and the design of the front anchoring part, channel steels, and rear anchoring part cannot be considered. From the force analysis of Fig. 9b–e, the axial force diagram is shown as Fig. 9f. The axial components should be designed according to strength, when the brace is under tension:

Figure 8. Schematic of the internal structure of the FZFDRB.

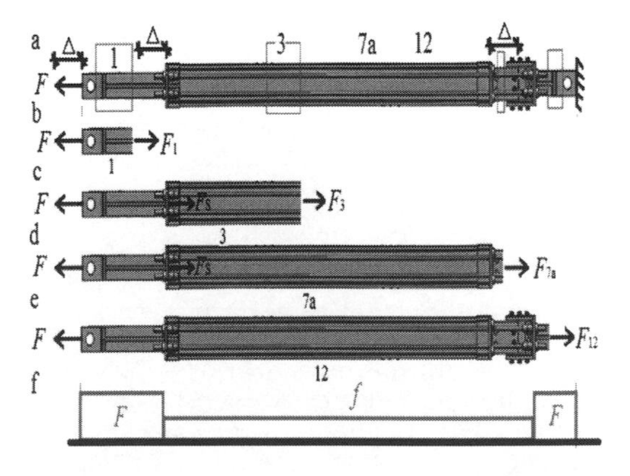

Figure 9. Schematic of the internal structure of the FZFDRB under tension.

$$1.\ \text{Loading end:}\left\{\begin{matrix}F_1 = F = F_s + f\\ A_1 > (F_s + f)/\sigma\end{matrix}\right\} \quad (7)$$

$$3.\ \text{Dowel bar}\left\{\begin{matrix}F = F_s + F_3 = F_s + f\\ A_3 > f/\sigma\end{matrix}\right\} \quad (8)$$

$$7a.\ \text{Guard plate}\left\{\begin{matrix}F = F_s + F_{7a} = F_s + f\\ A_{7a} > f/2\sigma\end{matrix}\right\} \quad (9)$$

$$12.\ \text{Fixed end:}\left\{\begin{matrix}F = F_{12} = F_s + f\\ A_{12} > (F_s + f)/\sigma\end{matrix}\right\} \quad (10)$$

Under compression: as shown in Fig. 10a, the loading end, dowel bar, and guard plates have the same displacement Δ to the right. The axial forces of structural parts are shown in Fig. 10b–e. And so, the design equation can be obtained by using the force equilibrium equation. From the force analysis, the axial force diagram is shown as Fig. 9j. The axial compression of the brace should be designed according to the strength and stability control:

$$6.\ \text{Fixed channel steel:}\left\{\begin{matrix}F_6 = F_s\\ A_6 > F_s/2\sigma\\ A_6 > F_s/2\psi_6\sigma\end{matrix}\right\} \quad (11)$$

$$1.\ \text{Loading end:}\left\{\begin{matrix}F_1 = F = f + F_s\\ A_1 > (f + F_s)/\sigma\\ A_1 > (f + F_s)/2\sigma\psi_1\end{matrix}\right\} \quad (12)$$

$$3.\ \text{Dowel bar:}\left\{\begin{matrix}F = F_3 + F_s = f + F_s\\ A_3 > f/\sigma\\ A_3 > f/\sigma\psi_3\end{matrix}\right\} \quad (13)$$

$$7a.\ \text{Guard plate:}\left\{\begin{matrix}F = F_{7a} + F_s = f + F_s\\ A_{7a} > f/2\sigma\\ A_{7a} > f/2\sigma\psi_{7a}\end{matrix}\right\} \quad (14)$$

$$12.\ \text{Fixed end:}\left\{\begin{matrix}F = f + F_s\\ A_{12} > (f + F)/\sigma\\ A_{12} > (f + F)/\sigma\psi_{12}\end{matrix}\right\} \quad (15)$$

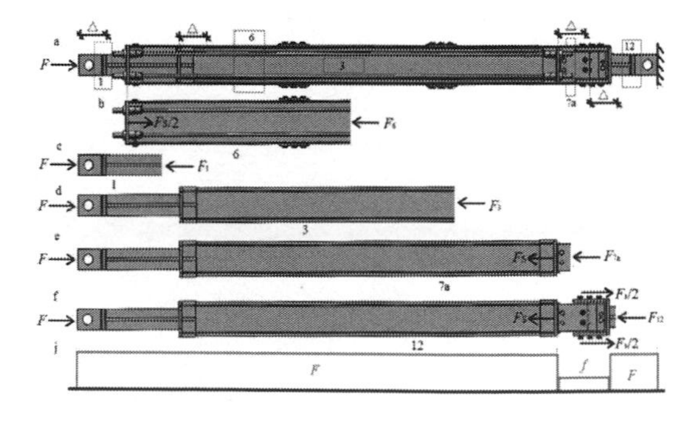

Figure 10. Schematic of the internal forces of the FZFDRB during compression.

The connection between the channel steel and rear anchoring part can be designed according to the shear of friction type high strength bolts:

$$N_v^b > F_s/n \quad (16)$$

Because there is only contact pressure between the channel steel and front anchoring part, the high strength bolts can be used in the mechanical design.

F_1 denotes the internal forces of the loading end section;

F_3 denotes the internal forces of the dowel bar section;

F_6 denotes the internal forces of the one channel steel section;

F_{7a} denotes the internal forces of the one guard plate section;

F_{12} denotes the internal forces of the fix end section;

F_s denotes the internal forces of one group cable;

f denotes the friction force;

A_1 denotes the cross-sectional area of the loading end;

A_3 denotes the cross-sectional area of the dowel bar;

A_{12} denotes the cross-sectional area of one channel steel;

A_{7a} denotes the cross-sectional area of one group cable;

A_{12} denotes the cross-sectional area of the fixed end;

ψ_1 denotes the stability coefficient of loading end;

ψ_3 denotes the stability coefficient of dowel bar

ψ_{7a} denotes the stability coefficient of one channel steel;

ψ_{12} denotes the stability coefficient of the fix end;

N_v^b denotes the shear bearing capacity of one single high-strength bolt,

$$N_v^b = 0.9 \cdot n_f \cdot \mu \cdot P \quad (17)$$

n is the number of high-strength bolts;

n_f denotes the number of friction surfaces: single shear: 1 and double shear: 2;

μ denotes the slip resistance coefficient of the friction surface, according to the literature [11];

P denotes the pretension design value of high-strength bolts, according to the literature [11];

A_{il} denotes the cross-sectional area of each part of the brace in tension; and

A_{is} denotes the cross-sectional area of each part of the brace in compression.

In conclusion, the size of the cross-sectional area of each part in the FZFDRB should meet $\{A_{il}, A_{is}\}_{\max}$

4 THEORETICAL ANALYSIS OF MECHANICAL PROPERTIES OF THE FZFDRB

The selection of specimen Φ14, Φ20 galvanized steel wire rope, S10.9, M16 and M20 high-strength bolts to test the theoretical analysis of seismic performance for the FZFDRB. The parameters of the galvanized steel wire rope and high-strength bolt are given in Table 1. According to the controlling target load method of taking the inter-laminar displacement angle of 0.00375, 0.005, 0.0075, 0.01, 0.015, and 0.02 rad in ANSI/AISC341 [12], the control target of the FZFDRB is the displacement. At each displacement amplitude (i.e., 10 mm, 15 mm, 20 mm, 25 mm, 30 mm, and 35 mm), two loading cycles are carried out. The theory of hysteresis curves and energy dissipation coefficient β_E under each amplitude are shown in Figure 11 and Table 2.

As shown in Figure 11, two hysteresis curves are all in the shape of quadrilateral. The stiffness remains the same in the loading process.

Table 1. List of energy dissipation coefficients at each displacement amplitude.

Test	Specimen 1	Specimen 2
Diameter of the cable/mm	20	14
Area of the cable/mm^2	152	89.6
Length of the cable/mm	3000	3000
Elasticity modulus/10^{5}Mpa	110	110
Class of the bolt	10.9	10.9
Diameter of the bolt/mm	16	20
Theory friction/kN	104	161.2
a/mm	4.67	12.27

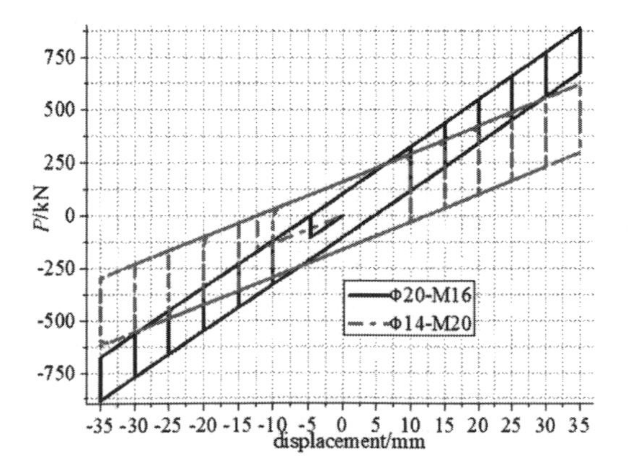

Figure 11. Theory hysteresis curve of the FZFDRB under amplitudes.

Table 2. List of energy dissipation coefficients at each displacement amplitude.

Displacement mm	β_E Φ20-M16	β_E Φ14-M20
10	1.77	2.20
15	1.38	1.80
20	1.13	1.52
25	0.96	1.32
30	0.84	1.16
35	0.74	1.04

In the process of re-centering, when the cable force is equal to the friction force (the residual displacement is a), the brace stops re-centering. Red and black dashed lines mean relaxing high strength bolts at this place, thereby eliminating the friction, and the brace continues re-centering to zero displacement under the re-centering force provided by the cables. By using M20 high strength bolt, the friction is bigger, the corresponding hysteresis loop back area is larger, the energy dissipation ability is stronger, and the re-centering force is greater than using M14 high strength bolts. By using the Φ20 steel wire rope, the load stiffness of the brace is bigger, and the bearing capacity is stronger. And so, the stiffness is proportional to the cable diameter, energy dissipation capacity is proportional to the high strength bolt's pre-tightening force. The residual displacement of brace Φ14-M20 is 12.27 mm, which is more than the 4.67 mm of brace Φ20-M16. Therefore, to make the structure have a better effect of energy dissipation and greater re-centering force, the reasonable collocation should be selected from cable force and high-strength bolts to meet the basic requirements of repairable structures after earthquakes.

As shown in Table 2, the energy dissipation coefficient β_E decreases with an increase in the displacement amplitude, and under each amplitude, β_E meets the requirements of more than 0.3 [13]. This shows that the brace has a good energy dissipation effect. Under the same displacement amplitude, the high-strength bolt model means the greater the friction, greater is the energy dissipation coefficient which shows the better energy dissipation effect.

Figures 12 and 13 show the force of Φ14 and Φ20 varies with the loading process of the brace. The load rule of the steel rope is always the same: one group of the cable force is under tension and the cable force of the other group is zero at the same time. Two groups of cable forces alternate in tension. The structure that uses two groups of cables alternating stress to avoid pre-stressed relaxation is feasible.

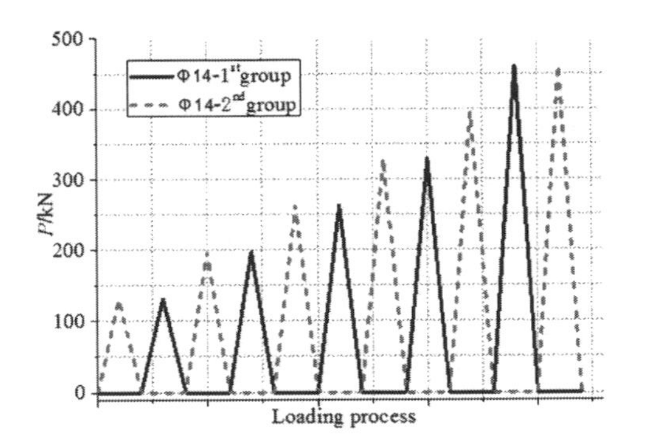

Figure 12. Graph showing the alternative variation of Φ14 cables force.

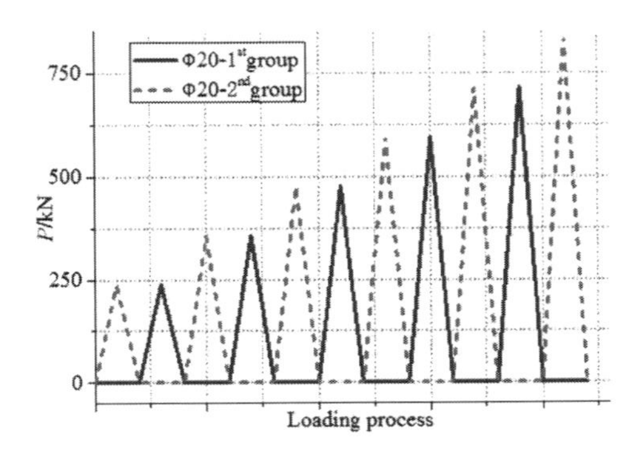

Figure 13. Alternative variation of Φ20 cables force.

5 CONCLUSION

In this paper, a fabricated zero initial cable force friction dissipation re-centering brace is proposed, its structure, working mechanism and design method of structure parts are expounded and theoretical analysis of the seismic performance is also performed to arrive at the following conclusion:

1. The FZFDRB has a simple structure. The initial cable force of two groups of the galvanized steel wire rope is zero. When the brace is under tension or compression, one group of the cables is always tensioned, the internal force of the other group of cables is zero, and natural relaxation occurs. This structure can simplify the complexity process that an existing re-centering brace needs to pre-stress in advance, and solve the question of pre-stressed relaxation.
2. The hysteresis curve of the FZFDRB is in the shape of a quadrilateral. The energy dissipation law is stability. In the process of loading, the stiffness remains the same; there is no stiffness degradation phenomenon. The energy dissipation coefficient β_E decreases with an increasing in the displacement amplitude, and under each amplitude, β_E meets the requirements of more than 0.3.
3. The load stiffness is associated with the cables and energy consumption is associated with high-strength bolts. The ability of energy consumption is proportional to the friction, but by increasing the friction, the corresponding residual deformation will also increase at the same time. And so, with different design requirements, different cables and high-strength bolts should be selected, thereby making the brace with better energy dissipation effects and a higher re-centering force, under the basic precondition of repairable structures after earthquake.

REFERENCES

ANSI/AISC 341 2005. Seismic provisions for structural steel buildings, *American Institute of Steel Construction.*

Christopoulos, C., Pampanin, S. & Priestley, M.J.N. 2003. Performance based seismic response of frame structures including residual deformations. Part 1: single-degree of freedom systems. *Journal of Earthquake Engineering. 7(1)*, 97–118.

Chen, Y. & Chen, Y.B. 2014. Research progress in self-centering energy dissipating braces. *Earthquake Engineering and Engineering Dynamic, 34(5)*, 239–246.

Erochkol, J. & Christopoulos, C. 2011. Residual drift response of SMRFs and BRB frames in steel buildings designed according to ASCE 7–05. *Journal of Earthquake Engineering. 137(5)*, 589–599.

GB 50017 2003. Code for design of steel structure. *China Planning Press.*

Liu, L., Wu, B & Li, W. 2012. Cyclic tests of novel self-centering buckling-restrained brace. *Journal of Southeast University (Natural science edition). 42(3)*, 536–541.

Miller, D.J., Fahnestock, L.A. & Eatherton, M.R. 2011. Self-centering buckling-restrained braces for advanced seismic performance. *ASCE,* 960–970.

Wakabayashi, M. & Nakamura, T. 1973. Experimental study of elastic-plastic properties of PC wall panel with built-in insulating brace. *Architectural Institute of Japan.* 1041–1044.

Wolski, M.E., Ricles, J.M. & Sause, R. 2009. Experimental study of a self-centering beam column connection with bottom flange friction device. *Journal of Structural Engineering. 135(5)*, 479–488.

Yoshino, T. & Kano, J. 1971. Experimental study on shear wall with braces: part2. *Architectural Institute of Japan. 11*, 403–404.

Zhang, A.L. & Zhao, L. 2014. Design Method of Initial Tension of Cables in Prestress-braced Steel Frame Structure. *Journal of Beijing University of Technology. 40(12)*, 1804–1809.

Zhang, A.L., Ye, Q.X., Zhan, X.X. & Liu, X.C. 2016. Research Outlook of Re-centering Function Brace. *Journal of Beijing University of Technology. 42(9)*, 1–7. doi: 10.11936/bjutxb2015120032.

Zhu, S.Y. & Zhang, Y.F. 2006. Seismic analysis of steel framed buildings with self-centering friction damping braces. *Structural Engineering & Mechanics*, 12–13.

Civil, Architecture and Environmental Engineering – Kao & Sung (Eds)
 ISBN 978-1-138-02985-9

An optimal allocation model of residual control rights in the hydraulic PPP project

Yunhua Zhang
Business School, Hohai University, Nanjing Jiangsu, China

Jingchun Feng, Song Xue & Ke Zhang
Business School, Hohai University, Nanjing Jiangsu, China
Institute of Project Management of Hohai University, Nanjing Jiangsu, China
Jiangsu Provincial Collaborative Innovation Center of World Water Valley and Water Ecological Civilization, Nanjing Jiangsu, China

ABSTRACT: The reasonable allocation of control rights of PPP is closely related to the efficiency of PPP. In this paper, an optimal allocation model of the residual control right of PPP is established from the perspective of the incentive constraint theory and incomplete contract theory, which is according to the ability difference and the goal difference between the public sector and the private sector and the uncertainty of the hydraulic PPP project. The relationship between the allocation proportion of residual control rights and many factors, such as the amount of investment of the private sector and public sector coordination, is analyzed based on the model, and a series of incentives and constraints are put forward based on optimization allocation of residual control rights. The research results of this paper not only provide theory and application basis for the optimal allocation of the residual control rights of PPP, but also perfect and supplement the contract design theory of the PPP model.

1 INTRODUCTION

As the innovation mode of the supply in the infrastructure, Public–Private Partnership (PPP) has been paid attention by the theory and industry widely, especially the co-operation efficiency between public and private sectors has attracted more and more attention. However, the efficiency of cooperation is closely related to the allocation of control rights (Hu 2012). The control rights between the public and private parties are assigned and transferred by using the concession agreement in PPP (Ye et al. 2011). The concession agreement drawn by the two parties is an incomplete contract on account of the limited rationality, the long-term cooperation, the incomplete information, and the uncertainty of the project (Rasmussen & Baird 2001). Who has the authority to make decisions when the contract is not expected to occur when there is a situation or decision-making problems? Although there is a principle for "who invests, who makes decisions, who gains, who bear the risk", but the main investments of hydraulic PPP projects are numerous, and their interests are not consistent. At this point, who has the right to make a decision? What should be done to improve the efficiency of the supply of the hydraulic infrastructure? The key to the problem lies in the allocation of control rights. The allocation of control rights can attract private sector participation and take advantage of the private sector's management and technical advantages. On the other hand, the allocation of control rights should enable the public sector to obtain regulatory projects and ensure public interest. Based on the consideration of the above-mentioned two aspects, we can come to the conclusion that rational allocation of control rights should combine the characteristics of the hydraulic PPP project and the requirements of supply efficiency, thereby giving full play to the advantages of the public and private sectors, and coordinate the interest conflicts of both sides.

2 LITERATURE ON CONTROL RIGHTS ALLOCATION

A series of theoretical models about control rights have been constructed from the 1980s to the present in the abroad, which is shown in Table 1. Those theories have researched on control rights allocation gradually from privates' cooperation to co-operation between public and private sectors, and from private goods to impure public goods.

Table 1. Theoretical model and influencing factors of control rights in foreign countries.

Influence factor	GHM (Grossman & Hart 1986, Hart & Moore 1990)	HSV (Hart et al. 1997)	BG (Besley & Ghatak 2001)	FM (Francesconi & Muthoo 2006)
	Theoretical model			
Partner type	Private and private	Private and public	Private and public	Private and public
Product degree of public	Private goods	Pure public goods	Pure public goods	Impure public goods
Level of product value evaluation	—	—	√	√
Investment type	—	√	—	—
Investment importance	√	—	—	√

The scope of research is constantly expanding and the applicability of the conclusion is rising too, which laid a solid theoretical foundation for the study of the rational allocation of control rights in China.

The research on control rights in foreign countries is systematic, but its research conclusion is not completely applicable to the current situation of China's hydraulic PPP project. The research on PPP control rights in our country has made some achievements, although it started relatively late. The domestic studies mainly focus on the theoretical model of the control rights allocation, which is derived from the incomplete contract theory or the principal agent theory. The control right is considered as that which is continuous from 0 to 1. The influencing factors of residual control rights are analyzed through the cooperation surplus maximization (Sun et al. 2011, Zhang et al. 2009). However, the existing research results not only simplify the private sector's output variables, but also ignore the differences of different enterprises' ability levels and different types of projects regardless of the mathematical model, game model, experimental research, case studies, and empirical research, and most of them dominated are by the static analysis (Chen 2010, Du & Wang 2013, Hu 2012, Sun et al. 2011, Zhang & Jia 2012). Although there are a few scholars who have discussed the influence of the ability to control the co-operation between the two sides (Cao et al. 2014), but they did not carry out a detailed and in-depth analysis of the differences in capacity. Some scholars have studied the control rights allocation of the public transportation service project, but the research on the control rights allocation of PPP is lacking practicality at this stage, whether it is applied in hydraulic facilities, municipal facilities, transportation facilities and environmental engineering, and especially lack of an in-depth study of control rights allocation of the hydraulic PPP project. At the same time, the research of control rights is not related to the interests' demand of the public and the private sectors, and the research on how to ensure the goal of the private sector's economic efficiency and the realization of the social benefits of the public sector is still lacking, in particular.

An optimal allocation model of residual control rights of PPP was constructed from the perspective of an incentive constraint theory and incomplete contract theory, which is according to the ability difference and the goal difference between the public sector and the private sector and the uncertainty of the hydraulic PPP project. The relationship between the allocation proportion of residual control rights and many factors, such as the amount of investment of the private sector and public sector coordination is analyzed based on the model, and a series of incentives and constraints are put forward based on optimization allocation of residual control rights.

3 OPTIMIZING ALLOCATION MODEL

3.1 *Model assumptions*

- Assuming that the public sector (G) and the private sector (E) jointly work together in hydraulic PPP projects, both sides invested money in I. The public sector invested money in I_G, the private sector invested money in I_E, so that the initial stake of the public sector is $\pi_0 = I_G / (I_G + I_E) = I_G / I, 0 \le \pi_0 \le 1$, and the initial stake of the private sector is $1 - \pi_0 = I_E / (I_G + I_E) = I_E / I$.
- Existing research over-simplified the PPP projects' output or the influence variables of the project income (Zhang et al. 2009, Du & Wang 2013, Cao et al. 2014), instead of distinguishing the ability difference of the public sector and private sector. Based on classic Cobb–Douglas production function in Economics, in this paper, specific investment A_i of both the public and the private parties involved in

hydraulic PPP projects is assumed, in addition to the labor and capital, including inputs of special technology and management of the private sector and inputs of special information, coordination, and guarantee of the public sector. L_E^a and L_G^{b+c} are used to distinguish the level of human capital investment of the private sector, public sector, and social public. The output (R) of the hydraulic PPP project is given as follows: $R = A_i(L_E^a + L_G^{b+c})I^\beta$.

- Assuming that λ and ω are respectively the economic and social benefits of the output factor of hydraulic PPP projects, and then the economic and social benefits of expectation values are given as follows: $R_r = \lambda R$ and $R_r = \omega R$.
- Assuming the level of effort of the private sector is η. If $\eta > 0$, then the cost of efforts to $\frac{1}{2}\eta a^2$; suppose the joint supervision cost factor in the public sector and the private sector is $\xi, \xi > 0$, then both monitoring costs is given by the following relation: $\frac{1}{2}\xi(b^2 + 2kbc + c^2)$.
- Assuming the benefits of residual control rights include non-monetary benefits m and monetary benefits n, where $m > 0$ and $n > 0$, among which the non-monetary benefits m are only owned by the public sector and the general public; the private sector can get part-monetary benefits. Residual control rights for the private sector is supposed as $\pi, 0 < \pi < 1$; the private sector can obtain the monetary income of $\pi \cdot n$. The residual control right of the public sector is $1 - \pi$.

3.2 *Model construction*

In the hydraulic PPP mode, while in both the public and the private parties, uncertain events occur or the contract has not been agreed upon decisions and problems in the process of co-operation, the two sides will make the choice based on the maximization of their own interests, with the premise of ensuring the maximization of the project's total revenue. Therefore, according to the above-mentioned assumptions, the total income of the project can be obtained by using the following equation:

$$\max[A_i(L_E^a + L_G^{b+c})I^\beta + m + n - I - \frac{1}{2}\eta a^2 - \frac{1}{2}\xi(b^2 + 2kbc + c^2)] \quad (1)$$

According to the efficiency theory, when combining the characteristics of hydraulic PPP projects, under the premise of satisfying the need of maximizing the total revenue, the private sector can attract social capital only in the case that the output is greater than the investment involved in hydraulic PPP projects; meanwhile, the public sector is also expected to obtain greater output than the expected social benefits. Hence, the following incentive constraints are observed:

$$(1-\pi_0)[A_i(L_E^a + L_G^{b+c})I^\beta] + \pi \cdot n - \frac{1}{2}\eta a^2 \geq I_E \quad (2)$$

$$(1-\pi)\pi_0[A_i(L_E^a + L_G^{b+c})I^\beta] + m + (1-\pi)\cdot n - \frac{1}{2}\xi(b^2 + 2kbc + c^2) \geq R_S \quad (3)$$

From the above-mentioned Equations (1)–(3), the Lagrange function is constructed as follows:

$$\begin{aligned} L = &\left[A_i(L_E^a + L_G^{b+c})I^\beta + m + n - I - \frac{1}{2}\eta a^2 - \frac{1}{2}\xi(b^2 + 2kbc + c^2)\right] \\ &+ \lambda_1\left[(1-\pi_0)[A_i(L_E^a + L_G^{b+c})I^\beta] + \pi \cdot n - \frac{1}{2}\eta a^2 - I_E\right] \\ &+ \lambda_2\left[(1-\pi)\pi_0[A_i(L_E^a + L_G^{b+c})I^\beta] + m + (1-\pi)\cdot n - \frac{1}{2}\xi(b^2 + 2kbc + c^2) - R_S\right] \end{aligned} \quad (4)$$

Suppose $\frac{\partial L}{\partial \lambda_1} = 0$, then the optimal controls of the private sector are defined as follows:

$$\pi^* = \frac{I_E - (1-\pi_0)A_i(L_E^a + L_G^{b+c})I^\beta - \frac{1}{2}\eta a^2}{n} = \frac{I_E - (1-\pi_0)R - \frac{1}{2}\eta a^2}{n} \quad (5)$$

Or suppose $\frac{\partial L}{\partial \lambda_2} = 0$, then the optimal control rights of the private sector are defined as follows:

$$\pi^* = \frac{\pi_0 A_i(L_E^a + L_G^{b+c})I^\beta - \frac{1}{2}\xi(b^2 + 2kbc + c^2) + n + m - R_S}{\pi_0[A_i(L_E^a + L_G^{b+c})I^\beta + n} \quad (6)$$

3.3 *Model analysis*

By using the proposed analysis of the allocation of the residual control rights mechanism and model assumptions in hydraulic PPP projects, obtaining mathematical equations that enable allocation of residual control rights are defined as equation (5) or equation (6) in the private sector. Based on the equation (5) or equation (6), further analysis is put forward for both parties such that the factors, such as capacity and goal difference, have an impact on the residual control rights and incentive constraint mechanism designing.

a. From equation (5), we can obtain the following equation:

$$\frac{\partial \pi^*}{\partial I_E} = \frac{1}{n} > 0, \quad \frac{\partial \pi^*}{\partial \pi_0} = \frac{A_i(L_E^a + L_G^{b+c})I^\beta}{n} = \frac{R}{n} > 0,$$

Whereby the public sector should be based on the amount of investment and equity in the private sector to determine the allocation of residual control rights and incentive programs, the incentive program is given as follows: When the amount of financing is large in the private sector, it should be allocated with more residual control rights, to encourage private sector participation in hydraulic PPP projects and contribute to the implementation of the ultimate realization of the PPP project financing; the greater the initial stake possessed by the private sector is, the greater residual control rights is allocated, to encourage the private sector to create more economic benefits.

b. From equation (5) or equation (6), the following equation is obtained: $\frac{\partial \pi^*}{\partial A_i} = -\frac{(L_E^a + L_G^{b+c})I^\beta}{n} < 0$, $\frac{\partial \pi^*}{\partial a} = -\frac{A_i I^\beta L_E^a \ln L_E}{n} < 0$

Whereby we receive the allocation of residual control rights and the incentive constraint program on the basis of capacity difference between the private sector and public sector, and the program is given as follows: when the private sector has advantages, such as the level of technical expertise, management ability, and the human capital investment is higher, the probability of hydraulic PPP projects succeeding and being controlled is greater. In this case, we can appropriately reduce the allocation of residual control over the private sector, to constrain the private sector, thereby keeping the level of good technical expertise and strong management and to achieve benefits; when the public sector has more advantages of the co-ordination and support ability, then the public sector should possess more residual control rights.

c. From equation (6), the following equation is obtained: $\frac{\partial \pi^*}{\partial R_S} = -\frac{1}{\pi_0 [A_i (L_E^a + L_G^{b+c}) I^\beta + n} < 0$

Whereby we can receive the allocation of residual control rights and the incentive constraint program in the public sector, and the program is given as follows: when the social efficiency exception is high, the public sector should enhance the proportion of residual control rights, which strengthen the monitor rights to hydraulic PPP projects, so as to ensure the interests of the public.

4 CONCLUSIONS

Residual control rights are continuous variables from 0 to 1, not simply by either the public sector or the private sector owning alone. In order to attract social capital to participate in hydraulic PPP projects, to utilize the governance effect of control rights, and to stimulate specificity investment from the private sector, the public sector should share one part of the residual control rights with the private sector. In the whole life cycle of the hydraulic PPP projects, according to the type of uncertainty events, the degree and range of influence, the residual control should be dynamically adjusted between public and private sectors.

In the process of hydraulic PPP projects, investment amounts and equity ratios are positively correlated with the allocation of residual control rights in the private sector; the ability advantages of the private sector, such as technical level, management ability, and effort level are in negative correlation with its allocation of residual control rights; The ability advantages of coordination, supervision, as well as the expectations values of the social benefits of the PPP project are positively correlated with its allocation of residual control in the public sector.

Due to the incompleteness of the hydraulic PPP project contract and the uncertainty of public–private cooperation, allocation of control rights is a very complex system process. In this paper, influence factors that the proposed model has suggested is insufficient, such as we did not consider how the policy support and the trust degree of both public and private sectors affect the allocation of residual control rights. It will require further research to solve the problem.

ACKNOWLEDGMENTS

The authors gratefully acknowledge the National Social Science Foundation of China (No. 14AZD024 and No. 15CJL023), National Science & Technology Pillar Program during the Twelfth Five-year Plan Period (2015BAB07B01) and the Fundamental Research Funds for the Central Universities (2015B08214) for financial support.

REFERENCES

Besley, T. & M. Ghatak (2001). Government versus private ownership of public goods. *J. The Quarterly Journal of Economics. 116(4)*, 1343–1372.

Cao, H. D., L. LI & J. L. Zheng (2014). Allocation of public project's control authority. *J. Journal of Industrial Engineering /Engineering Management. 02*, 55–63.

Chen, F. (2010). Governance mechanism study on PPP project based on contract relationship. *D. Central South University*.

Du, Y. L. & J. Y. Wang (2013). Research on the proper allocation of rights of control in the projects under BT mode—A multi-case comparative study. *J. Soft Science. 27(5)*, 56–61.

Francesconi, M. & A. Muthoo (2006). Control rights in Public-Private Partnerships. *R. IZA Discussion Papers*.

Grossman, S. J. & O. D. Hart (1986). The costs and benefits of ownership: a theory of vertical and lateral integration. *J. The Journal of Political Economy. 94(4)*, 691–719.

Hart, O. & J. Moore (1990). Property rights and the nature of the firm. *J. Journal of Political Economy. 98(6)*, 1119–1158.

Hart, O., A. Shleifer & R. W. Vishny (1997). The proper scope of government: theory and an application to prisons. *J. The Quarterly Journal of Economics. 112(4)*, 1127–1161.

Hu, Z. (2012). The decision model of the allocation of control rights in public-private partnerships projects. *J. Xi' an University of Architecture & Technology (Natural Science Edition). 44(01)*, 90–96.

Rasmussen, R. K. & D. G. Baird (2001), Control rights, priority rights, and the conceptual foundation of corporate reorganizations. *J. Virginia Law Review. 87(5)*, 921.

Sun, H., Z. Q. Fan & Y. Shi (2011). Research on the optimal ownership structure of an ex expressway under a PPP scheme. *J. Journal of Industry Engineering and Engineering Management. 3(1)*, 154–157.

Ye, X. S., P. C. Yi & S. X. Wu (2011). The discussion of PPP project control rights' essence. *J. Science & Technology Progress and Policy. 28(13)*, 67–70.

Zhang, Z. & M. Jia (2012). A experimental study on allocation of control rights in PPPs. *J. Journal of Systems & Management. 21(2)*, 166–179.

Zhang, Z., M. Jia & D. F. Wan (2009). Theoretical study on the efficient allocation of control rights in the public-private partnership (PPP). *J. Journal of Industrial Engineering and Engineering Management. 23(03)*, 23–29.

Civil, Architecture and Environmental Engineering – Kao & Sung (Eds)
© 2017 Taylor & Francis Group, ISBN 978-1-138-02985-9

A study on the optimized hydraulic circulation system of the artificial lake

Qiannan Jin, Aiju You & Haibo Xu
Zhejiang Institute of Hydraulics and Estuary, Hangzhou, Zhejiang, China

ABSTRACT: Constructing an optimized hydraulic circulation system is an effective means to improve the lake water's quality. The hydraulic circulation of the artificial lake in Shangyu was designed by including the external circulation system and the internal circulation system. The flow field of the lake was simulated and analyzed based on MIKE21 software, and the results showed that under the same water supplement and wind conditions, the internal circulation system combined with the morphological characteristics of the lake could eliminate the area of the dead water zone obviously.

1 INTRODUCTION

In recent years, the artificial landscape lakes come forth continuously in cities, for improving the living environment. However, artificial lakes face high risk of water pollution due to their poor flow-ability and low self-purification ability. And the lakes are always designed to be irregular in shape to meet the people's demand of landscape, and that is prone to create the "dead water zone". In the zone, the flow is slow, the water cannot be replaced, and a variety of pollutants are deposited, which leads to the deterioration of water quality, and spreading of pollutants to the entire lake.

The important measure to prevent dead water zone creation and to make water in the lake flow is constructing an optimized hydraulic circulation system. It means that, besides the strict control of pollution into the lake, good water should be supplied to the lake to update the lake water (Fu & Yu 2011, Kang et al. 2012), and the hydraulic circulation should be optimized and designed to prevent the adverse effects of irregular shape (You et al. 2014).

In this paper, as for the artificial lake in Shangyu, by combining with the morphological characteristics of lake, the hydraulic circulation system was proposed, based on the numerical simulation and comparative analysis of the flow field. We hope to provide references for similar projects.

2 PROJECT PREFACE

The planning artificial lake is located in the core area of the coastal town in Shangyu City. The area of the lake is about 80.54 hectares, the average water depth is 3.1 m and the normal water level is 3.1 m. The lake morphology is East–West, the distance from east to west is 2.8 km, and the widest point in the north–south direction is 850 m while the narrowest is 90 m. The original rivers on the south side of the lake were retained and formed with the lake in series and the width of the rivers is 30–70 m, and the total length is about 3 km. The layout of the artificial lake is shown in Figure 1. The rivers on the south side are winding prone to create a dead water zone.

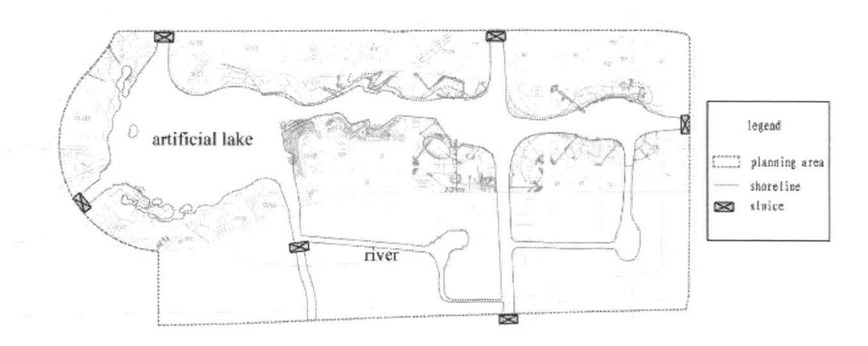

Figure 1. Layout of the artificial lake.

3 HYDRAULIC CIRCULATION SYSTEM DESIGN

Hydraulic circulation could promote the lake water flow, thereby having three benefits: (1) play the role of aeration, (2) prevent the dead water zone formation of the rivers on the south side, and (3) improve water quality combined with the ecological system in the rivers, where the pollutants could be better purified. The hydraulic circulation system was designed including two parts: the external circulation system and the internal circulation system, which is shown in Figure 2.

3.1 *External circulation system*

The external circulation system was constructed to supply good water to the lake to update the lake water. The supplement water requirement was mainly used for: (1) supplement for water losses due to evaporation and seepage to maintain the water level fluctuation in an acceptable range; (2) enhancing the water flow rate and reducing water retention time to prevent lake eutrophication. A variety of factors should be considered to determine the requirement, such as, the water quality condition of the supplement water source, the pollution source into the lake, and the lake target water quality. Through analysis, the water diversion scale was determined as 30,000 t/d (0.35 m^3/s). The details of water requirement analysis could be found elsewhere (You et al. 2012). The supplement water source was supplied into the lake through three water inlets in the west side of the lake. The three inlets' flow rates were 0.09 m^3/s, 0.17 m^3/s, and 0.09 m^3/s.

3.2 *Internal circulation system*

The internal circulation system was constructed to accelerate the water flow to prevent the adverse effects of irregular shape and the dead water zone of the rivers on the south side. The internal circulation system was composed of three weirs, a pump, an ecological system, and two adjustable outlets. Its working principle is as follows: 60,000 t/d water was pumped from the eastern lake through weir 1 and filled into the ecological rivers; the water treated by using the plant purification system in the rivers were then filled into the lake through inner outlets 1 and 2 (weir 2 and 3).

4 NUMERICAL SIMULATION

4.1 *Numerical model*

MIKE21 software (DHI, Denmark) was used to simulate the hydraulic circulation of the artificial lake, and the flow field of the lake whether the internal circulation system and whether the constant wind (southeast, 3 m/s) effect were analyzed. A 2-D flow model of the lake region was built by using on MIKE21 software.

4.1.1 *Computational grid generation*

A 10-m (x = y = 10 m) rectangular grid was used to divide the topography of the lake region. The grid-based terrain is shown in Figure 3.

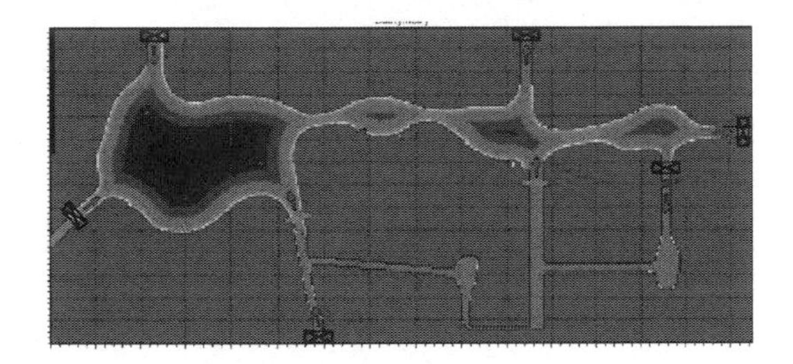

Figure 3. Schematic diagram of grid-based terrain.

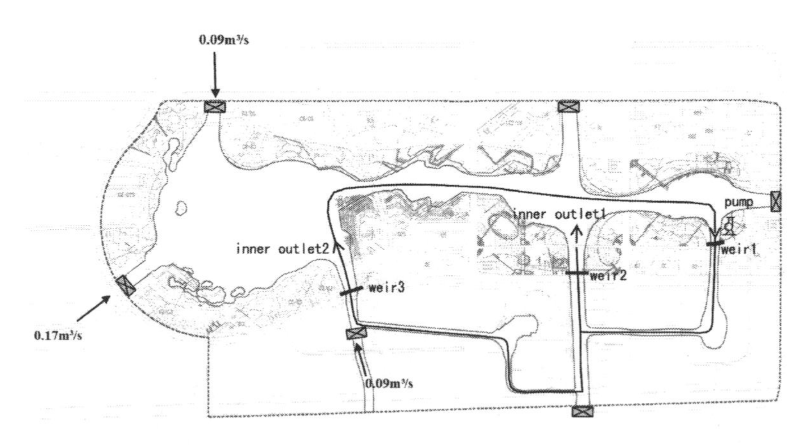

Figure 2. Schematic diagram of the hydraulic circulation system.

4.1.2 *Definite solution condition*

Initial conditions: the elevation of the lake was 3.1 m and the average water depth was about 3.0 m, according to the initial calculation of concentration.

Boundary conditions: the flow rate of the total amount of supplement water into the lake was 0.35 m^3/s, thereby maintaining the balance of import and export.

4.1.3 *Other conditions and parameters*

The simulation time step was 0.5 s, the diffusion coefficient was 0.5 m^2/s, and the continuous water exchange method was adopted.

4.2 *Result analysis*

The velocity variation of the lake based on different plans is listed in Table 1 and shown in Figure 4. The results show that wind played a leading role in the lake current (by comparison of plan 3 and plan 1 and plan 4 and plan 2). The wind-driven current played a dominant role while the inflow–outflow caused by water diversion made a relatively small impact on the lake current. However, the latter one still improved the area of relatively high flow velocity significantly, especially in the circulation path (by comparison of plan 3 and plan 5). The effect of the wind field on the lake was greater than that on the river, so that the velocity of the river was much slower than the lake, and was prone to create the dead water zone. Therefore, with the internal circulation system combining with the lake morphology, the dead water zone of the southern river could be eliminated (by comparison of plan 2 and plan 1 and plan 4 and plan 3). Moreover, the water

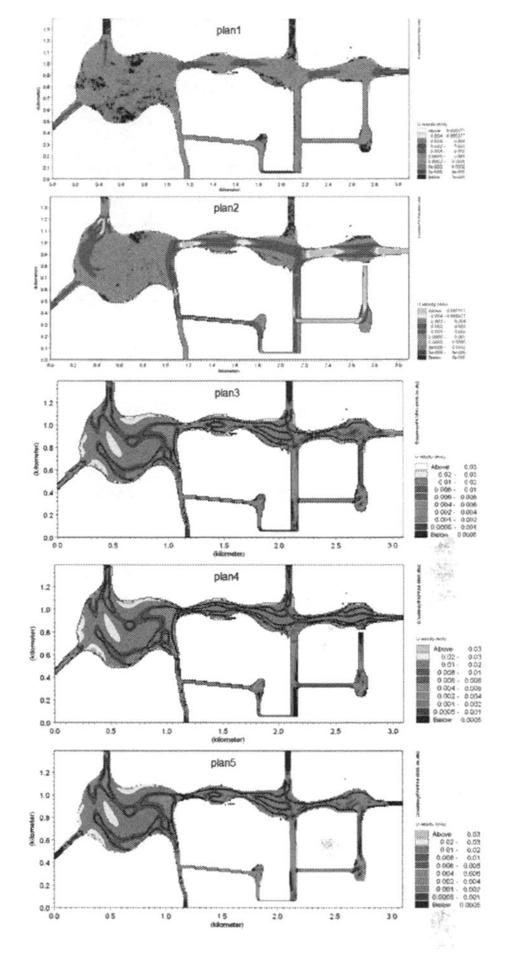

Figure 4. Velocity of the lake based on different plans.

Table 1. Area percentage of different grades of velocity in the artificial lake.

Plans	Area percentage of greater than a certain velocity, %							
	3 cm/s	2 cm/s	1 cm/s	0.8 cm/s	0.6 cm/s	0.5 cm/s	0.4 cm/s	0.3 cm/s
1: no wind, without internal circulation, and with external circulation	0	0	0	0	0	0	0.05	0.62
2: no wind, with internal circulation, and with external circulation	0	0	0	0.32	2.85	4.02	6.06	10.79
3: southeast wind, 3 m/s, without internal circulation, and with external circulation	0.25	6.96	40.52	52.36	65.83	72.5	79.42	85.72
4: southeast wind, 3 m/s, with internal circulation, and with external circulation	0.5	8.76	45.01	56.54	69.03	74	79.8	84.93
5: southeast wind, 3 m/s, without internal circulation, and without external circulation	1.07	8.12	27.11	34.44	44.88	51.39	58.52	65.89
2-1: compare the effects of internal circulation	0	0	0	0.32	2.85	4.02	6.01	10.17
4-3: compare the effects of internal circulation	0.26	1.8	4.49	4.18	3.2	1.5	0.38	–0.79
3-1: compare the effects of wind	0.25	6.96	40.52	52.36	65.83	72.5	79.37	85.1
4-2: compare the effects of wind	0.5	8.76	45.01	56.22	66.18	69.98	73.74	74.14
3-5: compare the effects of external circulation	–0.82	–1.16	13.41	17.92	20.95	21.11	20.9	19.83

of the lake could be further purified by using the river ecosystem.

5 CONCLUSION

The artificial lakes are conducive to improve the city landscape and ecological environment. The artificial lake in Shangyu was designed to be irregular in shape with winding rivers to meet the people's demand of the landscape that is prone to create the dead water zone, thereby leading to the deterioration of water quality. By combining with the morphological characteristics of the lake, the hydraulic circulation system was proposed including the external circulation system and the internal circulation system.

The hydraulic circulation of the artificial lake was simulated by MIKE21 software, and the flow field of the lake was analyzed. Under the same water supplement and wind conditions, the internal circulation system could achieve the following: the flow field distribution tends to be uniform, the state of the water flow is greatly improved, the area of the dead water zone is eliminated obviously, and the fluidity of water in the southern river increased. In general, the optimized hydraulic circulation system could effectively maintain and improve the water quality, and MIKE 21 software is an effective tool to construct an eco-design of artificial lakes.

REFERENCES

Fu Zongfu & Yu Guoqing 2011. Method of water quality improvement for urban lake based on concentrated and distributed water diverting. *Journal of Tianjin University* (3):51–56.

Kang Ling, Guo Xiaoming, Wang Xueli 2012. Study on water diversion schemes of large urban lake group. *Journal of Hydroelectric Engineering* (3):65–69.

You Aiju, Wu huaan, Jin Qiannan 2012. Study on the water requirement of man-made lake based on water pollution risk analysis. *Environmental Pollution & Control* 34(8):1–4.

You Aiju, Jin Qiannan, Han Zencui 2014. Water purification system design for man-made lake based on the source water quality. *Resources, Environment and Engineering* (3):3–8.

Civil, Architecture and Environmental Engineering – Kao & Sung (Eds)
© 2017 Taylor & Francis Group, ISBN 978-1-138-02985-9

Development of new elastic constant estimation method using laser ultrasonic visualization testing

T. Saitoh & A. Mori
Department of Civil and Environmental Engineering, Gunma University, Japan

ABSTRACT: This paper presents a new elastic constant estimation method using the laser ultrasonic visualization testing for anisotropic elastic solids such as CFRP (Carbon Fiber Reinforced Plastics). In this paper, first, EFIT (Elastodynamic Finite Integration Technique) formulations for general anisotropic elastodynamics are described. Next, the procedure of our developed elastic constant estimation method is described with anisotropic elastodynamic theory. As numerical examples, elastic constants of a CFRP specimen are predicted by the proposed method. In addition, simulation of ultrasonic wave propagation in the CFRP with estimated elastic constants is implemented by using the EFIT and the results are compared with those obtained by the laser ultrasonic visualization testing to validate our proposed method.

1 INTRODUCTION

In recent years, anisotropic materials have been attracted lots of interest for various engineering fields. CFRP (Carbon Fiber Reinforced Plastics) is known as one of the anisotropic materials and has been used as a main material of aircraft and as a seismic strengthening member of civil and architectural structures. In general, the Ultrasonic nondestructive Testing (UT) is widely used for material inspection because the UT is portable for ready to use and transport compared to the Radiographic Testing (RT) which can be practiced by only suitably trained and qualified personnel. However, anisotropic materials sometimes make it difficult for inspectors to detect flaws because the phase and group velocities of elastic waves in anisotropic materials depend on their propagation directions. Therefore, a better knowledge of anisotropy is a great help for accurate the UT. In general, it is essential to determine the elastic constants of the anisotropic materials should be inspected in order to understand the anisotropic properties. Several researches about elastic constant estimation methods have been done since decades ago. A problem of conventional elastic constant estimation methods that need to be overcome is that we have to cut the test specimen to investigate the phase velocity of waves propagating in an oblique direction. This disadvantage is not efficient and economic, and should be resolved from the point of view of environmental aspect of waste generation.

Therefore, in this research, a new elastic constant estimation method for anisotropic materials is developed. A laser ultrasonic visualization technique is incorporated to measure ultrasonic wave phase velocities which are required for elastic constant estimation.

As numerical examples, elastic constants of a CFRP specimen are predicted by the proposed method. In addition, simulation of ultrasonic wave propagation in the CFRP with estimated elastic constants is implemented by using the EFIT (Elastodynamic Finite Integration Technique) and the results are compared with those obtained by the laser ultrasonic visualization testing to validate our proposed method.

2 PROBLEMS

2.1 *Fundamental equations*

Let us consider a transversely anisotropic and homogeneous elastic material with density ρ as shown in Figure 1. The geometry of the elastic material is cuboid with width w, depth, d and height h. Elastodynamic field is excited by a laser ultrasonic non-contacting system with the central frequency f as shown in Figure 1. The displacement field $u_i(\boldsymbol{x}, \boldsymbol{t})$ on the field point $\boldsymbol{x}$ at time t satisfies the equation of motion and the constitutive equation as follows:

$$\rho\frac{\partial \dot{u}(\boldsymbol{x},t)}{\partial t} = C_{ijkl}u_{k,l}(\boldsymbol{x},t) \quad (1)$$

$$\sigma_{ij}(\boldsymbol{x},t) = C_{ijkl}u_{k,l}(\boldsymbol{x},t) \quad (2)$$

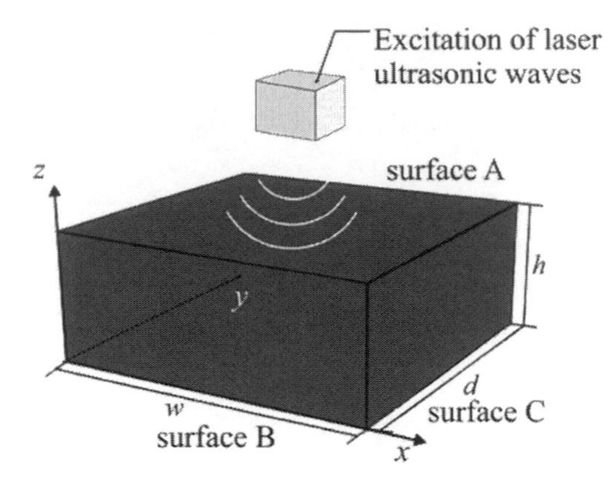

Figure 1. A model for the laser ultrasonic visualization testing for an anisotropic specimen.

where $\sigma_{ij}(\boldsymbol{x}, t)$ is the stress field and C_{ijkl} is the elastic constant. In addition, $(\dot{\cdot})$ and $()_{,i}$ denote the partial derivative with respect to time t and spatial variable x_i, respectively. For simplicity, 2-D formulation is considered hereafter. Taking time derivative of equation (2) and using engineering notation, which uses x, y, z for x_1, x_2, x_3, for the resulting equation, the relation between stress and particle velocity can be obtained as follows:

$$\frac{\partial}{\partial t}\begin{pmatrix} T_1 \\ T_3 \\ T_5 \end{pmatrix} = \begin{pmatrix} C_{11} & C_{13} & 0 \\ C_{31} & C_{33} & 0 \\ 0 & 0 & C_{55} \end{pmatrix} \begin{pmatrix} \dfrac{\partial \dot{u}}{\partial x} \\ \dfrac{\partial \dot{w}}{\partial z} \\ \dfrac{\partial \dot{w}}{\partial x} + \dfrac{\partial \dot{u}}{\partial z} \end{pmatrix} \tag{3}$$

where T_1, T_3 and T_5 are defined by $T_1 = \sigma_{11}, T_3 = \sigma_{33}$ and $T_5 = \sigma_{31}$, respectively. In addition, $u_1 = u$, $u_3 = w$, and $C_{\alpha\beta}(\alpha, \beta = 1, \ldots, 6)$ is the Voigt form of the elastic constant C_{ijkl}. Equation (1) can be rewritten using the engineering notation as follows:

$$\rho \frac{\partial \dot{u}}{\partial t} = \frac{\partial T_1}{\partial x} + \frac{\partial T_5}{\partial z} \tag{4}$$

$$\rho \frac{\partial \dot{w}}{\partial t} = \frac{\partial T_5}{\partial x} + \frac{\partial T_3}{\partial z} \tag{5}$$

Normally, Equations (3), (4) and (5) are solved by using appropriate numerical methods such as FEM (Finite Element Method), BEM (Boundary Element Method) or FDM (Finite Difference Method). In this research, EFIT (Elastodynamic Finite Integration Technique) (Nakahata et al., 2011, Fellinger et al., 1995) is used to obtain elastic wave fields in anisotropic materials.

2.2 *EFIT formulation*

A brief description of the EFIT formulation is given in this section. The starting point of the EFIT for elastodynamics is the integral form of equations (3), (4) and (5). The EFIT requires integrations over certain control surfaces S of the cells as defined in Figure 2. As seen in Figure 2, a square cell is used as an integral surface S. For each cell, the normal stresses T_1 and T_3 are located at the center of the cell. Considering a spatial staggered grid as shown in Figure 2, the particle velocities $\dot{u}$ and $\dot{w}$, and the shear stress T_5 are naturally arranged at the midpoints of the edges and the corners of the cell. Therefore, the stresses T_1, T_3 and T_5 are calculated as follows:

$$\begin{aligned} T_1^{n+\frac{1}{2}}(i,k) = T_1^{n-\frac{1}{2}}(i,k) &+ \gamma C_{11}\left\{\dot{u}^n\left(i+\frac{1}{2},k\right) - \dot{u}^n\left(i-\frac{1}{2},k\right)\right\} \\ &+ \gamma C_{13}\left\{\dot{w}^n\left(i,k+\frac{1}{2}\right) - \dot{w}^n\left(i,k-\frac{1}{2}\right)\right\} \end{aligned} \tag{6}$$

$$\begin{aligned} T_3^{n+\frac{1}{2}}(i,k) = T_3^{n-\frac{1}{2}}(i,k) &+ \gamma C_{31}\left\{\dot{u}^n\left(i+\frac{1}{2},k\right) - \dot{u}^n\left(i-\frac{1}{2},k\right)\right\} \\ &+ \gamma C_{33}\left\{\dot{w}^n\left(i,k+\frac{1}{2}\right) - \dot{w}^n\left(i,k-\frac{1}{2}\right)\right\} \end{aligned} \tag{7}$$

$$\begin{aligned} T_5^{n+\frac{1}{2}}\left(i-\frac{1}{2},k-\frac{1}{2}\right) = T_5^{n-\frac{1}{2}}\left(i-\frac{1}{2},k-\frac{1}{2}\right) &+ \gamma C_{55}\left\{\dot{w}^n\left(i,k-\frac{1}{2}\right) - \dot{w}^n\left(i-1,k-\frac{1}{2}\right)\right\} \\ &+ \gamma C_{55}\left\{\dot{u}^n\left(i-\frac{1}{2},k\right) - \dot{u}^n\left(i-\frac{1}{2},k-1\right)\right\} \end{aligned} \tag{8}$$

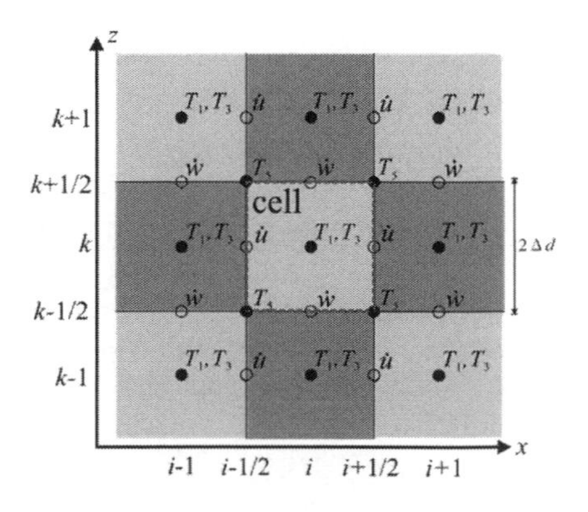

Figure 2. A spatial staggered grid of the 2D EFIT.

where $\gamma = (\Delta t / \Delta d)$, and Δt and Δd are time increment and the half length of the unit cell size, respectively. The particle velocities $\dot{u}$ and $\dot{w}$ are calculated as well as the stresses T_1, T_3 and T_5 as follows:

$$
\begin{aligned}
\dot{u}^{n+1}\left(i-\frac{1}{2},k\right) &= \dot{u}^{n}\left(i-\frac{1}{2},k\right) \\
&+\frac{\gamma}{\rho}\left\{T_1^{n+\frac{1}{2}}(i,k)-T_1^{n+\frac{1}{2}}(i-1,k)\right\} \\
&+\frac{\gamma}{\rho}\left\{T_5^{n+\frac{1}{2}}\left(i-\frac{1}{2},k+\frac{1}{2}\right)-T_5^{n+\frac{1}{2}}\left(i-\frac{1}{2},k-\frac{1}{2}\right)\right\}
\end{aligned}
\tag{9}
$$

$$
\begin{aligned}
\dot{w}^{n+1}\left(i,k-\frac{1}{2}\right) &= \dot{w}^{n}\left(i,k-\frac{1}{2}\right) \\
&+\frac{\gamma}{\rho}\left\{T_3^{n+\frac{1}{2}}(i,k)-T_3^{n+\frac{1}{2}}(i,k-1)\right\} \\
&+\frac{\gamma}{\rho}\left\{T_5^{n+\frac{1}{2}}\left(i+\frac{1}{2},k-\frac{1}{2}\right)-T_5^{n+\frac{1}{2}}\left(i-\frac{1}{2},k-\frac{1}{2}\right)\right\}
\end{aligned}
\tag{10}
$$

For time-domain, the stresses T_1, T_3 and T_5 are allocated at half time steps, while the particle velocities $\dot{u}$ and $\dot{w}$ are at full-time steps. The detail of the EFIT formulation is skipped due to the limitation of the spaces.

3 ELASTIC CONSTANT ESTIMATION

3.1 *Laser ultrasonic visualization testing*

Laser ultrasonic visualization testing is one of the non-contacting ultrasonic methods, which use lasers to generate ultrasonic waves and detect flaws in materials. The most important advantage of this method is that we can directly see the state of ultrasonic wave propagation in real time. The ultrasonic visualization testing is utilized to obtain the phase velocities of existing waves in anisotropic materials.

Figure 3 shows a system of the laser ultrasonic visualization testing. In this research, the system developed by Tsukuba Technology Co. Ltd. in Japan, which is called LUVI, is used. In this testing, only the ultrasonic waves propagating on the surfaces of test specimens can be visualized. No visualization data of wave propagation inside specimens can be obtained. Therefore, some numerical simulation techniques are used in conjunction with the laser ultrasonic visualization inspector to understand wave propagation phenomena inside specimens. Unfortunately, wave fronts seen in visualization data obtained by LUVI are not always clear. Consequently, the edge detection which is one of image processing methods is applied to the snapshots obtained by the laser ultrasonic visualization testing in order to enhance the wave fronts. Figures 4(a) and (b) show an example of snapshot obtained by the laser ultrasonic visualization testing and its edge detection result. As seen in Figure 4(b), we can clearly confirm the wave front of quasi P-wave (qP wave) and S-wave (qS wave) due to the image processing. The wave velocities of the qP and qS waves can be obtained by the time difference between some edge-detected pictures. It is possible to estimate elastic constants if qP and qS wave velocities are determined as explained in the following section.

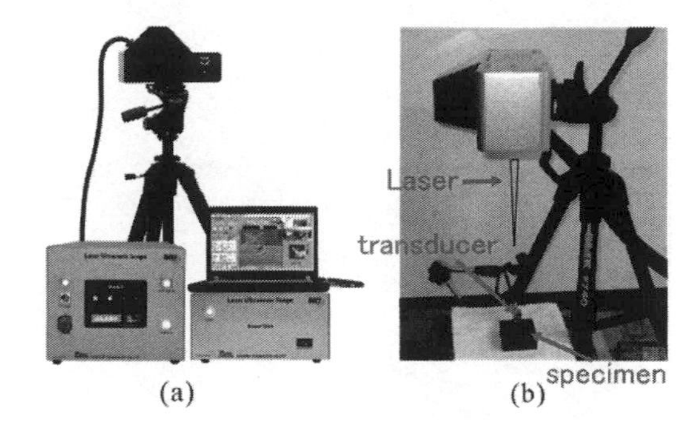

Figure 3. Laser ultrasonic visualization inspector (a) system of LUVI (b) excitation of laser ultrasonic waves.

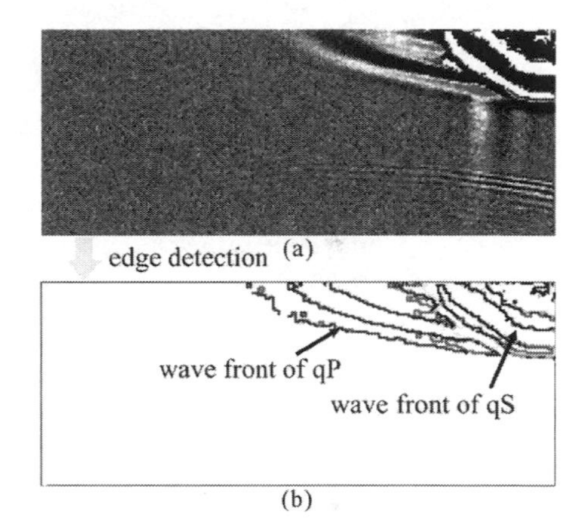

Figure 4. Laser ultrasonic visualization testing (a) a snapshot obtained by the Laser ultrasonic visualization testing (b) its edge detected-result.

3.2 *Phase velocities and group velocities in anisotropic elastic materials*

The phase velocity is equal to the group velocity for isotropic materials. However, the phase velocity is not consistent with the group velocity for anisotropic materials. Therefore, this fact should be taken into account when elastic constants of an anisotropic materials are evaluated. If the phase velocity is known, elastic constants can be obtained using the following equations derived from Christoffel equation (Auld, 1990):

$$C_{11} = \rho V_{L-L}^2 \tag{11}$$

$$C_{33} = \rho V_{L-Z}^2 \tag{12}$$

$$C_{55} = \rho V_{T\perp ZC-Z}^2 = \rho V_{T\perp LC-L}^2 \tag{13}$$

$$C_{13} = -C_{55} + \frac{1}{2} \times \sqrt{\left(4\rho V_{L-ZL}^2 - C_{11} - C_{33} - 2C_{55}\right)^2 - \left(C_{11} - C_{33}\right)^2} \tag{14}$$

where $V_{L\text{-}L}$ and $V_{L\text{-}Z}$ are the phase velocities of qP waves traveling to the x_1 and x_3 direction, respectively. In addition, $V_{T\perp LC-L}$ and $V_{T\perp ZC-Z}$ are the phase velocities of qS waves polarized in the z direction propagating to the x direction, and in the x direction propagating to the z direction, respectively. $V_{L\text{-}ZL}$ is the phase velocity of qP wave which propagates with an angle of 45° in the x-z plane. $V_{L-L}, V_{L-Z}, V_{T\perp LC-L}$ and $V_{T\perp ZC-Z}$ can be deduced from the results of the laser ultrasonic visualization testing. However, the derivation of the phase velocity $V_{L\text{-}ZL}$ is somewhat troublesome because the wave propagation seen in snapshots as shown in Figure 4(b) shows the group velocity. Therefore, in this research, the phase velocity $V_{L\text{-}ZL}$ was derived from the fact that the wave vector must always be normal to the ray surface.

4 NUMERICAL EXAMPLES

4.1 *Laser ultrasonic testing*

Laser ultrasonic visualization testing was carried out for the CFRP specimen, Toray T800S-2592. The geometry parameters of the CFRP were set to be width w = 5 cm, depth d = 5 cm, and height h = 2 cm. The density ρ was given by ρ = 160 kg/m³. The 45° angle probe with the central frequency f = 1 MHz was used as ultrasonic wave receiver. Figure 5(a) shows a snapshot of the ultrasonic waves obtained by using the laser ultrasonic visualization testing at surface B as shown in Figure 1. As seen in Figure 5(a), the qP wave propagating parallel to the x-direction shows the fastest speed of all directions. The tendency of wave propagation in CFRP is similar to that in graphite epoxy (Furukawa et al., 2014). Estimated elastic constants obtained by using the proposed method are shown in Table 1. As predicted, the component of

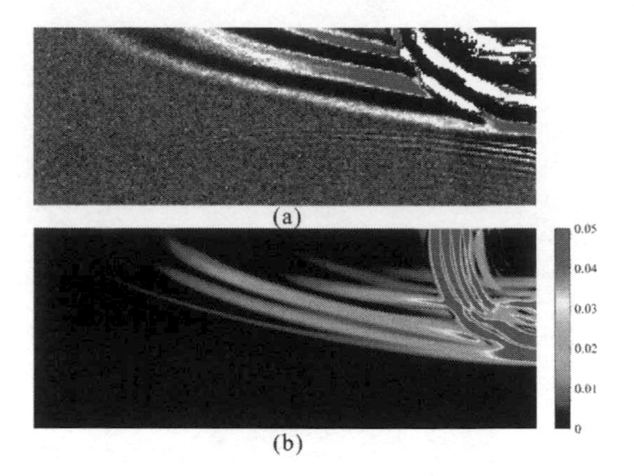

Figure 5. Numerical results obtained by (a) laser ultrasonic visualization testing and (b) EFIT simulation with estimated elastic constants shown in Table 1.

Table 1. Estimated elastic constants.

C_{11}	C_{33}	C_{13}	C_{55}
161.4	13.4	9.036	7.109 × 10⁹ GPa

the elastic constant C_{11} shows large value rather than other components.

4.2 *EFIT simulation using the estimated elastic constants*

Numerical examples using the 2-D EFIT were carried out for the ultrasonic wave propagation problem corresponding to the laser ultrasonic testing demonstrated in previous section. Estimated elastic constants as shown in Table 1 were used for the 2-D EFIT simulation in this section. In this analysis, the following time-domain incident wave with central period $T(=2\pi/\omega)$ and angular frequency ω is considered:

$$\dot{w}(\boldsymbol{x},t) = \begin{cases} \sin\left(\dfrac{2\pi}{T}\right) & \text{for} \quad 0 \le t \le T \\ 0 & \text{for} \quad \text{otherwise} \end{cases}$$

The time-domain incident wave is given as the boundary condition at the point (x, z) = (4.9, 2.0) cm. In addition, the stress free boundary condition is imposed over the surface of CFRP specimen except for the source point. Figure 5(b) shows the total particle velocity $\sqrt{\dot{u}^2 + \dot{w}^2}$ at the surface B of the CFRP specimen obtained by the EFIT simulation, which is corresponding to the result shown in Figure 5(a). As can be observed in Figure 5(b), the ultrasonic waves propagate at different speeds for different directions due to the influence of anisotropic property. We can confirm qP wave traveling with the wave front of ellipsoidal form. Although some of the difference can be seen in the shape of qS wave front, the numerical result obtained by using estimated elastic constants is agree with that from the laser ultrasonic visualization testing.

5 CONCLUSIONS

Elastic constant estimation method using the laser ultrasonic visualization testing was presented. As a numerical example, elastic constants of a CFRP specimen were calculated by the proposed method. The results from EFIT simulation using estimated CFRP elastic constants were consistent with the corresponding those from laser ultrasonic visualization testing. In near future, the proposed method will be extended to more complicated CFRPs with different type of laminations.

REFERENCES

Auld, B.A. 1990. Acoustic fields and waves in solids, vol. 1, 2, R.E. Krieger.

Fellinger, P., Marklein, M., Langenberg, K.J. and S. Klaholz, S. 1995. Numerical modeling of elastic wave propagation and scattering with EFIT-elastodynamic finite integration technique, wave motion, vol. 21:47–66.

Furukawa, A, Saitoh, T. and Hirose, S. 2014. Convolution quadrature time-domain boundary element method for 2-D and 3-D elastodynamic analyses in general anisotropic elastic solids, *Engineering Analysis with Boundary Elements,* vol. 39:64–74.

Nakahata, K., Ichikawa, S., Saitoh, T. and Hirose, S. 2011. Acceleration of the 3D image-based FIT with an explicit parallelization approach, *Review of progress in quantitative nondestructive evaluation*, American Institute of Physics: 769–776.

Civil, Architecture and Environmental Engineering – Kao & Sung (Eds)
© 2017 Taylor & Francis Group, ISBN 978-1-138-02985-9

Selection for the frequency control point of the VAV air supply system in large underground spaces

Kedi Xue, Zhihua Zhou & Yufeng Zhang
School of Environmental Science and Engineering, Tianjin University, Tianjin, China

Yimiao Sun
School of Mathematics, South China University of Technology, Guangzhou, China

ABSTRACT: Ventilation is an important part of the mine and other underground spaces to ensure personal safety. Previous rough design and managements lead to larger system energy consumption. A control method is proposed in this paper for the current status of the ventilation system in large underground spaces not having been taken seriously and its operation being uneconomical, where the main duct airflow rate is variable, while the branch pipe airflow rate is constant. Based on this, an experimental system was built in a laboratory, taking practical engineering as the model. Through experiments, the system's total energy consumption, control effect, and actual airflow rate of each branch under different constant pressure points in constant static pressure control mode were studied, and were then compared with designed values. Results showed that the optimal constant pressure point location of the ventilation system in large underground spaces was where the static pressure was equaled to the average value of static pressures at the front and end of the main air duct, where it can be better controlled leading to increased energy-conservation.

1 INTRODUCTION

Several major incidents and accidents related to the presence of methane in underground coal mines with fatalities have been reported (Torano et al. 2009). Ventilation is a very important part of underground mines. Studies have shown that depending on the types of mines, the power consumption of the ventilation system is also different. Some mines reveal as high as 25%–40% of the total underground electricity consumption (Karacan 2007, LI & WANG 2009). Currently, large-scale underground mines usually adopt steady ventilation. The mine ventilation system (Parra et al. 2006, Sahay et al. 2014) can be divided into many small regions, and meet the regional ventilation requirements by setting branch pipes. Since the thermal and moisture load in underground engineering undergo very little change over time, each region requires a certain air volume during operation, namely, but some regions may sometimes open and sometimes close, thereby resulting in the total airflow rate being mutative at any time. In order to guarantee the airflow rate of each branch pipe while reducing total power consumption, an appropriate Variable Air Volume (VAV) system is clearly required.

The constant static pressure method was the commonly used fan control method. In most cases, a static pressure sensor is generally located two-thirds downstream in a main supply air duct (Shim et al. 2014, Zhang et al. 2010) although it is no longer recommended by ASHRAE. However, for mines and other large-scale underground areas, which have long distance pipe networks, more complex systems and changeable operation conditions, it is difficult to determine where and how many static pressure sensors should be placed, and the energy-saving effect is also difficult to guarantee (ASHRAE 2011).

In this paper, an experimental system was built in a laboratory by taking practical engineering as the model. And then, different constant pressure points were set. Finally, by the analyzing airflow rate and stability of each branch and also taking energy consumption into account, the best location of the constant pressure point under constant static pressure control mode was determined.

2 METHODS

2.1 *Experimental system*

Based on an engineering model in practice, we designed a VAV air supply system, which comprised a main air duct and four branch pipes (I, II, III, and IV), as shown in Fig. 1. A main fan was installed on the main air duct, and three box-type centrifugal fans on II, III, and IV branches. In addition, there was also one fixed resistance valve, marked as 1, 2, 3, and 4, and one adjustable damping valve, marked as 5, 6, 7, and 8 on each branch system, respectively. Furthermore, a power meter was used to collect the electromotor input power. In order to test system pressure and control the main fan effectively, three measurement points were chosen, namely P1, P2, and P3 (see Fig. 1). The P1 constant pressure point was located at the main fan outlet, where air flow was stable. The distance from the P2 constant pressure point to the main fan was 2/3 length of the main air duct. The P3 constant pressure point was set at the end of the main air duct. In addition, a virtual static control parameter named as the P4 constant pressure point was set, which took the average value of P1 and P3 static pressure points.

2.2 *Parameter test*

2.2.1 *Static pressure*

A pressure sensor was installed at P1, P2, and P3 constant pressure points, respectively, and the static pressure value of the P4 constant pressure point was calculated by taking the average of P1 and P3 constant pressure point values.

2.2.2 *Airflow rate*

The method of testing the duct's airflow rate was as follows: measure the duct dynamic pressure at some point and convert it into wind speed at this position by using equation (1), and then calculate the duct airflow rate by using equation (2). In addition, for the circular pipe, we should arrange measuring points based on the equal area method (Zhao et al. 2009).

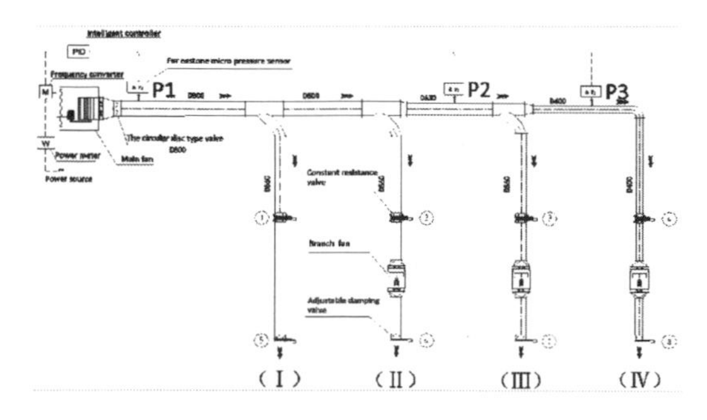

Figure 1. Schematic diagram of a VAV air supply system.

$$v_a = \sqrt{\frac{2P_{da}}{\rho}} \tag{1}$$

$$Q = 3600 \times v_a \times S \tag{2}$$

In equation (1), P_{da} was calculated based on equation (3); ρ can be obtained by measuring the absolute static pressure, temperature, and humidity within 20 m from the wind pressure measurement position in the tunnel, as given in equation (4).

$$P_{da} = \left(\frac{\sqrt{P_{d1}} + \sqrt{P_{d2}} + \ldots \sqrt{P_{dm}}}{m} \right)^2 \tag{3}$$

$$\rho = 3.484 \times 10^{-3} \times \frac{p_0 - 0.3779\varphi p_{sat}}{273 + t} \tag{4}$$

2.3 *Static pressure set point value*

Set the PID controller and the frequency converter to the manual working state, and open the fixed resistance valve of each branch pipe fully. Turn on the main fan, and maintain the frequency converter at 50 Hz. Next, turn on the branch pipe fan sequentially, and adjust the adjustable damping valve of branch IV, III, II, and I sequentially till the airflow rate of each branch system reached its design value. Record the static pressures of each constant pressure point, which were taken as static pressure set point values.

3 RESULTS

3.1 *Set point values*

The set point values at each constant pressure point were measured, as given in Table 1.

3.2 *Actual airflow rate*

Maintain the static pressure at P1, P2, P3, and P4 points at its set point values sequentially, and the actual airflow rates are listed in Tables 2–5.

Table 1. Static pressure set point values at each constant pressure point.

Constant pressure point	P1	P2	P3	P4
Pressure (Pa)	730	565	530	640

Table 2. Actual airflow rate of each branch while P1 was the constant pressure point (m^3/h).

Experimental condition	Branch I	Branch II	Branch III	Branch IV	Total
Open four branches	5451	4840	4835	2853	17979
Open three branches	0	4908	4863	2916	12687
	5483	0	4861	2909	13253
	5546	4960	0	2900	13406
	5504	4906	4859	0	15269
Open two branches	0	5010	0	2935	7945
	5617	0	4934	0	10551
Open one branch	5634	0	0	0	5634
	0	5091	0	0	5091
	0	0	5083	0	5083
	0	0	0	2946	2946

Table 3. Actual airflow rate of each branch while P2 was the constant pressure point (m^3/h).

Experimental condition	Branch I	Branch II	Branch III	Branch IV	Total
Open four branches	5415	4812	4812	2805	17844
Open three branches	0	4820	4823	2811	12454
	5403	0	4836	2845	13084
	5413	4835	0	2847	13095
	5409	4824	4828	0	15061
Open two branches	0	4852	0	2866	7718
	5428	0	4864	0	10292
Open one branch	5476	0	0	0	5476
	0	4879	0	0	4879
	0	0	4871	0	4871
	0	0	0	2886	2886

Table 4. Actual airflow rate of each branch while P3 was the constant pressure point (m^3/h).

Experimental condition	Branch I	Branch II	Branch III	Branch IV	Total
Open four branches	5407	4803	4808	2803	17821
Open three branches	0	4800	4770	2823	12393
	5268	0	4776	2825	12869
	5135	4761	0	2827	12723
	5112	4770	4857	0	14739
Open two branches	0	4745	0	2817	7562
	5162	0	4805	0	9967
Open one branch	5159	0	0	0	5159
	0	4711	0	0	4711
	0	0	4800	0	4800
	0	0	0	2885	2885

Table 5. Actual airflow rate of each branch while P4 was the constant pressure point (m^3/h).

Experimental condition	Branch I	Branch II	Branch III	Branch IV	Total
Open four branches	5420	4833	4818	2829	17900
Open three branches	0	4835	4828	2827	12490
	5421	0	4870	2833	13124
	5423	4840	0	2829	13092
	5420	4837	4857	0	15114
Open two branches	0	4855	0	2863	7718
	5438	0	4855	0	10293
Open one branch	5444	0	0	0	5444
	0	4863	0	0	4863
	0	0	4828	0	4828
	0	0	0	2889	2889

4 DISCUSSION

In this experiment, due to the different requirements for airflow rates among every region in the system, the rated airflow rate of branch I, II, III, and IV was 5400 m^3/h, 4800 m^3/h, 4800 m^3/h, and 2800 m^3/h, respectively. As can be seen from Section 3, the actual air supply volume of branch I fluctuated between 5000 m^3/h and 5500 m^3/h, only except for a few points; nevertheless, it gave about 7%–10% deviation when compared with the rated value of 5400 m^3/h. Similarly, all deviations of the four branches were within ±15%, thereby meeting the standard requirements.

4.1 *Operation stability*

For each branch pipe, we calculated the standard deviation of ratios between the actual and rated air supply volumes under different operating conditions, and the results are summarized as Table 6.

When the constant pressure point was located at P4, the standard deviation of each branch reached to the minimum values, which illustrated their best air supply stability.

4.2 *Energy consumption*

Fig. 2 illustrates the inverter frequency and input power of different constant pressure point positions.

When the constant pressure point was located at P3, the inverter frequency reached to the minimum value of 39.05 Hz, which was significantly smaller than that of 50 Hz under full load operation conditions. What is more, the input power has also decreased from 12 kW to 4.8 kW. Likewise, when the constant pressure point was located at P1, P2, or P4, the input power has also decreased by 49.20%, 58.30%, and 57.50%, respectively.

Table 6. Standard deviation of ratios between actual and rated air supply volumes for each branch pipe (%).

Constant pressure point	P1	P2	P3	P4
Branch I	0.012	0.005	0.019	0.002
Branch II	0.017	0.005	0.007	0.002
Branch III	0.018	0.004	0.006	0.004
Branch IV	0.011	0.010	0.009	0.008

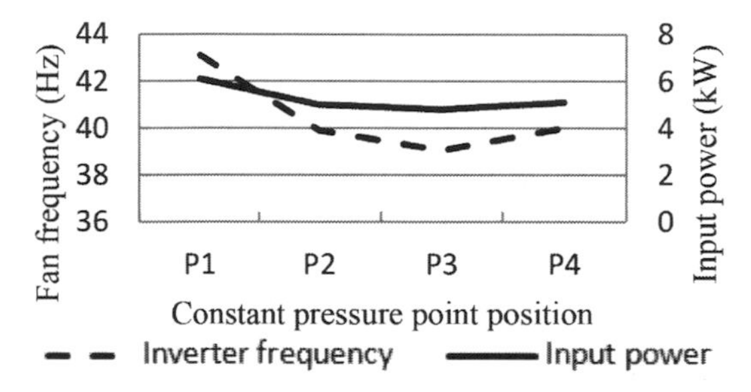

Figure 2. Comparison of fan frequency and input power for different constant pressure point positions.

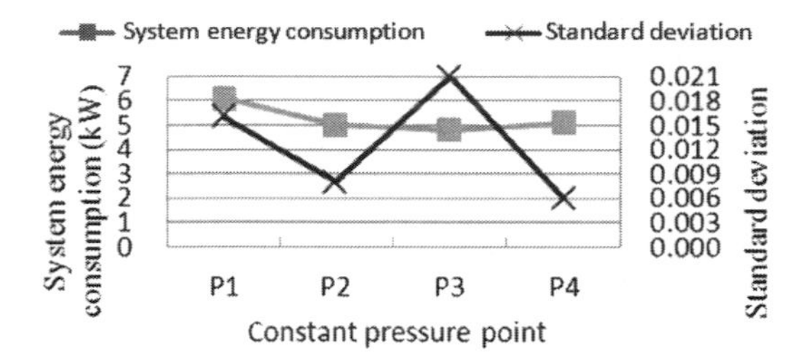

Figure 3. Comparison of system stability and system energy consumption among different constant pressure point positions.

4.3 *Determination of optimal constant pressure point location*

Fig. 3 illustrates the comparison of system stability and system energy consumption among different constant pressure points positions.

As shown in Fig. 3, the P1 constant pressure point possessed the poor air supply stability and the largest energy consumption. P2 possessed a better air supply stability, but not the most energy-efficient point. P3 possessed the lowest energy consumption but the worst air supply stability. The biggest advantage of the P4 constant pressure point is that it possessed the best air supply stability on the basis of meeting the rated requirements of each branch system. What is more, its energy consumption was slightly higher than the P2 constant pressure point, but decreased by 16.39% compared with P1 constant pressure point. Therefore, the P4 point was the optimal constant pressure point location in our experimental system.

5 CONCLUSIONS

In order to reduce the ventilation system energy consumption of the mines and other large underground spaces, on one hand, an appropriate ventilation system should be chosen; on the other hand, a suitable system operation control method should be adopted. In this paper, a frequency ventilation control program was proposed, taking an underground engineering for study. By the testing airflow rate, air pressure, fan input power, and other parameters under different operation conditions, different constant pressure points, and analyzing and comparison, the variable main duct air volume and constant branch pipe air volume control mode was adopted. Among the four constant pressure points, the point at where the static pressure is equaled to the average value of static pressures at the front and back of the main air duct showed the best results.

REFERENCES

ASHRAE 2011. HVAC application [Chapter 46] In Design and application of controls. *Atlanta: American Society of Heating, Refrigerating and Air-conditioning Engineers, Inc.*: 7–8.

Karacan CO. 2007. Development and application of reservoir models and artificial neural networks for optimizing ventilation air requirements in development mining of coal seams. *Int. J. Coal Geol. 72*: 221–239.

Li Man & Wang Xue-rong 2009. Performance evaluation methods and instrumentation for mine ventilation fans. *Mining Science and Technology 19*: 0819–0823.

Parra M.T., Villafruela J.M., Castro F. & Méndez C. 2006. Numerical and experimental analysis of different ventilation systems in deep mines. *Building and Environment 41*: 87–93.

Sahay N., Sinha A., Haribabu B. & Roychoudhary P.K. 2014. Dealing with open fire in an underground coal mine by ventilation control techniques. *The Journal of The Southern African Institute of Mining and Metallurgy 114*: 445–453.

Shim Gyujin, Song Li & Wang Gang 2014. Comparison of different fan control strategies on a variable air volume systems through simulations and experiments. *Building and Environment 72*: 212–222.

Torano, J., Torno, S., Menendez, M., Gent, M. & Velasco, J. 2009. Models of methane behaviour in auxiliary ventilation of underground coal mining. *Int. J. Coal Geol. 80*: 35–43.

Zhang Jie, Pan Yi-qun & Huang Zhi-zhong 2010. Study on static pressure reset control of VAV system [J]. *Building Energy & Environment 29(5)*: 25–29+10.

Zhao Rong-yi, Fan Cun-yang & Xue Dian-hua 2009. *Air Conditioning [M]*. Beijing: *China* Architecture & Building Press: 262–263.

Civil, Architecture and Environmental Engineering – Kao & Sung (Eds)
© 2017 Taylor & Francis Group, ISBN 978-1-138-02985-9

The innovations of the ship lock design for the Tugutang Hydropower Station in the Xiangjiang river

Huying Liu & Chongyu Wang
Hunan Provincial Communications Planning, Survey and Design Institute, Changsha, China
Changsha University of Science and Technology, Changsha, China

ABSTRACT: According to XJMWD and HNWTDP, the Tugutang Hydropower Station navigation-power hub was the last dam on the Xiangjiang River. And so, the construction and development of Hunan inland waterways is very important. Not only the general layout and engineering characteristics of its ship lock were analyzed, but also the innovations of the ship lock design are summarized. The innovations of the lock head, construction diversion program by structural optimizations, and considering the second ship lock in the general layout were adopted in this Hydropower Station. A design reference for similar hubs or ship locks was provided by using the above innovations.

1 BACKGROUND AND INTRODUCTION

According to "Xiangjiang River Mainstream Waterway Development Plan" (XJMWDP, 2007), the channel from the stream outlet to Hengyang was planned for the level II, and the channel from Hengyang to Yongzhou was planned for the level III. With the implementation of Changsha hydro-junction and the first phase of Xiangjiang River waterway improvement, the channel from the stream outlet to Zhuzhou will be considered as level II. Xiaoxiang, Jinweizhou, WuXi, and XiangQi are four hydro-junctions that had been built on the Xiangjiang River. The water level between Jinweizhou and Dayuandu hydro-junction is not yet joined. Therefore, the level III waterway in Xiangjiang River can be extended to the Jinweizhou hydro-junction because of the construction of the Tugutong navigation-power junction, which will promote the economic development of the middle reaches and the upper reaches of the basin of Xiangjiang River.

In addition, a modernization of the water transport system will be built in Hunan according to "Hunan Water Transport Development Plan" (HNWTDP, 2011). The system relies on the Yangtze River in the region and Dongting Lake. Moreover, "One vertical and five horizontal" channel–net will be formed. The backbone is the "one vertical" channel i.e. Xiangjiang River. The construction of a 2000-tons ship-lock will soon start in the Dayuandu hub and in the Zhuzhou hub located downstream of the Tugutang hub. The expansion of the 1,000-tons ship-lock of four hydro-junctions is planned upstream. The Xiang-Gui canal will be constructed, which can connect the Yangtze River basin and the Pearl River basin.

Therefore, the Tugutang hub is the last dam in Xiangjiang River. An important role will be played by this hub in the development of the Xiangjiang River channel and Hunan water transport system.

The Tugutang hub is located in the town of Yunji, Hengnan County, 39 km upstream of the city of Hengyang. The hub is a project for the ship, as well as taking into account hydropower generation, transportation, irrigation, water supply, aquaculture, and so on.

The navigation structure is a 1,000-tons ship-lock whose annual capacity is 15 million tons. The effective measure of the lock bay is 180 × 23 × 4.0 m, and the position of the second-tier ship lock is reserved.

2 THE OVERALL SCHEME AND INNOVATIONS OF THE SHIP LOCK

2.1 *The general layout scheme of the ship lock*

According to the general layout scheme of the hub, the ship lock is arranged in the right-hand side bank of the river, and the axis of the ship lock is perpendicular to the axis of the dam. The main

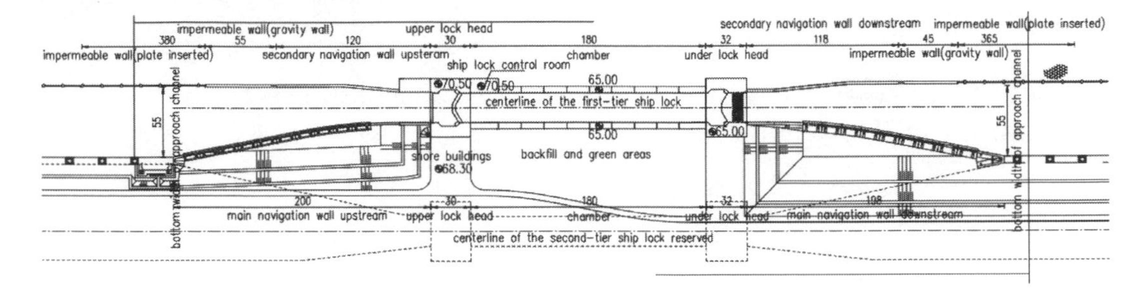

Figure 1. The layout of the ship lock of the Tugutang hub.

buildings of the ship lock include the ship lock head, bay, navigation wall, boat pier, separating wall, and so on. Figure 1 shows the overall layout of the ship lock.

2.2 *Innovations of the design*

Three innovations were embedded in the design of the ship lock as follows, according to the overall layout of the hub, engineering features, construction plans, and long-term plans:

1. According to the construction diversion scheme of the hub, part of the ship lock shall be in construction area above the water level. In addition, during the construction process, a small cofferdam for the ship lock needs to be formed, and part of the main buildings near the river may be used to be the longitudinal cofferdam as well. In order to meet the above-mentioned demand, the structure of buildings must implement innovation and optimization.
2. According to PHWTD, the 1000-ton channel extension in the upper reaches of the Xiangjiang River and Xianggui canal construction will be constructed in the long term. It should be reserved for construction conditions of the Tugutang second-tier ship lock to meet with the plan of regional economic development. Therefore, not only positions are reserved for the construction of the second-tier ship lock in the design, but also buildings landside of one line lock gets structural innovations in order to accommodate the construction of the second-tier ship lock, which can avoid duplication and demolition of buildings.
3. The dam of this project is located in Hengnan city; after the completion of the hub, it will become a part of the scenery in the city. Taking into account the landscape requirements, we need to fully optimize structural arrangement and control the structure height in the design of the upper and lower lock head. The requirements of the scenery will be met in the ship lock area after completion of the hub.

3 CONSTRUCTION DIVERSION, DESIGN INNOVATION, AND STRUCTURAL OPTIMIZATION

3.1 *Scheme of construction diversion*

The project includes a ship lock, 17.5 sluice gates, turbine generators, and other units. It is divided in three phases. The first phase is around the ship lock and 7.5 sluice gates at the right-hand side of the bank. The second one is separately around the powerhouse at the left-hand side bank. A small cofferdam is constructed before removing the cofferdam at the first phase. Three phases closure is carried out after the ship lock, and the rest of the sluice gates are fenced. Figures 2 and 3 show layouts of construction diversion.

The innovations of the ship lock design are mainly reflected as follows:

1. Cofferdam in the first phase includes only lock heads, bay, and other primary and secondary navigation walls; the rest of the buildings of the ship lock shall be constructed above the water level.
2. After the formation of the cofferdam in the second phase, the transverse cofferdam in the first phase must be removed. And then, the ship lock is not completed and therefore, we shall continue to build the small cofferdam of ship locks. And the buildings by the river side shall be considered to be the longitudinal cofferdam as well.

3.2 *The pile–pillar–slab separation wall*

The main buildings out of the cofferdam include berthing piers, separation walls, diversion piers, which all need to be constructed above the water level. The general pile cap foundation is used in berthing piers and diversion piers. The separation wall must have the function of water diversion, which may be the structure with slab between the piers. The pier is designed innovatively in this

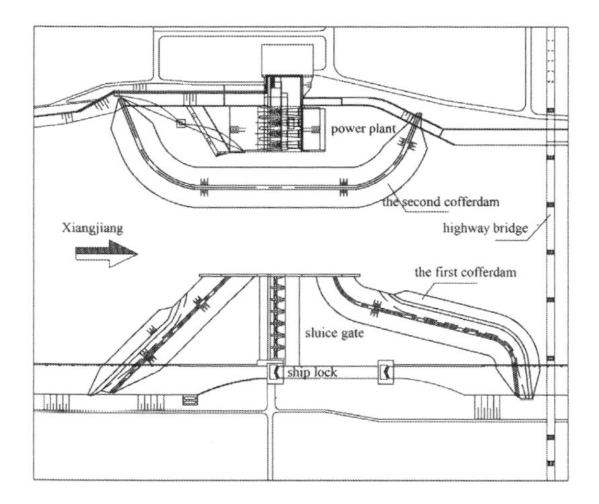

Figure 2. The layout of construction diversion in the first and second phases.

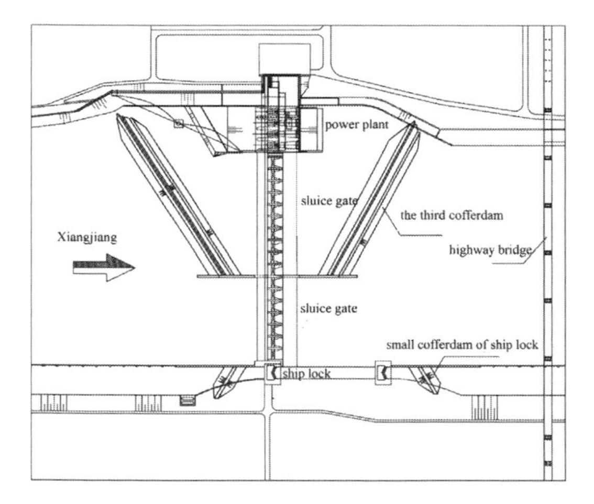

Figure 3. The layout of the small ship lock cofferdam and construction diversion in the third phase.

project as follows: "pile foundation + pillar structure". Figure 4 shows the separation wall's structure.

The pile–pillar–slab separation wall indicates the structure with pile foundation base, the pillar as the upper structure, and with slab between the pillars, whose pile is bored piles with a diameter of 2.5 m. In order to simplify the mechanical characteristics, the cylinders with the same diameter of 2.5 m are used as pillars, with 15 m as the column spacing. Mounting slots are reserved by the upstream and downstream ends of the cylinder. The precast slab with a thickness of 0.6 m is inserted. The riprap is used to closure it between the slab and the bed.

Advantages of the pile–pillar–slab separation wall are mainly given as follows: 1) deep foundations are facilitated through the construction above

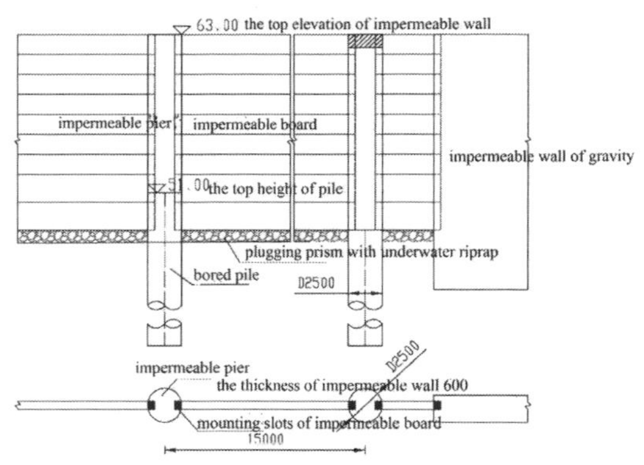

Figure 4. Schematic of an impermeable wall with piles plate inserted.

the water level; 2) the pile and pillar are designed with the same size for simple construction; 3) slabs are prefabricated firstly for a short time.

There are issues that need attention in this program: 1) the reserved position of the mounting groove must be precise in order to ensure a smooth installation of the slab; 2) when the force was calculated, the water level difference between both sides of the wall should be taken into account.

3.3 *Chamber wall*

According to the scheme of construction diversion, after the completion of the first phase of the sluice gate, the transverse cofferdam will be removed. The construction of the small cofferdam of the ship lock was proceeded, and the small cofferdam of the ship lock was used only to construct the lateral cofferdam. One part of the main structure of the ship lock caters the longitudinal cofferdam, including brake head, river-side chamber wall, and so on. The standard flow rate is 8000 m^3/s and the water level is 58.10 m.

According to the arrangement and the construction needs, it can be observed that it is different for structural forms between the shore side and the river side of the chamber wall. Figure 5 shows a cross-sectional view of the chamber wall.

Approximately, an "L"-type gravity structure is constructed in the chamber wall at the river side. The weight of the trailing floor, upper backfill and water are fully used to improve the stability of the river-side wall against sliding. It also meets requirements during construction as a cofferdam.

The weight-balanced structure with an unloading-board is carried out in the shore-side chamber wall. The bottom width of the structure is 13 m. The unloading-board is added at the balancing platform location to lower the soil pressure of the backfill.

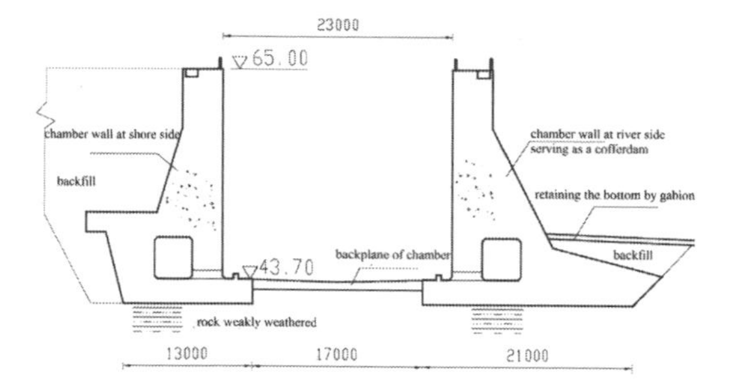

Figure 5. The cross-sectional view of the chamber wall.

With optimized design as mentioned above, both sides of the chamber wall can meet the overall stability requirements of the various conditions.

4 THE OPTIMIZATION OF THE DESIGN TAKING INTO ACCOUNT THE SECOND SHIP LOCK

According to the plan of the general layout, the second ship lock will be laid on the right-hand side of the ship lock. And so, buildings related to the second ship lock mainly include the navigation wall, boat pier, and so on.

4.1 *Main navigation wall with holes*

According to the layout of the first and the second ship locks (see Figure 1), the main navigation wall of the first and the second ship lock was splayed. The outer surface of the main navigation wall is constructed by using a boat segment. As an operational experience of multi-line locks, it will affect the approach channel's flow condition of the other ship lock when the water of the first ship lock flows in or out. Even great negative effects on the operation of the miter gate of the other by big water head are reversed.

To avoid the situation mentioned above, the main navigation wall with holes in the bottom was designed. It will improve the flow properties of approach channels. Figure 6 shows the structure drawing of the main navigation wall.

The broadened base is used in foundation of the main navigation wall. The counterfort wall was used as the upper structure, and the plate was used as the boat wall. There was a hole at the bottom of the plate. The hole size is determined by using the approach channel bottom elevation and the lowest navigable water level. The two lines of the ship lock water level are balanced by using holes. The reason for this structure in the first ship lock is that it can save the amount of concrete work at the present stage firstly, and also it is used to avoid concrete demolition projects in the second ship lock construction.

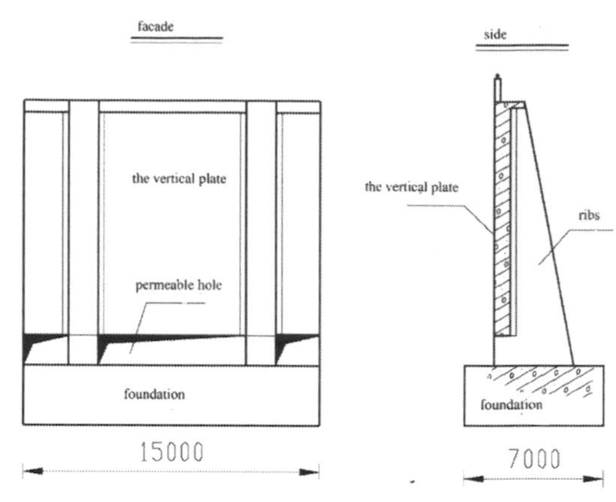

Figure 6. The chart of the main navigation wall.

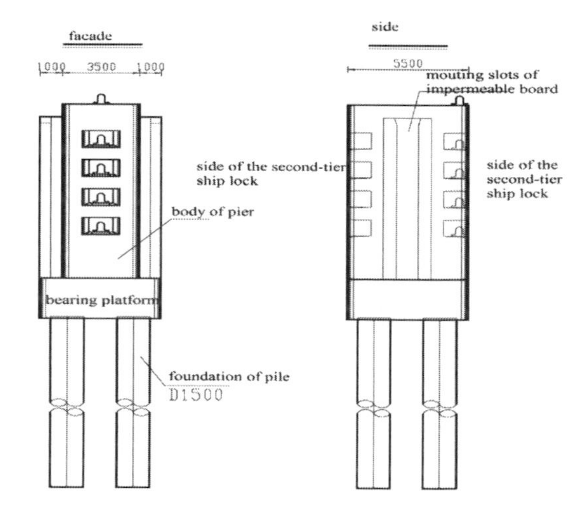

Figure 7. Schematic of the bidirectional berthing pier.

4.2 *Two-way boat pier*

A boat pier was constructed above the water level out of the cofferdam. It is concreted on the pile platform. According to the layout of the first and the second ship locks, boat piers of the first and the second ship lock are laid in the same location.

In order to avoid duplication and adverse effects on the first ship lock by the second ship lock's construction, the two-way berthing pier for the first tier ship lock was designed, taking into account the berthing capacity of the first and the second ship locks at the same time. But, the bollard installation slot was reserved by the side of the second ship lock. Figure 7 shows the structure drawing of a two-way berthing pier.

The plane size of the pier along the flow direction is 3.5 m, and vertical size is 5.5 m. The slab slots were reserved by the upstream and downstream surfaces of the pier for installation of the riser board when we constructed the second ship lock. To reduce the transverse flow velocity of the parking section, the boat pier can meet the requirements in two ways. It should be noted that the structure shall be calculated considering the combination of the impact force and the mooring force in both sides at the same time.

5 DESIGN OPTIMIZATION OF THE LOCK HEAD

5.1 *Upper lock head*

According to operation requirements of the ship lock, a hoist room needs to be arranged on both sides of the lock head. But the control center of the ship lock should only be arranged on one side, in order to facilitate the operators. Thus, the pipe line channels must be arranged over the ship lock for cables and water pipes. A road bridge or pedestrian bridge was practiced regularly over the lock.

However, taking into account the security of operation, the Xiangjiang River Bridge was arranged 500 m downstream of the river between the hub. And so, the pipe line channel cannot be laid by the bridge. If a footbridge over the lock was constructed, the investment will be added. In view of the factors mentioned above, the pipe line channel crossing the ship lock was arranged within the upper ship lock head.

Because the upper lock head was an entity concrete dock, the vertical shaft from the hoist room to the chamber's bottom and a horizontal tunnel across the chamber bottom were designed. A U-shaped pipeline corridor was formed, as shown in Figure 8.

5.2 *Under lock head*

Similarly with the upper lock head, the built-in hoist room is also used in the under lock head. Based on experience in similar projects, this elevation of the bottom of the hoist room may be determined by the flood control level of the hydraulic control device. It often leads to the elevation of the lock head much higher than the elevation of lock wall. It should be disharmony in visual. If we take measures to heighten the elevation of the lock wall, the cost of the lock wall would greatly increase.

To solve the problem mentioned above, the built-in hoist room of Tugutang is separately furnished. The room of the miter gate's hydraulic hoist and the room of the hydraulic control are separated. A concrete wall is set up between these, with waterproof casing as the communicating pipe

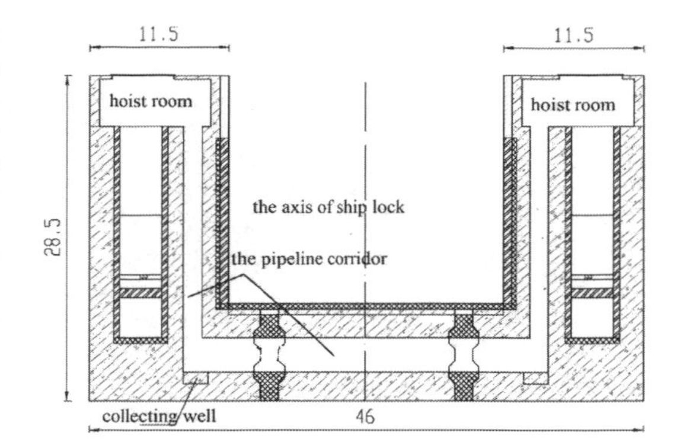

Figure 8. The cross-sectional view of the upper lock head. Not only was the cost of the pedestrian bridge over the lock saved, but also an open vision of the lock area was protected by this design.

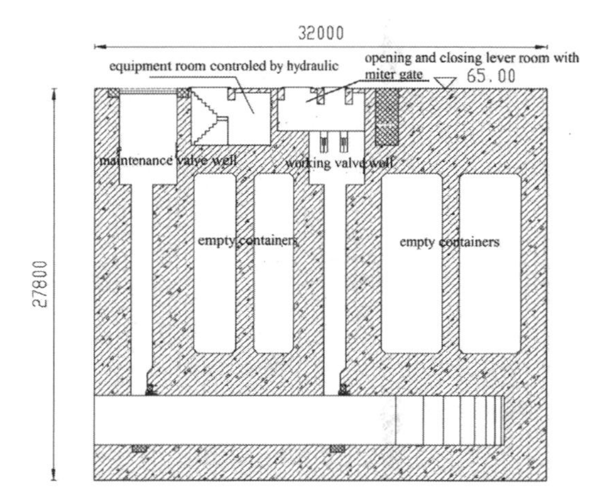

Figure 9. The profile view of an under lock head.

to connect these rooms. Meanwhile, the bottom elevation of the hydraulic control room is reduced to 61.00 m (the highest navigable water level is 62.50 m). The sunken design scheme was carried out. Four concrete walls were arranged around the hydraulic control rooms. A drain pipe was set in the hydraulic room. A control valve was set at the end for drainage of the hydraulic room at a low water level. When the water level was above 61.00 m, the valve was shut off to form an enclosed room, and the water was pumped, as shown in Figure 9.

The top elevation of the under lock head had been reduced from 67.50 m to 65.00 m. Taking into account the overall coordination of elevation in the lock region, the elevation of the lock wall had been increased from 64.3 m to 65.00 m, which is consistent with the elevation of the chamber wall of the under lock head, thereby avoiding the lock area that is visually impaired and formed by projections of the under lock head.

6 CONCLUSION

1. The following innovations of the construction diversion program and structural optimizations were carried out: the pile–pillar–slab separation walls, the lock walls as the cofferdam, and so on. All these decreased the investment, and produced a significant reduction in construction time.
2. The innovations taken into account the second ship lock were carried out: the main navigation wall with holes, the two-way boat piers and so on. Taking into account all conditions of building the second ship lock, these will save the cost of the second ship lock, reduce construction difficulty and adverse effects.
3. The following innovations of the lock head were carried out: the U-shaped pipeline corridor, the built-in hoist room, the waterproof hydraulic control room, and so on. The elevation of the under lock head was reduced and the landscape area of the lock was improved by these innovations.

The above innovations were not only unique, but also versatile. These can provide a reference for similar hub or ship lock designs.

REFERENCES

Department of Transportation of Hunan Province (2007). Plan of Xiangjiang Waterway Development. *R.*

Department of Transportation of Hunan Province (2011). Plan of inland waterway development of Hunan Province. *R.*

Overall design specifications of drainage engineering hub (JTS 182-1-2009). *S.*

Transportation Planning Survey and Design Institute of Hunan Province (2012). Preliminary design report of Tugutang of Xiangjiang Navigation Project. *R.*

Wu Xin (2012). Key Technology of Guiping Second Line Ship lock for Xijiang Navigation Route. *J. Journal of Hydraulic Engineering*. 04, 108–114.

Zhai Huijuan, Li Hao, Zhang Linjiang (2008), Avionics construction of hub and development of inland water transport. *J. Transport Construction and Management*. 07, 68–72.

Zhang Chunyang, Sun Yimin (2001). Discussion of Water Control Landscape Planning Features-Area planning design of Guangdong Feilaixia Project. *J. South China University of Technology (Natural Science)*. 07, 79–82.

Civil, Architecture and Environmental Engineering – Kao & Sung (Eds)
© 2017 Taylor & Francis Group, ISBN 978-1-138-02985-9

Knowledge management in construction—the framework of high value density knowledge discovery with graph database

Yong Jiang
School of Architecture, Tsinghua University, Beijing, China

Ying-chu Wang & Zuo Wang
UNIS Software Co. Ltd., Beijing, China

ABSTRACT: With the high speed development of information technology, the use of big data technology has been deepened into more and more areas. The outstanding problems of applying big data in construction industry are large scale, multi dimensions, strong supervision and serious information island phenomenon. How to obtain the useful business information (high value density knowledge) efficiently during the acquisition, storage and batch processing of large volume data is an important subject. Here we report a method of combining the graph algorithm and the traditional data warehouse concept to obtain high value density knowledge. This method uses the subjective judgment of human beings and the objective data analysis of the machines together to get more accurate results.

1 CONSTRUCTION INFORMATION CLASSIFICATION STANDARDS AND KNOWLEDGE TAG SYSTEM

The construction industry is the pillar industry of the national economy and the first major industry. It covers all aspects of the building life cycle, closely related to a variety of urban construction projects. It also relates to the construction of the second industry and design, management, financial services and other related contents of the third industry. How to use big data, Internet of things, cloud computing, mobile Internet and other modern information technology to improve efficiency is an important subject relating to the development of the construction industry, and it is also a prerequisite link in the future construction of smart cities.

The data processing during the whole life cycle of a construction project is a typical nonstructural and big data use case, fully embodies the big data's 4V (Volume, Velocity, Variety and Value) character. The future development and intelligence of the construction industry is relying on the accumulation and continuing the refinement of the construction project data, digging out the relationship between the various data. So as to eliminate the information island phenomenon and break information asymmetry. According to the characteristics and contents of the construction industry, building a unified data platform (see Figure 1) based on the classification of construction industry information and knowledge management is the most basic and important step for interconnecting information and making full use of high value data. And the establishment of information classification method and keyword tag system is the key technology for this data platform.

The definition of information classification method and keyword tag system: A graph theoretic database structure based on IFC standard faceted classification. It stores all the information in the quality supervision platform of the construction project; uses data reduction and retrieves the relationship between different data on the subsequent development and evolution.

2 USING GRAPH DATABASE AS THE CORE INFRASTRUCTURE OF THE HIGH VALUE DENSITY KNOWLEDGE DISCOVERY SYSTEM IN CONSTRUCTION INDUSTRY

The main way to find high value density knowledge in the field of construction industry is analyzing, inquiring and exploring the relationship between different information classification tags attached on construction project data. Graph theory model is a priori advantage technology for handling this kind of "many to many to many" relation inquiries and the traverse computation as well as the quantitative analysis.

Graph theory is a branch of Mathematics, which is mainly about the study of graphs. The graph in graph theory is a shape which composes of a number of given points and lines (one line

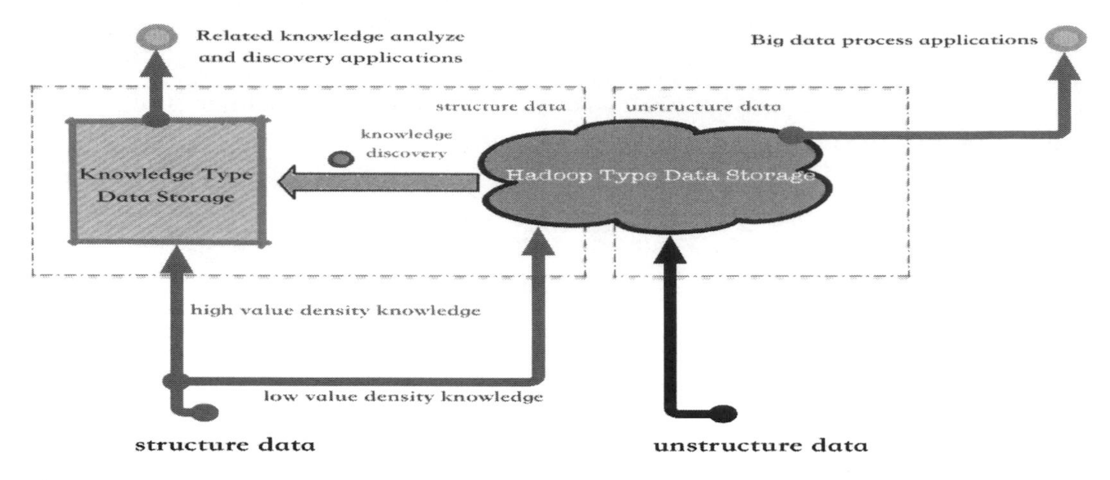

Figure 1. The data processing flow of the high value density knowledge discovery system.

joints two different points), it is usually used to describe certain relations of some things, with points representing things, by connecting the two points lines represent relationship between two things. In the most common sense of the term, a graph is an ordered pair G = (V, E) comprising a set V of vertices or nodes or points together with a set E of edges or arcs or lines, which are 2-element subsets of V (i.e. an edge is associated with two vertices, and that association takes the form of the unordered pair comprising those two vertices). Usually, describing a graph is to draw points as a small circle, if the corresponding vertex has an edge, connect the two small circles with a line. How to draw these small circles and connections is not important, it is important to correctly reflect which vertices are connected by edges, which vertices are not connected by edges. Most areas in the real world can be modeled as graphs, such as social system, the recommended system and the association discovery system can be conveniently modeled as graphics. And the high value density knowledge discovery system discussed in this article is also a great scene for using graph theory model.

Graph databases are based on graph theory, they use graph theory to construct data access model, the whole data set in graph databases is modeled as a large and dense network structure, so that graph databases allow simple and rapid retrieval of complex hierarchical structures. Most graph databases based on NoSQL structures such as key-value or document-oriented store. These storage engines have the concept of tags(properties) for data elements, it naturally fits the technic requirement of information classification method and keyword tag system and allows data elements to be categorized for easy retrieval.

3 DESIGNING HIGH VALUE DENSITY KNOWLEDGE DISCOVERY SYSTEM BASED ON GRAPH DATA MODEL AND TRADITIONAL DATA WAREHOUSE STAR SCHEMA CONCEPT

The value density of knowledge data isn't an objective value. It depends on the subjective needs of different users. In order to obtain the required high value density knowledge data, users must input a number of constraints (classification keyword tag information) as the basic filter for the selection of knowledge data. The more flexible users can set the constraint conditions; the higher accurate knowledge data users can get from system. On the other hand, the knowledge data stored in the system are combination of some scalar value, which is an objective fact. Since the knowledge data in system are limited objective value, and shared by all users, so that the constraints (classification keyword tag information) used for filter information should also be a set of describable measures within a limited scope. In order to implement these characteristics in graphic data model logically and systematically, we can learn the concept of star schema from the traditional data warehouse (based on the principle of relational database storage technology). The following is a brief definition of several concepts related to the traditional data warehouse and star schema:

Fact: The fact is the data unit in the data warehouse, and is also a unit in the multi-dimensional space, which is restricted by the analysis unit. Each fact includes the basic information about the facts (such as income, value, satisfaction, etc.), and is related to the dimension.

Dimension: The dimension is the space axis of the coordinate system. The coordinate system in the data warehouse defines the data unit, which contains the facts. In the data warehouse, the time is always one of the dimensions, other descriptive information with business implications are also main attributes used for defining dimension.

Star Schema: Star schema is a model that uses relational database to realize multidimensional analysis space (see Figure 2). It is a multidimensional data relationship, which is composed of a fact table (Fact Table) and a set of dimension table (Dimension Table). Each dimension table has a dimension as the primary key, all of these dimensions' primary key composite the primary key of fact table. The non-primary key attribute of the fact table is called Fact, which are generally numeric or other data that can be calculated, and dimensions are mostly text, time or other data types. By organizing data in this way, fact can be clustered and computed according to the different dimensions (part or all of the primary key of the fact table). By using this method, users can analyze the situation of business topics from different points of view.

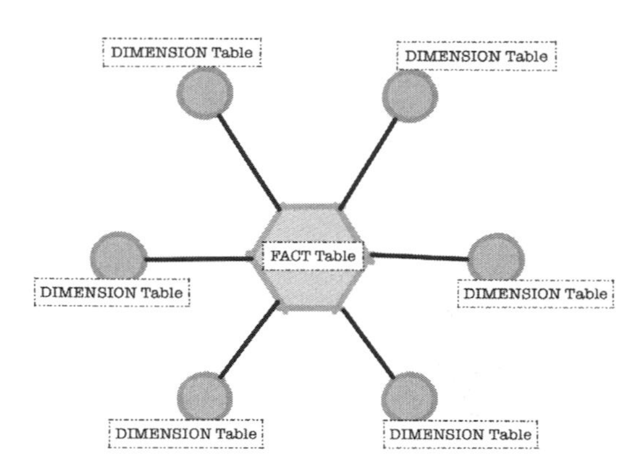

Figure 2. The internal database tables relationship of the star schema model.

On the macro, the internal components of star schema of the traditional data warehouse are essentially characterized by a graph. It just uses relational model in the specific data query and access operation. Inspired by this, we can use the graph data model to design an "improved" version of the star schema based on the graph database (see Figure 3) to serve the needs of the high value density knowledge discovery system.

The following is the main technical features of this new star schema model:

1. On design logic level, the model uses Fact to represent the concept of knowledge data and Dimension to represent the concept of constraint conditions.
2. On system implementation level, the model uses the vertex in the graph theory model to create the fact. The fact vertex is used to identify the objective scalar value of the knowledge data.

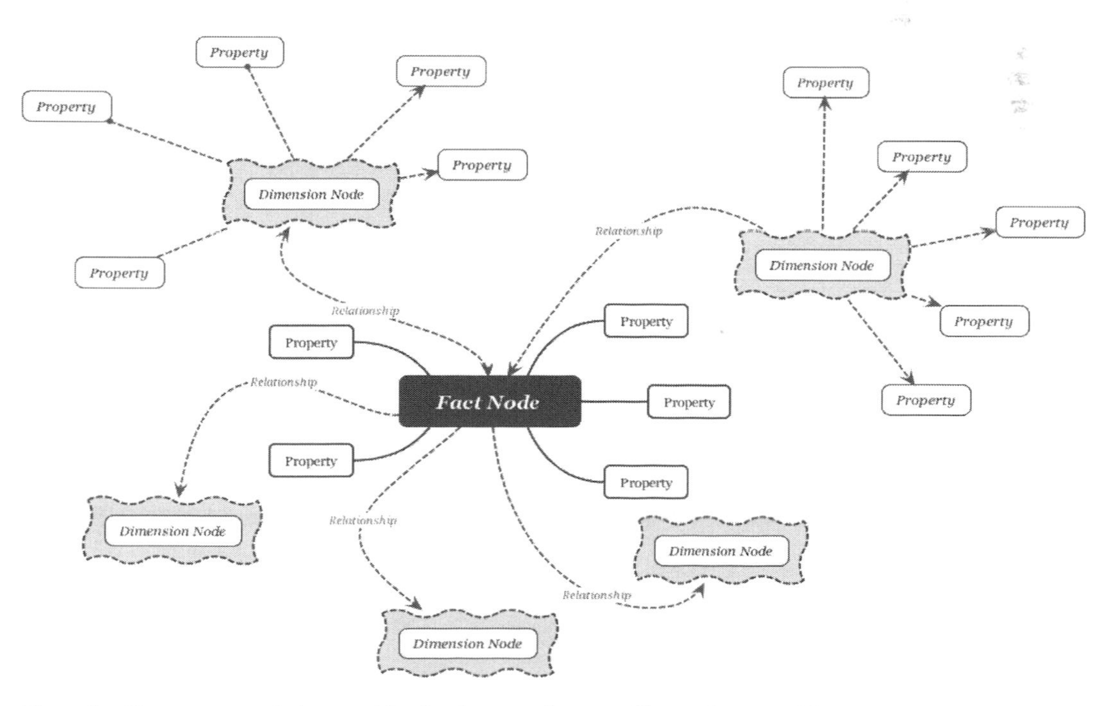

Figure 3. How to use graph data model to implement a "improved" star schema model.

These values are stored in the fact vertex as attributes.

3. On system implementation level, the model uses the vertex in the graph theory model to create the dimension. The dimension vertex is used to identify the constraint conditions users used to obtain high value density knowledge data. The dimension vertex only stores information associated with the constraint conditions. It does not store any data related to the knowledge data.
4. On system implementation level, the model uses the edge of graph theory to describe the relationship between Fact and Dimension. By setting the direction and different attributes of the edges, users can use the graph theory algorithm to screen the complex relationship between the fact vertices and the dimension vertices concisely.
5. Users can perform associated information discovery operations (multi node and multiple path correlation calculation algorithm in graph theory) on a particular dimension vertex to obtain the required high value density knowledge data (the fact vertices returned earlier by graph theoretic algorithms have higher value density).
6. Users can perform attributes query against fact vertices to select required knowledge data (similar to the traditional relational database data query).
7. When a specific fact vertex is obtained, according to their own subjective judgment, users can adjust and optimize the dimension vertices which are related to improve the accuracy of the value density of knowledge data under specific constraints.

4 CONCLUSIONS

The high value density knowledge discovery system which implemented on the framework we reported in this paper has the support of graph theory data model and algorithm. It can provide more powerful features and faster performance than the traditional relational model on discovering the association information under the multi dimension condition (the process of obtaining the high value density knowledge data). And by referencing the star schema concept in data warehouse field that has already been validated by IT industry, we can reduce the theoretical and practical risks of system design and implementation. At the same time, using the terminologies and operational methods that are familiar to the industry to obtain high density of knowledge data is conducive to the large-scale promotion and the use of the system.

REFERENCES

Bondy J.A. & U.S.R. Murty. Graph Theory by Bondy and Murty. Springer, 2008.

Ian Robinson, Jim Webber & Emil Eifrem. Graph Databases. O'Reilly, 2015.

Inmon W.H. Building the Data Warehouse. Wiley, 2005.

Jiawei Han, Micheline Kamber & Jian Pei. Data Mining Concepts and Techniques. Morgan Kaufmann, 2011.

Len Silverston, W.H. Inmon & Kent Graziano. The Data Model Resource Book: A Library of Logical Data Models and Data Warehouse Designs. Wiley, 1997.

Richard J. Trudeau. Introduction to Graph Theory (Dover Books on Mathematics). Dover Publications, 2013.

"From Relational to Graph Databases". Neo4j.

Civil, Architecture and Environmental Engineering – Kao & Sung (Eds)
© 2017 Taylor & Francis Group, ISBN 978-1-138-02985-9

Research of cold region green hospital design specification and practice

Yingzhi Gao & Hongyuan Mei
School of Architecture, Harbin Institute of Technology, Harbin, China

Jianfei Dong
School of Architecture, Hanyang Harbin Institute of Technology, Harbin, China

ABSTRACT: Presently, the green hospital design concept has attracted much attention, but the green hospital design work in cold regions involves more difficulties, so new technologies and methods for realizing "Green" hospital standards and effective design strategies for realizing the green hospital engineering in cold regions are also the research focus and difficulty in the medical design field. In this study, the design strategies for the green hospitals in cold regions that comply with the evaluation standards for green hospital buildings in cold regions are discussed, with the cold region hospitals that have been completed in recent years or are under research as the research objects, with the *Standard for Evaluation of Green Hospital Buildings* and *Heilongjiang's Standard for Evaluation of Green Hospital Buildings* as the benchmark, through discussion about the cold adaptability and land-saving strategies involved in the green hospital planning and design, the energy conservation involved in green hospital construction measures and material selection, the sustainability of the green hospital water-saving measures and the humanization and environmental protection performance standard index in terms of indoor acoustic environment, light environment and thermal environment of green hospitals.

1 INTRODUCTION

In China's building climate regionalization, the climate in cold regions is cool in summer and cold in winter. The energy consumption for heating in winter occupies the main part of annual energy consumption. Viewing the current building thermal insulation in China, compared with developed countries with similar climatic conditions, the unit building energy consumption in cold regions in China is 4~5 times that in developed countries, 2.5~3 times for roofing, 1.5~2.5 times for exterior windows, and 3~6 times in terms of air permeability of doors and windows. In terms of architectural design, blind pursuit of beauty is the main cause of increased building energy consumption. Besides, apparently, the division of building energy consumption types in cold regions is not thorough enough in many aspects in China. For example, in terms of space requirements for and habits of using medical buildings, the use of most office buildings peaks during the day, but the density of using such buildings involves uncertainty at night. Such use features can directly influence the energy consumption of buildings. Therefore, the issue concerning the ecological design of public buildings calls for detailed research and discussion.

The current architectural development trend is for green buildings, and architects in various countries are exploring ways to apply green technology to buildings. However, the question of how to create green buildings is an issue which is hard to solve in cold regions. Nowadays, most medical buildings involve large energy consumption, serious environmental pollution and lack of care for patients. With the cold regions in North China as the research object in this paper, based on the special climatic conditions in such regions and numerous green building theories at home and abroad, coupled with basic research on the medical environment within hospitals in cold regions, to further the practice of gradually meeting green hospital standards and characteristics of medical buildings, design strategies are presented which are suitable for green medical buildings in cold regions.

2 BASIC PRINCIPLES OF GREEN HOSPITAL BUILDINGS

Green hospital buildings refer to hospital buildings that provide people with healthy, useful and efficient spaces and which co-exist with nature harmoniously, featuring maximum resource conservation, environmental protection and pollution reduction in the whole life cycle of each building. Green hospital buildings belong to a "Green Building" system and involve special requirements for hospital buildings. A green building is the result of the incorporation and introduction of "Ecological and Sustainable Development and Intelligent and Efficient Energy Conservation" into the hospital building field and is a new development trend for future hospital buildings.

According to GB/T50378-2014 Standard for Evaluation of Green Buildings, whose trial implementation commenced in July 2011, GB/T51153-2015 (Standard for Evaluation of Green Hospital Buildings), which was formally implemented as a global standard in August 2016, and relevant evaluation standards for green hospitals, the space treatment of green hospital buildings should be based on several important principles, including matching the site, land conservation, tree conservation, energy conservation, indoor and outdoor design efficiency, economy and environmental comfort. The matching site and land conservation issues can be solved through environmental optimization design involving underground space utilization, site ventilation, noise, sunlight and daylight, as well as site ecological environment creation (native plants, stratified greening, roof greening, permeable ground, etc.).

2.1 *Site planning*

The problem considered in this study was a hypothetical 10-m-diameter tunnelling situation. The Standard for Evaluation of Green Hospital Buildings provides that site environmental noise shall meet the National Standard GB3096 (Environmental Quality Standard for Noise) in terms of planning. Such buildings intended as ward buildings and dormitories shall not be adjacent to urban trunk roads, and additional sound insulation measures shall be taken. In order to realize human design factors, 5% of the land planned for construction shall be provided for patients, visitors and hospital staff as relaxation space directly linked to the natural environment. A wind speed of less than 5 m/s shall be guaranteed in the pedestrian areas around the above buildings, and in cold and severely cold and windy regions, consideration shall be given to providing windproof waiting facilities at the main hospital entrances and exits intended for patients. In the case of flat roofing, the area of greening or high-reflectivity material when used for roofing shall be at least 50% of the calculable roofing area. Room entry design for ambulances shall be provided at the entrances and exits of emergency departments of hospitals in cold and severely cold regions.

2.2 *Architecture design*

In terms of architecture, the thermal performance of building envelopes should meet building energy conservation design standards, while heating, ventilation and air conditioning system design should comply with the National Standard GB50189 chapter 5 *Standard for Energy Efficiency Design of Public Buildings*; and it is acceptable for such special areas as operating rooms and ICUs not to comply with Article 5.3.26. Hospital buildings should not have Line-Of-Sight (LOS) interference, especially between ward buildings and surrounding buildings, with spacing greater than or equal to 20 m. Full use of underground spaces of buildings should be made, and such spaces should occupy at least 20% of the floor area.

2.3 *Indoor environment designing*

In terms of indoor environment, the main functional rooms in a building should have good vision, and the day-lighting coefficient should meet the requirements in the existing National Standard GB50033 (Standard for Day-lighting Design of Buildings), i.e. at least 60%. The required illumination, unified glare value and general color rendering index of indoor lighting are three major factors that influence lighting quality and they need to meet the relevant provisions of GB50034 (Standard for Lighting Design of Buildings) and JGJ49 (Code for Design of General Hospital Buildings).

Adjustable sunshade and thermal insulation measures should be taken. For the transparent parts of exterior windows and curtain walls, the area of controlled sunshade and thermal insulation adjustment measures should be up to a proportion of 25%.

As for the rooms with no special requirements for air pollution control, it is preferable to adopt natural ventilation measures and suitable air filtration systems, while it is desirable to select purification and filtration equipment with rational filter resistance, dust filtering efficiency and germ filtering efficiency.

The indoor noise level in the main functional rooms should not exceed the standard average value in the existing National Standard GB50118 (Code for Sound Insulation Design of Civil Buildings).

3 DESING STRATEGIES FOR GREEN HOSITALS IS COLD REGIONS

The design work for green hospitals in cold regions involves many difficulties, and I will summarize and analyze the green design strategies for many completed cold region hospitals involved in my participation or research.

3.1 *Attention to cold adaptation and land conservation-based green hospital planning and design*

The national land saving policies are adhered to and implemented in the design, and the functional relationships between buildings and the general layout of each building are rationally organized, culminating in layout relations between buildings which are mutually independent but closely linked. Based on land use, supporting engineering like landscape roads are rationally designed, and apart from the main buildings, land used for development is reserved, and at the same time, a large area of green space landscape is reserved as relaxation space, thereby realizing the dual purposes of land conservation and full utilization.

The square, buildings and green spaces in the entire courtyard, are subject to unified planning ideas, and under the premise of providing sufficient car parking spaces in the hospital, it is preferred to reserve as many sunny precincts as possible, with surrounding buildings and spatial gradients used as wind barriers and for wind induction, with more green vegetation for climate regulation. A south facing layout is adopted for buildings to ensure better sunshine. Underground parking arrangements, ambulance indoor entry design, etc. can help to shield hospital patients from the cold in winter. Attention is paid to the addition of indoor-outdoor transition areas, such as foyer arrangements at main entrances and exits, and hallways or antechamber transitions at secondary entrances and exits. Such areas as staircases, balconies, basements and roof terraces should match the thermal buffer area design

In terms of landscape, a green and ecological medical care environment is obtained by means of landscape gardens, much afforestation, and idyllic scenes with water body and micro-topography designs.

3.2 *Attention to the construction measures and material selection of green energy-saving hospitals*

Economical and practical thermal insulation materials are selected to ensure the thermal insulation of exterior walls of buildings. For example, thermal insulating rock wool boards with good thermal insulation and energy-saving and fire prevention performance can be selected. An extruded benzene board-based thermal insulation layer is adopted for roofing, while heat-insulating profile UPVC plastic steel windows are used as exterior windows, and single-frame three-layer double-hollow glass is adopted, having good air tightness and thermal insulation performance.

All stairwells are heated; for heating and non-heating rooms, relevant energy-saving structure standard atlas are selected for such detailed structures as floor slabs, walls, parapets, plinths, canopies and deformation joints, and for thermal insulation between exterior doors & windows and walls, which can meet any restriction requirements. Energy-saving equipment is selected for supporting facilities involving power supply, heat supply and water supply, and green environmentally-friendly materials are used as pipe materials of ventilation systems, thermal insulating materials, bonders, etc.

3.3 *Attention to sustainable development-based energy-saving measures for green hospitals*

Recycled water is provided for greening, landscape water bodies, washing and underground water sources through establishment of a rainwater collection system, comprehensive utilization of rainwater resources, and water conservation. In addition, in the water supply mode adopted in the Panjin City Central Hospital Project, water is directly supplied to some water points from municipal facilities and frequency-convertible feed water booster pumps are used as booster pumps in the other parts; energy-saving sanitary ware and water distribution components are selected in this practice, and at the same time, each building monomer and water consumption departments in each building monomer uses water meters for charge collection, thereby limiting the waste of water. Also, water pond and water tank overflow level alarms are provided, which is a good water-saving strategy.

3.4 *Attention to humanistic care-based indoor environmental design of green hospitals*

In the hospital street and medical care port mode, consulting space design is integrated and optimized to improve the extent of streamline identification, and consulting and medical treatment efficiency. For example, Panjin City Central Hospital uses a hospital street-based consulting mode for its clinic and medical technical building, with the consulting rooms of each department distributed on both sides of the hospital street to reduce cross interference and facilitate identification. At the same time, the top day-lighting skylight design of the large-scale atrium, the hospital street and the consulting space design combined with natural landscape are also important design methods for improving the comfort of the consulting environment.

In terms of air conditioning, based on the room function zoning and the principle of separation of clean and dirty material, air conditioners, exhaust fans and induction units-based combination of a mechanical exhaust system and natural ventilation is adopted. This ensures meeting the required degree of air purification in each functional space of the hospital, with the minimum per capita fresh air volume being 20 cubic meters/hour.

The table below shows the room ventilation plan we made while designing the "Jinzhou Central Hospital" Program.

In terms of the indoor acoustic environment, light environment and thermal environment, corresponding environmental protection measures are also taken to realize effective control, with basic indoor indicators basically meeting or exceeding the requirements of the *Standard for Evaluation of Green Hospital Buildings*. As for the acoustic environment, besides the use of sound-insulating materials for building envelopes, super low-noise water drainage plastic pipes are used for indoor water drainage systems, HDPE double-wall corrugated pipes are used for outdoor water drainage systems, and soundproofing plus anti-vibration measures are provided for fans. In the main tool room, the interior noise level meets the existing National Building Standard *GB50118* (Code for Sound Insulation Design of Civil Buildings), with all standard values met.

In terms of light environment adjustment, high-efficiency high-color rendering three-band straight tubular fluorescent lamps are used, with noise-free electronic ballasts used together with them, and anti-reflection lamps used in the special medical treatment rooms, creating a good light environment, with lower electricity consumption and reduced line loss.

Electric hot air screens are provided at the entrance for indoor heating of hospitals, and a ground radiation system (electric heating floor film) plus radiator-based heating mode is used in public spaces for winter heating featuring low energy consumption. In summer, an air conditioning system is combined with refrigeration. The table below shows the indoor environment indicators of the Cancer Hospital Affiliated to Harbin Medical University, which exceeds the specific data requirements in the Standard for Evaluation of Green Hospital Buildings, indicating a comfortable hospital medical treatment environment.

Table 1. Room Air Changes of Jinzhou Central Hospital (Design Program).

Room name	Number of air changes	Room name	Number of air changes
Consulting Room	5	Bathroom	10
Cleaning Room	10	Soiled Linens Room	5
Pharmacy	2	Lift Motor Room	10
Underground garage	6	Kitchen	30
Decontamination Room	10	Underground Restaurant	2
Disinfection Room	10	Treatment Room	5

4 CONCLUSIONS

In this paper, based on the writer's practical experience, through in-depth thinking and research in three aspects, i.e. adaptation to cold in special climate environments; use of green structure technology and new materials to realize the energy conservation and material conservation of buildings; attention to medical patients' psychological and physiological feelings and multi-aspect regulation of acoustic and photothermal effects to realize high-comfort indoor physical environment design, energy conservation, high efficiency and perfect function. A preliminary method for design of green medical buildings that fully adapts to the climatic conditions in cold regions has been constructed.

Research on improving the design of medical buildings in cold regions, a complex public building type, will be an important undertaking to improve the medical treatment environment in cold regions, and to achieve environmental protection and low energy consumption of medical buildings in cold regions.

REFERENCES

David Seamon, Robert Mugerauer. Dwelling, Place & Environment: Towards a Phenomenology of Person [M]. Krieger Pub Co, 2000. GB50118 Code for Sound Insulation Design of Civil Buildings.

GB50033 Standard for Delighting Design of Buildings.

GB50034 Standard for Lighting Design of Building.

GB/T50378-2014 Standard for Evaluation of Green Buildings, pp. 11, 21–24, 29, 32.

GB/T51153-2015 Standard for Evaluation of Green Hospital Buildings, pp. 4, 27.

JGJ49 Code for Design of General Hospital Buildings.

JiYan, StelliosPlainiotis. Sustainable building design practice [M], China Building Industry Press, 2006.

LiuHuang. International green ecological building evaluation method is introduced andanalyzed. Journal of Architecture [J]. 2003(2), pp. 58–60.

Luomeng. New Model Discussion of Green hospital energy management [M]. International Symposium on Green Hospital solution, pp. 207–222.

Mei Hong yuan. Cold Region Architecture [M]. China building industry press, 2012.

Richard L. Miller, Earl S. Swensson. Hospital and Healthcare Facility Design. W W Norton & Co Inc, 2002.

Civil, Architecture and Environmental Engineering – Kao & Sung (Eds)
© 2017 Taylor & Francis Group, ISBN 978-1-138-02985-9

Development of intelligent residual current devices based on the PIC micro-controller

Hao Chen
Metrology Center of Hunan Electric Power Research Institute in Hunan Electric Power Company, Changsha City in Hunan Province, China

Jie Xiang
National Electrical Equipment Inspection and Engineering Efficiency Evaluation Center (WUHAN), Wuhan City in Hubei Province, China

Rui Huang, Liman Shen, Fusheng Chen, Kai Li, Weineng Wang, Longmin Bu, Xing He & Xianyong Xu
Metrology Center of Hunan Electric Power Research Institute in Hunan Electirc Power Company, Changsha City in Hunan Province, China

ABSTRACT: In order to solve the problem of traditional residual current devices, a scheme of intelligent residual current devices was developed. In this paper, the basic principle of intelligent residual current devices is introduced, and then the hardware part of the PIC16F886 MCU system and the software architecture are described. The data showed that the intelligent residual current devices could monitor the real-time residual current and cut down the power supply quickly when there was a residual current caused by the fault, so that the devices could solve the residual current problem in the low-voltage distribution network.

1 INTRODUCTION

In low-voltage distribution networks, especially in the rural low voltage power grids, due to improper use of electrical appliances, electrical accidents, such as electric shock or short circuit, occur frequently. In order to protect people's lives and to ensure safety in low-voltage power supply networks, the installations of residual current devices are performed in the distribution network. But there are still some problems in traditional residual current devices which are installed in the system, such as the circuit breaks frequently, because of which the application of residual current devices in the distribution network is limited (Deng Xiaolin (2007), Gu Huimin (2006)). In order to solve the residual current problem in the low-voltage distribution network, it is important to develop an intelligent residual current device which can adapt to changes in low-voltage distribution networks.

2 PRINCIPLES OF RESIDUAL CURRENT DEVICES

In residual current circuit devices, the residual current is detected as well as the value of residual current is compared with a reference value. When the residual current value exceeds the reference value, the main circuit will be cut down. Some residual current devices experience an overload or short circuit protection and over-voltage protection (Gu Qiaoli (2006), Wu Shaoqin (2010)).

Figure 1 shows the schematic of residual current devices. It is a three-phase, four-wire, and low-voltage distribution network; according to circuit theory, when the system is operating normally, the three-phase circuit is symmetrical. In this case, the output of the zero sequence current transformer TA is zero, that is,

$$I_a + I_b + I_c + I_o = 0 \tag{1}$$

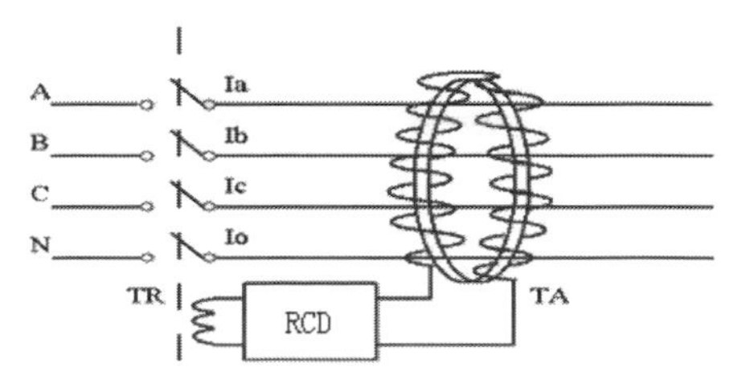

Figure 1. Principles of residual current devices.

By the principle of electromagnetic induction, each phase current in the line and the neutral line current generated in the zero-sequence current transformer's magnetic flux phase is equal to zero, that is,

$$\varphi_a + \varphi_b + \varphi_c + \varphi_o = 0 \qquad (2)$$

In this case, the output of the zero-sequence current transformer TA is zero, and current protection does not act. The normal power supply will not be cut down.

When short circuit or other electric shock occurs, the three-phase current will be no longer symmetrical. Residual current occurs, and the amount will no longer be equal to zero, that is,

$$I_a + I_b + I_c + I_o = I_r \qquad (3)$$

The amount of each phase lines' current and neutral current and residual current is I_r. In this case, the zero-sequence current transformer magnetic field will have the potential existence. The magnetic flux of the magnetic field generated by the potential of each phase is actually the current in the line and the neutral line current in the zero-sequence current transformer's iron ring phase generator's flux core and that,

$$\varphi_a + \varphi_b + \varphi_c + \varphi_o = \varphi_r \qquad (4)$$

Therefore, the output of the zero-sequence current transformer's secondary winding will not be zero. The voltage signal is proportional to the size of the residual current; when there is a more serious system failure, the larger the residual current is, the higher the zero-sequence current transformer's output voltages will be. The voltage is compared to a threshold voltage of the zero sequence current transformer's output; when the fault current reaches the setting value, the zero sequence current transformer output voltage exceeds the threshold voltage, the comparison circuit outputs a logic drive level start relay, and the relay circuit breaker will cut off the power supply quickly to achieve the residual current protection (Zhang lixu (2009)).

3 HARDWARE DESIGN OF INTELLIGENT RESIDUAL CURRENT DEVICES

3.1 *Hardware design of the overall program*

The hardware system of intelligent residual current devices includes the following: a micro-controller for signal conditioning parts, keyboard circuit, display circuit signal processing module, programming interface, operating circuit, and the power supply section. The structure of this hardware is shown in Figure 2.

Zero sequence current transformer, Current Transformer (CT), voltage transformers (PT) respectively collect the residual current, load current and bus voltage signals, and then output the voltage with a linear relationship to the amplitude of the current signal.

The signal processing module with the filtering and amplification processing to the signal obtains a suitable subsequent stage in the signal processing circuit.

In order to achieve an intelligent residual current protection, the micro-controller with internal hardware and software resources is used to achieve a residual current automatic switching gear, which is to meet the requirements of the operating environment adaptation. Meanwhile, the single-chip analog signal-conditioning circuit's signal sent was sampled and digitally processed, thereby displaying real-time voltage and current, in order to achieve good interactive features.

An action execution unit receives the I/O port of the MCU, to achieve the operation of the relay driver circuit breaker and cut down the fault circuit.

A digital display and status indicator with LED digital display and light-emitting diodes are controlled by using the micro-controller.

The program is downloaded via ICD2 with a 5-pin port, which connects the MCU to the PC's USB port, and then it will be programmed into the micro-controller FLASH memory by using the MPLAB integrated development environment.

3.2 *Design of the MCU module*

As shown in Figure 3, PIC16F886 micro-controller's clock source modes can be classified as follows: the external or internal. The external clock modes rely on external circuitry for the clock source, an oscillator module (EC mode), quartz crystal resonators or ceramic resonators (LP, XT, and HS modes) and Resistor–Capacitor (RC) mode circuits. The

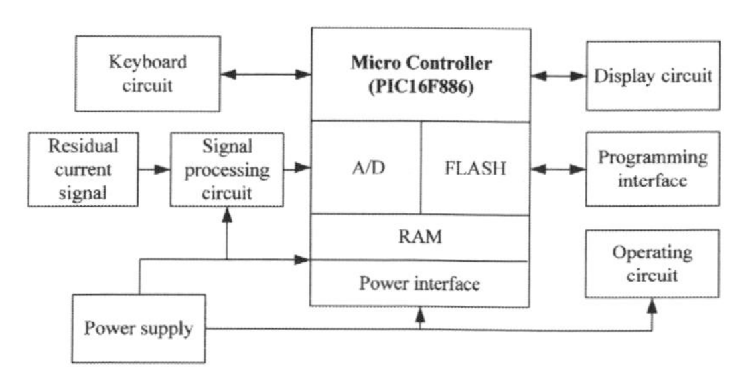

Figure 2. Hardware structure of the residual current devices.

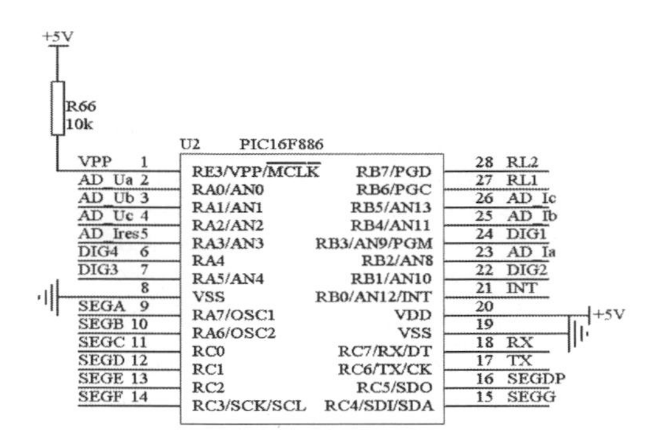

Figure 3. Schematic of the micro-controller system.

internal clock source module has two internal oscillators: 8 MHz High-Frequency Internal Oscillator (HFINTO-SC) and 31 kHz Low-Frequency Internal Oscillator (LFINTOSC); external or internal clock sources can be selected via the bit-select of the System Clock Select OSCCON register (SCS) (Microchip Technology (2005, 2002, 20-07)).

Since the external oscillator module requires a more complex I/O port and the peripheral circuit, the micro-controller internal oscillator is used as a clock source in this design; the clock frequency is selected to 8 MHz, and the corresponding bit is simply set in the OSCCON register.

3.3 *ADC conversion*

PIC16F886's ADC module can achieve an accuracy of 10-bit in digital–analog conversion. An analog input channel device uses a common sampling and holding circuit. The sampling and holding circuit is connected to the output of the converter input. A full approximation ADC generates a 10-bit binary result and saves the result in the ADC result registers (ADRESL and ADRESH). Software reference voltage conversion is connected to VDD or an external voltage reference pin is provided. The power supply voltage VDD is used as the ADC reference voltage in this design.

Conversion relation from the principle and the ADC reference voltage magnitude (+5V) can be drawn from the AD converter input voltage and by converting binary values i.e.,

$$NADC = 1023 \times (Vin - VR-) / (VR+ - VR-) \tag{5}$$

where V_{in} is the input voltage, V_{R+} is the reference voltage (+5V), and V_{R-} is a ground voltage, i.e. $V_{R-} = 0$. The conversion relationship is given as follows:

$$NADC = 204.6 \times Vin \tag{6}$$

3.4 *Design of the residual current signal processing and amplifying circuit*

Figure 4 shows the schwmatic of the signal rectifying and amplifying circuit. The amplifier circuit is divided into two parts, by using two integrated operational amplifiers, which are used in the design of an integrated four operational amplifier TL084. The operational amplifier with low signal distortion has a lot of advantages, such as the low input offset voltage, low offset current, low zero drift, and less noise (National Semi-conductor (2001)).

Since the voltage signal after the amplification stage after the AC signal is bipolar, while the micro-controller AD can be treated as a DC voltage signal 0~+5V; therefore, the AC signal should be rectified into a DC signal. In order to improve the measurement accuracy of small signals, the integrated operational amplifier amplification and depth of the negative characteristics of the feedback generated are used to overcome errors due to non-linear diode voltage and dead brought. The circuit diagram is shown in Figure 4.

In this circuit, the first part is composed of a high-precision half-wave rectifier, by using diodes D3 and D4 to achieve the half-wave rectifier, in which R3 and R4 should be equal, R3 = R4, otherwise the positive and negative half-wave will be asymmetrical, and thus the rectified waveform will be bad.

When $Vi > 0$ and $Vd < 0$, D4 works and D3 is turned off, so that,

$$V_{o1} = -V_i \tag{7}$$

When $Vi < 0$ and $Vd > 0$, D4 is turned off and D3 works, so that,

$$V_{O1} = 0 \tag{8}$$

The second part is the adder circuit made by using an operational amplifier, seen from the circuit parameters:

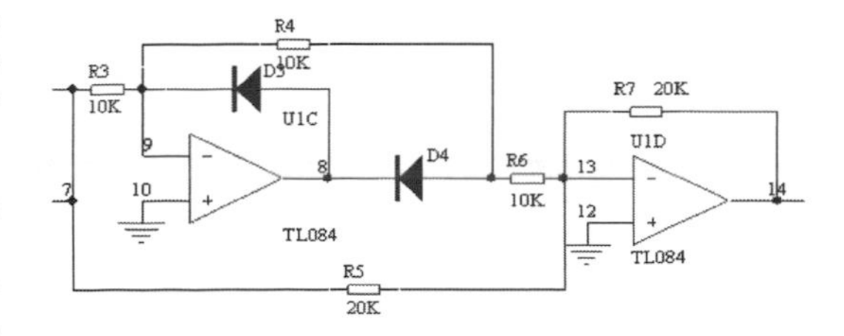

Figure 4. Schematic of the precision-rectifying and amplifying circuit for signal processing.

$$V_o = -(V_i + 2V_{o1}) \tag{9}$$

When $V_i > 0$,

$$V_{o1} = -(V_i - 2V_i) = V_i \tag{10}$$

When $V_i < 0$,

$$V_{o1} = -(V_i + 0) = -V_i \tag{11}$$

So that,

$$V_o = |V_i| \tag{12}$$

This circuit can achieve the full-wave rectification and transform AC residual current signals into DC signals.

4 SOFTWARE DESIGN OF INTELLIGENT RESIDUAL CURRENT DEVICES

The functional requirements of intelligent residual current protection are as follows: the devices can display the real-time value of residual current and other electrical parameters, when electric shock or short circuit occurs, and it can cut down the power supply quickly. The devices can be set automatically and the residual current and the load current can be operated manually through the key position, depending on the size of the over-current delay characteristics. The design based on the PIC16F886 micro-controller is used to control the internet module. The MCU control software is written by using C language and then it is developed by utilizing the microchip's MPLAB integrated development environment. Finally, the program is downloaded to the micro-controller FLASH through the compiler and the ICD2 machine code is also used to achieve intelligent protection.

According to the functional requirements of the system with a modular design idea, the software part is divided into the following modules: system initialization module, data acquisition module, data processing module, digital display and status display module, and movement gear switching module operation logic module. When the system works, it should make an initialization, and then it goes into the main loop, in which some tasks are operated, such as data acquisition and processing, current and voltage display, and the switch gear operation, in order to meet real-time requirements to ensure the normal operation and data processing of each module, by using a timer to interrupt and to control the execution of each module. Design and implementation of each module should be carried out with a lot of debugging.

5 TEST OF THE RESIDUAL CURRENT DEVICES

5.1 *Residual current measurement experiment*

The system hardware and software were debugged successfully, and then the IDB-1B residual current protection tester produced by Shanghai Siyuan equipment production company was used to generate a corresponding AC current source as a standard for residual current devices. The residual current RMS experiment was made to obtain a lot of experimental data in Table 1. According to the table, the test data show that the residual current devices could achieve an accuracy of at least 2.0% in measuring the residual current.

5.2 *Protection action characteristic test*

The residual current devices operation has inverse time operation characteristics, i.e. with an increase in the measured fault current, the delay acts more quickly, which can quickly cut off a serious fault. According to this requirement, it is necessary to protect the operating characteristics of the test, such as measurement data given in Table 2.

From Table 2, it can be observed that the residual current protection is achieved with inverse time characteristics in the action; the greater the fault

Table 1. Residual current measurement data.

Input residual current (mA)	Show value of the residual current devices (mA)	Absolute errors (mA)	Relative errors (%)
10	9.8	–0.2	–2.0
50	49.3	–0.7	–1.4
100	98.9	–1.1	–1.1
150	148.6	–1.4	–0.9
200	199.5	–0.5	–0.3
300	302.4	2.4	+0.8
400	404.8	4.8	+1.2
500	504.4	4.4	+0.9
600	605.7	5.7	+0.9

Table 2. Protection action characteristic test data (Tset = 0.3S).

Times of overload residual current	The maximum value of cut-off time (ms)	Test value of the residual current devices (ms)
1	300	240
2	200	159
5	40	32

current was, the lesser the break time would be. The test results were given within the required time limit, so that the devices could solve the residual current problem in the low-voltage distribution network.

6 CONCLUSIONS

In this paper, the principle of the residual current devices is introduced; it described the development scheme of intelligent residual current devices based on the PIC16F886 micro-controller. And then, the circuit design of the hardware and software, which make parts of the program were also discussed. The hardware and software were debugged, the performance indicators of the system were also tested, and the experimental results showed that the device exhibited a good performance and high reliability to ensure electricity supply in the low-voltage distribution network.

REFERENCES

Deng Xiaolin (2007). The analysis on difficulties of residual current devices. Rural Electrification.

Gu Huimin, Zhou Jigang (2006). Situation and development trend of the next generation of MCCB at home and aboard. Electrical Manufacturing, (10): 26–28.

Gu Qiaoli (2006). Research on intelligent leakage protection. Hebei University of Technology, Master's degree thesis.

Microchip Technology, PIC16F882/883/884/886/887 Data-sheet, 2007.

Microchip Technology, PIC16F872 Data Sheet, 2002.

Microchip Technology, Product Select Guide, 2005.

National Semiconductor (2001). LM78XX Series 3-Terminal Negative Regulator.

National Semiconductor (2001). LM79XX Series 3-Terminal Negative Regulator.

Wu Shaoqin (2010). Development and application of residual current devices. Transportation Technology of Hei longjiang province, (11).

Zhang lixu, Fu zhouxing (2009). Development of single phase residual current protector. Xi'an University of Science and Technology. Master's degree thesis.

Civil, Architecture and Environmental Engineering – Kao & Sung (Eds)
© 2017 Taylor & Francis Group, ISBN 978-1-138-02985-9

Design of Control Moment Gyro equivalent sound source based on transfer function matrix inversion

Wei Cheng & Wei-Qing Sun
Beijing University of Aeronautics and Astronautics, Beijing, China

Guang-Yuan Wang
Beijing Institute of Spacecraft System Engineering, Beijing, China

Jiang-Pan Chen, Xiong-Fei Li & Jiao Jia
Beijing University of Aeronautics and Astronautics, Beijing, China

ABSTRACT: The micro-vibration induced by a CMG (Control Moment Gyro) is very important for spacecraft. Vibration validating testing before launch is always necessary. However the boundary conditions of the validation measurement on the ground are different from space. One of the main differences is the presence of an atmosphere versus vacuum in space. Therefore, one should consider the structural vibrations due to the CMG acoustic excitation on the ground, which doesn't exist in space because there is no acoustic medium in vacuum. To research the micro-vibration due to CMG acoustic excitation, an equivalent acoustic source was designed in this paper, which can reproduce the CMG acoustic performance and therefore can be used to evaluate the contribution of the CMG acoustic excitation to the total vibration in the future. FRF matrix inversion based on PCA was used to derive the drive voltage of the speakers on the equivalent sound source. A comparison of the acoustic energy level and directional characteristics was done between the CMG and the equivalent sound source. Moreover, a new assurance criterion based on the microphone auto power linear spectrum was used to simplify the accuracy validation. It was shown that the equivalent sound source can accurately represent the CMG acoustic excitation from 80 Hz to 800 Hz.

1 INTRODUCTION

Because of the large torque amplification capability and good momentum storage performance of a CMG (Control Moment Gyro) attitude control system, it has been widely used on satellites for moving target tracking or imaging capturing, which need rapid attitude manoeuvrability and high attitude precision and stability. (Bhat, 2009).

However, the micro-vibration caused by the CMG;s broadband excitation can decrease pointing accuracy and the tracking control's stability for optical inter-orbit communication and astronomical telescope devices. (Kamesh, 2010; Luo, 2013) These disturbances are attracting more and more attention, because they are significant in the mid-high frequency range and excite flexible modes of the spacecraft, which cannot be controlled or reduced by the attitude and orbit control systems. (Steven, 1995) Therefore, spacecraft that use Control Moment Gyroscopes (CMG) for attitude control tend to have high sensitivity to pointing and jitter, creating a need for isolation. In recent years, some researchers have studied the methods to evaluate CMG disturbance forces/torques by detecting the disturbance displacements or accelerations. Many other researchers have tried to build analytical CMG dynamic models, and then based on the model, many different kinds of passive or active isolators were designed to minimize the load of the CMG acting on structures. (Belvi, 1995; Keith, 1995; Zhang, 2014; Tomio, 2006).

Anyhow, satellite micro-vibration must be measured and evaluated thoroughly to implement appropriate designs for the pointing and tracking systems. The micro-vibrations on LANDSAT-4 (Sudey, 1985), OLYMPUS (Manfred, 1990), and ETS-VI (Morio, 2001) satellite in orbit were reported from 1984 to 2001. Moreover, micro-vibration comparisons between in orbit and on ground measurements were conducted on ARTEMIS (Munoz, 1997; Galeazzi, 1996) in 1997, on the SPOT-4 (Privat, 1999) satellite in 1999, and on OICETS (Morio, 2010) in 2010. During the data comparison between in orbit and on ground, it was found that the satellite micro-vibration performance in space is different from on the ground. Morio Toyoshima (Morio, 2010) found that the in-orbit micro-vibration is greater, and thought it was caused by energy dissipation into air as heat or sound, which will not happen in space. It was also considered that the contribution of acoustical noise in the ground test might not be negligible. Acoustical noise does not exist in orbit but these effects might be included in the ground based measurement results.

In a ground based experiment, the vibration of a structure excited by a disturbance source can be separated into two parts. One part is structure borne vibration due to its structural force excitation. The other part is acoustic borne vibration due to its acoustic excitation. (Gajdatsy, 2010)

To evaluate the real vibration performance in space, the acoustic-borne vibration which will not exist in space needs to be subtracted from the ground testing result because there is no atmospheric medium and therefore no acoustic excitation in space. However, during operation of the CMG, the two kinds of excitation will act simultaneously. It's not possible to separate them directly. Therefore, if we could design an equivalent sound source, the acoustic energy of which is the same as the CMG, but which does not generate vibration, it could be put into the cabin structure to measure the vibration purely caused by acoustic excitation, then the difference of the micro-vibration performance in space and on the ground due to acoustic influence can be analysed.

In this paper, in order to design an equivalent sound source to simulate the CMG acoustic excitation, based on the acoustic signals from the microphones surrounding the CMG, a matrix inversion with PCA will be used to derive the drive voltages for the speakers on the equivalent sound source. The reproduction accuracy will be discussed for both energy level and directional characteristics aspects.

2 METHOD AND PRINCIPLE

2.1 *Identify the acoustic characteristics of the original source*

First of all, the acoustic characteristics of the original source need to be identified. Put the original acoustic source under a microphone array and get each microphone's sound pressure signal and then do Fast Fourier Transform (FFT) transform to get each microphone's spectrum, p_i ($i = 1, 2,\ldots, m$, the number of microphones), from which its sound power and its directional characteristics can be obtained.

2.2 *Frequency Response Function (FRF) testing from the speakers on equivalent source to microphones*

According to the original source's directional characteristics, appropriate numbers of speakers will be used to build up an equivalent sound source. Put the equivalent sound source under the same microphone array, the FRF from each speaker's drive voltage to each microphone, $h_{i,j}$ ($i = 1,2,\ldots, m$, the number of microphones; ($j = 1,2,\ldots, n$, the number of speakers), will be measured.

2.3 *Derive the drive voltage for the equivalent sound source*

To simulate the acoustic excitation from the original source, the appropriate drive voltage of the speakers on equivalent sound source need to be determined. The derivation of the voltage for the equivalent sound source can be considered a dynamic load identification, which is a kind of inverse problem. There are many kinds of methods for dynamic load identification, but roughly they can be divided into two types, frequency domain methods and time domain methods. (Gursoy, 2009; Yan, 2009) Because time domain methods are theoretically complex and are very sensitive to boundary conditions, currently for engineering applications, frequency domain methods are still widely used for stationary load identification.

However, due to the frequency response matrix being ill-conditioned, the numerical precision and stability of the frequency domain approach are often not as good as expected. (Mao, 2010) In this paper, PCA technology was used to handle the matrix conditioning problem.

Given, p_j ($j = 1, 2,\ldots, n$, the drive voltages for the equivalent sound source.

Therefore, the equation below can be obtained:

$$p_i = \sum_j h_{i,j} \cdot d_j \tag{1}$$

$$\mathrm{P} = \mathrm{H} \cdot \mathrm{D} \tag{2}$$

where p_i, $h_{i,j}$ and d_j represent Fourier spectra for their respective frequencies. Calculations are conducted in the same way for all target frequencies.

If the number of microphones, m, is equal to the number of drive voltages, n, the transfer function matrix H becomes a square matrix. Then, both sides of Eq. (2) can be multiplied by the inverse matrix of H to enable the drive voltage D to be calculated, as shown in Eq. (3).

$$\mathrm{D} = \mathrm{H}^{-1} \cdot \mathrm{P} \tag{3}$$

However, the accuracy of the calculation of the drive voltage D declines markedly in this method when noise is present in the measured transfer function matrix H, or when there is a high correlation among the transfer functions. Therefore principal component analysis was used here to provide a solution for these issues.

By means of principal component analysis, singular value decomposition is first applied to the transfer function matrix H measured between each microphone and drive voltage. The matrix will be decomposed into singular values, which will reveal the correlations between the principal components. The singular values in the matrix, S express the magnitude of the principal components in its diagonal elements.

$$H = U \cdot S \cdot V^{-1} \tag{4}$$

Before matrix pseudo-inversion, the principal components that are much smaller in comparison to the others are considered noise and should be discarded. If not, the small singular values will become big after matrix inversion, which will make phenomena with low importance become important and therefor the result will be unreliable.

After the small principal components have been discarded, the new regressed singular value matrix S_r, and unitary matrices U_r and V_r are formulated, and Eq. (5) is used to calculate the pseudo-inverse matrix of the transfer function matrix H, from which the noise components are removed.

$$H^{-1} = V_r^{-1} \cdot S_r^{-1} \cdot U_r \tag{5}$$

It is desirable to set the number of microphones, *m*, higher than the number of drive voltages, *n*, ($m \geq 3\,n$) to exclude the noise components included in the transfer functions and the microphone signals by means of principle component analysis.

$$D = V_r^{-1} \cdot S_r^{-1} \cdot U_r \cdot P \tag{6}$$

2.4 *Validation of the equivalent sound source*

We can put the equivalent sound source under the microphone array and drive the speakers by the voltages obtained from 2.3, measure the microphone signals to compare it to the original acoustic source to validate whether the equivalent sound source can replicate the acoustic emission of the original source.

3 EXPERIMENTAL TEST

Because of CMG's size not small enough compared to the cabin structure, and its directional acoustic properties, CMG cannot be considered a point acoustic source. By reason that CMG is always mounted on a plate, it can be approximated by 5 separated sources. Therefore, an equivalent sound source with five speakers was designed to simulate CMG acoustic radiation (see Fig. 1).

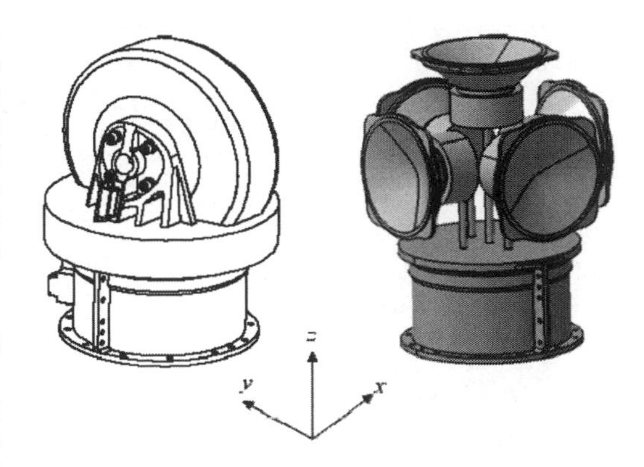

Figure 1. Sketch of the CMG (left) and equivalent sound source (right).

The size and mass distribution of the equivalent sound source is similar to the CMG. To validate whether the equivalent sound source can really represent the CMG's acoustic performance, a measurement based on ISO 3744 was done. If the signals of the 20 response microphones around the CMG and equivalent sound source are the same, we can conclude their sound power and directional characteristic are the same.

3.1 *CMG noise measurement*

The CMG was put under a microphone array (see Fig. 3). The signals of 20 microphones were measured to be used to calculate the five drive voltages of the equivalent sound source. Moreover, its sound power level and directional characteristics were also obtained to be compared with the equivalent sound source.

3.2 *FRF measurement*

For each of the five speakers, a broadband random signal with 800 Hz bandwidth was output separately to an independent amplifier, and then the FRFs between the output voltages from the signal generator and the 20 microphones surrounding the equivalent sound source were measured (see Fig. 2 and Fig. 3) while H_v estimator was used.

3.2.1 *Signal/noise ratio check*

To verify the speakers' frequency range, the coherence between speakers and microphones needs to be checked. For each speaker, the coherence curves of 20 microphones were averaged. The averaged coherence results of five speakers are shown in Fig. 4. It can be found that the signal/noise ratio is good from 80 Hz to 800 Hz. However, because of the speaker's limited size, the sound below 80 Hz can-

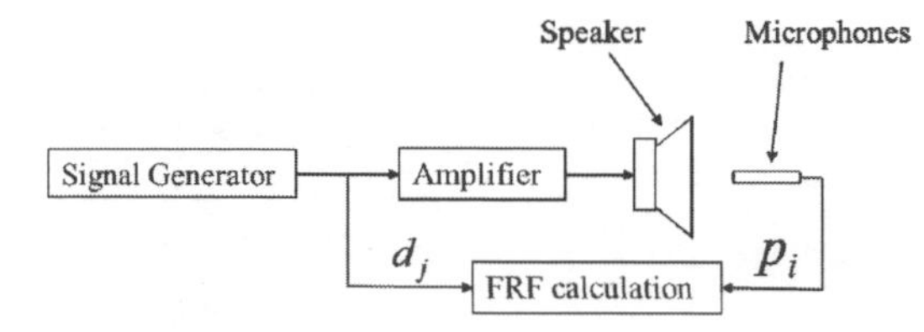

Figure 2. Principle of FRF measurement from drive voltage to microphones.

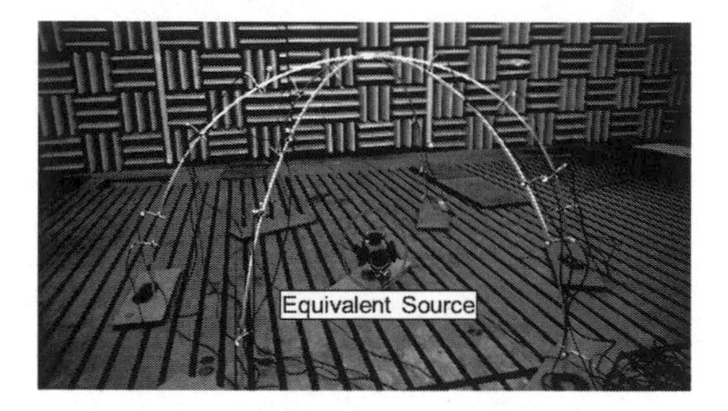

Figure 3. Noise identification and transfer function measurement in Anechoic room.

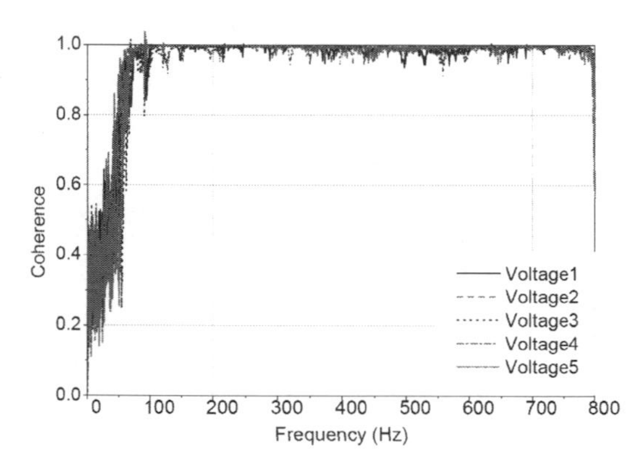

Figure 4. Averaged coherence results for the five speakers.

not be radiated. Therefore, the equivalent sound source can be used to simulate the CMG noise from 80 Hz to 800 Hz.

3.2.2 *System linearity check*

Because the sound pressure level of the microphones during FRF measurements could be different from the CMG, the linearity of the system needs to be checked. White noise signals with three different RMS levels (0.01 V, 0.05 V, 0.2 V) were output. As an example, the amplitude and also the phase of the FRF between the first microphone and drive voltage of one side speaker from 80 Hz to 800 Hz are compared in Fig. 5. We can see the

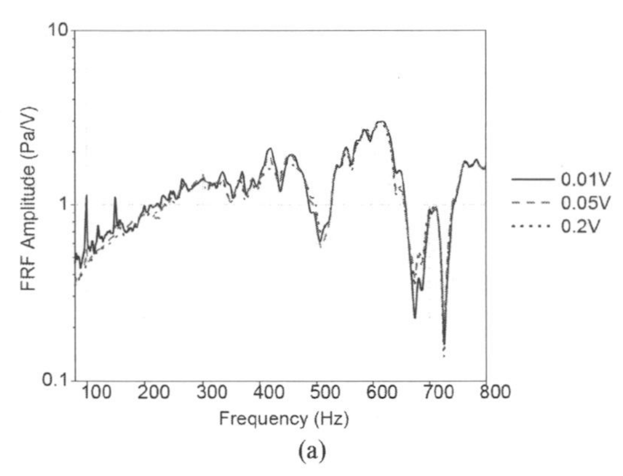

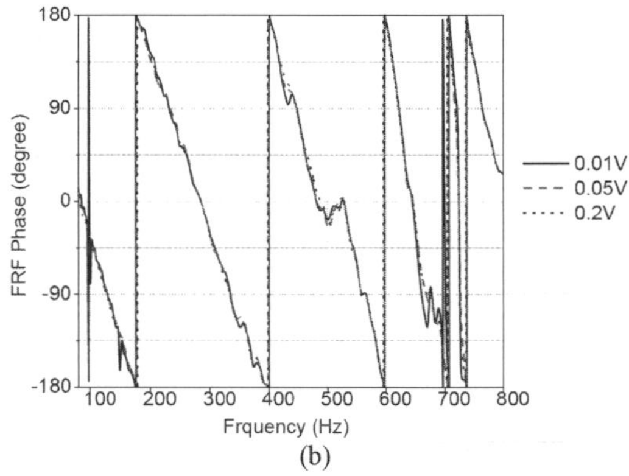

Figure 5. FRF amplitude and phase comparison for different of level acoustic excitation: (a) FRF Amplitude Spectrum; (b) FRF Phase Spectrum.

amplitude and phase are coincident, so the system is linear in the range 80–800 Hz.

3.2.3 *FRF measurement*

After system linearity and signal/coherence checks, FRFs between each of the 20 microphones and the 5 speakers were measured. Totally, there are 100 FRFs measured between each of the 5 source drive voltages and each of the 20 microphones as Eq. (7).

$$H_{20\times 5} = \begin{bmatrix} h_{1,1} & h_{1,2} & h_{1,3} & h_{1,4} & h_{1,5} \\ h_{2,1} & h_{2,2} & h_{2,3} & h_{2,4} & h_{2,5} \\ \vdots & \vdots & \vdots & \vdots & \vdots \\ h_{20,1} & h_{20,2} & h_{20,3} & h_{20,4} & h_{20,5} \end{bmatrix} \tag{7}$$

3.3 *Equivalent sound source noise measurement*

As per 2.3, the five speakers drive voltage spectrums were derived through FRF matrix inversion,

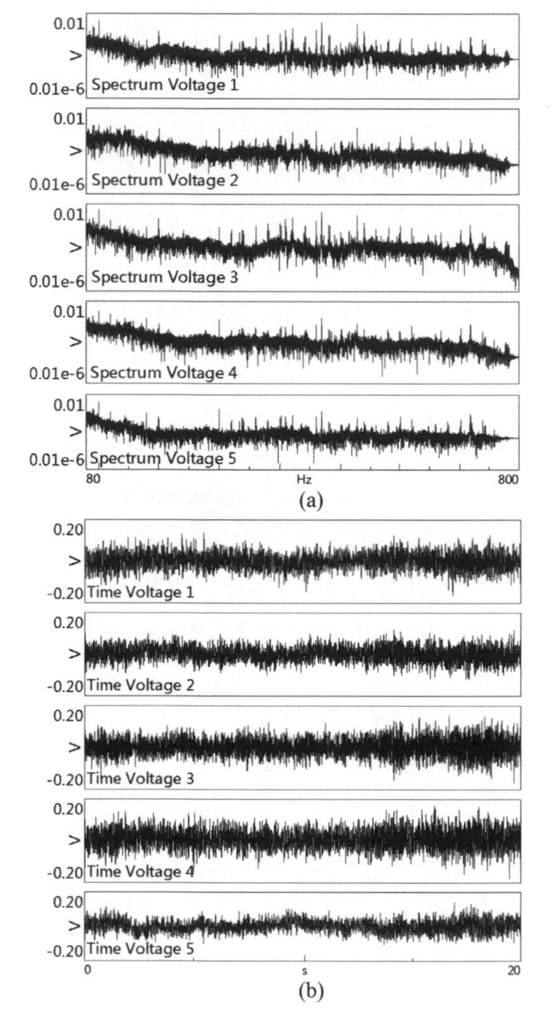

Figure 6. Five speakers' drive voltage: (a) Frequency Spectrum; (b) time Signal.

as below. Subsequently, the output time signals were obtained by inverse FFT. (See Fig. 6) Finally, the equivalent sound source was put under the microphone array again while the derived voltage signal was applied on the five speakers. The equivalent sound source can be validated by comparing the microphone signals with the signals measured when the CMG was present.

4 VALIDATION

Based on the 20 microphone signals, a validation based on energy level and also directional characteristics is done below.

4.1 *Sound power comparison*

According to ISO 3744, the sound power of an acoustic source can be derived as Eq. (8),

$$L_w = \overline{L_p} + 10\log\left(\frac{S}{S_0}\right) \tag{8}$$

where $\overline{L_p}$ is the sound pressure level averaged over the measurement surface; S is the area, in square meters, of the measurement surface; $S_0 = 1\ m^2$.

The sound power of the CMG and the equivalent sound source was calculated separately as shown in Fig. 7. It can be seen that the equivalent sound source can reproduce the energy level and spectrum characteristics of the CMG. On each peak frequency, the level difference was less than 1 dB. Moreover, the CMG's total sound power level is 58.1dB, compared to 57.4dB for the equivalent sound source, the difference still less than 1dB.

However, the equivalent sound source cannot reproduce the background signal of the CMG which is the reason of 1dB total sound power difference. The reason is that the background signal level is too low. Therefore, compared to the anechoic room background noise, the signal/noise ratio is not good enough.

4.2 *Each microphone signal comparison*

To validate whether the equivalent sound source can also reproduce the CMG's acoustic directional characteristics, each microphone's signal was compared.

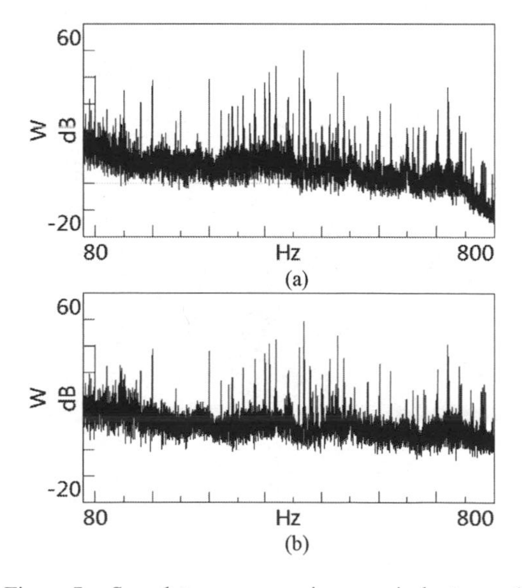

Figure 7. Sound power comparison: equivalent sound source (a) and CMG (b).

The results of all 20 microphone spectra from the CMG and the equivalent sound source are shown in Fig. 8. It proves that the distribution of the acoustic energy not only over frequency but also spatially is substantially the same. Since the 20 microphones were positioned on every direction of the test objects, the equivalent sound source can be deemed to accurately reproduce the acoustic directional characteristics.

4.3 *Auto Power Based Assurance Criterion (APAC)*

To simplify the validation of the reproduction accuracy, a new metric, Auto Power Based Assurance Criterion (APAC) was designed as per Eq. (9), which will represent an overall correlation result for the entire frequency range for each microphone point.

$$APAC(j)=\frac{\mathrm{AP}_{\mathrm{CMG}}(\omega_i)_j\cdot\mathrm{AP}_{\mathrm{EQS}}^{T}(\omega_i)_j}{\left|\mathrm{AP}_{\mathrm{CMG}}(\omega_i)_j\right|\cdot\left|\mathrm{AP}_{\mathrm{EQS}}(\omega_i)_j\right|} \tag{9}$$

where $\mathrm{AP}(\omega_i)_j$ is a row vector, which represent the auto power linear spectrum at microphone j for

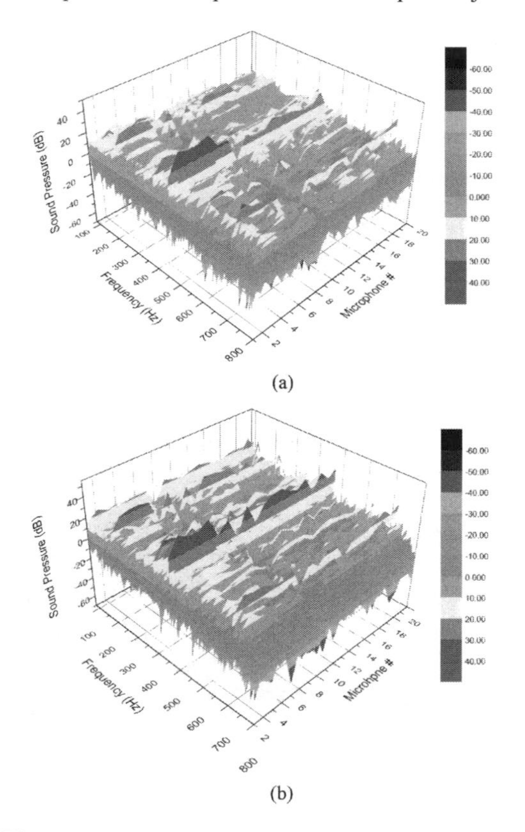

Figure 8. The microphone signals from the CMG and equivalent sound source: (a) CMG; (b) equivalent source.

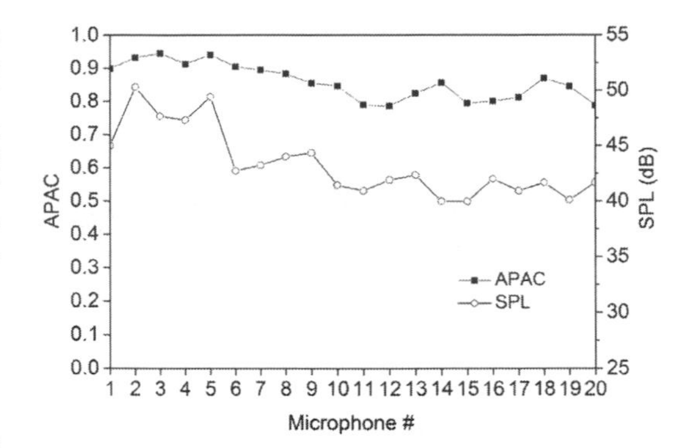

Figure 9. Auto Power Based Assurance Criterion (APAC) and SPL of each microphone.

a range of frequency ω_i; EQS means equivalent sound source; T means vector transverse; |.| means norm of vector.

An APAC close to 1 means good reproduction accuracy. The APAC for each microphone point is shown in Fig. 9. It can be seen that the APAC of all microphones are above 0.78, and the averaged APAC is 0.86, which means the reproduction accuracy should be acceptable.

Moreover, there is a trend: the higher the microphone number, the lower the APAC value, which is similar to the Sound Pressure Level (SPL) variation. The reason for that could be, because the higher number microphones were put closer to the top of the CMG, due to the CMG directivity characteristic, the signal/noise ratio was worse, therefore the reproduction accuracy was decreased.

5 CONCLUSIONS

In this paper, a method to design an acoustic source to each microphone simulate the CMG was proposed. Based on the acoustic signal from the microphones around the CMG, and the FRFs from the drive voltages of the speakers on the acoustic source to the microphones, an FRF matrix inversion by PCA was applied to calculate the required speaker drive voltages to simulate the CMG acoustic performance.

The accuracy of the reproduction was validated for both energy level and directivity aspects. We proved that the equivalent sound source can generally reproduce the acoustic excitation of the CMG.

Because of the frequency range of the speakers, the equivalent sound source in this paper cannot be used below 80 Hz. As a matter of fact, for sound below 80 Hz, the speaker size will be too large so that its geometry and weight cannot be similar to the CMG, which makes it not suitable for measuring the CMG acoustic borne micro-vibration in the cabin. Therefore, this method is currently limited to the frequency range above 80 Hz.

ACKNOWLEDGEMENTS

The authors would like to express their appreciation to the members of the CMG team at Beijing Institute of Spacecraft System Engineering for their kind help.

REFERENCES

Belvin, W. Sparks Dean, Horta Lucas, and Elliott Kenny, in 36th Structures, Structural Dynamics and Materials Conference (American Institute of Aeronautics and Astronautics, 1995).

Bhat S. and P. K. Tiwari, Automatic Control, IEEE Transactions on **54** (3), 585 (2009).

Gajdatsy, P., K. Janssens, Wim Desmet, and H. Van der Auweraer, Mechanical Systems and Signal Processing **24** (7), 1963 (2010).

Galeazzi, C., P.C. Marucchi-Chierro, and LV Holtz, in ESA International Conference on Spacecraft Structures, Materials and Mechanical Testing, Noordwijk, Netherlands (1996), pp. 997.

Gursoy E. and D. Niebur, IEEE Trans. Power Deliv. **24** (1), 285 (2009).

International Organization for Standardization, 2010. Acoustics—determination of sound power levels and sound energy levels of noise sources using sound pressure—engineering methods for an essentially free field over a reflecting plane.

Kamesh, D., R. Pandiyan, and Ashitava Ghosal, Journal of Sound and Vibration **329** (17), 3431 (2010).

Keith Belvin, W. Report No. NASA-TM-111574, 1995.

Luo, Q., D. X. Li, W. Y. Zhou, J. P. Jiang, G. Yang, and X. S. Wei, Journal Of Sound And Vibration **332** (19), 4496 (2013).

Manfred E. Wittig, L. van Holtz, D. E. Tunbridge, H. C. Vermeulen, and Bernard D. Seery, in Free-Space Laser Communication Technologies II, edited by David L. Begley (1990), Vol. 1218, pp. 205.

Morio Toyoshima, Yoshihisa Takayama, Hiroo Kunimori, Takashi Jono, and Shiro Yamakawa, OPTICE **49** (8), 083604 (2010).

Morio Toyoshima and Kenichi Araki, OPTICE **40** (5), 827 (2001).

Munoz L., E. M. Marchante, presented at the Proceeding 48th International Astronautical Congress, Torino, Italy, (1997).

Privat, M. presented at the 8th European Space Mechanisms and Tribology Symposium, (1999).

Steven P. Neeck, Thomas J. Venator, Joseph T. Bolek, and Brian J. Horais, in Platforms and Systems, edited by William L. Barnes (1995), Vol. 2317, pp. 70.

Sudey Jr J. and J. R. Schulman, Acta Astronautica **12** (7–8), 485 (1985).

Tomio Kanzawa, Daigo Tomono, Tomonori Usuda, Naruhisa Takato, Satoru Negishi, Shinji Sugahara, and Noboru Itoh, in Society of Photo-Optical Instrumentation Engineers (SPIE) Conference Series (2006), Vol. 6267.

Wentao Mao, Dike Hu, and Guirong Yan, International Journal of Applied Electromagnetics & Mechanics **33** (3/4), 1001 (2010).

Yan G. and L. Zhou, Journal Of Sound And Vibration **319** (3–5), 869 (2009).

Zhang Yao and Zhang Jingrui, Aerospace and Electronic Systems, IEEE Transactions on **50** (2), 1017 (2014).

Civil, Architecture and Environmental Engineering – Kao & Sung (Eds)
© 2017 Taylor & Francis Group, ISBN 978-1-138-02985-9

A study of the magnetic field distribution with saddle-shaped coil of the electromagnetic flow meter

Hua Jia, Yueming Wang, Wentao Li & Yaoyao Zhang
School of Information Engineering, Inner Mongolia University of Science and Technology, Baotou, China

ABSTRACT: The saddle-shape excitation coil is available in the electromagnetic flow meter usually. To discuss the saddle-shaped excitation coil, firstly the magnetic fields in the space generated respectively by straight, elliptical, and circular excitations of the coil segment are analyzed by using the Biot–Savart law. On this basis, the magnetic field distribution inside the electromagnetic flow measurement sensor produced by the saddle-shaped excitation coil is investigated and discussed. Finally, a comparative analysis between magnetic field distributions excited by rectangular and saddle-shape coils inside the instrument is suggested.

1 INTRODUCTION

The electromagnetic flow meter is an important flow measurement instrument, and the excitation coil design is one of the key technologies used in anelectromagnetic flow meter. Tingxiang Chen and Baosen Huang have calculated the magnetic field excited by several kinds of inhomogeneous type coils (Chen, 1982); Xutong Qiao and Lijun Xu gave the design of exciting Coils for multi-electrode electromagnetic flow meters (Qiao, 2002); Ningde Jin built simulations about magnetic field distributions of outward flow meters (Jin, 2009); Yueming Wang and Lingfu Kong carried out an optimization analysis for the excitation coil of concentrating flow meters (Wang, 2011; Kong, 2012; Li, 2012); and Chen Zhao and Bin Li suggested a line approximation algorithm to analyze the magnetic field distribution of saddle excitation coils in EM flow meter sensors (Zhao, 2008). Due to restrictions with installation space, the excitation coil of the saddle-shape structure is frequently used in the flow meter sensor, especially catering to a new sort of downhole flow meter. In this paper, the electromagnetic flow meter sensor's internal magnetic induction with saddle-shaped excitation coil is studied by using the Biot–Savart law, thereby providing a consulting reference to the magnetic field distribution inside the electromagnetic flow meter with saddle-shape excitation coil.

2 BIOT–SAVART LAW

From the Biot–Savart law, the magnetic induction generated by current element $I\vec{dl}$ at point P in space is decided by using the following equation:

$$dB = \frac{\mu_0}{4\pi}\frac{I\vec{dl}\times\vec{r}}{r^2} \tag{1}$$

where $\vec{r}$ is the radius vector that the current element $I\vec{dl}$ directs to point P, r is the distance from point P to the current element $I\vec{dl}$, and μ_0 is the permeability of vacuum.

The magnetic induction at point P in the space which any current carrying conductor generates could be obtained by using the superposition principle.

$$\vec{B} = \oint dB = \frac{\mu_0}{4\pi}\oint\frac{I\vec{dl}\times\vec{r}}{r^2} \tag{2}$$

3 INTERAL MAGNETIC FIELD INVESTIGATION ON THE SADDLE-SHAPED COIL

Different shapes of coil excite different magnetic fields; thus, the theoretical analysis of saddle-shaped coils is carried out to provide theoretical basis to the design of the electromagnetic flow meter.

3.1 *Magnetic distribution calculation of saddle-shaped coils*

Roughly speaking, a saddle-shaped coil is formed by two straight segments and two arc segments. And so, according to the magnetic induction superposition principle, the magnetic induction at point P in the space can be acquired by adding magnetic induction generated by using two straight segments and two arc segments at point P in the space.

Figure 1 shows the sketch map of the saddle-shaped coil. In Figure 1, the sub-graph (a) shows flow sensor schematic diagram with saddle-shaped excitation coil, and sub-graph (b) shows the cartesian coordinate system's schematic diagram with one of saddle-excitation coil. As shown in figure, L1 and L3 are the two straight segments and L2 and L4 are the two arc segments (L2 and L4 could be part of the ellipse or circle). Four pieces are connected each other into a form of coil circuit. To study the problem in universality, in this work, we define L2 and L4 ellipses (for circle is a special type of ellipse when the long and short radii are equal; we can study on elliptic coil and circle coil is also solved as which is pre-digested of the elliptic model). The x-axis and the y-axis are perpendicular to the z-axis; and so, x-axis, y-axis, and z-axis constitute a rectangular coordinate system. In electromagnetic flow meters, there are two identical coils that are symmetrically distributed. We can get the magnetic induction at one point by using the principle of superposition, and then the magnetic field distribution inside the instrument is described.

The Biot–Savart law is used to calculate the value of magnetic induction at point P in space. The function of L1 and L3 mentioned above are described as follows:

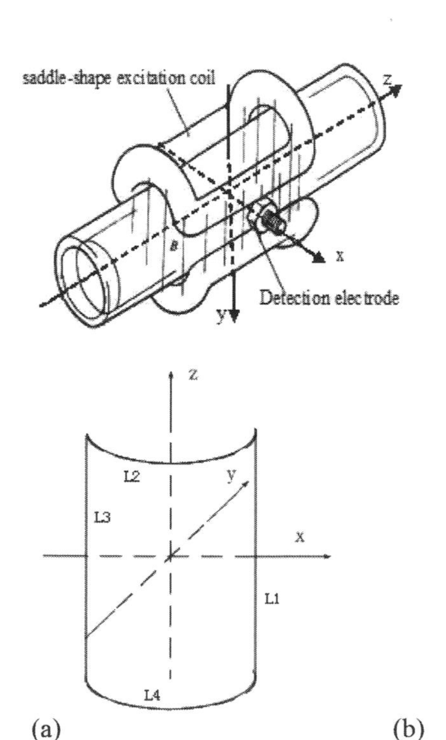

Figure 1. The schematic diagram of the saddle-shaped coil excitation.

$$\begin{cases} \mathrm{L1: x = a, y = b}, z \in [-c, c]; \\ \mathrm{L3: x = \text{-}a, y = b}, z \in [c, -c]; \end{cases} \tag{3}$$

In this work, for universality, the description of coordinates for both L1 and L3 consist of a, b, and c.

The measuring electrodes are completely covered by using the magnetic field of the excitation coil, and so the study focuses on the inside space of the coverage only. If a current element Idl in the coil whose coordinate position is (ς, η, ζ), and one arbitrary point P in the space whose coordinate position is (x1, y1, z1) are defined, the distance from p to any current element Idl in coil could be defined as follows:

$$r = \sqrt{(x_1 - \varsigma)^2 + (y_1 - \eta)^2 + (z_1 - \xi)^2} \tag{4}$$

Substituting equation (3) in equation (4), we can calculate the distance from P to any point in the excitation coil.

According to the Biot–Savart law, the magnetic induction and its direction excited at point P in space by excitation coils of L1 and L3 are solved as follows:

Point P's magnetic induction and direction caused by excitation coils of L1 is given as follows:

$$\begin{cases} B1x = \dfrac{\mu_0 I}{4\pi}\displaystyle\int_{L1}\dfrac{dl \times r1}{r1^2} = \dfrac{\mu_0 I(b - y_1)}{2\pi((a - x_1)^2 + (b - y_1)^2)}k1 \\ B1y = \dfrac{\mu_0 I}{4\pi}\displaystyle\int_{L1}\dfrac{dl \times r1}{r1^2} = -\dfrac{\mu_0 I(a - x_1)}{2\pi((a - x_1)^2 + (b - y_1)^2)}k1 \\ B1z = 0 \end{cases} \tag{5}$$

where

$$k1 = \frac{c + z_1}{\sqrt{(a - x_1)^2 + (b - y_1)^2 + (c + z_1)^2}} + \frac{c - z_1}{\sqrt{(a - x_1)^2 + (b - y_1)^2 + (c - z_1)^2}}$$

Point P's magnetic induction and direction caused by excitation coils of L3 is given as follows:

$$\begin{cases} B3x = \dfrac{\mu_0 I}{4\pi}\displaystyle\int_{L3}\dfrac{dl \times r1}{r1^2} = -\dfrac{\mu_0 I(b - y_1)}{2\pi((-a - x_1)^2 + (b - y_1)^2)}k3 \\ B3y = \dfrac{\mu_0 I}{4\pi}\displaystyle\int_{L3}\dfrac{dl \times r1}{r1^2} = -\dfrac{\mu_0 I(-a - x_1)}{2\pi((-a - x_1)^2 + (b - y_1)^2)}k3 \\ B3z = 0 \end{cases} \tag{6}$$

where,

$$k3 = \frac{c - z_1}{\sqrt{(-a - x_1)^2 + (b - y_1)^2 + (c - z_1)^2}} + \frac{c + z_1}{\sqrt{(-a - x_1)^2 + (b - y_1)^2 + (-c - z_1)^2}}$$

We can also use the Biot–Savart law to solve the magnetic induction at point P in space excited by two arc segments (L2 and L4) and then acquire the final magnetic induction distribution by using the superposition principle.

The functions for L2 and L4 (parts of ellipse) are shown in equation (7). In the equation, the projection of L2 and L4's long and short radii along the x-axis is e and along the y-axis is f, and they have a projection respectively along the z-axis –c and c.

$$\begin{cases} \dfrac{x^2}{e^2} + \dfrac{y^2}{f^2} = 1 \\ z = \pm c \end{cases} \tag{7}$$

r stands for the distance from the center to the current element $I\overrightarrow{dl}$ (ς, η, ζ) in the ellipse, according to elliptic equation in this work is given as follows:

$$\begin{cases} \varsigma = e\cos\theta \\ \eta = f\sin\theta \\ r = \sqrt{e^2\cos^2\theta + f^2\sin^2\theta} \end{cases} \tag{8}$$

It should be noted that the θ in equation (8) is an eccentric angle.

Figure 2 shows the vector calculation diagram of an elliptic excitation coil. As shown in the figure, the point P(x1, y1, z1) could be any point in the sensor space, the current element $I\overrightarrow{dl}$ belongs to the elliptic equation (L2 or L4), the center point of the elliptic equation is point O, $\vec{r}$ is the vector that connects point P to the current element $I\overrightarrow{dl}$'s initial point, $\overrightarrow{Pr}$ is the vector that connects the center point O of the elliptic equation (L2 or L4) to the point P, $\overrightarrow{r01}$ is the vector that connects point O to the current element $I\overrightarrow{dl}$'s initial point, and $\overrightarrow{r02}$ is the vector that connects point O to the current element $I\overrightarrow{dl}$'s terminal point, $\overrightarrow{dl}$ is the vector that connects current element $I\overrightarrow{dl}$'s initial point to the current element $I\overrightarrow{dl}$'s terminal point, and dθ is the angle between $\overrightarrow{r01}$ and $\overrightarrow{r02}$. And so, we can obtain the following vector conjunction with Figure 2.

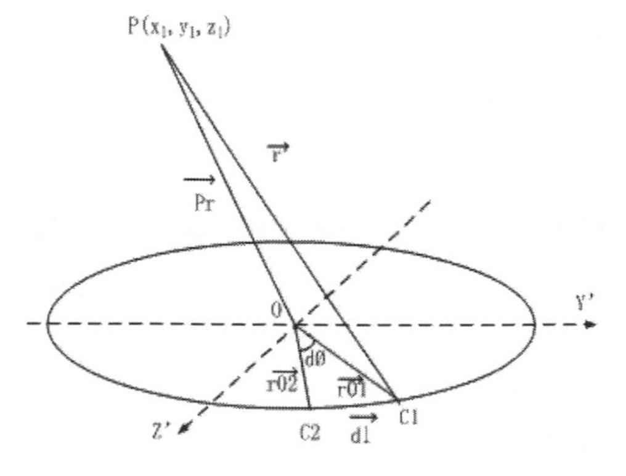

Figure 2. Vector diagram of the elliptic excitation coil.

The equation (9) has expressed the vector $\overrightarrow{Pr}$ as follows:

$$\overrightarrow{Pr} = x_1 i + y_1 j + (z_1 - c)k \tag{9}$$

where i, j, and k are the direction vectors; equation (10) expresses the vector $\overrightarrow{r01}$ and equation (11) expresses vector $\overrightarrow{r02}$ (because dθ is quite small, which means the approximate treatments such as $\cos d\theta \approx 1$ and $\sin d\theta \approx d\theta$ are possible).

$$\overrightarrow{r01} = e\cos\theta i + f\sin\theta j \tag{10}$$

$$\begin{aligned} \overrightarrow{r02} &= e\cos(\theta + d\theta)i + f\sin(\theta + d\theta)j \\ &= e(\cos\theta - \sin\theta \cdot d\theta)i + f(\sin\theta + \cos\theta \cdot d\theta)j \end{aligned} \tag{11}$$

And we can also obtain vector $\overrightarrow{dl}$, as given in equation (12), and the vector $\vec{r}$ is given in equation (13)

$$\overrightarrow{dl} = \overrightarrow{r02} - \overrightarrow{r01} = -e\sin\theta d\theta i + f\cos\theta d\theta j \tag{12}$$

$$\vec{r} = \overrightarrow{Pr} - \overrightarrow{r01} = (x_1 - e\cos\theta)i + (y_1 - f\sin\theta)j + (z_1 - c)k \tag{13}$$

Equation (9)–(13) are substituted into equation (2); the magnetic induction and direction caused by the excitation of two arc segments coils (L2 and L4) at point P in space is expressed as equation (14).

$$\begin{cases} Bx = \dfrac{\mu_0 I}{4\pi}\displaystyle\int_{\theta 1}^{\theta 2}\left(\frac{1}{\sqrt{r1^3}}(z_1 - c) + \frac{1}{\sqrt{r2^3}}(z_1 + c)\right) f\cos\theta d\theta \\ By = \dfrac{\mu_0 I}{4\pi}\displaystyle\int_{\theta 1}^{\theta 2}\left(\frac{1}{\sqrt{r1^3}}(z_1 - c) + \frac{1}{\sqrt{r2^3}}(z_1 + c)\right) e\sin\theta d\theta \\ Bz = \dfrac{\mu_0 I}{4\pi}\displaystyle\int_{\theta 1}^{\theta 2}\left(\frac{1}{\sqrt{r1^3}} + \frac{1}{\sqrt{r2^3}}\right)(ef - x_1 f\cos\theta - y_1 e\sin\theta)\, d\theta \end{cases} \tag{14}$$

It should be noted that θ is the eccentric angle of the elliptic equation. In the equation, $\theta 1$ and $\theta 2$ in the integrator are the initial and terminal points of the eccentric angle.

For the case of the circular shape coil, we can make out magnetic induction and direction by just substituting e = f in equation (14).

According to the superposition theorem, the gross magnetic induction and direction generated by each part of the excitation coil at the point P is given as follows:

$$\begin{cases} Bx = \sum_{i=1}^{8} Bix \\ By = \sum_{i=1}^{8} Biy \\ Bz = \sum_{i=1}^{8} Biz \end{cases} \tag{15}$$

where, i is defined as the identity number of each edge of the excitation coil; x, y, and z, respectively stand for the axial component of magnetic induction of point P on the x-axis, y-axis, and z-axis.

Similarly, the magnetic induction and direction of any point inside the sensor space can be calculated by using the above-mentioned method.

3.2 *Magnetic induction distribution simulation for saddle and other shape coil*

In the simulation experiment, substitute a = 0.6R, b = 0.6* $\sqrt{3}$ R and c = 5R, where R is the radius of the sensor. Numerical simulation of the magnetic induction distribution of the saddle-shaped excitation coil is given as follows: Figure 3 shows the distribution of By (magnetic induction along direction of y-axis) inside the measuring tube where z = 0. The sub-map (a) is the distribution of By inside the measuring tube where z = 0 and the sub-map (b) is the potential diagram of By inside the measuring tube where z = 0.

Obviously, the magnetic induction distribution changes along with the variation of position and shape (different values of a, b, c, e, and f); however, under a new condition (set different values of a, b, c, e, and f), the magnetic induction distribution can also be acquired by using the above-mentioned method.

3.3 *Comparison between the saddle and rectangular excitation coil*

Table 1 shows the magnetic induction distribution performance comparative analysis between rectangle and saddle excitation coils.

In accordance with the method described above, more simulation of the magnetic induction distribution generated by different shapes and parameters (set different values of a, b, c, e, and f) has been established for comparison. In this section we built a simulation for rectangular excitation coils which has the same size as saddle-shaped excitation coils mentioned in the previous section (a, b, and c are respectively 1.2R, 1.2R, and 5R). Table 1 gives a comparative analysis of the magnetic induction distribution performance between rectangle and saddle excitation coils inside the measuring tube where z = 0.

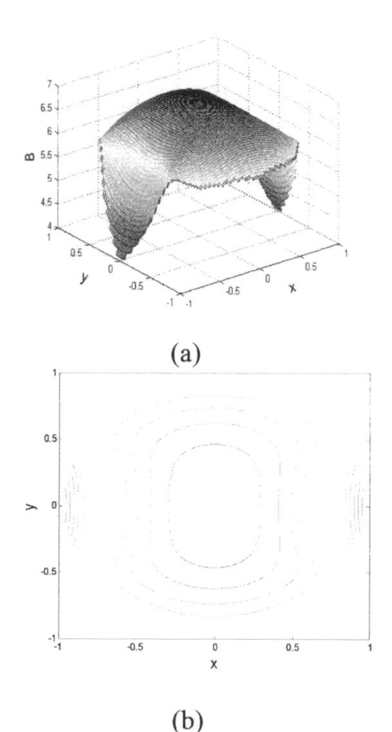

Figure 3. The distribution diagram of By (Y-direction of magnetic induction) with the saddle-shaped excitation coil in the measuring tube, where z = 0.

Table 1. The magnetic induction distribution performance comparative analysis.

Performance coil	Mean value	Standard deviation	Coefficient of variation	Uniformity of By
Rectangular	5.072	0.6345	0.1251	12.19%
Saddle-shaped	6.187	0.5888	0.0952	24.38%

The evaluation index given in the literature (WANG, 2011) is provided in Table 1. It is clearly seen that the saddle-shaped excitation coil has more advantages than the rectangular one under the same parameters by using the evaluation index from Table 1. In addition, the excitation coil of the saddle-shaped structure can save the sensor's installation space, which is suitable for installation in a relatively small position.

4 CONCLUSION

In this paper, the magnetic induction distribution in the space is analyzed by using the Biot–Savart law which produces straight line segments, oval arc and circular arc, and the magnetic induction distribution of saddle-shaped coil is investigated by calculation and simulation on this basis. Finally, the magnetic induction distribution performance is compared between rectangle and saddle-shape excitation coils inside the measuring tube in the same position.

In this paper, a study on the excitation coils of different shapes with different parameters is carried out, to provide a reference method of calculation of the magnetic field distribution inside the sensor, which provides technical support for the design of the sensor's excitation coil with different shapes, parameters, and positions. Meanwhile, through the analysis of the internal magnetic field of the sensor, the approach may help for the optimization in the design of coil parameters.

ACKNOWLEDGMENTS

This work is supported by the National Natural Science Foundation of China (61463042) and the Inner Mongolia Autonomous Region Natural Science Foundation (2016MS0611).

REFERENCES

Chen Tingxiang, Tai yachuan, Huang Baosen. Journal of Shanghai Jiaotong **01**: 83–94 (1982).

Jin Ningde, Zong Yanbo, Zheng Guibo. Acta Petrolei Sinica, **02:** 308–311 (2009).

Qiao Xutong, Xu Lijun, Dong Feng. Chinese Journal of Scientific Instrument, **23**: 867–869 (2002).

Wang Yue-ming, Kong Ling-fu. Journal of Inner Mongolia University (Natural Science Edition). **01**: 77–80 (2012).

Wangyueming, Konglingfu, Liyingwei. Journal of Computational Information Systems **8**: 1573–1580 (2012).

Yueming Wang, Lingfu Kong. Journal of Computational Information Systems, **8(7)**: 2779–2786 (2011).

Zhao Chen, Li Bin, Chen Wen-jian. Journal of Shanghai University (Natural Science Edition) **14**: 31–3 (2008).

Civil, Architecture and Environmental Engineering – Kao & Sung (Eds)
© 2017 Taylor & Francis Group, ISBN 978-1-138-02985-9

A study on a localizable CMOS temperature sensor based on wireless transmission

Liying Chen, Jiaqi Li & Yaoxian Li
School of Electronic and Information Engineering, Tianjin Polytechnic University, China

ABSTRACT: A CMOS temperature sensor is proposed, which is used in the wireless sensor and transmission network based on RFID technology. The current target's information can be tested timely and efficiently by using the intelligent electronic equipment in the Wireless Sensor Network. Meanwhile, the waste of human resources will be reduced. In this paper, the principle and hardware structure of the system are described. The results show that the CMOS temperature sensor's performance is good by being used in the RFID wireless sensor network. The accuracy of temperature is under 0.015°C in the range between –40 and 125°C. The error of the location system is within 0.4 meter in 5 × 5 meters range. Real-time transmission and data display can be achieved by using this sensor.

1 INTRODUCTION

With the rapid development of the Internet of Things (IOT), and with temperature sensors being widely used in many fields, such as the biological medical industry, vehicle-mounted system, the temperature measurement of industry, agriculture and fire-alarm systems, it becomes very increasingly important to trace the temperature of the target in real time. However, the current temperature tracking system lacks the characteristic of quick rhythm, which is convenience and veracity.

Radio Frequency Identification (RFID) technology has been considered as a way to increase the efficiency of the location in the temperature sensor field. There are many benefits in the RFID system, such as using small tools, real-time data transmission etc., which can allow data transmission for a long time (Amador, 2009). There are many location methods of the RFID, in which the most classic one is the LANDMARC algorithm. The LANDMARC algorithm is a very simple one with high accuracy of location and low complexity of hardware. And so, the LANDMARC algorithm is the foundation in most current location algorithms; in other words, the location algorithm originated from LANDMARC (Kung, 2015).

Two kinds of sensors are widely used in the area of temperature sensors. One is liquid-filled thermometer, the fluid material of which is hydrargyrum or kerosene. The liquid-filled thermometer with big volume is mainly used in hospitals and homes, which is only fit to measure a target's temperature in a specific time; sometimes it may have a large error. The other is the electronic thermometer which receives the temperature signals through a thermistor. This kind of thermometer has the same disadvantages as that with liquid-filled thermometers. What is more, both thermometers cannot be integrated into other electronic equipment. Comparatively speaking, the CMOS temperature sensor has so many advantages, such as small volume, low cost, and low-power dissipation. The most important is that it can integrate into other electronic equipment. Because of that, CMOS temperature sensors are used worldwide and will gradually replace other thermometers.

In this paper, research on combining temperature sensors with RFID orientation technology has been carried out for many years worldwide. In 2009, Cecilia Amador, Jean-Pierre Emond, and Maria Cecilia do Nascimento Nunes published papers about RFID technology being used in receiving the data of Pineapple temperature change from temperature sensors. In 2016, Andrea Luvisia, Alessandra Panattonib, and Alberto Materazzib discovered the relationship between the temperature change and the change of microorganisms in soil by RFID technology (Luvisi, 2016; Feng, 2014).

A design is proposed in this paper to improve the insufficiency mentioned above. The design is a system which combines the RFID location technique and CMOS temperature sensor, and it realizes both location and temperature data received in real time. In addition, it also displays the information on the screen. The design could be used in many occasions, such as hospitals, markets, factories, etc. To solve problems, first, the design can be used in fire prediction in factories and markers. Second, the design can track the temperature of every patient anywhere at hospital to determine its position timely, and then lead to finding and curing. Third, wireless sensing of biomedical equipment can become true by substituting the temperature sensor with any other physiology signs testing equipment.

2 SYSTEM STRUCTURE

The temperature sensor system includes mainly two blocks, which are the temperature sensing block and wireless communication system (Yue, 2012). The temperature sensing block includes reset circuit, CLK, PTAT voltage source, and digital transmission port. The wireless communication structure includes active frequency identification label, RF receiving block, location block, and PC data processing. The diagram of the whole system structure is shown in Figure 1.

The temperature sensing block collects the analog temperature data of the current environment; meanwhile, the digital transmission port can turn analog signals into digital signals. It sends the digital data to the active frequency identification label. And then, the label sends the digital data to the RF receiving block through the antenna of the label. Finally, the data will be processed and shown in the PC screen.

Simultaneously, the readers of the location block receive the RSSI of the active frequency identification label and then send the RSSI to the PC terminal. The RSSI data will be processed by using a specific location algorithm. In the end, the location information is received and also shown on the screen.

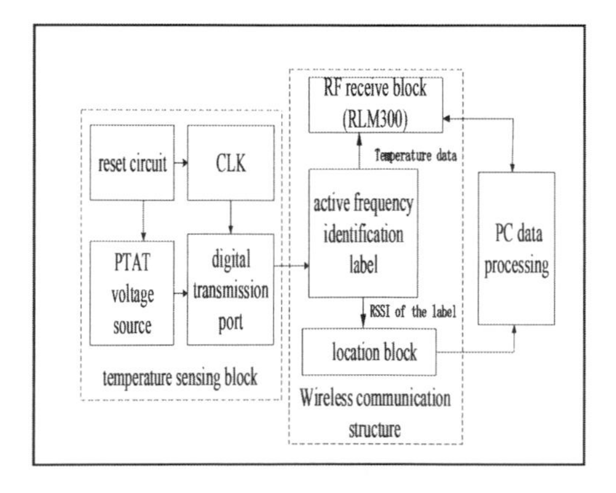

Figure 1. Diagram of the system structure.

3 TEMPERATURE SENSOR BLOCK

The temperature sensor is designed with 0.18 μm UMC CMOS technology. The circuit structure includes reset circuit, CLK, PTAT voltage source, and digital transmission port (Ohzone, 2007; Kim, 2013). After the circuit began to work, first, the circuit is reset. Second, the PTAT voltage source produces an analog voltage signal which is proportional to absolute temperature. Third, the digital transmission port outputs a series of parallel digital signals, when it turns the analog signal into digital one. Finally the digital signals will be sent to the RF label. In addition, the function of CLK is to provide a clock signal for the digital transmission port and RF label.

If that I1, I2, and I3 are the electricity of pnp1, pnp2, and R2 (in the figure of this article, the letter P represents PMOS transistors and the letter N represents NMOS transistors). The differential operational Amplifier (AMP) works on deep negative feedback mode, which compels to its positive and negative ports' voltage equally. When the circuit is stable, the relationship of each parameter is shown as follows:

$$I_2R_2 + V_{EB2} = V_{EB1} \tag{1}$$

$$V_{EB1} = V_T \ln(\mathrm{I}_1/\mathrm{I}_{S1}) \tag{2}$$

$$V_{EB2} = V_T \ln(\mathrm{I}_2/\mathrm{I}_{S2}) \tag{3}$$

$$V_T = \mathrm{k}T/\mathrm{q} \tag{4}$$

where IS1 and IS2 are the reverse saturation currents of pnp1 and pnp2. k is the Boltzmann's constant, T is the absolute temperature, and q = 1.60 × 10^{-19} Coulomb. When the voltages of P1 and P2's grids are equal, the expression for the output voltage is given as follows:

$$V_{out} = I_3R_2 = \left(\frac{\beta_3}{\beta_2}\right)I_2R_2 = \left(\frac{\beta_3}{\beta_2}\right)\left(\frac{R_2}{R_1}\right)\left(\frac{\mathrm{k}T}{\mathrm{q}}\right)\ln\left[\left(\frac{\beta_1}{\beta_2}\right)\left(\frac{A_2}{A_1}\right)\right] \tag{5}$$

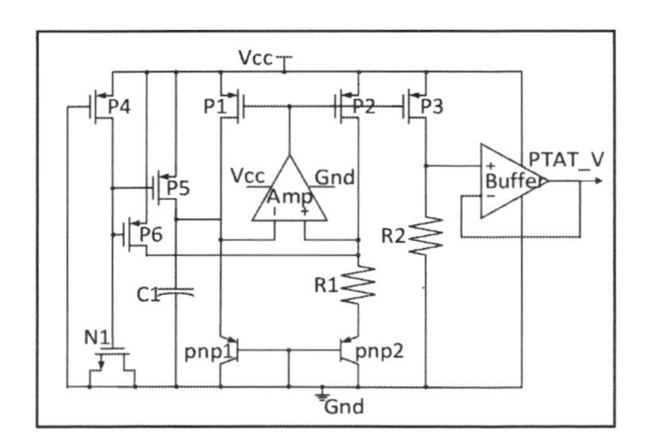

Figure 2. Circuit diagram of the PTAT voltage source.

When the voltages of P1 and P2's grids are equal, the expression for the output voltage is shown as follows: $\beta_3/\beta_2 = I_3/I_2$, $\beta_1/\beta_2 = I_1/I_2$, where A1 and A2 are the emitter areas of the pnp transistor. Other parameters are all constant except T. And so, Vout is a voltage which is proportional to the parameter T.

The starting circuit is constituted by using P4, P5, P6, N1, and C1. During the time that VCC changes from 0V to supply voltage, N1 amounts to a certain capacitance. When the power starts to work, the grid of N1 is under low voltage conditions. Thus, P4, P5, and P6 start to work, and the electric current will help to enable P1 and P2. Finally, the circuit starts to work. However, with the grid of N1 being charged, the gird will remain under high voltage conditions. Meanwhile, P5 and P6 will be cut-off, and the starting circuit will end the job (Crepaldi, 2010).

The output voltage and error of CMOS temperature sensor are simulated by using the Spectre of Cadence software. The simulation curve of the error is shown in Figure 3. From Figure 3, 1650 points are tested. The error of output is within ±0.015°C when the change of temperature is between –40 and 125°C. The simulation curve of the output voltage is shown in Figure 4. From Figure 4, there are also 1650 points are tested. The relationship between temperature and the voltage shows that when the change of temperature is between –40 and 125°C, the output is between 350 mV and 570 mV. The voltage range of output can be processed easily. In addition, when the supply power is 1.8V, the power consumption is under 1.5 uW. Therefore, this design can be used to satisfy the demand of application. Finally, the layout is designed and displayed, as shown in Figure 5.

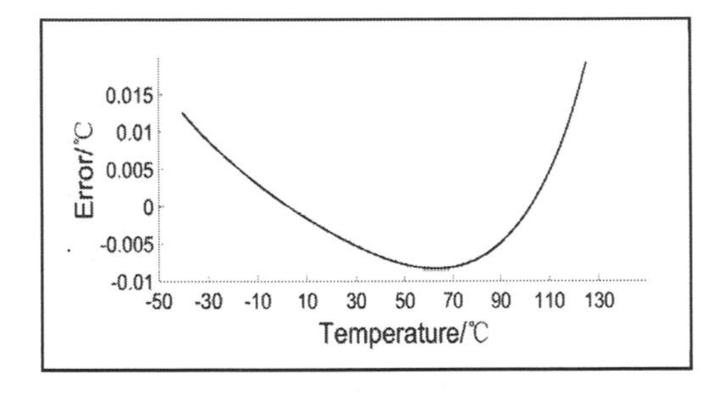

Figure 3. Simulation curve of errors.

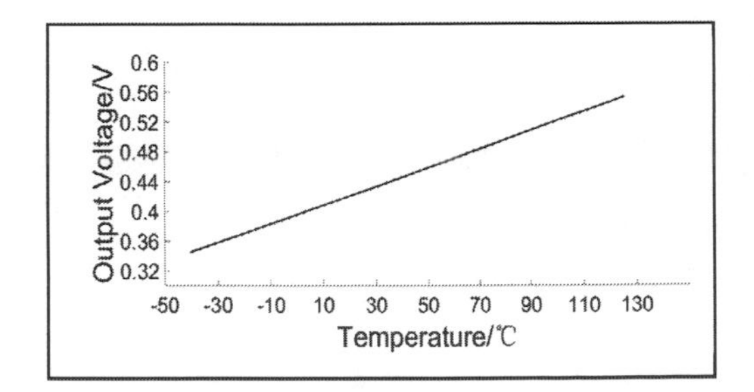

Figure 4. Simulation curve of output voltages.

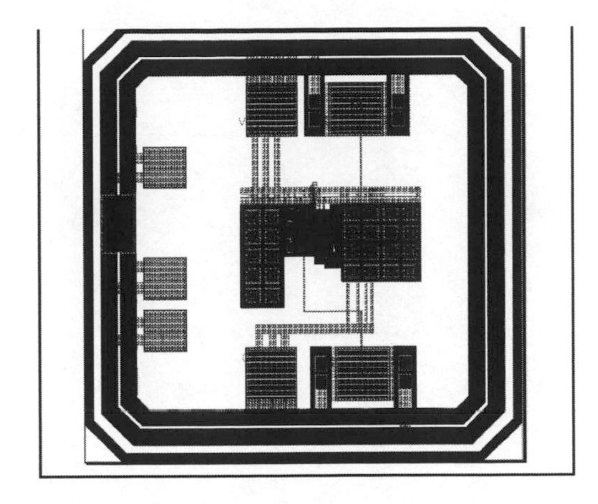

Figure 5. Layout of the CMOS temperature sensors.

4 WIRELESS COMMUNICATION STRUCTURES

The function of the wireless communication system is to connect the user terminal and the sensor for increasing the transmission distance. In this paper, a wireless communication system is proposed which is set up with advanced equipment based on RFID. This system includes RF receiving block and location block. And an active tag in the system can be used to simplify the equipment.

4.1 *RF receiving block*

The RLM300 UHF RFID reader–writer is used in the RF receiving block, which has the advanced RFID analog circuits. It has the power processing chip with DSP technique. The protocol standard of this block is ISO 18000–6C and the working frequency is between 840 MHz and 960 MHz. RLM300 provides the API of Windows OS, the function of which is to receive the digital signals sent by using an active frequency identification label in this block and then the RLM300 sends it to the PC directly to process the data. From Figure 6, which shows the structure of RLM300, the data are the data-flow between PC and RLM300. And the antenna receives the temperature data sent by using the active frequency identification label (Guédon, 2015).

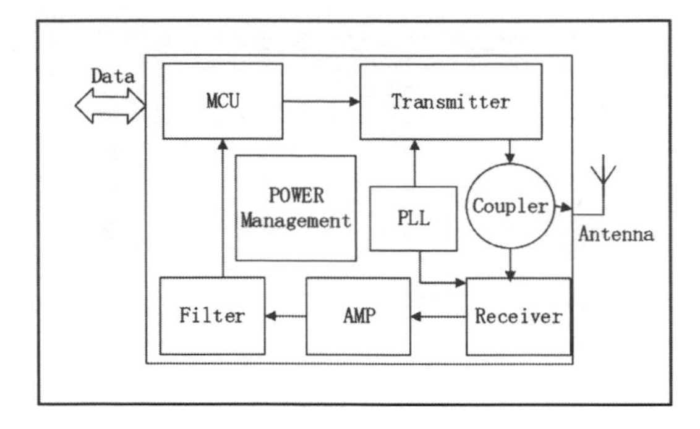

Figure 6. Schematic of the structure of RLM300.

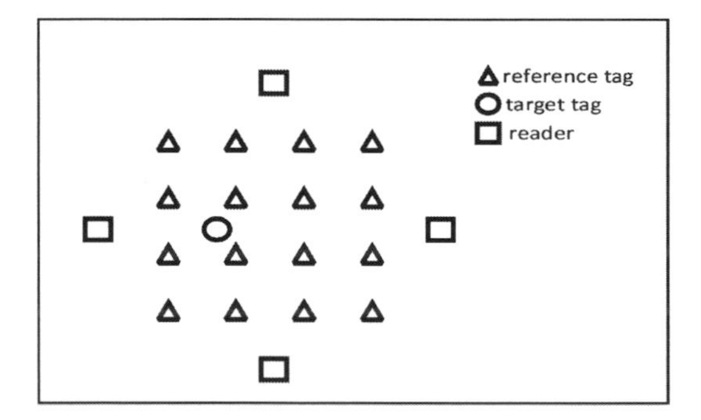

Figure 7. Schematic of the block of the location block.

4.2 *Location block*

The basis of the location block in this article is RFID technology. The method is that the readers measure the RSSI of THE target tag, and locate the position of the tag by using the LANDMARC algorithm. The model founding includes 16 reference tags, a target tag, and four readers, which is shown in Figure 7. The data from readers will be transmitted to the processor and then sent to the PC. The reader equipment of this system is Intel R1000, which is produced by Intel. The RSSIs of 16 reference tags and a target tag are read with the LANDMARC location principle. And then, the RSSIs are sent to the PC. Finally, the localization is realized.

The LANDMARC algorithm is given as follows:

Assume that there are n readers, m reference tags, and one target tag. And the coordinates of target tag are (x, y). We define the RSSI matrix of target tag as $S=[S_j]$ $(j=1\cdots n)$ where S_j is the RSSI of the jth reader. The matrix $Q=[Q_{ij}]$ $(i=1\cdots n)$ $(j=1\cdots n$ is the RSSI of the ith reference tag read by the jth reader. The matrix $E=E_i$ is the Euclidean distance between the target tag and the ith reference tag. Gett the minimum of $E_i(i=1\cdots k)$, where k is the number of nearest tags from the reference tag to the target tag. The coordinates of the reference tags are $(x_i, y_i)(i=1\cdots k)$. The coordinates of the target tag are defined as $(x,y)=\sum_{r=1}^{k} w_r \times (x_r', y_r')$, where w_r is the weight of the kth reference tag. The expression of w_r is $w_r = 1/E_r^2 \Big/ \sum_{i=1}^{k} 1/E_{ir}^2$. In addition, the systematic error estimation is given by the following relation: $e_i = \sqrt{(x_i - x_0)^2 + (y_i - y_0)^2}$.

5 RESULTS AND ANALYSIS

Based on the principle mentioned above, we tested the performance of the system in the temperature sensor block. When comparing the CMOS temperature sensor and other electronic thermometers, we tested 50 groups of different temperature values. The consequence is that the maximum error between them is 0.25°C. And the average one is about 0.1°C. Figure 8 shows the chart of the errors.

In the side of the location, the 50 groups' data are tested in a 5 m*5 m indoor area. The environment of the area is complicated. The consequence is that the maximum error is 0.6 m and the errors of the regional center are about 0.3 m. The details are shown in Figure 9.

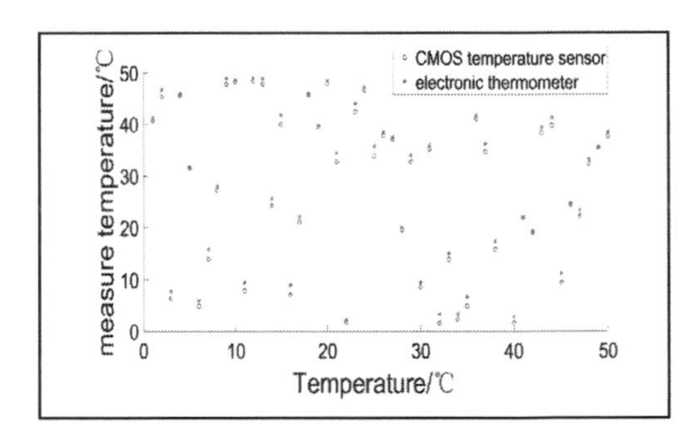

Figure 8. Chart of the errors.

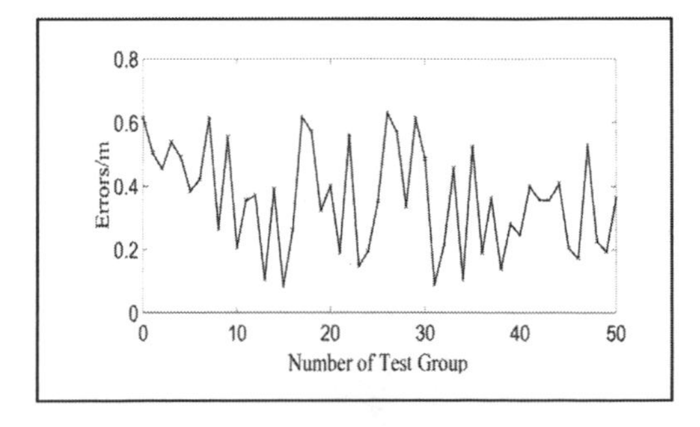

Figure 9. Graph showing the errors of location.

6 CONCLUSION

The CMOS temperature sensor represents easy-to-use and cheap tools, which can solve the problems of current thermometers such as large volume, high loss of power, and the problem about the temperature data cannot be transmitted on time. A CMOS temperature sensor system is proposed, which is used in wireless sensor networks. The system can receive and transmit data immediately and timely. And it also can locate the temperature sensor. The consequence of the test is that the error of temperature measurement under 0.015°C and location is under 0.6 m. These can be shown on screen simultaneously. This successful design can be a basis of researching other human signal testing equipment.

ACKNOWLEDGMENTS

This research work is funded by the Tianjin Research Program of Application Foundation and Advanced Technology (No. 15 JCYBJC16300) and The Science and Technology Plans of Tianjin (16 JCTPJC45500).

REFERENCES

Amador, C., J.P. Emond, M. Cecilia do N. Nunes. Application of RFID technologies in the temperature mapping of the pineapple supply chain [J]. Sensing and Instrumentation for Food Quality and Safety, 2009, 31.

Crepaldi, P.C., T.C. Pimenta, R.L. Moreno. A CMOS low-voltage low-power temperature sensor. [J]. Microelectronics Journal, 2010, 41.

Feng, J., M. Onabajo. Wide Dynamic Range CMOS Amplifier Design for RF Signal Power Detection via Electro-Thermal Coupling [J]. Journal of Electronic Testing, 2014, 301.

Guédon, A.C.P., L.S.G.L. Wauben, D.F. Korne, M. Overvelde, J. Dankelman, J.J. Dobbelsteen. A RFID Specific Participatory Design Approach to Support Design and Implementation of Real-Time Location Systems in the Operating Room [J]. Journal of Medical Systems, 2015, 391.

Kim, Y., P. Li. A 0.003-mm 2, 0.35-V, 82-pJ/conversion ultra-low power CMOS all digital temperature sensor for on-die thermal management [J]. Analog Integrated Circuits and Signal Processing, 2013, 751.

Kung, H.Y., S. Chaisit, N.T.M. Phuong. Optimization of an RFID location identification scheme based on the neural network [J]. Int. J. Commun. Syst., 2015, 284.

Lee, H., K. Kim, S. Jung, J. Song, J.K. Kim, C. Kim. A 0.0018 mm 2 frequency-to-digital-converter-based CMOS smart temperature sensor [J]. Analog Integrated Circuits and Signal Processing, 2010, 642.

Luvisi, A., A. Panattoni, A. Materazzi, RFID temperature sensors for monitoring soil solarization with biodegradable films, Computers and Electronics in Agriculture, Volume 123, April 2016, Pages 135–141, ISSN 0168–1699, http://dx.doi.org/10.1016/j.compag.2016.02.023.

Ohzone, T., T. Sadamoto, T. Morishita, K. Komoku, T. Matsuda, H. Iwata. A CMOS Temperature Sensor Circuit. [J]. IEICE Transactions, 2007, 90-C.

Yue, X., P. Hong-Bin, H. Shu-Zhuan, L. Li. A highly sensitive CMOS digital Hall sensor for low magnetic field applications. [J]. Sensors, 2012, 122.

Civil, Architecture and Environmental Engineering – Kao & Sung (Eds)
© 2017 Taylor & Francis Group, ISBN 978-1-138-02985-9

Finite element analysis of the roller coaster wheel bridge based on virtual prototype technology

Yongming Wang, Jun Ye, Xuetan Xu & Song Song
School of Mechanical Engineering, Anhui University of Technology, Maanshan, China

ABSTRACT: The wheel bridge of the roller coaster is a key component, which connects the wheel frame and vehicle frame, and so its security is related to the entire roller coaster running safety. When the roller coaster is running in high speed, the forces of the wheel bridge are changeable instantaneously. The traditional static analysis method cannot meet the demand of the security assessment. Therefore, based on virtual prototype technology, a dynamics analysis was carried out on the roller coaster in this paper, and then the force curves of the wheel bridge were obtained. The finite element analysis model of the wheel bridge was established in ANSYS Workbench, and then the forces were applied. Finally, the displacement diagram and stress nephogram of the wheel bridge were obtained. The results indicate that the strength safety factor of the wheel bridge meets the amusement facilities safety specification stated in national standard GB8408-2008.

1 INTRODUCTION

The roller coaster is deeply loved by young people for its high speed and strong irritation. The roller coaster belongs to high-speed running equipment, and the wheel bridge is its key component. The wheel bridge connects the wheel frame and vehicle frame, and so its security is very important. However, the forces of the wheel bridge are very complicated, which are influenced by the orbital shape, the running condition of the carrying car, the weight of the carrying car, the gravity of the tourists, the running speed, and so on. Because the stresses of the wheel bridge are changeable instantaneously, an ideal analysis result cannot be obtained by using the traditional static method (Wang, 2010). The virtual prototype technology provides an effective way to solve this problem (Liang, 2008; Shen, 2007).

2 KINEMATICS AND DYNAMICS SIMULATION OF THE WHEEL BRIDGE

The structure of the roller coaster wheel bridge is shown in Figure 1.

According to design requirements, the total length of the roller coaster orbit is 567 m, the maximum height of the roller coaster orbit is 30 m, the distance of two orbits is 960 mm, and the maximum running speed is 80 km/h. As shown in Figure 2, two space curves were used to simulate the roller coaster orbit in the global coordinate system of ADAMS. The orbit is mainly composed of the platform, the lifting section, the transition section, the floating ring, the vertical ring, and the spiral ring. The corresponding constraints of the components are listed in Table 1. The corresponding loads were applied on the components, such as the gravity, the friction between the wheel and orbit,

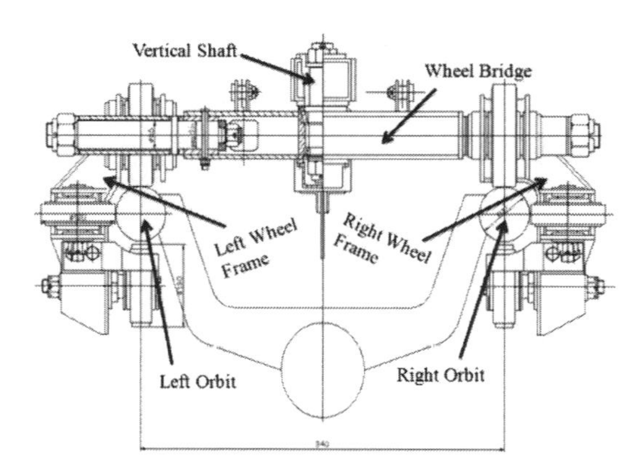

Figure 1. Schematic of the structure of the roller coaster wheel bridge.

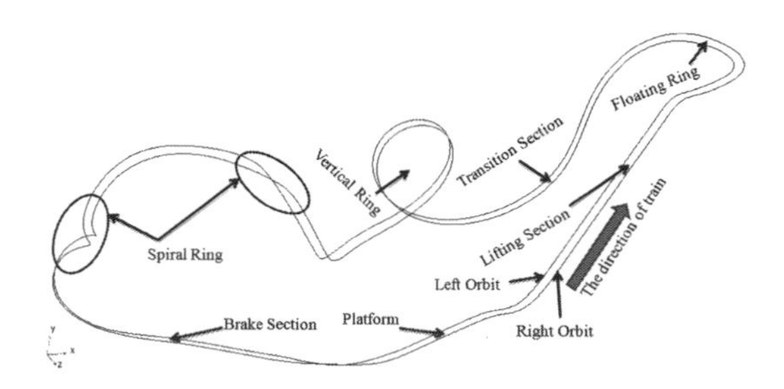

Figure 2. The orbital space curve of the roller coaster.

Table 1. List of the corresponding constraints of the components.

No.	Component	Constraints
1	Orbit and ground	Fixed
2	Left wheel and orbit	PTCV
3	Right wheel and orbit	PTCV
4	Left wheel frame and wheel bridge	Fixed
5	Right wheel frame and wheel bridge	Fixed
6	Wheel bridge and frame	Rotation
7	Wheel bridge and tail connector	Rotation
8	Connecting rod and clevis	Rotation

the traction of the carrying car, and the wind loads (Li, 2009; Tian, 2006; He, 2006). In order to ensure the accuracy of the simulation data collection, the sampling frequency must be more than 20 Hz according to the sampling theory, i.e. the simulation step's length must be less than 0.05. Considering the computer performance and the simulation time, the simulation step length was set to 0.01.

Based on the ADAMS kinematics and dynamics simulation, the reference system of the former wheel bridge was selected as the measuring marker, and then the forces of the wheel bridge imposed by the wheel frame and frame body were measured. According to the simulation results, the force of the wheel bridge imposed by the right wheel frame in the Z-direction was almost zero, and so it was ignored. The force curves of the wheel bridge imposed by the left wheel frame in the X, Y, and Z directions are shown in Figure 3.

As shown in Figure 3, the force curves of the wheel bridge in X, Y, and Z directions are almost constant during the time of 0 s–26.33 s. It is because the roller coaster is running in the chain-lifted segment, and the carrying car is rising at a constant speed by the traction. When it reaches the highest point, the traction hook unhooks the carrying car, and the carrying car is going to run under the influence of gravitational potential energy. With the change of the radius, height, and the lateral angle of the orbit, the wheel bridge begins to bear the complex stress. During the time of 26.33 s–44.95 s, the changes of the radius, height, and the lateral angle of the orbit are very little, and so the forces of the wheel bridge change with a small fluctuation. As the carrying car goes ahead, the orbit shape is complex and changeable, and so the forces of the wheel bridge fluctuate greatly.

Figure 3(a) shows that the roller coaster is beginning to pass through the second spiral ring segment at the time of 55.93 s. Except for some individual cusps, the maximum force of the wheel bridge in the X direction is 18114.25 N at that time.

Figure 3(b) shows that the roller coaster is beginning to pass through the second spiral ring

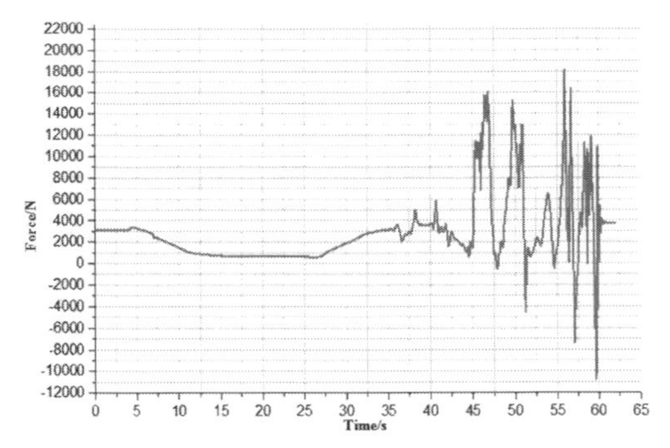

(a) The force curve of the wheel bridge in the X direction

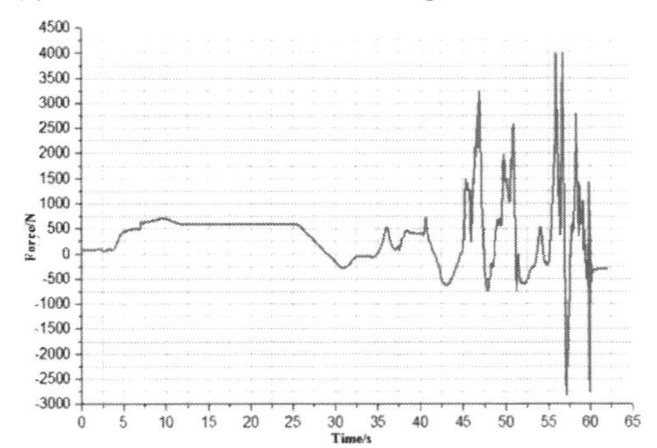

(b) The force curve of the wheel bridge in the Y direction

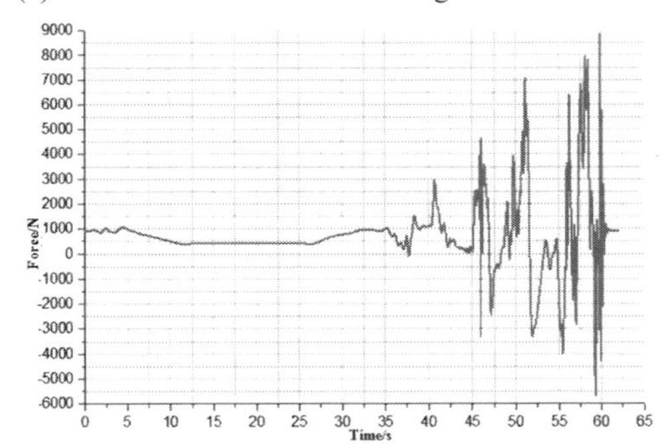

(c) The force curve of the wheel bridge in the Z direction

Figure 3. The force curves of the wheel bridge imposed by the left wheel frame.

segment at the time of 56.74 s. Except for some individual cusps, the maximum force of the wheel bridge in the Y direction is 3983.52 N at that time.

Figure 3(c) shows that the roller coaster is also beginning to pass through the second spiral ring segment at the time of 58.14 s. Except for some individual cusps, the maximum force of the wheel bridge in the Z direction is 7936.31 N at that time.

The force curves of the wheel bridge imposed by the right wheel frame in X and Y directions are shown in Figure 4. Figure 4 (a) shows that the roller coaster is beginning to pass through the second

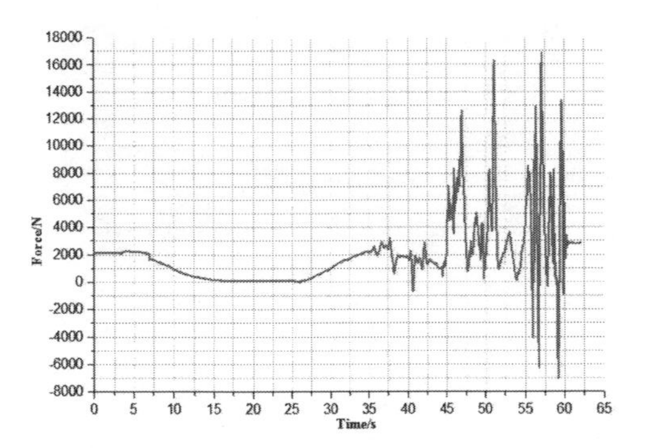

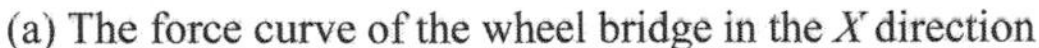
(a) The force curve of the wheel bridge in the *X* direction

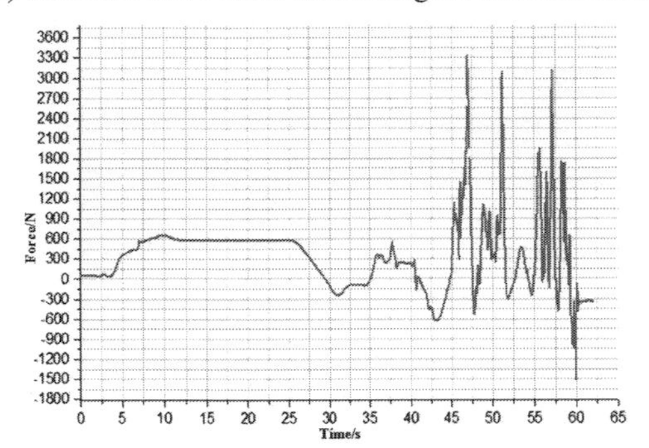

(b) The force curve of the wheel bridge in the *Y* direction

Figure 4. The force curves of the wheel bridge imposed by the right wheel frame.

spiral ring segment at the time of 57.13 s. Except for some individual cusps, the maximum force of the wheel bridge in the *X* direction is 16873.95 N at that time. Figure 4(b) shows that the roller coaster is also beginning to pass through the second spiral ring segment at the time of 57.13 s. Except for some individual cusps, the maximum force of the wheel bridge in the *Y* direction is 3109.28 N at that time.

The force curves of the wheel bridge imposed by the vehicle frame in *X*, *Y*, and *Z* directions are shown in Figure 4.

Figure 5(a) shows that the roller coaster is beginning to pass through the second spiral ring segment at the time of 59.26 s. Except for some individual cusps, the maximum force of the wheel bridge in the *X* direction is 5197.32 N at that time. Figure 5(b) shows that the roller coaster is beginning to pass through the vertical ring segment at the time of 46.98 s. Except for some individual cusps, the maximum force of the wheel bridge in the *Y* direction is 5866.7 N at that time. Figure 5(c) shows that the roller coaster is also beginning to pass through the vertical ring segment at the time of 46.98 s. Except for some individual cusps, the

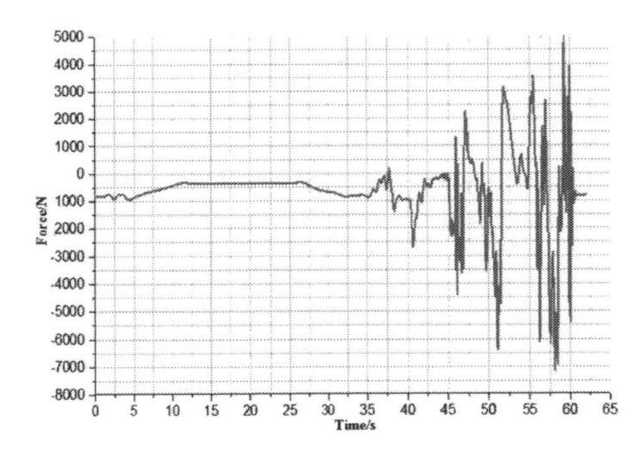

(a) The force curve of the wheel bridge in the *X* direction

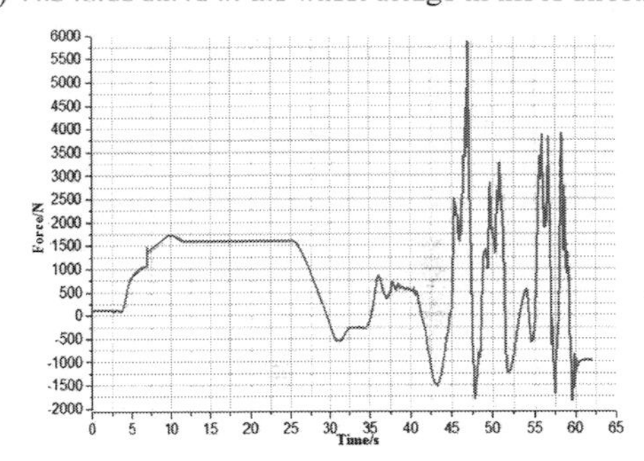

(b) The force curve of the wheel bridge in the *Y* direction

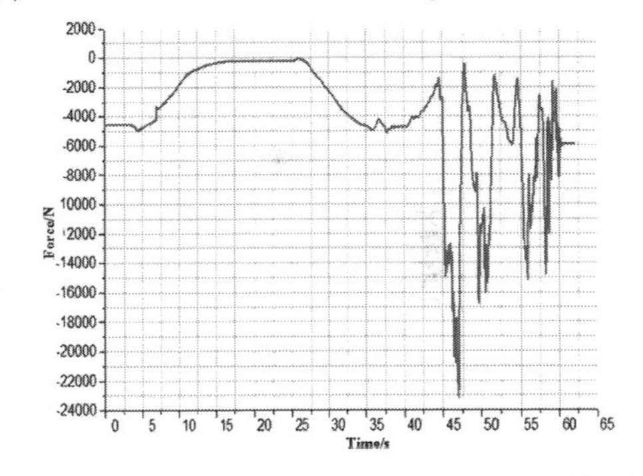

(c) The force curve of the wheel bridge in the *Z* direction

Figure 5. The force curves of the wheel bridge imposed by the vehicle frame.

maximum force of the wheel bridge in the *Z* direction is –23151.53 N at that time. Based on the data obtained from the above-mentioned simulation, the strength analysis of the wheel bridge was carried out in the following.

3 STATIC ANALYSIS OF THE WHEEL BRIDGE

According to the structure of the wheel bridge shown in Figure 1, its simplified 3D model was established and imported into ANSYS

Workbench. The first step was to divide the mesh, and then exert the constraints and loads. The finite element mesh model of the wheel bridge is shown in Figure 6.

According to its location on the roller coaster, the wheel bridge mainly bears the loads exerted by the wheel frame and the vehicle frame. According to ADAMS simulation, the loads are given in Table 2. The constraints and loading positions of the wheel bridge are shown in Figure 7.

FEM results show that the maximum deformation of the wheel bridge is 0.0056–0.0063 mm, which mainly occurred near the contact area between the wheel shaft and wheel bridge. The wheel bridge mainly bears bending deformation, and its deformation nephogram is shown in Figure 8(a). The stress nephogram of the wheel bridge is shown in Figure 8(b). Figure 8 shows that there is stress concentration at the connection part of the wheel bridge. According to the Von Mises equivalent stress analysis, the maximum stress value is 24.31 MPa.

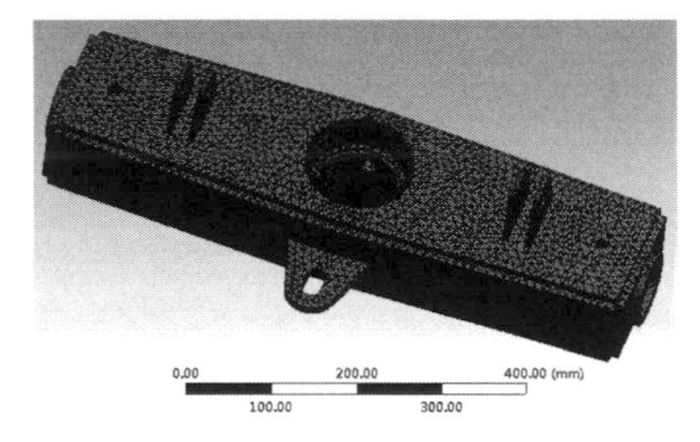

Figure 6. The finite element mesh model of the wheel bridge.

Table 2. List of force parameters of the wheel bridge.

Direction	Time/s	Force/N
The left wheel frame in the X direction	55.93	18114.25
The left wheel frame in the Y direction	56.74	3983.52
The left wheel frame in the Z direction	58.14	7936.31
The right wheel frame in the X direction	57.13	16873.95
The right wheel frame in the Y direction	57.13	3109.28
The wheel frame in the X direction	59.26	5197.32
The wheel frame in the Y direction	46.98	5866.71
The wheel frame in the Z direction	46.98	−23151.53

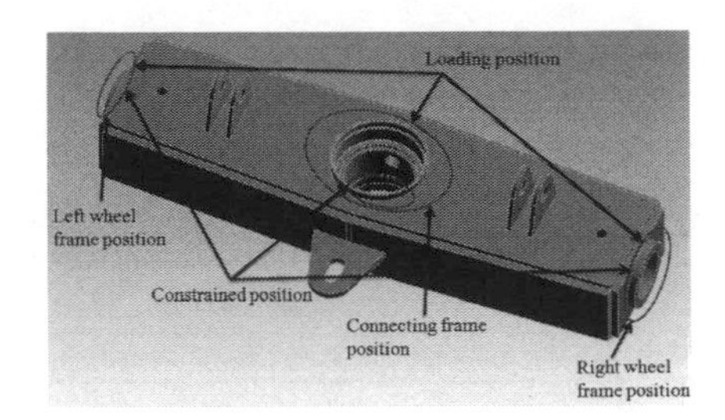

Figure 7. Constraints and loading position of the wheel bridge.

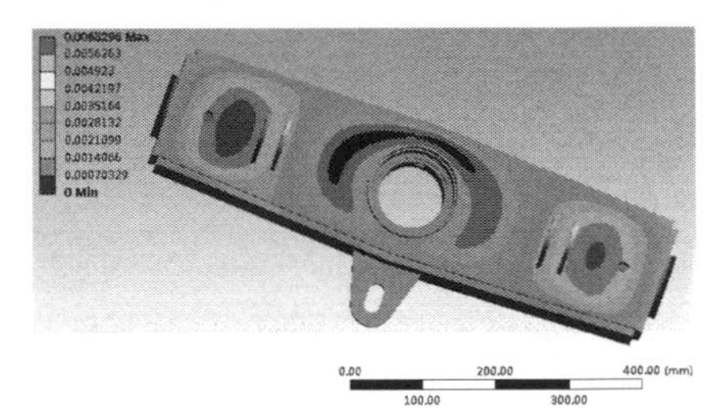

(a) Deformation nephogram of the wheel bridge

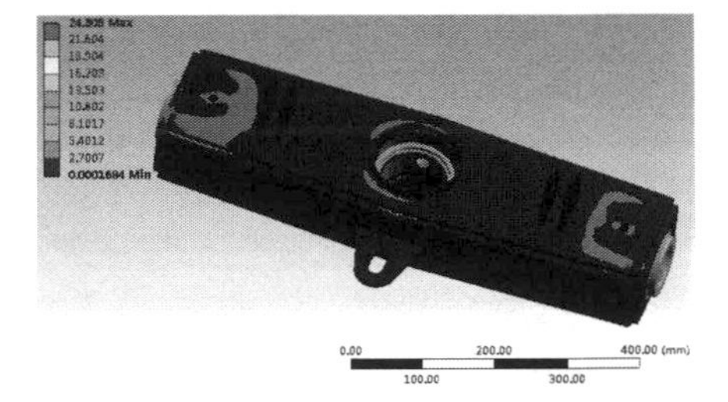

(b) Stress nephogram of the wheel bridge

Figure 8. Finite element analysis results of the wheel bridge.

The material of the wheel bridge is Q235-B, and its tensile strength $\sigma_b = 420\ MPa$ and yield strength is $\sigma_s = 235\ MPa$. According to GB8408-2008, the safety factor of the wheel bridge is selected as [n] = 5, and the impact coefficient $K = 2$. According to the design requirements, the safety factor of the wheel bridge n should be greater than the allowable safety factor [n].

$$n = \frac{\sigma_b}{K\sigma_{\max}} = \frac{420}{2 \times 24.31} = 8.64 > [n] \tag{1}$$

According to equation (1) calculation results, the safety factor of the wheel bridge meets the requirements of the strength design.

4 CONCLUSIONS

Based on virtual prototype technology, kinematics and dynamic simulation analysis of the wheel bridge was carried out. On this basis, the static analysis of the wheel bridge was completed. According to the simulation results, the maximum stress of the wheel bridge is 24.31 MPa and the strength safety factor of the wheel bridge is 8.64. The strength safety factor of the wheel bridge meets the amusement facilities' safety specification given in national standard GB8408-2008. And so, the structure of the wheel bridge meets the requirements of safety evaluation.

ACKNOWLEDGMENTS

The authors acknowledge the Anhui Province Education Department of China for providing financial support through its major project (No. JK2015ZD11).

REFERENCES

GB8408-2008, Amusement facilities safety specification standards, Beijing: Standardization Administration of China, 2008.

He, D.X. *Wind engineering and industrial aerodynamics*. National Defence Industry Press, Beijing, China (2006).

Li, Z.G. *ADAMS detailed introduction and examples*. National Defence Industry Press, Beijing, China (2009).

Liang, Z.H., Y. Shen, L.J.E. Analysis and criterion on the g-acceleration of amusement device. China Safety Science Journal, **18** (2008) 31–35.

Liang, Z.H., Y. Shen, P.Y. Qin. Dynamic modeling and simulation for amusement device as roller coasters. China Safety Science Journal, **17** (2007) 14–20.

Tian, H.Q., D. Zhou, P. Xu. Aerodynamic performance and streamlined head shape of train. *China Railway Science*, **27** (2006) 47–55.

Wang, H.Q. & J.R. Zheng. Analysis on the transient stress of the joints of roller coaster based on virtual prototype simulation Journal of Machine Design, **27** (2010) 25–28.

Civil, Architecture and Environmental Engineering – Kao & Sung (Eds)
© 2017 Taylor & Francis Group, ISBN 978-1-138-02985-9

Multi-axis coordination control based on siemens S7-CPU315T

Cunhai Pan, Yanlong Li & Mingwei Qin
College of Mechanical Engineering, Tianjin University of Science and Technology, Tianjin, China

ABSTRACT: Hot billets automatic marking equipment is an important part of the casting production lines, which is used to mark a group of particular characters on the billet side or its end surface according to the manufacturer, steel materials, production date and other information, and when quality problems arise, it can promptly trace back to the billet production information, in order to solve quality problems in the enterprise quickly and efficiently. In this paper, a multi-axis coordinated motion control system has been established by using Siemens CPU315T-2DP motion controller. The virtual cam technology of the Technology CPU is used to control two shaft (or multi-axis) coordinated motion, which can guarantee the trajectory curve coordinate movement between the two real axes. Experiments show that Siemens CPU315T-2DP with motion control functions can be designed a multi-axis coordinated control system flexibly, by means of poly nomial fitting can easily design the trajectory curve or CAM. The Siemens PLC control system itself has the wealth of motion control information, such as the state of motion of each axis, the axis configuration parameters, limits setting information. etc., and the real axis motion control accuracy can be up to 1 μm.

1 INTRODUCTION

Any iron and steel enterprises all need to mark a group of particular characters on the billet side or its end surface according to the manufacturer, steel materials, production date and other information, and when quality problems arise, it can trace back to the billet production information timely, in order to solve quality problems in the enterprise quickly and efficiently (Pan, 2010; Chen, 2011). Billet production line is in a high temperature, large dust and moisture environment, using the automatic identification device instead of manual operation, the marking information can be more specific and clear, it is not only to improve the efficiency, but also to lighten the labor intensity of workers, which is an inevitable trend of development.

The Technology CPU integrates Motion Control functions in a SIMATIC CPU in order to combine the function of a SIMATIC CPU S7-300 with PLC open-compliant Motion Control functions. The control unit of the Technology CPU performs the tasks known from a standard CPU of the S7-300 family, and S7-Technology is an options package used to configure Motion Control functions. The Technology CPU operates the DP (DRIVE) PROFIBUS interface in synchronous mode. The integrated technology controls, evaluates and monitors all hardware components at DP (DRIVE) and the cam track is configured and activated by calling the technology function blocks. The switching state of the individual output cams can also be read and edited.

In this paper, using Siemens PLC as the core controller, it aims to research a set of billet automatic identification equipment, in order to control the virtual axis to lead two real axes synchronous forward according to pre-designed virtual cam.

2 SYNCHRONIZATION AXIS AND VIRTUAL CAM

A synchronization axis is a following axis which follows a leading axis. Synchronization axis technology object is to use the motion and position values of a leading axis as a master set-point. The synchronization axis contains all functions of the speed-controlled and positioning axes. The master set-points and slave values are coupled without physical conversion in the relevant programmed unit. For example, if the system operates with a linear leading axis (in mm units) and a slave rotary axis (in degree units), then one millimeter corresponds to one degree at a conversion ratio of 1:1.

A following axis can be interconnected with multiple master set-points using the synchronization object. However, only one master set-point can be evaluated actively at any time. The master set-point can be returned by positioning axes, or by a synchronization axes (real or virtual), or by external encoders.

Interpolation of a cam disk is a basic requirement for using it in synchronous operation. The continuity in the definition ranges of leading axis values and in the ranges of the following axis is

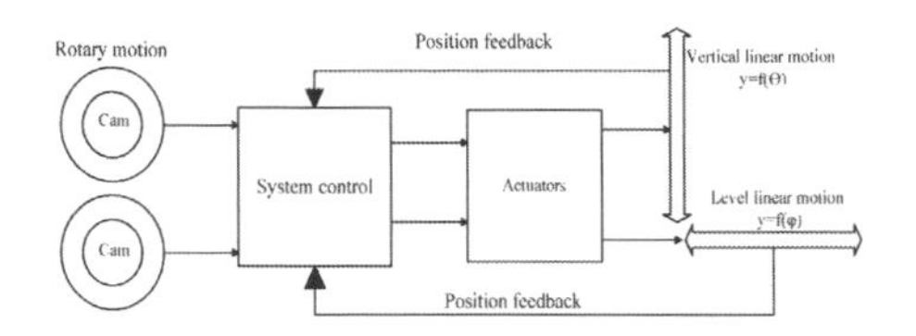

Figure 1. Cam control motion trajectory principle diagram.

checked. This check prevents redundant assignment of values to a definite value. These areas are padded according to the interpolation mode and cam disk type.

Virtual cam takes advantage of the flexible programming function of controller to modify the cam motion so as to adapt the needs of different occasions. This not only improves the adaptability of the organization, but also effectively avoids the problem in the process of cam manufacturing to prevent from movement distortion (Yu, 2000). As shown in Figure 1 is the principle diagram of using virtual cam to operate two axes to realize plane trajectory control. According to the rotation angle θ and φ of cam curves, which correspond to different plane positions, they can control indirectly motors to complete trajectory control.

In general, cam curve is often composed of straight line, sine curve and other lines. The function relation between the cam rotation angle and follower distance is:

$$y = F(\varphi) \tag{1}$$

The inverse function is:

$$\Phi = f(y) \tag{2}$$

In addition, using polynomial to express cam curve as follows:

$$y = C_0 + C_1\varphi + C_2\varphi^2 + C_3\varphi^3 + \dots + C_n\varphi^n \tag{3}$$

Therefore, according to equation 3, follower distance can be got by cam angle and reversed cam angle can also be obtained by the follower motion. The realization of virtual cam mainly includes three parts, which is to set the master and slave axis, the cam curve design and the cam path control implement.

3 BUILDING CONTROL SYSTEM

3.1 *CPU315T configuration and control system*

Using S7-CPU315T unique master and slave control mode, it needs to generate two real axes connected to S7-CPU315T by cam disk synchronization. Then, the virtual main axis drives motor shaft corresponding virtual cam disk, so that the axis can coordinate motion according to the preset trajectory. System control scheme is shown in Figure 2.

The motor real axis and virtual control axis are configured by S7-Technology software. For virtual axis, since it has no end actuators, and reference of other axis' position, so should be configured as the virtual main axis. The second is to create two linear real axes *X*, *Y* to drive servo motor. At the same time, according to the requirements of the cooperative move, the two axes configure for synchronous linear axis, then proper communicate parameter and type of encoder are configured; and the corresponding technical system data is also generated. The control relationship between axes is as shown in Figure 3. System network configuration is as shown in Figure 4. The control system is also designed with the WinCC operator interface HMI.

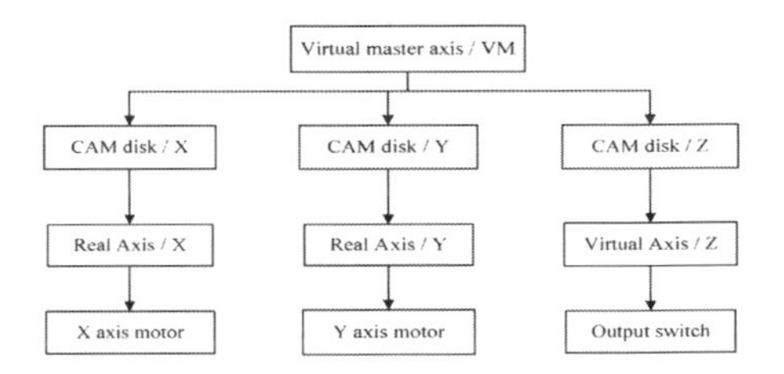

Figure 2. The control system scheme of S7-CPU315T.

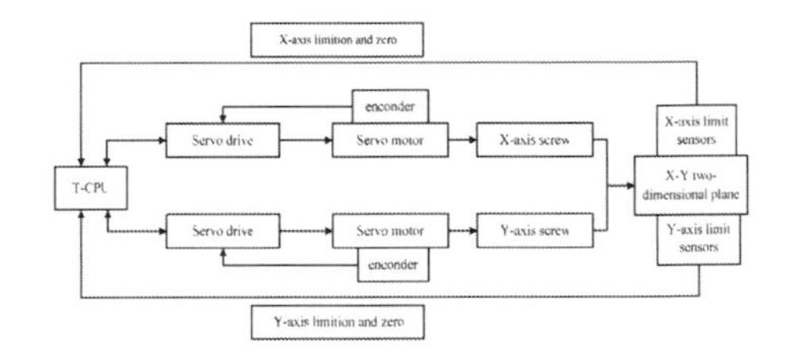

Figure 3. Relationship between servo controller, motor and S7-CPU315T control system.

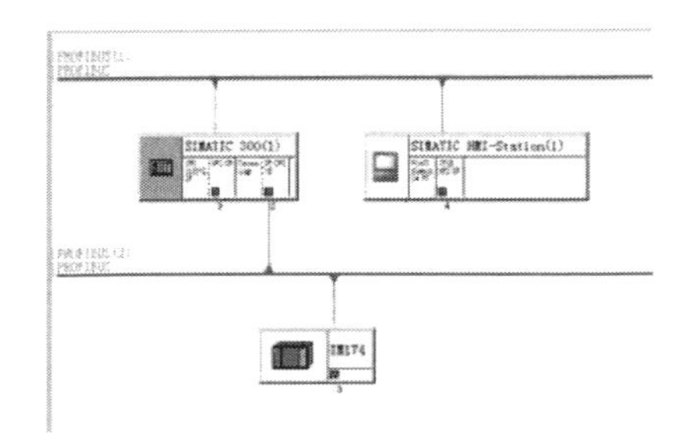

Figure 4. Network configuration system.

3.2 *Virtual cam based on polynomial fitting*

S7-Tech can be used to plan cam curve by interpolation points and polynomial function. According equation 3, this system mainly uses the method of polynomial to define cam curve. Equation 4 is polynomial cam curve formula based on Siemens S7-CPU315:

$$y = a_0 + a_1x + a_2x^2 + a_3x^3 + a_4x^4 + a_5x^5 + A\sin(\omega x + \varphi) \quad (4)$$

In order to avoid tedious coefficient calculation, cam curve can be planned for linear and circular arc form as far as possible. In this way, the equation 4 is simplified as:

$$y = a_0 + a_1x + A\sin(\omega x + \varphi) \quad (5)$$

After defining cam curve according to equation 5 in x,y cam disk, different position coordinates of X-Y plane can be got. Then, reversed rotation angle can be achieved by equation 3 to drive indirectly servo motor.

Cam configuration and parameter settings, coupling relation between master and slave axis through the cam plate are built by S7-Tech software and completed by configuration and axis synchronized window. Figure 5 defines cam plate polynomial curve of *X* axis.

S7-Tech provides the uniaxial, multiaxial and complex move control functions. For example, MC-CamIn function block can be called to control the virtual axis leading the real *X*-axis and Y-axis in accordance with the preset CAM curve synchronous movement. Furtherly it also need to set the master, slave and cam table of the modules according to DB block given by configuration. In addition, the parameters such as speed, acceleration, deceleration can be set according to the requirements of system.

3.3 *Programming design*

The program flow chart is as shown in Figure 6.

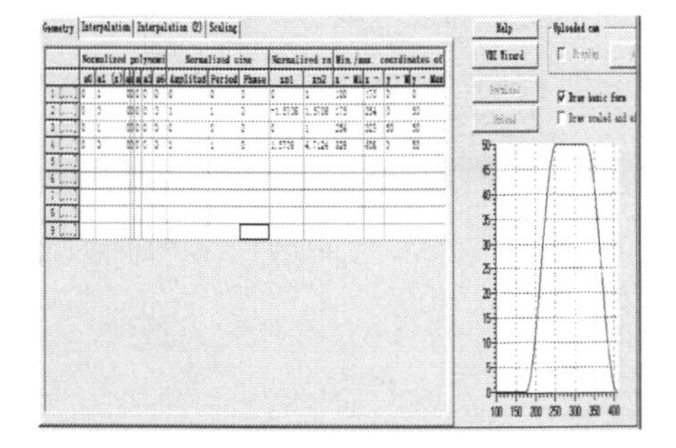

Figure 5. Configuration interface of *X*-axis cam curve.

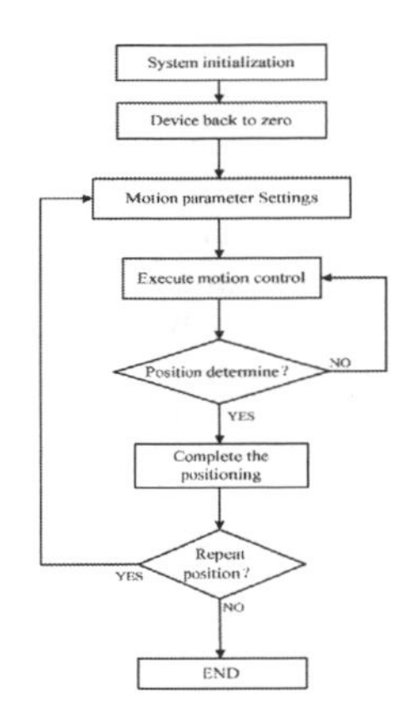

Figure 6. Programming flow chart.

Table 1. Name and function of functions.

Name	Function
MC-Power	Start/Stop axis
MC-Home	Zero/Setting axis
MC-Moveabsolute	Absolute position
MC-Halt	Normal stop
MC-Stop	Emergency stop
MC-CamIn	Start cam
MC-CamOut	Stop cam

The system requirement of S7-CPU315 is to control servo motor to do coordinated motion in the two-dimensional plane, in which includes system initialization, zero point setting, motion parameters set and motion process control, etc. Initialization process mainly call the MC-Power function block to start axis and the MC-Home function block to make the axis back to zero. Motion process need to call MC-Moveabsolute, MC-CamIn and MC-CamOut function blocks to realize the main axis absolute location and slave axis coordinate move. Furthermore the MC-Halt and MC-Stop function blocks are called to halt or stop move process. Program flow chart is shown in Figure 6 and function modules used in the movement process are shown in Table 1.

4 THE EXPERIMENTAL RESULTS AND ANALYSIS

The S7T-Config Trace tool always uses to analysis the dynamic response of an axis and its positioning error. This tool not only records the actual position and the command position, and its error, but also records the actual speed and the command speed, and many other parameters which can be used to analyze the stability of a motion axis (Li, 2013).

Figure 7 is the virtual axis to lead the real servo *X*-axis synchronous trajectory curve; it clearly shows the synergistic effect between the virtual axis and the *X*-axis.

Siemens T—CPU can use PID control to realize the closed loop control of the system. Figure 8 can clearly see that PID rapidly adjustment the speed of a real *X*-axis, but when the speed and the acceleration parameter of a virtual main axis and a real axis are set, it must consider the facts such as synchronous demand and system stability. It's because the bigger acceleration, the greater the impact, and the more unfavourable to system stability.

Figure 9 is the set position curve and the actual position curve of the real *X*-axis, it's clearly that cam curve defined by mathematical polynomial method can meet the marking location needs rapidly and precisely, and the following error of the control system is up to 0.001 mm, this high accuracy no doubt can be achieved for accurate positioning requirements of the identification system. Of course, *Y*-axis also can be achieved.

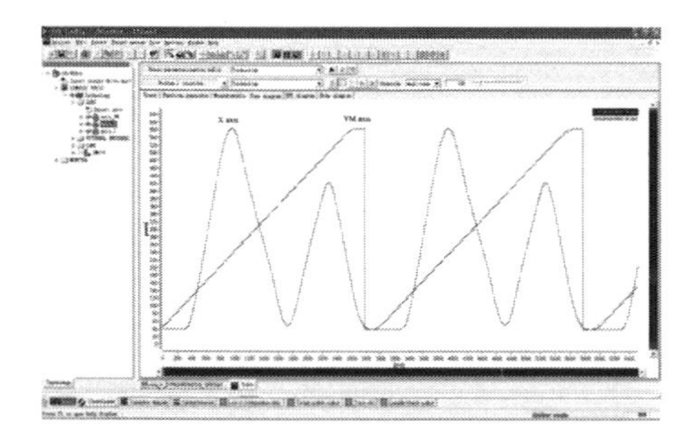

Figure 7. Synchronization of the virtual master axis and a real *X*-axis.

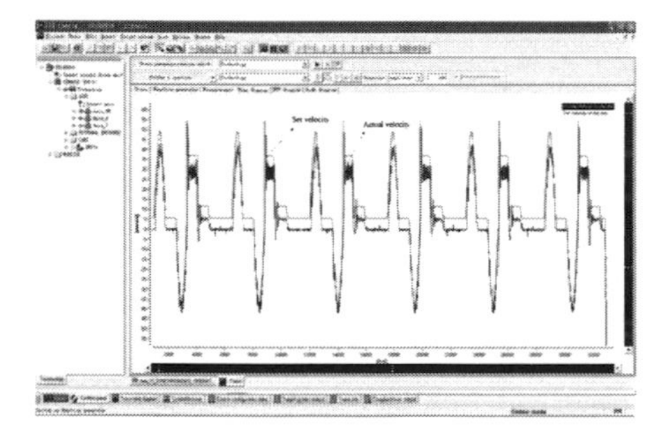

Figure 8. Relation between the set velocity and the actual velocity of a real *X*-axis.

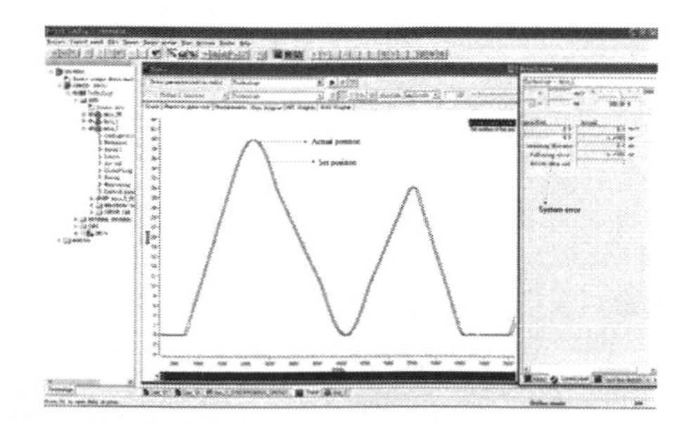

Figure 9. Relation between the set position and the actual position of a real *X*-axis.

According to the relationship between the command position and the actual position of an axis, it can analyze the response speed. And in accordance with the acceleration process of an axis, it can also determine that the system is subjected to impact severity, in other words, determine the dynamic stability of a system.

5 CONCLUSIONS

In this paper, a multi-axis coordinated motion control system has been established by using Siemens S7-CPU315 motion controller. The virtual cam technology of the Technology CPU is used to control two shaft (or multi-axis) coordinated motion, which can guarantee the trajectory curve coordinate movement. Experiments show that Siemens CPU315T-2DP with motion control functions can be flexibly designed a multi-axis coordinated control system, by means of a polynomial fitting can easily design the trajectory curve or CAM, and the real axis motion control accuracy can be up to 1 μm.

The dynamic response of an axis and its positioning error can be analysed by the S7T-Config Trace tool. This tool not only records the actual position and the command position, but also records the actual speed and the command speed, and many other parameters which can be used to research the stability of a motion axis.

According to the relationship between the command position and the actual position of an axis, the response speed can be analyzed, and also in accordance with the acceleration process of an axis, the dynamic stability of a system can be determined.

REFERENCES

Cunhai Pan, Ming Chen. Two-axises coordination motion based on cam mechanism [J]. Journal of Tianjin university of science and technology. **6** (2010) 42–45.

Junfeng Li. Research on coordinated motion control system based on programming logic controllers [D]. Tianjin: Tianjin university of science and technology, 2013.03.

Ming Chen. Research on two-dimensional plane control system based on Siemens T-CPU[D]. Tianjin: Tianjin university of science and technology, 2011.03.

SIEMENS. *Simotion CamTool Configuration Manual*. 2004.

SIEMENS. SIMATIC *S7-Technology, Function Manual, A5E00251798-06*. 2008.

Zhigen Yu. Realize and Application of the Model Cam Mechanism [J]. Mechatronics. **5** (2000): 39–40.

Civil, Architecture and Environmental Engineering – Kao & Sung (Eds)
© 2017 Taylor & Francis Group, ISBN 978-1-138-02985-9

Short-term photovoltaic power forecasting based on the human body amenity and least squares support vector machine with fruit fly optimization algorithm

Huabao Chen
School of Physics and Electronic Electrical Engineering, Huaiyin Normal University, Huai'an, China
Jiangsu Province Key Construction Laboratory of Modern Measurement Technology and Intelligent System, Huai'an, China

Wei Han
State Grid Huaian Power Supply Company, Huai'an, China

Ling Chen
College of Energy and Electrical Engineering, Hohai University, Nanjing, China

ABSTRACT: Accurate PhotoVoltaic (PV) power forecasting can provide reliable guidelines for power dispatching and construction planning of PV power generation, which is also important to the sustainable development of solar utilization. According to the characteristics of PV power generation and the factors impacting PV output power, the human body amenity is proposed. The forecasting performance of the Least Squares Support Vector Machine (LSSVM) model largely depends on the parameters of the kernel function and penalty factor. In this paper, a hybrid PV power forecasting model combining Fruit fly Optimization Algorithm (FOA) with LSSVM is proposed to solve this problem, where the FOA is used to automatically select the appropriate parameter values for the LSSVM PV power forecasting model. The numerical results verify the effectiveness and accuracy of the proposed model and improved algorithm.

1 INTRODUCTION

Along with our country paying more attention to environment and energy sources, PhotoVoltaic (PV) power generation becomes more important in theory and actual use because of little pollution and high energy utilization. Accurate PV power forecasting can relieve the conflict between electricity supply and demand. Short-term forecasting will contribute to arrange the output of the conventional electric power supply by using the electric power dispatching system (Damousis, 2004; Ai-Hamadi, 2005; Almonacid, 2009). Lu and Qin (Lu, 2011) applied metrical information including temperature, pressure, humidity, and solar irradiance as inputs to predict the day type, and then calculated the power output of PV power generation under day types. However, the cycle of this forecasting model is very shorter, only in per hour.

Moreover, owing to the rising of intelligence techniques, many new intelligence forecasting algorithms were used for PV power generation forecasting. Keerthi and Lin (Keerthi, 2003) proposed a hybrid model combining support vector regression and differential evolution algorithm to forecast the output power of PV, and this method was proved to outperform the SVR model with default parameters, regression forecasting model, and back propagation artificial neural network. Hsu and Chen (Hsu, 2003) implemented a knowledge-based expert system to support the choice of the most suitable forecasting model, and the usefulness of this method was demonstrated by using a practical application.

The human body amenity, which refers to take no effective measures under the premise of protection, can describe the degree of comfort by the person in the natural environment (Wu, 2003). This index can effectively use the various meteorological factors, reduce the input of the network, and improve the precision of forecasting. The Least Squares Support Vector Machine (LSSVM) extended by using the Support Vector Machine (SVM) is a powerful regression tool with a dynamic network structure (Cortes, 1995). The Fruit fly Optimization Algorithm (FOA) proposed by the scholar Pan (Pan, 2012) is a novel evolutionary computation and optimization technique. This new optimization algorithm has the advantages of being easy to understand and to be written into program code, which is not too long when compared with other algorithms. Therefore, in this paper, an attempt is made to use the FOA to automatically select the parameters' value of the LSSVM for improving the LSSVM's forecasting accuracy in the short-term PV power forecasting.

2 HUMAN BODY AMENITY

As we all know, the output power of PV generation is in positive correlation with the solar irradiance. When compared with the solar irradiance, temperature, humidity, and wind speed are easier to obtain by the devices (Li, 2010; Qin, 2006). Therefore, the human body amenity is adopted to replace so many influenced factors, reduce the input of the network, and simplify the network model in this paper.

Generally speaking, the meteorological factors such as temperature, humidity, and wind speed have a great effect on the human body. Hence, the human body amenity is a non-linear equation based on these elements. Usually it is described as follows:

$$D = f(T) + g(U) + h(V) \tag{1}$$

where D is the human body amenity, T is the temperature (°C), U is the average daily humidity (%rh), and V is the wind speed (m/s), which always assumes it as the average value of the maximum and minimum wind speed.

According to the empirical formula, the equation (1) can be represented as follows:

$$D = 1.8T + 0.55(1-U) - 3.2\sqrt{V} + T_N \tag{2}$$

Where T_N is the reference temperature (°C), which slightly varies with the region. In this paper, we take $T_N = 30$ (°C) as the reference temperature of Jiangsu province in China.

3 LEAST SQUARES SUPPORT VECTOR MACHINE WITH FRUIT FLY OPTIMIZATION

3.1 *Least Squares Support Vector Machine (LSSVM)*

The Support Vector Machine (SVM) is a trainable and machine learning method, which is used to solve the problem of pattern recognition and has better generalization abilities (Gestel, 2001; Du, 2008).

Set $\{x_i, y_i\}(i = 1, 2 \cdots l)$ as the training sample set; among these, $x_i \in R^n$ as the n-dimensional input vector, y_i as the output value of one-dimensional input vector, and l is the sample number.

The LSSVM model has two parameters, such as kernel function σ and penalty coefficient γ that need to be determined, which are very important in using the LSSVM model for forecasting. The parameters of σ and γ determine the squared bandwidth of the Gaussian RBF kernel and the trade-off between the training error minimization and smoothness, respectively. Many researchers selected the parameters by priori knowledge or individual experience, which may be un-efficient for forecasting. Therefore, we should develop an automatically efficiently method for selecting the appropriate parameters in the LSSVM model.

In order to achieve this goal, in this paper, the Fruit fly Optimization Algorithm (FOA) is used to automatically determine the parameter values of the LSSVM model.

3.2 *Fruit fly Optimization Algorithm*

Fruit fly Optimization Algorithm (FOA) is a new swarm intelligence method, which was proposed by Pan (Pan, 2012). The FOA is a new method for finding global optimization based on the food finding behavior of the fruit fly. The fruit fly is superior to other species in vision and osphresis. The food finding process of the fruit fly is as follows: firstly, it smells the food source by the osphresis organ, and flies towards that location; and then, after it gets close to the food location, the sensitive vision is also used for finding food and other fruit flies' flocking location, and it flies towards that direction. Figure 1 shows the food-finding iterative process of the fruit fly swarm.

The FOA belongs to a kind of interactive evolutionary computation, and also is a part of the artificial intelligence. Therefore, it has been widely applied to solve the problems in military, engineering, management, and financial affairs.

3.3 *Fruit fly optimization algorithm for the parameters selection of the LSSVM model*

Selecting appropriate values of the kernel function σ and penalty coefficient γ are particularly important.

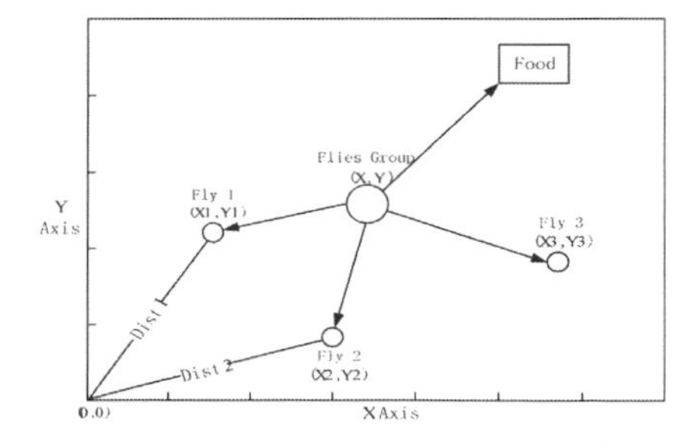

Figure 1. Graph showing the food-finding iterative process of the fruit fly swarm.

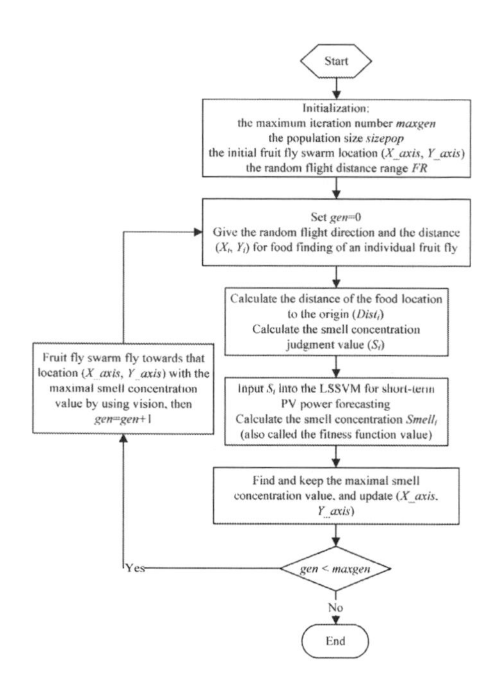

Figure 2. Flowchart of the FOALSSVM model.

In this paper, the FOA was used for selecting the suitable parameter values of the LSSVM model in order to effectively improve the short-term photovoltaic power forecasting accuracy.

The flowchart of the FOA for the parameters of the LSSVM model (abbreviated as FOALSSVM) is shown in Figure 2.

4 EXAMPLES COMPUTATION AND COMPARISON ANALYSIS

In this paper, the FOALSSVM algorithm was simulated by the MATLAB and one demonstration project of PV power generation in Jiangsu province was selected to prove the effectiveness of the proposed FOALSSVM algorithm.

The capacity of the PV power generation system is 380 kW. It was chosen as the local meteorological data and historical output power of PV generation as the training sample in July, and then forecasted the output power of the PV generation in step size of 15 minutes. As shown in Figure 3, it has provided the output power curves of PV power generation from July 13th to July 18th.

The operation curves of PV generation show that the effective working period of PV generation is from about 6 o'clock in the morning to 19 o'clock in the afternoon, while the output power of the PV generation is zero in the rest time of a day. Therefore, the output power forecasting of the PV generation can be only restricted in the working period.

The procedures for applying the proposed FOALSSVM algorithm to forecast the output power of the PV generation are described as follows:

Step 1: The process of sample data.
In this paper, the sample data were normalized to obtain data in the range from 0 to 1 by using the following formula:

$$X = \frac{x_i - x_{i\min}}{x_{i\max} - x_{i\min}}, i = 1, 2, \cdots l \quad (3)$$

where $x_{i\max}$ and $x_{i\min}$ are the maximal and minimal value of each input factor, respectively.
Step 2: Train the LSSVM model.
In the FOALSSVM model, the parameter values of the LSSVM are dynamically tuned by the FOA. The initial parameter values of the LSSVM were set up in the range of [0.01, 1]. After 100 times of iterative evolution, the optimal parameter values of the LSSVM model are obtained in Table 1.
Step 3: Forecast the output power of PV generation.
According to the result of FOA tuning of the parameters of the LSSVM model, the parameters of kernel function σ and penalty coefficient γ are chosen as 0.18 and 5.48 to forecast the output power of PV generation, respectively. The final forecasted output power of PV generation is shown in Figure 4.

The historical output power data and the predictive meteorological data of the day are fed as input to forecast the output power of July 19th in 15 minutes, and compared with the actual output power of the PV generation station. The Absolute Percentage Error (APE) and Mean Absolute Percentage Error (MAPE) are adopted to analyze and measure the performance of the forecasting

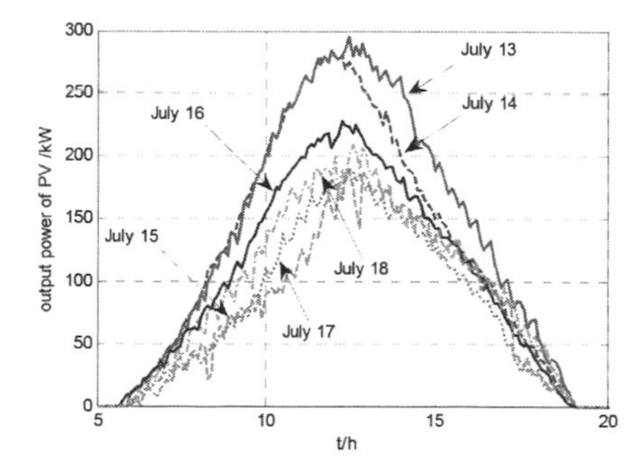

Figure 3. Curves of PV power generation data.

Table 1. The parameter values of the three algorithms.

Parameter	SVM	LSSVM	FOALSSVM
σ	0.2	0.5	0.18
γ	1	10	5.48

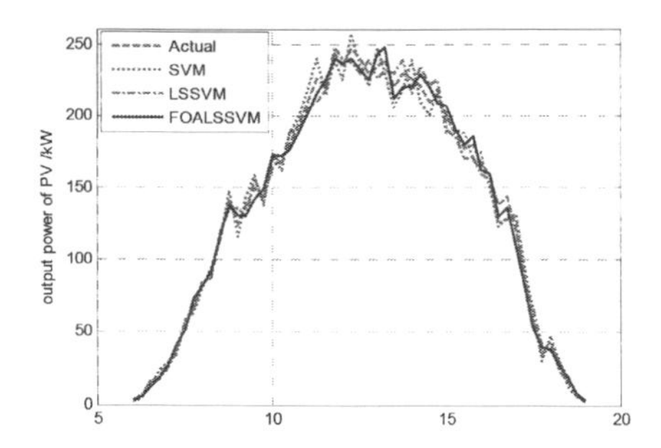

Figure 4. Graph showing the PV power forecasting results.

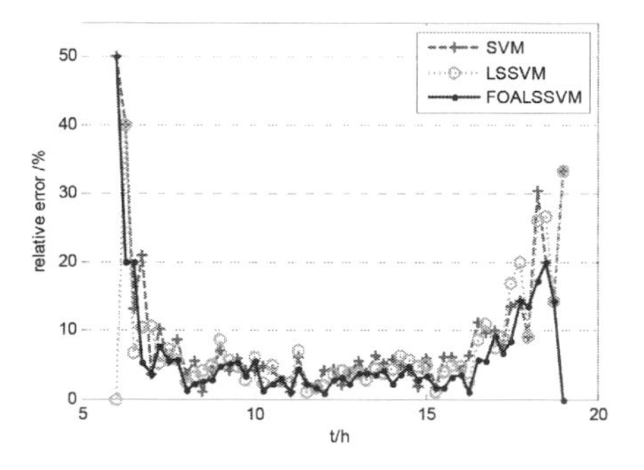

Figure 5. Error curves based on different algorithms.

models in the PV power forecasting system, which can be calculated by using the following equation:

$$\varepsilon_{APE} = \frac{|Y_r - Y_e|}{Y_r} \times 100\% \tag{4}$$

$$\varepsilon_{MAPE} = \frac{1}{n}\sum_{i=1}^{n} \varepsilon_{APE} \tag{5}$$

Table 1 shows the forecasting results of PV generation by the SVM, LSSVM, and FOALSSVM algorithms in 15 minutes, respectively. Among these, the MAPE of short-term PV power forecasting based on FOALSSVM is 6.32%, while the values based on SVM and LSSVM are 8.93% and 7.67%, respectively. Figure 4 gives the forecasting results with the SVM, LSSVM, and FOALSSVM algorithms. Figure 5 describes the error analysis of the three forecasting algorithms.

From the forecasting results, we can arrive at the following conclusions:

1. It can be clearly seen that all the three forecasting algorithms capture the output trends of the PV generation, but the performance of the FOALSSVM algorithm is better than the others.
2. From 6 to 8 o'clock in the morning, the forecasting errors are relatively larger. Because humidity has an effect on the PV power generation system, the curve of output power is unstable. However, there is an obvious correlation between the human body amenity and humidity, which causes the good forecasting performance by the FOALSSVM in this time quantum.
3. In the afternoon, the PV generation is covered by a part of cloud, which leads the output power of PV generation in high volatility. Therefore, there exist larger deviations between the forecasting results and the actual output.

Generally speaking, the proposed human body amenity and FOALSSVM algorithm can narrow the deviation between the forecasting results and the actual values, and it outperforms the SVM and LSSVM algorithms in the short-term PV power forecasting. When compared with the LSSVM algorithm, the FOALSSVM which uses the FOA to select the parameter values of the LSSVM can improve the forecasting accuracy effectively.

5 CONCLUSIONS

The LSSVM algorithm has been widely used in a variety of fields, but it is rarely found that the LSSVM has been applied to the PV power forecasting. In this paper, a hybrid model based on the human body amenity and LSSVM with FOA is proposed for the short-term PV power forecasting. The human body amenity can effectively use the various meteorological factors, reduce the input of the network, and improve the precision of forecasting. Meanwhile, the FOALSSVM model uses the FOA to automatically select the appropriate parameter values of the LSSVM in order to improve the forecasting accuracy. For comparison, another two models such as SVM and LSSVM are selected. The simulating results show that the FOA can select the appropriate parameter values of the LSSVM model, which could effectually improve the forecasting accuracy of PV output power. When compared with the SVM and LSSVM algorithm, the values of APE and MAPE are obviously smaller than that of other two algorithms.

REFERENCES

Ai-Hamadi, H.M., S.A. Soliman. Long-term/mid-term electric load forecasting based on short-term correlation and annual growth. Electric Power Systems Research, Vol. 74, No. 3, pp. 353–361 (2005).

Almonacid, F., C. Rus, P.J. Perez, et al. Estimation of the energy of a PV generator using artificial neural network. Renewable Energy, Vol. 34, No. 12, pp. 2743–2750 (2009).

Cortes, C., V. Vapnik. Support-vector networks. Machine Learning, Vol. 20, No. 3, pp. 273–297 (1995).

Damousis, I.G., A.G. Bakirtzis, P.S. Dokopoutos. A solution to the unit commitment problem using inter-coded genetic algorithm. IEEE Trans. on Power Systems, Vol. 19, No. 2, pp. 1165–1172 (2004).

Du, Y., J.P. Lu, Q. Li, et al. Short-term wind speed forecasting of wind farm based on least square-support vector machine. Power System Technology, Vol. 32, No. 15, pp. 62–66 (2008).

Hsu, C.C., C.Y. Chen. Regional load forecasting in Taiwan — applications of artificial neural networks. Energy Conversion and Management, Vol. 44, No. 12, pp. 1941–1949 (2003).

Keerthi, S.S., C.J. Lin. Asymptotic behaviors of support vector machines with Gaussian kernel. Neural Computation, Vol. 15, No. 1, pp. 1667–1689 (2003).

Li, R., Q. Chen, H.R. Xu. Wind speed forecasting method based on LS-SVM considering the related factors. Power System Protection and Control, Vol. 38, No. 21, pp. 146–151 (2010).

Lu, N., J. Qin, K. Yang. A simple and efficient algorithm to estimate daily global solar radiation from geostationary satellite data. Energy, Vol. 5, No. 1, pp. 3179–3188 (2011).

Pan. W.T. A new fruit fly optimization algorithm: taking the financial distress as an example. Knowledge-Based Systems, Vol. 26, No. 2, pp. 69–74 (2012).

Qin, H.C., W. Wang, H. Zhou, et al. Short-term electric load forecast using human body amenity indicator. Proceedings of the CSU-EPSA, Vol. 18, No. 2, pp. 63–66 (2006).

Van Gestel, T., J.A.K Suykens, B. Baesens, et al. Benchmarking least squares support vector machine classifiers. Machine Learning, Vol. 54, No. 1, pp. 5–32 (2001).

Wu. D. Discussion on various formulas for forecasting human comfort index. Meteorological Science and Technology, Vol. 31, No. 6, pp. 370–372 (2003).

Civil, Architecture and Environmental Engineering – Kao & Sung (Eds)
© 2017 Taylor & Francis Group, ISBN 978-1-138-02985-9

An RFID-enabled CEP model for cloud manufacturing shopfloor

Keqiang Chen
The Fifth Electronics Research Institute of the Ministry of Industry and Information Technology, Guangzhou, China
School of Information Engineering, Guangdong University of Technology, Guangzhou, China

Yingyou Qiu
Guangdong Polytechnic Normal University, Guangzhou, China

Meilin Wang
School of Information Engineering, Guangdong University of Technology, Guangzhou, China

Qingyun Dai
Guangdong Polytechnic Normal University, Guangzhou, China

Ray Y. Zhong
Department of Industry and Manufacturing Systems Engineering, The University of Hong Kong, Hong Kong

Chuangmian Huang
The Fifth Electronics Research Institute of the Ministry of Industry and Information Technology, Guangzhou, China
Guangdong Provincial Key Laboratory of Electronic Information Products Reliability Technology, Guangzhou, China

ABSTRACT: Cloud manufacturing is a new manufacturing paradigm developed from existing advanced manufacturing models and enterprise information technologies under the support of Internet of Things (IoT) such as Radio Frequency Identification (RFID) technology. This paper presents an RFID-enabled Complex Event Processing (CEP) model for the Cloud manufacturing shopfloor by defining the basic and complex event to describe the typical operations and behaviors in the Cloud manufacturing environment. Firstly, RFID technology is used for creating the Cloud manufacturing shopfloors where typical events could be presented by CEP. Secondly, by defining different events, this paper categorizes various management such as workers, materials etc, so that the CEP model could be used for demonstrating various typical production logic. Two demonstrative cases are given in this paper for displaying the implementation of the proposed model in an RFID-enabled Cloud manufacturing shopfloor.

1 INTRODUCTION

Cloud manufacturing is a new manufacturing paradigm developed from existing advanced manufacturing models and enterprise information technologies under the support of cloud computing, Internet of Things (IoT), virtualization and service-oriented technologies, and advanced computing technologies so as to transform manufacturing resources and capabilities into various services, which can be managed and operated in an intelligent and unified way to enable the full sharing and circulating of them (Xu, 2012). Cloud manufacturing can provide safe and reliable, high quality, cheap and on-demand manufacturing services for the whole lifecycle of manufacturing that refers to big manufacturing that includes the whole lifecycle of a product (e.g. design, simulation, production, test, and maintenance) (Tao,2011).

The Cloud manufacturing shopfloors play a very important role in supporting the sharing of various resources and capacities (Zhong, 2013). Figure 1 shows a typical Cloud manufacturing shopfloor production. In this environment, traditional resources are converted into Smart Manufacturing Objects (SMOs) by using the Radio Frequency Identification (RFID) technology (Qu, 2012; Dai, 2012; Zhong, 2015). However, there are large number of complexities and varieties such as differences of workers' skill, large quantities of

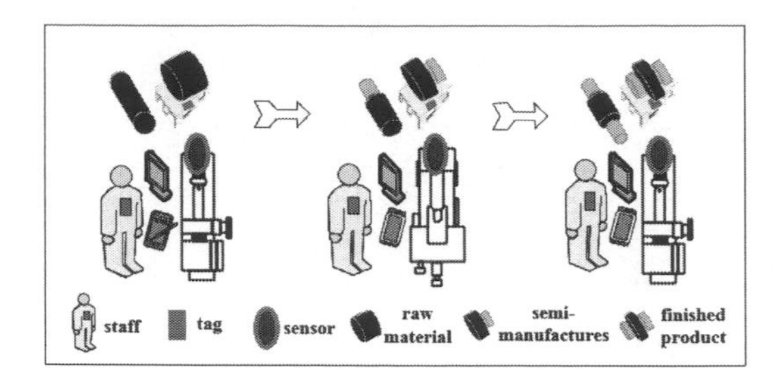

Figure 1. An RFID-enabled cloud manufacturing shopfloor.

materials, machine deviations, technical challenges, and the control of various SMOs (Wang, 2012).

Within the RFID-enabled Cloud manufacturing shopfloors, great myriad of events will occur. An event is an interaction of two SMOs such as a reading operation. So that all the predefined events will be constructed by large number of SMOs' basic data. How to present the events is an practical question with great significance in Cloud manufacturing.

Complex Event Processing (CEP) is a method of tracking and analyzing streams of data about things that happen (events) and deriving a conclusion from them that combines data from multiple sources to infer events or patterns that suggest more complicated circumstances (Zhang, 2009). CEP is able to identify meaningful events and respond to them as quickly as possible (Zhong, 2008). Therefore, this paper introduces an RFID-enabled CEP model for the Cloud manufacturing shopfloor according to the production logic from real-life case. This model includes the RFID processing mechanism and models to deal with large number of events happened from SMOs. A logic CEP presentation is then proposed to reflect the requirements from different end-users.

The rest of this paper is organized as follows. Section 2 introduces the basic idea of CEP. Section 3 presents an RFID-enabled CEP model. Section 4 illustrates the implementation of this model. Section 5 concludes this paper by highlighting our contributions and future research directions.

2 COMPLEX EVENT PROCESSING (CEP)

The CEP has roots in discrete event simulation, the active database area and some programming languages (Zhong, 2015). The activity in the industry was preceded by a wave of research projects in the 1990s and the first project that paved the way to a generic CEP language and execution model was the Rapide project in Stanford University, which is directed by David Luckham (Lan, 2015). In parallel there have been two other research projects: Infospheres in California Institute of Technology, directed by K. Mani Chandy, and Apama in University of Cambridge directed by John Bates (Huang, 2013). The commercial products were dependents of the concepts developed in these and some later research projects and community efforts started in a series of event processing symposiums organized by the Event Processing Technical Society, and later by the ACM DEBS conference series. One of the community efforts was to produce the event processing manifesto (Tao, 2014; Zhang, 2014; Zhong, 2014).

Based on the RFID-enabled Cloud manufacturing shopfloor production mode, the CEP in this paper is constructed as shown in Figure 2. From the bottom to the top, there are several layers such as hardware abstraction, event and data management, and application abstract. Each layer has different functionalities to enable the CEP in the Cloud manufacturing shopfloor work well. Hardware Abstraction Layer (HAL) is used for connecting the physical environment and abstract components. HAL manages the RFID readers, hardware connections and deletes the duplicated or error RFID data. Additionally, it provides the standard RFID data for the Event And Data Management Layer (EDML).

EDML is responsible for analyzing and processing the RFID data from HAL. It picks up the basic event from enormous data and interprets them so as to providing more rich information to

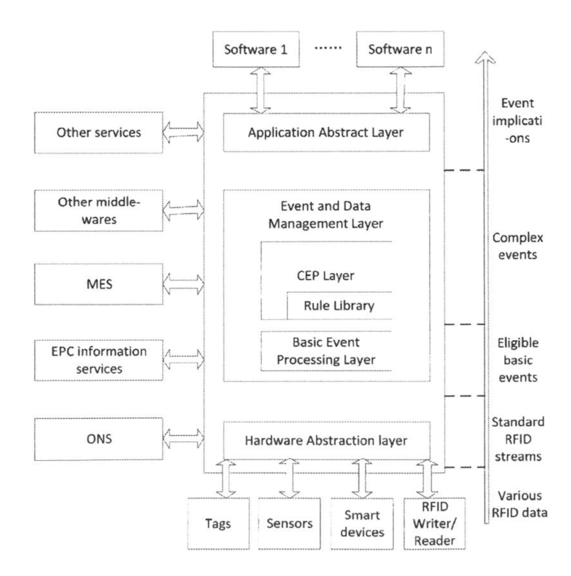

Figure 2. CEP mechanism for RFID-enabled Cloud manufacturing shopfloor.

Table 1. RFID CEP basic logic.

Logical relation	Representation method	Relation description	Application example
and	&	Two or more basic events occurring at the same time.	Employees tag and at least a material tag should be identified by PDA at the same time when employees receive materials.
or	$\vert$	At least one event occur in two or more basic events.	The same process can be processed in any of the same type of equipment.
period requirements	X < Min { time1, time2,..., time n} < Max { time1, time 2,..., time n} < Y	Basic events must occur within a certain period of time.	To complete a manufacturing order in a prescribed period of time.
time sorting constraint	time 1 < time 2 < ... < time n	Basic events must occur in a particular order.	There is a time sequence relation in the process of a product.
enumeration	Count {e1, e2,..., en}	Calculate the number of tags.	Statistics on the number of components of processing equipment.

the upper level like Application Abstract Layer (AAL). There are two sub-layers: basic event processing layer and CEP layer. The previous layer is responsible for sending the standard RFID data to Object Name System (ONS) so as to get the real-time status of the SMOs. The latter is for processing the obtained basic events according to the time and production logic.

AAL mainly focuses on various applications in distributed environments so as to providing reliable data communications. Moreover, the interpreted events could be issued to different parties so that they can be widely used.

3 AN RFID-ENABLED CEP MODEL

Basic event is the message presenter which carries the united information. It could be presented by a four tuple set e = <terminal Id, antenna Id, tag Id, time>. Where, terminal Id is the reader ID which indicates the readers' type so as to differentiate each device, antenna Id is the label of each antenna. Each reader may have several antennas. Tag Id presents a tag which contains the bound objects' information so as to differentiate the SMOs. Time is the time stamp recording the event. Basic event includes the worker event, material event, and machine event.

Complex event is the meaningful messager which is consisted by basic events. E(e1, e2, e3,..., en) = {operator (terminal Id1, terminal Id2,..., terminal Id n), operator (antenna Id1, antenna Id2,..., antenna Id n), operator (tagId 1, tagId 2,..., tagId n), operator (time 1, time 2,..., time n)}. Operator is an event code. Thus, based on the definition, the CEP model for RFID-enabled Cloud manufacturing shopfloor will use the following logic (as show in Table 1) to present various events in the production environment.

4 IMPLEMENTATION

This model is used in a typical RFID-enabled Cloud manufacturing shopfloor which is equipped with some digital devices for creating an intelligent environment. Different users and departments are defined by the CEP following the Table 2.

Based on Table 2, various CEP could be used for guiding the operations. Figure. 3 demonstrates a typical implementation of this model using a demonstrative case. An operator will enter into the RFID-enabled Cloud manufacturing shopfloor. He/She will be sensed by the antenna 1~4. Assume the basic events are e1~e4. Thus, we can get the entry and exit complex events:

E1 = (e1| e4) & (e2| e3), Max {time 1, time 4} < Min { time 2, time 3}

E0 = (e1| e4) & (e2| e3), Min {time 1, time 4} > Max {time 2, time 3}

Another implementation as shown in Figure 4 which demonstrates an inspector carries an RFID mobile PDA to do the quality check from three machine operators. Assume that the no. i product is sensed by the PDA which is an event ei. Then the complex event is:

E = e1 & e2 & ... ei & ... & en, time e1<Min{time e2 ... time en}.

Table 2. RFID CEP for different management.

<table>
<thead>
<tr><th>Task</th><th>CEP class</th><th>Task description</th><th>Event model</th></tr>
</thead>
<tbody>
<tr><td>Material management</td><td>Material receiving events</td><td>To monitor and control the use of material</td><td>E = e’1 & e’2 … & e’i… & e’n, time e’1 < Min {time e’2, … time e’i,… time e’n}. e’i = ei1|ei2…|eij…|ein. eij represents the basic event that tag i is identified by the j antenna of the same RFID reader.</td></tr>
<tr><td></td><td>Material flowing events</td><td>To monitor the flow of material</td><td>E = (e11| e14) & (e12| e13), Max { time e11, time e14} <Min{time e12, time e13}. e11 and e14 represent that the same tag is identified by two different antennas, both at one end of the shop floor aisle, of the same RFID reader. On the contrary, the rest two basic events are both identified by different antennas at the other end</td></tr>
<tr><td>Production management</td><td>Production schedule control events</td><td>To monitor the processing status of all parts.</td><td>E = e’1& e’2 … & e’i … & e’n, time e’1 <time e’2 … <time e’i… <time e’n. e’i = ei1|ei2…|eij…|ein. eij represents the basic event that tag i is identified by the j antenna of the same RFID reader.</td></tr>
<tr><td></td><td>Assembly events</td><td>To monitor the assembly of products</td><td>E = e’1 & e’2…&e’i…&e’n, time e’1 < Min{time e’2,…time e’i,…time e’n}. e’i = ei1|ei2…|eij…|ein. eij represents the basic event that tag i is identified by the j antenna of the same RFID reader.</td></tr>
<tr><td></td><td>Production efficiency events</td><td>To monitor the production capacity and equipment processing capacity</td><td>E = Count{e’1, e’2, e’3,..., e’n}. e’i = ei1|ei2…|eij…|ein. eij represents the basic event that tag i is identified by the j antenna of the same RFID reader.</td></tr>
<tr><td></td><td>Product quality events</td><td>To monitor inspect work and completion acceptance</td><td>E = e’1& e’2…&e’i…&e’n, time e’1 <Min{time e’2,…time e’i,…time e’n}. e’i = ei1|ei2…|eij…|ein. eij represents the basic event that tag i is identified by the j antenna of the same RFID reader.</td></tr>
<tr><td>Staff management</td><td>Staff attendance events</td><td>To monitor staff attendance and staff access to production areas</td><td>E = (e11| e14) & (e12| e13), Max{time e11, time e14} <Min{time e12, time e13}. e11 and e14 represent that the same tag is identified by two different antennas, both at one end of the shop floor aisle, of the same RFID reader. On the contrary, the rest two basic events are both identified by the different antennas at the other end.</td></tr>
<tr><td>Device management</td><td>Device management events</td><td>Equipment working status and machining quantity</td><td>E = e11|e12…|e1j…|e1n, X <Min{time e11, time e12,…time e1j,…time e1n} <Max{time e11, time e12,… time e1j,…time e1n} <Y. e1j represents the basic event that devices tag is identified by the j antenna of the same RFID reader.</td></tr>
</tbody>
</table>

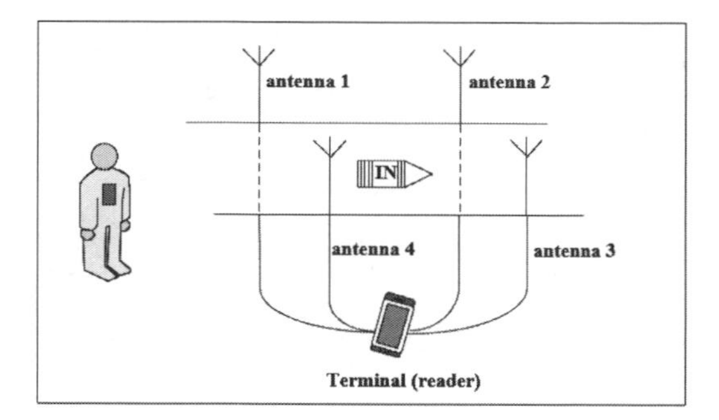

Figure 3. A demonstrative case for entry and exit.

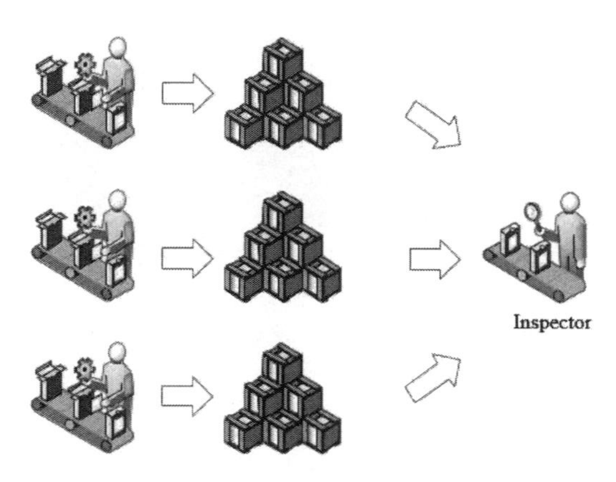

Figure 4. A demonstrative case for inspection.

5 CONCLUSIONS

This paper introduces an RFID-enabled CEP model for the Cloud manufacturing shopfloor. This model uses basic and complex event to present the typical operations and behaviors in the Cloud manufacturing environment. There are several contributions or innovations in this research. Firstly, RFID technology is used for creating the Cloud manufacturing shopfloors where typical events could be presented by CEP concept. Secondly, by defining different events, this paper categorizes various management such as workers, materials etc, so that the CEP model can be used for demonstrating various typical production logic.

Future research will be carried out as follows. First of all, this approach will be validated by using real-life data to present the real-life operations and logics in the Cloud manufacturing shopfloors. Secondly, more theoretical aspects will be focused in the near future to enhance the feasibility and practicality of the proposed CEP model. Finally, a prototype system will be worked out by using the model to support different visibility or traceability of various materials or SMOs.

ACKNOWLEDGEMENTS

This work is financially supported by the Science and Technology Plan Project of Guangdong Province (2013A061401003).

REFERENCES

Dai, Q. Y., R. Y. Zhong, G. Q. Huang, T. Qu, T. Zhang and T. Y. Luo. Radio frequency identification-enabled real-time manufacturing execution system: a case study in an automotive part manufacturer. International Journal of Computer Integrated Manufacturing, 2012, 25(1): 51–65.

Qu, T., H. Luo, G. Q. Huang, N. Cao, J. Fang, R. Y. Zhong, L. Y. Pang and X. Qiu. RFID-Enabled Just-In-Time Logistics Management System For "SHIP"–Supply Hub In Industrial Park. Proceeding of 42nd International Conference on Computers & Industrial Engineering (CIE 42), 16–18 July, 2012, Cape Town, South Africa.

Tao, F., L. Zhang, V. Venkatesh, Y. Luo and Y. Cheng. Cloud manufacturing: a computing and service-oriented manufacturing model. Journal of Engineering Manufacture, 2011, 225(10): 1969–1976.

Tao, F., Y. Zuo, L. Da Xu and L. Zhang. IoT-based intelligent perception and access of manufacturing resource toward cloud manufacturing." Industrial Informatics, IEEE Transactions on, 2014, 10(2): 1547–1557.

Wang, M. L., T. Qu, R. Y. Zhong, Q. Y. Dai, X. W. Zhang and J. B. He. A radio frequency identification-enabled real-time manufacturing execution system for one-of-a-kind production manufacturing: a case study in mould industry. International Journal of Computer Integrated Manufacturing, 2012, 25(1): 20–34.

Xu, X. From cloud computing to cloud manufacturing. Robotics and computer-integrated manufacturing, 2012, 28(1): 75–86.

Zhang, Y. H., Q. Y. Dai and R. Y. Zhong. An Extensible Event-Driven Manufacturing Management with Complex Event Processing Approach. International Journal of Control and Automation, 2009, 2(3): 1–12.

Zhang, Y., G. Q. Huang, S. Sun and T. Yang. Multi-agent based real-time production scheduling method for radio frequency identification enabled ubiquitous shopfloor environment. Computers & Industrial Engineering, 2014, 76: 89–97.

Zhong, R. Y., C. Xu and C. Chen. Big Data Analytics for Physical Internet-based Logistics Data from RFID-enabled Intelligent Shopfloors. International Journal of Production Research DOI:10.1080/00207543.2015.1086037.

Zhong, R. Y., G. Q. Huang, Q. Y. Dai and T. Zhang. Mining SOTs and Dispatching Rules from RFID-enabled Real-time Shopfloor Production Data. Journal of Intelligent Manufacturing, 2014, 25(4): 825–843.

Zhong, R. Y., G. Q. Huang, S. L. Lan, Q. Y. Dai, C. Xu and T. Zhang. A Big Data Approach for Logistics Trajectory Discovery from RFID-enabled Production Data. International Journal of Production Economics, 2015, 165: 260–272.

Zhong, R. Y., Q. Y. Dai, K. Zhou and X. B. Dai. Design and Implementation of DMES Based on RFID. 2nd International Conference on Anti-counterfeiting, Security and Identification, Guiyang, 20–23 Aug. 2008, 475–477.

Zhong, R. Y., Q. Y. Dai, T. Qu, G. J. Hu and G. Q. Huang. RFID-enabled Real-time Manufacturing Execution System for Mass-customization Production. Robotics and Computer-Integrated Manufacturing, 2013, 29(2): 283–292.

Zhong, R. Y., S. Lan, C. Xu, Q. Dai and G. Q. Huang. Visualization of RFID-enabled shopfloor logistics Big Data in Cloud Manufacturing. The International Journal of Advanced Manufacturing Technology, 2015: 1–12.

Civil, Architecture and Environmental Engineering – Kao & Sung (Eds)
© 2017 Taylor & Francis Group, ISBN 978-1-138-02985-9

Optimization of extraction of polysaccharides from mulberries by using high-intensity Pulsed Electric Fields

Shenglang Jin
College of Tourism, Huangshan University, Huangshan, China

ABSTRACT: The high intensity Pulsed Electric Fields (PEFs) was used to extract polysaccharides from mulberries. The factors affecting the extraction yield, such as electrical field intensity, extracting temperature, pulse number, and the ratio of the material to liquid, were analyzed under specific conditions and the optimal extracting parameters were obtained as follows: the electrical field intensity is 40 kV/cm, the extracting temperature is 80°C, pulse number is 10, and the ratio of the material to liquid is 1:30. Under these conditions, the extraction yield of total polysaccharides was 82%. The PEF and conventional extraction method were compared. The extraction yield of polysaccharides by using the conventional method was 54%, and the microwave-assisted extraction yield of polysaccharides was 67%. The study demonstrates that the PEF is a reliable and great efficiency tool for the fast extraction of polysaccharides from mulberries. Extraction of these compounds from mulberries is a crucial step for use of these compounds in the food. The PEF-assisted extraction by pressing of polysaccharides from fresh mulberry stands as an economical and environmentally friendly alternative to conventional extraction methods, which require the product to be dried, use large amounts of organic solvents, and need long extraction times.

1 INTRODUCTION

The mulberry tree, a typical plant of genus Morus, has been widely cultivated for its leaves that serve as an indispensable food for silkworm. There are about 3,000 accessions of mulberry germplasm resources in China. Of which, over 60 accessions can be used as the mulberry fruit (Li, 2005). Some varieties introduced from mid-Asia have white fruit. Mulberry has been used as medicine from ancient times. The root bark and mulberry leaves have been used as herbal medicine (Machii, 1989). The mulberry fruit contains polysaccharides, organic acids, free amino acids, vitamins, micronutrients and other components (Elmaci, 2002).

Polysaccharides play a substantial role in the food industry, especially as the health ingredient in healthy food (Li, 2006). Polysaccharides have been proposed as the first biopolymers formed on Earth. These are classified on the basis of their main monosaccharide components and the sequences and linkages between them, as well as the anomeric configuration of linkages, the ring size (furanose or pyranose), the absolute configuration (D- or L-), and any other substituents present. Later work has shown that polysaccharides are involved in a number of important biochemical functions, such as cell–cell interaction and communication, the attachment for infectious bacteria, viruses, toxins, and hormones. Some researchers now focus on the polysaccharides of mulberries. Generally, traditional extraction methods are very time-consuming and require large quantities of solvents (Chen, 2011). Consequently, the demand is increasing for extraction techniques that improve the yield, shorten the extraction time, and reduce the use of organic solvents. Pulsed Electric Field electro-technology (PEF) is an emerging technology in the field of food preservation. PEF can be widely used to analyze the functional components of natural products, with advantages of non-thermal, fast, efficient, low-power, and low-pollution. PEF-assisted extraction has shown promise as a technology for obtaining valuable compounds from soft vegetal materials.

Up to now, however, the extraction of polysaccharides from mulberries by PEF-assisted extraction technology has not been reported. In this study, the influence of the different factors on the extracting yield was investigated.

2 MATERIALS AND METHODS

2.1 *Materials*

The ripe mulberry fruits were sampled in Huangshan County, Anhui Province, China.

After sampling, the fruits were immediately carried to the laboratory, placed in plastic bags, and stored below 4°C in a refrigerator. Their size was about 1 cm long and their color was purple.

2.2 *Reagents and chemicals*

The reference standard of rutin was purchased from the National Institute for Control of Pharmaceutical and Biological Products (purity ≥ 98%). Ethanol, sodium nitrite, aluminum nitrate, sodium hydroxide, and other reagents were obtained from Tianjin Haiguang Chaemical Co., Ltd., respectively. All solvents were distilled prior to use.

2.3 *Instruments and equipments*

An AEU-210 electronic analytical balance (Xiangyi Balance Instrument & Equipment Co., Ltd.), FW100 high-speed universal grinder (Tianjin City Taisite Instrument Co., Ltd.), SHZ-D (III) circulating water vacuum pump (Gongyi Yingyu Yuhua Instrument Factory), LD5-2A centrifuge (Beijing Medical Centrifuge Factory), and GZX-9140M E digital blast oven (Shanghai Xunda Industrial) were employed in this research.

2.4 *Drawing standard curve*

Suck the control solutions 0, 0.1, 0.2, 0.3, 0.4, and 0.5 mL precisely and place them in 10 mL-centrifuge tubes, respectively. The method of anthrone is adopted and the absorbance is detected (K = 620 nm). Construct the standard curve by plotting the mean absorbance obtained from each standard against its concentration with absorbance value on the horizontal (y) and the vertical (x) axis. Calculate the regression equation of the standard curve:

$$y = 6.4609x - 0.0024, \quad r = 0.9997 \tag{1}$$

2.5 *Statistical analysis*

Results reported in this paper are the average of at least two measurements found in two independent experiments. To determine the influence of the electric field strength, the results were analyzed by multiple regressions starting from a second-order polynomial model using the software Statgraphics Plus 5.1 (Statistical Graphics Corporation, USA). A backward regression procedure was used to determine the parameters of the models. This procedure systematically removed the effects that are not significantly associated ($P > 0.05$) with the response until a model with only a significant effect was obtained.

3 RESULT AND DISCUSSION

3.1 *Characterization of PEF-induced damage in the cells of mulberries*

Zp was used to select the optimum PEF treatment conditions to permeabilize the mulberry cells. This index, which is based on the changes of the conductivity of intact and PEF-permeabilizated tissue, has been used previously for this purpose in diverse vegetable tissues. Fig. 2 shows the influence of PEF treatment times on the *Zp* of the mulberry tissue at different electric field strengths. The increase in the electric field strength and treatment time results in an increment of the *Zp* to the highest value of 0.33 for the most intense treatment conditions tested. Independent of the applied electric field strength, the *Zp* value increased significantly with an increase in the treatment time to approximately 60 μs. Above these values, the increment of the treatment time scarcely affected the *Zp* value. The general trend of the influence of the electric field strength and treatment time on the *Zp* value observed in this research is in good agreement with previously reported data for other vegetable tissues, such as potatoes, apples, and onions. However, in these tissues higher values (>0.8) for the *Zp* were obtained even by using lower electric field strengths (≤30 kV/cm). Therefore, the results obtained in this investigation seem to indicate that the permeabilization of mulberry tissue cells

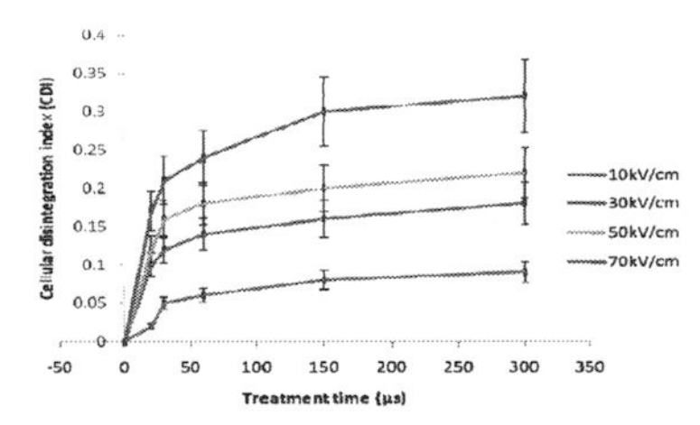

Figure 1. Influence of the electric field strength and treatment time on the cell disintegration index (Zp) of mulberries.

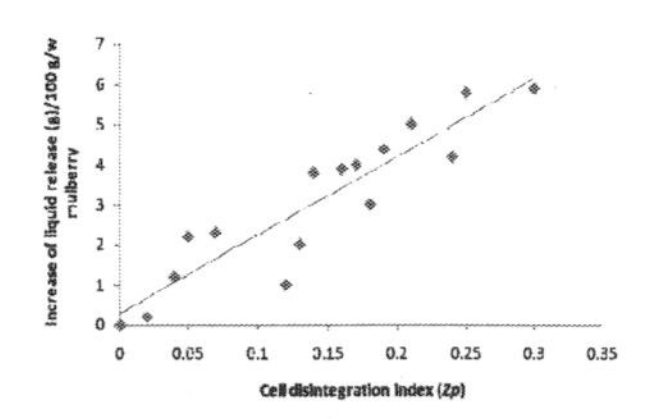

Figure 2. Relationship between the cell disintegration index (Zp) of mulberries treated by PEF and liquid release by centrifugation.

requires applying more intense PEF treatments than those for other vegetable tissues.

Due to the low disintegration index determined for the mulberry tissue even at electric field strengths as high as 70 kV/cm, the other experimental method based on the release of the intracellular liquid from the PEF-treated tissue was used. The increment of the liquid leakage in the PEF-treated samples dependent on the treatment intensity can be interpreted as a consequence of the PEF-cell permeabilization. Quantification of the liquid released from a PEF-treated material subjected to a given centrifugation has been proven to be useful in the testing cell membrane's permeability. Fig. 2 shows that a solid correlation ($r = 0.97$) was obtained between the Zp and the quantity of the liquid released by centrifugation when the same treatment conditions (electric field strength and treatment time) were applied. This good correlation indicates that the Zp index measuring the degree of cell permeabilization over a short period of time is a suitable procedure to select the PEF treatment conditions that induce the highest permeabilization of mulberry cells.

3.2 *Effect of electric field intensity on the extraction yield*

The effect of different electric field intensities on the extraction yield was detected at a pulse number of 4, at the ratio of the material to liquid (ratio of mulberry to absolute alcohol) of 1:10, and at a temperature of 40°C. The extraction yield was the highest at an electric field intensity of 40 kV/cm and then decreased appreciably with an increase in the electric field intensity (Fig. 3). This is due to the fact that the phenomenon of cavities increases along with the enhancement of the electric field intensity, which accelerates the polysaccharides in mulberries. With an increase in the electric field intensity, from the micro-perspective, the potential difference between the inside and outside of the cell membrane became larger, and the electroporation of the cell membrane occurred. A higher electric field intensity could cause more solvent to enter the cell membrane and permeate the cell membrane. Thus, increasing the electric field intensity could increase the extraction yield. However, when the electric field intensity increased from 20 to 40 kV/cm, the extraction yield of polysaccharides decreased. The reason might be that some of polysaccharides were decomposed in higher-intensity electric fields.

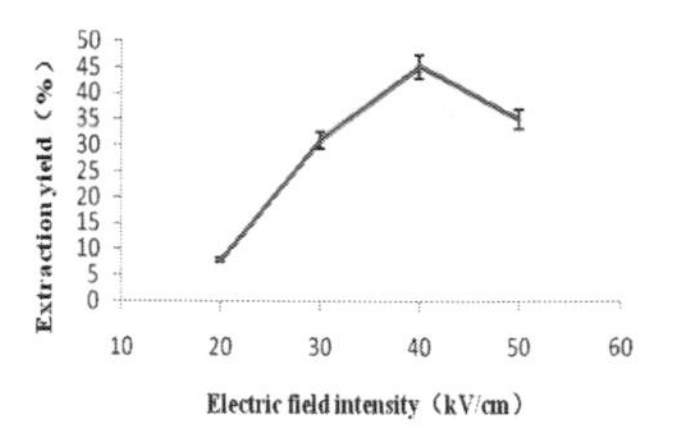

Figure 3. Influence of the electric field intensity on the extraction yield of polysaccharides from mulberries.

3.3 *Effect of pulse number on the extraction yield*

The effect of different pulse numbers on the extraction yield was detected at an electric field intensity of 30 kV/cm, at a ratio of the material to liquid of 1:10, and at a temperature of 40°C. The result showed that extraction yield increased with an increase of the pulse number, but after a certain time when osmotic-pressure of the cell reached a balance, the extraction yield maintained invariableness (Fig. 4). Therefore, pulse number of 8 was chosen in the following experiments.

3.4 *Effect of the ratio of material to liquid on the extraction yield*

The effect of different ratios of the material to liquid on the extraction yield was detected at an electric field intensity of 30 kV/cm, pulse number of 4, and temperature of 40°C. The extraction yield increased with an increase in the ratio of the material to liquid (Fig. 5). The extraction yield at the ratio of the material to liquid of 1:40 was 6.8% higher than that at a ratio of 1:30. The reasons might be that, at lower material to liquid ratios, the concentration of polysaccharides in the solution would be low, and the polysaccharides' exchange between the solid and solvent had reached equilibrium. Since a big ratio of the mulberry to absolute

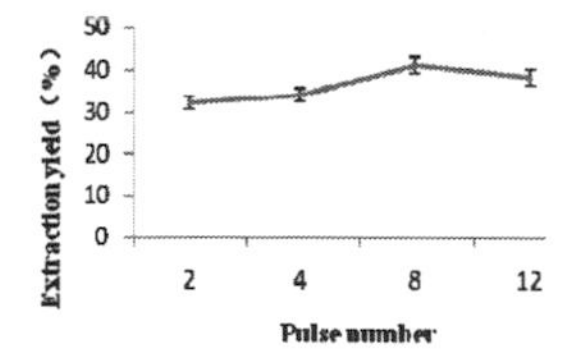

Figure 4. Influence of pulse number on the extraction yield of polysaccharides from mulberries.

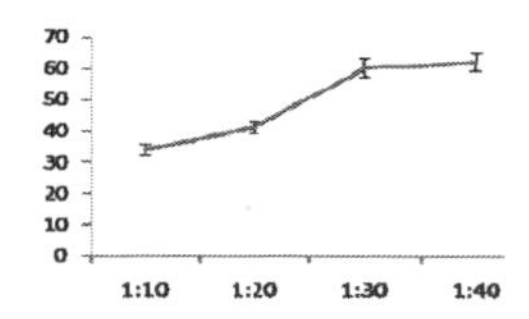

Figure 5. Influence of the material liquid on the extraction yield of polysaccharides from mulberries.

alcohol might increase the difficulty in reclaiming alcohol, the ratio of the material to liquid at 1:40 is suggested, according to the experimental results.

3.5 *Effect of temperature on the extraction yield*

The effect of different temperatures on the extraction yield was investigated at an electric field intensity of 30 kV/cm, at a pulse number of 4, and at a ratio of material and liquid of 1:10. The extraction yield increased with an increase of temperature (Fig. 6). The yield of extraction at 80°C increased by 4.2% when compared with that at 70°C. The result showed that it is not necessary to maintain higher temperature conditions during the extraction process. Too high temperatures will be a disadvantage for the stability of polysaccharides from mulberry, and thus the extraction temperature should be maintained under 80°C.

3.6 *Synthetic factors*

On the basis of experiment and analysis of single factor, electric field intensity, temperature, pulse number, and the ratio of the material to liquid were selected separately to set up three levels of experiments (Table 1), and the influence of each factor on the extraction yield was observed. Design and results of orthogonal experiments are shown in Table 2.

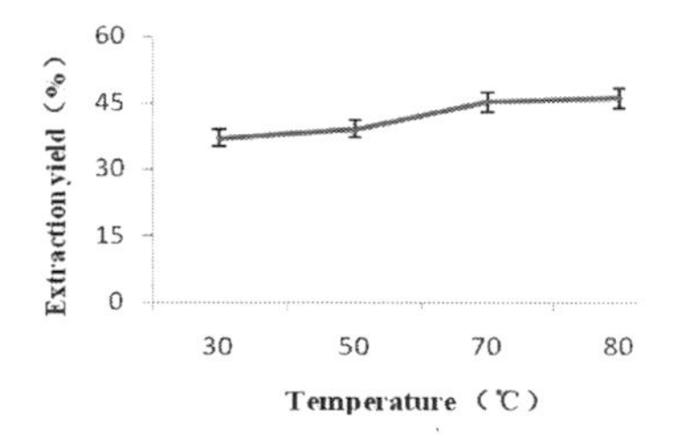

Figure 6. Influence of temperature on the extraction yield of polysaccharides from mulberries.

Table 1. Factors and levels of orthogonal experiments.

Levels	Factors Temperature (°C)	Pulse number	Ratio of the material to liquid (g:mL)	Electrical field intensity (kV/cm)
1	60	6	1:20	35
2	70	8	1:30	40
3	80	10	1:40	45

Table 2. The results of orthogonal experiments.

No. of test	Factors A	Br	C	D	Yield/ %
1	1	1	1	1	47
2	1	2	2	2	53
3	1	3	3	3	49
4	2	1	2	3	56
5	2	2	3	1	45
6	2	3	1	2	65
7	3	1	3	2	49
8	3	2	1	3	58
9	3	3	2	1	70
K_1	49.667	50.667	56.667	54.000	
K_2	55.333	52.000	59.667	55.667	
K_3	59.000	61.333	47.667	54.333	
R	9.333	10.666	12.000	1.667	

Note: A—temperature; B—pulse number; C—the ratio of the material to liquid; D—electrical field intensity.

Table 3. ANOVA analysis results.

Source of variance	SS	F	F_a	Significance
A	132.667	28.427	28.427	*
B	202.667	43.426	43.426	*
C	234.000	50.139	50.139	*
D	4.667	1.000	1.000	
Error	4.67			

Note: A—temperature; B—pulse number; C—the ratio of the material to liquid; D—electrical field intensity. All freedom degrees are equal to 2, $a = 0.05$. * represents significant difference.

From Table 2, we can draw a conclusion that the order of influence of different parameters on the extraction yield is as follows: the ratio of the material to liquid > pulse number > temperature > electric field intensity. The optimum condition is $A_3B_3C_2D_2$, which means that the extracting temperature is 80°C, pulse number is 10, the ratio of the material to liquid is 1:30, and the electric field intensity is 40 kV/cm. Variance analysis showed that extracting temperature, pulse number, and the ratio of the material to liquid also had remarkable influences on the extraction yield of polysaccharides from mulberries (Table 3). This was consistent with the results of the single factor test, indicating that the optimized extraction conditions were reliable. The verification experiment showed that under the optimal conditions, the extraction yield of polysaccharides from mulberries can reach up to 82%.

3.7 *Comparison of PEF and the conventional extraction method*

The yield of PEF-assisted extraction was higher than the conventional extraction method. Under the same experimental conditions (80°C, material-solvent ratio of 1:30) as well as the same filtering, concentration, and determination procedures, the extraction yield of flavonoids was 54%, and the microwave-assisted extraction yield of flavonoids was 67%.

4 DISCUSSIONS

PEF technology has been applied widely in the extraction of effective components of many materials. Some experiments have been carried out to develop the PEF extraction method. PEF extraction has been regarded as a fast, reliable, and inexpensive technique, and it is more applicable in the extraction of materials than the conventional techniques of extraction. Therefore, materials would produce huge energy by themselves and accelerate chemical reactions. Extracting the active materials from natural products by PEF-assisted extraction technology had the advantages of high extraction efficiency, strong selectivity, good repeatability, little time consumption, few solvents, low energy consumption, environmental benefits, and so on. Therefore, PEF-assisted extraction is expected to offer a new way for the production and analyses of the plant extractions and the modernization of pharmaceutical engineering.

Our study demonstrates that the PEF is a reliable and greatly efficient tool for the fast extraction of polysaccharides from mulberries, and the optimized PEF-assisted extraction parameters are as follows: extracting temperature of 80°C, pulse number of 10, the ratio of the material to liquid of 1:30, and the electrical field intensity of 40 kV/cm. The highest extraction yield is about 82%.

ACKNOWLEDGMENTS

This work is supported by 2014 Educational Commission of Anhui Province of China (No. KJ2014A243), 2014 Huangshan Municipal Science and Technology Project (No. 2014Z-04) and 2016 The Key Program in the Youth Elite Support Plan in Universities of Anhui Province (No. gxyqZD2016304).

REFERENCES

Chen, G.Y., C.X. Zhang, Y.H. Luo, Extraction of Polysaccharides from Lycopus lucidus, Chinese Journal of Experimental Traditional Medical Formulae, **17**, 38–40. (2011).

Elmaci, Y., T. Altug, Flavour evaluation of three black mulberry (Morus nigra) cultivars using GC M/S, chemical and sensory data. Journal of the Science of Food & Agriculture, **82**, 632–635. (2002).

Li, H.S., H.F. Wang, YX Sun, Prelmiinary study on extraction and characteristics of mulberry pigment, Acta Sericologica Sinica, **31**, 175–181. (2005).

Li, J.W., X.L. Ding, Study on ultrasonically assisted extraction of polysaccharides from Chinese Jujube, Chemistry and Industry of Forest Products, **26**, 73–76. (2006).

Machii, H. Varietal differences of nitrogen and amino acid contents in mulberry leaves, Acta Seric Entomol, **1**, 51–61. (1989).

Civil, Architecture and Environmental Engineering – Kao & Sung (Eds)
© 2017 Taylor & Francis Group, ISBN 978-1-138-02985-9

Stress analysis of the interference fit in helical gear transmission

Tao Zhao
Zhonghuan Information College Tianjin University of Technology, Tianjin, China

ABSTRACT: The material mechanics method is important for structure analysis and for reducing the volume and weight of the driving parts in gear transmission. In this paper, a mathematical model is presented for the calculation and analysis of the interference fit joint. The formulation for the gear and shaft are provided in the paper. The calculation includes two aspects: first, calculating the contact stress in the matching surface and the minimum interference required under a known load; second, calculating the maximum allowable interference for the connecting parts in which no deformation or rupture occur. This method may also be applied to other interference fit joints.

1 INTRODUCTION

The interference fit joint, which is also known as the tight fit connector, uses the interference fit between the parts to achieve the connection result. Stress analysis is a commonly used research method for the preliminary design of the structure size. Because the assembly interference quantity between the gear and shaft directly affects the working performance of gear transmission, a precisely mathematical model is important.

Many analytical techniques have been proposed and successfully applied in the interference fit joint, for instance, the analysis of the preload and magnitude of the interference fit joint, the retting analysis for the relative sliding mode and stress distribution in the shaft and thick wall cylinder model.

The goal of this paper is to present a mathematical model that is aimed at achieving the computation of the amount of interference. The actual installation interference should be between the minimum interference and the maximum. The damage reason of the pinion is obtained by theoretical calculation, and the analysis validation is conducted by the finite element analysis software ANSYS.

2 THEORETICAL ANALYSIS OF INTERFERENCE FIT

2.1 *The interference fit joint analysis*

The interference fit is a way to realize the tight joint through the interference between enveloping parts (hole) and enveloped parts (shaft). It is mainly used for assembling hole and shaft parts tightly to pass the axial force or torque. After being assembled, a surface pressure is created between the enveloping parts and enveloped parts due to the elastic deformation of the material. Work is dependent on the pressure produced by the friction to transmit torque, axial force, or both of the two complex loads. As a result, only if the contacted pressure on the surface of the interference fit joint is precisely calculated, the allowable load will be accurately determined. In this article, gear and shaft are used as interference fit joints. Figure 1 describes the geometric model of the interference fit of helical gears and shaft, and the theoretical model is simplified as a thick wall cylinder model.

The enveloped parts with an inside diameter of d_i is pushed into an enveloping part with an outside diameter of d_a. Nominal diameters of the hole and shaft are within their tolerance zone. The difference between the two parts is the magnitude of interference δ, and the coupled diameter value is d_f.

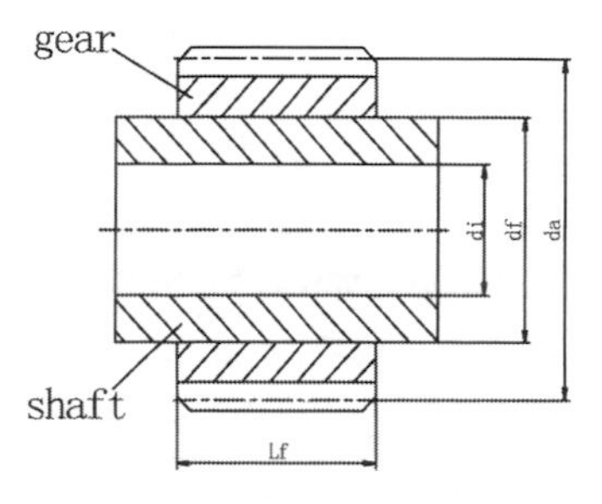

Figure 1. Schematic of the geometric model for the interference fit.

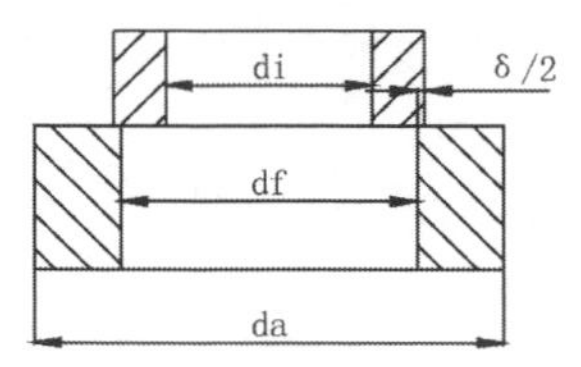

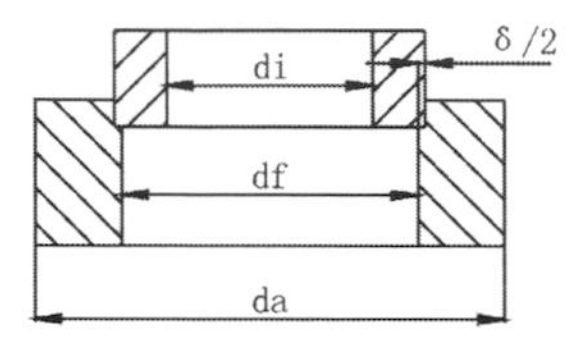

Figure 2. Schematic of the thick wall cylinder model.

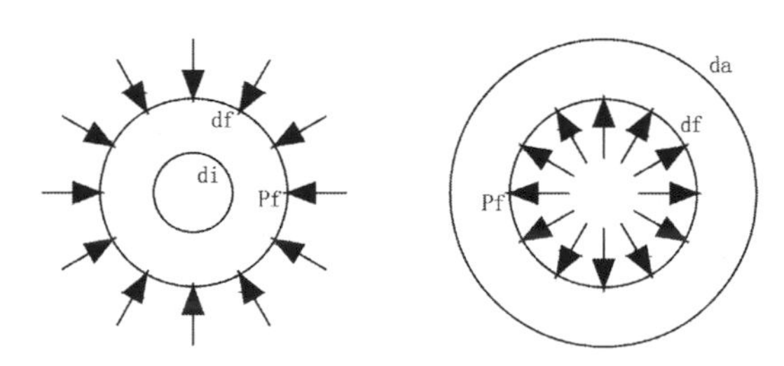

Figure 3. Schematic of the enveloping parts and enveloped parts.

As is shown in Figure 3, the combination of compaction pressure will be produced in the interface that forms the interference fit fastening because of the magnitude of interference δ. When the connection sustains an axial force or torque, the frictional resistance and the frictional torque are generated for transmitting the external load. The pressure P_f of the joint surface is not only related to the extruding magnitude of interference δ and material property (elasticity modulus E and Poisson's ratio μ), but also with relation to the physical dimension (the coupled diameter d_f and copulate length L_f), structural dimension (inside diameter d_i of the enveloped parts and outside diameter d_a of enveloping parts).

2.2 *Computation method of the interference fit*

Calculate the former model with the theory of mechanics of materials. The elasticity modulus E_a and Poisson's ratio μ_a are calculated for the enveloping parts, the elasticity modulus E_i and Poisson's ratio μ_i are calculated for the enveloped parts.

The diameter variable quantity of the enveloping parts e_a is given as follows:

$$e_a = \frac{p_f d_f}{E_a}\left(\frac{d_a^2 + d_f^2}{d_a^2 - d_f^2} + \mu_a\right) \tag{1}$$

The diameter variable quantity of the enveloped parts e_i is given as follows:

$$e_i = \frac{p_f d_f}{E_i}\left(\frac{d_f^2 + d_i^2}{d_f^2 - d_i^2} - \mu_i\right) \tag{2}$$

The effective magnitude of interference δ is given as follows:

$$\delta = e_a + e_i \tag{3}$$

The minimum pressure of the enveloping parts for the transmitting load $p_{f\,\min}$, is given as follows:

$$p_{f\min} = \frac{\left[(2T/d_f)^2 + F^2\right]^{\frac{1}{2}}}{\pi d_f L_f f} \tag{4}$$

In the equation, the transmitting torque is T, the transmitting axial force is F, and the friction coefficient is f.

The minimum diameter variable quantity of the enveloping parts $e_{a\,\min}$ is given as follows:

$$e_{a\min} = \frac{p_{f\min} d_f}{E_a}\left(\frac{d_a^2 + d_f^2}{d_a^2 - d_f^2} + \mu_a\right) \tag{5}$$

The minimum diameter variable quantity of the enveloped parts $e_{a\,\min}$ is given as follows:

$$e_{i\min} = \frac{p_{f\min} d_f}{E_i}\left(\frac{d_f^2 + d_i^2}{d_f^2 - d_i^2} - \mu_i\right) \tag{6}$$

The minimum effective magnitude of the interference for transmitting load $\delta_{e\,\min}$ is given as follows:

$$\delta_{e\min} = e_{a\min} + e_{i\min} \tag{7}$$

According to the analysis of the thick wall cylinder model by using the mechanics of materials, Figure 4 describes the distribution of stress and the relationship between the magnitude of interference δ and the pressure of joint surface P_f.

The interference of p_f can be calculated from the following equation:

$$p_f = \frac{\delta}{d_f} / \left[\frac{1}{E_a}\left(\frac{d_a^2 + d_f^2}{d_a^2 - d_f^2} + \mu_a\right) + \frac{1}{E_i}\left(\frac{d_f^2 + d_i^2}{d_f^2 - d_i^2} - \mu_i\right)\right] \tag{8}$$

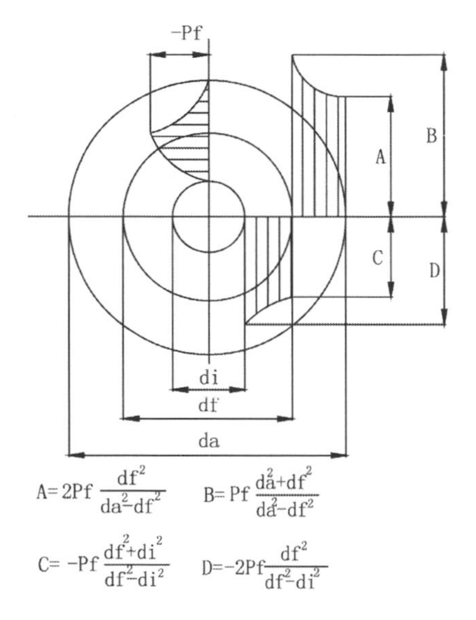

Figure 4. Schematic of the stress distribution of the interference fit joint.

Table 1. Geometrical parameters of helical gear transmission.

Name	Symbol	Pinion	Main gear
Number of teeth	z	21	101
Tooth width	b	169	167
Normal module	m_n	9	9
Normal pressure angle	a_n	20	20
Helical angle	β	8	8
Reference diameter	d	191	918
Contact ratio	ε	2.33	2.33

3 NUMERICAL EXAMPLE

The calculation of the numerical example is based on the above-mentioned mathematical model. Geometrical parameters of helical gear transmission are shown in Table 1. The basic parameters and machine settings of the example are adopted from the literature. The maximum stress of the gear hub is $\sigma_k = 551.85\ Mpa$, which exceeds the yield limit of the pinion that is $\sigma_s = 500\ Mpa$. And so, the gear wheel may undergo a fracture when it is working.

4 FINITE ELEMENT ANALYSIS

Authenticate the calculation of the model by using finite element analysis software ANSYS. Fig. 5 and 6 show the installation of the property of the contact surface such that normal penalty stiffness is 0.1; friction coefficient is 0.125; and the contact surface offset is 0. 18. Fig. 7 imposed the boundary conditions and the load.

It can be seen from Fig. 8 that 2–3 teeth participate in the gear mesh of the transmission, coinciding with the operation circumstance. The stress

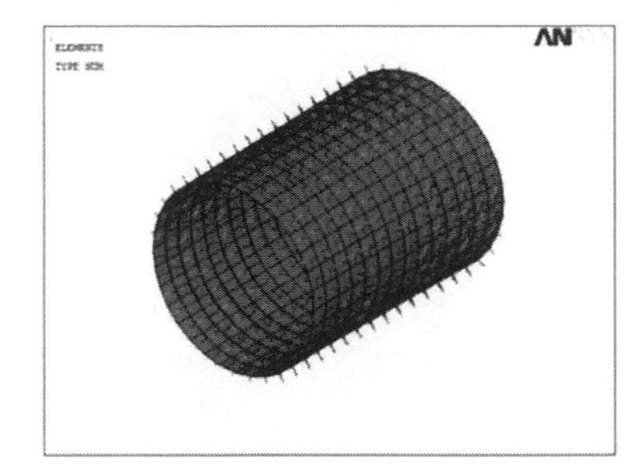

Figure 5. Stress direction of the pinion and shaft.

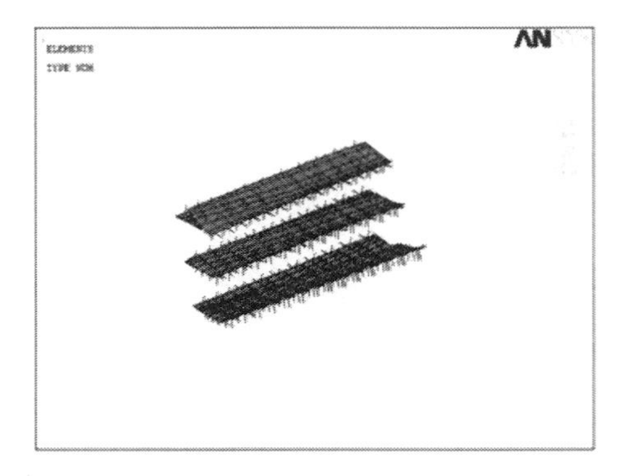

Figure 6. Stress direction of the mesh surface of the gear.

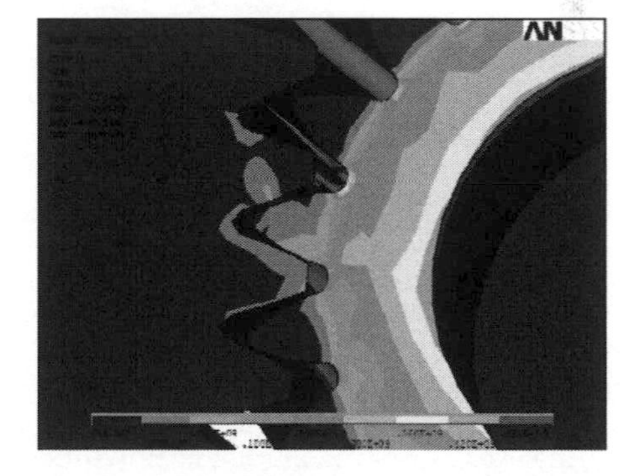

Figure 7. Graph showing the contact stress distribution.

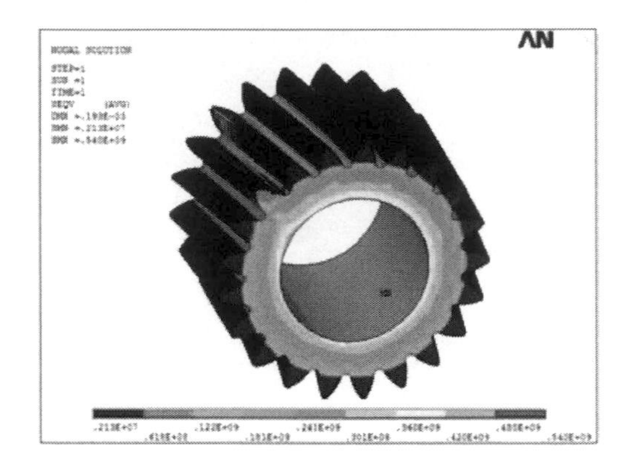

Figure 8. Results of finite element analysis.

of the root is 540 MPa. Exceeding the yield limit of the pinion may create the tooth root fracture failure.

5 OPTIMIZATION DESIGN OF THE DRIVE SHAFT BASED ON RELIABILITY

In order to avoid the fracture in the root of the helical gear tooth, the reliability design method is adopted in the structure optimization design. There are two ways to improve the structure: increase the root circle diameter of the helical gear tooth, or reduce the assembly interference. Because there is a corresponding function relationship between the root circle diameter and assembly limit interference quantity; any single change in variables affect the value of another, and so the use of the amount of interference δ and root circle diameter d_f of small gear tooth as design variables.

$$X=[x_1,x_2]^T=[\delta,d_f]^T \tag{9}$$

Failure modes of the interference fit including the relative slide in the mating surfaces of the shaft and pinion, plastic deformation of the pinion, or plastic deformation of the shaft. These are random events. As long as one occurs, it will lead to the abnormal working of the entire system. With series relationship of all, the reliability of the entire system is given as follows:

$$R(X)=R_1(X)\times R_2(X)\times R_3(X) \tag{10}$$

The goal is to make the reliability of the interference fit system maximum, and the objective function is given as follows:

$$\min f(x)=1-R_1(X)\times R_2(X)\times R_3(X) \tag{11}$$

If the three random events are assumed to be normally distributed, we can come to the conclusion that

$$\begin{aligned}R_1(X)&=\Phi\left[(\bar{\delta}-\bar{\delta}_{\min})/(C_{v\delta}\sqrt{\bar{\delta}^2+\bar{\delta}^2{}_{\min}})\right]\\&=\Phi\left[\frac{x_1-f_1(x_2)}{0.1\sqrt{x_1{}^2+f_1^2(x_2)}}\right]\end{aligned} \tag{12}$$

$$\begin{aligned}R_2(X)&=\Phi\left[(\bar{\delta}_{ea\max}-\bar{\delta})/(C_{va}\sqrt{\bar{\delta}^2_{ea\max}+\bar{\delta}^2})\right]\\&=\Phi\left[\frac{f_2(x_2)-x_1}{0.08\sqrt{f_2^2(x_2)+x_1{}^2}}\right]\end{aligned} \tag{13}$$

$$\begin{aligned}R_3(X)&=\Phi\left[(\bar{\delta}_{ei\max}-\bar{\delta})/(C_{vi}\sqrt{\bar{\delta}^2_{ei\max}+\bar{\delta}^2})\right]\\&=\Phi\left[\frac{f_3(x_2)-x_1}{0.08\sqrt{f_3^2(x_2)+x_1{}^2}}\right]\end{aligned} \tag{14}$$

Φ is the inverse function of the standard normal distribution function, δ, $\delta_{\min}$ are the actual interference amount and the minimum quantity, $\delta_{eq\max}$, $\delta_{ei\max}$ are the maximum effective interference amount of the small gear and the shaft which will not occur plastic deformation, $C_{v\delta}, C_{va}, C_{vi}$ are the variation coefficients of the yield strength gear and the shaft and the interference quantity. Given the system reliability requirements is R = 0.9988, according to the reliability allocation method of serial system, three random events of reliability is calculated respectively as R1 = 0.9997674, R2 = 0.99998665, and R3 = 0.9990324. The resulting target constraint conditions are as follows:

$$G_1(X)=\bar{\delta}-\bar{\delta}_{\min}-C_{v\delta}\sqrt{\bar{\delta}^2+\bar{\delta}^2{}_{\min}}\,\Phi^{-1}(R_1)\geq 0 \tag{15}$$

$$G_2(X)=\bar{\delta}_{ea\max}-\bar{\delta}-C_{va}\times\sqrt{\bar{\delta}^2_{ea\max}+\bar{\delta}^2}\times\Phi^{-1}(R_2)\geq 0 \tag{16}$$

$$G_3(X)=\bar{\delta}_{ei\max}-\bar{\delta}-C_{vi}\times\sqrt{\bar{\delta}^2_{ei\max}+\bar{\delta}^2}\times\Phi^{-1}(R_3)\geq 0 \tag{17}$$

$$G_4(X)=d_{f\max}-d_f\geq 0 \tag{18}$$

$$G_5(X)=d_f-d_f^*\geq 0 \tag{19}$$

The objective function is given as follows:
function
f = myfun(x)

```
f = 1-normcdf((x(1)–27.7/(1–x(2)^2)–48.8)/
     (0.1*(x(1)^2+(27.7/(1–x(2)^2)+48.8)^2)
*normcdf((–x(1)+575/(3+x(2)^4)^0.5)/
     (0.8*(x(1)^2+575^2/(3+x(2)^4))^0.5))
*normcdf((–x(1)+207/(1–x(2)^2))/
     (0.08*(x(1)^2+(207/(1–x(2)^2))^2)^0.5))
```

The constraint conditions:

```
function[c,ceq] = mycon(x)
c=[–x(1)+27.7/(1–x(2)^2)+48.8+0.35*(x(1)^2
     +(27.7/(1–x(2)^2)+48.8)^2)^0.5;
     x(1)–575/(3+x(2)^4)^0.5+0.36*(x(1)^2
     +575^2/(3+x(2)^4)^0.5;
     x(1)–207/(1–x(2)^2+0.248*(x(1)^2
     +(207/(1–x(2)^2))^2)^0.5;
     0.6865–x(2);
     x(2)–0.7005];
ceq = [];
```

Call the objective function and constraint conditions

```
x0 = [220, 0.70];
lb = [103.24, 0.6865];
ub = [352.27, 0.7005];
options = [];
[x,fval,exitflag,output] = fmincon('myfun',x0,[],[],[],
[],lb,'mycon',options)
```

The results:

```
x = 0.1875 186.7155
```

By using "MATLAB" software to calculate, the root circle diameter obtained is 178.8 mm. The optimization of the gear size and coupling interference quantity improves the defects of the initial design and achieve the expected purpose.

6 SUMMARY

In this paper, a mathematical model is presented for the calculation and analysis of the interference fit joint. The theoretical analysis of the interference fit joint is captured by the formulation for the gear and shaft provided in the paper. The calculation is in accordance with the finite element analysis results. Through the optimization design, the interference fit system achieved the highest reliability. Finite element analysis was carried out on the optimization model validation, and the results show that the optimized structure can avoid the small gear tooth root fracture and meet the strength requirements. This method may also be applied to other interference fit joints.

REFERENCES

Ding H C, Xiao L J, Zhang H Y, et al. Interference Fit Calculation and Stress Analysis for Rotor Sleeve of High-speed Permanent Magnet Electric Machine Design and Research, 2011, 27(5): 95.

Han C J, Zhang J, Huang G. Temperature Difference Method-based Analysis of Interference Fit between Gear and Shaft. China Petroleum Machinery, 2013, 41(1): 6.

He Y F, Yang X B, Gan W M. Resarch of the interference Fit of Gear and Axis in Press-fit Process. Journal of Mechanical Transmission, 2014, 38(12): 33.

Li S J, Le G, Lin G W, et al. The Research on interference fit between Missile Adapter and Launch Canister. Engineering Mechanics, 2011, 28(4): 245.

Tong J, Yuan D N, Liu H Z, et al. The Reliability Analysis of Motorized Spindle and Motor Rotor with Interference Fit Mechanical Science and Technology for Aerospace Engineering, 2013, 32(6): 845.

Zhang J P, Li C Y. Contact Pressure and Pullout Force Caused by interference Fit Mechanical Engineering & Automation, 2011, (1): 195.

Zhang J S, Yin Y F, Zhao X M, et al. Contact analysis of interference fit between shaft and shaft sleeve based on ANSYS. Journal Of Machine Design, 2014, 31(5): 21.

Civil, Architecture and Environmental Engineering – Kao & Sung (Eds)
© 2017 Taylor & Francis Group, ISBN 978-1-138-02985-9

A review of transformerless DC–DC boost converters for renewable energy sources

Jun Luo
School of Electrical and Electronic Engineering, Huazhong University of Science and Technology, Wuhan, China

ABSTRACT: The renewable energy sources and energy storage devices deliver output voltage at a range of around 12–125 VDC. Before being connected to the grid, such a low voltage level should be stepped up to a sufficient level according to different electrical network standards among countries. Boost DC–DC converters are used for the voltage adjustment, and have become the key parts in the energy system power chain as the overall performance of the system is affected by their efficiency. The major consideration in DC–DC conversion is often associated with high efficiency, reduction of stresses involving semiconductors, low cost, simplicity, and robustness of the involved topologies. In this review, we first analyze the conventional boost converter and then focus on the kinds of transformerless boost converters. Advantages and disadvantages of each boost converter topology are presented to help the designer with structure selection. A comparison and discussion of different DC–DC boost topologies are carried out across a number of parameters and presented in the end of this paper.

1 INTRODUCTION

In the last few decades, renewable energy sources like PhotoVoltaic (PV) modules, fuel cells and energy storage devices, such as super capacitors have been widely developed. However, the majority of these commonly used renewable energy sources deliver electric power at the output voltage range of 12V–125V, which cannot be directly connected to the grid for use. Typically, such low voltages are necessary to be stepped up to a voltage of around 300V–400V, varying with the different electrical network standards among countries (Abusara, 2013).

In the past, a centralized inverter was responsible for connecting several modules or other renewable energy sources into the grid. At the moment, centralized technology has been replaced. Boost DC–DC converters are now applied to adjust the voltage to a sufficient level, so that the attached cascade DC–AC inverter, which is coupled to the grid (1-or 3-phase), can be supplied (Kjaer, 2005). Therefore, the boost DC–DC converters have become the key part of a renewable energy system as the overall performance of the system is affected by their efficiency.

In applications where galvanic insulation is not a must, non-isolated DC–DC converters can be used to achieve voltage step-up or step-down, with consequent reduction of size, weight, and volume associated to the increase of efficiency because of the lack of a high-frequency transformer (Hu, 2014).

The classical or conventional DC–DC boost converter is widely employed in voltage step-up, and is also being studied in many books on power electronics (Mohan, 2007; Erickson, 2007). But for high-power applications, the boost converter is not a feasible solution because the load power is processed by using only two semiconductors, while appreciable current and/or voltage stresses exist. Particularly, in high-current applications, conduction losses lead to the significant reduction of efficiency because these increase with the square of the RMS current through the semiconductors. Although the parallelism of switches or even converters is possible, current sharing is compromised because of the intrinsic differences of the involved elements (Klimczak, 2008).

Considering the limitations of the classical boost converter, several studies have been carried out in the literature to improve key issues, such as the static gain, voltage stress across the semiconductors, efficiency, power capacity, and many other aspects of the original topology. The simultaneous search for the terms 'DC–DC' and 'boost converter' in IEEEXPlore® Digital Library demonstrates the relevance of this subject, revealing that more than 5000 papers have been published in conferences and journals over the last few decades.

In Session II, a topology of the conventional boost converter is presented and analyzed. And then, in Session III, some of the most popular improved transformerless DC–DC boost converters available in the literature are introduced and classified. The classification of these differentiations is based on their topologies. Session IV gives a comparison and discussion of these different DC–DC boost converter topologies across numbers of parameters. Finally, a conclusion is drawn in Session V.

2 CONVENTIONAL DC–DC BOOST CONVERTERS

A conventional boost converter is shown in Fig. 1 (Kim, 2009), where the intrinsic series resistance of the inductor is represented by R_L. By using the volt-second balance principle and considering the operation in continuous-conduction mode, it is easy to demonstrate that the static gain G_v is given by equation (1).

$$G_v = \frac{V_o}{V_i} = \frac{1}{1-D} \cdot \frac{1}{\left(1 + \left(R_L / \left((1-D)^2 \cdot R_o\right)\right)\right)} = \frac{1}{1-D} \cdot \eta, \tag{1}$$

where V_o is the output voltage, V_i is the input voltage, R_o is the load resistance, D is the duty cycle, ranging from 0 to 1, and η is used to represent the theoretical efficiency of the boost converter.

As shown in equation (1), both static gain and efficiency depend on R_L, D, and R_o. Some interesting conclusions can be led with the analysis of equation (1). Firstly, the static voltage gain G_v theoretically is infinite when the duty cycle D reaches 1. But switch turn-on period becomes long as D increases thereby causing conduction losses to increase. The power-rating of the conventional single switch boost converter is limited to switch-rating. Secondly, if $R_L = 0$, the static gain is the same as that of an ideal boost converter and the converter is lossless as its efficiency is 100%. In practical terms, $R_L \neq 0$ and both static gain and efficiency decrease as R_L is increased. Finally, the conventional boost converter is a simple and adequate approach for applications with medium power requirements, as a significant reduction of efficiency may be caused by the equivalent series resistance of the output capacitor, reverse-recovery losses, and the rectifier diode.

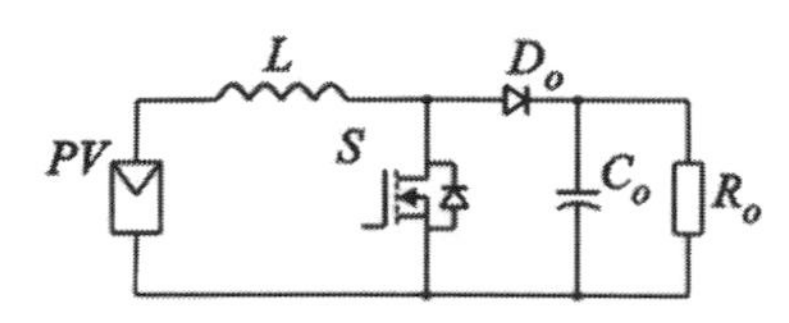

Figure 1. Topology of the conventional boost converter. R_L is the equivalent resistance of the inductor.

3 IMPROVED TRANSFORMERLESS DC–DC BOOST CONVERTERS

Due to the effect of power switches, rectifier diodes, equivalent series resistance of inductors and capacitors and high voltage stresses on the power switches, conventional DC–DC boost converters are unable to provide high step-up voltage gain. In this part, some topologies of improved transformerless DC–DC boost converters are introduced.

3.1 *Soft switching boost converters*

In conventional designs, hard switching leads to switching losses, which, especially for high values of the switching frequency, cause a reduction in the energy conversion efficiency. Converters with soft switching technology have a slightly improved voltage gain in comparison to conventional boost converters. As shown in Fig. 2, the converter operates in ZVS (Zero Voltage Switching) mode (Park, 2010), while there is another kind of soft switching boost converter operating in ZCS (Zero Current Switching) mode. This topology dramatically reduces switching losses and thus achieves better efficiency. The disadvantage of that topology is the complexity of the circuit, as some resonant components are added including a switch and an extra inductor. Though the driving sequence is a bit more complex, both switches operate at the same ground potential, so that additional separation at the driver side is needless.

3.2 *Interleaved boost converters*

A four-phase interleaved boost converter is shown in Fig. 3 (Chen, 2009). As discussed in the literature (Crews, 2008), this topology has the advantages such as simplicity, smaller input and output current ripples, lower switching losses, higher efficiency, and smaller inductor size. The input current ripples are small because each interleaved cell shares the input current, and they help in increasing the life of PV modules.

Considering that a generic number of phases N can be used, N represents the number of cells. By

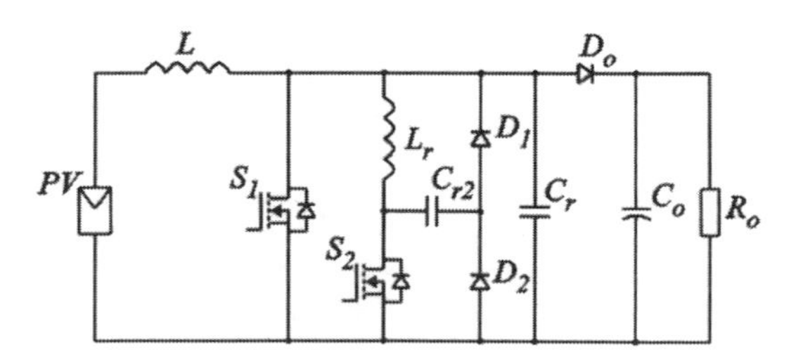

Figure 2. Topology of a soft switching boost converter.

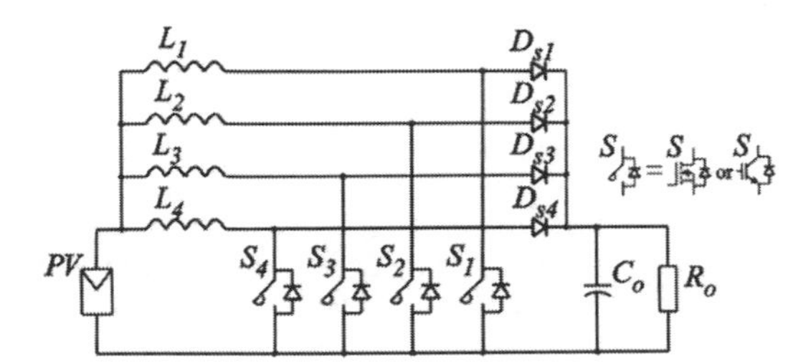

Figure 3. Topology of an interleaved boost converter.

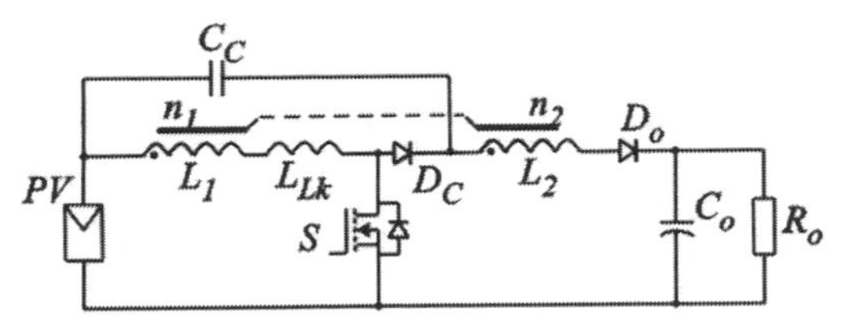

Figure 4. Topology of a boost converter with coupled inductors.

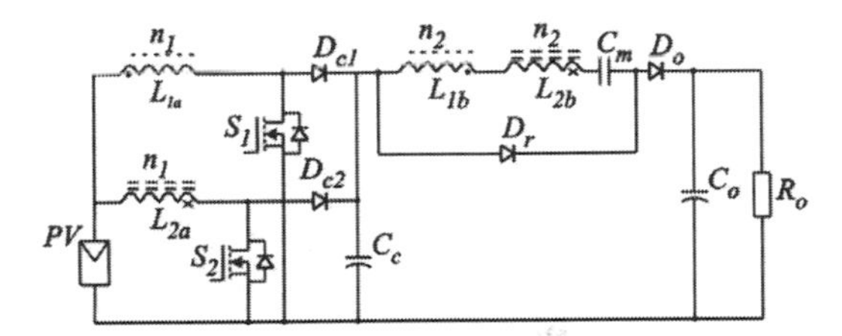

Figure 5. Schematic of the topology of a boost converter with voltage multiplier cell.

splitting the current into N power paths, conduction losses (I^2R) can be reduced, by increasing the overall efficiency when compared to a conventional converter. Moreover, the duty cycle of a single switch does not exceed 1/N, and smaller inductors can be used along with the decrease in the power rating of switches and diodes. Switches are turned on and off one by one when driving sequentially, thereby enabling low output voltage ripples. When the ripples have already been small, further reductions may be unnecessary. And then, it is possible to make some different trade-offs, for instance, reducing the switching frequency by a factor N (to increase the conversion efficiency) and reducing the inductance per cell by the same factor N (reducing the converter size). As a result, the system will present a per-cell ripple that is N^2 times larger than that of a single-cell converter, but the net interleaved ripple will remain the same. Therefore, interleaving is recommended to increase the conversion efficiency and power density associated with the reduction of the ripple current (Garth, 1971).

The main disadvantage of this topology is relatively low voltage gain, usually not higher than two. To improve the voltage gain, interleaved structures can be mixed with transformers (Li, 2010) or the inductors can be coupled (Yu, 2009).

3.3 *Boost converter with coupled inductors*

In non-isolated DC/DC boost converters, coupled inductors can act as a transformer to avoid extreme duty cycles and to reduce the current ripple in high-gain boost conversion (Li, 2009). Fig. 4 shows a high-voltage gain boost converter by using coupled inductors, whose voltage gain is in proportion to the winding turns ratio and can be properly adjusted (Zhao, 2003). These converters can easily achieve high voltage gain by using low R_{DS-on} switches working at a relatively low level of voltage. However, the leakage inductance may lead to high-voltage spikes, which will increase the switch voltage stress resulting in serious EMI (Electro-Magnetic Interference) noise and reducing efficiency. The input current is also pulsating in this case.

Thanks to the active clamp circuit used, the leakage energy can be recycled. A clamping circuit is used to reduce the voltage stress across the main switch because of the leakage inductance of the coupled inductor. This is an approach recommended to low-input voltages and high-voltage step-up since efficiency is typically low because of the voltage ringing, while EMI levels are appreciable because of the pulsating input current (Silva, 2009).

At the kilowatt power level, the power dissipation within the components becomes an important issue especially in case of inductive components. Interleaved solutions can tackle that problem as the input current is shared between the cells. The coupled inductors can be integrated into a single magnetic core to reduce the passive component size.

3.4 *Converter with voltage multiplier cell*

A Voltage Multiplier Cell (VMC) is an electrical circuit that converts electrical power from a lower voltage to a higher voltage, typically by using a network of capacitors and diodes. In Fig. 5, an interleaved boost converter with VMC is presented (Li, 2010). A voltage multiplier cell is inserted in the conventional interleaved boost converter structure that is composed of two diodes, capacitor, and secondary windings of coupled inductors. In such a topology, a high voltage gain can be achieved by adjusting the turns ratio of the two same coupled inductors instead of extreme duty cycles (Li, 2010). It allows good trade-offs between the number of VMCs and the turns ratio of the coupled inductor. By utilizing current-sharing technique by interleaving at the input, this topology allows the use of smaller inductors and lower power-rated switches. Low R_{DS-on} switches can be used to improve the converter performance. The presented circuit works in the turn-on ZCS mode, which can reduce switch losses and EMI noise as well.

3.5 *Non-inductive boost converter*

Without the inductive components, meaning no transformers are applied, there are remarkable benefits in circuit size, cost, weight reduction, and thus the reduction of overall complexity of the converter. What is more, the non-inductive boost converter is able to work at higher temperatures than inductor-based counterparts. In Fig. 6, a new converter without inductive components is presented (Qian, 2011), developed for the requirements of high efficiency and ability to work in high temperatures. The voltage gain is accomplished by VMCs that operate basing on the switching capacitor principle. The disadvantage of this topology lies in the need of the relatively big number of switches, which is 12 in this case. Moreover, due to the capacitive load, the switches are exposed to high current stress. However, the possibility to use low-voltage rated switches and the lack of inductors make it possible to achieve the compact and cost effective solution in the DC–DC boost converter design.

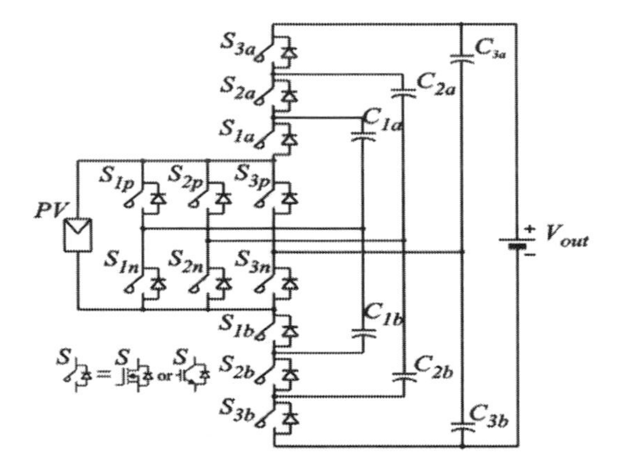

Figure 6. Schematic of the Topology of a non-inductive boost converter.

4 COMPARISON AND DISCUSSION

A brief comparison of the transformerless DC-DC boost converters mentioned in Session III is presented in Table 1 with performance proven in hardware prototype. Those highlighted are the best efficiencies and voltage gains reported in this work. The solution chosen by the designer depends on particular design constraints. The demands should be confirmed first, and then it comes to determine the most robust and best performance topology.

As shown in Table 1, the output powers of these converters are no more than 2.50 kW, which is quite small. The reason may be concluded as that the output power is processed by only two semiconductors and losses become appreciable, especially those regarding the intrinsic resistance of the filter inductor. In this case, most transformerless DC–DC boost converters are not recommended for high-power or high-current applications. In practical projects, interleaving of several converters may be a solution to high-power requirement and reduce the size of filter elements.

For high voltage gain and high efficiency applications, topologies of converters with coupled inductors and VMC are what should be focused on. As given in Table 1, their voltage gains are nearly 10, which are even competitive among converters with transformers.

When compared to non-inductive converters, converters with coupled inductors show great advantages among the aspects listed in Table 1.

Table 1. Comparison of transformerless DC–DC boost converter hardware prototypes.

Topology	Figure	η [%]	P_{MAX} [kW]	*Gain* [V/V]	f_s [kHz]	V_I [VDC]	V_{BUS} [VDC]	Number of switches	Number of diodes
Soft-switching boost converter	2	96.2	0.60	3.0	30	130 to 170	**400**	2	3
Interleaved boost converter	3	**97.3**	2.50	1.3	25	250	320	4	4
Converter with coupled inductors	4	92.3	1.00	**8.0**	100	48 to 75	380	**1**	2
Converter with voltage multiplier cell	5	94.7	1.00	**9.5**	100	40 to 56	380	4	2
Non-inductive boost converter	6	94.0	0.45	5.6	100	12	68	12	**0**

Multiple inductors can be integrated into a single magnetic core in order to reduce the very dimensions of the power circuit. However, it does not mean that non-inductive converters are useless. There is an important parameter, called working temperature, which is not presented in this comparison. As non-inductive boost converters can work in high temperatures, it is worth thinking about for designers when it comes to high temperature applications.

5 CONCLUSION

Start from the basis of the conventional DC–DC boost converter, a series of improved transformerless DC–DC boost converters is introduced and discussed in this paper. These improved topologies are mostly proposed to solve some key issues, such as efficiency, voltage gain, and power handling capacity. Some of the most important techniques typically step the voltage up without the need of extreme duty cycles and may use interleaving of multiple cells to increase the output power levels.

High efficiency of boost DC–DC converters can be achieved by decreasing the duty cycle (lower conduction losses) and reducing voltage stress on switches (cheaper and lower $R_{DS\text{–}on}$ switches) by applying the soft switching technique (minimizing switching losses) and utilizing active clamp circuits (recycling the energy stored in parasitic inductances).

Coupled inductor topology provides a compact design with the features rarely to be found in other topologies. First of all, the energy stored in parasitic leakage inductance of coupled inductors is recycled there. Avoiding the use of electrolytic capacitors improves the reliability of that solution. An active clamp circuit decreases the voltage stress on the switch. And finally, only one switch at a low-side is needed and so the driving scheme is simple.

Topology of using VMC technology is a compact and robust solution. The leakage energy is recycled by utilizing a passive clamp. Due to the voltage multiplier cell, the switch voltage stress is reduced and voltage gain extended. Input current ripples are low and input current sharing is obtained due to an interleaved input structure thereby allowing a lower duty cycle to be used.

REFERENCES

Abusara, Mohammad A., and Suleiman M. Sharkh. "Design and control of a grid-connected interleaved inverter." IEEE Transactions on Power Electronics 28.2 (2013): 748–764.

Chen, Chunliu, Chenghua Wang, and Feng Hong. "Research of an interleaved boost converter with four interleaved boost convert cells." 2009 Asia Pacific Conference on Postgraduate Research in Microelectronics & Electronics (PrimeAsia). IEEE, 2009.

Crews, Ron, and Kim Nielson. "Interleaving is Good for Boost Converters, Too." Available in: www. powerelectronics.com (2008).

Erickson, Robert W., and Dragan Maksimovic. Fundamentals of power electronics. Springer Science & Business Media, 2007.

Garth, Dean R., et al. "Multi-phase, 2-kilowatt, high-voltage, regulated power supply." Power Electronics Specialists Conference, 1971 IEEE. IEEE, 1971.

Hu, Xuefeng, and Chunying Gong. "A high voltage gain dc–dc converter integrating coupled-inductor and diode–capacitor techniques." IEEE Transactions on Power Electronics 29.2 (2014): 789–800.

Kim, E-H., and B-H. Kwon. "High step-up resonant push-pull converter with high efficiency." IET Power Electronics 2.1 (2009): 79–89.

Kjaer, Soeren Baekhoej, John K. Pedersen, and Frede Blaabjerg. "A review of single-phase grid-connected inverters for photovoltaic modules." IEEE transactions on industry applications 41.5 (2005): 1292–1306.

Klimczak, Pawel, and Stig Munk-Nielsen. "Comparative study on paralleled vs. scaled dc-dc converters in high voltage gain applications." Power Electronics and Motion Control Conference, 2008. EPE-PEMC 2008. 13th. IEEE, 2008.

Li, Weichen, et al. "A non-isolated high step-up converter with built-in transformer derived from its isolated counterpart." IECON 2010–36th Annual Conference on IEEE Industrial Electronics Society. IEEE, 2010.

Li, Wuhua, et al. "Application summarization of coupled inductors in DC/DC converters." Applied Power Electronics Conference and Exposition, 2009. APEC 2009. Twenty-Fourth Annual IEEE. IEEE, 2009.

Li, Wuhua, et al. "Interleaved converter with voltage multiplier cell for high step-up and high-efficiency conversion." IEEE Transactions on Power Electronics 25.9 (2010): 2397–2408.

Mohan, Ned, and Tore M. Undeland. Power electronics: converters, applications, and design. John Wiley & Sons, 2007.

Park, Sang-Hoon, et al. "Analysis and design of a soft-switching boost converter with an HI-bridge auxiliary resonant circuit." IEEE Transactions on Power electronics 25.8 (2010): 2142–2149.

Qian, Wei, et al. "A multilevel dc-dc converter with high voltage gain and reduced component rating and count." Applied Power Electronics Conference and Exposition (APEC), 2011 Twenty-Sixth Annual IEEE. IEEE, 2011.

Silva, Felinto SF, et al. "High gain DC-DC boost converter with a coupling inductor." 2009 Brazilian Power Electronics Conference. IEEE, 2009.

Yu, Wensong, et al. "High efficiency converter with charge pump and coupled inductor for wide input photovoltaic AC module applications." 2009 IEEE Energy Conversion Congress and Exposition. IEEE, 2009.

Zhao, Qun, and Fred C. Lee. "High-efficiency, high step-up DC-DC converters." IEEE Transactions on Power Electronics 18.1 (2003): 65–73.

Civil, Architecture and Environmental Engineering – Kao & Sung (Eds)
© 2017 Taylor & Francis Group, ISBN 978-1-138-02985-9

The linear fatigue cumulative damage analysis of the welding seam

Yuyu Wang
Beijing Luoxin Technology Company, China

ABSTRACT: The welding seam's fatigue failure under random vibration is considered; therefore, based on the Miner' rule and rain flow counting method, in this paper, a random vibration analysis of the welding seam is performed by ANSYS Workbench with an analytical model of the vehicle exhaust pipe. Finally, the linear fatigue cumulative damage of the welding seam was concluded. And the analysis method in this paper provides guidance for the welding seam design.

1 INTRODUCTION

The reliability analysis of components under random vibration mainly includes two methods: time domain and frequency domain (Andrew, 1998). The analysis based on time domain features, such as uncertain amplitude, unrepeated vibration waveform, and large volume of data to be processed, namely, for a certain time *t*, the vibration amplitude and frequency are stochastic. However, because the amplitude and frequency have certain statistical laws, the frequency domain analysis method based on the power spectral density is simple in calculation and does not require cyclic counting. The frequency domain analysis method has been broadly applied to estimate the fatigue life of the dangerous location of components in automobile, aerospace, machine manufacturing industries, etc. (Li, 2005).

The welding structure is located where is most vulnerable to fatigue failure, which is stemmed from the defects, such as incomplete penetration, slag inclusion, undercut, cracks, etc. in the welding seam. This "innate" fatigue crack source can directly transverse the fatigue crack initiation stage, thereby dwindling the process of fatigue fracture. The severe concentrated stress and higher residual welding stress in the welding seam will make the welded structure more vulnerable to the fatigue fractures, thanks to the fatigue cracks (Qin, 2013).

The random vibration of the vehicle exhaust pipe during the movement was simulated by using the ANSYS Workbench random vibration analysis module based on the Miner's rule and rain flow counting method. The simulative analysis gave the inherent frequency of the vehicle exhaust pipe and the cumulative damage value of linear fatigue in the welding seam of the vehicle exhaust pipe to provide the ground for the weld design of the vehicle exhaust pipe.

2 MINER'S RULE

In 1945, M A Miner popularized a rule that had first been proposed by A. Palmgren in 1924. The rule, variously called as the Miner's rule or the Palmgren–Miner linear damage hypothesis states that, where there are k different stress magnitudes in a spectrum, S_i ($1 \le I \le k$), each contributing ni(Si) cycles, then if Ni(Si) is the number of cycles to failure of a constant stress reversal Si (determined by uni-axial fatigue tests), failure occurs when:

$$C = \sum_{1}^{k} n_1 / N_i(l, ti = 1, 2, \ldots, k) \tag{1}$$

C is the fraction of life consumed by exposure to the cycles at a different stress level; in general, C is experimentally found to be between 0.7 and 2.2. Usually for design purposes, when the damage fraction reaches 1, failure occurs (Zhang, 2013). This can be thought of as the proportion of life is consumed by using a linear combination of stress reversals at varying magnitudes.

3 S–N CURVE

For fatigue analysis, a test-based S–N curve is needed. Most common equations for representing the S–N curve are power function (Zhao, 1993; Chen, 2005), as given below:

$$\sigma_a = \sigma'_f (2N_f)^b \tag{2}$$

$$\sigma_a = C + \mathrm{D}\log N_f \tag{3}$$

The equation gives a straight line on log-linear coordinates. The commonly used S–N curves are described under GB50017; BS7608-1993 Code of Practice for Fatigue Design and Assessment of Steel Structural; Fatigue Design Recommendations for Steel Structures; AWS D1.1/D1.1M-2006 Structural Welding Codes Steel; The NEW IIW Fatigue Design Recommendations Newly Revised and Expanded (Hobbacher, 2007); Manual of Standard and Recommended Practices Section C Part II: Specifications for Design, Fabrication, and Construction of Freight Cars.

4 RANDOM VIBRATION ANALYSES

4.1 *Radom vibration load*

The random vibration of the vehicle at driving is affected by various factors, including roughness of the road surface, characteristics of tires, characteristics of the suspension system, travel speed, engine operating conditions, the driver's operating level, etc. (Zhang, 2007). The vehicle exhaust pipe mounted on the frame is connected with the supercharger at one end, and connected with the post-processor of the engine at the other end. The exhaust pipe must withstand its own gravity as well as the random vibration load while driving the vehicle.

4.2 *3D model*

Before carrying out the finite element analysis of random vibration, 3D models of parts, such as the vehicle exhaust pipe, square flange, clamp, bracket, and bellows, were built by using Creo software, as illustrated in Fig. 2.

4.3 *Finite element model*

A 3D model of the vehicle exhaust pipe was exported into ANSYS Workbench to set up the fillet weld model as illustrated in Fig. 2.

The material properties of the vehicle exhaust pipe, square flange, clamp, and bracket are defined in Table 1. The bellows connected to the exhaust pipe is simulated by using the spring unit and its property is shown in Table 2.

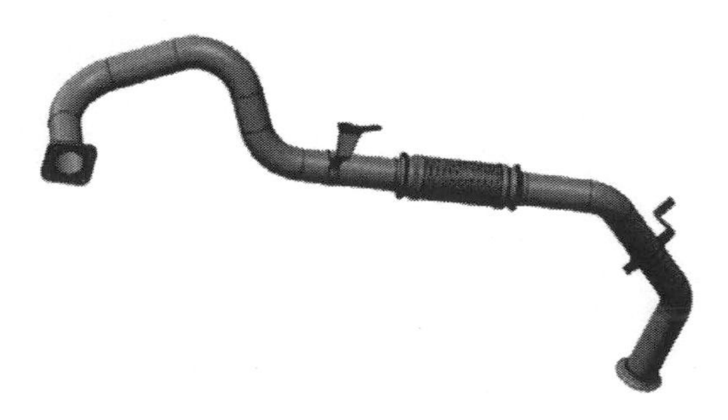

Figure 1. 3D model of the vehicle exhaust pipe.

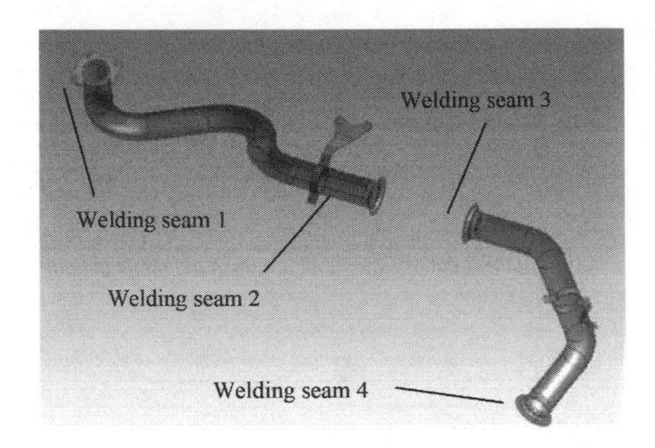

Figure 2. Schematic of the welding seam model of the vehicle exhaust pipe.

Table 1. Material properties.

Material	Young modulus	Poisson ratio	Density kg/m^3
304	200000	0.3	7700

Table 2. Bellow properties.

Axial stiffness	Radius stiffness
18 N/mm	5 N/mm

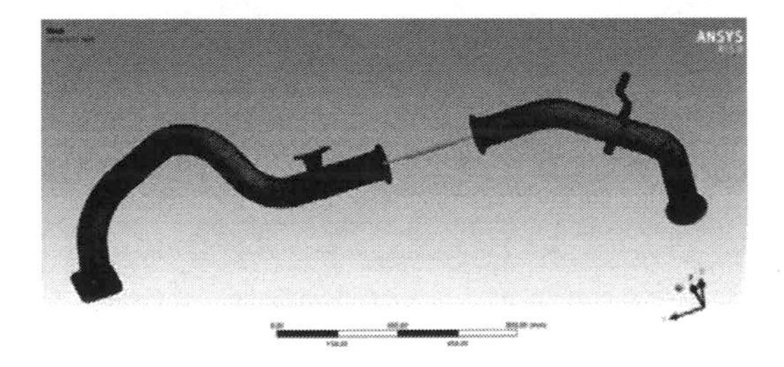

Figure 3. Schematic of the finite element model of the vehicle exhaust pipe.

The exhaust pipe of the vehicle, which is a kind of the thin-plate part, is grid-classified by shell units; The mesh is divided by using the shell unit. The square flange, clamp, and bracket are grid-classified by using the solid units; the number of elements is 35446 and the number of nodes is 42341. The finite element model of the vehicle exhaust pipe is shown in Fig. 3.

4.4 *Modal analysis*

The modal analysis, which is modern method to study the dynamic characteristics of structures, is an important branch in the vibration theory. It is also a system identification method that is applied in the engineering vibration field (Zhang, 2011). The boundary condition of the actual loading condition is defined in the modal analysis for the vehicle exhaust pipe. The bracket of the vehicle exhaust pipe is connected with the frame, the square flange is connected with the supercharger, and the clamp is connected with the vehicle muffler. In the finite element analysis, the mounting holes of the mounting bracket, the bolt holes of the square flange, and the end face of the clamp are set as 0 in the offset value in all directions. The coordinate system is defined as follows: in the X direction that is the vehicle advancing direction corresponds with the braking condition; in the Y direction that is the vehicle's transverse direction corresponds with the camber condition; in the Z direction that is the vertical direction corresponds with the bumpy conditions. The appropriate frequency is given by modal analysis for modalities of all orders at 0–2000 Hz. The inherent frequency of the former six orders in the modal analysis is indicated in Table 3 for the exhaust pipe of vehicles. The vibration plot of the former six orders is shown in Fig. 4.

4.5 *Random vibration analysis*

The modal analysis results are input data for random vibration analysis. The bracket of the exhaust pipe is connected with the frame and its power spectral density function described under ISO16750-3 VII 4.1.3.2.3.2 is used as its load input condition, as shown in Table 4. The square flange of the exhaust pipe is connected to the supercharger and its power spectral density function described under ISO16750-3 VI 4.1.3.2.2.3 is used as its load condition.

The S–N curve at the G level under BS7608 is used as an input, as shown in Fig. 5.

According to the defined coordinate system, the power spectral density function is applied in X, Y, and Z axles at the vibration time of 94 h in each direction, the life at 500,000 km was simulated and

Table 3. Natural frequencies of the vehicle exhaust pipe.

Modal	Natural frequency/Hz
1	54.07
2	87.803
3	110.85
4	128.47
5	156.08
6	172.75

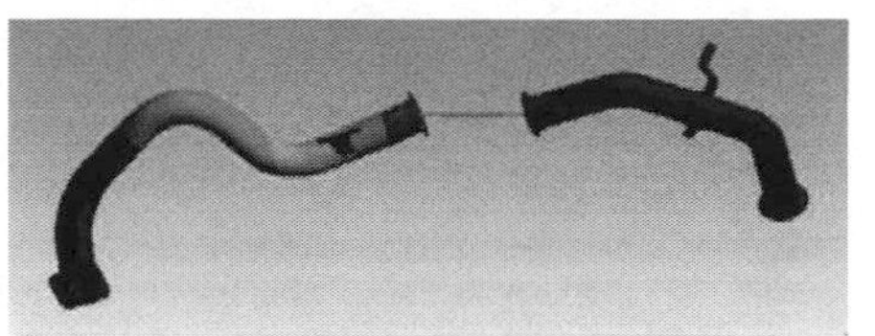

4.1 The first modal shape

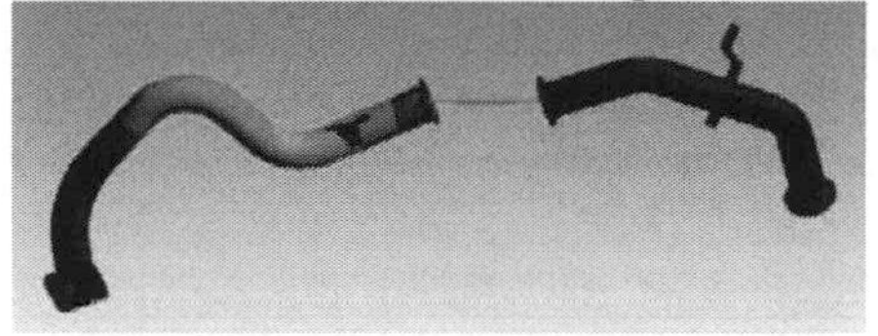

4.2 The second modal shape

4.3 The third modal shape

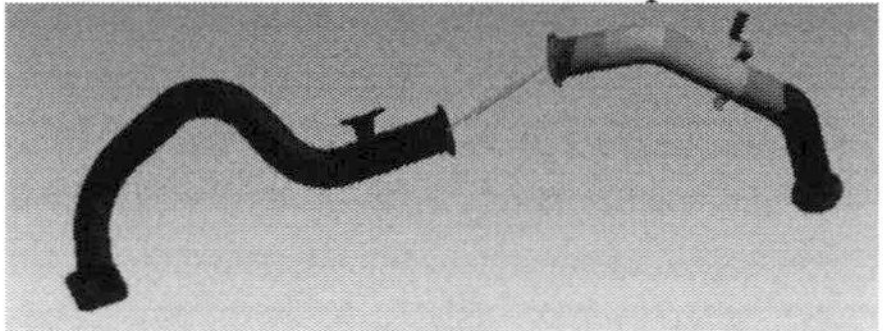

4.4 The fourth modal shape

4.5 The fifth modal shape

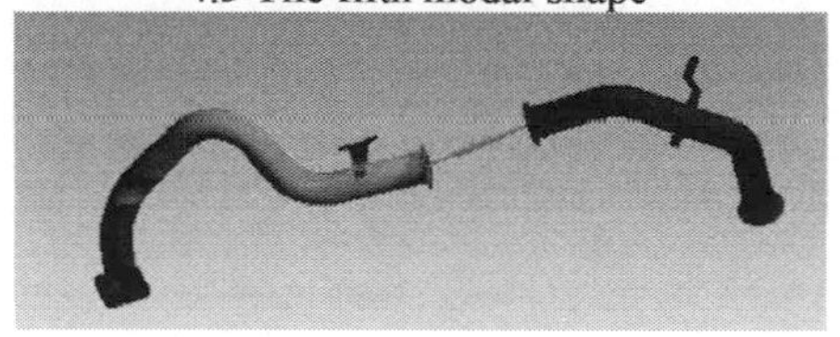

4.6 The sixth modal shape

Figure 4. Schematic of the modal shapes of the vehicle exhaust pipe.

Table 4. ISO16750-3 VII 4.1.3.2.3.2.

Frequency (Hz)	Power spectrum density (G^2/H)
10	18
20	36
30	36
180	1
2000	1

the cumulative damage value at X, Y, and Z directions was calculated. With the finite element analysis, the stress value σ_i of all the welding seams can be given. With comparison of S–N curves of the welding seam, the fatigue cyclic life N_i of the welding seam as appropriate to each stress value can be given. With the rain flow counting method, the times σi under the stress value n_i can be given during the random vibration analysis of the welding seam. With Miner's rule, the accumulative damage of linear fatigue can be given for the exhaust pipe, as shown in Table 6.

The cumulative damage values of linear fatigue of welding seam 1, 2, 3, and 4 are all less than one in the braking direction, camber, and bumpy directions. The maximum cumulative damage value of the linear fatigue appears in the camber direction 0.414 at weld 4. The distribution plot of the equivalent stress of the weld 4 is shown in Fig. 6; the rain flow plot of weld 4 is shown in Fig. 7. The cumulative damage value meets the design requirements as it is less than what is described under the acceptance rule 1.

Table 5. ISO16750-3 VII 4.1.3.2.2.3.

Frequency (Hz)	Power spectrum density (G^2/H)
10	14
20	28
30	28
180	0.75
300	0.75
600	20
2000	20

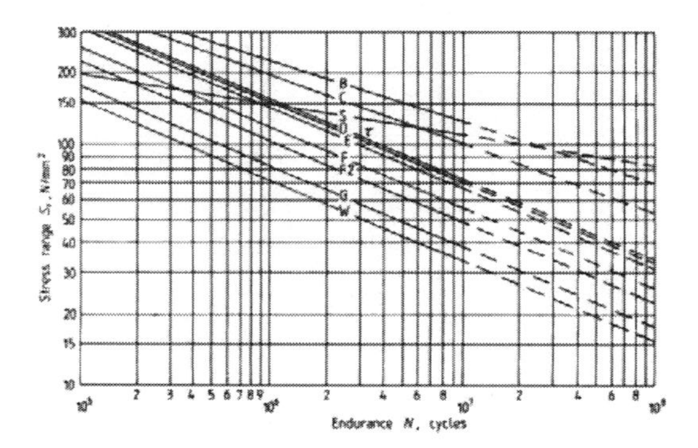

Figure 5. Graph showing the BS7608 S-N curve.

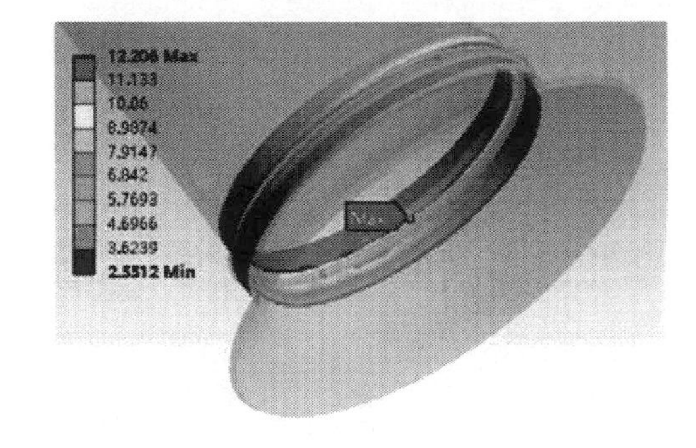

Figure 6. Graph showing the equivalent stress plot.

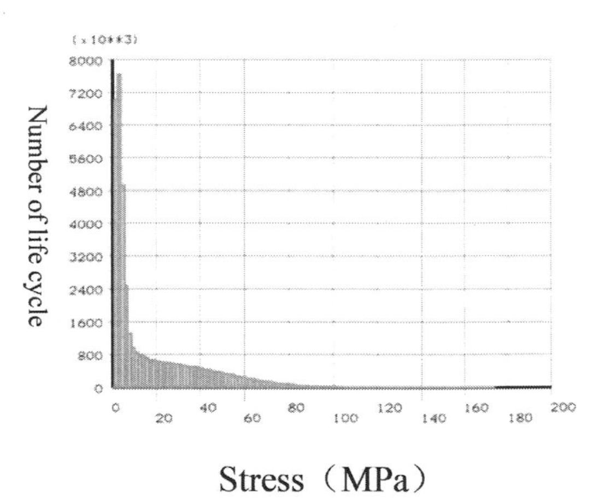

Figure 7. Graph showing the rain flow plot.

Table 6. Linear cumulative damages and frequencies of the welding seams.

	Braking condition x		Camber condition y		Bumpy condition z	
Welding seam	Linear cumulative damage	Frequency (Hz)	Linear cumulative damage	Frequency (Hz)	Linear cumulative damage	Frequency (Hz)
Welding seam 1	7.86e-2	54.1	2.79e-3	54.1	0.139	54
Welding seam 2	7.57e-10	54.1	1.41e-9	156.1	1.29e-6	128.4
Welding seam 3	4.60e-5	110.8	8.45e-5	87.8	2.36e-5	87.8
Welding seam 4	0.25	110.8	0.414	87.8	0.165	87.8

5 CONCLUSIONS

In this paper, a 3D model of the vehicle exhaust pipe is built by using Creo. The modal analysis and random vibration analysis are performed by using ANSYS workbench for the exhaust pipe and the accumulative damage of linear fatigue at the welding seam is given. In this paper, a method is provided for the welding seam design.

REFERENCES

AAR. Manual of Standard and Recommended Practices Section C Part II: Specifications for Design, Fabrication and Construcion of Freight Cars [S].

Andrew Halfpenny. Fatigue Lifecycle Estimation base on Power Spectrul Density signal [J]. China Mechanical Engineering, 1998, 9(11):16–19.

AWS. AWS D1.1/D1.1M-2006 Structural Welding Codes Steel [S].

BSI. BS7608-1993 Code of Practice for Fatigue Design and Assessment of Steel Structural [S].

Chen chuanyao, Fatigue and Fracture [M], Wuhan: Huazhong University of Science & Technology Press, 2005:17–21.

GB 50017-2003, Code for design of steel structures [S].

Hobbacher A F. The NEW IIW Fatigue Design Recommendations Newyly Revised and Expanded [J]. Welding in the World, 2007, 51(1):243–254.

ISO16750-3-2012, Road Vehicles—Environmental conditions and testing for electrical and electronic equipment—part 3: mechanical loads [S].

JSSC, Fatigue Design Recommendations for Steel Structures [S].

Li Chao. A Approach Based on Power Spectral Densiyt for Fatigue Life Estimation [J]. Machine Design and Research, 2005, 21(5):6–8.

Qin datong, Xie liyang. Modern handbook of mechanical design [M]. Beijing, Chemical Industry Press, 2013.

Zhang Li, Lin Jianlong, Xiang huiyu. Modal Analysis and Test [M]. Beijing, Tsinghua University Press, 2011.

Zhang Lijun, He Hui. Vehicle Radom Vibration [M]. Shenyang, Dongbei University Press, 2007.

Zhang yanhua. Welded structural fatigue analysis [M]. Beijing, Chemical Industry Press, 2013.

Zhao shaobian, Wang zhongbao, Fatigue Analysis [M]. Beijing, China Machine Press, 1993:13–17.

Civil, Architecture and Environmental Engineering – Kao & Sung (Eds)
© 2017 Taylor & Francis Group, ISBN 978-1-138-02985-9

Application of wavelet analysis and resonance demodulation in mechanical fault diagnosis

Chenhui Shen & Wei Wang
Ordnance Engineering College, Shijiazhuang, Hebei, China

ABSTRACT: The vibration signal of the mechanical failure is always submerged in the noise signal of the whole system, and it is unable to accurately separate the fault characteristic frequency by frequency domain analysis. The traditional resonance demodulation method is widely accepted as a powerful tool in the fault diagnosis of rotating machinery. The method of wavelet decomposition combined with resonance demodulation is described in this paper, which is used to extract the fault feature. The test data of the fault bearing is collected, and the feasibility of the method is verified.

1 INTRODUCTION

In the technology of condition monitoring and fault diagnosis of mechanical equipment, the research on the theory and method of vibration signal processing is the basis of extracting fault feature and effectively developing fault diagnosis. However, in the running process of rolling bearings, signal acquisition is inevitably interfered by a large number of non-monitoring site vibration, thereby resulting in effective information of submerged bearings; this kind of phenomenon can be observed in the early stage of rolling bearing fault behavior (Su, 2012). Rolling bearing is one of the most widely used mechanical parts in all kinds of rotating machinery, and its running state often directly affects the performance of the whole machine (including precision, reliability, and life span). Rolling bearing is the wearing part; according to statistics, 30% of the rotating machinery fault is caused by the rolling bearing; fault monitoring and diagnosis have been the subject of domestic and foreign technical fields (Wang, 2012). When the bearing undergoes local damage or defect, in the operation process, the damaged parts and other parts collide. This impact will produce attenuation impulse, and the impact force pulse frequency range is relatively wide, so as to stimulate the high frequency vibration of bearings. This natural vibration of high frequency is the carrier of the bearing vibration, and its amplitude should be modulated by the pulse excitation force caused by these defects, so that the ultimate bearing vibration waveform performance is the complex amplitude modulation wave (Wang, 2011). Therefore, the fault characteristic signal can be obtained by using the resonance demodulation method.

2 THEORETICAL METHODS

2.1 *Wavelet multi resolution analysis*

Wavelet analysis is a time–frequency analysis method, in which the window size is fixed but the shape can be changed, namely time and frequency windows may change (Zhang, 2010). The spectrum of the fault signal is refined by multi-resolution analysis, which makes full use of the superior performance of time–frequency localization, which makes the fault feature information more obvious. The low frequency part of the discrete signal is further decomposited by multi-resolution analysis, and the high frequency part is not considered. A three-layer wavelet decomposition tree structure is shown in Figure 1.

2.2 *Resonance demodulation technique*

The collected signals are processed to obtain the fault characteristic frequency based on the principle that the impact force generated by the collision of the faulty parts cause the inherent vibration of the

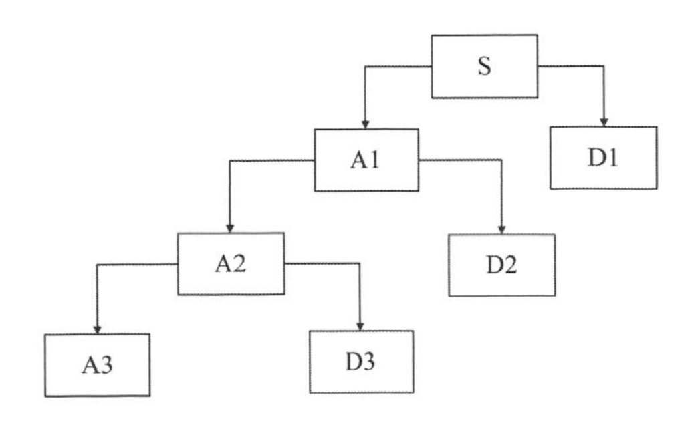

Figure 1. Three-structured of the three-layer wavelet decomposition.

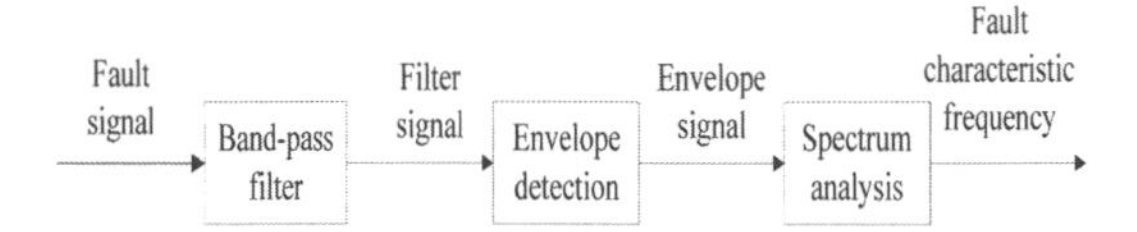

Figure 2. The basic principle of resonance demodulation.

system. The steps of the resonance demodulation technique of the signal are as follows: selecting a frequency band pass filter which is equal to the natural frequency of the selected component. The initial vibration signal collected by the vibration sensor is processed by the band-pass filter, and the natural vibration frequency component is extracted as the research object. Next, we detect the high frequency signal through the envelope detector, to eliminate the high frequency attenuation vibration frequency components, which contains only the rolling bearing fault feature of the low-frequency envelope signal. Finally, the envelope signal is obtained by using the traditional Fourier spectrum analysis, the location of the rolling bearing fault can be determined by using the observation of the spectrum (Wang, 2011; Wang, 2001). The basic principle of the resonance demodulation technique is shown in Figure 2.

2.3 *The tombination of wavelet analysis and resonance demodulation*

The high frequency vibration signal is extracted by using a digital filter through the traditional resonance demodulation method. Due to the fact that the signal collected by the actual sensor is flooded with a lot of noise, the result is not satisfactory. The method of wavelet decomposition and reconstruction can replace the digital filter to extract high frequency signals to achieve better results. Wavelet multi-resolution analysis of the signal is decomposed into a number of frequency signals, through the requirements of the selection of a frequency achieved by the work of filtering.

3 BEARING FAILURE FREQUENCY

The frequency of the rolling element passing the outer ring is given as follows:

$$f_o = \frac{z}{2} f_s \left(1 - \frac{d}{D} COS\alpha\right) \tag{1}$$

The frequency of the rolling element passing the inner ring is given as follows:

$$f_i = \frac{z}{2} f_s \left(1 + \frac{d}{D} COS\alpha\right) \tag{2}$$

The frequency of the rolling element passing on the holder is given as follows:

$$f_b = \frac{D}{2d} f_s \left(1 - \frac{d^2}{D^2} COS^2\alpha\right) \tag{3}$$

d is the diameter of the ball, D is the diameter of the ball center's rotary track, f_s is the active shaft rotation frequency, and α is the contact angle.

4 BEARING FAULT EXPERIMENT

To conduct the single-stage gearbox vibration test, the driving gear Z_1 is set at 36, driven gear Z_2 is set at 28, and the meshing frequency is set at 688 Hz, Bearing parameters are as follows: number of rolling elements $Z = 6$, $d = 10$, $D = 37$, and $\alpha = 60°$, and the active shaft rotation frequency f_s is 19.11 Hz. The outer ring on the active shaft bearing was damaged, as shown in Fig. 4. The acceleration sensor is attached to the bearing seat.

Figure 3. Picture of the single-stage gearbox vibration test.

Figure 4. Picture showing the damage of the outer ring.

Experimental procedure:

- acquisition of the vibration signal.
- the wavelet is used to decompose the collected signal, decompose the high frequency vibration component, and carry out the wavelet reconstruction.
- using the Hilbert transform to extract the signal envelope demodulation.
- spectral analysis of the obtained envelope signal to obtain the fault characteristic signal.

5 EXPERIMENTAL RESULTS AND ANALYSIS

Calculation result: the frequency of the rolling element passing the outer ring f_o is 49.5856 *Hz*, and the frequency of the rolling element passing the inner ring f_i is 65.0811 *Hz*; the frequency of the rolling element passing on the holder f_b is 34.7099 *Hz*.

The Fourier transform of the signal is collected by the frequency spectrum, as shown in Figure 5, from which the fault feature cannot be found directly. As can be seen from the chart, at a frequency of 688 Hz, there is an obvious peak, according to the experimental setup that is that the gear meshing frequency, the useful components required should be in the vicinity of the following two peaks. The useful signal should be in the second layer of the high-frequency frequency band, after the three-layer wavelet decomposition.

The collected signal is decomposed by using the db4 wavelet, and the wavelet is extracted and reconstructed. The reconstructed signal is shown in Fig. 6, and the 2-layer high frequency d2 is extracted as the useful frequency component.

By carrying out the envelope of high-frequency signal by Hilbert transform, the envelope spectrum analysis can be a significant fault characteristic frequency. As shown in Figure7, the fault characteristic frequency is 48 Hz, which is almost equal to calculation 49.5856 Hz, which verifies the feasibility of the method in this paper.

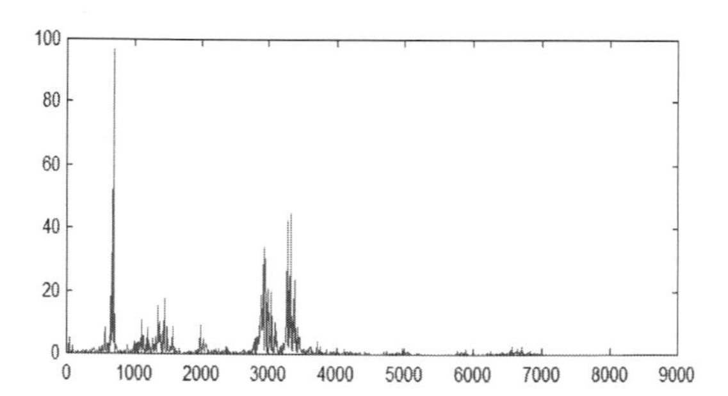

Figure 5. Graph showing the spectrum curve.

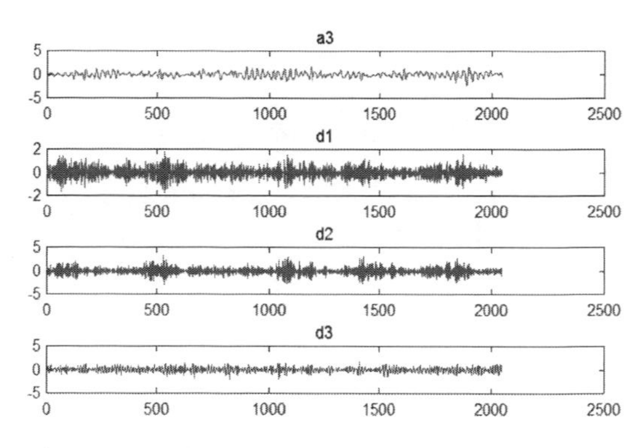

Figure 6. Each layer reconstruction signal after wavelet decomposition.

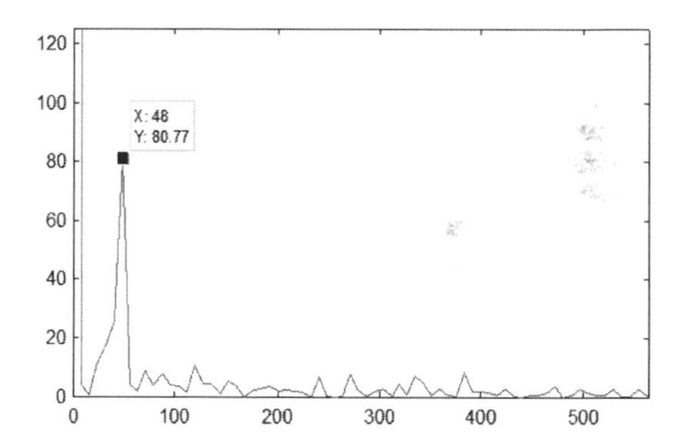

Figure 7. Fault feature spectrum curve of wavelet analysis and resonance demodulation.

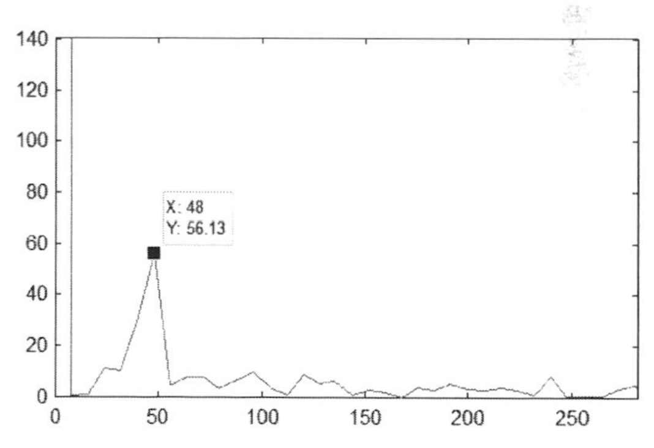

Figure 8. Fault feature spectrum curve of resonance demodulation without wavelet decomposition.

If the signal does not carry out the wavelet decomposition, the direct use of the original signal is to carry on the back of the step; the signal obtained by the curve is shown in Fig. 8. This situation can provide accurate results, but the signal wave crest is smaller than the wavelet decomposition, which fail easily in a lot of interferences.

The different wavelet decomposition coefficients are chosen to carry out the following steps:

Fig. 9 shows the results obtained from the reconstructed signal of the selected d3 coefficients. Fig. 10 shows the results obtained from the reconstructed signal of the d1 coefficients. It can be seen that the choice of the frequency segment has a great influence on the correctness of the results of the resonance demodulation.

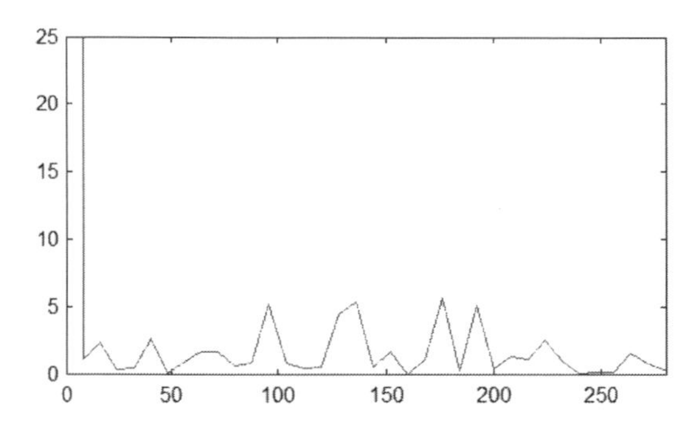

Figure 9. Fault feature spectrum curve obtained from the reconstructed signal of the selected frequency d3.

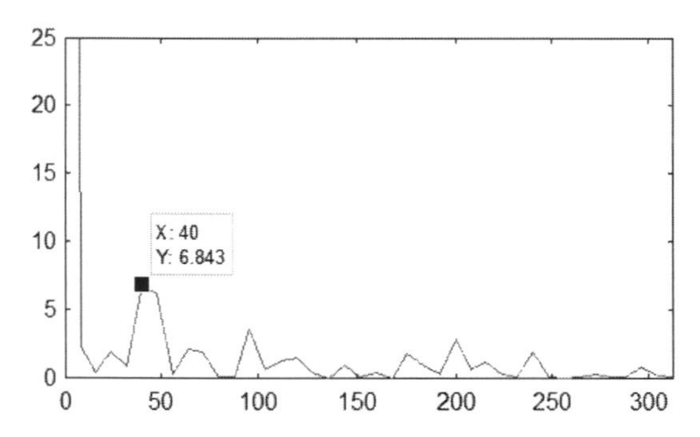

Figure 10. Fault feature spectrum curve obtained from the reconstructed signal of the selected frequency d1.

6 CONCLUSION

In this paper, the feasibility of the method is verified by the method of gear box experimental analysis. However, the selection of the wavelet frequency segment mainly depends on experience and observation, which cannot be accurately obtained. This will lead to inaccurate results. The method needs to be improved to find the frequency segment of the transient impact.

ACKNOWLEDGMENTS

The author thanks Zhichuan Liu for help with the theory and the experiment. The method used in this work was obtained by referring Wang Ziyu's article.

REFERENCES

Wang Ziyu, Kong Fangrang. The extraction of fault feature of rolling bearing-based on resonance demodulation and wavelet analysis [J]. Modern Manufacturing Engineering, 2012, 1, 117–121.

Wang Ziyu. Rolling Bearing's Breakdown Feature Extraction Technology Based on Wavelet Analysis [D], He Fei: University of Science and Technology of China, 2011.

Wensheng Su. Research on Rolling Element Bearing Vibration Signal Processing ang Fearure Extraction Method [D]. Da Lian: Dalian University of Technology, 2012.

Wenyi Wang. Early Detection of Gear Tooth Cracking Using the Resonance Demodulation Technique [C]. Mechanical Systems and Signal Processing, 2001:887–903.

Zhang Xiong-xi, Liu Zhen-xing. Application of resonance demodulation and wavelet denosingin fault diagnosis of induction motors [J]. Electric Machines and Control. 2010, 14(6), 66–70.

Civil, Architecture and Environmental Engineering – Kao & Sung (Eds)
© 2017 Taylor & Francis Group, ISBN 978-1-138-02985-9

A study on the modeling of the slider-crank mechanism based on parameter identification

Xiangjiang Wang, Xiaohuan Tang & Jiang Shen
School of Mechanical Engineering, University of South China, Hengyang, Hunan, China

ABSTRACT: In order to solve the problem of the interval in the slider–crank mechanism and improve its position accuracy, this article will establish a simplified model of the slider–crank mechanism based on its simple mechanism, and make full use of the sensor to measure the movement parameters of each component of the slider–crank mechanism. And then, in this article, the system parameter of the slider–crank mechanism is identified via the parameter identification method, and finally the slider–crank mechanism with interval will be set up and analyzed through the simulation experiment. The comparison results show that the model building via the parameter identification possesses high accuracy and a simple modelling process. Meanwhile, this kind of modelling method can afford a new thinking for the modelling of the complex and non-linear electromechanical system.

1 INTRODUCTION

The slider–crank mechanism is an evolution form of the hinged four-bar linkage and is commonly used in the transformation of the rotary motion of the crank to the reciprocating linear motion of the slider. It has been widely used in engineering practice because of its simple structure, ease of manufacturing, and reliable operation (LIU, 2008). But in the process of design, manufacture, installation, and employment, it has clearance between kinematic pairs in the slider–crank mechanism (Bai, 2011). Meanwhile, because of the external factors, the clearance will produce friction among each component and eventually lead to a decrease in the position accuracy of the slider–crank mechanism and its reliability. Therefore, to improve the position accuracy of the slider–crank mechanism, it is necessary to study and solve the gap problem of the slider–crank mechanism. For now, the common method to solve this problem is to establish its mathematical model and then improve the position precision of the slider–crank mechanism through the relevant method of control theory, such as Ai Zhihao who used the matrix method to establish the dynamics equation of the slider–crank mechanism and its mathematical model is based on Lagrange equation and Hamilton principle (Ai, 2008). Luo Jiman and Sun Zhili established the mathematical model of the slider–crank mechanism from the perspective of kinematics based on the theory of reliability engineering and probability design of the mechanism (Luo, 2002). Zhang You took advantage of the Newton–Euler method to build the mathematical model of the slider–crank mechanism and carry out a simulation analysis (Zhang, 2012). But all of these methods have a lot of disadvantages, such as the complex modeling process, heavy computation, involving many disciplines and so on. Meanwhile, it is difficult to set up their simulation model.

In this paper, the slider–crank mechanism is regarded as the research object, and the motion and connection relationship between its module component and other components are analyzed. At the same time, in this paper, the movement physical quantities of its each component will be measured by using the sensor and then analyze the measuring curve according to the parameter identification method and identify the system parameters, and eventually establish a model of the slider–crank mechanism with clearance.

2 SIMPLIFIED MODELING AND IDENTIFICATION PRINCIPLE

2.1 *Simplified model of the slider–crank mechanism*

Figure 1 shows the experiment equipment of this article, in which the crank length of the experimental device L_1 is 35.8 mm, quality m_1 is 0.32 kg, the center of mass moment of inertia $I_1 = 1185.4\ kg \cdot mm^2$, the length of the hole center on both ends of the link L_2 is 245.88 mm, quality m_2 is 0.175 kg, the center of the mass moment of inertia relative to the link I_2 is 4751.3 $kg \cdot mm^2$, slider quality m_3 is 0.895 kg, the difference semi-diameter of the hinge pin and shaft sleeve between the link and slide block e is 0.5 mm, and the shaft pin radius R is 4 mm.

The whole slider–crank mechanism was driven by using a servo motor. Therefore, the rotation of the servo motor stimulated the whole mechanism. In order to further reveal the motion law of the mechanism under this stimulation, we have to accurately set up its mathematical model.

Figure 2 shows the diagram of the slider–crank mechanism with no clearance, and the crank angle is θ_1, link angle is θ_2; the length of the crank and link are L_1 and L_2 respectively; the slider displacement is L_3.

The closed vector loop equation of the slider–crank mechanism with no clearance is given as follows:

$$\vec{L}_1 + \vec{L}_2 = \vec{L}_3 \tag{1}$$

The vector equation is decomposed on the axis X and Y as following:

$$L_1 \cos\theta_1 + L_2 \cos\theta_2 = L_3 \tag{2}$$

$$L_1 \sin\theta_1 + L_2 \sin\theta_2 = 0 \tag{3}$$

Figure 1. Picture of the experiment equipment.

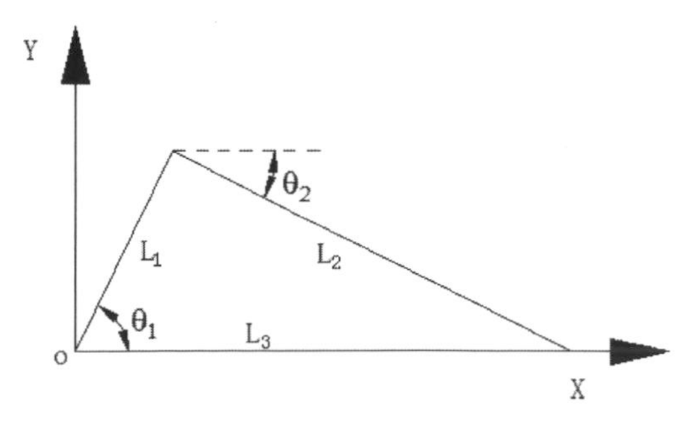

Figure 2. Diagram of the slider–crank mechanism with no clearance.

Figure 3 shows the diagram of the slider–crank mechanism with clearance. The clearance in the location C represents the clearance between the link and slider. Moreover, the clearance of the location C is the biggest in the experiment mechanism and has the greatest influence on the mechanism. While the location A and B also has clearance in the actual mechanism. If considering these clearances, the model of the mechanism, which is established by using the method of mechanism modeling, is too complicated for practical applications. And so, the parameter identification method is beneficial to establish a more precise mechanism model.

Because of the clearance on locations A, B, and C, we assume that the clearances in those three locations are disassembled to the link and the crank respectively in the process of movement, so that the length of the link and the crank not only is a real-time change value but also a function of time. If we can gain the law in which the length of the crank and link is changed with the time, we can receive the movement regulation of the slider through equation (2), and the laws we get must be consistent with the motion rule of the actual slide. Thus, this means that the accurate motion model of the mechanism has been built. This kind of modeling method has great significance for rapidly setting up a mathematic model of the mechanism and understanding its movement principle. Moreover, its modeling process is no longer as complex as the mechanism modeling, and it improves the efficiency and accuracy for the modeling of the complex electromechanical system.

From what we have mentioned above, we can know that the length of the crank L_1 and the link L_2 is the known quantity in the equations (2) and (3), so that the change rule of slider displacement L_3 can be deduced. If we regard L_1 and L_2 as variables and want to get their change rule, we have to rewrite the equations (2) and (3) to equations (4) and (5):

$$[\cos\theta_1 \quad \cos\theta_2]\begin{bmatrix} L_1 \\ L_2 \end{bmatrix} = L_3 \tag{4}$$

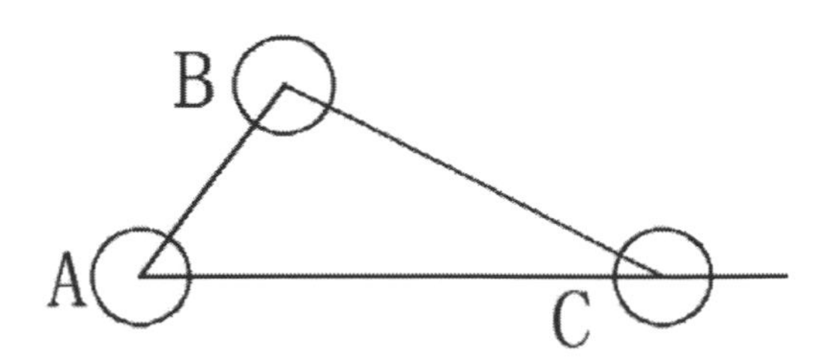

Figure 3. Diagram of the slider–crank mechanism with clearance.

$$[\sin\theta_1 \quad \sin\theta_2]\begin{bmatrix} L_1 \\ L_2 \end{bmatrix} = 0 \tag{5}$$

Comprehensive equations (4) and (5) to equation (6):

$$\begin{bmatrix} \cos\theta_1 & \cos\theta_2 \\ \sin\theta_1 & \sin\theta_2 \end{bmatrix}\begin{bmatrix} L_1 \\ L_2 \end{bmatrix} = \begin{bmatrix} L_3 \\ 0 \end{bmatrix} \tag{6}$$

2.2 *The principle of parameter identification*

Parameter identification is generally defined as that if the input and output of the system is the only thing we know, we can get the important parameters of the system by solving the unknown parameters in the system, so that the original unknown parameters of system can become clearer. From its identification, we can know that we need the specific data of the input and output of the system in order to identifying a system. And then, we can be aware of the types of the model according to the known conditions and also need to understand the identification algorithm; Figure 4 shows the principle of identification.

Figure 4 shows that the input signal of the system P is u(k), the output signal is y(k); v(k) is the disturbance signal, z(k) as the measuring error of the output signal of the actual system, *P* is the error, which is gained from a comparative study between the output of the identification model on the basis of the input and output of the actual system and the measured output of the actual system; e(k) is the input of the identification method. The result is used to improve the model of the identification system.

The method of the least squares identification is commonly used in the identification algorithm; the principle of the least squares identification is shown as follows:

For a general linear system, if we set its variable as y, x_1, x_2, ..., x_n, the parameters needed to be identified are listed as the θ_1, θ_2, ..., θ_n, and the following equation is arrived at (Ding, 2011; Xiao, 2002):

$$y = \theta_1 x_1 + \theta_2 x_2 + ... + \theta_n x_n \tag{7}$$

If the variable y, x_1, x_2, ..., x_n is a function of time, each sampling time is a fixed equation:

$$y(t) = \theta_1 x_1(t) + \theta_2 x_2(t) + ... + \theta_n x_n(t) \tag{8}$$

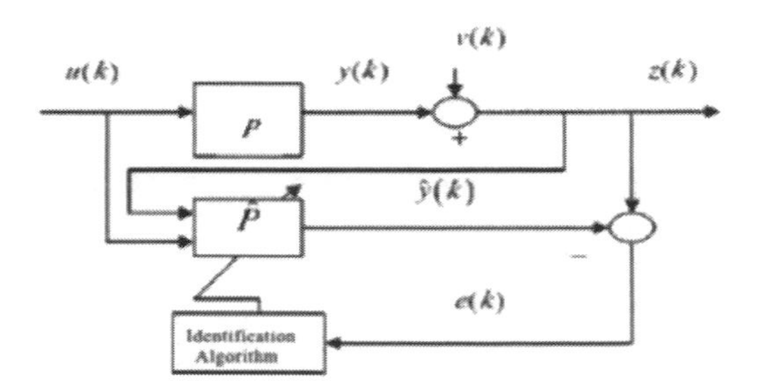

Figure 4. Schematic of the principle of the parameter identification.

3 ESTABLISHMENT AND VALIDATION OF THE PARAMETER IDENTIFICATION MODEL

3.1 *Analysis of the measured results*

To accurately get the law of the length of the crank and link, which change with time, the related sensor is particularly applied to measure the quantity of the corresponding components, as shown in Figure 5, and the data acquisition is shown in Figure 6.

Figure 6 shows the platform of experimental data acquisition, three analog input modules are analog input signals, which is used to collect the output data of sensor for the crank angle, link angle, and slider displacement. Before collecting, the interval time needs to be set, when the parameters of the

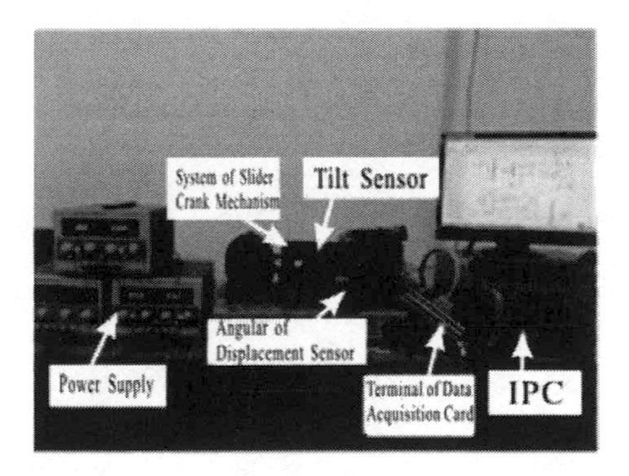

Figure 5. Picture of the object of the experiment device.

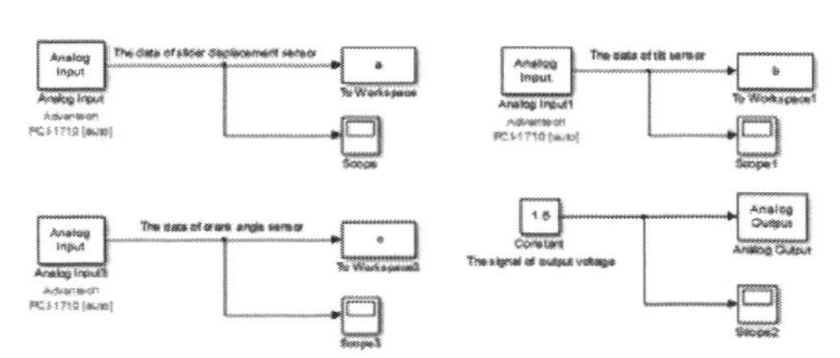

Figure 6. Schematic of the RTW platform of data acquisition.

module in each data point have been received. It is usually set as 0.005 s. Moreover, the voltage of the output analog signal has been set at 0.3V, the speed of the crank is 5 r/min; and acquisition time is 60 s. After successfully connecting the compile platform, the power of the experimental platform and sensor has been switched on. And then, the experiment was started. Finally, the experimental data were saved after the test.

The experimental data obtained from measurement are the voltage signals of the sensor. Therefore, it must be converted into the angle signal by using equation (9) and then fitted of the data by Originpro 8, as shown in Figures 7–9.

$$\theta_2 = \pi(0.25V - 0.625) \tag{9}$$

In this equation, θ_2 is the angle of the link and V is the output voltage of the angle sensor.

3.2 *Modeling parameter identification and verification*

The sine and cosine value of the crank angle and link angle are calculated separately through measuring the crank angle, link angle, and the slider position. After that, the length of the crank and link can be worked out according to equation (8), and the identification results are shown in Figures 10 and 11.

After obtaining the change rule of the length of the link and crank, it can be put into the ideal

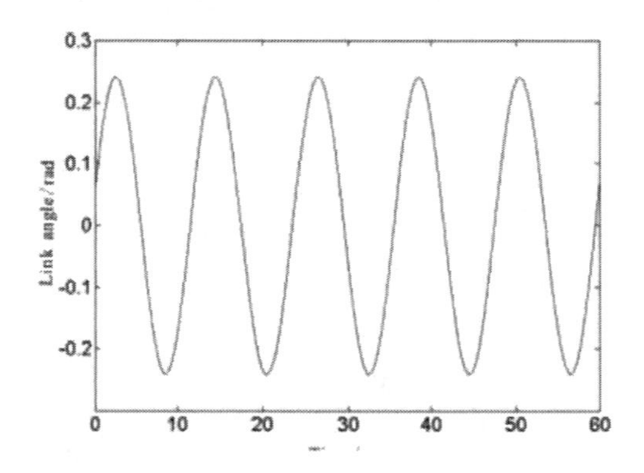

Figure 7. Graph showing the angle curve of the link.

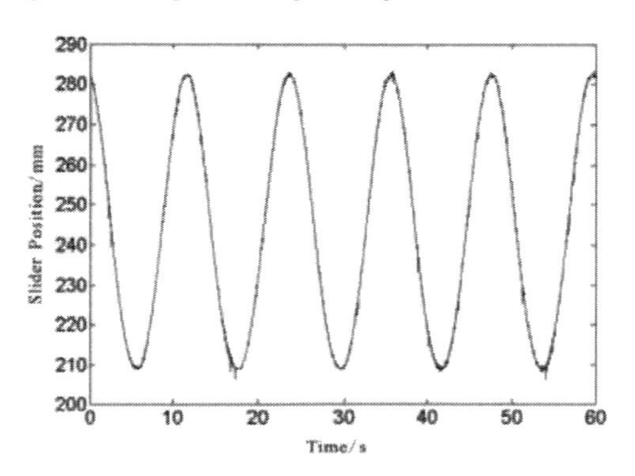

Figure 8. Graph showing the position curve of the slider.

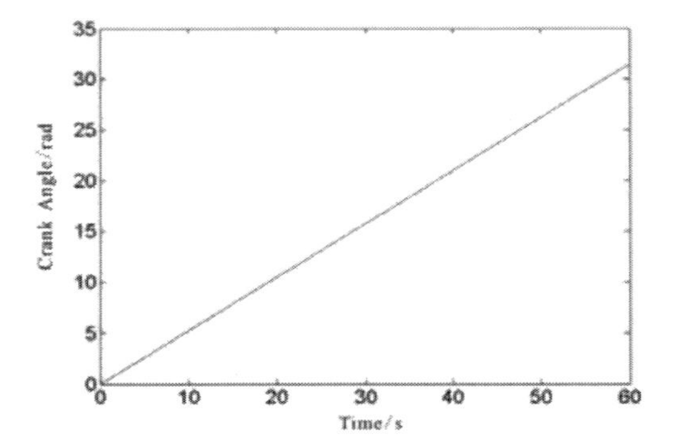

Figure 9. Graph showing the angle curve of the crank.

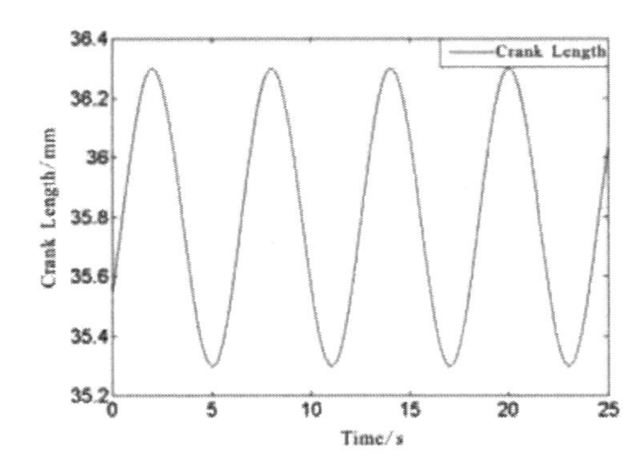

Figure 10. Graph showing the length curve of the crank after parameter identification.

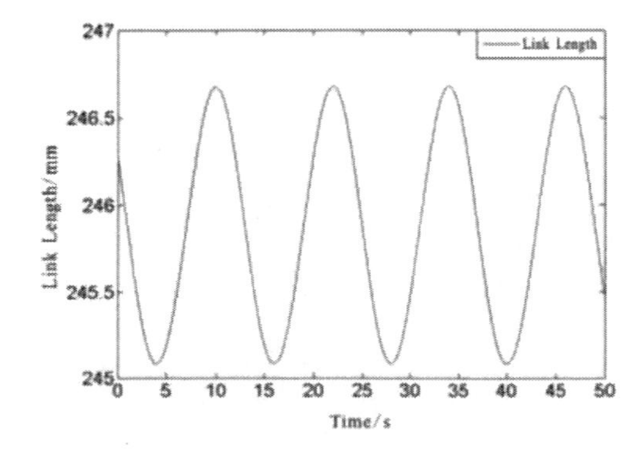

Figure 11. Graph showing the length curve of the link after parameter identification.

model. And then, this kind of law has been integrated into the model calculation. However, before integrating, the initial speed of the servo motor should be set as 5 r/min, and the fixed step of the simulation should be 0.005 s. Moreover, the frequency of the input model in the Workspace also should be 0.005 s. and the simulation time has been set as 60 s. Thus, the simulation has been conducted after setting all the parameters. The simulation platform is shown in Figure 12.

The simulation results should be conformed to the slider displacement in the experiment according to the theory of parameter identification. Figure 13 shows the comparison results between the calculated results and the measured results of the slider displacement.

From Figure 13, we can know that the slider displacement, which is gained from the simulation model after joining the parameter identification, basically agrees with the change trend of the slider displacement in the actual measured graph. However, the error between the farthest position and the nearest position of the slider is big and even the mean square error arrives at 0.520067 mm,

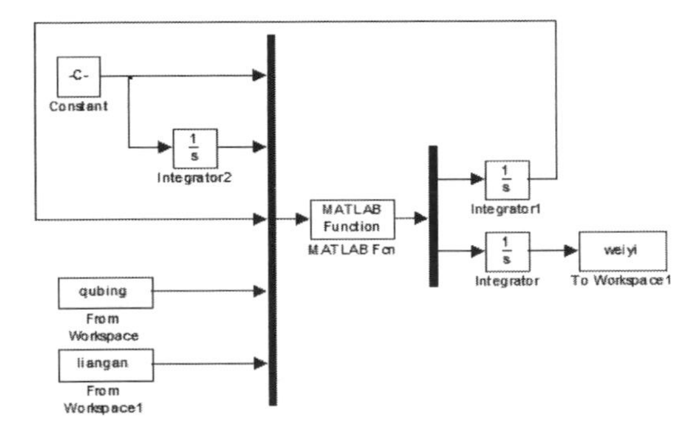

Figure 12. Schematic of the simulation platform of the model after identifying the parameters.

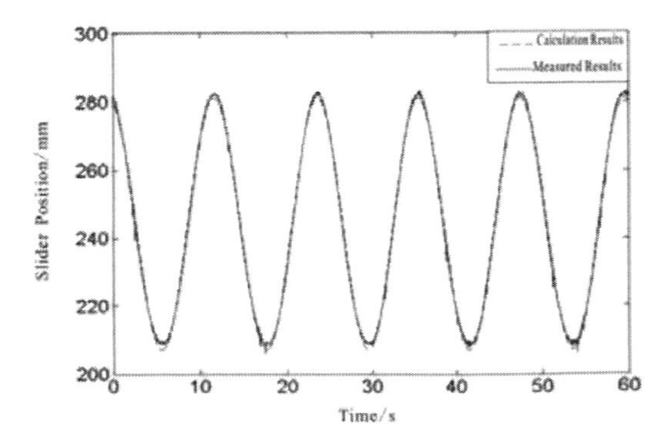

Figure 13. Graph showing the simulation results and measured results of the slider displacement.

which belongs to a reasonable range. The main reasons of this phenomenon are: firstly the too slow rotation speed that makes the shaft sleeve slower and need more time while the shaft sleeve reversely passes through the gap between the shaft pin and the shaft sleeve; secondly, due to the long process of the direction, the farthest position and the nearest position have been in an unstable transient state, so that the data which are collected in the slider displacement are not so accurate; thirdly, even though the calculable details collected from the link was fitted in order to facilitate the calculation, the fitting data still have differences with the original data, which leads to the result of calculation inaccuracy; fourthly, the types of error will be accumulated and even overlaid in the calculation, so that there will be errors between final simulation results and the actual measurement result.

To sum up, the slider–crank mechanism model building by using the parameter identification has more advantages than that of by the method of the mechanism model. Although the mechanism modeling method can be used in the establishment of the model on the basis of the laws and theories of the corresponding subjects, these also have some shortcomings, such as the process of the mechanism modeling is more complicated and the larger amount of calculation by comparing Figure 2 with Figure 14 (Shen, 2015) and the building simulation model is more complex by comparing Figure 12 with Figure 15 (Shen, 2015). This modeling method is not applicable for the complex system and not convenient for the computational simulation, and so the application of this method is limited. However, the method of the parameter identification all depends on the input and output of the system. Thus, the speed of the establishing model is quick and the modeling process is simpler, as shown in Figures 2 and 12. If we just know the simplified model of the mechanism in advance, we can carry on the parameter identification for the system, and get rid of the excessive consumption of the computational expense and time for the complex modeling. This method not only provides a new idea for the mathematical model of the complex electromechanical system, especially the mathematical model system with strong non-linear behavior, but also can be used more widely.

4 CONCLUSIONS

Through the analysis of the connection condition between the components status and the movement principle of the slider–crank mechanism, the clearance position in the slider–crank mechanism has been determined. Meanwhile, the simplified model of the system has been established. Moreover, in this paper, the sensor is used to introduce the measurement of the corresponding physical quantities and obtain its characteristic curve, analyze the characteristic curve of the system by using the method of the parameter identification, so that the model of the slider–crank mechanism is established and the simulation experiment is carried out. This study shows that the results obtained by using the model, which is built by using the parameter identification method, is different from the results gained in the actual measurement. It has errors while the error is in a reasonable range. When compared with the mechanism modeling method, this modeling method is simpler and easier to realize industrialization and has certain engineering application value, thereby providing a new thought for the modeling of the complicated and non-linear mechanical and electrical system.

REFERENCES

Ai Zhihao. Dynamic modeling and parameters estimation of slider-crank mechanism [D]. Hunan University of Technology, 2008.

Bai Zhengfeng. Research on dynamic characteristics of mechanism with joint clearance [D]. Harbin Institute of Technology, 2011.

Ding Feng, Ding Tao, Xiao DeYun, Yang Jiaben. Founded convergence of finite data window least squares identification for time-varying systems [J]. Act automatic science, **05**, 754–761(2002).

Ding Feng. System identification. Part C: Identification accuracy and basic problems [J]. Journal of Nanjing University of Information Science and Technology: Natural Science Edition, **03**, 193–226(2011).

Liu Shanlin, Hu Penghao. A kinetic characteristic analysis of slider-crank mechanism and simulation realization [J]. Machinery Design & Manufacture, **05**, 79–80(2008).

Luo Jiman, Sun Zhili. Research on the motion reliability design model of slider-crank mechanism [J]. Mechanical science and technology, **21(6)**, 959–960(2002).

Shen Jiang, Wang Xiangjiang. Dynamic modeling and error compensation of slider crank mechanism withclearance [J]. Modern Machinery, **06,** 7–8(2015).

Shen Jiang. Modeling and precise motion control considering the gap slider crank mechanism [D]. University of South China. 2015.

Yang Dan. Modeling and control for nonlinear systems with hysteresis [D]. University of south china, 2014.

Zhang You. Dynamic modeling and analysis of crank-slider mechanism with kinematic pair clearance [D]. Harbin Institute of Technology, 2012.

Civil, Architecture and Environmental Engineering – Kao & Sung (Eds)
© 2017 Taylor & Francis Group, ISBN 978-1-138-02985-9

Angle measurement by comparing the amplitude technique of the sum-difference pattern radiometer

Hong Wang & Jing Xiao
School of Electronic and Optical Engineering, Nanjing University of Science and Technology, Nanjing, China

Bo Bi
Key Laboratory of Science and Technology on Millimeter-wave Remote Sensing, Beijing, China

ABSTRACT: The glint problem existing in amplitude comparison of the mono-pulse radar leads to angle-measuring errors. However, the radiometer detects the target only by using the radiation energy, which is not affected by such a problem. In this paper, the design principle of the sum–difference patterns of the three-channel millimeter wave radiometer on the W-band, and the angle measurement technique of the radiometer are mainly discussed. By carrying out the experiment of detecting airborne metal targets, the radiometer recognizes the target radiation and pinpoints it. The center of the target is identified from the sum channel signal, and the pitch and azimuth angle are measured by comparing the signals of two difference channels to that of the sum channel.

1 INTRODUCTION

One of the most widely used mono-pulse directional method is amplitude sum–difference patterns (Sherman, 1984). The biplane mono-pulse radar of the amplitude sum–difference patterns tracks the target effectively by using the three channel receiver's in a unique manner with high angle measuring accuracy and strong anti-interference performance (Zhou, 2014). But in the near field, the glint problem affects the angle measurement accuracy, which cannot be overcome for the tracking radar (Liu, 2013; Ma, 2009). The fast median filtering algorithm of the small window can make some suppression in the glint problem (Guo, 1996; Dunlop, 1981). In this paper, the use of the millimeter wave radiometer to measure the angle can completely compensate the glint problem in the near field. In this paper, the design and identification principle of the sum–difference patterns of the three-channel millimeter wave radiometer are introduced, and verifies it by carrying out an experiment of detecting airborne metal targets. Finally by carrying out digital signal processing to the received signal, we can identify the center of the target and get the pitch and azimuth angle of the target.

2 THE PRINCIPLE OF SUM–DIFFERENCE PATTERNS MILLIMETRE WAVE RADIOMETER

2.1 *The design principle of the millimetre wave radiometer*

The radiometer receives the radiation energy of the natural material (Sato, 2008). The detection system of the millimeter wave receives the millimeter wave thermal radiation energy of the electromagnetic spectrum object emitted by the radiometer of high sensitivity. And then, target recognition can be realized by the energy difference of the target and background (Li, 2009). In this paper, we design an AC all-power radiometer of sum–difference patterns at 3 mm.

The all-power radiometer can be divided into two major categories of DC coupling and the AC radiometer according to the structure characteristics. The DC coupling radiometer makes the DC level after detection as the signal of identify target directly. The AC radiometer makes the DC level through a blocking capacitor, what is used to determine the target is the exchange coupling signal. An alternating current radiometer is more

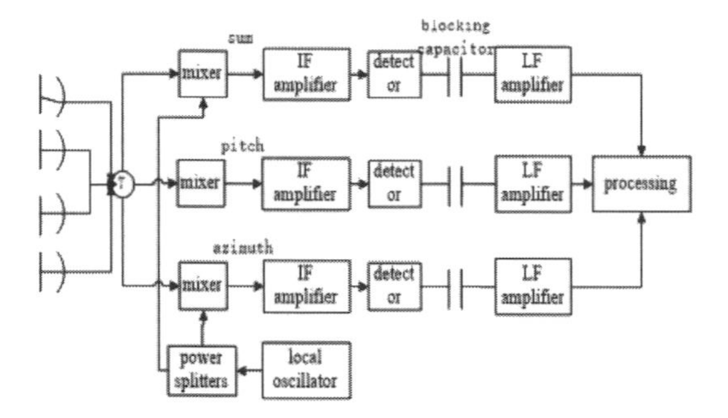

Figure 1. The schematic diagram of the AC all-power radiometer of sum–difference patterns.

likely to achieve as eliminating the calibration of the DC level when compared with direct current radiometer. The body of the sum–difference pattern radiometer is a three-channel radiometer, which is implemented by the antenna and the sum–difference network of the wave-guide. The schematic diagram of the AC all-power radiometer of sum–difference patterns is shown in Figure 1.

The AC radiometer of sum–difference patterns connects three radiometers by magic T, shares a local oscillator through the power splitters. The received electromagnetic wave of four antennas enters into the magic T. After sum–difference conversion, sum and differential signals can be obtained and then be put into three radiometers, respectively. In the radiometer of a single channel, the mixer converts the RF signal down to the IF signal. And then, by amplification and detection, the AC component can be extracted through the blocking capacitor, which is the required signal processing analysis.

The output voltage after low amplification of a single channel radiometer is given by the following equation:

$$U_d = C_d G k B \Delta T_a \tag{1}$$

where C_d is the detector sensitivity, G is the total gain, k is the Boltzmann constant, B is the bandwidth, and ΔT_a is the apparent temperature of the target and background temperature difference in the antenna. What we monitor and process is just the output voltage of the radiometer.

The antenna beams, respectively, scan the target and background back and forth. And then, the sum channel radiometer outputs the energy changes of four beams and the difference channel radiometers respectively output the energy changes in the pitch and azimuth plane. The recognition method of the sum channel of the sum–difference radiometer is the same as the ordinary radiometer. But it is different from the conical scan radiometer, which is pinpoint target by the conical scanning. The sum–difference radiometer obtains the orientation of the target in a plane through two differential energies of the beam.

2.2 *The principle of identification*

The radiation energy of metal objects detected by using the millimeter wave detection system mainly comes from the environmental radiation energy reflected by the metal. The reflectivity (Kim, 2013) and the environmental radiation energy are related to the detection angle, and so the metal objects radiation energy is a function of the angle.

The objects emissivity ε is in a linear relationship with the reflectivity Γ and transmittance τ.

$$\varepsilon + \Gamma + \tau = 1 \tag{2}$$

Because the transmittance of most objects is small, it can be approximately regarded as zero, and then

$$\varepsilon + \Gamma = 1 \tag{3}$$

As shown in Fig. 2, assuming that the four antenna beams respectively are 1, 2, 3, and 4, the square represents the metal objects, the sky temperature is 50 k, and the ground temperature is 300 k (room temperature is 23°). The metal target can be seen as a non-emitter. And so, in the analysis, we can approximately think that the emissivity is $\varepsilon_T \approx 0$, and the reflectivity is $\rho_T \approx 1$.

When surface-to-air metal target identification occurs, assuming air without a cloud, when to detect a target, the temperature near the antenna is given by the following equation:

$$T_{BT} = \varepsilon_T \cdot T_T + \rho_T \cdot T_g = \rho_T \cdot T_g \tag{4}$$

Not to detect a target, the temperature near the antenna is given by the following equation:

$$T_{Bs} = T_s \tag{5}$$

The temperature contrast of the metal target and sky is given by the following equation:

$$\Delta T_T = T_s - T_g \tag{6}$$

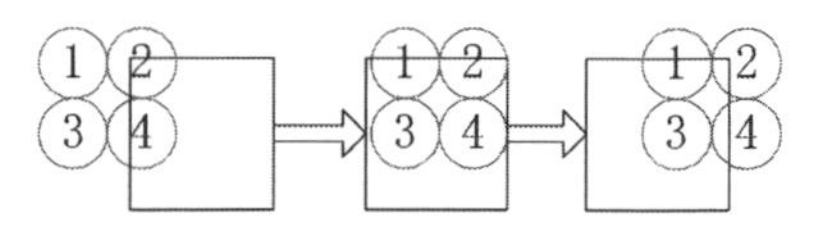

Figure 2. Schematic of the process diagram of the antenna scanning the target.

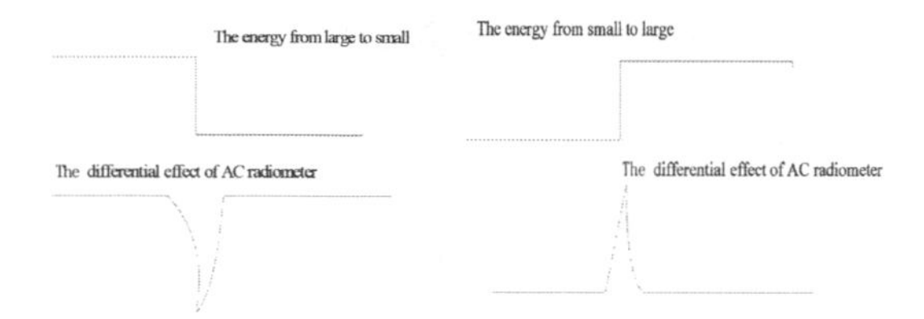

Figure 3. The differential effect scheme of energy change.

Figure 2 is the process diagram of the antenna scanning the target. The sum channel energy is $\Sigma = 1+2+3+4$, azimuth difference energy is $\Delta_a = (1+3)-(2+4)$, and pitch difference energy is $\Delta_p = (1+2)-(3+4)$. When compared to the metal, the sky is cold. Hence, in the process of the antenna close to or far away from the goal, Σ is from low to high and high to low, the target center can be obtained from its peak. Δ_a is obtained from the negative energy gradually to the positive energy. Δ_p is always negative energy, but the size is changing. Therefore, azimuth and pitch angle can be obtained respectively by using the sum–difference amplitude radio.

The energy change presented on the AC radiometer is a differential process, and so the result is a positive or negative pulse, as shown in Figure 3.

3 RADIATION SIGNAL MEASUREMENT AND ANALYSIS

The experiment with radiometer and metal plane mode is carried out to verify the identification ability of the sum–difference patterns millimeter wave radiometer. As shown in Figure 4.

First, by hanging the metal plane model in the air, the radiometer detects the target after aiming at the plane. Due to the antenna followed by magic T, the received sum and difference energy inputs into three receiver channels, respectively. After the frequency conversion and amplification, the signal waveform can be monitored by using an oscilloscope. The radiometer detects the energy difference between the metal plane and the sky. We can monitor the signal waveform of three channels by using an oscilloscope, which is shown in Figure 5. Figure 5(a) is the sum channel signal, (b) is the azimuth difference channel signal, and (c) is the pitch difference channel signal.

We can see from Fig. 5 that, once the radiometer detects the metal plane, it will output a signal. And the output voltages of three channels are different. The peak value of the sum channel signal is $V_\Sigma = 2.4V$, the peak value of the azimuth channel signal is $V_a = 500mV$, and the peak value of the pitch channel signal is $V_p = 700mV$. Therefore,

Figure 4. Picture of the experimental scheme.

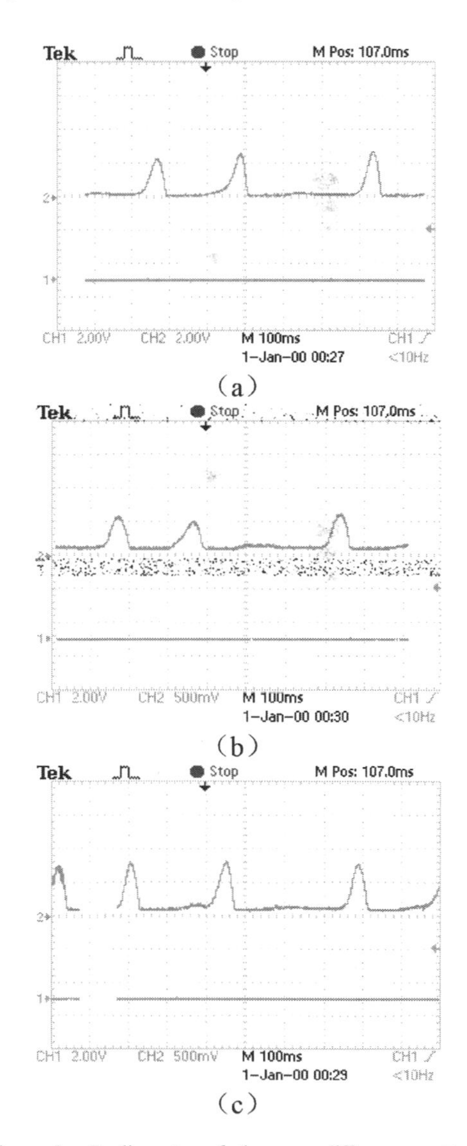

Figure 5. Radiometer of the sum–difference patterns signal waveform.

the amplitude of the sum channel signal is greater than the difference, and the amplitude difference between two signals reflect the angle target deviation from the center of the antenna. To measure the angle, we make the peak time of the AC energy as a reference point. According to the principle of the sum–difference radiometer angle measurement, the target angle can be calculated and determined.

After the target radiation signal required, we perform digital signal processing, thereby choosing DSP F2812 (Han, 2009) chip as the core chip. The total signal processing board is composed of DSP F2812 and its peripheral circuit. First, signal sampling is carried out by the AD module of 2812 to transform the analog signal to the digital signal. In the AD sampling, the choice of the sampling rate is a key problem.

The calculated equation of the sampling rate is given as follows:

$$f_s = \frac{30M \times (DIV/2) \div (HISPCLK \times 2)}{(T1PR + 1) \times 2^{TPS}} \tag{7}$$

where, DIV is the setting of the PLLCR register to change the PLL double-frequency coefficients of the CPU. HISPCP is the setting of the SysCtrlRegs register to change the prescaler standard of the high speed clock. T1PR is the setting of the EvaRegs register to change the value of the cycle register T1PR. TPS is the setting of the T1CON register to change the clock frequency of the timer. The T1PR is often changed to alter the AD sampling frequency as it is simple and easy. And the sampling frequency is set according to the input analog signal frequency.

After AD sampling, the digital value corresponds to the analog voltage and it can be calculated as follows:

$$digital = 4095 * (\text{anolog} - \text{ADCLO}) / 3 \tag{8}$$

For once experimental signals, the sampling results of sum, azimuth, and pitch channel are respectively displayed on CCS and are shown in Fig. 6(a)–(c). Also, the sum channel signal is greater, while the two difference signals is smaller and have some difference. We can see from Figure 5 that there is some noise in the signal.

After digital sampling, signal processing can be performed on the digital signal. First, for the sum channel signal, we find the first signal by the rising edge. And then, we continuously compared the signal value to find the maximum value until the fall edge. The fall edge is determined by the consecutive declining three signal values. Finally, if the maximum value reaches or exceeds the threshold value, the peak value can be required. Also, we can

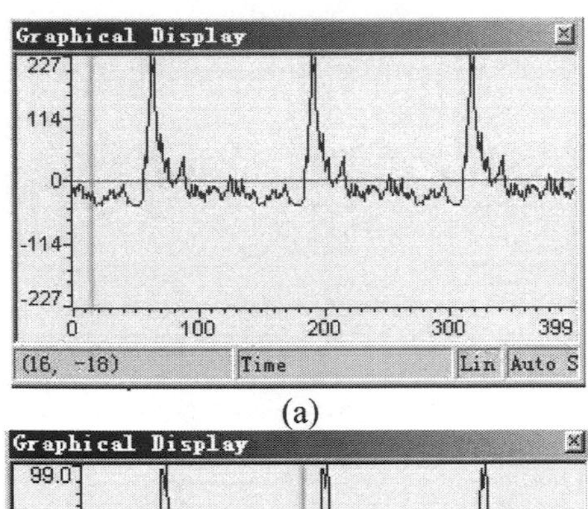

(a)

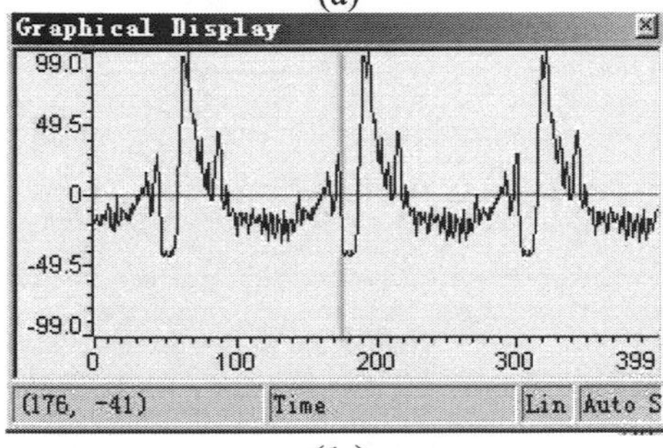

(b)

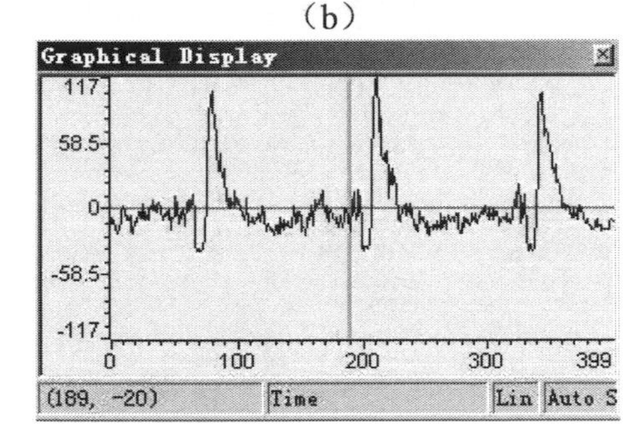

Figure 6. Sum–difference channels' sampling signal.

output a pulse signal in each peak position, and then the corresponding recognition signal is generated, as shown in Figure 7. We know that, when the antenna aims at the center of the target, the received energy is the maximum. Therefore, the target center can be determined by the maximum energy of the sum channel.

And then, we need to determine the deviation angle of the antenna. We can get the peak value of two difference channels the same as the sum channel. And then, the sum channel signal as a reference signal, respectively contrasted with the signal amplitudes of two difference channels, we can obtain the amplitude ratio. And then, azimuth and pitch angle can be calculated according to the corresponding relationship between the amplitude ratio and angle. The angle measuring result is shown in Figure 8. The azimuth angle a = 5°

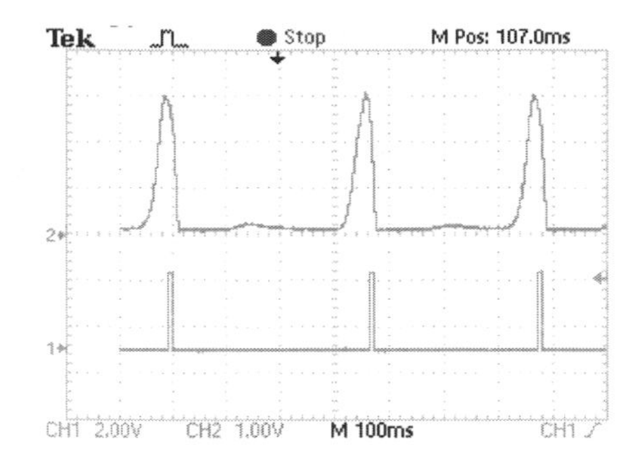

Figure 7. The recognition signal in the sum channel.

Name	Value
x	494
anglea	5
anglep	9

Watch Locals | Watch 1

Figure 8. Angle measuring results.

and the pitch angle p = 9°. Figure 6 shows that the peak value of the azimuth channel is a little less than the pitch channel, and so the azimuth angle deviation is less than the pitch angle in the result.

4 CONCLUSION

The millimeter wave radiometer of sum–difference patterns eliminate the influence of the mono-pulse glint problem and improve the angle measuring precision. The experiment of the radiometer detecting the air metal target verified the directional ability of the system.

REFERENCES

Dunlop, A.J. AGC response and target glint. Communications, *Radar and Signal Processing, IEE Proceedings F.* **182**, 83 (1981).

Guoying C. A statistical model of radar target glint based on discrete differenced gaussian noise. *Antennas and Propagation Society Int. Symp.* 1418 (1996).

Han F. *TMS320F281xDSP Principle and application technology*. (Beijing, Qinghua university press, 2009).

Kim, S., J. So, et al. Characterization of material emissivity using 4-Stokes W-band radiometer. IGARSS, 2013 IEEE International. 3014 (2013).

Liu H., Yang X. and Jiang H. The study of mono-pulse angle measurement based on digital array radar. *Radar Conference* 2013, IET Int. 1 (2013).

Li X., Li Y. *The basement of millimeter wave near sensing techniques*. (Beijing, Beijing institute of technology press, 2009).

Ma Z., Cui W. Angle measurement performance analysis and simulation of sum and difference phase-comparison monopulse radar. *Radar Conference, 2009* IET, 1(2009).

Sato, M. Compact receiver module for a 94 GHz band passive millimetre-wave imager. Microwaves, Antennas & Propagation, IET. **2**, 848 (2008).

Sherman S M. *Mono pulse principles and techniques*. (Dedham, MA: Artech House, 1984).

Zhou, W., J. Xie, et al. Angle measurement accuracy analysis of sum-difference amplitude-comparison monopulse in onshore or shipborne ISAR. *DSP 2014 19th Int. Conf. on IEEE.* 531 (2014).

Civil, Architecture and Environmental Engineering – Kao & Sung (Eds)
© 2017 Taylor & Francis Group, ISBN 978-1-138-02985-9

Tension control methodology of active continuous discharge and intermittent reciprocating work based on master-slave synchronous motion

Lei Han & Hong Hu
Harbin Institute of Technology Shenzhen Graduate School, Shenzhen, China

Wenhua Ding
Shenzhen Colibri Technologies Co. Ltd., Shenzhen, China

Jiyu Li
Harbin Institute of Technology Shenzhen Graduate School, Shenzhen, China

ABSTRACT: Tension control is the key methodology for the lamination process of lithium batteries. However, the isolation film is easily broken when the tension is too great, and when the tension is too small it is prone to wrinkle retraction resulting in an irregular phenomenon. The paper describes the establishment of a model according to the structural characteristics of the lithium battery lamination machine, and the building of an experimental platform. The control strategy was designed by using active unwinding methodology, combined with the classic PID control method for experimental verification. The motion of the platform was used as the main axis, and the motor of the storage material axis and the other servo motors were used as the slaver axis. Compared with the existing passive unwinding methods, this novel active continuous unwinding and master-slave synchronization method realizes high speed continuous unwinding, and keeps the tension stability in the process of intermittent reciprocating lamination.

1 INTRODUCTION

Tension control methods are applied to various fields of industry, and over the years many advanced control algorithms have been proposed. The first such one to appear in the literature (Lee, 2012) proposed an advanced taper tension control method. This established the mathematical model of stress changes that the material belt section went through in the transmission process, and designed a pair of rollers to control strip force in the process of transmission by adjusting the single stick angle: the system of tension was controlled by a passive tension swing rod on the winding production line to enhance performance. Later studies (Chen, 2004) proposed an algorithm of tension control for the web feed machine based on the sliding mode control, and a further study (Park, 2014) used different fuzzy adaptive gain control methods to control the winding tension.

Synchronization motion control commonly used in master-slave synchronization technology for the design and study of chaotic systems can also be successfully applied. The literature (Wua, 2006) proposed two non-autonomous horizontal platform systems of master-slave synchronization control, by proving the criteria of a general coupling matrix to derive the special coupling matrices by a linear state error feedback controller. Another study (Ding, 2011) proposed two horizontal master-slave synchronization platforms based on time delay feedback control, and in yet another study (Salarieh, 2009), the master slave synchronization technique was extended to the multiple synchronization problems, and the algorithm was designed by a feedback control strategy. Later studies (Yang, 2015) established the thrust equation of the linear motor. Other researchers (Albert, 2007) established the dynamic mathematical model of the traditional AC synchronous servomotor and linear motor dynamic mathematical model, and found that the linear motor has better positioning accuracy of the profile than a ball screw. The rationality of trajectory planning directly affects the vibration of the whole system. Studies (Liang, 2014; Li, 2009) have optimized the trajectory of the linear motor. The main function of the feedforward control is to predict the development trend of the signal, and the master axis is often ahead of the slave axis in the master-slave synchronous motion because of the time lag. In the high-speed motion of the linear motor, it is called the tracking error. A study (Chen, 2007) introduced the feedforward gains of speed and acceleration, which can forcefully enhance the tracking ability of the repetitive controller and improve on the errors of the system. The feedforward control rate is introduced into the industrial system to improve the overall performance of the system (Kang, 2010; Zheng, 2016).

This paper studies the tension control method of the lithium battery lamination machine, in which there is a continuous active unwinding and an intermittent reciprocating motion of the laminated platform. Subsequent chapters are arranged as follows: second section analyzes the lamination machine structure characteristics, and establishes the kinematics model of the whole system. The third section focuses on the design and study of the control strategy, using the traditional PID control algorithm to achieve the variable speed unwinding. The fourth section describes the experimental platform for verifying the accuracy of control strategy and the mathematics model. The research results from this study have been applied in practice for production equipment.

2 SYSTEM MATHEMATICAL MODEL

Figure 1 shows the laminate machine experimental platform working principle diagram for this study. It mainly comprises the unwinding roller, the tension roller and its storage shaft, the laminate platform and its lifting part, and the pressure knife and an isolating membrane. The main goal of the research is to control the tension of the isolation membrane so that it is stable when the laminate platform is working at high speed. This section of the paper mainly analyzes the structural characteristics of the lamination machine and establishes its kinematics mathematical model as well as part of the dynamic mathematical model. Software is used to simulate the system's motion.

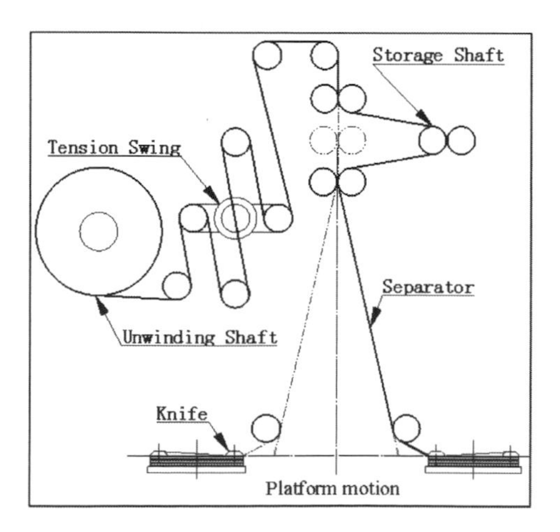

Figure 1. Working principle of the experiment platform.

2.1 *Unwinding mathematical model*

Changes in the unwinding diameter need to be analyzed because the changes occur in real time in the process of unwinding. The unwinding diameter and isolation membrane motion diagram is shown in Figure 2, while the flattened isolation membrane is shown in Figure 3.

The total length of the initial isolation membrane can be calculated according to the diameter of the coil and the thickness of the isolation film:

$$L_0 = (\pi[(D/2)^2 - (D_0/2)^2])/m \quad (1)$$

The relationship between the volume and the linear velocity of the film:

$$D = 2\sqrt{(L_0 - v_2 t)m/\pi + (D_0/2)^2} \quad (2)$$

Under the premise of ensuring uniform speed, the relationship between the volume change of the roll rate and the roll rate is further deduced:

$$n = 60v_1/\pi \cdot D$$
$$= 60v_1/\left(2\pi\sqrt{(L_0 - v_2 t)m/\pi + (D_0/2)^2}\right) \quad (3)$$

2.2 *The mathematical model of tension swing rod*

The tension swing rod is mainly used to control the tension of the whole system at a constant tension with the help of PID adjustments and the initial unwinding diameter calculation. The force diagram is shown in Figure 4.

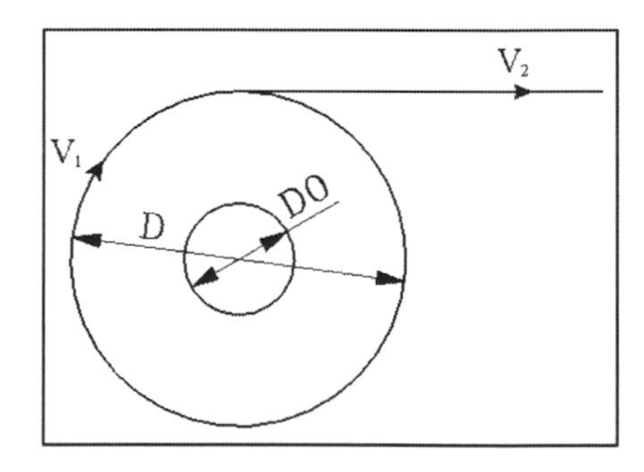

Figure 2. Roll dimension and membrane motion diagram.

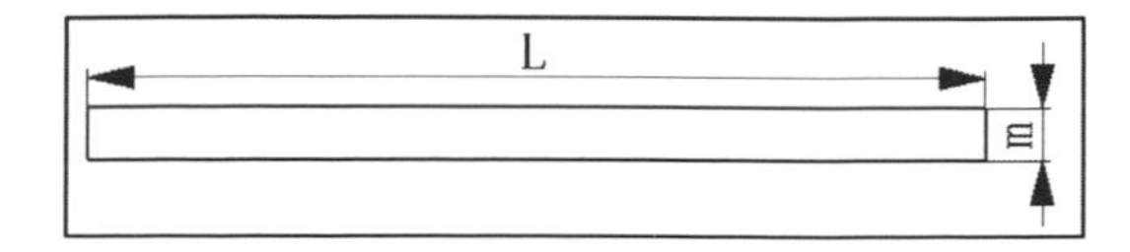

Figure 3. Isolation film flattening size diagram.

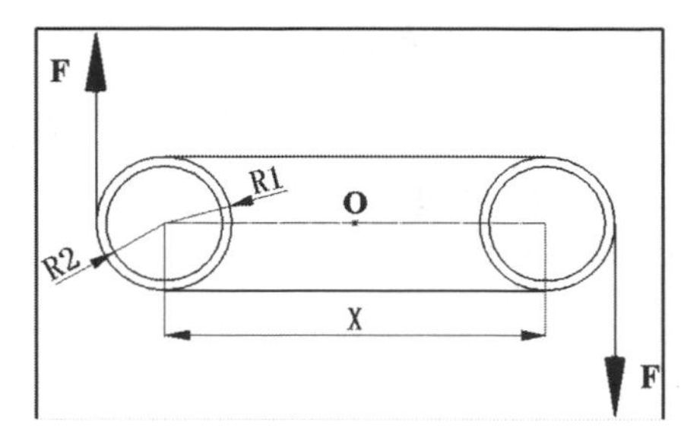

Figure 4. Force diagram of tension swing rod.

The tension swing is driven by a rotary motor, O, as a rotation center, and f the isolation membrane tension. In order to ensure the isolation membrane tension is constant, it is necessary to apply a driving force in the rotary motor, which is mainly affected by the isolation membrane tension and tension pendulum inertia effect, for the establishment of the dynamic mathematical model. Tension swing rod total moment of inertia is:

$$J = m_1(x+2R_1)^2/12 + 2[0.5m_1(R_1^2 - R_2^2) + m_2(0.5x+R_1)^2] \quad (4)$$

where m_1 is the tension swing rod quality and m_2 is the roll quality on the tension swing rod. In the constant tension control system, the driving force of the motor is balanced with the tension of the isolation membrane. The driving torque required by the tension swing drive motor is calculated as:

$$T = F(x+2R_1) + 0.5J\beta^2 \quad (5)$$

where β is the angular acceleration of the current moment of tension swing rod; in the practical application it is possible to get feedback from the servo system. From the above formula, in the ideal case, the tension swing angular acceleration is 0, and the motor drive force is only related to the tension of the isolated film.

2.3 *Mathematical model of laminated platform and mathematical model of linear motor*

According to a recent study (Liang, 2014) the dynamic state space equation of linear motor can be expressed as:

$$\begin{bmatrix} \frac{di_{d1}}{dt} \\ \frac{di_{q1}}{dt} \end{bmatrix} = \begin{bmatrix} -\frac{R_1}{L_{d1}} & \frac{\pi}{\tau_1} v_{p1} \frac{L_{q1}}{L_{d1}} \\ -\frac{\pi}{\tau_1} v_{p1} \frac{L_{d1}}{L_{q1}} & -\frac{R_1}{L_{q1}} \end{bmatrix} \begin{bmatrix} i_{d1} \\ i_{q1} \end{bmatrix} + \begin{bmatrix} v_{d1}/L_{d1} \\ (v_{q1} - \pi v_{p1} c\lambda_1/\tau_1)/L_{d1} \end{bmatrix} \quad (6)$$

where i_{dl} and i_{ql} indicate the armature currents, and v_{dl} and v_{ql} are the armature voltages for d-axis and q-axis, respectively. Similarly, L_{dl} and L_{ql} represent the armature inductance of each axis respectively. λ_1, τ_1, M, B represent maximum magnetic flux of PM, poles pitch of linear motor, mass of armature, and friction coefficient, respectively. Disturbance loading discussed here includes friction and propulsion ripple, in which for simplicity the ripple propulsion can be considered as a periodic sine wave.

There are some large impacts in the system when the laminated platform is in high speed reciprocating motion. In order to avoid the impacts and vibrations this study used a dissymmetrical S curve to plan a trajectory and added a dynamic model of the linear motor to simulate the system. The motor simulation curve is shown in Figure 5 and the tracking error curve in Figure 6.

2.4 *Mathematical modeling of the storage shaft*

The kinematic mathematical modeling of the storage shaft is mainly on the isolation membrane length change. The mathematical relationship of isolation membrane length change includes all three parts of the movement: the lamination platform, the pressure knife, and the unwinding shaft. The simulation trajectory of the linear motor combined

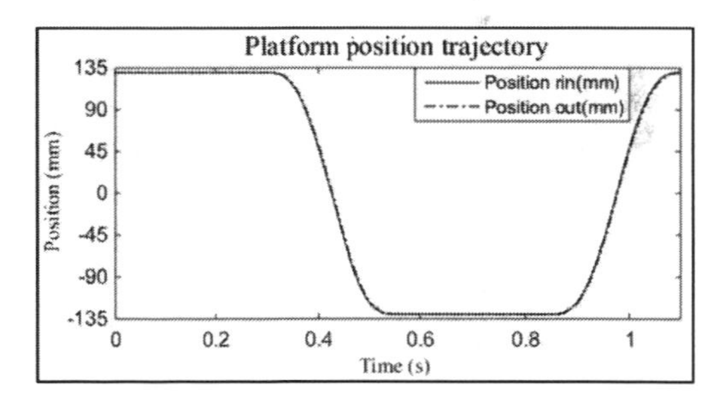

Figure 5. Running track of the lamination platform.

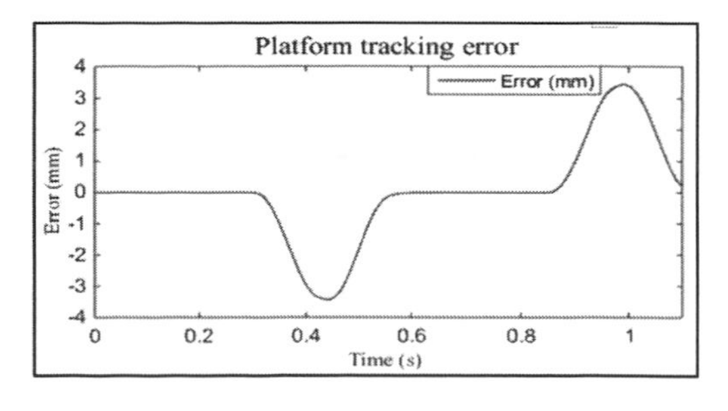

Figure 6. Simulation tracking error curve.

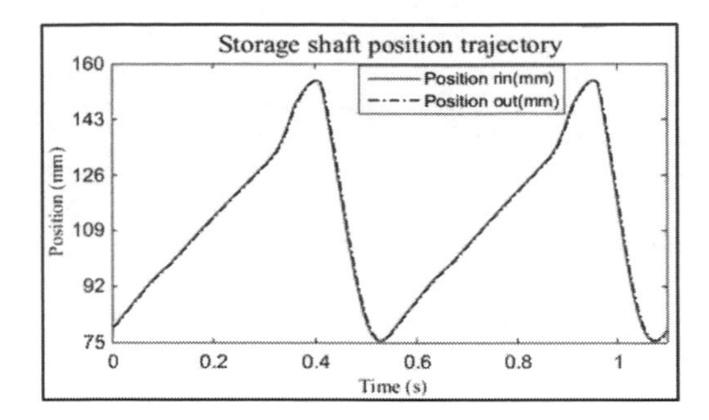

Figure 7. Running track of the storage shaft.

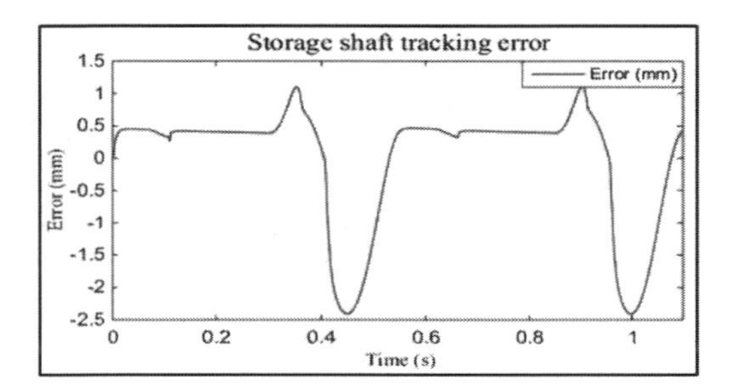

Figure 8. Simulation tracking error curve.

with the linear motor model is shown in Figure 7 and the tracking error curve in Figure 8.

3 CONTROL STRATEGY

Master-slave synchronization is defined as setting one as the master axis and the other as the slave axis in a multi axis control system; the velocity of the master axis changes the slave velocity. The storage shaft position is calculated according to the lamination platform and pressure knife position in this study, so the movement of laminated platform is set as the spindle.

Combined with the mathematical model established in the previous section, the control strategy diagram of master-slave synchronization is shown in Figure 9. The motion of the platform is used as the master axis, and the movement of the storage master is calculated according to the position of the master.

It is found that the increase of the current loop gain can improve the tracking performance of the linear motor. So we put feed forward on the current loop. A diagram of the feed forward control of the linear motor is shown in Figure 10.

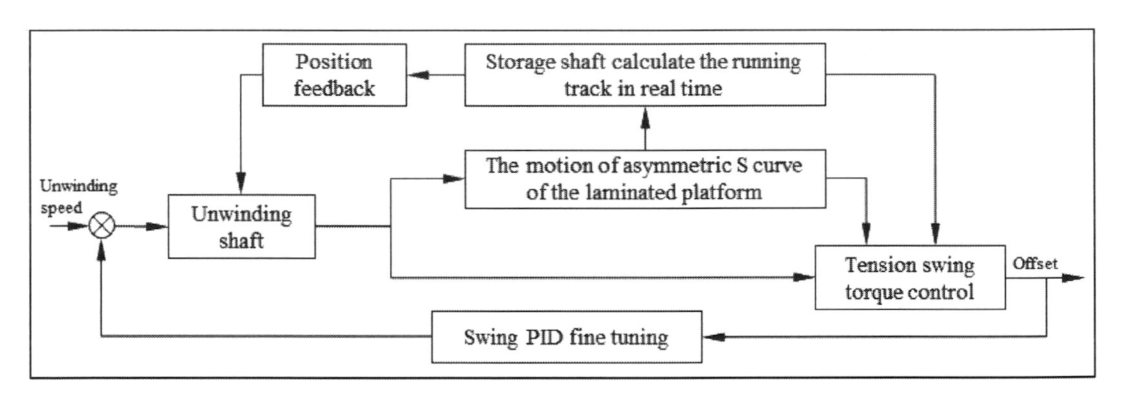

Figure 9. Lamination machine control strategy.

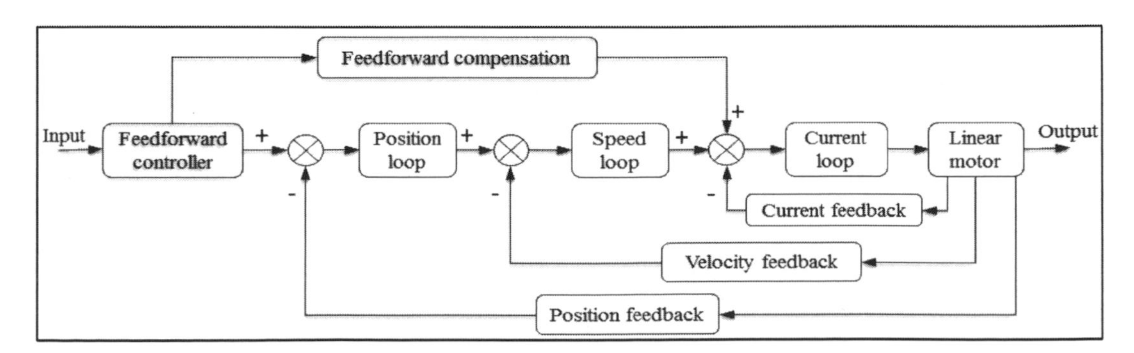

Figure 10. Diagram of feed forward control of the linear motor.

4 EXPERIMENT

The experimental platform is shown in Figure 11. The tension swing rod provides constant torque control, and so with the passive state of change of angular displacement it will move when the tension fluctuates. So the displacement of the tension swing rod reacts to the stability of tension fluctuations.

The actual curve of the displacement of the tension swing rod is shown in Figure 12. The tension swing angle range is only 0.15 degrees, which can be determined by the system tension being almost at a constant state throughout the whole process.

In the transmission process of the isolation membrane, the line velocity of the isolation membrane at a constant is based on the premise that tension must be stable. In the unwinding velocity curve is shown in Figure 13, the unwinding suddenly accelerates at the beginning followed by small-scale fluctuations. The fluctuations are impacted by PID rules, which dictate that the unwinding velocity is basically uniform.

The experiment of the overlapping separation membrane was carried out in the experimental platform with a set cycle time of 0.55 S, The effect of overlapping in the microscope observation chart is shown in Figure 14.

Figure 11. Experimental platform.

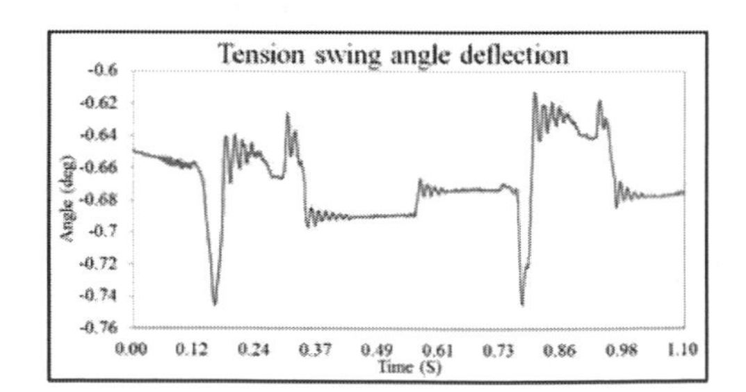

Figure 12. Displacement wave curve of tension swing rod.

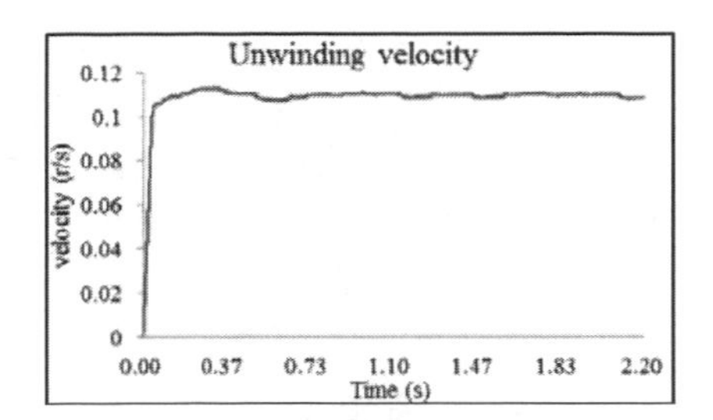

Figure 13. Velocity curve of the unwinding.

Figure 14. Effect of overlapping isolation membrane.

Table 1. Regularity data table of isolation membrane.

Itemz	Left OH Max (um)	Right OH Max (um)	Alignment metric
Spec	1.25	1.25	0
USL	1.75	1.75	0.25
LSL	0.75	0.75	–0.25
Max	0.95	1.74	0.09
Min	0.79	1.60	–0.02
Mean	0.85	1.67	0.02
Range	0.17	0.14	0.11
Sigma	0.043	0.035	0.026
CP	3.876	4.762	3.252
CPK	0.800	0.761	2.936

According to the observed data used to calculate other indices, the positive and negative deviation and process capability index is a very important indicators. For computing these data, some important parameters are summarized in Table 1. In the production process of lithium batteries, it is required that the isolation membrane CPK > 1.3, so as this study's CPK = 2.936 the performance requirements is met.

5 CONCLUSION

This paper describes the model of the kinematics and dynamics of active continuous unwinding and laminated platform intermittent reciprocating motion. It also explains the master-slave synchronization control method of a platform for experiment and analysis, with a lamination cycle of 0.55 S. The experiment has achieved very good results.

ACKNOWLEDGMENT

The authors gratefully acknowledge the support provided by the National Nature Science Foundation of China under NSFC Grant 51375118, as well as the support from Shenzhen Government Fund GJHZ20140419142750948.

REFERENCES

Albert W.-J. Hsue, M.-T. Yan, S.-H. Ke, JMPT *Comparison on linear synchronous motors and conventional rotary motors driven Wire-EDM processes*, **192–193** 478–485 (2007).

Chen, C.L., K.M. Chang, C.M. Chang, AMM, *Modeling and control of a web-fed machine*. **28** *863–876*(2004).

Chen, S.L., T.H. Hsieh, IJMTM, *Repetitive control design and implementation for linear motor machine tool*. **47** 1807–1816 (2007).

Ding, K., Q.L. Han, JSV, *Master-slave synchronization criteria for horizontal platform systems using time delay feedback control*. **330** 2419–2436 (2011).

Hast, M., T. Hägglund, Journal of Process Control, *Low-order feedforward controllers: Optimal performance and practical considerations*. **24** 1462–1471 (2014).

Kang, H.K., C.W. Lee, K.H. Shin, Journal of Process Control, *A novel cross directional register modeling and feedforward control in multi-layer roll-to-roll printing*. **20** 643–652 (2010).

Lee, C., H. Kang, K. Shin, IJMS, *Advanced taper tension method for the performance improvement of a roll-to-roll printing production line with a winding process*. **59** 61–72 (2012).

Li, H.Z., M.D. Le, Z.M. Gong, W. Lin, IEEE-ASME T MECH, *Motion Profile Design to Reduce Residual Vibration of High-Speed Positioning Stages*, **14-2** 264–270 (2009).

Liang, Z., J. Huang, P I MECH ENG C-J MEC, *Design of high-speed cam profiles for vibration reduction using command smoothing technique*, **228(18)** 3322–3328 (2014).

Park, J.C., S.W. Jeon, K.S. Nam, L. Liu, J. Sun, C.H. Kim, JJAP, *Tension control of web of winder span using adaptive gain control method*. **53**, 05HC11-1–05HC11-7 (2014).

Salarieh, H., M. Shahrokhi. JCAM, *Multi-synchronization of chaos via linear output feedback strategy*. **223** 842–852 (2009).

Wua, X.F., J.P. Cai, M.H. Wang, JSV, *Master-slave chaos synchronization criteria for the horizontal platform systems via linear state error feedback*. **295** 378–387 (2006).

Yang, X.J., D. Lu, J. Zhang, W.H. Zhao, IJMTM, *Dynamic electromechanical coupling resulting from the air-gap fluctuation of the linear motor in machine tools*, **94** 100–108 (2015).

Zheng, S.Q., R. Feng, JSV *Feedforward compensation control of rotor imbalance for high-speed magnetically suspended centrifugal compressors using a novel adaptive notch filter*. **366** 1–14 (2016).

Civil, Architecture and Environmental Engineering – Kao & Sung (Eds)
© 2017 Taylor & Francis Group, ISBN 978-1-138-02985-9

Design and kinematics analysis of a novel liquid-brewing steamer feeding robot

Jiashuang Zhang, Bin Li, Tongchen Zhang & Xinhua Zhao
Tianjin Key Laboratory for Advanced Mechatronic System Design and Intelligent Control, Tianjin University of Technology, Tianjin, China

ABSTRACT: In this paper, a novel steamer feeding robot is designed and studied. When compared to the currently used steamer feeding robot, the feeding process of the proposed robot is continuous, and this way will greatly improve the efficiency and quality of the feeding process. Firstly, the mechanical structure and kinematics of the novel steamer feeding robot are introduced and analyzed. And then, the control system of the robot is introduced. Finally, experiment evidences are given to prove the efficiently of brewing by using this robot.

1 INTRODUCTION

Currently, robotics is one of the world's most popular fields of the study; many mature robot products have been widely applied in many aspects such as space flight, aviation, busywork in sea and land, undergroundexploitation,manufacturing,computer-aided medical instrument, biological engineering and micromechatronics system, and so on. More and more experts and producers have been concerned about the robotics studies and applications. On one hand, the studies on the robotics have become systematic and deep, in terms of the robot theory, the design, kinematics, dynamitic, workspace, dimensional synthesis, etc., are the hot points of robotics studies. On the other hand, the applications of robotics have extended into more extensive domains, such as the Chinese traditional labor-intensive industry (Gu, 2006).

Liquor-making is a traditional labor-intensive industry in China. The whole process of liquor-making is mainly composed of cereals fermentation, feeding, steam distillation, filling, packaging, and transportation. Among these links in the production chain, the feeding operation is a very important link in the process of liquor-brewing. Its operation quality will directly determine the liquor yield and liquor quality. During the brewing process, we must spread the brewing raw materials as a uniform and loose layer in the wine barrel to ensure that brewing raw materials do not "run steam" and "pressure steam" in the process of distillation. Because of the high viscosity of brewing raw materials and because the barrel is large, round, and small, it is difficult to obtain the charging equipment with the operation (Qian, 2013). The current brewing industry generally uses the traditional artificial steamer, through which the workers spread the materials layer by layer. And so, the operator labor intensity and the liquor stability is poor. The feeding link of most winery is still manually operated at present. Heavy manual labor is the characteristic of this process. Changing the traditional way of feeding to upgrade the liquor industry is imminent. And after years of research, we developed some feeding robots that can be practically applied. These kinds of robots use the existing joint type robot as the main operational body, and installed the feeding mechanism at the end of the manipulator. We complete the feeding operation by controlling the robot's anthropomorphic operation. But this kind of robot in the operation process is the interrupted type of carrying on the material and then spreading. If steam is released during this period, robots will not be able to feed in time, and this will greatly reduce the liquor out rate. In addition, due to the limited operation of the robot body, it is difficult for the part of the wine barrel edge to achieve uniform material. "Babel gap" will appear (Wang, 1992; Qian, 2013; Serdar, 2014; Liu, 2009).

This article is aimed at the feature of the feeding operation with "light, loose, uniform, thin, accurate, and flat"; and design an intelligent feeding robot combined with infrared vision technology and PLC control technology. In this paper, its mechanical structure and kinematics are introduced and analyzed firstly. And then, the introduction of its control system is carried out too. Finally, the actual test of liquor-brewing with this robot is taken to prove the possibility of brewing.

2 MECHANICAL DESIGN AND DESCRIPTIONS

As we know, the robot must keep moving to the designated position in the circular wine barrel when feeding. And so, it is necessary to ensure that the robot can rotate around the "Z" axis, with the accuracy and stability of motion at the same time. The rotation of the mechanical arm is realized by installing the servo motor to control the gear rotation; the arm needs to rise for a distance for the next round of feeding after finishing a layer of material, and so in order to ensure that the feeding mechanism and the material layer are not in contact with each other, we must provide a vertical direction for the robot's translational degrees of freedom. And so, we use the servo motor to drive the rotation of the column and the ball–screw mechanism to realize the rise of the mechanical arm, and thinking of the global stability, we have designed four columns; one of them installed for the screw is for rising, and another is for support.

In the way of spreading, the feeding robot has been applied actually and is commonly used the hopper. When the material in the hopper is used up, the robot moves to the material to supplement, and then continues to feed. In the process of the supplement, the robot cannot give a remedy to the "leakage" phenomenon in the barrel. Therefore, in this paper, the robot manipulator is specially designed. Each of them is hollow inside and provided with a feed screw. It drives the continuous movement of the feeding screws to realize the transfer of the material in the mechanical arm through the auger motor. At the same time, the two mechanical arms are connected. The raw material can be passed through the two mechanical arms without leakage. There is a feed inlet that is directly connected with the feeding conveyor belt installed at the robot's forearm to ensure the continuity of feeding.

The feed operation needs to be "light, loose, uniform, thin, accurate, and flat". But the material out of the rear arm flows fast and spreads non-uniformly. And so, we design a spreader by using the leaf structure to make the material enter the barrel lightly, loosely, and uniformly. And it ensures the quality of the feeding process.

According to the above-mentioned structure requirements, the robot is built and assembled by Solidworks. Add the motor and other parts into the assembly; and we can get the three-dimensional model of the feeding robot, as shown in Figure 1. It is composed of a chassis, four upright posts, two mechanical arms three servo motors that control chassis and mechanical arm rotation and two auger motors that control feed screws rotate. The chassis not only plays the role of supporting the

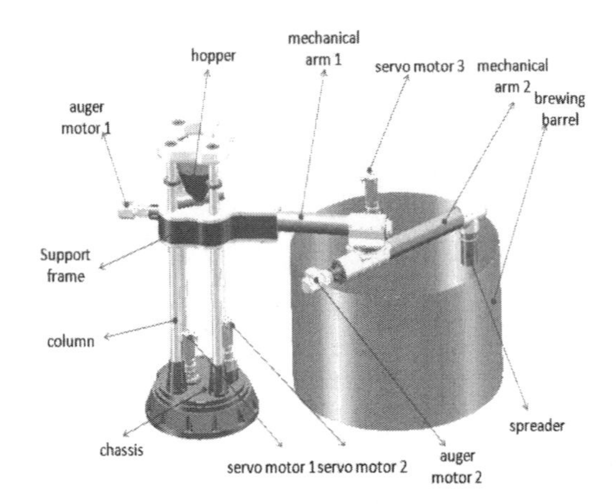

Figure 1. Schematic of the steamer feeding robot system.

whole robot, but also can be driven by the servo motor 1 to achieve the rotation on the horizontal plane; the mechanical arm 1 is connected with the chassis through the upright post, it controls the mechanical arm to achieve the vertical up and down movement driven by using servo motor 2, the two mechanical arms are connected through a gear mesh, the mechanical arm 2 rotary is driven by using servo motor 3.

3 INVERSE AND FORWARD POSITION ANALYSES

3.1 *Establishment of the direction matrix*

When we are studying the feed movement of the robot, upward movement is not a decisive factor. And so, we ignore the upward mobility, and focus on the rotary motion of the spreader. And then, simplify the structure of the robot and we obtained the structure diagram, as shown in Figure 2.

It consists of a base S and two arms (PR and RS). Among these, the base and mechanical arm and the two mechanical arms are connected through a rotating pair (S and R), and the two axes of rotation are parallel to each other, respectively, thereby establishing the coordinate system with the base consolidation (x_0, y_0, z_0) and the end of arms (x_1, y_1, z_1), (x_2, y_2, z_2). The clockwise joint angle is negative, and the counter clockwise angle is positive, as shown in Figure 2. And it is expressed with D–H table in Table 1.

3.2 *Analysis of positive kinematics*

Assume that the Z axis is perpendicular to the direction of the paper. And according to Table 1,

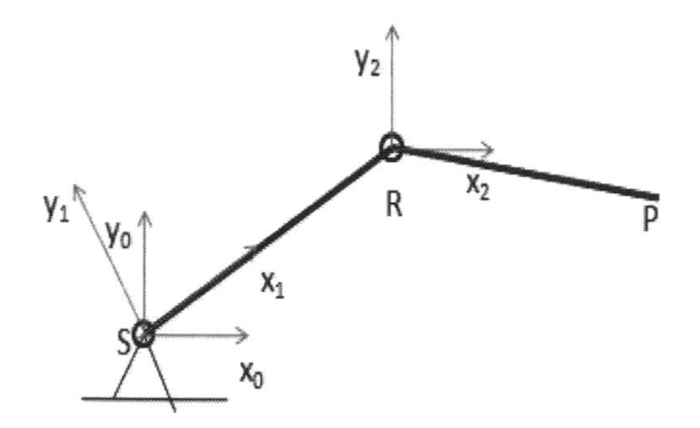

Fgure 2. Schematic of the structure.

Table 1. D-H parameters of the robot.

Connecting rot	α	A	D	θ
1	0	l_1	0	θ_1
2	0	l_2	0	θ_2

we obtain the homogeneous rotation transformation matrix from the coordinate system (x_0, y_0, z_0) to (x_2, y_2, z_2) as given in Equation (1).

$$
\begin{aligned}
{}_2^0T &= {}_1^0T \cdot {}_2^1T \\
&= \begin{bmatrix} \cos\theta_1 & -\sin\theta_1 & 0 & 0 \\ \sin\theta_1 & \cos\theta_1 & 0 & 0 \\ 0 & 0 & 1 & 0 \\ 0 & 0 & 0 & 1 \end{bmatrix} \begin{bmatrix} \cos\theta_2 & -\sin\theta_2 & 0 & l_1 \\ \sin\theta_2 & \cos\theta_2 & 0 & 0 \\ 0 & 0 & 1 & 0 \\ 0 & 0 & 0 & 1 \end{bmatrix} \\
&= \begin{bmatrix} \cos(\theta_1+\theta_2) & -\sin(\theta_1+\theta_2) & 0 & l_1\cos\theta_1 \\ \sin(\theta_1+\theta_2) & \cos(\theta_1+\theta_2) & 0 & l_1\sin\theta_1 \\ 0 & 0 & 1 & 0 \\ 0 & 0 & 0 & 1 \end{bmatrix}
\end{aligned} \tag{1}
$$

By dot multiplying the position vector ($p = (l_2, 0, 0, 1)$) and Eq. (1), we can obtain the position vector of the end of the connecting rod (P) in the base coordinate system (x0, y0, z0) as follows:

$$
\begin{aligned}
{}^0P &= {}_0^2T \bullet {}^2P \\
&= \begin{bmatrix} \cos(\theta_1+\theta_2) & -\sin(\theta_1+\theta_2) & 0 & l_1\cos\theta_1 \\ \sin(\theta_1+\theta_2) & \cos(\theta_1+\theta_2) & 0 & l_1\sin\theta_1 \\ 0 & 0 & 1 & 0 \\ 0 & 0 & 0 & 1 \end{bmatrix} \begin{bmatrix} l_2 \\ 0 \\ 0 \\ 1 \end{bmatrix} \\
&= \begin{bmatrix} l_1\cos\theta_1 + l_2\cos(\theta_1+\theta_2) \\ l_1\sin\theta_1 + l_2\sin(\theta_1+\theta_2) \\ 0 \\ 1 \end{bmatrix} = \begin{bmatrix} x_p \\ y_p \\ z_p \\ 1 \end{bmatrix}
\end{aligned} \tag{2}
$$

And it can be rewritten as follows:

$$x_p = l_1\cos\theta_1 + l_2\cos(\theta_1+\theta_1) \tag{3}$$

$$y_p = l_1\sin\theta_1 + l_2\sin(\theta_1+\theta_1) \tag{4}$$

3.3 *Analysis of inverse position*

After establishing the kinematic equations mentioned above, if the end position of the manipulator is known, we can obtain the joint angle of the robot arm (θ_1 and θ_2) by using the kinematic equations. This is the inverse kinematics of the manipulator. Inverse kinematics can be used to control the joint angle and the end position of the robot arm. For this feeding robot, its inverse kinematics is the process of known x_p, y_p for θ_1, θ_2.

Firstly, based on Equations (3) and (4), and performing the proper expansion and simplification yields the following equation:

$$x_p^2 + y_p^2 + l_1^2 - l_2^2 = 2l_1(x_p\cos c + y_p\sin\theta_1) \tag{5}$$

In order to simplify the calculation, assume

$$\tan\theta_p = \frac{x_p}{y_p}, \theta_p = \arctan\frac{x_p}{y_p} \tag{6}$$

Substituting Equation (6) in Equation (3) yields the following equation:

$$x_p^2 + y_p^2 + l_1^2 - l_2^2 = \frac{2l_1x_p}{\cos\theta_p}(\cos\theta_1\cos\theta_p + \sin\theta_1\sin\theta_p) \tag{7}$$

$$x_p^2 + y_p^2 + l_1^2 - l_2^2 = \frac{2l_1x_p}{\cos\theta_p}\cos(\theta_1 - \theta_p) \tag{8}$$

According to Equation (8), we can obtain θ_1 which is as follows:

$$\theta_1 = \arccos\frac{\cos\theta_p(x_p^2 + y_p^2 + l_1^2 - l_2^2)}{2l_1x_p} + \arctan\frac{x_p}{y_p} \tag{9}$$

In the same way as solving θ_1, Equation (10) and (11) are obtained, and at last completed the calculation of θ_2 as Equation (12).

$$x_p^2 + y_p^2 + l_2^2 - l_1^2 = 2l_2[\cos(\theta_1+\theta_2) + \sin(\theta_1+\theta_2)] \tag{10}$$

$$x_p^2 + y_p^2 + l_2^2 - l_1^2 = \frac{2l_2x_p}{\cos\theta_p}\cos(\theta_1+\theta_2-\theta_p) \tag{11}$$

$$\theta_2 = \arccos\frac{\cos\theta_p(x_p^2 + y_p^2 + l_2^2 - l_1^2)}{2l_2x_p} + \arctan\frac{x_p}{y_p} - \theta_1 \tag{12}$$

4 ANALYSIS OF THE WORKSPACE

The feed operation requires the robot to reach any point within the barrel. But according to the results of the positive and inverse position analysis mentioned above, we find that when the two arms are in the same line, the robot is in singular configuration. Except this, many other factors could influence the working space of the robot. And so, we respectively introduce these interferences and give the corresponding constraints.

1. Input angle: when the robot operates, the forearm may interfere with the conveyor or other equipment. And so, the swing angle of the forearm is often restricted to a certain angle. And so, the robot's working space is limited, so that the original part of the work space cannot be achieved. In this paper, the research on the robot forearm's swing angle sensor range is limited in the range of –60o–80o by using a photoelectric limit sensor. When the swing angle of the current arm reaches the upper and lower limit positions, the trigger sensor stops the robot. When the robot works regularly, the robot will not trigger the limit sensor.
2. Singular configuration: when the singularity occurs, the robot is in a state of "sickness", and the working space will show some holes. According to above-mentioned equations, when the two arms are in the same line, the robot is in a singular position. But due to the spreader and the limitation of the chassis we set, the robot will not arrive at the singular position and will clear up the effect of the singular configuration.

5 REALIZATION AND TESTING OF THE ROBOT

According to the mechanical structure and structure parameters designed above, the factory processes the robot and connects the conveyor belt system to the upper end of the feed inlet. And on the basis of the mechanical system, in order to realize the function of feeding, we add the control system and the infrared vision system to the robot (Tian, 2013; Hu, 2001), as shown in Figure 3.

5.1 *Working process of the robot*

Before feeding, the control system resets the manipulator to the initial position. And then, when the robot begins to feed, the control system controls the mechanical arm's movement circle and the circle around the inside barrel to spread layer by layer; and after finishing two material layers, the infrared vision system starts to leak and displays an uneven position picture according to the actual situation in the barrel, and then send the feeding information to the control system for controlling the robot's body to fill these positions. After the completion of the feeding of one layer, the control system controls the robot's rising by a layer distance, and then continues to carry on spreading the material layer by layer until the robot finishes feeding all of the entire barrel (Li, 2015; Zhang, 2015).

5.2 *Experiment and summary*

In order to detect the actual operation effect of the robot, the factory is powered on the spot to make the liquid. In the process of brewing, the robot can spread uniformly layer by layer; the vision system can detect the location of leakage accurately; and control the spreader to the accurate leak position. The leakage photos are taken by using the visual system, as shown in Figure 4.

According to the actual brewing results, this new feeding robot can simulate the artificial feeding process. During the operation, only one person will be required to shovel the material into the conveyor belt. It reduces the labor intensity of workers and saves labor cost. The yield and quality of the liquor brewed by using the robot is similar to traditional artificial feeding.

Figure 3. Picture of the steamer feeding robot.

Figure 4. Picture of the leakage.

6 CONCLUSIONS

In this paper, the urgent need of the feeding robot in the liquor-brewing industry is to design a new feeding robot, and carry out the related theoretical research work. Firstly, the mechanical structure design of the robot is carried out; and then, on this base, the forward and inverse solution analysis is carried out, and the working space is also analyzed. In the last part of this work, the control system of the robot is designed, and the field practice is carried out to verify the practicability of the robot. The experiment shows that the design of this robot is helpful for improving and perfecting the liquor production process, solving the poor working environment, recruitment difficulties, and other problems that the liquor industry faces. And there are also some guidelines for the further application of robot technology in the liquor-brewing industry.

REFERENCES

Gu, Z.Y. Status and trend of global industrial robot industry, *Mechanical and electrical integration*, **2**, (2006).

Hu, H.Q., H.R. Fang, J.B. Peng, Y.Z. Yang, H. Chen, Singularityanalysis of robot, *Robot Technology and Application*, **6**, (2001).

Li, H., Y. Hua, Design of mechanical steaming bucket feeding control system based on PLC, *Food and mechinery*, **31**, 114–118. (2015).

Liu, S.Z., Y.Q. Yu, J.X. Yang, L.Y. Su, Kinematic and dynamic analysis of a three-degree-of-freedom parallel manipulator, *Journal of Mechanical Engineering*, **45** (2009).

Qian, C., Y.H. Qian, X. Zhang, H.B. Ma, J. Xu, X. Fan, Y. Zhang, Application progess of mechanical automation liquor-production technologies, *China brewing*, **32**, 5–9 (2013).

Serdar Kucuk, Zafer Bingul, Inverse kinematics solutions for industrial robot manipulators with offset wrists, *Mathematical Modeling*, **38**: 1983–1999 (2014).

Tian, H.B., H.W. Ma, J. Wei, Workspace and structural parameters analysis for manipulator of serial robot, *Chinese Journal of agricultural machinery*, (2013).

Wang, J.Z., S.Q. Liu, The technique of solid liquor feeding, *Liquor-making ScienceTechnology*, **2**, 18–21, (1992).

Zhang, L.D., T.D. Meng, J.P. Tian, Y.Hu, H.S. Yang, automatic steamer distillation apparatus for Nongxiang liquor by solide-state fermentation, *Liquor-making ScienceTechnology*, **9**, 94–99, (2015).

Civil, Architecture and Environmental Engineering – Kao & Sung (Eds)
© 2017 Taylor & Francis Group, ISBN 978-1-138-02985-9

Design and simulation analysis of a new type of multi-gear pump

Xuhui Zhang, Pingyi Liu, Wenjun Wei & Haitao Li
College of Engineering, China Agricultural University, Beijing, China

Guoping Tian
Beijing YIHE ZHONGWEI Precision Machine Co. Ltd., Beijing, China

ABSTRACT: Aiming at solving problems of great influence on hydraulic cylinders' flow rate caused by load pressure and poor synchronization at composite action, a new type of parallel multi-gear pump is put forward in this paper. According to the relationship between the number of center gear teeth and driven gears, the pump can be divided into three kinds. And then, the expressions for outlets' instantaneous flow rate of the three kinds of multi-gear pumps are derived. By using the dynamic mesh technique in Fluent software, transient simulation on flow characteristics of the multi-gear pump under various working conditions is carried out. The instantaneous and mean flow velocities of the three kinds of multi-pumps are analyzed. The results show that the mean flow rate is in a linear relationship with outlet pressure, and the flow rate of a sub-pump outlet is only related to its pressure but not that of the other sub-pumps. Sub-pumps have no influence on each other. The working states of every sub-pump of Class I are completely synchronous. The working states of sub-pumps of Class II are divided into groups, and the working form is packets that are synchronous and cooperative among groups.

1 INTRODUCTION

Hydraulic transmission is widely used in agricultural machinery because of its advanges such as large power, flexible installation, easy adjustment, low breakdown rate, convenient to realize continuously variable transmission, and so on (Li, 2012; Saleh, 2015). The actions of multiple actuators are simultaneous in most modern agricultural machinery (Zhao, 2014; Yang, 2005; Zhang, 2015; Liang, 2012; LV, 2014), which are controlled by using a multi hydraulic circuit system using one pump. The hydraulic system has problems that the flow rate is greatly influenced by load pressure and synchronization of the composite action is poor (Dong, 2010).

In this paper, a new type of parallel multi-gear pump is proposed, which can avoid the interference of pressure and flow among actuating cylinders. The design and classification of the multi-gear pump are carried out, and the expressions of each outlet are derived. By taking the Fluent software as a platform, the flow characteristics of the multi-gear pump under various working conditions are studied. The results provide theoretical basis for the multi-gear pump application in the multi-hydraulic circuit system.

2 DESIGN AND CLASSIFICATION OF THE NEW TYPE OF MULTI-GEAR PUMP

As shown in Figure 1, the new type of multi-gear pump is composed of one center gear and several driven gears. Driven gears are uniformly distributed around the center gear and mesh with it. Being powered by using the center gear, driven gears rotate with high speed. And then, a series of sub-pumps are formed. Let z_1 be the number of center gear teeth and N be the number of driven gears. The multi-gear pump is composed of N sub-pumps. Every sub-pump has one outlet and one inlet. And then, the N inlets converge to one inlet. Therefore, the multi-gear pump has N outlets and one inlet.

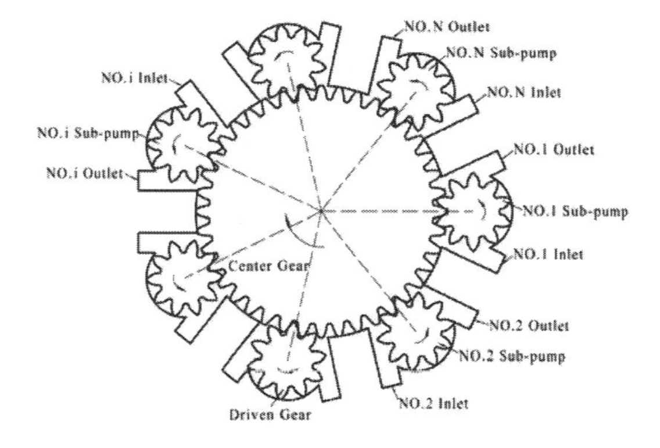

Figure 1. Schematic illustration of the new type of multi-gear pumps.

Sub-pumps are clockwise named as NO. 1, NO. 2, ..., NO. N. Draw the line which connects centers of the center gear and the NO. 1 driven gear in a horizontal manner, and also the symmetry line of the first tooth of the center gear. And then, the two lines coincide. When the line which connects centers of the center gear and NO. i driven gear coincides with the symmetry line of any tooth of the center gear, the instantaneous states of the NO. i and NO. 1 are completely synchronous. But if two lines do not coincide anywhere else, the instantaneous states of all sub-pumps contained in the multi-gear pump are not synchronous.

Let k be the greatest common divisor of z_1 and N. Based on the analysis mentioned above, the multi-gear pump is divided into three kinds.

1. $k = N$, that z_1 is divisible by N, and then the instantaneous states of N sub-pumps are fully synchronous. Define this kind of multi-gear pump as Class I.
2. $1 < k < N$, such that z_1 is not divisible by N, but z_1 and N is divisible by k simultaneously; and then, the working states of N sub-pumps are divided into groups. The working states of sub-pumps are packets that are synchronous, but cooperative among groups. The number of groups is N/k, and there are k sub-pumps in every group. The number of intervals among the k sub-pumps is (N/k-1), that is the working states of NO. i, NO. ($N/k + i$), NO. ($2N/k + i$), …, NO. [(k-1)$N/k + i$] sub-pumps are synchronous. Define this kind of multi-gear pump as Class II.
3. $k = 1$, the instantaneous states of all the sup-pumps contained in the multi-gear pump are different. Define this kind of multi-gear pump as Class III.

According to the analysis mentioned above, the classification of the new type of multi-gear pump is related to the number of center gear teeth and driven gears, but not the number of driven gear teeth.

3 OUTLETS' INSTANTANEOUS FLOW RATE OF THE NEW TYPE OF MULTI-GEAR PUMPS

3.1 *Position analysis of meshing points*

The meshing of the center gear and a driven gear is shown in Figure 2. Point A is the beginning mesh point. Point B is the ending mesh point. Point C is the meshing pitch point which is fixed. Point D is the mesh point when the next pair of teeth begin to mesh at point A. Point M is the instantaneous mesh point of the two gears for oil expulsion. f_i is the distance between point M and point C.

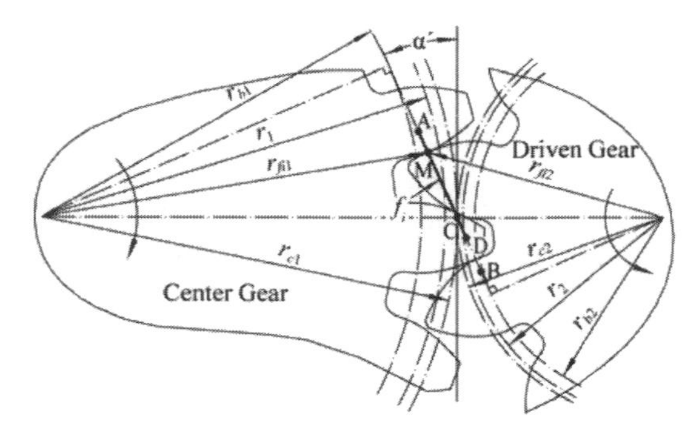

Figure 2. Schematic showing meshing of the center gear and a driven gear.

1. The instantaneous mesh point positions relationship of sub-pumps contained in the Class I multi-gear pump can be determined as follows:

$$f_1 = f_2 = \cdots = f_i = \cdots = f_N \tag{1}$$

2. The instantaneous mesh point positions relationship of sub-pumps contained in the Class II multi-gear pump can be determined as follows:

$$f_i = f_{(N/k+i)} = f_{(2N/k+i)} = \cdots = f_{[(k-1)N/k+i]} \quad (i = 1,2,\cdots,N/k) \tag{2}$$

$$f_i + \frac{[b_i \cdot N - (i-1)c]a}{k} = f_1 \tag{3}$$

wherein, define b_i as the distribution factor, and c as the tooth number difference factor.

$$\begin{cases} b_i = \text{ROUNDUP}\left[\dfrac{(i-1)c}{N}\right] \\ c = \text{MOD}\left(\dfrac{z_1}{N}\right) \end{cases} \tag{4}$$

wherein, define ROUNDUP( ) as the round up and integral function, and MOD( ) as the taking remainder function.

3. The instantaneous mesh point positions relationship of sub-pumps contained in the Class III multi-gear pump can be determined as follows:

$$f_i + [b_i \cdot N - (i-1)c]a = f_1 \tag{5}$$

wherein, b_i and c can be obtained from equation (4).

After collecting and analyzing equations (1), (2), (3), and (5) and their related parameter equations, the general formula of the instantaneous mesh point positions of sub-pumps for the N-link pump is obtained as follows:

$$\begin{cases} f_i = f_1 - \dfrac{\left[b_i \cdot N - (i-1)c\right]a}{k} \\ a = \dfrac{k\pi m \cos\alpha}{N} \\ b_i = \text{ROUNDUP}\left[\dfrac{(i-1)c}{N}\right](i = 1,2,\cdots,N) \\ c = \text{MOD}\left(\dfrac{z_1}{N}\right) \end{cases} \tag{6}$$

where the value of f_1 is expressed as follows: with point C as the zero point, to the left of point C (AC direction) is negative and to the right (direction CD) is positive. And then, the value range of f_1 is given as follows:

$$f_1 \in \left[r_2 \cos\alpha \tan\alpha' - \sqrt{r_{a2}^2 - r_2^2 \cos^2\alpha}, 0\right] \cup \left[0, \pi m \cos\alpha + r_2 \cos\alpha \tan\alpha' - \sqrt{r_{a2}^2 - r_2^2 \cos^2\alpha}\right] \tag{7}$$

The meshing point positions of the three kinds of multi-gear pumps show different characteristics along with different numbers of center gear teeth and driven gears.

3.2 *Instantaneous flow rate equations of outlets*

Based on the Finite Volume Method (FVM) (Huang, 2009), the instantaneous flow rate equations of N outlets can be obtained as follows:

$$\begin{cases} Q_{ex1} = \dfrac{B\omega_1}{2}\left[r_{a1}^2 + \dfrac{r_1}{r_2} r_{a2}^2 - r_1(r_1 + r_2)\dfrac{\cos^2\alpha}{\cos^2\alpha'} - \left(1 + \dfrac{r_1}{r_2}\right) f_1^2\right] \\ \vdots \\ Q_{exi} = \dfrac{B\omega_1}{2}\left[r_{a1}^2 + \dfrac{r_1}{r_2} r_{a2}^2 - r_1(r_1 + r_2)\dfrac{\cos^2\alpha}{\cos^2\alpha'} - \left(1 + \dfrac{r_1}{r_2}\right) f_i^2\right] \\ \vdots \\ Q_{exN} = \dfrac{B\omega_1}{2}\left[r_{a1}^2 + \dfrac{r_1}{r_2} r_{a2}^2 - r_1(r_1 + r_2)\dfrac{\cos^2\alpha}{\cos^2\alpha'} - \left(1 + \dfrac{r_1}{r_2}\right) f_N^2\right] \end{cases} \tag{8}$$

where,

B is the width of the center gear and driven gears.

ω_1 is the angular velocity of the center gear.

r_1 is the graduated circle radius of the center gear.

r_2 is the graduated circle radius of the driven gear.

r_{a1} is the addendum circle radius of the center gear.

r_{a2} is the addendum circle radius of the driven gear.

α is the pressure angle of gears.

α' is the meshing angle of the gear pair.

According to above-mentioned equations, when geometric and motion parameters of the gear pair are all constant, the instantaneous flow rate equations of outlets have their only variable that is f_i. According to equation (6), the instantaneous flow rate equations share only one variable, that is f_1. Just for Class II and Class III multi-pumps, the instantaneous flow rate phases of outlets are different.

4 SIMULATION AND ANALYSIS OF OUTLETS FLOW CHARACTERISTICS OF THREE KINDS OF MULTI-PUMPS

After building models for three kinds of multi-pumps, the simulation of flow characteristics is carried out by using CFD ICEM, Fluent and CFD-Post modules in ANSYS software.

4.1 *Models building and data post-processing of three kinds of multi-pumps*

The procedures of models building and data post-processing are as follows:

1. Building geometric models.

Building geometric models as three cases for three kinds of multi-pumps. Module m of center gears and driven gears is three. The teeth number of the driven gear z_2 is 10. The modification coefficient x_2 of the driven gear is 0.5. The number of driven gears N is 6. The teeth number of the three center gears are selected as 42, 43, and 44 for the three cases respectively, and the modification coefficients x_1 are all 0. Among these, the case of $z_1 = 42$ is considered as Class I, $z_1 = 44$ is considered as Class II, and $z_1 = 43$ is considered as Class III. The dimensions of the inlet and outlet are 14 mm and 12 mm, respectively. The radial gap between the housing and addendums of the center gear and driven gear is designed to be 0.1 mm.

2. Building calculation models.

After transferring the geometric models built as mentioned above to the CFD ICEM module, the mesh of flow domains (as shown in Figure 3) is divided properly by using triangular unstructured meshes globally. By using the dynamic mesh technique in Fluent, transient numerical simulation on the flow characteristics will be carried out. The calculation model is set for transient flow and k-ε turbulence mode, and the pressure–velocity coupling method is called as PISO. The density of oil is 866 kg/m^3. The viscosity is 0.0414 Pa•s. The gauge pressure values of six inlets are all set to 0 MPa. By using the UDF macro command, the rotating speed of the driven gear is 209.44 rad/s, and rotating speeds of the center gear of Class I, II, and

Figure 3. Schematic of the flow domain.

III are 49.867 rad/s, 47.6 rad/s, and 48.707 rad/s, respectively. The time step size is set as 1e-5, and the data files are auto saved every 25 steps.

3. Computing data post-processing.

After transferring the data files generated by using the Fluent module to the CFD-Post module, data post-processing is carried out. Due to the fixed dimension of outlets, the instantaneous flow characteristics can be expressed by using flow velocity, and the displacement characteristics can be expressed by using the mean flow velocity. By setting up monitoring points at the center of six outlets, the variation law of flow velocity with time is obtained.

4.2 *Simulation analysis of Class I multi-pumps*

By setting different gauge pressure values of six outlets in the Fluent module, flow characteristics simulation under different conditions are carried out.

1. The pressure values of six outlets are diagonally the same and different with adjacent outlets. The condition 1 is as follows: gauge pressure values of NO. 1, NO. 3 and NO. 5 outlets are all 3.5 MPa, and NO. 2, NO. 4 and NO. 6 outlets are all 2 MPa. The instantaneous flow velocity curves of six outlets are shown in Figure 4.
   From Figure 4, we can learn that the three outlet instantaneous velocity curves of 3.5 MPa coincide, and also the three curves of 2 MPa. The six curves form two groups. The greater the pressure, the lower is the flow velocity. The phases of six outlet instantaneous flow velocity curves are the same.
2. The pressure distribution of six outlets present offset loads. The condition 2 is as follows: gauge pressure values of NO. 1, NO. 2 and NO. 3 outlets are all 3.5 MPa, and NO. 4, NO. 5 and NO. 6 outlet are all 2 MPa. The instantaneous flow velocity curves of six outlets are shown in Figure 5.
   From Figure 5, we can learn that the curves of condition 2 exhibit the same characteristics as condition 1.
3. The pressure values of six outlets are different. The condition 3 is given as follows: gauge pressure values of NO. 1, NO. 2, NO. 3, NO. 4, NO. 5 and NO. 6 outlet are 4.5 MPa, 4 MPa, 3.5 MPa, 3 MPa, 2.5 MPa, and 2 MPa respectively. The instantaneous flow velocity curves of six outlets are shown in Figure 6, and the mean flow velocity trend curve under different pressure values is shown in Figure 7.
   From Figures 6 and 7, we can learn that, the six outlet instantaneous flow velocity values are degressive and the mean flow velocity decreases linearly along with an increasing in pressure.
4. Under different conditions, the instantaneous velocity values of the outlets whose pressure values are 2 MPa are compared. By collecting the outlets data of conditions 1, 2, and 3, the outlet instantaneous velocity curves are shown in Figure 8.
   From Figure 8, we can observe that the instantaneous flow velocity curves of six outlets are coincident under different conditions. The result shows that the instantaneous flow velocity of an outlet is only related to its own pressure, but not the other sub-pumps.

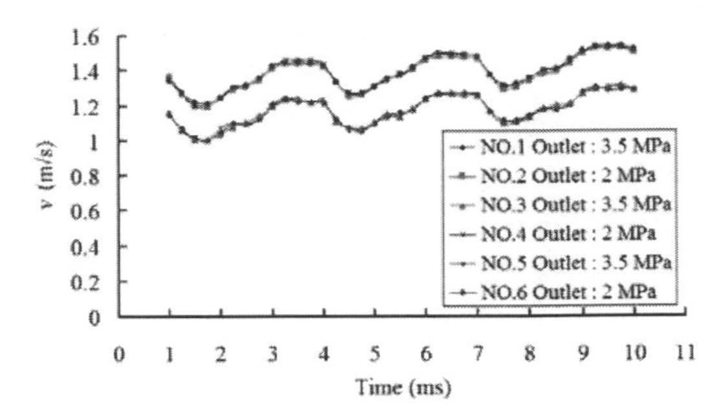

Figure 4. The outlet instantaneous flow velocity curves of Class I multi-pump under condition 1.

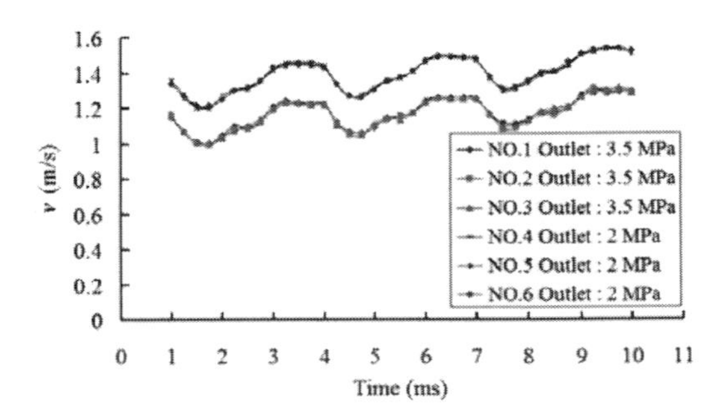

Figure 5. The outlet instantaneous flow velocity curves of Class I multi-pumps under condition 2.

4.3 *Simulation analyses of Class II and Class III multi-pumps*

Class II and III multi-gear pumps are simulated and analyzed under conditions which are the

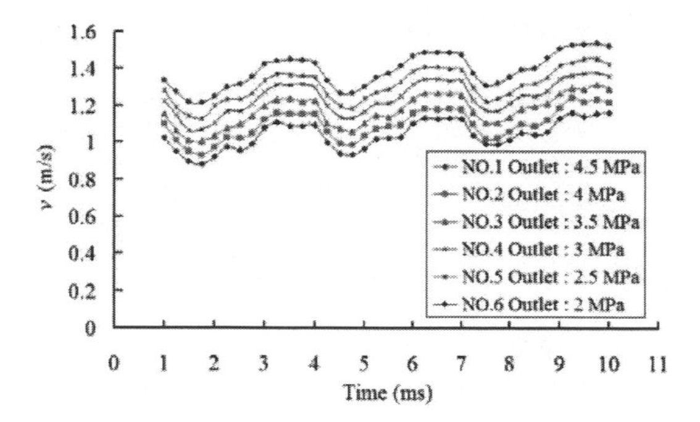

Figure 6. The outlet instantaneous flow velocity curves of Class I multi-pumps under condition 3.

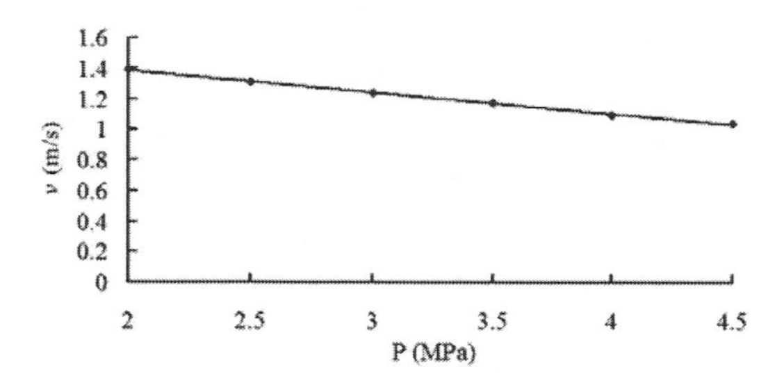

Figure 7. The outlet mean flow velocity trend curve of Class I multi-pumps under condition 3.

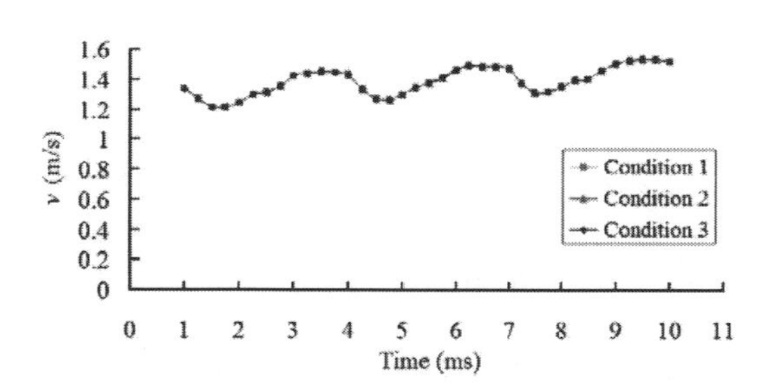

Figure 8. The outlet instantaneous flow velocity curves of Class I multi-pumps under different conditions.

same as that of Class I. The rest characteristics of flow rate are the same as that of Class I, except the phases of the outlet's instantaneous velocity curves. Therefore, we list only the conditions such that the six outlet pressure values are the same for Class II and Class III multi-pumps.

1. The condition of Class II multi-pumps is as follows: gauge pressure values of outlets are all 2 MPa. The instantaneous flow velocity curves of six outlets are shown in Figure 9, and the mean flow velocity values are given in Table 1. From Figure 9 and Table 1, it can be observed that the six outlet instantaneous flow velocity curves form three groups as follows: NO. 1 and NO. 4 outlets as a set, NO. 2 and NO. 5 outlets as a set, and NO. 3 and NO. 6 outlets as a set. The wave amplitudes and periods of the three group curves are the same, but there is a 1 ms phase difference between each other. The six outlet mean velocity values are almost the same. The result shows that the states of six sub-pumps are divided into three groups, and each group has two sub-pumps.
2. The condition of Class III multi-pumps is as follows: gauge pressure values of outlets are all 2 MPa. The instantaneous flow velocity curves of six outlets are shown in Figure 10, and the mean flow velocity values are given in Table 2. From Figure 10 and Table 2, it can be observed that wave amplitudes and periods of the six instantaneous flow velocity curves are the same, but there is a 0.5 ms phase difference between each other. The six outlet mean velocity values are almost the same. The result shows that the six sub-pumps are in different states.

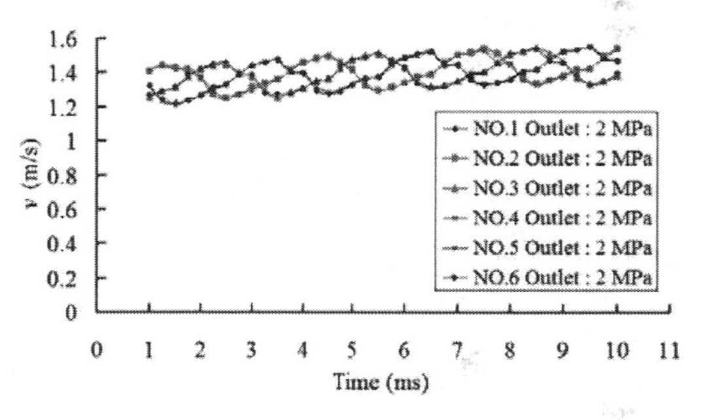

Figure 9. The outlet instantaneous flow velocity curves of Class II multi-pumps.

Table 1. The outlet's mean flow velocity of Class II multi-pumps.

Outlet	NO. 1	NO. 2	NO. 3	NO. 4	NO. 5	NO. 6
Mean flow velocity (m/s)	1.396	1.405	1.396	1.395	1.403	1.394

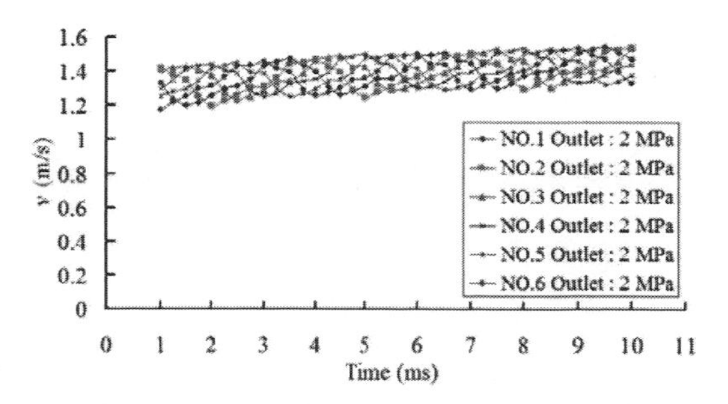

Figure 10. The outlet instantaneous flow velocity curves of Class III multi-pumps.

Table 2. The outlet's mean flow velocities of Class III multi-pumps.

Outlet	NO. 1	NO. 2	NO. 3	NO. 4	NO. 5	NO. 6
Mean flow velocity (m/s)	1.388	1.382	1.396	1.397	1.383	1.384

5 CONCLUSION

In this study, a new type of multi-gear pumps is put forward. The design of the new type of multi-gear pump and transient numerical simulation on the flow characteristics are carried out. The results of the study are as follows:

1. According to the relationship between the number of center gear teeth and the driven gears, the pump can be divided into three kinds. The working states of six sub-pumps of Class I multi-pumps are completely synchronous. The working states of six sub-pumps of Class II multi-pumps are divided into groups, and the working form is packets that are synchronous and cooperative among groups. The working states of six sub-pumps of Class III multi-pumps are different from each other.
2. The outlet mean flow velocity decreases linearly with an increase in pressure. The mean flow velocity values are the same when the outlet pressure values are the same. It indicates that the flow velocity of an outlet is only related to its own pressure value but not the other sub-pumps. The sub-pumps contained in one multi-gear pump has no effect on each other.
3. The characteristics of the multi-pump indicate that the pressure and flow interference among actuating cylinders can be avoided by using the new type of multi-gear pump in a multi-loop hydraulic system, and finally construct cylinders according to their respective stroke and movement time.

ACKNOWLEDGMENT

This research is supported by the National Natural Science Foundation of China (grant no. 51405494).

REFERENCES

Dong, D. S., H. CH. Deng, W. X. Ma, Power allocation analysis of hydraulic system with single pump and multiple motors of multifunctional snow plough, Transactions of the Chinese Society of Agricultural Engineering, **26**(7): 140–146 (2010)

Huang, K. J., W. C. Lian, Kinematic flowrate characteristics of external spur gear pumps using an exact closed solution, Mechanism and Machine Theory, **44**(6): 1121–1131 (2009)

Liang, R. Q., Z. Kan, Ch. S. Li, Applied Research of Hydraulic Drive Technology in Harvest Machinery, Machine Tool & Hydraulics, **40**(20): 152–155 (2012)

Lv, X. R., W. M. Ding, X. L. Lv, Design of hydraulic system of small multi-function chassis in hilly area, Journal of Huazhong Agricultural University, **33**(2): 128–132 (2014)

Saleh A. Al-Suhaibani, Mohamed F. Wahby, Farm tractors breakdown Classification, Journal of the Saudi Society of Agricultural Sciences, 1–5 (2015), http://dx.doi.org /10.1016/j.jssas.2015.09.005

Sh. M., Li, Zh. X. Zhu, E. R. Mao, Design of Proportional Raise Valve in Electro-hydraulic Lifting Mechanism of Big-power Tractor, Transactions of the Chinese Society for Agricultural Machinery, **43**(10): 31–35 (2012).

Yang, H. Y., J. Gao, B. Xu, G. M. Wu, Progress in The Evolution of Directional Control Valves and Future Trends, Chinese Journal of Mechanical Engineering, **41**(10): 1–5 (2005)

Zh. Q. Zhang, Zh. K. Cui, J. Y. Liu, Hydraulic System Design of 4YX-4 Type Full—hydraulic Self-propelled Corn Harvester, Journal of Agricultural Mechanization Research, 10: 97–101 (2015)

Zhao, J. J., ZH. X. Zhu, ZH. H. Song, Simulation and Experiment on Multi-directional Valve of Heavy Tractor Electro-hydraulic Hitch, Transactions of the Chinese Society for Agricultural Machinery, **45**(S1):1–9(2014)

Civil, Architecture and Environmental Engineering – Kao & Sung (Eds)
© 2017 Taylor & Francis Group, ISBN 978-1-138-02985-9

A study on the torsional fatigue fracture location of steel wires

Wenjie Wei
China Productivity Center For Machinery Beijing, Beijing, China

Decheng Wang
China Academy of Machinery Science and Technology Beijing, Beijing, China

Peng Cheng & Chenxi Shao
University of Science and Technology Beijing, Beijing, China

ABSTRACT: In the torsional fatigue test, the fracture location of steel wires is a critical factor influencing the effectiveness of the test. Based on the actual force of steel wires, in this paper, an attempt is made to carry out the finite element analysis, thus obtaining the stress variation of the cylinder busbar of steel wires. The additional stress influencing area of the clamping force is labeled, and the fracture location of steel wires in the actual test is adopted for verification. Finally, the run-length testing method shows that the fracture location of steel wires in the test area is randomly distributed.

1 INTRODUCTION

In previous torsional tests of steel wires, the clamping area is usually designed to be thicker than the test area, so as to avoid the influence of additional stress in the clamping area. However, due to the surface stress state, the drawing degree and dimension of the spring material are generally different from other steel wires, and it is hard to be made into test samples that are thin in the middle for fatigue tests. In this paper, through the torsional fatigue test of spring steel wires, the high-stress spring fatigue performance can be obtained quickly. The steel wire is directly clamped to the test machine in the torsional fatigue test of steel wires. Since there is additional stress in the clamping area, the obtained fatigue data cannot reflect the real fatigue life of steel wires. But the authenticity of the fatigue data can be judged by using the fracture location of steel wires. (Shao, 2013; Zhang, 1997)

2 PRINCIPLE (SONG, 2012; XIE, 2006)

The stress of the object caused by the load distributed in a small area (or volume) of the elastomer has no effect on the area that is away from the loading area, and its stress only makes a difference to the resultant force and the moment. The load can only affect the stress distribution near the loading zone. In the torsional test of steel wires, the clamping force is even with the moment and resultant force is 0. Thus, the additional force only affects the stress distribution near the clamping area.

Through the finite element analysis of steel wires, the stress distribution that is affected near the clamping area can be obtained.

3 FINITE ELEMENT ANALYSIS

3.1 *Geometric model*

The solidworks is applied in this paper to establish the geometric model of steel wires, as shown Fig. 1.

For the convenience of applied torque and clamping force, a 10 mm-long area is left on the steel wire whose diameter is 4.0001 mm, which is slightly larger than elsewhere. However, it cannot affect the result of finite element simulation (the unit size divided below is far larger than 0.0001 mm).

3.2 *Meshing and boundary conditions (Li, 2012; Xia, 2015; Lv, 2001)*

Considering the precision and model, the torsional steel wire belongs to the long and thin model, but the stress variation of the round section is huge, and so the bar unit cannot be adopted. The spatial solid ele-

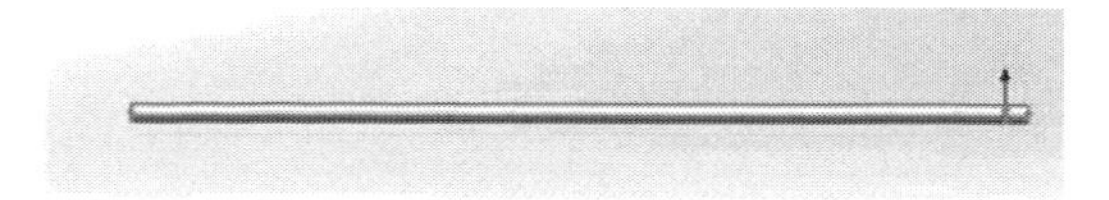

Figure 1. Schematic showing the geometric model of a steel wire.

ments of the four-nodes tetrahedron and eight-nodes hexahedron can be used together, and the automatic meshing can be used. In fact, the automatic switch between the tetrahedron mesh and hexahedron mesh can be conducted with this method. When the sweeping is available, the sweeping meshing will be executed; otherwise, the tetrahedron can be applied.

Since the diameter of the clamping area is slightly larger than that of other areas in this paper, the sweeping method cannot be applied for division in the whole model. In the arrows of Figure 1, a tetrahedron element is used. Its advantage is that its strong geometric adaptability and various irregular geometric models can be divided by it. And so, the four-node tetrahedron element unit becomes an indispensable unit type of the finite element analysis, and it can be used in the arrows of Figure 1. Because the tetrahedron element is a constant-strain element, many units are required to make up for the weaknesses of the function by itself in the big strain gradient area. Therefore, in the finite element analysis of the torsion shaft, the four-nodes tetrahedron element is used together with the eight-nodes hexahedron element, in order to ensure the precision of results. Furthermore, it can help to save the computing resources and reduce the calculation cost. In the local coordinate system, the standard unit of the eight-node hexahedral element is an eight-node cube. The coordinate origin is in the core of the cube with the edge length being 2 mm. The finite element model of the torsion shaft after meshing is shown in Figure 2.

3.3 *Finite element calculation and results*

Due to the constantly changing torque of the torsional steel wire in the test, the static analysis for the steel wire is meaningless. Instead, based on the actual situation, the transient dynamic analysis with the alternate torque is applied. According to the fact, a section of load and time process is applied to the steel wire in this work. Seven load steps have been applied to the steel wire within 1/8 s, as shown in Table 1. Through the application of alternating loads, the transient dynamic analysis is carried out (Cai, 2012).

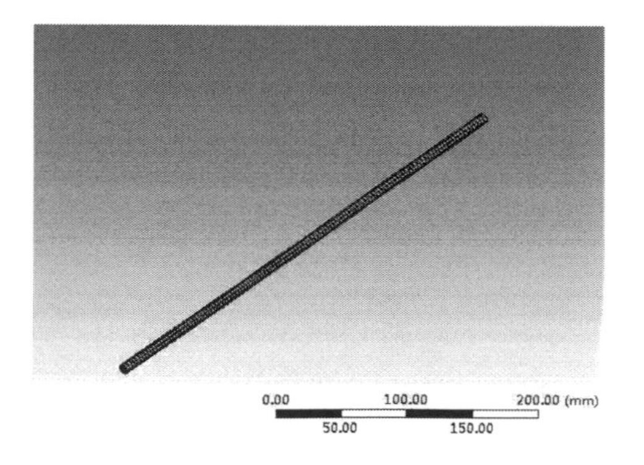

Figure 2. Schematic of the finite element model after meshing.

Because the left side of the steel wire is clamped on the static end in the actual test (as shown in Figure 1), six free degrees can be locked. The right-hand side cylinder diameter in Figure 1 is slightly larger than that of other areas. The alternating load can be applied as well as the 334 MPa additional force, so that the force of the steel wire will be completely consistent with that in the actual test.

According to the transient dynamic analysis of the finite element analysis for the torsional steel wire and the load–time process that is applied, the steel wire strain nephogram in a period is obtained, as shown in Figure 4. At the moment of the largest torsion angle, the maximum shear stress shows near the clamping area, but no dramatic change occurs due to the clamping force, and only 18 MPa is increased.

The stress diagram in the bus direction of the steel wire is extracted from the result. In the bus direction, one point of stress is taken every 1 mm to create the curve, as shown in Figure 5. It can be seen from the figure that, the additional stress influence is significant within 11 mm near the steel wire clamping area.

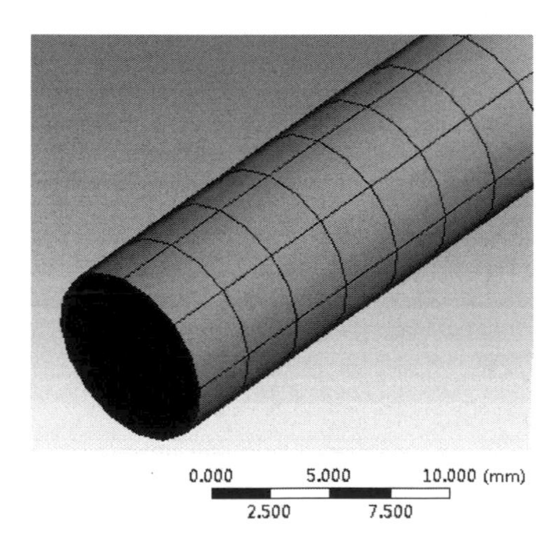

Figure 3. Schematic of the finite element model after amplification.

Table 1. The load–time process extracted according to the actual test.

Time/s	0	1/128	3/128	8/128	13/128	15/128	16/128
Load/MPa	0	500	1000	1000	1000	500	0

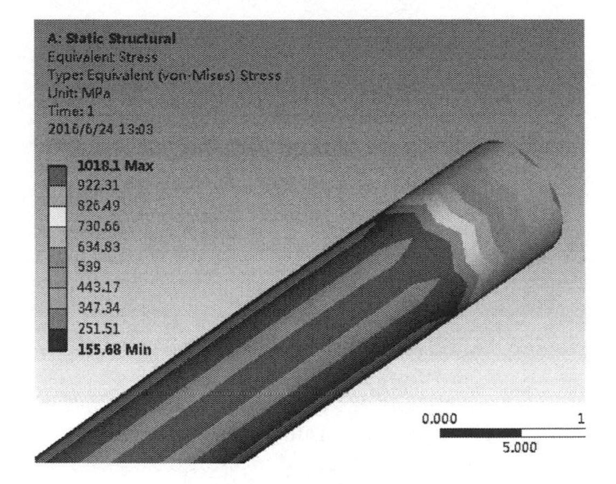

Figure 4. Schematic of the diagram of applying 1000 MPa stress.

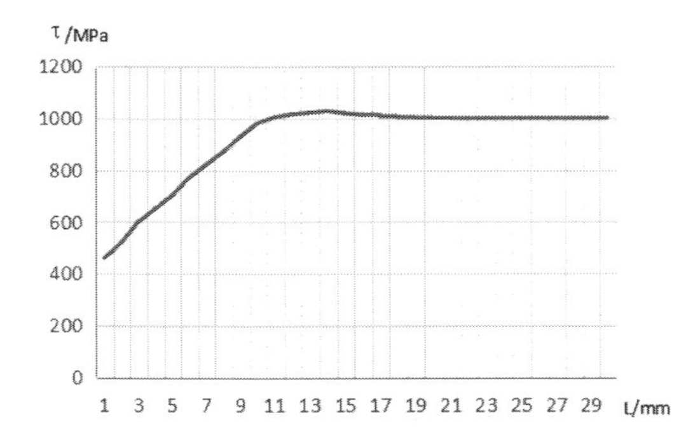

Figure 5. The stress distribution in the bus direction of the steel wire.

4 LABELING OF THE STEEL WIRE FRACTURE LOCATION IN THE TEST

4.1 *Test materials and methods*

The 55CrSi steel wire is adopted by this experiment as the test material. Its specific components are shown in Table 2, and others are Fe.

Since there's no specific test specification for the torsional fatigue test of steel wires, some labeling tests can be carried out in the early stage to determine the test specification of one diameter standard. The oil-quenching steel wire made by 55CrSi steel is used in this paper for the torsional fatigue test. Specific parameters are shown in Table 3.

Parameters confirmed in the test include the test frequency f (referring to ref. 1) shown in equation (1), and the clamping force F. The additional force needs to ensure that there is no slip of the steel wire affected by the alternating torque, which should not be too large to make the influence of the additional force significant.

According to the restrictive frequency f_n calculated in ref. 1, a safe value $\zeta = 0.8$ can be selected to obtain the following equation:

$$f_n = \frac{1}{4}\sqrt{\frac{\sqrt{T_n^2 - \frac{1}{384}\left(\tau_{\max}\pi d^3\right)^2}}{\left(J_M + J_L\right)a}} \tag{1}$$

And so, $f < f_n$. The clamping force can be measured by using the work pressure of the hydraulic power pack and labeled by the test, as shown in Table 4.

In the experiment, it is found that the torque test can proceed smoothly within the clamping force scope, the obtained fatigue life is similar, and that the variance fluctuation is within 5%. The mid-value of the clamping force is selected in the test, because the work pressure in the hydraulic power pack is an adjustable mid-value with small error.

Table 2. Setting mass fraction of the chemical components of the 55CrSi steel wire.

C	0.46–0.54	Si	0.17–0.37
Mn	0.50–0.80	Cr	0.80–1.10
V	0.10–0.20	S	< 0.30
Cu	< 0.25	P	< 0.30
Ni	< 0.35		

Table 3. Parameters of the steel wire.

Steel diameter	3.5 mm	Young's modulus E	200 GPa
Tensile strength	(1952–1972) MPa	Poisson's ratio	0.3

Table 4. Selection of the clamping force.

Length–diameter ratio	Clamping force	Clamping force for in the test
40	(4.0–5.0) MPa	4.5 MPa
50	(4.0–5.0) MPa	4.5 MPa
60	(5.0–6.0) MPa	5.5 MPa

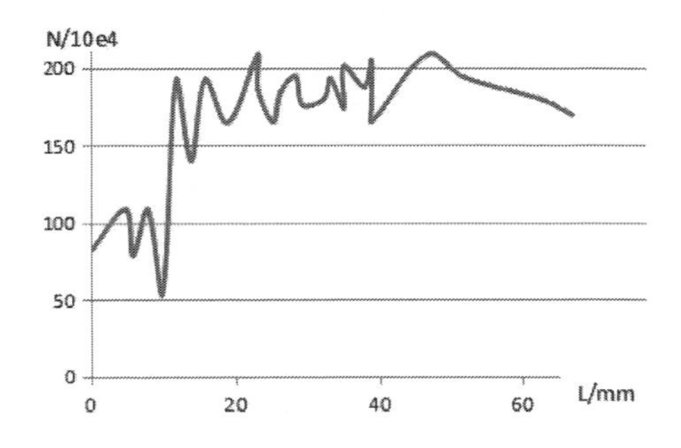

Figure 6. Curve of steel wire fracture locations and cycling times.

4.2 *Analysis of test results*

The steel wire whose length–diameter ratio is 50 and it is selected by this test. With the method mentioned in Section 3.1, the test-restrictive frequency f_n is obtained, so as to test with the frequency as large as possible. The selected clamping force is 4.5 MPa, and the 1000 MPa stress level is selected by the test. The torque test with the cycling performance being 0 is conducted. The cycling times and fracture locations of the test are recorded to create the fitting curve, as shown in Figure 6. As seen from the fitting curve, it can be divided into two parts.

When the fracture location is close to the clamping area, the cycling times will be smaller than 1 million, and when it is far away from the clamping area, the cycling times will be around 1.7 million. Obviously, the influence of the clamping force on the former is more significant.

As can be seen clearly from the curve, when the fracture location is 11 mm away from the clamping area, there are sudden changes for the cycling times. It is also the boundary of the two cycling times levels, which can also be regarded as the demarcation line of whether it is affected by the additional force. The experiment is basically consistent with the result of finite element analysis.

5 THE RUN TEST (ZHANG, 2014; KONG, 2011; LI, 2010; ZENG, 2008)

When combined with the finite element analysis and the test verification, the demarcation point of determining whether the fracture location is effective is labeled. No matter the finite element analysis or the test verification, system errors or measurement errors do exist all the time. Thus, verification for the above conclusions is necessary.

Within the effective fracture location, the steel wire stress is only relevant to the diameter according to the torque theory, and its stress is the maximum at the steel surface. Therefore, the fracture location of the steel wire should be a random variable. In this way, only the randomness of the fracture location shall be measured.

The runs test is a method specially used to check the data randomness. The test area is divided into several small areas, and the numbers are given in turn from the left to the right. In case there is steel wire fracture in the area, it is recorded as 1; otherwise, it is recorded as 0. As a result, the issue will turn to the check of the randomness of one 0–1 sequence. The test area is divided into 50 small areas evenly along with number provisions in turn. The obtained data are shown in the following table

It is assumed that H_0 denotes that the fracture in the table shows the random distribution, and H_1 denotes that the fracture in the table does not show the random distribution.

n_0 represents the number of 0 s, n_1 represents the number of 1 s, and R is the run length, and the statistics of the Runs test.

In cases when n_0 and n_1 are not larger than 20, the calculation test can be conducted by referring to the table, but n_0 and n_1 are both larger than 20 in this paper. Under the condition of H_0, R conforms to the normal distribution and its expectation and variance are given as follows:

$$E(R)=\frac{2n_0n_1}{n_0+n_1}+1 \qquad V(R)=\frac{2n_0n_1(2n_0n_1-n_0-n_1)}{(n_0+n_1)^2(n_0+n_1-1)}$$

Its form of standard normal distribution is given as follows:

$$Z=\frac{R-E(R)}{\sqrt{V(R)}}=\frac{R-\dfrac{2n_0n_1}{n_0+n_1}-1}{\sqrt{\dfrac{2n_0n_1(2n_0n_1-n_0-n_1)}{(n_0+n_1)^2(n_0+n_1-1)}}}\propto N(0,1)$$

Therefore, the critical value of the rejection region can be obtained by referring to the table in the given confidence level α.

Table 5. List of fracture locations of the steel wire.

No.	Fracture	No.	Fracture	No.	Fracture	No.	Fracture	No.	Fracture
1	0	11	0	21	0	31	0	41	0
2	0	12	0	22	1	32	0	42	1
3	1	13	1	23	1	33	0	43	0
4	1	14	0	24	0	34	0	44	0
5	0	15	1	25	1	35	0	45	0
6	1	16	1	26	0	36	0	46	0
7	0	17	0	27	0	37	1	47	1
8	0	18	1	28	1	38	0	48	0
9	1	19	1	29	1	39	0	49	0
10	0	20	0	30	0	40	0	50	0

Thus, the random test issue of the original series can be solved by using the normal distribution. In the above test data table, it can be obtained that $n_0 = 33$, $n1 = 17$, 23.44, and 21.72, and $Z = 0.00718$, because Z is in the range (–1.96, 1.96) (1.96 is the critical value of the standard normal distribution when the significance is 0.95).

H0 is accepted and the fracture in the table shows the normal distribution.

6 CONCLUSIONS

1. Based on the restriction of steel wires in the actual test, finite element analysis is carried out in this paper. The stress distribution curve of the cylinder bus of steel wires is elaborated and the demarcation point of the effective fracture location of steel wires is obtained. Based on the existing experimental data, the curve of fatigue life and fracture location are obtained, which is basically consistent with the theoretical demarcation point of the effective location.
2. The random test of the sample data of the steel wire fracture location is conducted, and the results show that the fracture location of steel wires in the test area is random, which further verifies the rationality of the demarcation point division of the effective location.

REFERENCES

Cai, L.G., S.M. Ma, Y.S. Zhao, Journal of Mechanical Engineering, 48, 165–168(2012).

Kong, K.L., J.J. Li, Journal of Applied Statistics and Management 32, 145–148(2011).

Li, H.F., J.C. Wu, J.B. Liu, Y.B. Liang, China Mechanical Engineering 3, 368–371(2012).

Li, X., J.M. Liu, Y.H. Jin, Statistical Research 17, 43–46 (2010).

Lv, J., Z.J. Wang, Z.R. Wang, Journal of Harbin Institute of Technology 33, 485–488(2001).

Shao, C.X. Study on the Torsional Fatigue Test Method and Test Prototype of Spring Steel Wires, 23, (2013).

Song, S.Y., F. Yin, Machinery Design & Manufacture 8, 63–66(2012).

Xia, S.L., J.W. He, Aircraft Design 28, 10–13(2015).

Xie, X.L., W.L. peng, R. qin, China Journal of Highway Transport 14, 33–36(2006).

Zeng, G. Science Technology and Industry 8, 73–76(2008)

Zhang, L.F. Value Engineering 13, 208–211(2014).

Zhang, Y.H., H.H. Liu, D.C. Wang, Spring Manual (1997).

Civil, Architecture and Environmental Engineering – Kao & Sung (Eds)

The lightweight design of the support arm

Yaqin Tian
Taiyuan University of Science and Technology, Taiyuan, China

Dan Meng & Xiaoyi Lin
JMC Heavy Duty Vehicle Co., Ltd, Taiyuan, China

ABSTRACT: The model of the arm was established for supported kids' sickbeds under the bed board by using Creo software in this paper. The relationship of the arm width, plate thickness, stress, and the total weight was analyzed through topological optimization under the ANSYS Workbench, and a lightweight design is implemented. The weight of arm is reduced about 33% by topological optimization under the ANSYS Workbench and the production cost of beds is also reduced.

Keywords: topological optimization, light weight, structure optimization

1 INTRODUCTION

With the development of the medical industry, more and more medical bed series were designed. There were different structure designs for different age levels of sickbeds. A lightweight design is an important method to reduce the cost of the large quantity production. The materials' design of lightweight (Cho, 2009), shape and size optimization, structure optimization, and multi-objective optimization design (Leite, 2015) are the main directions of lightweight research. The weight can be reduced by 48% after applying the weight loss design for some structures (Raedt, 2014). The lightweight design of the sandwich structure can reduce weight 7.5 kg/m^2 for the trailer (Michael, 2013). The lightweight design of a continuous variable cross-section roll plate is used to design the hybrid column, which makes the weight of the car reduced by about 20% (Thomas, 2015). The lightweight design method based on the impact of the car body, and the body weight can be reduced by 5.2% from the sensitivity analysis and optimization calculation to the thickness of the main structure (Lan, 2010). Lightweight structures for interior and exterior as well as car bodies, which can be produced in a flexible manner, are in the center of discussion. In order to satisfy the performance requirement of high dynamic and static characteristics and light weight of the activity crossbeam, the mass of the activity crossbeam was decreased by 12.2% through a method of multi-objective optimization (Sun, 2015). It also can reduce the structure weight by 21.4% with the improved response surface method (Pan, 2010). An arm's model was established which supported kids' sickbeds under the bed board by using Creo software in this paper. The lightweight design is implemented to the arm under the ANSYS Workbench, which contains the Modeler Design and three modules of the Mechanical and Xplorer Design.

2 TOPOLOGY OPTIMIZATION BASED ON THE VARIABLE DENSITY METHOD

The topology optimization method considers the material distribution as an object. The optimal distribution scheme can be found in the design space of uniformly distributed material, and then the final shape can be determined. Material utilization can greatly be improved (Kirsch, 1989). The main research methods of the continuum topology optimization are level set method, evolutionary structural optimization, (Kim, 2000), homogenization method (Nikos, 2015), variable density method etc. In addition, there is the discrete topology optimization (Cheng, 2014) and the method of topology optimization (Krister, 2009) based on the genetic algorithm (Kawamura, 2002) which is also being studied. ANSYS topology optimization simulation is the combination of the variable density method and finite element method. The variable density method includes SIMP (Solid Isotropic Microstructures with Penalization) and RAMP (Rational Approximation of Material Properties). These models are established by the penalty factor, which is a non-linear relationship between the elastic modulus of the material and the relative density of the element. Its effect is to punish the intermediate density value, so that the intermediate density value is gradually gathered to the 0/1 level. When the values of the design variables are in the range of (0, 1), in this way, the continuous variables can be well-approximated by the model of 0–1 discrete variables. At this time, the intermediate density unit corresponds to a very small elastic modulus, and the influence of the structural stiffness matrix will become very small and can be ignored. For the RAMP density stiffness interpolation model, the intermediate density of the penalty is carried out by the following equation:

$$E_i = \frac{x_i}{1+p(1-x_i)} E_0$$

where, Ei is the elastic modulus of the No. i unit. E0 is the elastic modulus of the material when the unit is filled, and p is the penalty factor.

Whether the intermediate density material can be effectively eliminated or not and obtain clear optimization results are affected by the penalty factor p. The general penalty factor p is in the range of 3–31.

In general, the optimization model of the variable density method is shown in Fig. 1. In the optimization model equation, xi is the design variable and it is the relative density of materials; n is the number of finite elements in the design domain; C(x) is the objective function, and it is the structural compliance; K is the total stiffness matrix of the structure; U is the general displacement vector of the structure; F is the load vector of the structure; V is the volume after the optimized structure; Vi is the unit volume; f is the volume fraction of a given material; V0 is the volume of an initial structure; V* is the volume cap; xmin is the minimum relative density. The optimization design process under the ANSYS Workbench is shown in Figure 1.

The Design Modeler is used to build a model module. It is the module to realize the interaction between CAD and CAE and it is the first step in product development and processing too. Simulation is the second step. The third is the Design Xplorer. It can obtain clear optimization results by

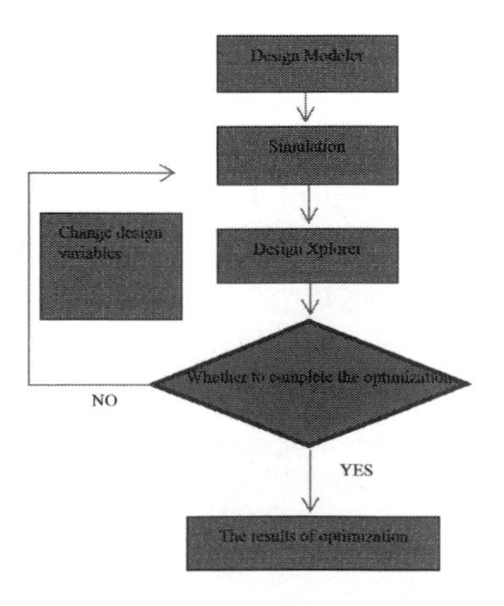

Figure 1. The process of optimization design on ANSYS Workbench.

choosing appropriate penalty factors and topology optimization.

3 THE MODELING PROCESS BY USING THE ANSYS FINITE ELEMENT

The specific steps of product optimization design by using the Ansys workbench is as follows: modeling, add dimension parameters, add material properties, meshing, add the loads and constraints, finite element analysis, topology optimization, shape optimization design, defined target parameters and status parameters, optimization, comparison, and validation. Parameter optimization is also solved directly for fixed shape.

The arm structure is the most important structure of the connection bed plate and hand screw pipe. In order to analyze the stress and deformation, the model must be simplified. The seamless steel pipe welded together with the arm is ignored. The model and the key parameters are imported to the Design Modeler Workbench platform through the Creo and Workbench ANSYS seamless connection and the parameters can be fully controlled by using the Workbench.

The material of the arm is ordinary carbon steel Q235. Its properties are as follows: the yield limit is 235 MPa, Poisson's ratio is 0.3, the elastic modulus is 2 × 105 MPa, and the density is 7.85 × 10–9 T/mm^3. The original size of the arm is as follows: the length is 180 mm, width is 50 mm, and thickness is 2 mm. As the arm is stamping, the free divides of the function in meshing is selected. The grid size is 1 mm, the number of discrete nodes is 136431, and the unit number is 74359.

4 STATIC ANALYSIS

4.1 *The boundary conditions and loading*

The screw is connected with the support arm by using the pin. On the other end, the supporting arm is welded in a seamless steel tube of Φ32 × 2. The arm has been in force when the bed is under working condition, and can be regarded as a two-force bar. The direction of the force is consistent with diameter connection of two different holes of the support arm. Children's weight is generally not more than 80 kg under 15 years of age, and the safety factor is 1.5. Thus, the putting weight is 120 kg. The maximum force is in the arm which lies on the left-hand side of the bed frame, and the bed component effects are neglected. And so, the applied force is 1200 N. The displacement constraint is 0 mm in the small hole in the X, Y, and Z directions. The solution result for the total mass is 0.27913 kg, the maximum stress is 34.517 MPa (Figure 2), and the maximum total deformation is 0.043639 mm (Figure 3).

Figure 2. Graph of arm stress.

Figure 3. Graph of arm deformation.

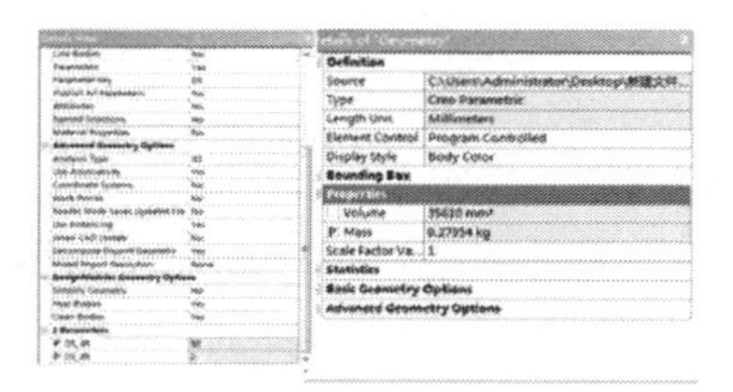

Figure 4. The setting process of parameter.

Figure 5. Schematic of the results of topology optimization.

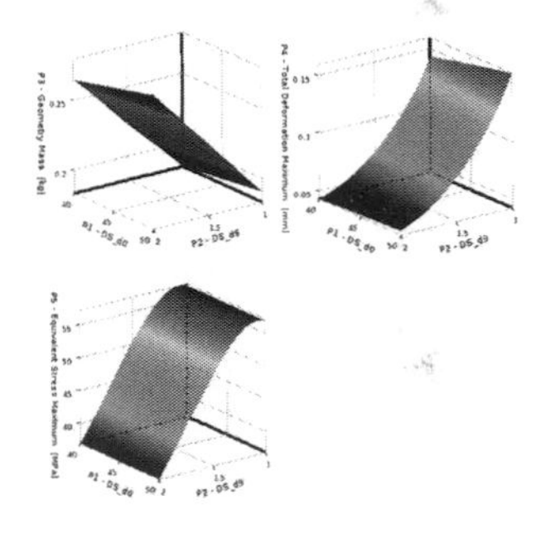

Figure 6. Graph showing the effect of mass, deformation, and stress by size.

4.2 *The size optimization of the arm*

Before importing Workbench, the variable name must be modified in Creo. The width of the supporting arm of the variable name is changed to DS_d0, the thickness is modified to DS_d9, so that these two dimensions can be identified in Workbench Ansys. At the same time, the model type and the parameters of the model are selected. "P" will appear before the size (Figure 4), which indicates that the size is optimized at this time. Finally, the Modeler Design module is exited. In the same way, the quality, the maximum total deformation, and the maximum equivalent stress are parameterized. After the parameter setting is completed, the Experience Design module is entered. The width dimension (DS-d0) is between 40–50 mm and the plate thickness (DS-d9) is in the 1–2 mm range at the Input Parameters. Taking into account the extreme use of the situation, a certain degree of safety is set aside. The result requires that the maximum stress is less than the Q235 yield stress of 55 MPa, and the solution of the target is the minimum of the mass and the deformation.

4.3 *Topology optimization design of the ANSYS Workbench*

There are three ways to display the results in the Workbench ANSYS topology optimization module. First is to show that the topological iteration process through the list. The second is to display the topology of the iterative process with graphics. And the other topology optimization results are shown by using different colors on the entity. The penalty factor is set to reduce weight by 20%, and then the supporting arm shape topology optimization is carried out. The most intuitive method of topology optimization is selected, and the results are shown in Figure 5.

The darker parts of the map can be removed in the region. The lighter part of the color indicates a less important part and can be removed or retained as needed. The gray-colored region is the retained region. After optimization, the effect of mass, deformation, and stress by size are shown Figure 6. The change of thickness has little influence on the quality of the arm while has a great influence on the total deformation and stress. The change of the width has little effect on the stress, as shown in Figure 6.

Ten groups of optimization design results (Figure 7) are compared. Stress over 55 MPa of the optimized groups 4, 6, 7, was excluded. The final optimization plan is the lightest of the quality of the group 2 in the remaining 7 groups. The arm width is 40 mm and thickness is 1.5 mm after optimization.

The arm size of the 2nd group was calculated by using the finite element method. The arm mass is

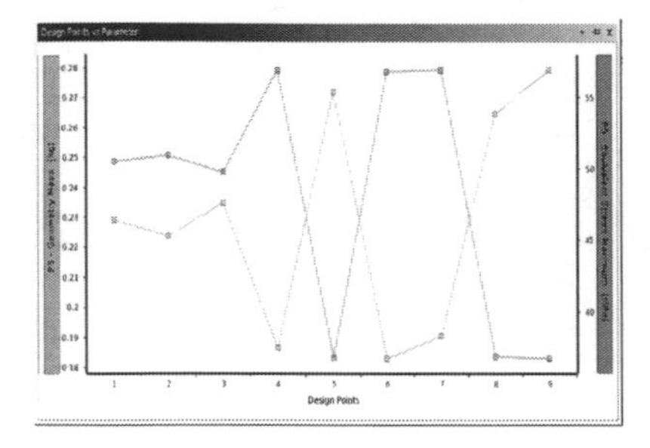

Figure 7. Graph showing the quality and stress of each scheme.

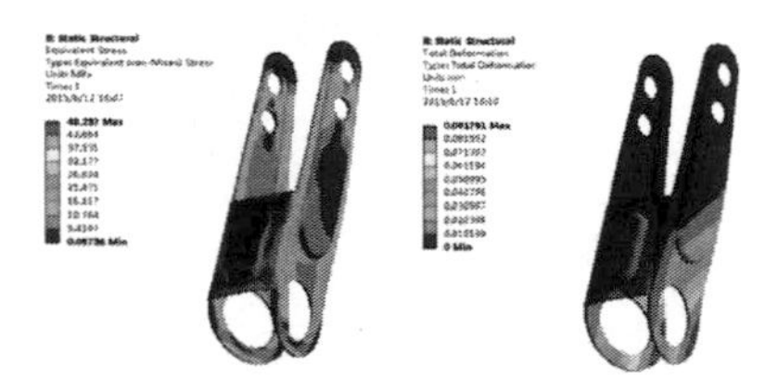

Figure 8. Graphs showing deformation–stress after optimization.

0.18709 kg after calculation. The maximum deformation is 0.09179 mm. The maximum stress is 48.287 MPa (Figure 8). It is less than the strength requirement of the material yield stress which is 55 MPa. After making a quality comparison before and after optimization, it can be observed that the weight reduced by 33% after optimization.

5 CONCLUSION

The arm's optimization design was carried out for the initial width of 50 mm and thickness of 2 mm for supported kids' sickbeds under the bed board by using ANSYS workbench in this paper. The width of the arm is 40 mm and the thickness is 1.5 mm after optimization. Based on the strength requirements, the optimized quality of the arm is reduced about 33%.

ACKNOWLEDGMENTS

This work was financially supported by the Doctoral Project of the Taiyuan University of Science and Technology of Shanxi Province in China (grant no. 20122003), the Funding of Science and Technology Activities of the Ministry of Human Resources and Social Security for International Students (the year 2015 startup), and Fund Program for the Scientific Activities of Selected Returned Overseas Professionals in Shanxi Province.

REFERENCES

Cheng Chang, Airong Chen, The gradient projection method for structural topology optimization including density-dependent force[J]. Structural and Multidisciplinary Optimization, 2014.10, Vol. 50(4), pp: 645–657.

Hans-Willi Raedt, Frank Wilke, Christian-Simon Ernst, The Lightweight Forging Initiative Automotive Lightweight Design Potential with Forging[J] ATZ worldwide, March 2014, Volume 116, Issue 3, pp: 40–45.

Kawamura, H., H. Ohmori, N. Kito, Truss topology optimization by a modified genetic algorithm [J]. Structural and Multidisciplinary Optimization, 2002.07, Vol. 23(6), pp: 467–473.

Kim, H., O.M. Querin, G.P. Steven, Y.M. Xie, A method for varying the number of cavities in an optimized topology using Evolutionary Structural Optimization[J]. Structural and Multidisciplinary Optimization, 2000.04, Vol. 19(2), pp: 140–147.

Kirsch, U., Optimal topologies of structures [J]. Applied Mechanics Review, 1989, 42:223–249.

Koo*, J.S. and H.J. Cho, Theoretical method for predicting the weight reduction rate of a box-type car body for rolling stock by material substitution design, International Journal of Automotive Technology, 2009, Vol. 10 No. 3, pp: 355−363.

Krister Svanberg, Mats Werme, On the validity of using small positive lower bounds on design variables in discrete topology optimization[J]. Structural and Multidisciplinary Optimization, January 2009, Volume 37, Issue 4, pp: 325–334.

Lan Fengchong, Zhuang Liangpiao, Zhong Yang, Chen Jiqing & Wei Xingmin, Study and Practice of Car Body Struture Lightweight Design[J]. Automotive Engineering, 2010.09 Vol. 32, pp: 763–768, 733 (In Chinese).

Leite, M., A. Silva, E. Henriques, J.F.A. Madeira, Materials selection for a set of multiple parts considering manufacturing costs and weight reduction with structural is performance using direct multisearch optimization[J]. Structural and Multidisciplinary Optimization, 2015.10, Vol. 52 (4), pp: 635–644.

Michael Hamacher, Lutz Eckstein, Birger Queckenstedt, Klaus Holz, Intelligent Trailer in lightweight Design[J], ATZ worldwide, May 2013, Vol. 115(5), pp: 22–25.

Nikos T. Kaminakis, Georgios A. Drosopoulos, Georgios E. Stavroulakis, Design and verification of auxetic microstructures using topology optimization and homogenization[J]. Archive of Applied Mechanics, 2015.09, Vol. 85 (9), pp: 1289–1306.

Pan F, Zhu P, Zhang Y. Met model-based lightweight design of B-pillar with TWB structure via support vector regression [J]. Computers and Structures, 2010, 88(1–2), 36–44.

Stolpe M, Svanberg K. An alternative interpolation scheme for minimum compliance topology optimization. Structural and Multidiscipline Optimization, 2001, 22: 116–124.

Sun Kang, Jin Chaofeng, Wu Xin, Dynamic and static multi-objective optimization of the activity crossbeam based on response surface method[J]. Manufacturing Technology & Machine Tool, 2015.12, pp: 71–76 (In Chinese).

Thomas Muhr, Johannes Weber, Andreas Theobald, Martin Hillebrecht, Economically Viable Lightweight Design Concept for a Hybrid B-pillar [J]. ATZ worldwide, March 2015, Vol.117(3), pp: 4–9.

Civil, Architecture and Environmental Engineering – Kao & Sung (Eds)
© 2017 Taylor & Francis Group, ISBN 978-1-138-02985-9

An analysis of the thermal elastohydrodynamic lubrication performance of the slope of the thrust bearing

YuLei Liu, YongHai Li, TianYu Niu, JueFei Huang & YuXin Wang
School of Mechanical and Engineering, Harbin University of Science and Technology, Harbin, China

ABSTRACT: A mathcmatical model of the thermal elastohydrodynamic lubrication of fixed and inclined thrust bearings is established according to the lubrication theory and the finite element numerical analysis method and the preparation of the calculation program and calculation are compared. The results show that the two calculation methods are in line with the general rules of the lubrication theory, and thermal elastohydrodynamic lubrication's calculation results tend to be safe. This conclusion can provide reference for the design and application of the thrust bearing.

1 INTRODUCTION

These are very important and the key problem in the design of the bearing is to improve the lubrication performance of the bearing, reduce the fiction of the bearing, and reduce the failure of the thrust bearing. In order to improve the lubrication performance and reduce the thrust bearing's burning tile to bring losses, many scholars studied variable viscosity and bearing deformations of the tilting pad's thrust bearing. The results showed that the bearing deformation had a larger effect on the lubrication performance of journal bearings (Wu, 2007; Zhu, 1990; An, 2011; Liu, 2010; Ge, 1983). But there were no studies on the fixed pad bearing, which is still under constant temperature and viscosity conditions of the simple calculation, to account for the variable viscosity and deformation of thc shell of the fixed pad thrust-bearing's properties and lubrication properties' analysis. In this paper, the lubrication mechanism of the fixed and inclined thrust bearings is revealed by using the knowledge of fluid lubrication, included in the viscosity, temperature change, bearing deformation etc. These are factors identified on the basis of the above-mentioned research results. The results have theoretical significance and application value to the design of the thrust bearing.

2 MATHEMATICAL MODEL

2.1 *Shape of the oil film*

The fixed pad's thrust-bearing inclined plane is used in the structure of the horizontal axis. The bearing will form a dynamic pressure oil film, and the film shape is formed between the tile surface and thrust plate (mirror plate) gap's oil film, and set the tile surface deformation as a parabolic and the thrust disc (mirror plate) as a rigid plane; the oil film shape is as shown in Figure 1. The governing equation of the oil film shape is given as follows (Ding, 1986; Chen, 1980; Pang, 1981; Wang, 2004; Han, 2004):

$$h_i = h_0 + \frac{c}{l} r_i \sin\left(\theta_i - \frac{\theta_0}{2}\right) + \frac{e^2}{e_A^2}\delta_{\max} \tag{1}$$

where *hi* is the oil film thickness of the surface; h_0 is the center of the surface of the oil film thickness; *c* is the elevation of the slope of the tile; *l* is the tile length; r_i is the tile surface at any point of the pole radius; Q_i is the tile surface at any point of the polar angle; Q_0 is tile angle; *e* is the distance to the tile surface at any point in the heart; e_A is the largest deformation point on the surface to the center distance; and δ_{max} is the largest deflection deformation of bearing.

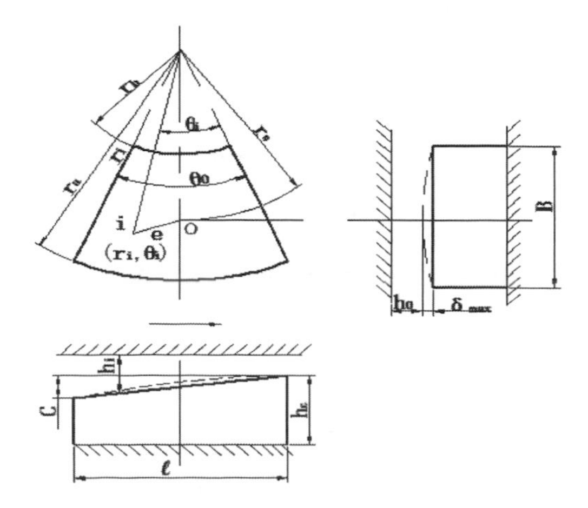

Figure 1. Schematic of the oil film's shape.

$$e^2 = r_i^2 + r_0^2 - 2r_i r_0 \cos\left(\theta_i - \frac{\theta}{2}\right) \tag{2}$$

$$e_A^2 = r_A^2 + r_0^2 - 2r_A r_0 \cos\frac{\theta_0}{2} \tag{3}$$

where r_A is the outer radius of the tile and r_0 is the polar radius of the center point of the tile surface.

2.2 *Oil film pressure equation*

$$\frac{\partial}{\partial r}\left(\frac{rh^3}{\mu}\frac{\partial p}{\partial r}\right) + \frac{\partial}{\partial \theta}\left(\frac{h^3}{\mu}\frac{\partial p}{r\partial \theta}\right) = 6\omega r\frac{\partial h}{\partial \theta} \tag{4}$$

where P is the pressure of the fluid; μ is the lubricating oil's viscosity; h is the thickness of the oil film; ω is the angular velocity; and r and θ are respectively the radial and circumferential coordinates. The boundary condition can be written as zero pressure along the surface of the tile.

$$\bar{p}|_{\Sigma} = 0 \tag{5}$$

where Σ is the boundary.

2.3 *Oil film temperature equation*

The temperature field of the bearing can be solved by using the energy equation. The energy equation of the thrust bearing of the inclined surface tile is given as follows:

$$JC_p\rho g\left[\left(\frac{\omega rh}{2} - \frac{h^3}{12\mu}\cdot\frac{\partial p}{r\partial \theta}\right)\frac{1}{r}\frac{\partial T}{\partial \theta} - \frac{h^3}{12\mu}\left(\frac{\partial p}{\partial r} - \rho\omega^2 r\right)\frac{\partial T}{\partial r}\right] - Jkh\left[\frac{\partial^2 T}{\partial r^2} + \frac{\partial^2 T}{r^2\partial \theta^2}\right] + JK_B(T - T_0) = \frac{\mu\omega^2 r^2}{h} + \frac{h^3}{12\mu} \times\left[\left(\frac{\partial p}{\partial r}\right)^2 + \left(\frac{\partial p}{r\partial \theta}\right)^2\right] \tag{6}$$

where J is the mechanical equivalent of heat; K and C_p are respectively the thermal conductivity of the oil film and specific heat capacity; T and ρ are respectively temperature and density; T_0 is the oil temperature; K_B is the heat conduction coefficient; and P is the solution of pressure values.

Boundary condition: $T=T_0$:

Temperature initial condition:

$$\varphi = 0, T = T_0, \frac{\partial T}{\partial R} = 0, r = R_2, \frac{\partial T}{\partial R} = 0.$$

2.4 *Viscosity temperature equation*

The viscosity temperature equation is obtained by using the Lagrange interpolation method.

$$\mu = \sum_{i=1}^{n}\left[\prod_{\substack{j=1\\ i\neq j}}^{n}\frac{T - T_j}{T_i - T_j}\mu_i\right] \tag{7}$$

where T_i and U_i are the temperature and the viscosity value of the reference point; n is the number of the reference point; T is the interpolation temperature point; and μ_T is the response viscosity.

The oil temperature boundary conditions of the bearing can be carried through the heat equation to be determined.

$$T_{in} = \left(\frac{1-K}{1-K/2}\right)T_s + \left(\frac{K}{2-K}\right)T_{out} \tag{8}$$

where T_s is the oil supply temperature; T_{out} is the oil outlet temperature; and K is the hot oil carrying factor, whose value is 0.5–0.9; depending on the speed of the bearing surface of the selected bearing, $K = 0$ is the carrying oil calculation.

$$T_{in} = T_{out}$$

2.5 *Single tile oil film control equation*

The general properties of the equations in a class of similar phenomena are taken into account, and marked as the dimensionless quantity:

$$\begin{cases} r = R\bar{r}, h_{min} = \min\{h\}, h = h_{min}\bar{h}, \mu = \mu_0\bar{u}, p = p_\rho\bar{p}, \\ T = T_\rho\bar{T}, \theta = \bar{\theta}, \mu_0 = \left(\sum_{i=1}^{N}\mu_i\right)\Big/N, p_\rho = 6\mu_0\omega R^2 / h_{min}^2, \\ T_\rho = p_\rho / JC_\rho\rho g = 6\mu_0\omega R^2 / JC_\rho g h_{min}^2; D = 2\lambda / C_\rho\rho g h_{min} \\ \omega d, M_R = m_r R / h_{min}, M_\theta = m_\theta R / h_{min}, A = \delta_{max} / h_{min}; e_A^2 = 1 \\ +\overline{R_0}^2 - 2\overline{R_0}\cos\left(\bar{\theta}_r/2\right), \bar{e}^2 = \bar{r}^2 + \bar{R}^2 - 2\overline{R_0}\bar{r}\cos\left(\bar{\theta} - \bar{\theta}_r/2\right) \end{cases} \tag{9}$$

The lubrication performance of the thrust bearing of the inclined surface tile can be solved by using the above equation. Because most of the mathematical models are composed of two partial differential equations, the finite element method is used to solve the problem.

3 EXAMPLE AND COMPARATIVE ANALYSIS

3.1 *Calculation method of the lubrication performance*

In the past, for the inclined plane thrust bearing's lubrication performance calculation with the traditional approximate calculation method, an approximate calculation equation was used, regardless of the lubricating oil viscosity changes and bearing deformation, through the look-up table and the curve of rough calculation and there are many kinds of calculation methods. The thermal EHL numerical calculation method, finite element method, control equation of the simultaneous solution of the oil film shape equation, Reynolds equation, energy equation, viscosity temperature control equation, in the meantime, considering the change of oil viscosity and the bearing deformation (Zhang, 2014; Fan, 2013; Li, 2014; Yu, 2014; Fan, 2014; Ting, 1971; Sharma, 2002; Yoshimoto, 1993; Younes, 1993; Adam, 2003).

3.2 *Calculation results and comparative analysis*

The calculation of bearing parameters respectively is performed for the following parameters: the bearing's outer radius is 0.19 M, bearing's inner radius is 0.09 m, tile is 10 mm thick, number of tiles is 3, the lubricating oil brand is L-TEA46, lubrication is oil-immersed lubrication, and the oil temperature on the inside is 33°C.

3.2.1 *Influence of speed on lubrication characteristics and comparative analysis*

Figures 2–5 show the relationship between the velocity and the minimum oil film thickness, the maximum film temperature, the maximum film pressure, and the power consumption of the two calculation methods. The sign of ▲ shows approximate calculation results. The sign of ■ shows thermal EHL calculation results.

It can be seen from the figures that, the relationship between two methods of calculating speed and minimum oil film thickness, maximal oil film temperature, maximum oil film pressure, and power consumption, in line with the general rules of the lubrication theory, oil film pressure and power consumption of the two calculation methods changed little, the minimum oil film thickness and maximum oil film temperature is slightly different, and the thermal EHL calculation tends to be more conservative. From the figures, the relationships between calculating speed and the minimum oil film thickness of two methods, the maximal oil film temperature, the maximum oil film pressure and power consumption can be obtained.

3.2.2 *Effect of specific pressure on lubrication characteristics and comparative analysis*

Figures 6–9 show the relationship between the two calculation methods, the specific pressure and the minimum film thickness, the maximum oil film temperature, the maximum oil film pressure, and power consumption. Because of the limited space, only the speed of the calculation is 18 m/s.

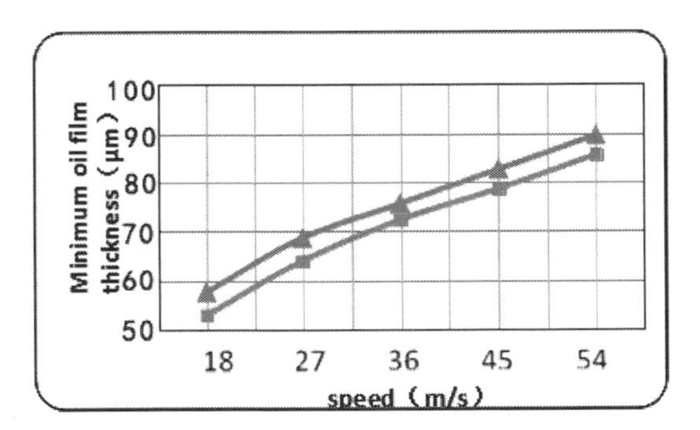

Figure 2. Variation curve of the minimum oil film thickness with speed.

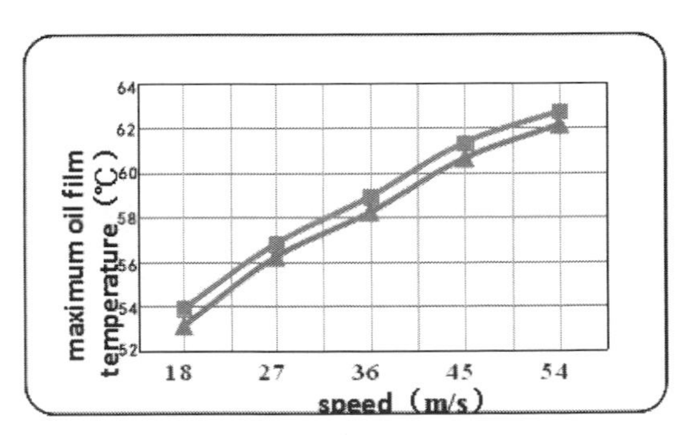

Figure 3. Variation curve of the maximum oil film temperature with speed.

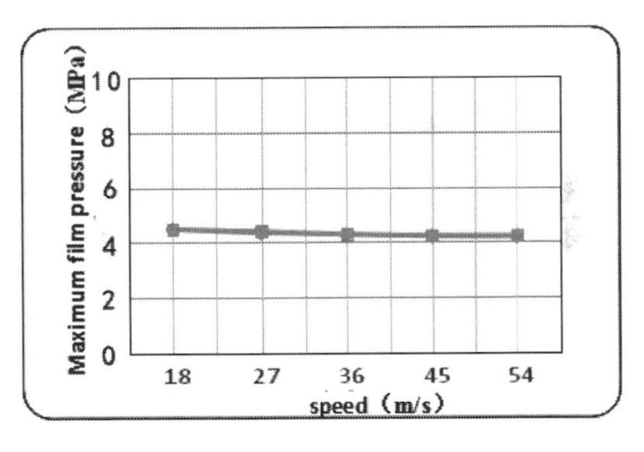

Figure 4. Variation curve of the maximum film pressure with speed.

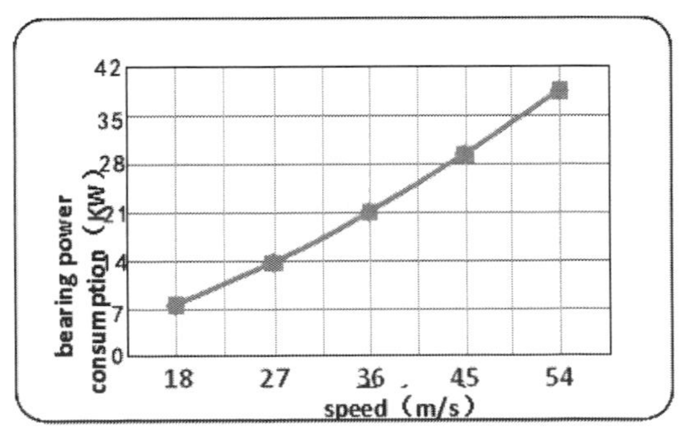

Figure 5. Variation curve of the bearing's power consumption with specific pressure.

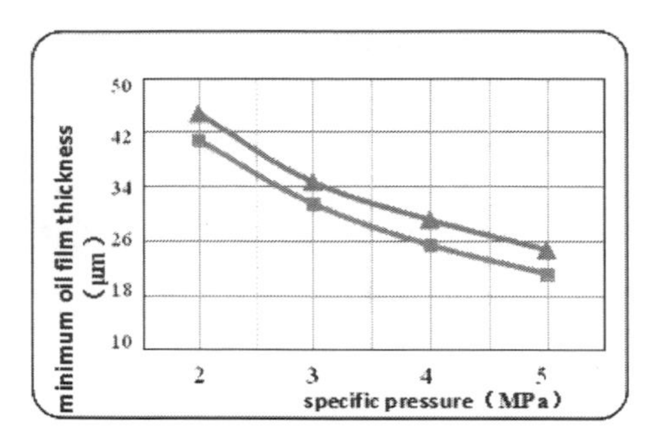

Figure 6. Variation curve of the minimum oil film thickness with specific pressure.

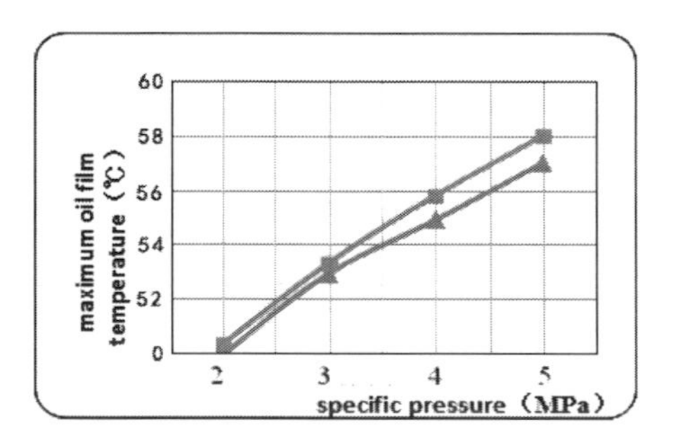

Figure 7. Variation curve of the maximum oil film temperature with specific pressure.

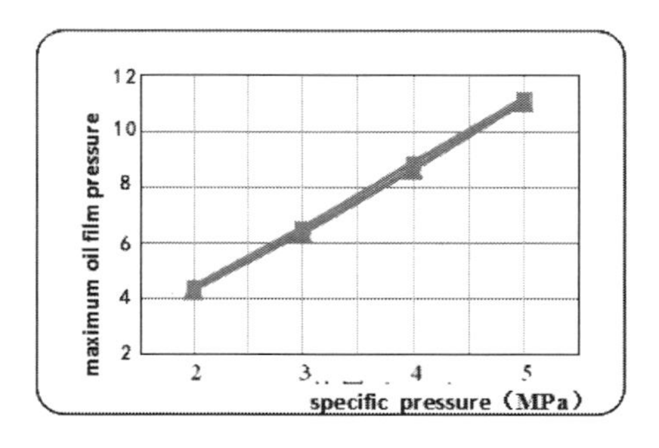

Figure 8. Variation curve of the maximum oil film pressure with specific pressure.

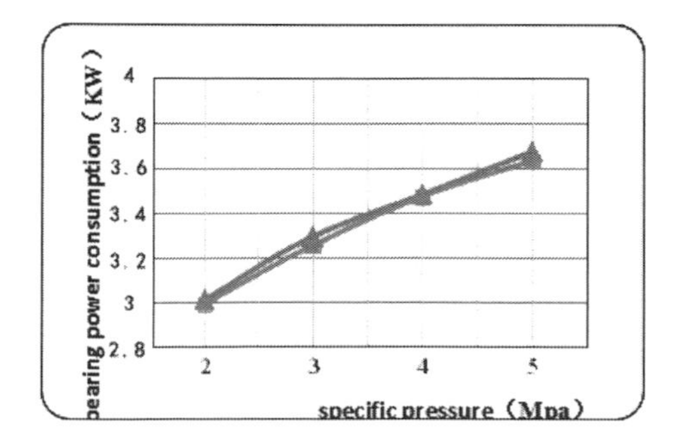

Figure 9. Variation curve of the power consumption with specific pressure.

From the graphs, we can see that the relationship between the ratio of the two calculation methods, the minimum oil film thickness, the maximum oil film temperature, the maximum oil film pressure and power consumption is similar to that mentioned in Section 3.2.1.

4 CONCLUSIONS

The maximum oil film temperature, the maximum oil film pressure and power consumption is computed by using the traditional method (regardless of the viscosity temperature change and the bearing deformation) and the thermal elastohydrodynamic lubrication method for computing (included in the viscosity temperature change and deformation of the shell), and the results show that calculation results of the two calculation methods are in line with the general rules of the lubrication theory, and thermal EHL calculation results tend to be more safe. These conclusions can provide reference for the design and application of the thrust bearing.

REFERENCES

Adam Stansfield, F.M. Application of Hydrostatic Bearing in Machine Tools, Beijing, China Machine Press, 2003.

Chen Yansheng: Designs and Principle of the Liquid Static Pressure Supporting, Beijing, National Defense Industry Press, 1980.

Ding Zhenkun: Designs of the Liquid Static Pressure Supporting, Shanghai, Shanghai Science and Technology Press, 1986.

Ge Zhi Qi. Mechanical Parts Design Handbook [M]. Beijing: Metallurgical Industry Press, 251–253. 1983.

Han Zhanzhong, Wang Jing, Lan Xiaoping: The FLUENT Fluid Project Simulates Calculating an Example and Applies, Beijing, Press of Beijing Institute of Technology, 2004.

Liu Jun, An Qi, Gao Lei. Numerical Study on the Thrust Bearing of the Inclined Plane [J]. Bearing, 2011 (9): 6–9.

Liu Yu, Liu Zhenglin, Wu Zhuxin, et al. The Influence of the Structural Parameters of the Thrust Bearing on the Elastic Deformation of the Slope of the Thrust Bearing [J]. Lubrication and Seal, 2010, 35 (11):62–68.

Pang Zhicheng: Liquid Gas Hydrostatic Technology, Harbin, Heilongjiang The People Press, 1981.

Sharma, S.C., Jain, S.C., and Bharuka, D.: Influence of Recess Shape on the Performance of a Capillary Compensated Annular Thrust Pad Hydrostatic Bearing, Tribol. Int., 35, 347, (2002).

Ting, L.L., and Mayer, J.E., Jr.: The Effects of Temperature and Inertia on Hydrostatic Thrust Bearing Performance, ASME J. Lubr. Technol., 93(2), 307, (1971).

Wang Fujun: Calculation Flow Mechanics Analyses FLUENT Software Principle And Applies, Beijing, Tsinghua University Press, 2004.

Wu Zhong De, Zhang Hong, Development of the Thrust Bearing Technology of Hydroelectric Generating Set [P]. Electric Appliance Industry, 2007.1(Supply):32–36.

Yan Qin Zhang, Wei Wei Li, Li Guo Fan: Simulation and Experimental Analysis of Influence of Inlet Flow on Heavy Hydrostatic Bearing Temperature Field, Asian Journal of Chemistry, 26(17), 5478, (2014).

Yanqin Zhang, Weiwei Li, Zeyang Yu: Flow Criterion Research on Fluid In Hydrostatic Bearing From Laminar to Turbulent Transition, International Journal of Hybrid Information Technology, 7(3), 369, (2014).

Yanqin Zhang, Yao Chen, Zeyang Yu: Influence of Strengthening Rib Location on Temperature Distribution of Heavy Vertical Lathe Worktable, International Journal of u- and e- Service, Science and Technology, 7(1), 65, (2014).

Yoshimoto S, Anno Y, Fujimora M: Static Characteristics of a Rectangular Hydrostatic Thrust Bearing With a Self-Controlled Restrictor Employing a Floating Disk. ASME J. of Tribo. 115, 307, (1993).

Younes YK: A revised design of annular hydrostatic bearings for optimal pumping power. Tribo. Int. 26(3), 195, (1993).

Zhang Yan-Qin, Fan Li-Guo: Simulation and Experimental Analysis on Supporting Characteristics of Multiple Oil Pad Hydrostatic Bearing Disk, J. of Hydrodynamics, 25(2), 236–241, (2013).

Zhang Yanqin, Yu Zeyang, Chen Yao: Simulation and Experimental Study of Lubrication Characteristics of Vertical Hydrostatic Guide Rail, High Technology Letters, 2014, 20(3), 315–320.

Zhu Jun. Failure Analysis of Thrust Bearing for Hydraulic Turbine Units [J]. Electrical Technology, 1990 (3): 13–17.

Civil, Architecture and Environmental Engineering – Kao & Sung (Eds)
© 2017 Taylor & Francis Group, ISBN 978-1-138-02985-9

Simulation of submarine underwater motion with an external weapon module

Yongqing Lian
School of Marine Science and Technology, Northwestern Polytechnical University, Xi'an, China
Department of Weaponry Engineering, Naval University of Engineering, Wuhan, China

Chunlai Li & Yang Liu
Department of Weaponry Engineering, Naval University of Engineering, Wuhan, China

ABSTRACT: To study the motion performance of submarines carrying an external weapon module, in this paper, a method of approximation is used to calculate the hydrodynamic coefficients of submarines carrying the weapon module, the Matlab mathematical model of six-freedom-motion is built, and a comparative simulation analysis of five kinds of typical motions of submarines before and after carrying the external weapon module is performed. The result of the simulation indicates that carrying an external weapon module has an effect on a submarine's motion performance to some extent, thereby resulting in the reduction of its maneuverability and steering quality. The simulated model can reflect motion features of submarines with external weapon module, which provides the theoretical foundation for further feasibility analysis.

1 INTRODUCTION

Submarines carrying an external weapon module (the module can be divided into UUV, missile, torpedo, etc.) play an important role in promoting the underwater combating ability in current and future warfare. Referring to the literature, studies on motion modeling and simulation of submarines carrying external weapons are rather rare. Applications about submarines carrying external weapons are mainly about the mine layer box and UUV (Fan, 1995; Qian, 2003; Sirmalis, 2001); on the aspect of study of submarines' external carrying, the domestic research is mainly about conducting calculations of the hydrodynamic coefficient of the submarine-carrying pod (Yang, 2009). While on the aspect of submarine space operation simulation, foreign scientists have carried out much work in the theory of submerged maneuverability, security, attack elusion effect validity, onshore manipulation, simulation training, and comprehensive display centralized control (Fu, 1992; Xia, 1992; John, 1978); domestic research has also made many achievements (Hu, 2006). Based on the achievements mentioned above, in this paper, modeling simulation and initial analysis for the motion of the submarine carrying an external weapon module are conducted.

2 MODELING AND SIMULATION OF THE SUBMARINE MOTION

2.1 *Modeling and solving*

Establish the fixed coordinate system $E - \xi\eta\zeta$ and moving coordinate system $G - xyz$, as is shown in Figure 1. Assuming that the speed relative to the earth of the submarine barycenter is V, the projection of V on $G - xyz$ is u (longitudinal speed), v (horizontal speed), and w (vertical speed); the submarine rotates with an angular velocity of Ω, and the projection of Ω on $G - xyz$ is p (heel angular velocity), q (trim angular velocity), and r (yawing rate). And set that right-leaning of the heel angle as φ, right-turning of the heading angle as ψ, and stern-leaning of the pitch angle as θ, the right rudder of the rudder angle δ_r is

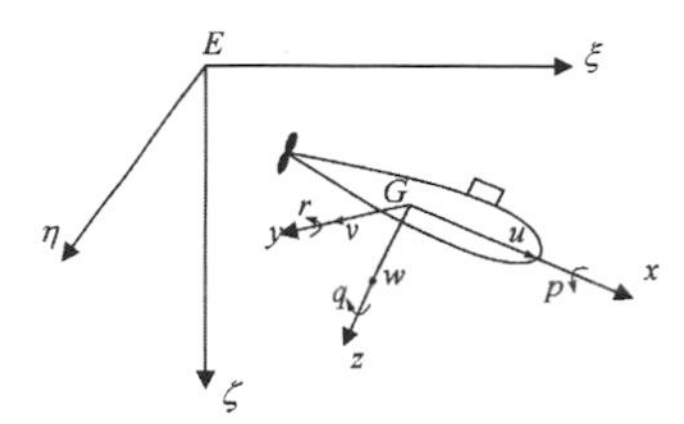

Figure 1. Schematic of the fixed coordinate system and moving coordinate system.

positive, and the sign of the head rudder angle δ_b and stern rudder angle δ_s is determined with right-hand rule, and then the heading rudder angle δ_b is the positive rudder angle, and the stern rudder angle δ_s is the positive diving angle.

On an established coordinate system, according to the standard motion equation for the submarine simulation study carried out by Gertler (Martlon, 1967), the influence of the propeller burden can be ignored and the differentiation is moved to the left-hand side and the rest to the right-hand side of the equation, and the equation of the six-freedom-motion of submarines is obtained as follows:

Longitudinal force equation:

$$a_{11}\dot{u}+a_{15}\dot{q}+a_{16}\dot{r}=f_1 \tag{1}$$

Horizontal force equation:

$$a_{22}\dot{v}+a_{24}\dot{p}+a_{26}\dot{r}=f_2 \tag{2}$$

Vertical force equation:

$$a_{33}\dot{w}+a_{34}\dot{p}+a_{35}\dot{q}=f_3 \tag{3}$$

Heel torque equation:

$$a_{42}\dot{v}+a_{43}\dot{w}+a_{44}\dot{p}+a_{45}\dot{q}+a_{46}\dot{r}=f_4 \tag{4}$$

Trim torque equation:

$$a_{51}\dot{u}+a_{53}\dot{w}+a_{54}\dot{p}+a_{55}\dot{q}+a_{66}\dot{r}=f_5 \tag{5}$$

Yaw torque equation:

$$a_{61}\dot{u}+a_{62}\dot{v}+a_{64}\dot{p}+a_{65}\dot{q}+a_{66}\dot{r}=f_6 \tag{6}$$

In which,

$$a_{11}=m-\frac{1}{2}\rho L^3X'_{\dot{u}},\ a_{15}=\frac{1}{2}\rho L^3mZ_G,\ a_{16}=-\frac{1}{2}\rho L^3my_G,$$
$$a_{22}=m-\frac{1}{2}\rho L^3Y'_{\dot{v}},\ a_{24}=-mZ_G-\frac{1}{2}\rho L^4Y'_{\dot{p}},$$
$$a_{26}=mx_G-\frac{1}{2}\rho L^4Y'_{\dot{r}},$$

$$a_{33}=m-\frac{1}{2}\rho L^3Z'_{\dot{w}},\ a_{34}=my_G,\ a_{35}=-mx_G-\frac{1}{2}\rho L^4Z'_{\dot{q}},$$
$$a_{42}=-mz_G-\frac{1}{2}\rho L^4Z'_{\dot{v}},\ a_{43}=my_G,\ a_{44}=J_x-\frac{1}{2}\rho L^5K'_{\dot{p}},$$
$$a_{45}=-J'_{xy},$$
$$a_{46}=-J_{zx},\ a_{51}=-mz_G,\ a_{53}=mx_G-\frac{1}{2}\rho L^4M'_{\dot{w}},$$
$$a_{54}=J_{xy},\ a_{55}=J_y-\frac{1}{2}\rho L^5M'_{\dot{q}},\ a_{56}=-J_{yz},\ a_{61}=-my_G,$$
$$a_{62}=mx_G-\frac{1}{2}\rho L^4,\ a_{65}=J_{yz},\ a_{64}=-J_{zx}-\frac{1}{2}\rho L^5N'_{\dot{p}},$$
$$a_{66}=J_z-\frac{1}{2}\rho L^5N'_{\dot{r}};$$

$$\begin{aligned}f_1=&\left[mx_G+\frac{1}{2}\rho L^4X'_{qq}\right]q^2+\left[mx_G+\frac{1}{2}\rho L^4X'_{rr}\right]r^2\\&-my_Gpq+\left[\frac{1}{2}\rho L^4X'_{qq}-mz_G\right]pr+\left[\frac{1}{2}\rho L^3X'_{vr}+m\right]vr\\&+\left[\frac{1}{2}\rho L^3X'_{wq}-m\right]wq+\frac{1}{2}\rho L^2\left[X'_{uu}u^2+X'_{vv}v^2+X'_{ww}w^2\right]\\&+\frac{1}{2}\rho L^2u^2[X'_{\delta_r\delta_r}\delta_r^2+X'_{\delta_b\delta_b}\delta_b^2+X'_{\delta_s\delta_s}\delta_s^2]\\&+\frac{1}{2}\rho L^2\left[a_Tu^2+b_Tuu_c+c_Tu_c^2\right]-(W-B)\sin\theta\end{aligned}$$

$$\begin{aligned}f_2=&\left[\frac{1}{2}\rho L^4Y'_{pq}-mx_G\right]pq+my_G\left[p^2+r^2\right]-mz_Gqr\\&+\left[\frac{1}{2}\rho L^3Y'_r-m\right]ur+\frac{1}{2}\rho L^3Y'_pup+\left[\frac{1}{2}\rho L^3Y'_{wp}+m\right]wp\\&+\frac{1}{2}\rho L^2\left[Y'_0u^2+Y'_vuv+Y'_{v|v|}v\left|(v^2+w^2)^{1/2}\right|\right]\\&+\frac{1}{2}\rho L^2\left[Y'_{vw}vw+Y'_{\delta_r}u^2\delta_r\right]+(W-B)\cos\theta\sin^{-1}\varphi\end{aligned}$$

$$\begin{aligned}f_3=&mz_G\left[p^2+q^2\right]-mx_Grp-my_Grq+\frac{1}{2}\rho L^4Z'_{rr}r^2\\&+\left[\frac{1}{2}\rho L^3Z'_q+m\right]uq-\left[m-\frac{1}{2}\rho L^3Z'_{vp}\right]vp+\frac{1}{2}\rho L^3Z'_{vr}vr\\&+\frac{1}{2}\rho L^2\left[Z'_0u^2+Z'_wuw+Z'_{w|w|}w\left|(v^2+w^2)^{1/2}\right|\right]\\&+\frac{1}{2}\rho L^2\left[Z'_{|w|}u|w|+Z'_{ww}\left|w(v^2+w^2)^{1/2}\right|\right]\\&+\frac{1}{2}\rho L^2\left[Z'_{vv}v^2+Z'_{\delta_b}u^2\delta_b+Z'_{\delta_s}u^2\delta_s\right]+(W-B)\cos\theta\cos\varphi\end{aligned}$$

$$\begin{aligned}f_4=&\frac{1}{2}\rho L^5\left[K'_{qr}qr+K'_{pq}pq+K'_{p|p|}p|p|\right]\\&+\frac{1}{2}\rho L^4\left[K'_pup+K'_rur+K'_{wp}wp\right]\\&+(J_y-J_z)qr-J_{xy}pr-J_{yz}(r^2-q^2)+J_{zx}pq\\&+\frac{1}{2}\rho L^3\left[K'_0u^2+K'_vuv+K'_{v|v|}v\left|(v^2+w^2)^{1/2}\right|\right]\\&+\frac{1}{2}\rho L^3\left[K'_{vw}vw+K'_{\delta_r}u^2\delta_r\right]+(y_GW-y_cB)\cos\theta\cos\varphi\\&-(Z_GW-Z_CB)\cos\theta\sin\varphi-T_N\end{aligned}$$

$$\begin{aligned}f_5=&(J_z-J_x)rp+J_{xy}qr+(r^2-p^2)J_{zx}-J_{yz}pq\\&+\frac{1}{2}\rho L^5\left[M'_{rr}r^2+M'_{rp}rp+M'_{q|q|}q|q|\right]+\frac{1}{2}\rho L^4M'_quq\\&+\frac{1}{2}\rho L^3\left[M'_0u^2+M'_ww+M'_{w|w|}w\left|(v^2+w^2)^{1/2}\right|\right]\\&+\frac{1}{2}\rho L^3\left[M'_{|w|}u|w|+M'_{ww}w\left|(v^2+w^2)^{1/2}\right|\right]\\&+\frac{1}{2}\rho L^3\left[M'_{vv}v^2+M'_{\delta_b}u^2\delta_b+M'_{\delta_s}u^2\delta_s\right]\\&-(x_GW-x_CB)\cos\theta\cos\varphi-(z_GW-z_CB)\sin\theta\end{aligned}$$

$$\begin{aligned}f_6=&(J_x-J_y)pq+(p^2-q^2)J_{xy}+J_{yz}rp-J_{zx}rq\\&+\frac{1}{2}\rho L^5\left[N'_{pq}pq+N'_{|r|r}|r|r\right]+\frac{1}{2}\rho L^4\left[N'_pup+N'_rur\right]\end{aligned}$$

$$+\frac{1}{2}\rho L^3\left[N_0' u^2 + N_v' uv + N_{v|v|}' v\left|\left(v^2+w^2\right)^{1/2}\right|\right]$$
$$+\frac{1}{2}\rho L^3\left[N_{vw}' vw + N_{\delta_r}' u^2\delta_r\right] + \left(x_G W - x_C B\right)\cos\theta\sin\varphi$$
$$+\left(y_G W - y_C B\right)\sin\theta$$

The meaning of notes in the above-mentioned equation can be referred from the literature (Martlon, 1967).

By solving equation (1)–(6) simultaneously, the acceleration matrix can be obtained as follows:

$$\begin{bmatrix}\dot{u}\\ \dot{v}\\ \dot{w}\\ \dot{p}\\ \dot{q}\\ \dot{r}\end{bmatrix} = \begin{bmatrix}f_1\\ f_2\\ f_3\\ f_4\\ f_5\\ f_6\end{bmatrix} \Big/ \begin{bmatrix}a_{11} & \cdots & a_{16}\\ \vdots & \ddots & \vdots\\ a_{61} & \cdots & a_{66}\end{bmatrix} \tag{7}$$

In which the value of unvalued a_{ij} is 0.

Integrate the acceleration matrix in equation (7) and obtain speed and angular velocity, integrate again with the converted relation of the fixed and moving coordinate systems, and obtain the position and attitude angle of any time.

2.2 *Verification of modeling*

To testify the accuracy of the established model, verification is conducted according to the modeling simulation program in five typical motions of the submarine.

1. Horizontal steady rotation

Horizontal steady rotation is the rotation motion conducted when the submarine is on a direct route and the rudder is inclined at a fixed angle. Conduct a simulation study for the submarine on this motion with rudder angles of 15°, 25°, and 35° and the obtained barycentre locus curves are shown in Figure 2.

2. Horizontal Z-type steering motion

On practical voyage, the steering maneuver of the submarine is conducted on the basis of steady heading, the maneuverability of which can be studied with Z-type steering, and the focus is the response ability to the rudder. Conduct a simulation study for the submarine with the z-type steering motion with the obtained variation curves of the rudder angle and heading angle, as shown in Figure 3.

3. Vertical steady steep submerging and surfacing motion

The submarine will conduct the vertical steady steep submerging and surfacing motion when elevators are fixed on an angle in a fixed depth voyage. This motion reflects the depth maneuverability of the submarine. Conduct a simulation study of this motion with elevators at the angles of 10° and 20° and the obtained barycenter locus curves are shown in Figure 4, and pitch angle variation curves are shown in Figure 5.

4. Vertical surpass maneuver motion

Adjust elevators regularly in the vertical plain and the submarine will conduct the vertical surpass maneuver motion. The response ability to elevators can be studied in this motion. Conduct a simulation study for the submarine with this motion with $\delta_s/\theta = 20°/10°$, the obtained rudder angle and pitch angle variation curves are shown in Figure 6 and the depth variation curves are shown in Figure 7.

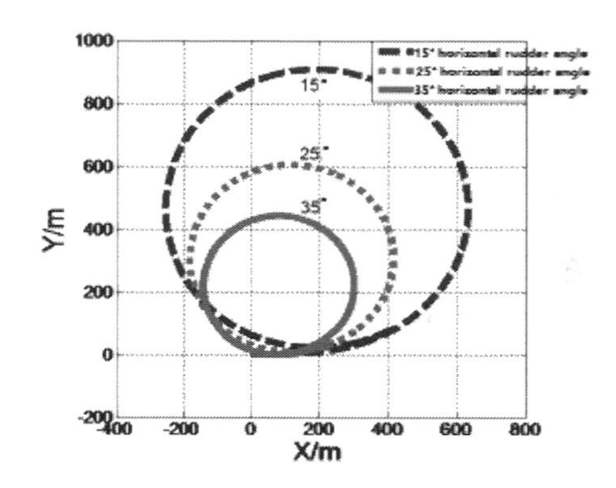

Figure 2. Locus curves of the motion in the horizontal plane.

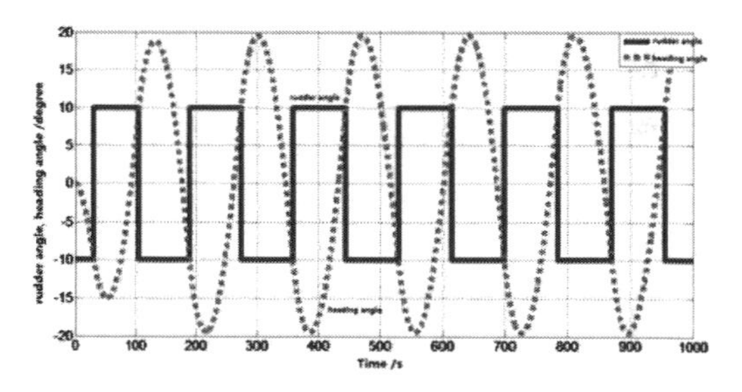

Figure 3. Variation curves of the rudder angle and the heading angle of motion in the horizontal plane.

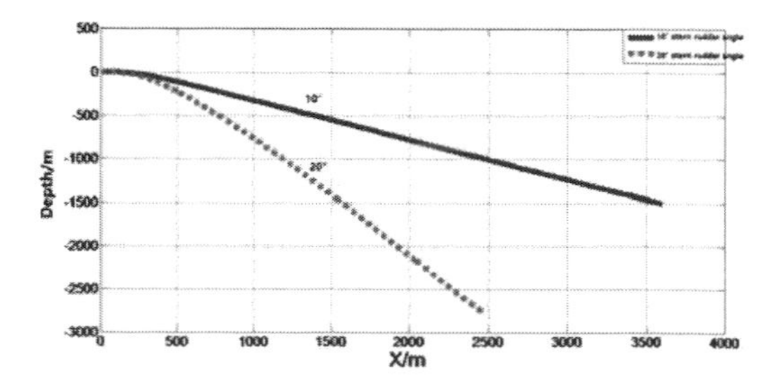

Figure 4. Locus curves of the motion in the vertical plane.

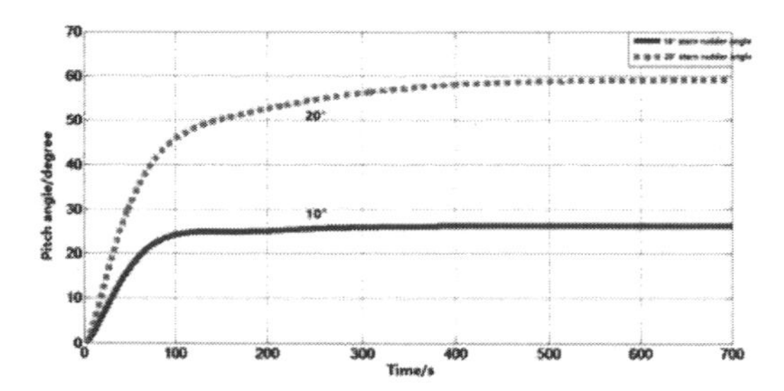

Figure 5. Variation curves of the pitch angle of the motion in the vertical plane.

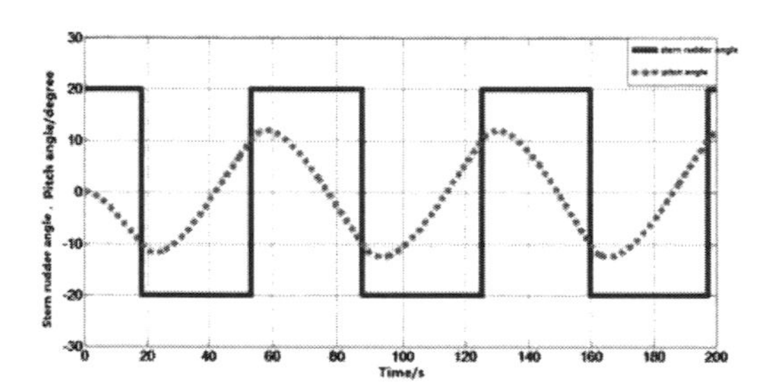

Figure 6. Variation curves of the rudder angle and the pitch angle of the motion in the vertical plane.

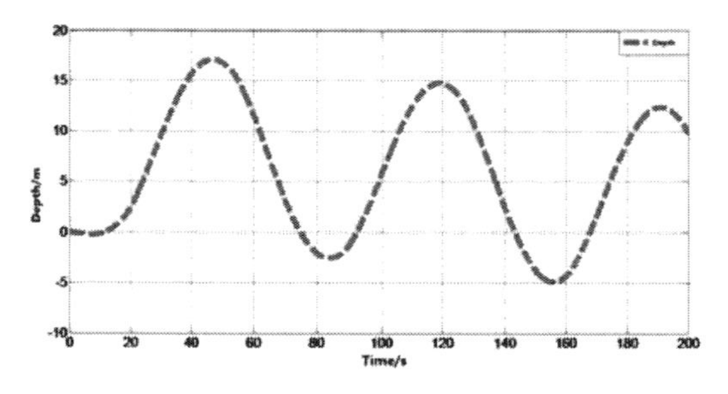

Figure 7. Variation curves of the diving depth of the motion in the vertical plane.

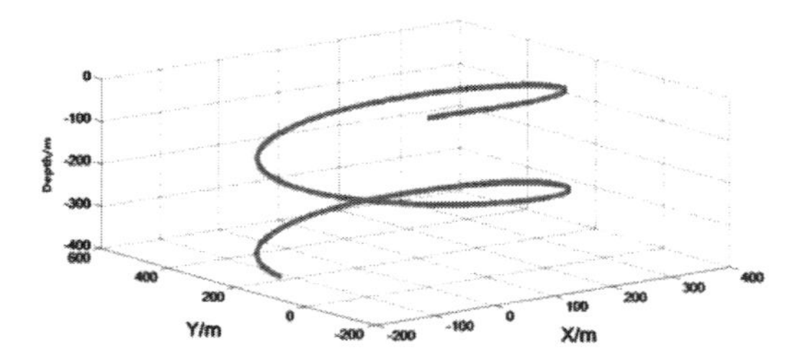

Figure 8. Locus curves of the spatial helical motion.

5. Spatial steady helical motion

When the directions of the rudder angle and the elevator angle are fixed or only the direction of the rudder is fixed, the submarine will rotate around axis $E\zeta$ and the barycenter's 3D motion path is a spiral line, and this motion is called spatial steady helical motion, which can be used to study the spatial maneuverability. Conduct a simulation study for the submarine with this motion with $\delta_r = 30°$, $\delta_b = 10°$, and $\delta_s = 0°$, and the obtained barycentre locus curves are shown in Figure 8.

The simulation results of five typical motions shows that the character of the motion obtained by simulation results satisfies the common practice, and the established submarine six-freedom model is correct and can be used in predicting the motion of the submarine that carries an external weapon module.

3 APPROXIMATION OF THE SUMARINE CARRYING EXTERNAL WEAPON MODULE'S HYDRODYNAMIC COEFFICIENTS

The hydrodynamic coefficients will change when the external weapon module is loaded, and thus it must be confirmed. The method of obtaining coefficients includes experiment, theoretical calculation, approximation, etc. In this paper, the primary demonstration phase is presented and exhibits the estimating method. An approximation method (Li, 2008) is carried out by sorting masses of experimental results and estimation, which can obtain the approximate value rapids if certain deviations are allowed. This method assumes that the hydrodynamic coefficients of the submarine carrying an external weapon module are equal to the sum of that of the submarine and the external weapon module.

3.1 *Approximation of weapon module acceleration*

There can be a few external weapon modules in a single submarine. In this paper, the work is based on the condition of one single external weapon module (as shown in Figure 9). The length of the module is L, which is set to 10 m, width B and height H are set to 1.5 m, and the total underwater volume of the displacement ∇ is set to 8 m^3.

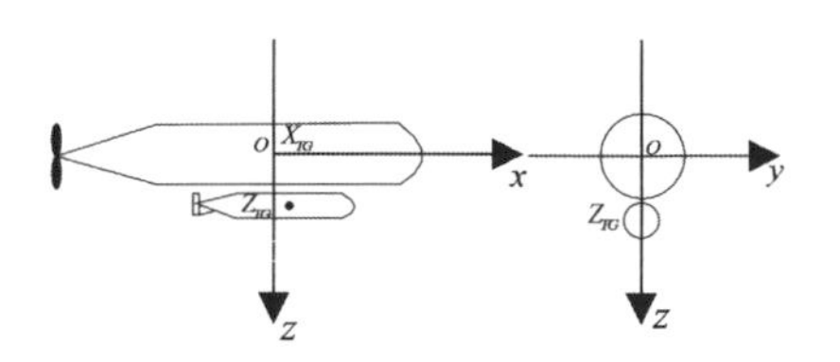

Figure 9. Schematic diagram of submarines carrying the weapon module.

The weapon module of streamline shape can be replaced with an ellipsoid with three axes, take the main criterion L,B,H as three axes $2a,2b,2c$. Since in an ellipsoid, $\lambda_{ij} = 0(i \neq j)$, the added weight that needs calculation is $\lambda_{ij}(i = j)$.

The added mass coefficients of the ellipsoid is obtained by Lamb through the potential-flow theory (Li, 2008). When the semi-major axis of the ellipsoid is a,b,c, the added weight of it can be obtained from Figure 10. K_{11}, K_{22}, K_{33} is the relative value of the added weight λ_{11}, λ_{22}, λ_{33}; K_{55}, K_{66} is the relative value of added rotational inertia λ_{55}, λ_{66}.

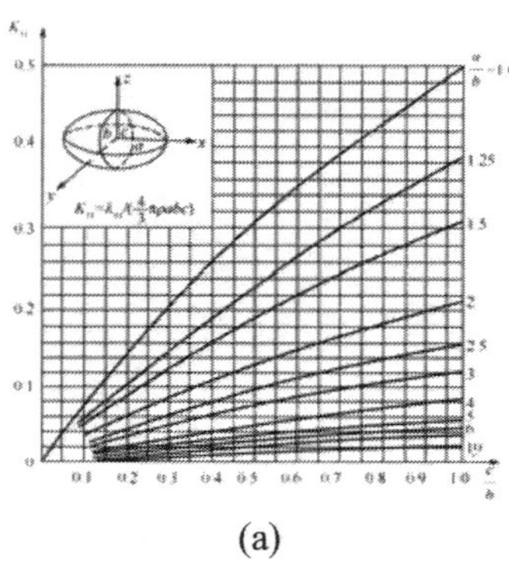

(a)

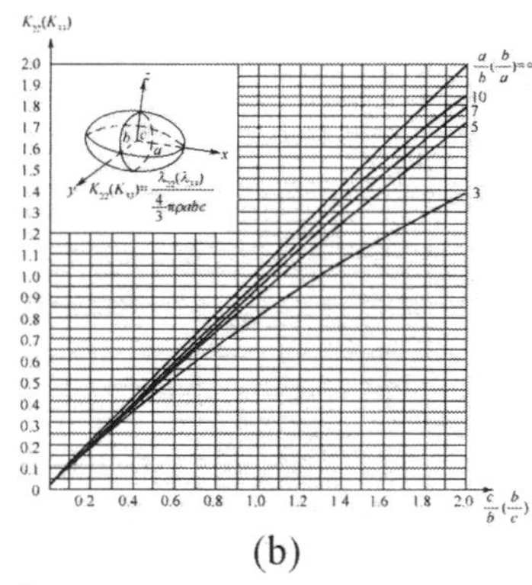

(b)

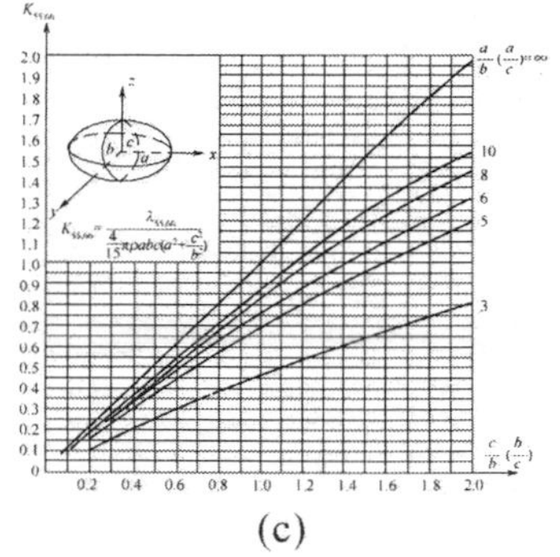

(c)

Figure 10. Graphs showing the added mass coefficient curves of the ellipsoid.

The added mass coefficient can be obtained by referring to Figure 10 with $L/B = 10.0$, $H/B = 1.0$, $K_{11} = 0.02$, $K_{22} = 0.96$, $K_{33} = 0.96$, $K_{55} = 0.87$, $K_{66} = 0.87$, through calculation, and the acceleration of the weapon module is given as follows:

$$(X'_{\dot{u}})_{wq} = -\frac{\pi}{3}\frac{B}{L}\frac{H}{L}K_{11} = -2.09\times10^{-4}$$

$$(Y'_{\dot{v}})_{wq} = -\frac{\pi}{3}\frac{BH}{L^2}K_{22} = -1.005\times10^{-2}$$

$$(Z'_{\dot{w}})_{wq} = -\frac{\pi}{3}\frac{BH}{L^2}K_{33} = -1.005\times10^{-2}$$

$$(M'_{\dot{q}})_{wq} = -\frac{\pi}{60}\frac{BH}{L^2}\left[1+\left(\frac{H}{L}\right)^2\right]K_{55} = -4.60\times10^{-4}$$

3.2 *Approximation of the weapon module velocity coefficient*

Refer to the attack angle derivation curves of ellipsoid (as is shown in Figure 11) with $L/\nabla^{\frac{1}{3}} = 5.0$ and $H/B = 1.0$,

$$\frac{\partial F'_L}{\partial \alpha} = 1.05\times10^{-2} \quad \frac{\partial M'}{\partial \alpha} = 1.03\times10^{-2}$$

The resistance coefficient of ellipsoid is given as follows:

$$F'_D = 0.05\times\frac{\pi B^2}{4L^2} = 0.39\times10^{-3}$$

And the velocity hydrodynamic coefficient of the weapon module can be calculated as follows:

$$(Y'_v)_{wq} = (Z'_w)_{wq} = -\left(\frac{\partial F'_L}{\partial \alpha} + F'_D\right) = -1.09\times10^{-2}$$

$$(N'_v)_{wq} = (M'_w)_{wq} = \frac{\partial M'}{\partial \alpha} = 1.03\times10^{-2}$$

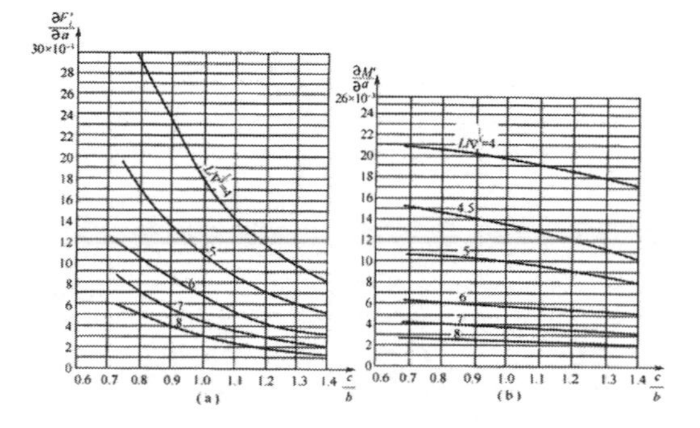

Figure 11. Attack angle derivation curves of the ellipsoid.

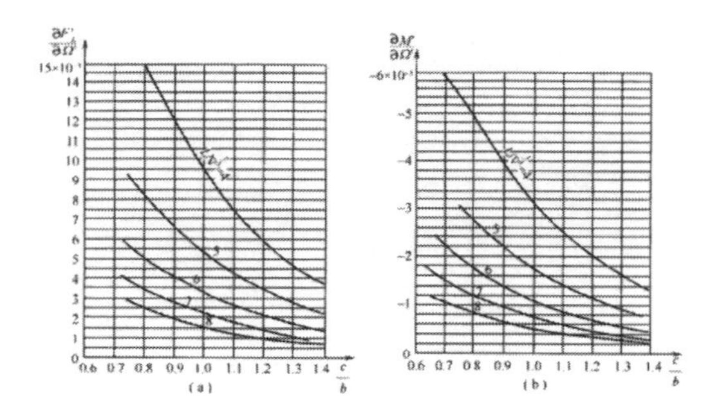

Figure 12. Angular velocity derivation curves of the ellipsoid.

3.3 *Approximation of the weapon module angular velocity*

Refer to the angular velocity derivation curves of the ellipsoid with $L/\nabla^{\frac{1}{3}} = 5.0$ and $H/B = 1.0$,

$$\frac{\partial F_L'}{\partial \Omega} = 5.3\times10^{-3} \quad \frac{\partial M'}{\partial \Omega} = -1.66\times10^{-3}$$

Assuming that the barycenter of the weapon module is in the very center, that is the correction factor of buoyant centre position is 1, and then the angular velocity coefficient of the weapon module is given as follows:

$$\left(Y_r'\right)_{wq} = -\left(Z_q'\right)_{wq} = \frac{\partial F_L'}{\partial \Omega} = 5.3\times10^{-3}$$

$$\left(N_r'\right)_{wq} = \left(M_q'\right)_{wq} = \frac{\partial M'}{\partial \Omega} = -1.66\times10^{-3}$$

4 MOTION SIMULATION OF SUBMARINES CARRYING THE EXTERNAL WEAPON MODULE

According to the obtained hydrodynamic coefficient of the submarine carrying the weapon module, the simulation of five typical motions can be conducted with the established model.

4.1 *Comparison of the horizontal steady rotation*

Conduct the simulation to the submarine with the rudder at 15°, 25°, and 35° separately, and the barycentre locus curves of the submarine before and after loading the weapon module with the rudder at 15° is shown in Figure 13.

After calculation, the variation of maneuverability of the submarine is shown in Table 1.

From Table 1, the conclusion can be drawn that, after loading the weapon module, rotation

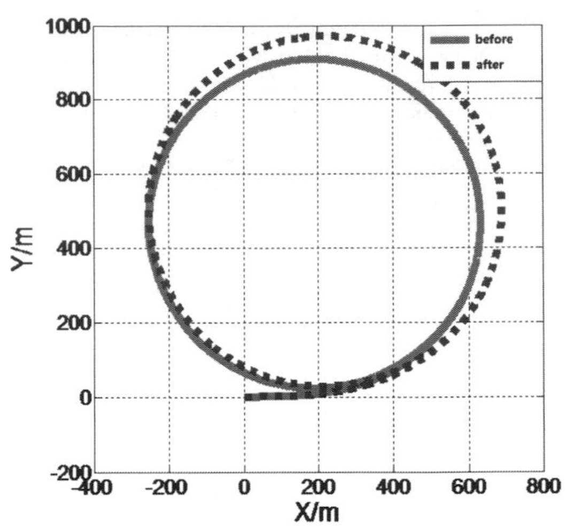

Figure 13. Comparison curves between motions in the horizontal plane.

Table 1. Contrast on performance indexes of maneuverability in the horizontal plane before and after carrying the weapon module.

		Rotation diameter/m	Steady time/s	Rotation period/s	Rotation angular velocity/ rad·s^{-1}
15°	Before	886.2	48.4	567.0	0.0111
	After	945.7	49.7	600.0	0.0105
25°	Before	596.2	127.3	454.4	0.0138
	After	624.8	129.3	473.4	0.0133
35°	Before	444.5	132.8	401.8	0.0156
	After	462.6	133.0	416.0	0.0151

diameter, steady time (the time needed to adjust the submarine to steady conditions), and rotation period, the rotation angular velocity and other maneuverability indexes are bigger than before, which means that the maneuverability has been reduced. Besides, through analysis of the absolute variation before and after carrying the weapon module, the conclusion can be drawn that the variation of maneuverability is bigger in a small rudder angle and smaller in a large angle.

4.2 *Comparison of the horizontal z-type steering motion*

The heading angle comparison curves obtained before and after carrying the weapon module obtained by using simulation is shown in Figure 14.

The calculation result shows that before carrying the weapon module, the primary rotation time is 31.0 *s*, the surpass heading angle is 11°; after carrying the weapon, the rotation time changes to

33.7 s and the surpass heading angle changes to 12°. The absolute variation of the primary rotation time before and after carrying the weapon module is 8.71% and the surpass heading angle is 4.76%. The conclusion can be drawn that response ability to the rudder is reduced.

4.3 *Comparison of vertical steady steep submerging and surfacing motion*

Conduct a simulation study for the submarine (stern elevator at 20°) before and after carrying the external weapon module in the vertical steady steep submerging and surfacing motion, and the obtained barycentre locus comparison curves

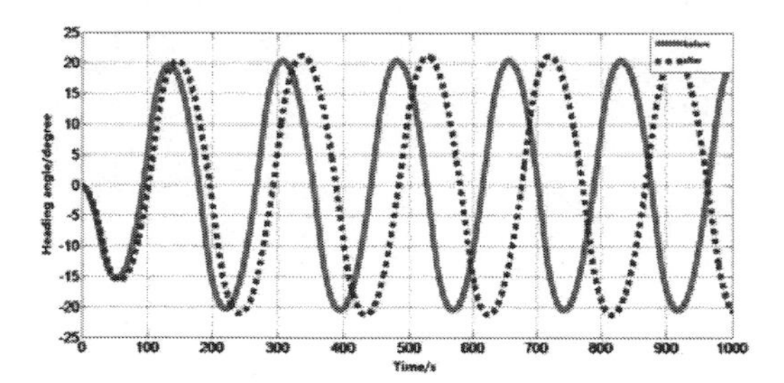

Figure 14. Comparison curves between heading angles in the horizontal plane.

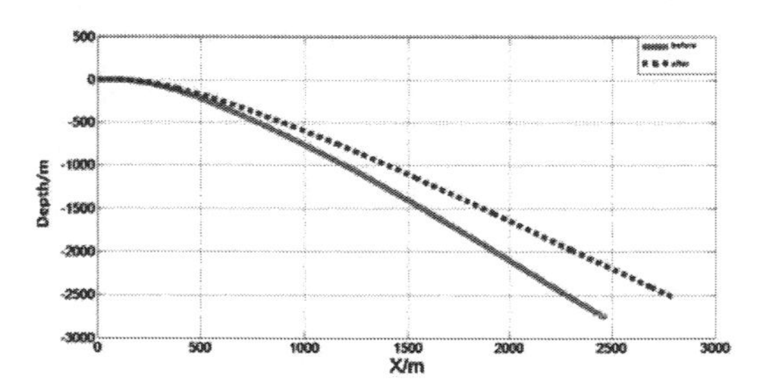

Figure 15. Comparison curves of the motion in the vertical plane.

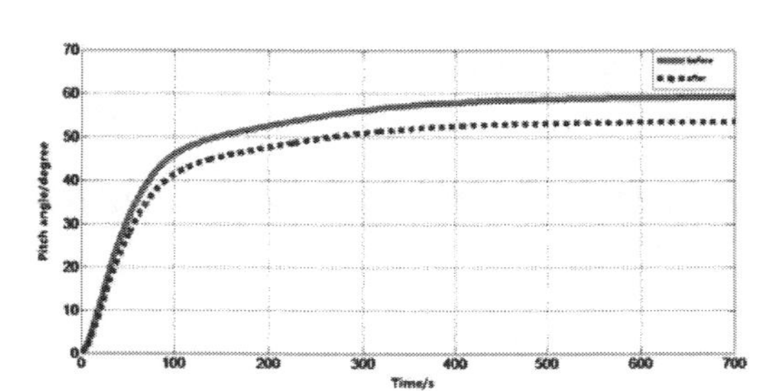

Figure 16. Comparison curves between pitch angles of the motion in the vertical plane.

and pitch angles comparison curves are shown in Figures 15 and 16.

Through analysis we can know that the steady time and maximum pitch angle before carrying the weapon is 46.2 s and 59°, while that of the after carrying condition is 50.1 s and 52°; the carrying weapon module has a rather significant influence on maneuverability especially for the pitch angle.

4.4 *Comparison of the vertical surpass maneuver motion*

Conduct a simulation study for the submarine before and after carrying the weapon module with the surpass maneuver motion with $\delta_s / \theta = 20° / 10°$, and the depth and pitch angle comparison curves are shown in Figures 17 and 18.

The result shows that, before carrying the weapon module, the primary rotation time is 17.8 s, surpass depth is 0.4 m, and surpass pitch angle is 2.5°, and after that, the primary rotation time is 20.0 s, surpass depth is 0.5 m, and surpass pitch angle is 3.2°. Through calculation, the absolute variation of these factors are determined as 15.7%, 25.0%, and 28.0%. The conclusion can be drawn that the carrying weapon module has a bigger influence on the vertical maneuverability that that of the horizontal maneuverability. The loss of maneuverability in the vertical plain is greater.

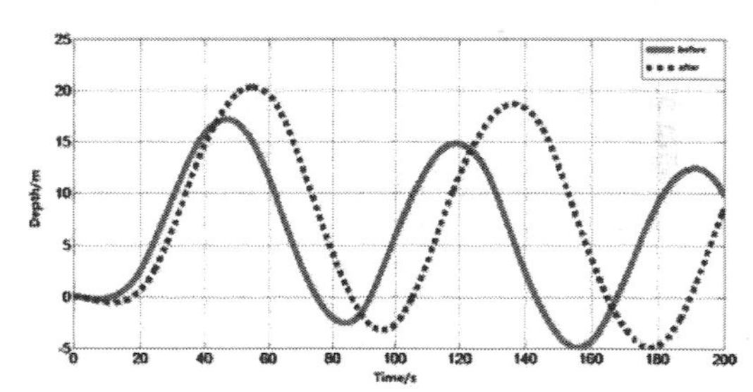

Figure 17. Comparison curves of the diving depth of the motion in the vertical plane.

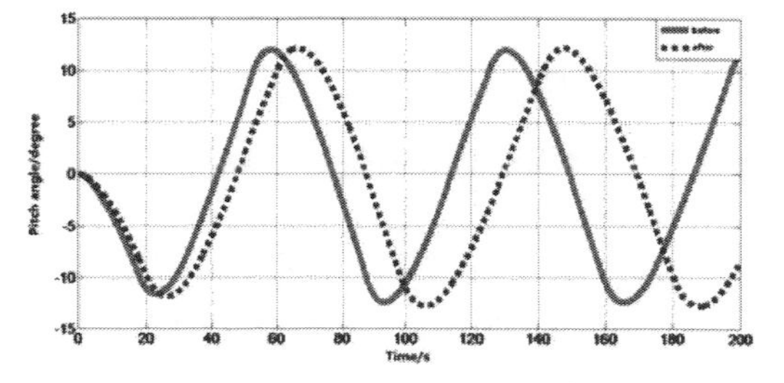

Figure 18. Comparison curves of pitch angles of the motion in the vertical plane.

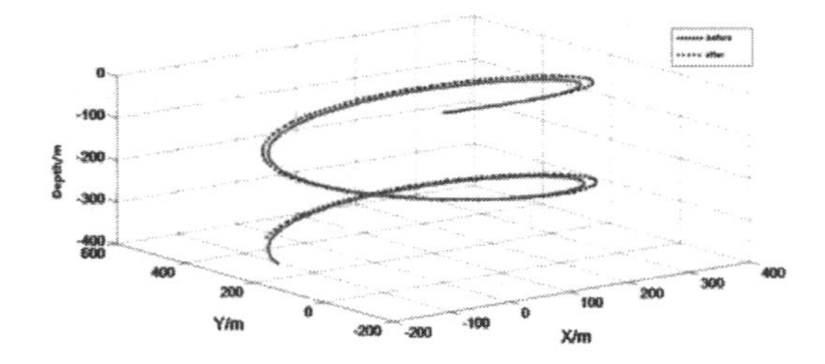

Figure 19. Comparison locus curves of the spatial helical motion.

Table 2. Contrast on performance indexes of spatial maneuverability before and after carrying the weapon module.

	Rotation diameter/m	Rotation period/s	Lift distance/m	Relative lift velocity/ $m \cdot s^{-1}$
Before	460.1	402.8	226.6	0.5626
After	485.1	421.4	230.1	0.5462

4.5 *Comparison of spatial, steady, and spiral helical motions*

Conduct a simulation study for the submarine before and after carrying the weapon module with the spatial steady spiral helical motion with $\delta_r = 30°, \delta_b = 10°$, and $\delta_s = 0°$, and the comparison locus curves are shown in Figure 19.

The rotation diameter, time, lift, relative lift velocity, and other main indexes obtained in simulation are shown in Table 2.

From Table 2, it can be observed that the submarine's spatial maneuverability before and after carrying the weapon module changes heavily, especially for the rotation diameter and time, which means that after carrying the weapon module, the submarine's spatial maneuverability is reduced.

5 CONCLUSION

By combining five typical motions, in this paper, a simulation study is conducted for the submarine carrying an external weapon module. The result shows that the carrying weapon module reduces the maneuverability indexes. The result satisfies the real submarine motion characteristics, indicating that the model is reasonable. The simulation in this paper lays the theoretical foundation for the study of the validity of submarines carrying the external weapon module.

REFERENCES

Chen Houtai, Xu Yuru, Huang Xirung. On the Measurement and Reckoning of Hydrodynamic Coefficients of Deep Submerged Vehicles with six-freedom-motion[J]. 1980, (1).

Fan Naizhong. External Mine Laying Equipment of Modern Submarine[J]. Modern Vessel, 1995, 6(2): 32–34.

Fu Yanhui, Wan Yancheng. Development of Submarine Spatial Motion Research [A]. Demonstration for Domestic and Home Research of Submarine Maneuverability [C]. Hydrodynamic Profession Group, National Defence Scientific Industry Council. Beijing, 1992: 13–31.

Hu Kun, Zhang Honggang and Xu Yifan. Simulation Study and Analysis on Underwater Space Motion of Submarine[J]. Computer Simulation. Beijing, 2006, 23(5): 10–13.

John Thomas Hammond. Development of a six degee of freedom motion Simulation model for use in submarine design analysis[R]. ADA078410. USA: DTNSRDC, 1978: 1–20.

Li Dianpu. Ship Motion and Modeling-Second edition[M]. National Defend Industy Press. Beijing, 2008.

Martlon Gertler and Grant P. Hagen. Standard Equations of motion for submarine simulation. AD653861 June 1967, Naval Ship Research and Development Center.

Qian Dong. The Large-scale Unmanned Undersea Vehicles by the US in the Future[J]. Torpedl Tochnology, 2003(01) 41–50.

Sirmalis J E. Pursuing the MANTA vision: recent at-sea technology demonstration results[C]//UDT 2001.

Xia Fei, Chen Yuan. Commentary of Submarine [A]. Demonstration for Domestic ang Home Research of Submarine Maneuverability [C]. Hydrodynamic Profession Group, National Defence Scientific Industry Council. Beijing, 1992: 1–12.

Yang Luchun. Study of the Calculation Approach of Hydrodynamic Derivatives of Submarine with Outer Carry[D]. Harbin Engineering University. Harbin, 2009.

Civil, Architecture and Environmental Engineering – Kao & Sung (Eds)
© 2017 Taylor & Francis Group, ISBN 978-1-138-02985-9

A multi-objective NC drilling parameter optimization model to achieve low energy consumption and costs

Wei Yan, Hua Zhang & Zhigang Jiang
School of Machinery and Automation, Wuhan University of Science and Technology, Hubei, China

ABSTRACT: Machining parameters are the important factors to energy conservation and emission reduction in the manufacturing industry. The drilling parameters' optimization problem for low energy consumption and costs is studied in this paper. Two objectives including energy consumption and costs are considered in this optimization model with two independent variables, namely drilling speed and feed rate per turn. Additionally, the constraints of the model are also considered. And then, the model is solved with the weighted method and the Adaptive Particle Swarm Optimization (APSO) algorithm. Finally, a case study is conducted for the verification of the proposed model and method.

1 INTRODUCTION

With the recent continuous increase in energy demand and constraints in carbon emissions, energy conservation and emission reduction have become priorities for the manufacturing industry (Li, 2015). In China, the manufacturing sectors consumed over 50% of the entire electricity produced, but the energy efficiency is low (Tang, 2006).

Drilling is a common processing method, which accounts for about 30% of the total machining methods. Energy consumption from drilling is huge. The drilling parameter plays a very important role in quality, efficiency, costs etc. (Dietmair, 2009). Some scholars have carried out much research on drilling parameter optimization. Zhang et al. (Zhang, 2006) proposed a minimum drilling time optimization method of cutting parameters based on fuzzy mathematics. Li et al. (Zhang, 2013) conducted an energy model for cutting processes of machine tools, and studied the energies of turning, drilling operations, etc. Zhao et al (Zhao, 2012) developed a parameter optimization system for costs and efficiency in the hole machining processes. Zhang (Zhang, 2011) studied a high-speed drilling parameter optimization method for small and deep-hole-processing of stainless steel. Zhang (Cui, 2007) proposed a drilling parameter fuzzy optimization method based on the machining reliability. Most of the above literature addressed the parameter optimization with the traditional single objective model such as quality, efficiency, or costs. Some are concerned with energy conservation and environmental emissions. However, there is hardly any reported research on the correlation between energy and costs of drilling.

Based on the above review, a multi-objective parameter optimization model of drilling for achieving low energy and costs is presented in this paper. Energy consumption and cost are the optimization objectives and drilling speed and feed rate per turn are taken as independent variables. The APSO (Adaptive Particle Swarm Optimization algorithm) is applied as the solution method. Finally, a hole-processing example is given for the verification of the proposed model and methodology.

2 ESTABLISHMENT OF THE DRILLING MULTI-OBJECTIVE OPTIMIZATION MODEL

Drilling parameters include drilling speed, feed rate per turn, and cutting depth. However, the cutting depth can be calculated with diameters and allowances of the hole to be processed. Therefore, the drilling speed and feed rate per turn are selected as the independent variables and minimizing energy consumption and costs are considered as the two optimization objectives.

2.1 *Multi-objective optimization functions*

2.1.1 *Costs function*

The costs of drilling mainly include machining costs, tool change costs, other ancillary costs etc. Because the artificial costs are greatly influenced by the geographical and processing environment, it has not been considered in this paper. The function of drilling costs objective is shown as follows:

$$C_p = T_m X + T_c \frac{T_m}{T} X + \frac{T_m}{T} C_t + T_o X \tag{1}$$

where, C_p represents the cost of processing, X represents the production cost per unit time, T_m is the drilling time, T_c is the tool change time, T_o is the other auxiliary time, C_t represents the tool change cost, T represents the tool durability, and the calculation methods for T and T_m are as follows:

$$T_m = \frac{60H}{nf} \tag{2}$$

$$n = \frac{1000v_c}{\pi D} \tag{3}$$

$$T = \left(\frac{60C_v D^{Z_v} K_V}{f^{Y_v} V_c} \right)^{\frac{1}{m}} \tag{4}$$

where, H is the drilling depth, f is the feed rate per turn, n represents the spindle speed, v_c is the drilling speed, D represents the drill diameter, C_v, Z_v, K_v, Y_v, and m represent the tool durability coefficients.

From Equation (1)–Equation (4), the cost function of drilling is shown in Equation (5).

$$Cp = 0.06 \times \frac{\pi DHX}{v_c f} + (T_c X + C_t) \frac{60^{\left(1-\frac{1}{m}\right)} \pi H D^{\left(1-\frac{Z_v}{m}\right)} f^{\left(\frac{Y_v}{m}-1\right)} v_c^{\left(\frac{1}{m}-1\right)}}{1000 C_V^{\frac{1}{m}} K_V^{\frac{1}{m}}} + T_o X \tag{5}$$

2.1.2 *Energy function*

The energy consumption of drilling is mainly composed of five parts, namely drilling energy, no-load energy, feed system energy, additional load energy, and auxiliary system energy.

1. Drilling energy: it can be calculated with drilling force. The mathematical expression is shown in Equation (6).

$$E_m = P_m \times T_m = 2 \times 9.81 C_m D^{X_m - 1} f^{Y_m} K_m T_m \tag{6}$$

where, P_m is the drilling power and C_m, X_m, Y_m, and K_m are the correlation coefficients, which can be obtained in the cutting quantity manual.

2. No-load energy: it continued along the whole drilling process, which is shown in Equation (7).

$$E_u = P_u \times T_m = \left(K_1 n^2 + K_2 n + K_3\right) T_m \tag{7}$$

where, P_u is the no-load power and K_1, K_2, and K_3 are the correlation coefficients with the spindle speed (Liu, 2012).

3. Additional load energy: it is related to the load and drilling force and the mechanism is very complicated. The energy consumption of additional load is commonly proportional to the energy consumption of the drilling process.

$$E_t = bE_m \tag{8}$$

where, b is the additional load factor of drilling; it is often 0.15–0.25 and depends on the experience.

4. Feed and auxiliary system energy: the energy consumption of the feed system is usually 1.5% of the main drive system, and the energy consumption of the auxiliary system has little impact on drilling parameters. Therefore, the energy consumption of the feed system and auxiliary can be regarded as a constant in Equation (9).

$$E_g + E_a = Q \tag{9}$$

Therefore, the energy consumption during the drilling operation is expressed as follows:

$$\begin{aligned} E_p &= E_m + E_u + E_t + E_g + E_a \\ &= 1.177(1+b)\left(C_m D^{X_m} f^{Y_m - 1} K_m\right) \\ &\quad + \left[K_1 \left(\frac{1000V_c}{\pi D} \right)^2 + K_2 \frac{1000V_c}{\pi D} + K_3 \right] \cdot \\ &\quad \frac{\pi DH}{1000V_c f} + Q \end{aligned} \tag{10}$$

2.2 *Constraints*

The parameters of drilling are constrained by the machine tools, drilling process, quality etc. Therefore, all operating parameters shall be within the constraints in the optimized solution.

In this paper, the constraints of the spindle speed, feed rate, cutting force, drilling power, spindle torque, and tool life are considered.

The constraints of drilling parameters are shown in Equation (11).

$$s.t. \begin{cases} g_1(v_c, f) = n_{\min} - n \le 0 \\ g_2(v_c, f) = n - n_{\max} \le 0 \\ g_3(v_c, f) = f_{\min} - f \le 0 \\ g_4(v_c, f) = f - f_{\max} \le 0 \\ g_5(v_c, f) = C_f D^{X_f} f^{Y_f} K_f - F_{\max} \le 0 \\ g_6(v_c, f) = P_m - P_{\max} \le 0 \\ g_7(v_c, f) = M - M_{\max} \le 0 \\ g_8(v_c, f) = T_m - T \le 0 \end{cases} \tag{11}$$

Therefore, the energy consumption and costs parameter optimization of NC drilling is a typical constrained multi-objective optimization problem, which can be expressed as Equation (12).

$$\begin{aligned} &\min F(v_c, f) = \left(\min C_p, \min E_p\right) \\ &s.t. g_i(v_c, f) \le 0 \quad i = 1, 2, \cdots, 8 \end{aligned} \tag{12}$$

3 SOLVING THE MODEL

3.1 *Transformation*

For most multi-objective optimization problems, it is often difficult to achieve all the multiple objectives at the same time. In order to facilitate the solution, the weighted sum method is introduced to construct in this paper.

$$F(v_c,f)=\lambda_C\frac{C_p-C_{\min}}{C_{\max}-C_{\min}}+\lambda_E\frac{E_p-E_{\min}}{E_{\max}-E_{\min}} \tag{13}$$

where, C_{max}, C_{min}, E_{max}, and E_{min} are the minimum and maximum values of costs and energy consumption respectively, and λ_C and λ_E are the weighted coefficients of costs and energy consumption, and $\lambda_C+\lambda_E=1$.

3.2 *Solution procedure of APSO*

PSO (Particle Swarm Optimization) is an evolutionary algorithm, which searches for the optimal solutions through the iterative operation based on random solution, and evaluates the quality of the solution with fitness. Because of the ease of implementation and the fact that it does not need the gradient information of the objectives, PSO is widely used in solving constrained optimization problems.

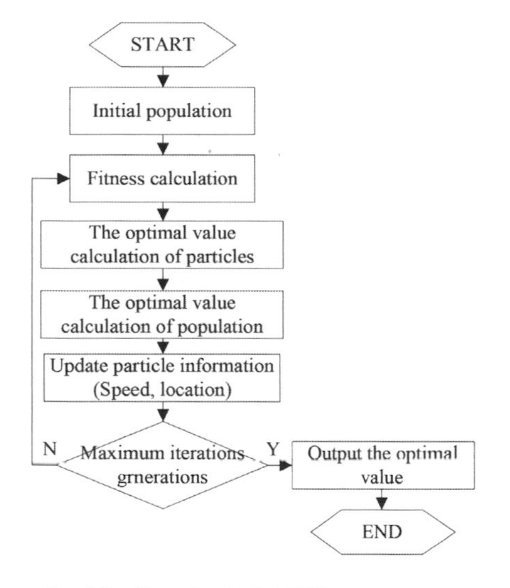

Figure 1. The flow chart of APSO.

However, the flight time of particles are fixed in the traditional PSO, which may lead to oscillation and reduce the convergence speed (Xia, 2006).

Therefore, an improved algorithm called APSO (Adaptive Particle Swarm Optimization) was proposed in this paper. The global optimal value was added in the APSO to adjust the inertial weight and particle flight time adaptively. The equation of APSO is given as follows:

$$\begin{cases} v_{id}^{(t+1)}=w^t\cdot v_{id}^{(t)}+c_1r_{1d}\left[p_{id}^{(t)}-x_{id}^{(t)}\right]+c_2r_{2d}\left[p_{gd}^{(t)}-x_{id}^{(t)}\right] \\ x_{id}^{(t+1)}=x_{id}^{(t)}+v_{id}^{(t+1)}\times T_0\times\left(1-\dfrac{k_0t}{I_{\max}}\right) \end{cases} \tag{14}$$

where, v_{id} is the particle speed, and $v_{id}\in[-v_{\max},v_{\max}]$, c_1 and c_2 are the study factors, r_{1d} and r_{2d} are the random numbers, and $r_{1d},r_{2d}\in[0,1]$, x_{id} is the current location of the particles, $w^t=\exp\left(-\frac{F_b^t}{F_b^{t-1}}\right)$ is the inertial factor, and F_b^t, F_b^{t-1} is the global optimal value of the particles of t and t-1 generation, T_0 is the initial flight time, K_0 is the adjust parameter, and $I_{\max}$ is the maximum evolution generation.

The flow chart of APSO is shown in Figure 1.

4 CASE STUDIES

In this paper, a hole-machining case is studied to verify the effectiveness of the optimization model. The equipment and machining information are shown in Table 1.

The program code of APSO was compiled with Matlab 2013b, and the algorithm parameters were set as follows: the initial population size was 30, the maximum generation was 150, the adjusting parameter was 0.9, and the initial flight time was 0.4. The APSO iteration process and optimization results are shown in Figure 2 and Tablc 2, respectively.

In this paper, the optimization results of cost ($\lambda_C=1$), energy consumption ($\lambda_E=1$) and cost and energy ($\lambda_C=\lambda_E=0.5$) are also considered, and the results are listed in Table 2.

Table 1. The information of the drilling case.

Minimum spindle speed (r/min)	Maximum spindle speed (r/min)	Minimum feed speed (mm/r)			Maximum feed speed (mm/r)			Maximum torque (Nm)		Maximum cutting force (N)		Maximum power (kw)		Power factor	
20	4000	0.1			3.4			28.5		5500		7.5		0.8	
Diameter (mm)	Hole depth (mm)	C_v	Z_v	Kv	Y_v	m	T_b(s)	C_F	X_F	Y_F	K_F	C_m	X_m	Y_m	K_m
20	35	11.1	0.4	0.87	0.5	0.2	1100	61.2	1.0	0.7	0.17	0.03	2	0.8	1.15
C_t	X	$T_c(s)$	$T_o(s)$		B		K_1		K_2			K_3		$Q(J)$	
4	0.05	16	60		0.2		0.00002		1.243			90.115		17115	

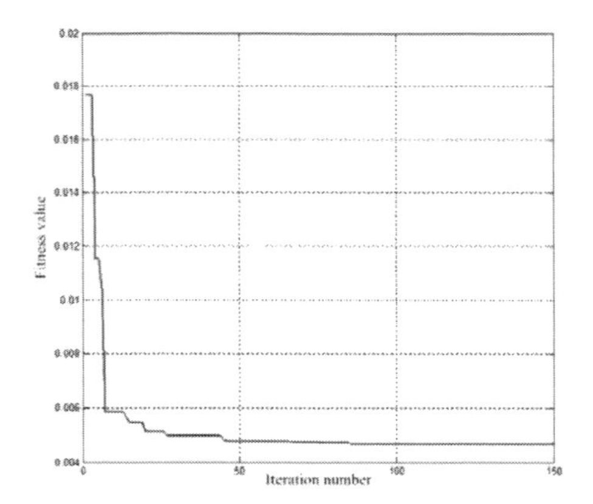

Figure 2. Graph showing the iterative convergence process of APSO.

Table 2. The optimization results of APSO.

Results	$\lambda_C = 1$	$\lambda_E = 1$	$\lambda_C = \lambda_E = 0.5$
Drilling speed (m/min)	24.05	27.09	26.18
Feed rate per turn (mm/r)	0.24	0.24	0.24
Time (s)	13.72	12.17	12.60
Cost (yuan)	0.881	0.918	0.901
Energy consumption (kJ)	207.80	207.63	207.66

When compared with the results of low costs or energy, the costs and energy consumption of drilling processes are changed with the drilling speed. When the goal is chosen as low cost and energy consumption, the results are between the above-mentioned values. It impacted in such a manner that the chosen parameters have important effects on energy consumption and costs, and the proposed model and method are effective for drilling processes optimization.

5 CONCLUSIONS

1. A multi-objective optimization model was established in this paper. The energy and cost are the optimization objectives, drilling speed, and feed rate per turn as independent variables, and the constraints of spindle speed, feed rate, drilling torque, etc. are also considered.
2. An improved algorithm (APSO) of PSO is proposed in this paper, and the optimization model was solved with this algorithm.
3. The case of this paper is aimed at one drilling condition, but the energy of the drilling process is not only affected by machining parameters, but also by the processing environment. Therefore, the next research work may focus on the various factors of the energy consumption.

REFERENCES

Congbo, Li, Tang Ying, Cui Longguo, Li Pengyu. A quantitative approach to analyze carbon emissions of CNC-based machining systems [J]. Journal of Intelligent Manufacturing, 26(5): 911–922 (2015).

Dietmair, A., A. Verl. Energy Consumption Forecasting and Optimization for Tool Machines [J]. Modern Machinery Science Journal 3: 62–67 (2009).

Liu Fei, Liu Shuang. Multi-period energy model of electro-mechanical main driving system during the service process of machine tools [J]. Journal of mechanical engineering, 48(21): 132–140 (2012).

Tang D, Du K, Li L. On the development path of Chinese manufacturing industry based on resource restraint [J]. Jiangsu Social Sciences 4, 51–58 (2006).

Xia Xiaohua, Liu Bo, Luan Zhiye, Jin Yihui. APSO-based Nonlinear Predicted Control and Its Application for the pH Neutralization Reactor Control [J]. Control and instruments in chemical industry, 33(1): 24–27 (2006).

Zhang Wenquan, Wei Wenshu. Parameter Optimization of High Speed Drilling in Small and Deep Hole on Stainless Steel [J]. Coal mine machinery, 32(4): 119–121 (2011).

Zhang Xinming, Cui Zhenshan. A study on fuzzy optimization of drilling parameters based on reliability analysis [J]. Journal of plasticity engineering, 14(5): 150–153 (2007).

Zhang Xinming, Qiu Jianjie, Zhou Zanxi, Cui Zhenshan. Study on Optimum Selection of Drilling Parameter Based on Fuzzing Mathematics [J]. Tool Engineering, 40(7): 27–29 (2006).

Zhang Zhe, Li Haolin. Quantitative analysis and comparison method for cutting process scheme energy consumption of machine tools [J]. Manufacturing technology & machine tool, 26(12): 21–23 (2013).

Zhao Pengfei, Liu Tingting, Wang Huifen. Research and Development on the optimization system of cutting parameters in the hole processing [J]. Aviation Precision Manufacturing Technology, 59(1): 34–37 (2012).

Civil, Architecture and Environmental Engineering – Kao & Sung (Eds)
© 2017 Taylor & Francis Group, ISBN 978-1-138-02985-9

An experimental study of the dynamic equivalence leakage of rolling-piston compressors

Kuihua Geng, Wei Wei, Xiaobo Ma, Xiao Wang, Xiaogang Han, Taibai Zhao, Chongli Chen & Han Sun
School of Mechanical Engineering, Guangxi University, Nanning, Guangxi, China

ABSTRACT: Based on the rolling-piston compressor as the research object, in this work, a new type of dynamic equivalence testing device for the compressor is put forward, which simulates the working condition of the inside compressor cavity and accomplishes dynamic adjustable speed, controllable differential pressure, and variable clearance. More groups of experiments with dynamic changes of multi-factors are carried out and finally the quantity and regularity about the leakage change with speed, differential pressure, and clearance are obtained.

Keywords: Rolling-piston compressor; dynamic equivalence testing device; dynamic leakage; experimental study

1 INTRODUCTION

A rolling piston compressor owns a good property and has been widely applied in many domains (Ma, 2001). However, due to the clearance of kinetic fitting pieces inside, its leakage problem is also really prominent, which has a major negative influence on the volumetric efficiency of the compressor (Yue, 2011). In view of this, the leakage study on the rolling-piston compressor by far has become a research hotspot in the compressor's academic field and engineering field both within the country and abroad (Hao, 2004). When the compressor is running, each motion pair is under a dynamic condition, such as the position, temperature, clearance, pressure, and even the quantity and distribution of the lubrication medium is constantly changed. Thus, the leakage characteristics are essentially associated with multiple factors and changes with work process (Li, 2006). The leakage inside the compressor can be collected and processed under the condition that it can accomplish dynamic differential pressure, dynamic clearance, and dynamic components of working medium, which makes comprehending leakage regularities of the compressor at a higher level. It is not only meaningful but it is also necessary and important that the research model with dynamic variable parameters can be concentrated instead of the original limited one with static parameters. However, there are large numbers of leakage channels and most of them could be extremely small (Jin, 1986; Zhou, 2007; Shen, 2015); the working conditions are complex and changing; it is really hard to complete the dynamic leakage tests inside the compressor. For the above-mentioned reason, in this work, a dynamic equivalence testing device is designed and fabricated for the rolling-piston compressor, by which there could be leakage data from the compressor under dynamic conditions.

2 EXPERIMENTAL TEST PLAN

The equivalent method is a kind of common scientific thinking method, which means, in a sense, unknown, complicated, and intractable issues could be converted into known, simple, and tractable ones on the premise that effects are same. It has been widely applied in mechanics, kinematics, electricity, optics, etc. which really means complicated practical problems are turned into simple familiar ones under the same effect. It will be easy to highlight the main factors, seize the essence and bring out laws if the team uses the simpler factors instead of the complicated ones. In this work, the testing about the leakage of the working medium was conducted inside compressors by using an equivalent device as the core. The selected dynamic equivalence leakage testing scheme in this work is shown in Fig. 1.

This experiment scheme consists of an air compressor, equivalent device, glass rotameter, precision pressure gauge, and so on. On one hand, the gas exhausted by using the air compressor enters the high pressure holder through the three-link control valve, and then turns into the precision pressure gauge after passing the control valve, and finally enters the high pressure chamber of the equivalent device through the inlet control valve. On the other hand, the leakage gas from the low pressure chamber of the equivalent device threads the glass rotameter and then passes the vent control valve and the precision pressure gauge, and finally enters into the constant tin through the control valve, and between this and the high pressure holder there is a balance valve. During the experiment, the balance valve is kept closed; meanwhile, the vent control valve and the control valve comprise the constant tin maintaining stability of the pressure. In the experiment system, the precision pressure gauge is used to display the differential pressure between two chambers of the equivalent device; the DC-infinitely adjustable-speed motor is used to change the speed of the device; and the precision micrometer head is used to adjust and display the leakage gap.

Based on the experiment's schematic shown in Fig. 1, the built experimental platform about the dynamic equivalence leakage of rolling-piston compressors is shown in Fig. 2. The type of the glass rotameter in the figure is LZB-10, in which the test medium is gas, and the test range is 250 L/h–2500 L/h.

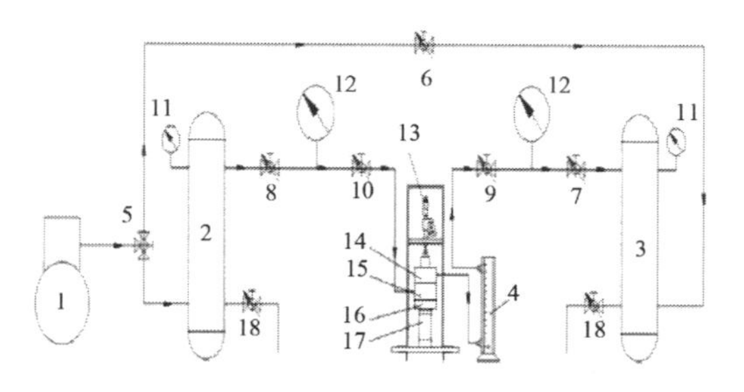

Figure 1. Schematic of the dynamic equivalent leakage testing setup.
1-Air compressor, 2-high pressure holder, 3-constant tin, 4-glass rotameter, 5-three-link control valve, 6-balance valve, 7-control valve, 8-control valve, 9-vent control valve, 10-inlet control valve, 11-pressure gauge, 12-precision pressure gauge, 13-precision micrometer head, 14-equivalent device, 15-high pressure chamber, 16-low pressure chamber, 17-DC infinitely adjustable-speed motor, and 18-drain valve.

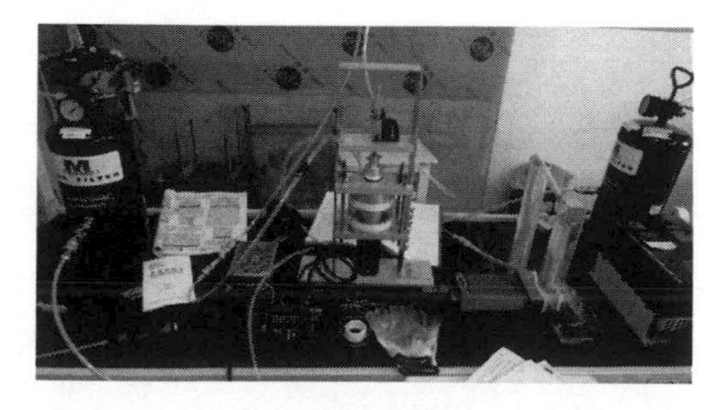

Figure 2. Picture of the dynamic equivalent leakage experiment's platform.

The experiment proceeded under the condition that three sets of factors were kept different from each other. Based on the orthogonal experiment, there would be 3^3 results, thus the change of clearance was the first factor to be studied by experiments in different groups, which could show the change of leakage by other factors when the clearance was kept specific.

3 EXPERIMENTAL EQUIVALENT DEVICE

3.1 *Basic structure and working principle*

The equivalent device's internal structure is shown in Fig. 3 and its dynamic equivalent device assembly drawings are shown in Fig. 4. The device consists of an adjustable-speed motor, static cavity, dynamic cavity, rotor, precision micrometer head, deep groove ball bearing, transparent cavity, gasket, and so on. The rotor is driven by using the the motor; the clearance is made by using the adjustable device on the right-hand side; the connection axis between the adjustable device and the bearing uses the interference fit. The viewing window is made of organic glass; the gasket is placed on each side to complete the clearance trimming.

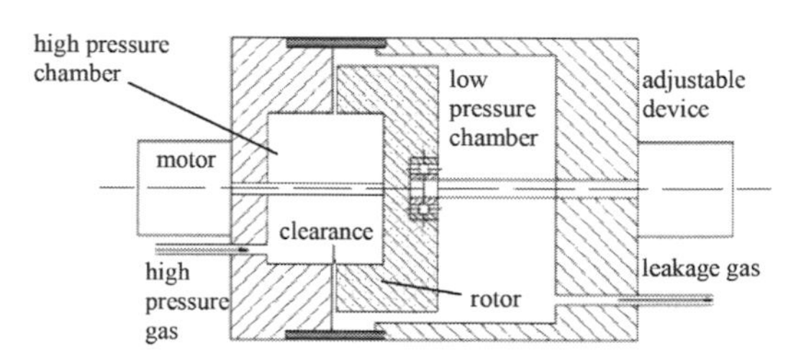

Figure 3. Schematic of the equivalent device's internal structure.

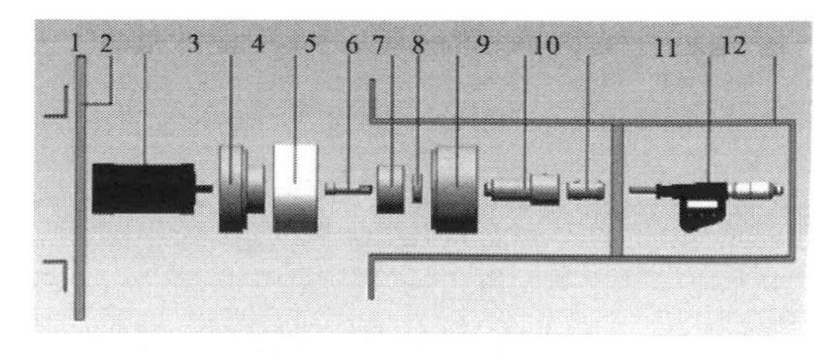

Figure 4. Schematic showing the dynamic equivalent device assembly drawings.
1-Pedestal, 2-adjustable-speed motor, 3-lower cavity, 4-transparent cavity, 5-motor transmission axis, 6-rotor, 7-bearing, 8-upper cavity, 9-rotor connection axis, 10-probe connection axis, 11-precision micrometer head, and 12-holder.

The leakage clearance forms by assembly between the rotor and the contact area of the static cavity on the left-hand side, in which there is a high pressure chamber. The low pressure chamber is formed among the rotor's outer surface, the dynamic cavity on the right-hand side and the transparent cavity's inner wall. The high pressure gas enters into the high pressure chamber from outside, flows into the low pressure chamber through the leakage clearance and then vents, which realizes an equivalence leakage course of compressors.

In the equivalent device, the motion of the compressor is controlled by the DC-infinitely adjustable-speed motor, which is placed under the device. The motor's rated power is 210 W, its rated speed is 3000 rpm, it is driven by using a 24-V DC power drive and controlled by using DC sensor-less and brushless controllers (ZM-6615). The speed is adjusted by using the speed-control switch inside, and its number is present on digital display. The precision micrometer head on the top of the device controls and adjusts the leakage clearance, which is produced by a company named Mast Correct. Its measurement range is 0–25 mm and its measurement accuracy is 0.001 mm, which could enable micron-sized control. The gas enters into or out through the tube of the device, and its difference value is controlled by using the high pressure holder, constant tin, and the pressure gauge.

The following is the test principle of the dynamic equivalence device: there are two cavities in the device: the left-hand side one is placed in high pressure gas during the experiment to simulate the high pressure chamber of compressors, and the right-hand side one is connected with the atmosphere to simulate the low pressure chamber of compressors; there is an adjustable leakage clearance between the rotor and the static cavity, which is equivalent to the leakage channels of compressors. The differential pressure of the high pressure chamber and the low pressure chamber is controlled by using a precision pressure gauge; the motor speed is controlled by using the machine controller; the leakage clearance is adjusted by using the precision micrometer head; the leakage is measured by using the float flowmeter in the vent-pipe. When the differential pressure and the reading from the flowmeter become stable, the leakage can be directly read. During the experiment, the speed can be adjusted alone, and so do the clearance and the differential pressure, which can make adjustable single-factor tests come true. Finally, the quantity and regularity about the leakage change with speed, and the differential pressure and clearance were obtained.

3.2 *Design and process*

Based on the design thought mentioned above, parts of the equivalence device are designed. Aluminium alloy and organic glass are chosen as work materials, which means that the upper and lower cavities, rotor, spindle, pedestal, and holder are made of aluminium alloy and the transparent cavity is made of organic glass. The SNS air hose is chosen as the gas tube. The finished parts are shown in Fig. 5.

3.3 *Installing and debugging*

Connect the assembled equivalence device, motor controller, tachometer, and the DC power supply. And then, connect the intake-tube and the gas supply system, and connect the vent-pipe, flowmeter, and the back pressure system. Check the pressure tightness of the sealing washer and working parts. As for the debugging process to increase the speed, turn the speed-control switch to the minimum, turn on the device, turn the switch upwards slowly to check the stationarity and check if it could reach the speed of the compressor working condition. The debugging experimental setup to increase the speed is shown Fig. 6, and it is found that the result is normal, which means the experiment could proceed.

As for the debugging to adjust clearance, adjust the micrometer head manually to make the bottom

Figure 5. Picture showing the dynamic equivalent device assembly and parts.
1-General assembly drawing, 2-rotor, 3-upper cavity, 4-transparent cavity and clearance, and 5-lower cavity.

Figure 6. Picture of the dynamic equivalent device that performs debugging to increase the speed.

of the rotor touch the superface of the lower cavity, and then lock the head and put in the pressure gas, check the pressure tightness and reset to the initial position. Turn on the self-locking switch, rotate the microdrum, and observe the reading from the micrometer head; turn off the self-locking switch when the reading reaches the default value. By this time, the clearance is well-adjusted, which could be lubricated to approximate the real compressor working conditions. The debugging experimental setup to adjust clearance is shown Fig. 7.

As for the debugging to maintain differential pressure, first blow up gasholders on both sides of the tube to reach the default value. Turn on the inlet control valve and observe the change of the pressure gauge, turn on the vent control valve when the reading is stable, and continue observing the pressure gauge. After debugging for a few times, the pressure could be maintained stable. The debugging experimental setup to maintain differential pressure is shown Fig. 8.

There are three major factors that affect the working medium leakage inside rolling-piston compressors: the clearance leakage (δ), the speed change (Δn) from the starting value to the default value and the differential pressure change (ΔP) between two chambers. The dynamic equivalence experiment of compressors below will proceed with these three factors.

Figure 7. Picture of the dynamic equivalent device's debugging experimental setup to adjust clearance.

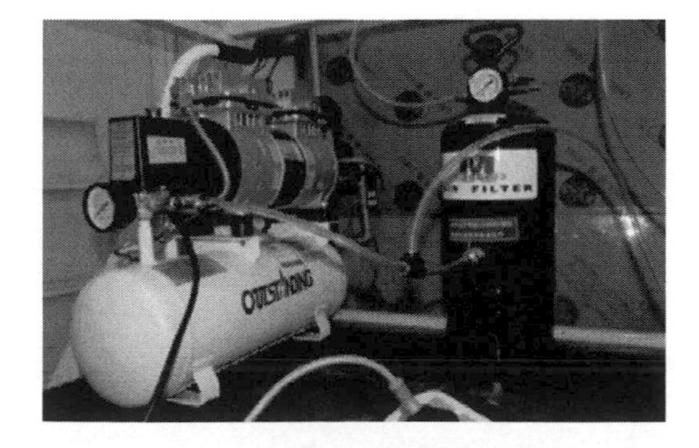

Figure 8. Picture of the dynamic equivalent device's pressure difference debugging.

4 EXPERIMENTAL PROCEDURE

The dynamic equivalence experiment of compressors includes the following three steps:

Firstly, turn on the self-locking switch of the precision micrometer head, turn the clearance leakage to the default value, and then lock the micrometer head. Turn on the power supply of the equivalence device, controller and tachometer, turn on the adjustable-speed motor, and adjust the speed to the default value.

Secondly, turn on the vent control valve next to the glass rotameter, turn on the control valve, and turn off the balance valve and the inlet control valve. After that, turn on the air compressor, turn on the three-link control valve to inflate the high pressure holder; when the reading from the pressure gauge reaches the default value, turn on the control valve. After the reading from the precision pressure gauge is maintained as a stable value, turn on the inlet control valve and fill the gas in the equivalence device to check the pressure-tightness of the device.

Thirdly, based on the two steps mentioned above, after maintaining the differential pressure and speed as stable values, record readings of the clearance, differential pressure, and rotameter. If the floater waggles seriously during the experiment, adjust the lower rotary knob of the rotameter and record the reading when it is maintained stable again.

Adjust the clearance, differential pressure, and speed for a few times, repeat the three steps above mentioned until the experiment is finished.

5 RESULTS AND DISCUSSION

5.1 *The effect of differential pressure*

The clearance value (δ1) of the equivalence device is set to 15 μm and the speed (n) is set to 1000 r/min, and then the test proceeded in the case that the differential pressure is maintained at 0.1, 0.2, and 0.3 MPa, respectively. The data are shown in Table 1 and the processing result is shown in Fig. 9.

Figure 9 shows that the corresponding leakage is about 0.41 g/s when the differential pressure is 0.1 MPa; it is about 0.50 g/s when the differential pressure is 0.2 MPa; it is about 0.59 g/s when the differential pressure is 0.3 MPa.

The clearance value (δ1) of the equivalence device is set to 15 μm and the speed (n) is set to 3000 r/min, and then the test proceeds in the case that the differential pressure is maintained at 0.1, 0.2, and 0.3 MPa, respectively. The processing result is shown in Fig. 10.

Figure 10 shows that the corresponding leakage is about 0.44 g/s when the differential pressure is

Table 1. The ΔPi experimental data under conditions such as δ1 = 15 μm and n1 = 1000 r/min (MFlow means mass flow).

Test number		1	2	3	4	5	6	7	8	9	10	11	12
Flow (m^3/h)	0.1 MPa	1.29	1.31	1.25	1.26	1.27	1.23	1.22	1.21	1.2	1.27	1.26	1.25
MFlow (g/s)		0.4254	0.432	0.4122	0.4155	0.4188	0.4056	0.4023	0.399	0.3957	0.4188	0.4155	0.4122
Flow (m^3/h)	0.2 MPa	1.58	1.57	1.53	1.52	1.52	1.51	1.51	1.52	1.53	1.53	1.52	1.53
MFlow (g/s)		0.521	0.5177	0.5045	0.5012	0.5012	0.4979	0.4979	0.5012	0.5045	0.5045	0.5012	0.5045
Flow (m^3/h)	0.3 MPa	1.8	1.81	1.82	1.8	1.79	1.79	1.8	1.8	1.79	1.78	1.79	1.79
MFlow (g/s)		0.5936	0.5968	0.6001	0.5936	0.5903	0.5903	0.5936	0.5936	0.5903	0.587	0.5903	0.5903

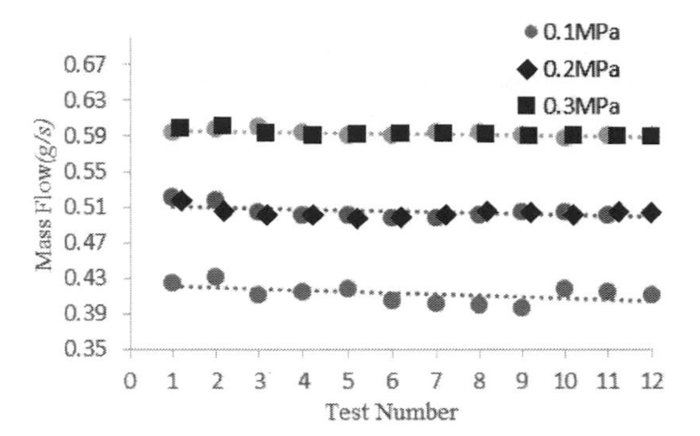

Figure 9. The ΔPi experimental results under conditions such as δ1 = 15 μm and n1 = 1000 r/min.

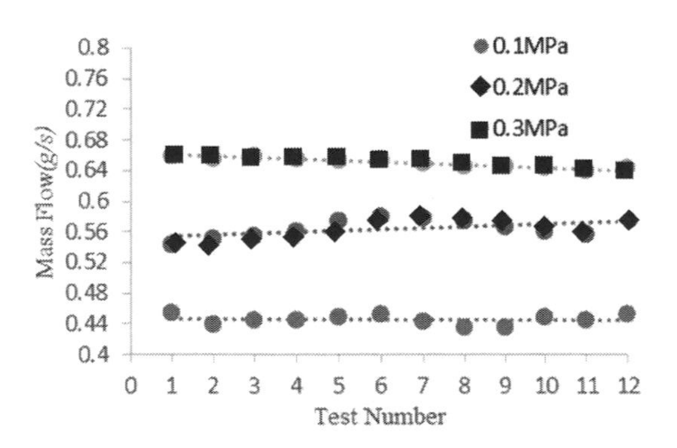

Figure 10. The ΔPi experimental results under conditions such as δ1 = 15 μm and n3 = 3000 r/min.

0.1 MPa; it is about 0.56 g/s when the differential pressure is 0.2 MPa; it is about 0.65 g/s when the differential pressure is 0.3 MPa. It is clear that when the speed and the clearance is maintained as constant, the leakage is greatly affected by the differential pressure. The more the differential pressure, the more is the leakage, and every 0.1 MPa differential pressure change could bring about 0.1 g/s leakage change.

Table 2. Δδi & ΔPi experimental data.

Constant condition	Clearance (Δδ)	Leakage ($\bar{m}_1$ (g/s))
$n_1 = 1000$ r/min $\Delta P_1 = 0.1$ MPa	$\Delta\delta_1 = 15$ μm	0.4127
	$\Delta\delta_2 = 25$ μm	0.4347
	$\Delta\delta_3 = 35$ μm	0.4443
$n_1 = 1000$ r/min $\Delta P_2 = 0.2$ MPa	$\Delta\delta_1 = 15$ μm	0.5048
	$\Delta\delta_2 = 25$ μm	0.5598
	$\Delta\delta_3 = 35$ μm	0.5641
$n_1 = 1000$ r/min $\Delta P_3 = 0.3$ MPa	$\Delta\delta_1 = 15$ μm	0.5925
	$\Delta\delta_2 = 25$ μm	0.6045
	$\Delta\delta_3 = 35$ μm	0.6400

5.2 *The effect of clearance*

The speed of the equivalence device is set to 1000 r/min and the differential pressure is set to 0.1, 0.2, and 0.3 MPa, respectively. And then, the test proceeds in the case that the clearance is maintained at 15, 25, and 35 μm, respectively. The processing results are shown in Table 2. As can be seen in Table 2, when the speed and the differential pressure is kept as constant, the more clearance makes more leakage, and every 10 μm clearance change could bring about a 0.02 g/s leakage change.

5.3 *The effect of speed data*

The clearance value of the equivalence device is set to 15 μm and the differential pressure is set to 0.1, 0.2, and 0.3 MPa, respectively. And then, the test proceeds in the case that the speed is maintained at 1000, 2000, and 3000 r/min, respectively. The processing results are shown in Table 3. As can be seen in Table 3, when the clearance and the differential

Table 3. The Δni and ΔPi experimental data.

Constant condition	Speed (Δn)	Leakage ($\bar{\dot{m}}_1$ (g/s))
$\delta_1 = 15$ μm	$\Delta n_1 = 1000$ r/min	0.4127
$\Delta P_1 = 0.1$ MPa	$\Delta n_2 = 2000$ r/min	0.4344
	$\Delta n_3 = 3000$ r/min	0.4452
$\delta_1 = 15$ μm	$\Delta n_1 = 1000$ r/min	0.5048
$\Delta P_2 = 0.2$ MPa	$\Delta n_2 = 2000$ r/min	0.5576
	$\Delta n_3 = 3000$ r/min	0.5644
$\delta_1 = 15$ μm	$\Delta n_1 = 1000$ r/min	0.5925
$\Delta P_3 = 0.3$ MPa	$\Delta n_2 = 2000$ r/min	0.6076
	$\Delta n_3 = 3000$ r/min	0.6507

pressure are maintained as constant, the more speed leads to more leakage, and an every 1000 r/min speed change could bring about a 0.02 g/s leakage change.

6 CONCLUSION

This work utilizes the self-designed dynamic equivalence testing device for the compressor, which simulates the working condition inside the compressor chamber; and builds the experimental platform, which accomplishes dynamic adjustable speed, dynamic controllable differential pressure, and dynamic variable clearance. After testing, that the following conclusions can be drawn:

1. The leakage is affected by the differential pressure. The more differential pressure leads to more leakage, and every 0.1 MPa differential pressure change could bring about a 0.1 g/s leakage change.
2. The leakage is affected by the clearance. The more clearance leads to more leakage, and every 10 μm clearance change could bring about a 0.02 g/s leakage change.
3. The leakage is affected by the speed. The more speed leads to more leakage, and every 1000 r/min speed change could bring about a 0.02 g/s leakage change.

ACKNOWLEDGMENTS

This work was supported by the China Natural Science Foundation (grant no. 51565004), Guangxi (China) Natural Science Foundation (Grant no. 2013GXNS FAA019309), and the Key Laboratory of Manufacturing System and Advanced Manufacturing Technology Fund (grant no. 15-140-30S001 and no. 10-0046-07S03).

REFERENCES

Hao, J., Y.F. Chang, W. Guo, Cmpr. Tech. J. **2** (2004).
Jin, G.X., Q. Lin, Flu. Mach. J. **1** (1986).
Li, X.M., A.N. Geng, K.H. Geng, Flu. Mach. J. **34**, 1 (2006).
Ma, G.Y., H.Q. Li, *Rotary Compressor* (China Machine Press, Beijing, 2001).
Shen, L.L., W. Wang, Y.T. Wu, Cryog. Supercond. J. **12** (2015).
Yue, X.J., D.C. Ba, Z. Lin, Eng. Mech. J. **28**, 9 (2011).
Zhou, H., Z.C. Qu, B.F. Yu, Chin. Mech. Eng. J. **18**, 2 (2007).

Civil, Architecture and Environmental Engineering – Kao & Sung (Eds)
© 2017 Taylor & Francis Group, ISBN 978-1-138-02985-9

The structure and transmission efficiency analysis research of a new pattern F2C-T pin–cycloid planetary transmission

Liang Xuan
School of Mechanical and Architectural Engineering, Jianghan University, Wuhan, China

Heng Jiang & Tianmin Guan
School of Mechanical Engineering, Dalian Jiaotong University, DaLian, China

ABSTRACT: F2C-T transmission is a new K-H-V type of two-stage cycloid transmission device. It has high transmission efficiency advantages. In this paper, the structure and transmission efficiency of the new high precision type of F2C-T transmission are analyzed. The drive ratio calculation model of F2C-T transmission is deduced by using the solving method of the differential gear train transmission ratio. And the calculation method of the system's meshing transmission efficiency is deduced with the transmission ratio method. The transmission efficiency mathematical model of F2C-T transmission is established. And the transmission efficiency of the F2C-T455 model is calculated and analyzed.

1 INTRODUCTION

F2C-T transmission is a new pattern with a two-stage closed type cycloid transmission device. It exhibits properties, such as high transmission precision, small rotation error, high transmission efficiency, stable transmission, and so on. And so, it has an extremely widespread application prospect in the field of transmission. While studying the F2C-T drive, its structure requires analysis at first. The shape of the components and the transmission relationship between the parts can be grasped. And then, the power loss of the system in the transmission process can be found. The transmission efficiency of F2C-T transmission is calculated. Transmission efficiency is one of the important indexes to evaluate the performance of mechanical transmission devices, and its level also directly affects the research value and development prospects of the device. Therefore, it is of great significance to study the transmission efficiency of F2C-T transmission.

2 SYSTEM STRUCTURE AND TRANSMISSION PRINCIPLE OF F2C-T TRANSMISSION

2.1 *Structure and characteristics of F2C-T transmission*

The structure of F2C-T transmission is more complex, and it is composed with a stage of involute gear planetary transmission and a stage of cycloid-pin planetary gear transmission as well as the bearing in the meshing transmission process. The structure diagram is shown as Figure 1.

According to the meshing transmission relationship between each part in the transmission process of the F2C-T transmission, the F2C-T transmission is divided into the following four parts:

1. The transmission part of the involute gear
The transmission part of the involute gear consists of the input shaft, the key gear, and three planetary gears. And the input shaft and the sun gear are fixedly connected together to form the input gear shaft. And three planetary gears are distributed at angles that are 120 degrees away according to the structure of the planet.
2. Eccentric transmission part
Each eccentric body is composed of a spline shaft, two eccentric shaft necks, and a rotary arm

Figure 1. Schematic of the F2C-T transmission structure.

bearing. Two eccentric shaft necks are 180 degrees away in distribution. In order to balance the inertia force, the two pieces of the cycloid gear should be also evenly arranged on the two eccentric axes at angles that are 180 degrees away from each other.
3. The part of the cycloid–pin gear transmission

It consists of a cycloid gear, a pin gear, and a pin gear housing. Different from the traditional two fulcrums and the three-fulcrums structure, the pin gear of F2C-T transmission chooses the horizontal pillow type installation method. It directly lies in the pin gear housing. In this way, the bending deformation of the pin gear can be greatly reduced, and the bending strength can be improved.
4. Output mechanism

It consists of output side flange, eccentric support bearing, spindle bearing, the supporting pin and the input side of the flange. The input and output side flanges are connected together with the supporting pin to form a simply-supported beam structure. It can greatly improve the carrying capacity of the mechanism body.

The structure of F2C-T transmission has the following characteristics:

1. The two stages of the transmission structure are used. Due to the transition of the first stage involute transmission part, the cycloid transmission part of the second stage is made more stable. It can greatly improve its life; (2) the cycloid gear adopts the two-tooth-difference structure and increases the carrying capacity; (3) the pin gear adopts the horizontal pillow type structure. In this way, when the pin gear and cycloid gear undergo meshing, the bending deformation and even bending damage can be avoided; (4) the output mechanism adopts the rigid flange output structure with bilateral support. When compared with the output mechanism of the traditional cycloid transmission cantilever beam structure, it has greater stiffness and better impact resistance performance.

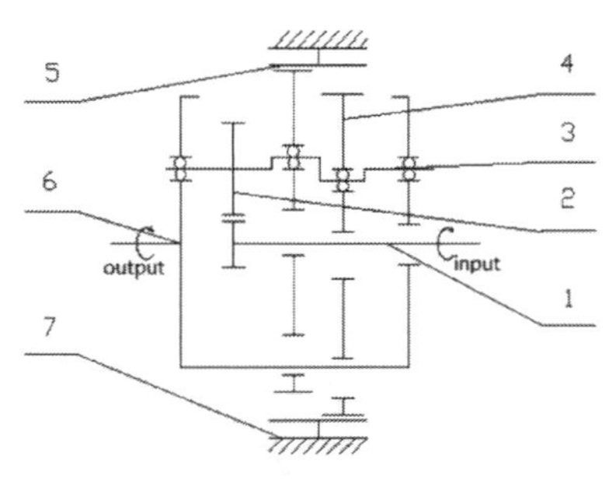

1) Input gear shaft, 2) planet gear, 3) eccentric body, 4) cycloid gear, 5) pin gear, 6) output shaft, and 7) pin gear housing

Figure 2. Transmission diagram of F2C-T transmission.

2.2 *The working principle of F2C-T transmission*

F2C-T transmission is a new pattern with two-stage closed type cycloid planetary transmission device. It is composed of the involute gear planetary transmission of the first stage and the cycloid–pin gear planetary transmission of the second stage. The transmission diagram is shown in Figure 2.

3 THE CALCULATION OF THE TRANSMISSION RATIO

According to Figure 2, the transmission diagram of the new pattern high precision F2C-T transmission, F2C-T transmission consists of two stages. The first stage transmission is composed of the input gear shaft 1 and the planet gear 2. The second stage transmission is composed of cycloid gear 4, the pin gear 5, and the pin gear housing 7. When the eccentric H(3) is fixed, the transformed transmission ratio can be obtained by using the following equation:

$$i_{12}^{H} = -\frac{Z_2}{Z_1},\ i_{45}^{H} = \frac{Z_5}{Z_4} \tag{1}$$

In this type, Z_1 is the tooth number of the input gear shaft, Z_2 is the tooth number of the planet gear, Z_4 is the tooth number of the cycloid gear, Z_5 is the tooth number of the pin gear.

The pin gear is mounted on the pin gear housing during the installation. And so, the pin gear and the pin gear housing can be treated as a whole part in the calculation process. If the angular velocity of the pin gear 5 is ω_5, the angular velocity of the eccentric body H is ω_H, the angular velocity of the cycloid gear 4 is ω_4, the angular velocity of the output flange 6 is ω_6, then $\omega_4 = \omega_6$ and the angular velocity of the eccentric H is 0, the relative angular velocity of each component is given as follows:

$$\begin{cases} \omega_5^H = \omega_5 - \omega_H \\ \omega_H^H = \omega_H - \omega_H = 0 \\ \omega_6^H = \omega_6 - \omega_H = \omega_4 - \omega_H \end{cases} \tag{2}$$

When the eccentric H is fixed, the transmission ratio of the mechanism can be obtained by using the following equation:

$$i_{65}^{H} = \frac{\omega_6 - \omega_H}{\omega_5 - \omega_H} = \frac{Z_5}{Z_4} \tag{3}$$

When the output shaft 6 is fixed, from the general equation $i_{ab}^{c}=\frac{1}{1-i_{bc}^{a}}$ of the transmission ratio calculation it can be obtained:

$$i_{H5}^{6}=\frac{1}{1-i_{56}^{H}}=\frac{Z_5}{Z_5-Z_4} \tag{4}$$

The output shaft is fixed, when the pin gear and the pin gear housing are obtained as the output, the transmission ratio is given as follows:

$$i_{15}^{6}=i_{12}^{H}\cdot i_{H5}^{6}=-\frac{Z_2Z_5}{Z_1(Z_5-Z_4)} \tag{5}$$

And according to the general calculation equation of transmission ratio $i_{ab}^{c}=1-i_{ac}^{b}$, when the pin gear housing is fixed, the system transmission ratio can be obtained as follows:

$$i_{16}=i_{16}^{5}=1-i_{15}^{6}=1+\frac{Z_2Z_5}{Z_1(Z_5-Z_4)} \tag{6}$$

The cycloid–pin gear transmission part of F2C-T transmission uses two-tooth-difference structure. And so, the result is $Z_5-Z_4=2$; then,

$$i_{16}=1+\frac{Z_2Z_5}{2Z_1} \tag{7}$$

4 TRANSMISSION EFFICIENCY ANALYSIS

The transmission efficiency of F2C-T transmission is mainly composed of two parts, which are meshing transmission efficiency and bearing efficiency. The meshing efficiency is the main part of the transmission efficiency. It is composed of two parts, which are involute gear meshing transmission efficiency and cycloid–pin gear transmission efficiency.

4.1 *Meshing transmission efficiency analysis*

The equation of meshing transmission efficiency is derived by using the method of transmission ratio. According to the transmission ratio equation of F2C-T transmission, it can be re-written in the following form:

$$i_{16}=\frac{-i_{12}^{H}\,i_{45}^{H}}{i_{45}^{H}-1}+1 \tag{8}$$

It turns out that the transmission ratio i_{16} of the differential gear transmission F2C-T transmission consists of the transmission ratio function of its various transmissions. (When the eccentric body is fixed, the transmission ratio of the corresponding transmission can be obtained.) That is,

$$i_{16}=\frac{\omega_1}{\omega_6}=f(i_{12}^{H},i_{45}^{H})=\frac{-i_{12}^{H}i_{45}^{H}}{i_{45}^{H}-1}+1 \tag{9}$$

In this type, ω_1 is the angular velocity of the input shaft (rad/min) and ω_6 is the angular velocity of the output shaft (rad/min).

Similarly, the force transmission ratio $\tilde{i}_{16}$ is also the function, which is composed of each corresponding force transmission ratio $\tilde{i}_{12}^{H},\tilde{i}_{45}^{H}$. That is,

$$\tilde{i}_{16}=\frac{T_6}{T_1}=f\left(\tilde{i}_{12}^{H},\tilde{i}_{45}^{H}\right)=\frac{-\tilde{i}_{12}^{H}\,\tilde{i}_{45}^{H}}{\tilde{i}_{45}^{H}-1}+1 \tag{10}$$

In this type, T_1 is the input torque $(N\cdot m)$, T_6 is the output torque $(N\cdot m)$, $\tilde{i}_{16}$ is the ratio of the output to the input torque.

Therefore, the meshing transmission efficiency of F2C-T transmission is given as follows:

$$\eta_{16}=\frac{P_6}{P_1}=\frac{T_6\omega_6}{T_1\omega_1}=\frac{\tilde{i}_{16}}{i_{16}}=\frac{f\left(\tilde{i}_{12}^{H},\tilde{i}_{45}^{H}\right)}{f\left(i_{12}^{H},i_{45}^{H}\right)} \tag{11}$$

In this type, P1 is the input power (kw), P6 is the output power (kw). $\tilde{i}_{12}^{H}=i_{12}^{H}(\eta_{12}^{H})^{x_1}$, $x_1=\pm 1$; $\tilde{i}_{45}^{H}=i_{45}^{H}(\eta_{45}^{H})^{x_2}$, $x_2=\pm 1$;

η_{12}^{H} is the meshing transmission efficiency of the involute gear; η_{45}^{H} is the meshing transmission efficiency of the cycloid–pin gear.

The value representation of the index x_1, x_2 is +1, which indicates that the motion transmission ratio and the power flow are in the same direction; –1 indicates that the motion transmission ratio is opposite to the power flow. According to the literature (Yang, 1986),

$$x_1=sign\frac{i_{12}^{H}}{i_{16}}\cdot\frac{\partial i_{16}}{\partial i_{12}^{H}} \tag{12}$$

While $\frac{i_{12}^{H}}{i_{16}}\cdot\frac{\partial i_{16}}{\partial i_{12}^{H}}<0, x_1=-1$. On the contrary, $x_1=+1$.

According to equation (9), function i_{12}^{H} is derived as the partial derivative and multiply $\frac{i_{12}^{H}}{i_{16}}$ to obtain the following equation:

$$\frac{i_{12}^{H}}{i_{16}}\cdot\frac{\partial i_{16}}{\partial i_{12}^{H}}=\frac{i_{12}^{H}}{i_{16}}\cdot\frac{-i_{45}^{H}}{i_{45}^{H}-1}=\frac{-i_{12}^{H}i_{45}^{H}}{i_{16}(i_{45}^{H}-1)}$$

Since $i_{12}^{H}<0, i_{45}^{H}>0$. Substitute in the above equation to obtain the following equation: $\frac{i_{12}^{H}}{i_{16}}\cdot\frac{\partial i_{16}}{\partial i_{12}^{H}}>0$, that $x_1=+1$.

Similarly, according to equation (8), the function i_{45}^{H} is derived as the partial derivative and multiply $\frac{i_{45}^{H}}{i_{16}}$, to obtain the following equation: $\frac{i_{45}^{H}}{i_{16}} \cdot \frac{\partial i_{16}}{\partial i_{45}^{H}} < 0$, that $x_2 = -1$.

The above equations are substituted into equation (9), and the following equation can be obtained:

$$\tilde{i}_{16} = \frac{-i_{12}^{H}(\eta_{12}^{H})^{+1} i_{45}^{H}(\eta_{45}^{H})^{-1}}{i_{45}^{H}(\eta_{45}^{H})^{-1} - 1} + 1 = \frac{-i_{12}^{H} i_{45}^{H} \eta_{12}^{H}}{i_{45}^{H} - \eta_{45}^{H}} + 1 \tag{13}$$

Equation (12) is substituted into the transmission efficiency parameter of equation (10). The equation for calculating the meshing transmission efficiency of F2C-T transmission is given as follows:

$$\eta_{16} = \frac{\tilde{i}_{16}}{i_{16}} = \frac{(i_{45}^{H} - 1)(i_{45}^{H} - \eta_{45}^{H} - i_{12}^{H} i_{45}^{H} \eta_{12}^{H})}{(i_{45}^{H} - \eta_{45}^{H})(i_{45}^{H} - 1 - i_{12}^{H} i_{45}^{H})} \tag{14}$$

4.2 *Involute gear transmission efficiency calculation*

The meshing transmission efficiency calculation equation of the involute gear transmission part (Zhu, 1983) is given as follows:

$$\eta_{12}^{H} = 1 - \pi\mu\left(\frac{1}{Z_1} + \frac{1}{Z_2}\right)(\varepsilon_1^2 + \varepsilon_2^2 + 1 - \varepsilon_1 - \varepsilon_2) \tag{15}$$

In this type, μ is the friction coefficient, $\mu = 0.05 \sim 0.1; \varepsilon_1 = \frac{Z_1}{2\pi}(tg\alpha_{a1} - tg\alpha); \varepsilon_2 = \frac{Z_2}{2\pi}(tg\alpha_{a2} - tg\alpha)$;

α is the meshing angle of the involute gear, and $\alpha = 20°$; α_{a1} is the addendum circle pressure angle of the key gear, $\alpha_{a1} = \arccos\frac{mZ_1\cos\alpha}{2r_{a1}}$, (°); r_{a1} is the addendum circle radius of the key gear, $r_{a1} = \frac{mZ_1}{2} + mh_a^*$, (mm); α_{a2} is the addendum circle pressure angle of the planetary gear, $\alpha_{a2} = \arccos\frac{mZ_2\cos\alpha}{2r_{a2}}$, (°); r_{a2} is the addendum circle radius of the planetary gear, $r_{a2} = \frac{mZ_2}{2} + mh_a^*$, (mm).

4.3 *Efficiency calculation of cycloid–pin gear planetary transmission*

The F2C-T pattern cycloid–pin gear planetary transmission is a type of K-H-V planetary transmission. The meshing transmission efficiency loss mainly comes from the loss of the cycloid gear and the pin gear meshing process. And so, the transmission efficiency calculation equation of the cycloid–pin gear transmission part can be derived according to the following equation:

$$\eta_{45}^{H} = 1 - \eta_{K}^{H} \tag{16}$$

In this type, η_{K}^{H} is the meshing efficiency loss of the cycloid gear and the pin gear.

4.4 *Bearing total efficiency calculation*

According to the F2C-T transmission structure, it can be known that the whole system is a total including three sets of bearings: spindle bearing, support bearing of the eccentric body, and rotary arm bearing. The former two are tapered roller bearings. The third one is a single row cylindrical roller bearing without outer ring. Therefore, the total efficiency η_B of bearing is given as follows:

$$\eta_B = \eta_{B1}\eta_{B2}\eta_{B3} \tag{17}$$

In this type, η_{B1} is the efficiency of the spindle bearing; η_{B2} is the efficiency of the eccentric body support bearing; and η_{B3} is the efficiency of the rotary arm bearing.

4.5 *System transmission efficiency calculation*

The F2C-T transmission efficiency is composed of two parts, which are the meshing transmission efficiency and the bearing total efficiency. Therefore, the transmission efficiency of F2C-T transmission is as follows:

$$\eta = \eta_{16} \cdot \eta_B \tag{18}$$

In this type, η_{16} is the meshing transmission efficiency and η_B is the bearing total efficiency.

5 CALCULATION EXAMPLE

Based on the above analysis and research as well as the transmission efficiency mathematical model of the F2C-T transmission, taking the F2C-T455 prototype as an example, the transmission efficiency is calculated. When the F2C-T455 prototype input shaft rated speed is $n = 1500$ r/min, the basic parameters are shown in Table 1.

The "transmission efficiency calculation" program is written by using the VB programming language, and the calculation program of F2C-T transmission efficiency can be obtained.

The transmission efficiency of the T455 prototype can reach 91.2%, thereby fully meeting the requirements that the transmission efficiency should be more than 90%.

Table 1. Basic parameters of the F2C-T455 prototype.

Parameter	Symbol	Numerical (mm)	Parameter	Symbol	Numerical (mm)
Eccentricity	α	2	Width of the cycloid gear	b	14
Tooth number of the cycloid gear	Z_c	27	Input gear's shaft teeth number	Z_1	12
Pin gear's teeth number	Z_p	28	Planet gear's teeth number	Z_2	36
Pin gear's center circle radius	r_p	90	Transmission ratio	i_{16}	81
Pin gear's radius	r_{rp}	3.5	Tooth difference	c	2
Module	m	2			

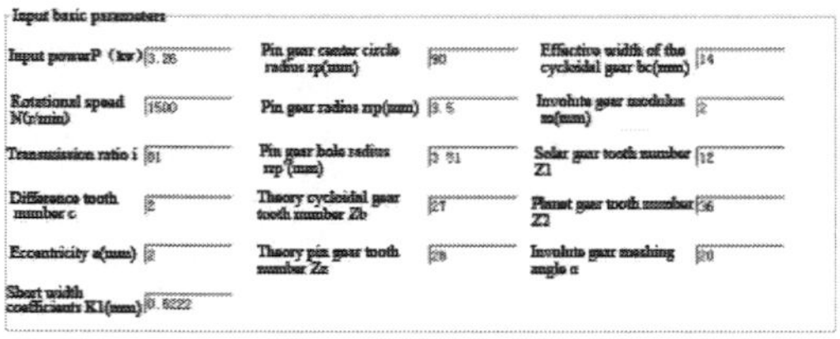

Figure 3. Basic parameter interface.

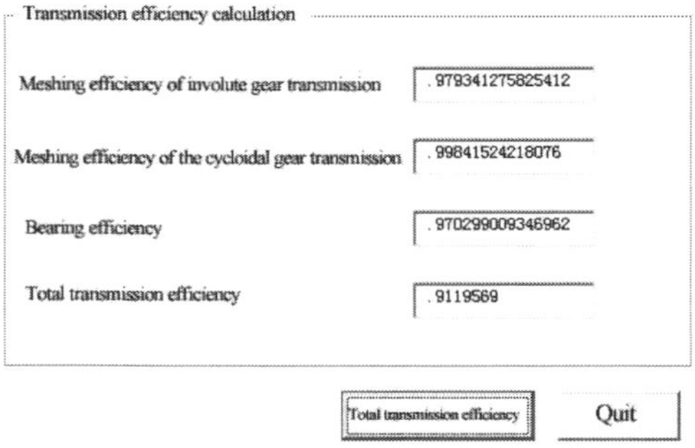

Figure 4. Transmission efficiency calculation interface.

6 CONCLUSIONS

In this paper, the structure and transmission efficiency of the new type of high precision F2C-T transmission are analyzed. The structure and characteristics of F2C-T transmission are mainly analyzed. And on those bases, according to the solving method of the differential gear train transmission ratio, the transmission ratio calculation equation of F2C-T transmission is deduced. The calculation equation of the system's meshing transmission efficiency is derived by using the method of transmission ratio. The transmission efficiency mathematical model of the F2C-T transmission is established. And the transmission efficiency calculation for the F2C-T455 prototype models is carried out. The transmission ratio distribution model of the new pattern can be further optimized by using the calculation of transmission efficiency. It has a great significance.

REFERENCES

Figliolini Giorgio, Stachel Hellmuth, Angeles Jorge. On Martin Disteli's spatial cycloidal gearing. Mechanism and Machine Theory. 2013, V60, P73–89.

Kumar Naren, Kosse Vladis, Oloyede Adekunle. A novel testing system for a cycloidal drive. International Conference on Intelligent Systems Design and Applications, 2012, P503–507.

Kumar Naren, Kosse Vladis, Oloyede Adekunle. Analysis of the damping properties of high-transmission—ratio cycloidal drives. Applied Mechanics and Materials, 2012 V215–216, P1038–1042.

Liu Yuting. Transmission error analysis of RV reducer. (Master's degree thesis). Dalian Jiaotong University. 2012.

Meng Yunhong. Research on the performance theory of 2 K-H type cycloidal-pin gear planetary transmission. Doctoral Dissertation of Huazhong University of Science and Technology. 2007.

Yang Yandong. Involute gear planetary transmission. Chengdu University of Science and Technology press. 1986:47–54.

Zhu Jingzi. Research on the efficiency of gear transmission. Journal of Mechanical Engineering. 1983, 19(3): 61–70.

Civil, Architecture and Environmental Engineering – Kao & Sung (Eds)
© 2017 Taylor & Francis Group, ISBN 978-1-138-02985-9

A study on surface zinc plating and heat treatment of welding gears

Rui Zeng
College of Field Engineering, PLA University of Science and Technology, Nanjing, China
Department of Airfield Engineering, Air Force Logistics College, Xuzhou, China

Junhong Luo & Zhenrong Lin
Department of Airfield Engineering, Air Force Logistics College, Xuzhou, China

Lulu Wang
College of Field Engineering, PLA University of Science and Technology, Nanjing, China

ABSTRACT: In order to improve the comprehensive mechanical properties and zinc coating qualities of welding gears, the new zinc plating solution was used to electroplate after heat treatment of the welding gear. The results showed that after quenched–tempered heat treatment, the weld area organization of the welding gear was small and uniform tempered sorbite, which led to achieving excellency and stability of mechanical properties. The gears which were processed and manufactured by the way of pre-stack low carbon steel were tested and found to be qualified. The main composition of the coating was Zn, and the Fe content was less. The mass fraction of Zn was increasing along the base metal to coating, and Fe was just the opposite. The thickness of the plating layer was about 6–8 μm, and the zinc coating was closely attached to the gear. The minimum deformation heat treatment process was used to weld the gear, which ensured the dimensional accuracy, reduced the processing difficulty, and was conducive to the production of the workpieces. The mechanical performances of the weld area were improved significantly by using the process of bead welding of the transition layer. The national standard test showed that the electroplated coating had uniform thickness, reasonable distribution, good density, adhesion, and corrosion resistance.

1 INTRODUCTION

When the welding gear was working, it was affected by the large alternating impact load. The gear was required to have good comprehensive mechanical properties and corrosion resistance. In this paper, the gear ring material used was 34CrNiMo, and the shaft material was 42CrMo, which adopted the whole forging and the intermediate frequency quenching process (ZHENG, 2004). But in the practical application, it was found that the tooth part and the end face underwent cracking easily, which was not conducive to the production of many workpieces. The gear and the shaft of the transmission with ordinary galvanized steel plating method is used to prevent rust, and this caused serious hydrogen evolution, easy peeling and foaming of the coating, longer plating time, and loss of life (Barbosa, 2006; El, 2008; Kavitha, 2008). The coating was easy to burn. In order to improve the comprehensive mechanical properties and zinc coating qualities of the welding gear, the new zinc plating solution was used to electroplate after heat treatment of the welding gear.

2 THE TECHNOLOGICAL SCHEME AND PROCESS

2.1 *Heat treatment*

The welding structure of the gear should focus on solving the following problems in heat treatment:

1. Mechanical properties of the welded HAZ zone decreased.

The content of high carbon martensite in the HAZ area increased after welding. In the welding process, it was difficult to be heated evenly, and the high temperature tempering zone would appear in the near-crack zone, which decreases the whole structure strength, thereby resulting in the brittleness and hardening of the gear.

2. Crack and deformation of the weld zone

In the heating and cooling process of the welding structure of the gear, the thermal conductivity of different gear surfaces and internal temperatures changed to thermal expansion and contraction, and thermal stress could make the gear crack and lead to deformation.

Besides, the phase transition points of different materials were different, and the microstructure transformation was not the same, which increased the stress and the cracking and deformation tendency of the gear-welded zone.

In view of the above-mentioned problems, the following heat treatment process is developed:

1. The gear is set aside for heat treatment or heat treatment of gear teeth. Although the size precision was high, the hardness of heat treatment was high, and it was difficult to obtain the gear. Considering the processing properties, the gear tooth surface was set aside by 0.4 mm for heat treatment, and the deformation of gear teeth must be less than 0.4 mm after heat treatment.
2. Individual parts were pre-heat-treated, and the hardness of the quenched and tempered gear was adjusted to 230~270 HBW to achieve the grain refinement, uniform structure, and reduce the amount of deformation after welding.
3. After welding, heat stress was eliminated in the well-type furnace to reduce the cracking tendency and deformation of the weld zone.
4. In order to reduce the deformation caused by the weight of the parts, the welding gear was mounted in a vertical furnace in a well-type furnace, which was carried out by quenching and tempering. First, the high temperature of 900 degrees–860 degrees is set for the rapid heating furnace; quenching is carried out after which, the heating time is reduced, and oxidation and decarburization deformation are prevented. And then, oil quenching was carried out in the vertical parts of the gear, parts were cooled to MS and when the temperature fell below MS, stirring was stopped, slow cooling was carried out, and tissue stress was reduced. Finally, it was tempered to eliminate the internal stress generated during quenching.

2.2 *Plating process plan*

The pre-treatment of the gear with heat treatment before electroplating is as follows:

First degreased in 12% sodium hydroxide, 5% sodium carbonate, 7% sodium phosphate, 0.5% OP-10 emulsifier solution at 60~65°C for 20 min. After washing with water, it was pickled with 18% hydrochloric acid, 0.2% hexamethylenimine and 0.2% OP-10 emulsifier solution at 20~25°C for 15 min. And then into the 5% hydrochloric acid solution activation of about 8 s, activated washed.

The formulation of the zinc plating solution is as follows: 20 g/L of ammonium chloride, 3 g/L of malic acid, 1 g/L of OP-20 emulsifier, 0.1 g/L of cinnamic acid, 0.2 g/L of the main brightener, and 1 g/L of polyethylene glycol, pH 1.8 to 2.4.

In the experiment, vanadium aldehydes, o-nitrobenzaldehyde, and benzylidene acetone were used as the main brighteners for galvanizing the welding gear parts. The comparison test showed that benzylideneacetone was more suitable for zinc coating preparation of the heat-treated parts (Gao, 2001; Liang, 2007).

The electroplating process is as follows:

The gears were placed in the prepared hook on the galvanized gear, to maintain the current intensity of 0.5 A/dm^2, pre-plated for 5 min. And then, the current intensity is adjusted to 1 A/dm^2, by adding a mixture of the main brightener and cinnamic acid solution; adjust the current density; the pH value of zinc plating solution control is 1.8–2.4, and the plating time is 20–30 min for the gear surface to form a uniform zinc coating. After the galvanized gear parts are immersed in the mixture of 1 g/L sulfuric acid, 0.2 g/L nitric acid, and 60 g/L chromic anhydride, passivation is carried out for 8 s, hot water washing is carried out for 10 s, and finally, the gear is placed at 60°C for drying in the oven for 15 min.

In order to verify the effectiveness of the heat treatment process, comparison tests were carried out on the post-weld, quenched, and tempered gear and the original gear metal's performances.

An LDW200-4XB-type table-type optical microscope was used to observe the structures of the original gears and quenched and tempered gear, as given in Table 1. A wew-1000b hydraulic universal testing machine was used to measure the mechanical properties of the original gears and quenched and tempered gears, as given in Table 2.

From Tables 1 and 2, it could be seen that the residual stress was large because the heat treatment of the original gear structure had the upper bainite, so that the comprehensive mechanical properties decline. The mechanical properties of the weld zone were fine and homogeneous. The mechanical properties of the HAZ zone were softened, brittle, and hardened. The mechanical properties of the pre-heaped low carbon steel were the best after welding. According to the above-mentioned experimental results and analysis, in accordance

Table 1. A comparison of the metallographic structures.

Test number	After welding	Weld metal area	Heat-affected zone
1		B+M (trace)	B+F+M (few)
2	860°C quenching 500°C tempering	Tempering S	Tempering S
3	860°C quenching 500°C tempering	Tempering S + F + P	Tempering S + F + P (few)

Table 2. A comparison of mechanical properties.

Test number	After welding	Tensile strength/N·mm^{-2}	Extensibility/%	Impact ductility/J
1		973.2	7.1	37.1
2	860°C quenching 500°C tempering	1218.2	6.5	40.6
3	860°C quenching 500°C tempering	1369.4	5.5	66.3

Table 3. Gear test results.

	Radial pulsation of the gear/mm	End face pulsation of the gear/mm	Tooth axis bending/mm	Hardness/ HBW
Technical requirement	<0.40	<2.0	<2.5	350~380
Detection result	0.30~0.40	1.0~1.1	1.5~1.8	370~380

with the surfacing of the welding layer after the quenching and tempering process gear production, all the gears produced qualified the tests, as given in Table 3.

3 TEST AND ANALYSIS

3.1 *Metal performance test*

3.2 *Electroplating layer quality test*

1. Distribution of iron and zinc contents in coating

A JSM-6360 LV Scanning Electron Microscope (SEM) is used to test the metal content of the plating layer area, as shown in Figure 1. The composition of the electroplating layer in the scanning area was analyzed by using the Genesis 2000XM60S EDAX spectrometer, as shown in Figure 2. The coating chemical compositions of elements are given in Table 4.

From Figures 1 and 2 and Table 4, it could be seen that the main component of the electroplating layer was Zn, and a minor component was Fe. The mass fraction of Zn increased along the base metal due to the coating, while Fe decreased. This is because, in the initial stage of electroplating, Fe atoms on the based metal and Zn atoms formed an alloy, and with the continuous electroplating process, the main form of Zn coating formed polymer complexes, the substrate metal surface was covered by the Zn layer, so far away from the base metal outside the coating was mainly Zn, which could significantly enhance the corrosion resistance of the base metal.

2. Densification and thickness of the coating

The surface morphology of the plating layer magnified 5000 times is shown in Figure 3. The lower part of the base metal was galvanized for the middle layer and the upper part was galvanized for the

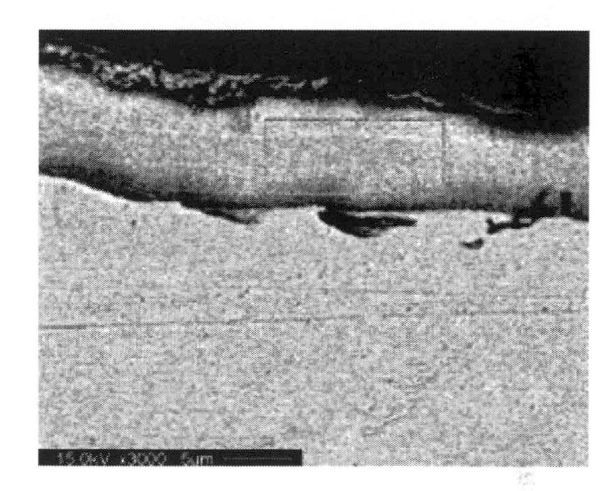

Figure 1. The coating area to be measured.

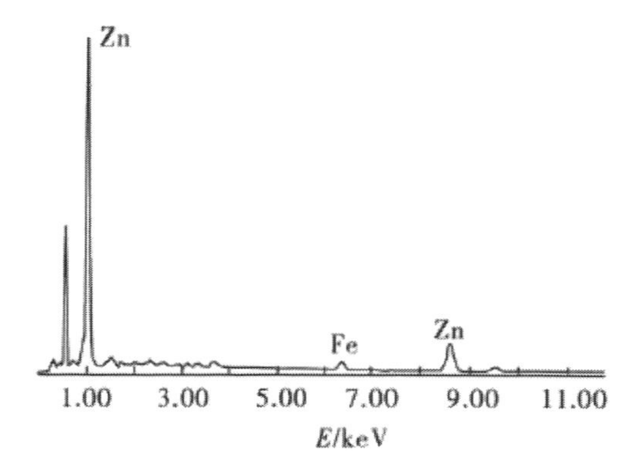

Figure 2. The energy spectrum of the coating components' analysis.

mosaic test SEM. The plating layer of good density was obtained to ensure adhesion and excellent corrosion resistance. By measuring the thickness

Table 4. List of coating components.

Element	Mass fraction/%	Atomic number fraction/%
Zn	95.17	94.61
Fe	4.83	5.39

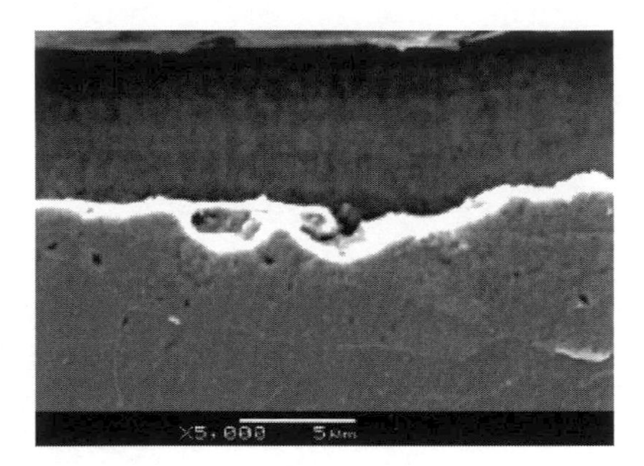

Figure 3. SEM image of the coating.

of different parts of the galvanized layer, it could be seen that, the thickness of the layer was about 6–8 μm, thickness exhibited uniformity, reaching up to the national standard.

3. Coating adhesion

In accordance with GB/T5270-2005 adhesion strength test methods and standards, the adhesion strength of the galvanized layer was tested.

A 25 mm 3M fiber tape was applied to the cover of the test coupon with a 2 kg roller, and 10 seconds after the removal of the bubble, the tape was peeled off with a quasi-constant tensile force perpendicular to the surface of the overlay. No peeling phenomenon was found in the test, which indicated that the adhesion between the covering layer and the substrate was strong.

The standard sharp blade of the hard steel knife was used to draw six parallel lines, with a spacing of 2 mm, when the parallel line through the standard cuts through the cover to the substrate. No peeling of the cover layer was observed, indicating that the covering layer of the part was closely attached to the substrate.

4. Appearance and corrosion resistance

In accordance with GB/T 9799-1997 standards, the zinc plating layer on the standard appearance of the coating was detected. According to GB/T10125-1997 standards for the salt spray test and no corrosion point in 96 h indicated that the surface of the galvanized layer exhibited uniform thickness; it was compact with excellent corrosion resistance.

4 CONCLUSIONS

1. The welding gear tooth surface reserved for grinding and the use of the minimal deformation heat treatment process ensured the dimensional accuracy and reduced the difficulty of processing.
2. After welding and tempering, the mechanical properties of the weld zone could be improved, and the gears produced by processing could meet the testing results.
3. Scanning electron microscopy and Energy Dispersive Spectrometry (EDS) showed that the main component of the electroplated layer was Zn and the fractions of Fe and Zn were increasing along the base metal of the coating, while Fe was decreasing. The plating thickness was about 6–8 μm, which indicated that the thickness exhibited uniformity, reasonable distribution, and good density.
4. The results showed that the galvanized coating was close to the substrate and had excellent corrosion resistance. The coating had no roughness, blister, crack, porosity, or local coating. In this paper, the galvanized solution formulation and electroplating process worked well.

REFERENCES

Barbosa L L, Carlos I A, Ass, **201**, 1695 (2006).

El hajjami A, Gigandet M P, De Petris-Wery M, Ass, **255**, 1654 (2008).

GAO Meilan, Ntnp, **12**, 26 (2001).

Kavitha B, Santhosh P, Renukadevi M, Sct, **201**, 1438 (2006).

Liang Jun, Hu Litian, Hao Jingcheng, Ass, **253**, 939 (2007).

Zheng Ruiting, Ieic, **12**, 56 (2004).

Civil, Architecture and Environmental Engineering – Kao & Sung (Eds)
© 2017 Taylor & Francis Group, ISBN 978-1-138-02985-9

Development and application of a test platform for a battery management system

Jie Xiong
State Key Laboratory of Automotive Safety and Energy, Tsinghua University, Beijing, China

Zhenhua Jin
State Key Laboratory of Automotive Safety and Energy, Tsinghua University, Beijing, China
Collaborative Innovation Center of Electric Vehicles in Beijing, China

Zhucheng Li
Department of Mechanical Systems Engineering, Chung Ang University, Seoul, Korea

ABSTRACT: A hardware-in-the-loop simulation platform is developed to test the performance and function of a BMS. Hardware is constructed based on a PXI real time controller. Veristand software is adopted to manage the I/O channel configuration and user interface. The HIL test system can realize battery cell voltage simulation, current sensor simulation and CAN bus monitoring. It has been used for the static and dynamic test of the BMS product.

1 INTRODUCTION

In recent years, new energy source vehicles have become a hot area of research and development due to petroleum energy depletion and environment protection requirements. There are different technology options for new energy vehicles, such as electric vehicles, hybrid electric vehicles and fuel cell vehicles. A battery package is one of the key components of electric vehicles as the energy storage system, which can output energy to the electric motor during the vehicle driving state and recover energy from regenerative braking. The battery management system is responsible for state monitoring and safety management to realize suitable charging and discharging of the battery package. To reduce costs, shorten the development cycle. In the early development stage of the control system, Hardware in the Loop (HIL) simulation technology is often used (Ye, 2009). In the model-based development process, the test bed system plays a key role, which can be developed to test different kinds of ECU (Electronic Control Unit) (Ouyang, 2008; Haupt, 2013; Jason, 2012). In this paper, design of a BMS HIL test system is introduced and then application of this test system in the BMS performance test is described in detail.

2 BATTERY MANAGEMENT SYSTEM

A distributed structure is often adopted for BMS and Figure 1 shows a star topology commonly used in product BMS.

A slave board is used for data acquisition of battery cell voltage and package temperature and equalization control. Data are sent to the master board through a CAN bus. The master board is responsible for SOC calculation, charging and discharging management and thermal management. One master can be connected with several slave boards according to the

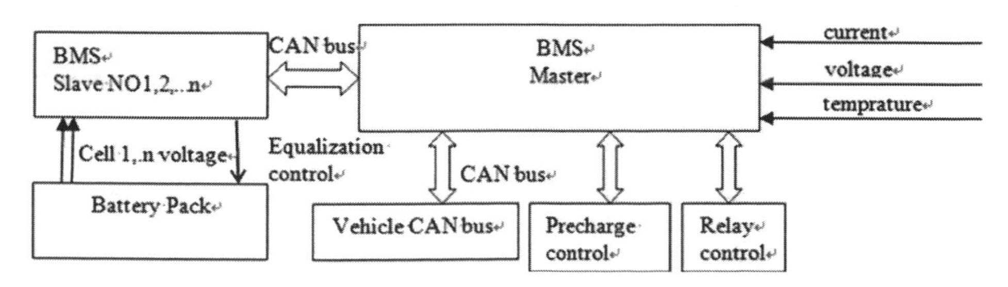

Figure 1. BMS structure.

number of cells in the battery package. A BMS must rule and check the energetic car's behavior, reporting to the user all the relevant information about the battery. It must maximize the runtime per discharge and the number of life cycles. Usually a BMS is connected to the remaining electronic systems via a CAN bus (Buccolini, 2016).

Generally speaking, the function of BMS includes several aspects:

1. Battery state monitoring. Battery package operating parameters such as voltage, current, temperature, etc. are acquired. Based on these details, suitable algorithms are developed which includes cell equalization control, thermal management and safety management.
2. SOC and SOH calculation. SOC is an important parameter for the energy distribution algorithm of the vehicle control unit.
3. Bus communication. The BMS communicates with the vehicle control unit, information system and battery charger through the CAN bus.

3 DESIGN OF THE BMS TEST SYSTEM

3.1 *Hardware structure*

The design of the test system is shown in Figure 2. It consists of a host computer and a real time controller.

A notebook with Windows7 is adopted as host computer and Veristand software is used on it. The host computer has the function of operating panel design, I/O board configuration, model management and test sequence management. The host computer is connected with a real time controller through a LAN and TCP/IP protocol.

The real time controller is constructed with PXI hardware. PXI 1062Q is selected as the chassis which provides 4 PXI slots and 3 PXIe slots. PXIe4322 is adopted to simulate the cell voltage. It has 8 isolated analog output channels which can output maximum 20 mA current and 16 V voltage. PXIe6361 is selected to simulate the package voltage and current sensor. It has 16 analog input channels, 2 analog output channels and a 24-channel I/O signal. PXI 2722 is adopted to simulate the temperature sensor. It has a 5-channel programmable resistor output channel and the maximum resistor value is 16 kilohm. PXI 8512 provides a 2-channel CAN interface to connect with the BMS CAN bus.

3.2 *Battery model*

The Rint model is adopted for battery voltage calculation, which is shown in Eq. 1 to Eq. 3. Vbat and Ibat are the battery package voltage and current. Rint is the inner resistor of the battery, which can be obtained through a two-dimensional lookup table. Voc is the open circuit in the model, which is the function of SOC.

$$V_{bat} = V_{oc} - I_{bat} R_{int} \tag{1}$$

$$R_{int} = f(I_{bat}, SOC) \tag{2}$$

$$V_{oc} = f(SOC) \tag{3}$$

3.3 *HIL test process with Veristand*

The hardware-in-the-loop test process with Veristand software is shown in Figure 3. A project in Veristand contains a system definition file and a

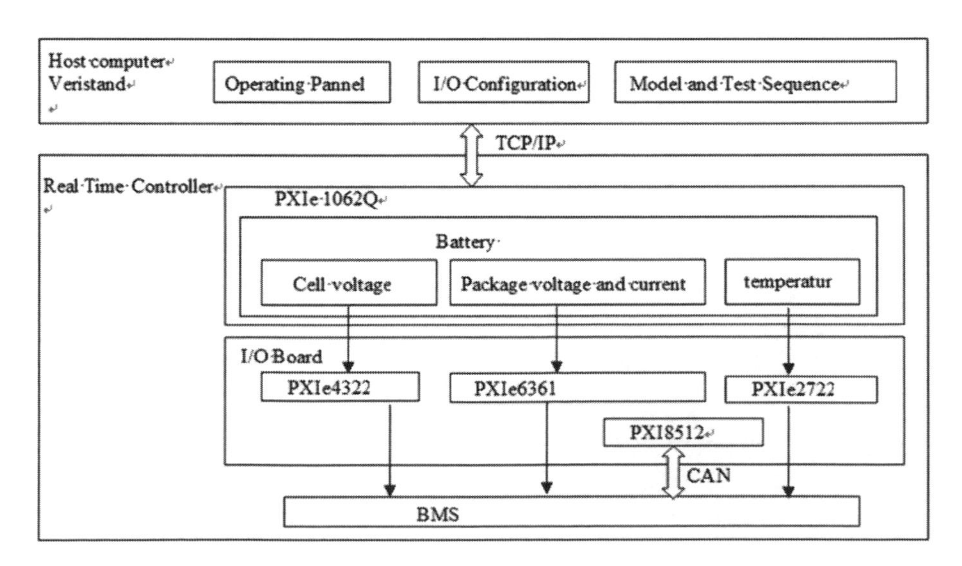

Figure 2. BMS test system structure.

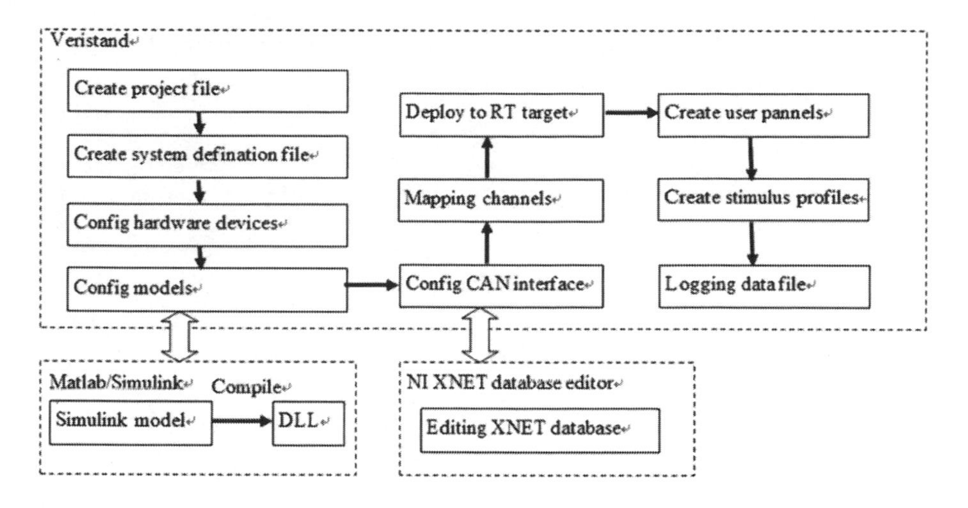

Figure 3. HIL test process with veristand.

workspace which consists of user panels to display data and curves. The system explore window is used to configure a system definition file which includes hardware configuration, model configuration, CAN communication information and mapping information. The battery model and power train model in Matlab/Simulink can be compiled into a DLL file and loaded into Veristand. The XNET database editor can create and edit the CAN frame definition file which can be loaded into Veristand to resolve CAN frame information. After the system definition file is defined, it can be deployed to the real time target, the PXI hardware. User panels contain display indicators and graphs can be designed on line to monitor test data. A stimulus profile can be created to realize test automation.

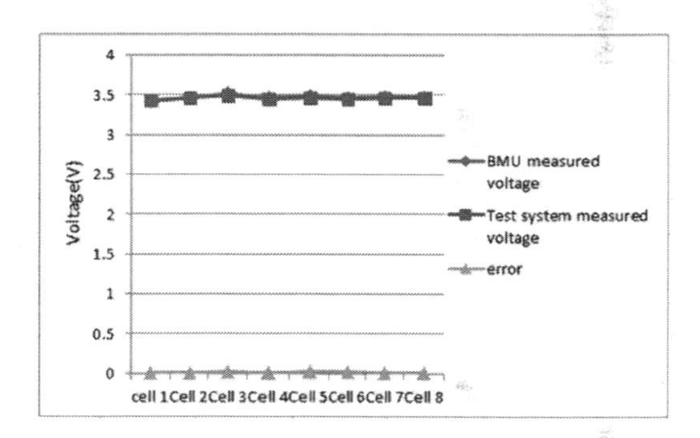

Figure 4. Cell voltage acquisition accuracy.

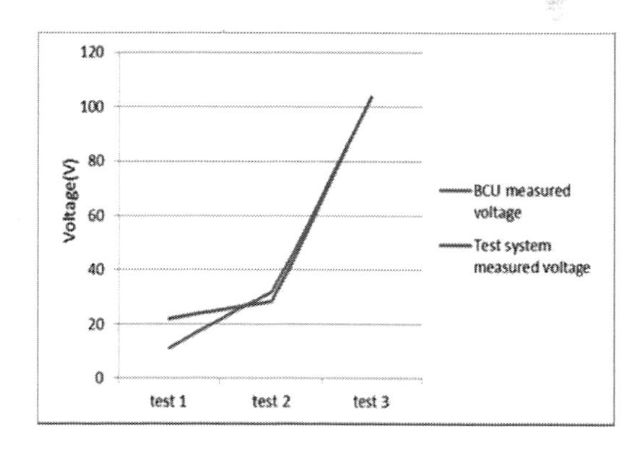

Figure 5. Total voltage acquisition.

4 APPLICATION OF THE TEST SYSTEM

4.1 *Testing of BMU cell voltage acquisition*

The PXI NI 4322 board provides 8 cell voltage output channels, using a tandem structure, each cell voltage is set to 3.500 V, and the total voltage of the battery pack is 28 V. The PC can monitor the battery voltage through the CAN data using Veristand software. Its detection accuracy is less than 10 mv through resolving the CAN bus data. Figure 4 shows that the BMU can collect voltage with low error.

4.2 *Testing of BCU total voltage acquisition function*

The total voltage acquisition range of the BCU is 0–500 V We can input a different analog voltage signal to the measurement port of the BCU using the HIL system. Figure 5 shows that the test accuracy will be increased by the increase of the total voltage.

4.3 *Testing of BCU pack current acquisition function*

The BCU pack current acquisition range is 0–500 A. This study uses a NI PXI-6221 IO board card to simulate primary current by outputting different voltage whose range is 0–5 V to the BCU current acquisition port. Figure 6 shows that the pack current measured

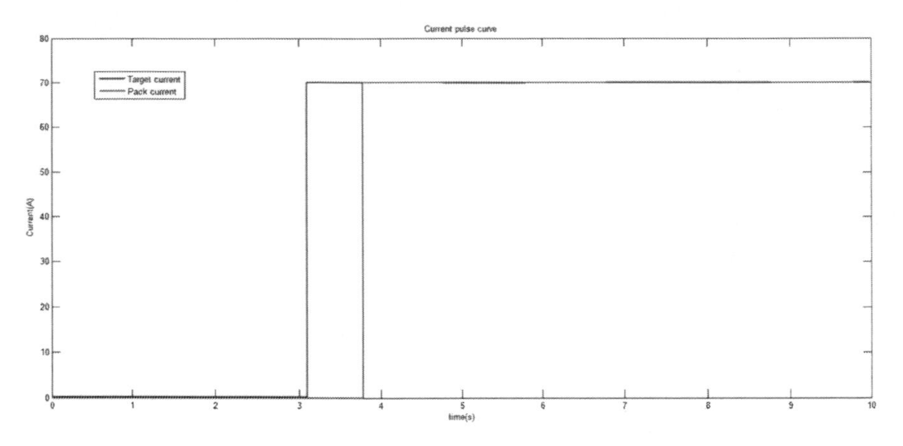

Figure 6. Current pulse curve.

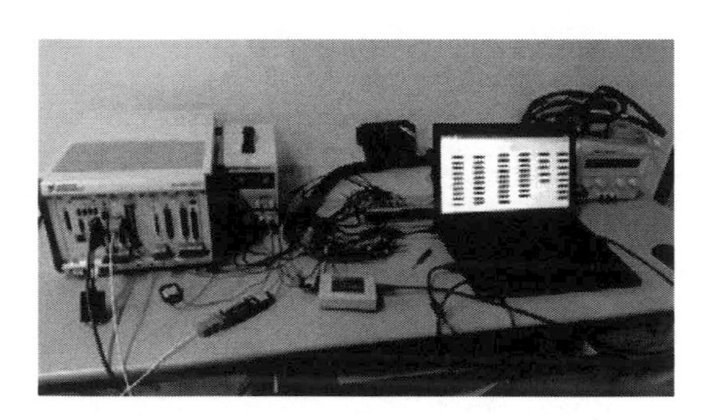

Figure 7. Photo of a BMS HIL test platform product.

by BMS delays for less than 1 second, compared with the target current, which is a current pulse up to 70 A, and we can obtain BMS measurement error less than 0.5 A from the acquisition data.

The real BMS HIL wiring diagram is shown in Figure 6. It includes a host computer, a real time controller and a BMS.

5 CONCLUSION

In this study, we developed a test bed for a BMS. It can assess the completeness and effectiveness of the BMS function, the accuracy of data acquisition, and the accuracy of the tested BCU and calculated SOC. This platform is a very powerful tool for testing and developing a BMS.

ACKNOWLEDGEMENTS

This study was supported by the International Collaborative R&D Project between PMGROW and Tsinghua University.

REFERENCES

Buccolini, L., A. Ricci, C. Scavongelli, G. DeMaso-Gentile, S. Orcioni and M. Conti, *Battery Management System (BMS) simulation environment for electric vehicles, 2016 IEEE 16th International Conference on Environment and Electrical Engineering (EEEIC)*, Florence, Italy, (2016).

Hagen Haupt, Joerg Bracker, Markus Ploeger. *Hardware-in-the-Loop Test of Battery Management Systems*. SAE International, (2013), **01**-1542.

Ouyang Minggao, Li Jianqiu, Yang Fuyuan, Lu Languang, *Automotive New Powertrain: Systems, Models and Controls*, Tsinghua University Press, Beijing, (2008).

Poon Jason J, Kinsy Michel A, Pallo Nathan A, et al. *Hardware-in-the Loop Testing for Electric Vehicle Drive Applications.* //Conference Proceedings – IEEE Applied Power Electronics Conference and Exposition, (2012).

YE X, JIN Z, *Hybrid vehicle motor test platform dcvelopment and application*, Proceedings of 9 th lnternational ConfeFence Oil Electronic Measurem ent & Instruments, Beijing, (2009).

Civil, Architecture and Environmental Engineering – Kao & Sung (Eds)
© 2017 Taylor & Francis Group, ISBN 978-1-138-02985-9

Experimental study on yarn friction slip in the yarn pull-out test

Weikang Li, Zhenkun Lei, Fuyong Qin, Ruixiang Ba & Qingchao Fang
State Key Laboratory of Structural Analysis for Industrial Equipment, Dalian University of Technology, Dalian, China

ABSTRACT: The single yarn pull-out test is a model experimental method to research the mechanical properties of fabric under impact. This paper aims to understand the yarn pull-out mechanism of Kevlar49 plain fabric by the single yarn pull-out test. The load–displacement curve contains the typical physical phenomena such as crimp extension, crimp swap, frictional slip of yarn, and fabric deformation behavior. During static friction, the pull-out load increased nonlinearly with the displacement and the crimp extension of the pulled warp yarn occurred gradually. Then, the kinetic friction started when the friction load reached its maximum, and the load–displacement curve decreased undulately until the wrap was pulled out. During kinetic friction, the crimp swap and frictional slip were observed. Further analysis showed that in the pull-out test process, the friction force is not uniformly distributed along the pulled yarn and the friction dissipation is nonlinear.

1 INTRODUCTION

Kevlar fiber is a kind of aramid composite developed by DuPont in the 1960s, and it has high specific strength and modulus, good toughness, high temperature resistance, and high energy storage capacity and energy transfer efficiency. Kevlar fabric is weaved by Kevlar yarn, and each yarn consists of hundreds to thousands of Kevlar fibers. It is an ideal composite fiber for manufacturing bullet clothing (Carr, 1999).

In the impact process, there are three kinds of transfer of the kinetic energy of a bullet: the kinetic energy of yarn, the strain energy of yarn, and the friction dissipation caused by the frictional slip (Duan, 2005). The main role of friction force during impact is to strengthen the interaction of yarns, and the remote yarns can join into energy absorption.

Because of the high speed of the bullet impact, the fabric properties were studied by comparing the change of fabric state and bullet speed before and after the impact, which makes it difficult to monitor the deformation process of fabric during the impact and is impossible to understand deeply the friction mechanism during the impact. Some researchers developed the yarn pull-out test to simulate the process of yarn pulled out from fabric under impact (Bazhenov, 1997). This method gives a good look to the fabric deformation process under impact, and is helpful to understand the friction mechanism and energy absorbing mechanism.

Some studies show that yarn pull-out force is positive correlation with impact resistance, and a fabric with high yarn pull-out force behaves better in the impact test (Bilisik, 2010; Dong, 2009). Zhu (2011) modeled the yarn pull-out process into the process of fiber pull out from the matrix, and established the analysis model of force-displacement. The model has important reference values because it takes crimp extension and contact area into consideration. Bilisik et al (Bilisik, 2010) studied the influence of fabric size and pull-out yarn mount as follows: the maximum of yarn pull-out load is proportional to the fabric longitudinal dimension, while the fabric transverse size has little effect on the maximum pull-out load. Bislisik et al (Bilisik, 2011) studied the effects of yarn mount and width/length ratio on the shear strength by the yarn pull-out test. Nilakantan's (2013) result shows that transverse preload is in positive correlation with the maximum pull-out force. Dolatabadi (2009) researched the deformation behavior of fabric under bias extension using the 2D-FFT method, and the result shows that there is a critical shear angle during bias extension which may correlate with the geometry parameter. However, the above studies are statistical research on the relationships between load and displacement, and load and fabric geometric size. There are few studies focusing on the distribution of friction force and the energy absorption mechanism at mesoscopic level.

In this paper, the process was measured by the DIC method, and the load–displacement curve was obtained by the yarn pull-out test. Combined with the mesoscopic geometry of the fabric structure, the pull-out load, frictional force distribution and energy absorption were investigated, and the yarn crimp extension, crimp swap, and frictional slip were discussed.

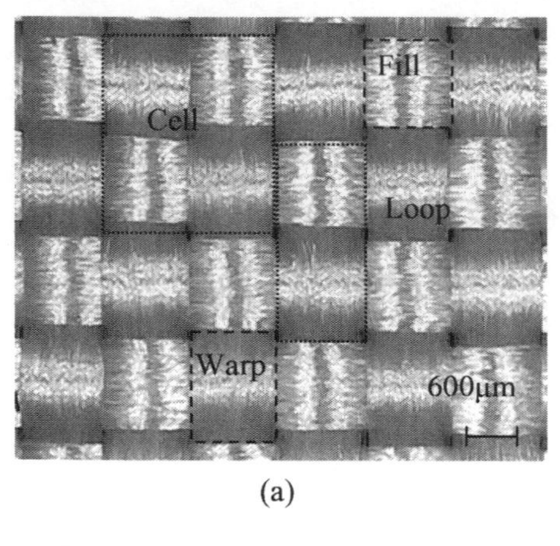

(a)

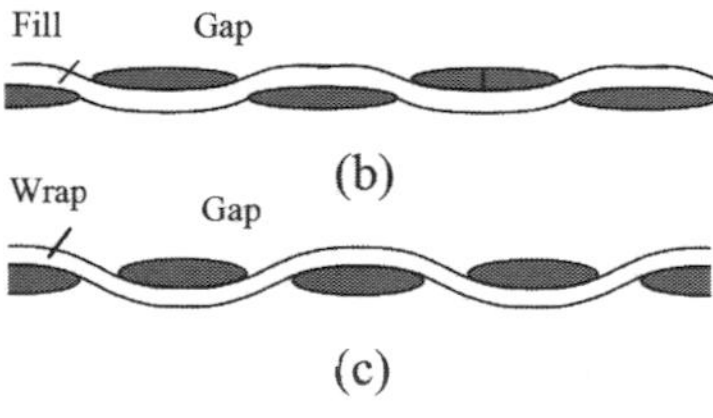

(b)

(c)

Figure 1. (a) Geometric dimension of plain woven fabric and (b) schematic diagram of warp trajectory with cross-fill yarns and a schematic diagram of fill trajectory with cross-warp yarns.

2 GEOMETRIC PROPERTIES OF FABRIC

The micrograph of Kevlar 49 plain fabric is shown in Figure 1(a), which is woven alternately by warp yarn in the vertical direction and fill (weft) yarn in the horizontal direction. There is a period cell structure consisted of 2 × 2 intersections of wrap and fill yarns. Because of the interaction of warp and fill yarns formed during the process of weaving, the yarn wavy crimp exists in the fabric under relaxation conditions. The abridged view of warp yarn trajectory with cross-fill yarns and the fill yarn trajectory with cross-warp yarns is shown in Figure 1(b) and 1(c), respectively.

3 SINGLE YARN PULL-OUT TEST

3.1 *Sample and experimental process*

The test equipment used to conduct the experiment is shown in Figure 2(a). Two clamps are placed on a sliding rail to clip the two sides of the fabric. Each clamp is comprised of three parts: the front U-shaped groove, the back U-shaped groove, and the board between them. Before the test, the fabric was placed in the middle of the board and the back groove and fixed by the bolts installed in the front groove. So, the fabric was clamped tightly between the board and the back groove. This design reduces the stress concentration of the boundary and improves the accuracy of the experiment.

Before the experiment, the direction of warp yarn is along the tension direction and the upper end of warp yarn is loaded. The width and length of the sample are both 15 cm. There are totally 65 cells (130 intersections) along the warp direction and same number of cells along the fill direction. There is no preload in the sample before the experiment. The middle wrap yarn is pulled out using a universal testing machine (Instron 3345) and the pull-out speed is 100 mm/min.

3.2 *Pull-out load and displacement curve*

The load–displacement curve obtained during the yarn pull-out test is shown in Figure 2(b). There are two main distinct regions: the static friction region

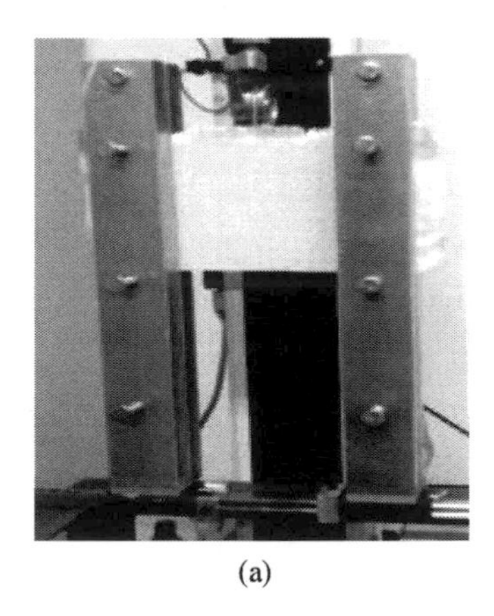

(a)

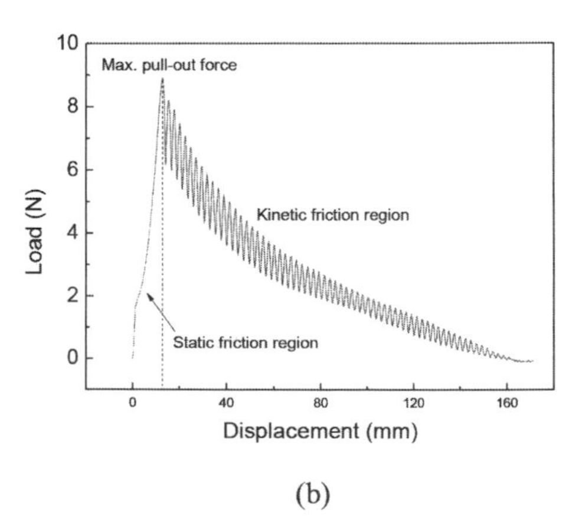

(b)

Figure 2. (a) Experimental setup and (b) typical curve of load and displacement during the single yarn pull-out test.

and kinetic friction region. In the static friction region, the pulled yarn had crimp extension and produced the static friction force among the intersections, which led to fabric deformation. When the pull-out load reached the maximum, the pulled yarn began to slip in the kinetic friction region.

In the kinetic friction region, the load–displacement curve shows a classic "sine wave" feature corresponding to a yarn crimp swap, where the pull-out load first increased and then decreased.

The increase of pull-out load indicates that the bottom end of the pulled wrap yarn slips into the adjacent intersection, while the decrease of pull-out load indicates the bottom end slipping out of the current intersection. A sine wave corresponds to 2 intersections. When the pulled wrap yarn gets through 2 intersections, there is a "slip in-out" process corresponding to produce a sine wave in the load–displacement curve. Therefore, the load–displacement curve comprehends the fabric shear deformation at macroscopic scale and the yarn deformation at mesoscopic scale, which reflects a coupled physical phenomenon of yarn crimp extension, crimp swap, and friction slip.

4 RESULTS AND DISCUSSION

4.1 *Friction distribution during kinetic friction*

For simplicity, taking the first wave valley after the maximum static friction force as the starting point, the first wave number is 1, and so on, and the last wave number is 65. Extracting the load data of the wave peak and wave valley, the distribution of pull-out load during kinetic friction is shown in Figure 3(a).

In Figure 3(a), the horizontal axis represents the wave number, while the vertical axis represents the pull-out load. The pull-out load curves of the wave peak and wave valley have the same trend, and are exponentially decreased. The pull-out loads decreased sharply at beginning and then decreased slowly. At last, the curves coincided together.

4.2 *Energy absorbing mechanism*

In Figure 3(a), the area under the pull-out load and displacement curve is the total energy of the test machine. Each wave area of the curve during kinetic friction can be calculated by the sum of several sections, as shown in Figure 3(b), which shows that there is the relationship between the absorbed energy and the wave number. Every point on the curve represents the sum of energy absorption for all intersections. The absorbed energy decreases exponentially with the intersection number as the yarn is pulled out.

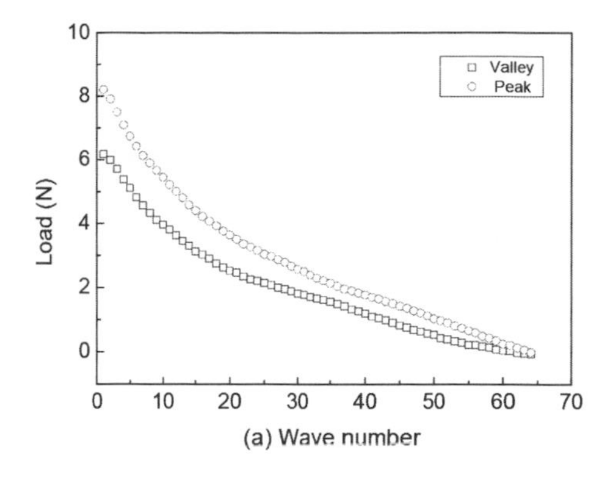

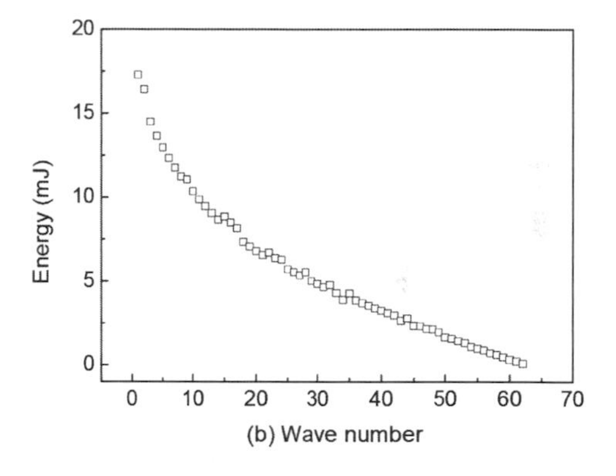

Figure 3. (a) Pull-out load curves and (b) absorbed energy curve.

5 CONCLUSIONS

The yarn pull-out test results indicate that the pull-out load and displacement curve well describe the macroscopic fabric shear deformation and mesoscopic yarn deformation, and reflects the yarn extension crimp, crimp swap, and friction slip. During static friction, the pulled yarn produced the crimp extension so that the overall shear deformation of fabric occurred under the pull-out load. During kinetic friction, when the pulled wrap yarn and the cross-fill yarns were a convex to convex contact, the pull-out load increased; when the pulled wrap yarn and the cross-fill yarns were a convex to concave contact, the pull-out load decreased. Further analysis showed that the friction force on the pulled wrap yarn was not uniformly distributed, which decreased exponentially with the pull-out intersections of fabric.

ACKNOWLEDGEMENTS

This work was supported by the National Basic Research Program of China (No. 2014CB046506), the National Natural Science Foundation of China (Nos. 11472070, 11572070), the Research Subject of State Key Laboratory of Structural Analysis for Industrial Equipment (Nos. S14207, S15202), and the Natural Science Foundation of Liaoning Province of China (No. 201502014).

REFERENCES

Bazhenov S. Dissipation of energy by bulletproof aramid fabric. Journal of Materials Science. 1997, 32(15): 4167–4173.

Bilisik K, Korkmaz M. Single and multiple yarn pull-outs on aramid woven fabric structures. Textile Research Journal. 2010, 81(8): 847–864.

Bilisik K. Experimental determination of fabric shear by yarn pull-out method. Textile Research Journal. 2011, 82(10): 1050–1064.

Carr DJ. Failure mechanisms of yarns subjected to ballistic impact. Journal of Materials Science Letters. 1999, 18(7): 585–588.

Dolatabadi MK, Kovař R, Linka A. Geometry of plain weave fabric under shear deformation. Part I: measurement of exterior positions of yarns [J]. Journal of the Textile Institute. 2009, 100(4): 368–380.

Dong Z, Sun CT. Testing and modeling of yarn pull-out in plain woven Kevlar fabrics [J]. Composites Part A: Applied Science and Manufacturing. 2009, 40(12): 1863–1869.

Duan Y, Keefe M, Bogettic TA, Cheeseman BA. Modeling friction effects on the ballistic impact behavior of a single-ply high-strength fabric. International Journal of Impact Engineering. 2005, 31(8): 996–1012.

Nilakantan G, Gillespie JW. Yarn pull-out behavior of plain woven Kevlar fabrics: Effect of yarn sizing, pullout rate, and fabric pre-tension [J]. Composite Structures. 2013, 101(15): 215–224.

Zhu D, Soranakom C, Mobasher B, Rajan SD. Experimental study and modeling of single yarn pull-out behavior of Kevlar 49 fabric [J]. Composites Part A: Applied Science and Manufacturing. 2011, 42(7): 868–879.

Civil, Architecture and Environmental Engineering – Kao & Sung (Eds)
© 2017 Taylor & Francis Group, ISBN 978-1-138-02985-9

Analysis and research into the effect of dithering technology on ADC's SNR

KunKun Jin & Xiao Long Chen
School of Mechano-Electronic Engineering, Xidian University, Xi'an, China

ABSTRACT: The dithering technique that has been researched frequently in the current ADC application field is often used to improve the dynamic performance of ADC. However, adding a dither in the input signal will have a bad influence on ADC's SNR. In this paper, the characterization of the SNR and the factors that affect the ADC's SNR are analyzed theoretically. By establishing a simulation system on the Matlab platform, it is found that dithering technology increased the ADC's SFDR from 26.40 dB to 33.71 dB, but at the same time it reduced the ADC's SNR from 18.49 dB to 16.95 dB.

1 INTRODUCTION

With the development of modern science and technology, digital analogy conversion technology-ADC, as the most commonly used data acquisition technology, has played a more and more important role in military, medical, civil and other fields. However, ADC's dynamic range is greatly influenced by the existence of non-ideal characteristics of ADC. Dithering technology is an effective means to improve the non-ideal characteristics and increase the dynamic range of ADC. But, with the improvement of ADC's non-ideal characteristics, the dithering technique has bad influence on the ADC's SNR.

In this paper, firstly, we study the concept of ADC's SNR. Then we get the main factors that affect the ADC's SNR. Finally, the simulation is implemented to verify the theoretical result.

2 THE CHARACTERIZATION OF ADC'S SNR

The SNR refers to the ratio of the effective value of the fundamental component of the input signal and the effective value of the other noise except the harmonic component. We can express it by the mathematical expression given below:

$$SNR = 20\lg\frac{V_{in}}{V_n} \tag{1}$$

Among the expression, V_{in} represents the effective value of the fundamental component of the input signal, and V_n indicates the effective value of the other noise except the harmonic component.

Considering the above formula (1), in order to calculate the signal-to-noise ratio, first, we need to know the effective value of the fundamental component and the noise component. Assuming that the input signal is a sine wave, thus, the effective value of the fundamental signal is equal to the ADC's full-scale input signal divided by $\sqrt{2}-(2^{N-1}\times LSB)/\sqrt{2}$ where LSB means least significant bit. The quantization noise amplitude range of ADC is $\pm\frac{1}{2}LSB$, and the quantization error is caused by the analog input signal. According to the relationship between the effective value and the peak value of the triangle wave, we know that the effective value of the quantization noise is equal to its magnitude divided by $\sqrt{3}-(LSB/2)/\sqrt{3}=LSB/\sqrt{12}$. Then, we reorganize the formula (1):

$$\begin{aligned} SNR &= 20\lg\frac{V_{in}}{V_N} \\ &= 20\lg\frac{(2^{N-1}\times LSB)/\sqrt{2}}{LSB/\sqrt{2}} \\ &= 20\lg 2^N + 20\lg\sqrt{6} - 20\lg 2 \\ &\approx 6.02N + 1.76 \end{aligned} \tag{2}$$

In practical applications, the ADC's SNR can be roughly calculated by the following method: take the conversion accuracy of ADC and multiply by 6. For example, for a 10 bit analog to digital converter, the SNR should be greater than 60 dB, and for a 12 bit, the SNR should be greater than 72 dB.

3 INFLUENCE OF THE DITHERING TECHNIQUE ON ADC'S SNR

3.1 *SNR under ideal conditions*

According to the analysis above, under ideal conditions, the SNR of ADC refers to the ratio of the effective value of the full range single frequency ideal sine wave input and the total effective value of the Nyquist bandwidth's all other frequency components' output of ADC (not including the DC component and harmonic components). In this condition, the ideal ADC's noise is only generated by its inherent quantization error, and we can know that the ADC's SNR is just about N through expression (2). Then, what is the relationship between N and quantization error? Since the conversion accuracy (N) of the ADC cannot be infinite in practical application, we can only use a limited conversion accuracy to represent an input signal. The limited conversion accuracy implies a finite quantization interval of q, and the finite quantization interval cannot accurately quantify each voltage value. Therefore, the voltage value of the input signal and the quantized voltage value usually have error which is called the quantization error. That is to say, N and ADC's quantization error has a direct relationship. The relation between the quantization error and the input signal can be described by the following picture:

3.2 *SNR under non-ideal conditions*

The ADC device we use in practice is non-ideal usually, there always is a deviation between the actual conversion curve and ideal transfer curve, and it appears as varieties of errors, such as zero error, full scale error, gain error, integral nonlinearity error-INL and differential nonlinearity error-DNL. Among them, the zero error, the full degree error and the gain error are constant errors, and they affect only the absolute accuracy of ADC but do not affect the SNR of ADC. INL refers to the maximum deviation between the actual conversion curve of ADC and the ideal conversion curve of ADC based on the calibration of the above constant error. While DNL refers to the maximum deviation between the actual quantization interval of ADC and the ideal quantization interval of ADC, we change the quantization error of ADC, which can directly calculate the influence of the deviation of the actual conversion curve of ADC and the ideal conversion curve of ADC on the ADC's SNR.

Non-ideal ADC, in addition to the above error, has all kinds of noise, such as thermal noise, aperture jitter and peripheral circuit noise. The thermal noise is caused by the internal molecular thermal motion of the semiconductor device. The aperture jitter is caused by the uncertainty of the ADC's aperture delay. While the peripheral circuit of ADC will also bring noise, such as the thermal noise of the ADC's input circuit, the power/ground clutter, the space electromagnetic interference and the external clock's instability (leading to the ADC's sampling clock edge's uncertainty that brings the aperture jitter), it can be all equivalent to the above two kinds of internal noise of ADC.

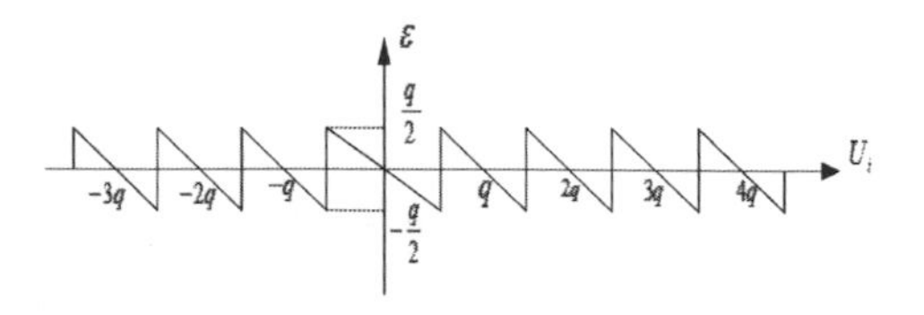

Figure 1. The relation between quantization error and input signal.

In general, the non-ideal ADC's quantization noise, NDL, aperture jitter and thermal noise are independent of each other. Considering the influence of these four factors, the SNR calculation formula of ADC is as follows:

$$
\begin{aligned}
SNR &= \frac{P_s}{Pn} = 10\lg\left(\frac{V_s^2}{V_n^2}\right) \\
&= 10\times\lg\left(\frac{V_s^2}{V_{DNL}^2 + V_{\Delta_{tj}}^2 + V_{tn}^2}\right) \\
&= -10\lg\left[8\times\left(\frac{1}{12}+\frac{\varepsilon^2}{4}\right)+\left(2\pi f_{in}\sigma_{\Delta_{tj}}\times 2^N\right)^2 + 8\sigma_{tn}^2\right] \\
&\quad + 6.02N(dB)
\end{aligned}
\tag{3}
$$

In formula (3), N represents the quantization bits of ADC. Δ_{tj} represents the aperture jitter. ε represents the effective value of the actual quantization interval and the ideal quantization interval error of ADC—the unit is LSB, f_{in} represents the frequency of the ADC's input signal—the unit is Hz. $\sigma_{\Delta_{tj}}$ represents the effective value of the ADC's aperture jitter—the unit is s. $\sigma_{\Delta_{tn}}$ represents the input noise of the ADC—the valid value is LSB.

According to formula (3), in the case of the ADC's inherent quantization error, DNL and aperture jitter are fixed, due to the minus sign in front of the formula. If the input jitter noise increases, $\sigma_{\Delta_{tn}}$ is large, and the SNR of ADC will be smaller. That is to say, the dithering technique will deteriorate the ADC's SNR.

4 SIMULATION AND VERIFICATION

Through the theoretical analysis, we can conclude that the dithering technique deteriorates the ADC's SNR while improving some of its non-ideal properties.

4.1 *Simulation conditions*

We construct a simulator through the Simulink module in MATLAB. The simulation condition is as follows: a sine wave generation module in Simulink is used to produce a sine wave whose frequency is f = 3 MHZ, and amplitude is $V_{PP} = 7$ V. We keep the sine wave signal through the sampling module and use a quantizer whose conversion precision is 3 bits and the maximum input voltage is 8V to quantify the signal. We set the sampling holder's sampling frequency to $f_s = 100$ MHZ. For the dither signal, we use the noise module to generate the dither noise signal whose amplitude is 0.3 V less than 1 LSB (1V). The block diagram of the simulation system is shown in Figure 2.

Figure 2 is the block diagram of the simulation system without dithering. Figure 3 shows the block diagram of the simulation system with dithering. We can see that the difference between the two pictures is just with or without dithering.

4.2 *Simulation result*

We observe the oscilloscope module's output in the Simulink and compare the quantized output results between without dithering and with dithering. The result is shown in Figure 4.

Figure 4 is the ADC's quantization output graph without dithering. We can see from the figure that the ADC's harmonic interference is serious, and the Spurious Free Dynamic Range (SFDR) is 26.40 dB and the SNR is 18.49 dB. Figure 5 shows the ADC's quantization output graph with dithering. We can see from the figure that the ADC's SFDR is improved, reaching 33.71 dB, but the substrate of the noise increases from the original 100 dB to 90 dB, and the Signal-to-Noise Ratio (SNR) is reduced to 16.95 dB.

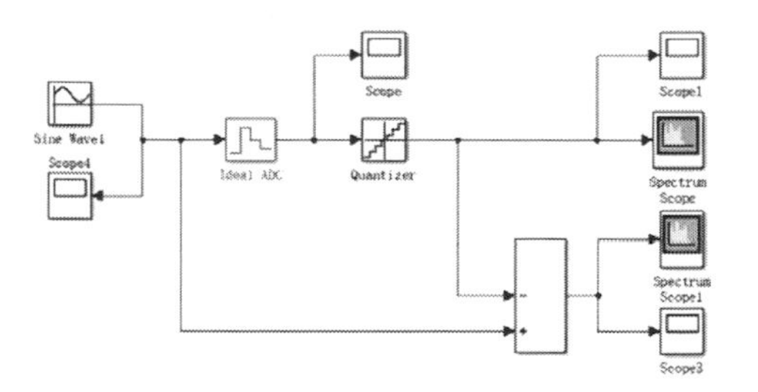

Figure 2. The block diagram of the simulation system without dithering.

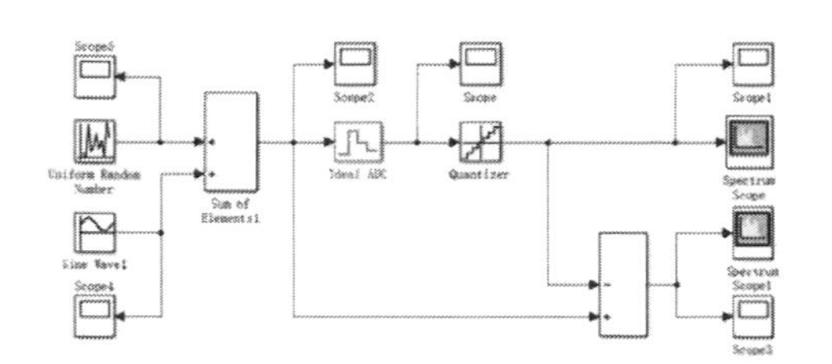

Figure 3. The block diagram of the simulation system with dithering.

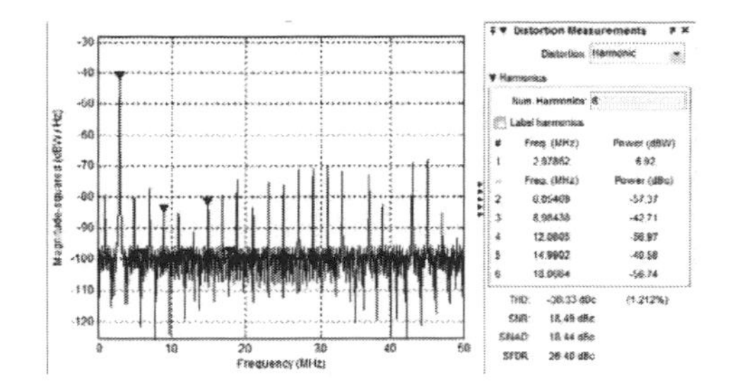

Figure 4. The ADC quantization output graph without dithering.

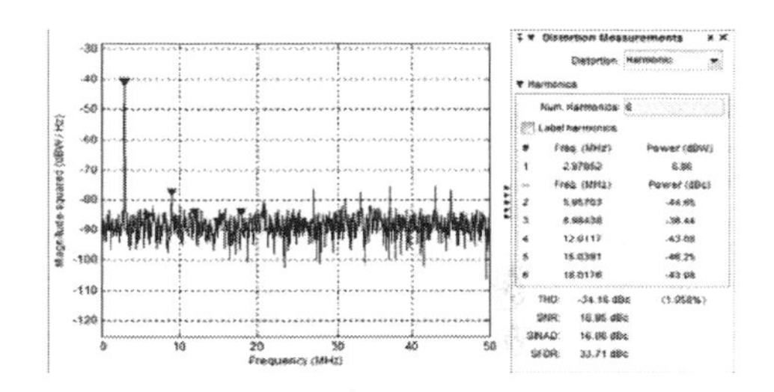

Figure 5. The ADC quantization output graph with dithering.

5 CONCLUSION

From theoretical analysis and simulation, it can be concluded that dithering technology can indeed improve some non-ideal characteristics of ADC, but at the same time it will deteriorate the ADC's SNR. So, when using the dithering technology to improve the non-ideal characteristics of ADC, we should pay more attention to the requirement of the SNR.

ACKNOWLEDGEMENT

This work was supported by the National Natural Science Foundation of China under Grant No. 61471282.

REFERENCES

Goodall, WM. *Television by Pulse Code Modulation*. BSTJ, **30**(1):33–49.(1951).

Graybeal, TD. *The nature of vibration in electric machinery. EE*, **63**(10):1458–1459. (1944).

Wagdy, M.F. *Simulation results on A/D converter dithering*. IMTC, **98**. 1998:78–83 vol.1, (1998).

Xu, J., Y. Lu, Z. Ding. *Analysis of signal to noise ratio of ADC and implementation of high speed and high resolution ADC circuit*. EAT, (2004).

Yang, S., J. Cheng, P Wang. *Variable-amplitude dither-based digital background calibration algorithm for linear and high-order nonlinear error in pipelined ADC*. MJ, **41**(7):403–410, (2010).

Civil, Architecture and Environmental Engineering – Kao & Sung (Eds)
© 2017 Taylor & Francis Group, ISBN 978-1-138-02985-9

Design and implementation of a CANopen master stack for servomotor controllers in a 6-DoF manipulator

Junhui Ji
Institute of Plasma Physics, Chinese Academy of Sciences, Hefei, China
University of Science and Technology of China, Hefei, China

Zihang Tao & Tao Zhang
University of Science and Technology of China, Hefei, China

Minzhong Qi
Institute of Plasma Physics, Chinese Academy of Sciences, Hefei, China

ABSTRACT: The CANopen master stack is used for network management and communication with CANopen slave nodes. This paper presents the design and implementation of a specialized CANopen master stack for servomotor controllers in a 6-DoF manipulator. Based on the hardware and general CANopen master stack, additional profile configuration for the 6-DoF manipulator is defined. It is shown by the real-time performance that the specialized CANopen master stack satisfies the timing constraint of control.

1 INTRODUCTION

CANopen is an application-layer communication protocol that includes the layer above the data link layer in the OSI reference model (Farsi, 1999). The low layer under it can be implemented by Controller Area Network (CAN), Ethernet for Control Automation Technology (EtherCAT) and so on (CAN, 2016). By the application of various standardized device profiles (e.g. CiA 401, CiA 402) and a communication profile (i.e. CiA 301), devices with different specialization can be connected to one network, allowing interoperation with each other (Jeon, 2001; Portillo, 2006; Li, 2013; Duan, 2007; Zhe, 2010), as shown in Figure 1 (H, 2015).

The communication profile CiA 301 serves as the basis of various device profiles. It specifies the object dictionary as well as the CANopen communication objects. In addition, it specifies the CANopen network management services and protocols. Following the way of how CiA 301 specified, a CANopen master stack can be realized.

As for a high-level flexible communication protocol, CANopen is widely used in many different areas nowadays, showing good compatibility and performance (Jeon, 2001; Portillo, 2006; Li, 2013; Duan, 2007; Zhe, 2010). In this paper, according to the hardware of the 6-DoF manipulator, additional definition of CiA DSP-402 for servomotor controllers is provided on the basis of the master stack, together with the definition of the connection set and procedure for setting and using modes

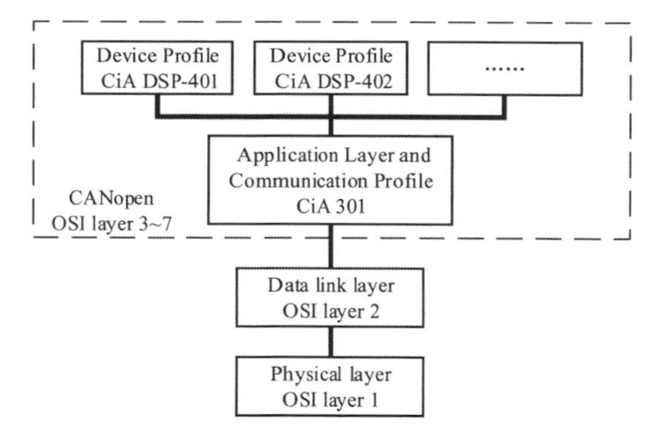

Figure 1. Schematic overview of CANopen and the low layer in the OSI reference model.

of operation. As a result, a specialized version of the CANopen master stack for a 6-DoF manipulator is implemented. The real-time performance testing result shows that the worst-case response time is less than 1.67 ms, which meets the demand of the controlling period of 50 ms for 6 servomotor controllers all in interpolated position mode.

2 HARDWARE CONFIGURATION

The hardware of the manipulator consists of several parts, of which some are highly concerned with CANopen, as illustrated in Figure 2.

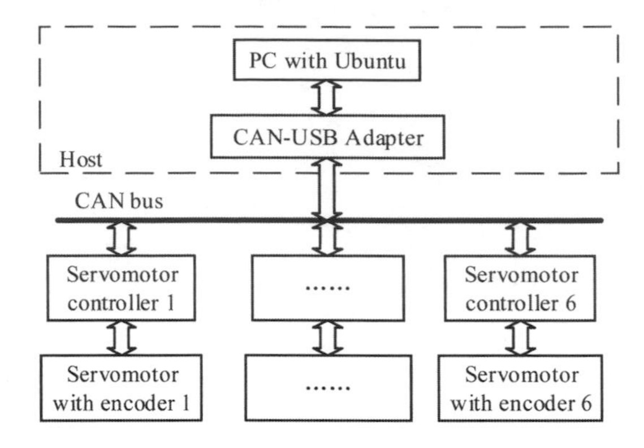

Figure 2. Network structure.

- PC with Ubuntu: the version of Ubuntu is 14.04 LTS; PC equipped with USB interface
- CAN-USB adapter: PCAN-USB adapter from PEAK system, of which the bit rates vary from 5 kbit/s up to 1 Mbit/s
- Servomotor controller: Gold Twitter from Elmo, with CAN and standard RS232 communication interface provided

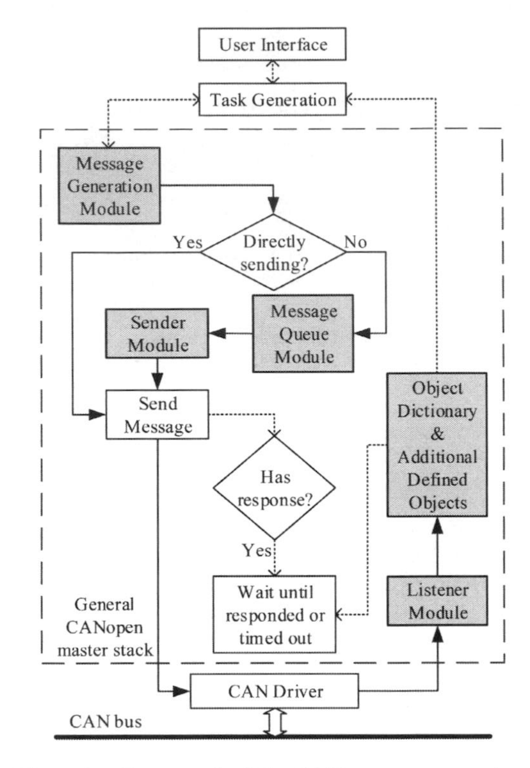

Figure 3. Framework of the CANopen master stack.

3 SOFTWARE ARCHITECTURE OF THE GENERAL CANOPEN MASTER STACK

As mentioned before, the general CANopen master stack is instructed by CiA 301. According to the modularity, it can be roughly divided into several modules, as shown in Figure 3. All the programs mentioned are written in C++ 11.

Generally, there are two kinds of tasks that are generated by UI: one is to send messages to the devices and the other one is to get data from the object dictionary or additional defined objects. To meet the requirement of these two types of tasks, modules illustrated in Figure 3 are designed and implemented.

3.1 *Message generation module*

As its name prompts, this module generates different kinds of messages that are requested by the task from the host, including node guarding and node control from the NMT protocol, upload and download from the SDO protocol, RPDO from the PDO protocol, syncWrite from the SYNC protocol, etc. For the various protocols mentioned above, one same structure can be defined to describe the messages, consisting CAN-ID, message type (standard frame or extended frame), data length and data.

After the message to be sent is generated, there comes two ways of sending it. One is to send messages directly to the CAN bus in the thread that is the same as the one where the message is generated. This approach aims at the tasks that do not require real-time communication. The other one is to push the message into a circular queue, which is called the message queue. In contrast to the former way, this method aims at providing a mechanism that guarantees real-time communication as well as sequentiality and periodicity of sending messages.

3.2 *Message queue module*

The message queue is a FIFO circular buffer with a fixed maximum size. It deals perfectly with the producer-consumer problem. The basic idea is to use a front pointer and a rear pointer to make the queue perform as if it was connected from front to rear, in a ring form.

There are two functions that will change the number of element stored in the queue: push and pop. The diagram of the push function is presented in Figure 4. The pop function is almost the same except that it checks whether the queue is empty, gets the element from the front position and moves the front pointer to the next.

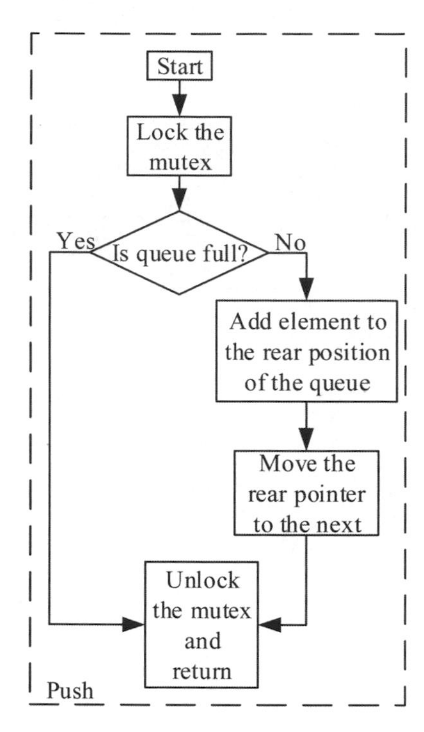

Figure 4. Push function of the message queue.

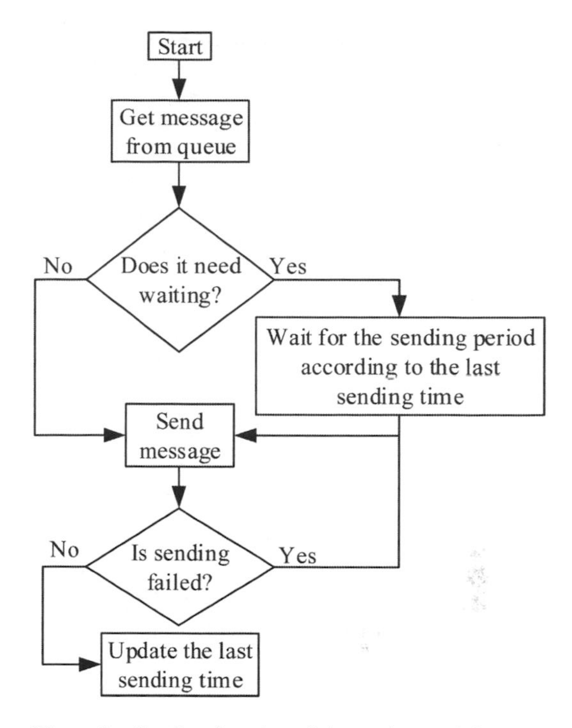

Figure 5. Sending function of the sender module.

3.3 *Sender module*

The sender module gets the message from the queue using the pop function and then sends it to the CAN bus. This module lies in a separate thread that is different from the thread where messages are generated or stored.

The sending function in this module is illustrated in Figure 5. It is located in a loop that can be paused or exited until a corresponding command is delivered. The waiting operation after getting the message is the approach to keep the sending period of the "same message" according to the last sending time. Here, the last sending time only makes sense to the "same message", and the "same message" has the same node ID and CAN ID.

3.4 *Listener module*

The listener module is responsible for getting the messages on the CAN bus from devices. Like the sender module, it is located in another thread. The function of the listener is in a loop that can be exited until the related command is delivered. Some of the code is listed in Figure 6, which shows that the process is switched by the function code of the coming CAN message. The processed result

```
switch(getFunctionCode(pcan_rd_msg)){

 case(3)://PDO1 (tx)
 case(5)://PDO2 (tx)
 case(7)://PDO3 (tx)
 case(9)://PDO4 (tx)
   pdo.processTpdo(pcan_rd_msg);
   break;

 case(11)://SDO (tx)
   sdo.processTsdo(pcan_rd_msg);
   break;

 case(14)://NMT error control
   nmtErrCtrl.processMsg(pcan_rd_msg);
   break;
}// switch
```

Figure 6. Part of the code in the listener mode function.

will be stored in the object dictionary or additional defined objects.

4 CONNECTION SET AND PROCEDURE FOR SETTING MODES OF OPERATION

After the implementation of the CANopen master stack mentioned above, more work should be done to make it a specialized version for controlling a 6-DoF manipulator with 6 servomotor controllers. The main work goes to the additional description of servomotor controllers profiled in CiA DSP-402 and the process of various tasks.

4.1 *Connection set*

For the sake of convenience and less effort to use devices supporting CANopen, a pre-defined connection set is provided in CiA 301. This pre-defined connection set supports 1 NMT, 1 SYNC, 1 TIME, 1 Emergency, 4 TPDOs, 4 RPDOs, 1 TSDO, 1 RSDO and 1 NMT error control, as shown in Table 1. Note that some of these items are needed for sending messages by the master stack, and the others should only be used for received messages.

4.2 *Procedure for setting modes of operation*

The 6-DoF manipulator uses several modes of operations, including the profile position mode, profile torque mode, interpolated position mode, etc. The main task of the specialized CANopen stack is to process the right procedure of setting these modes of operation.

An example of setting the interpolated position mode is given in Figure 7. Note that not all the modes of operation should be strictly set in this way. Modification is necessary in some situations. For example, when setting the profile torque mode, it is not allowed to set the motor state to OPERATION_ENABLE at the end because this operation may release the brake immediately and may lead to unwanted movement.

Table 1. Connection set.

COB	Function code	CAN-IDs
NMT	0000b	000h
SYNC	0001b	080h
TIME	0010b	100h
Emergency	0001b	081h~0FFh
TPDO1	0011b	181h~1FFh
RPDO1	0100b	201h~27Fh
TPDO2	0101b	281h~2FFh
RPDO2	0110b	301h~37Fh
TPDO3	0111b	381h~3FFh
RPDO3	1000b	401h~47Fh
TPDO4	1001b	481h~4FFh
RPDO4	1010b	501h~57Fh
TSDO1	1011b	581h~5FFh
RSDO1	1100b	601h~67Fh
NMT error control	1110b	701h~77Fh

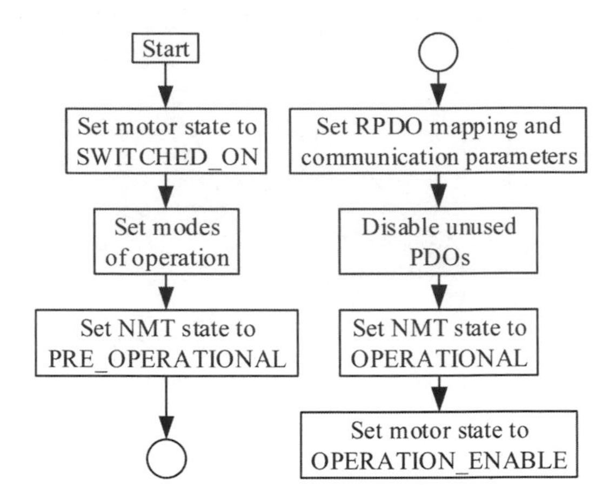

Figure 7. Procedure for setting the interpolated position mode.

Table 2. Worst-case response time experiment.

Platform	Clock frequency	Worst-case response time
Intel Core i5-480 M	2.66 GHz	1.361 ms
Intel Core i7-6700	3.40 GHz	1.025 ms

5 REAL-TIME PERFORMANCE

For the servomotor controllers working in the interpolated position mode, the response time is very important, because this mode of operation is highly time-sensitive. If the servomotor controller cannot receive the corresponding message within the required sending period, serious faults would happen immediately. For one servomotor controller in the interpolated position mode, 5 related messages are regarded as one sending unit. As a result, if all 6 servomotor controllers are in the interpolated position mode, considering that the sending period is limited to 50 ms, the response time of one message can only be within 1.67 ms.

The worst-case response time is tested using the SDO upload protocol, the result of which is listed in Table 2. For these two platforms, both of the worst-case response times are less than 1.67 ms, which meets the aforementioned requirement.

ACKNOWLEDGEMENT

Special thanks to Yongfeng Xia from HFUT for his help in guidance and participation in programming. Additional thanks to Qi Wang, Kun Wang and Xuanchen Zhang for their help in testing the program.

REFERENCES

Boterenbrood, H. CANopen: high-level protocol for CAN-bus (Version 3.1). (2015). at <http://www.nikhef.nl/pub/departments/ct/po/doc/CANopen31.pdf>

CANopen wikipedia. *Wikipedia Free Encycl.* (2016). at <https://en.wikipedia.org/w/index.php?title=CANopen&oldid=734892394>

CiA 301 version 4.2.0. at <http://www.canopen.org/standardization/specifications/>

Duan, J., Xiao, J. & Zhang, M. Framework of CANopen Protocol for a Hybrid Electric Vehicle. in *2007 IEEE Intell. Veh. Symp.* 906–911 (2007). doi:10.1109/IVS.2007.4290232

Farsi, M., Ratcliff, K. & Barbosa, M. An introduction to CANopen. *Comput. Control Eng. J.* **10,** 161–168 (1999).

Gold Board Level Module Hardware Manual Ver. 1.002. (2015). at <http://www.elmomc.com/products/gold-twitter-servo-drive.htm>

Jeon, J.M., Kim, D.W., Kim, H.S., Cho, Y.J. & Lee, B.H. An analysis of network-based control system using CAN (Controller Area Network) protocol. in *IEEE Int. Conf. Robot. Autom. 2001 Proc. 2001 ICRA* **4,** 3577–3581, vol. 4 (2001).

Li, W., Lijin, G. & Zheyuan, L. Design of STM32-Based CANopen Motion Control Master in Transfer Robot. In *2013 Third Int. Conf. Instrum. Meas. Comput. Commun. Control IMCCC* 1609–1612 (2013). doi:10.1109/IMCCC.2013.357

PCAN-USB User Manual version 2.5.0. (2016). at <http://www.peak-system.com/PCAN-USB.199.0.html>

Portillo, J., Estevez, E., Cabanes, I. & Marcos, M. CANopen Network for μcontroller-based Real Time Distributed Control Systems. In *IECON 2006—32nd Annu. Conf. IEEE Ind. Electron.* 4644–4649 (2006). doi:10.1109/IECON.2006.348148

Zhe, X. & Shifeng, D. The Design and Implementation of a CANopen Slave Stack for Powertrain Controller in Hybrid Electric Vehicle. In *2010 Int. Conf. Intell. Comput. Technol. Autom. ICICTA* **3,** 755–758 (2010).

Civil, Architecture and Environmental Engineering – Kao & Sung (Eds)
© 2017 Taylor & Francis Group, ISBN 978-1-138-02985-9

Design of a crane intelligent control system

Cancan Zhang, Shuguang Liu & Huawei Xie
School of Electronic and Information Engineering, Xi'an Polytechnic University, Xi'an, China

ABSTRACT: At present, the majority of cranes are still managed by manual control. The working efficiency is lower, so the method cannot meet the requirement of industrial production. Therefore, we design a crane intelligent control system, which is based on PLC and frequency converter. Particularly, we expound the design of hardware and software of the electric control system. The successful debugging proves that the system is feasible. Crane, as a transport machinery, is widely applied to the metallurgical industry, construction industry, etc. With the continuous expansion of modern industrial production, industrial production demands increasingly higher working efficiency of cranes. But currently, most of the cranes are still managed by manual control, to a certain extent, which restricts the working efficiency of the crane. We designed the intelligent control system, which is based on PLC and frequency converter. PLC serves as the kernel controller. It accomplishes the data acquisition of the field sensor and the output of the control signal. PLC connects the frequency converter with the CANopen bus, whose aim is to realize the variable frequency control of motor speed of the system. The communication between industrial PC and PLC is transmitted by a wireless transmission module.

1 OVERALL STRUCTURE AND WORKING PRINCIPLE OF THE SYSTEM

The system of crane intelligent control is mainly composed of industrial PC, PLC, frequency converter, wireless transmission module and different kinds of sensors. The overall structure is shown in Figure 1.

The intelligent control system is PLC and frequency converter-based design (Sun, 2014). PLC serves as the kernel controller of the crane intelligent control system, which accomplishes the data acquisition of the field sensor. After the logical operation, PLC will efferent the control signal to the frequency converter and then it will achieve the goal of managing the work of the crane. The communication between industrial PC and PLC is transmitted by the wireless transmission module. IPC can read the storage data of the PLC register and IPC can real-time display it at the software interface. At the same time, IPC can send the controlling signal to the PLC, and then PLC controls the frequency converter and the frequency converter controls the movement of the crane.

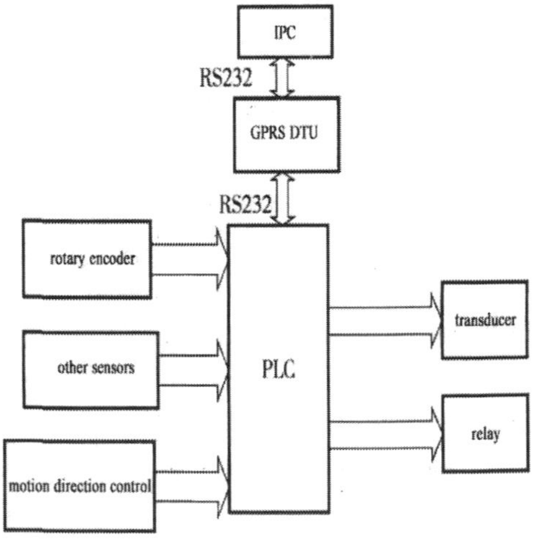

Figure 1. Overall structure block.

2 HARDWARE OF THE SYSTEM

The hardware of the intelligent control system is mainly composed of industrial PC, PLC, frequency converter, GPRS wireless module and some kinds of sensors. The structure diagram is shown in Figure 2.

2.1 *Composition of PLC hardware*

The signal, which the load module of PLC mainly receives, comes from 3 parts: the flashlight door signal (manual operation), data signals of the field sensor and the state of the system running. The signals of the flashlight door mainly include the signals of the system starting, stopping, the switching of the remote control/computer control and the quick signal. The quick signal controls the direc-

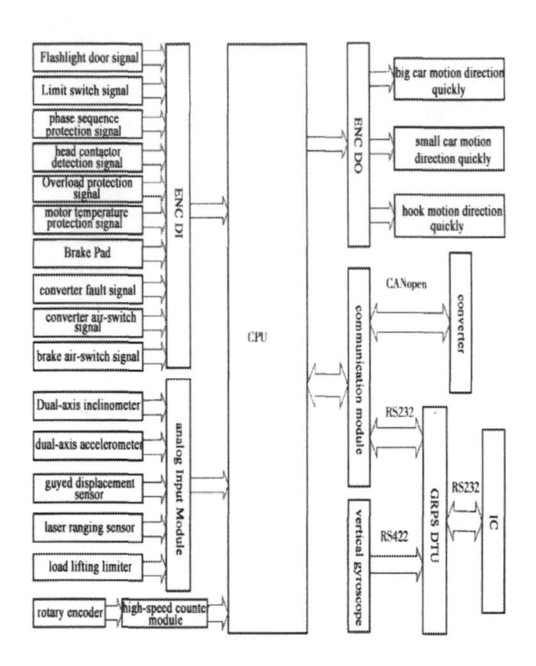

Figure 2. System hardware structure drawing.

tion of the big/small cars running institutions and lifting mechanism. It needs 12 input ports of digital quantity. Field sensors include three rotary encoders, which need to connect the high speed counting module (Jiang, 2011). The output of the other sensors is all analog. The state signals of the system include limit signals, deceleration limit signals, converter fault signals and guard signals. The limit signals come from the big car, small car and hook. The deceleration limit signals come from the hook. The converter fault signals come from the mechanism of the big car running and lifting. The guard signals mainly come from the phase order protection, the head contact detection, brake pad detection, overload protection, and motor temperature protection. All the above totally need 24 input terminals of digital quantity.

The output terminals of PLC mainly export the starting signal of the system and quick signal, which controls the direction of the big/small cars running institutions and lifting mechanism. It needs 20 output terminals of digital quantity.

2.2 *Communication between PLC and frequency converter*

The PLC, which the system adopts, comes from the series of Schneider M258 and the converters come from the series of ATV31 and ATV71. The PLC and converters are equipped with the CANopen communication protocols. So, they can be connected with the networking through the CANopen bus to communicate, so as to realize E-control of the PLC to the converter. Through the CANopen bus, PLC can set the output frequency of the frequency converter and read the real-time output frequency value of the frequency converter (Xu, 2008).

The network is composed of PLC and frequency converter. In the network, PLC is set to the master station and the frequency converter is set to the slave station. The converter receives the controlling signal that is sent by PLC. In the software of So Machine, the ID of the master station node of PLC is set to 127 and the ID of the slave station node of the lifting mechanism converter is set to 1(Song, 2008). The ID of the slave station node of the traveling mechanism converter of the big car is set to 2 and the ID of the slave station node of the traveling mechanism converter of the small car is set to 3.

Among the settings of the CANopen network, there will be a crucial parameter setting, the baud rate setting, to distinguish the different slave nodes, in addition to setting the address of slave nodes. Before the data exchange, the exchanging data comes from the master station and slave station in CANopen (Xu, 2009). The information of the baud rate needs checking. If the baud rate is different, they will not be able to do connection of communication, and hence the baud rate of the master station must be strictly consistent with that of the slave station. In the software of So Machine, the baud rate of PLC of the master station could be set and we set the baud rate to 125 kbit/s. Through the operation of the frequency converter, we could set the baud rate of the converter. Taking the lifting mechanism converter for example, in the communication menu, we could set the address of slave station 1, baud rate 125 kbit/s. The setting means that we set COM-/CnO-/AdCO = 1 (the address of the slave station) in the COM—of communication settings of the converter. The COM-/CnO-/bdCO is set to 125bit. The settings of two parameters must be strictly consistent with the settings of the lifting mechanism converter in the software of So Machine. The aim is to make sure that the communication between the PLC and the converter could go well.

2.3 *Wireless transmission modules*

There are two main functions of the wireless transmission module in the system: one of them is to realize the wireless communication of PLC and industrial control and the other is that industrial PC receives the data signal of vertical gyro.

2.3.1 *Hardware connection*

The ST-05 wireless transmission module, which the system chooses, can adopt the RS232 or RS485 bus to connect and communicate with others. When the RS232 bus is connecting, the wireless transmission distance is up to 100 m. It can guarantee the normal work of the wireless transmission module.

The system needs two channels to do wireless transmission. One is for the wireless communication between PLC and industrial control, and the other is for the wireless data transmission from vertical gyro to industrial PC. Each channel needs two wireless modules, so we need two couples for four modules. One of them connects with the PLC, the other connects with the vertical gyro, and the remaining two connect with the industrial PC. The PLC, which the system chooses, contains communication port RJ45 that can be used to connect with RS232 or RS485 serial links. PLC can connect with the wireless transmission module through the RJ45 port then turning to the RS232 bus. The communication follows the Modbus—ASCII or the RTU protocol. The output signal of vertical gyro is a digital signal, and through the RS422 bus, the signal outputs. Then, the signal should be shifted from RS422 to RS232 to connect with the wireless transmission module. The wireless transmission module connects with the industrial PC by RS232. After the connection between each device and the wireless transmission module, we need to do some relative configuration of the module of each channel. The goal is that the communication from PLC to vertical gyro and industrial works normally (Zheng, 2004).

2.3.2 *Software configuration*

The system chooses the ST—05 wireless communication module and it installs a matching set of X-CTU software to configure the wireless transmission module. The main interface is shown in Figure 3.

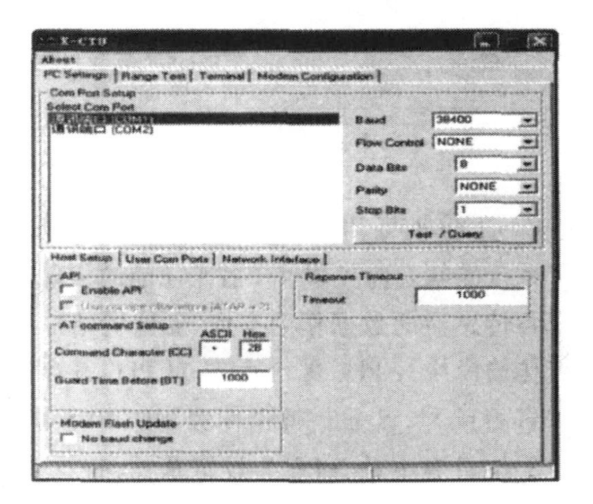

Figure 3. Wireless module configuration main interface.

In Figure 3, the Communication port 1 (COM1) is the communication port of PLC and industrial PC. The Communication port 2 (COM2) is the communication port of industrial PC. Industrial PC receives the data that comes from the vertical gyro (Wu, 2007). When it is configuring, firstly, we should select the communication port 1. Secondly, in the option of "PC Setting", we should choose the communication channel which needs configuring. Then, we should set the "Baud" (Baud rate) 9600, and the rest of the options are the default settings. Clicking "Test/Query", we test the module, because we do not know that the communication between module and software is normal. After the successful test, we should set the corresponding parameters to go on communicating.

3 SOFTWARE OF THE SYSTEM

The design of system software mainly includes 2 parts: the design of upper computer program and the design of PLC program. The program design of the upper computer mainly includes the system running state monitoring program and the communication with PLC program. PLC program is mainly the redact of the motion control program (Huang, 2009).

3.1 *Communication between upper computer and PLC*

The upper computer communicates with PLC through the wireless transmission module. It reads and writes the data of the PLC register, whose communication follows the standard MODBUS protocol. Among them, the upper computer is the master station, the PLC is the slave station, and the address is 01 (set in the PLC).

When the system is communicating, the upper computer, as the master station, sends instructions to PLC agreeing the standard MODBUS protocol (Xiong, 2008). PLC, as a slave station, responds the instructions of the upper computer.

3.2 *PLC program*

The PLC control program adopts modular programming, mainly including three parts: the sensor signal acquisition module, flashlight door control module and computer control module. By doing that, we can not only reduce the program statements, but also shorten the processing time and increase the responsiveness of the system so as to realize the rapid debugging of the program. The program flow chart of the PLC control system is shown in Figure 4.

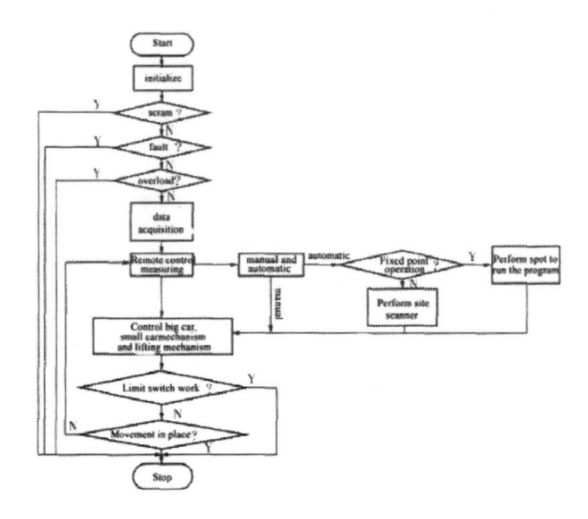

Figure 4. Flow chart of the PLC control system.

4 CONCLUSION

PLC is the core of the crane intelligent control system. The system communicates with industrial PC through the wireless module, which can realize the wireless remote monitoring to system status. PLC and frequency converter are managed by CANopen bus control, which not only saves the complex connection ports, but also improves the reliability of data transmission. For software design, we use the idea of modular to compile the program. Program 1 is acquisition data of PLC. Program 2 is motion control of the system. After scene debugging and testing, the system runs in good condition. We achieved the design goal and have provided the basis for future related design.

REFERENCES

Di Wu. The well data acquisition system Based on PC based PLC study [J]. Industrial control computer, **15** (2007).

JiaJing Song, Zhigang Liu. Based on the doors of the CAN_open train communication network control system design [J]. Mechanical and electrical engineering, **25** (2008).

Jiang Yong. Common fault analysis and processing industrial converter [J]. Mechanical and electrical information, **86** (2011).

Mingzhu Huang. The application of frequency converter and PLC in crane [J]. Mechanical and electrical technology, **3** (2009).

Qian Xu, Jie Zhang, Bi Wen. Bridge gantry crane intelligent monitoring system design [J]. Chinese test, **122** (2009).

Shu Zheng. Computer control system of crane design [J]. Electrical automation, 26 (2004).

Xuecheng Xiong. Of gantry crane based on PLC frequency control of motor speed control system [J]. China water transport, **12** (2008).

Yinlong Sun. Shearer fault analysis and processing of frequency converter [J]. Journal of technology and market, **43** (2014).

Zhenbang Xu. Warping machine control system based on CANopen bus [J] equipment manufacturing technology, **79** (2008).

Civil, Architecture and Environmental Engineering – Kao & Sung (Eds)
 ISBN 978-1-138-02985-9

Valve clearance fault diagnosis of an internal combustion engine based on wavelet packets and k-nearest neighbors

Rui Tan, Yu Zhang, Taixiong Zheng, Bin Yang & Yanjun Wang
School of Advanced Manufacturing Engineering, Chongqing University of Posts and Telecommunications, Chongqing, China

ABSTRACT: Considering the valve clearance fault that cannot be detected by a vehicle's on-board diagnostic system, the time-domain characteristics based on nearest-neighbor classifiers are proposed by using engine vibration acceleration signals. Especially, the diagnosis method consists of feature extraction using discrete wavelet packet decomposition and fault classification using the k-Nearest Neighbors (kNN) algorithm. Based on the feature extraction, the classifier is evaluated in terms of running time and identification rate compared with the Back Propagation Neural Network (BPNN), Support Vector Machine (SVM) and Least Square Support Vector Machine (LSSVM). The experimental results indicate that the proposed method has higher accuracy and the running time is shorter than that of the other methods.

1 INTRODUCTION

Internal Combustion Engine (ICE) is the power source of many machines in our daily life. It generates the necessary drive power to overcome the resistance loads by burning fuel and converting the energy content of the inlet mixture to mechanical motions. The valve-train is mainly a part of ICE, and it usually operates in high temperature, high pressure and gas shock environment, which can result in intermittent faults including valve head corrosion, broken spring and abnormal valve clearance. The abnormal valve clearance is difficult to be detected and it will bring serious damage to the engine, thereby increasing emission and fuel consumption. In this condition, the OBD systems as well as other monitoring and control systems may not detect the fault due to their adaptation to the technical condition change which has occurred (Mohammadpour, 2011; Dziurdź, 2010; Dąbrowski, 2011). Thus, it's very important to diagnose the fault of valve clearance when it occurs.

Several useful feature extraction techniques have been presented in abnormal valve clearance, such as Empirical Mode Decomposition (EMD) (Li, 2011; Wang, 2007; Si, 2012), short time Fourier transform (STFT) (Lei, 2011), Wigner-Ville Distribution (WVD) (Cai, 2012; Tang, 2010), and Wavelet Transform (WT) (Tang, 2010; Geng, 2010; Figlus, 2014; Pons-Llinares, 2011; Yang, 2014; Kankar, 2011). Although, WVD and STFT methods have a better performance in time-frequency analysis, the time window is fixed. WT is used to decompose the vibration signal, but the high-frequency resolution of signals is not very accurate. For most random complicated signals, the high-frequency features are very important to analyze the signal. To this point, WT will lead to be inadequate feature extraction of signals. Wavelet packets can efficiently overcome this disadvantage and improve the high-frequency analytical ability of complicated signals. Furthermore, wavelet packets have a good solution simultaneously in both frequency and time domains, and they can also extract more information in the time domain at different frequency bands (Wu, 2012).

At present, in order to improve the diagnostic accuracy, many machinery diagnosis systems with intelligent classification have been developed, such as Support Vector Machine (SVM) (Yang, 2014; Kankar, 2011; Wu, 2012; Wu, 2012) and the Artificial Neural Network techniques (ANN). They, especially SVM, are powerful classification tools for small sampling and nonlinear data. However, SVM is sensitive to missing data and has no certain solution for nonlinear problems. ANN needs a large number of parameters and its learning process cannot be observed, which will affect the credibility and acceptable degree of the results.

In this paper, according to the engine vibration signal, a fault diagnosis method of valve clearance based on the discrete wavelet packet decomposition frequency band energy for feature extraction and classification using the kNN algorithm is proposed. Then, the simulation is conducted, compared with SVM (Yang, 2014), LSSVM (Wu, 2012) and BPNN, showing that the running time and classification accuracy of kNN are better.

2 PRINCIPLES OF FAULT DIAGNOSIS

In 1986, Stephane G. Mallat found the wavelet function in the field of harmonic analysis. Since then, some of the physical and mathematical earthquake researchers have begun to pay close attention to this issue. Recently, the wavelet theory, which mainly includes Wavelet Transform (WT) and Wavelet Packet Decomposition (WPD), has been applied in various fields, such as data compression, signal de-noising and feature extraction, etc.

The discrete wavelet packet decomposition is described as follows.

$$\begin{cases} d_{i,2j}(t) = \sqrt{2}\sum_k g(k)d_{i-1,j}(2t-k) \\ d_{i,2j-1}(t) = \sqrt{2}\sum_k h(k)d_{i-1,j}(2t-k) \\ d_{0,0}(t) = f(t) \end{cases} \tag{1}$$

where, $f(t)$ is the original signal, $h(k)$ is the high-pass filter, $g(k)$ is the low—pass filter, and $d_{i,j}(t)$ is the reconstruction signal of wavelet packet decomposition at the i th level for the j th node.

Based on the feature of the vibration signal, three times decomposition of the original signal is conducted. To this point, the energy $E_{i,j}$ of every frequency band is worked out, which can be defined as follows.

$$E_{i,j} = \int \left|d_{i,j}(t)\right|^2 dt = \sum_{k=1}^{M} \left|d_{i,j}(k)\right|^2 \tag{2}$$

where M is the number of discrete points of the reconstruct signal $d_{i,j}$.

It has great influence on the energy signal of each frequency band in the diagnosis system. Therefore, each frequency band energy based vector F is constructed in the third layer.

$$F = [E_{30}, E_{31}, E_{32}, E_{33}, E_{34}, E_{35}, E_{36}, E_{37}] \tag{3}$$

During the process of data analysis, the value of E_{3i} is so high that it is hard to compute. Thus, normalization processing of E_{3i} is necessary. And the feature vector F^* of the original signal is established.

$$F^* = F_i \Big/ \sum_{j=0}^{7} E_{3j} \tag{4}$$

This paper measures four fault samples through the experiment and finds that the feature of each fault has a prominent performance in frequency band 5. Therefore, we define a feature vector F_{i5}^* of sample X_i. Based on the previous theory, the valve clearance faults can be classified using the kNN algorithm.

The kNN algorithm is one of the simplest and most commonly used machine learning methods. It is an important approach for nonparametric classification. The rules of kNN classification need to assign a category label for unknown samples by estimating their k-nearest neighbors based on known samples. A is the eigenvalues of known samples with respect to label B.

$$\begin{cases} A = (F_{1,5}^*, F_{2,5}^*, \ldots, F_{m,5}^*) \\ B = (1,1,\ldots,1,2,2,\ldots,2,3,3,\ldots,3,4,4,\ldots,4) \end{cases} \tag{5}$$

First, define a test F_0^*, and kNN can find the k training examples that are most similar to its k-nearest neighbors from $d_i(F_0^*, A_i)$ by measuring the similarity of the standard Euclidean distance between each training example A_i and the query point F_0^*:

$$d_i(F_0^*, A_i) = \left\|F_0^* - A_i\right\| \tag{6}$$

The label that appears the maximum times in k training examples is regarded as the prediction category label of the test point. For a discrete-valued target function, kNN returns the most common target function value among the neighbors of the test point. The kNN method gathers the k-nearest neighbors and lets them vote the class of most neighbors wins. k is the parameter of the kNN algorithm and its value should be selected between 1 and the total number of examples. The higher values of k provide smoothing that reduces vulnerability to noise in the training data. Note that if k is chosen as the total number of samples in the training set, then for any new instance, all the examples in the training set become the nearest neighbors. In this case the predicted response for each new or test case becomes just the frequent response variable in the training set.

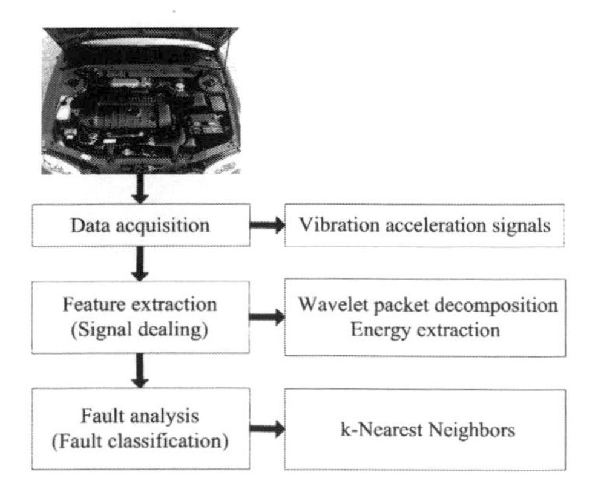

Figure 1. Block diagram of engine valve clearance fault diagnosis.

3 EXPERIMENTAL STUDIES

In the experiment, in order to achieve the proposed discrete wavelet packet decomposition using the kNN in the valve clearance fault diagnosis, an engine platform is used to measure the engine vibration signal of each valve clearance fault. Therefore, the engine vibration signal is regarded as the original data for the fault diagnosis process. And the process is divided into three parts: data acquisition, feature extraction, and fault analysis, which is shown in Figure 1.

The tests are conducted using two four-cylinder spark ignition engines of passenger cars. The vibration signal is extracted by an accelerometer, and the signal is recorded using a serial debugging assistant and the sampling frequency is chosen to be 20 kHz.

Additionally, the work condition of the engine valve-train can be divided into four parts: (i) ideal operation condition, (ii) exhaust valve clearance is 0.3 mm, (iii) exhaust valve clearance is 0.5 mm, and (iv) exhaust valve clearance is 0.7 mm. The ICE is operated under normal conditions and the speed is 3000 rpm. The accelerometer is mounted on the first cylinder head to collect the vibration signals generated during the operation of the engine.

4 RESULTS AND ANALYSIS

According to previous study, the exhaust valve clearance fault introduced to the engine is increased to 0.3 mm (positive), 0.5 mm (positive) and 0.7 mm (positive), respectively, which represents minor fault, moderate fault and serious fault. The vibration acceleration signal is collected from each fault condition, and the data of the time domain and frequency domain have been treated.

As the valve clearance is increased, the speed of the valve seat is increased. And the impact knock on the cylinder is also enhanced, which increases the signal energy in the fault frequency band. Therefore, this study decomposes sample signals three times and calculates the energy of each reconstructed signal in the third layer. Simultaneously, the frequency band energy normalized is regarded as the significant feature of the valve clearance fault. A sample of each band can be shown in Table 1 and the diagram is shown in Figure 2.

Figure 2 presents the changes of energy in each band which has been decomposed from different valve clearance vibration acceleration signals in an ICE. The data in Figure 2 show that the response value of frequency in band 5 increases from 0.1161 to 0.1577 with the exhausts valve clearance increasing from 0.2 mm to 0.7 mm. And the response value of frequency in band 6 is the same, but it is not obvious. As the condition of valve clearance changes, the normalized values of frequency energy in band 5 and band 6 have a regular variation from 5 Hz to 7.5 Hz. Therefore, the conclusion can be drawn that the feature frequency of the vibration acceleration signal ranges from 5 Hz ~ 7.5 Hz.

Furthermore, the energy ratio of frequency in band 5 is regarded as the fault feature that prepares for the latter fault classification. The paper randomly divide the total 40 samples of each condition into two parts, one of 30 for training to get a classier and the other of 10 for testing the classier to obtain the diagnostic yield. If the classifier has an ideal result, it can be used to predict the new or unknown data. In the course of experiment, 30 measured values of each fault are applied to the input of the k NN classier and we define the number (1, 2, 3, and 4, respectively) as fault label. The selection of k is based on experimental tests, where k is selected from a small integral value to increase gradually.

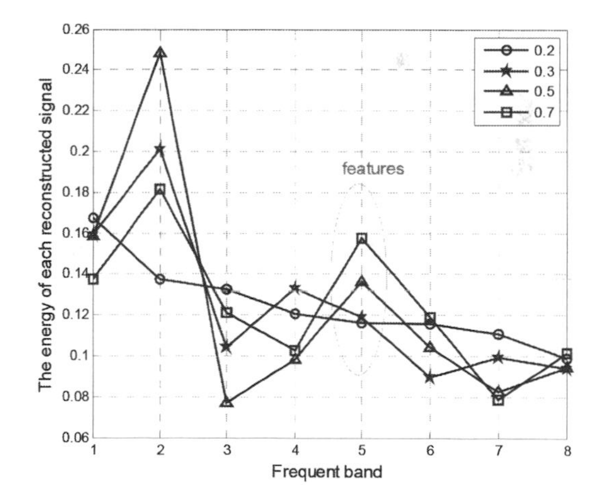

Figure 2. Block diagram of engine valve clearance fault diagnosis.

Table 1. The energy of each frequency band.

Band/Hz Gap/mm	0.00~1.25	1.25~2.50	2.50~3.75	3.75~5.00	5.00~6.25	6.25~7.50	7.50~8.75	8.75~10.00
0.20	0.1673	0.1373	0.1327	0.1208	0.1161	0.1158	0.1108	0.0991
0.30	0.1591	0.2012	0.1044	0.1331	0.1191	0.0897	0.0994	0.0941
0.50	0.1587	0.2483	0.0771	0.0982	0.1365	0.1043	0.0825	0.0944
0.70	0.1377	0.1815	0.1213	0.1026	0.1577	0.1192	0.0786	0.1014

By comparing the accuracy of classification and the running time taken to build a classier, we try to observe the efficiency and effectiveness of classifiers created by different values of k in the experiment. The optimal value of the nearest neighbors is acquired by comparing the results. Therefore, the value of k is selected from 2 to 17 and the results are shown in Table 2.

From Table 2, it can be seen that the accuracy of classification is 95.0% when k chooses 4, 5, 8, 9, 10, 17, respectively. However, the longer the running time, the larger the value of k. It shows that the accuracy of classification is always over 90.0% and the time response is about 2.3 seconds. The accuracy of classification do not increase with the increase of k in kNN classifiers, such as $k = 13$. Therefore, the optimal value of k in the experiment is selected as $k = 9$ and the accuracy of the classifier can reach to 95.0%.

The running time and accuracy are also recorded in the same simulation platform compared with BPNN, SVM and LSSVM. The data in Table 3 show that the identification accuracy of those algorithms is relatively high based on wavelet packet decomposition. The identification rate of kNN is better than BPNN and SVM. Under same conditions, the identification rate of kNN is 95.0% and the running time is 2.3108 s shorter than LSSVM.

Table 2. The comparison of accuracy and running time for different k.

k	Time (s)	Accuracy
2	2.2581	90.0%
3	2.2596	92.5%
4	2.2624	95.0%
5	2.2773	95.0%
6	2.2819	92.5%
7	2.2822	92.5%
8	2.2974	95.0%
9	2.3108	95.0%
10	2.3161	95.0%
13	2.3303	92.5%
17	2.4062	95.0%

Table 3. Comparison of diagnosis data for four algorithms.

Algorithm	Time (s)	Accuracy
BPNN	1.6356	90.0%
SVM	2.4973	92.5%
LSSVM	2.7926	95.0%
kNN	2.3108	95.0%

5 CONCLUSIONS

This paper presents a fault diagnostic method of valve clearance of ICE based on discrete wavelet packet decomposition and kNN that is applied to feature extraction and classification, respectively. Compared with BPNN, SVM and LSSVM, kNN achieves high accuracy. Moreover, the running time is rather short. The results indicate that the proposed method can also be used to detect the engine misfire fault. Furthermore, it has the following features:

1. Discrete wavelet packet decomposition overcomes the shortcoming of inadequate feature extraction of high-frequency signals.
2. For multiple classification problems, the effect of kNN is better than that of SVM.
3. kNN classification is easy to understand and the running time is short. In addition, the classification accuracy is high, and thus it is useful for other applications such as text classification, image-processing and predictive analysis.

ACKNOWLEDGMENTS

This work was jointly supported by the Scientific and Technological Innovation Program of Chongqing (grant no. CYS16168).

REFERENCES

Cai, Y.P., A.H. Li, L.S. Shi, P. Xu and W. Zhang, "IC Engine Fault Diagnosis Method Based on EMD-WVD Vibration Spectrum Time-Frequency Image Recognition by SVM," *Chinese Internal Combustion Engine Engineering*, vol. 33, no. 2, pp. 73–78, Apr. 2012.

Dziurdź, J. "Transformation of Nonstationary Signals into Pseudostationary Signals for the Needs of Vehicle Diagnostics," *Acta Physica Polonica A*, vol. 118, no. 1, pp. 49–53, Jul. 2010.

Dąbrowski, Z., M. Zawisza. "Investigations of the Vibroacoustic Signals Sensitivity to Mechanical Defects Not Recognised by the OBD System in Diesel Engines," *Solid State Phenomena*, vol. 180, no. 4, pp. 194–199, Nov. 2011.

Figlus, T., Liščák Štefan, A. Wilk, et al. "Condition monitoring of engine timing system by using wavelet packet decomposition of a acoustic signal," *Journal of Mechanical Science & Technology*, vol. 28, no. 5, pp. 1663–1671, May 2014.

Geng, X.J., Y. Cheng, "Research on Vibration Signal Characteristic Parameters of Diesel Engines by the Wavelet Technique," *Chinese Internal Combustion Engine Engineering*, vol. 31, no. 4, pp. 100–104, Aug. 2010.

Kankar, P.K., S.C. Sharma, S.P. Harsha. "Rolling element bearing fault diagnosis using wavelet transform," *Neurocomputing*, vol. 74, no. 10, pp. 1638–1645, May. 2011.

Lei, Y., J. Lin, Z. He, et al. "Application of an improved kurtogram method for fault diagnosis of rolling element bearings," *Mechanical Systems & Signal Processing*, vol. 25, no. 5, pp. 1738–1749, Jul. 2011.

Li, G., D.Y. Cai, S. Wang and H. Bai, "Application of EMD and SOM Neural Network in Gas Engine Fault Diagnosis," *Compressor Technology*, vol. 1, no. 2, pp. 31–34, Feb. 2011.

Mohammadpour, J., M. Franchek, K. Grigoriadis. "A survey on diagnostics methods for automotive engines," *Proceedings of the American Control Conference* 47(3): 985–990, May. 2011.

Pons-Llinares, J., J.A. Antonino-Daviu, M. Riera-Guasp, et al. "Induction Motor Diagnosis Based on a Transient Current Analytic Wavelet Transform via Frequency B-Splines," *IEEE Transactions on Industrial Electronics*, vol. 58 no. 5, pp. 1530–1544, May. 2011.

Si, J.P., J.H. Liu, L.N. Guo and J.C. Ma, "Application of EEMD and SVM in Engine Fault Diagnosis," *Vehicle Engine*, Serial no. 1, pp. 81–86, Feb. 2012.

Tang, B., W. Liu, T. Song. "Wind turbine fault diagnosis based on Morlet wavelet transformation and Wigner-Ville distribution," *Renewable Energy*, vol. 35, no. 12, pp. 2862–2866, Bec. 2010.

Wang, Z.P., W. Wang, X.Y. Li and J. Zhang, "Fault Diagnosis of Engine Valve Based on EMD and Artificial Neural Network," *Transactions of the Chinese Society for Agricultural*, vol. 38, no. 12, pp. 133–136, Dec. 2007.

Wu, J.D., J.B. Chain, C.W. Chung, et al. "Fault Analysis of Engine Timing Gear and Valve Clearance Using Discrete Wavelet and a Support Vector Machine," *International Journal of Computer Theory & Engineering*, vol. 4, no. 489, pp. 386–390, Apr. 2012.

Wu, J.D., J.B. Jian, C.W. Chen and Y. Hao, "Fault Analysis of Engine Timing Gear and Valve Clearance Using Discrete Wavelet and a Support Vector Machine," *International Journal of Computer Theory & Engineering*, vol. 4, no. 3, pp. 386–390, June. 2012.

Yang, L., H.S. Kang, Y.C. Zhou, et al. "Intelligent Discrimination of Failure Modes in Thermal Barrier Coatings: Wavelet Transform and Neural Network Analysis of Acoustic Emission Signals," *Experimental Mechanics*, vol. 55, no. 2, pp. 321–330, Oct. 2014.

Civil, Architecture and Environmental Engineering – Kao & Sung (Eds)
 ISBN 978-1-138-02985-9

RV reducer dynamic and static performance test system design

Wen-bo Fu & Chang-sheng Ai
School of Mechanical Engineering, University of Jinan, Jinan, China

Cheng-long Tang & Yong-shun Yang
Shandong Shuaike Machinary Ltd., Weifang, China

Guo-ping Li & Hong-hua Zhao
School of Mechanical Engineering, University of Jinan, Jinan, China

ABSTRACT: The RV reducer performance test system based on an open power flow structure is proposed to test the dynamic and static performance of a RV reducer. The test system can be used for the measurement of the RV reducer's transmission efficiency, transmission angle error and hysteresis error. The system consists of three parts: mechanical structure, measurement and control unit, and data processing software. The magnetic powder brake is used to apply different load torques. The IPC combined with the data acquisition card is used for parameter and load control. The Labview platform is used for related software development, which can accurately measure the RV reducer relevant parameters such as motion characteristics and transmission performance.

1 INTRODUCTION

RV (Rotate Vector) reducers have been widely used in the modern industry as the main joints of industrial robot precision components, with their high load, large transmission range and high precision (L, 2011). To ensure the accuracy of RV reducers and transmission performance, it is particularly important to perform the RV reducer comprehensive performance test.

2 THE MEASURING PRINCIPLE AND METHOD

A RV reducer is a two stage high-precision reducer using a cycloid reducer mechanism. The structure uses a way of closed planetary transmission combined with a planetary gear drive and cycloid wheel drive (Gola C, 2008).

The main parameters affecting the transmission performance of the RV reducer include rated input speed, input power, output torque, transmission ratio, transmission efficiency, transmission error, torsion stiffness, backlash clearance, etc. Among them, the main indicators of dynamic efficiency include transmission accuracy, and static indicators mainly include hysteresis error, torsional stiffness and so on (Z, 2014). It is better to improve the transmission performance of the reducer for the purpose of these parameters to detect the study. In this test system, the static test mainly carries on the hysteresis error test, and the dynamic test mainly carries on the transmission efficiency and the angle transmission error measurement (L, 2016).

2.1 *Mechanical efficiency test*

Mechanical efficiency, namely the ratio of the output of the mechanical work (amount of useful work) and the ratio of input power (power work quantity). For the RV reducer, the mechanical efficiency is the ratio between the output power and the input power. The reducer input power and output power are as follows:

$$P = \frac{T \cdot n}{9550} \tag{1}$$

The mechanical efficiency of reduction is as follows:

$$\eta = \frac{P_2}{P_1} \tag{2}$$

Among them, P is the output of the reducer or the input power, and the unit is Kw; T is the reducer input shaft and output shaft torque, and the unit is N·m; n is the reducer input shaft and output shaft speed, and the unit is r/min.

In the actual test, both ends of the reducer installed the torque sensor. Considering the coupling influence, in calculating the input and output power is to take into account the coupling mechanical efficiency η_1 (Predrag, 2007). The power formula becomes:

$$\eta = \frac{P_2}{P_1} = \frac{T_2 * n_2}{T_1 * n_1 * \eta_1} \tag{3}$$

Among them, n_1 is the input speed of the reducer, and the unit is r/min. n_2 is the output speed, and the unit is r/min.

The test method is as follows: the torque speed sensor installs in both ends of the measured RV reducer, real-time gathers the input end and the output end rotational speed and the torque signal of the measured RV decelerator. The signal is transmitted through the data acquisition card which connects with the IPC (Industrial Personal Computer) to the test program, and it is converted to the actual data through the program calculation. Then, the mechanical efficiency test result is obtained.

2.2 *Transmission angle error test*

Angle transmission error θ_{er}, namely reducer drive error. For the gear unit, the difference value between the theoretical output angle and the actual output angle θ_{out} at any input rotation angle θ_{in} of the gear unit can be calculated by formula (4):

$$\theta_{er} = \frac{\theta_{in}}{i} - \theta_{out} \tag{4}$$

The test method is as follows: during the test, the control motor to maintain low-speed rotation, the measured RV reducer input and output are installed around the grating encoder to capture the angular signal. The signal is transmitted through the data acquisition card which connects with IPC transfer to the test program, and converted to the specific test data for the program calculation. When the output shaft is rotated by 5°, a set of data is collected, and the output shaft is rotated by 360° for one cycle. 73 theoretical input angles $\theta_{in0}/i \sim \theta_{in72}/i$ and 73 actual output angles $\theta_{out0} \sim \theta_{out72}$, according to the formula (4) to calculate all of the actual test angle transmission error value $\theta_{er0} \sim \theta_{er72}$, and calculated according to the formula (5) of the gearbox angle transmission error θ_{er}.

$$\theta_{er} = \{\theta_{ern}\}_{max} - \{\theta_{ern}\}_{min} \tag{5}$$

2.3 *Hysteresis error test*

Hysteresis error in the gear transmission is abbreviated as hysteresis, which is defined as when the reducer input shaft steering angle changes, the amount of hysteresis on the output shaft also changes.

The hysteresis error measured in the hysteresis test of the reducer generally refers to the output shaft rotational angle caused by geometrical factors such as flank clearance in the drive chain, bearing clearance, etc., where the components are in good contact. Normally ± 3% of rated torque is applied during the process to overcome internal friction and oil film resistance. Formula (6) is used for the calculation of hysteresis error:

$$j_T = [(\theta_{in})_{max} - (\theta_{in})_{min}]/i - [(\theta_{out})_{max} - (\theta_{out})_{min}] \tag{6}$$

The hysteresis error is mainly due to two reasons. First, it is the output shaft gear backlash hysteresis error caused by gear backlash. The second is the elastic deformation of hysteresis error caused by the geared rotor system and gear elastic deformation under the action of the external torque.

The test method is as follows: when measuring the hysteresis error, the output shaft of the measured RV reducer is locked by the magnetic powder brake. The motor applies the positive torque to the input shaft, gradually loads from 0 to the rated gear torque M, and then unloading to 0. The input rotary angle grating corresponding to each torque value is acquired at the same time during the loading and unloading processes. The data are transferred to the acquisition card and uploaded to the IPC. And then, the software analyzes the data and draws the hysteresis curve of the RV reducer.

3 DESIGN OF THE TEST BED STRUCTURE

3.1 *The general system structure*

The reducer static and dynamic performance test system test bed can be divided into two types such as open power flow and closed power flow, according to whether the energy is recycled in the experimental process of transmission.

The open power flow test stand structure is shown in Figure 1. In the entire measurement and control system, the drive unit for the measurement and control system to provide the power source is the general choice of synchronous motor and other devices. The load unit can be effectively loaded on the device under test to provide the RV reducer of different stress state and enhance system reliability

(Y, 2015). The measurement and control system with sensors is used to measure the performance of the reducer parameters. In the open power flow measurement and control system, the output power of the motor and other drive device flows through the measured reducer and the transmission to reach the loading unit. The load unit uses a variety of loading methods to consume energy, and to achieve effective loading.

Compared to the open power flow structure, closed power flow can realize energy cycles. For larger energy consumption and a longer run-time test, its operating costs are lower and more efficient. But for the system, the structure of the closed power flow test-bed is more complicated, and the open power flow is more mature and reliable. In this test, the test object is the small and medium RV reducer, the power requirements are low, and the dynamic transmission efficiency, transmission error and the static under the hysteresis error test does not require long-running time. Considering comprehensive test bench energy consumption and the general structure, an open power flow test system is designed. The functional block diagram is shown in Figure 2.

3.2 *Mechanical structure design of the test system*

The test bench of the RV reducer static and dynamic performance test system mainly composed of the base, reducer bracket, measuring unit, mobile platform, etc. It mainly plays a role in the fixed supporting function of the components and the measuring elements in the system, and also has a key role in ensuring the measurement accuracy of the measurement and control system. The measuring unit consists of circle grating, torque speed sensor, magnetic powder brake and other components, which is used to complete the measurement and control in the test system.

The test bench adopts a horizontal structure, which is easy to ensure the coaxial degree. The mechanical institutions are as compact as possible, and they use coupling connection or key connection between devices. Input components can be moved in the horizontal direction, for different types of gear reducer, and it can be installed and tested normally. A three dimensional schematic diagram is shown in Figure 3.

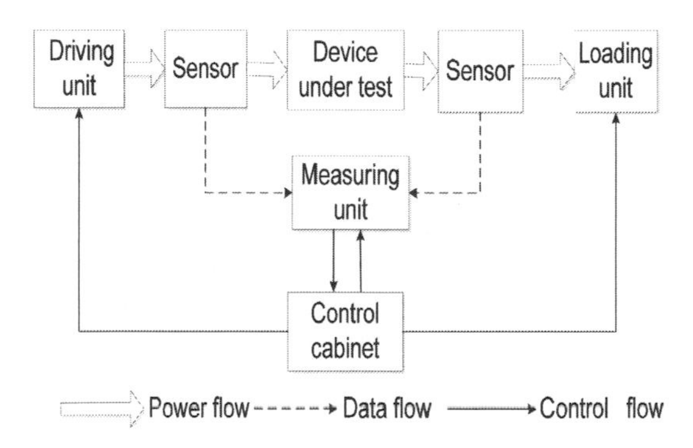

Figure 1. Schematic diagram of an open power flow test rig structure.

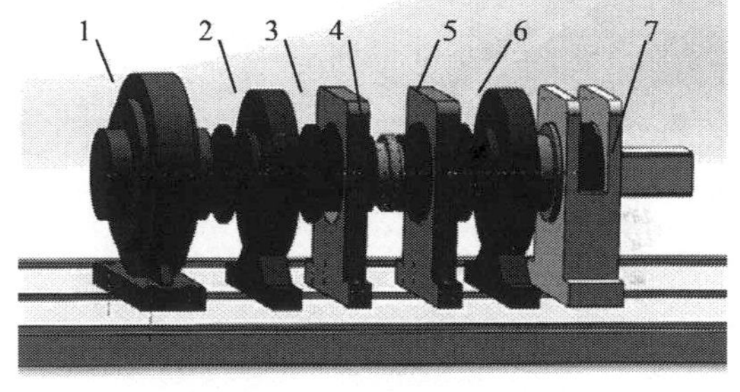

Figure 3. The diagram of the RV reducer test bed structure. 1-magnetic powder brake 2-output shaft torque and speed sensor 3-output shaft circular grating 4-measured RV reducer 5-input shaft round grating 6-input shaft torque and speed sensor 7-drive motor.

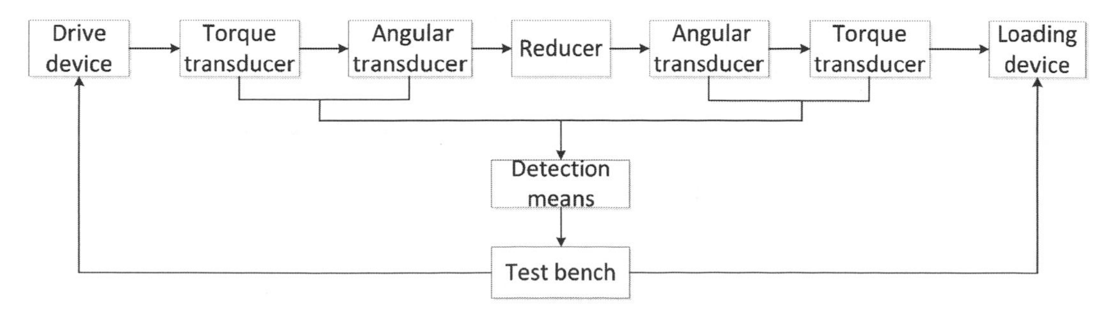

Figure 2. The principle diagram of the RV reducer test system.

4 MEASUREMENT AND CONTROL SYSTEM

4.1 *Operating principle*

The static and dynamic performance measurement and control system of the RV reducer is shown in Figure 4. When the RV reducer is tested for dynamic or static performance, the measured RV reducer is installed on the test platform. The input shaft and output shaft of the reducer are connected with the measuring spindle, the output shaft and torque sensor respectively. The two ends of the input and output of the reducer are equipped with high-precision circular gratings to measure the angle signals at both ends of the gearbox. The torque and speed sensors connected at both ends of the gearbox are used to output the speed and torque signals at both ends of the reducer.

The system uses motor to drive and magnetic brake to load. The IPC is equipped with a dedicated acquisition card for collecting the round grating signals of input and output as well as the torque and speed sensor signals of input and output.

When measuring, the operator sent control commands to the controller through the IPC. The driver receives the command and then drives the motor rotation to provide the power for the system. At the same time, the IPC connected with the program-controlled power, for the control loading of the magnetic particle brake. The measured RV reducer parameters are received by the sensor in the test, then collected by the acquisition card and uploaded to the IPC. The data are processed by the software for the output display to complete the measurement.

4.2 *Drive control unit*

A permanent magnet synchronous AC servo motor is selected, which is provided with a drive control unit, and it can realize the control and drive of the motor so as to provide energy sources for the system.

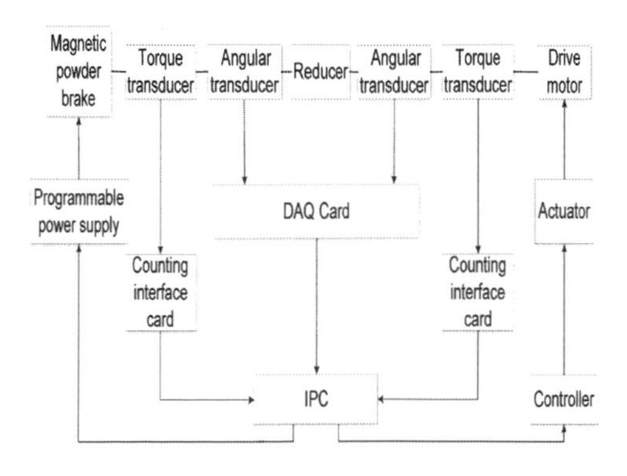

Figure 4. The diagram of the test system structure.

4.3 *Signal acquisition unit*

The signal acquisition system mainly includes torque sensors and circular gratings, which combines the acquisition card and other equipment to achieve the acquisition and collection of the signal. Then, the signal is uploaded to the IPC to achieve data processing.

4.4 *Load unit*

The magnetic particle brake acts as a load device in the system. The excitation current of the magnetic particle brake and transmission torque is basically a linear relationship, and as long as the value of the exciting current changes, the torque value can be controlled. So, the system can control the braking torque size according to the characteristics of the magnetic particle brake.

5 DATA PROCESSING SOFTWARE

The signal processing is carried out based on the hardware measurement. The software platform of the measurement and control system is set up in view of Labview. The measurement process and data processing method of the static and dynamic performance parameters (transmission efficiency, transmission error and hysteresis error) are analyzed, so as to realize the automatic detection of the RV reducer measurement and control system.

RV reducer dynamic and static performance test system software needs to achieve three parts of the target with motor control, data acquisition, data analysis and processing, so that the measurement and control system has become an organic whole.

1. Motor control: the return difference experiment in the static test needs to achieve the reverse load and unload for the reducer, and so it needs the measurement and control software to control the loading value of motor torque accurately. In the test of dynamic transmission error and transmission efficiency, it is necessary to make the reducer at different speeds, which requires measurement and control system software to accurately control the motor speed.
2. Data acquisition: the software needs to set up the data acquisition card. It can complete data acquisition with the input and output angle and torque of the RV reducer, and communicate normally with the controller and programmable power supply.
3. Data analysis and processing: the signal is collected for analysis and processing, according to the performance parameters of the system,

and the required output data and performance curves are obtained.

6 SUMMARY

On the basis of the dynamic and static performance parameters of the RV reducer, the principle and method of mechanical efficiency, angular drive error and return error of the RV reducer are deeply researched. With these methods, a RV reducer static and dynamic performance test system is designed. The system can quickly and accurately test the performance of different types of RV reducer. It is important to verify and improve the design theory and manufacturing method of the RV reducer, and its batch design and manufacture.

ACKNOWLEDGEMENT

This work was financially supported by the Shandong Province R&D Key Projects (2014GJJS0401).

REFERENCES

Fengshou Z, Linlin Z, Jianting L, Peng Z, Journal of Mechanical Transmission, 08(2014).

Gola C, Davoli P, Rosa F, Journal of Mechanical Design, **130**, 11(2008).

Predrag Z, Journal of Machine Design, (2007).

Qi L, Xinhui Wu, Weridong H, Journal of Mechanical Transmission, 04(2016).

Shidang Y, Zhongming L, Journal of Modern Manufacturing Engineering, 11(2015).

Xichang L, Hongzhan L, Journal of Mechanical Engineering, 07(2011).

Civil, Architecture and Environmental Engineering – Kao & Sung (Eds)
© 2017 Taylor & Francis Group, ISBN 978-1-138-02985-9

Analysis of automotive suspension compliance based on the main effects of rubber bushing

Bao Chen, Jiang-hua Fu & Zhe-ming Chen
College of Vehicle Engineering, Chongqing University of Technology, Chongqing, China

ABSTRACT: Taking the rubber bushing of a twist beam suspension as the research object, based on the bushing's static characteristic test results, the main effects of the six bushing's stiffness were determined by using the orthogonal experimental design method. The objective of this study is the case that the convergence of the suspension kinematics' evaluation index response is close to the target value range and the convergence of the suspension compliance' evaluation index response is smaller. The final values of the three axial bushing's stiffness were determined by using a full factorial experimental design method. The two sets of bushing stiffness data before and after optimization were substituted into the Motion-Solver for suspension compliance analysis. The results indicate that the optimization of the bushing parameters had almost zero effect on each evaluation index of compliance under the aligning torque, and the camber compliance loaded lateral and longitudinal force is also very small, but the compliance of the toe angle and wheel displacement was changed.

1 INTRODUCTION

Due to its simple structure, light weight and low cost, the twist beam suspension is widely used in the rear suspension of various economical cars. The main components connected with the sub-frame are the rubber bushings of a trailing arm. The lateral force, longitudinal force and aligning torque at the tire contact point are transmitted through the bushing to the vehicle body. It directly or indirectly influences the correct positioning of the suspension, vehicle ride comfort and NVH characteristics through its axial stiffness, torsional stiffness and its own viscoelastic properties (Liu, 2012). In addition, for the suspension with rubber bushings, its compliance and roll stiffness also impact on the vehicle handling stability (Chen, 2011). Therefore, it is important to study the Kinematics and Compliance (K&C) of a suspension with rubber bushings in order to improve the suspension and vehicle performances.

2 STATIC CHARACTERISTIC TEST OF A RUBBER BUSHING

The static characteristic test refers to the mechanical behaviour of a rubber bushing under static load. That is, the curve between the loading force and the displacement in the slower loading state, including three axial stiffness and three torsional stiffness (Wen, 1999).

Figure 1 a) shows the position of the rubber bushing of the trailing arm studied in the paper. Figure 1 b) shows a computer controlled universal electronic material test rig (RGM-4300).

The corresponding measurement principle is shown in Figure 2 a). The bushings installed in the fixture are fixed to the U-shaped fork by a screw to measure the radial stiffness (Y/Z direction). The corresponding measurement principle is shown in Figure 2 b).

The test loading method is unidirectional pressure. The force is increased linearly from zero to the set value. The setting range of loading force in the Z direction is 0~3000 N, and the setting range of loading force in X and Y direction is 0~5000 N

a) Rubber bushing b) Material test rig

Figure 1. Twist beam suspension and a test rig.

and 0~9000 N, respectively. The Z-direction is designed with a radial shrinkage hole. The test procedure starts from zero time, and the running speed of the test rig is 1.0 mm/min. Its computer control system automatically records the load and the deformation of bushing at each moment. The exported data format is a .PSD file inputted to EXCEL. The three axial stiffness curves, as shown in Figure 3, are generated by using the insertion point pattern. Based on the three axial stiffness curves of bushing, the coefficients of its constitutive equation are optimized (Zhai, 2009), and finally the three torsional stiffness curves are shown in Figure 4.

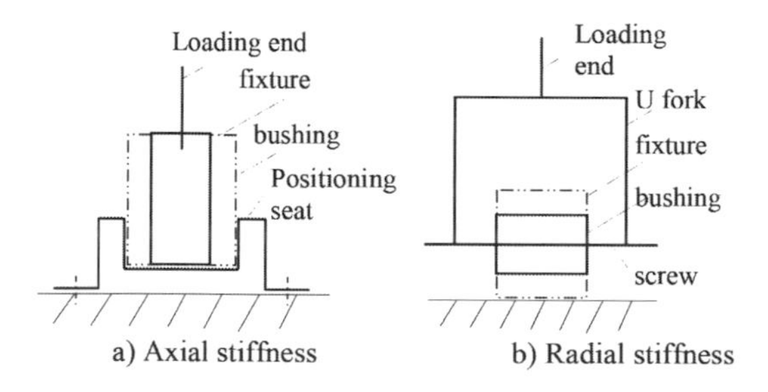

Figure 2. Clamping diagram of the bushing's axis and radial direction.

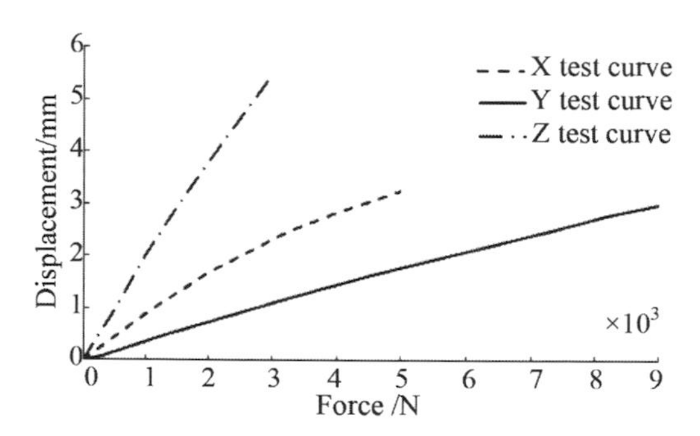

Figure 3. Three axial stiffness curves of bushing.

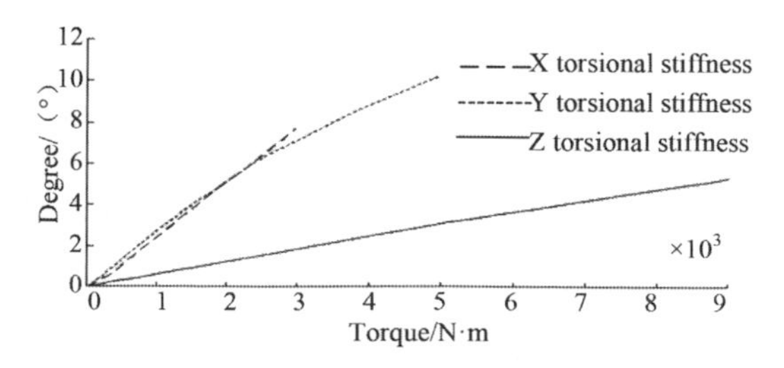

Figure 4. Three torsional stiffness curves of bushing.

3 MAIN EFFECT ANALYSIS OF THE BUSHING STIFFNESS

The six bushing's stiffness was used as the variable to optimize the evaluation index of suspension K&C, and the coordinate system of the optimization variables was consistent with the local coordinate system of the rubber bushing. In order to reduce the optimal parameters, shorten the development cycle, before the optimization of the bushing stiffness, the main effects of the bushing stiffness were analysed by using the experimental design of Hyper-study (Pei, 2014).

The range of bushing stiffness was set, that is, the upper and lower limits of design variables range of increase and decrease were 10% of the initial value, as the main effect of the bushing stiffness analysis of the upper and lower levels. Six design variables were just the six bushing's stiffness. It was designed as L8 (2^7) orthogonal experimental table (Table 1), in which the number of experiments required is 8. The optimization targets are the out-of-ideal range of each index response in the suspension kinematic analysis, such as the toe angle, camber angle and longitudinal displacement. The constraint range was the interval value of each characteristic index in the suspension kinematic analysis, where "–1" indicates the lower level and "1" indicates the upper level.

Submitted to the orthogonal experimental table to calculate, listed 6 design variables impacted on the main effect of each index response shown in Figure 5. In this study, the toe angle is taken as an example. And after analysis, it is found that the three axial stiffness has a certain influence on the toe angle under the two-wheeled reverse excitation, in which the positive main effect of the radial stiffness Kx and axial stiffness Kz is larger, the positive main effect of the radial stiffness Ky is smaller, and it shows that when the stiffness increases in the upper and lower limits of the interval, the corresponding toe angle is also gradually increased.

The toe angle increases in the range of about 0.1321°~0.1366°, 0.1323°~0.1363°, and 0.1340°~0.1348° of K_x, K_y, and K_z, while the other three torsional stiffness has no effect of the toe angle. The six design variables on the evaluation index response of the suspension kinematics shows that the three axial bushing's stiffness has a great effect on the index response, and the three torsional stiffness effect on the index response is very small, and so the three axial bushing's stiffness was used as the object of further optimization research.

Table 1. Orthogonal experimental table.

No.	K_x	K_y	K_z	K_{rx}	Y_{ry}	Z_{rz}
	Axial stiffness (N/mm)			Torsional stiffness N·mm/(°)		
1	1(1375)	1(2933)	1(550)	1(501)	1(440)	1(1870)
2	1(1375)	1(2933)	−1(450)	1(501)	−1(360)	−1(1530)
3	1(1375)	−1(2400)	1(550)	−1(410)	1(440)	−1(1530)
4	1(1375)	1(2400)	−1(450)	−1(410)	1(360)	1(1530)
5	−1(1125)	1(2933)	1(550)	−1(410)	−1(360)	1(1870)
6	−1(1125)	1(2933)	−1(450)	−1(410)	1(440)	−1(1530)
7	−1(1125)	−1(2400)	1(550)	1(501)	−1(36)	−1(1530)
8	−1(1125)	−1(2400)	1(550)	1(501)	1(440)	1(1870)

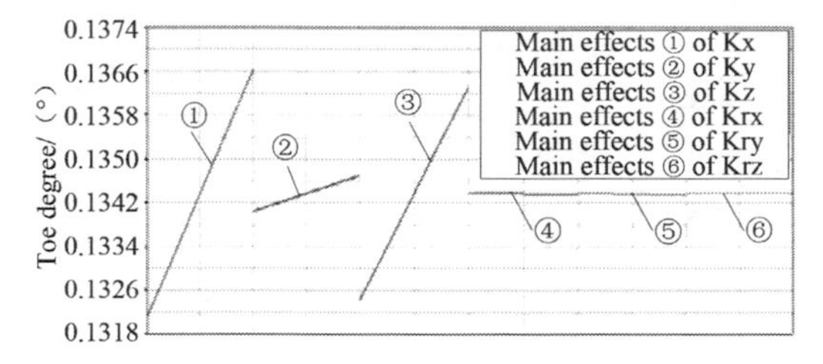

Figure 5. The diagram showing the main effects of the design variables.

Figure 6. Test rig of a twist beam suspension simulation.

4 THE IMPACT ANALYSIS OF BUSHING FOR SUSPENSION COMPLIANCE

Suspension kinematics is the relationship between the evaluation index and the wheel jump. Suspension elastic kinematics refers to the change of the evaluation index under the external forces such as lateral force, longitudinal force, aligning torque, etc. (Lin, 1992), which is also called suspension compliance (Sun, 2012).

4.1 *Establishment of virtual test rig of a twist beam suspension with rubber bushings*

Firstly, a subsystem model of the twist beam suspension was built. The main method was basing on the modal synthetic method and adopted motion/ FLEX module to input the modal file for building a flexible body's twist beam. The rubber bush characterization data use the six bushing's stiffness at the zero level, as shown in Table 1.

Other components such as coil springs, shock absorbers, hubs, etc., were linked in Motion View/ Solver modules according to the positioning and physical parameters of the actual suspension model. All hard points were created in a no-load state. The twist beam suspension simulation test rig is shown in Figure 6.

The main function of this simulation test rig is executing a simulation experiment of suspension K&C, wherein the outputs were set to toe angle, camber angle, and wheel displacement as evaluation index. In the simulation load process, the calculation was automatically carried out, and the relationship between output evaluation index and wheel jump and load was obtained by HyperView or HyperGraph.

4.2 *Full factorial experimental design of the bushing*

Among the six design variables, only the three axial stiffness of the rubber bushing had a significant effect on the indexes of suspension performances. So the three axial stiffness was chosen as the optimization variable. The upper and lower limits of the optimization variable were set in the range of 50% to 150% of the initial value.

The optimization objectives are determined as follows. The response convergence of the evaluation index of the suspension kinematics tends to be within the target value range, and the evaluation index's response of the suspension compliance is small.

Through the full factorial experimental design, the upper and lower limits of the three axial bushing's stiffness were divided into five equal parts, the number of levels was set to 5, which was a 3-factor and 5-level full factorial experimental design. The number of calculations was 5 of the 3 power, a total of 125 times (Gao, 2014). The data of 125 sets of test combinations could be read from EXCEL, which analysed and selected the 46th group that accorded with the optimization objectives as the optimization result, which shows that the corresponding stiffness of the X direction is 937.5 N/mm, the stiffness of the Y direction is 4000 N/mm, and the stiffness of the Z direction is 250 N/mm.

4.3 *Analysis on the main evaluation index's change rule of suspension compliance*

Through various experiments and theoretical research, it is found that the rubber bushing has certain influence on the compliance of the twist beam suspension, but the contribution rate to the various sub-performance indexes are different.

After the optimization of bushing, the simulation was carried out again by using the simulation test rig shown in Figure 6. In the simulation of the lateral force loading, the vertical height of the wheel and the degrees of freedom of the axle were fixed, and only the results of one side of the tire were used; the load range was set to ± 4000 N. Then by using the Motion-Solver's calculation, the compliance results of the toe angle and lateral displacement are shown in Figures 7 and 8.

From the comparison of the dotted and solid line in Figure 7, it is found that the inclination of the right wheel's toe angle under the lateral force is smaller, which indicates that the optimization has greater influence on the toe angle under the lateral force, that is, when the vehicle turns left, it is possible to reduce the tendency of excessive steering. At the same time, as the lateral force changes, as shown in Figure 8, the comparison of the dotted and solid line shows that the slope of the lateral displacement increases, which will lead to a larger wheel slip, increasing rolling resistance.

From the comparison of the dotted solid line in Figure 9, it is found that the inclination of the toe angle under the longitudinal force becomes large, which reduces the tendency of understeer. As can be seen from the comparison of the dotted and solid line in Figure 10, the longitudinal displacement slope under the longitudinal force increases,

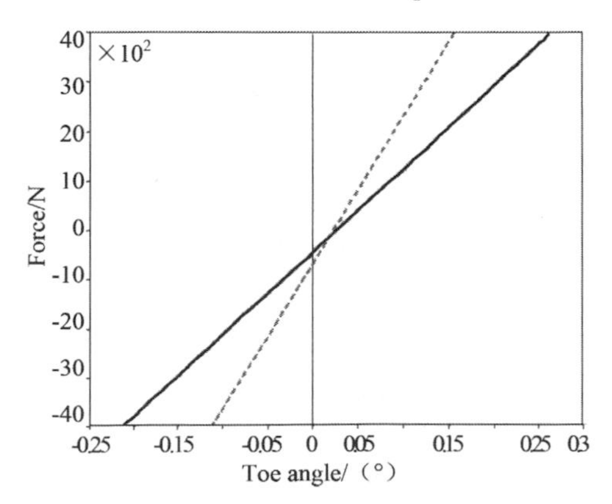

Figure 7. Toe angle's compliance (lateral force).

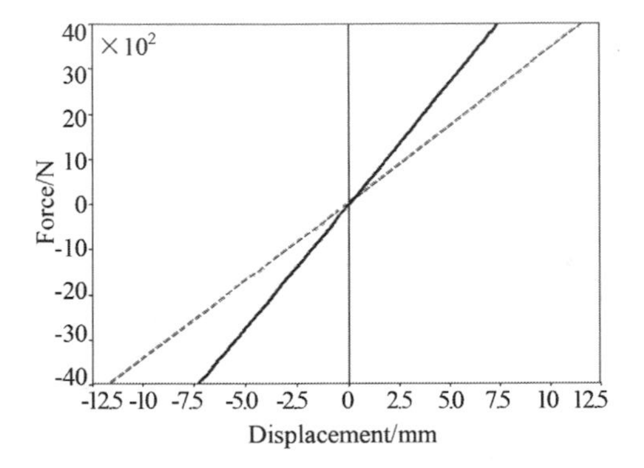

Figure 8. Lateral displacement's compliance (lateral force).

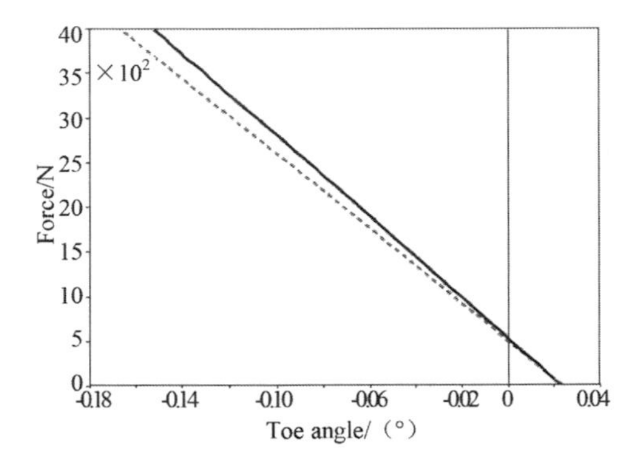

Figure 9. Toe angle's compliance (longitudinal force).

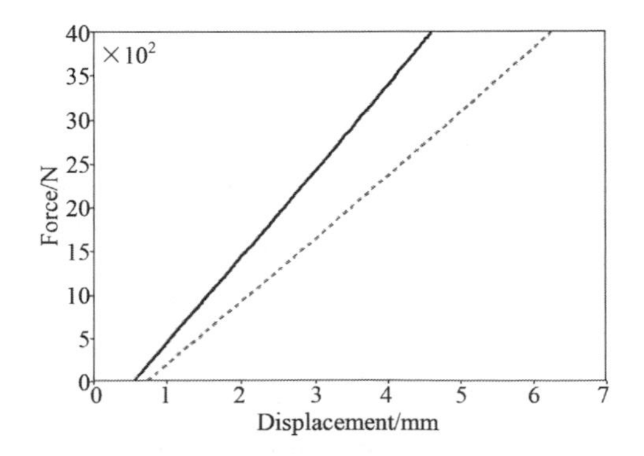

Figure 10. Longitudinal displacement's compliance (longitudinal force).

Table 2. Compliance change of the suspension performance indexes.

Compliance with lateral force			Compliance with longitudinal force			Compliance with aligning torque		
Toe angle (°)/kN	Camber angle (°)/kN	Lateral displacement mm/kN	Toe angle (°)/kN	Camber angle (°)/kN	Lateral displacement mm / kN	Toe angle (°)/kN	Camber angle (°)/kN	Lateral displacement mm/kN
0.05875	0.03442	1.875	−0.0435	−0.03575	1.00	0.0002765	0.00000485	0.00101
0.03406	≈0.03442	2.958	−0.0480	≈−0.0358	1.37	—	—	—
−42.02%	≈0	57.73%	10.34%	≈0	37.00%	—	—	—

which is helpful to improve the vehicle ride comfort. In the longitudinal force loading test, the wheel vertical height and axle freedoms were also fixed, and only the test results of one-sided tires were studied. The load range was set between 0 and 4000 N. After the calculation in Motion-Solver, the compliance results of the toe angle and longitudinal displacement are shown in Figures 9 and 10.

In addition, the variation of each evaluation index's compliance under the lateral force and longitudinal force and the compliance index's change under the aligning torque was very small. So, the study ignores the corresponding change results. Table 2 shows the change value of the suspension performance index's compliance before and after optimization of the bushing stiffness.

From Table 2, it can be seen that by optimizing the bushing stiffness, the change of the toe angle and displacement compliance under the lateral force and longitudinal force is large. The absolute values of the change rate of the toe angle compliance are 42.02% and 10.34%, respectively, and the absolute values of the displacement compliance rate are 57.73% and 37.00%, respectively. But the camber compliance is significant under lateral force and longitudinal force basically unchanged. Under the aligning torque, it can be seen that the compliance indexes of the toe angle, camber angle and lateral displacement have been very small before optimization, that is, the aligning torque has little effect on the changes of compliance indexes.

5 CONCLUSIONS

Based on the test results of the static characteristics of rubber bushings, the coefficient of constitutive equation of rubber bushing can be optimized and the accurate bushing model can be constructed. Furthermore, the three torsional stiffness of bushing can be obtained by simulation, and the six bushing's stiffness are reconstructed.

Through the analysis of the main effect of the rubber bushing and the compliance analysis of the twist beam suspension, and compared with the main evaluation index, it can be found that the compliance of the toe angle, lateral and longitudinal displacement is greatly affected by the change of the bushing stiffness under the lateral and longitudinal force, and the camber compliance is not affected by this. In particular, under the effect of the aligning torque, the change in each evaluation index's compliance is close to zero.

ACKNOWLEDGEMENTS

This research was supported by the Chongqing Research Program of Basic Research and Frontier Technology (No. cstc2013 jcyjA60005) and the National Natural Science Foundation of China (No. 51205433).

REFERENCES

Chen Wu-wei, Li Xin-ran, Chen Xiao-xin, Wang Lei, Middle-high frequency vibration transfer analysis of vehicle suspension and optimization of rubber bushings, Transactions of the Chinese Society of Agricultural Machinery, **42**(10), 25–29(2011).

Gao Jin, Yang Xiu-jian, Niu Zi-ru, Robust optimization and sensitivity analysis of hardpoints on suspension characteristics and full vehicle handling performance, Journal of Jiangsu University, **35**(3), 249–256(2014).

Lin Yi, Zhang Hong-xin, Wen Wu-fan, A study of the Compliance of Automotive Suspension Automotive Engineering, **14**(3), 175–180(1992).

Liu Wei, A study on multi-objective optimization and bus performance affected by suspension rubber bushing, Jilin University, 1(2012).

Pei Wei-chi, Zhang Wen-ming, Qiao Chang-sheng, The Optimization of micro tourism electric vehicle ride comfort based on ADAMS/Car Ride, Science Technology and Engineering, **14**(6), 251–255(2014).

Sun Hai-yang. Research of evaluation for suspension K&C characters based on vehicle handling and stability, Jilin University, 3(2012).

Wen Qian, Yu Zhuo-ping, Zhang Li-jun, A testing study of static and dynamic rigidity characteristics of rubber bushings, Shanghai Auto, **8**, 7–9(1999).

Zhai Yuan, Zhang Lin-bo, Wu Shen-rong, Parameter identification of rubber bushing based on the axial and radial stiffness, Proceedings of the Hyperworks Technology Conference, 1–6(2009).

Civil, Architecture and Environmental Engineering – Kao & Sung (Eds)
© 2017 Taylor & Francis Group, ISBN 978-1-138-02985-9

Experimental and simulation research on the friction and wear properties of surface textured friction pairs in CST

Shaoni Sun & Liyang Xie
School of Mechanical Engineering and Automation, Northeastern University, Liaoning, Shenyang, China

Risheng Long
College of Mechanical Engineering, Taiyuan University of Technology, Taiyuan, Shanxi, China

ABSTRACT: In order to realize the reliable soft starting and intelligent running of heavy duty scraper conveyors, the Controlled Start Transmission (CST) equipment is introduced to provide a feasible way to realize the soft starting, overload protection, and multi-driver power balance of scraper conveyors. The Hydro-viscous Drive Unit (HDU), which consists of multiple sets of friction pairs, is the core component of CST. The wear resistance and reliability of friction pairs are the decisive factors for the performance. The experimental models, with non-smooth surface as well as smooth surface having regular pit distribution, were established based on the typical sample size of the friction tester. The stress and wear resistant performance of the models with different non-smooth morphology parameters were studied through LS-DYNA analysis and wear tests. The results indicate that the wear resistant performance of biomimetic non-smooth samples is better than that of smooth ones; the pits can improve the contact stress distribution in the contact area of friction pairs, and the Von Mises stresses of biomimetic non-smooth samples are less than those of smooth samples; the wear resistant performance is the best when the pit distance equals 1 mm and the pit diameter equals 0.8 mm. This study would provide a basis for the future optimal design of the HDU of CST.

1 INTRODUCTION

As an effective method for soft starting, overload protection as well as multi-driver power balance, Controlled Start Transmission (CST) equipment has been increasingly applied to heavy duty scraper conveyors. The largest type of heavy duty scraper conveyor has reached to 3 × 1600 kW until now. The Hydro-viscous Drive unit (HDU) is the core component of CST. It consists of multiple sets of friction pairs, which are the decisive factors for the performance and reliability of the equipment. The bionic non-smooth surface morphologies provide a feasible way to improve the wear resistant performance of the friction pair.

The surface characteristics, such as reducing adhesion and wear resistance, have a great relationship with their non-smooth surface morphologies. The typical non-smooth surface morphologies, including synapses, pits and scales etc., are mainly from natural creatures. Some typical bionic microscopic non-smooth surface morphologies are shown in Figure 1. These bionic surface misconstructions have been successfully applied to

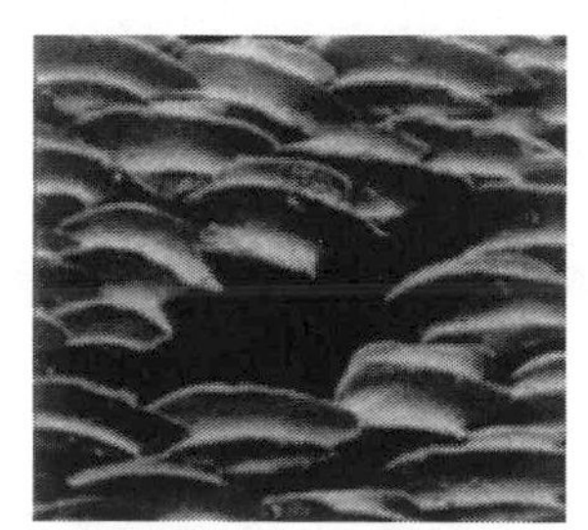
(a) White shark epidermis surface

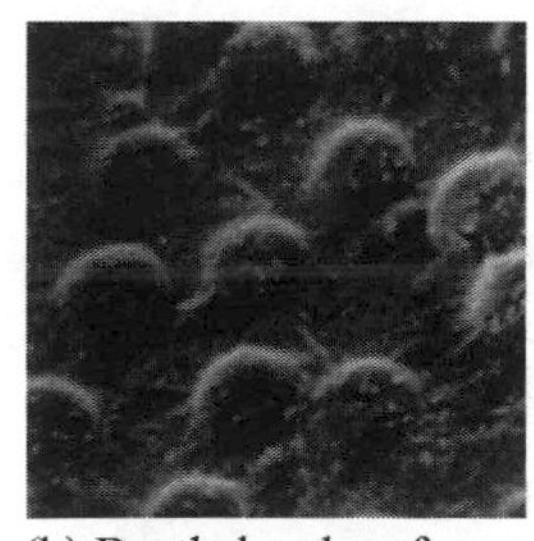
(b) Beetle head surface

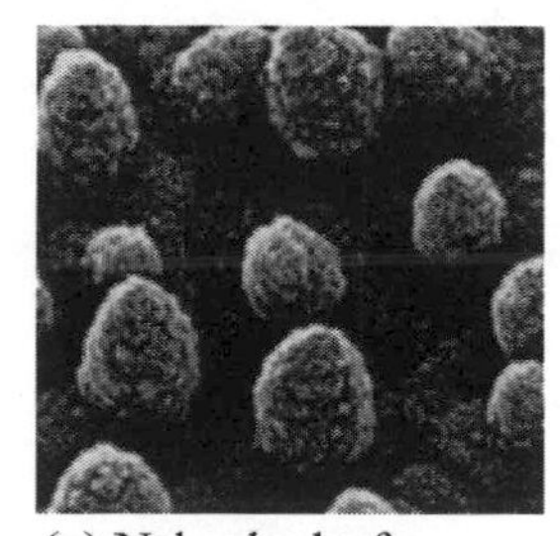
(c) Nelumbo leaf

Figure 1. Some SEM photos of the typical bionic non-smooth surface morphology.

Figure 2. Laser processed samples with regularly distributed pits.

the following fields: surface resistance enhancing of airplanes, stick reducing of the plough surface and wear performance improvement on the engine piston surface (Wang, 2001; Etsion, 2004; Etsion, 2004). At the same time, the wear-resisting performances of the biological surface morphology are becoming the research hotspot of tribology, and getting more and more attention (Wakuda M, 2003; Kligerman, 2001).

The recent studies show that the non-smooth technology can improve surface wear resistance significantly (Borghi, 2008; Song, 2006; Chen, 2008; Huang, 2008). L. Q. Ren et al. (2005) studied the wear resistances of different forms of bionic non-smooth surface morphology through micro-friction and wear tests. B. Y. Jia (2008) explored the freestyle abrasive wear performances of seven kinds of bionic non-smooth surface morphology by using a JMM wear-test machine. Yang (2005) studied the friction and wear behavior of W9Gr4V non-smooth samples with different pit-diameter and pit-distance concaves using a pin-on-disk wear tester.

In order to ensure the reliable starting and running performances of the heavy-duty scraper conveyor, it is necessary to further improve the wear resistance of friction pairs used in CST. The non-smooth surface morphologies are the future research direction of high performance friction pairs. Using laser cutting and drilling methods, nine kinds of samples are fabricated with different morphology parameters (see Figure 2). The wear resistance tests of different samples are conducted on a universal pin-on-disk wear tester. Based on the Finite Element Analysis (FEA) theories and the LS-DYNA module of ANSYS (Jovicic, 2009; Rusinski, 2013), the stress behaviors of pin-on-disk models during the wear testing process were studied. The results would provide a basis for the future improvement and optimization of CST.

2 WEAR EXPERIMENTS OF SAMPLES WITH NON-SMOOTH AND SMOOTH SURFACES

According to the actual installation requirement of the pin-on-disk wear test machine, the samples were machined to a standard disk shape (see Figure 2). Through laser cutting and drilling methods, the upper surfaces of samples were processed with regularly distributed pits of different pit diameters and pit distances. A MMW-1 universal friction testing machine was used to test the friction and wear resistance of these samples, while an L-200 photoelectric analytical balance was used to measure their wear loss. The test conditions are listed here: load F, equals 0.98 MPa; testing time T, equals 5 min; rotating speed R, equals 480 r/min.

2.1 *Wear loss analysis of non-smooth samples*

The wear losses of non-smooth samples with different pit diameter and pit distance were analyzed.

Table 1. Wear loss and kinetic friction coefficient of non-smooth and smooth samples.

	Smooth	pad1	pad2	pad3	pad4	pad5	pad6	pad7	pad8	pad9
Pit distance/mm	0	1	1	1	1.25	1.25	1.25	1.5	1.5	1.5
Pit diameter/mm	0	0.4	0.6	0.8	0.4	0.6	0.8	0.4	0.6	0.8
Wear loss/mg	6.8	5.2	5.3	4.4	5.8	5.1	4.5	6.1	5.7	5.2
Kinetic friction coefficient	0.391	0.382	0.373	0.368	0.39	0.377	0.371	0.396	0.385	0.38

The results are listed in Table 1. From the wear loss data, it can be easily seen that the surface morphology parameters have great influence on the wear resistance of non-smooth samples. Theoretically, under the same test conditions, the wear resistance of the sample has a close relationship with its wear loss. If the mass loss of one sample is more than others, it means that its wear resistant performance is poorer than others.

Similarly, as shown in Table 1, it is evident that the wear loss of pad 3 is the minimum. So, under the experiment conditions, pad 3 possesses the best wear resistant performance. Its pit distance equals 1 mm and its pit diameter equals 0.8 mm. On the contrary, pad 7 possesses the worst wear resistant performance. Its pit distance equals 1.5 mm and its pit diameter equals 0.4 mm.

The influence of pit distance on wear loss is shown in Figure 3. When the pit diameter is fixed, it can be found that the wear loss of the sample increases gradually when the pit distance becomes larger. When the pit distance equals 1.5 mm, the wear losses of pad 7, pad 8 and pad 9 are all beyond 5 mg. Therefore, it can be concluded that the wear resistances of non-smooth samples take on opposite variation with the increase in pit distance.

The influence of pit diameter on wear loss is shown in Figure 4. From Table 1 and Figure 4, when the pit diameter equals 0.8 mm, it can be easily seen that the wear loss is significantly reduced. When the pit distance is constant, the larger the pit diameter is, the smaller the value of the wear loss is. Therefore, under the same experiment conditions, the wear resistant performance of the sample is enhanced when the pit diameter becomes larger. Instead, when the pit diameter gets smaller, the wear resistance of the sample will be accordingly reduced.

In addition, to compare the wear resistance of the non-smooth sample with that of the smooth sample, the wear test of smooth samples were performed under the same conditions. The average wear loss of smooth surface samples is 6.8 mg, as shown in Table 1.

It is assumed that the increment ratio of wear resistance is Q,

$$Q = (K_2 - K_1)/K_2 \tag{1}$$

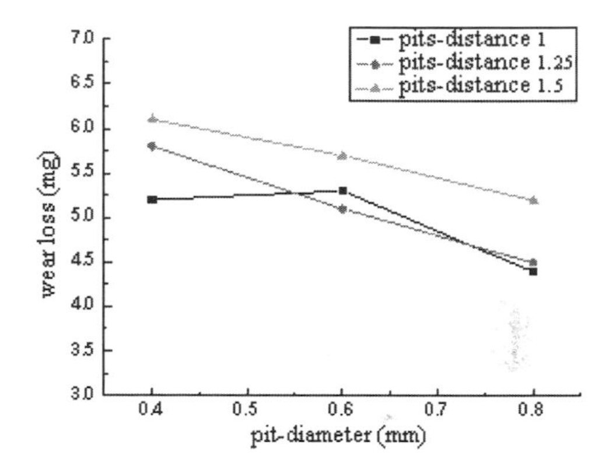

Figure 3. Wear loss versus pit distance.

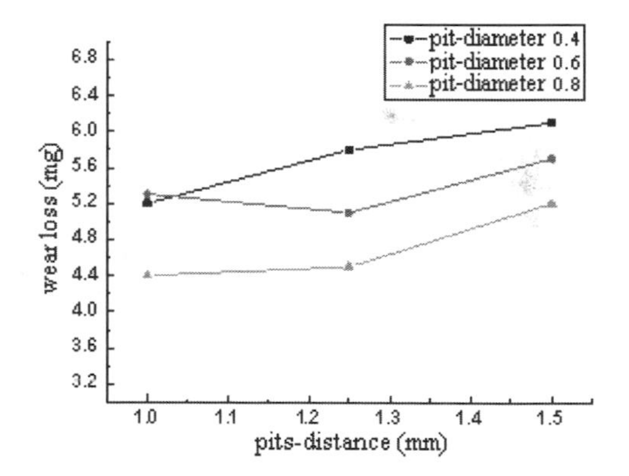

Figure 4. Wear loss versus pit diameter.

Wherein, K_1 means the average wear loss of non-smooth samples; K_2 means the average wear loss of smooth samples. K_1 and K_2 are gained under the same test conditions mentioned before. It can be found that the average wear loss of smooth samples is evidently bigger than the average mass loss of non-smooth samples. That is, the wear resistance of the non-smooth sample with regular pit distribution is obviously higher than that of the smooth sample. Obtaining K_1 and K_2 from Table 1, the wear resistance increment ratio Q of pad3 equals 35.29%.

2.2 *Kinetic friction coefficient analysis*

The kinetic friction coefficient is an important parameter which cannot be ignored during the process of friction and wear test. The kinetic friction coefficients of the non-smooth sample with different surface morphology parameters and the average friction coefficient of smooth samples are both listed in Table 1.

As shown in Table 1, the average friction coefficient of the smooth samples equals 0.391. Almost all the kinetic friction coefficients of non-smooth samples are less than the average friction coefficient of the smooth samples. The decrease of the friction coefficient does not affect the HDU driving performance. Inversely, it is helpful to improve the wear resistance of the friction pair to some degree. The friction coefficient of pad 3 is the minimum, which equals 0.368. In general, the friction coefficients of samples with different surface morphology parameters have the same change tendencies with their wear losses.

3 FEA OF SAMPLES WITH BIONIC NON-SMOOTH MORPHOLOGY

The 3D pin-on-disk model was constructed in Pro/E, and the pit diameter and pit distance could be adjusted according to the actual morphology parameters of different samples. The model with a non-smooth surface is shown in Figure 5(a). The meshed Finite Element (FE) model is shown in Figure 5(b). The other parameters are the same with the test samples.

According to actual experimental conditions of the pin-on-disk friction and wear tester, the surface-surface contact type and automatic contact algorithm was used in the following simulations. During the FE analysis process, the upper surface of the disk sample was chosen as the target surface and the lower surface of the coupling pin was chosen as the contact surface. The friction coefficient of the contact pair was set to 0.38 (Ren, 2005; Jia, 2008; Yang, 2005; Jovicic, 2009; Rusinski, 2013). All the constraints of the sample and pin are imposed according to actual work conditions. The upper non-smooth surface of the disk sample was applied with surface loads and a rotating speed.

The Von Mises stress curves of all non-smooth samples are shown in Figure 6. Wherein, Figure 6 (a)-(i) respectively corresponds to the Von Mises stress curves of nine kinds of non-smooth samples. In each sub-picture, the blue curve is the Von Mises stress of the current non-smooth sample, while the red curve is the Von Mises stress of the smooth sample. The average Von Mises stress of each non-smooth sample and the smooth sample are also embodied in every sub-picture.

As shown in Figure 6 (g), the Von Mises stress of pad 7 is slightly higher than the stress of the smooth sample, while the average Von Mises stress of the other non-smooth samples was less than the average of the smooth sample. Furthermore, the volatility of Von Mises stress of non-smooth samples is also significantly less than that of the smooth sample. From Figure 6(c), it is easy to find that the average Von Mises stress of pad 3 is significantly less than the average stress of the smooth sample. The volatility of Von Mises stress and its stress value was both significantly lower than other non-smooth samples too.

In addition, the Von Mises stresses of non-smooth samples have a certain relationship with the key parameters including pit diameter and pit distance. Overall, when the pit distance is fixed, the Von Mises stress of non-smooth samples will become small with the increase of pit diameter. For example, as shown in Figure 6 (a), (b), and (c), the pit distance equals 1 mm, the Von Mises stress of pads 1–3 decreased when the pit diameter varied from 0.4 mm to 0.8 mm. Similarly, Figure 6(d)-(f) and Figure 6 (g)-(i) show the same stress variation trends.

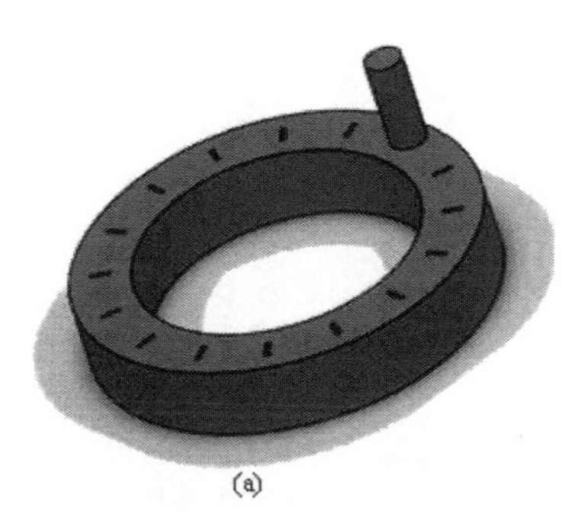
(a)

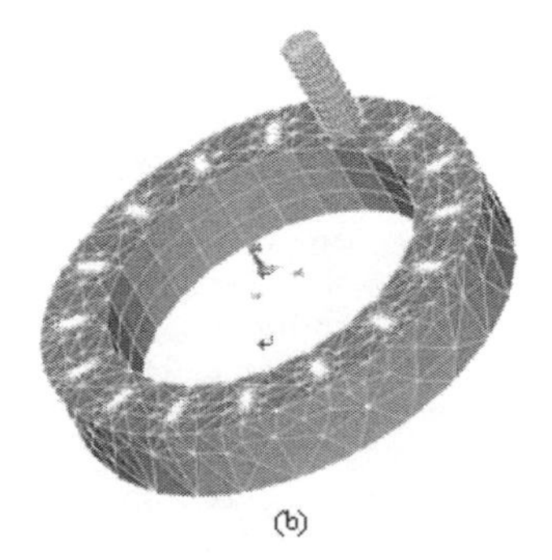
(b)

Figure 5. 3D model of a sample with non-smooth morphology (a) 3D model; (b) meshed FE model.

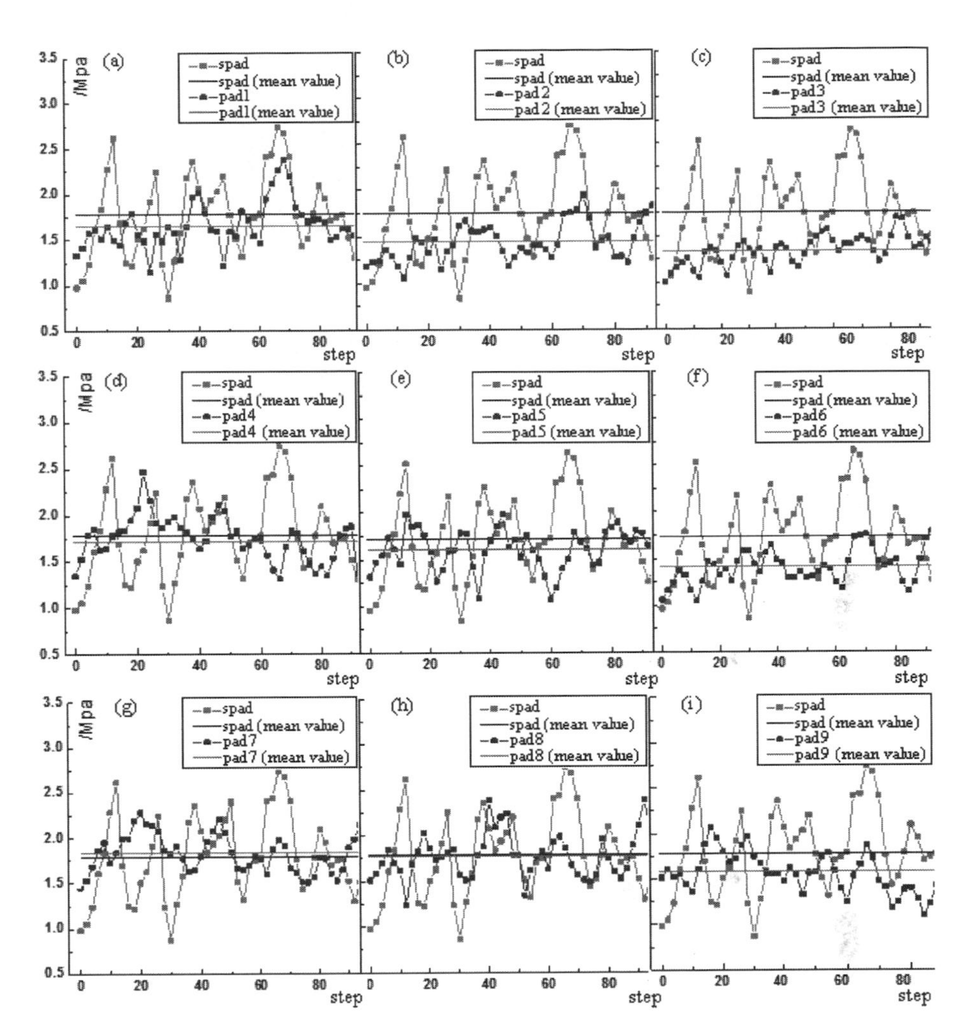

Figure 6. Von Mises stress contrast curve of all non-smooth disk samples (a) pad1; (b) pad2; (c) pad3; (d) pad4; (e) pad5; (f) pad6; (g) pad7; (h) pad8; (i)pad9.

4 CONCLUSIONS

The bionic non-smooth surface morphology is a feasible way to improve the wear resistant performance of friction pairs used in CST. Using laser cutting and drilling methods, nine kinds of samples are fabricated with different surface morphology parameters. The wear performance experiments of different samples are also conducted on a pin-on-disk wear tester. Based on the FEA theories and the LS-DYNA module of ANSYS, the stress behavior of the pin-on-disk model during wear test processing was also studied.

The experimental and FEA results are listed below:

1. The pit distance and pit diameter have a great influence on the wear loss of non-smooth samples. The smaller the pit distance is, the smaller the value of wear loss is; the larger the pit diameter is, the smaller the value of wear loss is.
2. When the pit spacing equals 1 mm, and the diameter equals 0.8 mm, the non-smooth sample has the best wear resistant performance.
3. Through the comparison and analysis between the non-smooth and smooth samples, it can be concluded that the wear resistance of the non-smooth sample is obviously better than that of the smooth sample.
4. The Von Mises stresses of non-smooth samples are less than those of smooth samples. The non-smooth surfaces with regular pit distribution can reduce the stress concentration in the contact area, and the non-smooth samples have better wear resistant performance than smooth samples.

The above results will provide a basis for the future product improvement and optimization of the HDU of CST.

REFERENCES

Borghi, A., E. Gualtieri, D. Marchetto, et al. Wear, **265**, 1046–1051(2008).

Chen, L., H. Zhou, Y. Zhao, et al. Journal of Mechanical Engineering, **44**, 173–176(2008).

Etsion, I., G. Halperin, V. Brizmer, Y. Kligerman. Tribology Letters, **2**, 295–300(2004).

Etsion. I. Tribology Letters, **17**, 733–737(2004).

Huang, J.M., Ch. H. Gao, X. Sh. Tang, et al. Journal of Mechanical Engineering, **44**, 145–151(2008).

Jia. B.Y. Master diss., Jilin University, (2008).

Jovicic, G., Zivkovic, M., Jovicic, N. Strojniski Vestnik-Journal of Mechanical Engineering, **55**, 549–554(2009).

Kligerman, Y., I. Etsion. Tribology Transactions, **44**, 472–478(2001).

Ren, L.Q., Zh. J. Yang, Zh. W. Han. Transactions of the Chinese Society for Agricultural Machinery, **36**, 144–147(2005).

Rusinski, E., Moczko, P., Pietrusiak, D., et al. Strojniski Vestnik-Journal of Mechanical Engineering, **59**, 556–563(2013).

Song, Q.F., H. Zhou, Y. Li, et al. Tribology, **26**, 24–27(2006).

Wakuda M, Yamauchi Y, Kanzaki S, et al. Wear, **254**, 356–363(2003).

Wang, X.L., K. Kato, K. Adachi, et al. Tribology International, **34**, 703–711(2001).

Zh. J. Yang, Zh. W. Han, L.Q. Ren. Tribology, **25**, 374–378(2005).

Civil, Architecture and Environmental Engineering – Kao & Sung (Eds)
© 2017 Taylor & Francis Group, ISBN 978-1-138-02985-9

Simulation and analysis of MPPT used in photovoltaic grid-connected systems based on the improved disturb-and-observe method

Junfeng Wang & Hairu Zhao
School of Physics and Electronic Engineering of Yuxi Normal University, Yuxi, China

ABSTRACT: Based on the analysis of the output characteristics of photovoltaic cells and the theory of the disturb-and-observe method, a new improved disturb-and-observe method is proposed in this paper. This method first sets a time threshold, and then establishes the simulation model using MATLAB/Simulink. Simulation results indicate that the model has better dynamic performance and allows fast, accurate tracking of the maximum power point, which validates the proposed method to be feasible and reliable.

1 INTRODUCTION

In the solar photovoltaic power generation system, the output characteristics of photovoltaic power generation is nonlinear due to the effects of many conditions from outside world such as light intensity, temperature and load characteristic and their own factors (Lu, 2011). The main problem of photovoltaic power generation is that the output characteristics of photovoltaic cells are greatly affected by the external environment. The change of temperature and light intensity will lead to great changes to the output characteristics of photovoltaic power generation. Therefore, making full use of the energy produced by photovoltaic cells is the basic requirement of photovoltaic power generation systems (Zhao, 2004). In order to solve this problem, Maximum Power Point Tracking (MPPT) devices are required to be added between the photovoltaic power generation system and load, which makes photovoltaic cells' output maximum power so that the solar energy can be fully used (Min, 2006).

In order to obtain the maximum power output unceasingly under any conditions, maximum power point tracking control is needed for photovoltaic cells. MPPT is a self-optimizing process essentially. That is to say, photovoltaic cells can output maximum power in the environment of a variety of light intensities and temperatures by controlling terminal voltage or physical quantities (Jiang, 2013). The research of maximum power point tracking is a relatively mature technology. The common methods include the incremental conductance algorithm, three-point comparing algorithm, quadratic interpolation method, disturb-and-observe method and so on. An improved disturb-and-observe method is proposed in this paper, which sets a time threshold.

2 EQUIVALENT CIRCUIT AND OUTPUT CHARACTERISTICS OF PHOTOVOLTAIC CELLS

2.1 *Equivalent circuit of photovoltaic cells*

The common ideal equivalent circuit of photovoltaic cells is shown in Fig. 1. *VD* represents an ideal diode, namely the semiconductor PN junction. One part of photo-electromotive force offsets the electric field within the PN junction and recharges the diode. The other part of photo-electromotive force connects to external circuit through the internal resistance in series and in parallel so that the terminal voltage of load R_L is U.

According to Kirchhoff's current theorem, the volt-ampere relation of photovoltaic cells can be expressed as Eq. 1.

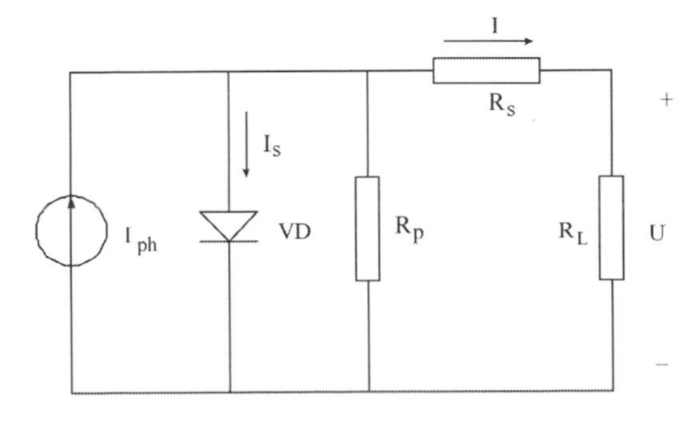

Figure 1. Equivalent circuit of photovoltaic cells.

$$I = I_{ph} - I_s[e^{\frac{q(U+IR_s)}{AKT}} - 1] - \frac{U + IR_s}{R_p}. \qquad (1)$$

In Eq. 1, I is the output current of photovoltaic cells and U is the output voltage. I_{ph} is the current of the photoproduction current source and Is is the saturation current of the diode. q is the electronic charge and its unit is 1.6×10^{-19}C. K is Boltzmann's constant and its unit is 1.38×10^{-23}J/K. T is the absolute temperature and A is the ideal factor of the PN junction. R_p is the parallel resistance of photovoltaic cells and R_s is the series resistance.

2.2 *Output characteristics of photovoltaic cells*

According to Eq. 1, we know that the volt-ampere relation of photovoltaic cells is a transcendental exponential equation and it can't be expressed by linear equation. The output characteristics of photovoltaic cells are nonlinear functions and affected by the environmental temperature and light intensity. Based on short-circuit current I_{sc}, open-circuit voltage U_{oc}, the current of maximum power point I_m and the voltage of maximum power point U_m provided by solar battery manufacturers, the output characteristic curves of photovoltaic cells under different light conditions are shown in Figure 2 (Chen, 2013).

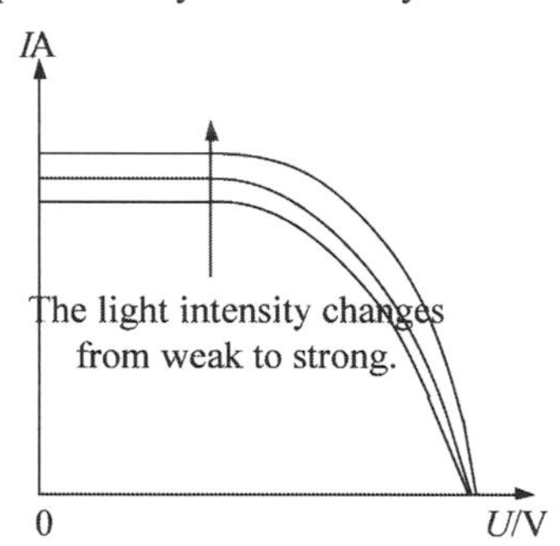

(a) *U-I* curves under different light intensities

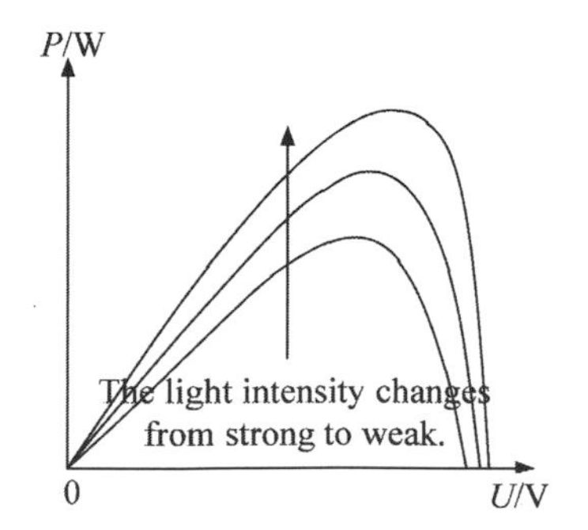

(b) *P-U* curves under different light intensities

Figure 2. Output characteristic curves of photovoltaic cells under different light conditions.

As is shown in Figure 2, in the case of other conditions unchanged, the output power of photovoltaic cells will increase as light intensity enhances (Lu, 2011). There is a working point with maximum power output at any time under a certain light intensity, namely the maximum power point. The maximum power output of photovoltaic cells will change when the external environment changes. In order to obtain maximum power output of photovoltaic cells, the Maximum Power Point Tracking (MPPT) device is required to be added between photovoltaic cells and load (Liu, 2010).

3 IMPROVED DISTURB-AND-OBSERVE METHOD

3.1 *Principle of the disturb-and-observe method*

The disturb-and-observe method is also called the hill climbing method. Since its principle is easy and hardware implementation is economic, it has been the most common method. The principle of the disturb-and-observe method is shown in Figure 3. Its principle is described as follows. First, initiatively apply a small voltage disturbance to a certain photovoltaic operating voltage. Then, compare the output power after disturbance with the output power before disturbance. If the output power after disturbance increases, it states that the disturbance can enhance the output power and the disturbance in the same direction should be applied to the output voltage of photovoltaic cells next time. On the contrary, if the output power of the photovoltaic array decreases, the disturbance in the reverse direction should be applied (Liu, 2011). If $P/(dU) > 0$, then apply on the left side of the maximum power point. If $P/(dU) = 0$,

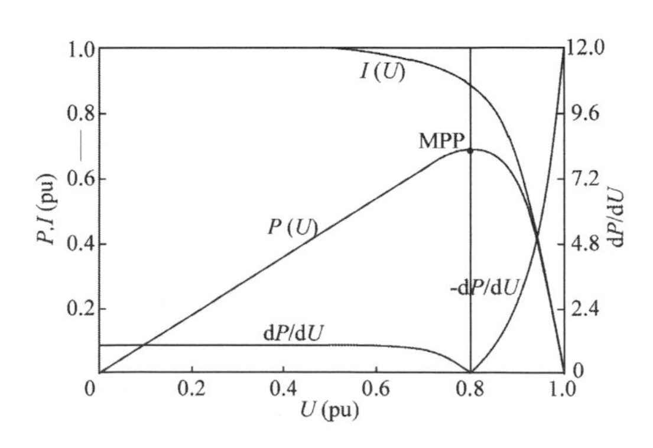

Figure 3. Curves of *I-U*, *P-U* and |*dP/dU*|–*U* after calibrating the photovoltaic array.

then apply at the maximum power point. If $P/(dU) < 0$, then apply on the right side of the maximum power point.

The voltage disturbs the output voltage of photovoltaic cells which can be expressed as Eq. 2.

$$U_{ref} = U_{ref} + \alpha\frac{dP}{dU} = U_{ref} + \alpha\frac{P(k)-P(k-1)}{U(k)-U(k-1)} \quad (2)$$

In Eq. 2, α is a positive number. It represents the variable step size speed factor and is used to adjust the tracking speed. When the operating point of the photovoltaic array is far away from the maximum power point, the tracking step size is big. Otherwise, the tracking step size is small. When the operating point of the photovoltaic array is close to the maximum power point, the tracking step size approaches zero (Huang, 2011). α can be determined by Eq. 3.

$$\alpha \leq \frac{U_{step-\max}}{\left|dP/dU\right|_{\max}}. \quad (3)$$

In Eq. 3, $U_{step\text{-}max}$ is the maximum step size allowed by the disturb-and-observe method with fixed step size. $|dP/dU|_{max}$ can be calculated based on the photovoltaic array, and can also be estimated according to Eq. 4.

$$\left|\frac{dP}{dU}\right| \approx \left|\frac{P|_{U_{ref}=mU_{oc}} - P|_{U_{ref}=U_{oc}}}{mU_{oc} - U_{oc}}\right| = \left|\frac{mI|_{U_{ref}=mU_{oc}}}{m-1}\right| \quad (4)$$

In Eq. 4, m is a positive number approaching one and Uoc is the open-circuit voltage of the photovoltaic array. The range of variable step size speed factor α is first calculated by Eq. 3, and then the final value is determined by experiments (Wang, 2014).

3.2 *Process of the improved disturb-and-observe method*

The flow chart of improved MPPT is shown in Fig. 4. First, set a time threshold Δt, and the time threshold is about the time of the output power of the photovoltaic electrical source from zero to the maximum power point. Track the output power using normal disturbance within the time threshold so that the output power can be up to the maximum. Second, set a variation threshold of output power ε. When the output power reaches the maximum power point (namely $|P(k)-P(k-1)| \leq \varepsilon$), stop the disturbance to make the output voltage remain the same. That is to say, the output voltage lies near the maximum power point.

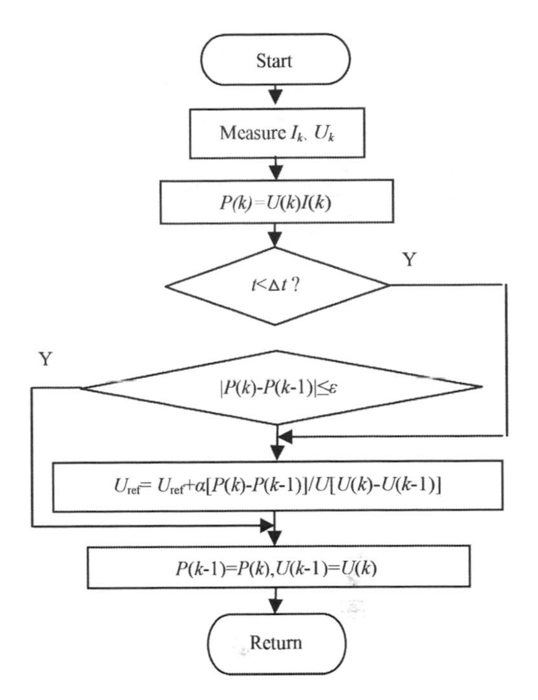

Figure 4. Flow chart of the improved disturb-and-observe method with variable step size.

If $U(k)-U(k-1) \neq 0$, automatically adjust the disturbance step size based on Eq. 4. Otherwise, end it and return.

4 MODEL AND SIMULATION OF THE PHOTOVOLTAIC GRID-CONNECTED SYSTEM

4.1 *Simulation model*

According to the above design, the simulation model of the overall system is set up in the simulation environment of Matlab/Simulink, and is shown in Figure 5. The model is composed of photovoltaic cells (PV modules), a controller module, a pulse-width modulation module and the Boost DC-DC circuit. In this model, the values of parameters are set as follows: load R is 30Ω; inductance L is 6×10^{-4}H; $C1$ is 1.3×10^{-6}F; $C2$ is 4.7×10^{-6}F. Two constant input modules are used as the parameter T of photovoltaic cells and the parameter S of light intensity.

4.2 *Results and simulation analysis*

According to the simulation results of Figure 6, it can be known that each parameter of the photovoltaic output is stable after 0.1 second since there is a disturbance. Analysis of the results under the

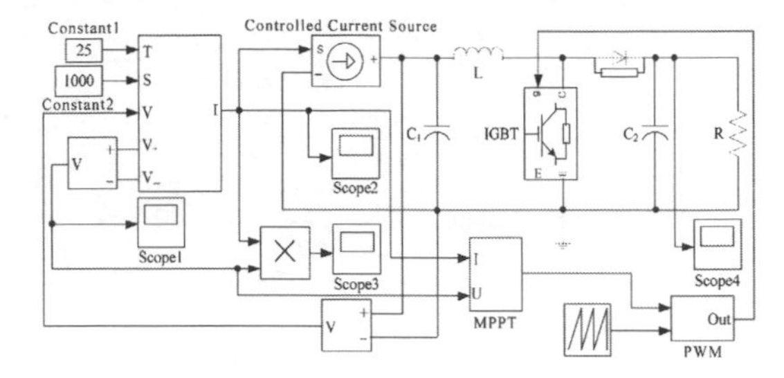

Figure 5. Simulation model of the overall system.

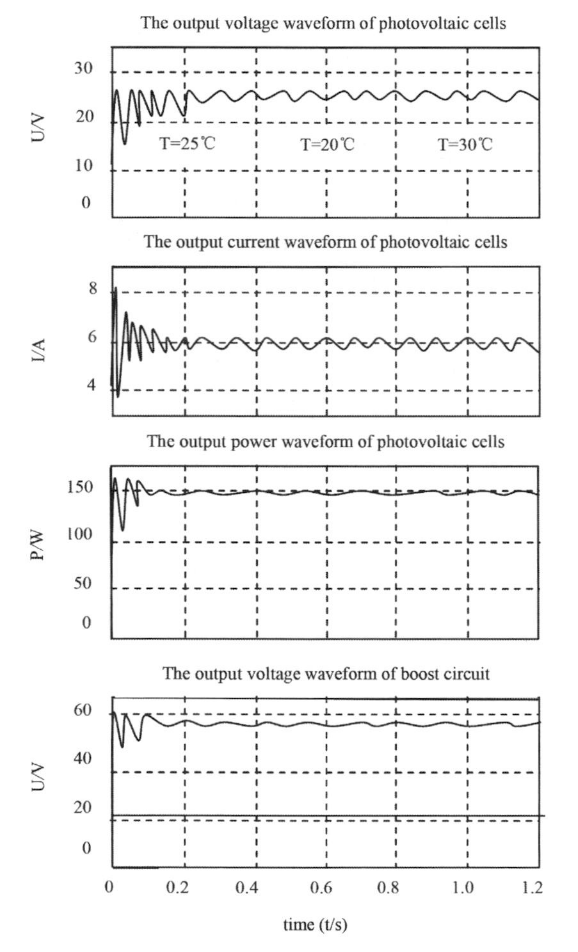

Figure 6. Simulation waveforms under the same light intensity and variable temperatures.

same light intensity and variable temperatures is shown in Table 1.

It can be stated based on Table 1 that when the temperature changes, the output voltage, current, power of photovoltaic cells and the output voltage of boost circuit only have little fluctuation and always keep the same value. That is to say, the change of temperature has little influence on the output parameters of photovoltaic cells. Photovoltaic cells can fast track the maximum power point when temperature changes, and the adjustment time is short.

It can be known based on the simulation results of Figure 7 that when the temperature changes,

Table 1. Analysis of the results under the same light intensity and variable temperatures.

T(°C)	Output voltage (V)	Output current (A)	Output power (W)	Output voltage of boost circuit	Analysis of results
25	24~25.5	5.8~6.08	140~155	53~57	Change in temperature has a little influence on the output parameters
20	24~25.4	5.8~6.07	140~154	53~56.9	
30	24~25.6	5.8~6.09	140~156	53~57.1	

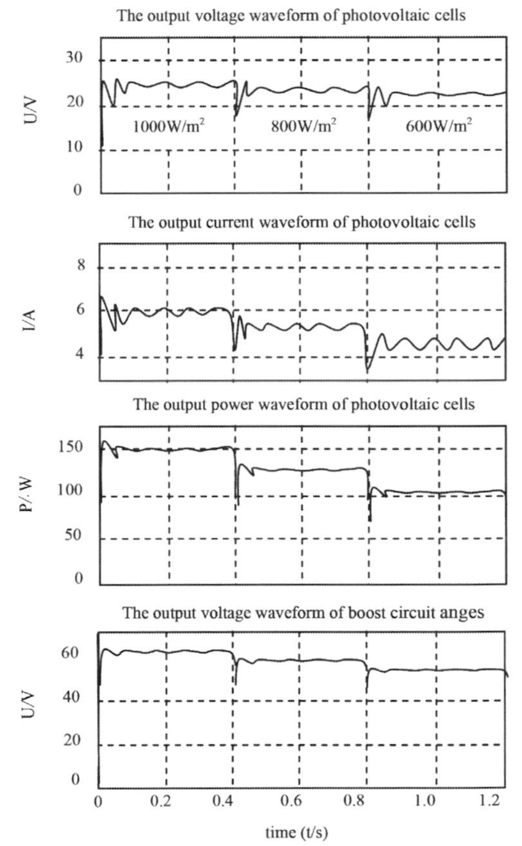

Figure 7. Simulation waveforms under the same temperature and variable light intensities.

Table 2. Analysis of the results under the same temperature and variable light intensities.

Light intensity (W/m^2)	Output voltage (V)	Output current (A)	Output power (W)	Output voltage of boost circuit	Analysis of results
1000	24~25.5	5.8~6.08	140~155	61~63	Each parameter of photovoltaic output will change when the light intensity changes
800	23.5~24	5.4~5.6	127~135	59~61	
600	23~23.5	5.0~5.2	115~122	57~59	

each parameter of the photovoltaic output can be stable after 0.05 second. Analysis of the results under the same temperature and variable light intensities is shown in Table 2.

It can be stated according to Table 2 that the output voltage, current, power of photovoltaic cells and the output voltage of boost circuit all reduce when the light intensity declines (Wu, 2014). Each parameter of the photovoltaic output will change when the light intensity changes. The photovoltaic cells can track the change of external environment quickly, and also can quickly realize the maximum power point tracking and the steady accuracy is high.

The above simulation results show that the temperature has little influence on the output characteristics of photovoltaic cells, while the light intensity has great influence on the output characteristics of photovoltaic cells. This is consistent with the characteristics analysis of photovoltaic cells. The improved disturb-and-observe method prose in this paper can fast track the maximum power point of photovoltaic cells, and the tracking accuracy is high, which verifies that the proposed method is feasible.

5 CONCLUSION

An improved disturb-and-observe method was proposed in this paper, and the simulation model of the photovoltaic grid-connected system was built in Matlab/Simulink. Simulation results indicate that the model can fast track the maximum power point and has a higher tracking accuracy and a better dynamic performance than the incremental conductance algorithm, three-point comparing algorithm, quadratic interpolation method and traditional disturb-and-observe method, which proves that the proposed method is feasible and reliable.

REFERENCES

Chen Huanglin, MPPT algorithm research based on mathematical model of PV module, Mechanical & Electrical Technology. 5 (2013) 27–29.

Huang Shuyu, Mou Longhua, Shi Lin, Adaptive variable step size MPPT algorithm, Proceedings of the CSU EPSA. 5 (2011) 26–30.

Jiang Zheng, Simulation of three-phase photovoltaic grid-connected system based on matlab/simulink, Power System and Clean Energy. 10 (2013) 59–65.

Liu Li, Zhang Yanmin, Application of perturbation and observation method in MPPT of photovoltaic power generation, Chinese Journal of Power Sources. 34 (2010) 186–189.

Liu Lihong, Chen Qizheng, MPPT simulation research for PV module based on boost circuit, Electrical Engineering. 5 (2011) 26–30.

Lu Xiao, Qin Lijun, Application and simulation of adaptive perturbation and observation method in PV MPPT, Modem Electric Powar. 28 (2011) 80–84.

Min Jiangwei, Duan shanxu, et al, Photovoltaic maximum power point tracking system based on push-pull converter, Telecom Power Technologies. 2 (2006) 7–9.

Wang Liqiao, Sun Xiaofeng, Photovoltaic power generation technology in the distributed generation system, Mechanical Industry Press, Beijing, 2014.

Wu Yuwei, Shibin, et al, Photovoltaic array maximum power point tracking based on improved variable step P&O, Electrical Engineering. 6 (2014) 23–25.

Zhao Hong, Pan Junmin, Photovoltaic maximum power point tracking system using boost converter, Telecom Power Technologies. 3 (2004) 55–57.

Civil, Architecture and Environmental Engineering – Kao & Sung (Eds)
© 2017 Taylor & Francis Group, ISBN 978-1-138-02985-9

Automatic control loop optimization method of thermal power plant based on internal model control

Song Gao, Xiangkun Pang & Jun Li
State Grid Shandong Electric Power Research Institute, Jinan, Shandong, China

Changwei Li & Huadong Li
Shandong Zhongshiyitong Co. Ltd., Jinan, Shandong, China

ABSTRACT: Automatic control loop is one of the important subsystems in the automatic control technology of thermal power plant. It is difficult to achieve satisfactory control effect when setting PID controller parameters in actual production process. This paper uses typical three impulse control of boiler water level as an example, introduces the closed-loop steady state time, gives the optimization strategy of PID controller parameters based on internal model control, and determines the design of feed-forward controller. Finally, the effectiveness and practicability of the control strategy are verified by the application on a typical 300MW unit.

1 INTRODUCTION

Automatic control loop is one of the important subsystems in the automatic control technology of thermal power plant. PID controller is used in more than 90% industrial process control loops, and therefore the PID controller parameter is one of the important factors that affect the quality of the control loop. PID regulator parameter optimization is generally determined by trial and error, and it is difficult to reach the satisfied result. This paper used the typical three impulse water control loop as an example, introduced an optimization method of PID controller parameters based on internal model control strategy and the concept of closed-loop steady state time.

The paper is organized as follows. In Section 2, the typical three impulse water control loop is introduced and some definitions are given. In Section 3, the control model based on internal model control is presented and Section 2 presents the automatic control strategy for three impulse water control loop. In Section 4, the application result is given to show the efficiency of the strategy. Finally, we conclude our paper in section 5.

2 THREE IMPULSE FEED-WATER CONTROL LOOP

Three impulse feed-water control loop is typical cascade control loop, and the control target is to maintain the constant value of drum water level through adjusting the feed-water flow under different work conditions (Babykina, 2013; Rajkumar, 2013; Wu, 2012).

Three impulse water control loop is shown in Fig. 1, where, *SP* is the set-point value of drum water level; *PV* is the process value of drum water level; *OP* is the set-point value of feed-water flow; *PV1* is the process value of feed-water flow; *OP1* is the instruction of feed-water control valve.

The cascade control loop is divided into inner control loop and outer control loop. The inner water supply control loop is a single feedback control loop, and can be regarded as a fast follow-up system. The output value of the inner PID controller changes the water supply valve instruction, and then changes the feed water flow. The aim of the outer control loop is to maintain the drum water level as a constant value (always be considered as 0 mm), and the output value of the outer PID controller is the set-point value of the inner control loop (W, 2013; Huang, 2013; Chakraborty, 2014). Since the inner control loop is a fast follow-up system, the performance of the inner control loop

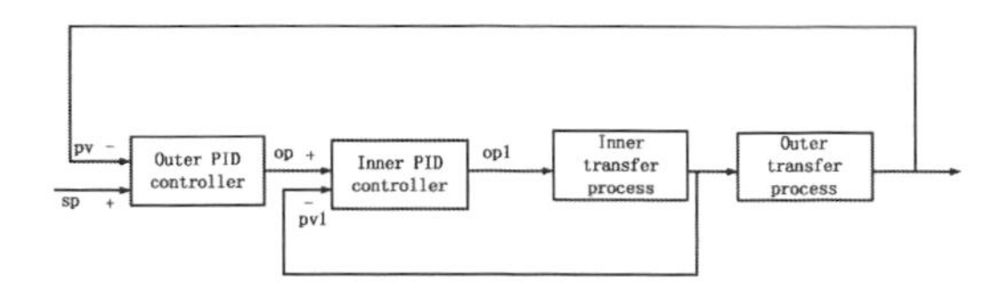

Figure 1. Three impulse feed-water control loop.

directly determines the performance of the outer control loop.

In the cascade loop, the closed-loop steady state time of the inner loop is about 1/4 of the steady state time of the outer loop. In this paper, the outer control loop are maintained unchanged, and the control structure and parameters of the inner loop are optimized only. The main purpose is to improve the performance of the inner control loop and reduce the steady-state time of the inner control loop.

3 AUTOMATIC CONTROL LOOP OPTIMIZATION METHOD

3.1 *Internal model control*

For the typical three impulse water supply control loop of thermal power plant, the first-order plus pure delay model is used to describe the characteristics of the transfer process of inner control loop. In fact, a large number of research results show that the first-order model is suitable for most industrial processes.

Internal Model Control (*IMC*) has advantages of high control accuracy, good robustness and anti-interference (Liu, 2013; Jahanshahi, 2014; Garrido, 2014; Jahanshahi, 2013). Therefore, it is widely used in the large-lag system in many industrial processes.

Consider a *SISO* feedback control loop depicted in Fig. 2. Here $P(s)$ and $C(s)$ are the process and the PID controller, respectively; $r(t)$, $u(t)$ and $y(t)$ are the set-point, the control signal, and the process output, respectively; $e(t)$ is the error between $r(t)$ and $y(t)$.

$P(s)$ is first-order plus delay time model,

$$P(s)=\frac{K}{T_1 s+1}e^{-\theta s} \tag{1}$$

Here, K is the gain of transfer function, T_1 is first-order lag time constant, and θ is the pure delay time.

$C(s)$ is the PID controller,

$$C(s)=K_p(1+\frac{1}{T_i s}+T_d s) \tag{2}$$

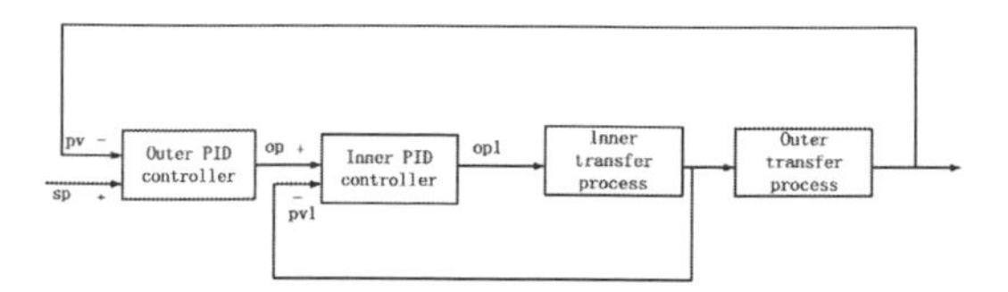

Figure 2. A SISO feedback control loop.

Here, K_p is the proportional gain of PID controller, T_i is the integration time of PID controller, and T_d is the derivative time of PID controller.

The PID controller setting for first-order plus pure delay time model as

$$K_p=\frac{T_1}{K(\tau_c=\theta)},T_i=T_1,T_d=0 \tag{3}$$

By taking the IMC tuning rule, the desired closed-loop response will be

$$G_{cl}(s)=\frac{1}{(\theta+\tau_c)s+e^{-\theta s}}e^{-\theta s}\approx\frac{1}{(\theta+\tau_c)s+1-\theta s}e^{-\theta s}=\frac{1}{\tau_c s+1}e^{-\theta s} \tag{4}$$

Here, the user-selected parameter τ_c stands for the desired close-loop time constant.

Based on the IMC tuning rule, the PID parameters can be determined by equation (3), with the desired closed-loop time constant τ_c is selected by users to adjust the performance of PID controller. The closed-loop time constant is shown in Fig. 3.

3.2 *Object model and parameter optimization*

In this section the method of optimization strategy is introduced based on a typical industrial case.

The corresponding relationship between the *OP-PV* (output value of PID controller to the process value of feed water flow) is established, which is the open-loop model of the control object. Data trends are shown in Figure 4, there, *PV* is process value of feed water flow, *SP* is set-point value of the feed water flow, and *OP* is the output value of

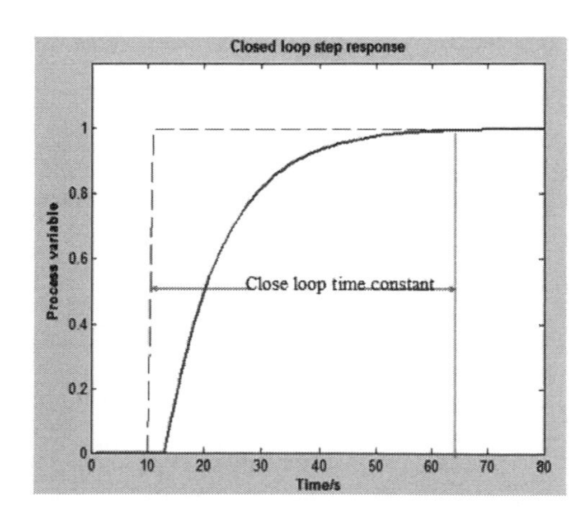

Figure 3. Closed-loop time constant τ_c.

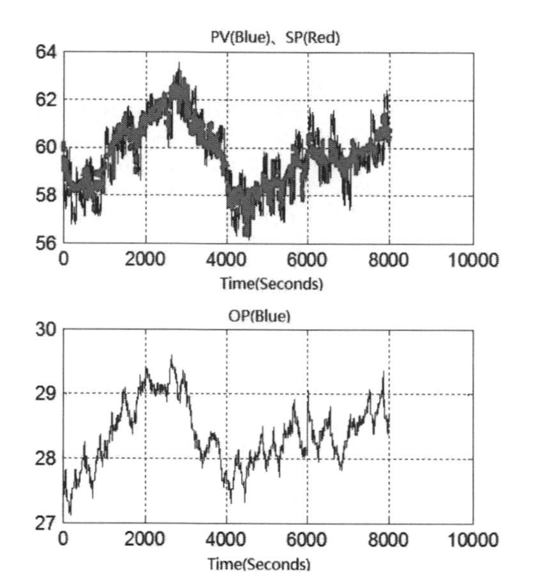

Figure 4. Data trends of *PV/SP/OP*.

PID controller, which then gives the instructions of feed-water control valve.

The data of Fig. 4 can be identified based on the least-square method, and the process transfer function of the open-loop model is

$$\frac{3.143}{89.12s+1}e^{-1s}$$

Based on the above process transfer function, *PV* and *OP* value can be estimated. The matching degree between estimated value and the actual value is 84.9%, so the open-loop model can be considered reliable to describe the characteristics of the control object.

Based on the IMC algorithm, the performance of the control loop can be evaluated. It is found that the performance is good, but the robustness is poor. That is, once the external disturbance occurs, the control performance will drop significantly. Hence, the original closed-loop steady time of 202 seconds is maintained, and parameters of PID controller is adjusted: K_p is updated from 0.12 to 0.55, T_i is updated from 100 seconds to 160 seconds.

3.3 *Feed-forward controller design*

After designing the parameters of the feedback controller, it is expected that the closed-loop steady state time can be further reduced, so the feed-forward controller is introduced. If the closed-loop steady state time reduced 50%, from 200 seconds to 100 seconds, we can design a feed-forward controller based on the model, $K_l = 0.156$, $K_2 = 0.394$, $T_f = 25.25$. The feed-forward controller is shown in Fig. 6.

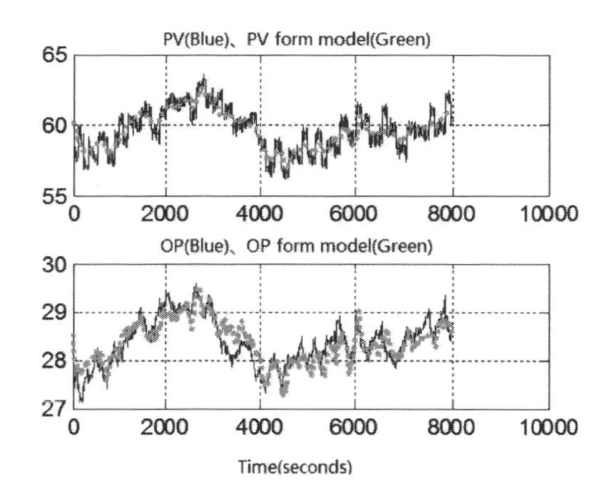

Figure 5. Matching degree between estimated value and the actual value.

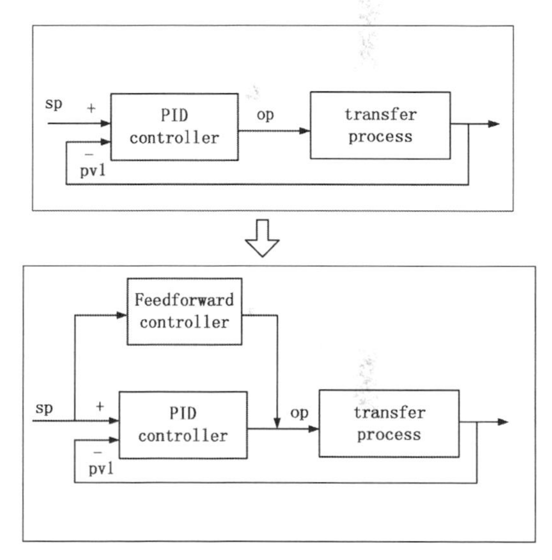

Figure 6. Feed-forward controller.

The feed-forward controller is

$$C_f(s) = K_1 + K_2\left(1 - \frac{1}{T_f s + 1}\right)$$

4 APPLICATION

In this section, the automatic control loop optimization method is used in an example of 300 MW

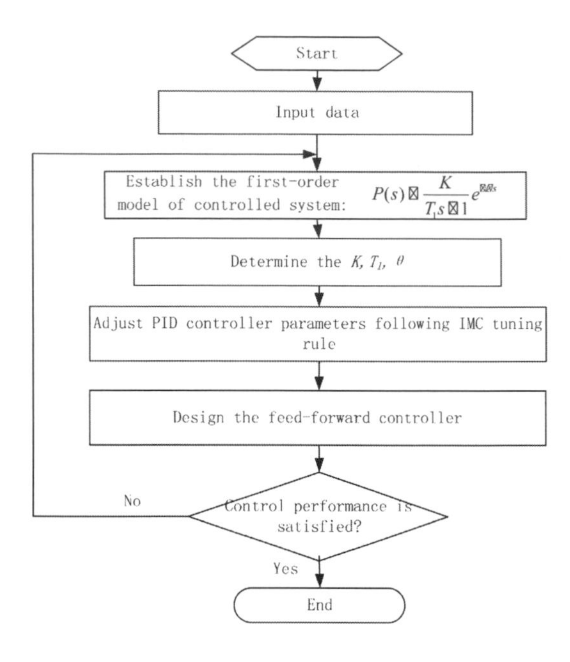

Figure 7. Control flow of optimization method.

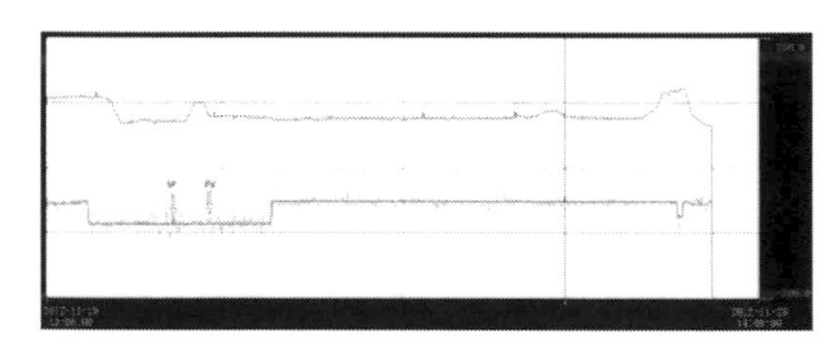

Figure 8. Data trend of feed-water control loop after optimization.

thermal power plant. The implementation process is completed by the following steps in Fig. 7.

The actual test results of this control strategy which is used in a typical 300 MW thermal power plant are shown in Fig. 8, where SP is the set-point value of drum water level, and PV is the process value of drum water level.

Before optimization the standard deviation of the drum water level is 8.7628, and the standard deviation of the drum water level is 4.8612 after optimization, which is 55% times better than the original.

5 CONCLUSION

This paper introduces the typical three impulse feed-water control loop of thermal power plant, and establishes optimization strategy of PID controller based on internal model control. Finally, the effectiveness and practicability of the control strategy are verified by the application on a typical 300 *MW* thermal power unit.

REFERENCES

Babykina G, Brinzei N, Aubry J F, et al. Modelling a feed-water control system of a steam generator in the framework of the dynamic reliability[C]//Annual Conference of the European Safety and Reliability Association, ESREL 2013: 3099–3107.

Chakraborty S K, Manna N, Dey S. Importance Of Three-Elements Boiler Drum Level Control And Its Installation In Power Plant [J]. International Journal of Instrumentation and Control Systems (IJICS), 2014, 4(2): 1–12.

Garrido J, Vázquez F, Morilla F. Inverted decoupling internal model control for square stable multivariable time delay systems [J]. Journal of Process Control, 2014, 24(11): 1710–1719.

Huang Z, Yan F. Improvement of Boiler Feed Water Control Strategy in Coal mine Owned Power Plant [J]. Coal Technology, 2013, 6: 031.

Jahanshahi E, De Oliveira V, Grimholt C, et al. A comparison between Internal Model Control, optimal PIDF and robust controllers for unstable flow in risers [J]. IFAC Proceedings Volumes, 2014, 47(3): 5752–5759.

Jahanshahi E, Skogestad S. Comparison between nonlinear model-based controllers and gain-scheduling Internal Model Control based on identified model[C]//52nd IEEE Conference on Decision and Control. IEEE, 2013: 853–860.

Liu G, Chen L, Zhao W, et al. Internal model control of permanent magnet synchronous motor using support vector machine generalized inverse [J]. IEEE Transactions on Industrial Informatics, 2013, 9(2): 890–898.

Pizhou W, Junjie G, Dafei Q, et al. Analysis of the Two Feed Water Control System of 600 MW Supercritical Once-through Boiler [J]. Electric Power Science and Engineering, 2013, 4: 014.

Rajkumar T, Priyaa V M R, Gobi K. Boiler drum level control by using wide open control with three element control system [J]. International Monthly Referred Journal of Research in Management and Technology, 2013, 2.

Wu J P, Zhu J F, Yao J, et al. Analysis and Optimization of Feed Water Control System in 1000 MW Ultra-Supercritical Units [J]. East China Electric Power, 2012, 40(6): 1075–1078.

Civil, Architecture and Environmental Engineering – Kao & Sung (Eds)
© 2017 Taylor & Francis Group, ISBN 978-1-138-02985-9

Multidisciplinary optimization and parameter analysis of the in-wheel permanent magnet synchronous motor of a tram

X.C. Wang & Z.G. Lu
Institute of Rail Transit, Tongji University, Shanghai, China

ABSTRACT: Based on the theory of electromagnetic field and thermodynamics, this paper puts forward a multidisciplinary optimization model for In-wheel Magneto Synchronous Motors (I-PMSMs). On the condition of satisfying the limits for output torque and motor temperature, this model sets motor efficiency at rated condition and cogging torque ripple as the optimization objectives, designs the motor gap length, permanent magnet thickness, pole arc coefficient, pole-slot combination, and stator slot width as variables, and subjects to the constrains of flux density in the stator teeth, the stator yoke flux density, tank full rate and no-load back EMF. The results indicate that the optimized model not only meets the requirements of outline integrated parameter and temperature limits, but also reduces cogging torque and volatility of output torque, decreases the losses and improves the motor efficiency.

1 INTRODUCTION

Tram is a kind of light rail transport, driven by electric power and driving on tracks. Due to its features of construction with less upfront investment cost, energy saving, convenience and passenger comfort, many medium and small sized cities apply the tram widely. Wheel motor-driven trams have been running in foreign countries for many years, such as R3.1. Trains are wheel-motor driven low-floor trams produced jointly by Duewag and Siemens, which began running in Frankfurt, Germany in 1993. The domestic I-PMSM driven trams are also in rapid development. In 2014, 100% low-floor tram-like vehicles in CSR Qingdao Sifang Locomotive Co. rolled off the production line. That car was the first one to drive with a magneto synchronous motor, featured with high efficiency, low noise, environmental protection, good curving performances and so on. Low-floor bogies directly driven by permanent magnet motors don't have a traditional gear transmission system, which approves light truck weight, compact design and high transmission efficiency.

With further studies in energy saving and cost control of PMSM, lots of scholars focus on motor performance optimization. Kim et al. (1984) focus on energy consumption study of motors. Yue-jun et al. (2011) proposed the use of oblique pole rotors, a stator chute or changing the shape of permanent magnet (points segment or unequal thickness) and other methods to reduce motor output torque ripple and other issues. Haodong et al. (2011) proposed optimization of the pole arc coefficient and pole-slot combination to study vibration noise caused by unbalanced magnetic force generated during operation of the motor and present optimization analysis method of the line size in the cooling system to solve the problem of the motor system temperature rise.

However, the optimized methods mentioned above can only concern several parameters or single discipline, which cannot analyze a set of parameters in different structures and multiple disciplines for optimization. In order to study the motor performance with different parameters in different disciplines, it is necessary to build a multi-parameter model with multidisciplinary factors when designing motors.

2 ANALYSIS OF THE I-PMSM MODEL

Considering the motor system consistency in the axial direction, the short axial length and limited axis end heat accumulation phenomenon, this paper simplifies the 3D electromagnetic field and thermodynamic models into two-dimensional plane models to simplify the model in multidisciplinary design optimization and improve computational efficiency. As shown in Figure 1,

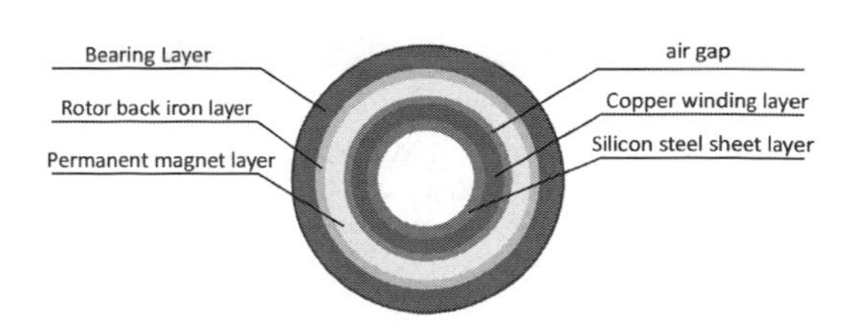

Figure 1. Simplified cross-section model of I-PMSM.

the cross-section of the two-dimensional model presents isotropic in the circumferential direction when neglecting process structure impacts in the motor section.

2.1 *Electromagnetic finite element analysis*

The basic structure parameters of a permanent magnet motor can be determined according to the formula. But for many structure variables, including slot numbers, the stator slot shape, slot full rate, rotor poles, stator and rotor air gap, etc., the adjustments of which will considerably change the power density of the motor and further influence the output efficiency and output capability. So, it is essential to specifically analyze those variables and select a preferred structure to meet the needs of train traction power.

In Ansoft electromagnetic finite element software, the solve domain is generally divided into a discrete triangular network, which means the magnetic vector potential in an arbitrary unit will be divided into three node vectors by (1):

$$A = N_i A_i + N_j A_j + N_m A_m \tag{1}$$

Transient magnetic field analysis can solve the voltage, non-sinusoidal energizing current source, or there is motion in model. In the Ansoft transient magnetic field solver, magnetic vector potential A satisfies the following field (2):

$$\nabla \times \nu \nabla \times A = J_S - \sigma \frac{\partial A}{\partial t} - \sigma \nabla \nu - \nabla \times H_C + \sigma \nu \times \nabla \times A \tag{2}$$

where H_C = coercive force of the permanent magnet; v = speed of moving objects; A = magnetic vector potential; J_S = source current density.

Maxwell 2D uses a reference frame when carrying on transient magnetic field analysis, fixed to a part of the model so that the speed is zero. Moving objects are fixed to their own coordinate system, and the partial time derivative turns into a full-time derivative. The motion equation is

$$\nabla \times \nu \nabla \times A = J_S - \sigma \frac{\partial A}{\partial t} - \sigma \nabla \nu - \nabla \times H_C \tag{3}$$

Thus, the magnetic vector potential can be calculated at every point of the finite element model in each time period. Based on the above principle, this paper establishes the electromagnetic finite element analysis model for I-PMSM by Ansoft software. The 2D transient electromagnetic analysis model produced by Ansoft is a 1/2 model, which can speed up the analysis, shown in Figure 2; Figure 3 is the electromagnetic analysis process.

2.2 *Thermodynamic analysis model*

In order to enhance the optimization efficiency, the thermodynamic modeling of the I-PMSM system is based on the basic theory of heat transfer. The element internal energy equations on the 3D cylindrical coordinates are simplified and integrate to get the one-dimensional temperature distribution of the cylindrical wall with heat source inside as (4). According to (4), if the inside heat source power is constant, it is easy to deduce that the temperature distribution is parabolic.

$$T = -\frac{q_v}{4\lambda} r^2 + c_1 \ln r + c_2 \tag{4}$$

In the temperature curve, there is always a point of maximum temperature, and heat energy flows from this point to the sides, so this point is the starting heat point of the cylindrical wall. Similarly, according to the general expression of the thermodynamic equation, this point can be regarded as the heat insulating point. When the motor system is under normal operating conditions, the highest temperature part is located at the copper wire windings, so the heat insulating point, which also means the starting heat point, should be located

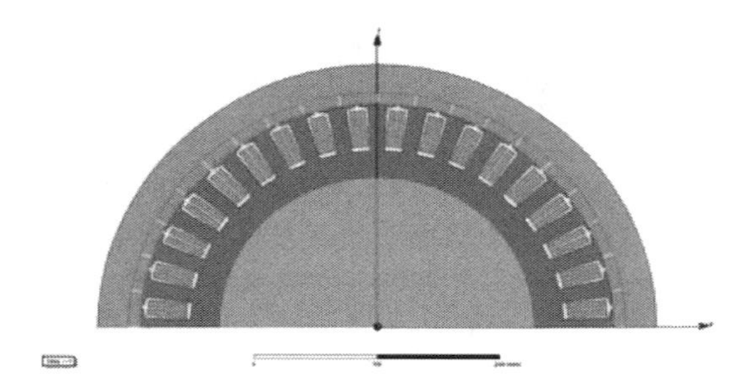

Figure 2. Motor 1/2 finite element model.

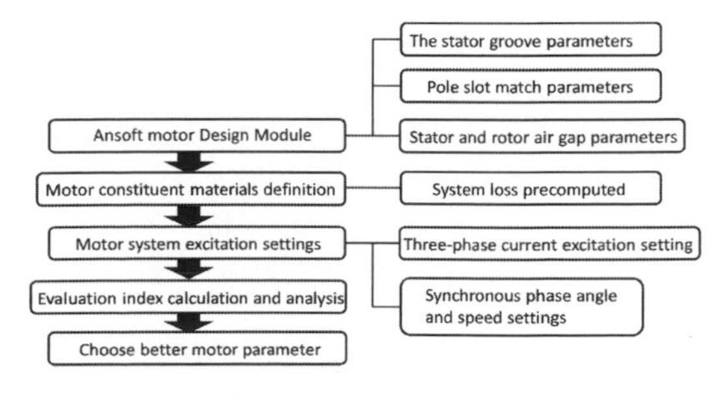

Figure 3. PMSM analysis process based on Ansoft.

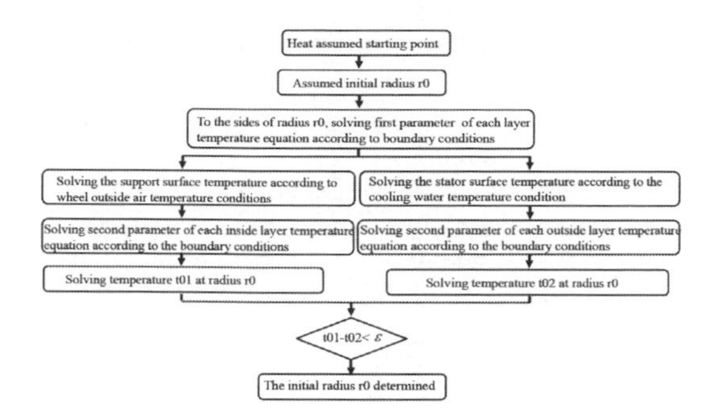

Figure 4. Algorithm of the thermodynamic analysis model.

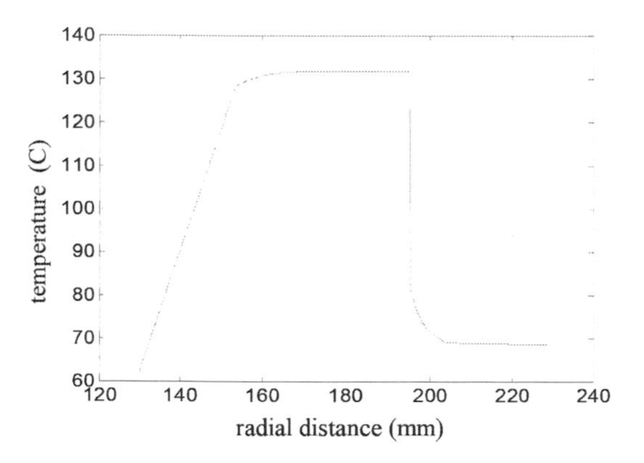

Figure 5. Temperature distribution of the motor.

at some radial position on the copper windings. Two beams of heat transfer to the sides from the starting point. One inward transfers to the cooling water, and the other outward transfers to the air outside of the wheel support.

This paper sets up a thermodynamic mathematical model by MATLAB software. The algorithm flow is given in Figure 4. With the application of the above model and on the condition that the convection heat transfer coefficient of the stator inside cooling water is 800 $W/K \cdot m^2$, and the convection heat transfer coefficient of the external environment is 30 $W/K \cdot m^2$, Figure 5 shows the simulation result of motor internal temperature distribution along the radial direction at rated power output.

3 I-PMSM MULTIDISCIPLINARY OPTIMIZATION MODEL AND IMPLEMENTATION

3.1 *Motor electromagnetic analysis and selection of evaluation parameters*

Easy to know the efficiency of the motor rated conditions is an important parameter to evaluate the performance of the motor, and hence this paper takes it as an optimization objective. Meanwhile, since the permanent magnet interacts with the slotted stator armature, it will produce cogging torque, which will not only fluctuate the motor output torque and generate additional vibration and noise, but also cause more difficulties to control the system. Therefore, cogging torque is another optimization objective of this paper.

From section 1.1, it is known that some parameters such as the gap length, the thickness of permanent magnets in the magnetization direction, the pole arc coefficient, the number of pole pairs and stator slot width have important effects on the motor operating performance. So, this paper selects these parameters as design variables. Meanwhile, the area of the stator slots, the width of stator teeth and stator slots, the air gap flux density, stator teeth magnetic density, the stator yoke magnetic density, motor slot full rate, the motor no-load back EMF and the rotor outer surface temperature must meet the I-PMSM constraints.

3.2 *Design of I-PMSM multidisciplinary optimization model*

The purpose of multidisciplinary optimization is to take advantage of synergies through the interaction of various disciplines (subsystem) to obtain the overall optimal solution. By enabling concurrent design, the design cycle is shortened, so that the developed products are more competitive. Multidisciplinary optimization can be simplified as a mathematical model and expressed as following:

find: x

minimize: $f = f(x, y)$

constrain: $hi(x, y) = 0 \ (i = 1, 2... m)$

$gj(x, y) \leq 0 \ (j = 1, 2... n)$

where f = objective function; x = design variables; y = state variable; $hi(x, y)$ = equality constraints; $gj(x, y)$ = inequality constraints. The calculation of state variables y, constraints hi, gj and objective function f involves multiple disciplines.

Through the above analysis, by meeting the traction torque and lower temperature limits, the I-PMSM multidisciplinary optimization model is designed as follows (Fenlian et al. 2012):

$$\min_{z_{i1} \leq z_i \leq z_{in}} \{f_1(z_1, z_2, ..., z_n), f_2(z_1, z_2, ..., z_n), ...f_i(z_1, z_2, ..., z_n)\}(i = 1, 2, ...n) s.t. \{c_1(z_1, z_2, ...z_n) \leq 0, c_2(z_1, z_2 ..., z_n) \leq 0, ...c_j(z_1, z_2 ..., z_n) \leq 0\} \quad (5)$$

where $z_{il} = z_i$ upper limit value; $z_{in} = z_i$ lower limit value;

$f_1(z_1, z_2\ldots z_n), f_2(z_1, z_2\ldots z_n)\ldots f_i(z_1, z_2\ldots z_n)$, are objective functions;

$c_1(z_1, z_2\ldots z_n) \le 0, c_2(z_1, z_2\ldots z_n) \le 0\ldots c_i(z_1, z_2\ldots z_n) \le 0$ are constraint functions;

$z_1, z_2\ldots z_n$ are design variables.

This paper uses the OPTIMUS multidisciplinary design optimization platform to set up the I-PMSM optimization model. This software can integrate CAE, Matlab, ADAMS and other simulation tools. It can achieve the automation of the simulation process, including test design, single or multiple objective optimizations, robust and reliability of the design module.

Electromagnetic analysis of the optimization model is based on Ansoft electromagnetic finite element software, while the thermodynamic analysis of the model is based on MATLAB. The implementation process on computer is shown in Figure 6.

Corresponding to optimization model in Figure 6, Figure 7 shows the flow structure at OPTIMUS.

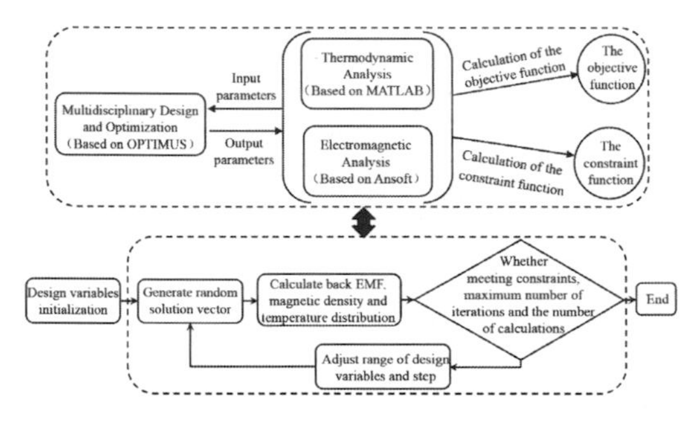

Figure 6. The design schematic of the motor multidisciplinary optimization model.

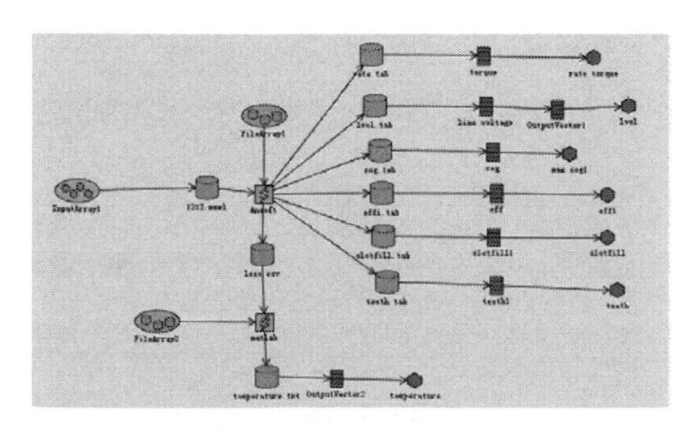

Figure 7. Workflow interface of PMSM performance optimization at OPTIMUS.

3.3 *Multidisciplinary optimization algorithm*

The multi-objective optimization algorithm has a variety of options, including the weight sum method, the weighted Chebyshev algorithm, and many other species. This paper uses the weighted Chebyshev aggregation algorithm, which is an improved method based on the weight sum method and Chebyshev method. Parameter ρ is added in that method. The weighted Chebyshev aggregation algorithm adjusts parameter ρ to control the ratio of two polymerizations, trying to combine the high speed of the weight sum method and reasonable distribution of the Chebyshev method. The mathematical description of the method is as follows:

$$\min g^{AT} \times \lambda, z^* = \max_{i=1,\ldots,n}\{\lambda_i \,|\, f_i(x) - z^* \,|\} + \rho \sum_{j=1}^{m} | f_j(x) - z_j^* | \tag{6}$$

where $\lambda = (\lambda_1 \ldots\ldots \lambda_m)$ = a set of weights vector, for all $i = 1\ldots\ldots n$, $\sum_{i=1}^{n} \lambda_i = 1 \,\&\&\, \lambda_i \ge 0$.

$z^* = (z_1^* \ldots\ldots, z_i^*)^T$ = the reference points, for all $i = 1\ldots\ldots n\, z^* = \min\{f_i(x) \,|\, x \in \Omega\}$

4 OPTIMIZATION MODEL ANALYSIS AND RESULTS

4.1 *The sensitivity (contribution) analysis of main parameters*

Take I-PMSM shown in Figure 2 for an example. In Ansoft software, the motor speed is set to 253 rpm, and the energizing current is set to 97 A. Calculate the I-PMSM multidisciplinary optimization model by Ansoft solver. The optimization process and results are shown in Figures 8 to 10.

Figures 8 and 9 show the contribution analysis of design variables to optimization objectives

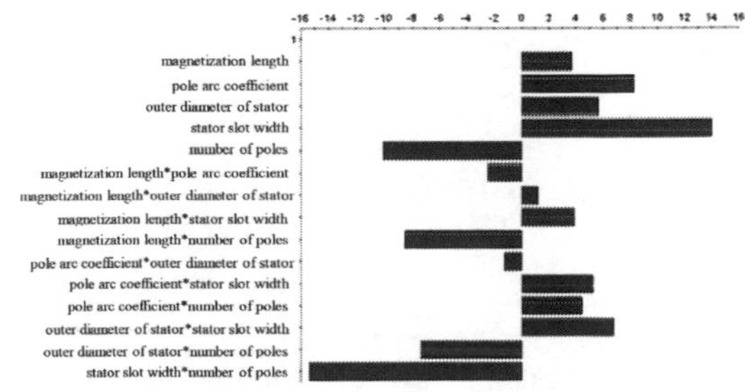

Figure 8. Contribution of design variables to cogging torque.

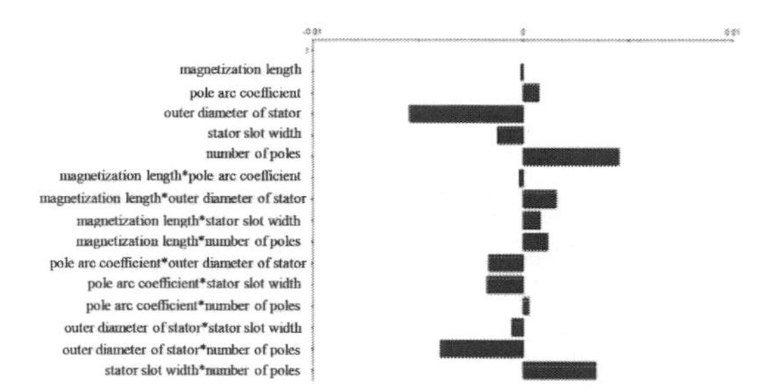

Figure 9. Contribution of design variables to motor efficiency.

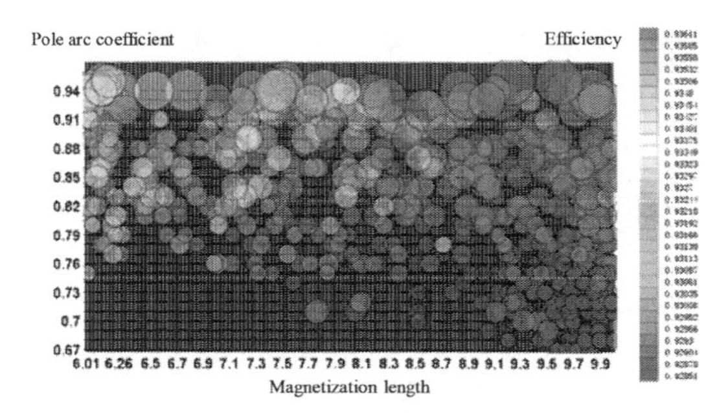

Figure 10. Bubble chart of pole arc coefficient and permanent magnet thickness contribution to motor efficiency.

including cogging torque and motor efficiency. The abscissa shows the impact of each variable on the optimization objective, wherein the numerical size indicates the impact level, and sign indicates that the design variable is in positive or negative correlation with the optimization objective. As shown in Figure 8, increasing the number of pole pairs makes maximum contribution to reducing cogging torque, and increasing the outer diameter of the stator will decrease the gap length and then lead to torque fluctuation increase. At the same time, the change of stator slot width will cause the change of effective gap length between stator and rotor, and then affect the cogging torque. As shown in Figure 9, the decrease of gap length, caused by the increase of stator outer diameter, has a maximum effect on the decrease of motor efficiency. As is known, reducing the gap spacing will result in increased motor losses, thereby reducing its efficiency, which also verifies the validity of the model results.

Figure 10 is the Bubble chart for the workflow test. The size of the bubble represents the size of cogging torque, while the color of the bubble shows the change in efficiency. In practical application, it is hoped to get small cogging torque and meanwhile ensure a certain degree of efficiency. So, in the bubble chart, it helps to find a set of parameters with smaller bubble and brighter color. As shown in Figure 9, the set of parameters (the magnet thickness ranges from 9.5–9.9 mm and pole-arc coefficients around 0.7–0.76) in the bottom right corner gets better results in the objective function.

4.2 *The optimized results of motor performance parameters*

Figure 11.a and 11.b illustrates the response surface in a 3D diagram for cogging torque and electric efficiency to pole-arc coefficients and the permanent magnet thickness in magnetization direction on the given slot-pole combination. It can be seen that the influence of improving motor efficiency by increasing the permanent magnet thickness is greater than that by increasing the pole-arc coefficients. Figure 11.c Figure 11.d shows the response surface for cogging torque and electric efficiency to stator slot width and outer diameter on the given slot-pole combination. The stator outer diameter has the same impact on cogging torque and efficiency, which can simultaneously decrease the torque and efficiency if it is too large or small. Similar to the stator slot width, when widening the slot, the electric efficiency and cogging torque both decline, but cogging torque shows a relatively more effective decline.

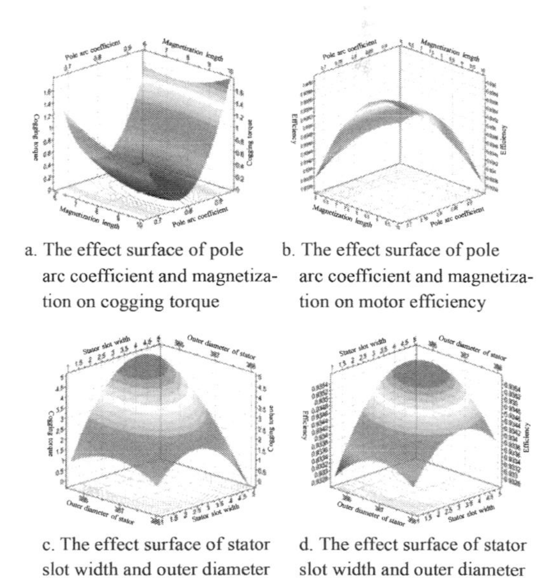

a. The effect surface of pole arc coefficient and magnetization on cogging torque

b. The effect surface of pole arc coefficient and magnetization on motor efficiency

c. The effect surface of stator slot width and outer diameter of stator on cogging torque

d. The effect surface of stator slot width and outer diameter of stator on efficiency

Figure 11. The effect surfaces of design variables on cogging torque and motor efficiency.

5 ANALYSIS OF MOTOR PERFORMANCE AFTER OPTIMIZATION

At rated conditions (speed: 253 rpm, power: 50 kW), the multidisciplinary optimized results of cogging torque and efficiency values on different weights are stated as Table 1.

Table 1. Comparisons of parameters before/after optimization.

Construction parameters	Before optimization	After optimization
Slot-pole combination	44/48	34/30
Number of conductors per slot	26	38
Gap length (mm)	1.5	1.5
Magnetization length (mm)	9	9.6
Stator slot width (mm)	3	4.6
Pole arc coefficient	0.85	0.75

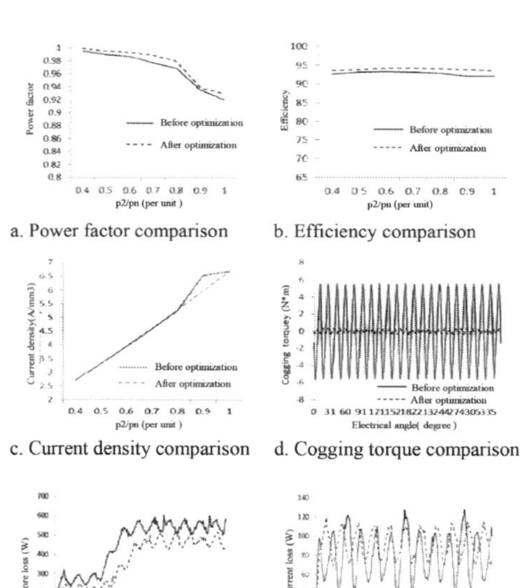

a. Power factor comparison

b. Efficiency comparison

c. Current density comparison

d. Cogging torque comparison

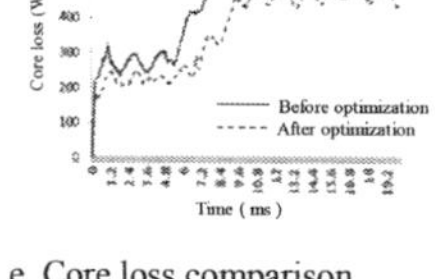

e. Core loss comparison

f. Eddy current loss comparison

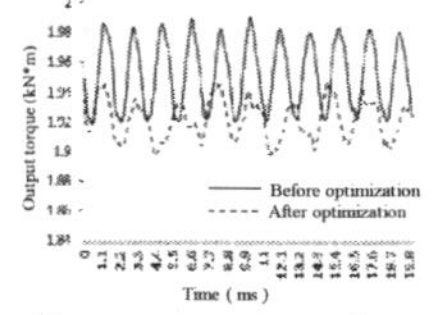

g. Output torque comparison at rated condition

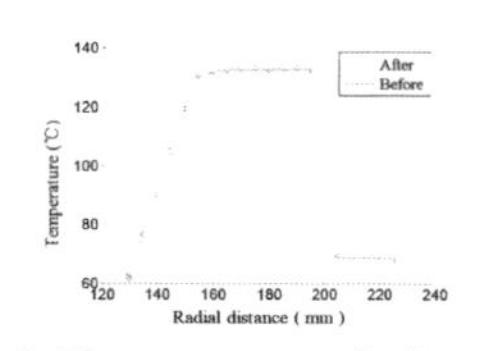

h. Motor temperature distribution comparison

Figure 12. Comparison of motor characteristics before / after optimization.

Table 1 lists the comparisons of parameters before and after optimization. In order to testify and prove the optimized results, this paper uses the corresponding finite element model to analyze the difference between motor performance parameters (power factor, efficiency, stator current density) before and after optimization. The comparison is shown in Figure 12, which can manifest that the power factor and efficiency relatively get improved. In addition, the optimized parameters greatly decrease the cogging torque and consequently reduce the torque ripple during the motor running process.

From Figures 12 to 14, we can find that the copper loss and iron loss fade to some extent after optimization, both of which arise from the decrease of current density. From Figure 12, it is easy to understand that the optimized output torque can satisfy the demands on specific conditions and meanwhile decreases the frequency of torque ripple. For the outer surface temperature of the motor rotor, there is no big difference in optimization. However, the temperature around stator winding decreases a little due to the decrease of copper loss; the sine of air-gap flux density gets improved due to the possibility of the reduced slots in stator slotting, which further weaken the influence on effective air-gap fluctuations. Meanwhile, the decline of fundamental amplitude avoids the over saturation of air-gap flux density.

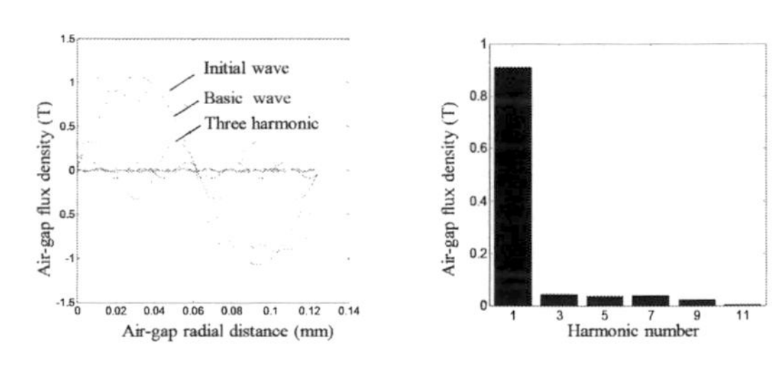

Figure 13. Analysis of air-gap flux density and harmonic number before optimization.

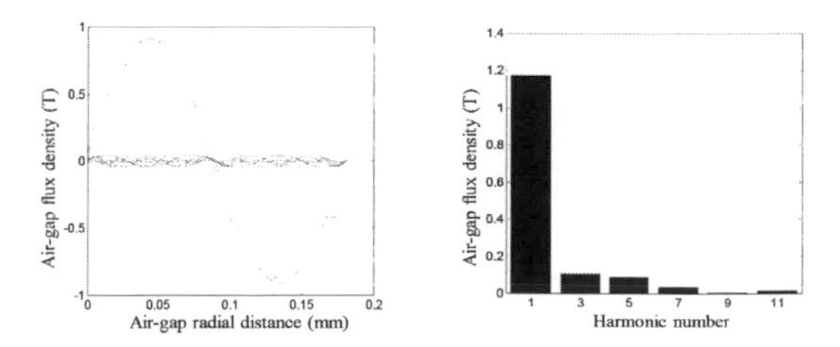

Figure 14. Analysis of air-gap flux density and harmonic number after optimization.

6 CONCLUSIONS

Based on the theory of electromagnetic fields and thermodynamics, this paper establishes a multidisciplinary optimization model for tram I-PMSM, which aims at optimizing the motor torque ripple and electrical efficiency based on the OPTIMUS platform. Emphasis is put on analyzing some design variables such as slot-pole combinations, pole-arc coefficient, gap length, the permanent magnet thickness in the magnetization direction, and stator slot width. The main conclusions are as follows:

1. Increasing the number of permanent magnet poles contributes the most to smoothing the cogging torque, while reducing the gap length and increasing the thickness of the permanent magnet will intensify the cogging torque ripple.
2. On the condition of equally adding the permanent magnets and the same output torque, compared with the method of increasing the pole-arc coefficient, increasing the permanent magnet thickness will produce smaller current density, which further means a less copper loss of stator windings and a higher efficiency.
3. Slot-pole combination affects the operating performance greatly. However, it is not a simple correlation due to the complexity of influence. The model proposed in this paper can set the range for pole pairs and slots of the permanent magnet, automatically achieve the slot-pole combinations and provide the results under specific conditions. Overall, it is a good solution for optimum simulation.

REFERENCES

Alexande, K., & Donald, G. (1983). Control means for minimization of losses in AC and DC Moto drives. *J. IEEETrans.onIA.*, 13(4):561–57.

AN Yue-jun, Zhao Wen-qing, Xue Li-ping, & Li Yong (2011). Pole width modulation method and its experimental research for reduce cogging torque of permanent magnet motor. *J. Electric Machines and Control,* 2011(10).

Chen Yiguang, Pan Yuling, & He Xin. (2010). Magnetomotive Force in Permanent Magnet Synchronous Machine With Concentrated Fractional-Slot Winding. *J. Transactions of China Electrotechnical Society.* 30–36.

Huang Fenlian, Ji Wei, & Zhang Lu. (2012). Optimization of valve timing and injection advance angle of diesel engine based on OPTIMUS. *J. Transactions of the CSAE.* 28(15): 27–32.

Kim, H.G., Sul, S, K., & Park, M.H. (1984). Optimal efficiency drive of a current source inverter fed induction motor by flux control. *J. IEEETrans. OnIA.*, 20(6):1453–1459.

Wang Aimeng (2010). Optimal IPM Machine Design and Flux-weakening Control. *D. North China Electric Power University.*

Yang Haodong, & Chen Yangsheng. (2011). Electromagnetic Vibration Analysis and Suppression of Permanent Magnet Synchronous Motor with Fractional Slot Combination. *J. Proceedings of the CSEE.* 83–86.

Zhou Jun-jie, Fan Cheng-zhi, Ye Yun-yue, & Lu Qin-fen (2011). Method for reducing cogging torque based on magnet skewing in disc-type permanent magnet motors. *J. Journal of Zhejiang University (Engineering Science).* 2011(10).

Civil, Architecture and Environmental Engineering – Kao & Sung (Eds)
 ISBN 978-1-138-02985-9

Dynamic simulation and muzzle vibration analysis of vehicle-mounted automatic mortar

Yu Fang, Jun-qi Qin & Chang-chun Di
Department of Guns Engineering, Ordnance Engineering College, Shijiazhuang, China

ABSTRACT: The vibration of gun muzzle is an important factor that affects the firing dispersion. In this paper, the dynamic model of vehicle-mounted automatic mortar considering suspension and tire parameters is established, and the vibration characteristics of the system are solved by the Lagrangian equation method of the Multi-Degree-of-Freedom (DOF) vibration system. Based on ADAMS software, the model of the tire-road system and the virtual prototyping model of vehicle-mounted automatic mortar are established. The dynamic characteristics of the mortar and the vibration of the muzzle are analyzed, which can provide reference for further research on vehicle-mounted weapon systems.

1 INTRODUCTION

The firing dispersion is one of the important indexes to evaluate the performance of artillery, and the gun muzzle vibration is an important factor that affects the firing dispersion. The position of the muzzle and the velocity of the muzzle are greatly changed when the gun fired, which leads to the change of the initial flight parameters of the projectile, and then affects the final bullet spread. As for vehicle-mounted weapons, there are more links between barrel and the ground. On the one hand, the barrel and the cradle of the weapon will be deformed, and on the other hand, the vibration of the vehicle will lead to muzzle vibration.

At present, there are some pieces of literature on the muzzle vibration of towed gun and vehicle-mounted weapons. CAI studied the impact of the mass of the muzzle brake, the position of the barrel support and other parameters on the muzzle vibration, while LIANG studied the effect of the arrangement of the recoil mechanism.

In this paper, vehicle-mounted automatic mortar is studied, which shoots without a rigid support. The body is connected to the ground through the suspension and the tire. At present, there are few research articles on this special case. The vehicle-mounted automatic mortar is used as the object of study, and the vibration characteristics of the system are solved by using the Lagrangian equation of the multi-degree-of-freedom vibration system. The dynamic model and the virtual prototyping model of the vehicle-mounted automatic mortar are established, and the dynamic characteristics of the vehicle are numerically simulated. The variation law of the muzzle is given, which will provide reference for further research of vehicle-mounted weapon systems.

2 DYNAMIC MODEL OF VEHICLE-MOUNTED AUTOMATIC MORTAR

2.1 *Establishment of the dynamic model*

The vehicle-mounted automatic mortar mainly includes a driving system and a fire system. Among them, the driving system is the modified Dongfeng warrior EQ2050A chassis, composed of tires, suspension, body and other components, and the front and rear suspensions are double cross-arm independent suspension.

The fire system is the mortar system with an automatic mechanism. The fire system is mainly composed of the larger automaton, balancing machine, shelves, machine, car body, height direction machine and so on. The fire system is connected to the driving system through the trunk. The trunk can be rotated horizontally relative to the chassis, the barrel can rotate up and down relative to the trunk, in order to achieve a different direction angle and pitch angle of the mortar to fire.

In order to show the main movement and force of the mortar, according to the structural characteristics of the vehicle-mounted automatic mortar, the design of the mortar is simplified to the following components: the mass under suspension, body of car, cradle of mortar and recoil device.

The chassis of vehicle-mounted is a common two-axle vehicle, which can be simplified as a 1/2 dynamics model of vehicle-mounted automatic mortar with 6-DOF, which is shown in Figure 1.

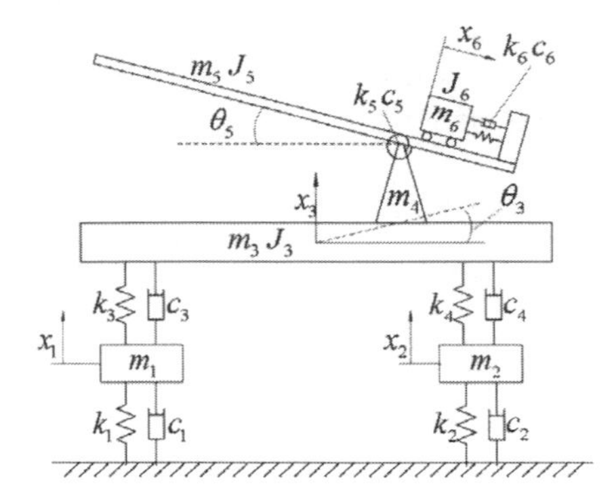

Figure 1. 1/2 Model of vehicle-mounted automatic mortar with 6-DOF.

In Figure 1, m_1, k_1, c_1, and x_1, respectively, represent the mass of the front wheel, the equivalent stiffness and damping coefficient of the front wheel, and the vertical displacement of the front tire; m_2, k_2, c_2, and x_2, respectively, represent the mass of the rear wheel, the equivalent stiffness and damping coefficient of the rear wheel, and the vertical displacement of the rear tire; k_3, k_4, c_3, and c_4, respectively, represent the front and rear suspension equivalent stiffness and front and rear suspension damping coefficient; M_3, J_3, x_3, and θ_3 are the total mass and moment of inertia of the body and the vertical displacement and pitch angle of the body center of mass; M_5, J_5, m_6, and J_6, respectively, are the mass and moment of inertia of the barrel and mass and moment of inertia of the recoil part; m_4 and J_4, respectively, are the mass and moment of inertia of the cradle; θ_5 and x_6, respectively, are the elevation of the barrel and displacement of the recoil part. The stiffness coefficient and damping coefficient between barrel and cradle are denoted by k_5 and c_5.

2.2 *Assumptions and calculations of the model*

The model assumes the following:

1. The body contacts with the ground through four double cross-arm independent suspension and four tires. Only vertical and pitching movement of the body and barrel are considered; there is no lateral force and movement.
2. The suspension system is equivalent to the spring damping system. The elastic connection between barrel and cradle forms into a coil spring damping system.
3. In this model, the barrel, the body of the car, the cradle and other parts are rigid; only consider the movement and stress, regardless of deformation factors.
4. The tire will not slip; the contact point between the tire and the ground is always fixed. The contact with the ground is simplified as a vertical movement of the spring damping system, that the stiffness coefficient and damping coefficient are determined by the tire-surface coupling characteristics.

Based on the above assumptions, the Lagrangian equation of the multi degree-of-freedom system is used to construct the vibration equations of a 6-DOF model:

$$\frac{d}{dt}\left(\frac{\partial T}{\partial \dot{q}_i}\right)-\frac{\partial T}{\partial q_i}=Q_i-\frac{\partial V}{\partial q_i} \tag{1}$$

where T is the kinetic energy of the system:

$$T=\sum_{i=1}^{n} T_i \tag{2}$$

Q_i is the generalized force corresponding to non-dominated forces. In this model, the damping exists. It is necessary to determine the energy dissipation coefficient D of the system, that is, the energy dissipated by damping:

$$D=\frac{1}{2}c\dot{q}_i^2(t)=\frac{1}{2}\{\dot{q}(t)\}^T[c]\{\dot{q}(t)\} \tag{3}$$

The generalized damping force is:

$$R_i=-\frac{\partial D}{\partial \dot{q}_i}=-\sum_{j=1}^{n} c_{ji}\dot{q}_i \tag{4}$$

The resulting multi-degree-of-freedom system equations of motion are:

$$[M]\{\ddot{q}(t)\}+[C]\{\dot{q}(t)\}+[K]\{q(t)\}=\{Q(t)\} \tag{5}$$

In the formula,

$$\{q(t)\}=[x_1,x_2,x_3,\theta_3,\theta_5,x_6]^T \tag{6}$$

The above formulas are vibration equations of the model.

3 ESTABLISHMENT OF THE VIRTUAL PROTOTYPE MODEL

3.1 *Virtual prototype model of the tire–road system*

According to the characteristics of the vehicle in an actual weapon system, the UA tire model is a commonly used model at present. Using the SAE coordinate system, considering the unsteady effect, side slip and longitudinal slip interaction, the model can only be used on flat roads.

MSC.ADAMS software was used to create the model of the vehicle. The model mainly includes a vehicle chassis model, double cross arm independent suspension model, tire model and road. The parameters of the UA tire model used in this paper are shown in Table 1.

3.2 *Virtual prototype model of vehicle-mounted automatic mortar*

Before the establishment of the model of vehicle-mounted automatic mortar, the mortar model was simplified and the cradle model was deleted, which has little influence on the simulation results. The mass of the cradle is equivalent to the car body and the sphere model is used to simulate the mortar projectile with the same mass as the real projectile. On stimulation at firing, the ADAMS software comes up with a STEP function to apply the force in bore, acting between the projectile and the breechblock, pointing along the barrel axis. During the firing process, the chassis is in the braking state, and the tire is not rotated. In order to reduce the computational cost, the component can be regarded as a rigid body with larger stiffness and smaller deformation, and the equivalent mass and moment of inertia can be processed for the components which have no influence on the motion and force.

In the modeling process of the paper, the vehicle is considered as a rigid body to be analyzed, the suspension damping system of the vehicle is standardized designed according to the spring damping system, and the virtual prototyping model of the vehicle-mounted automatic mortar, the model consists of 30 motion components, 22 revolute joints, 9 fixed joints and 9 translation joints. The virtual prototype model is shown in Figure 2.

Table 1. UA tire model parameters.

Parameter	Value	Parameter	Value
Tire unload radius (m)	0.48	Tire width (m)	0.28
Vertical stiffness (n/m)	1.9e5	Minimum friction coefficient	0.9
Vertical damping (n·s/m)	280	Maximum friction coefficient	1.1

Figure 2. Virtual prototype model of vehicle-mounted automatic mortar.

4 SIMULATION OF MUZZLE VIBRATION

Based on the virtual prototype model, the corresponding tire and road parameter files were imported, and physical parameters and structural parameters such as the fire line height and mass of body were imported. The linear and angular displacements of the muzzle point in the global coordinate system are calculated. The model is symmetrical, and only the forward firing condition is studied. The horizontal displacements in the muzzle parameters, the vertical displacement in the vertical direction, and the angular displacement in the pitch angle are the dependent variables. Table 2 shows the parameters of suspension.

The figures presented below show the muzzle dynamic simulation results in the single-shot condition. The angle of fire is 12 degrees, shooting straight ahead.

The curve in the figure can be divided into two parts: before firing and after firing. Since the model is not in the equilibrium position at the beginning of the simulation, from t = 0 s to t = 3.99 s, the vibration of the body and the gun body is small, and gradually stabilizes. When t = 3.99 s, the STEP function is triggered, resulting in force in the bore of mortar, making the projectile flow out of the barrel, while the breech recoils.

Starting from the moment of firing, the muzzle center point parameters began to change greatly. Muzzle has produced the front and back direction, the up and down direction vibration as well as pitching rotation. Among them, the amplitude of the pitch rotation angle is 1.6 degree, the displacement amplitude in the vertical direction is 71.08 mm, and the displacement amplitude in the front-back direction is 84.23 mm. From the firing time to the projectile flying out of the muzzle interval, the time taken is only 0.016 s. After 2 seconds, the muzzle center position tends to be stable, back to the initial launch position.

As can be seen from the figure, although there is muzzle vibration when shooting, the projectile from firing to the time between the muzzle time

Table 2. Suspension parameters.

Parameters	Value
Front suspension stiffness (n/mm)	200
Front suspension damping (n·s/mm)	100
Rear suspension stiffness (n/mm)	380
Rear suspension damping (n·s/mm)	100

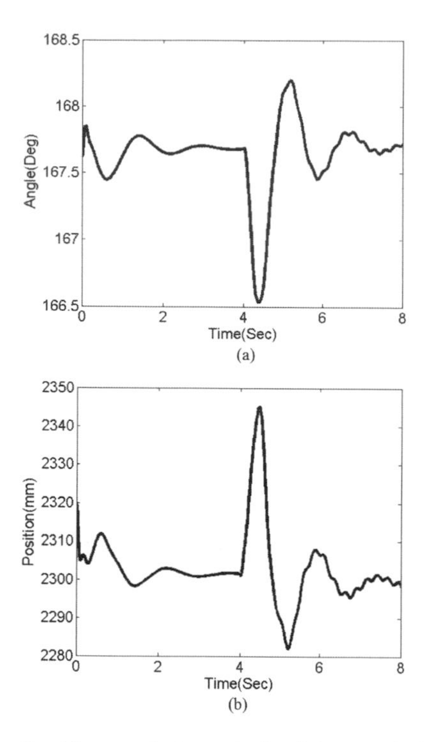

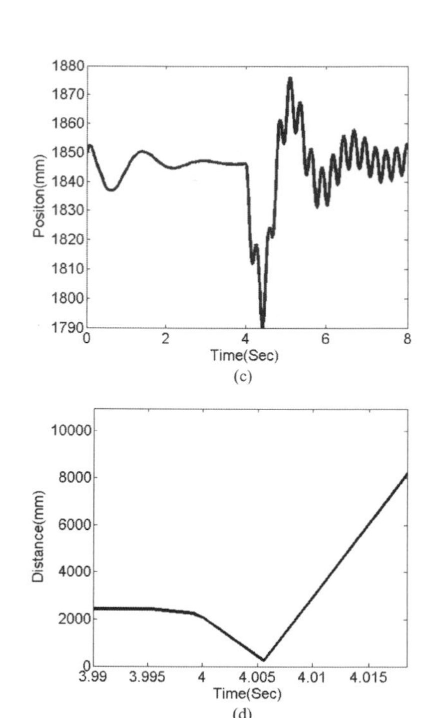

Figure 3. The muzzle dynamic simulation results in the single-shot condition. Every curve means:
(a) Change of rotation angle of muzzle axis in the vertical plane.
(b) Displacement of the muzzle center in the vertical direction.
(c) Displacement of the muzzle center in the front-back direction.
(d) The distance between projectile and the muzzle center.

interval is very short, only 0.016 s. When the projectile is out of the muzzle, the change in the muzzle parameters is still very small, almost no change; so, the intensity of a single-shot has little effect. Therefore, it can be concluded that for the vehicle-mounted automatic mortar in the single-shot condition, the muzzle vibration is harmless.

5 CONCLUSIONS

Based on the theory of the multi-degree-of-freedom vibration system, the dynamic characteristics of vehicle-mounted automatic mortar are modeled and analyzed, which are then simulated with MSC.ADAMS software. This modeling method is simple, efficient and accurate, which provides reference for further modeling research on vehicle-mounted weapons.

This type of vehicle-mounted automatic mortar, which shoots without a rigid support, indicates that the muzzle vibration is harmless when the mortar is fired in a single-shot condition.

Future research can be performed on the effects of muzzle vibration in continuous firing of the vehicle-mounted automatic mortar, to further explore the characteristics of muzzle vibration of vehicle-mounted weapons.

REFERENCES

Chen Yanhui, Guo Min, He Zongying. 2010. Muzzle vibration test method and practice. *Journal of Gun Launch & Control* 3(1): 80–83.

Cai Wen-yong, Chen Yun-sheng, Yang Guo-lai. 2005. Impacts of Structural Parameters on Muzzle Disturbance for a Vehicle Mounted Howitzer. *Journal of Nanjing University of Science and Technology* 29(6):658–661.

Liang Chuan-jian,Yang Guo-lai, Ge Jian-li, Wang Xiao-feng, Zhang Zhen-hui. 2013. Study on Influence of Structural Arrangement of Recoil Mechanism on Muzzle Vibration of Gun. *Acta Armamentarii* 34(10):1210–1214.

Li Tao, Wang Rui-lin, Zhang Jun-nuo. 2013. Simulation of Coupled Rigid and Flexible Multi-body Dynamics on Gatling Gun. *Journal of System Simulation* 25(6): 1382–1387.

Mao Bao-quan, Yu Zi-ping, Shao Yi. 2009. *Introduction to Vehicle-Mounted Weapon Technology*. Beijing: National Defense Industry Press.

Civil, Architecture and Environmental Engineering – Kao & Sung (Eds)

 ISBN 978-1-138-02985-9

Numerical simulation of aspheric glass lens during the non-isothermal molding process

Hongjie Duan & Zhiling Xiao
College of Mechanical and Electrical Engineering, Zhengzhou University of Light Industry, Zhengzhou, China

Junyi Tao
School of Management, Huazhong University of Science and Technology, Wuhan, China

ABSTRACT: Defects such as low productivity and short service life of the mold occur in the isothermal molding process of aspheric lens. On the basis of these problems, a new method of non-isothermal molding was introduced and the fundamental principle of non-isothermal molding was carefully analyzed. Non-isothermal molding is formed at different temperature conditions between glass blank and mold. The deformation flow of the glass is accompanied by heat transfer. Thus, non-isothermal molding of aspheric lens was simulated by the thermo-mechanical coupling 3D finite element method. The internal temperature and stress field of the glass and mold were analyzed. And then the simulation results of isothermal and non-isothermal molding were compared. The results indicated that the non-isothermal molding process was feasible and superior.

1 INTRODUCTION

Optical glass lenses are widely used in optical engineering, automotive engineering, biomedical, aerospace and other fields, which have huge market demand (Chen et al. 2010). The lens manufacturing not only requires ensuring the quality and use of optical performance, but also meets the high efficiency and low cost. At present, the isothermal molding method is widely used in the production of aspheric lens, which involves employing chunk glass into the mold at room temperature, and after a specified time, the mold is prepared when the temperature reaches molding temperature (Yan et al. 2009). The processing lenses have high surface precision using this method, but only after the end of the previous production cycle, we can start the next production cycle. Therefore, the production efficiency is quite low. In addition, the mold has a shorter service life because of repeated heating and cooling processes, and a large temperature range (Ma et al. 2009, Bai et al. 2015).

In this paper, a new method for manufacturing aspheric lens with non-isothermal molding is presented. The advantages and disadvantages of non-isothermal and isothermal molding methods were compared by means of a finite element simulation method, aiming at exploring the efficient non-isothermal molding process under the premise of ensuring the quality of lenses.

2 THE BASIC PRINCIPLES OF NON-ISOTHERMAL MOLDING

2.1 *Non-isothermal molding process*

The glass blank was usually heated to a temperature dozens of degrees above the conversion point in a preheat device at first. Then, it was put into the mold with temperature, maintained at a lower level by a mechanical transport device, molded and annealed in the mold. When the inner stress was reduced to a certain extent, the lens was removed from the molding chamber through a mechanical conveying device, and then further cooled to room temperature in a cooling device. The whole process must be performed in a vacuum environment to ensure that the high-temperature glass is not oxidized by air.

2.2 *Thermo-mechanical coupled finite element method*

The process involving non-isothermal molding of aspheric glass lenses included a flow-forming process for softened glass at high temperatures and a heat transfer process with a strong coupling effect. For the coupling problem covering temperature and displacement, the accurate analysis method was to follow the solution of the thermo-mechanical coupling field and process the two different field equations containing heat conduction and force balance simultaneously. In this study, Marc software was used to conduct the coupling analysis of aspheric glass lens molding. Marc software has strong structure and contact analysis capability, which is suitable for large deformation analysis of viscoelastic materials under isothermal and non-isothermal conditions.

3 ESTABLISHMENT OF THE FINITE ELEMENT MODEL

3.1 *Mesh model*

Figure 1 shows the geometric model of aspheric lens during the simulation process.

The curve of the upper and lower aspheric surface satisfies the equation (Zhou, 2009):

$$y = \frac{x^2}{R\left(1+\sqrt{1-\frac{(1+k)x^2}{R^2}}\right)} + B_4x^4 + B_6x^6 + B_8x^8 + B_{10}x^{10} \quad (1)$$

where R is the vertex radius of the curve, x is the horizontal coordinate, y is the vertical coordinate, k is the curve constant and B is the aspheric surface coefficient. Among the aspheric parameters:

$$R = -10.75431\,mm \quad (2)$$

$$K = 0 \quad (3)$$

$$B_4 = 8.621796 \times 10^{-3} \quad (4)$$

$$B_6 = -2.56179 \times 10^{-3} \quad (5)$$

$$B_8 = -3.037114 \times 10^{-4} \quad (6)$$

$$B_{10} = -1.82822 \times 10^{-5} \quad (7)$$

The parameters can be determined similarly.

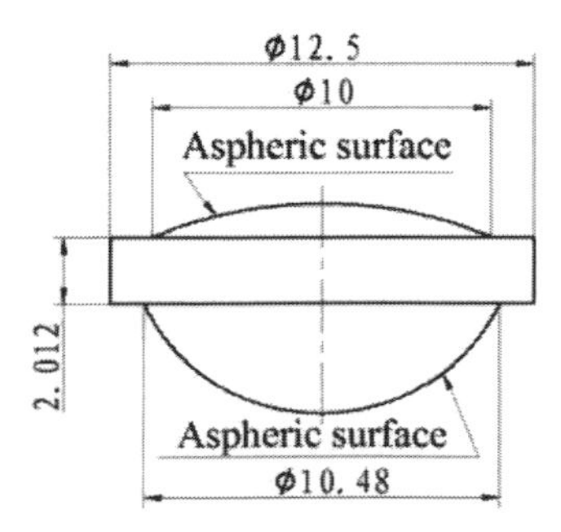

Figure 1. Double aspheric lens geometry model.

Table 1. Glass characteristic parameters.

Performance parameter	Size
Young's modulus E/MPs	15800
Density ρ/(g/cm³)	3.05
Poisson's ratio	0.247
CTE α/ × 10–6/°C8.8 (100~300°C)	
Thermal conductivity k/(W/(m.°C)	10.28
Specific heat Cp/(J/(kg.°C)	750

3.2 *Material parameters*

The prefabricated part of glass molding was glass ball blank which is L-BAL42. The material characteristic parameters are shown in Table 1. The three-dimensional model of glass ball blank was established by Solid Works software, and the solid model of upper and lower mold with an aspheric curve was established by Solid Works software too. Then the upper mold, lower mold and glass ball were fitted together according to the actual working position. The assembly was imported into the Marc software. The finite element meshing was divided by the auto mesh function. The assembly model was divided into tetrahedral elements.

4 ANALYSIS OF SIMULATION RESULTS

In the actual non-isothermal molding process of the aspheric surface of glass lens, the temperature was always not less than the molding temperature and not higher than its softening temperature Sp. When the temperature is higher than the softening point Sp, the viscosity of glass is very low, and its behavior is similar to the viscous fluid. It also can form bond to the mold (Zhao et al., 2009; Ni et al., 2015). The temperature of the mold should not be too low, and the surface and interior thermal contraction of glass will inconformity if there is a large gap between the temperature of glass and mold, so it comes into being a lot of internal stress and ever crack inside of lens. Therefore, aiming to the L-BAL42 optical glass, we defined the initial temperature of the glass as 590°C, the initial temperature of tungsten carbide WC mold and die holder was 560°C, and the temperature of molding was 580°C.

4.1 *Analysis of the temperature field*

The temperature field of aspheric lens in the molding process is shown in Figure 2. The spherical glass blank with 590°C was put into the mold by mechanical devices. When the lower mold moved upward, the extrusion was produced between the upper surface of the glass blank and the upper mold, and the precise aspheric surface of the upper

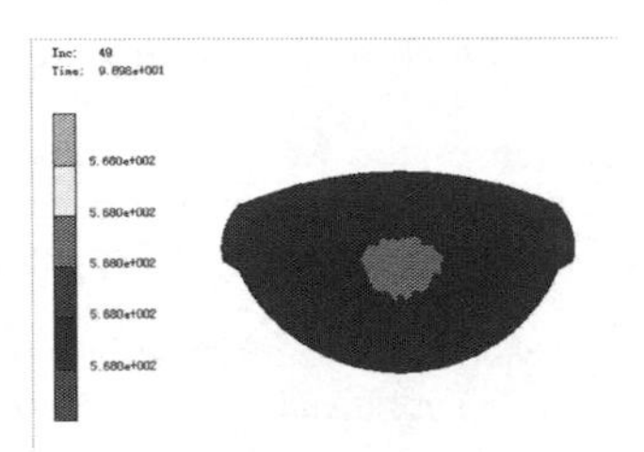

(a) Internal temperature distribution of the glass with molding 98s

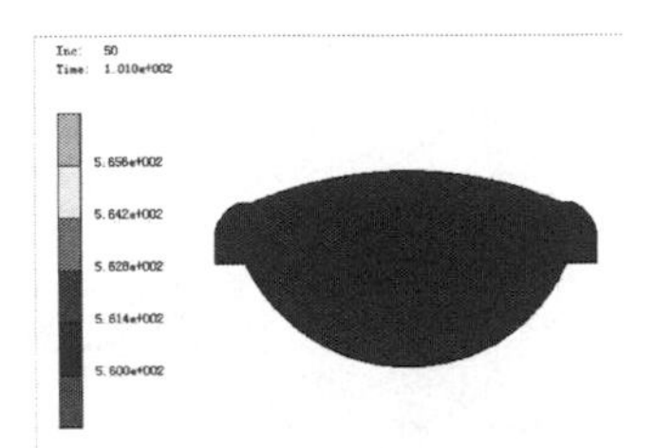

(b)Internal temperature distribution of the glass with molding 101s

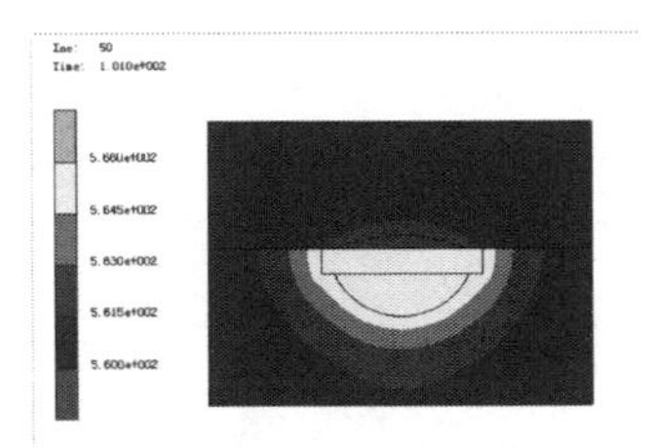

(c) Internal temperature distribution of the mold with molding 101s

Figure 2. The temperature field of the aspheric lens glass molding process.

mold started to copy on the lens surface. The lowest temperature area inside the lens was contacted with the mold at this moment. With the mold pressing, the lowest temperature area on the upper surface of the glass lens moved towards the edge with the contact surface of the glass and mold enlargement. The temperature of the lower surface edge was relatively low, because it contacted with the die holder for a long time, and heat transfer was more fully performed.

After molding, glass was almost filled into the closed space formed by the mold and mold base. The upper and lower mold contacted with the glass, and the precision aspheric surface of the mold was copied on the lens surface. The internal temperature gradient of aspheric lens was clearly visible, and the highest temperature area occurred in the interior of the aspheric lens. The further the distance, the lower the temperature. The upper and lower surfaces were in direct contact with the mold surface (Jain 2006), so the lowest temperature region appeared at the side edge of the mold holder. Its temperature was 560°C, which was close to the initial temperature of the mold. In the whole forming process, the internal heat of the high-temperature glass was transferred to the mold, mold holder and the environment filled with nitrogen gas through two ways, which are solid heat conduction and convection heat transfer.

4.2 *Analysis of the stress field*

Fig. 3 shows the stress field of the aspheric lens glass molding process. When molding runs at 78 seconds, the glass blank started to contact with the upper mold, and the equivalent stress of the glass ball surface center increased rapidly because the extrusion stress of the upper die and the stress of the edge of the mold began to increase. With the molding proceeding, two maximum stress points

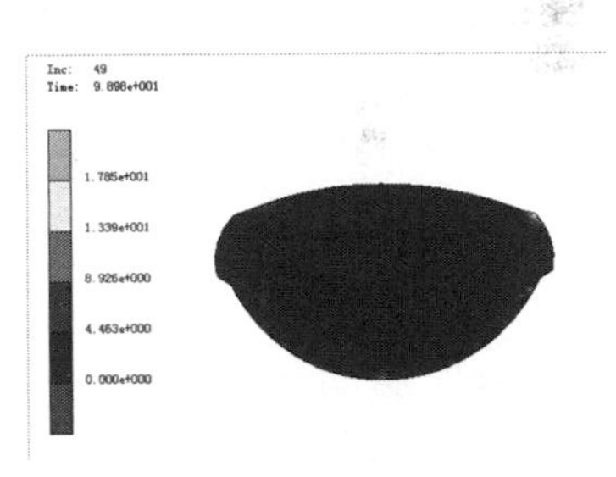

(a) Internal stress distribution of the glass with molding 98s

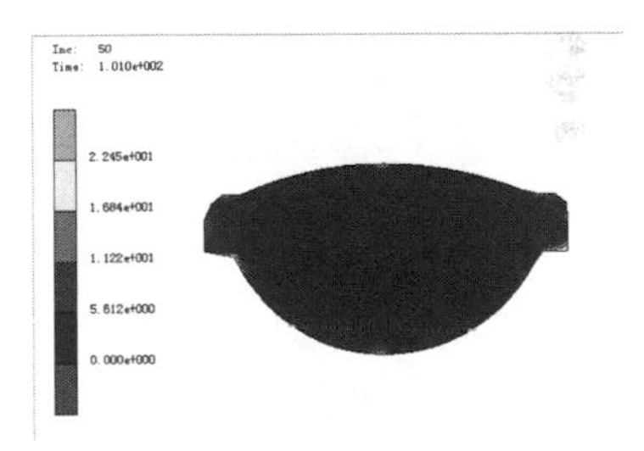

b) Internal stress distribution of the glass with molding 101s

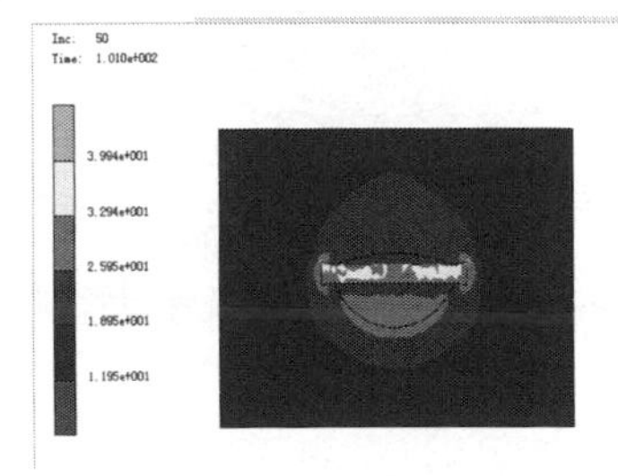

(c) Internal stress distribution of the mold with molding 101s

Figure 3. Stress field of the spherical glass lens molding process.

Table 2. The stress field of the isothermal and non-isothermal molding.

Type of lens/MPa of mold/MPa	Maximum stress	Maximum stress
Isothermal molding	16.56	51.54
Non-isothermal molding	22.45	39.94

appeared inside the glass. One is in the center of the upper surface with the lowest temperature, which moved outward with the contact area with the upper surface and the mold enlarging. Another is the edges of the lower surface, namely the intersection of the aspheric surface and the plane of the glass edge. The glass blank bears a large shearing force, so a larger stress is produced. When molding proceeds to 98 seconds, the glass is close to the final shape, and the maximum stress area moves further to the edge. The stress of the upper and lower surfaces increases with deformation increasing. The stress distribution has no change but an increase in the value.

4.3 *Contrastive analysis of non-isothermal and isothermal molding*

Comparing with isothermal molding, non-isothermal molding has two main advantages such as high production efficiency and longer mold life. However, two factors involving strong heat transfer and internal temperature gradient can affect inner stress and surface accuracy of after-mold lens. In this paper, the simulation model of non-isothermal molding was used to compare non-isothermal molding (glass 590°C, mold 560°C) with isothermal (molding temperature 580°C). The comparison results are shown in Table 2.

It can be found that the lens have a small average stress when conducting the isothermal molding. The average stress within the mold is also small as the non-isothermal molding proceeds. A smaller mold stress will help to prolong the mold life, which confirms the advantage of the non-thermal molding method. Nevertheless, the non-isothermal molding method has a bigger stress of lens, which contributes to glass shrinkage during the process of deformation. The inner stress in the next step can be reduced by the annealing and cooling processes.

5 CONCLUSIONS

1. The reasons for the temperature gradient and the stress gradient inside the mold were explained. The spherical lenses had a higher internal temperature during the non-isothermal molding process. Moreover, the average temperature and stress in the upper die were higher than those in the lower die, indicating that the upper die had a short service life.
2. The comparison between the non-isothermal molding and isothermal molding processes showed that the mold had good optical performance due to a small internal stress within the lens during isothermal molding, while it had a long service life because of a small internal stress within the mold during non-isothermal molding.

ACKNOWLEDGEMENT

This work was supported by the International Science and Technology Cooperation Project (2012DFG70640) and Key Scientific Research Projects of Colleges and Universities, Henan (15A460012).

REFERENCES

Bai Daiping; Li Yanqin; Duan Hongjie; et al. 2015. Microstructure evolution numerical simulation law research of turbine blade in the high temperature deformation. Modern Manufacturing Eng. 6, 104–109.

Chen Fengjun; Yin Shaohui; Huang He et al. 2010. Profile Error Compensation in Ultra–precision Grinding of Aspheric Surfaces with On–machine Measurement. Int.J. Mach. Tools & Manufacture. 50, 480–486.

Jain A. 2006. Experimental study and numerical analysis of compression molding process for manufacturing precision aspheric glass lenses. Columbus: The Ohio State University.

Ma Tao; Yu Jingchi & Wang Qinhua. 2011. Precision glass molding process small aperture thin lens. Infrared and Laser Eng. 40(1):87–90.

Ni Jiajia; Fan Yufeng & Chen Wenhua. 2013. Simulation Study of Molding of Spherical Optical Glass Lens. Laser & Optoelectronics Progress. 50, 032201.

Yan Jiwang; Zhou Tianfeng; Masuda Jun; et al. 2009. Modeling high-temperature glass molding process by coupling heat transfer and viscous deformation analysis. Precision Eng. 33, 150–159.

Zhang Xiaobing; Yin Shaohui; Zhu Kejun; et al. 2013. Simulation technology of compression molding spherical glass lens. Optical Tech. 39(3), 204–211.

Zhao Yanzhao & Yin Hairong. 2006. Chemical Industry Press. Beijing: Chemical Industry Press.

Zhou Tianfeng. 2009. Research on high-precision glass molding press for optical elements. Tokyo: Tohoku University.

Civil, Architecture and Environmental Engineering – Kao & Sung (Eds)
© 2017 Taylor & Francis Group, ISBN 978-1-138-02985-9

Temperature field analysis of a porous modified concrete composite pavement

Yonghong Wang
School of Transportation, Wuhan University of Technology, Wuhan, China

Ximing Tan
CCCC Second Highway Consultants Co. Ltd., China

Xiongjun He
School of Transportation, Wuhan University of Technology, Wuhan, China

Tao Liu
Xianning Anda Highway Maintenance Engineering Co. Ltd., China

Yu Wang
Xingda Luqiao Hubei Company, China

Cheng Wang
CCCC INVESTMENT Co. Ltd., China

ABSTRACT: The pavement structure is usually under the impact of frequent changes in the temperature when it is completely exposed to the external environment. The solar and atmospheric radiation energy is part of the energy that can be reflected from the pavement surface. But the remaining energy will be stored in the form of heat in the pavement. This will lead to high road surface temperatures, and the pavement structure will be affected by the dynamic changing environmental conditions. It is known from the long highway maintenance experience that the road surface temperature can seriously damage the pavement structure strength and pavement performance. Therefore, it is very important to analyze the influence of the temperature field.

1 INTRODUCTION

The solar radiation and atmospheric radiation energy can be reflected from the pavement surface. As is known, the remaining energy will be stored in the form of heat that is stored in the pavement. It will cause high road surface temperature and will be affected by the dynamic changing environmental conditions (Elmer C. Hansen, 2013). It is known from the long highway maintenance experience that the road surface temperature can seriously damage the pavement structure strength and pavement performance, so it is significant to analyze the influence of the temperature field.

2 ESTABLISHMENT OF THE TEMPERATURE FIELD CALCULATION MODEL

2.1 *Unit selection*

In the process of analyzing the temperature field, the 3D solid model is used to simulate the temperature field of the pavement structure (Bhutta m, Tsurutak, 2013). It usually analyses the temperature field of the pavement structure in the calculation by the finite element software ANSYS, and put the SOLID70 THERMAL unit of the three-dimensional solid thermal unit as the thermal unit (Bentz D P, 2007). In the experiment, put it to be equivalent to structural unit SOLID45 in the calculation of temperature stress (Huang Hui, 2010). The SOLID70 unit structure model is shown in Figure 1 and the SOLID45 structure model is shown in Figure 2.

SOLID70 is to be used as a three-dimensional solid element for the analysis of three-dimensional heat conduction. The element is defined by 8 nodes, as well as the characteristics of orthotropic materials (Tennis P D, 2004). This unit is applied to three-dimensional, steady-state or transient thermal analysis for analysis. In view of the characteristics of this unit, the temperature field model of PMC is established by using the element.

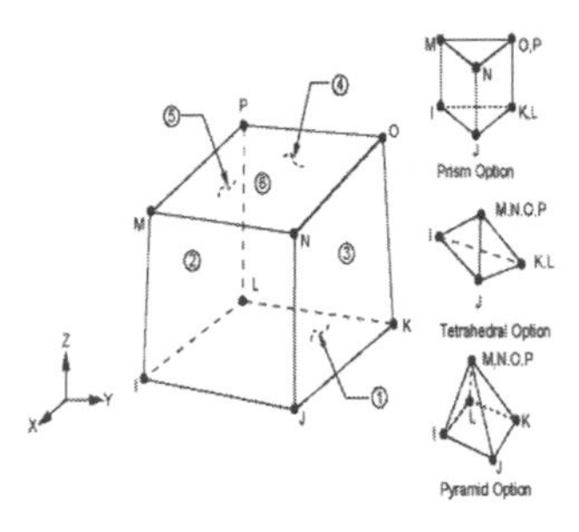

Figure 1. SOLID70.

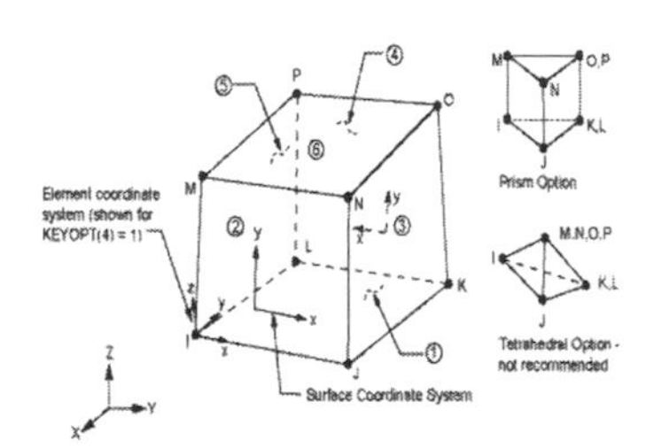

Figure 2. SOLID45.

2.2 *Computational model*

According to the characteristics of porous modified concrete and regional characteristics, it can determine the modified porous cement concrete pavement structure in the cross-section that is shown in Figure 3. The road width is 12 meters and the top layer is the asphalt layer with a thickness of 4 cm. The top layer is regarded as a function of the asphalt layer that provides the structure of road roughness. The bottom layer is a porous modified cement concrete layer with a thickness of 30 cm. The sub-base is a cement stabilized macadam sub-base and its thickness is 18 cm. The gravel slope is 1:1.5. When the temperature reaches a certain depth, the temperature remains the same. Consulting from the literature, the depth of the foundation is 3m. The calculation of temperature field and temperature stress is based on the model.

In the calculation and analysis by ANSYS software, it usually employs the cross-section as the basis. Use the expansion foundation form to establish the finite element model of temperature field, then to divide the grid and the temperature field model of the grid. This is shown in Figure 4.

2.3 *Boundary conditions and calculation parameters*

In the pavement structure, the temperature change is relatively small. The pavement structure parameters can be regarded as constant. Thermal

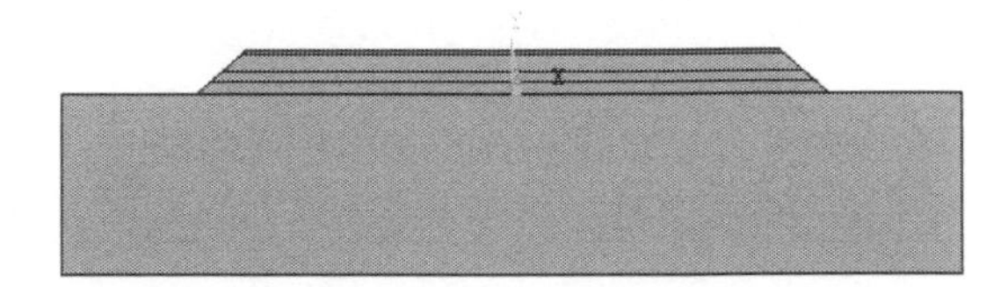

Figure 3. Typical road cross-section.

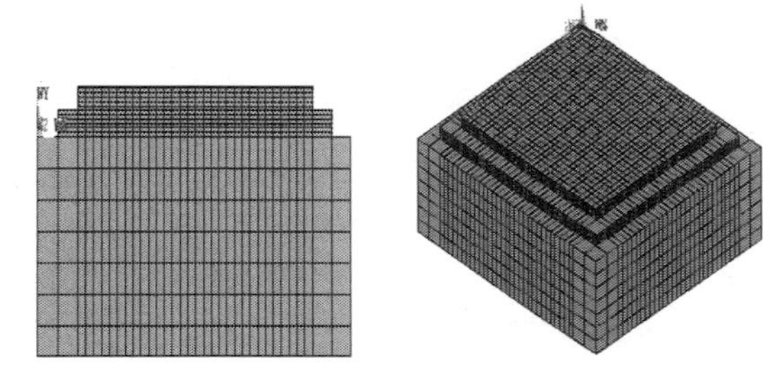

Figure 4. Grid division of the temperature field model.

Table 1. Thermal physical parameters of the road structure.

Material type	Temperature coefficient α (1/°C)	Thermal conductivity (W/m.°C)
Cement concrete	0.6×10^{-5}	1.1
Crushed stone	0.8×10^{-5}	1.2
Cement grave	1.4×10^{-5}	1.2
Subgrade	5×10^{-5}	1.3

Table 2. Thermal physical parameters of the road structure.

Temperature (°C)	−20°C	0°C	20°C	60°C
Thermal conductivity (W/m.°C)	1.84	1.22	1.56	2.36

parameters of the cement concrete layer are little affected by temperature, which can be seen as constant. According to the study on the pavement structure layer, the thermal parameters are shown in Table 1.

The thermal conductivity coefficient of different temperatures is shown in Table 2, and the temperature coefficient is 2×10^{-5} (1/°C). The heat exchange coefficient is selected by the different climate.

3 PMC TEMPERATURE FIELD ANALYSIS

3.1 *Climatic condition analysis of different natural zoning*

In the calculation of temperature stress, the changes of temperature field are mainly caused by the temperature change of climate and it is mainly affected by solar radiation. According to the specification and China atlas of climate division and other related materials, it is classified as the temperature characteristics of different natural regionalization that is shown in Table 3.

According to the different climatic conditions, it usually calculates the temperature field distribution under the different natural conditions; the climatic zoning is only considered from the first level division.

3.2 *Analysis of temperature field distribution in different natural zoning*

According to the model and calculation parameters, the temperature field distribution of different natural zones in different seasons can be obtained. The maximum temperature stress is mainly occurred in the hot summer season about the analysis in pavement temperature. It chooses the representative weather to analyze the temperature field that is the unfavorable situation. According to the model, it can get the distribution of temperature field in a moment of pavement structure and it is shown in Figure 5.

Table 3. Temperature characteristics of different natural regionalization.

Natural zoning	II, V	III	IV, VI	VII
Maximum temperature (°C)	34~37	30~35	36~40	27~31
Minimum temperature (°C)	−10~−5	−12~−8	3~7	−14~−19

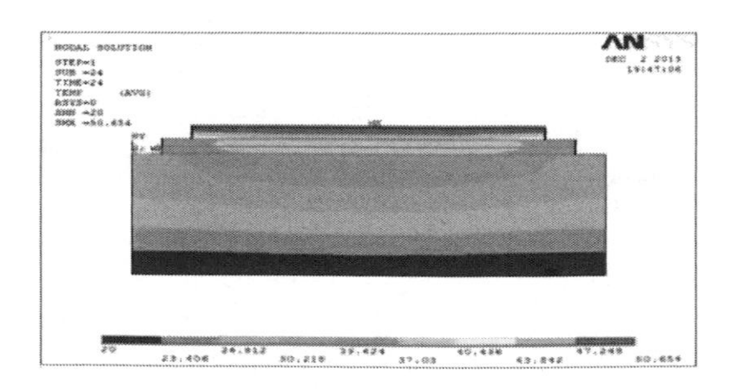

Figure 5. Temperature distribution of the pavement structure in summer 14:00.

According to the different temperature schedules, it chooses 6:00, 14:00, and 22:00 in the summer. It calculates the distribution of road structure temperature in different times by ANSYS. From the metabolic diagram, it can be seen that the temperature changes with depth and time. The specific value is shown in Figures 6, 7, 8 and 9.

It can be seen from Figure 6 that the temperature difference in the pavement structure is the

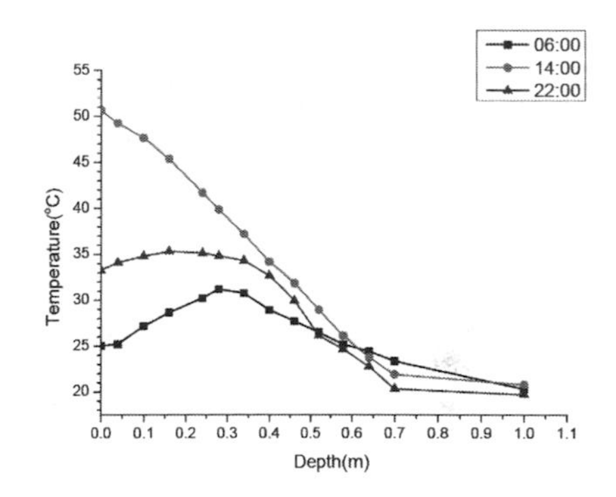

Figure 6. PMCCP temperature changes with depth in area III.

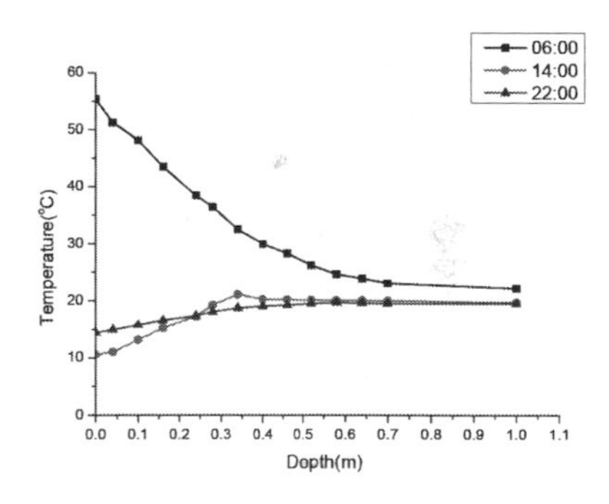

Figure 7. Temperature changes of PMCCP in area V.

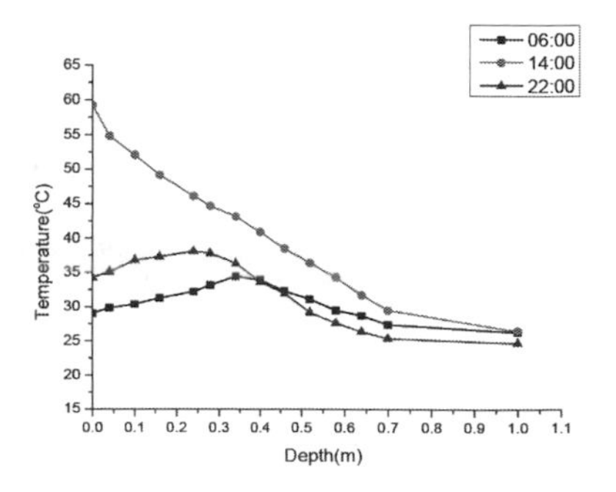

Figure 8. IV, VI PMCCP temperature changes with depth.

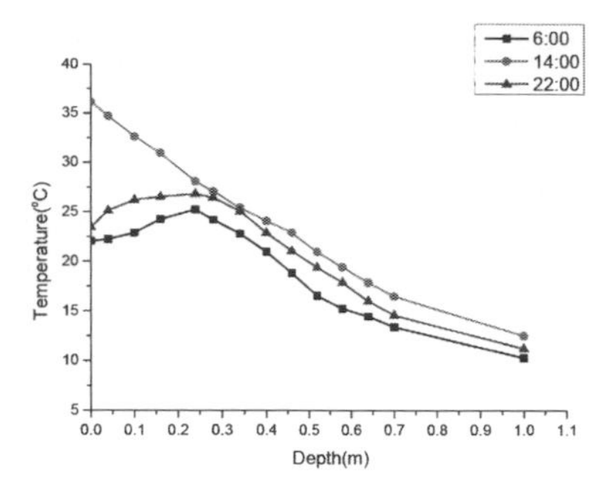

Figure 9. VII PMCCP temperature changes with depth.

largest at 14:00 in the natural division III district of summer. The porous cement concrete layer will reach the maximum positive temperature gradient with depth from the road extending to the bottom of the pavement. The temperature changes from 50 centigrade down to 20 centigrade and the temperature changes obviously. The pavement structure increases first, then show a decreased trend in the surface layer when the time is 6:00 and 22:00. The temperature reaches the maximum at the bottom layer, and then decreases with the depth change. When the depth is 1 m, the temperature variation is small and basically unchanged.

It can be seen from Figure 7 that the pavement structure temperature gradient reached the maximum temperature at 14:00. It reaches the highest temperature on the road surface. The temperature at the bottom of the road surface increases by 45% at 6:00 in the morning. The temperature in the concrete layer can reach the maximum. The temperature at the bottom of the road surface increases by 25% at 22:00 in the evening. When the depth increases, the temperature will not change with the depth.

We can see from Figure 8 that the summer temperature of the pavement structure has a sinusoidal variation in divisions IV and VI with the change of time. The concrete surface reached maximum at 14:00. In the hot season of summer, the pavement temperature can change from 25 to 60 centigrade. The change range can get to 55%. The middle and the bottom surface changes in the amplitude are not obvious for the relatively low temperature.

It can be seen from Figure 9 that the pavement structure temperature decreases with depth at 14:00. The road surface temperature can reach to 37 centigrade when the depth reaches to 70 cm. But the temperature will be reduced to 16 centigrade when the depth is 70 cm. When the temperature is 12 centigrade, the depth is 1 m. The temperature change is not very obvious.

Table 4. Maximum temperature gradients.

Natural zoning	II, V	III	IV, VI	VII
$T_{g,m}$ (°C/m)	63~68	71~76	67~73	75~81

3.3 *Temperature gradient analysis of different natural zoning*

The pavement temperature changes with the depth. The temperature gradient of the modified porous cement concrete slab is the ratio of the top temperature, fingerboard thickness and the bottom plate. The temperature gradient plays an important role in the temperature field analysis. The temperature gradient can reflect the temperature change of the pavement structure more intuitive.

According to the calculation result of the temperature field, the temperature gradient value is recommended, and the results are shown in Table 4.

It can be seen from Table 4 that the maximum temperature gradient of the porous cement concrete slab is 60~80°C/m, but the ordinary concrete maximum temperature gradient is up to 80~100°C/m. The temperature gradient of porous concrete is less than common concrete.

3.4 *Influence of the asphalt layer thickness on the temperature field*

Because of the temperature sensitivity of asphalt materials, the temperature of the porous cement concrete surface layer will change when the asphalt surface layer is added to the cement concrete surface layer. It is proposed to be the appropriate temperature gradient correction factor for analyzing the influence of the asphalt layer thickness on the temperature field.

3.4.1 *Infuence of the asphalt layer on the temperature field*

Taking the temperature distribution of III area in summer as an example, the influence of the asphalt layer on the temperature field is analyzed in the table. The porous modified concrete layer temperature change condition is shown in Figure 10.

It can be seen from the Figure 10 that the asphalt road surface temperature is slightly small when the layer that does not pave asphalt compared to a thin layer. The reason is that the asphalt material absorbs a lot of heat in solar radiation and convection. If the PMC layer depth is deeper, the temperature change is smaller. When it reaches the bottom, the temperature basically remains unchanged. The reason is that the heat has certain loss with the heat delivering to the bottom layer. This will result in a little difference when heat reaches the bottom.

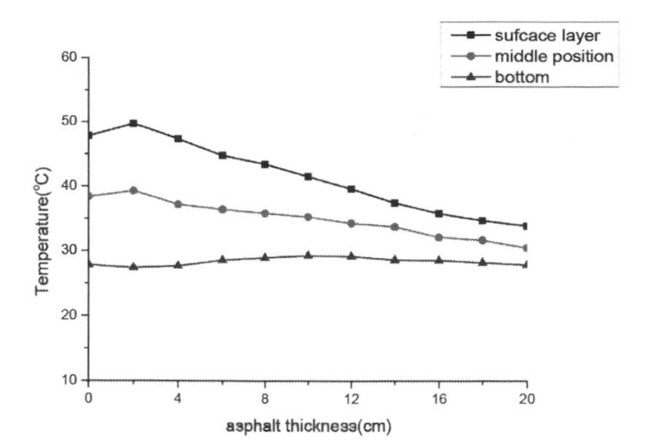

Figure 10. Temperature distribution of porous layer with different asphalt layers in summer.

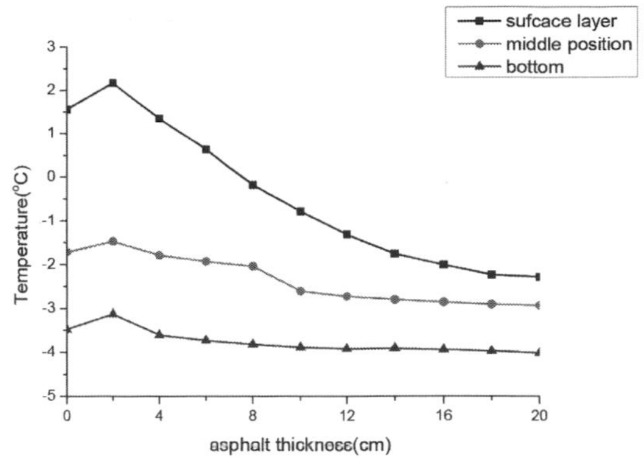

Figure 11. Temperature distribution of the PMC layer in different asphalt layers during winter.

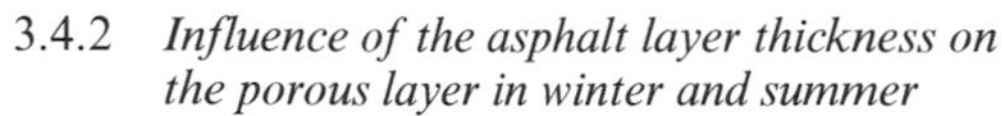

3.4.2 *Influence of the asphalt layer thickness on the porous layer in winter and summer*

The asphalt mixture is a kind of typical temperature sensitive material, in which its mechanical properties and road performance will be affected prominently with the change of temperature. It analyses the influence of asphalt layer thickness on the temperature field during the summer and winter. In winter, the temperature is 2.4 centigrade, the layer temperature is 0 centigrade, and the other parameters are consistent with the calculation method in the same summer. The data can be calculated by the finite element analysis, and the obtained data is shown in Figure 11.

It can be seen from Figure 11 that with the increase of asphalt layer thickness, the PMC layer temperature increased first and then decreased. The law is similar to summer. The temperature reaches the maximum when the PMC layer thickness is 2 cm. Then, when the asphalt layer thickness increases, the PMC layer temperature decreases. The PMC surface temperature change is obvious with the asphalt layer thickness increase. The middle and the bottom temperature changes are not big.

3.4.3 *Temperature gradient correction coefficient of different asphalt layer thicknesses*

The change of temperature gradient has great influence on the temperature stress. The temperature gradient is influenced by temperature, solar radiation and pavement material characteristics. Due to the thermal conductivity of different materials, the effects of heat transfer and diffusion are different. The asphalt layer gets the heat from the atmosphere by the action of radiation and convection. It transfers the heat to the pavement structure by conduction. So, the change of thickness affects the temperature gradient of the concrete slab.

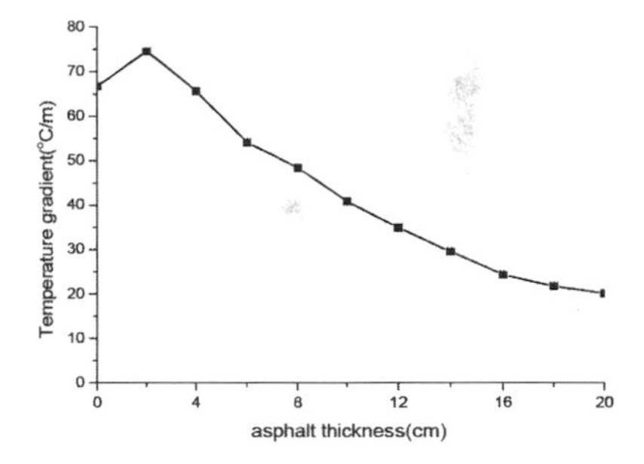

Figure 12. Temperature gradient of PMC layer in different asphalt layers.

The temperature gradient of the PMC layer in different asphalt layers is shown in Figure 12.

It can be seen from Figure 12 that the PMC layer temperature gradient is the maximum in summer. The reason is that the maximum positive temperature gradient is generated at high temperature. At the beginning of the asphalt layer at 0~2 cm, the temperature gradient increases from 67°C/m to 73°C/m. With the increase of the thickness of the asphalt layer to greater than 2 cm, the temperature gradient decreases gradually.

4 CONCLUSION

The main conclusions by analyzing the formation of the temperature field, the selection of thermal parameters and the temperature field of the porous modified concrete pavement can be summarized as follows:

1. The distribution of temperature field of the road structure under different conditions by the temperature field model was calculated.

The distribution of the temperature field in different zoning conditions in summer was obtained by analyzing the temperature characteristics of different natural divisions. The temperature gradient in different regions was recommended by the calculation results.

2. The temperature gradient was found to be highest in summer and temperature stress was found to be in the most unfavorable conditions.
3. The appropriate temperature gradient values were recommended in different natural regions by analyzing the change in the temperature gradient of the PMC layer under different natural regionalization and analysis results.

REFERENCES

Bentz D P, Martys N S. A Stokes permeability solver for three-dimensional porous media[M]. Gaithersburg, MD, USA: US Department of Commerce, Technology Administration, National Institute of Standards and Technology, 2007.

Bhutta m, Tsurutak, Mirza J. Evaluation of high performance porous concrete propoties [J]. Contruction and Building Material, 2013, 31(6):67–73.

Elmer C. Hansen. Field effects of water pumping beneath concrete pavement slabs[J]. Journal of Transportation Engineering, 2013, 117(6):679–697.

Huang Hui. Study on thermal properties and testing methods of asphalt and asphalt mixture [D] Changsha: Changsha University of Science and Technology 2010.

Tennis P D, Leming M L, Akers D J. Pervious concrete pavements[M]. Skokie, IL: Portland Cement Association, 2004.

Civil, Architecture and Environmental Engineering – Kao & Sung (Eds)
© 2017 Taylor & Francis Group, ISBN 978-1-138-02985-9

A retrieval method based on a CBR finite element template

Qinyi Ma, Dapeng Xie, Lihua Song & Maojun Zhou
Department of Mechanical Engineering, Dalian Polytechnic University, Dalian, China

ABSTRACT: A framework of the finite element analysis template case database is systematically constructed based on Case-Based Reasoning (CBR). The analysis information, feature information, and attribute information are used as indices to achieve the reasoning. The information in the finite element analysis case is structurally described by means of ontology representation. The conceptual similarity computation method based on semantic information is used to compute the similarity between the case analysis information, the characteristic information and the attribute information in the process of using the basic information to retrieve the case. The priority of retrieval information is defined, and the different weights are assigned in the process of similarity calculation. On this basis, the global similarity is computed to obtain the comparison between the cases.

1 INTRODUCTION

Case-Based Reasoning (CBR) is a reasoning method adjusting the solution of previous problem to solve the current problem (Reisbeck C K, 1989). The basic process of CBR mainly includes four aspects: case retrieval, case reuse, case modification and case study. Among them, the main content of case retrieval is to retrieve the same or similar historical case set from the existing case base, and solve the current problems with the knowledge experience of historical cases. CBR is mainly based on the "similar problem with similar solution", and it is one of the key technologies of case based reasoning to retrieve the target case which is similar to the source case.

The template case of finite element analysis includes the relevant knowledge and experience of analysis information, modeling characteristic information and attribute information. Because of the lack of analytical thinking, the new users will try to find some similar historical analysis cases and learn from them to find suitable method. It is the key for case retrieval to establish the finite element analysis template case library.

Finite element analysis template retrieval based on CBR can be divided into four steps: firstly, retrieving similar cases; then, solving the current problem with the case of information and knowledge; revising the solution; finally, learning experience knowledge in the process of case solving and coping with the needs of the case retrieval in the future. The case retrieval process is illustrated in Figure 1. The main standard of case retrieval is that the source case and the target case are labeled with the same semantic structure tree, so that the structure and content are comparable in form. For example, source case A is indexed according to the basic information of the case. Case basic information includes analysis information, modeling characteristic information and attribute information. These details are respectively marked using the semantic annotation, and the corresponding semantic annotation file is the input for the case base. A higher matching target case Y will be found by the search algorithm and reach the solution of case A.

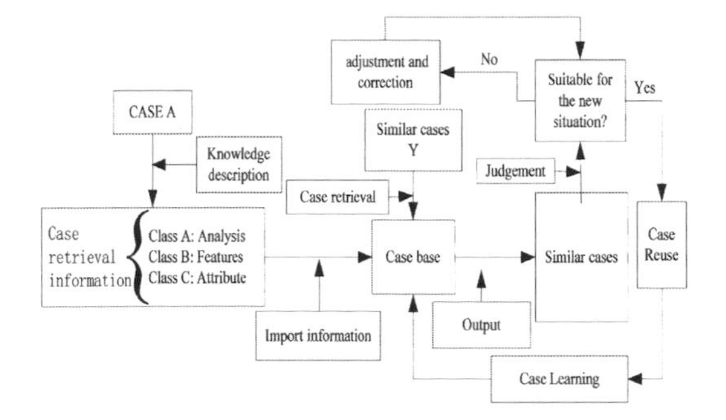

Figure 1. Framework of CBR.

2 RESEARCH STATUS

CBR is an important technology in the field of artificial intelligence. As early as the end of 1970s, some scholars have studied CBR, and the early CBR thought comes from the Rog Schank (Sha Zhongyong, 2014). In 1983, Janet Kolodner adopted the idea of Rog Schank, and developed the first CBR system CYRUS. With the development of CBR technology, the model of CBR has been gradually maturing. In the early stage of the development of the technology, Aamodt (Aamodt A, 1994) and Hunt (Hunt J, 1995) proposed the process oriented CBR model. Subsequently, combined with other technologies in the field of artificial intelligence, BeddoeG R proposed GACBR (Beddoe G R, 2006), and Kobti Z and others proposed a CBR model based on ontology (Kobit Z, 2010). The keyword matching was used in these CBR models, case information covered by the keyword for the whole case is limited, and it easily ignored the key semantic information, so the system is difficult to get the accurate matching case by keyword matching. This method has certain limitations to a great extent. In order to solve the problems of keyword matching, Li Botao and Niu Qinzhou proposed a CBR retrieval method based on FCA semantic understanding and the similarity distance between the domain ontology knowledge and knowledge element was taken as retrieval basis. This paper mainly studies the application and the realization of CBR in the field of finite element analysis.

3 FRAMEWORK OF THE MODEL CASE

The key of case retrieval is how to correctly and objectively define and quantify the degree of similarity between cases. In the process of long-term using CAE, people will accumulate a lot of analysis design experience and knowledge. If the method of CBR can be used in the finite element analysis field, and the useful information is extracted from the case analysis as the index information and reasoning application, the threshold and efficiency of finite element analysis will be greatly improved in the future.

According to the understanding and research of case knowledge of finite element analysis, a specific FEA model should include the analysis information, basic characteristic information, material attribute information and so on. The case analysis information can include the analysis of types, boundary conditions, load conditions, constraints types, element types, mesh division and so on. The basic feature information of the model case includes model specifications, the use of the field, functional information, modeling information, etc. Model attribute information mainly includes material properties, such as material name, material density information, Poisson's ratio, elastic modulus and other basic information. According to the relevant information of case analysis, application of domain ontology knowledge and expert experience and knowledge, this paper constructs

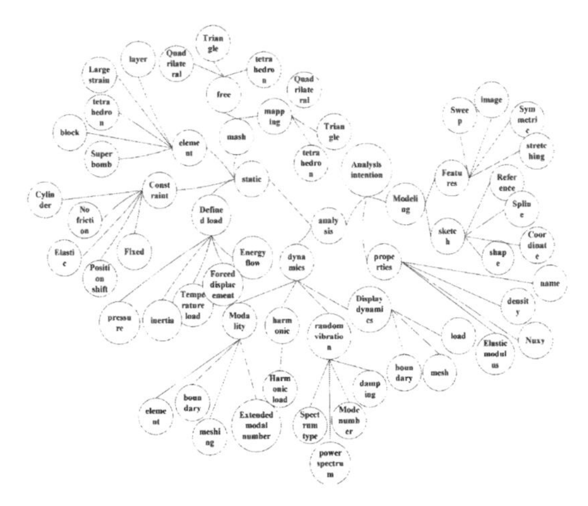

Figure 2. Ontology semantic tree.

an ontology semantic tree based on model information, and semantic annotations for cases using the semantic tree structure and specification. The basic form of an ontology semantic tree is shown in Figure 2. Each node in the tree has corresponding semantic knowledge, and the sub node and the root node have some knowledge of inheritance. For example, the root node represents the domain knowledge of A in the semantic tree, and its sub node represents the domain knowledge of B, namely $B \cap A = B$, $A > B$.

4 SIMILARITY CALCULATION OF THE ONTOLOGY SEMANTIC TREE

Usually, the case is retrieved by the single information. This method will lead to a low retrieval accuracy, because single information can't fully express the case. In the literature, Lu Ying (Lu Ying, 2015) put forward the application of precursory information as the accident case retrieval index case; although the argument is reasonable, the information type is too simple. This paper presents a retrieval method of the finite element template based on CBR, according to the model analysis information, feature information and attribute information, one by one to match by the priority order, the search range will be gradually reduced, and the retrieval accuracy is improved. This process is shown in Figure 3.

In the process of case retrieval using model information, we often think that some information of cases does not have intersection, but in fact there is a logical relationship between the information in the same semantic tree. So, this paper presents a calculation method for the similarity of the

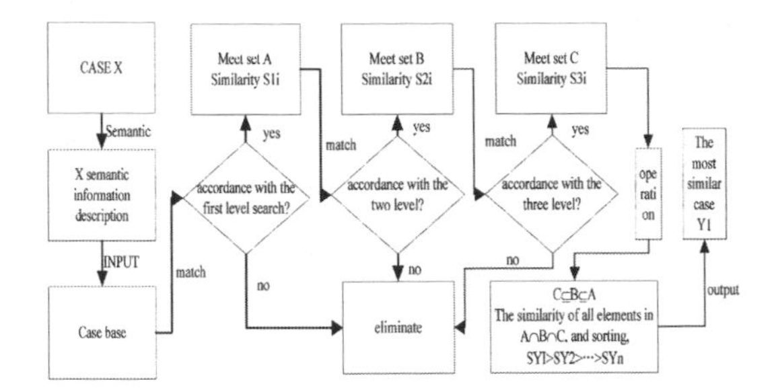

Figure 3. Three-level retrieval.

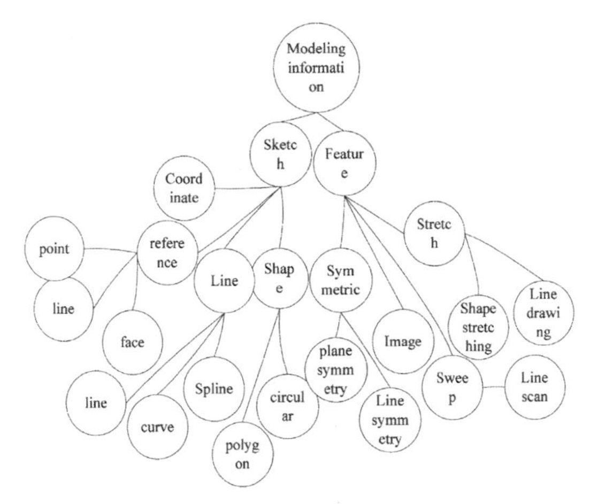

Figure 4. Feature semantic tree.

ontology semantic tree. The knowledge ontology is constructed according to the relevant field knowledge of FEA, case information is conceptualized and semantic, so that each case has the same semantic standard. So, there is a semantic hierarchy and semantic distance relationship between cases.

Taking the modeling feature information of the case as an example, the semantic relation of model feature information is constructed, as shown in Figure 4. Semantic similarity can be divided into two parts: direct similarity and indirect similarity. Direct similarity can be obtained according to the depth and distance between the semantic nodes; indirect similarity is a similar degree between the upper root nodes of the semantic node. This paper calculates the semantic similarity based on the inheritance relationship.

4.1 *Calculation of direct similarity*

Definition 1. The level of the semantic node in the semantic ontology tree is called the semantic depth, and it is denoted as f. The maximum depth of the semantic tree is denoted as m. In the semantic tree, the number of line segments that are connected to the shortest path of the two semantic nodes is called the semantic distance, denoted as d.

The ontology semantic tree is constructed by model modeling information; its each node represents certain semantic information. The ontology semantic tree describes the relationship of the position and the information between semantic. In the model information semantic tree, the total depth of the tree is m = 4, and the direct similarity can be obtained from the depth and the semantic distance of the semantic node in the tree structure.

Definition 2. In the semantic tree, when the semantic distance is larger, the similarity is smaller; when the depth difference is smaller, the similarity is bigger. The direct similarity between C1 and C2 is defined as,

$$Sc1, c2 = \frac{a}{d(c1, c2) + a * \max\left[\left|f(c1) - f(c2)\right|, 1\right]} \tag{1}$$

Among them, f (c) represents the semantic depth of node c; d(c1, c2) represents the semantic distance of node c1 and c2; a represents the coordination coefficient, generally take a = 2.

Basically, formula (1) is used to calculate the similarity of the information in Figure 4. Comparison of the similarity between the linear correlation information and the spline information is S_{t1}. Comparison of the similarity between the linear correlation information and the reference plane is S_{t2}. Comparison of the similarity between the linear correlation information and the face is S_{t3}. The calculation procedure is as follows:

1. The semantic depth of the linear correlation information and the spline is 4, that is $f(C_{11}) = f(C_{12}) = 4$; the semantic distance is $d(c_{11}, c_{12}) = 2$.
2. The semantic depth of linear correlation information and reference plane information is 4, that is, $f(C_{21}) = f(C_{22}) = 4$; the semantic distance is $d(c_{21}, c_{22}) = 4$.
3. The semantic depth of linear correlation information and the face of the information is 4, that is, $f(C_{31}) = f(C_{32}) = 4$; the semantic distance of $d(c_{31}, c_{32}) = 6$.
4. a = 2.

By formula (1), direct similarity is calculated.

$$S_{t1} = \frac{a}{d_{(c_{11}c_{12})} + a \times \max\left[\left|f(c_{11}) - f(c_{12})\right|\right]} = \frac{2}{2+2} = 0.5$$

$$S_{t2} = \frac{a}{d_{(c_{21},c_{22})} + a \times \max\left[\left|f(c_{21}) - f(c_{22})\right|\right]} = \frac{2}{4+2} \approx 0.33$$

$$S_{t3} = \frac{a}{d_{(c_{31},c_{32})} + a \times \max\left[\left|f(c_{31}) - f(c_{32})\right|\right]} = \frac{2}{6+2} = 0.25$$

4.2 *Indirect similarity calculation*

Definition 3. The upper node of the semantic node is called as the parent node. Indirect similarity is the similarity of upper nodes, recorded as S_f. The specific algorithm can be used to calculate S_f according to formula (1).

1. The semantic depth of the linear correlation information and the spline is 3, that is, $f(C_{11}) = f(C_{12}) = 3$; the semantic distance is $d(c_{11}, c_{12}) = 0$.
2. The semantic depth of linear correlation information and reference plane information is 3, that is, $f(C_{21}) = f(C_{22}) = 3$; the semantic distance is $d(c_{21}, c_{22}) = 2$.
3. The semantic depth of linear correlation information and the face of the information is 3, that is, $f(C_{31}) = f(C_{32}) = 3$; the semantic distance of $d(c_{31}, c_{32}) = 4$.

$$S_{f1} = \frac{a}{d_{f(c_{11},c_{12})} + a \times \max\left[\left|f_f(c_{11}) - f_f(c_{12})\right|\right]} = \frac{2}{0+2} = 1$$

$$S_{f2} = \frac{a}{d_{f(c_{21},c_{22})} + a \times \max\left[\left|f_f(c_{21}) - f_f(c_{22})\right|\right]} = \frac{2}{2+2} = 0.5$$

$$S_{f3} = \frac{a}{d_{f(c_{31},c_{32})} + a \times \max\left[\left|f_f(c_{31}) - f_f(c_{32})\right|\right]} = \frac{2}{4+2} \approx 0.33$$

4.3 *Semantic similarity computation*

Definition 4. The semantic similarity should not only satisfy the direct similarity, but also meet the indirect similarity. According to the probability principle, the probability that two conditions are met is equal to the product of both probability. So, the semantic similarity can be obtained by comprehensive direct similarity and indirect similarity, Sc representation. So, we define the semantic similarity between two semantic C_1 and C_2 as follows:

$$Sc(c_1, c_2) = S_t(c_1. c_2) * S_f(c_1, c_2) \tag{2}$$

From formula (2), the semantic similarity can be calculated:

$$S_{c1} = S_{t1} \times S_{f1} = 0.5 \times 1 = 0.5$$

$$S_{c2} = S_{t2} \times S_{f2} = 0.5 \times 0.33 = 0.165$$

$$S_{c3} = S_{t3} \times S_{f3} = 0.25 \times 0.33 = 0.0825$$

According to the above results, the position relation between a line and a spline in the semantic tree is that two nodes have the same parent node, so the similarity between the two is largest in line with the actual situation; linear relationship with the rest of the two are consistent in common sense. This method considers the semantic relation from the semantic level and structure, which makes the case matching more practical.

In the actual situation, the case similarity calculation includes the local similarity computation and the global similarity computation. The global similarity is obtained based on the local similarity according to the different weight influence [10]. There are two FEA cases X and Y; $X = \{P_x, Q_x, F_x\}$ and $Y = \{P_y, Q_y, F_y\}$; P is the case analysis information, Q is the model characteristics information, F is the case attribute information. P_x is the case of X analysis information, $P_x\{P_{x1}.... P_{xi}...P_{xm}\}$; P_y is the case of Y analysis information, $P_y\{P_{y1}...P_{yi}...P_{yn}\}$. In the same way, model characteristic information and attribute information are expressed. The case X and Y global similarity is S (X, Y), that is the case similarity.

$$S(X,Y) = \left[\sum_{i=1}^{\min(m,n)} w_i \max Sc(XPi, YPi) + \sum_{j=1}^{\min(r,z)} w_j \max Sc(XQi, YQi) + \sum_{k=1}^{\min(e,u)} w_k \max Sc(XFi, YFi)\right] \Big/ \left(\sum_{i=1}^{\min(m,n)} w_i + \sum_{j=1}^{\min(r,z)} w_j + \sum_{k=1}^{\min(e,u)} w_k\right) \tag{3}$$

1. In formula (3), w_i is the weight of the analysis information; w_j is the weight of the feature information; w_k is the weight of the attribute information. According to the priority of the case retrieval, provisions wi >wj > wk.

2. Sc (XPi, YPi) is the local similarity that is i class analysis information of X and Y.
3. Sc (XQi, YQi) is the local similarity that is i class feature information of X and Y.
4. Sc (XFi, YFi) is the local similarity that is i class attribute information of X and Y.

5 CONCLUSION

The ultimate goal of the CBR finite element template is to extract the solution of the matching case, and obtain a correct solution to achieve the case retrieval according to the related parameter information of the source case. This paper attempts to use semantic to express all analysis cases by integrating an ontology semantic tree in the case-based reasoning system, and establish the semantic structure and the relationship of case information. In this way, the similarity algorithm based on priority is proposed. The system is easy to determine the most matching case by comparing the similarity value, and the case retrieval is achieved.

ACKNOWLEDGEMENT

This work was financially supported by the National Natural Science Foundation of China (No. 51305051), the program for Liaoning Excellent Talents in University (No. LJQ2015007), the Liaoning Province Natural Science Foundation (No. 201602060), and the Liaoning Province Natural Science Foundation (No. 2014026006).

REFERENCES

Aamodt A, Plaza E. Case-based reasoning: Foundation issues, methodological variations, and system approaches[J]. AI Communications, 1994, 7(1): 39–59.

Beddoe G R, Petrovic S. Selecting and weighting features using a genetic algorithm in a case-based reasoning approach to personnel rostering[J]. European Journal of Operational Research, 2006, 175(2):649–671.

CBR database organization and retrieval method based on semantic understanding and FCA [J]. Journal of Guilin University of Technology, vol. 35, No. 2. 5, 2015. 383–390.

Hunt J. Evolutionary Case Based Design[M]//Progress in Cased-Based Reasoning. Berlin, Heidelberg: Springer, 1995: 17–31.

Kobit Z, Chen D, Baljeu A. A Domain Ontology Model for Mould Design Automation[M]//Advances in Artificial Intelligence. Berlin, Heidelberg: Springer, 2010:336–339.

Li feng, Wei ying. Approach to Case Retrieval in Distributed Environment Based on Semantic Similarity Caculation [J]. Computer Engineering. Vol. 33, No. 9. May 2007.

Lu Ying, Li Qiming. Construction of accident case database for subway operation based on case-based reasoning[J]. JOURNAL OF SOUTHEAST UNIVERSITY(Natural Science Edition), 2015, 9.45(7): 990–995..

Reisbeck C K, Schank R C. Inside Case Based Reasoning[M]. Hillsdale:Lawrence Erlbaum Associates, 1989.

Sha Zhongyong, Shi Zhongxian. Case Retrieval Mathods of Public Crisis Based on Semantic Similarity[J]. Information and Documentation Services, 2014, 6: 5–11.

Civil, Architecture and Environmental Engineering – Kao & Sung (Eds)
© 2017 Taylor & Francis Group, ISBN 978-1-138-02985-9

Improved reduction of graphene oxide through femtosecond laser pulse trains

Ruyu Yan, Pei Zuo & Xiaojie Li
Laser Micro/Nano Fabrication Laboratory, School of Mechanical Engineering, Beijing Institute of Technology, Beijing, China

ABSTRACT: We report a novel and effective reduction method of graphene oxide films using femtosecond laser double-pulse trains. Femtosecond laser reduction of graphene oxide films under ambient air by using three laser processing approaches: (1)1 KHz laser pulses, (2) 76 MHz laser pulses, and (3) 76 MHz double-pulse trains were investigated for comparison. As characterized by X-ray photoelectron spectroscopy, scanning electron microscopy and Raman techniques, femtosecond laser double-pulse trains with repetition rates of 76 MHz and a delay time of 3 picosecond could decrease the content of oxygen atoms in graphene oxide films from ~33.3% to ~15.1% and maintain the integrity of reduced graphene oxide sheets, which demonstrated the most effective reduction. This research adequately reveals the high-efficient reduction and patterning ability of fs laser double-pulse trains on graphene oxide for further large-area applications in various fields.

1 INTRODUCTION

Since the discovery of graphene in 2004, extensive studies regarding the synthesis of this promising material have been reported for widespread applications (Zhu et al. 2010). The production of graphene-based materials with various dimensions, quality and patterns is crucial for the market of graphene (Novoselov et al. 2012, Xiong et al. 2014). Significant advances in this field have been achieved, including growth by chemical vapor deposition, thermal or liquid phase exfoliation of graphite, and reduction of Graphene Oxide (GO) (Compton et al. 2010). GO is an electrical insulator which contains numerous Oxygen-Containing Groups (OCGs). Altering the OCGs of GO through chemical, thermal or photo reduction could restore its electrical properties and make it close to graphene (Chen et al. 2012). Among these reduction methods, photo-reduction, especially laser reduction, displays particular superiority because of its flexible, low-cost, eco-friendly and high-efficiency properties (Strong et al. 2012).

Femtosecond (fs) laser has been broadly adopted for high precision, high quality, flexible, and three-dimensional micro/nanoscale fabrication (Deng et al. 2015). Recently, femtosecond (fs) laser reduction and patterning of GO films has been reported. For instance, graphene microcircuits were created on GO films by direct fs laser reduction with 80 MHz repetition rates according to preprogrammed patterns (Zhang et al. 2010). In addition, fabrication of highly conductive graphene layers on flexible substrates were realized by fs laser nonthermal reduction of GO films (Kymakis et al. 2013). However, the reduction degree and the integrity of the graphene lattice in reduced Graphene Oxide (rGO) films via fs laser were still restricted when compared with those reduced by continuous (El-Kady et al. 2012) or long pulse (Guo et al. 2012) lasers. Hence, further study on the improvement of reduction degree and integrity of rGO layers using fs laser is necessary.

Fs laser pulse trains technique which is based on electron dynamics control by using temporally shaped fs laser, has been reported to hold great advantages in the fabrication of ripples (Shi et al. 2013), high-quality concave microarrays (Zhao et al. 2015), and many other micro/nanostructures. In this work, we proposed an effective reduction method by using fs laser double-pulse trains with 76.08 MHz repetition rates (2M-fs laser), which has not been reported previously. Conventional fs laser with 1 kHz (K-fs laser) and 76.08 MHz (M-fs laser) repetition rates were studied for comparison. We investigated 2M-fs laser reduction of GO films by changing the delay time (Δt) between two subpulses. Experiment results demonstrated obvious improvement of reduction by 2M-fs laser with a Δt of 3 picosecond (ps).

2 MATERIALS AND METHODS

2.1 *Preparation of graphene oxide films*

In our experiments, the freestanding GO films with ~30 μm thickness were prepared by vacuum

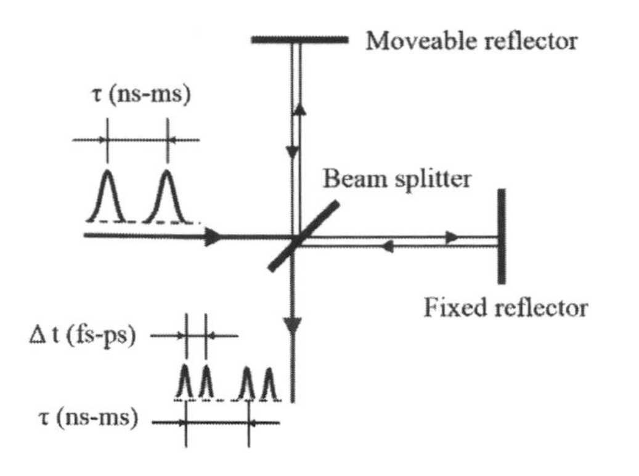

Figure 1. Schematic diagram of the Michelson interferometer.

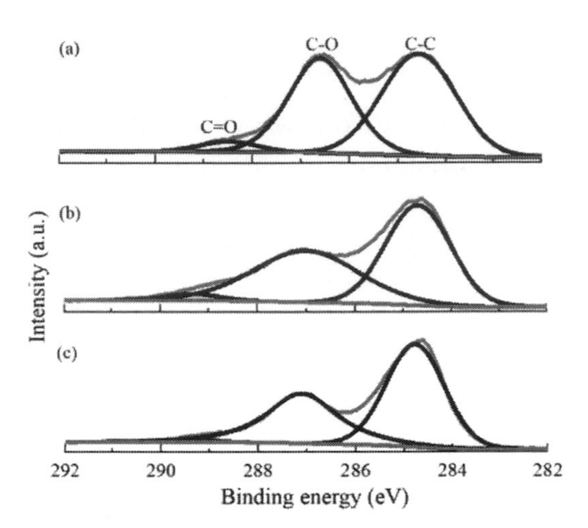

Figure 2. XPS spectra of (a) pristine GO, (b) K-rGO under 1 overlapping K-fs pulses reduction, and (c) K-rGO under 100 overlapping K-fs pulses reduction.

filtration of GO suspensions (XFNANO Advanced Materials Supplier, XF008). The dried GO films were peeled off from the filter membrane directly.

2.2 *Experimental set-up*

A regeneratively amplified Ti:Sapphire laser system (Spectra Physics, Inc.) was used to deliver 50 fs linearly polarized laser pulses with 800 nm central wavelength. Three kinds of laser sources: (1) K-fs laser, (2) M-fs laser and (3) temporally shaped 2M-fs laser were investigated. The reduced GO samples were respectively identified as K-rGO, M-rGO and 2M-rGO by the three types of laser sources we used. The fs laser double-pulse trains were generated via a Michelson interferometer which consisted of a beam splitter, a fixed reflector and a moveable reflector (Fig. 1). The delay time between the two subpulses had an accuracy of 66.7 fs. The splitting ratio for the double pulses was 50:50.

The GO samples were mounted on a computer-controlled, six-axis motion stage (M-840.5DG, PI, Inc.) with a maximum adjustable speed of 2 mm/s. The three kinds of laser beams were normally incident to the surfaces of GO samples under ambient air by a microscope objective lens (5×, N.A. = 0.15). The energy of fs laser beam was measured before the objective lens by a power meter (Newport Corporation). Surface morphology was studied by Scanning Electron Microscopy (SEM). X-ray photoelectron Spectroscopy (XPS) and micro-Raman spectroscopy (632.8 nm light source) were performed to analyze surface chemical and physical properties.

3 RESULTS AND DISCUSSION

3.1 *K-rGO*

Differences between pristine GO and rGO could be observed clearly in the XPS spectra. As shown in Figure 2, the C1s spectra of GO and rGO were decomposed into three peaks, which corresponding to C-C (284.6 eV), C-O (286.6 eV) and C = O (288.5 eV) peaks, respectively (Guo et al. 2012). For pristine GO, the content of oxygen atoms was as high as ~33.3% and the percentage of C-C bond was ~53.0% (Figure 2a). After 1 overlapping K-fs pulses reduction with laser energy of 0.5 μJ, the content of oxygen atoms slightly decreased to ~28.9%. As shown in Figure 2b and Table 1, the contents of C-O and C = O bonds were partially reduced. After 100 overlapping K-fs pulses irradiation, the content of oxygen atoms could further decrease to ~23.0% (Fig. 2c), and the percentage of C-C bond was increased to ~59.7%. The oxygen content of K-rGO could be reduced by increasing the overlapping number of pulses and tuning the laser energy. However, the effect of laser energy was weak. Hence, the reduction efficiency of K-fs laser was relatively low because multi-pulse overlap was necessary for a better reduction degree.

Table 1. XPS components for different samples, normalized to the C-C peak intensity.

Sample	C-C	C-O	C = O
GO	53.0%	42.8%	4.2%
K-rGO (1)	55.2%	41.5%	3.3%
K-rGO (100)	59.7%	39.3%	1.0%
M-rGO	63.1%	34.7%	2.2%
2M-rGO (0 ps)	63.9%	34.1%	2.0%
2M-rGO (3 ps)	70.1%	29.3%	0.6%

Additionally, K-fs laser reduction of OCGs in GO films was accompanied with the destruc-

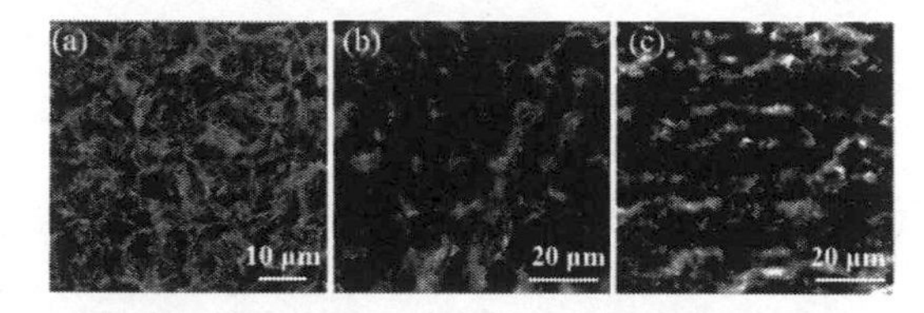

Figure 3. SEM images of typical morphologies for K-rGO respectively under (a) 1, (b) 10 and (c) 100 pulses overlapping K-fs laser reduction.

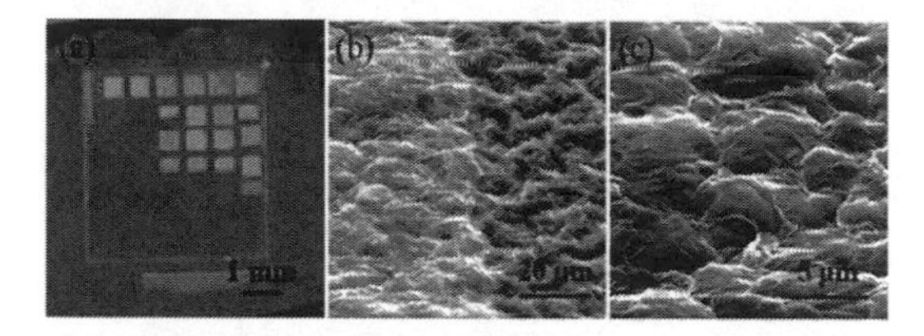

Figure 4. SEM images of M-fs laser ablation of GO films. (a) M-rGO area with silver gray color; (b) comparison between M-rGO surface (left) and pristine GO surface (right); (c) enlarged image of M-rGO surface. The laser energy was ~0.13 nJ, corresponding to a measured laser power of ~10 mW.

tion of graphene sheets, which was a huge drawback for its electronics applications. Figures 3a-c showed the morphologies of K-rGO films respectively under 1, 10 and 100 pulses overlapping K-fs pulses reduction with laser energy of 0.5 μJ. SEM images indicated that the structural integrity of rGO sheets obtained by K-fs laser reduction was rather poor. The conductivity of K-rGO would be greatly restricted even though the OCGs could be removed partly. Lower laser energy could benefit the intergrity of K-rGO layers, however, the reduction degree and reduction efficiency would become much lower.

3.2 *M-rGO and 2M-rGO*

To achieve effective reduction and maintain the integrity of rGO sheets, we studied the reduction of GO films by M-fs laser, which had ultrahigh repetition rates and ultralow single pulse energy. The overlapping number of pulses was limited by the translational speed of the six-axis motion stage. High laser energy could arouse obvious ablation of GO films, which was not conducive to maintain its structural integrity. Meanwhile, low laser energy was insufficient for effective reduction. Therefore, after carefully studying the effect of laser energy and overlapping number of pulses on GO films, laser energy of ~0.13 nJ, scanning speed of 2000 μm/s, and line interval of 5 μm (~1.1×10^6 overlapping number of pulses) were adopted for further study.

Figure 4b showed the comparision between M-rGO surfaces (silver gray area) and GO surfaces (dark area). Compared with K-rGO (Fig. 3), the reduction region in M-rGO could maintain intact. Furthermore, the reduction degree and efficiency by using M-fs laser were much higher than those by using K-fs laser. The content of oxygen atoms could decrease to ~20.2%, and the content of C-C bond could increase to ~63.1%. The accumulation of maxi-multiple pulses with ultralow single pulse energy by using M-fs laser induced effective reduction of GO films without surface damage to M-rGO layers.

Based on the good reduction effect of M-fs laser on GO films, we further investigated the reduction effect of 2M-fs laser, which has not been studied previously. The same laser energy (~0.13 nJ) was adopted for clearer comparison. As illustrated in Table 2, the content of oxygen atoms decreased gradually from ~19.9% when $\Delta t = 0$ ps to ~15.1% when $\Delta t = 3$ ps. Specifically, the proportion of C-C band increased from ~63.9 to ~70.1% (Fig. 5). Some residual OCGs were known to be advantages for numerous electrochemical redox reactions because OCGs could aid electron transfer (Griffiths et al. 2014). The morphologies of 2M-rGO with different delay times were almost the same as those of M-rGO with the same laser energy.

Raman spectra of GO and 2M-rGO with different delay times were analyzed to investigate the reduction effect of 2M-fs laser (Fig. 6). Raman spectra of GO films displayed obvious *D* and *G* peaks at ~1350 and ~1591 cm^{-1}, respectively. The *D* peak was induced by structural disorder, and the *G* peak was the in-plane

Table 2. The content of oxygen atoms in 2M-rGO samples with increasing delay times.

Delay time	0 ps	0.5 ps	1 ps	3 ps
Oxygen atoms	19.9%	19.7%	18.2%	15.1%

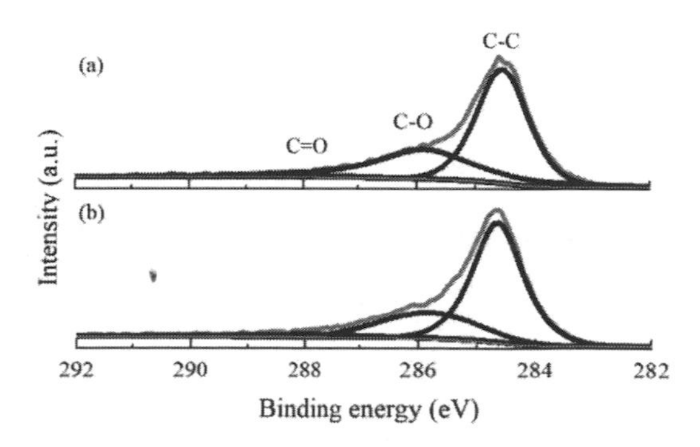

Figure 5. XPS spectra of 2M-rGO reduced by 2M-fs laser with delay time of (a) 0 ps and (b) 3 ps.

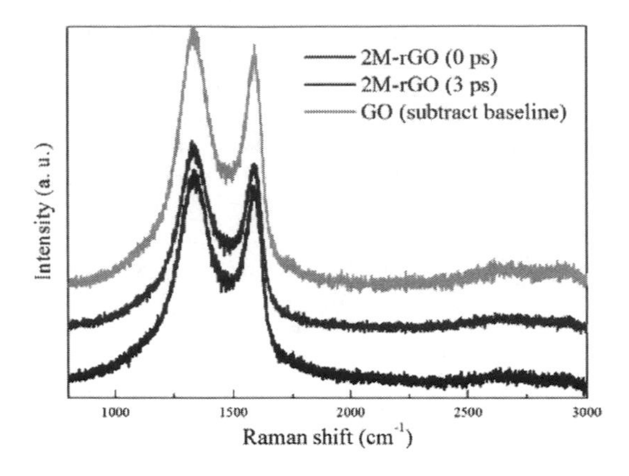

Figure 6. Raman spectra of GO and 2M-rGO with different delay times.

vibration of the C atoms sp^2 network (Kymakis et al. 2013). GO displayed strong fluorescent background. The background fluorescence in the Raman spectra of 2M-rGO was significantly quenched (Senyuk et al. 2015), indicating effective reduction of GO. There was no significant shift of *D* and *G* bands or change of full width at half maximum of the G band among GO, 2M-rGO ($\Delta t = 0$ ps) and 2M-rGO ($\Delta t = 3$ ps). The ratio between *D* and *G* peak intensities (I_D/I_G) increased from 2.41 of pristine GO to 2.65 of 2M-rGO ($\Delta t = 0$ ps), which indicated a slight increase of defects after reduction. After 2M-fs ($\Delta t = 3$ ps) laser reduction, the I_D/I_G ratio of 2.52 was smaller than that of 2M-rGO ($\Delta t = 0$ ps). The more effective removal of OCGs by using 2M-fs ($\Delta t = 3$ ps) laser was responsible for this phenomenon, as we verified by using XPS.

4 CONCLUSIONS

In summary, we studied fs laser reduction of GO films by using three laser processing approaches: (1) 1 KHz fs pulses, (2) 76 MHz fs pulses, and (3) 76 MHz fs laser double-pulse trains. A direct and robust comparison for reduction effect of the three methods was performed. As characterized by XPS, SEM and Raman techniques, 76 MHz fs laser double-pulse trains with delay time of 3 ps was the most effective reduction condition in our experiments, which could decrease the content of oxygen atoms from ~33.3% to ~15.1% and maintain the integrity of rGO films. In addition, the reduction efficiency of 2M-fs laser was much higher than K-fs laser, demonstrating its high-efficient and large-area reduction as well as patterning ability. This study opens up alternative routes for fs laser reduction of GO with superior reduction effect and none surface damage, indicating predictable widespread applications of fs laser pulse trains for processing GO materials.

REFERENCES

Chen, D., Feng, H., & Li, J. 2012. Graphene oxide: preparation, functionalization, and electrochemical applications. *Chem. Rev.* 112(11), 6027–6053.

Compton, Owen, C., & SonBinh, T. N. 2010. Graphene oxide, highly reduced graphene oxide, and graphene: wersatile building blocks for carbon-based materials. *Small* 6, 711–723.

Deng, Z., Yang, Q., Chen, F., Meng, X., Bian, H., Yong, J., Shan, C., & Hou X. 2015. Fabrication of large-area concave microlens array on silicon by femtosecond laser micromachining. *Opt. Lett.* 40(9), 1928–1931.

El-Kady, M. F., Strong, V., Dubin, S., & Kaner, R. B. 2012. Laser scribing of high-performance and flexible graphene-based electrochemical capacitors. *Science* 335(6074), 1326–1330.

Griffiths, K., Dale, C., Hedley, J., Kowal, M. D., Kaner, R. B., & Keegan, N. 2014. Laser-scribed graphene presents an opportunity to print a new generation of disposable electrochemical sensors. *Nanoscale* 6(22), 13613–13622.

Guo, L., Jiang, H. B., Shao, R. Q., Zhang, Y. L., Xie, S. Y., Wang, J. N., & Sun, H. B. 2012. Two-beam-laser interference mediated reduction, patterning and nanostructuring of graphene oxide for the production of a flexible humidity sensing device. *Carbon* 50(4), 1667–1673.

Kymakis, E., Savva, K., Stylianakis, M. M., Fotakis, C., & Stratakis, E. 2013. Flexible organic photovoltaic cells with in situ nonthermal photoreduction of spin-coated graphene oxide electrodes. *Adv. Funct. Mater.* 23(21), 2742–2749.

Novoselov, K. S., Fal, V. I., Colombo, L., Gellert, P. R., Schwab, M. G., & Kim K. 2012. A roadmap for graphene. *Nature* 490(7419), 192–200.

Senyuk, B., Behabtu, N., Martinez, A., Lee, T., Tsentalovich, D. E., & Smalyukh, I. I. 2015. Three-dimensional patterning of solid microstructures through laser reduction of colloidal graphene oxide in liquid-crystalline dispersions. *Nat. Commun.* 6.

Shi, X., Jiang, L., Li, X., Wang, S., Yuan, Y., & Lu, Y. 2013. Femtosecond laser-induced periodic structure adjustments based on electron dynamics control: from subwavelength ripples to double-grating structures. *Opt. Lett.* 38(19), 3743–3746.

Strong, V., Dubin, S., El-Kady, M. F., Lech, A., Wang, Y., Weiller, B. H., & Kaner, R. B. 2012. Patterning and electronic tuning of laser scribed graphene for flexible all-carbon devices. *ACS nano* 6(2), 1395–1403.

Xiong, W., Zhou, Y. S., Hou, W. J., Jiang, L. J., Gao, Y., Fan, L. S., Jiang, L., Silvain, J. F., & Lu, Y. F. 2014. Direct writing of graphene patterns on insulating substrates under ambient conditions. *Sci. Rep.* 4.

Zhu, Y., Murali, S., Cai, W., Li, X., Suk, J. W., Potts, J. R., & Ruoff, R. S. 2010. Graphene and graphene oxide: synthesis, properties, and applications. *Adv. Mater.* 22(35), 3906–3924.

Zhang, Y., Guo, L., Wei, S., He, Y., Xia, H., Chen, Q., & Xiao, F. S. 2010. Direct imprinting of microcircuits on graphene oxides film by femtosecond laser reduction. *Nano Today* 5(1), 15–20.

Zhao, M., Hu, J., Jiang, L., Zhang, K., Liu, P., & Lu, Y. 2015. Controllable high-throughput high-quality femtosecond laser-enhanced chemical etching by temporal pulse shaping based on electron density control. *Sci. Rep.* 5.

Civil, Architecture and Environmental Engineering – Kao & Sung (Eds)
 ISBN 978-1-138-02985-9

A nanoparticle-decorated silicon substrate for SERS via circularly polarized laser one-step irradiation

Ge Meng, Peng Ran & Yongda Xu
Laser Micro/Nano Fabrication Laboratory, School of Mechanical Engineering, Beijing Institute of Technology, Beijing, P.R. China

ABSTRACT: A simple and efficient approach to obtain nanostructured silicon substrates that are uniformly decorated with silver nanoparticles for surface enhanced Raman scattering by using circularly polarized femtosecond laser one-step irradiation in water is reported. The silver nanoparticles are produced by femtosecond laser photoreduction, with a diameter of 40–80 nm in the majority, thereby accounting for 75.6% and the mean diameter is around 66 nm. The irradiated surfaces exhibit excellent SERS performances with a maximum enhancement factor of up to 7.69×10^7 when Rhodamine 6G is used as the probe molecule, which presents a huge potential for trace detection. The nanostructured silicon surfaces decorated with silver nanoparticles can confine and enhance the Raman excitation laser through localized surface plasmon resonance, which played a determined role in such signal enhancement.

1 INTRODUCTION

In 1974, Fleischmann et al. (1974) acquired the Raman spectra of the monolayer pyridine for the first time by using a roughened silver electrode. Since then, Surface Enhanced Raman Scattering (SERS) has been widely researched for several decades (Jeanmaire et al. 1977, Ding et al. 2016, Wei H & Xu H 2013). The following two primary mechanisms are established to explain the enhanced Raman scattering: electromagnetic theory and chemical theory. In terms of electromagnetic theory, the localized surface plasmon resonance is regarded as the main reason for enhanced Raman scattering. In terms of chemical theory, the charge transfer between the chemisorbed species and the metal surface play a significant role in signal enhancement (Cialla et al. 2011). In this work, a roughened silicon surface covered with silver nanoparticles is investigated to enhance the Raman scattering primarily via the electromagnetic mechanism.

Nanostructures enhance the scattering signals via increasing the surface roughness and facilitating the generation of "hot spots", which usually exist between the adjacent noble metal nanoparticles and could greatly enhance the electromagnetic field around them. Femtosecond (fs) laser direct writing has been proven to be an efficient approach to acquire such nanostructures for SERS (Yang et al. 2014, Jiang et al. 2012, Zhang et al. 2013, Diebold et al. 2009). In our previous study, the nanoripple-roughened surface and nanopillar-roughened surface for SERS via linearly polarized fs laser have been fully studied, which exhibit a good SERS performance (Lin et al. 2009, Yang et al. 2015). However, to the best of our knowledge, the circularly polarized fs laser that is used to fabricate the SERS-active substrate has not been reported by far. And as we know, the morphologies and properties of the surface that is fabricated by using circularly polarized laser will be definitely different from that fabricated by using linearly polarized laser (Yasumaru et al. 2003, Varlamova et al. 2006, Huang et al. 2008, Zhu et al. 2006).

In this study, we present a one-step method for fabrication of silver nanoparticles decorated on the structured silicon surface by using circularly polarized fs laser irradiation in silver nitrate solution ($AgNO_3$) and demonstrate its ability for SERS for the first time. The silver nanoparticles are uniformly formed on the nanostructured silicon surface with a mean diameter of 66 nm, which is suitable for exciting localized surface plasmon at the visible light range for related applications. When compared to the multi-step approaches, the one-step approach we proposed to obtain the silver nanoparticle-covered SERS-active substrates is much more simple and efficient.

2 EXPERIMENTAL SETUP

The fs laser processing system consists of a Spectra Physics Spitfire regenerative amplifier (800 nm, 35 fs, 1 KHz), a computer-controlled six-axis motion stage, and a Charge-Coupled Device (CCD) camera. The Gaussian profile laser focused by using a plano-convex lens (f = 100 mm) was incident normally to the sample surface with a focus diameter ($1/e^2$) of 31.6 μm. A quarter-wave plate was applied to obtain a circularly polarized laser. The morphologies were characterized by using a Scanning Electron Microscope (SEM) and the surface components were analyzed by using energy dispersive X-ray spectroscopy. The Raman spectra were recorded with a Renishaw Raman microscope by using a 50× objective, and the maximum exciting power was 6 mW.

Silicon wafers (111) were cleaned in an ultrasonic bath with acetone for 15 min, rinsed with distilled water for 1 min, and then air-dried at room temperature. The cleaned silicon wafers were fixed in the bottom of a cuvette (25 mm × 25 mm × 30 mm) and submerged in a liquid. The height of the liquid layer was about 5 mm. Two silicon samples were raster-scanned in water and silver nitrate solution with circularly polarized laser, respectively. For comparison, two additional silicon samples were prepared. One was raster-scanned in air with circularly polarized laser and another was raster scanned in water with linearly polarized laser. The adjacent distance and the scan speed were 6 μm and 40 μm/s, respectively. The dimension of fabrication areas was 100 μm × 100 μm.

3 RESULTS AND DISCUSSION

Figure 1 shows the morphologies of the silicon samples that were obtained under different processing conditions. With circularly polarized laser irradiating in water at a lower laser energy level (E = 0.3 μJ), nanoparticles are uniformly formed on the silicon surface (Fig. 1a), which is very different from the high spatial frequency LIPSS (HSFL) structures (mean period of 120 nm) that are obtained with linearly polarized laser (Fig. 1d). The nanoparticles are distributed on the silicon surface tightly and uniformly with a mean diameter below 100 nm, which reveals a potential for high sensitive SERS. However, when the laser energy increases to 0.6 μJ, nanoparticle-covered microholes appear (Fig. 1b). The mean diameter of these nanoparticles (covered on microholes) is around 150 nm and the nanoparticles are distributed unevenly. When the silicon is irradiated in air with circularly polarized fs laser at a laser energy of 0.6 μJ, distorted microstructures covered by floccules are obtained (Fig. 1c).

To demonstrate the ability of the nanoparticle morphology for SERS, we irradiated the silicon sample in 0.01 M $AgNO_3$ solution with circularly polarized fs laser. As shown in Figure 2a, the

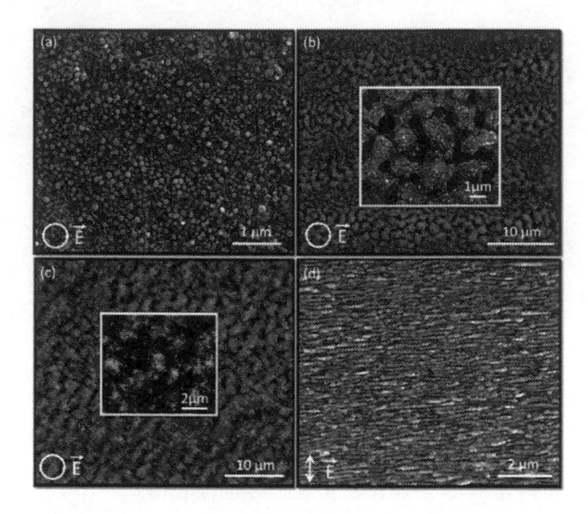

Figure 1. SEM images of the irradiated silicon surfaces by fs laser: (a) circularly polarized, in water, E = 0.3 μJ; (b) circularly polarized, in water, E = 0.6 μJ; (c) circularly polarized, in air, E = 0.6 μJ; and (d) linearly polarized, in water, E = 0.3 μJ. The scan speed is 40 μm/s. The insets are local enlarged pictures.

nanostructures are covered by using silver, which is acquired due to photoreduction. Besides, silver nanoparticles are also observed, which create "hot spots" and cause enhanced Raman scattering. In fact, fs-laser-ablation-induced chemical reduction dominates the synthesis of silver nanoparticles. When an fs laser source irradiates the silicon surface, silicon plasma plumes with high temperature and pressure are formed and cooled quickly to generate silicon nanoparticles. And then, the reactive silicon nanoparticles are easily oxidized, thereby contributing to the reduction of Ag (I) to Ag through the following process (Jiménez et al. 2010):

$$Si(NP) + 4AgNO_3 + 2H_2O \rightarrow SiO_2(NP) + 4Ag(NP) + 4H^+ + 4NO_3^- \quad (1)$$

Figure 2b shows the components of the area 1 in Figure 2a. The weight ratio of Ag is 4.98% and the atom ratio of Ag is 1.35%, which demonstrates that silver is reduced definitely.

The size of the nanostructures has a significant effect on the SERS. And so, we analyzed the particle size distribution, as shown in Figure 3. The mean diameter of the silver nanoparticles is 66 nm. And the number of particles with sizes of around 40–80 nm is high, thereby accounting for 75.6% of the total number of particles, which is suitable for localized surface plasmon resonance at the visible light range (Kelly et al. 2003). The interparticle spacing is almost below 100 nm.

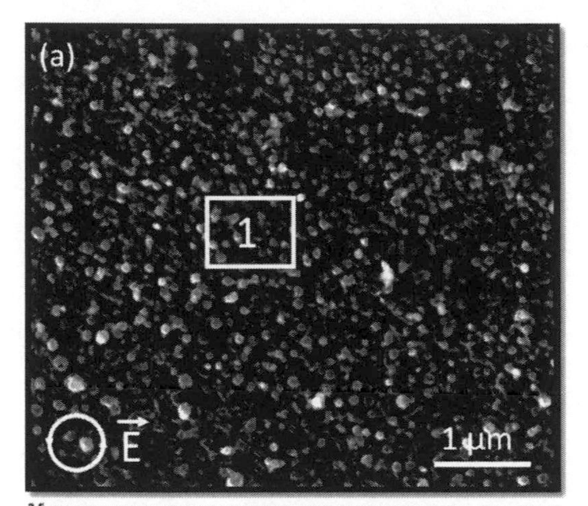

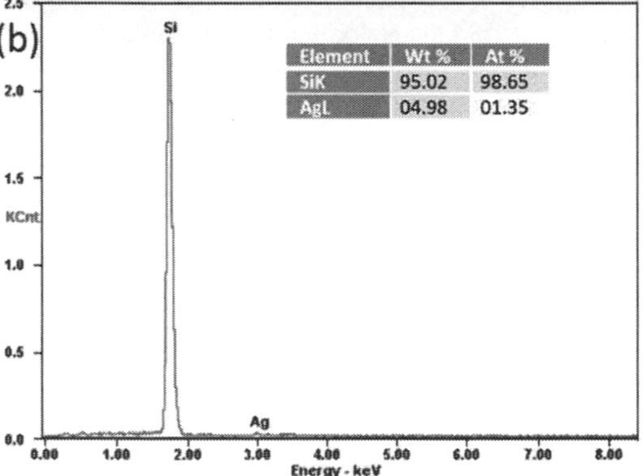

Figure 2. Silicon sample that is obtained by using circularly polarized fs laser irradiating in 0.01 M $AgNO_3$ solution: (a) SEM image of the silicon surface; (b) EDX spectrum of the area 1 in (a).

All of these may contribute to the electromagnetic field enhancement.

Raman spectra of Rhodamine 6G molecules acquired from the SERS substrates, which covered silver nanoparticles, are shown in Figure 4. Four different spots of the substrates were detected. The samples were immersed in 10^{-6} M Rhodamine 6G solution for 90 minutes in order to ensure that more Rhodamine 6G molecules are adsorbed on the roughened surface. Raman spectra of the four different spots are represented by using a red curve (for spot 1), blue curve (for spot 2), magenta curve (for spot 3), and green curve (for spot 4), respectively. The lowest black curve represents the normal Raman spectra of 10^{-2} M Rhodamine 6G outside the processed region, and a weak Rhodamine 6G signal (almost a flat line) is seen. For SERS and normal Raman spectra, the exciting laser powers of the Raman spectrometer were 0.03 mW and 6 mW; and the accumulating times were 10 s and 20 s, respectively. The Enhancement Factor (EF) was evaluated by Baia et al. 2006:

$$EF = \frac{I_{surf}}{N_{surf}} \bigg/ \frac{I_{vol}}{N_{vol}} \tag{2}$$

where I_{vol} and I_{surf} are the normal Raman and SERS intensities in the unit of mW^{-1} s^{-1}; and N_{vol} and N_{surf} represent the number of molecules detected in the reference sample and on the SERS-active substrate, respectively. The accumulating N_{vol} and N_{surf} were estimated by using the solution concentration method. Table 1 presents the calculated EFs of the four different spots at the Raman peaks of Rhodamine 6G. The Raman scattering is magnified up to 10^7 and the highest EF is 7.69×10^7, which occurs at the Raman peak 1360 cm^{-1} in the fourth spot.

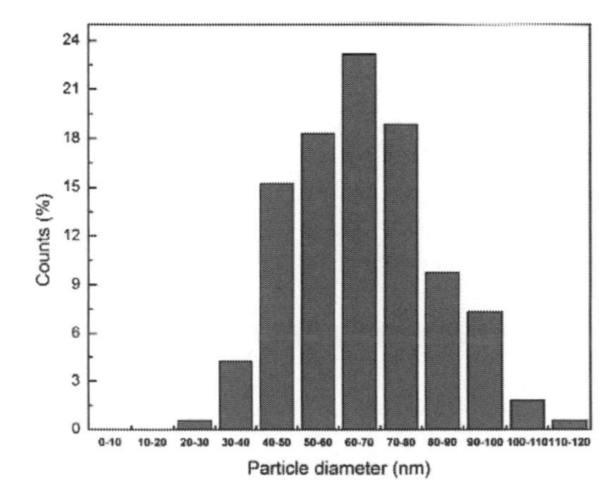

Figure 3. Size distribution of the silver nanoparticles fabricated by using circularly polarized fs laser.

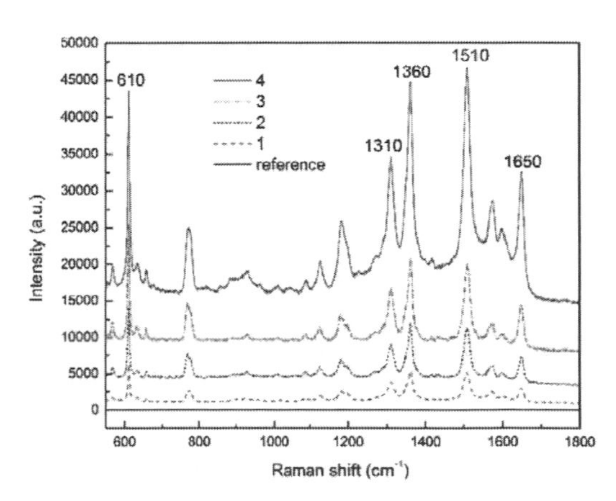

Figure 4. SERS spectra of different spots on the silver nanoparticle-covered substrate.

Table 1. Enhancement factors at different spots.

Raman shift (cm^{-1})	Spot 1	Spot 2	Spot 3	Spot 4
1650	9.89E+06	1.46E+07	2.33E+07	5.96E+07
1510	1.09E+07	1.61E+07	2.56E+07	6.56E+07
1360	1.71E+07	2.10E+07	2.88E+07	7.69E+07
1310	1.30E+07	1.81E+07	2.57E+07	6.03E+07
610	8.09E+06	1.07E+07	1.49E+07	2.54E+07

Our experiments results demonstrate that the nanoparticle-decorated substrates that are fabricated by using circularly polarized fs laser have a great potential for high-sensitive SERS.

4 CONCLUSIONS

In summary, disparate surface nanostructures on the silicon surface have been obtained by using circularly polarized fs laser one-step irradiation in the water/$AgNO_3$ solution. For the first time, uniformly distributed Ag-coated nanoparticles are fabricated on the silicon surface, and the mean diameter of the nanoparticles is below 100 nm, which are quite different from the LIPSS structures that are obtained by using linearly polarized fs laser. Therein, fs-laser-ablation-induced chemical reduction of silver ions and structuring silicon synchronously play a determined role in such hybrid structures' formation. The unique structures exhibit excellent SERS performance such that the maximum enhancement factor is up to 7.69×10^7 when Rhodamine 6G is used as the probe molecule, which presents a great potential for trace detection and sensing applications.

REFERENCES

Baia, M., Baia, L., Astilean, S., et al. 2006. Surface-enhanced Raman scattering efficiency of truncated tetrahedral Ag nanoparticle arrays mediated by electronic couplings. Applied Physics Letters: 88(14):143121-143121-3.

Cialla, D., März, A., Böhme, R., et al. 2011. Surface-enhanced Raman spectroscopy (SERS): progress and trends. Analytical & Bioanalytical Chemistry: 403(1):27-54.

Diebold, E.D., Mack, N.H., Doorn, S.K., et al. 2009. Femtosecond Laser-Nanostructured Substrates for Surface-Enhanced Raman Scattering. Langmuir the Acs Journal of Surfaces & Colloids: 25(3):1790-4.

Ding, S.Y., Yi, J., Li, J.F., et al. 2016. Erratum: Nanostructure-based plasmon-enhanced Raman spectroscopy for surface analysis of materials. Nature Reviews Materials: 1.

Fleischmann, M., Hendra, P.J. & Mcquillan, A.J. 1974. Raman spectra of pyridine adsorbed at a silver electrode. Chemical Physics Letters: 26(2):163–166.

Huang, M., Zhao, F., Cheng, Y., et al. 2008. Large area uniform nanostructures fabricated by direct femtosecond laser ablation. Optics Express: 16(23):19354-65.

Jeanmaire, D.L., Duyne, R.P.V., Jeanmaire, D.L., et al. 1977. Surface raman spectroelectrochemistry: Part I. Heterocyclic, aromatic, and aliphatic amines adsorbed on the anodized silver electrode. Journal of Electroanalytical Chemistry & Interfacial Electrochemistry: 84(1):1–20.

Jiang, L., Ying, D., Li, X., et al. 2012. Two-step femtosecond laser pulse train fabrication of nanostructured substrates for highly surface-enhanced Raman scattering. Optics Letters: 37(17):3648–50.

Jiménez, E., Abderrafi, K., Abargues, R., et al. 2010. Laser-ablation-induced synthesis of SiO2-capped noble metal nanoparticles in a single step. Langmuir: 26(10):7458–63.

Kelly, K.L., Coronado, E., Lin, L.Z., et al. 2003. The Optical Properties of Metal Nanoparticles: The Influence of Size, Shape, and Dielectric Environment. Cheminform: 34(16):668–677.

Lin, C.H., Jiang, L., Chai, Y.H., et al. 2009. One-step fabrication of nanostructures by femtosecond laser for surface-enhanced Raman scattering. Optics Express: 17(24):21581–9.

Varlamova, O., Costache, F., Reif, J., et al. 2006. Self-organized pattern formation upon femtosecond laser ablation by circularly polarized light. Applied Surface Science: 252(13):4702–4706.

Wei, H. & Xu, H. 2013. Hot spots in different metal nanostructures for plasmon-enhanced Raman spectroscopy. Nanoscale: 5(22):10794–805.

Yang, J., Li, J., Du, Z., et al. 2014. Laser Hybrid Micro/nano-structuring of Si Surfaces in Air and its Applications for SERS Detection. Scientific Reports: 4:6657–6657.

Yang, Q., Li, X., Jiang, L., et al. 2015. Nanopillar arrays with nanoparticles fabricated by a femtosecond laser pulse train for highly sensitive SERRS. Optics Letters: 40(9):2045–8.

Yasumaru, N., Miyazaki, K. & Kiuchi, J. 2003. Femtosecond-laser-induced nanostructure formed on hard thin films of TiN and DLC. Applied Physics A: 76(6):983–985.

Zhang, N., Li, X., Jiang, L., et al. 2013. Femtosecond double-pulse fabrication of hierarchical nanostructures based on electron dynamics control for high surface-enhanced Raman scattering. Optics Letters: 38(18):3558–61.

Zhu, J.T., Shen, Y.F., Li, W., et al. 2006. Effect of polarization on femtosecond laser pulses structuring silicon surface. Applied Surface Science: 252(252):2752–2756.

Civil, Architecture and Environmental Engineering – Kao & Sung (Eds)
© 2017 Taylor & Francis Group, ISBN 978-1-138-02985-9

An experimental study on rotation ultrasound-assisted drilling for high-strength engineering of mechanical connection components

S. Hong
Mechanical Engineering, Dalian Polytechnic University, China

ABSTRACT: To solve the serious wear problem of the drill tool in the drilling process of high-strength engineering of mechanical components with carbon fiber-reinforced composite material, in the conventional drilling and rotary ultrasound-assisted drilling tests have been conducted. The effects to the drilling tool's wear of both guide hole and rotary ultrasonic vibration-assisted drilling, and the axial force and drill torque of the work piece are studied. It is shown in the results that, with an increase in the drilling hole, the wear rate and axial force of the drill tool are increased. The guide hole can be added to effectively reduce the drilling axial force. Rotary ultrasonic vibration-assisted drilling effectively reduces the wear rate, the axial force, and torque of the drill tool. These tests can provide an important preference for researching the high-strength engineering mechanical connectors and carbon fiber-reinforced composite material drilling and drilling tool-processing characteristics.

1 INTRODUCTION

The carbon fiber-reinforced composite material has the advantages of light weight, high strength, high temperature resistance, corrosion resistance, good fatigue resistance, and good anti-vibration performance and mechanical properties. It has been widely used in military, aerospace, engineering machinery, automotive, and other fields (Xun Liu, 2016 & Liu Yang, 2015). With an increase in the high strength and good performance requirements on the connection components of engineering machinery, the new composite difficulty processing materials has been widely used in high-strength connection components of engineering machinery (Ren Shunan, 2013 & Jinyang Xu, 2016). Because the carbon fiber material has a unique high hardness and high strength, which make it more difficult to process, the carbon fiber material has the performance characteristics of long heating time, long processing time, strong axial force and big torque, and easy to cause tool wear (Julian M, 2015 & Shivraman Thapliyal, 2016). The wear of the drilling tool will lead to problems such as tearing, fluffing, roughness, and delamination in hole-drilling. At the same time, the drill tool will be scrapped too quickly, the efficiency will be low, and the economic cost will be high (Jun Cheng, 2015). In view of these problems, many scholars put forward the ultrasonic vibration-assisted method to process the material, and some scholars have carried out the corresponding experimental study. Jing Liu (2012) believes that the effect of RUEM can be beneficial to improve the cutting force during processing and can increase the use life of cutting tools and improve the accuracy of drilling holes. Pei Z J (2005) studied the influence of rotary ultrasonic machining on the characteristics of brittle materials. Wang Jinming (2006) studied the three kinds of super-hard material on the different carbon fiber-milling tools to obtain a good performance in processing.

In this paper, the high-strength engineering of mechanical connection components of the carbon fiber-reinforced composite material has been studied in Dalian Shenglong Machinery Co., Ltd and our university. The experimental method has been used to analyze the effect of guide hole and rotary ultrasonic vibration on high-strength engineering of mechanical connection components. The experiment can show the axial force and torque of the work piece during the drilling process. The experiment can provide the theoretical guidance for processing of high-strength engineering of mechanical connection components and the carbon fiber-reinforced materials, and it also plays a positive promotion for the special processing of composite materials.

2 DRILLING TEST

2.1 *Test equipment*

The main test equipment for ultrasonic-assisted drilling is the Ultrasonic 65 device in the WMG high-tech equipment machining center. The other test equipment includes the YE5850 charge amplifier,

A/D converter, and computer. The measurement equipments of the drilling tool's wear include the stereomicroscope, the image scale software called Image Measure, and the JEOLJSM-6380 LV scanning electron microscope.

The ultrasonic driver is mounted on the drill tool holder and it generates a certain frequency of ultrasonic vibration through the piezoelectric effect (see Figure 1). The ultrasonic vibrator is installed in the tool feed direction. In this test, the vibration frequency and amplitude of the ultrasonic vibrator are determined by using the tool. Once the tool is fixed on the holder, the vibration frequency and amplitude of the tool system are fixed. In the application of ultrasonic vibration, the vibration frequency and amplitude will change correspondingly in different drilling processes, and different drilling processes will lead to different surface qualities finally.

2.2 *Test materials and methods*

The drilling test material is the isotropic, epoxy resin-based, carbon fiber-reinforced composite material; the volume fraction of the carbon fiber is 65%, and the material properties are given in Table 1. The work piece is a connection plate of the carbon fiber-reinforced composite material in the high-strength engineering machinery.

In order to obtain the variation of wear characteristics of the drill tool and the axial drilling force under different conditions, the 1.5 mm diameter guide hole is drilled uniformly on the carbon fiber-reinforced composite work piece, and the main purpose of the guide hole is to reduce the axial thrust in twist drilling. Similarly, in order to test the impact of the guide hole on the cutting edge, and ensure the integrity of the test, the same test has been conducted with no guide hole.

During the test, the cutting speed is set at 150 m/min and the feed rate is 0.06 mm/rev, and the conventional drilling and rotary ultrasonic-assisted drilling tests are carried out, respectively. The work piece numbers are 1#–10# in the drilling process; when all the hole-drilling processes are completed, a new drill is exchanged for conducting the same drilling processes. An average of five trials is taken for the test results to obtain the final calculated value. The parameters of the rotational ultrasonic-assisted drilling test are respectively given in Tables 2 and 3.

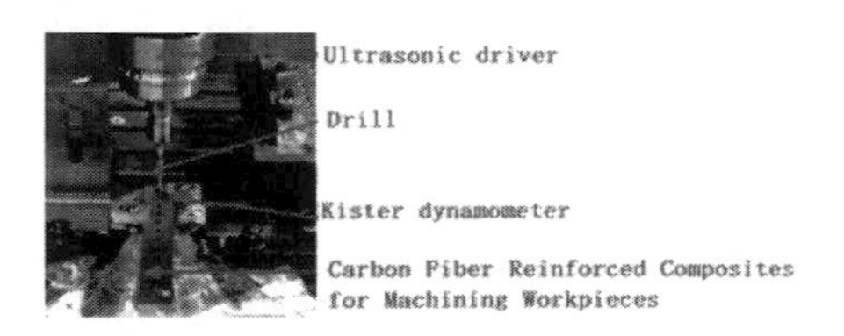

Figure 1. Picture of the test parts and major test equipment.

Table 1. Mechanical properties of carbon fiber-reinforced composites.

Setting	Tensile strength/MPa	Tensile modulus/GPa	Density/ g.cm^{-3})
Value	4300	145	1.44

Table 2. List of drilling bit tools.

Drill bit	Material	Coat	Diameter/mm
Guide hole drilling	Carbide alloy	TiAlN	1.5
Concentric main hole drilling	High speed steel	TiN	6

Table 3. Vibration parameters of the rotational ultrasonic-assisted drilling process.

Vibration parameters	No hole-A	No hole-B	Hole-A	Hole-B
Frequency/*Hz*	40360	40030	40130	39980
Peak-to-peak amplitude/*um*	5.4	4.8	5.3	4.4

3 TEST RESULTS

3.1 *Drilling tool wear*

In order to compare the wear effect of the drill tool and wear region, the drill tool is captured by using a ZEISS Axiocam micro-camera. The abrasion area and the wear region are calculated with Matlab.

With an increase in the number of holes in the drilling process, the wear area of the drill tool's flank surface is increased (see Figure 2). For example, in the traditional without guide hole drilling process, the 1# hole wear area is 0.2 mm^2; when the all 10 holes are completed, the wear area of the 10# hole is increased to 1.6 mm^2. However, the tool wear area is 1.19 mm^2 in the rotary ultrasonic-assisted drilling without guide holes process. In the traditional drilling process, the tool wear area is 1.61 mm^2.

When compared to traditional drilling, the edge wear of the drilling tool has been greatly improved in the rotary ultrasonic-assisted drilling process. When all the 10 holes' drilling processes are completed, the wear of the drilling tool's cutting edge in rotary ultrasonic-assisted drilling is 0.031 mm^2, while the same wear in traditional drilling is 0.058 mm^2 (see Figure 3).

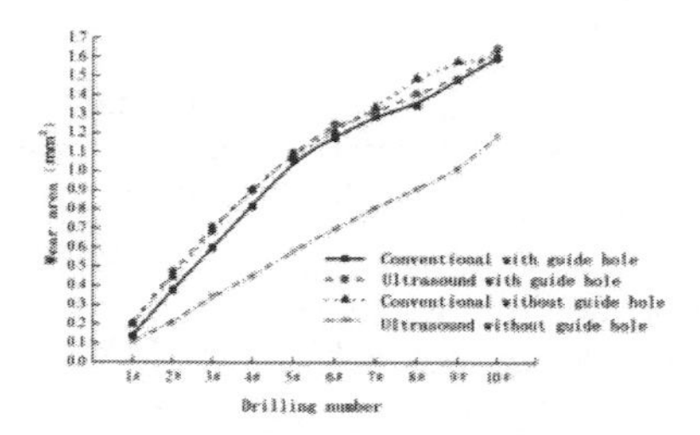

Figure 2. Graph showing the tool flank wear.

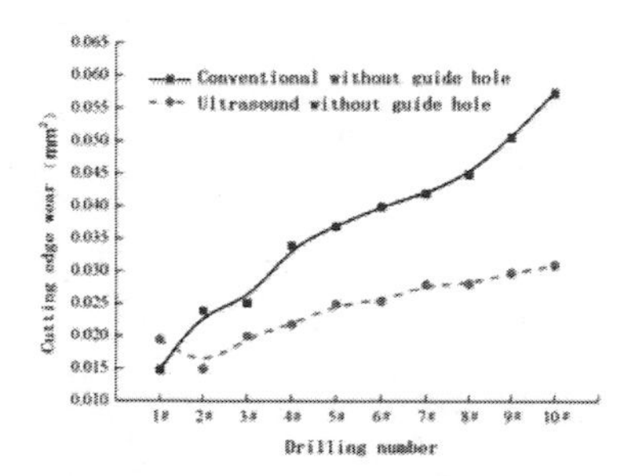

Figure 3. Graph showing the tool's cutting edge wear.

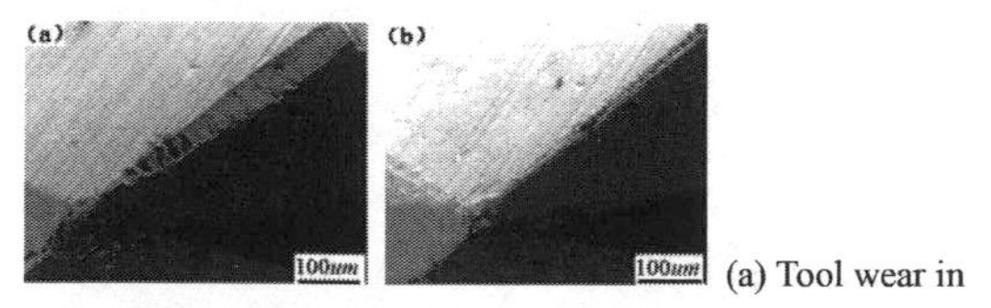

(a) Tool wear in conventional drilling and (b) tool wear in rotary ultrasonic vibration-assisted drilling

Figure 4. SEM images showing the tool wear.

The conventional drilling and the rotary ultrasonic-assisted drilling are conducted under the without guide hole condition. After all the 10 holes' drilling processes are competed, the wear area of the drill tool is observed by using Scanning Electron Microscopy (SEM). The cutting edge wear of the drilling tool in the traditional drilling process is more severe, while under the same conditions of rotary ultrasonic-assisted drilling, the cutting edge wear of the drilling tool is less (see Figure 4).

3.2 *Axial force*

The axial force of the drill tool under different conditions are measured and stored by using the Kistler dynamometer with the test instrument. With an increase in the number of drilling holes, the drill tool's axial force increases. The drill tool's axial force in the traditional drilling process is different for different holes. For example, the axial force in hole 1# is 49 N, while the axial force in hole 10# is increased to 721 N (see Figure 5). The axial forces are almost the same under the with guide hole condition in the conventional drilling and rotary ultrasonic-assisted drilling process. The magnitude of the axial force is obviously higher in the traditional drilling tool than that in the rotary ultrasonic auxiliary drilling under the no guide hole condition. The axial force in the traditional drilling process is 874 N, while the magnitude of the axial force in the rotary ultrasonic-assisted drilling process is only 550 N. Under the with guide hole condition, the ultrasonic vibration in the axial direction is useless; the vibration cannot effectively reduce the drill tool's axial force. Under the no guide hole condition, the ultrasonic vibration can lead to reduction in the magnitude of the drilling tool's axial force, the ultrasonic vibration can reduce drill tool wear and increase the drill tool's service life.

3.3 *Torque*

The drill torque can be greatly fluctuated in the drilling process, with an increase in the number of drilling holes. The torque has an increasing tendency (see Figure 6). The drill torque is less in the rotary ultrasonic vibration-assisted drilling process than that under the other conditions (the drill torque in the 1# hole is only 28.2 N.cm, while the drill torque in the conventional drilling process is 37 N.cm). The drill torque of the 10# hole in the ultrasonic drilling process is 36.5 N.cm, while the drill torque in the traditional drilling process is 41.4 N.cm.

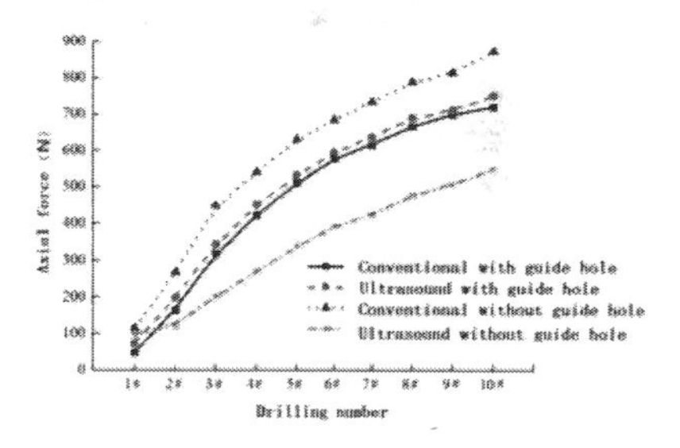

Figure 5. Graph showing the drilling tool's axial force distribution in the drilling process.

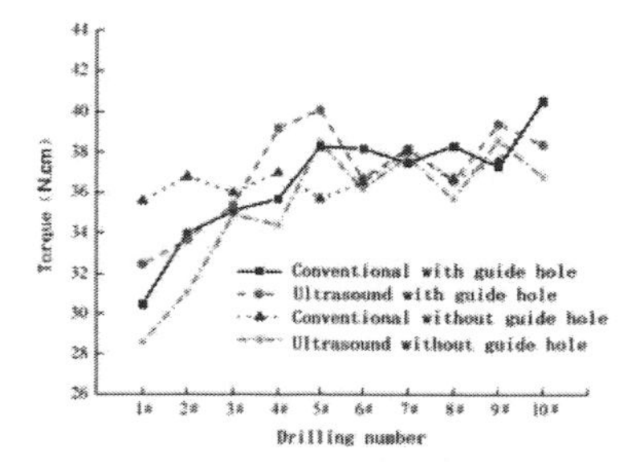

Figure 6. Graph showing the torque changes in the drilling process.

4 CONCLUSIONS

In order to reduce the magnitude of the drilling axial force, the ultrasonic vibration-assisted drilling process combined with the pre-processed guide hole has been used to study the high-strength engineering of machinery connection components of the carbon fiber-reinforced composite material. It is shown in the results that the wear rate of the drilling tool increased with an increase in the number of drill holes; the ultrasonic vibration-assisted drilling process can effectively improve the drilling tool's wear. The magnitude of the drilling tool's axial force in the ultrasonic-assisted drilling process is less than that of the traditional drilling process.

ACKNOWLEDGMENTS

The author gratefully acknowledges the following projects for financial assistance: Research on the Engineering Application of Rotating Ultrasonic Assisted Drilling Technology (DZ2016B-01); ASDR Pump Cover-Drilling Long Super-long Skew (DZHX-13-05); YOX1000 Hydraulic Coupling Spindle Machining Process Design and Structural Design (DZHX-12-02); and High-pressure Injection Pump Level Transition between the Hole Processing Rotation Positioning Tire Design and Processing (DZHX-12-01)

Author: Sun Hong (1963,12-) female, associate professor, Mechanical design and manufacturing, mechanical materials. Mobile phone number: 15840815533.

REFERENCES

Jing Liu, Deyuan Zhang & Longgang Qin (2012). Feasibility study of the rotary ultrasonic elliptical machining of carbon fiber reinforced plastics. J. Machine Tools and Manufacture, 53, 141–150.

Jinyang Xu & Mohamed El Mansoria (2016). Experimental study on drilling mechanisms and strategies of hybrid CFRP/Ti stacks. J. Composite Structures. 157, 461–482.

Julian M. Allwood & A. Erman Tekkaya (2015). Writing a good review for the Journal of Materials Processing Technology. J. Materials Processing Technology. 215, 5–7.

Jun Cheng, Guoqiang Yin, Quan Wen, Hua Song & Yadong Gong (2015). Study on grinding force modelling and ductile regime propelling technology in micro drill-grinding of hard-brittle materials. J. Materials Processing Technology. 223, 150–163.

Liu Yang, Li Pengnan, Chen Ming, Tang Siwen & Yang jin (2015). Comparison of cutting forces and cutting temperatures of two drills in high speed drilling carbon fiber composite. J. Materials for Mechanical Engineering. 39, 36–40.

Pei Z J, Ferreira P M & Haselkon M (2005). Plastic flow in rotary ultrasonic machining of ceramics. J. Materials Processing Technology, 48, 771–777.

Ren Shunan, Wu Dan & Chen Ken (2013). Thrust force on the main cutting edge when cutting carbon fiber reinforced plastics. J. Tsinghua University (Science and Technology). 53, 487–492.

Shivraman Thapliyal & Dheerendra Kumar Dwivedi (2016). Microstructure evolution and tribological behavior of the solid lubricant based surface composite of cast nickel aluminum bronze developed by friction stir processing. J. Materials Processing Technology. 238, 30–38.

Wang Jinming, Yang Zhixiang, AoMing, Yang Donjun, Ai huanzhi & zhaoFuling (2006). Study on Machining Processing and Evaluation of Carbon Fiber Reiforeed Composites Cutting Suafrees. C. The 14th National Composites Conference, 1094–1095.

Xun Liu, Yong Chae Lim, Yongbing Li, Wei Tang, Yunwu Ma, Zhili Feng & Jun Ni (2016). Effects of process parameters on friction self-piercing riveting of dissimilar materials. J. Materials Processing Technology. 237, 19–30.

Civil, Architecture and Environmental Engineering – Kao & Sung (Eds)
 ISBN 978-1-138-02985-9

Synthesis of PbO nanosheet-based thin films for all-solid-state supercapacitors

Jun Zhang & Haiyan Xie
Department of Physics, College of Science, Hohai University, Nanjing, P.R. China

Zhaojun Chen
College of Mechanics and Materials, Hohai University, Nanjing, P.R. China

ABSTRACT: In this work, the PbO nanosheet-based thin film was synthesized directly on the indium tin oxide-coated polyethylene terephthalate substrate through the electrodeposition method. The morphology and composition of the products were characterized systematically. The nanosheet-based film was used as as building block to fabricate flexible, All-Solid-State (ASS) SuperCapacitors (SCs) by using the PVA/H_3PO_4 gel as the solid electrolyte. Cyclic Voltammetry (CV), galvanostatic charge–discharge, and electrochemical impedance spectroscopy were carried out to study the electrochemical performance of PbO-based ASS thin film SCs. The results demonstrated that the device deposited at –0.7 V (*vs.* SCE) exhibited an areal capacitance of 173 μF cm^{-2} at 2.0 μA cm^{-2}, which suggested its potential application as an electrode-active material for SCs.

1 INTRODUCTION

Lead oxide is one kind of semiconductor with a direct band gap energy of ~1.9 eV that has important applications in storage batteries, glass industry, and pigments (Shi et al. 2008, Li et al. 2012). PbO nanoparticles had been used as electrode materials for the valve-regulated lead acid battery and had shown improved electrochemical performances, such as excellent discharge capacity and prolonged cyclic life (Wang et al. 2001). The results implied that nanosized PbO would be advantageous as electrode materials in designing high performance SCs. So far, various PbO nanostructures, such as nanoplates, nanostars, nanorods (Ghasemi et al. 2008) and nanotubes (Behnoudnia et al. 2012) had been synthesized by using thermal deposition (Behnoudnia et al. 2012), laser-assisted deposition (Tuncheva & Baleva 1995), solvothermal route (Gao et al. 2013), sonochemical (Sadeghzadeh et al. 2009), and hydrothermal methods (Wang et al. 2013). However, most of these techniques are complex, expensive, and time-consuming, which are against its application in SCs.

In this paper, we report the synthesis of PbO nanostructures by using the simply electrodeposition method and its application as active materials to build ASS thin film SCs. The detailed synthetic procedure is shown in the Experimental section. The results show that the PbO nanosheets deposited at a potential of –0.7 V were almost vertical to the substrate and formed large open channels between the nanosheets. The nanosheets-based film was used to fabricate ASS SCs by using the PVA/H_3PO_4 gel as the solid electrolyte. Since the configuration avoids the stocking of PbO nanosheets and fully utilizes the surfaces and edges for the Faradaic redox reaction, the PbO-based ASS thin film SCs exhibit an improved electrochemical performance with a high areal capacitance of 173 μF cm^{-2} at 2.0 μA cm^{-2}.

2 EXPERIMENTAL SECTION

2.1 *Synthesis of PbO-based thin film*

Electrodeposition was performed by using an electrochemical workstation (Kest Wuhan CS300) in a conventional three-electrode mode with a platinum electrode as the counter electrode, a Saturated Calomel Electrode (SCE) as the reference electrode and a piece of ITO-coated PET film (1.0×2.0 cm^2) as the working electrode. For the deposition of PbO nanostructures, the electrolyte used was an aqueous solution of 5 mM $Pb(CH_3COO)_2$, 5 mM EDTANa (EthyleneDiamine TetraAcetic acid disodium salt), and 0.1 M KCl. The pH value was adjusted to 3 with 0.2 M HCl. Electrodeposition was carried out under potentiostatic conditions with the electrolyte temperature being kept at 60°C. After deposition, the working electrodes were rinsed in distilled water several times and dried naturally in the atmosphere.

2.2 *Fabrication and electrochemical measurements of ASS thin film SCs*

ASS thin film SCs were fabricated by stacking two identical PbO thin films adhered with PVA/H_3PO_4 gel electrolyte. Firstly, two pieces of electrodes were painted with PVA/H_3PO_4 gel electrolyte in order to form a thin layer of the electrolyte and were then dried for 24 h at room temperature to remove the redundant water. Secondly, the electrodes were painted with the gel again. After that, the two electrodes were sandwiched with a piece of cellulose membrane to form the ASS thin film SC device. Furthermore, the fabricated device was pressed under definite pressure for 5 min to enhance the interfacial contact of the gel electrolyte with PbO electrodes. The device was dried for another 24 h at room temperature to obtain the ASS thin film SC. The Cyclic Voltammetry (CV), galvanostatic charge–discharge and Electrochemical Impedance Spectroscopy (EIS) properties of prepared devices were evaluated by using a CHI 660D Electrochemical Workstation (Chenhua Co. Shanghai).

The PVA/H_3PO_4 gel electrolyte was prepared as follows (Ma et al. 2014): 6 g of H_3PO_4 was added to 60 mL of deionized water, and then 6 g of PVA powder was added. The whole mixture was heated to 85°C under stirring until the solution became clear.

3 RESULTS AND DISCUSSION

3.1 *XRD results*

The XRD patterns of the deposited thin film are shown in Figure 1. At –0.7 V (vs. SCE), the phase of the samples is very complicated, the weak peaks are located at 24.707°, 26.602°, and 35.352° corresponding to the tetragonal phase of α-PbO·H_2O (JCPDS No.77-1895), the strong reflections are located at 29.082°, 30.309°, and 35.948° thereby confirming the orthorhombic phase of β-PbO (JCPDS No.77-1971 and JCPDS No.78-1663), respectively. The main reason for multi-phase coexistences in one sample is that the lattice constant of different phase-structured PbO is little different. Small changes of the micro-environment (small changes of the electrolyte concentration near the surface of the deposits or the temperature of the electrolyte) would change the atomic distances and formed the second phase during electrodeposition. The phenomenon is very common during the electrodeposition process (Didier et al. 2014). At –0.8 V (vs. SCE), a new peak appeared at 31.305°, which corresponds to (111) of metallic Pb. While decreasing the deposition potential to –0.9 V (vs. SCE) or –1.0 V (vs. SCE), all diffraction peaks at 31.305°, 36.266°, 52.228°, 62.119°, and 65.236°

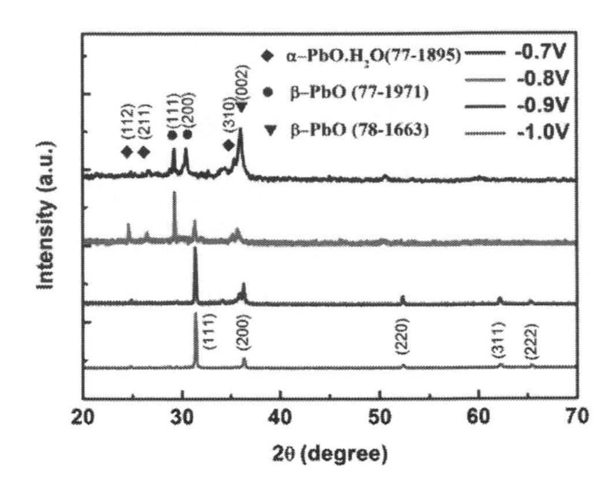

Figure 1. XRD patterns of the samples at different electrodeposition voltages: –0.7 V (vs. SCE); –0.8 V (vs. SCE); –0.9 V (vs. SCE); and –1.0 V (vs. SCE), respectively.

matched well with the cubic Pb (JCPDS card no. 04-0686). The results show that –0.7 V (vs. SCE) is the optimal deposition voltage for the formation of PbO nanostructures.

3.2 *Surface morphology*

Generally, the deposition potential is the most important parameter in the composition and morphology of electrodeposited nanostructures. In order to investigate the morphology evolution along with the electrodeposition potentials, SEM images of the samples are obtained. SEM images of the samples obtained at different voltages show drastic change in their morphologies, as shown in Figure 2. At –0.7 V (vs. SCE) (Figure 2a), the as-prepared products are nanosheets, and most of the nanosheets are perpendicular to the substrates with an edge length of about 10 μm. When decreasing the deposition potential to –0.8 V (vs. SCE) (Figure 2b), some nanowires with diameters of about several hundred nanometers are laid flatly on the substrate, and mixed with some Pb octahedron nanoparticles (proved by XRD results). While decreasing the deposition potential to –0.9 V (vs. SCE) or –1.0 V (vs. SCE) (Figure 2c and d), the products present sparsely distributed nanosheets; meanwhile, a layer of compact Pb nanoparticles appeared under the PbO nanosheets, which is consistent with XRD results. To optimize the ASS thin film SCs device properties, all PbO nanosheets thin films were synthesized under the deposition potential of –0.7 V (vs. SCE). Based on previous reported results, the vertically aligned PbO nanosheets film configuration is beneficial for fabricating the ASS SCs device for several reasons: 1) the vertically oriented nanosheet-based film formed large open channels between the nanosheets; the configura-

Figure 2. SEM images of the samples electrodeposited at various deposition potentials: (a) –0.7 V (vs. SCE); (b) –0.8 V (vs. SCE); (c) –0.9 V (vs. SCE); and (d) –1.0 V (vs. SCE).

tion avoids the stocking of PbO nanosheets and fully utilizes the surfaces and edges for the Faradaic redox reaction. 2) The solid electrolyte contacted with all the surfaces of the nanosheets, which is beneficial for the electronic transport during the charge–discharge process, thereby enabling these to be directly used as electrode materials without adding carbon black and binder.

3.3 *TEM and EDS*

TEM was used to study the morphologies and microstructures of the sample deposited at –0.7 V (vs. SCE). Although the sample displayed a uniform nanosheet-based structure in the SEM image, it is clearly observed that the sample presents two different morphologies (nanorods and nanosheets), as shown in Figure 3a and b. The reason for this is that the amount of nanorods is much smaller than that of nanosheets, and these were interleaved among the nanosheet-based film. High Resolution (HR)-TEM was also performed to study PbO microstructures. Figure 3c shows the HR-TEM image of the nanorods and clear lattice fringes indicated its good crystallinity. The measured d-spacings are 0.31 nm, 0.31 nm, and 0.28 nm as labeled on the image, which were corresponding to the (202), (103), and (220) planes of α-PbO·H_2O. Figure 3d shows the HR-TEM images of the nanosheets, and the measured d-spacings are 0.29 nm and 0.49 nm, which matched well with the (111) and (001) planes of the orthorhombic phase of β-PbO (JCPDS No. 77-1971). The results illustrated that the nanosheets are β-PbO, while the nanorods are α-PbO. It is clear that the size of nanosheets is much bigger than that of nanorods

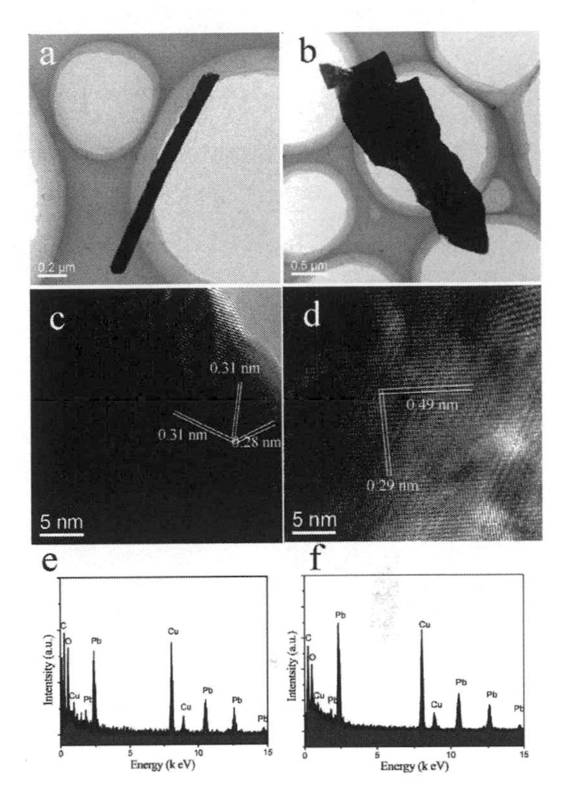

Figure 3. (a) and (b) TEM images of the sample deposited at –0.7 V (vs. SCE); (c) and (d) HR-TEM images of the nanorod and nanosheet, respectively; (e) and (f) EDS spectra of the nanorod and nanosheet, respectively.

from the SEM image in Figure 2a, and this may explain the diffraction intensity of β-PbO such that it is much stronger than that of α-PbO·H_2O as shown in Figure 1. Furthermore, EDS was also performed to investigate PbO composition. The EDS spectra of the nanorod and nanosheet are shown in Figure 3e and f, respectively. All of these are composed of Pb and O elements (Cu and C signals are obtained from the supporting film). The quantitative EDS results show that the ratio of Pb:O is about 3:4 (Figure 3e) and 1:1 (Figure 3f), which are further evidence that the nanosheets are β-PbO while the nanorods are α-PbO, respectively.

3.4 *Supercapacitive studies*

ASS thin film SCs were fabricated by stacking two pieces of PbO nanosheet-based films adhered with PVA/H_3PO_4 gel electrolyte. The electrochemical properties of the devices were investigated systematically. Figure 4a shows CV curves of the ASS thin film SC device at scan rates of 5-200 mV s^{-1} within a potential window of 0–0.5 V, which shows nearly rectangular shape without obvious redox

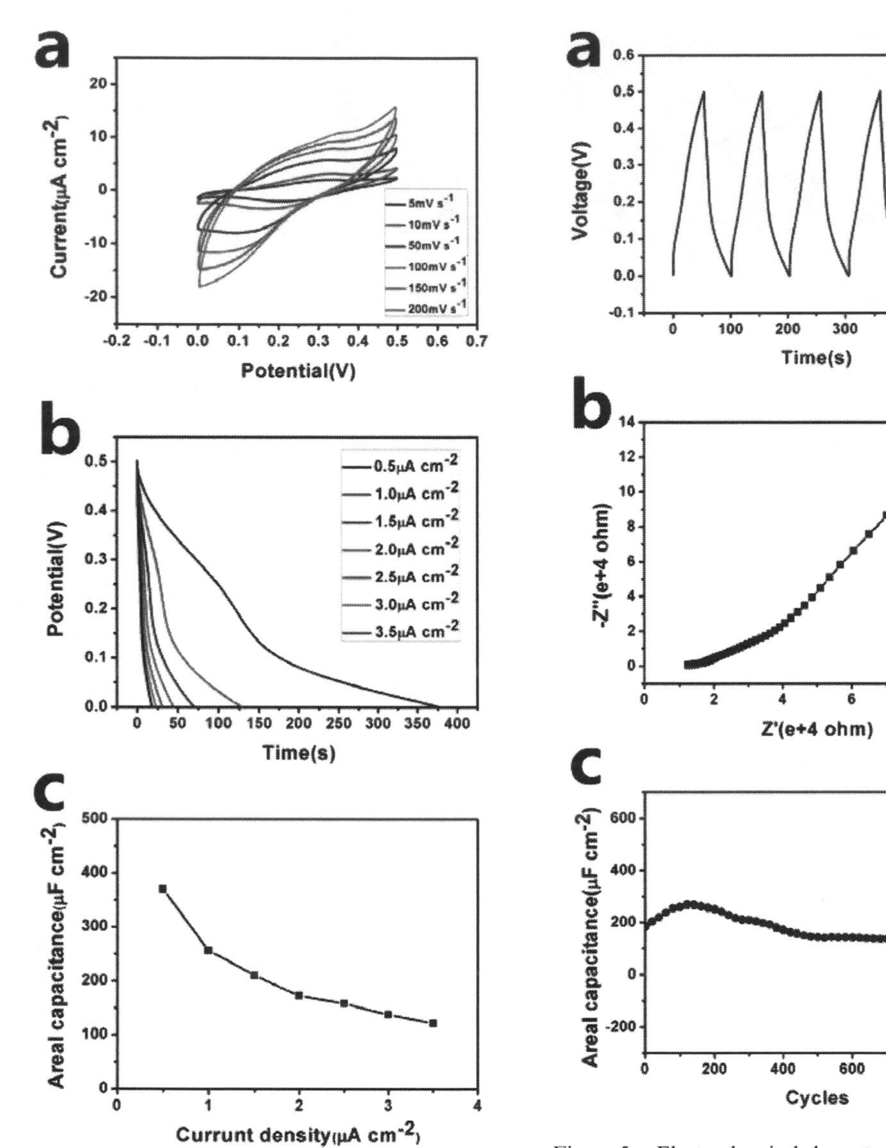

Figure 4. Electrochemical characterization of the PbO-based all-solid state device: (a) CV curves of PbO-SC in a potential range of 0–0.5 V at scan rates of 5–200 mV s^{-1}. (b) Galvanostatic discharge curves of PbO-SC at current densities of 0.5–3.5 μA cm^{-2}. (c) Areal capacitance at different current densities.

Figure 5. Electrochemical characterization of the PbO-based all-solid state device: (a) Five charge-discharge curves; (b) Nyquist plot of the PbO-SC; and (c) cycling performance at 2.0 μA cm^{-2}.

peaks. With an increase in the scan rate, the scan current is increased; the feature reveals the ideal capacitive and fast charge–discharge behavior of PbO SCs. The CV curves maintained their original shape and reversibility even at high scan rates, thereby suggesting excellent capacitive behavior of PbO SCs. Galvanostatic discharge curves of PbO nanosheet-based thin film ASS SC at various current densities are presented in Figure 4b; the barely IR drop during the discharge process indicates the good capacitive and fast charge–discharge behavior of PbO SCs. Figure 4c shows the corresponding areal capacitances; the calculated areal capacitances of the device are 370, 256, 210, 173, 158, 138, and 122 μF cm^{-2} at 0.5, 1.0, 1.5, 2.0, 2.5, 3.0, and 3.5 μA cm^{-2}, respectively. Figure 5a shows five charge–discharge curves, the symmetry of the charge and discharge characteristics declare its excellent capacitive behavior, and the slope of the curves changing at a high cell voltage reveals the occurrence of the redox process. Figure 5b shows the Nyquist plots of PbO nanosheets-based SC ranging from 100 kHz to 0.01 Hz. The Nyquist plot consists of a flat base in high frequency regions and a linear shape in low frequency regions. This plot features a high phase-angle (exceeding 45°) impedance plot, thereby indicating a fast ion transfer behavior in PbO nanosheets (Liu et al. 2010).

Excellent working stability under thousands of cycles is expected for SCs in practical applications. Further studies on the cycling stability of the PbO nanosheets-based SCs were performed by using the galvanostatic charge–discharge technique at a current density of 2.0 μA cm^{-2}. Figure 5c shows the areal capacitance as a function of the cycle number, the areal capacitance increased during the initial 150 cycles, which is attributed to the gradual increase of the electrochemically active surface area. After 150 cycles, the capacitance decreases gradually and tends to stabilization. The PbO SCs exhibit capacitance retention with 94 μF cm^{-2} after 1000 cycles, which maintains only 50.8% of the initial value. The PbO SCs has an unsatisfactory cycling stability performance. The problem may be that the stability of the reversible electron transfer process between PVA/H_3PO_4 and PbO is not good enough.

4 CONCLUSIONS

In summary, we demonstrated a facile electrodeposition method for the synthesis of PbO nanosheet-based thin film directly on the ITO-coated PET substrate. Through a series of structural characterizations, it is confirmed that the sample deposited at −0.7 V (vs. SCE) mainly has the phase of β-PbO nanosheets. ASS thin film SCs were fabricated by using two identical PbO thin films as the working electrode materials and PVA/H_3PO_4 gel as the solid electrolyte. The PbO nanosheet-based thin film SC device exhibited an areal capacitance of 173 μF cm^{-2} at 2.0 μA cm^{-2}. The good electrochemical properties of the PbO nanosheet-based thin film SC device are attributed to its vertically aligned configuration, which formed large open channels between the nanosheets and avoided the stocking of PbO nanosheets and fully utilized the surfaces and edges for the Faradaic redox reaction. Although the long life stability of the device needs to be improved in the future, the vertically aligned design for electrode materials may provide a general route for high-performance ASS SCs.

REFERENCES

Behnoudnia, F. (2012). Synthesis and characterization of novel three-dimensional-cauliflower-like nanostructure of lead(II) oxalate and its thermal decomposition for preparation of PbO. *Inorganic Chemistry Communications* 24: 32–39.

Didier, M. (2014). Method development for evaluating the redox state of Callovo-Oxfordian clayrock and synthetic montmorillonite for nuclear waste management. *Applied Geochemistry* 49: 184–191.

Gao, P. (2013). Solvothermal synthesis of alpha-PbO from lead dioxide and its electrochemical performance as a positive electrode material. *Journal of Power Sources* 242: 299–304.

Ghasemi, S. (2008). Sonochemical-assisted synthesis of nano-structured lead dioxide. *Ultrasonics Sonochemistry* 15(4): 448–455.

Li, L. (2012). Preparation and characterization of nano-structured lead oxide from spent lead acid battery paste. *Journal of Hazardous Materials* 203: 274–282.

Liu, C. G. (2010). Graphene-Based Supercapacitor with an Ultrahigh Energy Density. *Nano Letters* 10(12): 4863–4868.

Ma, G. F. (2014). High performance solid-state supercapacitor with PVA-KOH-K-3 Fe(CN)(6) gel polymer as electrolyte and separator. *Journal of Power Sources* 256: 281–287.

Sadeghzadeh, H. (2009). New Reversible Crystal-to-Crystal Conversion of a Mixed-Ligand Lead(II) Coordination Polymer by De- and Rehydration. *Inorganic Chemistry* 48(23): 10871–10873.

Shi, L. (2008). Controlled growth of lead oxide nanosheets, scrolled nanotubes, and nanorods. *Crystal Growth & Design* 8(10): 3521–3525.

Tuncheva, V. & Baleva, M. (1995). The application of the pulsed laser deposition for producing of PbO coatings. *Elevated Temperature Coatings: Science and Technology I*: 145–153.

Wang, J. (2001). Electrochemical performance of nanocrystalline lead oxide in VRLA batteries. *Journal of Alloys and Compounds* 327(1–2): 141–145.

Wang Y. G. (2013). Selected-control hydrothermal growths of alpha- and beta-PbO crystals and orientated pressure-induced phase transition. *Crystengcomm* 15(18): 3513–3516.

Civil, Architecture and Environmental Engineering – Kao & Sung (Eds)
© 2017 Taylor & Francis Group, ISBN 978-1-138-02985-9

A large-scale SERS substrate fabricated by using femtosecond laser with superhydrophobicity

Yongda Xu, Ge Meng & Peng Ran
Laser Micro/Nano Fabrication Laboratory, School of Mechanical Engineering, Beijing Institute of Technology, Beijing, P.R. China

ABSTRACT: In this study, a simple method is proposed to fabricate surface-enhanced Raman scattering substrates on Cu surfaces by using the crossed femtosecond laser direct-writing technique. As the scanning speed increases, the nanostructured surfaces exhibit superhydrophobic properties with the contact angle showing a generally rising tendency. With the speed of 2000 μm/s, we can acquire an area of 2500×2500 μm^2 in just 4 minutes at a static contact angle of 151°, which makes it a potential for large-area fabrication. After being modified with Au nanoparticles, the Cu substrates display high SERS sensitivity with an enhancement factor up to 3.60×10^9. Possible reasons for the high sensitive SERS involve: 1) the Au nanoparticles with the size of 60 nm are uniformly distributed on the nanostructure-roughened Cu surfaces which induce strong localized surface plasmon resonance; and 2) due to the superhydrophobicity of the substrates, the detected molecules are confined into a small area, which makes the concentration increase equivalently.

1 INTRODUCTION

Superhydrophobic surfaces attract much attention, because of its practical applications such as self-cleaning (Ding et al. 2011, Sas et al. 2012), oil-water separation (Tai et al. 2014, Singh A.K. & Singh J.K. 2016), and anticorrosion (Leon et al. 2012, Shenton 2009). In the past few decades, many methods were successfully used to fabricate superhydrophobic surfaces, such as chemical etching (And & Shen 2005, Luo et al. 2011), spin coating (Xu et al. 2012), and laser fabrication (Baldacchini et al. 2006). The superhydrophobic Surface-Enhanced Raman Scattering (SERS) substrate is another important research field due to its ability of breaking through the detection limit by shrinking the aqueous solution to a small area. However, the fabrication of the superhydrophobic SERS substrate lacks controllability and consumes much time just as chemical etching would, which is not suitable for industrial applications. Femto-Second (fs) laser surface nanotexturing has been proved to be an efficient and controllable technology for fabricating various materials, which has a wide range of applications in modifying physicochemical properties of the surfaces.

In this study, we have obtained high sensitive SERS Cu substrates with superhydrophobicity by using fs laser one-step irradiation. As the scanning speed increases, the contact angles of the roughened Cu surfaces show an increasing tendency, thereby reaching up to 151° at a scanning speed of 2000 μm/s. Au nanoparticle colloids with the diameter of 60 nm are used to acquire "hot spots" on the roughened surfaces and strong Raman scattering signals are detected with an enhancement factor of 3.60×10^9, due to the combination of strong localized plasmon resonance and superhydrophobicity.

2 EXPERIMENTAL SETUP

A copper sample with a size of 10 mm × 10 mm × 1 mm is cleaned in an ultrasonic bath with acetone for 30 minutes and then rinsed by using deionized water. The cleaned Cu wafer is fixed on the six-axis motion stage. The fs laser is incident normally to the Cu surface and this is achieved by using a plano-convex lens (f = 100 mm) with a spot size of ~30 μm. The total power of the fs laser used to process is 100 mW. After processing, the morphologies of the processed Cu wafer are characterized by using Scanning Electron Microscopy (SEM) and its superhydrophobicity is measured by using an optical contact angle measuring device. And then, the nanostructured Cu sample is immersed in Au nanoparticle colloids for assembly. Rhodamine 6G (R6G) with the concentration of 10^{-7} M is employed for the Raman signal detection. Raman spectra are obtained by using a Ramanscope (inVia, Reflex, Reinshaw) with an exciting power of 6 mW, a wavelength of 632.8 nm, and a 50 × objective.

3 RESULTS AND DISCUSSION

The Cu sample is scanned line-by-line by using the laser in the horizontal direction (x direction) and then in the vertical direction (y direction). Figure 1 shows the typical SEM micrograph of the nanostructures formed on the Cu surface with a laser power of 100 mW. Figure 1a–d present the surface morphologies of Cu substrates, which are processed at different scanning speeds of 500 μm/s, 900 μm/s, 1300 μm/s, and 2000 μm/s, respectively. After processing, tidy ravines appear on the Cu surface. In the case of low scanning speeds, large irregular microstructures are produced on the surface. With an increase in the scanning speed, the depths of the ravine become shallow intuitively and a large number of nanoscale spherical protrusions are present on the surface.

The wettability of the processed Cu surface is measured by using static water contact angle measurement, as exhibited in Figure 2. The Contact Angle (CA) for the roughened Cu shows a generally rising tendency with an increase in the scanning speed. When the scanning speeds are 1300 μm/s and 2000 μm/s, the CAs of processed areas are beyond 150°, which reveals superhydrophobic properties. The apparent increasing of CAs can be ascribed to the increasing number of the nanoscale spherical protrusions, which prevents the water droplet from penetrating into the surfaces, as is shown in Figure 3. It is consistent with the Cassie model (Guo et al. 2007), which described that droplets are suspended across the large amount of protrusions and air is trapped between them, thereby resulting in a higher CAs.

Figure 1. Surface morphologies of the Cu substrates processed by using fs laser at different scanning speeds: (a) 500 μm/s, (b) 900 μm/s, (c) 1300 μm/s, and (d) 2000 μm/s.

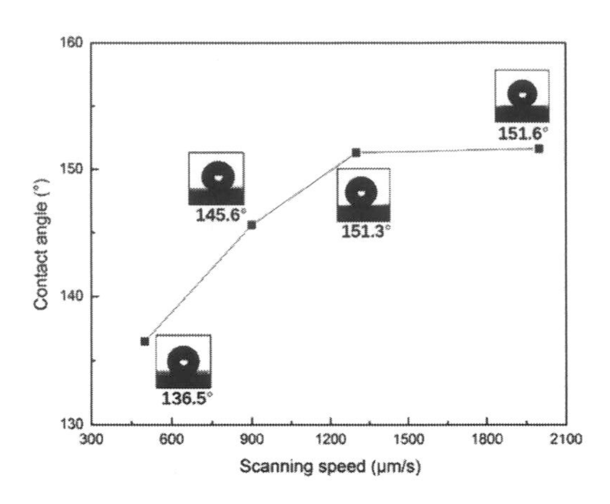

Figure 2. CAs of the laser-processed areas on the Cu substrate at different scanning speeds.

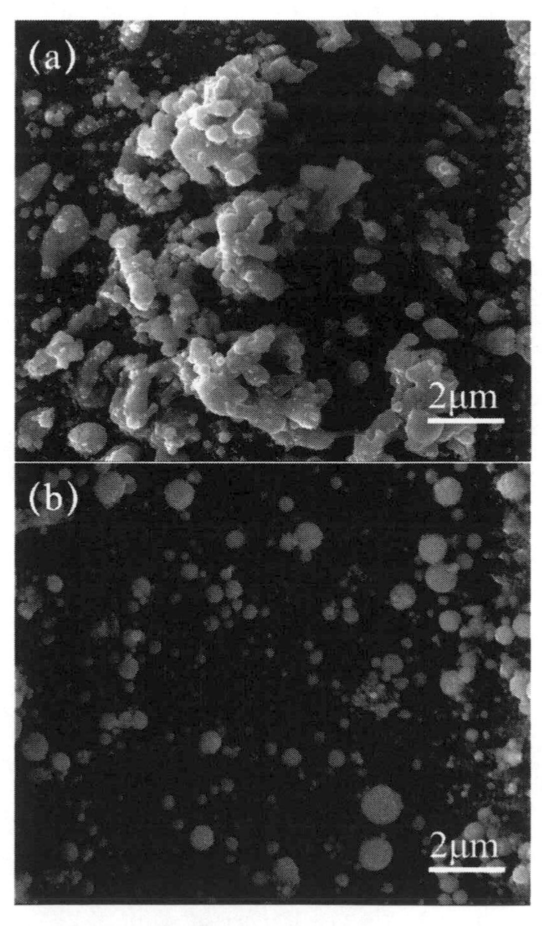

Figure 3. Enlarged surface morphologies of the Cu substrate processed by using fs laser at different scanning speeds: (a) 500 μm/s and (b) 2000 μm/s.

It was confirmed that noble metal nanoparticles, such as gold, silver, and copper, can exhibit localized Surface Plasmon Resonance (SPR), which produces strong extinction and scattering spectra for SERS enhancement (Haynes et al. 2005, Kneipp et al. 2006). And then, a simple method that requires no additional equipment is used to obtain the SERS substrate. We chose the substrate that was obtained at the scanning speed of 2000 μm /s as the sample, thereby looking forward to enhancing the Raman signal via superhydrophobic properties. The Cu substrate was decorated with Au nanoparticles by immersing the sample in Au nanoparticle colloids for 2 hours to enable the self-assembly of Au nanoparticles on the roughened Cu surface. The Au nanoparticles with the diameter of 60 nm adhere on the nanostructures, which cause "hot spots", as shown in Figure 4. Some studies have indicated that the nanogaps between nanoparticles play an important role in SERS enhancement (Tong & Käll 2014). As shown in Figure 4b, the nanogaps of Au nanoparticles are small enough to exhibit high localized plasmon resonance. Consequently, the proper diameter of the nanoparticles and the small nanogaps between the Au nanoparticles give rise to strong electromagnetic field enhancements, thereby leading to a high enhancement factor.

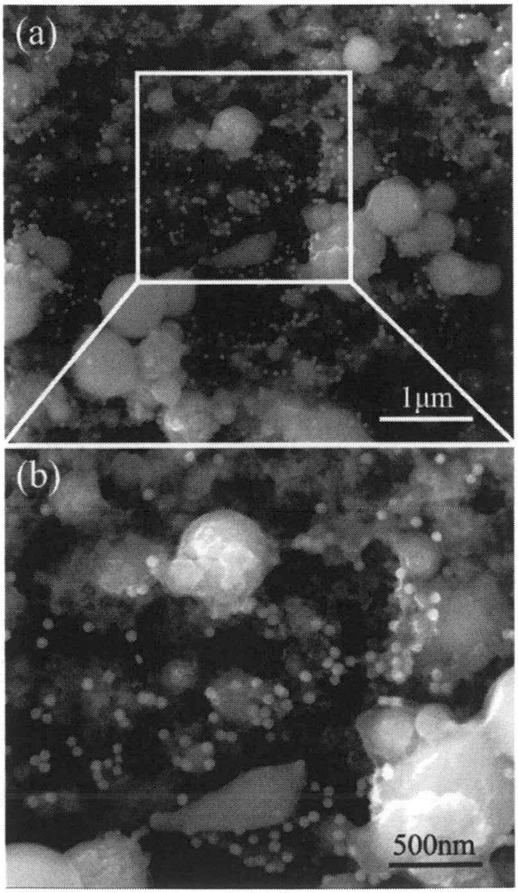

Figure 4. SEM pictures of the Cu substrate modified with Au nanoparticles. The local enlarged picture is shown in (b).

A droplet of R6G solution is added to the Au nanoparticles-decorated superhydrophobic Cu surface and natural evaporation is allowed to occur. The concentration of the R6G is 10^{-7} M. After solvent evaporation, the droplet is limited to a small area on the surface of the Cu substrate, which equivalently increases the concentration of R6G. Raman spectra of R6G molecules obtained from the Cu-nanotextured substrate decorated with Au nanoparticles are shown in Figure 5. Three spots were randomly selected in the area where R6G shrinks and evaporates. The purple curve of spot 1 shows the highest enhancement. The black curve represents the measured SERS spectra of 10^{-2} M R6G in the plain Cu surface. Obviously, the Cu substrates fabricated by fs laser and then modified by Au nanoparticles exhibit strong Raman signal intensity.

For obtaining Raman spectra in the fabricated regions and normal Raman spectra in the plain Cu surface, the Raman spectrometer laser powers were 0.6 mW and 6 mW; and the accumulated time was 10 s. The SERS EFs can be evaluated by the following equation (Hamad et al. 2014):

$$EF = \frac{I_{SERS}}{N_{SERS}} \Big/ \frac{I_{Raman}}{N_{Raman}} \tag{1}$$

where I_{SERS} is the Raman signal intensity of a Cu substrate at 10^{-7} M R6G, N_{SERS} is the number of

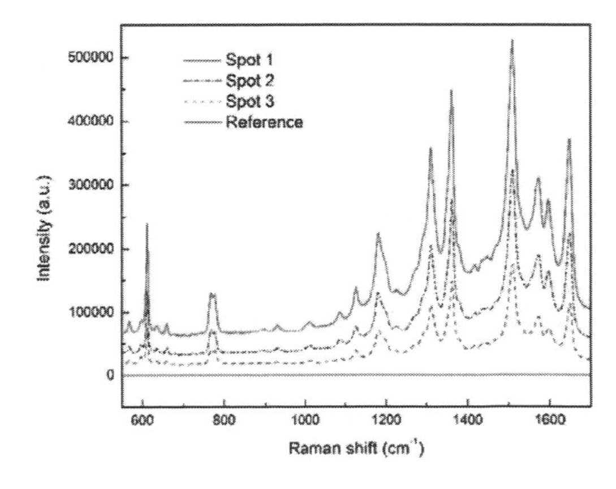

Figure 5. SERS spectra of Cu substrates with 10^{-7} M R6G.

molecules detected in the Cu SERS substrate at 10^{-7} M R6G, I_{Raman} is the normal Raman signal intensity of a plain Cu surface at 10^{-2} M R6G, and N_{Raman} is the number of molecules detected in the plain Cu surface at 10^{-2} M R6G. The highest EF, 3.60×10^9, occurs at Raman peak 1510 cm^{-1} of the spot 1, which is very promising for detecting a single molecule (Lim et al. 2011, Zheng et al. 2015) and demonstrates high sensitivity.

4 CONCLUSIONS

In summary, superhydrophobic SERS substrates have been fabricated in one step by fs laser irradiation in open air. The maximum static Contact Angle (CA) of the processed Cu substrate can reach up to 151°, which demonstrates superhydrophobic properties. With Au nanoparticle decoration, the nanotextured Cu substrate presents a strong potential for SERS with an enhancement factor of 3.60×10^9, when R6G is used as detection molecules. The superhydrophobic Cu surface confines the R6G droplet into a small area, which increases the concentration equivalently. And the uniformly distributed 60 nm Au nanoparticles with small nanogaps cause enough "hot spots", which can excite strong localized plasmon resonance, thereby resulting in electromagnetic enhancement. Both of these contribute to the high intensity of Raman scattering signals. In addition, the approach to fabricate the SERS substrate is simple and efficient (fabricate an area of $2500 \times 2500\ \mu m^2$ in just 4 minutes), which shows its potential application in industrial production.

REFERENCES

And, B.Q. & Shen, Z. 2005. Fabrication of Superhydrophobic Surfaces by Dislocation-Selective Chemical Etching on Aluminum, Copper, and Zinc Substrates. Langmuir the Acs Journal of Surfaces & Colloids: 21(20):9007–9.

Baldacchini, T., Carey, J.E., Zhou, M., et al. 2006. Superhydrophobic surfaces prepared by microstructuring of silicon using a femtosecond laser. Langmuir the Acs Journal of Surfaces & Colloids: 22(11):4917–9.

Ding, X., Zhou, S., Gu, G., et al. 2011. A facile and large-area fabrication method of superhydrophobic self-cleaning fluorinated polysiloxane/TiO_2 nanocomposite coatings with long-term durability. Journal of Materials Chemistry: 21(17):6161–6164.

Guo, Z., Jian, F., Wang, L., et al. 2007. Fabrication of superhydrophobic copper by wet chemical reaction. Thin Solid Films: 515(18):7190–7194.

Hamad, S., Podagatlapalli, G.K., Mohiddon, M.A. et al. 2014. Cost effective nanostructured copper substrates prepared with ultrafast laser pulses for explosives detection using surface enhanced Raman scattering. Applied Physics Letters: 104, 263104–263105.

Haynes, C.L., McFarland, A.D. & Duyne, R.P.V. 2005. Surface-Enhanced Raman Spectroscopy: Analytical Chemistry: 77, 338 A-346 A.

Kneipp, K., Kneipp, H. & Kneipp, J. 2006. Surface-enhanced Raman scattering in local optical fields of silver and gold nanoaggregates-from single-molecule Raman spectroscopy to ultrasensitive probing in live cells. Accounts of Chemical Research: 39, 443–450.

Leon, A.C.C.D., Pernites, R.B. & Advincula, R.C. 2012. Superhydrophobic Colloidally Textured Polythiophene Film as Superior Anticorrosion Coating. Acs Applied Materials & Interfaces: 4(6):3169–3176.

Lim, D.K., Jeon, K.S., Hwang, J.H., et al. 2011. Highly uniform and reproducible surface-enhanced Raman scattering from DNA-tailorable nanoparticles with 1-nm interior gap: Nature Nanotechnology: 6, 452–460.

Luo, Y.F., Lang, H.Y., Liang, J., et al. 2011. Fabrication of Superhydrophobic Surfaces on Aluminum Alloy by Simple Chemical Etching Method. Advanced Materials Research: 239–242:2270–2273.

Sas, I., Gorga, R.E., Joines, J.A., et al. 2012. Literature review on superhydrophobic self-cleaning surfaces produced by electrospinning. Journal of Polymer Science Part B Polymer Physics: 50(12):824–845.

Shenton, C. 2009. Fabrication and anti-corrosion property of superhydrophobic hybrid film on copper surface and its formation mechanism. Surface and Interface Analysis: 41(11):872–877.

Singh, A.K. & Singh, J.K. 2016. Fabrication of zirconia based durable superhydrophobic−superoleophilic fabrics using nonfluorinated materials for oil-water separation and water purification. Rsc Advances: 6, 103632–103640.

Tai, M.H., Gao, P., Tan, B.Y., et al. 2014. Highly efficient and flexible electrospun carbon-silica nanofibrous membrane for ultrafast gravity-driven oil-water separation. Acs Applied Materials & Interfaces: 6(12):9393–9401.

Tong, L. & Käll, H.X.M. 2014. Nanogaps for SERS applications. Mrs Bulletin: 39(2):163–168.

Xu, L., Karunakaran, R.G., Guo, J., et al. 2012. Transparent, superhydrophobic surfaces from one-step spin coating of hydrophobic nanoparticles. Acs Applied Materials & Interfaces: 4(2):1118–1125.

Zheng, Y., Soeriyadi, A.H., Rosa, L., et al. 2015. Reversible gating of smart plasmonic molecular traps using thermoresponsive polymers for single-molecule detection: Nature Communication: 6 (2015).

Civil, Architecture and Environmental Engineering – Kao & Sung (Eds)
ISBN 978-1-138-02985-9

A study on the ship model resistance test and numerical simulation

H.C. Pan, S.N. Yu & Y.J. Yang
School of Mechanical Engineering, Hangzhou Dianzi University, Hangzhou, Zhejiang, China

ABSTRACT: The ship resistance is a very important parameter, which can directly affect the performance of the ship. In order to study the resistance of the ship, in this paper, a model S-175 is considered as a research object and combines theoretical analysis, numerical analysis, and test methods to study on it. A comprehensive test platform for the ship model has been built, and also carried out some related model tests. The resistance data of both CFD and experiments under different ship speeds have been obtained for comparison. The results show that the numerical calculated resistance data are closed to model experiment results, which can satisfy the engineering applications of ship resistance prediction. Simulating the flow field around the hull-based on CFD shows that, with an increase in the Froude number, the wave resistance increases, and the bow and stern wave height also increase. However, the height increase of the stern wave is not obvious when compared with the bow. Under low speed conditions, the wave-making was not obvious. As the speed increased to a certain range, the stern creates the stern wake. And its baseline will be stretched when the speed is increased.

1 INTRODUCTION

The main performance parameters of the ship include heavy resistance, buoyancy, stability, maneuverability and speed, etc. (Yu 2009). Among these, the ship's speed has a significant influence on the economy and utility, which also makes greater contribution to the maneuverability and operational capability of the warship. Basically, the speed is mainly affected by ship resistance. Hence, studies on the resistance of ship sailing are always the research hot spot in the ship design field.

At present, there are three ways to discover the ship resistance: theoretical research method, ship model test method, and numerical simulation method (Sheng & Liu 2014). The theoretical research method creates simple analysis, reasoning, and conclusion according to corresponding fluid mechanical knowledge and mathematical tool, and then the approximate ship's resistance can be evaluated. However, only depending on the approximate resistance evaluation method and the empirical equation to forecast the ship's resistance has no longer satisfied our request on accuracy. The ship model test method is the major and the most effective method in ship's resistance forecasting (Sun et al. 2003). However, it takes a lot of time to make the ship model and easily produce error, which also costs a large number of human and material resources. The numerical simulation method can reflect the real condition of the flow field around the ship and make contribution to the research of speed, which decreases the human and material resources' costs in experiments. It also has great reference value in the aspect of optimization design and type selection (Deng et al. 2013). And so, studying on the ship's CFD method is essentially significant and meaningful (Xu 2012, Gao et al. 2014, Liu et al. 2001). At present, the numerical calculation method has been widely used in ship design and analytical calculation. There are three main numerical methods to simulate viscous flow: DNS (Piller et al. 2002, Wissink 2003, Versteeg & Malalaseker 1995), LES, and RANS (Michelassi et al. 2003, Stephane 2003). With the development and rapid progress of the modern computer technology, fluid researchers constantly create new and effective numerical methods to make it convenient for use. However, the ship structure is too complex to find out the real flow field around it; and so, the numerical simulation method can only solve the part of problems. As a result, the three typical research methods of ship resistance need to be combined to support each other.

Based on the theoretical research and aiming to achieve S-175 ship model resistance conditions, the vertical circulating tank experiment and numerical simulation research have been operated in this paper. The numerical simulation results were compared with the experimental results. In this paper, the gas and liquid distributions are also analyzed and the free surface condition with different Froude numbers (different speeds) are explored in detail.

2 SHIP MODEL RESISTANCE TEST

2.1 *Model preparation*

The S-175 model with length between perpendiculars of 2 m is made of fiber glass reinforced plastics, which has some advantages such as light unit weight and non-deformation after long-term use. The main particulars of the ship are given in Table 1.

2.2 *Model water tank test*

The model test is performed at the vertical circulating tank in Hangzhou Dianzi University. The length, width, and depth of the vertical circulating tank are respectively 4000 mm, 1500 mm and 1300 mm. The tank consists of a barrel, rectifier, honeycomb, porous plate, propelling plant, AC variable frequency motor, X–Y working platform, electronic speed control system, electronic control cabinet, control panel, IPC, etc. The arrows in Figure 1 show the flow direction of the tank.

The vertical circulating tank is adopted for resistance experiments with the velocity of flow in the range of 0.2–2.0 m/s, and the ship model size is in the range of 1.5–2.5 m. Merchant ship and fishing boat are used as the experiment objects.

Table 1. Main particulars of S-175.

S-175	Ship	Model
Lpp (m)	175	2
B (m)	25.7	0.292
H (m)	15.3	0.175
T (m)	9.08	0.104
Displacement	23900 (T)	35.67 (kg)
Cb	0.568	0.568
S (m2)	5420	0.734
Scale	1	2/175

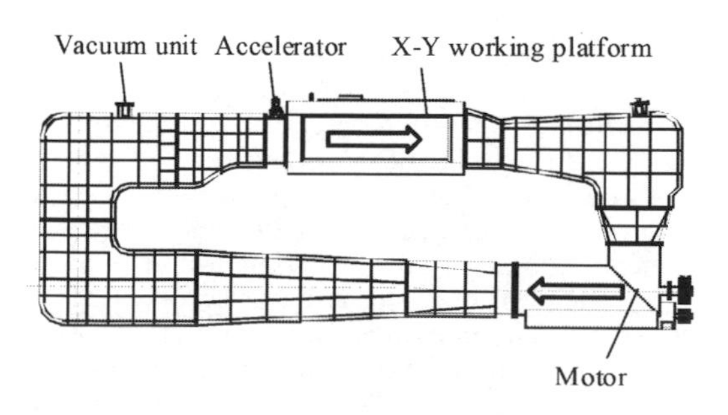

Figure 1. Schematic of the vertical circulating tank.

2.3 *Resistance experiment and results*

The ship model resistance experiment operation system is shown in Figure 2.

The experiment system can achieve following functions:

1. The generator's frequency and velocity of flow are adjusted by using an electronic control system.
2. The current meter monitors the real-time velocity of flow in the tank, which can be seen on the electronic control panel velocity of the flow measure system.
3. The ship model which is affected by flow can produce resistance that makes a tiny deformation on the steel wire. The S-type strain sensor on the ship model converts this deformation into an electrical signal. The data collector receives this signal and then converts it into resistance.

To perform the test, a ship model should be installed in the vertical circulating tank, according to Figure 3, and the ship model's plimsoll line should be located on the calm water surface by adjusting sand bags. After completion of the above preparations, the resistance meter and current meter are installed and the velocity of flow is adjusted by controlling the motor's frequency. Eight points were measured from low to high speed. The data were acquired and the experimental resistance values were obtained in Table 2.

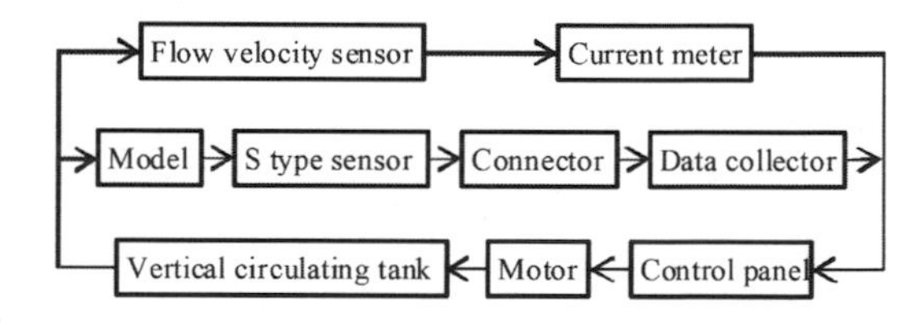

Figure 2. Schematic of the ship model resistance experiment operation system.

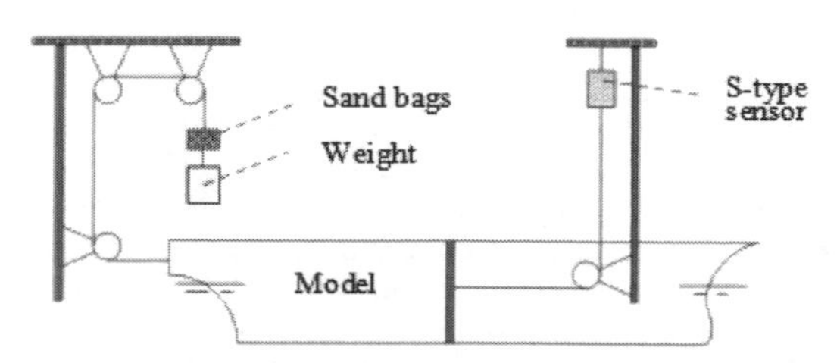

Figure 3. Schematic of the installation process.

Table 2. Experimental results of S-175.

V(m/s)	0	0.21	0.39	0.61	0.81	1.02	1.21	1.42
Fr	0	0.047	0.088	0.14	0.18	0.23	0.27	0.32
R_{tt}(N)	0	0.1000	0.3569	0.6129	1.2140	1.6955	2.5173	3.4352

Figure 4. Picture of the ship model of S-175.

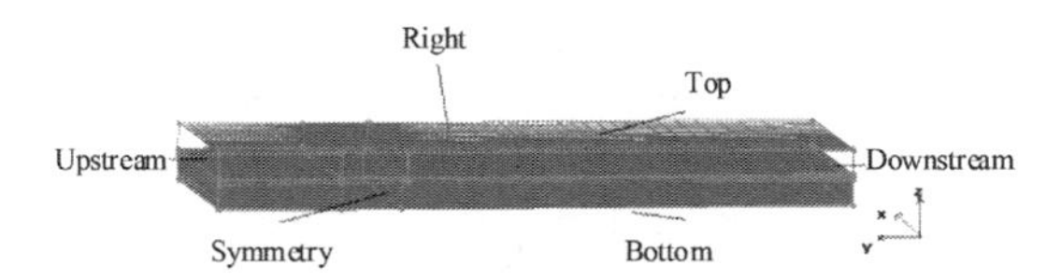

Figure 5a. Schematic of the computational domain and boundary conditions.

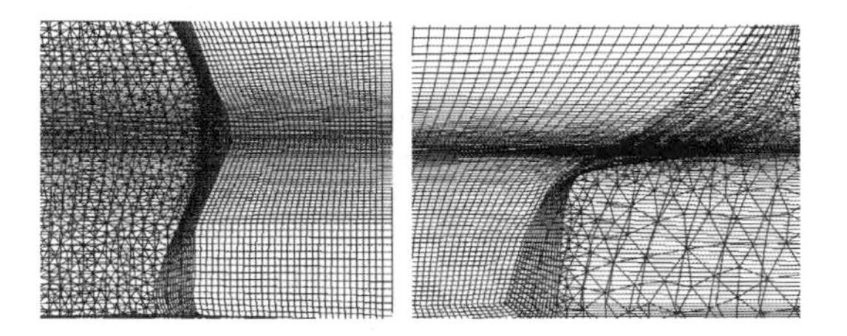

Figure 5b. Unstructured grids of the bow and stern.

3 MODEL CONSTRUCION AND MESH GENERATION

In this paper, the calculation model and the experimental model are in the same measure. In view of the symmetrical structure, only half of the ship hull is used in the numerical simulation. Figure 4 shows the hull geometry of S-175.

The computational domain and boundary conditions are shown in Figure 5a. The length of the computational domain extends 6 Lpp in the y-direction, the width extends 10B in the x-direction, and the height extends -3H~3H from the free surface in the z-direction.

As shown in Figure 5b, the mesh near the bow and stern is an unstructured grid. Other computational domains are discretized by using a H-type structural mesh. The total numbers of meshes are 3771338. In order to well-reflect the complex flow around the hull and ship motion, the meshes near the bow, stern, plimsoll line, and around the hull surface are refined locally.

In the numerical simulation, the boundary conditions are set as follows:

1. Inlet boundary: the direction of flow is set to the Y-axis negative direction. The model test gives the flow velocity and the upstream is set to be the velocity entrance.
2. Outlet boundary: the downstream and top are prescribed to be an open outlet. And the relative static pressure and fluid direction are set in this work.
3. Wall boundary: hull, bottom, and right are set to be the non-slip wall.
4. Symmetry boundary: the symmetry is assumed to be under symmetrical boundary conditions.

4 ANALYSIS OF NUMERICAL SIMULATION RESULTS

4.1 *Numerical simulation*

In this paper, the gas–liquid flow simulation is involved, which is needed to set the fluid volume fraction under inlet and outlet conditions. The free surface was set to be no slip, and the turbulence simulation was the SST model. The wall approach was automatic, and the simulation solving the format was of higher order. The physical timescale was 0.02 s and the maximum iteration step number was 1000.

4.2 *Computational results*

In this paper, the S-175 standard ship model fluid field distribution conditions are simulated under seven speeds: 0.21 m/s, 0.39 m/s, 0.61 m/s, 0.81 m/s, 1.02 m/s, 1.21 m/s, and 1.42 m/s. And the simulation resistance was compared with experimental results to make the analysis. The comparison results are shown in Table 3.

From Figure 6, the increasing trend of the total resistance calculated with CFD method is the same as the increasing trend of the total resistance measured in the experiment. From Table 3, we can observe that when $0.088 \leq Fr \leq 0.27$, their deviation decreased while the fluid number was increasing. And the deviation achieved a minimum value of 0.954%, when the Froude number was 0.27. However, when the Froude number was greater then 0.27, the deviation was enlarged further.

In the reality model test, when Fr = 0.32, waves can be observed on the deck, which had serious effects on the ship posture and made the

Table 3. List of simulation results and experimental results.

Fr	V(m/s)	R_{tl} (N)	R_{tc} (N)	$\Delta R_t/R_{tl}$
0	0	0	0	0
0.047	0.21	0.100028	0.0958242	−1.78%
0.088	0.39	0.356989	0.311562	−12.72%
0.14	0.61	0.612916	0.6734	9.86%
0.18	0.81	1.21406	1.144382	5.73%
0.23	1.02	1.69556	1.787076	5.39%
0.27	1.21	2.51736	2.54138	0.954%
0.32	1.42	3.43526	4.19744	22.18%

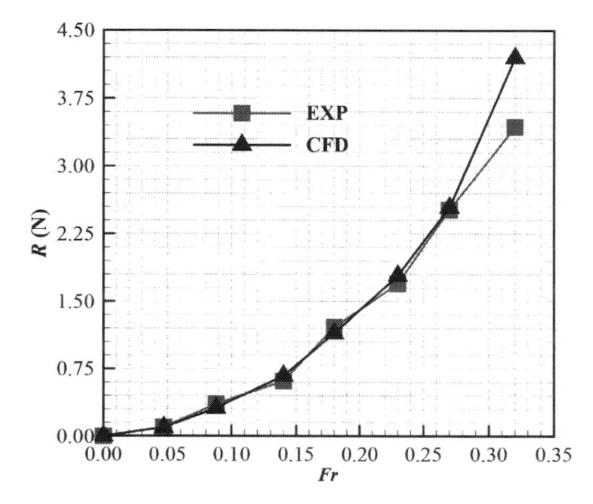

Figure 6. Graph showing the comparison between simulation results and experimental result.

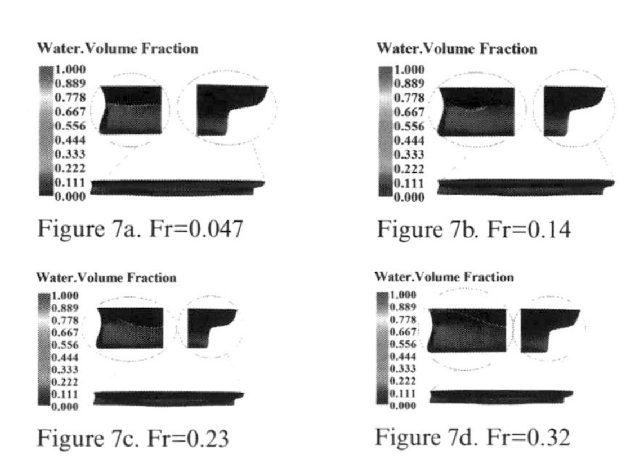

Figure 7a. Fr=0.047

Figure 7b. Fr=0.14

Figure 7c. Fr=0.23

Figure 7d. Fr=0.32

Figure 7. Graphs showing the gas and liquid distributions for different Fr values.

ship extremely unstable in the water. Similarly, in numerical simulation, from this certain speed (Fig. 7d), the phenomenon of the water wave on the deck can be also observed clearly. And so, it was proved that the numerical calculating method can accurately simulate the flow field distribution and ship posture of the ship model.

4.3 *Ship model analysis of the viscous flow field*

1. Analysis of the hull surface's gas–liquid distribution:

Figure 7 indicates the hull surface's gas and liquid distribution for different Froude numbers. From the figures, it can be observed that the water wave at the bow and stern were more obvious in which the wave of the bow was the highest. The wave decreased at the middle of the hull and further increased at the stern. With an increase in the ship speed, the wave height of the bow changed obviously, the wave height of the stern also increased but not as obvious as the bow. And the wave resistance was also increased. When Fr = 0.32, it can be observed that the wave was on the deck.

2. Simulation analysis of the free surface:

In this paper, the VOF method is used to simulate the flow of the free surface. The VOF method is based on the fluid in the grid cell, in every moment, thereby occupying volume fraction function F to create and track the free surface. Figure 8 shows the free surface conditions for different Froude numbers. From these figures, the wave of the water and the ship wave for the ship movement can be observed, which are the superposition results of the transverse wave and longitudinal wave. In low speed cases, the wave was not obvious. However, when the speed increased to a certain range, the stern created the stern wake, and its baseline will be stretched when the speed is added. From the figure, it can be observed that the flow field condition of the ship stern was complex.

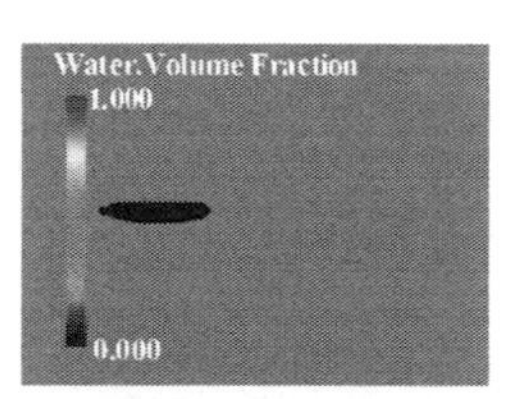

Figure 8a. Fr=0.047

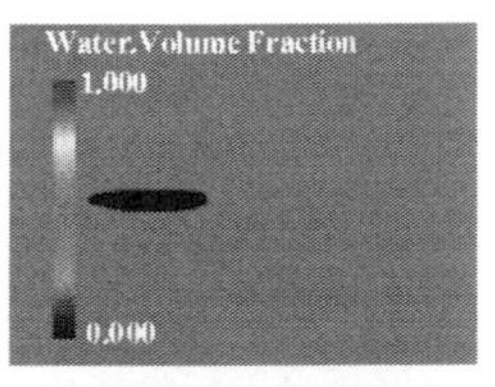

Figure 8b. Fr=0.14

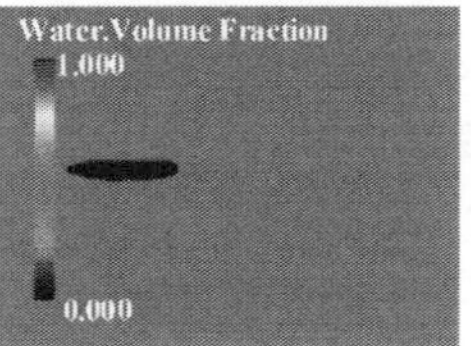

Figure 8c. Fr=0.23

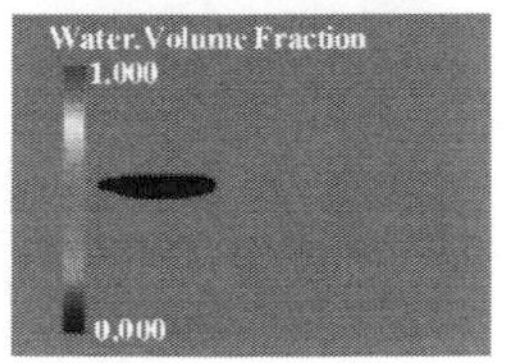

Figure 8d. Fr=0.32

Figure 8. Graphs showing free surface conditions for different Fr values.

In the design of the propeller, the corresponding wave on the stern will affect the efficiency of the propeller. Therefore, the analysis for flow fields on the ship stern is a necessary work.

5 CONCLUSION

According to the comparison of the ship resistance numerical simulation and experimental results of the S-175 ship model, the following conclusions can be drawn:

1. Although the CFD calculation data and the experimental data had a little deviation, the data fit better enough in a certain range. It means that the method used in this paper to forecast the ship resistance is reasonable under certain condition. To forecast the resistance efficiently, the numerical simulation method can be combined with the ship model test.
2. When the ship model operates at a low speed, a tiny disturbance has a considerable effect on the resistance. Therefore, the accuracy of the experimental measurement and numerical simulation of the resistance is affected by test conditions, operational details, parameter settings, etc.
3. Different types of ships have different designed speeds. When the testing speed was set to 1.42 m/s in this paper, the phenomenon of the water wave on the deck can be observed. And so, the ship model test should be less than this speed.
4. When compared with the model test, CFD has obvious advantages to simulate the model flow field and capture the local details and the characteristics of the flow field. With an increase in the Froude number, the wave resistance increases, and the bow and stern wave heights also increase. However, the height increase of the stern wave is not obvious when compared with the bow. Also, with an increase in the speed, the pressure of the bow and the stern has increased obviously. Due to the effect of the water viscosity, under the effect of the hull longitudinal pressure gradient distribution, the stern starts a boundary layer separation, which makes the pressure of the stern lesser than that of the bow. At low speeds, the wave-making was not obvious. When the speed increased to a certain range, the stern creates the stern wake, and its baseline will be stretched when the speed is increased.

REFERENCES

Deng, R. & Huang, D.B. & Zhou, G.L. 2013. Investigation on some factors effecting ship resistance calculation with CFD code FLUENT. *Journal of Ship Mechanics* 17(6): 616–624.

Gao, Z.L. & Pang, Q.X. & Zhang, R.B. 2014. Calculation of ship resistance in muddy navigation area based on CFD method. *SHIPBUILDING OF CHINA* 35(4): 104–111.

Liu, Y.Z. & Zhang, H.X. & Li, Y.L. 2001. Calculations of the ship performances and solving of RANS equations in the 21st century. *Journal of Ship Mechanics* 5(5): 66–84.

Marzio Piller. & Enrieo Nobile. & J. Thomas. 2002. DNS study of turbulent transport at low Prandtl Numbers in a channel flow. *Journal of Fluid Meehanies* (458): 419–441.

Michelassi, V. & J.G. Wissink. & W. Rodi. 2003. Direct numerical simulation, large eddy simulation and Unsteady Reynolds-averaged Navier-Stokes simulations of Periodic unsteady flow in a low-Pressure turbine cascade: A comparison. *Journal of Power and Energy* 217(4): 403–412.

Sheng, Z.B. & Liu, Y.Z. 2014. *Principle of ship.* Shanghai: Shanghai Jiao Tong University Press.

Stephane, V. 2003. Local mesh refinement and penalty methods dedicated to the Direct Numerical Simulation of incompressible multiphase flows. *Proceedings of the ASME/JSME Joint Fluids Engineering Conference*: 1299–1305.

Sun, X.J. & He, Y.P. & Tan, J.H. 2003. Study of 2500 units PCTC resistance performance with experiment and CFD calculation method. *Journal of East China Shipbuilding Institute (Natural Science Edition)* 17(6): 1–5.

Versteeg, H.K. & W. Malalasekera. 1995. *The Finite Volume Method.* New York.

Wissink, J.G. 2003. DNS of separating low Reynolds number flow in a turbine cascade with incoming wakes. International Journal of Heat and Fluid Flow 24(4): 626–635.

Xu, L. 2012. Optimization of ship hull resistance based on CFD method. *Shanghai Jiao Tong University*: 101.

Yu, W.J. 2009. Research on calculation and prediction for ship resistance based on CFD theory. *Shanghai Jiao Tong University*: 98.

Civil, Architecture and Environmental Engineering – Kao & Sung (Eds)
© 2017 Taylor & Francis Group, ISBN 978-1-138-02985-9

Magnetic memory testing signal analysis of Q345 steel welding parts under the static tension condition

Miao Lou, Xiang-Hui Lv, Fei Lu & Bo Li
North West Institute of Nuclear Technology, Xi'an, China

ABSTRACT: The paper is based on the research object of the Q345R steel butt welding specimen under static tension and analyzes the characteristics and change rule of magnetic memory testing signals in the load process. Results show that the welded specimen can determine the weld parts before magnetic memory testing. In the elastic stage, magnetic field gradient values in weld parts decrease obviously, which are consistent with those of the basic metal. After analysis, it is noted to be caused by the welding residual stress. In the yield phase, local magnetic field gradient values in the weld parts increase slightly and then decrease obviously, which is thought to be due to the fact that after welding, metallographic structures and mechanical properties change, and welding and heating affected the zone display performances, which are inconsistent with those of the basic metal in the tensile process. In the strengthening phase, welded specimens gradually show the same features as the basic metal specimens.

1 INTRODUCTION

Metal magnetic memory testing was put forward by Russian professor Dubois in the late 20th century (Dubov A A, 1999), which can accurately and reliably detect dangerous parts of iron work pieces that are characterized by the area of stress concentration, becoming the only effective nondestructive testing method in the early diagnosing of work pieces on the verge of injury and damage statuses. After nearly 20 years of research and development, the method now has been widely used in boiler pressure vessels, pipelines, turbine blades, engine crankshafts, bridges, railways, and other tastings (Ren Jilin, 2000 & Leng Jiancheng, 2010). Part of structural components in the Q345R steel container implement welding and assembly at the scene; the container is put into use without removing the welding residual stress, and the regular safety inspection is indispensable. In this paper, we carry out an experimental study on whether the ability of early warning in the metal magnetic memory method can conduct tastings and evaluations for the site welding parts of the container, obtain the magnetic memory signal characteristics of the basic metal, and weld assemblies and under the tensile loading process so as to provide channels for security evaluation of the container through a comparative analysis.

2 TEST SPECIMEN PREPARATIONS

The Q345R steel, basic metal and weld tensile specimens are designed by loading in a statically tensile way and are closed down to uninstall in the elastically tensile stage, yield tensile stage, strengthening phase, and shrunken phase. The magnetic memory tastings are carried out on the specimens so as to obtain the magnetic field intensity and its gradient changes of the testing lines. The material tensile strength (σ_b) is 510–640 MPa, yield point (σ_s) is 345 MPa, and elongation rate (S) is 21%. In accordance with the (GB/T 228.1-2010 standard (Roskosz M, 2011), the specimen has the shape of a plate, with a thickness of 6 mm, containing three pieces. Among them, the second and third are the butt welding specimens after longitudinal cuttings, with V-type weld groove of the welding specimens for manually electric arc welding, and the welding materials are J507 welding rods, with welded width of 10 mm. After passing the radiographic testing, it belongs to welding grade I without excessive defects (Long Feifei, 2012), and its welded location is shown in Figure 1.

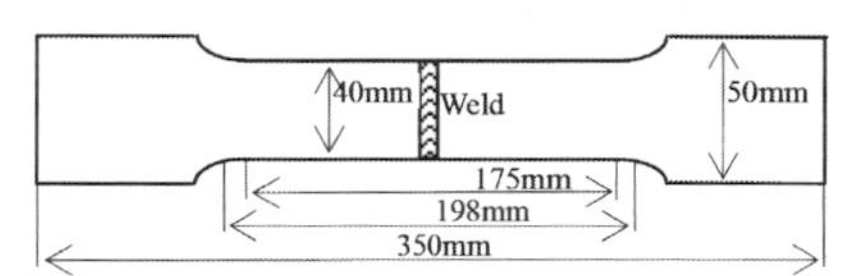

Figure 1. Butting welded tensile specimens sketch map.

3 TESTING TIME AND INSTRUMENTS

The aim of closing down the specimens to detect is to test before the tensile loading of the specimens and obtain the initial state signal. The mean stresses of the elastic stage are 110, 220, and 330 MPa. We selected five points according to the different sizes in the yield stage, randomly selected testing points in the strengthened and necking stages of the specimens, and measured after the specimen are fractured. The specific testing points are shown in Table 1. The positive and negative sides of the specimens are tested respectively, and the magnetic memory test lines are perpendicular to the welded joints, which are located in the middle of the specimens, as shown in Figure 2.

Magnetic memory testing equipment include the TSC-1M-4 metal magnetic memory detectors produced from Russia Dynamic Diagnostic Company. The loading equipment adopt the RAS-250 type of tensile testing machine manufactured by German Scheck.

4 EXPERIMENTAL RESULTS

4.1 *Initial state*

The test results before loading the specimens are shown in Figure 3.

Table 1. Downtime testing point.

Testing times	Experimental stage		Testing times	Experimental stage	
1	Not loaded		8	Yield	Strains 1.6%
2	elastic	Stresses 110 MPa	9	Intensive	Strains 2.2%
3		Stresses 220 MPa	10		Strains 6%
4		Stresses 330 MPa	11		Strains 10%
5	yields	Strains 0.2%	12		Strains 15%
6		Strains 0.3%	13	Necking	
7		Strains 0.8%	14	Sheared	

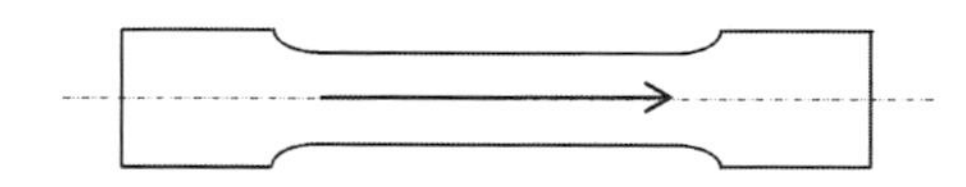

Figure 2. Sketch maps of the magnetic memory test lines.

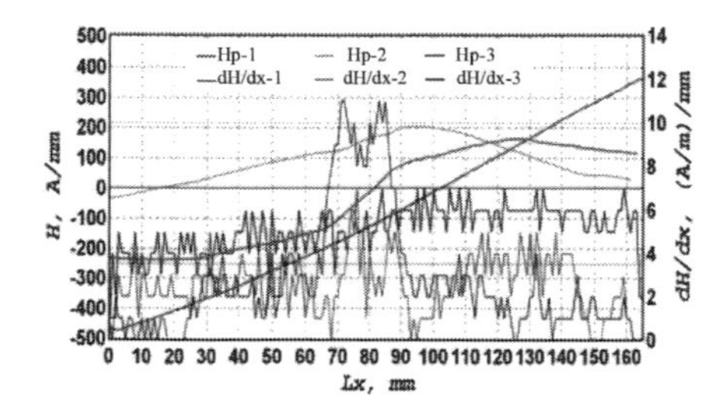

Figure 3. Test results before loading.

When not loaded, the magnetic field strength of the first specimen is almost a diagonal line, with its magnetic field gradient values between 4 and 7; the magnetic field intensity values of the first and second specimens in the welded area have mutations, and the magnetic ladder number has obvious peak value compared with the basic metal.

4.2 *Elastic stage*

In the elastic stage, test results that load the stresses as 110, 220, and 330 MPa are shown in Figures 4–6, respectively.

From Figures 4 to 6, the magnetic field strength is still the oblique line with the loading of the first specimen, the change levels off, and the slope decreases; the change scope of the gradient values continues to narrow, the gradient value is almost a constant until 330 MPa, with a numerical value about 4. The slope rate of the magnetic field intensity for the second and third specimens decreases, the left-hand side is turned into a positive value from a negative value, and the right-hand side is turned into a negative value from a positive value, which leads to the emergence of the deflection phenomenon. The magnetic field gradient value of welded parts decreases to 330 MPa, which is consistent with the basic metal parts, almost being a constant.

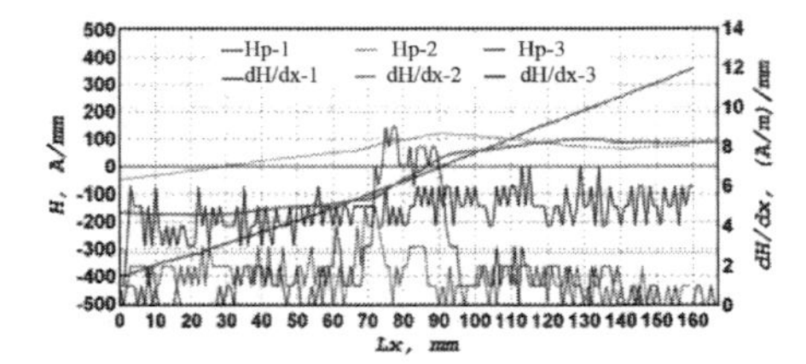

Figure 4. Specimen test results under 110 MPa.

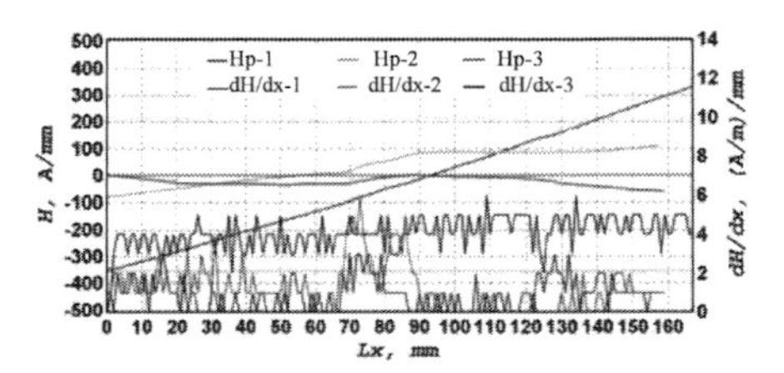

Figure 5. Specimen test results under 220 MPa.

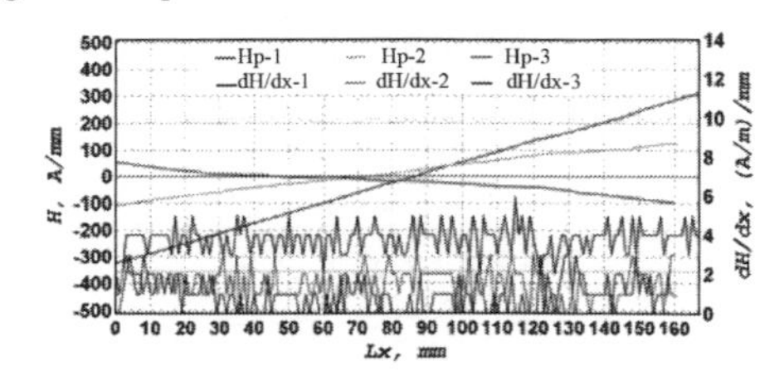

Figure 6. Specimen test results under 330 MPa.

4.3 *Yield stage*

Four measurements are conducted in the yield phase, with the test results of the loading stresses of 0.2%, 0.38%, 0.8%, and 1.65%, which are shown in Figures 7–10, respectively.

From Figures 7 to 10, it can be seen that the magnetic field intensity of the first specimen is almost an oblique line, with its gradient value keeping decreasing; in the yield stage, the magnetic field gradient values of the second and third specimens have significantly local change characteristics compared with the elastic stage, namely the gradient values in the welded parts increase; in the weld seam position, the slope rate of the magnetic field strength is increases obviously.

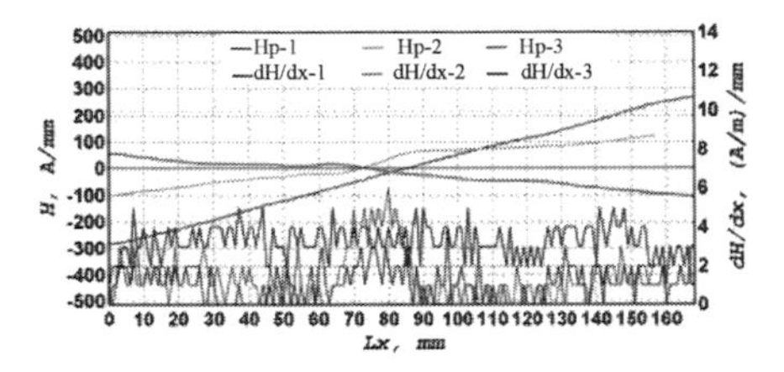

Figure 7. Specimen test result when the stress is 0.2%.

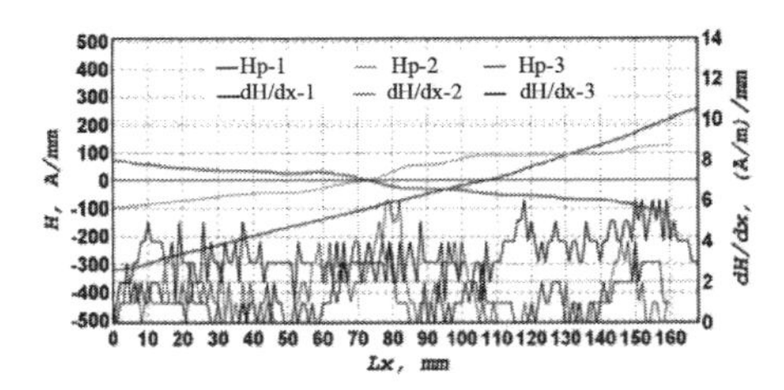

Figure 8. Specimen test result when the stress is 0.38%.

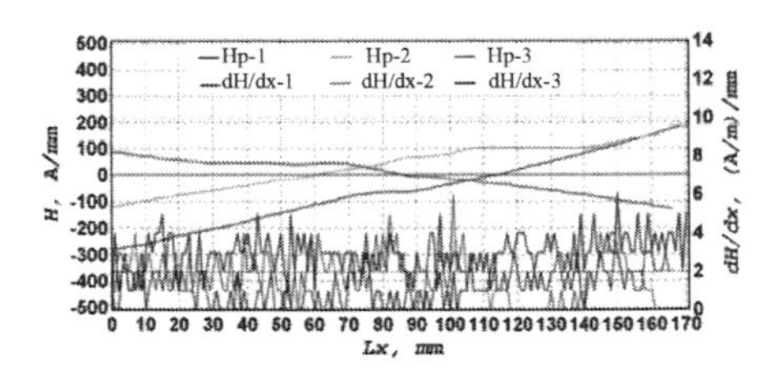

Figure 9. Specimen test result when the stress is 0.8%.

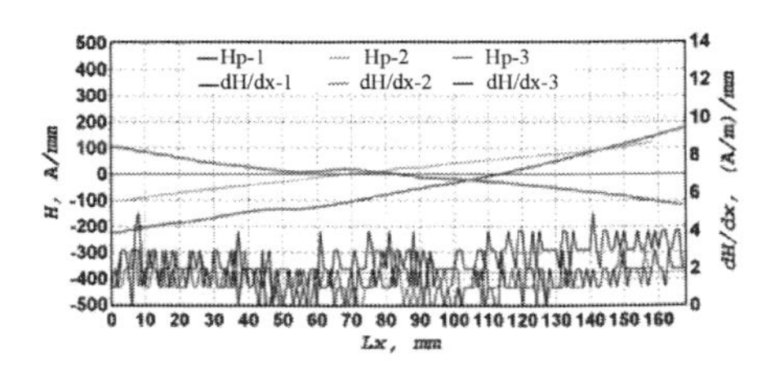

Figure 10. Specimen test result when the stress is 1.65%.

4.4 *Reinforcement, constriction, and fracture of specimens*

The test results of the specimen stresses are loaded to 2.2%, 6%, 10%, and 15%, which are shown in Figures 11–14, respectively.

With the increase of load, the magnetic field gradient numerical values become smaller, with the numerical values reduced to 0–2 as well as the gradual reduction of the slope rates of the magnetic field intensity. The magnetic field strength of the second and third specimens is almost a diagonal line again, and the numerical value of the magnetic field gradient is almost a constant.

The specimen continues to be loaded, and the apparent constriction appears in the basic metal

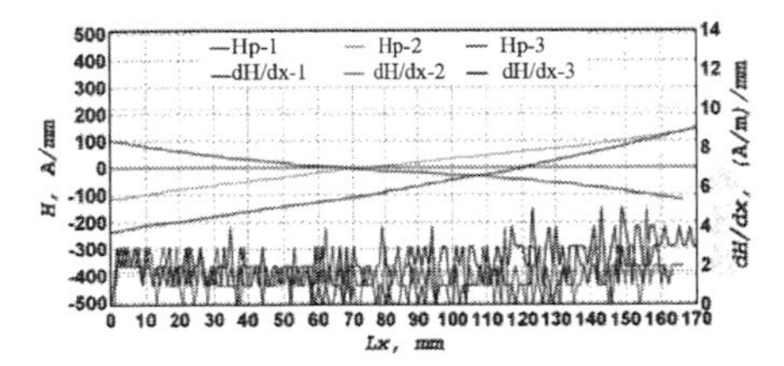

Figure 11. Specimen test result when the stress is 2.2%.

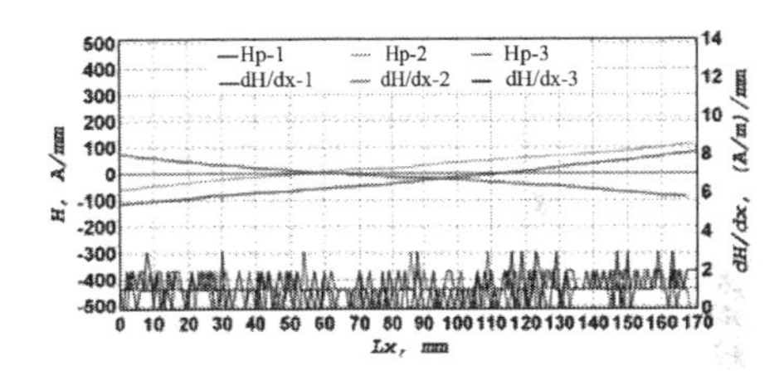

Figure 12. Specimen test result when the stress is 6%.

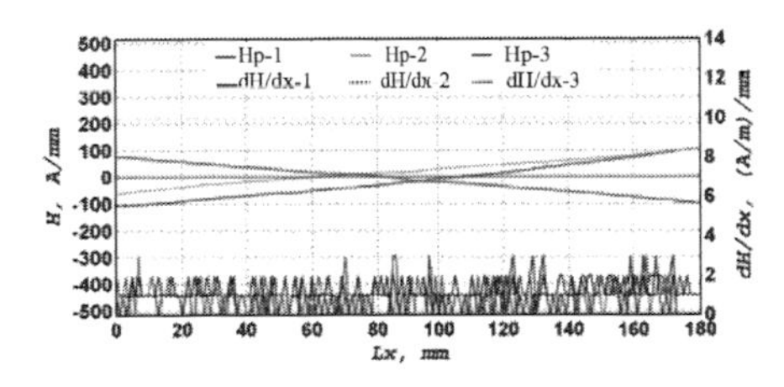

Figure 13. Specimen test result when the stress is 10%.

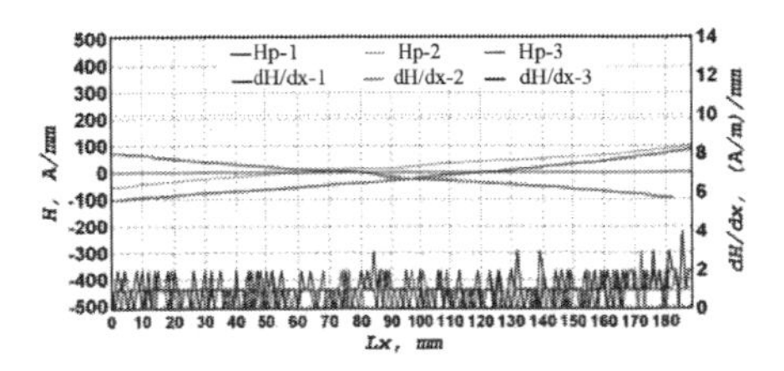

Figure 14. Specimen test result when the stress is 15%.

parts when the stress is 23.75%, as shown in Figure 15. Then, the specimen continues to be loaded to fracture; magnetic memory scanning and testing are conducted for the fractured specimen, as shown in Figure 16.

It is can be seen from Figures 15 and 16 that the specimen enters the strengthening phase to necking until fracture in the end, and the magnetic field intensity value of the specimen exceeds zero, which show the phenomenon of moving to the right-hand side and left-hand side, respectively, with its numerical values instantaneously jumping for change and the magnetic field gradient of the fracture specimens making a big difference. Magnetic memory testing signals show no obvious reaction to the constriction but demonstrate the most obvious reaction to the fracture.

4.5 *Over-zero value point*

Over-zero value points of many times of the magnetic field strength values for all specimens are shown in Figure 17, which displays the repeatedly

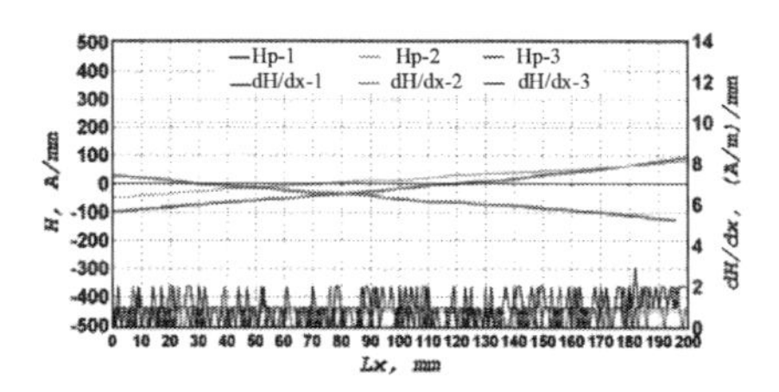

Figure 15. Specimen test result when the stress is 23.75%.

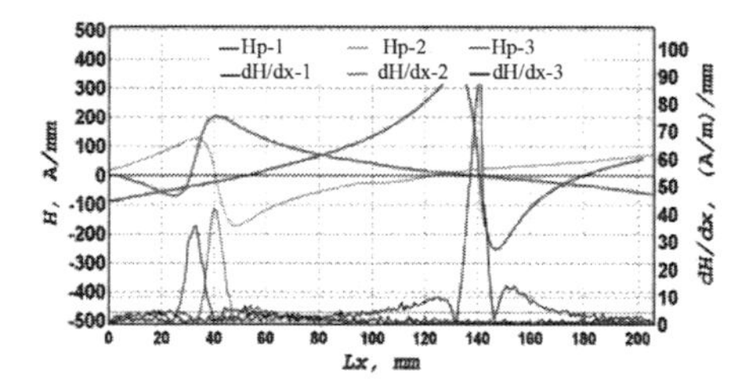

Figure 16. Testing result of the fractured specimen.

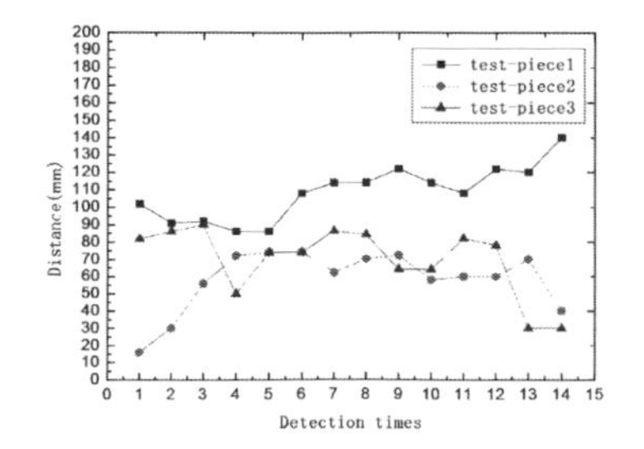

Figure 17. Over-zero value point of the magnetic field strength value.

moving process of over-zero value points, with all specimens' movements having no significant rules to follow. Compared with the actual tensile states of the specimens, the magnetic field intensity value in the fractured part is zero, and its directions change as well as the gradient value obtains the maximum value.

5 CONCLUSIONS

1. The test results can directly reflect the weld positions before loading the specimens.
2. In the elastic stage, along with the increase of load, the magnetic field gradient values in the weld parts of the welding specimens decrease. When loaded to 330 MPa, the detection result gradually converges with that of the basic metal specimen, with its magnetic field gradient value almost a constant. The feature is thought to be caused by being gradually eliminated through the tensile process of welding residual stresses.
3. In the yield stage, the welded parts of welding specimens change obviously locally, and the magnetic field gradient values increase again and then decrease until the value is consistent with that of the basic metal. In the strengthening and necking stages later, the magnetic field gradient values decrease, which are consistent with those of the basic metal of specimens.
4. After the specimen fractures, the magnetic field strength value of the fracture position shows an over-zero value, with the location showing no unified law with the previous zero value points that have been tested.

REFERENCES

Dubov A A. Diagnostics of metal items and equipment by means of metal magnetic memory[C]. Proceedings of CHSNDT 7th Conference on NDT and International Research Symposium. Shantou, China: Non-Destructive Testing Institution, CMES, 1999: 181–187.

Leng Jiancheng, Xu Minqiang, Xing Haiyan. The Research Progress of Ferromagnetic Component Magnetic Memory Testing Technology [J]. Journal of Materials Engineering, 2010, 11: 88–92.

Long Feifei, Gao Guozhu, Zhang Xiaoyong, etc. Fracture Magnetic Memory Testing Technology Research in the Plates and Weld Joints [J]. Chinese Journal of Safety Science. 2012, 22 (7): 102–107.

Ren Jilin, Lin Junming. Metal Magnetic Memory Testing Technology [M]. Beijing: China power press, 2000: 44–69.

Roskosz M. Metal magnetic memory testing of welded joints of ferritic and austenitic steels[J]. 2011, 44: 305–310.

Civil, Architecture and Environmental Engineering – Kao & Sung (Eds)
© 2017 Taylor & Francis Group, ISBN 978-1-138-02985-9

Research on reliability assessment of success or failure product based on small sample

Han-peng Li, Zhao-long Xuan & Yao-dong Wang
Ordnance Engineering College, Shijiazhuang, China

ABSTRACT: Based on the problem of intelligent ammunition reliability assessment on small sample, in this paper, the three type reliability evaluation methods for success or failure product are introduced, and their applications in small sample cases are compared. The effects of prior test information on reliability with three methods are studied by simulation analysis. The results show that the hybrid Bayesian method has more advantages than the maximum likelihood method and the traditional Bayesian method in the reliability of the small samples of the success or failure components.

1 INTRODUCTION

With the rapid development of information technology, intelligent ammunition came into being. It is able to automatically search, detect, capture and attack targets, due to a variety of optoelectronic components, precision machinery parts of the large number of applications. It is showed a technology intensive, complex structure, high cost and short design life characteristics. For a number of expensive, high-risk ammunition or parts, it is impossible to do a lot of field tests for reliability assessment, which faced a small case. There are many success or failure of products in smart ammunition. Success or failure of the product usually means that its use each time are independent of each other, and only "success" or "failure" of the two possible outcomes. A variety of one-time throwing or firing of ammunition, missiles and launch vehicles (Zhang & Cai 2003), etc. are typical success or failure product. In the development process of their reliability assessment, due to cost, time and other factors, they are often unable to conduct a large number of field tests. Bayes method (Mao et al. 2009) can fully integrate pre-test information such as field test information and historical information, so it can greatly reduce the number of trials, save test products, shorten the test cycle. It has been widely used in recent years.

2 THE CLASSICAL EVALUATION METHOD OF RELIABILITY

Using the classical statistical method to calculate the reliability expressed as "R" of the success or failure product, there is the maximum likelihood estimation (Bar & Lavi & Reiser 1992) method. With the test data $(n-f, n)$, regardless of the pre-test information of the product ("n" is the number of trials, "f" is the number of failures), likelihood function of the sample for the test data is as follows:

$$L(R) = P\{S = s\} = C_n^{n-f} R^{n-f} (1-R)^f \tag{1}$$

Where s is number of successful trials.

To facilitate the calculations formula (1), it often is taken the logarithm on both side yields:

$$\ln L(R) = \ln C_n^{n-f} + (n-f)\ln R + f\ln(1-R) \tag{2}$$

To derive R, let the equation result be zero, the likelihood equation is obtained:

$$\frac{d\ln L(R)}{dR} = \frac{n-f}{R} - \frac{f}{1-R} = 0 \tag{3}$$

The maximum likelihood estimate of the reliability R is obtained by solving the likelihood equation:

$$\hat{R} = \frac{n-f}{n} \tag{4}$$

When the total number of tests "n" is constant, the reliability R of the component is monotonous with the test success number expressed as "s". For the test data $(n-f, n)$, the lower credible limit of product reliability which is expressed as "R_L" can be obtained by the formula (5) when the confidence is γ.

$$\sum_{k=0}^{f} C_n^k R_L^{n-k} (1-R_L)^k = 1-\gamma \tag{5}$$

At present, this method is the most used in many military product reliability assessment. The method is simple and intuitive. And for some relatively low value and production of ammunition products it is very applicable. But for some high-value, high reliability success or failure products there is not a lot of field tests. The method is difficult to meet the acceptance requirements. Bayes method to evaluate the reliability of products of success or failure has been developed.

3 BAYES EVALUATION METHOD FOR RELIABILITY

3.1 *Traditional bayesian method*

Bayes method in the binomial distribution of the success or failure of the product reliability evaluation, is commonly used conjugate prior method to determine the pre-test distribution, the parameter that the reliability of R conjugate prior for the Beta distribution:

$$\pi(R) = Beat(a,b) = \frac{R^{a-1}\cdot(1-R)^{b-1}}{\beta(a,b)}, 0 < R < 1 \tag{6}$$

Where $\beta(a,b) = \Gamma(a)\Gamma(b)/\Gamma(a+b)$.

The "a" and "b" are the hyperparameter of pre-test distribution ($a > 0$, $b > 0$). Given a total of N number of development phase, the hyperparameter a_i and b_i are the key to determine the pre-test distribution ($i = 1, 2, \ldots, N$). In the development number i stage, there are number m_i test information, l_{ij} is the number of trials in each batch ($j = 1, 2, \ldots, m_i$). R_{ij} is the expected estimate of product reliability in each batch test ($R_{ij} = (l_{ij}-f_{ij})/l_{ij}$). Then according to the empirical Bayes method, the a_i and b_i are determined by following principles (Martz & Waller 1976).

1. When tests are enough that m_i are larger:

$$(a_i+b_i) = \frac{m_i^2\left(\sum_{j=1}^{m_i} R_{ij} - \sum_{j=1}^{m_i} R_{ij}^2\right)}{m_i\left(m_i\sum_{j=1}^{m_i} R_{ij}^2 - H_i\sum_{j=1}^{m_i} R_{ij}\right) - (m_i - H_i)\left(\sum_{j=1}^{m_i} R_{ij}\right)^2} \tag{7}$$

$$a_i = (a_i+b_i)\overline{R}_i \tag{8}$$

Where $H_i = \sum_{j=1}^{m_i} l_{ij}^{-1}$, $\overline{R}_i = \sum_{j=1}^{m_i} R_{ij}/m_i$.

2. When the number m_i is small, the value of $(a_i + b_i)$ may be negative because of sampling error in Eq. (7). In this case, the following corrections can be made:

$$(a_i+b_i) = \left(\frac{m_i-1}{m_i}\right)\frac{m_i\sum_{j=1}^{m_i} R_{ij} - \left(\sum_{j=1}^{m_i} R_{ij}\right)^2}{m_i\sum_{j=1}^{m_i} R_{ij}^2 - \left(\sum_{j=1}^{m_i} R_{ij}\right)^2} - 1 \tag{9}$$

3. When $m_i = 1$, and no other pre-experience information such as expert experience, then $a_i = l_i - f_i$, $b_i = f_i$, which is equivalent to take no information before the distribution $\beta(0,0)$. If according to Bayes assumption that when there is no prior information $\pi(R)$ is uniformly distributed on [0, 1], that is $\beta(1,1)$. It is desirable $a_i = l_i - f_i + 1$, $b_i = f_i + 1$. If according to the Jeffreys criterion, $\pi(R)$ is $\beta(0.5, 0.5)$ when no pre-test is performed, then $a_i = l_i - f_i + 0.5$, $b_i = f_i + 0.5$. $\beta(0, 0)$ is superior to the latter two distribution in terms of theoretical support and engineering practice, especially in the reliability evaluation of high reliability products. To reduce subjectivity, it is generally recommended $a_i = l_i - f_i$, $b_i = f_i$ (Zhou 1980).

When a_i and b_i are determined, the pre-test distribution $Beta(a_i, b_i)$ of the development phase i is obtained. At the same time, the traditional Bayesian method is based on the Bayes succession rate, that is, the pre-posterior distribution of the previous stage is the pre-test distribution of the next stage. Finally, combining with the field test information (field test n times, failure f times), the post-mortem distribution can be obtained using Bayes theorem:

$$\pi(R|D) = Beat\left(\sum_{i=1}^{N} a_i + n - f, \sum_{i=1}^{N} b_i + f\right) \tag{10}$$

The traditional Bayes evaluation method does not take account of the difference between the historical test sample with the sample of the field test. In the case of equation (6), the posterior expectation of the reliability R can be obtained.

$$\hat{R}_E = E(R\,|\,x) = \frac{a+f}{a+b+n} \tag{11}$$

Where $a = \sum_{i=1}^{N} a_i$, $b = \sum_{i=1}^{N} b_i$.

The lower credible limit R_L of the product is got under given confidence γ from formula (12).

$$\int_{R_L}^{1} \pi(R\,|\,D)dR = \gamma \tag{12}$$

It can be seen from the posteriori expression (11) of the parameter R. This traditional method of pre-distribution construction based on likelihood function actually treats pre-test information and field test information without discrimination. But the historical information and field test information may not be compatible (that means the same population). Even if compatible, it can only be in a certain degree of confidence. Therefore, the correctness of the above method is questionable. In general, the number of $(a + b)$ is much larger than n. It shows that if the two information directly is mixed without distinction of integration, often leads to historical information annihilation of field information and field test results on the final assessment of almost no impact with analysis of formula (11). The number of historical information is often relatively large and the number of field trials is relatively small, so the traditional Bayesian method is obviously very unreasonable.

3.2 *Hyprid prior bayesian method*

In order to make good use of pre-test information and distinguish the differences between pre-test and on-site information, a mixed beta pre-test distribution (Ming et al. 2008) has been put forward. After obtaining the pre-test distributions for each stage, the pre-mix distribution of the tectonics is as follows:

$$\pi(R)=\sum_{i=1}^{N}\left[\rho_i\frac{R^{a_i-1}\cdot(1-R)^{b_i-1}}{\beta(a_i,b_i)}\right]+(1-\rho) \tag{13}$$

Where $\rho=\sum_{i=1}^{N}\rho_i, 0\le R\le 1, 0\le\rho_i\le 1, 0\le\rho\le 1.$

The ρ and ρ_i are called inheritance factors, and (1-ρ) is called update factors. The ρ_i reflects the experimental stage i information and field test information from the overall similarity. The ρ reflects the current product in the reliability of the overall historical information on the degree of succession and the (1–ρ) is described the uncertainty of the current product in terms of reliability. In $\rho = 1$, it's considered that the two populations are the same, and the hybrid priori $\pi_\rho(R)$ is conjugate priori *Beta*(a, b). In $\rho = 0$, the two populations are completely different, that is, the historical test data have no reference value, and the a priori distribution is conservatively chosen as *Beat*(1, 1). In $0 < \rho < 1$, the use of mixed a priori is more reasonable. After pretest to determine the distribution, according to Bayes theorem, substitute site information, rearranging the posterior distribution as follows:

$$\pi(R|D)=\frac{(1-\rho)R^{n-f}(1-R)^f+\sum_{i=1}^{N}\rho_i\frac{R^{n-f+a_i-1}(1-R)^{f+b_i-1}}{\beta(a_i,b_i)}}{(1-\rho)\beta(n-f+1,f+1)+\sum_{i=1}^{N}\rho_i\frac{\beta(n-f+a_i,f+b_i)}{\beta(a_i,b_i)}} \tag{14}$$

With (14), the product confidence lower limit R_L and product moments μ_k of the product reliability can be obtained under given confidence γ by follow:

$$\int_{R_L}^{1}\pi_\rho(R\,|\,D)dR=\gamma \tag{15}$$

$$\mu_k=E(R^k)=\int_{R_L}^{1}\pi_\rho(R\,|\,D)dR=\gamma \tag{16}$$

3.3 *Determination of inheritance factors ρ and ρ_i*

To sum up, the ρ and ρ_i have a great influence on product reliability evaluation, so their values must be careful. Therefore, the expert should give the value of ρ according to the improvement degree of the product. When the value of ρ cannot be given accurately, then ρ can be treated as a random variable, and the probability distribution and range of ρ are given by the hierarchical Bayes method. However, the probability distribution and range of ρ are also determined by expert experience. The method is subjective, and it is difficult to deal with mathematical problems when the probability distribution of N ρ_i is given. If the value of ρ is directly determined by the goodness-of-fit test of the two samples from the population, the calculation is simple and its rationality is also well explained. Since ρ is a measure of the overall population of the pre-test information and the overall sample similarity from the field sample, it is necessary to determine ρ before determining ρ_i. Determine ρ using the overall goodness of fit with the field test method. That is required to test information on all stages of synthesis. The so-called synthesis is not a simple addition, but the weight and then sum, so the weight must be first determined of each stage of information. The "i" stage historical test sample $\left(\sum_{j=1}^{m_i}(l_{ij}-f_{ij}),\sum_{j=1}^{m_i}l_{ij}\right)$ belongs to population Y, and the field test sample $(n–f, n)$ belongs to population X. The following tables are obtained.

$$K_i=\frac{\left[\left(\sum_{j=1}^{m_i}(l_{ij}-f_{ij})f\right)-\left(\sum_{j=1}^{m_i}f_{ij}(n-f)\right)\right]^2\left(n+\sum_{j=1}^{m_i}l_{ij}\right)}{n\sum_{j=1}^{m_i}l_{ij}\left[(n-f)+\sum_{j=1}^{m_i}(l_{ij}-f_{ij})\right]\left(f+\sum_{j=1}^{m_i}f_{ij}\right)} \tag{17}$$

In the above equation, the K_i is the person statistic and its distribution converges to the χ^2

Table 1. Joint list of two binomial populations.

	Number of success	Number of failures	Sum
X	$n-f$	f	n
Y	$\sum_{i=1}^{m_i}(l_{ij}-f_{ij})$	$\sum_{j=1}^{m_i} f_{ij}$	$\sum_{j=1}^{m_i} l_{ij}$
Sum	$n-f+\sum_{j=1}^{m_i}(l_{ij}-f_{ij})$	$f+\sum_{j=1}^{m_i} f_{ij}$	$n+\sum_{j=1}^{m_i} l_{ij}$

distribution with 1 degree of freedom. It is required that the number of successes and failures of the two samples be greater than 5 in (17). When this condition is not satisfied, it is corrected as follows (Fisher & Yates 1957).

$$K_i = \frac{\left[\left|\left(\sum_{j=1}^{m_i}(l_{ij}-f_{ij})f\right)-\left(\sum_{j=1}^{m_i} f_{ij}(n-f)\right)\right| - \frac{1}{2}\left(n+\sum_{j=1}^{m_i} l_{ij}\right)\right]^2 \left(n+\sum_{j=1}^{m_i} l_{ij}\right)}{n\sum_{j=1}^{m_i} l_{ij}\left[(n-f)+\sum_{j=1}^{m_i}(l_{ij}-f_{ij})\right]\left(f+\sum_{j=1}^{m_i} f_{ij}\right)} \tag{18}$$

K_i also approximates the χ^2 distribution of 1 degrees of freedom in the above equation. At a given test level γ, K_i can be used as the test statistic to test whether X and Y are the same population. Even if the compatibility test, two parts of the test information cannot be simply mixed together, it still need to determine each stage of the test information and the similarity of the field information measure T. The goodness of fit $Q_i = P(\chi_1^2 > K_i)$. Q_i represents the degree of similarity between X and Y in probability, and it is related to T_i. But it is difficult to give the function relation of Q_i and T_i exactly. It is desirable that $T_i = Q_i^{1/2}$. Then the number of test L and the success number S can be obtained (Wang & Guo 2006).

$$S = \sum_{i=1}^{N}\sum_{j=1}^{m_i}(l_{ij}-f_{ij})\frac{T_i}{T_1+T_2+\cdots T_N} \tag{19}$$

$$L = \sum_{i=1}^{N}\sum_{j=1}^{m_i} l_{ij}\frac{T_i}{T_1+T_2+\cdots T_N} \tag{20}$$

The value of K is obtained by Substitute (S, L) and field test sample $(n\text{-}f, n)$ into formula (17) or (18), and then the value of Q can be obtained by using the following formula.

$$Q = P(\chi_1^2 > K) \tag{21}$$

Taking $\rho = Q_i^{1/2}$, ρ_i can be obtained as follows from the value of ρ.

$$\rho_i = \rho\frac{T_i}{T_1+T_2+\cdots T_N} \tag{22}$$

3.4 *Simulation calculation and discussion*

In this paper, three methods of reliability evaluation are introduced, and their merits and demerits in the reliability evaluation of the success or failure product are studied. The following is a simulation study of the number of pre-test information on the reliability of the three methods of reliability evaluation of the impact of the one-sided lower credible limit. Because the multi-development stage Bayes evaluation method is an extension of the two overall Bayesian evaluation methods, in order to reduce the computational complexity, the historical test information is set as the same population and the same batch. Let history test as $(n_0\text{-}f_0, n_0)$. The simulation parameter is set to $\gamma = 0.8$, $n = 10$, $p = 0.9$, $p_0 = 0.7$ and let n_0 changes (Where p and p_0 are the probabilistic truth values of field test information and historical test information, $p_0 = n_0 - f_0/n$, $p = n - f/n$). Under the maximum likelihood estimation method, the traditional Bayesian method and the hybrid method, the estimated value R_L of the lower credible limit of the unilateral reliability varies with n_0/n, as shown in Fig. 1.

The simulation parameter is set to $\gamma = 0.9$, $n = 20$, $p = 0.95$, $p_0 = 0.95$ and let n_0 changes. Under the maximum likelihood estimation method, the traditional Bayesian method and the hybrid method, the estimated value R_L of the lower credible limit of the unilateral reliability varies with n_0/n, as shown in Fig. 2.

In Fig. 1, since the classical reliability assessment method does not take into account the historical

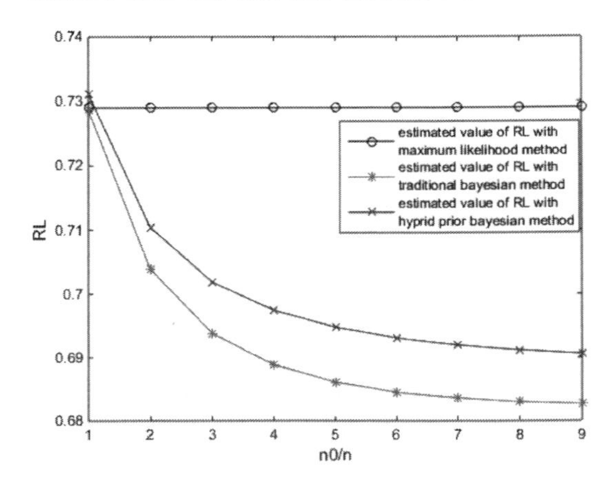

Figure 1. The value of RL with $p = 0.9$, $p_0 = 0.7$.

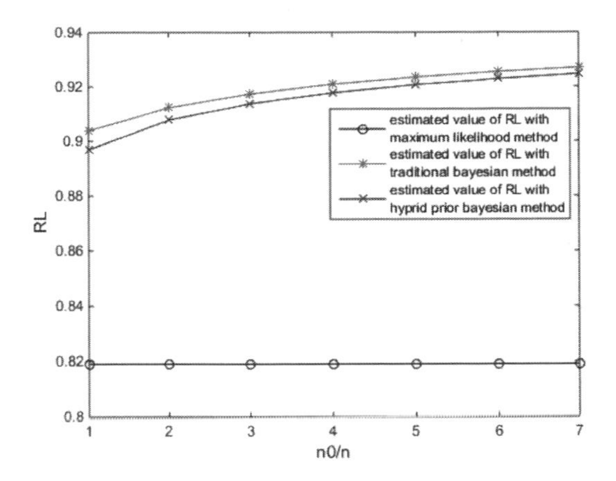

Figure 2. The value of R_L with $p = 0.95$, $p_0 = 0.95$.

test information, it leads to the lower credible limit R_L is too aggressive. Traditional Bayes method does not consider the difference between historical information and on-site information, which leads to the results evaluation too conservative. And as n_0/n increases, the influence of field test information on R_L evaluation results is getting smaller and smaller. The hybrid method takes into account both the historical information and the historical and field information differences, its evaluation results are more reliable.

In Fig. 2, the results of the classical reliability assessment are too conservative due to less field test information. Traditional Bayes method does not take into account the difference between historical information and on-site information, leading to the results of the evaluation too aggressive. The results of the hybrid method evaluation are between the above two methods. It can be seen from the figure 2 that as n_0/n increases, the difference between the traditional Bayes method and the hybrid method to evaluate the results is reduced.

From the comprehensive analysis of Fig. 1 and Fig. 2, it is found that the evaluation result of the traditional evaluation method is not applicable due to the less information of field test. Traditional Bayesian method does not consider the difference between historical information and field information, that leads to the prior data inundate the posteriori data. The results indicate that the hyprid prior approach can avoid the problem of that the prior data inundate the posteriori data effectively. And the evaluation results are more reliable.

4 CONCLUSION

This paper presents three reliability evaluation methods for the success or failure product, analyzes and compares the advantages and disadvantages with their assessment methods. Bayes evaluation method for the reliability of success or failure product based on the test information of all development stages not only make full use of historical information, but also consider the different stages of the test sample heterogeneity. The inheritance factor can be determined directly based on goodness of fit without additional information. Its practicality is strong.

REFERENCES

Bar S.K., Lavi I., Reiser B. 1992. Bayesian Inference for the Power Law Process. *Ann. Inst. Statist. Math* 44(4):623–639.

Fisher R.A., Yates F. 1957. Statistical Tables for Biological Agricultural and Medical Research. Edinburgh: Oliver & Boyd.

Mao Zhao-yong, Song Bao-wei, Hu Hai-bao, et al. 2009. The Bayes reliability evaluation method based on Amsaa model for torpedo system. *Acta Armamentarii* 30(10):1401–1404.

Martz H.F., Waller R.A. 1976. The Basics of Bayesian Reliability Estimation from Attribute Test Data. *Los Alamos Scientific Laboratory*, Report UC-79p.

Ming Zhi-mao, Tao Jun-yong, Chen Xun, et al. 2008. A Bayes plan of reliability qualification test based on the mixed Beta distribution for success/failure product. *Acta Armamentarii* 29(2):204–207.

Wang Jian-xia, Guo Bo. 2006. Bayes Assessment for Reliability of Success or Failure Product Based on Fusing Test Information of Development Phases. *Advanced Manufacture and Management* 25(15):31–33.

Zhang Shi-feng, Cai Hong. 2003. Bayesian approach of performance reliability assessment of solid rocket motors. *Journal of Solid Rocket Technology* 26(4): 9–11.

Zhou Yuanquan. 1980. Pre-test distribution without prior knowledge. *Acta Mathematica Sinica*, 23(3): 359–371.

Civil, Architecture and Environmental Engineering – Kao & Sung (Eds)
© 2017 Taylor & Francis Group, ISBN 978-1-138-02985-9

Numerical simulation of hydroforming of a stainless steel sink

Qihan Li, Pei Zhu, Xueqiang Guo, Jianwen Hou & Xiaomei Li
School of Mechanical Engineering Changchun University of Technology, Changchun, Jilin, China

ABSTRACT: Box-shaped stainless steel sinks are more difficult to shape. In the stamping process, the factors affecting the flow of material are more complex. On the basis of the technological difficulties and the shape characteristics, this paper uses the Dynaform software to simulate the first drawing of the stainless steel flume by using the technology of hydroforming, which can effectively solve problems such as low pass rate, low surface quality, and rupture of parts. The influence of the loading path and liquid chamber pressure on the result of hydroforming during the forming process was studied. The optimum process parameters were established by simulation. This effectively controls the sidewall wrinkles and the excessive thinning of the corner of punch. The results provide a theoretical guidance and reference for the actual production. Compared with the general deep drawing of the kitchen sink, the quality of the hydroforming sink is much better.

1 INTRODUCTION

1.1 *Type area*

Low-end manufacturing industries such as resource-intensive, labor-intensive, high-energy-consuming, and high-polluting industries have experienced sluggish growth. However, recently, computer hardware and software technology and related control technology have been widely promoted and used. A number of new technologies and new crafts are widely used in traditional manufacturing. They promote the transformation and upgrading of production methods (Zhang. S. H & Jensen. M. R., 2000). Because of the lightweight and high quality of the parts, the technology of hydroforming has attracted more attention in sheet metal forming.

Hydroforming technology is the use of liquid as a force transmission medium of a processing technology. In the sheet metal stamping, it uses liquid instead of a rigid punch or a die (Dursun T & Soutis C., 2014; Lang Lihui & Xu Nuo, 2013). In the liquid filling process, the sheet will be pressed into the liquid chamber by the punch so that the liquid is pressurized and forms a blank around the punch. The friction generated between the blank and the punch may allow the use of higher pressures. The sheet metal forming performance will be improved, which is called the "friction retention" effect. The liquid pressure will make a blank between the punch and die shoulder to form a raised structure similar to the drawbead, which can help prevent wrinkling of parts (Shihong Zhang & Lixin Zhou, 2003; Hartl C., 2005). In this flexible manufacturing process, the hydraulic chamber can also be used repeatedly to adapt to the different shapes of parts, greatly saving mold costs. In this paper, the numerical simulation of the hydroforming of the stainless steel sink is carried out. The optimal process parameters are determined, and the failure problem in the forming process is solved. The theoretical guidance is provided for the production.

2 MATERIAL AND DIMENSIONS OF STAINLESS STEEL SINKS

2.1 *Shape of the stainless steel sink*

Figure 1 shows the three-dimensional digital model parts diagram of a stainless steel sink. Its size is 380 * 330 * 165 mm. With the development

Figure 1. 3D model of sink.

Table 1. Material performance parameters.

Young's modulus/GPa	207
Yield stress, MPa	205
Strain-hardening index, N	0.498
Anisotropic index, r	1.2
Poisson's ratio	0.28
Strain hardening coefficient k	1426

of science and technology and improvement of the quality of life, the performance and quality requirements of the sink has also increased. The most important part of the production process is the drawing process. Because the parts forming has relatively large stretch, and drawing effect is usually a low pass rate, the use of traditional drawing needs repeated mold debugging (Lang L H & Danckert J., 2004). The forms of failure are difficult to control. Through the study of numerical simulation of key process parameters, reasonable control of sheet metal flow in the forming process helps to produce qualified parts.

2.2 *Sheet performance parameters*

Because of its high temperature, corrosion resistance, and good performance, SS304 steel is used in a wide range of applications in the sink. Table 1 shows the performance parameters.

3 FINITE-ELEMENT MODEL OF LIQUID-FILLED FORMING

3.1 *Establishment of finite-element model*

The numerical simulation of sheet metal forming is carried out by the ETA Company and LSTC Company. In order to improve the simulation precision, the four-node Belytschko-Tsay shell element is used to mesh, and the maximum grid is set to 2. The die, the punch, and the blank holder are regarded as the rigid body. The rigid four-node element is used to divide the mesh. In the automatic setting module of the software, the tools are defined respectively. Figure 2 shows the finite-element model of the punch, blank, blank holder, and die.

3.2 *Setting of simulation conditions*

In the simulation of the stainless steel sink, the single-action drawing was adopted. Blank material selects the SS304 anisotropic material model. The punch speed is set at 2000 mm. Then, the contact parameters are defined. The contact type of the tool is surface contact. For the parts forming effect and improving the drawing ratio, the billet and punch friction factor is set to 0.3, and the friction

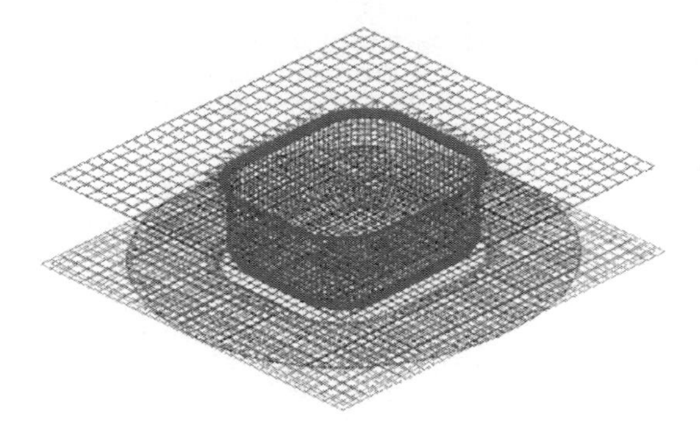

Figure 2. Sink stamping model.

coefficient between the sheet metal blank holder and the die is set to 0.005. According to Dynaform software calculations and experience, the blank size is 700 mm*700 mm and the thickness is 1 mm.

4 NUMERICAL SIMULATION ANALYSIS OF THE PROCESS OF LIQUID–FILLED FORMING

In the process of hydroforming of stainless steel sink, the process parameters influence the forming quality and control failure form of the sheet metal. They include the size and performance of the sheet metal, the arrangement of the drawbead, the blank holder force, the gap, the pressure of the liquid chamber, the liquid chamber pressure loading curve, and so on (Peng Chengyun and Zou Qiang, 2011; Yuan Shijian & Wang Zhongren, 2000). The key process parameters are liquid chamber pressure and its loading path, which have important influence on the failure forms such as excessive thinning of parts and wrinkling of forming surface. In this paper, the numerical simulation and analysis focus on the influences of liquid chamber pressure loading path in the process of hydroforming of stainless steel sink.

4.1 *Liquid pressure loading path and forming effect*

In the process of hydroforming, the way of loading liquid chamber pressure is mainly of two types (the passive natural pressurization and active forced pressurization). The former is the punch running into the liquid chamber, thus the liquid in the cavity is compressed to form a reaction. Because of the small initial forming pressure, the forming limit of the plate is low. The latter is the use of valve to control the liquid chamber pressure, which greatly increases the molding limit.

The load path of the liquid chamber pressure has a great influence on the forming quality of the part. The pressure value will cause the sink to have a different degree of instability (Wang Jianmin & Hu Mi, 2009). "Friction retention" and "fluid lubrication" resulting from the proper liquid chamber pressure can improve the stress and strain in the drawing process. These effects improve the drawing limit of the material and effectively suppress the rupture and wrinkling in the traditional drawing. In order to study the influence of the pressure loading path in the drawing process, eight different loading paths were chosen to simulate the study. The gap was 1.1 mm. In this paper, the simulation test of AUTOSETUP in Dynaform 5.9.1 was used. In line with the actual production and to determine the appropriate pressure, the maximum pressures of liquid chamber were 10, 20, 25, 28, 30, 32, 35, and 50 MPa, as shown in Fig. 3.

The overflow pressure was the maximum liquid pool pressure. Between 0 and 0.005 s, the pressure of the liquid chamber was zero. Until the punch was down to 0.005 s, the liquid chamber pressure started to load. When the punch was down to 0.008 s, the pressure of the liquid chamber reached 1 MPa and kept 0.002 s. At 0.01 s, the pressures of the liquid chamber linearly increased to 10, 20, 25, 28, 30, 32, 35, and 50 MPa respectively, keeping the hydraulic pressure until the forming end.

4.2 *Analysis of pressure loading path of liquid overflow*

The overall thickness of the stainless steel sink should be uniform thickness after forming. The thickness was generally 0.7–1.0 mm. In order to improve the service life and enhance the strength, the sink should not be much thin. From the drawing process requirements, the maximum thinning rate should not exceed 35% so that the sheet metal thickness of the minimum is not less than 0.65 mm in the process.

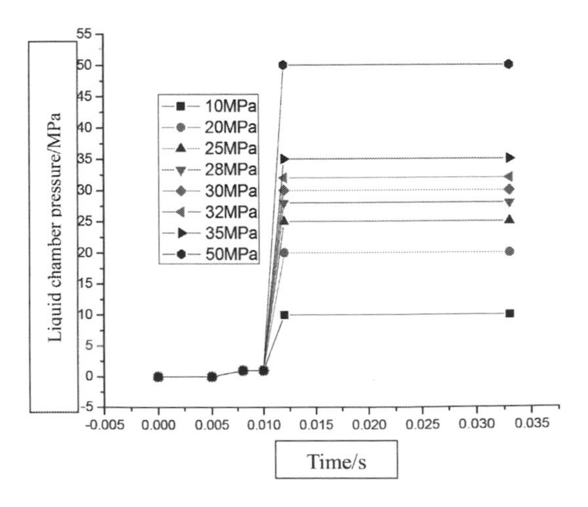

Figure 3. Liquid overflow pressure.

When using the load curve No. 1, the overflow pressure was 10 MPa, which is not sufficient to reduce the initial tensile stress of the flange part. Therefore, the sheet was cracked at the beginning of drawing, and the rupture position was close to the punch fillet.

When using the load curve No. 8, the relief pressure was 55 MPa. As the overflow pressure setting was too high, the formed part was no longer deformed under the action of favorable frictional force. Under the action of the hydraulic pressure, the unfavorable friction of the flange portion was increased. The radial tensile stress of the sheet near the die cavity in the late stage of forming was too high, and then rupture occurs.

When using the load curve No. 2–No. 7, the overflow pressures were 20, 25, 28, 30, 32, and 35 MPa. The sink could be formed. The change of the thickness distribution of the final shape is shown in Figures 4–8. Observing the changes of color in these figures, as the relief pressure increases, we can find the change of the quality of the formed stainless steel sink.

When the overflow pressure was 20 MPa, as shown in Figure 4, the parts were able to shape.

Figure 4. Thickness of the cloud with 20 MPa.

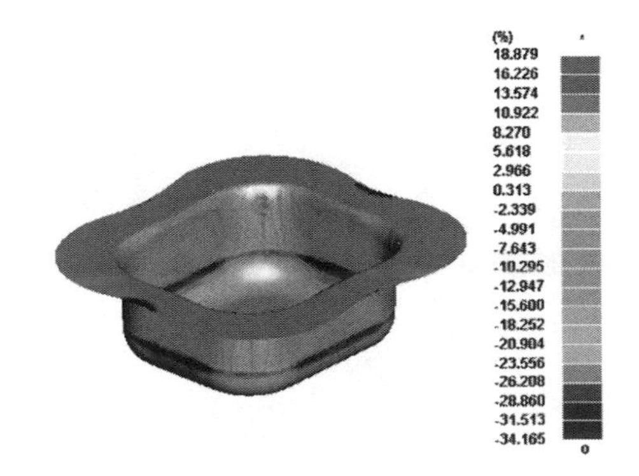

Figure 5. Thickness of the cloud with 25 MPa.

The thickness changes more uniformly. The largest thinning area occurs in the straight wall near the bottom and edge corner.

The maximum thinning rate is 19.239%, not exceeding the limit, to meet the requirements. As the overflow pressure increases to 25 MPa, as shown in Figure 5, the thickness of the edge corner increases significantly. The maximum thinning area occurs in the straight wall near the bottom. The maximum thinning rate is 18.879% in the extent specified. When the overflow pressure increases to 28 MPa, the maximum thinning rate and the maximum thickening rate are about the same as that at the relief pressure of 25 MPa.

When the relief pressure was 30 MPa, it is clear from Fig. 6 that the distribution of the thickness of the part is not uniform. The straight wall of the bottom part is still the largest thinning area. And the rest is well formed. However, compared with the above thickness distribution, the maximum thinning rate increases significantly. The maximum thinning rate was 31.746%, not exceeding the allowable range, to meet the technical requirements. However, a minimum thickness of less than 0.7 mm did not meet the quality requirements of parts.

When the overflow pressure was 32 MPa, it can be seen from Fig. 7 that the forming quality of the parts was good. The thickness distribution was uniform, and the maximum thinning rate was reduced to 18.826%, which satisfies with the technological requirements and the quality requirements. When the overflow pressure was 35 MPa, it can be seen from Fig. 8 that the maximum thinning rate increases to 31.922%. The overall forming quality was generally as the same as that at 30 MPa, which is not good.

It can be seen that when the relief pressure was 32 MPa, the largest thickness was 1.329 mm. The smallest thickness was 0.812 mm, which is called uniform thickness. The maximum thinning rate

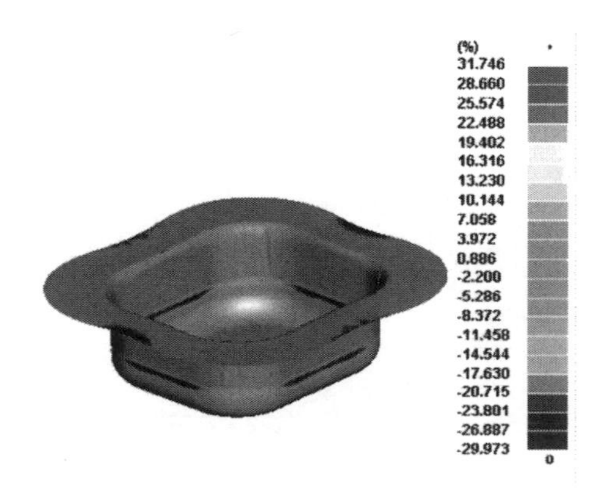

Figure 6. Thickness of the cloud with 30 MPa.

Figure 7. Thickness of the cloud with 32 MPa.

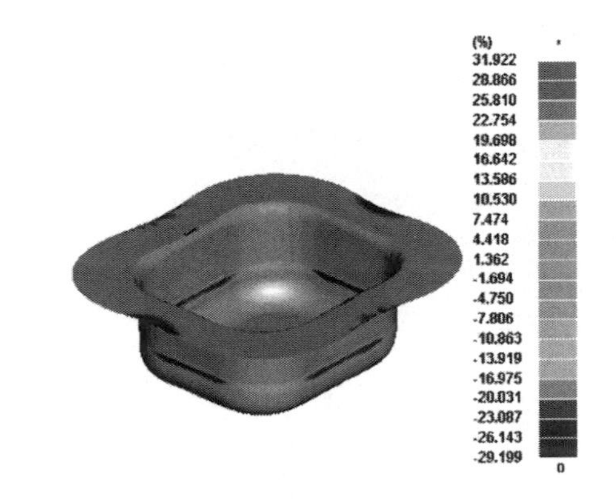

Figure 8. Thickness of the cloud with 35 MPa.

was 18.826%. When the maximum pressures of the sink were 20, 25, 28, 30, 32, and 35 Mpa, the sink forming quality was best under the relief pressure of 32 MPa.

5 COMPARISON OF COMMON FORMING AND HYDROFORMING

We can analyze the forming results of the sink in the ordinary deep drawing process and the hydroforming process. The FLD diagram of the sink was used as the evaluation standard to compare the effects of the two forms on the quality. In this simulation, 304 stainless steel was selected as the sheet metal. The thickness was 1.0 mm, and the friction coefficient between the blank and the blank holder was 0.005. The friction coefficient between the punch and the sheet was 0.3. The friction coefficient between sheet metal and die was 0.005. The blank gap was 1.1mm. The punch speed was 2000 mm/s. The loading curve 1 was selected as the loading form. In the ordinary drawing simulation,

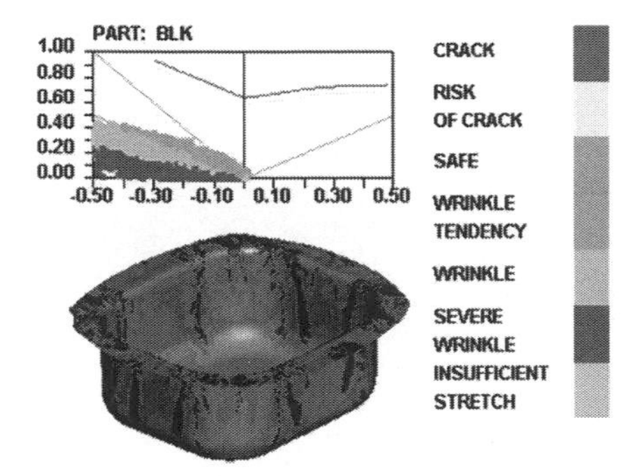

Figure 9. FLD of ordinary deep drawing.

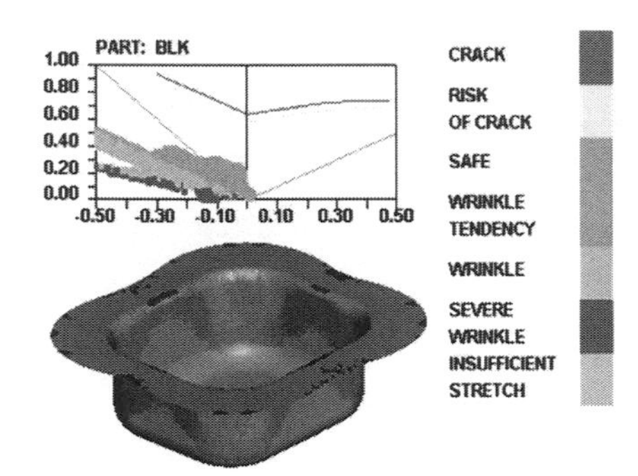

Figure 10. FLD of hydroforming.

the friction coefficient between the sheet metal and the punch and blank holder was 0.125, and all the other process parameters were the same.

From the forming effects of the sink in the general drawing, we can see that the drawing of the straight wall part was not fully adequate, which was due to uneven flow of sheet metal. Form the forming effects of the hydroforming process, although wrinkled at the flange, the quality of the sink was not affected. The wrinkling area would be trimmed in the subsequent process, and the other parts were formed in good quality.

It could be seen from Figures 9 and 10 that the formability of the sink using the hydroforming was much higher than that in the ordinary deep drawing. Compared with the ordinary deep drawing, the problem of defects, such as insufficient drawing, was avoided in hydroforming. The forming quality of the parts is remarkably improved.

6 CONCLUSIONS

The advantages of the hydroforming process were mainly in the hydraulic retention effect. The layout of a reasonable hydraulic loading path was the key to the success of qualified parts (Wang Jianmin & Qian Chunmiao, 2012). It was very important to select the appropriate relief pressure for the hydroforming process (Chen Xuguo & Li Jiguang, 2015). Only when the appropriate relief pressure was determined, the sheet metal forming can proceed smoothly. In the numerical simulation of the sink, when the safe pressure value was 32 MPa, the wall thickness was the most uniform. If the overflow pressure was not appropriate, there will be rupture phenomenon.

By comparing the ordinary deep drawing of the sink with the liquid chamber pressure loading curve No. 1, it was obvious that the quality of the sink under the hydroforming process was better than that of the ordinary deep drawing process.

REFERENCES

Chen Xuguo & Li Jiguang (2015). Numerical simulation of liquid—Filled deep drawing of 2A12 aluminum alloy flat—bottom cylindrical. J. Journal of Net shape Forming Engineering. 6, 86–91.

Drawing without draw die. J. Machinery Design & Manufacture. 8, 230–232.

Dursun T & Soutis C. (2014). Recent developments in advanced aircraft aluminum alloys. J. Materials & Design. 56, 862–871.

Hartl C. (2005). Research and Advances in Fundamentals and Industrial Applications of Hydroforming. J. Journal of Materials Processing Technology. 167, 383–392.

Lang L. H. & Danckert J. (2004). Investigation into Hydrodynamic deep drawing assisted by radial pressure part I Experimental observations of the forming process of aluminum alloy. J. Journal of Materials Processing Technology. 148, 119–131.

Lang Lihui & Xu Nuo. (2013). Research on liquid filling technology of deep cavity box. J. Forging & Stamping Technology. 38, 21–25.

Nakamura K. & Nakagawa T. (1987) Sheet metal forming with hydraulic counter pressure in Japan. J. Annal Cirp. 36, 191–194.

Peng Chengyun & Zou Qiang (2011). Study of Variable Draw Bead Technology for Rectangular Box Drawing Based on Numerical Simulation. J. Journal of Netshape Forming Engineering. 3, 4–6.

Shihong Zhang & Lixin Zhou (2003). Technology of sheet hydroforming with a movablefemakle die. J. International Journal of Machine Tools & Manufacture. 43, 781–785.

Wang Jianmin & Hu Mi (2009). Study on liquid pressure loading paths of square cup in Hydro-mechanical deep

Wang Jianmin & Qian Chunmiao. (2012). Numerical simulation of sheet metal pulling multiple times deep forming. J. Thermal Processing. 41, 136–139.

Yuan Shijian & Wang Zhongren (2000). Research and application of internal hydroforming technology. J. Journal of Harbin Institute of Technology. 32, 60–63

Zhang. S. H. & Jensen. M. R. (2000). Analysis of the Hydrometrical deep drawing of cylindrical cups. J. Mater. Process. Technol. 103, 367–373.

Civil, Architecture and Environmental Engineering – Kao & Sung (Eds)
© 2017 Taylor & Francis Group, ISBN 978-1-138-02985-9

The fabrication and measurement of a bi-axial thermal convection acceleration sensor

Huan Wang, Yigui Li, Ping Yan & Yuan Huang
School of Science, Shanghai Institute of Technology, Shanghai, China

Susumu Sugiyama
Department of Micro System, Ritsumeikan University, Japan

ABSTRACT: In this paper, a bi-axial (x and y axial) thermal convection acceleration sensor based on MEMS is described. The sensing principal, sensor design, fabrication process and the performance is illustrated. The sensor is fabricated by MEMS process using the SOI wafer. A thermistor with p-type slightly doped silicon beam is fabricated by micro fabrication technology and one end of the beam is fixed and the other end of the beam was free to increase the sensitivity of the sensor because the free end of the beam has a large Temperature Coefficient of Resistance (TCR) value.

1 INTRODUCTION

1.1 *MEMS accelerometer*

There are different kinds of acceleration sensors, such as thermal accelerometer, piezoelectric accelerometer, semiconductor piezo-resistive accelerometer, electrostatic capacitance accelerometer etc. (MA Ming-jun & RAN Yingling, 2015). In general, an accelerometer sensor uses an inertial mass to detect the acceleration (GUO Wei, 2012 & LI Li-Jie, 2001). Considering the fragility and complexity, a thermal convection chamber can be used instead of the inertial mass to overcome the disadvantages of the usual mechanical accelerometer (LI Li-Jie, 2001; WAN, Cai-xin, 2012; Stephen B. Pope, 2000; Alan J. Chapman, 1987 & S.M. Sze, 1985).

1.2 *Gas thermal accelerometer*

The accelerometer is based on free thermal convection principle. The heat source can be created by a heater inside a packaged cap. The accelerometer includes two parts, a sensing part and the packaged cap.

The sensing part is consist of an arc thermistor and its center part is fixed on the beam. The sensor may simultaneously detect acceleration at the direction of X and Y axial. The silicon thermistor is slightly p-type doped by doping boron. Further, the thermistor structures have been optimized in thermal stress.

2 PRINCIPLE

2.1 *Temperature coefficient of resistance*

The detection principle of the device is based on the thermal resistance effect of p-type silicon. The thermal resistance effect is an effect that the lattice scattering of a substance changes due to a temperature change. In general, the metal has a property that its resistance rises almost linearly with scattering of electrons as the temperature rises. If the resistance value at the reference R_0 temperature is T_0, the resistance value R_1 at temperature T_1 can be expressed by the following equation:

$$\mathrm{R}_1 = R_0\left\{1 + \alpha(T_1 - T_0)\right\} \tag{1}$$

In equation (1) α is the temperature coefficient of resistance. The same as metals, semiconductors also have the property that remarkable change of resistance value when temperature changes. And the semiconductors are more sensitive than the metals.

2.2 *Detection principle*

The sensing chip of the sensor is packaged in an enclosed chamber. The sensing chip is consist of a heater in the center and four micro thermistors surrounding it. The thermistors are designed as free arc-shape, and its center part is fixed on the beam.

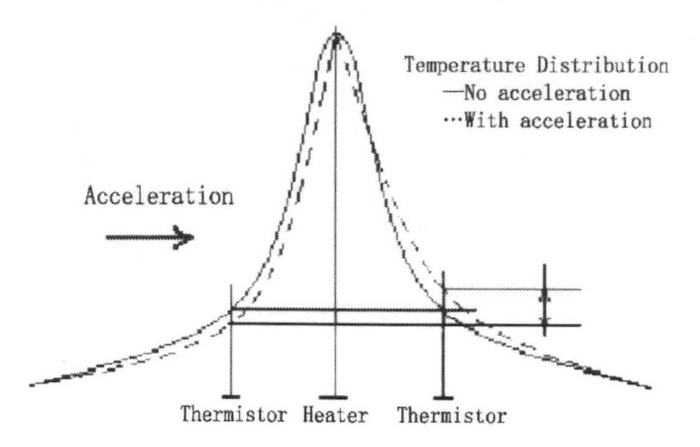

Figure 1. Schematic view of the accelerator.

The heater creates a symmetric temperature distribution in the chamber. When no acceleration, the temperature is no difference on the thermistor. When the sensor is given an acceleration in one direction, the symmetrical distribution is broken, a set of the opposite of the thermistor has a temperature rise while the other is falling. Due to thermal effects, therefore resistance changes with temperature. The calculation formula is as following:

$$R_{a,b} = R(1 + 0.5\alpha\Delta t) \tag{2}$$

Here, R_a, R_b represent the resistance of the opposite thermistor, respectively. R is the resistance of the thermistor at room temperature. The resistance change is converted into an output voltage change by Wheatstone bridge and written equation:

$$V_{out} = \{(R_a - R_b)/2(R_a + R_b)\}V_{in} \tag{3}$$

$$V_{out} = 0.25\alpha \cdot \Delta t \cdot V_{in} \tag{4}$$

And the schematic diagram of sensor was shown in Figure 1.

3 DESIGN AND ANALYSIS

3.1 *Basic analysis*

The mean temperature of the thermistor is an important factor and it directly affects to the performance of the sensor. The high value of the mean temperature means the high sensitivity of the sensor. The temperature of the thermistor is determined by the heat balance equation. In solving this equation, the effect of heat conduction between the tiny thermistor and the huge support arms is ignored.

Temperature distribution along the two thermistor is analyzed by ANSYS, and the result shows that temperature at the ends of the thermistor is higher than the ambient temperature. The dimensions of each thermistor is designed as $160 \times 6 \times 2.5\ \mu m^3$. Four kinds of materials are used: aluminum wire, light doped-silicon thermistor, high doped-silicon contact hole, and silicon dioxide insulator layer. The size of packaged cap and cavity are studied in the simulation.

3.2 *Simulation*

Using analysis software ANSYS, the three-dimensional model is built and the temperature distribution of the heat flow in the sensor is simulated.

Cross-sectional view of the sensor is shown in Figure 2 and the influence of cavity size on temperature difference is shown in Figure 3.

The L is the length of the cap, d is diameter of the cavity, the x is the position of the thermistor along the x axial and the x/L means the relative position of the thermistor.

The simulation results show that the length of the cavity L is 2400 μm and it is most sensitive. It was determined the size of the sensing chip to some extent. Through simulation it shows the change relationship between the sizes of the

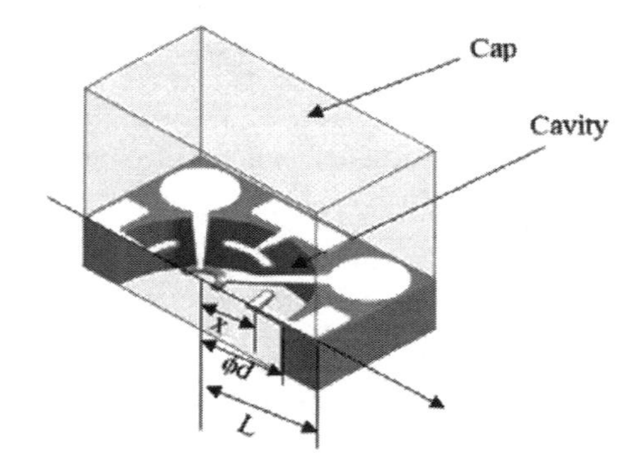

Figure 2. Cross-sectional view of the sensor.

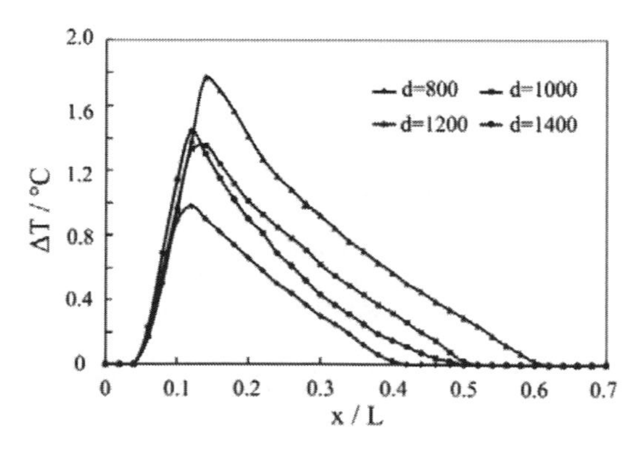

Figure 3. Influence of cavity size on temperature difference.

packaged cap and the heater. The size of the cap cause different sensor sensitivity. The sensitivity of the sensor is increased with the volume of the cap. In other words, the width and the height of the cap increases, the sensitivity increases. However, bigger size makes the frequency characteristic of the sensor worse and that limit the application. Therefore, a suitable packaged cap size was selected to improve the sensitivity of the sensor. Finally the size was $6000 \times 6000 \times 500\ \mu m^3$.

4 FABRICATION PROCESS

SOI wafer is selected for the fabrication of the accelerometer. The fabrication process is shown in Figure 4.

Firstly, a 0.3 um-thick SiO_2 insulation layer is formed by thermal oxidation on an SOI wafer. Next, photoresist is coated on the wafer and photo-lithography is carried out. Thereafter, to etch the oxidation layer, wet etching is performed with BHF, and 3 um × 3 um contact holes are made after the removal the resist. When the contact hole is opened, boron is added through the contact hole. Then, in order to improve the metal semiconductor of the contact, p+ diffusion is carried out by putting it into a diffusion furnace. In this process, first, it is placed in a diffusion furnace at 1000°C for 1 hour, then wet etching is carried out by BHF, and then diffusing is carried out again by placing it in a diffusion furnace at 1200°C for 30 minutes. Aluminum wires (0.15 μm-in-thickness) are formed by the process of vacuum vapor deposition, photolithography, and aluminum etching. By placing the material in an annealing furnace at 450°C for 30 minutes, Ohmic contact between aluminum and silicon is formed by sintering. The thermistor is formed by photo-lithography and deep reactive ion etching (Deep-RIE). Deep-RIE is performed only in the front wafer for 3 minutes, and etching is performed. The front surface of the thermistor is coated with photoresist. A layer of polyimide is deposited on the surface to provide the chip from mechanical damage. The rear surface is also etched by deep-RIE for about 5 hours and etching is performed to create the thermistor structure. Then, the HF (hydrofluoric acid) vapor removes the oxide buried to form the supports of the beams. The protective layer of polyimide and photoresist is finally peeled off by removing the underlying photoresist. The fabricated sensor is shown in Figure 5.

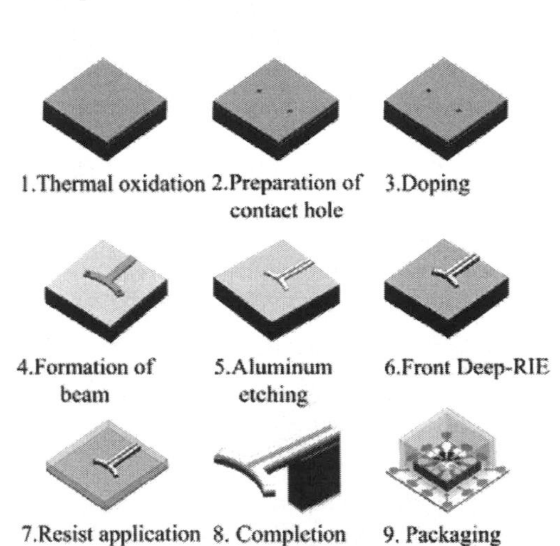

Figure 4. Fabrication process of the sensor.

Figure 5. The fabricated sensor.

5 PERFORMANCE OF THE SENSOR

The output voltage of the sensor is measured at a given acceleration from –6 g to +6 g. The measurement result is shown in Figure 6.

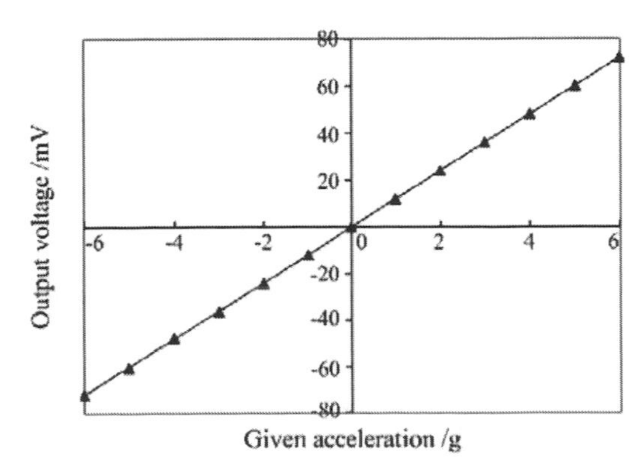

Figure 6. The sensor sensitivity.

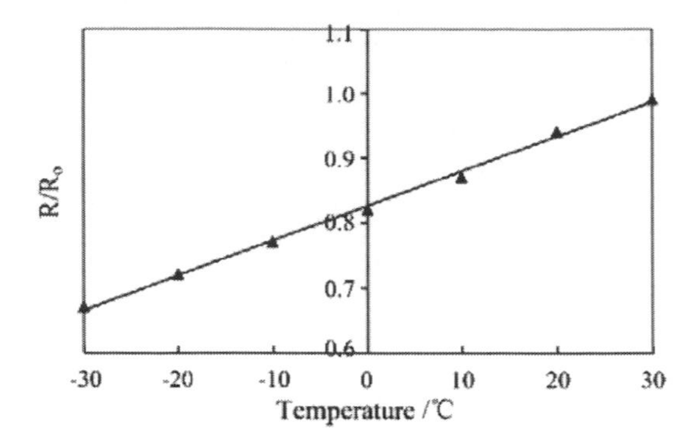

Figure 7. The TCR value of P-type silicon.

The power of the heater is 14.7 mW and the input voltage is 1 V. The sensor sensitivity is 12 mV/g and the nonlinearity of the full-span output is 0.32% obtained from the measured sensitivity curve.

And the TCR value of the thermistor is tested by placing the sensing chip in an oven and increasing temperature from −30°C to 30°C. The P-type silicon TCR value is shown in Figure 7. The TCR value is 6000 ppm/°C.

6 CONCLUSION

In this paper, bi-axial thermal convection acceleration sensor based on micro-machining is described. The sensing principal, structure design, fabrication process and performance test is illustrated in detail. Based on heat convection simulation results, the sensor was designed and manufactured. The new type structure of the sensor reduces the 93% thermal stress of the thermistor compared with the commonly thermistor.

The convective sensor performance is strongly influenced by the size of cavity and cap of the package. The simulation result shows the sensitivity change with the sizes of the packaged cap and the cavity. By compromising the accelerometer with the sensitivity and the bandwidth of the frequency, an appropriate size was estimated. The performance of the sensor was tested by experiments, and result shows that a linear relationship between the acceleration and the output signal is obtained.

One end of thermistors is free due to considering thermal stress is the merit of the sensor while that the load concentrates on the fixed end is its disadvantage. Since the large portion of aluminum layer causes very large thermal stress, the future work is to reduce aluminum wire size in order to lower the thermal stress.

REFERENCES

Alan J. Chapman, "Fundamentals of Heat Transfer", MacMillan Express, 1987.

GUO Wei, WANG Rong-Qing. Thermal analysis of single-axis MEMS convective accelerometer[J]. China measurement & test, 2012, 38(2): 42–45.

LI Li-Jie, LIANG Chun-guang. Micromachined Convective Accelerometer[J]. Chinese journal of semiconducors, 2001, 22(4): 465–468.

Ming-jun MA, JIN Zhong-he. Noise behaviors of a closed-loop micro-electromechanical system capacitive accelerometer [J]. J. Cent. South Univ, 2015, (12): 4634–4644.

RAN Yingling, JIN Bin. Design and research of micro thermal conductivity detector structure[J]. Electronic components and materials, 2015, 34(10): 64–67.

Stephen B. Pope, "Turbulent flows", Cambridge Univ. Express 2000, 96–178.

Sze S.M., "Semiconductor Devices", John Willey and Sons, 1985.

WAN, Cai-xin, [1], FU, Li-ping, [2], XUE, Xu, [1], LI, Dan-dong. A Probe Into Actualities of the Microaccelerometer Development[J]. Navigation and Control, 2012, (2): 73–771.

Civil, Architecture and Environmental Engineering – Kao & Sung (Eds)
© 2017 Taylor & Francis Group, ISBN 978-1-138-02985-9

Innovative green design of the frog ramming machine based on TRIZ/FRT

Z.G. Xu, J.L. Li & Z.Y. Zhu
School of Mechanical Engineering, Shandong University, Jinan, Shandong, China

ABSTRACT: The green innovation design of the frog ramming machine is done based on FRT (Future Reality Tree) and TRIZ. First, the traditional 39 parameters are classified into four major classes and 16 small classes, which are put into the green conflict parameters. The functions are analyzed for green innovation design, targets are solved, green conflicts are collected, and the subtractive model of the frog ramming machine is obtained by the FRT method. The product design process verified and exemplified the effectiveness of the cooperative TRIZ and FRT model.

1 INTRODUCTION

Yamamoto (1999) pointed out that green product design can improve environmental efficiency five times. Meanwhile, Hsiang-Tang Chang (2004) developed a new ecological design tool, which is proved to be effective in the application of TRIZ theory into green design, and five functional tools are developed: ecological retrieval tool, product evaluation tool by AHP, TRIZ engineering parameters describing the ecological factors, TRIZ invention tools, and invention retrieval tools. S. Vinodh (2014) proposed a quality function deployment on the basis of the awareness of environmental protection (ECQFD), the theory of inventive problem solving (TRIZ), and the Analytic Hierarchy Process (AHP) integration model for computer-aided innovation of auto parts and sustainable design.

The Future Reality Tree (FRT) is mainly used to verify whether the injection scheme can improve the system and achieve the desired results or adverse results after the injection (Qian, 2011). FRT describes the future prospects, the bottom-up approach is used, and the solution injection effect is predicted. FRT systems have two important roles: one is to make a preliminary evaluation scheme to remove the fault scheme, which does not meet the environmental requirements; the other is to find the negative effects through the solution injection and taking measures to prevent or eliminate the negative effects (Liu, 2008). Innovation programs are injected to insert the corresponding problem, and the corresponding FRT chart is illustrated.

2 GREEN CONFLICT RESOLUTION

The optimization goal of green design is to improve the green conflicts of a product. There are three kinds of parameters: green parameters, general parameters, and half-general half-green parameters. To find out the potential conflicts, we first determine the innovation goal of a green product and then combined the functional model with the substance-field analysis. A detailed analysis of the relationship between subsystems and components determines the potential problem of the product, and then, the problem standardization and TRIZ conflict are found and solved.

Similar forms of substance field analysis model and optimization should be selected to cope with the following five models: complete function model, perfect incomplete function model, noneffective complete function model, excessive functions, and harmful function. The following two need to be removed.

When the improvement of noneffective complete function model increases the harmful functions to the existing system, conflicts will result. Table 1 shows the correspondence between green design factors with the engineering parameters.

The green design conflict resolution model is shown in the following steps.

1. Determine design goals: Subsystems are determined.
2. Establish the substance-field analysis model: According to the relationship between the subsystems, the existing product substance-field analysis model is established.
3. Find the existing problems: According to the product substance-field function analysis model, determine the deficiency function, excess function, harmful function, and the problem of standardization and choose the appropriate TRIZ tool to solve the problem and obtain the final solutions.

Table 1. Green product engineering parameters.

Green design para types	Green design factors	Corresponding engineering parameters
Green environmental factors	Low toxicity of materials	No.31
	Compatibility of materials	No.35
	Reusability of materials	No.23
	Easy degradation of materials	No.31
	Types of materials	No.26
	Reduction of resource consumption	No.1, No.2, No.7, No.8, No.23, No.26, No.32
Green energy factor	Energy use efficiency	No.22, No.23
	Reduction of energy consumption	No.15, No.21, No.22, No.25
	Energy cleanliness	No.30, No.31
Green environmental factors	Vibration noise	No.31
	Emission of pollutants	No.31
	Cleanliness of production	No.30, No.31
Green technology factor	Security	No.13, No.27, No.30
	Degree of intelligence	No.34, No.37, No.38
	Life	No.15, No.16
	Removable property	No.13, No.25, No.33, No.34, No.36

4. Determine the green conflict: Use FRT to determine whether there is a conflict between the target solution and the environment. If yes, product finding needs to optimize the green design parameters and deterioration of parameters and then determine the product of the green conflict.
5. Green conflict resolution: According to Table 1, the green conflict into the corresponding engineering parameters, find the conflict matrix, according to the corresponding principle of the invention to get the target solution.
6. Innovative program generation: According to the above-mentioned five steps, get the target solution, and the design of personnel knowledge and experience, get the product's green innovation program.

3 GREEN INNOVATION DESIGN OF FROG RAMMING MACHINE

Ramming machine in the work process generates severe vibration and makes a large noise, which can easily cause deviation from the original direction in the process of advancing. Therefore, it needs to be pulled back when advancing, and hard labor works are needed as well as the security of the whole apparatus is enormously reduced.

It is necessary to analyze ramming functions by TRIZ tools to find the existing problem of the ramming machine. TRIZ tools are used to get the target solution. FRT tools are also introduced to find the target solution, and the conflicts between the machine and the environment are classified. The conflicting problems are solved by the conflict matrix.

3.1 *Functional model*

According to the relationship between the parts of the actual situation and the ramming machine in the process, model function in the process is shown in Fig. 1. Ramming machine is analyzed to find the functional model, where four problems are described as the follows:

Question 1: when reinforcing the soft soil foundation, each time the frog rammer tamps the foundation, there will exist excess soil at the bottom part of the front plate of the compaction;

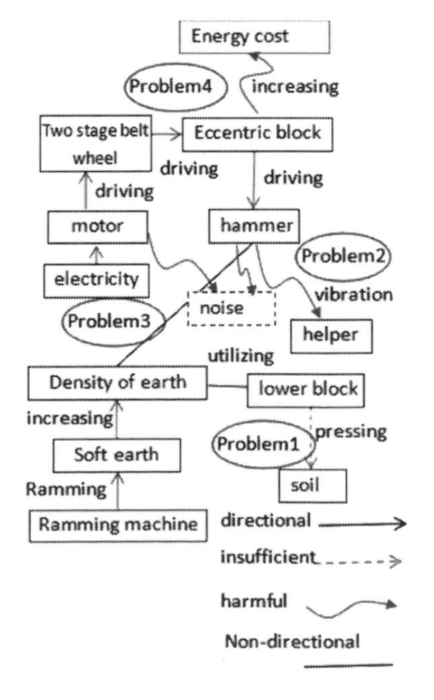

Figure 1. Functional model of the ramming machine.

inhomogeneity of soil distribution caused ramming machine offset line direction; and the pulling back to its original direction needs intensity of manual labor.

Question 2: horizontal centrifugal force generated by the rotation of the eccentric block drives the ramming machine move forward, causing the transverse vibration; rammer lifting caused vertical vibration; in addition, the rotation of the motor caused a certain vibration.

Question 3: tremendous noise is produced in the process of work, by the rotation of the motor, moving forward of the rammer, and the lifting of the hammer.

Question 4: because of the heavy eccentric block, the ramming machine starts very slowly, consuming huge amount of energy.

3.2 *Target solution*

Problem 1 came from the ill functional structure; substance-field model is used to solve this problem. It is a kind of noneffective complete system. Problem 1 and its substance-field analysis model are shown in Fig. 2. The possible solutions are shown in the following three schemes.

Scheme 1: a small hammer is added in the dynamic consolidation, which is located in the front of the floor, in the large ram compaction foundation, while the small ram compaction of the soil is in front of the floor.

Scheme 2: a "sweeping structure" is installed in the front floor, which is consisted of a cam and rack gear. A cam is arranged on both ends of the intermediate shaft; the cam and spring acted together to drive the reciprocating rack. The drive gear rotated repeatedly, and the gear and a swing rod are made into an integrated structure to remove the front soil.

Scheme 3: A small motor is installed in front of the floor; the small motor drives the fan to blow away surplus soil.

Problems 2 and 3 are very similar, which can be seen as a technical conflict. At the same time, the rammer increases the blow strength as well as the vibration and noise. If parameter 10 is improved, parameter 31 gets deteriorated, the principles of 13, 3, 36, and 24 are obtained, and the following schemes are suggested:

Solution 4: introduce agent in the motor to absorb vibration caused by the motor.

Plan 5: Principle of acting conversely: adding a bias offset from the existing eccentric block to offset horizontal centrifugal force, thereby offsetting lateral vibration.

Plan 6: Principle of intermediary: installation of vibration damping material in the arm absorbs the transmitted vibration.

Plan 7: Principle of intermediary: add a motor cover, and reduce the noise caused by the motor rotation.

Problem 4 is regarded as a group of physical conflicts; the centrifugal force of eccentric block mass provides large force intensity, but will lead to slow starting of the ramming machine and high energy consumption. Spatial separation principle is used to solve the problems. The following schemes are suggested.

Scheme 8: Separation: the eccentric block is designed to be removed. When the eccentric block up-rotated, it moved upward; when rotated downward, it moved downward.

Plan 9: Local quality: eccentric block is made of hollow magnetic steel balls, with a magnet block installed on the eccentric block near the shaft hole.

4 GREEN DESIGN OF FROG RAMMING MACHINE

Scheme 1 is shown in Fig. 2. The difference between scheme 1 and scheme 2 is that the former has no the "pre-scanning structure", where a small hammer (10) is utilized to remove the front soil; the view is shown in Fig. 2. Compared with scheme 2, scheme 1 has the following properties: simple structure, material saving, and better effects. Therefore, we adopt scheme 1 in this paper.

The front view of scheme 2 is shown in Figure 3; the down view of scheme 2 is shown in Figure 4.

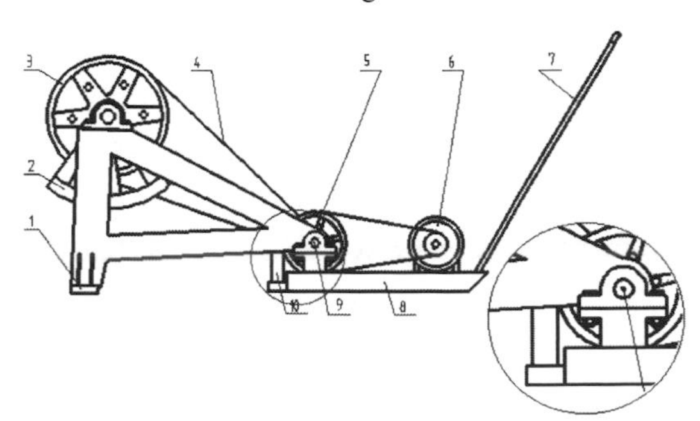

Figure 2. Structure view of scheme 1.

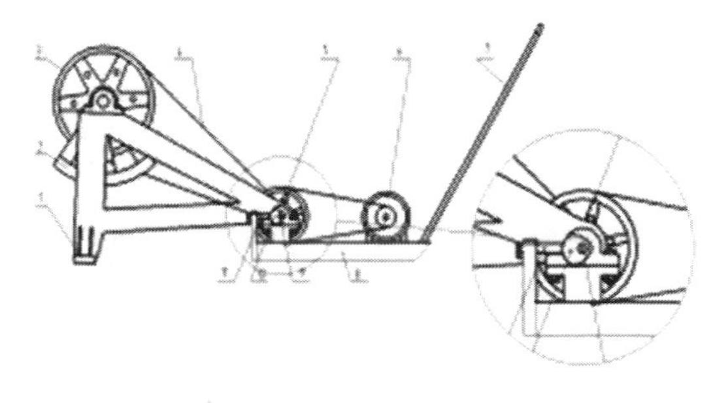

Figure 3. Front view of scheme 2.

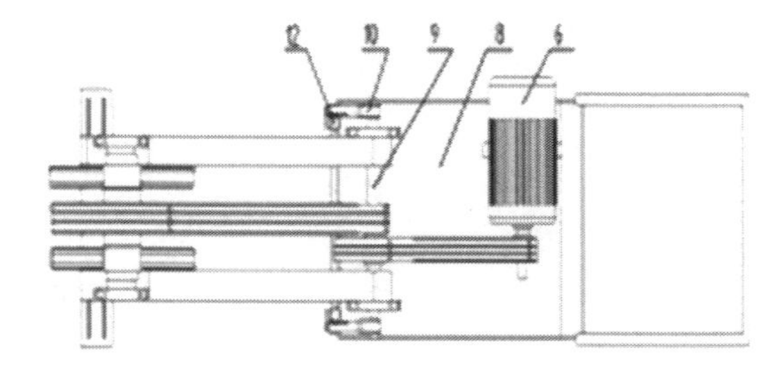

Figure 4. Down view of scheme 2.

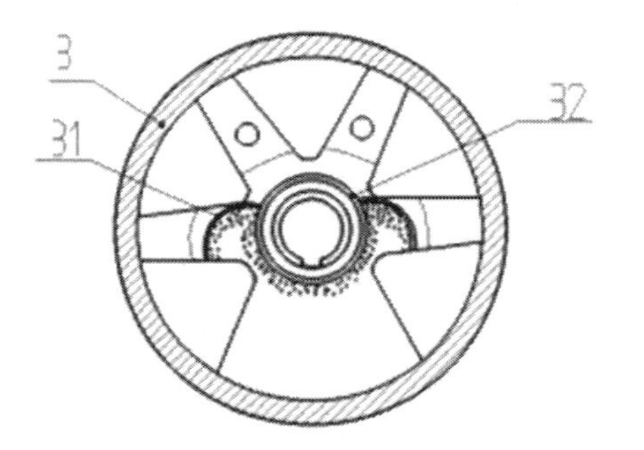

Figure 5. Static state view of pulley 3.

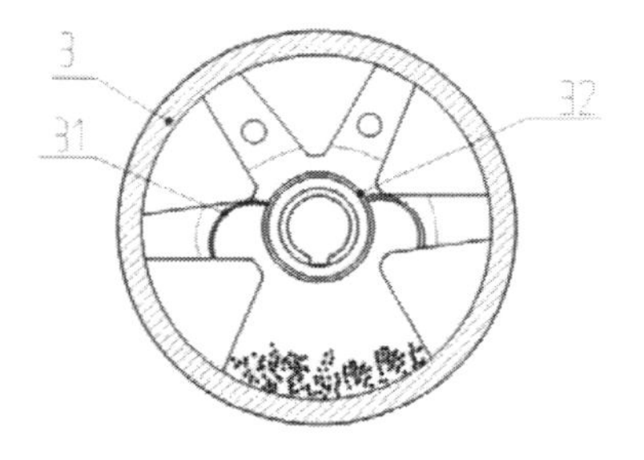

Figure 6. Performance view of pulley 3.

A "front sweeping structure" is installed at the bottom front, which is consisted of a cam (10), gear (12), and rack cam (11). The cam is installed in both ends of the intermediate shaft; cam and spring (not marked in the figure) acted together to drive the reciprocating movement of the rack and the gear. The reciprocating rotation gear and a swing rod are made into an integrated structure to remove the front soil.

Compared with scheme 8, scheme 9 has properties of simple structure, material saving, easy to wear, long service life, and so on. Scheme 9 changed the structure of the output belt pulley 3, which is made of a hollow structure containing magnetic steel balls designed near the eccentric block.

In addition, the four ribs have a semi-solid structure. Near the center axis, where the belt wheel is made into a hollow structure and permanent magnet balls are put inside, it consisted of an arc magnet (31) and a cylindrical magnet (32). The section view of pulley 3 is shown in Figure 4. Figure 4 is the state view of pulley 3; when the ramming machine does not work, the magnetic balls are put inside the arc magnet (31) and (32). Because of the lower weight of the eccentric block, the machine can start work quickly.

When the ramming machine works, the magnetic steel balls move toward the eccentric block; when the hammer was landed, the magnetic ball moved downward, as shown in Figure 6, and crackdown power increased. This scheme reduces the material consumption, improves the starting efficiency, and increases the blowing efforts.

5 CONCLUSIONS

The green conflict resolution model based on TRIZ/FRT is put forward in this paper for the ramming machine green design. Four types of parameters are put forward: green resources, green energy, green environment, and green technology. To sum up, these four aspects of engineering parameters corresponding to green design factors. The design process is verified by the green innovative design of a frog ramming machine, which becomes more complied with environmental requirements.

ACKNOWLEDGMENTS

This work was supported by the Chinese NSFC, 61272017, P.R. China, and the Key Laboratory of High-efficiency and Clean Mechanical Manufacture at Shandong University, Ministry of Education.

REFERENCES

Hsiang-Tang Chang, Jahau Lewis Chen. The conflict-problem-solving CAD software integrating TRIZ into eco-innovation. Advances in Engineering Software 2004, 35: 553–566.

Liu Xiaomin, Tan Runhua, Theory of constraints in the current implementation of tree and conflict resolution chart drive. Innovation design of Chinese mechanical engineering, 2008, (12): 1442–1445.

Qian Weimiao, Research on product improvement design based on functional analysis, theory of constraints and TRIZ, Dissertation of master degree, Hangzhou: Zhejiang University, 2011.

Vinodh S, Kamala V, Jayakrishna K. Integration of ECQFD, TRIZ, and AHP for innovative and sustainable product development. Applied Mathematical Modelling, 2014, 38(11): 2758–2770.

Yamamoto R. Closing Remarks of Eco-design'99: Manifesto of Eco-design. Ecodesign'99: First International Symposium on Environmentally Conscious Design and Inverse Manufacturing, Tokyo, Japan, 1999.

Civil, Architecture and Environmental Engineering – Kao & Sung (Eds)
© 2017 Taylor & Francis Group, ISBN 978-1-138-02985-9

Rotor fault diagnosis based on SVM and improved D–S theory

Zhan-hong Yan & Ping Liao
College of Mechanical and Electrical Engineering, Central South University, Changsha, China

ABSTRACT: In order to overcome the shortcomings of the D–S (Dempster–Shafer) evidence theory, that is, the basic probability distribution is difficult to determine and the conflicting evidence is poorly treated, a diagnosis method based on improved evidence theory was proposed and applied to rotor fault diagnosis. First, using the pairwise coupling method and the one-to-one support vector machine to obtain the basic probability distribution function, the problem of how to obtain the original evidence was solved. Second, the theory of evidence was improved. The conflicting evidence was judged by the distance function of evidence. Then, the evidence was modified and combined to get fusion results. A variety of rotor fault signals were collected and analyzed in the experiments. The results showed that the diagnosis method was effective.

1 INTRODUCTION

Rotor system always plays an important role in electric power, aviation, and other industries. Its condition is related to equipment safety; therefore, it is necessary to perform research on rotor fault diagnosis. The D–S (Dempster–Shafer) evidence theory can fuse evidence without prior knowledge, so it is widely used in fault diagnosis of rotor system (Chen 2005). However, the D–S theory faces two problems (Sima 2012): (1) how to quantify the sensor information to get the original Basic Probability Assignment (BPA). The main solutions to this problem include the neural network method and the expert experience assignment method. However, neural network training requires a lot of historical data, and there are over fitting, local minimum phenomena in it; the expert experience assignment method relies on subjective judgment, and its scope is limited (Jiang 2012); (2) Uncertainties in measuring systems and error accumulation may bring conflicting evidence, but the D–S theory will cause unreasonable conclusions when dealing with highly conflicting evidence. Although some researchers proposed some improved methods to solve this problem, there are still slow convergence, low accuracy, and some other shortcomings (Yager 1987, Liu 2014).

Therefore, to overcome the aforementioned shortcomings of the D–S theory, we proposed a rotor fault diagnosis method on the basis of Support Vector Machine (SVM) and improved evidence theory. The pairwise coupling (Hastie 1998) and one-to-one SVM were used to obtain the BPA function. Then, the D–S evidence theory was improved to judge the conflicting degree of evidence, and the evidence was modified. Finally, data fusion was performed to obtain the diagnostic results. The method was applied to the rotor system in unbalance, misalignment, and other conditions. The results showed that the diagnostic accuracy was improved.

2 THEORETICAL BASIS OF THE EVIDENCE THEORY

Supposing that frame of discernment $\Theta = \{A_1, A_2, \dots A_n\}$ consists of evidence E_1 and E_2, the BPA functions are $m_1(A_i)$ and $m_2(A'_j)$, the focal elements are A_i and A'_j, and the D–S rule of combination is:

$$\begin{cases} m(\Phi) = 0 \\ m(A) = \dfrac{\sum\limits_{\substack{A_i, A'_j \subset 2^\Theta \\ A_i \cap A'_j = A}} m_1(A_i) m_2(A'_j)}{1-k}, \forall (A \neq \Phi) \end{cases} \tag{1}$$

where k is the normalization constant:

$$k = \sum_{\substack{A_i, B_j \subset 2^\Theta \\ A_i \cap B_j = \Phi}} m_1(A_i) m_2(A'_j) \tag{2}$$

K represents the conflicting degree of evidence. The larger the k, more serious is the conflict of evidence. If we combine the evidence by using equations (1) and (2), it will lead to incorrect results.

3 OBTAINING BPA BASED ON SVM

As the general form of BPA is not given by the D–S theory, in application, obtaining BPA requires a detailed analysis. SVM is a new machine learning method proposed by Vapnik. Compared with neural network, SVM can achieve a good accuracy with only a small number of training samples. It also has better generalization ability and no local minimum phenomenon (Ai 2005). Meanwhile, SVM has strict theoretical and mathematical basis, so the BPA function obtained from SVM overcomes the shortcomings, which the expert experience method lacks in (Zhou 2014).

However, the output of standard SVM can only judge whether the input samples belong to or not to a certain class. Platt (2000) proposed a method to convert the standard SVM output value f to probability p_i by sigmoid function:

$$p_i = \frac{1}{1+e^{Af+B}} \tag{3}$$

$$F(A,B) = \min - \sum_{i}^{K} [t_i \log(p_i) + (1-t_i)\log(1-p_i)] \tag{4}$$

$$t_i = \begin{cases} t_+ = \dfrac{N_+ + 1}{N_+ + 2} \\ t_- = \dfrac{1}{N_- + 2} \end{cases} \tag{5}$$

where p_i is the probability of sample x_i, K is the number of training samples, $t_i = t_+$ or t_- when the classification result is 1 or −1, respectively, and N_+ and N_- are the number of samples classified into 1 or −1, respectively. By solving (4) and (5), we can gain A, B in (3) and then get a function of f and p_i.

The method proposed by Platt is only suitable for dealing with two classification problems and cannot be applied to multiclassification problems directly. Therefore, we use the pairwise coupling classification proposed by Hastie (1998) to convert two classes probabilities into multiclassification probability. By solving the optimization problem of (6), the probability p_i of the class i is obtained as the original BPA function in the D–S rule:

$$\begin{cases} \min\limits_{p} \sum\limits_{i=1}^{k} \sum\limits_{j:j\neq i} (r_{ji}p_i - r_{ij}p_j)^2 \\ \sum\limits_{i=1}^{k} p_i = 1, p_i \geq 0, i = 1, \ldots, k \end{cases} \tag{6}$$

where $j=1,\ldots, k$, k is the number of classes, $i\neq j$; r_{ij} represents the probability that the sample x_i belongs to class i when pairing the ith class and the jth class. Equation (6) can be rewritten as:

$$\min_{p} 2p^T Qp = \frac{1}{2} \min_{p} p^T Qp \tag{7}$$

$$Q_{ij} = \begin{cases} \sum\limits_{s:s\neq i} r_{si}^{2}, i = j \\ -r_{ji}r_{ij}, i \neq j \end{cases} \tag{8}$$

Where $\boldsymbol{p}$ should satisfy the following equation:

$$\begin{bmatrix} Q & e \\ e^T & 0 \end{bmatrix} \begin{bmatrix} p \\ b \end{bmatrix} = \begin{bmatrix} 0 \\ 1 \end{bmatrix} \tag{9}$$

Where $\boldsymbol{p} = [p_1, p_2, \ldots, p_k]^T$ is the solution of (6), $\boldsymbol{e}$ is a $k \times 1$ vector, whose elements are all one, $\boldsymbol{0}$ is a $k \times 1$ vector, whose elements are all zero, and b is the Lagrangian multiplier of equality constraint $\sum p_i = 1$.

Wu (2004) gave the following algorithm to compute $\boldsymbol{p}$:

1. Give initial value to $\boldsymbol{p}$, which satisfies:

$$\sum_{i=1}^{k} p_i = 1, p_i > 0 \tag{10}$$

2. Repeat ($t = 1, \ldots, k$) (11) and (12):

$$p_t \leftarrow \frac{1}{Q_{tt}} \left[-\sum_{j:j\neq t} Q_{tj}p_j + p^T Qp \right] \tag{11}$$

Normalize $\boldsymbol{p}$ (12)

until (9) is satisfied.

4 IMPROVEMENT OF THE D–S THEORY

In practice, the error factors lead to conflicting evidence, but the D–S theory is not suitable for dealing with it. Hence, we use the distance function of evidence as a criterion to determine whether evidence is conflicting to other or not. Then, the evidence will be amended and combined by the D–S rules (Pan 2013, Zhang 2009).

The distance function (d_{ij}) of evidence E_i and E_j in the frame of discernment is defined as:

$$d_{ij} = \sqrt{\frac{1}{2}\left(\| m_i \|^2 + \| m_j \|^2 - 2\langle m_i, m_j \rangle\right)} \tag{13}$$

where $\| m_i \|^2 = < m_i, m_i >$,$< m_i, m_j >$ is the inner product of m_i, m_j, $d_{ij} \in [0, 1]$ and $d_{ij} = d_{ji}$. d_{ij} describes the distance of evidence. The greater the value, the higher the conflicting degree of evidence E_i and E_j.

If there are X pieces of evidence, the distance function matrix formed by them is:

$$S=\begin{pmatrix} 0 & d_{12} & \cdots & d_{1X} \\ \vdots & & \ddots & \vdots \\ d_{X1} & d_{X2} & \cdots & 0 \end{pmatrix}_{X\times X} \quad (14)$$

Introducing the parameter c_i represents the conflicting degree of E_i and other evidence:

$$c_i=\frac{1}{X}\sum_{\substack{j=1\\ i\neq j}}^{X} d_{ij},\ i,j=1,2,\ldots X \quad (15)$$

As $d_{ij}\in[0,\ 1]$ and the diagonal elements of the matrix $\boldsymbol{S}$ are all zero, there should be $0\le c_i\le [(X\ 1)/X]$.

Thus, we introduce the discounted factor, ω_i, as the weight of the evidence. The smaller the c_i, indicating that E_i is more similar to other evidence, the greater the value of ω_i. Otherwise, ω_i should be smaller.

Let $\omega_i=f(c_i)$, ω_i and c_i should meet the following conditions:

a. $c_i\to[(X\text{-}1)/X], f(c_i)\to 0;\ c_i\to 0, f(c_i)\to 1$
b. $0\le f(c_i)\le 1$, $f(c_i)$ should be monotonically decreasing in its interval.
c. When c_i small, ω_i should decrease slowly. As c_i increases, ω_i should decrease rapidly to zero.

Thus, $f(c_i)$ should be an exponential function. A similar creditable factor was defined by Pan (2013) and Zhang (2009). We define:

$$\omega_i=\left(1-c_i\cdot\frac{X}{X-1}\right)\beta^{c_i\cdot\frac{X}{X-1}} \quad (16)$$

Where β is an unknown parameter to be determined. Supposing $t=c_i\cdot[X/(X-1)]$, then $0\le t<1$. Taking derivation of t, we have:

$$f'(t)=\beta^t(\ln\beta-t\ln\beta-1)<0 \quad (17)$$

Rewriting (17), we get $\beta<\mathrm{e}$. When β decreases, $f'(t)$ will decrease slowly, which do not meet the condition c. Therefore, replacing t with c_i, we have:

$$\omega_i=\left(1-c_i\cdot\frac{X}{X-1}\right)e^{c_i\cdot\frac{X}{X-1}} \quad (18)$$

After the discounted factor ω_i is determined, if the BPA function before modification is $m_i(A_j)$ and that after modification is $m_i'(A_j)$, we use (19) to modify the evidence:

$$m_i'(A_j)=m_i(A_j)\cdot\omega_i,\ A_j\neq\Theta$$
$$m_i'(\Theta)=1-\sum_{j=1}^{X} m_i'(A_j) \quad (19)$$

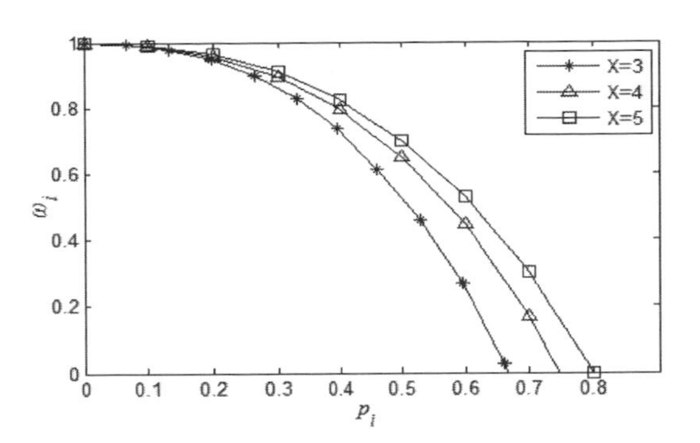

Figure 1. Relationship between discounted factor and conflicting degree when $X=3, 4, 5$.

Finally, the D–S rule is used to combine the modified evidence and then we can obtain the results of evidence fusion.

5 APPLICATION IN ROTOR FAULT DIAGNOSIS

5.1 *Diagnostic model*

Large rotor common faults include loose bearings, unbalance, misalignment, and rubbing. In order to fully and accurately capture these faults, it is necessary to set multiple measurement points to collect vibration signals in different locations on the rotor system. The vibration signals were used to extract features and form diagnostic eigenvector. Those eigenvectors formed by each measuring point data were input to SVM for training and testing, and the BPA function was generated. The BPA function produced by each measuring point was regarded as a piece of evidence. Then, we can calculate the conflicting degree and modify evidence to eliminate the impact of uncertain information. Finally, the D–S evidence theory can be used to fuse the evidence and get the diagnosis result.

On the basis of the common faults of the large rotor, a rotor fault simulation platform was built. The platform consists of multistage pull rod rotor, motor, eddy current sensors, and brake (Fig. 2). The fault diagnosis system was established by applying the SVM and the improved evidence theory as shown in Figure 3.

5.2 *Rotor fault signal feature extraction*

The vibration signals generated during the operation can reflect the fault characteristics. Because of error accumulation, analysis of single sensor data is difficult to accurately identify the fault.

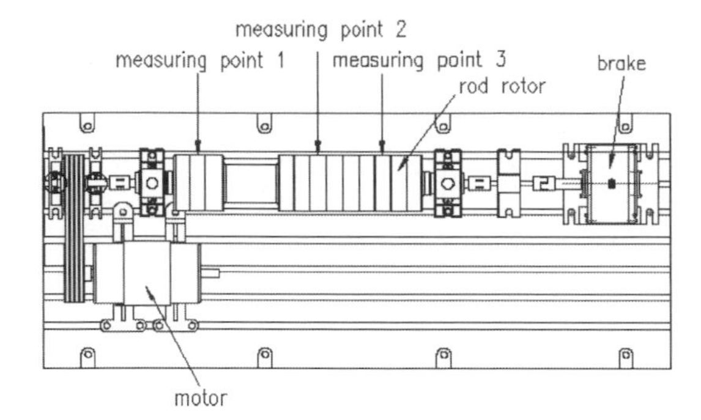

Figure 2. Fault simulation platform.

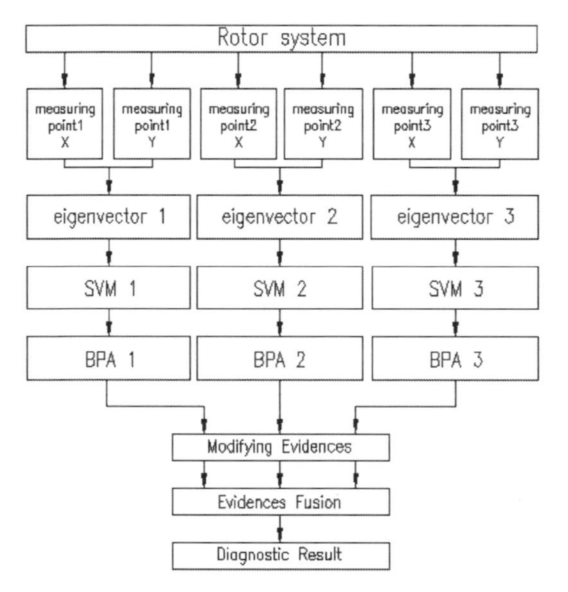

Figure 3. Rotor fault diagnosis system.

To solve this problem, we collected the vibration signals in different positions of the rotor and used the wavelet packet to process the signals to get the diagnostic eigenvector.

The rotor radial vibration signal was measured by setting eddy current sensors in three positions of the rotor. As only the signals measured by two sensors perpendicular to each other can fully describe vibration information (Han 2009), one eddy current sensor was set in horizontal direction and another one in vertical direction. The signals were recorded as x_1, y_1, x_2, y_2, x_3, and y_3, which represented horizontal and vertical signals in three points. During the experiments, rotor speed was 1460 rpm and sampling frequency was 1 kHz. Some of the fault vibration signals are shown below (unbalanced fault as example).

The frequency band energy distribution of the vibration signal reflects the fault characteristics,

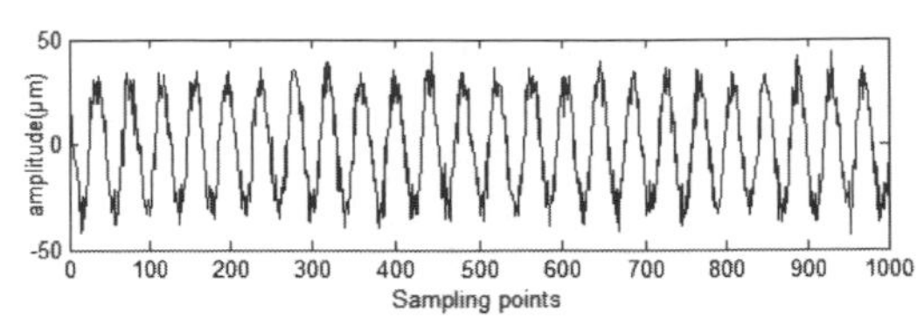

Figure 4. Unbalanced fault, sensor x_2 signal.

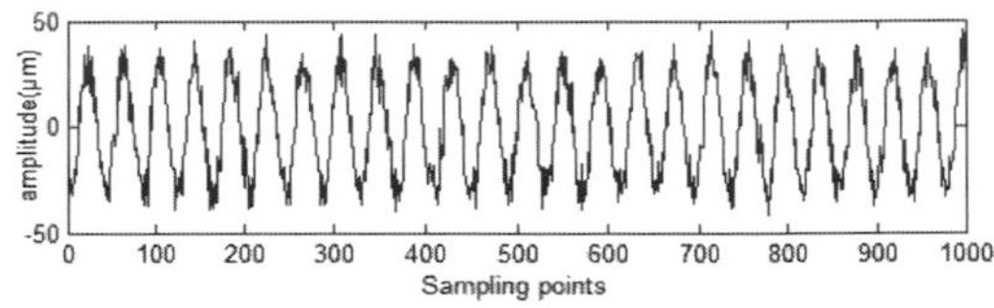

Figure 5. Unbalanced fault, sensor y_2 signal.

so we can form the eigenvector of diagnosis by decomposing band energy spectrum. We used the wavelet packet to decompose the signal and obtain the energy spectrum and wavelet energy entropy. The eigenvector, which is the full description of the fault information characteristics (Han 2009), was obtained by combing energy spectrum and the energy entropy. Meanwhile, noticing that $1/2f$, $2f$, $3f$, and so on (f = rotor speed) are fault characteristic frequencies (Zhou 2014), in order to capture the fault information more accurately, we decomposed the signal as shown in Table 1.

db12 was selected as the wavelet basis. The x, y signals from one measuring point were filtered and decomposed by wavelet packet. Wavelet reconstruction coefficients obtained from eight nodes were numbered as $S_{x1}(n)$, $S_{y1}(n)$, $S_{x2}(n)$,...$S_{y8}(n)$ in descending order of frequency. Then, the energy spectrum and energy entropy of the jth frequency band were:

$$E_{zj} = \sum_{n=0}^{N}(|S_{xj}(n)|^2 + |S_{yj}(n)|^2) \tag{20}$$

$$E_{zTotal} = \sum_{j=1}^{8} E_{zj} \tag{21}$$

where N = the number of sampling points. Normalizing E_{zj}, we can obtain the vibration signal wavelet relative energy spectrum $q_z = [q_{z1}, \ldots q_{zi}, \ldots q_{z8}]$

The wavelet energy entropy, H_{zW}, is:

$$H_{zW} = -\sum_{i=1}^{8} q_{zi} \log q_{zi} \tag{22}$$

In conclusion, the eigenvector of rotor vibration signal is: $W = [q_{z1}, \ldots q_{zi}, \ldots q_{z8}, H_{zW}]$

5.3 *BPA function design*

With the eigenvector, now we can establish the frame of discernment $\Theta = \{A_1, A_2, \ldots A_5\}$ representing misalignment, imbalance, rubbing, loose bearings, and normal five conditions. Several sets of vibration signals were collected under five conditions and some eigenvectors are shown in Table 2:

It can be seen that the decomposition according to Table 1 showed the signal energy distribution under different faults. But considering the noise and error, single sensor information may still cause misjudgment.

According to the algorithm shown in Section 2, 75 sets of data from the measuring point 1 (15 groups for each case) were input as training samples. SVM was set with RBF kernel, and the cross-validation method was used to obtain SVM-related parameters (Wang 2014) to form SVM1. Similarly, SVM2 and SVM3 were formed. For each of the three SVMs, 175 sets of test samples (35 groups for each case) were input. The results of BPA are shown in Table 3.

Table 1. Wavelet packet decomposition nodes.

Node	Frequency range	Node	Frequency range
(1,1)	0–0.3125f	(4,3)	2.5–3.75f
(2,1)	0.3125–0.625f	(5,1)	3.75–5f
(4,1)	0.625–1.25f	(6,0)	5–10f
(4,2)	1.25–2.5f	(6,1)	10–20f

Table 2. Fault eigenvectors (from point 2).

	A_1	A_2	A_3	A_4	A_5
q_{z1}	0.0157	0.0095	0.0082	0.0120	0.0021
q_{z2}	0.0113	0.0072	0.0151	0.0211	0.0053
q_{z3}	0.3487	0.8515	0.5873	0.7137	0.8605
q_{z4}	0.5071	0.0633	0.2087	0.1650	0.0541
q_{z5}	0.0593	0.0074	0.0731	0.0227	0.0054
q_{z6}	0.0054	0.0055	0.0283	0.0089	0.0061
q_{z7}	0.0289	0.0247	0.0353	0.0273	0.0258
q_{z8}	0.0495	0.0461	0.0470	0.0478	0.0455
H_{zW}	0.5534	0.2995	0.5629	0.4535	0.2700

Table 3. BPA function under unbalanced fault (from SVM2).

Sample number	$m(A_1)$	$m(A_2)$	$m(A_3)$	$m(A_4)$	$m(A_5)$
1	0.028	0.429	0.028	0.028	0.487
2	0.031	0.886	0.030	0.030	0.023
3	0.029	0.490	0.029	0.029	0.423
4	0.028	0.567	0.028	0.028	0.349
5	0.029	0.463	0.029	0.029	0.450
…	…	…	…	…	…

The accuracy of the diagnosis result was judged according to the following rule (Geng 2006):

$$
\begin{aligned}
&m(A_r) = \max(m(A_i)) \\
&m(A_r) > \eta
\end{aligned}
\tag{23}
$$

5.4 *Evidence fusion*

As can be seen in Table 4, relying on information from a single sensor will reduce the accuracy. Therefore, we combined the multisensor information and improved the diagnostic accuracy. According to the improved method shown in Section 2, 175 groups of samples were combined. Taking unbalanced fault as an example, Contrast Yager (1987), Liu (2014, $\tau = 0.5$), D–S and improved D–S theory in Section 4, the following results are obtained.

In Table 5, the probabilities of unbalance fault from SVM1, SVM2, and SVM3 were 0.869, 0.553, and 0.025, respectively. Because of the high conflict, the probability of normal condition (0.584) was higher after treatment by the D–S rule; The uncertainty degree of Yager's rule was 0.971, which means it is difficult to determine the fault; Only Liu's (0.582) and the improved method (0.603) produced a result that the rotor was more likely in unbalanced fault, which was consistent with the actual situation.

In Table 6, the BPA functions provided by SVM1, SVM2, and SVM3 were slightly different from each other (the probabilities were 0.799, 0.717, and 0.553, respectively). After the fusion, the uncertainty of Yager's rule was 0.673, and still

Table 4. SVM diagnosis results ($\eta = 0.8$).

SVM number	Number of correct predictions	Number of test samples	Accuracy
SVM1	134	175	76.6%
SVM2	109	175	62.3%
SVM3	78	175	44.6%

Table 5. Conflicting evidence fusion.

	$m(A_1)$	$m(A_2)$	$m(A_3)$	$m(A_4)$	$m(A_5)$	Θ
SVM1	0.027	0.869	0.026	0.027	0.051	0
SVM2	0.024	0.553	0.024	0.024	0.375	0
SVM3	0.026	0.025	0.026	0.028	0.895	0
D–S	0.001	0.413	0.001	0.001	0.584	0
Liu	0.019	0.582	0.018	0.019	0.362	0
Yager	0	0.012	0	0	0.017	0.971
Improved method	0.011	0.603	0.011	0.011	0.331	0.033

Table 6. Comparison of convergence.

	$m(A_1)$	$m(A_2)$	$m(A_3)$	$m(A_4)$	$m(A_5)$	Θ
SVM1	0.025	0.799	0.025	0.026	0.125	0
SVM2	0.025	0.717	0.025	0.025	0.208	0
SVM3	0.024	0.553	0.024	0.024	0.375	0
D–S	0	0.970	0	0	0.030	0
Liu	0.016	0.791	0.016	0.016	0.161	0
Yager	0	0.317	0	0	0.010	0.673
Improved method	0	0.965	0	0	0.035	0

Table 7. Accuracy of different algorithms.

Methods	Number of correct predictions	Number of test samples	Accuracy
D–S	140	175	80.0%
Liu	146	175	83.4%
Yager	27	175	15.4%
Improved method	162	175	92.3%

it is difficult to determine the type of fault. Liu's method correctly diagnosed the rotor condition but with slow convergence (0.791). Only the results of the D–S method and improved method showed that the probability of the unbalanced fault was more than 0.96, which showed better convergence efficiency as well.

Let the threshold $\eta = 0.9$. 175 groups of test samples be fused. The accuracy of diagnosis is shown in Table 7, according to (23):

It can be seen that all the methods improved the diagnostic accuracy. Among them, the improved method was the best; while Yager's method had the worst result. Liu's method and the D–S method had a certain improvement in the correct rate; however, the effect was not good as that of the improved method.

6 CONCLUSION

The pairwise coupling was combined with SVM to solve the problem of evidence generation and overcame the shortcomings of the expert assignment method and neural networks.

The D–S evidence theory was improved. The conflicting degree of evidence was judged by the distance function of evidence. Then, the evidence was modified and combined. Experiments showed that the improved method could deal with conflicting evidence well and had a high convergence speed.

The fault characteristics of rotor system were analyzed. Combined with the signal characteristics, a rotor fault diagnosis method was established by using wavelet, SVM, and improved evidence theory. Experiments showed that the accuracy of diagnosis was improved, indicating that the method was valuable in actual rotor fault diagnosis.

REFERENCES

Ai, N., Wu Z.W. & Ren J.H. 2005. Support vector machine and artificial neural network. *J. Journal of Shandong University of Technology (Sci & Tech).* 19(5), 45–49.

Chen, L.Y. 2005. *Multisensor data fusion and its application in induction motor fault diagnosis*, Hangzhou, China: Zhejiang University.

Geng, J.B., Huang S.H. Jin J.S., Chen F., Shen S. & Liu W. A rotational machinery fault diagnosis method Based on close degree of information entropy and evidence theory. *J. Mechanical Science and technology.* 25(6):664–667.

Han, J. & Xie K. 2009. Study of full information wavelet energy entropy and its application in rotating machine condition monitoring. *J. Journal of Mechanical Strength Engineering* 39(1), 290–293.

Hastie, T. & Tibshirani R. 1998. Classification by Pairwise Coupling. *J. The Annals of Statistics*, 26(1), 457–471.

Jiang, W.L. & Wu S.Q. 2010. Multi-data fusion fault diagnosis method based on SVM and evidence theory. *J. Chinese Joumal of Scientific Instrument* 31(8), 1738–1743.

Liu, X.L., Chen G.M. & Li F.X. 2014. Research on fault diagnosis approach of gear pump based on improved evidence theory. *J. Mechanical Science and Technology for Aerospace Engineering* 33(2), 184–188.

Pan, K., Li H. & Xing G. 2013. Conflict Evidence Synthesis Method Based on Confidence Distance. *J. Computer Engineering* 39(1), 290–293.

Platt, J.C. 2000. *Probabilistic outputs for support vector machines and comparison to regularized likelihood methods.* J. Cambridge, MA, USA: MIT Press: 61–74.

Sima, L.P., Shu N.Q., Li Z.G., Huang Y. & Luo X.Q. 2012. Identification of interior fault position based on SVM and D-S evidence theory for electric power. *J. Electric Power Automation Equipment* 32(11), 72–76.

Wang, J.F., Zhang L. & Chen G.X. 2012. A parameter optimization method for an SVM based on improved grid search algorithm. *J. Applied Science and Technology* 39(3), 28–31.

Wu, T.F., Lin C.J. & Weng R.C. 2004. Probability Estimates for Multi-class Classification by Pairwise Coupling. *J. Journal of Machine Learning Research* 5(4), 975–1005.

Yager, R.R. 1987. On The Dempster-Shafer Framework and New Combination Rules. *J. IEEE Trans. On System*, 41(2), 93–137.

Zhang, S.G., Li W.H. & Ding K. 2009. A novel approach to evidence combination based on the evidence credibility. *J. Control Theory & Application*, 26(7), 812–814.

Zhou, Y. 2014. *Rotor faults diagnosis based on vibration signal analysis*, Nanjing, China: College of Energy and Power Engineering.

Civil, Architecture and Environmental Engineering – Kao & Sung (Eds)
© 2017 Taylor & Francis Group, ISBN 978-1-138-02985-9

A study of the relationship between pre-tightening force and torque in precision instrument assembly

Shuang Huang, Xin Jin, Zifu Wang, Zhijing Zhang, Muzheng Xiao & Can Cui
Beijing Institute of Technology, Beijing, China

ABSTRACT: The screw thread connection is widely used in precision instruments and the quality of threaded connections directly affects the assembly quality and reliability of the product. (Yang X, 2011) With the development of micro precision mechanical and electrical products, the requirement of controlling on micro-thread connection is becoming much higher. The present work aims to embark from theoretical analysis and experimental research to explore the rules and characteristics of high precision threaded connections. During the research, the accurate torque-preloading force mapping relationship of the miniature titanium screw in the aerospace field is measured. (Nassar SA, 2005) The experimental method as well as all the results in the present work can be utilized to improve the accuracy and the dynamic stability of the precision instruments and provide guidance for the assembly process of precision instrument.

1 INTRODUCTION

One of the most important reasons why threaded connection is widely used is that it can provide considerable connection force, besides, it can be repeatedly disassembled and interchanged. In the field of aerospace precision instruments, there are higher requirements for the threaded connections. The precise controlling of the pretightening force has a great influence on the precision and the stability of the instrument, so that the accurate measurement of torque-preload relationship for realizing the quantitative control of assembly has the vital significance. As early as 1953, the screw torque controlling has strict and precise assembly requirements in the assembly process standard of Sidewinder missile. In the field of assembly, researches have done a lot of work in the design of threaded connection form, but rarely involves the study of the torque-preload force accuracy control, especially small size high precision thread connection. (Zaki AM, 2008) Consequently, the aim of the present work is to explore the torque—preload force mapping relationship of the miniature titanium screws in the aerospace field through theoretical analysis and experimental research.

2 MECHANICAL MODEL OF THE TORQUE METHOD IN THE TIGHTENING PROCESS

In the process of uniform tightening bolts as shown in figure 1, there are four kinds of torque. The tightening torque T is initiatively applied, and the friction torque Tb between the bolt head and the supporting surface which is a passive torque. The friction torque Tt between the screw thread surface and the internal thread surface is another passive torque, and the torque which is provided by lead angle generate the preload force. This is constantly increasing in the process of tightening and has the following relationship:

$$T = T_b + T_t + T_\alpha \tag{1}$$

The yellow ring in figure 2 is the joint surface of threaded connection. The torque of the yellow joint surface is called the face friction torque T_b. If

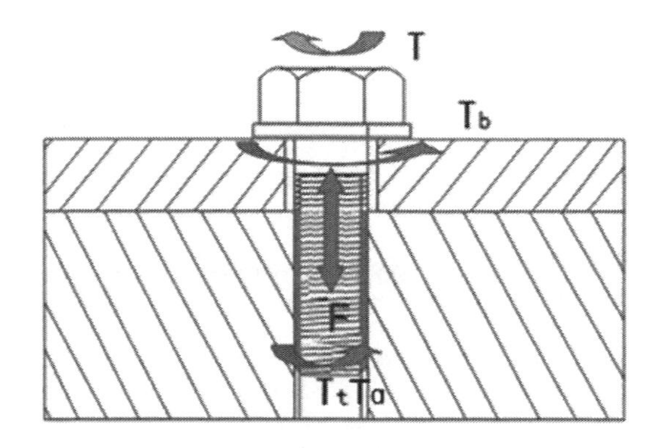

Figure 1. The mechanical model in the process of tightening.

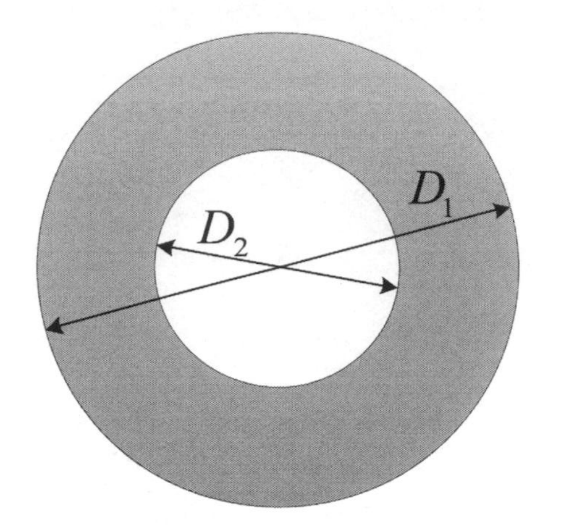

Figure 2. The schematic diagram of the friction area under the surface of the nut.

the equivalent friction diameter is d_b, the friction torque can be expressed as follows:

$$T_b = \frac{1}{2} d_b F \mu_b \tag{2}$$

F is screw preload force, μ_b is the friction coefficient of the relative sliding surface.

The equivalent friction diameter can be calculated by

$$d_b = \frac{D_1 + D_2}{2} \tag{3}$$

Where D_1 is the washer outer diameter or the outer diameter of bolt head circular contact area, which is the large diameter of contact area, D_2 is thread through hole internal diameter or washer internal diameter, which is the small diameter of contact area.

Take the rectangular bolt nut connecting as an example as shown in Figure 3 (a). The nut is regarded as a slider, and it is applied on the positive pressure F which is equal to the bolt interior preload (parallel to the bolt axis), the thrust F_t (perpendicular to the bolt axis) can be equivalent as the friction torque on the thread surface. In figure 3 (a), spread along pitch diameter of thread to draw the figure 3 (b). If the friction on the supporting surface is ignored, the tightening process can be converted into the oblique plane-block mechanism. In that case, there remain a horizontal force F_t and a vertical pressure F. And if we proceed the orthogonal decomposition, then we can get figure 3 (b), the mechanical relationship of the forces can be expressed as follows:

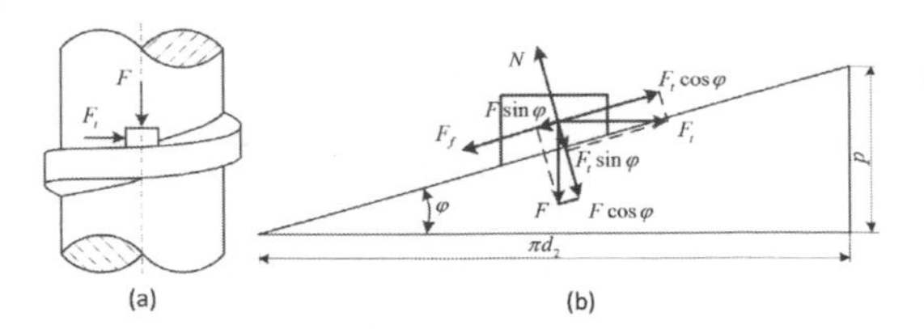

Figure 3. Rectangular thread force analysis.

$$\begin{aligned} F_f &= F_t \cos\varphi - F \sin\varphi \\ N &= F \cos\varphi + F_t \sin\varphi \\ F_f &= \mu_t N = \mu_t (F \cos\varphi + F_t \sin\varphi) \\ &= F_t \cos\varphi - F \sin\varphi \end{aligned}$$

$$F_t = \frac{\mu_t + \tan\varphi}{1 - \mu_t \tan\varphi} F \tag{4}$$

Where ϕ is the lead angle, μ_t is the thread surface friction coefficient, N is the support force of the bevel to the sliding block, F is the equivalent preload force, F_t is the equivalent thrust, F_f is the equivalent thread surface friction.

Setting thread surface friction angle as ρ, and put it into the formula (4), the equivalent thrust can be expressed as follows:

$$F_t = \frac{\tan\rho + \tan\varphi}{1 - \tan\rho \tan\varphi} F = F \tan(\rho + \varphi) \tag{5}$$

Setting equivalent friction diameter value is the thread pitch diameter d_2, so in the tightening process, the friction torque T_t can be expressed as follows:

$$T_t = \frac{d_2}{2} F \tan(\rho + \varphi) \tag{6}$$

The equivalent friction angle ρ' of the triangular thread can be expressed as follows:

$$\tan\rho' = \frac{\mu_t}{\cos\alpha'} \tag{7}$$

Where α' is the included angle between the thread flank and the vertical thread axis.

Put ρ' into the formula (6), T_t can be expressed as follows:

$$T_t = \frac{d_2}{2} F \tan(\rho' + \varphi) \tag{8}$$

Where d_2 is pitch diameter of thread, at the same time it is also the equivalent friction diameter.

For the average metric triangular thread:

$25° < \alpha' < 30°$

$\varphi < 3°$

Then:

$$1.103 < \frac{1}{\cos\alpha'} < 1.155 \tag{9}$$

$$\tan\varphi < 0.0524 \tag{10}$$

However the scope of the friction coefficient of the general metal materials:

$$0.1 < \mu_t < 0.3 \tag{11}$$

Synthesize the formula (7) to (11), we can get

$$\tan\rho' \tan\varphi \approx 0 \tag{12}$$

Put the formula (12) into the formula (8)

$$T_t = \frac{d_2}{2} F(\tan\rho' + \tan\varphi) \tag{13}$$

For

$$\tan\varphi = \frac{p}{\pi d_2} \tag{14}$$

Putting the formula (2), the formula (7), the formula (13) and the formula (14) into the formula (1), the ordinary triangular thread tightening torque T can be calculated as follows:

$$T = Fd\left(\frac{d_b\mu_b}{2d} + \frac{d_2\mu_t}{2d\cos\alpha'} + \frac{p}{2\pi d}\right) = KFd \tag{15}$$

From the formula (15), we know the use of empirical the formula (1) can describe the tightening process of the torque method, but there are many factors which will more or less affect the number of K, they will enlarge the dispersion degree of the preload force.

When the thread is loosened, there are still two friction torques. The friction torque on the support surface has an exactly opposite direction, yet the algorithm does not change. In that case, the thrust in the force model of the screw thread reverses at the moment, and the direction of the preload force do not change. Therefore, breakaway torque T' can be calculated as follows:

$$T' = Fd\left(\frac{d_b\mu_b}{2d} + \frac{d_2\mu_t}{2d\cos\alpha'} - \frac{p}{2\pi d}\right) = K'Fd \tag{16}$$

3 EXPERIMENTAL STUDY

In order to study the rule of F/M mapping in the tightening process of small size screws, an experimental device for screw tightening is designed. Screw tightening conditions can be changed to detect the torque coefficient of the screw in different conditions, then the torque curve and the tension curve in the process of screw tightening can be obtained.

The schematic diagram of the screw tightening experimental apparatus is shown in figure 4. We use the torque tester and high precision sensor to display in real time and collect the tightening torque and the preload. The experimental bolt is fixed on both sides of the support surface by the experiment nut, a replaceable gasket is fixed on the supporting surface, which can simulate the contact between the connected part support surface and the screw lower surface. The experiment nut is fixed on the connecting column by two pressing plates, and the bottom end of the connecting column is connected with the tension-compression sensor which is fixed to the bottom plate. The four corners of the bottom plate have four studs, which are used to support the supporting surface and adjust the distance between the upper and the lower boards. In the tightening process, when the screw rotates under the action of torque, due to friction of the contact area, the nut will rotate to some extent along with the screw. The tighter the bolt screws, the greater the friction and the nut torque is. In order to prevent the tension-compression sensor from the excessive torque damaging, a limit device is designed to limit the rotation of the connecting column.

A torque tester is used to make a real-time detection of the screw tightening torque and a tension-compression sensor is utilized to measure the size of the screw preload force. The experiment is carried out by using the connecting column, which can replace the nut. Besides, each screw corresponding to a nut. The upper gasket will be

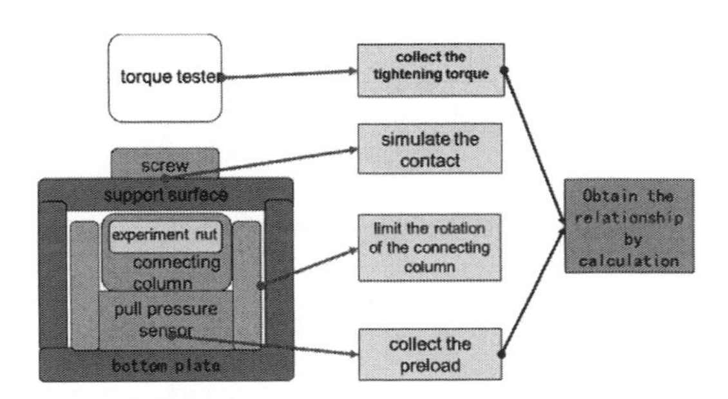

Figure 4. The schematic diagram of the screw tightening experimental apparatus.

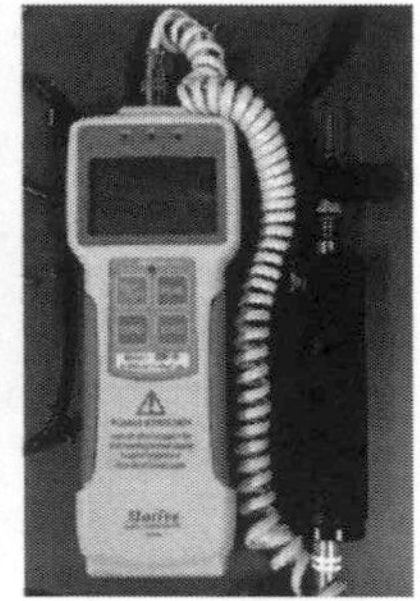

Figure 5. The experiment device.

changed to satisfy different materials and different hole sizes.

4 THE MEASUREMENT OF THREAD PITCH DIAMETER

The machine vision detection is the main method that used in precision measurement of micro-miniature device. It is a measurement method which combines optical microscopy techniques and computer vision technology. In this paper, we use a CCD sensor to measure the pitch diameter of the screw. The basic principle is: Place the screw in the field of the CCD view, and then light the screw using a backlight, screw images are optical lens imaging on the light-sensitive chips, light sensitive chip converts the collected image into image signal. With the basis of pixel brightness and pixel color information, we put them into digital signals after image processing system receiving the light-sensitive chips transmit to image signal. Image system takes a variety of operations in order to change these signals into the characteristics of the target, so that realize precision measurement of the tested pieces.

The rapid detection system based on machine vision is shown in figure 6.

The image processing methods is as follow:

First, we take five points in the thread path in order to fit a straight line. Then a contour line on the other side of the thread should be taked. We find the closest point of contour line to the straight line and get the distance between which names L1. Second, we take five points on the large diameter using the same method of screw thread in order to fit a straight line. Then we find the closest point of contour line to straight line, and name the distance between them as L2. Then we can calculate the thread of the large diameter using L1 and L2. Since the M2 screw pitch P is 0.4 mm, and the M3 screw pitch P is 0.5 mm, we can obtain the screw pitch diameter by:

Figure 6. Device rapid detection system based on machine vision.

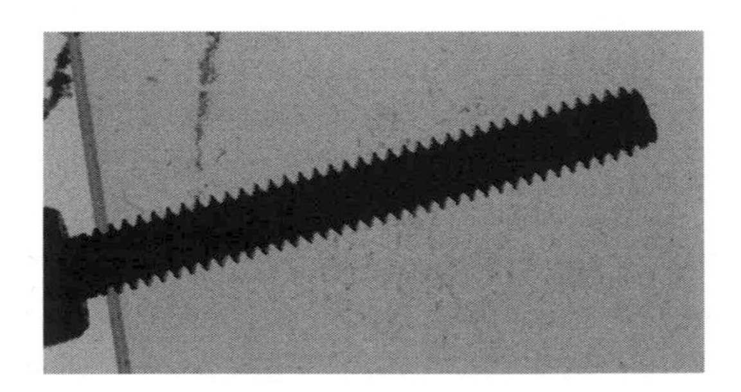

Figure 7. Experimental sample image.

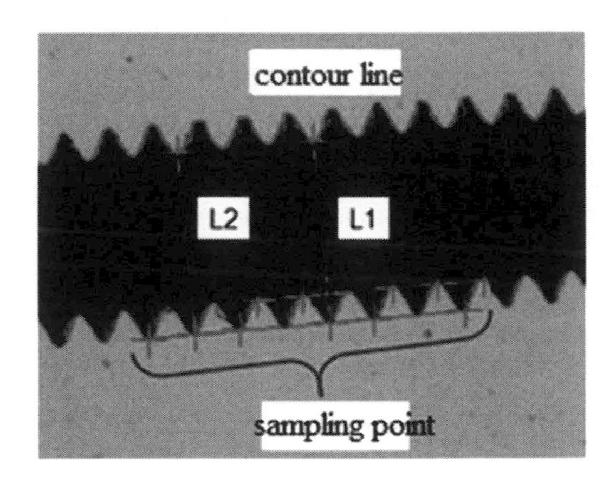

Figure 8. Measuring the thread.

$$D_2 = 2L_2\text{-}L_1\text{-}0.6495P \quad (17)$$

The measured results are shown in table 1.

5 RESULTS AND DISCUSSION

Experienced workers of the aviation instrument assembly process are invited to help the experiment, each worker loads and releases respectively

Table 1. Screw pitch diameter.

	M1/mm				M2/mm		
	L2	L1	D2		L2	L1	D2
A1	2.527	2.283	2.449	B3	1.638	1.38	1.573
A8	2.539	2.291	2.462	B7	1.632	1.373	1.568
B2	2.517	2.243	2.469	B9	1.628	1.37	1.563
C4	2.500	2.208	2.469	C5	1.637	1.376	1.575
D5	2.555	2.269	2.518	D9	1.634	1.373	1.572
Average	2.527	2.259	2.473	Average	1.634	1.374	1.570
Standard deviation	0.021	0.034	0.026	Standard deviation	0.004	0.004	0.005

Table 2. The preload results.

	Master1				Master2				Master3				Master4			
NO.	Ist	2nd	3rd	4th	Ist	2nd	3rd	4th	Ist	2nd	3rd	4th	Ist	2nd	3rd	4th
1	542	615	569	447	529	564	523	264	530	626	335	405	630	653	528	358
2	518	537	493	490	564	758	623	483	537	739	671	355	556	473	519	382
3	569	609	561	516	521	534	535	582	606	606	540	414	608	576	512	335
4	607	566	586	519	518	514	499	474	568	528	539	407	569	595	607	412
5	622	661	645	590	501	546	444	469	467	499	607	348	585	536	516	422
6	793	605	622	427	454	565	479	343	610	664	620	342	563	584	489	360
Average	571	598	579	498	514	580	517	436	550	594	548	372	585	569	528	371
	583				537				567				561			
Standard deviation	43	42	53	58	36	89	61	113	52	86	107	32	32	60	45	33
	45				69				86				44			
Coefficient of variation	7%	7%	9%	11.6 %	7%	15%	11%	26%	9%	14%	19%	8.5%	5%	10%	8%	8.8%
	7.7%				12.9%				15.15%				7.86%			

with the load wrench of experimental torque tester and his customary production assembly tools. After the statistical analysis of a number of experimental results, the conclusions and results are shown in table 2.

As shown in table 2, each worker in this experiment loads and unloads for four times. The first three times are loaded with the load wrench of the experimental torque tester, and the last one is loaded with the tools customary used in the assembly. While the front is intended to accurately measure the mapping relationship of torque-preload force of precision aerospace screws, and the latter is intended to measure the magnitude and consistency of torque and preload for threaded fasteners in existing production processes.

Results of the worker shows that the process in the production of bolt pre-tightening force distri-

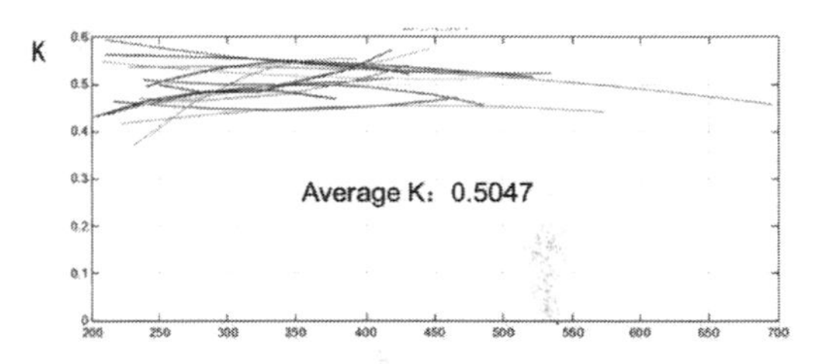

Figure 9. The average K of all experimental results.

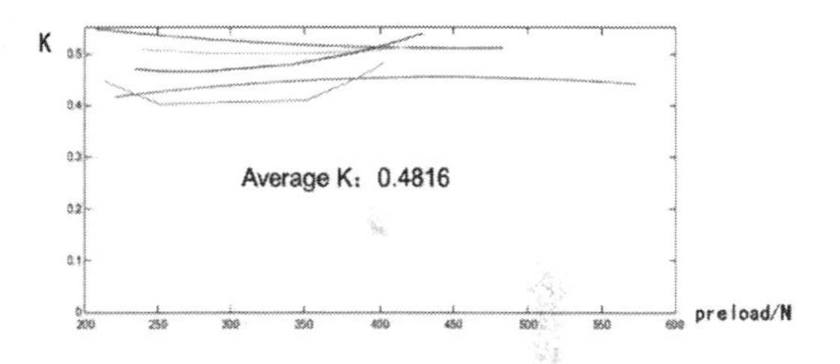

Figure 10. Tightening for the 1st time.

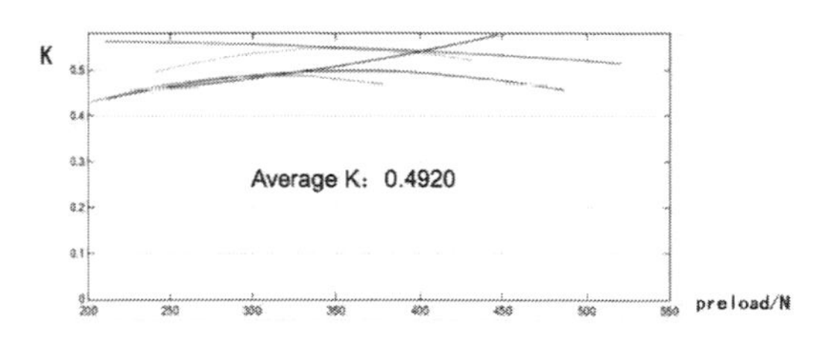

Figure 11. Tightening for the 2nd time.

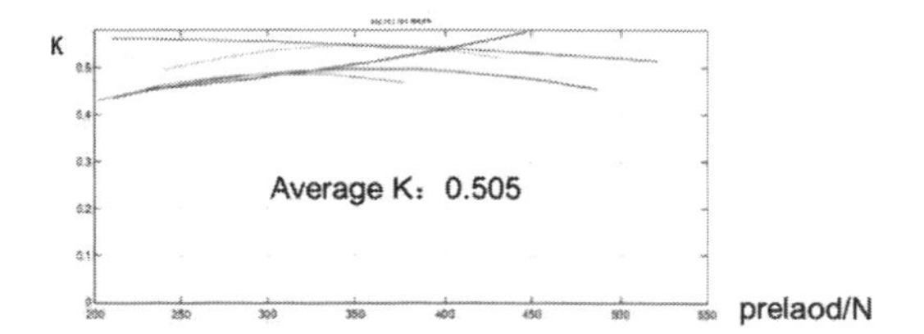

Figure 12. Tightening for the 3rd time.

bution is around 400 N, but the consistency is not ideal because the variation coefficient is extremely high, therefore, to optimize the assembly process and improve assembly standard is very important.

Figure 9 illustrates the torque coefficient K that a worker tighten a lot of screws three times, and figure 10 to 12 are the results of the individual time. It is found that the average of K is above 0.5, and this does not conform with engineering experience value 0.3, it shows that precision of threaded fasteners in the field of aerospace torque coefficient K have their own different rules.

As is shown in the pictures above, among three times of tightening screw, average torque coefficient K reduce with the increase of the crew tight times. This is because the tight make the thread much smoother, and these can lead to reduction of the torque coefficient, which means less preload can achieve the same torque. Other tighten the supervision of the process have similar laws.

6 CONCLUSION

In this article, we analyzed the research status of precision instrument at home and abroad, and then delve into the influence of threaded fasteners which is precise assembled in mechanical and electronic products in the field of aerospace. A theory research and experimental study on miniature threaded fastener tighten process is implemented, further, the distribution law of coefficient K is summarized and the mapping relationship of preload force is explored. The whole work of the present study makes a foundation for instrument assembly process optimization in the field of aerospace.

ACKNOWLEDGMENTS

This research was supported by National Natural Science Foundation of China (grant number 51375054) and National Natural Science Foundation of China (grant number E051002).

REFERENCES

Housari BA, Nassar SA, (2007) "Effect of thread and bearing friction coefficients on the vibration-induced loosening of threaded fasteners", Journal of vibration and acoustics, Vol. 129, No. 4, pp484–94.

Nassar SA, Barber G, Zuo D, (2005) "Bearing friction torque in bolted joints", Tribology Transactions, Vol. 48, No. 1, pp69–75.

Nassar SA, Ganeshmurthy S, Ranganathan RM, Barber GC, (2007) "Effect of tightening speed on the torque-tension and wear pattern in bolted connections", Journal of pressure vessel technology, Vol. 129, No. 3, pp426–40.

Nassar SA, Housari BA, (2006) "Effect of thread pitch and initial tension on the self-loosening of threaded fasteners", Journal of pressure vessel technology, Vol. 128, No. 4, pp590–8.

Nassar SA, Housari BA, (2007) "Study of the effect of hole clearance and thread fit on the self-loosening of threaded fasteners", Journal of mechanical design, Vol. 129, No. 6, pp586–94.

Nassar SA, Matin PH, (2006) "Clamp load loss due to fastener elongation beyond its elastic limit", Journal of pressure vessel technology, Vol. 128, No. 3, pp379–87.

Nassar SA, Veeram AB, (2006) "Ultrasonic control of fastener tightening using varying wave speed", Journal of pressure vessel technology, Vol. 128, No. 3, pp427–32.

Nassar SA, Yang X, (2007) "Novel formulation of the tightening and breakaway torque components in threaded fasteners", Journal of Pressure Vessel Technology, Vol. 129, No. 4, pp653–63.

Yang X, Nassar SA, Wu Z, (2011) "Criterion for preventing self-loosening of preloaded cap screws under transverse cyclic excitation", Journal of vibration and acoustics, Vol. 133, No. 4, pp041013.

Zaki AM, Nassar SA, (2008) "Effect of Coating Thickness on the Friction Coefficients and Torque-Tension Relationship in Threaded Fasteners", ASME 2008 Pressure Vessels and Piping Conference, Vol. No. pp253–61.

Zaki AM, Nassar SA, Yang X, (2012) "Effect of conical angle and thread pitch on the self-loosening performance of preloaded countersunk-head bolts", Journal of Pressure Vessel Technology, Vol. 134, No. 2, pp021210.

Zou Q, Sun T, Nassar S, Barber G, Gumul A, (2007) "Effect of lubrication on friction and torque-tension relationship in threaded fasteners", Tribology Transactions, Vol. 50, No. 1, pp127–36.

Civil, Architecture and Environmental Engineering – Kao & Sung (Eds)
 ISBN 978-1-138-02985-9

Thermal characteristic analysis of a hollow screw based on the fluid structure thermal coupling

X.S. Li, W. Xiong & J.N. Xu
School of Mechanical Engineering, Shenyang Ligong University, Shenyang, P.R. China

ABSTRACT: Thermal deformation problems of the hollow screw of a machine tool were analyzed during the working process. The mathematical models of the screw deformation were established. On the basis of the heat transfer theory, the fluid structure thermal coupling problem of screw was solved under different coolant flow rates and temperatures. According to the temperature field and thermal deformation under different working conditions, the rules of thermal deformation about hollow screw were obtained under different flow rates and coolant temperatures. The analyzing results show the best flow and optimum temperature of cooling fluid.

1 INTRODUCTION

As the ball screw is an important transmission part of a machine tool, its performance directly affects the running state and precision of the tool (Ramesh et al. 2000). If a machine tool works at high speed and heavy loading, the screw-nut and bearing will generate a lot of heat. If the heat could not be spread out in time, then it would lead to a higher temperature and thermal deformation. Therefore, low temperature and narrow thermal deformation are very important to reduce machining errors (Wang et al. 2009).

Much research has been done on the problem of thermal deformation of screw. Kim et al. (1997) obtained the temperature distribution of screw at different speeds and running times using the method of finite-element analysis (FEM). Cao et al. (2011) analyzed the optimum coolant flow value for the hollow screw by the ANSYS thermal analysis module. However, few people have studied the fluid structure thermal coupling effect on the deformation of screw.

The Y-axis hollow screw of a boring-milling machining center named TX1600G is taken as the research object. In this paper, temperature field and thermal deformation were analyzed combining with the actual working conditions, that is, mean different cooling fluid flow and temperature. This paper will provide a theoretical basis for decreasing the thermal deformation of hollow screw.

2 MATHEMATICAL MODEL OF TEMPERATURE FIELD

In the process of heat generation, the heat quantity is directly proportional to the temperature gradient and the section area in unit time. The direction of heat transfer is opposite to the temperature rise, which is the Fourier law (Guo 2002):

$$q = -\lambda \frac{\partial T}{\partial n} \tag{1}$$

$$-\lambda \frac{\partial T}{\partial n} = \lambda_x \frac{\partial T}{\partial x} n_x + \lambda_y \frac{\partial T}{\partial y} n_y + \lambda_z \frac{\partial T}{\partial z} n_z \tag{2}$$

where q is the heat flux; λ is the coefficient of thermal conductivity; T is the object temperature; λ_x, λ_y, and λ_z are the isothermal surface coefficient thermal conductivities along the x, y, and z directions, respectively; n_x, n_y, and n_z are the isothermal surface components along the x, y, and z directions, respectively; and $\partial T/\partial n$ is the temperature gradient.

Because inner heat source is absent in the working process of the screw, the load form is axisymmetric. The heat conduction differential equation can be expressed as (Sun et al. 2015):

$$\frac{\partial^2 T}{\partial r^2} + \frac{\partial T}{r\partial r} + \frac{\partial^2 T}{\partial z^2} = \frac{\partial T}{\alpha \partial \tau} \tag{3}$$

where α is the thermal diffusivity, T is the temperature, and τ is the time. Its initial temperature was $T_o = 22°C$.

3 BOUNDARY CONDITIONS

The heat source of hollow screw mainly includes the bearing rotating friction heat and nut moving frictional heat. The methods of heat exchange with the outside include the cooling liquid convection heat transfer, the hollow screw and air convection heat transfer, the hollow screw, and the lubricant convection heat transfer.

3.1 *Calorific value analysis*

3.1.1 *Screw–nut pair frictional heat*

The equation of calorific value screw per unit time is shown as (Sun et al. 2015)

$$Q = 1.047 \times 10^{-4} M \cdot n \tag{4}$$

where Q is the calorific value in per unit time, N is the rotating speed, and M is the friction torque of screw. Friction torque of screw is mainly caused by the friction torque, which is caused by the pre-tightening force (Liu et al. 2006). The friction torque is calculated as:

$$M = 2Z\left(M_e + M_g\right)\cos\beta \tag{5}$$

where Z is the rolling element number, M_e is the group of friction torque, M_g is the geometry sliding friction torque, and β is the rolling spiral angle of the screw.

3.1.2 *Friction heat of bearing*

Bearing calorific value calculation formula is the same as that of the screw–nut pair, and it also can be calculated using equation (4). The friction torque of bearing is shown as (Chen et al. 1999):

$$\begin{cases} M = M_0 + M_1 \\ M_0 = 10^{-7} f_0 (\nu n)^{2/3} D_m^3 & (n \geq 2000 \text{ rpm}) \\ M_0 = 160 \times 10^{-7} f_0 D_m^3 & (n < 2000 \text{ rpm}) \\ M_1 = f_1 P_1 D_m \end{cases} \tag{6}$$

where f_0 is a factor related to bearing type and lubrication method, v is the kinematic viscosity of the lubricant, n is the rotating speed of screw, f_1 is a factor related to the bearing type and load, P_1 is the bearing load, and D_m is the mean diameter of the bearing.

3.2 *Heat transfer calculation of screw*

3.2.1 *Screw exchanges heat with air*

In the working process, the type of heat exchange between the air and screw is forced convective heat transfer. Therefore, the heat transfer coefficients can be given as follows:

$$h_a = \frac{\lambda Nu}{L} \tag{7}$$

Table 1. Screw thermal boundary conditions.

Parameter	Value
Heat production rate of bearing Q_b	197623 W/m³
Heat flow density (screw) q_s	2100 W/m²
Heat flow density (nut) q_n	2784 W/m²
Heat transfer coefficient (air) h_a	54 W/m²°C
Heat transfer coefficient (lubricating oil) h_o	288.5 W/m²·°C
Heat transfer coefficient (grease) h_g	235 W/m²·°C

$$Nu = 0.133 \mathrm{Re}^{\frac{2}{3}} P_r^{\frac{1}{3}} \tag{8}$$

$$\mathrm{Re} = \frac{\omega \times d^2}{\nu} \tag{9}$$

where λ is the coefficient of thermal conductivity, L is the feature size, Nu is the Nusselt number, Re is the Reynolds number, P_r is the Prandtl number, ω is the angular velocity of screw, d is the diameter of screw, and v is the kinematic viscosity of air.

3.2.2 *SCRW exchanges heat with lubricating oil*

The hollow screw nut–pair is lubricated by oil, and the heat transfer coefficient is calculated as:

$$h_o = 0.11 \frac{\lambda}{d} (0.5 \mathrm{Re}^2 P_r)^{0.35} \tag{10}$$

3.2.3 *Bearing exchange heat with grease*

Bearing is lubricated by grease, and its heat transfer coefficient is expressed as:

$$h_g = 0.332 \frac{\lambda}{L} P_r^{\frac{1}{3}} R_e^{\frac{1}{2}} \tag{11}$$

The thermal boundary conditions could be calculated while the rotation speed of the screw is 1500 rpm, and the results are shown in Table 1.

4 SCREW HEAT FLOW SIMULATION ANALYSIS

The method of one-way steady-state analysis is used in this paper. This means that the result of fluid analysis is loaded in the static heating analysis and static forces analysis.

4.1 *Analysis and calculation of the fluid*

The flow field was analyzed with CFX module of Workbench. The analysis results of the flow field can be obtained after setting the fluid inlet temperature, initial temperature, and cooling fluid velocity. In addition, the fluid outlet pressure is equal to the atmospheric pressure. The coolant flow rate is 3L/min, and the inlet temperature is 20°C. The results are shown in Figure 1.

4.2 *Thermal analysis in steady state*

At the steady state of thermal analysis, the screw is simplified as a hollow cylinder, and the bearing is simplified as a ring. The boundary conditions of the steady-state thermal analysis are shown in Table 1. The temperature distribution of the screw system is shown in Figure 2.

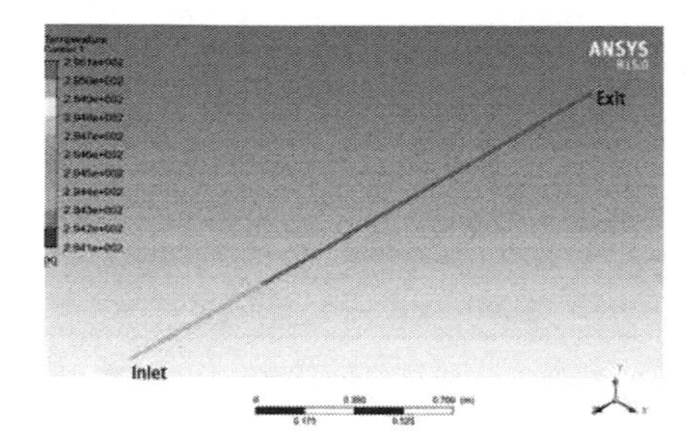

Figure 1. Temperature distribution of hollow screw.

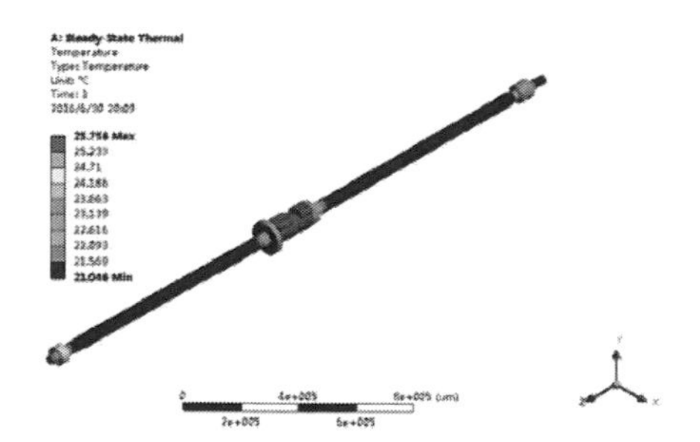

Figure 2. Temperature distribution of screw system.

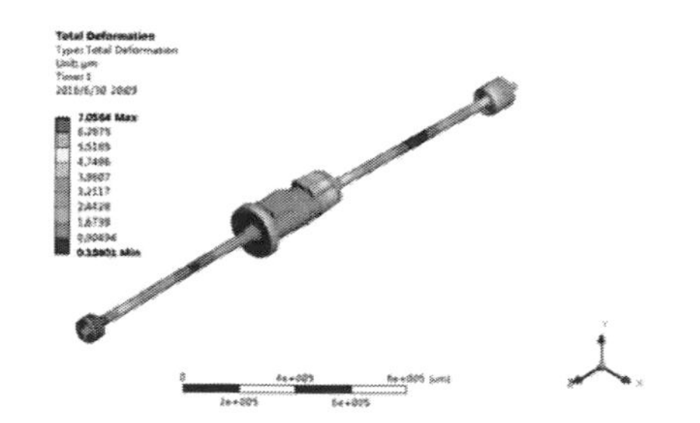

Figure 3. Deformation distribution of screw system.

4.3 *Coupling analysis*

The results of the temperature field were loaded into the model as the thermal load and pre-tightening force. It was also loaded on screw bearing. By setting constraints and the deformation condition, the coupling results are shown in Figure 3.

5 RESULTS OF FINITE-ELEMENT ANALYSIS

5.1 *Results under different temperatures of cooling fluid inlet*

In thermal characteristic analysis, the rotating speed of the milling shaft screw is 1500 rpm, coolant flow rate is 3 L/min, and the coolant inlet temperature is 22°C. The highest temperature of screw and the temperature difference of cooling fluid are shown in Figure 4.

The relationship between the maximum temperature of the screw and the inlet temperature difference of the cooling liquid is shown in Figure 4. The results show that the lower the inlet temperature of the coolant, the greater the temperature difference between the inlet and outlet.

The screw axial deformation law and total deformation law are always similar in Figure 5. The axial deformation is the main deformation. With the increasing of cooling fluid inlet temperature, the deformation decreases first and then increases. Because the coolant inlet temperature is too low, there is a large temperature difference between the two ends of the lead screw, which produces a large thermal deformation.

Combining with the cooling effect and the deformation of screw, best results are obtained when the inlet temperature of cooling fluid is slightly lower than the temperature of the environment.

5.2 *Results of simulation analysis under different cooling fluid flow*

Heat flow and solid coupling were analyzed when the rotating speed of the milling shaft screw is 1500 rpm, cooling fluid inlet temperature is 21°C, and cooling fluid flows are 1, 3, 5, 8, and 10 L/min. The highest temperature of the screw and the temperature of cooling fluid between inward and outward are shown in Figure 6.

Figure 6 shows that the temperature of the coolant becomes smaller while the coolant flow rate is higher. Lead screw temperature is significantly reduced first, and then decreased slowly. The results show that the cooling fluid flow is too small and the lead screw undergoes insufficient cooling.

Figure 7 shows that coolant flow rate increases, and the total deformation of the screw decreases

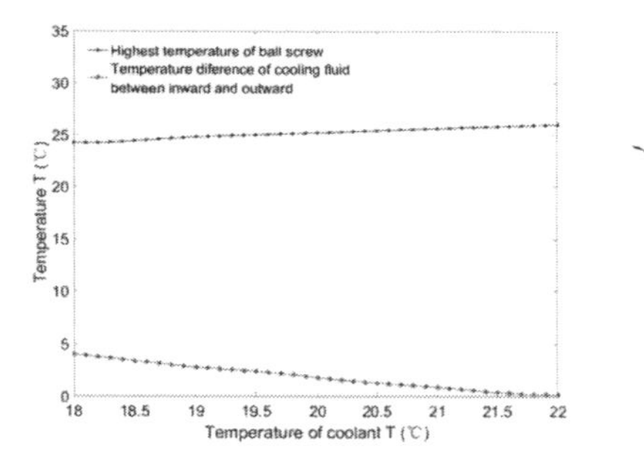

Figure 4. Coolant inlet temperature with its exit temperature, screw temperature curves.

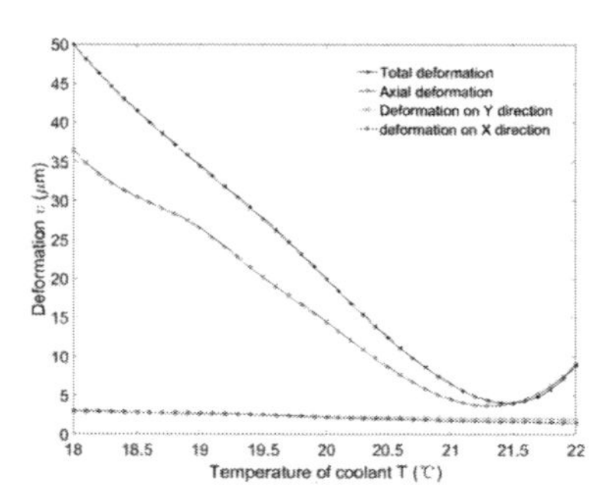

Figure 5. Deformation of screw in all directions under different cooling fluid flow.

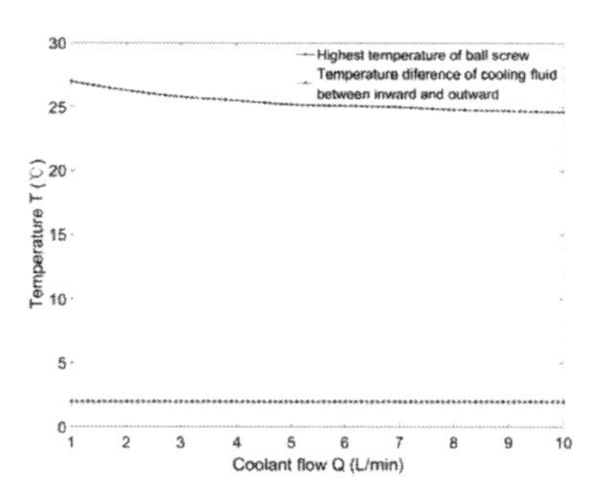

Figure 6. Highest temperature of screw and the temperature difference between inward and outward.

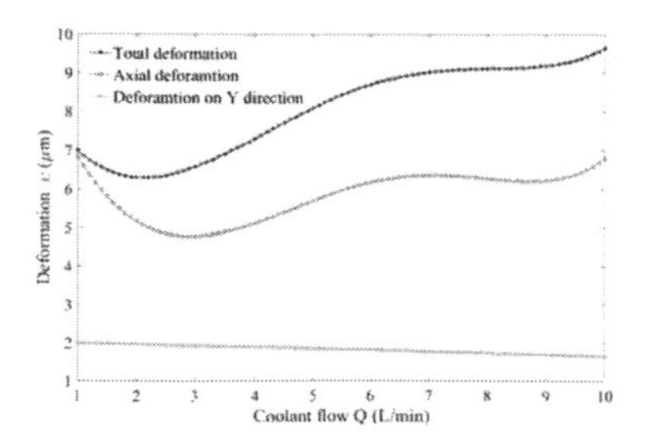

Figure 7. Relation curves of deformation and coolant flow.

first and then increases. The deformation of Y-axis direction decreases gradually. It also shows that the coolant flow should be appropriate, or it will cause a larger deformation. Considering the cooling effect, the suitable coolant flow should be 2–3 L/min.

6 CONCLUSION

In this paper, the thermal deformation of the hollow screw rod of the boring and milling center is studied. When the lead screw is at a different temperature and flow rate, the heat, flow, and solid coupling analysis has been carried out in this paper. The influence of cooling fluid on the temperature and deformation of screw is studied. The coolant inlet temperature is lower, the coolant temperature will be higher, and the cooling effect will be better. If the coolant flow is too small, the cooling effect of the screw will be insufficient. If coolant inlet temperature is lower, the temperature difference of both ends of the screw will be higher. Then, a larger thermal deformation will appear.

REFERENCES

Cao, J. J., Long-Gang, L. I., Liu, Y. S., & Sun, J. G. (2011). Research on thermal characteristics of high speed hollow ball screw in different working conditions. *J. Modular Machine Tool & Automatic Manufacturing Technique. (3)*, 30–35.

Chen, T. Y., Wei, W. J., & Tsai, J. C. (1999). Optimum design of headstocks of precision lathes. *J. International Journal of Machine Tools & Manufacture. 39(12)*, 1961–1977.

Guo, D. (2002). Heat transfer. *M. Higher education press.*

Kim, S. K., & Cho, D. W. (1997). Real-time estimation of temperature distribution in a ball-screw system. *J. International Journal of Machine Tools & Manufacture. 37(4)*, 451–464.

Liu, X. H & Song, X. C. (2006). Influence factor and test method research of ball screw pair friction moment. *J. Tool Engineering. 40(6)*, 59–61.

Ramesh, R., Mannan, M. A., & Poo, A. N. (2000). Error compensation in machine tools—a review: part ii: thermal errors. *J. International Journal of Machine Tools & Manufacture. 40(9)*, 1257–1284.

Sun, J., Qin, X., Qian, B., & Huang, Y. (2015). Thermal characteristics analysis of ball screw of tx1600g cnc boring and milling machining center. *J. Machinery & Electronics. (9)*, 32–36.

Wang, D. A., Liu, Y. A., Zhang, L. A., Tian, X. A., & Qinyun b, L. I. (2009). Temperature field and thermal distortion simulation in feeding process of ball screw based on finite element method. *J. Computer Aided Engineering. (2)*, 29–35.

Civil, Architecture and Environmental Engineering – Kao & Sung (Eds)

ISBN 978-1-138-02985-9

Research on intelligent exhaust device design of the breaker in the substation

Xiao Yu
School of Electrical Engineering, Fuzhou University, China
State Grid Fujian Electric Power Co. Ltd., Zhangzhou Power Supply Company, Fujian, China

Wu Wang
School of Electrical Engineering, Fuzhou University, China

ABSTRACT: To understand the frequent bulge fault of hydraulic operating mechanism of the breaker in a substation, oil pump failure is analyzed, and liquid level and gas sensors are used to design an intelligent exhaust device, which weakens the influence of the environment and improves the safety and reliability of the equipment. This device, with neural network self-learning height measurement algorithm as its core, processes the data obtained by barometric sensor and liquid-level sensor by detecting the breaker, to get an accurate oil-level height and reach real-time control over solenoid valve exhaust device as well as automatic record of system working condition, thus realizing the intelligent detection and automatic identification.

1 INTRODUCTION

Recently, the breaker hydraulic operating mechanism is finding wide application for its small size, large output power, reliability and maintainability (Li Haibo, 2014). High-voltage circuit breaker adopts a hydraulic operating mechanism, but in practical operation, the probability of hydraulic operating mechanism fault is relatively high, in which the frequent oil pump bulge fault accounts for a relatively large part. If breaker is used for an extended time, there will be some gas aggregated in the oil pump's low-pressure section of hydraulic system as well as inside of some high-pressure oil system (energy storage canister and high-pressure oil cylinder) (Zhou Xuan & Zhao Chenguang, 2011). The gases prevent the oil pump from effectively sending the hydraulic oil from the low-pressure section to the high-pressure section, thus resulting in shorter oil pump bulge interval time, and the oil pressure does rise with continuous running of the oil pump.

When the pressure of the oil pump drops below the monitoring working pressure P1 (32 MPa) or less, the pump is started and it pumps the hydraulic oil from the tank storage cylinder. When the pressure reaches P1 (32 MPa), the pump is stopped automatically after about 3 minutes. Under normal circumstances, each is weighed one time during 24–48 h (Li Jianming, 2011).

Pressure pump breaker failure occurs once every minute or tens of minutes depending on the severity. Each time it lasts for only about 10s. The hydraulic system over time gradually after stabilization suppresses longer intervals, and each time the interval is extended to 1 h to 10 hours gradually. Through the rapid build pressure relief method, the situation has changed (Li Zhiyong, 2010).

To overcome the breaker operating mechanism fault, intelligent exhaust device is designed in the present paper, in which liquid level and gas sensors are applied to detect the breaker, automatically identify the oil pump's gas content and liquid level, control breaker solenoid valve, to realize intelligent detection and automatic identification.

2 INTELLIGENT EXHAUST DEVICE OPERATING PRINCIPLE

When there is gas in the oil pump of the breaker operating mechanism, oil pump cannot effectively send hydraulic oil from the low-pressure section to the high-pressure section, thus resulting in shorter oil pump bulge interval time, and also the oil pressure does rise with the continuous running of the oil pump (Fan Shouxiang, 2000). To make breaker resume its normal operation, gas in the oil pump and high-pressure oil system need be exhausted in time.

A joint detection of liquid-level sensor and barometric sensor is employed in an intelligent exhaust device. When the liquid level sensor is only used, due to its multiple problems, such as mechanical fraction, short service life, narrow measurement range, it cannot measure oil-level height accurately. In practical measurements, oil temperature within the oil pump and its interior pressure are obtained by barometric sensor first. Then a neural algorithm is used to calculate the network space to be measured, so that the relative oil-level height at current measuring point is obtained indirectly. Finally comparing this measured value to the oil-level height obtained by liquid-level sensor, an accurate oil-level height is obtained. When accurate oil-level height is acquired, oil-level height is alternated to control the solenoid valve to the exhaust. The exhaust when oil level is under the lowest oil-level height set by system, and close to solenoid valve when it is above set oil level, shown as Figure 1.

The exhaust device is equipped with a working condition indicator, shown as Figure 2, a fault indicator and a buzzer alarm. If abnormal working condition or malfunction is detected by system device through a self-checking program, fault indicator will glow and buzzer will raise an alarm, and it will enter the system protection state. When system resumes the normal working condition, the alarm will disappear and the working indicator will return to normal. The exhaust device is also equipped with system data record and data transmission module to record real-time exhaust valve working condition, which helps to understand working characteristics of master device and benefits maintenance.

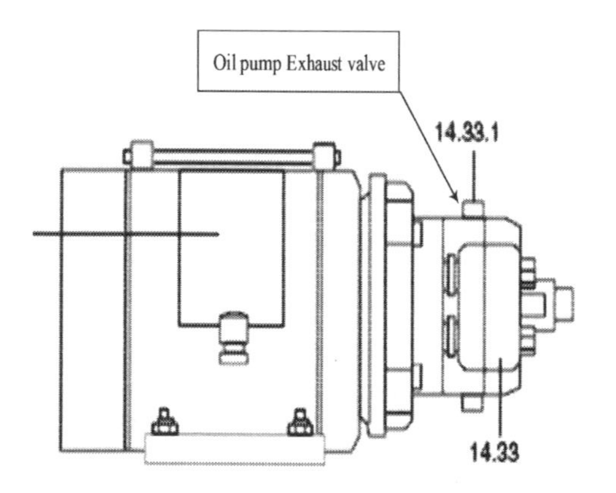

Figure 1. Vent valve of oil pump.

Figure 2. Intelligent exhaust device detection module.

3 INTELLIGENT EXHAUST DEVICE HARDWARE DESIGN

The measurement system process of intelligent exhaust device is as shown in Figure 3. The whole measurement system mainly includes a sensor module, analog signal modulation circuit, A/D switching circuit, single chip microcomputer processing unit with STM32F103 as the core and solenoid valve exhaust device.

3.1 *Sensor module*

The sensor module is composed of BMP085 digital barometric sensor and liquid-level sensor, shown in Figure 4. BMP085 is a digital barometric sensor with high precision and low power consumption, which adopts powerful ceramic leadless chip bearing ultrathin packaging, made up of resistance-type pressure sensor, AD convertor and control unit.

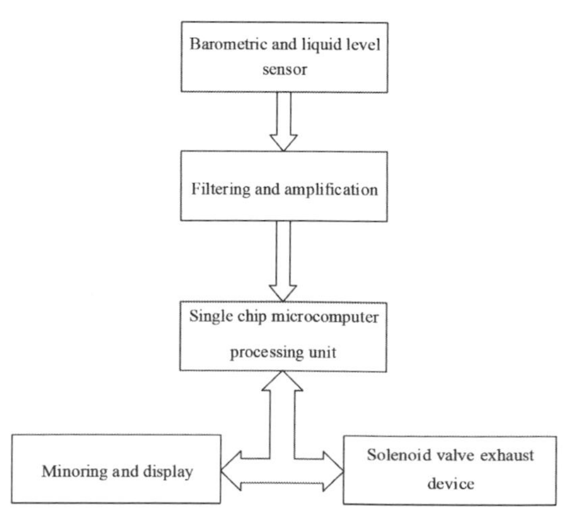

Figure 3. System flow chart.

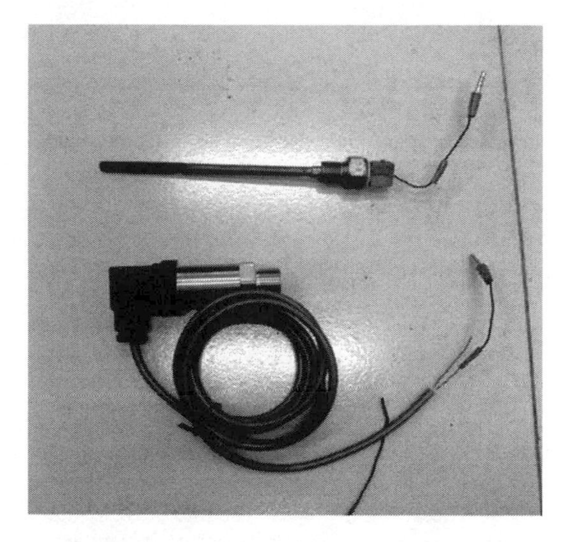

Figure 4. BMP085 digital barometric sensor and liquid level sensor.

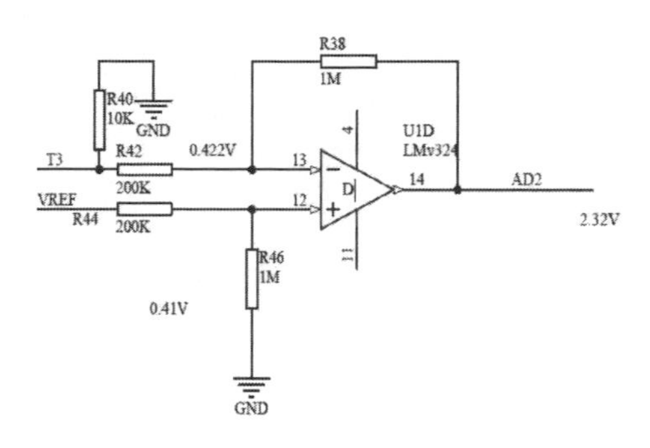

Figure 5. Analog signal conditioning circuit diagram.

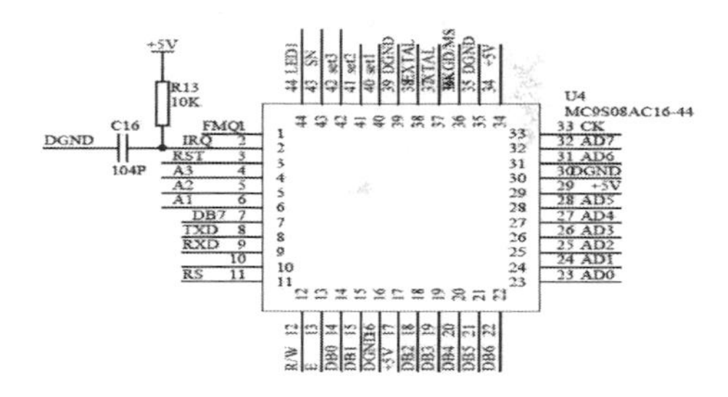

Figure 6. Microcontroller processing block diagram.

The internal temperature compensation in the BMP085 barometric sensor is always limited. When its measurement is combined with the liquid level sensor, it can reduce pressure height measuring error due to environmental factors such as temperature, reducing fault operation and improving measurement accuracy.

3.2 *Analog signal modulation circuit*

To guarantee that sensor output signal does not exceed voltage range of analog-digital conversion under a relatively high voltage grade, and is higher than the lowest resolution ratio of the analog-digital conversion, a proportional amplifying circuit is adopted in the design to adjust the input signal amplitude, to make its signal within the optimal range of AD conversion, shown in Figure 5. In consideration of that input signal lower than zero potential, a DC component is superpositioned in input signal in the design.

3.3 *Single-chip microcomputer processing unit*

A single-chip microcomputer processing unit, shown in Figure 6, takes neural network self-learning height measurement algorithm as its core, processes data obtained by barometric sensor and liquid level sensor, to get an accurate oil-level height and realize real-time control over the solenoid valve module. Single-chip microcomputer processing unit is equipped with circuit to automatically record the working condition while the exhaust valve can automatically conduct relevant operation according to received data.

4 IMPLEMENTATION PLAN

After powered on, the exhaust device begins to self-check, with working condition indicator and fault indicator lit as along with the one-second buzzer. When self-check is completed, it immediately enters monitoring. When the first entry is made into the working condition, as oil level is under the lowest oil level set by system, system-state indicator is lit on with solenoid valve opened and oil level raised. When oil level reaches the highest oil level set by the system, the system closes the solenoid valve with working green light flashing. When abnormal working condition or fault is detected by the system through the self-checking program, fault light will flash and buzzer will raise an alarm, and device will enter system the protection state. After the system resumes its normal working condition, the alarm will disappear with working green light flashing. Intelligent exhaust device working flow chart is shown in Figure 7.

The exhaust breaker intelligent detection device is shown in Figure 8. When the exhaust solenoid valve system data record and data transmission module are working, all kinds of system working states will be recorded, and system will also receive

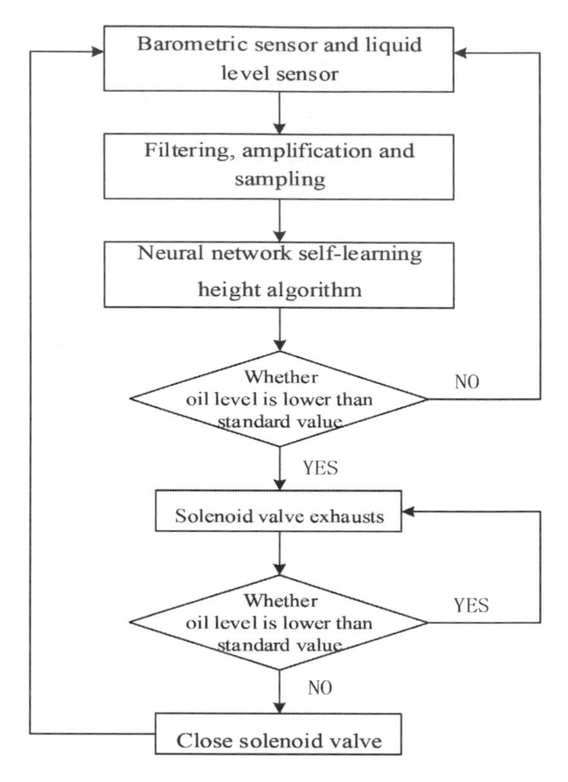

Figure 7. Intelligent exhaust device working flow chart.

Figure 8. Exhaust breaker intelligent detection device.

the command by handheld data recorder, and send the recorded files to it. The exhaust valve data transmission module and portable data recorder can send command and data through photoelectric coded signal, avoiding mutual interference between high-voltage electromagnetic field and other devices, which is secure and reliable. Data can be stored on the computer room with portable data recorder connected by a wired connection. Exhaust valve serial number responds to equipment serial number one-to-one. A comparison of exhaust valve and master equipment's working state helps to understand some working characteristics of the master equipment, which also benefits master equipment maintenance.

5 CONCLUSION

Intelligent exhaust device processes the data obtained by barometric sensor and liquid level sensor to get an accurate oil-level height, realizing real-time control over solenoid valve exhaust device as well as automatic record of system working condition and intelligent identification.

Combined measurement of gas and liquid level sensors is adopted in an intelligent exhaust device, which weakens environmental influence, avoids regular blind manual exhaustion, reduces the danger of the maintainer touching high-risk equipment during live working, improves equipment working security and reliability, and significantly improves working efficiency.

REFERENCES

Fan Shouxiang, Yang Maojun, "500kVSF_6 Breaker Hydraulic Mechanism Faults Analysis," East China Electric Power., vol. 2, 2000.

Li Haibo, Yu Jun, "The Impact of Air Content in Hydraulic System on Hydraulic Mechanism," High-Voltage Apparatus., vol. 1, 2014.

Li Jianming, Wang Zhiyong, Wang Shuchen, "3AQ/3AT Breaker Hydraulic Mechanism Oil Pump Frequent Bulge Fault Analysis and Treatment," High-Voltage Apparatus., vol. 5, 2011.

Li Zhiyong, "High-Voltage Oil-Minimum Breaker Hydraulic Mechanism Faults Diagnosis and Treatment," Chinese High-Tech Enterprises., vol. 25, 2010.

Zhao Chenguang, Kang Guojun, "Breaker Hydraulic Mechanism Common Faults Analysis and Treatment," Electric Appliance Industry., vol.9, 2011.

Zhou Xuan, "High-Voltage Circuit Breaker Hydraulic Operating Mechanism Common Faults Analysis," Electrotechnical Application., vol. 19, 2011.

Civil, Architecture and Environmental Engineering – Kao & Sung (Eds)
© 2017 Taylor & Francis Group, ISBN 978-1-138-02985-9

An improved calibration method for a nondestructive testing system based on infrared thermopile sensors

Wei He, Tong Zhou, Bo Jiang, Yin Wan & Yan Su
School of Mechanical Engineering, Nanjing University of Science and Technology, Nanjing, China

ABSTRACT: This paper presents an improved calibration method for a nondestructive testing system based on infrared thermopile sensors. The method uses reference source radiation to get the calibration parameters. A new two-point with multi-step calibration algorithm serving as the calibration of infrared thermopile sensors was proposed after researching the two-point calibration algorithm used for the calibration of infrared detector array pixels. The experiment verified that this method has advantages of simple operation, is easy to implement and has good calibration effect when compared with the commonly used two-point calibration algorithms.

1 INTRODUCTION

The infrared nondestructive testing technology is currently one of the hot research topics in the field of nondestructive testing, which has broad development prospects in the industry and national defense field. The principle of infrared nondestructive testing is based on thermal radiation characteristics of the object. Any object above the absolute zero kelvin would launch the infrared radiation outward. When heating the surface of the tested sample, the heat will flow inside the material and if there is a defect inside the sample, then the heat will accumulate in the defect, which will lead to the difference of heat distribution between the surface corresponding to defects and the surface corresponding to other no-defects area. Finally, we can detect the defects and their location by using infrared sensors to perceive the difference.

The detection module of nondestructive testing system based on infrared thermopile sensors is composed of a number of thermopile sensors and different sensors tend to have difference at the offset and gain coefficients, which will influence the outcome of the nondestructive testing. So a nondestructive testing system needs to be calibrated before it is used. There are a variety of calibration methods of sensors such as hardware circuit compensation and linear interpolation compensation. However, the hardware circuit compensation will enlarge the size of the system and increase the factors affecting the effects of the calibration. The linear interpolation compensation not only increases the complexity of the calibration but also has the shortcoming of bad calibration when there is a large difference between the two adjacent sensors. Generally speaking, for the calibration of the sensors, software processing is relatively simple and can accurately correct the nonuniformity error of the system. In this paper, a new two-point with multi-step calibration algorithm used for the calibration of infrared thermopile sensors was proposed based on the two-point calibration algorithm that was used for the calibration of infrared detector array pixels. This method uses reference source radiation to get the calibration parameters. Then the parameter was used to calibrate the thermopile sensors with two-point calibration algorithm and improved two-point with multi-step calibration algorithm. It was demonstrated through experiments that this method is easy to operate and has prior calibration effect than the commonly used two-point calibration algorithms.

2 DESIGN OF AN INFRARED NONDESTRUCTIVE TESTING SYSTEM BASED ON INFRARED THERMOPILE SENSORS

The infrared nondestructive testing system based on infrared thermopile sensors is mainly composed of pulse excitation source, linearity array infrared sensors group, signal conditioning and acquisition module, data output and display module and calibration module for sensors, etc. The system block diagram is shown in Figure 1.

The linear array infrared sensing module is composed of six TPS334L55 infrared thermopile sensors and its working principle connects several thermocouples together.

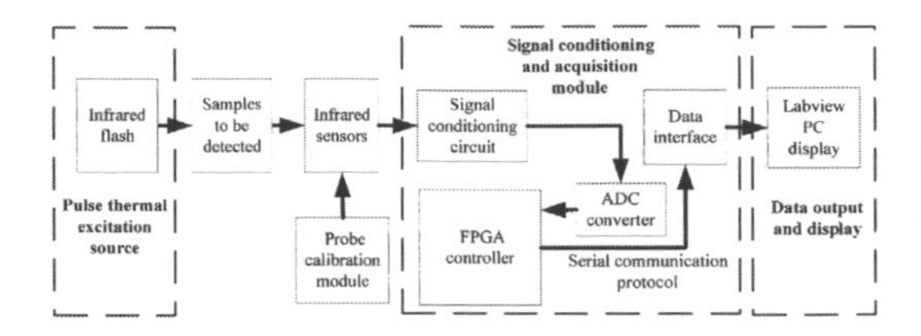

Figure 1. Block diagram of the infrared nondestructive testing system.

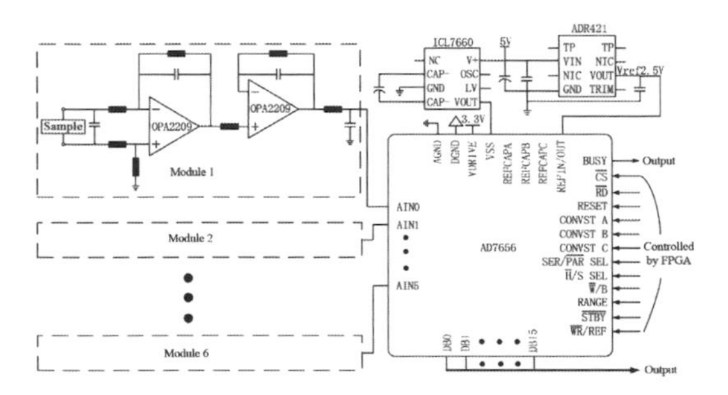

Figure 2. Circuit diagram of signal conditioning and acquisition module.

The signal conditioning and acquisition module is mainly responsible for enlarging and filtering the signal voltage transformed by thermopile sensors and then the signal is input into the AD converter modulus. The circuit diagram of this module is shown in figure 2.

The signal was imported by a high common mode rejection ratio differential amplifier circuit and it was then changed to digital signals by a high-speed sampling ADC AD7656 after being enlarged by a single-ended operation amplifier and being filtered by a passive low-pass filter. The system uses FPGA that contains rich IO interface as the main controller. The sampling data was transferred through serial port to the Labview for display and preservation.

3 SELF-CALIBRATION REFERENCE SOURCE EXPERIMENT

The reference source experiment is designed to obtain infrared response data of each sensor, which provides data support for selecting the calibration algorithm. The experiments used temperature-controlled plane array blackbody to radiate the linear array infrared thermopile sensor group uniformly. The physical and schematic diagrams of the system are shown in Figure 3.

The self-calibration experiment system is composed of temperature-controlled plane array

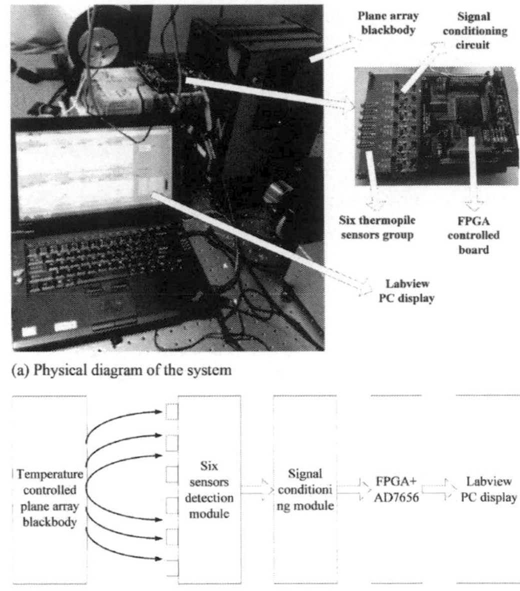

Figure 3. Physical diagram and schematic diagram of the system.

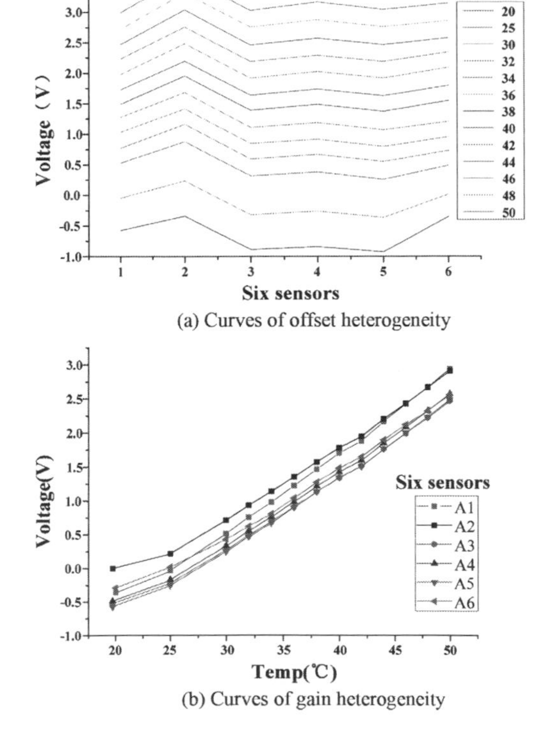

Figure 4. Curves of offset heterogeneity and gain heterogeneity.

blackbody, six-sensor probe group, signal conditioning module, data acquisition and control module and PC labview data display and save module. First, the six-sensor probe group was placed to the center position of plane array blackbody, then the array blackbody is opened and the temperature set in the working state. The system is powered and labview opened to save the data when the temperature of the blackbody tended to be stable and sampling data is saved every 30 s. The working temperature of blackbody is kept from 20°C to 50°C and we tested the output of six sensors under 20°C, 25°C and from 30°C to 50°C at a 2°C interval.

According to the tested data, the curves of offset heterogeneity and the curves of gain coefficient heterogeneity is shown in Figure 4.

As shown from the above figure, the linear array infrared thermopile sensor group has the large offset heterogeneity and gain coefficient heterogeneity. Therefore, the thermopile sensor group needs to be calibrated in order to ensure the accuracy of detection.

4 TWO-POINT CALIBRATION ALGORITHM

4.1 *Theory of two-point calibration algorithm*

The two-point calibration algorithm is a kind of calibration method that selects two irradiance as punctuation and it takes the output of the sensors' response and incident irradiance as a linear relationship. First, irradiance ψ_1 and ψ_2 are selected as the punctuation within the dynamic response scope of the sensor. Then the average response output $V(\psi_1)$ and $V(\psi_2)$ under ψ_1 and ψ_2 of all N sensors can be obtained according to equation 1:

$$V(\psi)=\sum_{i=1}^{N} V_i(\psi)/N \tag{1}$$

Provided that any sensor's gain heterogeneity calibration coefficient is m_i and its offset heterogeneity calibration coefficient is n_i, then equation 2 can be obtained:

$$\begin{cases} V(\psi_1)=\mathrm{m}_i V_i(\psi_1)+n_i \\ V(\psi_2)=\mathrm{m}_i V_i(\psi_2)+n_i \end{cases} (i=1,2,3\ldots N) \tag{2}$$

$V_i(\psi_1)$ and $V_i(\psi_2)$ are the actual output under the calibration irradiance ψ_1 and ψ_2. $V(\psi_1)$ and $V(\psi_2)$ means the calibration output under the calibration irradiance ψ_1 and ψ_2. According to equation 2, there is:

$$\begin{cases} \mathrm{m}_i=\dfrac{V(\psi_2)-V(\psi_1)}{V_i(\psi_2)-V_i(\psi_1)} \ (i=1,2,3\ldots N) \\ n_i=V(\psi_1)-\mathrm{m}_i V_i(\psi_1) \end{cases} \tag{3}$$

The offset heterogeneity calibration factor and gain coefficient heterogeneity calibration factor of each sensor can be calculated according to equation 3. Then the calibration response output of any sensor under any irradiance can be represented as the following equation:

$$V_i^{"}(\psi)=\mathrm{m}_i V_i(\psi)+n_i (i=1,2,3\ldots N) \tag{4}$$

4.2 *Calibration of infrared thermopile sensors based on the two-point calibration algorithm*

According to the two-point calibration algorithm, we selected 30°C and 44°C as the fixed point for irradiance. According to equation 1, The average response voltage under 30°C and 44°C is 0.48469V and 2.29675V respectively. The calibration factor of six sensors can be calculated according equation 3. Table 1 shows the gain coefficient nonlinear calibration factor and offset coefficient nonlinear calibration factor of six sensors.

The offset nonlinear calibration parameters and gain nonlinear calibration parameters are loaded from above table in each sensor channel and the same temperature points' response voltage again under the uniform radiation of plane array blackbody is sampled. Then the offset nonlinear curve and gain nonlinear curve can be obtained as shown in Figure 5.

It can be seen from the picture above that the two-point calibration method improved the offset nonlinearity and gain nonlinearity of infrared thermopile sensor group greatly in a certain temperature range. However, in the range of 20°C to 30°C, the offset nonlinearity and gain nonlinearity has not improved, which fit the theory that two-point calibration algorithm is only suitable for a small temperature range.

As we all know that the larger the temperature difference is, the easier the defect can be detected.

Table 1. Calibration parameters of six sensors within 20°C to 50°C.

Sensors No.	m_i	n_i
A1	1.09796	–0.07975
A2	0.98583	–0.37031
A3	0.98695	0.16826
A4	0.96735	0.10177
A5	0.97105	0.20052
A6	1.00195	–0.03405

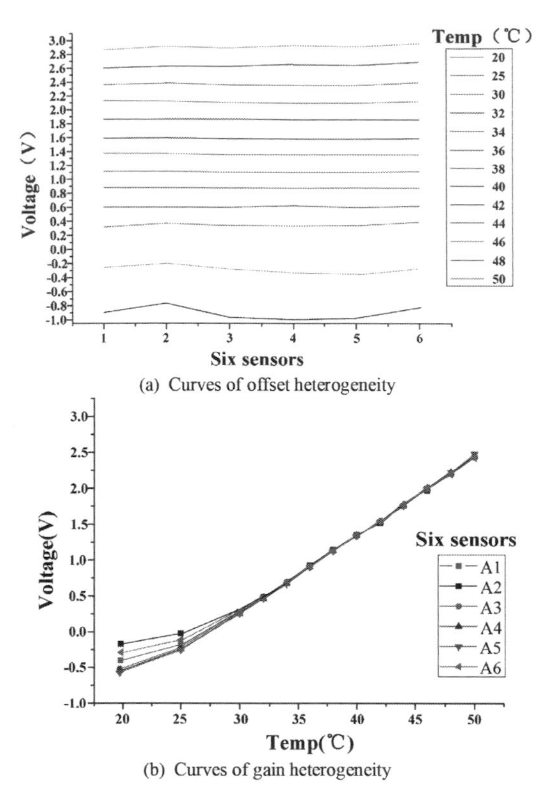

Figure 5. Curves of offset heterogeneity and gain heterogeneity after two-point calibration.

So the infrared nondestructive testing system needs to adapt to the detection in wide dynamic temperature range and in this case, the two-point calibration method of infrared thermopile sensor group can't adapt to the work environment of the system. In order to meet the calibration in a wide temperature range, the infrared thermopile sensor group needs a new self-calibration algorithm for calibration.

5 AN IMPROVED TWO-POINT WITH MULTI-STEP CALIBRATION ALGORITHM

5.1 *Theory of two-point with multi-step calibration algorithm*

The two-point calibration algorithm cannot calibrate the infrared thermopile sensors in a wide dynamic temperature range and multipoint interpolation calibration algorithm has the shortcoming of large amount of computation. By referring to the thermal response curve of the thermopile infrared sensors and the idea of multi-point approximation, an improved two-point with multi-step calibration algorithm was proposed based on two-point calibration algorithm.

The improved two-point with multi-step calibration algorithm divides the dynamic temperature range of the system into a number of intervals, which can be determined according to the practical test response output curves of the sensors. Then the two-point calibration method is used for the calibration again in each interval. As the infrared non-destructive testing system only needs to detect the relative infrared response between surface with defects and surface with no defects, it can minimize the system error as long as it can ensure that each sensor's gain coefficient and offset coefficient changes are synchronized.

The theory of two-point with multi-step calibration is as follows. First, select M+1 irradiation punctuation $\psi_i (i = 0,1,2,3\ldots m)$, under which the radiation for all N sensors is calibrated respectively, and then the average response $V(\psi_i)$ under corresponding irradiance according to equation 1 is calculated. All irradiance is divided into M section (ψ_{j-1}, ψ_j) (j = 1,2,3...M) based on the level of irradiance. Then, the sensors are calibrated according to the two-point calibration method in each interval and the gain heterogeneity calibration coefficient $m_{j,i}$ and the offset heterogeneity calibration coefficient $n_{j,i}$ are calculated. As equation 5 shows.

$$\begin{cases} m_{j,i} = \dfrac{V(\psi_i - 1) - V(\psi_i)}{V_i(\psi_i - 1) - V(\psi_i)} \\ \qquad (i = 0,1,2,3\ldots m), (j = 1,2,3\ldots M) \\ n_{j,i} = V(\psi_i) - \mathrm{m}_{j,i} V_j(\psi_i) \end{cases} \tag{5}$$

Therefore, in any irradiance, the sensors calibration method can be summarized as follows. First, determine which subinterval the sensors' infrared response belongs to, namely:

$$V_j(\psi_i - 1) < V_j(\psi) \leq V_j(\psi_i) \tag{6}$$

Then, the calibration output $V_j''(\psi)$ can be corrected as follows after judging the irradiance ψ_i according to the interval:

$$V_j''(\psi) = m_{j,i} V_j(\psi_i) + b_{j,i} \tag{7}$$

$V_j(\psi_i)$ is the actual response output, $m_{j,i}$ is the gain calibration parameter and $b_{j,i}$ is the offset calibration parameter.

It can be seen from the derivation process above that the two-point with multi-step calibration algorithm generally includes two steps including judge interval and two-point calibration algorithm is employed at every interval. Compared with multipoint interpolation correction algorithm, the two-point with multi-step calibration algorithm reduced the amount of calculation greatly and the defect of narrow calibration temperature that two-point calibration algorithm has been overcome. So it is a simple and effective calibration method.

5.2 *Calibration of infrared thermopile sensors based on two-point with multi-step calibration algorithm*

It can be seen from figure 4 that the infrared thermopile sensors has good offset linearity and gain coefficient linearity after being calibrated by two-point calibration method at the temperature range from 30°C to 50°C. However, the calibration effect is not ideal at the temperature range from 20°C to 30°C, so the irradiance range is supposed to be divided into two intervals of 20°C to 30°C and 30°C to 50°C. The nonlinear calibration coefficient within 30°C to 50°C still uses the result in Section 3.2 and the nonlinear calibration within 20°C to 30°Cis described as follows:

First of all, select 20°C and 30°C as irradiance point. According to equation 1, the average response at corresponding irradiance is –0.36079V and 0.48469V. Then according equation 3, the offset nonlinearity calibration factor and gain nonlinearity calibration factor can be calculated as follows.

Loading the calibration parameters from 20°C to 30°C and 30°C to 50°C respectively into the infrared sensors response curves of six sensors under radiation of plane array blackbody are obtained. The offset nonlinear curves and gain nonlinear curves can be obtained as shown in Figure 6.

It can be seen from the above figure that the two-point with multi-step calibration algorithm has good correction effect, which meets the design requirements of the system. The two-point with multi-step calibration algorithm can not only solve the problem that the sensor group cannot be calibrated in the whole working temperature range by two-point calibration algorithm but also improved the offset linearity and gain linearity of six sensors. In addition, this algorithm has advantages of simple operation, easy implementation and involves a small amount of calculation. Its application in the calibration of linear array infrared thermopile sensors laid a solid foundation to the normal working of the infrared nondestructive test system.

Table 2. Calibration parameters of six sensors within 20°C to 30°C.

Sensors No.	m_i	n_i
A1	1.07483	–0.06785
A2	0.87621	–0.27521
A3	1.01094	0.16057
A4	0.94068	0.11233
A5	1.03936	0.18053
A6	1.09409	–0.08175

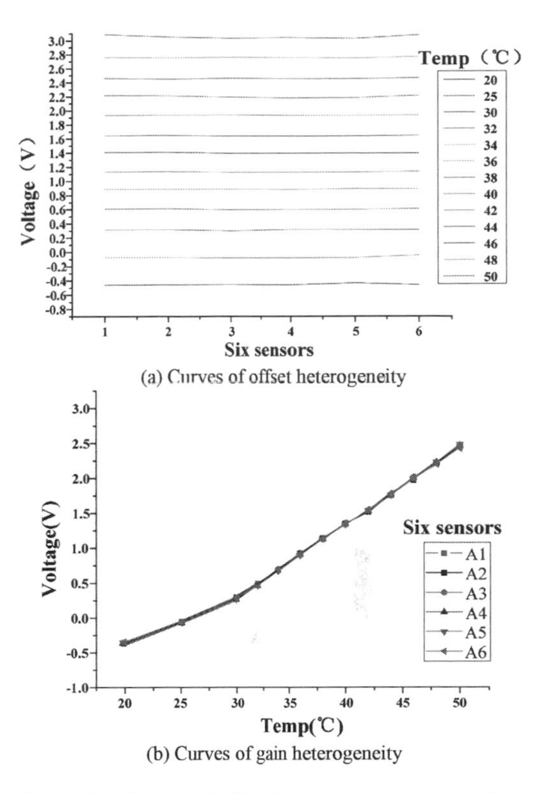

(a) Curves of offset heterogeneity

(b) Curves of gain heterogeneity

Figure 6. Curves of offset heterogeneity and gain heterogeneity after two-point with multi-step calibration.

6 CONCLUSION

This article introduced a calibration algorithm of an infrared nondestructive test system based on thermopile sensors. This method obtains calibration parameters by reference source radiation method and it is improved based on the two-point calibration method, which is commonly used in the calibration of infrared sensor array pixels. Finally, compared with experiment, the method demonstrated that the improved two-point with multi-step calibration algorithm has advantages of simple operation, easy implementation and has a good calibration effect.

REFERENCES

Blanco M, Cueva-Mestanza R, Peguero A. NIR analysis of pharmaceutical samples without reference data: improving the calibration [J]. Talanta, 2011, 85(4).

Chen Xiangbo. Research on Infrared sensors' temperature drift and nonlinear compensation technology [J]. Journal of Beijing industrial vocational and technical college, 2007, 6(1):1–8.

Ferraro Simona, Marano Giuseppe, Ciardi Laura et al. Impact of calibration fitting models on the clinical value of chromogranin A [J]. Clinical Chemistry and Laboratory Medicine, 2009, 47(10).

Li Xiangdi, Huang Ying, et al. Infrared imaging system and its application [J]. Laser and Infrared, 2014, 44(3):229–234.

Nie Yanwei. The development and research of infrared array thermopile platform [D]. Taiyuan: North University of China.

Scriloner D A, et al. Physical limitations to non-uniformity correction in IR focal plane arrays[C]// Technology and Applications, SPIE, 1987, 865: 185–202.

Torres, S N, J E Pezoa, M M Hayat. Scene-based non-uniformity correction for focal plane arrays by the method of the inverse covariance form [J]. Applied Optics, 2003, 42(29):5872–5881.

Zhang Jun, Qian Weixian, Liu Zewei. Analysis and correction to four quadrant detector's output heterogeneity [J]. Infrared technology, 2016, (7):565–570.

Zhao Min, Chen Xiaoping, et al. Implementation of sensor's real-time self-calibration/self-compensation [J]. Chinese Journal of Scientific Instrument, 1999, 20(4):432–434.

Zhou T, Dong T, Su Y, et al. High-precision and low-cost wireless 16-channel measurement system for multi-layer thin film characterization [J]. Measurement, 2013, 46(9):360–361.

Civil, Architecture and Environmental Engineering – Kao & Sung (Eds)
© 2017 Taylor & Francis Group, ISBN 978-1-138-02985-9

Evaluation of wind turbine power generation performance based on a multiple distribution model

Shuangquan Guo, Jiarong Yang & Hui Li
Central Academe, Shanghai Electric Group Co. Ltd., China

ABSTRACT: The evaluation of the performance of wind turbine power generation generally focuses on the wind power curve. The power value of wind turbine fluctuates within a reasonable range and follows a different distribution at different speeds caused by wind speed variation, anemometer accuracy, operation control and so on. One evaluation method of wind turbine power generation performance based on a multiple distribution model, which is validated on the history data of wind turbine, is proposed to calculate the performance of the power generation and power loss by data preprocessing steps, training the multi-distribution model and performance evaluation. The result reveals the trend of turbine's power generation performance well and several typical faults are detected before they happen.

1 INTRODUCTION

With the increasing of wind power's installed capacity in recent years, China has become the world's largest country in this field. According to the wind power development plan in the "13th Five-Year" period formulated by the National Energy Administration, the installed capacity of wind power in 2020 will reach to 200 million kilowatts, which is almost twice as much as the current installed capacity, and the market prospects and scale are very large (Xiao Q, 2014). The increase in single turbine capacity and wind farm turning gradually from land to sea are the two major trends in wind power field in the future (Guo S.Q. 2016). At the same time, the maintenance costs are increasing, according to the literature statistics. The maintenance cost of onshore wind turbine accounts for 10% of the total cost, while the maintenance cost of offshore wind turbine is up to 30% (M.I. Blanco, 2009). Therefore, research on the assessment of wind turbine generation performance is of great significance to reduce costs by optimizing operation and maintenance.

At present, the evaluation of the performance of wind turbine power generation mainly focuses on the wind power curve, obtained by general methods include spline curve fitting, self-organizing map, artificial neural network and Gaussian mixture model and so on. However, these methods have some limitations on describing the distribution characteristics of wind power points. In the actual operation process, the fluctuation of wind speed and anemometer accuracy often lead changing the power curve (Chu Zheng, 2014). Meanwhile, each turbine has different control strategy in different conditions, so the wind-power points are always scattered around the wind-power curve. When the turbine is in normal state, its power fluctuates in a reasonable range and satisfies a certain distribution in a certain wind speed interval. In this paper, a model for the performance of wind turbine power generation based on a multiple distribution model is proposed to evaluate its generation performance and calculate the power loss.

2 WIND-POWER MULTIPLE DISTRIBUTION MODEL

As shown in Figure 1, the wind-power points scattered around the power curve, usually wind turbine has its own design parameters, such as cut-in wind speed v_{In}, cut-out wind speed v_{Out} and rated wind speed v_N. Since the turbine has different control strategy at different speeds, and the power satisfies different distributions, here we partition the whole wind speed range by minimal interval Δv (like 0.2m/s), which is much less than the value of $v_{Out} - v_{In}$, to obtain different operation conditions. We take the average speed of each minimal interval, v_i, as the speed of all points in the interval. Then several common distributions (beta, exponential, gamma, generalized extreme value, logistic, normal, Rayleigh, Weibull, Poisson and so on) are applied to fit the power distribution in each minimal interval of wind speed by maximum likelihood estimation, and the best distribution is selected by the principles such as negative of the log likelihood (NLogL), Bayesian Information Criterion (BIC),

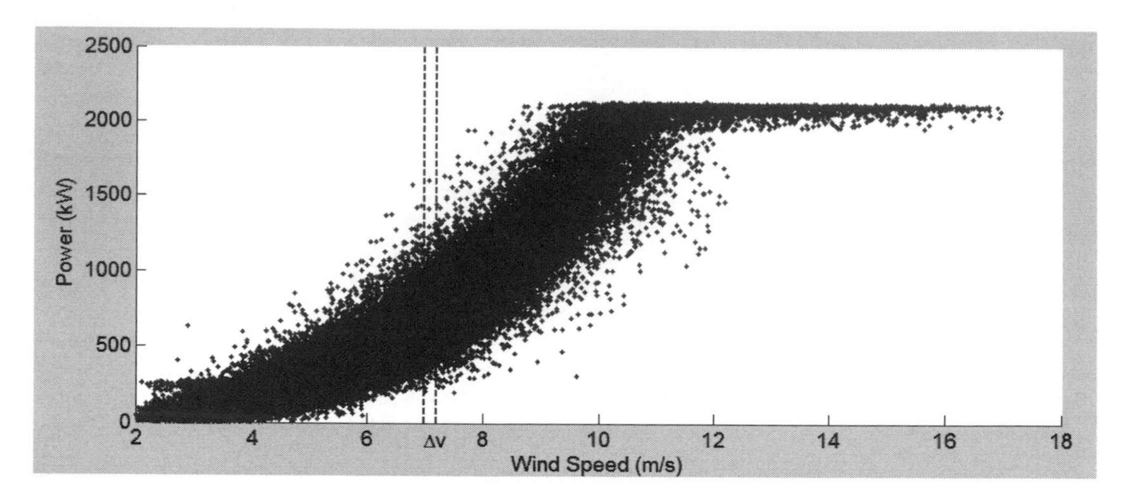

Figure 1. The wind-power scatter diagram, and the minimal interval Δv (like 0.2 m/s), is much less than the value of $V_{Out} - V_{In}$.

or Akaike Information Criterion (AIC) (Liao L, 2010), shown in Equation 1.

$$f(P \mid v = v_i) = f_{\text{best}}(P) \text{ with } \min(NLogL, BIC \text{ or } AIC) \quad best \in (1, 2, 3, \dots N_d) \tag{1}$$

The sign N_d is the number of common distributions.

For a certain period, if the turbine has worked in wide wind speed ranging from the cut-in speed to the cut-out speed, we can build its wind-power multiple distribution model during this period in Equation 2.

$$f(P \mid v) = \sum_{i=1}^{N_v} \omega_i f(P \mid v = v_i) \tag{2}$$

The sign N_v is the number of partitions of the whole range, ω_i is the weight of the power point number in the speed interval of $v = v_i$ accounted for the total number of power points.

3 POWER GENERATION PERFORMANCE MODEL

To evaluate the performance of the power generation of wind turbine in a certain period (like two-day period), we should build the wind-power multiple distribution models in the normal-power period and current period respectively, and quantify the difference between the two models as the index of power generation performance of wind turbine. If $f_B(P \mid v)$ represents the multiple distribution model of wind turbine in normal or ideal period, namely baseline model, and $f_C(P \mid v)$ represents the distribution model in current period, namely current model, The power generation performance model of wind turbine in this period, which is given by Equation 3, can be built through evaluating the weighted average of overlapping areas of corresponding distributions between the baseline model and the current model.

$$PGP = Difference\left(f_B(P \mid v), f_C(P \mid v)\right) = \sum_{i=1}^{N_v} \omega_{i,B} S\left(f_B(P \mid v = v_i), f_C(P \mid v = v_i)\right) \tag{3}$$

In Equation 3, $\omega_{i,B}$ is the weight of each wind speed interval in the baseline model, $S(f_B, f_C)$ is the overlapping area between baseline distribution and current distribution where $v = v_i$. As shown in Figure 2, in the minimal interval corresponding $v = v_i$, the average and variance of baseline distribution $f_B(P \mid v = v_i)$ and current distribution $f_C(P \mid v = v_i)$ are (μ_B, σ_B) and (μ_C, σ_C) respectively. The overlapping area of two distributions is expressed by Equation 4.

$$S\left(f_B(P \mid v = v_i), f_C(P \mid v = v_i)\right) = \frac{\int_{\min(\mu_B - \sigma_B, u_C - \sigma_C)}^{\max(\mu_B + \sigma_B, u_C + \sigma_C)} \min\left(f_B(P \mid v = v_i), f_C(P \mid v = v_i)\right) dP}{\int_{\min(\mu_B - \sigma_B, u_C - \sigma_C)}^{\max(\mu_B + \sigma_B, u_C + \sigma_C)} f_C(P \mid v = v_i) dP} \tag{4}$$

For a more intuitive understanding of power generation performance, we could also evaluate

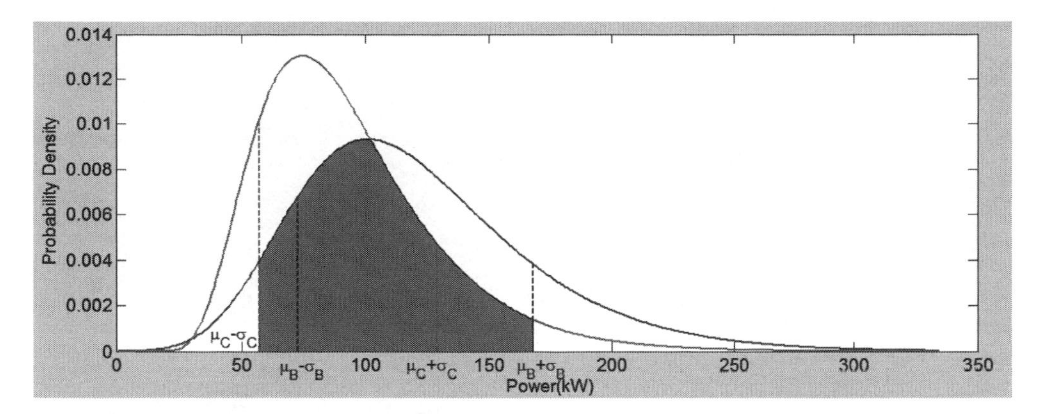

Figure 2. Schematic of overlapping area between two distributions, and the averages and variances are (μ_B, σ_B) and (μ_C, σ_C).

the expectation of power generation based on the baseline model in the current wind condition, and calculate the difference with the actual power generation, namely power loss, shown in Equation 5.

$$P_{loss} = \sum_{i=1}^{N_v} \omega_{i,C} E_B \left(P \mid v = v_i\right) * N_C * \Delta t - \sum_{i=1}^{N_C} P_i * \Delta t \quad (5)$$

The sign $\omega_{i,C}$ is the weight of each wind speed interval in current period, $E_B\left(P \mid v = v_i\right)$ is the expected power at each speed, N_C is the total sample number of this period, and Δt is the sampling interval of power points.

In addition, when partitioning the sample points during the selected period by wind speed intervals, if there are too few points to fit the distributions in some intervals, it can be considered that the distribution characteristic remains unchanged, and take the same distribution parameters of last period, then combine these samples with the samples of next period to evaluate the distribution.

4 APPLICATION EXAMPLE

4.1 *Data preprocessing*

To verify the practicability of the proposed method, we choose the history sample data of one wind turbine, including the parameters such as wind speed, wind direction, power, pitch angle and wind turbine state. Before training the model, we should filter the data by the principles as follows:

1. Remove the points where power is less than 0.
2. Remove the points where wind speed is out of the range between cut-in speed and cut-out speed.
3. Remove the pitching points where the power is less than the rated power.
4. Remove the points in non-generating state.
5. Filter out the outliers by the principle of Grubb's test.

4.2 *Training the baseline model*

The data for training the baseline model should be selected from the appropriate period when the turbine works in a normal state without major maintenance or failure records. Here the cut-in speed and cut-out speed of turbine are 3.5 m/s and 17 m/s, and the rated speed is 14 m/s. If we take 0.2 m/s as the minimal interval to partition the whole range, put all the data below the speed 3.5 m/s in the first interval and put all the data beyond the speed 14 m/s in the last interval. Finally, the whole sample would be partitioned to 54 intervals corresponding to the representative speeds (3.5, 3.6, 3.8, …, 13.6, 13.8, 14.0). Through training, the distribution of each interval can be fitted by Equation 1. Several typical distributions were shown in Figure 3. The fitted distributions at speeds of, 7, 9.6, and 12 m/s are generalized extreme value distribution, log-logistic distribution, normal distribution and extreme value distribution, respectively.

4.3 *Power generation performance and power loss*

After training the baseline model, we can evaluate the turbine's power generation performance and power loss in different periods. Through partitioning the half-year testing data into different periods every two days and filtering data of each period by the principles in Section 4.1, finally we could obtain the wind turbine's Power Generation Performance (PGP) curve and power loss curve shown in Figures 4 and 5 from Equation 3–5.

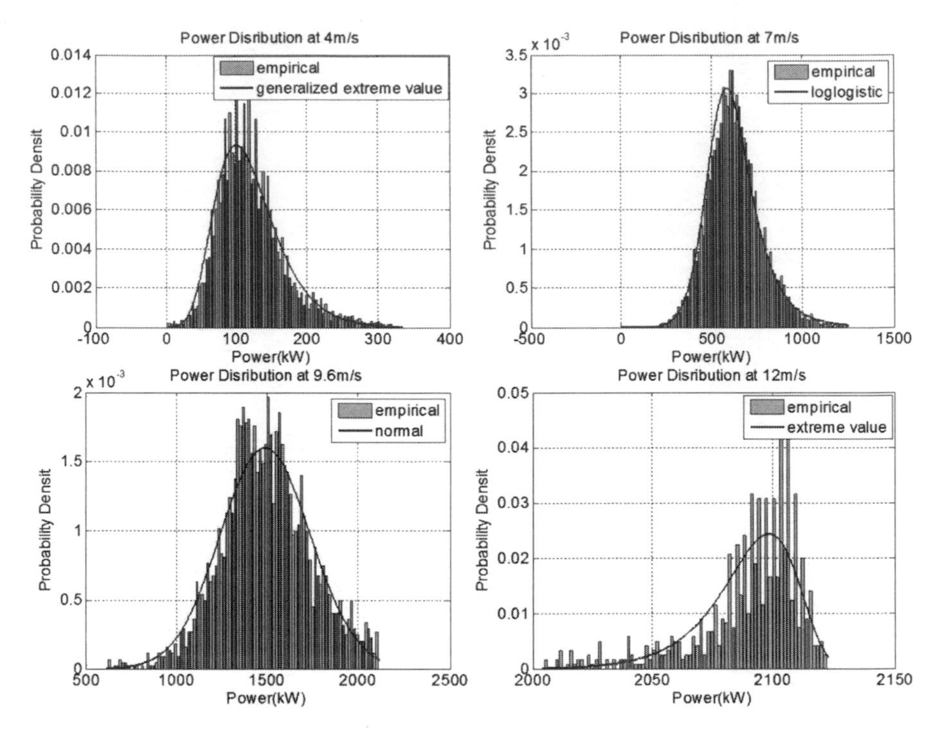

Figure 3. Several typical distributions at different wind speed intervals such as 4 m/s, 7 m/s, 9.6 m/s, and 12 m/s.

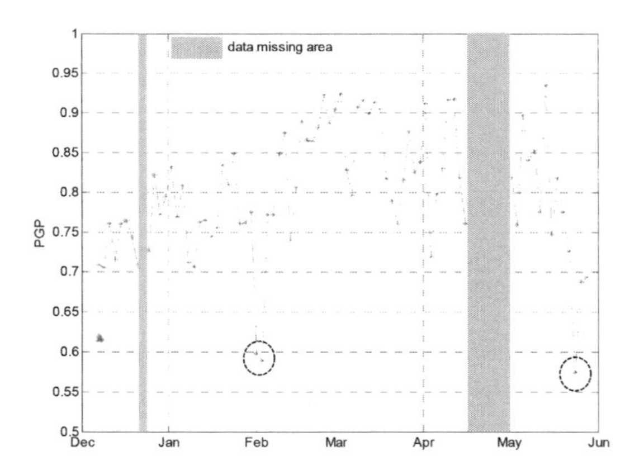

Figure 4. Wind turbine's power generation performance curve.

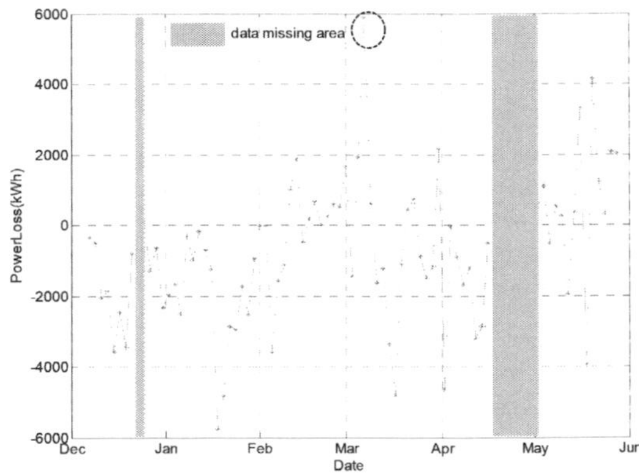

Figure 5. Wind turbine's power loss curve.

From the PGP curve in Figure 4, we find that the turbine works mostly in the condition of PGP > 0.7, but the value drops rapidly in early February and late May. The real operation records show that there were respectively an anemometer looseness fault on February 7 and ash cleaning maintenance operation in late May, and PGP value's significant degradations happen earlier than the actual records. Moreover, in Figure 5, we can find that the wind turbine's power loss are mostly negative during 12 months, namely the late winter and early spring, which is related to good wind condition in this time or the training data selection for the baseline model. At last, a maximum of the power loss value appeared at early March before the pitch lubrication fault alert happened.

5 CONCLUSIONS

In this paper, one evaluation method of performance of wind turbine power generation based on a multiple distribution model has been proposed and the whole process including data preprocessing, baseline training and performance evaluation has been implemented well on the wind turbine's history data. Results show that the PGP value and power loss value are good characterizations of the generation status of wind turbine and can identify several typical faults before the records happen.

The method can also be tentatively applied to the component-level performance evaluation of wind turbine, such as drive train, anemometer, and pitch system. However, there are still some aspects need to improve in the future: (1) take consideration of season factor in the model training and evaluation; (2) partition the wind range by dynamic interval instead of the fixed minimal interval.

REFERENCES

Blanco M.I. 2009. The economics of wind energy. *Renewable and Sustainable Energy Reviews*, 2009(13):1372–1382.

Chu Zheng & Li Nan. 2014. Air density self-adapting wind turbine optimal torque control. *Journal of Shanghai Electric Technology*, 2014(2):41–44.

Guo S.Q. 2016. Method study on health performance evaluation of wind turbine based on gray correlation. *Magazine on Equipment Machinery*:7–11.

Lapira E, Brisset D, Davari H, Siegel D. & Lee J, Wind turbine performance assessment using multi-regime modeling approach. International Journal of Renewable Energy, vol. 45, pp. 86–95, 2012.

Liao L. An adaptive modeling for robust prognostics on a reconfigurable platform. Department of Mechanical Engineering. Cincinnati: University of Cincinnati; 2010.

Xiao Q. 2014. The early sight of the wind power development framework in "13th Five-Year". *Energy Research & Utilization*, 2014(6):4–5.

Civil, Architecture and Environmental Engineering – Kao & Sung (Eds)
© 2017 Taylor & Francis Group, ISBN 978-1-138-02985-9

Research on the acoustic emission testing of the tank truck and on the defect acoustic emission source location method

Z.W. Ling, M.L. Zheng, D. Bin & X.L. Guo
Zhejiang Provincial Special Equipment Inspection and Research Institute, Hangzhou, China

ABSTRACT: The tank truck is an important type of transportable pressure vessel, which transports poisonous and harmful media. Due to erosion by the natural environment and poisonous and harmful media, corrosion perforation, crack propagation, and fracture of the tank are caused, which can lead to failure of the tank structure resulting in huge economic losses and serious environmental and ecological pollution. In this paper, acoustic emission testing technology carried out to test the overall structure of the tank truck and the location of defect acoustic emission signals were researched. Based on the comparison of several common defect acoustic emission source location methods, a defect acoustic emission source location method combining arbitrary planar triangle location and local scope linear location was proposed. Then, the non-open online detection of the tank truck was realized.

1 INTRODUCTION

Tank truck is an important type of transportable pressure vessel, which transports poisonous and harmful media. Due to erosion by poisonous and harmful media, corrosion perforation, crack propagation and fracture of the tank are caused, which can lead to failure of the tank structure resulting in huge economic losses and serious environmental and ecological pollution.

Currently, the main testing methods for tank truck and other transportable pressure vessels are conventional testing methods such as visual inspection, wall thickness measurement, and surface nondestructive testing and ultrasonic testing if necessary (TSG R0005-2011 2011). However, the conventional testing needs replacement, cleaning, and other pretreatments. If the tank truck contained inflammable, explosive, toxic and harmful media, the replacement, cleaning, and waste liquid discharge is put forward as higher safety and environmental protection requirements. When compared with conventional non-destructive testing, Acoustic Emission (AE) testing can avoid the pre-processing stage of the tank truck. At the same time, without replacement, cleaning and other pretreatments can reduce inspection of the environmental impact and personal risk (Wang 2009).

In this paper, acoustic emission testing technology is carried out to test the overall structure of the tank truck and location of defect acoustic emission signals were researched. Based on the comparison of several common defect acoustic emission source location methods, a defect acoustic emission source location method combining arbitrary planar triangle location and local scope linear location is proposed. Then, the non-open online detection of the tank container is realized.

2 ACOUSTIC EMISSION TESTING OF THE TANK TRUCK

Acoustic emission testing technology is a kind of dynamic non-destructive testing technology which can realize rapid detection of the overall structure of large size components. The acoustic emission testing of tank trucks is based on structural characteristics of the tank truck and acoustic emission sensors are placed in a specific location on the outer surface of the tank truck. Then, a load is applied to the tank truck. The elastic wave generated by the acoustic emission source is propagated to the surface of the tank material and received by the acoustic emission sensor through the coupling interface, thus causing the displacement vibration of the acoustic emission sensor surface. These sensors convert the mechanical vibration of the materials into electrical signals, which are then amplified and acoustic emission signals filtered. The acoustic emission signals generated by the defects are analyzed and studied by acoustic emission testing instruments. This is followed by deducing the location, state level, and development trend of the internal defect of the material or structure, and then safety status of the tested material or component evaluated.

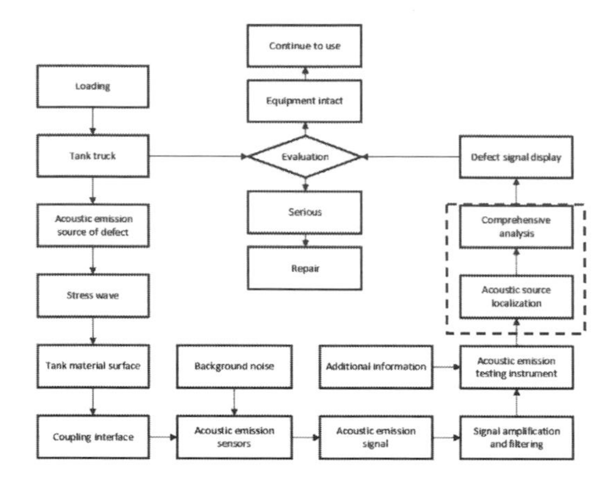

Figure 1. The acoustic emission testing process.

The acoustic emission testing process is shown in Figure 1. In this paper, signal processing and comprehensive analysis, especially the defect acoustic source location, are the most important research contents.

A. Leakage of safe valve nozzle

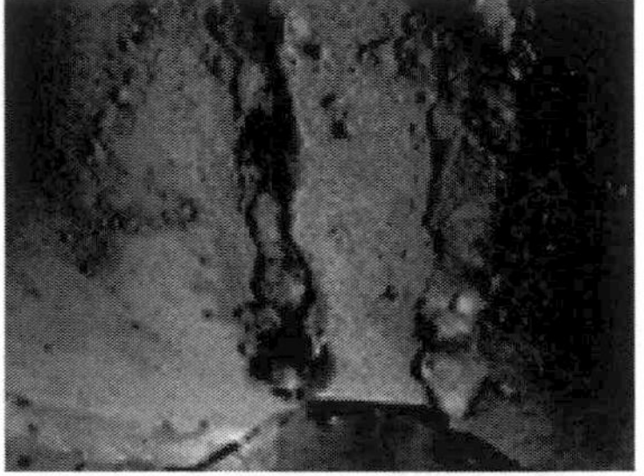

B. The crack of tank reinforcing ring

Figure 2. Defect of the tank truck.

3 LOCATION OF DEFECT ACOUSTIC EMISSION SOURCES

Currently, as shown in Figure 2, during tank truck regular inspection, it has been found that typical defects that have a great impact on safety of the overall structure of the tank truck are as follows: the tank joint crack, the wave preventer fillet weld crack, and tank and nozzle surface corrosion or leakage.

Therefore, the study of acoustic emission characteristics of these defects and accurately locating defects in acoustic emission source is considered important. In engineering practice, the correlation between acoustic emission signal characteristics and extent of damage caused by defects is particularly critical to the acoustic emission testing and structural safety evaluation of the tank truck. Furthermore, accuracy of defect acoustic emission source location is an important indicator in the acoustic emission testing and safety assessment of the tank truck.

In practical engineering, acoustic emission testing is used in a passive manner for defect dynamic testing and acoustic emission signals are usually accompanied with interference noise. Thus, in order to obtain much more accurate location of defects of the acoustic emission source, it is necessary to amplify, filter, and noise-reduce the acoustic emission testing signals first. Currently, there are many location methods for defects of acoustic emission source with the linear location and arbitrary planar triangle location being the most commonly used.

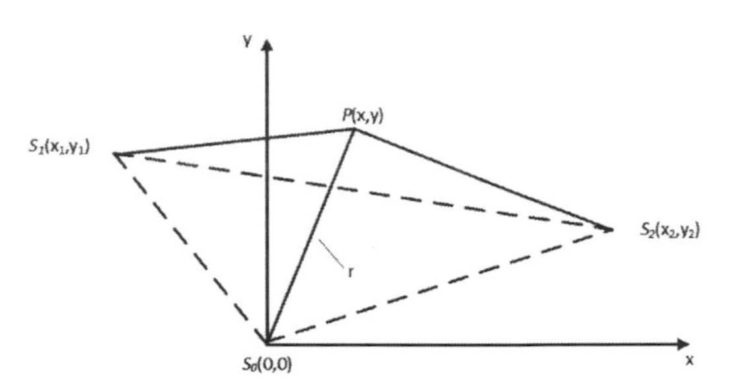

Figure 3. The method of arbitrary planar triangle location.

The tank truck is a cylindrical container, but the tank truck outside is usually covered with a thermal insulation layer, which can make an enormous impact on the acoustic emission source location. So based on comprehensive analysis and comparison of several common defect acoustic emission source location methods (Wang 2005, Wang 2008 & Zhang 2012), a defect acoustic emission source location method combining arbitrary planar triangle location and local scope linear location is proposed (Long 2002).

As shown in Figure 3, three acoustic emission sensors are assumed to form an arbitrary triangle, S_0 (0, 0), S_1 (x_1, y_1), and S_2 (x_2, y_2); at the point P (x, y) it has a defect acoustic emission source, r is the distance from S_0 to P, there are:

$$\delta_1 = \Delta t_{10} \cdot v = (t_1 - t_0) \cdot v \quad (1)$$

$$\delta_2 = \Delta t_{20} \cdot v = (t_2 - t_0) \cdot v \quad (2)$$

Where: Δt_{10} is the time difference of the defect acoustic emission source signal when it reaches S_1 and S_0;

Δt_{20} is the time difference of the defect acoustic emission source signal when it reaches S_2 and S_0; v is the sound velocity.

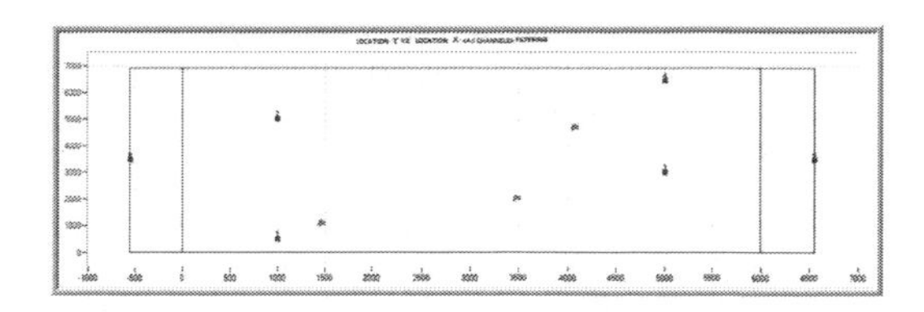

Figure 5. Acoustic emission signal location of the tank truck.

4 ENGINEERING APPLICATION OF ACOUSTIC EMISSION TESTING OF THE TANK TRUCK

Generally, six acoustic emission sensors are arranged on the overall structure for the standard acoustic emission testing of the tank truck, with four sensors arranged on the cylinder body, and a sensor placed on each of the vessel heads as shown in Figure 4. Through these six acoustic emission sensors, the overall structure of acoustic emission testing of the tank truck can be realized.

In this paper, the acoustic emission testing instrument used in testing is the type of EXPRESS 32 produced by PAC Company, USA and the sensor type of DP 151.

The acoustic emission sensors of the tank truck and the defect acoustic emission signal location are shown in Figure 5. The blue marking points in the figure are acoustic emission sensors and the red ones are acoustic emission signal source locations. The acoustic emission signal location point appears from the beginning stage up to the holding stage. At this stage, there are several acoustic emission events near the sensor number 1 and number 3, which disappeared in subsequent pressure stable stage. Therefore, the judgment of acoustic emission signal may not be caused by defects in acoustic emission events. It is possible that the interference signal during the pressure loading stage, the impact or friction between the loading medium and the tank internal components, and then at the pressure stable stage, were responsible for the acoustic emission signals to disappear.

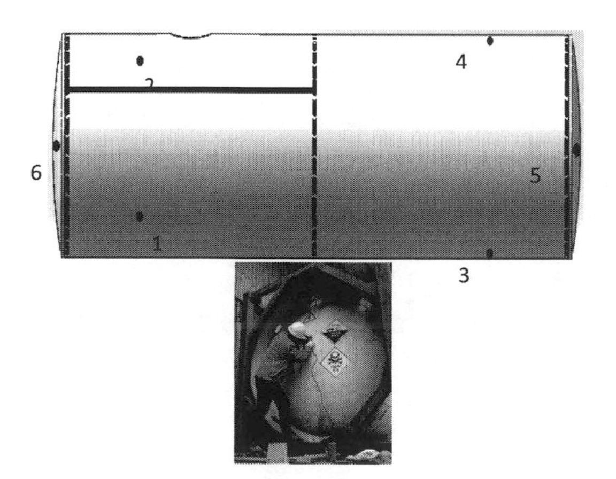

Figure 4. Acoustic emission sensor arrangement of the tank truck.

5 CONCLUSION

1. In the process of acoustic emission testing of the tank truck, at the pressure loading stage, due to the impact or friction of the loading medium and wave preventer inside the tank, it may produce some interference acoustic emission signals, easy to produce some miscalculation or make the true defect signals submerged in the interference signal such that the defect location is not accurate.
2. Acoustic emission testing is only effective for active defects and as such the result of acoustic emission testing relative to conventional non-destructive testing methods poses a slightly higher risk. Therefore, it is proposed to use both the acoustic emission testing and conventional non-destructive testing to reduce the risk of tank container inspection and ensure the safety of the equipment.

ACKNOWLEDGMENT

This paper is supported by the Science and Technology project of Zhejiang Provincial Administration of Quality and Technology Supervision (NO. 20160227).

REFERENCES

Long, F. F., (2002), Study of a New Type of Acoustic Emission Testing System and Technology of AE Sources Orientation. Daqing Petroleum Institute.

TSG R0005-2011 Supervision Regulation on Safety Technology for Transportable Pressure Vessel (2011).

Wang, H. L., M. C. Ge. (2008), Acoustic emission/microseismic source location analysis for a limestone mine exhibiting high horizontal stresses. International Journal of Rock Mechanics and Mining Sciences, 45(5), 720–728.

Wang, Q., G. X. Zhang, Z. K. Zhou, et al. (2005), Research on acoustic emission location method for oil pipeline breakage. Journal of Zhejiang University: Engineering Science, 39(3), 322–329.

Wang, Y. D., Y. T. Xu, et al. (2009), Application of Acoustic Emission Technology onto Container Tank Testing. NONDESTRUCTIVE TESTING, 3(6), 499–500, 508.

Zhang, W. G., Y. F. Huang, R. Y. YE, et al. (2012), Background noise on accuracy of line localization for acoustic emission detection. Journal of China University of Metrology, 23(4), 326–331.

Civil, Architecture and Environmental Engineering – Kao & Sung (Eds)
© 2017 Taylor & Francis Group, ISBN 978-1-138-02985-9

Nonlinear dynamic analysis of a cantilever plate for generating electricity based on electromagnetic induction

J.H. Yang, K. Ye, G. Zhu & H. Guo
Beijing Vocational College of Agriculture, Beijing, China

ABSTRACT: The nonlinear dynamic responses of a rectangular cantilever plate to power performance based on electromagnetic induction are studied in this paper. At the end of the rectangular cantilever plate, an electromagnet is placed and a set of coil is wound around the electromagnet. The rectangular cantilever plate is forced by the base excitation which is assumed to be a harmonic load. Based on the von Karman type equations and Reddy's classic plate theory, the nonlinear partial differential governing equations of motion for the cantilever plate are derived using the Hamilton's principle. Numerical simulations are used to investigate the effects of parameters on the steady-state responses of the cantilevered plate. Substituting the physical parameters of cantilever plate into the ordinary differential equation about transverse displacement and the equation of the output voltage, the bifurcation diagram of the cantilevered plate for the output voltage via the base excitation frequency is obtained.

1 INTRODUCTION

With the rapid development of MEMS, wireless communication technology, and computer technology, wireless sensor network technology has become a research hotspot. At present, a vast majority of wireless sensor nodes are battery-powered (Li & Gao 2008). However, the traditional chemical batteries have many shortcomings such as limited life span, the requirement of periodical replacement, environmental pollution, etc. In order to solve this problem, research about energy harvesting devices and spontaneous electrical systems are increasing rapidly (Culler et al. 2004).

In daily production and life, vibration is a common energy form. However, it is not used reasonably and wasted in most cases. If vibrational energy can be converted into electricity, not only can the efficiency of energy utilization be improved, but also a new green method of energy access is provided (Beeby et al. 2007). Thus vibrational energy harvesting technology has become a research hotspot.

The most common energy harvesting mechanisms are electrostatic, electromagnetic, and piezoelectric. Among them, electromagnetic energy harvester has been paid more attention because of small volume, powerful power generation-higher frequency sensing without power supply, and working in harsh environments (Vinod et al. 2009).

2 EQUATIONS OF MOTION

As shown in Figure 1, a rectangular cantilever plate clamped at edge Ob whose edge length and width in the x and y directions are, respectively, a and b and the thickness is h, the harmonic translation $F\cos(\Omega t)$ is subjected at the base of the plate. An electromagnet is placed at the end and a set of coil is wound around the electromagnet. A Cartesian coordinate $Oxyz$ is located on the middle surface of cantilevered rectangular plate.

According to Reddy's classic deformation plate theory and von Karman type equations for geometric nonlinearity, the displacement field of the cantilevered plate is assumed to be (Lim et al. 1998):

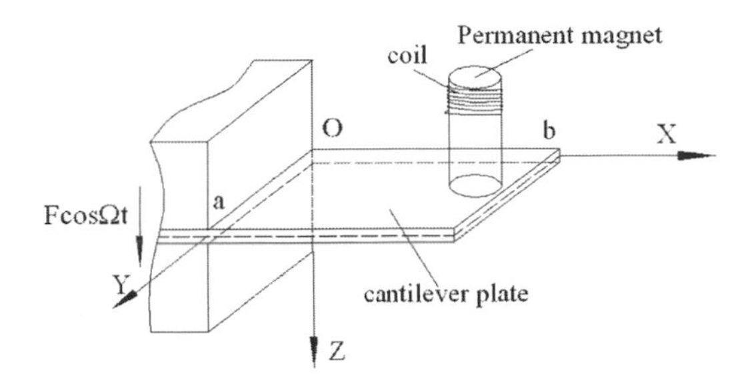

Figure 1. The simplified mathematical model of a cantilevered plate structure.

$$u_1 = u(x,y,z,t) = u_0(x,y,t) - z\frac{\partial w_0}{\partial x}$$
$$u_2 = v(x,y,z,t) = v_0(x,y,t) - z\frac{\partial w_0}{\partial y} \quad (1)$$
$$u_3 = w(x,y,z,t) = w_0(x,y,t)$$

where (u_1,u_2,u_3) are the displacement of an arbitrary point of the cantilever plate in the x, y, and z directions; (u_0,v_0,w_0) is the deflection of a point on the middle plane $(z = 0)$.

The nonlinear strain-displacement relations are given as follows:

$$\varepsilon_{xx} = \frac{\partial u}{\partial x} + \frac{1}{2}(\frac{\partial w}{\partial x})^2, \varepsilon_{yy} = \frac{\partial v}{\partial y} + \frac{1}{2}(\frac{\partial w}{\partial y})^2,$$
$$\varepsilon_{xy} = \frac{1}{2}(\frac{\partial u}{\partial y} + \frac{\partial v}{\partial x} + \frac{\partial w}{\partial x}\frac{\partial w}{\partial y}) \quad (2)$$

The stress–strain relationship of the cantilever plate is given as follows:

$$\begin{Bmatrix} \sigma_{xx} \\ \sigma_{yy} \\ \sigma_{xy} \end{Bmatrix} = \begin{Bmatrix} Q_{11} & Q_{12} & 0 \\ Q_{21} & Q_{22} & 0 \\ 0 & 0 & Q_{66} \end{Bmatrix} \begin{Bmatrix} \varepsilon_{xx} \\ \varepsilon_{yy} \\ \gamma_{xy} \end{Bmatrix} \quad (3)$$

$$Q_{11} = \frac{E_1}{1-\nu_{12}\nu_{21}}, Q_{12} = \frac{\nu_{12}E_2}{1-\nu_{12}\nu_{21}}, Q_{21} = Q_{12},$$
$$Q_{22} = \frac{E_2}{1-\nu_{12}\nu_{21}}, Q_{66} = G_{12} \quad (4)$$

where $Q_{ij}(i, j = 1,2)$ is the elastic stiffness coefficient; $E_i(i = 1,2)$ is the elastic modulus; G_{12} is the shear modulus; v_{12} and v_{21} are Poisson's ratio.

The Hamilton's principle is given as follows:

$$\delta\int_{t_1}^{t_2}(T-U)dt + \int_{t_1}^{t_2}\delta W dt = 0 \quad (5)$$

The variation of kinetic energy is shown as follows:

$$\delta\int_{t_1}^{t_2} T dt = \delta\int_{t_1}^{t_2}\left(\int_V \frac{1}{2}\rho\dot{u}_i\dot{u}_i dV + \frac{1}{2}m\dot{u}_i^2\Big|_{x=a}\right) dt, (i = 1,2,3) \quad (6)$$

The variation of potential energy is shown as follows:

$$\delta\int_{t_1}^{t_2} U dt = \int_{t_1}^{t_2}\int_V (\sigma_{xx}\delta\varepsilon_{xx} + \sigma_{yy}\delta\varepsilon_{yy} + \sigma_{xy}\delta\varepsilon_{xy}) dV dt \quad (7)$$

The virtual work is shown as follows:

$$\delta\int_{t_1}^{t_2} W dt = \delta\int_{t_1}^{t_2}\int_{x_1}^{x_2}\int_{y_1}^{y_2} F w_0 dx dy dt$$
$$-\delta\int_{t_1}^{t_2}\int_{x_1}^{x_2}\int_{y_1}^{y_2} c_3\dot{w}_0 w_0 dx dy dt \quad (8)$$

Substituting Equation (6) (7) (8) into Equation (5) yields the nonlinear governing equations of motion for the cantilever plate:

$$\frac{\partial N_{xx}}{\partial x} + \frac{\partial N_{xy}}{\partial y} = I_0\ddot{u}_0 - I_1\frac{\partial \ddot{w}_0}{\partial x},$$
$$\frac{\partial N_{yy}}{\partial y} + \frac{\partial N_{xy}}{\partial x} = I_0\ddot{v}_0 - I_1\frac{\partial \ddot{w}_0}{\partial y},$$
$$\frac{\partial N_{xx}}{\partial x}\frac{\partial w_0}{\partial x} + \frac{\partial N_{xy}}{\partial x}\frac{\partial w_0}{\partial y} + \frac{\partial N_{xy}}{\partial y}\frac{\partial w_0}{\partial x} \quad (9)$$
$$+ \frac{\partial N_{yy}}{\partial y}\frac{\partial w_0}{\partial y} + F\cos(\Omega t) - c_3\dot{w}_0$$
$$= I_0\ddot{w}_0 + I_1\left(\frac{\partial \ddot{u}_0}{\partial x} + \frac{\partial \ddot{v}_0}{\partial y}\right) - I_2(\frac{\partial^2 \ddot{w}_0}{\partial x^2} + \frac{\partial^2 \ddot{w}_0}{\partial y^2})$$

where the dot represents the partial differentiation with respect to time t; c_3 is the damping coefficient.

$$N_{\alpha\beta} = \int_{-h/2}^{h/2}\sigma_{\alpha\beta}dz, I_i = \int_{-h/2}^{h/2}\rho(z)^i dz, (i = 0,1,2) \quad (10)$$

where $\alpha = x, y; \beta = x, y$.

The stress-strain relations are given as follows:

$$\begin{Bmatrix} N_{xx} \\ N_{yy} \\ N_{xy} \end{Bmatrix} = \begin{Bmatrix} A_{11} & A_{12} & 0 \\ A_{21} & A_{22} & 0 \\ 0 & 0 & A_{66} \end{Bmatrix} \begin{Bmatrix} \varepsilon_{xx} \\ \varepsilon_{yy} \\ \gamma_{xy} \end{Bmatrix} \quad (11)$$

where $A_{ij} = \int_{-h/2}^{h/2} Q_{ij} dz, (i, j = 1,2,6)$.

Substituting Equation (10) and Equation (11) into Equation (9), the governing equations of motion in terms of generalized displacement for the cantilever plate are as follows:

$$A_{11}\frac{\partial^2 u_0}{\partial x^2} + A_{66}\frac{\partial^2 u_0}{\partial y^2} + (A_{12} + A_{66})\frac{\partial^2 v_0}{\partial x \partial y}$$
$$+(A_{12} + A_{66})\frac{\partial w_0}{\partial y}\frac{\partial^2 w_0}{\partial x \partial y} + A_{11}\frac{\partial w_0}{\partial x}\frac{\partial^2 w_0}{\partial x^2} \quad (12a)$$
$$+A_{66}\frac{\partial w_0}{\partial x}\frac{\partial^2 w_0}{\partial y^2} = I_0\frac{\partial^2 u_0}{\partial t^2} - I_1\frac{\partial^3 w_0}{\partial x \partial t^2},$$

$$A_{22}\frac{\partial^2 v_0}{\partial y^2} + A_{66}\frac{\partial^2 v_0}{\partial x^2} + (A_{21} + A_{66})\frac{\partial^2 u_0}{\partial x \partial y}$$
$$+(A_{21} + A_{66})\frac{\partial w_0}{\partial x}\frac{\partial^2 w_0}{\partial x \partial y} + A_{22}\frac{\partial w_0}{\partial y}\frac{\partial^2 w_0}{\partial y^2} \quad (12b)$$
$$+A_{66}\frac{\partial w_0}{\partial y}\frac{\partial^2 w_0}{\partial x^2} = I_0\frac{\partial^2 v_0}{\partial t^2} - I_1\frac{\partial^3 w_0}{\partial y \partial t^2},$$

$$
\begin{aligned}
&-D_{11}\frac{\partial^4 w}{\partial x^4}+(-D_{21}-4D_{66}-D_{12})\frac{\partial^4 w}{\partial x^2\partial y^2}\\
&-D_{22}\frac{\partial^4 w}{\partial y^4}+\frac{3A_{11}}{2}\left(\frac{\partial w}{\partial x}\right)^2\frac{\partial^2 w}{\partial x^2}\\
&+(\frac{1}{2}A_{21}+A_{66})\left(\frac{\partial w}{\partial x}\right)^2\frac{\partial^2 w}{\partial y^2}\\
&+(\frac{1}{2}A_{12}+A_{66})\left(\frac{\partial w}{\partial y}\right)^2\frac{\partial^2 w}{\partial x^2}\\
&+\frac{3}{2}A_{22}\left(\frac{\partial w}{\partial y}\right)^2\frac{\partial^2 w}{\partial y^2}+2\left(\frac{1}{2}A_{21}+2A_{66}+\frac{1}{2}A_{12}\right)\\
&\frac{\partial w}{\partial x}\frac{\partial w}{\partial y}\frac{\partial^2 w}{\partial x\partial y}+A_{11}\frac{\partial u}{\partial x}\frac{\partial^2 w}{\partial x^2}+A_{22}\frac{\partial v}{\partial y}\frac{\partial^2 w}{\partial y^2}\\
&+A_{21}\frac{\partial u}{\partial x}\frac{\partial^2 w}{\partial y^2}+A_{12}\frac{\partial v}{\partial y}\frac{\partial^2 w}{\partial x^2}+2A_{66}\frac{\partial u}{\partial y}\frac{\partial^2 w}{\partial x\partial y}\\
&+2A_{66}\frac{\partial v}{\partial x}\frac{\partial^2 w}{\partial x\partial y}+(A_{21}+A_{66})\frac{\partial w}{\partial y}\frac{\partial^2 u}{\partial y\partial x}\\
&+A_{11}\frac{\partial w}{\partial x}\frac{\partial^2 u}{\partial x^2}+A_{66}\frac{\partial w}{\partial x}\frac{\partial^2 u}{\partial y^2}+A_{66}\frac{\partial w}{\partial y}\frac{\partial^2 v}{\partial x^2}\\
&+(A_{12}+A_{66})\frac{\partial w}{\partial x}\frac{\partial^2 v}{\partial y\partial x}+A_{22}\frac{\partial w}{\partial y}\frac{\partial^2 v}{\partial y^2}\\
&+F\cos(\Omega t)=c_3\dot{w}_0+I_0\ddot{w}_0-I_2\frac{\partial^2\ddot{w}}{\partial x^2}\\
&-I_2\frac{\partial^2\ddot{w}}{\partial y^2}+\left(I_1\frac{\partial\ddot{u}}{\partial x}+I_1\frac{\partial\ddot{v}}{\partial y}\right)
\end{aligned}
\tag{12c}
$$

where $D_{ij}=\int_{-h/2}^{h/2}Q_{ij}z^2dz,(i,j=1,2,6)$

The Galerkin method is a numerical analysis method, which can transform partial differential equations to ordinary differential equations (Nayfeh & Mook 1979). Through the experimental study, the first order modal vibration plays a key role in the vibration of the nonlinear system (Yu 2009). So we choose a suitable mode function to satisfy the boundary condition of the cantilever plate and consider nonlinear dynamics of the cantilever plate in the first mode of u_0,v_0,w_0 and F. Thus, we write them in the following forms (Oh & Nayfeh 1996):

$$
\begin{aligned}
&u_0(x,y,t)=u_1(t)\sin(\frac{\pi x}{2a})\cos(\frac{\pi y}{b}),\\
&v_0(x,y,t)=v_1(t)\sin(\frac{\pi x}{2a})\cos(\frac{\pi y}{b}),\\
&w_0(x,y,t)=w_1(t)\alpha_1\beta_1,\\
&F(x,y,t)=F_1(t)\alpha_1\beta_1,\\
&\alpha_1=\cosh\frac{k_1x}{a}-\cos\frac{k_1x}{a}\\
&-\frac{\sinh k_1-\sin k_1}{\cosh k_1+\cos k_1}(\sinh\frac{k_1x}{a}-\sin\frac{k_1x}{a}),\\
&\beta_1=\sqrt{3}(1-\frac{2y}{b}),\\
&\cos k_1\cosh k_1+1=0,k_1{}^4=\omega^2\frac{\rho A}{EJ}
\end{aligned}
\tag{13}
$$

For the cantilever thin plate, transverse displacement is more obvious than the other directions, so all inertia terms in Equation (12) about u_0,v_0 are neglected (Narita 1985). By using the Galerkin method and substituting Equation (13) into Equation (12), we yield the governing ordinary differential equations of transverse motion:

$$
\begin{aligned}
&\ddot{w}_1+\mu_1\dot{w}_1+q_1w_1+q_2w_1{}^3=f_1\cos\Omega t\\
&\mu_1=-\frac{c_3a^4b^2}{a^4b^2I_0-a^2b^2I_2},q_1=\frac{10D_{11}b^2}{a^4b^2I_0-a^2b^2I_2},\\
&q_2=\frac{1}{a^4b^2I_0-a^2b^2I_2}\\
&\left(\begin{aligned}&-\frac{A_{11}A_{66}a^3b^2}{8A_{11}ab^2+30A_{66}a^3}\\&-\frac{A_{11}^2ab^4}{1600A_{11}ab^2+6000A_{66}a^3}\\&+\frac{A_{11}^2ab^4}{160A_{11}ab^2+600A_{66}a^3}-1.5A_{11}b^2\end{aligned}\right),\\
&f_1=\frac{a^4b^2F}{a^4b^2I_0-a^2b^2I_2}
\end{aligned}
\tag{14}
$$

According to Faraday's law of electromagnetic induction, the output voltage is equal to the change rate of flux through the closing coil.

$$
V=-\frac{d\Phi}{dt}=-\frac{dNBs}{dt}=-\frac{dNBl\dot{w}_1(t)}{dt}=-NBl\ddot{w}_1(t) \tag{15}
$$

where ϕ denotes magnetic flux; N denotes coil number of turns; B denotes the magnetic induction intensity; l denotes the effective length of the coil; $\ddot{w}_1(t)$ represents the ordinary differentiation with respect to time t.

3 NUMERICAL SIMULATION

The fourth-order Runge–Kutta algorithm is used to numerically analyze the nonlinear dynamic

Table 1. Physical parameters of the cantilevered plate.

Physical quantity	Value
a	2 m
b	1 m
h	0.008 m
E_1	125 Gpa
E_2	7.2 Gpa
G_{12}	4.0 Gpa
V_{12}	0.33
ρ	1570 kg/m³
c_3	2.83×10^{-5}
B	3300 Gs
N	800

responses of the cantilever plate to the output voltage based on the base excitation.

In Table 1, the physical parameters chosen are as follows: a denotes the length, b denotes the width, h denotes thickness of the plate, E_1, E_2 denotes Young's moduli, G_{12} denotes the shear moduli, v_{12} denotes the Poisson's ratios, ρ denotes the density, c_3 denotes the damping coefficient, N denotes the turns of the coil, B denotes the magnetic induction intensity.

The Equation (14) and Equation (15) are chosen for numerical simulation. The base excitation frequency Ω is used as the controlling parameter to investigate the periodic and chaotic responses of the cantilever plate to generating electricity based on electromagnetic induction.

As shown in Figure 2, the abscissa denotes the frequency of the base excitation while the ordinate denotes the output voltage of the plate. When the base excitation frequency Ω is located in the interval 30 Hz–300 Hz, the bifurcation diagram is obtained.

Analyzing the bifurcation diagram in Figure 2, it is found that the motions of the rectangular cantilever plate are as follows: the periodic motion $\rightarrow$ the multiple periodic motions $\rightarrow$ the chaotic motion $\rightarrow$ the multiple periodic motions $\rightarrow$ the chaotic motion with an increase in the base excitation frequency Ω.

Based on the bifurcation diagram and using the same physical parameters, the base excitation frequency Ω is changed to obtain different waveforms of the cantilever plate, including periodic motion, multiple periodic motion, and chaotic motion.

When the base excitation frequency $\Omega = 60$ Hz, Figure 3(a) shows the existence of periodic motion for the cantilever plate. When the base excitation frequency $\Omega = 170$ Hz, Figure 3(b) shows the existence of multiple periodic motions for the cantilever plate. When the base excitation frequency changes to $\Omega = 230$ Hz, the chaotic motion of the cantilever plate is observed, as shown in Figure 3(c).

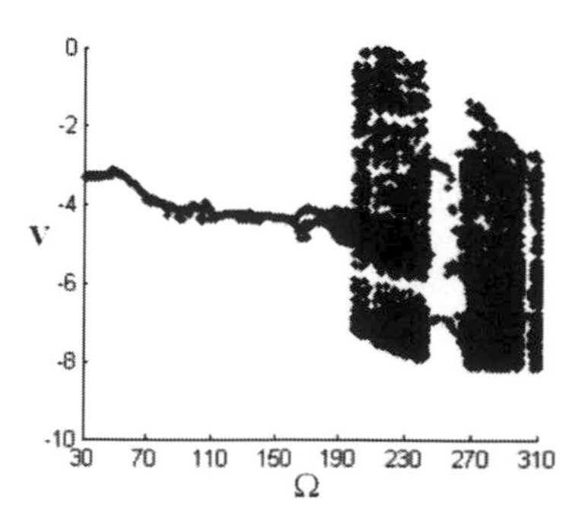

Figure 2. The bifurcation diagram of the output voltage V via the base excitation frequency Ω.

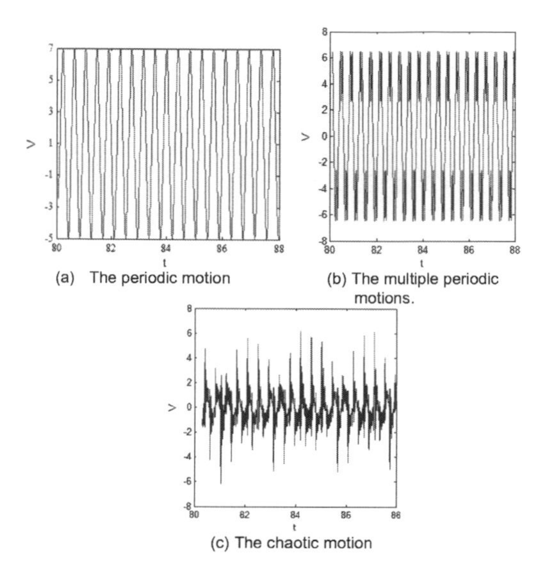

Figure 3. The motion of the cantilever plate.

4 CONCLUSIONS

The nonlinear dynamic responses of a rectangular cantilever plate based on electromagnetic induction to generating electricity subjected to the base excitation were investigated. Based on Reddy's classic deformation plate theory and von Karman type equations, the governing equations of motion for the cantilever rectangular plate was derived by using Hamilton's principle. Numerical simulation is provided by using the fourth-order Runge–Kutta algorithm. The numerical results show that there exist periodic, multiple periodic, and chaotic motions of the cantilever plate under certain conditions.

The bifurcation diagram of the cantilever plate for the output voltage V via the base excitation frequency Ω is obtained. Analyzing the bifurcation diagram, the periodic motion appears firstly, and the multiple periodic motion and chaotic motion appear alternately when the base excitation frequency Ω exceeds a certain value.

The results show that the output voltage of the rectangular cantilever plate is different with a change in the base excitation frequency Ω. In order to get the optimal efficiency of power generation, the appropriate interval of the base excitation frequency Ω must be chosen.

ACKNOWLEDGMENTS

The authors gratefully acknowledge the support of the Research Project from Beijing Vocational

College of Agriculture (project number: XY—YF—14-42).

REFERENCES

Beeby, S.P., R.N. Torah, M.J. & P. Tudor. (2007). A Micro Electromagnetic Generator for Vibration Energy Harvesting. *Journal of Micromechanics and Microengineering 17(7)*, 1257–1265.

Culler, D., D. Estrin, & M. Srivastava. (2004). Guest Editors' Introduction: Overview of Sensor Networks. *Computer*, 41–49.

Li, J.Z. & Gao, H. (2008). The research progress of wireless sensor network. *Research and development of the computer (01),* 1–15.

Lim, C.W., K.M. Liew, & S. Kitipornchai. (1998). Vibration of cantilevered laminated composite shallow conical shells. *Int J Solids Struct (35),* 1695–1707.

Narita Y. (1985). The effect of point constraints on transverse vibration of cantilever plates. *J Sound Vib (102)*, 5–13.

Nayfeh, A.H. & D.T. Mook (1979). Nonlinear oscillations. New York: Wiley.

Oh, K. & A.H. Nayfeh (1996). Nonlinear combination resonances in cantilever composite plates. *Nonlinear Dynamics (11)*, 143–169.

Vinod, R., M.G. Challa, & T. Prasad. (2009). A Coupled Piezoelectric–Electromagnetic Energy Harvesting Technique for Achieving Increased Power Output through Damping Matching. *Smart Materials and Structures 18(9)*, 095029.

Yu S.D. (2009). Free and forced flexural vibration analysis of cantilever plates with attached point mass. *J Sound Vib (321),* 27–85.

Civil, Architecture and Environmental Engineering – Kao & Sung (Eds)
© 2017 Taylor & Francis Group, ISBN 978-1-138-02985-9

Influence of inductance on an interleaved parallel boost PFC circuit

Zhonghua Kong, Ligang Wu & Zaifei Luo
Ningbo University of Technology, Ningbo, China

ABSTRACT: With power rating enhancing unceasingly, the increasing demand for high-power PFC (Power Factor Correction) converters is becoming stronger, but the traditional boost PFC converter is only suitable for small- and medium-sized power applications. By using the interleaved technology, the current stress of switches can be decreased, the ripple of the input current and switching losses can also be reduced, and the power rating of the PFC converter can be improved. Interleaved boost PFC is very suitable for high power applications. Inductance influence on the characteristics of interleaved parallel boost PFC circuit was studied. So the different material inductance and inductance values are designed, and the characteristics of the interleaved parallel boost PFC circuit are studied in this paper. Experimental results are presented for a prototype boost converter converting universal AC input voltage (85–265 V) to 400 V DC output at up to 1 kW load.

1 INTRODUCTION

With the rapid development of power electronics technology, power electronic devices have been widely used. This may cause severe harmonic pollution posing a serious threat to the health of the grid. Power Factor Correction (PFC) technology is a good way to solve the problem of harmonic pollution caused by power electronic devices. The most commonly used PFC converter in an onboard charger is a single-phase diode bridge rectifier followed by a boost converter. With power rating enhancing unceasingly, the increasing demand for high-power PFC converters is becoming stronger, but the traditional boost PFC converter is only suitable for small- and medium-sized power applications. By using the interleaved technology, the current stress of switches can be decreased, the ripple of the input current and switching losses can also be reduced, and the power rating of the PFC converter can be improved. Interleaved boost PFC is very suitable for high-power applications. Two-phase interleaved power factor correction boost converters in single phase AC-DC converter can reduce input current ripple for the EMI filter, the RMS ripple current for output capacitor and current stress for switches and inductors (L. Huber, 2000; C. Wang, 2008; Choudhury, 2005; T. Grote, 2009; B. A. Miwa, 1992; C. H. Chan, 1997). So this article will use interleaved boost PFC converter as the research object to study the characteristics of interleaved technology. Different material inductance and inductance values are designed, and the characteristics of interleaved parallel boost PFC circuit are studied in this paper. Experimental results are presented for a prototype boost converter converting universal AC input voltage (85–265 V) to 400 V DC output at up to 1 kW load.

2 DESIGN OF AN INTERLEAVED BOOST CONVERTER

The boost interleaved PFC converter topology is presented in Figure 1. It uses a dedicated diode bridge to rectify the AC input voltage to DC, which is then followed by two boost PFC converters in parallel, operating 180° out of phase. The boost interleaved PFC converter has the advantage of paralleled semiconductors. The input current is the sum of two inductor currents; therefore, it reduces output capacitor high-frequency ripple, but still has the prob-

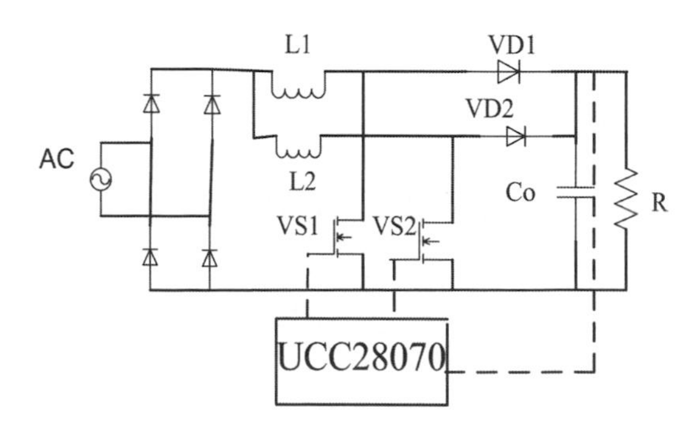

Figure 1. Interleaved PFC boost converter.

lem of heat management for the input diode bridge rectifiers. The PFC controller adopted is UCC28070. The UCC28070 is an advanced power factor correction device that integrates two pulse-width modulators operating 180° out of phase. This interleaved PWM operation generates a substantial reduction in input and output ripple currents, and the conducted-EMI filtering becomes easier and less expensive. A significantly improved multiplier design provides a shared current reference to two independent current amplifiers that ensure matched average current mode control in both PWM outputs while maintaining a stable, low-distortion sinusoidal input line current.

The interleaved inductors are designed to meet the requirement of high power factor over a wide range of AC line voltages. Two non-coupled inductors are used. The inductance value of inductors should be high enough to avoid Discontinuous Current Mode (DCM) operation of each boost leg and acquire high power factor, low Total Harmonic Distortion (THD), and low Electromagnetic Interference (EMI). As the inductance value of an inductor decreases at higher operating currents, the nominal value of inductance is designed at maximum operating current, corresponding to full load conditions (1 kW) at a lower line voltage (85 Vac). The inductance value is determined by the switching frequency, which is set at 80 kHz and the current ripple of each boost leg, which is selected to be equal to 20% (Δi) of the maximum AC current. The boost inductors (L1 and L2) are selected based on the maximum allowable input ripple current. In universal applications (e.g., 85 V to 265 V RMS input), the maximum input ripple current occurs at the peak of low line, and for this design, the maximum input ripple current was set to 30% of the peak nominal input current at low line.

The following calculations are used to select the appropriate inductance for L1 and L2, where variable D is the converter's duty cycle at the peak of low line operation, the max duty cycle D_{max} is as follows:

$$D_{max} = \frac{(V_o - \sqrt{2} \times V_{in\min})}{V_o} = \frac{400 - \sqrt{2} \times 85}{400} = 0.7 \quad (1)$$

where V_o is the output voltage, V_{inmin} is the minimum AC input voltage

Variable K(D) is the ratio of input current to inductor ripple current at the peak of low line operation as follows:

$$k = \frac{2D_{max} - 1}{D_{max}} = 0.57 \quad (2)$$

ΔI_L is the boost inductor ripple current at the peak of low line based on the converter input ripple current requirements as follows:

$$\Delta I_L = \frac{P_{out} \times \sqrt{2} \times 0.4}{V_{in\min} \times \eta \times k} = 6.49A \quad (3)$$

The inductance is calculated as follows:

$$L_1 = L_2 = \frac{V_{in\min} \times \sqrt{2} \times D}{\Delta I_L \times f} = 131\mu H \quad (4)$$

where f is the switching frequency at 80 kHz.

3 EXPERIMENTAL RESULTS

In order to study the influence of different inductance material and inductance values on the characteristics of the interleaved parallel boost PFC circuit, TDK ferrite N97 and Magnetics Kool Mμ 77090 materials are selected as the magnetic core of the inductor. The ferrite core adopted is EE42/21, the boost inductance (L1 and L2) values are 265, 600, and 835 μH, the Kool Mμ core adopted is ring cores, and the boost inductance (L1 and L2) values are 290 and 1100 μH. Input voltage and current waveform are shown in Figure 2. The yellow waveform is the voltage, the blue waveform is the current, and the current waveform is close to the sinusoidal waveform. Current waveform of the inductor and voltage waveform of switching are shown in Figure 3. The green waveform is voltage and blue waveform is current, the underpart waveform is a partially enlarged view form of the upper part waveform.

Figure 4 to Figure 6 show the relation between the input voltage and power factor and efficiency, where B is the power factor curve, C is the efficiency curve, and TDK ferrite N97 material is adopted. The values of boost inductors (L1 and L2) are 265,

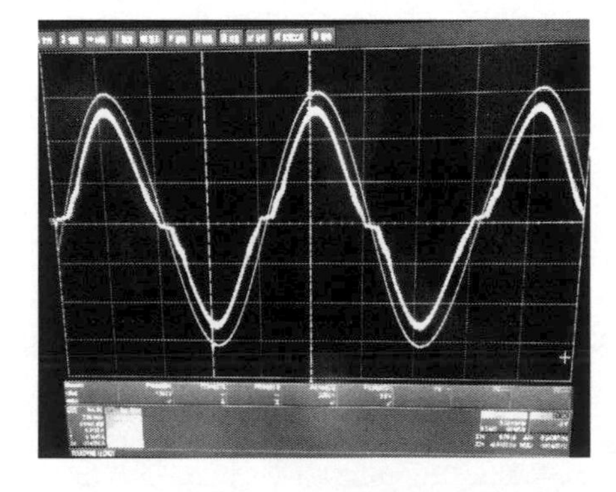

Figure 2. Input voltage and current waveform.

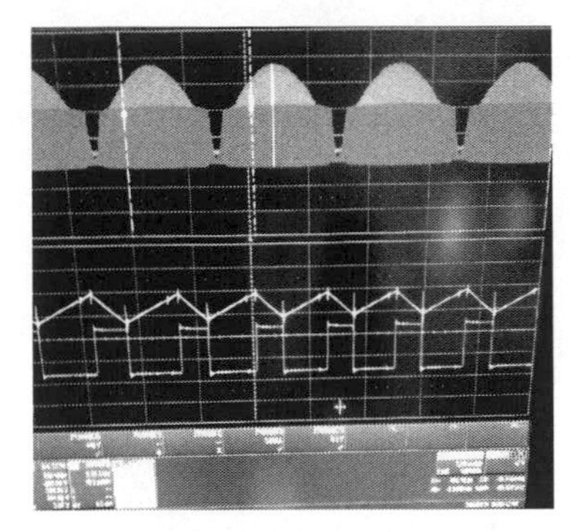

Figure 3. Current waveform of the inductor and voltage waveform of switching.

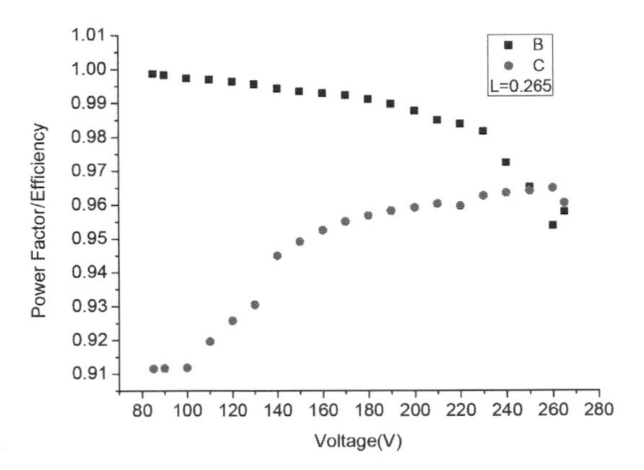

Figure 4. Relation between the input voltage and power factor and efficiency.

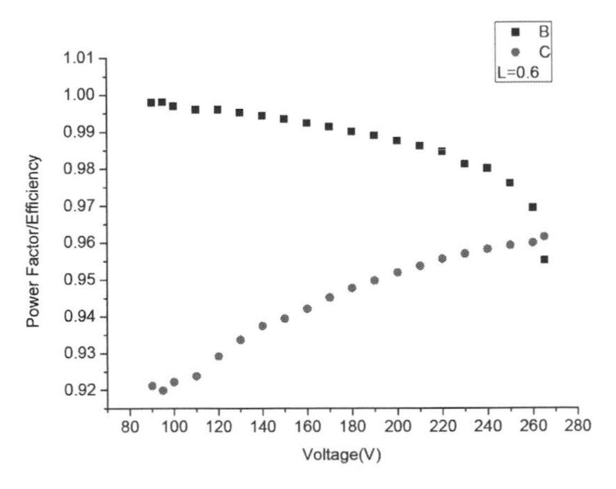

Figure 5. Relation between the input voltage and power factor and efficiency.

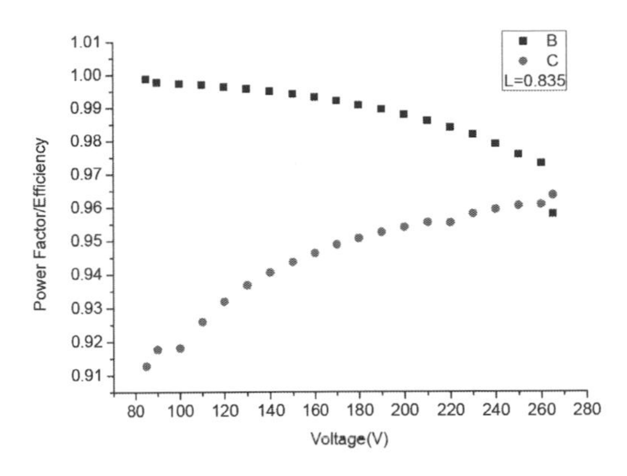

Figure 6. Relation between the input voltage and power factor and efficiency.

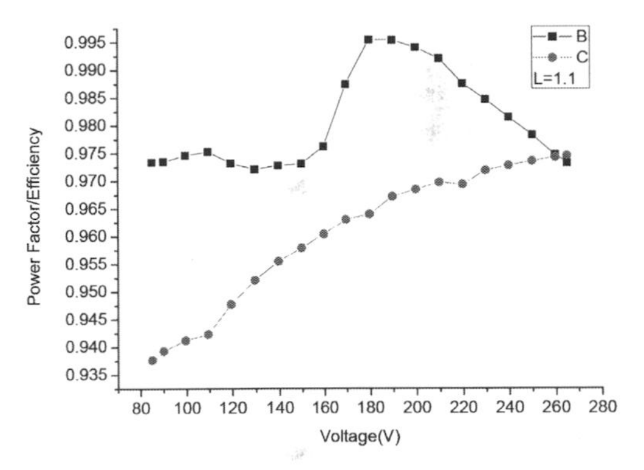

Figure 7. Relation between the input voltage and power factor and efficiency.

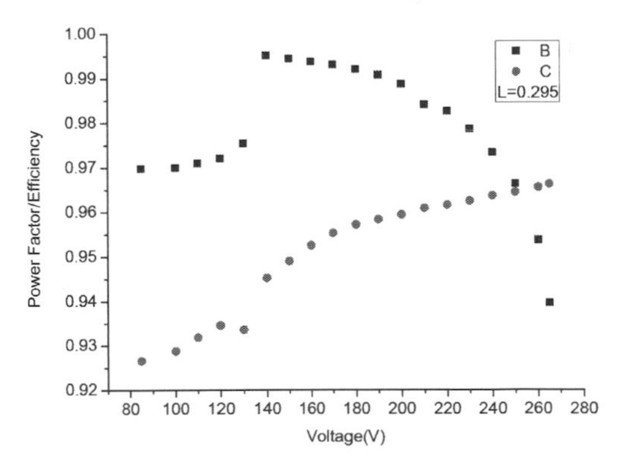

Figure 8. Relation between the input voltage and power factor and efficiency.

600, and 835 μH. Power factor decreases with the increase in input voltage, but efficiency increases with the increase in input voltage. The input current decreases with increase in input voltage, so power loss of the power device increases with increase in input current. Overall, considering the power factor and efficiency, the boost inductor with 600 μH is better than the other inductors.

Figure 7 and Figure 8 show the relation between the input voltage and power factor and efficiency, B is the power factor curve, C is the efficiency curve, and the Magnetics Kool Mμ 77090 is the adopted material, the boost inductor (L1 and L2) values are 295 and 1100 μH. Efficiency increases with increase in input voltage, power factor is nearly stable under low voltage and decreases with increase in input voltage. So, the characteristics of Magnetics Kool Mμ must be considered in the design of switching power supply.

4 CONCLUSION

The relation between the input voltage and power factor and efficiency is studied in this paper. When ferrite material is adopted for power supply, the power factor decreases with increases in input voltage, the efficiency increases with the increase in input voltage, and the boost inductor with 600 μH is better than the other inductors. When the Magnetics Kool Mμ material is adopted for power supply, the efficiency increases with the increase in input voltage, and the power factor is nearly stable under a lower voltage and then decreases with the increase in input voltage. So, the characteristics of Magnetics Kool Mμ must be considered in the design of switching power supply.

ACKNOWLEDGMENTS

This paper was supported by the Ningbo ZD-A01—Major Industrial Projects Foundation (2012A610111) and the Ningbo Natural Science Foundation (2014A61008).

REFERENCES

Chan C.H. and M.H. Pong, Input Current Analysis of Interleaved Boost Converters Operating in Discontinuous-Inductor-Current Mode[C]. in Proc. IEEE PESC 1997, pp. 392–398.

Choudhury and J.P. Noon. A DSP Based Digitally Controlled Interleaved PFC Converter, in Proc[C]. IEEE APEC 2005, pp. 648–654.

Grote, T., H. Figge, N. Frohleke, W. Beulen, F. Schafmeister, P. Ide and J. Bocker. Semi-Digital Interleaved PFC Control with Optimized Light Load Efficiency[C]. in Proc. IEEE APEC 2009, pp. 1722–1727.

Huber, L., B.T. Irving, and M.M. Jovanovic. Closed-Loop Control Methods for Interleaved DCM/CCM Boundary Boost PFC Converters[C]. in Proc. IEEE APEC 2009, pp. 991–997

Miwa, B.A., D.M. Otten and M.E. Schlecht. High Efficiency Power Factor Correction Using Interleaving Techniques[C]. in Proc. IEEE APEC 1992, pp. 557–668.

Wang, C., M. Xu, F.C. Lee and Z. Luo. Light Load Efficiency Improvement for Multi-Channel PFC[C]. in Proc. IEEE PESC 2008, pp. 4080–4085.

Civil, Architecture and Environmental Engineering – Kao & Sung (Eds)
© 2017 Taylor & Francis Group, ISBN 978-1-138-02985-9

Author index